The Chemistry of Food

The Chemistry of Food

Second Edition

Jan Velíšek
University of Chemistry and Technology Prague

Richard Koplík
University of Chemistry and Technology Prague

Karel Cejpek
University of Chemistry and Technology Prague

WILEY Blackwell

Registered Office(s)
John Wiley & Sons Ltd, The Atrium, Southern Gate, Chichester, West Sussex, PO19 8SQ, UK

Editorial Office
The Atrium, Southern Gate, Chichester, West Sussex, PO19 8SQ, UK

For details of our global editorial offices, customer services, and more information about Wiley products visit us at www.wiley.com.

Wiley also publishes its books in a variety of electronic formats and by print-on-demand. Some content that appears in standard print versions of this book may not be available in other formats.

Library of Congress Cataloging-in-Publication Data
9781119537649

Cover Image: © oxygen/Getty Images
Cover Design: Wiley

Set in 9.5/12pt MinionPro by SPi Global, Chennai, India
Printed and bound by CPI Group (UK) Ltd, Croydon, CR0 4YY

10 9 8 7 6 5 4 3 2 1

Contents

Preface

Preface to the First Edition

During the 15 years that have elapsed since the first Czech edition of the textbook *The Chemistry of Food* was published, many important monographs and scientific articles from various areas of food science have appeared in the literature, thanks to developments in analytical instrumentation and advances in the knowledge in the field of food technology. All these, plus the interest expressed by the readers, were the impetus that led to the third Czech edition of the book in 2009. This first English edition of *The Chemistry of Food* has remained essentially unchanged in terms of the basic structure of the 12 chapters and the majority of the text, tables, figures and formulae. Certain specific parts, however, have, necessarily, been revised, supplemented and updated.

The Chemistry of Food is the result of many years of experience by workers at the Department of Food Chemistry and Analysis (now Department of Food Analysis and Nutrition), Institute of Chemical Technology in Prague, through the teaching of this subject, plus other related topics taught in the Faculty of Food and Biochemical Technology. The reader will first be introduced to the chemical composition of foods, the important properties of food components and their functions. After these descriptions, subsequent sections deal with the changes and reactions that occur, or may occur, in foods under certain conditions. It is necessary for the reader to be able to understand the context and complexity of all the processes taking place in foods. As in the previous Czech editions, the authors have tried to reduce to a minimum the amount of text, as is typical for textbooks on biochemistry, microbiology, organic, inorganic and physical chemistry, so that it should not be necessary to consult other specialised textbooks too often, but, at the same time, ensuring the text is clear to both students and professionals.

The textbook contains an introductory chapter and then 11 chapters dealing with the main and accessory nutrients that determine the nutritional and energy value of food raw materials and foods. There is a chapter describing amino acids, peptides and proteins, a chapter dealing with fats, oils and other lipids, as well as chapters on carbohydrates, vitamins, mineral substances and water. Another chapter considers compounds responsible for the aroma, taste and colour attributes that determine the sensory quality of food raw materials and foods. The remaining chapters discuss substances that affect, or may affect, the hygienic–toxicological quality of food raw materials and foods, including antinutritional, toxic and other biologically active food components, food additives and food contaminants.

The third Czech edition featured the work of many workers from the Department of Food Chemistry and Analysis, along with colleagues from other departments of the Faculty of Food and Biochemical Technology and also external authors. Jan Pánek acted as a co-author of the chapters dealing with the main nutrients, and Helena Čížková and Michal Voldřich participated in these and other chapters. Co-authors of the chapter on amino acids, peptides and proteins were Karel Cejpek and Roman Kubec, of the chapter on fats, oils and other lipids Jana Dostálová and Vladimír Filip, of the chapter on carbohydrates Karel Cejpek and Kamila Míková, of the chapter on minerals Richard Koplík, of the chapter on flavour-active substances Jan Šavel, of the chapter on pigments and other colourants Jana Dostálová, of the chapter of antinutritional and toxic substances Pavel Kalač and Přemysl Slanina, and of the chapter on contaminants Jaroslav Dobiáš, Marek Doležal and Jana Hajšlová. The co-authors of the sections dealing with food legislation were Vladimír Kocourek and Kamila Míková. Most of these co-authors were also involved in the first and second Czech editions of this publication. In addition, co-authors of the first and second Czech editions included Jan Pokorný (the chapter on fats, oils and other lipids), Jiří Davídek (the chapter on vitamins and contaminants), Tomáš Davídek (the chapter on carbohydrates), Karel Hrnčiřík (chapters on vitamins and antinutritional and toxic compounds) and Helena Valentová (the chapter on pigments and other colourants). I would like to take this opportunity to thank all of my co-authors and other colleagues for their thorough work. Considerable thanks are also due to the reviewers of the first Czech edition, Prof. Alexander Pribela, DSc (Slovak University of Technology in Bratislava, Slovak Republic) and Assoc. Prof. Jan Staněk, PhD, who also did a great deal of hard work on the project.

The co-authors of the Czech editions were Assoc. Prof. Karel Cejpek, PhD, Helena Čížková, PhD, Prof. Jiří Davídek, DSc, Assoc. Prof. Jaroslav Dobiáš, PhD, Assoc. Prof. Marek Doležal, PhD, Prof. Jana Dostálová, PhD, Prof. Vladimír Filip, PhD, Prof. Jana Hajšlová, PhD, Prof. Vladimír Kocourek, PhD, Prof. Richard Koplík, PhD, Assoc. Prof. Kamila Míková, PhD, Assoc. Prof. Jan Pánek, PhD, Prof. Jan Pokorný, DSc, Assoc. Prof. Kateřina Riddellová, PhD, and Prof. Michal Voldřich, PhD, staff of the Faculty of Food and Biochemical Technology, Institute of Chemical Technology in Prague. External co-authors were Tomáš Davídek, PhD (Nestlé, Orbe, Switzerland), Karel Hrnčiřík, PhD

(Unilever, Vlaardingen, The Netherlands), Prof. Pavel Kalač, PhD (University of South Bohemia, České Budějovice), Assoc. Prof. Roman Kubec, PhD (University of South Bohemia, České Budějovice), Prof. Přemysl Slanina, DVM, PhD (National Food Administration, Uppsala, Sweden), Assoc. Prof. Jan Šavel, PhD (Budweiser Budvar n.p., České Budějovice) and Helena Valentová, PhD (Hügli Food Ltd, Zásmuky).

Jan Velíšek
Prague, February 2013

Preface to the Second Edition

This second English edition of *The Chemistry of Food* appears 6 years after the first. The number of chapters remains the same, but all of the text has been carefully checked, revised, and updated with new knowledge gained in the field of food chemistry from 2013 to 2018.

Jan Velíšek, Richard Koplík, and Karel Cejpek
Prague, April 2019

1

Introduction

The term **food** refers to any consumed material, eaten or drunk, that contains nutrients; typically, this means proteins, fats, carbohydrates, vitamins, and minerals, but it actually includes a wide range of many other chemical substances. Most food is of plant or animal origin but, to a lesser extent, it may also come from other sources, such as insects, algae, and microorganisms. Organisms use food to perform one or more of four essential functions: (i) to generate energy, (ii) to build, repair, and maintain body tissue, (iii) to supply chemical substances that regulate psychosomatic processes, and (iv) to provide chemical substances that protect the organism. For many centuries, human food was obtained directly through agriculture, but today the food consumed by the majority of the global population is either supplied by the food industry or homegrown.

The human diet, typically including meat, poultry, fish, milk, cereals, vegetables, and fruit, is diverse and, consequently, chemically very complex. According to rough estimates, fresh foods and foodstuffs contain around half a million different chemical compounds. A much larger number of compounds result from the biochemical (enzymatic) and chemical (non-enzymatic) reactions that occur during the storage of food raw materials and their subsequent industrial and culinary processing.

Food science is a major branch of the life sciences. It covers every aspect of food, incorporating and integrating concepts and information from many different fields, including the natural, technical, and social sciences. Primarily, it is based on chemical science (biochemistry; organic, inorganic, physical, and analytical chemistry), but it also draws heavily on some areas of physics (such as the mechanics of solids and liquids), biology (especially microbiology), biotechnology, some branches of medical science (human nutrition, human physiology, pharmacology, toxicology), and agricultural science (crop and livestock production, post-harvest plant physiology, and post mortem muscle physiology). It also utilises the expertise of technical engineering disciplines, such as agricultural and food engineering, particularly in the form of genetic food engineering. And, from time to time, it calls on knowledge from economics, sociology, psychology, and other branches of social science.

The most important building blocks of food science are **food chemistry** and **food technology**. Food chemistry deals not only with the composition of food raw materials and final food products, but also with the interactions and reactions of food components during the production, storage, and processing of food. Food technology runs the entire gamut from procuring food raw materials, through processing them into food products, to preserving, packaging, and distributing them. The shared goals and objectives of both disciplines are to enhance desirable positive changes and to prevent unwanted ones. Producing nutritious, safe, and, if possible, attractive food for human and animal consumption is the principal aim of both food chemistry and food technology.

This book covers all aspects necessary for a comprehensive understanding of how food works. The 11 chapters that follow provide detailed knowledge of and modern insights into different classes of food constituents and their corresponding changes. Each details a particular class of food components, in virtually all cases covering the following: structure, classification, sources and occurrence, nutrition and physiology, physical and organoleptic properties, and reactions and interactions during storage and processing.

Chapters 2–4 cover the macronutrients (primary, basic, or main nutrients), the **proteins**, **lipids**, and **carbohydrate**s that primarily provide structural (building) material and energy, but also have other functions. They show how some constituents of macronutrients may act as detrimental bioactive substances (e.g. toxic amino acids and antinutritive proteins), as beneficial bioactive substances (e.g. certain oligopeptides and oligosaccharides), or as precursors of both (e.g. amino and fatty acids). Other macro- and micronutrients act as sensory-active substances or their precursors.

Chapters 5 and 6 describe the chemistry, interactions, and reactions of the micronutrients: the **vitamins** and dietary **minerals** often referred to as accessory nutrients or essential nutritional factors.

Chapter 7 focuses on water, which is formed *in vivo* in small amounts by the oxidation of primary nutrients, but is obtained in much larger amounts from foods. Water is necessary for the absorption of nutrients. The chapter includes essential information on drinking water, on water activity in foods, on the interaction of water with other food components, and on the role of water in food dispersed systems.

Chapters 8 and 9 turn the spotlight on the main substances that affect the sensory qualities and organoleptic properties of food. **Sensory-active compounds** determine the sensory value of food; namely, the olfactory (aroma, odour, smell), gustative (taste), visual (colour), haptic (tactile), and auditory (sound) perceptions it induces. Chapter 8 introduces flavour- and odour-active organic compounds, classifying them according to both their structure and their occurrence in particular foods. Haptic and auditory sensations (e.g. crispiness) are mainly affected by proteins and polysaccharides (hydrocolloids), and, thus, are mostly covered in earlier chapters. Chapter 9 is devoted to pigments and other natural colourants.

Chapter 10 concentrates on three classes of undesirable natural compounds: (i) **antinutrients**, which impair the utilisation of nutrients by biochemical mechanisms, (ii) **natural compounds toxic to some individuals**, in whom they induce food allergies or intolerance, and (iii) **toxic substances** or toxins, which are toxic to all individuals.

Chapters 11 and 12 are concerned with **food additives** and **contaminants**, respectively. Additives are intentionally added to foods to protect them against spoilage and oxidation, as well as to maintain or improve their safety, freshness, taste, texture, or appearance. Contaminants are toxic products of microorganisms and harmful chemical substances that are unintentionally present in food mostly as a result of human agricultural and industrial activities. The chapter outlines the health risks associated with certain contaminants, together with techniques aimed at preventing contamination and ensuring food safety.

This book takes you on a tour of food chemistry, an incredibly fascinating field of study that inspires a constant sense of discovery. For better or worse, food chemistry is all around us all the time, and everything in food involves chemistry. Because food touches almost every aspect of our existence in some way, it is not surprising that the 'magic' of food composition has been attracting inquisitive minds from time immemorial. So, let us wade into this information soup and immerse ourselves in the chemistry of food, glorious food.

2

Amino Acids, Peptides, and Proteins

2.1 Introduction

Aminocarboxylic acids occurring in nature have vital functions in living organisms. These amino acids are found as free substances or higher-molecular-weight compounds, where the amino acid building units are connected to one another by amide bonds, —C(=O)—NH—, termed **peptide bonds**. Depending on the number of bound amino acids, these compounds are divided into two main groups:

- **peptides**, usually composed of 2–100 monomers;

- **proteins**, which contain more than 100 monomers, but also hundreds or even thousands of amino acids.

Proteins and peptides may even contain some other compounds in addition to amino acids. Proteins are undoubtedly the most important amino acid derivatives. They are the basic chemical components of all living cells and therefore are also part of almost all food raw materials and foods of plant, animal, and microbial origin. In organisms, proteins perform a number of unique and extraordinary functions. Along with ribonucleic acids (RNA) and deoxyribonucleic acids (DNA), polysaccharides, some lipids, and other macromolecules, peptides and proteins are often known as biological polymers or biopolymers. Nucleic acids (RNA and DNA) have almost no significance as food components in human and animal nutrition, although they play fundamental roles in living systems. Plants and some microorganisms are capable of synthesising proteins from basic substrates such as carbon dioxide, water, and inorganic nitrogen compounds, but animals rely on getting their necessary vegetable or animal protein from their food. In the process of digestion, the food proteins are enzymatically broken down (hydrolysed) to their building blocks, from which animals may synthesise their own proteins or use them (along with carbohydrates and lipids) as a source of energy. Therefore, proteins, together with carbohydrates and lipids, belong to the category known as **main (primary) nutrients**.

The following sections deal with food-related important amino acids, peptides, and proteins, their structure, occurrence, properties, and fate in the human organism. Nutritional aspects and important interactions and reactions that fundamentally affect the nutritional value, organoleptic properties (odour, taste, colour, and texture), and toxicological quality of food commodities are also discussed.

2.2 Amino acids

Plants, animals, and other organisms have been shown to contain more than 700 different amino acids. Some of these are spread quite generally throughout nature, whilst others occur only in certain species of plants, animals, and other organisms. According to their origins, therefore, the following two groups of amino acids are recognised:

- amino acids found in all living organisms (bound in proteins or peptides or occurring as free amino acids);

- amino acids found only in some organisms (bound in peptides or present as free compounds) that are not protein constituents.

The Chemistry of Food, Second Edition. Jan Velíšek, Richard Koplík and Karel Cejpek.
© 2020 John Wiley & Sons Ltd. Published 2020 by John Wiley & Sons Ltd.

The amino acids bound in proteins (22 compounds) are called proteinogenic, encoded, basic, standard, or primary amino acids; 21 of them are constituents of proteins in food raw materials and foods. Of the 22 proteinogenic amino acids, 20 are encoded by the universal genetic code. The remaining two amino acids (selenocysteine and pyrrolysine) are incorporated into proteins by unique synthetic mechanisms. For example, pyrrolysine occurs as a component of enzymes involved in the production of methane in some methanogens, members of a group of single-celled microorganisms of the Archaea domain. Amino acids bound in peptides and free amino acids have the same nutritional value as proteins, but their importance in nutrition is usually negligible due to the small amounts in which free amino acids and peptides commonly occur in foods.

The process of protein biosynthesis is called translation. Posttranslational oxidation, alkylation, and esterification of some amino acids that are bound in proteins yield modified proteinogenic amino acids (see Section 2.2.1.1.2). Non-proteinogenic (non-encoded, non-standard, or secondary) amino acids do not function as building blocks of proteins, as they have other roles in organisms. Amino acids also have an influence on the organoleptic properties of food, especially on their taste. Products of reactions of amino acids are often important compounds influencing the odour, taste, and colour of foods.

2.2.1 Structure, terminology, classification, and occurrence

Amino acids are organic compounds that contain at least one amino group, in most cases primary —NH$_2$, together with at least one carboxyl group, —COOH, and various side groups in the molecule. They can also be defined as carboxylic acids substituted by an amino group. According to the distance of the amino group from the carboxylic group, amino acids (**2-1**) are generally divided into:

- **2-aminocarboxylic acids** (α-aminocarboxylic acids), which have the amino and carboxyl groups attached to the same carbon atom (such as α-alanine);

- **3-aminocarboxylic acids** (β-aminocarboxylic acids), which have the amino and carboxylic groups attached to adjacent carbon atoms (such as β-alanine);

- **4-aminocarboxylic acids** (γ-aminocarboxylic acids), which have three ($n = 2$) carbon atom (group) between the amino and the carboxylic groups (such as γ-aminobutyric acid, GABA);

- **5-aminocarboxylic acids** (δ-aminocarboxylic acids), which have four ($n = 3$) carbon atoms (groups) between the amino and the carboxylic groups (such as δ-aminolaevulinic acid);

- **6-aminocarboxylic acids** (ε-aminocarboxylic acids), which have five ($n = 4$) carbon atoms (groups) between the amino and the carboxylic groups (such as ε-aminocaproic acid).

2-1, 2-amino acid ($n = 0$)
3-amino acid ($n = 1$)
4-amino acid ($n = 2$)
5-amino acid ($n = 3$)
6-amino acid ($n = 4$)

The amino acids also include carboxylic acids, which contain a secondary amino group —NH— in the molecule that is a part of three-, four-, five-, or six-membered rings. These amino acids are actually derivatives of the saturated nitrogen heterocycles aziridine, azetidine, azolidine (pyrrolidine), and azinane (piperidine), or of other more complex heterocyclic compounds. The only proteinogenic amino acid with a secondary amino group is proline, which comprises a pyrrolidine ring.

In most foods, about 99% of amino acids are bound in proteins and peptides. The rest (about 1%) are free amino acids. More free amino acids are found in foods in which proteolytic enzymes or chemical agents have hydrolysed proteins during manufacture or storage. Larger amounts of free amino acids can be found in some cheeses, beer, and wine. The enzymatic hydrolysates of proteins (such as soy sauce) or acidic protein hydrolysates (used as soup condiments) contain only free amino acids with small amounts of peptides, but no protein.

2.2.1.1 Amino acids of proteins

2.2.1.1.1 Proteinogenic amino acids

Proteinogenic amino acids bound in all proteins are exclusively 2-amino (α-amino) acids that have the primary or secondary amino and carboxyl groups attached to the same carbon atom in position 2 (α) to the carboxyl group.

The proteins in most organisms contain only 21 basic amino acids, of which 20 have the primary amino group, whilst one is an alicyclic amino acid with a secondary amino group (**2-2**). All amino acids except glycine (which is achiral) are optically active (chiral) compounds that contain at least one asymmetric centre (chiral carbon atom) and thus can occur in at least two non-superimposable mirror-image forms, D- and L-forms, known as optical isomers or enantiomers. Proteinogenous amino acids almost exclusively belong to the L-series, and thus have the L-configuration (see Section 2.2.3.2).

aliphatic α-amino acid   alicyclic α-amino acid

2-2, basic structures of amino acids

One or more hydrogen atoms of the substituent R of L-α-amino acids, known as a side chain, specific to each amino acid (**2-2**), may be substituted by a carboxyl group (called a distal carboxyl group), an amino group (distal amino group), or other functional groups, such as a hydroxyl group —OH, a sulfhydryl (mercapto) group —SH, a sulfide group —S—CH_3, a guanidino group —NH—C(=NH)—NH_2, or a phenyl group —C_6H_5. The methylene groups or nitrogen atom of the secondary amino group of α-amino acids can also be substituted (**2-2**).

The proteinogenic amino acids (**2-3**) are often known by their trivial names, which are derived from their properties or the source from which they were first isolated. Systematic names of the proteinogenic amino acids are rarely used, but are more frequent in the case of non-protein amino acids. Each amino acid has both a three-letter code (mostly the first three letters of the trivial name) and a single-letter code, which is used for the registration of long sequences of amino acids in proteins (Table 2.1).

Each classification of amino acids is somewhat imprecise and conforms to certain purposes. The most common way of sorting the proteinogenic amino acids in food chemistry is classification according to the structure of their side chain and functional groups present therein. This will distinguish the following groups of amino acids:

- aliphatic amino acids (monoaminomonocarboxylic acids) with an unsubstituted side chain, which include the simplest amino acid **glycine** (sometimes also classified as the only amino acid without a side chain), its higher homologue **alanine**, and the branched-chain amino acids **valine**, **leucine**, and **isoleucine**;

- aliphatic hydroxyamino acids, which include **serine** and **threonine**;

- aliphatic sulfur-containing amino acids **cysteine** and **methionine**;

- selenoanalogue of cysteine, called **selenocysteine**;

- amino acids with carboxyl groups in the side chain (monoaminodicarboxylic acids), **aspartic acid** and **glutamic acid**;

- their monoamides (amino acids with carboxamide groups in the side chains): **asparagine** and **glutamine**;

- amino acids with basic functional groups in the side chain; that is, diaminomonocarboxylic acid **lysine**, lysine derivative with a 1-pyrroline ring known as **pyrrolysine**, **arginine** with a guanidino group, and the derivative of imidazole 1*H*-isomer called **histidine**;

- amino acids with aromatic and heterocyclic side chains, which include **phenylalanine**, its hydroxyl derivative **tyrosine**, and **tryptophan** with an indole ring;

- **proline**, the amino acid in which the functional group is involved in the ring structure.

For clarity, formulae of amino acids are shown in the non-ionised form. Non-ionised molecules are, however, virtually absent in aqueous solutions and in animal and plant tissues. Amino acids are ionised and form inner salts (**2-2**), where carboxyl groups exist as —COO^- anions and amino groups as —NH_3^+ cations. The amino acid molecule simultaneously carries positive and negative charges. When the amino acid contains one positive and one negative charge, it is a neutral molecule called a **zwitterion**. These forms are the predominant ionic forms of amino acids under neutral conditions (about pH 7 in solution). Under these conditions, the distal carboxyl groups of aspartic and glutamic acids, distal amino group of lysine, guanidine group of arginine, and imidazole cycle of histidine are also more or less ionised. The nitrogen atoms of the imidazole ring of histidine are denoted by *pros* ('near', abbreviated π) and *tele* ('far', abbreviated τ) to show their position relative to the side chain.

In biochemistry, the most common and perhaps most practical classification of proteinogenic amino acids is according to the side-chain polarity and its ionic forms occurring in neutral solutions, which are related to non-bonding interactions in proteins (see Section 7.6.2.2). The following groups of amino acids are recognised:

Table 2.1 Trivial and systematic names of proteinogenic amino acids and their codes.

Trivial name	Systematic name	Three-letter code	One-letter code
Glycine	Aminoacetic acid	Gly	G
L-Alanine	L-2-Aminopropionic acid	Ala	A
L-Valine	L-2-Amino-3-methylbutanoic acid	Val	V
L-Leucine[a]	L-2-Amino-4-methylpentanoic acid	Leu	L
L-Isoleucine[a]	L-2-Amino-3-methylpentanoic acid	Ile	I
L-Serine	L-2-Amino-3-hydroxypropanoic acid	Ser	S
L-Threonine	L-2-Amino-3-hydroxybutanoic acid	Thr	T
L-Cysteine	L-2-Amino-3-mercaptopropanoic acid	Cys	C
L-Selenocysteine	L-2-Amino-3-selanylpropanoic acid	Sec	U
L-Methionine	L-2-Amino-4-methylthiobutanoic acid	Met	M
L-Aspartic acid[a]	L-Aminosuccinic acid	Asp	D
L-Glutamic acid[a]	L-2-Aminoglutaric acid	Glu	E
L-Asparagine[a]	L-2-Amino-4-carbamoylbutanoic acid	Asn	N
L-Glutamine[a]	L-2-Amino-5-carbamoylpentanoic acid	Gln	Q
L-Lysine	L-2,6-Aiaminohexanoic acid	Lys	K
L-Pyrrolysine	(2R,3R)-N6-[3-Methyl-3,4-dihydro-2H-pyrrole-2-ylcarbonyl]-L-lysine	Pyl	O
L-Arginine	L-2-Amino-5-guanidylpentanoic acid	Arg	R
L-Histidine	L-2-Amino-3-(4-imidazolyl)propionic acid	His	H
L-Phenylalanine	L-2-Amino-3-phenylpropionic acid	Phe	F
L-Tyrosine	L-2-Amino-3-(4-hydroxyphenyl)propionic acid	Tyr	Y
L-Tryptophan	L-2-Amino-3-(3-indolyl)propionic acid	Trp	W
L-Proline	L-Pyrrolidine-2-carboxylic acid	Pro	P

[a] In addition to the specific amino acid codes, three-letter code, and one letter-code, placeholders are used for ambiguous amino acids, e.g. Asx and B for aspartic acid or asparagine, Glx and Z for glutamic acid and glutamine, Xle and J for leucine or isoleucine, and Xaa (Unk) and X for an unspecified (unknown) amino acid.

- **hydrophobic amino acids** with non-polar side chains, which include valine, leucine, isoleucine, methionine, phenylalanine, tyrosine, and proline; sometimes the hydrophobic amino acids also include glycine, alanine, and tryptophan, even though these amino acids are rather amphiphilic and form a transition between the hydrophobic amino acids and the following group;

- **hydrophilic amino acids** with polar side chains, which include serine, threonine, cysteine, selenocysteine, aspartic, and glutamic acids and their amides asparagine and glutamine, as well as lysine, pyrrolysine, arginine, and histidine.

Hydrophilic amino acids are classified according to the ionic form in which they occur in living organisms into:

- **neutral** (polar side chain has no electric charge in neutral solutions), which includes most amino acids;

- **acidic** (polar side chain has a negative charge in neutral solutions), which includes aspartic acid and glutamic acid;

- **basic** (polar side chain has a positive charge in neutral solutions), such as lysine, pyrrolysine, arginine, and histidine.

According to the significance in human nutrition, proteinogenic amino acids are divided into:

- **essential** (or indispensable: valine, leucine, isoleucine, threonine, methionine, lysine, phenylalanine, and tryptophan), which cannot be synthesised by the human body;

- **semi-essential** (arginine and histidine);

- **non-essential** (or dispensable: other amino acids that are synthesised by the human body *de novo*; e.g. tyrosine is formed by hydroxylation of the essential amino acid phenylalanine).

Essential amino acids are routine constituents of most dietary proteins and are fairly readily available in a reasonably well-balanced diet. However, there are some amino acids that are present in some protein-based food at lower concentrations than in a high-quality protein (e.g. lysine in wheat versus egg protein). These amino acids are called **limiting amino acids**, because if a person's diet is deficient in one of them, this will limit the usefulness of the others (as well as the extent of protein synthesis in the body), no matter how much of the others is present. To enhance the nutritional value of food and animal feed, limiting amino acids are sometimes used as additives.

In rapidly growing organisms (such as infants), some non-essential amino acids (arginine and histidine), which the young organism cannot synthesise in sufficient quantities (rates), become essential amino acids. These amino acids are sometimes called **semi-essential amino acids**. Histidine was initially thought to be essential only for infants, but longer-term studies established that it may be also essential for adult humans. Cysteine and tyrosine are also sometimes referred to as semi-essential (conditionally indispensable), because if they are not adequately supplied in the diet, indispensable methionine and phenylalanine, respectively, are required in large amounts to synthetise them.

The amino acids that the body can synthesise are called **non-essential amino acids**. They are biosynthesised from intermediates of the citric acid cycle used by all aerobic organisms to generate energy and through other metabolic pathways (glycolysis and the pentose cycle).

Essential and semi-essential amino acids
- **Valine** Valine occurs in animal and plant proteins (meat, cereals) in amounts of 5–7% (average content is 6.9%). Egg and milk proteins contain 7–8% valine. The structural protein elastin has the highest amount (16%).

- **Leucine** Leucine occurs in all common proteins, usually in amounts of 7–10% (average content is 7.5%). Cereals contain a variable amount of leucine, wheat proteins contain about 7%, and maize proteins about 13%. Free leucine is formed in larger amounts during cheese ripening due to bacterial activity.

- **Isoleucine** The largest amounts of isoleucine are found in milk and egg proteins (6–7%), whilst meat and grains contain 4–5% of leucine (average content is 4.6%).

- **Threonine** A rich source of threonine is meat and brewer's yeast. The content in animal proteins (meat, eggs, and milk) is around 5%. In cereals, threonine content is lower (often around 3%), meaning that it sometimes becomes the limiting amino acid.

- **Methionine** Animal proteins contain 2–4% of methionine, whereas plant proteins contain only 1–2% (average content is 1.7%). Methionine is the limiting amino acid in legumes. Methionine (and cysteine) is present only in small amounts in histones and is completely lacking in protamines.

- **Lysine** The average content of lysine in proteins is 7%. High amounts of lysine are found in most animal proteins; meat, eggs, and milk proteins commonly contain 7–9% of lysine, fish and shellfish proteins contain 10–11%. In contrast, vegetable proteins, such as proteins of cereals (especially gliadins but not glutelins) and cereal products contain only 2–4% of lysine, which is the limiting amino acid.

- **Phenylalanine** Good sources of phenylalanine are meat and fish (4–5%, average content is 3.5%). In some individuals, its presence in the diet causes phenylketonuria, an inherited defect in which phenylalanine is incompletely and abnormally metabolised (see Section 10.3.1.2).

- **Tryptophan** The average tryptophan content in proteins is 1.1%. Animal proteins contain 1–2% of tryptophan, except for histones and collagen, which do not contain tryptophan at all. Therefore, tryptophan is not present in gelatine or in acid protein hydrolysates used as soup seasonings. The content of tryptophan (as well as 3-methylhistidine and creatinine) in meat products can then serve as an indicator of the quality of the meat used. Cereals contain <1% of tryptophan; the glutenine fraction of gluten has a somewhat higher content of tryptophan. In fruits and fruit juices, N^1-(β-D-glucopyranosyl)-L-tryptophan is also found. Its amount in juices ranges from 0.1 to >10 mg/l. The highest amount of this glycoconjugate was found in pear juices (13.5 mg/l), whilst it was low in other juices (<0.7 mg/l); it may thus be helpful in the authentication and characterisation of fruit origin. Similarly, a striking difference in the glucoside content was found between the green bean of robusta (40–50 mg/kg) and arabica (below 1.0 mg/kg) coffee species. Humans can partly use tryptophan for biosynthesis of nicotinic acid (see Section 5.8.3) and the neurotransmitter serotonin (see Section 10.3.4). In ruminants, if present in excessive amounts in the diet, tryptophan causes pulmonary oedema (fluid accumulation in the lungs), which may lead to respiratory failure. The cause is the heterocyclic compound 3-methylindole (skatole), which is a fermentation product of tryptophan in the rumen.

- **Arginine** Arginine occurs in all proteins in amounts of 3–6% (average content is 4.7%). Basic proteins protamines from fish roes have particularly high levels. Peanuts and other oilseeds are also rich sources (arginine content of up to 11%).

- **Histidine** Common proteins contain 2–3% of histidine (average content is 2.1%), and blood plasma proteins contain up to 6% of histidine. The flesh of some fish (especially mackerel and tuna) contains from 0.6 to 1.3% (and sometimes more than 2%) of free histidine. The free histidine content in the flesh of other fish is only 0.005–0.05%.

Non-essential amino acids

- **Glycine** Glycine is contained in a significant amount (25–30%) in collagen and gelatine; in the majority of albumins, it is not present at all (average content is 7.5%).

- **Alanine** Alanine occurs in almost all proteins in amounts of 2–12% (average content is 9.0%). The prolamine protein of maize – zein – and animal protein gelatine contain about 9% of alanine.

- **Serine** Serine is found in many proteins, generally in an amount of 4–8% (average content is 7.1%).

- **Cysteine** The highest amount of this amino acid and its oxidation product cystine (**2-4**) is found in keratin (up to 17%); it occurs in many other proteins in smaller amounts (average content is 2.8%). In the organism, cysteine can partially replace the essential amino acid methionine.

- **Aspartic acid and asparagine** The average content of aspartic acid in proteins is 5.5%; the average content of asparagine is 4.4%. Aspartic acid is the major amino acid of animal proteins known as globulins and albumins (6–10%). Vegetable proteins contain 3–13% of aspartic acid, mainly in the form of asparagine (e.g. wheat proteins contain about 4% and maize proteins about 12%).

- **Glutamic acid and glutamine** The average content of glutamic acid and glutamine in proteins is 6.2 and 3.9%, respectively. Glutamic acid is the most abundant amino acid in the nervous tissue. In conventional proteins, both amino acids are usually found in larger quantities (especially in globulins) in cereal and legume proteins (18–40%). Wheat gluten (in its component gliadin) contains about 40%, soy protein about 18%, and milk proteins about 22% of glutamic acid.

- **Selenocysteine** In most foods of both vegetable and animal origin, selenocysteine is the main form of selenium bound in proteins. The content of this amino acid, as well as the contents of other amino acids and peptides containing selenium (L-selenocystine, Se-methyl-L-selenocysteine, L,L-selenocystathionine, L-selenomethionine, and γ-glutamyl-Se-methyl-L-selenocysteine), is unknown. Selenocysteine is typically located in a small number of active centres of proteins of Archaea, bacteria, and eukaryotes (in glutathione peroxidase, thioredoxin reductase, formate dehydrogenase, glycine reductase, and some hydrogenases). The presence of selenocysteine provides much better kinetic and thermodynamic properties of redox-related proteins than cysteine. Usually only one molecule of selenocysteine is bound in the peptide chain; however, proteins containing multiple molecules of this amino acid are also known.

- **Tyrosine** Tyrosine accompanies phenylalanine in most proteins in an amount of 2–6% (average content is 3.5%). Gelatine contains only traces of tyrosine.

- **Proline** Proline is present in most proteins in amounts of 4–7% (average content is 4.6%). Its content in the component of wheat gluten, gliadin, is about 10%, and about 12% of proline is contained in casein. Proline content in gelatine can be up to 13%. In bacteria, plants, and animals, proline also plays a role as an osmoprotectant that helps stabilise proteins and cell membranes against the damaging effect of high osmotic pressure.

2.2.1.1.2 Modified proteinogenic amino acids

Post-translational modification, which extends the range of functions of the protein, is one of the later steps in the biosynthesis of many proteins. Post-translational modification of proteins occurs by methylation, hydroxylation, acetylation, and phosphorylation of the protein functional groups, by attaching various lipids and carbohydrates to the protein molecule, and by making structural changes such as the formation of disulfide bonds from cysteine residues by oxidation.

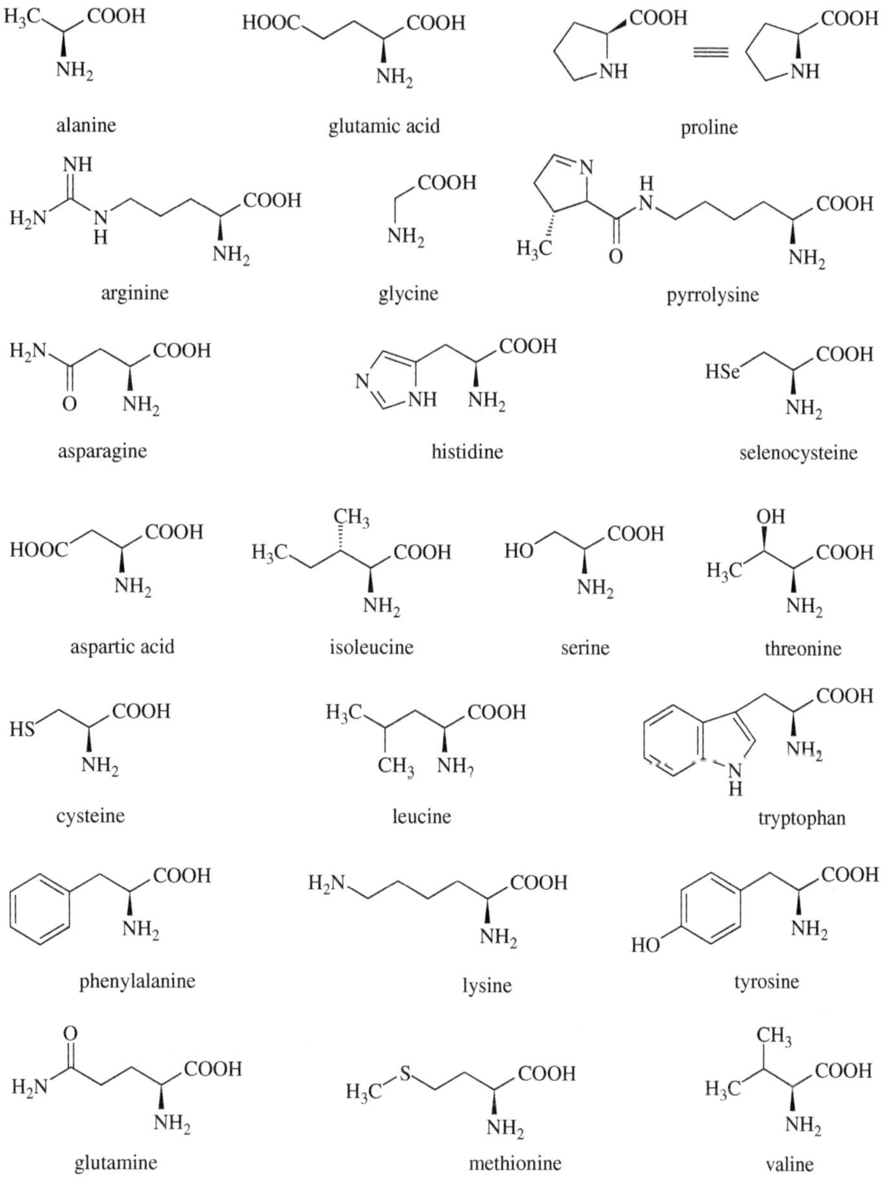

alanine

glutamic acid

proline

arginine

glycine

pyrrolysine

asparagine

histidine

selenocysteine

aspartic acid

isoleucine

serine

threonine

cysteine

leucine

tryptophan

phenylalanine

lysine

tyrosine

glutamine

methionine

valine

2-3, proteinogenic amino acids

Post-translational oxidation of the thiol groups of cysteine residues in proteins yields L-**cystine** (Cys-Cys, **2-4**). The disulfide bond (bridge) plays an important role in the structure of many proteins, as it can combine two different polypeptide chains or two molecules of cysteine in the same peptide chain. For example, β-lactoglobulin, the major globular protein of milk, contains two disulfide bridges, and ovomucoid, the glycoprotein of egg white, has three domains connected by disulfide bridges.

Another common post-translationally modified amino acid is the proline derivative L-4-**hydroxyproline**, (2S,4R)-4-hydroxyproline, also known as L-4-hydroxypyrrolidine-2-carboxylic acid (abbreviated Hyp, **2-4**), which is an important structural component of collagen, gelatine (about 12% of content), and the polypeptide (glycopeptide) of plant cell walls known by the trivial name extensin (see Section 4.5.6.7.2). Its content is low in most other proteins. The amount of 4-hydroxyproline in meat products therefore correlates with the quality of the meat, for instance where products contain skin, wherein collagen is the major protein. In plants, 4-hydroxyproline occurs in smaller amounts. For example, concentrations ranging from 1.0 to 4.2 mg/kg have been found in the edible part of bergamot fruits (*Citrus bergamia*, Rutaceae).

4-Hydroxyproline in collagen is accompanied by a small amount of its isomer, L-3-hydroxyproline (L-3-hydroxypyrrolidin-2-carboxylic acid), and by a hydroxy derivative of lysine, also known as L-5-**hydroxylysine**, (2S,5R)-5-hydroxylysine, L-5-hydroxy-2,6-diaminohexanoic acid (abbreviated Hyl, **2-4**). In glycoproteins, hydroxylysine is bound as *O*-glycoside. Small amounts of free hydroxylysine occur in plants, for example in alfalfa forage (*Medicago sativa*, Fabaceae).

The minor amino acid typically present in the meat myofibrillar protein actin (also in some myosin isoforms and in dipeptide anserine, see Section 2.3.3.1.3) is L-3-**methylhistidine** (**2-4**), which is formed by methylation of histidine bound at the 73rd position of the protein chain. The functional significance of this modification of actin is not known; it is probably related to the metabolism of phosphates, with which the side chain of methylhistidine interacts. Methylhistidine does not occur in other protein-rich foods, such as milk, eggs, and soybeans. Its contents might therefore serve as a criterion for determining the quality of meat products.

Another modified amino acid is L-serine. It can be esterified with phosphoric acid. This ester, **O-phosphoserine** (**2-4**), occurs in many proteins, such as glycophosphoprotein phosvitin (also known as phosphovitin) from egg yolk (see Section 2.4.5.3.2). Phosvitins are one of the most phosphorylated (10%) proteins in nature, and are important for sequestering cations of calcium, iron, and other metals for the developing embryo. Serine in phosvitin represents nearly 50% of all amino acids, and about 90% of it is phosphorylated. Phosphoserine also occurs in the glycoprotein α_{S1}-casein (eight phosphoserine residues) and β-casein (five phosphoserine residues). Phosphoserine (as phospatidyl-L-serine) is also a component of phospholipids, distributed widely amongst animals, plants, and microorganisms. It usually constitutes less than 10% of total phospholipids, the greatest concentration being in myelin from brain tissue and in wheat germ. Threonine residues in proteins are esterified fairly frequently, with phosphoric acid yielding **O-phosphothreonine**.

cystine   4-hydroxyproline   5-hydroxylysine

3-methylhistidine   *O*-phosphoserine

2-4, modified proteinogenic amino acids

2.2.1.2 Other amino acids

Plants and microorganisms are able to generate all 21 amino acids necessary for food protein synthesis, and can additionally synthesise many more. Foods contain numerous less common amino acids in addition to proteinogenic and modified proteinogenic amino acids. It is estimated that there are around 700 amino acids known in nature, of which at least 300 are found in plants. They are often bound in peptides (see Section 2.3) or are present as free amino acids. In biochemistry, these **non-protein amino acids** are often classified as the so-called secondary metabolites, as they are the products of three major routes: modification of an existing (often proteinogenic) amino acid, modification of an existing pathway, and a novel pathway. These amino acids become the biosynthetic precursors of many biologically active nitrogenous compounds, such as signalling molecules, vitamins, alkaloids, bile acids, and pigments. Some of these amino acids also have specific functions in organisms: they act as neurotransmitters, stimulants, and hormones. For example, phenylalanine (or 3,4-dihydroxyphenylalanine) is a precursor of adrenal hormones catecholamines (see Section 10.3.4) and the thyroid hormone thyroxine (see Section 2.2.1.2.5). Other amino acids increase plant tolerance to abiotic stress factors, act as toxic substances that protect plants against invading viruses, microorganisms, other plants, and predators (e.g. the arginine analogue canavanine found in some legumes), and serve as a storage and transport form of nitrogen and sulfur. Some of the more unusual amino acids are formed secondarily during storage of raw materials and food processing via the activities of microorganisms and by chemical transformation of proteins (e.g. lysinoalanine), peptides, and free amino acids.

The following sections present some of the most important non-protein amino acids in foods and feeds. For clarity, they are sorted according to the structure in which they are present in living organisms, and not by function (their functions are often not well known).

2.2.1.2.1 Neutral aliphatic and alicyclic amino acids

3-Amino acids and 4-amino acids In addition to α-amino acids, β- and γ-amino acids can also be found in food. Naturally occurring β-alanine (3-aminopropionic acid, **2-5**) acts in the biosynthesis of pantothenic acid, acetyl coenzyme A, other acyl coenzymes A (see Section 5.9.1), and some histidine dipeptides. β-Alanine is synthesised in several different ways, including by aspartic acid decarboxylation (in bacteria), by the transformation of propionic acid (in bacteria and some plants), by degradation of the polyamines spermine and spermidine (in yeasts and many plants, such as tomato, soybean, and maize), and via the pyrimidine derivative uracil that occurs in all animals and in some plants, such as wheat.

The higher homologue of β-alanine is γ-aminobutyric acid (4-aminobutanoic acid), known by the acronym of GABA (**2-5**), which is formed almost exclusively by enzymatic decarboxylation of glutamic acid. It is mainly present in the brain tissue of animals, where it acts as an inhibitor of nerve impulse transmissions. It is also found in plants and microbial cells, where it has a role as a signal molecule, and is a constituent of some peptides, such as nisine. High levels of GABA accumulate in plant tissues under various stresses.

β-alanine γ-aminobutyric acid

2-5, the most common 3-amino acid and 4-amino acid

N-Substituted amino acids N-Alkylsubstituted amino acids are commonly found in foods. The simplest is N-methylglycine (sarcosine, **2-6**), which is a product of the catabolism of choline (see Section 3.5.1.1.1) and creatine (see Section 2.2.1.2.4); it is also formed by N-methylation of glycine. Sarcosine probably has a regulatory function in methylation reactions. The subsequent N-methylation of sarcosine gives N,N-dimethylglycine (**2-6**), which is a component of pangamic acid. N,N,N-Trimethylglycine (N,N,N-trimethylammonioacetate, **2-6**) was named glycine betaine or just betaine due to its discovery in sugar beet (*Beta vulgaris,* Amaranthaceae), where it occurs in the amount of about 2.5 g/kg in the root. Glycine betaine is a very effective osmolyte of microorganisms and higher plants. It is formed from choline or by N-methylation of glycine or sarcosine, and acts in various transmethylation reactions. It is present in higher amounts in sugar beets and molasses, and accumulates at high levels in some higher fungi; for example, its content in horse mushroom (*Agaricus arvensis*) is 7.8% dry matter.

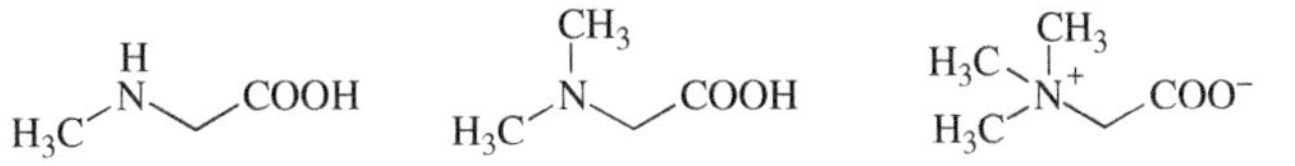

N-methylglycine N,N-dimethylglycine N,N,N-trimethylglycine

2-6, examples of N-methylsubstituted amino acids

The term **betaine** includes not only glycine betaine but also all zwitterionic ammonium compounds derived from other amino acids. By extension, betaines are also neutral molecules having charge-separated forms with an onium atom that bears no hydrogen atoms (e.g. phosphonium, sulfonium, arsonium ions) and that is not adjacent to the anionic atom. Examples of the most common onium ions are given by the general formulae **2-7**. The attached groups are typically organic substituents, such as methyl groups. The most common betaines are quaternary ammonium compounds that occur in foods of animal and plant origin. An arsenic analogue of betaine, arsenobetaine, accumulates in the body of some fish (see Section 6.2.3.1). Betaines are particularly ubiquitous in plants. They are generally referred to as osmolytes, as they, like their amino acid precursors, tend to accumulate in the cytoplasm and intercellular fluids, where they exert protective functions for proteins, nucleic acids, and cell membranes in response to abiotic stresses such as low, freezing, or high temperatures, the presence of toxic metals, the reduced availability of water, and high salinity. The presence and accumulation of betaines in plants is a species-specific process. For example, the homostachydrine content in robusta coffee beans (31 mg/kg) is much higher than in arabica beans (1.5 mg/kg); it can thus be considered an authentication marker. Some of the major betaines are listed in Table 2.2. Their structures are given by formulae **2-8**.

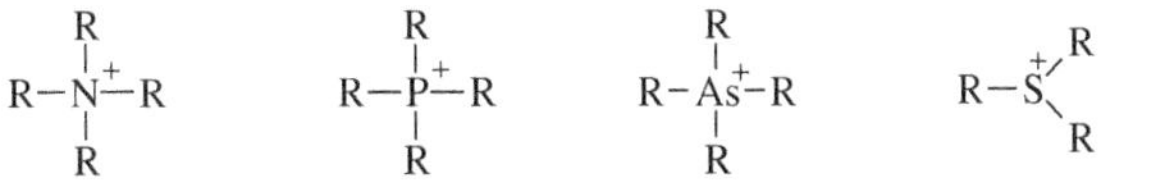

ammonium ion phosphonium ion arsonium ion sulfonium ion

2-7, general structures of some onium ions

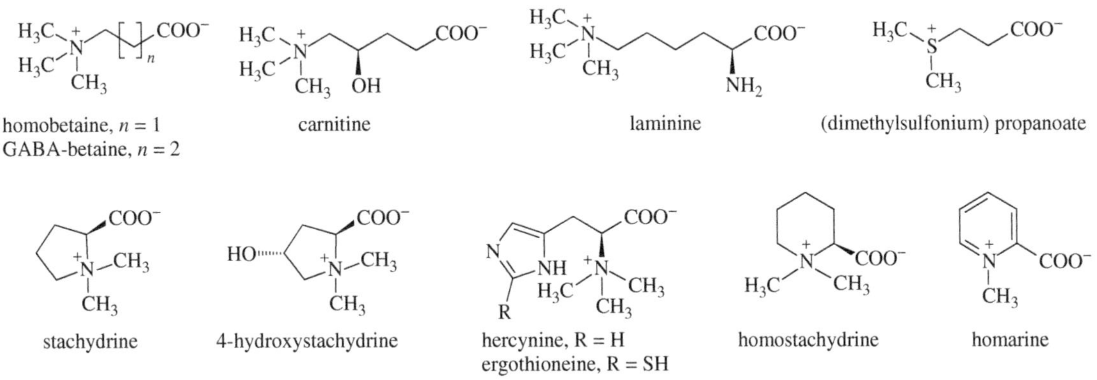

homobetaine, n = 1 carnitine laminine (dimethylsulfonium) propanoate
GABA-betaine, n = 2

stachydrine 4-hydroxystachydrine hercynine, R = H homostachydrine homarine
ergothioneine, R = SH

2-8, structures of selected betaines

Table 2.2 Some important betaines.

Betaine	Initial amino acid	Occurrence
β-Alaninebetaine (homobetaine)	β-Alanine	Plants of the Plumbaginaceae family, citrus species, meat
GABA-betaine	γ-Aminobutyric acid	Molluscs, citrus species
L-Carnitine	Lysine and methionine	Meat, dairy products, legumes, vegetables, higher fungi
Laminine (lysine betaine)	Lysine	Higher plants
Stachydrine (cadabine, proline betaine)	Proline	Higher plants, citrus species, vegetable oils, fungi
Homostachydrine (pipecolic acid betaine)	Pipecolic acid	Coffee beans, citrus species, some higher plants
Betonicine (4-hydroxystachydrine, 4-hydroxyproline betaine)	4-Hydroxyproline	Plants of the Lamiaceae family
Hercynine (histidine betaine)	Histidine	Higher fungi, citrus species
L-Ergothioneine (thiol form)	2-Mercaptohistidine	Higher fungi, higher plants
Homarine	Pyridine-2-carboxylic acid	Higher plants
(Dimethylsulfonium)propanoate	Methionine	Marine phytoplankton, seaweeds

A group of *N*-substituted amino acids is the *N*-acylamino acids. An important metabolite is hippuric acid, also known as *N*-benzoylglycine (**2-9**). Hippuric acid arises from glycine as a result of detoxification of benzoic acid and some other aromatic acids in humans and animals. In higher concentrations, it occurs in the urine of herbivores and reportedly inhibits some pathogenic bacteria in the urinary tract. Benzoic acid is found (usually in small amounts) in many plant materials, including fruit, vegetables, and forage crops (see Section 8.2.6.1.6), and is used as a food preservative (see Section 11.2.1.1.1). Cows partially excrete hippuric acid into their milk, up to a concentration of 60 mg/kg. Microorganisms used in the production of fermented dairy products hydrolyse hippuric acid to glycine and benzoic acid. For example, the content of benzoic acid in yoghurts can reach, on average, 15 mg/kg.

N-Phenylprop-2-enoyl amino acids (**2-9**) have been identified as the key contributors to the astringent taste of non-fermented cocoa beans and cocoa products. Besides the already known (*E*)-*N*-[3′,4′-dihydroxycinnamoyl-3-hydroxy-L-tyrosine (known as clovamide), (*E*)-*N*-(4′-hydroxycinnamoyl)-L-tyrosine (deoxyclovamide), and (*E*)-*N*-(3′,4′-dihydroxycinnamoyl)-L-tyrosine, seven additional amides derived from cinnamic, 4-coumaric, caffeic, and ferulic acids were recently identified. At light exposure, (*E*)-isomers of *N*-phenylpropenoyl amino acids rapidly convert into the corresponding (*Z*)-isomers.

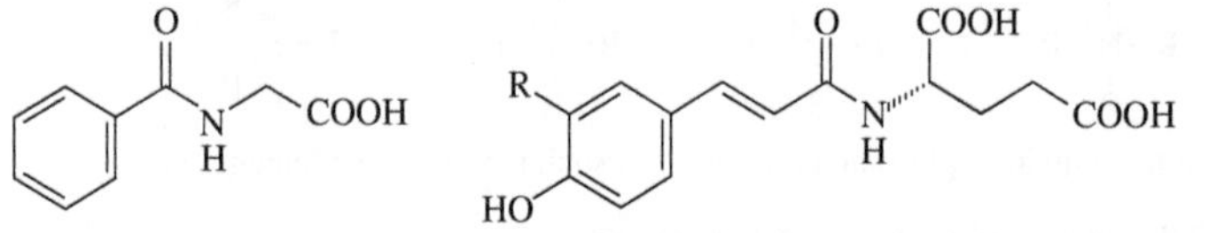

hippuric acid

(*E*)-*N*-(4′-hydroxycinnamoyl)-L-glutamic acid, R = H
(*E*)-*N*-(3′,4′-dihydroxycinnamoyl)-L-glutamic acid, R = OH

2-9, *N*-acylamino acids

Alicyclic amino acids A common neutral amino acid derived from cyclopropane is 1-aminocyclopropane-1-carboxylic acid. The precursor of this carboxylic acid is *S*-adenosyl-L-methionine, which is derived from methionine (Figure 2.1). 1-Aminocyclopropane-1-carboxylic acid is present in apples, pears, and other fruits and serves as a precursor of ethylene (ethene) in essentially all tissues of vascular plants and also in fungi and algae. Ethylene is a signalling molecule in plants (a hormone) that regulates seed germination, plant growth, flowering, ripening of fruits, leaf abscission, and ageing. Plants also synthesise ethylene in various stressful situations (mechanical damage, frost damage, flooding, pathogen attack, and the presence of heavy metals). Production of ethylene is connected with the formation of hydrogen cyanide, a process known as cyanogenesis (see Section 10.3.2.3).

The lychee seed and fruit (*Litchi chinensis*) from the soapberry family (Sapindaceae), native to southern China and Southeast Asia, contains an unusual amino acid L-α-(methylenecyclopropyl)glycine, also known as (2*S*,3*S*)-2-(methylenecyclopropyl)glycine (**2-10**).

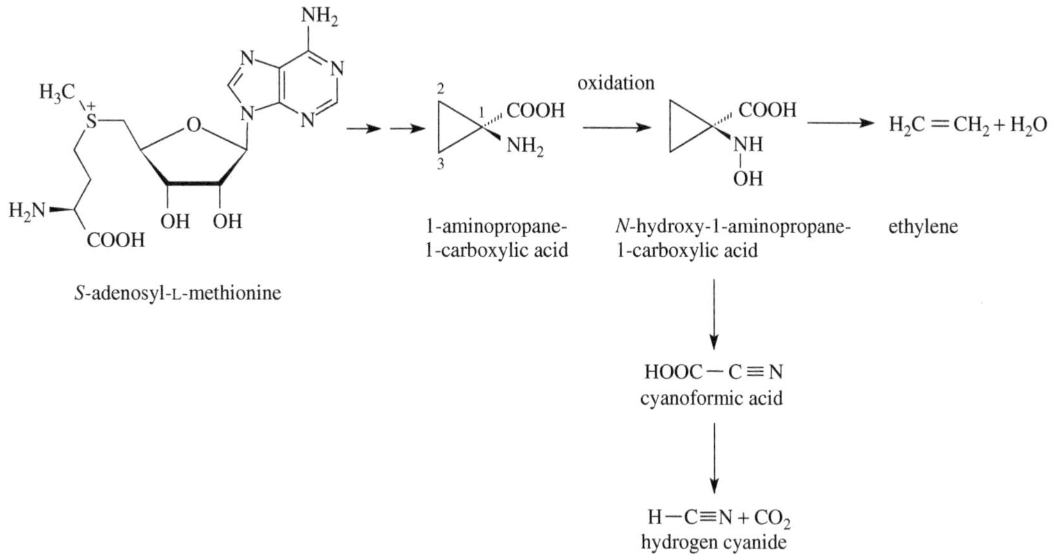

Figure 2.1 Formation of ethylene and hydrogen cyanide in plants.

Lychee seeds contain 1.8 mg/kg 2-(methylenecyclopropyl)glycine, whilst the unripe and ripe arils (the edible white parts of fruits) contain 82–220 mg/kg and 45–68 mg/kg (f.w.) 2-(methylenecyclopropyl)glycine, respectively. Lychee fruits also contain its higher homologue L-3-(methylenecyclopropyl)alanine, better known as hypoglycin A or hypoglycin (**2-10**). The concentrations of hypoglycin in the unripe and ripe arils are 19–152 and 12–74 mg/kg, respectively. The active principle in mammalian digestion after ingestion of hypoglycin is methylenecyclopropylacetyl-CoA, which interferes with the metabolism of branched-chain amino acids. The same metabolite also occurs in lychee seeds.

2-(Methylenecyclopropyl)glycine (**2-10**), together with derived γ-glutamyl dipeptide and (2S,1′S,2′S)-2-(2′-carboxycyclopropyl)glycine (**2-10**), occurs in the seeds of ackee fruit (*Blighia sapida*) from the Sapindaceae family as lychee, native to tropical West Africa. Hypoglycin A (**2-10**) occurs in this fruit and its seeds as a mixture of (2S,4R)- and (2S,4S)-diastereomers. The former is the dominant compounds in the arils. Ackee fruit also contains the γ-glutamyl dipeptide of hypoglycin A, known as hypoglycin B (**2-10**). The amount of hypoglycin A in green arils is 6.7–7.9 g/kg and decreasess to 0.27–0.55 g/kg in the ripe fruit. On the contrary, the content of hypoglycin B in the seeds of unripe fruit is 1.6–3.1 g/kg, whilst that in seeds of ripe fruit ranges from 11.7 to 12.6 g/kg. The strong inverse relationship demonstrated that hypoglycin B in the seeds serves as a sink for hypoglycin A from the ripening arils and is thereby involved in the detoxification mechanism of the fruit. Consumption of improperly ripened ackee often results in fatalities; the toxic effect is known as Jamaican vomiting sickness (see Section 10.3.5.3), manifested by vomiting, drowsiness, fatigue, and hypoglycaemia. Higher concentrations of hypoglycins A and B also occur in seeds of the common sycamore (*Acer pseudoplatanus*, Aceraceae). For example, the U.S. Food and Drug Administration and Health Canada set the maximum permissible level of hypoglycine A to 100 mg/kg.

An important alicyclic amino acid in some genera of tropical plants of the Flacourtiaceae, Passifloraceae, and Turneraceae families is cyclopentenylglycine, occurring as a mixture of two stereoisomers, predominant (2S,1′R)-2-(cyclopent-2′-en-1′-yl)glycine (**2-10**) and (2S,1′S)-2-(cyclopent-2′-en-1′-yl)glycine. Cyclopentenylglycine is a precursor of unusual fatty acids (see Section 3.3.1.4.3) and cyanogenic glycosides occurring in these plants (see Section 10.3.9.1).

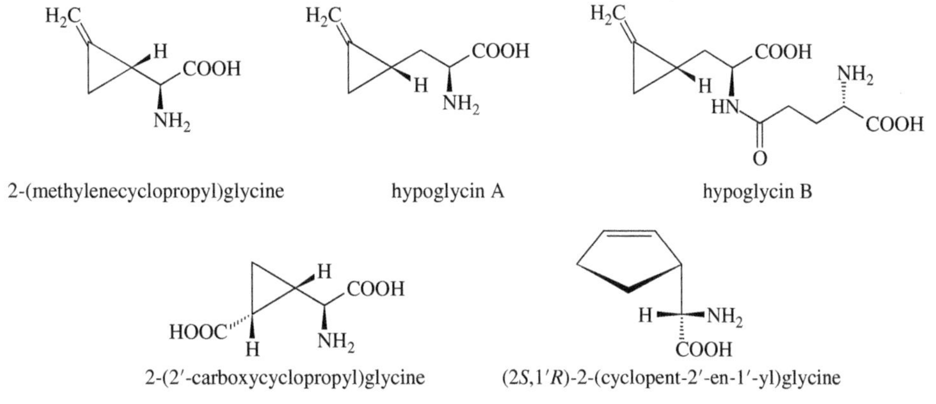

2-(methylenecyclopropyl)glycine　　hypoglycin A　　hypoglycin B

2-(2′-carboxycyclopropyl)glycine　　(2S,1′R)-2-(cyclopent-2′-en-1′-yl)glycine

2-10, alicyclic amino acids

Hydroxyamino acids Various legumes (seeds of the Fabaceae family plants) are rich sources of free aliphatic hydroxyamino acids. Common amino acids such as 4-hydroxyleucine, 4-hydroxyisoleucine, 4-hydroxynorvaline, and 5-hydroxynorleucine are hydroxyderivatives of branched-chain amino acids, L-leucine (2-3) and L-isoleucine (2-3), and their homologues L-norvaline (2-11) and L-norleucine (2-11). Another common amino acid in legumes is L-homoserine (2-11), which is derived from 2-aminobutyric acid, the intermediate in the biosynthesis of threonine from aspartic acid. *O*-Acyl- and *O*-amino derivatives of homoserine are also common. An example of an *O*-aminoderivative of homoserine is a toxic amino acid known as canaline (see Section 2.2.1.2.4). The most common source of this amino acid is the jack bean (*Canavalia ensiformis*).

Peptides of some *Amanita* species (see Section 10.3.7.2) contain (2*S*,4*R*)- and (2*S*,4*S*)-isomers of 4-hydroxyleucine and some other hydroxysubstituted amino acids, including amino acids with unsaturated side chains. 4-Hydroxyisoleucine is found in fenugreek (*Trigonella foenum-graecum*, Fabaceae) seeds in an amount of 5–12 g/kg (represents about 80% of total free amino acids). These seeds are known in traditional medicine for their antidiabetic properties. The absolute configuration of the major isomer was shown to be (2*S*,3*R*,4*S*)-2-amino-4-hydroxy-3-methylpentanoic acid, also known as (4*S*)-4-hydroxy-L-isoleucine (2-11), whilst the minor isomer configuration is (4*S*)-4-hydroxy-D-isoleucine. In the fruiting body of an edible mushroom *Hypsizygus marmoreus* (white beech mushroom), (2*R*,3*S*)-2-amino-3,4-dihydroxybutanoic acid was found in relatively high concentrations (1.3 mg/g).

2-11, homologous branched amino acids and hydroxyamino acids

2.2.1.2.2 Sulfur and selenium amino acids

Cysteine derivatives Foods of plant origin contain numerous sulfur amino acids, mostly formally derived from cysteine and their higher homologues, such as L-homocysteine (2-12). The unusual cysteine derivative L,L-cystathionine (2-12) is an intermediate of the biosynthesis of methionine and other amino acids. The toxic amino acid L,L-djenkolic acid, also known as 3,3′-(methylenedithio)di-L-alanine, consists of two cysteine residues joined by a methylene group between the sulfur atoms (2-12). It was first identified in the urine of Indonesians consuming tropical legumes known as jengkol or jering (*Archidendron pauciflorum*, syn. *Pithecellobium lobatum*, Fabaceae). Cysteine (or cystine) is also the precursor of dehydroalanine (2-aminoacrylic acid), from which the amino acid lanthionine (see 2-33) is formed. Dehydroalanine is also a component of the microbial peptide nisin (see Sections 2.3 and 11.2.1.2.1) and a precursor of the other unusual amino acid lysinoalanine. Lysinoalanine and its analogues can be generated as cross-links on heating of proteins and, especially, in proteins after treatment in alkaline solutions (as occurs, for example, in soybean processing). Digestive enzymes do not hydrolyse these cross-links. Other sulfur-containing amino acids include *N*-acetyl-*S*-substituted cysteines (or mercapturates, 2-12), which are formed as products of metabolic detoxification of xenobiotics, with glutathione following by hydrolysis of the intermediates.

2-12, less common sulfur amino acids and cysteine derivatives

Plants also synthesise a number of *S*-cysteine (and *S*-glutathione) conjugates that are released by the action of L-cysteine-*S*-conjugate thiol-lyases (deaminating) to produce various volatile flavour-active thiols in fruits and vegetables (see Section 8.2.9.1.2).

Very important cysteine derivatives are *S*-alk(en)yl-L-cysteines (2-12), which accompany, in small amounts, the predominant *S*-alk(en)yl-L-cysteine *S*-oxides, also called *S*-alk(en)yl-L-cysteine sulfoxides (2-13). Owing to the free electron pair on the sulfur atom, the latter amino acids may occur in two optical isomers. In vegetables, only (+)-*S*-isomers, also known as (*S*$_S$)-isomers, occur. However, isoalliin is an exception. It occurs as an (*R*$_S$)-isomer because of the priority of the other substituent. The bond between the sulfur and oxygen atoms in *S*-alk(en)yl-L-cysteine sulfoxides differs from the double bond between carbon and oxygen. As a result, the molecule has a dipolar character, with the negative charge centred on the oxygen and the sulfur atom being a chiral centre. These amino acids originate from γ-glutamyl S-alk(en)yl-L-cysteine storage compounds via S-alk(en)yl-L-cysteines (**2-12**) and are the precursors of many biologically and flavour-active compounds (see Section 8.2.9.1.4). Particularly rich sources of these amino acids are some genera of the plant families Brassicaceae (the genus *Brassica*), Amaryllidaceae (subfamily Allioideae, genera *Allium* and *Tulbaghia*), and Fabaceae (*Phaseolus*, *Vigna*, and *Acacia*). S-Alk(en)ylcysteine derivatives also occur in plants belonging to the families Olacaceae (*Scorodocarpus*) and Phytolaccaceae (*Petiveria*), in some higher fungi such as mushrooms in the genera *Marasmius*, *Collybia*, and *Lentinula* (known as shiitake), and in brown (*Undaria*) and red (*Chondria*) seaweeds.

alk(en)ylcysteine sulfoxides   marasmin

S-(pyridin-2-yl)cysteine sulfoxide   *S*-(pyrrol-3-yl)cysteine sulfoxide

2-13, *S*-substituted L-cysteine sulfoxides

The basic member of the homologous series, *S*-methylcysteine and its sulfoxide, (R_C,S_S)-*S*-methylcysteine sulfoxide, also known as (+)-*S*-methyl-L-cysteine sulfoxide or methiin (**2-13**, R = CH$_3$), occurs in cabbage, garlic, onion, and other *Brassica* and *Allium* vegetables, and also in beans. The concentration is roughly 0.2–0.7 g/kg in fresh Brassica vegetables. The methiin levels in *Allium* vegetables are given in Table 2.3. Note that methiin is a toxic amino acid to ruminants (see Section 10.3.5.1). (R_C,S_S)-*S*-Carboxymethylcysteine sulfoxide (**2-13**, R=CH$_2$COOH) occurs, for example, in radishes (*Raphanus* spp., Brassicaceae). Some species of *Brassica* plants and some members of the genus *Allium* (Amarillidaceae), such as elephant garlic, onion, shallot, leek, and chive, contain traces of ethiin; that is, (R_C,S_S)-*S*-ethylcysteine sulfoxide (**2-13**, R=CH$_2$CH$_3$) (Table 2.3).

Other related amino acids are (R_C,S_S)-*S*-allylcysteine (deoxyalliin) and (R_C,S_S)-*S*-(prop-2-en-1-yl)cysteine sulfoxide, which is known as alliin (**2-13**, R=CH$_2$CH=CH$_2$). Alliin is the major sulfur-containing amino acid of garlic (Table 2.3). Its isomer, ($R_C,R_S,1E$)-*S*-(prop-1-en-1-yl)cysteine sulfoxide, or isoalliin (**2-13**, R=CH=CHCH$_3$), is the major free amino acid of onions. It is also found in smaller amounts in garlic (Table 2.3). (R_C,S_S)-*S*-Propylcysteine sulfoxide (propiin, **2-13**, R=[CH$_2$]$_2$CH$_3$) occurs in onions, shallots, and chives and in trace amounts in some garlic species, for example in bulbs of *Allium vineale* (0.01 mg/kg f.w.), *A. ursinum* (0.02 mg/kg f.w.), and *Allium triquetrum* (0.09 mg/kg f.w.). Its higher homologue butiin, (R_C,S_S)-*S*-butylcysteine sulfoxide, R=[CH$_2$]$_3$CH$_3$ (**2-13**), occurs in some species of the genus *Allium*, in bulbs of *A. siculum* (0.3 g/kg f.w.) and *A. tripedale* (5.7 g/kg f.w.), together with ($R_C,S_S,1E$)-*S*-(but-1-en-1-yl)cysteine sulfoxide, R=CH=CHCH$_2$CH$_3$ (**2-13**), known as homoisoalliin. The concentrations found in fresh bulbs of the ornamental plants *Al. siculum* and *A. tripedale* were 3.5 and 1.2 g/kg, respectively. The important cysteine sulfoxide (R_S,R_C)-*S*-(methylthiomethyl)cysteine-4-oxide, known as marasmin (**2-13**), which carries a methylthiomethyl moiety as its aliphatic residue, was found in South African society garlic (*Tulbaghia violacea*) and in wild (woodland) garlic (*Tulbaghia alliacea*), in *A. stipitatum* and *A. suworowii* (*Allium* subgenus *Melanocrommyum*). Its γ-glutamyl derivative, γ-glutamyl-(S_S,R_C)-marasmin, was found in various mushroom species belonging to the genus *Marasmius* (such as *M. alliaceus*). Marasmin is the precursor of the thiosulfinate marasmicin, which shows antifungal and tuberculostatic activities. (R_C,S_S)-*S*-(Pyridin-2-yl)cysteine sulfoxide (**2-13**) and several pyridyl compounds, which were formed from this amino acid by the action of

Table 2.3 Concentrations of alk(en)yl-L-cysteine sulfoxides in *Allium* vegetables.

Substituent	Concentration in g/kg fresh weight			
	Garlic (*Allium sativum*)	Onion (*Allium cepa*)	Leek (*Allium porum*)	Chive (*Allium schoenoprasum*)
Methyl	0.05–4.24	0.22	0.04	0.32
Ethyl	0–trace	0.05	trace	0.01
Propyl	0–trace	0.06	trace	0.07
Prop-2-en-1-yl (allyl)	1.10–10.5	Trace–0.05	trace	0.02
Prop-1-en-1-yl	Trace–1.11	0.50–1.31	0.18	0.31

alliinase, were identified in bulbs of the drumstick onion *A. stipitatum* (*Allium* subgenus *Melanocrommyum*), which is used as a crop plant and in folk medicine in Central Asia. (R_C,S_S)-*S*-(Pyrrol-3-yl)cysteine sulfoxide is found in another member of the subgenus *Melanocrommyum*, *A. giganteum*.

To date, six *S*-alk(en)ylcysteine sulfoxides, namely *S*-allyl-, (*E*)-*S*-(prop-1-en-1-yl)-, *S*-methyl-, *S*-propyl, *S*-ethyl, and *S*-butylcysteine sulfoxides, have been found in common vegetables of the genus *Allium*; the latter two are present only in trace amounts. It has been well documented that the total contents and relative proportions of the individual *S*-alk(en)ylcysteine sulfoxides in alliaceous plants are significantly affected by a number of genetic and environmental factors (e.g. plant species and variety, climatic conditions, sulfur content in the soil, irrigation, fertilisation, and harvest date, amongst other things) but are usually around 0.1–0.8%, which represents more than 10% of the crude protein content. The contents also vary considerably over the growth period. For example, in garlic, the highest amounts are found in spring, in the green parts of the plant, which are the most important sites of cysteine sulfoxide biosynthesis, although some biogenesis may also occur in the bulbs. The cysteine sulfoxide levels in the bulbs start to increase dramatically approximately five weeks before harvest. This accumulation in bulbs may be associated with the possible role of cysteine sulfoxides as storage compounds for sulfur and nitrogen during dormancy.

Some vegetables also contain alk(en)ylthio-substituted L-cysteine sulfoxides in small amounts. For example, (R_C,S_S)-*S*-methylthiocysteine sulfoxide (**2-13**, R = SCH_3), (R_C,S_S)-*S*-propylthiocysteine sulfoxide (**2-13**, R = $SCH_2CH_2CH_3$), and (R_C,S_S,1*E*)-*S*-(prop-1-en-1-ylthiocysteine sulfoxide (**2-13**, R = SCH=CHCH$_3$) occur in onions at concentrations of 0.19, 0.01, and 0.56 g/kg (f.w.), respectively. These amino acids and the corresponding peptides (see Section 2.3.3.1.2) are formed by the reaction of thiosulfinates (see Section 8.2.9.1.4) with cysteine residues.

An interesting sulfur amino acid derived from cysteine is 2-amino-4,6,8,10,10-pentaoxo-4,6,8,10-tetrathiaundecanoic acid, also known as deglutamyllentinic acid (**2-13**, R=CH$_2$S(=O)CH$_2$S(=O)CH$_2$SO$_2$CH$_3$, which is formed from the corresponding γ-glutamyl peptide known as lentinic acid (see Section 8.2.12.9.7) through the action of γ-glutamyl transpeptidase and becomes the precursor of the characteristic sulfur compound lenthionine in shiitake mushrooms (*Lentinula edodes*).

Methionine derivatives A common amino acid in *Brassica* vegetables, *S*-methyl-L-methionine, was formerly classified as vitamin U (see Section 5.15). Higher homologues of methionine (**2-14**), such as L-homomethionine (also known as L-5-methylthionorvaline), L-dihomomethionine, and L-trihomomethionine up to L-hexahomomethionine, are starting compounds for the biosynthesis of many important glucosinolates (see Section 10.3.10). For example, homomethionine is the precursor of sinigrin and glucoibervirin, whilst dihomomethionine is the precursor of progoitrin and gluconapin.

L-homomethionine   L-dihomomethionine   L-hexahomomethionine

2-14, selected methionine homologues

Selenium amino acids *Allium* vegetables and various other plants contain small amounts of amino acids derived from cysteine and methionine, in which sulfur is replaced by selenium. These amino acids represent the main organic form of selenium and, together with *Se*-containing proteins, represent the main selenium source in foods (see Section 6.2.3.1). Examples are *Se*-alk(en)ylcysteines and their γ-glutamyl derivatives.

2.2.1.2.3 Acidic amino acids and their amides

Most non-protein acidic amino acids and their amides are structurally related to L-glutamic acid and L-glutamine (**2-15**). For example, sea pea seeds (*Lathyrus maritimus*, Fabaceae) contain L-4-methylglutamic acid, that is, (2*S*,4*R*)-2-amino-4-methylpentanedioic acid, whilst the seeds of peanuts (*Arachis hypogaea*, Fabaceae) contain L-4-methyleneglutamic acid together with L-4-methyleneglutamine. (2*S*,3*R*)-2-Amino-3-hydroxyglutaric (*threo*-3-hydroxy-L-glutamic) acid is a precursor of ibotenic acid and other toxic compounds in fly agaric (*Amanita muscaria*) and panther cap (*Amanita pantherina*). It is also a precursor of the tricholomic acid found in some agarics (*Tricholoma muscarium*) (see Section 10.3.7.3.1). The so-called false rhubarb (*Rheum rhaponticum*), red currant (*Ribes silvestre*, Grossulariaceae), garden cress (*Lepidium sativum*, Brassicaceae), and other plant materials contain L-3,4-dihydroxyglutamic acid, that is, (2*S*,3*S*,4*S*)-2-amino-3,4-dihydroxypentanedioic acid (**2-15**). Plants also contain other stereoisomers of these amino acids and many other related compounds that exhibit a range of biological effects.

N-Ethyl-L-glutamine, also known as L-theanine or L-ethanine (**2-15**), is a unique major free amino acid in tea leaves. L-Glutamic acid, a precursor of L-theanine, is present in most plants, whilst ethylamine, another precursor of L-theanine, specifically accumulates in *Camellia* species, especially *C. sinensis*. Depending on the variety of Chinese tea (*C. sinensis*, Theaceae), its content in the leaves is normally around 1.3% dry matter (0.7–2.0%); this is higher in green tea and tea of higher quality (0.8–2.7% in green tea, 0.6–2.1% in black tea, 2.0–9.2%

in oolong tea, 0.7–1.8% in decaffeinated black tea). Tea extracts contain about 3% of theanine. The theanine content is used to detect tea extracts in products based on tea (instant and iced tea). Theanine has neuromuscular sedative effects with no side effects and is used in some special food supplements.

4-methylglutamic acid   4-methyleneglutamic acid   3-hydroxyglutamic acid

3,4-dihydroxyglutamic acid   theanine

2-15, derivatives of glutamic acid and glutamine

The fruiting bodies of common mushrooms, *Agaricus bisporus* (also known as button mushrooms, white mushrooms, table mushrooms, and champignon mushrooms), contain a mutagenic hydrazine derived from glutamic acid, viz. β-N-(γ-L-glutamyl)-4-(hydroxymethyl) phenylhydrazine, known as L-agaritine, at concentrations of 100–1700 mg/kg. The metabolic fate of agaritine has been linked with the carcinogenity of this mushroom (see Section 10.3.7.4.1).

2.2.1.2.4 Basic amino acids and related compounds

Plants contain various basic diaminomonocarboxylic acids that are homologues of lysine or derivatives of these homologues. Arginine derivatives and their homologues are also relatively common.

The lowest member of the homologous series of diaminomonocarboxylic acids is L 2,3-diaminopropionic acid (**2-16**), occurring in plants of the genus *Mimosa* of the legume family Fabaceae. The toxic N^3-methyl derivative occurs in the cycads (plants of the cycas family Cycadaceae), whilst the N^2-oxalyl derivative is found in vetchlings (*Lathyrus* spp.) and vetches (*Vicia* spp.) that belong to the legume family Fabaceae. Together with L-2,4-diaminobutyric acid (**2-16**), this compound is the originator of the neurodegenerative disease neurolathyrism (see Section 10.3.5.1).

A common basic amino acid is the lower homologue of lysine, L-ornithine (L-2,5-diaminovaleric acid, **2-16**), an intermediate in the biosynthesis of arginine and an important amino acid of the ornithine (urea) cycle, which has the function of converting toxic ammonia into less toxic urea in mammals. Ornithine formed in dough through the action of yeast (*Saccharomyces cerevisiae*) is the main precursor of the typical aroma of bread crust, for which 2-acetyl-1-pyrroline, together with 6-acetyl-1,2,3,4-tetrahydropyridine and its isomer 6-acetyl-2,3,4,5-tetrahydropyridine, is responsible (see Section 8.2.12.4.1).

The reaction of L-lysine with reducing sugars (known as the Maillard reaction) during the processing and storage of food produces the so-called bound lysine, which is present in the form of unavailable derivatives. Examples of these compounds are the amino acids furosine and pyridosine (see Section 4.7.5.12.3). Plants often contain derivatives of lysine, such as 4-hydroxylysine, which is found in some *Salvia* species of the mint family (Lamiaceae), commonly referred to as sage. N^6-Acetyllysine occurs in beet (*Beta vulgaris*, Amaranthaceae). The metabolic precursor of lysine, L-2-aminoadipic (α-aminoadipic) acid, occurs in many legume seeds, such as lentils (0.17 g/kg) and garden peas (0.18 mg/kg). In commercially available seedlings of leguminous plants sold for human consumption, aminoadipic acid also occurs, but in lower amounts.

A non-protein amino acid related to arginine is the carbamoyl derivative of ornithine, L-citrulline (**2-16**), which is an intermediate in the urea cycle. It was first identified in watermelon (*Citrullus lanatus*, Cucurbitaceae), where it is present at concentrations of 640–890 mg/kg (f.w.). In plants, citrulline forms part of the nitrogen reserves and is a means of transporting nitrogen. It can also be formed by the decomposition of arginine, for example by yeast *S. cerevisiae*. The seeds of some legumes, such as vetchlings (*Lathyrus* spp.) and vetches (*Vicia* spp.), contain its higher homologue, L-homocitrulline (**2-16**), and a number of amino acids containing the guanidino group; examples are L-4-hydroxyarginine, L-homoarginine (2-amino-6-guanidinohexanoic acid, **2-16**), and L-4-hydroxyhomoarginine. For example, homoarginine occurs in the seeds of red peas (*Lathyrus cicera*) and grass peas (*Lathyrus sativus*) and in lentils (*Lens culinaris*).

The subtropical legume jack bean (*Canavalia ensiformis*, Fabaceae), which is grown commercially as a source of the enzyme urease, contains two unusual amino acids derived from homoserine, known as L-canaline, or L-2-amino-4-(aminooxy)butanoic acid, and L-canavanine, or L-2-amino-4-(guanidinooxy)butanoic acid (**2-16**). Toxic canaline, with a unique aminooxy group, is a structural analogue of ornithine and a lysine antagonist. Canavanine is a structural analogue of arginine and its highly potent antagonist (competitor) that blocks the binding of arginine (agonist) at a receptor molecule. By acting as an antagonist of arginine, canavanine inhibits the growth and development of other organisms and thus has an allelochemical effect. Its toxicity is based upon its degradation to canaline (by arginase) and the subsequent reaction of canaline with aldehydes, for instance with pyridoxal 5′-phosphate in molecules of decarboxylases and aminotransferases yielding stable oximes. Both amino acids show toxic effects to bacteria, insects, fungi, higher plants, and predators. The canavanine concentration in seeds can be as high as 10% of dry matter. Canavanine has also been tested experimentally for cytotoxicity against cancer cells. A specific amino acid, L-indospicine (**2-16**), is a hepatotoxin, and is present along with canavanine in seeds of the *Indigofera spicata* tree, which belongs to the same plant family (see Section 10.3.5.1). Other species of the genus *Indigofera* have found use as sources of a natural blue dye known as true indigo (*I. tinctoria* and *I. suffruticosa*).

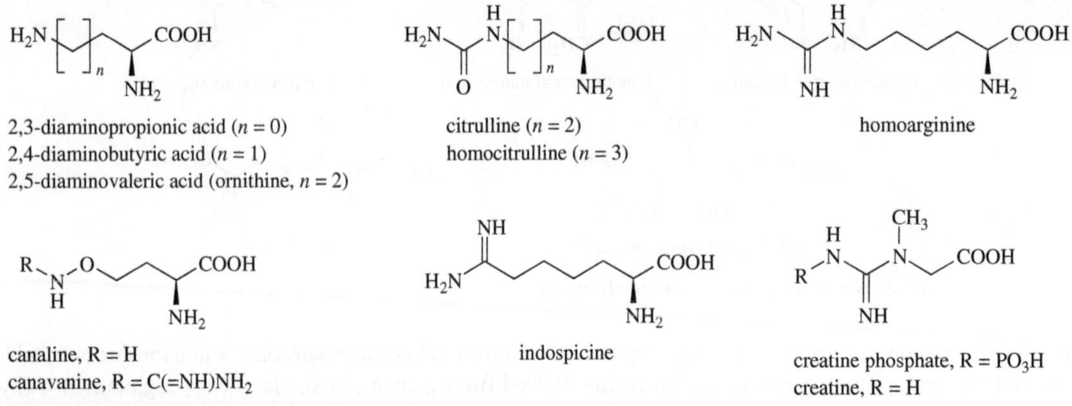

2,3-diaminopropionic acid (*n* = 0)
2,4-diaminobutyric acid (*n* = 1)
2,5-diaminovaleric acid (ornithine, *n* = 2)

citrulline (*n* = 2)
homocitrulline (*n* = 3)

homoarginine

canaline, R = H
canavanine, R = C(=NH)NH$_2$

indospicine

creatine phosphate, R = PO$_3$H
creatine, R = H

2-16, basic amino acids

The muscle tissues of warm-blooded animals and of fish contain an unusual amino acid, creatine phosphate (or phosphocreatine, **2-16**), in relatively high concentrations (3–6 g/kg). Creatine phosphate acts as a phosphate donor (an energy reserve that can be rapidly mobilised) for anaerobic formation of ATP (adenosine triphosphate) from ADP (adenosine diphosphate) in skeletal muscles and the brain following an intense muscular or neuronal effort. Creatine is also a scavenger for methylglyoxal under physiological conditions, suppressing carbonyl stress *in vivo* (see Section 4.7.5.6). It is biosynthesised from glycine and arginine. Natural hydrolysis of phosphocreatine post mortem yields creatine (**2-16**), which occurs in raw meat. During thermal processing of meat and meat products, creatine dehydrates spontaneously to form creatinine, which may become the precursor of toxic imidazopyridines, imidazoquinolines, and imidazoquinoxalines in some processed meats and meat products (see Section 12.2.1). Crustaceans employ arginine instead of creatine. Creatine is often used as a food supplement for athletes and bodybuilders. It is claimed that it helps stimulate and maintain instantaneous muscle power.

Cyanoamino acids The production of hydrogen cyanide in plants, called cyanogenesis (see Section 10.3.2.3), requires effective mechanisms, which eliminate this toxic substance. One of these mechanisms is the biosynthesis of neurotoxic β-substituted alanine derivative β-cyano-L-alanine (L-3-cyanoalanine) from cysteine, which takes place in all vascular plants under the catalysis of β-cyanoalanine synthase (Figure 2.2). β-Cyanoalanine is then converted into asparagine, which is involved in normal metabolism. This reaction is catalysed by β-cyanoalanine hydratase. Seeds of some vetches (*Vicia* spp.) and vetchlings (*Lathyrus* spp.) contain β-cyanoalanine (together with γ-glutamyl-β-cyanoalanine) at high concentrations. The seeds of common vetch (*V. sativa*) contain up to 1.5 g/kg β-cyanoalanine and 0.6 g/kg of the γ-glutamyl derivative. Decarboxylation of β-cyanoalanine and other reactions produce β-aminopropionitrile, which is considered a cause of the disease lathyrism (see Section 10.3.5.1).

Other enzymes involved in cyanide catabolism are cyanide hydratase and rhodanase. Cyanide hydratase, found in many fungi, catalyses the hydration of cyanide to formamide (H$_2$N—CH=O). Rhodanase found in bacteria, plants, and animals catalyses the conversion of cyanide into thiocyanate (rhodanide, R—S—C≡N).

2.2.1.2.5 Aromatic amino acids

Most non-protein aromatic amino acids are derived from L-phenylglycine (**2-17**), L-phenylalanine (**2-3**), and L-tyrosine (see **2-3**). For example, various plants contain 3-carboxyphenylalanine, which is accompanied by 3-carboxyphenyltyrosine. Legumes of the genus *Vigna* (Fabaceae) contain 4-aminophenylalanine, and (*R*)-3-amino-3-phenylpropionic acid (β-phenyl-β-alanine) (**2-17**) occurs in different types of beans (*Phaseolus* spp.).

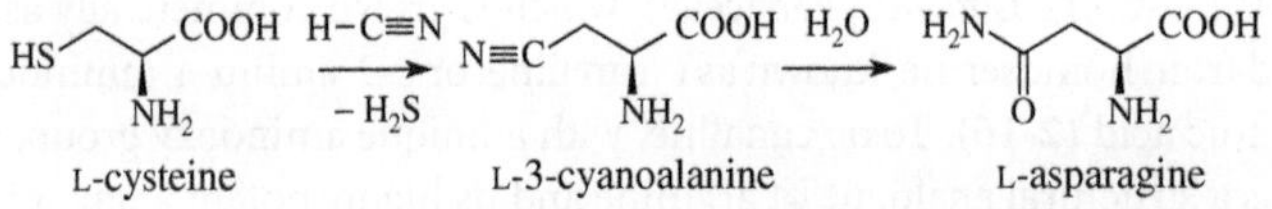

L-cysteine   L-3-cyanoalanine   L-asparagine

Figure 2.2 Formation of 3-cyanoalanine and asparagine.

An important aromatic hydroxyamino acid is L-3,4-dihydroxyphenylalanine, known by the acronym DOPA (from the name dioxyphenylalanine, **2-17**). It is formed by enzymatic oxidation of tyrosine and becomes the precursor of brown and black pigments termed melanins (see Section 9.3.1.1). Melanins are formed from DOPA by enzymatic browning reactions that proceed *in vivo*.

Tyrosine is additionally a precursor of the thyroid hormones 3,5,3′,5′-tetraiodothyronine, also known as thyroxine (**2-17**), and 3,5,3′-triiodothyronine, which was previously called iodogorgic acid. Thyronine is a deiodinated form of thyroxine.

phenylglycine   β-phenyl-β-alanine   3,4-dihydroxyphenylalanine   thyronine, $R^1 = R^2 = R^3 = R^4 = H$
                                                                                             thyroxine, $R^1 = R^2 = R^3 = R^4 = I$

2-17, less common aromatic amino acids

2.2.1.2.6 Heterocyclic amino acids

Foods of plant and animal origin often contain amino acids structurally related to proline and its lower and higher homologues, amino acids derived from histidine, and various other nitrogen heterocycles. These amino acids are mainly derivatives of azetidine, pyrazole, pyridine, piperidine, pyrimidine, purine, isoxazole, and isoxazoline. Oxygen heterocycles (such as pyran-2-ones) and sulfur heterocycles (thiazanes) are less frequent precursors. The side chains are formally derived from glycine, alanine, and 2-aminobutyric acid, although the heterocyclic amino acid skeletons are synthesised from other biochemical precursors. Many heterocyclic amino acids have significant biological effects. They often exhibit neurotoxicity, hallucinogenic, insecticidal, and other toxic effects.

Heterocyclic derivatives formed from proline, lysine, and cysteine Examples of proline derivatives, which reportedly occur in apples and loquat (Japanese plum; *Eriobotrya japonica*, Rosaceae), are (*S*)-4-methyl-L-proline (**2-18**), (*S*)-4-hydroxymethyl-L-proline, and 4-methylene-L-proline. *N*-Amino-D-proline (linatin, **2-18**), together with the corresponding dipeptide (γ-glutamyl derivative), occurs in linseed (*Linum usitatissimum*, Linaceae); it has antinutritional effects, as it acts as an antivitamin B$_6$. Cucurbitin, (*R*)-3-aminopyrrolidine-3-carboxylic acid (**2-18**), is an unusual compound that is found in the seeds of various squashes. It is responsible for their antihelminthic (vermifugal) effects, which are destructive to parasitic helminth worms. In *Cucurbita pepo* (Cucurbitaceae) squash seeds, the concentration ranges from 1.7 to 6.6%, and in *Cucurbita maxima* squash seeds it can reach up to 19.4%.

The higher homologue of proline, L-pipecolic (L-2-piperidinecarboxylic) acid (**2-18**), is commonly found in many plants. It is a metabolite of L-lysine, which arises from the oxidative deamination of its α-amino group. Pipecolic acid is found at high concentrations in beans (17–44 mg/kg) and cruciferous vegetables (12–21 mg/kg). Lower concentrations are found in potatoes (2.5 mg/kg) and fruits (1.3 mg/kg). Some legumes, such as those of the genus *Inga* (originating in tropical America), contain 4-hydroxy-, 5-hydroxy-, and 4,5-dihydroxypipecolic acids. Pipecolic acid is also found in animal tissues. In the mammalian brain, it acts as a stimulator of GABA receptors. Abnormally high amounts of pipecolic acid in blood plasma (known as hyperpipecolataemia or hyperpipecolic acidaemia) are found in chronic liver diseases and some genetic disorders.

4-methylproline   linatin   cucurbitin   pipecolic acid   cycloalliin

2-18, amino acids derived from pyrrolidine, piperidine, and 1,4-thiazane

A special sulfur-containing heterocyclic amino acid is cycloalliin, (1*S*,3*R*,5*S*)-3-carboxy-5-methyl-1,4-thiazane-*S*-oxide (**2-18**), which is formed in onion and garlic from the amino acid isoalliin. The cycloalliin content in raw onion bulbs and garlic cloves ranges from 0.5 to 1.5 g/kg.

Substituted glycines Amanitas, such as fly agaric (*Am. muscaria*), contain a toxic derivative of isoxazole known as ibotenic acid (**2-19**), which is biosynthesised from glutamic acid. Hallucinogenic effects are produced by dihydroderivative tricholomic (dihydroibotenic) acid (**2-19**), which is a component of *T. muscarium* mushrooms (see Section 10.3.7.3.1).

ibotenic acid   tricholomic acid

2-19, glycines substituted with isoxazole derivatives

Substituted alanines In addition to the previously mentioned 3-methylhistidine derived from 3*H*-imidazole isomer (see Section 2.2.1.1.2), the imidazole skeleton occurs in many other amino acids. Acetylhistidine, diacetylhistidine, and triacetylhistidine (**2-20**) are present as free amino acids in spinach (*Spinacia oleracea*, Asteraceae). The free amino acid N^1-methyl-4-mercaptohistidine, known as ovothiol A (**2-20**), can be found in many marine animals. It is one of the mercaptohistidines isolated from the eggs and ovaries of marine invertebrates. Owing to the imidazole ring, ovothiol A shows a higher reactivity of the thiol group (the active form is thiolate) towards free radicals, including reactive oxygen species, compared to glutathione. An effective antioxidant is 2-thiohistidine betaine, known as L-ergothioneine (**2-8**), which is synthesised in Actinomycetales bacteria (such as *Mycobacterium* ssp.) and in various non-yeast-like fungi, including edible genera of the Basidiomycota division, such as *Boletus* ssp. In contrast, no ergothioneine synthesis occurs in higher plants or any animal species. It has been suggested that the incorporation of ergothioneine in plants may result from the absorption of the ergothioneine produced by microorganisms in the soil and in animals through ergothioneine-containing plant and animal foods. Allegedly, king bolete (*Boletus edulis*) has the highest amount of this amino acid (528 mg/kg f.w.). Concentrations in other fungi are lower (e.g. 119 mg/kg in oyster mushroom, *Pleurotus ostreatus*, 0.46 mg/kg in button mushrooms, *A. bisporus*, 0.06 mg/kg f.w. in chanterelle, *Cantharellus cibarius*, and undetectable amounts in shiitake, *Le. edodes*). Concentrations of ergothioneine in pork liver were 8.71 mg/kg, in pork meat (loin) 1.33 mg/kg, in ham 1.12 mg/kg, in whole-grain wheat bread 0.53 mg/kg, and in broccoli 0.24 mg/kg.

There is a relatively large group of substituted alanines that are derived from isoxazoline, isoxazolidine, and oxazoline. They can be found mainly in germinating seeds and in higher fungi. The donor of the alanyl residue is usually *O*-acetylserine.

An example of these compounds is β-(isoxazolin-5-on-2-yl)-L-alanine, also known as 2-(2-amino-2-carboxyethyl)isoxazolin-5-one (**2-20**), which occurs in the seeds and seedlings of many legumes (Fabaceae), vetchlings (*Lathyrus* spp.), vetches (*Vicia* spp.), garden peas (*Pisum* spp.), and lentils (*Lens* spp.). For example, the amount of β-(isoxazolin-5-on-2-yl)-L-alanine in the seedlings of *Lathyrus odoratus* is 0.7–0.8 g/kg (f.w.), that in seedlings of lentils is 0.17 g/kg, and that in seedlings of garden peas is 1.31 g/kg. Seedlings of lentils also contain a γ-glutamyl derivative of β-(isoxazolin-5-on-2-yl)-L-alanine (0.40 g/kg). β-(Isoxazolin-5-on-2-yl)-L-alanine shows antimycotic effects and, as a structural analogue of glutamic acid (an antagonist), has a neuroexcitatory effect. In vetchlings and vetches, but not in sweet peas, garden peas, or lentils, β-(isoxazolin-5-on-2-yl)-L-alanine is a metabolic precursor of 3-aminopropionitrile, L-2,4-diaminobutanoic acid, and neurotoxic 3-(*N*-oxalyl)-L-2,3-diaminopropionic acid (β-*N*-oxalyl-L-alanine), which occur in some species of these plants and are responsible for the crippling human disease lathyrism (see Section 10.3.5.1). The seeds of the vine *Quisqualis indica* (Combretaceae), known as Rangoon creeper and native to tropical Asia, contain L-quisqualic acid (**2-20**), which is a stimulant of glutamic acid receptors and has neuroexcitatory activity. Oil extracted from these seeds is used in traditional Chinese medicine as a nematocide and in the treatment of ulcers of the stomach and duodenum. Its leaves, shoots, and pods are popular vegetable plants.

3-Pyrazol-1-yl-L-alanine (**2-20**), derived from pyrazole, occurs in some plants of the Cucurbitaceae family. It was first isolated from the seeds of watermelon (*Citrullus lanatus*), where it represents approximately 0.1% of dry matter. It also occurs in the seeds of many other melons, squashes, and gourds, together with cucurbitin and other heterocyclic amino acids.

The pyridine (dihydropyridine)-derived amino acid β-(3-hydroxy-4-pyridon-1-yl)-L-alanine, L-mimosine (**2-20**), is thyreotoxic (causes overactivity of the thyroid gland) to non-ruminant animals. It occurs in the subfamily Mimosoideae of the legume family Fabaceae, in plants of the genera *Mimosa* and *Leucaena*. *Leucaena leucocephala* (syn. *Mimosa glauca*) is native to the tropical and subtropical Americas, known as white leadtree, and used as livestock fodder. The mimosine content in seeds can reach up to 5% of dry matter (see Section 10.3.5.1). The amino acid pyridosine arising from lysine in the Maillard reaction can also be considered a dihydropyridine derivative (see Section 4.7.5.12.3).

Several other free amino acids derived from purines and pyrimidines have roles as defence agents. Plants of the genus *Mimosa* synthesise 5-aminouracil, which blocks mitotic division and the incorporation of guanosine into nucleic acids. The product of its catabolism is the amino acid L-albizziine (**2-20**), the precursor of L-2,3-diaminopropionic acid, which accumulates in plants.

The neuroactive pyrimidine derivative L-willardiine, β-(uracil-1-yl)-L-alanine (**2-20**), occurs in peas (*Pisum sativum*, Fabaceae), the seeds of European beech (*Fagus sylvatica*, Fagaceae), and, together with mimosine, some acacia (*Acacia* spp., Fabaceae). In subtropical legumes of the genus *Crotalaria* (young pods and leaves of certain species are used as vegetables, others are used as fodder and green manure), it is accompanied by L-isowillardiine, β-(uracil-3-yl)-L-alanine (**2-20**). Amounts of isowillardine in garden pea seedlings, for example, have been reported to be 0.67 g/kg. The amino acid L-lathyrine (**2-21**) occurs in various types of vetchlings (*Lathyrus* spp., Fabaceae). For example, lathyrine makes up 2% of dry matter in seeds of Tangier peas (*Lathyrus tingitanus*). Unlike other N-heterocyclic β-substituted alanines, the alanyl residue is derived from L-alanine and not from *O*-acetyl-L-serine.

The fly agaric (*Am. muscaria*), other Amanitas, and some higher plants contain stizolobic acid (L-2-amino-3-(6-carboxy-2-oxo-2*H*-pyran-4-yl)propionic acid, **2-20**) and stizolobinic acid (L-2-amino-3-(6-carboxy-2-oxo-2*H*-pyran-3-yl)propionic acid, **2-20**). Stizolobic acid is a precursor of the orange betaxanthine muscaaurine II, which is, along with other pigments (see Section 9.3.1.2), responsible for the orange–red colour of the hat of this toadstool.

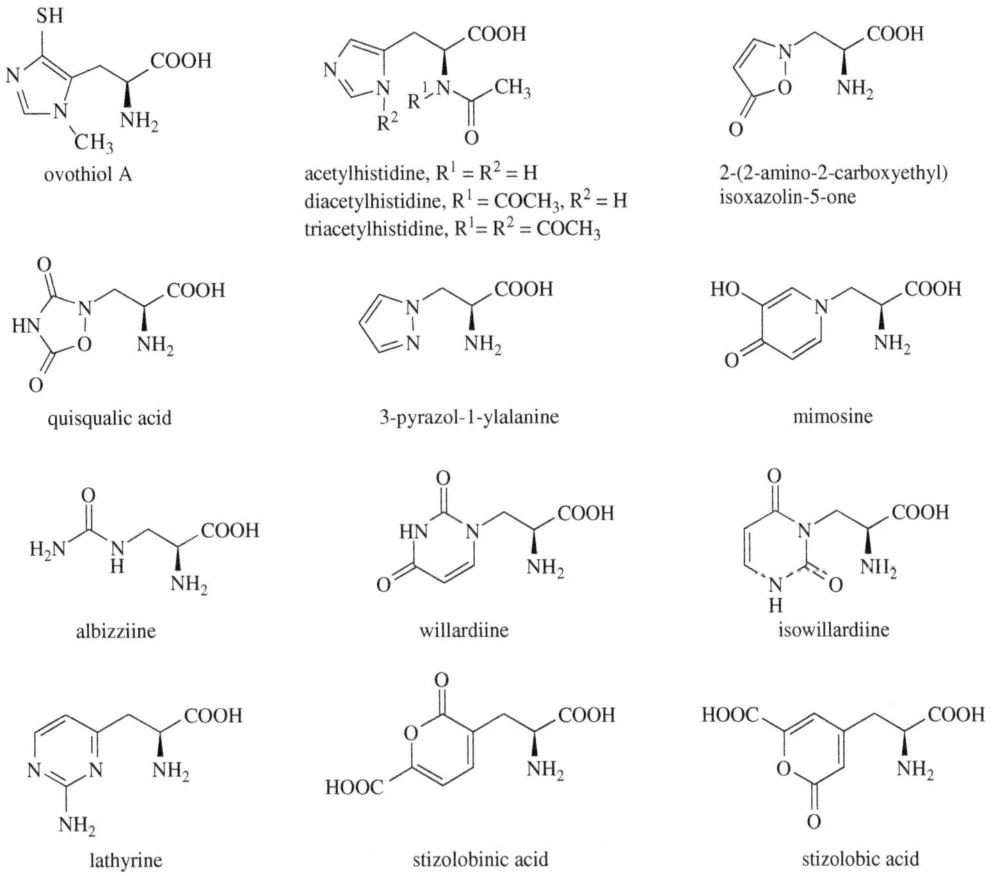

ovothiol A

acetylhistidine, R¹ = R² = H
diacetylhistidine, R¹ = COCH₃, R² = H
triacetylhistidine, R¹= R² = COCH₃

2-(2-amino-2-carboxyethyl)
isoxazolin-5-one

quisqualic acid

3-pyrazol-1-ylalanine

mimosine

albizziine

willardiine

isowillardiine

lathyrine

stizolobinic acid

stizolobic acid

2-20, substituted alanines

Substituted aminobutanoic acids Higher homologues of heterocyclic β-substituted alanines, with another carbon atom in their side chain, can be considered as compounds derived from homoserine or as γ-substituted 2-aminobutanoic acids.

The lower homologue of proline, azetidine-2-carboxylic acid (**2-21**), has antimicrobial properties associated with its incorporation into proteins in place of proline. A small amount of this amino acid is found in some legumes (Fabaceae), plants of the Lily family (Lilliaceae), plants of the Amaryllidaceae family, and red algae (Rhodophycae). Some other amino acids with the azetidine ring act in graminaceous and other plants as metal ligands, known as siderophores. They are excreted under conditions of iron deficiency as part of a strategy of solubilising Fe from the root environment for uptake by the plant (see Section 6.3.6.1). The basic member of this series is *N*-(3-amino-3-carboxypropyl)azetidine-2-carboxylic acid (**2-21**), identified, for example, in beechnuts (*Fagus sylvatica*, Fagaceae). Important siderophores are nicotianamine (**2-21**, found in tobacco, *Nicotiana tabacum*, Solanaceae), its hydroxylated derivative mugineic acid, 3-hydroxymugienic acid (**2-21**), and 2′-deoxymugienic acid. Related compounds without the azetidine ring are avenic acid (**2-21**, found in oat, *Avena sativa*, Poaceae) and distichonic acid (**2-21**, produced by barley, *Hordeum distichon*).

Tobacco also contains the pyridine derivative L-nicotianine (**2-21**). A number of heterocyclic substituted aminobutanoic acids occur in legumes. The sweet pea (*L. odoratus*) contains 3.5% (dry matter) neurotoxic β-(isoxazolin-5-on-2-yl)alanine and a higher homologue known as α-amino-β-(isoxazolin-5-on-1-yl)butanoic acid or 2-(3-amino-3-carboxypropyl)isoxazolin-5-one (**2-21**). Poisoning shows similar symptoms to poisoning by neurolathyrogenic L-2,4-diaminobutyric acid (see Section 10.3.5.1). Other toxic amino acids with a 3-amino-3-carboxypropyl residue are canavanine and canaline (see Section 10.3.5.1).

azetidine-2-carboxylic acid

N-(3-amino-3-carboxypropyl)-
azetidine-2-carboxylic acid

nicotianamine

3-hydroxymugineic acid

nicotianine

2-(3-amino-3-carboxypropyl)-
isoxazolin-5-one

avenic acid

distichonic acid

2-21, azetidine-2-carboxylic acids and substituted aminobutanoic acids

2.2.2 Physiology and nutrition

All tissues have some capability to biosynthesise non-essential amino acids, to modify amino acids, and to convert non-amino acid carbon skeletons into amino acids. At times of dietary surplus, the potentially toxic nitrogen of amino acids is eliminated via transamination, deamination, and urea formation to yield various compounds, which are then used for the biosynthesis of glucose or fatty acids, or which may be converted into carbon dioxide and water in the citric acid cycle, which releases energy during times of starvation.

People with a sufficiently varied and balanced diet (whose energy intake from protein is in the range of 10–15% of their total energy intake and in whom the ratio of animal and vegetable proteins is 1 : 1) usually have a good supply of essential amino acids. Complete proteins contain a balanced set of essential amino acids. Some countries allow enrichment of foods with essential amino acids, especially limiting amino acids: lysine is usually the limiting amino acid in cereals and vegetable proteins; methionine, possibly with cysteine, is limiting due to low content in beans, meat, and dairy proteins; threonine is limiting due to low content in wheat and rye proteins; and tryptophan is limiting in milk caseins and maize and rice proteins. In Japan and other East Asian countries, for example, lysine and threonine are used to fortify rice; lysine is used for bread fortification, whilst methionine is added to soy products. Problems with the intake of essential amino acids may occur where there are significant nutritional restrictions that lead to a rather monotonous diet. This applies, for example, to vegans and those on macrobiotic diets. Such diets are inappropriate for children, because the lack of some essential amino acids can lead to serious developmental defects.

The content of essential amino acids in livestock feed should be monitored, as the lack of limiting amino acids can significantly cause weight loss. Essential amino acids are commonly used as feed additives at levels of 0.05–0.2%.

Certain risks to human nutrition, and especially to livestock nutrition, may be posed by some toxic amino acids.

2.2.3 Properties

2.2.3.1 Acid–base properties

The common amino acids are weak polyprotic acids. The ionisable groups do not show a strong tendency to dissociate, and the degree of dissociation thus depends on the pH of the medium. In the physiological range of pH values, the α-carboxyl groups of amino acids are usually fully dissociated, and the α-amino groups are protonated. Compounds with these properties are called **amphoteric** or **ampholytes** (amphoteric electrolytes). Ampholytes form inner salts that are dipolar (ambiguous or amphoteric) ions called amphions (the German equivalent is zwitterion). An amphion carries a positive and a negative electrical charge (**2-2**), so the resulting electrical charge of the molecule is zero. The predominant forms of almost all amino acids in the physiological pH of animal and plant tissues are amphions. In addition to amphions, in aqueous solutions amino acids may also (depending on pH) form cations (as they react with acids and behave as bases) or anions (when they act as acids) (Figure 2.3). Therefore, amino acids are very polar compounds, being fairly soluble in water and polar solvents. They have high dipole moment values and high melting points (they often melt with decomposition around 200 °C).

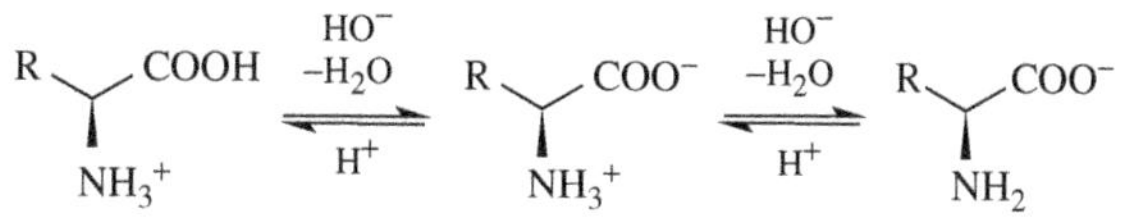

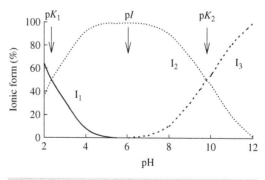

Figure 2.3 Influence of pH on the dissociation of L-amino acids.

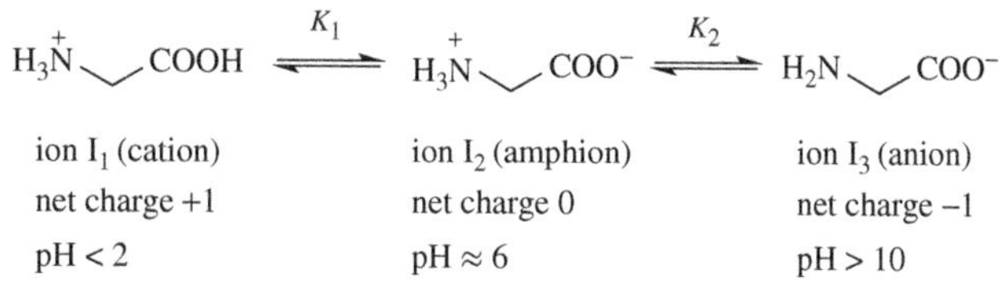

$$H_3\overset{+}{N}\diagdown COOH \underset{K_1}{\rightleftharpoons} H_3\overset{+}{N}\diagdown COO^- \underset{K_2}{\rightleftharpoons} H_2N\diagdown COO^-$$

ion I_1 (cation)	ion I_2 (amphion)	ion I_3 (anion)
net charge +1	net charge 0	net charge −1
pH < 2	pH ≈ 6	pH > 10

Figure 2.4 Influence of pH on the dissociation of glycine.

2.2.3.1.1 Glycine

The dissociation dependence of the simplest monoaminomonocarboxylic acid, glycine, on pH is shown schematically in Figure 2.4. In an acidic solution (pH about 2 and below), the predominant form is the ion I_1 (cation) and the net amino acid charge is +1. In a neutral solution (pH around 6), the ion I_2 (amphion) dominates and the net charge is zero. In alkaline medium (pH approximately 10 and higher), glycine is predominantly present as an I_3 ion (anion) and the net charge of the molecule is −1.

Measures of acidity and basicity of functional groups of amino acids are known as **dissociation constants**. The dissociation constant K_1 describes the acidity of the α-carboxyl group of glycine, as shown by the following reaction:

$$^+NH_3CH_2COOH + H_2O = {}^+NH_3CH_2COO^- + H_3O^+$$

It is therefore defined by the equation (the activities and concentrations of ions and neutral molecules, respectively, are given in parentheses):

$$K_1 = \frac{[^+NH_3CH_2COO^-][H_3O^+]}{[^+NH_3CH_2COOH][H_2O]}$$

The dissociation constant K_2 describes the basicity of the α-amino group:

$$^+NH_3CH_2COO^- + H_2O = NH_2CH_2COO^- + H_3O^+$$

and is defined by the equation:

$$K_2 = \frac{[NH_2CH_2COO^-][H_3O^+]}{[^+NH_3CH_2COO^-][H_2O]}$$

The values of the dissociation constants of glycine at 25 °C are: $K_1 = 4.47 \times 10^{-3}$, $K_2 = 1.67 \times 10^{-10}$. Most often these values are shown as negative logarithms, that is as pK_i ($-\log K_i$) values. Then, $pK_1 = 2.35$ and $pK_2 = 9.78$.

The pH of solution at which the maximum concentration of amphions is reached is called the isoelectric point (denoted pI). The value of pI may be calculated as the arithmetic mean of the pK_1 and pK_2 values. The pI value for glycine is:

$$pI = \frac{pK_1 + pK_2}{2} = 6.1$$

At pH = pK_1 the concentrations of the cation and the amphion are equal, and so each amounts to 50% of the total amino acid. At pH = pK_2 the concentrations of the amphion and the anion are equal. The dependences of ionic forms of glycine on pH are shown in Figure 2.5.

Figure 2.5 Dependence of ionic forms of glycine on pH (see Figure 2.4). –, cation (I_1); ... ..., amphion (I_2); -----, anion (I_3).

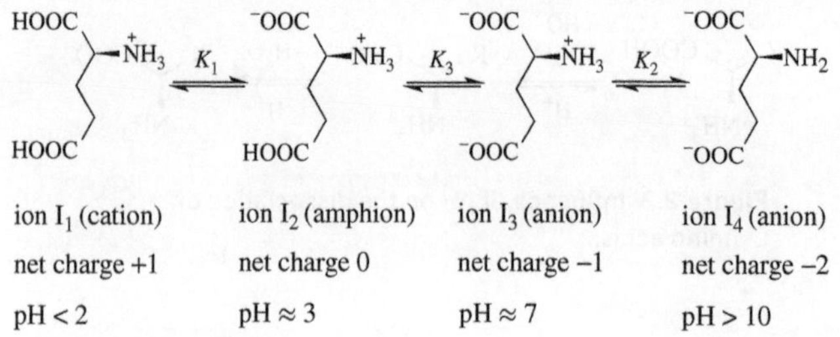

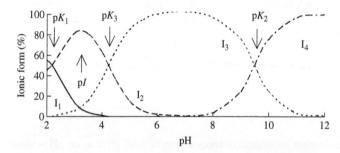

ion I_1 (cation)   ion I_2 (amphion)   ion I_3 (anion)   ion I_4 (anion)

net charge +1   net charge 0   net charge −1   net charge −2

pH < 2   pH ≈ 3   pH ≈ 7   pH > 10

Figure 2.6 Influence of pH on the dissociation of glutamic acid.

Figure 2.7 Dependence of ionic forms of glutamic acid on pH (see Figure 2.6). –, cation (I_1); -----, amphion (I_2); ... ..., anion (I_3); --, anion (I_4).

2.2.3.1.2 Acidic amino acids

A somewhat different situation is found with aspartic and glutamic acids. Owing to the presence of two carboxyl groups in the molecule, these amino acids form four types of ions: cations, amphions, and two types of anions. The dissociation of glutamic acid, depending on the pH value of the solution, is given as an example in Figure 2.6. The dependence of its ionic forms on pH is shown in Figure 2.7.

The amphion again has charge on the α-amino group and on the α-carboxyl, which is more acidic than the distal carboxyl. The value pI (the mean of the pK_1 and pK_3 values) is 3.1. At pH = pI, unlike glycine, the amphion I_2 is not the only ion in solution; it makes up about 80% of the total amount of glutamic acid present. Cation I_1 is also present at a level of about 10% (this is the main ion in an acidic medium at pH < pK_1), and anion I_3 occurs in the same level. In a neutral solution at pH 7, virtually the only ion present is anion I_3. This ion is a sensory active form of glutamic acid, and is responsible for its unique organoleptic properties. The anion I_3 is the dominant form of glutamic acid in solutions at pH values ranging from pK_2 to pK_3 (pH 4.3–9.5). Therefore, foods to which sodium hydrogenglutamate is added as a flavour enhancer (see Section 11.3.5) should have pH in this range. At pH values around 10 and above, the predominant form of glutamic acid is the anion I_4, which has a net charge of −2. Under suitable conditions (pH and temperature), there is equilibrium between glutamic acid and its lactam, L-5-oxopyrrolidine-2-carboxylic acid. This lactam is in equilibrium with its anion.

2.2.3.1.3 Basic amino acids

Lysine also has four types of ions (Figure 2.8). In an acidic solution, two types of cations may occur. At pH < pK_1 (a pH lower than 2), the predominant form is a cation with a net charge of +2, whilst in a solution around pH 7, the monocation predominates. Its maximum concentration is in solutions of pH equal to the average values of pK_1 and pK_2. The maximum concentration of the amphion is found in solutions of pH around 10, where the pI value is equal to the average values of pK_2 and pK_3. In solutions of pH > pK_3 (pH about 11 and higher), lysine anion predominates (Figure 2.9).

Four ionic forms can also exist with arginine (Figure 2.10) and histidine (Figure 2.11), depending on the pH. The predominant forms of both amino acids at pH 7 are monocations.

2.2.3.1.4 Other amino acids

Other monoaminomonocarboxylic acids behave in an analogous way to glycine. These include alanine, valine, leucine, and isoleucine, as well as amino acids substituted in the side chain (serine, threonine, methionine, phenylalanine, and proline). The cysteine SH-group can also be dissociated at higher pH; therefore, there are four ionic forms. In solutions of pH 7, the main form is the amphion (Figure 2.12),

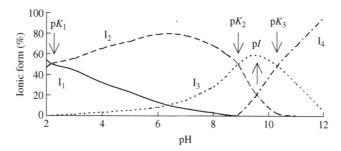

ion I_1 (cation)   ion I_2 (cation)   ion I_3 (amphion)   ion I_4 (anion)

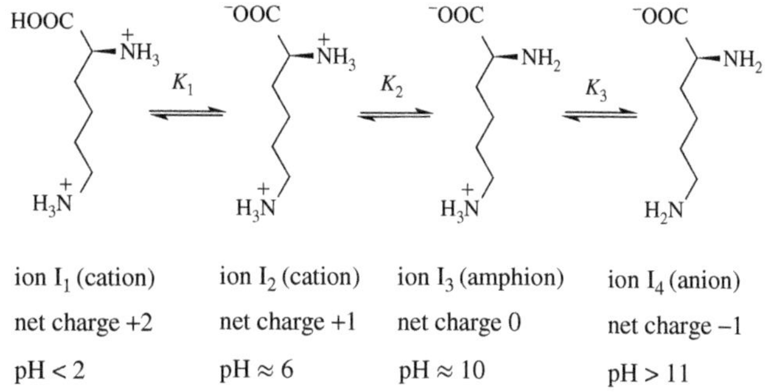

net charge +2   net charge +1   net charge 0   net charge −1

pH < 2   pH ≈ 6   pH ≈ 10   pH > 11

Figure 2.8 Influence of pH on the dissociation of lysine.

Figure 2.9 Dependence of ionic forms of lysine on pH (see Figure 2.8). –, cation (I_1); -----, cation (I_2); ...., amphion (I_3); –·–, anion (I_4).

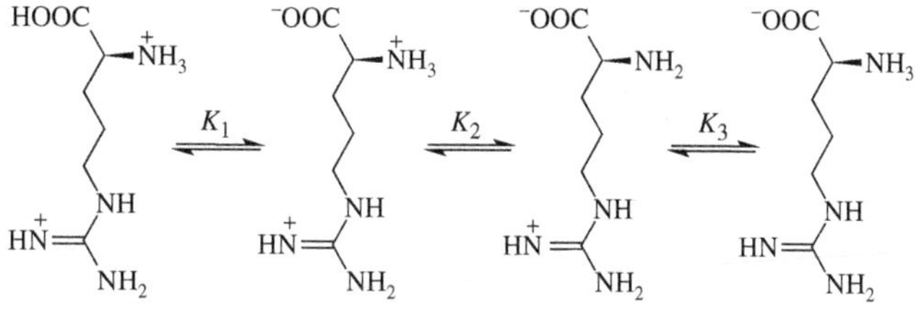

Figure 2.10 Influence of pH on the dissociation of arginine.

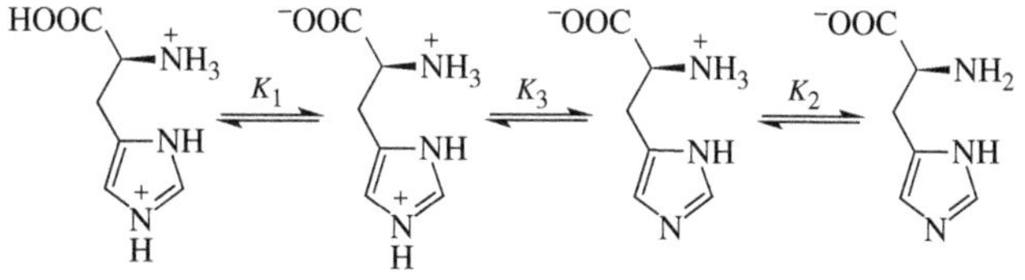

Figure 2.11 Influence of pH on the dissociation of histidine.

which also prevails in tyrosine solutions at pH 7 (analogous to phenylalanine). In an alkaline solution, the hydroxyl group of tyrosine can be deprotonated, forming a phenoxide anion (Figure 2.13). An overview of the pK_i and pI values of proteinogenic and some other common amino acids is given in Table 2.4.

2.2.3.2 Optical activity

With the exception of glycine, all amino acids contain a chiral carbon atom in the position α to the carboxylic group, which is called a C_α carbon. Therefore, all exist in two optical isomers (enantiomers). The proteins contain almost exclusively the amino acids of the L-configuration

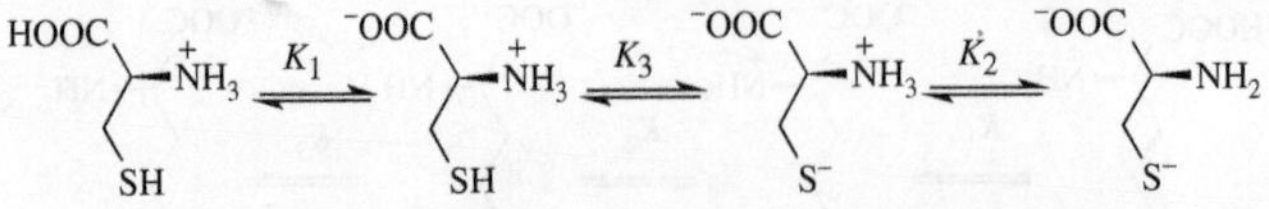

Figure 2.12 Influence of pH on the dissociation of cysteine.

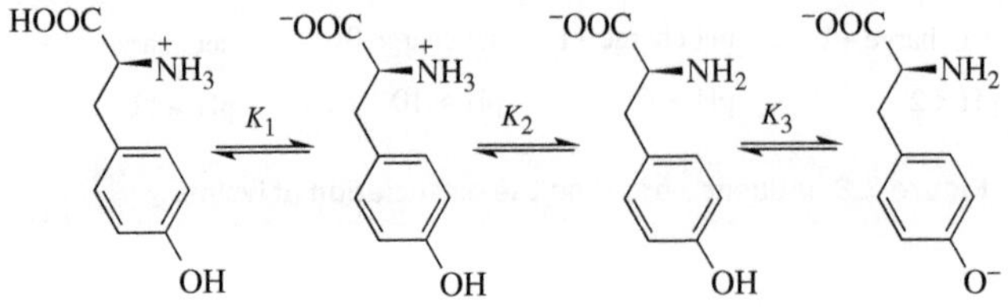

Figure 2.13 Influence of pH on the dissociation of tyrosine.

Table 2.4 pK and pI values of amino acids.

Amino acid	pK_1	pK_2	p$K_3{}^a$	pI	Amino acid	pK_1	pK_2	p$K_3{}^a$	pI
Proteinogenic					Asparagine	2.10	8.84	–	5.5
Glycine	2.34	9.78	–	6.1	Glutamine	2.17	9.13	–	5.7
Alanine	2.35	9.87	–	6.0	Lysine	2.16	9.18	10.79	10.0
Valine	2.29	9.74	–	6.0	Arginine	1.82	8.99	12.48	10.9
Leucine	2.33	9.74	–	6.0	Histidine	1.80	9.33	6.04	7.7
Isoleucine	2.32	9.76	–	6.0	Phenylalanine	2.16	9.18	–	5.7
Serine	2.19	9.21	–	5.7	Tyrosine	2.20	9.11	10.13	5.7
Threonine	2.09	9.10	–	5.6	Tryptofan	2.43	9.44	–	5.9
Cysteine	1.92	10.78	8.33	5.1	Proline	1.99	10.60	–	6.3
Selenocysteine	1.27	8.42	5.24		Other				
Methionine	2.28	9.21	–	5.7	4-Hydroxyproline	1.82	9.65	–	5.7
Aspartic acid	1.99	9.90	3.90	3.0	β-Alanine	3.55	10.24	–	6.9
Glutamic acid	2.10	9.47	4.07	3.1	γ-Aminobutyric acid	4.03	10.56	–	7.3

^{a}The dissociation constant pK_3 of cysteine (selenocysteine) is the dissociation constant of the -SH (-SeH) group. The dissociation constants of both carboxyl and amino groups of cystine have values 1.04 and 2.10 and 8.02 and 8.71, respectively. The pI value is 6.1. The dissociation constant pK_3 of aspartic acid and glutamic acid is the dissociation constant of the distal carboxyl group. The dissociation constant pK_3 of lysine, arginine, histidine, and tyrosine is the dissociation constant of the distal amino group (ε-amino group), the guanidino group, the imino group of imidazole ring, and the hydroxyl group, respectively.

(2-22); these are (S)-stereoisomers. The spatial arrangement of substituents around the asymmetric carbon in L-amino acids is the same as in L-glyceraldehyde (see Section 4.2.1.1). The exceptions are L-cysteine and L-selenocysteine, which are (R)-stereoisomers (the priority of substituents R=CH$_2$SH and R=CH$_2$SeH is higher than the priority of a carboxyl group COOH). S-Substituted cysteine sulfoxides have two chiral centres, at the C$_\alpha$ carbon and the sulfur atom (2-13).

Amino acids of the D-series are derived from D-glyceraldehyde and are (R)-stereoisomers of amino acids (2-23). They are a mirror image of the L-series of amino acids. Compared with L-amino acids, D-amino acids occur relatively rarely in nature. They are mainly bound in a variety of biologically active plants and animal peptides, occuring in peptidoglycans of microbial cell walls or as free compounds (such as linatine in flax seeds). For a long time, it was believed that the proteins of all living organisms contained only L-amino acids and that D-amino acids were not present in mammals. Recently, D-aspartic acid (synthesised by aspartate racemase) and D-serine (synthesised by serine racemase) were found in proteins of various tissues, where they play important roles in living organisms.

2-22, L-amino acid or (*S*)-amino acid

Free D-amino acids found in fresh or heat-processed foods are mostly of microbial origin. They are formed by hydrolysis of peptidoglycans in the cell walls of microorganisms or by the action of microbial racemases on L-amino acids. Therefore, they mainly occur in foods produced by lactic acid fermentation such as yoghurt and are termed **biogenic** D-amino acids. D-Amino acids may also form through non-enzymatic isomerisation of L-amino acids in foods processed at high temperatures, and especially in foods treated with alkaline reagents. These amino acids are bound in proteins, from which they are released by hydrolysis. They are called **abiogenic** D-amino acids.

Isoleucine, threonine, 4-hydroxyproline, and 3-hydroxyproline also have an asymmetric carbon atom in the side chain (two chiral centres in total). They can therefore occur in four stereoisomeric forms (this number is equal to $2^n = 2^2 = 4$, where n = number of asymmetric carbon atoms) known as L-, D-, L-*allo*-, and D-*allo*- according to the configuration of the carbon carrying the amino group. The unique *R/S* system is not widely used in biochemistry. The stereoisomers of isoleucine are thus called L-isoleucine or (2*S*,3*S*)-isoleucine (**2-3**), D-isoleucine or (2*R*,3*R*)-isoleucine (**2-24**), L-*allo*-isoleucine or (2*S*,3*R*)-isoleucine (**2-24**), and D-*allo*-isoleucine or (2*R*,3*S*)-isoleucine (**2-24**). Proteins contain only L-isoleucine. The product of enzymatic or thermal isomerisation of L-isoleucine is D-allo-isoleucine. Likewise, proteins contain L-threonine or (2*S*,3*R*)-threonine (**2-3**) and L-4-hydroxyproline or (2*S*,4*S*)-4-hydroxyproline (**2-4**). Cystine (as well as lanthionine, see **2-32**) contains two identical asymmetric carbons in the molecule and therefore only exists in three different isomers. Proteins contain just L-cystine (**2-4**); its isomer is D-cystine (**2-24**) and the mesoform *meso*-cystine (**2-24**) is a symmetrical molecule.

The sorting of amino acids into L- or D-series is not linked to their optical rotation. For example, aqueous solutions of natural alanine and glutamic acid are dextrorotatory, and are denoted (+)-L-Ala and (+)-L-Glu, respectively. Dextrorotation (turning of polarised light by solutions of certain optically active substances to the right, i.e. clockwise) shows a total of 12 proteinogenic amino acids; the single remaining amino acid shows laevorotation. For example, aqueous solutions of leucine and histidine are laevorotatory, and are denoted (−)-L-Leu and (−)-L-His, respectively. Optical rotation depends greatly on pH. For example, leucine in acidic and alkaline solutions is dextrorotatory, whilst histidine is dextrorotatory in acidic and laevorotatory in neutral and alkaline solutions.

2-23, D-amino acid or (*R*)-amino acid

2.2.3.3 Organoleptic properties

Some amino acids are sensorially active substances, which can therefore have some impact on the organoleptic properties of food. For example, glycine has a sweet taste (70% of the sweetness of sucrose) and is sometimes mixed with saccharin as a sweetening agent. According to their organoleptic properties, the proteinogenic amino acids can be classified as:

- **sweet** (glycine, alanine, threonine, proline, and hydroxyproline);

- **acidic** (aspartic and glutamic acids);

- **bitter** (amino acids with hydrophobic side chains, i.e. valine, leucine, isoleucine, phenylalanine, tyrosine, and tryptophan);

- **indifferent** (other amino acids).

The taste of the L- and D-enantiomer of an amino acid differs in many cases, demonstrating correlation between stereochemistry and flavour perception. For example, D-forms of most bitter proteinogenic amino acids show more sweetness than bitterness.

Free amino acids are usually found in small amounts in foods. Therefore, they influence flavour only in foods where significant proteolysis occurs (some cheeses, but also meat and fish). Enzymatic hydrolysates of proteins (such as soy sauces) or hydrolysed vegetable proteins (acid hydrolysates), which contain only amino acids, are used extensively as seasonings.

Glutamic acid and its monosodium salt, sodium hydrogen L-glutamate (**2-25**), have unique organoleptic properties. They are salty, but also have the taste known as **umami** (see Section 8.3.3.1). They are thus used as food additives, intensifying the flavour of meat and vegetable dishes.

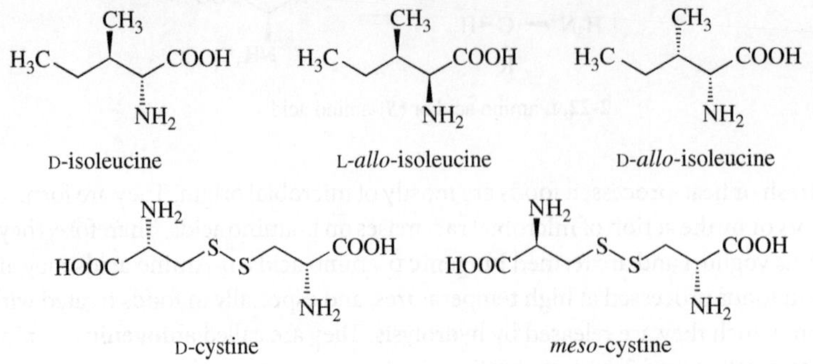

D-isoleucine   L-*allo*-isoleucine   D-*allo*-isoleucine

D-cystine   *meso*-cystine

2-24, isoleucine and cysteine stereoisomers

2.3 Peptides

Peptides have a number of different functions in organisms and exhibit a variety of biological effects. Some are widely distributed in nature, occurring in diverse organisms, whereas others have a restricted occurrence. Many oligopeptides and polypeptides act like hormones (e.g. the linear peptides secretin, insulin, and thyroliberin and the cyclic peptides oxytocin and vasopressin) or antibiotics (gramicidine, polymyxin, bacitracin, and other peptide antibiotics). Many microbial peptides, collectively known as **bacteriocins**, have antibacterial properties. In practice, the polycyclic peptide nisin is used as a food preservative (see Section 11.2.1.2.6).

Furthermore, peptides include a number of mycotoxins such as botulinum toxin produced by the bacterium *Clostridium botulinum* (see Section 12.3.2.1), cyclic toxins of higher fungi such as fallotoxins and amatoxins of toxic agarics of the genus *Amanita* (see Section 10.3.7.2), insect toxins such as apamine and mellitine in bee venom, and toxins of many other insect species (spiders, scorpions, reptiles, and amphibians).

Some peptides that are present in foods significantly influence their organoleptic properties, especially the taste.

2.3.1 Structure, nomenclature, and classification

Peptides are amino acid oligomers and polymers in which the carboxyl group of one amino acid is bound to the amino group of another via an **amide bond**. The amide bond thus created is called a **peptide bond** because of its specific nature (amide bonds resulting from the condensation of amino groups of α-amino acids with carboxylic groups of organic acids, e.g. in pantothenic acid and folic acid, are not classified as peptides). The combination of two α-amino acids yields a dipeptide, three molecules of amino acids produce a tripeptide, and so on. The formation of a dipeptide from two amino acids is accompanied by the loss of a water molecule. Each linear peptide contains one amino acid with a free carboxyl group, which is a **C-terminal amino acid**, and one amino acid with a free amino group, an **N-terminal amino acid**. In addition to linear peptides, the condensation of amino acids can also lead to cyclic structures, which have no free carboxyl group and no free amino group. Such compounds are called cyclic peptides. Examples of simple cyclic dipeptides are 3,6-disubstituted 2,5-dioxopiperazines (Figure 2.14).

The peptide bond is essentially a rigid planar structure. In dipeptides, six atoms lie in the same plane: the α-carbon atom and C=O group from the first amino acid and the NH group and α-carbon atom from the second amino acid. The peptide bond shows substantial double-bond characteristics, which prevent rotation about it (Figure 2.15). Owing to the rigidity of the bond, peptide groups may be present in *cis* or *trans* conformations (Figure 2.16). The *trans* form is strongly favoured; therefore, natural peptides and proteins contain almost exclusively the energetically preferable *trans* conformers, where the neighbouring C$_a$ atoms are situated on the opposite sides of the peptide

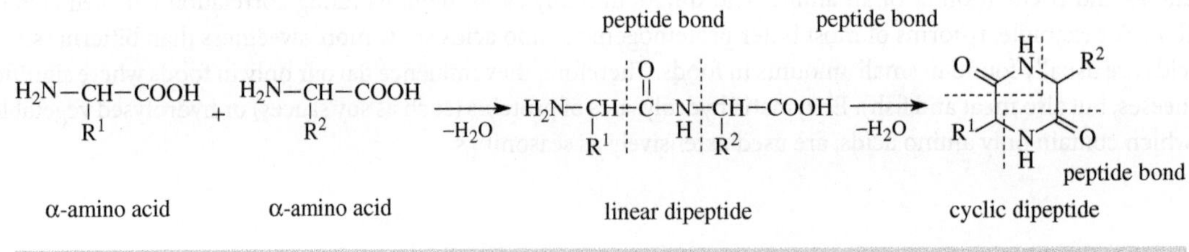

Figure 2.14 Formation of linear and cyclic dipeptides.

bond. Typically, the *cis* conformation occurs only when the nitrogen atom in the peptide bond is from proline. In the peptide bond region, only the substituents on C_α carbons can freely rotate, and they thus determine the conformational structure of the peptide chain. The C_α carbon attachment into the peptide chain involves two single bonds, a C_α—N bond and a C_α—C bond. Other groups of the peptide chain can freely rotate around these bonds. The degree of rotation can be specified by dihedral angles of rotation around the bonds C_α—N (angle φ; phi) and C_α—C (angle ψ; psi), which are called torsion angles. The different values of these angles give rise to a large number of conformations of peptides and proteins.

Figure 2.15 Peptide bond (substituents at the C_α are not marked).

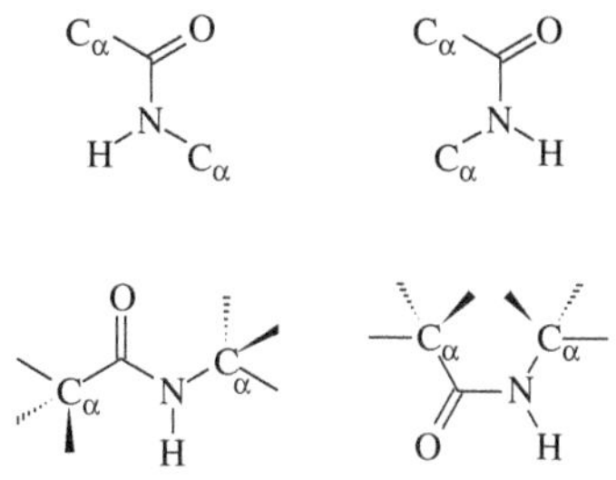

Peptides are usually classified according to the molecule size (number of bound amino acids), the shape of the chain, and the type of links in the chain. According to the number of bound amino acids (monomers), they are divided into:

- **oligopeptides** (containing typically 2–10 amino acid molecules in the chain);

- **polypeptides** (containing typically 11–100 amino acid molecules or having relative molecular weights of about 10 kDa; higher polypeptides were formerly known as **macropeptides**).

Figure 2.16 Peptide bond *trans* and *cis* conformations.

According to the chain type, peptides are divided into:

- **linear peptides** (which represent the clear majority);

- **cyclic** peptides.

Peptides containing only amino acids (often the less common amino acids) are called **homeomeric peptides**, whilst those that also contain other compounds are **heteromeric peptides** or **peptoids**. This group includes:

- nucleopeptides;

- lipopeptides;

- glycopeptides;

- phosphopeptides;

- chromopeptides;

- metalopeptides.

Peptides in which all of the covalent linkages between the constituent amino acids are peptide bonds are termed **homodetic peptides**, whereas those that contain peptide bonds and other covalent linkages between certain amino acid residues are called **heterodetic peptides**. Examples of the latter covalent linkages are disulfide bonds (R—S—S—R′), thioester bonds (R—CO—S—R′), and ester bonds (R—CO—O—R′). The relevant peptides with one or more ester bonds are **depsipeptides**. The peptide bonds usually contain bound α-amino groups and α-carboxyl groups. Unlike proteins, the peptide bonds in peptides sometimes contain the distal functional groups, such as the γ-carboxyl group of glutamic acid in the so-called γ-peptide bond. Also unlike proteins, peptides frequently contain D-amino acids and unusual amino acids, for example β-alanine, GABA, pyroglutamic acid, and others.

The terminology considers linear peptides as acylated amino acids. Therefore, the amino acid at the carboxyl terminus of the chain is what gives it its name; for the other amino acids, the ending -ine is replaced by the suffix -yl. The practice uses one or three letters to label peptide chains for amino acids. The simplest dipeptide, composed of two molecules of glycine, is called glycylglycine (**2-26**), or Gly–Gly or G–G for short. The important tripeptide with the trivial name glutathione (**2-26**) is γ-L-glutamyl-L-cysteinylglycine. Using the three letter symbols, glutathione is:

$$\underset{\lfloor}{\overset{\text{Glu}}{}}\ \text{Cys-Gly}\quad\text{or}\quad \gamma\underset{\lfloor}{\overset{\text{Glu}}{}}\ \text{Cys-Gly}$$

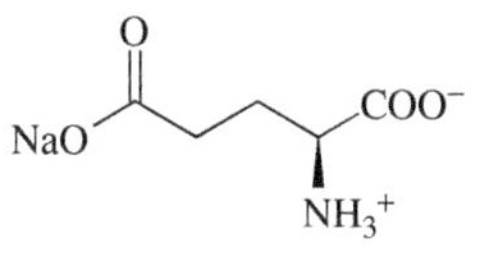

2-25, sodium hydrogen glutamate

The symbol ⌐ indicates the distal carboxyl group linkage. Using one-letter amino acid symbols, it is E-C-G. The nomenclature of cyclic peptides is based on either the peptide or the heterocycle. The cyclic dipeptide derived from glycine is cyclo(Gly–Gly) or 2,5-dioxopiperazine (Figure 2.14, R=H). Most naturally occurring cyclic peptides contain a higher number of amino acids.

glycylglycine   glutathione

2-26, examples of simple linear peptides

2.3.2 Biochemistry

Peptides are biosynthesised by both ribosomal and non-ribosomal processes from a wide range of amino acids. Ribosomal peptide biosynthesis leads to peptide hormones and many other physiologically active substances. Many structures biosynthesised by ribosomal processes are post-translationally modified. Glycopeptides are produced by adding sugar residues via *O*-glycoside linkages to the hydroxyls of serine and threonine residues or via *N*-glycoside linkages to the amino group of asparagine. With phosphopeptides, the hydroxyl groups of serine or threonine are esterified with phosphoric acid. Many peptides contain a pyroglutamic acid residue at the N-terminus, which is a consequence of glutamine intramolecular cyclisation between the γ-carboxyl and the α-amino group. The C-terminal carboxylic acids may also frequently be converted into an amide. Non-ribosomal processes synthesise many natural peptides that occur in foods via a sequence of enzyme-controlled reactions. They are responsible for the formation of glutathione, histidine dipeptides of skeletal muscles, peptide toxins, and peptide antibiotics.

2.3.3 Occurrence

Peptides occurring in foods are either products of metabolism, a result of the genetic dispositions of the animal or plant organisms, or products of hydrolysis of diverse food proteins during processing and storage. The process is known as **proteolysis**. Peptides can also arise when amino acids are heated to higher temperatures, but this reaction has no significance in practice.

2.3.3.1 Natural peptides

2.3.3.1.1 Glutathione

Tripeptide glutathione, γ-L-glutamyl-L-cysteinylglycine (**2-26**), is a common peptide that is found in animal and plant tissues and microorganisms. It occurs in two forms: a reduced form with free sulfhydryl group (denoted G-SH) and an oxidised form with disulfide bond (G-S-S-G), which are part of an important redox system, similar to cysteine and cystine:

$$2\text{G-SH} \rightleftharpoons \text{G-S-S-G} + 2\text{H}$$

Glutathione is the most abundant non-protein thiol in cells and is also a cofactor of some enzymes, such as glyoxalase. In animal cells, glutathione is present in high amounts of 300–1500 mg/kg. Higher plants contain lower amounts of glutathione (wheat flour 10–15 mg/kg), whilst some microorganisms, such as yeast *S. cerevisiae*, occur at about 5000 mg/kg dry cell weight. The levels of glutathione in common species of mushrooms, where ergothioneine (**2-8**) is the characteristic thiol antioxidant, are relatively low (0.11–2.41 mg/g d.w.).

The glutathione redox system of reduced and oxidised glutathione allows it to act as an effective electron acceptor and donor for numerous chemical and biochemical reactions. Glutathione also modifies protein sulfhydryl groups via a number of reactions: reduction of protein sulfenic acids, formation of protein mixed disulfides, and the subsequent reduction of them. For example, glutathione in flour affects the rheological properties of dough (see Section 5.14.6.1). The nucleophilic thiol group of glutathione allows reactions with reactive oxygen and reactive nitrogen radicals, hydrogen peroxide, and hydroperoxides of fatty acids, formation of mercaptides with metals, and reactions with various electrophiles. Thus, glutathione is an ideal substance to protect organisms against oxidative stress, poisoning by metal ions, and poisoning by a number of exogenous and endogenous toxic organic substances, including carcinogens.

2.3.3.1.2 Other glutamyl peptides

Both animal and plant tissues contain many different γ-L-glutamyl peptides that have a number of important functions. γ-Glutamyl peptides are formed in reactions catalysed by the enzyme L-γ-glutamyl transferase, which catalyses the splitting of the γ-glutamyl linkages in reduced

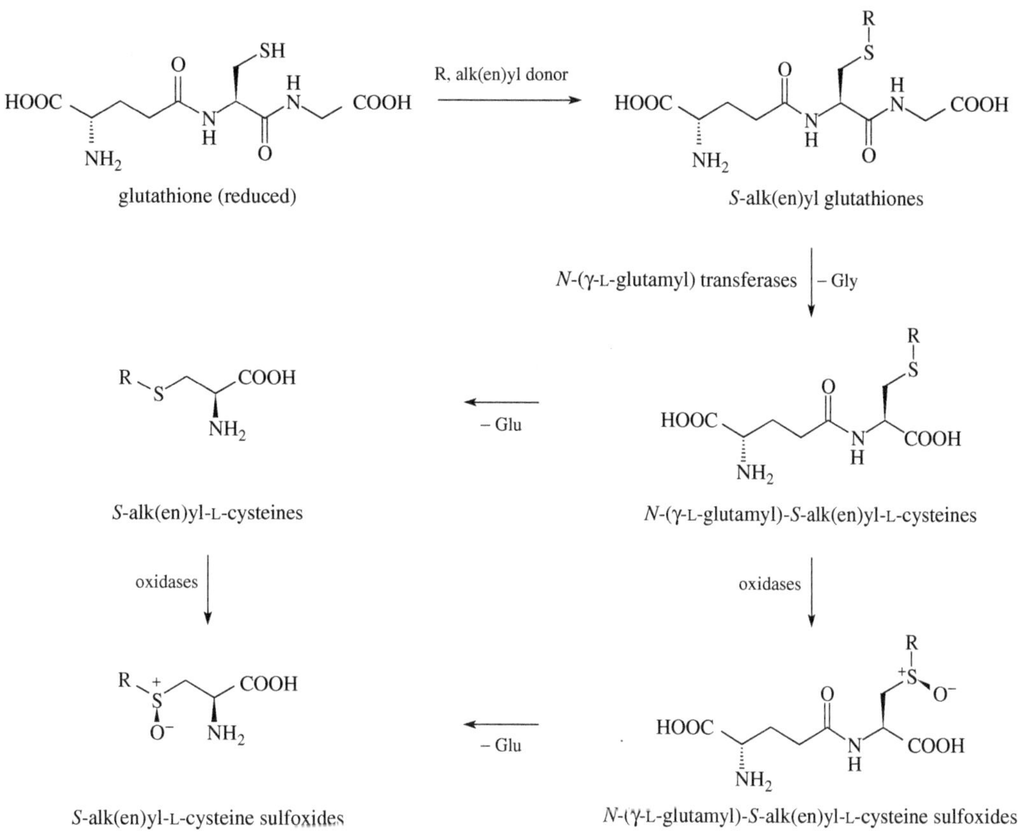

Figure 2.17 Formation of S-alk(en)ylcysteine sulfoxides in *Allium* vegetables.

glutathione (G-SH) and the transfer of the glutamate moiety to amino acid or peptide acceptors. For example, plant γ-glutamyl peptides play a role in the transport of amino acids through membranes, protecting plant cells against the effects of phytotoxic heavy metals (by acting as phytochelatins; see Section 6.2.2.2). In *Allium* and *Brassica* plants, they have a key role as a reserve for nitrogen, sulfur (they contain up to 50% organically bound sulfur), and selenium compounds, and they are a part of the defence mechanisms of plants against elicitors and predators. For some peptides, such as γ-glutamyl-S-allyl-cysteine, an antiglycation effect was shown.

Plants of the genus *Allium* typically contain more than 20 γ-glutamyl peptides, such as S-alk(en)yl glutathiones, N-(γ-glutamyl)-S-alk(en)yl cysteines, and N-(γ-glutamyl)-S-alk(en)yl cysteine sulfoxides. These peptides occur as intermediates in the biosynthesis of S-alk(en)ylcysteine sulfoxides in *Allium* vegetables (Figure 2.17). The main γ-glutamyl peptides in garlic cloves are N-(γ-glutamyl)-S-(prop-2-en-1-yl)-L-cysteine (R=CH$_2$—CH=CH$_2$) and (E)-N-γ-glutamyl-S-(prop-1-en-1-yl)-L-cysteine (R=CH=CH$_2$—CH$_3$), which are present at levels of approximately 2.9–4.6 and 4.6–5.4 g/kg f.w., respectively. They are followed by N-(γ-glutamyl)-S-(2-carboxypropyl)-L-cysteinyl glycine, R=CH$_2$—CH(COOH)CH$_3$, known as S-(2-carboxypropyl)glutathione (0.8–1.2 g/kg f.w.), N-(γ-glutamyl)-S-methyl-L-cysteine (0.04–0.4 g/kg f.w.), and N-(γ-glutamyl)-S-(prop-2-en-1-yl)-L-cysteine sulfoxide (R=CH$_2$—CH=CH$_2$; 0.04–0.08 g/kg f.w.). S-(2-Carboxypropyl) glutathione is the main γ-glutamyl peptide of onion bulbs (0.1–0.9 g/kg f.w.), where it is accompanied by (E)-N-γ-glutamyl-S-(prop-1-en-1-yl)-L-cysteine sulfoxide (R=CH=CH$_2$—CH$_3$; 0.2–2.3 g/kg f.w.), (E)-N-γ-glutamyl-S-(prop-1-en-1-yl)-L-cysteine (up to 0.1 g/kg f.w.), and N-(γ-glutamyl)-S-methyl-L-cysteine (up to 0.1 g/kg d.w.). In pre-bulbing onions, levels of (E)-γ-glutamyl-S-(prop-1-en-1-yl)cysteine sulfoxide and S-(2-carboxypropyl)glutathione were found to be below 0.05 g/kg (f.w.), and at the bulbing onion stage they amounted to 2.1 and 4.0 g/kg (f.w.), respectively. The main γ-glutamyl peptide in beans is N-(γ-L-glutamyl)-S-methyl-L-cysteine, and an important γ-glutamyl peptide of shiitake mushrooms (*Le. edodes*) is lentinic acid (see Section 8.2.12.9.7).

2.3.3.1.3 Histidine dipeptides

In addition to glutathione, the muscle tissue of animals contains a group of dipeptides (**2-27**) derived from L-histidine (its *1H*- and *3H*-isomers, respectively). These dipeptides include carnosine (β-alanylhistidine), anserine (β-alanyl-1-methylhistidine), balenine (also known as ofidine, β-alanyl-3-methylhistidine), and homocarnosine (γ-aminobutyrylhistidine). The biological role of these dipeptides is not yet fully known. It is assumed that they are involved in the contraction of skeletal muscles and are associated with some enzymes containing copper in the molecule that show the activity of neurotransmitters, vasodilators, and modulators. Anserine and carnosine can inhibit lipid

Table 2.5 Content of histidine dipeptides in fresh meat.

Meat	Content (mg/kg) Carnosine	Anserine	Balenine	Meat	Content (mg/kg) Carnosine	Anserine	Balenine
Pork	1040–4190	70–160	180	Horse	3820–4023	30–48	0
Beef	1520–3650	110–552	17	Rabbit	497	4536	0
Sheep	670–1898	430–1992	24	Chicken	100–1117	550–3350	–
Goat	520–1030	750–2016	0	Turkey	1600–2400	6150	–

oxidation through a combined mechanism. Carnosine can react with carbonyls, suppressing protein glycation (see Section 4.7.5.6). The buffering capacity of these dipeptides at physiological pH values is also important (they are present as cations with an ionised primary amino group, imidazole ring nitrogen, and carboxyl group). Carnosine has recently appeared in some food supplements for athletes, but its use is not yet supported by credible scientific results.

carnosine ($n = 1$)
homocarnosine ($n = 2$)

anserine

balenine

2-27, histidine dipeptides

Because of the variations in different types of meat, the dipeptide contents can serve as criteria for determining the origin of meat and meat products (Table 2.5). For example, anserine is the dominant dipeptide in meat from birds, whilst both anserine and carnosine are the main dipeptides in animal tissues (human tissues contain only carnosine). The content of individual dipeptides varies in different muscles and depends on many factors, such as the age of the animal and the type of processing. Breast muscles of poultry, for example, contain about 5–10 times higher amounts of carnosine and 5 times higher amounts of anserine than leg muscles (white breast muscles are involved in intense anaerobic metabolism and therefore require a higher buffering capacity). Determination of these two peptides has been used to evaluate the amount of chicken meat in pork meat products. Balenine is usually present in small amounts in meat from snakes. In marine mammals such as whales, balenine is the main histidine dipeptide.

The sensory properties of histidine dipeptides and their reaction products may affect the oral sensation of meat products as they contribute to the umami taste (see Section 8.3.3.1) and enhance the flavouring properties of sodium hydrogen glutamate. The key contributors to the typical chicken broth flavour (taste) are, for example, anserine, carnosine, and β-alanylglycine. During the processing of meat, histidine dipeptides are partly hydrolysed by endogenous peptidases (e.g. during curing of ham about one-half of these dipeptides are hydrolysed) and non-enzymatically. They also partly react with protein-bound asparagine and glutamine and are incorporated into proteins during heating; in beef soup stock solution, hydrolysis of proteins with incorporated carnosine yields β-aspartyl-β-alanylhistidine and γ-glutamyl-β-alanylhistidine tripeptides, respectively.

2.3.3.2 Proteolytic products

Proteolytic enzymes in virtually all raw food materials hydrolyse proteins during the storage of these materials and during food production, which leads to a mixture of different peptides with proper organoleptic peculiarities (see Section 8.3.4). In some technologies, spontaneous proteolysis is very desirable or necessary, such as during meat curing and in cheese and malt production.

Skeletal muscles in post mortem state have to undergo a number of biochemical changes in order to became the final product: meat. Changes of tissue proteins are catalysed by multiple intracellular proteolytic systems, including lysosomal (thiol proteases cathepsins), calcium-dependent, and other proteases. Proteolysis of myofibrillar proteins presumably plays a major role in post mortem meat tenderisation. Intense proteolysis has been reported to occur during the processing of dry-cured ham. This gives rise to the formation of free amino acids and short peptides, especially from actin.

In cheese production (see Section 2.4.5.2.1), proteolysis has three phases: proteolysis in milk before cheese manufacture due to indigenous milk protease (plasmin) activity, enzymatically induced coagulation of the milk in rennet cheeses (hydrolysis of κ-casein by rennin), and proteolysis during ripening of most cheeses. The latter is the most important reaction, having a major impact on flavour and texture. Dozens of different peptides have been identified in cheeses. Most of them arise from α_{S1}- and β-caseins, and a few come from α_{S2}- and κ-caseins. The proteinases involved in hydrolysis of α_{S1}-casein are mainly cathepsin D originating from milk and cell-envelope proteinase from thermophilic starters, whilst β- and α_{S2}-caseins are mainly hydrolysed by plasmin. Moreover, peptidases from starters are also active throughout the ripening process, presumably like those from non-starter lactic acid bacteria (see Section 8.3.5.1.4).

Some proteins (especially lipid transfer protein and protein Z, see Section 2.4.5.4.1) and hydrophobic polypeptides, resulting from barley proteins in malt via endogenous proteases and yeast autolysis, are important foam stabilisers in beer.

2.3.3.2.1 Linear peptides

Many food products (including soy sauce and other East Asian sauces) that are produced by enzymatic hydrolysis of proteins or protein-rich materials practically contain only peptides, instead of proteins. Other materials that contain variable amounts of peptides include yeast autolysates and blood, whey, and casein hydrolysates, which are used as food additives of high nutritional value.

During the partial enzymatic hydrolysis of proteins, bitter-tasting peptides are often released, which limits the application of these hydrolysates in food processing. Casein and soy protein hydrolysates in particular have a tendency towards bitterness. The bitter taste is caused by peptides containing hydrophobic amino acids (valine, leucine, isoleucine, phenylalanine, tyrosine, and tryptophan). Proline (which commonly occurs in the penultimate position) and pyroglutamic acid are also often present in bitter peptides, whilst cysteine and methionine are usually absent (see Section 8.3.5.1.4). For example, four main bitter peptides, YGLF, IPAVF, LLF, and YPFPG-PIPN, which originated from α-lactalbumin, β-lactoglobulin, serum albumin, and β-casein, respectively, were found in a whey protein hydrolysate. The compounds with the highest perceived bitterness intensity in aged sharp Cheddar cheese were peptides GPVRGPFPIIV, YQEPVLGPVRGPFPI, MPFPKYPVEP, MAPKHKEMPFPKYPVEPF, and APHGKEMPFPKYPVEPF; all originated from β-casein. Whilst MAPKHKEMPFPKYPVEPF was found in young mild Cheddar at similar levels, APHGKEMPFPKYPVEPF was reported only in the sharp one. The concentrations of the other compounds increased during maturation of the sharp cheese, especially GPVRGPFPIIV (about 30-fold).

Methods for eliminating unintended bitter peptides in partial protein hydrolysates are known, but they often cause a significant loss of essential amino acids. These procedures usually include additional enzymatic hydrolysis under controlled conditions (a shorter time for the hydrolysis leads to higher peptides that are not bitter) and a selection of suitable proteases, such as aminopeptidases, carboxypeptidases, and some other types. Enzymes of plant and microbial origin have been successfully used for this purpose. For example, the intracellular peptidases from *Lactococcus lactis* ssp. *cremoris* and *Brevibacterium linens*, which have high proteolytic activity, successfully hydrolyse bitter peptides in cheeses. The bitterness of hydrolysates can be reduced considerably by the plastein reaction. Under certain conditions (temperature, pH, peptidase concentration, and other factors), proteases (such as endopeptidases) can synthesise new larger peptides from undesirable bitter peptides; these larger peptides are known as plasteins and are not bitter. The plastein reaction involves a transpeptidation or condensation mechanism. Plastein synthesis can also be used for the incorporation of nutritionally valuable amino acids (methionine, lysine, and tryptophan) into proteins of low biological value, for the removal of undesirable amino acids from proteins (phenylalanine, see Section 10.3.1.2), as well as for the recovery of protein from non-traditional and waste proteinaceous materials.

The bitter taste-lowering potential of β- and γ-cyclodextrin was shown for a rice protein hydrolysate. It was found that the hydrophobic side chain of amino acids penetrates directly into the cavity of the cyclodextrins, and β-cyclodextrin in particular reduces bitterness significantly. Microencapsulation of hydrolysates together with incorporation of masking agents is believed to be favourable in reducing bitterness in hydrolysates. For example, bitter taste decreased when casein hydrolysates were spray-dried and mixed with gelatin and soy protein isolate as carriers.

Another impact on food quality may be increased susceptibility of protein hydrolyzates to non-enzymatic browning reactions.

Some peptides with particular amino acid sequences that are inactive in the intact protein may exert biological functions after their release from the intact protein molecule. Such bioactive peptides have positive impacts on a number of physiological functions in living beings. The search for bioactive peptides has thus increased significantly in the last 20 years. They have been described as peptides with hormone- or drug-like activity that eventually modulate human physiological functions by binding to specific receptors on target cells. They can be used as part of a functional food or as therapeutic peptides. Not all peptides in a protein hydrolysate exhibiting an effect *in vitro* exert the same effect *in vivo*; only those resisting intestinal digestion will do so. The discovery of some food-derived prolyl and pyroglutamyl peptides with 2–3 amino acids in the blood of humans indicates that the biological activities of food-derived peptides in the body, rather than in food, are crucial to understanding the mechanism of the beneficial effects of orally administered peptides. Bioactive peptides (protein fragments), generally consisting of 3–20 amino acids, but sometimes more, arise through hydrolysis of many dietary and other proteins (e.g. those from food-industry byproducts) during:

- fermentation processes using proteolytic starter cultures;

- the manufacture of protein hydrolysates;

- digestion of proteins;

- the action of digestive enzymes on proteins *in vitro*.

Bioactive peptides display a broad scope of functions and can be divided into many groups, the best known of which are:

- opioids;

- antihypertensive peptides;

- mineral-binding peptides;

- antioxidative and anti-inflammatory peptides;

- hepatoprotective peptides;

- antimicrobial peptides;

- immunomodulating and cytomodulating peptides.

Unlike endorphins (peptide hormones found mainly in the brain that bind to opiate receptors and reduce the sensation of pain and affect emotions), exorphins are found in food, mainly in milk, dairy products, and gluten-rich wheat. The most important are β-casomorphins, which are released by the digestion of β-casein in bovine milk and dairy products. β-Casomorphins were detected in raw cow's milk, human milk, and a variety of commercial cheeses, but not in commercial yoghurts. It is suggested that they could form in yoghurt, but are degraded during processing. They are denoted by a numeral indicating the number of amino acids. The first β-casomorphin, heptapeptide (also known as casomorphin 7, Tyr-Pro-Phe-Tyr-Gly-Tyr-Ile), was isolated from a casein hydrolysate in 1970. In 2009, the European Food Safety Agency (EFSA) published a comprehensive review on casomorphins, which are classified as substances with opioid effects. This is why milk, for instance, can help people fall asleep, because it contains opioid peptides. Exorphins also arise from other milk proteins (α_{S1}-casein, κ-casein, α-lactalbumin, β-lactoglobulin, and serum albumin), cereal proteins (such as wheat gluten and gliadin), soybeans (α-protein), meat and poultry proteins (albumin, haemoglobin, γ-globulin), eggs (ovalbumin), and other proteins.

Another group of bioactive peptides can inhibit the activity of the angiotensin-converting enzyme (ACE inhibitors) and thus reduces blood pressure. ACE inhibitory peptides with a chain length of 2–10 amino acids were first obtained from milk proteins (caseins and serum proteins), although they have also been found in a number of other animal and plant proteins. For example, one of about 150 identified bioactive peptides from α_{S1}-casein (of the genetic variant B, see Section 2.4.5.2.1) is a hexapeptide, comprising Lys-Thr-Thr-Met-Pro-Leu. An antihypertensive hexapeptide Val-Pro-Phe-Gly-Val-Gly isolated from wheat sourdough, traditionally used in bread making, is a product of glutenin hydrolysis. Fermentation enables the broader formation of bioactive peptides. For example, lactic-fermented buckwheat (*Fagopyrum esculentum*) led to the formation of new, highly potent blood pressure-lowering peptides and increased levels of GABA and tyrosine in buckwheat sprouts. Analyses of amino acid sequences of chia seed (*Salvia hispanica* L.) showed peptides with bioactive potential, including dipeptidyl peptidase-IV inhibitors, ACE inhibitors, and antioxidants.

Similarly to mineral-binding proteins, casein-derived phosphopeptides can form salts with calcium due to the binding properties of the phosphoserine residue. These peptides are involved in the increased absorption and bioavailability of calcium and other minerals (zinc, copper, manganese, and iron) in the intestine. An example is a heptapeptide Pro-Val-Ala-Leu-Ile-Asn-Asn derived from κ-casein. Atlantic salmon (*Salmo salar* L.) ossein and tilapia (*Oreochromis niloticus*) protein hydrolysate are also good sources for calcium-binding peptides. The amino acid sequence of the tilapia calcium-binding peptide is Trp-Glu-Trp-Leu-His-Tyr-Trp, with Glu and His as principal binding residues.

Peptides with antioxidant properties are effective against enzymatic and non-enzymatic oxidation of lipids, as free radical scavengers, and in chelation of metal ions. An example is a soybean pentapeptide with the sequence Leu-Leu-Pro-His-His. Bioactive peptides Ala-Glu-Glu-Arg-Tyr-Pro and Asp-Glu-Asp-Thr-Gln-Ala-Met-Pro isolated from hydrolysed chicken egg white showed strong oxygen radical absorbance capacities. Peptides that show *in vitro* antioxidant activity have to be resistant to gastrointestinal (GI) digestion in order to exert antioxidant activity *in vivo*. For example, amongst many *in vitro*-active low-molecular-weight peptides of casein hydrolysis with alcalase, only Ala-Tyr-Pro-Ser might strongly contribute to antioxidant action *in vivo*.

In Japanese rice wine (sake), 19 pyroglutamyl peptides have been identified. The major peptide is a hepatoprotective peptide pyroGlu-Leu. Lunasin, a soybean seed-derived bioactive peptide, is an example of the high-molecular bioactive peptide, since it is composed of 43 amino acid residues with a molecular weight of 5.5 kDa. Lunasin possesses inherent antioxidative, anti-inflammatory, and anticancerogenic properties, and plays a role in regulating cholesterol biosynthesis in the body.

Antimicrobial peptides are effective against various bacteria and yeasts. An example is lactoferricin, which includes residues 17–41 from lactoferrin. Other peptides with antimicrobial properties are isracidin, derived from α_{S1}-casein (residues 1–23), and casocidin-I, formed by hydrolysis of α_{S2}-casein (residues 150–188). Some of the bacteriocins are used as food preservatives (see Section 11.2.1.2.1). Lactoferricin also has immunomodulating and cytomodulatory properties, which similarly show many peptides arising by hydrolysis of milk caseins and serum proteins. Many immunomodulatory peptides have been identified in soybeans, honey, and other food materials.

In addition to the presence of bioactive peptides, consumption of protein hydrolysates results in more rapid uptake of amino acids compared with intact protein or free amino acid mixtures and better hyppoallergenic characteristics compared with the intact proteins.

2.3.3.2.2 Cyclic peptides

Cyclic dipeptides, 2,5-dioxopiperazines, are mainly formed by dehydration of linear dipeptides in acidic media and partly through stepwise cyclisation from free amino acids at high temperatures. They have been identified in a wide range of fermented and thermally treated foods. These smallest cyclic peptides are reported to demonstrate antibacterial, antifungal, and many anthroprotective effects. They also display an array of chemesthetic effects (bitter, astringent, metallic, and umami) that can sometimes significantly contribute to the perceived bitterness taste of a variety of foods. 2,5-Dioxopiperazines are partly responsible for the bitterness of coffee, cacao, roasted cereal grains, roasted malt, beer, aged sake, and other foods. For example, the following proline-based diketopiperazines have been identified in roasted coffee: cyclo(Ile-Pro), cyclo(Leu-Pro), cyclo(Phe-Pro), cyclo(Pro-Pro), and cyclo(Val-Pro). Seven proline-based diketopiperazines, namely cyclo(Ala-Pro), cyclo(Val-Pro), cyclo(Ile-Pro), cyclo(Leu-Pro), cyclo(Met-Pro), cyclo(Phe-Pro), and cyclo(Pro-Pro), have been identified in beer, and (Z)-cyclo(Leu-Pro) and (Z)-cyclo(Phe-Pro) have been found in wheat dough and bread. The levels of the latter cyclic dipeptides increased from approximately 1–35 and 25 µg/kg of dough, respectively, after 48 hours of incubation. Almost 100 and 2000 times higher levels of the dipeptides were found in the crumbs and crust, respectively, compared to the dough prior to baking. Acid protein hydrolysates used as seasonings contain about 25 mg/kg of dioxopiperazines derived from the hydrophobic amino acids valine, leucine, isoleucine, and phenylalanine. The main sensory active products are cyclo(Val-Val), cyclo(Val-Ile), cyclo(Val-Phe), cyclo(Leu-Phe), and cyclo(Ile-Ile).

Bitter, off-tasting cyclopeptides can also arise during the storage of cold pressed vegetable oils. For example, fresh linseed oil provides a nutty flavour, but a bitter off-taste is developed on storage at room temperature. The key bitter compound responsible for the bitter taste is octapeptide cyclo(Pro-Leu-Phe-Ile-MetO-Leu-Val-Phe), known as cyclolinopeptide E, which contains methionine sulfoxide (MetO). Other hydrophobic cyclolinopeptides found in flaxseed and flax oil show immunosuppressive activity.

2.3.4 Properties

2.3.4.1 Physical–Chemical properties

As with amino acids, peptides dissociate in aqueous solutions and form internal salts (zwitterions). The values of the dissociation constants (pK values) and isoelectric points (pI) of some simple peptides are shown in Table 2.6. It is clear (by comparing the values for Gly, Gly-Gly, and Gly-Gly-Gly) that the increase in the molecular weight of a peptide reduces the acidity and basicity of the functional groups. The pI values of peptides are generally different from those of the corresponding amino acids. The pK and pI values of peptides containing the same amino acids are different depending on the order of the amino acids (compare Gly-Asp and Asp-Gly).

Table 2.6 pK and pI values of peptides.

Peptide	pK_1	pK_2	pK_3	pK_4	pK_5	pI
Gly-Gly	3.12	8.17	–	–	–	5.6
Gly-Gly-Gly	3.26	7.91	–	–	–	5.6
Ala-Ala	3.30	8.14	–	–	–	5.7
Gly-Asp	2.81	4.45	8.60	–	–	3.6
Asp-Gly	2.10	4.53	9.07	–	–	3.3
Lys-Ala	3.22	7.62	10.70	–	–	9.2
Asp-Asp	2.70	3.40	4.70	8.26	–	3.0
Lys-Lys	3.01	7.53	10.05	11.01	–	10.5
Lys-Lys-Lys	3.08	7.34	9.80	10.54	11.32	10.9

2.3.4.2 Organoleptic properties

Peptides, like amino acids, can taste bitter, sweet, salty, or indifferent. Some peptides reveal umami taste and show umami-enhancement (see Section 8.3.4). Most natural and synthetic oligopeptides have a bitter taste (see Section 8.3.5.1.4). The best-known sweet-tasting peptide is the methyl ester of L-aspartyl-L-phenylalanine, which has long been used as the sweetener aspartame (see Section 11.3.2.1.2). A sweet taste also have other compounds derived from L-aspartic acid (advantame, alitame, and neotame) and its lower homologue, aminomalonic (aminopropanedioic) acid.

The hydrochlorides of some amides (peptides) have a salty taste, which suggests the possibility of using them as a replacement of sodium chloride for people who need to cut down their salt intake for health reasons (see Section 8.3.2). Hydrochlorides of L-ornithyl (**2-28**) and L-lysyltaurine exhibit a salty taste of similar quality and intensity as sodium chloride. A weaker salty taste is found in the hydrochloride of L-ornithyl-β-alanine (**2-28**) and the hydrochloride of L-ornithyl-γ-aminobutyric acid.

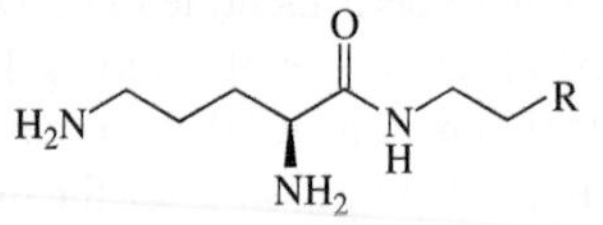

ornithyltaurine, R = SO$_3$H
ornithyl-β-alanine, R = COOH

2-28, structures of salty peptides (as hydrochlorides)

2.4 Proteins

The process of protein synthesis creates polymers from amino acids. Proteins commonly contain more than 100 amino acids bound into linear chains by covalent peptide bonds. Their relative molecular weight ranges from 10 000 to several million daltons. In addition to peptide bonds, some other covalent bonds (such as disulfide bonds between bound cysteine side chains and ester bonds, which allow the connection of, for example, serine and threonine through phosphoric acid) can be important determinants of protein structure. Apart from covalent bonds, various electrostatic interactions can also be important to the protein structure (ionic, hydrogen, and hydrophobic interactions; see Section 7.6.1). Hydrogen bonds (bridges) can exist between a variety of compounds (functional groups), for example between two different amino acid side chains or between an amino acid side chain and a water molecule. Hydrophobic bonds occurring in the hydrophobic protein core are the major forces driving the correct folding of the protein. Protein molecules are also bound with water and various inorganic ions. Some proteins contain physically or chemically bound organic compounds, such as lipids, sugars, nucleic acids, and various coloured organic compounds.

Apart from water, proteins make up the major portion of the mass of living organisms. According to the biological functions they perform in biochemical processes, proteins are often distinguished as:

- **structural** proteins (occurring primarily as structural cell components of animal and plant tissues);

- **catalytic** proteins (enzymes, hormones);

- **transport** proteins involved in the movement of ions, small molecules, or macromolecules (e.g. haemoglobin);

- **mobility** proteins participating in muscle contraction (actin, myosin, and actomyosin);

- **defence** proteins (antibodies and immunoglobulins);

- **storage** proteins (e.g. ferritin);

- **sensory** proteins (e.g. rhodopsin);

- **regulation** proteins (histones and hormones).

Proteins ultimately perform **nutritional** functions, together with lipids, carbohydrates, and minor nutrients, and are amongst the most important components of human diet. They are the main source of nitrogen, essential amino acids, and other material needed for the growth and repair of muscles, bones, skin, and other tissues, and are a source of energy. For human nutrition, proteins are derived from various foods. Foods of animal origin (meat, milk, and eggs) provide on average about 25% of the protein in the human diet worldwide, whilst proteins of vegetable origin (mainly cereals and pulses, but also fruits, vegetables, and root crops) contribute to about 65% of the per capita supply of protein on a worldwide basis. Of course, these numbers vary greatly between different countries and geographic regions. Non-conventional sources such as various seeds, plant waste, and algae can also potentially provide protein for human consumption.

In food raw materials, which are, with some exceptions, post mortem animal tissues and post-harvest plant tissues, the majority of proteins do not possess their original biological functions. Furthermore, animal and plant tissues of raw materials are often damaged during food processing or cooking, which can result in desirable or undesirable activity of various enzymes. Apart from the enzymes that occur naturally in raw materials, foods may also contain enzymes produced by microorganisms (both naturally occurring and those whose properties are used in food technology procedures), as well as enzymes added for various reasons during processing. In many cases, foods processing leads to a number of physical and chemical changes in proteins, which is known by the general term **denaturation**. Thus, proteins are often classified according to the state in which they are found in foods as either **native** (natural) or **denatured**. Native proteins preserve all the biological functions they have in living organisms, whilst denatured proteins do not, as the quaternary, tertiary, and secondary structures present in their native state are lost. The intentional modification of food protein side chains with chemical reagents, currently applied only to a limited extent, produces chemically **modified** proteins with altered properties. The modified proteins, mostly used as food additives for specific purposes, can have an improved nutritional quality, physical state (such as texture), and functional properties (e.g. whipping capacity). Though many such modifications offer opportunities for improving food proteins and extending their availability from non-conventional sources, careful consideration of their safety and acceptability is required.

2.4.1 Classification and nomenclature

In protein nomenclature, trivial names are predominantly used. Proteins can be classified from a variety of perspectives, which reflect their importance in nutrition, origin, structure, chemical and biochemical properties, and other attributes.

The amount of protein that the body is able to absorb and use from a certain protein source varies because a protein from a food may be absorbed better than a protein from another source, even if the same amounts are provided. According to their bioavailability (see Section 2.4.4.2), the following protein categories can be identified:

- **fully usable**, which contain all essential amino acids at levels required for human nutrition (e.g. egg and milk proteins);

- **almost fully usable**, in which some essential amino acids are present in somewhat lower amounts than required (e.g. meat proteins);

- **partly (insufficiently) usable**, in which essential amino acids are present in much lower amounts than required (e.g. all vegetable proteins and animal proteins from connective tissues).

Depending on their structure (or the presence of non-proteinous components), native proteins are divided into:

- **simple proteins**;

- **conjugated proteins**.

Simple proteins that consist of amino acids only are also traditionally separated into two basic groups according to the shape of the molecule (to a certain extent, this classification coincides with a further classification into soluble and insoluble proteins):

- **globular** proteins or spheroproteins (e.g. albumins and globulins), where the molecules have a round to spherical shape, non-polar functional groups are found within the molecule, and the polar functional groups form the outer core and bind to water molecules; these are generally soluble in water or dilute salt solutions, forming colloidal solutions (see Section 7.8.1.2), and many act as enzymes;

- **fibrillar** proteins or scleroproteins (e.g. practically insoluble structural proteins collagen, keratin, and elastin), the molecules of which have the shape of macroscopic fibres.

Depending on the kind of prevailing non-proteinous compounds covalently bound to the proteins, conjugated proteins are divided into several groups:

- **nucleoproteins**, which contain ester-bound nucleic acids;

- **lipoproteins**, proteins conjugated with neutral lipids, phospholipids, steroids (such as cholesterol), which mostly occur in egg yolk and blood plasma;

- **glycoproteins** containing bound saccharides, mostly oligosaccharides (e.g. the structural protein collagen in meat, κ-casein in milk, and some egg white proteins); saccharides are bound by an *O*-glycosidic bond to serine, threonine, and 5-hydroxylysine;

- **phosphoproteins** containing bound phosphoric acid (α- and β-caseins in milk and the egg yolk protein phosvitin);

- **chromoproteins** containing bound porphyrin or flavin derivatives (haemoglobin, myoglobin, ferritin, peroxidases, catalases, and dehydrogenases with NAD and FAD cofactors);

- **metalloproteins**, which contain metals bound by coordinate bonds (examples include the protein ferritin, which is the storage form of iron in the liver, and ceruloplasmin, which contains bound copper; these proteins represent the most important sources of iron and copper in human nutrition).

In the past, proteins were classified according to their solubility, but the use of this classification is now limited. However, many trivial names of proteins are still based on this classification:

- **soluble proteins;**

- **insoluble proteins**, which include fibrillar proteins (scleroproteins).

The soluble proteins are divided into the following six groups:

- **albumins** are neutral proteins that are soluble in water, salted out from aqueous solutions by ammonium sulfate at a saturation of 60% at 75 °C, irreversibly coagulating; albumins include lactalbumin in milk, ovalbumin and conalbumin in egg white, leucosin in wheat, legumelin in lentils and peas, and phaseolin in kidney beans;

- **globulins** are weakly acidic proteins that are insoluble in water but soluble in dilute salt solutions, e.g. in 5% sodium chloride, in acids and bases, salted out by ammonium sulfate at a saturation of 40% and coagulating on heating; globulins soluble in salt-free water are pseudoglobulins, those insoluble in salt-free water are euglobulins; examples include the muscle proteins myosin and actin, lactoglobulin in milk, ovoglobulin in egg white, edestin in wheat, avenalin in oats, legumin and vicilin in lentils, peas, and other legumes, conglycinin and glycinin in soybeans, tuberin in potatoes, the almond protein amandin, the Brazil nut protein excelsin, and the hemp protein edestin;

- **prolamins** are plant storage proteins that are insoluble in water but soluble in dilute solutions of salts, acids, and bases and in 70% ethanol, which do not coagulate on heating; they contain higher amounts of bound proline and glutamine and primarily occur in cereal grains, e.g. gliadin in wheat and rye, secalin in rye, hordein in barley, oryzin in rice, zein in maize, and kafirin in sorghum;

- **glutelins** are insoluble in water but soluble in dilute solutions of salts, acids, and bases; they differ from prolamins in that they are insoluble in ethanol and coagulate on heating, and contain a considerable amount of glutamic acid; the most common glutelin is glutenin in wheat, which is responsible for some of bread's baking properties; examples of other glutelins are rye secalinin, barley hordenin, rice oryzenin, and maize zeanin;

- **protamines** are basic proteins that are soluble in water, dilute acids, and ammonium hydroxide and do not coagulate on heating; they contain large amounts of basic amino acids (about 80% arginine) and occur in fish sperm (e.g. cyprimine in carp, salmine in salmon, sturine in sturgeon, clupeine in herring, and scombrine in mackerel);

- **histones** are basic proteins that are soluble in water and dilute acids but not in ammonium hydroxide and do not coagulate on heating; they contain high amounts of lysine, arginine, and histidine and are present at high levels in the nuclei of animal and plant cells, where they are bound to nucleic acids.

2.4.2 Structure

Knowledge of the occurrence of specific amino acids in individual proteins is very important in nutrition, food technology, and other disciplines. In many cases, detailed knowledge of the protein structure is also required. Four levels of protein structure are recognised: **primary**, **secondary**, **tertiary**, and **quaternary**. Detailed descriptions of the structure of proteins can be found in biochemistry textbooks.

2.4.2.1 Primary structure

The primary structure of proteins includes the data on the structure of covalently bound amino acid molecules, the number and order (sequence) of amino acids in the peptide chain, the character of basic peptide bonds, and the number, character, and position of other covalent bonds (disulfide bridges and other bonds).

2.4.2.2 Secondary structure

The secondary structure of a segment of a polypeptide chain is the local spatial arrangement of its main chain atoms, without considering the conformation of its side chains or its relationship to other segments. The polypeptide chains of native proteins have particular secondary structures in different parts of the chain. A certain spatial arrangement (conformation) of the chain is given by its primary structure (amino acid sequence), fixed by noncovalent interactions of functional groups of the amino acids. Hydrophobic interactions are present when the amino acids have hydrophobic side chains. The amino acids with an electric charge in the side chain (acidic and basic hydrophilic amino acids) are involved in electrostatic interactions. The other hydrophilic and amphiphilic amino acids can form hydrogen bonds through their functional groups. There are three common secondary structures in proteins, namely **helices**, **sheets**, and **turns**.

The most important elements of the secondary structure (conformation of the polypeptide chain) are helical structures arising from the coiling of peptide chains (or parts of a chain) in the neighbourhood of the C_α into a spiral known as a helix. Helices have characteristic torsion angles, numbers of amino acid residues per turn, and helix heights. These characteristics are described by values of n_m, where n is the number of amino acid residues per turn and m is the number of atoms in the ring, including the hydrogen atom forming a hydrogen bond. Helices are chiral and can be right-handed or left-handed. Only right-handed helices are found in natural proteins (with some exceptions, for instance in collagen).

The major secondary structure of proteins is the right-handed **α-helix** (Figure 2.18). An α-helix is a rigid arrangement of a polypeptide chain, in which intramolecular hydrogen bonds play a significant role. If the helix has a translation of 0.15 nm along the helical axis and a pitch height (the vertical distance between one consecutive turn of the helix) of 0.54 nm, then it has 3.6 amino acid residues per turn (0.54/0.15 = 3.6). The average size of an α-helix is 11 amino acid units, which corresponds to about three turns. The C=O group of the peptide bond of an amino acid with a sequence number n is bound by hydrogen bonding to the N—H group of the peptide bond of an amino acid with a sequence number $n + 4$. This α-helix is also known as a 3.6_{13} helix as the hydrogen bond is between atoms 1 and 13 of the polypeptide chain. This secondary structure can be found in at least part of any polypeptide chain in the majority of proteins (in myoglobin, collagen, and other proteins).

β-Structures are somewhat less common regular secondary structures of proteins, also known as β-pleated sheets or just β-sheets (Figure 2.19). These are characterised by long extended polypeptide chains as compared to the compact helices. They are formed by combining parallel- or antiparallel-oriented polypeptide chains (β-strands), which line up and form bridges of extramolecular hydrogen bonds (unlike the helices, which are bound by intramolecular hydrogen bonds), creating a stable β-sheet. On average, a β-sheet consists of six β-strands, and structures with fewer than five β-strands are rare. Each polypeptide chain contains up to 15 amino acid residues. Some globular proteins rarely produce a parallel structure; an antiparallel structure occurs more frequently, and polypeptide chains containing proline, such as the collagen polypeptide chain, never form β-structures. A large closed β-sheet (typically having an antiparallel structure) is called a β-barrel. β-Barrels are found in proteins known as porins that occur in the cytoplasmic membrane of Gram-negative bacteria and in the outer membranes of mitochondria and plastids.

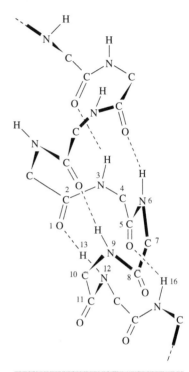

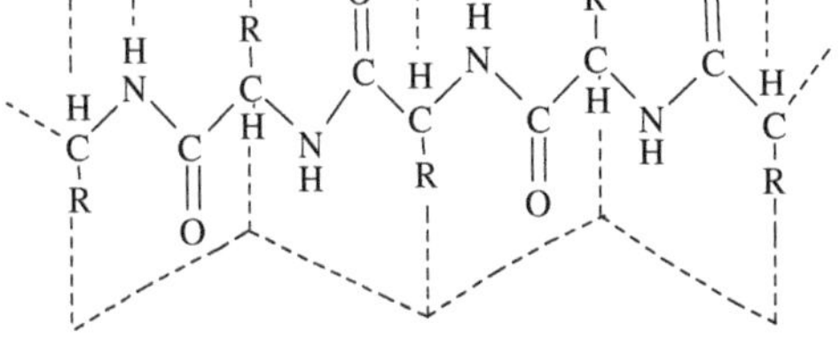

Figure 2.18 α-Helix (C_α atoms are shown without substituents).

Figure 2.19 Antiparallel β-sheet consisting of two β-strands.

There are many structural motifs that are formed by the combination of helices and sheets, but the common types of protein secondary structures (helices and pleated sheets) represent about half of the common globular proteins structures. These combinations require turns in the protein structure that reverse the direction of the peptide chain. Such so-called β-turns are considered a third common secondary structure motif. Approximately one-third of all the residues in globular proteins are found in turns. Turns are almost always found on the surface of proteins and often contain Pro or Gly. The remaining segments of the molecule have a so-called non-repetitive structure (random coils, turns, and bulges), which is also highly organised and characterised in terms of the backbone conformation and hydrogen bonding.

Fibrillar proteins have a specific secondary structure. For example, the basic secondary structure of keratin consists of two pairs of closely linked right-handed α-helices (a superhelix) that are coiled into a left-handed helix. The basic secondary structure of collagen is a triple helix in which the left-handed polypeptide helix is coiled into the right-handed superhelix.

2.4.2.3 Tertiary and quaternary structures

The tertiary structure of a protein molecule is the arrangement of all its atoms in space (conformation) without regard to its relationship with neighbouring molecules or subunits. It describes the packing of individual sections of the peptide chains (α-helices, β-sheets, and random coils) in the whole polypeptide chain with respect to one another, as these secondary structure units are not planar, but may be bent, coiled, or connected together in different ways. Various types of covalent bonds (such as disulfide bridges) and non-covalent interactions between functional groups of amino acids (hydrophobic and electrostatic interactions, see Section 7.6.1) participate in the binding of individual protein sections and in the full fixation of the tertiary structure.

The tertiary structures of many proteins are now known. One example is the myoglobin molecule, which is composed of 153 amino acids, about 80% of which are involved in α-helical structures (Figure 2.20). Its tertiary structure is that of a typical water-soluble globular protein. The polypeptide chain is coiled into eight separate right-handed α-helices (designated A through H) that are connected by short non-helical regions. The size of each helix ranges from 7 (helix D) to 26 amino acid residues. In total, 121 amino acids out of 153 occur in helical structures. The last coils of helices A, C, E, and G are reinforced by helices 3_{10}. The hydrophobic cleft (hydrophobic pocket), containing one haem prosthetic group with an iron atom, is formed largely by helices E and F, and is in contact with helices B, C, G, and H and turns CD and FG. The fifth ligand of the coordination bond of Fe^{2+} is histidine F8, the eighth amino acid in the α-helix F, which is known as proximal histidine. The so-called distal histidine (histidine E7) bonds oxygen to oxymyoglobine via a hydrogen bond (see Section 9.2.1.1). In deoxymyoglobine, the sixth binding position of Fe is vacant.

Many protein molecules consist of several different or identical polypeptide chains, which form a complex structure; this is known as the quaternary structure of a protein. The quaternary structure is the arrangement of protein subunits in space and the association of its inter-subunit contacts and interactions, without regard to the interior geometry of the subunits. The subunits in a quaternary structure must be in non-covalent association. Therefore, a quaternary structure exists only if there is more than one polypeptide chain present in a complex protein. For example, avidin in eggs (68 300 kDa) and haemoglobin in blood (64 500 kDa) consist of four subunits, whilst β-lactoglobulin

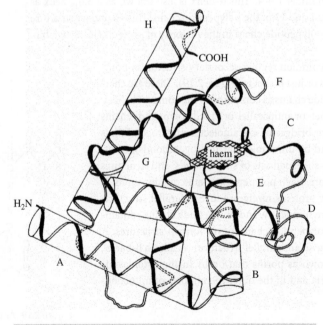

Figure 2.20 Structure of a muscle-tissue protein myoglobin.

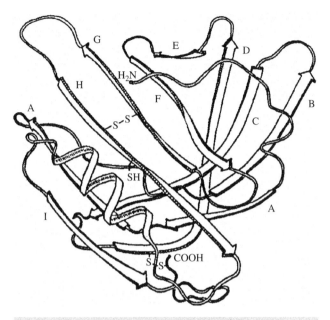

Figure 2.21 Structure of β-lactoglobulin of cow's milk.

in milk (36 000 kDa) is composed of two subunits (monomers). Each subunit of a haemoglobin tetramer has a haem prosthetic group with Fe^{2+}, which is identical to that described for myoglobin. The common peptide subunits are designated α, β, γ, and δ. Both the α and the β subunits have structural characteristics similar to that of myoglobin.

The subunit structure of β lactoglobulin (18 kDa) of cow's milk is shown in Figure 2.21. In a solution of pH 5–7.5 it is generally a dimer, in a solution of pH 3.5–5 it is an octamer, and in a solution of pH 3.5 it is a monomer. The monomer polypeptide chain consists of 162 amino acid residues. It contains two disulfide bridges (Cys 66-Cys 160 and Cys 106-Cys 119) and one free thiol group (Cys 121); the *N*-terminal amino acid is leucine, and the *C*-terminal amino acid is isoleucine (milk of other mammals contains genetic variants of β-lactoglobulin, which differ in the structure of certain amino acids). The polypeptide skeleton is composed of nine β-strands (marked with arrows pointing to the *C*-terminus of the molecule in the figure), which form a cylindrical β-barrel. The strands contain the following amino acid residues: A 16–27, B 39–44, C 47–58, D 62–76, E 80–84, F 89–97, G 102–109, H 115–124, and I 145–150. The connection between strands H and I is made through an α-helix, which contains amino acid residues 130–140. Strand I participates in dimer formation, which is based on an antiparallel hydrophobic interaction with another monomer, between isoleucine 29 of one monomer and isoleucine 147 of the other.

2.4.3 Properties

2.4.3.1 Dissociation and hydration

Globular proteins are soluble in polar solvents such as water and in aqueous solutions of acids and bases. Structural fibrous proteins are insoluble in water. Prolamins also dissolve in less polar ethanol. In addition to the protein structure, solubility depends on the relative permittivity of the solvent, the pH value of the solution (the minimum solubility is at or in the vicinity of the isoelectric point), the ionic strength of the solution (a low concentration of salts increases the solubility through the salting-in effect, whilst higher concentrations reduce the solubility through the salting-out effect; see Section 7.8.3.1), temperature, and other factors.

Proteins dissociate in solutions, as do amino acids and peptides, to form macromolecular polyions. Therefore, proteins are polyampholytes. Depending on the pH of the solution, protein molecules are either positively or negatively charged ions due to the dissociation of the functional groups of the various amino acids, especially basic amino acids (such as lysine) and acidic functional groups (aspartic and glutamic acids) in the side chains. The net charge (the difference between the number of positive and negative charges) is therefore positive or negative depending on the pH value. The pH at which the protein has no net charge is the isoelectric point (the negative logarithm is the p*I* value). It depends on the type of protein and lies in the pH range 2–11. It should be noted that the protein still contains charged side chains at its isoelectric point; however, at the isoelectric point, the number of positively charged side chains is equal to the number of negatively charged side chains.

Hydration dictates a range of processes, including protein folding, ligand binding, macromolecular assembly, and enzyme kinetics. For example, solutions of globular proteins are colloidal dispersions. The protein molecule has the character of a molecular micelle with hydrophobic amino acids arranged towards the interior of the molecule, whereas hydrophilic amino acids are bound outwards. The polar core is hydrated in aqueous solutions, which explains the solubility of globulins. The amount of bound water (monomolecular layer) is about 0.2–0.5 g per 1 g of protein. Dispersions of proteins are therefore hydrophilic colloids. Other layers of water are now bound less tightly. Proteins in aqueous solutions are usually molecular dispersions as the dispersed phases are neutral molecules or ions. Most protein solutions are monodisperse colloidal systems since they contain single molecules, but some proteins are oligodisperse colloidal systems that contain dimers and higher oligomers (see β-lactoglobulin in Section 2.4.2.3). Some proteins form aggregates of molecules, known as micellar colloids (e.g. α-, β-, and κ-casein in milk). The mechanical, kinetic, optical, and electrical properties of disperse systems of proteins (sols and gels) are described in Section 7.8.4.

2.4.3.2 Denaturation

Native proteins are precisely folded polypeptidic chains. The denaturation of proteins involves the disruption and possible destruction of the secondary, tertiary, and quaternary structures under the mild effects of various physical factors and in the presence of chemical agents. Since denaturation reactions are not strong enough to break the peptide bonds, the primary structure (sequence of amino acids) remains the same after the denaturation process. However, the long-term and severe effects of some factors that led to the denaturation can also cause changes in the primary structure (degradation of the polypeptide chain to shorter units and possibly to free amino acids). The conformational changes may be reversible or irreversible. Native and denatured proteins are in equilibrium, and the original conformations can sometimes be restored, returning the protein to its original condition. This process is called renaturation (Figure 2.22). Most changes in protein conformation, however, are irreversible. Loss of biological activity and the original function of the protein is thus a result of these changes.

Denaturation disrupts the normal α-helices and β-sheets in a protein and uncoils the protein into less organised and random shape structures. New (previously unavailable) functional groups can interact with water, and therefore many proteins in the denatured state have an increased ability to bind water (30–45%). Denaturation is often accompanied by protein coagulation, due to the aggregation of protein molecules caused by mutual reactions of functional groups of the proteins, reducing the number of functional groups available to enter into interactions with water. Hence, the ability of proteins to bind water decreases (see Section 7.3.2). For example, the proteins in eggs denature and coagulate during cooking. Denatured proteins can also react with other food components, for example with reducing saccharides in the Maillard reaction.

Various physical factors can cause denaturation (mostly changes in temperature, pressure, exposure to ultrasound, and penetrating electromagnetic radiation). Denaturation as a result of chemical agents occurs, for example, in the presence of salts, acids, bases, and surfactants. For example, heat can be effectively used to disrupt hydrogen bonds and non-polar hydrophobic interactions. This occurs because heat increases kinetic energy and causes molecules to vibrate so rapidly and violently that their bonds are disrupted. The activation energy of denaturation by heat is dependent on temperature and the amount of water present, amongst other factors, and is very high compared with enzymatic reactions. For example, the activation energy of the thermal denaturation of the milk proteins β-lactoglobulin and α-lactalbumin during pasteurisation (at 80 °C) is approximately 270 kJ/mol. During sterilisation (at >100 °C), the activation energy of denaturation is about 60 kJ/mol. The denaturation of fibrillar proteins in meat requires a significantly higher activation energy and proceeds in a different way. An example is the thermal denaturation of collagen in the production of gelatine, which requires a long, intense heating procedure in the presence of water.

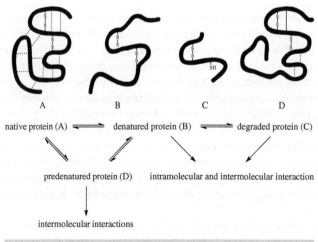

Figure 2.22 Thermal denaturation of proteins.

The denaturation of proteins does not only occur during thermal processing of foods, but also can be the result of low temperatures (when food is exposed to freezing temperatures of between 0 and $-15\,°C$). As a result, crystals of ice are produced, which can break the cell membranes, followed by denaturation due to surface phenomena at the interface of the two phases (protein solution and ice crystals), or possibly due to the freezing of the water required to maintain the native conformation of the protein. Loss of water also leads to an increase in salt concentration and osmotic pressure, which can accelerate protein denaturation. Lipoproteins are particularly sensitive to denaturation (e.g. those in egg yolk). However, the extent of the denaturation is limited in rapid freezing at very low temperatures, as smaller ice crystals are formed.

Denaturation of globular proteins also occurs at the interface of a protein with another phase (solid, liquid, or gaseous). As a result of conformation changes, the molecule unfolds, the hydrophilic groups are oriented into the aqueous phase, and hydrophobic groups into the less polar phase. This leads to partial denaturation of proteins in the presence of surfactants, for example in emulsions and foams (see Section 7.8.3.1). When whipping egg whites, partial denaturation occurs due to the mechanical stress, through the action of shear forces.

From a nutritional point of view, denaturation is usually desirable, since denatured proteins are more amenable to digestive enzymes. Denaturation, therefore, generally increases the bioavailability of proteins (e.g. sulfur-containing amino acids in cereals and legumes). Some antinutritional and natural toxic substances are also denatured, such as protease inhibitors (see Section 10.2.1.1), lectins (see Section 10.3.3), enzymes, and other unwanted proteins and undesirable microorganisms.

2.4.4 Physiology and nutrition

2.4.4.1 Fate of proteins in human organism

The human organism is not able to use dietary proteins as such. They must be hydrolysed into single amino acid molecules before they can be absorbed. The hydrolysis of proteins (mostly denatured proteins) is catalysed by proteolytic enzymes called proteases (proteinases or peptidases), which have relatively high substrate specificity. They catalyse the hydrolysis of interior peptide bonds to form peptides of different sizes (endopeptidases such as pepsin, trypsin, and chymotrypsin) or attack the terminal amino acids (exopeptidases). Hydrolysis of the N-terminal amino acids is catalysed by aminopeptidases, whilst carboxypeptidases hydrolyse amino acids from the C-terminus.

Protein digestion begins in the stomach. The gastric juices contain the endopeptidase pepsin, created in the peptic cells of the gastric mucosa. Pepsin is released from the proenzyme pepsinogen in the acid medium of the stomach (pH 1–2), which is provided by hydrochloric acid also secreted by the gastric mucosa. The major component of pepsin in adult mammals is pepsin A. Rennin, also known as chymosin, is another proteolytic enzyme secreted by the stomach of newborn mammals. Protein hydrolysis by pepsin just gives smaller units, polypeptides. Digestion of polypeptides continues in the upper part of the small intestine, the duodenum, through the action of the proteolytic enzymes trypsin and chymotrypsin, which are produced in the pancreas as the inactive proenzymes known as trypsinogen and chymotrypsinogen, respectively. They act in an alkaline medium in the small intestine and break down polypeptides into smaller peptides, which are hydrolysed to their component amino acids by a mixture of aminopeptidases and dipeptidases (known as erepsin) released by the mucosa. Free amino acids are absorbed across the intestinal mucosa into the lymphatic circulatory system and from there into the bloodstream.

The bloodstream provides a readily available pool of amino acids, which can be taken up by the liver and cells of various organs. The liver is the major site for synthesis of tissue proteins and for the conversion of excess amino acids into many other products. After deamination, the carbon skeleton enters the citric acid cycle and is used for the production of glucose (gluconeogenesis) and fatty acids and to generate energy through the oxidation of acetate into carbon dioxide and water. The resulting ammonia is converted into urea and excreted by the kidneys. D-Amino acids and peptides that occur in dietary proteins are eliminated from the body in urine (10–20%), and the remainder are catabolised by the flavoenzyme D-amino acid oxidase to 2-oxo acids in the liver, kidneys, and brain.

2.4.4.2 Nutrition

Dietary proteins are essential nutrients that provide adequate amounts of amino acids to allow the body to synthesise its own proteins and other nitrogen-containing compounds and serve as a source of energy. On average, proteins contain 16% nitrogen by weight.[1] Dietary protein requirements are determined by the need for essential amino acids and nitrogen, intake of other nutrients (carbohydrates and fat for energy intake), body weight, and physical activity. Specific needs include growth, pregnancy, lactation, illness, and injury. The Recommended Dietary Allowances (RDA values) are: 1.5 g/kg/day for infants, 1.1 g/kg/day for 1–3-year-old children, 0.95 g/kg/day for 4–13-year-old children, 0.85 g/kg/day for 14–18-year-old adolescents, 0.8 g/kg/day for adults, and 1.1 g/kg/day for pregnant and lactating women. On average, based on the RDA data, a male who weighs 70 kg should consume approximately 56 g of protein per day, whilst a female who weighs 50 kg should consume approximately 40 g of protein per day. Many nutritional experts feel the RDA for protein is far too low and is only suitable for sedentary adults. For teenagers, active adults, and some other people, it should be 1.2–1.8 g/kg. Research does not support protein

---

[1] The nitrogen contents of pure proteins and particular foods are in fact slightly different: meat and eggs have a nitrogen content of 16%, egg white 15%, gelatine 18%, milk and dairy products 15.7%, and cereals (wheat, rye, barley, and oats) 17.2%.

intake in excess of 2.0 g/kg body weight. Excess protein intake is associated with body dehydration and may be related to excessive urinary calcium losses and inadequate carbohydrate intake. Inadequate protein intake can lead to disturbances in mental and physical development, decreased resistance to infection, deterioration of wound healing after injury, and other problems.

The **energy yield** from proteins is 17 kJ/g (4 kcal/g), roughly the same as for carbohydrates (with the exception of sugar alcohols). Nutrition recommendations indicate how much protein is needed by considering daily energy intake and the percentage of energy that should come from protein. Although carbohydrates are the body's preferred fuel source, approximately 20–30% of energy requirements should be derived from proteins. The U.S National Academy of Sciences has recommended that between 45 and 65% of energy should be derived from carbohydrates, 20–35% from fat, and 10–35% from protein, with no more than 25% of the total energy being from added sugars. Recommendations by the American Heart Association and the American Diabetes Association fall well within these broad guidelines: approximately 50% of energy from carbohydrates, 30% from fat, and 20% from protein. In other words, if the average male who weighs 70 kg consumes 9350 kJ (2200 kcal) per day and 20% of the energy is derived from protein, this corresponds to 110 g of protein daily (1.6 g/kg body weight). The ratio of animal and vegetable protein should be about 1 : 1.

2.4.4.3 Protein nutritional value

The nutritional value of a food is mainly determined by the amount of protein it contains and the quality of that protein. Dietary protein quality depends on how well the essential amino acid content matches the requirements of the human body, the availability of peptide bonds to digestive enzymes (protein digestibility), and how efficiently amino acids are absorbed into the body for utilisation. Proteins from animal sources provide all of the essential amino acids in adequate amounts and are thus called **complete (whole) proteins**. Proteins of plant origin tend to be deficient in one or more of the essential amino acids and are thus termed **incomplete proteins**. The human body uses dietary essential amino acids in specific ratios, so if a person does not get enough of one essential amino acid, it can only use the others at that level.

Proteins have different biological availabilities in the human body and a number of methods have been introduced to evaluate and measure protein utilisation and retention. Two types of measurements are used to estimate protein quality: biological assays and chemical analysis. The most common measures of protein quality include Biological Value (BV), Protein Efficiency Ratio (PER), Net Protein Utilisation (NPU), Amino Acid Score (AAS; also known as chemical score, CS), Essential Amino Acid Index (EAAI), Protein Digestibility Corrected Amino Acids Score (PDCAAS), and Digestible Indispensable Amino Acid Score (DIAAS), amongst other procedures and modifications.

The BV is the proportion of nitrogen retained in the body compared with the total nitrogen absorbed (taking no account of digestibility). A dietary protein that is completely utilisable (a complete or whole protein), that is, one that can be completely ingested, absorbed, and incorporated into body proteins (such as eggs and human milk), has BV = 90–100, whilst gelatine has BV = 0 as it does not contain the essential amino acid tryptophan (Table 2.7). The PER is determined by dividing the weight gain of a test subject who had been given certain proteins by the intake of that particular protein during the test period. The NPU is the ratio of nitrogen retained (converted into proteins) to nitrogen supplied. Methods like BV, PER, and NPU, together with nitrogen balance, may not reveal much about the amino acid profile and digestibility of the protein source, but can still be considered useful in that they determine other aspects of protein quality not taken into account by PDCAAS and DIAAS.

Table 2.7 Comparison of various methods for the evaluation of nutritional value.

Food	BV	NPU	PER	PDCAAS	DIAAS
Beef meat	74–80	67–73	2.7–2.9	0.92–0.94	
Fish	76–83	78–80	2.7	0.9–1.0	
Milk (cow)	84–91	81–82	2.5	1.00	124–139
Eggs (whole)	93–100	88–94	3.1–3.9	1.00	
Whey protein	96	92	3.0	1.00	139
Casein	77–80	76	2.5	1.00	124
Wheat	54–64	67	0.8	0.37–0.54	29–77[a]
Rice	83	40	1.5	0.54–0.78	
Soybeans	73–74	48–61	2.3	0.91	100
Potatoes	73	60	1.8	0.71–0.82	
Gelatine	0	0	0	0	

[a]Raw cereal grains in ascending order of sorghum, maize, wheat, rye, rice, barley, and dehulled oats.

Table 2.8 Essential amino acid contents in a reference protein (in mg/g protein) and their RDI in mg/kg body weight.

Amino acid	Protein FAO/WHO	RDI	Amino acid	Protein FAO/WHO	RDI
Valine	32	26	Threonine	27	15
Leucine	55	39	Lysine	51	30
Isoleucine	25	20	Phenylalanine/tyrosine	47	25
Methionine/cysteine	25	15	Tryptofan	7	4

The amino acid (chemical) score is expressed in milligrammes of an amino acid in 1 g of the test protein divided by milligrammes of the same amino acid in a reference protein. The essential (limiting) amino acid that has the lowest AAS value thus determines the nutritional value of a protein. The reference protein is a protein of high nutritional value (such as β-lactalbumin) with balanced amounts of all essential amino acids. The AAS value for each of them is 100%. The composition of the reference protein is shown in Table 2.8, which also gives the recommended daily intake (RDI) of essential amino acids in adult humans. These values depend on nutritional and health status, age, and other individual factors, and have undergone considerable revision in recent years. The AAS value only takes into account one amino acid. More accurate data on the nutritional value of proteins are provided by the index of essential amino acids, which includes the contributions of all the essential amino acids as the geometric mean of the AAS values. The calculated AAS and EAAI values for meat, chicken, and fish are given in Table 2.9, for milk in Table 2.10, for eggs in Table 2.11, for cereals and pseudocereals in Table 2.12, and for legumes, oilseeds, and nuts in Table 2.13. The PDCAAS is a more recent ranking standard for the evaluation of protein quality. It is based on both the amino acid requirements of humans and their ability to digest them. It is expressed in milligrammes of limiting amino acid in 1 g of test protein divided by milligrammes of the same amino acid in 1 g of the reference protein (which corresponds to AAS), and multiplied by faecal digestibility percentage. The limitations of PDCAAS (e.g. it underestimates the value of high-quality proteins and overestimates the protein quality of products containing antinutrients) can be overcome by the use of DIAAS, proposed in 2013 by the US Food and Agriculture Organization (FAO). This method improves on the usefulness of the AAS and provides a more accurate measure of the contribution of proteins to human amino acid and nitrogen requirements. It is defined as follows: DIAAS% = 100 × [(mg of digestible dietary indispensable amino acid in 1 g of the dietary protein)/(mg of the same dietary indispensable amino acid in 1 g of the reference protein)] (Table 2.7).

In human nutrition, which assumes a varied and well-balanced diet, the evaluation of the nutritional quality of proteins is not particularly important. An exception may be some extreme diets (such as those of vegans), which might not supply some essential amino acids in sufficient quantities. The evaluation of protein quality in feeds for livestock and poultry is important in cases requiring rapid muscle growth or high yields of milk or eggs. In these cases, feed is often enriched (fortified) with limiting amino acids, mostly lysine and methionine, which increase its nutritional value.

2.4.5 Occurrence, composition, and changes

Foods differ considerably in their protein contents, and the protein contents vary in their amino acid compositions and quality. Protein contents can effectively vary anywhere between 0 and 100% of dry matter. The richest sources of protein are mainly foods of animal origin (Table 2.14). For example, the amount of protein in dried egg whites approaches 100%, whereas refined vegetable oil does not contain any protein at all. However, some products of plant origin, especially legumes (peas, beans, lentils) and oilseeds (soybeans, peanuts, poppy seeds, nuts) are also good protein sources. Cereals and cereal products have medium protein content, whilst vegetables, fruits, and root crops have a low protein content (Table 2.15). Vegetable oils, vinegar, and sugar contain only traces or no protein at all.

2.4.5.1 Meat, meat products, poultry, and fish

Animal cells produce four main types of tissue:

- **epithelial** tissue;

- **connective** (supporting) tissue;

- **muscle tissue;**

- **nervous tissue.**

Table 2.9 Amino acid content of meat, chicken, and fish (in g per 16 g nitrogen).

Amino acid	Beef	Pork	Pork offal	Mutton	Horse	Chicken	Fish
Ala	5.8	5.5	6.1	6.6	5.4	3.4	6.0
Arg	6.3	6.4	6.4	6.9	7.2	5.6	5.7
Asx	9.0	8.9	8.2	8.8	8.3	9.2	10.4
Cys	1.3	1.1	1.4	1.3	1.3	1.3	1.2
Glx	15.3	14.5	11.7	14.8	12.2	15.0	14.1
Gly	4.9	5.7	–	5.9	4.3	5.3	4.8
His	3.4	3.3	2.6	2.7	2.8	2.6	3.5
Ile	4.8	5.1	6.1	5.0	6.5	5.3	4.8
Leu	8.1	7.6	8.3	7.7	9.5	7.4	7.7
Lys	8.9	8.1	8.5	8.2	10.0	8.0	9.1
Met	2.7	2.7	2.5	2.5	2.8	2.5	2.9
Phe	4.4	4.2	4.8	4.0	3.8	4.0	3.9
Pro	3.8	4.6	5.3	4.7	4.0	4.1	3.7
Ser	4.0	4.2	4.7	4.2	4.2	3.9	4.3
Thr	4.6	4.9	4.5	4.7	3.9	4.0	4.6
Trp	1.1	1.4	1.3	1.3	1.0	1.0	0.6
Tyr	3.6	3.6	3.4	3.3	3.7	3.3	3.7
Val	5.0	5.2	6.0	5.1	5.0	5.1	6.1
Total EAA	44.5	43.8	46.8	42.9	47.2	41.9	45.0
Total AA	97.0	96.8	98.5	97.4	95.7	91.0	97.5
EAAI (%)	80	81	78	81	69	79	80
AAS (%)	69	69	71	67	63	64	70
Limiting AA	Val	Ser	Ser	Ser	Trp	Trp	Trp

EAA, essential amino acids; AA, amino acids; EAAI, Essential Amino Acid Index; AAS, Amino Acid Score (for limiting amino acids).

Epithelial tissues cover the whole surface of the body and the digestive tract. Connective tissues are present in cartilages and bones; adipose tissue is also a connective tissue. Muscle tissue covers the bones and is the building material of the gut. Nervous tissue is a major part of the nervous system, which is divided into the central nervous system (nervous tissue of the brain and spinal cord) and the peripheral nervous system. Blood is often classified as an additional tissue. Almost all tissues of animals are used as food for humans, especially the meat of mammals, poultry, and fish. The term 'meat' generally means the skeletal muscle and associated fat and tissues, but it may also describe other edible tissues such as organs and offal of mammals, poultry, and fish. A typical consumer connects the very vague term 'meat' with a product that, besides muscle tissue, contains a certain proportion of epithelial and connective tissue (such as skin, fat, cartilage, and bones). The main types of muscles that are used as food are skeletal and smooth muscles of the digestive tract and internal organs and cardiac muscle tissue (also called heart muscle or myocardium).

In addition to water and proteins, meat commonly contains about 1.5% fat (although in some meats this can be much more, such as pork belly, which can contain >50% fat), about 1% minerals, and small or trace amounts of carbohydrates. The glycogen content in muscle tissue is 1–2%, but its amount decreases very rapidly post mortem and meat usually contains only a small part of the original amount; a typical concentration is about 0.2%. Meat also contains sugar phosphates (mainly glucose 6-phosphate, see Section 4.2.2.1) in small amounts (0.05–0.2%) and some free sugars (mainly glucose).

Muscle proteins in typical mammalian muscle tissue constitute around 20% of the muscle weight. The major proportion of muscle is made up of muscle fibre proteins (elongated, threadlike cells) called **myofibrillar proteins**. Smaller amounts of soluble **sarcoplasmatic proteins** and insoluble **structural proteins** from connective tissue are also present (Table 2.16). Myofibrillar and sarcoplasmatic proteins are almost complete (whole) proteins, whilst the nutritional value of structural proteins is very low as they are almost indigestible. Table 2.17 gives the amino acid composition of some pure animal proteins; Table 2.9 presents the amino acid compositions for the main types of proteins.

Table 2.10 Amino acid content of milk proteins (in g per 16 g nitrogen).

Amino acid	Milk[a]				Amino acid	Milk[a]			
	Cow	Goat	Sheep	Human		Cow	Goat	Sheep	Human
Ala	3.5	2.7	3.2	3.9	Pro	9.1	8.3	10.8	7.1
Arg	3.3	1.3	2.9	3.9	Ser	5.8	4.1	5.6	4.7
Asx	7.7	7.6	7.0	8.5	Thr	4.5	4.4	3.7	4.5
Cys	0.8	1.6	1.4	1.3	Trp	1.4	1.3	1.9	1.8
Glx	22.2	18.3	22.1	15.2	Tyr	4.8	3.2	5.0	3.3
Gly	2.0	1.4	1.9	2.5	Val	5.8	6.5	6.2	4.5
His	2.7	3.6	2.5	2.5	Total EAA	47.2	41.7	45.8	40.1
Ile	4.7	5.2	4.6	4.1	Total AA	103.5	89.0	101.8	88.5
Leu	9.5	9.2	9.3	8.8	EAAI (%)	100	99	60	69
Lys	7.8	5.2	7.2	6.8	AAS (%)	75	73	53	53
Met	2.5	1.3	1.6	1.6	Limiting AA	Sulfur	Sulfur	Sulfur	Sulfur
Phe	5.4	3.8	4.9	3.5		and Ile	and Lys	and Ile	and Ile

[a]Colostrum (milk produced by the mammary glands in late pregnancy) has a different composition.

EAA, essential amino acids; AA, amino acids; EAAI, Essential Amino Acid Index; AAS, Amino Acid Score (for limiting amino acids).

Table 2.11 Amino acid content of hen's egg proteins (in g per 16 g nitrogen).

Amino acid	Egg			Amino acid	Egg		
	Whole	White	Yolk		Whole	White	Yolk
Ala	5.9	6.1	5.5	Phe	5.7	6.0	4.2
Arg	6.1	5.7	7.5	Pro	4.2	3.6	4.3
Asx	9.6	11.0	10.6	Ser	7.6	7.3	9.0
Cys	2.4	2.4	1.7	Thr	5.1	4.8	5.5
Glx	12.7	12.2	14.0	Trp	1.6	1.4	1.9
Gly	3.3	3.6	3.1	Tyr	4.2	3.5	4.0
His	2.4	2.4	2.5	Val	6.8	4.8	7.2
Ile	6.3	6.1	5.1	Total EAA	51.3	47.9	48.4
Leu	8.8	8.3	8.5	Total AA	107.1	99.8	104.9
Lys	7.0	6.6	7.7	EAAI (%)	100	94	95
Met	3.4	4.0	2.6	AAS (%)	100	71	74

EAA, essential amino acids; AA, amino acids; EAAI, Essential Amino Acid Index; AAS, Amino Acid Score (for limiting amino acids).

In addition to proteins, the other nitrogen compounds present in meat are free amino acids. The level of the individual amino acids is about 0.005% (their total amount is 0.1–0.3%); alanine (about 0.01%), glutamic acid (about 0.05%), and taurine (0.02–0.1%) occur in somewhat higher concentrations, followed by histidine dipeptides and guanidine compounds such as creatine and creatinine. Important groups of nitrogenous compounds are purine and pyrimidine nucleotides, nucleosides, and free bases (0.1–0.25%). Rigor mortis usually sets in between two and four hours after death. ATP becomes the main nucleotide in meat within approximately one hour post mortem. As rigor mortis dissipates, 5′-inosinic acid (5′-inosine monophosphate, 5′-IMP) predominates at a level of 0.02–0.2%, being formed by the enzymatic decomposition of ATP. The inosinic acid then gradually decomposes during the storage of meat to inosine and hypoxanthine. In addition to IMP, ATP and its breakdown products, ADP and AMP, and other purines and pyrimidines (mainly NAD, 5′-GMP, 5′-CMP,

Table 2.12 Amino acid content of cereal and pseudocereal proteins (in g per 16 g nitrogen).

Amino acid	Wheat	Rye	Barley	Oats	Rice	Maize	Millet	Buckwheat	Amaranth
Ala	3.6	4.3	4.0	4.5	6.0	7.5	7.9	4.7	3.4
Arg	4.6	4.6	4.7	6.3	8.3	4.2	5.3	9.8	7.4
Asx	4.9	7.2	5.7	7.7	10.3	6.3	8.0	8.9	8.3
Cys	2.5	1.9	2.3	2.7	1.1	1.6	2.4	2.4	1.4
Glx	29.9	24.2	23.6	20.9	20.6	18.9	18.6	17.3	15.4
Gly	3.9	4.3	3.9	4.7	5.0	3.7	3.8	5.0	8.7
His	2.3	2.2	2.1	2.1	2.5	2.7	2.4	2.1	2.3
Ile	3.3	3.5	3.6	3.8	3.8	3.7	4.1	3.4	3.6
Leu	6.7	6.2	6.7	7.3	8.2	12.5	9.6	5.9	5.3
Lys	2.9	3.4	3.5	3.7	3.8	2.7	3.4	3.8	5.0
Met	1.5	1.5	1.7	1.7	2.3	1.9	2.5	1.5	1.8
Phe	4.5	4.4	5.1	5.0	5.2	4.9	4.8	3.8	3.6
Pro	9.9	9.4	10.9	5.2	4.7	8.9	6.1	4.3	3.6
Ser	4.6	4.3	4.0	4.7	5.4	5.0	4.9	5.0	7.1
Thr	2.9	3.3	3.3	3.3	3.9	3.6	3.9	3.6	3.5
Trp	0.9	1.0	0.9	1.1	0.8	0.7	2.0	1.4	1.5
Tyr	3.0	1.9	3.1	3.3	3.5	3.8	3.2	2.4	3.4
Val	4.4	4.8	5.0	5.1	5.5	4.8	5.5	6.7	4.3
Total EA	32.8	31.6	35.8	37.1	38.5	40.2	41.1	34.8	28.4
Total AA	96.5	92.0	94.6	93.3	101	97.5	98.1	93.3	89.4
EAAI (%)	68	75	78	79	76	55	67	76	76
AAS (%)	44	46	54	57	57	41	53	51	54
Limiting AA	Lys	Trp, Ile	Lys, Leu	Ile, Lys	Ile, Lys	Lys	Lys	Lys, Ile	Lys Ile

EAA, essential amino acids; AA, amino acids; EAAI, Essential Amino Acid Index; AAS, Amino Acid Score (for limiting amino acids).

and 5′-UMP) occur in small quantities. The other compounds present in significant amounts are numerous organic acids. The main acid is lactic acid, the content of which ranges from 0.2 to 0.8% in fresh meat. Glycolic acid (about 0.1%) and succinic acid (about 0.05%) are also present at relatively higher levels; the other acids of the citric acid cycle occur in lower amounts. Another important group of nutrients present in meat is the vitamins. In general, meat is a good source of all vitamins, particularly water-soluble vitamins.

2.4.5.1.1 Myofibrillar proteins

The basic structural unit of skeletal muscle is a muscle fibre composed of long, cylindrical cells of 10–100 μm diameter and several tens of millimetres length (usually 20–30 mm). Each muscle fibre is actually a muscle cell containing 100–200 nuclei and normal cell organelles. The muscle fibre is covered by a thin, extensible, semi-permeable membrane, the sarcolemma (or myolemma), which transmits electrical impulses from the nerves. It consists of a cell membrane (plasma membrane) and an outer coat made up of a thin layer of polysaccharide material with numerous thin collagen fibrils. The sarcoplasm of a muscle fibre is comparable to the gel-like cytoplasm of other cells.

Muscle cells contain contractile elements, **myofibrils**, which are essentially bundles of proteins found in the sarcoplasm (Figure 2.23). Muscle myofibrils contain isotropic sections (I-bands about 0.8 μm long) and anisotropic sections (A-bands about 1.5 μm long). The I-band

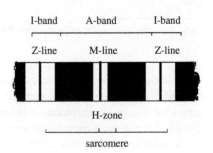

Figure 2.23 Muscle myofibril.

Table 2.13 Amino acid content of legumes, oilseeds, and nuts (in g per 16 g nitrogen).

Amino acid	Soybeans	Lentils	Peas	Beans	Sunflowerseed	Peanuts	Sesame seed	Walnuts	Hazel nuts
Ala	4.3	4.3	4.1	4.2	4.2	3.9	4.5	4.1	4.2
Arg	7.2	8.7	9.5	5.7	8.0	11.2	12.1	12.3	15.0
Asx	11.7	11.6	11.0	12.0	9.3	11.4	8.2	8.3	7.2
Cys	1.3	0.9	1.1	0.8	1.5	1.2	1.8	0.5	0.4
Glx	18.7	16.6	16.1	14.8	21.8	18.3	19.4	20.1	20.5
Gly	4.2	4.2	4.0	3.8	5.4	5.6	4.9	7.0	8.7
His	2.5	2.7	2.3	2.8	2.3	2.4	2.4	2.0	1.8
Ile	4.5	4.3	4.3	4.2	4.3	3.4	3.6	3.9	6.2
Leu	7.8	7.6	6.8	7.6	6.4	6.4	6.7	7.5	6.2
Lys	6.4	7.2	7.5	7.2	3.6	3.5	2.7	1.6	2.9
Met	1.3	0.8	0.9	1.1	1.9	1.2	2.8	1.3	0.8
Phe	4.9	5.2	4.6	5.2	4.4	5.0	4.4	4.1	3.6
Pro	5.5	4.3	3.9	3.6	4.5	4.4	3.7	4.7	5.6
Ser	5.1	5.3	4.3	5.6	4.3	4.8	4.7	6.1	9.6
Thr	3.9	4.0	4.1	4.0	3.7	2.6	3.6	2.7	2.7
Trp	1.3	1.5	1.4	1.4	1.4	1.0	1.0	1.0	1.1
Tyr	3.1	3.3	2.7	2.5	1.9	3.9	3.1	3.1	3.7
Val	4.8	5.0	4.7	4.6	5.2	4.2	4.6	4.4	6.4
Total EAA	39.3	39.8	38.2	38.6	34.1	32.4	34.8	26.5	33.1
Total AA	98.5	97.4	93.4	90.9	93.9	94.2	94.7	94.5	106.7
EAAI (%)	62	41	50	47	93	69	63	60	35
AAS (%)	47	31	37	34	56	43	43	24	22
Limiting AA	Sulfur, Val	Sulfur, Trp	Sulfur, Trp	Sulfur, Trp	Lys, sulfur	Sulfur, Ile	Lys, Ile	Sulfur, Lys	Sulfur, Lys

EAA, essential amino acids; AA, amino acids; EAAI, Essential Amino Acid Index; AAS, Amino Acid Score (for limiting amino acids).

is interrupted by a Z-line about 80 nm wide. The low-density central part of the A-band is the H-zone. The M-line is situated in the centre of the H-zone. The entire structural unit of myofibrils, which is the span between two Z-lines, is called the sarcomere. In a relaxed state (relaxed muscle), it is about 2.5 µm long. During muscle contraction, it gets shorter, at 1.7–1.8 µm long.

Myofibrils are composed of two types of protein microfilaments (microfibers). Strong microfilaments have a diameter of 12–16 nm and a length of 1.5 µm. They are composed of the protein myosin, which exhibits ATPase activity. Thin filaments, with a diameter of 8 nm and a length of 1 µm, are predominantly composed of the protein actin (Figure 2.24). Other components of microfilaments are proteins with regulatory functions such as tropomyosin, troponin, and other proteins important for the stability of the sarcomer structure (Table 2.16).

Myosin has a relative molecular weight of about 470 kDa. The molecule contains a fibrillar section, formed by two identical polypeptide chains, largely found in the α-helix conformation. The fibrillar part of the molecule is connected to a globular head consisting of four chains with a relative molecular weight of about 20 kDa, responsible for the ATPase activity of the protein.

Actin is a one-chain globular protein with a relative molecular weight of 43.5 kDa (G-actin, where G stands for globular). This monomeric form polymerises to yield a fibrous form of actin (F-actin, where F stands for fibrous). The units of G-actin are organised into a helix composed of two monomers. This polymer binds to the protein tropomyosin (relative molecular weight 70 kDa), which has a similar structure

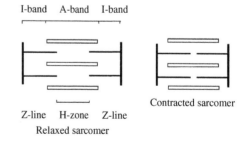

Figure 2.24 Arrangement of actin and myosin. –, actin; ▭, myosin.

Table 2.14 Protein content in foods of animal origin.

Food	Content (%) (from–to)	Content (%) (mean)	Food	Content (%) (from–to)	Content (%) (mean)
Meat, meat products			Deer	22.7–23.2	23.0
Beef meat	13.1–27.0	20.8	*Fish*	16.0–29.0	18.7
Veal meat	18.3–28,0	21.8	Carp	17.7–17.9	17.8
Pork meat	9.1–20.2	15.5	Trout	20.2–20.8	20.5
Sheep meat	14.9–18.0	16.4	Cod	17.8–17.9	17.8
Rabbit meat	19.8–20.3	20.1	*Milk, dairy products*		
Offal	10.4–22.7	17.2	Cow's milk	3.0–3.4	3.2
Pork liver	21.1–21.7	21.4	Curd	18.0–20.6	19.4
Beef liver	20.2–20.5	20.4	Soft cheeses	12.5–20.2	15.0
Sausages	12.8–28.0	20.8	Hard cheeses	23.8–40.6	24.8
Poultry			Butter	0.4–0.6	0.5
Chicken	21.2–21.4	21.3	Sheep milk	3.3–8.1	6.1
Turkey	19.2–19.8	19.5	Goat milk	1.1–2.8	1.9
Duck	11.2–11.8	11.5	*Eggs*	12.5–12.6	12.6
Goose	15.1–16.7	15.9	Egg white	10.8–11.0	10.9
Game	20.8–24.3	22.8	Egg yolk	15.8–16.0	16.0

Table 2.15 Protein content in foods of plant origin.

Food	Content (%) (from–to)	Content (%) (mean)	Food	Content (%) (from–to)	Content (%) (mean)
Cereals, cereal products			*Vegetables*		
Wheat flour	8.1–12.8	10.1	Fruit vegetables	0.7–1.7	1.2
Rye flour	5.1–12.0	9.6	Brassica vegetables	0.7–1.8	1.4
Rice (peeled)	7.0–7.3	7.1	Leaf vegetables	1.3–3.9	2.6
Bread (wheat)	6.7–11.4	6.7	Root vegetables	1.0–3.3	2.0
Bakery products	7.3–9.7	8.5	*Fruits*		
Pastry	3.5–7.8	5.6	Raw	0.3–1.5	1.0
Pasta	9.8–12.5	11.8	Dried	1.4–4.0	2.3
Legumes, oilseeds, nuts	21.4–44.7	24.2	*Other foods*		
Soybeans		44.7	Mushrooms		2.6
Lentil		25.8	Yeast		10.6
Poppy seed		18.0	Cocoa		18.0
Potatoes		2.0	Chocolate	4.9–8.1	6.8

to the fibrillar part of myosin (two unequal polypeptide chains, essentially curled into an α-helix). The actin is further connected to the regulatory troponins proteins (troponin complexes C, I, and T) with relative molecular weights of 18, 24, and 37 kDa, respectively.

In muscle, actin and myosin molecules are able to associate to form a protein complex, actomyosin. Calcium ions lead to this association, and ATP causes the dissociation of the actomyosin complex. These are the fundamental processes required for muscle activity (contraction and relaxation). The primary process during muscle contraction is the release of calcium ions (e.g. from the sarcoplasmic

Table 2.16 Muscle proteins.

Protein	Proportion (%)	Protein	Proportion (%)
Myofibrillar proteins	60.5	*Sarcoplasmatic proteins*	29.0
Myosin	29	Enzymes	24.5
Actin	13	Myoglobin	1.1
Connectin	3.7	Haemoglobin and other extracellular proteins	3.3
Tropomyosin	3.2	*Structural proteins and proteins of organelles*	10.5
Troponin (C, I, T)	3.2	Collagen	5.2
Actinin (α-, β-, γ-)	2.6	Elastin	0.3
Myomesin, desmin	5.8	Mitochondrial proteins	5.0

Table 2.17 Amino acid content in pure animal proteins (in g per 16 g nitrogen).

Amino acid	Actin[a]	Myosin[a]	Collagen[a]	Elastin[a]	Keratin[b]	α-Casein[c]	α-Lactalbumin[c]	β-Lactoglobulin[c]	Ovalbumin[d]
Ala	6.1	9.3	11.0	21.1	5.0	4.9	2.6	9.2	8.9
Arg	6.3	5.4	4.9	1.2	7.2	2.8	0.7	1.9	3.9
Asx	10.4	8.6	5.0	1.0	6.0	7.6	15.0	9.9	8.2
Cys	1.3	1.5	0.0	0.3	11.2	0.5	5.7	1.9	1.8
Glx	14.2	19.3	7.6	2.4	12.1	17.9	9.3	15.4	13.2
Gly	4.8	3.2	31.4	25.5	8.2	3.4	4.6	2.2	4.8
His	2.8	2.0	0.5	0.1	0.7	2.4	2.0	1.2	1.8
Hyl	0.0	0.0	0.6	0.0	0.0	0.0	0.0	0.0	0.0
Hyp	0.0	0.0	10.1	1.5	0.0	0.0	0.0	0.0	0.0
Ile	7.2	5.3	1.2	3.7	2.8	6.1	5.6	6.2	6.3
Leu	7.9	10.0	2.8	8.6	6.9	7.6	9.4	13.7	8.3
Lys	7.3	10.4	2.6	0.5	2.3	8.0	8.4	9.3	5.1
Met	4.3	2.9	0.5	trace	0.5	2.2	0.7	2.5	4.1
Phe	4.6	3.4	1.6	5.9	2.5	3.5	2.9	2.5	5.5
Pro	4.9	2.1	11.8	11.6	7.5	9.0	1.8	5.2	3.7
Ser	5.6	5.3	3.8	0.9	10.2	7.6	4.8	4.1	9.1
Thr	6.7	5.5	2.0	1.1	6.5	4.4	4.9	4.9	4.0
Trp	2.0	0.5	0.0	0.0	1.2	1.2	3.7	1.5	0.7
Tyr	5.6	2.4	0.3	1.3	4.2	5.0	3.2	2.5	2.4
Val	4.7	2.8	2.1	16.5	5.0	6.0	4.3	6.0	7.1

[a]Beef meat protein.
[b]Sheep wool protein.
[c]Cow milk (whey) protein.
[d]Hen egg (white) protein.

reticulum) initiated by a nerve impulse. Calcium ions bind to troponin, changing the formation of the molecule, which causes a formational change of tropomyosin and consequently of the actin molecule. Actin then reacts with myosin, forming actomyosin, which results in a shortening of the sarcoma, that is, contraction of the whole muscle. During the process of muscle relaxation, the actomyosin complex dissociates through the action of ATP to give actin and a myosin-ATP complex, which hydrolyses to myosin, ADP, and inorganic phosphate. Under normal conditions, ATP is synthesised from pyruvic acid through the process of glycolysis. In extreme cases (during short, intense exercise or post mortem), the formation of ATP is less efficient, as it is a product of anaerobic glycolysis (from glucose or glycogen) via lactic acid.

2.4.5.1.2 Sarcoplasmic proteins

The sarcoplasm of muscle tissue in the dry state contains on average 1% myoglobin (Table 2.16). Its main role is to facilitate the transport of oxygen by diffusion in muscle *in vivo*. In the muscles of aquatic mammals, where its content is about 10 times higher than in the muscles of terrestrial mammals, myoglobin also acts as an oxygen reserve. The levels of myoglobin generally depend on the type and origin of the muscle. Besides myoglobin, sarcoplasm contains various enzymes, especially glycolytic enzymes and enzymes from the pentose cycle, along with numerous other compounds, such as glycogen, ATP, ATP degradation products, and so on.

2.4.5.1.3 Structural proteins

The structural proteins form a specific group of extracellular proteins with protective and supporting roles. They include fibrillar proteins, the most important of which are collagens, elastins, and keratins. Their nutritional value is usually low (collagen) or almost nil (elastin, keratin) due to their poor digestibility and inappropriate composition of amino acids.

Collagens Collagens are insoluble extracellular glycoproteins that can be found in all animals. They are the most abundant proteins in the human body (29 different types can be identified). Collagens occur in nearly all connective tissues (skin, cartilage, tendons, filaments, and bones). Type I collagen is the main component of tendons, ligaments, skin, and bones. Type II collagen is the main component of cartilage.

The fundamental structural unit of collagen is the tropocollagen molecule (relative molecular weight about 300 kDa, thickness 1–2 nm, length ≤300 nm), which consists of a helical structure of three polypeptides, each with molecular weight of about 95 kDa. Type I collagen is the most abundant and consists of two chains of α_1 collagen and one chain of α_2 collagen, each having a left-handed helix formation. These three helices are twisted together into a unique right-handed triple helical structure (a heterotrimer known as a super helix), which is stabilised by hydrogen bonds. With type I collagen, each triple helix of tropocollagen naturally aggregates with other triple helices to form a collagen microfibril (thickness < 5 nm, length ≤ 500 nm), which is a right-handed super-super-coil. Cross-linking of collagen microfibrils (catalysed by lysyl oxidase) produces collagen fibres (thickness ≤ 500 nm, length ≤ 10 mm). Type I collagen molecules are behind the mechanical strength of tissues, and provide the major scaffolding units for the attachment of cells and macromolecules (such as integrins, fibronectin, fibromodulin, and decorin). In bones and dentin, the type I collagen is mineralised with hydroxyapatite crystals.

The amino acid composition of collagen is anomalous in comparison with other proteins. Collagen contains high amounts of glycine (30%) and proline (about 12%), and also contains hydroxyproline (10%) and 5-hydroxylysine (about 0.5%, Table 2.17). The amino acid sequence is composed of repeating units of Gly-X-Y, where X is often proline and Y is hydroxyproline or hydroxylysine. The free hydroxyl groups of the peptide chains derived primarily from hydroxylysine contain glycosidically bound glucose or galactose. For example, one fragment of the molecule is 2-*O*-α-D-glucosyl-*O*-β-D-galactosylhydroxylysine. Sugars represent between 0.4 and 12% of collagen molecules by weight. Collagens in connective tissues are accompanied by proteoglycans.

During ageing, the structure of collagen is stabilised by covalent cross-links. These are formed not only by disulfide bridges (collagen contains low amounts of cysteine, unlike keratin), but also by cross-links from the side chains of lysine, hydroxylysine, and histidine. For example, the enzyme lysyl oxidase[2] catalyses the oxidation of lysine (hydroxylysine is oxidised analogously) to the corresponding ω-aldehyde, that is, adipic acid monoaldehyde, or allysine (**2-29**). Two molecules of allysine generate an aldol, which then reacts with the side chains of lysine and histidine to produce imine; these compounds are then stabilised by a reduction or an oxidation. For example, the reaction with lysine followed by imine reduction yields lysinonorleucine (**2-29**). Condensation of three side chains of allysine with the side chain of bound lysine produces cyclic structures derived from pyridine (substituted in positions 1, 3, 4, and 5 of the nucleus), such as desmosine (**2-29**), dihydrodesmosine, and tetrahydrodesmosine, as well as pyridines substituted in other positions. The formation of these

---

[2]Lysyl oxidase is inhibited by β-aminopropionitrile formed in some vetchling seeds (*Lathyrus* spp.), which leads to disturbances in the collagen structure, with characteristic distortion of limbs and damage to blood vessels. These diseases are known as osteolathyrism and angiolathyrism, respectively (see Section 10.3.5.1).

structures is one of the causes of the toughening of the meat in older animals. Desmosine is also present in elastin.

allysine   lysinonorleucine   desmosine

2-29, cross-link structures in collagens

Native collagen is hydrolysed by proteases known as collagenases. Collagen denatured in muscle by lactic acid is hydrolysed during the maturation of meat by other enzymes, such as cysteine proteinase cathepsin B_1. Heat-denatured collagen is hydrolysed by pepsin and trypsin.

Collagen does not dissolve in cold water, salt solutions, or dilute solutions of acids and bases. Its characteristic feature is contraction (shortening of the molecule) when heated to a certain temperature, which is also observed on cooking meat. This form is sometimes referred to as collagen B, whilst native collagen is known as collagen A. The temperature at which collagen shrinkage occurs depends on its origin. For example, the shrinkage of fish collagen starts at about 45 °C, but mammalian collagen shrinkage occurs at 60–65 °C. At higher temperatures (approximately 90 °C), the structure of the molecule breaks, the bonds between the polypeptide chains disrupt, and each of the tropocollagen molecules is released to form a soluble sol of gelatine. The sol consists of mixtures of the original polypeptides together with their oligomers and breakdown products. Upon cooling (depending on the concentration of the gelatine and the temperature gradient), a more or less organised structure reforms. The new bonds are restored randomly between different parts of the same string or between two strings. The coil–helix transition, followed by aggregation of the helices, leads to the formation of collagen-like right-handed triple-helical proline/hydroxyproline-rich junction zones, similar to the original collagen structure, composed of three polypeptide chains. These structures capture large amounts of water and the gelatine sol forms a gel. Higher concentrations of the proline/hydroxyproline-rich junction zones form stronger gels (see Section 7.8.3.1).

Gelatine can also be produced from collagen during cooking, baking, and other thermal processing of meat. The extent of gelatinisation depends on the number of cross-links present in the collagen, and therefore also on the age of the animal and the parameters of the thermal process (temperature, time, pressure). Prolonged exposure to high temperatures leads to partial hydrolysis of collagen or gelatine.

Edible gelatine is made from collagen after partial extraction with water and partial acid (type A gelatine) or alkaline (type B gelatine) hydrolysis. The acid treatment uses pig skin, whereas the alkaline treatment makes use of cattle hides and bones. This gives rise to products with different molecular weights and properties (the alkaline treatment converts asparagine and glutamine residues into their respective acids and results in gelatine with a higher viscosity). Gelatine is primarily used as a gelling agent, forming transparent elastic thermoreversible gels on cooling. In addition, the amphiphilic nature of the molecules endows them with useful emulsifying (e.g. in whipped cream) and foam-stabilising properties (as in mallow foam). On dehydration, irreversible conformational changes take place that may be used in the formation of surface films. Gelatine is also used as a clarifiyng agent for wine and fruit juices.

Collagen from the skin dermis is used for the production of collagen casings used in the edible packaging material of sausages. Collagen casings are mainly produced from the corium layer (splits) of cattle hides, which consist essentially of collagen (although it can also be derived from pigs, poultry, and fish). The hides are chopped in water and mixed with lactic acid, which causes them to swell and form a slurry. The acid-swollen slurry is homogenised and filtered to tease the collagen fibres apart. The resultant slurry containing separated collagen fibres is extruded into the casing form. The proteins are then coagulated in a solution of an inorganic salt, plasticised with glycerol, dried, partially rehumidified, wound on reels, and shirred.

Elastins Elastins are insoluble rubber-like proteins conferring elasticity on tissues and organs. Elastin accompanies collagen in a number of connective tissues, undergoing reversible deformations in tendons, blood vessel walls, lung parenchyma, ligaments, skin, tissue membranes, and cartilage; it is not hydrolysed by common proteolytic enzymes, but only by the action of specific elastases.

Unlike collagen, elastin is a very flexible net-like structure found in organisms in the form of fibres. The fundamental building component of elastin is the soluble protein tropoelastin, which has a relative molecular weight of about 70 kDa and is formed from the soluble precursor, proelastin. Tropoelastin accumulates on the cell surface in small particles that grow into larger spherules (about 1 μm in diameter). The net-like structure of insoluble elastin is based on the association of many cross-linked tropoelastin molecules that are oxidised by lysyl oxidase enzymes at bound lysine molecules, which subsequently undergo aldol condensation and imine formation. One of these cross-links is desmosine (**2-29**) and its isomer, isodesmosine (unlike in desmosine, the side chains are in positions 1, 2, 3, and 5 of the pyridine ring). The process of elastic fibre formation includes a number of other molecules. The elastin is then introduced into microfibrils in the extracellular matrix as the central part and interacts with proteins known as fibulins, which form the sheath of these fibres.

Keratins Keratins are products of epithelial cells. They occur in the outer skin layer (epidermis) and in some skin structures (fur, horns, and hooves). The main group are the so-called α-keratins, based on a right-handed polypeptide with an α-helical structure and molecular weight of 10–50 kDa and stabilised by disulfide bridges and hydrogen bonds. Three polypeptides form left-handed helices called protofibrils, and 11 protofibrils (two inside and nine outside) are needed to form microfibrils. Several hundred microfibrils combine to make up macrofibrils, which finally give the keratin fibre (e.g. hairs and wool fibres).

Keratins have limited uses in the food industry. In combination with vegetable protein-rich materials, keratin is sometimes used for the production of acid protein hydrolysates used as food seasonings. Outside the food industry, alkaline hydrolysis of keratin is used for the production of glues.

2.4.5.1.4 Blood proteins

Blood is a distinct liquid tissue of higher animals, ensuring the transfer of nutrients and metabolic products, gas exchange, and heat balance. It is also important in organisms' defence reactions, such as the immunological response and coagulation involved in clotting. Blood represents about 5% of live weight of cattle (cows, calves), 3.3% of pigs, and 8% of poultry. In fresh weight, it contains about 80% water (80.5% in beef blood, 79.2% in pork blood) and around 18% protein (17.8% in beef blood, 18.5% in pork blood). About 40% of blood consists of blood cells, of which about 94% are erythrocytes containing the blood pigment haemoglobin, 6% thrombocytes, and 0.1% leukocytes. About 60% of blood is plasma, which has a dry matter content of 10%. About 6.5–8.5% of this dry matter is albumins and globulins, present in a ratio of 1.5 : 1; 0.3–0.4% is fibrinogen. Blood contains about 0.1% lipids, 1% minerals (mainly phosphates and chlorides of sodium and potassium; calcium, magnesium, iron, and other ions are present at lower concentrations), and small amounts of sugars (0.06–0.07%), free amino acids, vitamins, and other low-molecular-weight nitrogenous substances (urea, creatine, and creatinine).

Blood from farm animals is an important raw material for food, feed, and industrial use. Clotting is prevented by binding the calcium ions using citric or phosphoric acid. Blood drawn aseptically from slaughtered pigs is used to produce some speciality products (e.g. blood sausages and black puddings). Dried blood plasma is used to replace egg whites in baking and in the confectionery industry, whilst protein fibrin resulting from fibroin can be used in making soup spice preparations (such as acid protein hydrolysates and stock cubes). Dried or cooked blood is also used as feed for livestock, in the production of various veterinary and human medicines, and for other technological purposes.

2.4.5.1.5 Changes during storage and processing

After death (post mortem), a sequence of important biochemical, structural, and functional changes occur in the bodies of slaughtered animals when muscle tissue is converted into meat. Although the post mortem changes proceed in a relatively orderly fashion, a variety of external factors and intrinsic characteristics may accelerate or retard meat maturation and have a marked effect on the quality of the meat. Other important changes occur during thermal processing of meat and during the manufacturing of meat products.

Post mortem changes and meat maturation In contrast to tissues *in vivo*, only anaerobic glycolysis takes place in muscles post mortem, until the muscle glycogen store is depleted and the glycolytic enzymes are inactivated. The production of lactic acid causes the pH of the muscle tissue to decrease from the physiological value of 6.8–7.4 in warm-blooded animals to the ultimate post mortem pH of around 5.3–5.8. In such an acidic medium, the glycolytic enzymes are inhibited, and therefore a small proportion of glycogen remains preserved in the muscle. Even after the death of an animal, calcium ions still lead to the interaction of actin with myosin and hydrolysis of ATP to ADP and inorganic phosphate; however, ATP for dissociation of the resulting actomyosin is no longer available. The muscles stay shortened (contracted) due to actomyosin, and the posthumous stiffening (rigor mortis) begins. A muscle that is soft immediately after the death of the animal becomes stiffened within a few hours. In cattle at normal room temperatures, rigor mortis takes 10–24 hours post mortem; it takes 4–18 hours in pigs and 2–4 hours in chickens. Fish generally exhibit a shorter rigor mortis period commencing 1–7 hours after death. Rigor mortis then gradually subsides, for example after two to three days post mortem in beef, or more slowly at lower temperatures. Immediately after killing the animal, the meat is dry, since own water and any added water is bound well, meaning it has a high water binding capacity. In rigor mortis, when lactic acid acidifies the meat, the muscle proteins are present in an environment with a pH around their isoelectric points. The meat is moist-to-wet and its binding capacity is often half or lower than immediately after killing the animal. It is therefore advantageous to produce some meat products before the onset of rigor mortis (in the so-called warm state), otherwise the meat can only be processed after rigor mortis subsides and after maturation, when the pH and binding capacity will increase slightly. To increase the binding capacity of meat, different food additives are used (see Section 11.3.3.1). In practice, meat in rigor mortis cannot be cooked as it remains rigid and does not have the desirable organoleptic properties. During maturation, the cleavage of actomyosin and other meat proteins is due to the action of specific proteases that include a variety of endogenous enzymes of sarcomers, calcium-dependent calpains, and lysosomal enzymes (cathepsin B, D, and L), whilst colagenases break down collagen fibres. More recent studies suggested also a possible role for proteosomes. During tenderization, desirable properties of the meat are acquired, its texture is improved, and it becomes suitable for industrial and culinary processing.

Meat defects   Meat quality defects known as PSE (Pale-Soft-Exudative) and DFD (Dark-Firm-Dry) occur in pork as a result of sensitive animals being subjected to stress. There are a variety of physiological and environmental conditions that can cause stress in animals, including extremes in temperature, humidity, light, sound, and confinement, as well as excitement, fatigue, pain, hunger, and thirst. The two defects, PSE and DFD, can occur simultaneously in different types of muscle in the same animal. In beef, although the PSE defect is rare, with far less severe effect on quality than in pork, a defect similar to DFD is seen, known as DCB (Dark-Cutting-Beef), where the meat is dark on the cut.

PSE meat is more acidic than normal meat since the rapid drop in pH within one hour following death leads to a pH of around 5.6 instead of more than 5.8 whilst the muscle temperature is still above 36 °C; this value lasts much longer than 24 hours. This is explained by faster and more extensive hormonally controlled glycolysis, resulting in a faster and higher production of lactic acid. In the acidic environment, the muscle proteins are partially denatured and their ability to bind water decreases. The affected meat is abnormally pale due to the tissue with the small myofibrillar volume possessing a high light-scattering ability. The fluid is released from the tissue and may drip from the surface, which leads to weight loss during storage and thawing of frozen meat.

In DFD and DCB meats, the acidification one hour post mortem results in a pH around 6.5 instead of 5.8 (the pH at 24 hours post mortem is about 6.3), due to the loss of lactic acid via blood at bleeding out of the animal. The high pH result in negligible denaturation of proteins so that the water remains tightly bound with little or no exudate. The result is meat with a high water binding capacity, stiff consistency, and dark colour. An additional problem with this type of meat is that it is more susceptible to spoiling since it lacks lactic acid, which normally impedes the growth of microorganisms after slaughter.

Appropriate handling of animals before and during slaughter can greatly reduce their discomfort and stress. This includes proper feeding and rest, as well as the use of suitable techniques for moving and transporting animals.

Thermal processing   During heat processing of meat, the first conformation changes in proteins occur at temperatures around 35 °C, when proteins in the sarcoplasm associate into unstable structures reducing the water binding capacity of the meat and increasing its rigidity. The first visible changes occur at around 45 °C as a result of shortening of the muscles caused by myosin denaturation. Actomyosin is denatured at 50–55 °C, whilst the sarcoplasmic proteins are denatured at between 55 and 65 °C, including globins bound to myoglobin and haemoglobin. Depending on the intensity and duration of thermal processing, more severe changes in the structure of haematin occur (the central iron atom splits off and the porphyrin skeleton is degraded). As a result, the red colour of meat changes to reddish-brown and finally to grey-brown. Myoglobin in meat and meat products is often stabilised by adding sodium nitrite (see Section 9.2.1.5.3).

During denaturation of actomyosin and sarcoplasmatic proteins, stable associated structures and a firm, solid gel are formed. Partial denaturation may also occur during local overheating, such as during meat grinding. Changes in meat colour are caused by oxidation of haem to haematin. The coagulation of proteins reaches a maximum at temperatures of 60–65 °C, which is followed by conformational changes (a shrinkage to about a third or a quarter of the original length) of the collagen molecules (for fish, this occurs even at 45 °C). At temperatures above 80 °C, virtually all myofibrillar and sarcoplasmic proteins are coagulated, and free thiol groups of actomyosin are oxidised to disulfide groups. Around 90 °C, collagen gelatinises and the water binding capacity of meat increases. At temperatures around 100 °C (during boiling of meat), there are also some chemical changes in the proteins. Significant reactions include desulfuration and deamination of proteins yielding hydrogen sulfide and ammonia, which have an important impact on the development of meat flavour. These reactions are accompanied by a certain loss of cysteine/cystine and lysine.

Thermal processing of foods at temperatures around 150 °C (e.g. during baking) is much more complex than at 100 °C. The result is a certain loss of all amino acids, but also the formation of typical flavour substances. Roasting, frying, and grilling of meat at temperatures around 200 °C and higher lead to the extensive isomerisation of native protein-bound L-amino acids in the surface layers to the corresponding D-amino acids, and also to less usual cross-links such as lysinoalanine. Some amino acids (such as tryptophan) and other nitrogenous compounds (such as creatinine) can produce toxic products (see Section 12.2.1.1), which are present at significant concentrations in grilled meat.

When drying meat and meat products (such as durable salami), a significant aggregation and partial denaturation of actomyosin molecules occurs, and the toughness of the meat or meat products is increased. The freezing process also induces significant changes in the structure of meat proteins. Myosin gradually denatures and associates with actin. This greatly strengthens the structure and decreases the water binding capacity of meat. However, these reactions do not significantly affect the nutritional value of meat. The texture and nutritional value of long-term stored frozen meat is, however, considerably influenced by the reactions of proteins with oxidised lipids. Significant changes in proteins occur in the surface layers of smoked meat due to the reactions with reactive aldehydes, such as formaldehyde and glyoxal, which lead to cross-links between polypeptide chains. Smoked products (e.g. meat and sausages) thus obtain a smooth and rigid surface.

Table 2.18 Content of nutrients in milk.

Component	Content (%)			
	Cow's milk	Goat's milk	Sheep milk	Human milk
Total protein	3.2	3.2	4.6	0.9[a]
Caseins	2.6	2.6	3.9	0.4
Whey protein	0.6	0.6	0.7	0.5
Fat	3.9	4.5	7.2	4.5
Saccharides	4.6	4.3	4.8	7.1
Minerals	0.7	0.8	0.9	0.2

[a]Increases during lactation to 1.6%.

2.4.5.2 Milk and dairy products

The main component of milk is water. Its content can vary considerably according to the origin. Cow's milk contains 87–91% water. The basic chemical composition of cow's, goat's, and sheep milk in comparison with human milk is shown in Table 2.18. The amino acid composition of milk is given in Table 2.10.

Milk is a very complicated dispersed system in which casein molecules form a micellar dispersion, globular whey proteins form a colloidal dispersion, fat droplets (milk microsomes) form an emulsion, particles of lipoproteins form a colloidal suspension, and polar low-molecular-weight substances (lactose and other sugars, amino acids, minerals, and water-soluble vitamins) form a true solution. Milk's typical white opaque colour is related to the scattering and absorption of light by the fat particles and casein micelles. The yellowish colour that is sometimes seen is caused by carotenoid compounds dissolved in the fat phase (cream and butter), whilst the greenish colour of whey is caused by riboflavin.

The protein composition of cow's milk is shown in Table 2.19. Milk proteins are a mixture of two main types of proteins:

- **caseins** (account for approximately 80% of milk proteins);

- **whey (serum) proteins** (about 20% of milk proteins).

2.4.5.2.1 Caseins

Caseins are categorised according to the homology of their primary amino acid sequences into four genetic families: α_{S1}-, α_{S2}-, β-, and κ-caseins. The individual casein fractions differ from one another in their behaviour towards calcium ions. The main components of the casein fraction in bovine milk are α_S-caseins. Bovine milk contains phosphoproteins α_{S1}-casein and α_{S2}-casein, both occurring in four genetic variants, identified by the letters A, B, C, and D, which differ somewhat in their primary structure. The most common variant is B, which is a polypeptide chain composed of 199 amino acids (23.6 kDa, pI = 4.92–5.35). α_S-Caseins comprise from 44 to 55% of the total caseins and α_{S1}-casein comprises 75–80% of the total α_S-caseins. The polypeptide chain contains eight phosphoserine residues, located mainly at positions 43–80, which makes this part of the molecule polar. Non-polar side chains of amino acids are located in positions 100–199. α_{S1}-Casein forms insoluble calcium salts with calcium ions. Fragments of α_{S1}-casein are considered to be λ-casein. α_{S2}-Casein has a similar structure (25.2 kDa), but it is less sensitive to the presence of calcium ions.

Table 2.19 Protein composition of cow's milk.

Protein	Proportion (%)	Content (g/l)	Protein	Proportion (%)	Content (g/l)
Total caseins	80	25.6	*Total whey proteins*	20	6.4
α_S-Casein	42	13.4	α-Lactalbumin	4	1.3
β-Casein	25	8.0	Serum albumin	1	0.3
γ-Casein	4	1.3	β-Lactoglobulin	9	2.9
κ-Casein	9	2.9	Immunoglobulins	2	0.6
			Polypeptides (proteoses, peptones)	4	1.3

β-Casein comprises from 25 to 35% of the total casein. Polypeptide chains of β-casein consist of 209 amino acid residues (24 kDa, pI = 5.20–5.85). Each variant fits into one of two main categories, A1 and A2. In cattle, they differ in the amino acid at position 67 – histidine is engaged in the A1-type, whereas proline is engaged in the A2-type β-caseins. They are also phosphoproteins, as they contain five phosphoserine residues (in positions 1–40). Non-polar side chains of amino acids are concentrated in positions 136–209. At temperatures <1 °C, β-caseins and calcium ions form soluble salts, which are insoluble at higher temperatures. In human milk, β-casein is the major protein.

γ-Caseins (pI = 5.8–6.0, 3–7% of the total casein) are degradation products of β-caseins as a result of the action of indigenous milk proteinase (plasmin). Plasmin causes hydrolysis of β-casein, especially in late lactation and prior to milking. Splitting off the residue with amino acids 1–28 yields γ1-casein (20.5 kDa) and proteose-peptone-labelled PP8F; γ2-casein (11.8 kDa) is created by splitting off the amino acid residues 1–105, and γ3-casein (11.6 kDa) is formed by splitting off the amino acid residues 1–107.

κ-Caseins occur in two genetic variants (A and B) in cow's milk and represent about 8–15% of caseins. The molecule of the more common variant B is composed of 169 amino acid residues (18 kDa, pI = 5.37). The molecules occur as trimers and higher oligomers connected by disulfide bridges. Unlike caseins α and β, κ-caseins are glycoproteins containing bound oligosaccharides composed of D-galactopyranose (D-Galp), N-acetyl-D-galactosamine (D-GalpNAc), and N-acetylneuraminic acid (NeuAc). The main components of κ-caseins (56.0%) include branched tetrasaccharide α-NeuAc-(2 → 3)-β-D-Galp-(1 → 3)-β-D-GalpNAc-(6 → 2)-α-NeuAc, branched trisaccharide β-D-Galp-(1 → 3)-β-D-GalpNAc-(6 → 2)-α-NeuAc (18.5%), linear trisaccharide α-NeuAc-(2 → 3)-β-D-Galp-(1 → 3)-β-D-GalpNAc (18.4%), disaccharide β-D-Galp-(1 → 3)-β-D-GalpNAc (6.3%), and N-acetyl-D-galactosamine (0.8%). Sugars are bound to the protein via threonine at position 133 of the peptide chain through the glycosidic bond of N-acetyl-β-D-galactosamine (2-30). Calcium ions form soluble salts with κ-caseins that stabilise αS1- and β-caseins in milk.

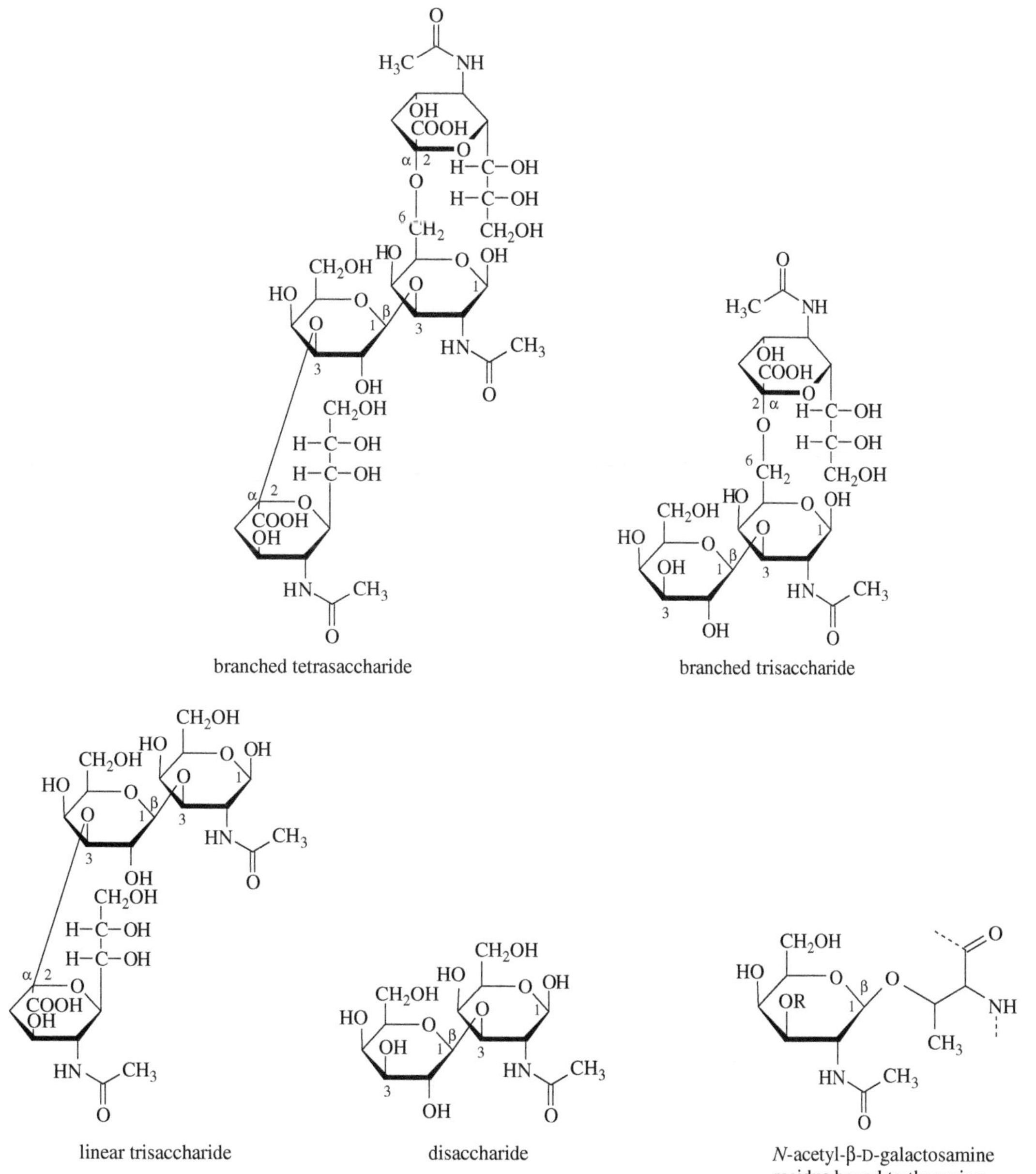

branched tetrasaccharide

branched trisaccharide

linear trisaccharide

disaccharide

N-acetyl-β-D-galactosamine residue bound to threonine

2-30, saccharides bound in κ-caseins

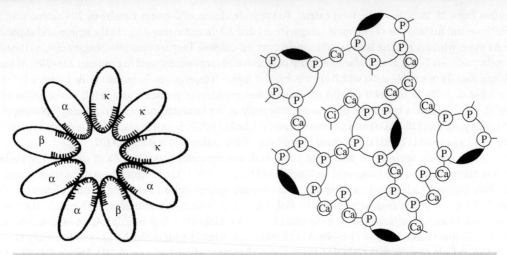

Figure 2.25 Left: Cross-section of a typical submicelle (hydrophobic parts of molecule are indicated by dashes). Right: Interconnection of submicelles through phosphate (P), calcium ions (Ca), and citrate (Ci) (whole of the non-bonding areas with molecules of κ-casein are indicated). Source: Coultate (1989), p. 144. Reproduced by permission of the Royal Society of Chemistry.

Caseins in milk do not occur as monomers, but are aggregated into casein complexes and spherical particles called **micelles**. Caseins β and γ can easily associate into polymeric structures at about 20 °C, whilst at temperatures <8 °C they dissociate back to the monomers. The aggregation of molecules of α_S-, β-, and κ-caseins into micelles occurs at temperatures >5 °C. Molecules of α_S-, β-, and κ-caseins are first arranged into particles called **submicelles**, which have the shape of a rotating ellipsoid and contain 25–30 protein molecules (Figure 2.25). The non-polar parts of caseins are oriented into the submicelle centre and are bound through hydrophobic interactions. The polar molecules of caseins, that is, phosphoserine residues of α_S- and β-caseins and the threonine residue with bound oligosaccharides, interact with calcium ions and water. The individual submicelles are connected to each other via phosphate (phosphoserine) groups of α_S- and β-caseins (the κ-casein molecule has no more than a single phosphoserine) and calcium ions, either directly or through free phosphates and citrates (nanoclusters of colloidal calcium phosphate; Figure 2.25). On the micelle surface, hydrophilic C-terminal ends of κ-casein predominate.

A typical micelle of cow's milk contains approximately 20 000 casein molecules. In addition to caseins (about 93% d.w.), the micelles contain about 3% calcium ions, 3% inorganic phosphates (free), 2% phosphates bound to phosphoserine, 0.4% citrates, and 0.5% sodium, potassium, and magnesium ions. The micelles have an open, porous structure, and their size distribution is 50–400 nm, with the peak at around 200 nm depending on the α_S- and κ-caseins content. The smallest particles contain about 50% of α-casein and 15% of κ-casein, whilst the largest particles contain about 42% α-casein, 26% κ-casein, and 30% β-casein. Milk contains about 1.10^{12} micelles/ml.

2.4.5.2.2 *Whey proteins*

The globular protein β-lactoglobulin (Figure 2.21) constitutes about 50% of whey (serum) proteins. Its polypeptide chain consists of 162 amino acids (18 kDa, $pI = 5.35–5.41$). β-Lactoglobulin occurs in three genetic variants, the main ones being labelled A and B. In milk, it is present as a dimer that denatures irreversibly on heating, and also in the presence of high concentrations of calcium ions in solutions of pH >8.6. The partially denatured protein reacts with other milk proteins (κ-casein and α-lactalbumin) through one accessible thiol group to form dimers linked by disulfide bonds.

Immunoglobulins, which are globular glycoproteins, are biologically important proteins with antibody efficacy, occurring in whey at low concentrations. They include IgG1 (162 kDa, $pI = 5.5–6.8$), IgG2 (152 kDa, $pI = 7.5–8.3$), IgA (400 kDa as a dimer), and IgM (950 kDa as a pentamer). The clustering of fat globules in raw milk, which results in the formation of larger particles and finally in the formation of a layer of cream on the surface of the milk, is not caused by the coalescence of fat globules, but is due to the action of a specific minor protein, **macroglobulin**, which creates cross-links between the globule membranes. Heating for several minutes to a temperature higher than 100 °C causes coagulation of this protein and prevents formation of a cream layer on the surface of pasteurised and otherwise heat-treated milk. Another important whey protein is α-lactalbumin, which has a biological function as a component of some enzymes. This protein occurs in two genetic variants (A and B, 14 kDa, $pI = 4.2–4.5$) and constitutes 20–30% of whey proteins. Whey or serum albumin (66 kDa, $pI = 5.13$) is present in smaller amounts. From the other minor proteins, lactoferritin with specific iron-binding properties displays anti-inflammatory and antiviral activity. Peptides and low-molecular-weight proteins called peptoses and peptones have relative molecular weights of 4–41 kDa and $pI = 3.3–3.7$.

2.4.5.2.3 Changes during storage and processing

Changes in the composition, structure, and properties of milk proteins (caseins and whey proteins) occur during pasteurisation, sterilisation, and especially fermentation, drying, and thickening of milk and in cheese making.

Thermal processing Amongst milk proteins, the whey protein immunoglobulins, serum albumin, β-lactoglobulin, and α-lactalbumin are particularly labile. During milk pasteurisation[3] at a temperature of 72–74 °C (20–40 seconds), about 50–90% of whey proteins are denatured and most enzymes present are inactivated. At temperatures >75 °C, the disulfide bonds in whey protein molecules are partly reduced to sulfhydryl groups and hydrogen sulfide is eliminated (mainly from β-lactoglobulin). The partial degradation of methionine yields thiols and disulfides, which also contribute to the so-called cooked flavour of milk. This defect is most obvious immediately after heating, but dissipates within one or two days. During sterilisation and ultra-heat treatment at 140 °C (for four seconds), 100% of the milk proteins are denatured. Denatured milk proteins have a slightly higher nutritional value compared with the proteins of raw milk. The denatured β-lactoglobulin in canned milk can interact with κ-casein, which results in increased viscosity and sometimes even flocculation of sterilised milk.

In pasteurised and sterilised milk, caseins do not denature, but thermal processes lead to their partial dephosphorylation (phosphoserine hydrolysis), proteolysis, and aggregation of molecules. Under extreme conditions, when incubated at 37 °C with 1% sodium hydroxide, casein can be completely dephosphorylated within 24 hours. During heating of sodium caseinate at 120 °C for one hour, about 50% of phosphoproteins are dephosphorylated; after five hours of heating, the dephosphorylation reaches 100%, and the extent of the caseinate hydrolysis is 10–20%.

Lactose, the main milk sugar, reacts with whey proteins to cause the partial loss of available lysine. This reaction is particularly intense during milk evaporation or drying (see Figure 7.37).

Precipitation and proteolysis of caseins The pH of fresh milk is 6.50–6.75. When the pH decreases to 4.6 due to lactic acid formation during fermentation of lactose to lactic acid, caseins precipitate, leaving a yellowish solution of whey (serum). The acidification is not only the spontaneous activity of contaminating microorganisms during storage of milk, but also a result of the desirable activity of a range of cultural microorganisms used in dairy industry.

Partial precipitation of caseins occurs in the production of yoghurt and cottage cheese using appropriate strains of bacteria (such as *Streptococcus thermophillus* and *Lactobacillus bulgaricus*). Yoghurt gets its characteristic gel texture as a result of the associations of casein micelles.

In the production of most cheeses, coagulation (clotting reaction) of milk is brought about by an enzymatic method. In the production of some hard and semi-hard rennet cheeses (e.g. Edam, Gouda, and Cheddar), the action of various cultural bacteria belonging to the genera *Streptococcus* and *Lactobacillus* partly transforms lactose into lactic acid, and thus acidify milk to a pH of around 5.5. At this stage, the setting of proteolytic enzymes, rennet (its key component is chymosin, also called rennin) is added, which accomplishes the coagulation of milk. Rennet is extracted from the stomach of calves, but today it is often replaced by proteases from fungi of the genus *Mucor* and by pepsin obtained from the stomach of chickens or pigs. κ-Casein, which mosty resides on the surface of micelles, is hydrolysed by chymosin almost exclusively at the phenylalanine (105)-methionine (106) bond and the molecule splits into two parts:

- ***para*-κ-casein;**

- κ-casein **macropeptide.**

The resulting *para*-κ-casein (protein composed of 105 amino acids) contains the original hydrophobic part of the κ-casein molecule and therefore remains in the micelles, but unlike the native κ-casein, it has no stabilizing function in the micelles. Loss of the stabilising layer cause micelles to became unstable and, with the participation of Ca^{2+} ions, the coagulation (curdling) of *para*-casein is initiated. κ-Casein macropeptide is a glycopeptide composed of 64 amino acid residues corresponding to the C-terminal 106–169 residues of κ-casein. Owing to the bound oligosaccharides, κ-casein macropeptide is hydrophilic and passes into the whey.

After several hours of storage, the casein curds reach the desired firmness and are cut into small pieces, which causes the curds to shrink and allows the whey to be released. Heat drives the shrinking (syneresis) process even further. The increase in temperature causes the protein to shrink due to increased hydrophobic interactions and increases the rate of fermentation of lactose to lactic acid. The increased acidity

---

[3] Pasteurisation, in order to extend shelf-life and inactivate pathogenic microorganisms, is carried out at temperatures of 85 °C (2 seconds), 71–74 °C (14–40 seconds), or 62–65°C (30 minutes). ESL (extended shelf-life) milk is obtained by indirect heating at 120–130 °C (<1–4 seconds). UHT (ultra high-temperature) milk is obtained by indirect heating at 135–140 °C (6–10 seconds) or by direct steam heating at 140–150 °C (2–4 seconds).

also contributes to the shrinkage of the curd particles. When the curds have reached the desired moisture and acidity, they are separated from the whey. Curd handling from this point on is very specific for each type of cheese. After salting (salting of brine is used in Gouda, salting in a vat in Cheddar, and salting of the cheese surface in Feta) and other operations (such as pressing and milling), the curds are left to mature (ripen) at low temperature until the characteristic flavour and texture profile is achieved. This final stage varies from weeks to years according to the particular cheese. During maturation, proteins are partly hydrolysed by chymosin and microbial proteases, and milk fat is partly hydrolysed by milk lipases. Proteolysis is probably the most important biochemical reaction during the ripening of most cheeses, having a major impact on their flavour and texture.

Use of whey and casein   Whey is the cloudy yellowish liquid that is left after milk has curdled. Sweet whey (pH 6.5–6.7) is a byproduct when milk is curdled with chymosin in cheese production. Acid or sour whey (pH 4.3–4.6) results from milk to which an acid has been added within the curdling process (e.g. in the production of Mozzarella cheeses). Whey has often been used to feed livestock, but the presence of proteins and other nutritionally important constituents led the food industry to take advantage of this in many utilizations of whey and whey constituents. For example, acid whey is used as a raw material for the production of lactose and other sugars (see Section 4.4.3.1.2). Sweet whey is a raw material in the production of some cheeses (including the Italian cheese ricotta). Dried whey is added to some food products (infant formulae, breads, pancakes, and biscuits) to enhance their nutritional value and functional properties. Whey proteins isolate is used as edible films and coatings after addition of some structuring agents, such as plasticizers, polysaccharides, emulsifiers, and lipids. Whey protein concentrate is used to improve athletic performance and as an alternative to milk for people with lactose intolerance.

About 50 years ago, the major uses of casein were in technogical applications, but nowadays casein products are regularly used as food additives, for example as ingredients that enhance some physical properties of foods, such as whipping, foaming, water binding, thickening, and emulsification.

Insoluble sweet casein (pH 6.5–6.7) obtained by precipitation of milk with chymosin is still used for technological purposes, for instance in the production of imitation tortoise shell and ivory (combs, hairclips, knife handles, piano keys, and other products). The insoluble acid casein (pH 4.3–4.6), obtained by precipitation of milk using mineral acids or lactic acid, is a raw material for the production of **caseinates**. Caseinates are produced by the reaction of acid casein curd or dry acid casein with dilute alkali (e.g. NaOH). The resulting solution can be spray dried to produce a caseinate powder. The pH of sodium caseinate ranges from 6.5 to 6.9. Soluble caseinates (sodium, potassium, and ammonium caseinates; solubility 100%) and dispersible calcium (solubility 90–98%) and magnesium caseinates are used in various branches of the food industry as water-binding additives and emulsifying agents. For example, sodium caseinate is used as an emulsifying and foaming agent in the production of mayonnaise, ice creams, processed cheeses, coffee whiteners, instant breakfasts and beverages, and meat products; potassium caseinate is used in the manufacture of confectionery. Mixtures of sodium, potassium, and calcium caseinates are employed in the production of confectionery products, and to stabilise and fortify meat products, yoghurts, and breads. Technological uses of acid casein include the production of synthetic fibres, paints, adhesives for wood, paper and foil laminates, coatings for paper, and cardboard.

The so-called insoluble **co-precipitates** contain all of the milk proteins (caseins and serum proteins). They are obtained from milk by precipitation with acids and salts during heating. Similarly to caseinates, co-precipitates are used as food additives for some meat products.

2.4.5.3 Eggs

The eggs of birds and reptiles have been eaten by mankind since time immemorial. The most commonly consumed eggs are those of hens, ducks, geese, quails, and ostriches.[4] Eggs contain a considerable amount of protein in a concentrated form (about 13% of the edible part), which has high nutritional value. The basic chemical composition of a hen's egg (average weight is around 58 g) is summarised in Table 2.20. The egg white proteins constitute about 53% and the yolk ones about 47% of the total protein. The levels of the individual proteins in egg white and yolk are shown in Table 2.21. The amino acid composition of egg proteins is shown in Table 2.11.

2.4.5.3.1 Egg white proteins

The egg white constitutes about two-thirds of the weight of a whole fresh hen's egg without eggshell, and contains more than 40 different proteins, which are classified as globulins, glycoproteins, and phosphoproteins. Glycoproteins contain various oligosaccharides composed of galactose (Gal), mannose (Man), acetylderivatives of glucosamine (GlcNAc), galactosamine (GalNAc), and neuraminic acid (NeuAc) (Table 2.22).

The main protein of egg white is a group of related compounds known as ovalbumin A, or ovoalbumin A (44.5 kDa, pI = 4.6–4.8; coagulation temperature ≥57.5 °C depending on pH). In addition to sugars, ovalbumin A contains up to two molecules of bound phosphoserine (two are in the main fraction A1) and six cysteine residues (two of which form a disulfide bridge). The thiol groups are buried within the

---

[4] Also eaten, although rarely, are the eggs of some wild birds, such as lapwings (e.g. in Great Britain and the Netherlands). Nowadays the large-scale collection of such eggs for food is prohibited as it has contributed to a decline in the numbers of the birds.

Table 2.20 Average content of nutrients in hen's egg.

Component	Content (%)		
	Shell	Egg white	Egg yolk
Proteins	6.4[a]	9.7–10.6	15.7–16.6
Fats	0.03	0.03	32.0–35.0
Carbohydrates	–	0.4–0.9	0.2–1.0
Minerals	91.9[b]	0.5–0.6	1.1
Water	1.6–1.7	88	49
Total weight in %	9.5[c]	63	27.5

[a]Complex of proteins with mucopolysaccharides (50 : 1).

[b]Inorganic salts: calcium carbonate (98.4%), magnesium carbonate (0.8%), tricalcium phosphate (0.8%).

[c]Including shell membrane.

Table 2.21 Composition of proteins in the white and yolk of hen's eggs.

Proteins	Content		Proteins	Content	
	%	g/kg		%	g/kg
Total egg white proteins	100	106	*Total egg yolk proteins*	100	166
Ovalbumin	54	57	*Granules*	47	78
Ovotransferrin (conalbumin)	12	13	Lipovitellenins (I–VII)	37.3	62
Ovomucoid	11	12	Lipovitellin apoproteins (HDL)[a]	40	66
Globulin G$_1$ (lysozyme)	3.4	4	α-Lipovitellin		
Globulin G$_2$	4	4	β-Lipovitellin		
Globulin G$_3$	4	4	Phosvitin (phosphovitin)	13.4	22
Ovomucin	3.5	4	LDL[b]	12	20
Ovoflavoprotein	0.8	1	*Plasma*	53	88
Ovoglycoprotein	1.0	1	Lipovitellenin (LDL)[b]	16	27
Ovomacroglobulin (ovostatin)	0.5	1	Livetins	15	25
Ovoinhibitor	1.5	2	α-Livetin (serum albumin)		
Avidin	0.05	<1	… β-Livetin α$_2$-glycoprotein)		
Cystatin	0.05	<1	… γ-Livetin (γ-globulin)		

[a]Lipoprotein fractions (see Section 3.6.1) of high density (high-density lipoprotein, HDL).

[b]Lipoprotein fractions of low density (low-density lipoprotein, LDL).

protein core. During heating, these thiol groups become exposed and react with other thiol groups to form disulfide bonds, which are crucial in the formation of protein networks and the texture of egg white gels. During storage of eggs, the conformational changes of ovalbumin A due to the isomeration of several serine residues into their D-amino acid forms lead to ovalbumin S, which is more thermoresistant and coagulates at 92.5 °C (at pH 7). In other words, ovalbumin S exposes thiol groups later during heating than ovalbumin A, and thus delays the incorporation of both ovalbumin and an important fraction of other proteins into the protein network. However, when the egg white is whipped, ovalbumin S denatures relatively easily. As for the stability of whipped foam, ovoglobulins G$_2$ (40 kDa, pI = 5.5) and G$_3$ (58 kDa, pI = 4.8) are of particular importance.

Ovotransferrin, also known as conalbumin (76 kDa, $pI = 6.1$), which is identical with serum transferrin in hens, shows antimicrobial effects. This protein coagulates at a lower temperature than ovalbumin (coagulation temperature is 53 °C) and forms complexes with divalent and trivalent metal ions. Complexes with iron can cause a pink discolouration of products containing egg white.

Ovomucoid and ovomucin are the proteins responsible for the viscosity and gel-like consistency of egg white. Ovomucoid (28 kDa, $pI = 4.4–4.6$) has three components, which differ in carbohydrate composition on the N-terminus of the molecule. Ovomucoid inhibits trypsin and other proteolytic enzymes and is a strong allergen. The high stability to thermal denaturation (in an acidic medium) is associated with nine disulfide bonds in the molecule. Ovomucin is thermostable and exists in two forms differing in the content of N-acetylneuraminic acid (α-form 220 kDa, β-form 720 kDa, $pI = 4.5–5.0$).

The basic protein ovoglobulin G_1, known as lysozyme, contains four disulfide bonds (14.3 kDa, $pI = 10.7$). It occurs in many animal tissues and secretions (e.g. tears), in the latex of some plants, and in microorganisms. This protein possesses the activity of N-acetylmuramidase and hydrolyses the β-glycosidic linkage between N-acetylmuramic acid and N-acetyl glucosamine in the peptidoglycan murein, which is the construction material of cell walls in Gram-positive bacteria (see Section 4.3.3.7). Lysozyme is therefore used as an antimicrobial agent.

Similarly to ovomucoid, native ovoflavoprotein (49 kDa, $pI = 5.1$) has certain antinutritional effects, as it inhibits serine proteases (trypsin, chymotrypsin and also microbial proteases). Ovomacroglobulin (ovostatin) is an inhibitor of serine, cysteine, thiol, and metalloproteases and shows antimicrobial activity. In raw egg white, the basic glycoprotein avidin (15.6 kDa, $pI = 9.5$) also shows some antinutritional effects. It contains four identical subunits, each of which binds one molecule of biotin to give an unavailable biotin complex. However, the denatured avidin, for example in hard-boiled eggs, does not interact with biotin. The interaction of riboflavin with flavoprotein (32 kDa, $pI = 4.0$), on the other hand, a positive influence on vitamin stability. Cystatin acts as a cysteine protease inhibitor and shows antimicrobial, antitumor, and immunomodulating activities.

2.4.5.3.2 Egg yolk proteins

The egg yolk is an emulsion of fat in water, which makes up about 36% of the weight of a whole fresh hen's egg. Rougly one-third of the dry matter consists of protein and the two-thirds of lipid. Egg yolk contains a range of droplets (20–40 μm in diameter), similar to fat droplets, which are enveloped by a lipoprotein membrane composed mainly of low-density lipoproteins (LDL) (Table 2.21). It also contains granules (1–1.3 μm in diameter) consisting of proteins, lipids, minerals, and a plasma. Egg yolk proteins are various glycoproteins, lipoproteins, glycophosphoproteins, and glycophospholipoproteins.

Granule proteins The most important lipoprotein is lipovitellin, which contains neutral triacylglycerols (35%), polar phospholipids (about 60%), and cholesterol and its esters (5%). α-Lipovitellin and β-lipovitellin differ in their phosphorus content. In the yolk, lipovitellin forms complexes with phosvitin. This name includes two aggregates of related molecules, α-phosvitin (160 kDa) and β-phosvitin (190 kDa). α-Phosvitin is composed of three basic units (37.5, 42.5, and 45 kDa, respectively) whilst β-phosvitin has only one unit (45 kDa). Phosvitin is a glycophosphoprotein (Table 2.22) with a high content of phosphoric acid (10%) bound in the form of phosphoserine, which provides

Table 2.22 Composition of saccharides in important glycoproteins of egg white and yolk.

Protein	Content (%)	Content (mol/mol of protein)				
		Gal	Man	GlcNAc	GalNAc	NeuAc
Ovalbumin	3	–	5	3	–	–
Ovotransferrin	–	–	4	8	–	–
Ovomucoid	23	2	7	23	–	1
α-Ovomucin[a]	13	21	46	63	6	7
Ovoglycoprotein	31	6	12	19	–	2
Ovoinhibitor A	9	10[b]	–	14	–	0.2
Avidin[c]	10	–	4 (5)	3	–	–
Phosvitin	–	3	3	5	–	2

[a]Sulfuric acid bound as ester is contained.

[b]Sum of galactose and mannose content.

[c]Basic unit (16 kDa); Gal, D-galactose; Man, D-mannose; GlcNAc, N-acetyl-D-glucosamine; GalNAc, N = acetyl-D-galactosamine; NeuAc, N-acetylneuraminic acid.

efficient metal-binding sites for calcium, iron, and other cations. This sequestering activity of phosvitin is related to its biological functions (the availability of metal ions for the developing embryo).

Plasma proteins The name lipovitellenin refers to lipoproteins of the LDL type, where the lipid portion represents approximately 84–90% of the molecule weight. This lipid portion is composed of triacylglycerols (74%) and phospholipids (26%). Livetins are water-soluble globular proteins including α-livetin (albumin), β-livetin (glycoprotein), γ-livetin (globulin), and δ-livetin (polypeptide). These proteins are identical to serum proteins of hens, namely serum albumin, $α_S$-glycoprotein, and γ-globulin, respectively. Immunoglobulin IgY (167 Da, pI = 5.7–7.6) represents the main antibody of egg plasma and blood serum and is analogous to mammalian immunoglobulin G (IgG). It is transmitted from the blood into the yolk and contributes to the development of immunity in the chicken in its early stages of life. The isolated protein is used in human and veterinary medicine.

2.4.5.3.3 Changes during storage and processing

Due to their chemical composition, eggs have been associated with a variety of functionalities and both fresh and refrigerated eggs are widely used in the food industry. Further industrial processing produces a variety of egg products, such as pasteurised, frozen, and dried egg whites, egg yolks, and the whole egg content. Smell and taste defects affect the long-term storage of raw eggs. Sulfur and nitrogen compounds, such as hydrogen sulfide, thiols, sulfides, and amines arising from cysteine and methionine, and indoles derived from tryptophan, are particularly responsible for these changes.

The mechanical denaturation of proteins occurs, for example, when whipping egg whites. Denatured proteins located at the interface with the air have a positive effect on foam stability. During the heat treatment of eggs, white protein denaturation begins at temperatures around 57 °C. Most egg white proteins (with the exception of ovomucoid, ovomucin, and complexes of avidin with biotin) are denatured in the temperature range of 60–65 °C, whilst most of the yolk protein (except for phosvitin) is not denatured until 65–70 °C. Egg proteins play a major role in the development of the typical aroma of boiled and fried eggs. When drying egg whites, yolks, or whole egg contents, glucose in egg white reacts with lysine bound in proteins and with phosphatidylethanolamine bound in egg yolk lipoproteins, which leads to undesirable colour and flavour changes in the product. Before drying, the content of glucose in egg whites is therefore removed by fermentation using bacteria of the genera *Aerobacter* or *Streptococcus* or *S. cerevisiae* yeasts. Another possibility is to oxidise glucose to gluconic acid using glucose oxidase (in the presence of catalase). Cold storage of pasteurised egg whites slightly increases the viscosity of the material, but the functional properties remain practically unchanged. For yolks and whole egg contents, cold storage leads to a substantial increase in viscosity due to the formation of products of a gel consistency. Slow freezing (critical temperature is −6 °C) leads to changes in conformation of the lipoprotein complex and probably irreversible partial dehydration of the protein part of lipoproteins.

2.4.5.4 Foods of plant origin

Nowadays plant protein sources provide less than 60% of the world's supply of protein for human nutrition, with cereal grains (41%) and pulses (5%), nuts and oilseeds (3–4%) as the major sources. Other limited sources of plant protein are fruits, leaves, tubers, and other parts of plants included under the terms 'fruits', 'vegetables', or 'root crops'. Plant protein sources can differ from animal protein sources in terms of digestibility, amino acid composition, and the presence of antinutritional (such as enzyme inhibitors) and toxic factors (e.g. saponins, cyanogens, and lectins), which adversely influence protein digestibility, nutritional value, and food safety.

Seed proteins usually contain a large amount of aspartic and glutamic acids and their amides, but some essential amino acids occur in lower amounts and become the limiting amino acids. With some exceptions (such as soybeans), lysine levels are uniformly lower in wheat, rice, maize, potatoes, and other crops. Soybeans have lower levels of the sulfur amino acids, maize is low in tryptophan, and cassava in threonine. As a consequence, the nutritional value of plant protein is lower than that of animal protein; nevertheless, plants can cover all human protein needs. An appropriate combination of plant materials gives a mixture of proteins with high nutritional value, which can provide protein for the diet of vegetarians. The question is whether this is only achievable with the carefully selected diets consumed by vegetarians, or whether it could also be provided by cereal or other staple diets available to poor or developing communities. In addition to protein, the body receives a number of other nutritionally valuable components such as fibre, carbohydrates, vitamins, and minerals from plants.

2.4.5.4.1 Cereals and pseudocereals

Cereals are grasses of the Poaceae family cultivated for seeds – grains. Cereal grains have been the principal components of human nutrition for thousands of years, and even today, most of the energy intake of the human population is derived from their consumption. The highest production of all cereals is of maize (36%), which is an important staple food in many countries, is used as an animal feed, and has many industrial applications. The second largest cereal production is of rice, which is a staple food for over half the world's population. Wheat is probably the third most produced cereal and it is the staple food used in a wide variety of products in Europe and North America. Wheat is also used to produce animal feedstuffs, starch, and ethanol. Barley grain ranks fourth in production in Europe and in arid and semi-arid areas of Asia, the Middle East, and North Africa. It is predominantly used as groats and flour for human consumption, in animal feed, and as malt in alcoholic beverages. The fifth most produced cereal is sorghum (*Sorghum bicolor*), an important crop worldwide, used for food as grain and syrup. It is grown in Africa, Central America, and South Asia and is also used as animal fodder and to produce alcoholic beverages and biofuel. The most widely grown millet is pearl millet (*Pennisetum glaucum*), which is an important crop in India and parts of Africa.

Some cereals with increased protein content have been introduced relatively recently. An example is the interspecies hybrid of wheat and rye called triticale (the name comes from Latin *Triticum*, wheat and *Secale*, rye). The grains contain 15–20% protein. The protein content in the interspecies hybrid of wheat and barley, known as tritordeum (the Latin name for barley is *Hordeum*), is 19–22%.

Edible grains of other plant families than Poaceae are referred to as pseudocereals. Increasingly, some locally important, previously little-known or almost forgotten pseudocereal grains are appearing as regular components of nutrition. Amaranth (*Amaranthus caudatus, Amaranthus cruentus, and Amaranthus hypochondriacus* of the amaranth family Amaranthaceae) is an ancient pseudocereal, formerly a staple crop of the Aztec Empire and now widely grown in Africa. Buckwheat of the knotweed family (Polygonaceae) has two main species, common buckwheat (*F. esculentum*) and tartary buckwheat (*Fagopyrum tartaricum*). Buckwheat was domesticated and first cultivated in Southeast Asia, and from there spread to Central Asia and later to Europe. The biggest current producers are Russia and China. Chia (*Sa. hispanica* of the mint family Lamiaceae) is a pseudocereal originated in Latin and Central America. In the past, the seeds belonged to the staple crop of the legendary Aztecs. Quinoa (*Chenopodium quinoa*, Amaranthaceae) is a pseudocereal native to South America, an important staple food for the Incas. Today it is also grown in North America, and for decades it has been produced in Europe.

The whole-grain cereals and pseudocereals contain the germ (the embryo rich in enzymes and lipids), endosperm (the nutritious part containing starch granules in a protein matrix), and bran (the grain coat) in variable proportions. For example, in brown rice, germ, endosperm, and bran fractions are 2, 70, and 8%, respectively. The rest (i.e. 20%) is the outer seed coating, called the husk or hull. White rice is milled rice that has had its germs and brans removed. In wheat, germ, endosperm, bran, and husk represent 2, 83, 15, and 0%, respectively. The refined grains retain only the endosperm. Wheat endosperm is ground into white flour for bread, but the whole flour is derived by milling the whole grain of wheat. The protein content in the outer layers of the grain is higher than in the inner layers. Therefore, the protein content in flour significantly depends on the degree of milling (flour extraction rate) and, of course, on the plant species, varieties, and some other factors. The dark wholemeal wheat flour has a higher protein content than white flour; the difference is about 4%.

The amino acid composition of the common cereals and pseudocereals is shown in Table 2.12. Their basic chemical composition is shown in Table 2.23.

Wheat proteins   Wheat usually contains 7–13% protein, but can contain up to 15%, with lysine as the limiting amino acid. Distribution of proteins within the wheat kernel differs significantly. About 72% of total protein is in the endosperm, 15.5% in the aleurone layer, 4.5% in the pericarp, 4.5% in the scutellum, and 3.5% in the embryo, whilst starch only occurs in the endosperm. The major factors influencing the protein content of wheat are genetics (species and variety) and environmental conditions (timing and amount of growing season precipitation, temperature during the growing season, soil nitrogen reserve levels, and applied nitrogen fertilisers). Wheat proteins are composed of albumins known as leucosin (about 14%), globulins known as edestin (8%), prolamins known as gliadin (33%), and glutelins known as glutenin (46%). In other words, about 20–25% of wheat proteins are represented by water-soluble cytoplasmic proteins that predominantly have structural, metabolic, and protective functions (e.g. enzymes with activity of α- and β-amylases, proteases, lipases, phytases, and lipoxygenases), and 75–80% of proteins are water-insoluble storage proteins prolamins and glutelins. The total number of wheat proteins is about 40.

The water-insoluble storage proteins are the major proteins of wheat. Prolamins (gliadins) and glutelins (glutenins) are represented by a number of related proteins, each with slightly different amino acid compositions (there are several dozen gliadins in each of the wheat varieties). Gliadins have a relative molecular weight of between 30 and more than 100 kDa (usually 60–80 kDa). Three basic fractions are identified: high-molecular-weight gliadins (>100 kDa), ω-gliadins (60–80 kDa), and low-molecular-weight gliadins (α- and γ-gliadins and other low-molecular proteins, 30–40 kDa). Gliadins contain a large amount of glutamine (36–45%) and proline (14–30%), somewhat less aspartic and glutamic acid, and unusually low amounts of the basic amino acids arginine, lysine, and histidine. In contrast to ω-gliadin with no cysteine, α-, β-, and γ-gliadins have less proline, glutamine, and phenylalanine, but 2–3% cysteine and methionine. The relatively low content of acidic and basic amino acids with polar side chains is linked to low gliadin solubility. Glutenins have a higher relative molecular weight, which typically ranges from 40 to 20 000 kDa (usually about 2000 kDa), as they are made up of polypeptide chains linked by disulfide bridges.

The high proline content virtually prevents the formation of helical secondary structures in both types of proteins. Gliadin and glutenin molecules therefore contain only short helical sections associated with segments of the unorganised secondary structure. The ω-gliadin molecules (stabilised by strong hydrophobic interactions) are rich in segments known as β-turns. Their number increases with temperature. In contrast, the α-, β-, and γ-gliadins contain α-helices (30–35%) and β-sheets (about 10% in α-gliadins) stabilised by disulfide bridges and hydrogen bonds. Heating causes an increase in the unorganised structures and partial loss of the α-helical structures.

Bread flour is obtained from wheat varieties with higher protein content (12–14%) and is also known as strong flour. This term relates to the properties of the dough, which is elastic and stiff (and therefore requires intense mixing), retains the carbon dioxide produced by yeast and air well, and provides voluminous products. The milling and baking properties of flour are linked not only to its protein content,

Table 2.23 Average basic chemical composition of important cereals and pseudocereals (%).

Raw grains[a]	Water	Protein	Total lipid	Total sugars	Carbohydrate[b]	Dietary fibre	Minerals (ash)
Cereals							
Barley (pearled)[c]	10.1	9.9	1.2	0.8	77.7	15.6	1.1
Barley	11.7	10.6	2.1	–	63.5	9.8	2.3
Maize	12.5	9.2	3.8	0.6	62.9	9.7	1.3
Maize (sweet, yellow)	76.1	3.3	1.4	–	18.7	2.0	0.6
Millet	8.7	11.0	4.2	–	72.9	8.5	3.3
Oats	8.2	16.9	6.9	–	66.3	10.6	1.7
Rice (brown, long-grain)	11.8	7.5	3.2	0.7	76.3	3.6	1.2
Rice (white, long-grain)	11.6	7.1	0.7	0.1	80.0	1.3	0.6
Rye	10.6	10.3	1.6	1.0	75.9	15.1	1.6
Sorghum	12.4	10.6	3.5	2.5	72.1	6.7	1.4
Wheat (hard red winter)	13.1	12.6	1.5	0.4	71.2	12.2	1.6
Wheat (soft red winter)	12.2	10.4	1.6	0.4	74.2	12.5	1.7
Pseudocereals							
Amaranth	11.3	13.6	7.0	–	65.3	6.7	2.9
Buckwheat	9.8	13.3	3.4	–	71.5	10.0	2.1
Chia	5.8	16.5	30.7	–	42.1	34.4	4.8
Quinoa	13.3	14.1	6.1	–	64.2	7.0	2.4

[a]Values from National Nutrient Database for Standard Reference, U.S. Department of Agriculture (average values from various sources, therefore the sum may not be 100%).

[b]Carbohydrate content has been calculated from the sum of the contents of other components (protein, fat, water, ash) subtracted from the total weight of the food. This fashion includes fibre as well as some components that are not carbohydrates. The main available carbohydrate is mostly starch.

[c]Grains that have been processed to remove the husk and bran.

but also to its composition. Flours referred to as weak flours usually contain less than 10% protein. They are suitable for the production of biscuits, pastries, confectionery, and other products (along with a baking powder), but not for making bread.

Wheat flour mixed with water gives a dough composed of starch and a viscoelastic sticky material called **gluten**, two-thirds of which is water and one-third hydrated gliadins and glutenins. The formation of the quaternary structure of gluten during dough mixing and bread making is extremely complex and not well characterised. The typical viscoelastic properties of gluten are determined by the glutenin subunits, as their molecules are able to generate a three-dimensional network stabilised by different types of bonds. The most important are hydrogen bonds mediated mainly by glutamine residues, ionic and hydrophobic interactions of amino acids, and primarily disulfide bonds present in the gluten structure that contribute to the process of dough formation through a disulfide/sulfhydryl exchange. Tyrosine bonds (bityrosine and isobityrosine cross-links) formed in wheat dough during the processes of mixing and baking also contribute to the structure of the gluten network (see Section 2.5.1.1.3). The main factor that determines the quality of flour is the ratio between glutenins and gliadins. Gliadin molecules have a modifying effect on the viscoelastic properties of dough. The amino acid composition of wheat protein can be seen in Table 2.12.

Gluten proteins cause an autoimmune disorder of the small intestine in some individuals, called coeliac disease (see Section 10.3.2.3).

Proteins of other cereals Rye albumins (44% of rye proteins) and globulins (10%) have similar properties to albumins and globulins of wheat. The other rye proteins are prolamins, known as secalin (21%), and glutelins, known as secalinin (25%). The trivial names of insoluble proteins of rye and other grains are seldom used, but they have become known by the collective name gluten, as in the case of wheat proteins. However, this so-called rye gluten differs from wheat gluten in the proportion of prolamins and glutelins, in the content of some amino acids, and, especially, in its viscoelastic properties. Wheat gluten is ductile, whilst rye gluten tears, but rye is still the only cereal grain other than wheat from which a typical bread can be produced. Nevertheless, rye bread has a completely different texture and consistency (it is compact and sticky) from wheat bread, and also a different content of other flour components, in particular pentosans (see Section 4.5.1).

The main components of barley and oat proteins are prolamins (hordein in barley, 25%; gliadin in oats, 15%) and glutelins (hordenin in barley, 55%; avenin in oats, 54%), albumins (12% in barley, 20% in oats), and globulins (8% in barley, 12% in oats; known as avenalin). Important constituents of barley albumins are the lipid transfer protein (about 9 kDa), responsible for the shuttling of phospholipids and other fatty acid esters within cell membranes, and protein Z (43 kDa), a major barley endosperm albumin. Both survive the malting and brewing processes and participate in the formation and stability of beer foam. Rice proteins are mainly composed of glutelins known as oryzenin (about 80%), whilst prolamins known as oryzin (2–5%) and other proteins (albumins 11%, globulins 10%) are minor components. Rice and, to a lesser extent, oats differ from other cereals in containing little prolamin. Thus, these cereals have much more lysine, making them nutritionally superior. Maize proteins are composed of small amounts of albumins (4%), globulins (3%), gliadins called zein (about 50%), and glutelins (20–45%). The typical limiting amino acids in maize gluten are lysine and tryptophan.

2.4.5.4.2 Legumes and oilseeds

Legumes and oilseeds are important sources of proteins, lipids, and other nutrients in the diet. Legumes are plants of the Fabaceae family that are cultivated for their seeds, used for human and animal consumption and as forage crops (e.g. alfalfa and clover). The most important edible legume seeds (pulses) are soybeans, peas, beans, lentils, and peanuts, in addition to some other relatively less frequently used types (e.g. chickpeas, vigna, and lupin seeds). Legumes are high in protein (a typical protein content is 20–45%) and the limiting amino acids are usually sulfur amino acids.

Proteins in soybeans, peas, and other legumes are mostly composed of globulins, whilst minor components are low-molecular-weight albumins. For example, soybean protein contains about 80% globulins. Globulins in many legumes are further divided into legumins and vitilins, but soybean globulins are known as glycinin and conglycinin. Legumins have higher relative molecular weight, are less soluble, and have better thermal stability than vitilins. Peanuts contain about the same amount of globulin arachin and related conarachins, 15% of albumins and 10% of glutelins. Pea protein contains about 70% globulins (legumin and vicilin), 20% albumins, and 10% glutelins.

A recent, more precise characterisation of legume proteins is based on their sedimentation coefficients during ultracentrifugation. In this way, soybean proteins can be separated into four fractions, designated 2S, 7S, 11S, and 15S (S denotes the so-called Svedberg's units in which the sedimentation rate is expressed). The 2S fraction termed α-conglycinin contains 8–22% of soybean protein and consists of a number of enzymes (the Bowman-Birk and Kunitz trypsine inhibitors and cytochrome c). The 7S globulin fraction, known as β- and γ-conglycinins, accounts for 35–37% of the total protein (containing haemaglutinin, lipoxygenase, β-amylase, and globulins labelled a, a_1, and b). The 11S fraction, which accounts for 31–52% of the total protein, is known as glycinin, and the 15S fraction, which has no trivial name, accounts for 5–11% and is poorly characterised; it is thought to be composed of polymers of other soy proteins.

The typical basic chemical composition of some legumes and oilseeds is summarised in Table 2.24. The content of the individual nutrients and other components, however, varies broadly and depends on the seed origin (plant variety), ripeness, and many other factors. For example, in ripe common pea seeds, the protein content varies between 14.2 and 36.1%, the dietary fibre content ranges from 16.7 to 25.5%, and the water content is 11%. Green peas, however, contain 78.9% water, 5.4% protein, and 5.1% dietary fibre. Furthermore, oilseeds accumulate oil rather than starch. The amino acid composition of important legumes is given in Table 2.13.

Some legume seeds that contain significant amount of lipids, such as soybeans and peanuts, are used as oilseeds. The protein content of the oilseeds usually ranges from 20 to 50%. Other important oilseeds, including rapeseed, sunflower seed, and many other plant seeds, are described in Section 3.4.3.4.1. Almonds and nuts have high protein contents (about 20%). Owing to their low consumption, these

Table 2.24 Average basic chemical composition of some legumes and oilseeds (%).

Mature seeds[a]	Water	Proteins	Total lipids	Total sugars	Starch	Total dietary fibre	Minerals (ash)
Soybeans	8.5	40.2	20.5	7.3	0.5–2.0	9.3	4.5
Peanuts	6.5	27.6	49.6	4.0	1.0	6.0	2.8
Peas	11.3	26.3	2.2	15.8	40	15.9	2.9
Beans (kidney)	11.8	27.2	1.0	2.1	40	11.9	3.8
Chickpeas	11.5	20.7	5.7	10.7	47	13.9	3.6
Lentils	10.4	27.7	2.1	2.0	44	12.2	2.9

[a]Average values, from various sources, therefore their sum may exceed 100%. Non-starch carbohydrates are calculated by difference.

commodities cannot be considered as significant sources of protein. The amino acid composition of major oil seeds and nuts is given in Table 2.13.

Legumes and oilseeds often contain various antinutritional factors or even toxic substances. Examples of such substances are various allergens, protease inhibitors, lectins, saponins, indigestible sugars such as raffinose and its homologues, and phytic acid (see Section 10.3.2.6).

2.4.5.4.3 Changes during storage and processing

Denaturation of proteins can occur during grain drying and milling, but on a very small scale. The extent of protein denaturation depends on water activity, temperature, and time of drying and milling. When completely denatured, gluten lacks its typical viscoelastic properties. Bread flour improves its baking properties during storage of up to about a year; with longer storage, the properties get worse again. The improved baking properties of flour are caused by natural bleaching, which is based on autoxidation of unsaturated fatty acids (see Section 3.8.1.8.2) and their enzymatic oxidation by lipoxygenases. The resulting hydroperoxides of fatty acids are strong oxidising agents that oxidise the free thiol groups of gluten proteins. Their polypeptide chains are connected by disulfide bridges to high-molecular-weight units, which have a positive effect on the volume of the bread loaf and the texture of the crust. At the same time, degradation of carotenoid pigments (see Section 9.9.5.1) results in a more attractive flour of lighter colour. Bread flour is sometimes conditioned with ascorbic acid, which increases the bread volume and creates a better texture. The mechanism of this reaction is described in Section 5.14.6.1. Attempts to simulate the ageing process have led to the use of flour additives (oxidising agents) in some countries, such as chlorine dioxide, dibenzoyl peroxide, bromates, and iodates of alkali metals (see Sections 11.4.2.2.1, 5.2.6.2, and 11.4.2.2).

During the mixing of wheat flour with water, yeast, and other ingredients, the dough is processed mechanically (in a process known as kneading). The flour proteins gliadin and glutenin are hydrated, and air and carbon dioxide (produced by yeast) are incorporated, which is the basis for the bread structure. On baking, water evaporates from the crust, starch begins to gelatinise at about 50 °C, and proteins are denatured at about 70 °C, releasing water, which is absorbed by the starch and increases the loaf volume. As the temperature increases, the loaf shape is fixed, and gluten creates the typical three-dimensional net structure, which surrounds all the other ingredients. Baking bread and other cereal products results in a relatively intensive Maillard reaction, which is desirable in order to develop the typical colour and flavour. However, it has a negative effect on the nutritional value of proteins, because the bound essential amino acid lysine primarily enters this reaction and thus becomes unavailable (blocked).

Significant changes in the structures of proteins also occur in other methods of thermal processing of cereals. For example, extrusion leads to texturised proteins that lose their prevailing globular structure, gain a fibrous structure, and are denatured, accompanied by the disruption of existing non-covalent interactions and disulfide bridges and the emergence of new bonds between filamentous molecules.

2.4.5.5 Non-traditional protein sources

Since the middle of the last century, both unicellular and multicellular organisms have been studied as possible non-traditional sources of protein for human nutrition and as animal feed. Special interest has been paid to yeast proteins (e.g. the genera *Candida* or *Torula*), to bacteria, fungi, and algae (e.g. the green sweet water algae of the genus *Chlorella*, which contain about 45% protein d. w.), and to higher plants (oilseed meals, the leaves of certain plants, and some vegetables). These protein sources have not found significant use in human nutrition, but are often used as animal feed.

Novel protein sources like insects, microalgae, seaweed, duckweed, and rapeseed are expected to enter feed and food market as replacers for animal-derived proteins. Their use is limited by potential hazards including a range of contaminants such as heavy metals, mycotoxins, pesticide residues, toxins, and pathogens, amongst other factors.

Potential insects for application in food in the Euro-Atlantic area are *Gryllodus sigillatus* and *Acheta domesticus* (crickets), *Alphitobius diaperinus* (lesser mealworm), and *Tenebrio molitor* (yellow mealworm). In general, the protein content (20–76% d.w.) varies according to the species and the development stage. The content of crude protein of giant mealworm larvae (*Zophobas morio*), larvae of the yellow mealworm, and nymphs of the field cricket (*Gryllus assimilis*) ranges from 46 to 56% (d.w.). The green algae (Chlorophycea) *Chlorella vulgaris*, *Haematococcus pluvialis*, and *Dunaliella salina* and the cyanobacteria *Spirulina maxima* are used for human consumption, mainly as nutritional supplements, and as animal feed additives. Seaweeds can be both harvested from the sea and cultivated. Several seaweed species are used for direct consumption, such as *Monostroma* sp., *Enteromorpha* sp., *Laminaria* sp., and *Ulva rigida. Laminaria* sp. (kelp weed) are mainly used in food supplements.

Some traditional food and feed can be employed as a novel source of proteins due to an improvement of the availability and technological functionality of the original proteins via novel technologies, and owing to the valorization of waste streams. Thus, legumes (e.g. lupine), grains (e.g. rice and maize), mushrooms, and potatoes have already been used in food and food technology. Some proteins are intended to replace the original proteins in many applications for health, economical, and technological reasons. For example, potato proteins can be used for a variety of functional effects: as water binders in meat and sausage, as foaming aids in confectionery, bakery, and dairy products, and as emulsifiers in spreads, desserts, and dressings. Patatin, a storage and enzymatically active glycoprotein recovered from potato aqueous byproducts, can be used as an alternative to animal proteins in fining red wine.

Various technologies have been developed that utilise oilseed meals as the raw material for the production of 'flours', protein isolates, and protein concentrates. These technologies are mainly used in the processing of soybeans and, to a lesser extent, of peanuts, cotton, lupine, and other meals. Additional sources of protein are whey and fishmeal, amongst others. The final products are mixtures rich in proteins (often enriched by minerals and vitamins).

Protein concentrates generally tend to contain around 65–80% protein, whereas in **protein isolates**, a higher protein content (≥90%) is found. Concentrates are mostly prepared from oilseed meals (such as soybean and rapeseed meals) and from pulses and cereals. The proteins are first denatured by heating, and soluble substances are extracted with water. Protein isolates are prepared by extracting proteins and other substances soluble in dilute aqueous solutions of sodium or calcium hydroxides. The extract obtained is purified and its pH is adjusted to the value of the isoelectric point, when the proteins precipitate. This procedure gives a relatively clean proteinaceous material, which can then be used as a dietary supplement or as an intermediate for further processing. The texture of protein preparations can be further modified to resemble meat. Well-known examples are so-called soy meat, soy curd (tofu), and soy cheese (sufu). Modifications include addition of fat, flavouring agents, dyes, and binders, and the finished products are mostly processed by extrusion. Another meat-resembling example with high-quality protein is a texturised pure egg white (Proteiner) enriched by nutritious substances such as fibre and unsaturated fatty acids.

2.4.5.6 Modified proteins

The modification of native protein structures is based on nutritional, hygienic and toxicological, and technological requirements. It can be carried out chemically or enzymatically, with the aim of improving:

- desirable physico-chemical properties (solubility, dispersibility, elasticity, viscosity, adhesivity, cohesivity, emulsification, foaming, gelation, stabilisation of the dispersion systems, and water-binding ability);

- nutritional value (by inactivation of antinutritional and toxic substances, through improved availability of essential amino acid content);

- organoleptic properties (texture and taste);

- the use of non-traditional raw materials for food purposes (e.g. yeast proteins).

Functional groups of bound amino acids can be chemically modified in various ways. Amino groups can be derivatized by acylation or methylation (reaction with formaldehyde and reduction of the resulting hydroxymethyl derivatives). Carboxyl and hydroxyl groups can be modified by esterification, amide bonds (including peptide bonds) by hydrolysis, and thiol groups via oxidation to disulfides and vice versa. For example, succinylated casein is, unlike native caseins, practically insoluble in solutions of pH 2–3 but soluble in solutions of pH 4.5. Succinylated yeast protein, soluble in solutions of pH 4–6, is more resistant to heat denaturation and has a higher emulsifying strength. Alkylated proteins, for example, do not react with reducing sugars in the Maillard reaction. The enzyme modifications are mainly based on the protein dephosphorylation and plastein reaction. For example, in the presence of calcium ions, the dephosphorylated β-casein is more soluble than the native protein. The plastein reaction (see Section 2.3.3.2) has found use in the debittering of protein hydrolysates and in the incorporation of nutritionally valuable essential amino acids into proteins of low nutritional value (e.g. of tryptophan, threonine, and lysine into maize zein). Another approach to modifying protein structure is to use physical food processing technologies, such as extrusion, high-pressure processing, and high-power ultrasound treatment. Recently, recombinant technology and genetic engineering approaches have also been used, for example, to increase both total protein and lysine content in cereal crops and to alter gluten and soybean protein structure in order to provide better technological properties such as solubility, emulsification, gelation, and foaming properties.

2.5 Reactions

In food raw materials, which mainly include animal tissues post mortem and plant tissues during post-harvest storage, free amino acids and amino acids bound in peptides and proteins take part in a variety of both biochemical and chemical reactions. Other changes of these compounds are caused by the action of various physical factors, such as heat, mechanical force, hydrostatic pressure, and radiation, and of chemical agents, such as acids, bases, salts, and surfactants, during processing and cooking. These reactions often result in oxidation of side chains of amino acids, peptide chain fragmentation, aggregation, enzymatic inactivation, conformational changes known as denaturation, and even changes in the primary structure of proteins due to hydrolysis of peptide bonds. Proteins, peptides, and amino acids react with one another and with the other food components, particularly with oxidised lipids, reducing sugars, oxidised phenolic compounds, some food additives, and some contaminants. The extent and type of the reactions depend on the particular food (its chemical composition), conditions during storage and processing such as water activity, temperature, pH, and oxygen, and the reaction partners present. All the reaction may significantly affect the nutritional, sensory, technological, and hygienic and toxicological quality of foods.

The positive consequences of these changes, interactions, and reactions are the inactivation of undesirable enzymes and microorganisms, denaturation of protein antinutritional factors and toxins, generally higher protein digestibility, production of desired flavour-active compounds and colourants, and increased food shelf-life.

Negative consequences also occur, such as some reduction in the nutritional value of the food following from the reactions of essential amino acids, essential fatty acids, and vitamins, and sometimes reduced digestibility of proteins. The formation of unwanted flavour compounds and undesirable discolouration may also occur. Some reactions may negatively affect the hygienic-toxicological quality of food products, because certain reaction products show specific biological effects and are classified as contaminants.

2.5.1 Intramolecular and intermolecular reactions

Amino acids in living organisms undergo oxidation during normal metabolism, provided that the diet is rich in protein, and during starvation, when they serve as an energy source. In tissues post mortem or after harvesting, both free and peptide-bound amino acids undergo enzymatic and chemical oxidation along with many other elimination, addition, isomeration, and other reactions, including with other food constituents, which produce a variety of products.

2.5.1.1 Oxidation

Proteins and free amino acids contain numerous reactive functional groups and can become substrates for oxidoreductases or may react with reactive oxygen species, such as hydroxyl radicals ($HO^\bullet$), superoxide radicals ($O_2^{\bullet-}$), singlet oxygen ($^1O_2$), fatty acid hydroperoxides (R—O—OH), alkoxyl ($RO^\bullet$) and peroxyl ($ROO^\bullet$) radicals, atmospheric oxygen, and other oxidising agents, often under the catalysis of transition metal ions. The reaction of proteins with the most reactive oxygen species, hydroxyl radicals, results in abstraction of the hydrogen atom from the α-carbon, but also from the polypeptide chain and the side chains of hydrophobic amino acids residues, and leads to the formation of various oligomers and hydroxy derivatives of proteins (Figure 2.26).

Free peroxyl radicals react with proteins with the formation of free radicals of proteins, which react with other protein radicals to yield protein dimers, or with free radicals of lipids, yielding copolymers (lipoproteins with covalent bonds). The formation of protein oligomers and lipoproteins during oxidation of proteins (PH) by alkoxyl ($RO^\bullet$) and peroxyl ($ROO^\bullet$) radicals is shown in the following equations. A protein radical ($P^\bullet$) is formed most frequently by splitting off the labile hydrogen atom on C_α, an alkoxyl radical yields a hydroxy acid, and a peroxyl radical yields hydroperoxide:

$$P-H+R-O^\bullet \quad \rightarrow \quad P^\bullet + R-OH$$

$$P-H+R-O-O^\bullet \quad \rightarrow \quad P^\bullet + R-O-OH$$

Recombination of protein radicals subsequently yields protein oligomers:

$$P^\bullet + P \quad \rightarrow \quad P^\bullet - P$$

$$P-P^\bullet + P \quad \rightarrow \quad P-P-P^\bullet \quad \text{etc.}$$

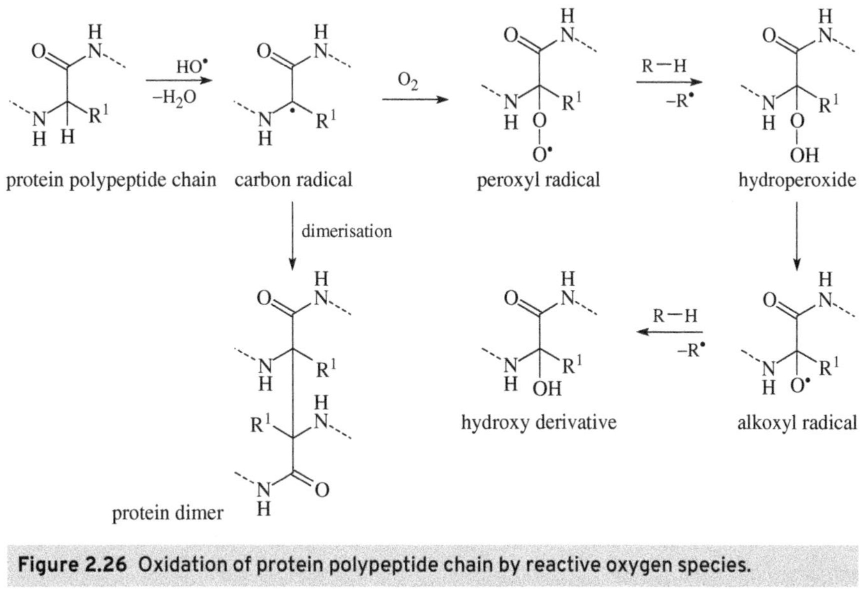

Figure 2.26 Oxidation of protein polypeptide chain by reactive oxygen species.

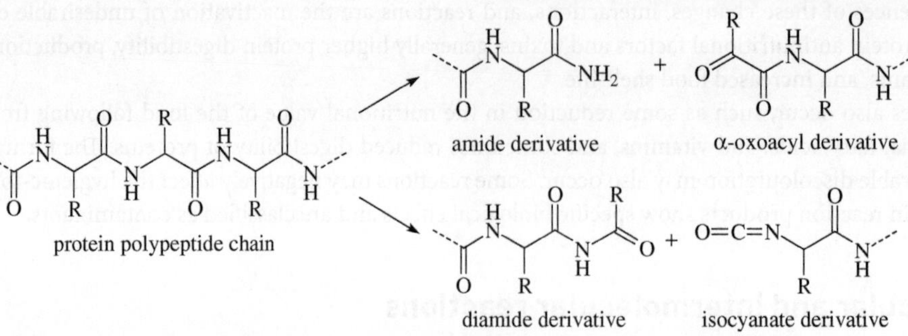

Figure 2.27 Oxidative cleavage of protein polypeptide chain. Source: Stadtman and Levine (2003). With kind permission from Springer Science and Business Media.

$$P^\bullet + P^\bullet \quad \rightarrow \quad P - P$$

$$P - P^\bullet + P^\bullet \quad \rightarrow \quad P - P - P \quad \text{etc.}$$

Cross-links of the lipid–protein type are produced mostly by the following reactions:

$$P^\bullet + R - O - O^\bullet \quad \rightarrow \quad P - R + O_2$$
$$P^\bullet + R - O^\bullet \quad \rightarrow \quad P - O - R$$
$$P^\bullet + R^\bullet \quad \rightarrow \quad P - R$$

Oxidation of the polypeptide chains of proteins can also lead to the cleavage of peptide bonds. Two different mechanisms are proposed, α-amidation and diamide pathways (Figure 2.27). In both, the polypeptide chain is shortened and the *N*- and *C*-amino acid residues of the resulting fragments are modified (aspartyl, glutamyl, and prolyl residues produce specific *N*- and *C*-terminal fragments).

Sulfur-containing amino acids (cysteine and methionine) and aromatic and heterocyclic amino acids (phenylalanine, tyrosine, histidine, and tryptophan) are relatively oxylabile compounds (both free and peptide-bound). Other side chains in some proteins are also readily oxidised. Formation of some oxidation products of amino acids is irreversible in food and *in vivo*, leaving the amino acids unavailable for nutrition. Some advanced products of these reactions are undesirable, leading to off-flavours and hygienic and toxicological issues.

Oxidation of proteins *in vivo*, which modifies their structure and affects their function, results from oxidative stress and is related to regular ageing and pathogenic changes in some diseases.

2.5.1.1.1 Cysteine and cystine

Oxidation of the thiol group (—SH) to the disulfide group (—S—S—) proceeds typically through the action of dehydrogenases, reactive oxygen species such as singlet oxygen, fatty acid hydroperoxides, and even atmospheric oxygen (autoxidation). Cysteine (free or bound in peptides and proteins) yields cystine (Figure 2.28). The reaction is reversible, and the cleavage of the disulfide bond in cystine may proceed, according to conditions, by homolytic (photolysis and other reactions) or heterolytic mechanisms.

$$2\,R{-}SH \xrightleftharpoons[\qquad]{1/2\,O_2,\ -H_2O} R{-}S{-}S{-}R$$

Figure 2.28 Oxidation of cysteine to cystine, R=CH$_2$CH(NH$_2$)COOH.

It is assumed that the first stage of cysteine autoxidation by homolytic cleavage is the formation of the alkylthiolate (RS⁻ anion) through the reaction of the thiol with a hydroxyl ion (the reaction is therefore faster in alkaline media). Alkylthiolate reacts with oxygen to form a thiyl radical (RS•) and anion of superoxide radical (O$_2^{-\bullet}$) (see Section 3.8.1.13.2). Homolytic cleavage of thiols by reactive oxygen species also produces thiyl radicals. Volatile thiols, which are components of many food flavours, are oxidised in the same way (see Section 8.2.9.1.2):

$$R\text{-}SH + HO^- \rightleftharpoons R\text{-}S^- + H_2O$$

$$R\text{-}S^- + O_2 \rightleftharpoons R\text{-}S^\bullet + O_2^{-\bullet}$$

$$2\,R\text{-}S^\bullet \longrightarrow R\text{-}S\text{-}S\text{-}R$$

$$R\text{-}SH + O_2^{-\bullet} \longrightarrow R\text{-}S^\bullet + HO_2^-$$

$$R\text{-}SH + HO^\bullet \longrightarrow R\text{-}S^\bullet + H_2O$$

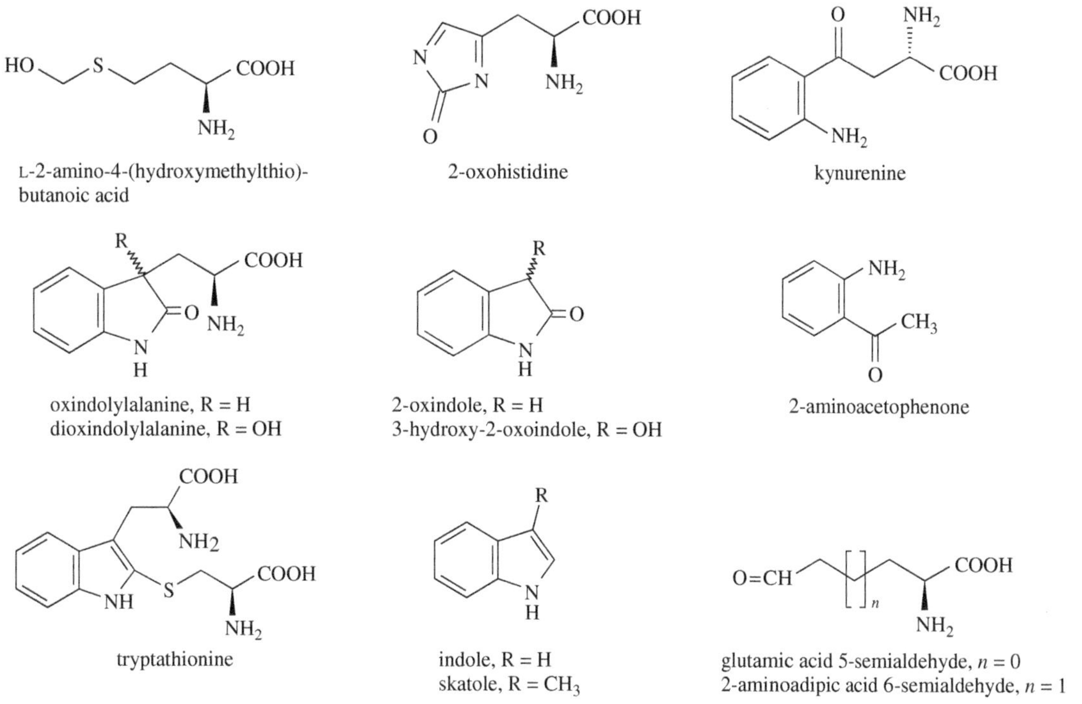
L-2-amino-4-(hydroxymethylthio)-butanoic acid
2-oxohistidine
kynurenine
oxindolylalanine, R = H
dioxindolylalanine, R = OH
2-oxindole, R = H
3-hydroxy-2-oxoindole, R = OH
2-aminoacetophenone
tryptathionine
indole, R = H
skatole, R = CH3
glutamic acid 5-semialdehyde, n = 0
2-aminoadipic acid 6-semialdehyde, n = 1

2-31, selected oxidation products of amino acids

Cysteine is oxidised to cystine more effectively by fatty acid hydroperoxides, hydrogen peroxide, and reactive oxygen species. Fatty acid hydroperoxides are able to oxidise simultaneously the sulfur atoms; thus, in addition to cystine, the reaction yields thiosulfinate (monooxide), disulfoxide, thiosulfonate (dioxide), sulfoxidosulfone (trioxide), and disulfones (tetraoxides) (Figure 2.29). In foods, the degree of oxidation does not usually exceed the level of thiosulfinate and dioxide. Both the latter compounds are the main products of cysteine and cystine oxidation. Their bioavailability is about 20–50% that of cysteine. The products of cysteine oxidation by the more powerful oxidising agents are, in addition to cystine, the unstable cysteine sulfenic acid, cysteine sulfinic acid, and cysteine sulfonic acid, known as cysteic acid (Figure 2.30). Both the latter acids are completely unavailable as a source of cysteine.

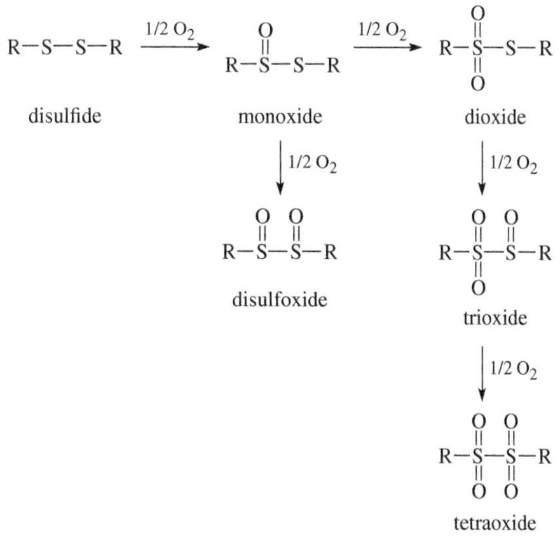
disulfide
monoxide
dioxide
disulfoxide
trioxide
tetraoxide

Figure 2.29 Oxidation of cystine to oxides, R=$CH_2CH(NH_2)COOH$.

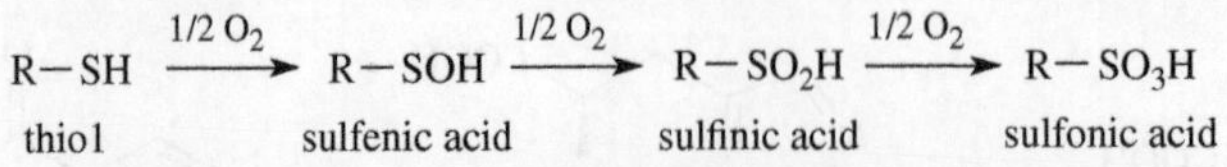

$$R-SH \xrightarrow{1/2\,O_2} R-SOH \xrightarrow{1/2\,O_2} R-SO_2H \xrightarrow{1/2\,O_2} R-SO_3H$$

thiol   sulfenic acid   sulfinic acid   sulfonic acid

Figure 2.30 Oxidation of cysteine to acids, $R=CH_2CH(NH_2)COOH$.

2.5.1.1.2 Methionine

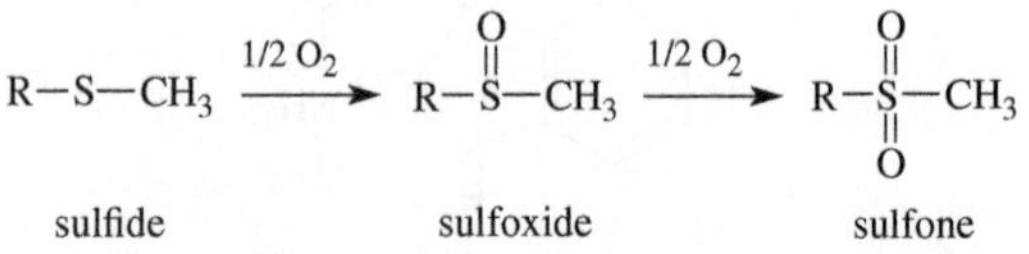

Methionine can be oxidised by reactive oxygen species, fatty acid hydroperoxides, and oxidised polyphenols. The primary oxidation product, methionine sulfoxide, is a frequent component of various proteins. The contents of methionine sulfoxide in protein sources used for foods and feeds usually vary from 0 to 30% of the total methionine content (e.g. up to 11% methionine sulfoxide apparently occurs in milk powder, 13% in lean beef, 16–28% in wheat gluten, and around 50% in orange juice). Methionine sulfoxide is fully available as a source of methionine as it can be reduced back to methionine by sulfoxide reductases, but the final oxidation product, methionine sulfone, is not at all (Figure 2.31).

$$R-S-CH_3 \xrightarrow{1/2\,O_2} R-\overset{\displaystyle O}{\underset{\displaystyle ||}{S}}-CH_3 \xrightarrow{1/2\,O_2} R-\overset{\displaystyle O}{\underset{\displaystyle O}{S}}-CH_3$$

sulfide   sulfoxide   sulfone

Figure 2.31 Oxidation of methionine, $R=CH_2CH_2CH(NH_2)COOH$.

The oxidation of the methyl group to a hydroxymethyl group by hydroxyl radicals gives L-2-amino-4-(hydroxymethylthio)butanoic acid (**2-31**). Another oxidation product, the so-called hydroxymethionine, (R,S)-2-hydroxy-4-(methylthio)butanoic acid, which also can be formed by the reduction of the corresponding oxoacid (see Section 2.5.1.3.2), is widely used in poultry nutrition as a precursor of dietary methionine. It is also used in cosmetic preparations intended to prevent the signs of cutaneous ageing, such as sagging of the cutaneous and subcutaneous tissue and loss of cutaneous elasticity.

2.5.1.1.3 Phenylalanine and tyrosine

Phenylalanine residues in proteins and free phenylalanine are oxidised by various reactive oxygen species to 2- and 3-hydroxy derivatives and other products. The most important reaction of tyrosine is the direct post-translational enzymatic oxidation of tyrosine to 3,4-dihydroxyphenylalanine (DOPA) in animals *in vivo*, which is catalysed by tyrosine hydroxylase. Subsequent reactions then lead to melanins, an important group of pigments widely distributed in all living organisms (see Section 9.3.1.1). Tyrosine is also a substrate of oxidoreductases in the enzymatic browning reactions (see Section 9.12).

Oxidation catalysed by peroxidases (Figure 2.32) converts the tyrosine residues in proteins into bityrosine and isobityrosine derivatives. Bityrosine has been found in animal proteins from resilin, a protein found in the cuticle of insects and arthropods, to elastin and collagen, found in vertebrates. Bityrosine is also formed by oxidation with reactive oxygen species and oxidising agents (ascorbic acid, azodicarbonamide, and potassium bromate) in wheat flour, dough, and bread. It has been suggested that it is a stabilising cross-link in the wheat gluten structure in addition to disulfide bonds.

The oxidation of tyrosine residues by singlet oxygen suggests the initial formation of an unstable endoperoxide via 1,4-cycloaddition of singlet oxygen to the aromatic ring. The endoperoxide undergoes ring opening to produce a hydroperoxide, which decomposes to the corresponding alcohol. The hydroperoxide derived from free tyrosine yields a hydroperoxide of an indole derivative by ring closure, which is transformed to the corresponding alcohol and other heterocyclic products (Figure 2.33).

2.5.1.1.4 Histidine

Oxidation of histidine by singlet oxygen results in a complex mixture of products. The initial 1,4-cycloaddition of singlet oxygen, yielding one or more endoperoxides, is followed by subsequent ring opening, which produces unstable hydroperoxides that decompose to a variety of products via free radicals. Some intermediates can form adducts with various nucleophilic reagents, dimers with another histidine-derived molecule, and a range of other products. The final reaction products are aspartic acid, asparagine, and urea. The reaction is schematically illustrated in Figure 2.34. Recent studies revealed the formation of 2-oxohistidine (**2-31**) from the histidine residues in proteins. The oxo group can be hydrated or can undergo reaction with a lysine side chain amine group to form a protein cross-link.

2.5.1.1.5 Tryptophan

Tryptophan (free and bound in proteins) is a very oxylabile compound, especially in acid solutions. It is oxidised by reactive oxygen species, hydrogen peroxide, fatty acid hydroperoxides, sulfoxides, peroxyacids, and other oxidation agents, and can also react via physical quenching, which results in energy transfer and de-excitation of singlet oxygen.

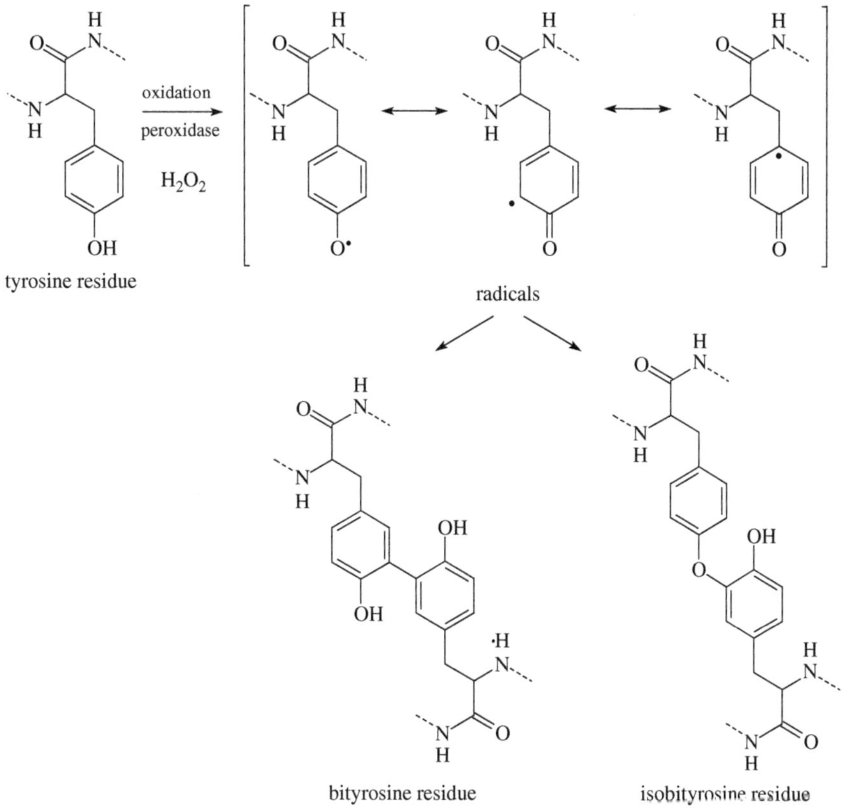

Figure 2.32 Oxidation of tyrosine by peroxidases.

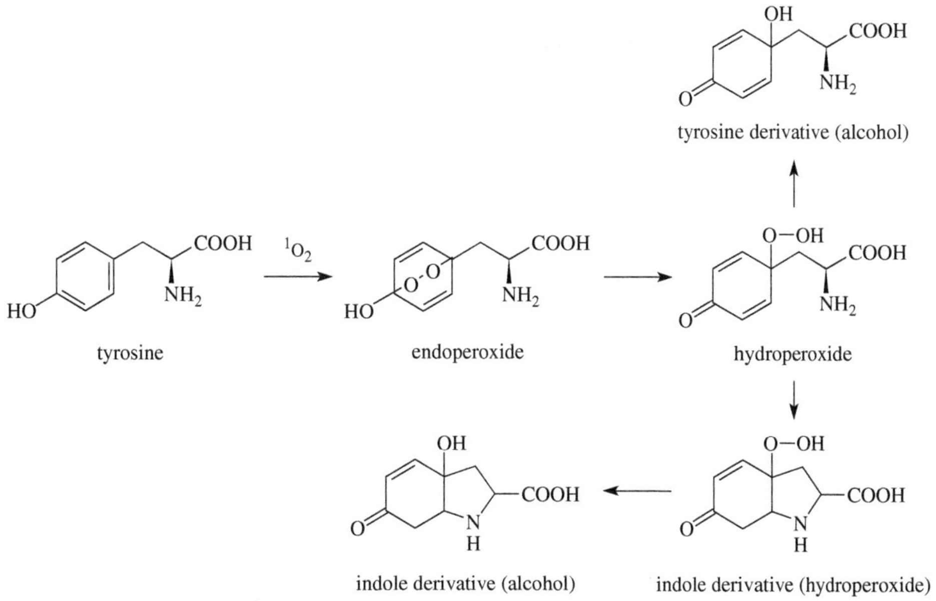

Figure 2.33 Oxidation of tyrosine by singlet oxygen. Source: Wright et al. (2002), Scheme 1. Reproduced by permission of John Wiley and Sons.

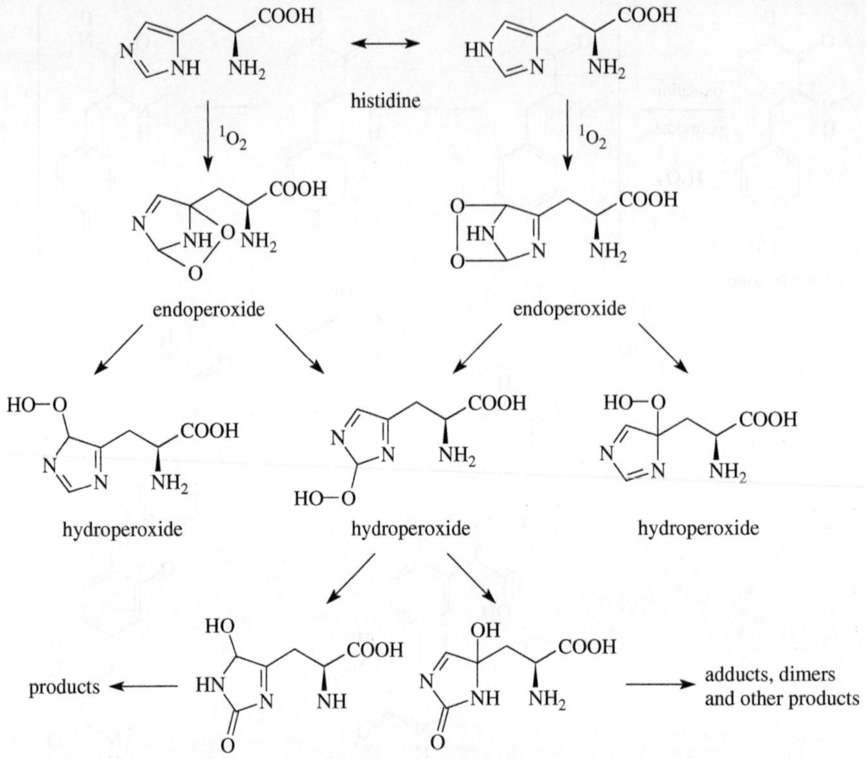

Figure 2.34 Oxidation of histidine by singlet oxygen. Source: Agon et al. (2006). Reproduced by permission of Elsevier.

The main product of tryptophan oxidation with singlet oxygen is a dioxethane derivative formed via 1,2-cycloaddition and a hydroperoxide at C-3. Subsequent decomposition of these intermediates yields *N*-formylkynurenine, whereas their ring closure leads to *cis*- and *trans*-isomers of 3α-hydroxypyrroloindoles and 3α,8α-dihydroxypyrroloindoles. The simplified reaction sequences are given in Figure 2.35.

The oxidation of tryptophan with hydrogen peroxide yields a number of oxygen-containing products via *cis*- and *trans*-isomers of hydroperoxides and free radicals. The main degradation products are the physiological metabolites *N*-formyl-L-kynurenine and L-kynurenine (3-anthraniloyl-L-alanine, **2-31**), unavailable as a source of tryptophan. Other reaction products include 3α-hydroxypyrroloindole derivative (Figure 2.35), 2,3-dihydro-2-oxotryptophan, and 2-hydroxytryptophan (keto–enol tautomerism), also known as oxindolylalanine (**2-31**) and dioxindolylalanine (**2-31**); that is, 2,3-dihydro-3-hydroxy-2-oxo-tryptophan and 5-hydroxytryptophan.

The main intermediate of tryptophan degradation occurring in acid hydrolysates of proteins, where tryptophan is completely degraded, is oxindolylalanine (**2-31**), followed by simple indoles, such as 3-hydroxy-2-oxindole (**2-31**) and 2-oxindole (**2-31**). The reaction of tryptophan with cystine in acid solutions yields cysteine and α-amino-2-[(2-amino-2-carboxyethyl)thio]-1*H*-indole-3-propionic acid, known as tryptathionine (**2-31**). The tryptathionine cross-link formed between tryptophan and cysteine is characteristic of the toxic bicyclic peptides amatoxins and phallotoxins (see Section 10.3.7.2). 2-Aminoacetophenone (**2-31**) (together with its *N*-formyl derivative) is known as the character impact compound responsible for the atypical gluey and glutinous off-flavour of milk powders and sulfurated wines. In non-fat dry milk, it is responsible for a strong flavour characterised as potato-like or cocoa-like. In wine, where it is formed by oxidative degradation of the phytohormone indole-3-acetic acid triggered by sulfuration, its flavour resembles naphthalene. Its odour threshold is 1 μg/l in wine. The products of microbial degradation of tryptophan, indole (**2-31**) and 3-methylindole known as skatole (**2-31**), are responsible for the off-flavour of some foods, such as stored eggs and boiled tripe (see Section 8.2.11.1.4).

2.5.1.1.6 Other amino acids

In addition to the side chain of bound cysteine that is oxidised to bound cystine, other amino acid side chains in proteins can also react with reactive oxygen species such as hydroxyl radicals, frequently mediated by transition metal ions. These oxidations lead to the formation of hydroxyl, oxo, and formyl protein derivatives and peptide fragments. For example, the side chain of bound leucine is oxidised to 3-, 4-,

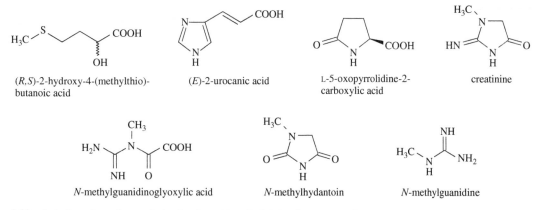

Figure 2.35 Oxidation of tryptophan by singlet oxygen. Source: Gracanin et al. (2009). Reproduced by permission of Elsevier.

or 5-hydroxyleucine, the hydroxyl group of threonine is oxidised to an oxo group (yielding 2-amino-3-oxobutanoic acid on hydrolysis), the side chain of arginine yields a glutamic acid 5-semialdehyde residue (**2-31**), whilst the side chain of lysine produces 2-aminoadipic acid 6-semialdehyde residue (**2-31**).

(R,S)-2-hydroxy-4-(methylthio)-
butanoic acid

(E)-2-urocanic acid

L-5-oxopyrrolidine-2-
carboxylic acid

creatinine

N-methylguanidinoglyoxylic acid

N-methylhydantoin

N-methylguanidine

2-32, selected transformation products of amino acids and related amino compounds

2.5.1.2 Isomerisation

Conventional processing of food raw materials as well as culinary methods used during food preparation do not usually cause a significant isomerisation (racemisation) of L-amino acids (Table 2.25). L-Aspartic acid and L-serine undergo racemisation relatively easily. L-Isoleucine (isomerises to D-*allo*-isoleucine), L-proline, L-threonine (isomerises to D-*allo*-threonine), and L-valine yield smaller amounts of D-isomers.

Table 2.25 Content of D-amino acids in some foods.

Food	Content of D-amino acids as a % of total amount of both optical isomers							
	Ala	Asp	Glu	Leu	Met	Phe	Pro	Val
Hamburger (outer layer)	2.8	5.5	2.4	3.2	2.9	2.7	1.8	1.5
Original beef meat	3.2	6.2	4.9	3.1	2.4	2.8	2.0	1.6
Pasteurised milk	1.8	7.3	5.1	–	–	–	–	–
Condensed milk	2.5	–	–	–	–	–	–	–
Original milk	1.9	7.3	4.8	–	–	–	–	–
Sour milk[a]	38.6	14.1	4.0	18.4	–	–	–	6.8
Kefir[a]	37.4	17.6	4.9	22.6	–	–	–	5.6
Yoghurt[a]	61.3	20.9	12.4		–	–	–	–
Toast (outer layer)	2.8	10.5	3.2	2.7	1.7	2.4	2.1	1.1
Originál bread (white)	2.4	5.6	2.8	3.2	2.3	2.3	0.9	0.9
Extruded soybean flour	2.7	7.6	3.9	2.7	–	2.4	1.6	0.8
Original soybean flour	2.5	4.4	3.1	1.4	–	2.8	2.3	1.0
Roasted almonds	7.2	4.7	–	6.5	–	1.3	–	1.1
Original almonds	0.0	2.6	–	3.0	–	1.1	–	0.0

[a] Amino acids are mainly released by hydrolysis of peptidoglycans of microbial cell walls that contain D-amino acids.

Free amino acids are roughly 10 times more stable against racemisation than amino acids bound in proteins. An extensive racemisation of amino acids does however occur, even at relatively low temperatures, in alkaline media used to inactivate enzymes, microorganisms, and microbial toxins and for the extraction of nucleic acids, removal of residual meat from bones, production of protein isolates, peeling of fruits and vegetables, or debittering of olives. In sterilised alkali-treated olives, significant amounts of D-Asx, D-Glx, D-Ser, D-His, D-Arg, D-Ala, D-Tyr, D-Val, D-Phe, D-Ile, and D-Leu were found. D-Lys was not detected. The most-abundant D-amino acids were D-Asx, D-Glx, D-Ser, and D-Leu. The total content of D-amino acids in olives averaged 260 mg/kg in the edible portion, which is 4.2% of the total content of amino acids (L- and D-enantiomers, 6170 mg/kg edible portion).

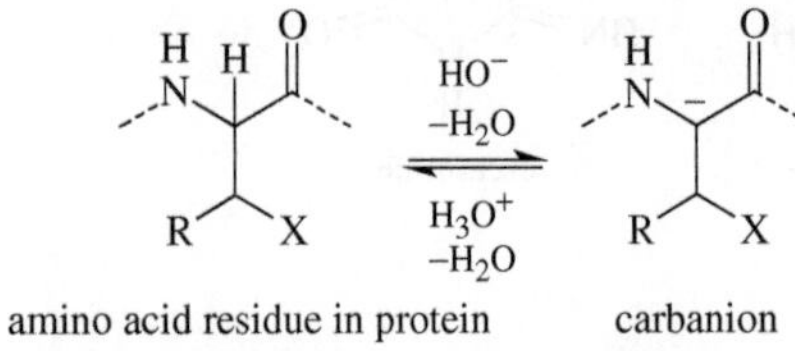

amino acid residue in protein　　　carbanion

Figure 2.36 Racemisation of amino acids.

The racemisation of amino acids bound in proteins is accompanied by β-elimination of certain functional groups in side chains of amino acids, which yields **dehydroproteins** and subsequently cross-links in protein molecules. Racemisation begins with the α-proton elimination of protein-bound amino acids, giving rise to an immediately isomerisable intermediate carbanion. This carbanion reacts with a proton (hydronium ion), which gives an equimolar mixture of the D-enantiomer and the original L-amino acid bound in protein (Figure 2.36). Proteolysis then releases the free D-amino acid.

The formation of D-amino acids reduces the nutritional value of a protein, because proteins containing D-amino acids have reduced digestibility and hence lower availability. The reason is the different method of absorption; whilst L-amino acids are absorbed very efficiently in the digestive tract by an active transport, D-amino acids are absorbed by passive diffusion, the effectiveness of which is limited. The toxicity of D-isomers is insignificant, with the exception of D-serine, which is reportedly nephrotoxic.

Racemisation of L-amino acids (elimination of an α-proton of protein-bound amino acids) occurring *in vivo* during ageing and some diseases (Alzeheimer's disease, cataracts, and diabetes) is facilitated by reducing sugars. An example is the reaction of D-glucose, shown in Figure 2.37. Briefly, an Amadori product formed in the Maillard reaction (see Section 4.7.5) is isomerised and then hydrolysed to a mixture of L- and D-amino acids under the formation of 3-deoxy-D-*erythro*-hexos-2-ulose or 1-deoxy-D-*erythro*-hexo-2,3-diulose.

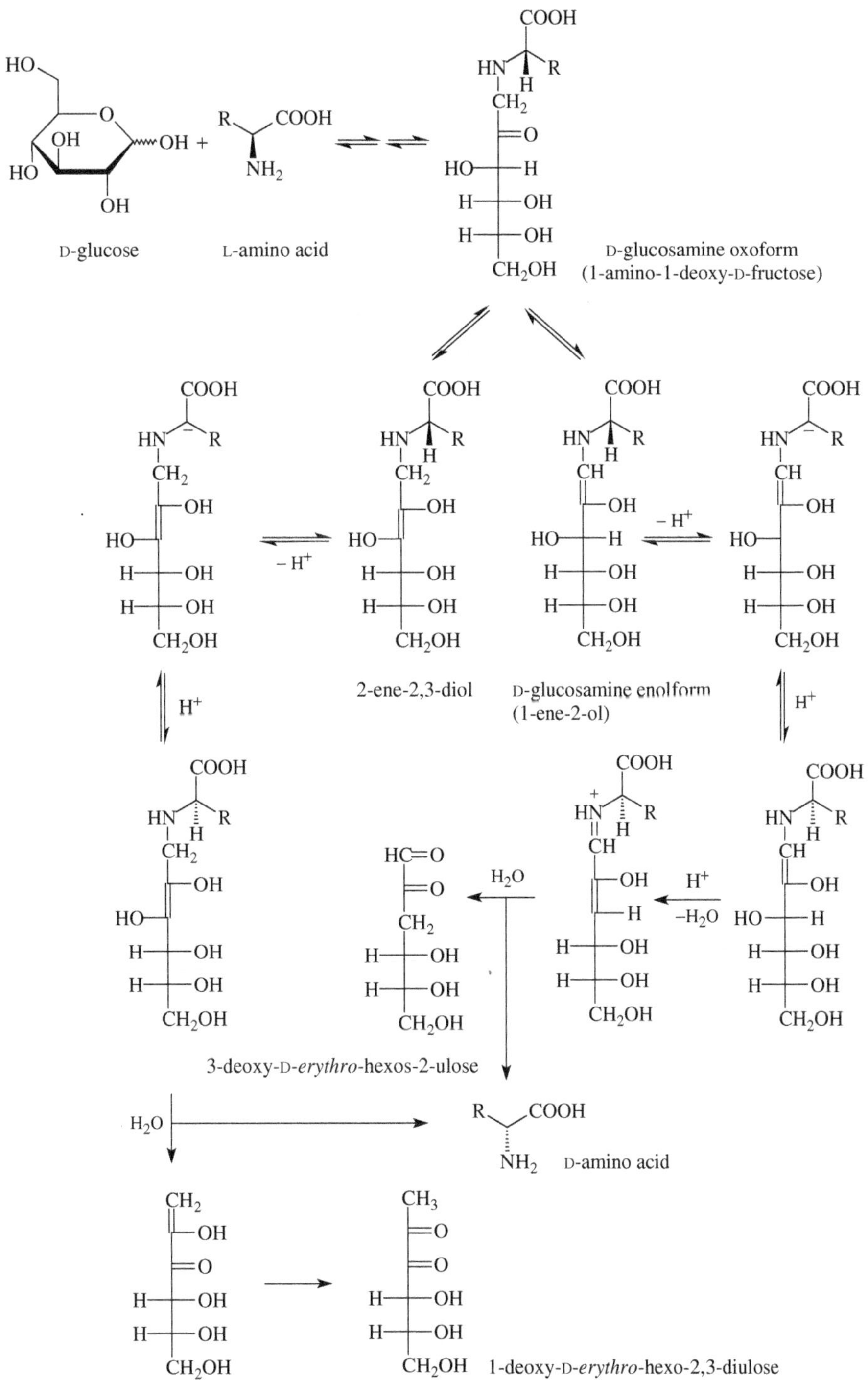

Figure 2.37 Isomerisation of L-amino acids catalysed by reducing sugars.

2.5.1.3 Elimination and other reactions

2.5.1.3.1 Decarboxylation

Amino acids eliminate carbon dioxide under the action of specific decarboxylases and, to a lesser extent, during heating to temperatures around their melting point (at around 200 °C). The products of decarboxylation of amino acids are the corresponding amines (Figure 2.38). Amines can be also formed by transamination of aldehydes catalysed by transaminases (see Section 8.2.10.1.2). Aliphatic amines (e.g. methylamine formed from alanine and isobutylamine arising from valine) are flavour-active compounds of non-acidic foods, such as some cheeses. Amines derived from aromatic, heterocyclic, and basic amino acids (such as histamine or tryptamine) are biologically active compounds; therefore, they are collectively termed biogenic amines (see Section 10.3.4).

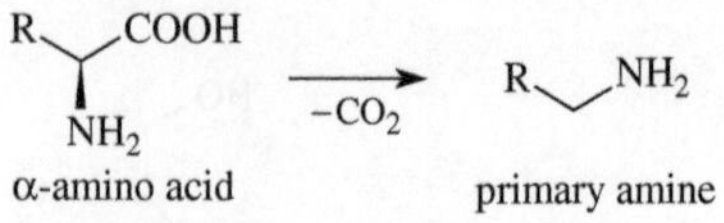

Figure 2.38 Decarboxylation of amino acids.

2.5.1.3.2 Transamination and oxidative deamination

Enzymatic reactions   Transamination catalysed by aminotransferases and oxidative deamination catalysed by oxidases transform α-amino acids into the corresponding 2-oxoacids (Figure 2.39). Dehydrogenases transform 2-oxoacids into the corresponding secondary alcohols. For example, the oxidative deamination product of glutamic acid is 2-oxoglutaric acid, serine yields pyruvic acid, and 2-oxobutanoic acid arises from threonine. Transamination of 2-oxobutanoic acid gives 2-aminobutyric acid. Certain bacteria, yeasts (including *S. cerevisiae*), and fungi produce 2-hydroxy-4-(methylthio)butanoic acid (**2-32**) (see Section 2.5.1.1.2) from methionine via reduction of the corresponding oxoacid. It is therefore a natural component (in amounts of up to 60 mg/kg) of foods and feeds produced by fermentation processes, such as fermented milk products, beer, bread, and silage.

Under the catalysis of ligases of the acid-thiol type, 2-oxoacids are transformed into acyl-CoA esters, or by the action of decarboxylases (carboxylases), into aldehydes having one carbon atom less than the original amino acid. For example, decarboxylation of pyruvic acid yields acetaldehyde, whilst decarboxylation of 2-oxobutanoic acid gives propanal. The same aldehydes are formed by the Strecker degradation of amino acids, which is carried out without enzyme catalysis. Reduction of aldehydes by aldehyde dehydrogenases yields the corresponding primary alcohols. These can react with acyl-CoA esters in the reaction catalysed by alcohol acyltransferases to form carboxylic acid esters. Oxidation of aldehydes by aldehyde dehydrogenases provides carboxylic acids, and the reaction with alcohols catalysed by carboxylesterases yields esters of carboxylic acids.

Aldehydes, alcohols, carboxylic acids, their esters, and other transformation products of amino acids are important food volatiles. Aldehydes and fusel alcohols formed during alcoholic fermentation from 2-oxoacids through Ehrlich pathway are flavour-active compounds of alcoholic beverages (see Sections 8.2.4.1.1 and 8.2.2.1.1). Esters of carboxylic acids are flavour constituents of many foods, especially fruits, alcoholic beverages, and dairy products. Aldehydes are also formed in the Maillard reaction during cooking, baking, frying, and other thermal operations.

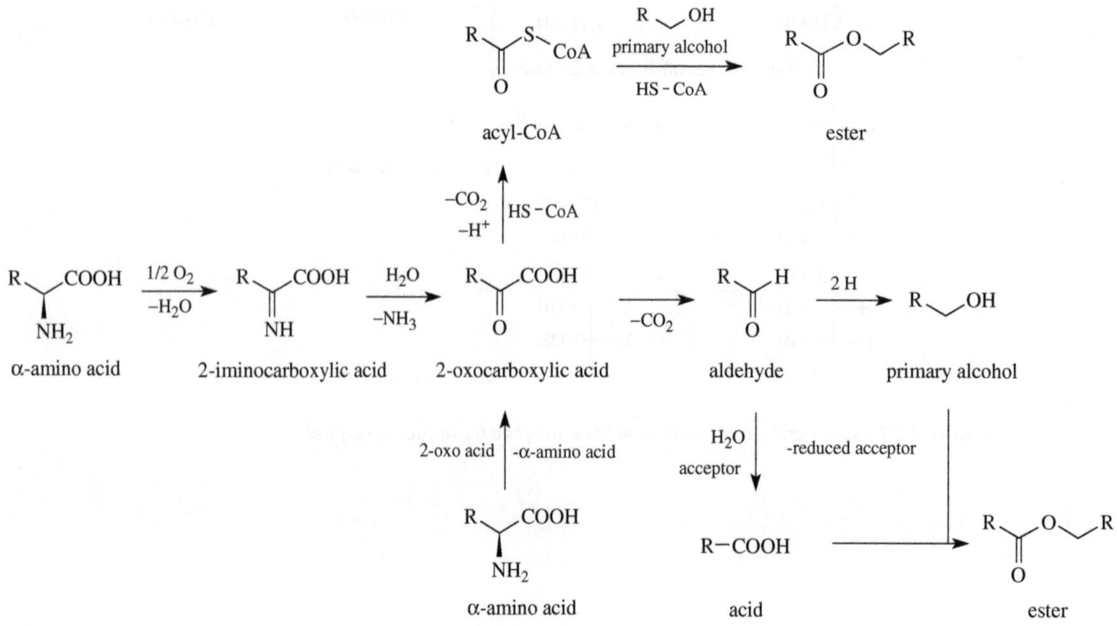

Figure 2.39 Formation of aldehydes, alcohols, carboxylic acids, and their esters.

Table 2.26 Strecker aldehydes produced from α-amino acids.

Amino acid	Degradation product	Amino acid	Degradation product
Glycine	Formaldehyde (methanal)	Cysteine	2-Mercaptoethanal
Alanine	Acetaldehyde (ethanal)	Methionine	3-Methylthiopropanal (methional)
Valine	2-Methylpropanal	Aspartic acid	3-Oxopropionic acid
Norvaline	Butanal	Glutamic acid	4-Oxobutyric acid
Leucine	3-Methylbutanal	Lysine	5-Aminopentanal
Norleucine	Pentanal	Ornithine	4-Aminobutanal
Isoleucine	2-Methylbutanal	Phenylalanine	2-Phenylethanal (phenylacetaldehyde)
Serine	Glycolaldehyde	Tyrosine	4-Hydroxyphenylacetaldehyde
Threonine	2-Hydroxypropanal	3,4-Dihydroxyphenylalanine	3,4-Dihydroxyphenylacetaldehyde

Strecker degradation (oxidative decarboxylation and deamination) Oxidation of amino acids by oxidising agents, which generates carbon dioxide, ammonia, and a carbonyl compound containing one carbon atom less than the starting amino acid, is known as the **Strecker degradation** of amino acids (Table 2.26).

Monoaminomonocarboxylic α-amino acids with a primary amino group produce sensory active aldehydes called **Strecker aldehydes**. Strecker degradation of β-amino acids yields alkan-2-ones known as methylketones (see Section 8.2.4.1.2). By analogy, alkane-3-ones (ethylketones) are formed from γ-amino acids. The general reaction is schematically indicated in Figure 2.40. The reaction mechanism, however, varies considerably depending on the type of oxidant and amino acid. In some cases, 2-imino acids and 2-oxoacids can be apparently formed as intermediates, analogous to those arising in enzymatically catalysed transamination and oxidative deamination of amino acids (see Section 2.5.1.3.2). Some Strecker aldehydes readily decompose, such as methional, or yield cyclic products, such as 5-aminopentanal, which dehydrates to 2,3,4,5-tetrahydropyridine.

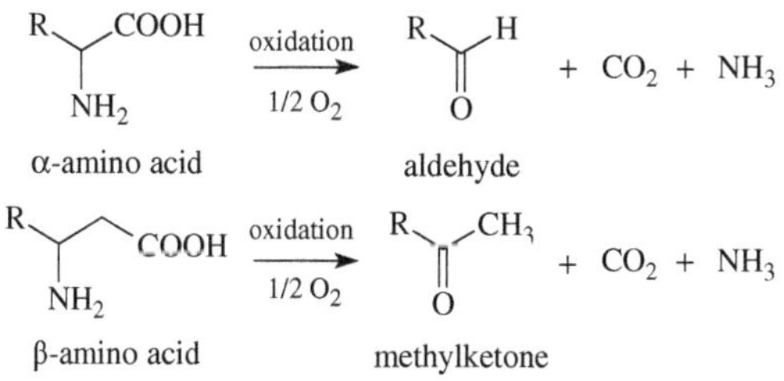

Figure 2.40 General scheme of the Strecker degradation of amino acids.

Inorganic oxidising agents, such as hypochlorites found in some sanitation and disinfection agents, may oxidise amino acids only in exceptional circumstances. The important oxidising agents are mainly organic compounds, which contain electronegative functional groups in the molecule. The active substances in the Strecker degradation of amino acids in foods are therefore fatty acid hydroperoxides, unsaturated aldehydes and ketones, derivatives of furan-2-carbaldehyde, α-hydroxyaldehydes and α-hydroxyketones, sugars (aldoses and ketoses), and sugar-derived α-dicarbonyl compounds, such as (deoxy)glycosuloses or methylglyoxal. Pyruvic acid, L-dehydroascorbic acid, and quinones such as benzoquinones, naphthoquinones, and anthraquinones, as well as some compounds belonging to vitamins K or E (oxidised to tocopheryl quinones), are also effective oxidising agents. Some unsaturated lipid oxidation products, such as epoxyoxoene fatty acids and their products (e.g. 4,5-epoxy-2-decenal, see Section 3.8.1.8.2), are able to convert amino acids to 2-oxo acids (see Section 2.5.1.3.2) as intermediates that react with an amino acid yielding the corresponding Strecker aldehyde. Autoxidation of the tautomeric forms of a Strecker aldehyde produce the corresponding peroxides, which, after elimination of formic acid, are converted into the corresponding aldehyde having two carbons less than the initial amino acid; thus, for example, benzaldehyde can be formed in addition to phenylacetaldehyde from phenylalanine. Some transition metal ions with high oxidation potential, such as Cu^{2+} and Fe^{3+}, are also amongst the active substances. They are able to induce formation of Strecker aldehydes during thermal degradation of metal–amino acid complexes in the absence of sugars and sugar-derived α-dicarbonyl compounds.

The Strecker degradation of amino acids is an extremely important reaction that occurs during storage and particularly thermal processing of food. The main products of this reaction (Strecker aldehydes) are important volatiles in a variety of foods, and many other aromatic and flavouring substances are formed by the subsequent reactions of Strecker aldehydes and other Strecker degradation products such as α-aminocarbonyl compounds, ammonia, amines, and various sulfur compounds. The Strecker degradation of amino acids also has its negative side, which is a loss of some essential amino acids.

2.5.1.3.3 Elimination of ammonia, hydrogen sulfide, and water

α-Aminocarboxylic acids When heated to higher temperatures (above the melting points of amino acids) and in acidic solutions, α-amino acids condense through the formation of linear and cyclic dipeptides, cyclic six-membered amides known as 3,6-disubstituted 2,5-dioxopiperazines (see Section 2.3.3.2 and Figure 2.14).

β-Aminocarboxylic acids   Heating β-amino acids leads to the elimination of ammonia and the formation of alk-2-enoic acids (α,β-unsaturated) acids (Figure 2.41). For example, aspartic acid and histidine behave as β-amino acids. Aspartic acid produces (E)-but-2-enedioic acid, known under the trivial name fumaric acid. The reverse reaction, in which fumaric acid gives aspartic acid, occurs in living organisms and is catalysed by the enzyme aspartase. Analogously, elimination of ammonia from asparagine produces fumaric acid monoamide, known as fumaramic acid. Its decarboxylation could theoretically produce toxic acrylamide. However, acrylamide is produced in the reactions of asparagine with reducing sugars (known as the Maillard reaction) that enable its decarboxylation (see Section 12.2.2).

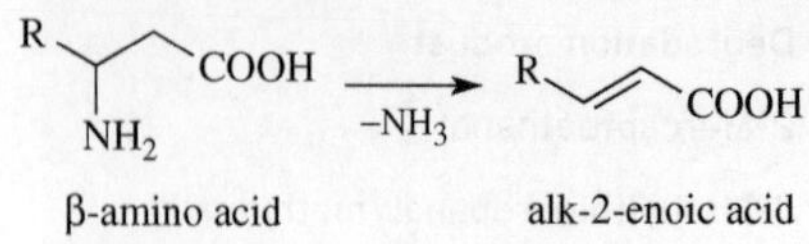

Figure 2.41 Formation of alk-2-enoic acids from β-amino acids.

An intermediate in the catabolism of L-histidine is urocanic acid (**2-32**), 3-(1H-imidazol-4-yl)prop-2-enoic acid, which is formed by the elimination of ammonium through the action of histidine ammonialyase (also known as histidase or histidinase). Together with histamine and other biogenic amines (see Section 10.3.4), urocanic acid is formed in the flesh of fish due to inappropriate handling during storage or processing. The highest levels of (E)- and (Z)-urocanic acids were found in the meat of frigate tuna (*Auxis thazard*) during storage for 15 days at 3 °C (23.7 and 21.8 mg/kg, respectively).

γ-Aminocarboxylic acids and δ-aminocarboxylic acids   The heating of γ- and δ-amino acids leads to intramolecular condensation and the formation of N-analogues of lactones, termed lactams. γ-Lactams (butane-4-lactams or pyrrolidine-2-ones) are formed from γ-amino acids (Figure 2.42), whilst δ-amino acids yield δ-lactams (pentane-5-lactams or piperidine-2-ones) (Figure 2.43).

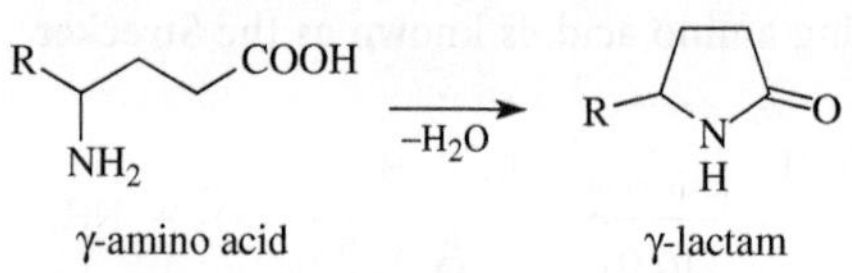

Figure 2.42 Formation of γ-lactams from γ-amino acids.

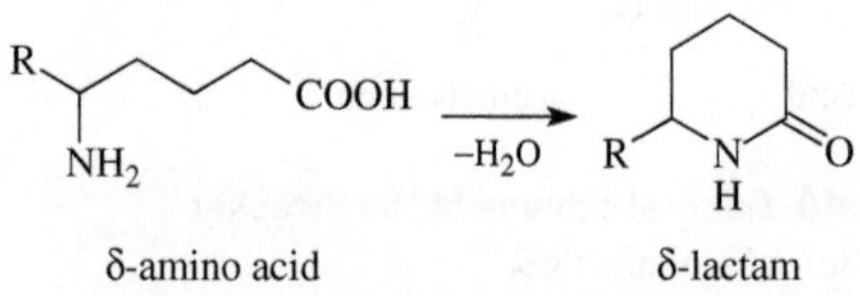

Figure 2.43 Formation of δ-lactams from δ-amino acids.

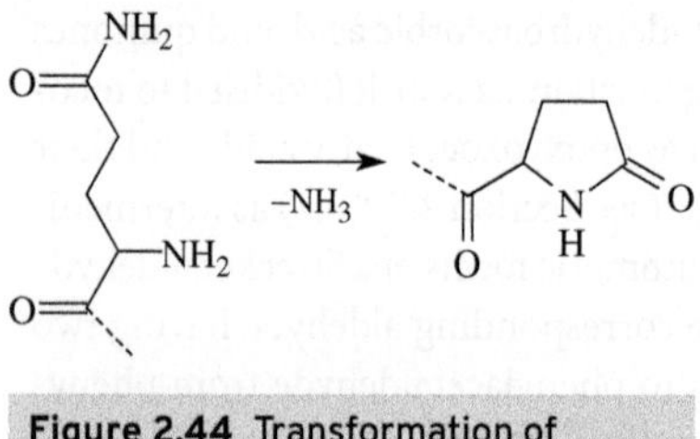

Figure 2.44 Transformation of N-terminal glutamine.

During the thermal processing of food, glutamic acid and glutamine behave as γ-amino acids, as they are easily transformed into L-5-oxopyrrolidin-2-carboxylic acid (also known as pyroglutamic acid, glutiminic acid, or 5-oxoproline, **2-32**) and its ammonium salt, respectively. Pyroglutamic acid reportedly exhibits umami-like taste. The free acid occurs at levels of up to 300 mg/kg in fresh vegetables and fresh and preserved fruits, and in concentrations of 500–3000 mg/kg in canned vegetables such as pickled beets. The pyroglutamic acid content can serve as a qualitative index to estimate the proportion of tomatoes in tomato products, for example in tomato paste (ketchup). Fresh tomatoes typically contain less than 10 mg/kg pyroglutamic acid. During processing, its content increases by one to two orders, and the final concentration of pyroglutamic acid is 25–40 mg/kg in tomato juice, 150–220 mg/kg in ketchup, and 630–820 mg/kg in puree, depending on the tomato content and the process used.

In some proteins (e.g. κ-casein and para-κ-casein), during thermal processing, transformation (spontaneous cyclisation) of the N-terminal glutamine into N-terminal pyroglutamate also occurs, especially in acidic media (Figure 2.44). Under the same conditions, aspartic acid only undergoes interconversion of α- and β-carboxylic groups (Figure 2.45).

When heated, the γ-amino acid creatine (**2-16**) yields an imidazolidine derivative known as creatinine (**2-32**). For example, fresh beef muscle contains creatine in amounts ranging from about 0.30 to 0.60% (according to the muscle type), whilst the creatinine content is only 0.020–0.040%. The creatine content of pork meat is lower than that of beef, at 0.25–0.37%, whilst its creatinine content is only 0.003–0.009%. However, when the meat is cooked, the creatine content decreases whilst the creatinine content increases: the proportion of these two substances might therefore be a useful indicator for quantifying the heat treatment applied in the processing of meat products. For example, cooked ham contains 0.29–0.32% of creatine and 0.032–0.073% of creatinine.

The quality criteria for beef extract specify the minimum creatinine content as 8.5% (d.w., added salt excluded) for use in the production of bouillon cubes. Creatine and creatinine levels may be therefore used as a measure during the detection of beef extracts in food products. Serum creatinine concentration is the most frequently used clinical estimate of renal function.

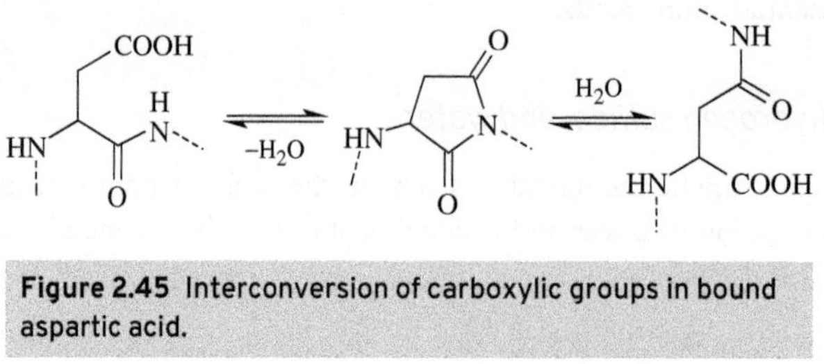

Figure 2.45 Interconversion of carboxylic groups in bound aspartic acid.

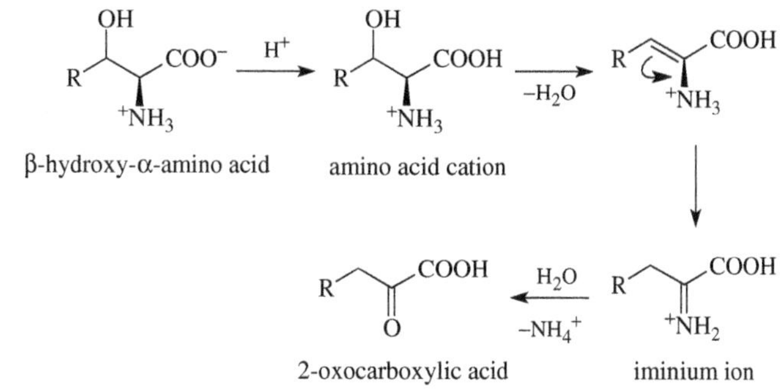

Figure 2.46 Formation of 2-oxocarboxylic acids from β-hydroxy-α-amino acids.

The product of creatine degradation via *N*-methylguanidinoglyoxylic acid (creatone, **2-32**) is oxalic acid and *N*-methylguanidine (**2-32**), which is found in fresh pork, beef, chicken, and fish at levels of 1–10 mg/kg. *N*-Methylguanidine occurs in larger amounts (20–180 mg/kg) in some Japanese dishes prepared from smoked and dried fish. In the production of beef and whale meat extracts, which are used in soup seasoning preparations, creatinine produces small amounts of the amino acid sarcosine (*N*-methylglycine, **2-6**) and urea. Deamination of creatinine by bacteria or in alkaline solutions can yield the toxic *N*-methylhydantoin (**2-32**). Creatinine is also a precursor of aminoimidazoazaarenes found mainly in grilled meat and fish (see Section 12.2.1.1).

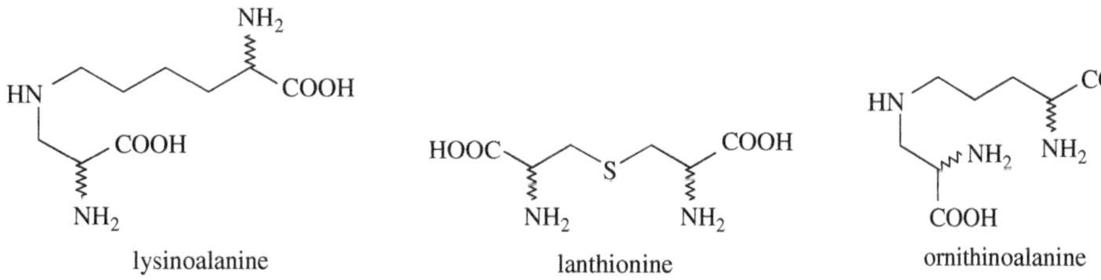

2-33, amino acids hydrolysed from the proteins cross-linked via dehydroprotein structures

Hydroxyamino acids Under acidic conditions, and especially during acid hydrolysis of proteins, serine and threonine isomerise to imino acids via β-elimination of water molecule, followed by hydrolysis of the iminium ion to give the corresponding 2-oxoacids (Figure 2.46). Thus, serine produces pyruvic acid (R=H) and threonine yields 2-oxobutanoic (2-oxobutyric) acid (R=CH$_3$). Decarboxylation of these oxocarboxylic acids at elevated temperatures yields acetaldehyde and propanal, respectively.

Thermal degradation of serine has been reported to produce a wide range of products. The initial thermal degradation products include pyruvic acid, glycine, and formaldehyde (formed by retro aldol condensation), 2-aminoethanol (formed by decarboxylation), and acetaldehyde (formed by elimination of ammonia from 2-aminoethanol). A number of products then arise in reactions of these initial products. Reaction of pyruvic acid with 2-aminoethanol produces alanine and glycolaldehyde, reaction of acetaldehyde with 2-aminoethanol yields glycolaldehyde, ethylamine, and 2-methyl-2-oxazoline, aldol condensation of formaldehyde to glycolaldehyde forms glyceraldehyde, and reactive α-hydroxycarbonyl compounds are involved in the formation of heterocyclic compounds such as substituted pyrazines and pyrroles and some other heterocyclic compounds.

Amino acid amides At temperatures around 100 °C, asparagine and glutamine side chains react with the side chain of bound lysine (Figure 2.47), with the elimination of ammonia (the reaction is known as deamidation). Thus, the peptide chains are connected by intermolecular and intramolecular transverse covalent bonds. These bonds, known as isopeptide bonds, consist of dipeptides ε-*N*-(β-aspartyl)-L-lysine and ε-*N*-(γ-glutamyl)-L-lysine, respectively. Depending on the protein type, up to 15% of lysine residues can react in this manner. The digestive proteases of some animals (e.g. chickens and rats) can split these bonds, but the lysine bound in this manner is not available for humans. The result is the certain, but usually not very significant, reduction of the nutritional value

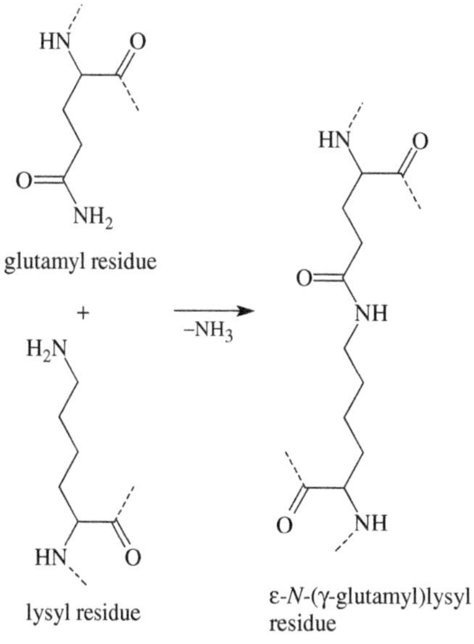

Figure 2.47 Deamidation of proteins.

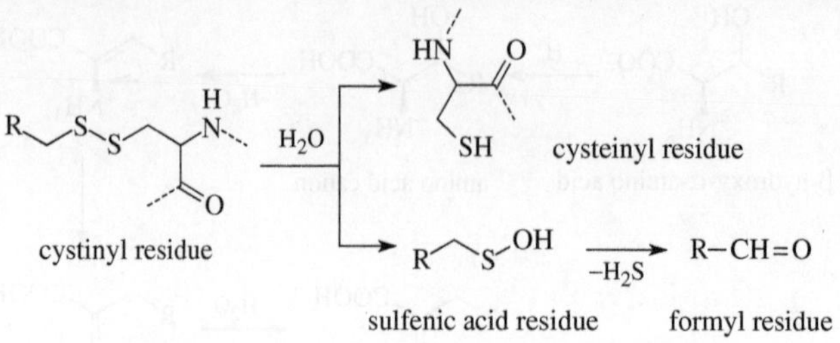

Figure 2.48 Desulfuration of proteins.

of proteins. Reduction of the nutritional value can be important in the cases where a diet is low in protein and the limiting amino acid is lysine. The Maillard reaction often leads to higher loss of lysine.

Sulfur amino acids At temperatures around 100 °C, free cystine is relatively stable. The reaction that is observable during thermal processing of foods at temperatures of approximately 100 °C is partial elimination of hydrogen sulfide from bound cystine, which is often called protein desulfuration (Figure 2.48). This reaction splits (hydrolyses) the disulfide bridges in protein with the formation of cysteine and sulfenic acid residues. The unstable sulfenic acid intermediate is decomposed with the formation of hydrogen sulfide, yielding a reactive molecule containing a free carbonyl group. This compound reacts with the ε-amino group of lysine, which results in a clear loss of cystine (cysteine) and lysine. The collagen-bound lysine reacts similarly (see Section 2.4.5.1.3). At higher temperatures or in alkaline solutions even under milder conditions, bound cysteine eliminates hydrogen sulfide, yielding dehydroprotein as an intermediate (Figure 2.49). The resulting hydrogen sulfide is involved in the formation of many flavour-active compounds and colourants in thermally processed foods. Similarly to protein bound cysteine, some other protein-bound amino acids can eliminate small molecules, yielding dehydroprotein. For example, protein-bound methionine eliminates dimethyl sulfide.

Dehydroproteins and cross-linked amino acids If allowed by the structure of protein-bound amino acids, the heat treatment of protein or reaction in an alkaline solution produces carbanion, an intermediate in the pathway to D-amino acids (see Section 2.5.1.2) and dehydroprotein (Figure 2.49). The most frequently eliminated groups are the cysteine thiol group (X=SH), the hydroxyl group of serine (X=OH), and the phosphorylated hydroxyl group of phosphoserine (X=OPO_3H_2). When R=H, the structure formed is a dehydroprotein containing protein-bound 2-aminoacrylic acid, known as dehydroalanine (R=H). The reactive side chains of bound amino acids, such as side chains of lysine, ornithine, and cysteine, can add to the double bond of the dehydroprotein, which leads to cross-linking of protein polypeptide chains (Figure 2.50).

During proteolysis of cross-linked proteins, unusual amino acids are formed. Bound lysine yields lysinoalanine (**2-33**), cysteine lanthionine (**2-33**), and ornithine ornithinoalanine (**2-33**). Bound ornithine forms in proteins (together with urea) through alkaline hydrolysis of arginine. In the same way, many other amino acids (arginine, histidine, threonine, serine, tyrosine, and tryptophan), as well as amines and ammonia, react with dehydroprotein.

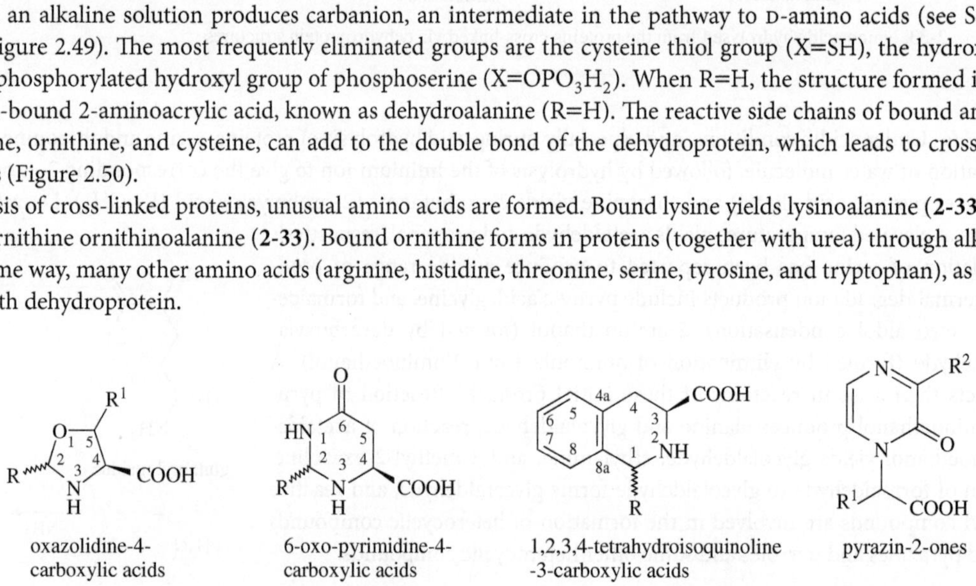

oxazolidine-4-carboxylic acids

6-oxo-pyrimidine-4-carboxylic acids

1,2,3,4-tetrahydroisoquinoline-3-carboxylic acids

pyrazin-2-ones

2-34, selected heterocyclic products of carbonyl compounds with amino acids and peptides

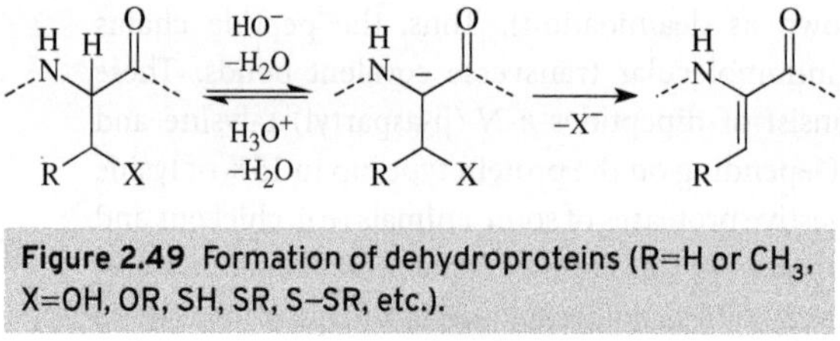

Figure 2.49 Formation of dehydroproteins (R=H or CH_3, X=OH, OR, SH, SR, S–SR, etc.).

Table 2.27 Lysinoalanine content in some foods.

Food	Content (g/kg protein)	Food	Content (g/kg protein)
Soybean meal (defatted)	1.5	Sausage (boiled)	0.05
Soy protein (texturised)	0.3	Sausage (baked)	0.2
Soy protein (isolate)	0.4–11.6	Chicken (raw)	0.0
Milk evaporated	0.4–0.9	Chicken (baked)	0.1
Casein (acidic)	0.1–0.2	Egg white (original)	0.02
Sodium caseinate	0.3–6.9	Egg white (boiled)	0.1–0.4
Calcium caseinate	0.2–4.3	Egg white (baked)	0.4–1.1
Sausage (original)	0.0	Egg white (dried)	0.2–1.8

These unusual amino acids are not bioavailable and therefore reduce the nutritional value of proteins. More than 50 xenobiotic amino acids are known. Some products are even toxic, especially lysinoalanine. In theory, the hydrolysis of protein containing bound L-lysine produces the same amount of free L,L- and L,D-lysinoalanine. Furthermore, the isomerisation of bound L-lysine produces bound D-lysine, and protein hydrolysis then gives free D,L- and D,D-lysinoalanine. It is suggested that alkali treatment of protein might have important toxicological implications. Free lysinoalanine is partly excreted in the urine and partly catabolised in the kidney. Feeding lysinoalanine to rats induces changes in kidney cells characterised by enlargement of the cytoplasm designated as nephrocytomegaly and of the nucleus (karyomegaly). The toxicity of the individual isomers is different. The least toxic is D,D-lysinoalanine, about three times more toxic is L,L-isomer, 10 times more toxic is D,L-isomer, and about 30 times more toxic is L,D-lysinoalanine. Lysinoalanine and related amino acids of protein cross-links are present in many foods, especially those processed at high temperatures or in alkaline media (Table 2.27). The peptide nisin, for example, contains *meso*-lanthionine (see Section 11.2.1.2.6).

To prevent lysinoalanine and other cross-link amino acid formation, ammonia was used when reacting with dehydroprotein to form 2,3-diaminopropionic acid (β-aminoalanine, **2-16**).

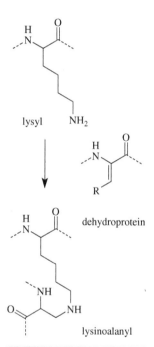

Figure 2.50 Formation of lysinoalanyl cross-link in proteins via dehydroprotein (R=H).

2.5.1.4 Hydrolysis and protein hydrolysates

Proteolytic enzymes are present in all raw food materials, and spontaneous hydrolysis of proteins therefore occurs during storage in a number of foods of animal and plant origin. Proteases in microbial, plant, or animal proteolytic preparations have found practical application in various food and many technologies; for instance, rennin is used in the dairy industry for the manufacture of hard cheeses, whilst plant proteases, such as papain, are employed in the prevention of protein hazes in beer. Actinidin is commercially useful as a meat tenderiser and in coagulating milk for dairy products. Plant proteases such as papain, bromelain, actinidin, and zingibain show the potential for targeting specific meat tenderising applications. Whilst actinidin, found in many fruits, including kiwifruit, is most effective at hydrolysing beef myofibril protein, zingibain, a cysteine protease in ginger, is most effective at hydrolysing connective tissue proteins.

Autolysis of fish meat, oysters, and other marine animals is employed for the production of special products used as condiments, such as oyster sauce and yeast autolysates (autolysates of waste brewery yeast). Partial hydrolysis of microbial proteins is used for the production of soy sauces. In the production of hydrolysates from vegetable proteins, which are used as seasonings, mineral acids are utilised to achieve almost complete hydrolysis of the proteins.

Soy sauce (or soya sauce or shoyu) is the best-known representative of the **enzymatic protein hydrolysates**. Soy sauce was invented in China between the third and fifth centuries AD and introduced to Japan in the seventh century. Approximately 80% of all Japanese soy sauce is brewed using a method known as 'honjozo', in which roasted wheat and steamed soy beans are combined with 'koji' mould (*Aspergillus oryzae* or *Aspergillus soja*) then placed in tanks with brine water, making an unfermented soy sauce solution called 'moromi'. The moromi is left to ferment for six to eight months before being pressed. The liquid is then pasteurised and filtered before being aged or bottled and sold. Variations of this standard production method occur depending on what type of soy sauce is being produced. The five types of Japanese soy sauces are 'Common' (Koikuchi Shoyu), 'Light Color' (Usukuchi Shoyu), 'Tamari' (Tamari Shoyu), 'Refermented' (Sai-Shikomi Shoyu), and 'Extra Light' Colour (Shiro Shoyu). Soy sauces of different origins have different flavours.

Acid hydrolysates of proteins, also called hydrolysed plant proteins or hydrolysed vegetable proteins, are typically produced by hydrochloric acid (HCl) hydrolysis of various proteinaceous vegetable raw materials (defatted oilseed meals, wheat gluten, maize and rice protein, and occasionally some animal proteins such as casein and keratin). The hydrolysis is usually carried out with 20% (6 M) HCl at temperatures exceeding 100 °C for up to about eight hours. After cooling, the hydrolysate is neutralised by sodium hydroxide or sodium carbonate, generally to pH 4.5–7. A product of this neutralisation is sodium chloride; its content in the hydrolysate is about 20%. The filtered raw hydrolysate can be bleached and fortified with flavour-active compounds, and is then ripened for several weeks. The ripe product is provided as a liquid seasoning; alternatively, it can be concentrated to a paste or dried to powder. Acid hydrolysis, however, leads to complete degradation of oxylabile tryptophan, whilst glutamine and asparagine are hydrolysed to aspartic acid and glutamic acid, respectively. Approximately 5–10% of the hydroxyamino acids serine and threonine are degraded with the formation of 2-oxoacids, and the cysteine is desulfurated to a large extent.

2.5.2 Reactions with food components

2.5.2.1 Reactions with minerals

Polypeptide chains of proteins and peptides and free amino acids contain basic functional groups (typically amino or imino groups) and acidic functional groups (usually carboxyl groups or phosphoric acid residues in phosphoproteins), and therefore react with acids and bases. Salts of amino acids and proteins with acids and bases are stable only in a certain range of pH values. High-molecular-weight complexes can also be formed between basic and acidic proteins and between proteins and some polysaccharides. The most important reaction is the formation of salts and complexes with metal ions. Metal ions, however, often act as cofactors of enzymes and constitute a natural part of many complex proteins, such as milk caseins. Formation of salts and complexes of proteins with various metal cations (sodium, potassium, calcium) is employed in various food technologies, for example in the production of caseinates and soybean curd (tofu). The formation of protein salts with metal ions affects the solubility of proteins. Sodium, potassium, and ammonium salts are predominantly soluble, whilst calcium salts are less soluble, and salts of heavy metals are mostly insoluble. Particularly strong complexes are formed by cysteine, tyrosine, and the basic amino acids lysine and histidine. Such properties imply that proteins, peptides, and amino acids often act as natural antioxidants of lipids and other oxylabile substances, when complexing ions of heavy metals such as Fe^{3+} and Cu^{2+} that act as active catalysts in the oxidation of lipids. A few simple examples of the structures of metal complexes are given in Section 6.2.2.1.

In some cases, however, the formation of salts and complexes is an unwanted phenomenon, as protein salts and complexes with polyphenols form hazes and sediments in preserved fruit, fruit juices, and beer. Protein hazes can also form in white wines after bottling, where the main factor is inorganic sulfate ions. Protein complexes with transition metals are often coloured, and their formation in processed foods is generally undesirable. An example is the protein conalbumin occurring in egg white, which readily forms coloured complexes with metal ions (pink with Fe^{3+}, yellow with Cu^{2+} and Mn^{3+}) in media of pH $\geq$ 6 through tyrosine and histidine residues of the polypeptide chain. Complexes with iron ions often cause discolouration of egg products, but in media with pH < 4, these complexes dissociate to the original colourless compounds.

In addition to the oxidation reactions summarised in Section 2.5.1.1, the side chains of amino acids residues in proteins are readily oxidised through the catalysis of metal ions.

2.5.2.2 Reactions with oxidised lipids

Oxidised lipids react with one another and with non-lipid food components under the conditions used in culinary and technological processing and even during storage and *in vivo*. Hydroperoxyl free radicals and epoxy and aldehyde functional groups of oxidised lipids are mainly involved in reactions with proteins, peptides, and free amino acids. They react chiefly with sulfhydryl, amino, guanidyl, and imidazole functional groups of proteins, and to a small extent with their peptide bonds. Reactions of proteins with oxidised lipids result in products of decreased solubility and digestibility, and often lead to unwanted discolourations, especially in fish.

Reactive aldehydes derived from oxidised lipids, such as acrolein, (*E*)-4-hydroxynon-2-enal and malondialdehyde, react with lysine, arginine, and other amino acids to give the reaction products described in Section 4.7.5.6. These products, ALE (advanced lipoxidation end products), formed *in vivo*, are markers of oxidative stress in the organism. The reaction mechanisms are discussed in Section 3.8.1.12.1. As the final reaction products of proteins and oxidised lipids, dark insoluble macromolecular products are formed containing variable proportions of lipid and protein fractions. In particular, such products include protein oligomers, proteins with oxidised sulfur amino acids, proteins containing imine bonds (C=N) formed mostly by the reaction of ε-amino group of bound lysine with aldehydes or hydroperoxides, and lipoproteins, in which lipids and proteins are bound by covalent bonds in contrast to the native lipoproteins.

The amino groups of free amino acids and *N*-terminal amino groups of proteins also react with free fatty acids and triacylglycerols yielding the corresponding fatty acid amides. The latter may be precursors of non-volatile nitroso compounds (nitrosamides) classified as food contaminants (see Section 12.3.7).

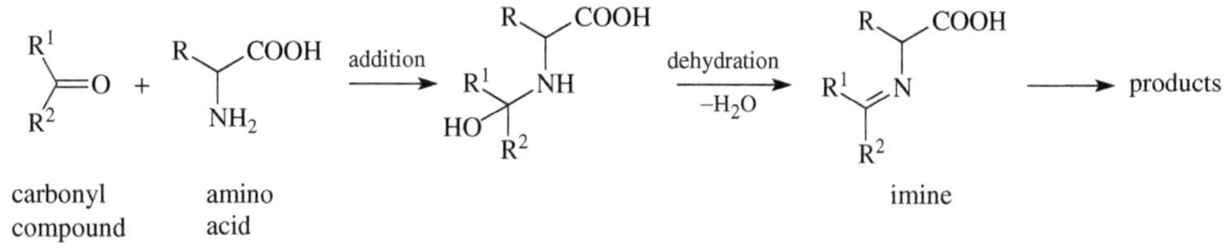

Figure 2.51 Reaction of amino acids with carbonyl compounds.

2.5.2.3 Reactions with aldehydes, sugars, and dicarbonyl compounds

2.5.2.3.1 Reactions with aldehydes

α-Amino groups of free amino acids, N-terminal α-amino groups of amino acids bound in peptides and proteins, and, amongst others, the ε-amino group of bound lysine react with carbonyl compounds such as aldehydes, ketones, and reducing sugars. The reaction starts with the addition of an amino group of an amino acid to a carbonyl group of a carbonyl compound to form imines, formerly known as Schiff bases (Figure 2.51). Subsequent reactions of the imines lead to a variety of reaction products. The reaction with reducing sugars, the Maillard reaction, is the most important in food processing. An important factor influencing the reactivity of amino acids and carbonyl compounds is pH. The reactivity of amino acids also depends greatly on the basicity, or more precisely on the nucleophilicity of the amino group, but also depends on other factors described in detail in Section 4.7.5.

Aromatic and heterocyclic amino acids react with carbonyl compounds with the formation of a variety of heterocyclic products. Common carbonyl food components that are present, added, or formed during food production are aldoses and various aliphatic, aromatic, and heterocyclic aldehydes, such as formaldehyde, acetaldehyde, benzaldehyde, and 2-furancarbaldehyde, as well as vanillin and anisaldehyde, which are added to foods as flavouring agents. With formaldehyde, cysteine gives thiazolidine-4-carboxylic acid, also known as thioproline. Higher aldehydes yield C-2 substituted (2RS,4R)-thiazolidine-4-carboxylic acids (Figure 2.52). 2-Methylthiazolidine-4-carboxylic acid, a condensation product of cysteine and acetaldehyde, occurs even in human blood as a consequence of ethanol consumption. Serine ($R^1 = H$) and threonine ($R^1 = CH_3$) analogously produce C-2-substituted (2RS,4S)-oxazolidine-4-carboxylic acids (**2-34**). Heterocyclic products, C-2-substituted (2RS,4S)-pyrimidine-4-carboxylic acids, are also produced in the reaction of aldehydes with asparagine (**2-34**). Phenylalanine yields C-1-substituted (1RS,3S)-tetrahydroisoquinoline-3-carboxylic acids (**2-34**), and analogous products arise from tyrosine. Tryptophan reacts with aldehydes under the formation of β-carboline derivatives, which are classified as indole alkaloids (see Section 10.3.6.6). Of special importance are thiazolidine-4-carboxylic acids and β-carboline derivatives that may react with nitrogen oxides or nitrites with the formation of nitroso compounds, which can occur in foods as contaminants (see Section 12.3.7). Most of the reactions that occur in amino acids also take place in peptides and proteins. Heterocyclic products are formed when the N-terminal amino acid is involved in the ring formation. Products such as N-3 substituted 2-alkylpyrazin-2-ones form, for example, in the reaction of dipeptides with glyoxal (**2-34**, R^1 and R^2 = amino acid residues).

2.5.2.3.2 Reactions with sugars

Reducing sugars are very reactive compounds, and their reactions with amino acids as compounds assisting in the transformation of sugars are described in detail in Section 4.7.5.1. Here, only the reactions where sugars act as oxidising agents towards amino acids are described (see also Section 2.5.1.3.2); this is about the oxidative degradation of amino acids by the action of sugars, α-dicarbonyl compounds, and other conjugated dicarbonyl compounds formed from sugars, lipids, and Amadori products. In addition to the Strecker aldehydes, ammonia, and carbon dioxide, this reaction provides various reducing compounds essential for the development of the Maillard reaction (Figure 2.53).

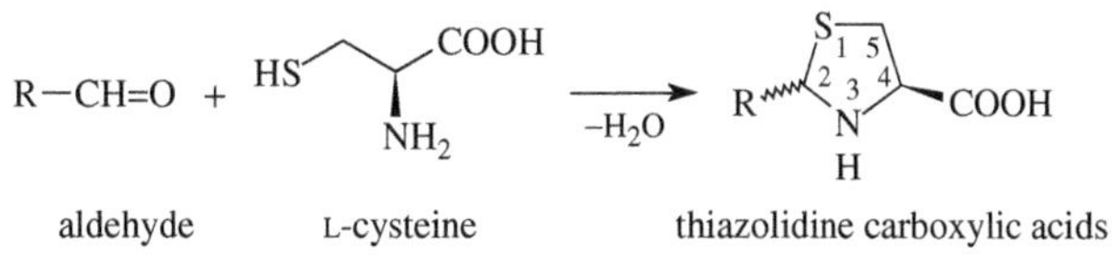

Figure 2.52 Reaction of cysteine with aldehydes.

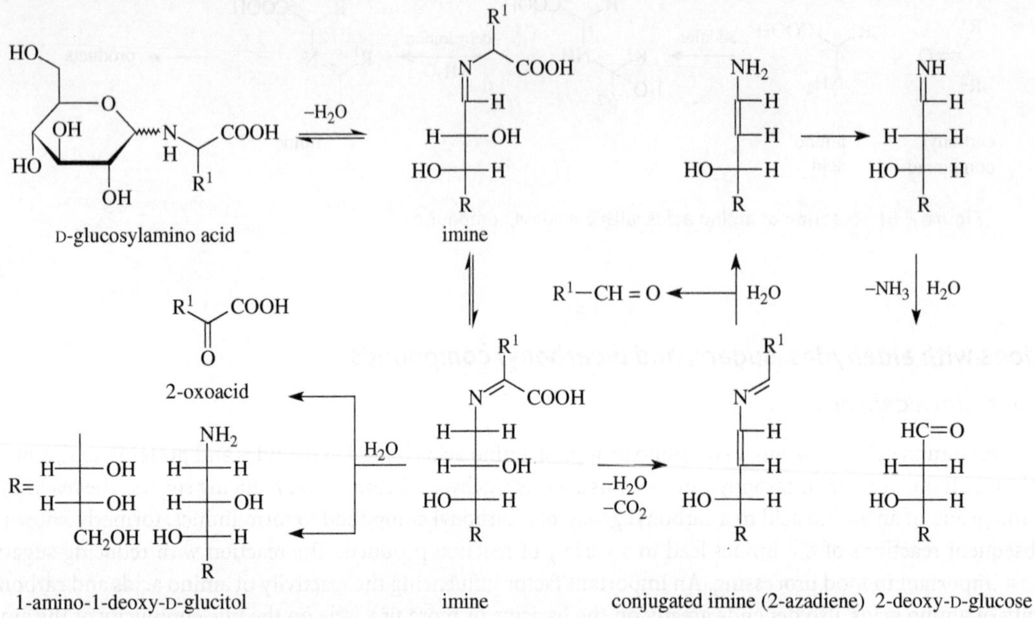

D-glucosylamino acid $-H_2O$ imine

2-oxoacid

$R^1-CH=O$ H_2O $-NH_3$ H_2O

$R=$ H_2O $-H_2O$ $-CO_2$

1-amino-1-deoxy-D-glucitol   imine   conjugated imine (2-azadiene)   2-deoxy-D-glucose

Figure 2.53 Strecker degradation and oxidative deamination of amino acids by aldoses.

2.5.2.3.3 Reactions with dicarbonyl compounds

α-Dicarbonyl compounds, such as glyoxal and methylglyoxal, and their higher homologues, glycosuloses, as well as deoxyglycosuloses, are potent Strecker oxidants. The mechanism of oxidative decarboxylation and deamination of amino acids by α-dicarbonyl compounds is in fact more complex than that shown in Figure 2.54. As in the reaction of other carbonyl compounds with amino acids, the first step is the addition of an amino acid on the carbonyl group of an α-dicarbonyl compound, with the formation of imine and elimination of water. The resulting imine then eliminates carbon dioxide, and hydrolysis of the product yields Strecker aldehyde (preferably in a weakly acidic solution). Reduction of an α-dicarbonyl compound (transamination) produces a reactive aminocarbonyl compound. Other products include ammonia, amines, α-hydroxycarbonyl compounds, and carboxylic acids.

The carboxylic acids are formed in the presence of traces of transition metals (Me^{n+}), preferably in weakly alkaline solutions, and by oxidation with α-dicarbonyl fragments (glycosuloses in their prevailing cyclic semiacetal form do not have free enaminol moiety to bind the transition metal). The formation of the Strecker carboxylic acids is not very significant, but some of them are important aroma components of cooked foods. The levels of amines formed by Strecker degradation, such as during roasting of cocoa beans, exceed by several orders the number of identical amines that are formed during fermentation and are comparable to the level of Strecker aldehydes generated.

The Strecker degradation of ornithine proceeds analogously, but the final product, 1-pyrroline, is formed by cyclisation of intermediate 4-aminobutanal (Figure 2.55). This reaction is important for the development of the characteristic aroma of bread crust (see Section 8.2.12.4.1), the aroma of some other cereal products, fragrant rice varieties (such as Basmati rice), and the biosynthesis of pyrrolidine alkaloids in plants (see Section 10.3.6.2). It was proposed that pyrrolidine and 1-pyrroline also results from the Strecker-type reaction of proline with α-dicarbonyl compounds.

In some cases, Strecker aldehydes are unstable and readily decompose; for example, aspartic acid produces 3-oxopropionic acid, which provides acetaldehyde by decarboxylation. Other products are pyruvic acid and fumaric acid. Glutamic acid produces 4-oxobutyric acid (4-oxobutanoic acid), which eliminates carbon dioxide and yields propionaldehyde and other products, such as 2-oxobutanoic acid and 5-oxopyrrolidine-2-carboxylic acid.

More complex is the Strecker degradation of hydroxyamino acids (serine and threonine) and sulfur amino acids (cysteine and methionine). The Strecker degradation product of serine is glycolaldehyde, whilst threonine produces 2-hydroxypropanal (lactic acid aldehyde). Both aldehydes are highly reactive compounds that enter a number of other reactions. Decomposition of the imine intermediate of decarboxylated cysteine yields 2-mercaptoacetaldehyde, as well as vinylamine as a minor degradation product (Figure 2.56).

Methionine via the corresponding imine yields an unstable volatile aldehyde, methional (3-methylthiopropanal). Methional further decomposes to acrolein and methanethiol (methylmercaptan), which spontaneously oxidises to dimethyldisulfide in the presence of oxygen (Figure 2.57). Oxidation of methional yields the corresponding sulfoxide, which decomposes via methanesulfenic acid to acrolein. Methanesulfenic acid reacts with methanethiol or other thiols producing dimethyldisulfide or mixed disulfides, respectively.

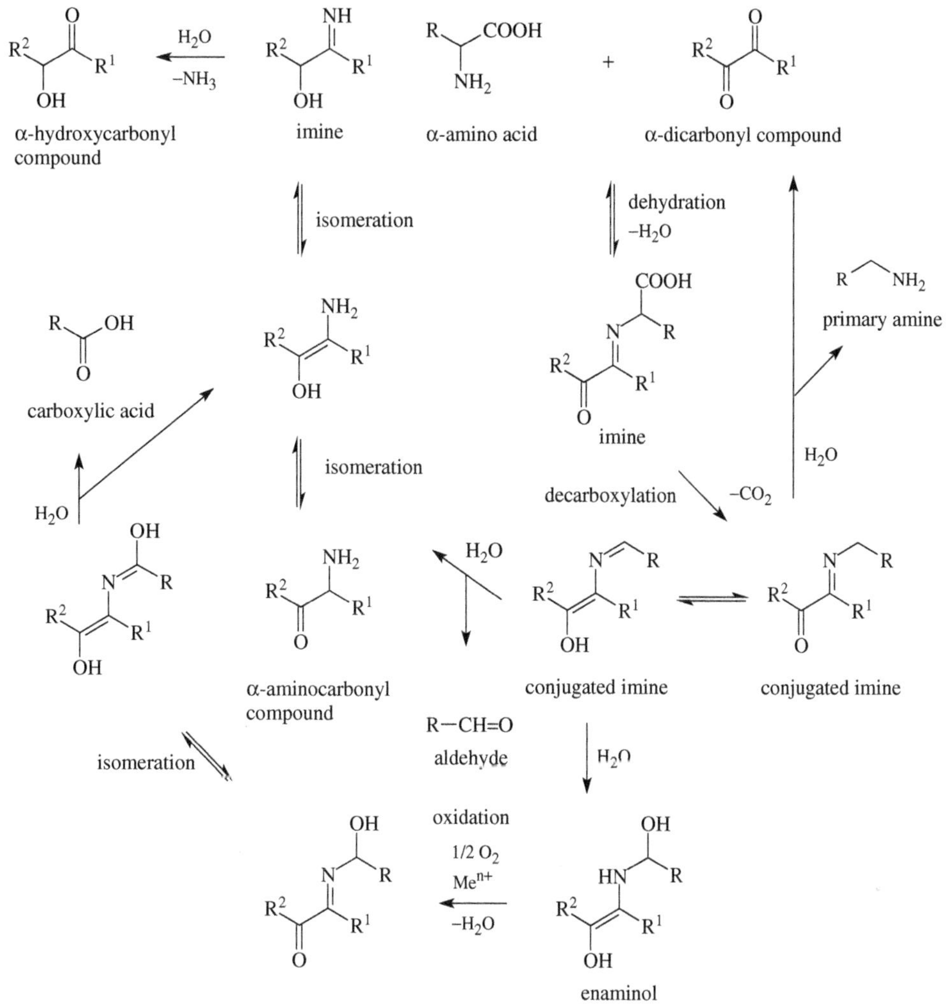

Figure 2.54 Strecker degradation of α-amino acids by α-dicarbonyl compounds.

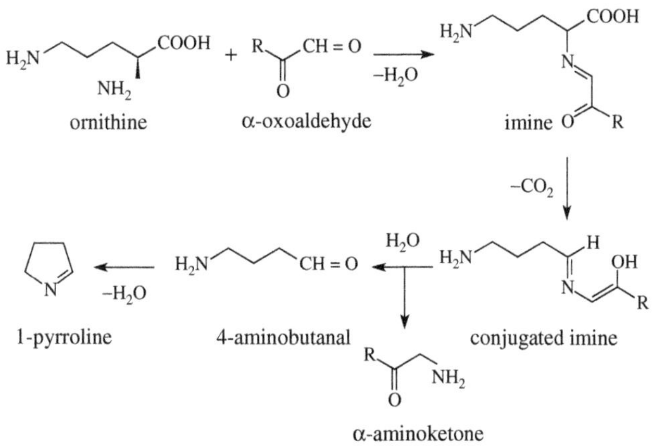

Figure 2.55 Strecker degradation of ornithine by α-dicarbonyl compounds.

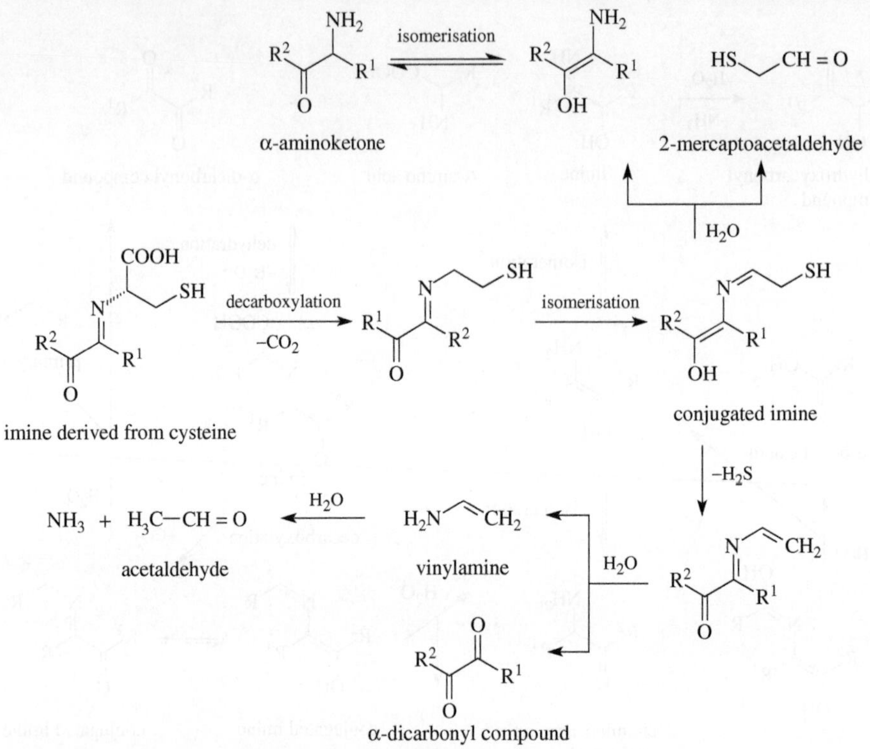

Figure 2.56 Strecker degradation of cysteine.

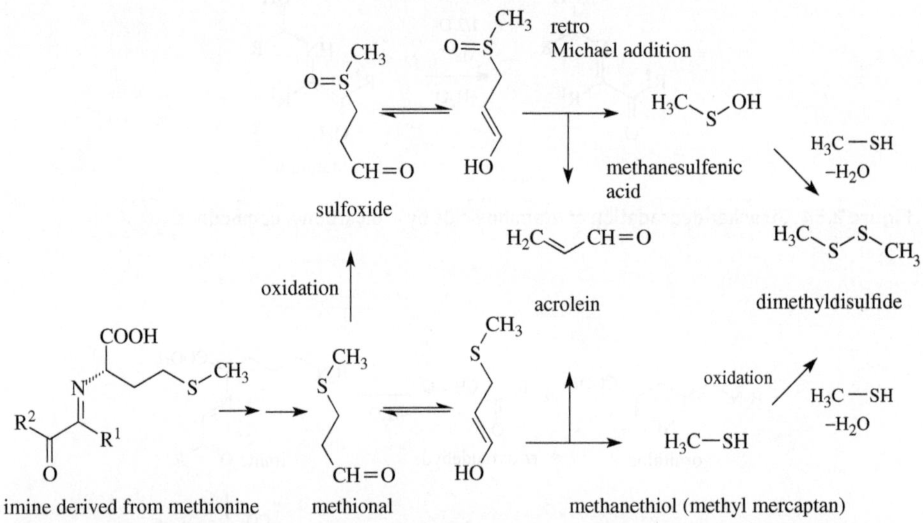

Figure 2.57 Strecker degradation of methionine.

2.5.2.4 Reactions with isothiocyanates

Isothiocyanates that are formed by the degradation of glucosinolates in cruciferous vegetables (see Section 10.3.10.2) react with proteins, peptides, and amino acids under the formation of a variety of products. Proteins and peptides react with isothiocyanates via amino groups, mercapto groups, or other nucleophilic functional groups (hydroxyl groups). Their reactivity depends on their nucleophilicity, pK_a, and pH values. Reactions of the α-amino groups of amino acids with isothiocyanates in weakly alkaline media produce N,N'-disubstituted thioureas (N-thiocarbamoyl derivatives of amino acids) as primary products. The thioureas cyclise under acidic conditions or on heating to 2-thiohydantoins (Figure 2.58), which are considered responsible for some of the biological effects of isothiocyanates. In neutral and alkaline media, 2-thiohydantoins yield symmetric dehydrodimers as a result of cleavage of the acidic hydrogen atom in position C-5. The

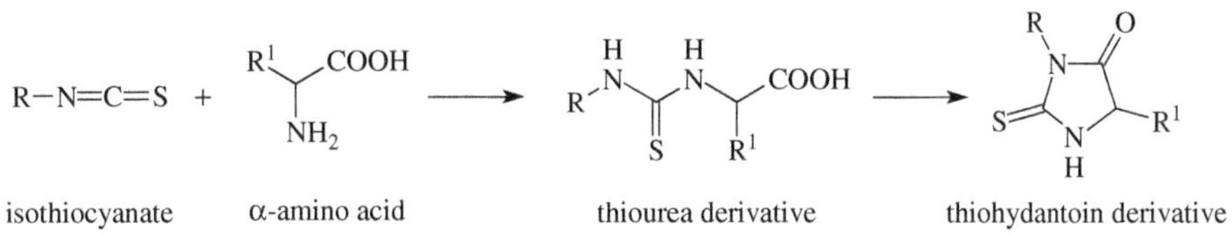

Figure 2.58 Reaction of α-amino acids with isothiocyanates.

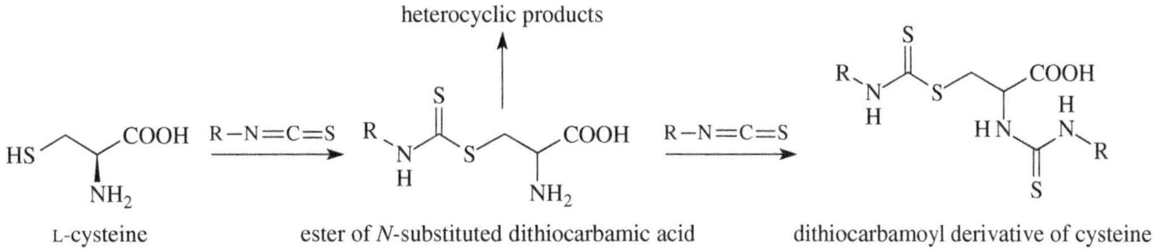

Figure 2.59 Reaction of cysteine with isothiocyanates.

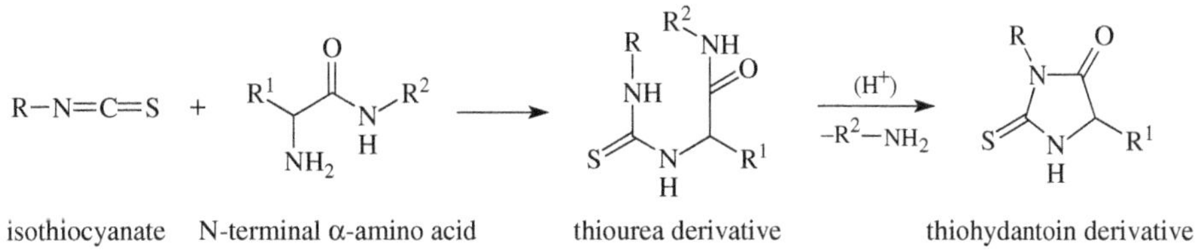

Figure 2.60 Reaction of peptides with isothiocyanates.

dehydrodimers of 2-thiohydantoins with the methylene group at C-5 (R^1 = H), derived from glycine or N-substituted glycines, are of yellow, red, or blue colour. Similarly to amino acids, isothiocyanates are supposed to react to heterocyclic aromatic amines or their precursors, yielding stable thioureas, and thus decrease the mutagenicity of meat processed with brassicaceous vegetables or condiments.

Depending on pH, the reaction of isothiocyanates with cysteine may give two types of primary products. In weakly acidic solutions (pH 5–6), isothiocyanate reacts exclusively with the thiol group of cysteine, producing N-substituted esters of dithiocarbamic acids and heterocyclic products. The same products (and cysteine sulfenic acid) are formed from cystine. Detoxification of isothiocyanates in the body is based on their reaction with the reduced glutathione (G-SH) mercapto group. The resulting mercapturate is excreted in the urine. In weakly alkaline media, a surplus of isothiocyanate can also react with the α-amino group of cysteine (Figure 2.59).

Isothiocyanates react with the N-terminal amino group of peptides to form a substituted thiourea, which under acidic conditions spontaneously cyclises to a thiohydantoin (imidazolidine) derivative; the other reaction product is the original peptide without the N-terminal amino acid (Figure 2.60). Reactions of isothiocyanates with the N-terminal amino group and other nucleophilic sites (thiol, disulfide, ε-amino, and hydroxyl groups) of proteins can cause changes in the conformation of the proteins, which is associated with the antimicrobial and antithyroid activity of isothiocyanates.

2.5.2.5 Reactions with plant phenols

Reactions of plant phenols catalysed by oxidoreductases are known as **enzymatic browning reactions**. The oxidation products of o-diphenols (1,2-dihydroxybenzenes) are o-quinones (1,2-benzoquinones), highly reactive compounds that react with amino acids, peptides, and proteins (N-terminal amino groups, ε-amino groups of lysine and thiol groups), which leads to adducts that are further transformed into a variety of different products described in detail in Section 9.12.4.

2.5.2.6 Other reactions and interactions

Food processing at high temperatures during baking, frying, grilling, and roasting leads to the degradation of amino acids and the production of many different compounds that significantly affect the smell, taste, and colour of the food. Reactions are usually complex.

Pyrolysates of amino acids and various foods processed at high temperatures (around 200 °C and above) contain some mutagenic and carcinogenic products, such as the technological contaminants acrylamide (the decomposition product of asparagine; see Section 12.3.2) and furan (formed from alanine, serine, and other precursors; see Section 12.3.3), as well as aminoimidazoazaarenes (see Section 12.3.1), formed from creatinine and some Maillard reaction products. Aromatic amines are generated by pyrolysis of amino acids such as tryptophan, glutamic acid, lysine, ornithine, and phenylalanine. Structurally related heterocyclic products (derivatives of imidazopyrazines) were also found in the pyrolysates of aliphatic amino acids. Pyrolysis of amino acids and proteins at very high temperatures in burnt foods and in burning tobacco (around 800 °C) generates a number of different products, including polycyclic aromatic hydrocarbons (see Section 12.3.5).

Amino acids react with some food additives (such as sulfur dioxide), resulting in the loss of cysteine and formation of non-available products. Carcinogenic nitroso compounds are formed during the reaction of nitrites, nitrous acid, or nitrogen oxides with amino acids containing a secondary amino group (N-alkylamino acids or imino acids). Toxic products are also formed in the reaction of cysteine with chlorinated solvent residues and of methionine with some oxidising agents (e.g. nitrogen trichloride NCl_3) used as food additives.

Proteins have a potential role in the encapsulation, protection, and delivery of bioactive components such as fatty acids, vitamins, and phenolic compounds. Alternatively, the controlled reactions and interactions of proteins can follow a different approach, such as modifying technological functionalities (see Section 2.4.5.6) or producing hypoallergenic food. For example, complexation with bovine β-lactoglobulin significantly increases the hydrosolubility of anthraquinones and the stability of cyanidin-3-O-glucoside, promoting its application as a natural pigment in the food industry. The covalent conjugation with (−)-epigallocatechin-3-O-gallate and chlorogenic acid reduces the allergenicity and improves the functional properties of β-lactoglobulin. Recently, shell-like proteins have been used for the encapsulation of bioactive compounds in nutrition. For example, ferritin, which is distributed widely in animals, plants, and bacteria, becomes apoferritin upon removal of iron from its cage and is used as a pH-controlled vehicle for the protected delivery of bioactive nutrients such as carotenoids, curcumin, rutin, and calcium. Overall, reactions and interactions change the bioactivity and bioavailability of different components during food processing, and are helpful in maintaining the health benefits of functional foods.

3
Fats, Oils, and Other Lipids

3.1 Introduction

Lipids are one of the major components in foods and their role in human nutrition is one of the most important areas of concern in the field of nutritional science. Interestingly, however, lipids are such a diverse range of compounds that no single agreed definition exists; the main criterion for inclusion in this group of compounds is their hydrophobicity and not their chemical properties.

Lipids are usually defined as natural compounds containing bound fatty acids that contain more than three carbon atoms in the molecule. This definition is used by most biologically oriented workers, mainly physicians. However, ethyl esters of fatty acids with 4–12 carbon atoms, which are commonly found as odorous food components, are not classified as lipids. Some difficulty is caused by the inclusion of free fatty acids because they do not contain a bound alcohol residue. However, they are generally regarded as lipids and form a separate group together with their salts, which are known as soaps. Foods also frequently contain derivatives of fatty acids arising from industrial or other human activities (such as sugar esters and esters of sugar alcohols with higher fatty acids) that are not natural substances, but are still usually classified as lipids. Today, lipids are often defined as fatty acids and their derivatives, which are derived biochemically and related by their solubility in nonpolar solvents and general insolubility in water.

In practice, the non-volatile lipophilic compounds that accompany true lipids in both natural and manufactured products are also classified as lipids. They are termed **compounds accompanying lipids** (formerly called lipoids) in food chemistry, and 'unsaponifiable lipophilic substances' in food analysis. Their chemical structure is different and often these compounds do not even contain bound fatty acids. This group includes a large number of lipophilic compounds, for example some **terpenoids**, especially steroids (see Section 3.7.4) and carotenoids (see Section 9.9). Monoterpenoids, sesquiterpenoids, and diterpenoids, despite their structural similarity with higher terpenoids, are not classified as lipids, although they are lipophilic. Compounds accompanying lipids also include lipophilic vitamins (see Chapter 5), some pigments (see Chapter 9), natural antioxidants (see Section 11.2.2), and other lipophilic compounds.

The following sections describe fatty acids and the individual classes of lipids and lipid accompanying substances, their structure, occurrence, importance in nutrition and metabolism, and production of lipid derivatives. A large part of the chapter is devoted to the interactions and reactions of lipids with other food components that positively or negatively affect the nutritional value and organoleptic properties of foods. Attention is also paid to reactions of lipids under normal physiological conditions, such as during ageing of living organisms and reactions linked with pathogenesis of certain diseases.

3.2 Classification

Recent classification systems define lipids as hydrophobic or amphipathic compounds biosynthesised solely or partially by condensation of thioesters (fatty acids and polyketides) or isoprenoid units (prenols, sterols, and other compounds). Lipids are sometimes divided into three large groups: simple (or neutral) lipids, complex lipids, and proteolipids. Simple lipids provide at least two types of hydrolysis products (such as fatty acids and sterols or fatty acids and acylglycerols). Lipids containing sugar (glycolipids) are excluded from that list and are classified

The Chemistry of Food, Second Edition. Jan Velíšek, Richard Koplík and Karel Cejpek.
© 2020 John Wiley & Sons Ltd. Published 2020 by John Wiley & Sons Ltd.

as 'complex lipids', even if they contain two compounds. The complex lipids are divided into three main groups: phospholipids, arsonolipids, and glycolipids, as well as lipoamino acids. Complex lipids provide three or more types of products through hydrolysis. Glycolipids include glycoglycerolipids, glycosphingolipids, lipopolysaccharides, steryl glycolipids, and fatty acid glycosides, alcohols, or aminoalcohols.

This classification is fairly logical in basic chemical or biochemical research of lipids, but in practice, this system is not often used. More frequently we come across a totally illogical designation such as 'neutral lipids' and 'polar lipids'. The neutral lipids include esters with glycerol, steroids and their esters, but also free fatty acids, although they are not neutral. Polar lipids include phospholipids, as well as many other heterolipids that often do not contain phosphorus or even fatty acids. This classification system of lipids is usually seen in biologically oriented areas, and is based mainly on the behaviour of compounds during chromatographic separation. In food technology and oleochemistry, the name 'lipids' is not commonly used. Instead, only fats, oils, fatty acids, waxes, and lecithin are recognised, since only these components have industrial relevance.

In this chapter, we use a very similar global and fairly comprehensive lipid classification system that is clear and logical but remains only a guide. Lipids are classified into three major groups:

- **homolipids**;

- **heterolipids**;

- **complex lipids**.

Homolipids are compounds that contain fatty acids and alcohols. They are divided further according to the structure of the bound alcohol. Heterolipids are lipids that also contain, in addition to the fatty acids and alcohols, other covalently bound compounds, such as phosphoric acid in phospholipids or D-galactose in some glycolipids. Complex lipids are macromolecular substances whose lipid constituent is bound to the non-lipid components by hydrogen bonds, hydrophobic interactions, and other physical bonds, but some may also be bound by covalent bonds. The non-lipid components are mostly proteins (in lipoproteins), polysaccharides (in mucolipids), mixtures of proteins with polysaccharides, lignin, and other substances.

3.3 Fatty acids

3.3.1 Structure and occurrence

Fatty acids are the building blocks of lipids and nutritionally the most important lipid components. The term fatty acid usually means a saturated or unsaturated carboxylic acid usually having an even number of carbon atoms in an unbranched chain of four or more carbon atoms. Fatty acids bound in lipids can also, however, be branched, alicyclic, or even aromatic compounds. In nature, and therefore in foods, lipids contain the following groups of fatty acids:

- saturated fatty acids;

- unsaturated fatty acids with one double bond (monoenoic);

- unsaturated fatty acids with several double bonds (polyenoic);

- fatty acids with triple bonds and with various substituents (alkynoic, branched-chain, and cyclic acids, acids with oxygen-, nitrogen-, or sulfur-containing functional groups).

Free fatty acids are only found in plants and animal organisms in small quantities. Most of fatty acids are bound as esters or amides in homolipids and heterolipids. Some fatty acids, such as palmitic or oleic acids, are common constituents of lipids of all natural materials. Other fatty acids are specific only to microorganisms, plants, or animals – some only to certain genera – and therefore have particular importance in taxonomy. For simplicity, the occurrence of free fatty acids and fatty acids bound in lipids is described collectively.

3.3.1.1 Saturated fatty acids

Saturated fatty acids are common constituents of natural lipids. They normally contain from 4 to 38 carbon atoms, but there are also higher fatty acids with about 60 carbon atoms (normally an even number), which are mostly linear (with unbranched chains). In most natural lipids, saturated fatty acids constitute between 10 and 40% of the total fatty acids. Depending on the number of carbon atoms (chain length),

Table 3.1 Saturated fatty acids occurring in lipids.

Fatty acid	Number of carbon atoms	Trivial name	Fatty acid	Number of carbon atoms	Trivial name
Butanoic	4	Butyric	Eicosanoic	20	Arachidic
Hexanoic	6	Caproic	Docosanoic	22	Behenic
Octanoic	8	Caprylic	Tetracosanoic	24	Lignoceric (carnaubic)
Decanoic	10	Capric (caprinic)	Hexacosanoic	26	Cerotic
Dodecanoic	12	Lauric	Octacosanoic	28	Montanic
Tetradecanoic	14	Myristic	Triacontanoic	30	Melissic
Hexadecanoic	16	Palmitic	Dotriacontanoic	32	Lacceric
Octadecanoic	18	Stearic	Tetratriacontanoic	34	Gheddic (geddic)

saturated (as well as unsaturated) fatty acids can be divided into short-chain fatty acids (C_4 and C_6), medium-chain fatty acids (C_8—C_{12}), long-chain fatty acids (C_{14}—C_{18}), very-long-chain fatty acids (C_{20}—C_{26}), and ultra-long-chain fatty acids (C_{28}—C_{38}).

An overview of important fatty acids with an even number of carbon atoms in the molecule is given in Table 3.1. Besides the systematic names derived from the corresponding hydrocarbons, trivial names are also given, and these are used predominately in routine practice, especially in the nomenclature of common fatty acids. In the literature, for brevity, various short designations predominate, such as an $N : M$ ratio, where N is the number of carbon atoms in the molecule and M the number of double bonds. The most abundant saturated fatty acids in animal and plant tissues are palmitic (hexadecanoic, $16 : 0$; **3-1**) acid and stearic (octadecanoic, $18 : 0$; **3-2**) acid.

$$H_3C \diagup\diagdown\diagup\diagdown\diagup\diagdown\diagup\diagdown\diagup\diagdown\diagup\diagdown\diagup COOH \equiv H_3C \left[CH_2 \right]_{14} COOH$$

3-1, palmitic acid

$$H_3C \diagup\diagdown\diagup\diagdown\diagup\diagdown\diagup\diagdown\diagup\diagdown\diagup\diagdown\diagup\diagdown COOH \equiv H_3C \left[CH_2 \right]_{16} COOH$$

3-2, stearic acid

An overview of the occurrence of saturated fatty acids in fats and oils is given in Table 3.2. Saturated short-chain fatty acids, such as butyric and caproic acids, and the group of medium-chain fatty acids acids with 8 and 10 carbons in the molecule, typically occur in milk fat (Table 3.3). Butyric acid constitutes 4–8% (being found exclusively at the *sn*-3 position of triacylglycerols) and caproic acid 1–3% (also found mostly at the *sn*-3 position of triacylglycerols) of the weight of all fatty acids in cow's milk fat. These fatty acids are also present in milk fat triacylglycerols of other ruminants, such as sheep and goats, but they are not found in the fats of ruminants other than in the milk fats. Caproic acid is also a minor component of some seed fats (e.g. of palm seeds), which mainly contain fatty acids with medium chain length.

Medium-chain fatty acids (C_8—C_{12}) are present, usually as minor components, in milk fat triacylglycerols (Table 3.3). They are not found in the majority of vegetable oils, with a few exceptions. A high content of these fatty acids, especially of lauric acid, which is accompanied by other acids, is found in palm seed oils, such as coconut and palm kernel oils (Table 3.4), and seed oils of the Laurel family of plants (Lauraceae). Coconut oil is derived from the seeds (kernels) of coconuts. Fat extracted from the pulp (mesocarp) of the oil palm is palm oil, which has a completely different composition from palm kernel oil, the fat derived from the kernel (endosperm). Its fractions (palmolein and palmstearin) are commercially available, and are often added to food products.

Myristic acid is a common lipid component of most living organisms. It usually represents about 1–2% of total fatty acids and is present at higher levels in milk fat (10%) and coconut and palm kernel oils (14–21%). The fat of farm animals, especially of pigs and ruminants, contains long-chain saturated fatty acids: palmitic and stearic acids. The content of these fatty acids is lower in the depot fat of commercially reared birds (Table 3.5). Palmitic acid (**3-1**) is the most common saturated fatty acid. Its content in cow's milk ranges from 26 to 41%, and in breast milk from 6.1 to 12.1%. Most animal tissue lipids contain 20–30% palmitic acid; its content in vegetable oils from seeds is 5–30%, whilst in palm oil it reaches 40% or more. Stearic acid (**3-2**) is found in the highest amount in ruminant fats (milk fat in Table 3.3, tallow in Table 3.5). In vegetable fats, its highest levels are found in cocoa butter and shea butter (Table 3.6), and then in shortenings. In gangliosides,

Table 3.2 Saturated, mono-, and polyenoic fatty acids contents of some fats and oil (% of total fatty acids).

Type of fat	Fatty acids			Type of fat	Fatty acids		
	Saturated	Monoenoic	Polyenoic		Saturated	Monoenoic	Polyenoic
Beef tallow	47–86	40–60	1–5	Olive oil	8–26	54–87	4–22
Pork lard	25–70	37–68	4–18	Rice oil	19–35	42–50	16–37
Chicken lard	27–30	42–47	20–24	Cotton seed oil	24–33	15–23	46–59
Milk fat	53–72	26–42	2–6	Wheat germ oil	12–24	24–42	40–62
Carp fat	22–25	46–50	23–28	Soybean oil	14–20	18–26	55–68
Cod liver oil	14–25	35–68	20–45	Sunflower oil	9–17	13–41	42–74
Herring oil	17–29	36–77	10–24	Sesame oil	13–18	36–44	42–48
Coconut oil	88–94	5–9	1–2	Safflower oil	7–13	8–23	68–84
Palm kernel oil	75–86	12–20	2–4	Peanut oil	14–28	40–68	15–45
Palm oil	44–56	36–42	9–13	Rapeseed oil	5–10	52–76	22–40
Cocoa butter	58–65	33–36	2–4	Linseed oil	10–12	18–22	66–72

Table 3.3 Composition of major fatty acids of milk fat (% of total fatty acids).

Fatty acids	Cow's milk	Breast milk	Fatty acids	Cow's milk	Breast milk
Butyric	2.8–4.0	4–8	Σ saturated acids	61	38
Caproic	1.4–3.0	1–4	Σ isoacids	1	0.2–0.6
Caprylic	0.5–1.7	2–4	Σ anteisoacids	0.8	0.2
Capric	1.7–3.2	2–6	Σ branched acids	1.8	0.4
Lauric	2.2–4.5	4–9	Σ trans-C18:1 isomers	6.3	0.5
Myristic	5.4–14.6	8–14	Σ trans-C20:1 isomers	0.1	trace
Palmitic	26–41	18–35	Σ trans-monounsaturated	6.8	–
Stearic	6.1–12.1	7–15	Σ cis-C16:1 isomers	1.9	2.2
Arachidic	0.95–2.4	0–1	Σ cis-C18:1 isomers	23	38
Oleic	18.7–33.4	18–28	Σ cis-monounsaturated	26	41
Linoleic	0.9–3.7	2.0–5.2	Σ polyunsaturated	3.5	20
Linolenic	0.1–1.4	0.1–1.1	Σ C18:2 isomers	3.0	17
Arachidonic	0.8–3.0	0.4–1.5	Conjugated linoleic acid	0.7	0.4

stearic acid may constitute up to 80% of the total fatty acids. Arachidic acid and other very-long-chain fatty acids (behenic and lignoceric) are usually present in traces in animal fats; only in some oils do they occur in relatively high concentrations. For example, peanut oil contains 5–7% of C_{20}—C_{24} acids, of which arachidic acid represents about a third. Around 2% of behenic acids are found in canola oil. These and higher fatty acids are found in larger quantities in waxes.

Saturated fatty acids with an odd number of carbon atoms (C_{13}—C_{25}) are relatively rare, and occur only as trace components in triacylglycerols, for example in the milk of mammals. The particularly common acids are pentadecanoic (15 : 0) and heptadecanoic (also known as margaric or daturic acid, 17 : 0) acids, with pentanoic (valeric, 5 : 0), heptanoic (enanthic, 7 : 0), and nonanoic (pelargonic, 9 : 0) acids, and many others, being present at low levels. At higher concentrations, these acids are found as esters in microbial lipids, in skin lipids, and in depot fat and the milk of ruminants (5% or more), from which they enter into the human depot fat.

Table 3.4 Composition of major fatty acids in the fat of palm seeds (% of total fatty acids; range of values and average value are reported).

Fatty acid	Coconut oil[a]	Palm kernel oil[b]	Babassu oil[c]	Palm oil[d]	Palmolein	Palmstearin
Caproic	0–0.6	0.0–0.8/0.5	–	–	–	–
Caprylic	4.6–9.4/8	2.4–6.2/4	2.6–7.3	–	–	–
Capric	5.5–8.0/6	2.6–5.0/4	1.2–7.6	–	–	–
Lauric	43.0–51.0/47	41.0–55.0/47	40.0–55.0	0–0.4	0.1–0.2	0.1–0.6
Myristic	16.0–21.0/18	14–18.0/16	11.0–27.0	0.5–2.0/1	0.9–1.4	1.1–1.9
Palmitic	7.5–10.0/9	6.5–10.0/9	5.2–11.0	40–47/45	39.5–41.7	47.2–73.8
Palmitoleic	–	–	–	0–0.6	0–0.4	0.05–0.2
Stearic	2.0–4.0/2.5	1.3–3.0/2.5	1.8–7.4	3.5–6/4.5	4.0–4.8	4.4–5.6
Oleic	5.4–8.1/7	12–19/15	9.0–20.0	36–44/38	40.7–43.9	16.6–37.0
Linoleic	1.0–2.5/2	1.0–3.5/2.5	1.4–6.6	6.5–12.0/10	10.4–13.4	3.2–9.8
Linolenic	0.0–0.2	0.0–0.7	0	0–0.5	0.1–0.6	0.1–0.6
Arachidic	0.0–0.2	0.0–0.3	0	0–1.0	0.2–0.5	0.1–0.6
Eikosenic	0.0–0.2	0.0–0.5	0	0.1	–	–

[a]From the kernel of coconut palm seed (*Cocos nucifera*, Arecaceae).
[b]From the kernel of oil palm fruits (*Elaeis guineensis*).
[c]Derived from the kernel of the fruit of several varieties of the palm *Attalea* spp. In Brazil, from Babassu palm kernels (*Attalea speciosa*). In Mexico and Honduras, related cohune oil is pressed from the endosperm of cohune palm seeds (*A. cohune*).
[d]From the pulp (mesocarp) of oil palm fruits.

Table 3.5 Composition of major fatty acids in the fat of farm animals and birds (% of total fatty acids).

Fatty acid	Beef tallow	Sheep tallow	Pork lard	Rabbit lard	Chicken lard	Goose lard	Emu lard[a]
Lauric	1.0	0.8	trace	trace	0.1	–	–
Myristic	1.4–7.8	2–5	0.5–2.5	4	0.9	0.5	0.3
Palmitic	17–37	20–27	20–32	32	22	21	20
Palmitoleic	0.7–8.8	1.4–4.5	1.7–5.0	6	6	3	3
Stearic	6–40	23–34	5–24	8	6	6	1
Oleic	26–50	30–42	35–62	29	37	54	51
Linoleic	0.5–5.0	1.9–2.4	3–16	19	20	10	14
Linolenic	<2.5	0.6	<1.5	2	1	0.5	1
Arachidic	<0.5	1	<1.0	–	–	–	0.2
Eicosenoic	<0.5	0.2	<1.0	–	1	0.1	0.5

[a]*Dromaius novaehollandiae* (Casuariidae).

3.3.1.2 Unsaturated fatty acids

3.3.1.2.1 Unsaturated fatty acids with one double bond

Unsaturated fatty acids with one double bond, with the trivial name of monoenoic fatty acids, differ from one another by the number of carbon atoms, the double bond position in the hydrocarbon chain, and the spatial organisation of molecules (*cis-trans*-isomerism, also known as *E/Z* isomerism or geometric isomerism). The most important of these are listed in Table 3.7. Many monoenoic fatty acids have trivial names, which are commonly used. A typical example is *cis-* or (*Z*)-octadec-9-enoic acid, better known as oleic acid (**3-3**). Oleic acid

Table 3.6 Fatty acid composition of plant butters (% of total fatty acids).

Fatty acid	Cocoa butter	Shea butter	Illipé butter
Lauric	–	0.4	0.2
Myristic	0.1	0.3	0.3
Palmitic	23–30	4–8	23
Palmitoleic	0.1–0.4	–	0.2
Stearic	30–36	36–41	23
Oleic	33–36	45–50	34
Linoleic	1–4	4–8	14
Linolenic	0–0.5	0–0.4	0.2
Arachidic	0.2–1.0	1.2	0.2

Table 3.7 Major monoenoic fatty acids occurring in lipids.

Fatty acid	Number of carbon atoms	Double bond position	Isomer	Trivial name
Decenoic	10	4	cis	Obtusilic
Decenoic	10	9	cis	Caproleic
Dodecenoic	12	4	cis	Linderic
Dodecenoic	12	9	cis	Lauroleic
Tetradecenoic	14	4	cis	Tsuzuic
Tetradecenoic	14	5	cis	Physeteric
Tetradecenoic	14	9	cis	Myristoleic
Hexadecenoic	16	6	cis	Sapienic
Hexadecenoic	16	9	cis	Palmitoleic
Hexadecenoic	16	9	trans	Palmitelaidic
Octadecenoic	18	6	cis	Petroselinic
Octadecenoic	18	6	trans	Petroselaidic
Octadecenoic	18	9	cis	Oleic
Octadecenoic	18	9	trans	Elaidic
Octadecenoic	18	11	cis	Asclepic (cis-vaccenic)
Octadecenoic	18	11	trans	trans-Vaccenic
Eicosenoic	20	9	cis	Gadoleic
Eicosenoic	20	11	cis	Gondoic
Docosenoic	22	11	cis	Cetoleic
Docosenoic	22	11	trans	Cetelaidic
Docosenoic	22	13	cis	Erucic
Docosenoic	22	13	trans	Brassidic
Tetracosenoic	24	15	cis	Nervonic (selacholeic)
Hexacosenoic	26	17	cis	Ximenic
Triacontenoic	30	21	cis	Lumequeic

is the most widespread unsaturated fatty acid, which occurs, at least in a small amount, in virtually all animal and plant lipids. Oleic acid constitutes 30–40% of the total fatty acids in adipose tissue of animals and 20–80% of the total fatty acids in vegetable fats and oils. For example, olive oil contains up to 78% oleic acid.

Systematic names are often used for monoenoic fatty acids, especially when it is necessary to describe precisely the location and spatial configuration of the double bond in the molecule. For example, the major (Z)-octadec-9-enoic (oleic) acid is accompanied by a small amount of (Z)-octadec-11-enoic acid known as (Z)-vaccenic acid or asclepic acid. (Z)-Octadec-12-enoic acid appears to exist in hardened oils as a result of hydrogenation of linoleic acid so that only the double at the C-9 position is hydrogenated, whereas the double bond in the C-12 position remains unchanged. The short designation, 18 : 1, for all of these acids says nothing about the position of the double bonds. If we wish to indicate the position of the double bond, for example in oleic acid, the short designation 9–18 : 1 or 18 : 1Δ9 can be used, where the number 9 indicates that the double bond is at the ninth carbon atom from the carboxyl. The spatial configuration of natural fatty acids is usually cis (Z) or, more rarely, trans (E). Systematic names now increasingly use the newer designations Z and E, but far more common (and still correct) is the designation cis and trans. Oleic acid is then (Z)-octadec-9-enoic acid or cis-octadec-9-enoic acid, also abbreviated as c-octadec-9-enoic acid or 9c-18 : 1 or 18 : 1 cis-9.

If the trivial name is used instead of the systematic name, it should be noted that this generally refers only to an isomer with a specific location of the double bond and steric configuration. Trivial names should therefore not be used when the individual isomers are not known, for example in the data on fat composition. In other words, the group of octadecenoic acids cannot be generally referred to as oleic acid, and it is not possible to speak of 11-oleic acid, which does not exist. The designation isooleic acids has been applied to all isomers of octadecenoic acids, which have the double bond of the 18th carbon chain at some other place in the chain, instead of at C-9.

Milk fat contains some other monoenoic fatty acids, such as caproleic, palmitoleic (about 4%), and asclepic (cis-vaccenic, 1–5%; commonly present in vegetable oils, where it accompanies oleic acid) (Table 3.3). Larger quantities of myristoleic, palmitoleic (approximately 12–14%), and selacholeic acids are found in fish and whale fats. Palmitoleic acid is present at high levels in the fruits of sea buckthorn (*Hippophae rhamnoides,* Elaeagnaceae) and the oil of macadamia nuts (*Macadamia integrifolia,* Proteaceae). Petroselinic acid, in quantities of 50% and higher, occurs in the seeds of plants of the Apiaceae family, to which carrots, parsley, and coriander belong. Ximenic acid is present in the seeds of tropical trees and shrubs of the genus *Ximenia* (Olaceae), from which technical as well as cooking oils are obtained.

Many less common monoenoic fatty acids are only found in large amounts in inconsequential sources of lipids, but in foods they are usually present only in traces. For example, (Z) hexadec-6-enoic acid (known as sapienic acid) is the most abundant fatty acid in human sebum, as amongst hair-bearing animals it is restricted to humans.

Natural unsaturated fatty acids are mostly of a cis (Z) configuration. Acids with trans (E) configuration occur mainly in the depot fat and milk of ruminants (Table 3.3), resulting from conversion of the digested vegetable fats through biohydrogenation by microorganisms in the rumen, the first and the largest stomach of ruminants. These fatty acids are ingested by humans as a part of their food and therefore occur in human depot fat and in breast milk. The main trans-fatty acids are (E)-octadec-9-enoic (elaidic) acid (**3-4**) and (E)-octadec-11-enoic (trans-vaccenic) acid (**3-6**), which arises from linoleic or α-linolenic acids.

Trans-fatty acids are also formed during industrial catalytic hydrogenation (see Section 3.8.1.7) of plant or fish oils. This hydrogenation improves the thermal stability and prevents oxidation of unsaturated oils. Unsaturated fatty acids with a trans-configuration also form during heating of polyenoic fatty acids (at temperatures over 240 °C); for example, they form during deodorisation of oils (see Section 3.4.3.7).

$$H_3C \diagdown\diagup\diagdown\diagup\diagdown\diagup \overset{9}{=}\diagdown\diagup\diagdown\diagup\diagdown\diagup COOH \equiv H_3C \left[CH_2 \right]_7 CH = \overset{9}{CH_2} \left[CH_2 \right]_7 COOH$$

3-3, oleic acid

elaidic acid

trans-vaccenic acid

3-4, main *trans*-fatty acids

3.3.1.2.2 Unsaturated fatty acids with two double bonds

Fatty acids with two isolated double bonds separated by one methylene group (methylene-interrupted double bonds) are very important in nutrition. Theoretically, there should be far more of these acids than of monoenoic fatty acids in natural lipids, but only a few are found in significant amounts (Table 3.8). The most important dienoic acid is linoleic acid (**3-5**). Linoleic acid belongs to a group of fatty acids called **essential fatty acids** (see Section 3.3.2.2) It is present, at least in traces, in all fats. Sunflower and soybean oils usually contain 50–60% of

Table 3.8 Dienoic, trienoic, and other polyenoic fatty acids occurring in lipids.

Fatty acid	Number of carbon atoms	Positions of double bonds	Isomer	Trivial name
Dienoic				
Hexadecadienoic	16	9,12	*cis, cis*	—
Octadecadienoic	18	9,12	*cis, cis*	Linoleic
Octadecadienoic	18	9,12	*trans, trans*	Linolelaidic
Octadecadienoic	18	12,15	*cis, cis*	—
Eicosadienoic	20	11,14	*cis, cis*	—
Dokosadienoic	22	13,16	*cis, cis*	—
Trienoic				
Hexadecatrienoic	16	6,10,14	all *cis*	Hiragonic
Octadecatrienoic	18	6,9,12	all *cis*	γ-Linolenic
Octadecatrienoic	18	8,10,12	*trans, trans, cis*	α-Calendic
Octadecatrienoic	18	8,10,12	*trans, trans, trans*	β-Calendic
Octadecatrienoic	18	9,11,13	*cis, trans, trans*	α-Eleostearic
Octadecatrienoic	18	9,11,13	*trans, trans, trans*	β-Eleostearic
Octadecatrienoic	18	9,11,13	*cis, trans, cis*	Punicic (trichosanic)
Octadecatrienoic	18	9,12,15	all *cis*	α-Linolenic
Eicosatrienoic	20	8,11,14	all *cis*	Dihomo-γ-linolenic
Tetraenoic				
Octadecatetraenoic	18	6,9,12,15	all *cis*	Stearidonic (moroctic)
Octadecatetraenoic	18	9,11,13,15	*cis, trans, trans, cis*	α-Parinaric
Octadecatetraenoic	18	9,11,13,15	all *trans*	β-Parinaric
Eicosatetraenoic	20	5,8,11,14	all *cis*	Arachidonic
Docosatetraenoic	22	7,10,13,16	all *cis*	Adrenic
Pentaenoic				
Eicosapentaenoic	20	5,8,11,14,17	all-*cis*	Timnodonic (EPA)
Docosapentaenoic	22	4,7,10,13,16	all-*cis*	Osbond or (n-6) DPA
Docosapentaenoic	22	4,8,12,15,19	all-*cis*	Clupadonic
Docosapentaenoic	22	7,10,13,16,19	all-*cis*	Clupanodonic or (n-3) DPA
Hexaenoic				
Docosahexaenoic	22	4,7,10,13,16,19	all *cis*	Cervonic (DHA)
Tetracosahexaenoic	24	6,9,12,15,18,21	all *cis*	Nisinic

linoleic acid; safflower oil contains 75% linoleic acid. In the fat of animals, where this essential fatty acid gets from plant food, its content is typically 15–25%, but may be higher (cardiolipin of heart muscle contains 75% linoleic acid).

Again, positional as well as geometric isomers of dienoic fatty acids exist. Linoleic acid is a (9Z,12Z)-isomer or 9*cis*,12*cis*-octadeca-9,12-dienoic acid, abbreviated 9*c*12*c*-18 : 2 or 18 : 2 *cis*-9,*cis*-12. Isomers of linoleic acid with conjugated double bonds occur in milk fat and in lower amounts (2.9–11.3 g/kg) in dairy products and in the depot fat of ruminants. The main fatty acids are (9Z,11E)-octadeca-9,11-dienoic acid (formerly known as conjugated linoleic acid or bovinic acid; the recommended trivial name is rumenic acid, **3-6**) and (10E,12Z)-octadeca-10,12-dienoic acid. Rumenic acid is reported to exhibit antiatherogenic effects and (10E,12Z)-octadeca-10,12-dienoic acid provides anticarcinogenic and other beneficial effects. The ratio of these two fatty acids is about 1 : 3. Fatty acids with conjugated double bonds differ significantly from fatty acids with isolated double bonds in their reactivity and have distinct physiological effects.

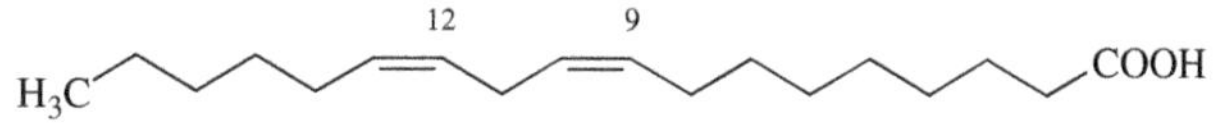

3-5, linoleic acid

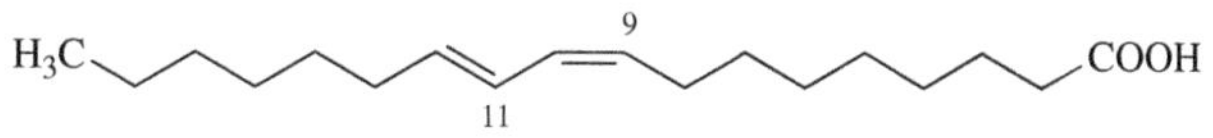

3-6, (9Z,11E)-octadeca-9,11-dienoic acid

3.3.1.2.3 Unsaturated fatty acids with three or more double bonds

The number of naturally occurring fatty acids with three *cis*-double bonds is considerably smaller than would be expected from the possibilities of isomerism (Table 3.8). The most important representative of these fatty acids is linolenic acid, an essential fatty acid with three isolated double bonds (**3-7**). Linolenic acid is the 9*cis*,12*cis*,15*cis*-isomer, that is, (9Z,12Z,15Z)-octadeca-9,12,15-trienoic acid, abbreviated to 9c12c15c-18 : 3 or 18 : 3 *cis*-9, *cis*-12,*cis*-15. In biological texts, it is often referred to as α-linolenic acid, which is a member of the *n*-3 series, unlike the isomeric γ-linolenic acid (octadeca-6,9,12-trienoic acid), a member of the *n*-6 series (**3-8**), which has different physiological effects. Linolenic acid is the main component of leaves, especially in the photosynthesising apparatus of algae and higher plants. It is present in linseed oil in amounts of up to 65%. Soybean and rapeseed oils only contain up to 10% linolenic acid. In animal tissues, it is usually a minor component (up to 1%), although the adipose tissue of horses contains up to 10% of this essential fatty acid.

Conjugated polyenoic fatty acids with three and four double bonds are natural components of oils obtained from the seed (endosperm) of some plants. (8E,10E,12Z)-octadeca-8,10,12-trienoic (α-calendic) acid (**3-9**) is the major fatty acid (50–60%) of oil from the seeds of marigold (*Calendula officinalis*, Asteraceae). (9Z,11E,13Z)-Octadeca-9,11,13-trienoic (punicic or trichosanic) acid (**3-10**) occurs at the level of about 60% in oil from the seeds of pomegranate fruit (*Punica granatum*, Lythraceae). (9Z,11E,13E)-octadeca-9,11,13-trienoic (α-eleostaric) acid is the major fatty acid (38%) of white mahleb (*Prunus mahaleb*, Rosaceae) seed oil. α-Eleostearic acid and its all-*trans*-isomer β-eleostearic acid are found in significant quantities in tung oil (Chinese wood oil) and abrasin oil, formerly used for lighting and now used for biodiesel production. Tung oil is obtained by pressing the nut of the tung tree (*Vernicia fordii*, Euphorbiaceae), which is native to southern China, Burma, and northern Vietnam. Abrasin oil, used in technological applications similarly to tung oil, is obtained from the *Vernicia montana* tree.

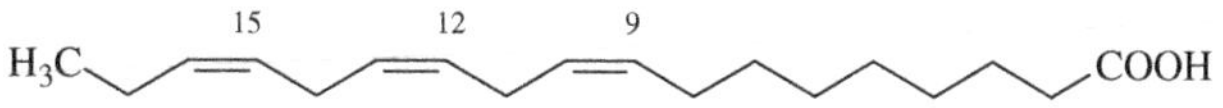

3-7, α-linolenic acid

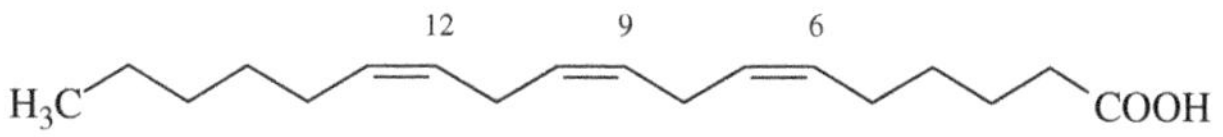

3-8, γ-linolenic acid

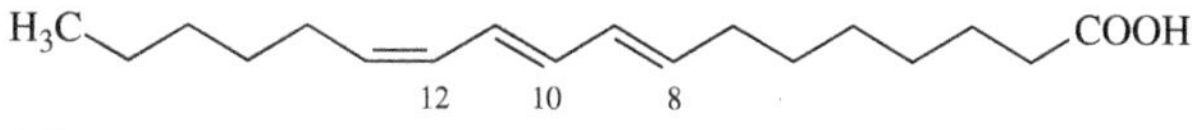

3-9, α-calendic acid

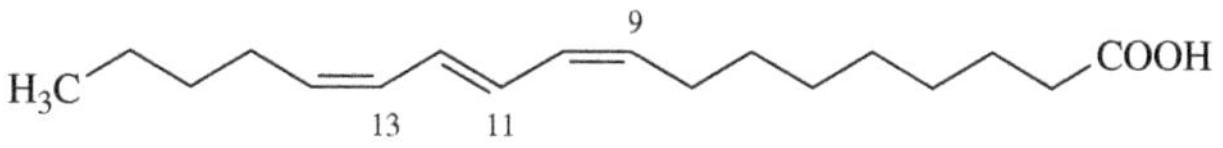

3-10, punicic acid

Highly unsaturated fatty acids with four to six isolated *cis*-double bonds in the molecule (Table 3.8) occur relatively rarely in nature. They belong to the *n*-3 and *n*-6 series. The most important of these fatty acids are those with four and five double bonds. The acid with four all-*cis* double bonds (*n*-6 series) is arachidonic (5,8,11,14-eicosatetraenoic) acid (**3-11**); all-*cis* acids with five double bonds are represented by clupanodonic (7,10,13,16,19-docosapentaenoic) acid (**3-12**) of the *n*-3 series. Polyenoic acids with isolated double bonds, such as

Table 3.9 Composition of major fatty acids of fish oils (% of total fatty acids).

Fatty acid	Herring oil[a]	Menhaden oil[b]	Pilchard oil[c]	Cod liver oil[d]
Myristoic	3–10	6–12	4–12	3–5
Myristoleic	–	0.2–0.4	–	–
Pentadecanoic	–	–	–	0.3–0.5
Palmitoic	13–25	14–23	9–22	10–14
Palmitoleic	5–8	7–15	6–13	6–12
Hexadecadienoic	–	3–6	–	0.5–1.6
Stearic	1–4	2–4	2–7	1.0–4.0
Oleic	9–22	6–16	7–17	19–27
Linoleic	1–2	1–2	1–3	1.0–2.0
Linolenic	0.6–2	1–2	0.4–1	0.2–1.0
Octadecatetraenoic	1–5	1–5	2–3	0.4–2.0
Eicosenoic	9–15	0.5–2	1–8	7–15
Eicosadienoic	0.5–0.7	–	–	0.1–0.4
Arachidonic	0.3–0.5	1–4	1–3	–
Eicosapentaenoic	–	12–18	9–35	8–14
Docosenoic	12–27	0.2–0.4	–	4–13
Docosadienoic	0.4–1	–	–	–
Docosatetraenoic	–	–	1–3	–
Docosapentaenoic	0.5–1.3	2–4	1–4	1.1–3.8
Docosahexaenoic	4–10	4–15	4–13	6–17

[a]Atlantic herring (*Clupea harengus*; the herring family Clupeidae).
[b]Atlantic menhaden (*Brevoortia tyrannus*; Clupeidae).
[c]European pilchard (*Sardina pilchardus*; Clupeidae).
[d]Atlantic cod (*Gadus morhua*; Gadidae).

hiragonic (hexadecatrienoic), stearidonic (octacosatetraenoic), clupanodonic (docosapentaenoic, commonly known as DPA), and nisinic (tetracosahexaenoic) acid, are found primarily in fats of marine fish (Table 3.9).

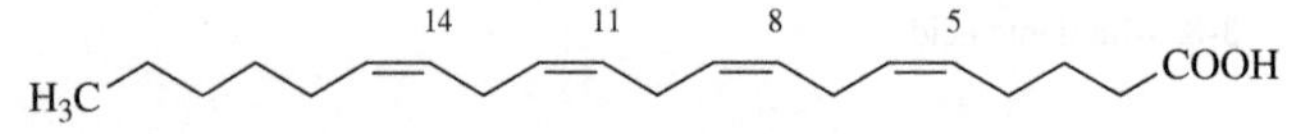

3-11, arachidonic acid

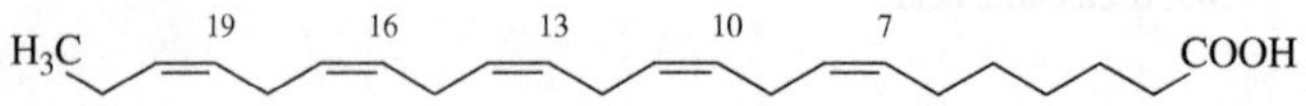

3-12, clupanodonic acid

A representative of polyenoic fatty acids with four conjugated double bonds is (9Z,11E,13E,15Z)-octadeca-9,11,13,15-tetraenoic (α-parinaric) acid. It is present in the seeds of the plant *Atuna racemosa* (Chrysobalanaceae), originally from the island of Fiji and other Pacific islands, where it is present at about 46% (34% of the fatty acids of the oil consists of α-eleostearic acid).

In addition, categorising polyunsaturated fatty acids according to the position of the double bond away from the terminal methyl group is common in biological documents. The nomenclature *n-x* (or ω-*x*) means that the first double bond is located on carbon number *x* counting from the terminal methyl group. The omega notation is common in popular nutritional literature. Linoleic acid is an example of a polyunsaturated fatty acid of the *n*-6 series (**3-13**). A short notation for linoleic acid is then $18:2n$-6 (or $18:2\omega$-6). Similarly fatty acids of *n*-3 (or ω-3) series are found in nature (**3-14**). Sometimes, *n*-9 (*ω*-9) fatty acids are also recognised. Unlike the 3's and 6's, n-9s are not 'essential' fatty acids. For example, n-9 fatty acids include oleic acid and erucic acid.

$$H_3C \left[CH_2 \right]_4 \left[CH{=}CH{-}CH_2 \right]_n \left[CH_2 \right]_\omega COOH$$

3-13, polyenoic fatty acids of *n*-6 series (ω=1–8, *n*=2–5)

$$H_3C{-}CH_2 \left[CH{=}CH{-}CH_2 \right]_n \left[CH_2 \right]_\omega COOH$$

3-14, polyenoic fatty acids of *n*-3 series (ω= 1–8, *n*=3–6)

Animal fats and vegetable oils mostly contain unsaturated straight-chain fatty acids with 10–36 carbon atoms. The most common fatty acids are monoenoic and polyenoic acids with 16–18 carbon atoms. The unsaturated fatty acids content of fats and oils ranges widely, from more than 90% of total fatty acids in rapeseed oil to less than 10% in coconut oil. Unsaturated fatty acids in animal fats occur in a much smaller concentration range, usually between 50 and 70% (Table 3.2). The only exception is fish oil, because it contains fatty acids with 20–22 carbon atoms and 4–6 double bonds (Table 3.9). The fat of freshwater fish differs in composition of the fatty acids in the fat of marine fish. Fish do not synthesise these fats themselves, but acquire them through their food (they are present in plankton, such as crustaceans and algae). Therefore, aquatic mammals (such as whales) that feed on small crustaceans have a similar fatty acid composition to that of fish. The content of *trans*-unsaturated fatty acids in animal fats is given in Table 3.10.

There is far greater diversity in the composition of unsaturated fatty acids in plants than in animals. Vegetable fats and oils are divided into groups according to their related fatty acid composition. Information about the fat content of major vegetable oil raw materials is listed in Tables 3.4 and 3.11–3.13. The following groups are recognised:

- Oils from palm seeds, which generally contain a small amount of oleic acid and only traces of acids with more double bonds (Table 3.4); typical examples are coconut oil and palm kernel oil.

- Plant butters,[1] where the content of unsaturated fatty acids is similar to that in animal fats and consists mainly of oleic acid (Tables 3.6 and 3.12); a typical example is cocoa butter (today, there are objections to the term 'plant butter' as it might be confused with cows' butter).

Table 3.10 Distribution of *trans*-unsaturated fatty acids in animal fats according to double bond position (% of total *trans*-unsaturated fatty acids).

Double bond position	Milk fat[a]	Butter[a]	Depot fat (beef tallow)[a]
8	1–3	1–2	1–2
9	7–15	5–16	8–14
10	4–13	4–7	5–7
11	28–55	51–68	64–69
12	4–9	3–6	2–3
13	4–9	3–6	2–3
14	4–10	4–7	3–4
15	4–8	3–5	2–3
16	5–10	4–7	3–4

[a] Total amount of *trans*-unsaturated fatty acids in milk fat and butter is 2–8% and in beef tallow is 2–3% of total fatty acids.

---

[1] The name 'plant butter' is derived from the fact that in their countries of origin, these fats have a consistency like cow's butter; in temperate climates, they are hard fats, but melt in the mouth. A typical example is cocoa butter from cacao beans (*Theobroma cacao*, Sterculiaceae). It is obtained by pressing and is used unrefined in the chocolate and pharmaceutical industries. Because it is expensive, it is often replaced by other plant butters. The most common is shea butter, extracted from fruits of a traditional African food plant commonly known as shea tree, shi tree, vitellaria, or karité (*Vitellaria paradoxa*, syn. *Butyrospermum parkii*, Sapotaceae). Throughout Africa, it is used extensively for food and for medicinal purposes. Butter illipé comes from an Indian plant *Madhuca longifolia* from the same family, butter mowrah from a similar plant (*Madhuca latifolia*), Borneo tallow from the East Asian plant *Shorea stenoptera* (Dipterocarpaceae), Chinese vegetable tallow from the plant *Stillingia sebifera* (Euphorbiaceae; it is often regarded as a wax due to its rigid consistency), and nutmeg (*Myristica fragrans*, Myristicaceae) fat from plants

Table 3.11 Composition of major fatty acids of vegetable oils with a predominance of oleic acid (% of total fatty acids; range of values is reported).

Fatty acid	Olive oil	Almond oil	Hazelnut oil	Avocado oil	Sunflower oil	Peanut oil	Safflower oil	Argan oil
Myristic	0.0–0.1	–	–	–	–	–	<0.1–0.1	0.15
Palmitic	7.5–20	4–13	5.1–7.2	9–18	3–5	4–7	3–6	12–13
Palmitoleic	0.3–3.5	0.2–0.6	0.1–0.3	3–9	<0.2	<0.4	<0.2	<0.12
Stearic	0.5–5.0	2–10	1.5–2.4	0.4–1.0	3–5	1–4	1.5–5.0	5–7
Oleic	55–83	43–60	71.9–84.0	56–74	70–87	70–90	74–82	43–49.1
Linoleic	3.5–21.0	20–34	5.7–22.2	10–17	3–20	7–15	7–18	29.3–36
Linolenic	0–1.5	0	0.0–0.2	0–2	0	<0.5	<0.2	<0.1
Arachidic	0–0.8	0.1–0.5	–	–	–	2	<1.0	0.3–0.5
Eicosenoic	0	0.0–0.3	0.1–0.3	–	–	2	<0.2	0.4–0.5
Behenic	0–0.2	–	–	–	–	4	<0.2	<0.2
Lignoceric	0–1.0	–	–	–	–	2	–	–

Table 3.12 Composition of major fatty acids of oils with moderate to high content of linoleic acid (% of total fatty acids; range of values and average value are reported).

Fatty acid	Peanut oil[a]	Sesame oil	Maize germ oil	Sunflower oil[a]	Safflower oil[a]	Cottonseed oil	Poppyseed oil
Lauric	0.0–0.1	0	0.0–03	0.0–0.1	0	0.0–0.2	0
Myristic	0.0–0.1	0.0–0.1	0.1–0.3	0.0–0.2	0.0–0.2	0.6–1.0	0.1–0.7
Palmitic	8.3–14.0/10	7.9–10.2/8.5	10.7–16.5	5–8/6.5	5.3–8.0	21.4–26.4/25	7–11
Palmitoleic	0.0–0.2	0.1–0.2	0.0–0.3	<0.5	0.0–0.2	0.1–1.2	0.8–1.6
Stearic	1.9–4.4/3.5	3.5–6.1/4.5	1.6–3.3	2.5–7.0/5	1.9–2.9	2.1–3.3	1–4
Oleic	36.4–67.1/59	35–50/42	24.6–42.2	13–40/24	8.4–21.3	14.7–21.7/18	16–30
Linoleic	14.0–43.0/20	35–50/44	39.4–60.4	40–74/63	67.8–83.2	46.7–58.2/52	62–73
Linolenic	0.0–0.1	0.3–0.4	0.7–1.3	pod 0.3	0.0–0.1	0.0–0.4	–
Arachidic	1.1–1.7	0.3–0.6	0.3–0.6	pod 0.5	0.2–0.4	0.2–0.5	–
Eicosenoic	0.7–1.7	0.0–0.3	0.2–0.4	pod 0.5	0.1–0.3	0.0–0.1	–
Behenic	2.1–4.4	0.0–0.3	0.1–0.5	0.5–1.0	0.2–0.8	0.1–0.6	–
Docosenoic	0.0–0.3	0	0–0.1	0.0–0.2	0.0–1.8	0.0–03	–
Lignoceric	1.1–2.2	0.1–0.3	0.1–0.4	0.2–0.3	0.0–0.2	0.0–0.1	–

[a]Corresponds to traditional varieties of the oil plant.

- Oils with predominant oleic acid and small amounts of polyunsaturated fatty acids; the most common representative is olive oil (today, a number of plants containing seeds with oils high in linoleic acid have been bred so that the fatty acid composition resembles the composition of olive oil; examples include sunflower, safflower, and peanut oils (Tables 3.11 and 3.12); palm oil can also be included, as it contains about 10% linoleic acid and has a high content of palmitic acid (Table 3.4)).

- Oils with a medium content of linoleic acid, but no linolenic acid; these oils include the traditional peanut oil (also known as arachis oil or groundnut oil) (Table 3.12).

related to nutmeg. Plant butters have a specific composition of triacylglycerols with palmitic, stearic, and oleic acids as the main components, which causes a narrow interval of melting points of these fats. Sometimes, plant butters are replaced with artificial fat prepared by fractionation of palm oils.

Table 3.13 Composition of major fatty acids of oils containing linolenic acid (% of total fatty acids; range of values and average value are reported).

Fatty acid	Rapeseed oil (traditional)	Rapeseed oil (low-erucic acid)	Mustard seed oil	Crambe oil	Soybeen oil	Wheat germ oil	Linseed oil
Lauric	0.1	0	Traces	Traces	0.0-0.1	–	–
Myristic	0.2	0.0-0.2	0-1	Traces	0.0-0.2	–	–
Palmitic	1.5-6	3.3-6.0/4.5	0.5-4.5	2	8.0-13.3/10	12-19	4-7/6.5
Palmitoleic	0-3	0.1-0.6	0.0-0.5	0.3	0.0-0.2	0.3-0.5	–
Stearic	0.5-3.1	1.1-2.5/1.5	0.5-2	1	2.4-5.4/4	0.3-3.0	2-5/3.5
Oleic	8-60	52.0-66.9/56	8-23	12-15	17.7-25.1/21	14-23	12-34/18
Linoleic	11-23	16.1-24.8/21	10-24	8-10	49.8-57.1/56	50-56	7-27/14
Linolenic	5-13	6.4-14.1/10	6-18	6-7	5.5-9.3/8	3.5-7.0	35-65/52
Arachidic	0-3	0.2-0.8	0-1.5	1-2	0.1-0.6	0.3	–
Eicosenoic	3-15	0.1-3.4	5-13	3-4	0.0-03	0.3	–
Eicosadienoic	0-1	0.0-0.1	0-1	–	–	–	–
Behenic	0-2	0.1-0.5	0.2-2.5	0.2	0.3-0.7	0.1	–
Docosenoic	5-60	0-2.0	22-50	55-60	0-0.3	0.3	–
Docosadienoic	0-2	0.0-0.1	0-1	1	–	–	–
Lignoceric	0-2	0.0-0.2	0-0.5	–	0.1-0.4	0.0-0.1	–
Tetracosenoic	0-3	0.1-0.4	0.5-2.5	–	0	–	–

- Oils high in linoleic acid, but not containing linolenic acid, such as traditional sunflower, cottonseed, safflower, poppy seed, germ, or sesame oils (Table 3.12).

- Oils with a medium content of linolenic acid, such as soybean oil and rapeseed 00 oil (Table 3.13); this group also includes oils from seeds of Brassica plants, which originally contained erucic acid[2]; linseed oil is the traditional oil that is rich in linolenic acid.

- Oils containing some specific fatty acids such as γ-linolenic acid (Table 3.14), which occurs in oil from the seeds of evening primrose (*Oenothera biennis,* Oenotheraceae), borage (*Borago officinalis*, Boraginaceae), and seed oil from currants and gooseberries (*Ribes* spp.; Grossulariaceae); erucic acid is found in seed oils of cruciferous plants (Brassicaceae) and plants belonging to the family Tropaeolaceae; petroselinic acid is found in oils of carrot parsley and celery seeds, plants of the Apiaceae family, and plants of the Araliaceae family.

3.3.1.3 Alkynoic, branched-chain, and cyclic fatty acids

3.3.1.3.1 Alkynoic acids

Fatty acids with one or more triple bonds (alkynoic, ethynoic, or acetylenic acids) and fatty acids containing both double and triple bonds occur rarely in nature. Alkynoic acids are distributed mainly in the lipids of mosses, of mushrooms, and of some tropical plants.

---

[2] The traditional rapeseed oil (known as colza oil, turnip rape oil, ravison oil, sarson oil, or toria oil) was produced from seeds of *Brassica napus, Brassica campestris, Brassica juncea*, and *Brassica tournefortii* and contained up to 60% of erucic acid (C22:1), which is toxic to humans in large doses. This oil was mainly used for domestic lighting, soap making, high-temperature lubricating oils, and plastics manufacturing. Since 1991, virtually all rapeseed production in the European Union has shifted to edible rapeseed 00 oil (double-zero oil), low in both erucic acid and glucosinolates, as opposed to colza oil. In the 1970s, Canadian plant breeders developed through conventional plant breeding a premium-quality rapeseed oil called canola with erucic acid level below 2%. The name was derived from 'Canadian oil, low acid'. Monola oil is obtained from a high-oleic-acid canola cultivar. European rapeseed production is conventional (non-GMO). The maximum permissible concentration of erucic acid in vegetable oils, fats, and foods with more than 5% fat is 5% of total fatty acid content. Rapeseed oil has also become the primary feedstock for biodiesel in Europe.

Table 3.14 Composition of fatty acids of oils for special dietary purposes (% of total fatty acids).

Fatty acids	Evening primrose oil	Borage oil	Blackcurrant oil
Myristic	0.07	0.1	0.1
Palmitic	6–10	9.4–11	6–8
Palmitoleic	0.04	0.4	<0.2
Stearic	1.5–3.5	2.6–5.0	1–2
Oleic	5–12	14.6–21.3	9–13
Linoleic	65–80	36.5–40.1	45–50
α-Linolenic	0.2	0.2	12–15
γ-Linolenic	8–14	17.1–25.4	14–20
Octadecatetraenoic	–	0.2	2–4
Arachidic	0.3	0.2	0.2
Eicosenoic	0.2	2.9–4.1	0.9–1.0
Behenic	0.1	–	0.1
Docosenoic	–	1.8–2.8	–
Lignoceric	0.05	–	0.1
Tetracosenoic	–	1.2–4.5	–

Various taste-modulating C18 octadecadien-12-ynoic acids were identified in golden chanterelles (*Cantharellus cibarius*). The lowest threshold concentrations (kokumi enhancement) had the pungent 14-oxo-octadeca-9,15-dien-12-ynoic acid, the astringent 14,15-dehydrocrepenynic acid methyl ester (occurring also as ethyl ester and free acid), and the bitter 9-hydroxyperoxyoctadeca-10,14-dien-12-ynoic acid (**3-15**).

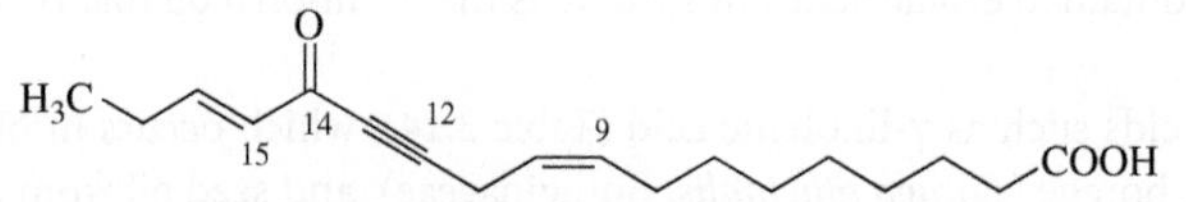

14-oxo-octadeca-9,15-dien-12-ynoic acid

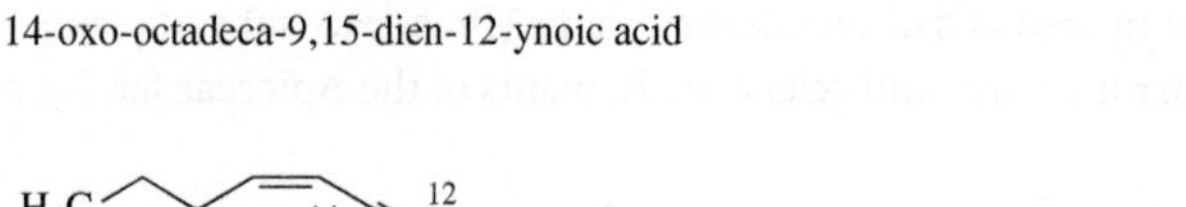

14,15-dehydrocrepenynic acid

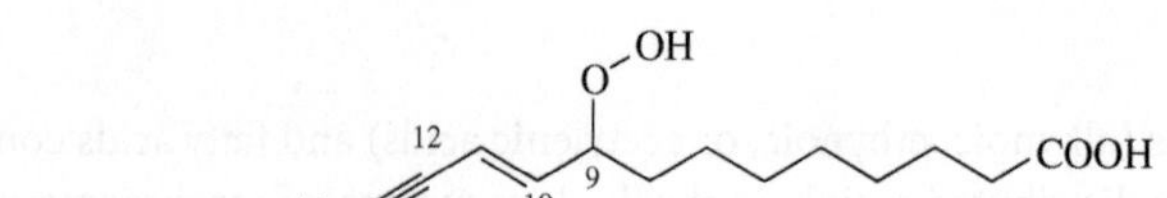

9-hydroxyperoxyoctadeca-10,14-dien-12-ynoic acid

3-15, octadecadien-12-ynoic acids

The occurrence of alkynoic acids in plants is largely limited to species of the family Santalaceae, where the seeds of some plants of the genera *Exocarpus*, *Santalum*, and *Ximenia* contain stearolic (octadec-9-ynoic) acid (**3-16**); plants of the Simaroubaceae family contain its positional isomer tariric (octadec-6-ynoic) acid, whilst plants of the Olaceae family contain isanic (octadec-17-en-9,11-diynoic) acid, also known as bolecic acid (**3-17**), in short 9a-18:1 (or 18:1, 9a) and 17E,9a,11a-18:3 (18:3, 17E,9a,11a), respectively. Boleka oil, containing tariric acid, and isano oil, containing isanic acid, are used in technological applications.

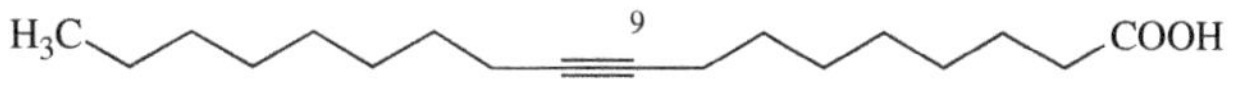

3-16, stearolic acid

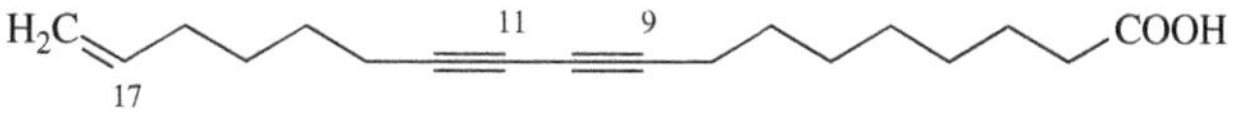

3-17, isanic acid

Alkynoic fatty acids are toxic. It is assumed that they interfere with the metabolism of lipids and fatty acids and inhibit lipoxygenases and cycloxygenases.

3.3.1.3.2 Branched acids

A large number of fatty acids (usually saturated) have a side chain consisting of one carbon. The methyl group is usually bound to the penultimate carbon atom (n–1). These acids are generally referred to as **isoacids** (**3-18**). If the methyl group is bound to the third carbon atom from the end (n–2), these fatty acids are called **anteisoacids** (**3-18**). Rare **neoacids** (**3-18**) have two methyl groups on the penultimate carbon atom; therefore, their terminal group is a tertiary butyl group (Table 3.15).

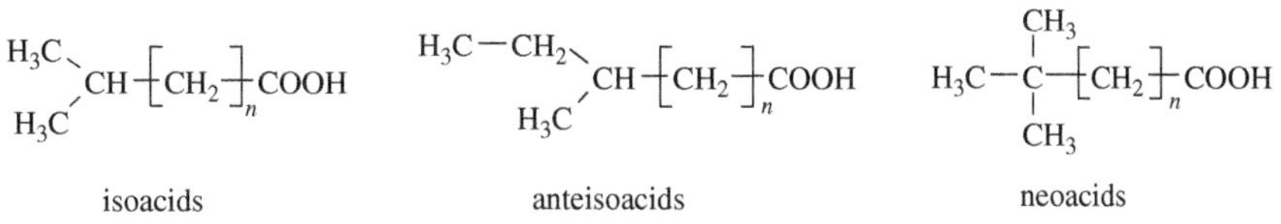

isoacids anteisoacids neoacids

3-18, fatty acids methylated in the carbon chain

Table 3.15 Fatty acids with a branched hydrocarbon chain.

Fatty acid	Number of carbon atoms	Trivial name
3-Methylbutanoic	5	Isovaleric
2,2-Dimethylpropanoic	5	Pivalic
4-Methylpentanoic	6	Isocapric
8-Methylnonanoic	10	Isocapric
10-Methylundecanoic	12	Isolauric
12-Methyltridecanoic	14	Isomyristic
12-Methyltetradecanoic	15	Isopentadecanoic (sarcinic)
14-Methylpentadecanoic	16	Isopalmitic
13,13-Dimethyltetradecanoic	16	Neopalmitic
14-Methylhexadecanoic	17	Anteisoheptadecanoic (methylpalmitic)
15-Methylhexadecanoic	17	Isoheptadecanoic (methylpalmitic)
16-Methylheptadecanoic	18	Isostearic
10-Methyloctadecanoic	19	Tuberculostearic
16-Methyloctadecanoic	19	Anteisononadecanoic (methylstearic)
17-Methyloctadecanoic	19	Isononadecanoic (methylstearic)
2,6,10,14-Tetramethylpentadecanoic	19	Pristanic
3,7,11,15-Tetramethylhexadecanoic	20	Phytanic
3,12,19-Trimethyltricosanoic	26	Phthioic
2,4,6-Trimethyloctacosanoic	31	Mycoceranic

Table 3.16 Contents of branched-chain fatty acids in animal fats (% of total fatty acids).

Number of carbon atoms	Milk fat (cow)	Beef tallow	Sheep tallow	Pork lard
13	0.06	trace	trace	0
14	0.05	<0.3	0.1	0
15	0.77	<1.5	1.2	<0.1
16	trace	<0.5	0.5	<0.1
17	0.42	<0.5	1.6	trace
18	trace	trace	0.2	0

The basic member of the homologous series of branched-chain carboxylic acids is 3-methylbutanoic (isovaleric) acid, which is ranked amongst the hemiterpenoids. Triacylglycerols containing isovaleric acid are important components of fat in European sturgeon, known as beluga (*Huso huso*), a species of anadromous fish in the sturgeon family (Acipenseridae). Isovaleric acid occurs in larger amounts in valerian root (*Valeriana officinalis*, Valerianaceae), in the form of esters of various monoterpenoids (such as borneol) and some less common sesquiterpenoids.

A number of higher branched acids is a typical characteristic of lipids in the cell walls of certain bacteria (including pathogens). Higher isoacids and anteisoacids of plants are the main component of the surface waxes. For example, pine seeds (*Pinus* spp., Pinaceae) contain 14-methylhexadecanoic acid at a level of 0.5–1%. Higher isoacids and anteisoacids of animals are predominantly found in the depot fat and milk of ruminants (Table 3.16) at levels of 1–2%. The dietary branched-chain fatty acids come from milk, dairy products, and the meat of ruminants. Several mechanisms appear to be involved in their formation. Some branched-chain fatty acids arise from acetyl-CoA elongation of 2-oxoacids acting as the precursors of amino acids (such as valine and isoleucine), without the involvement of fatty acid synthase-mediated reactions, suggesting integration of amino acid and fatty acid metabolism. Methyl side chains can also be introduced when methylmalonyl-CoA replaces malonyl-CoA as the chain-extending unit in fatty acid biosynthesis.

Common branched-chain fatty acids are anteisoheptadecanoic acid (14Me 16 : 0), anteisononadecanoic acid (16Me 18 : 0), anteisopentadecanoic acid (12Me 14 : 0), and anteisotridecanoic acid (10Me 12 : 0). These occur in higher levels in the lipids of mammal hairs. Lanolin (wool fat), for example, contains isoacids and anteisoacids C_{10} to C_{34}. One of these, the anteisoacid 18-methyleicosanoic acid, represents up to 60% of fatty acids bound by thioester bonds in sheep wool. Fish oils contain 1–2% C_{14}—C_{18} isoacids and anteisoacids.

More rarely, methyl groups are bound in the middle of the hydrocarbon chain, and some fatty acids may have a higher number of methyl group. (R/S)-4-Methyloctanoic (4-methylcaprylic or hircinoic), 4-ethyloctanoic, and 4-methylnonanoic acids have been implicated as primary contributors to the so-called 'soo' odour of mutton, which is an unpleasant fatty-goaty type note in goat subcutaneous tissue, as well as milk.

(2R,4R,6R,8R)-2,4,6,8-Tetramethyldecanoic acid occurs in the sebaceous glands of mammals and birds, for example in the glands of geese, producing an oily and waxy secretion, called sebum, which lubricates and waterproofs the feathers. Microbial degradation of isoprenoids in the rumen of ruminants, for example of the chlorophyll side chain formed by (2E,3,7R,11R,15)-3,7,11,15-tetramethylhexadec-2-en-1-ol (phytol), yields some more branched acids, such as (2RS,6R,10R)-tetramethylpentadecanoic (pristanic) or (3RS,7R,11R,15)-3,7,11,15-tetramethylhexadecanoic (phytanic) acid (**3-19**). Phytanic acid is therefore found in dairy products such as butter and in the fat of ruminants, as a mixture of (3RS)-epimers. Phytanic acid is normally present in small amounts in human tissues. Many defects in the α-oxidation pathway, including Refsum's disease, result in an accumulation of phytanic acid leading to neurological distress, deterioration of vision, deafness, loss of coordination, and eventually death. Branched-chain fatty acids obtained by oxidation of paraffins are used for technical purposes, for example in cosmetics.

Neoacids occur in some microorganisms, algae, higher plants, and marine animals. For example, 13,13-dimethyltetradecanoic acid (**3-19**) is a minor component of conifer resins. Further examples are presented in Table 3.15.

phytanic acid   13,13-dimethyltetradecanoic acid

3-19, examples of fatty acids with branched hydrocarbon chains

3.3.1.3.3 Alicyclic acids

Alicyclic fatty acids have saturated or unsaturated three-membered, five-membered, and six-membered rings. These naturally occurring acids usually have the cyclopropane or cyclopropene ring in the middle of the hydrocarbon chain. Alicyclic saturated fatty acids with cyclopropane and cyclopropene rings are relatively rare in natural fats and oils. They are found at trace levels in the lipids of bacteria, other lower organisms, and plants. An example is *cis*-11,12-methyleneoctadecanoic acid, also known as (11*R*,12*S*)-10-(2-hexylcyclopropyl)decanoic (IUPAC), lactobacillic, or phytomonic acid (**3-20**). It was first isolated as a constituent of triacylglycerols from the bacterium *Lactobacillus arabinosus*, and was later found in many other bacteria. This acid is accompanied by *cis*-9,10-methylenehexadecanoic acid and higher homologues. Some microorganisms contain *cis*-9,10-methyleneoctadecanoic, that is, (9*R*,10*S*)-8-(2-octyl-1-cyclopropyl)octanoic (dehydrosterculic) acid.

(*Z*)-8-(2-Octyl-1-cyclopropenyl)octanoic (sterculic) acid (**3-20**) occurs at low concentrations (1.0–2.3 g/kg) in cottonseed (*Gossypium* spp., Malvaceae) oil, accompanied by (*Z*)-7-(2-octyl-1-cyclopropenyl)heptanoic (malvalic) acid (2.4–3.4 g/kg) and dihydrosterculic acid (1.0–1.8 g/kg). In unroasted cottonseed meal, concentrations of these acids are about 10 times lower. This fatty acid is found in a significantly higher amount in seed oils of *Sterculia* plants in the mallow family (Malvaceae), originating from tropical South America, where it is also accompanied by malvalic and dihydrosterculic acids. Fruits known as tropical chestnuts have a flavour reminiscent of pistachios. The sterculic and malvalic acid contents in oils from *Sterculia foetida* are about 49 and 6%, respectively, whilst in *Sterculia monosperma* they are about 19 and 0.6%. These oils have special properties because they polymerise at normal temperatures and form gels at elevated ones (around 250 °C). Gum karaya is extracted from *Sterculia* species and used as a thickener and emulsifier in foods. Dihydrosterculic acid is found at a level of 35–48% in the oil from seeds of Chinese lychee (*Litchi chinensis*) and longan fruits (*Dimocarpus longan*) of the Sapindaceae family.

Cyclic unsaturated fatty acids with a cyclopentene ring at the end of the chain, such as (13*R*)-13-(cyclopent-2-ene-1-yl)tridecanoic (chaulmoogric) acid (**3-20**) and (11*R*)-11-(cyclopent-2-en-1-yl)undecanoic (hydnocarpic) acid, are present in the seeds of evergreen trees known as chaulmoogra (*Hydnocarpus*, syn. *Taraktogenos*) of the Achariaceae (formerly Flacourtiaceae) family, which grow in the rainforests of Southeast Asia (Table 3.17). Poisonous fruits of *Hydnocarpus wightiana* contain about 35% oil with a specific odour that is used in folk medicine for the treatment of leprosy and skin diseases (see Section 10.3.9.1). The oil of related species *Hydnocarpus anthelminticus* contains about 68% chaulmoogric, 9% hydnocarpic, and 1% gorlic acids. The content of these acids in the oil of *Hydnocarpus kurzii* is about 35, 23, and 23%, respectively.

Fatty acids with a cyclohexane ring at the end of the hydrocarbon chain, such as 11-cyclohexylundecanoic (**3-20**) and 13-cyclohexyltridecanoic acids, are minor constituents of milk, butter, sheep's tallow, and rumen bacteria. The amount of 11-cyclohexylundecanoic acid in cow's and goat's milk is 1–2 g/kg total fat. It was found that this acid had inhibitory effects on the bacteria *Bacillus cereus* and

Table 3.17 Fatty acids with cyclic residue in chain.

Systematic name	Number of carbon atoms	Number of double bonds	Trivial name
(5*R*)-5-(Cyclopent-2-en-1-yl)pentanoic	10	1	Aleprestic
(1*R*,2*S*,2*Z*)-2-(3-Oxo-2-pent-2-en-1-ylcydopentyl)ethanoic	12	1	Jasmonic
(7*R*)-7-(Cyclopent-2-en-1-yl)heptanoic	12	1	Aleprylic
(9*R*)-9-(Cyclopent-2-en-1-yl)nonanoic	14	1	Alepric
(11*R*)-11-(2-Cyclopent-2-en-1-yl)undecanoic	16	1	Hydnocarpic
(13*R*)-13-(2-Cyclopent-2-en-1-yl)tridecanoic	18	1	Chaulmoogric
(*Z*)-7-(Octylcycloprop-2-en-1-yl)heptanoic	18	1	Malvalic
(15*R*,9*Z*)-15-(Cyclopent-2-en-1-yl)pentadec-9-enoic	18	2	Hormelic
(13*R*,6*Z*)-13-(Cyclopent-2-en-1-yl)tridec-6-enoic	18	2	Gorlic
(*Z*)-8-(Octylcycloprop-2-en-1-yl)octanoic	19	1	Sterculic
(9*R*,10*S*)-8-(2-Octylcyclopropyl)octanoic	19	1	Dihydrosterculic
(11*R*,12*S*)-10-(2-Hexylcyclopropyl)decanoic	19	0	Lactobacillic (phytomonic)

Escherichia coli and on the fungus *Fusarium culmorum*. Cyclic fatty acids are also found in the ordinary oils, where their formation is induced by heating.

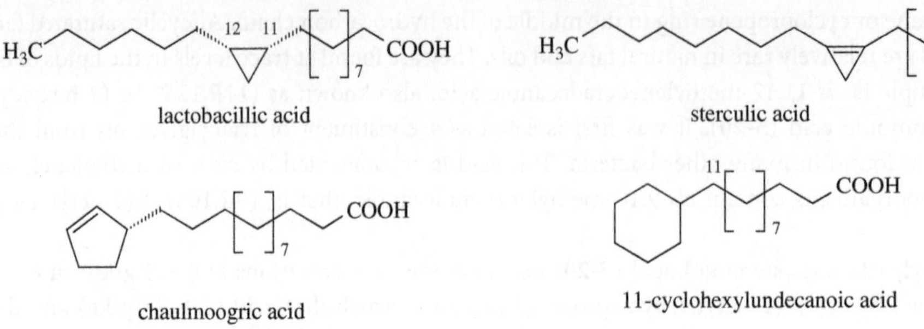

lactobacillic acid

sterculic acid

chaulmoogric acid

11-cyclohexylundecanoic acid

3-20, examples of alicyclic fatty acids

3.3.1.4 Fatty acids with additional oxygen functional group

3.3.1.4.1 Epoxyacids

Epoxy fatty acids occur either as 1,2-epoxy compounds derived from ethylene oxide (oxirane) or 1,4-epoxides derived from furan (furan acids).

Natural 1,2-epoxides are saturated and unsaturated compounds with 18 carbon atoms, which are found mainly in small amounts as components of cutin and oils from plant seeds. An example of a *cis*-epoxide is (+)-(9Z,12S,13R)-epoxyoctadec-9-enoic acid, known by the trivial name vernolic acid (**3-21**). Vernolic acid occurs in (+)- and (−)-optical isomers in the seeds of trees of the genus *Vernonia* of the Asteraceae family, originating from tropical Asia and America. For example, seeds of *Vernonia anthelmintica* and of *V. galamensis* contain 62–78% of vernolic acid. Related optical isomers of (12Z)-9,10-epoxyoctadec-12-enoic acid, known as coronaric acid, are found at levels of 8–18% in the seeds of garland, also known as chrysanthemum greens or edible chrysanthemum (*Chrysanthemum coronarium*) from the same plant family. The leaves of these plants are often eaten as a salad. Small amounts of 9,10-epoxystearic (about 2.5% of total fatty acids), vernolic, and coronaric acids occur in germ peanut oil.

Furan fatty acids, known as F-acids (**3-22**), are present in plants, fishes, amphibians, reptiles, and mammals. In some fish, F-acids can typically represent 1–6% of the fatty acids in the liver lipids and in some freshwater fish they may represent up to 25% of the total fatty acids. They also occur in butter, meat, vegetable oils, shrimps, and mushrooms in small amounts. Their concentration in soybean oil (250–400 mg/kg) is much higher than in rapeseed oil and corn oil (13–40 mg/kg) or in extra virgin olive oil (1–3 mg/kg). The basic member of the homologous series of non-olefinic furan fatty acids is (8Z,11Z)-8,11-epoxy-9,10-dimethylhexadecanoic acid (abbreviated as F_0-acid; m = 6, n = 4, R = CH_3). The most common compound is (12Z,15Z)-12,15-epoxy-13,14-dimethyleikosa-12,14-dienoic acid (abbreviated as F_6-acid, m = 10, n = 4, R = CH_3), as well as (10Z,13Z)-10,13-epoxy-11,12-dimethyloctadeca-10,12-dienoic acid (F_3-acid, m = 8, n = 4, R = CH_3) in vegetable oils. Common F-acids are olefinic fatty acids with one unsaturation on the side chains conjugated with the furan ring. Furan fatty acids also arise as products of lipid oxidation. Autoxidation or photoxidation of furan fatty acids yields α, β-unsaturated γ-lactones known as bovolides (see Figure 8.54).

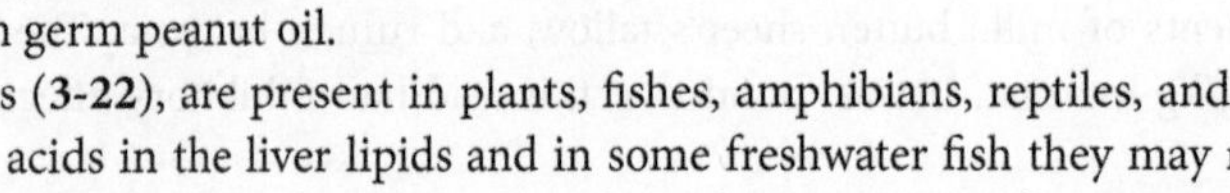

3-21, vernolic acid

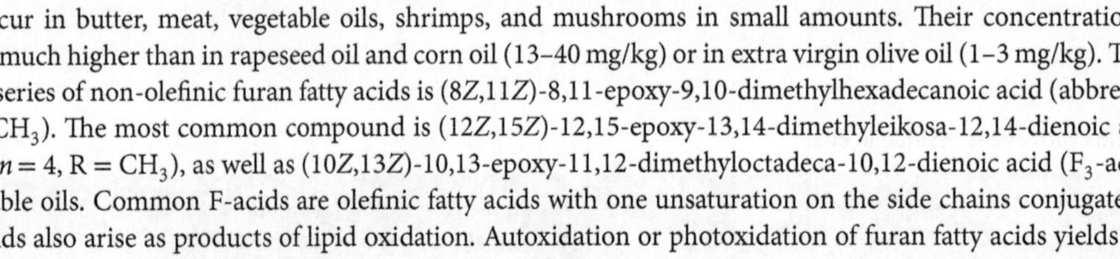

non-olefinic furan fatty acids

olefinic furan fatty acids

3-22, furan fatty acids

3.3.1.4.2 Hydroxyacids

Hydroxy acids are found naturally mostly as minor fatty acids. 2-Hydroxy fatty acids with 16–26 carbon atoms are important components of animal sphingolipids and plant waxes occurring on the surface of leaf vegetables. In addition, 2-hydroxy fatty acids and 3-hydroxy fatty acids are characteristic compounds of the lipopolysaccharides, which are located in the outer membrane of Gram-negative bacteria. Even-numbered 2- and 3-hydroxyfatty acids (with chain lengths of from 8 to 16 carbons) were identified in bovine milk fat. 2-Hydroxyoctanoic acid (45 mg/kg) was present in the highest amount, followed by 3-hydroxyhexadecanoic (42 mg/kg), 3-hydroxydodecanoic (34 mg/kg), 3-hydroxydecanoic (32 mg/kg), and 3-hydroxytetradecanoic (25 mg/kg) acids. Other 2- and 3-hydroxy fatty acids were present in lower amounts. Both (*R*)-enantiomers (D-hydroxy fatty acids) and (*S*)-enantiomers (L-hydroxy fatty acids) of 2- and 3-hydroxy acids were detected, with the (*R*)-enantiomer being predominant.

4-Hydroxycarboxylic acids and 5-hydroxycarboxylic acids occur in the form of corresponding γ- and δ-lactones in many fruits, especially apricots and peaches. Many other hydroxy fatty acids are also found in seed oils of plants. A palmitic acid derivative (+)-(*S*)-11-hydroxyhexadecanoic acid, known as (*S*)-jalapinolic acid (**3-23**), occurs in lipophilic ester-type dimers of acylated heterotetrasaccharides (operculinic acid C) and pentasaccharides derived from L-rhamnose (simonic acids B) in sweet potato tuberous roots (*Ipomoea batatas*, Convolvulaceae), which are known as batatins. The unsaturated hydroxy acid (9*Z*,12*R*)-12-hydroxyoctadec-9-enoic acid, known as ricinoleic acid (**3-24**), is the best-known representative of the series of (+)-D-acids. Ricinoleic acid occurs in castor oil, where it represents about 90% of the total fatty acids. Castor oil is extracted from the seeds of the castor oil plant (*Ricinus communis*) of the Euphorbiaceae family, and is used only for technical purposes as it has purgative properties. Ricinoleic acid is accompanied by 9,10-dihydroxy- and 9,10,12-trihydroxystearic acids, which can also be found in some edible fats as secondary oxidation products.

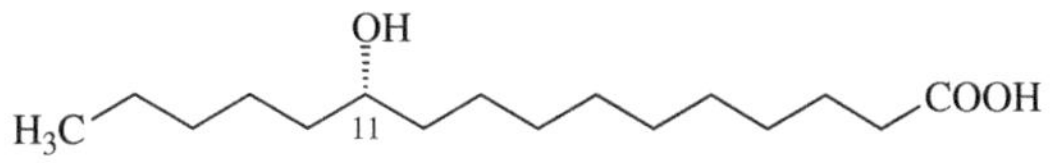

3-23, jalapinolic acid

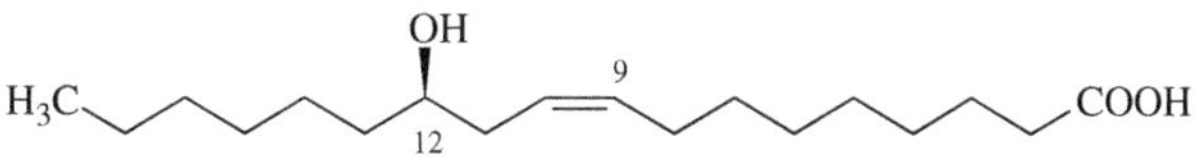

3-24, ricinoleic acid

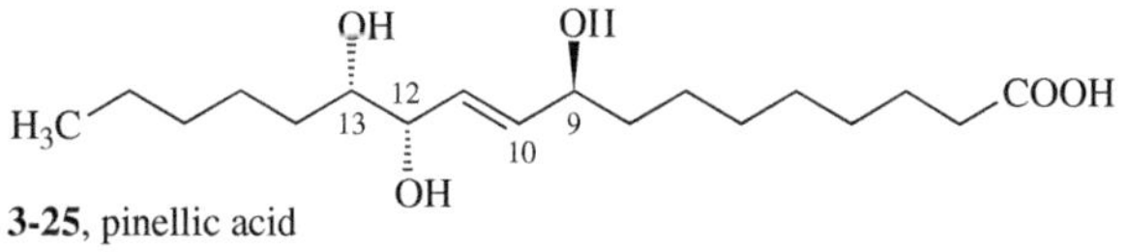

3-25, pinellic acid

An important polyhydroxy acid derived from linoleic acid is (9*S*,10*E*,12*S*,13*S*)-trihydroxyoctadec-10-enoic acid, known as pinellic acid (**3-25**). It can be found in alcoholic beverages, garden onion, cereals, and cereal products (see Section 8.3.5). Other metabolites of linoleic acid are steric isomers of 9,10,12,13-tetrahydroxyoctadecanoic acids, known as sativic (sativinic) acids. Oxidation of linolenic acid analogously produces 9,10,12,13,15,16-hexahydroxyoctadecanoic (linusic) acids. An overview of common hydroxy fatty acids is given in Table 3.18.

3.3.1.4.3 Oxoacids

Oxoacids (ketoacids) are less common than hydroxy acids. Saturated oxoacids with 10–24 carbon atoms and unsaturated oxoacids with 14–18 carbon atoms and the carbonyl group at C-5 to C-13 represent about 1% of fatty acids in milk fat. The so-called oiticica oil is extracted from the seeds of South American shrubs and trees of the genus *Licania* (Chrysobalanaceae), such as *Licania rigida* and *Licania arborea*. This oil typically contains (9*Z*,11*E*,13*E*)-4-oxooctadeca-9,11,13-trienoic acid, which is known under the trivial name α-licanic (couepinic) acid (**3-26**).

Avocado contains acetates of unsaturated fatty acid that contain both oxo and hydroxyl groups. Examples of these fatty acid esters are (2*R*,12*Z*,15*Z*)-2-hydroxy-4-oxoheneicosa-12,15-dien-1-yl acetate, (2*R*,5*E*,12*Z*,15*Z*)-2-hydroxy-4-oxoheneicosa-5,12,15-trien-1-yl acetate, known as persenone A (**3-27**) and (2*R*,5*E*)-2-hydroxy-4-oxononadec-5-en-1-yl acetate (persenone B, **3-28**), which act as antioxidants, preferentially suppressing radical generation.

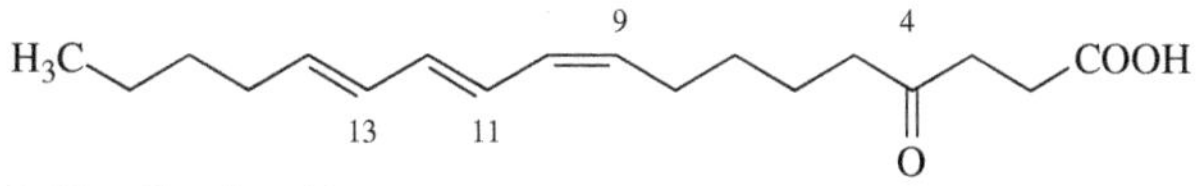

3-26, α-licanic acid

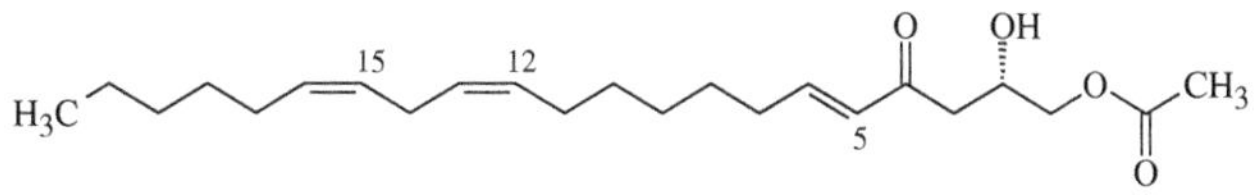

3-27, persenone A

Table 3.18 Some important hydroxy acids, oxoacids, and epoxy acids occurring in lipids.

Fatty acid	Number of carbon atoms	Positions of double bonds	Positions of functional groups	Trivial name
Hydroxy acid				
Hydroxydocosanoic	22		2	Fellonic
Hydroxytetracosanoic	24		2	Hydroxynervonic
Hydroxyhexadecenoic	16	7	16	Ambrettolic
Hydroxyoctadecenoic	18	9	12	Ricinoleic
Hydroxyoctadecenoic	18	12	9	Isoricinoleic
Hydroxytetracosenoic	24	15	2	Cerebronic
Hydroxyoctadecadienoic	18	9,15	12	Densipolic
Hydroxyoctadecadienoic	18	10,12	9	Dimorphecolic
Hydroxyoctadecatrienoic	18	9,12,13	2	Cemolenic
Hydroxyoctadecatrienoic	18	9,11,15	18	Hydroxylinolenic
Dihydroxyoctadecanoic	18	–	9,10	Dihydroxystearic
Dihydroxytriacontanoic	30	–	9,10	Lanoceric
Tetrahydroxyoctadecanoic	18	–	9,10,12,13	Sativic
Hexahydroxyoctadecanoic	18	–	9,10,12,13,15,16	Linusic
Oxo acid				
Oxooctadecatrienoic	18	9,11,13	4	Licanic
Oxooctadecatetraenoic	18	9, 11,13,15	4	Parinaric
Epoxy acid				
Epoxyoctadecenoic	18	9	12, 13	Vernolic
Epoxyoctadecenoic	18	12	9,10	Coronaric
Epoxyoctadecadienoic	18	9,12	15,16	Epoxylinoleic

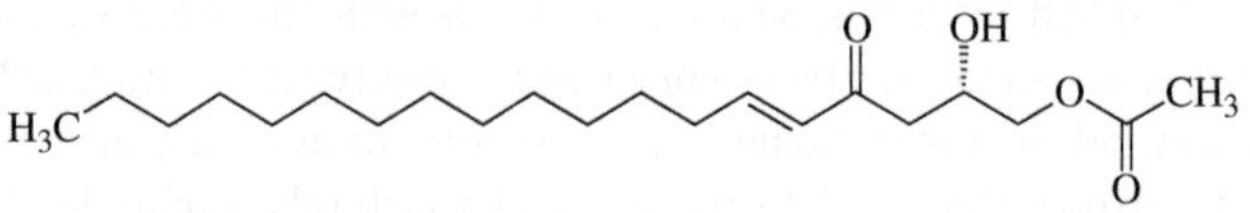

3-28, persenone B

3.3.1.5 Other fatty acids

Fatty acids are composed primarily of carbon, hydrogen, and oxygen, but may also contain nitrogen, sulfur, and chlorine.

Nitro fatty acids are one of the forms used to link the signalling pathways of the two major groups of cell-function regulators: eicosanoids and reactive forms of nitrogen, derived from nitric oxide (NO). Their interaction with the oxidised fatty acids can be crucial in the regulation of pro-oxidative and antioxidative processes in the immune responses and development of inflammatory processes. The most abundant nitro fatty acids are 9-nitrooleic (**3-29**), 10-nitrooleic, and 10- and 12-nitrolinoleic acids. Nitro derivatives of linolenic and icosapentaenoic acids have also been described.

Sulfur-containing fatty acids are rarely encountered. The lower homologue of (*R*)-5-(1,2-dithiolan-3-yl)pentanoic acid (known as α-lipoic acid, see Section 5.15) is 1,2-dithiolane-3-carboxylic (tetranorlipoic or tetranorthiooctic) acid. It was recently isolated from garlic, together with sulfur-substituted C_{18} fatty acids with thiophene, tetrahydrothiophene, and tetrahydrothiopyran rings. An example is

8-(5-hexyltetrahydrothiophen-2-yl)octanoic acid (**3-30**, R = H). A derivative of this tetrahydrothiophene fatty acid with a methyl group in the β-position (**3-30**, R = CH$_3$) and two other position isomers (the corresponding C$_{18}$ hexanoic and heptanoic acids) were found as sulfur-bearing components of unprocessed canola oil.

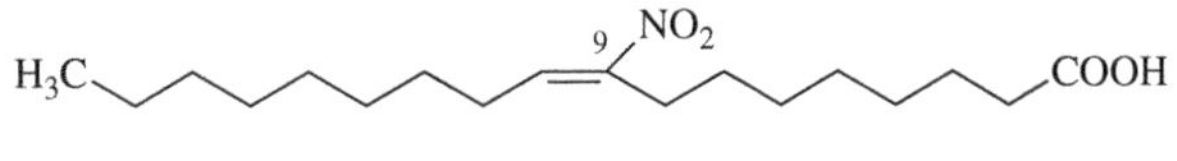

3-29, 9-nitrooleic acid

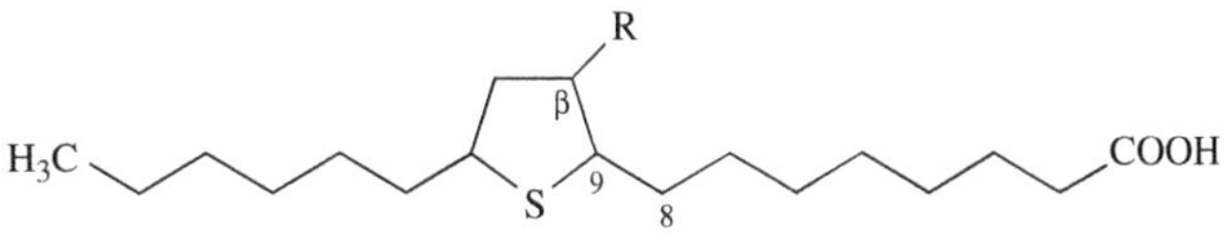

3-30, 8-(5-hexyltetrahydrothiophen-2-yl)octanoic acid

Chlorine-containing fatty acids arise as contaminants in reactions of unsaturated fatty acids with chlorine or chlorine dioxide (see Section 11.4.2.2.2).

3.3.2 Biochemistry, physiology, and nutrition

3.3.2.1 Biosynthesis of fatty acids

In mammals, the biosynthesis of fatty acids starts with the breakdown of excess dietary carbohydrates (glycolysis) and their transformation into acetic acid (C$_2$) units. These units are acetyl coenzyme A (acetyl-CoA) formed from coenzyme A (HS-CoA) and the glycolytic product pyruvic acid in the citric acid cycle. Fatty acid biosynthesis *de novo* occurs primarily in the liver, adipose tissue, central nervous system, and lactating mammary gland and is catalysed by multienzyme complex, known as type I fatty acid synthase. In prokaryotes and plants, the process is carried out by type II fatty acid synthase.

The molecules of fatty acids and other polyketides are theoretically formed by coupling of acetyl-CoA via condensation reactions in the so-called acetate pathway. Acetyl-CoA is first converted into more reactive malonyl-CoA. The formation of the polyketide chain can be envisaged as a series of Claisen condensation reactions. During each cycle, the fatty acid chain is extended by two carbon atoms and the resulting chain is reduced before connecting another molecule of acetyl-CoA. Therefore, the fatty acids with an even number of carbon atoms are present in lipids far more often than fatty acids with an odd number of carbon atoms. Biosynthesis stops at 16–18 carbon atoms and palmitic and stearic acids become the precursors for the biosynthesis of unsaturated fatty acids (palmitoleic and oleic). Fatty acids with an odd number of carbon atoms are formed by incorporation of propionyl-CoA instead of acetyl-CoA or by the loss of one carbon atom through α-oxidation. Branched fatty acids with methyl-substituted side chains are produced from acetyl-CoA precursors and methylmalonyl-CoA chain extender units.

In addition to fatty acids obtained in the diet, the human organism is able to synthesise some saturated and unsaturated fatty acids in the same way as other animals and plants do (Figure 3.1). Unlike plants, however, polyene fatty acids of *n*-6 (linoleic acid) and *n*-3 series (linolenic acid) cannot be synthesised, although they are strictly necessary. Therefore, these so-called essential fatty acids must be obtained through food in sufficient quantities. In the body, linoleic acid can be extended by two carbon atoms and α-linolenic acids by six carbon atoms in elongation reactions catalysed by elongases. New double bonds are created by desaturation, which is catalysed by desaturases. These reactions lead to fatty acids with 20–24 carbon atoms and from three to six double bonds. Some people have less active Δ^6-desaturase catalysing the desaturation of *n*-6 and *n*-3 fatty acids, so the transformations of linoleic and linolenic acids are difficult. The main factors that negatively affect the activity of Δ^6-desaturase are age (in the elderly the enzyme activity is lower), nutrition (ethanol has an inhibitory effect, and there are negative impacts resulting from deficiencies of vitamin B$_6$, biotin, Zn, Mg, and Ca,

n-6 Fatty acids	*n*-3 Fatty acids
linoleic	α-linolenic
(18:2 Δ9,12)	18:3 Δ9,12,15)
↓ Δ^6-desaturase	↓ Δ^6-desaturase
γ-linolenic	stearidonic
(18:3 Δ6,9,12)	(18:4 Δ6,9,12,15)
↓ elongase	↓ elongase
dihomo-γ-linolenic	eicosatetraenoic
(20:3 Δ8,11,14)	(20:4 8,11,14,17)
↓ Δ^5-desaturase	Δ^5-desaturase
arachidonic	eicosapentaenoic (EPA)
(20:4 Δ5,8,11,14)	(20:5 Δ5,8,11,14,17)
	↓ elongase
	docosapentaenoic (DPA)
	(22:5 Δ7,10,13,16,19)
	↓ elongase
	tetracosapentaenoic
	(24:5 Δ9,12,15,18,21)
	↓ Δ^6-desaturase
	tetracosahexaenoic
	(24:6 Δ6,9,12,15,18,21)
	↓ β-oxidation
	docosahexaenoic (DHA)
	(22:6 Δ4,7,10,13,16,19)

Figure 3.1 Biosynthesis of higher essential fatty acids.

a higher intake of *trans*-unsaturated fatty acids, and positional isomers of unsaturated acids), stress, and viral infections. Today, various products containing γ-linolenic acid, dihomo-γ-linolenic acid, eicosapentaenoic acid (EPA), and other substances are available as dietary supplements.

The oxygenated derivatives of C_{20} are pivotal signalling molecules in animals. In the bodies of animals, these essential fatty acids have an indispensable role as precursors of biologically active compounds called **eicosanoids** (prostaglandins, leucotrienes, prostacyclins, thromboxanes, and lipoxins). In animal systems, eicosanoids regulate cell differentiation, immune responses, and homeostasis. In human metabolism, the most important substance is arachidonic acid (Figure 3.1), which is stored in biological membranes as the C-2 ester of phosphatidylinositol and other phospholipids. In terrestrial plants, the biological role of eicosanoids is played by other biologically active substances, derivatives of C_{18} and C_{16} fatty acids, known as **octadecanoids** and **hexadecanoids** that act as developmental or defence hormones. The oxygenated derivatives of fatty acids are collectively known as **oxylipins**.

3.3.2.2 Metabolism of fatty acids

Fatty acids yield energy through a multi-step process called β-oxidation, which takes place in the mitochondria of most body tissues and is the reverse of the process involved in the biosynthesis of fatty acids. Less common is α-oxidation, which splits the carboxyl group, shortening the chain by one carbon atom, whilst generating a fatty acid with an odd number of carbon atoms. ω-Oxidation (oxidation of the methyl group at the end of a chain) can also occur, leading to the formation of dicarboxylic acid.

The saturated fatty acids oxidised during the process of β-oxidation are degraded to acetyl-CoA. The end product of fatty acids with an odd number of carbon atoms is propionyl-CoA, which is transformed to succinyl-CoA ready to enter the citric acid cycle. β-Oxidation of unsaturated fatty acids takes place via similar reactions, but requires additional enzymes. Fatty acids with *trans*-double bonds and hydroxy acids are difficult to break down and are a burden for the organism if they are present in large quantities in the diet. *Trans*-unsaturated fatty acids also adversely affect the ratio of low- and high-density lipoproteins (LDL/HDL) in serum and incorporate instead of saturated fatty acids into the *sn*-1 position of the membrane and nervous tissue phospholipids. *Trans*-isomers of fatty acids deteriorate the insulin resistance and thus increase the risk of the metabolic disorder *diabetes mellitus* type 2. Certain pro-inflammatory and adverse effects on foetal development have also been described. Recommendations of the WHO/FAO therefore limit the intake of *trans*-unsaturated fatty acids to 1% of energy intake.

3.3.2.3 Nutrition

The human diet contains only small amounts of free fatty acids. Dietary fat is mostly composed of neutral triacylglycerols accompanied by polar phospholipids and other fatty acid esters, which cannot be absorbed by the intestine. Digestion and absorption of fat requires that the complex fat molecules are broken down enzymatically into smaller, more manageable molecules. Dietary fat should constitute 30–35% of a person's total energy intake, and should contain saturated, monoenoic, and polyenoic fatty acids in a ratio of <1 : 1.4: >0.6. Saturated acids should contribute <10%, polyenoic acids of the *n*-6 series 4–8% (on average about 5%), and acids of the *n*-3 series about 1% of the energy provided by food. It is often stated that the ratio *n*-6 : *n*-3 should be no greater than 5 : 1. At least 0.5% of the energy intake should come from EPA, DHA, and other higher polyunsaturated acids of the *n*-3 series (Figure 3.1).

3.3.3 Properties

Fatty acids are colourless viscous thixotropic liquids (see Section 7.8.3.2.2) or solids. Lower saturated fatty acids are liquid, whilst caprinic (capric) acid and higher fatty acids are solid at room temperature. Their melting points depend on the number of carbon atoms, but when this number is higher than 20, the melting point does not change significantly (Table 3.19). Fatty acids with an odd number of carbon atoms have a slightly lower melting point than fatty acids with an even number with one less carbon atom. Unsaturated fatty acids with a double bond in the middle of the chain have significantly lower melting points than saturated fatty acids. Considerable influence on the melting point (the conformation of the molecule) comes from the *cis*-double bond, which causes bending of the straight chain by about 42°; therefore, *cis*-derivatives (**3-31**) have melting points a few tens of degrees lower than the corresponding *trans*-derivatives (**3-31**). The shift of the double bond towards the carboxyl in *cis*-monoenoic acids leads to an increase in the melting point. A higher number of double bonds in the all-*cis* configuration results in lowering of melting points (Table 3.19).

Melting point also depends on the polymorphic form (crystalline modification) of fatty acids. On cooling, fatty acids pass from the liquid phase first to a solid unstable phase, which is gradually transformed into more stable modifications. Modifications are labelled A, B, and C

Table 3.19 Melting points of important fatty acids.

Fatty acid Saturated	Melting point (°C)	Fatty acid Unsaturated	Melting point (°C)
Butyric	−7.9	Petroselic	33.0
Caproic	−3.4	Oleic	16.0
Caprylic	16.7	Elaidic	45.0
Capric	31.6	Asclepic	15.0
Lauric	44.2	Vaccenic	43.8
Myristic	54.1	Linoleic	−5.0
Palmitic	62.7	Linolelaidic	28.5
Stearic	69.6	Linolenic	−10.6
Arachidic	75.4	Erucic	33.5
Behenic	79.9	Brassidic	61.5
Lignoceric	84.2	Arachidonic	−49.5

in acids with an even number of carbon atoms, and A′, B′, and C′ in acids with an odd number of carbons in the molecule (Figure 3.2).

cis-monoenoic acid   *trans*-monoenoic acid

3-31, chain conformation of monoenoic fatty acids

Forms A and A′ are triclinic crystals (one of the seven crystal systems), whilst the other forms are orthorhombic crystals. Fatty acids can generally be present in any form (considering the conditions of crystallisation). Saturated fatty acids usually form the stable modification rather quickly, whilst unsaturated fatty acids are converted more slowly (because e.g. oleic acid has two modifications melting at different temperatures, 14 and 16 °C). The crystals of fatty acids form layers, in which carboxyl groups are oriented to one side and the methyl groups of the chain end to the other. The chains are parallel to each other (Figure 3.2). This arrangement of fatty acid molecules leads to soft crystals. Single-crystal modifications differ by the slope of fatty acids to the plains of individual layers. Even above the melting point, the fatty acids preserve the partially oriented structure, which causes the already mentioned thixotropy of these compounds.

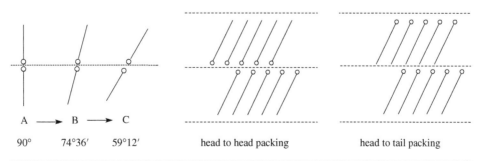

A ⟶ B ⟶ C

90°   74°36′   59°12′   head to head packing   head to tail packing

Figure 3.2 Left: Slope of saturated fatty acid chain in different crystal modifications: −, hydrocarbon chain; o, carboxyl; ------, crystal plane. Right: Simplified packing patterns of fatty acids in crystals (solid phase).

Fatty acids are weak acids that exist almost exclusively in non-dissociated form in solutions. Lower fatty acids are volatile at atmospheric pressure, whilst higher fatty acids are non-volatile. Their boiling point increases with the increasing number of carbon atoms, but double bonds have little influence. Fatty acids with short carbon chains are miscible with water, whilst others are basically soluble in water, but solubility decreases rapidly with an increasing number of carbon atoms in the molecule and a two-phase system is created. It is also necessary to consider the water solubility in fatty acids, which rapidly decreases with an increasing number of carbon atoms. Higher fatty acids dissolve in water only slightly, but can form monomolecular films (a closely packed layer of molecules) in which fatty acid molecules are oriented towards each other, with carboxyl groups to the aqueous phase and the methyl groups to the gaseous phase. When reducing the surface area, the monomolecular films collapse, creating a multimolecular layer (see Section 7.6.2.3).

3.4 Homolipids

Homolipids, which consist of fatty acids linked to various alcohols, are categorised exclusively by the structure of the bound alcohol. The most common alcohol in natural lipids is glycerol. Rarely occurring alcohols are glycols, higher monohydric aliphatic and aromatic alcohols, aliphatic, and alicyclic terpenoid hydroxy compounds, such as xanthophylls (see Section 9.9.1.2) and various steroid compounds (see Section 3.7.4).

3.4.1 Esters of monohydric alcohols

Esters of fatty acids with monohydric alcohols are known by the trivial name of **waxes**. The most important natural compounds are fatty acids esters with:

- aliphatic alcohols, formerly known as **cerides**;

- alicyclic alcohols, formerly known as **sterides**.

The term wax, however, is used for pure compounds and for a variety of natural products of animal and vegetable origin, components of which are fatty acid esters with monohydric alcohols, but also other lipophilic compounds, such as higher hydrocarbons, free alcohols, ketones, and fatty acid. These compounds, with the exception of fatty acids, are classified as lipid-accompanying substances (see Section 3.7). Esters of fatty acids with monohydric alcohols may even be minority constituents in these natural mixtures. It may therefore be more appropriate to use the term wax just for natural products rather than for pure compounds that are often called **wax esters**.

The waxy material on the surface of the aboveground parts of plants, such as the skins of fruits and the leaves of vegetables, seeds, and pollen grains, is called surface or **epicuticular wax**. The next hydrophobic layer of the cuticle is **cutin**, which is a polymeric material built from hydroxy fatty acids. Wax and cutin together constitute the **epicuticular lipids**. In most plants, wax is not associated with cutin. The building units of cutin are cutin acids, which include mono-, di-, and trihydroxycarboxylic fatty acids and also higher dicarboxylic, dihydroxydicarboxylic, and other acids. The characteristic monomers are 18-hydroxyoctadeca-9,12-dienoic, 10,16-dihydroxyhexadecanoic, and *threo*-9,10,18-trihydroxyoctadecanoic acids.

On the surface of the underground parts of plants, wax does not occur. Polymeric lipid material on the surface of the underground parts of plants and on any injured tissues is called **suberin**. Suberin has a qualitatively similar composition to cutin, but the amounts of the individual compounds differ. The prevailing acids are C_{16}—C_{22} dicarboxylic acids (such as octadec-9-ene-1,18-dioic acid), C_{20}—C_{26} fatty acids, and alcohols, which constitute about one third of the suberin weight. The remainder is composed of phenolic compounds.

Cutin and suberin are not classified as lipids.

3.4.1.1 Structure and nomenclature

Esterified fatty acids bound in waxes are usually saturated acids with long and ultra-long chains having 20–34 carbon atoms. Fatty acids with hydroxyl group often occur, for example, in position C-4. These then easily form δ-lactones, which are always present in industrial waxes in small amounts.

The common name for a mixture of high-molecular-weight (12–36 carbon atoms) saturated aliphatic primary alcohols, which are constituents of plant epicuticular waxes, is polycosanol. Alcohols bound in waxes include lauryl (C_{12}), myristyl (C_{14}), cetyl (C_{16}), stearyl (C_{18}), arachinyl (C_{20}), docosyl (C_{22}), lignoceryl or carnaubyl (C_{24}), ceryl (C_{26}), tetraenanthyl (C_{28}), myricyl or melissyl (C_{30}), and lacceryl (C_{32}) alcohol. The cuticular wax of wheat, for example, contains lignoceryl, ceryl, and myricyl alcohol as the major polycosanol components. Myristoleyl alcohol (C_{14}) and oleyl alcohol (C_{18}) are derived from unsaturated fatty acids. Less frequently, secondary aliphatic alcohols can be found.

Alcohols are usually esterified with fatty acids of similar structure. The characteristic structure is expressed by formula **3-32**. The compound name is formed by the radical names of the alcohol and the bound acid. For example, the main wax ester on the surface of sunflower seeds is ceryl cerotate, the ester of cerotic acid with ceryl alcohol. The main surface wax esters of apples and cabbage leaves are ceryl palmitate and ceryl stearate. The constituents of the epicuticular wax of apple fruits are also C_{16}–C_{26} fatty acid esters of (E)- and (Z)-p-coumaryl alcohol.

Fatty acids can also be bound to various secondary aliphatic alcohols such as nonacosane-15-ol, to carotenoids containing hydroxyl groups (known trivially as xanthophylls, see Section 9.9.1.2), and to flavonoids and some terpenoids (steroids) (see Section 3.7.4). Other wax constituents, in particular the epicuticular wax components, include alkanes, aldehydes, ketones, and free fatty acids.

3-32, general structure of aliphatic waxes

cetyl palmitate, $n = 14$, $\omega = 15$
ceryl cerotate, $n = 24$, $\omega = 25$
lacceryl laccerate, $n = 30$, $\omega = 31$

3.4.1.2 Biochemistry, physiology, and nutrition

Waxes are widely distributed in nature in both the animal and plant kingdoms, but are usually only seen in foods at low levels. In animals they occur in the skin, hair, fur, or feathers, whilst in plants they are found in the surface layers of leaves, fruits, and other aerial (above-ground) parts. The main significance of waxes is that they form a hydrophobic layer on the surface of organisms. They act mainly to control transpiration, preventing water loss by evaporation and protecting the organism against environmental influences, insects, and parasites.

Waxes are synthesised by reduction of fatty acids to primary alcohols via aldehydes. Primary alcohols react with acyl-CoAs to form esters, whilst aldehydes eliminate carbon monoxide (by decarbonylation), giving rise to hydrocarbons (alkanes). Oxidation of alkanes yields secondary alcohols, and oxidation of secondary alcohols gives rise to ketones.

Waxes do not decompose easily in the body, but when they do, the mechanism is similar to that of fatty acids in fats and oils with similar chain lengths. In human nutrition, waxes have virtually no significance.

3.4.1.3 Occurrence

3.4.1.3.1 Animal waxes

The wax found in the cranial cavity of the marine mammal the sperm whale (*Physeter macrocephalus*, syn. *Physeter catodon*) and two related species (the pygmy sperm whale and dwarf sperm whale) from the sperm whale superfamily has industrial importance. The milky-white waxy substance, or spermaceti, is composed mainly of cetyl palmitate (**3-32**). The part of spermaceti with a lower melting point, containing predominantly unsaturated fatty alcohols, is called sperm oil. Another animal wax is beeswax, which is used by honey bees to build honeycomb cells in which their larvae are raised and honey and pollen are stored. Its major component is ceryl cerotate (**3-32**), but a diverse mix of its homologues is also present. The composition of animal waxes is shown in Table 3.20.

The industrially important animal waxes include the wax on the surface of wool, which is obtained as a byproduct of wool scouring; after bleaching and purifying, this wax is termed lanolin. The alcohol components are mainly triterpenic (alicyclic) alcohols, such as lanosterol, cholesterol, and related steroids (see Section 3.7.4). Lanolin is used in the manufacture of cosmetics and pharmaceuticals.

3.4.1.3.2 Plant waxes

An example of the composition of cuticular waxes is the cuticular wax of thale cress, also known as mouse-ear cress or Arabidopsis (*Arabidopsis thaliana*, Brassicaceae), which is a plant used as a model organism in molecular biology and genetics. The major components of this wax are alkanes (38%), followed in quantity by ketones (30%), primary alcohols (12%), secondary alcohols (10%), aldehydes (6%), free fatty acids (3%), and their esters (1%). The main alkane is nonacosane (C_{29}), the main ketone is ditetradecylketone (C_{29}), the main primary alcohols are myricyl alcohol (melissyl alcohol, C_{30}) and ceryl alcohol (C_{26}), the main secondary alcohols are C_{29}, C_{31}, and C_{27} compounds, and the major fatty acids are mellisic acid (C_{30}) and montanic acid (C_{28}). The epicuticular wax of wheat and related species contains alkanes (5–10%), esters (10–30%), aldehydes ($\leq$5%), alcohols (15–55%), and acids ($\leq$5%).

The plant waxes that are of industrial importance include a hard, white wax known as Chinese wax, which is secreted by Indian Wax Scale insects (*Ceroplastes ceriferus*, syn. *Coccus ceriferus*) on injury of Chinese ash (*Fraxinus chinensis*, Oleaceae) and laid on tree branches. Japan wax is actually not the true wax, but a pale-yellow fat obtained from the fruits of the sumac wax tree (*Rhus succedanea*, Anacardiaceae) that

Table 3.20 Main wax acids and alcohols.

Wax	Fatty acid	Fatty alcohol
Animal		
Beewax	Palmitic, cerotic	Ceryl alcohol
Spermaceti	Palmitic	Cetyl alcohol
Sperm oil	Lauric, myristic	Oleyl alcohol, myristoleyl alcohol
Plant		
Chinese	Cerotic, montanic	Ceryl alcohol
Carnauba (Brazil, palm)	Cerotic	Ceryl alcohol, myricyl alcohol
Jojoba	Eicosenoic, docosenoic	Elcosanol, docosanol
Shellack	Cerotic, lacceric	Ceryl alcohol, lacceryl alcohol

grows in China and Japan. The seeds of jojoba shrubs (*Simmondsia californica*, Buxaceae), native to the desert of Sonora and adjacent areas of the United States and Mexico, contain a liquid wax (with melting point of 7 °C) called jojoba oil (wax). Candelilla wax is a yellowish-brown, hard wax located on the leaves and shoots of the small Candelilla shrub (*Euphorbia antisyphilitica*, Euphorbiaceae) growing in northern Mexico and the southwestern United States. Carnauba wax (also known as Brazil wax or palm wax) is a hard yellow-brown wax secreted on the surface of leaves of *Copernicia cerifera* (Arecaceae) palms from tropical South America. A similar palm wax is obtained from Andean palm (*Ceroxylon andicola*) and related species. Myrica wax comes from the California Wax Myrtle growing in North America (*Myrica cerifera*, Myricaceae). It is used to manufacture candles and perfumes. Similar waxes are produced by other species of the genus *Myrica*. Shellac is a product obtained after removal of dyes from lacks, resinous materials arising from various fig trees (*Ficus* spp., Moraceae), flowering plants of the genus *Butea* (Fabaceae), and other plants after injury caused by the lac insect (*Coccus lacca*). An overview of fatty acids and alcohols bound in important plant waxes is given in Table 3.20.

Esters of fatty acids with monohydric alcohols can also be produced by industry. Esters of fatty acids with methanol are components of so-called biodiesel; esters of fatty acids with ethanol and fusel oil alcohols are common components of alcoholic beverages.

3.4.1.3.3 Mineral waxes

Some waxes are of mineral origin. They come from the breakdown products of natural plant waxes. These include montan wax (also known as lignite wax) and the refined product ozokerite. Their alcoholic component and bound fatty acids have longer hydrocarbon chains than plant waxes.

3.4.1.4 Properties

Waxes are usually hard materials with melting points between 60 and 110 °C. They are insoluble in water and poorly soluble in organic solvents. Some waxes are liquid and also hydrophobic, but they dissolve well in organic solvents, can leave greasy stains, and easily form emulsions. Esters of lower alcohols produced by industry also have some of the properties of liquid waxes, but due to their lower molecular weights they have a lower viscosity and better solubility in alcohols.

3.4.1.5 Use

In the food industry, waxes are used as agents to repel water (hydrophobicity) on the surface of fruits and other food products, and to improve their appearance (fruits and some sweets). The waxes that are allowed for use in foods are legally limited to white and yellow beeswax, candelilla wax, carnauba wax, and shellac.

Spermaceti was historically used primarily for the manufacture of cosmetic products such as ointments, cosmetic creams, and pomades. Later, it was used for the manufacture of fine wax candles, in textile finishing, and as an industrial lubricant. Natural or bleached beeswax is used in the cosmetics and pharmaceutical industries, and practical uses include the making of candles (today only for luxury and church candles) and various dressings and polishes. Lanolin is an important raw material in the cosmetics and pharmaceutical industries.

Chinese wax is used in perfumery and skin polishing. Japan wax is used for the manufacture of polishing creams for floors, footwear, and furniture and for the manufacture of candles and matches and in the cosmetics industry. Jojoba oil is a liquid; it is therefore used for the production of fat emulsions and moisturisers in the cosmetics and pharmaceutical industries, as a carrier for speciality fragrances, and as a lubricant for delicate mechanisms. It has also been considered as a low-energy replacement for edible fats. The wax prepared from the oil is used as a polishing wax, in the manufacture of carbon paper and linoleum. It also has a potential use as a biodiesel fuel.

Candelilla wax and carnauba wax are mainly used for the production of candles. Carnauba wax can produce a glossy finish and as such is used in automobile waxes, shoe polishes, and waxing skin. Carnauba wax is also used for the preparation of polishing waxes, carbon paper, and many other purposes. Palm wax is the raw material for the production of wax candles and matches. Myrica wax and related waxes have found application in the pharmaceutical and food industries. Shellac is used to make paints, varnishes, sealing waxes, adhesives, and insulating materials.

As waxes are a relatively expensive raw material, they have been replaced by mineral hydrocarbon fractions (paraffin), solid triglycerides (tristearin, also known as stearin), and various synthetic materials (polyethylene and polypropylene) in the production of cosmetics and candles. Mineral waxes are used almost exclusively for non-food purposes. The so-called montanic acid esters and oxidised polyethylene wax can also be used to modify the surfaces of fresh citrus fruits. Montanic acid esters are used pharmaceutically to improve the slowdown of drug release from tablets.

3.4.2 Glyceryl ethers

Monoalk(en)ylethers of glycerol or 1-O-alk(en)yl-*sn*-glycerols or 1-alkoxy-*sn*-glycerols (**3-33**) are compounds that have bound fatty alcohol in the *sn*-1 position of the glycerol. They are related to plasmalogenes (see Section 3.5.1.1.2), which are glycerylethers of phospholipids and classified as heterolipids.

The main monoalkylethers of glycerol are 1-O-hexadecyl-*sn*-glycerol (chimyl alcohol), 1-O-octadecyl-*sn*-glycerol (batyl alcohol), and (*Z*)-1-O-octadec-9-en-l-ylglycerol (selachyl alcohol, **3-33**). Trivial names are derived from the names of the animals from which the fats were first isolated. Glyceryl ethers are not hydrolysed by lipases. In nature, glyceryl ethers are mostly found as esters of higher fatty acids. For example, 2-O-arachidonoylglyceryl ether derived from arachidonic acid, trivial name noladin ether, occurs in the brain of pigs. The most common compounds are diesters of glyceryl ethers, 1-O-alk(en)yl-2,3-diacyl-*sn*-glycerols.

3-33, glyceryl ethers (glyceryl 1-O-alkylethers)
chimyl alcohol, R = $(CH_2)_{15}CH_3$
batyl alcohol, R = $(CH_2)_{17}CH_3$
selachyl alcohol, R = $(CH_2)_8CH=CH(CH_2)_7CH_3$

Glyceryl ethers are present in animal and plant lipids in amounts that normally do not exceed 1%. They are found at higher levels in microbial lipids and lipids of some marine animals (cartilaginous fish, starfish, urchins, and clams), especially in lipids of cartilaginous fishes of the class *Chondrychthyes*, subclass Elasmobranchii, which includes sharks (Selachii) and rays (Batoidea). Shark liver oils commonly contain about 53% triacylglycerols, 45% diesterified glyceryl ethers, 1.5% cholesteryl esters, and 0.5% squalene. In the liver oils of some species of these animals, depending on the season, glyceryl ethers can constitute about 10% (in spiny dogfish, *Squalus acanthias*), but also up to 50–80% of the unsaponifiable fraction (in squaliform sharks of the genus *Centrophorus* and in kitefin sharks of the genus *Scymnorhinus*).

3.4.3 Glyceryl esters

Glyceryl esters are the most important food lipids. They are generally known by the names of their physical state as **fats** or **oils**, even if the name fats should be only be used for natural products containing triacylglycerols that are usually solid at ambient temperature, rather than for individual compounds. If fats are liquid at ambient temperature, they are called oils. In the past, vegetable oils were categorised according to the behaviour of a thin film of the oil when exposed to the air. The categories that could be distinguished were:

- **non-drying oils** (such as olive, peanut, coconut, palm, palm kernel, and castor oils) that do not harden when exposed to air;

- **semi-drying oils** (soybean, sunflower, poppyseed, sesame, and cottonseed oils);

- **drying oils** (linseed, safflower, and perlila oils; see Section 8.2.1.1.1) that harden after exposure to air.

However, these groups do not have precise boundaries. Today, the division of triacylglycerols into fats and oils has only historical significance, and the term 'fat' commonly refers to the entire group, regardless of consistency.

Table 3.21 Composition of glycerol esters in refined edible oils.

Type of esters	Content (%)	
	Rapeseed oil	Sunflower oil
1-Monoacylglycerols	0.6	0.2
2-Monoacylglycerols	0.1	0.05
1,3-Diacylglycerols	1.9	0.9
1,2-Diacylglycerols	0.2	0.1
Triacylglycerols	96.5	97.8

3.4.3.1 Structure and nomenclature

The glycerol molecule can be bound to one, two, or three fatty acids. The resulting monoesters are then either 1-**monoacylglycerols** or 2-monoacylglycerols (**3-34**). These esters were formerly called monoglycerides, and this name is often still used, especially amongst experts from industry. The ratio of both isomers in natural materials depends on the stereospecifity of lipase synthesising monoacylglycerols, and the major compounds in food lipids are the more stable 1-monoacylglycerols (Table 3.21). Two fatty acids bound on the glycerol molecule yield 1,2-**diacylglycerols** and 1,3-diacylglycerols (**3-34**); the latter predominate in natural fats. Diacylglycerols were previously called diglycerides. Monoacylglycerols and diacylglycerols are collectively referred to as **partial esters of glycerol**. In nature, the most frequently occurring glyceryl esters contain three fatty acids bound to the glycerol molecule. They are known as **triacylglycerols** (formerly known as triglycerides). Nearly all commercially important fats and oils of animal and plant origin consist almost exclusively of triacylglycerols (Table 3.21).

All three fatty acids bound in triacylglycerols can be the same, and in this case we may speak about **simple triacylglycerols** (**3-35**). For example, the triacylglycerol with three identical acyls, derived from palmitic acid, is 1,2,3-tripalmitoylglycerol or simply tripalmitoylglycerol, or tripalmitin to give it its trivial name. Three acyl residues derived from oleic acid are found in triolein and three stearic acid residues are found in tristearin. Alternatively, the two or three fatty acids bound in triacylglycerols can be different, and in this case we speak about **mixed triacylglycerols** (**3-36**).

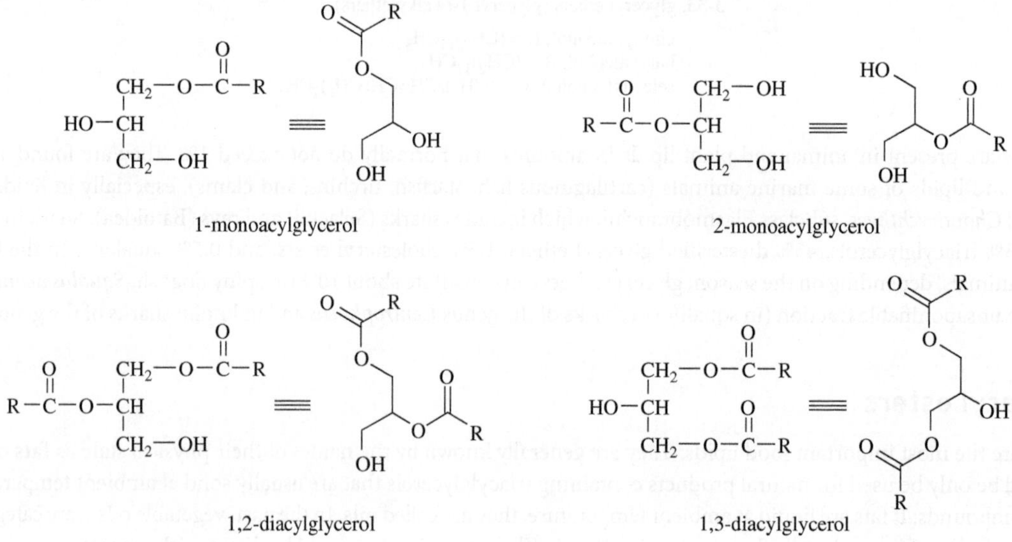

1-monoacylglycerol   2-monoacylglycerol

1,2-diacylglycerol   1,3-diacylglycerol

3-34, partial glycerol esters

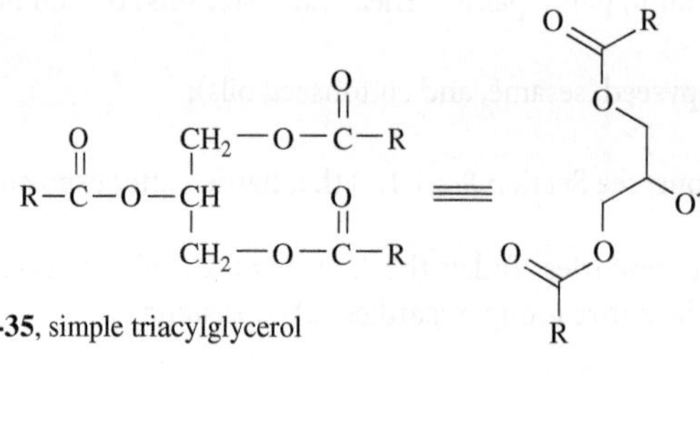

3-35, simple triacylglycerol

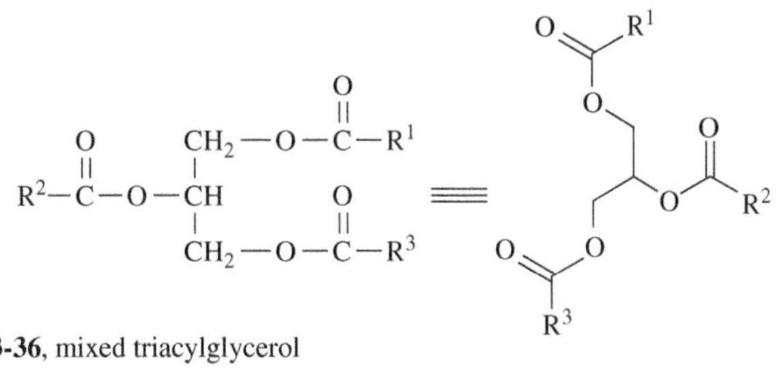

3-36, mixed triacylglycerol

For example, 1-palmitoyl-2-stearoyl-3-oleoylglycerol (formerly also known as l-palmito-2-stearo-3-oleine) is different from l-palmitoyl-2-oleoyl-3-stearoylglycerol and other positional isomers. These positional isomers lead to great diversity in the composition of natural fats and oils. When the two primary hydroxyl groups (in positions C-l and C-3) of glycerol are esterified with different fatty acids, the resulting triacylglycerol is asymmetric and optically active and the diversity of composition of natural fats and oils thus increases. Natural fats contain optically active triacylglycerols because they are formed by esterification through the catalytic action of hydrolases that are stereospecific. Since the bound fatty acids have very similar properties, these stereoisomers do not differ significantly in their optical rotations. Nevertheless, an exact definition of the composition of fats should also mention their steric structure. To express the exact composition and stereo configuration of triacylglycerols, the so-called stereospecific numbering system (*sn*-system) is used, as recommended by an IUPAC-IUB commission, which defines the numbering of atoms in prochiral glycerol ('L-glycerol' derivatives) as that shown in formula **3-37**. The conventional D/L or *R/S* systems are not used. The secondary hydroxyl group is shown to the left of C-2, and the carbon atom above this then becomes C-l, whilst that below becomes C-3 and the prefix *sn* is placed before the stem name of the compound. The term triacyl-*sn*-glycerol should then be used instead of triacylglycerol.

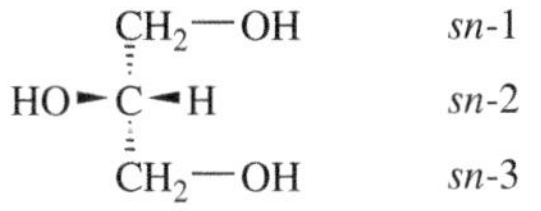

3-37, *sn* numbering of glycerol in Fischer projection

Three different fatty acids can occur from 27 different mixed triglycerides. To simplify the description, one- or two-letter abbreviations of fatty acids are used, such as B = butanoic (butyric), D = decanoic (capric or caprinic), H = hexanoic (caproic), L = linoleic, La = lauric, Ln = linolenic, M = myristic, O = oleic, Oc = octanoic (caprylic), P = palmitic, Po = palmitoleic, S = saturated, St = stearic, U = unsaturated, and V = vaccenic. For example, the mixed triacylglycerol esterified with palmitic acid in position *sn*-l, with stearic acid in the position *sn*-2, and with oleic acid in the position *sn*-3 of glycerol is called l-palmitoyl-2-stearoyl-3-oleoyl-*sn*-glycerol (abbreviated to *sn*-PStO). Racemate is a mixture of *sn*-PStO and *sn*-OStP in a molar ratio of 1 : 1 (or rac-PStO), where the fatty acids in positions *sn*-1 and *sn*-3 are reversed. Triacylglycerol PStO can generally be a mixture of up to six optically active substances.

3.4.3.2 Biochemistry and physiology

3.4.3.2.1 Biosynthesis

Synthesis (lipogenesis) of triacylglycerols occurs primarily in the liver, adipose tissue, central nervous system, and lactating mammary gland. The necessary fatty acids are obtained from food or synthesised through the oxidation of fats, glucose, and some amino acids via acetyl-CoA. In animals, the liver and intestines are the most active, although most of the body stores of triacylglycerols are in adipocytes, the fat-storing cells of adipose tissue. Two main biosynthetic pathways are known. The most important routes are the *sn*-glycerol 3-phosphate pathway (also known as the Kennedy pathway), which predominates in the liver, adipose tissue, and skeletal muscle, followed by the 3-monoacyl-*sn*-glycerol pathway in the intestines. In some animal tissues, a third, less well-known biosynthetic pathway has been recognised. In this pathway, triacylglycerols are synthesised by transacylation between two racemic diacylglycerols.

The building block for the biosynthesis of glycerol is the glycolytic product (*R*)-phosphoglyceric acid.[3] Glycerol is activated by phosphorylation with glycerol kinase to prochiral *sn*-glycerol 3-phosphate, which is esterified by fatty acid residues in acyl-CoA with catalysis by glycerol 3-phosphate *O*-acyltransferase. The first reaction product is l-acylglycerol 3-phosphate (3-*sn*-lysophosphatidic) acid, which is

---

[3]To a lesser extent, 1,3-dihydroxyacetone phosphate (formed as a product of glycolysis and photosynthesis in plants) is employed for the biosynthesis of glycerol. In the depot tissue (in adipocytes), 1,3-dihydroxyacetone phosphate is the only precursor of triacylglycerols. It can either be reduced to glycerol 3-phosphate by glycerol 3-phosphate dehydrogenase or first acylated to 1-acylglycerol, which is reduced by acylglyceronphosphate reductase to 3-*sn*-lysophosphatidic acid.

esterified to 2-diacylglycerol 3-phosphate (3-*sn*-phosphatidic acid) by 1-acylglycerol 3-phosphate *O*-acyltransferase. Hydrolysis of the phosphate by phosphatidate phosphatase yields 1,2-diacylglycerol. The last esterification of 1,2-diacylglycerol to triacylglycerol is catalysed by diacylglycerol *O*-acyltransferase. Phosphatidic acid also serves as a precursor of glycerophospholipids.

3.4.3.2.2 Fat digestion, absorption, transport, and mobilisation

The human fat-digestive enzymes include triacylglycerol lipases and phospholipases. The triacylglycerol digestive enzymes comprise the preduodenal lingual, gastric, and extra-duodenal pancreatic lipase. Lipases catalyse the hydrolysis of dietary triacylglycerols primarily to fatty acids (their sodium or potassium salts) and 2-monoacyl-*sn*-glycerol via 1,2-diacyl-*sn*-glycerol or 2,3-diacyl-*sn*-glycerol. The hydrolytic attack is regiospecific and occurs at the *sn*-1 or *sn*-3 positions, but not at the *sn*-2 position of triacylglycerols. To some extent, 2-monoacylglycerol is partly spontaneously racemised to 1-monoacylglycerols, which may be subsequently completely hydrolysed to fatty acid and glycerol.

Glycerol is transported to the liver or kidneys and is either converted into glucose via glycerol 3-phosphate (gluconeogenesis) or used to help the breakdown of glucose into energy (glycolysis). Fatty acids with a chain length of less than 14 carbon atoms are absorbed directly into the blood via intestine capillaries as complexes with albumin, enter into the portal vein (which can also act as an absorptive route for dietary long-chain fatty acids), and are transported to the liver. Monoacylglycerols and fatty acids with 14 or more carbons are not released directly into the intestinal capillaries and need to be re-esterified to triacylglycerols. The re-esterification proceeds by the 3-monoacyl-*sn*-glycerol pathway, which produces more than 75% of triacylglycerols. In addition to the 3-monoacyl-*sn*-glycerol pathway, the intestine can also synthesise triacylglycerols via the glycerol 3-phosphate pathway, which is the dominant triacylglycerol synthetic pathway in other tissues, such as adipose and liver. Beginning in the endoplasmic reticulum, the main site of lipid synthesis, and continuing in the organelle called Golgi apparatus, triacylglycerols are assembled into a lipoprotein particle called chylomicron, which consists of triacylglycerols (85%), phospholipids (6–12%), cholesterol and cholesteryl esters (1–3%), and proteins (apolipoproteins, 1–2%). Cholesterol esters are hydrolysed by pancreatic steryl ester acylhydrolase (cholesterol ester hydrolase) to give cholesterol and fatty acids.

Chylomicron is transported into the bloodstream for transport to tissues where triacylglycerols are stored or metabolised for energy. In the target tissues, the chylomicron triacylglycerols are hydrolysed by lipoprotein lipase to fatty acids and 2-monoacylglycerols. Most fatty acids and 2-monoacylglycerols are absorbed by the cells (adipose cells or muscle fibres) and employed for triacylglycerol re-synthesis. A significant proportion of digested fat is typically stored as body fat in the adipocytes found mostly in the abdominal cavity and subcutaneous tissue as cytoplasmic lipid droplets called adiposomes, organelles enclosed by a monolayer of phospholipids and hydrophobic proteins, such as the perilipins in adipose tissue. The remnants of the chylomicron are transported to the liver and hydrolysed.

Animals also accept fats in the diet in very small amounts through the mechanism of pinocytosis (the process of taking fluid, together with its contents, into the cell by forming narrow channels through the cell membrane that divide off into vesicles).

Triacylglycerols represent a form of energy storage for most organisms (in animal adipose tissue and plant seeds), but also act as membrane components and as signalling molecules. The fat reserves in humans are usually 10–30 kg, which is sufficient to cover the body's energy needs for several months. When energy is required, triacylglycerols in the form of hydrophilic lipoproteins are supplied to the relevant cells, where they are hydrolysed and fatty acids are oxidised, mainly by a series of reactions termed β-oxidation, to produce energy and building blocks for growth. The use of fatty acids for energy production and re-synthesis of triacylglycerols is predominantly performed in the liver.

3.4.3.3 Nutrition

The human diet contains fats of animal and vegetable origin. Some fats are obvious (such as those added during cooking), but roughly the same amount are hidden fats already present in the raw materials (mainly of animal origin). Triacylglycerols derived from the diet of the populations in industrialised countries deliver 30–40% of the body's energy requirements. In developed countries, this energy intake is usually excessive (more than 100% of the recommended amount). In Europe, it would be prudent ('it is recommended') to reduce ('to lower') the contribution of fat to total energy intake below 30% to prevent the major public health problems associated with fat inadequate intake (such as the risk of developing high blood pressure, diabetes, cardiovascular diseases, and certain forms of cancer). However, the proportion of fat must not fall below 20% of energy supplied, otherwise various disorders may occur, mainly due to an inadequate supply of lipophilic vitamins and essential fatty acids.

3.4.3.4 Occurrence and composition

3.4.3.4.1 Occurrence

The total content of lipids in common foods is given in Table 3.22. The dietary fats occur almost exclusively as triacylglycerols. However, they may contain 1–10% of the partial esters of glycerol, a smaller amount of phospholipids, and about 1% of accompanying compounds. Humans mainly receive fats by eating plant tissues and animal reserve tissues in which they are stored. According to their origin, fats are divided into:

Table 3.22 Fat contents in common foods.

Foods	Lipid content in %		Food	Lipid content in %	
	Fresh weight	Dry weight		Fresh weight	Dry weight
Foods of animal origin			Bread (white)	0.8–1.1	1.3–1.7
Pork meat (lean)	18	51	Bakery products (white)	2–3	8–12
Pork meat (fatty)	41	75	Confectionary products	14–34	22–44
Beef meat	2–36	9–63	Chocolate	32–38	33–40
Calf meat	3–7	4–10	Fruits[a]	0.2–0.7	1.0–2.8
Poultry (fowl)	1–35	5–50	Leafy vegetables	0.2–1.0	2.8–6.2
Poultry (waterfowl)	17–33	40–65	Root vegetables	0.3–0.4	1.3–1.8
Offal	2–15	8–45	Potatoes	0.2	0.8
Sausages	25–48	60–65	Walnuts	64	66
Fish	0.4–16	2–44	Almonds	54	57
Milk (full fat)	3.8	30	Soybeans	13–20	14–22
Milk (reduced fat)	1.8	12	Beans	1.6	1.8
Milk (condensed)	9.5	12	Pea	1.4	1.6
Milk (dry, full fat)	27	28	Margarine	80–82	98–99
Cream	34	87	reduced fat content	62–71	98–99
Butter	81	99	low fat content	41–61	98–99
Cheeses	12–28	20–68	very low fat content	31–41	98–99
Egg yolk	33	66	Shortening	80–100	100
Egg white	0.02	0.15	Vegetable oil	100	100
Egg (whole, dried)	42	44	*Other foods*		
Foods of plant origin			Mayonnaise	80	95
Wheat flour (whole-grain)	2.5	2.8	Yeast	0,4	1.3
Wheat flour (white)	1.0–1.5	1.1–1.7	Mushrooms	0.4	3.3

[a]The average fat content of avocado cultivars is 14.6% (the fat content in seeds is given in Table 3.25).

- plant fats;

- animal fats;

- other fats.

In plants, fats are mainly found in seeds, but also in the pericarp layers, the outside exocarp layer (peel), the middle layer (pith), and the inner endocarp layer. Fat is also present in the germs of seeds, in which starch is the main reserve substance (e.g. the germ of cereals). Fat composition is due mainly to the fatty acid content, but specific properties of lipases also play a role.

Animal fats are further divided into the fats of terrestrial animals (including milk and depot fats) and seafood (including the fat of marine mammals, such as whales, and fish oils). In animal products, it is mainly the subcutaneous adipose tissue that is consumed, but fat is also stored in the muscle and offal, and in the liver of some fish (such as cod).

Fats are also present in microorganisms, higher fungi, algae, and other organisms, but these sources are of negligible significance in human nutrition.

In recent decades, a belief in the need to reduce fat intake in the diet has become widespread, and this is reflected in the availability of food products. Data on the fat content of meat in Table 3.22 relate to traditional products, but new breeds of pigs have very low fat content in muscle (only 2–3%). Also, new breeds of dairy cows have somewhat reduced fat content of milk (only 3.5–3.6%). Specially processed cheeses can have fat contents significantly lower than 20%. Palatability in such products is achieved by increasing their viscosity using soluble proteins or polysaccharides. The fat content in fish depends on their type and the time of year at which they are caught; fatty fish include eel, carp, and herring, whilst cod is an example of a lean fish. Emulsified fats often now have only 20–70% fat, whilst margarine-type products and butter spreads tend to have only 30–40% or even lower amounts of fat. Reduced-fat mayonnaise and yoghurts are also available, along with many other reduced-fat products.

3.4.3.4.2 Composition

Previously, the belief was that fats were mixtures of simple triacylglycerols. However, it was then found that mixed triacylglycerols exist and that they are the main components of natural fats. It was also thought that fatty acids in the triacylglycerol molecules occurred randomly, but neither this assumption was correct. According to another theory, the composition of fatty acids depends on the distribution of saturated (S) and unsaturated (U) fatty acids in triacylglycerols; therefore, four possible types of triacylglycerols (G = glycerol) with random distributions of fatty acids were possible: SSS (GS_3), SSU (GS_2U), SUU (GSU_2), and UUU (GU_3). Saturated fatty acids (mainly palmitic and stearic acids) and unsaturated fatty acids (mainly oleic and linoleic acids) might substitute for each other. Because the experimental data did not correspond to this assumption, it was proposed that each fatty acid tends to occur in as many molecules of triacylglycerols as possible. According to this theory, a fatty acid that occurs in less than 33% of total fatty acids can be represented in each triacylglycerol molecule only once. If its ratio is between 33 and 67%, it can occur in the triacylglycerol molecule once or twice. Only if it is present in amounts higher than 67% can single triacylglycerols be formed in which this fatty acid occurs three times. Experimental data did support this theory for a long time, but it was necessary to revise it based on the assumption that the first compounds formed are specific 2-monoacylglycerols and their positions C-l and C-3 in the glycerol molecule are then randomly occupied by the remaining fatty acids.

Today, these strict principles are not asserted because acyltransferases are not quite specific and frequent deviations from this scheme occur. There is a high variation in stereoisomers in oils and fats of different biological origin, typically with strong species specificity. Lard is unique amongst animal depot fats because it has a strong predominance of palmitic acid in the *sn*-2 position (90–100%), most stearic acid is found in the *sn*-1 position, and position *sn*-3 is rich in oleic acid (Table 3.23). In beef tallow and other bovine adipose tissues, nearly 50% of the fatty acids in the *sn*-2 position are oleic acid; other acids, in decreasing order, are palmitic and stearic acids. Palmitic, stearic, and oleic acids are the major fatty acids in the *sn*-1 and *sn*-3 positions. In eggs, a high degree of asymmetry between positions *sn*-1 and *sn*-3 is found. Palmitic acid is mainly in the *sn*-1 position, whereas >60% of *sn*-2 fatty acids and *sn*-3 fatty acids is oleic acid. Linoleic acid predominates in the *sn*-2 position. Seed oils tend to have polyunsaturated fatty acids in the *sn*-2 position, but relatively little difference can be found between positions *sn*-1 and *sn*-3, although less abundant fatty acids are often concentrated in the *sn*-3 position. Generally, saturated fatty acids (mainly palmitic acid) are in the *sn*-1 position and unsaturated fatty acids (mainly oleic acid) are in the *sn*-2 position. In triacylglycerols of marine mammals, highly unsaturated fatty acids, such as DHA and EPA, are mainly located at the *sn*-1,3 positions. In

Table 3.23 Typical composition of triacylglycerols of pork lard and cocoa butter.

Glycerol position[a]			Content (mol %)		Glycerol position[a]			Content (mol %)	
sn-1	*sn*-2	*sn*-3	Pork lard	Cocoa butter	*sn*-1	*sn*-2	*sn*-3	Pork lard	Cocoa butter
3 saturated fatty acids					S	S	U	1.2	0.1
P	P	P	0.3	0.3	P	U	P	0.1	14.1
P	P	S	1.7	0.9	P	U	S	0.7	39.3
S	P	S	2.4	0.6	S	U	S	1.1	27.4
P	S	P	0.2	0.2	*2 unsaturated fatty acids*				
P	S	S	0.1	0.4	U	P	U	36.5	0.0
S	S	S	0.2	0.3	U	S	U	2.4	0.0
1 unsaturated fatty acid					P	U	U	2.9	6.4
P	P	U	6.5	0.1	S	U	U	8.3	8.9
S	P	U	18.9	0.2	*3 unsaturated fatty acids*				
P	S	U	0.4	0.1	U	U	U	16.1	0.7

[a] P, saturated fatty acid C_{16} and lower; S, saturated fatty acid C_{18} and higher; U, unsaturated fatty acid.

Table 3.24 Main groups of triacylglycerols in some fats and oils (in mol %).

Fat or oil	Type[a]			
	SSS	SSU	SUU	UUU
Coconut[b]	82	18	0	0
Palm kernel[b]	61	38	1	0
Beef tallow	18	42	32	8
Pork lard	5	29	50	16
Cocoa butter	2	82	15	<1
Palm	7	51	33	9
Maize germ	0	4	34	62
Cotton	<1	18	48	34
Peanut	0	9	44	47
Olive	0	5	35	60
Sunflower	0	3	26	71
Soybean	0	5	29	66
Linseed	0	5	43	52
Rapeseed 00	0	0	20	80

[a]S, saturated fatty acid; U, unsaturated fatty acid.
[b]Fat contains mainly lauric and myristic acids.

phospholipids, however, they are primarily bound to the *sn*-2 position where they are used as a starting substrate for eicosanoid synthesis via Lands' cycle. In contrast, the preference of *sn*-1,3 positions for the highly unsaturated fatty acid in triacylglycerols is much lower in marine fish. The simplified composition of triacylglycerols of some other fats and oils is shown in Table 3.24.

The milk fat of mammals is very different from the fat depot, particularly in that it contains much higher amounts of triacylglycerols of fatty acids with shorter hydrocarbon chains and lower amounts of triacylglycerols of unsaturated fatty acids (Table 3.3). Characteristic triacylglycerols of milk fat have a total carbon number of three bound fatty acids C_{44} (7.4%), C_{42} (7.6%), C_{34} (6.7%), and C_{32} (2.9% of triacylglycerols), which can be used for the detection of vegetable and animal fats in milk fat. Other triacylglycerols are: C_{54} (3.3%), C_{52} (7.8%), C_{50} (10.0%), C_{48} (9.1%), C_{46} (7.9%), C_{40} (9.7%), C_{38} (12.9%), C_{36} (12.1%), C_{30} (1.3%), C_{28} (0.7%), C_{26} (0.3%), and C_{24} (0.04%). The last of these triacylglycerols can contain, for example, three bound caprylic acids or one butyric acid, one caprylic acid and one lauric acid, and so on.

The composition of triacylglycerols of plant seeds is usually very different from the composition of triacylglycerols in the pericarp, which is of practical importance for olive oil and palm oil. The difference is shown in Table 3.4. The composition of the storage fat in animals varies considerably, too (Table 3.5). Intermuscular fat of fish and chondrichthyans (sharks and rays) is fairly different from the liver oil (Table 3.9). Subcutaneous fat of mammals usually contains more triacylglycerols with unsaturated fatty acids than visceral fat (found around the viscera, especially the kidneys) in the abdominal cavity, which is known as soft fat and leaf lard. However, these differences are not very significant.

3.4.3.5 Use

Most fats are used for nutritional purposes or as animal feed, either directly or after isolation from the food raw materials. Part of the fat that is consumed is added to the food during cooking or frying or used as a spreadable fat. A further proportion of fat is not visible, and the consumer does not realise that the food contains hidden fats. The ratio of the two types in human diet is approximately 1 : 1.

Oils and fats are also used for non-food purposes in oleochemistry and cosmetics. Some oils are obtained specifically for technological purposes, such as castor oil or tung oil. Oleochemicals from oils and fats manufacturing include fatty acids, fatty alcohols, and other derivatives for the production of surfactants and subsequently detergents, paints, plastics, adhesives, building materials, and many other products. A typical example of the use of oils as fuel for diesel engines is the production of fatty acid methyl esters, especially from rapeseed oil.

3.4.3.6 Procedures for obtaining fats and oils

The various methods for obtaining fats differ according to the raw material, and whether it is of animal or vegetable origin. Special procedures exist for bacterial fats.

3.4.3.6.1 Procedures for vegetable fats and oils

On an industrial scale, the most commonly processed raw material is oil seeds. A summary of the important typical raw materials is given in Table 3.25. In terms of world oil production, the major oils are soybean, palm, rapeseed, sunflower, cotton, peanut, coconut, palm kernel, sesame, and olive oils. A large number of other oils have only local significance, and a number of other oils are obtained solely for

Table 3.25 Fat contents in some raw materials.

Oil	Latin name of plant	Processed part of plant	Lipid content (%)
Coconut	Cocos nucifera	Seed (copra)	63–68
Palm	Elaeis guineensis	Pericarp	44–53
Palm kernel	Elaeis guineensis	Seed (kernel)	50–60
Olive	Olea europea	Pericarp	35–70
Olive kernel	Olea europea	Seed (kernel)	30–45
Almond	Prunus amygdalus	Seed	45–53
Hazelnut	Corylus avellana	Seed	50–65
Avocado	Persea americana	Seed	10–30
Sunflower	Helianthus annuus	Seed	22–36[a]
Peanut	Arachls hypogaea	Seed	45–55
Safflower (carthamus, kurdee)	Carthamus tinctorius	Seed	25–37
Sesame	Sesamum indicum	Seed	44–54
Cotton	Gossyplum hirsutum	Seed (unpealed)	15–24[b]
Poppyseed	Papaver somniferum	Seed	36–50
Rapeseed (traditional)	Brassica napus[c]	Seed	38–45
Rapeseed (low erucic acid)	Brassica napus[d]	Seed	30–40
Mustard	Sinapis alba[e]	Seed	30–42
Soybean	Soja max	Seed	17–22
Linseed	Linum usitatissimum	Seed	35–45
Perllla	Perllla frutescens	Seed	42–51
Hempseed	Cannabis sativa	Seed	30–35
Maize	Zea mays	Germ	12–20
Wheat	Triticum aestivum	Germ	8–14
Rice	Oryza sativa	Bran	15–20

[a]Kernels contain up to 55% of oil.

[b]Kernels contain 30–38% of oil.

[c]Also known as turnip rape oil, colza oil, ravison oil, sarson oil, and toria oil; produced from seeds of Brassica napus, Brassica campestris, Brassica juncea, and Brassica tournefortii species.

[d]Also known as low erucic acid turnip rape oil, low erucic acid colza oil, and canola oil; produced from low erucic acid oil-bearing seeds of varieties derived from the Brassica napus, Brassica campestris, and Brassica juncea species.

[e]Derived from the seeds of white mustard (Sinapis alba), brown and yellow mustard (Brassica juncea), and black mustard (Brassica nigra).

pharmaceutical, cosmetic, and oleochemical purposes (linseed, castor, and perilla oils in particular). Oils from the pericarp (pulp) of fruits, such as palm and olive oils, are obtained by specific procedures, unlike the oils from seeds.

The vegetable oils from seeds and beans are typically obtained by both physical extraction (mechanical expeller pressing) and chemical extraction, usually with hexane as a solvent (a petroleum-derived fraction without aromatic hydrocarbons), or by a combination of both processes. If the seed oil content is lower than 25%, the oil is obtained by direct extraction. From seeds with higher oil content, most of the oil is obtained by pre-pressing, reducing the fat content to 17–20% in the meal. Pre-treatment is necessary to produce leakage of the oil from the seeds, which consists of milling followed by exposure to water at elevated temperatures. This causes tissue disruption and then the subsequent disruption of cell walls and membranes. The remaining oil in the meal is usually obtained by extraction, which lowers the residual fat content in meal to 1.5–2%. The solution obtained by extraction contains 30–40% of the oil in hexane, the so-called miscela, from which the solvent is evaporated under reduced pressure, and any residual solvent is then removed by steam distillation under reduced pressure. The so-called **crude oils** contain about 95–97% of triacylglycerols, have a high content of phospholipids (up to 2.5% in canola oil and up to 3.5% in soybean oil), contain about 2% unsaponifiable compounds (such as hydrocarbons, steroids, tocopherols, and various pigments), have a higher content of free fatty acids due to the hydrolytic activity of enzymes, and have a higher content of mineral elements. The organoleptic characteristics of crude oils make them unsuitable for consumption in the natural state, and they must therefore be refined.

The so-called **virgin oils**, suitable for direct consumption, are obtained by grinding the material and extracting the oil by mechanical means (pressing). The most important virgin oils are oils from the fruit of the olive tree (*Olea europea*, Oleaceae). According to the International Olive Oil Council (IOOC), there are four commercial types of olive oils: extra virgin olive oil, virgin olive oil, olive oil, and pomace oil. Virgin oils are obtained entirely through physical methods (pressing and heating) under conditions that do not lead to alteration of the oil (see Section 3.4.3.7). Virgin olive oils are only filtered and not refined. Extra virgin olive oil and virgin olive oil contain up to 0.8% and 2.0% of free fatty acids (expressed as oleic acid), respectively. Olive oil is a mixture of refined oil (oils of lower quality are refined by filtration through activated carbon) with virgin oil. Pomace oil (obtained from pomace, which is the ground pulp and seeds, by solvent extraction and other physical methods, followed by refining of the oil thus obtained) is also a blend of refined and virgin olive oil. The oleic acid content of the last two types is up to 1.0%.

3.4.3.6.2 Procedures for animal fats

Animal fats are mostly obtained by the action of hot water, when the fat is washed from the tissue and then separated from the aqueous phase. Previously, animal fats were obtained using hot steam and by direct heating (rendering), as is still practiced in the home. Rendered fats have a distinctive flavour due to the pyrolytic products of proteins contained in adipose tissue, whilst the unrendered fat obtained using hot water is virtually odourless and tasteless.

Milk fat is usually obtained from milk in the form of butter. The milk fraction rich in fat (cream) is separated from the aqueous phase poor in fat (skimmed milk) by centrifugation. The cream is an oil-in-water (o/w) emulsion from which the fat (butter) is mechanically separated by inversion to a water-in-oil (w/o) emulsion, in which milk proteins act as emulsifiers. The residual liquid that remains is buttermilk. Butter contains at least 80% fat and about 2% non-fat portion (proteins, carbohydrates, and other substances), with the remainder being water. It generally has a pale yellow colour, which is dependent on the animal's feed. Some food colourings, most commonly annatto and β-carotene, are used in the commercial manufacturing process. Butter is most frequently made from cow's milk, but can also be manufactured from the milk of other mammals, including sheep, goats, buffalo, and yaks. For specific applications, anhydrous milk fat can be obtained. In India, Pakistan, and the Middle East, anhydrous milk fat, known as *ghee*, is prepared by heating butter to evaporate the water and separate the precipitated proteins.

3.4.3.7 Vegetable oil refining

Unlike animal fats and virgin oils, which are usually used without further treatment, vegetable oils obtained from seeds by mechanical expelling and solvent extraction are refined in order to make them more acceptable to consumers. The refining process includes:

- **degumming** (hydration);

- **deacidification** (neutralisation);

- **bleaching**;

- **deodorisation** or **physical refining**.

The crude oil is first centrifuged or filtered to remove solid particles (impurities, seeds, or parts of the cellular tissue), which are the main causes of undesirable enzyme activities. The next step is hydration, which separates out a substantial portion of the heterolipids (mainly phospholipids) as hydrated gums that are highly insoluble in oil and can be quickly separated out as a hydration sludge. Plant mucilages, carbohydrates, and proteins are also separated. Hydration is based on heating crude oil with water or, in modern processes such as superdegumming and total degumming, with concentrated hydrochloric, phosphoric, or citric acid and then separating the hydration sludge. The hydration sludge is used in the production of lecithin.

In the next step, free fatty acids are separated from the oil by neutralisation (deacidification or alkaline refining) with sodium hydroxide solution (which is also effective in the removal of toxic gossypol from cottonseed oil; see Section 9.11). The free fatty acid content varies widely within 0.5–1.5% in crude canola oils and within 0.3–0.7% in crude soybean oils. After neutralisation, the fatty acid content falls below 0.1%. The resulting soaps are separated as soap stock and typically used in animal feed. Tropical oils, such as palm and coconut oils, contain 3–7% free fatty acids. They are therefore separated together with volatiles by distillation in the end stage of physical refining process.

Bleaching is achieved by adsorption on activated bleaching clays (sorbents of the aluminum silicate type) or in combination with other adsorbents (such as activated carbon, also known as activated charcoal) to remove oil soluble pigments (such as carotenoids and chlorophylls), residual phospholipids, and eventually soap residues resulting from the deacidification process. Volatile substances are removed by deodorisation via steam distillation under reduced pressure. The volatile compounds are mainly responsible for the unpleasant smell and aftertaste of crude oil, so this process provides organoleptically neutral, indifferent oils.

During the deodorisation process, there is a partial loss of tocopherols and sterols. They pass into the distillate, or deodorisation condensate, from which they may be recovered. An undesirable phenomenon in deodorising (physical refining), in addition to partial loss of sterols and tocopherols, is partial geometrical isomerisation of double bonds of polyenoic fatty acids. By adjusting the process parameters, these adverse effects can be minimised. The refining process simultaneously leads to decomposition of hydroperoxides, and other oxidation products of fatty acids and volatile secondary oxidation products are distilled off. The vitamin and carotenoid pigments content is partly reduced, but the nutritional value of the oil is not altered significantly. The resulting refined oil is virtually a pure mixture of triacylglycerols with small amounts of partial esters of glycerol, residual free fatty acids, and some desirable accompanying substances (in particular phytosterols, tocopherols, fat-soluble vitamins, and carotenoids). For example, the refined soynbean oil contains triacyglycerols (>99%), phospholipids (0.003–0.5%), unsaponifiable matter (0.3%), free fatty acids (<0.05%), and traces of other compounds.

Winterisation is the process of removing components with high melting points (such as waxes) from some vegetable oils, for example sunflower, rice bran, and cotton seed oils or partially hydrogenated soybean oil. Vegetable oils then remain clear liquids even when stored at low temperatures for extended periods of time.

3.4.3.8 Crystallisation and emulsification of fats

Vegetable oils are often eaten directly (100% cooking fats or oils). The basic problem is that most vegetable oils are liquid, but what is usually required is solid fats of suitable consistency that have similar properties to traditionally used animal fats, such as butter and lard. Therefore, vegetable oils are processed further to fats of suitable consistency.

Vegetable oil used for the production of margarines and other w/o emulsified spreads and shortenings for use as dough fat or filling fat is called **fat blend**. Fat blend is made by mixing the so-called **structural fat** and liquid oil. Structural fat contains a mixture of triacylglycerols of higher melting points, which crystallise in the blend. The fat blend for the manufacture of shortenings requires the formation of a dispersion of crystals in oil (β-crystalline modification; see Section 3.4.3.9), whilst that for the manufacture of margarines requires the formation of a spatial network (β'-crystalline modification). Liquid triacylglycerols (liquid oil) are adsorbed (immobilised) on the surface of the crystals or in the spatial network of crystals. Structural fat can be produced by partial catalytic hydrogenation, fractionation, or ester interchange (transesterification). The main process of structural fats production is transesterification of triacylglycerols.

A classic emulsified fat is **margarine**, a w/o emulsion, which contains at least 80% fat, with the remainder being the aqueous phase (minimum 16% of total emulsion content by weight). It is obtained by mechanically emulsifying fat blends, containing an emulsifier (usually monoacylglycerols with the addition of diacylglycerols, often mixed with lecithin), with an aqueous phase, and by subsequent crystallisation and emulsification Margarines are produced either as spreadable fats, often with reduced fat content, or in a harder consistency for baking purposes.

Another type of emulsified fat is **mayonnaise**, an o/w emulsion, where the emulsifier is egg yolk (the phospholipids present in it). Today, many similar products with lower oil contents are produced in abundance. A similar composition is found in tartar sauce and some salad dressings, but the fat content is usually lower in these products.

Non-food fat emulsions are used for paints, cosmetics, pharmaceuticals, and other purposes.

3.4.3.9 Composition, crystallisation, and properties of fats and oils

Fats can be hard and brittle substances (cocoa butter), hard substances (beef tallow), gooey solids (pork lard, palm oil), or viscous liquids (vegetable oils). The melting and solidification points of some fats are listed in Table 3.26. The range of values is mainly related to the variability of the composition of fatty acids and triacylglycerols.

Table 3.26 Melting and solidification points of solid fats.

Fat	Melting point (°C)	Solidification point (°C)	Fat	Melting point (°C)	Solidification point (°C)
Pork lard	28–40	22–32	Coconut oil	20–28	18–23
Beef tallow	40–50	30–38	Palm oil	30–37	27–43
Cow's milk fat	28–38	15–25	Cocoa butter	32–36	21–27
Breast milk fat	30–32	22–23	Hydrogenated vegetable oils	28–42	23–36

The molecules of triacyglycerols in solid fats display polymorphism, having different crystalline forms with different melting points. Saturated monoacid triacylglycerols crystallise in three fundamental polymorphic forms, namely α (the least stable form), β', and β (the most stable form). Compared with the saturated monoacid triacylglycerols, polymorphism of unsaturated monoacid triacylglycerols is more complicated, because of diverse variations in the number and the position of double bonds of their acyl chain moieties. Triacylglycerols containing unsaturated fatty acid moieties or saturated diacid moieties often exhibit multiple β' or β forms. The basic polymorphism of saturated–unsaturated mixed acid triacylglycerols, which are the main components of vegetable fats and fish oils, is more complicated than that of the saturated monoacid triacylglycerols. For example, liquid tristearin generates on cooling first the less stable α-polymorphic form, which is, on further cooling, transformed via the β'-form into the most stable β-form. Heterogeneous fats containing different fatty acids of different chain lengths and fats with asymmetric distribution of fatty acids (SSU or UUS types) tend to occur in the β'-modification and are transformed to β-modification slowly. Homogeneous fats containing fatty acids of roughly the same length and fats with a symmetrical distribution of fatty acids (SUS type) are transformed into the β-modification very quickly. Crystallisation of saturated–unsaturated mixed acid triacylglycerols leads from α to β_1 through γ, β', and β_2-forms of highest density with the highest melting points (Table 3.27). Except for the composition of the lipid phase, the arrangement of the molecules into the crystalline state also depends on such factors as the cooling rate and the agitation rate.

The polymorphic forms differ in the arrangement of fatty acid hydrocarbon chains in the crystal lattice. All hydrocarbon chains are packed in one of a few possible ways. The α-modifications of triacylglycerols form crystals in a hexagonal, β'-modifications in an orthorhombic and β-modifications in a triclinic crystal lattice. In the lattice, triacylglycerol molecules are oriented in a chair or tuning fork configuration. Crystals of triacylglycerols in the α-modification containing residues of the same fatty acids tend to form tuning fork configurations with the fatty acid chains perpendicular to the crystal planes. Crystals in the β'-modification have tuning fork configuration with angled fatty acids chains. Chains of fatty acids in positions C-l and C-3 of glycerol are located opposite to the acid chain in position C-2 and inclined to the crystal planes. Crystals of triacylglycerols in the β-modification have chair configuration. The molecules are packed in such a way that two chair-like molecules build up a dimeric unit. The bilayer structure then consists of four triacylglycerol molecules or two dimers (type β_2). The triple chain layer (type β_3) occurs in saturated–unsaturated mixed acid triacylglycerols when one fatty acid chain in a triacylglycerol differs significantly in length by four or more carbons or where one fatty acid is unsaturated (Figure 3.3).

Table 3.27 Melting points of polymorphic forms of triacylglycerols.

Triacylglycerol	Melting point of polymorphic forms (°C)				Triacylglycerol	Melting point of polymorphic forms (°C)			
	γ	α	β'	β		γ	α	β'	β
Tricaprylin	–	–	–21.0	8.3	Trilinolein	–	–45.0	–13.0	–10.0
Tricaprin	–	–15.0	18.0	31.5	Trilinolenin	–	–44.6	–	–24.0
Trilaurin	–	15.0	35.0	46.5	1,2-Dipalmitoyl-3-oleoylglycerol	18.5	28.9	–	34.8
Trimyristin	–	33.0	46.5	57.0	1,3-Dipalmitoyl-2-oleoylglycerol	18.0	26.5	–	38.5
Tripalmitin	–	45.0	56.5	65.5	1,2-Dipalmitoyl-3-linoleylglycerol	–	–	–	26.5
Tristearin	–	55.0	65.0	72.5	1,2-Distearoyl-3-oleoylglycerol	30.4	43.5	–	–
Triolein	–	–32.0	–12.0	5.5	1,3-Distearoyl-2-oleoylglycerol	22.4	37.0	41.5	44.3

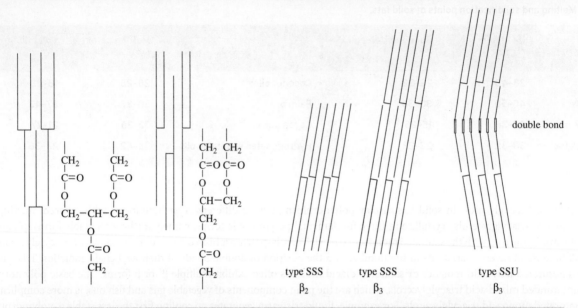

Figure 3.3 Left: Tuning fork (β'-modification) and chair configuration (β-modification) of triacylglycerols in crystal lattices. Right: Arrangement of polymorphic forms of triacylglycerols in double- and triple-chain layers.

Coconut and palm kernel oil, palm oil, traditional rapeseed oil and cottonseed oil, beef tallow, and butter tend to preferentially crystallise in the metastabile β'-modification, which has characteristic small needle-like crystals. These triacylglycerols characteristically form clusters and aggregates to create the crystal spatial network; ideally, they form a gel structure, which shows thixotropy. Fats crystallising in this modification are suitable for the production of margarines and shortenings. In dough and pastry, these fats help incorporate air bubbles and suspended particles of flour and sugar.

Olive, peanut, sunflower, soybean, low erucic acid rapeseed, and safflower oils, cocoa butter, and pork lard crystallise in the stable β-modification, for which large fat crystals conferring grainy texture to the fats are characteristic. Such crystals arise, for example, in pork lard during storage in a refrigerator. Association of crystals in the β-modification is significantly weaker than in the β'-modification, and therefore fats in this configuration of the crystal lattice create not a crystalline network, but a simple dispersion of fat crystals in liquid triacylglycerols, which is a sol (see Section 7.8.3.1).

In many food products, it is important to control lipid crystallisation in order to obtain the desired number, size distribution, polymorph, and dispersion of the crystalline phase. Proper control of the crystalline microstructure leads to products with the desired textural properties and physical characteristics. For example, tempering of chocolate prior to moulding or enrobing is designed to control crystallisation of the cocoa butter into a large number of very small crystals that are all in the desired polymorphic form. When controlled properly, the cocoa butter crystals in chocolate contribute to the desired appearance (the shine or gloss), meltdown rate upon consumption, and stability during shelf life (prevention of fat bloom, the grey film of fat on the surface). Natural cocoa butter has a total of six polymorphs (γ, α, β'$_2$, β'$_1$, β$_2$, and β$_1$), the melting points of which range from 17.3 to 36.3 °C. Only the fifth form (β$_2$), with a layer corresponding to three lengths of the fatty acid chains (the main triacylglycerols are StOSt, POSt, and POP; Table 3.23) and melting at 33.8 °C, has the desirable organoleptic properties. To ensure that this form is created in the production of chocolate, tempering is required. Liquid chocolate is allowed to cool, and when the fat begins to crystallise, it is heated to a temperature just below the melting point of the desired polymorph, and the undesired polymorphic forms melt. Chocolate is stirred at this temperature for some time to ensure that a greater proportion of fat is crystallised into small fine crystals. Inappropriately tempered chocolate, or chocolate exposed to fluctuations in temperature, creates a fat bloom as a result of the fat transformation into more stable polymorphic forms, which then crystallise on the chocolate surface.

3.4.4 Esters of polyhydric alcohols

This group includes industrially produced substances. Esters of sorbitol or saccharose with one or two molecules of fatty acids are used as emulsifiers. Sucrose esters with 5–8 molecules of fatty acids are used as fat substitutes (see Section 3.8.2.2.2), with the advantage that they have physical properties similar to fat and give the same sensory impression, but they are non-utilisable in the human body and therefore do not supply the body with energy (see Section 11.5.2.1.5).

3.5 Heterolipids

Unlike homolipids, heterolipids contain not only bound fatty acids and alcohols, but also other components. They are further subdivided into:

- **phospholipids;**
- **glycolipids;**
- **sulfolipids.**

The most important group of phospholipids are **glycerophospholipids**. Homolipids derived from glycerol (such as mono-, di-, and triacylglycerols), glycerophospholipids, and **glyceroglycolipids** together constitute one class of lipids known as **glycerolipids**. Another major group of lipids are **sphingolipids**, which are not derived from glycerol and are not very important in nutrition. Sphingolipids include fatty acid amides derived from nitrogen-containing alcohols **sphingosines**. Sphingolipids are further divided into **sphingohomolipids** (*N*-acylsphingosines called **ceramides**) and **sphingoheterolipids** (including **sphingophospholipids** and **sphingoglycolipids**). Glycerophospholipids (formerly known as phosphoglycerides), glyceroglycolipids, and sphingolipids are structural lipids of biological membranes.

3.5.1 Phospholipids

3.5.1.1 Structure and nomenclature

The most important group of phospholipids are glycerophospholipids, which are glycerol derivatives esterified with both fatty acids and phosphoric acid. Glycerophospholipids are divided into three groups according to the number of bound fatty acids, other types of bonds, and the structure of the bound components:

- **phosphatides;**
- **lysophosphatides;**
- **plasmalogenes.**

3.5.1.1.1 Phosphatides and lysophosphatides

Phosphatides are derivatives of phosphatidyl residues based on 1,2-diacyl-*sn*-glycerol, which contains bound phosphoric acid on the C-3 hydroxyl. The basic compound is 3-*sn*-phosphatidic acid (1,2-diacyl-*sn*-glycerol 3-phosphoric acid) (**3-38**). If only position C-1 is occupied by an acyl, the resulting residue derived from 1-acyl-*sn*-glycerol is called lysophosphatidyl. The corresponding acid is 3-*sn*-lysophosphatidic acid (1-acyl-*sn*-glycerol 3-phosphoric acid) (**3-39**).

In phosphatides and lysophosphatides, phosphoric acid is further esterified by some hydroxy compounds, namely choline (*N,N,N*-trimethylethanolamine), ethanolamine (2-aminoethanol), L-serine, *myo*-inositol, and glycerol.

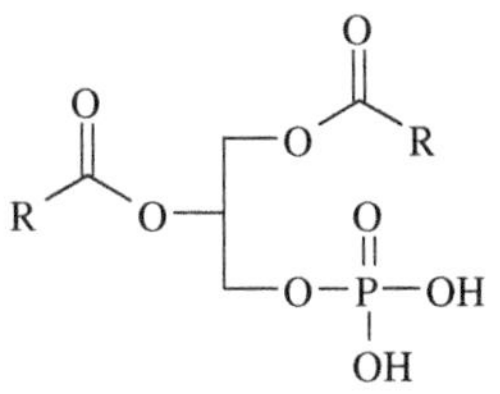

3-38, phosphatidic acid

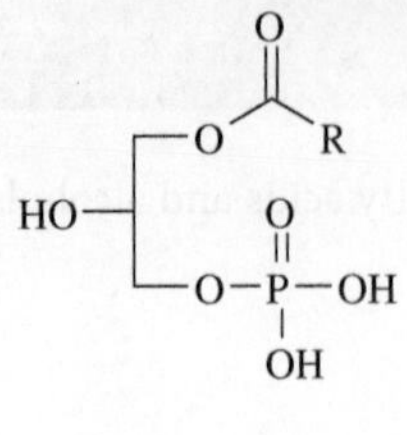

3-39, lysophosphatidic acid

The most frequently occurring phosphatide is the ester with amino alcohol choline, which is known as (3-*sn*-phosphatidyl)choline or l,2-diacyl-*sn*-glycero-3-phosphocholine, or phosphatidylcholine for short (**3-40**). Previously, this ester was known as lecithin, but this name should no longer be used for the pure compound, but only for the industrial concentrate of phospholipids. If ethanolamine (colamine) is bound to the phosphatidyl residue, the resulting phospholipid is called (3-*sn*-phosphatidyl)ethanolamine or l,2-diacyl-*sn*-glycero-3-phosphoethanolamine (phosphatidylethanolamine, **3-40**). Previously, the fraction containing phosphatidylethanolamine, together with other phospholipids, was known as cephalin. This name has been used only rarely in non-chemical literature, and should not be used as it is a poorly defined mixture of compounds. Another phospholipid is (3-*sn*-phosphatidyl)-L-serine or l,2-diacyl-*sn*-glycero-3-phospho-L-serine (phosphatidylserine, **3-40**).

The compound of this group that does not contain nitrogen is (3-*sn*-phosphatidyl)-*myo*-inositol or 1,2-diacyl-*sn*-glycero-3-phospho-*myo*-inositol (or phosphatidyl-*myo*-inositol for short, **3-40**), which may have the second phosphoric acid bound to *myo*-inositol via the C-4 hydroxyl group (see Section 4.3.1.1.2). Another compound of this group that does not contain nitrogen is (3-*sn*-phosphatidyl)-*sn*-glycerol or 1,2-diacyl-*sn*-glycero-3-phospho-*sn*-glycerol (phosphatidylglycerol, **3-40**). Phosphatidyl-glycerol is found as a minor component in various animal and plant tissues, particularly in chloroplasts. The structurally related compound is cardiolipin (**3-40**), an acyl derivative of 1′,3′-di-*O*-(3-*sn*-phosphatidyl)-*sn*-glycerol, which was first isolated from cardiac muscle, hence its name.

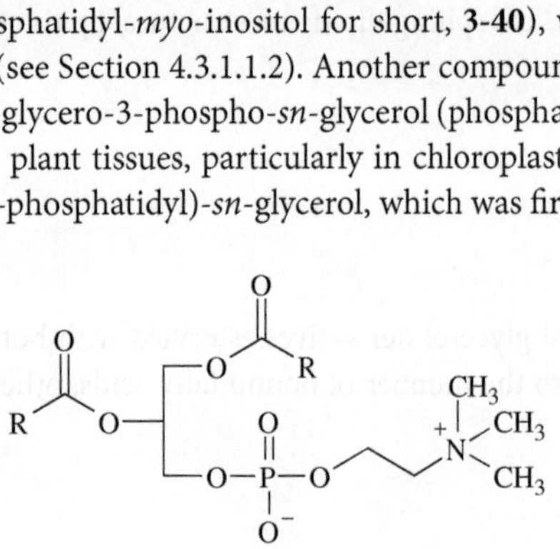

phosphatidylcholine

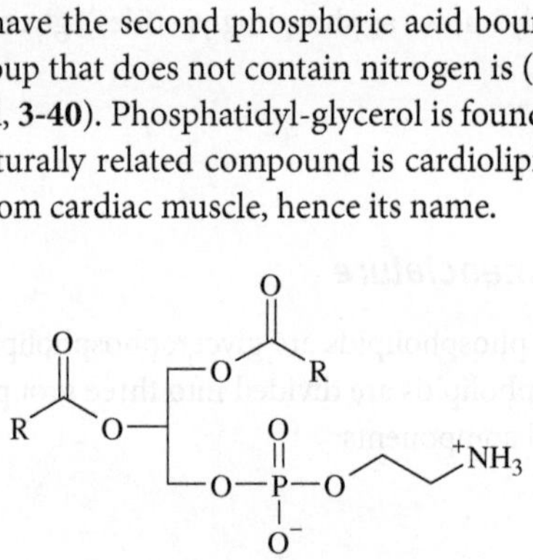

phosphatidylethanolamine

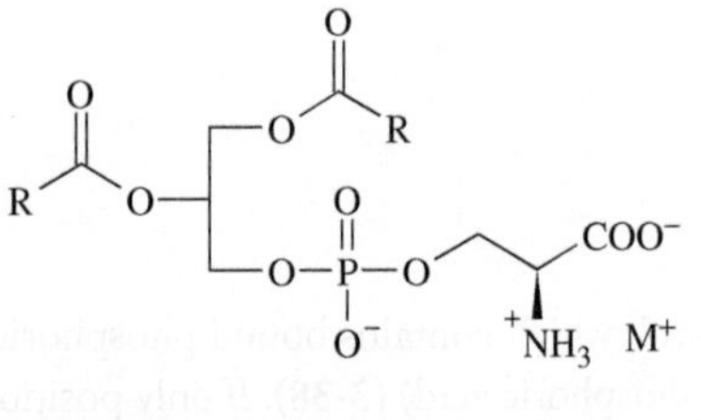

phosphatidylserine (M⁺ = counter cation)

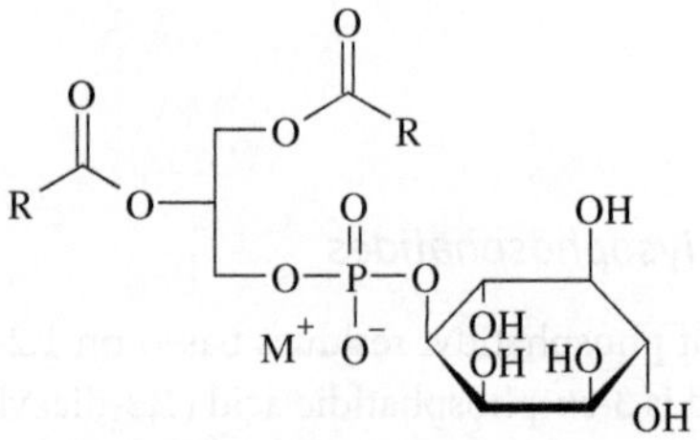

phosphatidyl-*myo*-inositol (M⁺ = counter cation)

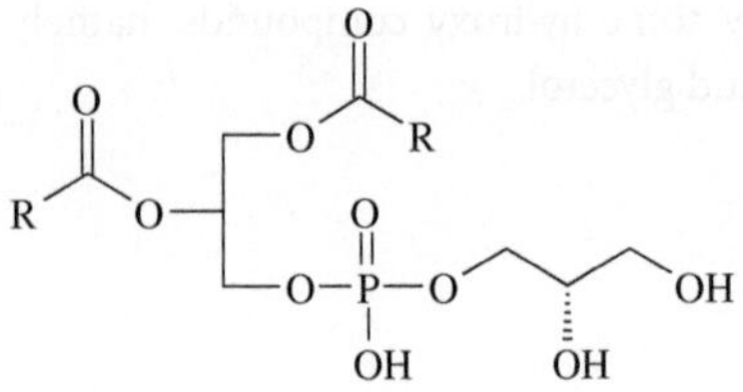

phosphatidylglycerol

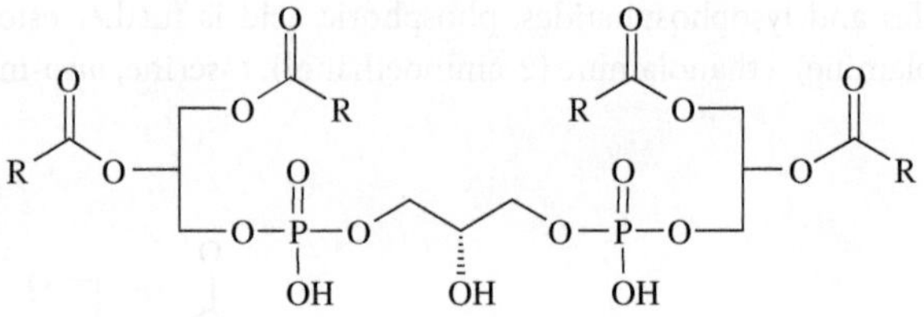

cardiolipin

3-40, phosphatides and derived compounds

3.5.1.1.2 Plasmalogens

The acyl in the *sn*-1 position of glycerol in plasmalogenes (phosphate esters of glycerylethers) is replaced by a higher aliphatic aldehyde bound as a hemiacetal. It usually occurs in a dehydrated form, but may be also present as a hydrate (**3-41**). If the *sn*-3 position of glycerol is esterified with phosphoric acid then this plasmalogen is formally derived from 3-*sn*-plasmenic acid, the systematic name for which is 2-acyl-l-(alk-1'-en-1'-yl)-3-*sn*-glycerol 3-phosphate (**3-41**). Depending on the nitrogenous bases bound in plasmalogenes, 2-acyl-1-(alk-1'-en-1'-yl)-3-*sn*-phosphocholine (for short plasmenylcholine), 2-acyl-1-(alk-1'-en-1'-yl)-3-*sn*-phosphoethanolamine (plasmenylethanolamine), and 2-acyl-l-(alk-1'-en-1'-yl)-3-*sn*-phospho-L-serine (plasmenylserine) are recognised. The corresponding compounds with a saturated chain in the *sn*-1 position also exist. They are derived from 3-*sn*-plasmanic acid; that is, 2-acyl-1-alkyl-*sn*-glycerol 3-phosphate (**3-41**). Such a compound is, for example, 2-acyl-1-alkyl-3-*sn*-phosphocholine (briefly plasmanylcholine). Plasmalogenes probably show cytostatic effects based on the inability of tumour cells to metabolise their molecules.

glyceryl ether
dehydrated form

glyceryl ether
hydrated form

plasmenic acid

plasmanic acid

3-41, glycerylethers and their phosphates

3.5.1.1.3 Sphingophospholipids

An important group of phospholipids, the sphingophospholipids, do not contain glycerol, but the N-analogues of glycerol generally known as sphingosines. Sphingosines (2-amino-1,3-dihydroxyalkanes) are bases with a long chain substituted on the nitrogen atom by acyls of fatty acids with 14–26 carbon atoms. In mammals, this base is usually (*E*)-sphing-4-enine, (2*S*,3*R*,4*E*)-2-aminooctadec-4-ene-1,3-diol, also known as (*E*)-D-*erythro*-2-amino-1,3-dihydroxyoctadec-2-ene or sphingosine (**3-42**). Plants contain up to eight sphingoid-C$_{18}$ bases derived from D-*erythro*-sphinganine, also known as (2*S*,3*R*)-sphinganine. As a result of *cis*- or *trans*-dehydrogenation at C-8, the prevalent regioisomers are (*E/Z*)-sphing-8-enine (C18:1), (4*E/*8*E/Z*)-sphinga-4,8-dienine (C18:2), and (8*E/Z*)-4-hydroxysphing-8-enine. In bacterial lipids, a common component is 2-aminopropane-1,3-diol derived from L-serine, which is called serinol (**3-42**). In the yeast *Saccharomyces cerevisiae*, 4-hydroxysphinganine predominates, which is known by its systematic name of (2*S*,3*S*,4*R*)-2-aminooctadecan-1,3,4-triol or phytosphinganine, C18:0 (**3-42**).

The amino group of sphingosine linked to a fatty acid via an amide bond occurs in the *N*-acylsphingosines known as ceramides (**3-42**), which belong to the homolipids. If the primary hydroxyl group of ceramide (sphingosine) is esterified with phosphoric acid, the resulting phospholipid is called ceramide phosphate (**3-42**). Ceramide phosphate esterified with choline is known as sphingomyelin (or ceramide phosphorylcholine) (**3-42**). Sphingomyelin is thus the sphingolipid analogue of phosphatidylcholine.

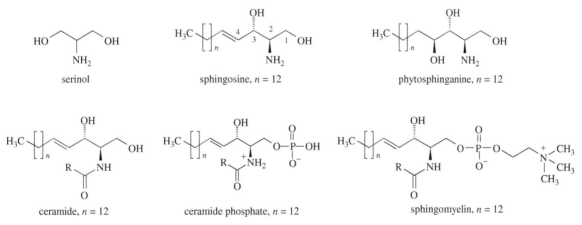

serinol

sphingosine, *n* = 12

phytosphinganine, *n* = 12

ceramide, *n* = 12

ceramide phosphate, *n* = 12

sphingomyelin, *n* = 12

3-42, sphingosines and sphingophospholipids

The composition of fatty acids linked to phospholipids differs from the fatty acid composition of lipid reserves of the same organism. Phospholipids typically contain higher amounts of palmitic acid, usually bound in the *sn*-1 position, and linoleic acid, which is most often bound in the *sn*-2 position. Some phospholipids contain large amounts of higher aliphatic hydroxy acids. The fatty acid composition of phospholipids of different origins is given in Table 3.28. The composition of fatty acids in individual soybean phospholipids is shown in Table 3.29. Phospholipids, particularly phospholipids of seafood, contain small amounts of arsenic analogues (see Section 6.2.3.1).

Table 3.28 Fatty acid composition of phospholipids of different origin (% of total fatty acids).

Fatty acid	Phospholipids				
	Soybean	Sunflower	Rapeseed	Egg	Milk
Palmitoic	14	15	12	32	20
Stearic	3	6	5	4	16
Oleic	14	19	18	46	60
Linoleic	58	46	52	4	2
Linolenic	4	0	1	0	0

Table 3.29 Composition of fatty acids bound in soy phospholipids (% total fatty acids).

Type of phospholipid	Fatty acid				
	Palmitic	Stearic	Oleic	Linoleic	Linolenic
Phosphatidylcholine	14–20	4–6	8–14	58–65	2–6
Position *sn*-1	34	8	19	33	4
Position *sn*-2	1	1	10	78	10
Phosphatidylethanolamine	15–30	3–5	5–9	53–65	3–6
Phosphatidic acid	34	8	12	45	1
Phosphatidylinositol	25–48	8–12	8–9	36–45	2–7

3.5.1.2 Biochemistry, physiology, and nutrition

Phospholipids are essential components of animal and vegetable organisms, in which they are present as a part of the cellular and intracellular membranes and as components of lipoproteins that are stabilised due to the emulsifying ability of phospholipids. The phospholipids content also affects the solubility of LDL in blood.

Phospholipids are formed from precursors by the action of several phospholipases (Table 3.30), which catalyse the connection of individual components to glycerol. The same enzymes catalyse the cleavage of phospholipids. Phospholipids are not an essential food component for humans because the body can synthesise their basic building blocks through at least two different mechanisms. The first uses an appropriate alcohol (such as choline and ethanolamine), activated by binding to cytidine 5′-monophosphate (CMP), which is then connected to phosphatidic acid. The second uses the reaction of an alcohol with 1,2-diacylglycerols (activated by binding to CMP). In the absence of the basic building units, a third possibility is the direct synthesis of phospholipids. Decarboxylation of glycerophosphatidylserine yields glycerophosphatidylethanolamine and, subsequently, produces glycerophosphatidylcholine by methylation of the amino group

Table 3.30 Phospholipases.

Phospholipase	Activity	Origin and characteristics
A	A$_1$: splits fatty acids bound in the *sn*-1 position; A$_2$: splits fatty acids bound in the *sn*-2 position	Pancreas, intestinal mucosa, microorganisms (e.g. *Escherichia coli* and *Saccharomyces cerevisiae*)
B	Splits acyl in the *sn*-2 position, then acyl in the *sn*-1 position	Microorganisms (e.g. *Penicillium notatum*)
C	Splits bonds between the residue of glycerol and bound phosphoric acid	Microorganisms (e.g. *Bacillus cereus*, *Clostridium perfringens* and *Clostridium welchii*)
D	Splits bonds between phosphoric acid and nitrogen base	Vegetables, nuts, legumes

Table 3.31 Contents of phospholipids in foods.

Food	Content (% dry matter)	Food	Content (% dry matter)
Animal materials		Pork lard	0.01–0.1
Brain	5–6	Butter	0.6–1.4
Liver	3–4	*Plant materials*	
Heart	2.6–3.0	Vegetable oils	0.02 (refined)
Kidney	1.6–3.0		0.6–3.2 (raw)
Egg yolk	28	Fruits, vegatables, cereals	0.5–1.5

of the ethanolamine residue. Of course, it is advantageous if the diet contains adequate amounts of phospholipids, because the body has sufficient building material for their re-synthesis. For this reason, various phospholipid preparations are currently recommended for dietetic purposes.

Phospholipids in food are hydrolysed by pancreatic phospholipase A_1 (phosphatidylcholine 1-acylhydrolase) and phospholipase A_2 (phosphatidylcholine 2-acylhydrolase), present as a zymogen that requires activation by trypsin. The products of 1,2-diacyl-*sn*-glycero-3-phosphocholine breakdown are fatty acids and lysophospholipids; that is, 2-acyl-*sn*-glycero-3-phosphocholine and 1-acyl-*sn*-glycero-3-phosphocholine, respectively.

3.5.1.3 Occurrence

Phospholipids are found in all plants and animals as a part of the cellular and intracellular membranes. Their content is about 1% of dry matter (Table 3.31) and is not increased even in depot fat. Some animal tissues are particularly rich in phospholipids, especially the nervous tissue and egg yolk. Egg yolk contains about 28% of phospholipids and constitutes an appropriate source of phospholipids for pharmaceutical purposes. Soybeans are also a relatively rich source of phospholipids, but their phospholipid composition is very different to that in eggs (Table 3.32). Soy and egg phospholipids are used as emulsifiers.

When obtaining oil from oilseeds, a considerable amount of the phospholipids is extracted into the lipid phase, from which it is possible to isolate the phospholipid fraction by adding water or acidic solutions, such as phosphoric or citric acids that release phospholipids from their salts. The phospholipids are hydrated, become less lipophilic, and therefore are excluded from the lipid phase. The obtained dark brown phospholipid concentrate is called **lecithin**.

3.5.1.4 Lecithin

Food lecithin is made from specially selected high-quality crude oils, because its refining is difficult and bleaching using hydrogen peroxide is not permitted in Europe. The most important product for industrial purposes is soy lecithin. Its composition is shown in Table 3.32.

Table 3.32 Distributions of individual phospholipids in the phospholipid fraction (% of total phospholipids).

Compound	Phospholipid				
	Rapeseed	Soybean	Milk	Egg	Liver
Phosphatidylcholine	26–38	24–46	20–29	66–83	43–55
Phosphatidylethanolamine	19–30	21–34	28–36	8–24	23–28
Phosphatidylinositol	12–26	13–21	0–1	0–1	6–9
Phosphatidylserine	trace	5–6	0–8	1–3	3–4
Lysophospholipids	3–9	1–5	trace	3–7	0–2
Sphingolipids	–	–	15–29	1–3	3–5
Phosphatidic acid	0–3	0–14	–	–	0–1

Rapeseed lecithin and sunflower lecithin have a quite different phospholipid fraction composition. Lecithin contains 30–50% of homolipids (acylglycerols), free fatty acids, sterols and tocopherols, and various other substances (e.g. sugars, which represent about 6.5% of soy lecithin, amino acids, and metal ions), as well as chlorophylls and carotenoid pigments. In addition to glycerophospholipids, lecithin contains other heterolipids, of which the major group is glyceroglycolipids (6–7% of soy lecithin).

For some purposes, lecithin is fractionated to increase the phosphatidylcholine content, which is more valued than other phospholipids. Another way to increase the phosphatidylcholine content in lecithin is through enzymatic transesterification after the addition of choline. Alternatively, modification can be achieved by the addition of free fatty acids, which make the product more fluid. Chemical modification of lecithin (hydrogenation and hydroxylation of the double bonds of fatty acids) is used to increase its stability against oxidation. For specific purposes, a phospholipid concentrate (>90% of phospholipids) is produced by a selective extraction that separates neutral lipids, especially triacylglycerols and free fatty acids.

3.5.1.5 Use

The majority of lecithin is used as feed (as extraction meals dipped in hydration sludge), and only a smaller part of a better-quality product is used in the food industry, mostly in the bakery industry. Lecithin is used as a substance for improving dough properties, as it increases the amount of gas retained in the rigid dough foam and reduces the rate of amylose retrogradation. It is also used as an emulsifier in the production of mayonnaise and emulsified fats of the margarine type (it allows the formation of both o/w and w/o emulsions), and to reduce the viscosity of the chocolate mass in chocolate production. The phospholipid concentrate is used in the production of instant powdered products such as dried milk drinks. The use of lecithin for non-food purposes is very diverse. Previously, a substitute for lecithin was manufactured by phosphorylation of diacylglycerols and subsequent neutralisation of the phosphatidic acids formed (see Section 11.5.2.1.1).

3.5.2 Glycolipids

Glycolipids are fatty acid derivatives that contain bound sugars. If they also contain bound glycerol, they are called glyceroglycolipids, whereas if they contain bound sphingosine, they are called sphingoglycolipids. The most common sugar bound in glycolipids is D-galactose (sometimes several galactose units are linked to one another in one glycolipid molecule), and rarely D-glucose or D-fructose, as well as some other sugars. Glycolipids sometimes contain bound phenolic acids and may then act as antioxidants. Glycolipids accompany phospholipids as a part of cell structures and are also bound in lipoproteins. Reactions of glycolipids are similar to reactions of other heterolipids, but they also exhibit the typical reactions of sugars. The hydrolysis of glycolipids in lecithin yields free sugars that react with the free amino groups of amino compounds (in the Maillard reaction), which leads to browning of lecithin and formation of undesirable odours.

3.5.2.1 Glyceroglycolipids

Galactolipids (monogalactosyldiacylglycerols and digalactosyldiacylglycerols) are, along with sulfolipids, major components of lipid membranes of chloroplasts and related organelles in photosynthetic organisms (higher plants, algae, and some bacteria). The predominant compounds are 1,2-di-O-acyl-3-O-β-D-galactopyranosyl-sn-glycerol (**3-43**) and 1,2-di-O-acyl-3-O-(6′-O-α-D-β-galactopyranosyl-D-galactopyranosyl)-sn-glycerol (**3-43**). Higher homologues also exist. For example, trigalactosylglycerol has been found in potatoes and tetragalactosylglycerol occurs in oats. In addition to these glyceroglycolipids, a number of plants and some bacteria contain linear galactolipids in which two to four galactose units are linked by the β-(1 → 6) glycosidic bond. Oats and rice sprouts contain interesting diacylgalactosylglycerols with an estolide bond composed of 15-hydroxylinoleic acid esterified with linoleic acid and bound to the glycerol sn-2 hydroxyl group. The rice germ contains galactolipids together with 1,2-di-O-acyl-3-O-β-D-glucopyranosyl-sn-glycerol; rice seeds also contain triglycosyldiacylglycerols that contain bound galactose and glucose.

The main fatty acids of galactolipids in photosynthetic tissues of higher plants are polyene fatty acids, of which α-linolenic acid dominates. Its concentration is up to 95% of total fatty acids. In pea (18 : 3 plant), α-linolenic acid is practically the only fatty acid bound in positions sn-1 and sn-2 of monogalactopyranosyl-diacylglycerol. The model plant A. thaliana (16 : 3 plants) contains exclusively hexadecatrienoic acid bound in the position sn-2. Palmitic acid is only found in digalactosyldiacylglycerols, and only in small quantities. Other tissues (tubers, roots, and seeds) contain fatty acids with lower numbers of double bonds. Wheat flour, for example, contains 1.5–2.5% of lipids, of which the smaller part (25%) is bound to starch as the so-called starch lipids, whilst the rest are non-starch lipids. The composition of the two types of lipids differs significantly. The main components of starch lipids are glycerophospholipids (89.4%) and glyceroglycolipids (5.0%), whilst triacylglycerols (1.4%) are minor components. The nonstarch lipids contain as the major components triacylglycerols (46.7%); other major components are glyceroglycolipids (26.9%) and glycerophospholipids (14.7%). The detailed composition of wheat flour glyceroglycolipids is shown in Table 3.33.

3.5.2.2 Sphingoglycolipids

In addition to N-acylsphingosines (ceramides) and sphingophospholipids, sphingosine also occurs in the form of glycosides. These compounds belong to the sphingoglycolipids group. The common sugar bound in sphingoglycolipids is D-galactose. An example of sphingosine

Table 3.33 Fatty acid composition of galactolipids of wheat flour (in % of total fatty acids).

Position in galactolipid	Fatty acid				
	16:0	18:0	9c-18:1	9c12c-18:2	9c12c15c-18:3
Monogalactosyldiacylglycerol					
sn-1	11	1	5	81	1
sn-2	trace	trace	9	83	7
Digalactosyldiacylglycerol					
sn-1	26	2	4	63	4
sn-2	2	trace	7	83	7

galactoside is psychosine (1-*O*-β-D-galactopyranosylsphingosine, **3-43**). Glycosides of ceramides (amides of mostly saturated fatty acids and sphingosine) are called **cerebrosides** (1-*O*-β-D-galactopyranosylceramides, **3-43**). Also, sphingophospholipids may still contain bound sugar, again mostly galactose.

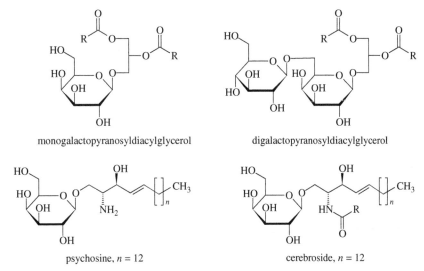

monogalactopyranosyldiacylglycerol   digalactopyranosyldiacylglycerol

psychosine, *n* = 12   cerebroside, *n* = 12

3-43, structures of selected galactolipids and sphingoglycolipids

3.5.3 Sulfolipids and lipid sulfates

Some heterolipids contain bound sulfuric acid, including the **sulfoglycosylsphingolipids**, formerly called sulfatides. Members of this lipid group are glycosylceramide sulfates, exemplified by 1-*O*-β-D-galactopyranosylceramide sulfate (**3-44**). Sulfur in some lipids is also bound as sulfonic acid. An examples of these compounds is 2,3-diacyl-1-(6-deoxy-6-sulfo-β-D-galactopyranosyl)glycerol (**3-44**). Sulfolipids and lipid sulfates accompany phospholipids in nature, and are constituents of some complex lipids.

glycosylceramide sulfates   sulfolipids

3-44, sulfolipids and lipid sulfates

3.5.4 Sialolipids

A physiologically important group of lipids is the sialoglycosphingolipids, known as gangliosides, which contain bound sialic acid or several sialic acid residues (see Section 4.3.3.7). The amount of gangliosides can reach up to 6% by weight of the brain lipids, where they constitute 10–12% of the total lipids of neuronal membranes. These lipids are mostly oligoglycosphingolipids.

3.5.5 Other heterolipids

Vegetable oils often contain a variety of natural phenolic antioxidants, but usually only as admixtures. Phenolic acids with antioxidant effects, such as caffeic (R = H) and ferulic (R = CH$_3$) acids, occur in lipids relatively rarely, usually bound to the glycerol residue (**3-45**).

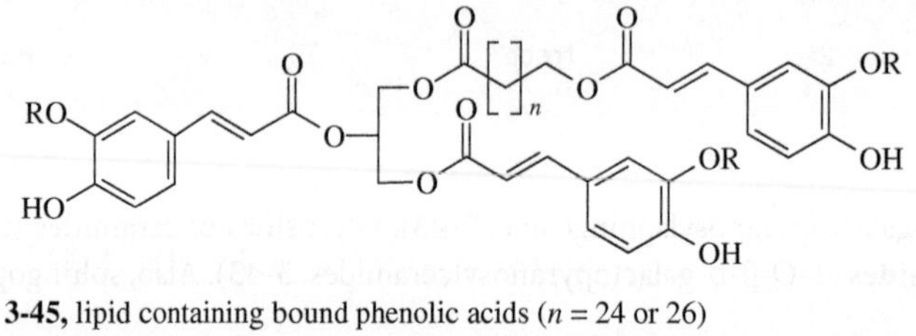

3-45, lipid containing bound phenolic acids (*n* = 24 or 26)

3.6 Miscellaneous simple and complex lipids

3.6.1 Lipoamino acids and fatty acid amides

Many different simple fatty acyl-amino acids, also known as lipoamino acids, are present in animal tissues and as constituents of bacterial lipids. Examples of simple lipoamino acids are *N*-palmitoylglycine (**3-46**) and *N*-oleoylglycine, which have roles in sensory neuronal signalling and regulation of body temperature and locomotion, respectively. *N*-Arachidonylglycine has been shown to suppress inflammatory pain.

Biologically active compounds are also long-chain *N*-acylethanolamides, such as *N*-palmitoylethanolamide and *N*-oleoylethanolamide. For example, *N*-arachidonylethanolamide (anandamide, **3-47**) is a ubiquitous trace constituent of animal and human cells, tissues, and body fluids. Anandamide is an endogenous cannabinoid neurotransmitter that occurs naturally in the brain and in some foods (as chocolate), and which binds to the same brain receptors as the cannabinoids (as tetrahydrocannabinol). The name is taken from the Sanskrit word *ananda*, which means bliss or delight.

N-Alkanoyltryptamides and 5-hydroxytryptamides derived from higher fatty acids occur in wax layers of cocoa shells and coffee beans, respectively (see Section 10.3.4.1.2).

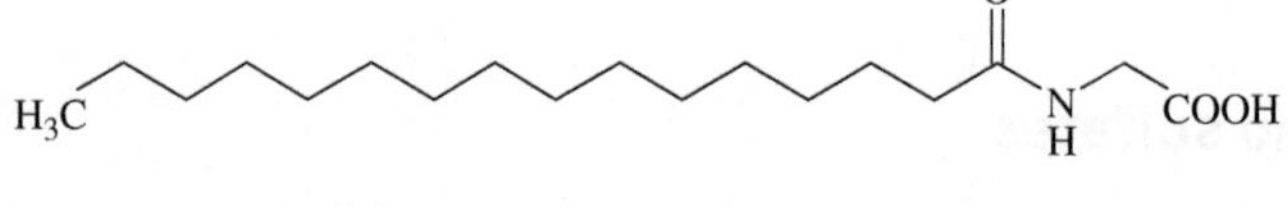

3-46, *N*-palmitoylglycine

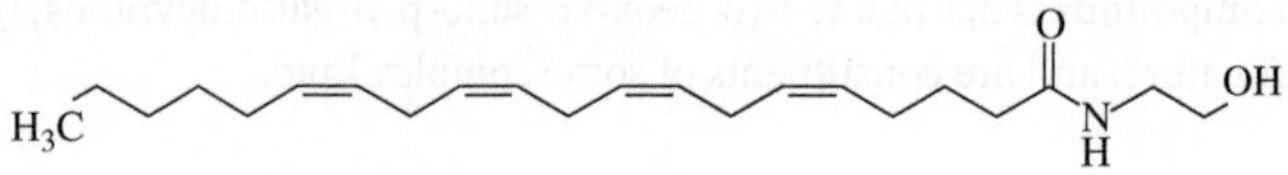

3-47, anandamide

3.6.2 Complex lipids

3.6.2.1 Lipoproteins

Lipoproteins are the most important complex lipids, composed of proteins and lipids. Lipoprotein particles consist of a central core containing non-polar lipids (cholesterol esters and triacylglycerols), surrounded by polar lipids (free cholesterol, phospholipids), which facilitate the link between lipids and proteins. The best examined lipoproteins are serum lipoproteins, whose primary purpose is the transport of

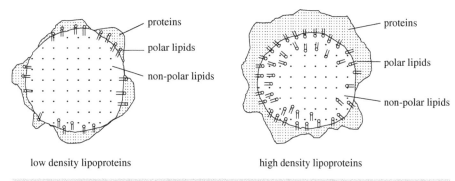

low density lipoproteins   high density lipoproteins

Figure 3.4 Schematic composition of lipoproteins.

Table 3.34 Overview of serum lipoproteins.

Lipoprotein type[a]	Density (kq/l)	Triacylglycerols(%)	Cholesterol/its esters (%)	Phospholipids (%)	Proteins (%)
Chylomicrons	<0.95	85–88	1/3	8	1–2
VLDL	0.95–1.006	50–55	8–10/12–15	18–20	7–10
IDL	1.006–1.019	25–30	8–10/32–35	25–27	10–12
LDL	1.019–1.063	10–15	8–10/37–48	20–28	20–22
HDL	>1.063	3–15	2–10/15–30	32–43	33–57

[a]VLDL, very low-density lipoprotein; IDL, intermediate-density lipoprotein, LDL, low-density lipoprotein; HDL, high-density lipoprotein.

non-polar lipids in blood and extracellular fluids. Historically, serum lipoproteins were divided according to their specific gravity (density). When lipoproteins contain more non-polar lipids, their density is lower, and vice versa. The lower the density of lipoproteins, the weaker and less complex is the protein coat, which maintains lipoproteins in an aqueous solution (Figure 3.4). To date, five major groups of lipoproteins are recognised (Table 3.34):

- **chylomicrons**;

- very low-density lipoproteins (VLDL);

- intermediate-density lipoproteins (IDL);

- **low-density lipoproteins (LDL)**;

- high-density lipoproteins (HDL).

Chylomicrons are particles (100–1000 nm in diameter) that carry ingested lipids (triacylglycerols) from the intestinal wall into the tissues (adipose tissue and muscles), where they are stored, whilst VLDL (30–80 nm) carry the newly synthesised triacylglycerols from the liver to adipose tissue. Milk fat globules (see Section 2.4.5.2) have a similar structure to chylomicrons. IDL (the degradation products of VLDL; 25–50 nm) circulate through the body and transport cholesterol. They also have the ability to promote plaque formation in vessels. LDL (18–28 nm) carry cholesterol from the liver to the body cells, whilst HDL (5–15 nm) carry cholesterol from the body cells to the liver. The LDL are much less stable than the HDL. Lipids from lipoproteins are readily cleared and temporarily stick to the walls of blood vessels.

Lipoproteins are also components of membranes, forming structures with bilayers of oriented molecules of polar lipids (Figure 3.5). Owing to their polar character, induced by the bilayer (double layer) of lipids with an intermediate layer of protein, they are a suitable barrier for the transport of substances between cells or within individual intracellular structures. These properties are also used in the pharmaceutical industry for the emulsification of fats with proteins to form liposomes, which facilitate the transfer and absorption of certain medications. The formation of the liposomes can explain some changes in the properties of foods during processing. Reactions of lipoproteins involve all lipoprotein components, lipids and proteins. Protein reactions with free radicals yield modified structures, where the lipid part is tightly bound to protein. These structures are thought to be the major components of aterosclerotic deposits in blood vessels. Reactions with peroxyl radicals, which are derived from unsaturated lipids, lead to insoluble macromolecular brown-coloured compounds called **ceroids**

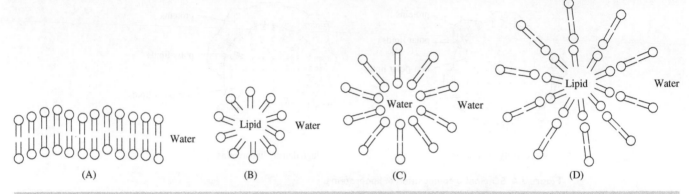

Figure 3.5 Schematic illustrations of lipid membranes in food materials. A) Double lipid cell membrane. B) Nanoparticles bound with simple lipid membrane with lipid phase within the particles. C) Liposome bound by a double lipid membrane with the aqueous phase inside the particle. D) Fat particles stabilised by a lamella structure bound by a double lipid membrane and by a simple lipid membrane with fat phase inside the particle.

Table 3.35 Composition of gangliosides.[a]

Sialic acid units	Oligosaccharide	Ceramide	Sialic acid units	Oligosaccharide	Ceramide
II^3-NeuAc	Lac	Cer	V^3-NeuAc	GgO_5	Cer
II^3-NeuAc	triaose	Cer	II^3NeuGc	$nLacO_4$	Cer
IV^3-NeuAc$_2$	GgO_4	Cer	V^3NeuGc	$iGgO_5$	Cer

[a]NeuAc, acetylneuraminic acid; NeuGc, glycolylneuraminic acid (Roman numerals at the beginning = bond position in sugar; exponent = to which oligosaccharide unit sialic acid is bound; index at the end = number of sialic acid units bound in a row); Lac, lactose; triaose (three sugar units); GgO, globoose, oligosaccharide with any sugar units; n, neobond (1 → 3); i, isobond (1 → 4) (index at the end = number of sugar units); Cer, residue of any sphingoid base (generally identified as ceramide).

or **stichosterins** that are deposited in the walls of blood vessels and other places in the body such as the kidneys or uterus. Similar structures are found in deposits in nerve tissue, and are called age pigments (see Section 4.7.5.6), or **lipofuscin**.

3.6.2.2 Mucolipids

Significant mucolipids are gangliosides containing bound sialic acid, which are present in neural tissues. The individual components are bound by covalent bonds, but also by physical links. Examples of gangliosides are listed in Table 3.35.

3.6.2.3 Lipid clathrates

Clathrates (cage or inclusion compounds) are compounds that consist of a lattice of one type of molecule (a host molecule) trapping another type (a guest molecule) by intermolecular interactions. A clathrate is therefore a mechanical structure without chemical bonds. Such compounds include protein complexes containing β-carotene or lipids (fatty acids) bound in the starch macromolecules. The most famous clathrates are those of long unbranched saturated fatty acids with urea.

3.7 Substances accompanying lipids

Food raw materials and products contain, in addition to lipids, many lipophilic compounds sometimes called **lipoids**, which, during the isolation of lipids, pass into the lipid fraction due to its low polarity. They are therefore called lipid-accompanying substances, although this does not necessarily mean that they accompany lipids in the original material or that they are somehow functionally related to lipids. The lipid-accompanying substances include higher hydrocarbons, higher primary and secondary alcohols, ketones and diketones, various steroids, lipophilic vitamins and pigments, and other lipophilic compounds specific to certain materials.

Table 3.36 Contents of alkanes in sunflower and virgin olive oils.

Alkane	Name	Content (mg/kg oil)		Alkane	Name	Content (mg/kg oil)	
		Sunflower	Olive			Sunflower	Olive
C_{16}	Hexadecane	0.13	0.06	C_{25}	Pentacosane	1.52	17.98
C_{17}	Heptadecane	0.16	0.12	C_{26}	Hexacosane	0.41	2.04
C_{18}	Octadecane	0.90	0.08	C_{27}	Heptacosane	11.19	15.72
C_{19}	Nonadecane	0.12	0.13	C_{28}	Octacosane	2.38	1.84
C_{20}	Eicosane	0.02	0.08	C_{29}	Nonacosane	49.63	12.38
C_{21}	Heneicosane	0.04	0.81	C_{30}	Triacontane	5.52	1.70
C_{22}	Docosane	0.04	1.24	C_{31}	Hentriacontane	47.96	9.41
C_{23}	Tricosane	0.15	18.54	C_{32}	Dotriacontane	1.79	1.54
C_{24}	Tetracosane	0.17	9.54	C_{33}	Tritriacontane	3.60	5.66

Lipophilic polycyclic aromatic hydrocarbons (see Section 12.3.5) and other lipophilic compounds formed in processed foods or compounds derived from external sources are classified as food contaminants.

3.7.1 Hydrocarbons

Higher hydrocarbons are found mainly in waxes, where they can form as much as a few percent of the weight of the wax, but they can also be found in small quantities in common edible fats and oils. They come mainly from the waxy surface layer of seeds. The majority of hydrocarbons have 15–35 carbon atoms in the molecule. The most frequently occurring hydrocarbons are alkanes (*n*-alkanes) with an odd number of carbon atoms. Less common are isoalkanes (mainly with an odd number of carbon atoms) and ante-isoalkanes (mainly with an even number of carbon atoms). Sometimes, higher alkenes are also present. The qualitative and quantitative composition of hydrocarbons is often characteristic of certain foods. For example, the surface waxes of fruits and oilseeds, and therefore also most food oils, contain alkanes C_{27}, C_{29}, and C_{31} as major components, but olive oil is dominated by alkanes C_{23}, C_{25}, C_{27}, and C_{29} (Table 3.36).

The total alkane content in vegetable oils ranges from about 10 to 200 mg/kg. Their amount in sunflower oil is 105–170 mg/kg, and in olive oil approximately 30–100 mg/kg. The amount of alkanes in the cuticular wax of apples is about 33% of the lipid accompanying substances. About 97% of all hydrocarbons in the cuticular wax are represented by nonacosane (C_{29}), about 2% by heptacosane (C_{27}), and the remaining 1% mainly C_{26}, C_{28}, C_{30}, and C_{31}. The cuticular wax of grapefruits is composed primarily of oxygenated compounds, whilst the content of alkanes is only around 1%. Fats of marine organisms (e.g. fish liver oils) contain saturated and unsaturated hydrocarbons C_{15}, C_{17}, and C_{21} as the main components. The total content of alkanes is about 10–30 mg/kg.

Edible oils also contain alkenes and monoterpenic and sesquiterpenic hydrocarbons. For example, the amount of alkenes in olive oils ranges from 0.5 to 2 mg/kg and includes a series of alk-9-enes from C_{22} to C_{27}, heptadec-8-ene, and 6,10-dimethylundec-1-ene. An important compound is the linear triterpenic (C_{30}) hydrocarbon squalene (**3-48**), all-*trans*-2,6,10,15,19,23-hexamethyltetracosa-2, 6,10,14,18,22-hexaene, which is the universal precursor of all triterpenoids and steroids. It was given this name because it was discovered in the liver of sharks (Squalidae). Shark liver oil contains about 30% squalene, about 7% pristane (2,6,10,14-tetramethylpentadecane), and smaller amounts of phytane (2,6,10,14-tetramethylhexadecane). Squalene occurs in small amounts in edible oils and especially in olive oil, where its content is 1–7 g/kg.

In addition to the previously mentioned less volatile higher hydrocarbons, hydrocarbons with a short chain, such as pentane and hexane, may also be found in fats. They are formed by cleavage of fatty acid hydroperoxides produced by oxidation of unsaturated fatty acids. These hydrocarbons, not ranked amongst the lipid-accompanying compounds, are easily removed by heating, so they do not occur in freshly refined oils.

3.7.2 Aliphatic alcohols

Aliphatic primary alcohols C_{12}–C_{36} are found in waxes, but may also occur in trace amounts in ordinary cooking oils, where they pass from the surface layers during seed extraction. Their structure therefore corresponds to the alcohols bound in waxes. For example, the major components of the wax on the surface of sunflower seeds (and of apple wax), along with a number of minor alcohols, are the alcohols C_{24},

C_{26}, C_{28}, and C_{30}. In addition to these higher alcohols, called fatty alcohols, vegetable oils and animal fats also contain lower primary alcohols with 3–11 carbon atoms, which are formed from hydroperoxides by the hydrocarbon chain cleavage. In addition to primary alcohols, higher secondary alcohols also occur in vegetable oils. Their chain length corresponds to the alkanes from which they arise by oxidation. Their hydroxyl group is located near the centre of the hydrocarbon chain. A common secondary alcohol is nonacosan-15-ol. Vegetable oils may also contain traces of the diterpenic alcohol phytol (**3-48**), (2*E*,7*R*,11*R*)-3,7,11,15-tetramethylhexadec-2-en-1-ol, which is, for example, bound in chlorophyll pigments (see Section 9.2.2.5). Free alcohol arises during catabolic processes catalysed by chlorophyllase. Diterpenes, however, are not included in the lipid-accompanying substances.

squalene   phytol

3-48, selected hydrocarbons and alcohols accompanying lipids

3.7.3 Aliphatic ketones

Higher ketones, usually C_{24}—C_{33} compounds, arise as products of oxidation of alkanes via secondary alcohols. Higher β-diketones, mainly C_{31} and C_{33} compounds (**3-49**), accompany higher ketones in smaller amounts.

3-49, structures of aliphatic ketones accompanying lipids

3.7.4 Triterpenoids and steroids

Triterpenoids and steroids are the major group of lipid-accompanying substances in all natural lipids. They belong to a large group of compounds known as **terpenoids** or **isoprenoids** (through the oxidation or rearrangement of modified terpenes). Six isoprene (2-methylbuta-1,3-diene) units give **triterpenes** with 30 carbon atoms, and their modifications yield **triterpenoids**. Several thousand natural triterpenoids include compounds with nearly 200 types of skeletons. **Steroids** are a group of terpenoids arising from triterpenoid precursors. Some steroids play a number of vital functions in living organisms. Examples of such compounds are the steroid hormones of animals, the plant steroid hormones brassinins (**10-12**), steroid glycoalkaloids (see Section 10.3.6.11), and various phytoanticipins and phytoalexins (see Section 10.1). Some compounds have found use in medicine (such as cardioactive steroid glycosides).

3.7.4.1 Structure and nomenclature

Triterpenoids are mostly C_6—C_6—C_5—C_6 tetracyclic or C_6—C_6—C_6—C_6—C_5 and C_6—C_6—C_6—C_6—C_6 pentacyclic compounds, but there are also acyclic, monocyclic, bicyclic, tricyclic, and hexacyclic triterpenoids. Triterpenoids with a β-hydroxyl group at C-3 and a skeleton derived theoretically from the hydrocarbon α-cholestane (**3-50**) are structures with a (20*R*)-configuration and are called **steroids**. Almost all steroids are, in addition to 5α-cholestan-3β-ol (**3-51**), secondary alcohols with a perhydro-1,2-cyclopentanophenanthrene C_6—C_6—C_6—C_5 skeleton, which consists of three six-membered rings A, B, and C in a nonlinear arrangement, with the circle C connected to a five-membered ring D.[4] The steroid skeleton contains a secondary hydroxy group in the C-3 position of ring A, therefore sterols are

---

[4]Numbers 28, 29, and 30 are reserved for methyl groups of triterpenoids at C-4 and C-14. The basic structure has eight chiral centres (carbon atoms C-3, C-5, C-8, C-9, C-10, C-13, C-14, and C-17) and another in the side chain at C-17. Most natural steroids have the same basic structure. The circles AB, BC, and CD are always in the *trans*-position. The steroid molecule is nearly a planar unit and individual substituents lie above or below this plane. The orientation point is a methyl group

actually alicyclic alcohols. The positions C-10 and C-13 always contain methyl groups and the position C-17 contains a side chain with 8–10 carbon atoms. All of these substituents are in the β-position. Individual steroids vary in carbon number and the number and location of double bonds in the side chain attached to C-17. In addition, steroids differ in substituents at C-4 (hydrogen, one or two methyl groups), the number and location of double bonds in ring B, and the stereochemistry of some asymmetric centres. Some related compounds contain the fifth ring E formed by cyclisation of the side chain that is connected to ring D. These compounds are found, for example, as aglycones of many saponins (see Section 10.3.2.2).

Most steroids have trivial names, usually created by the Latin or Greek name of the material from which they were isolated. For less common steroids, systematic names are used, which are derived from basic structures. In addition to 5α-cholestane (**3-50**), other principal structures of hydrocarbons are 5α-lanostane (**3-52**), 5α-ergostane (**3-53**), 5α-campestane (**3-54**), 5α-poriferastane (**3-55**), and 5α-stigmastane (**3-56**), which differ in the structure of the side chain at C-17. The classification reflecting the biochemical origins distinguishes three groups of steroids by the number of methyl groups on carbon C-4. The recognised groups are:

- **4.4-dimethylsterols;**

- **4-methylsterols;**

- **4-demethylsterols** (as methyl groups in position 4 are missing), simply called sterols.

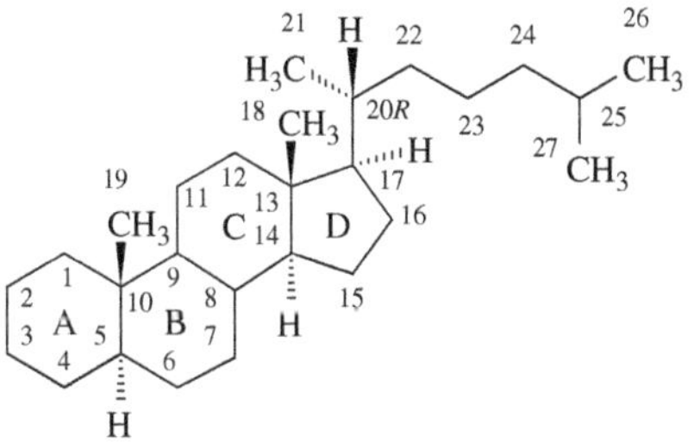

3-50, 5-cholestane

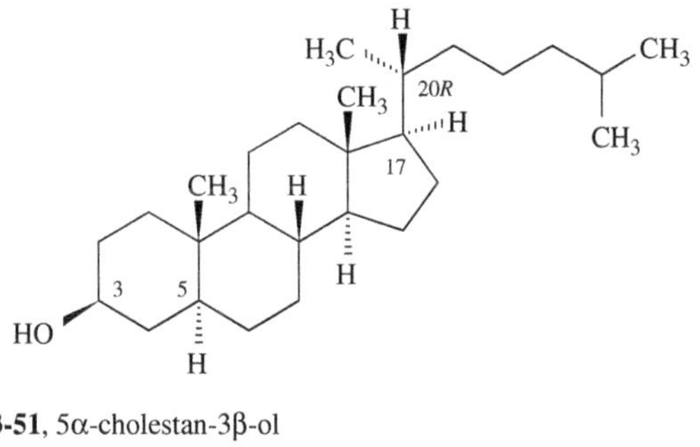

3-51, 5α-cholestan-3β-ol

Steroids of the first two groups that contain 30 carbon atoms are also known as **triterpenic alcohols**. This name is often used for all 4,4-dimethylsterols and 4-methylsterols. Both types of steroids may, however, similarly to 4-demethylsterols, contain less or more than 30 carbon atoms in the molecule.

An overview of significant compounds is given in Table 3.37. Lanosterol, for example, is 4,4,14-trimethyl-5α-cholesta-8,24-diene-3β-ol or 5α-lanosta-8,24-diene-3β-ol. Lanosterol is derived from the hydrocarbon lanostane (**3-52**, 4,4,14-trimethyl-5α-cholestane). Analogously, cycloartenol is 4,4,14-trimethyl-9,19-cyclo-5α,9β-cholesta-24-en-3β-ol or 9,19-cyclo-5α-lanost-24-en-3β-ol (**3-58**), campesterol is (24R)-24-methylcholesta-5-en-3β-ol or campest-5-en-3β-ol (**3-82**), stigmasterol is (24S)-24-ethylcholesta-5,22-dien-3β-ol or

---

(C-19) bound at C-10, which is always above the plane. Substituents that are in *cis*-position to this methyl group (they also lie above the plane) are β-substituents, whilst those in *trans*-position are called α-substituents. The name must always indicate the position of hydrogen on C-5, which in this case is in 5α position.

(*E*)-stigmasta-5,22-dien-3β-ol (**3-85**), β-sitosterol is (24*R*)-24-ethylcholest-5-en-3β-ol or stigmasta-5-en-3β-ol (**3-84**), and ergosterol is (22*E*,24*R*)-24-methyl-cholesta-5,7,22-trien-3β-ol or (*E*)-ergosta-5,7,22-trien-3β-ol (**3-90**).

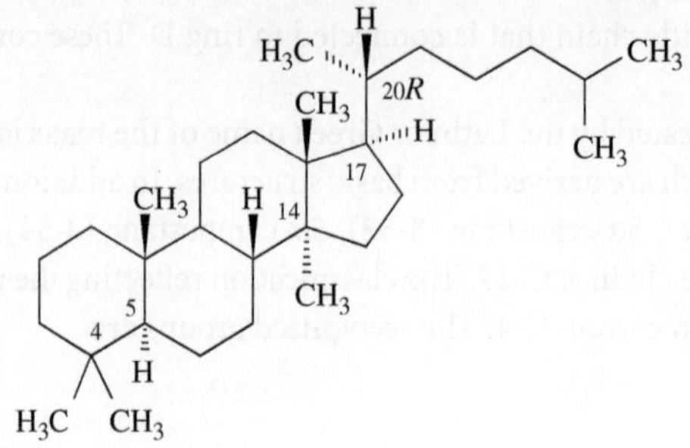

3-52, 5-lanostane

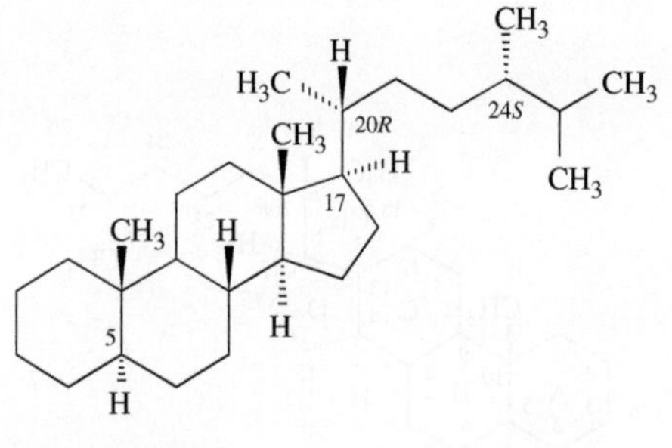

3-53, 5-ergostane

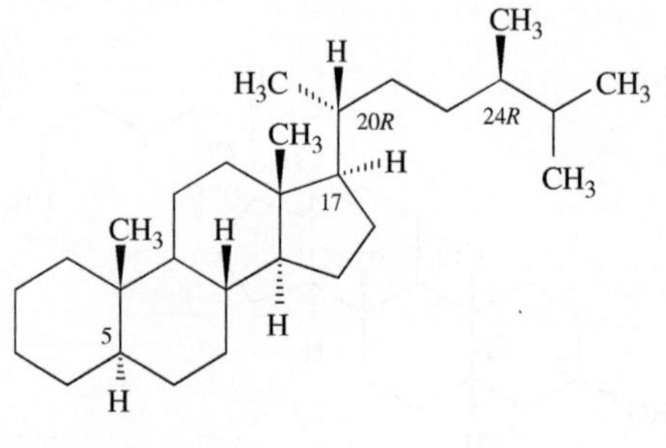

3-54, 5-campestane

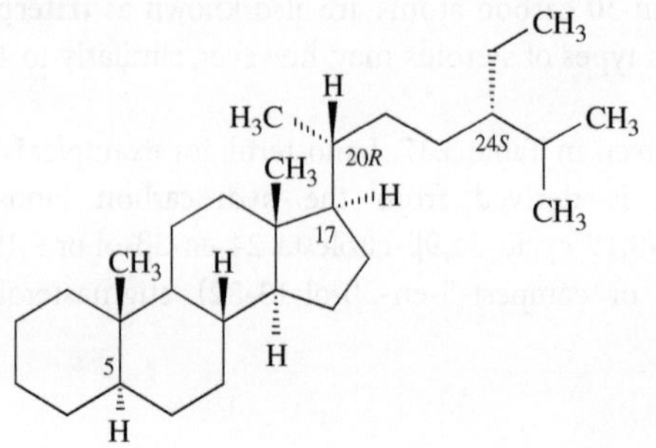

3-55, 5-poriferastane

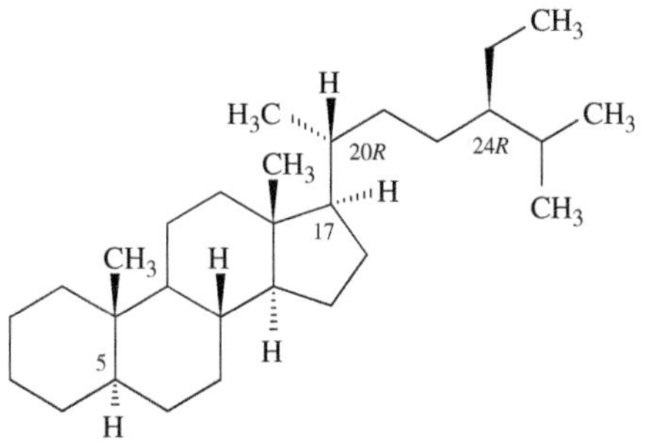

3-56, 5-stigmastane

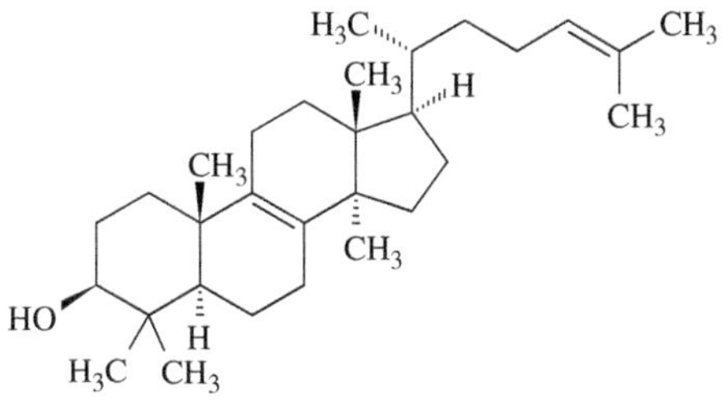

3-57, lanosterol

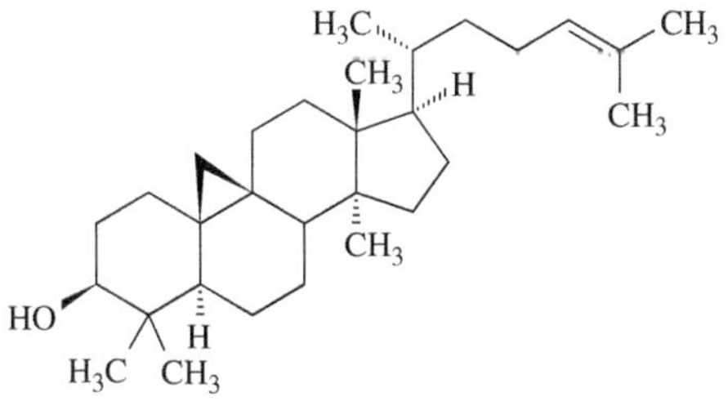

3-58, cycloartenol

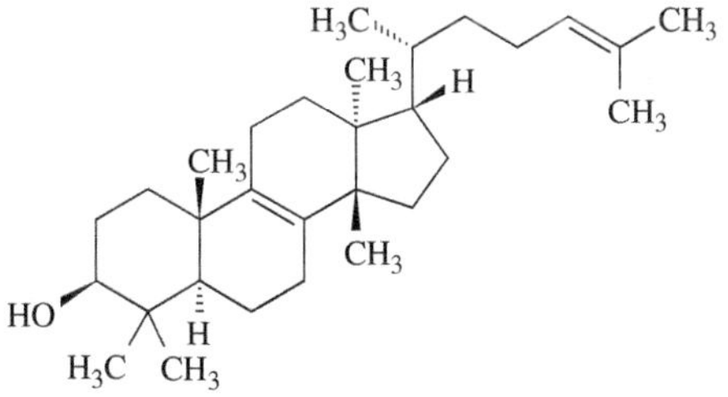

3-59, euphol

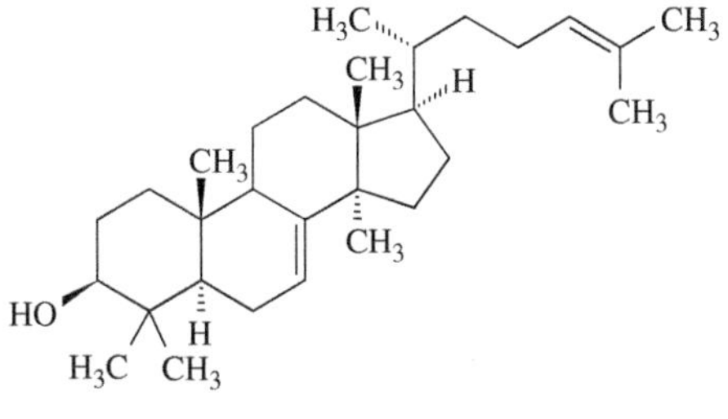

3-60, butyrospermol

Table 3.37 Overview of trivial and systematic names of selected steroids.

	Steroid name
Trivial	**Systematic**
4,4-Dimethylsterols	
α-Amyrin	5α-Urs-12-en-3β-ol
β-Amyrin	5α-Olean-12-en-3β-ol
Betulinic Acid	5α-Eupha-7,24-dien-3β-ol-28-carboxylic acid
Butyrospermol	5α-Eupha-7,24-dien-3β-ol
Cycloartenol	9,19-Cyclo-5α-lanost-24-en-3β-ol
Cyclobranol	24-Methyl-9,19-cyclo-5α-lanost-24-en-3β-ol
Cyclolaudenol	(24S)-24-Methyl-9,19-cyclo-5α-lanost-25-en-3β-ol
Cyclosadol	24-Methyl-9,19-cyclo-5α-lanost-23-en-3β-ol
Erythrodiol	5α-Olean-12-en-3β,28-diol
Euphol	5α-Eupha-8,24-dien-3β-ol
Lanosterol	5α-Lanosta-8,24-dien-3β-ol
Lupeol	5α-Lup-20(29)-en-3β-ol
24-Methylenecycloartanol	24-Methylene-9,19-cyclo-5α-lanostan-3β-ol
Oleanolic acid	5α-Olean-12-en-3β-ol-28-carboxylic acid
Parkeol	5α-Lanosta-9(11),24-dien-3β-ol
Ursolic acid	5α-Urs-12-en-3β-ol-28-carboxylic acid
Uvaol	5α-Urs-12-en-3β,28-diol
4-Methylsterols	
Citrastadienol	4α-Methyl-24-ethylidene-5α-cholest-7-en-3β-ol
Cycloeucalenol	4α,14α-Dimethyl-24-methylene-9β,19-cyclo-5α-cholestan-3β-ol
Gramisterol	4α-Methyl-24-methylene-5α-cholest-7-en-3β-ol
Lophenol	4α-Methyl-5α-cholest-7-en-3β-ol
Obtusifoliol	4α,14α-Dimethyl-24-methylene-5α-cholest-8-en-3β-ol
4-Demethylsterols	
Δ⁵-Avenasterol	(Z)-Stigmasta-5,24(28)-dien-3β-ol
Δ⁷-Avenasterol	(Z)-Stigmasta-7,24(28)-dien-3β-ol
Brassicasterol	(R)-24-Methylcholesta-5,22-dien-3β-ol
Desmosterol	Cholesta-5,24-dien-3β-ol
Fucosterol	(E)-Stigmasta-5,24(28)-dien-3β-ol
Cholesterol	Cholest-5-en-3β-ol
Campesterol	(R)-24-methylcholest-5-en-3β-ol
Clerosterol	(Z)-Stigmasta-5,25(26)-dien-3β-ol
Lathosterol	Cholest-7-en-3β-ol
Lichesterol	(Z)-Ergosta-5,8,22-trien-3β-ol
Poriferasterol	(E)-Stigmasta-5,22-dien-3β-ol
β-Sitosterol	(R)-24-Ethylcholest-5-en-3β-ol or stigmasta-5-en-3β-ol
Spinasterol	(Z)-Stigmasta-5,22-dien-3β-ol or (24R)-24-ethylcholesta-5,22-dien-3β-ol
Stigmasterol	(S)-24-Ethylcholesta-5,22-dien-3β-ol or (22E)-stigmasta-5,22-dien-3β-ol
Ergosterol	(E)-Ergosta-5,7,22-trien-3β-ol or (24R)-24-methylcholesta-5,7,22-trien-3β-ol

3.7.4.1.1 Dimethylsterols

The basic 4,4-dimethylsterols are the C_{30} steroids lanosterol (**3-57**) and cycloartenol with a cyclopropane ring (**3-58**), derived from lanostane (4,4,14-trimethylcholestane). Lanosterol is the building block for the biosynthesis of all other zoosterols (including cholesterol); the biosynthesis of many other phytosterols is based on cycloartenol. Common compounds also incude euphol (**3-59**) and its isomer butyrospermol (Δ^7-euphol or Δ^7-tirukallol, **3-60**), derived from the hydrocarbon euphane.

Triterpenoids (C_{30} steroids) with four C_6 cycles and one C_5 cycle E, derived from the hydrocarbon lupane, are lupeol (**3-61**) and a product of its oxidation on C-28 known as betulinic acid (**3-62**). Also common are pentacyclic triterpenes with five C_6 cycles. The most important of these compounds are α-amyrin (**3-63**), derived from the hydrocarbon ursane, and β-amyrin (**3-64**), derived from the hydrocarbon oleane. Also common is taraxasterol (**3-65**), which is structurally related to α-amyrin. Fruit waxes very often contain triterpenoid diols and acids formed by oxidation of the C-28 methyl group. Determination of uvaol (**3-66**) and erythrodiol (**3-67**) concentrations can be used to distinguish olive oils from other vegetable oils. Widespread in cuticular wax of apples, pears, grapefruits, and other fruits is ursolic acid (**3-68**), which is structurally related to α-amyrin. Some citrus fruits contain in significant quantities oleanolic acid (**3-69**), which is related to β-amyrin. Its 2α-hydroxy derivative, known as maslinic acid, occurs in olives. Many saponins derived from C_{30} triterpenoids (see Section 10.3.3.2) have been reported to exert beneficial effects on health, including anticarcinogenic activity.

The common C_{31} 4,4-dimethylsterols are represented by isomeric compounds with a cyclopropane ring, such as cycloartenol (**3-58**) and 24-methylenecycloartanol (**3-70**), cyclosadol (**3-71**), cyclobranol (**3-72**), and cyclolaudenol (**3-73**), which are derived from the hydrocarbon lanostane.

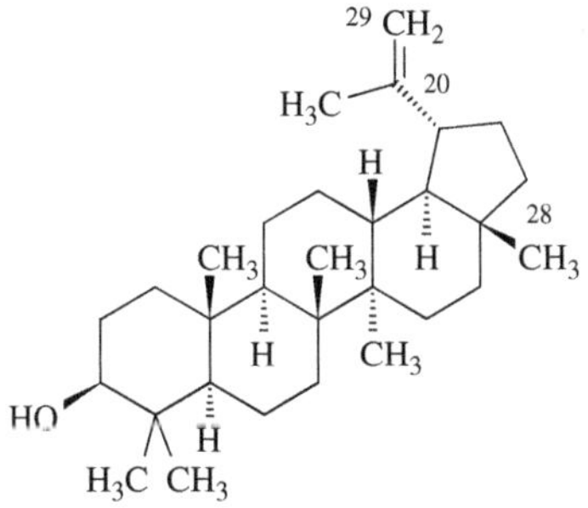

3-61, lupeol

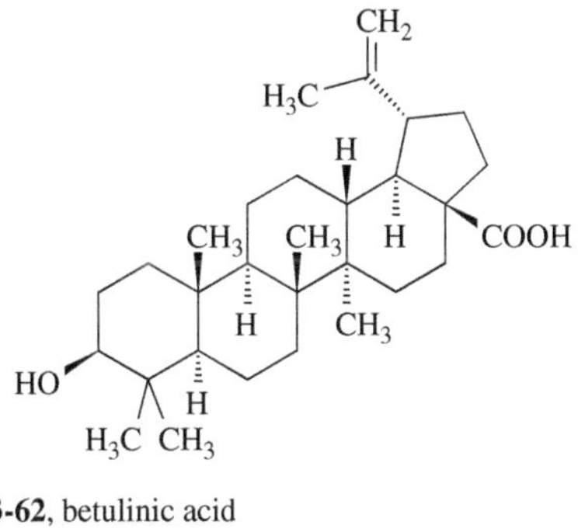

3-62, betulinic acid

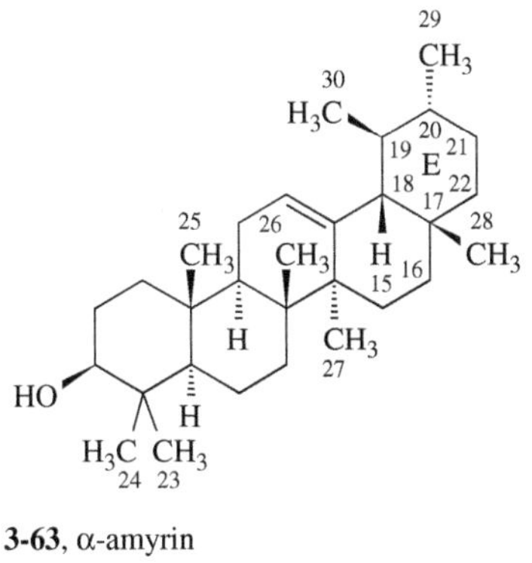

3-63, α-amyrin

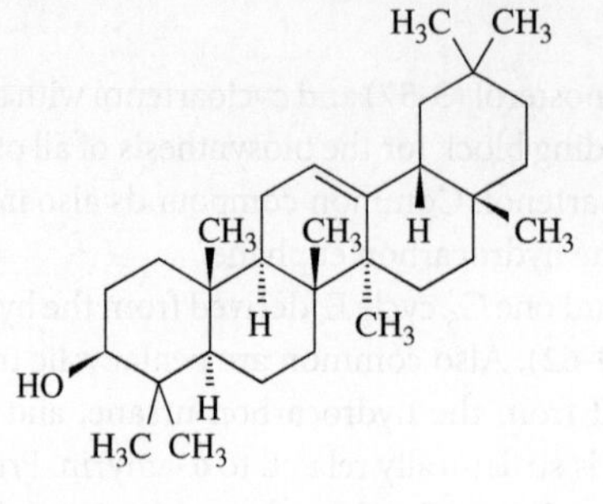

3-64, β-amyrin

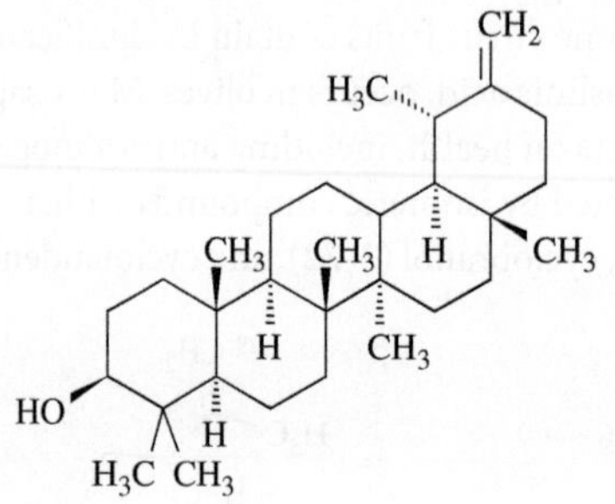

3-65, taraxasterol

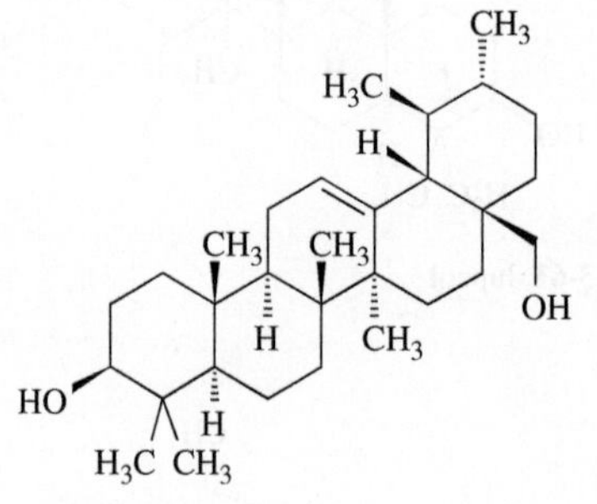

3-66, uvaol

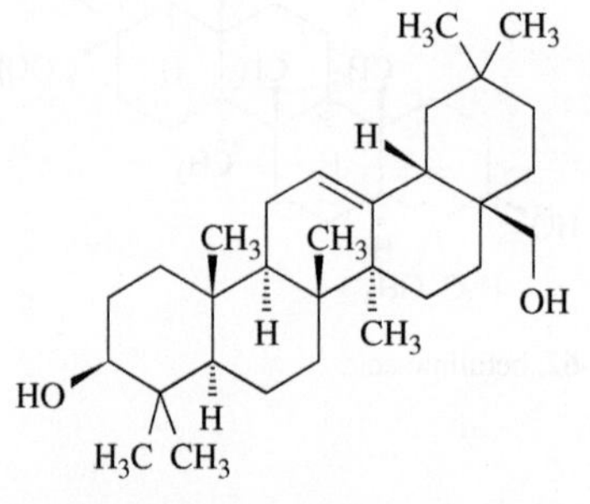

3-67, erythrodiol

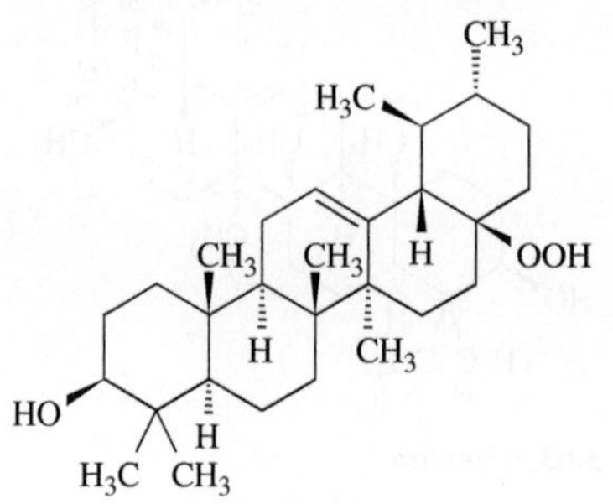

3-68, ursolic acid

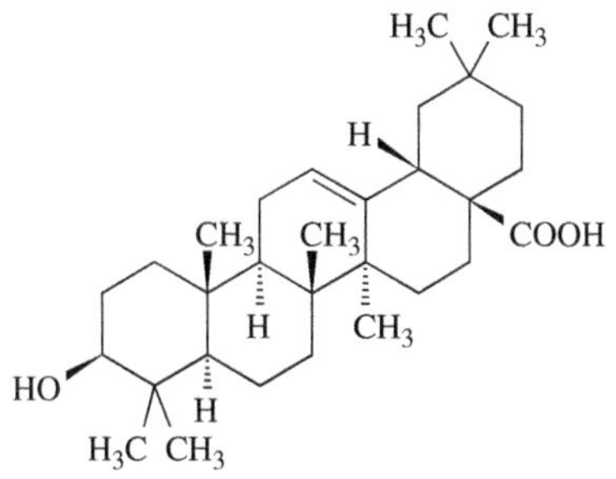

3-69, oleanolic acid

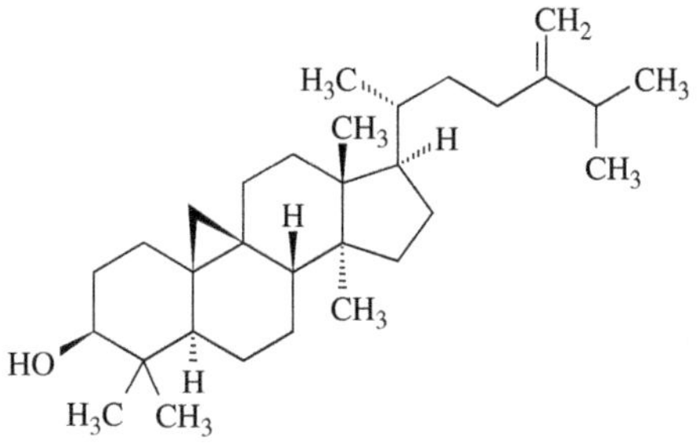

3-70, 24-methylenecycloartanol

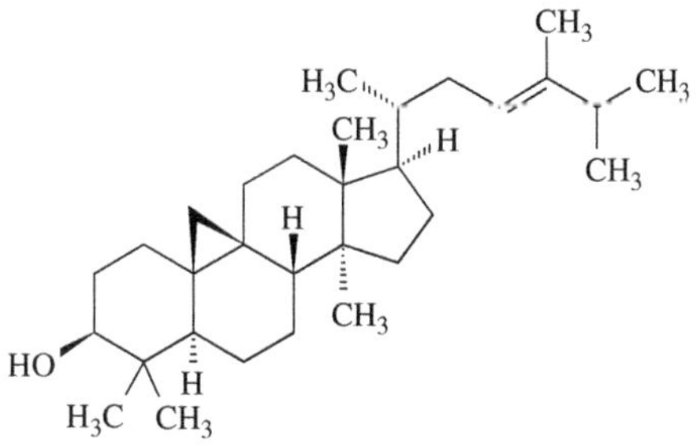

3-71, cyclosadol

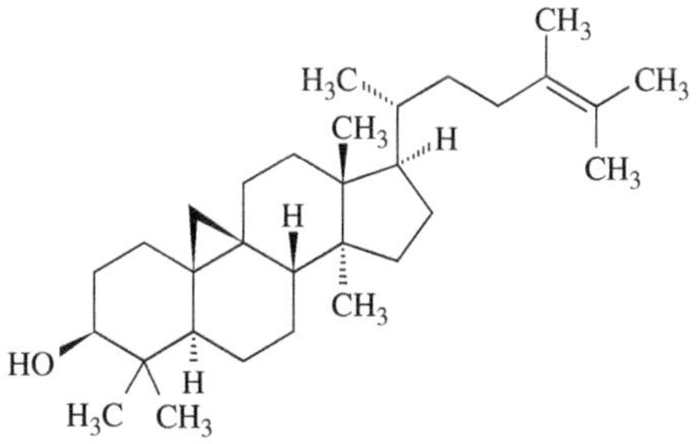

3-72, cyclobranol

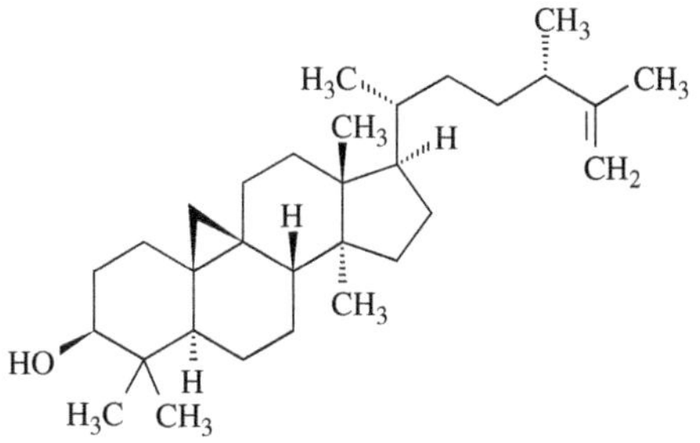

3-73, cyclolaudenol

3.7.4.1.2 Methylsterols

Typical 4-methylsterols are the C_{30} compounds cycloeucalenol (**3-74**) and obtusifoliol (**3-75**), which are intermediates in the biosynthesis of phytosterols from cycloartenol. Other C_{30} compounds include citrastadienol (24-ethylidenelophenol, **3-76**), which is the intermediate in the biosynthesis of β-sitosterol and stigmasterol. The C_{29} compounds are represented by gramisterol (episterol, **3-77**) and the C_{28} compounds by lophenol (**3-78**).

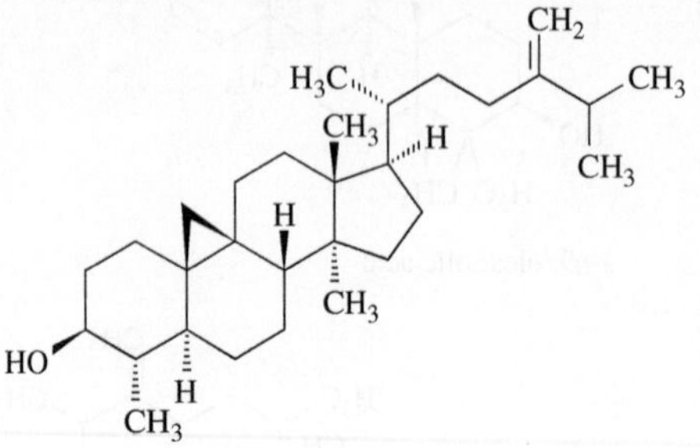

3-74, cycloeucalenol

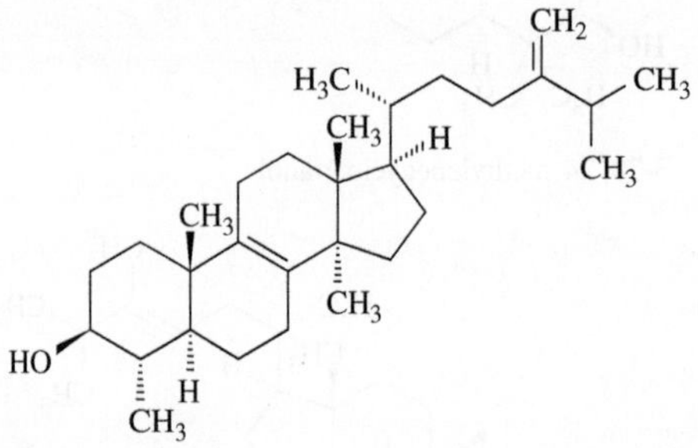

3-75, obtusifoliol

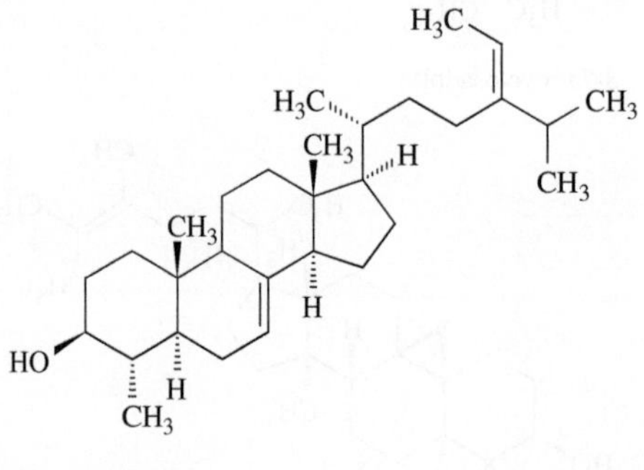

3-76, citrastadienol

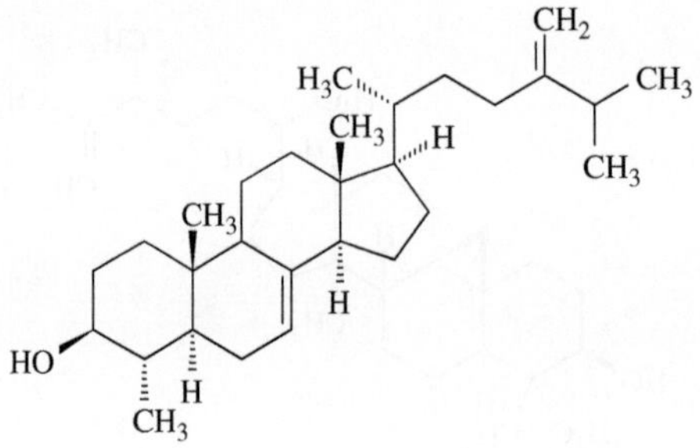

3-77, gramisterol

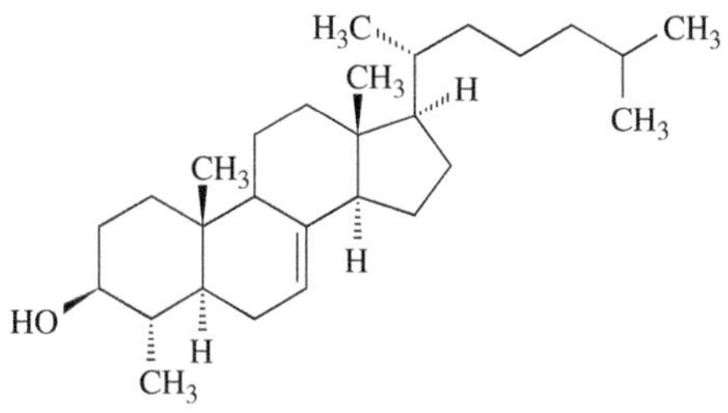

3-78, lophenol

3.7.4.1.3 Sterols

Sterols (demethylsterols) are commonly divided into categories according to their occurrence in nature. The recognised categories are:

- **zoosterols** (animal sterols);

- **phytosterols** (plant sterols);

- **mycosterols** (sterols of fungi).

This classification is somewhat imperfect. Some bacteria, for example, accept the zoosterol cholesterol from the host animals and use it as a component of membranes, some types of prokaryotic organisms even synthesise sterols *de novo*, and some eubacteria (species of the genus *Methylobacterium* and *Methylosphaera*) synthesise 4-methylsterols and 4,4-dimethylsterols (including lanosterol). In many species of bacteria, the so-called hopanoids – pentacyclic triterpenoids with the C_6—C_6—C_6—C_5—C_6 skeleton of the hydrocarbon hopane (with cyclopentane ring E) – play the role of sterols.

Modern classification is based on the structure of the sterols. The most frequently occurring compounds are sterols with a double bond in position C-5 in ring B (Δ^5-sterols) and a saturated or unsaturated side chain at position C-17; less common are Δ^7- and $\Delta^{5,7}$-sterols. In plants, these sterols are accompanied by small quantities of saturated sterols known as phytostanols that are analogues of Δ^5-sterols without the double bond in ring B. They occur in cereals in relatively high amounts.

The basic C_{27} sterol is cholesterol, cholesta-5-en-3β-ol or (3β)-cholest-5-en-3-ol (**3-79**), with saturated C_8 side chain. The same side chain also occurs in its precursor lathosterol (**3-80**); the unsaturated C_8 side chain has its other precursor, desmosterol (**3-81**). Normally, there are also sterols with 28 carbon atoms in the molecule. Their representatives are the Δ^5-sterols campesterol (**3-82**) and brassicasterol (**3-83**). The most common C_{29} Δ^5-sterol is sitosterol (also called β-sitosterol, **3-84**). Also found, at lower levels, are stigmasterol (**3-85**), avenasterol (or Δ^5-avenasterol or 5-avenasterol, **3-86**), clerosterol (**3-87**), poriferasterol, fucosterol (**3-88**), and Δ^7-sterols such as spinasterol (**3-89**), Δ^7-campesterol, Δ^7-avenasterol, and Δ^7-stigmasterol. An important $\Delta^{5,7}$-sterol is ergosterol (**3-90**).

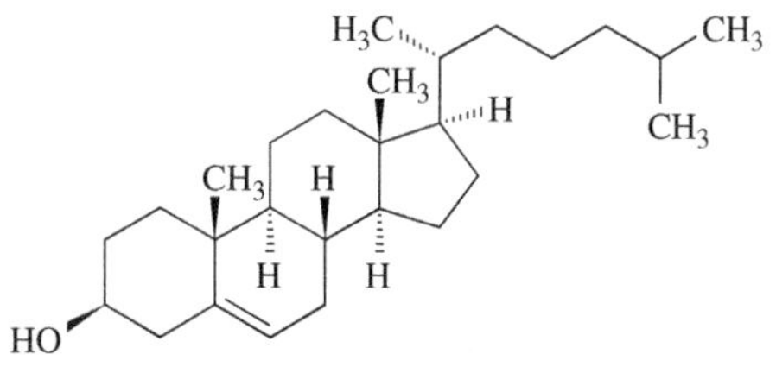

3-79, cholesterol

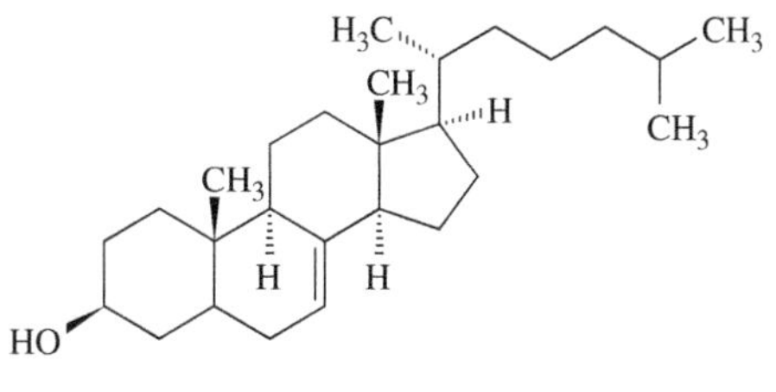

3-80, lathosterol

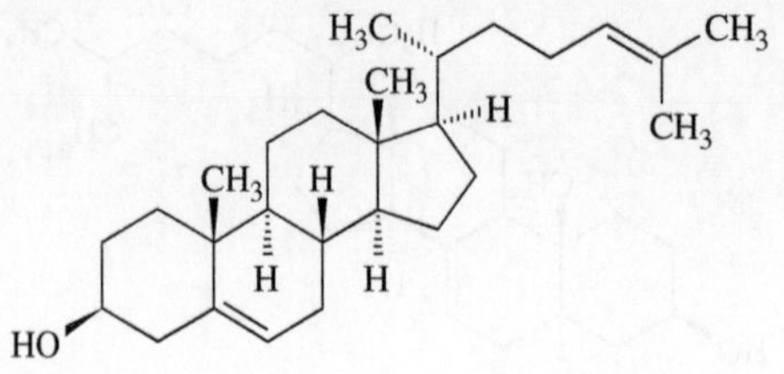

3-81, desmosterol

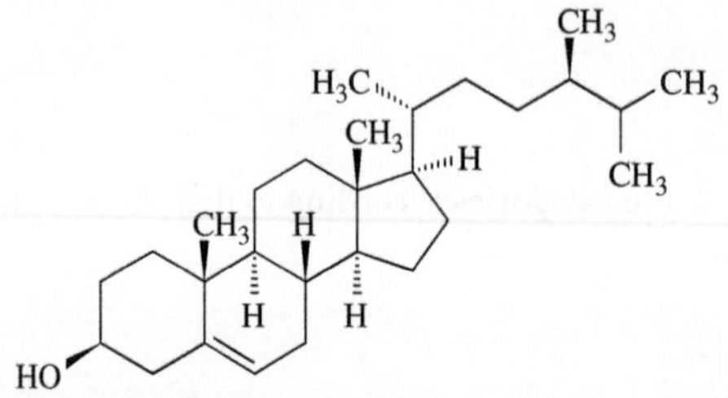

3-82, campesterol

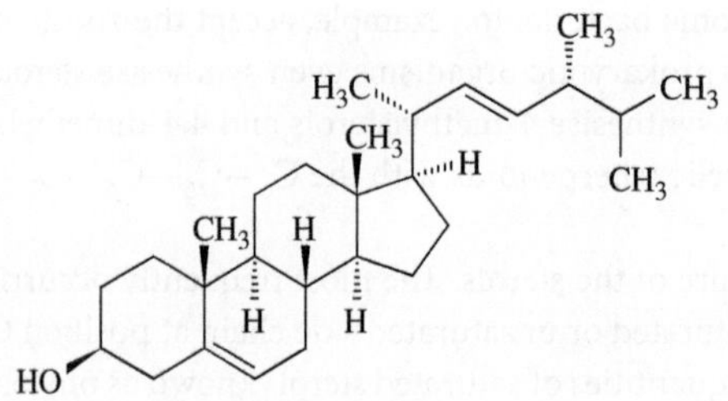

3-83, brassicasterol

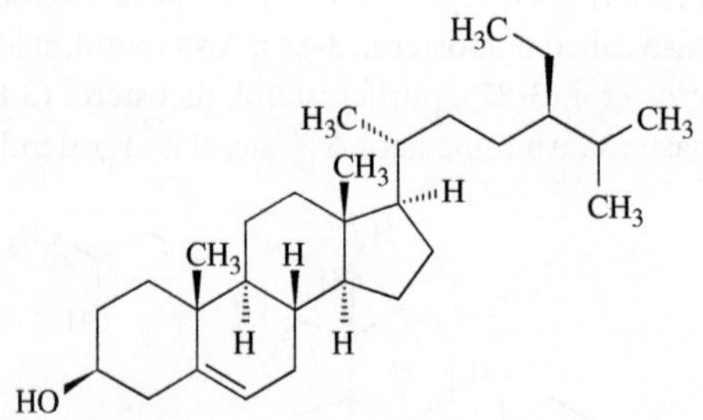

3-84, β-sitosterol

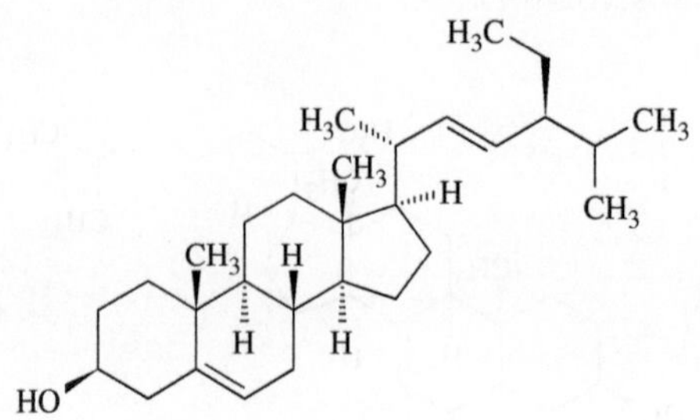

3-85, stigmasterol

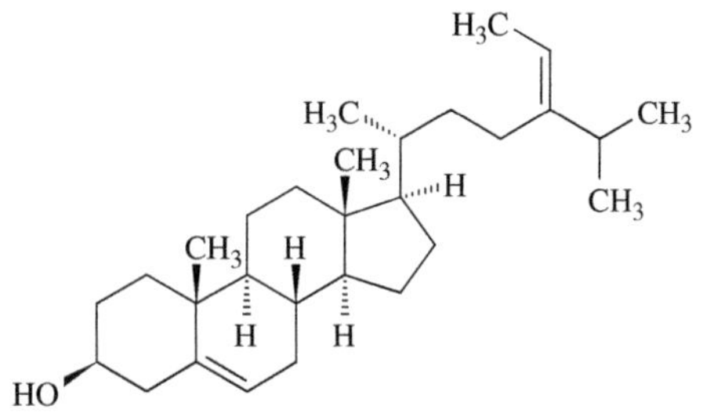

3-86, Δ^5-avenasterol

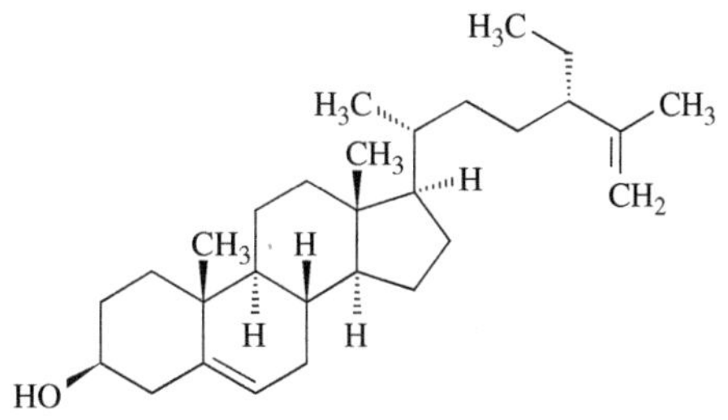

3-87, clerosterol

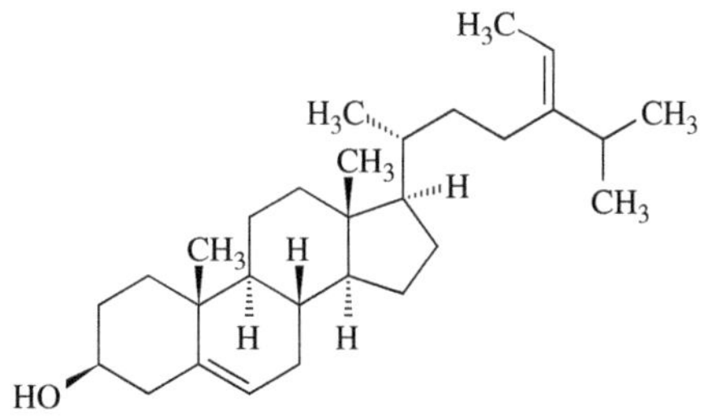

3-88, fucosterol

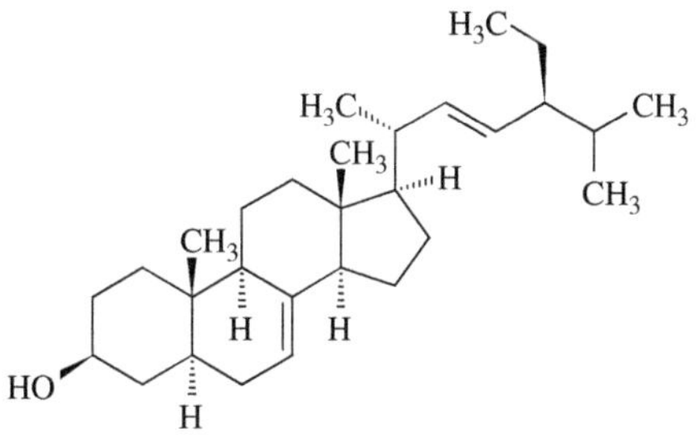

3-89, spinasterol

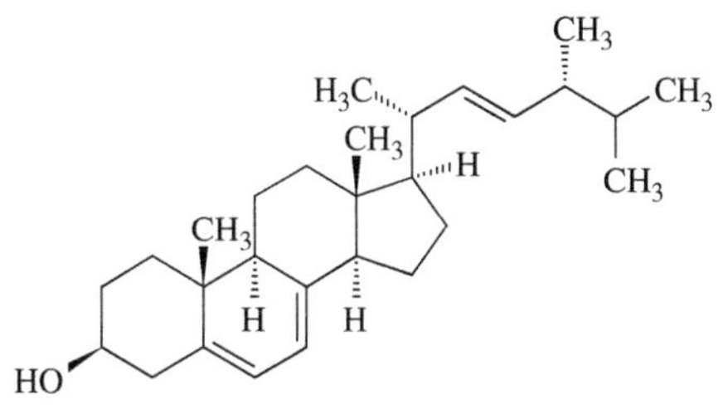

3-90, ergosterol

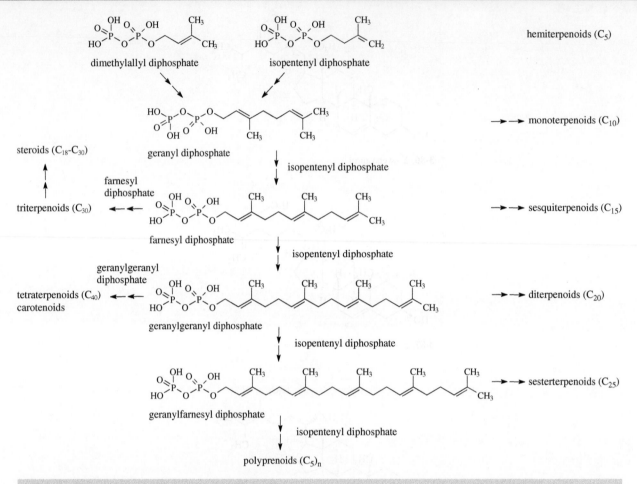

Figure 3.6 Biosynthesis of steroids and other terpenoids.

3.7.4.2 Biochemistry, physiology, and nutrition

Steroids are synthesised in organisms via complex mechanisms from the isoprene units isopentenyl diphosphate and dimethylallyl diphosphate, which first yield geranyl diphosphate. Reaction with another molecule of isopentenyl diphosphate gives an important intermediate, farnesyl diphosphate (Figure 3.6). Two molecules of farnesyl phosphate give rise to triterpenic hydrocarbon squalene, which in the body of animals yields triterpenic alcohol lanosterol and in plants yields the triterpenic alcohol cycloartenol. Plants synthesise a number of steroid substances from cycloartenol, including 4,4-dimethylsterols, 4-methylsterols, and demethylsterols. Cycloartenol is a precursor of many other steroids that are aglycons of saponins, steroidal glycoalkaloids, and other compounds.

Lanosterol in animals is a precursor for the biosynthesis of the most important zoosterol cholesterol. An intermediate in the biosynthesis of cholesterol is 7-dehydrocholesterol, which is a precursor of vitamin D_3. Cholesterol in the body is used for the biosynthesis of steroid hormones and bile acids (**3-91**).

3-91, bile acids and their conjugates

chenodeoxycholic acid, $R^1 = H$, $R^2 = OH$
glycochenodeoxycholic acid, $R^1 = H$, $R^2 = NHCH_2COOH$
taurodeoxycholic acid, $R^1 = H$, $R^2 = NHCH_2CH_2SO_3H$
cholic acid, $R^1 = OH$, $R^2 = OH$
glycocholic acid, $R^1 = OH$, $R^2 = NHCH_2COOH$
taurocholic acid, $R^1 = OH$, $R^2 = NHCH_2CH_2SO_3H$

The main bile acids are cholic acid and chenodeoxycholic acid (**3-91**). They are excreted from the liver as glycine salts or taurine conjugates (see Section 5.15) and into the gallbladder and then the small intestine. In the duodenum, bile acids act as emulsifiers in the absorption and digestion of fats and fat-soluble vitamins. The recycling system allows bile acids to re-enter the bloodstream and return to the liver for reuse. They are partially (some 1000 mg per day) metabolised in the colon by microorganisms, which cleave peptide bond of conjugates and bile acids, and converted to deoxycholic ($R^1 = OH$, $R^2 = OH$) and lithocholic ($R^1 = H$, $R^2 = OH$) acids (the hydroxyl group at C-7 is replaced by an H atom in both compounds), which is excreted in the faeces from the body. This is the only way of removing cholesterol indirectly from the body. The amount of excluded bile acids may be increased due to their binding to the soluble fibre, such as pectin, fructooligosaccharides, or arabinoxylans.

Sterols are essential components of lipoproteins and lipid membranes in animals. They are particularly important in nerve tissues and in the transport of lipids (fatty acids), which are bound in lipoproteins. In humans, dietary cholesterol intake is lower than the daily requirement, so the body synthesises the majority of the cholesterol that it needs (a larger intake in the diet decreases the amount synthesised by the body). Cholesterol in the diet is easily absorbed, but problems may occur during its transport from the intestinal wall as part of lymph and blood circulation. Excessive cholesterol transport in LDL may cause health problems. It is therefore recommended that the intake of dietary cholesterol not exceed 300 mg per day. However, it can often be more than twice as high as that in developed countries, so a reduction of dietary fat is recommended.

Dietary phytosterols have some influence on the biosynthesis of cholesterol in the body, although they are not used for making membranes, and in the body they break down. To reduce the amount of cholesterol in blood plasma, the Recommended Daily Intake of phytosterols is about 250 mg. The importance of triterpene alcohols in the diet is not known.

3.7.4.3 Occurrence

Steroids in food of animal and plant origin appear as:

- **free compouds**;

- **esters** with higher fatty acids and cinnamic acids;

- **glycosides** (common sugars are D-glucose and D-mannose);

- **glycoside** esters of higher fatty acids (called acylsterylglycosides).

Examples of these compounds are β-sitosterylpalmitate (**3-92**), β-D-glucopyranosyl-β-sitosterol (**3-93**), and its ester with palmitic acid (**3-94**). In sunflower oil, for example, the amount of steroids is 0.32%; 0.07% of steroids occur as esters, 0.03% as glycosides, and the rest are free sterols. Fatty acids bound in sterol esters are mainly palmitic, stearic, oleic, linoleic, and linolenic acids, but the fatty acid composition of sterol esters may not match the composition of fatty acids in triacylglycerols.

An example of steroids esterified with cinnamic acids is the 3-*O*-caffeoyloleanolic acid, which, together with other esters containing the (*E*)- or (*Z*)-caffeic acid moiety, has been isolated from the peel of pear fruit (*Pyrus pyrifolia*, Rosaceae) grown throughout East Asia and known by many names, including Asian pear and nashi. Esters of steroids with ferulic acid have been identified in many cereal grains and are valued for their antioxidant properties. The so-called γ-oryzanol is a mixture of ferulic acid esters of sterols and triterpene alcohols that occurs, for example, in brown rice (260–630 mg/kg) and rice bran oil (1–2%), where it serves as a natural antioxidant. The major components of γ-oryzanol are ferulic acid esters of 4-demethylsterols (campesteryl ferulate, campestanyl ferulate, stigmasteryl ferulate, and β-sitosteryl ferulate, **3-95**) and of 4,4′-dimethylsterols (cycloartenyl ferulate and 24-methylenecycloartanyl ferulate).

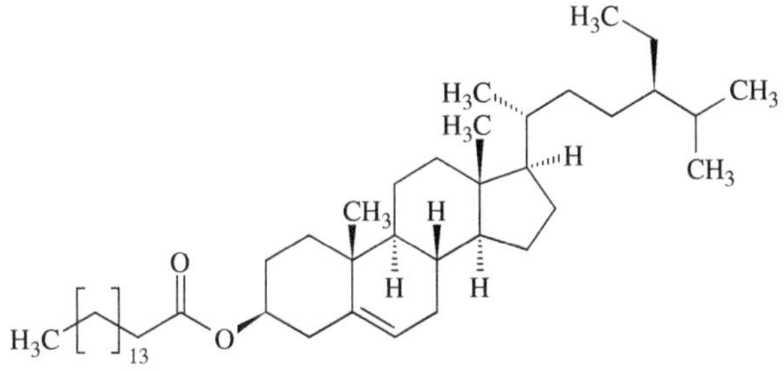

3-92, β-sitosteryl palmitate

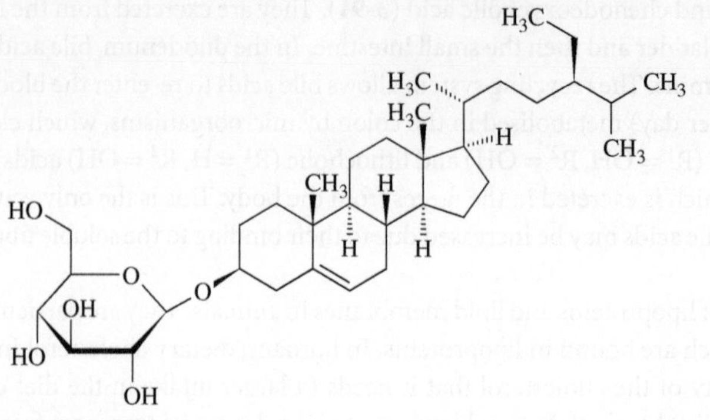

3-93, β-D-glucopyranosyl-β-sitosterol

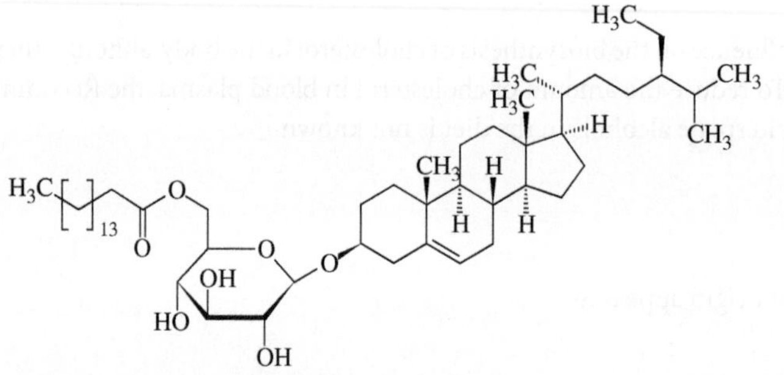

3-94, 6-O-palmitoyl-β-D-glucopyranosyl-β-sitosterol

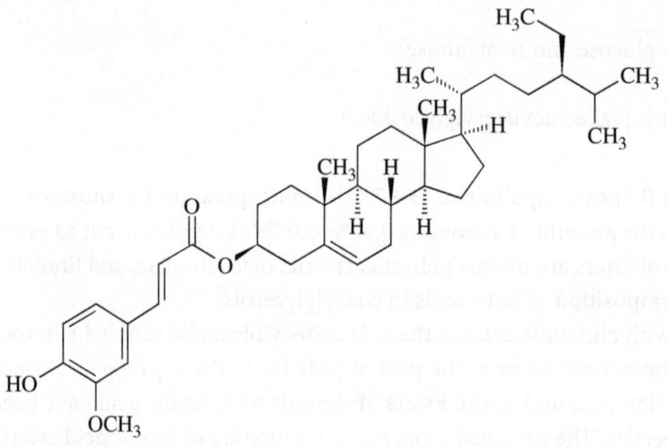

3-95, β-sitosteryl ferulate

3.7.4.3.1 Animal fats

In practice, cholesterol is mainly found in animal fats and human tissues (Table 3.38). In lower animals, other sterols, known collectively as zoosterols, may also be present.

Cholesterol and its esters are present in all membranes and in blood lipids, but particularly rich sources are nervous tissues, especially the brain. Therefore, the cerebellum would be the food with the highest cholesterol content. Egg yolks also contain high amounts of cholesterol. Other important sources include meat, milk, and cheeses, but also animal fats, lard, and butter, to a greater extent. Together with cholesterol and other steroids, wool fat contains 4,4-dimethylsterol lanosterol as the main steroid.

3.7.4.3.2 Vegetable oils

A variety of steroid compounds are usually present in plants and other organisms, but the major steroids are phytosterols. Those found in higher plants (β-sitosterol being the most predominant) and some other organisms are shown in Table 3.39.

Table 3.38 Cholesterol contents in foods of animal origin.

Food		Content (mg/kg edible portion)
Meat, meat products, poultry, fish	Beef	590–670
	Mutton	700–720
	Veal	650–700
	Pork	600–760
	Sausages	470–1150
	Poultry	650–900
	Fish	420–1500
Milk and dairy products	Milk	120–140
	Yoghurt	40–100
	Cream	190–1050
	Cheeses	290–1050
	Curd	50–130
Fats	Beef tallow	1000
	Butter	2400
	Pork lard	940
Eggs	Egg yolk	8400–13100
	Egg white	0
	Eggs (whole)	2000–3540
	Mayonnaise	1100
Pastry[a]	Pastry (various products)	150–2800

[a]Almost exclusively from eggs and animal fats.

Table 3.39 Main sterols in plants and other organisms.

Organism	Main sterols	Organism	Main sterols
Bacteria	Cholesterol, β-sitosterol	Brown algae	Fucosterol, 24-methylenecholesterol
Lower fungi	Ergosterol	Red algae	Cholesterol, desmosterol
Higher fungi	Ergosterol	Mosses	β-Sitosterol
Lichens	Ergosterol, lichesterol, poriferasterol	Ferns	β-Sitosterol
Green algae	Various Δ^5-, Δ^7-, and $\Delta^{5,7}$-sterols	Higher plants	β-Sitosterol, other Δ^5- and Δ^7-sterols

The composition of major sterols found in several vegetable oils is shown in Table 3.40. The total contents of phytosterols in vegetable oils are given in Table 3.41. Usually, vegetable oils contain a mixture of phytosterols and related compounds, which are characteristic of a particular oil. Cholesterol is also a phytosterol as it is found in many vegetable oils, but is usually present at very low levels that have no practical importance in the nutritional balance. Vegetable oils also contain numerous 4,4-dimethylsterols and 4-methylsterols in small quantities. The relative abundance of these compounds and other steroids in maize oil is given in Table 3.42 as an example.

Table 3.40 Sterol composition of vegetable oils (% of total sterols).

Sterol	Rapeseed oil	Soybean oil	Sunflower oil	Palm oil
Cholesterol	0.5–1.3	0.6–1.4	<0.5	2.6–2.7
Brassicasterol	5.0–13.0	0.0–0.3	<0.1	0
Δ^5-Campesterol	24.7–38.6	15.8–24.2	8–12	18.7–27.5
Stigmasterol	0.0–0.7	15.9–19.1	7–11	8.5–13.9
β-Sitosterol	45.1–57.9	51.7–57.6	50–62	50.2–62.1
Δ^5-Avenasterol	3.1–6.6	1.9–3.7	1.5–7	0–2.8
Δ^7-Stigmasterol	0.0–1.3	1.4–5.2	20	0.2–2.4
Δ^7-Avenasterol	0.0–0.8	1.0–4.6	3–6.5	0–5.1
Δ^7-Campesterol	–	–	2–3	–
Clerosterol	–	–	0.7–1.0	–

Table 3.41 Total contents of sterols in some vegetable oils.

Oil	Phytosterols content (mg/kg)	Oil	Phytosterols content (mg/kg)
Olive	>1000	Sunflower	2437–4545
Peanut	901–2854	Rapeseed	4824–11 276
Cotton	2690–6425	Rice	10 550
Maize germ	7950–22150	Palm	376–627
Wheat germ	5500	Palm kernel	792–1406
Sesame	401–18 957	Almond	2660
Safflower	2095–2647	Hazelnut	1200
Soybean	1837–4089	Hempen	3700

3.7.4.3.3 Other lipids

In yeast and fungi (moulds and higher fungi), the main sterol is ergosterol (**3-90**), which belongs, along with related compounds (precursors and their metabolites), to the C_{28} mycosterols. Ergosterol usually represents 60–70% of the total sterols present. Other sterols are primarily derived from ergosterol, such as ergosta-5-en-3β-ol, ergosta-7-en-3β-ol, ergosta-5,7-dien-3β-ol (dihydroergosterol), ergosta-5,22-dien-3β-ol, ergosta-7,22-dien-3β-ol, ergosta-5,24(28)-dien-3β-ol, ergosta-7,24(28)-dien-3β-ol, and ergosta-5,7,9(11),22-tetraen-3β-ol.

Other minor steroids include C_{27} sterols, such as cholesterol and desmosterol, and C_{29} sterols, such as β-sitosterol and stigmasterol. Sterols are found in fungi as free compounds, as fatty acid esters (mainly esters of linoleic, oleic and palmitic acids), and as glycosides, in which the sugar component is D-glucose.

3.7.4.4 Properties

Sterols are solid crystalline substances with high melting points, which are insoluble in water, poorly soluble in alcohol, but soluble in non-polar solvents. Pure sterols are relatively stable in air during storage, but they are easily oxidised in solutions. Catalytic hydrogenation of oils transforms sterols into the corresponding saturated compounds.

Table 3.42 Steroid composition of maize oil (% of total steroids).

Steroids	Composition (%)	Steroids	Composition (%)
4,4-Dimethylsterols		*4-Demethylsterols*	
Cycloartenol	0.6	Cholesterol	trace
24-Methylenecycloartanol	1.4	24-Methylenecholesterol	1.3
Cyclosadol	trace	Stigmasterol	7.3
4-Methylsterols		Isofucosterol	9.0
Obtusifoliol	1.2	Campesterol	9.3
24-Methylenelophenol	1.0	24-Epicampesterol	13.9
Citrastadienol	0.5	β-Sitosterol	52.5

3.7.4.5 Use

Sterols are a regular part of the diet. The most common is cholesterol, which is used in the cosmetics and pharmaceutical industries as an emulsifier and as a starting material for the synthesis of various biologically active derivatives, such as vitamin D_3 and steroid hormones. During deodorising of vegetable oils, a mixture of phytosterols can be obtained, which can become a viable addition to a variety of dietary products such as margarines.

3.7.5 Lipophilic vitamins

Edible vegetable oils regularly contain varying amounts of tocopherols (vitamin E) or, rarely, their esters. Synthetic all-rac-α-tocopheryl acetate (D,L-α-tocopheryl acetate) is often added to oils and other products as a food additive. The fat products often contain synthetic retinyl acetate (vitamin A), which is more stable than retinol. Emulsified fats sometimes contain vitamin D. Animal fats contain only small amounts of lipophilic vitamins, except for fish oils, which are sometimes very rich in vitamins A (see Section 5.2) and D (see Section 5.3).

3.7.6 Lipophilic pigments

Plant lipids contain various natural pigments, particularly carotenoids (see Section 9.9.2.3.3) and chlorophylls (see Section 9.2.2). Of the animal fats, only butter has an important natural carotenoid pigment content. Carotenoid pigments can be added to margarines as additives (see Section 11.4.1.5).

3.7.7 Natural antioxidants

In addition to tocopherols, vegetable oils, especially olive, sesame, and soybean oils, contain natural antioxidants that are described elsewhere (see Section 11.2.2.4).

3.8 Reactions

3.8.1 Reactions of fatty acids

Chemical reactions of fatty acids involve the carboxyl group, as well as the hydrocarbon residue of the molecule, particularly the double bonds and neighbouring methylene groups. The reactions of the hydrocarbon chain are common to free fatty acids and to their esters.

3.8.1.1 Formation of salts

Although fatty acids are relatively weak acids, they readily form salts with metal cations or organic bases, most easily with alkali metal cations and with ammonia. The former compounds are called **soaps**. Soaps of alkali metals are surfactants. Sodium soaps of unsaturated fatty acids and medium-chain saturated fatty acids are very soluble in water (this corresponds to a high critical micellar concentration), but soaps of saturated acids with long chains (>C_{14}) dissolve better if the temperature is raised (this corresponds to a very low critical micellar concentration). Technologically, this is utilised specifically in the formation of sodium soaps in alkaline neutralisation (refining) of vegetable oils, when free fatty acids are soluble in oil but their soaps are soluble in water. Sodium salts of fatty acids are important for the production of toilet soaps, detergents, and cleaning agents. Potassium soaps are more alkaline and have a significantly higher detergent effect, but are now limited in their production.

Soaps of divalent cations and aluminum soaps are called **metal soaps**. They have a hydrophobic character and therefore are of great significance in non-food applications in industry and construction. They are seldom found in foods. Calcium and magnesium soaps commonly arise from alkaline soaps in hard water, but they are also natural components of foods at low levels. The concentrations at which soaps occur naturally have no significant effect on the sensory and nutritional value of a food or its functional properties. However, higher concentrations of soaps may lower the acidity and negatively affect some functional properties of the food and cause diarrhoea. Calcium soaps may be used as a source of calcium in the diet of dairy cows.

Soaps only form true solutions in water at a high dilution. Concentrated solutions of soaps, after exceeding the so-called critical micellar concentration, form micelles, which are agglomerations of many molecules where the polar groups are directed to the surface and the hydrocarbon chains into the interior of micelles. Solutions of alkali soaps hydrolyse in water and thus produce an alkaline solution. Salts of divalent cations, mainly of calcium and magnesium, are practically insoluble in water, and in organic solvents they only dissolve with difficulty.

3.8.1.2 Esterification reactions

One of the main reactions of the carboxyl group of fatty acids is esterification. This is an ionic reaction, which is usually catalysed by acids, that proceeds similarly to the esterification of other organic carboxylic acids. Esterification of higher fatty acid usually takes place at high temperatures (autocatalysis) or under catalysis of strong acids (Figure 3.7). In fatty acids in foods, the process is often catalysed by lipases. Esterification or interesterification of fatty acids, which is used in specific instances, is catalysed by microbial lipases.

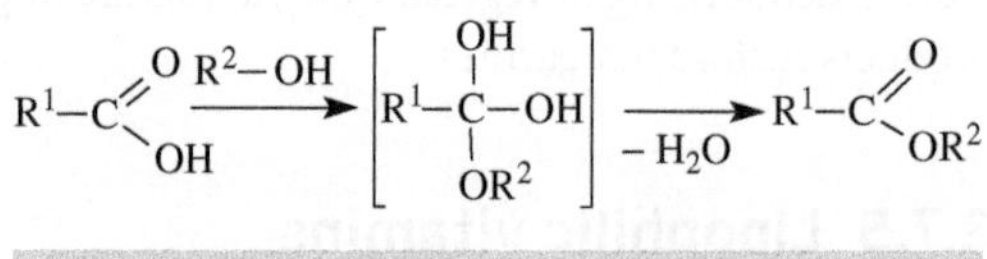

Figure 3.7 Esterification of fatty acids.

In food technology, esterification is usually employed in the production of monoacylglycerol emulsifiers, obtained by esterification of glycerol with fatty acids. The main reaction products are mono- and diacylglycerols. The same products can be obtained by interesterification (glycerolysis). Similarly, various other emulsifiers that are used as food additives (see Section 11.5.2) are produced from glycols, hydroxy acids, and sugars. Esterification is also utilised in the manufacture of particular dietary fats (triacylglycerols) with medium-chain fatty acids (caproic and capric). A wide range of esters that act as wax analogues, such as 2-propyltetradecanoate (2-propylmyristate), are manufactured by the pharmaceutical and cosmetics industries. Long-chain esters formed from hydroxy acids (e.g. ricinoleic acid) by reaction of two molecules either of the same or of different acids are known as **estolides** (Figure 3.8). Estolides can be free acids or esters, or can occur within a triacylglycerol structure.

3.8.1.3 Formation of amides

Naturally occurring fatty acid amides are described in Section 3.6.1. Amides of fatty acids are also formed in foods from fatty acids during thermal treatments by reaction with ammonium salts, amines, or amino acids. In non-food uses of lipids, fatty amides are produced synthetically in industry for use as ingredients of detergents, lubricants, inks, and many other products. Fatty primary amines are used as reactants for the synthesis of cationic and amphoteric surfactants.

3.8.1.4 Isomerisation reactions of unsaturated acids

Double bonds of unsaturated fatty acids can isomerise so that they change their steric configuration (*cis–trans* or geometrical isomerisation) or their double bond shifts in the hydrocarbon chain (positional isomerisation).

3.8.1.4.1 Geometrical isomerisation

Double bonds of unsaturated fatty acids in natural lipids are almost exclusively in the *cis* configuration, although *trans* double bonds are thermodynamically more stable. Under equilibrium conditions, the ratio of *cis* to *trans* double bonds is approximately 30 : 70. Therefore, under appropriate conditions (if free radicals are temporarily created and electrons can move to the carbon with an sp^2 hybridisation), an interconversion between *cis*- and *trans*-isomers occurs. This happens during heating, autoxidation, and hydrogenation.

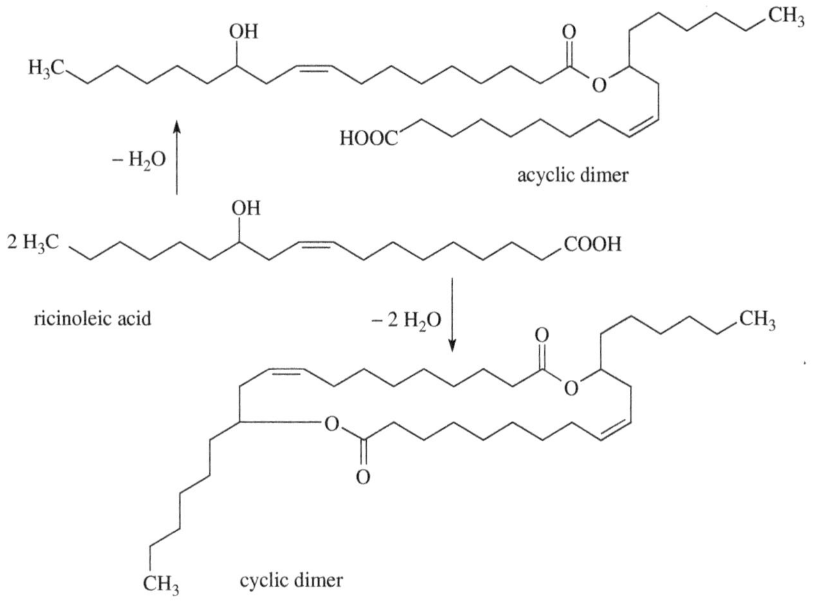

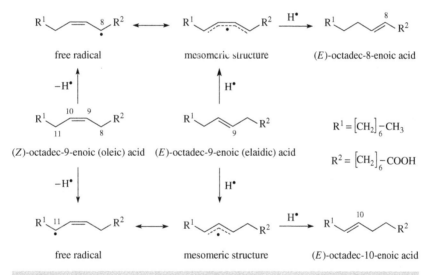

Figure 3.8 Formation of estolides from ricinoleic acid.

Figure 3.9 Isomerisation of monoenoic fatty acids.

Thermally induced geometrical isomerisation is a nonequilibrium reaction, which means that the isomerisation proceeds until exhaustion of the *cis* double bonds. The degree of isomerisation achieved is thus a function of temperature and time of heating. This isomerisation occurs in particular during deodorising and physical refining of oils and fats where the temperatures reach 200–260 °C.

Monoenoic fatty acids (such as oleic acid) require heating to temperatures around 270 °C at which *cis–trans*-isomerisation of the allylic system of one double bond proceeds at a reasonable rate (Figure 3.9).

Pentadienoic systems of isolated double bonds in dienoic acids, which have two double bonds separated by a methylene group, are far less stable (Figure 3.10). The splitting off of a hydrogen radical from the methylene group adjacent to both double bonds creates a free radical transformed into the mesomeric form. The reaction of the mesomeric form with a hydrogen radical forms the most stable system possible, which is a system with one or two *trans* double bonds. At temperatures 240 °C and higher, *cis–trans*-isomers and *trans–trans*-isomers of linoleic acid form at a significant rate and deodorised edible oils always contain small amounts of isomerised dienoic acids.

Acids with three double bonds have an even greater tendency to isomerise, because each acyl contains two labile methylene groups. Linoleic acid yields *cis–trans*-, *trans–cis*-, and *trans–trans*-isomers. Linolenic acid mainly produces *cis–cis–trans*-, *trans–cis–cis*-, *cis–trans–cis*-, and *trans–cis–trans*-isomers at temperatures of 210–220 °C and higher. Oils with a higher content of trienoic, tetraenoic, pentaenoic, and hexaenoic acids are very easily isomerised at temperatures above 200 °C, which means that in practice these oils cannot be deodorised.

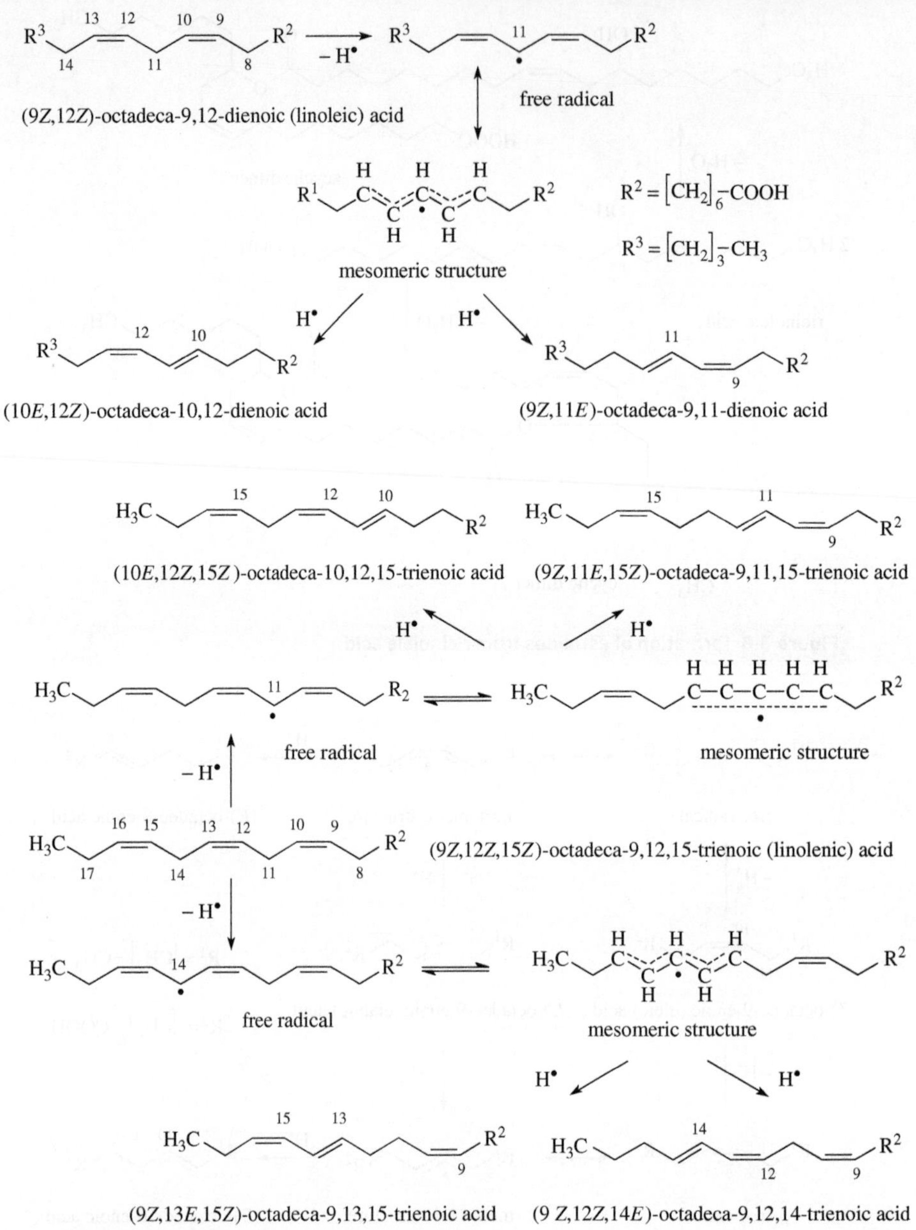

Figure 3.10 Isomerisation of polyenoic fatty acids.

In parallel with saturation of the double bonds during the catalytic hydrogenation of oils, geometrical isomerisation of fatty acids via intermediate free radicals also occurs. Isomerisation using a nickel catalyst proceeds to equilibrium; therefore, partially hydrogenated fats can contain 50–60% of *trans*-isomers of fatty acids. The main isomer is usually elaidic acid. Similarly, *trans*-isomers of octadecenoic acid are formed by biohydrogenation in the composite stomach of ruminants, where the main *trans*-isomer is vaccenic acid. The process, in this case, is far from equilibrium, so the *trans*-isomers content in the milk fat of ruminants is about 3–6%.

The fatty acid double bonds also isomerise during oxidation, which begins with the formation of a free radical at the position adjacent to the double bond. In this case, the amount of *trans*-isomers is proportional to the amount of formed hydroperoxides.

3.8.1.4.2 Positional isomerisation

In unsaturated fatty acids, positional isomerisation is also possible, where the double bond shifts by one carbon atom either away from or towards the carboxyl. Positional isomerisation is usually associated with *cis–trans*-isomerisation and takes place under similar conditions to that isomerisation, on heating to temperatures higher than about 240 °C. Isolated double bonds are relatively stable, but the pentadiene systems easily isomerise to conjugated systems. Two conjugated isomers are formed from linoleic acid, whereas linolenic acid yields four conjugated isomers (Figure 3.10).

In deodorised oils, where the oil temperature reaches 220–260 °C for several tens of minutes, conjugated dienes are always present in small amounts. The isomerisation is catalysed by bleaching clays and other materials with an active surface. Conjugated triene systems do not accumulate in oils, as they are quickly cyclised or polymerised.

Conjugated unsaturated fatty acids also form at the beginning of hydrogenation and during autoxidation of unsaturated fatty acids. In these processes, even isolated double bonds isomerise. The partially hydrogenated oils, which originally contained linoleic and linolenic acids, in addition to isomers with double bonds at positions C-9, C-12, and C-15, thus also contain other isomers of octadecenoic acids with double bonds in positions C-5 to C-16.

Geometric and positional isomers arising from biohydrogenation of unsaturated fatty acids in the rumen of ruminants become components of lipids of milk, butter, and meat. Linoleic acid yields a mixture of positional and geometric isomers called conjugated linoleic acid. The main constituents of this mixture are (9Z,11E)- and (9E,11Z)-octadecadienoic acids. The first acid represents about 90% of conjugated acids of cow's milk and more than 75% of the conjugated acids of tallow. Also formed in small amounts are (8E,10Z)-, (9Z,11Z)-, (9E,11E)-, (10Z,12Z)-, and (10E,12E)-octadecadienoic acids. Conjugated linoleic acid has some beneficial effects, including anticarcinogenic, antiatherogenic, antidiabetic, and immunomodulatory effects.

3.8.1.5 Cyclisation of unsaturated acids

The unsaturated fatty acids are also transformed, via the radicals formed during heating to higher temperatures (e.g. during frying), into cyclic fatty acids with five- and six-membered rings. Cyclic acids derived from oleic acids are saturated compounds. The products formed from linoleic acid include cyclopentene acids and cyclopentane and cyclohexane acids with one *cis* double bond in the side chain. Linolenic acid forms products with two double bonds, such as cyclopentene and cyclohexene acids with one *cis* double bond in the side chain.

After deep frying, oils contain cyclic fatty acids bound in triacylglycerols (free acids evaporate during frying) at a concentration of 0.1–7 g/kg. There are approximately twice as many acids with five-membered rings as with six-membered rings. The radical mechanism for the formation of cyclic fatty acids from oleic acid is given in Figure 3.11. Recent suggestions assume also thermally induced [1,6]- and [1,7]-prototropic migrations and the formation of a ring by a pericyclic rearrangement. The structures of products derived from oleic acid cyclisation are represented in formulae **3-96**, two of the numerous products of linoleic acid are given in formulae **3-97**, whilst formulae **3-98** show two of the many products of linolenic acid.

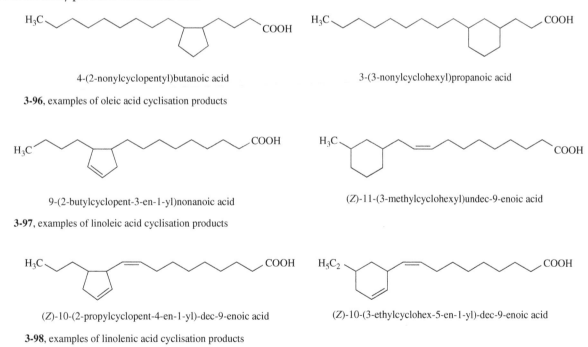

4-(2-nonylcyclopentyl)butanoic acid

3-96, examples of oleic acid cyclisation products

3-(3-nonylcyclohexyl)propanoic acid

9-(2-butylcyclopent-3-en-1-yl)nonanoic acid

3-97, examples of linoleic acid cyclisation products

(Z)-11-(3-methylcyclohexyl)undec-9-enoic acid

(Z)-10-(2-propylcyclopent-4-en-1-yl)-dec-9-enoic acid

3-98, examples of linolenic acid cyclisation products

(Z)-10-(3-ethylcyclohex-5-en-1-yl)-dec-9-enoic acid

3.8.1.6 Polymerisation of unsaturated acids

Polymerisation refers to processes in which the overall composition of a compound does not substantially alter, but the molecular weight increases by a multiple of the weight of the monomer. Polymers usually arise when a system can create free radicals. Therefore, polymerisation

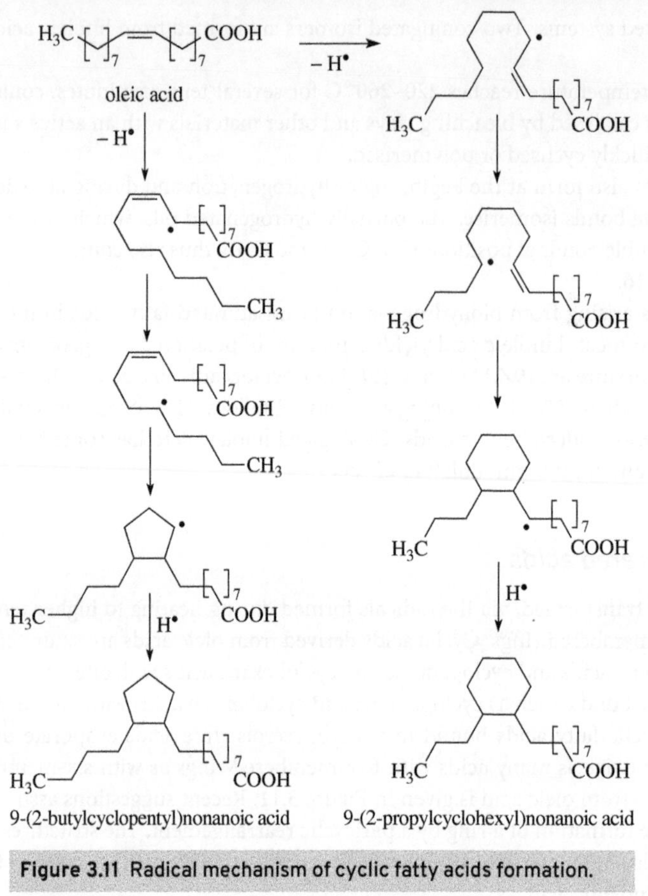

Figure 3.11 Radical mechanism of cyclic fatty acids formation.

reactions usually accompany isomerisation reactions and the formation of cyclic fatty acids under extreme heating. In the fully refined (and thus also deodorised) edible oils, polymers represent several tenths of a percent, but their content increases during heating. Oils with a polymer content higher than 10% are not recommended for use.

Under a low partial pressure of oxygen and at higher temperatures (such as during frying), cyclic and linear polymers are formed, in which the original fatty acids are bound together mainly by C—C bonds. When an adequate supply of air is present (in an environment with a high partial pressure of oxygen) and at lower temperatures, the oxidation of polyenoic fatty acids produces a large number of free radicals containing oxygen. In addition to cyclic and linear polymers of the type C—C, their recombination yields linear polymers with ether bonds (C—O—C) and linear polymers with peroxide bonds (C—O—O—C). Therefore, oxidation is always accompanied by polymerisation, especially in more advanced stages of the reaction. Such a polymerisation is called **oxypolymerisation**. Oxypolymerisation is associated with secondary polymeric products of autoxidation of fatty acids. Under typical conditions encountered during food processing, monoenoic fatty acids barely polymerise and dienoic fatty acids usually only form dimers. Trienoic fatty acids, however, polymerise very readily, and can also form higher oligomers.

Cyclic dimers, which are actually substituted cyclohexene derivatives, arise during polymerisation by the Diels–Alder reaction. This reaction takes place via a cyclic transition state and is therefore also referred to as cycloaddition. In this reaction, the so-called dienophile, a substituted alkene (olefin), which is a monoenoic fatty acid, reacts with a conjugated *cis,cis*-dienoic fatty acid (Figure 3.12). If the dienophile has two different substituents, various stereoisomers may be formed. If the substituents in the dienophile are in the *cis* position, the reaction product (substituted cyclohexene) also has substituents in the *cis* position, and vice versa.

The reaction can take place intermolecularly, between two fatty acids bound in two different triacylglycerol molecules, or intramolecularly, within a single triacylglycerol. An example of a cyclic dimer of the cyclohexene type is the reaction of linoleic acid with 10,12-octadecadienoic acid (Figure 3.13). An example of a cyclohexene structure resulting from the intramolecular reaction of two unsaturated fatty acids bound in a triacylglycerol molecule is shown in Figure 3.14. Cyclic dimers of the cyclohexene type containing an additional double bond and also a system of conjugated double bonds may react further to form tricyclic dimers (Figure 3.15). If the trienoic fatty acids forming a dimer still contain a system of two conjugated double bonds in the molecule, there is the possibility of further reaction to form a trimer (Figure 3.16).

In addition to the cyclic dimers of the cyclohexene type, some cyclic cyclopentane C—C dimers and acyclic C—C dimers may also be formed. The mechanism of their formation (Figure 3.17) is demonstrated for reactions of one of the four main free radicals generated from

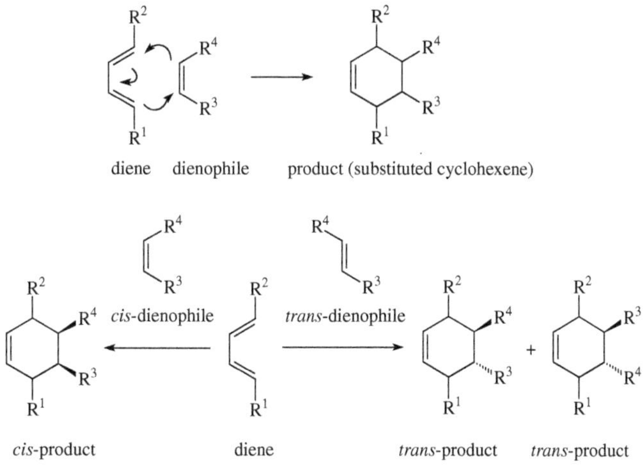

Figure 3.12 Diels–Alder mechanism of polymerisation of polyenoic fatty acids. R¹, R³, residues of hydrocarbon chains; R², R⁴, residues of chains with carboxylic groups).

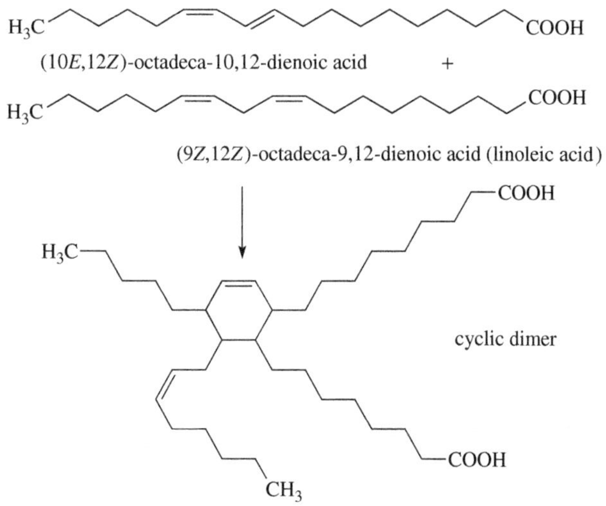

Figure 3.13 Dimerisation of dienoic fatty acids to cyclohexene derivatives.

oleic acid. An acyclic dimer forms by recombination of two radicals (reaction A in Figure 3.17), but also by addition of a radical to the double bond of another radical (reactions B and C in Figure 3.17) and by a termination reaction with a hydrogen radical (reaction D in Figure 3.17). The formation of bicyclic derivatives is illustrated in Figure 3.18.

In practice, mixed linear polymers in fats arise from radicals of different fatty acids. Free radicals can also be formed by splitting the chain around the double bond. They therefore have a smaller number of carbon atoms, and their combinations with free radicals of fatty acid that have not split yield a dimer with the number of carbon atoms being between a monomer and a dimer. In oils used for frying, cyclohexene and cyclopentane structures, bicyclic derivatives, acyclic dienes, and other products have been identified. Polymerisation mainly occurs in fats exposed to higher temperatures for a long time, such as frying oils. It manifests itself through colour changes and deterioration of fat functional properties, mainly by foaming and increased viscosity.

For example, oleochemistry produces specific so-called dimeric fatty acids for the production of adhesives.

3.8.1.7 Hydrogenation reaction of unsaturated acids

Hydrogenation of double bonds of fatty acids, commonly known as hardening of oils and fats, is a chemical modification of oils and fats in the double bonds of unsaturated fatty acids. The common confusion between the terms 'hydrogenation' and 'hardening' reflects the fact that

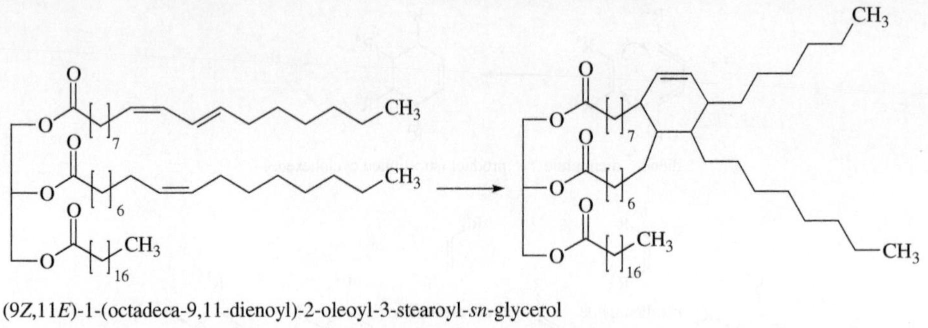
CH3
O
O
CH3
7
O
O
O
O
7
O
CH3
O
6
O
6
O
O
CH3
O
CH3
O
16
16
CH3
(9Z,11E)-1-(octadeca-9,11-dienoyl)-2-oleoyl-3-stearoyl-sn-glycerol

Figure 3.14 Structure of cyclic dimers in a triacylglycerol molecule.

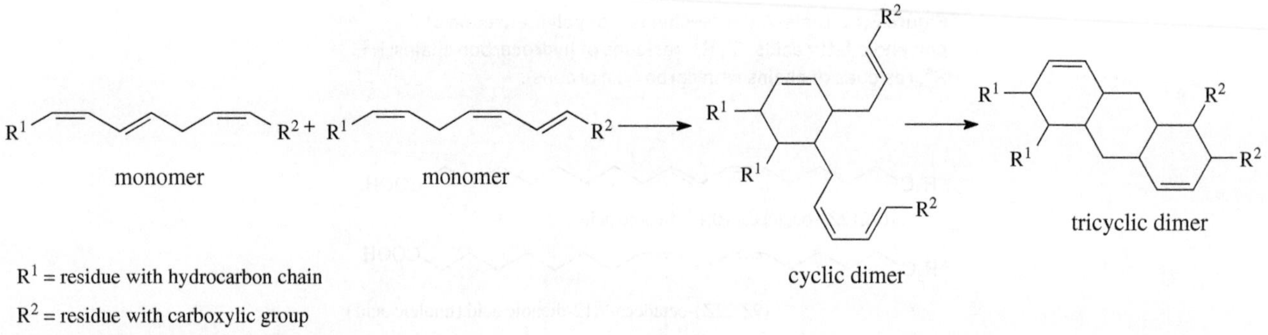
monomer
monomer
cyclic dimer
tricyclic dimer
R1 = residue with hydrocarbon chain
R2 = residue with carboxylic group

Figure 3.15 Formation of condensed tricyclic dimers.

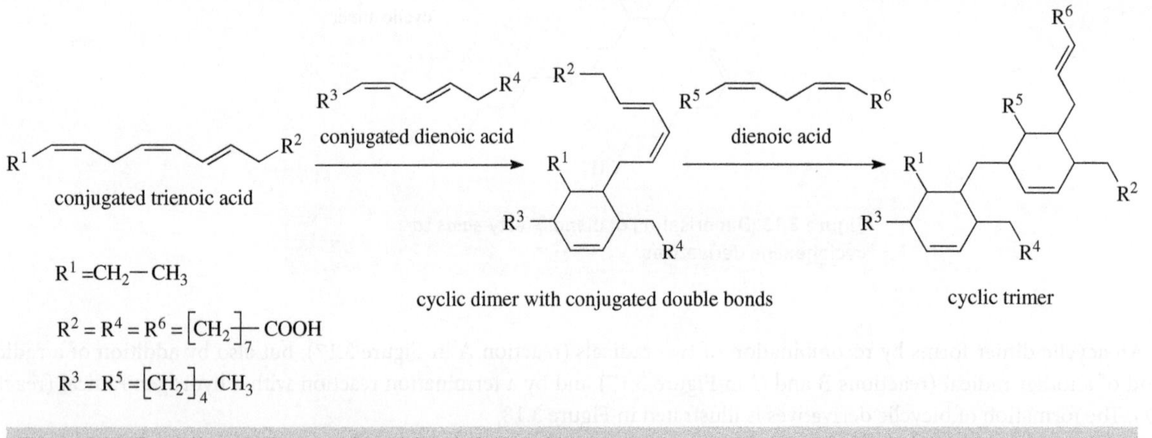
conjugated dienoic acid
conjugated trienoic acid
dienoic acid
cyclic dimer with conjugated double bonds
cyclic trimer
R1 =CH2—CH3
R2 = R4 = R6 = [CH2]7 COOH
R3 = R5 =[CH2]4 CH3

Figure 3.16 Formation of oligomers from trienoic fatty acids.

the primary objective when introducing this technology was to change the consistency of liquid oils and fats into viscoplastic substances of a required consistency. Hydrogenation was an important technological development in the fat industry. Owing to the growth of human population in the nineteenth and twentieth centuries, the traditional animal fats, lard, and butter that were suitable for baking, for addition to dough, for use as spreadable fats, and for other purposes, became scarce. At the same time, there was a relative abundance of fish oils and whale blubber in the world, but these fats were not suitable for direct use because of their adverse physical and organoleptic properties. Therefore, suitable procedures were sought to produce solid fats (correctly known as viscoplastic thixotropic substances) from liquid oils. The best procedure seemed to be the industrial hydrogenation of double bonds, which was introduced in 1902–1910. Paradoxically, in Europe, specifically oils and fats of animal origin were originally hydrogenated, whilst in the United States, soybean and cottonseed oils

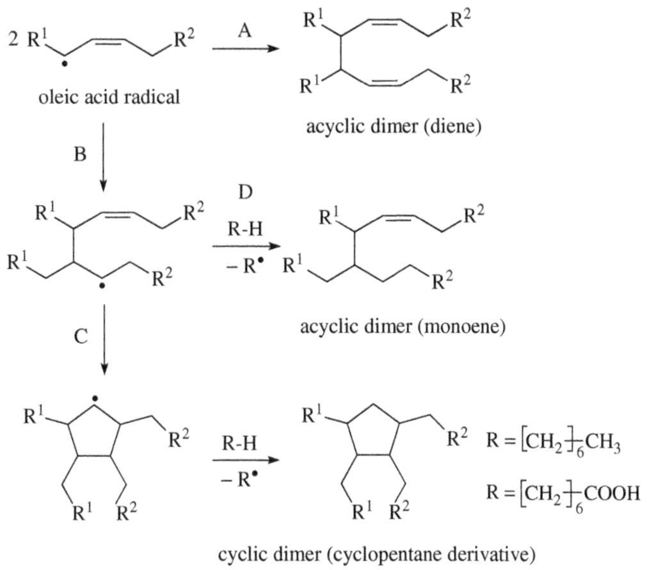

Figure 3.17 Formation of dimers from free radicals of oleic acid.

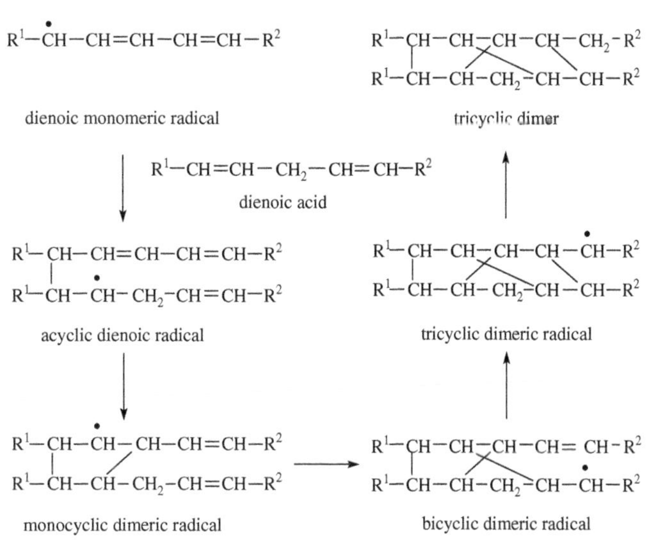

Figure 3.18 Formation of oligocyclic dimers in heated dienoic fatty acids.

began to be hydrogenated gradually before World War II. In Europe, vegetable oils became increasingly hydrogenated when the whaling era ended.

3.8.1.7.1 Hydrogenation of double bonds

Oils can be either partially or fully hydrogenated through the hydrogenation process. Partially hydrogenated oils, such as those found in salad dressings, produce hardened – but not solid – fats. Fully hydrogenated oils, such as shortenings, are solid fats. Food items that contain hydrogenated oils can remain solid or semisolid at room temperature and have a comparatively long shelf life, because they are more stable against autoxidation than the original oils.

The hydrogenation of unsaturated organic compounds, that is treatment with gaseous hydrogen in the liquid phase on solid metal catalysts (usually nickel, palladium, or platinum) to decrease the number of double bonds, was developed at a time when there was demand for solid fats. The common procedure is based on nickel catalyst (pre-reduced supported Ni-catalysts) formed by depositing partially reduced nickel from nickel oxide (NiO) on a diatomaceous earth. Molecules of hydrogen and triacylglycerols are adsorbed, through the double

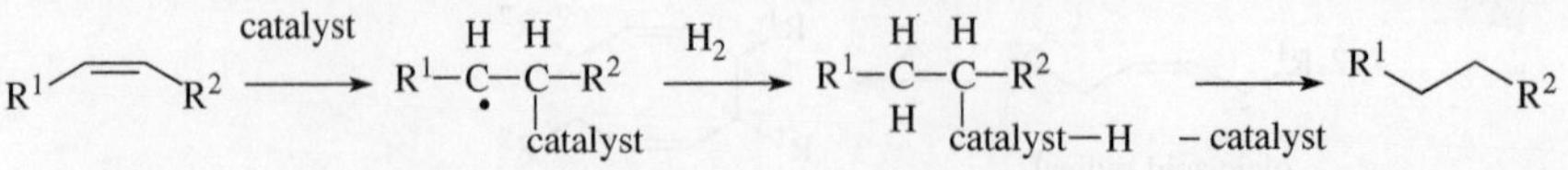

Figure 3.19 Scheme of hydrogenation of unsaturated fatty acids and their esters.

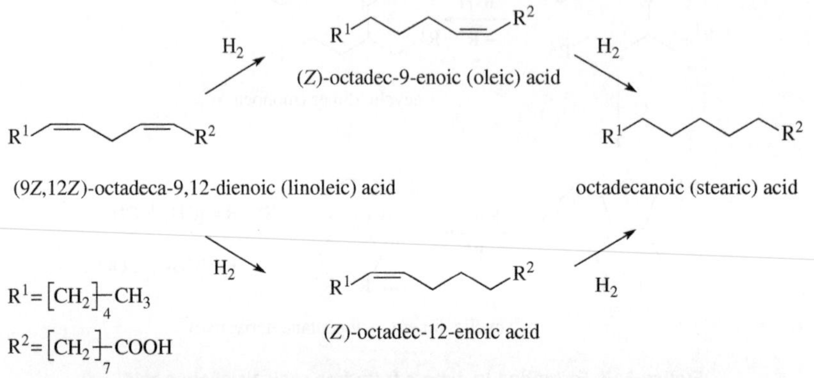

Figure 3.20 Hydrogenation of linoleic acid.

bond, onto the surface of the active metal. The hydrogenation is then conducted via the adsorbed free radicals of the fatty acids bound in the triacylglycerols that form a complex with the catalyst. This complex reacts with hydrogen and dissociates to give a hydrogenated fatty acid and the catalyst (Figure 3.19). The reaction scheme for hydrogenation of linoleic acid into stearic acid is shown in Figure 3.20. In addition to the reduction of the double bond, positional isomerisation (migration of the double bond) and, in particular, geometric isomerisation (*cis–trans*-isomerisation) also proceed. Unsaturated fatty acids in the *trans* configuration have melting points much higher than the corresponding *cis* unsaturated fatty acids, thus contributing to the consistency of solid hydrogenated oils. The hydrogenation of double bonds is usually incomplete and only 25–40% of double bonds are reduced and the extent of *cis–trans*-isomerisation and positional isomerisation is often even greater than that of the hydrogenation reactions. In partially hydrogenated fats, up to 70% of double bonds of fatty acids can be in the *trans* configuration so the ratio of *cis*- and *trans*-isomers is about 30 : 70. It is therefore more accurate to describe this situation by the terms **partial catalytic hydrogenation** or **hardening**, which includes both hydrogenation and isomerisation reactions. Examples of the composition of hydrogenated oils are listed in Table 3.43. By a mechanism similar to industrial catalytic hydrogenation, microbial biohydrogenation proceeds in ruminants.

The negative health effect of *trans* fatty acid isomers has stimulated efforts to achieve their full elimination, which for industry means producing no partially hydrogenated fats. Major manufacturers have actually abandoned this technology and have been forced to replace it by another process, which is usually a transesterification procedure in combination with fractionation of triacylglycerols (fractional crystallisation). Hydrogenation of oils for food purposes is now only used for the production of **fully hydrogenated fats**, where the residual content of monoenoic fatty acids is 1–1.5% and the residual content of *trans*-monoenoic fatty acids is 0.7–1.0%.

Table 3.43 Fatty acid composition of partially hydrogenated fats (% of total fatty acids).

Fatty acid types	Hydrogenated sunflower oil	Hydrogenated soybean oil	Hydrogenated rapeseed oil
Saturated	18–28	20–30	12–20
cis-Monoenoic	15–20	10–20	10–23
trans-Monoenoic	35–45	22–45	20–40
Dienoic	1–15	1–10	1–8
Trienoic	0	trace–0.1	trace–0.4

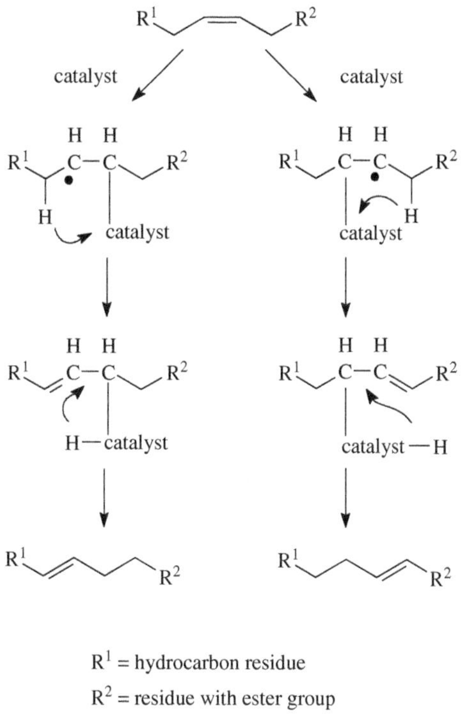

R^1 = hydrocarbon residue

R^2 = residue with ester group

Figure 3.21 Positional isomerisation of a double bond during unrealised hydrogenation.

3.8.1.7.2 Side reaction during hydrogenation

After adsorption on the surface of the catalyst, the fatty acid is not necessarily hydrogenated because it is often desorbed just before the adsorbed double bond reacts with hydrogen. The double bond adsorbed on the surface of the catalyst can also migrate to different positions in the chain (Figure 3.21). For example, partially hydrogenated oleic acid contains isomers with double bonds in positions Δ7–Δ11. The *cis*-unsaturated fatty acids content is 42% of the total unsaturated fatty acids, in which the main geometrical isomer is elaidic acid (13%), followed by other positional isomers: Δ8 (10%), Δ10 (10%), Δ7 (7%), and Δ11 (7%). In natural menhaden oil, mainly *cis*-Δ11 isomers occur, but in hydrogenated oil, positional isomers are present with double bonds in positions Δ3–Δ17, amongst which *cis*-Δ6, *cis*-Δ9 and *cis*-Δ11, *trans*-Δ11 and *trans*-Δ13 isomers predominate. The representation of individual isomers may vary depending on the fatty acid composition of the starting oil. Distribution of *trans*-octadecenoic acids in hydrogenated fats is very different from their distribution in cow's milk fat (Table 3.44). The nutritional value of various positional isomers has not been fully investigated.

3.8.1.7.3 Hydrogenation of carboxyl or ester groups

At high temperatures, under pressure, and using metal complexes prepared from chromium, copper, nickel, or platinum as catalysts, the fatty acid carboxyl groups are also hydrogenated to primary hydroxyl groups and the hydrogenation of esters gives two types of alcohols (Figure 3.22). Alcohol R^1–OH is usually methanol. These hydrogenations are very important in oleochemistry for the production of fatty alcohols as hydrophobic substances for the production of surfactants (e.g. dodecanol and tetradecanol from coconut and palm kernel oils).

3.8.1.8 Oxidation reactions

3.8.1.8.1 Classification

In foods, the following types of oxidation reactions of lipids may occur:

- autoxidation by triplet oxygen;

- oxidation by hydrogen peroxide or hydroperoxides;

Table 3.44 Relative representation of *trans*-octadecenoic fatty acids in margarine blends containing partially hydrogenated fats in comparison with cow's milk fat (% of total *trans*-octadecenoic acids; ranges of values and median values are indicated).

Double bond position	Partially hydrogenated vegetable oil	Cow's milk fat
Δ4	0.0–1.6/0.3	0.9–1.7/1.2
Δ5	0.3–1.3/0.7	0.7–1.4/1.0
Δ6–8	9.4–24.1/18.5	3.8–5.0/4.5
Δ9	9.1–34.2/23.7	4.3–6.9/5.8
Δ10	8.7–26.1/20.7	2.7–5.3/4.3
Δ11	6.1–21.1/13.4	42.8–59.0/49.2
Δ12	5.1–18.8/10.9	4.2–6.1/5.4
Δ13–14	3.4–32.5/9.4	9.7–14.2/12.6
Δ15	0.5–5.2/1.5	6.9–8.7/7.5
Δ16	0.3–7.8/1.1	7.2–9.3/8.5
trans-C18:1 of all fatty acids (%)	0.2–25.9/9.3	3.2–5.9/4.3

- oxidation by singlet oxygen (mostly photooxidation);

- oxidation catalysed by enzymes (lipoxygenases);

- oxidation by metals in higher valency;

- oxidation by quinones and related compounds.

Oxidation reactions of the fatty acid hydrocarbon chains are common to free fatty acids and their esters. Carboxyl groups of free fatty acids, however, accelerate the decomposition of fatty acid hydroperoxides and can react with some oxidation products.

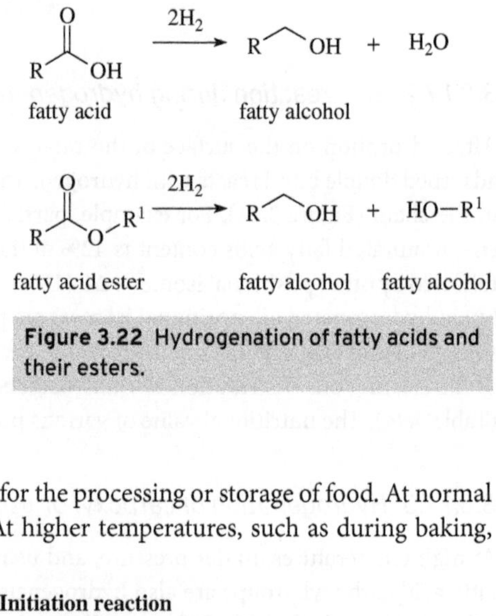

Figure 3.22 Hydrogenation of fatty acids and their esters.

3.8.1.8.2 Autoxidation

Autoxidation of fatty acids is the most common type of oxidation under conditions suitable for the processing or storage of food. At normal temperatures, only unsaturated fatty acids are oxidised by atmospheric (triplet) oxygen. At higher temperatures, such as during baking, frying, and roasting, autoxidation of saturated fatty acids also proceeds.

Autoxidation of the hydrocarbon chain of fatty acids, and of many other hydrocarbons, is a radical chain reaction that proceeds in three stages. The simplified reaction mechanism is shown in Figure 3.23. The first stage of the reaction is the formation of a free **hydrogen radical** (H$^\bullet$) and a free fatty acid radical (R$^\bullet$) formed by homolytic cleavage of the covalent bond H—C of the hydrocarbon chain. The energy needed to break down the H—C bond of fatty acids may be obtained from various sources. It may be heat energy (from a heating source), ultraviolet or radioactive radiation, or visible radiation (in the presence of photosensitisers in the case of a two-electron oxidation by singlet oxygen). The dissociation energy of the **H—CH—** bond in the middle of the saturated fatty acid chain is 410 kJ/mol, that of the —CH—**H** bond adjacent to COOH end of its chain is 376 kJ/mol, that of the methylene group adjacent to the double bonds, —**H**—CH—CH=CH—, is 322 kJ/mol, but that of the methylene group between two double bonds is only 272 kJ/mol. The fission of the fatty acid molecule also occurs by reaction with another free radical (immediate reaction of the hydrocarbon chain

Initiation reaction
　R–H (fatty acid residue in lipid) → R$^\bullet$ (free radical)

Propagation reaction
　formation of peroxyl radical
　R$^\bullet$ + O$_2$ → R–O–O$^\bullet$
　formation of hydroperoxide
　R–O–O$^\bullet$ + R–H → R–O–OH + R$^\bullet$

Termination reaction
　2R$^\bullet$ → R–R
　R$^\bullet$ + R–O–O$^\bullet$ → R–O–O–R
　2R–O–O$^\bullet$ → R–O–O–R + O$_2$

Figure 3.23 Autoxidation chain reaction of lipids.

directly with oxygen is thermodynamically difficult) or by reaction with transition metals. This first stage is called the **initiation stage** (initiation) of an autoxidation reaction.

The resulting fatty acid (lipid) free radical ($R^{\bullet}$) is very reactive, so it easily combines with an oxygen molecule, which is actually a biradical. This results in the formation of a peroxyl radical ($R—O—O^{\bullet}$). The peroxyl radical is also very reactive and splits off a hydrogen atom from another unsaturated fatty acid molecule with the formation of a **hydroperoxide** ($R—O—OH$) and another fatty acid free radical ($R^{\bullet}$). This second stage of the autoxidation reaction is called the **propagation stage** (propagation). The sequence of these two reactions in the propagation stage may be repeated just once or many more times; chains having more than 1000 steps have been observed. Therefore, autoxidation is called a radical chain reaction. The reaction of a fatty acid free radical with oxygen is much faster than the reaction of a peroxyl radical with a hydrocarbon chain of fatty acid. As a peroxyl radical reacts with a fatty acid molecule relatively slowly, this reaction therefore determines the reaction rate of the autoxidation.

If the concentration of free radicals in the reaction system is quite high, it is likely that two free radicals react together to form a relatively stable product, and the chain reaction ends. This third stage is called the **termination stage** (termination) of an autoxidation reaction.[5] With a limited supply of oxygen, when the rate of autoxidation depends on its partial pressure, the main radicals in the system are fatty acid radicals ($R^{\bullet}$) and the main termination reaction is their recombination. With an adequate oxygen supply, the reaction rate is independent of its partial pressure. Higher amounts of peroxyl radicals ($R—O—O^{\bullet}$) are formed, and the main termination reactions are then recombination of radicals of the fatty acids ($R^{\bullet}$) with peroxyl radicals and mutual recombination of peroxyl radicals.

Unsaturated fatty acids Unsaturated monoenoic fatty acids split off hydrogen atoms relatively easily; at least, the hydrogen from the methylene group that is adjacent to the double bond does. Even less energy is needed to split hydrogen from dienoic and trienoic fatty acids. The primary reaction products of autoxidation are hydroperoxides. The number of double bonds remains unchanged, but they are usually shifted by one carbon atom to either the carboxyl or the methyl end; this is particularly easy in dienoic and trienoic fatty acids, because the free radicals arise between two double bonds and are stabilised by mesomerism. The reaction with oxygen yields a peroxyl radical at one end of the mesomeric system, and the double bond is simultaneously rearranged from the *cis* configuration to the more stable *trans* configuration.

Monoenoic acids Autoxidation of oleic acid is illustrated in Figure 3.24 as an example of autoxidation of monoenoic acids. The C—H bond cleavage in the initial stage of the reaction happens in the vicinity of the double bond, which occurs on carbon C-8 or C-11 (the dissociation energy is 322 kJ/mol). The resulting fatty acid radical, which is called allylic radical, may or may not isomerise, therefore reactions of fatty acid radicals with oxygen in the propagation stage of the reaction produce four hydroperoxides (isomeric hydroperoxyoctadecenoic acids). At room temperature, the amounts are approximately the same, but are dominated to a certain extent by 8- and 11-hydroperoxides. These two hydroperoxides that are generated without the double bond shift, are a mixture of *cis*- and *trans*-isomers, whilst hydroperoxides formed by positional isomerisation of double bonds are *trans*-isomers.

Dienoic acids The higher reactivity of the methylene group between two double bonds causes the oxidation process to take place almost exclusively at this carbon atom (dissociation energy is 272 kJ/mol). In the case of linoleic acid, the oxidation occurs at the C-11 methylene group and the main oxidation products are hydroperoxides with conjugated double bonds: (10*E*,12*Z*)- and (10*E*,12*E*)-9-hydroperoxyoctadeca-10,12-dienoic acids, and also (9*Z*,11*E*)- and (9*E*,11*E*)-13-hydroperoxyoctadeca-9,11-dienoic acids. At room temperature or lower, it is mainly *cis,trans*- and *trans,cis*-hydroperoxides that are formed, whilst at higher temperatures, mainly *trans,trans*-hydroperoxides arise.

11-Hydroperoxyoctadeca-9,12-dienoic acid (*cis,cis*-isomer) is formed at a low concentration (without double bond isomerisation). Radicals on carbons C-8 or C-14, adjacent to the double bond, are also formed to a lesser extent. As in the previous cases, these radicals give rise to the corresponding geometric isomers: 8-hydroperoxyoctadeca-9,12-dienoic and 10-hydroperoxyoctadeca-8,12-dienoic acids, and 12-hydroperoxyoctadeca-9,13-dienoic and 14-hydroperoxyoctadeca-9,12-dienoic acids. These reactions of linoleic acid are given in Figures 3.25 and 3.26.

Trienoic acids In the initial stage of the autoxidation reaction, a linolenic acid C—H bond is cleaved preferentially at C-11 and C-14; that is, at the methylene groups located between two double bonds. The main oxidation products of autoxidation are four hydroperoxides with a system of two conjugated double bonds and an isolated third double bond, with hydroperoxyl groups located in positions C-9, C-12,

---

[5] Termination of hydrocarbon radicals is not the only reaction. Other possibilities (depending on the structure of the radicals) are: (i) the termination by disproportionation, where hydrogen is eliminated, which yields an olefin ($R—CH_2—CH_2^{\bullet} \rightarrow R—CH=CH_2 + H^{\bullet}$); (ii) termination by transfer (e.g. in reaction with antioxidants); (iii) termination by recombination with a hydroxyl radical, which leads to an alcohol ($R—CH_2—CH_2 \bullet + HO^{\bullet} \rightarrow R—CH_2—CH_2—OH$); and (iv) further oxidation of the terminal hydroperoxide ($R—CH_2—CH_2 \bullet + O_2 \rightarrow R—CH_2—CH_2—O—O\bullet \rightarrow R—CH_2—CH_2—O—OH$). In the terminal hydroperoxide, the link between oxygen atoms is then cleaved, which gives an alkanal ($R—CH_2—CH_2—O—OH \rightarrow R—CH_2—CH=O + H_2O$). Alternatively, the link between carbon atoms can be cleaved, which results in a shorter alkyl radical ($R—CH_2^{\bullet}$). Some of these reactions are discussed in relation to the secondary reactions of hydroperoxides.

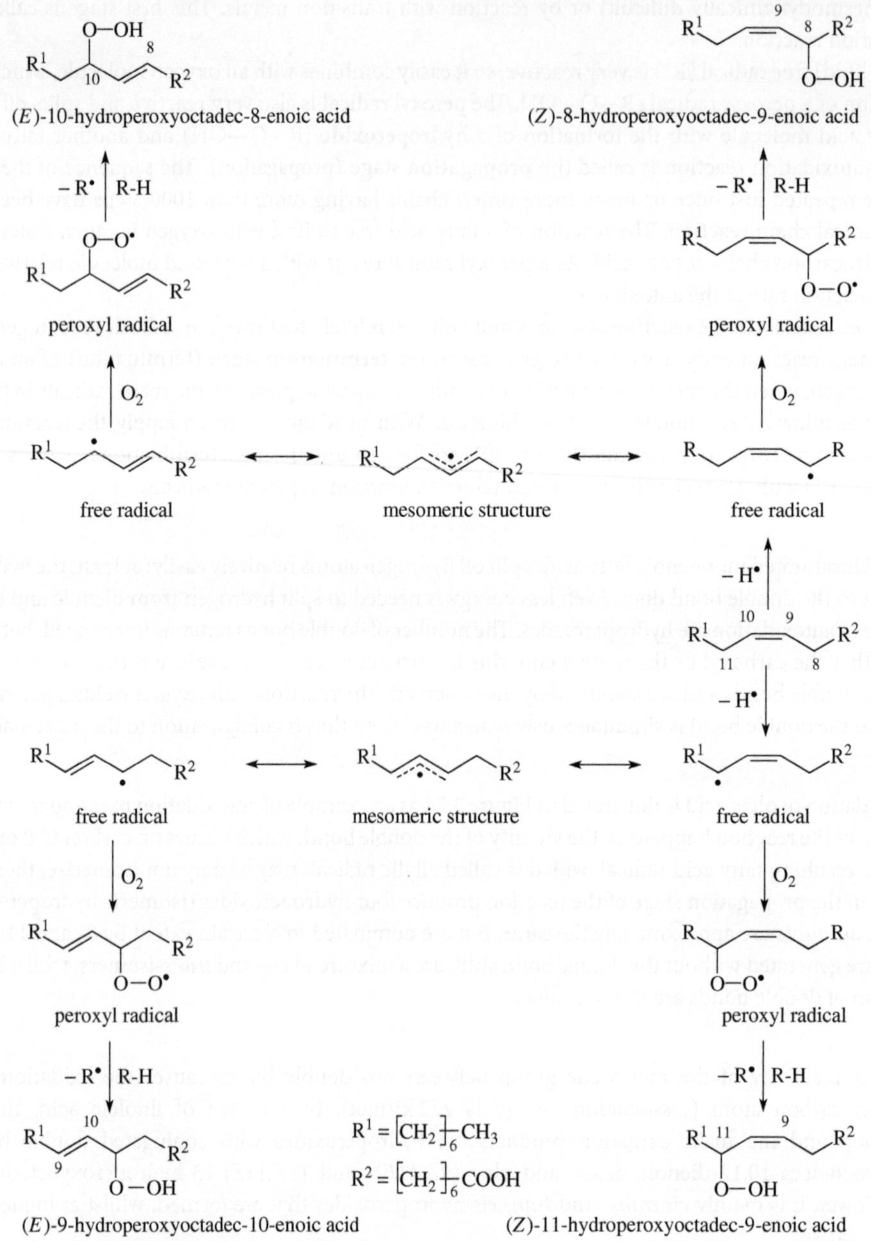

Figure 3.24 Autoxidation of oleic acid.

C-13, and C-16. In fact, these hydroperoxides are a mixture of *cis,trans*- and *trans,trans*-isomers and the isolated double bond has the *cis* configuration (Figure 3.27).

The predominant oxidation products are 9-hydroperoxide and 16-hydroperoxide, as 12-hydroperoxide and 13-hydroperoxide are less stable and tend to transform into cyclic products (similarly to the reaction catalysed by lipoxygenase that produces prostaglandins). As in the case of linoleic acid, many other hydroperoxides are formed as minor products.

Saturated acids   Cleavage of the carbon–hydrogen bond in a saturated hydrocarbon chain requires a significant level of activation energy, so this reaction does not come into practical consideration at normal temperatures. At higher temperatures, however (corresponding to temperatures during baking, frying, and roasting), it is possible. In addition to the last carbon atom, all the other carbon atoms are almost equally prone to the creation of free radicals by the C—H bond cleavage. The third and, to a lesser extent, the second and fourth carbons, are somewhat more reactive than the other carbon atoms of the hydrocarbon chain. The resulting free radicals bind oxygen through the formation of peroxyl radicals, and the reaction proceeds in a similar way to reactions of unsaturated acid peroxyl radicals, as already shown.

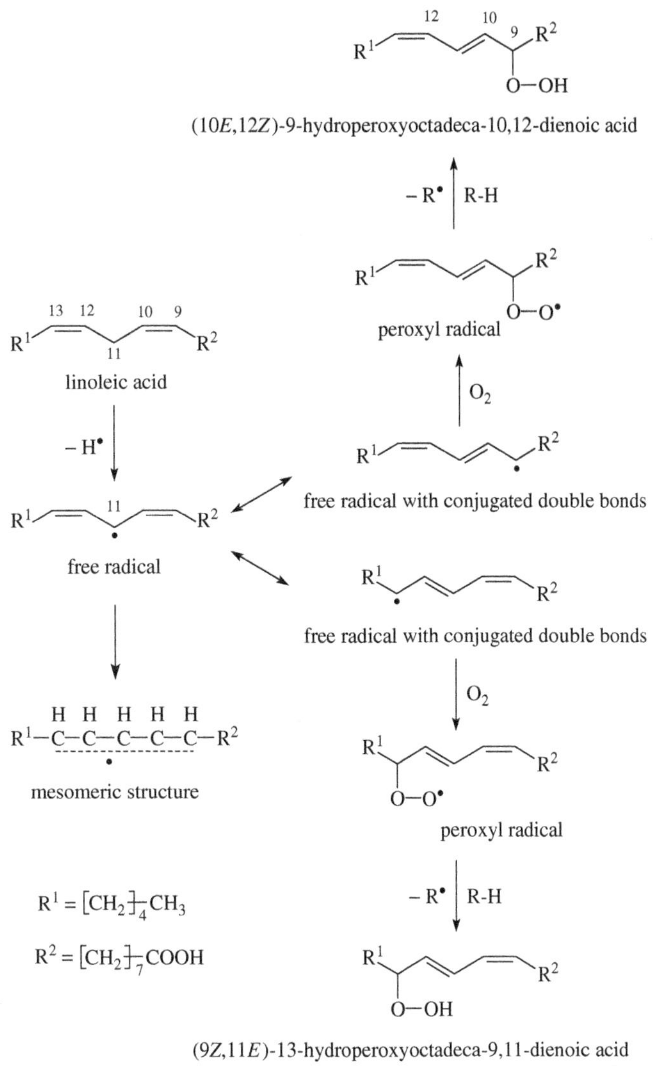

Figure 3.25 Main primary products of linoleic acid autoxidation.

Since saturated acids can form free radicals in many different positions along the hydrocarbon chain, the composition of the oxidation products is diverse.

Hydroperoxides The primary products of autoxidation are sensorially indifferent hydroperoxides of fatty acids; that is, products that have no taste and smell in comparison with other autoxidation products, such as aldehydes. Hydroperoxides, especially those of dienoic and trienoic acids, are very unstable compounds that decompose to either a hydrogen radical or a hydroxyl radical. In the former case, this decomposition yields a peroxyl radical, and in the latter case, the reaction produces an **alkoxyl radical** (Figure 3.28). The decomposition of hydroperoxides, especially at higher concentrations, takes place via dimeric intermediates associated by hydrogen bonds (Figure 3.29). Both of these free radicals can initiate the chain radical reaction of fatty acids (Figure 3.28).

According to current knowledge, formation of the alkoxyl radical is favoured, as the O—O bond dissociation energy is about 184 kJ/mol, whilst the O—H bond dissociation energy is about 377 kJ/mol. During oxidation (and particularly in its advanced stages), these reactions are the major initiation reactions. The reaction of a fatty acid with a peroxyl radical gives fatty acid hydroperoxide, whilst reaction with an alkoxyl radical produces a hydroxy acid.

The timeline of an autoxidation reaction is shown in Figure 3.30. At the beginning, the reaction is usually initiated by heat or radiation, so the initial reaction rate is low. This reaction stage is called the **induction period**. Hydroperoxides gradually accumulate in the system, which causes formation of other radicals, thus the initiating reaction rate increases with an increasing concentration of hydroperoxides. Therefore, this reaction is called an autocatalytic reaction. If there is sufficient oxygen, the reaction rate increases rapidly until, at some point, the rate of decomposition of hydroperoxides exceeds the rate of their formation, and then the amount of hydroperoxides gradually

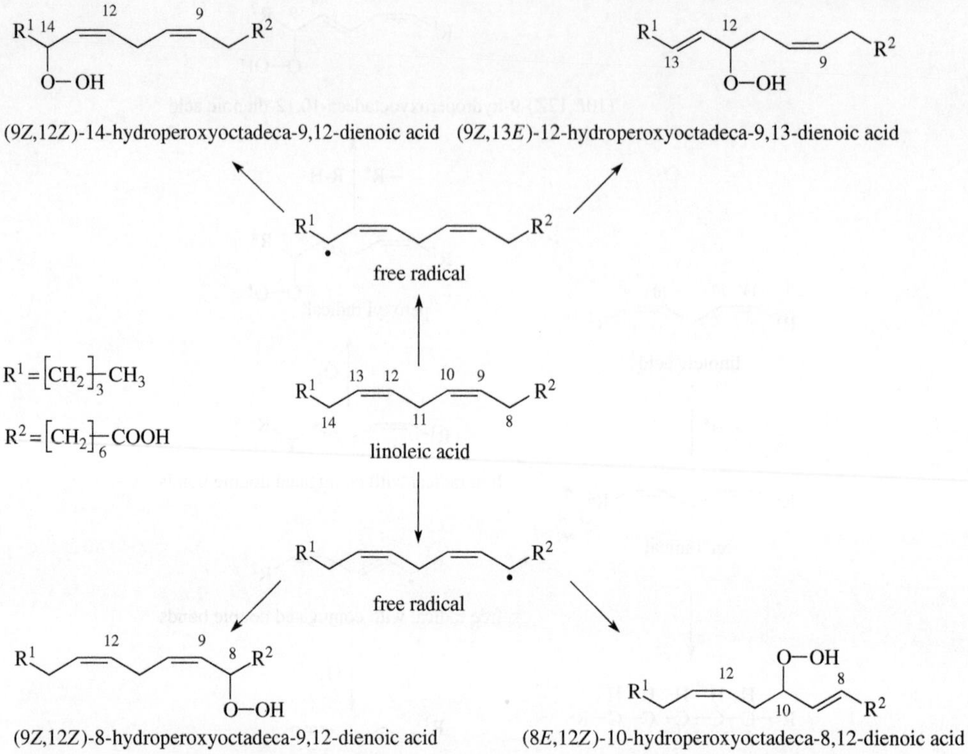

(9Z,12Z)-14-hydroperoxyoctadeca-9,12-dienoic acid   (9Z,13E)-12-hydroperoxyoctadeca-9,13-dienoic acid

free radical

$R^1 = \left[CH_2\right]_3 - CH_3$

$R^2 = \left[CH_2\right]_6 - COOH$

linoleic acid

free radical

(9Z,12Z)-8-hydroperoxyoctadeca-9,12-dienoic acid   (8E,12Z)-10-hydroperoxyoctadeca-8,12-dienoic acid

Figure 3.26 Minor primary products of linoleic acid autoxidation.

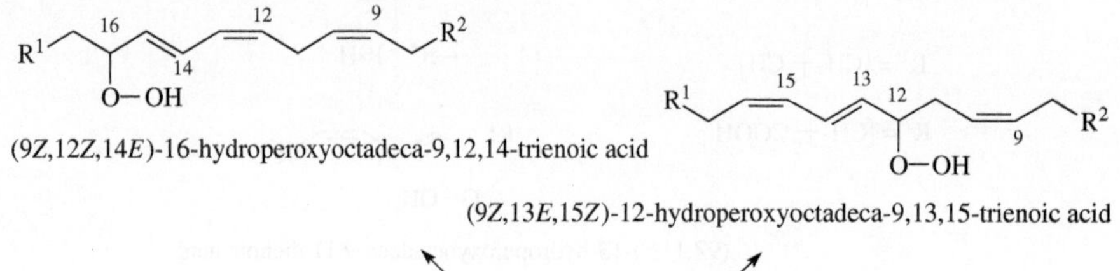

(9Z,12Z,14E)-16-hydroperoxyoctadeca-9,12,14-trienoic acid

(9Z,13E,15Z)-12-hydroperoxyoctadeca-9,13,15-trienoic acid

free radical

$R^1 = CH_3$

$R^2 = \left[CH_2\right]_6 - COOH$

linolenic acid

free radical

(9Z,11E,15Z)-13-hydroperoxyoctadeca-9,11,15-trienoic acid

(10E,12Z,15Z)-9-hydroperoxyoctadeca-10,12,15-trienoic acid

Figure 3.27 Main primary products of linolenic acid autoxidation.

$$R-O-OH \rightarrow R-O-O^{\bullet} + H^{\bullet} \qquad R-H + R-O-O^{\bullet} \rightarrow R^{\bullet} + R-O-OH$$

$$R-O-OH \rightarrow R-O^{\bullet} + HO^{\bullet} \qquad R-H + R-O^{\bullet} \rightarrow R^{\bullet} + R-O-OH$$

$$2\,R-O-OH \rightarrow R-O-O^{\bullet} + R-O^{\bullet} + H_2O$$

Figure 3.28 Hydroperoxides as initiators of autoxidation reactions.

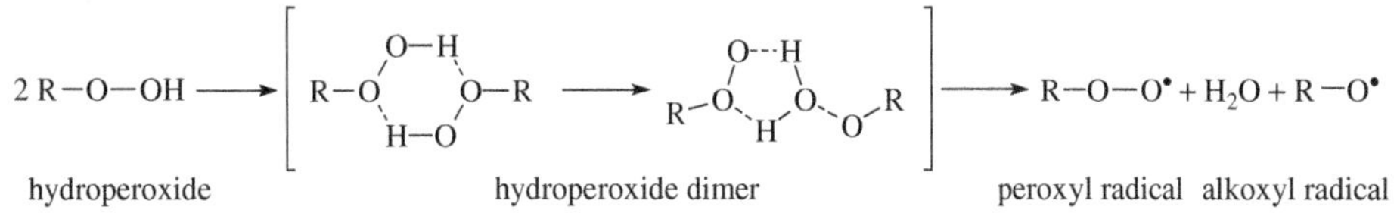

$$2\,R-O-OH \longrightarrow \left[\underset{\text{hydroperoxide dimer}}{R-O{\overset{O-H}{\underset{H-O}{\diagdown}}}O-R} \longrightarrow R-O{\overset{O\cdots H}{\underset{H}{\diagdown}}}O-R \right] \longrightarrow R-O-O^{\bullet} + H_2O + R-O^{\bullet}$$

hydroperoxide   hydroperoxide dimer   peroxyl radical   alkoxyl radical

Figure 3.29 Decomposition of hydroperoxides to free radicals.

decreases. During deceleration of the formation of additional hydroperoxides, their decomposition becomes more important. Decomposition of hydroperoxides does not depend on the presence of oxygen or of unreacted lipids. At higher temperatures, such as during deep-fat frying, hydroperoxides do not accumulate in fat, because they decompose at about 150 °C.

Secondary autoxidation products Hydroperoxides of fatty acids and their radicals can in principle react in three ways in secondary reactions:

- reactions that do not change the number of carbon atoms in the molecule (e.g. the formation of cyclic peroxides and endoperoxides, epoxy acids, hydroxy acids, and oxoacids);

- reactions involving decomposition of molecules, producing products with fewer carbon atoms (formation of aldehydes, hydrocarbons, or oxoacids);

- polymerisation reaction in which the number of carbons in the molecule increases.

Cyclic peroxides and endoperoxides Hydroperoxides of polyenoic fatty acids with three or more double bonds in the molecule, which have a system of conjugated double bonds adjacent to hydroperoxide group, are very unstable compounds. They tend to form more stable cyclic six-membered peroxides derived from 1,2-dioxane by 1,4-cyclisation, five-membered peroxides derived from 1,2-dioxolane by 1,3-cyclisation, and endoperoxides.

As an example, the formation of 1,2-dioxanes from 12-hydroperoxyoctadeca-9,13,15-trienoic acid derived from linolenic acid is shown in Figure 3.31. 1,2-Dioxanes are unstable compounds and decompose to low molecular weight flavour-active products. For example, 10-hydroperoxy-9,12-peroxyoctadeca-13,15-dienoic acid is decomposed into octa-3,5-dien-2-one. The mechanism of formation of 1,2-dioxolanes, cyclic peroxides, and cycloendoperoxides from 13-hydroperoxyoctadeca-9,11,15-trienoic acid is shown in Figure 3.32. Peroxohydroperoxides of the 1,2-dioxolane type are considered to be the main precursors of toxic malondialdehyde (see Section 3.8.1.12.1).

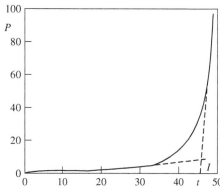

Figure 3.30 Course of autoxidation of sunflower oil at 40 °C. P, amount of fatty acid hydroperoxides in milliequivalents of oxygen per kg (peroxide number); t, duration of the autoxidation in days; I, induction period.

Epoxy acids, hydroxy acids, and oxoacids Hydroperoxides, or strictly peroxyl radicals, readily react with unsaturated double bonds of olefins to form epoxide derivatives. The addition of a peroxyl radical to the double bond of an olefin can proceed as an intermolecular reaction (Figure 3.33).

Similarly, epoxides are also formed during the reaction of olefins with hydrogen peroxide. Secondary reactions of oxidised polyenoic acids produce red-coloured cyclopentane derivatives (**3-99**) causing adverse colour changes to stored fish.

A particularly important reaction of unsaturated fatty acids is an intramolecular reaction (Figure 3.34), occurring when epoxide arises from an alkoxyl radical. The resulting radical of epoxy acid reacts with oxygen to form a hydroperoxyl radical, which is decomposed

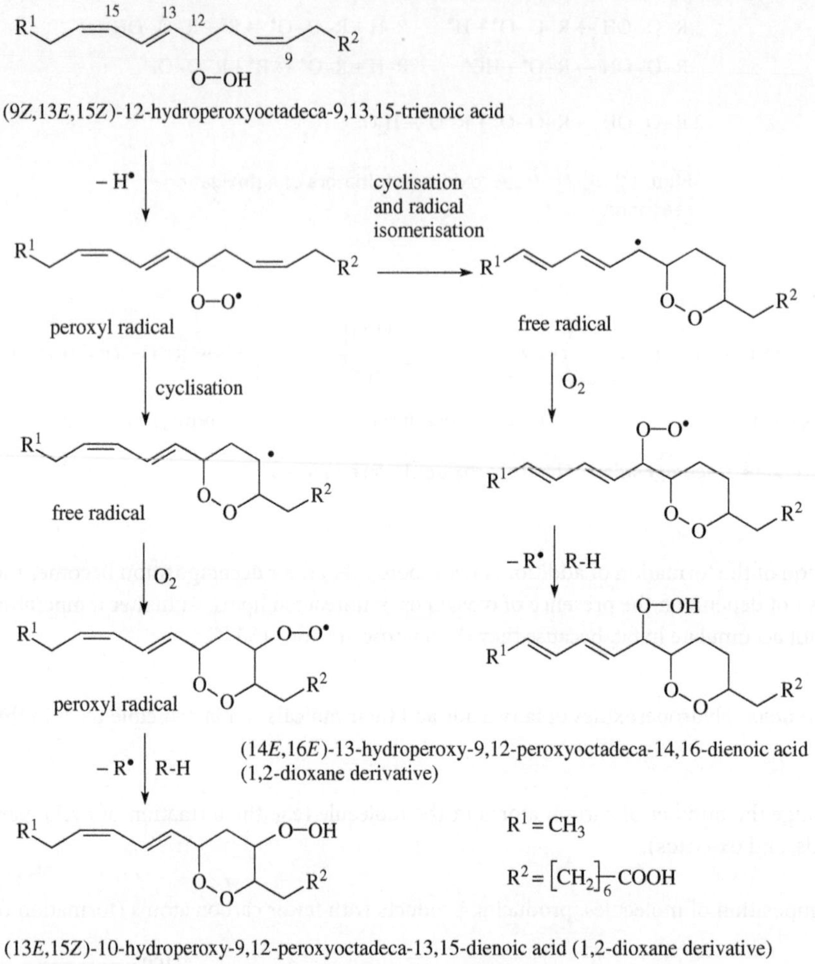

Figure 3.31 Formation of 1,2-dioxanes from 12-hydroperoxyoctadeca-9, 13,15-trienoic acid.

via hydroperoxide to an alkoxyl radical. This alkoxyl radical provides, by recombination with a hydrogen radical, the corresponding hydroxy acid and elimination of a hydrogen atom yields the corresponding oxoacid. Somewhat more complicated is the intramolecular reaction of hydroperoxides with conjugated double bonds. Conjugated dienes, such as 13-hydroperoxyoctadeca-9,11-dienoic and 9-hydroperoxyoctadeca-10,12-dienoic acids, arise as major autoxidation products of linoleic acid. This reaction is illustrated in Figure 3.35 in the example of 13-hydroperoxyoctadeca-9,11-dienoic acid. Decomposition of this hydroperoxide via alkoxyl radicals creates 13-hydroxyoctadeca-9,11-dienoic- and 13-oxooctadeca-9,11-dienoic acids. Oxidation and other reactions of the alkoxyl radical give rise to two isomeric hydroperoxy acids: 12,13-epoxy-11-hydroperoxyoctadec-9-enoic and 12,13-epoxy-9-hydroperoxyoctadec-10-enoic acids. Decomposition of these epoxyhydroperoxy acids produces, analogously, epoxyhydroxy acids and epoxyoxoacids.

Radicals of dihydrofuran derivatives are formed when the alkoxyl radicals resulting from the decomposition of hydroperoxides react with adjacent conjugated double bonds through 1,4-addition. These radicals are stabilised, in the same way as 1,2-epoxide radicals, by reaction with oxygen and subsequent reactions provide dihydrofuran derivatives. Radicals of dihydrofuran derivatives can also react with oxygen to form a peroxyl radical, which cleaves another fatty acid molecule and is transformed into the corresponding hydroperoxide from which, via an alkoxyl radical, hydroxy acids and oxoacids are formed (Figure 3.36).

Cyclic peroxides (1,2-dioxolane derivatives) provide substituted derivatives of tetrahydrofuran by ring cleavage and rearrangement (Figure 3.37). Dihydrofuran derivatives (so-called 1,4-epoxy compounds) can also be formed from unsaturated 1,2-epoxides by ring opening under acidic conditions and by cyclisation of the free radicals generated (Figure 3.38). This reaction is analogous to the formation of so-called 5,8-epoxides from 5,6-epoxides, which occurs in carotenoid pigments (see Section 9.9.1.2).

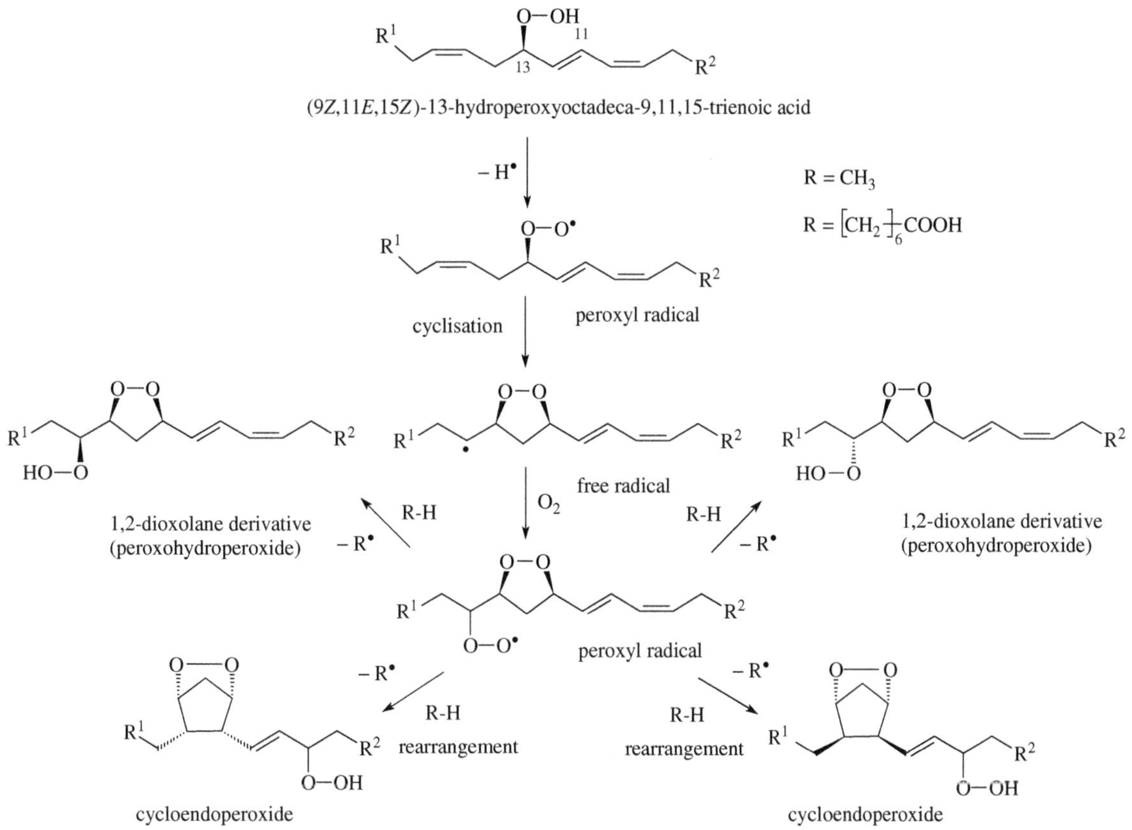

Figure 3.32 Formation of 1,2-dioxolanes from 13-hydroperoxyoctadeca-9,11,15-trienoic acid.

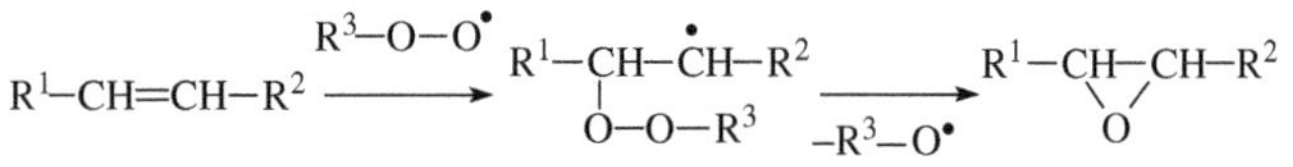

Figure 3.33 Formation of epoxide from peroxyl radical and olefins.

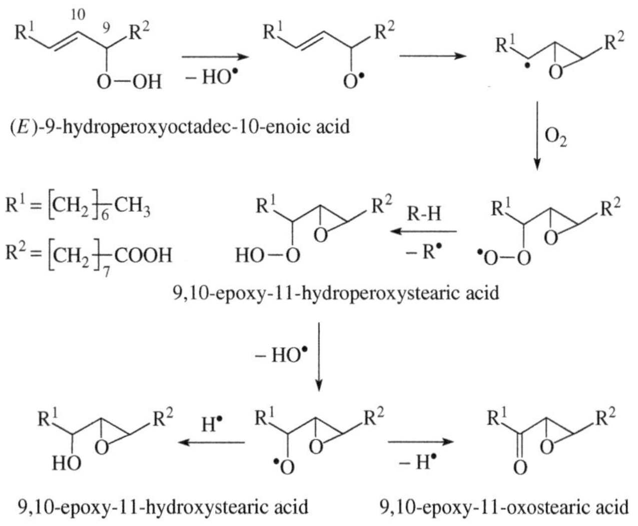

Figure 3.34 Formation of epoxy acids, hydroxy acids, and oxoacids from oleic acid 9-hydroperoxide.

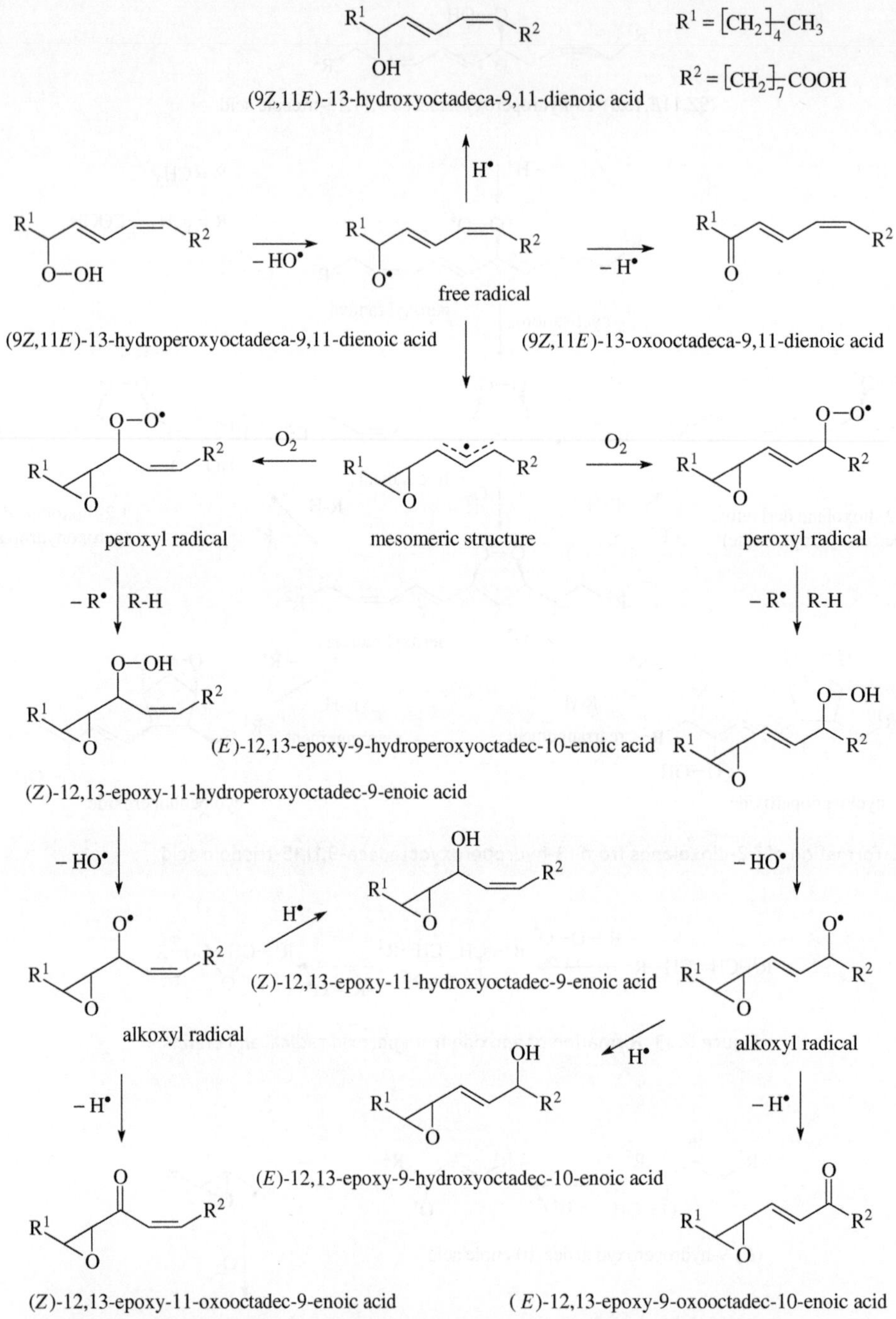

Figure 3.35 Formation of epoxy acids, hydroxy acids, and oxoacids from linoleic acid 13-hydroperoxide.

Epoxides are reactive compounds, especially at higher temperatures. They form dihydroxy derivatives with water, ethers with alcohols, esters with the carboxyl group of fatty acids, chlorohydrins (α-chlorohydroxy compounds) with hydrochloric acid and amino compounds with amines (Figure 3.39). Chlorinated hydroxy acids arise, for example, in polyvinyl chloride (PVC) by reaction of epoxidised soybean oil (used as a stabiliser of PVC) with hydrogen chloride (see Section 11.4.2.2.2).

Common oxidation products are derivatives with a hydroxyl group in the α-position to the oxo group, known as ketols. Ketols are formed in reactions of hydroperoxides with the neighbouring double bonds (Figure 3.40). They have significant reducing properties; their reactivity is close to the reactivity of aldehydes. The hydrolysis product of dihydrofuran derivative of 9-oxo-10,13-epoxyoctadec-11-enoic acid, which is 10,13-dihydroxy-9-oxooctadec-11-enoic acid, is also a ketol. The structure of this ketol is shown in formula **3-99**.

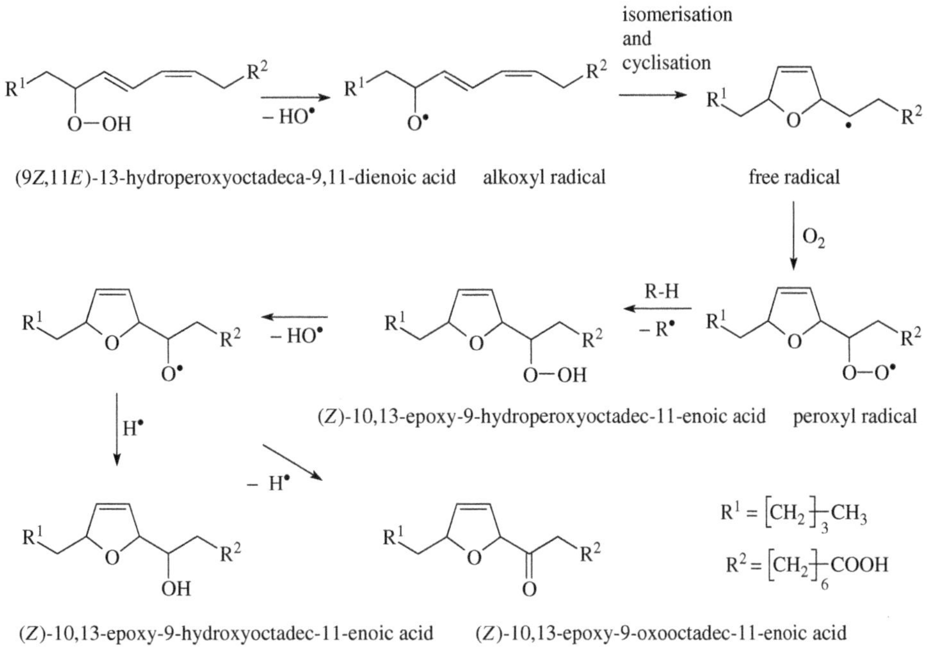

Figure 3.36 Formation of dihydrofuran derivatives from an alkoxyl radical of linoleic acid 13-hydroperoxide.

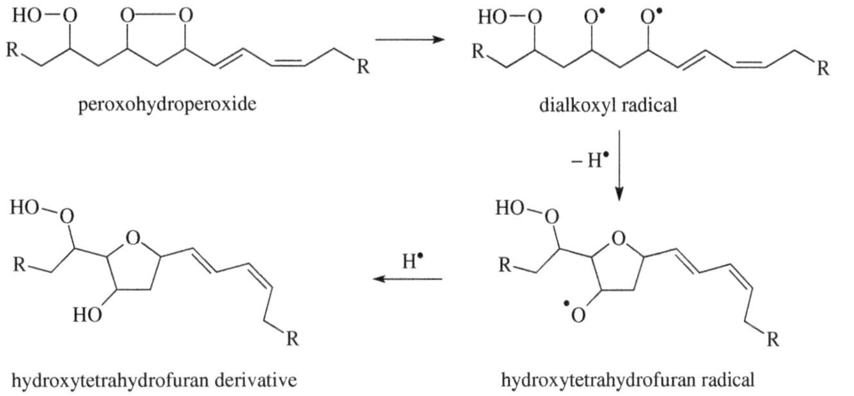

Figure 3.37 Formation of tetrahydrofuran derivatives from 1,2-dioxolanes.

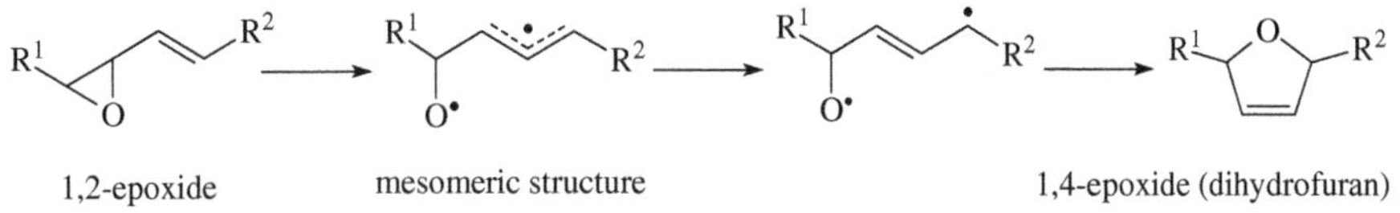

Figure 3.38 Formation of fatty acid 1,4-epoxides from 1,2-epoxides.

Volatile secondary products Another important group of reactions are those in which the hydrocarbon chain of the alkoxyl radicals cleaves to form low-molecular-weight products, mainly volatile and sensory active compounds. The cleavage takes place on both sides of the alkoxyl radicals (Figure 3.41). The composition of reaction products formed from alkoxyl radicals derived from unsaturated fatty acids depends on at which carbon, next to the hydroperoxide group, the double bond is located. In addition to non-volatile oxoacids and hydroxy acids, the main volatile sensorially active decomposition products include saturated and unsaturated aldehydes and saturated and unsaturated hydrocarbons. As an example, the cleavages of two major hydroperoxides of linoleic acid, 13-hydroperoxy-9-octadeca-9,11-dienoic (Figure 3.42) and

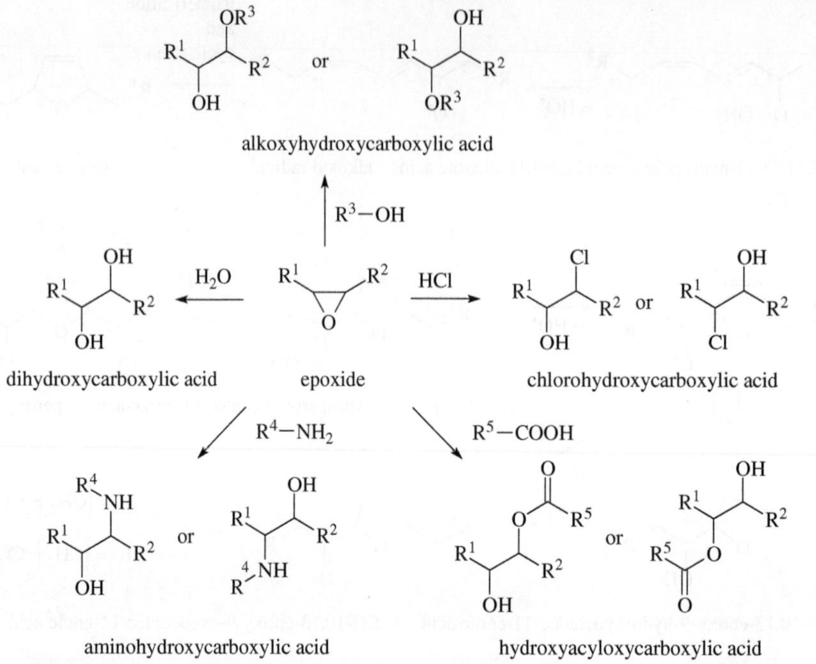

Figure 3.39 Important reactions of fatty acid epoxides.

9-hydroperoxyoctadeca-10,12-dienoic acids (Figure 3.43), are shown. Other hydroperoxides of unsaturated and saturated acids decompose in an analogous manner.

Hydrocarbons give a typical rancid flavour to slightly oxidised oils, whilst aldehydes carry a typical rancid flavour in the advanced stages of oxidation, and thus there is a close relationship between the aldehyde content and the intensity of the rancid flavour. The nature of the rancid odour changes according to the type of aldehyde. Alkanals and alk-2-enals impart a typical rancid flavour to oils but the sensory effects of alk-2-enals are more intense than those of alkanals. The oxidation of oils with higher linoleic acid contents produces alka-2,4-dienals, which are another important group of volatile oxi-

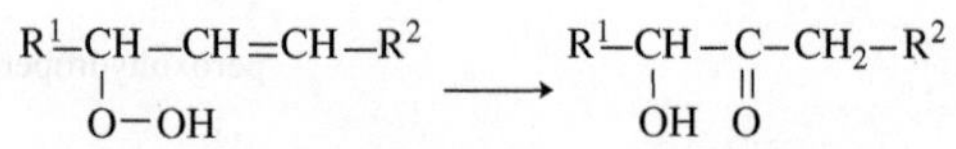

Figure 3.40 Formation of ketols by oxidation of fatty acids.

dation products. In small amounts, they impart a fried flavour to oils, whilst at slightly higher levels, the flavour resembles peanuts, and at even higher levels, it becomes unpleasant. Conjugated alkatrienals, formed during the oxidation of linolenic acid in oils, impart an off-flavour to oils, resembling varnish or fish oil. Substances that are less sensorially active than alkanals and alkenals, but nevertheless are important reaction products, are alkan-2-ones, the odour of which is reminiscent of mouldy cheese. An overview of the hydrocarbons and carbonyl compounds formed by decomposition of hydroperoxides of oleic, linoleic, and linolenic acids is given in Sections 8.2.1.1.2 and 8.2.4.1.1.

A specific type of rancidity, known as light-induced flavour reversion in the United States, occurs during the oxidation of soybean oil. This beany and green off-odour is attributable to the furan fatty acids present in the soybean oil in relatively high levels (see Section 3.3.1.4.1). Upon photooxidation, 3-methyl-2,4-nonanedione is formed as a key intermediate with an intense strawlike and fruity flavour that is further photooxidised to 3-hydroxy-3-methyl-2,4-nonanedione with the odour description of rubbery, earthy, and plasticlike (Figure 3.44), which is shown to be one of the compounds responsible for the odour reversion in light-exposed soybean oil.

Some other furan derivatives found in soybean oil include pentylfuran and (Z)- and (E)-2-(pent-2-en-1-yl)furan (**3-99**), as well as other decomposition products of hydroperoxides. Pentylfuran arises as a byproduct of the decomposition of 9-hydroperoxyoctadeca-10,12-dienoic acid derived from linoleic acid via the terminal radical. The content of linoleic acid in soybean oil is about 51% of the total fatty acids. Similarly, both isomers (*cis* and *trans*) of 2-(pent-2-en-1-yl) furan arise by decomposition of 9-hydroperoxyoctadeca-10,12,15-trienoic acid formed by autooxidation of linolenic acid. Soybean oils used at present contain 7–10% linolenic acid.

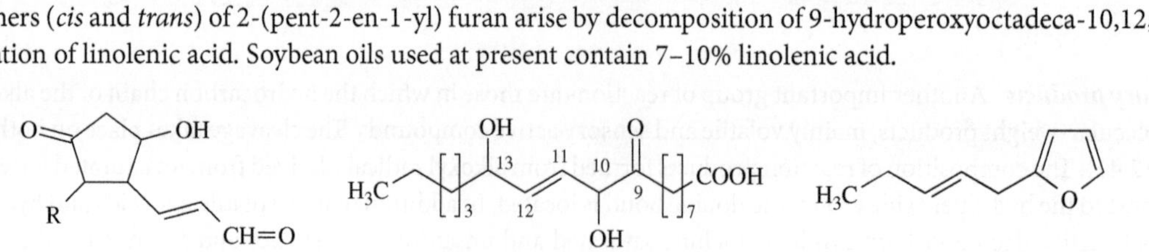

cyclopentanolone derivative

(*E*)-10,13-dihydroxy-9-oxooctadec-11-enoic acid

(*E*)-2-(pent-2-en-1-yl)furan

3-99, selected oxidation products of fatty acids

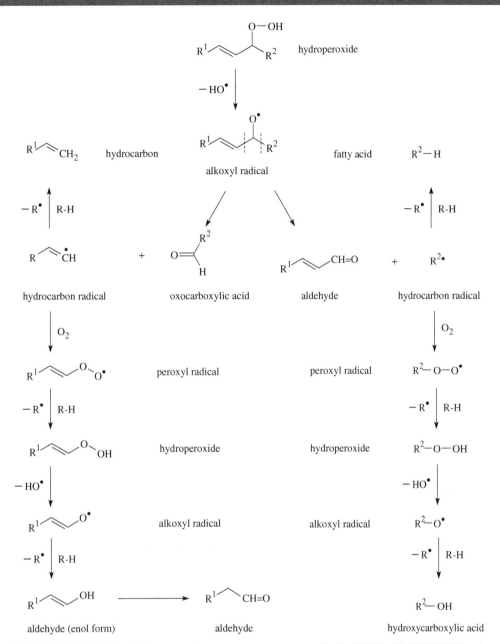

Figure 3.41 Fission of a fatty acid hydrocarbon chain during hydroperoxide decomposition.

Subsequent reactions of aldehydes Aldehydes are very reactive compounds that enter many other reactions in oxidised fats. Their carbonyl group is oxidised to a carboxyl group and, as a rule, they are further oxidised in the hydrocarbon chain. Common reactions are aldolisation, retroaldolisation, and reactions with other food components, such as proteins.

Alkanals Autoxidation of the hydrocarbon chain and of the carbonyl group of alkanals (via unstable peroxy acids and 2-hydroperoxy derivatives) leads to a mixture of products. The important final products are the corresponding fatty acids, formic acid and alkanals containing fewer carbon atoms per molecule than the parent compound (Figure 3.45). Lower alkanals, such as hexanal, also condense to trimeric 2,4,6-trialkyl-1,3,5-trioxanes.

Alk-2-enals The autoxidation of alk-2-enals leads to a number of products that are formed by mechanisms similar to the reactions of alkanals. An important product is highly reactive malondialdehyde (propanedial, also incorrectly known as β-hydroxyacrolein;

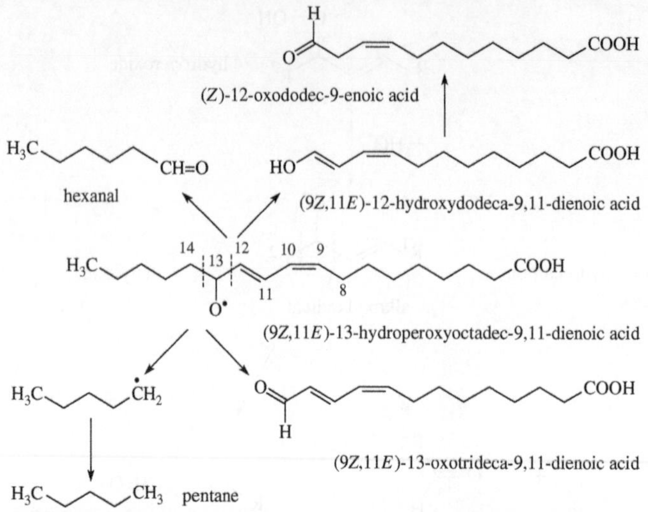

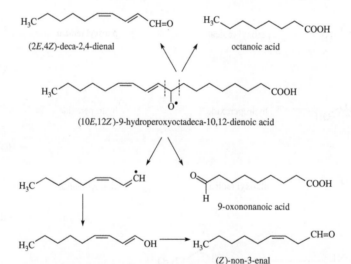

Figure 3.42 Fission of a 13-hydroperoxyoctadeca-9,11-dienoic acid hydrocarbon chain.

Figure 3.43 Fission of a 9-hydroperoxyoctadeca-10,12-dienoic acid hydrocarbon chain.

Figure 3.46). Malondialdehyde exists predominantly in enol form, as a *cis*-isomer in organic solvents and as a *trans*-isomer in water (Figure 3.47).

As with alkanals, the carbonyl group of alk-2-enals is oxidised to a carboxyl group via the corresponding peroxy acid. Acetaldehyde and an aldehyde with two less carbon atoms are formed by retroaldolisation, whereas aldolisation with acetaldehyde produces alka-2,4-dienals (Figure 3.48).

Alka-2,4-dienals Autoxidation of alka-2,4-dienals accompanied by retroaldolisation leads temporarily to epoxides and finally to relatively stable end products. Similarly to alk-2-enals, the main reaction products include alkanals, alkenals, dialdehydes (including malondialdehyde), and hydrocarbons. Some secondary reactions of alka-2,4-dienals are shown in Figure 3.49.

Secondary polymeric products The third group of secondary reactions of oxidised fats are reactions in which the number of carbons in the molecule increases. In specialised areas, these reactions are known under the term **oxypolymerisation** (oxidative polymerisation). Oxypolymerisation is one of the consequences of oxidative degradation of lipids. Free radical addition and sometimes cycloaddition reactions

(10Z,12Z)-10,13-epoxy-11,12-dimethyloctadecanoic acid

oxidation $^1O_2$

hydroperoxide

radical fragmentation
tautomerisation

oxidation

3-methyl-2,4-nonanedione   3-hydroxy-3-methyl-2,4-nonanedione

Figure 3.44 Formation of the odour principle of flavour reversion.

can be seen as the main causes of the production of oxidative polymers (oxypolymers) in triacylglycerols, especially if the polyunsaturated fatty acids content is high. Polymers are usually formed by reactions of two free radicals. If both are alkyl radicals, the formed dimers have the oxidised fatty acids bound by a single C—C bond between two carbon atoms. This reaction is described in more detail in Section 3.8.1.6. Alkyl radicals, of course, do not occur in high amounts and react slowly, so reactions between oxygen-containing radicals are more common (Figure 3.50). Most products contain ether bonds of the type C—O—C, although peroxide bonds of the type C—O—O—C can also occur. In theory, diperoxide bonds could also form, but they are unstable and their formation is accompanied by elimination of oxygen. Two fatty acids may also be linked to one another by two bonds, which yields tetrahydrofuran, tetrahydropyran, or dioxan derivatives (**3-100**). Usually, polymers are described by schematic formulae that indicate the type of link between the chains of original fatty acids (Figure 3.51).

tetrahydrofuran dimer   tetrahydropyran dimer   1,4-dioxan dimer

3-100, polymeric cyclic products of fatty acids

Polymers may also arise by non-radical mechanisms. For example, they are formed by reactions of epoxides with hydroxyl groups or by reactions of epoxides with carboxyl groups of free fatty acids (Figures 3.39 and 3.52). Many other types of polymerisation reactions have also been described.

Influence of reaction conditions The rate of formation and decomposition of hydroperoxides is very dependent on the structure and concentration of reactants and on the reaction conditions, mainly on the temperature, but also on the concentration of oxygen, the surface area exposed to air, water activity, presence of prooxidants or antioxidants, and other factors.

Structure of acids An important factor is the presence of reactive double bonds. Saturated fatty acids are practically stable during fat storage at normal temperatures. At temperatures around 20 °C, monoenoic fatty acids (such as oleic acid) are only oxidised very slowly, dienoic fatty acids (such as linoleic acid) are oxidised about 10 times faster, and trienoic fatty acids (such as linolenic acid) approximately 20 times faster. Arachidonic acid, as an example of a tetraenoic fatty acid, is oxidised about 30–40 times faster. At temperatures above 100 °C, all fatty acids

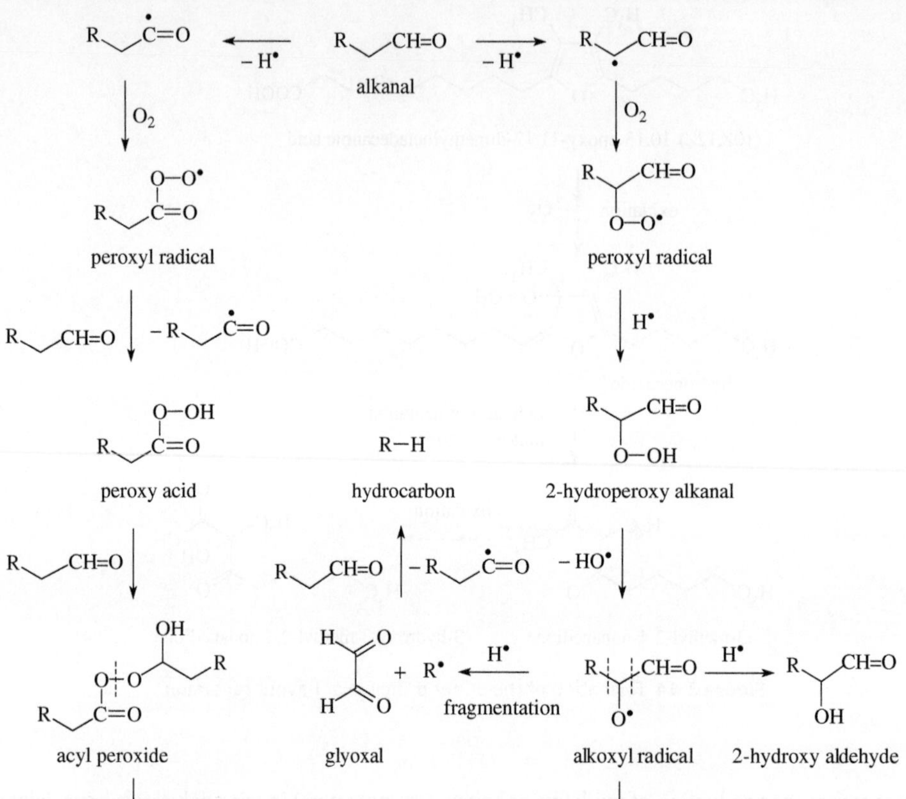

Figure 3.45 Oxidative fission of alkanals formed by decomposition of lipid hydroperoxides.

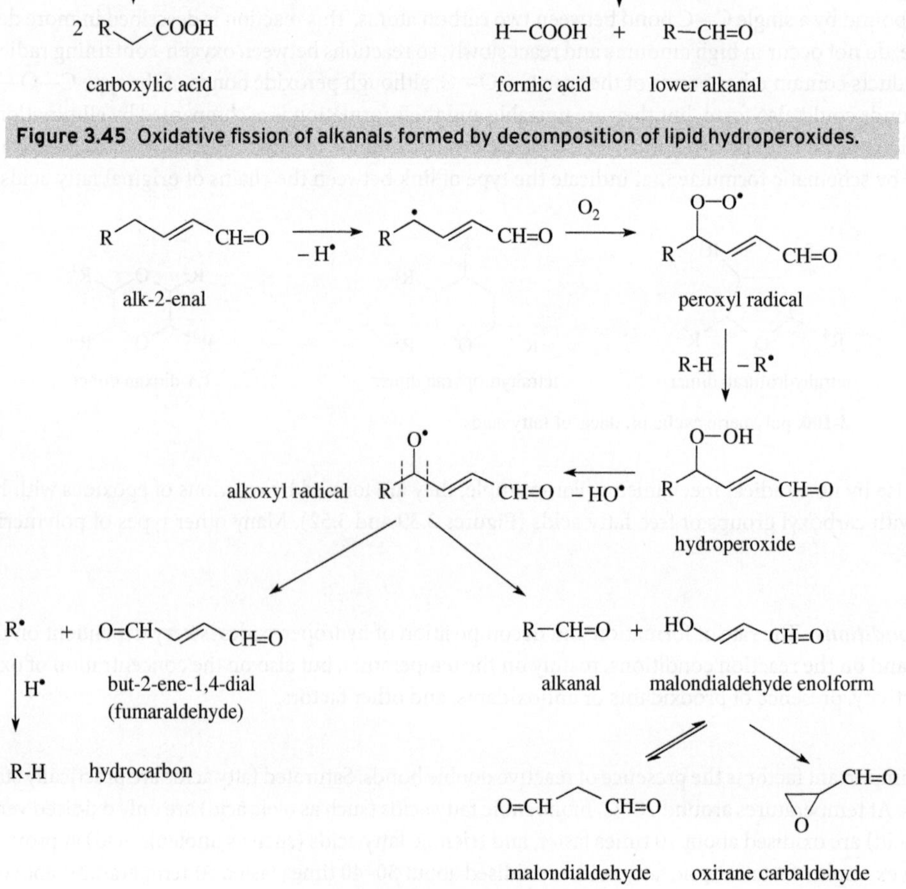

Figure 3.46 Oxidation fission of alk-2-enals formed by decomposition of lipid hydroperoxides.

are oxidised at about the same rate; *cis*-isomers are generally less stable than *trans*-isomers, conjugated dienes are less stable than dienes with isolated double bonds, and free fatty acids are less stable than fatty acids bound in triacylglycerols.

A fatty acid's inclination to oxidation also depends on the presence of other fatty acids, which is important for natural fats and oils, where fatty acids occur as mixtures of esters. When, for example, higher amounts of linoleic acid hydroperoxides are formed, their decomposition leads to radicals, which are able to initiate oxidation of monoenoic fatty acids. Vegetable oils that contain less than 10% of polyenoic fatty acids are therefore very stable against oxidation under typical storage conditions. This is the main reason why oilseeds are bred to have low linoleic acid and linolenic acid contents.

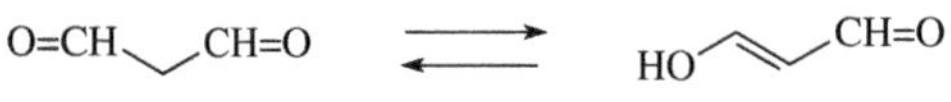

Figure 3.47 Isomerisation of malondialdehyde.

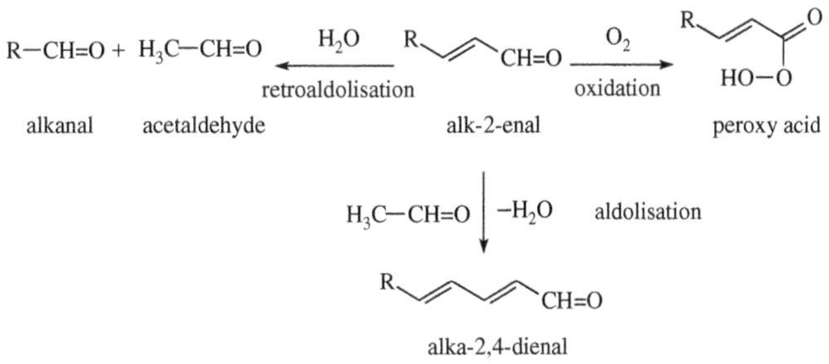

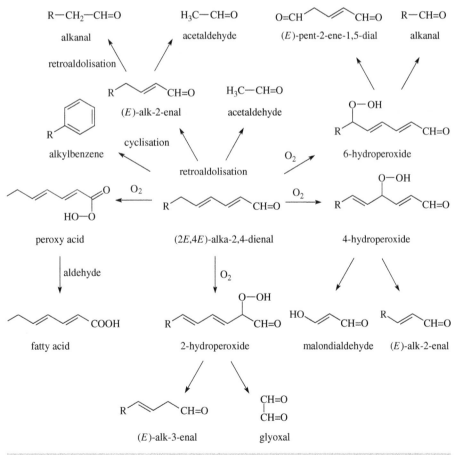

Figure 3.48 Oxidation and (retro)aldolisation of (*E*)-alk-2-enals.

Figure 3.49 Secondary reactions of alka-2,4-dienals.

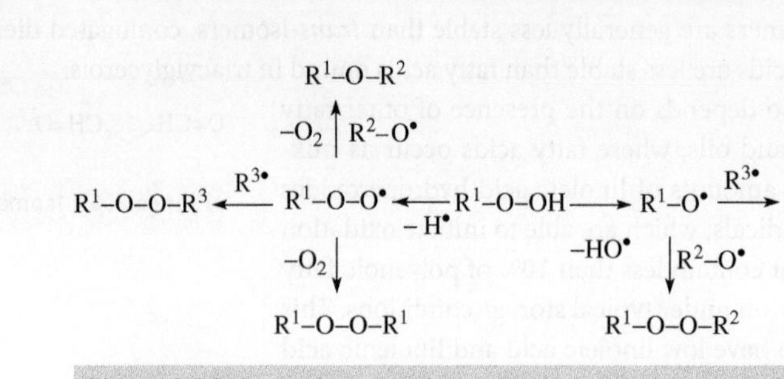

$$R^1\text{-}O\text{-}R^2$$

$$-O_2 \Big\uparrow \begin{array}{c} R^2\text{-}O^\bullet \end{array}$$

$$R^1\text{-}O\text{-}O\text{-}R^3 \xleftarrow{R^{3\bullet}} R^1\text{-}O\text{-}O^\bullet \xleftarrow[-H^\bullet]{} R^1\text{-}O\text{-}OH \longrightarrow R^1\text{-}O^\bullet \xrightarrow{R^{3\bullet}} R^1\text{-}O\text{-}R^3$$

$$-O_2 \Big\downarrow \qquad\qquad -HO^\bullet \Big\downarrow R^2\text{-}O^\bullet$$

$$R^1\text{-}O\text{-}O\text{-}R^1 \qquad\qquad R^1\text{-}O\text{-}O\text{-}R^2$$

Figure 3.50 Secondary reactions of fatty acid hydroperoxides.

C—C bonds

double C—C—bonds
(cyclopentane ring)

double C—C—bonds
(cyclohexene ring)

ether bonds

peroxide bonds

dioxan ring

tetrahydropyran ring

tetrahydrofuran ring

C—C and peroxide bonds

Figure 3.51 Schematic representation of bonds in dimers of oxidised fatty acids.

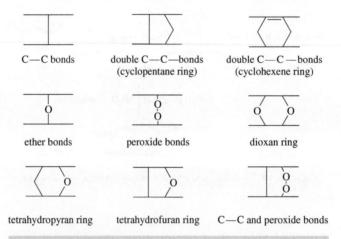

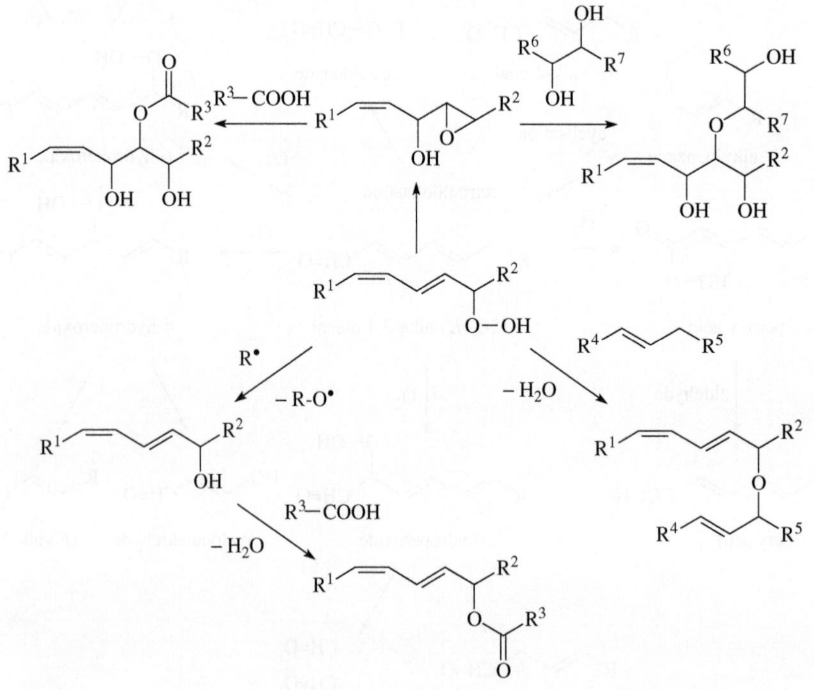

Figure 3.52 Examples of non-radical polymerisation during autoxidation of unsaturated fatty acids.

Temperature and oxygen concentration A rise in temperature accelerates the oxidation process and thus also increases the content of hydroperoxides, but hydroperoxides are very unstable at high temperatures. Particularly unstable are the hydroperoxides of trienoic and other polyenoic fatty acids. With a rise in temperature, the increase of the content of hydroperoxides is therefore soon compensated for by their faster decomposition. The maximum hydroperoxide content is achieved sooner, but it is lower and less distinct.

An important factor is the concentration of oxygen, as only oxygen molecules dissolved in the fat phase can react with free radicals. At higher temperatures, such as during frying, oxygen is consumed within a few minutes and the rate of oxidation is then determined only by the amount of oxygen that penetrates into the fat from the atmosphere by diffusion (diffusion of oxygen is also an important factor in autoxidation of fatty emulsions and dispersions). Under such conditions, the rate of formation of hydroperoxides and the rate of their decomposition are almost comparable, and their content remains approximately constant until the reactive unsaturated acids are oxidised. If a flow of air is introduced into fat, its oxidation is much faster, and the maximum content of hydroperoxides significantly increases.

Water activity The rate of fat oxidation in foods depends greatly on water activity (see Section 7.9). In dry foods, oxygen can more easily penetrate into the material than in normal foods. The minimum rate of oxidation occurs in foods with water activity values around 0.3. This situation is explained by the decrease in catalytic activity of metals, by the quenching of free radicals, and by the formation of antioxidants in the Maillard reaction. In foods with higher water activity, the rate of oxidation is again higher, probably due to increased mobility of metal ions, which catalyse the autoxidation.

Pro-oxidants and antioxidants Heavy metals with variable valencies have a significant influence on the oxidation of unsaturated fatty acids and their derivatives, as they can accelerate the reaction, as pro-oxidants. On the other hand, antioxidants will retard the oxidation of fatty acids.

3.8.1.8.3 Oxidation by hydrogen peroxide

Hydrogen peroxide, which arises naturally in food processes, can oxidise unsaturated lipids. The primary oxidation product is epoxide, which is immediately hydrolysed to form a fatty acid dihydroxy derivative. For example, the oxidation of oleic acid by hydrogen peroxide yields 9,10-dihydroxystearic acid. These reactions are also of great importance in the development of vascular diseases.

3.8.1.8.4 Oxidation by singlet oxygen

Atmospheric oxygen, O_2, exists in the triplet state ($^3O_2$), unlike many other compounds that exist in the singlet state. **Triplet oxygen** has 12 valence electrons, of which two electrons with higher energy and parallel spin are located in different orbitals. It is actually a biradical ($\bullet O\!\!-\!\!O\bullet$). Excitation of triplet oxygen creates singlet oxygen ($^1O_2$), whose electrons are paired, occuring in the same orbital and spins are antiparallel or the electrons occur in different orbitals and also have the opposite spins. These two of a total of five possible excited states of oxygen are known as $^1\Delta$ (delta) singlet oxygen and $^1\Sigma$ (sigma) singlet oxygen. The energy of $^1\Delta$ singlet oxygen is 93.8 kJ/mol higher and the energy of $^1\Sigma$ singlet oxygen is another 157 kJ/mol higher than the energy of the basic triplet state. The $^1\Sigma$ singlet oxygen is so reactive that it does not survive relaxation to the ground state, but the $^1\Delta$ singlet oxygen is sufficiently stable (its lifetime is several microseconds) and can therefore react with other molecules also occurring in the singlet state. Therefore, this excited state is simply called **singlet oxygen**. It reacts with compounds rich in electrons (such as unsaturated lipids) so that electrons are added to the free molecular orbital. With conventional unsaturated fatty acids, singlet oxygen reacts at least 1450 times faster than triplet oxygen. The relative reactivity of linolenate, linolate, and oleate are roughly in the proportions 3.5 : 3 : 1.

Singlet oxygen arises in chemical reactions, enzymatic reactions, by decomposition of hydroperoxides, and in some other ways. Under the conditions eligible for the processing and storage of foods, the most important formation pathway is by photochemical reactions with the participation of natural photosensitisers.

Singlet oxygen adds to the double bond of fatty acids with the formation of unstable peroxide (formerly called moloxide) or a six-membered ring. The resulting intermediates rapidly decompose to form hydroperoxides (Figures 3.53 and 3.54). These hydroperoxides differ from hydroperoxides resulting from autoxidation reaction in the ratio of the individual positional isomers, since the primary mechanism of the formation of free radicals is different. For example, oxidation of oleic acid by singlet oxygen results in an equimolar mixture of 9- and 10-hydroperoxide, since isomeric hydroperoxides with the hydroperoxide group in positions C-8 and C-11 cannot be formed. The products arising from linoleic acid by oxidation with singlet oxygen are a mixture that contains 66% of conjugated hydroperoxides (9- and 13-hydroperoxides) and 34% of unconjugated hydroperoxides (10- and 12-hydroperoxides). Linolenic acid oxidation by singlet oxygen produces a mixture containing 75% of conjugated 9-, 12-, 13-, and 16-hydroperoxides and 25% of unconjugated 10- and 15-hydroperoxides.

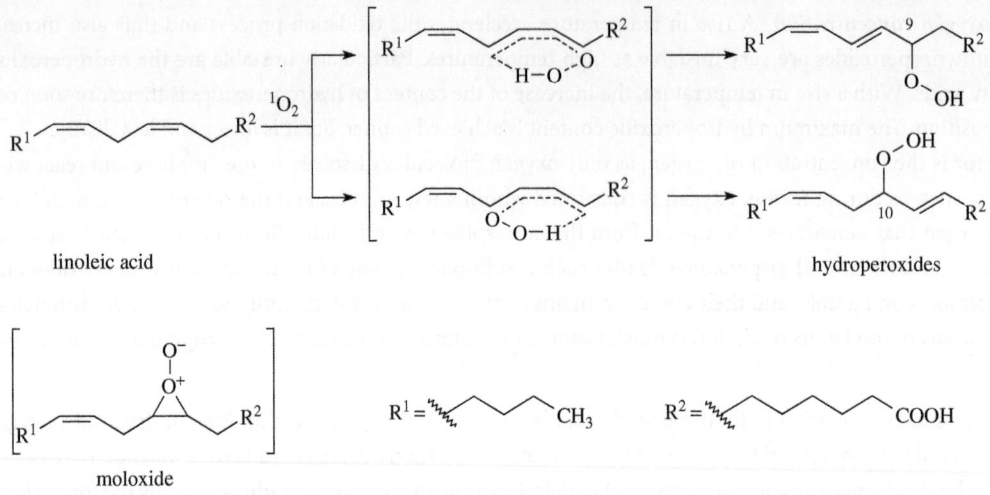

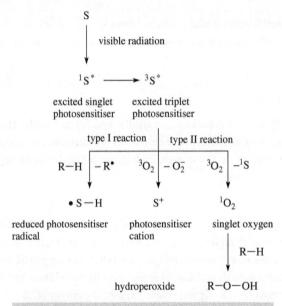

Figure 3.53 Oxidation of unsaturated fatty acids by singlet oxygen.

Figure 3.54 Photosensitised oxidation of unsaturated fatty acids.

Photosensitisers Photosensitisers are compounds that catalyse the oxidation of organic substances with oxygen when exposed to visible radiation. They act as carriers of absorbed energy that is transmitted to triplet oxygen transforming it into singlet oxygen.[6] The mechanism of their action is shown in Figure 3.54. Type I reactions occur, for example, in phenols, amines, and quinones (easily oxidisable or reducible compounds), via direct interactions of a triplet photosensitiser with other molecules (R—H), and involve a transfer of a hydrogen atom or electron, which produces free radicals (R$^•$ and $^•$SH). Triplet sensitisers are therefore photochemically activated free radical initiators. Reactions of type II describe the most common method of energy transfer from the excited triplet photosensitisers to triplet oxygen to

---

[6] Photolytic cleavage of fatty acid molecules in the initial autoxidation stage with ultraviolet exposure is negligible and has no practical significance in comparison with oxidation that occurs in the presence of photosensitisers during exposure to visible light. Photosensitisers contain a conjugated system of double bonds that easily absorbs light energy. The photosensitiser absorbs a photon (a quantum of light) and is transferred from the basic singlet state to an unstable singlet excited state. The molecule existing in this state may lose the energy obtained in three ways: (i) by internal conversion with the loss of thermal energy (transformation into another excited state); (ii) by light emission (through the process of fluorescence, the state where the emission begins and ends is a singlet state); and (iii) by conversion into a triplet excited state that is an intermediate in the photosensitised reaction and is followed by degradation to a lower energy triplet state and further to the basic singlet state with the emission of light. The process is called phosphorescence (a type of photoluminescence), the state where the emission begins is a triplet and where it ends is a singlet state.

form singlet oxygen. Less than 1% of triplet oxygen is simultaneously transferred to the superoxide anion (O_2^-) and oxidised form of the sensitiser (S^+).

Of the substances that are present in foods, the most important photosensitisers are chlorophylls, phaeophytins, haem pigments, and riboflavin. In vegetable oils, chlorophylls (and their degradation products, phaeophytins) act mainly as photosensitisers (type II reaction). The concentration of chlorophyll pigments in soybean oil is, for example, 1–1.5 mg/kg; their content in virgin olive oil is higher, around 6 mg/kg. In milk, the main photosensitiser is riboflavin; the main damaged amino acid is methionine. Riboflavin, on absorbing visible light, generates a metastable (triplet) excited state from the initially populated singlet state that may react with triplet oxygen to form singlet oxygen in the type II photooxidation pathway, or it may react directly with a substrate such as methionine (protein) by accepting hydrogen or electrons and producing radicals through the type I photooxidation mechanism. The significance of the type II reaction in milk has been described in several studies, whereas type I reactions in milk have received less attention. In meat and meat products, myoglobin and oxymyoglobin mainly act as photosensitisers, and the photooxidation product is metmyoglobin.

Other photosensitisers are cofactors of metalloproteins (enzymatic reactions produce singlet oxygen in some dairy products), pyridoxal and its derivatives (see Section 5.10), coumarins of the psoralene type (see Section 10.3.12.1.2), polycyclic aromatic hydrocarbons (such as anthracene; see Section 12.3.5), synthetic food colours (e.g. erythrosine; see Section 11.4.1.3), sulfides, and metal oxides (such as CdS, ZnS, and ZnO).

Scavengers of singlet oxygen The effects of photosensitisers can be neutralised by substances that serve as scavengers for singlet oxygen. In foods, the most important scavengers are carotenoids, tocopherols, and L-ascorbic acid. The mechanism of reactions of carotenoids with singlet oxygen can be simply described by the following reactions:

$$^1\text{carotenoid} + {}^1\text{O}_2 \rightarrow {}^3\text{carotenoid} + {}^3\text{O}_2$$
$$^1\text{carotenoid} + {}^3\text{S}^* \rightarrow {}^3\text{carotenoid} + \text{S}$$
$$^3\text{carotenoid} \rightarrow {}^1\text{carotenoid}$$

The reaction products of singlet oxygen with carotenoids occurring in the natural singlet state (^{1}carotenoids) are triplet oxygen and carotenoids in the excited triplet state (^{3}carotenoids). Other reaction sequences of carotenoids are described in Section 5.2.6.2.

3.8.1.8.5 Enzymatic oxidation

Oxidoreductase enzymes called lipoxygenases (linoleate: O_2 oxidoreductases), formerly known as lipoxidases, are widespread. Multiple forms or isozymes have been detected in both animal and plant species. Lipoxygenases are present in most raw materials and food products, unless they have been heated, as the enzymes are denatured on heating and lose their activity. Lipoxygenases are non-haem iron-containing dioxygenases involved in the catalysis of the oxidation of polyunsaturated fatty acids containing a (1Z,4Z)-penta-1,4-diene system of isolated double bonds. They catalyse the oxidation of essential fatty acids to hydroperoxides, but the other unsaturated fatty acids are not oxidised. Hydroperoxides formed by catalysis of lipoxygenases differ in their optical activity from hydroperoxides formed by autoxidation. Lipoxygenases have different specificity, so they catalyse the formation of hydroperoxide groups only on certain fatty acid carbon atoms and are therefore **regioselective**. Some examples are given in Table 3.45. They are also **stereoselective**, as they produce enantiomeric hydroperoxides, whilst the hydroperoxides arising by autoxidation of fatty acids are racemates. Hydroperoxides of fatty acids produced by the action of lipoxygenases can be further transformed into a variety of biologically active products collectively known as **oxylipins** (oxidised fatty acid derivatives) through the action of several participating enzymes.

The most common substrates for lipoxygenases in plants are linoleic and linolenic acids. Their oxidation, catalysed by lipoxygenases, yields 13- and 9-hydroperoxy fatty acids. Some lipoxygenases mainly catalyse the formation of (13S)-hydroperoxides, whilst less specific lipoxygenases produce a higher proportion of (13R)-hydroperoxides. Other lipoxygenases catalyse the formation of optically active 9-hydroperoxides. The 13-hydroperoxides of linoleic and linolenic acids, by the action of hydroperoxide lyases, produce aldehydes with six carbon atoms (Figure 3.55) that possess green (hexanal and hex-3-enal) grassy or beany (nona-3,6-dienal) flavours. Similarly, products with cucumber and melon flavours are formed from 9-hydroperoxides derived from linoleic and linolenic acids (Figure 3.56). For example, isomerisation of (3Z,6Z)-nona-3,6-dienal yields (2E,6Z)-nona-2,6-dienal that typically has a flavour resembling fresh cucumbers. Under the action of enal isomerases, *cis*-alkenals are isomerised to *trans*-alkenals or reduced by alcohol dehydrogenases to unsaturated alcohols (Figure 3.57). Both types of reaction products deliver characteristic flavour components to strawberries, bananas, cauliflower, tomatoes, and many other fruits and vegetables.

Hydroperoxides of linoleic and linolenic acids also react with sensitive food components, such as carotenes, to form colourless products (lipoxygenase, however, do not directly catalyse the decomposition of carotenes) and induce oxidative changes in sulfur amino acids and proteins. Secondary products of lipid oxidation initiate radical reactions of proteins that lead to a reduction in their nutritional value.

Champignon mushrooms and certain other fungi contain 10-lipoxygenase, which catalyses the formation of unconjugated hydroperoxide, (8E,10S,12Z)-10-hydroperoxyoctadeca-8,12-dienoic acid, from linoleic acid. From this hydroperoxide, lyases produce (E)-10-oxodec-8-

Table 3.45 Overview of properties of selected plant lipoxygenases.

Plant material	Isoenzyme	Regiospecificity[a]	Occurrence
Soybeans	L-1	95/5	Seeds, water stress
Soybeans	L-2	50/50	Seeds
Soybeans	L-3	50/50	Seeds, water stress
Lentils	Lox1	82/18	Seeds
Potatoes	H1	13-Hydroperoxides	Leaves, wounding
Potatoes	LOX5	9-Hydroperoxides	Tubers

[a]Ratio of 13- and 9-hydroperoxides.

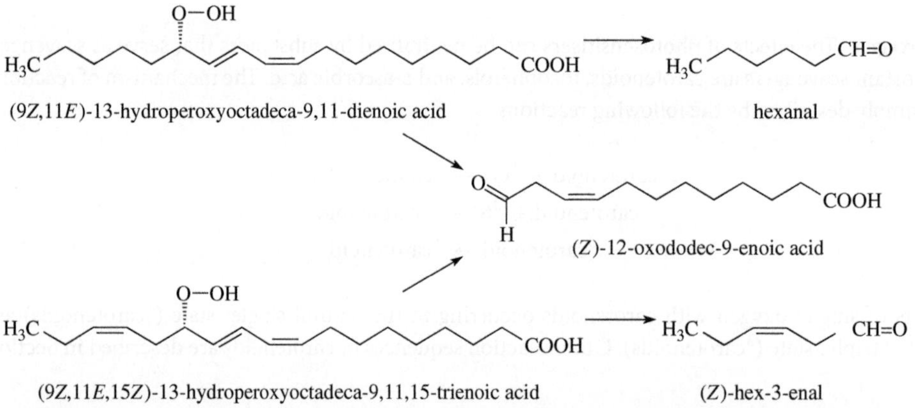

Figure 3.55 Mechanism of action of lyases to 13-hydroperoxides of unsaturated fatty acids.

enoic acid and (R)-oct-1-en-3-ol, which is the key component of mushroom aroma (Figure 3.58). Analogously, (8E,10S,12Z,15Z)-10-hydroperoxy-8,12,15-octadecatrienoic acid formed by oxidation of linolenic acid yields (3S,5Z)-octa-1,5-dien-3-ol (Figure 3.58).

Biosynthesis of biologically active oxygenated derivatives of C_{18} fatty acids in plants that are known as **octadecanoids** proceeds by similar mechanisms to the biosynthesis of eicosanoids (e.g. prostaglandins) in animals. For example, an important product formed from linolenic acid by the activity of 13-lipoxygenase, via (9Z,11E,13S,15Z)-13-hydroperoxyoctadeca-9,11,15-trienoic acid, is the plant hormone (–)-jasmonic acid (**3-101**), which plays a role in the inhibition of plant growth and leaf senescence. It is also responsible for the formation of potato tubers and bulbs of some root vegetables, such as onion. Volatile transformation products of jasmonic acid (such as *cis*-jasmone, **3-101**) occur in many other essential oils, such as bergamot and lemon essential oils, and are components of the aroma of tea.

Other accompanying enzymes of lipoxygenases are divinylether synthases that catalyse the rearrangement of fatty acid hydroperoxides into divinyl ether fatty acids, which have a role in plant defences towards pathogens. The volatile fission products of these ethers give products with green and grassy flavours. Hydroperoxide generated by 13-lipoxygenases from linolenic acid serves as a precursor of (1′E,3′Z,9Z,11E)-12-hexa-1′,3′-dien-1-yloxy-9,11-dodecadienoic acid, which is known as etherolenic acid (**3-101**). Similarly, 9-hydroperoxide of linolenic acid is transformed into another divinyl ether fatty acid, (1′E,3′Z,6′Z,8E)-9-nona-1′,3′,6′-trien-1-yloxynon-8-enoic acid, known as colnelenic acid (**3-101**). Analogously, 13- and 9-hydroperoxides of linoleic acid become precursors of colneleic and etheroleic acids, respectively. (7Z,10Z,13Z)-Hexadeca-7,10,13-trienoic acid, abundant in many plant species known as 16 : 3 plants, is a precursor of a range of biological active substances (oxylipins) called **hexadecanoids**.

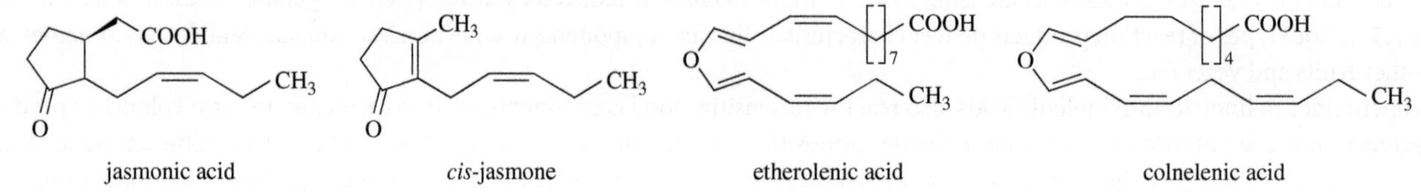

jasmonic acid   *cis*-jasmone   etherolenic acid   colnelenic acid

3-101, selected products of oxygenases in plants

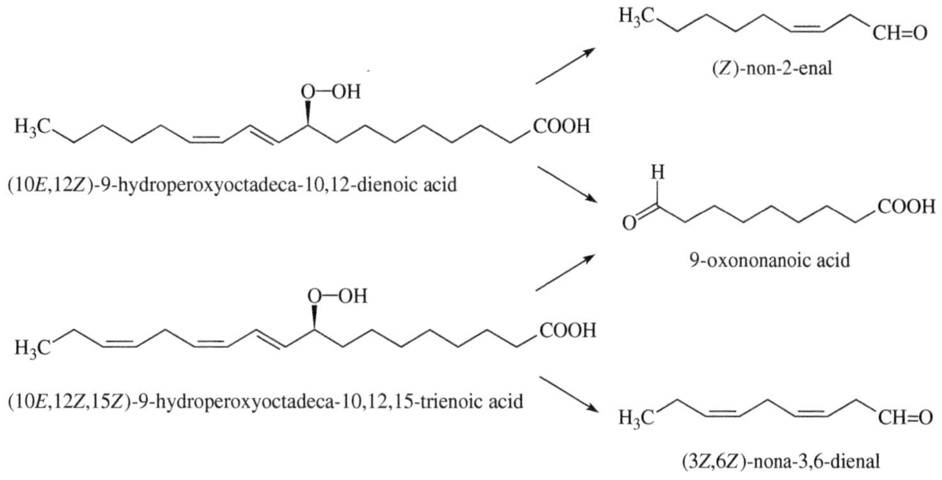

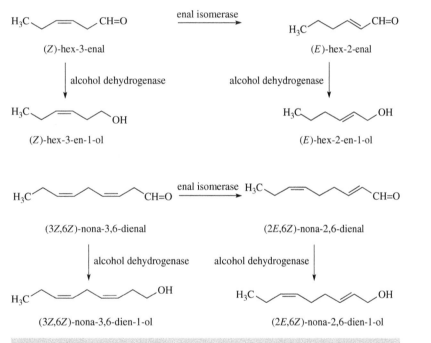

Figure 3.56 Mechanism of action of hydroperoxide lyases to 9-hydroperoxides of fatty acids.

Figure 3.57 Mechanism of action of isomerases and alcohol dehydrogenases to hydroperoxide fission products.

3.8.1.8.6 Oxidation catalysed by metals

Lipid oxidation is catalysed by transition metals that form compounds in many oxidation states, due to the relatively low reactivity of unpaired d electrons. These compounds, which include mainly iron,[7] copper, manganese, nickel, cobalt, and chromium, are reduced by adopting one electron. The last three elements are indeed fairly active, but the level of their active forms is so low that these metals are almost of no significance. Other metals, as free ions or as some undissociated salts or complexes, act directly or indirectly as catalysts in the initiation, propagation, and termination phase of autoxidation reaction.

---

[7] It is not yet quite clear whether, in addition to ferric (Fe^{3+}) and ferrous ions (Fe^{2+}), mixed complexes of both ions with oxygen (Fe^{3+}-O_2-Fe^{2+}) and the so-called hypervalent iron (ions with valences +4 or +6, for example ferryl cations FeO^{2+} and ferrate anions FeO_4^{2-}), which are the active forms of the enzymes containing haem cofactors (peroxidases, catalases, and cytochrome P-450), are also involved in the process of the oxidation of lipids.

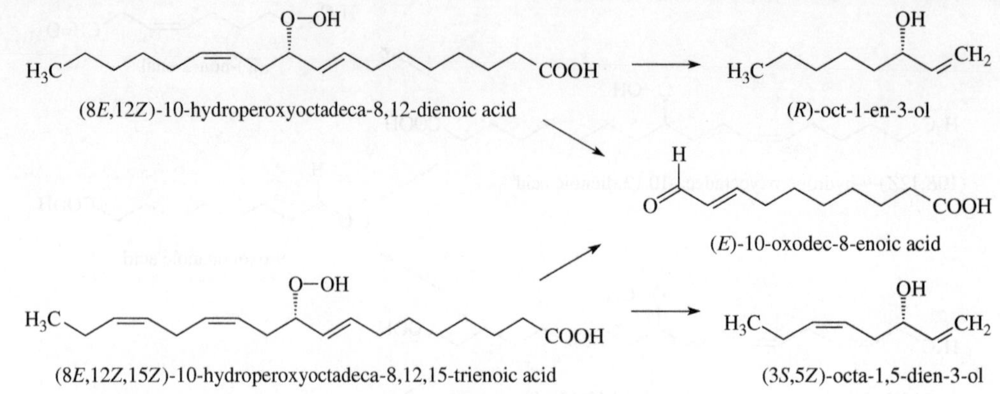

Figure 3.58 Mechanism of action of lyases in mushrooms.

Initiation reaction Metals in their higher valency $M^{(n+1)+}$ are initiators of autoxidation reactions. The initiation reaction takes place by electron transfer to form free radicals of hydrocarbons ($R^\bullet$), that is, $M^{(n+1)+} + R–H \rightarrow M^{n+} + R^\bullet + H^+$, and with ferric ions as $Fe^{3+} + R{—}H \rightarrow Fe^{2+} + R^\bullet + H^+$. The initiation phase is also indirectly catalysed by metals in the lower valency M^{n+} via their transient metal complexes with oxygen. The reaction products are hydrocarbon radicals ($R^\bullet$), metals in higher valences, reactive oxygen species such as superoxide radical anion ($O_2^{-\bullet}$), or the protonised form of the superoxide radical ($HO_2^\bullet$):

$$M^{n+} + O_2 \rightarrow [M^{(n+1)+} \ldots O_2] \xrightarrow{R-H} R^\bullet + [M^{(n+1)+} \ldots {}^-O_2H]$$
$$M^{n+} + O_2 \rightarrow [M^{(n+1)+} \ldots O_2] \xrightarrow{R-H} R^\bullet + M^{n+} + HO_2^\bullet$$
$$M^{n+} + O_2 \rightarrow [M^{(n+1)+} \ldots O_2] \xrightarrow{R-H} RO^\bullet + [M^{(n+1)+} \ldots {}^-OH]$$
$$M^{n+} + O_2 \rightarrow [M^{(n+1)+} \ldots O_2] \xrightarrow{(M^{n+} \cdots R-H)} HO_2^\bullet + [M^{(n+1)+}R^-] + M^{n+}$$
$$M^{n+} + O_2 \rightarrow [M^{(n+1)+} \ldots O_2] \rightarrow M^{(n+1)+} + O_2^{-\bullet}$$

The autoxidation reaction is also indirectly catalysed by reactive oxygen species (e.g. $O_2^{-\bullet}$, $HO_2^\bullet$, and H_2O_2), which are formed by the previously mentioned and other reactions:

$$O_2^{-\bullet} + H^+ \rightarrow HO_2^\bullet$$
$$O_2^{-\bullet} + H^+ + HO_2^\bullet \rightarrow H_2O_2 + O_2$$

The reaction of superoxide radical $HO_2^\bullet$ with unsaturated acids is very slow and, therefore, not too significant. Superoxide radical anion $O_2^{-\bullet}$ does not react with fatty acids. However, much more reactive are hydroxyl radicals generated by reduction of hydrogen peroxide by metals that directly initiate autoxidation reaction. Like the metals, these radicals are one-electron initiation agents:

$$M^{n+} + H_2O_2 \rightarrow M^{(n+1)+} + HO^\bullet + HO^-$$
$$R - H + HO^\bullet \rightarrow R^\bullet + H_2O$$

Propagation reactions Some metals (Fe, Cu, and Ni) catalyse the decomposition of hydroperoxides to alkoxyl radicals in their lower valency, which results in a change to a higher valency. Metals in the higher valency then catalyse the decomposition of hydroperoxides to peroxyl radicals and switch back to a lower valency:

$$R - O - OH + M^{n+} \rightarrow R - O^\bullet + HO^- + M^{(n+1)+}$$
$$R - O - OH + M^{(n+1)+} \rightarrow R - O - O^\bullet + H^+ + M^{n+}$$

The resulting alkoxyl and peroxyl radicals increase the autoxidation reaction rates of initiation and propagation phases, since the rate of cleavage of hydroperoxides by metal ions is much faster than the formation of radicals *ab inicio*. Metal ions and non-ionised salts may react in this way (Figure 3.59). Of the metals bound in complexes, some are effective, but some are ineffective. Metals may also become less effective in the presence of fats if micelles are formed. Another catalyst for the oxidation of lipids may be iron that is bound in complexes.

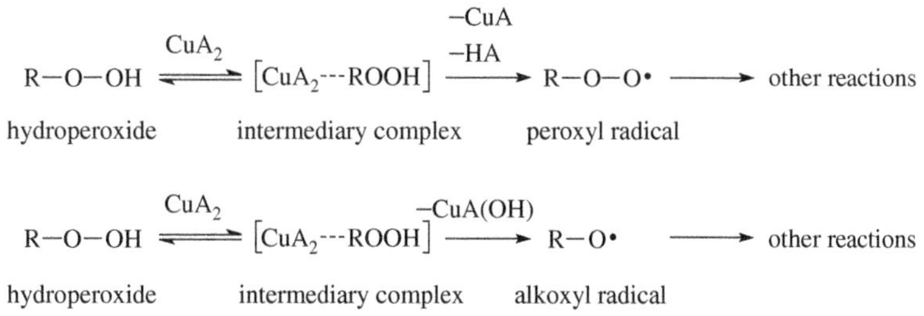

Figure 3.59 Catalytic activities of cuprous salts in autoxidation reactions. CuA_2, cuprous salt; CuA, cupric salt; HA, fatty acid; CuA(OH), hydrated cupric salt.

Iron bound in haem pigments has the same catalytic activity as the ions Fe^{2+} and Fe^{3+}; moreover, in aqueous solutions it is even more active, as it catalyses the cleavage of hydroperoxides as follows:

$$R - O - OH + Fe^{2+} \rightarrow R - O^{\bullet} + HO^{-} + Fe^{3+}$$
$$R - O - OH + Fe^{3+} \rightarrow R - O - O^{\bullet} + H^{+} + Fe^{2+}$$

In addition to these reactions, which result in direct electron transfer, four other mechanisms have been postulated that may be involved in catalysis of lipid oxidation by haem iron. The first of these mechanisms implies formation of complexes with hypervalent iron (with valency +4), which can oxidise lipids directly. The second mechanism is indirect. Haem iron catalyses the formation of hydroxyl radicals $HO^{\bullet}$, which (as one-electron agents) initiate the autoxidation reaction. Hydroxyl radicals are formed by reaction of Fe^{2+} ions bound in haem pigments via $O_2^{\bullet-}$ and hydrogen peroxide (H_2O_2), which is reduced to hydroxyl radicals with the formation of haematin:

$$H_2O_2 + haem - Fe^{2+} \rightarrow HO^{\bullet} + HO^{-} + haem - Fe^{3+}$$

The third mechanism of lipid autoxidation catalysis by haem pigments is a photosensitised reaction and oxidation of lipids by either singlet oxygen or free radicals. The fourth mechanism assumes hydroxyl radical attack, in which haem iron is released in an ionic form, which then catalyses the oxidation of lipids in the same way as non-haem iron.

The redox potentials of other metals, such as Mn, Co, and so on, are too low, so these metals cannot catalyse the cleavage of hydroperoxides in aqueous systems (e.g. in emulsions), but can catalyse the decomposition of lipid hydroperoxides in anhydrous fats through the transient complexes formed with hydroperoxides:

$$R - O - OH + Co^{2+} \rightarrow [R - O - OH \ldots Co]^{2+} \rightarrow R - O^{\bullet} + Co^{3+} + HO^{-}$$
$$R - O - OH + Co^{3+} \rightarrow [R - O - OH \ldots Co]^{3+} \rightarrow R - O - O^{\bullet} + Co^{2+} + H^{+}$$

Inhibition Heavy metals increase not only the amount of free radicals in the lipid phase, but also the rates of initiation, propagation, and termination reactions. Heavy metals therefore also change the composition of the reaction products. At high concentrations of free radicals, the termination reaction may dominate and metals then act as the inhibitors of autoxidation. Autoxidation reaction can also be inhibited by metals when they are present at higher concentrations. It is assumed that the reason is the oxidation and reduction of free hydrocarbon radicals to anions and cations by Fe and Cu ions and the formation of complexes of free radicals. Similar complexes are also formed with Co. All these reactions interrupt the radical chain autoxidation reaction. Reactions with Fe ions are given as examples:

$$R^{\bullet} + Fe^{2+} \rightarrow Fe^{3+} + R^{-}$$
$$R^{\bullet} + Fe^{3+} \rightarrow Fe^{2+} + R^{+} \rightarrow products$$
$$R^{\bullet} + Fe^{3+} \rightarrow [R^{\bullet} \ldots Fe^{3+}]$$

The formation of Co complex can be described by the following equations:

$$R^{\bullet} + CoA_3 \rightarrow R - CoA_2$$
$$R - O^{\bullet} + CoA_3 \rightarrow R - O - CoA_2$$
$$R^{\bullet} - O - O^{\bullet} + CoA_3 \rightarrow R - O - O - CoA_2$$

where CoA_3 is the undissociated salt of Co^{3+} with fatty acids.

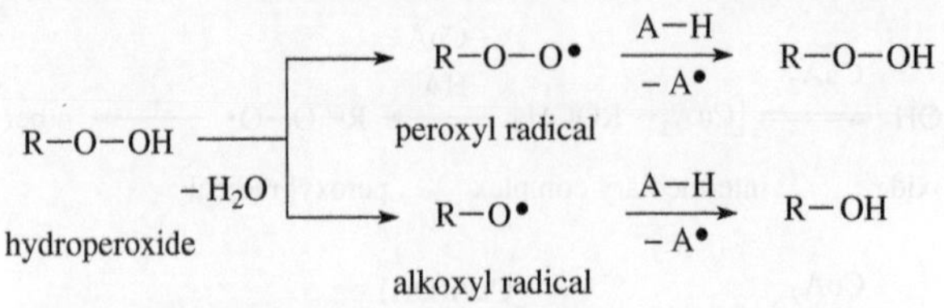

$$R-O-OH \xrightarrow{} \begin{array}{c} R-O-O^{\bullet} \xrightarrow[-A^{\bullet}]{A-H} R-O-OH \\ \text{peroxyl radical} \\ \\ R-O^{\bullet} \xrightarrow[-A^{\bullet}]{A-H} R-OH \\ \text{alkoxyl radical} \end{array}$$

hydroperoxide $-H_2O$

Figure 3.60 Reactions of antioxidants with free radicals formed by autoxidation of fatty acids. A–H, antioxidant; A•, antioxidant radical.

Crude oil contains mainly iron and copper at levels up to several milligrammes per kilogramme, but during degumming and subsequent refining the content of these metals is reduced to negligible amounts. In foods, however, these metals are present in considerable concentrations and can pass into the oil phase, although the original oil may have contained only traces of the metals.

3.8.1.9 Inhibitors of oxidation

The inhibitors of the oxidation reaction of fats are substances that reduce the oxidation rate, regardless of the mechanism of their action. These compounds include antioxidants, synergists, chelating agents, and compounds decomposing hydroperoxides through nonradical reactions. Also, agents stabilising hydroperoxides may reduce the reaction rate because they inhibit the formation of free radicals.

3.8.1.9.1 Antioxidants

Antioxidants (see Section 11.2.2) are substances that can react with free radicals of the autoxidation chain, especially peroxyl radicals (Figure 3.60). The reaction creates hydroperoxides or other non-radical lipid products. The antioxidant is transformed into a free radical, which, however, is fairly stable, so that it is unable to continue in the autoxidation reaction. The role of the antioxidant thus lies in shortening the autoxidation chain and increasing the rate of termination reactions. During the reaction, the antioxidant is consumed. When all molecules of the antioxidant have been consumed, the autoxidation reaction proceeds as if no antioxidant were present. Antioxidants therefore cannot completely stop the autoxidation reaction; they just slow it down.

Figure 3.61 shows the reaction without an antioxidant and the reaction in its presence. It is clear that antioxidants prolong the induction period (they slow autoxidation), but do not affect the rate of the subsequent rapid oxidation. The ratio of the length of the induction period of inhibited and uninhibited reactions is called the **protection factor**. It is usually expressed as a percentage of increased stability (extension of the induction period).

In addition to the inhibition reaction, other reactions run in parallel, such as oxidation of free antioxidant radical to peroxyl radical or the reverse reaction with the formation of a free lipid radical, especially at high antioxidant concentrations. The reaction mechanism is given in detail in Section 11.2.2.2.

The most commonly used antioxidants are synthetic phenol derivatives, which contain two (or three) hydroxyl groups in the *ortho-* or *para*-positions. Natural antioxidants are usually substituted in the *ortho*-position, so they are more efficient as antioxidants. Synthetic antioxidants are usually substituted in the *para*-position, so they are less toxic than the corresponding *ortho*-derivatives. Instead of one hydroxyl group, a methoxyl group or at least a branched alkyl substituent may be present. Substitution in a benzene ring by additional alkyl group or two alkyls increases the antioxidant efficiency. Far less frequently used, especially for their higher toxicity, are compounds with heterocyclic nitrogen (such as dihydropyridine or dihydroquinoline derivatives). Natural antioxidants, mainly tocopherols (see Section 5.4.1), occur in all natural fats and oils. Other common natural antioxidants include phenolic and numerous other compounds (see Section 11.2.2.4).

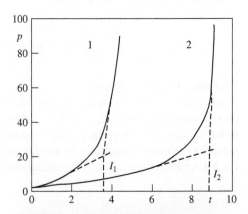

Figure 3.61 Impact of added antioxidant on the course of autoxidation reactions. p, amount of fatty acid hydroperoxides in milliequivalents of active oxygen per kg (peroxide number); t, duration of autoxidation in days. Antioxidant (BHA) concentration = 1, 0%, 2, 0.02%; I_1 and I_2, corresponding induction periods, protection factor $PF = (I_2 - I_1)/I_1$.

3.8.1.9.2 Synergists and sequestrants

Synergists are substances that do not have antioxidant activity themselves, but are able to increase the efficiency of antioxidants. The most common synergists include polyhydric acids, such as citric, tartaric, malic, ascorbic, and phosphoric acids. These substances are not very soluble in fats, which limits their activity. Therefore, their lipophilic derivatives, such as esters of ascorbic acid or phospholipids instead of phosphoric acid, are often used. The mechanism of synergism is not fully elucidated and is probably not uniform. It may be caused by the non-radical acceleration of decomposition of lipid hydroperoxides, partly by the regeneration of antioxidants, and partly by the binding of prooxidants (mainly heavy metals) into inactive complexes. This last mechanism is fairly widespread and the synergist is also a chelating agent or sequestrant, a chemical that promotes sequestration. Numerous substances, known as **sequestrants**, have the ability to bind heavy metals to complexes, such as pyrocatechol and pyrogallol derivatives, some carbohydrates, acidic phospholipids and lysophospholipids, oxalic acid, phytic acid, and other compounds. Some sequestrants are used as food additives, such as synthetic ethylenediaminetetraacetic acid (EDTA). Metal ions are inactivated due to formation of insoluble and undissociated salts, but also some soluble complexes may be inactive, whilst others may act as prooxidants.

3.8.1.9.3 Heterolytic decomposition of hydroperoxides

Lipid hydroperoxides are mostly decomposed into free radicals, but sometimes they can decompose into ions, for example in the presence of acids and alkalis (Figure 3.62) or by reactions with transition metals.

3.8.1.10 Oxidation and quality of lipids

3.8.1.10.1 Rancidity during storage

Oxidation reactions that proceed during storage of fats reduce their sensory quality. Such processes are known as rancidity. Rancidity is not just caused by oxidation of fatty acids, but also by other reactions. Therefore, several types of rancidity can be distinguished:

- **hydrolytic rancidity**;

- **oxidative rancidity**;

- **ketonic rancidity**;

- **flavour reversion**.

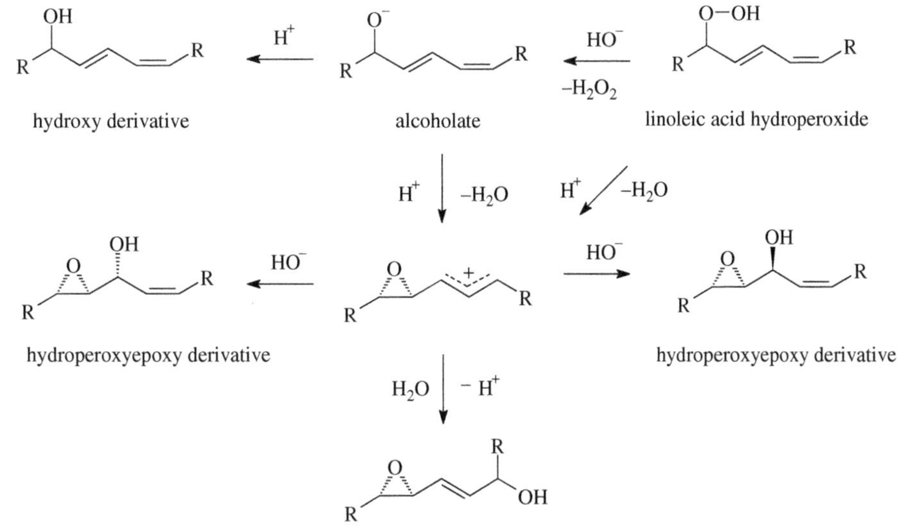

Figure 3.62 Heterolytic decomposition of hydroperoxides in acid and alkaline media.

Hydrolytic rancidity Fatty acids released during hydrolysis of fats do not bring about fat rancidity, because they are sensorially imperceptible in small amounts. The exceptions are fats containing bound fatty acids with shorter carbon chain (4–10 carbons). Butyric acid is released from butter and imparts a nasty, very pungent odour. Acids with 6–10 carbon atoms are also present to a large extent in coconut or palm kernel oils, as well as in milk fats. Native enzymes or enzymes released by moulds give the product a typical soapy off-taste.

Oxidative rancidity Hydroperoxides resulting from fat oxidation do not affect the sensory quality of fats and oils, but their decomposition products produce distinctive odours that depend on the concentration of the secondary products and their composition. These reactions are described in detail in other sections. A certain range of oxidation reactions, both enzymatic and non-enzymatic, is often desirable, because it leads to the formation of characteristic flavour active products in many foods (such as fruits, vegetables, and fried foods). However, during more extensive oxidation of polyunsaturated fatty acids with a high number of double bonds such as EPA or docosahexaenoic acid, the odour becomes unpleasant, for example with fishy taints caused mainly by *trans*-4,5-epoxy-(*E,Z*)-2,7-decadienal, (*Z*)-1,5-octadien-3-one, (*Z*)-3-hexenal, (*Z,Z*)-2,5-octadienal, (*Z,Z*)-3,6-nonadienal, and (*E,E,Z*)-2,4,6-nonatrienal.

Ketonic rancidity Ketonic or perfume rancidity typically occurs in butter, where it is undesirable. Fatty acids with 6–12 carbon atoms released from triacylglycerols by hydrolysis with lipases of microorganisms (e.g. of fungi of the genera *Penicillium* and *Aspergillus*) are enzymatically degraded largely by β-oxidation to 3-oxoacids. Their decarboxylation yields alkan-2-ones, known as methylketones, which can be reduced to the corresponding alkan-2-ols (Figure 3.63). The most common methylketones (see Section 8.2.4.1.2) are pentan-2-one, hexan-2-one, heptan-2-one, octan-2-one, and nonane-2-one, which have specific perfume odours. Pentane-2-one and hexane-2-one posses a fruit odour reminiscent of bananas, heptan-2-one has a floral and herbal odour, octane-2-one has a flowery odour, and nonan-2-one exhibits a floral and oily odour. The odour itself is not unpleasant, and in mould cheeses it is even desirable, but is unusual and atypical in edible fats.

Flavour reversion Flavour reversion is characteristic of soybean oil, which contain high levels of furan fatty acids (see Section 3.3.1.4.1). If the furan fatty acids are decomposed via photooxidation pathways, flavour-active secondary oxidation products such as 3-methyl-2,4-nonanedione and 3-hydroxy-3-methyl-2,4-nonanedione are formed (Figure 3.44), among other products. The off-flavour resemble grass and beans, but also rubber and soil. The off-flavour can be removed during subsequent refining of the oils in which this defect is apparent, but the defect will reappear after a certain amount of time, hence the name **flavour reversion**. Flavour reversion is especially off-putting to consumers in the United States as they require utterly tasteless cooking oil. Substances causing flavour reversion are thus actually present in small quantities that many other consumers do not mind, assuming they ever notice the flavour change.

3.8.1.10.2 Changes during frying

During frying, the oil or fat is preheated to temperatures around 150–200 °C. The frying process is commonly carried out in:

- a thin layer of fat;

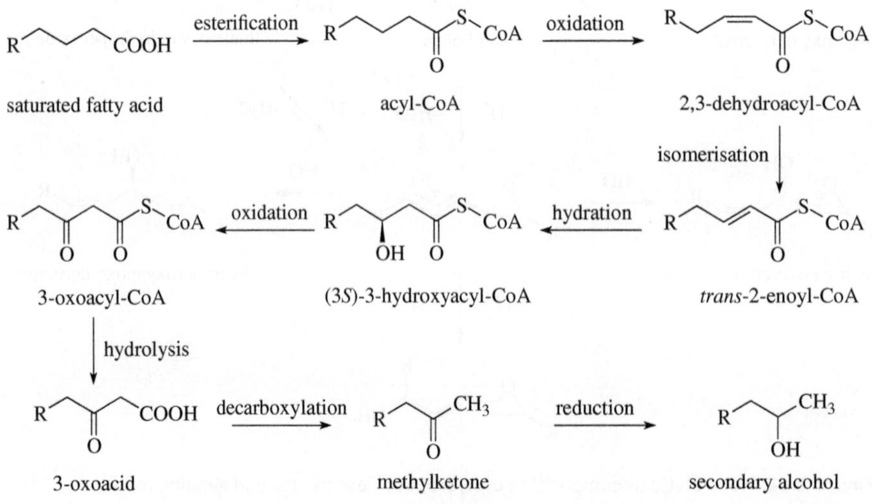

Figure 3.63 Formation of alkan-2-ones in ketone rancidity.

- a layer of fat more than 50 mm deep (usually between 100 and 200 mm), where the fried food is either submerged or floats on the surface.

The second approach is more common today and is called **deep frying** or **deep-fat frying**. After inserting the food, water evaporates upon contact with hot fat, which cools the fat. During the deep-frying process, which usually takes a few minutes, the temperature again increases by a few tens of degrees. In the processes taking place in the fat exposed to high temperature in the presence of air and moisture, several types of reactions can be distinguished:

- Hydrolytic processes induced by the action of hot water vapour (released from the fried food) on the hot fat are often the main processes occurring during frying; they lead to the hydrolysis of fatty acids, which are largely adsorbed on fried food or escape into the air.

- Oxidation processes that are very fast at frying temperatures; dissolved oxygen in fat is consumed, so that further oxidation is slow and depends on the rate at which oxygen diffuses from the air; the rate of diffusion, however, significantly increases if oil starts to foam, as foaming increases the interface between fat and air.

- Processes causing both polymerisation reactions between free radicals and reactions of carboxyl groups of free fatty acids with hydroxyl and epoxy groups of oxidised fatty acids of fat.

- Pyrolytic processes such as dehydration of oxidation products or their reactions with proteins and other components of fried foods; these processes yield significant sensorially active substances but also include the decomposition of glycerol (released by hydrolytic processes) to acrolein (see Section 8.2.4.1.1).

Under conditions employed in deep frying, a complex series of reactions takes place resulting in a loss of quality of both the frying oil and the fried food. The levels of polar compounds, polymers, free fatty acids, oxidised fatty acids, and the smoke point of the fat have become the most generally accepted indicators for quality evaluation of frying fats and are included in some of the current official regulations. In general, if the content of polar compounds generated by oxidative and hydrolytic reactions exceeds 25% and the amount of polymers exceeds 10% of the fat weight, it is recommended to replace the used fat with the fresh one.

3.8.1.11 Effects of oxidised lipids on human health

Oxidised lipids have only a low acute toxicity, so their effect on human health is often underestimated. The toxicity of fats used for frying has not been firmly established, therefore the limits for their use (25% of polar compounds and 10% of polymers) indicate that in reality the functional properties are deteriorating and the flavour of the fried food will get worse. Adversely reflected in terms of chronic toxicity are cyclic dimers and particularly cyclohexene derivatives. A higher content of hydroperoxides causes symptoms of vitamin E and essential fatty acid deficiency, which results in increased permeability of the skin to water. It is also difficult to enzymatically hydrolyse (digest) oxidised lipids. Recently, it has been shown that a higher intake of oxidised fats increases their levels in blood serum, and oxidised fatty acids or free radicals arising from oxidised fatty acids then react with certain proteins in blood serum and in the walls of blood vessels and form atherosclerotic deposits. Oxidised sterols are particularly active in this respect. Similar deposits are formed, for example, in the nerve tissue and some other important organs. Oxidation products of lipids, especially the reactive acrolein (E)-4-hydroxyalk-2-enals and malondialdehyde, also react with proteins and nucleic acids, where any alterations may facilitate development of malignant tumours. For these reasons, an increased intake of natural antioxidants, mainly tocopherols and carotenes, is recommended, which leads to an increased intake of easily oxidisable polyenoic fatty acids.

3.8.1.12 Oxidation of food constituents and other reactions of oxidised lipids

Oxidation products of lipids (hydroperoxides, free alkoxyl and peroxyl radicals, epoxides, and aldehydes) react with a number of food constituents during processing and storage. These reactions often lead to a reduction in the nutritional value of foods (such as reactions with proteins and vitamins) and a deterioration of their organoleptic properties (e.g. reactions with flavour active substances).

3.8.1.12.1 Reactions with proteins and deoxyribonucleic acids

Reactions of proteins with lipid hydroperoxides and other oxidised lipids give rise to different types of lipoproteins, in which lipids and proteins are bound by physical bonds (as in natural lipoproteins), but also by covalent bonds. Reactions with hydroperoxides, free radicals, epoxides, and aldehydes lead to the fission of some protein bonds, formation of protein radicals and oligomers, cross-links between protein chains, and oxidation of some sensitive functional groups of amino acids (see Section 2.5.1.1).

Cysteine and methionine are particularly sensitive to oxidation, and also histidine, arginine, tryptophan, and phenylalanine. In foods, oxidation mainly takes place in sulfur-containing amino acids (cysteine, cystine, and methionine). Cysteine can be oxidised to cystine or

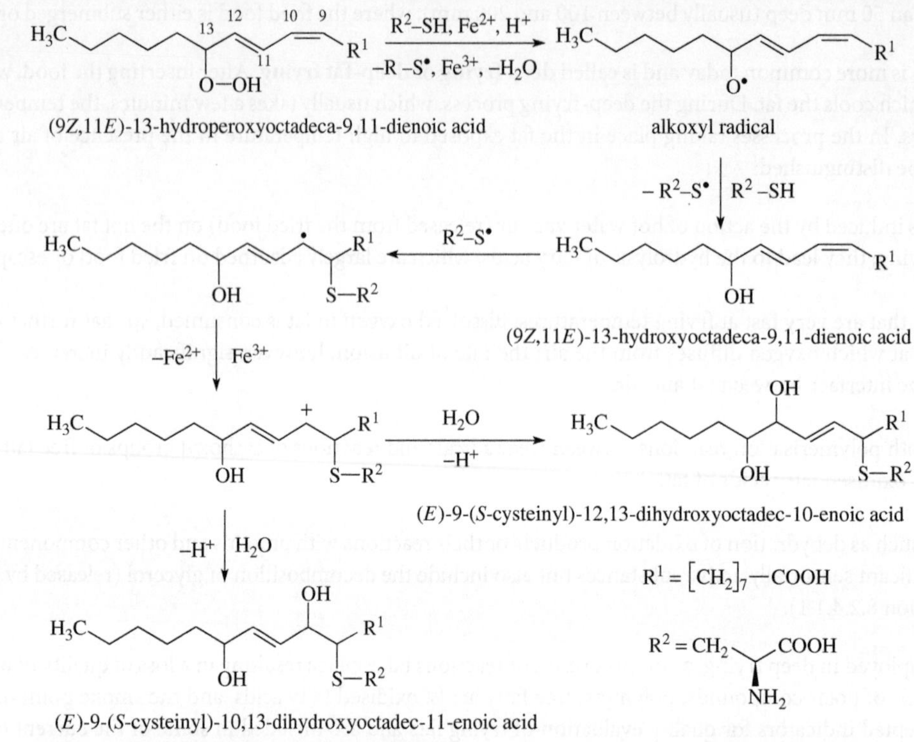

Figure 3.64 Reaction of cysteine with 13-hydroperoxyoctadeca-9,11-dienoic acid.

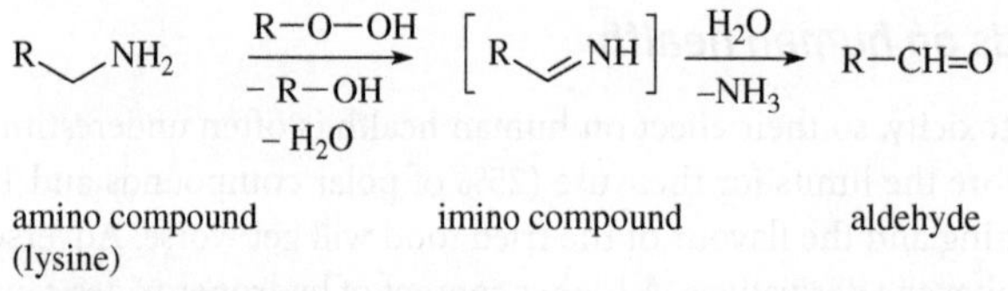

Figure 3.65 Reaction of hydroperoxides with amines or basic amino acids.

even to cysteic acid, but the oxidation by hydroperoxides in foods does not go so far. Methionine is oxidised to methionine sulfoxide, but not to methionine sulfone. Figure 3.64 gives as an example of a simplified reaction of cysteine with 13-hydroperoxyoctadeca-9,11-dienoic acid, which is formed by oxidation of linoleic acid. Further reactions of sulfur-containing amino acids are presented in Section 2.5.1.1. Thiols, sulfides, oligosulfides, and inorganic sulfides are also sensitive to oxidation. Selenium compounds are oxidised in the same way as sulfides. Furthermore, selenites can be oxidised to selenates.

The reaction of hydroperoxides with primary amino groups probably produces an imine, according to the mechanism indicated in Figure 3.65. Upon processing at acidic pH values, conversion of biogenic amines into Strecker aldehydes takes place in a similar way, through reaction with many secondary oxidation products (e.g. 4-4-oxonon-2-enal). Imines are also formed as reaction products of protein-bound amino acids with carbonyl compounds that arise as secondary products of lipid oxidation. In particular, alk-2-enals (α,β-unsaturated aldehydes), such as acrolein, (E)-4-hydroxyhex-2-enal, (E)-4-hydroxynon-2-enal and (E)-4,5-epoxydec-2-enal, and malondialdehyde, are very reactive compounds that accumulate in the forms that are covalently bound to proteins and nucleosides *in vivo* and can be found in foods. The covalent products of these aldehydes, **advanced lipoxidation end products** (ALEs), are used as markers of oxidative stress in the organism. Analogous reaction products of proteins with reactive products arising from sugars, known as AGEs, are described in Section 4.7.5.6. Dietary ALEs derived from acrolein and other alk-2-enals are found in the greatest amounts in heat-treated dairy and meat products.

Of all the alk-2-enals, acrolein is by far the strongest electrophile and, therefore, shows the highest reactivity with nucleophiles, such as the sulfhydryl group of cysteine, ε-amino group of lysine, and imidazole group of histidine. Acrolein is present in foods mainly as a decomposition product of methionine, arachidonic acid (see Figure 3.66), glycerol, hydroxyacetone (acetol), and spermine. With the amino group

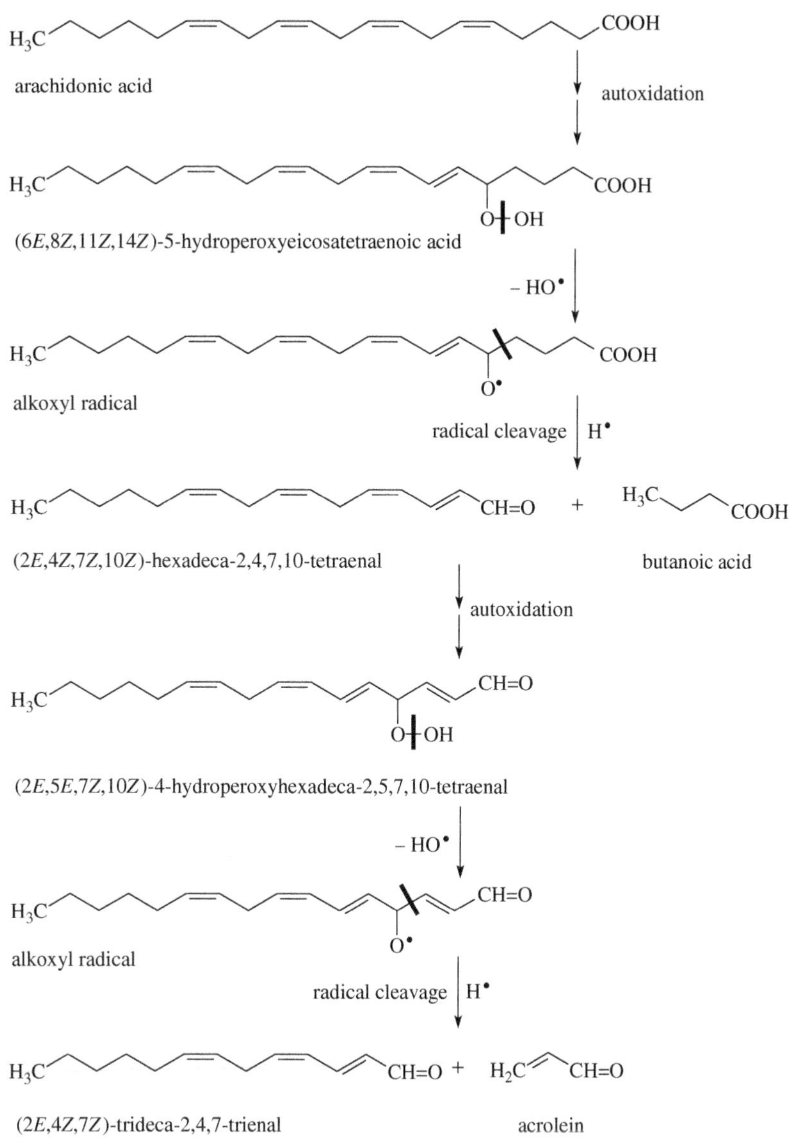

Figure 3.66 Formation of acrolein from arachidonic acid.

of amino acids, acrolein undergoes nucleophilic addition at the double bond (C-3) to form a derivative with the retention of the aldehyde group, resulting in the formation of the Michael addition-type acrolein–amino acid adducts. It has been shown that acrolein modifies lysine and histidine residues of human serum albumin and some proteinases. Although it has been proposed that, upon reaction with amino groups, acrolein forms β-substituted propanals (R—NH—CH$_2$—CH$_2$—CH=O) and imines (R—NH—CH$_2$—CH$_2$—CH=N—R), an adduct of two acrolein molecules with lysine, 2-amino-6-(3-formyl-1,2,5,6-tetrahydropyridine)hexanoic acid, also known as N^ε-(3-formyl-3,4-dehydropiperidino)lysine or FDP-lysine (**3-102**), has recently been identified as the major product.

(E)-4-Hydroxyalk-2-enals arising from lipid peroxidation in biological membranes *in vivo* and in foods have been proven to be involved in carcinogenesis. The most important compound is (E)-4-hydroxynon-2-enal, which arises from the ω-6 fatty acids (linoleic or arachidonic acids) (Figure 3.67). Linolenic acid analogously yields (E)-4-hydroxyhex-2-enal. 4-Hydroxynon-2-enal has also been found in many animal tissues, where it is produced in significant quantities during oxidative stress. In foods, it is formed in various meat products, such as frankfurter sausages (0.08–0.62 mg/kg), and in cooking oils used repeatedly in catering and at home. For example, the thermally oxidised soybean oil after intermittent heating at 185 °C for 1 hour contained 2.3 mg/kg of 4-hydroxynon-2-enal. 4-Hydroxynon-2-enal has various beneficial effects (such as stimulation of guanylate cyclase and phospholipase C) at low concentrations (below 0.1 mmol·l^{-1}), but at higher concentrations (1–20 mmol·l^{-1}) it inhibits protein and DNA synthesis, activates phospholipase A2, and is associated with many diseases, such as

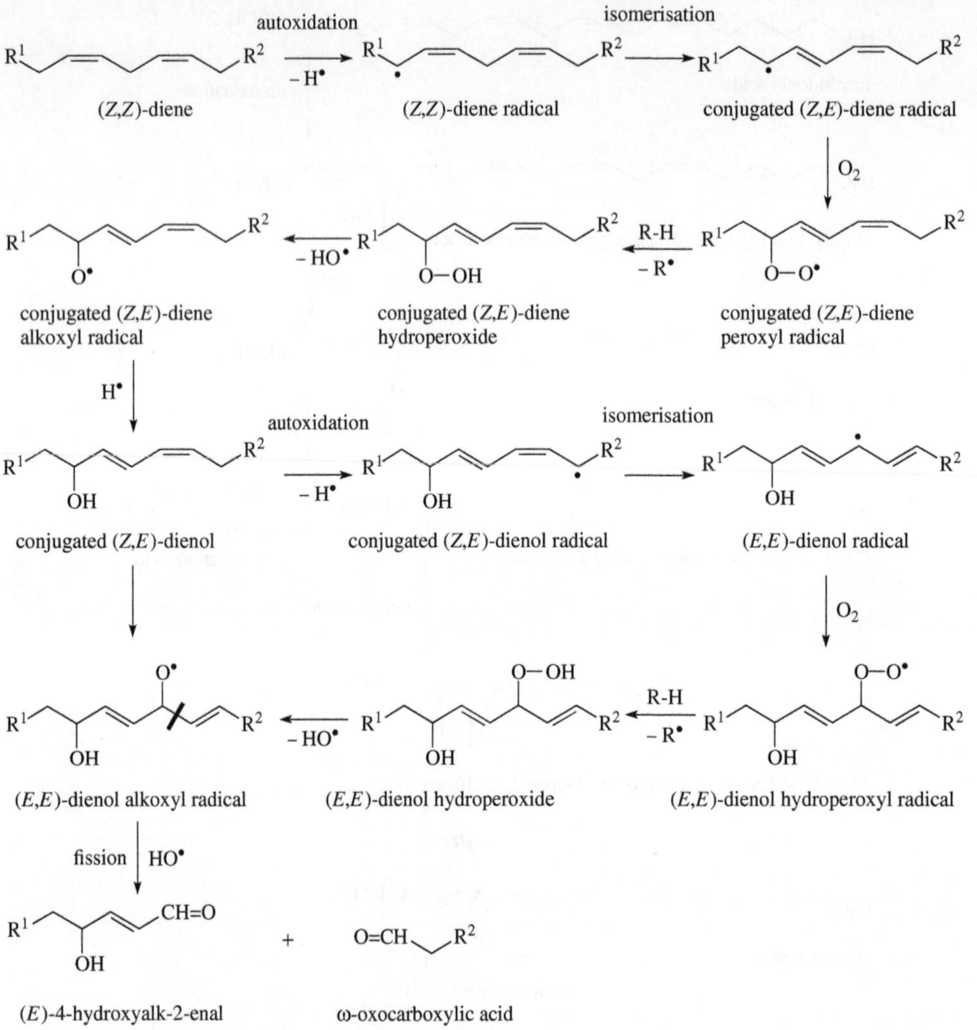

Figure 3.67 Expected mechanisms of formation of (*E*)-4-hydroxyalk-2-enals from ω-6 fatty acids.

chronic inflammation, neurodegenerative diseases, atherosclerosis, diabetes, and various types of cancer. Detoxification and elimination of 4-hydroxynon-2-enal is provided by several enzymes, including glutathione *S*-alkyltransferase, also known as *S*-hydroxyalkyl glutathione lyase (which catalyses the conjugation of aldehyde to glutathione with the formation of polar products), and aldehyde dehydrogenase (which reduces aldehyde to alcohol). An example of a product of 4-hydroxynon-2-enal with lysine is 3-(N^ε-lysino)-4-hydroxynonan-1-ol (cylic form), known as HNE-lysine (**3-102**), whilst a 2-aminopyrimidine derivative, ACR-arginine (**3-102**), results from a reaction of acrolein with arginine. Products analogous to HNE-lysine arise in reactions of 4-hydroxynon-2-enal with cysteine and histidine.

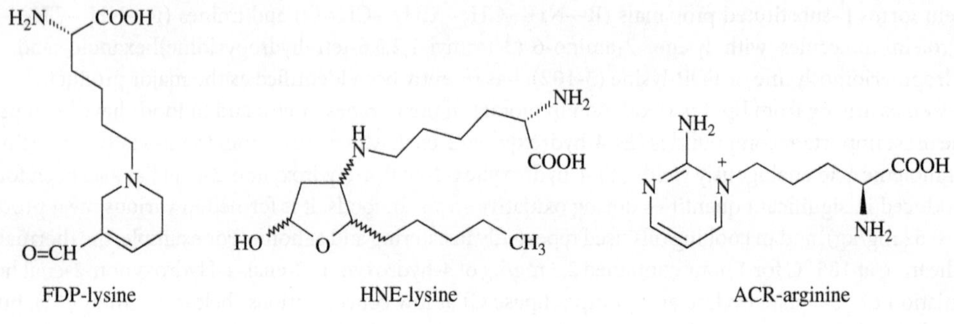

3-102, selected reaction products of acrolein and 4-hydroxynon-2-enal with amino acids

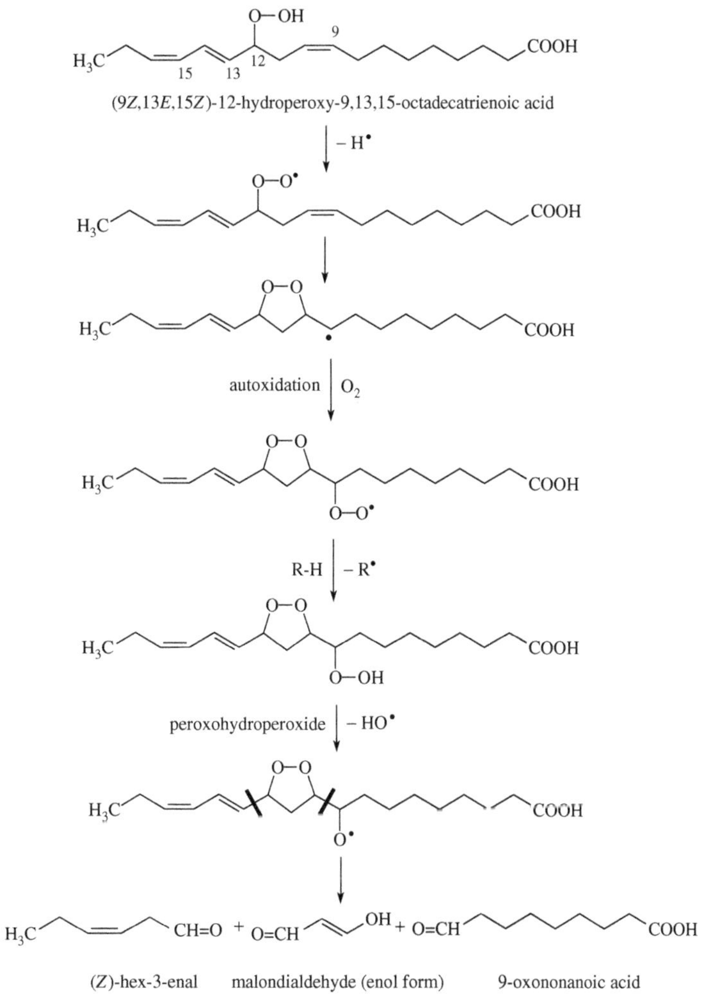

Figure 3.68 Formation of malondialdehyde from linolenic acid peroxohydroperoxide.

Considerable attention has also been paid to malondialdehyde, but the mechanisms of its formation from fatty acids are not yet fully understood. As shown in Figures 3.46 and 3.49, malondialdehyde arises as a secondary product by cleavage of alk-2-enals and alka-2,4-dienals. Its precursors are thus, under certain conditions (acidic pH, singlet oxygen), dienoic fatty acids, such as linoleic acid. Biochemically, the most important sources of malondialdehyde, however, are probably cyclic peroxides of fatty acids with three or more double bonds and similar structures (hydroperoxybisepidioxides, hydroperoxybiscycloendoperoxides, and dihydroperoxides) (Figure 3.32). Malondialdehyde in tissues may also arise from certain precursors of prostaglandins. For example, (9Z,13E,15Z)-12-hydroperoxy-9,13,15-octadecatrienoic acid formed by oxidation of linolenic acid can cyclise to a derivative of 1,2-dioxolane, oxidation of which yields a cyclic peroxohydroperoxide as a major product. Cleavage of bonds on both sides of peroxohydroperoxides gives malondialdehyde (Figure 3.68). This aldehyde readily reacts with functional groups present in proteins, nucleic acids, and phospholipids, especially amino groups.

Reaction of a malondialdehyde carbonyl group with amino groups leads to the formation of imines, but formation of these structures is of no nutritional concern, because they are hydrolysed at the acidic pH of the stomach. N-Prop-2-enals, which are absorbed from the gut, are also formed in neutral or acidic aqueous media, but most of the absorbed material is not metabolised. A third type of reaction products is unavailable 4-substituted 1,4-dihydropyridine-3,5-dicarbaldehydes, which arise in reactions of malondialdehyde with amino compounds, such as lysine, in the presence of alkanals. Examples of malondialdehyde reaction products with lysine are N^ε-(prop-2-enal)lysine, so-called malondialdehyde-lysine, N^α-(prop-2-enal)lysine, N,N'-di(prop-2-enal)lysine (**3-103**), and a conjugated cross-link in proteins termed lysine-malondialdehyde-lysine. An example of the reaction product of lysine with malondialdehyde and acetaldehyde is N,N'-di(4-methyl-1,4-dihydropyridine-3,5-dicarbaldehyde)lysine, which, for example, arises in the reaction of bovine serum albumin with malondialdehyde and acetaldehyde (**3-104**).

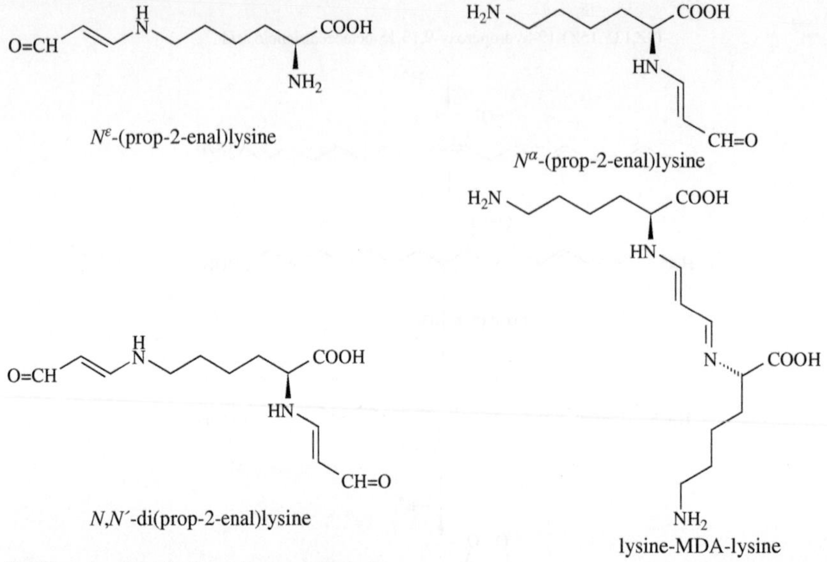

N^{ε}-(prop-2-enal)lysine

N^{α}-(prop-2-enal)lysine

N,N′-di(prop-2-enal)lysine

lysine-MDA-lysine

3-103, malondialdehyde conjugates with lysine

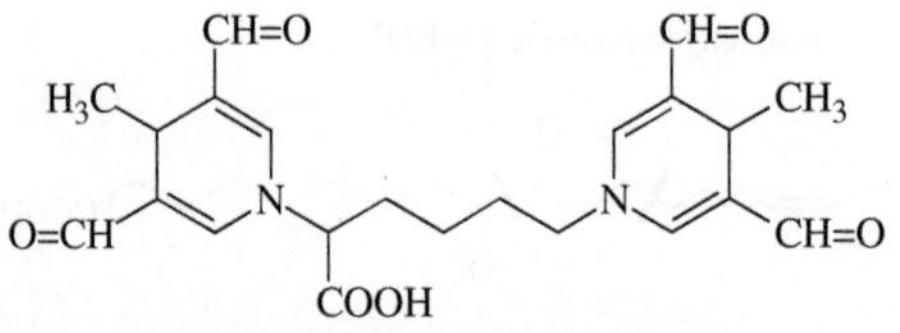

3-104, N,N′-di(4-methyl-1,4-dihydropyridine-3,5-dicarbaldehyde)lysine

ALEs can also be formed from prostaglandin H_2 (PGH$_2$), which rearranges non-enzymatically into levuglandin E$_2$. Levuglandin E$_2$ condenses with the ε-amino group of protein-bound lysine to form a pyrrole derivative levuglandin E$_2$-lysine (**3-105**), which has been identified in the blood plasma.

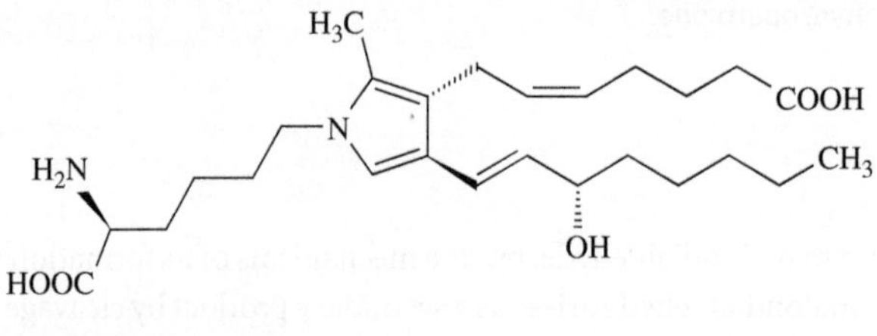

3-105, levuglandin E$_2$-lysine

Alk-2-enals also react to form adducts with 2′-deoxyguanosine and other bases in DNA. If the damaged DNA is not repaired, these adducts may be mutagenic. The proposed reaction mechanism involves a Michael addition of the N^2-amino group of deoxyguanosine at the C-3 position of the alk-2-enals. The Michael product then cyclises at the N-1 of deoxyguanosine to form two pairs of diastereomers. The simplest alk-2-enal, acrolein, reacts with 2′-deoxyguanosine in DNA to form 1,N^2-propanodeoxyguanosine adducts: (6R/S)-3-(2′-deoxyribos-1′-yl)-5,6,7,8-tetrahydro-6-hydroxypyrimido[1,2-a]purine-10(3H)ones (Figure 3.69) and (8R/S)-3-(2′-deoxyribos-1′-yl)-5,6,7,8-tetrahydro-8-hydroxypyrimido[1,2-a] purine-10(3H)ones. Other alk-2-enals, such as (E)-4-hydroxyhex-2-enal and (E)-4-hydroxynon-2-enal, react with deoxyguanosine analogously. Malondialdehyde possibly reacts with deoxyguanosine in a different way (and also with adenine) to form a variety of adducts, the most abundant of which is pyrimido[1,2-a]purin-10(3H)-one, known as M$_1$G (Figure 3.69). The M$_1$G adduct is mutagenic and has been detected in human liver cells and other products.

3.8.1.12.2 Reactions with saccharides

Oxidation of saccharides is usually initiated by one-electron oxidising agents, such as transition metal ions (especially Fe^{3+} and Cu^{2+}) and free radicals (mainly the hydroxyl radical HO• and alkoxyl radical RO•). The primary products of oxidation by free radicals (Figure 3.70),

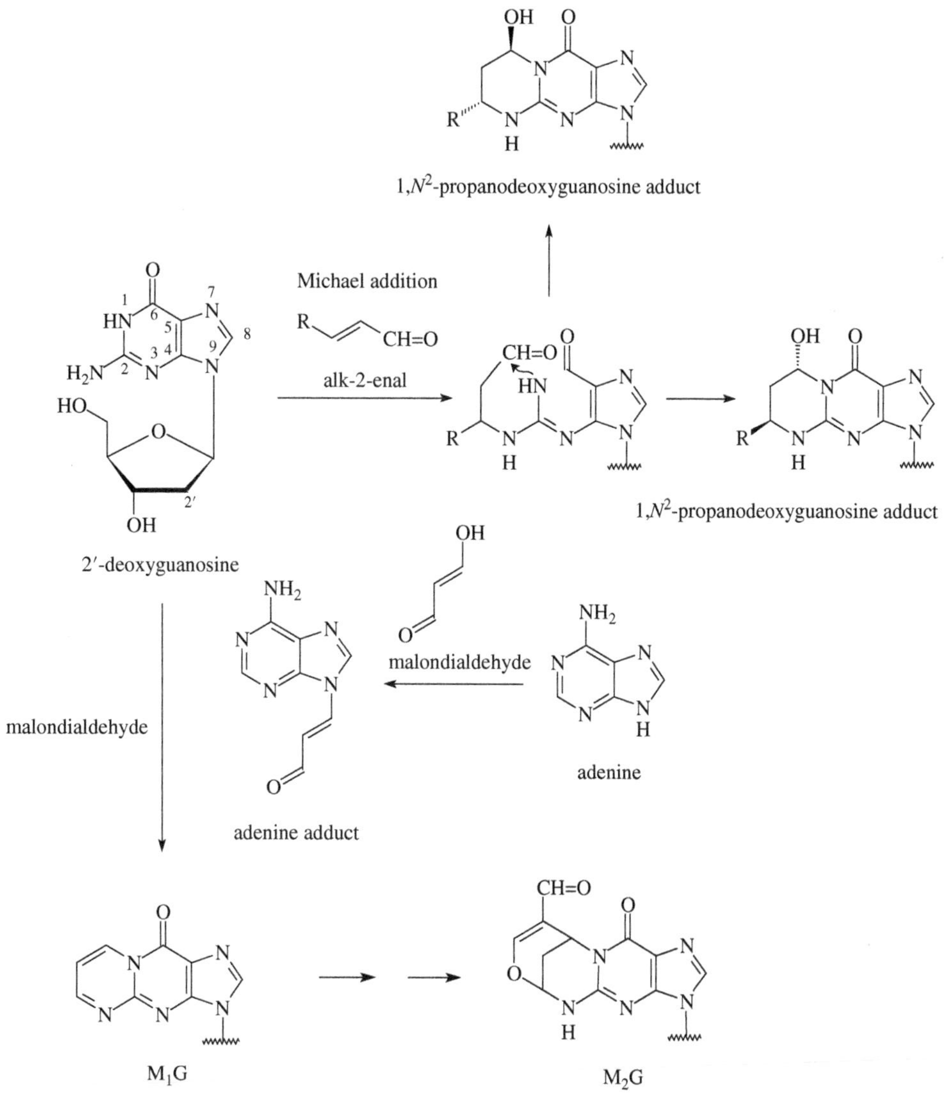

Figure 3.69 Formation of 2′-deoxyguanine adducts with alk-2-enals and malondialdehyde: R = H (acrolein), R = CH(OH)CH$_2$CH$_3$ (4-hydroxyhex-2-enal, R = CH(OH)[CH$_2$]$_4$CH$_3$ (4-hydroxynon-2-enal).

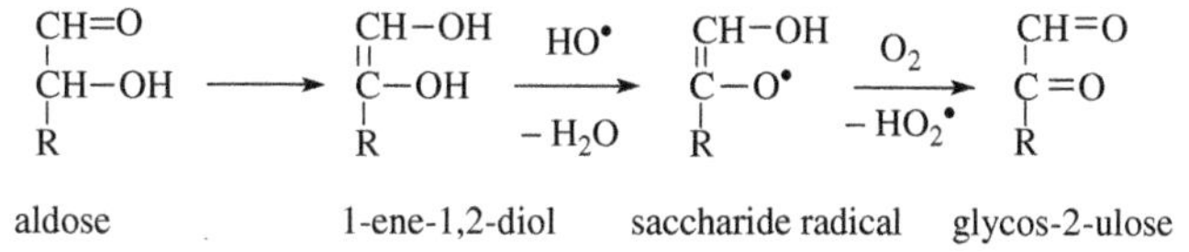

Figure 3.70 Oxidation of monosaccharides to glycosuloses by radicals.

which takes place after the isomerisation to 1-ene-1,2-diols, are α-dicarbonyl compounds containing the original number of carbon atoms, known as glycos-2-uloses. Glycos-2-uloses are further oxidised and decomposed. Radicals formed as intermediates and α-dicarbonyl compounds react further with proteins in the Maillard reaction.

3.8.1.12.3 Reactions with vitamins

Vitamins that have a role as antioxidants (vitamin E, ascorbic acid, and provitamins A) are easily oxidisable substances that react with free radicals or act as singlet oxygen scavengers. The reaction of tocopherols (T—OH) with hydroperoxyl radicals (ROO•) is also a part of the protective mechanisms in biomembranes of living tissues:

$$R - O - O^\bullet + T - OH \rightarrow R - O - OH + T - O^\bullet$$
$$(R - O - OH = \text{hydroperoxide}, \ T - O^\bullet = \text{tocopheryl radical})$$

At the phase interface, tocopheryl radicals are reduced back to tocopherols by ascorbic acid (H_2A), which is oxidised to ascorbyl radical ($HA^\bullet$):

$$H_2A + T - O^\bullet \rightarrow HA^\bullet + T - OH$$

Another one-electron oxidation of ascorbyl radical produces dehydroascorbic acid (A), which is reduced back to ascorbic acid by dehydroascorbate reductase:

$$2HA^\bullet + NADH + H^+ \rightarrow 2H_2A + NAD^+$$

Other vitamins are also oxidised. In folacin, for example, the links C-9 and N-10 are cleaved (see Section 5.12.2); vitamins D undergo photodegradation in the presence of photosensitisers (see Section 5.3.6).

3.8.1.12.4 Reactions with other compounds

Other important reactions are reactions of oxidised lipids with phenolic compounds, many of which are used as antioxidants (see Section 11.2.2).

3.8.1.13 Oxidation in biological systems

Oxygenic photosynthesis in plants proceeds according to the following formal equation:

$$A + H_2O + \text{light energy (photons)} \rightarrow AH_2 + 1/2 \, O_2$$

where A is the acceptor (carbon dioxide) and AH_2 is the carbohydrate unit ($H_2C{=}O$).

The reverse reaction, oxidation of nutrients in the respiratory chain of animals (e.g. the oxidation of sugars in the citric acid cycle and oxidation of fatty acids by β-oxidation), proceeds according to the reaction:

$$AH_2 + 1/2 \, O_2 \rightarrow A + H_2O + \text{ energy.}$$

Triplet oxygen is reduced to water during oxygenic photosynthesis in plants and in analogous reactions in the respiratory chain of animals:

$$^3O_2 + 4e^- + 4H^+ \rightarrow 2H_2O$$

At the same time, toxic forms of oxygen, free radicals, and covalent compounds are formed as byproducts (through one-, two-, and three-electron reduction), which oxidise lipids of biological membranes, and subsequently DNA, proteins, and other biomolecules (Figure 3.71). Fortunately, organisms are also equipped with detoxifying mechanisms, which include:

- enzymes whose cofactors are Mn, Zn, Cu, Se, and Fe;

- vitamins (vitamin E, provitamins A, and vitamin C).

3.8.1.13.1 Singlet oxygen

Singlet oxygen, which originates in photosensitised reactions (Figure 3.54), is very effectively scavenged by carotenes, tocopherols (vitamin E), and ascorbic acid (vitamin C) in living organisms and in edible oils and fats.

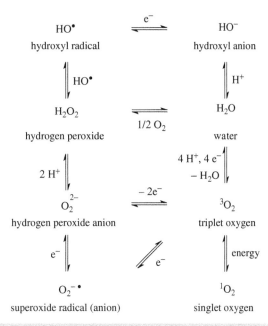

Figure 3.71 Metabolism of oxygen.

3.8.1.13.2 Superoxide anion

Superoxide anion O_2^- has one unpaired electron; therefore, it is a free radical, $O_2^{-\bullet}$. This superoxide radical arises by one-electron reduction of triplet oxygen. In plants, this reaction is catalysed by NADPH oxidase; in animals, superoxide radical is a byproduct of other oxidases:

$$^3O_2 + e^- \rightarrow O_2^{-\bullet}$$

The superoxide anion is removed by the activity of superoxide dismutase. The mitochondrial enzyme contains Mn as a cofactor; the cytosolic enzyme has Zn and Cu as cofactors. A donor of two hydrogen atoms might be ascorbic acid ($H_2A + O_2^{-\bullet} + H^+ \rightarrow HA^\bullet + H_2O_2$), for example:

$$O_2^{-\bullet} + e^- \rightarrow O_2^{2-}$$
$$O_2^{2-} + 2H^+ \rightarrow H_2O_2$$

Spontaneously, without the involvement of any enzymes, one-electron reduction or dismutation (the so-called Haber–Weiss reaction) takes place:

$$O_2^{-\bullet} + H_2O_2 \rightarrow HO^\bullet + HO^- + {}^3O_2$$
$$O_2^{-\bullet} + Fe^{3+} \rightarrow {}^3O_2 + Fe^{2+}$$

3.8.1.13.3 Hydrogen peroxide

Hydrogen peroxide is the result of two-electron reduction of triplet oxygen:

$$^3O_2 + 2e^- + 2H^+ \rightarrow H_2O_2$$

It is mainly decomposed by microsomal catalase (H_2O_2: H_2O_2 oxidoreductase), which contains Fe (haematin) as a cofactor, or by cytoplasmic or mitochondrial GSH peroxidase (H_2O_2: donor oxidoreductase), which contains Se as a cofactor:

$$2H_2O_2 \rightarrow 2H_2O + {}^3O_2$$
$$2G-SH + H_2O_2 \rightarrow G-S-S-G + 2H_2O$$

Otherwise, a single-electron reduction takes place, which proceeds as in the Haber–Weiss reaction or Fenton reaction, yielding a highly reactive hydroxyl radical (HO$^\bullet$):

$$H_2O_2 + e^- \rightarrow HO^\bullet + HO^-$$
$$H_2O_2 + Fe^{2+} \rightarrow HO^\bullet + HO^- + Fe^{3+}$$

3.8.1.13.4 Hydroxyl radical

The most dangerous toxic form of oxygen is the hydroxyl radical, which is generated by two-electron reduction of triplet oxygen:

$$^3O_2 + 3e^- + 3H^+ \rightarrow HO^\bullet + H_2O$$

Hydroxyl radicals damage the mitochondrial (chromoplastic), microsomal (peroxysomal), and endoplasmic reticulum membranes. The damage is associated with many degenerative processes (diseases) including certain forms of cancer and occurs even during the normal ageing of organisms. Hydroxyl radicals operate mainly at the place of formation, where they attack unsaturated fatty acid (RH) bound in the membrane phospholipids. Unsaturated fatty acids yield hydroperoxides (R–O–OH) through the previously described mechanisms:

$$R - H + HO^\bullet \rightarrow R^\bullet + H_2O$$
$$R^\bullet + O_2 \rightarrow R - O - O^\bullet$$
$$R - O - O^\bullet + R - H \rightarrow R - O - OH + R^\bullet$$

The protective mechanism of organisms lies in the reduction of hydroperoxyl radicals by vitamin E (tocopherols, TH):

$$R - O - O^\bullet + T - H \rightarrow R - O - OH + T^\bullet$$

Tocopheryl radicals T$^\bullet$ are stabilised by a number of reactions. They can be reduced back to tocopherols by ascorbic acid (H$_2$A), which is oxidised to dehydroascorbic acid (A):

$$2T^\bullet + H_2A \rightarrow 2T - H + A$$

Hydroperoxides are removed from bonds in damaged phospholipid by activated phospholipase A$_2$, and free hydroperoxide becomes a substrate for peroxidases. Otherwise, the oxidation of fatty acids results in the accumulation of lipid hydroperoxides and their secondary metabolites such as (E)-4-hydroxynon-2-enal and malondialdehyde, which react with proteins and DNA (see Section 3.8.1.12.1).

Proteins and deoxyribonucleic acids The main agents that damage proteins are hydroxyl radicals (HO$^\bullet$), singlet oxygen ($^1O_2$), and superoxide radicals (HO$_2^\bullet$). Reactions are analogous to those in foods and can result in loss of enzymatic activity, cell cytolysis, and cell death. Damaged proteins in biomembranes are often associated with oxidised lipids.

The main oxidising agent of DNA is the hydroxyl radical (HO$^\bullet$), which oxidises the purine and pyrimidine bases and deoxyribosyl residues. The main oxidation products are 5-(hydroxymethyl)uracil (**3-106**) and 8-hydroxyguanine (**3-106**), which are formed by the reaction of guanine with singlet oxygen. In the presence of transition metals, the main compound that is attacked is adenine. The reaction product is adenine-N^1-oxide (**3-106**).

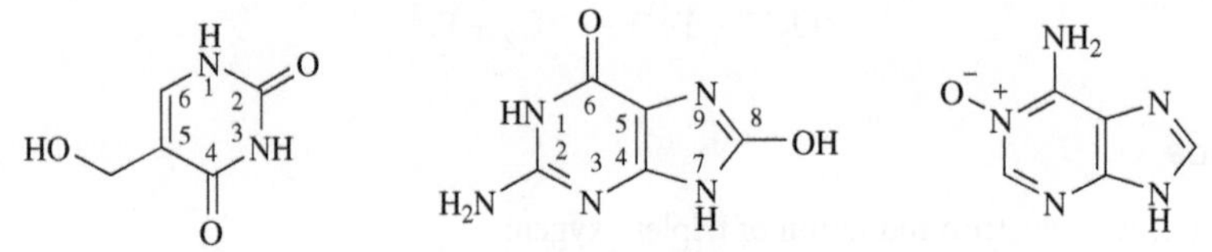

5-(hydroxymethyl)uracil   8-hydroxyguanine   adenine-N^1-oxide

3-106, oxidation products of pyrimidine and purine bases

3.8.1.14 Degradation reactions

At high temperatures, pyrolytic reactions of lipids can yield fatty acid anhydrides, and fatty acids may eliminate carbon dioxide by decarboxylation. These reactions take place during food processing only at trace levels.

3.8.2 Reactions of homolipids

3.8.2.1 Reactions of waxes

Waxes do not have specific chemical properties and their reactivities are similar to those of other fatty acid esters. The only difference is that waxes react more slowly due to their higher molecular weight and high melting point. Their most important chemical reaction is hydrolysis to the parent components, primary alcohols and fatty acids.

3.8.2.2 Reactions of fats and oils

3.8.2.2.1 Hydrolysis and saponification

Hydrolysis of fats and oils (lipolysis) in foods is primarily caused by the action of lipases. The sources of lipases can be crude oils (lipase from vegetable seeds, beans, and fruit pulps) or microorganisms. Some lipases show substrate specificity according to the length of the acyl group and regiospecificity (sn-non-specific and sn-1,3-specific lipases). Lipolysis in food technology may proceed on purpose, for example during ripening of certain cheeses, with specific types of moulds added during the cheesemaking process (surface-ripened Camembert or blue cheeses such as Roquefort and Gorgonzola with moulds growing in the curd), in cured and fermented dried sausages (by *Lactobacillus* bacteria and other bacteria added as starter cultures), or by the activity of natural microflora. Examples are given in Table 3.46.

Hydrolysis products are dependent on the specificity of lipase. In the case of pancreatic lipase, the final products are fatty acids and 2-monoacylglycerols; in the case of sn-non-specific lipases from oil seeds, di- and monoacylglycerols, glycerol, and free fatty acids gradually form upon hydrolysis of crude vegetable oils. A necessary condition for the enzymatic hydrolysis process is the presence of water. The solubility of water in refined oils is around 0.1% at 20 °C, and may be higher in crude oils due to the presence of heterolipids.

Spontaneous partial hydrolysis of fats occurs at high temperatures during deep frying of foods, when the released water causes hydrolysis of the ester bonds. The released fatty acids are distilled off from the oil bath to the surroundings. Particularly prone to this behaviour are fats containing medium-chain fatty acids, such as coconut or palm kernel oil and milk fat.

Fatty acids in oleochemistry (as well as fatty acids for food use) are obtained through autocatalysed hydrolysis (Figure 3.72) at high temperatures, where the byproduct is glycerol. The reaction proceeds in a two-phase system of oil–water, where the lipid phase is a dispersion medium, and at temperatures above 200 °C, in order to increase the solubility of water in the fat. Glycerol is soluble in water and is extracted from the fat phase into the aqueous phase; therefore, the reaction equilibrium is shifted in favour of hydrolysis and a degree of hydrolysis of up to 98–99% can be achieved. Enzymatic hydrolysis is used, for example, to obtain essential fatty acids containing C_{20}—C_{22} fatty acids, when high temperatures cannot be used.

Saponification of lipid ester bonds is based on the reaction of aqueous solutions of alkali metal hydroxides and the formation of soaps (alkali metal salts of fatty acids) and glycerol as byproducts. Intermediates of saponification of triacylglycerols are di- and monoacylglycerols.

A partial saponification of acylglycerols also occurs during deacidification (neutralisation) of raw oils by alkaline agents, which is a side reaction that leads to some loss of oil. A partial saponification also takes place during ester exchange, when sodium hydroxide is used as a catalyst. Saponification of fat blends is used deliberately to produce sodium soaps, toilet soaps (from a mixture of coconut oil and beef tallow), and detergent soap powders (usually from beef tallow only).

Table 3.46 Specific lipases from different sources.

Lipase source	Specificity to acyl length	Regiospecificity (sn)	Lipase source	Specificity to acyl length	Regiospecificity (sn)
Pancreatic lipase	S > M, L	1,3	*Penicillium camemberti*	MAG, DAG > TAG[a]	1,3
Stomach lipase	S, M ≫ L	1,3	*Penicillium roqueforti*	S, M ≫ L	1,3
Aspergillus niger	S, M, L	1,3 ≫ 2	*Rhizopus javanicus*	M, L > S	1,3 > 2
Candida lipolytica	S, M, L	1,3 > 2	*Rhizopus japonicus*	S, M, L	1,3 > 2
Candida cylindracea	S, M, L	1,2,3	*Rhizopus oryzae*	M, L > S	1,3 ≫ >2
Rhizomucor mihei	S > M, L	1 > 3 ≫ 2	*Pseudomonas fluorescens*	M, L > S	1,3 > 2
Mucor javanicus	M, L ≫ S	1,3 > 2	*Pseudomonas* spp.	S, M, L	1,3 > 2

[a] Substrate specificity to acylglycerols.

L, long-chain fatty acids; M, medium-chain fatty acids; S, short-chain fatty acids; MAG, monoacylglycerols; DAG, diacylglycerols; TAG, triacylglycerols.

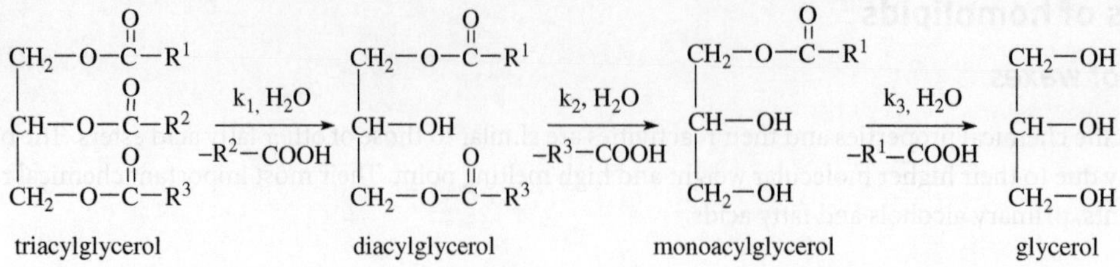

Figure 3.72 Hydrolysis (saponification) of fats (reaction rate constants $k_3 > k_2 > k_1$).

3.8.2.2.2 Interesterification reactions

Interesterification reactions in fats and oils composed of triacyglycerols include a set of reactions where the original esters reacts with fatty acids, alcohols, and other esters. The following reactions are then recognised:

- **acidolysis**;

- **alcoholysis**;

- **ester interchange (transesterification)**.

This set of reactions is catalysed either by enzymes or by acidobasic catalysts and is characterised by migration of acyls between the molecules of glycerol (or other alcohols). Interesterification reactions and esterification of fatty acids with glycerol are also known as re-esterification reactions. When using acid–base catalysts, acyl migration is statistically random. There is a growing interest in using enzymes (lipases or phospholipases) in technological applications, where it is possible to perform selective interesterification reactions catalysed by specific enzymes.

Acidolysis   Acidolysis is a reaction of triacylglycerols with carboxylic acid, which leads, under acid catalysis, to the exchange of acyl residues. The reaction is in dynamic equilibrium, and if the equilibrium is not shifted by removing reaction components, the product contains all components, including the starting reactants (Figure 3.73).

An example is the reaction of C_{16} and C_{18} fatty acids with triacylglycerols of coconut oil, resulting in the release of C_8—C_{14} fatty acids, which can be continuously distilled from the reaction mixture as the product. Acidolysis of fatty acids with diterpene abietic acid (**3-107**), the main component of the so-called resin acids, is used, for example, to produce varnishes, whilst the exchange of fatty acids for phthalic acid yields glyptals (polymers containing the ester functional group with properties of natural resins).

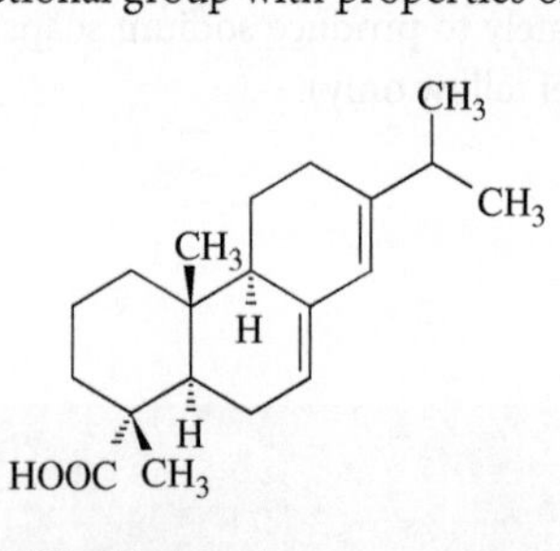

3-107, abietic acid

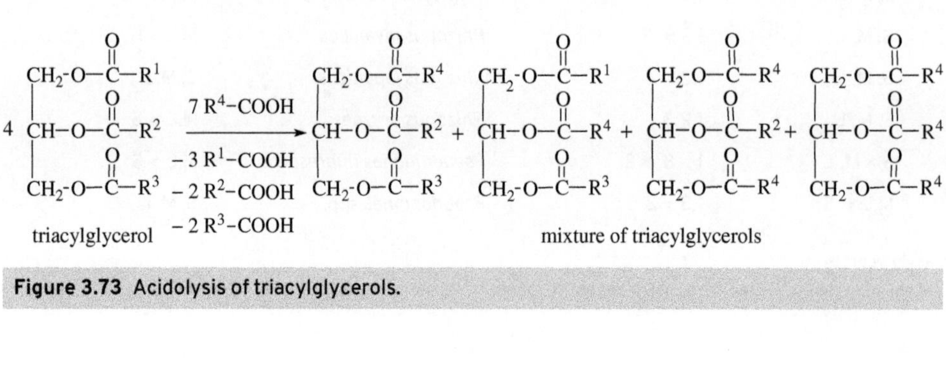

Figure 3.73 Acidolysis of triacylglycerols.

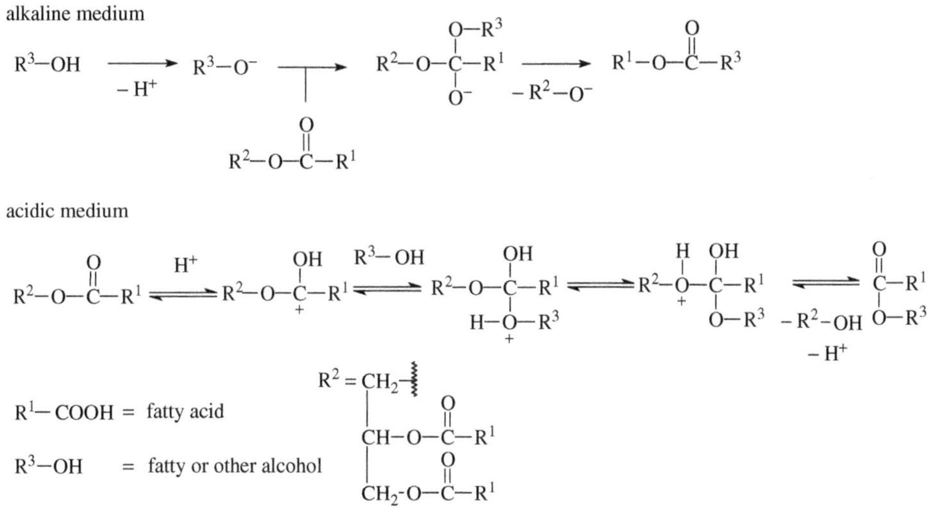

alkaline medium

$$R^3\text{--}OH \xrightarrow[-H^+]{} R^3\text{--}O^- \longrightarrow R^2\text{--}O\overset{\displaystyle O\text{--}R^3}{\underset{\displaystyle O^-}{\text{--}\overset{|}{\underset{|}{C}}\text{--}}}R^1 \xrightarrow{-R^2\text{--}O^-} R^1\text{--}O\overset{O}{\overset{\|}{C}}\text{--}R^3$$

$$R^2\text{--}O\overset{O}{\overset{\|}{C}}\text{--}R^1$$

acidic medium

$$R^2\text{--}O\overset{O}{\overset{\|}{C}}\text{--}R^1 \underset{H^+}{\rightleftharpoons} R^2\text{--}O\overset{OH}{\underset{+}{\text{--}\overset{|}{C}\text{--}}}R^1 \underset{R^3\text{--}OH}{\rightleftharpoons} R^2\text{--}O\overset{OH}{\underset{\underset{+}{H\text{--}O\text{--}R^3}}{\text{--}\overset{|}{C}\text{--}}}R^1 \rightleftharpoons R^2\overset{H\;\;OH}{\underset{O\text{--}R^3}{\overset{+}{\text{--}O\text{--}C\text{--}}}}R^1 \underset{-R^2\text{--}OH}{\rightleftharpoons} \overset{O}{\overset{\|}{C}}\overset{\text{--}R^1}{\underset{O\text{--}R^3}{}}$$

$$-H^+$$

$R^1\text{--}COOH$ = fatty acid

$R^3\text{--}OH$ = fatty or other alcohol

$$R^2 = CH_2\text{--}$$
$$CH\text{--}O\overset{O}{\overset{\|}{C}}\text{--}R^1$$
$$CH_2\text{--}O\overset{O}{\overset{\|}{C}}\text{--}R^1$$

Figure 3.74 Mechanism of alcoholysis of triacylglycerols.

An important application of enzymatic acidolysis, using *sn*-1,3-lipase, is the synthesis of structured triacylglycerols containing palmitic or stearic acid in the *sn*-1 and *sn*-3 positions and oleic acid in the *sn*-2 position, the so-called cocoa butter equivalent (CBE) fat, which serves as a substitute for cocoa butter. Acidolysis of 1,2-dipalmitoyl-2-oleoyl-*sn*-glycerol (obtained from palm oil) with stearic acid gives a mixture of 1-palmitoyl-2-oleyl-3-stearoyl-*sn*-glycerol and 1,3-distearoyl-2-oleoyl-*sn*-glycerol. The third major triacylglycerol contained in cocoa butter is obtained by acidolysis of trioleoylglycerol (triolein) with stearic acid.

Enzymatically catalysed acidolysis using specific *sn*-1,3-lipases is also used for the preparation of dietetically important structured triacylglycerols of the MUM type that contain medium-chain fatty acids (caprylic or capric) in the outer *sn*-1,3 positions and unsaturated essential fatty acids in the *sn*-2 position.

Alcoholysis The reaction of triacylglycerols with alcohols is called alcoholysis. It can be catalysed by acids (sulfuric acid or 4-toluenesulfonic acid) or bases (methoxides, hydroxides, and carbonates of alkali metals). Industrially produced alcoholysis products are methyl, ethyl, or propyl esters (Figure 3.74) of fatty acids.

Typically made by alcoholysis of vegetable oils (such as rapeseed, soybean, and palm oils) or animal fats (beef tallow) with methanol, ethanol or propan-1-ol is the so-called biodiesel, a diesel fuel consisting of long-chain methyl, ethyl, or propyl esters. Methyl and ethyl esters can also be obtained by enzyme-catalysed alcoholysis using non-specific lipases.

Glycerolysis Glycerolysis is a particular example of alcoholysis, being a reaction in which fat (triacylglycerols) reacts with glycerol (Figure 3.75) to give a mixture of monoacylglycerols and diacylglycerols. Glycerolysis takes place at high temperatures (200–250 °C) in the presence of an alkaline catalyst (usually either potassium hydroxide, calcium oxide, or sodium methoxide). The reaction mixture contains less than 55–60% of monoacylglycerols that are used as emulsifiers. This mixture can be used as such, but for more demanding applications it is subjected to molecular distillation, which gives the product (emulsifier) containing up to 95% of monoacylglycerols (Table 3.47). Emulsifiers produced in this way are used in the food, cosmetics, and pharmaceutical industries. Emulsifiers based on monoacylglycerols can also be produced by direct esterification of fatty acids with glycerol, but this method is only rarely used.

Glycerolysis can instead be catalysed by specific *sn*-1,3-lipases. A mixture of di- and monoacylglycerols is obtained in the first stage; further glycerolysis gives products with a reduced diacylglycerols content, which in practice is a mixture of 1- and 2-monoacylglycerols. 2-Monoacylglycerols can also be used for the synthesis of structured triacylglycerols.

A special case of alcoholysis is the use of sugars, usually saccharose and sugar alcohols. Direct esterification of sucrose is very difficult, because it leads to dehydration products and caramelisation (see Section 4.7.6). Alcoholysis may, depending on the initial fatty acid ester, give different products. The reaction of saccharose with a mixture of triacylglycerols gives a mixture of saccharose esters with mono-, di-, and triacylglycerols and glycerol. As saccharose has a total of eight hydroxyl groups, compounds ranging from saccharose mono to octa fatty acid esters can be produced. This reaction mixture is known by the name **sugar esters**. Sugar esters are mixed non-ionic surfactants (see Section 11.5.2.1.5) widely used as emulsifiers in foods and beverages, in detergents, industrial cleaners, agricultural chemicals, and as excipients in pharmaceuticals. Products consisting of sucrose monoesters, and to a small extent sucrose diesters, are manufactured by a transesterification reaction of sucrose and fatty acid methyl esters. Esterification gives a mixture in which hexa, hepta, and octa esters of sucrose predominate. The product, known as Olestra (or by its brand name, Olean) is a lipid-like substance that was developed as a non-energy (non-absorbable)

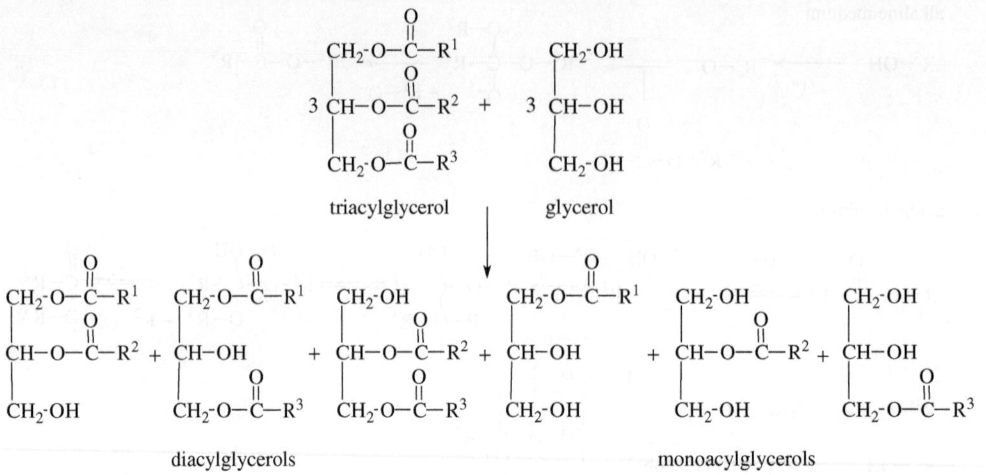

Figure 3.75 Glycerolysis of triacylglycerols.

Table 3.47 Compositions of monoacylglycerol emulsifiers.

Component (%)	Emulsifier[a]	Distilled emulsifier[b]
Monoacylglycerols	50–55	90–95
Diacylglycerols	30–35	2–5
Triacylglycerols	<5	0.2–1.0
Free fatty acids	1.0	1.0–2.0
Glycerol	<8	0.2–1.0

[a]Emulsifier composition corresponds to the composition of the fat phase after glycerolysis.
[b]Molecularly distilled emulsifier.

fat substitute, since the ester bonds are not hydrolysed by enzymes of the digestive tract. However, it was found that Olestra has some adverse gastrointestinal effects and depletes the levels of many valuable fat-soluble substances, including carotenoids, in blood.

Ester interchange The ester interchange (or transesterification) is catalysed by alkaline catalysts. For food uses, sodium hydroxide or sodium methoxide is usually used. The choice of catalyst is closely related to reaction temperature. In the resulting equilibrium mixture (Figure 3.76), the distribution of fatty acids in triacylglycerols is random, and thus very different from the fatty acid distribution in natural triacylglycerols. Therefore, this transesterification is often also known as **randomisation**. For example, the alkali-catalysed uncontrolled transesterification is used for the preparation of structural fats that do not contain trans-fatty acids.

In base-catalysed transesterification, some side reactions take place. For example, the reaction of water with sodium methoxide creates sodium hydroxide, which can cleave the ester bonds to form soaps. Oxidation of tocopherols and other antioxidants in alkaline media also occurs, which reduces the oxidative stability of transesterified fat.

Enzymatically catalysed transesterification The mechanism of acyl exchange during enzymatic transesterification of triacylglycerols is explained by the formation of the complexes enzyme–enzyme–acylglycerol and enzyme–fatty acid. A substantial benefit when using lipases that act as transesterification catalysts is their substrate specificity (depending on the acyl length) and especially the regiospecificity of sn-1,3-specific lipases. As a result of their sn-1,3-regiospecifity, the randomisation of acyls occurs only in the sn-1 and sn-2 positions of triacylglycerol molecules, and a substantial portion of unsaturated fatty acids remains in the sn-2 position, which is considered from a nutritional point of view an asset. Structural fats derived from enzyme catalysis have physical properties, including texture, different from the structural fats obtained by base-catalysed randomisation.

Enzymatic transesterification of triacylglycerols does not cause the oxidation of tocopherols, but the enzyme is sensitive to the presence of hydroperoxides and aldehydes, which may cause changes in its structure.

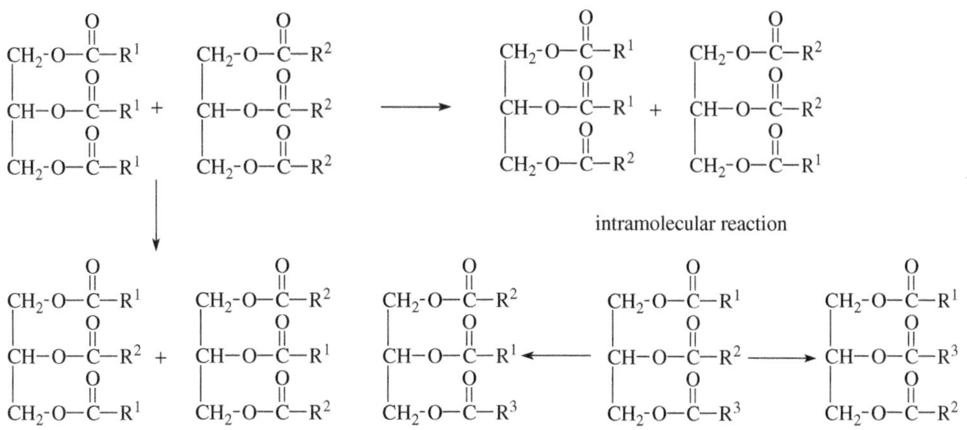

intermolecular reaction

Figure 3.76 Ester interchange in triacylglycerols.

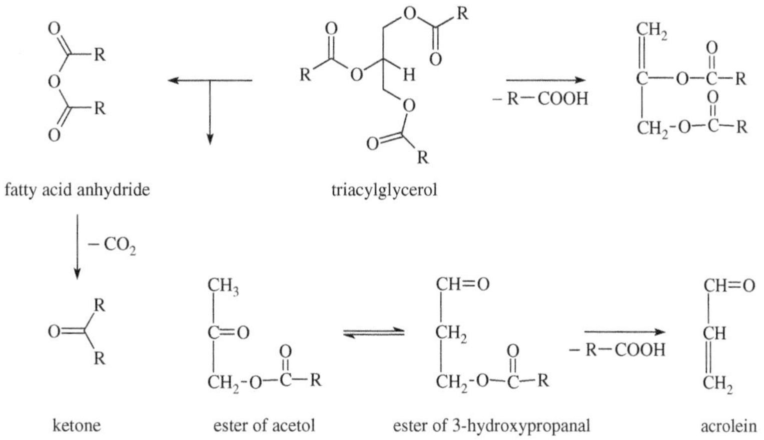

Figure 3.77 Formation of acrolein in triacylglycerol thermolysis.

3.8.2.2.3 Pyrolytic reactions

Heating fat to high temperatures causes its pyrolysis in parallel with oxidation, polymerisation, oxypolymerisation, and some other reactions. The presence of water leads to hydrolysis of fatty acids bound in triacylglycerols, and diacylglycerols, monoacylglycerols, glycerol, and free fatty acids are formed as products. Free fatty acids and other products are also formed during pyrolysis of triacylglycerols in the absence of water (Figure 3.77). Another important product of pyrolysis is acrolein, which is produced in anhydrous fats, for example during deep-fat frying, and which irritates the eyes and mucous membranes of operating personnel. Linoleic and linolenic acid hydroperoxides in particular are shown to be the key intermediates in acrolein formation, supposing a radical-induced reaction pathway. Although certain frying oils contain high amounts of acrolein after heating, deep-fried foods themselves are quite low in acrolein. Acrolein is also produced directly by dehydration of glycerol.

3.8.3 Reactions of heterolipids

Reactions of heterolipids are generally similar to those of other lipids, but the reactions of bound phosphoric acid and other less lipophilic moieties are specific to heterolipids. Phospholipids are hydrolysed by various phospholipases, an overview of which is given in Table 3.46, whilst the mechanism of their action is shown in Figure 3.78. Phosphatidylcholine is an in-demand product, and its concentration in lecithin can be increased by enzymatically controlled interesterification after the addition of choline. Phosphatidylcholine can also be produced directly from diacylglycerols or phosphatidic acids using immobilised enzymes.

Phospholipids can form salts, for example with metal ions, due to the presence of one free hydroxyl group of the bound phosphoric acid in phosphatidylethanolamine, phosphatidylserine, and phosphatidylinositol and of two free hydroxyl groups in phosphatidic acid. Mostly,

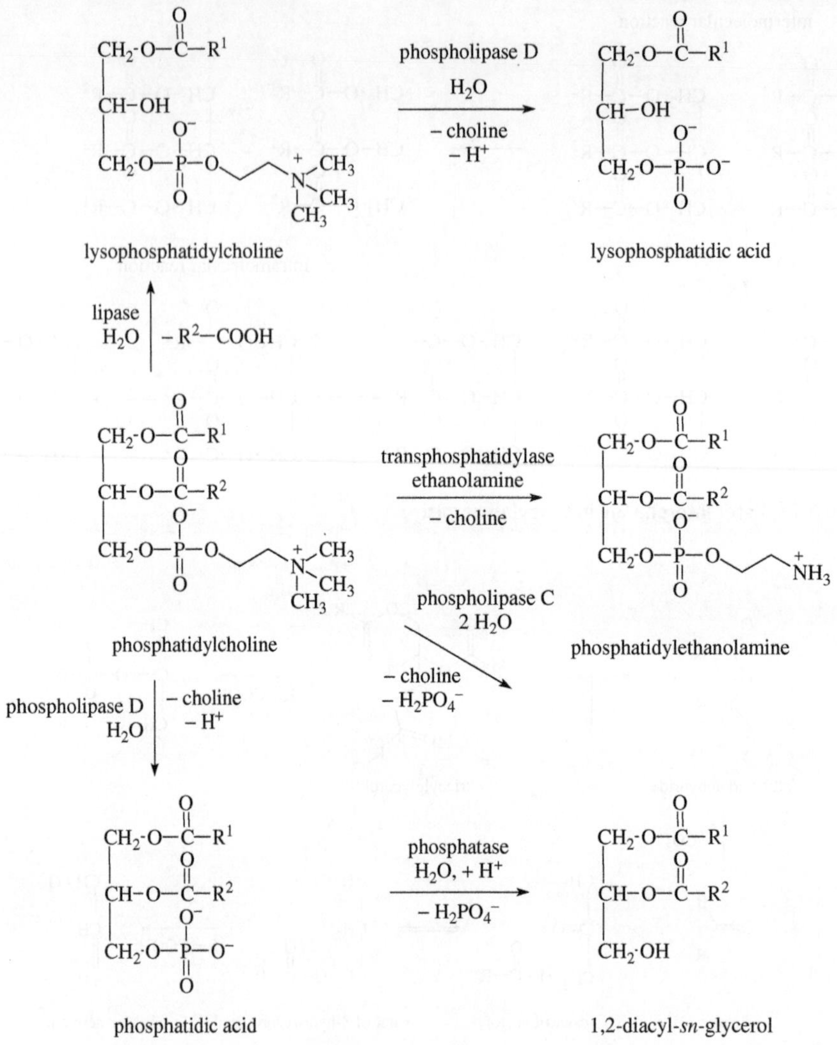

Figure 3.78 Degradation of phospholipids catalysed by phospholipases.

calcium and magnesium ions are involved in these reactions, but the resulting complexes have hydrophobic character, as in metal soaps. Cations of heavy metals (copper, manganese, and iron) bound in phospholipids catalyse autoxidation significantly less than free metal ions; phospholipids can thus become synergists of antioxidants.

Unsaturated fatty acids bound in phospholipids can be oxidised by oxygen, similarly to fatty acids in triacylglycerols. Phospholipid hydroperoxides may react with amino groups of other phospholipids and other amino compounds to form dark-brown macromolecular products similar to melanoidins (see Section 4.7.5.7) formed in the Maillard reaction. Therefore, phospholipids act as synergists of phenolic antioxidants. In addition to the oxidation of unsaturated fatty acids bound in phosphatidylcholine, the bound choline is also oxidised. Its oxidation product is trimethylamine N-oxide (Figure 3.79), which is degraded to dimethylamine.

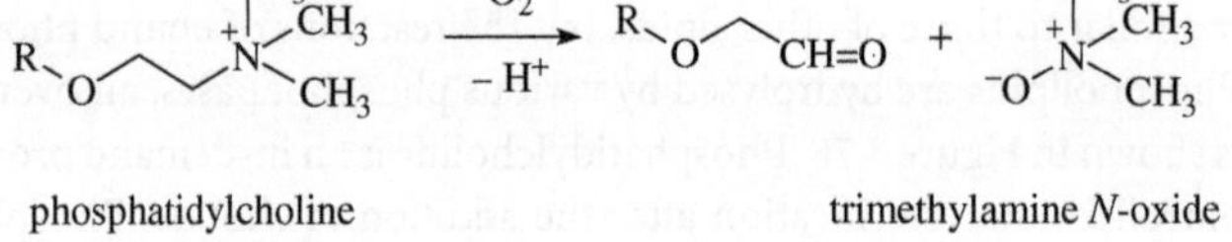

Figure 3.79 Oxidation of bound choline (R = residue of phospholipid molecule).

An important group of reactions of phospholipids containing free amino groups is the non-enzymatic browning reactions of the amino groups with reducing sugars or dehydroascorbic acid. The reaction mechanism is similar to that of sugars with other amines. Phosphatidylethanolamine and phosphatidylserine are more reactive than phosphatidylcholine.

3.8.4 Reactions of steroids

Important reactions of free steroids include esterification and hydrolysis. These reactions in food raw materials are catalysed by sterol esterases and glycosidases, respectively. Other important reactions of steroids include elimination and substitution reactions and oxidation. Hydrogenation of steroids is of industrial importance.

3.8.4.1 Elimination and substitution reactions

When sterols are oxidised at moderate temperatures ($\leq 100\,°C$), their reaction products are mainly derived from hydroperoxides, but at temperatures close to $200\,°C$ and above, thermal reactions such as dehydration and condensation become important. The secondary hydroxyl groups are eliminated as water molecules, and new double bonds of Δ^5 sterols arise preferentially at the Δ^3 position, which results in the formation of sterol dienes with conjugated double bonds. For example, stigmasta-3,5-diene is produced by dehydration of β-sitosterol. The concentration of stigmasta-3,5-diene in virgin olive oils is very low (<0.1 mg/kg), but its concentrations in refined olive oils (such as pomace oil) may reach ≤ 120 mg/kg. Refined oils also contain significant amounts of other steroidal hydrocarbons, including campesta-3,5-diene and stigmasta-3,5,22-triene, in addition to stigmasta-3,5-diene. The relative amounts of these steroidal hydrocarbons can be used to detect refined seed oils.

The elimination of water may also proceed intermolecularly, which yields disteryl ethers (Figure 3.80). Disteryl ethers are formed in low concentration (a few milligrammes per kilogramme) by dehydration of sterols during the conventional bleaching of fats and oils with acid-activated bleaching earths. Disteryl ethers are the main steryl dimers characterised in foods; other dimers may be formed by direct sterol ring linkages.

In the production of acid hydrolysates of proteins using hydrochloric acid, the residual sterols and their esters occurring in the starting materials yield small amounts of 3-chloro derivatives by replacement of the C-3 hydroxyl group with chlorine. Possible intermediates in the formation of 3-chlorosterols may be 3,5-dienes derived from sterols (Figure 3.80).

3.8.4.2 Addition reactions

Double bonds of sterols can be hydrogenated similarly to those of unsaturated fatty acids bound in triacylglycerols. Hydrogenation of oils and fats results in changes to the structure of steroids, especially in the molecules of 4,4-dimethylsterols and 4-methylsterols, and to a lesser extent in sterols (4-demethylsterols). Possible reactions are as follows:

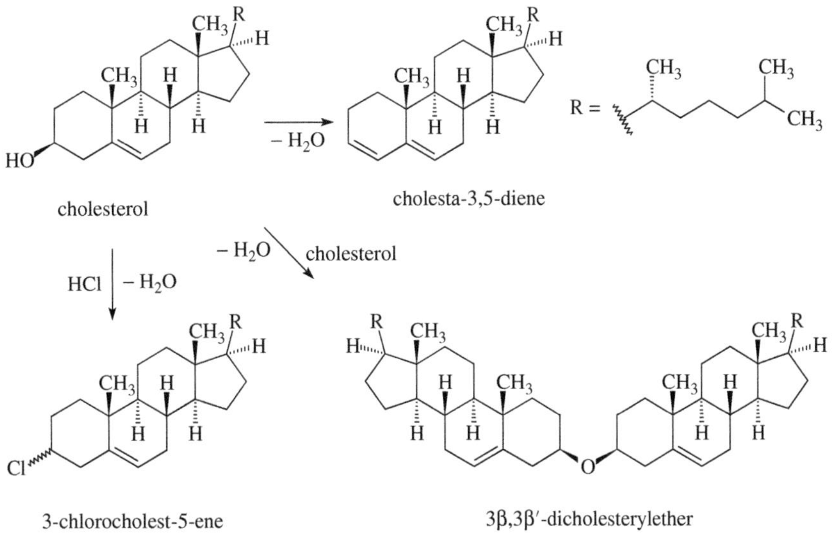

Figure 3.80 Dehydration and other reactions of sterols during heating.

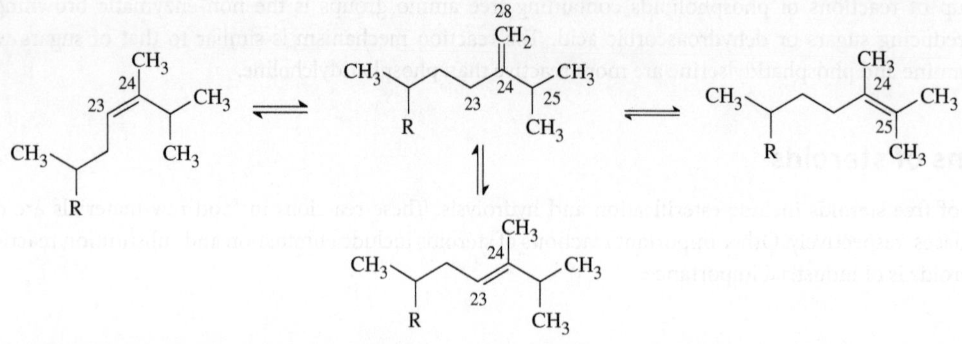

Figure 3.81 Isomerisation of steroids with a methylene group in the side chain (R = steroid residue).

- isomerisation of double bonds in the side chain, which gives rise to the geometric and positional isomers of the original steroids;

- hydrogenation of double bonds in the side chain, which leads to steroids with saturated side chains;

- opening of the cyclopropane ring;

- hydrogenation of all double bonds.

The predominant reaction is isomerisation of the double bond in position C-24 (C-28) to position C-24 (C-25), leading to thermodynamically stable isomers. For example, the 4,4-dimethylsterol 24-methylenecycloartanol yields a mixture of isomers consisting of, amongst other compounds, cyclosadol and cyclobranol. Similar isomerisation proceeds in 4-methylsterols such as obtusifoliol and others (Figure 3.81). The formation of geometric and positional isomers of 4-methylsterols (such as citrastadienol) and sterols (such as Δ^5-avenasterol) with ethylidene group in the side chain at C-24 is shown in Figure 3.82. Hydrogenation of double bonds in the side chain yields saturated derivatives. For example, 24-methylenecycloartanol is transformed to 24-methylcycloartanol, and stigmasterol yields β-sitosterol.

The opening of the cyclopropane ring during hydrogenation occurs in cycloartenol and 24-methylenecycloartanol. The hydrogenation of cycloartenol produces lanosta-9(11),24-dienol (known as parkeol) and lanosterol (lanosta-8,24-dienol). 24-Methylenecycloartanol yields 24-methylenelanost-9(11)-enol and 24-methylenelanost-8-enol. Changes of some steroids in hydrogenated sunflower and soybean oils are documented in Table 3.48.

For food purposes, phytosterols are isolated from tall oil, a byproduct of wood pulp manufacture from coniferous trees. Crude tall oil contains 5–10% of phytosterols that are selectively hydrogenated at the double bond in position C-5, yielding the corresponding **stanols**, the main component of which is sitostanol. It has been shown that if phytosterols (or their fatty acid esters) are consumed at higher levels (e.g. in margarines), they inhibit the absorption of exogenous and endogenous cholesterol in the gastrointestinal tract and reduce the total and LDL cholesterol concentrations; the commonly consumed amounts of phytosterol do not significantly affect cholesterol absorption.

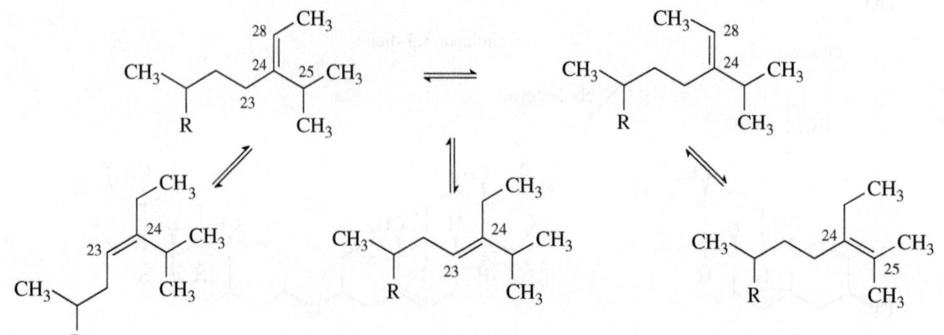

Figure 3.82 Isomerisation of steroids with an ethylidene group in the side chain (R = steroid residue).

Table 3.48 Changes to significant steroids during hydrogenation of vegetable oils.

Steroid	Sunflower oil hydrogenated		Soybean oil hydrogenated	
	Once	Twice	Once	Twice
4,4-Dimethylsterols[a]				
Cycloartenol	35	26	35	26
24-Methylenecycloartanol	25	4	16	0
Cyclobranol	3	18	2	11
24-Methylcycloartanol	2	10	4	9
4-Methylsterols[b]				
Obtusifoliol	25	6	10	2
Gramisterol and cycloeucalenol	12	2	7	2
Citrastadienol	33	15	38	12
Sterols[c]				
Stigmasterol	10	9	17	15
Δ^5-Avenasterol	3	1	1	0.2
Δ^7-Avenasterol	3	2	0.5	0.1

[a]In % of total 4,4-dimethylsterols (triterpenic alcohols).
[b]In % of total 4-methylsterols.
[c]In % of total sterols.

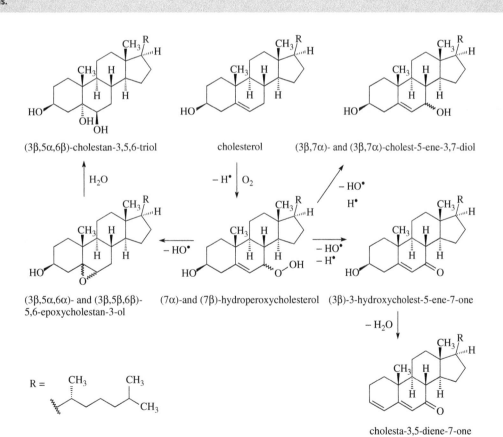

Figure 3.83 Main products of cholesterol oxidation.

Table 3.49 Oxidation products of phytosterols in potato chips fried in sunflower oil.

Phytosterol	Amount of oxidation product (mg/kg)[a]				
	7α-hydroxy-	7β-hydroxy-	7-oxo-	5α,6α- and 5β,6β-epoxy-	5α,6β-dihydroxy-
Sitosterol	4.7	9.7	13.5	5.4	2.8
Campesterol	1.4	1.8	9.2	3.6	1.6

[a] Total amount of phytosterol oxidation products is 53.7 mg/kg.

3.8.4.3 Oxidation

In vivo (in the biosynthesis of bile acids and some hormones), and also during heat processing and storage of foods, sterols are oxidised in the nuclei and side chain to form various oxygenated derivatives known by the general term **oxysterols**. Some of these compounds are synthesised *in vivo* from cholesterol, but they can also be formed in food during processing procedures and storage. For example, oxidation of cholesterol by oxygen via a C-7 radical yields 7α- and 7β-hydroperoxides. Analogous to the secondary oxidation reactions of fatty acids, hydroperoxides of sterols are transformed into the corresponding 5,6-epoxides (5α,6α-epoxides and 5β,6β-epoxides), and their hydrolysis gives 3,5,6-triols. Alkoxyl radicals formed by the decomposition of hydroperoxides give rise to the major cholesterol oxidation products, including epimeric 7-hydroxy derivatives and the 7-oxo derivative, which is further transformed into 3,5-diene-7-one by dehydration (Figure 3.83). In addition to 7-hydroperoxides of cholesterol, epimeric 22-hydroperoxides, 20,25-dihydroperoxides, 5α,6β-dihydroxy derivatives, and 2,4,6-trienes have also been identified as oxidation products with triplet oxygen. Oxidation of cholesterol with singlet oxygen produces epimeric 5-hydroperoxides that rearrange to 7-hydroperoxides via the corresponding hydroperoxyl radicals. Phytosterols and steryl esters are oxidised similarly. Plant sterols also undergo oxidative processesses comparable to those involved in cholesterol oxidation.

Sterol oxidation products are found in stored oils and especially in oils used for deep frying; therefore, these products are also present in fried potato chips (Table 3.49), fried potato crisps, and other fried food products, such as hamburgers (up to 30 mg/kg fat). Sterol oxidation products are also found in lipids of dried foods, such as dried milk (up to 10 mg/kg) and dried eggs (up to 700 mg/kg fat), and in butter oil (up to 30 mg/kg), refined plant oils (up to 110 mg/kg), and peanuts (50 mg/kg fat). Some food contains naturally high amounts of oxysterols (salami, parmesan).

Whilst dietary cholesterol does not correlate with its concentration in blood, rather good correlation exist between oxidised forms of cholesterol and atherosclerosis. Unlike phytosterols, oxyphytosterols are easily absorbed by humans. After absorption, oxysterols react with blood plasma lipoproteins to form complexes that can initiate formation of atherosclerotic deposits in the vascular wall. However, the important oxidation reaction from the physiological point of view is the transformation of provitamins D (ergosterol and 7-dehydrocholesterol) into vitamins D_2 and D_3, respectively.

Reactions of triterpene alcohols are probably related to the reactions of sterols, but they have not yet been adequately explored.

4

Saccharides

4.1 Introduction

4.1 Introduction

The term **saccharide** comes from the Greek word *zahari*, meaning **sugar**. It is frequently used in chemistry to describe polyhydroxyaldehydes $H-[CHOH]_n-CH=O$ and polyhydroxyketones $H-[CHOH]_n-C(=O)-[CHOH]_m-H$ that contain three or more aliphatic carbon atoms in the molecule. It also includes derived substances that are formed from these compounds by condensation reactions with the formation of acetal bonds, including monosaccharides, oligosaccharides, and polysaccharides, as well as substances derived from saccharides through the reduction of the carbonyl group, the oxidation of one or more terminal groups, or the replacement of one or more hydroxy group(s) by a hydrogen atom, an amino group, a thiol group, or similar heteroatomic groups. The term **carbohydrate** is actually a descriptor of what these molecules are theoretically composed of. It was applied originally to monosaccharides, in recognition of the fact that they are carbon hydrates in a ratio of one carbon atom to one water molecule, so that their empirical composition can be expressed as $C_n(H_2O)_n$. However, the term is now used generically in a wider sense, and also includes derivatives of these compounds. The term carbohydrate is most common in biochemistry, medicine, and nutrition, but not in chemistry. The former name, 'glycides', is now used less frequently and is not recommended for use in chemistry.

Depending on the number of carbon atoms in the molecule, saccharides are divided to **trioses**, **tetroses**, **pentoses**, **hexoses**, and higher sugars. Compounds with aldehyde functional group are called **aldoses** (e.g. aldopentoses and aldohexoses) and compounds with a ketone function are called **ketoses** (e.g. ketohexoses).

Depending on the number of sugar units bound in the molecule, saccharides are divided into:

- **monosaccharides**;

- **oligosaccharides**;

- **polysaccharides**, also known as **glycans**;

- **complex** or **conjugated saccharides**.

Monosaccharides consist of only one sugar unit. Oligosaccharides consist of between 2 and 10 of the same or different monosaccharides connected to each other by glycosidic (hemiacetal) bonds. Monosaccharides and oligosaccharides are sometimes referred to collectively by the name sugars, as they share many common properties and often have a sweet taste. Polysaccharides are composed of more than 10 of the same or different monosaccharides; commonly, they consist of multiple molecules of monosaccharides, which are often precisely determined. Complex carbohydrates also contain compounds other than saccharides, such as peptides, proteins, and lipids.

Saccharides arise naturally in cells of photoautotrophic organisms (green plants and photosynthetic bacteria) by assimilation of carbon dioxide in the air in the presence of water and by use of natural light (photosynthesis) converted by photosystems into chemical energy. Heterotrophic organisms obtain the necessary saccharides from autotrophic organisms or from non-saccharide carbon substrates, such as certain amino acids, hydroxy acids, glycerol, and other substances (the metabolic pathway is called gluconeogenesis), or else transform saccharides into different structures. Saccharides are common components of all cells. The level of saccharides in animal tissues is only a few percent, but in plant tissues, they commonly represent 85–90% of dry matter.

The Chemistry of Food, Second Edition. Jan Velíšek, Richard Koplík and Karel Cejpek.
© 2020 John Wiley & Sons Ltd. Published 2020 by John Wiley & Sons Ltd.

Saccharides have various functions in cells:

- They are used primarily as an energy source (e.g. polysaccharides, oligosaccharides, and monosaccharides); 1 g glucose provides 17 kJ (4 kcal), the energy yield of sugar alcohols is 10 kJ/g (2.4 kcal/g); saccharides are therefore main (primary) nutrients together with proteins and lipids.

- They are the basic building units of many cells, and protect cells against the effects of various external factors (e.g. some oligosaccharides, polysaccharides, and complex saccharides).

- They are biologically active substances (e.g. oligosaccharides in milk) or components of many biologically active substances, such as glycoproteins, some coenzymes, hormones, and vitamins.

Saccharides are highly reactive substances that are transformed into many different products, even without the participation of other reaction partners, during food storage and processing. The most common and important reactions of carbohydrates are their reactions with amino compounds and many other food components, which are known as non-enzymatic browning reaction. They also involve reactions of carbohydrates in the absence of amino and other compounds. Reactions of carbohydrates with amino compounds (proteins, amino acids) are referred to as the Maillard reaction. Among the products of non-enzymatic browning reactions we can found important flavour-active compounds, various yellow, brown, or black pigments, as well as compounds possessing health benefit (e.g. antioxidants); however, many of them also can show antinutritional or even toxic effects.

This chapter is divided into two main parts. The first describes major monosaccharides and their functional derivatives, oligosaccharides and polysaccharides. Their structure and nomenclature, occurrence in major food commodities, properties and importance in human physiology and nutrition, recommended intake, and use in food technology are all described. The second part is devoted to the reactions of saccharides that lead to the formation of products that influence the odour, taste, and colour of food raw materials and foods during storage and processing.

4.2 Monosaccharides

4.2.1 Structure and nomenclature

Monosaccharides, aldoses, and ketoses are divided – according to the number of carbons in the chain – into trioses, tetroses, pentoses, hexoses, and higher sugars. The carbon chain of monosaccharides present in foods is usually linear, but branched-chain monosaccharides also exist.

Monosaccharides occur as substances with a free carbonyl group (acyclic compounds) and as cyclic **hemiacetals**, also called **lactols**. Trioses are exclusively acyclic substances, whilst tetroses and higher monosaccharides exist predominantly in five- and six-membered and, exceptionally, seven-membered cyclic structures. They can therefore be regarded as substances derived from oxolane (tetrahydrofuran), oxane (tetrahydropyran), or oxepane (hexamethylene oxide), and are thus actually heterocyclic compounds. Acyclic forms, which exist in constitutional equilibrium with cyclic forms, occur in zigzag conformers, as are many alditols.

4.2.1.1 Aldoses

The simplest aldose is aldotriose, known as glyceraldehyde, which contains the aldehyde group in position C-1 and a chiral carbon atom in position C-2. It exists therefore in two ($2^1 = 2$; generally 2^n, where n is the number of chiral carbon atoms) configuration isomers, as D-glyceraldehyde (**4-1**) and L-glyceraldehyde (**4-2**). Solutions of D-glyceraldehyde rotate the plane of polarised light to the right (clockwise), and D-glyceraldehyde is therefore (+)-glyceraldehyde or (+)-D-glyceraldehyde (dextrorotary). L-Glyceraldehyde solutions rotate the plane of polarised light to the left (counter clockwise), so L-glyceraldehyde is (−)-glyceraldehyde or (−)-L-glyceraldehyde (laevorotary).

4-1, D-glyceraldehyde (D-*glycero*-triose)

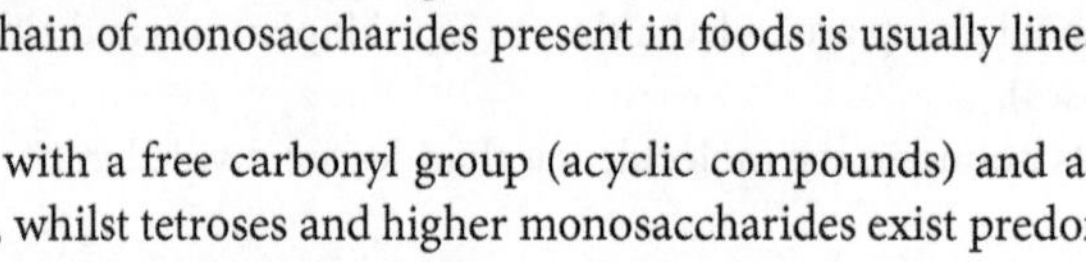

4-2, L-glyceraldehyde (L-*glycero*-triose)

The isomers are called optical isomers, antipodes, or enantiomers. Their equimolar mixture is optically inactive and is called a **racemate**. Symbols (affixes) *R* and *S* are used only rarely for labeling configurations of saccharides. D-Glyceraldehyde is thus (*R*)-glyceraldehyde and L-glyceraldehyde is (*S*)-glyceraldehyde. The former name for glyceraldehyde was glycerose. In biochemistry, the simplest two-carbon sugar glycolaldehyde (biose) and the simplest one-carbon sugar formaldehyde (monose) are also classified as saccharides (aldoses).

The prefixes D- and L-are also used with other monosaccharides. The formal insertion of clusters of atoms H–C–OH (or the mirror image of HO–C–H) between the first and second carbons of a D-glyceraldehyde molecule gives two optically active aldotetroses called D-erythrose and D-threose. L-Glyceraldehyde can yield L-erythrose and L-threose in the same way. The total number of **configuration isomers (stereoisomers)** is $2^2 = 4$ (**4-3**). The pairs of saccharides D-erythrose with L-erythrose and of D-threose with L-threose are **enantiomeric** substances (asymmetric carbons have the opposite configuration; one substance is a mirror image of the other). The pair from D-erythrose with D-threose and the pair from L-erythrose with L-threose are **diastereomeric** substances that differ in configuration at one chiral centre, whilst the configuration at the second chiral centre is identical.

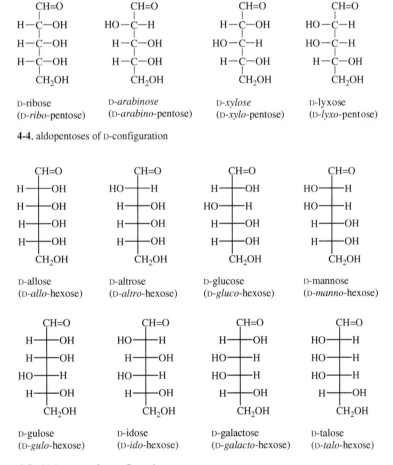

4-4, aldopentoses of D-configuration

4-5, aldohexoses of D-configuration

The affiliation of an aldose to the D- or L-series is determined by the conformity of the configuration of the chiral carbon atom with the highest sequence number in the molecule (such as carbon C-3 in aldotetroses, carbon C-4 in aldopentoses, and carbon C-5 in aldohexoses), with the configuration of the chiral carbon atom of either D-glyceraldehyde or L-glyceraldehyde. Whether the aldose belongs to the D- or L-series of sugars is independent of the configuration of the other carbon atoms. For example, (+)-D-glucose configuration at carbons C-5, C-4, and C-2 is D (R configuration), whilst carbon C-3 has an L configuration (*S* configuration). The change in configuration at other chiral atom provides a different aldose of the same series (e.g. changing the configuration at C-4 of D-glucose gives D-idose). The change in configuration of all chiral atoms, in the case of the D-isomer, gives the corresponding L-isomer. For example, D-glucose is transformed into L-glucose and vice versa. Aldoses that differ only in configuration at C-2 are called **epimers**. Examples of epimers are D-glucose and D-mannose.

Removing the suffix '-se' from the trivial names of trioses, tetroses, pentoses, and hexoses gives the configuration prefixes *glycero-*, *erythro-*, *threo-*, *ribo-*, *arabino-*, *xylo-*, *lyxo-*, *allo-*, *altro-*, *gluco-*, *manno-*, *gulo-*, *ido-*, *galacto-*, and *talo-*, which are, together with the configuration symbols D and L, the basis of systematic carbohydrate nomenclature. The systematic name of D-glyceraldehyde is then D-*glycero*-triose, D-erythrose is D-*erythro*-tetrose, D-ribose is D-*ribo*-pentose, and D-glucose is D-*gluco*-hexose. For practical reasons, however, the preferred names are trivial names, if the saccharide structure is expressed uniquely.

In addition to sugars with straight chains, branched-chain sugars also exist in nature, for example as components of pectin and other polysaccharides. Again, their preferred names are the trivial names. Aldopentose D-apiose with one chiral carbon atom has the systematic name 3-*C*-hydroxymethyl-D-*glycero*-tetrose (**4-6**).

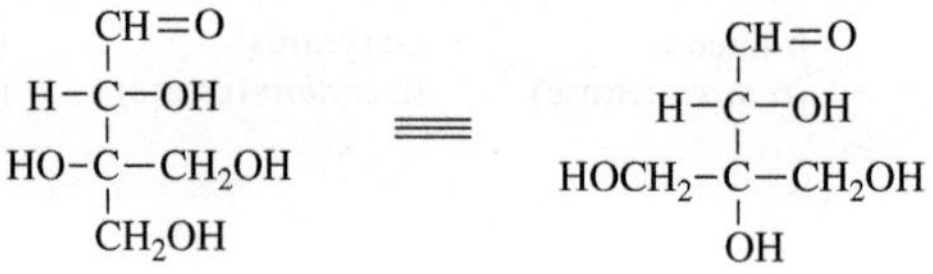

4-6, D-apiose

The abbreviated notations of monosaccharides are formed as they are for amino acids. An abbreviation of the monosaccharide name consists of the first three letters of the trivial name. An exception is glucose, for which the abbreviation Glc is used, as the symbol Glu is reserved for glutamic acid.

4.2.1.2 Ketoses

The simplest ketose (ketotriose) is optically inactive 1,3-dihydroxyacetone, also known as 1,3-dihydroxypropan-2-one or glycerone (**4-7**). Similar to aldotetroses and higher aldoses derived from glyceraldehyde, one D-ketotetrose (**4-7**), two D-ketopentoses (**4-7**), four D-ketohexoses (**4-8**), and the same number of ketoses of the L-series can be derived from 1,3-dihydroxyacetone. Trivial names are used for ketohexoses, but for higher ketoses systematic names are preferred. Ketoses with more than four chiral centres are formally divided into two parts so that the four chiral carbon atoms have the lowest serial numbers. The name starts from the centre with the highest number. A ketose **4-9**, for example, is called D-*manno*-hept-2-ulose, D-*altro*-hept-2-ulose is called trivially sedoheptulose, other sugar **4-9** is D-*glycero*-D-*manno*-oct-2-ulose, and another sugar in **4-9** is D-*erythro*-L-*gluco*-non-2-ulose. A branched ketohexose with the trivial name dendroketose (**4-9**) has the systematic name D-4-deoxy-4-*C*-hydroxymethyl-D-*glycero*-pent-2-ulose.

1,3-dihydroxyacetone | D-erythrulose (D-*glycero*-tetrulose) | D-ribulose (D-*erythro*-pent-2-ulose) | D-xylulose (D-*threo*-pent-2-ulose)

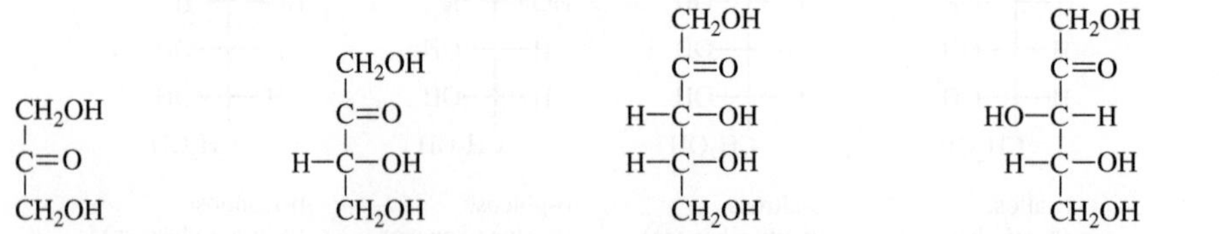

4-7, ketotriose, ketotetrose, and ketopentoses

D-psicose (D-*ribo*-hex-2-ulose) | D-fructose (D-*arabino*-hex-2-ulose) | D-sorbose (D-*xylo*-hex-2-ulose) | D-tagatose (D-*lyxo*-hex-2-ulose)

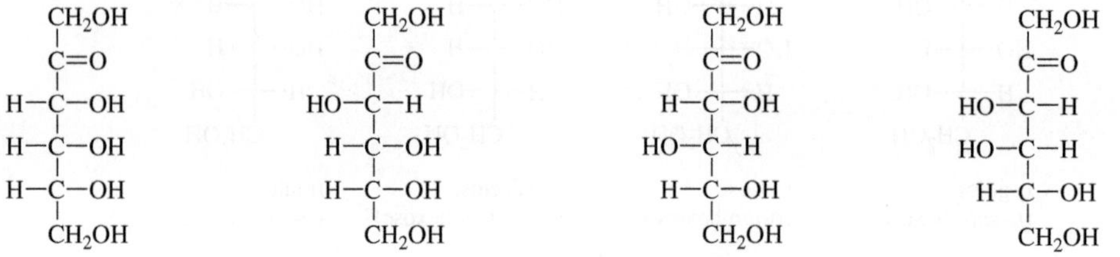

4-8, ketohexoses of D-configuration

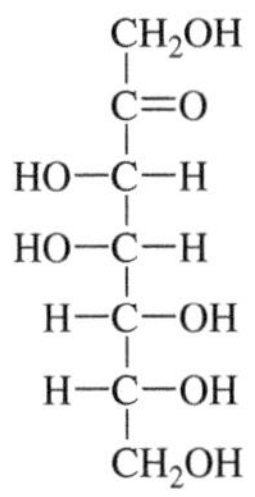

CH₂OH
C=O
HO−C−H
HO−C−H
H−C−OH
H−C−OH
CH₂OH

D-*manno*-hept-2-ulose

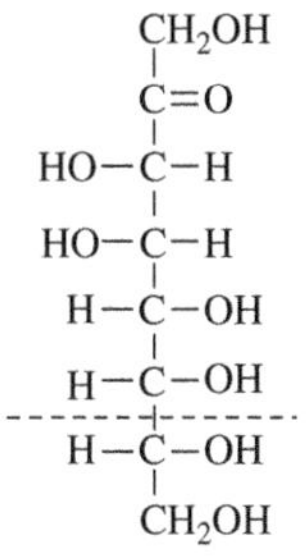

CH₂OH
C=O
HO−C−H
HO−C−H
H−C−OH
H−C−OH

H−C−OH
CH₂OH

D-*glycero*-D-*manno*-oct-2-ulose

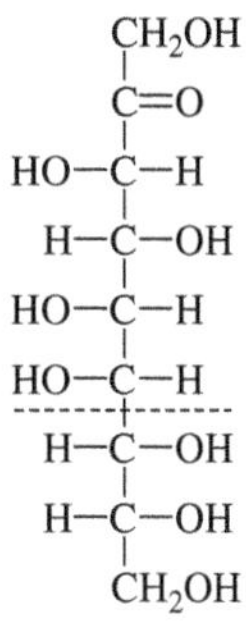

CH₂OH
C=O
HO−C−H
H−C−OH
HO−C−H
HO−C−H

H−C−OH
H−C−OH
CH₂OH

D-*erythro*-L-*gluco*-non-2-ulose

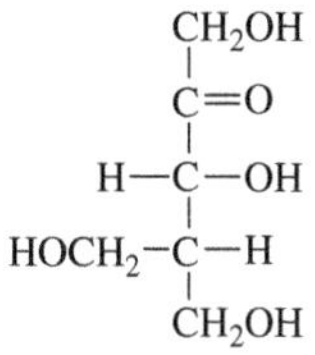

CH₂OH
C=O
H−C−OH
HOCH₂−C−H
CH₂OH

D-dendroketose

4-9, examples of heptuloses, octuloses, and branched ketohexoses

4.2.1.3 Cyclic structures

Aldehydes and ketones react with alcohols to form unstable hemiacetals. A reaction of hemiacetal with another molecule of alcohol yields stable acetals (see Figure 8.22). In contrast, saccharides spontaneously yield only stable cyclic hemiacetal by intramolecular addition of one hydroxyl group (primary or secondary hydroxyl group) to the carbonyl group. These cyclic forms of aldoses and ketoses are generally termed **lactols**. Six- and five-membered (and, exceptionally, seven-membered) rings are mainly formed. Structures with five-membered rings are called **furanoses**, those with six-membered rings are **pyranoses**, and those with seven-membered rings are **septanoses**.

On the carbon of the carbonyl group (C-1 carbon of aldoses and C-2 carbon of ketoses), a new chiral centre arises due to the ring formation. The carbon of the carbonyl group is called an **anomeric carbon**, the newly formed hydroxyl group at the anomeric carbon is an **anomeric hydroxyl group** (hemiacetal group), and the corresponding pairs of isomers are **anomers**. To mark the configuration of substituents on the anomeric carbon, configuration prefixes α and β, indicating the relative configuration to the chiral carbon atom with the highest number (which determines the affiliation of the sugar to the D- or L-series), are used. The α anomer has the same configuration, whilst the β anomer has the opposite configuration. Both anomers, which are special types of diastereomers, differ in their optical rotations.

The formation of cyclic structures from acyclic structures and the relationship between different types of formulae (Fischer, Tollens, Haworth, and Mills projection formulae) are shown in Figure 4.1 for the example of D-glucose. In the Fischer representation of a monosaccharide, the carbon chain is written vertically, with the lowest-numbered carbon atom at the top and the neighbouring carbon atoms below. The H and OH groups projects to the right or the left. A monosaccharide is assigned to the D or the L series according to the configuration at the highest-numbered centre of chirality. A sugar that belongs to the D series has the highest-numbered hydroxy group projected to the right in the Fischer projection. If a cyclic form of a sugar is to be represented by the Tollens formula in the Fischer projection, a long bond can be drawn between the oxygen involved in ring formation and the (anomeric) carbon atom to which it is linked. Cyclic forms of D-glucose and D-fructose can be represented by formulae **4-10** and **4-11**, respectively. Conventionally, the formula in the Haworth projection is drawn so that the planar heterocyclic ring lies perpendicular to the picture plane and the oxygen atom is behind the plane. Axial and equatorial positions are not distinguished; instead, substituents are positioned above or below the ring atom to which they are connected and are directed above or below the plane, if they lie in the Fischer projection on the left or the right side of the formula. The ring part directed to the front of the plane is usually drawn in bold. Formulae can rotate or flip in space if necessary. In some cases, formulae can be clarified by use of the Mills depiction, in which the main hemiacetal ring is drawn in the plane of the paper; dashed bonds denote substituents below this plane, and thickened bonds those above.

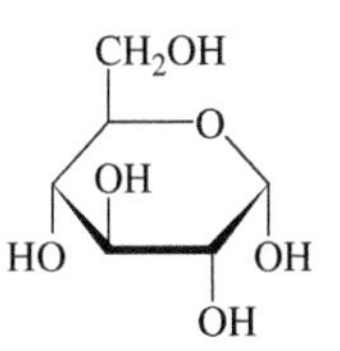

α-D-glucopyranose 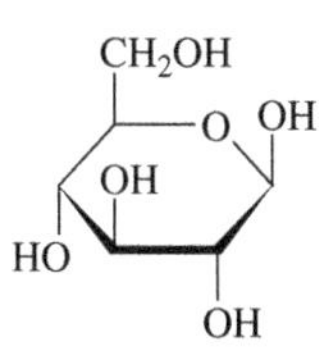   β-D-glucopyranose 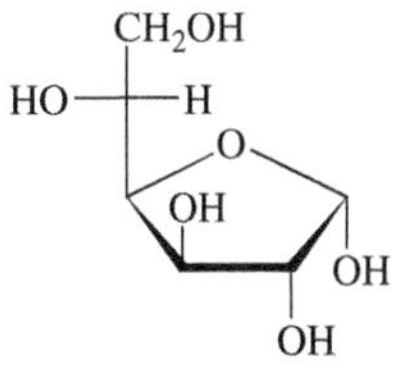   α-D-glucofuranose 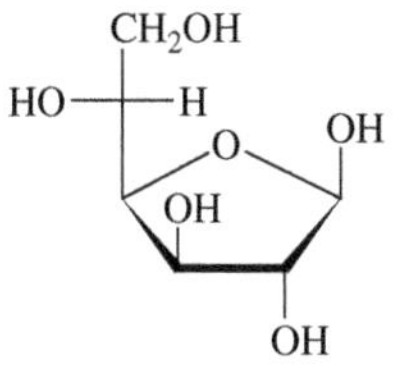   β-D-glucofuranose

4-10, cyclic forms of D-glucose

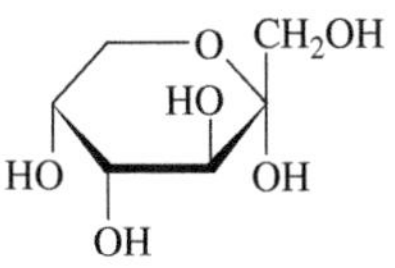

α-D-fructopyranose 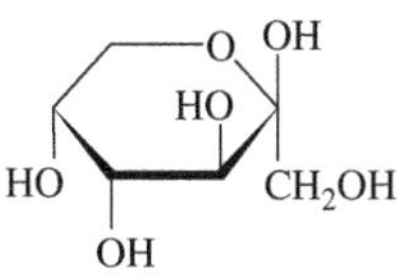   β-D-fructopyranose 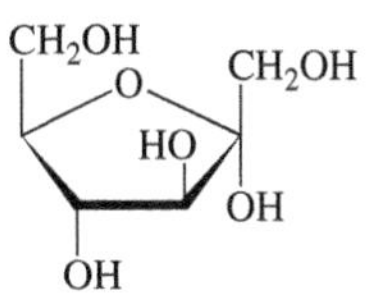   α-D-fructofuranose 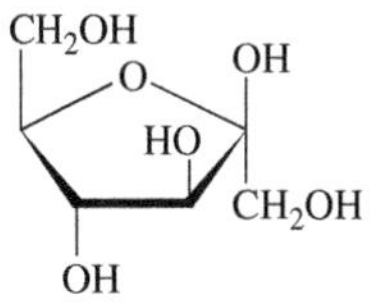   β-D-fructofuranose

4-11, cyclic forms of D-fructose

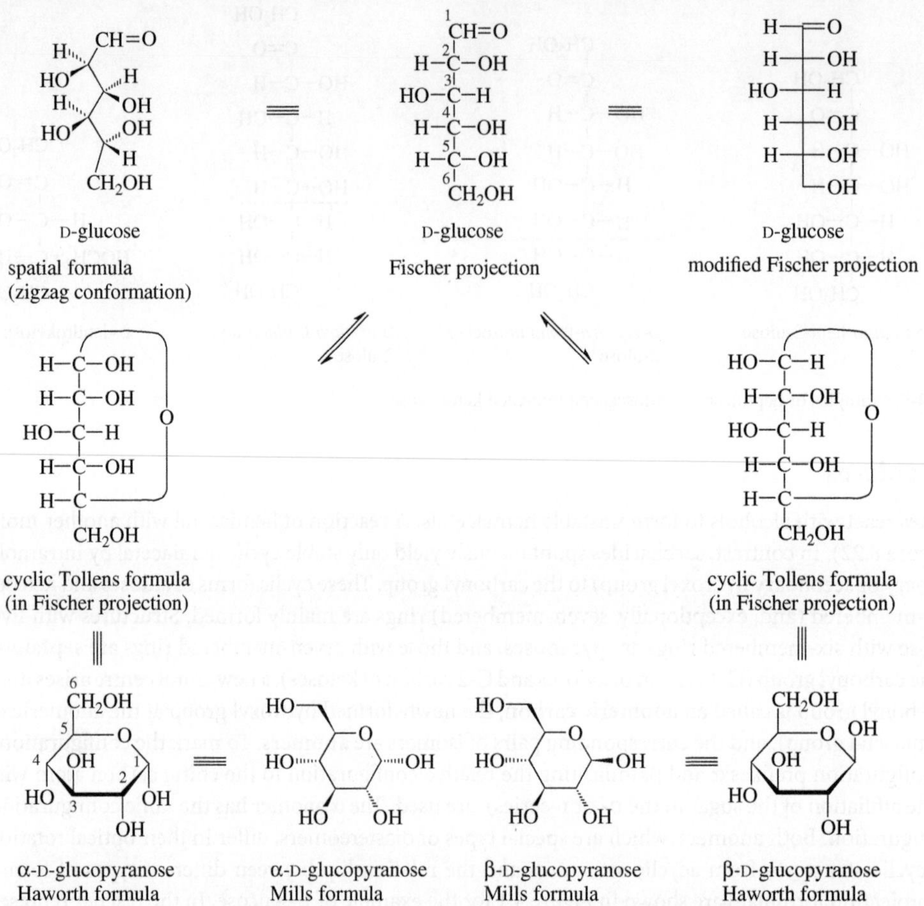

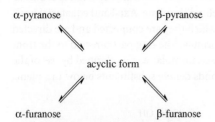

Figure 4.1 Acyclic and cyclic forms of D-glucose.

4.2.1.3.1 Mutarotation

α-pyranose   β-pyranose

acyclic form

α-furanose   β-furanose

Figure 4.2 Mutarotation of monosaccharides.

In the crystalline state, reducing monosaccharides exist exclusively in cyclic structures (α- or β-anomers). In solutions, an equilibrium between the α- and β-anomers and acyclic forms is established as the individual forms interconvert over time. This process, schematically shown in Figure 4.2, is termed **mutarotation**. The mechanism of mutarotation assumes cleavage of the saccharide cyclic forms to neutral acyclic saccharide or its ion that have free carbonyl groups. The equilibrium composition of the mixture (the amount of pyranoses, furanoses, or septanoses) depends on the type of sugar, solvent, pH (the reaction generally involves acid–base catalysis), and temperature.

Crystalline D-glucose, existing as an anhydrous substance or monohydrate, is α-D-glucopyranose. In weakly acidic and neutral aqueous solutions at ambient temperature, the establishment of equilibrium takes for about four hours, and in weakly alkaline solutions, the equilibrium is established immediately. The equilibrium aqueous solution at 20 °C contains about 33% of α-D-glucopyranose and 67% of β-D-glucopyranose; the solution at 40 °C contains 36% of the α-anomer and 64% of the β-anomer. The pyranose forms are thermodynamically more stable than the furanose form that are present at levels of up to 1%. The composition of the equilibrium solutions of ordinary aldoses and fructose are given in Table 4.1. In aqueous solutions (at 20 °C), the acyclic form of D-glucose is present at levels of 0.02–0.06%. Aqueous solutions of other aldohexoses and ketohexoses contain similar amounts of acyclic forms. Solutions of pentoses contain higher levels of acyclic forms than hexoses (e.g. D-ribose, 8.5%). Trioses are found exclusively in the acyclic forms. In enzymatically active plant and animal material, the mutarotation of saccharides containing bound glucose and galactose is catalysed by mutarotase (aldose 1-epimerase).

Table 4.1 Cyclic forms of some pentoses and hexoses in aqueous solutions at 40 °C.

Saccharide	Furanose (%)		Pyranose (%)		Saccharide	Furanose (%)		Pyranose (%)	
	α	β	α	β		α	β	α	β
D-Ribose	6	18	20	56	D-Gulose	<1	<1	21	79
D-Allose	5	7	18	70	D-Idose	16	16	31	37
D-Altrose	20	13	28	39	D-Galactose	<1	<1	27	73
D-Glucose	<1	<1	36	64	D-Talose	<1	<1	58	42
D-Mannose	<1	<1	67	33	D-Fructose	<1	25	8	67

4.2.1.3.2 Conformation

Pyranoses Glucose can exist in 38 distinct basic pyranose conformations: 2 chairs, 6 boats, 6 skew-boats, 12 half-chairs, and 12 envelopes. The thermodynamically most favourable and therefore most common conformations of pyranoses are chair (C) conformations. Boat (B), skew-boat (S), half-chair (H), and envelope (E) conformations are rarely found.

Each glucose anomer in the chair (C) conformation can exist in two chair structures, $^4C_1$ and $^1C_4$. Numbers 1 and 4 correspond to atoms C-1 and C-4, which are above (expressed as superscript) or below (expressed as subscript) the reference plane defined by carbons C-2, C-3, C-5, and the oxygen atom. Structures of both chair conformers of β- and α-D-glucopyranose are given in formulae **4-12**.

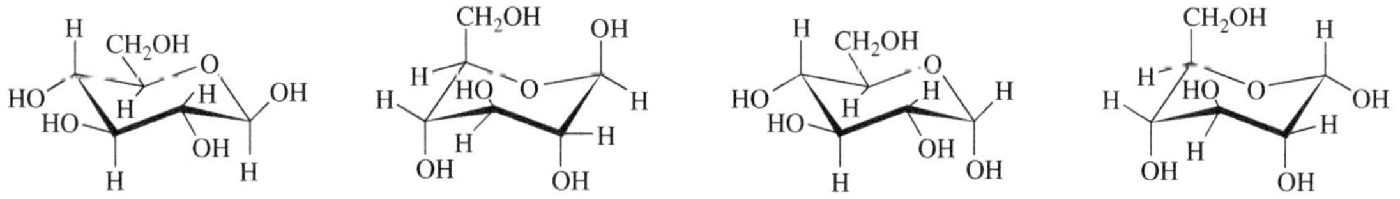

β-D-glucopyranose-$^4C_1$ β-D-glucopyranose-$^1C_4$ α-D-glucopyranose-$^4C_1$ α-D-glucopyranose-$^1C_4$

4-12, conformations of D-glucopyranose anomers

In solutions, conformers $^4C_1$ and $^1C_4$ of saccharides are in equilibrium, where the predominant conformer has a larger number of substituents (hydroxyl or hydroxymethyl groups) in the equatorial positions because (except for the substituent on the anomeric carbon atom) bulky axial substituents are thermodynamically unfavourable. Predominantly **1,3-*syn*-axial** interactions of hydroxyl and hydroxymethyl groups take place, such as interactions of the hydroxyl group in position C-4 and the hydroxymethyl group in $^1C_4$ conformer. The hydroxymethyl group, as well as the anomeric hydroxyl of aldohexoses, is always equatorial, which means that D-aldohexoses prefer the conformation $^4C_1$. From the conformational point of view, the most stable aldohexose is the $^4C_1$ conformer of β-D-glucopyranose because all five bulky substituents are in equatorial positions (Table 4.2). An aqueous solution of glucose, however, contains only about 64% of this anomer, and the rest is α-anomer. The reason is that the stability of the axial hemiacetal hydroxyl group of α-D-glucopyranose-$^4C_1$ is increased due to the **anomeric effect** manifested by the interaction with the oxygen atom of the pyranose ring, through electrostatic repelling of their non-bonding electrons. If the hydroxyl group is in the equatorial position (β-anomer), the angle between the dipole moments is

Table 4.2 Position of substituents (e, equatorial; a, axial) in $^4C_1$ conformers of aldohexoses.

Saccharide	Hydroxyl group			Saccharide	Hydroxyl group		
	C-2	C-3	C-4		C-2	C-3	C-4
D-Allose	e	a	e	D-Gulose	e	a	a
D-Altrose	a	a	e	D-Idose	a	a	a
D-Glucose	e	e	e	D-Galactose	e	e	a
D-Mannose	a	e	e	D-Talose	a	e	a

small and electrostatic repulsive forces are large. If the hydroxyl group is in the axial position, the repulsive forces are smaller. The axial position is therefore preferable. The 1,3-diaxial interaction between a C-1 hydroxyl group and the axial protons at C-3 and C-5 also operates in favour of β-anomers. The effect is that about 36% of D-glucose in equilibrium is present in the form of the α-anomer (at 20 °C). The conformation of glucose molecules explains the almost universal role of glucose in living systems as a structural unit of many biopolymers and an intermediate in a number of important reactions.

D-Fructose is found primarily as β-pyranose in conformation $^1C_4$, in which the hydroxyl groups at C-3 and C-4 are in the equatorial positions and the hydroxyl group at C-5 is in the axial position (4-13).

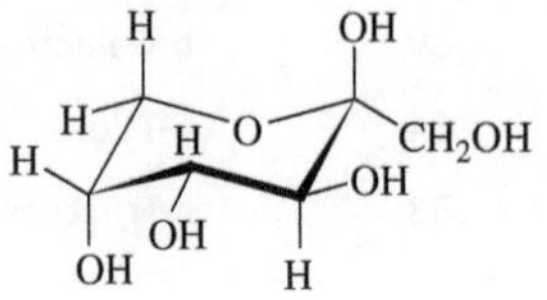

4-13, β-D-fructopyranose-$^1C_4$

Furanoses Furanose conformers differ from one another in their energy only slightly and can interconvert between the different conformations very quickly. The most frequently occurring are envelope (E) and twist (T) conformations. Both of these basic conformations have almost equal energy. In fact, there are 10 E conformations and 10 T conformations that interconvert. The interconversion between the many conformers is termed **pseudorotation**. Formulae 4-14 are examples of E and T conformers of β-D-glucofuranose. These types of conformer and levels of anomers in furanose solutions and in foods depend on the intramolecular non-bonding interactions.

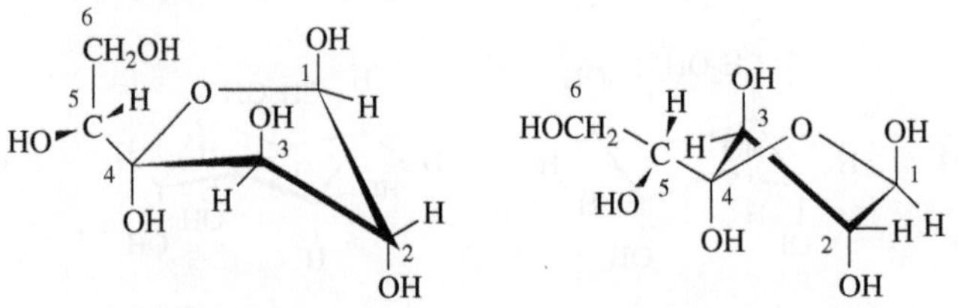

β-D-glucofuranose-E_2 β-D-fructofuranose-$^3T_2$

4-14, basic conformations of D-glucofuranose anomers

4.2.2 Occurrence

Monosaccharides are common components of almost all foods, but their content is highly variable. The most common monosaccharides in foods are hexoses and pentoses.

Monosaccharides are present in relatively large quantities in fruits, where their content increases during ripening. However, this varies greatly depending on the type of fruit, the level of maturity, the conditions of post-harvest storage, processing, and so on. In apples, for example, only traces of starch are present at the time of harvest, and during post-harvest ripening this starch is completely degraded. A partial decomposition of hemicelluloses and pectin also occurs, and the monosaccharide content rises.

D-Glucose, also known as dextrose, grape sugar, or starch sugar, is, together with D-fructose, known as laevulose or fruit sugar, the main monosaccharide of most foods. In the blood of animals, it occurs at concentrations of about 1 g/kg, whilst in diabetic patients (*diabetes mellitus*) it may occur in the urine, in extreme cases at concentrations up to 10%. Free D-mannose, D-galactose, and other hexoses and their derivatives are found in the small quantities in a variety of foods. Pentoses in foods are generally present in smaller amounts than hexoses. The main pentoses are D-ribose, L-arabinose, and D-xylose, which is also known as wood sugar.

Saccharides are deliberately added to a variety of food products to improve their organoleptic properties (taste, texture). Monosaccharides are usually added as invert sugar and in the form of glucose or fructose syrups.

4.2.2.1 Meat and meat products

Glycogen (also known as animal starch) occurs in the muscles of warm-blooded animals in concentrations ranging from 0.02 to 1% (0.3% in fish), depending on age and other factors, and is rapidly degraded post mortem. In the matured flesh, only monosaccharides and their phosphoric acid esters occur. They are commonly present at levels of 0.1–0.15%, of which glucose 6-phosphate constitutes about 0.1%; about a fifth of this amount represents (0.02%) glucose 1-phosphate and fructose 1,6-bisphosphate. The rest is mainly free glucose (0.009–0.09%), fructose, and ribose. Ribose arises primarily as a hydrolysis product of free nucleotides (NAD, NADP, ATP, and the corresponding nucleosides). A certain proportion of ribose is present in the form of ribitol in riboflavin (vitamin B$_2$) and other flavins.

4.2.2.2 Milk and dairy products

Monosaccharides, particularly glucose, are present in milk in insignificant quantities. The main sugar is disaccharide lactose, whilst other related oligosaccharides occur in milk in smaller amounts.

4.2.2.3 Eggs

The amount of saccharides in eggs (in dry matter) is about 10 g/kg. About 9 g/kg of sugars are present in egg white and 1 g/kg in the yolk. Protein-bound saccharides in the form of glycoproteins occur in the egg white at a level of about 5 g/kg and in the yolk at about 0.2 g/kg. The rest are free sugars, especially monosaccharides, but free oligosaccharides and polysaccharides do not occur in the egg white. About 98% of free monosaccharides are glucose, whilst mannose, galactose, arabinose, xylose, ribose, and 2-deoxyribose (2-deoxy-D-*erythro*-pentose) are present at concentrations of 2–20 mg/kg. In the protein-bound carbohydrates, galactose, mannose, glucosamine, galactosamine, and lactaminic acid predominate.

4.2.2.4 Honey

Honey is composed of approximately 17% water and 82% carbohydrates. The main components are glucose, fructose, and maltose (Table 4.3). Various oligosaccharides, whose composition is given in Table 4.12, occur to a lesser extent. The amount of other components (minerals, proteins, amino acids, vitamins, and pollen grains) is lower than 5%. In a 100 g serving, honey provides 1272 kJ (304 kcal).

4.2.2.5 Cereals and cereal products

Mono-, di-, tri-, and higher oligosaccharides resulting from the degradation of starch are present in low concentrations in cereals. Wheat flour contains 100–900 mg/kg glucose and 200–800 mg/kg fructose. The maltose content is 500–1000 mg/kg, saccharose and raffinose occur in amounts of 1000–4000 and 500–1700 mg/kg, respectively, and the amounts of other oligosaccharides ranges from 0.4 to 1.6%. The sugars content in cereal products is highly variable as it primarily depends on the degree of starch hydrolysis or the amount of added saccharides.

4.2.2.6 Fruits and vegetables

The main sugars in fruits are glucose (about 0.3–37%) and fructose (about 0.2–33%), and other monosaccharides are present to a lesser extent (Table 4.4).

Ripe grapes, for example, contain glucose and fructose in roughly the same amounts (about 8–11%), but fructose dominates in overripe grapes. The sugar content in wine musts is 120–250 g/l (expressed as glucose). In addition to glucose and fructose (their ratio is 0.5–0.9), wines contain relatively large quantities of arabinose, xylose, galactose, and small amounts of other monosaccharides and oligosaccharides (Table 4.5). According to the sugar content, wines are categorised as dry (up to 4 g of residual sugar per liter), semi-dry (4.1–12 g/l), semi-sweet (12.1–45 g/l), or sweet (minimum of 45 g residual sugar). Sparkling wines (champagne wines) are classified as brut nature (naturally dry, sugar content is lower than 3 g/l where sugar has not been added), extra brut (particularly dry, 0–8 g/l), brut (dry, <15 g/l), extra dry (especially dry, 12–20 g/l), sec (dry, 17–35 g/l), demi-sec (half dry, 33–50 g/l), or doux (sweet, >50 g/l).

Table 4.3 Basic composition of mixed floral honey from several US regions.

Constituent	Average content (%)	Range	Constituent	Average content (%)	Range
Water	17.2	13.4–22.9	Glucose	31.28	22.03–40.75
Proteins (enzymes)	0.3	0.1–0.6	Maltose	7.31	2.74–15.98
Minerals (ash)	0.17	0.02–1.03	Saccharose	1.31	0.25–7.57
Fibre	0.2	–	Higher sugars	1.50	0.13–8.49
Fructose	38.19	27.25–44.26	Other components	3.1	0–13.2

Table 4.4 Content of monosaccharides and saccharose in fresh fruits (% of edible portion).

Fruits	Water	Glucose	Fructose	Saccharose
Apple	84.0	1.8–2.6	5.0–8.7	2.0–2.4
Apricot	87.4	1.9	0.4	4.4
Avocado	73.2	0.5	0.2	0.1
Banana	73.6	5.6–5.8	3.5–3.8	6.6–13.9
Black currant	80.3	2.4	3.7	0.6
Blueberry	84.2	3.5	3.6	0.2
Cherry sweet	81.3	8.1	6.2	0.2
Date dried	12.9–16.1	35.7–37.1	28.6–33.3	0.2–2.2
Fig[a]	79.1	0.1–2.3	2.8–7.5	0.1–0.3
Grapefruit	88.6	2.0–2.7	1.2–2.7	2.1–2.2
Grape[b,c]	82.7	8.2–9.7	8.0–11.4	0.0–0.03
Kiwifruit	83.1	4.8	5.3	0.7
Lemon	88.3	0.5	0.5–0.9	0.1–0.2
Mandarin	85.2	2.2	2.4	0.9
Mango	83.5	0.3	2.5	4.5
Orange	87.0	1.8–2.4	1.9–2.4	4.7
Papaya	88.1	1.4	2.7	1.8
Peach	87.1	0.7–1.5	0.9–1.1	6.7–9.6
Pear	82.5	1.4–2.2	6.0–11.2	0.3–1.1
Pineapple	84.6	2.3–3.5	1.4–3.9	1.0–7.9
Plum	86.0	3.5	1.3	1.5
Raspberry	86.1	2.2–2.3	2.4–2.5	0.0–1.0
Red currant	83.6	2.3–3.7	1.0–4.5	0.0–0.2
Carambola (starfruit)	91.4	3.1	3.2	0.8
Strawberry	89.8	1.8–2.6	2.1–2.3	1.3–1.7

[a]The galactose content in fresh fruit is 1.9–7.9%. Dried figs contain 30.1% water and 47.9% total sugars.

[b]The saccharose content in varieties of common grape vine (*Vitis vinifera*, Vitaceae) is low, but in some varieties of Fox grape vine (*V.* labrusca) originating from North America and their hybrids it may account for up to 25% of total sugars.

[c]Raisins (golden, seedless) contain 14.8% water, 32.7% glucose, 37.1% fructose, and 0.8% saccharose.

The galactose content of fresh fruits ranges from about 11 mg/kg in strawberries to 70–160 mg/kg in grapes and 270 mg/kg in kiwifruit. Some fruits contain large amounts of less common sugars. Rowanberry wine, for example, contains L-sorbose, which is formed by oxidation of D-glucitol on carbon C-5 during fermentation. Higher ketoses (heptuloses) occur in small quantities in strawberries, grapes, and other fruits and in wines. Avocado has a particularly interesting composition of sugars, consisting of a number of heptuloses, octuloses, and nonuloses (0.2–5% of f.w.), such as D-*manno*-hept-2-ulose (**4-9**), D-*talo*-hept-2-ulose, D-*glycero*-D-*manno*-oct-2-ulose (**4-9**), D-*glycero*-L-*galacto*-oct-2-ulose, D-*erythro*-L-*gluco*-non-2-*ulose* (**4-9**), D-*erythro*-L-*galacto*-non-2-ulose, and others.

The main monosaccharides in vegetables, as well as in fruits, are glucose and fructose (Table 4.6). Other monosaccharides (such as galactose, arabinose, and xylose) are present at low levels. The galactose content in fresh vegetables ranges from about 22–26 mg/kg in butter lettuce to 33–108 mg/kg in potatoes and 140–400 mg/kg in red bell peppers. Free arabinose occurs in larger quantities in a dried byproduct, known as dried beet pulp, which is left after most of the sugar has been extracted from the sliced beets.

Hydrolysis of saccharose in harvested potato tubers results in the production of glucose and fructose. Changes in the reducing sugar level can be used as an indicator of stress (such as low storage temperature and physical damage) in tubers. For example, tubers stored at low temperature have a sweet taste and will produce French fries of darker colour.

Table 4.5 Content of saccharides in wines.

Saccharides	Content (mg/l)	Saccharides	Content (mg/l)
Monosaccharides		Fucose	2–9
Ribose	6.3–62	*Oligosaccharides*	
Arabinose	1.0–242	Trehalose	0–61
Xylose	0.6–146	Cellobiose	2–7
Glucose	56–25 000	Maltose	1–5
Mannose	2–37	Saccharose	0
Galactose	6.3–249	Lactose	1–5
Fructose	93–26 500	Melibiose	trace–1
Rhamnose	2.2–121	Raffinose	0–1

Table 4.6 Content of main sugars in fresh vegetables (% of edible portion).

Vegetable	Water	Glucose	Fructose	Saccharose	Vegetables	Water	Glucose	Fructose	Saccharose
Artichoke	84.9	0.02	0.01	0.02	Eggplant	92.3	0.96	0.83	0.15
Beetroot	87.6	0.22	0.14	10.70	Endive	93.8	0.03	0.59	0.03
Broccoli	89.0	0.49	0.68	0.10	Lettuce (iceberg)	95.6	0.49	0.39	0.02
Cabbage	92.2	0.69	0.61	0.25	Onion	89.1	1.54	1.76	0.22
Carrots	88.3	0.59	0.55	3.59	Pepper	93.9	1.16	1.12	0.11
Cauliflower	91.9	0.58	0.70	0.15	Radish	95.3	0.80	0.80	0.05
Chicory buds	94.5	0.23	0.35	0.13	Spinach	91.4	0.07	0.05	0.05
Celery	88.0	0.16	0.22	0.02	Tomato	94.5	1.25	1.37	0.00
Cucumber	95.2	0.76	0.87	0.03	Yam	69.6	0.91	2.62	0.00

4.2.2.7 Other foods

In legumes, the main monosaccharides are glucose and fructose. In beans, the content of glucose is 0.1–1.1% (f.w.), in peas around 0.3%, and in soybeans from 0.04 to 0.2%. Fructose content in beans is 0.1–1.2%, whilst in peas and soybeans it is about 0.2 and 0.5–3.2%, respectively. Saccharose and other oligosaccharides are also present in higher amounts in legumes (see Table 4.16).

The monosaccharide and sugar alcohol content of higher fungi is generally less than 1% of dry matter. The main monosaccharide is glucose, but the content of sugar alcohols is generally higher.

Glucose, fructose, mannose, galactose, and other monosaccharides are the building blocks of many oligosaccharides, polysaccharides, and heteroglycosides. A less common monosaccharide is apiose, which is an important component of pectin and, in the form of glycosides, occurs in vegetables of the Apiaceae family, commonly known as the carrot or parsley family, where the main carbohydrate starch is found in the root vegetables and root crops. Some gourmet vegetables belonging to the Asteraceae family, such as globe artichoke (*Cynara cardunculus*), common chicory (*Cichorium intybus*), and black salsify (*Scorzonera hispanica*), contain, in addition to starch, a reserve of polysaccharide inulin, composed of fructose units. Like in fruits, other main polysaccharides are cellulose, hemicelluloses, and pectin.

4.2.3 Physiology and nutrition

The key compound of carbohydrate metabolism, and a source of energy in animals and plants, is glucose. Heterotrophic organisms obtain energy for endergonic reactions by oxidation of glucose and other primary nutrients. Carbohydrates should account for more than 55% of energy. About 80–90% of the energy intake provided by carbohydrates should come from polysaccharides, with up to 20% coming from oligosaccharides and monosaccharides.

The ingested polysaccharides and oligosaccharides must be hydrolysed to their component monosaccharides before being absorbed. The digestion of starch starts by the action of salivary α-amylase and continues with pancreatic α-amylase in the small intestine. The hydrolysis yields α-dextrins, maltose oligomers, maltose, and small amounts of glucose. Maltase and isomaltase from the lining of the small intestine (so-called brush-border hydrolases) split the products produced by amylases into glucose. Saccharose is digested in the lining of the small intestine by saccharase (sucrase) into glucose and fructose. Lactose is hydrolysed to glucose and galactose by lactase, which is also found in the intestinal lining, but is absent in most adult humans. Other carbohydrates, for example pectins, pass undigested into the large intestine and are partly hydrolysed by intestinal bacteria. Some carbohydrates, such as cellulose, are not digested at all, as humans lack the enzyme cellulase. After digestion, the small intestine (mainly duodenum) absorbs the available glucose and other monosaccharides and carries them into enterocytes. Only glucose and galactose are actively absorbed (by co-transport with sodium using the same hexose transporter), whilst fructose passes into the enterocyte via diffusion through another hexose transporter. Glucose, galactose, and fructose are tranported out of the enterocyte through another hexose transporter, and then diffuse into blood capillaries within the villus.

The concentration of glucose in blood (glycaemia) is controlled by three hormones: insulin, glucagon, and epinephrine (adrenaline). If the concentration of glucose in the blood is too high, insulin is secreted by the pancreas and stimulates the transfer of glucose into the liver, muscles, and other organs that are able to metabolise glucose by the metabolic process called **glycolysis** to give pyruvic acid and ATP. Since glycolysis releases relatively little ATP, pyruvic acid is converted into acetyl-CoA, which enters the citric acid cycle. Glucose oxidation yields simple organic compounds and the final products of oxidation are carbon dioxide and water. During strenuous muscular activity, pyruvic acid is converted into lactic acid, which is transformed during the resting period back into pyruvic acid. Pyruvic acid in turn yields glucose by a process called **gluconeogenesis**. Excess glucose is stored in the liver and muscles as polysaccharide glycogen, which is formed by the process of **glycogenesis**. Glycogen is stored in the liver and muscles until the blood glucose levels are low. Then, epinephrine and glucagon hormones are secreted to stimulate the transformation of glycogen into glucose by the process called **glycogenolysis**.

The blood glucose level induced by carbohydrates in food can be measured by the **glycaemic index** (GI), which evaluates the biological value of dietary carbohydrates and is a measure of the effects of food carbohydrates on the blood glucose level. It is defined as the glycaemic response elicited by a 50 g portion of carbohydrate food and expressed as a percentage of that elicited by 50 g of D-glucose (GI = 100). For comparison, D-fructose has a GI value of 19, whilst D-glucitol 9, xylitol 13, and D-mannitol have a value of 0. Rice (white) has a GI value of 81, milk (whole) 39, apples 52, and banana 83.

The main pathway of galactose metabolism is the Leloir pathway, in which, in the liver, β-D-galactose is converted into UDP-α-D-glucose (a precursor of glycogen and a building block of nucleotide metabolism) via α-D-galactose, α-D-galactose 1-phosphate, and UDP-α-D-galactose under the action of mutarotase, galactokinase, galactose 1-phosphate uridyltransferase, and galactose-4′-epimerase, respectively.

Ingested fructose is phosphorylated in the liver to D-fructose 1-phosphate by fructokinase. Fructose phosphate is split by aldolase B into dihydroxyacetone phosphate and glyceraldehyde. Under the action of triokinase, glyceraldehyde is transformed into glyceraldehyde 3-phosphate. The resulting triose phosphates can enter the gluconeogenic pathway, and can be used for glucose or glycogen synthesis, or else further catabolised.

Exogenous polyols are absorbed slowly and metabolised mainly by the hepatic enzymes. D-Glucitol (sorbitol) and xylitol are metabolised completely after oxidation. Glucitol is oxidised to fructose by glucitol dehydrogenase. Xylitol is oxidised to D-xylulose by xylitol dehydrogenase and xylulose is phosphorylated by xylulose kinase to D-xylulose 5-phosphate, which is an intermediate of the pentose phosphate pathway. D-Mannitol is poorly utilised because of the low affinity for glucitol dehydrogenase. Exogenous erythritol is very poorly metabolised, being almost completely excreted in urine.

4.2.4 Use

Glucose and glucose syrups are raw materials for the production of D-glucitol, D-mannitol, D-fructose, D-gluconic acid and its lactones, and various fat and sugar substitutes, including an indigestible synthetic polymer of glucose called polydextrose with predominantly α-(1 → 6) bonds. Polydextrose is obtained by fusion of glucose with glucitol and citric acid. Glucose is still the most important raw material for ethanol production by fermentation.

4.3 Derivatives of monosaccharides

4.3.1 Sugar alcohols

Reduction of carbonyl groups of aldoses and ketoses gives rise to acyclic polyhydroxy derivatives of hydrocarbons called sugar alcohols, polyols, or alditols (formerly also known as glycitols). Reduction of aldoses produces only one sugar alcohol, whilst reduction of ketoses creates a new asymmetric carbon, and the products are two diastereomeric polyols. Polyols formed by reduction of some sugars have a plane of symmetry that bisects a molecule into halves that are mirror images of each other and the polyols are therefore achiral meso forms.

Through the hydrophilic hydroxyl groups, sugar alcohols are involved in hydration of macromolecules. They also act in the prevention of various stress factors and lipid peroxidation in plant cells.

Sugar alcohols also include cyclitols (alicyclic polyols), because they have similar properties, although they are not derived from sugars by simple reduction. Cyclitols are cycloalkanes in which at least three-ring carbon atoms are substituted (each of them only once) by hydroxyl or alkoxyl groups.

4.3.1.1 Structure and nomenclature

4.3.1.1.1 Alditols

The simplest alditol, which is derived from D-glyceraldehyde, L-glyceraldehyde, or 1,3-dihydroxyacetone, is glycerol (propane-1,2,3-triol).[1] The name of an optically active higher alditol is derived from the name of sugar from which the alditol is derived and to which the suffix -itol is connected (e.g. erythritol and threitol). From the four tetroses (aldoses of D- and L-series), only three tetritols can be derived by reduction. Erythritol (**4-15**), which is a meso form, arises from the D- or L-erythrose; D-threitol (**4-15**) is formed from D-threose; and L-threitol is formed from L-threose. Aldopentoses of D- and L-series (total of eight aldoses) provide four pentitols. D-Arabinitol (**4-15**) is formed from D-arabinose or D-lyxose, L-arabinitol from L-arabinose or L-lyxose, ribitol (**4-15**) from D- or L-ribose, and xylitol (**4-15**) from D- or L-xylose. The reduction of the 16 aldohexoses of D- and L-series yields 10 alditols: allitol (**4-16**), D-altritol (**4-16**), L-altritol, D-glucitol (**4-16**, formerly sometimes incorrectly called L-gulitol or more often D-sorbitol, from the Latin name *sorbus* for the rowans or mountain-ash trees in which glucitol was first identified in the fruits), L-glucitol, D-mannitol (**4-16**), L-mannitol, D-iditol (**4-16**), L-iditol, and galactitol (**4-16**, formerly called dulcitol). For example, D-glucitol is also formed by reduction of D-glucose and L-gulose.

erythritol D-threitol D-arabinitol ribitol xylitol

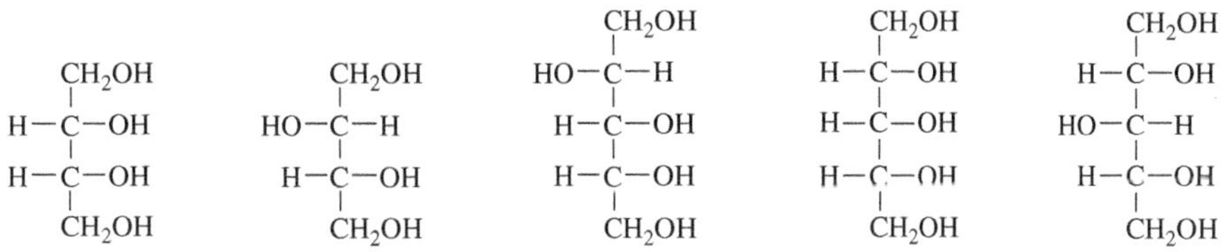

4-15, structures of selected tetritols and pentitols

allitol D-altritol D-glucitol D-mannitol D-iditol galactitol

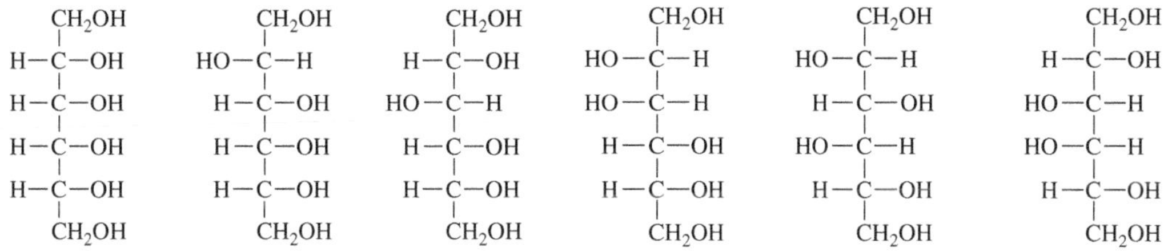

4-16, structures of selected hexitols

The systematic name of D-glucitol is (2*R*,3*S*,4*S*,5*S*)-hexane-1,2,3,4,5,6-hexol, that of D-mannitol is (2*R*,3*R*,4*R*,5*R*)-hexane-1,2,3,4,5,6-hexol, and that of galactitol is (2*R*,3*S*,4*R*,5*S*)-hexane-1,2,3,4,5,6-hexol.

The reduction of ketoses yields both C-2 isomers. For example, the reduction of D-*glycero*-tetrulose creates erythritol and D-threitol. Reduction of D-fructose yields D-glucitol and D-mannitol.

Conformation Alditols in solutions most frequently occupy the zigzag conformation with carbon atoms in one plane. The planar conformer of xylitol is shown in stereoformula **4-58** (using the Newman projection, diagrammatic formula). As in the cyclic forms of aldoses and ketoses, non-bonding *syn*-axial interactions also exist in planar zigzag conformers of alditols, which are the most important destabilising factor determining the relative abundance of different conformers. As a result of interactions of hydroxyl groups on C-2 and C-4 of xylitol, displaying the internal stress, the functional groups rotate around bonds C-3–C-4, resulting in the thermodynamically most stable conformer (**4-17**), whilst the least stable (**4-17**) conformer with the hydroxyl group at C-2 interacts with bulky hydroxymethyl group at C-4. The prevailing conformer is therefore the most stable anticlinal (also called gauche or bent) conformer.

[1] Reduction of glycolaldehyde or glyoxal yields ethylene glycol (ethane-1,2-diol) and the reduction of formaldehyde gives methanol. However, these compounds are not ranked amongst the alditols.

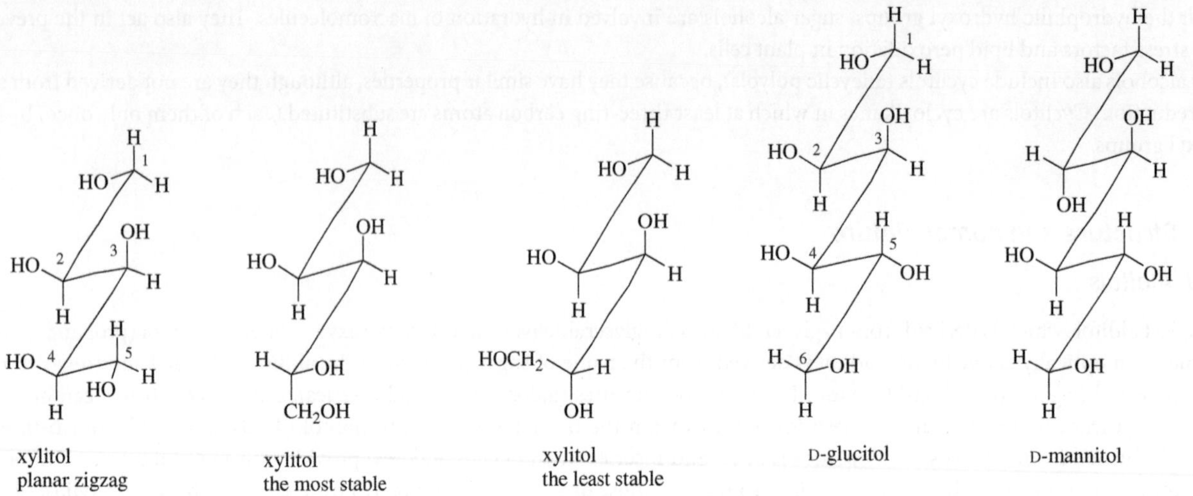

xylitol
planar zigzag

xylitol
the most stable

xylitol
the least stable

D-glucitol

D-mannitol

4-17, selected conformers of alditols

The *syn*-axial interactions also occur between the hydroxyl groups at C-2 and C-4 of ribitol, but not with D-arabinitol. Generally speaking, alditols in the configuration *xylo* largely exist as conformers of the most stable xylitol type, and alditols in the configuration *arabino* primarily exist as planar zigzag conformers. Analogously, the *syn*-axial interactions in the zigzag conformation also exist in hexitols allitol, altritol, D-glucitol (**4-17**) and D-iditol, but not with D-mannitol (**4-17**) and galactitol. Only hexitol in the *galacto* and *manno* configuration therefore occurs in the planar zigzag conformations in solutions, and the predominant conformation of other hexitols is gauche with other non-planar conformations.

4.3.1.1.2 Cyclitols

An important group of polyols are cycloalkanes, with three to six hydroxy groups bound to carbon atoms of the ring. The most important cyclitols are polyhydroxyalcohols formally derived from cyclohexane, which are called cyclohexitols. In addition to hexahydroxy derivatives, penta-, tetra-, and trihydroxycyclohexanes and their derivatives are commonly found in plants. The most important cyclohexitols are hexahydroxycyclohexanes (cyclohexane-1,2,3,4,5,6-hexols), known under the generic name **inositols** (formerly called **cycloses**). Altogether, there are eight stereoisomers, which are distinguished by affixes or by numerical prefixes (locants) indicating the individual atoms of the ring (**4-18**). As a result of the symmetry of the molecules, seven inositols are achiral substances, optically inactive meso forms, and *chiro*-inositol is the only cyclohexitol that occurs as the (+)-D- or (−)-L-enantiomer. Formerly, this substance was called inositol, without the affix. The affix *chiro* was chosen so as to underline that *chiro*-inositol is the only optically active inositol.

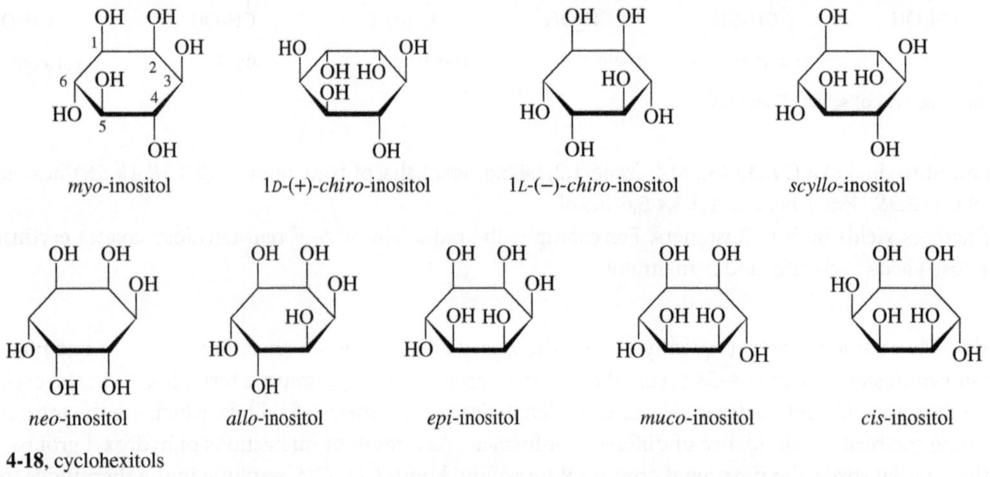

myo-inositol

1D-(+)-*chiro*-inositol

1L-(−)-*chiro*-inositol

scyllo-inositol

neo-inositol

allo-inositol

epi-inositol

muco-inositol

cis-inositol

4-18, cyclohexitols

The most important compound is *myo-inositol*, (1Z,2Z,3Z,5Z,4E,6E)-cyclohexane-1,2,3,4,5,6-hexol, formerly also called *meso*-inositol (or *i*-inositol, phaseomannitol, nucitol, bios I, mouse antialopaecia factor, or vitamin B_8). The affix *myo* was preferred to the affix *meso* as it defines a certain configuration of hydroxyl groups above and below the plane of the ring (1,2,3,5/4,6-), whilst the second affix has a general meaning and is used to label achiral, optically inactive compounds with the same number of identically bound enantiomeric groups.

Table 4.7 Main alditols in wines.

Alditol	Content (mg/l)	Alditol	Content (mg/l)
Glycerol	4000–10 000	Xylitol	4–11
Erythritol	35–292	D-Glucitol	9–277
D-Arabinitol	10–577	D-Mannitol	6–152

Conformation Analogously to pyranoses, cyclohexitols and derived carboxylic acids occur in sterically stable chair conformations. The lowest energy has *scyllo*-inositol with all substituents in equatorial positions.

4.3.1.2 Occurrence

4.3.1.2.1 Alditols

Alditols occurring in foods as natural components are the result of biochemical reactions, but they can also arise as products of Cannizzaro reaction or other non-enzymatic browning reactions. Many synthetic alditols are used as food additives (as sweeteners; see Section 11.3.2.2).

The simplest alditol, glycerol, is a component of food lipids and also occurs as a byproduct of fermentation in addition to (2R,3R)-butane-2, 3-diol, also known as (−)-D-butane-2,3-diol. It is accompanied by a small amount of (2S,3S)-butane-2,3-diol, also known as (+)-L-butane-2, 3-diol, and *meso*-2,3-butane-2,3-diol. The glycerol and (2R,3R)-butane-2,3-diol contents can be used as an indicator of wine quality. Erythritol is found in significant quantities in some algae, and small amounts are present in wine musts (juices); hence, along with glycerol and other alditols, it is found in wines (Table 4.7). The content of alditols in wines is higher or the same as in the musts and depends on the type of yeast and other factors. Ribitol is a constituent of riboflavin; D-arabinitol and xylitol are found fairly frequently in fruits, vegetables, and mushrooms. Mushrooms (such as the common white button mushroom, *Agaricus bisporus*) contain D-arabinitol at levels of about 3.5 g/kg (d.w.) and xylitol at about 1.3 g/kg (d.w.).

Widely spread alditols in foods are hexitols, particularly D-mannitol, which is the most common hexitol. Other common hexitols are D-glucitol and galactitol. The accumulation of D-mannitol in plants increases their resistance to the high salinity of soil. Fruits and vegetables usually contain very low amounts of mannitol (up to 0.01–0.02%); a relatively high mannitol content is found in celery (0.48%). Its content in green (unroasted) coffee is around 0.05%, whilst that in soluble coffee ranges from 0.20 to 1.03%, and waste from the production of soluble coffee contains 1.61–2.03% mannitol. This waste is therefore used as a raw material for isolating mannitol. Higher amounts of mannitol in wines (1–30 g/l) may be caused by attack of the *Botrytis cinerea* fungus invading the grapes, or by activities of bacteria *Bacterium mannitopoeum*. The fermentation type is called mannitol fermentation or slimy fermentation of sugars. Higher fungi can have a very high mannitol content. The level of mannitol in fungi *Tricholoma portentosum*, commonly known as the charbonnier, is 1.0%, that in Saffron milk cap (*Lactarius deliciosus*) is 13.7%, that in horse mushroom (*Agaricus arvensis*) 6.5%, and that in the cultivated common mushroom (*A. bisporus*) 15.7% in dry matter of young fruiting bodies or 26% in dry matter of mature fruiting bodies. The mannitol content in brine of canned common mushrooms is 2.5–10.5 g/l.

The D-glucitol contents in some common fruits and vegetables are given in Table. 4.8. For example, rowan berries, in addition to the main hexitol D-glucitol, also contain D-mannitol and D-iditol. Prunes have a higher amount of glucitol, and for that reason, they are used in various laxative preparations.

Table 4.8 D-Glucitol in fruits and vegetables.

Fruits and vegetables	Content (% of edible portion)	Fruits and vegetables	Content (% of edible portion)
Apples	0.2–0.8	Cloudberries	0.01
Pears	1.2–2.8	Red raspberries	0.03
Plums	0.6–13.9	Rowan berries	.3.4–5.3
Cherries	0.1	Grapes	0.01
Peaches	0.03–0.48	Cucumbers	0.01
Apricots	0.05–0.46	Tomatoes	0.01

A higher concentration of galactitol has been found in fungi and fermented milk products. Mushrooms have a galactitol content of about 0.5 g/kg (d.w.), whilst about 9 g/kg (d.w.) can be found in yoghurts.

4.3.1.2.2 Cyclitols

myo-Inositol is found as a free compound in fruits and vegetables, where its content ranges from 0.01 to 0.05%; 0.02–0.30% of *myo*-inositol is present in legumes, cereals, and nuts, particularly in bound forms, and is accompanied by a minor cyclitol 1D-*chiro*-inositol and other cyclitols. In legumes, cereals, and nuts, *myo*-inositol is mainly present as *myo*-inositol-1,2,3,4,5,6-hexakisdihydrogenphosphoric acid, known as phytic acid, occurring in the form of a mixed K^+, Mg^{2+}, Ca^{2+}, Fe^{2+}, and Zn^{2+} salt called phytin (see Section 6.3.4.2.2). Phytin is the main bound form of phosphorus used during seed germination. It also occurs in other parts of plants, such as pollen, roots, and tubers. Wheat contains about 10 g/kg of phytin. Approximately 70–80% of phytin in dough is hydrolysed by the yeast enzyme phytase to *myo*-inositol. *myo*-Inositol also commonly occurs in the form of phospholipids (phosphatides), known as phosphoinositols.

myo-Inositol is a precursor for the biosynthesis of other cyclitols, *O*-methylinositols, and other derivatives (**4-19**). Methylation of *myo*-inositol and isomeric inositols (*scyllo*-, *chiro*-, *muco*-, and *neo*-inositol) provides *O*-methylinositols bornesitol (1D-1-*O*-methyl-*myo*-inositol), ononitol (1D-4-*O*-methyl-*myo*-inositol), sequoitol (1D-5-*O*-methyl-*myo*-inositol), pinitol (1D-4-*O*-methyl-*chiro*-inositol), quebrachitol (1L-2-*O*-methyl-*chiro*-inositol), and others that are involved in a plant's adaptability to various forms of stress and are used as seed storage substances and building blocks for biosynthesis of glycosides. D-Sequoitol occurs as a minor component in beans and many other plants. A common methyl ether of legumes, derived from D-*chiro*-inositol, is D-pinitol.

Cyclitols (hexahydroxycyclohexanes) and their ethers also occur in a variety of α-D-galactosides, which are known as **pseudooligosaccharides**. Conjugation of *myo*-inositol and galactose yields galactinol, known by its systematic name as 3-*O*-α-D-galactopyranosyl-D-*myo*-inositol, and also as 1-*O*-α-D-galactopyranosyl-L-*myo*-inositol (**4-20**), the well-known and first identified representative of pseudooligosaccharides. It occurs in all plants containing raffinose and other higher α-galactooligosaccharides (stachyose, verbascose, and ajugose), and therefore occurs in all legumes, as it is employed as the starting compound for the biosynthesis of galactooligosaccharides. Higher α-D-galactopyranosyl homologues of galactinol are also present. The highest concentration of these compounds has been found in seed germs.

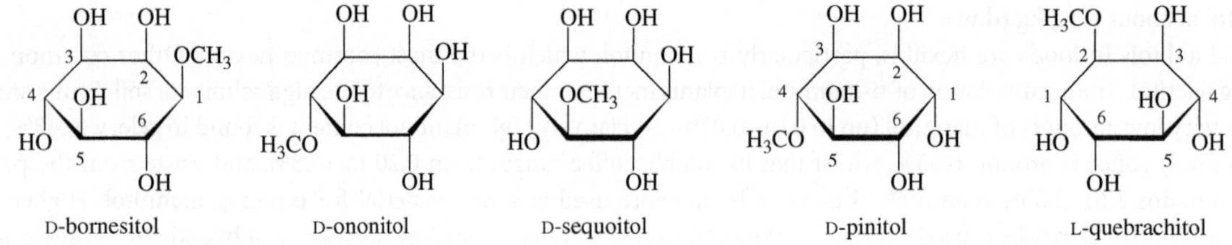

4-19, selected cyclitol ethers

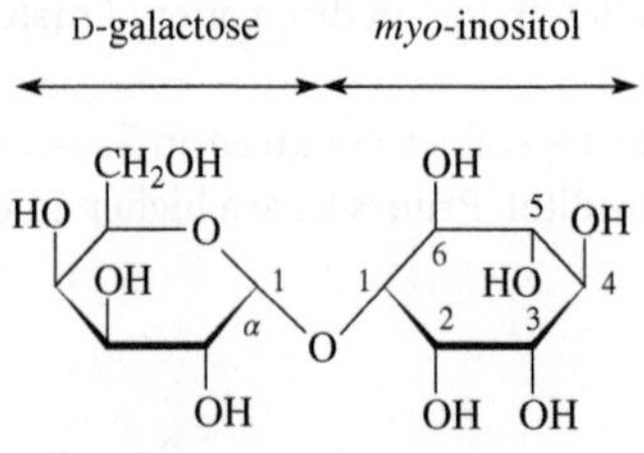

4-20, galactinol

The reaction of 1D-*chiro*-inositol with D-galactose provides fagopyritols (**4-21**), which are present in legumes. Fagopyritols occur at higher concentrations in the seeds and especially the germ of buckwheat (*Fagopyrum esculentum*, Fabaceae). Fagopyritols of the A series are 3-*O*-α-D-galactopyranosyl-1D-*chiro*-inositols, whilst those of the B series are 2-*O*-α-D-galactopyranosyl-1D-*chiro*-inositols.

Galactinol in soybeans and other legumes is accompanied by α-D-galactosides of pinitol, known as galactopinitols. A common compound is α-D-digalactopyranoside of pinitol, called ciceritol or *O*-α-D-galactopyranosyl-(1 → 6)-*O*-α-D-galactopyranosyl-(1 → 2)-4-*O*-methyl-1D-chiro-inositol (**4-21**). Contents of major cyclitols and their galactosides occurring in legumes are given in Table 4.9.

Table 4.9 Important cyclitols and their galactosides in legumes (% of dry matter).

Legume	Latin name	*myo*-Inositol	D-Pinitol	Galactinol	Galactopinitols[a]	Ciceritol
Beans	*Phaseolus vulgaris*	0.02–0.06	0.08–0.20	0.04–0.05	0.00–0.04	trace
Peas	*Pisum sativum*	0.10–0.17	0.05	0.07	0.00	0.00
Lens	*Lens culinaris*	0.07–0.11	1.11–0.40	0.10–0.12	0.36–0.39	1.60
Soybeans	*Glycine max*	0.03–0.10	0.20–0.90	0.00	0.35–0.70	0.08
Chickpeas	*Cicer arietinum*	0.10–0.30	0.40–0.45	0.08–0.20	0.50–0.80	2.80

[a]Sum of 2-*O*-α-D-galactopyranosyl-1D-*chiro*-inositol (fagopyritol B1) and D-pinitol derivatives, 2-*O*-α-D-galactopyranosyl-4-*O*-methyl-1D-*chiro*-inositol and 5-*O*-α-D-galactopyranosyl-4-*O*-methyl-1D-*chiro*-inositol.

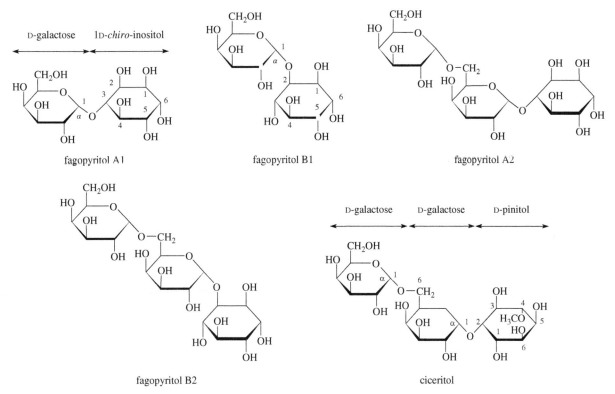

4-21, examples of galacto-*chiro*-inositols and galactopinitols

4.3.1.3 Physiology and nutrition

4.3.1.3.1 Alditols

The most important hexitols are D-glucitol and D-mannitol. The most important pentitol is xylitol. The relative sweetness of both hexitols (in 10% solutions) and of xylitol, compared with saccharose, is about 60 and 100%, respectively. All three polyols have very little effect on blood glucose level and are therefore used as sweeteners for diabetics. However, their content must be included in the total energy intake. They exhibit mild laxative effects.

The anticariogenic effect of xylitol is under discussion; xylitol exposure may have an inhibiting effect on the growth of cariogenic bacteria, and this may create a permanent change in their oral population. The sweetness and pleasant cooling effect of xylitol sweetened products in the mouth (such as mints and chewing gums) create an increase in salivary flow. The cooling effect of xylitol is produced in its contact with saliva due to its negative heat (enthalpy) of dissolution ($-149\,\text{kJ/kg}$, which corresponds to $-27\,\text{kJ/mol}$). D-Glucitol and D-mannitol show only weak cariogenicity.

In individuals with galactosaemia (type 1), a rare disorder that affects galactose metabolism, galactose is accumulated in the body tissues together with galactitol, which is an alternative product of degradation. Galactosaemia, which is especially dangerous for children, is caused by a lack of the enzyme galactose-1-phosphate uridyltransferase. Galactitol is toxic to the liver (which may lead to failure), brain (mental retardation), and kidney and to the eye lens (cataract).

4.3.1.3.2 Cyclitols

In nature, the most common cyclitol is *myo*-inositol. It is the key compound in the metabolism of microorganisms, plants, and animals, therefore it was formerly considered one of the vitamins (see Section 5.15), but the *myo*-inositol derivative phytin (see Section 10.2.3.1) is classified as an antinutritional food constituent.

4.3.1.4 Use

Polyols xylitol, D-glucitol, and D-mannitol have found extensive use primarily as sweeteners in foods for diabetics. They are used in many bakery and confectionery products (mainly xylitol and D-glucitol) to reduce water activity, as substances suppressing crystallisation of saccharose, and as humectants improving hydration of dry goods.

Alditols are mainly produced from aldoses and ketoses by hydrogenation, reduction with amalgams or complex hydrides, or by electrolytic reduction. Xylitol is prepared by reduction of D-xylose obtained by hydrolysis of natural xylems (hemicelluloses); D-glucitol is prepared from D-glucose and D-mannitol via hydrogenation of D-fructose obtained from either starch or invert sugar.

Some other polyols may find some application in nutrition (as sweet substances). Such examples include L-arabinitol (sweetening power of about 100%) obtained by hydrogenation of L-arabinose, galactitol (40%) obtained from D-galactose, and polyols derived from disaccharides.

4.3.2 Sugar acids

Sugar acids are saccharides with a carboxyl group(s) that commonly occur in foods as free substances and components of many oligosaccharides, polysaccharides, heteroglycosides, and other food constituents. Frequently, these acids are formally derived from monosaccharides by oxidation of aldehyde groups or primary alcohol groups. Sugar acids are usually produced by enzymatic reactions, but some sugar acids derived from both hexoses (such as saccharinic acids) and lower sugars are formed by chemical reactions during storage and processing of foods. Of particular importance in this respect is the Maillard reaction.

Some carboxylic acids, hydroxy acids, and oxoacids produced by glycolysis, or their modifications, arising in the citric acid or glyoxylate cycle, can thus also be considered sugar derivatives. An example of a C_6 substance is L-ascorbic acid and other forms of vitamin C; 2-oxoglutaric acid is a C_5 substance, tartaric and malic acids are C_4 substances, glyceric, lactic, and pyruvic acids are C_3 substances, and glyoxylic acid is a C_2 substance. Sugar acids also include carboxylic acids derived from alicyclic sugar alcohols (cyclitols) (**8-71**).

4.3.2.1 Structure and nomenclature

Oxidation of aldehyde groups of aldoses yields aldonic (generally glyconic) acids. Free acids readily (especially in acidic solutions) lactonise to give relatively stable five-membered γ-lactones or less stable six-membered δ-lactones. The nomenclature of aldonic acids and their lactones is based on the substitution of the suffix -ose in the aldose name with the suffix -onic acid and -onolactone, respectively. For example, D-glucose gives rise to D-gluconic acid, (2R,3S,4R,5R)-2,3,4,5,6-pentahydroxyhexanoic acid, and its dehydration yields the corresponding gluconolactones (Figure 4.3).

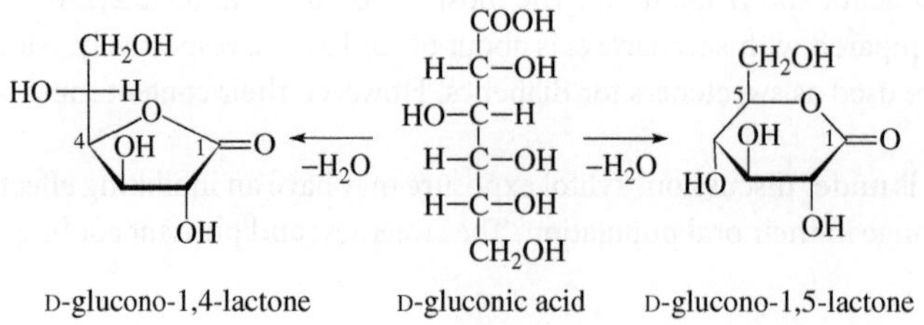

D-glucono-1,4-lactone   D-gluconic acid   D-glucono-1,5-lactone

Figure 4.3 Formation of gluconic acid lactone.

Another common type of sugar acid is the uronic (alduronic) acids, which are derived by oxidation of primary hydroxyl groups. The name of a uronic acid is formed by replacing the suffix -ose with the suffix -uronic acid. Uronic acids easily produce 6,3-lactones of furanose or pyranose forms. D-Glucose thus forms D-glucuronic acid, (2S,3S,4S,5R,6R)-3,4,5,6-tetrahydroxyoxane-2-carboxylic acid, occurring as pyranose, and its lactone (Figure 4.4), D-galactose yields D-galacturonic acid, and D-mannuronic acid is derived from D-mannose.

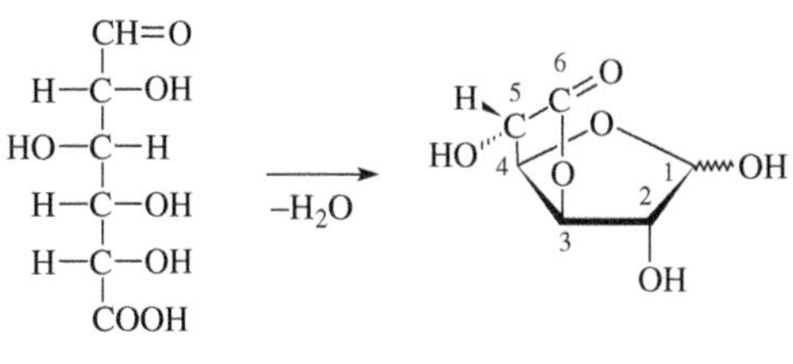

Under more drastic conditions (e.g. oxidation with nitric acid), the terminal hydroxymethyl group of aldonic acids is oxidised to carboxyl group, which produces dicarboxylic acids known as aldaric acids that are generally called glycaric acids. D-glucose yields D-glucaric acid, (2S,3S,4S,5R)-2,3,4,5-tetrahydroxyhexanedioic acid, also known as saccharic acid, which is a product of glucuronic-acid metabolism in the liver of mammals. From D-galactose arises achiral galactaric (mucic) acid.

The most important alicyclic acid derived from tetrahydroxycyclohexane is (3R,5R)-1,3,4,5-tetrahydroxycyclohexane-1-carboxylic acid, known as (−)-L-quinic acid. The trihydroxycyclohexene derivative is (3R,4S,5R)-3,4,5-trihydroxycyclohexene-1-carboxylic acid, known as (−)-L-shikimic acid (see Section 8.2.6.1.5).

D-glucuronic acid   D-glucurono-6,3-lactone

Figure 4.4 Formation of glucuronic acid lactone.

4.3.2.2 Occurrence

D-Gluconic and D-mannonic acids are present in relatively large amounts as natural components in many plant materials, as well as in those foods in which non-enzymatic browning reactions take place. Table 4.10 illustrates the composition and content of aldonic, deoxyaldonic, and some other acids in chicory root and malt.

Alduronic acids, especially D-glucuronic, D-galacturonic, D-mannuronic, and L-guluronic acids, occur chiefly as the building blocks of some polysaccharides. Constituents of rhamnogalacturonan chains in pectin (see Section 4.5.6.6) are unusual acids such as aceric acid (3-C-carboxy-5-deoxy-β-L-xylofuranose, β-L-AcefA, **4-22**), deoxyheptonic acid (3-deoxy-β-D-lyxo-hept-2-ulopyranaric acid, β-D-Dhap, **4-22**), and ketodeoxyoctonic acid (3-deoxy-β-D-manno-oct-2-ulopyranosonic acid, β-D-Kdop, **4-22**), which is also a component of lipopolysaccharides of gram negative bacteria.

Alicyclic acids, quinic acid, and shikimic acid (intermediates in the biosynthesis of phenylalanine, tyrosine, and tryptophan, numerous flavourings, natural colourings, and other compounds) are common compounds in foods. They occur as free compounds and also in the form of various derivatives, for example in depsides such as chlorogenic acids.

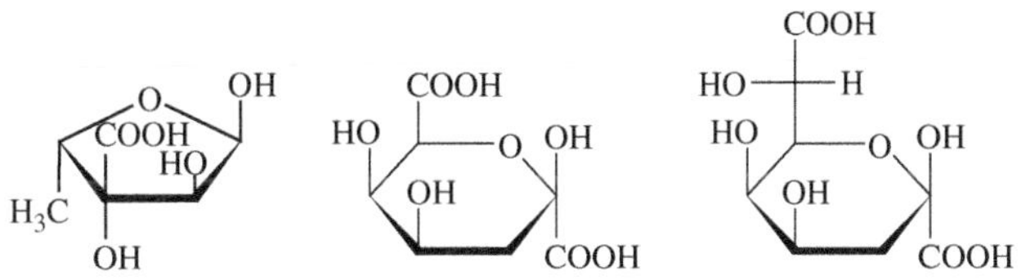

aceric acid   deoxyheptonic acid   ketodeoxyoctonic acid

4-22, unusual sugar acids in pectin

4.3.2.3 Physiology and nutrition

6,3-Glucuronolactone is used in the diet of athletes as a popular supplement for body building. It was found that it accelerates formation and decelerates degradation of muscle glycogen. Glycosides of γ-lactone of D-glucuronic acid, called D-glucuronides, act in the body in detoxification reactions.

4.3.2.4 Use

δ-D-Gluconolactone is added at levels of around 0.1% to some smoked meat products, such as fermented sausages and salami. Gradually, hydrolysed lactone produces free gluconic acid, which inhibits the growth of undesirable putrid microflora, especially at the beginning of product ripening, together with lactic and acetic acids produced by fermentation of gluconic acid by some bacteria of the genus *Lactobacillus*.

In dermatology, gluconolactone is capable of chelating metals and may also function by scavenging free radicals, thereby protecting the skin from some of the damaging effects of UV radiation.

Table 4.10 Sugar-derived and related carboxylic acids in chicory root and malt (g/kg of dry matter).

Carboxylic acids	Chicory root			Malt	
	Fresh	Dried	Roasted[a]	Dried	Roasted[a]
Aldonic					
Glycolic	0.46	0.15	0.94	0.03	0.10
Glyceric	0.15	0.05	0.24	0.04	0.11
Erythronic	0.00	0.03	0.09	0.00	0.04
Threonic	0.00	0.01	0.07	0.01	0.07
Ribonic	0.00	0.04	0.07	0.03	0.08
Arabinonic	0.00	trace	0.13	trace	0.02
Gluconic	0.46	0.03	0.08	0.58	0.30
Mannonic	0.30	0.03	0.05	0.20	0.15
Deoxyaldonic					
Lactic (2-hydroxypropanoic)	0.45	0.45	0.80	0.19	0.33
3-Hydroxypropanoic	0.00	0.12	0.31	0.08	0.21
2,4-Dihydroxybutanoic	0.00	0.00	0.06	0.04	0.08
3,4-Dihydroxybutanoic	0.00	0.00	0.15	0.05	0.20
β-D-Glucometasaccharinic (3-deoxy-D-*ribo*-hexonic)	0.00	0.02	0.32	trace	0.07
β-D-Glucometasaccharinic (3-deoxy-D-*arabino*-hexonic)	0.00	0.04	0.36	0.02	0.06
α-D-Isosaccharinic	0.00	trace	trace	0.00	trace
α-D-Glucosaccharinic	0.00	0.00	trace	0.00	0.02
Various					
Pyruvic (2-oxopropanoic)	0.30	0.11	0.43	0.49	1.06
Laevulinic (4-oxopentanoic)	0.00	0.00	trace	0.00	0.03
Furan-2-carboxylic (pyromucic, furoic)	0.00	0.03	0.07	0.00	0.03
5-Hydroxymethylfuran-2-carboxylic	0.00	0.00	0.15	0.00	0.06

[a] At 170 °C.

4.3.3 Other sugar derivatives

4.3.3.1 Glycosides

The reaction of the hemiacetal hydroxyl group of sugars with hydroxy compounds yields sugar glycosides, also known as *O*-glycosides. For example, glycosides derived from glucose are glucosides (**4-23**) and mannose gives rise to mannosides. The non-sugar part of the glycoside molecule is generally termed **aglycone** (formerly also referred to as genine). Aglycones are mostly phenols, alicyclic triterpenic alcohols, and other steroids, but can also be simple aliphatic alcohols, such as ethanol. An example is ethyl β-D-glucopyranoside, found in berries of common sea buckthorn (*Hippophaë rhamnoides*, Elaeagnaceae). Glycoside is then called a **heteroglycoside**. If the reacting hydroxy compound is another sugar, the resulting product is a **homoglycoside**. Homoglycosides are all oligosaccharides and polysaccharides. They are formed mainly in reactions catalysed by enzymes, but also in other reactions, for example by reversion of monosaccharides.

Glycosides can be also linked by an *N*-, *S*-, or *C*- glycosidic bond. Nitrogen analogues of glycosides are *N*-glycosides (glycosylamines, see Section 4.3.3.7). They occur as natural food compounds and are also formed in the Maillard reaction. *S*-Glycosides, known as thioglycosides (**4-23**), can also be found; such examples are glucosinolates. Glycosides also include *C*-glycosides (*C*-glycosyl compound), which are actually derivatives of anhydroalditols. Formally, *C*-glycosides resemble glycosides, but they are not hydrolysed by enzymes or acids to form the parent aglycone and sugar. Examples of these glycosides include some natural dyes, such as carminic acid, carthamin, mangiferin, and aloin. Some *C*-glycosides arise in the Maillard reaction.

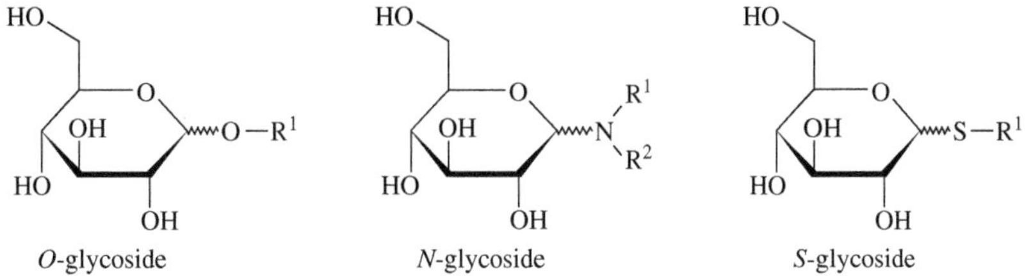

O-glycoside   *N*-glycoside   *S*-glycoside

4-23, glycosides derived from glucose

4.3.3.2 Ethers

Ethers of sugars occur in foods as minor structural units of some polysaccharides. For example, 4-*O*-methyl-D-glucuronic acid (**4-24**) is a component of hemicelluloses (arabinoxylans) and other saccharides, whilst 4-*O*-methyl-D-galacturonic acid, 2-*O*-methyl-D-xylose, and 2-*O*-methyl-L-fucose are components of pectin. Starch and cellulose ethers are used as food hydrocolloids.

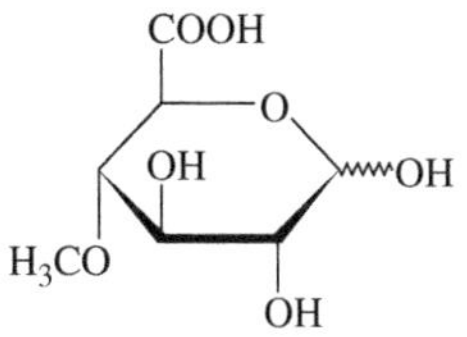

4-24, 4-*O*-methyl-D-glucuronic acid

4.3.3.3 Esters

Esters of saccharides are natural components of virtually all foods. The most common sugar esters are phosphoric acid esters (phosphates), which serve as key intermediates in the metabolism of animals, plants, and microorganisms. Examples are phosphates of D-glucose (**4-25**), D-glucose 1-phosphate, and D-glucose 6-phosphate, but many other sugar phosphates also take part in the metabolic processes. Phosphate esters also include metabolically active sugar nucleotides, such as adenosine diphospho-D-glucose (ADP-D-Glc), which has a role in the biosynthesis of starch. Of particular importance is D-ribose 5-phosphate, a component of pyrimidine and purine nucleotides in ribonucleic acid (RNA), free nucleotides (such as ATP), and many important cofactors of oxidoreductases and transferases, such as NADH and $FADH_2$.

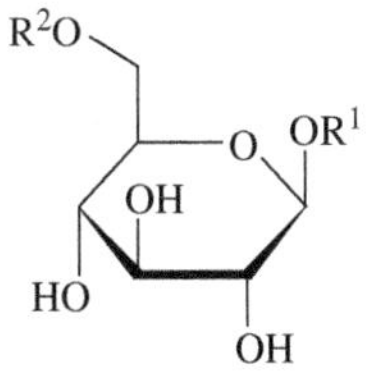

4-25, β-D-glucopyranose 1-phosphate, $R^1 = PO_3H_2$, $R^2 = H$
β-D-glucopyranose 6-phosphate, $R^1 = H$, $R^2 = PO_3H_2$

Esters with sulfuric acid are common components in proteoglycans (mucoproteins) of animal tissues as building blocks of mucopolysaccharides. They are also components of seaweed polysaccharides agar and carrageenan.

Acetates of sugars are often found in various glycosides (e.g. in saponins). Sweet–bitter-tasting hexose acetates, 6-*O*-acetyl-β-D-glucopyranose and 1-*O*-acetyl-β-D-fructopyranose, were identified in balsamic vinegar, where they are formed by esterification of the respective saccharides with acetic acid.

The esters of D-glucose with aromatic acids (**4-26**) are also widespread compounds. For example, a common metabolite in plants is 1-*O*-benzoyl-β-D-glucopyranose, which occurs, for example, in cranberries (*Vaccinium vitis-idaea*, Ericaceae), together with 6-*O*-benzoyl-β-D-glucopyranose, known as vaccinin. (*E*)-1-*O*-Cinnamoyl-β-D-glucopyranose has been found in strawberries (*Fragaria vesca*, Rosaceae) and cashew apples (*Anacardium occidentale*, Anacardiaceae). An example of a glycoside with antioxidant effects is (*E*)-1-*O*-sinapoyl-β-D-glucopyranose, which occurs in germinating seeds of cruciferous plants (Brassicaceae), where it is produced from sinapine. Various esters of sucrose with *p*-coumaric and ferulic acids, such as 1,3,6,6′-tetraferuloylsaccharose, known as taroside (**8-27**), have been found in tartary buckwheat (*Fagopyrum tataricum*, Polygonaceae). An effective antioxidant in olives (*Olea europea*, Oleaceae) and some other plants is verbascoside (**8-28**), a glycoside derived from 3,4-dihydroxyphenylethanol (hydroxytyrosol) and esterified with caffeic acid. The sugar component is the disaccharide 6-deoxy-β-D-glucopyranosyl-(1 → 3)-β-D-glucopyranose (chinovose), esterified by (*E*)-caffeic acid. Hydrolysable tannins are esters of glucose with gallic acid and its derivatives. Synthetic esters of some sugars and sugar alcohols with fatty acids are used as additives, primarily as emulsifiers.

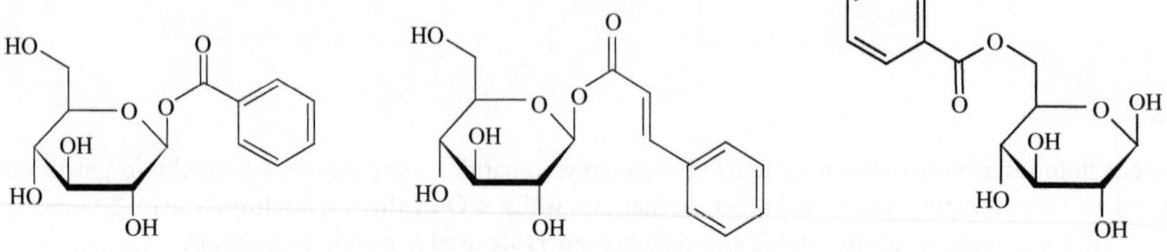

1-*O*-benzoyl-β-D-glucopyranose   (*E*)-1-*O*-cinnamoyl-β-D-glucopyranose   vaccinin

4-26, esters of glucose with aromatic carboxylic acids

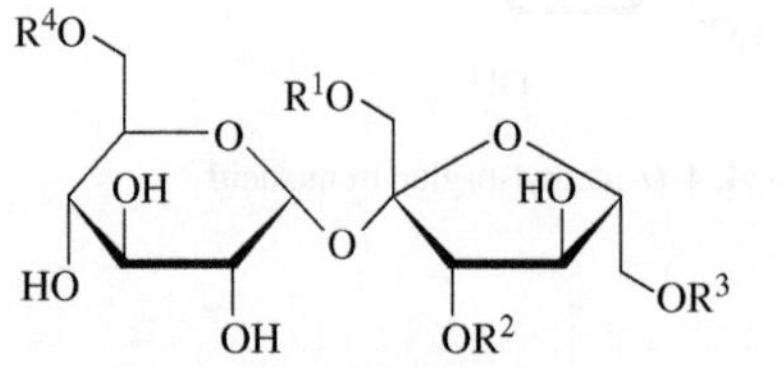

4-27, esters of saccharose with aromatic carboxylic acids

R^1 = feruloyl, R^2 = R^3 = R^4 = *p*-coumaroyl
R^1 = R^2 = feruloyl, R^3 = R^4 = *p*-coumaroyl
R^1 = R^2 = R^3 = feruloyl, R^4 = *p*-coumaroyl
R^1 = R^2 = R^3 = R^4 = feruloyl

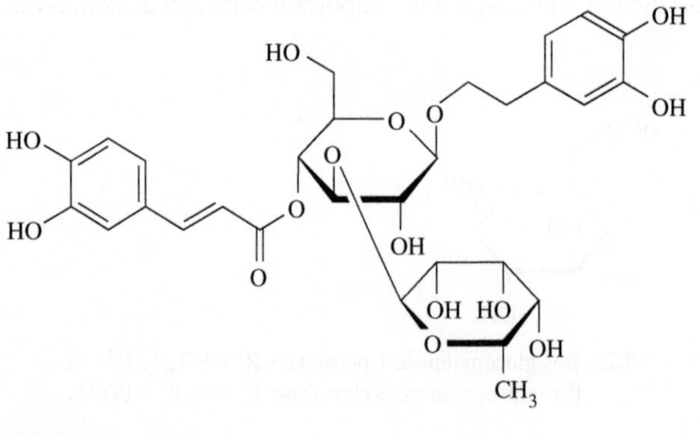

4-28, verbascoside

4.3.3.4 Ketoaldoses and diketoses

Monosaccharides containing both the aldehyde group and the keto group in the molecule are called **ketoaldoses** (aldoketoses or aldosuloses). If monosaccharides contain two keto groups, they are called **diketoses**. The names of ketoaldoses are formed by replacing the suffix -e in the name of aldose by the suffix -ulose and using a suitable prefix (e.g. pento- or hexo-). The hexosulose **4-29** is D-*arabino*-hexos-2-ulose (formerly also referred to as D-glucosulose or D-glucosone). The hexosulose arising from D-galactose during milk processing is D-*lyxo*-hexos-2-ulose (D-galactosulose or D-galactosone). The names of diketoses are formed by replacing the suffix -se by the suffix -diulose. Sugars containing three keto groups are triuloses. The diketose in formula **4-29** is hexodiulose, which has the systematic name 1-deoxy-D-*erythro*-hexo-2,3-diulose. Deoxyglycosuloses are the key Maillard reaction intermediates.

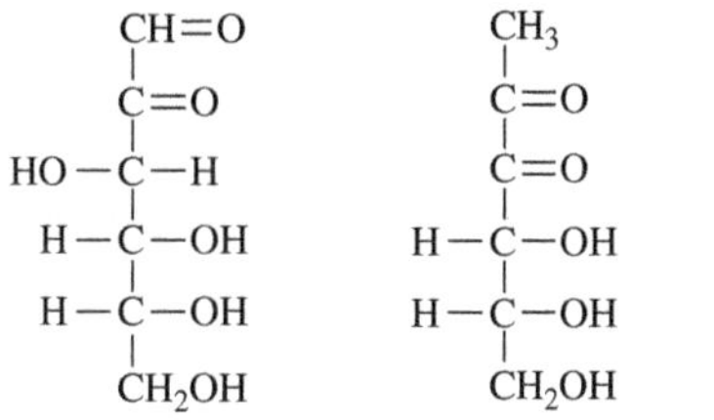

D-arabino-   1-deoxy-D-*erythro*-
hexos-2-ulose   hexo-2,3-diulose

4-29, acyclic forms of ketoaldoses and diketoses

Hydroxymethylglyoxal can also be considered a ketoaldose, methylglyoxal is deoxyketoaldose, and biacetyl is dideoxydiketose. All these compounds are formed as degradation products of sugars. Other products of sugar fragmentation include deoxyaldose 2-hydroxypropanal (lactic acid aldehyde), deoxyketose hydroxyacetone (acetol or pyruvic acid aldehyde), and dideoxyketose acetoin (3-hydroxybutan-2-one).

4.3.3.5 Anhydrosugars

Sugar anhydrides or **anhydrosugars** were previously generally termed **glycosans**. Anhydrosugars formed from glucose were known as glucosans, anhydrosugars derived from mannose were mannosans, and anhydrosugars arising from fructose were fructosans. Anhydrosugars are formed by intramolecular condensation of the hemiacetal hydroxyl group and any other hydroxyl group in the sugar molecule when it is heated in acidic solutions or in solid form. The anhydrosugar derived from D-glucose, 1,6-anhydro-β-D-glucopyranose (also known as β-glucosan or laevoglucosan), along with other anhydrosugars, is therefore a common constituent of caramel.

Polysaccharide agar is composed of 3,6-anhydro-α-L-galactopyranose, and polysaccharide carrageenan of 3,6-anhydro-α-D-galactopyranose.

4.3.3.6 Deoxysugars

Deoxysugars are sugar derivatives in which one or more hydroxyl groups (but not the hemiacetal hydroxyl) are replaced by a hydrogen atom. In foods, deoxysugars are bound in glycosides, glycoproteins, and bacterial lipids. An important deoxysugar occurring in foods is 2-deoxy-D-ribose (formerly known as thyminose or 2-deoxy-D-*erythro-pentofuranose*, **4-30**), which is a constituent of deoxyribonucleic acid (DNA). Another deoxysugar, L-fucose (6-deoxy-L-galactopyranose, **4-30**), occurs as a constituent of milk oligosaccharides. Deoxysugars L-rhamnose (6-deoxy-L-mannopyranose, **4-30**) and D-chinovose (6-deoxy-D-glucopyranose, **4-30**) are constituents of many glycosides. Various deoxysugars, especially deoxyglycosuloses, arise as intermediate products during degradation of monosaccharides in acidic media and as Maillard reaction intermediates.

The properties of deoxysugars are similar to those of sugars, but their glycosides, especially glycosides of 2-deoxysugars, are hydrolysed faster.

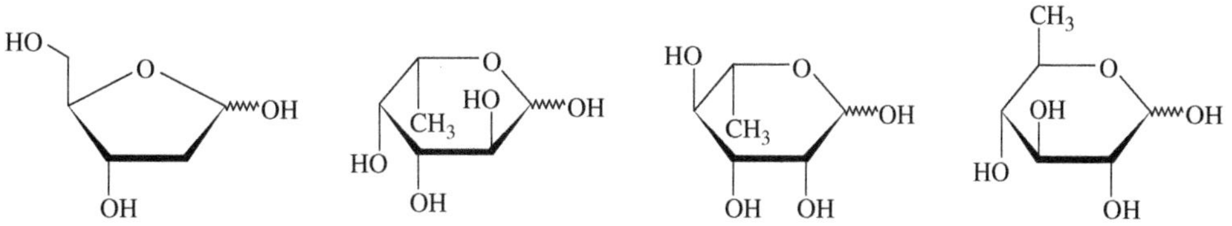

2-deoxy-D-ribose   L-fucose   L-rhamnose   D-chinovose

4-30, examples of deoxysugars

4.3.3.7 Amino derivatives

Hydroxyl groups of saccharides can be replaced by amino groups. Substitution of the hemiacetal hydroxyl group by an amino group yields **N-glycosides** called **glycosylamines**. Aldoses, such as glucose, produce aldosylamines (**4-31**, R = H), and ketoses, such as fructose, produce ketosylamines (**4-31**). Various glycosylamines are natural components of most foods. Common compounds are nucleosides, nucleotides (such as ATP), and their polymers RNA and DNA, which contain D-ribose and 2-deoxy-D-ribose, respectively. Glycosylamines are formed also as primary products of the Maillard reaction and in the reactions of sugars with ammonia. With amines, amino acids, and proteins, sugars yield *N*-substituted glycosylamines, and reactions of glycosylamines with another molecule of sugar give diglycosylamines. Glycosylamines are generally unstable basic compounds that are hydrolysed by acids to the parent sugars and amino compounds, rearranged to aminodeoxysugars or readily decomposed.

Substitution of a group other than the hemiacetal hydroxyl group for an amino group yields aminodeoxy derivatives of monosaccharides, which are commonly called **aminodeoxysugars**, or **amino sugars** for short. The prefix amino in the nomenclature indicates the replacement of a hydrogen atom, therefore aminosugars are derivatives of deoxysugars. Aminosugars are generally known as glycosamines, aminodeoxysugars derived from hexoses are called **hexosamines**, aminodeoxysugars derived from aldoses are **aldosamines**, and **ketosamines** are derived from ketoses. Aminosugars derived from aldoses commonly have the amino group on carbon C-2 and amino sugars derived from ketoses on carbon C-1. Amino sugars are unstable as free bases and decompose to a variety of products. Their salts (such as hydrochlorides) are more stable compounds.

Aldosamines and their *N*-acetyl derivatives are important components of various biologically active oligosaccharides and biopolymers. The most important aldosamines are D-glucosamine, D-galactosamine, and D-mannosamine. D-Glucosamine is also known as D-chitosamine (2-amino-2-deoxy-D-glucose, **4-31**, R = H). *N*-Acetyl-D-glucosamine (**4-31**, R = COCH₃) is a structural unit of milk oligosaccharides, of polysaccharide chitin, and of numerous heteropolysaccharides found in proteoglycans of connective tissue, in synovial fluids, in joints connecting bones, in respiratory tract mucus, in saliva, in glycoproteins of milk, eggs, and blood serum, and in peptidoglycans of bacterial cells. D-Galactosamine, also known as D-chondrosamine (2-amino-2-deoxy-D-galactose), is in the form of the *N*-acetate (*N*-acetyl-D-galactosamine) the building unit of milk oligosaccharides occurring in various connective tissue proteoglycans (mucoproteins) and in glycoproteins of milk and eggs. D-Mannosamine, like *N*-acetyl-D-mannosamine, is a precursor of acetylneuraminic acid.

β-D-glucosylamine   β-D-fructosylamine   β-D-glucosamine   β-D-fructosamine

4-31, examples of glycosylamines and glycosamines

Ketosamines, also known as 1-amino-1-deoxyketoses, have an analogous structure (**4-31**). Some important ketosamines are formed as intermediates of the Maillard reaction.

Biochemically important amino derivatives of sugars that are widespread in animal connective tissues (mainly in glycoproteins and gangliosides) and microorganisms are **sialic acids**. Sialic acids are a group of α-oxoacids that are formed by condensation of *N*-acetylhexosamines with three carbon-carboxylic acids. About 50 sialic acids are known. They are *N*- or *O*-substituted derivatives of neuraminic (5-amino-3,5-dideoxy-D-*glycero*-D-*galacto*-*non*-2-ulopyranosonic) acid (abbreviated as Neu), a monosaccharide with a nine-carbon backbone that does not occur in nature. The neuraminic acid amino group can be acylated by acetic or glycolic acid, and hydroxyl groups, except for the hemiacetal hydroxyl, can be esterified with lactic, sulfuric, or phosphoric acids, or else methylated. The most common compound, which is generally widespread, is *N*-acetylneuraminic acid (5-acetamido-3,5-dideoxy-D-*glycero*-D-*galacto*-non-2-ulopyranosonic acid, also known as lactaminic acid, or Neu5Ac for short; **4-32**), which is biosynthesised from *N*-acetyl-D-mannosamine and pyruvic acid. The cyclic form of this acid (α-anomer), commonly called *N*-acetyl-α-neuraminic acid, is a constituent of many minor milk oligosaccharides and, together with the corresponding *N*-glycolyl derivative, occurs in numerous glycoproteins (e.g. in milk and egg white glycoproteins). Examples of other sialic acids are *N*-glycolylneuraminic acid (Neu5Gc) and *N*-acetyl-9-*O*-acetylneuraminic acid (Neu5,9Ac2), which, however, do not occur in all living organisms.

Structurally related to sialic acids is *N*-acetylmuramic acid (**4-33**), the building unit of peptidoglycans of bacterial cell walls or **mureins**.

4-32, *N*-acetylneuraminic acid

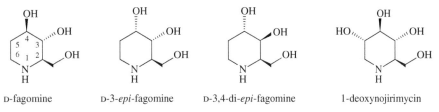

4-33, *N*-acetylmuramic acid

4.3.3.8 Iminosugars

Analogues of pyranoses or furanoses, where the oxygen atom in the ring is replaced by a nitrogen atom, are known as iminosugars, iminosaccharides, or formerly as azasugars. Iminosugars are common components of plants and may be responsible for some of their medicinal properties. The naturally occurring iminosugars also fall under the definition of the polyhydroxylated alkaloids and therefore can be classified as pyrrolidines, piperidines, indolizidines, pyrrolizidines, and nortropanes. Compounds structurally similar to iminosugars are hydroxylated cyclic amino acids, such as proline and pipecolic acid. To date, over 100 compounds have been isolated from natural sources, the majority of which have been from plants.

One of the most common iminosugars is D-fagomine, (2*R*,3*R*,4*R*)-2-(hydroxymethyl)piperidine-3,4-diol, which occurs in common buckwheat (*F. esculentum*, Polygonaceae), cultivated for its grain-like seeds. Its amount in buckwheat groats is 6.7–44 mg/kg. Fagomine (**4-34**) is biosynthesised upon sprouting and is stable during baking, frying, boiling, and fermentation. Buckwheat also contains its diastereomers D-3-*epi*-fagomine and D-3,4-di-*epi*-fagomine (**4-34**). Fagomine is capable of inhibiting the adhesion of potentially pathogenic bacteria to epithelial mucosa and reducing the postprandial (after a meal) blood glucose concentration. It can therefore be used as a dietary ingredient or functional food component to reduce the health risks associated with an excessive intake of fast-digestible carbohydrates or an excess of potentially pathogenic bacteria.

Another example of this class of compounds is 1-deoxynojirimycin (**4-34**), also called moranolin, (2*R*,3*R*,4*R*,5*S*)-2-(hydroxymethyl) piperidine-3,4,5-triol, which was found in herbal tea from mulberry (*Morus alba*) leaves. It acts as an α-glucosidase inhibitor with antidiabetic and antiviral activities. Mulberry leaves also contain D-fagomine and D-3-*epi*-fagomine as minor components.

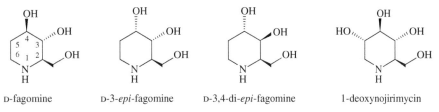

D-fagomine D-3-*epi*-fagomine D-3,4-di-*epi*-fagomine 1-deoxynojirimycin

4-34, selected iminosugars

4.4 Oligosaccharides

Oligosaccharides include those oligomers of monosaccharides in which 2–10 molecules of monosaccharides are connected with one another through glycosidic linkages. Oligosaccharides are therefore glycosides in which the aglycone is a molecule of another saccharide. They are thus homoglycosides. Depending on the number of monose units, oligosaccharides are divided into di-, tri-, tetra-, up to decasaccharides. Monosaccharides in oligosaccharides may occur in the form of pyranoses or furanoses. The most common monosaccharides found are hexoses.

Disaccharide can theoretically be formed by the condensation of the α- or β-anomeric hydroxyl group of a monosaccharide with any hydroxyl group of another monosaccharide. The biosynthesis of glycosides, oligosaccharides, and polysaccharides requires the activation of sugar (its 1-phosphate) through binding to nucleoside bisphosphate. The active forms of glucose, for example, are adenosine diphospho-α-D-glucopyranose, guanidine diphospho-α-D-glucopyranose, and uridine diphospho-α-D-glucopyranose. The condensation of two hemiacetal hydroxyls then yields a disaccharide that does not contain a free anomeric hydroxyl group and is therefore non-reducing. In any other cases, the condensation of two monosaccharides gives a reducing disaccharide, which, like the parent monosaccharides, exhibits mutarotation in solution and thus occurs as the α- or β-anomer. Connecting monosaccharides to disaccharides generates trisaccharides, and subsequently higher oligosaccharides.

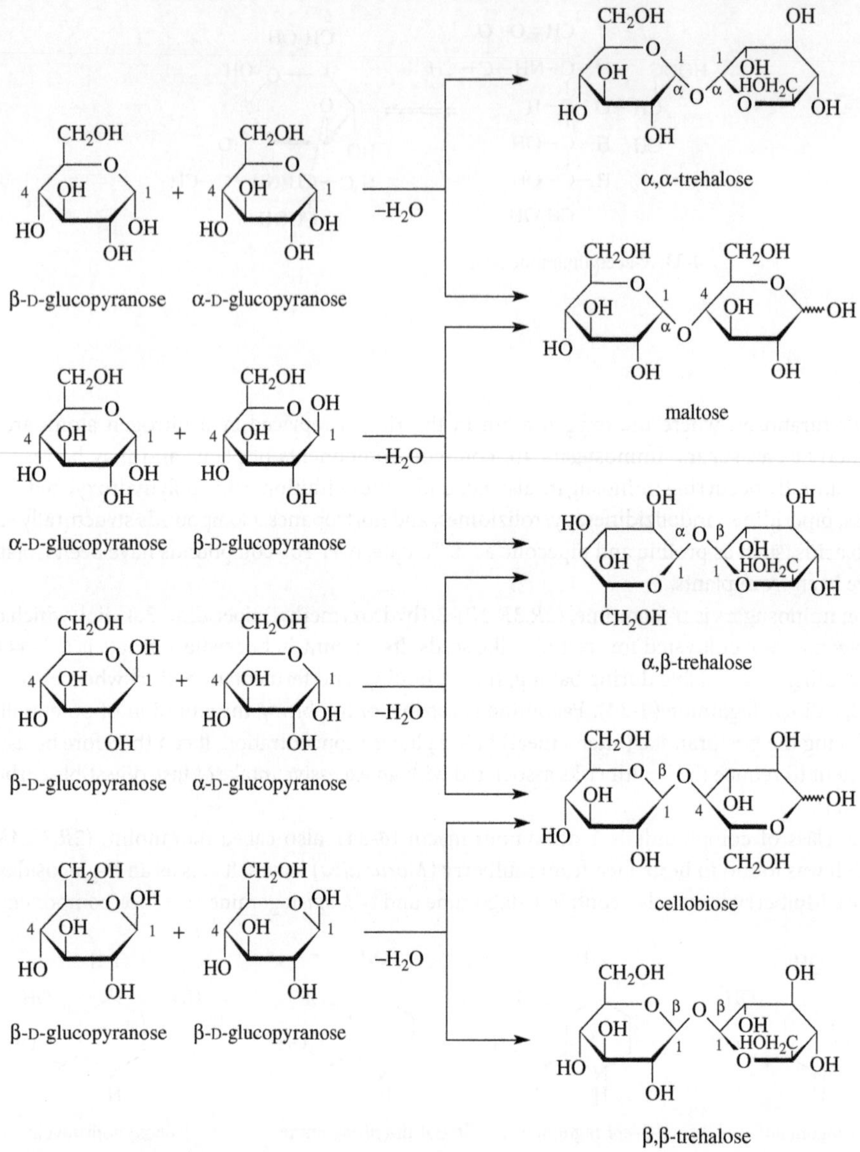

Figure 4.5 Condensation products of α-D-glucopyranose and β-D-glucopyranose.

As an example, Figure 4.5 lists all the hypothetical condensation products of α-D-glucopyranose and β-D-glucopyranose via the anomeric hydroxyl groups with hydroxyls in position C-4. The condensation of two molecules of α-D-glucopyranose through the hemiacetal hydroxyl groups yields non-reducing disaccharide known as α,α-trehalose (Figure 4.5). The condensation of the hemiacetal hydroxyl of one molecule with the hydroxyl group of the second molecule of α-D-glucopyranose at C-4 gives reducing disaccharide maltose, its α-anomer, which partly yields β-anomer by mutarotation.

Maltose is also produced by condensation of the anomeric hydroxyl of α-D-glucopyranose with the hydroxyl at C-4 of β-D-glucopyranose when mutarotation again gives the equilibrium mixture of α- and β-anomers of maltose. Condensation of both hemiacetal hydroxyl groups yields α,β-trehalose, known as neotrehalose. Condensation of the anomeric hydroxyl of β-D-glucopyranose with the hydroxyl at C-4 of α- or β-D-glucopyranose gives the reducing disaccharide cellobiose. Condensation of both anomeric hydroxyls of β-D-glucopyranose gives β-β-trehalose, known as isotrehalose. Eight other reducing sugars can be formed by condensation of the anomeric hydroxyl groups of α-D-glucopyranose and β-D-glucopyranose with hydroxyl groups at C-2, C-3, and C-6 of another molecule of D-glucose (see Table 4.11).

In the nomenclature of oligosaccharides, trivial names are still used for a number of substances (e.g. maltose, cellobiose, trehalose; Figure 4.5). In the case of reducing disaccharides, the systematic nomenclature is based on the name of the monosaccharide with a free hemiacetal hydroxyl group, which is preceded by the name of the substituting monosaccharide, and the corresponding anomeric configuration is

Table 4.11 Overview of important glucooligosaccharides.

Trivial name	Abbreviated notation	Occurrence
Disaccharides		
Trehalose (α,α-trehalose)	α-D-Glc$p$-(1 ↔ 1)-α-D-Glc$p$	Yeast, mushrooms, honey
Neotrehalose (α,β-trehalose)	α-D-Glc$p$-(1 ↔ 1)-β-D-Glc$p$	Honey, koji[a]
Isotrehalose (β,β-trehalose)	β-D-Glc$p$-(1 ↔ 1)-P-D-Glc$p$	Honey, glucose reversion
Kojibiose	α-D-Glc$p$-(1 → 2)-D-Glc$p$	Honey, koji[a]
Nigerose (sakebiose)	α-D-Glc$p$-(1 → 3)-D-Glc$p$	Honey, beer
Maltose	α-D-Glc$p$-(1 → 4)-D-Glc$p$	Starch structural unit, starch syrup, honey, malt, sugar beet
Isomaltose (brachiose)	α-D-Glc$p$-(1 → 6)-D-Glc$p$	Starch structural unit, starch syrup, honey
Sophorose	β-D-Glc$p$-(1 → 2)-D-Glc$p$	Glucose reversion, glycosides
Laminaribiose	β-D-Glc$p$-(1 → 3)-D-Glc$p$	Honey
Cellobiose	β-D-Glc$p$-(1 → 4)-D-Glc$p$	Cellulose structural unit, honey
Gentiobiose	β-D-Glc$p$-(1 → 6)-D-Glc$p$	Honey, glycosides
Trisaccharides		
Maltotriose	α-D-Glc$p$-(1 → 4)-α-D-Glc$p$-(1 → 4)-D-Glc$p$	Starch syrup, honey
Cellotriose	β-D-Glc$p$-(1 → 4)-β-D-Glc$p$-(1 → 4)-D-Glc$p$	Cellulose degradation
3-α-isomaltosylglucose	α-D-Glc$p$-(1 → 6)-α-D-Glc$p$-(1 → 3)-D-Glc$p$	Honey
Isomaltotriose (dextrantriose)	α-D-Glc$p$-(1 → 6)-α-D-Glc$p$-(1 → 6)-D-Glc$p$	Honey
Centose	α-D-Glc$p$-(1 → 4)-α-D-Glc$p$-(1 → 2)-D-Glc$p$	Honey
Panose	α-D-Glc$p$-(1 → 6)-α-D-Glc$p$-(1 → 4)-D-Glc$p$	Amylopectin degradation, honey
Isopanose	α-D-Glc$p$-(1 → 4)-α-D-Glc$p$-(1 → 6)-D-Glc$p$	Honey
Neapolitanose	β-D-Glc$p$-(1 → 2)-[β-D-Glc$p$-(1 → 6)]-D-Glc$p$	Glycosides
Tetrasaccharides		
Maltotetraose	β-D-Glc$p$-(1 → 4)-[α-D-Glc$p$-(1 → 4)]$_2$-D-Glc$p$	Starch syrup
Isomaltotetraose	α-D-Glc$p$-(1 → 6)-[α-D-Glc$p$-(1 → 6)]$_2$-D-Glc$p$	Honey

[a]Koji is a mixture of soybean meal and wheat meal fermented by the fungi *Aspergillus oryzae* and *A. soyae*. It is an intermediate in the production of soy sauce (shoyu).

indicated (α- or β-). Disaccharide maltose is generally **glycosylglycose** and is referred to as *O*-α-D-glucopyranosyl-(1 → 4)-D-glucopyranose. The previously used name was 4-*O*-α-D-glucopyranosyl-D-glucopyranose. The molecule of α-D-glucopyranose is bound by the hemiacetal hydroxyl group (at carbon C-1) and the *O*-glycoside to the hydroxyl group at C-4 (which is marked with numbers 1 → 4) to the second molecule of D-glucopyranose (α- or β-). In the case of the α-anomer of maltose, the trivial name α-maltose or the systematic name *O*-α-D-glucopyranosyl-(1 → 4)-α-D-glucopyranose can be used. The abbreviated name of maltose can be written as follows: α-D-Glc$p$-(1 → 4)-D-Glc$p$ (abbreviated names of oligosaccharides are formed from trivial names of monosaccharide units). The systematic name of cellobiose (Figure 4.5) is *O*-β-D-glucopyranosyl-(1 → 4)-D-glucopyranose, or β-D-Glc$p$-(1 → 4)-D-Glc$p$ for short.

Cellobiose, as well as maltose, is a reducing sugar and therefore occurs in the form of α- and β-anomers.

Non-reducing disaccharides are actually double glycosides, and are therefore called **glycosylglycosides**. Disaccharide α,α-trehalose (Figure 4.5) is α-D-glucopyranosyl-α-D-glucopyranoside, or α-D-Glc$p$-(1 ↔ 1)-α-D-Glc$p$ for short.

Disaccharides and higher oligosaccharides exist in the form of different conformational isomers, which differ in the energy state given by the rotation of molecules of monosaccharides around the glycosidic bond; therefore, the conformational isomers differ in their various spatial orientations. They can be characterised as peptides by torsion angles φ and ψ. Equilibrium at a given temperature exists for the populations of individual conformers. The predominant species are the conformers with the lower free energy content. Most oligosaccharides have the same stable and sterically most favourable conformers in both aqueous solutions and the solid (crystal) state. Conformers of disaccharides are commonly stabilised by hydrogen bonds between adjacent hydroxyl groups or between hydroxyl groups and the oxygen atom of pyran or furan rings.

Foods (such as fruits, vegetables, milk, and honey) contain a large number of free and bound oligosaccharides, which are natural food ingredients. Oligosaccharides are usually composed of a number of common monosaccharides of the D-series (glucose, fructose, galactose, and mannose) in various combinations. Aldoses are present in the form of pyranoses and fructose as furanose. Less common sugars, such as 6-deoxyhexose (methylpentose) L-rhamnose and the pentoses L-arabinose and D-xylose, are also present in some oligosaccharides.

The most common oligosaccharides present as natural components of foods are oligomers of D-glucose known as **glucooligosaccharides**, in which glucose is the sole or predominant monosaccharide. Another important group are the **fructooligosaccharides**, which contain only D-fructose or D-fructose and D-glucose. The dominant disaccharide of this type in foods is saccharose (sucrose). Equally important are the **galactooligosaccharides**, which consist of D-galactose, D-glucose, D-fructose, and sometimes also of other monosaccharides.

The human organism can hydrolyse some oligosaccharides, such as saccharose, lactose, maltose, and isomaltose-type oligosaccharides, to monosaccharides by the action of the hydrolytic enzymes maltase, isomaltase, saccharase (sucrase), and lactase (see Section 4.2.3). Utilisable oligosaccharides have a similar effect on blood glucose level as the corresponding monosaccharides. Saccharose and maltose exhibit strong cariogenic effects, whilst the effects of lactose are weaker. Lactose has some laxative effects.

In recent years, a number of oligosaccharides have been produced commercially and used as food additives. Incorporation of the non-digestible, low-glycaemic oligosaccharides into the regular diet provides a lot of health benefits. Currently, more than 10 different types of oligosaccharides are produced. The functional non-digestible oligosaccharides include lactulose, fructooligosaccharides, galactooligosaccharides, soybean oligosaccharides, lactosucrose, isomaltooligosaccharides, glucooligosaccharides, xylooligosaccharides, gentiooligosaccharides, arabinoxylan oligosaccharides, mannan oligosaccharides, pectin-derived acidic oligosaccharides, chitooligosaccharide, agarooligosaccharide, human milk oligosaccharide, cyclodextrins, xanthan-derived oligosaccharides, and alginate-derived oligosaccharide. Common manufacturing methods are hydrolysis of polysaccharides (starch, inulin, xylans, and other polysaccharides) and chemical or enzymatic polymerisation from naturally present disaccharide (e.g. saccharose and lactose) substrates. For example, saccharose is a raw material for the production of palatinose (isomaltulose), glycosylsaccharose, lactosaccharose, and fructooligosaccharides. Lactulose, lactosaccharose, and galactooligosaccharides are produced from lactose. Starch is the raw material for the production of maltooligosaccharides, isomaltooligosaccharides, cyclodextrins, and gentiooligosaccharides. Fructooligosaccharides are produced from inulin, and xylooligosaccharides from xylans. Feruloylated oligosaccharides are derived by hydrolysis of arabinoxylans in cereal bran. Mannan oligosaccharides can be obtained from *Gleditsia sinensis* (locust tree) gum using enzymatic hydrolysis. Hydrogenation of some disaccharides (lactose and maltose) yields the corresponding polyols. The oligosaccharides and polyols obtained are water-soluble substances having a slightly sweet taste. Their solutions usually have a higher viscosity than solutions of natural monosaccharides and disaccharides, and only 30–60% of the sweetness of saccharose. These sugars therefore find use as fillers for low-energy products (with reduced monosaccharide and saccharose content) and for products that require increased viscosity and increased water binding capacity as a means of preventing drying, and also have the desired physiological effects. They are not utilised by microflora of the mouth and are therefore non-cariogenic carbohydrates. Many of them are indigestible and can be used in the production of low-energy foods and foods for diabetics. Some oligosaccharides exhibit **prebiotic** effects, as they selectively stimulate the growth and metabolism of desirable microflora of the colon.

4.4.1 Glucooligosaccharides

Glucooligosaccharides include a large number of disaccharides, trisaccharides, and higher oligosaccharides, which commonly (but usually in small quantities) occur in many foods or are used as food additives (Table 4.11). The most important representative of these oligosaccharides is the disaccharide maltose.

4.4.1.1 Maltose

4.4.1.1.1 Structure and nomenclature

A molecule of the reducing sugar maltose (also called malt sugar, as it is the result of starch hydrolysis in malt), known by the systematic name *O*-α-D-glucopyranosyl-(1 → 4)-D-glucopyranose (Figure 4.5), is stabilised in aqueous solution by hydrogen bonds between the hydroxymethyl group of one molecule of glucose (C-6) and the hydroxyl group at carbon C-3 of the second molecule of glucose (**4-35**). In crystals of maltose (and also in solutions in non-aqueous media), maltose is present as a different conformer, which is stabilised by hydrogen bonds between the hydroxy group at the C-2 carbon of one glucose molecule and the hydroxyl at the C-3 of the second molecule of glucose (**4-36**).

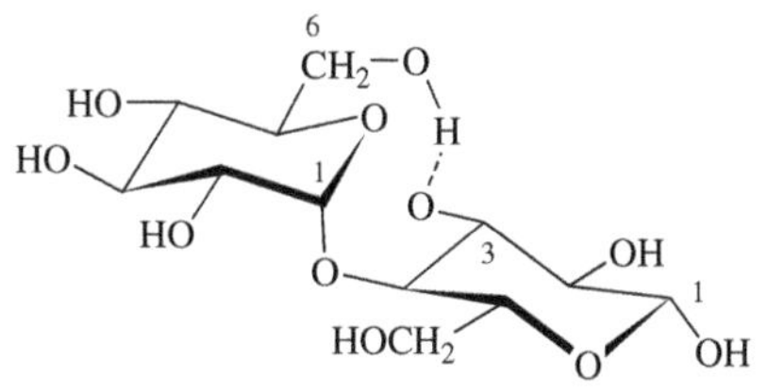

4-35, conformation of maltose in aqueous solutions

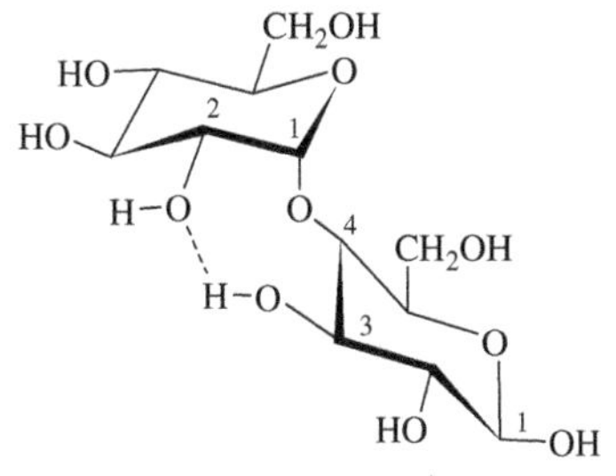

4-36, conformation of anhydrous crystalline maltose

4.4.1.1.2 Occurrence

Small (but sometimes also relatively large) amounts of maltose are typically found in most foods. Bread dough contains maltose formed by hydrolysis of starch by yeast (*Saccharomyces cerevisiae*) enzymes and by enzymes present in the flour. At the beginning of fermentation, its content increases due to the hydrolysis of starch. After the preferred fermentation of glucose and fructose, the maltose content decreases, as it is also partially fermented by yeast. The amount of maltose in bread is 1.7–4.3%.

Maltose, a product of the enzymatic hydrolysis of starch, is present in germinating seeds and thus also in germinating barley and barley malt (hence the name 'malt sugar'). Malt may be partly substituted by starch-rich materials, such as rice, maize, or wheat. Brewer's yeast is able to ferment about 95% of maltose and 80% of maltotriose.

Maltose is present at a relatively large concentration in honey (2.7–16%, 7.3% on average), together with fructose (27.3–44.3%), glucose (22.0–40.8%), saccharose (0.3–7.6%), and other oligosaccharides. Higher quantities of maltose are also found in cooked cereals (e.g. 4–20 g/kg in spaghetti), vegetables (e.g. 2 g/kg in raw broccoli and 17 g/kg in ketchup), and fruits (e.g. 1 g/kg in raw and 9–14 g/kg in canned peaches). Starch hydrolysates, such as maltose syrups, contain maltose at levels of up to 85%.

4.4.1.1.3 Physiology and nutrition

The relative sweetness of maltose is 30–60% of saccharose sweetness (see Table 8.37). After hydrolysis to glucose by maltase (α-D-glucoside glucohydrolase), maltose is a utilisable sugar. Maltose intake in foods has a significant effect on blood glucose levels and insulin secretion. It is a cariogenic sugar.

4.4.1.1.4 Use

Maltose and various products containing maltose and maltooligosaccharides (such as maltodextrins, glucose, and maltose syrups) are obtained from starch, often by the combined effects of mineral acids and amylolytic enzymes. Maltose is obtained from maltose syrups. A product of maltose hydrogenation (at the reducing end of the glucose molecule) is maltitol, O-α-D-glucopyranosyl-(1 → 4)-D-glucitol (**4-37**). Its sweetening power is about 90% of that of saccharose. Maltitol has a very little impact on blood sugar level and has a weak laxative effect. Isomerisation in an alkaline medium (glucose at the reducing end of the molecule isomerises to fructose) produces disaccharide maltulose (**4-37**).

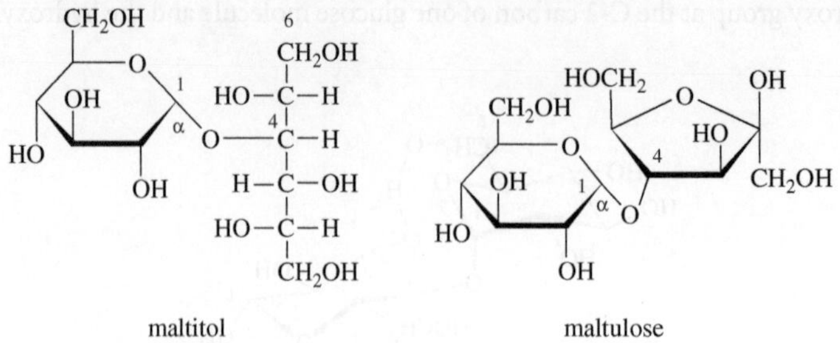

maltitol   maltulose

4-37, sugars produced from maltose

4.4.1.2 Other glucooligosaccharides

Many different glucooligosaccharides are present in honey as minor sugars (Table 4.12). Their structures are shown in Table 4.11.

An important non-reducing disaccharide is α,α-trehalose, which occurs in many plants and invertebrate organisms (bacteria, fungi, worms, crustaceans), where it serves as an energy source and has a protective function in stress induced by heat and drought. In fruits and vegetables, trehalose is found in very small amounts (see Table 4.5), but larger levels are located in fungi. In the cultivated common mushroom (*A. bisporus*), the content of α,α-trehalose is 0.5% in very young fruiting bodies and 0.09% in fruiting bodies at the end of their development, but it constitutes 21% (d.w.) in the charbonnier mushroom (*T. portentosum*). Trehalose, produced by enzyme-catalysed process from starch, is used as a multifunctional food additive for its aroma-stabilising effects. It has about half the sweetness of saccharose. As a non-reducing sugar, trehalose is a quite stable and non-reactive substance that does not enter the Maillard reaction during thermal processing of foods.

Numerous disaccharides and higher oligosaccharides containing glucose and other monosaccharides (e.g. arabinose, xylose, and rhamnose) are components of many glycosides. An overview of the major reducing disaccharides occurring in anthocyanins and other glycosides is shown in Table 4.13.

Table 4.12 Composition of minor disaccharides of honey.

Disaccharides	Content (g/kg)	Trisaccharides	Content (g/kg)
Kojibiose	3.0	Erlose	1.6
Turanose	1.7	Theanderose	0.99
Isomaltose	1.6	Panose	0.91
Saccharose	1.4	Maltotriose	0.69
Maltulose	1.1	1-Kestose	0.33
Nigerose	0.62	Isomaltotriose	0.22
Isotrehalose	0.40	Melezitose	0.11
Gentiobiose	0.15	Isopanose	0.09
Laminaribiose	0.03	Centose	0.02

Table 4.13 Important disaccharides occurring in glycosides.

Trivial name	Short notation	Trivial name	Short notation
Vicianose	β-L-Ara*p*-(1 → 6)-D-Glc*p*	Robinobiose	α-L-Rha*p*-(1 → 3)-D-Glc*p*
Sambubiose	β-D-Xyl*p*-(1 → 2)-D-Glc*p*	Chacobiose	α-L-Rha*p*-(1 → 4)-D-Glc*p*
Primeverose	β-D-Xyl*p*-(1 → 6)-D-Glc*p*	Rutinose	α-L-Rha*p*-(1 → 6)-D-Glc*p*
Neohesperidose	α-L-Rha*p*-(1 → 2)-D-Glc*p*	Solabiose	β-D-Glc*p*-(1 → 3)-β-D-Gal*p*

4.4.2 Fructooligosaccharides

4.4.2.1 Saccharose

4.4.2.1.1 Structure and nomenclature

Saccharose or sucrose (**4-38**), β-D-fructofuranosyl-α-D-glucopyranoside (sugars are given in alphabetical order), also called beet, cane, or table sugar, is the most important representative of non-reducing disaccharides. In aqueous solutions and in the solid state, the conformer stabilised by two hydrogen bonds predominates (**4-39**). The first bond is between the hydroxyl group on carbon C-2 of glucose ($^4C_1$-pyranose) and the hydroxyl group at C-1 of fructose ($^4T_3$-furanose); the second is between the pyranose ring oxygen and the C-6 hydroxyl group of the fructose.

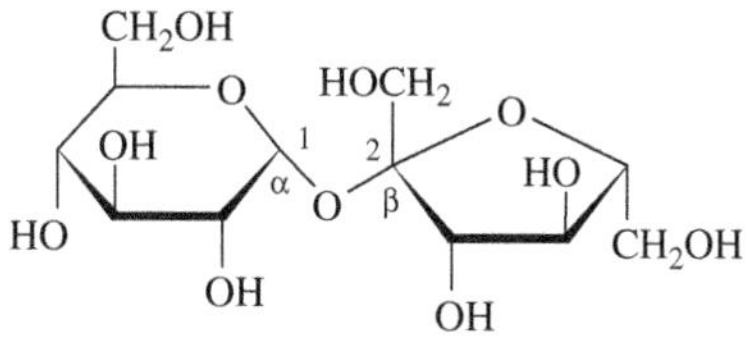

4-38, saccharose

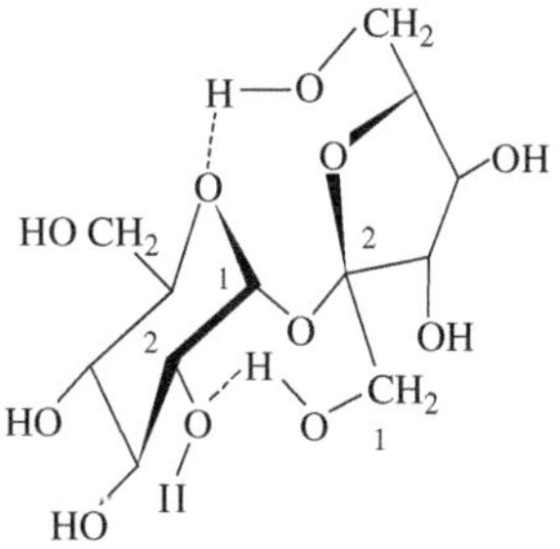

4-39, conformation of saccharose
in aqueous solutions

4.4.2.1.2 Occurrence and production

Saccharose is a very widespread sugar synthesised by most eukaryotic and some prokaryotic organisms. As the main product of photosynthesis, saccharose is commonly present in higher plants, where it has a role in transport and energy reserves and is used in signal transmission and in stress situations. As the energy reserve in plants, saccharose is mobilised under the catalysis of the enzyme invertase to fructose and glucose and is used for seed germination and plant growth and during fruit ripening. Another enzyme that plays a role in the catabolism of sucrose is sucrose synthetase, which splits sucrose into fructose and UDP-α-D-glucose, used for the biosynthesis of starch.

Saccharose is present in larger amounts in the vegetative parts of plants, such as leaves and stems (sugar cane contains 12–26%, sugar-rich varieties called sweet maize 12–17%, sweet millet 7–15% sucrose) and fruits (e.g. apples, oranges, apricots, peaches, pineapples, and papayas), which contain up to about 8% saccharose (Table 4.4). Some fruits, however, do not contain sucrose (cherries, grapes, figs), because it is hydrolysed during fruit ripening. Vegetables commonly contain 0.1–12% saccharose (e.g. onion 10–11% and sugar beet 3–20%). The content of saccharose in Jerusalem artichokes is 2–3%. The saccharose content in beans is given in Table 4.17. Sucrose is the transport form from the leaves to the tubers and is therefore the major sugar in immature potato tubers. The amount of saccharose in mature tubers varies between cultivars and is generally less than 0.3%. If saccharose is low in tubers at harvest, the concentration of all sugars remains low in healthy tubers under storage. Mustard seeds, rapeseeds, and other oilseeds contain about 4%, wheat flour 0.1–0.4%, green coffee 6–7% (roasted coffee about 0.2%), and groundnuts 4–12% saccharose.

The main industrial sources of sucrose are sugar cane (*Saccharum officinarum*, Poaceae) and sugar beet (*Beta vulgaris* group *Altissima*, Amaranthaceae, formerly Chenopodiaceae), the varieties of which today contain 15–20% saccharose (typically 16–17%). In addition to sugar cane and sugar beet, on a smaller scale and locally, other plants are sometimes used as a source of saccharose. In Algeria and Iraq, date sugar is obtained from fruits of date palm dates (*Phoenix dactylifera*, Arecaceae) that contain up to 81% saccharose in the dry state. In India, Cambodia, the Philippines, and elsewhere, table sugar is traditionally made from the juice of various palms of the genus *Borassus* (e.g. *Borassus flabelliformis*) and of some other species, such as arenga palm (*Arenga pinnata*), date palm, and coconut palm (*Cocos nucifera*). Maple sugar in Canada, the United States, and Japan is produced from the juice of sugar maple (*Acer saccharum*, Sapindaceae), which contains about 5% saccharose. Sweet sorghum syrup, traditionally used as a substitute for sugar, is known as molasses, especially in Southern US states, although it is not true molasses. It is obtained from the stalks of sweet sorghum (*Sorghum bicolor*, Poaceae), containing 12% saccharose.

The production of saccharose from sugar beet consists of several steps, including **diffusion**, juice **purification**, **evaporation**, and **crystallisation**. The processing of thoroughly washed sugar beet starts with slicing the beets into thin chips called **cossettes**. The cossettes are extracted with hot water at an elevated temperature (50–80 °C) in a process called **diffusion** (sulfur dioxide, chlorine, ammonium bisulfite, or commercial biocides are used as disinfectants). The sugar-enriched water, called **raw juice** (or mixed juice containing between 10 and 15% of saccharose) is filtered, heated to 80–85 °C, and cleaned by a process known as **carbonatation**, which removes non-sucrose impurities in the raw juice. Lime milk (a solution of calcium hydroxide) is added to adsorb or adhere to the impurities, and carbon dioxide gas is bubbled through the mixture to precipitate the lime as insoluble calcium carbonate crystals. After filtration (the liming processes with subsequent carbonation and filtration are known as **epuration**), sulfur dioxide is added to the juice (termed **thin juice**) to inhibit the Maillard reaction. The next stage of the process is the juice **evaporation** to **thick juice**, which contains 50–67% saccharose. The filtered juice with added pure sugar, or **standard liquor**, proceeds **crystallisation**. It is boiled under vacuum until it becomes supersaturated and the saccharose crystallises. Sugar crystals in the mixture with liquor, known as **massecuite** (or fillmass), are separated and refined. Raw sugar obtained by crystallisation contains about 96% saccharose, 1.0–1.2% organic compounds, 0.8–1.0% inorganic substances (so called non-sugars), and 1–2% water. It is light yellow to dark brown because it contains a certain amount of molasses. It can be used as raw brown sugar. The refining is mainly done with hot water, and the product is dried and cooled. The liquid separated from the sugar crystals, the **syrup**, is introduced back to the crystallisation operation; the process is repeated so that the yield of saccharose is 58–90%.

The final liquid obtained is **molasses**. Beet molasses contains about 50% sucrose and about 80–85% d.w., of which about 1% accounts for glucose and fructose and 1.2% (or sometimes more than 8%) for raffinose. In small quantities, some other oligosaccharides also occur. Approximately 20% of the molasses weight is accounted for by non-sugars consisting of organic acids and amino acids (particularly glutamic acid, 5-oxopyrrolidine-2-carboxylic acid) and *N,N,N*-trimethylammonioacetate (betaine), the content of which is generally 0.3%, and other substances. The rest (about 12%) is accounted for by inorganic substances such as potassium salts of sugars and other substances. Molasses can be further desugarised using an ion-exchange process called deep molasses desugarisation, used as a substrate for the production of yeast, ethanol, citric, and lactic acid, glycerol, butan-1-ol, acetone, and many other substances, or used in the production of livestock feed or for a range of other purposes. The pressed beet pulp (cossettes) is mixed with molasses that is not desugarised, then dried and used as a constituent of some animal feeds.

Sugar cane molasses contains about 80% d.w., of which 30–40% is saccharose and 30% glucose and fructose. The content of non-sugars is about 10%. The non-sugars include small amounts of raffinose, no betaine, but about 5% aconitic acid, which is not present in beet molasses. The content of inorganic substances is about 8%. Sugar cane molasses is mainly used for production of rum and arrack.

4.4.2.1.3 Physiology and nutrition

Saccharose occurring in foods is hydrolysed to glucose and fructose by invertase, β-D-fructofuranosidase; therefore, it has a large influence on the glucose content in plasma and insulin secretion. A high sucrose intake causes a significant excess of energy and the body synthesises higher amounts of fat. Saccharose is a cariogenic sugar.

4.4.2.1.4 Use

Saccharose is predominantly used as a versatile sweetener, and is also used in the production of invert sugar, fructooligosaccharides (also produced from inulin), palatinose and palatinitol, glycosylsaccharose, and lactosaccharose (also produced from lactose).

Invert sugar   Acid or enzymatic hydrolysis of sucrose, **inversion**, produces an equimolar mixture of D-glucose and D-fructose known as invert sugar. Invert sugar is used as a food additive substance, usually in the form of syrup (relative sweetness is 95–105% of saccharose sweetness). It also serves as a starting material for obtaining D-glucose and D-fructose, the sweeteners mannitol and glucitol, and other substances.

Invert sugar obtained by acid hydrolysis of saccharose contains trace amounts of reversion products of glucose (especially isomaltose and gentiobiose) and fructose (dianhydrides or laevulosans). Invert sugar obtained by enzymatic hydrolysis (due to the transglucosidase or transfructosidase activity of invertase) also contains trace amounts of some less common oligosaccharides.

Fructooligosaccharides   Fructooligosaccharides (**4-40**) are produced from saccharose by the action of microbial and plant fructosyltransferase (transfructosylation) or by the use of the transfructosylase activity of invertase. Fructooligosaccharides are oligomers of the type [β-D-Fruf-(2 → 1)]$_{n-1}$-β-D-Fruf-(2 ↔ 1)-α-D-Glcp, or GF$_n$ (G = glucose, F = fructose, n = degree of polymerisation) for short. The basic member of the homologous series is saccharose, and the degree of polymerisation is 2–4. This mixture of fructooligosaccharides is known as **neosugar**.

Fructooligosaccharides of somewhat different composition are produced by a controlled enzymatic hydrolysis of inulin using β-D-fructofuranosidase (invertase, saccharase) from the fungus *Aspergillus niger*. The hydrolysis products are fructooligosaccharides of the type [β-D-Fruf-(2 → 1)]$_{n-1}$-B-D-Fruf-(2 ↔ 1)-α-D-Glcp, abbreviated GF$_n$, or of the type [β-D-Fruf-(2 → 1)]$_n$, F$_n$, where n = 2–9.

These products are called **oligofructose** (**4-40**). The basic member of the fructooligosaccharides of the GF_n type is the non-reducing trisaccharide 1-kestose (**4-40**), abbreviated to GF_2. Higher oligosaccharides are tetrasaccharide 1,1-nystose (GF_3, **4-40**) and pentasaccharide 1-O-β-D-fructofuranosylnystose (1,1,1,-fructosylnystose, GF_4). The GF_n-type oligomers are non-reducing sugars. An example of a disaccharide consisting of two molecules of fructose is inulobiose, abbreviated F_2.

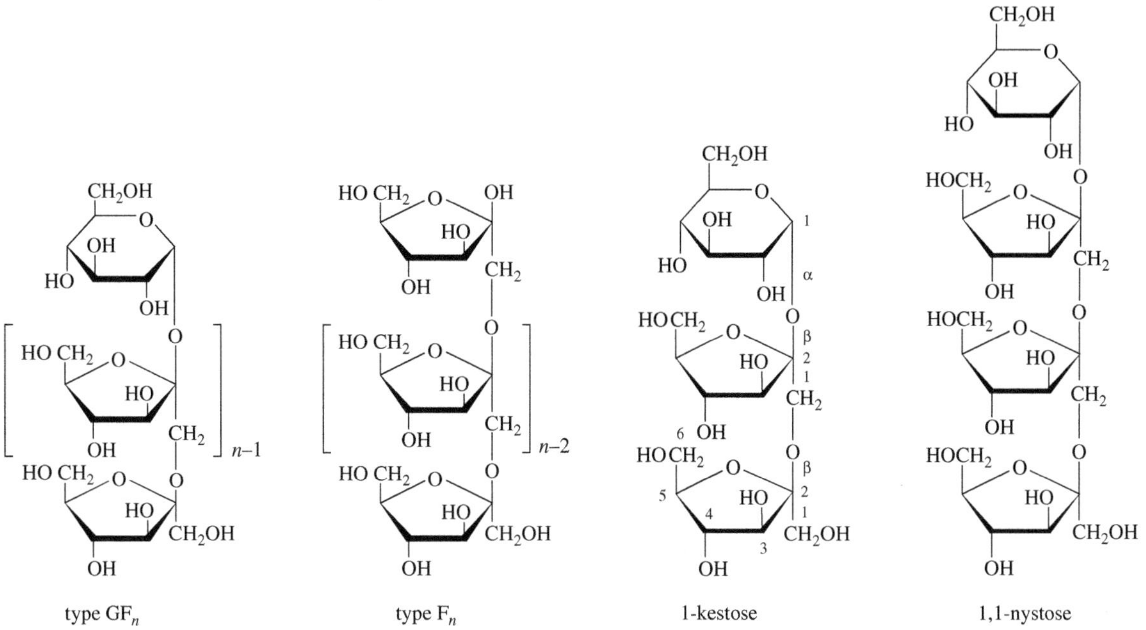

| type GF_n | type F_n | 1-kestose | 1,1-nystose |

4-40, fructooligosaccharides of GF_n and F_n types

Fructooligosaccharides are water-soluble sweet substances that show 40–60% of the sweetness of saccharose. They are not hydrolysed by saccharidases and therefore are classified as soluble fibre. In the colon, however, fructooligosaccharides are fermented by anaerobic bacteria to lower fatty acids (mainly acetic, propionic, and butyric acids), L-lactic acid, and gases (carbon dioxide, methane, and hydrogen). They are therefore also called intestinal food. The yield of energy of fructose bound in oligofructose is only about 25–35% of the energy yield of free fructose (about 4.2–6.3 kJ/g). Fructooligosaccharides are the growth factor of beneficial bifidobacteria (*Bifidobacterium bifidum*). They produce carboxylic acids (mainly acetic acid and lactic acid), and their lower pH is associated with the production of substances with antibiotic and immunomodulatory effects (so-called bifidin and other substances), which results in suppression of the growth of undesirable microflora, such as *Escherichia coli*, *Streptococcus faecalis*, *Streptococcus proteus*, and *Clostridium perfringens*, and also *Staphylococcus aureus*, *Salmonella typhosa*, and some other bacteria, which produce toxic fermentation products (ammonia, amines, nitrosamines, phenols, indoles, and so on). In addition, bifidobacteria produce thiamine, riboflavin, niacin, pyridoxine, folacin, and corrinoids. Fructooligosaccharides and the so-called soluble fibre bind bile acids in the intestine, and thus prevent their reabsorption, which is manifested by lowering of the cholesterol level in blood plasma.

Palatinose Isomerisation of saccharose using immobilised microorganisms *Leuconostoc mesenteroides* and *Protaminobacter rubrum* yields palatinose, also known as isomaltulose (**4-41**; Table 4.14). The ratio of α- to β-anomers of isomaltulose (usually 1 : 4) depends on the reaction conditions. Palatinose is accompanied by a small amount of α-D-fructofuranosyl-(1 ↔ 1)-α-D-glucopyranoside. Intermolecular condensation of these disaccharides yields higher oligosaccharides. Palatinose is a non-cariogenic sugar that stimulates the growth of bifidogenic microflora. The sweetening power of palatinose is about 40% of the sweetening power of saccharose.

Palatinitol Catalytic hydrogenation of palatinose yields palatinitol, which is known by the trade name of **isomalt**. Isomalt is a mixture of O-α-D-glucopyranosyl-(1 → 6)-D-glucitol (isomaltitol, **4-41**) and O-α-D-glucopyranosyl-(1 → 1)-D-mannitol in a ratio of approximately 1 : 1. The sweetening power of this non-cariogenic sugar is about 50% of that of saccharose. It has no effect on blood glucose level, but it is relatively effective laxative.

Glucosylsaccharose Trisaccharide glucosylsaccharose (**4-41**) is produced under the trade name **neosugar** from saccharose or maltose using the enzyme cyclomaltodextrin glucanotransferase. It has about 50% of the sweetening power of sucrose, is less digestible, and is non-cariogenic. It is appreciated for its technological properties, such as reduced crystallisation of saccharose, reduced retrogradation of starch, and low reactivity in the Maillard reaction.

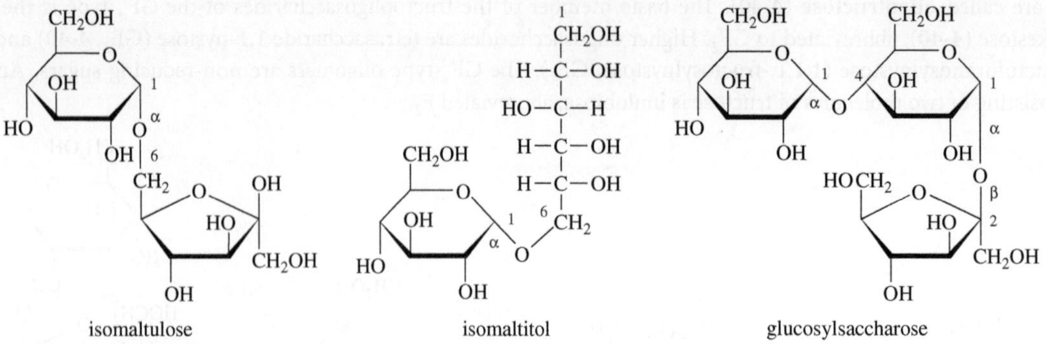

isomaltulose   isomaltitol   glucosylsaccharose

4-41, less common fructooligosaccharides produced from saccharose

4.4.2.2 Other fructooligosaccharides

In addition to saccharose, numerous other disaccharides and higher oligosaccharides derived from glucose and fructose are present in foods as natural components. An overview of relatively commonly occurring oligosaccharides is shown in Table 4.14. Turanose, maltulose, and isomaltulose are examples of reducing disaccharides (sucrose isomers).

Higher fructose oligomers linked by glycosidic bonds β-$(2 \rightarrow 6)$ are found in small amounts in foods. These sugars are mainly natural components of polysaccharides called fructosans (levans or phleines). The simplest substances of this type are the trisaccharides 6-kestose (**4-42**), neokestose (**4-42**), and kelose (**4-42**). 6-Kestose and neokestose occur, for example, as minor components in molasses and invert sugar produced by acid hydrolysis of saccharose (inversion). The reaction by which these oligosaccharides are formed from lower saccharides is called **reversion**.

Table 4.14 Overview of important fructooligosaccharides.

Trivial name	Abbreviated notation	Occurrence
Disaccharides		
Saccharose	β-D-Fruf-$(2 \leftrightarrow 1)$-α-D-Glcp	Honey, sugar beet, sugar cane plants
Turanose	α-D-Glcp-$(1 \rightarrow 3)$-D-Fruf	Honey
Maltulose	α-D-Glcp-$(1 \rightarrow 4)$-D-Fruf	Honey, maltose reversion
Isomaltulose (palatinose)	α-D-Glcp-$(1 \rightarrow 6)$-D-Fruf	Maltose reversion, saccharose isomerisation
Inulobiose	β-D-Fruf-$(2 \rightarrow 1)$-β-D-Fruf	Inulin degradation
Levanbiose	β-D-Fruf-$(2 \rightarrow 6)$-β-D-Fruf	Inulin degradation
Trisaccharides		
1-Kestose (isokestose)	β-D-Fruf-$(2 \rightarrow 1)$-β-D-Fruf-$(2 \leftrightarrow 1)$-α-D-Glcp	Honey, inulin, invert sugar, by saccharases from Suc
6-Kestose (kestose)	β-D-Fruf-$(2 \rightarrow 6)$-β-D-Fruf-$(2 \rightarrow 1)$-α-D-Glcp	Bananas, honey, invert sugar, molasses
Neokestose	β-D-Fruf-$(2 \rightarrow 6)$-α-D-Glcp-$(1 \rightarrow 2)$-β-D-Fruf	Inulin, invert sugar, by saccharases from Suc
Erlose	α-D-Glcp-$(1 \rightarrow 4)$-α-D-Glcp-$(1 \leftrightarrow 2)$-β-D-Fruf	Honey
Theanderose	α-D-Glcp-$(1 \rightarrow 6)$-α-D-Glcp-$(1 \leftrightarrow 2)$-β-D-Fruf	Honey
Gentianose	β-D-Glcp-$(1 \rightarrow 6)$-α-D-Glcp-$(1 \leftrightarrow 2)$-β-D-Fruf	Glycosides
Melezitose	α-D-Glcp-$(1 \rightarrow 3)$-β-D-Fruf-$(2 \leftrightarrow 1)$-α-D-Glcp	Nectars, honey
Kelose	α-D-Fruf-$(2 \rightarrow 6)$-β-D-Fruf-$(2 \leftrightarrow 1)$-α-D-Glcp	By saccharases from saccharose
Tetrasaccharides		
Nystose	β-D-Fruf-$(2 \rightarrow 1)$-β-D-Fruf-$(2 \rightarrow 1)$ Fruf-$(2 \leftrightarrow 1)$-α-D-Glcp	Inulin, saccharases from saccharose

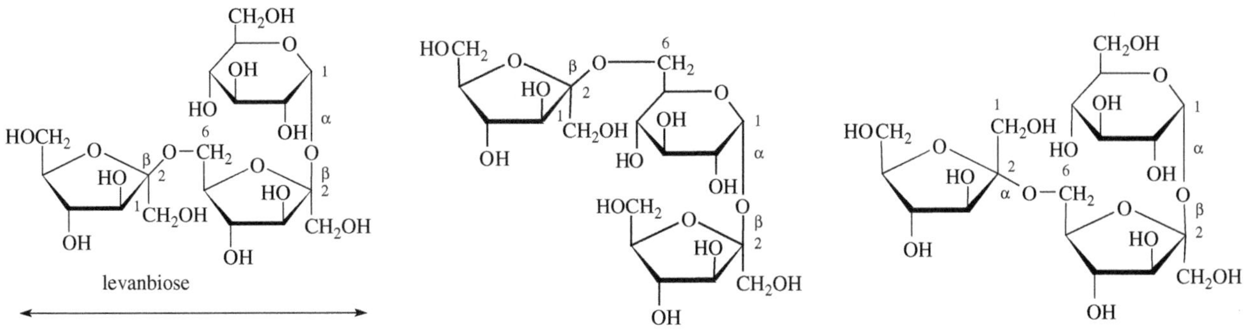

4-42, fructooligosaccharides in fructosans

4.4.3 Galactooligosaccharides

4.4.3.1 Lactose

4.4.3.1.1 Structure and nomenclature

Lactose, *O*-β-D-galactopyranosyl-(1 → 4)-D-glucopyranose (**4-43**), is a reducing disaccharide present in mammalian milk, and is therefore also referred to as milk sugar. The most stable form is α-lactose monohydrate (α-anomer). Lactose crystallises in this form from aqueous solutions at temperatures up to 93.5 °C. Drying of lactose in vacuum at temperatures above 100 °C yields hygroscopic α-anhydride. Crystallisation from aqueous solutions at temperatures above 93.5 °C gives anhydrous β-lactose (β-anhydride). An amorphous hygroscopic mixture composed of α- and β-lactose forms during rapid drying of lactose solutions and during drying of milk.

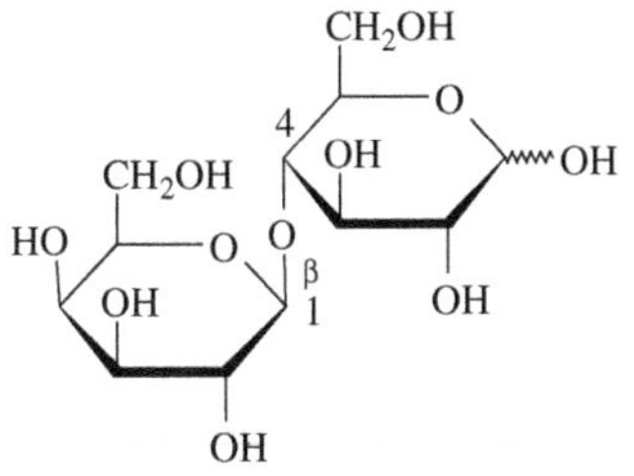

4-43, lactose

The mutarotation rate of lactose (as with other carbohydrates) depends on the pH. The minimum rate is found in solutions of pH 4–5. In solutions of pH <2 and pH >7, the mutarotation rate increases significantly. Conformation of lactose in aqueous solutions (**4-44**) resembles the conformation of cellobiose.

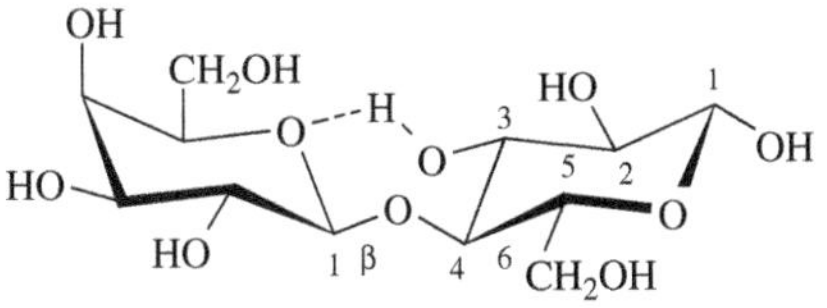

4-44, conformation of lactose in aqueous solutions

4.4.3.1.2 Occurrence and production

Lactose is only synthesised in the mammary glands of mammals. Cow's milk contains 4–5% of lactose, human milk 5.5–7% of lactose. In addition, milk contains smaller amounts of D-glucose and a wide variety of free oligosaccharides. Lactose is also present naturally in all milk-containing products (e.g. milk chocolates and ice creams). Its content in the products prepared using homofermentative lactic acid bacteria (such as yoghurt, acidophilic milk, and kefir) is lower than in fresh milk, generally about 1% or less.

Lactose is obtained by ultrafiltration of the whey of cow's milk or by crystallisation from whey concentrated to 55–65% of solids, which is called lactose syrup. Raw sugar is purified by recrystallisation.

4.4.3.1.3 Physiology and nutrition

Lactose is used as an energy source, as it is hydrolysed by lactase in the small intestine. A large part of the human population (as well as of other mammals) produces lactase only in childhood. In adulthood, a number of individuals have the enzyme activity reduced or completely absent, and consumption of milk is then problematic. Lactase is produced by lactic acid bacteria that break down lactose into lactic acid (L- or D-lactic acid or racemate). Fermented milk products, such as yoghurt and acidophilic milk, can therefore also be consumed by people with lactase deficiency. The relative sweetening power of lactose is about 40% (20–60%) of the sweetening power of saccharose. Intake of lactose (as well as galactose intake) leads to a significant increase in blood glucose level. Lactose has relatively low cariogenic and laxative effects.

4.4.3.1.4 Use

Lactose is used as a sweet substance and also serves as a raw material for the production of oligosaccharides and some alditols: lactulose, lactosaccharose, galactooligosaccharides, lactitol, and also galactose.

Lactulose   Lactulose (**4-45**) is obtained from lactose by isomerisation in alkaline solutions, where the main structure is β-pyranose, followed by α-furanose and β-furanose. Epilactose (**4-45**) is formed as a minor product. Lactulose is an indigestible disaccharide that is somewhat sweeter than lactose (about 60% of sweetening power of sucrose) and has a weak laxative effect. It stimulates the growth of bifidogenic microflora.

Lactosaccharose   Lactosaccharose (**4-45**) is obtained from lactose and sucrose under the action of the transfructosylase activity of microbial β-fructofuranosidase.

Higher galactooligosaccharides   Galactooligosaccharides (also called transgalactooligosaccharides) are obtained from lactose using the galactosyltransferase activity of the enzyme β-galactosidase, which is reflected in solutions containing higher concentrations of lactose. These oligosaccharides contain two to five molecules of galactose (usually three, so-called galactotriose), joined by β-$(1 \rightarrow 6)$ bonds (**4-45**).

Lactitol   Lactitol, O-β-D-galactopyranosyl-$(1 \rightarrow 4)$-D-glucitol or 4-O-β-D-galactopyranosyl-D-glucitol (**4-45**) is an alcoholic sugar produced by hydrogenation of lactose at the reducing end. Hydrogenation of lactulose gives a mixture of lactitol and 4-O-β-D-galactopyranosyl-$(1 \rightarrow 4)$-D-mannitol. Lactitol is currently used as a bulk sweetener in calorie-controlled foods. It has about 30–40% the sweetening power of sucrose. As an indigestible and non-cariogenic compound with slight laxative properties, lactitol has no effects on blood glucose levels or insulin secretion.

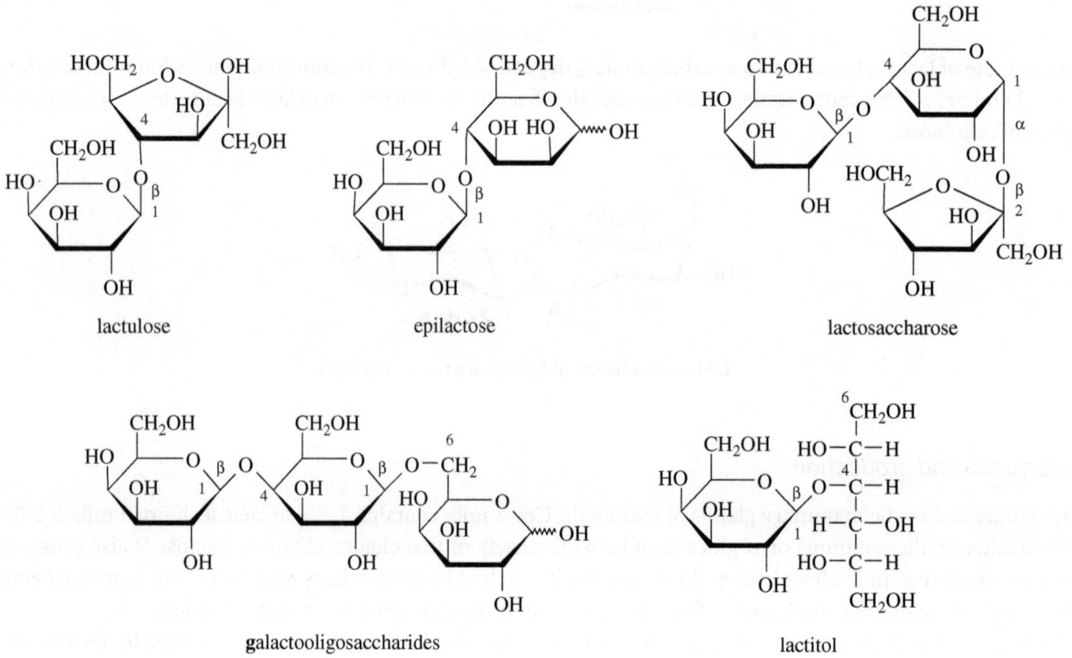

lactulose   epilactose   lactosaccharose

galactooligosaccharides   lactitol

4-45, oligosaccharides and other products derived from lactose

Table 4.15 Main oligosaccharides of breast milk and cow's milk.

Oligosaccharide	Abbreviated notation[a]	Content (g/l)	
		Breast milk	Cow's milk
Lactose	β-D-Galp-(1 → 4)-D-Glcp	55–70	40–50
Lacto-*N*-tetraose	β-D-Galp-(1 → 3)-β-D-GlcpNAc-(1 → 3)-β-D-Galp-(1 → 4)-D-Glcp	0.5–1.5	trace
Lacto-*N*-fucopentaose I	α-L-Fucp-(1 → 2)-β-D-Galp-(1 → 3)-β-D-GlcpNAc-(1 → 3)- β-D-Galp-(1 → 4)-D-Glcp	1.0–1.5	–
Lacto-*N*-fucopentaose II	β-D-Galp-(1 → 3)-β-D-GlcpNAc-(1 → 3)-β-D-Galp-(1 → 4)-D-Glcp 4 ↑ 1 α-L-Fucp	0.5–1.0	–
Sialyl-α-(2 → 6)-lactose	α-NeupAc-(2 → 6)-β-D-Galp-(1 → 4)-D-Glcp	0.3–0.5	0.03–0.06[b]
Sialyllactotetraose a	α-NeupAc-(2 → 3)-β-D-Galp-(1 → 3)-β-D-GlcpNAc-(1 → 3)-β-D-Galp-(1 → 4)-D-Glcp	0.03–0.2	trace
Sialyllactotetraose c	α-NeupAc-(2 → 6)-β-D-Galp-(1 → 4)-β-D-GlcpNAc-(1 → 3)-β-D-Galp-(1 → 4)-D-Glcp	0.1–0.6	trace
Disialyllacto-*N*-tetraose	α-NeupAc-(2 → 3)-β-D-Galp-(1 → 3)-β-D-GlcpNAc-(1 → 3)-β-D-Galp-(1 → 4)-D-Glcp 6 ↑ 2 α-NeupAc	0.2–0.6	trace

[a]Fuc, fucose; GlcNAc, *N*-acetylglucosamine; GalNAc, *N*-acetylgalactosamine; NeuAc, *N*-acetylneuraminic acid.
[b]Together with sialyl-α-(2 → 3)-lactose.

Galactose Galactose can be obtained from vegetable raw materials containing galactans, but the most common source is lactose or protein-free whey. The process involves acid or enzymatic hydrolysis (by lactase) of galactans to an equimolar mixture of glucose and galactose. This mixture has about 40–60% of the sweetening power of saccharose. Galactose can be separated from mixtures of galactose and glucose by selective fermentation of glucose to ethanol using various microorganisms or by glucose oxidase catalysed transformation of glucose to gluconic acid.

4.4.3.2 Other galactooligosaccharides

Breast milk and cow's milk contain, in addition to lactose, small amounts of mono- and oligosaccharides, such as D-glucose and D-galactose, *N*-acetyl-D-glucosamine, *N*-acetyl-D-galactosamine, L-fucose, and *N*-acetylneuraminic acid (lactaminic acid), and a number of oligosaccharides derived from lactose. More than 130 oligosaccharides have been identified in breast milk. Some of these, such as lacto-*N*-tetraose (**4-46**) and lacto-*N*-fucopentaose I (**4-47**), are present in relatively large amounts (at around 1–2 g/l). The total content of higher oligosaccharides in human milk is 3–6 g/l (Table 4.15). These amino sugars of milk are a growth factor for the microorganism

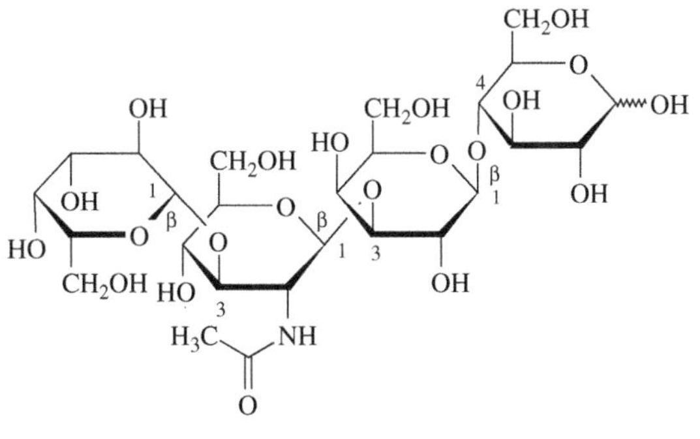

4-46, lacto-*N*-tetraose

Table 4.16 Overview of important galactooligosaccharides.

Trivial name	Abbreviated notation	Occurrence
Disaccharides		
Melibiose	α-D-Galp-(1 → 6)-D-Glcp	Cocoa beans, grapes
Lactose	β-D-Galp-(1 → 4)-D-Glcp	Milk, dairy products
Lactulose	β-D-Galp-(1 → 4)-D-Fruf	Milk, dairy products
Epilactose	β-D-Galp-(1 → 4)-D-Manp	Milk, dairy products
Solabiose	β-D-Glcp-(1 → 3)-D-Galp	Glycosides
Higher oligosaccharides		
Manninotriose	α-D-Galp-(1 → 6)-α-D-Galp-(1 → 6)-D-Glcp	Legumes
Raffinose	α-D-Galp-(1 → 6)-α-D-Glcp-(1 ↔ 2)-β-D-Fruf	Sugar beet, legumes, grapes
Umbelliferose (isoraffinose)	α-D-Galp-(1 → 2)-α-D-Glcp-(1 ↔ 2)-β-D-Fruf	Root vegetables
Planteose	α-D-Galp-(1 → 6)-β-D-Fruf-(2 ↔ 1)-α-D-Glcp	Cocoa beans, tobacco, spices
Stachyose	α-D-Galp-(1 → 6)-α-D-Galp-(1 → 6)-α-D-Glcp-(1 ↔ 2)-β-D-Fruf	Legumes, grapes, artichokes
Verbascose	α-D-Galp-(1 → 6)-[α-D-Galp-(1 → 6)]$_2$-α-D-Glcp-(1 ↔ 2)-β-D-Fruf	Legumes
Ajugose	α-D-Galp-(1 → 6)-[α-D-Galp-(1 → 6)]$_3$-α-D-Glcp-(1 ↔ 2)-β-D-Fruf	Legumes

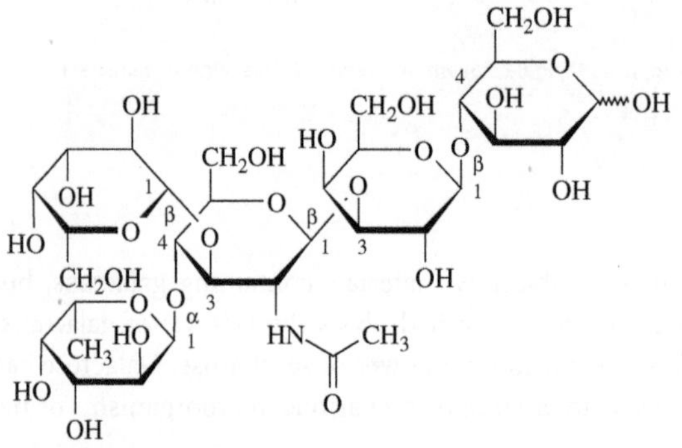

4-47, lacto-*N*-fucopentaose I

B. bifidum. They effectively inhibit the adhesion of pathogenic bacteria on cell walls, which is regarded as the initial phase of an infectious process. Therefore, they play a significant role in non-immunological protection of infants. Similar oligosaccharides composed of D-galactose, *N*-acetyl-D-galactosamine and *N*-acetylneuraminic acids, are components of κ-casein. An overview of other important galactooligosaccharides found in foods is given in Table 4.16.

Soybeans, other legumes, and other foods of plant origin contain the α-galactooligosaccharides raffinose (**4-48**), stachyose (**4-49**), verbascose, and ajugose, as well as higher homologues that do not have trivial names (see Table 4.17). The highest member of the α-galactooligosaccharide series is nonasaccharide. These oligosaccharides can be considered derivatives of saccharose or melibiose. They are accompanied by a series of cyclitols and galactocyclitols (Table 4.9). Table 4.17 shows the content of main α-galactoligosaccharides in legumes.

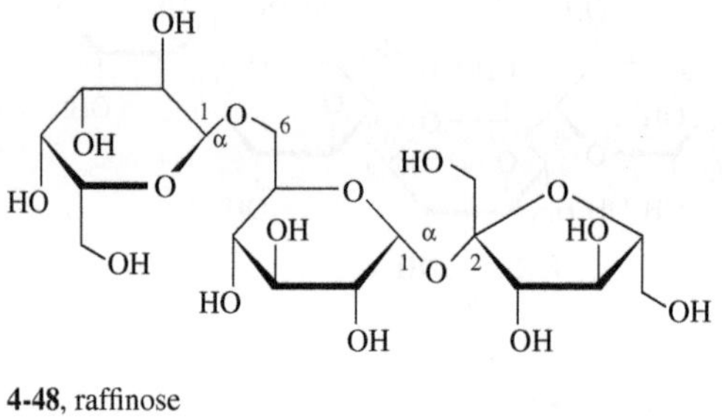

4-48, raffinose

Table 4.17 Contents of major oligosaccharides in legume seeds (% of dry matter).

Legumes	Latin name	Saccharose	Raffinose	Stachyose	Verbascose
Common bean	*Phaseolus vulgaris*	2.2–4.9	0.3–1.1	3.5–5.6	0.1–0.3
Vigna mungo	*Vigna mungo*	1.3	0.3	1.7	2.8
Pea	*Pisum sativum*	2.3–3.5	0.6–1.0	1.9–2.7	2.5–3.1
Lentil	*Lens culinaris*	1.3–2.0	0.3–0.5	1.9–3.1	1.2–1.4
Soybean	*Glycine max*	2.8–7.7	0.2–1.8	0.02–4.8	0.1–1.8
Chickpea	*Cicer arietinum*	2.0–3.5	0.7–0.9	1.5–2.4	0.0

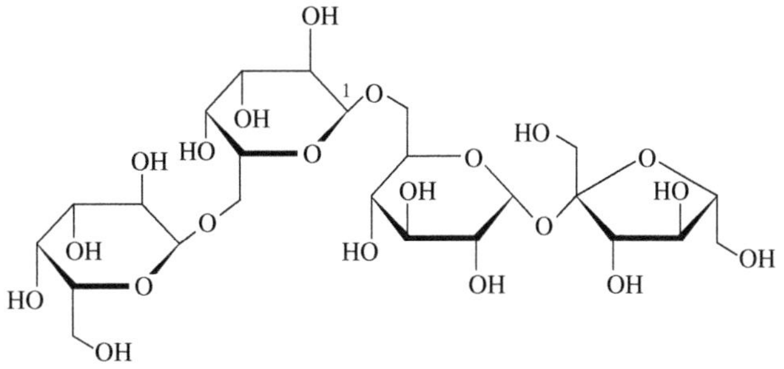

4-49, stachyose

α-Galactooligosaccharides are not digested by saccharases in the small intestine, but stimulate the growth of bifidobacteria and can be utilised by other bacteria of the colon that produce α-D-galactosidase and metabolise these sugars to form gases (carbon dioxide, methane. and hydrogen). α-Galactooligosaccharides are considered the main cause of bloating (flatulence) when consuming legumes. Enzymatic hydrolysis by α-D-galactosidase can reduce levels of these oligosaccharides. Use of this enzyme, which catalyses the hydrolysis of raffinose, can improve the yield of saccharose from sugar beet.

So-called soy whey, which is the waste from the production of soy protein isolates and concentrates, can be used for the isolation of a mixture of α-galactooligosaccharides, which also contains saccharose, glucose, and fructose. In the form of syrup, this mixture can be used as a proteinaceous food additive.

4.4.4 Other oligosaccharides

Hydrolysis of waste wheat flour containing arabinoxylans, using the enzyme *endo*-1,4-xylanase, yields xylooligosaccharides containing D-xylose molecules linked by β-$(1 \rightarrow 4)$ bonds. The degree of polymerisation of these oligosaccharides is $n = 2$–9. Enzymatic synthesis using D-mannose has been used for the preparation of mannooligosaccharides containing two to eight mannose molecules linked by α-$(1 \rightarrow 6)$ bonds. Both types of oligosaccharides can be used as food additives because they possess similar physiological effects to other unavailable oligosaccharides.

4.5 Polysaccharides

4.5.1 Structure and nomenclature

Polysaccharides or glycans consist of more than 10 monosaccharide units and can contain several thousand, hundreds of thousands, or even around a million structural units linked to one another by glycosidic bonds.

Polysaccharides are composed either of identical monomers (excluding terminal units) or, more often, of molecules of two or more different monosaccharides; alternatively, they may contain monosaccharide derivatives such as glycuronic acids, their esters, and deoxysugars. It is therefore possible to distinguish:

- **homopolysaccharides** (homoglycans);

- **heteropolysaccharides** (heteroglycans).

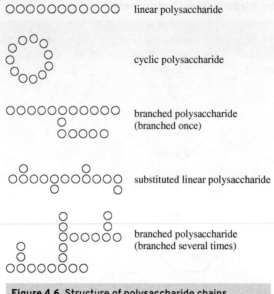

Figure 4.6 Structure of polysaccharide chains.

Homoglycans are, for example, components of starch, amylose and amylopectin, glycogen, and cellulose that are composed exclusively of D-glucose molecules. Most of the other polysaccharides are heteroglycans. Polysaccharide (and also oligosaccharide) chains (Figure 4.6) can be:

- **linear** (e.g. amylose, cellulose, and amylopectin);

- **cyclic** (e.g. higher cyclodextrins).

Linear chains can be unbranched (as in amylose and cellulose) or branched (as in amylopectin). The branched chains can be branched only once as in dextran, substituted as in guar gum, or branched several times as in amylopectin.

Names of homoglycans are derived from the monosaccharide name by substitution of the ending -ose by the ending -an. Some older trivial names have changed, and coincide with today's terminology (e.g. caragenin, carrageenin, or carrageenan); others are still in use (dextrin, pectin, and inulin, and also starch and cellulose).

The building blocks of homopolysaccharides are mostly pentoses, hexoses, and glycuronic (alduronic) acids. Their respective polymers are called:

- **pentosans**;

- **hexosans**;

- **glycuronans** (formerly also **polyuronides**).

The most common pentose bound in pentosans is D-xylose present in **xylans**. The most common building blocks of hexosans are D-glucose, D-mannose, D-galactose, and D-fructose. Homopolysaccharides, termed **glucans**, are composed entirely of glucose units (amylose, amylopectin, and cellulose). If glycosidic bonds link exclusively the α-anomers of a monomer, the glucans are called α-glucans (amylose). Cellulose is β-glucan, or, more precisely, **β-(1 → 4)-glucan**. **Fructans** are composed exclusively from fructose units, **mannans** of mannose units, and galactans of **galactose** units. Common components of many polysaccharides are alduronic acids, such as D-galacturonic acid in pectin and D-mannuronic and L-guluronic acids in alginates.

Heteroglycans having the main chain composed of a single type of monosaccharide have the names that end with the name of the homopolysaccharide that forms the main chain. Other sugar residues present in side chains are shown in alphabetical order before the basic name. For example, pentosans include **arabinoxylans**, whose main chain consists of D-xylose units and whose side chain contains L-arabinose. The sugars D-xylose and D-glucose occur in xyloglucans. The hexoses D-glucose, D-fructose, D-mannose, and D-galactose are common constituents of heteropolysaccharides known as **glucofructans** (inulins and phleins), **glucomannans** (konjac gum), and **galactomannans** (guar gum). If the main polysaccharide chain is not homopolymeric, all monosaccharide residues contained in the chain are shown in alphabetical order. Examples of such heteropolysaccharides are **arabinoglucuronoxylans**. An overview of the main building blocks of major polysaccharides is presented in Table 4.18.

Table 4.18 Main building monosaccharide units of important polysaccharides.

Monosaccharide	Abbreviated notation	Polysaccharides
α-L-Arabinofuranose	α-L-Araf	Plant gums (arabic, larch, ghatti), pectins, arabinoxylans
β-L-Arabinopyranose	β-L-Arap	Plant gums (arabic, larch, ghatti), pectins, arabinoxylans
α-D-Xylopyranose	α-D-Xylp	Arabinoxylans, xyloglucans
β-D-Xylopyranose	β-D-Xylp	Tragacanth
α-L-Rhamnopyranose	α-L-Rhap	Pectins, plant gums (arabic, karaya), okra, gellan gum
α-L-Fucopyranose	α-L-Fucp	Xyloglucans, pectins, tragacanth
α-D-Glucopyranose	α-D-Glcp	Starches, modified starches and their derivatives, starch hydrolysates, glycogen, dextran, elsin, pullulan
β-D-Glucopyranose	β-D-Glcp	Cellulose, modified celluloses, xyloglucans, glucomannans, β-glucans with mixed linkages, xanthan, gellan gum, curdlan, scleroglucan
β-D-Glucopyranuronic acid	β-D-GlcpA	Pectins, plant gums (arabic, karaya, ghatti)
α-D-Mannopyranose	α-D-Manp	Xanthan, plant gums (ghatti)
β-D-Mannopyranose	β-D-Manp	Galactomannans, glucomannans, xanthan
β-D-Mannopyranuronic acid	β-D-ManpA	Alginates
α-L-Gulopyranuronic acid	α-L-GulpA	Alginates
α-D-Galactopyranose	α-D-Galp	Galactomannans, pectins, plant gums (arabic, karaya), okra
β-D-Galactopyranose	β-D-Galp	Xyloglucans, pectins, plant gums (arabic, larch, ghatti, tragacanth), agar, carrageenans
α-D-Galactopyranuronic acid	α-D-GalpA	Pectin, plant gums (tragacanth), okra
3,6-Anhydro-α-D-galactopyranose	α-D-Galp3,6An	Carrageenans
3,6-Anhydro-α-L-galactopyranose	α-L-Galp3,6An	Agar
β-D-Fructofuranose	β-D-Fruf	Glucofructans and fructans (inulins and phleins)

Polysaccharides mostly have the reducing monosaccharide residue at one end of the chain, but some may have the non-reducing residues on both ends. The linear homopolysaccharide amylase has monosaccharide non-reducing residue at the beginning of the chain, and the end unit consists of a reducing unit with a hemiacetal hydroxyl group. Branched polysaccharides, such as amylopectin, have one reducing unit and $n + 1$ non-reducing units at the beginning of each of the n-molecule chain branches.

The **primary structure** of a polysaccharide (monosaccharide sequences) is regular in homoglycans (such as amylose and cellulose) and in some heteroglycans (arabinoxylans). The individual monosaccharides are also regularly alternating in the number of heteroglycans (carrageenan); alternatively, the regular structure may be disrupted and the order of monomers vary (e.g. in pectin) in certain chain sections.

The type of monosaccharide units, their conformation and type of linkages affect the conformation of macromolecules, the so-called glycan **secondary structure**. Linear macromolecules, such as cellulose, are often stabilised by hydrogen bonds between hydroxyl groups of one molecule of glucose and oxygen atoms of the second pyranose ring. Another type of conformation is known as egg box conformations, which have hydrogen bonds between hydroxyl groups of monomers and calcium cations (in pectins) or ionic bonds between dissociated carboxyl groups and calcium ions (e.g. alginates). For some polysaccharides, helical conformations are typical (e.g. in carrageenans).

Combinations of secondary structures result in **tertiary structures**, such as crystalline microfibrils in cellulose or double and triple helices in κ-carrageenan.

Polysaccharides are polydisperse substances, as they occur as mixtures of polymers with different degrees of polymerisation, and thus have an average molecular weight.

4.5.2 Classification

Polysaccharides are commonly divided according to their origin. Natural polysaccharides of plants have the greatest importance in human nutrition, whilst polysaccharides of animals and other natural polysaccharides have little or no importance. Many polysaccharides of plants (guar gum or locust bean gum), seaweed (such as agar, carrageenans, and alginates), and microorganisms (e.g. xanthan gum) are becoming part of various foods as additives in natural or modified forms (modified starches and celluloses, and modified chitin called chitosan).

According to the basic functions that are executed in the tissues of animals, tissues and cells of plants, algae, higher fungi, and microorganisms, polysaccharides are divided into:

- **storage or reserve polysaccharides**;

- **structural polysaccharides**;

- **polysaccharides with other functions**.

The storage polysaccharide in animals is glycogen. Structural functions are carried out by chitin, forming the exoskeletons of crustaceans, molluscs, and insects, and mucopolysaccharides occurring in proteoglycans of connective tissues.

Plant storage polysaccharides in seeds, tubers, rhizomes, bulbs, and roots are:

- **starches**, formerly also known as amyloids (e.g. in cereals, legumes, and potato tubers);

- **non-starch polysaccharides**, which include glucofructans and fructans (chicory root and cereal seeds), galactomannans that are known as seed gums (storage polysaccharides of some legumes, such as guar gum and locust bean gum), glucomannans (konjac tubers), and xyloglucans (e.g. rape and tamarind seeds).

Functioning as building materials in the walls of plant cells are:

- **cellulose**;

- **non-cellulose polysaccharides** associated with cellulose.

The non-cellulose polysaccharides include:

- **hemicelluloses** (xyloglucans in fruits, most vegetables, root crops and legumes, arabinoxylans and β-glucans in cereals, and galactomannans in some legumes);

- **pectins** (in fruits and vegetables).

The polymer of phenylpropanoid units is **lignin**, which is not composed of saccharide units but is associated with cellulose as well as with structural non-cellulose polysaccharides; it is also a structural material of plant cell walls. Lignin is accompanied by other phenolic compounds (tannins), proteins, and polymers of lipids.

Other plant polysaccharides seem to have different functions related to water management and protection of damaged tissues. These include:

- **plant exudates** or **gums** of some plants (such as gum arabic and tragacanth);

- **mucilages** (e.g. okra).

Some structural polysaccharides from algae are used for consumption as food additives, especially agar, carrageenans, and alginates. Some extracellular polysaccharides of microorganisms also have significance as food additives (such as xanthan gum). They have functions other than structural and reserve. Some extracellular and structural polysaccharides of microorganisms and higher fungi, such as β-glucans with a different structure to that of β-glucans of cereals, have found use as immunomodulators and anticarcinogenic substances.

4.5.3 Natural occurrence in foods

Pectin is the dominant polysaccharide in fruits, which also contain (in smaller amounts) some other polysaccharides, cellulose and hemicelluloses (the predominant components are xyloglucans), starch, and lignin.

Starch is the predominant polysaccharide of root vegetables and root crops (potatoes). Unlike fruits, its content increases during ripening, and a higher content of starch is found in overripe vegetables. In dicotyledonous (dicot) vegetable plants of the aster family (Asteraceae), such as black root, artichokes (grown mainly for the inflorescences, which are consumed fresh or preserved), Jerusalem artichokes (tubers),

Table 4.19 Content of main polysaccharides and lignin in wheat flour.

Polymer	Content (%)	Polymer	Content (%)
Starch	60–80	β-Glucans	0.5–2
Non-starch polysaccharides	3–11	Xyloglucans	0.2–0.4
Cellulose	0.2–3	Pectins	0.3–0.5
Hemicelluloses	2–7	Glucofructans (fructans)	1–4
Arabinoxylans	1–3	*Lignin*	0.7–12

and bulbs of monocotyledonous (monocot) plants of the lily family (Liliaceae), such as garlic and onion, the main reserve carbohydrates are glucofructans (fructans). Cellulose, hemicelluloses (predominantly xyloglucans), pectins, and lignin are important components of most vegetables.

Starch is also the major polysaccharide in all cereals. Of the non-starch polysaccharides, the predominant ones are hemicelluloses, whose main components in wheat and rye are arabinoxylans, whilst the so-called β-glucans are in oats and barley. Glucofructans (fructans), xyloglucans, cellulose, and lignin (the latter is mainly present in bran) are also found in significant quantities. The contents of main polysaccharides in wheat flour are given in Table 4.19. The content of polysaccharides varies widely, particularly since it depends on the degree of milling (known as flour extraction rate; the highly straight run flours are higher in non-starch polysaccharides) and other factors. Endosperm represents about 83% of grain by weight, bran 15%, and germ 2%. The starch content in the bran is about 15%, but the cellulose content is about 35%, and the content of hemicelluloses reaches 40–50%.

4.5.4 Physiology and nutrition

From the nutritional point of view, the following polysaccharides are recognised:

- **utilisable polysaccharides**;

- **non-utilisable polysaccharides** (formerly known as ballast polysaccharides), since the enzyme machinery for their digestion in humans and other monogastric animals is lacking and they are not cleaved by saliva, the pancreas, or small intestine saccharases.

Utilisable plant polysaccharides are starches (the main source of energy) and animal glycogen. Non-cellulose polysaccharides include hemicelluloses and pectin, as well as polysaccharides used as food additives (seaweed polysaccharides, microbial polysaccharides, vegetable gums and mucilages, and modified polysaccharides) and lignin and chitin of animal polysaccharides (and modified chitin, or chitosan). Some non-cellulose polysaccharides, such as pectin, can be relatively easily utilised (see Table 4.22), but these are known by the widespread and accepted term **fibre** or **dietary fibre**, although the name is misleading, inaccurate, and poorly definable. Some older definitions included plant polysaccharides (except starch) and lignin under the term dietary fibre, as materials resistant to hydrolysis by human digestive juices. Other definitions included all undigested and unabsorbed food components, including undigested proteins, minerals, and phytates, and possibly also endogenously excreted mucopolysaccharides. The EU definition of dietary fibre describes carbohydrate polymers with three or more monomeric units, which are neither digested nor absorbed in the human small intestine and belong to the following categories:

- edible carbohydrate polymers naturally occurring in the food as consumed;

- edible carbohydrate polymers that have been obtained from food raw material by physical, enzymatic, or chemical means, which have a beneficial physiological effect demonstrated by generally accepted scientific evidence;

- edible synthetic carbohydrate polymers that have a beneficial physiological effect demonstrated by generally accepted scientific evidence.

The current definition of dietary fibre is based on the chemical composition of polysaccharides and includes all non-utilisable carbohydrates, including polysaccharides and oligosaccharides used as food additives (e.g. plant gums and mucilages, algal and microbial polysaccharides, and modified starches, celluloses, and chitosan). Regarding lignin, it is stated that the carbohydrate polymers of plant origin that meet the definition of fibre may be closely associated in the plant with lignin or other non-carbohydrate components such as

Table 4.20 Fibre composition of fruits, vegetables, and cereals (in %).

Fibre source	Non-cellulose polysaccharides		Cellulose			Lignin
	Range	Average	Range	Average	Range	Average
Fruits	46–78	62.9	9–33	19.7	1–38	17.4
Vegetables	52–76	65.6	23–42	31.5	0–13	3.0
Cereals	71–82	75.7	12–22	17.4	0–15	6.7

phenolic compounds, waxes, saponins, phytates, cutin, and phytosterols. The definitions do not take into account the fact that fibre is not a static concept, as it does not pass through the digestive system completely intact.

A substantial component of dietary fibre (due to the amount consumed) is a certain percentage of not-completely-degraded starch known as **resistant starch**.

The gross composition of the dietary fibre of fruits, vegetables, and cereals is presented in Table 4.20. According to the solubility in water, the following two categories are recognised:

- **soluble dietary fibre;**

- **insoluble dietary fibre.**

Soluble dietary fibre includes a certain proportion of hemicelluloses. For example, about one-third of cereal structural arabinoxylans are soluble. Also, one-quarter to about one-half of barley β-glucans are soluble, and even a certain proportion of glucomannans and galactomannans of legumes. Other soluble polysaccharides are pectins, plant mucilages, seaweed polysaccharides, modified starches, and modified celluloses (Table 4.21).

The main components of insoluble fibre include cellulose, a certain proportion of hemicelluloses, and lignin. Higher lignin content is found in the bran and seeds of consumed fruits, such as garden strawberries, raspberries, and currants. Insoluble fibre increases the volume of food, reduces the time of its passage through the digestive tract, and increases peristalsis. Soluble fibre increases the viscosity of the contents of the stomach and intestines, slows down the mixing of their content, and limits the access of pancreatic amylase and lipase to substrates. As a result, the nutrient absorption by the intestinal wall slows down, faster passage rate of the intestinal content through digestive tract reduces the diffusion of bound nutrients and minerals (especially calcium, iron, copper, and zinc ions), and thus their availability is modified. Part of the bound ions is released during fermentation in the colon. Some minor components, such as accompanying tannins, also can partially inhibit digestive enzymes.

Dietary fibre is a beneficial material for constipation, gastric and duodenal ulcers, haemorrhoids, rectum and colon (colorectal) cancer (commonly known as bowel cancer), and other diseases. Eating foods high in fibre is recommended for the modulation of glucose levels

Table 4.21 Proportion of soluble and insoluble dietary fibre in selected foods.

Food	Dietary fibre (% of dry matter)			Food	Dietary fibre (% of dry matter)		
	Soluble	Insoluble	Total		Soluble	Insoluble	Total
Fruits				*Legumes*			
Apples	5.6–5.8	7.2–7.5	12.8–13.3	Beans	7.2–12.4	9.1–9.6	16.8–21.5
Peaches	4.1–7.1	3.4–6.4	7.5–13.5	*Potatoes*			
Strawberries	5.1–7.7	6.8–10.6	11.9–18.3	Raw	2.8–3.5	2.4–3.2	5.2–6.7
Oranges	6.5–9.8	3.9–5.2	10.4–15.0	Boiled	4.8	2.6	2.2
Vegetables				*Cereals*			
Carrot	4.4–14.9	10.4–11.1	14.8–26.0	Wheat flour (white)	2.0	1.2	3.2
Cabbage	13.5–16.6	4.2–20.8	27.6–37.4	Wheat flour (whole)	2.6	7.7	10.3
Tomatoes	0.8–3.5	3.2–12.8	6.7–13.6	Wheat bread	1.6–2.7	1.1–2.9	2.7–5.6
Peas	5.9	15.0	20.9	Rye bread	6.7	6.6	13.3

Table 4.22 Utilisation of soluble fibre components.

Polysaccharide	Utilisation (%)	Polysaccharide	Utilisation (%)
Hemicelluloses	19–85	Karaja gum	5
Pectins	65–97	Agar	21–28
Guar gum	76	Carrageenans	9–16
Locust gum	<15	Alginates	0–78
Gum arabic	71	Dextran	78–90

in blood serum and in some forms of diabetes (especially type 2 diabetes). This also has the effect of reducing serum cholesterol level, and can thus help to prevent cardiovascular diseases. The effect is explained by the reduced absorption of cholesterol from viscous foods and by cholesterol binding to fibre, which results in increased excretion of cholesterol in the faeces. Bile acids are also bound to dietary fibre and excreted. The result is a decrease of bile acid reserves in the liver. This deficit of bile acids has to be compensated by biosynthesis from cholesterol, which reduces the cholesterol concentration in the blood plasma. Cholesterol synthesis in the liver is also inhibited by lower fatty acids being formed during fibre fermentation by intestinal microflora.

Soluble fibre is partially hydrolysed by digestive enzymes already in the small intestine. Insoluble fibre is not hydrolysed in the small intestine, and along with soluble fibre is partly metabolised only by colon and appendix microorganisms (e.g. xanthan gum is not metabolised at all). Colon and appendix microorganisms assimilate on average 70% of dietary fibre polysaccharides (Table 4.22). The final products are gases (carbon dioxide and hydrogen, often methane) and lower fatty acids (acetic, propionic, and butyric acids). The amount of energy obtained in the form of these acids is around 3 kJ/g carbohydrate (compared with about 17 kJ/g obtained from starch). It is estimated that the consumption of 20–30 g of dietary fibre (the Recommended Daily Allowance of dietary fibre is >25 g per day in Europe) contributes to the total energy intake in the range of 0.5–1%; the rest is covered by the main nutrients. The ratio of insoluble to soluble fibre in the diet should be 3 : 1.

4.5.5 Properties and use

Polysaccharides in foods contribute to the formation of texture and also affect other organoleptic properties. Soluble polysaccharides are used in many food industry sectors and other fields as fillers, thickeners, viscosity increasers, dispersion stabilisers, thickening agents, and gelling substances.

The importance of polysaccharides has increased with the development of new technologies and of products with reduced fat and saccharose contents. Previously, the market was dominated by native starch, but its consumption is declining and consumption of modified starches is significantly increasing. Important substances are modified celluloses, plant gums, and polysaccharides of seaweeds and microorganisms. At the forefront of world consumption of non-starch polysaccharides are plant seed gums (guar gum and locust gum), followed by carrageenan, agar, gum arabic, pectins, alginates, modified celluloses (carboxymethyl cellulose), and xanthan gum. There have also been major increases in the sale of food supplements containing natural soluble fibre, which occurs mainly in the form of modified fructans, chitosan, and arabinoxylans. Arabinoxylans occur in dietary fibre supplements, known as *psyllium*, which is the name used for seeds of sand plantain (*Plantago psyllium*, Plantaginaceae).

4.5.6 Plant polysaccharides

4.5.6.1 Starch

Starch is the main plant storage polysaccharide, serving as a prompt supply of energy. Unlike structural polysaccharides, which are part of the plant cell walls, starch is found in organelles of cytoplasm called plastids, where its biosynthesis takes place. Storing of glucose obtained by photosynthesis in the form of starch strongly reduces the high intracellular osmotic pressure to which cells would otherwise be exposed. Chloroplasts of the specialised photosynthetic tissues produce so-called **transient starch**, which serves as a temporary sugar reserve. All transient starch granules synthesised during the day undergo nocturnal breakdown and serve as a major source for saccharose synthesis at night; this is required for metabolism in the whole plant. Saccharose is then transported to the storage organs of plants, such as seeds, fruits, tubers, and roots, where the specialised leucoplasts, known as amyloplasts, synthesise and store the so-called **reserve starch**. This starch is stored in insoluble micelles called **starch granules** or starch grains that have species-specific shapes (e.g. round and oval) and dimensions (Table 4.23).

Table 4.23 Basic characteristics of starch granules.[a]

Source	Diameter (μm)	Mean value (μm)	Source	Diameter (μm)	Mean value (μm)
Wheat	4–6; 15–25	15	Potatoes	38–50	33
Barley	3–5; 19–25	15	Cassava	6–36	20
Rice	3–9; 15–30	5	Amaranth	1.07–1.10	1.1

[a]Cereal starches have a bimodal distribution of granules. Large granules (A type granules) have a lens shape and a diameter of about 20 μm; small granules (B type granules) are spherical particles with a diameter of about 5 μm.

4.5.6.1.1 Structure

The majority of native starches are mixtures of **amylose** and **amylopectin**, two homopolysaccharides composed of α-D-glucopyranose molecules in $^4C_1$ conformation. Amylose and amylopectin usually occur in a weight ratio of 1 : 3. In some varieties of cereals (maize, barley, and rice) and other plants (potatoes), either amylose (starches of high amylose content or amylostarches) or amylopectin (starches high in amylopectin or waxy starches) predominates.

Amylose   Amylose is a linear α-D-(1 → 4)-glucan (**4-50**) and, therefore, is actually a polymer of the disaccharide maltose. To a limited extent, branching occurs at about 10 sites of the molecule. Amylose is partially esterified with phosphoric acid (wheat starch contains 0.055% and potato starch 0.07–0.09% of phosphorus), and in cereal starches forms complexes with lipids. An amylose molecule has one reducing monosaccharide residue.

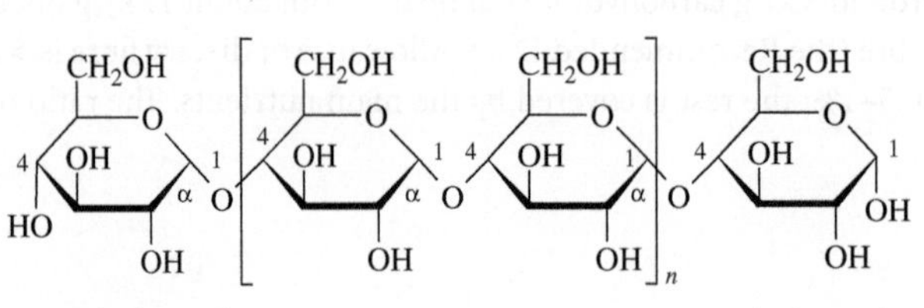

non-reducing end   reducing end

α-D-Glcp-(1→4)-[α-D-Glcp-(1→4)-α-D-Glcp-(1→4)$_n$-α-D-Glcp

4-50, amylose

Owing to the prevailing bonds 1 (axial) → 4 (equatorial), the amylose molecule can be randomly coiled in water and in neutral solutions, but generally tends to wind up into a rather stiff left-handed single helix or forms even stiffer parallel left-handed double-helical junction zones (**4-51**). Single helical amylose has hydrogen-bonded O-2 and O-6 atoms. In alkaline solutions, globular structures dominate. As with each polysaccharide, amylose is a mixture of polymers with different degrees of polymerisation. The average content of glucose units is 1000–2000 (in cereal starches), but can reach up to around 4500 (in potato starch). The molecular weight of amylose ranges between 180 and 1000 kDa.

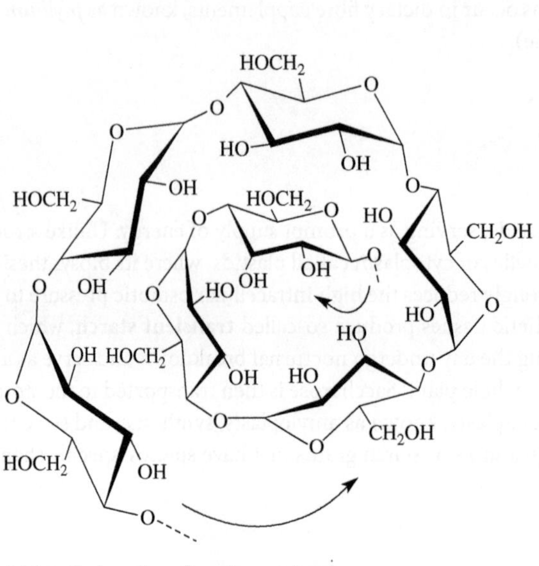

4-51, helical section of amylose molecule

Amylopectin An amylopectin molecule (**4-52**) consists of chains of D-glucose units linked by α-(1 → 4) linkages (maltose polymer), which are branched after 10–100 (average of 25) units by non-random α-(1 → 6) side chains (building unit is isomaltose). Occasionally, there may also be an α-(1 → 3) bond (building unit of this biose is laminaribiose). One in about 400 glucose residues is esterified with phosphoric acid.

···→4)-α-D-Glc*p*-(1→4)-α-D-Glc*p*

1

↓

6

···→4)-α-D-Glc*p*-(1→4)-α-D-Glc*p*-((1→4)-α-D-Glc*p*-((1→···

4-52, amylopectin

The degree of polymerisation is 50 000–1 000 000, so the molecular weight varies between 10 and 200 MDa. The amylopectin macromolecule is a many-times-branched structure consisting of three types of chain: outer chain A, inner chain B, and main chain C. About 5% of glucose molecules form the branch points. The inner and outer short side chains are called S chains and the internal long chains are L chains in modern terminology. The amylopectin molecule has one reducing end in the main chain (Figure 4.7).

Starch granules The ultrastructure of starch granules varies depending on the plant source, but they have a common general model (Figure 4.8). The amylopectin molecules are oriented radially (from the centre to the edge) in the starch granule. The non-reducing amylopectin ends are situated outside the granules and form their surface. In areas of non-reducing ends and in the central parts of the chains (corresponding to a length of about 15 glucose units), antiparallel double helices with an ordered (crystalline) three-dimensional structure are formed. The presence of amylose in starch tends to reduce the crystallinity of the amylopectin, and influences the ease of water penetration into the granules. In areas of chain branching, amylopectin and the accompanying amylose have an unordered amorphous structure. Crystalline and amorphous regions alternate regularly. Depending on the degree of crystallinity of the granules (related to the structure and chirality of double helices of amylopectin side chains in crystalline regions of granules), four polymorphic forms of starch have been identified, labelled A, B, C, and V. The most stable form, A, is found in cereal starches, with the exception of starches with a high content of amylose. A double spiral creates a central channel that is filled by another double helix, and in the space between the double helices is bound water (tighter than in form B). The least stable form is form B, where the central channel double helix is only filled by water molecules. This form occurs in starches of root vegetables, potatoes, and starches of cereals with high amylose content (>40%). Form C is present in legumes (it can be a combination of forms A and B). Retrogradation in gelatinised starches with amorphous structure first forms the less stable form B, which is subsequently transformed into form C, and then into the most stable form A. Form V occurs in gelatinised starch containing lipids, in which amylose interacts with fatty acids.

The amylopectin molecules are associated with amylose molecules, which form left-handed helices in certain parts of the molecule, oriented with non-reducing ends to the surface of the granules. They are located mainly in the amorphous zones, together with radially oriented lipid molecules whose fatty acids are immersed in helical parts of amylose molecules. These interactions create non-stoichiometric complexes called

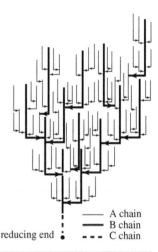

reducing end —— A chain
　　　　　　 —— B chain
　　　　　　 - - - C chain

Figure 4.7 Schematic structure of the amylopectin molecule. Robin et al. 1974, figure 11. Source: Reproduced by permission of AACC International.

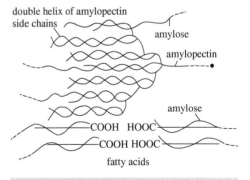

double helix of amylopectin side chains

amylose

amylopectin

amylose

COOH HOOC

COOH HOOC

fatty acids

Figure 4.8 Arrangement of starch grain components. Stephen 1995, figure 2.3. Source: Reproduced by permission of Taylor & Francis – Marcel Dekker.

Table 4.24 Content of lipids and proteins in starches.

Starch	Content (%)		Starch	Content (%)	
	Lipids	Proteins		Lipids	Proteins
Wheat	0.38–0.72[a]	0.3	Peas	0.1	0.7
Maize	0.02–1.09	0.3	Potatoes	0.05	0.06
Beans	0.1	0.9	Cassava	0.1	0.1

[a]Lipids in starch form 25% of total flour lipids; the main components are lysophospholipids.

inclusion compounds. An integral part of the granules is the small amounts of proteins with relative molecular weights of 5–97 kDa. For example, the hardness of the endosperm of cereals is attributed to the protein friabilin (15 kDa). Proteins are found mainly in the surface layers of granules (Table 4.24).

4.5.6.1.2 Occurrence and production

The starch content and its composition (amylose content) in some foods is given in Table 4.25; the amylopectin content makes up the remainder. The main sources of starch in foods and industrial sources of starch are potatoes and cereals (grasses of the Poaceae family cultivated for grains). Important sources of starch in foods are pseudocereals and ripe seeds of legumes (grain crops of other plant families). The starch content of cereal grains ranges from 40 to 90% (d.w.), legumes contain 30–70% of starch, and plant tubers 65–85%. Important sources of starch in many countries are sweet potato (*Ipomoea batatas*, Convolvulaceae), Jerusalem artichoke (also called topinambours, *Heliantus tuberosus*, Asteraceae), and tubers of *Manihot esculenta* (Euphorbiaceae), an important source of food carbohydrates in the Tropics, known in Asia and Africa by the name cassava and in South America as manioc, yucca, or tapioca. Another source of starch is sago (sago palm), obtained from the spongy pith of stems of various tropical palms, especially true sago palm (*Metroxylon sagu*, Arecaceae). The sago cycad (often also called sago palm) is obtained from the pitch of *Cycas revoluta* (Cycadaceae).

Ripe fruit, with some exceptions, does not contain starch. Starch is only found in immature fruits (e.g. about 2.5% in immature apples), and its content decreases during ripening. The exceptions are bananas, which even at full maturity contain at least 3% starch by weight and

Table 4.25 Content of starch and its composition in important food sources.

Source	Starch (%)	Amylose (%)
Amaranth	48–69	0–22
Barley[a]	52–62	38–44
Beans[b]	46–54	24–33
Cassava	28–35	17–19
Maize	65–75	24–26
Oats	40–56	25–29
Potatoes[c]	17–24	20–23
Rice	70–80	8–37
Rye	52–57	24–30
Sweet potatoes	10–30	19–25
Wheat	59–72	24–29

[a]Waxy barley contains 2–8% amylose, waxy maize about 1%, and barley amylostarch and maize amylostarch 60–70%.

[b]Lentils and ripe seeds of peas have similar levels of starch and amylose as beans. Green peas have about 4% starch, soybeans contain less than 1% starch.

[c]Industrial potato varieties have the starch content at the upper limit of the range.

about 1% glucofructans (fructans). Starch also occurs in sweet chestnuts, known as marrons (*Castanea sativa*, Fagaceae), and various nuts. A higher starch content is found in the seed of the cashew tree (*A. occidentale*, Anacardiaceae), commonly called the cashew nut.

Production of starch is relatively simple. Starch granules occurring in amyloplasts are neither chemically nor physically bound to other components of raw materials. Their density is high (around $1600\,kg/m^3$); therefore, pure starch can be separated from the crushed material by either washing or decantation on sieves, or by centrifugation.

4.5.6.1.3 Properties and changes

Gelatinisation At ambient relative humidity, starch granules take about 0.2 g of water per 1 g of dry starch from the atmosphere and contain about 17% water (wheat starch is about 13% water and potato starch 18–22% water), without changing the volume of the grains. The process is called imbibition (see Section 7.8.3.1). One molecule of glucose binds 1.5 water molecules. The structural units of glucans contain a total of five oxygen atoms that can interact with water (**4-53**).

Starch granules are insoluble in cold water and form a suspension. In cold water, the suspension of undamaged starch granules may absorb a small amount of water. Relatively large amounts of water are absorbed during heating without disturbing the integrity of the granules. Up to a certain temperature, called the initial **gelatinisation temperature** (T_o), at which swelling of granules occurs, this is a reversible process. The final gelatinisation temperature (when the gelatinisation process ends, T_e), the gelatinisation temperature range $T = T_e - T_o$, and the energy necessary to complete the process (gelatinisation enthalpy H_p) are also important starch characteristics. The initial gelatinisation temperature is generally 50–70 °C (Table 4.26) and depends on the starch origin, the ratio of starch to water, the pH, and the presence of other components (salts, sugars, lipids, and proteins). The gelatinisation temperature range is usually 10–15 °C. It refers to the temperature range over which all the granules are fully swollen.

4-53, interaction sites of glucose units with water

The changes to swollen starch granules in the process of gelatinisation are irreversible. Thermal motion of molecules interrupts the existing hydrogen bonds; water molecules penetrate the amorphous regions of granules and interact with free binding sites of the polymers. Hydrated polymer chains move away from one another, revealing new binding sites, which also interact with water; double helices of the side chains of amylopectin break down, which leads to the disappearance of crystalline zones, and the whole structure becomes disorganised and amorphous. Granules swell intensely, and their volume increases. For example, the size of wheat starch granules at this stage can be up to 30 μm (Table 4.23).

Upon further heating, some amylose and amylopectin molecules (originally located in the granule in the radial direction) reach the surface. Linear amylose molecules (less bulky than amylopectin molecules) penetrate this tangentially located sieve of molecules and are released into the extragranular environment (partly also broken down into shorter molecules), where they are fully hydrated. A small proportion of amylopectin molecules are also released to the extragranular environment.

As a consequence of hydration at elevated temperatures (e.g. at 70 °C), granules absorb water of about 25 times their weight. For example, 1 g of dry potato starch can take up a volume of about 200 ml after swelling. The release of amylose from the granules increases the viscosity,

Table 4.26 Gelatinisation temperatures of selected starches.

Starch source	Gelatinisation temperature (°C)			Starch source	Gelatinisation temperature (°C)		
	Initial	Medium	Final		Initial	Medium	Final
Wheat	52	58	64	Rice	66	72	78
Maize	62	67	72	Potatoes	50	60	68
Waxy maize	63	68	72	Cassava	61	66	71

Table 4.27 Typical viscosity of starch and other hydrocolloid dispersions at 25 °C.

Polysaccharide	Dispersion concentration (% w/w)	Dynamic viscosity (mPa/s)	Polysaccharide	Dispersion concentration (% w/w)	Dynamic viscosity (mPa/s)
Wheat starch	10	10	Tragacanth	1	54
Pectin	1	50	Agar	1	4
Guar gum	1	3025	Carrageenan	1	57
Locust gum	1	59	Alginate (Na)	1	214
Gum arabic	10	17	Xanthan	1	2000

and at sufficient concentration (roughly in a 1% suspension), starch yields viscous **starch paste**. It contains collapsed starch grains, but magnified many times. These contain mainly molecules of amylopectin and any remaining molecules of amylose (e.g. wheat starch contains about 8% of the original amount of amylose after heating to 90 °C). The branched amylopectin molecules give viscosity to the cooked paste. Their side chains and bulky shape keep them close enough to bond together.

Additional heating results in decreased viscosity and further loss of integrity of granules. Viscosity gradually increases again when the solution is chilled because of renewing hydrogen bonds between the macromolecules of amylose and amylopectin, and continuous chilling makes the solution cloudy. Left to stand, the solution becomes white, and in cases where there is a higher concentration of starch, it gels. The **starch gel** is a solid three-dimensional network that captures a large amount of water. Less concentrated starch suspensions form a viscous paste or viscous colloidal solution (Table 4.27). Starch gel is a complex system of gelatinised granules located in the matrix formed by amylose that does not contribute significantly to viscosity, but which does contribute to the gel formation as the linear chains can orient parallel to each other, moving close enough together to bond. Amylose forms a gel at room temperature at concentrations higher than 0.9–1.0%. Amylose gels are thermally stable even at a temperature of 120 °C. Amylopectin molecules do not usually contribute to gel formation. Amylopectin forms gels only at concentrations higher than 10% and temperatures lower than 5 °C. The gel formation is slow and the formed gels are thermoreversible (they melt when heated to 40–60 °C), like starch gels.

The rheological properties of starch gels depend on the origin of the starch, the degree of granule degradation, the ratio between the interacting polysaccharides (amylose and amylopectin), the temperature, the amount of water present, and the type and amounts of other compounds. Cereal starches, for example, generally form turbid, opalescent gels. Gels from amylose starches are formed faster and at higher temperatures than gels from waxy cereal varieties. They are stronger, and their strength increases with starch concentration, but they quickly undergo retrogradation and loss of water binding capacity caused by recrystallisation of amylose through cooling. Starches from waxy varieties of cereals, where amylopectin predominates, form gels with difficulty and only after cooling to low temperatures. They are clear, soft, and often have a paste-like consistency. The linear segments of amylopectin also have a tendency to associate over time. Prolonged storage at low temperatures causes retrogradation, as in amylose gels. Clear and moderately firm gels are formed from potato starch. The viscosities of diluted dispersions of major polysaccharides and starch are compared in Table 4.27.

Retrogradation Gelatinised starch is not in thermodynamic equilibrium; therefore, the structure and rheological properties of starch gels, pastes, and diluted dispersions change after several hours of storage. Intermolecular associations by hydrogen bonds between aligned chains of amylase lead to a loss of sites that bind water molecules, hence water binding capacity decreases. Gels and concentrated pastes get a rubbery texture and higher strength, dilute dispersions lose viscosity and precipitate, and bound water is released and eliminated, which gives rise to a solid–liquid two-phase system. These changes are related to the properties of amylose (its recrystallisation) and only very little to the properties of amylopectin. This process is generally called **syneresis**. Syneresis in starch gels is known as **retrogradation**. The process of retrogadation is actually the opposite of gel formation. The rate and extent of retrogradation depend on many factors, such as the origin of the starch (the amount and degree of polymerisation of amylose), the temperature, the water content, and the presence of other components. For example, maize starch undergoes retrogradation more easily than potato starch.

Starch retrogradation is often considered a major contributor to the staling of bread and other starch-rich foods, which can cause reduced shelf-life and consumer acceptance. On the other hand, modification of the structural, mechanical, and sensory properties via starch retrogradation may be desirable in some products, such as breakfast cereals, parboiled rice, and dehydrated mashed potatoes. It is also desirable in terms of nutrition, due to the slower enzymatic digestion of retrograded starch and moderated release of glucose.

During low-temperature storage (below about −5 °C), retrogradation of starch gels containing 45–50% water is strongly inhibited. At temperatures ranging from −5 °C to just below room temperature, the retrogradation rate is higher than its rate at room temperature. Higher temperatures (32–40 °C) effectively suppress retrogradation, and at temperatures around 65 °C and above, retrogradation does not occur at all. Therefore, waxy starches high in amylopectin, where the range of retrogradation is suppressed, are recommended for frozen foods. When stored at low temperatures, even gels of these starches lose clarity and water binding capacity, which is attributed to intermolecular associations of amylopectin side chains. Retrogradation is also influenced by water content in starch gel. It occurs in gels containing 20–90% water. The greatest tendency to retrogradation is seen in gels with 45–50% water. In the presence of salts (sodium chloride) or sugars, the degree of retrogradation is lower. The effect of individual sugars varies and depends on many factors, such as concentration. Retrogradation is also suppressed in the presence of lipids that form inclusion compounds with amylose.

Changes in foods Grain grinding mechanically damages about 5–10% of starch granules and during dough rising, amylases (also known as flour diastase) preferentially attack the damaged starch granules. Starch is partially hydrolysed by α-amylase (1,4-α-D-glucan glucanohydrolase) and β-amylase (1,4-α-D-glucan maltohydrolase) to maltose, which is hydrolysed to glucose by maltase (α-glucosidase, α-D-glucoside glucohydrolase). Endoglycosidase α-amylase randomly attacks α-(1 → 4) glycosidic bonds in amylose polymer, yielding glucose, maltose, and units of higher molecular weight, known as **linear dextrins** or **α-dextrins**, and is therefore known as dextrinogenic enzyme. Exoglycosidase β-amylase hydrolyses α-(1 → 4) glycosidic bonds of amylose molecules from the non-reducing end, splits off maltose molecules (if the chain has an odd number of glucose units, glucose and maltotriose are also formed), and is therefore known as saccharogenic enzyme. Amylopectin is attacked by both enzymes, but it is hydrolysed by about 50–60% because amylases only hydrolyse the molecule to the point of branching and do not hydrolyse α-(1 → 6) glycosidic bonds. β-Amylase hydrolyses amylopectin from the reducing end to the branching points, so that residues containing 2–3 glucose units remain unhydrolysed. Products of this concerted action of amylases are maltose and dextrins (maltodextrins). Highly branched dextrins of this type are called **limit dextrins** or **β-dextrins**. Limit dextrins are further hydrolysed gradually by pullulanase (1,6-amyloglycosidase or amylopectin 6-glucano hydrolase), which hydrolyses α-(1 → 6) bonds, and remnants of dextrins are further broken down by amylases.

The rheological properties of dough are mainly determined by the properties of gluten. The desired structure of the dough also arises partly as a result of interactions of swollen starch granules with gluten proteins, with pentosans, denatured proteins, and starch gelatinisation. If a dough contains a limited amount of water, starch is only partially gelatinised (raw starch with no water added does not undergo gelatinisation, but instead undergoes dextrinisation if heated). This condition exists in the production of bread and pastries (similarly in the cooking of potato tubers and opposed to cooking a pudding). Because the water content of the bread crumb stays virtually unchanged, the resulting degree of starch gelatinisation and crumb consistency depends on the amount of water used, but also on the origin of the flour (its chemical composition), the amount of fat, emulsifiers, and other factors.

During bread and thicker pastry baking, the enzymatic hydrolysis of starch by amylases proceeds for quite a long time (α-amylase is very thermostable, and the optimum temperature is around 90 °C). Bread and pastry therefore always contain less starch and more products of its fission (dextrins and lower sugars) than dough. Sugars bind a certain proportion of water, which can further contribute to the formation of gel. The decrease of water activity results in increased gelatinisation temperature and reduced rate of gel formation, and also reduces its viscosity and strength. Slow baking at lower temperatures produces higher amounts of lower sugars. For example, pumpernickel, a very heavy rye bread produced in Germany, is baked very slowly (24–100 hours at 100–150 °C). It is sweet, as it contains up to 20% reducing sugars; these undergo non-enzymatic browning reactions, and products of these reactions confer a bread crumb with a typical dark brown colour and distinctive flavour.

Starch in bread crust exposed to temperatures of 160–180 °C during baking is non-enzymatically hydrolysed into smaller units, which condense in larger molecules called **pyrodextrins** that contribute to bread flavour, colour, and crispness. Pyrodextrins contain glucose units bound by α-(1 → 6) glycosidic bonds or (6 → 6) ether bonds, which are not cleaved by saccharases present in the digestive tract.

Lipids (fats and oils) and monoacylglycerols used as emulsifiers form inclusion compounds with amylose, and slow down the swelling of starch granules and the extent of starch gelatinisation. For example, around 96% of starch is fully gelatinised in white bread, which is low in fat. Bakery products rich in fat, especially in the surface layers with lower water activity, contain a considerable proportion of ungelatinised starch. Small concentrations of sodium chloride have only limited impact on gel formation.

A typical symptom of retrogradation is ageing (hardening) of bread and pastries. Reheating of such bread or bakery products, for example in microwave ovens or during bread toasting, causes new starch gelatinisation, but such products harden very quickly.

Starch gelatinisation also occurs during hydrothermal processes, such as extrusion cooking. In some cases, the release of amylose from the granules into the environment is an undesirable phenomenon (e.g. during pasta and rice cooking), causing stickiness of the material. To prevent this phenomenon in pasta products, monoacylglycerols or other emulsifiers are added. In rice, cooking is helped by the decantation of starch and addition of oil to cooking water.

Preservation of unripe fruits and vegetables with higher starch content (e.g. apples and ripe green peas) often gives turbid and viscous brines. Acidic solutions can cause hydrolysis of starch to dextrins, which is manifested by less viscous solutions. Hydrolysis of starch also occurs in puddings containing acidic fruit juices. The liquefaction of dressings and mayonnaises containing starch is likewise caused by the action of enzymes (saccharases) of fresh vegetables and spices.

4.5.6.1.4 Physiology and nutrition

Starch is a utilisable polysaccharide that is easily digested in the small intestine. In the digestion of starch, the enzymes α-amylase (in saliva), pancreatic α-amylase, and isoamylase (isomaltase or glycogen 6-glucanohydrolase) cleave the α-(1 → 6) of amylopectin and glycogen.

Table 4.28 Classification and examples of digestible and resistant starches.

Starch type	Characteristics	Small intestine digestion	Source
Quickly digestible	Digestible	Total	Freshly cooked starchy foods
Slowly digestible	Digestible	Slow, but total	Majority of raw cereals
Resistent starch I	Physically inaccessible	Indigestible	Unprocessed grains, seeds, and legumes
Resistent starch II	Natural granules	Slow, partial	Raw potatoes and bananas, high-amylose maize
Resistent starch III	Retrograded	Partial or resistant	Boiled and stored potatoes, old bread, bread crust, legumes, cornflakes
Resistent starch IV	Chemically modified	Resistant	Not found in nature

Most starches belong to the category of quickly or slowly digestible starches and, under normal conditions, are completely hydrolysed in the small intestine. However, some starches are partially resistant to enzymatic hydrolysis or are unacceptable to the host amylolytic enzymes and rank amongst the non-utilisable polysaccharides that become the components of dietary fibre (e.g. cereal grains consumed raw and protected by cell walls, preventing access to amylases). Four types of resistant starch are recognised (Table 4.28). Resistance to digestion in the small intestine can be lowered by the method of preparation and storage of food. For example, retrogradation of starch increases the resistance to amylolytic enzymes by about 1%. Resistant starches represent about 1% of the total amount of starches consumed. They pass through the colon, where they are partially metabolised (also in the appendix) and used by the colon microflora. Resistant starch can therefore be considered a low-energy material that has a theoretical energy value of only about 8–12 kJ/g (2–3 kcal/g), whilst the energy value of digestible starch is 17 kJ/g (4 kcal/g).

4.5.6.1.5 Use

Starches are important natural components of many food commodities, which significantly affect or determine their texture and functional properties. Native and modified starches are used as additives in many products. The commercial and technological uses of starch are extremely diverse. Native starches are used as fillers and thickeners, gelling agents, water binders, fat substitutes, carriers of odour-active compounds, and stabilisers of foams or emulsions. Those applied directly include native starch granules, dispersed granules, extruded starch, and films obtained by drying starch dispersions. Starches are also an essential raw material for the manufacture of modified starches, sugars, and some sugar derivatives. Roughly half of the starches produced are used as natural or modified starches in food and feed production, the rest are used in many other industries (pharmaceutical, paper, textile, construction, cosmetics, and so on).

Modified starches   The use of native starches is limited due to their physical and chemical properties. Starch grains are insoluble in cold water, and in order to obtain starch dispersion it is necessary to cook the starch. Starch viscosity is often high, and confers to products a rubbery and cohesive texture. Amylose containing starches form rigid, turbid, and retrogradable gels; waxy starches form clear but soft gels, which become turbid during storage at low temperatures. Furthermore, native starches are easily hydrolysed in acidic solutions, for example in acidic sauces during prolonged heating. Native starches are therefore modified in various ways so that some of these undesirable properties are limited or the obtained products have desirable properties. Some modification of starches may also promote the reduction of glycemic index. Modified starches are divided into:

- **transformed** (converted, degraded) starches;

- **cross-linked** starches;

- **stabilised** starches;

- **other modified** starches.

Transformed starches   Transformed starches are obtained from native starches by acid hydrolysis (starches modified by acids), oxidation (bleached and oxidised starches), and heating (dextrinised starches).

During acid hydrolysis, amylose and amylopectin are partially hydrolysed in the amorphous regions of granules, and the product is a so-called **soluble starch**, as the damaged granules swell in cold water. Starches modified by acids are used primarily in the manufacture of confectionery (gum drops and candy, and slow-digestible cookies in the form of resistant-starch-rich powder) and as fillers (as saccharose

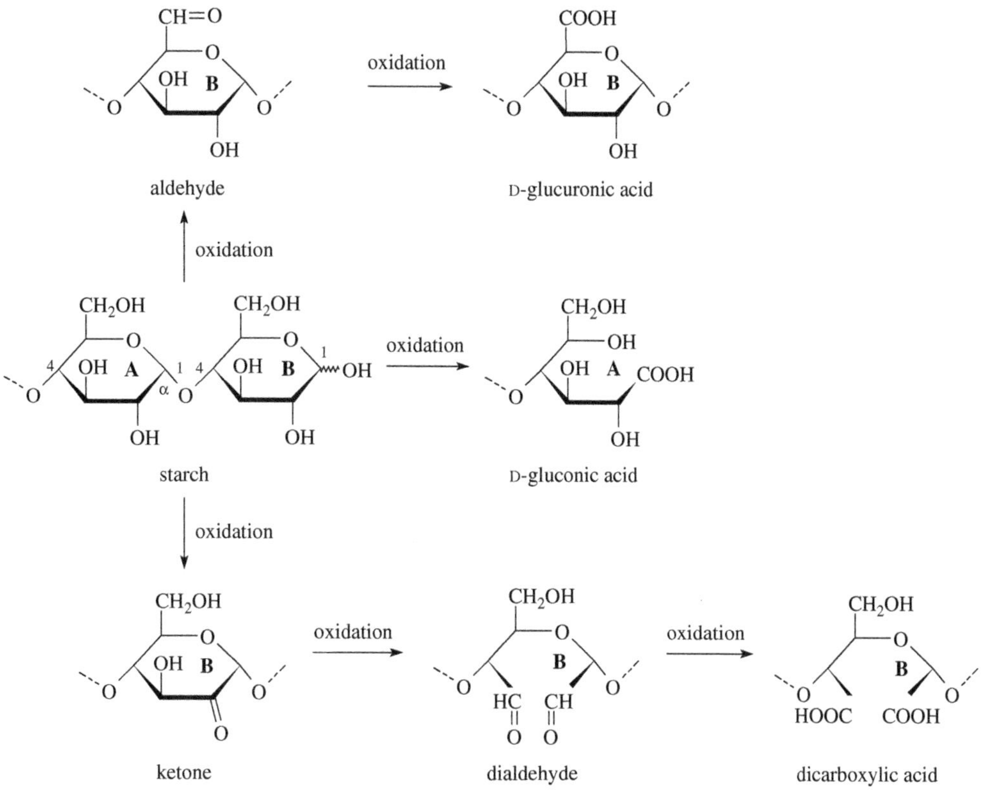

Figure 4.9 Major products of starch oxidation with sodium hypochlorite.

or fat replacers). Soluble starch in blends with native starch is used to prepare pudding powders. In addition to single acid modification, acid hydrolysis in combination with other types of modifications (e.g. hydrothermal, or autoclaving with subsequent β-amylolysis) on starch structure can greatly reduce the starch digestibility.

Oxidation is used to produce two types of starch, **bleached** and **oxidised**. Bleached starches are produced under milder conditions where starch oxidation is minimal, but accompanying colouring substances (especially carotenoids) are removed. Intense oxidation produces oxidised starches that contain carbonyl and carboxyl groups resulting from the oxidation of primary hydroxyl groups at C-6. The secondary alcohol groups at C-2 and C-3 are oxidised to formyl groups, through the ring opening between the C-2 and C-3 carbon atoms, and subsequent oxidation yields a dicarboxylic acid. Ring opening and oxidation of a formyl group in the reducing end of the molecule yields gluconic acid. To some extent, the secondary hydroxyl groups are oxidised to keto groups (Figure 4.9).

Oxidised starches form clear liquid sols and stable gels with a reduced tendency to retrogradation. Carboxylic groups with electric charges of the same sign repel one another, and the tendency of molecules to associate is lower. Oxidised starches have the same use as starches modified by acids. In addition, they are suitable for coating meat and fish, because the package adhesion is significantly higher in comparison with unmodified starches.

Heating of native dry starch or starch acidified by diluted mineral acids yields three basic types of products:

- **white dextrins** (in more acidic media by short heating at lower temperatures);

- **yellow dextrins**;

- **British gums** (in less acidic media).

The main reactions are hydrolysis (the products are mainly linear molecules branched at 2–3% of the glucose units) and transglucosylation (yielding branched dextrins resistant to amylase attack). In later stages of the reaction, high-molecular-weight dextrins are formed by repolymerisation of the hydrolysed products.

The dispersions of white dextrins (hydrolysis predominates) are viscous and have the highest tendency to retrogradation. Yellow dextrins (hydrolysis prevails at the beginning of the process; later, transglucosylation and finally polymerisation take place) contain branched molecules. British gums (the main reaction is transglucosylation) produce the most stable and least viscous dispersions containing higher amounts of branched molecules than yellow dextrins. Yellow dextrins and British gums are soluble in cold water; the least soluble products

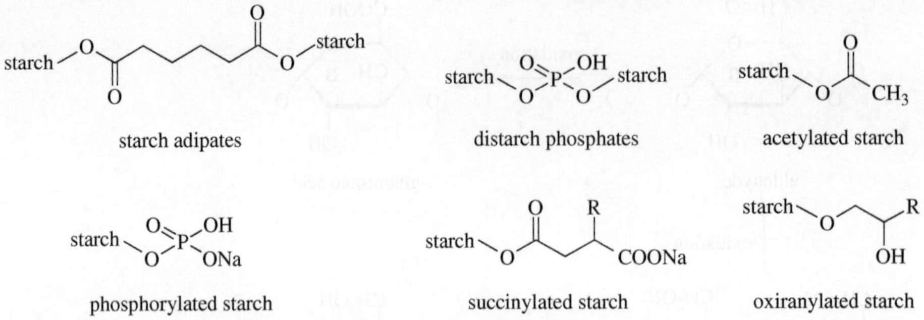

Figure 4.10 Schematic structures of cross-linked starches.

are white dextrins. Dextrins find use as adhesive substances for the preparation of shiny surfaces of sweets and tablets, as carriers for aromatic compounds, spices, and dyes, and for the encapsulation of oils and water-soluble flavours.

Cross-linked starches Cross-linked starches include two types of products, **adipates** and **phosphates** (Figure 4.10). The degree of starch cross-linking is relatively low (usually one cross-link in 1000–2000 glucose units), but the rheological properties of starch are modified significantly. Swollen granules retain their integrity, and the considerable viscosity of dispersions remains virtually unchanged. In contrast to cohesive pastes formed from unmodified starches, pastes of cross-linked starches are non-cohesive. Cross-linked starches are used for thickening, stabilising, and adjusting the texture of foods (pastry fillings, dressings, thickening of soups and sauces), but they are not suitable for products stored at low temperatures.

Stabilised starches The modification in stabilised starches consists in the substitution of some hydroxyl groups of polysaccharides. Stabilised starches are divided into **starch esters** (e.g. acetates, phosphates, and succinates) and **starch ethers** (e.g. hydroxyalkyl ethers). Stabilised starches are usually made from native starches, but also from starches modified by other means (acid hydrolysis, dextrination, and cross-linking). Acetylated starches contain up to 2.5% of acetyl groups. The degree of substitution, which represents the average number of substituted hydroxyl groups per glucose unit, is 0.1. The maximum number is 3 (Figure 4.10). The esterification occurs mainly at C-6 hydroxyl groups, less at C-3, and only partially at C-2. The modification is manifested by a reduced gelatinisation temperature of modified starches (especially of high-amylase starches), higher stability in acidic media, and higher stability against retrogradation during storage of products at low temperatures. The increasing degree of acetylation reduces the ability to form gels by cooling. Acetylated starches (especially pre-cross-linked starches) are used for similar purposes as cross-linked starches.

The degree of substitution in phosphorylated starches is usually <0.25. The decrease in gelatinisation temperature in comparison to non-esterified starches is so great that even phosphates with a degree of substitution of 0.07 swell in cold water. Phosphorylated starches provide non-gelatinisable dispersions with higher clarity, viscosity, and stability at low temperatures. Like other polyelectrolytes, they interact with polyvalent cations (e.g. in hard water) under flocculation. Phosphorylated starches are used for thickening and stabilising of unsalted, non-acidic (sodium salt), and stored refrigerated products, and for production of pudding powders soluble in cold milk.

Succinylated starches have a degree of substitution of 0.02 (one substituent to 50 units of glucose). Succinylated starches are used as thickeners. Some alk(en)ylsuccinylated starches have interesting properties, such as oct-1-enylsuccinated starch (R = $[CH_2]_6$–CH=CH$_2$; Figure 4.10), which is used as the sodium salt or acid. Some types are soluble in cold water and form stable dispersions of higher viscosity than the original starch, and do not tend to form pastes and gels. By the combination of hydrophobic substituent with ionised or non-ionised carboxyl groups, the modified starch has the functional properties of an emulsifier. It is used primarily to stabilise emulsions of oil-in-water in the pharmaceutical (e.g. encapsulation of vitamin A) and food industries (stabilisation of non-polar flavours in soft drinks, mayonnaises, and dressings), or as a substitute for gum arabic.

Examples of starch ethers are 2-hydroxyethyl (R = H; Figure 4.10) and 2-hydroxypropyl (R = CH$_3$; Figure 4.10). The degree of substitution tends to be <0.2. According to the reaction conditions, polyoxaalkyl starches substituted at positions O-2, O-3, and O-6 (**4-54**) can be also prepared.

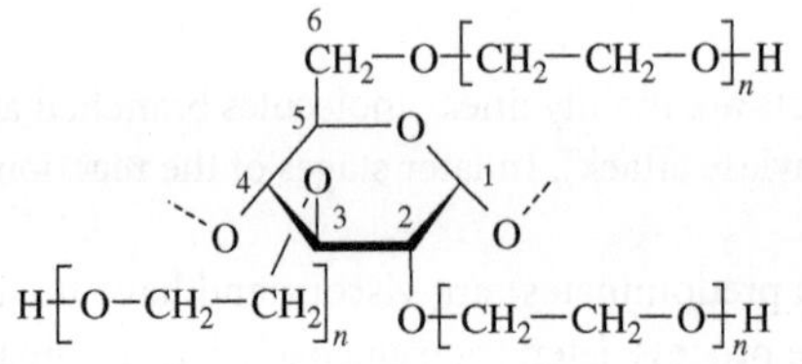

4-54, polyoxaalkyl substituted starch derivatives

In the food industry, hydroxypropylated distarch phosphates are mainly used, providing a series of appropriate functional properties. They have greater stability at low temperatures, and so are used mainly for frozen products. They are suitable for products of low acidity (with pH 5–6), but also for acidic salad dressings. In acidic conditions and at higher temperatures, they show a tendency toward partial hydrolysis, but generally, the ether bond is more stable than the ester bond.

Other modified starches The functionality of modified (cross-linked and stabilised) starches can be increased by additional modifications, which usually include a combination of acid hydrolysis and dextrination, dextrination and cross-linking, and other modifications. Another possibility for increasing the functionality of modified starches is the action of enzymes. For these purposes, pullulanase (isoamylase) is used, hydrolysing α-D-(1 → 6) bonds of amylopectin and dextrins. Such products are, for example, substitutes of caseinates in cheese imitations. Starches swelling in water or milk at ambient temperature (used for the production of puddings, pie fillings, and baby foods as a thickening agent) are prepared by heating the slurry above the gelatinisation temperature and subsequently drying swollen starches These are called **pre-gelatinised starches**. Techniques such as heating of starches in aqueous alcohol, high pressure and high temperature, and alcoholic-alkaline treatment may be used to prepare the granular cold water-soluble starches. Heat–moisture treatment and annealing are the hydrothermal processes in which the granular structure is preserved. Certain non-thermal processes, such as ultrasound, high hydrostatic pressure, and microwave treatments are also used.

Chemical, enzymatic, and physical treatments may additionally be used for the preparation of starch nanoparticles, which may serve as food additives (e.g. as emulsion stabilisers, fat replacers, or packaging components).

Physiology and nutrition Modified starches are considered normal food components, because analogous products are formed during technological operations and starch digestion *in vivo*. Their digestibility is therefore comparable to the digestibility of native starches. Starches stabilised by esterification have comparable degrees of digestibility. Somewhat lower digestibility is shown by cross-linked starches and starch ethers.

Starch hydrolysates In the past, starch hydrolysis was conducted exclusively by acids. At present, starch (native and less frequently modified, e.g. pregelatinised) is hydrolysed by acids, enzymes (commonly several different enzymes are used), or combined methods (partially acid-hydrolysed starch is hydrolysed enzymatically). Depending on the enzymes and technological procedures used, a number of products can be obtained. They are used as sweeteners, fat and sugar substitutes in low-energy products, and agents regulating the texture and other properties of foods. Starch hydrolysates are also used as raw materials for the production of other sugars and substances.

Acid hydrolysis Partial hydrolysis with acid is carried out at higher temperatures than the production of soluble starch (generally in 40% starch suspensions containing HCl at a concentration of 0.02–0.03 mol/l at temperatures of 135–150 °C for 5–8 minutes).

Enzymatic hydrolysis and other reactions Enzymatic hydrolysis provides similar products to acid hydrolysis (dextrins, maltooligosaccharides, maltose, and glucose), but the products are better defined and the process can be better controlled. Biotechnological procedures can be used to obtain products that cannot be provided by chemical hydrolysis (such as fructose and cyclodextrins). Amylolytic enzymes used for starch hydrolysis are exclusively of microbial origin. These enzymes hydrolyse α-(1 → 4) bonds in amylose and amylopectin with the formation of oligosaccharides of different chain lengths. The basic groups of enzymes are **endoamylases** (thermolabile and thermostable α-amylases) and **exoamylases** (maltogenic β-amylase and glucogenic amyloglucosidase, or glucoamylase).

Thermostable α-amylases derived from bacteria of the genus *Bacillus* (*B. subtilis* var. *amyloliquefaciens* and *B. licheniformis*) break down starch into maltooligosaccharides. For example, the primary intermediate generated by α-amylase from *B. licheniformis* is linear maltopentaose, which is gradually degraded to lower carbohydrates. The main degradation product is maltotriose followed by maltose, glucose, and maltotetraose. Maltotriose and maltose are the final products, which are not further hydrolysed. These sugars are also the main products of thermolabile maltogenic α-amylase derived from the fungus *Aspergillus oryzae*. Maltogenic α-amylase breaks down starch (amylose) to maltose almost exclusively. The end product of hydrolysis of amylopectin by β-amylase is called **maltodextrin**. This enzyme, which also partly cleaves the α-(1 → 6) bonds of amylopectin, can be obtained from bacteria of the genera *Bacillus* and *Pseudomonas* and from *Clostridium thermosulfurogenes*.

Glucogenic amyloglucosidase attacks α-(1 → 4) bonds of amylose and amylopectin molecules from the non-reducing end and splits off β-D-glucopyranose. It also partly cleaves the α-(1 → 6) bonds of amylopectin. At high concentrations of glucose, this enzyme acts as transglucosidase and catalyses the reversion of glucose to maltooligosaccharides and isomaltooligosaccharides.

The α-(1 → 6) bonds of amylopectin and dextrins are split by α-(1 → 6)-saccharidases known by the trivial names of pullulanases or 1,6-amyloglycosidases, or by the systematic name of amylopectin 6-glucanohydrolases. These bonds are also cleaved by isoamylases (isomaltases), known by the systematic name of glycogen 6-glucanohydrolases. Pullulanase and isoamylase are obtained from the bacterium *Klebsiella aerogenes*, and isoamylase from bacteria of the genus *Pseudomonas*. Conversion of glucose into fructose is catalysed by glucose isomerase, which is obtained from *Bacillus circulans*.

Hydrolysis of starch (amylodextrins) to non-reducing cyclomaltooligosaccharides called cyclodextrins is catalysed by cyclodextrin glycosyltransferase (also known as CGTase). This enzyme is produced by bacteria of the genus *Bacillus*. *B. macerans* synthesises mainly cyclomaltohexaoses (known as α-cyclodextrins), *B. stearothermophillus* produces cyclomaltohexaoses and cyclomaltoheptaoses (β-cyclodextrins), and the enzyme from *B. subtilis* yields cyclomaltooctaoses (γ-cyclodextrins).

Products of starch hydrolysis   The degree of hydrolysis of starch is characterised by the so-called **dextrose** (glucose) **equivalents** (DE) that indicate the percentage content of free glucose, respectively the content of terminal (reducing) glucose in maltose, maltotriose, and other reducing maltooligosaccharides after conversion to dry matter. Native starch has the value of DE = 0, whilst the hydrolysate containing only glucose has the value of DE = 100. Products with a DE value of about 20 or less form viscous solutions and have no sweet taste. The viscosity of starch hydrolysates decreases with the increased value of DE, whilst their sweetness (which ranges from 25 to 50% of saccharose sweetness) increases. A wide range of products with the values of DE within about 5–95 and more are produced. Depending on the extent of hydrolysis, products contain variable amounts of glucose, maltose, maltotriose, and higher glucooligosaccharides. The products are classified by the content of predominant components as **maltodextrins** and **starch syrups** (maltose syrups and glucose syrups).

Maltodextrins are products with a DE value of ≤20. They generally contain 0.3–1.6% of glucose, 0.9–5.8% of maltose, 1.4–11.0% of maltotriose, 1.4–6.1% of maltotetraose, and 75.5–96.0% of higher sugars. Hydrolysates containing maltodextrins are mostly dried or, exceptionally, prepared as syrups. They are able to increase the viscosity, smoothness, and surface gloss of confectionery products, prevent the formation of crystals in ice creams and frozen dairy products, and are used as carriers for flavours, pigments, and fats, as well as for replacement of gum arabic. Their main use is as fat substitutes.

According to their DE values, starch syrups are divided into type I (DE = 20–38), type II (DE = 38–58), and type III (DE = 58–73). Types II and III fall into the category usually called maltose syrups. Type IV syrups (DE > 73) are glucose syrups. Their basic chemical composition is shown in Table 4.29. Starch syrups are used in making sweets, soft drinks, fruit syrups, and jams, serve as stabilisers of consistency (in ice creams), and are used as substitutes for fat and as raw material for the production of caramel and other products. They are fermentable and serve in the biotechnological production of various substances (such as ethanol and citric acid). Maltose is isolated from syrups with a high maltose content. Isomerisation of maltose in maltose syrups in alkaline media is employed for the preparation of products that contain maltulose as the main component. Hydrogenation leads to sugar alcohol maltitol, to be more precise a mixture of alditols that contain as the essential components maltitol and maltotriitol. Glucose syrups are used for the isolation of glucose and are also used for the production of fructose syrups. Hydrogenation yields a mixture of sugar alcohols, in which the main component is D-glucitol.

Fructose syrups are obtained by the action of bacterial enzyme glucose isomerase (from *B. circulans*) on glucose syrups. Historically, the first products contain fructose in the amount of about 42% d.w. (content of glucose was 52%, contents of higher sugars and dry matter were 6 and 75%, respectively). Today, produced syrups contain 55% fructose (40% glucose, 5% higher sugars, 77% dry matter) and have a similar composition and similar sweetening power as invert sugar (about 100% relative to saccharose sweetening power). Using different physico-chemical methods, these syrups are used to produce high-fructose syrups containing up to about 90% fructose (9% glucose, 1%

Table 4.29 Composition of selected starch hydrolysates (% of dry matter).

DE value[a]	Glucose	Maltose	Maltotriose	Maltotetraose	Higher saccharides
Acid hydrolysate	10	9	8	7	66
28					
55	31	18	12	10	29
Enzymatic hydrolysate	0	1	1	1	97
5					
10	1	4	6	5	84
Acid and enzymatic hydrolysate					
15/22	2	6	8	–[b]	84
10/95	93	–[c]	–[c]	–[c]	–[c]

[a]The first number for combined acid and enzymatic hydrolysis is the DE value after acid hydrolysis, whilst the second number is the DE value after enzymatic hydrolysis.

[b]The content is included in the content of higher sugars.

[c]The total content of oligosaccharides approaches 7%.

higher sugars, 80% dry matter). Their sweetening power is 160–180% that of saccharose. Fructose syrups are used similarly to invert sugar and saccharose and are particularly suitable for sweetening soft drinks and confectionary.

Special products similar to maltose syrups are maltooligosaccharides containing two to seven glucose units linked to one another by α-D-(1 → 4) bonds. They are obtained by enzymatic hydrolysis of starch with isoamylase or pullulanase and by hydrolysis of the obtained products using α-amylase. Maltooligosaccharides are fully utilisable sugars, which are used similarly to maltose syrups.

Other special products are isomaltooligosaccharides containing two to five glucose units linked by α-D-(1 → 6) and partly by α-D-(1 → 4) bonds, which occur in the trisaccharide panose. They are obtained by hydrolysis of starch by α-amylase and by the hydrolysis of the thus obtained products using β-amylase and α-glucosidase (α-D-glucoside glucohydrolase) under controlled conditions. They are not utilisable, but stimulate the growth of bifidogenic bacteria in the colon.

Glucose syrups are also transformed by enzymatic transglucosylation (using amyloglucosidase activity) to gentiooligosaccharides that contain two to five glucose units linked by β-D-(1 → 6) glycosidic bonds which occur in the disaccharide gentiobiose (Table 4.11). They are used for the same purposes as other indigestible oligosaccharides, mostly as fillers in low-energy products. Like other indigestible oligosaccharides, gentiooligosaccharides also stimulate the growth of bifidogenic microflora in the colon.

Cyclodextrins Cyclodextrins (formerly called Schardinger dextrins) are cyclic maltooligosaccharides to maltopolysaccharides containing 6–12 glucose molecules joined by α-D-(1 → 4) bonds. They are obtained by hydrolysis of amylodextrins (DE < 10) by the action of cyclodextrin glucanotransferase. The main representatives are cyclomaltohexaose (formerly Schardiner α-dextrin or α-dextrin, **4-55**), cyclomaltoheptaose (β-cyclodextrin), and cyclomaltooctaose (γ-cyclodextrin), which are composed of six, seven, and eight glucose monomers, respectively.

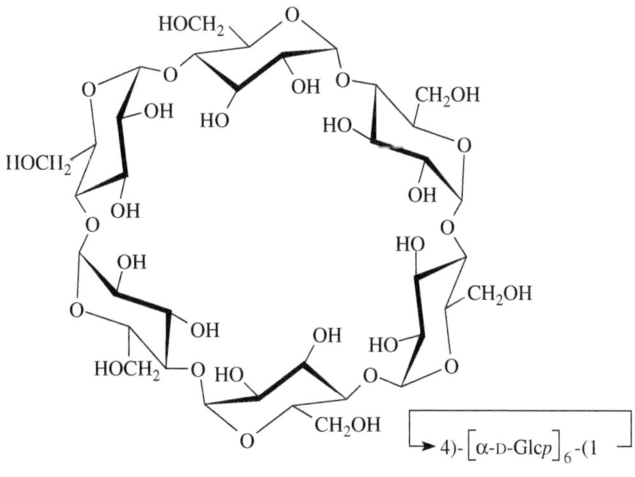

4-55, cyclomaltohexaose

Cyclodextrins form crystalline inclusion complexes with many organic compounds, including some gases, which are bound within the molecule. Formation of these complexes is called **encapsulation**. Encapsulation results in a change of the physico-chemical properties of encapsulated compounds (e.g. the volatility of flavour-active compounds and their increased stability against oxidation and photodegradation). Cyclodextrins are, therefore, of greatest use as carriers (encapsulators) of odoriferous substances, as emulsion stabilisers, and in the removal of bitter substances from citrus juices (see Section 8.3.5.1.1).

4.5.6.2 Other storage polysaccharides

Tubers, roots, seeds, and some vegetative parts of plants contain non-starch storage polysaccharides, which are involved in processes associated with germination and growth. Most of these polysaccharides are structurally similar to non-cellulose polysaccharides of cell walls that are classified as hemicelluloses and pectins. The most important representatives of this group of polysaccharides are:

- **heterofructans**;

- **heteromannans**;

- **heteroglucans**.

4.5.6.2.1 Heterofructans

Polymers and oligomers of D-fructose are called fructans or fructosans, but they often include D-fructose polymers containing one molecule of D-glucose as the terminal unit, which are correctly named glucofructans or glucofructosans. Fructans and glucofructans are synthesized as energy reserves by about 15% of higher plant species and by some microorganisms (some species of the mould genera *Aspergillus*, *Claviceps*, *Fusarium*, *Penicillium*, and yeasts, such as *S. cerevisiae* and others). In addition to being storage polysaccharides, they play a role in osmoregulation in plants and in their tolerance to some stress factors (cold and lack of water).

Natural fructans (glucofructans) are usually classified into:

- **inulins**;

- **levans**, otherwise called **phleins**.

Polymers referred to as inulins are composed of linear chains of D-fructofuranoses (fructans) usually containing a D-glucopyranose molecule (glucofructans) as a terminal unit. They are mutually bound by β-(1 → 2) glycosidic bonds. Typically, inulin is found in some plants of the Asteraceae family, for example in chicory roots (*C. intybus*), yacon roots (*Smallanthus sonchifolius*) native to the Andean region of South America, tubers of Jerusalem artichokes (*H. tuberosus*, also known as topinambours), and the fleshy flower heads of globe artichokes (*Cynara scolymus*).

Levans are branched polymers of D-fructofuranoses associated exclusively by β-(2 → 6) glycosidic bonds. In addition to inulin, they may also contain D-glucose as the terminal unit. Typical levans are synthesised, for example, by the bacterium *Bacillus subtilis* from saccharose or raffinose in sugar beet juice during sugar production. Levans with β-(2 → 6) glycosidic bond also predominate in oats.

Glucofructans that contain both types of bonds (1–4% of branched molecules) occur most frequently as components of numerous cereals (wheat, rye, triticale, and barley), fruits, and vegetables.

In the past, the name inulin was only used for polysaccharides of chicory, Jerusalem artichokes, and dahlias. Inulin of asparagus was called **asparogesin**. Trivial names were also used for some cereal fructans and glucofructans. Wheat levan with β-(2 → 6) bonds was called **pyrosin**, rye levan was **secalin**, wheat polysaccharide with both types of bonds was **triticin**, and the corresponding polysaccharide of rye was known as **graminin**.

If the molecule end is α-D-glucopyranose, the abbreviated notation is GF_n (G = glucose, F = fructose, n = degree of polymerisation). The prototype of glucofructans of inulin type is actually disaccharide saccharose (GF). The enzymatic hydrolysis of inulin by native plant endoglycosidases, known as 2,1-β-D-fructan fructanohydrolase or inulase, provides linear polymers of type GF_n or, after hydrolysis of glucose, F_n-type polymers. The latter type is a linear homopolymer of fructose, which always accompanies inulin. Low-molecular-weight sugars with a degree of polymerisation 2–10 are called fructooligosaccharides or oligofructose. Typical microbial levans (of bacteria *Bacillus subtilis* and other bacteria) are of type $G-F_n$ or F_m-G-F_n. The structure of a glucofructan of a mixed type occurring in wheat and other grains is shown in formula **4-56**.

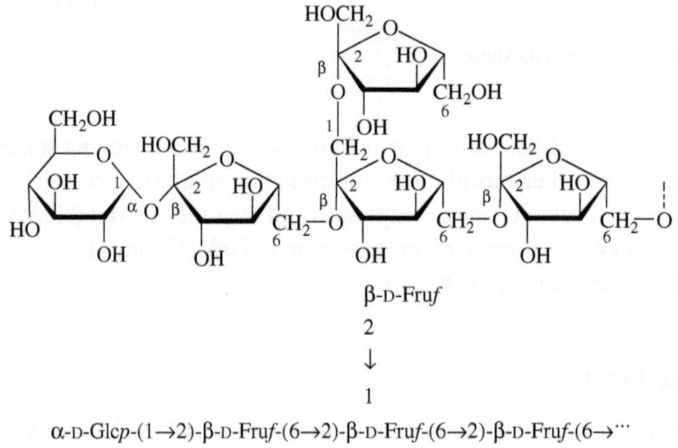

$$\alpha\text{-D-Glc}p\text{-}(1\rightarrow2)\text{-}\beta\text{-D-Fru}f\text{-}(6\rightarrow2)\text{-}\beta\text{-D-Fru}f\text{-}(6\rightarrow2)\text{-}\beta\text{-D-Fru}f\text{-}(6\rightarrow\cdots$$

4-56, glucofructan of mixed type

Glucofructans, together with oligofructose (oligosaccharides formed by partial enzymatic hydrolysis), occur in many foods of plant origin. They are highly polydisperse mixtures of related compounds with a degree of polymerisation (number of bound fructose molecules) $n = 2$–60 and sometimes more. Therefore, they can be classified simultaneously as oligosaccharides and polysaccharides. In some plants,

Table 4.30 Content of glucofructans and fructans in the fresh edible part of some crops.

Source	Latin name	Glucofructans (%)	Fructans (%)
Asparagus	*Asparagus officinalis*	2–3	2–3
Bananas	*Musa* sp.	0.3–0.7	0.3–0.7
Chicory (root)	*Cichorium intybus*	15–20	8–11
Garlic	*Allium sativum*	9–16	3.5–6.5
Globe artichokes	*Cynara scolymus*	3–10	0.3–1
Jerusalem artichokes	*Helianthus tuberosus*	16–20	12–15
Onion	*Allium cepa*	1.1–7.5	1.1–7.5
Leek	*Allium ampeloprasum*	3–10	2.5–8
Rye	*Secale cereale*	0.5–1	0.5–1
Wheat	*Triticum aestivum*	1–4	1–4
Yacon	*Smallanthus sonchifolius*	10–12	–

only oligosaccharides are present, while the higher polymers are absent. For example, levans of oats contain two to five monosaccharide molecules, the main glucofructan of onion is an oligomer with a degree of polymerisation $n = 5$, yacon roots are of a low degree of polymerisation ($n = 3–9$) containing as their main component the GF_3 oligosaccharide, inulin of Jerusalem artichokes (topinambours) has an average degree of polymerisation $n = 30$, and globe artichokes synthesise inulin containing up to 200 monosaccharide molecules (Table 4.30). Glucofructans and fructans are commonly accompanied by saccharose ($n = 1$), fructose, glucose, and other saccharides. Some aminoacyl sugars, e.g. saccharose acylated with amino acids (glycine, alanine, valine, threonine, tyrosine, tryptophan, and histidine), have also been identified at the C-2 of the glucose moiety in sweet potato (*Ipomoea batatas*, Convolvulaceae).

In the enzymatically active material, inulin is easily hydrolysed by endogenous inulase (inulinase), and the resulting monosaccharides are degraded during thermal processes. Such an example may be changes in the content of inulin during storage and thermal processing of chicory root, which is used for the manufacture of coffee substitutes. When chicory root is stored at 5 °C, inulin is completely enzymatically hydrolysed to oligofructose F_n ($n = 3–7$) within six weeks. Roasting of chicory root leads to extensive degradation of inulin, oligofructose, and higher reducing sugars, which caramelise and participate in the Maillard reaction with proteins and amino acids. The reaction is dependent on the temperature and roasting time. The optimum roasting time is about 55–60 minutes at 170 °C or 14–15 minutes at 175–200 °C. A well roasted product should then contain 13–20% of fructose and 28–32% of the original amount of inulin (a fraction with a lower degree of polymerisation than that of original inulin).

Glucofructans and fructans are physiologically active polysaccharides or oligosaccharides that are classified as dietary fibre because they are not hydrolysed in the upper gastrointestinal system by saliva and pancreas hydrolases or by intestinal hydrolases (α-amylase, saccharase, maltase, and other saccharidases). They also act as a growth factor for bifidobacteria in the colon. The daily intake of inulin and its hydrolytic products per capita in developed countries is estimated at 2–12 g.

Inulin extracted from Jerusalem artichokes or chicory roots was used as the raw material for the production of fructose syrups and fructooligosaccharides. At present, the main raw material for the production of fructose syrups is starch and the fructooligosaccharides are mainly produced from saccharose.

4.5.6.2.2 Heteromannans

In addition to starch and heterofructans, the storage polysaccharides of some seeds are heteromannans. Their main chain is homopolymeric and formed by D-mannose units linked to one another by β-($1 \rightarrow 4$) glycosidic bonds. Some of the mannose units are bound by ($1 \rightarrow 4$) bonds to α-D-galactose. These heteromannans are therefore called **galactomannans** (4-57).

Galactomannans that differ in their degree of substitution of the main chain by D-galactose (4–96% of mannose units are substituted) are found in the seeds of many plants. Palm seeds (e.g. seeds of dates) contain galactomannans in which about 4% of mannose units are substituted. Related galactomannans with a degree of substitution of 30–96% are located in the endosperm of coffee seeds and in some legumes. The most important representatives of galactomannans, which have found use as food hydrocolloids, are **guar gum** and **locust gum**.

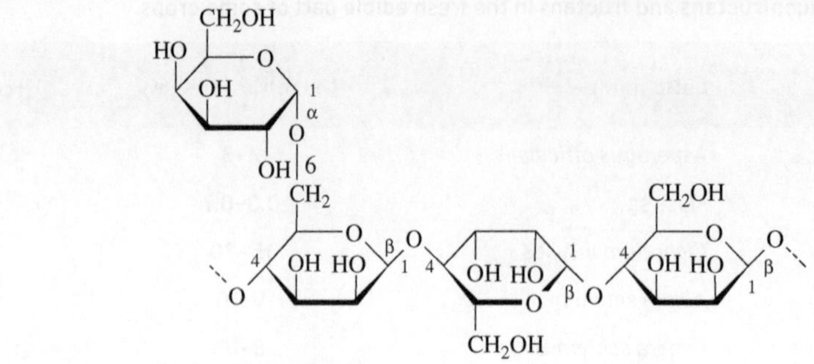

α-D-Gal$p$

1

$\downarrow$

6

····→4)-β-D-Man$p$-(1→4)-β-D-Man$p$-(1→4)-β-D-Man$p$-(1→····

4-57, basic structure of galactomannans

Related polysaccharides are linear copolymers in which some of the mannose units are randomly substituted by residues of D-glucopyranose bound by β-(1 → 4) bonds. These heteromannans are called **glucomannans**. The mannose to glucose ratio varies from 1.5 to 4. About 2–6% of sugars in the main chain are substituted by β-D-galactopyranose bound by (1 → 3) glycosidic bonds. Mannose residues (every 5th to 20th) may be acetylated in the C-2 and C-3 positions (**4-58**). Glucomannans are found in varying amounts in plant tubers, roots, and seeds. The only glucomannan that is used as a food hydrocolloid is **konjac gum**.

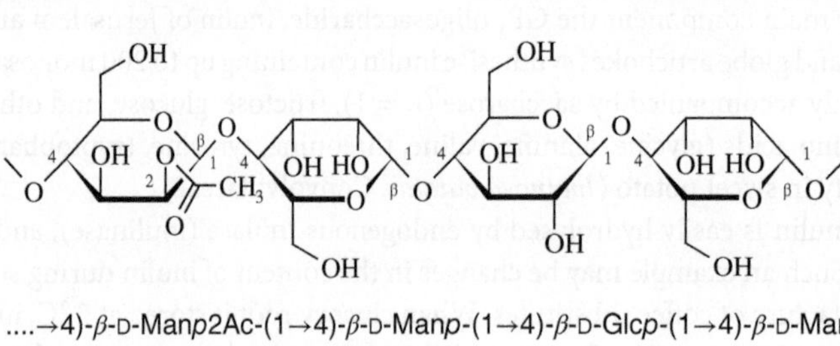

....→β-D-Man$p$2Ac-(1→4)-β-D-Man$p$-(1→4)-β-D-Glc$p$-(1→4)-β-D-Man$p$-(1→....

4-58, basic structure of glucomannans

Guar gum Guar gum (guaran) is obtained as the flour from the endosperm of guar bean seeds (*Cyamopsis tetragonoloba*, Fabaceae) after separating the germ and the surface layer. The plant is native to Central Africa, but today it is grown primarily in India, Pakistan, and the United States (Texas). Guar gum is a neutral galactomannan, whose primary structure is shown in formula **4-57**. Approximately every second residue of D-mannose is substituted by a D-galactose residue (the mannose to galactose ratio varies from 1.5 to 1.8). The minor sugars are D-glucose, D-xylose, and L-rhamnose (Table 4.31). The average molecular weight of guar gum polysaccharides is usually in the hundreds of kilodaltons.

Table 4.31 Basic composition of guar and locust gum monosaccharides.

Gum	Latin name of plant	Neutral sugars (% w/w)			
		L-Arabinose	D-Glucose	D-Mannose	D-Galactose
Guar	*Cyamopsis tetragonoloba*	1.3–2.2	1.5–4.5	45.6–56.5	28.6–37.2
Locust	*Ceratonia siliqua*	0.8–1.4	1.5–2.7	59.3–69.9	16.0–18.2

Guar gum is soluble in water, giving highly viscous solutions stable in the pH range 4–10. Gel is formed in the presence of small amounts of borates. Guar gum can be combined with almost all natural gums, starches, pectins, and cellulose and its derivatives, and very often is combined with xanthan gum, which increases (as a synergist) viscosity of dispersions.

Guar gum has a very wide range of applications and is the most commonly used gum of plant seeds. It is used as a thickener, viscosity modifier, and stabiliser of dispersions in foods and beverages. It is used not only in the food industry, but also in the paper industry as a glue, and often in cosmetics.

Locust gum Locust gum, also called locust bean gum, carob, carobin, or algarroba, is obtained as flour from the endosperm of seeds of the carob tree also known as St John's bread (*Ceratonia siliqua*, Caesapliniaceae). The tree comes from the Western Mediterranean region (Southern Europe, Northern Africa), but now grows mainly in Spain and the subtropical regions of the United States and Australia.

Locust gum is galactomannan (**4-57**), as is guar gum, and is substituted on approximately every fourth mannose unit by galactose residues (the ratio of mannose to galactose is 3.6–4.2; Table 4.31). The average molecular weight is approximately 310 kDa. The galactomannan content in the flour is around 88%.

The properties of locust gum are similar to those of guar gum, but the flour from locust beans is insoluble in cold water and only swells, although a viscous dispersion (of somewhat lower viscosity than a dispersion of guar gum) is obtained by heating. Locust gum is compatible with the majority of plant and microbial hydrocolloids and proteins (gelatin). Like guar gum, it does not form gels, but increases the elasticity and strength of agar and carrageenan gels, although a gel is formed with xanthan, which does not form a gel itself. In acidic media, locust gum is prone to hydrolysis.

Locust gum binding water is used as an emulsion stabiliser (e.g. in meat products), as a thickening agent for dairy products, and as fillings for frozen foods and bakery products, as it bounds water. Like guar gum, locust gum is not suitable for the preparation of clear dispersions as it contains insoluble residues of seeds. It is also used in the manufacture of cosmetic and pharmaceutical products. Locust gum is a relatively expensive hydrocolloid, so it is frequently substituted with cheaper guar gum.

Konjac gum The only commercially important glucomannan is konjac gum, also known as konjac mannan. Konjac flour is obtained from the starchy tubers of the plant *Amorphophallus konjac* (Araceae), grown in subtropical to tropical East Asia (Japan, China, and Indonesia).

The basic structure of konjac gum polysaccharides is represented by formula **4-58**. The dominating sequences are composed of one, two, and five molecules of mannose and one or two molecules of glucose. The mannose to glucose ratio is around 1.6, and the acetyl groups content is about 15%. The average molecular weight is high, around 6000 kDa.

Konjac mannan dissolves in water to give highly viscous dispersions, which are pseudoplastic systems. Heating with a small amount of alkaline agents gives (under deacetylation) longer blocks of an unsubstituted polymannose skeleton capable of mutual associations with the formation of thermostable elastic gels. Konjac flour is used mainly in East Asian countries as a thickener and gelling agent, especially in traditional Japanese cuisine in the preparation of noodles and gels, which are insoluble even in boiling water. It is also used for the preparation of edible films and coatings, which are semi-permeable to water vapour and oxygen. It acts synergistically with κ-carrageenan, xanthan gum, and starch and affects the viscosity of their dispersions and structure of the gels.

4.5.6.2.3 Heteroglucans

Other plant-storage polysaccharides are **xyloglucans**. Their primary structure consists of a chain of D-glucose units linked by β-(1 → 4) bonds. The chain branches consist of D-xylopyranose units linked by α-(1 → 6) bonds. The C-2 atoms of about one-half of the xylose units are connected to β-D-galactose molecules by glycosidic bonds (**4-59**). These xyloglucans are related to structural xyloglucans of cell walls (components of hemicelluloses), but they are not bound to cellulose and are extractable with hot water. The only xyloglucan that is used as a food hydrocolloid is **tamarind gum**.

Tamarind gum Tamarind gum is a xyloglucan (**4-59**) located in the endosperm of the seeds of the tamarind tree (*Tamarindus indica*, Fabaceae), probably originating from Africa, but now grown in tropical regions of other continents. The viscosity of dispersions is increased by heating. Under acidic conditions, however, the tamarind gum polysaccharides undergo rapid hydrolysis. In the presence of high concentrations of saccharose (>65%), and over a wide range of pH values, tamarind gum forms gels that are stronger than pectin gels. Tamarind gum is only rarely used in food production, except in some special applications (as a thickener and foam stabiliser). The main use of tamarind gum is in the textile industry.

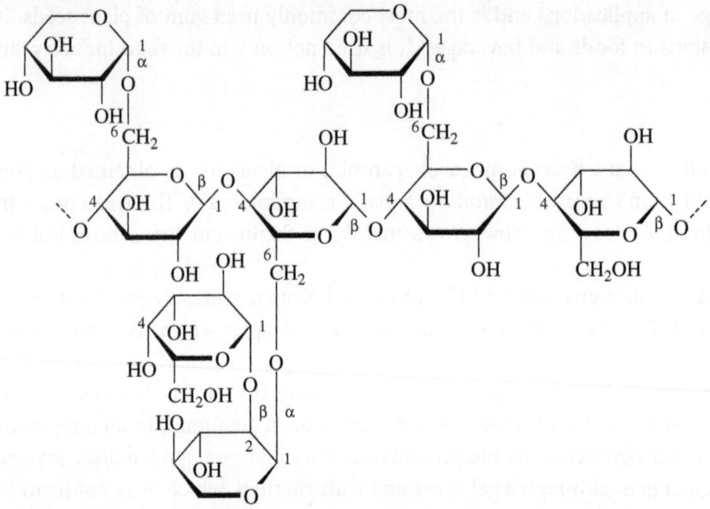

$$\cdots\rightarrow4)\text{-}\beta\text{-}D\text{-}Glc}p\text{-}(1\rightarrow4)\text{-}\beta\text{-}D\text{-}Glc}p\text{-}(1\rightarrow4)\text{-}\beta\text{-}D\text{-}Glc}p\text{-}(1\rightarrow4)\text{-}\beta\text{-}D\text{-}Glc}p\text{-}(1\rightarrow\cdots$$

```
        6              6              6
        ↑              ↑              ↑
        1              1              1
    α-D-Xylp       α-D-Xylp       α-D-Xylp
                       2
                       ↑
                       1
                   β-D-Galp
```

4-59, basic structure of storage xyloglucans

4.5.6.3 Cellulose

Cellulose is the most widespread natural organic compound. It occurs as a basic structural polysaccharide in the cell walls of higher plants. It is also found in green algae, fungi, and, exceptionally, cell walls of simple marine invertebrates (tunicates).

4.5.6.3.1 Structure

Homoglucan cellulose (**4-60**) is a high-molecular-weight linear polymer of D-glucose units linked by β-(1 → 4) glycosidic bonds. Each of the glucose units bound in the chain is turned toward the previous unit, and this position is maintained by intramolecular hydrogen bonds between the hydroxyl group at C-3 and oxygen of the pyranose ring and between the hydroxyl groups at C-2 and C-6 (**4-61**). The degree of polymerisation is as high as 15 000. Individual cellulose macromolecules interact with one another through hydrogen bonds to form more or less organised three-dimensional structures in the walls of plant cells called cellulose fibres or cellulose **microfibrils**, which have high tensile strength. They have a thickness of approximately 10–20 nm, a length of several μm, and contain about 30–100 cellulose macromolecules, arranged in parallel in microfibrils and forming a planar unit (sheet). Individual sheets are arranged so that they are alternately shifted by

$$\beta\text{-}D\text{-}Glc}p\text{-}(1\rightarrow4)\text{-}[\beta\text{-}D\text{-}Glc}p\text{-}(1\rightarrow4)\text{-}\beta\text{-}D\text{-}Glc}p\text{-}(1\rightarrow4)]_n\text{-}\beta\text{-}D\text{-}Glc}p$$

4-60, primary structure of cellulose

4-61, linear cellulose chain stabilised by hydrogen bonds

half the length of the glucose units. This arrangement is stabilised by intermolecular hydrogen bonds between the glucose units in adjacent sheets (between oxygens of the pyranose rings and hydroxy groups at C-6). Such areas of microfibrils with high numbers of intermolecular bonds are crystalline; less structured areas with a low degree of interaction are amorphous. The crystalline and amorphous regions are alternated in the microfibrils. A higher proportion of crystalline areas are in the microfibrils of secondary cell walls (e.g. in wood), whilst amorphous regions predominate in primary cell walls found in the flesh of fruits and vegetables. Compared with starch, cellulose is much more crystalline, and several crystalline structures (allomorphs) are recognised. Cellulose I is natural cellulose, cellulose produced by bacteria and algae is mainly a cellulose I_α allomorph, whilst cellulose I_β is mainly found in higher plants.

Metabolically active (primary) cell walls of plant tissues have a common structure consisting of randomly oriented cellulose microfibrils with a predominant amorphous structure. Microfibrils are bound through interactions and covalent bonds by a gel matrix composed of other structural polysaccharides, called non-cellulose polysaccharides (hemicelluloses and pectins), lignin, and polypeptides, such as extensin (Figure 4.11), to form a biocomposite. The most important hemicelluloses in fruits and vegetables are xyloglucans, whilst cereals contain mainly arabinoxylans and β-glucans.

Differences in the structures of cell walls of various plants and their parts are related to the differentiation of the primary cells, cellulose crystalinity, and the type and amount of non-cellulose polysaccharides. Deposition of additional layers of cellulose (in the form of clearly oriented parallel microfibrils), deposition of non-cellulose polysaccharides, and lignification of the polysaccharide network result in the formation of thick secondary cell walls, which have various special functions. For example, they ensure the rigidity of tissues and the transport of water and have protective and other functions.

4.5.6.3.2 Occurrence

Cellulose represents a significant proportion of non-starch polysaccharides in foods (Table 4.20) and is a substantial part of the so-called insoluble fibre (Table 4.21). Fruits and vegetables contain around 1–2% of cellulose, cereals and legumes 2–4%, and wheat flour only 0.2–3%, depending on the milling process (Table 4.19), because a large proportion of cellulose is found in the bran (30–35%). Cellulose also forms about 40–50% of wood mass, 80% of linen fibres, and 90% of cotton fibres.

4.5.6.3.3 Properties

Cellulose is insoluble in water, dilute acids, bases, and most solvents. Solvents, however, penetrate into the more accessible amorphous regions of microfibrils, which leads to swelling, but always less than in starches. Cellulose can be dissolved in concentrated acids, because, depending on the conditions (acid concentration, temperature), it is hydrolysed to soluble fragments with shorter chains, such as disaccharide cellobiose, or to glucose. Swelling in solutions of hydroxides is larger than that in water or acidic solutions. At elevated temperatures, cellulose is hydrolysed and oxidised.

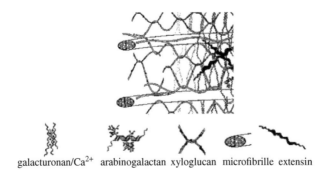

galacturonan/Ca^{2+} arabinogalactan xyloglucan microfibrille extensin

Figure 4.11 Schematic structure of plant cell walls. Carpita and Gibeaut 1993. Source: Reproduced by permission of John Wiley and Sons.

4.5.6.3.4 Physiology and nutrition

Cellulose is hydrolysed by a complex of cellulolytic enzymes of certain microorganisms (bacteria and moulds) and higher fungi called cellulases that break down plant tissues. Cellulase or *endo*-1,4-β-glucanase (1,4-β-D-glucan 4-glucanhydrolase) is an endoenzyme that splits β-(1 → 4)-D-glucosidic bonds in the amorphous regions of cellulose microfibrils with the formation of glucooligosaccharides and cellobiose. Cellobiohydrolase, also known as *exo*-β-D-1,4-glucanase (1,4-β-D-glucan cellobiohydrolase or avicelase), is an exogenous enzyme that splits β-(1 → 4) bonds from the non-reducing end of the chain to form oligosaccharides and cellobiose. Another enzyme, β-glucosidase or cellobiase (β-D-glucosid glucohydrolase), splits off cellobiose from the partially hydrolysed cellulose, and higher oligosaccharides are hydrolysed to β-D-glucose.

Vertebrates do not have their own cellulase, but the digestive tracts of herbivores contain symbiotic bacteria that produce cellulolytic enzymes. Cellulose is therefore a utilisable polysaccharide for polygastric animals. It is broken down into glucose, which is fermented by bacteria to give lower fatty acids that are absorbed and utilised by the animals. Monogastric animals, including human beings, do not have cellulolytic enzymes, and cellulose is a non-utilisable polysaccharide. Together with other polysaccharides, known as dietary fibre, cellulose is an important and beneficial food component.

4.5.6.3.5 Use

Native cellulose is added to some foods as a non-caloric thickener, causing turbidity, and to products processed by extrusion. However, **modified celluloses** have more applications in the food industry. Cellulose can be modified by either physical or chemical processes. Nanofibres of cellulose and cellulosic materials have good potential for use in pharmaceutical and food systems, as well as in food technology. The most important applications are immobilisation of bioactive substances such as enzymes, vitamins, and antimicrobials, nutraceutical delivery systems with controlled release of bioactive compounds, films, biosensors, and biodegradable food packaging materials.

Modified celluloses Physical modification (influence of high tangential stress and pressure) leads to cleavage of cellulose microfibrils; the product, with a high ability to bind water, is called **microfibrillar cellulose**. It can be used in food products as a non-caloric thickening agent and flavour carrier, in skin creams and paints, and as a carrier for medicines.

The two main groups of chemically modified celluloses are:

- **hydrolysed cellulose**;

- **derivatised cellulose**.

The only representative of hydrolysed cellulose is **microcrystalline cellulose**. It is obtained by partial acid hydrolysis of cellulose, which dissolves the amorphous zones, but leaves the crystalline zones unchanged. The product has properties of thixotropic and pseudoplastic systems. Viscosity is independent of temperature and pH. Functional properties remain constant even at high temperatures and in acidic media (during baking, microwave heating, and ultra high-temperature processes). Microcrystalline cellulose is used as a dietary fibre, low-energy bulking agent, carrier of flavourings, stabiliser of foams, and in extrusion technologies.

Of the many derivatives of cellulose, only some cellulose ethers have been used in food technologies. The most commonly used derivative is **carboxymethylcellulose** (its sodium salt, degree of substitution 1.0), followed by **methylcellulose** (degree of substitution 2.0) and **hydroxypropylcellulose** (degree of substitution 2.0, **4-62**).

4-62, cellulose ethers

R = CH_2COONa, R^1 = H, carboxymethylcellulose (sodium salt)
R = R^1 = CH_3, methylcellulose
R = R^1 = $CH_2\,CH(OH)CH_3$, hydroxypropylcellulose

In the polyelectrolyte carboxymethyl cellulose (sodium salt), the degree of substitution ranges from 0.4 to 1.5. In solutions of pH < 4, carboxymethyl cellulose is present mainly as a free acid of low solubility that produces turbid solutions with divalent cations, such as Fe and Zn (the salts are poorly soluble). In the presence of trivalent ions (Al and Fe), gels and precipitates of carboxymethyl cellulose are formed. Similar properties and uses have other cellulose ethers, such as methylcellulose, hydroxypropylcellulose, hydroxypropylmethylcellulose (hypromellose), and, less commonly, ethylmethylcellulose.

Unlike other gums, cellulose ethers show thermal gelatinization, as their colloidal solutions (at concentration of >1.5% by weight) form gels by heating to 50–85 °C. On cooling, these gels melt to viscous colloidal solutions, which are non-Newtonian pseudoplastic systems.

Cellulose ethers are used as thickeners (carboxymethylcellulose in cottage cheeses and cheese spreads), emulsion stabilisers (sauces, soups, and dressings), protein solubilisers (gelatin and casein), retarders of crystal formation (ice creams), and foaming agents (hydroxypropyl cellulose). They can be added to bakery and confectionery products to increase its water binding capacity and to reduce fat absorption by a product (e.g. doughnuts during frying), to slow the retrogradation of frozen products, and to produce edible films (gels) that protect, for example, frozen products against dehydration (desiccation).

4.5.6.4 Callose

Callose is an amorphous β-(1 → 3) glucan accompanying cellulose in plant cell walls, but it is less common than cellulose. Most plants synthesise callose in response to injury or only at certain stages of development of cell walls. The basic building unit of callose is disaccharide laminaribiose, β-D-glucopyranosyl-(1 → 3)-D-glucopyranose (**4-63**).

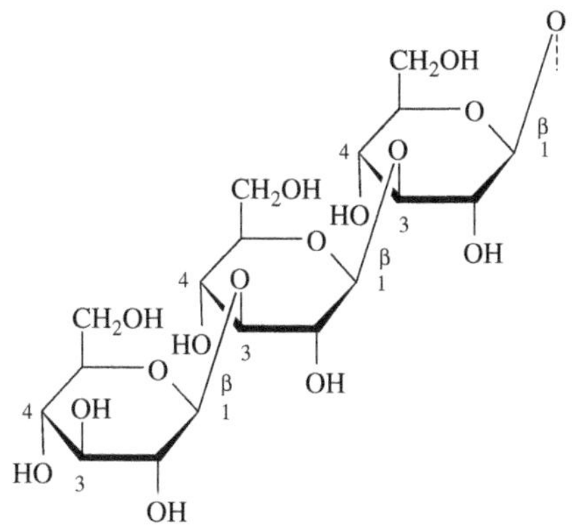

4-63, basic structure of callose

β-D-Glcp-(1→3)-β-D-Glcp-(1→3)-β-D-Glcp-(1→....

4.5.6.5 Hemicelluloses

The term hemicellulose is a common name for non-cellulose structural polysaccharides of plant cell walls that fill the spaces between the cellulose fibres (Figure 4.11). Hemicelluloses include two main groups of polysaccharides:

- **heteroglucans**;

- **heteroxylans**.

Plant cell walls contain a large number of other polysaccharides that can be included under the term dietary fibre. Other important hemicelluloses are **heteromannans** (galactomannans and glucomannans), which are present in lower amounts in plant cell walls, constituting about 12–15% of cell wall polysaccharides in the flowering plants (angiosperms). Their role as the building components of cell walls is still not well known. In larger quantities, heteromannans are present in some legume seeds, where they have the function of storage polysaccharides. Structures of heteromannans are illustrated by formulae **4-57** and **4-58**.

4.5.6.5.1 Heteroglucans

The main structural heteroglucans that are classified as hemicelluloses are:

- **xyloglucans**;

- **β-glucans**.

Xyloglucans The basic structural unit of the xyloglucan molecule is β-D-(1 → 4)-glucan (cellulose), with units of D-xylopyranose in side chains that are bound to glucose by α-(1 → 6) glycosidic bonds. Unlike the related reserve xyloglucans present in seeds of plants (**4-59**), structural xyloglucans contain the residue of D-galactose bound by β-(1 → 2) bonds to D-xylose near the reducing end of the polysaccharide. They may also contain L-fucopyranose linked to an L-galactose by an α-(1 → 2) bond (**4-64**).

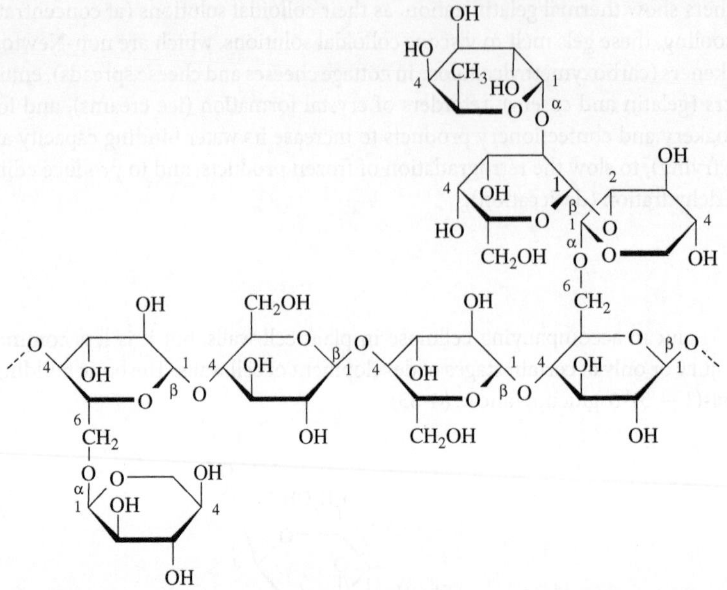

$$\text{α-L-Fuc}p\text{-}(1{\rightarrow}2)\text{-β-D-Gal}p\text{-}(1{\rightarrow}2)\text{-α-D-Xyl}p$$
$$1$$
$$\downarrow$$
$$6$$
$$\rightarrow 4)\text{-β-D-Glc}p\text{-}(1{\rightarrow}4)\text{-β-D-Glc}p\text{-}(1{\rightarrow}4)\text{-β-D-Glc}p\text{-}(1{\rightarrow}4)\text{-β-D-Glc}p\text{-}(1{\rightarrow}\cdots$$
$$6$$
$$\uparrow$$
$$1$$
$$\text{α-D-Xyl}p$$

4-64, basic structure of xyloglucans of hemicellulose type

Xyloglucans of the hemicellulose type, known as structural xyloglucans, are the dominant hemicelluloses of cell walls of dicotyledons plants (dicots) that include most vegetables, root crops, legumes, and fruit trees. The xyloglucan content is a characteristic of various types of fruits. For example, the content in strawberries is 14.6–15.9 g/kg, depending on the variety, cultivation, and degree of maturity. Xyloglucans are therefore a good indicator of the amount of fruit present in various products, for example in fruit yoghurts.

Xyloglucans are also present in small amounts in monocotyledon plants (monocots), including some vegetables (e.g. onion, garlic, and asparagus) and, especially, cereals. The major portion of xyloglucans is a component of insoluble fibre.

β-Glucans Polysaccharides known as β-glucans or β-$(1 \rightarrow 3)$, $(1 \rightarrow 4)$-D-glucans, or mixed-linkage β-glucans (formerly lichenins) are found in negligible amounts in the cell walls of dicot plants, but in larger quantities in those of cereals, where they constitute up to 30% of non-starch polysaccharides. Whilst their content in wheat and rye is only 0.2–2% by weight of grain and 1–2% of unhulled rice, the European and American varieties of oats contain 3.2–6.8% of β-glucans, and their content in barley is 3–7%. Some cultivars of barley even contain 14–16% of β-glucans.

Oat bran contains β-$(1 \rightarrow 3)$,$(1 \rightarrow 4)$-D-glucan, which is also called oat gum. Typical β-glucans for barley have two or more neighbouring $(1 \rightarrow 4)$ linkages (**4-65**). β-D-Glucans contain small quantities of arabinosyl and xylosyl residues. The relative molecular weights of β-glucans

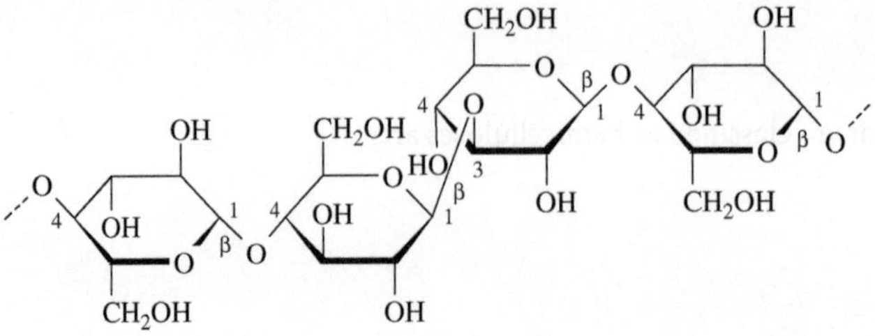

$$\cdots\rightarrow 4)\text{-β-D-Glc}p\text{-}(1{\rightarrow}4)\text{-β-D-Glc}p\text{-}(1{\rightarrow}3)\text{-β-D-Glc}p\text{-}(1{\rightarrow}4)\text{-β-D-Glc}p\text{-}(1{\rightarrow}\cdots$$

4-65, basic structure of mixed-linkage β-glucans containing (1→3), (1→4) bonds

vary over a wide range, according to their origin, from tens to thousands of kilodaltons. Similar polymers, also called β-glucans or β-(1 → 3), (1 → 6)-D-glucans, or mixed-linkage β-glucans, are synthesised by higher fungi, moulds, and yeasts.

β-Glucans are partly soluble and partly insoluble dietary fibre. Their solubility in water depends on their structure, which is related to their origin and decreases in the order barley > oats > wheat. The higher number of (1 → 4) bonds is in the molecule, the lower is the solubility of the polymers. More soluble polymers contain about 30% of (1 → 3) bonds and 70% of (1 → 4) bonds, and their chain is composed of two or three units of β-D-glucose linked by (1 → 4) bonds, between which is located a unit linked by a (1 → 3) bond. The solubility of β-glucans increases with temperature. For example, at 40 °C, only about 20% of barley β-glucans can be extracted, whilst at 65 °C, this amount increases to 30–70% (β-glucans of wheat are not extracted at this temperature). Native molecules do not form gels; however, a gel is formed after partial hydrolysis. β-Glucans bound to proteins are insoluble.

β-Glucans are particularly important in brewing technology. Their content in malting barley is 0.2–1%, and it varies according to a variety of climatic conditions, storage period, and other factors. During mashing, water-soluble fractions of β-glucans are extracted, the accompanying proteins are hydrolysed by carboxypeptidases, and formerly insoluble substances are dissolved. β-Glucan molecules are hydrolysed by *endo*-β-(1 → 3),(1 → 4)-glucanase, *endo*-β-(1 → 3)-glucanase, and β-glucosidase (laminarinase). The intermediates are soluble mixed-linkage β-glucan dextrins and dextrins with β-(1 → 3) bonds. The final products are cellobiose, laminaribiose, and glucose. The mashing temperature must not inactivate the enzymes hydrolysing the β-glucans, as native (undegraded) β-glucans can cause various problems in the production of beer. They prevent sedimentation of solid particles in the wort; beer filtration is complicated, and unfiltrable haze and sediments can be formed as a result of aggregation of β-glucan molecules amongst themselves or with other polymers, such as proteins and polyphenols.

β-Glucans of oats and barley reduce the bioavailability of feeds, which results in lower weight gain in poultry.

4.5.6.5.2 Heteroxylans

The main chain of heteroxylans consists of D-xylanopyranose units mutually linked by β-(1 → 4) bonds. The terminal unit is α-L-arabinofuranose. Most xylose chain units are not substituted, although there is some substitution by α-L-arabinofuranose linked by (1 → 3) bonds and less often by (1 → 2) bonds. Xylose is often substituted by two molecules of arabinofuranose (at C-2 and C-3), which form short side chains connected by (1 → 2), (1 → 3), or (1 → 5) bonds (**4-66**). Some heteroxylans, such as heteroxylans from cereal brans, are acetylated on C-2 of the xylose residues. Heteroxylan molecules also contain sections in which xylose units are substituted twice, once, or not at all. For example, wheat endosperm contains heteroxylans with xylose units substituted in positions C-2 and C-3 (about 12–20%), xylose units (19–31%) substituted once at C-3, and unsubstituted xylose units (55–69%). Rye heteroxylans contain chain segments exclusively substituted at C-2 and a smaller number of segments containing disubstituted xylose residues at C-2 and C-3.

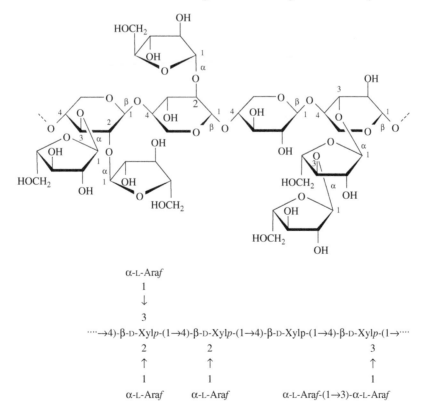

4-66, basic structure of arabinoxylans of cereals

With regard to the primary structure, these heteroxylans are called **arabinoxylans** and often also **pentosans**. The name is not very logical, because in addition to arabinose and xylose, they contain D-glucose and some other minor hexose building units, such as D-galactose, D-glucuronic acid and its 4-O-methyl derivative, and some other sugars that are rarely found as terminal units in the side chains. The average xylose content is 52–60%, the arabinose content 36–46%, and the glucose content 1.5–4.8%.

Arabinoxylans of various cereals differ not only in how the xylan chains are substituted, but also in their arabinose contents or ratios. The arabinose to xylose ratio in wheat arabinoxylans varies from 0.50 to 0.71, whilst in less substituted rye arabinoxylans it ranges from 0.48 to 0.55. D-Glucuronic acid and its 4-O-methyl ether are found mainly in arabinoxylans of husked rice, rice hulls (husks), sorghum, and maize brans. For example, these arabinoxylans, sometimes also referred to as **arabinoglucuronoxylans**, constitute about 4% of barley bran by weight.

Arabinoxylans serve a major structural role by binding to cellulose microfibrils and becoming cross-linked with other arabinoxylan moleculs by the means of hydroxycinnamic acid residues. A special feature is the presence of ferulic acid (*cis-* and *trans-*isomers in a ratio of about 1 : 1), which makes up 0.1–0.2% of arabinoxylans by weight. Ferulic acid is bound by an ester linkage at C-5 of the arabinose residue, which is connected to the C-3 of the xylose residue. In the secondary cell walls, the bound ferulic acid is oxidised and forms various C—C dimers (see Section 8.2.6.1.6), is bound to lignin in lignified cells through ether bonds, and forms addition products with thiol residues of proteins. All these reactions play an important role in the development of cross-links between the individual components of cell walls (Figure 4.12). In arabinoxylans, proteins are also present (0.9–3.9% by weight), like in a number of other polysaccharides.

The average molecular weight of arabinoxylans of wheat ranges from about 220 to 260 kDa, and that of rye arabinoxylans from 520 to 770 kDa. The arabinoxylan molecules are relatively rigid long chains resembling cellulose chains. There is a clear link with the considerable viscosity of the dispersion and its high ability to bind water.

Heteroxylans are the main polysaccharides of primary cell walls of vegetative parts of monocot plants and of lignified cells of monocot and dicot plants that are very important components of human diet. They occur in the stems of plants and in larger quantities in maize ears containing seeds (20–35%) and in wood mass (20–30% d.w.).

Arabinoxylans have a high water binding capacity (15–100 g of water per 1 g of dry matter). Some fractions are soluble in water and form extremely viscous dispersions. The differences in solubility depend on the degree of branching; therefore, branched molecules are more soluble. In the presence of oxidising agents, arabinoxylans form soft and elastic gels. Ferulic acid bound in arabinoxylans plays a key role in the gel formation.

In foods, heteroxylans are mainly present in cereals, where they are found in thin endosperm cell walls, the aleurone layer, and lignified bran cells. The endosperm cell walls of most cereals contain 60–70% arabinoxylans, with 20% in barley and 40% in rice. Glumes (husks) of wheat grains contain about 64% heteroxylans. Wheat grains contain on average 1.4–2.1% of heteroxylans, of which 0.8–1.5% represent water-soluble pentosans. Rice grains contain 7–8% heteroxylans. Soluble arabinoxylans are important components of wheat and particularly of rye flour. They have considerable influence on water uptake (hydration) by the flour and its distribution in the dough, and on the dough's viscosity and its rheological properties. Other desirable baking properties of flour are dependent also on the presence of arabinoxylans (e.g. larger bread volume as a result of carbon dioxide retention; reduced rate of starch retrogradation connected with ageing of bread and pastries; and impact on desirable organoleptic properties of bread crust). Reaction products of water-soluble pentosans, especially reactions of ferulic acid with cysteine residues of gluten proteins, significantly contribute to the desirable rheological properties of dough, but water-insoluble pentosans deteriorate the baking properties of flour. Like many other polysaccharides, arabinoxylans positively influence the composition of colon microflora.

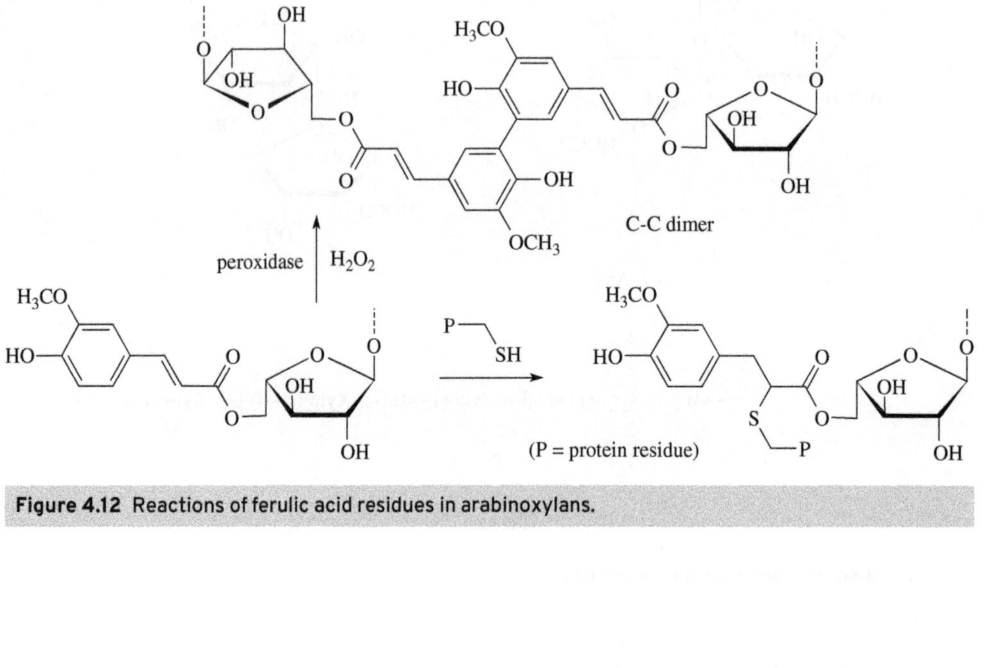

Figure 4.12 Reactions of ferulic acid residues in arabinoxylans.

4.5.6.6 Pectins

The general term **pectin**(s) or **pectic** (pectin) **substances** now refers to previously differentiated categories, which were polygalacturonates with higher amounts of methoxyl groups (**pectinic acids**), their salts **pectinans**, non-esterified polygalacturonates known as **pectic acids** and their salts **pectates**, and the accompanying neutral polysaccharides (arabinans and arabinogalactans of different structures). Native, insoluble cell-wall pectins associated with cellulose are known as **protopectins** (formerly pectoses). Enzymatic hydrolysis by the enzyme complex protopectinase converts protopectins into more or less soluble pectin substances with shorter chains, for example during fruit ripening.

Pectins are a group of highly polydisperse, complex, acidic polysaccharides of alternating chemical composition. They occur in all tissues of higher plants, except monocots, as a part of the primary cell walls and intercellular space (Figure 4.11). Pectins are formed and stored mainly in the early stages of growth, when the area of the cell walls increases. The presence of pectins and their changes during growth, maturation, storage, and processing have a major impact on the texture of fruits and vegetables.

4.5.6.6.1 Structure

A pectin molecule consists of three domains: homogalacturonan, rhamnogalacturonan I, and rhamnogalacturonan II. These three polysaccharides are bound together by covalent bonds.

The basic structure of homogalacturonan consists of a linear chain composed of 25–100 units of D-galacturonic acid linked by α-$(1 \rightarrow 4)$ bonds. This polymer is often called **polygalacturonic acid**. Galacturonic acid units are, in various degrees (70% in average), esterified with methanol so they do not have any electrical charge, whilst the remaining units contain dissociated carboxyl group and are negatively charged. These carboxyl groups located in different chains can be cross-linked by calcium ions. Some α-D-galactopyranuronates or methyl α-D-galactopyranuronates are acetylated at positions C-2 or C-3 (**4-67**), which increases the polysaccharide hydrophobicity. Pectins of the goosefoot family of plants (Amaranthaceae, formerly Chenopodiaceae), such as spinach, also contain small amounts of ferulic acid linked by ester bonds to neutral sugars (arabinose and galactose), similar to in arabinoxylans (Figure 4.12).

$\cdots \rightarrow 4)$-α-D-Gal$p$A6Me-$(1\rightarrow4)$-α-D-Gal$p$A-$(1\rightarrow4)$-α-D-Gal$p$A2Ac6Me-$(1\rightarrow4)$-α-D-Gal$p$A6Me-$(1\rightarrow\cdots$

4-67, basic structure of homogalacturonans

The degree of esterification (methylation) is defined as a percentage of esterified carboxyl groups. If the degree of methylation exceeds 50%, then the pectin is called **highly methoxylated** or **highly esterified** pectin. If the degree of methylation is lower than 50%, the pectin is called **low methoxylated** or **low esterified** pectin. The degree of acetylation is generally low, but sugar beet pectin contains a higher amount of acetyl groups (Table 4.32).

Rhamnogalacturonan I (degree of polymerisation is about 1000) contains chains of repeating units of α-D-galacturonic acid ($^{4}C_{1}$ conformers) with terminal α-L-rhamnopyranose linked by α-$(1 \rightarrow 2)$ bonds that correspond to the disaccharide residue α-D-Gal$p$A-$(1 \rightarrow 2)$-α-L-Rha$p$-$(1 \rightarrow 4)$-. The total content of rhamnose in pectin is 1–4%. Galacturonosyl and rhamnosyl residues are approximately in the ratio of 2 : 1 (**4-68**). Some units are methylated galacturonic acid molecules (4-O-methyl-D-galacturonic acid). About half of the rhamnosyl residues contain a galacturonic acid residue at C-4.

Table 4.32 Content of galacturonic acid (GalA), degree of methylation (DM), and degree of acetylation (DA) of selected pectins.

Pectin source	GalA (%)	DM (%)	DA (%)	Pectin source	GalA (%)	DM (%)	DA (%)
Apricots	64	57	8	Carrots	61	63	13
Peaches	90	79	4	Potatoes	40	53	15
Grapes	63	69	2	Sugar beet	65	62	35

Molecules of rhamnogalacturonans I also contain a significant amount of branched side chains with a considerable number of arabinose and galactose units (**4-69** and **4-70**). These chains have the structure of **arabinans** and **arabinogalactants**. The side chains are attached to the main chain through a rhamnose molecule at C-4 or C-3, or less frequently through C-2 or C-3 of galacturonic acid. Pectins with type I arabinogalactans have been found in a number of fruits and vegetables. Apple pectins have 25–100% of rhamnosyl residues substituted by arabinose and galactose units, whilst carrot pectin substitution is 10–50%. About 32% of galacturonosyl residues of potato pectin and up to 75% of galacturonosyl residues of rapeseed pectin are branched.

Arabinans and galactoarabinans of rhamnogalacturonan I create the so-called hair regions of the pectin molecule, whilst smooth (unsubstituted) regions consisting of polygalacturonic acids in homogalacturonan are binding zones. Rhamnose is a sugar that is incompatible with the regular conformation of polygalacturonates; therefore, its location in the chain determines the size of the binding zones, which play a crucial role in the formation of pectin gels.

R = COO⁻ or COOCH₃

···→4)-α-D-Gal*p*A-(1→4)-α-D-Gal*p*A-(1→4)-α-D-Gal*p*A-(1→2)-α-L-Rha*p*-(1→4)-α-D-Gal*p*A-(1→···

4-68, basic structure of rhamnogalacturonan I

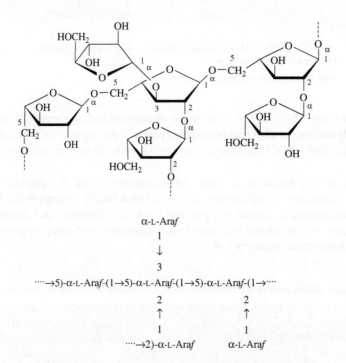

α-L-Ara*f*
1
↓
3
····→5)-α-L-Ara*f*-(1→5)-α-L-Ara*f*-(1→5)-α-L-Ara*f*-(1→····
2   2
↑   ↑
1   1
····→2)-α-L-Ara*f*   α-L-Ara*f*

4-69, general structure of arabinans in side chains of polygalacturonan I

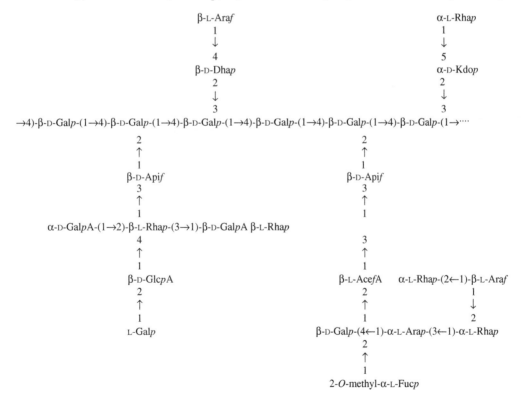

····→4)-β-D-Galp-(1→4)-β-D-Galp-(1→4)-β-D-Galp-(1→4)-β-D-Galp-(1→····

                        3   3
                        ↑   ↑
                        1   1

····→5)-α-L-Araf   ····→5)-α-L-Araf-(1→5)-α-L-Araf

4-70, general structure of arabinogalactans in side chains of polygalacturonan I

Rhamnogalacturonan II is a low-molecular-weight polymer (degree of polymerisation is about 60 and relative molecular weight is about 4.8 kDa) with a chain consisting of α-D-galacturonic acids linked by α-(1 → 4) bonds. Different side chains composed of α- and β-D-galactopyranuronic acids, α- and β-D-rhamnopyranoses, α-D-galactopyranose, α-L-fucopyranose, α-L-arabinopyranose, β-L-arabinofuranose, α-D-xylopyranose, and D-glucopyranuronic acid are connected to this main chain. The side chains also contain four unusual sugar residues (see Section 4.3.2.2) derived from 3-*C*-(hydroxymethyl)-β-D-erythrofuranose known as β-D-apiofuranose (β-D-Api*f*), 3-*C*-carboxy-5-deoxy-β-L-xylofuranose (β-L-Ace*f*A, aceric acid), 3-deoxy-β-D-*lyxo*-hept-2-ulopyranaric acid (β-D-Dha*p*), and 3-deoxy-β-D-*manno*-oct-2-ulopyranosonic acid (β-D-Kdo*p*). Xylose and fucose partly occur as 2-*O*-methyl ethers (**4-71**; the anomeric

β-L-Araf   α-L-Rhap
1   1
↓   ↓
4   5
β-D-Dhap   α-D-Kdop
2   2
↓   ↓
3   3
→4)-β-D-Galp-(1→4)-β-D-Galp-(1→4)-β-D-Galp-(1→4)-β-D-Galp-(1→4)-β-D-Galp-(1→4)-β-D-Galp-(1→····

2   2
↑   ↑
1   1
β-D-Apif   β-D-Apif
3   3
↑   ↑
1   1
α-D-GalpA-(1→2)-β-L-Rhap-(3→1)-β-D-GalpA β-L-Rhap
4   3
↑   ↑
1   1
β-D-GlcpA   β-L-AcefA   α-L-Rhap-(2←1)-β-L-Araf
2   2   1
↑   ↑   ↓
1   1   2
L-Galp   β-D-Galp-(4←1)-α-L-Arap-(3←1)-α-L-Rhap
                                                      2
                                                      ↑
                                                      1
                                                2-*O*-methyl-α-L-Fucp

4-71, basic structure of rhamnogalacturonan II

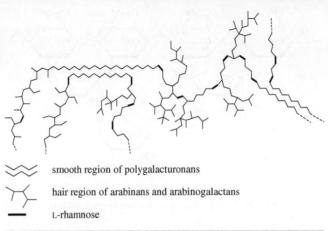

smooth region of polygalacturonans

hair region of arabinans and arabinogalactans

L-rhamnose

Figure 4.13 Pectin structure.

Table 4.33 Pectin content of fresh fruits and vegetables.

Pectin source	Pectin content (%)	Pectin source	Pectin content (%)
Apples	0.5–1.6	Grapes	0.1–0.9
Pears	0.4–1.3	Bananas	0.7–1.2
Peaches	0.1–0.9	Pineapple	0.04–0.13
Strawberries	0.6–0.7	Carrots	0.2–0.5
Gooseberries	0.3–1.4	Tomatoes	0.2–0.6
Red and black currants	0.1–1.8	Beans	0.5
Oranges	0.6	Onions	0.5
Peels of oranges	3.5–5.5	Potatoes	0.4

configuration of the terminal L-galactopyranose is not known). The rings of rhamnogalacturonan II are furthermore cross-linked by boric acid, $B(OH)_3$ (borate ions), covalently bound to the *cis*-hydroxyl groups at positions C-2 and C-3 of the apiose molecules located in different chains of rhamnogalacturonan II. Therefore, this part of the pectin molecule is present predominantly as a dimer (see Section 6.3.18). The overall arrangement of pectin chains, indicating the binding zones responsible for the formation of gels, is shown schematically in Figure 4.13.

4.5.6.6.2 Occurrence

Pectins are found in virtually all fruits and vegetables. Their content is not high; in fruit flesh, it varies around 1%. Higher amounts of pectin occur, for example, in apples, plums, damsons, currants, gooseberries, quinces, and cranberries, whilst lower amounts are found in cherries, sour cherries, blueberries, and elderberries ($\leq 0.5\%$). Vegetables generally contain low levels of pectin; the higher amounts are in tomatoes and carrots. Sugar beets also have a higher pectin level. The pectin content of some fruits and vegetables is given in Table 4.33.

4.5.6.6.3 Properties

Pectins are generally soluble in water and insoluble in most organic solvents. Water solubility decreases with increasing molecular weight and degree of esterification of the carboxyl groups (highly esterified pectins are soluble in hot water). Salts of polygalacturonic acids are generally more soluble than free acids (salts with monovalent ions are more soluble than calcium salts; salts of low esterified pectins are soluble in cold water).

Pectin dispersion has a relatively low viscosity, and therefore pectin is not used as a thickener. The viscosity of highly esterified pectins is increased by adding saccharose, and that of low esterified pectins increases in the presence of Ca^{2+} ions. Both types of pectins form gels under defined conditions.

Formation of gels The mechanism of gel formation depends on the degree of pectin esterification. Highly esterified pectins form gels with sugar in acidic solutions. Sugar binds water, thus reducing the degree of hydration of the pectin. Acids suppress the dissociation of carboxylic acid groups. The higher the degree of esterification, the smaller the amount of acid is necessary; therefore, totally esterified pectins form gels without addition of acid. Fast-gelling pectins form gels at pH $\cong$ 3.3, slowly gelling pectins form gels at pH $\cong$ 2.8. These gels are thermally irreversible. Low-esterified pectins (<50% of esterified carboxyl groups) form gels only in the presence of Ca^{2+} ions. Gelatinisation depends on the temperature, pH, ionic strength, and amount of added calcium. Under acidic conditions (pH $\cong$ 3.5), higher amounts of calcium ions are needed than in less acidic solutions. Gels of low-esterified pectins are thermoreversible.

Pectin molecules have a negative electrical charge in a neutral environment (p$K \cong$ 3.5), and therefore react with polymers carrying positive charges, such as proteins (in solutions of pH < pI). Under acidic conditions, pectin stabilises casein, which allows heat treatment of fermented dairy products. The interaction of pectin with plant proteins affects the consistency and texture of fruit products.

Firmer thermoreversible gels arise in the presence of sodium alginate in acidic solutions (pH < 3.5), in the presence of low amounts of sugar, and in the absence of calcium ions. The mechanical properties of these mixed gels depend on the ratio of pectin to alginate and the degree of pectin esterification. Alginates with a higher guluronic acid content form stable gels in combination with pectin with a degree of esterification of about 70%.

Changes and reactions Hydrolysis of pectin substances proceeds by the action of enzymes and in acidic or alkaline solutions. A number of native enzymes and enzymes produced by microorganisms are involved in the enzymatic hydrolysis of pectins of fruits and vegetables. Two groups of enzymes can be distinguished:

- **pectinesterases**;

- **pectin depolymerases**.

Pectinesterases (pectinmethylesterases or pectinpectylhydrolases) catalyse the hydrolysis of methyl esters to low-esterified pectins, also known as pectic acids (Figure 4.14). The resulting acids react with the naturally present divalent ions, which may lead to spontaneous gelling (in fruit juices) or hardening (in the processing of potatoes or cauliflower). Pectinesterases are present in various fruits and vegetables and are particularly active in cherries, citrus fruits, tomatoes, and carrots. Pectinacetylesterases can split off the acetyl groups in the smooth regions of pectin and rhamnogalacturonan acetylesterase catalyses the same reaction in the areas of branching. Together with rhamnogalacturonase, the latter enzyme also splits rhamnose-containing oligosaccharides.

Enzymes hydrolysing glycosidic bonds (pectin depolymerases) are glycosidases and lyases. Glycosidases, termed polygalacturonases or poly-α-1,4-D-galacturonid glycanohydrolases, hydrolyse α-(1 $\rightarrow$ 4) glycosidic and ester bonds. Exoenzymes split off monomers from the end of the chain, whilst endoenzymes operate within the chain. The products of hydrolysis are either monomers or oligomers with variable chain lengths. Highly esterified pectins are split by polymethylgalacturonases; polygalacturonases break down completely (or almost completely) de-esterified pectins (Figure 4.15). These enzymes are found in higher plants and microorganisms.

Pectate lyases (poly-α-1,4-D-galacturonid lyases) are also divided into exo- and endoenzymes. They degrade esterified or non-esterified pectins by a different mechanism to glycosidases, which is known as β-elimination (Figure 4.16). Pectate lyases are typical bacterial enzymes. Pectin lyases (poly-α-1,4-D-methoxygalacturonid lyases) cleave the glycosidic bond between esterified galacturonic acids by β-elimination (Figure 4.17). Only fungi produce pectin lyases.

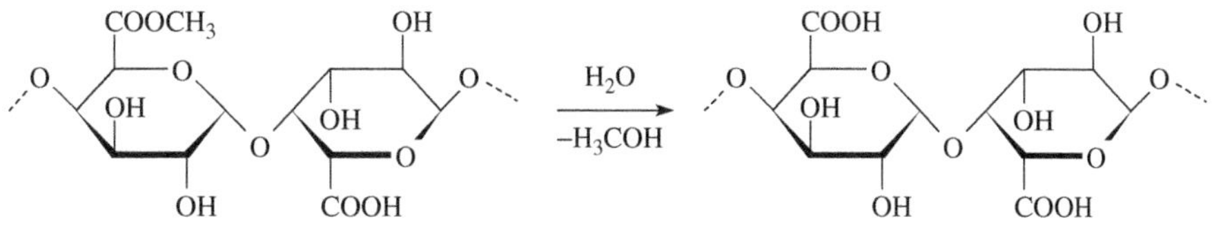

Figure 4.14 Reaction catalysed by pectinesterases.

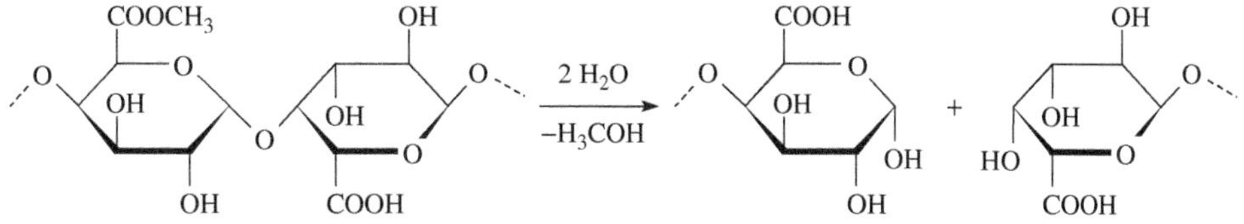

Figure 4.15 Reaction catalysed by polygalacturonases.

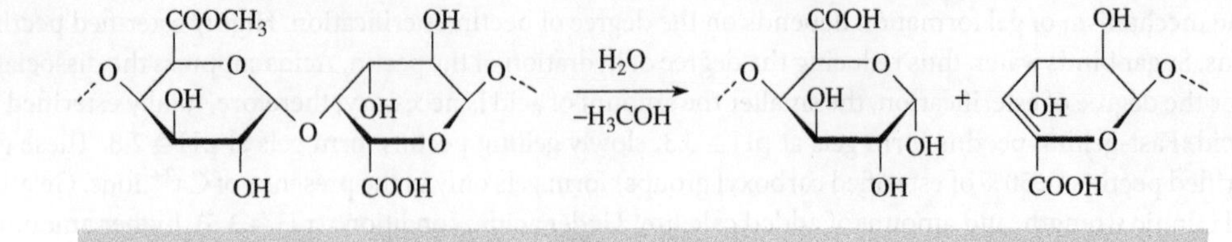

Figure 4.16 Reaction catalysed by pectate lyases.

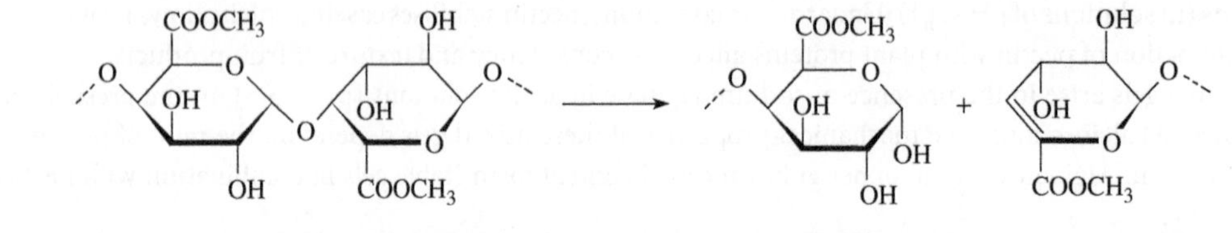

Figure 4.17 Reaction catalysed by pectin lyases.

Pectolytic enzymes usually act in combination. The enzyme exhibiting the activities of pectinesterase and polygalacturonase is sometimes referred to as pectinase. Industrially used pectinases are of bacterial origin, and often also have the activity of lyase, protease, cellulase, and other glycosidases.

Pectin is relatively stable under acidic conditions, and its stability is highest in the pH range of 3–4. Hydrolysis of methoxyl and acetyl groups and hydrolysis of glycosidic bonds occurs in strongly acidic solutions. Free D-galacturonic acid can degrade to furan-2-carbaldehyde and other products. At elevated temperatures, hydrolysis of pectin occurs even in weakly acidic solutions (pH > 5); the polysaccharide chain is cleaved by β-elimination. This reaction is also important in alkaline solutions, where ester bonds are mainly hydrolysed, and this reaction is accompanied by depolymerisation of the polysaccharide chains (β-elimination). The depolymerisation reaction occurs at the glycosidic bond of the non-reducing methoxylated galacturonic acid residue (Figures 4.16 and 4.17). Pectic acids are not depolymerised.

Modified pectin Pectin derivatives have found use in many fields taking advantage of novel functional properties of modified properties such as solubility, hydrophobicity, and other physicochemical, as well as biological characteristics. Pectin is capable of preparing a wide range of derivatives due to a number of hydroxyl and carboxyl groups on its backbone, as well as a certain amount of neutral sugars in side chains. The modification of pectin has been achieved by substitutions (e.g. alkylation, amidation, thiolation, and sulfation), chain elongations (cross-linking and grafting), and depolymerisations (e.g. acid or enzymatic hydrolysis, β-elimination, and mechanical degradation).

For example, treatment of a suspension of pectin in ethanol with ammonia yields **amidated pectin**, which contains –(C=O)-NH$_2$ functional groups. The degree of amidation is around 20%. Amidated pectin is thermoreversible and has a higher affinity for calcium ions and a reduced sensitivity against pH in comparison with low-esterified pectins. Alkylating the carboxyl group into an ester group is used to increase the hydrophobicity of pectin, whilst cross-linking can lead to improved swelling ability and other rheological properties. Another means of modifying the physico-chemical properties of pectin is through the synthesis of pectin graft (branched) copolymers. The mucoadhesive properties of natural pectins can be improved by their thiolation via the formation of either amide or ester bonds in pectin chains.

4.5.6.6.4 Physiology and nutrition

Pectin belongs to the polysaccharides that form dietary fibre, which affects glucose metabolism and decreases the amount of cholesterol in the blood serum. Pectin with a higher content of methoxyl groups is more effective.

4.5.6.6.5 Significance and use

Insoluble pectic substances produce the hardness and strength of immature fruits and vegetables. During maturation, post-harvest storage, and processing, these substances are subject to enzymatic and non-enzymatic degradation, which results in softening of fruits and the loss of the gelling ability of pectin. Pectins are released from the complex polysaccharides that form the cell walls, and this process continues after harvest during storage. Fruits containing active *endo*-polygalacturonases and pectinmethylesterases soften significantly and so quickly that this process often becomes an economic problem (e.g. in pears, cherries, kiwi, and tomatoes). Softening is less pronounced in fruits containing only *exo*-polygalacturonases (apples, peaches). Changes during processing are suppressed by heat inactivation of pectolytic enzymes and by the addition of monovalent (softening) or bivalent (texture hardening) cations. Bivalent ions (such as Ca^{2+} ions) protect pectins against depolymerisation, which results in a firmer texture of the tissues; monovalent cations displace divalent ions, which have the opposite effect.

Pectins are also responsible for the consistency of sterilised fruits and vegetables, the pressability of oilseeds, the filterability of fruit juices, and the formation of hazes in fruit juices. Some manufacturing processes, for example in the canning industry, use pectolytic enzymes of microbial origin to increase the yield in the production of fruit juices and to maintain their clarity. Pectolytic preparations have also found use in the oenological industry, sugar industry, and other sectors.

In industrial practice, pectins are mostly extracted from citrus fruit peels (their albedo), which contain about 20–40% of pectin. Another source of pectin is apple pomace, containing about 10–20% pectin. The isolation of pectin is based on extraction from an acidified aqueous slurry (pH 1.5–3) at temperatures of 60–100 °C. The extracts are then concentrated by evaporation or dried. Commercial pectin products are obtained by precipitation using metal ions that form insoluble salts with pectin (e.g. Al^{3+}), or by precipitation of pectin solutions with alcohols (ethanol or propan-2-ol).

4.5.6.7 Accompanying substances

Cellulose and other structural polysaccharides of cell walls are associated with different polymeric non-sugar materials that fix and firm the cell walls and also form their outer hydrophobic layers, which are impermeable to water. In nutrition, they are classified as dietary fibre. According to their chemical composition, substances accompanying polysaccharides are classified as lignin, phenolic compounds (tannins), proteins, and lipids.

4.5.6.7.1 Lignin

The structural polymer lignin is a major component of wood, where it makes up to about 25% of the biomass. Shelled nuts also have a similar amount of lignin. In lower amounts, lignin is a constituent of the dietary fibre of fruits, vegetables, and cereals (Table 4.20). Primary cell walls are virtually free of lignin. A high content is found in lignified secondary cell walls, such as aleurone and subaleurone cells of cereals (in bran), which contain around 8% lignin. Lignin occurs in small amounts (tens to hundreds of mg/l) in spirits aged in oak barrels, where it is extracted from wood.

The structural polymer lignin is a polymer of phenylpropanoid units known as **monolignols**, which may include, according to the type of plant, p-hydroxyphenyl, guaiacyl, and syringyl units. These units are hypothetically derived from p-hydroxyphenol, guaiacol, and syringol (see Section 8.2.8.1). Lignin is formed by lignification, a polymerisation process of three primary precursors, the monolignols p-coumaryl alcohol, coniferyl (ferulyl) alcohol, and 5-hydroxyconiferyl (sinapyl) alcohol (**4-72**). Dominating structures are (E)-isomers. Also involved in lignin biosynthesis are monolignol glucosides, aldehydes derived from monolignols, and glucosides of these aldehydes. The most common monolignol glucosides are 4-O-β-D-glucopyranoside of (E)-coniferyl alcohol, known as coniferin, and 4-O-β-D-glucopyranoside of (E)-sinapyl alcohol, known as syringin (**4-73**). In enzymatic reactions, these phenylpropanoid units produce radicals and couple with other monomer radicals to build up a polydisperse three-dimensional polymer lignin. Its composition is generally characterised by the relative abundance of p-hydroxyphenyl, guaiacyl, and syringyl units derived from the monolignols and by distribution of interunit linkages in the polymer. For example, grape stalk lignin is an HGS-type lignin with molar proportions of p-hydroxyphenyl (H), guaiacyl (G), and syringyl (S) units of 3 : 71 : 26, with a predominance of β-O-4′ structures (39% mol) and moderate amounts of β-5′, β-β, β-1′, 5–5′ and 4-O-5′ bonds (see Section 10.3.12.6). This type of lignin can also be associated with tannins. The catechol alcohols, caffeyl, and 5-hydroxyconiferyl alcohol may be also incorporated into lignin.

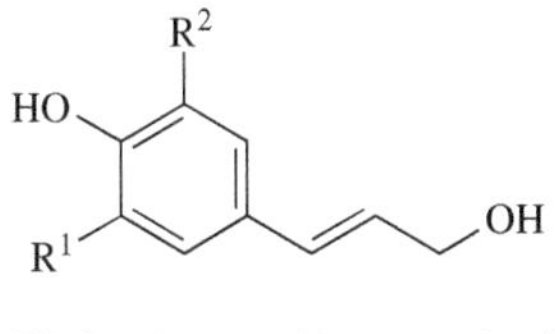

4-72, fenylpropanoid structural units of lignin

4-coumaryl (p-coumaryl) alcohol, $R^1 = R^2 = H$
coniferyl alcohol (ferulyl) alcohol, $R^1 = H, R^2 = OCH_3$
5-hydroxyconiferyl (sinapyl) alcohol, $R^1 = R^2 = OCH_3$

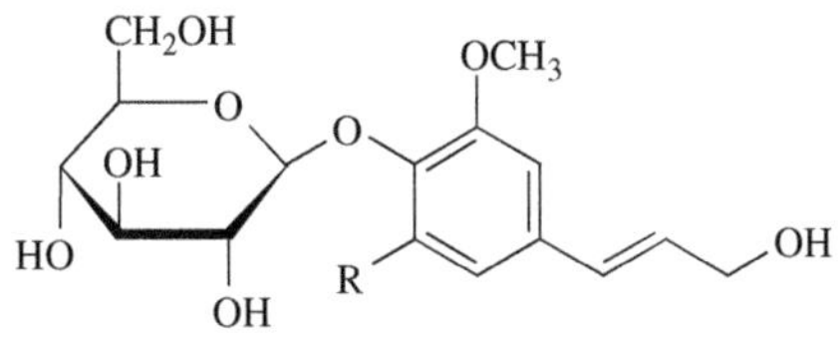

4-73, monolignol glucosides

coniferin, $R = H$
syringin, $R = OCH_3$

The digestive system does not decompose lignin, and only cleaves the bonds between lignin and other polymers.

During maturation of wines and spirits in oak barrels, lignin is decomposed to phenolic compounds that act as significant flavour-active components of these commodities. Phenolic components are also formed by pyrolysis of lignin in wood that is used in the process of smoking, and therefore are also found in smoked meat products and other smoked foods as well as in liquid smoke used for both food preservation and flavouring.

4.5.6.7.2 Other polymers

In addition to lignin, plants also contain structurally similar types of phenolic polymers, which are classified as **tannins** (see Section 8.3.6).

The best-known structural protein of cell walls of plants is **extensin**, which is present in cells at levels of about 0.5–5%. Extensin is a glycoprotein with an unusual amino acid composition. It contains about 40% hydroxyproline and a large proportion of lysine and serine. Octasaccharides and decasaccharides composed of arabinose and galactose are bound through hydroxyamino acids, and phenolic compounds are bound through tyrosine molecules, forming cross-links in the extensin molecule. Other structural proteins are present in smaller quantities, but may have an important function in building cell walls, as a connecting material with lignin.

Further components of dietary fibre are lipidic materials on the surface of cell walls that are composed of **waxes**, **cutin** (a cross-linked hydroxy fatty acid polyester), and **suberin** (a polyester composed of higher ω-hydroxy acids, α,ω-dicarboxylic acids, fatty acids, and fatty alcohols, which also contains ferulic acid).

4.5.6.8 Plant gums and mucilages

Plant exudates, known as **vegetable gums**, are usually sticky juices flowing spontaneously from plant tissues as a result of different stress factors, especially the attack of microorganisms and injury. They eventually solidify in a solid gummy mass on contact with air. **Mucilages** are slimy secondary metabolites occurring in different parts (such as the fruits and seeds) of some plants. The basic chemical composition of gums and mucilages of some plants is shown in Table 4.34.

The polysaccharides of most important vegetable gums and mucilages are acidic polysaccharides. According to their primary structure, they may be divided into:

- substituted **arabinogalactans** (gum arabic, larch gum);

- mixed **arabinogalactans** and **glycanogalacturonans** (gum tragacanth);

- **glycanorhamnogalacturonans** (gum karaya);

- **glycanoglucuronomannoglycans** (ghatti gum);

- **glycanorhamnogalacturonans** (okra).

Table 4.34 Basic monosaccharide composition of plant gums and mucilages (% w/w).

Gum or mucilage	Latin name of plant (family)	Neutral sugars					Uronic acids			
		Ara	Xyl	Rha	Fuc	Glc	Man	Gal	GlcA[a]	GalA
Arabic	*Acacia* spp. (Fabacaeae)	2–60	–	<24	–	–	–	28–80	2–23	–
Tragacanth	*Astragalus* spp. (Fabaceae)	22	13	6	4	–	–	12	–	23
Karaya	*Sterculia* spp. (Malvaceae)	1	–	20–24	–	–	–	33–40	5	28
Ghatti	*Anogeissus latifolia* (Combretaceae)	32–41	2	7	2	–	7–8	27–42	6	–
Okra	*Abelmoschus esculentus* (Malvaceae)			30				30	–	30[b]

[a]Total amount of D-glucuronic acid (GlcA) and 4-O-methyl-D-glucuronic acid (GlcA4Me).
[b]Other sugars are D-glucose, D-mannose, L-arabinose, and D-xylose; some sugars are partly acetylated.

Vegetable gums and mucilages are highly hydrophilic, water-soluble polysaccharides. They are significantly polydisperse, branched, and of very non-uniform structure. They are ranked amongst the hydrocolloids, but the low-molecular-weight fractions form true solutions. Dispersions or solutions are viscous, and in some cases gels may also arise. Plant gums often include non-starch storage polysaccharides of some seeds and tubers, such as guar, locust, tamarind, and konjac gums.

4.5.6.8.1 Gum arabic

Gum arabic (also called acacia gum) is hardened exudate from acacia tree species, in particular *Acacia senegal* and *Acacia seyal* (Fabaceae), growing mainly in Senegal, Nigeria, and West African countries. There are over 100 different acacia gums, but the structural differences between them are small.

Gum arabic is a substituted acid arabinogalactan (**4-74**). The basic building units are β-D-galactopyranose (the side chains also contain small amounts of α-D-galactopyranose), L-arabinose (α-furanose with small amounts of β-pyranose), and α-L-rhamnopyranose. β-D-Glucuronic and 4-*O*-methyl-β-D-glucuronic acids are present at lower levels. The gum from *A. senegal* contains around 44% galactose, 27% arabinose, 13% rhamnose, and about 16% uronic acids (glucuronic and 4-*O*-methylglucuronic acids). In various types of acacia, the individual sugars are present over a relatively wide range of concentrations (Table 4.34).

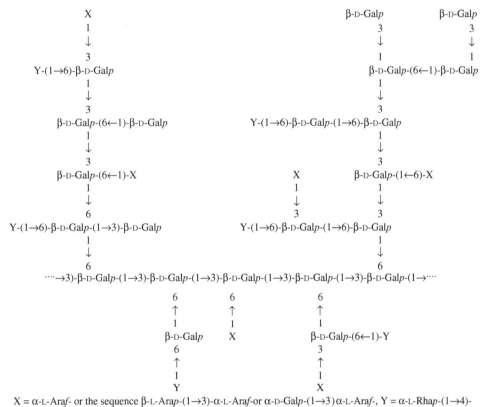

X = α-L-Araf- or the sequence β-L-Arap-(1→3)-α-L-Araf-or α-D-Galp-(1→3)α-L-Araf-, Y = α-L-Rhap-(1→4)-β-D-GlcpA- or β-D-GlcpA4Me-

4-74, basic structure of gum arabic

The main polysaccharide chain consists of β-D-galactopyranose units linked by (1 → 3) glycosidic bonds. Side chains, often repeatedly branched like amylopectin, occur in all residues of β-D-galactopyranoses of the main chain. β-D-Galactopyranose, linked to the main chain by β-(1 → 6) bonds and mutually by β-(1 → 6) and β-(1 → 3) bonds, predominates in the side chains. The abbreviated notation of a segment of gum arabic molecule originating from the plant *A. senegal* is given in formula **4-74**, which represents one of many possible structures. Some major segments of the molecule are shown in formulae **4-75** to **4-77**.

The relative molecular weight ranges from about 260–1200 kDa, but there are also polymers with molecular weights of several tens to 2300 kDa. The molecules are spherical globules with a twisted main chain. A protein (1.5–3%), associated with the gum, contains hydroxyproline, serine, and proline as the main amino acids, and is covalently bound to arabinose in the side chain via hydroxyproline hydroxyl group.

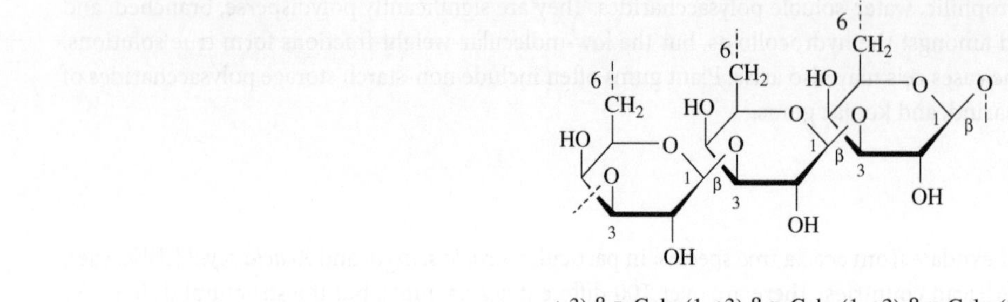

$\rightarrow$ 3)-β-D-Gal$p$-(1$\rightarrow$3)-β-D-Gal$p$-(1$\rightarrow$3)-β-D-Gal$p$-(1$\rightarrow$

4-75, main-chain fragment of gum arabic

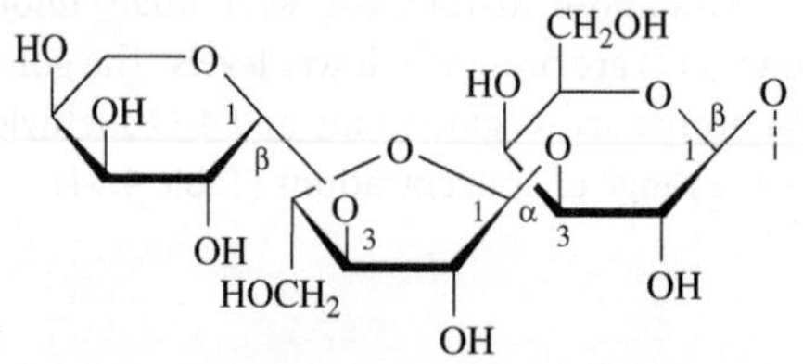

β-L-Ara$p$-(1$\rightarrow$3)-α-L-Ara$f$-(1$\rightarrow$3)-β-D-Gal$p$-(1$\rightarrow$

4-76, secondary-chain fragment of gum arabic

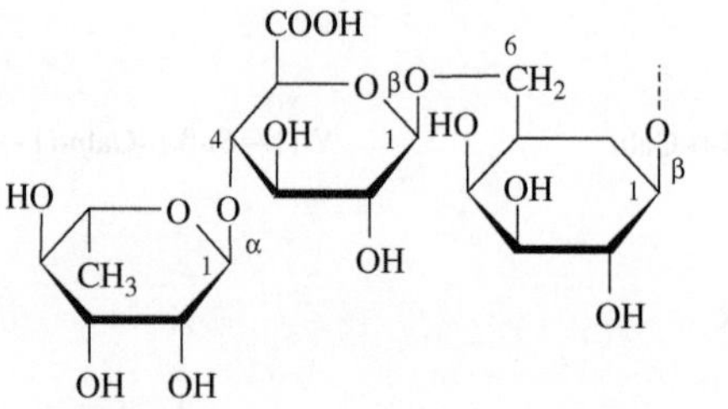

α-L-Rha$p$-(1$\rightarrow$4)-β-D-Glc$p$A-(1$\rightarrow$6)-β-D-Gal$p$-(1$\rightarrow$

4-77, secondary-chain fragment of gum arabic

The advantage of gum arabic is its very good solubility in water and the formation of low viscosity dispersions. Dispersions containing 40% gum behave as Newtonian fluids, and dispersions of higher concentration of gum behave as pseudoplastic systems (due to aggregation of the molecules). Viscosity is, as in all other polyelectrolytes (acidic polysaccharides), strongly influenced by the pH of the medium and the presence of electrolytes (salts). Maximum viscosity is achieved at pH 4.5–8.0, but it is decreased at the presence of electrolytes. In the presence of sugar, concentrated solutions (approximately 40–50%) of gum arabic form soft gels; however, gels are not formed at commonly used concentrations.

The ability of gum arabic to form concentrated dispersions (up to 50%), without significant increase of the viscosity, is used in stabilising and emulsifying various food systems. The gum stabilises emulsions of oil-in-water as it is firmly adsorbed on oil droplets due to proteins bound to polysaccharides. In ice cream, gum arabic helps to maintain a smooth consistency (formation of small crystals), and in confectionery it prevents the crystallisation of sugars and wetting of icings. Gum arabic can also be combined well with other gums, starches, gelatin, and sugars.

4.5.6.8.2 Larch gum

An industrial source of larch gum is the western larch (*Larix occidentalis*, Pinaceae), native to the mountains of western North America, whose wood contains up to 35% galactoarabinans; these are obtained by extraction with hot water. Some other types of larch trees (*Larix lyallii*, *Larix sibirica*, *Larix laricina*, and *Larix leptolepis*) have similar gums. Some species of pines, firs, and other conifers may also be alternative sources.

Larch gum resembles the composition of gum arabic (arabinose content is usually 10–20%), but contains a lower amount of uronic acids. It consists of two fractions of nearly neutral arabinogalactans of different relative molecular weights (about 16 and 100 kDa). The main chain consists, as in gum arabic, of D-galactopyranose units linked by β-(1 $\rightarrow$ 3) bonds. The C-6 positions contain short side chains composed of arabinose and galactose (**4-78**).

Larch gum forms highly concentrated aqueous solutions (60%). It is stable in the pH range of 1.5–10.5, even in the presence of electrolytes. Larch gum is used in the food industry as a substitute for gum arabic. It is used also as a thickener and surfactant.

β-L-Ara*p*-(1→3)-α-L-Ara*f*   β-D-Gal*p*-(1→6)-β-D-Gal*p*   β-D-Gal*p*-(1→6)-β-D-Gal*p*
1   1   1
↓   ↓   ↓
6   4   6
····· →3)-β-D-Gal*p*-(1→3)-β-D-Gal*p*-(1→3)-β-D-Gal*p*-(1→3)-β-D-Gal*p*-(1→3)-β-D-Gal*p*-(1→ ·····
6
↑
1
α-L-Ara*f*-(1→3)-β-D-Gal*p*

4-78, basic structure of larch gum

4.5.6.8.3 Tragacanth

The sources of tragacanth (also referred to as traganth, tragakanth, gum bassora, and bassorin) are shrubs of the genus *Astragalus* (the common name milkvetch includes most species) belonging to the legume family Fabaceae. The most important plants are *Astragalus gummifer*, *Astragalus microcephalus*, and *Astragalus kurdicus*, which grow in dry mountainous areas of Iran and Turkey.

Except for small amounts of starch and proteins (1.0–3.6%), tragacanth consists of two groups of polysaccharides. The first group is characterised by neutral arabinogalactans (so-called tragantin or tragacanthin), which represent about 60–70% of gum weight. Their structure is similar to that of larch gum. The main chain consists of units of D-galactopyranose linked by β-(1 → 4) bonds. Side chains contain other units of D-galactopyranose linked by β-(1 → 6) bonds with terminal arabinose units attached to β-D-galactopyranose through (1 → 2), (1 → 3), and (1 → 5) bonds.

The second group of acidic polysaccharides of the pectin type (so-called tragacanthic acid or bassorin) consist of α-D-galacturonic acid (or its potassium salt) units linked by (1 → 4) bonds. Tragacanthic acid also contains small amounts of α-L-rhamnopyranose. Units of the main chain are either unsubstituted or substituted by β-D-xylanopyranose residues, or by short chains formed by xylose and β-D-galactopyranose that can be replaced by α-L-fucose (**4-79**). The side chain also contains D-glucuronic acid. Some sugars are acetylated and α-D-galacturonic acid is methylated analogously to pectin. The average molecular weight is 840 kDa.

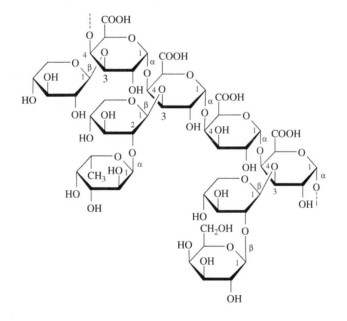

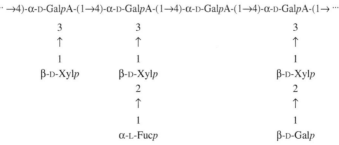

··· →4)-α-D-Gal*p*A-(1→4)-α-D-Gal*p*A-(1→4)-α-D-Gal*p*A-(1→4)-α-D-Gal*p*A-(1→ ···
3   3   3
↑   ↑   ↑
1   1   1
β-D-Xyl*p*   β-D-Xyl*p*   β-D-Xyl*p*
2   2
↑   ↑
1   1
α-L-Fuc*p*   β-D-Gal*p*

4-79, basic structure of tragacanth acidic component

The neutral fraction of tragacanth is soluble in water; acidic fractions only swell and form a thick slime or gel in the presence of Ca^{2+} ions. The viscosity of sols is high even at low concentrations, where the dispersion behaves as a Newtonian fluid (in the concentration range 0.3–0.5%). The pH value of solutions has little effect on viscosity and viscosity does not change under acidic conditions and even at pH 7–10. An important feature of tragacanth is its resistance to hydrolysis and mechanical stress.

Tragacanth is used as a thickening agent, emulsifier, and stabiliser in salad dressings, ice creams, pastry fillings, and other products.

4.5.6.8.4 Gum karaya

Gum karaya (sometimes referred to as Sterculia gum) is a bark exudate of trees of the genus *Sterculia* (Sterculiaceae), mainly of *Sterculia urens*, growing on the plateaus of central and northern India. Similar properties are found in some other species of the genus *Sterculia*, for example *Sterculia setigera*, *Sterculia caudata*, and *Sterculia villosa*, and in some species of the genus *Cochlospermum* of the Cochlospermaceae family.

Gum karaya is glycanorhamnogalacturonan. The main polysaccharide chain consists of alternating units of L-rhamnopyranose and D-galacturonic acid, which are connected by α-(1 → 4) and α-(1 → 2) glycosidic bonds. Units of the main chain are substituted by β-D-galactopyranose and β-D-glucuronic acid or remain unsubstituted. The end units are rhamnose, galactose, and glucuronic acid; some units are partially acetylated (**4-80**). Gum karaya contains calcium and magnesium ions bound to the carboxylic acid groups of glycuronic acids.

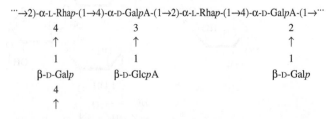

$$\cdots{\to}2)\text{-}\alpha\text{-L-Rha}p\text{-}(1{\to}4)\text{-}\alpha\text{-D-Gal}p\text{A-}(1{\to}2)\text{-}\alpha\text{-L-Rha}p\text{-}(1{\to}4)\text{-}\alpha\text{-D-Gal}p\text{A-}(1{\to}\cdots$$

4	3	2
↑	↑	↑
1	1	1
β-D-Galp	β-D-GlcpA	β-D-Galp
4		
↑		

4-80, basic structure of gum karaya

Gum karaya is very slightly soluble. At very low concentrations (<0.02% in cold water and 0.06% in hot water), it forms true solutions. Dispersions of a concentration of 0.3–0.5% show behaviour of Newtonian liquids as well as dispersions of tragacanth, guar, and locust gums. At higher concentrations (up to 5%), gum karaya forms viscous colloidal dispersions, which behave as non-Newtonian liquids.

Gum karaya is used as a thickener for soups, sauces, ketchups, and mayonnaise, as a substance that increases water binding capacity in processed cheese and meat products, and for stabilisation of foams produced from proteins (egg white whipped cream or whipped cream). It is also used in the pharmaceutical industry.

4.5.6.8.5 Ghatti gum

Ghatti gum (also known as Indian gum) is an exudate of *Anogeissus latifolia* trees (Combretaceae) growing in arid regions of India, Ceylon, and Africa (Ghana). The gum of *Anogeissus leiocarpus* has similar properties. Ghatti gum is not very significant commercially.

The molecule of glycanoglucuronomannoglycan (formerly called ghattic acid) is formed by periodically repeating residues of α-D-mannose and β-D-glucuronic acid, which are linked in the main chain through (1 → 4) and (1 → 2) bonds. The side chain contains mannose, to which are linked α-L-arabinofuranose by (1 → 6) bonds and β-L-arabinopyranose by (1 → 3) bonds; β-D-glucuronic acid or β-D-galactopyranose is bound to the β-L-arabinopyranose (**4-81**). The average molecular weight is about 12 kDa. The gum of the *A. leiocarpus* tree contains two mutually different polysaccharides of similar composition. The minor component is leiocarpan A and the major component is leiocarpan B.

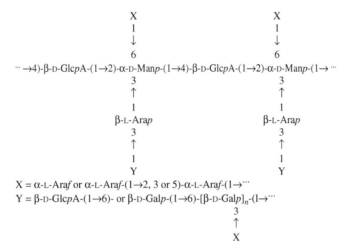

4-81, basic structure of ghatti gum

The functional properties of ghatti gum are more similar to those of gum arabic than to the viscous gums tragacanth and karaya. Ghatti gum has low solubility in water and does not form a gel. At concentrations lower than 5%, it forms true solutions, and at higher concentrations, viscous colloidal dispersions result.

Ghatti gum, thanks to its higher viscosity, has found use as a stabiliser of emulsions and suspensions.

4.5.6.8.6 Okra and other plant mucilages

Plant mucilages are used in many eastern Mediterranean, West Asian, and African countries, and are popular in Indian, Pakistani, Chinese, Malaysian, Japanese, Caribbean, and many other cuisines as ingredients in soups and sauces, to which they give their characteristic mucilaginous consistency. The main representative of this group of polysaccharides is **glycanorhamnogalacturonan**, which comes from unripe fruits of edible hibiscus (*Abelmoschus esculentus*) of the mallow family (Malvaceae), and is commonly known as okra (also bamie orgombo) or lady's fingers.

The basic structural unit of the main okra chain is acidic disaccharide →4)-α-D-GalpA-(1 → 2)-α-L-Rhap-(1→. Side chains contain the disaccharide D-galactobiose, β-D-Galp-(1 → 4)-β-D-Galp-(1→, which is linked by (1 → 4) bound to about one-half of the α-L-rhamnopyranose residues.

Okra polysaccharides are already viscous in diluted solutions (0.5%), and the viscosity increases rapidly with increasing concentration. Heating increases their solubility, but after some time it leads to irreversible loss of viscosity caused by hydrolysis. The maximum viscosity is achieved at pH 6–9. Okra mucilage is a typical polyelectrolyte. It has hypoglycaemic effects.

Other acid mucilages of composition similar to that of okra (glycanorhamnogalacturonans) are junsai originating from the leaves and stems of *Brasenia schreberi* (Cabombaceae), baobab mucilage obtained from the leaves of baobab *Adansonia digitata* (Malvaceae), and ruredzo from the plant *Dicerocaryum zanguebarium* (Pedaliaceae). Slimy substances are produced by many other plants, but these mucilages are usually important locally and are not used commercially.

4.5.7 Seaweed polysaccharides

The most important representatives of this group of polysaccharides are:

- **agar**;

- **carrageenans** and **furcellaran**;

- **alginates** (alginic acid salts).

4.5.7.1 Agar

Agar forms an intracellular matrix of a number of species of red seaweed (Rhodophyceae) that plays a similar role in algae to that of cellulose in higher plants. Algae, which are a source of agar (agarophytes), come mainly from the Gelidaceae, Gracilariaceae, and Pterocladiaceae families, growing on the coast of Portugal, South Africa, India, Japan, Mexico, Chile, and New Zealand. Agar is usually obtained from algae by extraction with hot neutral, acidic, or alkaline water (at a temperature higher than the melting point of the agar gel). Gels obtained from extracts are dried. Alkaline solutions cause partial hydrolysis of the sulfate groups, which results in modified agar properties.

The building units of linear polysaccharides agars are β-D-galactopyranose and 3,6-anhydro-α-L-galactopyranose, alternately linked by glycosidic $(1 \rightarrow 3)$ and $(1 \rightarrow 4)$ bonds. The basic neutral polysaccharide is often still called **agarose** (by analogy with starch amylose). Its building block is disaccharide agarobiose, 4-O-β-D-galactopyranosyl-3,6-anhydro-α-L-galactopyranose (**4-82**). The structure of agars is in fact much more complex. Some polysaccharide fractions contain α-L-galactose instead of 3,6-anhydro-α-L-galactose. Hydroxyl groups at C-6 of β-D-galactose or at C-2 of α-L-galactose can be partially methylated. Hydroxyl groups at C-4 of β-D-galactose and at C-6 of α-L-galactose are esterified with sulfuric acid (the SO_4^{2-} ion concentration is up to about 7% w/w). These fractions also contain bound pyruvic acid (as acetal in positions C-4 and C-6 of β-D-galactose) in 4,6-O-(1'-carboxyethylidene)-D-galactose, the concentration of which is 0.02–1%. These fractions also contain about 1% of D-galacturonic acid. Previously, these acidic agar fractions were known as **agaropectin** (by analogy with amylopectin). According to recent studies, up to 18% of total sugars in some agars are represented by D-xylopyranose molecules substituted in some units of the main chain.

Agars are largely polydisperse polysaccharides whose molecular weight ranges from 80 to 420 kDa. These molecules, similarly to other algal polysaccharides, form organised structures (double right-handed helices).

$\cdots \rightarrow$3)-β-D-Gal$p$-(1→4)-α-L-Gal$p$3,6An-(1→3)-β-D-Gal$p$-(1→4)-α-L-Gal$p$3, 6An-(1→ $\cdots$

4-82, basic structure of agarose

Agars are soluble in hot water at 85 °C and higher, and the dispersion yields gel by cooling. Commonly used concentrations range from 0.5 to 2.0%, but gels are formed even in 0.04% solutions. The sol–gel transition and the reverse process show a hysteresis. The melting-point temperature of the gel (at temperatures up to 95 °C) is higher than the temperature at which the gel is formed. During ageing, agar gels are subject to syneresis. Their resistance to deformation is improved when combined with vegetable gums, such as locust gum. Agars are polysaccharides of low acidity and, unlike carrageenans, the gel formation does not require the presence of neutralising cations.

The use of agar in food production is based on its ability to bind water and form thermoreversible gels. Its gelling properties depend on the proportion of the agarose fraction present. Owing to the high melting point of gels, agar is mainly used in bakery products, as well as in the production of jams and jellies, confectionery products, dairy products, meat, fish, and poultry products, and beverages. In some Asian countries (e.g. in Japan), agar is used not only as a food additive, but also as a separate food (in various types of flavoured gels and edible packagings).

4.5.7.2 Carrageenan

Carrageenan is an extract from red seaweed (Rhodophyceae), especially of the genera *Chondrus* and *Gigartina*. Carrageenans differ in their structure, which is largely related to their origin. Seaweed algae of the genus *Euchema* (*E. cottonii* and *E. spinosum*) are fibrous shrubs of about 0.5 m height that grow on coral reefs along the Philippines, Indonesia, and other tropical regions of the Pacific Ocean. They are also grown on marine farms. The algae *Chondrus crispus* (also known as Irish moss) are small, dark-red shrubs growing to a height of about 0.1 m along the coast of the North Atlantic, particularly in Canada, the British Isles, and France. Algae of the genus *Gigartina* grow to a height of up to 5 m in the cool coastal waters of South America (Chile).

Carrageenans are mainly extracted as sodium salts using alkaline hot water (Na_2CO_3 or NaOH solutions). On acidification (HCl), carrageenans are obtained. The final materials are produced by drying or by precipitation with solvents (such as propan-2-ol).

Carrageenans are linear polysaccharides with a structure similar to the structure of agars, but, unlike agars, the structural unit is only D-galactopyranose and not L-galactopyranose. The structure is a repetitive sequence of β-D-galactopyranose and 3,6-anhydro-α-D-galactopyranose, a disaccharide, known as carabiose (**4-83**). In fact, the primary structures of carrageenans are much more complex. They

Table 4.35 Basic structure of carrageenans.

Polysaccharide	Building unit	
	A	B
Agarose	β-D-**Galactose**	**3,6-Anhydro-α-L-galactose**
β-Carrageenan	β-D-**Galactose**	**3,6-Anhydro-α-D-galactose**
κ-Carrageenan	β-D-**Galactose-4-sulfate**	**3,6-Anhydro-α-D-galactose**
ι-Carrageenan	β-D-**Galactose-4-sulfate**	**3,6-Anhydro-α-D-galactose-2-sulfate**
μ-Carrageenan	β-D-Galactose-4-sulfate	α-D-Galactose-6-sulfate
θ-Carrageenan	β-D-Galactose-2-sulfate	3,6-Anhydro-α-D-galactose-2-sulfate
ξ-Carrageenan	β-D-Galactose-2-sulfate	α-D-Galactose-2-disulfate
ν-Carrageenan	β-D-Galactose-4-sulfate	α-D-Galactose-2,6-disulfate
λ-**Carrageenan**	β-D-**Galactose-2-sulfate**	α-D-**Galactose-2,6-disulfate**

can be expressed by notation $\cdots \to 3)\text{-}\beta\text{-A-}(1 \to 4)\text{-}\alpha\text{-B-}(1 \to 3)\text{-}\beta\text{-A-}(1 \to 4)\text{-})\text{-}\alpha\text{-B-}(1 \to \cdots$, where A and B are units of D-galactose; its derivatives are listed in Table 4.35 (for comparison, the structure of polysaccharide agarose is also given).

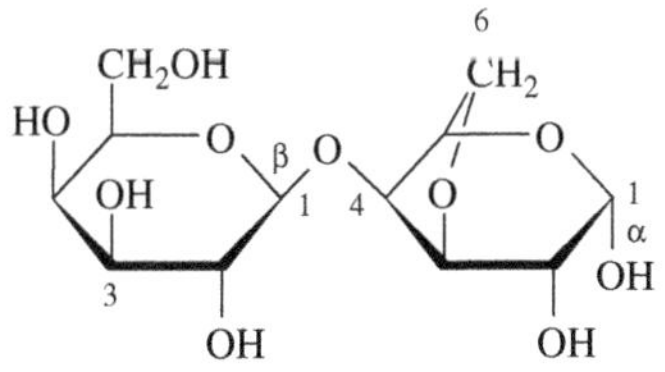

β-D-Gal*p*-(1→4)-α-D-Gal*p*3,6An

4-83, carabiose

In the carrageenan molecules, at least eight types of the sequences of monomers are known. They are indicated with lower-case Greek letters: β (beta), θ (theta), ι (iota), κ (kappa), λ (lambda), μ (mu), ν (nu), and ξ (xi). The same names are used for the individual carrageenans if these sequences predominate in their primary structures. In the food industry, attention is devoted to just three dominant species, known as ι-carrageenan (its precursor is ν-carrageenan), κ-carrageenan (its precursor is μ-carrageenan), and λ-carrageenan. These three carrageenans are highlighted in bold in Table 4.35.

Carrageenans obtained from algae are predominantly complex mixtures of polysaccharides. They can be fracionated by precipitation with potassium salts (a KCl solution of concentration 0.25 mol/l) to separate the two main components, insoluble κ-carrageenan and soluble λ-carrageenan. The potassium salt of κ-carrageenan is insoluble in cold water, whereas that of ι-carrageenan is slightly soluble in cold water and soluble at temperatures above 60 °C, and λ-carrageenan is soluble in the form of any salt. Virtually pure carrageenans can also be obtained from certain species of algae; for example, ι-carrageenan can be obtained from *Euchema cottonii* algae, κ-carrageenan from *E. spinosum* algae, and λ-carrageenan from *C. crispus* algae.

The structures of the repeating units A and B in κ-carrageenan ($R^1 = H$, $R^2 = SO_3^-$), ι-carrageenan ($R^1 = R^2 = SO_3^-$), and λ-carrageenan ($R = SO_3^-$) are provided in formulae **4-84** and **4-85**, respectively. In addition to sulfate groups, and similarly to agar, carrageenan macromolecules also contain other functional groups (methoxyl groups or pyruvic acid bound as acetal).

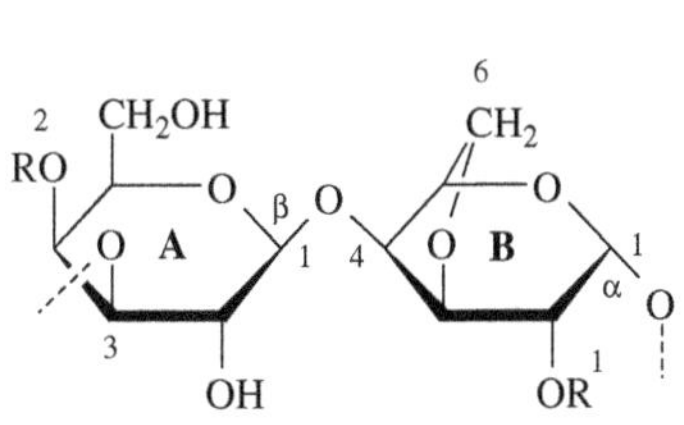

4-84, basic structure of κ- and ι-carrageenans

6

CH$_2$OH   CH$_2$OR

HO   O   β   O
     O   A   OH   B
                 4   α
  3   2   2   O
     OR   OR

4-85, basic structure of λ-carrageenan

The average molecular weights of all three very polydisperse polymers ranges from 100 to 1000 kDa. In the sequence shown in Table 4.35, the sulfate group contents of the polysaccharides increase, which is associated with conformations of the molecules and the carrageenan properties. In the same way as agar molecules, molecules of κ-carrageenan and ι-carrageenan form double helices, but molecules of λ-carrageenan occur in a zigzag conformation due to sulfuric acid bound at C-2 and C-6 of the B unit (disulfate is primarily found in the $^4C_1$ conformation, but the formation of helical structures requires a $^1C_4$ conformation of the 3,6-anhydro derivative). Gelling properties can be improved by intramolecular substitution of sulfate group at C-6 of the B units in alkaline solution, which yields 3,6-anhydrogalactose-2-sulfate (θ-carrageenan).

Carrageenans are anionic hydrophilic colloids. Their solubility in water depends on the type of carrageenan, ions present, temperature, and pH. The ratio of hydrophilic hydroxyl and sulfate groups to hydrophobic 3,6-anhydro-D-galactose residues also affects the solubility. Highly sulfated λ-carrageenan is very soluble and forms viscous dispersions (but does not form gels), κ-carrageenan containing higher amounts of hydrophobic groups is less soluble, and the solubility of ι-carrageenan is between the two. Carrageenans are stable in solutions of pH 5–10, but in more acidic solutions (pH < 4) they are hydrolysed and the viscosity of dispersions decreases.

An important feature is the formation of gels. They are formed, like agar gels, by cooling dispersions of κ- or ι-carrageenans of even 0.5%, where intermolecular associations of double helices lead to the formation of superhelical structures. κ-Carrageenan forms a firm, brittle gel that undergoes syneresis, whilst ι-carrageenan provides flexible and cohesive thixotropic gels where syneresis does not occur. The formation of strong but brittle gels requires the presence of neutralising ions (e.g. potassium or ammonium ions, but not sodium ions in κ-carrageenan or calcium ions in ι-carrageenan).

An important property is the ability to form complexes with milk proteins (caseins). They can also be combined with modified starches.

Commercial carrageenan is a mixture of all three types in which κ-carrageenan (gelling) mostly prevails over λ-carrageenan (non-gelling) in a ratio of about 3 : 2. Carrageenan is used as a thickener, gelling agent, stabiliser, and emulsifier in the production of dairy desserts, milk drinks, and ice creams and in the manufacture of canned meat. It also finds applications in cosmetics, in the stabilisation of industrial suspensions, and in the production of various pigments.

4.5.7.3 Furcellaran

Furcellaran is obtained mainly from red algae of the genus *Furcellaria* (*F. lumbricalis* and *F. fastigiata*). It is sometimes known as Danish agar, because its source is predominantly along the coast of Denmark. Furcellaran is a sulfated polysaccharide consisting of D-galactose units (46–53%), 3,6-anhydro-D-galactose (30–33%), and their sulfates (16–20%). Its structure and properties are similar to κ-carrageenan, and it is therefore often considered one of the carrageenans. The difference between furcellaran and κ-carrageenan lies in the fact that κ-carrageenan has a sulfate group linked to every second sugar unit, whilst furcellaran has one sulfate group linked to every third unit.

Furcellaran is soluble in warm water and forms soft, flexible, and thermoreversible gels. The addition of sugar positively affects the strength of the gel. Furcellaran forms very strong gels in the presence of K$^+$ and NH$_4$$^+$ ions, just like κ-carrageenan; the presence of Ca^{2+} ions has less effect, and with Na$^+$ ions, it does not produce gels. It is used in the production of milk puddings and desserts.

4.5.7.4 Algin

Algin is the name used to refer to **alginic acid** and its salts, **alginates**. Algin is an intercellular matrix (a gel containing ions of Na, Ca, Mg, Sr, and Ba) in the brown seaweed family Phaeophyceae growing off the coast of the North Atlantic, especially in the United States, Norway, France, and Great Britain. The main industrial sources are the algae *Macrocystis pyrifera*, *Laminaria hyperborea*, and of the genera *Ascophyllum*, *Lessonia*, and *Sarrgasum*. Alginate makes up about 40% of dry matter of algae.[2]

Algin, like agar and carrageenan, is obtained as a sodium salt (alginate) by extraction of algae with alkaline aqueous solutions (NaOH or Na$_2$CO$_3$). The extract is precipitated as calcium salts by adding CaCl$_2$ or as alginic acid by acidification using HCl. The calcium salt is

---

[2] A range of extracellular bacterial polysaccharides of the genera *Azotobacter* (*A. vinelandii* and *A. crococcum*) and *Pseudomonas* (*P. aeruginosa*, *P. putida*, *P. mendocina*, *P. fluorescens*, and *P. syringae*), known as bacterial alginates, have similar compositions and properties. They are almost exclusively polymannuronans. Some bacteria are considered as a future potential industrial source of bacterial alginate, which could replace production from algae.

converted into alginic acid, from which the final commercial product (sodium salt) is obtained by neutralisation with Na_2CO_3. A significant portion of native polysaccharides are hydrolysed during the isolation of alginates.

Alginates are linear copolymers of non-branched β-D-mannuronic acid (M) and α-L-guluronic acid (G) linked by (1 → 4) glycosidic bonds (**4-86**). Residues of mannuronic acid are in the $^4C_1$ conformation, whilst residues of guluronic acid are in the $^1C_4$ conformation. The chain contains alternating sections of different lengths containing only molecules M, sections formed exclusively of molecules G, and mixed sections G-M. Representation and alternation of both components is particularly variable, depending mainly on the origin of the alginate. The mannuronic acid content is usually 22–90%, and that of guluronic acid varies between 10 and 78%. The structures of alginate sections M-M, G-G, and M-G are shown in formulae **4-87**.

M−M−M−M−G−M−G−G−G−G−G−M−G−M−G−G−G−G−G−G−G−M−M−M−G−M−G−M−G−G−M

| section M | section G | section G | section M-G |

4-86, basic structure of alginates (M = β-D-mannuronic acid, G = α-L-guluronic acid)

section M-M   section G-G   section M-G

4-87, structures of alginate sections

Alginates of alkali metals, ammonium salts, amine salts, and magnesium salts are soluble, whilst those of calcium salts are insoluble. Solubility is affected by pH, ionic strength, and ion type. Alginic acid is precipitated in acidic solution; slow acidification gives a gel rather than a precipitate. Solutions of alkali metal alginates are highly viscous and behave as pseudoplastic systems. Viscosity is strongly dependent on ionic strength (they are polyelectrolytes); as the ionic strength increases, the viscosity decreases. Through heating and in the presence of reducing agents (such as ascorbic acid), the polymers are degraded and viscosity decreases.

An important property of alginates is the ready formation of gels and heat-resistant films on addition of calcium ions to sodium alginate dispersions. In the binding of calcium ions, not only do electrostatic forces participate, but their chelation is involved. The binding zones are the G sections that contain at least four units of guluronic acid. This creates a structure called the egg-box structure, as in pectin. Depending on the guluronic acid concentration (the amount in the binding zones), the formed gels can have different properties. The relative molecular weight of alginates is 32–200 kDa, and the chains (spiral to almost linear) contain 180–930 sugar units.

Alginates are used, in concentrations of 0.25–0.5%, as thickeners, stabilisers, and emulsifiers to improve the consistency of bread, sauces, dressings, ice creams, fruit juices, and many other foods. Their gelling properties are employed in the production of fruit and dessert jellies, puddings, and reconstituted fruit prepared from fruit pulp. With highly esterified pectins, thermoreversible gels are formed. Alginates also readily react with protonised amino acid residues in proteins to form precipitates, and may therefore also be used to remove proteins, for example from beer. Besides the native alginates, alginates modified by propylene oxide (propylene glycol alginates) are also used.

4.5.8 Polysaccharides of microorganisms and higher fungi

Microorganisms and higher fungi produce two basic types of polysaccharides:

- **extracellular**;

- **intracellular** (structural and storage polysaccharides).

Extracellular polysaccharides of bacteria are accumulated in the form of capsules, which remain part of the cell wall, or as an amorphous mucilaginous material surrounding the outer cell wall and diffusing into the growth medium. These **mucilages**, also known as **bacterial gums**, appear to have a barrier function to protect cells from infection by bacterial viruses (bacteriophages), prevent dehydration, and fix microorganisms to the environment (e.g. to soil particles). Bacterial gums have unique properties, for which they find use in the food industry, pharmaceutical industry, and elsewhere. The most important bacterial extracellular hydrocolloid used for food purposes is **xanthan (xanthan gum)**, whilst on a smaller scale **gellan** (gellan gum) is also used. In addition to xanthan and gellan, a number of other bacterial polysaccharides are obtained, but for various reasons (e.g. low production often on a semi-industrial scale, high cost, access to better resources) they are used only sporadically. Dextran and curdlan, for example, have limited uses. Bacteria of the genus *Azotobacter* are a potential source of bacterial alginates.

Kefiran, a water-soluble extracellular heteropolysaccharide with a molecular weight of 1.35 MDa and produced by lactic acid bacteria, *Lactobacillus kefiranofaciens*, is a part of kefir grains. Kefiran possesses a backbone of [→6)-β-D-Glc*p*-[1 → 2(6)]-β-D-Gal*p*-(1 → 4)-α-D-Gal*p*-(1 → 3)-β-D-Gal*p*-(1 → 4)-β-D-Glc*p* -(1→] with a branch attached to *O*-2 of galactose residues and terminated with glucose residues. The glucose to galactose molar ratio is 1.0 : 1.1. Due to its beneficial probiotic properties, kefiran has wide applications in pharmaceutical industries.

Yeasts, moulds, and higher fungi produce a large variety of extra- and intracellular polysaccharides. Certain applications have been found for glucans as hydrocolloids in the food industry, especially α-glucans with multiple α-(1 → 4), α-(1 → 3) bonds and some β-glucans with combined β-(1 → 3) and β-(1 → 6) bonds. Extracellular (excreted as mucilages) and intracellular (structural polysaccharides of cell walls) polysaccharides, which are commonly called **β-glucans**, or more precisely **β-(1 → 3)-D-glucans**, have unique properties. They are produced by some yeasts, moulds, and higher fungi. Their antibacterial, antiviral, anticoagulant, and anticarcinogenic effects are mainly exploited by the pharmaceutical industry and in human medicine.

4.5.8.1 Xanthan

The extracellular polysaccharide xanthan (xanthan gum) is produced by bacteria of the genus *Xanthomonas* (industrially, that most commonly used is *X. campestris*).

The main xanthan chain consists of β-D-(1 → 4) glucose units, which also occur in cellulose. Side chains (usually trisaccharides) consist of a D-glucuronic acid residue and two D-mannose residues (**4-88**). To the terminal D-mannose unit of the side chain, β-D-glucuronic acid through β-(1 → 4) glycosidic bond is attached. Glucuronic acid is further linked to α-D-mannose by α (1 → 2) bond. The structure is complicated by the presence of pyruvic acid linked, as an acetal, at positions C-4 and C-6 of the terminal β-D-mannose unit, which is 4,6-*O*-(1′-carboxyethylidene)-β-D-mannopyranose. The internal mannose unit in the side chain is acetylated in position C-6 (6-*O*-acetyl-α-D-mannopyranose). The structure may also vary in the degree of substitution depending on the bacterial strain used in the production. The relative molecular weight is about 15 000 kDa. Xanthan molecules form single or double helices stabilised by side chains.

Xanthan gum is water soluble; its dispersions are highly viscous, and show thixotropic behaviour even at low concentrations. The viscosity is strongly dependent on temperature. During heating, viscosity decreases at first, but then increases again with changes in the conformation of the molecules. Xanthan dispersions are stable in acidic and alkaline conditions and at elevated temperature (80 °C). In the presence of guar gum, the viscosity of dispersions increases, which is utilised in products that require a stable viscosity over a wide range of salt concentrations, pH, and temperatures.

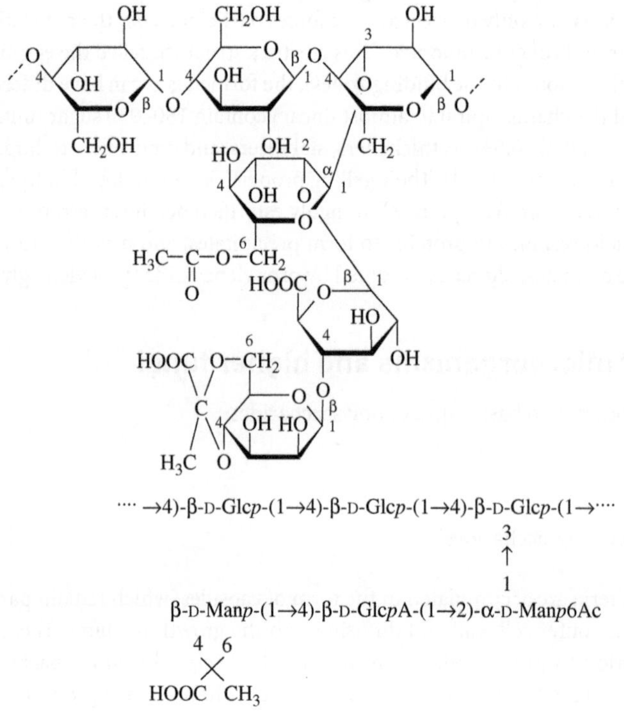

···· →4)-β-D-Glc*p*-(1→4)-β-D-Glc*p*-(1→4)-β-D-Glc*p*-(1→ ····

3

↑

1

β-D-Man*p*-(1→4)-β-D-Glc*p*A-(1→2)-α-D-Man*p*6Ac

4 6

✕

HOOC CH₃

4-88, basic structure of xanthan

Xanthan does not form gels, but thermoreversible gels are formed in mixtures with some polysaccharides, such as galactomannans (locust gum), glucomannans (konjac gum), and κ-carrageenan. Gel formation requires the interaction of xanthan molecules (arranged in a double helix) with the unbranched part of another polysaccharide molecule (with its binding zone). Better elastic and cohesive gels are formed from deacetylated xanthan.

Xanthan gum is used primarily as a thickener and emulsion stabiliser and, in combination with other hydrocolloids, as a gelling agent. The thermostability of xanthan is typically exploited in the preparation of instant soups and sauces and as binders in various canned foods.

4.5.8.2 Gellan

Gellan (gellan gum) is an extracellular polysaccharide produced commercially by aerobic submerged fermentation from bacteria *Sphingomonas elodea* (previously called *Pseudomonas elodea*). The gellan chain contains tetrasaccharide repeating structures κ-D-Glc*p*-(1 → 4)-β-D-Glc*p*A-(1 → 4)-β-D-Glc*p*-(1 → 4)-α-L-Rha*p*-(1 → 3) consisting of β-D-glucose esterified at position C-6 by acetic acid (the degree of substitution is about 50%), to which is bound L-glyceric acid at C-2 and D-glucuronic acid. D-Glucuronic acid is bound to the second molecule of D-glucose by a β-(1 → 4) bond, which is connected with α-L-rhamnose by a (1 → 4) bond (**4-89**).

···· →3)-β-D-Glc*p*6Ac-(1→ 4)-β-D-Glc*p*A-(1→ 4)-β-D-Glc*p*-(1→ 4)-α-L-Rha*p*-(1→ ····
            2
            ↑
        L-glyceric acid

4-89, basic structure of gellan

Gellan can also occur in a branched form, known as **welan**. The molecular weights range from thousands to tens of thousands of kilodaltons. Gellan gum forms coaxial triangular threefold double helices from two left-handed chains coiled around each other, with the acetate residues on the periphery and glyceryl groups stabilising the interchain associations. The molecule is further stabilised by hydrogen bonds.

Gellan is dissolved in water to form highly viscous solutions even at low concentrations. The gelling properties depend on the degree of acetylation. The native acetylated product forms soft, elastic gels, whereas gellan produces hard and brittle gels. Gellan is an acidic polysaccharide and in the presence of cations, even at low (0.1% w/w) to very low (0.005% w/w) concentrations, forms thermoreversible gels on cooling at about 50 °C. Gelatinisation depends on the cation type and its valence. Divalent cations form stronger and more elastic gels than monovalent cations.

Depending on their structure (degree of deacetylation), gellan gels are formed in either cold or warm water and are thermoreversible or thermoirreversible, which means they are perfect as broad-spectrum gelatinisation agents that find use in many food and non-food applications (texture modification and stabilisation of foams and emulsions). They are used mainly in Japan. Gellan is used also in combination with other hydrocolloids, for example xanthan gum, locust gum, and gelatin. In microbiology, it is used as an alternative to agar gels due to its thermal stability (up to temperatures of 120 °C) and is particularly suitable for the cultivation of thermophilic microorganisms.

4.5.8.3 Dextran

The extracellular polysaccharide dextran is produced by the bacteria *L. mesenteroides*, *Streptobacterium dextranicum*, *Streptococcus mutans*, and some others. A dextran molecule consists of about 95% of α-(1 → 6) linked D-glucose units. The remainder are D-glucose molecules linked by α-(1 → 3) bonds that form side chains (**4-90**). Some dextrans contain side chains in which D-glucose molecules are also partly bound by α-(1 → 4) and α-(1 → 2) bonds. The type and number of these glycosidic bonds depends on the origin of the dextran. The relative molecular weight of dextran is usually 60–90 kDa.

Dextran is soluble in water, but water-insoluble dextrans also exist. Dextran dispersions have lower viscosity in comparison with xanthan dispersions, but are relatively resistant to hydrolysis.

In food applications, dextrans (*L. mesenteroides* strains product) have been used, usually in combination with other polysaccharides such as gum arabic, for purposes similar to other hydrocolloids. They are highly effective emulsifiers and stabilisers of emulsions of oil in water.

They bind water and inhibit the crystallisation of saccharose. Much more extensive uses have been found for dextran in pharmacy and medicine (replacement of blood plasma) and in analytical chemistry, as a chromatographic material for gel filtration and other applications. Dextran is naturally formed in contaminated sugar beet and juice, where high viscosities can cause problems in sugar beet processing.

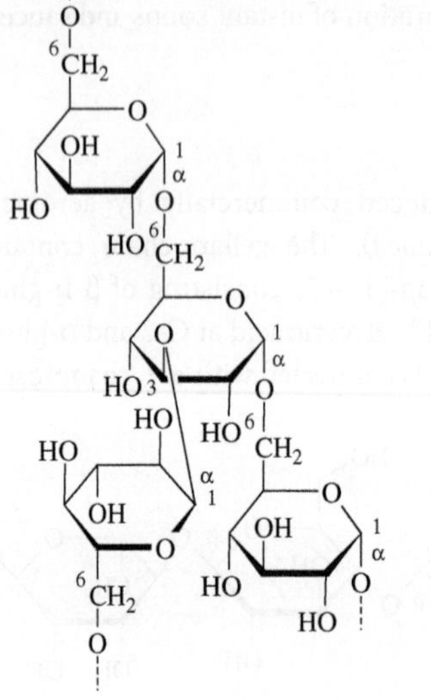

$$\cdots\rightarrow 6)\text{-}\alpha\text{-}D\text{-}Glc}p\text{-}(1\rightarrow 6)\text{-}\alpha\text{-}D\text{-}Glc}p\text{-}(1\rightarrow 6)\text{-}\alpha\text{-}D\text{-}Glc}p\text{-}(1\rightarrow \cdots$$
$$3$$
$$\uparrow$$
$$1$$
$$\cdots\rightarrow 6)\text{-}\alpha\text{-}D\text{-}Glc}p$$

4-90, basic structure of dextran

4.5.8.4 Curdlan

Curdlan is an extracellular polysaccharide produced by the bacterium *Alcaligenes faecalis* var. *myxogenes*. It is also found in many yeasts and fungi as a component of cell walls or as a storage polysaccharide.

Curdlan is a linear polymer of glucose units linked by β-$(1 \rightarrow 3)$ bonds (**4-91**). The basic unit is actually disaccharide laminaribiose. It can be partially esterified by succinic acid. The relative molecular weight of curdlan is 44–77 kDa.

Curdlan swells in water, and at 80 °C forms a turbid gel that is stable over a wide range of temperatures and pH values.

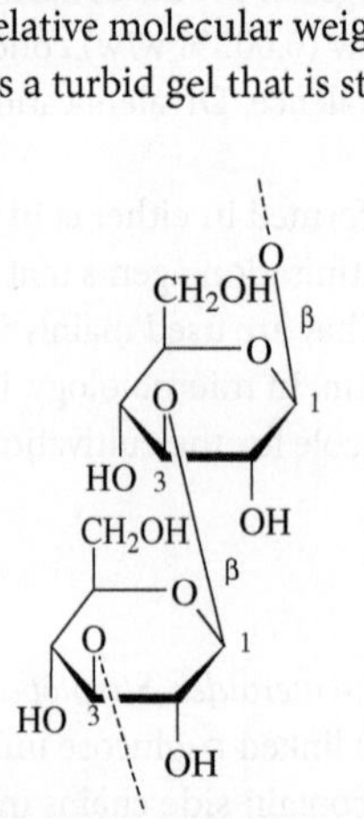

$$\cdots\rightarrow 3)\text{-}\beta\text{-}D\text{-}Glc}p\text{-}(1\rightarrow 3)\text{-}\beta\text{-}D\text{-}Glc}p\text{-}(1\rightarrow \cdots$$

4-91, basic structure of curdlan

Curdlan serves as a gelling agent, thickener, and stabiliser, improves water binding capacity and viscoelasticity, masks various odours, and has an excellent ability to form films that are insoluble in water, impermeable to oxygen, and biodegradable. It is mainly used in Japan.

4.5.8.5 Elsinan

Elsinan is an exocellular glucan produced by the fungus *Elsinoe lencospila*, which forms white patches on the leaves of tea plants (*Camellia sinensis*, Theaceae) whose leaves and leaf buds are used to produce Chinese tea.

The elsinan chain is composed of D-glucose units linked by approximately 70% α-(1 → 4) bonds and about 28% α-(1 → 3) bonds (**4-92**).

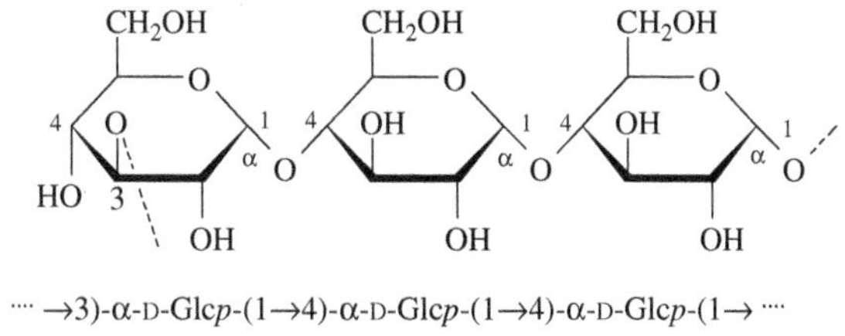

···· →3)-α-D-Glc*p*-(1→4)-α-D-Glc*p*-(1→4)-α-D-Glc*p*-(1→ ····

4-92, basic structure of elsinan

Elsinan is soluble in hot water. It is stable over a wide range of pH values and in the presence of salts, can be hydrolysed by some amylolytic enzymes (e.g. by α-amylase), can form highly viscous solutions even at low concentrations, and forms a gel at concentrations over 5%. An important feature is the formation of strong, flexible films during evaporation of elsinan solutions.

Elsinan is used as a low-energy filler. Films made from elsinan are suitable for use as edible food packagings that are impermeable to oxygen.

4.5.8.6 Pullulan

Extracellular polysaccharide pullulan is produced by the fungus *Aureobasidium pullulans* (syn. *Pullularia pullulans*). The molecule consists of D-glucopyranose units linked alternately by two α-(1 → 4) bonds and one α-(1 → 6) bond (**4-93**). The molecular weight varies widely, from 1.5 to about 800 kDa, according to the origin of the fungus and the cultivation conditions.

Pullulan is soluble in water. The pressed polysaccharide has similar properties to polystyrene. When mixed with sorbitol or glycerol, pullulan forms translucent films that are impermeable to oxygen and fats.

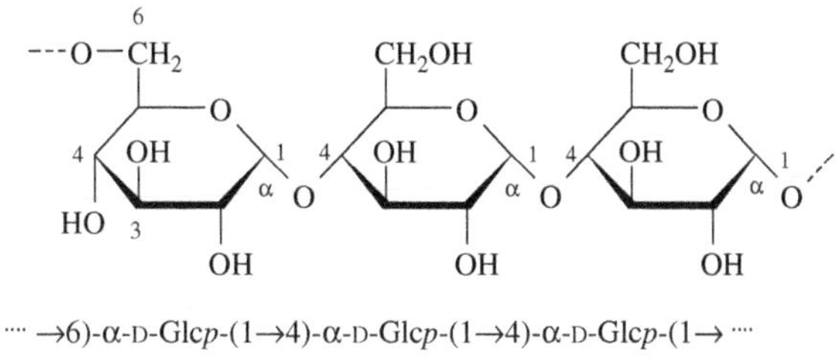

···· →6)-α-D-Glc*p*-(1→4)-α-D-Glc*p*-(1→4)-α-D-Glc*p*-(1→ ····

4-93, basic structure of pullulan

Pullulan is not used for food purposes. Its most widespread applications are in medicine (infusion solutions), pharmacy (coated tablets), and the paper industry (adhesives).

4.5.8.7 Scleroglucan

Scleroglucan is the capsular polysaccharide excreted by some lower fungi of the genus *Sclerotium*. The fungi *S. glucanicum* and *S. roefsii* serve as an industrial source. Other sources are fungi of the genera *Sclerotinia*, *Corticium*, *Botrytis*, and *Stromatinia* that produce polysaccharides similar to scleroglucan. Similar polysaccharides are also cell-wall components and storage polysaccharides of yeasts and higher fungi.

Scleroglucan produced by *P. glucanicum* has a linear chain composed of D-glucose residues linked by β-(1 → 3) bonds, with side glucose residues linked by β-(1 → 6) bonds that occur on approximately every third glucose unit of the main chain (**4-94**). The polymers contain between one hundred and several hundred molecules of glucose.

Scleroglucan is soluble in water and forms highly viscous dispersions that are stable over a wide range of pH values and at higher temperatures. Dispersions of concentrations higher than 1.5% form gels. In combination with glycerol, scleroglucan forms a strong and flexible film and emulsions with vegetable oils.

The food use of scleroglucan is minimal; greater use is made in cosmetics, pharmaceuticals, and the manufacture of porcelain, paper, and paints.

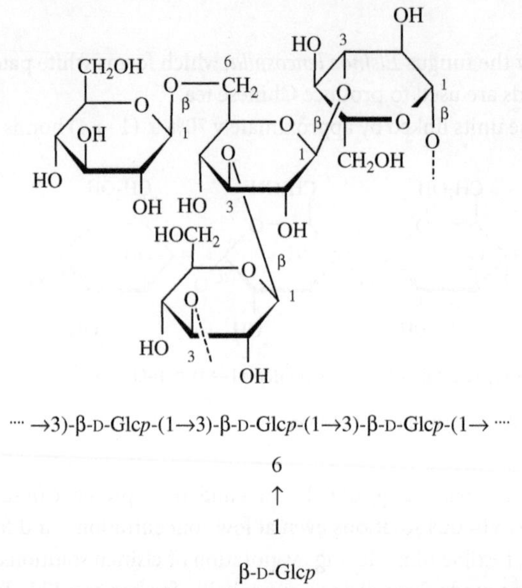

$\cdots \rightarrow 3)\text{-}\beta\text{-}\text{D-Glc}p\text{-}(1\rightarrow 3)\text{-}\beta\text{-}\text{D-Glc}p\text{-}(1\rightarrow 3)\text{-}\beta\text{-}\text{D-Glc}p\text{-}(1\rightarrow \cdots$

$$6$$
$$\uparrow$$
$$1$$

$\beta\text{-}\text{D-Glc}p$

4-94, basic structures of scleroglucan, schizophyllan, and lentinan

4.5.8.8 Other glucans

Polysaccharides structurally related to scleroglucan, which are β-glucans with glucose units in the main chain linked by $(1 \rightarrow 3)$ bonds and side chains attached by $(1 \rightarrow 6)$ bonds, are produced as structural and storage (extra- and intracellular) polysaccharides by many yeasts, moulds, and higher fungi. For example, the yeast *S. cerevisiae* contains as its major polysaccharides β-glucan with $(1 \rightarrow 6)$ bonds in the main chain and $(1 \rightarrow 3)$ bonds in the side chains and α-mannan with $(1 \rightarrow 6)$ bonds in the main chain and $(1 \rightarrow 2)$ and $(1 \rightarrow 3)$ bonds in the side chains. Cultivated yeast strains produce only β-glucan with $(1 \rightarrow 3)$ bonds in the main chain and $(1 \rightarrow 6)$ bonds in the side chains.

Polysaccharides composed of glucose units are also known as β-$(1 \rightarrow 3)$-D-glucans. Individual polysaccharides differ in substitution of the main chain and the number of glucose units in the side chains. The chains form triple helices. β-$(1 \rightarrow 3)$-D-Glucans are mostly isolated from so-called club fungi, belonging to the phylum Basidiomycota, and from so-called sac fungi of the phylum Ascomycota. Examples of these polysaccharides are given in Table 4.36.

The most important properties of β-$(1 \rightarrow 3)$-D-glucans are the abilities to strengthen the immune system and to inhibit tumour growth. Their activity depends on the molecular weight, frequency of branching, and conformation of molecules. The highest efficiency reported is shown by β-D-glucans with a degree of branching of 0.20–0.33 and higher relative molecular weight (100–200 kDa). Some β-D-glucans (Table 4.36), for example schizophyllan (extracellular polysaccharide of Basidiomycota fungi) and lentinan (structural cell-wall

Table 4.36 Some β-D-$(1 \rightarrow 3)$-glucans with significant anticarcinogenic activity.

Name	Source		Degree of branching
	Latin name	Other names	
AM-ASN	*Amanita muscaria*	Fly agaric (fly amanita) mushroom	0.3
β-Glucan I	*Auricularia auricula-judae*	Jew's ear (jelly ear) mushroom	0.75
Grifolan	*Grifola frondosa*	Hen of the woods (sheep's head) mushroom, maitake mushroom (in Japan)	0.33
HA	*Pleurotus ostreatus*	Oyster mushroom	0.25
Lentinan	*Lentinula edodes*	Shiitake mushroom	0.23–0.33
Pachyman	*Wolfiporia extensa* (syn. *Poria cocos*)	Fu Ling mushroom (in China)	0.015–0.02
Schizophyllan	*Schizophyllum commune*	–	0.33
Scleroglucan	*Sclerotium glucanicum*	–	0.3
Tylopilan	*Tylopilus felleus*	Bitter bolete mushroom	0.33
Zymosan	*Saccharomyces cerevisiae*	Baker's yeast (brewer's yeast)	0.03–0.2

polysaccharide of Oomycota fungi), are used in clinical medicine as immunotherapeutic agents in the treatment of cancer diseases, often in combination with radiotherapy. Schizofyllan also protects against bacterial infections. Lentinan is used against viral diseases. The structures of both compounds are similar to that of scleroglucan (**4-94**), differing only in the level of substitution of the main chain and the degree of polymerisation.

Schizofyllan is a product of the phytopathogenic fungus *Schizophyllum commune* (Table 4.36), which produces three types of β-(1 → 3)-D-glucans with molecular weights 10, 20–80, and 200 kDa. The first two types contain disaccharide laminaribiose in the side chains. Glucans with a relative molecular weight of 200 kDa show the highest antitumor and cytotoxic activities. Their side chains contain trisaccharide β-D-Glc*p*-(1 → 3)-β-D-Glc*p*-(1 → 3)-β-D-Glc*p*.

4.5.9 Animal polysaccharides

Amongst the polysaccharides of animal origin, only the storage homopolysaccharide **glycogen** and structural homopolysaccharide **chitin** are important. Animal tissues also contain many other heteropolysaccharides with structural, protective, and other functions.

4.5.9.1 Glycogen

The primary structure and function of glycogen resemble those of the starch component amylopectin, and therefore glycogen is also called animal starch. Glycogen is the main utilisable source of glucose (energy) present in all animal cells, but it is also found in cells of bacteria, moulds, and higher fungi.

Glycogen is an α-glucan that contains more than 10^6 D-glucose residues linked by α-(1 → 4) bonds. In every 8–12 glucose residues, side chains of other α-(1 → 4) glucans are bound, attached by α-(1 → 6) bonds, much like in amylopectin, although the glycogen molecule is more branched and structurally more compact. Glycogen contains also phosphoric acid in small amounts.

In mammals, glycogen is found mainly in the liver and muscles. The liver cells, hepatocytes, contain glycogen at a level of 2.9–8.1% of their weight (in total, 100–120 g in adults). It occurs in the form of granules in the cell cytoplasm. Approximately 1–2% of glycogen is also present in the skeletal muscles of animals (*in vivo*), so the total amount is higher than the content in the liver. Meat contains only about 0.15–0.18% of glycogen, with the exception of horse meat, which contains 0.9%.

Glycogen plays an important role in post mortem changes in muscle. Anaerobic glycolysis transforms glycogen into L-lactic acid, which lowers the pH of the meat, increases its tenderness during ripening, and has antimicrobial effects, thereby extending the shelf life during storage. The level of glycogen in the muscle of animals, poultry, and fish prior to slaughter has a major impact on the quality of meat during further processing (see Section 2.4.5.1.5). In fish, glycogen is present at levels of up to 0.3%. It is also found in higher fungi, moulds, yeasts, and bacteria. Higher fungi contain 5–10% glycogen. From a nutritional point of view, glycogen has little significance.

Glycogen is more soluble than starch and does not form gels. The large number of non-reducing ends of the glycogen molecule plays an important role in biochemical processes, as it allows rapid mobilisation of glucose during biodegradation of glycogen. In cells, glycogen is degraded for metabolic purposes by glycogen phosphorylase, which splits α-(1 → 4) bonds from the non-reducing end of the molecule with the formation of glucose 1-phosphate. Glucose 1-phosphate is transformed into glucose 6-phosphate, which produces glucose via the process of glycolysis (with the catalytic action of hexokinase), or other sugar phosphates. The α-(1 → 6) bonds are cleaved by α-(1 → 6)-glucosidase (isomaltase).

Glycogen is a utilisable polysaccharide. In food, it is hydrolysed by hydrolases (α-amylase and β-amylase) in the same way as starch. The branched dextrins formed are further broken down by isoamylase to maltose and glucose.

4.5.9.2 Chitin

Chitin is an almost linear copolymer of *N*-acetyl-β-D-glucosamine (70–90%) and β-D-glucosamine (10–30%) in which both monomers are bound to each other by β-(1 → 4) glycosidic bonds (**4-95**). The basic building block of chitin is the disaccharide **chitobiose**, composed of two molecules of *N*-acetyl-β-D-glucosamine (also known as chitosamine). Its relative molecular weight is about 1000 kDa.

····· →4)-β-D-Glc*p*NAc-(1→4)-β-D-Glc*p*N-(1→4)-β-D-Glc*p*NAc-(1→4)-β-D-Glc*p*NAc-(1→ ·····

4-95, basic structure of chitin

The chains are coiled along the longitudinal axis into a helix at certain distances (1–1.5 nm) and are oriented parallel or antiparallel. The most common conformation is an α-conformation, in which the individual chains are arranged in antiparallel. In the less frequent β-conformation, the chains are oriented in parallel.

Chitin is the second most abundant organic compound in nature after cellulose. It is found mainly in the animal kingdom, where it is the main structural polysaccharide in the exoskeleton (shells) of crustaceans, insects, and other invertebrates. It also occurs in some algae, fungi, yeasts, and bacteria, usually associated with proteins. Naturally occurring chitin is consumed only rarely. Some beetles and seafood with shells that are largely made of chitin (such as crabs) are eaten as delicacies. The chitin content in the exoskeleton of crabs is 61–77%. The composition and amount of chitin in insects varies from 1.2 to 12% according to the species and developmental stage.

The main source of chitin in the diet is the higher fungi, which contain about 1% chitin; such examples are cultivated common mushrooms (*A. bisporus*, 1.3–8.0% d.w.) and shiitake mushrooms (*Lentinula edodes*, 3.6–8.1% d.w.). Fermented soybeans and fermented rice used in the preparation of oriental foods contain chitin derived from moulds (*A. oryzae* and *A. sojae*). Fungi contain up to 42% chitin (e.g. moulds, *A. niger*). Baker's yeast (*S. cerevisiae*) contains around 2.9% chitin, which becomes part of the baked products.

Chitin is insoluble in water and barely soluble in acidic media (forming ammonium salts). Heating in acid or alkali solutions leads to the hydrolysis of bound *N*-acetyl-D-glucosamine with the formation of D-glucosamine and acetic acid. Heating under acidic conditions causes partial depolymerisation of chitin through the hydrolysis of β-glycosidic bonds. In alkaline solutions, depolymerisation occurs to a lesser extent.

The enzyme lysozyme cleaves β-(1 → 4) glycosidic bonds between *N*-acetylglucosamine and D-glucosamine. The more chitin is acetylated, the faster the reaction. The ability to decompose chitin is based on the antimicrobial effects of lysozyme. Some bacteria produce chitinases known as chitodextrinases or poly[β-1,4-(2-acetamido-2-deoxy-D-glucosidases)] and chitosanase (chitosan-*N*-acetylglucosamine hydrolases), which cleave chitin chains to form *N*-acetylglucosamine.

Chitin forms complexes with most transition metals. The complexes with copper ions are among the strongest ones. The rates of formation and stability of the metal complexes depend on the temperature, the solubility of the ions, their type, the pH, the degree of acetylation of chitin, and other factors. Complexes with toxic metals (mercury or lead) can cause poisoning, because they break down in the digestive tract.

Chitin is virtually indigestible, because no human intestinal microflora contains the chitin-cleaving enzymes. It is only partially hydrolysed by saliva (by lysozyme) and in the stomach (by lysozyme and hydrochloric acid). A significant proportion of chitin can be utilised by some animals (e.g. fish, fowl, and rabbits).

Chitin is obtained industrially mainly from the shells of marine molluscs. The technology involves the removal of accompanying proteins in dilute sodium hydroxide, calcium carbonate, and hydrochloric acid. Partial alkaline hydrolysis of acetyl groups with sodium hydroxide is used for the preparation of modified chitin, which is called **chitosan**.

Chitosan contains 5–25% of *N*-acetylglucosamine and 75–95% of D-glucosamine units. Although the structures of chitin and chitosan differ only slightly, their chemical reactivity and physical properties are quite different. Chitosan is soluble in water, acids, and organic solvents and insoluble in neutral and alkaline solutions (it is a stronger base than chitin). Chitosan molecules are polycations and coagulate when molecules or particles carrying negative charge (e.g. sodium alginate, sulfates, phosphates, and proteins) are added to chitosan solutions. Dispersions of chitosan are highly viscous.

Chitosan is, like chitin, hydrolysed by acids, alkalis, and enzymes and forms complexes with metals. Unlike chitin, it reacts with aldehydes to form imines. It is also indigestible, but significantly reduces the cholesterol level in the blood serum and liver. The mechanism of this effect is not fully understood. Apparently, coagulation and flocculation of chitosan play an important role. Bile acids form micelles containing cholesterol, which are absorbed through the intestinal wall. Chitosan forms a gummy coagulate that captures and binds these micelles, which reduces the level of emulsified cholesterol. In the colon, chitosan blocks the conversion of cholesterol into coprostanol, which increases faecal excretion of cholesterol and reduces its transfer into the blood serum. Chitosan also forms salts with fatty acids, which do not dissociate even in acidic gastric juice. These salts bind lipids (triacylglycerols), which are thus excluded from the body.

Chitosan is used as an emulsifier (for margarine) and emulsion stabiliser (it is added to hamburgers, ice creams, and cheeses), a thickener (viscosity increases), a foam stabiliser, a gelling agent, a clarification agent for fruit juices, and in the pharmaceutical industry and cosmetics.

4.6 Complex saccharides

In animal and plant tissues and in cells of microorganisms, saccharides are often part of more complex structures containing, simultaneously, proteins, peptides, lipids, and other non-saccharide components. These saccharides are called **conjugated** or **complex** saccharides.

Peptides and proteins form a wide range of conjugated compounds with carbohydrates. Carbohydrates are present mostly as linear or branched oligosaccharides. In some conjugated compounds, the properties of peptide (protein) predominate, therefore these complex saccharides include:

- **glycopeptides**;

- **glycoproteins**.

In other complex carbohydrates, the properties of saccharides predominate, and then these compounds are divided into:

- **peptidoglycans** (also known as **mucopeptides**);

- **proteoglycans** (also known as **mucoproteins**).

The linear polysaccharide components of peptidoglycans of bacterial cell walls, known as **mureins**, are made up of alternating units of *N*-acetyl-D-glucosamine and *N*-acetylmuramic acid that are linked by β-(1 → 4) glycosidic bonds. The residue of lactic acid in *N*-acetylmuramic acid is linked by an amide bond to L-alanine in tetrapeptide Ala-D-isoGlu-L-Lys-D-Ala. Saccharides in proteoglycans of higher animals are predominantly linear polymers, so-called **mucopolysaccharides**, composed of aldoses (D-galactose), glycuronic acids (D-glucuronic or L-iduronic acids), amino sugars (*N*-acetyl-D-glucosamine and *N*-acetyl-D-galactosamine), and their sulfates.

Saccharides also form a very diverse range of compounds with lipids. These compounds are categories of homolipids, heterolipids, and complex lipids. Glycolipids are homolipids that include naturally occurring esters of higher fatty acids (e.g. glycoglycerolipids such as diacylgalactosylglycerol), synthetic sugar esters (e.g. esters of sucrose), and various sugar derivatives (such as glucitol esters). Heterolipids include some phospholipids (phosphatides, such as phosphatidylinositol) and lipamides (e.g. lipamide glycosides called cerebrosides). For example, complex lipids include mucolipids and gangliosides.

4.6.1 Mucopolysaccharides

Mucopolysaccharides, also known as **glycosaminoglycans**, are acidic linear polysaccharides composed of repeating disaccharide units. Mucopolysaccharides are sugar components of proteoglycans, peptidoglycans, glycoproteins, and glycopeptides. Proteoglycans are essential components of the extracellular parts of the epithelial and connective tissues (skin, tendons, blood vessels, cartilages, and bones) and also have protective functions in cells. Glycoproteins are slimy and gelatinous substances in the synovial fluids of joints, liquids of eye sockets, and various secretions, such as respiratory tract mucus. They also occur in saliva, egg white, and other viscous materials.

Glycosaminoglycans are covalently bound to proteins through a tetrasaccharide bound to the serine residue (**4-96**). Elongation of this tetrasaccharide (consecutive addition of sugar units to gluconic acid), esterification with sulfuric acid, and various other reactions yield the individual glycosaminoglycans.

β-D-Glc*p*A-(1→3)-β-D-Gal*p*-(1→3)-β-D-Gal*p*-(1→4)-β-D-Xyl*p*-(1→

4-96, binding region of glycosaminoglycans

Hyaluronic acid is a common glycosaminoglycan that is not esterified with sulfuric acid. It consists of 250–25 000 alternating units of D-glucuronic acid and *N*-acetyl-D-glucosamine linked to one another by a β-(1 → 3) glycosidic bond (**4-97**). Hyaluronic acid is present in the skin (which prevents penetration of bacteria into the tissues), in cartilage, and in the aqueous humour of the eyes, resulting in the lubrication properties of synovial fluid. These properties are attributed to the formation of hydrogen bridges between the chains and their high degree of hydration. The enzyme hyaluronidase, hydrolysing β-(1 → 3) bonds, occurs in animal tissues, bacteria (where it is probably responsible for their invasiveness), and snake and insect poisons.

There are two types of sulfated glycosaminoglycans. The first includes galactosaminoglycans, known as chondroitin sulfate and dermatan sulfate, whilst the second are glucosaminoglycans, referred to as heparin sulfate and heparin. An example of a commonly occurring galactosaminoglycan is chondroitin sulfate, which contains alternating units of D-glucuronic acid and *N*-acetyl-D-galactosamine. The units of D-glucuronic acid can be esterified with sulfuric acid at position C-2, and the units of *N*-acetyl-D-galactosamine are esterified at positions C-4 or C-6 or are non-esterified (**4-97**). Chondroitin sulfate connects cells in animal epithelial and connective tissues.

Dermatan sulfate differs from chondroitin sulfate in that some units of D-glucuronic acid are replaced by units of L-iduronic acid (**4-97**) or its 2-sulfate and *N*-acetyl-D-galactosamine is esterified in position C-4 or C-6 (monoesters), simultaneously at C-4 and C-6 (diesters), or else is not esterified. Dermatan sulfate is found mostly in the skin (also in the blood vessels and tendons).

Instead of *N*-acetyl-D-galactosamine, keratan sulfate (kerato sulfate, **4-97**) contains *N*-acetyl-D-glucosamine, and instead of D-glucuronic acid, it contains D-galactose esterified at C-6 with sulfuric acid. *N*-Acetyl-D-glucosamine can also be esterified at C-6 with sulfuric acid. Other sugars are present in small amounts, such as L-fucose, which is found in type II keratan sulfate. Keratan sulfate is part of cartilages and the cornea.

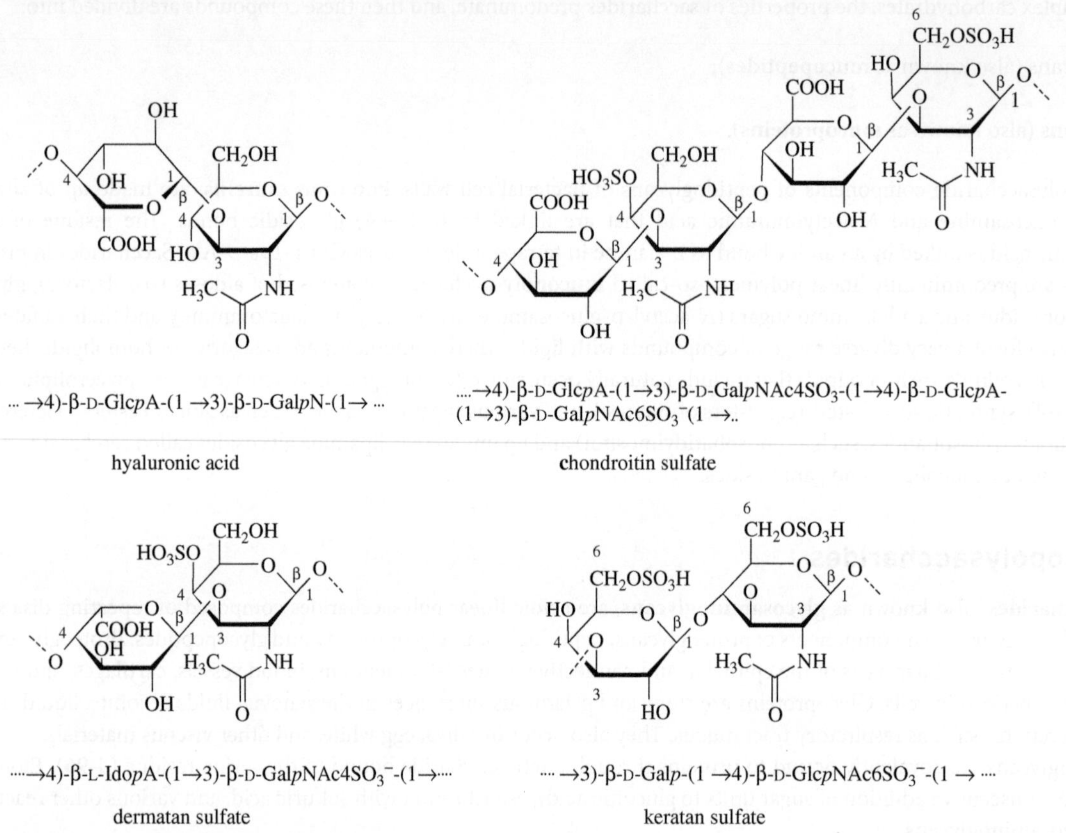

... →4)-β-D-GlcpA-(1 →3)-β-D-GalpN-(1→ ...

hyaluronic acid

....→4)-β-D-GlcpA-(1→3)-β-D-GalpNAc4SO₃-(1→4)-β-D-GlcpA-
(1→3)-β-D-GalpNAc6SO₃⁻(1 →..

chondroitin sulfate

·····4)-β-L-IdopA-(1→3)-β-D-GalpNAc4SO₃⁻-(1→····

dermatan sulfate

···· →3)-β-D-Galp-(1 →4)-β-D-GlcpNAc6SO₃⁻-(1 → ····

keratan sulfate

4-97, basic structures of mucopolysaccharides

A related acidic heteropolysaccharide is heparin. It consists primarily of D-glucuronic acid (and also contains L-idurono-2-sulfate) and N-acetyl-D-glucosamine, which can be esterified with sulfuric acid at nitrogen or C-6, or at both simultaneously. These basic units are joined by α-(1 → 4) glycosidic bonds. Heparin does not occur in connective tissues, but is found in granular form in specific cells (particularly cells of the liver, lungs, and skin) and is a part of the bloodstream. It has anticoagulant properties (prevents blood clotting). A very similar polysaccharide is heparan sulfate, which contains predominantly D-glucuronic acid (or L-iduronic acid and L-idurono-2-sulfate) and N-acetyl-D-glucosamine (or its sulfates as heparan). D-Glucuronic acid is linked by β-(1 → 4) glycosidic bonds with N-acetyl-D-glucosamine (its sulfates), which is linked in turn with an additional unit of D-glucuronic acid by an α-(1 → 4) glycosidic bond. Disaccharide units are organised into sulfate and non-sulfate domains.

4.6.2 Proteoglycans

Mucopolysaccharides are covalently and non-covalently bound to various proteins to form macromolecules called proteoglycans or mucoproteins. The basis of the molecule of proteoglycans is a linear molecule of hyaluronic acid, to which are non-covalently attached proteoglycan subunits through the globular N-terminus. The subunits are composed of proteins covalently bound to mucopolysaccharides, mainly to chondroitin sulfate and keratan sulfate through serine and threonine hydroxyl groups or the amide group of asparagine and the hemiacetal hydroxyl groups of N-acetyl-D-glucosamine or N-acetyl-D-galactosamine.

4.7 Reactions

4.7.1 Monosaccharides and oligosaccharides

The forms of sugars with free carbonyl groups and the cyclic forms are in equilibrium, and therefore most of the reactions of sugars involve the reactive carbonyl group and the anomeric hydroxyl group. Some reactions also involve primary and secondary hydroxyl groups of the chain (salt formation, formation of complexes with metals, and other reactions). The presence of the hydroxyl groups allows carbohydrates to interact with the aqueous environment and to participate in non-bonding interactions.

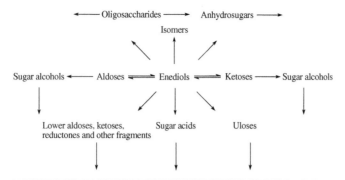

Figure 4.18 Main transformations of monosaccharides.

Saccharides in the enzymatically active food materials often become substrates for various enzymes that catalyse oxidoreduction reactions of sugars (oxidoreductases), their hydrolysis (hydrolases), and a range of other reactions.

The aldehyde group of aldoses is easily oxidised, yielding sugar acids (glyconic acids) that eliminate water and form lactones. Aldoses and ketoses isomerise under the catalysis of bases and break down to lower sugars and other compounds. Reduction of aldoses and ketoses yields sugar alcohols. In aqueous solutions (of pH 3–7) and in diluted solutions of acids, monosaccharides are relatively stable at normal temperatures. At higher temperatures, to a small extent, intermolecular condensation leads to disaccharides and higher oligosaccharides. The long-term effect of diluted acids or concentrated acids results in hydrolysis of polysaccharides and oligosaccharides, and monosaccharides undergo dehydration to furan and pyran derivatives. In alkaline solutions, monosaccharides are unstable, and even at normal temperature they undergo isomerisation and elimination reactions and rearrangements. At higher temperatures, sugar carbon chains can split to lower-molecular-weight products, which may be again recombined. Chemical changes of monosaccharides are shown schematically in Figure 4.18.

Reactions of sugars in foods are generally complex biochemical (enzymatic) and chemical (non-enzymatic) reactions. All the functional groups of sugar molecules participate in these reactions, which are influenced by pH, temperature, water content, and many other factors, especially the presence of other food components. In this respect, reactions of sugars themselves and reactions of sugars with other food components that result in the formation of brown pigments are generally classified as **non-enzymatic browning reactions**. The most important reaction is the reaction of reducing sugars with proteins, the **Maillard reaction**.

4.7.1.1 Reactions in acid media

4.7.1.1.1 Formation and hydrolysis of glycosides

Hemiacetal hydroxyl groups of pyranoses and furanoses are more acidic in comparison with other hydroxyl groups due to the negative inductive effect (–I effect) of the ring oxygen atom. Therefore, sugars tend to split these reactive hydroxyl groups in acidic media with the formation of mesomeric carbenium cations (Figure 4.19). This reaction is reversible. The existence of this cation explains a number of reactions that proceed on anomeric carbon, especially the formation and hydrolysis of glycosides.

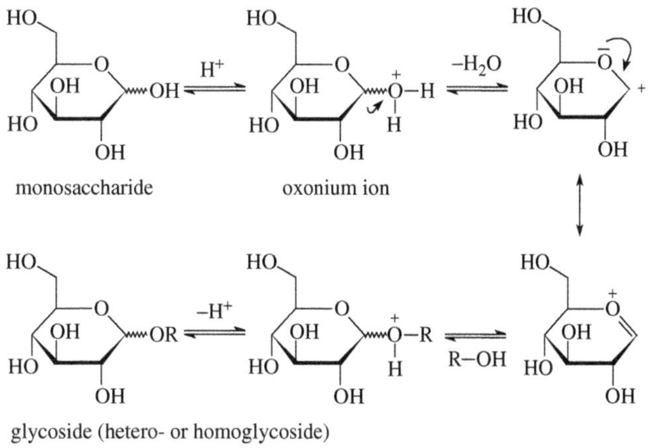

Figure 4.19 Mechanism of glycoside formation and hydrolysis.

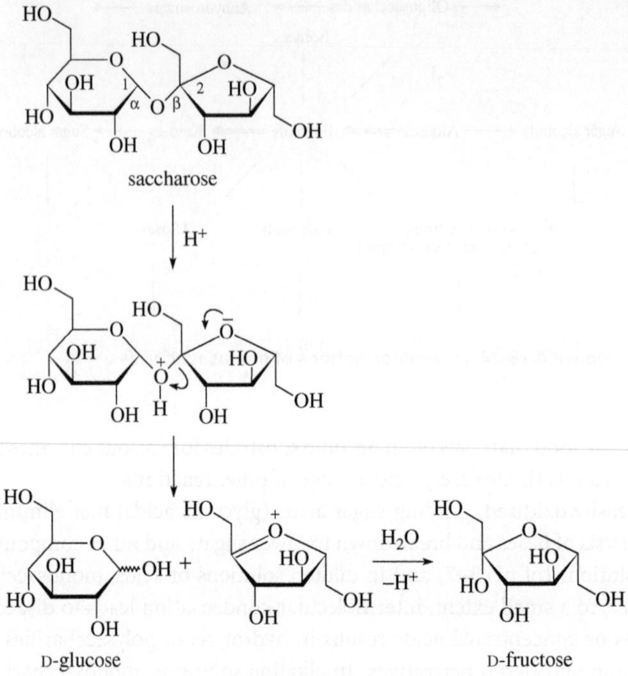

Figure 4.20 Mechanism of saccharose hydrolysis.

A typical example is the formation of oligosaccharides from monosaccharides in acidic solutions. The reaction is called **reversion**. Its mechanism is opposite to the hydrolysis of oligosaccharides, which is **inversion**. The main products of D-glucose reversion are the $(1 \rightarrow 6)$ disaccharides isomaltose and gentiobiose, because the primary hydroxyl group on the C-6 carbon is more reactive than the secondary hydroxyl groups. Of the total products of glucose reversion, isomaltose amounts to about 68–70% and the gentiobiose to 17–18%. Minor reaction products are $(1 \leftrightarrow 1)$, $(1 \rightarrow 2)$, $(1 \rightarrow 3)$, and $(1 \rightarrow 4)$ disaccharides and some higher oligosaccharides. Reversion products are commonly found in hydrolysates of saccharose, starch, and other polysaccharides, and arise also during caramelisation of sugars. For example, the reversion products in hydrolysates of starch are about 5–6%.

Glycosides have the typical properties of acetals. They are generally stable in alkaline solutions, but are hydrolysed in acidic solutions to the parent sugar and aglycone. Hydrolysis of glycosides by acids starts with protonation of the aglycone oxygen atom (Figure 4.19). β-Glycosides are generally hydrolysed faster than α-glycosides; the reaction rate depends considerably on the sugar moiety substituents. Hydrolysis of glycosides containing a 2-deoxysugar with hydrogen as a C-2 substituent is relatively rapid. For sugars containing either the electronegative hydroxyl group (which the majority of important sugars have) or the protonated amino group (in 2-amino-2-deoxysugars) in position C-2 hydrolysis is much slower.

One of the most stable disaccharides is α,α-trehalose, which contains two C-2 hydroxyl groups and is therefore about 10 times more stable than maltose or lactose. Increased stability determines, in particular, the negative inductive effect of the hydroxyl groups at C-2. The hydrolysis rate of the galactose-O-glucose bond in raffinose, and also of the glucose-O-fructose bond in the reducing disaccharide turanose (Table 4.14), is comparable to the hydrolysis rate of maltose or lactose.

Hydrolysis (inversion) of saccharose to glucose and fructose (Figure 4.20) is about 10^3 times faster than hydrolysis of other disaccharides, such as maltose and lactose (the rate constant of this first-order reaction in solutions of pH 1–3 at 100 °C is approximately 1×10^{-3}/s). Hydrolysis of saccharose takes place even at low temperatures, in the presence of diluted mineral or organic acids, and in the presence of even small amounts of water. The glycosidic bond in glucose-O-fructose in raffinose, which yields melibiose (Table 4.16), is hydrolysed at a similar rate to saccharose. The reason is that the mechanism of hydrolysis of saccharose is somewhat different from the mechanism of hydrolysis of other glycosidic bonds (such as glucose-O-glucose or galactose-O-glucose). It is assumed that the resulting cation is degraded to glucose and a stable five-membered cation of fructose. The reaction of this cation with water forms fructose.

4.7.1.1.2 Formation of anhydrosugars

In addition to extramolecular condensation to homoglycosides, aldohexoses also undergo intramolecular condensation of the hemiacetal and other hydroxyl groups to form the corresponding anhydrosugar (sugar anhydride). As well as the main product, 1,6-anhydropyranose, small amounts of 1,6-anhydrofuranoses and other anhydrosugars may also form.

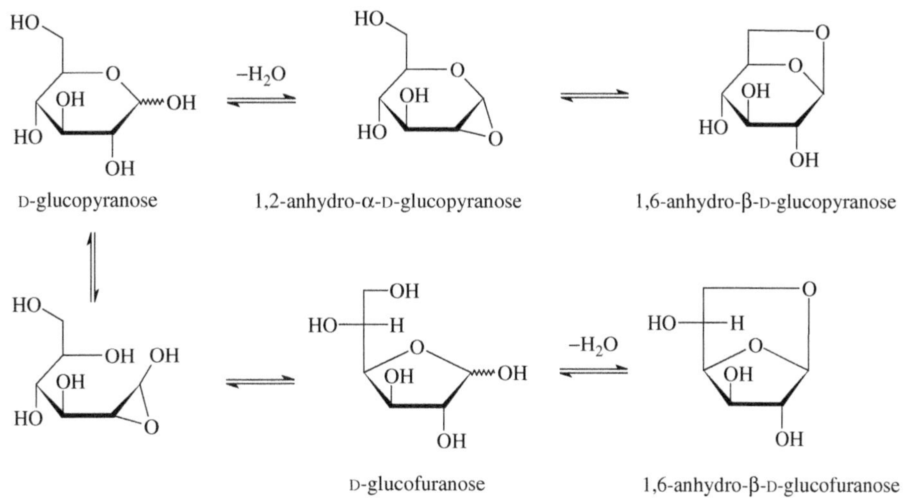

Figure 4.21 Formation of glucose anhydrides.

In acidic solutions, D-glucose yields about 1% of 1,6-anhydro-β-D-glucopyranose, known as β-glucosan or laevoglucosan. It is also produced when glucose syrup is heated to temperatures higher than 100 °C, and by pyrolysis of glucose, starch, and cellulose. It also can partially polymerise to branched polysaccharides. β-Glucosan is hydrolysed by acids to D-glucose. The mechanism of formation of anhydrosugars from D-glucose is shown in Figure 4.21. Under similar conditions, D-mannose yields 1,6-anhydro-β-D-mannopyranose (β-mannosan) and D-galactose yields 1,6-anhydro-β-D-galactopyranose (β-galactosan). The pyranose ring of 1,6-anhydrosugars is fixed in the $^1C_4$ conformation. Anhydrosugars of those monosaccharides that have a higher number of equatorial hydroxyl groups with appropriate conformation, such as β-D-idopyranose, are formed in larger amounts.

Ketoses do not form monomeric anhydrides, but intermolecular reactions yield tricyclic compounds. Heating of D-fructose (and also heating of saccharose or inulin) gives rise to various dianhydrides of the corresponding furanoses and pyranoses. Examples of these intermolecular anhydrides formed from fructose are the 1,2′:1′,2-dianhydride of α-D-fructofuranose with α-D-fructofuranose and 1,2′:1′,2-dianhydride of α-D-fructopyranose with β-D-fructopyranose (**4-98**). Anhydrides are also formed by dehydration of disaccharides and sugar alcohols.

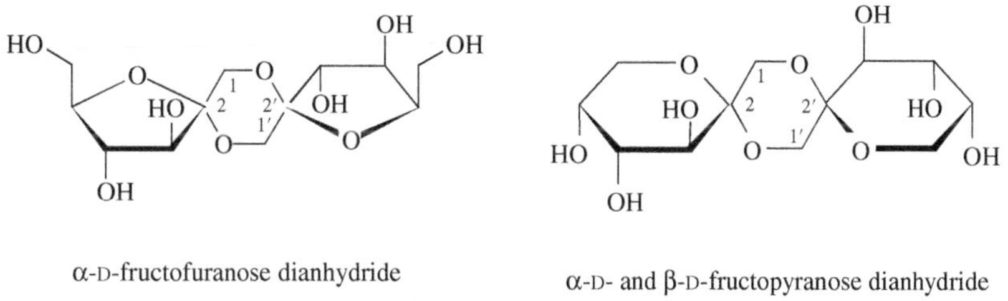

α-D-fructofuranose dianhydride

4-98, fructose anhydrides

α-D- and β-D-fructopyranose dianhydride

4.7.1.1.3 Formation of furan and pyran derivatives

The main reaction of monosaccharides in strong acids is dehydration, associated with the loss of one to three molecules of water and the formation of furan and pyran derivatives. Both types of compounds are also formed during caramelisation of sugars. In practice, dehydration of monosaccharides and the formation of furan and pyran derivatives also occur in slightly acidic solutions, such as during long-term storage of fruit compotes, especially at elevated temperatures. Higher levels of furans and pyrans are formed during hydrolysis of starch, and significant amounts of these compounds arises in the production of acid protein hydrolysates that are used as seasonings for soups, sauces, and other products.

The reaction mechanism of the formation of furans and pyrans from D-glucose and D-fructose (ketoses are more reactive than aldoses) is shown in Figure 4.22. The intermediate product of dehydration of both sugars is 1-ene-1,2-diol (1-ene-1,2-diolate) stabilised by hydrogen bonds (Figure 4.23). The formation of 1-ene-1,2-diolate is known as **1,2-enolisation**. The reaction involves acid/base catalysis, so endiolates are common products of acid-catalysed dehydration as well as isomerisation, saccharinic acid rearrangement, and fragmentation of sugars in alkaline media. All these reactions are often collectively called Lobry de Bruyn–Alberda van Ekenstein transformations of aldoses and ketoses.

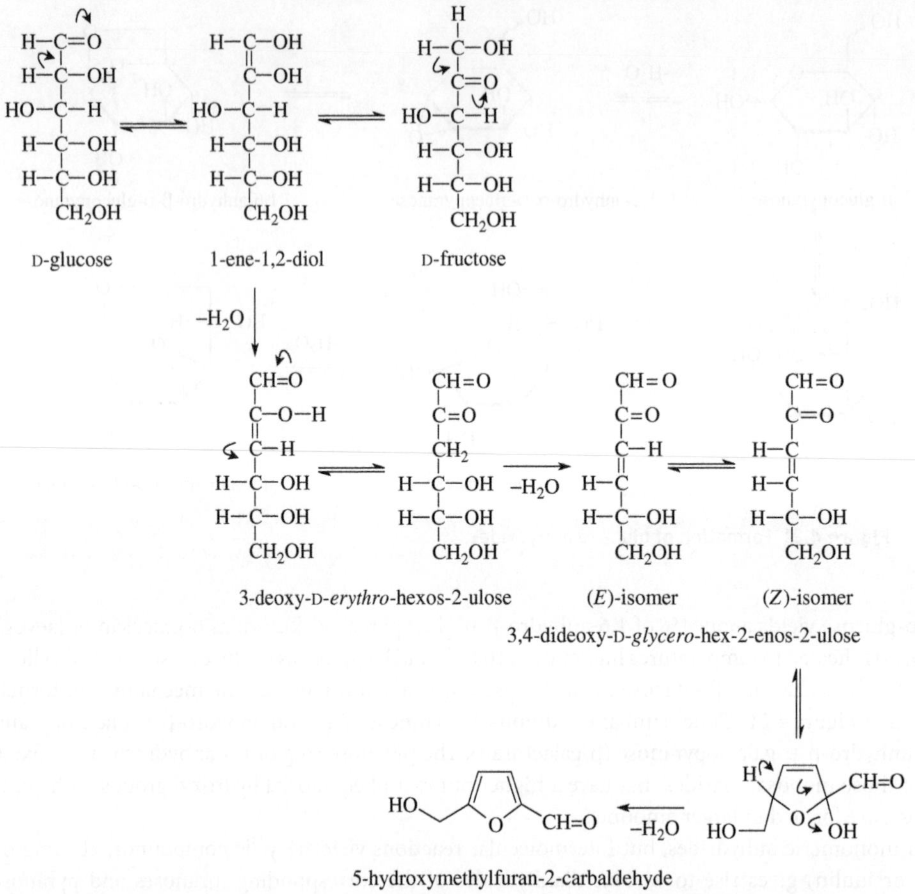

Figure 4.22 Dehydration of glucose and fructose.

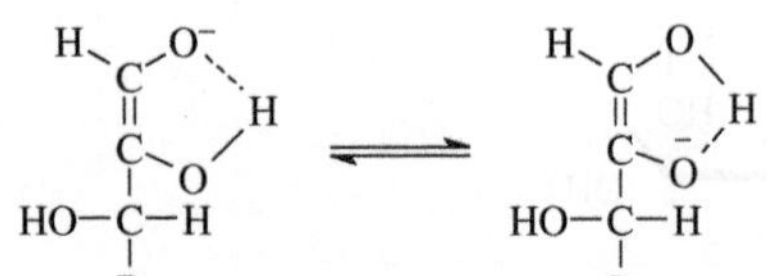

Figure 4.23 Structure of (Z)-1-ene-1,2-diolate.

The elimination of one molecule of water from D-glucose or D-fructose produces a 3-deoxyglycos-2-ulose known under the systematic name 3-deoxy-D-*erythro*-hexos-2-ulose (or by older name 3-deoxy-D-glucosone). In analogy, 3-deoxy-D-*threo*-hexos-2-ulose (3-deoxy-D-galactosone) results from D-galactose in milk products. Loss of the second water molecule creates 3,4-dideoxy-D-*glycero*-hex-3-enos-2-ulose (3,4-dideoxy-D-glucosulos-3-ene), and loss of the third then yields 5-hydroxymethylfuran-2-carbaldehyde, which can further decompose to laevulinic and formic acids. 5-Hydroxymethylfuran-2-carbaldehyde is one of the most common products of the dehydration of hexoses. It is virtually absent in fresh and untreated foods, but occurs in many carbohydrate-rich processed foods. For example, the concentration of 5-hydroxymethylfuran-2-carbaldehyde in honeys is 3.3–26.3 mg/kg, in apple juices 2.9–3.5 mg/kg, in orange juices 2.7–10.6, in jams 2.7–15.9 mg/kg, in orange marmalade 47.1 mg/kg, in toasts 41.7–591 mg/kg, and in roasted coffee 113–1093 mg/kg. When heated with hydrochloric acid, as in the production of acid hydrolysates of proteins (hydrolysis is performed at a temperature of 100–120 °C in about 20% acid), hexoses yield 5-chloromethylfuran-2-carbaldehyde (about 1–2 mg/kg) as a minor product.

The formation of 5-hydroxymethylfuran-2-carbaldehyde from saccharose proceeds via hydrolysis to fructofuranosyl cation (Figure 4.20) and glucose. Elimination of a proton and two molecules of water from the fructofuranosyl cation yields 5-hydroxymethylfuran-2-carbaldehyde. Under dry pyrolytic conditions and at temperatures above 250 °C, 90% of 5-hydroxymethylfuran-2-carbaldehyde originates from the fructose moiety and only 10% from glucose. Pentoses and L-ascorbic acid dehydrate in the same way as hexoses under acidic conditions, yielding furan-2-carbaldehyde (via reactive 3-deoxy-L-*threo*-pentos-2-ulose and 3,4-dideoxypentosulos-3-ene) as the main product (see Section 5.14.6.1). 6-Deoxyhexoses, such as L-rhamnose, yield, analogously, 5-methylfuran-2-carbaldehyde.

3-Deoxy-D-*erythro*-hexos-2-ulose occurs mainly as α- or β-pyranose in aqueous solutions. Owing to the presence of a keto group at C-2, other cyclic forms may also be present, such as furanoses (Figure 4.24). In weakly acidic and neutral solutions, 3-deoxy-D-*erythro*-hexos-2-ulose eliminates water and yields γ-lactone of metasaccharinic acid and other products (Figure 4.25). However, metasaccharinic and

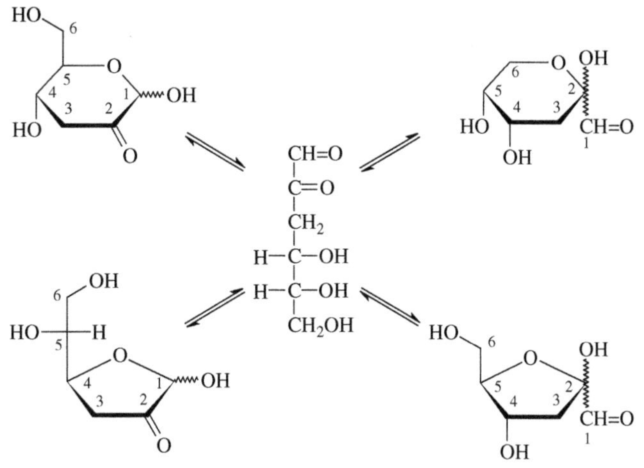

Figure 4.24 Cyclic forms of 3-deoxy-D-*erythro*-hexos-2-ulose in solutions.

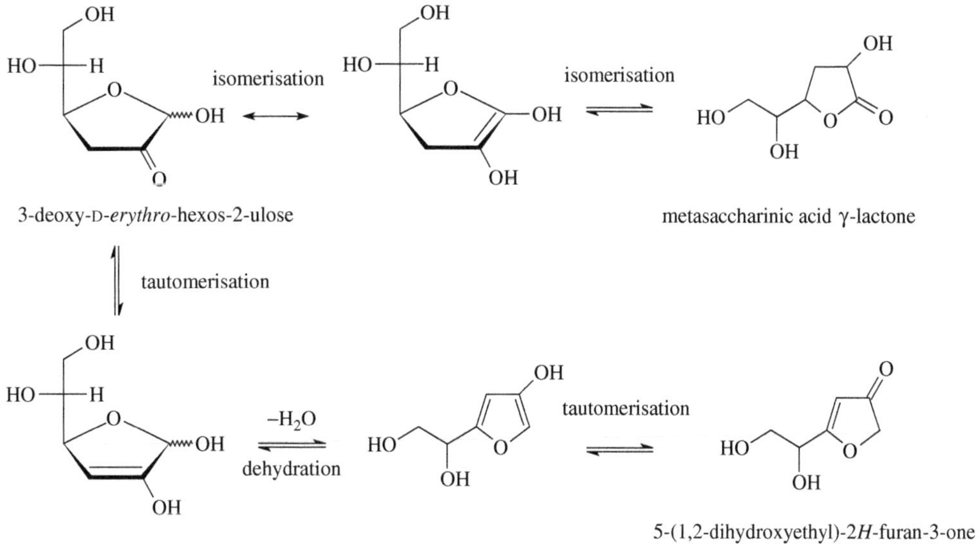

3-deoxy-D-*erythro*-hexos-2-ulose

metasaccharinic acid γ-lactone

5-(1,2-dihydroxyethyl)-2*H*-furan-3-one

Figure 4.25 Formation of metasaccharinic acids and furan-3-ones from 3-deoxy-D-*erythro*-hexos-2-ulose.

other sugar acids are formed preferably in alkaline solutions. Pyranoses also dominate in (*Z*)-3,4-dideoxy-D-*glycero*-hex-3-enos-2-ulose (**4-99**) solutions. The existence of a six-membered ring in (*E*)-3,4-dideoxy-D-*glycero*-hex-3-enos-2-ulose (**4-99**) is not possible; therefore, in aqueous solutions, this isomer occurs as a straight chain form or as its hydrate.

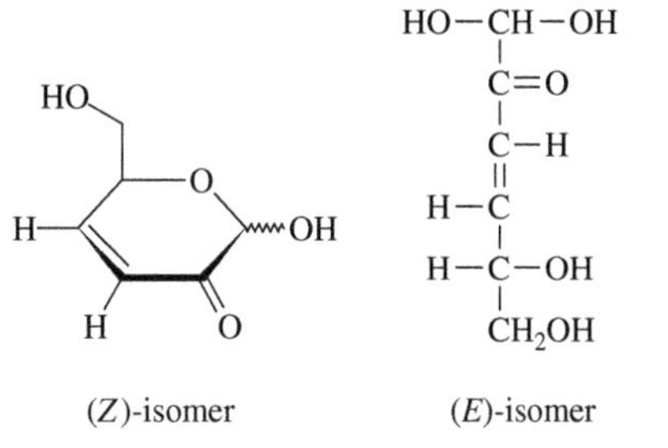

(*Z*)-isomer

(*E*)-isomer

4-99, 3,4-dideoxy-D-*glycero*-hex-3-enos-2-ulose

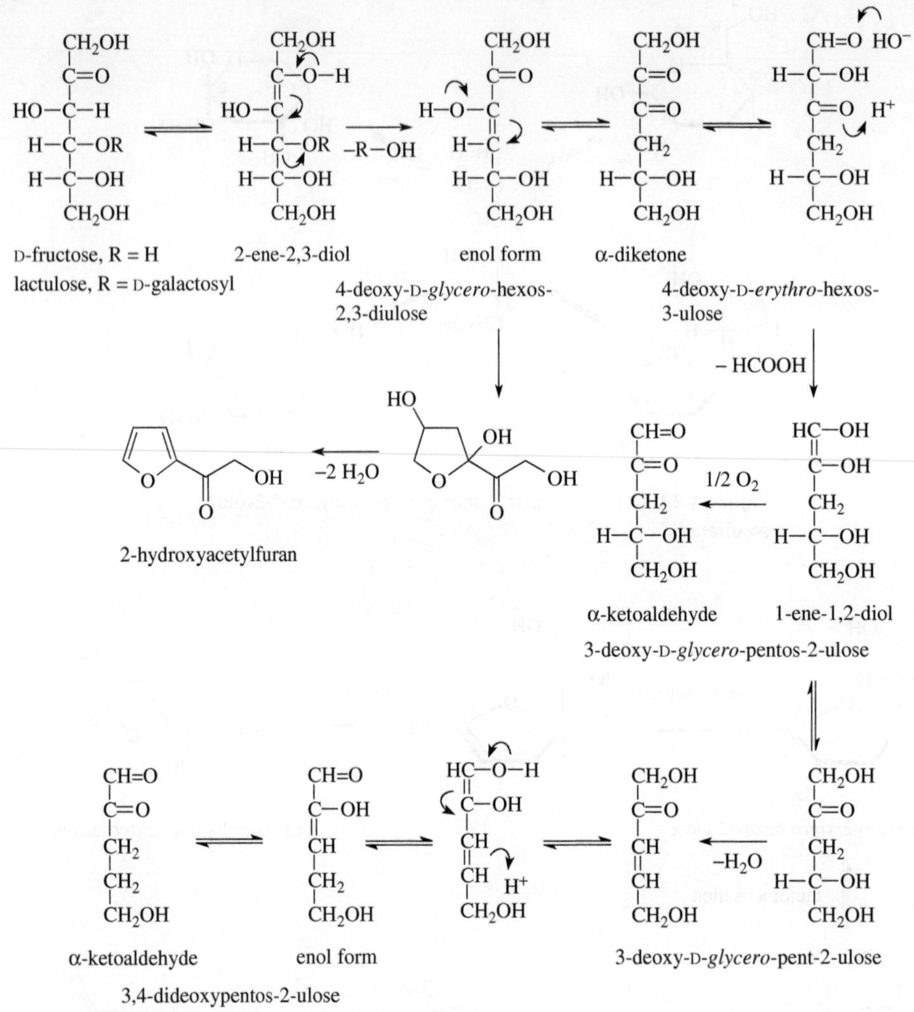

Figure 4.26 Formation of 2-hydroxyacetylfuran and 3-deoxy-D-*glycero*-pent-2-ulose from fructose and lactulose.

5-Hydroxymethylfuran-2-carbaldehyde is the main reaction product of glucose and fructose in strongly acidic solutions, but hexose degradation also produces some other furan derivatives. Enolisation of fructose may continue along the chain, and isomerisation via 2-ene-2,3-diol leads to a wider range of products than in the case of glucose. This isomerisation is known as **2,3-enolisation**. Splitting off the hydroxyl group at C-4 probably yields, by a sequence of isomerisation and dehydration reactions via 4-deoxy-D-*glycero*-hexo-2,3-diulose, other important degradation product of fructose-2-hydroxyacetylfuran, also known as furyl hydroxymethyl ketone (Figure 4.26). Another dehydration product of sugars, 2-acetylfuran, is formed via 1-deoxy-D-*erythro*-hexo-2,4-diulose (an isomerisation product of 1-deoxy-D-*erythro*-hexo-2,3-diulose) and 1,4-dideoxy-D-*glycero*-hexo-2,3-diulose. The latter is a product of oligosaccharide degradation (e.g. lactose) or it is arisen by a reduction of 1-deoxy-D-*erythro*-hexo-2,4-diulose or its isomers. 1-Deoxy-2,3-diulose derived from pentoses (and ascorbic and dehydroascorbic acids) reacts with formaldehyde (formed, for example, in the Strecker degradation of glycine) to a six-carbon unit, 1-deoxy-D-*erythro*-hexo-2,4-diulose. Sequence of further reactions again gives 2-acetylfuran (Figure 4.27).

4-Deoxy-D-*glycero*-hexo-2,3-diulose is a typical product of degradation and of 4-*O*-substituted derivatives of fructose (e.g. lactulose, which arises by isomerisation of lactose). 1,2-Enolisation of 4-deoxy-D-*glycero*-hexo-2,3-diulose yields, via 1-ene-1,2-diol, unstable 4-deoxyhexos-3-ulose, which splits into formic acid and 1,2-enediol of 3-deoxy-D-*glycero*-pentulose in neutral or alkaline solutions through a reverse Claisen condensation. This sugar can be oxidised to 3-deoxy-D-*glycero*-pentos-2-ulose or may dehydrate to form 3,4-dideoxypentosulose (Figure 4.26). Other reaction products are isosaccharinic acids.

Splitting off the hydroxyl group at C-1 of fructose yields a key intermediate, 1-deoxy-D-*erythro*-hexo-2,3-diulose (Figure 4.28). In addition to the straight chain form, also the corresponding furanoses and pyranoses are formed in solution (Figure 4.29). Isomerisation of 1-deoxy-D-*erythro*-hexo-2,3-diulose along the chain yields other important intermediates: 1-deoxy-D-*erythro*-hexo-2,4-diulose, 1-deoxy-D-*erythro*-hexo-3,4-diulose, 1-deoxy-D-*erythro*-hexo-3,5-diulose, 1-deoxy-D-*erythro*-hexo-4,5-diulose, and 1-deoxy-D-*ribo*-hexo-4,6-diulose (Figure 4.30). 1-Deoxy-D-*erythro*-hexo-2,3-diulose is considered a precursor of many important furanones and pyranones that

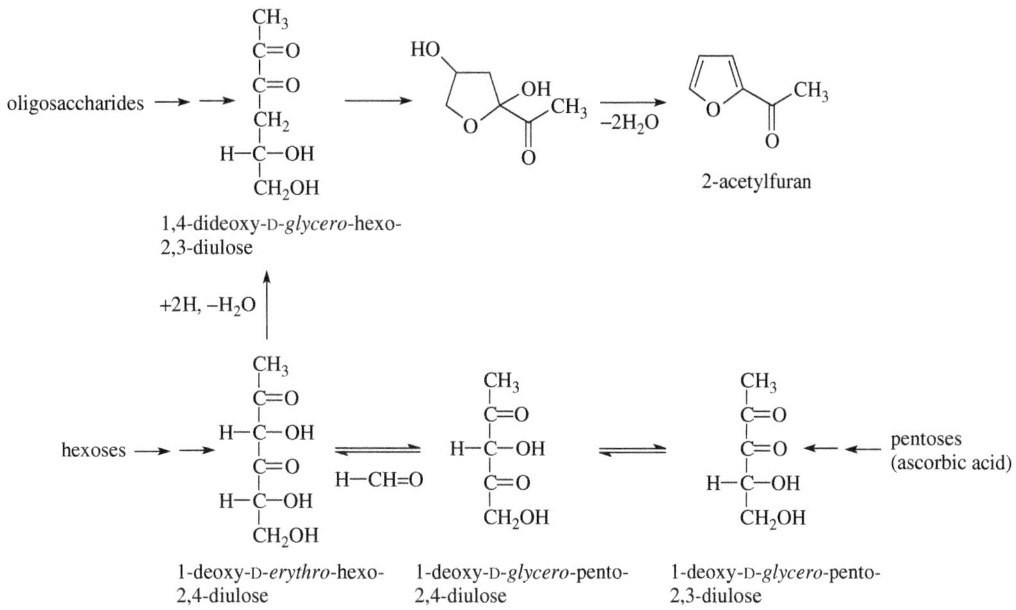

Figure 4.27 Formation of 2-acetylfuran from pentoses, hexoses, and oligosaccharides.

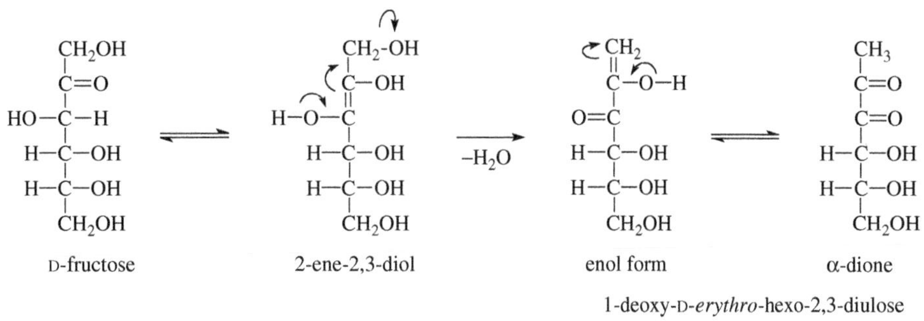

Figure 4.28 Formation of 1-deoxy-D-*erythro*-hexo-2,3-diulose from fructose.

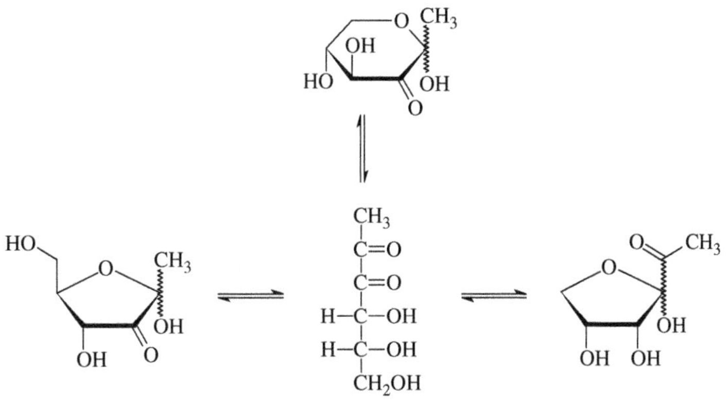

Figure 4.29 Cyclic forms of 1-deoxy-D-*erythro*-hexo-2,3-diulose.

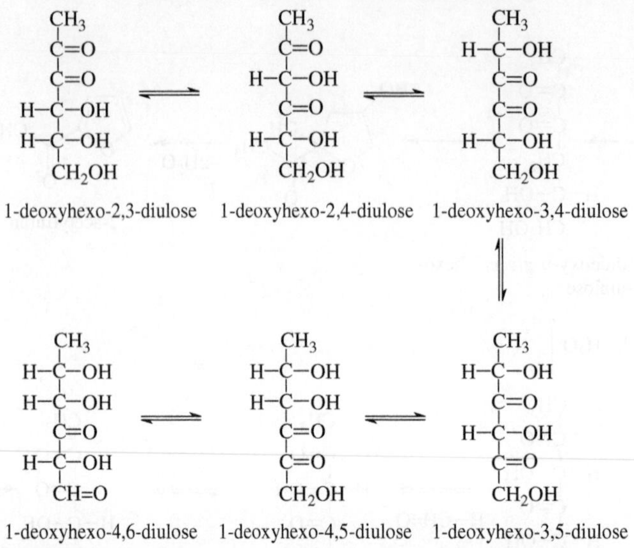

Figure 4.30 Products of 1-deoxy-D-*erythro*-hexo-2,3-diulose isomerisation.

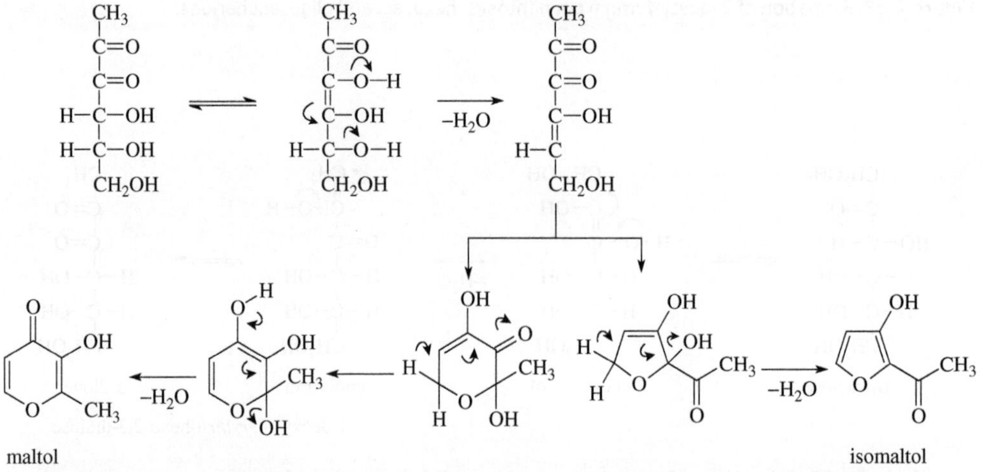

Figure 4.31 Formation of maltol and isomaltol from monosaccharides.

arise particularly in weakly acidic and neutral solutions. For example, this diulose is the precursor of 3-hydroxy-2-methylpyran-4-one, which is trivially termed maltol, and of its isomer 2-acetyl-3-hydroxyfuran, known as isomaltol (Figure 4.31). Maltol and isomaltol have a characteristic caramel odour and taste.

In comparison with hexoses, higher quantities of maltol and isomaltol are produced from 4-*O*-substituted glucose derivatives such as maltose and lactose. For example, in malt coffee (coffee substitutes), maltol occurs at a level of 300 mg/kg, in chocolates at approximately 3 mg/kg, and in dark beers at up to 3 mg/kg. The mechanism of degradation of 1-deoxyglycodiuloses derived from disaccharides is somewhat different than that of 1-deoxyglycodiuloses arising from monosaccharides (Figure 4.32). Elimination of the hydroxyl group in position C-5 of the cyclic diulose (pyranose) yields pyran-3-one. This compound, which was also found in heat-treated milk (R = β-D-galactosyl), is the key intermediate of the degradation of disaccharides. Its fate depends on the properties of glycosidically bound sugar. If that sugar is β-galactose (in the case of lactose), the main reaction product is galactosylisomaltol. If it is glucose (in the case of maltose), maltol is preferably formed, whilst glucosylisomaltol is found in lower amounts. For example, glucosylisomaltol was determined in pre-baked bread (heated at 190 °C for 30 minutes), where its amount increased from non-detectable to 20.9 mg/kg after 30 minutes of baking.

A compound with similar organoleptic properties to those of isomaltol or maltol is 2,4-dihydroxy-2,5-dimethyl-2*H*-furan-3-one, the cyclic form of so-called diacetylformoin. Diacetylformoin arises from 1-deoxy-D-*erythro*-hexo-2,3-diulose (Figure 4.33) via 1,6-dideoxy-D-*glycero*-hexo-2,4,5-triulose (acyclic form of diacetylformoin) and other intermediates (Figure 4.34). Diacetylformoin is considered a precursor of other important furans and pyrans and of many lower-molecular-weight degradation products. In addition to diacetylformoin, hexoses (cyclic forms of 1-deoxy-D-*erythro*-hexo-2,3-diulose) also give rise to 4-hydroxy-5-hydroxymethyl-2-methyl-2*H*-furan-3-one and to isomeric 2,3-dihydro-3,5-dihydroxy-6-methyl-4*H*-pyran-4-one. The latter is commonly formed in thermally processed

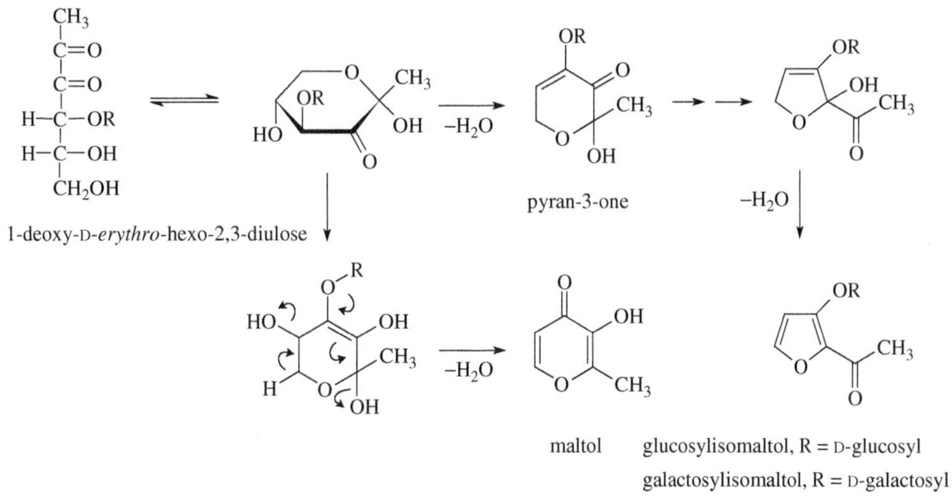

Figure 4.32 Formation of maltol and isomaltol from disaccharides.

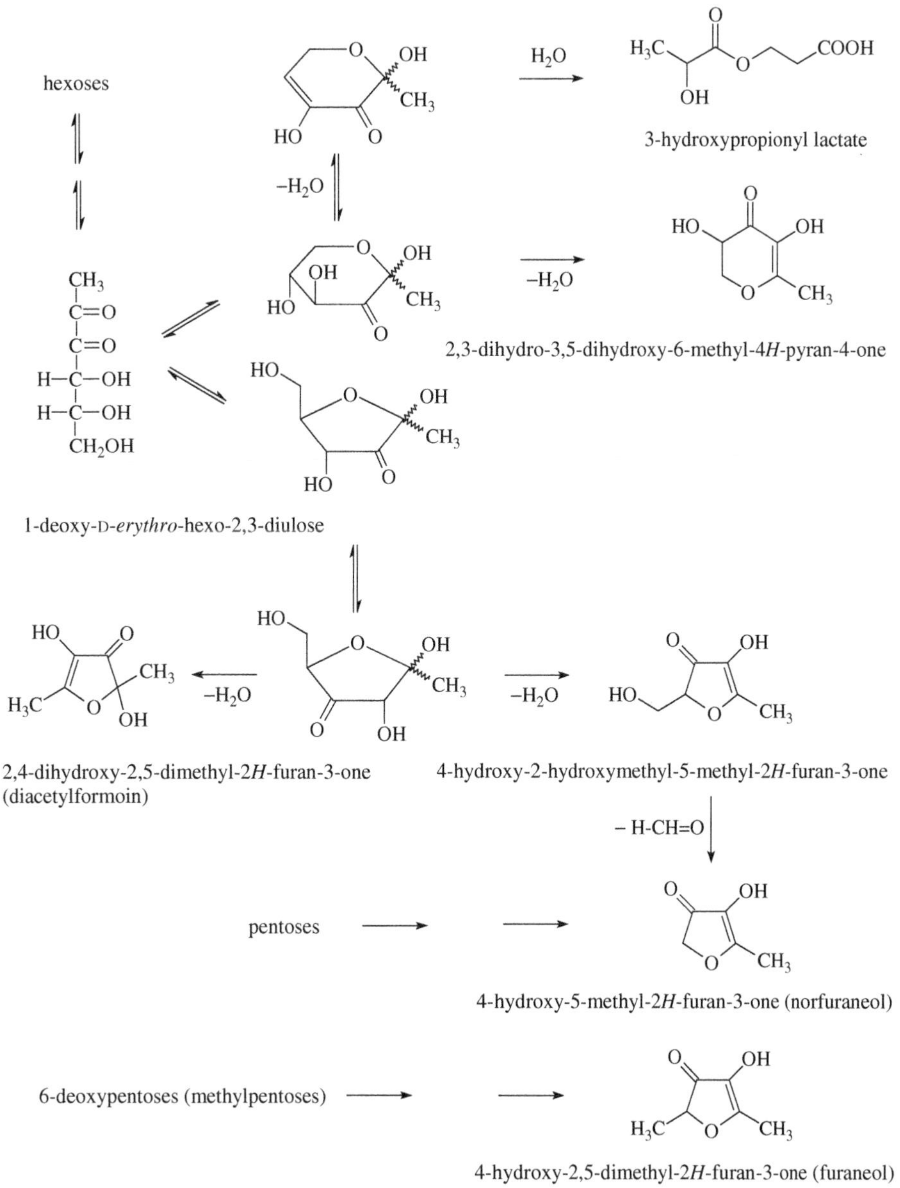

Figure 4.33 Formation of furan-3-ones and pyran-4-ones.

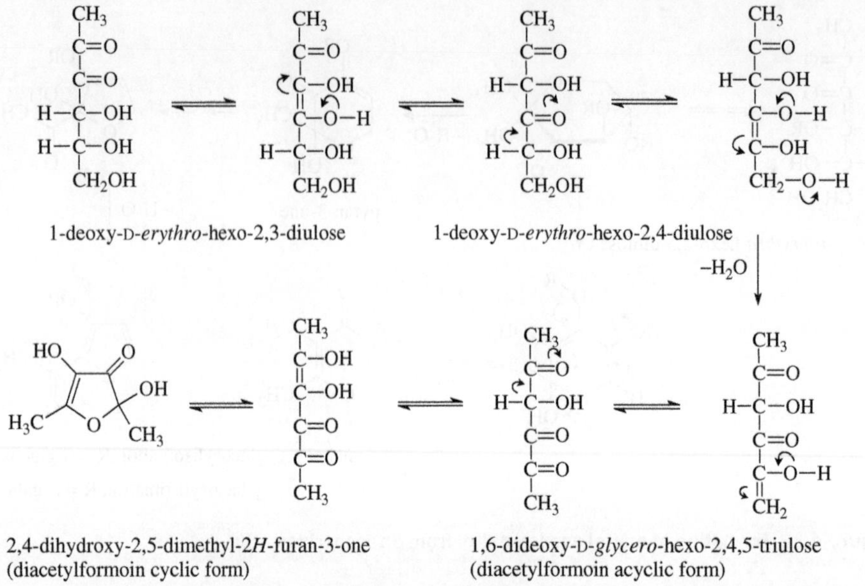

Figure 4.34 Formation of cyclic and acyclic forms of diacetylformoin.

1-deoxy-D-*erythro*-hexo-2,3-diulose

1-deoxy-D-*erythro*-hexo-2,4-diulose

−H₂O

2,4-dihydroxy-2,5-dimethyl-2*H*-furan-3-one
(diacetylformoin cyclic form)

1,6-dideoxy-D-*glycero*-hexo-2,4,5-triulose
(diacetylformoin acyclic form)

foods, and its presence indicates the extent of non-enzymatic browning reactions. This dihydropyranone is related to maltol, but maltol is not produced from this compound under the conditions of the Maillard reaction. Along with 2,3-dihydro-3,5-dihydroxy-6-methyl-4*H*-pyran-4-one, the corresponding pyran-3-one is also formed as an unstable isomer, which is rearranged to a lactic acid ester.

The reaction product of pentoses, hexoses (e.g. D-fructose via 1-deoxy-D-*erythro*-hexo-2,3-diulose), and alduronic acids is 4-hydroxy-5-methyl-2*H*-furan-3-one, which is known as norfuraneol, toffee furanone, or chicory furanone (Figure 4.33). 6-Deoxyhexoses (methylpentoses) produce, analogously, 4-hydroxy-2,5-dimethyl-2*H*-furan-3-one, which is known as furaneol, strawberry furanone, or pineapple furanone. Furaneol, norfuraneol, and homofuraneol are sensorially active substances with low threshold concentrations. They are synthesised and added to numerous aroma compositions with a pleasant sugary, jammy, fruity, and caramel flavour reminiscent of cooked strawberries and of pineapple – which is where furaneol was first identified (see Section 8.2.11.1.1).

4.7.1.1.4 Subsequent reactions of furans and pyrans

Under acidic conditions, 5-hydroxymethylfuran-2-carbaldehyde partially decomposes to laevulinic (4-oxopentanoic) acid and formic acid (Figure 4.35). Both acids are, amongst other things, important components of acid protein hydrolysates. The laevulinic acid content in acid protein hydrolysates can be up to about 2% w/w. Its cyclisation product, α-angelica lactone, was identified in acid hydrolysates of proteins, extracts from vanilla pods, raisins, bread, and other products.

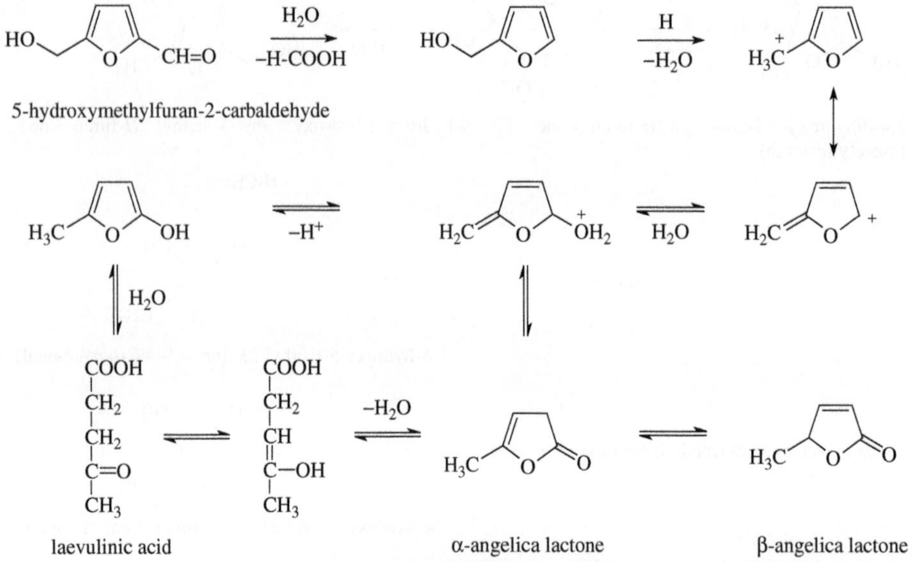

5-hydroxymethylfuran-2-carbaldehyde

laevulinic acid

α-angelica lactone

β-angelica lactone

Figure 4.35 Degradation of 5-hydroxymethylfuran-2-carbaldehyde.

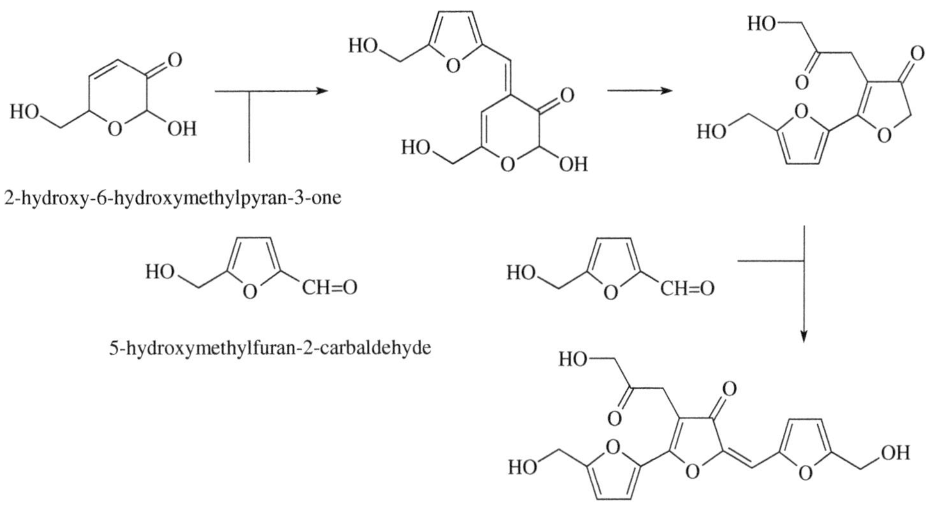

Figure 4.36 Condensation of 5-hydroxymethylfuran-2-carbaldehyde and 2-hydroxy-6-hydroxymethylpyran-3-one.

All derivatives of furan-2-carbaldehyde are reactive substances. They are easily oxidised by oxygen to the corresponding acids, which, together with the corresponding alcohols, are also formed from furan-2-carbaldehyde derivatives by the Cannizzaro reaction. Derivatives of furan-2-carbaldehyde condense together or with other carbonyl compounds and can enter advanced non-enzymatic browning reactions. For example, reaction of 5-hydroxymethylfuran-2-carbaldehyde with acetic acid in balsamic vinegar yields 5-acetoxymethyl-2-carbaldehyde (**4-100**), which acts as a sweetness modulator. Oxidation of 5-hydroxymethylfuran-2-carbaldehyde yields 5-hydroxymethylfuran-2-carboxylic acid (**4-100**). Condensation of two molecules of this aldehyde form the corresponding ether (**4-100**), whilst condensation with 3,4-dideoxy-D-*glycero*-hex-3-en-2-ulose (2-hydroxy-6-hydroxymethylpyran-3-one) yields the corresponding dimer and other compounds (Figure 4.36). The dehydration product of fructose 2-hydroxyacetylfuran gives a coloured trimer with furan-2-carbaldehyde by aldolisation (**4-100**). Condensation products are also formed in reactions of 4-hydroxy-5-methyl-2*H*-furan-3-one (**4-101**) or 3,5-dihydroxy-2-methyl-4,5-dihydropyran-3-one (**4-101**) with aliphatic aldehydes, and similar products arise with other furans and pyrans. Some of these products are coloured, and uncoloured compounds produce coloured polymers by condensation reactions.

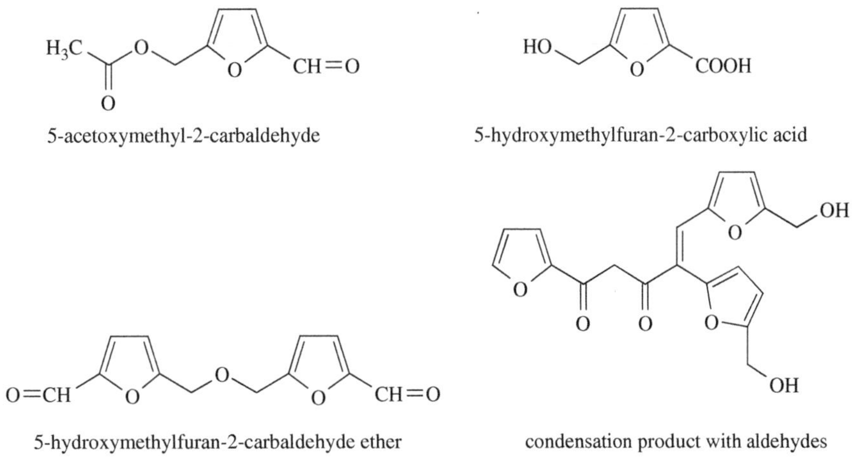

5-acetoxymethyl-2-carbaldehyde

5-hydroxymethylfuran-2-carboxylic acid

5-hydroxymethylfuran-2-carbaldehyde ether

condensation product with aldehydes

4-100, examples of 5-hydroxymethylfuran-2-carbaldehyde reaction products

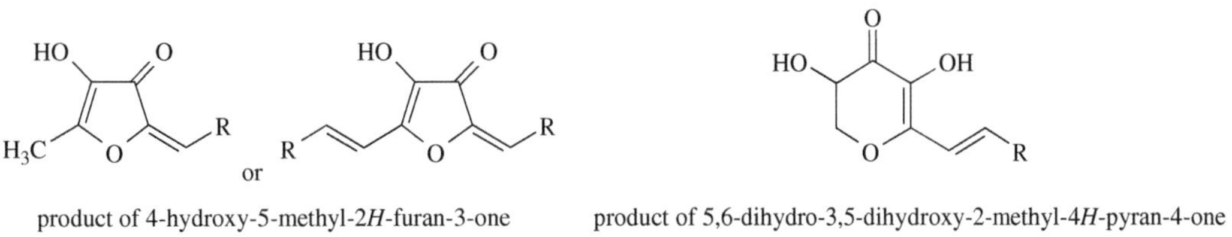

product of 4-hydroxy-5-methyl-2*H*-furan-3-one

product of 5,6-dihydro-3,5-dihydroxy-2-methyl-4*H*-pyran-4-one

4-101, condensation products of furan-3-ones and pyran-3-ones with aldehydes

4.7.1.1.5 Formation of reductones

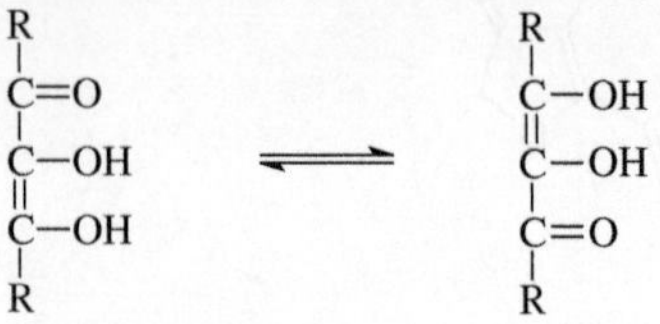

Figure 4.37 Structure of reductones.

Some of the degradation products of 1-deoxyhexo-2,3-diuloses are reductones, compounds that have the enediol group adjacent to a carbonyl group. In solutions of pH < 6, reductones (like enediolates) occur as monoanions (Figure 4.37) stabilised by resonance. In media of higher pH value, reductones form unstable dianions. In foods, reductones act as antioxidants, as in acidic solutions they can reduce a number of organic compounds and metal ions. Their oxidation products are triuloses (Figure 4.38).

Reductones with six carbon atoms in the molecule, formed in acidic solutions, include aliphatic compounds that form as intermediates in the transformation of 1-deoxy-D-*erythro*-hexo-2,3-diulose (Figure 4.31). Diacetylformoin is also one of the reductones (Figure 4.34). In alkaline media, aliphatic reductones with four and three carbon atoms in the molecule, such as *C*-methyltriosoreductone and alicyclic triosoreductone, are formed by fragmentation of sugars. Reductones containing nitrogen in the molecule are formed as products of the Maillard reaction.

4.7.1.2 Reactions in basic media

4.7.1.2.1 Isomerisation

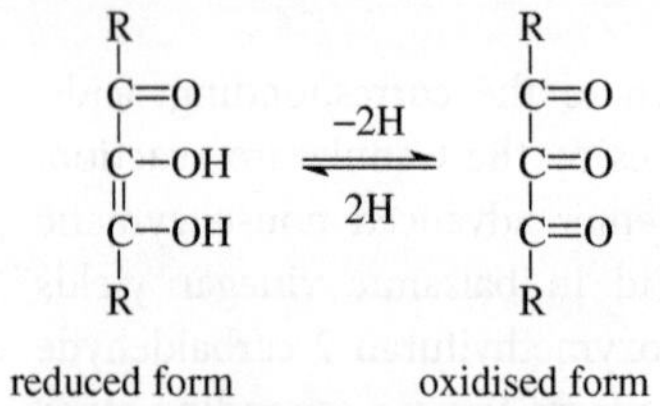

Figure 4.38 Reversible oxidation of reductones.

Enolisation of acyclic forms of reducing monosaccharides occurring in neutral and weakly alkaline media proceeds by a mechanism similar to the dehydration in acidic solutions. Enolisation of D-glucose or D-fructose produces 1-ene-1,2-diol (enediol anion species 1-ene-1,2-diolates). Enolisation is a reversible reaction, so D-glucose, its epimer D-mannose, and D-fructose produce a mixture of all three sugars as the main products (Figure 4.39). For example, an aqueous solution of D-glucose of pH 10 is still dominated by D-glucose after 10 days of heating to 35 °C and contains 63.5% D-glucose, 31% D-fructose, and 2.5% D-mannose. Isomerisation of glucose to fructose is also known as **aldose–ketose isomerisation** and the change of configuration on carbon C-2 is called **epimerisation** or **aldose–aldose** isomerisation. The corresponding pairs of aldoses (e.g. glucose and mannose) are called **epimers**. D-Fructose further produces a small amount of D-psicose by 2,3-enolisation. Enolisation can continue along the chain, and a small amount of D-sorbose is produced by 3,4-enolisation of fructose, whilst glucose can yield traces of D-allose and D-mannose gives rise to small amounts of D-altrose.

Isomerisation of pentoses, tetroses, and trioses, as well as of disaccharides, takes place analogously. For example, D-ribose produces ketose D-ribulose as the major product plus a smaller amount of aldose D-arabinose, glyceraldehyde yields ketose 1,3-dihydroxyacetone (Figure 4.40), and the major product of disaccharide lactose is lactulose, accompanied by smaller amount of epilactose.

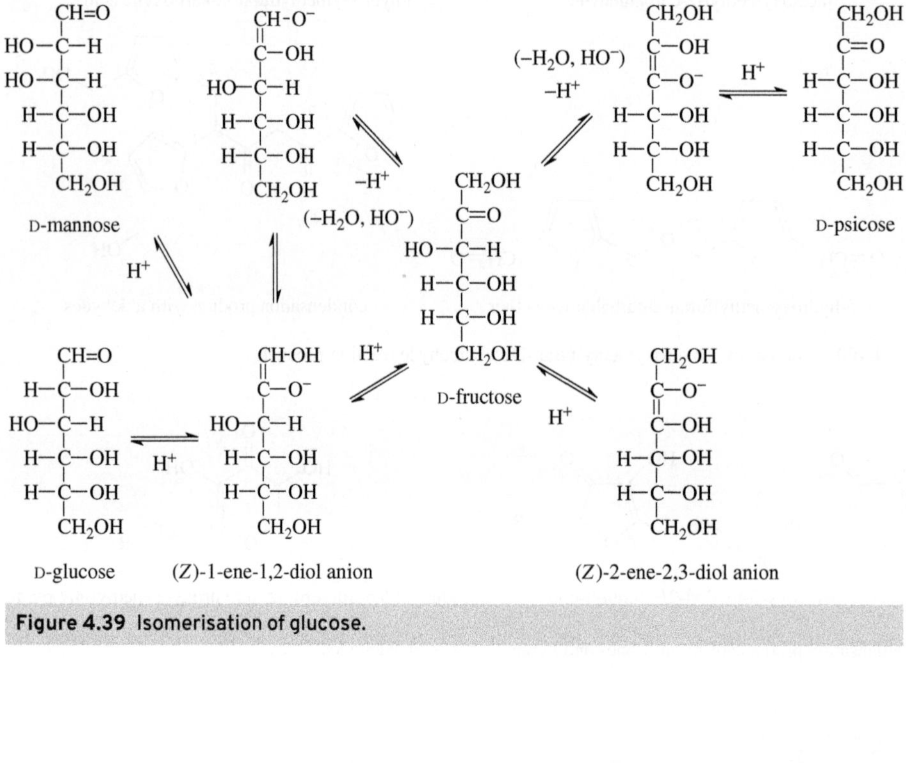

Figure 4.39 Isomerisation of glucose.

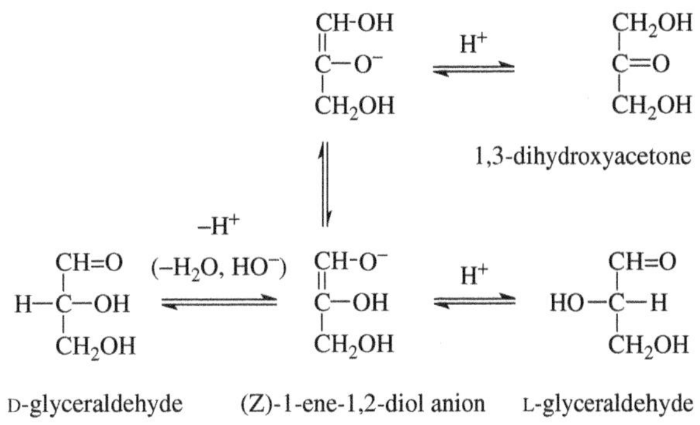

Figure 4.40 Isomerisation of trioses.

4.7.1.2.2 Rearrangement to acids

When sugars are heated in a solid state (caramelization) or in alkaline solutions, the enolisation of sugars with the formation of deoxyglycosuloses can also be followed by a rearrangement to sugar acids. The mechanism of this reaction, using the example of D-glucose and D-fructose, is shown in Figure 4.41. The 1-ene-1,2-diolate eliminates the hydroxyl group at C-3 producing the reactive α-dicarbonyl

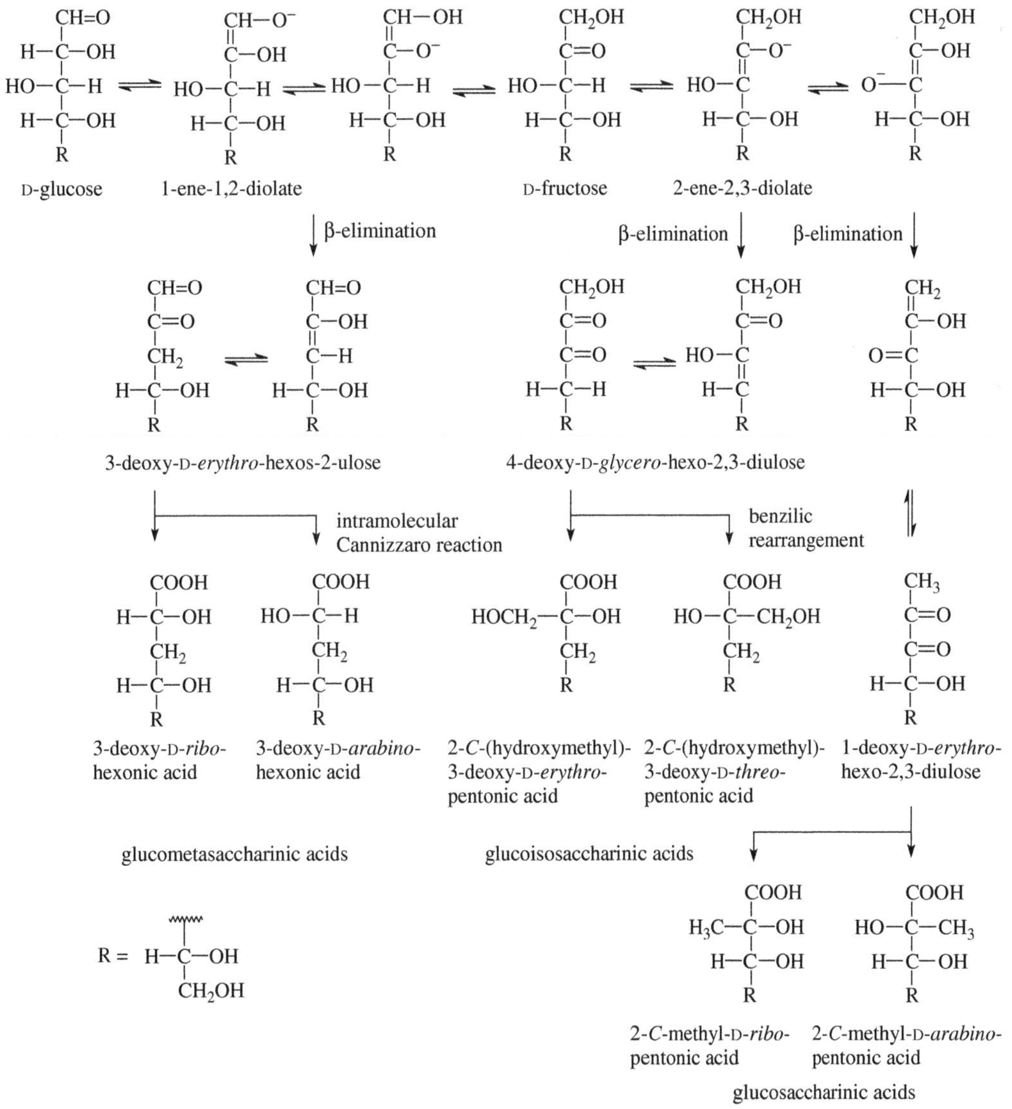

Figure 4.41 Formation of saccharinic acids from glucose and fructose.

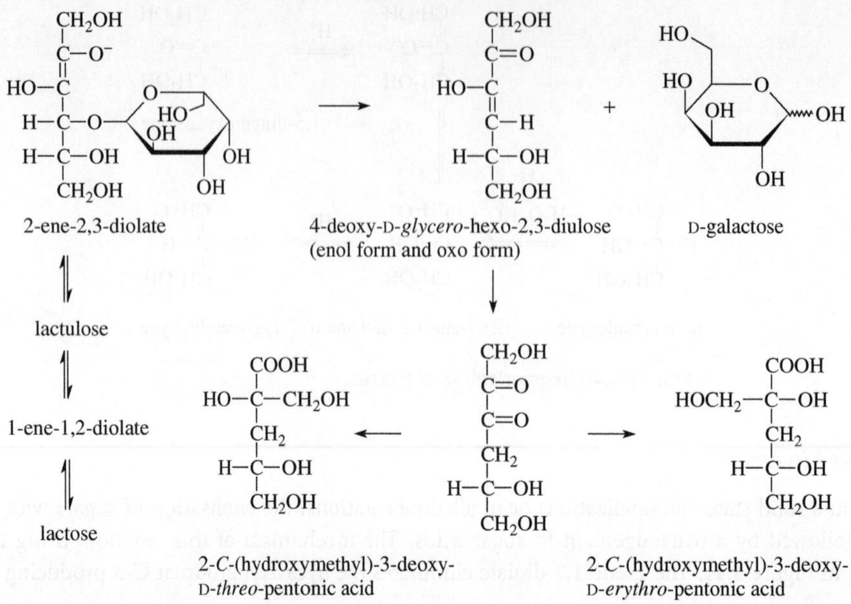

Figure 4.42 Formation of isosaccharinic acids from lactose.

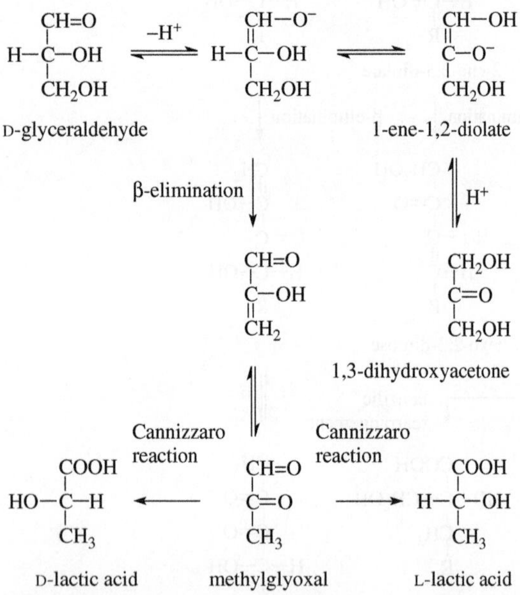

Figure 4.43 Formation of methylglyoxal and lactic acid from trioses.

compound (or 2-oxoaldehyde) 3-deoxy-D-*erythro*-hexos-2-ulose. Isomers of **metasaccharinic acids**, such as 3-deoxy-D-*ribo*-hexonic acid (3-deoxy-D-gluconic acid) and 3-deoxy-D-*arabino*-hexonic acid (3-deoxy-D-mannonic acid), are formed from 3-deoxy-D-*erythro*-hexos-2-ulose by the intramolecular Cannizzaro reaction. Branched acids, called **isosaccharinic acids**, such as 2-C-(hydroxymethyl)-3-deoxy-D-*erythro*-pentonic acid and 2-C-(hydroxymethyl)-3-deoxy-D-*threo*-pentonic acid, as well as isomers of saccharinic (aldonic) acids (2-C-methyl-D-*ribo*-pentonic and 2-C-methyl-D-*arabino*-pentonic acids), are formed via 2-ene-2,3-diols by benzilic acid rearrangement of α-dicarbonyl intermediates (2,3-diuloses).

The reaction is generally acid–base-catalysed, and small amounts of saccharinic acids (their lactones) are formed even in an acidic solution. D-Glucose is the precursor, via 3-deoxy-D-gluconic acid, of the corresponding γ-lactone (Figure 4.25). In the case of 3-O-substituted D-glucose, 4-O-substituted D-glucose, and 1-O-substituted D-fructose (and the corresponding disaccharides), the reaction mechanism is analogous and the reaction products are again isosaccharinic and saccharinic acids. The formation of isosaccharinic acids from lactose is shown in Figure 4.42 as an example.

The reaction of trioses takes place in a similar manner (Figure 4.43). D-Glyceraldehyde or 1,3-dihydroxyacetone dehydration yields 1,2-diulose methylglyoxal (also known as pyruvic acid aldehyde or pyruvaldehyde), whilst the intramolecular Cannizzaro reaction of methylglyoxal yields lactic acid (racemate).

4.7.1.2.3 Fragmentation and reactions of fragments

Fragmentation   Another important reaction in alkaline media is monosaccharide splitting, producing fragments with fewer carbon atoms. The cleavage occurs by oxidation of the molecule (after previous isomerisation and dehydration) or by retroaldolisation (opposite reaction to aldolisation). Most of the fragmentation pathways lead through glycosuloses, which are formed by enolisation and dehydration of sugars. Five major mechanisms are involved in dicarbonyl decomposition: retro-aldol fragmentation, hydrolytic α-dicarbonyl cleavage, oxidative α-dicarbonyl cleavage, hydrolytic β-dicarbonyl cleavage, and amine-induced β-dicarbonyl cleavage. The main products are acids and highly reactive carbonyl, hydroxycarbonyl, and dicarbonyl compounds, from which a large number of secondary products are formed

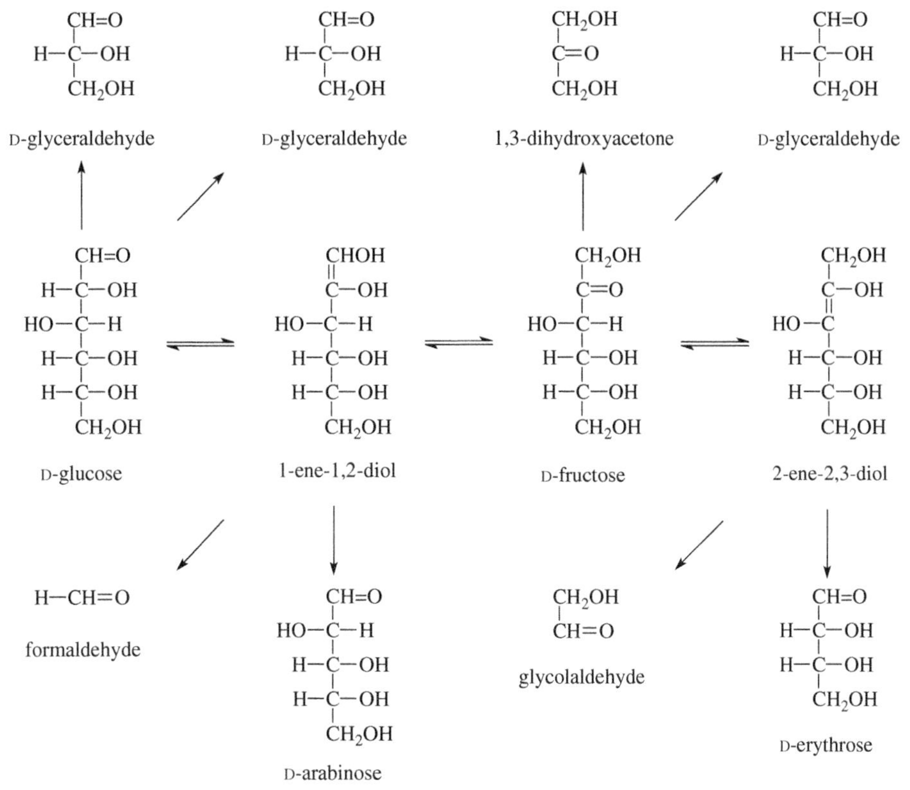

Figure 4.44 Fragmentation of glucose and fructose.

by subsequent reactions such as isomerisation, aldolisation, and the Cannizzaro reaction. Some of these products may be the final reaction products, whilst the others subsequently enter other reactions, for example with amino acids or proteins.

An example of a degradation product of sugars with one carbon atom in the molecule is formaldehyde. Acetaldehyde, glyoxal, and glycolaldehyde have two carbon atoms, whilst glyceraldehyde, 1,3-dihydroxyacetone (1,3-dihydroxypropan-2-one), hydroxy-acetone (1-hydroxypropan-2-one, also known as acetol), lactaldehyde (2-hydroxypropanal or lactic acid aldehyde), methylglyoxal (pyruvaldehyde), and hydroxymethylglyoxal have three. The four-carbon fragment consists of hydroxybiacetyl (1-hydroxybutan-2, 3-dione), biacetyl (butan-2,3-dione), and aldose D-erythrose, whilst the five-carbon fragment consists of pentoses such as D-arabinose. At the same time, aliphatic three- and four-carbon reductones are produced, such as the so-called triosoreductone, C-methyltriosoreductone, and alicyclic reductones. Monosaccharides produce more fragments than disaccharides, and glucose gives more fragments than fructose. Fragmentation of monosaccharides is illustrated by the fragmentation of D-glucose (Figure 4.44) and D-fructose (Figure 4.45).

Formation of carboxylic acids Fragmentation of sugars and glycosuloses leads not only to reactive α-dicarbonyl and α-hydroxycarbonyl products, but also to carboxylic acids. Carboxylic acids with short chains can be formed by several mechanisms. Acetic acid is a major degradation product of sugars, especially in neutral and alkaline solutions. As a product of degradation of hexoses, acetic acid is almost exclusively produced from C-1/C-2 atoms of their transformation products, 1-deoxy-2,3-diuloses. The most common pathway is the isomerisation of the reactive α-dicarbonyl intermediate to β-dicarbonyl compounds, followed by hydrolytic (β-dicarbonyl) cleavage (reverse Claisen condensation). For example, 1-deoxy-D-*erythro*-hexo-2,3-diulose, via 1-deoxy-D-*erythro*-hexo-2,4-diulose, and after the nucleophilic attack of hydroxyl anion to the C-2 carbonyl group and subsequent protonisation of C-4 carbonyl group, yields acetic acid and an 1-ene-1,2-diol intermediate, which isomerises to yield either C_4 ketose D-erythrulose or C_4 aldoses D-erythrose and D-threose (Figure 4.46). Oxidative (α-dicarbonyl) cleavage of 1-deoxy-D-*erythro*-hexo-2,4-diulose yields acetic acid and D-tetronic acids (D-erythronic and D-threonic acids). Hydrolytic cleavage of 1-deoxy-D-*erythro*-hexo-2,4-diulose leads to D-glyceric acid and hydroxyacetone (acetol). Oxidative cleavage of the 1-ene-1,2-diol intermediate leads to the formation of D-*glycero*-tetros-2-ulose, whilst dehydration of the 1-ene-1,2-diol yields 3-deoxytetros-2-ulose and (after isomerisation) 1-deoxytetro-2,3-diulose (Figure 4.47). Another possibility is the formation of acetic acid from carbon C-3/C-4 of 1-deoxy-D-*erythro*-hexo-2,4-diulose by hydrolytic cleavage of 1-deoxytetro-2,3-diulose. To a lesser extent, acetic acid results from carbon C-5/C-6 after dehydration of 1-deoxy-D-*erythro*-hexo-2,4-diulose to diacetylformoin

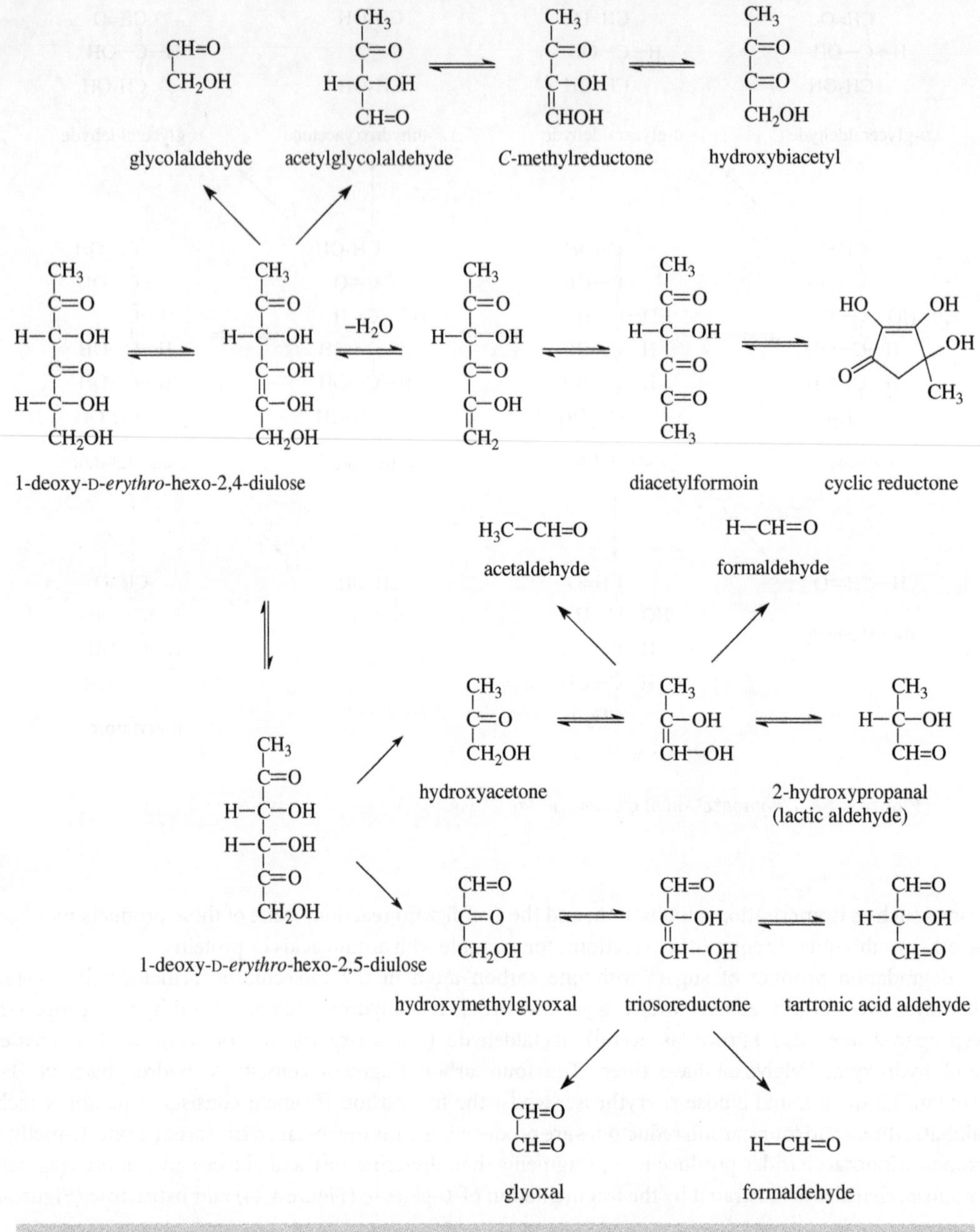

Figure 4.45 Formation of fragments and reductones from fructose 2,3-enolisation products.

and its hydrolytic β-cleavage. Similar mechanisms – cleavage of C-1/C-2 bonds of 1-amino-1,4-dideoxyhexo-2,3-diulose – lead to the formation of formic acid and 3-deoxypentos-2-ulose. 1-Amino-1,4-dideoxy-D-*glycero*-hexo-2,3-diulose arises almost exclusively from the transformation of oligosaccharides, such as maltose, linked by a $(1 \rightarrow 4)$ glycosidic bond, and hydrolysis of the amino compound yields 1,4-dideoxy-D-*glycero*-hexo-2,3-diulose.

In addition to formic, acetic, and glyceric acids, the transformation of sugars yields a number of other acids, from C_2 (glycolic acid) to C_6. The acids are formed at higher temperatures, especially in alkaline media and in the presence of oxygen. To a lesser extent, they may arise from α-dicarbonyl intermediates by oxidative cleavage. When oxygen binds to one of the carbonyl group carbons, electron transfer (Baeyer–Villiger rearrangement) and hydrolysis of the product yield two molecules of carboxylic acids (Figure 4.48). The options also include oxidation of the terminal aldehyde groups of sugars by molecular oxygen, which produces carboxylic acids occurring in acidic solutions in the form of lactones.

Reactions of fragments with water   Carbonyl compounds with electron-withdrawing groups on the C_α carbon can be hydrated in aqueous solution. Carbonyl compounds and their hydrates are in equilibrium with each other, and most hydrates (as well as hemiacetals) decompose

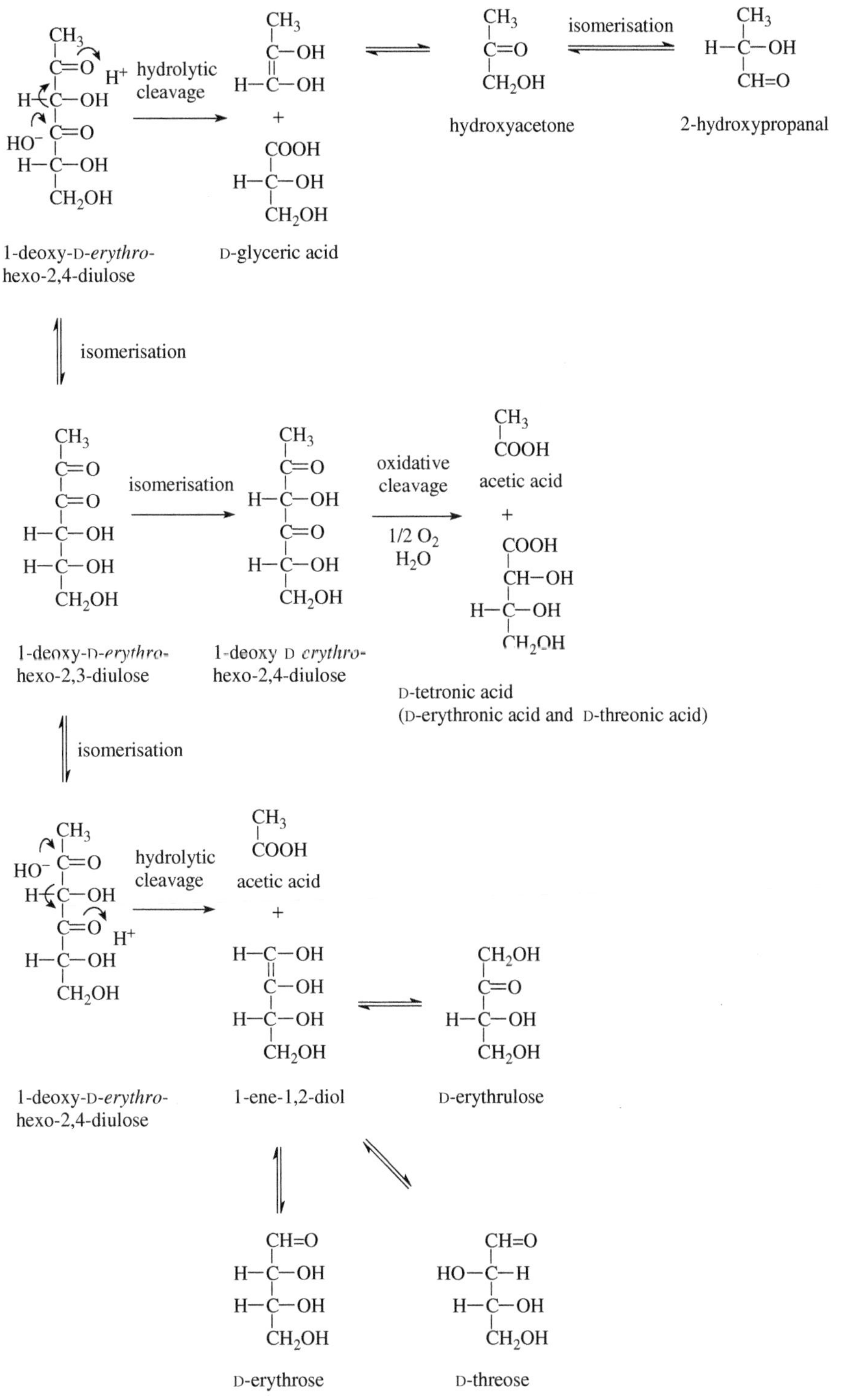

Figure 4.46 Formation of acetic acid, glyceric acid, and other products from
1-deoxy-D-*erythro*-hexo-2,3-diulose.

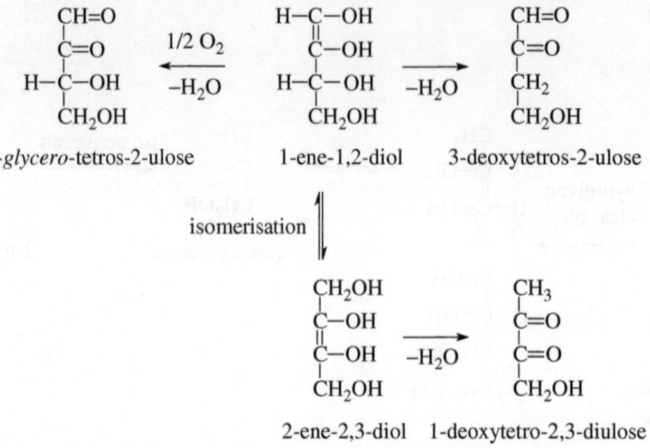

Figure 4.47 Formation of D-*glycero*-tetros-2-ulose,
3-deoxytetros-2-ulose, and 1-deoxytetro-2,3-diulose.

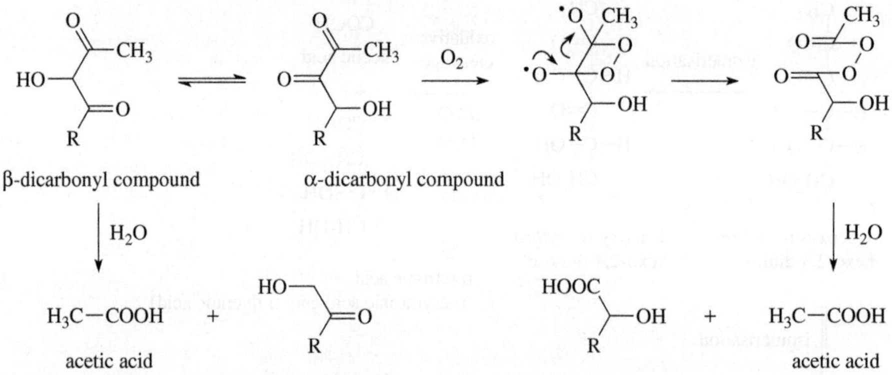

Figure 4.48 Formation of acetic acid by cleavage of α- and β-dicarbonyl compounds.

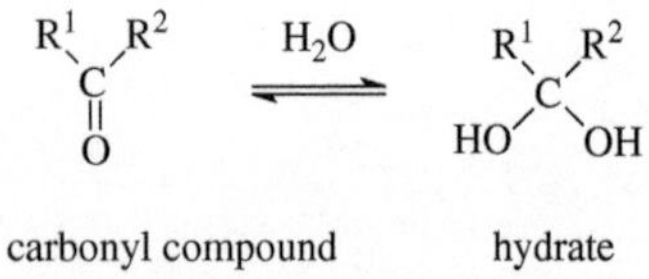

Figure 4.49 Addition of water
and formation of a hydrate.

spontaneously to the corresponding carbonyl compounds (Figure 4.49). Formaldehyde in dilute aqueous solutions is almost completely hydrated, acetaldehyde is hydrated to about 60%, whilst higher (fatty) aldehydes and acetone are not hydrated at all. The hydration of carbonyl compounds is the result of the positive inductive effect of alkyl substituents. In addition to the hydrate (methyleneglycol), concentrated solutions of formaldehyde also contain a series of oligomeric hydrates of general formula $H-(O-CH_2)_n-OH$. In neutral and especially acidic solutions, formaldehyde also forms a cyclic trimer, 1,3,5-trioxane (**4-102**), and polymeric hydrates known as paraformaldehyde (Figure 4.50). A cyclic trimer, 2,4,6-trimethyl-1,3,5-trioxane (paraldehyde, **4-102**), is formed from acetaldehyde.

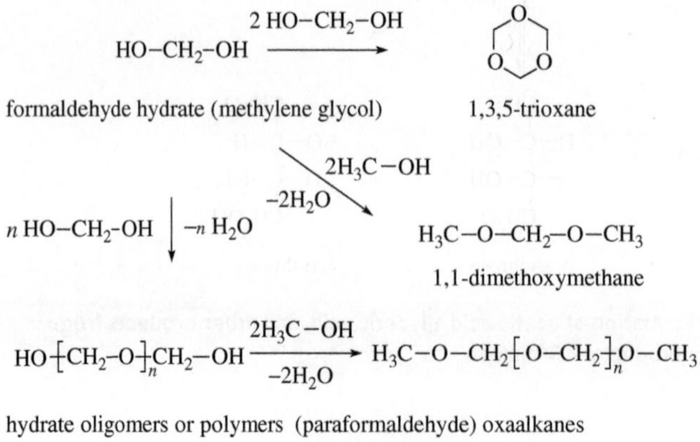

Figure 4.50 Reactions of formaldehyde hydrate.

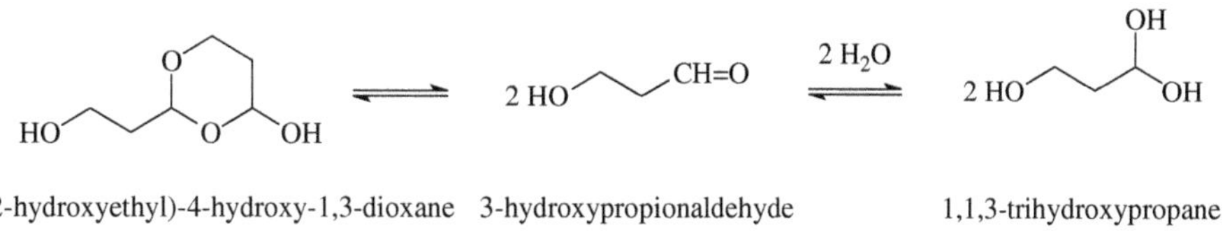

Figure 4.51 Formation of 3-hydroxypropionaldehyde hydrate and dimer.

The product of microbial transformation of glycerol, 3-hydroxypropionaldehyde, undergoes a reversible dimerisation and hydration (Figure 4.51), which results in an equilibrium of 3-hydroxypropionaldehyde, 1,1,3-trihydroxypropane (3-hydroxypropionaldehyde hydrate), and 2-(2-hydroxyethyl)-4-hydroxy-1,3-dioxane (3-hydroxypropionaldehyde dimer). The formation of hydrates is attributed also to α-hydroxycarbonyl compounds, such as glycolaldehyde (**4-102**), glyceraldehyde, and 1,3-dihydroxyacetone, and α-dicarbonyl compounds, such as glyoxal, methylglyoxal (**4-102**), and related substances. These hydrates are spontaneously transformed into various cyclic dimers with 1,4-dioxolane rings. Similar hydrates are formed by some higher α-hydroxycarbonyl and α-dicarbonyl compounds, such as L-dehydroascorbic acid and 2,3-dioxo-L-gulonic acid.

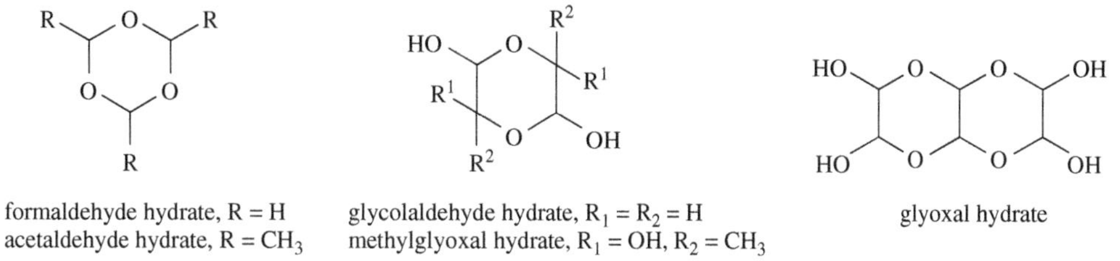

formaldehyde hydrate, R = H
acetaldehyde hydrate, R = CH$_3$

glycolaldehyde hydrate, R$_1$ = R$_2$ = H
methylglyoxal hydrate, R$_1$ = OH, R$_2$ = CH$_3$

glyoxal hydrate

4-102, oligomeric hydrates of selected carbonyl compounds

The hydrates of most carbonyl compounds, such as methylglyoxal hydrate (Figure 4.52), are unstable and exist only in solutions. Hydrates that are stable in the solid state arise, for example, from (E)-3,4-dideoxy-D-glycero-hex-3-enos-2-ulose. Oligomers that are stable in the solid state are also formed from hydrates of glycolaldehyde (**4-102**) or glyoxal (**4-102**), and thermal degradation of these hydrates produces a number of different products. For example, glyoxal hydrate trimer produces the corresponding dimer, monomer, carbon dioxide, and formic acid when heated to 250 °C (formic acid is a common degradation product of sugars and also the oxidation product of formaldehyde). Methylglyoxal hydrate yields about 40 different compounds during heating to 100 °C. The main decomposition products are acetic acid (30%), propane-1,2-diol (18%), hydroxyacetone (acetol, 7%), and 2,4-dimethyl-1,3-dioxolane, which is a cyclic acetal of formaldehyde and propane-1,2-diol (7%). Other products are formed in smaller quantities, for example 2-hydroxypropanal (lactic aldehyde) and formic acid. Dehydration of hydroxyacetone yields acrolein (Figure 4.52).

Other reactions of fragments Sugar fragments isomerise in alkaline solutions just like the original sugars. Glyceraldehyde, for example, isomerises to 1,3-dihydroxyacetone (Figure 4.40), hydroxyacetone isomerises to lactic aldehyde, hydroxymethylglyoxal isomerises to triosoreductone, and C-methyltriosoreductone isomerises to 3-oxobutanal (Figure 4.45).

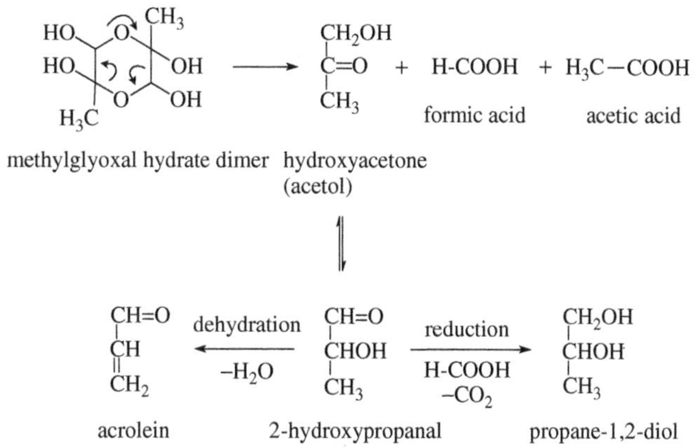

methylglyoxal hydrate dimer   hydroxyacetone
(acetol)

acrolein   2-hydroxypropanal   propane-1,2-diol

Figure 4.52 Degradation of methylglyoxal and hydroxyacetone.

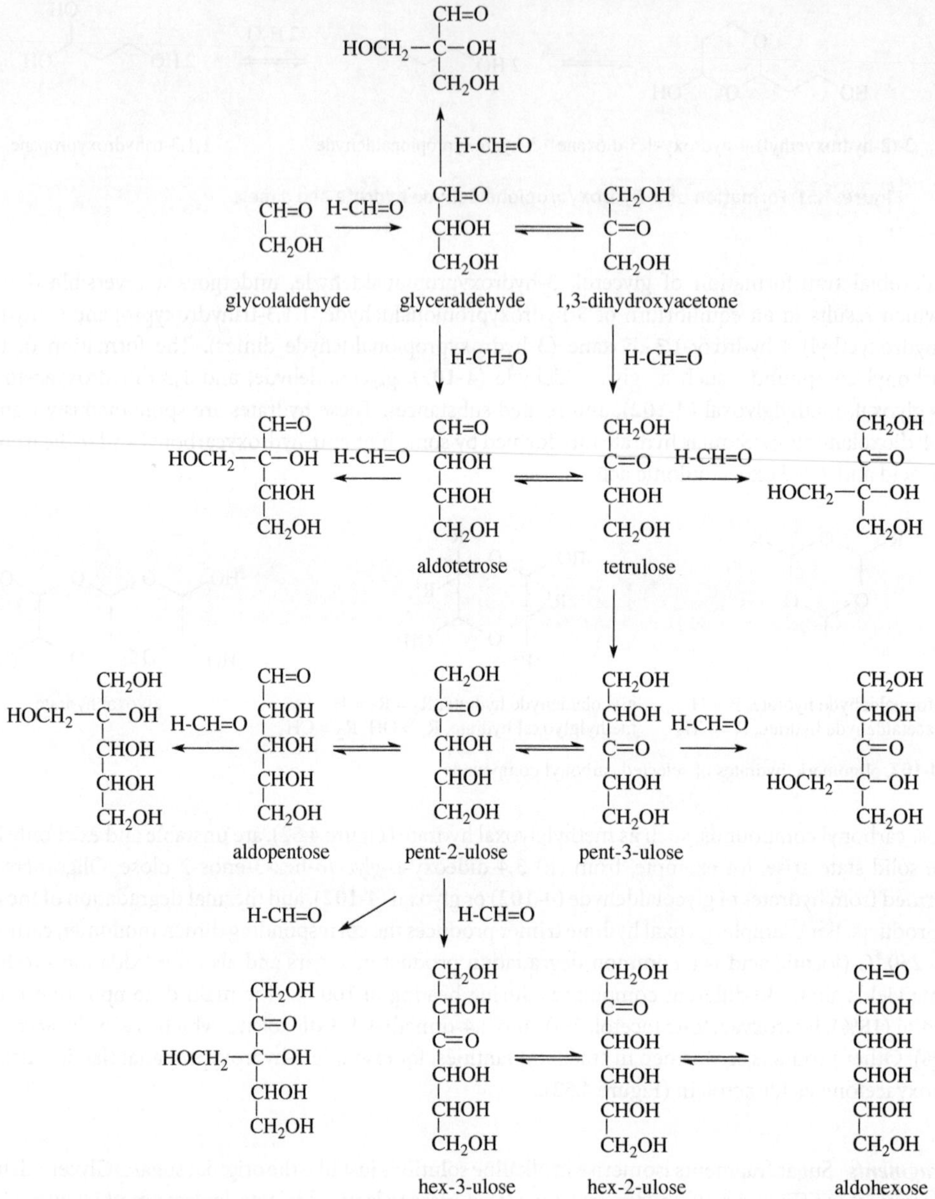

Figure 4.53 Formose reaction.

Sugar fragments can also be rearranged into acids. For example, dehydration of glyceraldehyde and 1,3-dihydroxyacetone yields methyl-glyoxal, which is rearranged to lactic acid (Figure 4.43). Aldehydes and ketones that have at least one hydrogen on the C_α carbon undergo aldolisation in alkaline solutions. The reaction of formaldehyde with the reactive degradation sugar product glycolaldehyde yields glycer-aldehyde, which isomerises to 1,3-dihydroxyacetone and then reacts, in the rate-determining step, with another molecule of formaldehyde. The reaction of formaldehyde in alkaline earth hydroxide solutions, leading to a mixture of optically inactive sugars (so-called formose), is called the formose reaction. Originally, it was assumed that glycolaldehyde, which catalyses this reaction, arises from formaldehyde. It was later shown that formaldehyde contained trace amounts of glycolaldehyde as an impurity. The formed tetroses break down relatively quickly by retroaldolisation to glycolaldehyde, and the cycle is repeated. In subsequent steps, tetroses can react with formaldehyde to yield pentoses or hexoses with branched or straight chains. At the level of hexoses, the reaction virtually stops because they are relatively stable compounds that react slowly with formaldehyde (Figure 4.53). Based on its kinetics, the formose reaction is considered a rare example of a non-radical chain process with degenerate branching. Knowledge of the formose reaction may be useful in the elaboration of catalytic methods for the synthesis of rare and non-natural monosaccharides and polyols.

The main aldolisation products from the reaction of glycolaldehyde with D-glyceraldehyde are D-arabinose and D-xylose. Similarly, D-glyceraldehyde and 1,3-dihydroxyacetone aldolisation produces a mixture of D-fructose and D-sorbose. The reaction is highly stereos-elective, as a diol arrangement of *threo*-configuration and not of *erythro*-configuration results on carbons C-3 and C-4. A similar highly

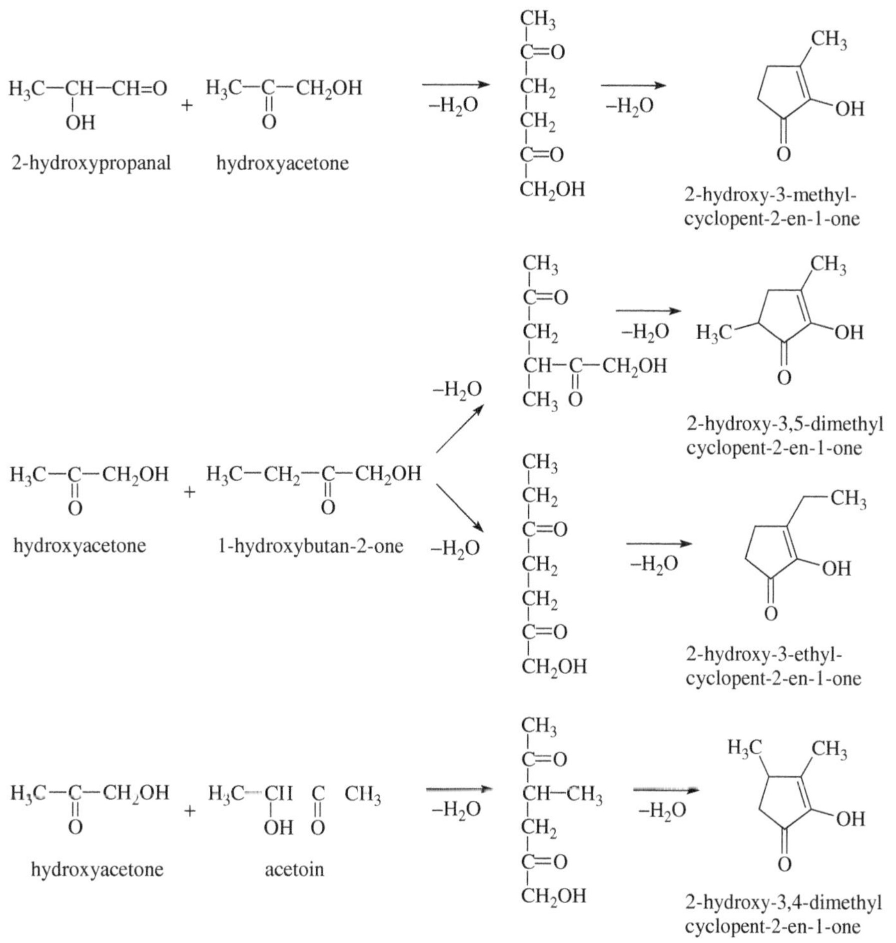

Figure 4.54 Formation of cyclopentenolones.

selective aldol reaction of 1,3-dihydroxyacetone phosphate is catalysed by the enzyme aldolase. Branched sugar D-dendroketose is formed as one of the minor products.

The mutual condensation of some degradation products of sugars yields important alicyclic substances called cyclopentenolones, which are characterised by a caramel flavour similar to maltol, and other secondary reaction products. For example, condensation of hydroxyacetone with lactic aldehyde yields the basic member of the homologous series 2-hydroxy-3-methylcyclopent-2-en-1-one, which is known as cyclotene (Figure 4.54). In aqueous solutions, cyclotene exists partly as the corresponding hydrate. Similarly, the condensation of fragmentary carbonyl degradation products can yield 2,6-dimethyl-1,4-cyclohexanedione, which contributes to the pungent odour of roasted coffee (Figure 4.55).

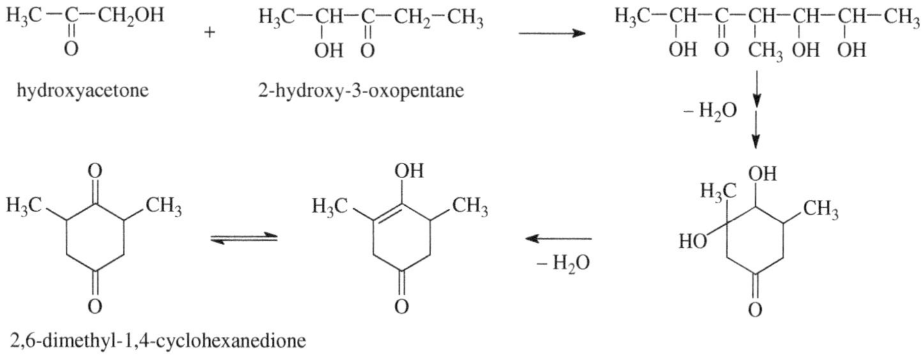

Figure 4.55 Formation of 2,6-dimethyl-1,4-cyclohexanedione.

In addition to Strecker-type reactions, biacetyl can undergo cyclocondensation through aldol type reaction to generate 4,5-dimethyl-1, 2-benzoquinone intermediate and its isomers under acidobasic catalysis. Under pyrolytic conditions, the o-benzoquinone can be converted by amino acids into o-diamine derivative. These all-redox-active moieties could play an important role in the browning associated with the Maillard reaction.

Condensation of sugar degradation products in alkaline, neutral, and weakly acidic media or a sequence of isomerisation, dehydration, and oxidation/reduction reactions of sugars can lead to aromatic compounds such as phenols. For example, minor degradation products of D-glucose and D-fructose in neutral media are catechol (benzene-1,2-diol) and pyrogallol (benzene-1,2,3-triol), but products with a higher number of carbon atoms can also be formed, such as 3,4-dihydroxybenzaldehyde, 3,4-dihydroxyacetophenone (1-acetylbenzene-2,4-diol), 3,8-dihydroxy-2-methyl-4H-chromen-4-one, and some other phenolic compounds. Even more reactive than sugars in this way is uronic acid (e.g. galacturonic acid); thus, polygalacturonic acid and pectin under heat treatment may also be considered as a source of phenolic compounds possessing condensation activity.

Phenolic compounds play a particularly important role in the non-enzymatic browning reactions of cellulose and cellulose products, which result in yellowing of paper and cotton textile products because phenolic compounds are easily oxidised to coloured quinones that further react with the original phenols and amino compounds with the formation of coloured reaction products. Quinones also react with sugars and may form glycosyl derivatives, better known under the older term C-glycosides.

4.7.1.3 Oxidations and reductions

4.7.1.3.1 Oxidation by oxygen, peroxides and peroxyl radicals

Aldehyde groups can be oxidised to carboxyl groups enzymatically by common oxidising agents in basic and acidic media and during caramelisation. The oxidation with atmospheric oxygen (autoxidation) is slow under neutral and faster in alkaline media. During autoxidation of D-glucose and D-fructose, unstable hydroperoxides are formed via 1-ene-1,2-diol. Their decomposition yields arabinonic and formic acids (Figure 4.56). Glucose and fructose are also oxidised in the presence of transition metal ions, such as Cu^{2+} and Fe^{3+}, via enediol complex with metal and oxygen, to D-$arabino$-hexos-2-ulose, which can be further oxidised and decomposed while reducing metal ions.

Similar products are formed by oxidation with hydrogen peroxide in alkaline media. Hydrogen peroxide is formed, e.g., by decomposition of the sugar enediol complex with metal and oxygen. In neutral and weakly acidic solutions, hydrogen peroxide itself is a weak oxidising agent. However, in the presence of Fe^{2+}, the decomposition of hydrogen peroxide generates free radicals (hydroxyl radicals $^{•}OH$), which also oxidise sugars to glycosuloses (Figure 4.57).

The mechanism of oxidation by lipid peroxyl radicals involves C—H bond cleavage at carbon C-2. The radical that is produced reacts with oxygen with the formation of a peroxyl radical that spontaneously decomposes to the corresponding glycosulose radical and hydrogen peroxide (Figure 4.58). This hydrogen peroxide radical, after disproportionation to hydrogen peroxide and oxygen ($2\,HOO• \rightarrow H_2O_2 + O_2$), may act in other oxidation reactions. The formation of a hydrogen peroxide radical ($HOO^{•}$) by this mechanism, which assumes transfer of hydrogen in the five-membered intermediate structure, is not the only way for the oxidation of sugar by peroxyl radicals. An alternative reaction is decomposition with hydrogen transfer in the intermediate six-membered structure. This generates lower aldonic acids (such as tetronic acid), hydrogen peroxide, and glyoxal (Figure 4.59).

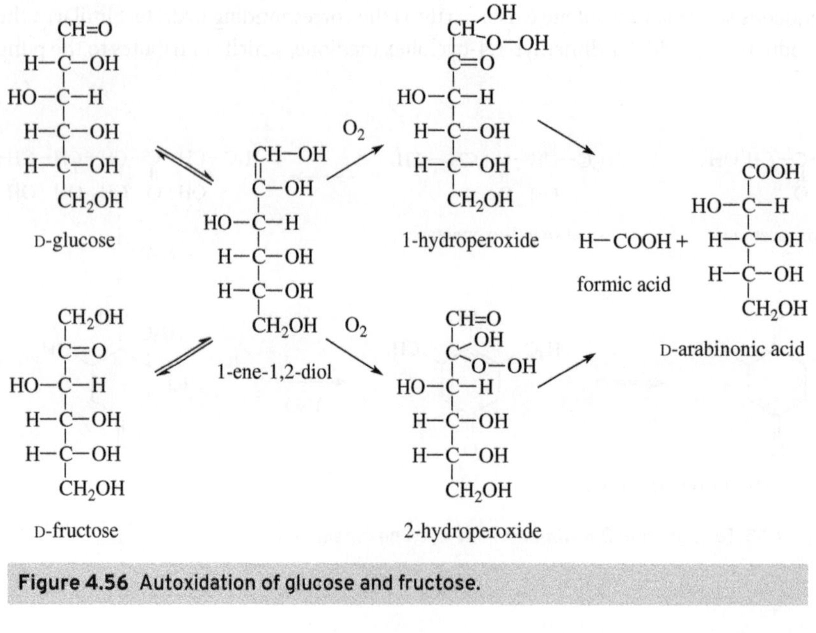

Figure 4.56 Autoxidation of glucose and fructose.

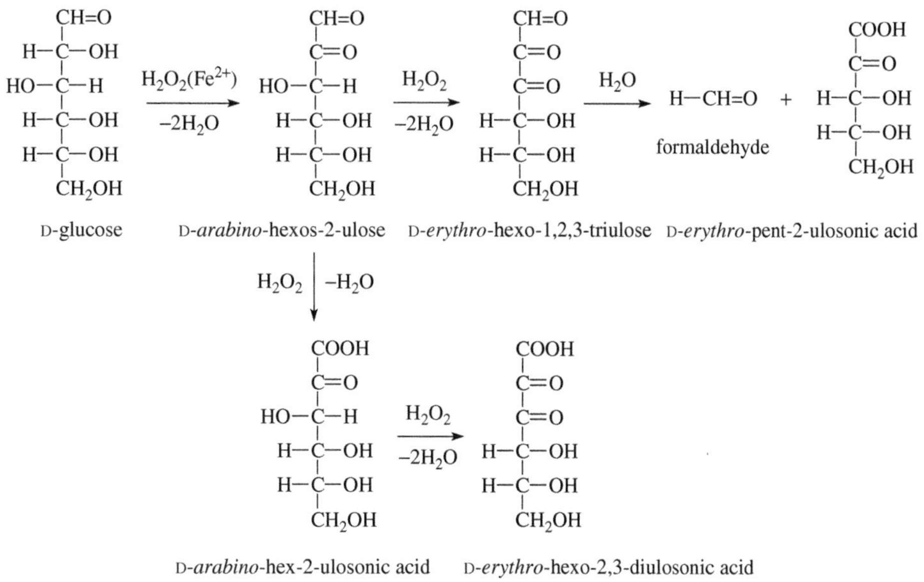

Figure 4.57 Oxidation of glucose to glycosulosonic acids by hydrogen peroxide/Fe²⁺ ions.

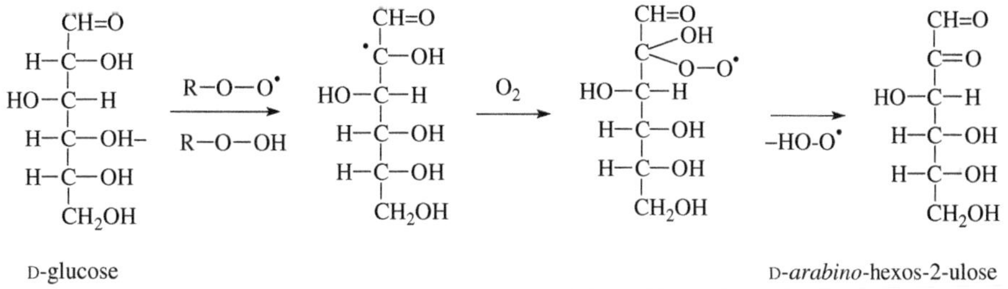

Figure 4.58 Oxidation of glucose to glycosulose by peroxyl radicals.

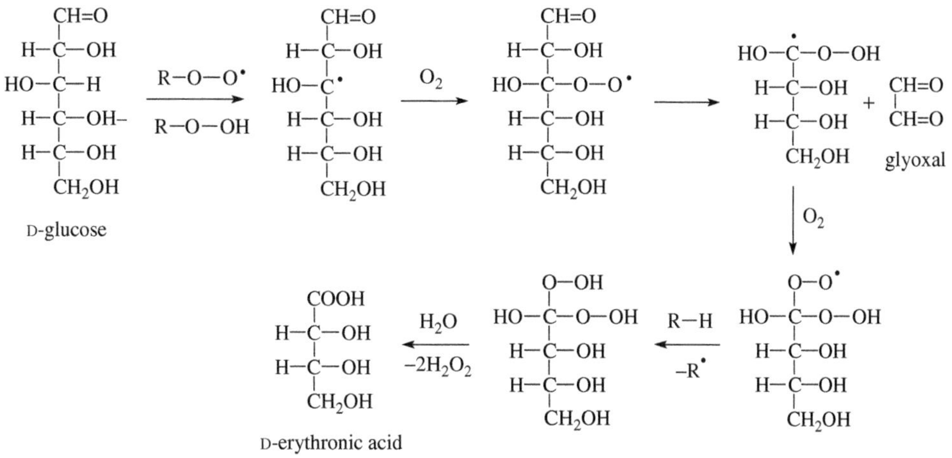

Figure 4.59 Oxidation of glucose to aldonic acid and glyoxal by peroxyl radicals.

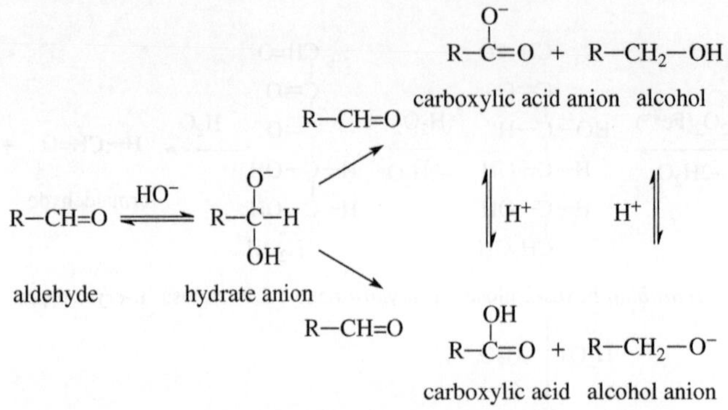

Figure 4.60 General mechanism of the Cannizzaro reaction.

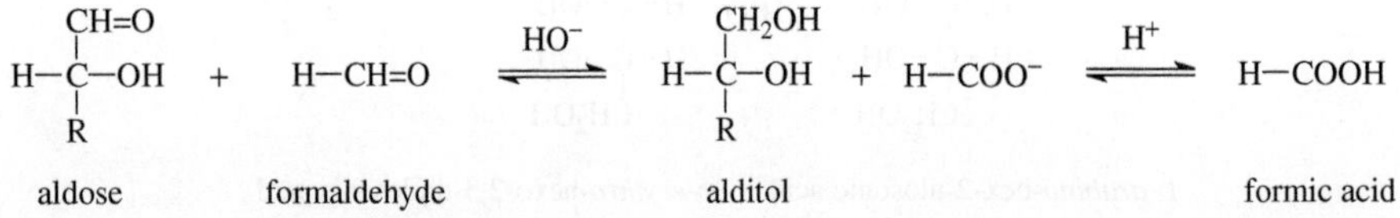

Figure 4.61 Cannizzaro reaction of formaldehyde with aldoses.

4.7.1.3.2 Cannizzaro reaction

A special kind of an oxidation–reduction (redox) reaction catalysed typically by alkalis is the Cannizzaro reaction. It occurs in aldehydes that do not have a hydrogen atom on the C_α carbon, which means that the usual aldol condensation cannot take place. One molecule of aldehyde is oxidised with the simultaneous reduction of the second molecule and the reaction product is a mixture of acid and primary alcohol (Figure 4.60). The Cannizzaro reaction of formaldehyde yields formic acid and methanol; furan-2-carbaldehyde gives rise to furan-2-carboxylic acid (also known as 2-furoic acid or pyromucic acid) and furfuryl alcohol, also known as (furan-2-yl)methanol.

Formaldehyde, as the most reactive aldehyde, undergoes Cannizzaro reaction even with aldehydes that have a hydrogen atom at the C_α carbon. Formaldehyde also reacts to some extent with aldoses that are reduced to sugar alcohols (alditols) and formaldehyde is oxidised to formic acid (Figure 4.61). Reaction of formaldehyde with D-glucose yields D-glucitol, reaction with D-glyceraldehyde yields glycerol, and reaction with glycolaldehyde gives ethane-1,2-diol (ethylene glycol).

With glyoxal, methylglyoxal, and other α-dicarbonyl compounds, the Cannizzaro reaction is an intramolecular redox reaction, which yields 2-hydroxy acids (Figure 4.62). Glycolic acid results from glyoxal and lactic acid is produced from methylglyoxal.

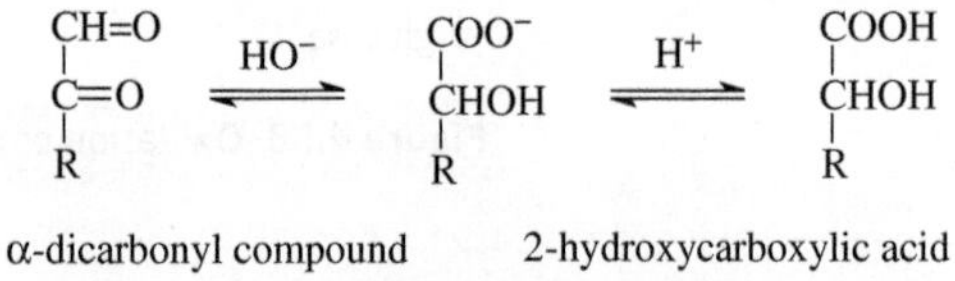

Figure 4.62 Cannizzaro reaction of α-dicarbonyl compounds.

4.7.1.3.3 Disproportionation of α-hydroxycarbonyl and α-dicarbonyl compounds

The enediol forms of α-hydroxycarbonyl compounds can react with α-dicarbonyl compounds through interconversion, in which α-hydroxycarbonyl compounds are transformed into α-dicarbonyl compounds and vice versa (Figure 4.63). This reaction explains a number of oxidation–reduction reactions that occur in degradation products of sugars. Within a complex of an α-dicarbonyl compound and an endiol, the reaction is accomplished either by the transfer of two protons in the basic singlet state or by the use of more advantageous biradical mechanism at the excited triplet state.

4.7.2 Derivatives of monosaccharides

The chemical reactions of polyols are similar to reactions of the hydroxyl groups of monosaccharides. For example, glycerol yields a mixture of different products containing diglycerol, oligoglycerols, and polyglycerols. Alditols can dehydrate to form monoanhydro and dianhydro derivatives in acidic solutions. Acid-catalysed dehydration of D-glucitol yields a mixture of cyclic anhydrides that are used as emulsifiers. Oxidation of the secondary alcoholic group of D-glucitol on C-2 or C-5 by the microorganisms *Acetobacter xylinum* or *Gluconobacter oxydans* produces L-sorbose, an intermediate used in the synthesis of L-ascorbic acid. During food processing and storage, alditols and

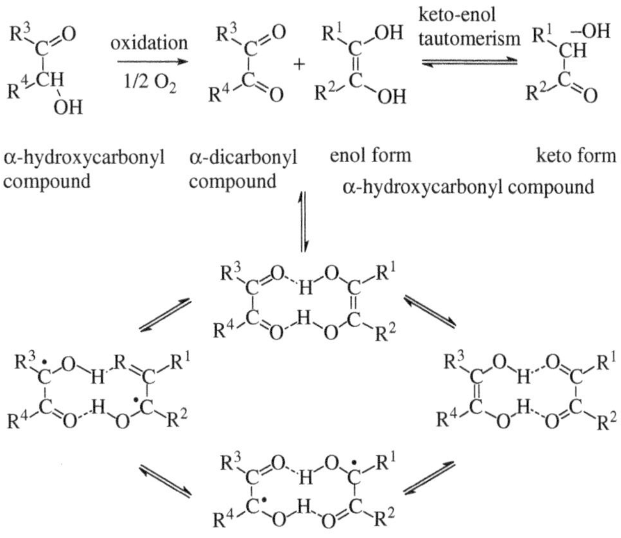

Figure 4.63 Disproportionation of α-hydroxycarbonyl and α-dicarbonyl compounds.

cyclitols are virtually stable. On the other hand, uronic acids, such as D-galacturonic acid, are highly reactive in non-enzymatic browning reactions.

4.7.3 Oligosaccharides

The properties and reactivity of reducing oligosaccharides are comparable to the properties and reactivity of monosaccharides; non-reducing oligosaccharides behave and react similarly to glycosides of monosaccharides. The reactivity and reaction pathways of reducing oligosaccharides may differ from those of monosaccharides. For example, maltose predominantly yields α-dicarbonyls that still carry a glucosyl moiety, and thus subsequent reactions to 5-hydroxymethylfuran-2-carbaldehyde and 2-acetylfuran are favored due to the elimination of D-glucose, which is an excellent leaving group in aqueous solution. Consequently, higher amounts of these heterocycles are formed from maltose than glucose.

Oligosaccharides are hydrolyzed to monosaccharides; at the same time intramolecular dehydration, which produces small amounts of anhydrosugars, occurs as a side reaction. The rate of hydrolysis (inversion) depends on the acidity and temperature, but also on other factors related to the structure of the oligosaccharides, such as inductive and steric effects. The hydrolysis of α-anomers of disaccharides is easier than hydrolysis of β-anomers, pyranosides are more stable than furanosides, and non-reducing oligosaccharides are more stable than the reducing sugars. The mechanism of hydrolysis of most disaccharides is similar to the mechanism of hydrolysis of O-glycosides. The hydrolysis of higher sugars is performed via disaccharides and other lower oligosaccharides. The α- and β-glycosidic bonds are also cleaved enzymatically under the action of α- and β-glycosidases, similarly to the heteroglycosides.

The opposite reaction to hydrolysis (inversion) is reversion, where di- and higher oligosaccharides are synthesised from monosaccharides and lower oligosaccharides in acid solutions. Reversion often proceeds together with acid catalysed hydrolysis and the glycosidic bonds can isomerise simultaneously, so that O-β-D-oligosaccharides are transformed into O-α-D-oligo-saccharides and vice versa.

During heating in solutions, fructooligosaccharides are less stable than glucooligosaccharides. For example, fructose decomposes already at 60 °C in neutral solutions, whilst glucose or saccharose solutions can be rapidly heated to temperatures up to 100 °C without decomposition.

Hydrolysis of saccharose is undesirable reaction during production of saccharose from sugar beet or sugar cane, which depends on the presence of hydrolases, pH, temperature, concentration of saccharose, and other factors. Invertases, which catalyse saccharose hydrolysis during the diffusion process, have to be inactivated by heating the raw juice to temperatures up to 80 °C. At this temperature, and at the relatively low pH of the juice (pH 5.2–5.8), however, saccharose is significantly lowered (e.g. in solutions of pH 5.2, and also at pH 11.4, the loss of saccharose is about 0.5% per hour); this can be prevented by an increase in pH through the addition of calcium hydroxide. The rate of hydrolysis of saccharose is minimal in solutions of pH around 8.4. Alkaline pH, on the other hand, initiates non-enzymatic browning reactions of reducing monosaccharides. The main manifestation of these reactions is the formation of yellow, brown, to black pigments called melanoidins, which, along with the pigments (known as melanins) derived from phenolic compounds in enzymatic browning reactions, cause discolouration of raw sugar and molasses.

Reducing oligosaccharides are somewhat more stable than monosaccharides in alkaline solutions. Under similar conditions as those for monosaccharides, isomerisation of the sugar unit bound at the reducing end of the oligosaccharide is performed, which leads to

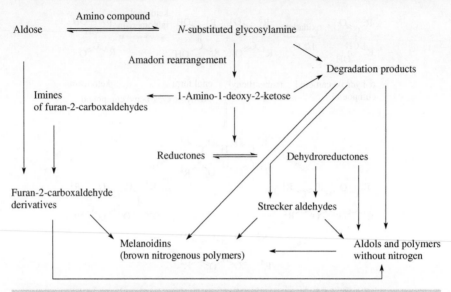

Figure 4.64 Schematic representation of the Maillard reaction according to Hodge.

the formation of the corresponding epimer (isomerisation of the type aldose–aldose) whilst bound aldoses isomerise to bound ketoses (isomerisation of the type aldose–ketose).

4.7.4 Polysaccharides

Functional properties of polysaccharides (formation of viscous dispersions and gels) are associated with the mutual interactions of their chains and interactions with other food components (especially water, proteins, and lipids).

The most important reaction of polysaccharides is hydrolysis. Hydrolysis by digestive enzymes or by colon microbial saccharases allows many polysaccharides, and especially starch, to be utilised. Spontaneous hydrolysis of polysaccharides by endogenous enzymes of food raw materials or microbial saccharases is an important reaction in many food technologies. It is related to the technological, functional, and organoleptic characteristics of a number of food products (including production of bread, post mortem changes in meat, fruit ripening, and other processes). Hydrolysis of some polysaccharides (e.g. starch and inulin) is used to obtain monosaccharides, oligosaccharides, and polysaccharide functional derivatives. Derivatives of some polysaccharides (e.g. starch and cellulose) are used as food additives (hydrocolloids).

4.7.5 Maillard reaction

The very important and widespread chemical reactions occurring during storage and processing of food are reactions of reducing sugars with amino compounds. These reactions produce a number of highly reactive carbonyl compounds that react with each other and also with the amino compounds present. This series of reactions is generally known as the Maillard reaction,[3] the consequence of which is the formation of brown pigments called **melanoidins**, and therefore these reactions belong to the group of **non-enzymatic browning reactions**. In addition to the Maillard reaction, three other major types of non-enzymatic browning reactions are recognised:

- **caramelisation**;

- **reactions of proteins with oxidised lipids**;

- **reactions of proteins with oxidised phenolic compounds**.

The Maillard reaction can therefore be regarded as a special kind of non-enzymatic browning reaction of sugars with proteins (amino acids).

---

[3]The reaction was named after the French chemist Louis-Camille Maillard, who first described the formation of brown pigments during the heating of glucose with glycine. The Maillard reaction has always attracted, and still attracts, the attention of many chemists, nutritionists, physiologists, and physicians.

Only a small number of compounds produced in these reactions have been characterised. These are mainly low-molecular-weight compounds, which are relatively stable during isolation and identification. Much less is known about the reactive intermediates that arise in very low concentrations and, moreover, are usually decomposed during isolation. Little is also known about the resulting free radicals. Knowledge of the structure of these compounds is very important because they play an important role in the formation of flavour-active substances and high-molecular-weight coloured pigments (melanoidins), and also in the reactions that occur *in vivo*. The attention of food chemists has mainly been focused on the following:

- Formation of a brown colour, which may be either desirable as a manifestation of the Maillard reaction (such as the colour of bread crust, roasted coffee and fried onions) or a negative phenomenon (e.g. in the production of dried foods, especially milk, as well as fruits and vegetables).

- Formation of flavour-active compounds including flavour-active compounds with adverse organoleptic properties.

- Nutritional and physiological aspects of the Maillard reaction (reduction of the nutritional value of foods mainly due to the reaction of sugars and other carbonyl compounds with lysine, an essential and often limiting amino acid).

- Antioxidant activities of the reaction products (mainly reductones and coloured melanoidins).

- Toxicity of some products (formation of mutagenic and carcinogenic substances).

The most important food saccharides involved in the Maillard reaction are the monosaccharides glucose and fructose, and in some cases ribose (e.g. in meat and meat products). The most important disaccharides are lactose (in milk and dairy products) and maltose (in cereal products, e.g. malt). Saccharides linked by glycosidic bonds in glycoproteins, glycolipids, heteroglycosides, and non-reducing sugars (such as saccharose) can participate in the Maillard reaction after hydrolysis to the parent monosaccharides.

Reaction partners of reducing sugars are mainly proteins and amino acids. Proteins react with reducing carbohydrates primarily through the ε-amino group of bound lysine. To a smaller extent, α-amino groups of *N*-terminal amino acids and other amino acid functional groups, such as the mercapto group of cysteine and guanidyl group of arginine, are also involved in the Maillard reaction. In addition to proteins and amino acids, biogenic and other amines contribute significantly to it in some foods (e.g. cheeses).

In addition to sugars and their degradation products, as well as the degradation products of amino acids (amines, ammonia, and aldehydes), these reactions include carbonyl compounds already present in foods as primary substances (e.g. ascorbic acid and aldehydes and ketones occurring in essential oils) and carbonyl compounds formed in foods from precursors other than carbohydrates (e.g. aldehydes resulting from fat oxidation and oxidised phenolics), making the reactions even more complex.

Owing to the complexity of the Maillard reaction, model systems containing only one reducing sugar and one amino acid are often chosen because they are simpler systems than foods. Research has shown that even in simple reaction systems, for example in glucose and glycine solutions, surprisingly numerous reaction products are formed. Therefore, even in such simple systems, the Maillard reaction mechanisms have not been fully elucidated and many of the reaction products have still not been identified. The modified reaction schemes of Hodge and Tressl[4] are given in Figures 4.64 and 4.65, respectively.

In the reaction of an aldose with an amino compound, such as that of glucose with amino acids, three basic phases are recognised:

- the **initial phase**, involving the formation of glycosylamine followed by Amadori rearrangement;

- the **middle phase**, involving dehydration and fragmentation of saccharides and Strecker degradation of amino acids;

- the **final phase**, involving reactions of intermediates leading to the formation of heterocyclic and other compounds, which are usually important as flavour-active compounds and coloured pigments called melanoidins that cause the yellow, brown, or black colouration.

Ketoses may also enter into the reaction, and then the formation of glycosylamine is followed by Heyns rearrangement. The initial reaction may also include not only carbonyl compounds derived from the degradation of sugars, but also Strecker aldehydes arising from amino acids and reactive aldehydes, and other compounds produced, for example, as secondary decomposition products of fatty acid hydroperoxides in lipid peroxidation.

In short, amino compounds play a dual role during the transformation of sugars and other carbonyl substances. They can act as both catalysts and reactants in various phases of the Maillard reaction. Therefore, the reaction cascade of sugars in the presence of amino compounds can be described as an amino-assisted transformation of sugars.

---

[4]The classification of individual reactions taking place within the Maillard reaction was first described by John Edward Hodge in 1953; it is still the most concise description of the Maillard reaction to this day. Exactly 40 years later, Roland Tressl supplemented this scheme.

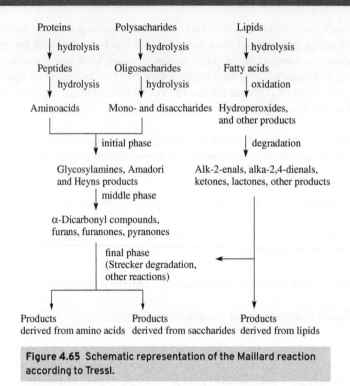

Figure 4.65 Schematic representation of the Maillard reaction according to Tressl.

4.7.5.1 Formation of glycosylamines and rearrangement to aminodeoxysugars

The Maillard reaction starts by the addition of a non-protonated amino group of an amino compound (amine, amino acid, or protein) to an electron-deficient carbon atom of a polarised carbonyl group of a reducing sugar (acyclic form of an aldose or ketose). The resulting addition product, N-substituted hemiaminal (formerly also known as monotopic carbinolamine), dehydrates with the formation of an imine (formerly also known as azomethine or a Schiff base). The electron density on the carbonyl group carbon, and thus the ease of addition, is affected by polar factors (mainly inductive effects of the carbonyl group substituents) and steric factors. Alkyl and aryl carbonyl group substituents have a positive inductive effect (electron repelling; $+I$). The most reactive substance is therefore formaldehyde, and less reactive compounds are higher aliphatic aldehydes. The introduction of the electronegative group exhibiting a negative inductive effect (electron attracting; $-I$) into the aliphatic substituent to a position adjacent to the carbonyl group, such as a hydroxyl group (in sugar) or other carbonyl group (in glycosuloses), decreases the electron density on the carbon of the carbonyl group, which results in a higher reactivity of such a derivative. The reactivity of carbonyl compounds, therefore, generally decreases in the following order:

- aldoses > ketoses;

- trioses > tetroses > pentoses > hexoses > disaccharides;

- α-dicarbonyl compounds > aldehydes > ketones > saccharides > oxoacids.

Alditols are believed not to enter the Mailard reaction, but the simplest one, glycerol, was proved to contribute to the formation of proline-specific compounds 2-acetyl-1-pyrroline and 2-acetyl-1(3),4,5,6-tetrahydropyridines, known for their roasty, popcorn aroma. In fact, during thermal processing of foods, glycerol may be oxidised into dihydroxyacetone, which then can lose a molecule of water to form methylglyoxal. The latter, and the hydroxy-2-propanone that might be formed by direct dehydratation of glycerol, are the ultimate reactants in the Maillard reaction.

The reactivity of sugars is mainly determined by the amounts of acyclic forms present. The reactivity of amino compounds is closely related to their basicity and therefore falls in the following order:

- ammonia > amines > amino acids;

- 6-amino acids > … > 3-amino acids > 2-amino acids.

Another very important factor influencing the reactivity of both reaction partners is pH. Protonisation of the carbonyl group in acidic solutions increases the reactivity of nucleophilic reagents, whilst protonised amino groups are less reactive as the nitrogen atom does not have a free electron pair (Figure 4.66). Acid–base properties of amino acids, peptides, and proteins (formation of cations, amphions, and anions) are described in Section 2.2.3.1.

The mutual reaction of a carbonyl compound with an amino compound is shown in Figure 4.67. The concentration of both reactants varies with pH value in the opposite direction: the cation concentration of the carbonyl compound increases, whilst the concentration of the unprotonised amino compound decreases with decreasing pH. Thus, the reaction rate maximum ranges from slightly acidic solutions (e.g. in reactions of aldoses with amines) to slightly basic solutions (in reactions of α-dicarbonyl compounds with amino acids). Schiff bases are usually unstable compounds that are stabilised by subsequent reactions. The exception is the Schiff base of an aromatic aldehyde with an aromatic amine, in which case the reaction equilibrium is shifted significantly in favour of the base. The Schiff bases formed from sugars are stabilised by a reversible cyclisation to N-substituted glycosylamines (Figure 4.68).

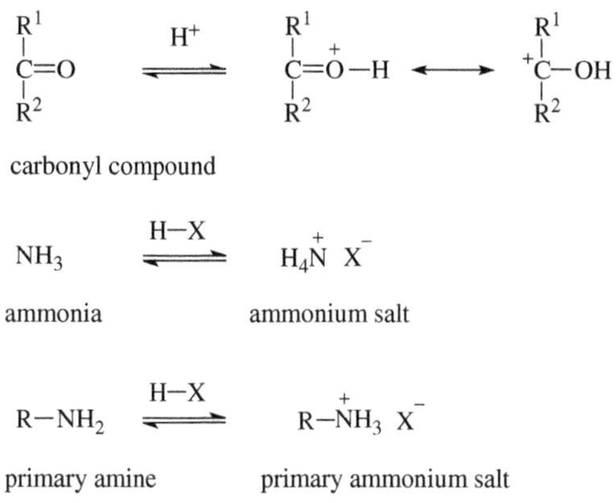

Figure 4.66 Formation of cations of carbonyl compounds and amino compounds.

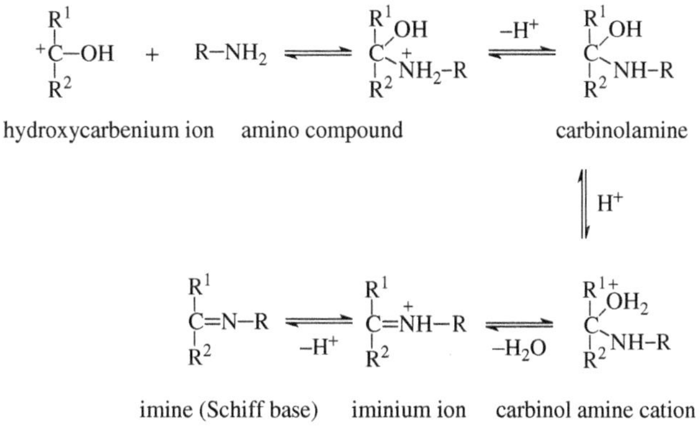

Figure 4.67 General scheme of the reactions of carbonyl compounds with amino compounds.

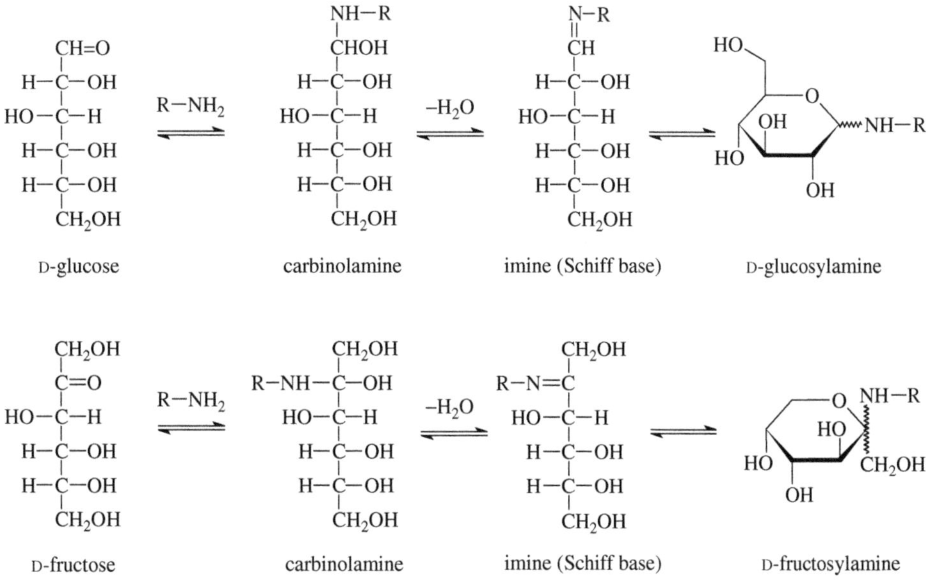

Figure 4.68 Formation of glycosylamines.

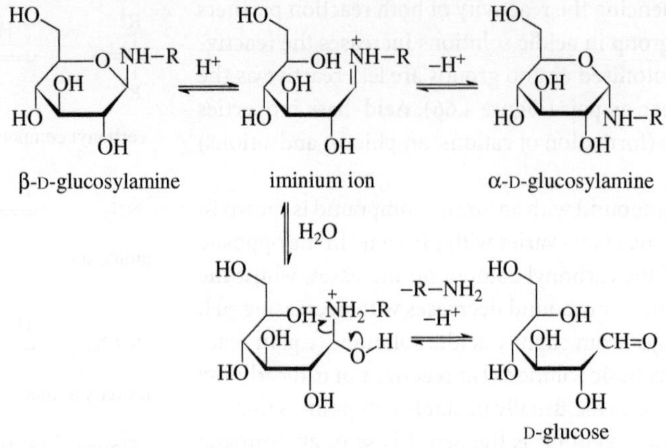

Figure 4.69 Mutarotation and hydrolysis of glycosylamines.

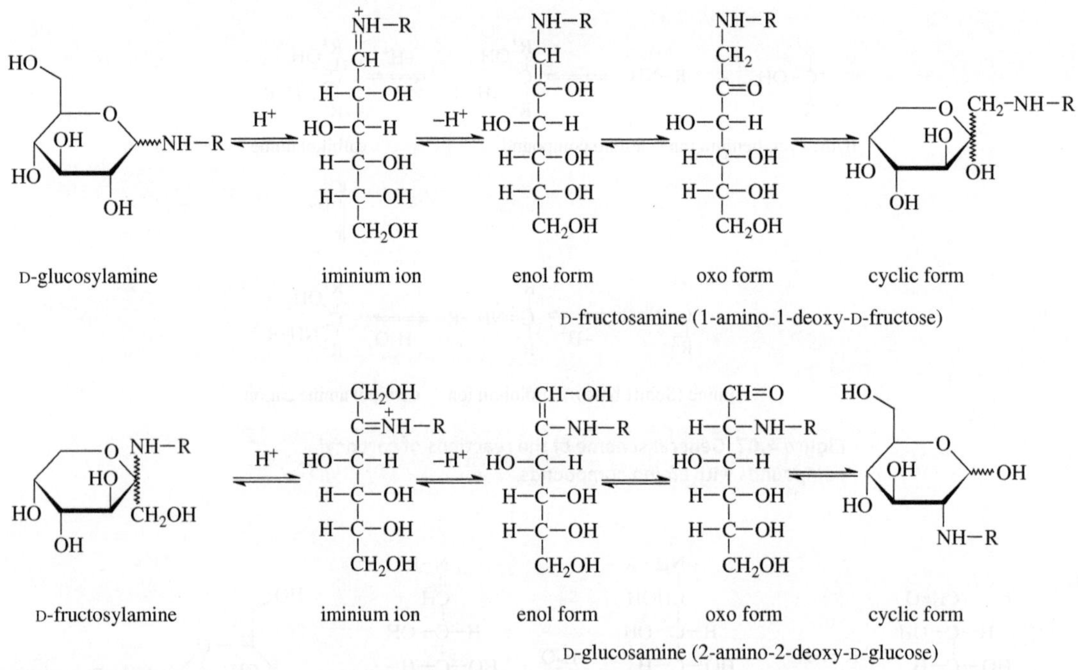

Figure 4.70 Amadori and Heyns rearrangements catalysed by acids.

In solutions, *N*-glycosylamines (as well as the parent sugars) occur mainly in the form of pyranoses or furanoses. They mutarotate as a result of the formation of an iminium intermediate (Schiff base cation) and are easily hydrolysed to the original sugar and amine (Figure 4.69). Glycosylamines derived from aromatic or heterocyclic amines are relatively stable, whilst those derived from aliphatic amines and especially amino acids (i.e. glycosylamino acids) are subjected to further reactions to give more stable products.

The fate of *N*-glycosylamines depends on water activity, pH, and temperature. In media with moderate to high water content and a pH range of 5–7, the Schiff bases are usually rearranged to substantially more stable compounds. Aldosylamines are transformed into **ketosamines**, also known as **1-amino-1-deoxyketoses** or **Amadori compounds**, by the **Amadori rearrangement**. Ketosylamines are similarly transformed into **aldosamines** (**2-amino-2-deoxyaldoses**, also called **Heyns compounds**) by the **Heyns rearrangement**. The reaction is equivalent to aldose–ketose isomerisation. The Amadori and Heyns rearrangements are generally acid–base-catalysed reactions (Figure 4.70).

The Amadori rearrangement starts by protonisation of aldosylamine, leading to ring opening and the formation of a hemiacetal iminium ion. This reaction is followed by proton cleavage on carbon C-2, and the arising enaminol is stabilised to the keto form by isomerisation and by cyclisation to form a hemiacetal. The reaction rate is significantly influenced by the form in which the glycosylamines occur. Aldosylamines present as furanoses rearrange about 10 times more quickly than pyranoses. The reacting amino acid can act simultaneously as

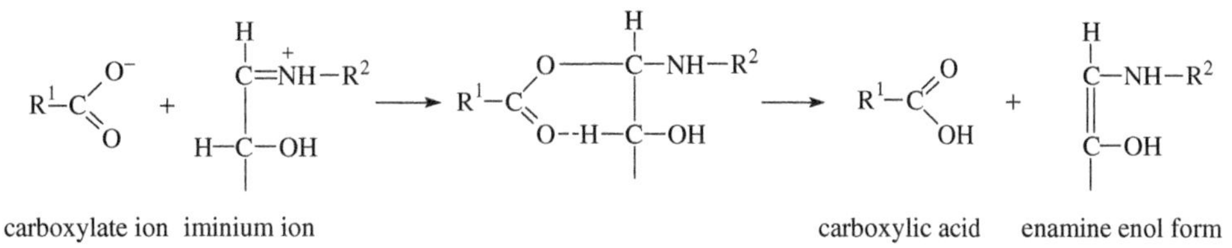

Figure 4.71 Amadori rearrangement catalysed by carboxylic acids.

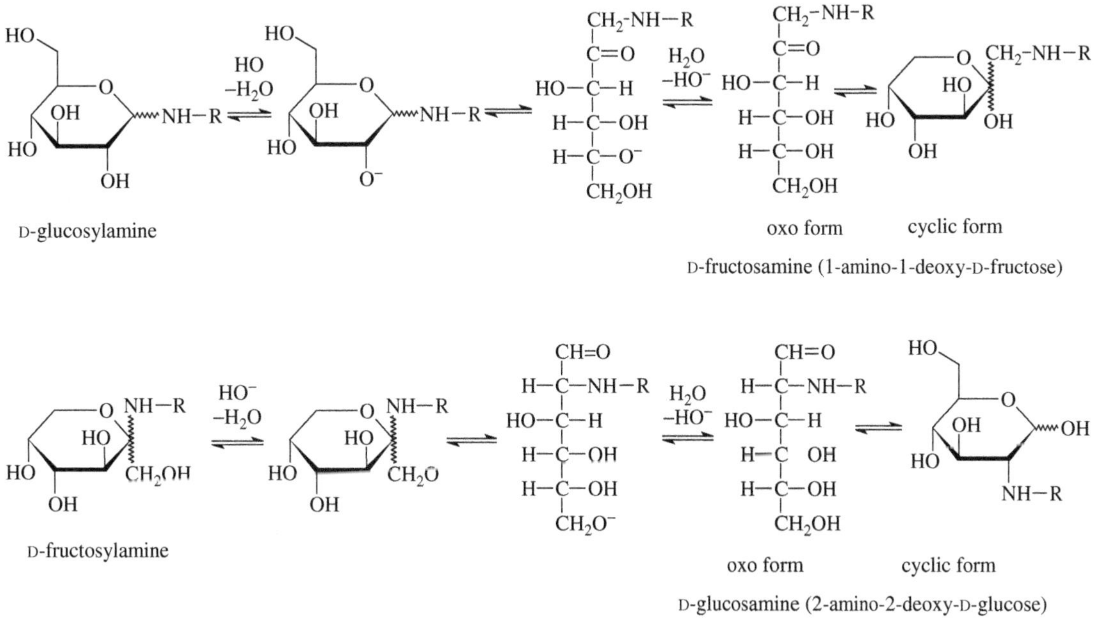

Figure 4.72 Amadori and Heyns rearrangements catalysed by hydroxyl ions.

a catalyst, because the amino acid carboxyl group provides the proton required. The reaction rate increases in the presence of carboxylic acids and phosphates. The catalytic effect of carboxylic acids is explained by reaction of the iminium ion with the carboxyl anion to form an intermediate, which is decomposed to enaminol and carboxylic acid (Figure 4.71).

The active ion of phosphates (the catalytic effect is optimal at pH 5–7) is the dihydrogenphosphate anion. This anion is a base that withdraws a proton from the carbon C-2 of glycosylamine during the Amadori rearrangement. The Amadori and Heyns rearrangements are favoured under acidic conditions, but they may also take place in a strongly alkaline medium, as they are generally acid–base-catalysed (Figure 4.72). The Amadori rearrangement has long been considered an irreversible reaction. Some studies in recent years, however, have admitted the possibility of a reversible reaction to form the parent sugar or its epimer.

Ketosamines have been shown in a number of stored and heat-processed foods, such as dried fruits, dried vegetables, dried milk, soy sauces, and other products. They arise similarly in infusions containing glucose and amino acids, and even in the human organism (particularly in diabetics), as intermediates of the Maillard reaction that take place *in vivo*. Some aminodeoxysugars occurring in foods, however, arise by biochemical (enzymatic) reactions, such as the D-glucosamine and D-galactosamine that are the building units of chitin.

4.7.5.2 Other reactions of glycosylamines

The Amadori and Heyns rearrangements are the main but not the only possible conversions of *N*-glycosylamines. Aldosylamines derived from primary amines (and also from amino acids) can react with another aldose molecule to form a dialdosylamine before the Amadori rearrangement. For example, the reaction of D-glucose with D-glucosylamine gives di-D-glucosylamine (**4-103**). Analogously to aldosylamines, dialdosylamines are rearranged to diketosamines; for example, di-D-fructosamine (**4-104**) is formed from di-D-glucosylamine.

In foods or parts of foods with low water activity, other transformation pathways can also occur. In alkaline solutions and at low temperatures, and even during pyrolysis, transamination can yield the corresponding oxoacid and 1-amino-1-deoxyalditol. In weakly acidic or

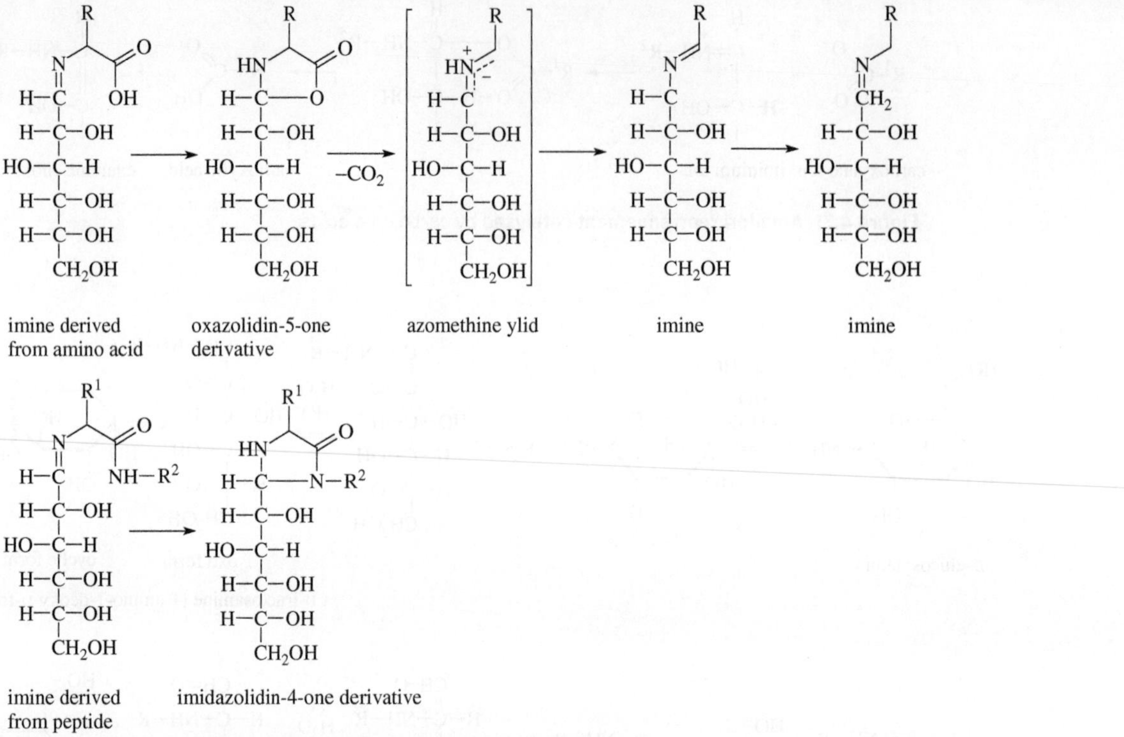

Figure 4.73 Formation of imines, imidazolidinones, and oxazolidinones.

neutral solutions and at elevated temperatures, Schiff bases can form cyclic products. The Schiff bases derived from amino acids can yield oxazolidinone derivatives, whilst those derived from peptides generate imidazolidinone derivatives. The former cyclic product decarboxylate easily to form isomeric imines (Figure 4.73). The Strecker degradation of bound amino acids in Schiff bases derived from sugars and amino acids yields the corresponding aldehyde, carbon dioxide, and 2-deoxyaldose through oxidation.

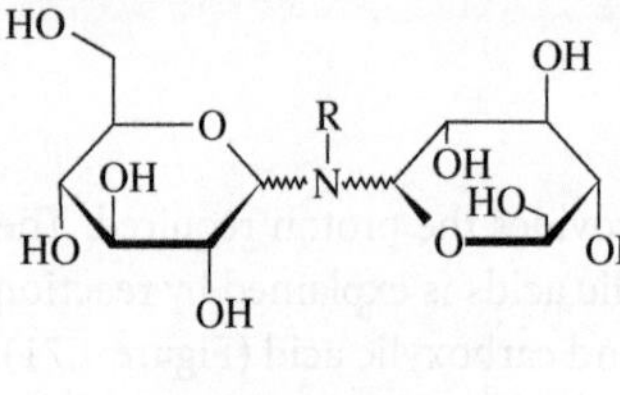

4-103, di-D-glucosylamine

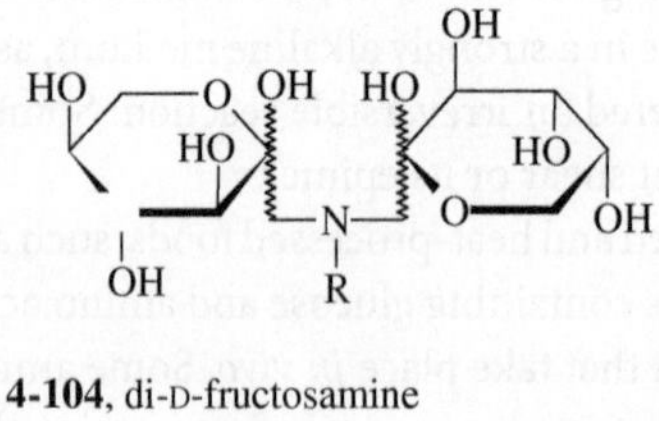

4-104, di-D-fructosamine

Schiff bases derived from the α-hydroxycarbonyl fragmentation products of sugars (such as glycolaldehyde), where the free hydroxyl group in the original glycolaldehyde is in the β-position to the nitrogen atom, may yield an imine by rearrangement of an azomethine ylide. In N-glycosylamines, the hydroxyl group is usually part of a pyranose or a furanose ring, and therefore their decomposition via Amadori product prevails. The imine is a decarboxylated analogue of an Amadori compound (Figure 4.74), transformed by isomerisation, elimination of water, and amine hydrolysis to the original α-hydroxycarbonyl compound. If the decarboxylated Amadori compound is in amine form, the C—N bond can be directly cleaved to form a 1-amino-2-oxo compound (1-amino-1-deoxyketose) and the original amino acid derivative with a vinyl group, as is the case with the formation of acrylamide from asparagine. Another possibility for acrylamide formation is oxidation of glycosylated asparagine to oxoimine and its decarboxylation, followed by hydrolysis to the immediate precursor of acrylamide, 3-aminopropionamide (see Section 12.2.2).

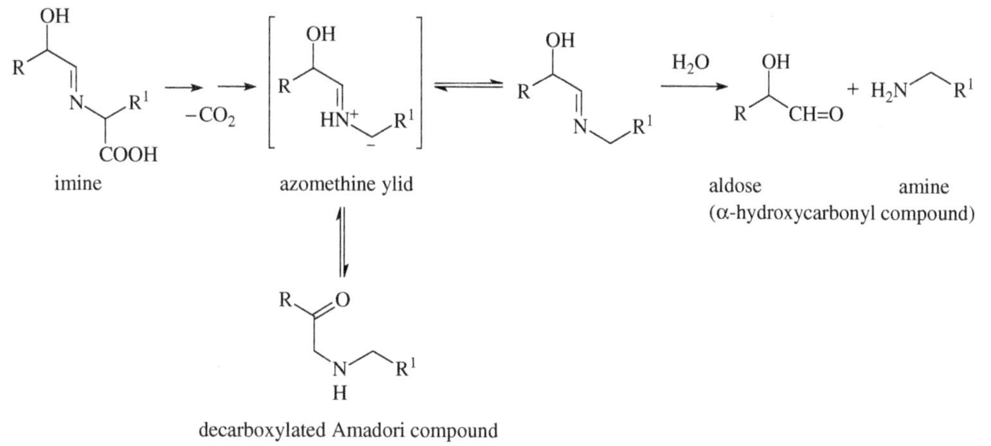

Figure 4.74 Imine degradation and formation of decarboxylated Amadori product. R, aldose or α-hydroxy compound residue; R¹, amino acid residue.

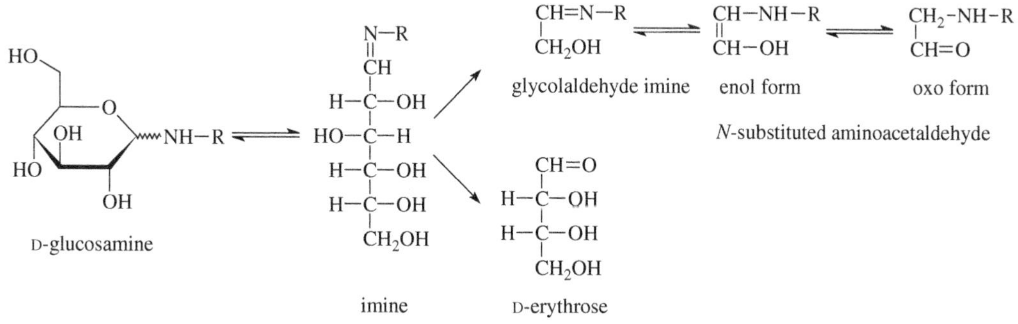

Figure 4.75 Retroaldolisation of glycosylamines.

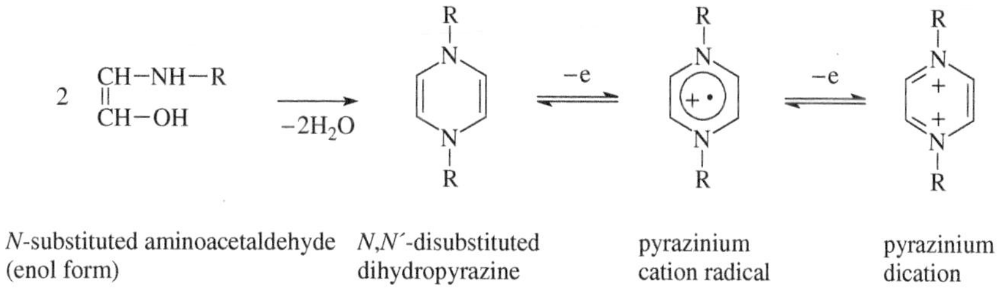

Figure 4.76 Formation of pyrazine derivative.

The just described 1-amino-1-deoxyketoses can be precursors of other compounds. For example, 1-amino-1-deoxyketopentoses can undergo dehydration and cyclisation reactions leading to furfuryl amine (**8-95**), which reacts with 3-deoxyribosulose under oxidative conditions to form aroma active 1-(furan-2-ylmethyl)-1*H*-pyrrole derivatives (furfuryl pyrroles), which possess a diverse range of flavours described as roasted, cocoa-like, green, horseradish-like, and mushroom-like (**4-118**). These were detected in various foods, such as coffee, chocolate, popcorn, and roasted chicken.

Another important reaction of *N*-glycosylamines is retroaldolisation, which generates the highly reactive C_2 intermediates, glycolaldehyde *N*-alkylimines (*N*-substituted aminoacetaldehyde derivatives), and D-erythrose (Figure 4.75). Glycolaldehyde alkylimines dimerise almost immediately to *N*,*N*′-dialkyldihydropyrazines, which are transformed into pyrazinium salts by two-electron oxidation (Figure 4.76). Imine hydrolysis yields glycolaldehyde, whilst glyoxal is formed by the oxidation of imine (Figure 4.77). Similarly, glyoxal is formed by retroaldolisation of hexos-2-uloses. Reactive hydroxycarbonyl and dicarbonyl compounds (glycolaldehyde and glyoxal) can thus be formed even in the early stage of the Maillard reaction; the original Hodge scheme did not expect their formation until the advanced stages, such as the decomposition of aminodeoxysugars via 2,3-enolisation. Glyoxal, however, is also a product of lipid oxidation.

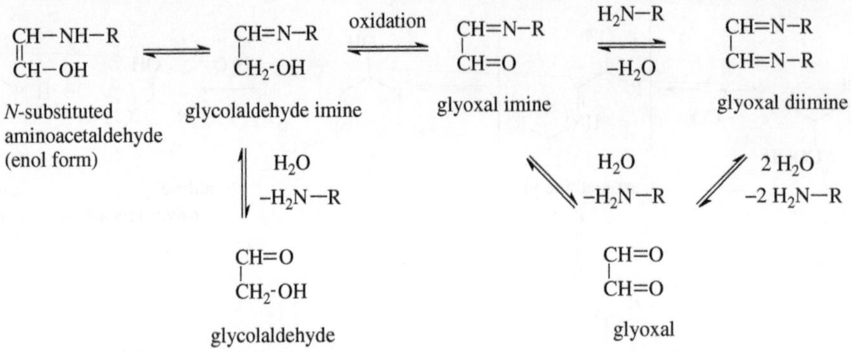

Figure 4.77 Formation of glycolaldehyde and glyoxal.

Reactions of sugars with amino acids, which have more reactive functional groups (such as sulfur, hydroxy, aromatic, and heterocyclic moieties), also bypass the Amadori rearrangement, and the reaction gives rise to various heterocyclic compounds. For example, glucose yields a thiazolidine derivative 2-(D-gluco-1,2,3,4,5-pentahydroxypentyl)thiazolidine-4-carboxylic acid (**4-105**) with cysteine and a β-carboline derivative 1,2,3,4-tetrahydro-1-(D-*gluco*-1,2,3,4,5-pentahydroxypentyl)-β-carboline (**10-35**) with tryptophan.

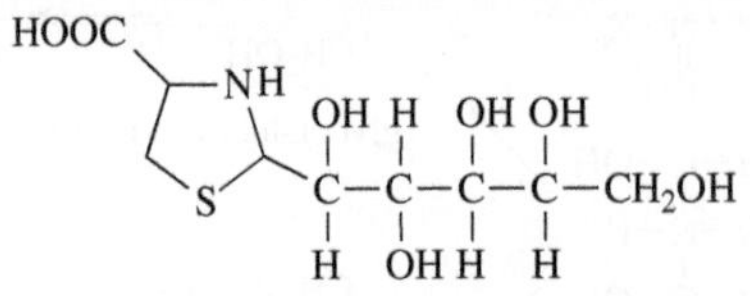

4-105, thiazolidine-4-carboxylic acid derivative

4.7.5.3 Decomposition of aminodeoxysugars

All aminodeoxysugars are strong reducing agents: less reactive than reductones, but more reactive than the original sugars. Their mutarotation in aqueous solutions leads to an equilibrium mixture of different forms of aminodeoxysugars.

Diketosamines are unstable compounds that decompose to ketosamines and many other compounds, even in aqueous solutions at room temperature. Ketosamines (Amadori compounds) are relatively stable in the solid state and in neutral aqueous solutions, but their stability is significantly lower than that of the original sugar. In acidic and alkaline solutions, ketosamines are decomposed. Their degradation in alkaline solutions is faster than in acidic solutions. In addition to pH, the rate of ketosamine degradation is considerably affected by different stabilities of the individual cyclic forms. Their decomposition is particularly fast when ketosamines are present in the less stable furanose and acyclic forms. For example, ketosamines formed in the reaction of glucose with amino acids (derivatives of 1-amino-1-deoxyfructose) at 20 °C are present in the form of β-pyranose (approximately 64%), α-furanose (15%), β-furanose (15%), and α-pyranose (6%). The equilibrium mixture contains the acyclic form in quantities up to 2%. The composition of the equilibrium mixture varies only slightly with pH. Decomposition of aldosamines proceeds in a similar way to that of ketosamines, with extra formation of some original ketose.

Decomposition of ketosamines starts, like most other reactions of monosaccharides, with 1,2-enolisation, which yields an amino analogue of 1-ene-1,2-diol: 1-ene-1-amino-2-ol (Figure 4.78). The reaction proceeds by elimination of the C-3 hydroxyl group, and hydrolysis of the bound amino compound provides 3-deoxy-D-*erythro*-hexos-2-ulose and, further, 3,4-dideoxy-D-*glycero*-hex-3-enos-2-ulose, 5-hydroxymethylfuran-2-carbaldehyde in addition to other products that are formed by degradation of glucose by 1.2-enolisation.

2,3-Enolisation of ketosamines (Figure 4.78) gives 2-ene-2,3-diol as a primary intermediate, thus decomposition of the bound amino compound yields 1-deoxy-D-*erythro*-hexo-2,3-diulose as the main intermediate. In the absence of amino compounds, the latter is formed from fructose only to a small extent. A minor reaction product is 4-deoxy-D-*glycero*-hexo-2,3-diulose. Both uloses are transformed into a variety of products by subsequent reactions, covering also those formed in the absence of amino compounds, but in different yields.

It is obvious that the decomposition of aminodeoxysugars by 1,2-enolisation and 2,3-enolisation gives rise to the same key intermediates that arise during the degradation of sugars in the absence of amino compounds (3-deoxy-D-eryihro-hexos-2-ulose, 1-deoxy-D-erythro-hexo-2,3-diulose, and 4-deoxy-D-*glycero*-hexo-2.3-diulose). In the absence of amine compounds, however, these and many other products of carbohydrates are formed at comparable rates only in strongly acidic or strongly alkaline media (pH lower than 3 or higher than 8) or at higher temperatures (e.g. during caramelisation). Amino compounds catalyse the degradation of saccharides, so the reaction rate is higher and degradation of aminodeoxysugars proceeds under normal temperatures and in virtually a neutral medium, which are typical for most foods (pH 4–7). Moreover, the significance of many particular pathways differs with the presence or absence of amino compounds.

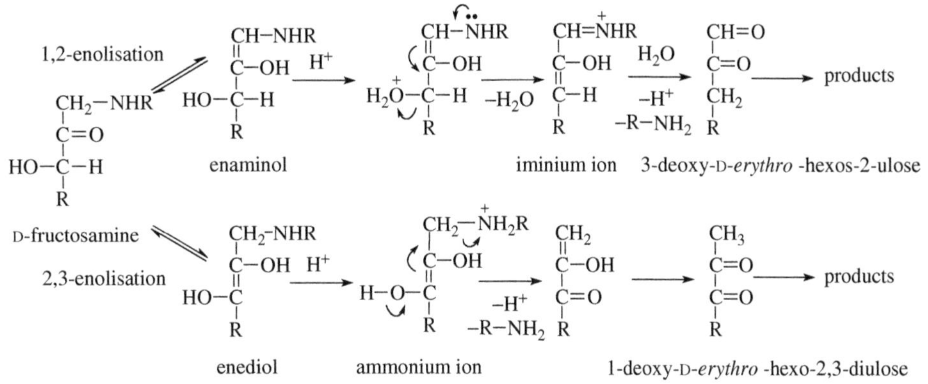

Figure 4.78 1,2-Enolisation and 2,3-enolisation of ketosamines.

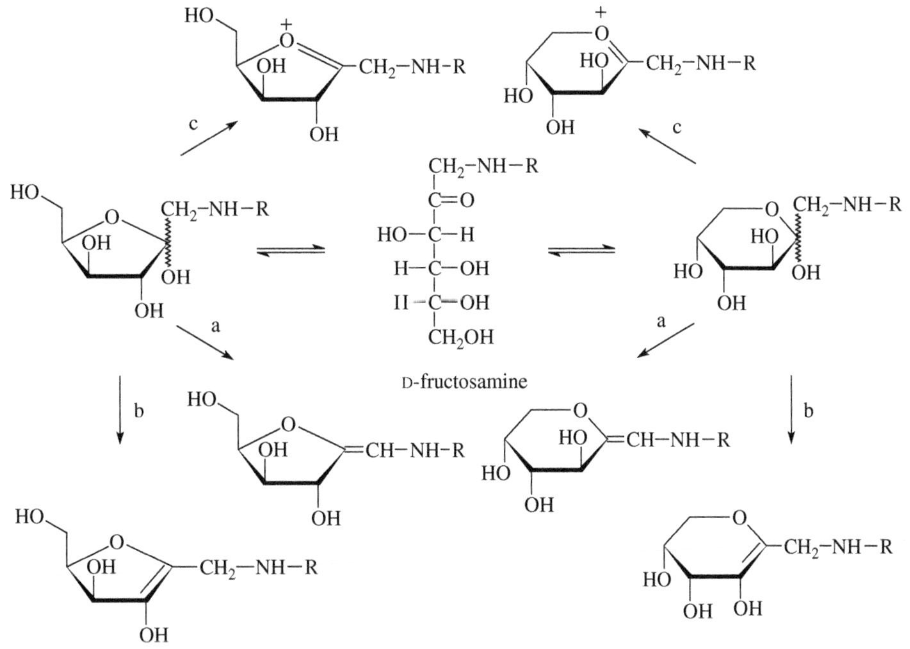

Figure 4.79 Ketosamine degradation by (a) 1,2-dehydration, (b) 2,3-dehydration, and (c) 2-dehydroxylation.

Without amino compounds, mainly 3-deoxyuloses are formed at sugar transformation, whilst the characteristic products in the presence of amino compounds are 1-deoxyuloses. A number of degradation products of sugars, including deoxy sugars, furans, pyrans, and fragments containing reactive carbonyl groups, also react with amino compounds to form products (or intermediates) containing nitrogen in the structures. If reactive sulfur amino acids and other sulfur compounds are present in reaction systems, sulfur is also involved in many products. The characteristic products of the Maillard reaction are, therefore, nitrogen- and sulfur-containing heterocyclic compounds.

Factors that determine whether aminodeoxysugars undergo 1,2-enolisation or 2,3-enolisation have been studied for decades, but the reaction mechanisms are still not fully understood. It is generally considered that 1,2-enolisation takes place in acidic media where the nitrogen atom of the aminodeoxysugars is protonated. The level of 1,2-enolisation products, such as derivatives of furan-2-carbaldehyde, also increases with increasing amino compound basicity. Aminodeoxysugar must be present as the free base in order to allow 2,3-enolisation. For this reason, 2,3-enolisation occurs especially in alkaline or non-aqueous media, for example during caramelisation.

According to recent surveys, ketosamines degrade preferentially via 1,2-enolisation and 2,3-enolisation under milder reaction conditions (in solutions at relatively low temperatures), as mentioned previously. At higher temperatures, however, ketosamine decomposition prevails by 1,2-dehydration, 2,3-dehydration, and the so-called 2-dehydroxylation, which is an elimination of the hemiacetal hydroxyl group (Figure 4.79). It is assumed that the 1,2-dehydration occurs primarily in furanoses, whilst the 2.3-dehydration mechanisms predominates in pyranoses. The 2.3-dehydration products of pyranoses are stabilised by hydrogen bonds between nitrogen atoms and C-3 hydroxyl groups.

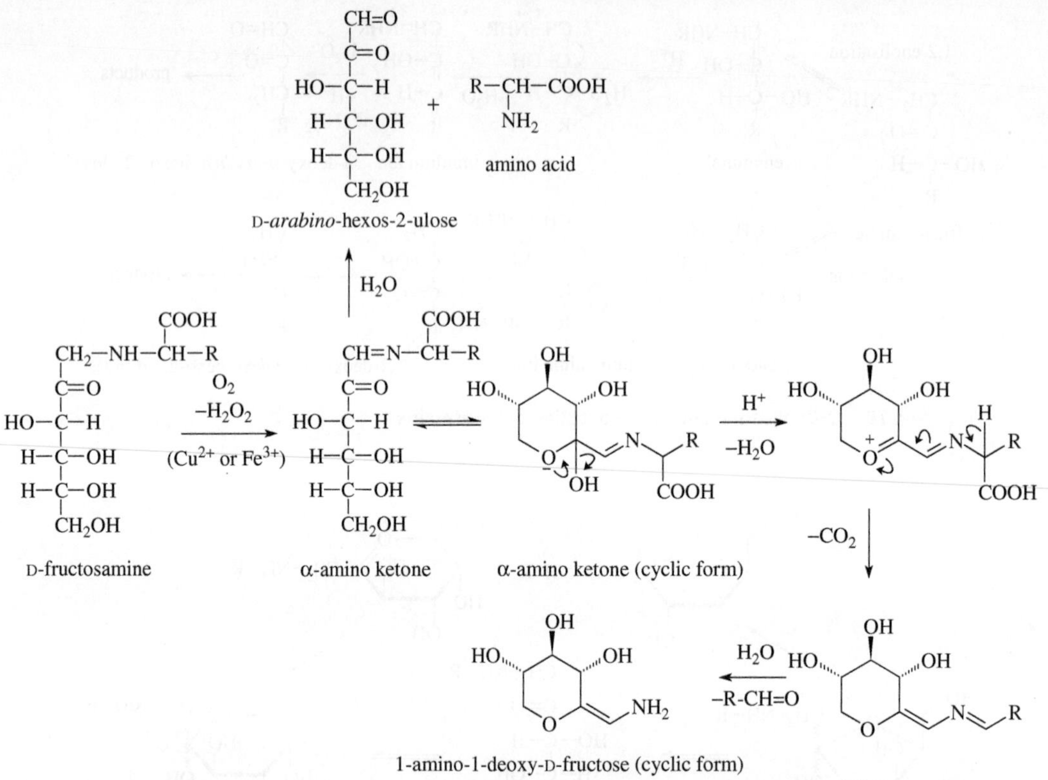

Figure 4.80 Oxidation of ketosamines to D-*arabino*-hexos-2-ulose and Strecker aldehydes.

Recently, the formation of 2-deoxy-D-*glycero*-hexo-3,4-diulose (2-deoxyglucosulose) was suggested in the Maillard reaction of fructose with γ-aminobutyric acid under mild conditions (50 °C) at low water contents (<50%). This indicates the formation of a 2,3-enaminol from the Schiff base of fructose and the formation of 2-amino-2-deoxy-3-ketose as an alternative to the Heyns product.

Another important reaction is decomposition of the ketosamines derived from amino acid in the presence of oxygen, which is catalysed by transition metal ions. Products of this decomposition are glycos-2-uloses. For example, D-fructosamine yields D-*arabino*-hexos-2-ulose and, in addition to other products, the corresponding Strecker aldehyde (Figure 4.80).

4.7.5.3.1 Decomposition of 3-deoxyglycosuloses

The main products of 3-deoxy-D-*erythro*-2-hexos-2-ulose degradation are (Z)-3,4-dideoxy-D-*glycero*-hex-3-enos-2-ulose (2-hydroxy-6-hydroxymethylpyran-3-one), the corresponding (E)-isomer, and 5-hydroxymethylfuran-2-carbaldehyde.

In addition to furan (pyran) derivatives arising at the decomposition of 3-deoxyglycos-2-ulose, considerable amounts of pyrrole derivatives, N-substituted 5-hydroxymethylpyrrole-2-carbaldehydes (**4-106**), and their isomeric pyridines (pyridinium betaines, **4-106**) are formed in a surplus of amine compounds. In the reaction of 5-hydroxymethylfuran-2-carbaldehyde (HMF) with amino compounds, the nitrogen products are formed in only trace amounts, which suggests that incorporation of amino group occurs preferentially in a reaction with an HMF precursor. If the reacting amino compound is ammonia, nitrogen atoms of the resulting pyrrole and pyridine derivatives are unsubstituted. In the presence of amino acids, the reaction products are carboxymethyl-substituted 1-pyrrole-2-carbaldehydes (**4-106**). Their cyclisation yields the corresponding lactones of N-carboxymethyl-5-hydroxymethylpyrrole-2-carboxaldehydes (**4-106**). In the presence of peptides, the lactams (dihydropyrrolopyrazinones, **4-106**) are formed.

The C-1/C-2 bond cleavage of 3-deoxy-D-*erythro*-hexos-2-ulose yields formic acid and 2-deoxy-D-*erythro*-pentose, which is transformed into furfuryl alcohol and, in the presence of amino compounds, N-substituted 2-hydroxymethylpyrroles (Figure 4.81). These substances tend to form polymers. Pyrrole and pyridine derivatives are formed also in reactions of amino compounds with 3-deoxyglycos-2-uloses derived from disaccharides. Another mechanism for the formation of furfuryl alcohol from glucose in aqueous systems involves the oxidation of glucose to gluconic acid, which is decarboxylated to a pentitol and then dehydrated and cyclised. In wort, *S. cerevisiae* yeast can reduce furan-2-carbaldehyde to furfuryl alcohol, but the main formation pathway seems to be the decomposition of the Amadori product derived from maltose via 3-deoxy-D-ribulose, which dehydrates to furfuryl alcohol during boiling.

Reactions of 3-deoxyhexos-2-uloses with ketosamines derived from ammonia yield polyhydroxyalkyl substituted pyrroles and reactions of ketosamines derived from amines produce N-substituted polyhydroxyalkyl pyrroles. When the aforementioned uloses react with

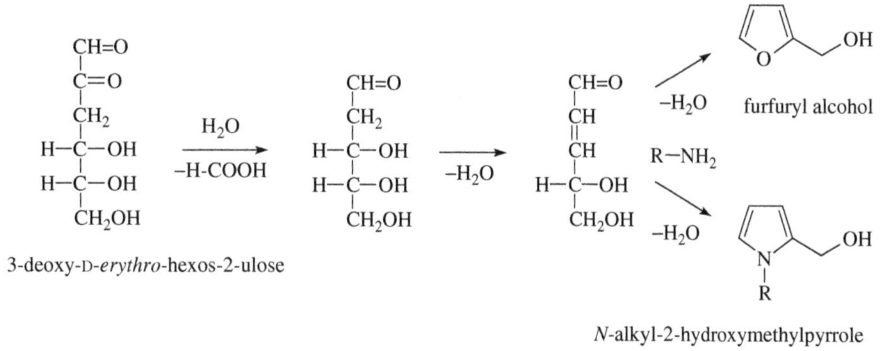

Figure 4.81 Degradation of 3-deoxy-D-*erythro*-hexos-2-ulose.

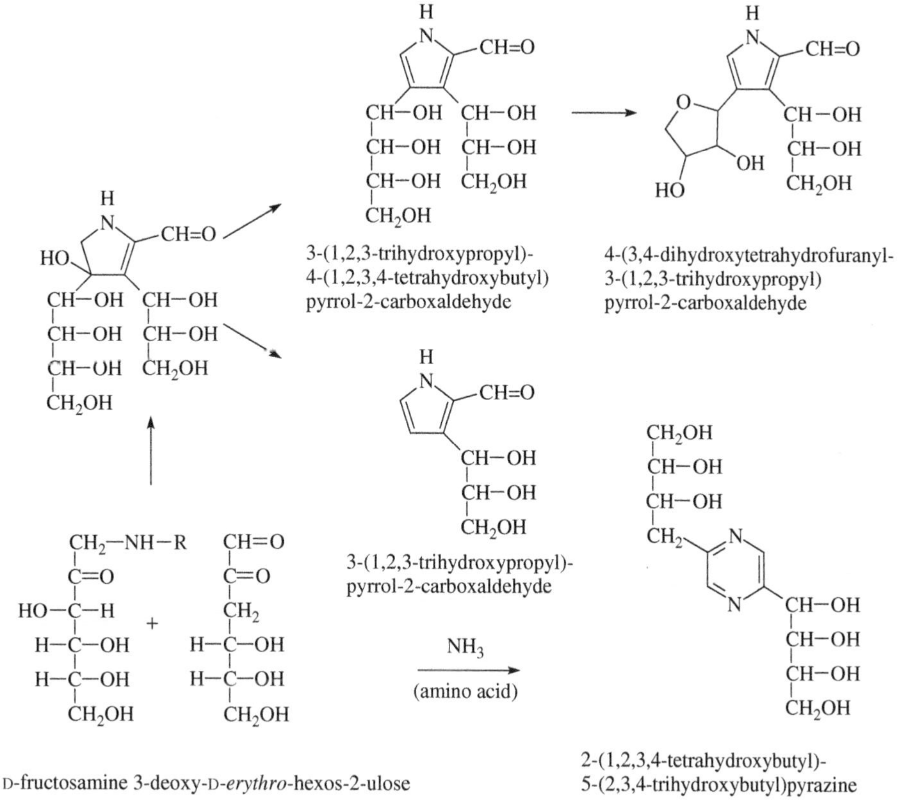

Figure 4.82 Formation of polyhydroxyalkyl substituted pyrrole-2-carbaldehyde and pyrazine derivatives (R = H).

ketosamines in the presence of ammonia, 2,5-polyhydroxyalkyl-substituted pyrazines are formed (Figure 4.82). These compounds also arise by dimerisation of ketosamines. 2,6-Disubstituted analogues of these compounds are formed by condensation of aldosamines with ketosamines. Polyhydroxyalkyl-substituted imidazoles (**4-106**) are formed in reactions of 3-deoxyglycos-2-uloses with ammonia. It is assumed that all polyhydroxyalkyl-substituted heterocyclic compounds can be transformed into simple alkyl-substituted analogues (such as dimethyl derivatives) during thermal operations.

The reaction course of 3-deoxyglycos-2-uloses with secondary amines is different from that with primary amines. Usually, significant amounts of cyclic compounds are formed, but 5-hydroxymethylfuran-2-carbaldehyde does not arise at all (or only in small amounts). For example, the reaction of proline with 3-deoxy-D-*erythro*-hexos-2-ulose yields the so-called maltoxazine, a flavour-active compound occurring in beer and malt (Figure 4.83).

Condensation of *N*-substituted pyrrole-2-carbaldehyde derivatives mainly leads to coloured products, examples of which are ethers and dimers (**4-106**). Derivatives of furan-2-carbaldehyde and many other furans are also significantly involved in non-enzymatic

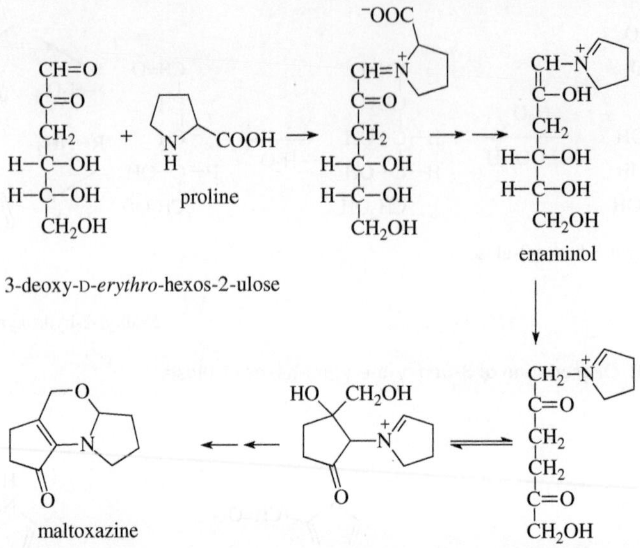

Figure 4.83 Formation of maltoxazine.

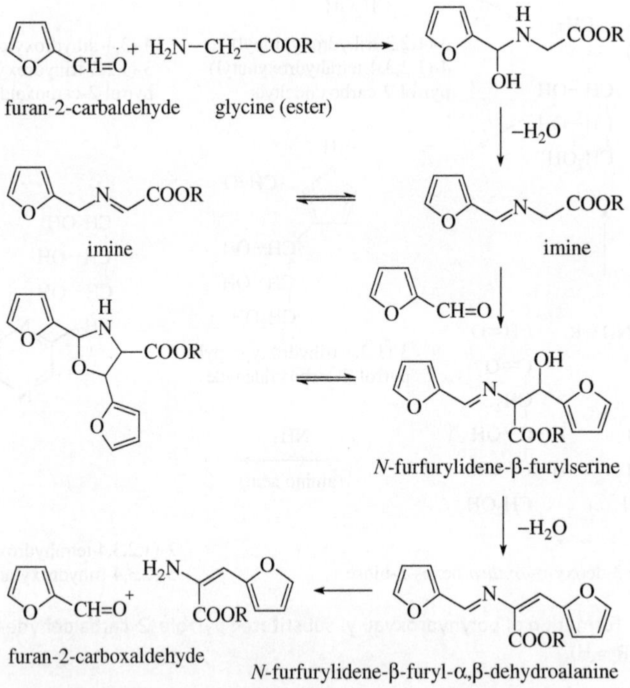

Figure 4.84 Reaction of furan-2-carbaldehyde with glycine.

browning reactions. Reactions of furan-2-carbaldehydes with amino acids produce common Maillard reaction products, such as imines. Subsequent reactions of imines produce more complex structures, mostly coloured compounds (Figure 4.84). The reaction of 4-hydroxy-5-methyl-2H-furan-3-one with primary and secondary amines also yields coloured products (**4-106**).

4.7.5.3.2 Decomposition of 1-deoxyglycodiuloses

1-Deoxyglyco-2,3-diuloses are highly reactive compounds that scientists have so far failed to isolate from either foods or model reaction systems. Their presence has been indirectly confirmed by analysis of stable derivatives. Important degradation products of 1-deoxyhexo-2,3-diuloses are furanones.

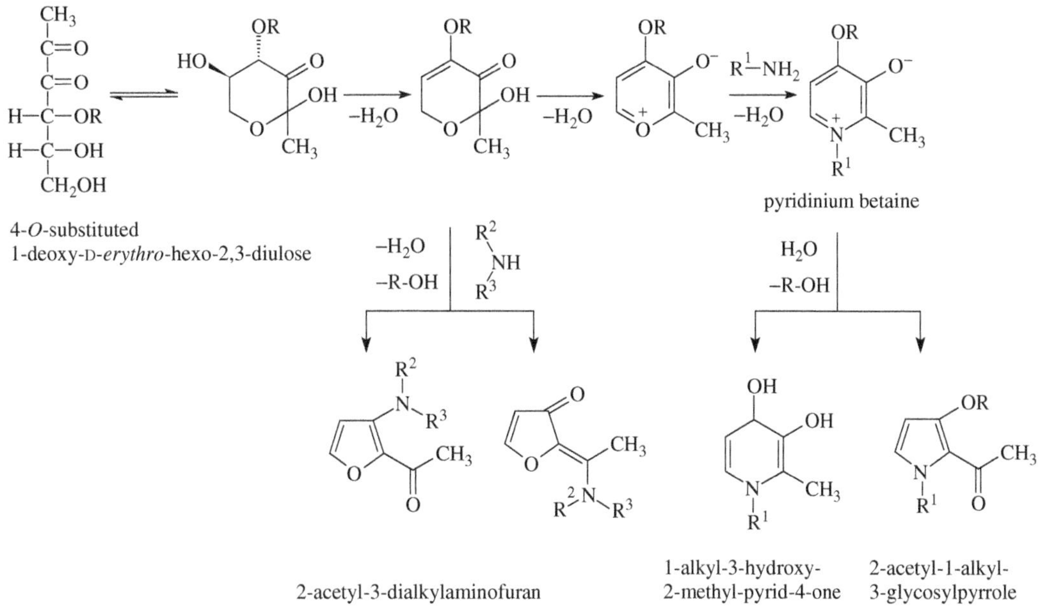

Figure 4.85 Reaction of amines with 1-deoxyglyco-2,3-diuloses derived from disaccharides.

In the presence of sulfur compounds (e.g. cysteine), pentoses and methylpentoses produce, in addition to 4-hydroxy-5-methyl-2*H*-furan-3-one and 4-hydroxy-2,5-dimethyl 2*H*-furan-3-one, respectively, related compounds containing sulfur in the molecule, some of them also via the formation of cysteine *S*-conjugates (see Section 8.2.9.1.2). Many of these compounds resemble the aroma of cooked or roasted meat or coffee. For example, odourless *S*-furfuryl-L-cysteine and *S*-(2-methyl-3-furyl)-L-cysteine are formed in the Maillard reaction of xylose and cysteine. Both compounds are cleaved by microbial oral β-lyases, releasing furfuryl thiol and 2-methyl-3-furanthiol, which smell coffee-like and meaty, respectively.

In the presence of primary amines, 1-dexyhexo-2,3-diuloses derived from disaccharides yield pyridinium betaines and, further, pyrid-4-ones and isomeric 2-acetylpyrroles. The reaction with secondary amines leads to *N*-substituted furan derivatives (Figure 4.85). 1-Deoxyhexo-2,3-diuloses also condense with carbonyl compounds and form coloured products (**4-106**). The reaction with proline and hydroxyproline results in cyclopentaazepin-8(1*H*)-ones and bispyrrolidinohexose reductones (**4-106**).

4.7.5.3.3 Decomposition of 4-deoxyglycosuloses and 1-amino-1,4-dideoxyglycosuloses

In the presence of amino compounds, the decomposition of 4-deoxyglycosuloses, such as 4-deoxy-D-*glycero*-hexo-2,3-diuloses (whose existence is expected during the transformation of lactulose and fructose into 2-hydroxyacetylfuran), yields the corresponding pyrroles (**4-106**). The other products are isomeric pyridinium betaines with a hydroxymethyl group in position C-2 of the pyridine nucleus, which differ from the pyridinium betaines (**4-106**) formed by decomposition of 3-deoxyglycos-2-uloses.

In the Maillard reaction of oligosaccharides with a 1,4-glycosidic bond (such as maltose, lactose, and lactulose) in media of low water activity, 1-deoxyhexo-2,3-diuloses, 3-deoxyhexos-2-uloses, and short-chain α-dicarbonyl compounds are formed. However, these represent only a minor portion of the dicarbonyl intermediates. A principal intermediate is the amino analogue of 4-deoxy-D-*glycero*-hexo-2,3-diulose; that is, 1-amino-1,4-dideoxy-D-*glycero*-hexo-2,3-diulose. Probably the most common product of decomposition of this diulose is 2-aminoacetylfuran, which arises (analogously to 2-hydroxyacetylfuran) by cyclisation and dehydration. 2-Aminoacetylfuran is stable in acidic media, but at pH higher than 5 it readily oxidises and condenses to form more complex structures (Figure 4.86). When the amino compound bound in aminoacetylfuran is a protein (e.g. bound by the ε-amino group of lysine), the resulting aminopyrrole forms cross-links, which connect the peptide chains of the proteins. Aminoacetylfurans react with ammonia with the formation of imidazole derivatives.

The main dicarbonyl product of the Maillard reaction with oligosaccharides that have 1,4-glycosidic bonds is 1,4-dideoxyhexo-2,3-diulose (Figure 4.87). 2,3-Enolisation of the Amadori compound is followed by splitting off the oligosaccharide residue by β-elimination to form 1-amino-1,4-dideoxy-D-*glycero*-hexo-2,3-diulose. Reduction and hydrolysis of the bound amino acid yields 1,4-dideoxy-D-*glycero*-hexo-2, 3-diulose. The residue of the reducing oligosaccharide with a 1,4-glycosidic linkage in the reducing end can be transformed in this way repeatedly. 1,4-Dideoxyhexo-2,3-diuloses may also result from 1-deoxyhexo-2,3-diuloses, which are reduced during the Strecker degradation of amino acids.

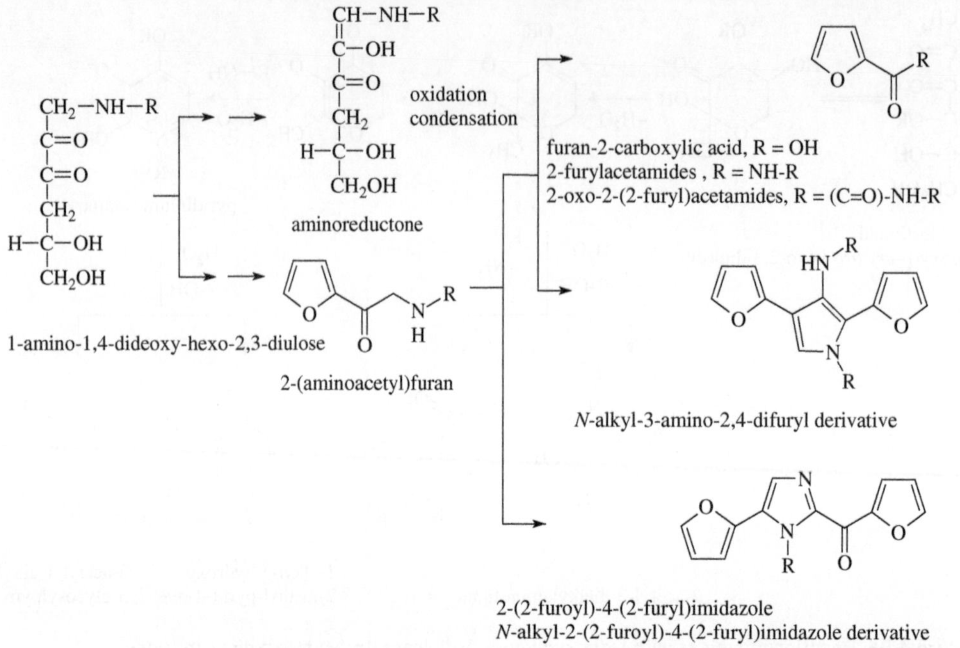

Figure 4.86 Important reactions of 1-amino-1,4-dideoxyhexo-2,3-diuloses.

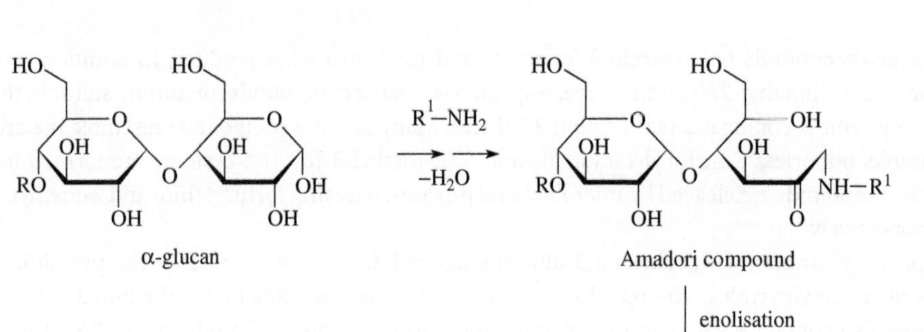

Figure 4.87 Formation of diuloses from maltooligosaccharides.

1,4-Dideoxyhexo-2,3-diuloses are unstable compounds, and their cyclisation and dehydration lead, for example, to 5-hydroxymethyl-2-methylfuran-3-one (**4-106**). The cyclisation and dehydration of 1-amino-1,4-dideoxyhexosulose produce amino-acetylfurans. The best-known of these products is furosin. In non-acidic media, maltooligosaccharides and other (1 → 4)-oligosaccharides give rise, via 1-amino-1,4-dideoxyhexosuloses, to α-dicarbonyl compounds with five carbon atoms, namely 3-deoxy-D-*glycero*-pentulose, 3-deoxy-D-*glycero*-pentos-2-ulose, and 3,4-dideoxypentosulose (Figure 4.26).

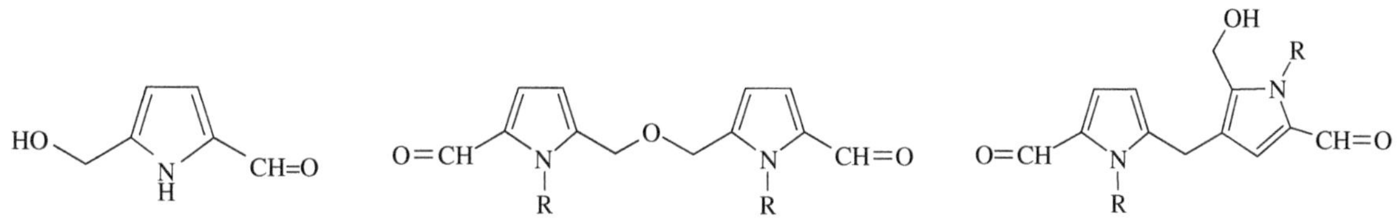

5-hydroxymethylpyrrole-2-carboxaldehyde, its *N*-substituted ethers and dimers

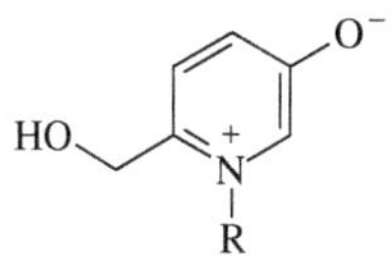

pyridinium betaines 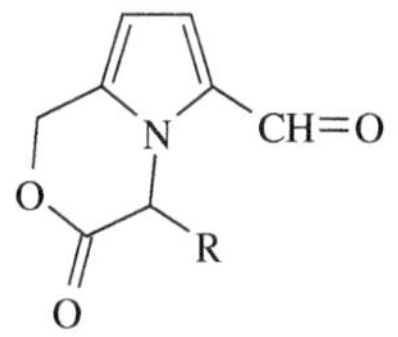   lactones 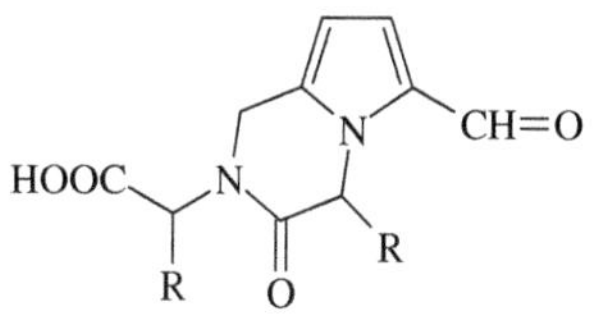   lactams

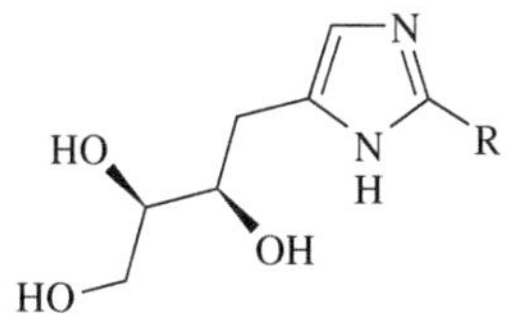

1,4-polyhydroxyalkylimidazoles 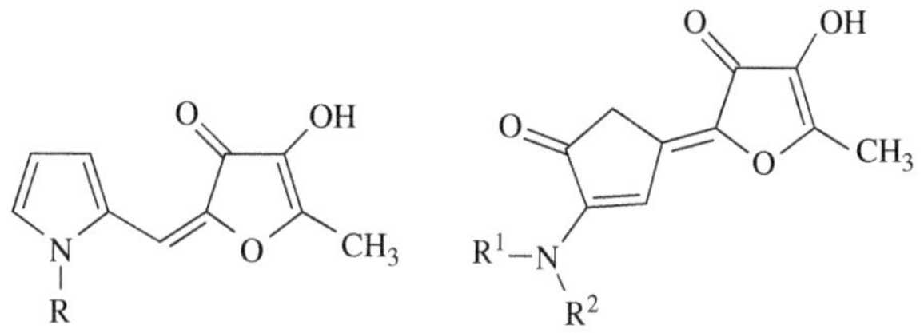   products derived from 4-hydroxy-5-methyl-2*H*-furan-3-one...

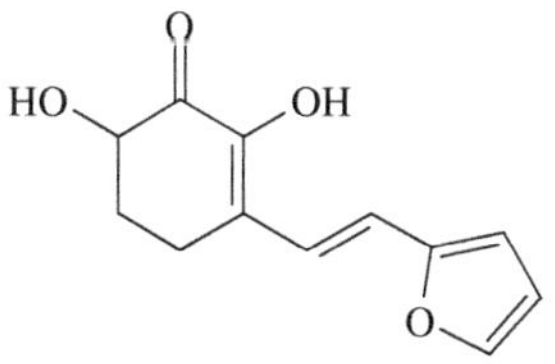

...with furan-2-carboxaldehyde

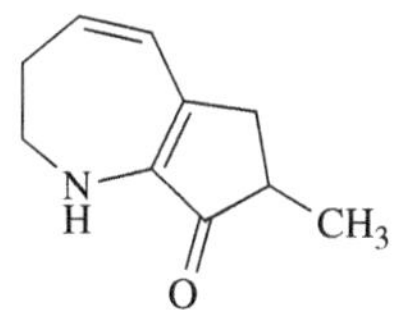

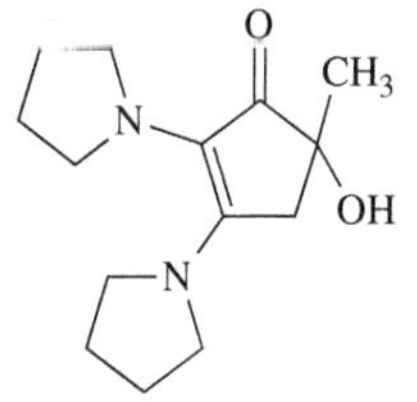

1-deoxyhexo-2,3-diulose products

with proline (1 molecule)   with proline (2 molecules)

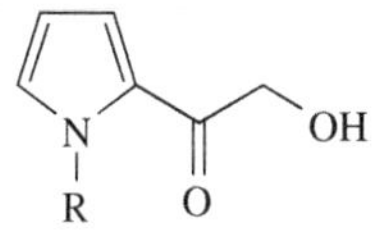

2-hydroxyacetylpyrroles

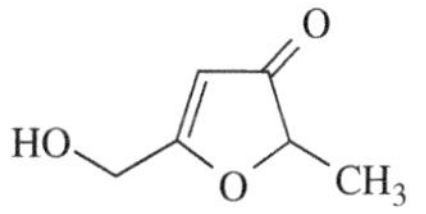

2-methyl-5-hydroxymethyl-2*H*-furan-3-one

4-106, simple heterocyclic Maillard reaction products

The key intermediates in the formation of furan and pyrrole derivatives from 3-deoxyglycos-2-uloses via β-dicarbonyl compounds are 3-deoxy-2,4-diuloses and 1-amino-1,3-dideoxy-2,4-diuloses. Cyclisation, dehydration, and keto-enol tautomerisation of diuloses yield 2-acetylpyrrole derivatives as the main products (Figure 4.88).

4.7.5.4 Formation of aminoreductones

1-Deoxyhexo-2,3-diuloses can be easily converted into aliphatic aminoreductones of the general structure given by formula **4-107**. Aliphatic aminoreductone derived from propylamine was the first assumed and, later, identified linear six-carbon chain aminoreductone (Figure 4.86). A very reactive compound, formed by the degradation of 1-deoxyhexodiuloses, is diacetylformoin (R = H, Figure 4.34). This key intermediate, existing in several tautomeric forms, reacts with primary amines to form pyrrolid-3-one derivatives that are transformed into methylene reductic acid (R = H, Figure 4.89). Analogously, reductic acid is formed from pentoses and ascorbic acid. Diacetylformoin reacts with

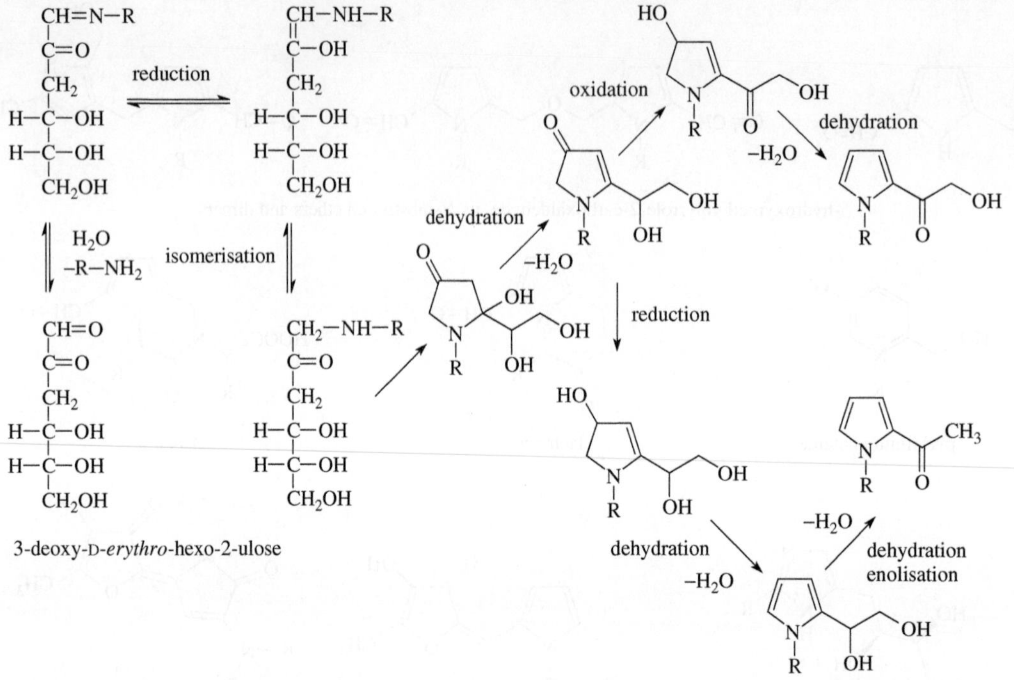

Figure 4.88 Transformation of 3-deoxy-D-*erythro*-hexos-2-ulose in the presence of amino compounds.

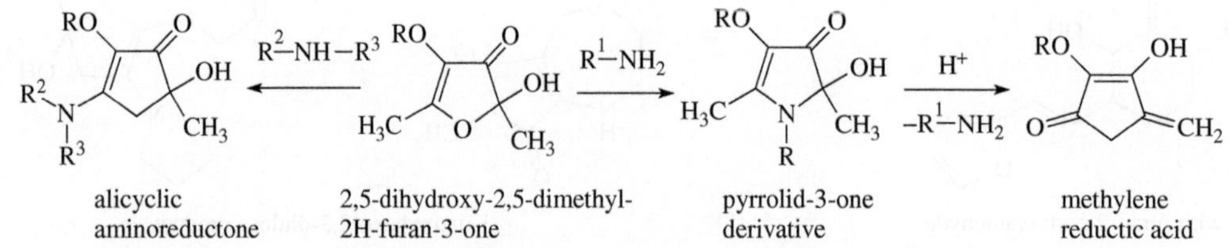

Figure 4.89 Reaction of diacetylformoin with amines.

secondary amines to form 5-dialkylamino substituted 2*H*-dihydrofuran-3-one derivatives. The reaction of diacetylformoin with the ε-amino group of lysine bound in proteins causes protein cross-linking. Aminohexose reductones and derivatives of pyrrolidone derived from disaccharides (R = glycosyl, Figure 4.89) are formed by similar reactions.

4.7.5.5　*Strecker degradation of amino acids*

The mechanism of this very important reaction is given elsewhere (see Sections 2.5.1.3.2 and 2.5.2.3). In addition to numerous other substances, the active compounds are aldoses, ketoses, derivatives of furan-2-carbaldehyde, α-dicarbonyl compounds derived from sugars such as glycos-2-uloses or glyco-2,3-diuloses, and, in particular, the simple decomposition products of sugars, such as glyoxal and methylglyoxal.

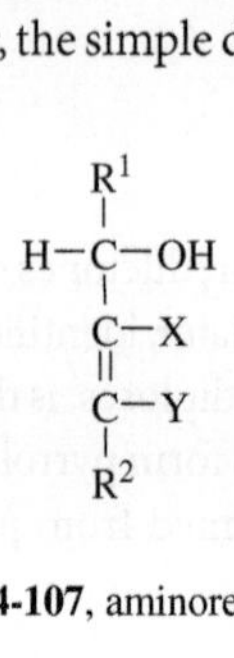

4-107, aminoreductones

X = OH, NR₂ or NHR
Y = OH, NR₂ or NHR

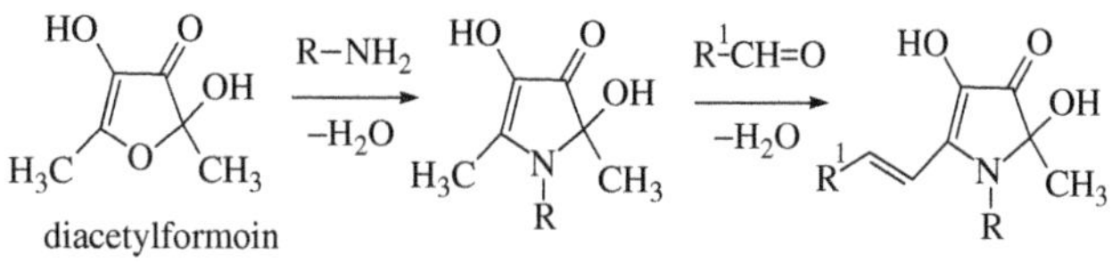

Figure 4.90 Formation of chromophores from diacetylformoin, aldehydes, and amino acids.

The importance of Strecker degradation is mainly in the formation of reactive and often sensory-active aldehydes and ammonia (free or bound in reactive α-amino carbonyl compounds). Strecker aldehydes, together with α-amino carbonyl intermediates, contribute significantly to the formation of many heterocyclic compounds, which are important flavour-active components of processed foods, and of the brown pigments melanoidins. Aldoses simultaneously produce the corresponding 2-deoxyaldoses or 1-amino-1-deoxyalditols and 2-oxoacids. In addition to free amino acids or proteins, the Amadori reaction products have been confirmed as direct precursors of Strecker aldehydes, without the need for an α-dicarbonyl compound as catalyst. Other Strecker aldehyde precursors are 2-substituted 5-methyl-3-oxazolines (see Section 8.2.11.1.8), which are stable in dry food but easily hydrolyse to form Strecker aldehydes in the presence of water during processing with water addition or during thorough mastication. In such a way, for example, the Strecker aldehyde of phenylalanine, phenylacetaldehyde, may be liberated from 2-isobutyl-5-methyl-3-oxazoline during consumption of chocolate.

4.7.5.6 Melanoidins

The compounds produced in the first two phases of the Maillard reaction are usually colourless compounds, which are often referred to as **premelanoidins**. Various coloured compounds are formed mainly in the final stage of this reaction. Some of these are low-molecular-weight substances (usually with a relative molecular weight below 1000 Da), but other coloured substances known as **melanoidins** have relative molecular weights higher than 1000 Da. They have characteristic physico-chemical properties, such as solubility and antioxidant activity.

In mixtures of pentoses and hexoses with amino acids, numerous low-molecular-weight chromophores have been characterised; for example, those formed from the transformation products of 3-deoxyglycos-2-uloses (e.g. furan-2-carbaldehyde) and from reactive sugars fragments (**4-108**). The significant degradation product of 1-deoxyhexo-2,3-diuloses diacetylformoin is another precursor of nitrogen-free chromophores, but only if primary amino acids are not present. Otherwise, its reaction with amino acids gives rise to pyrrolinone reductones that react via a methyl group with carbonyl compounds (such as furan-2-carbaldehyde) to form coloured compounds (Figure 4.90; for other structures, see Section 4.7.1.1.4). Yellow Maillard pigment, 7-(2-furanyl)-2,3,5,6-tetrahydrocyclopenta[1,4]thiazine-3-carboxylic acid (furpenthiazinate, **4-108**), was found after acid hydrolysis of soy protein in the presence of xylose, with cysteine and furan-2-carbaldehyde as intermediates.

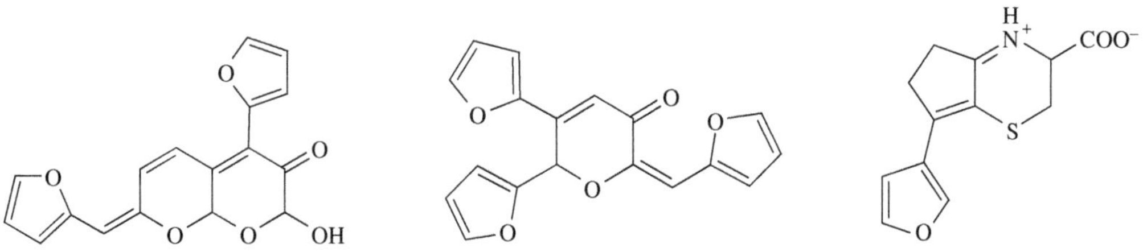

pyrano[2,3-b]pyran-3-one   6H-pyran-3-one   furpenthiazinate

4-108, coloured substituted furyl derivatives

Whilst low-molecular-weight chromophores arise mainly in systems with free amino acids, the coloured Maillard reaction products with proteins are almost exclusively macromolecular substances, unless there is subsequent proteolysis. Current hypotheses include two main types of high-molecular-weight pigments. The first type assumes the condensation of low-molecular-weight Maillard reaction intermediates to form the coloured substances. The structure of melanoidins resulting from the condensation of monomeric units (especially heterocyclic Maillard reaction intermediates) has not been described satisfactorily. For example, melanoidins may arise in the reactions of hydroxylated 1,4-dihydropyrazines formed by oxidation and hydration of pyrazinium radicals (Figure 4.76) or by their condensation with aldehydes (especially with furan-2-carbaldehyde). Also other degradation products of pentoses and hexoses, such as N-substituted pyrroles and

N-substituted pyrrole-2-carbaldehydes, easily form coloured products. Coloured oligomers formed from 2-hydroxymethyl-1-methylpyrrole (**4-109**) by condensation of 1-methylpyrrole with 1-methylpyrrole-2-carbaldehyde (**4-109**) or furan-2-carbaldehyde (**4-109**) have been identified. The other two types of polymers are formed by nucleophilic addition of amino groups of Amadori compounds to carbonyl compounds and by subsequent dehydration. The resulting polymer is oxidised with the formation of a system of conjugated double bonds. Decarboxylation and elimination reactions transform these polymers into other conjugated systems. Coloured pigments that do not contain nitrogen can be hypothetically formed, for example, by dehydration of the individual oligomers produced from 3-deoxy-D-*erythro*-hexos-2-uloses (the corresponding acids formed by their oxidation; **4-109**) or oligomers of 1-deoxy-D-*erythro*-hexo-2,3-diulose (**4-109**).

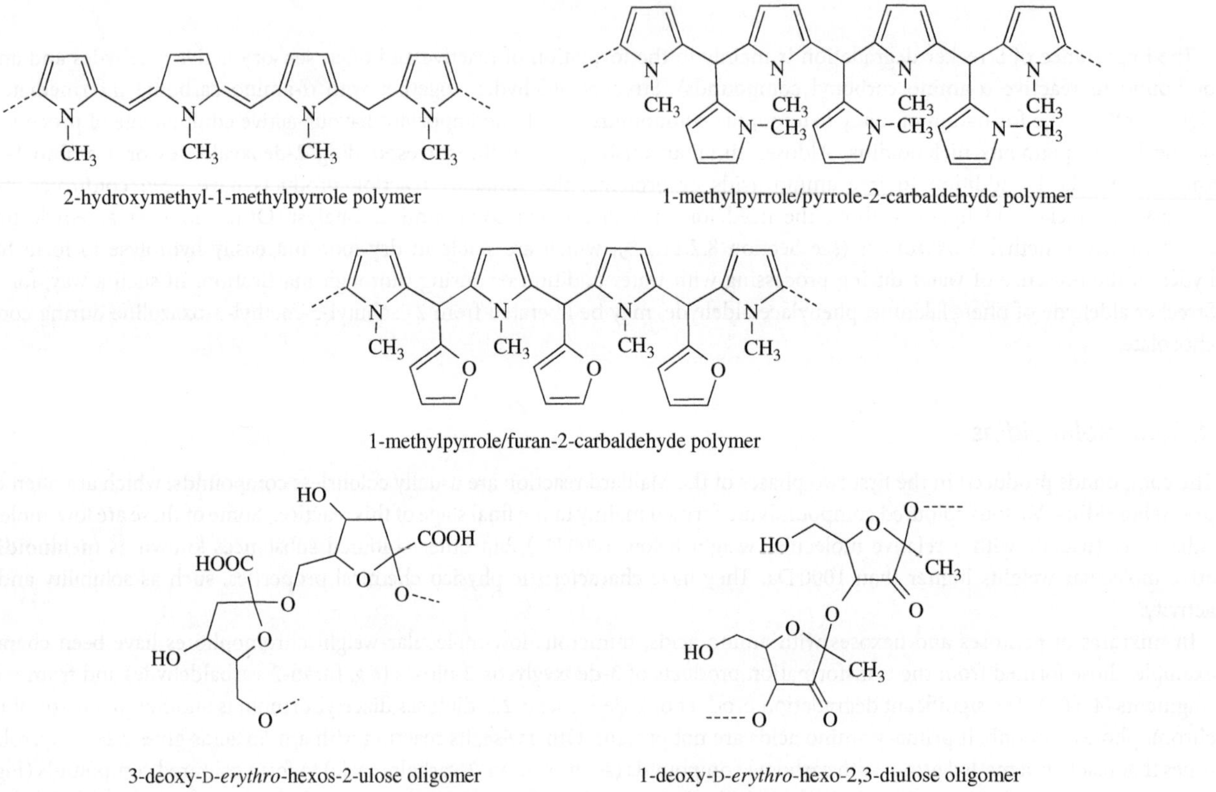

2-hydroxymethyl-1-methylpyrrole polymer

1-methylpyrrole/pyrrole-2-carbaldehyde polymer

1-methylpyrrole/furan-2-carbaldehyde polymer

3-deoxy-D-*erythro*-hexos-2-ulose oligomer

1-deoxy-D-*erythro*-hexo-2,3-diulose oligomer

4-109, examples of hypothetical homeomeric poly(oligo)mers of heterocyclic Maillard intermediates

The second method of melanoidin formation, which also leads to protein cross-linking, is much more probable in foods. This reaction assumes the covalent bond formation of transformation products of sugars on the side chains of proteins, especially of bound lysine, arginine, or cysteine, to form chromophore structures. Some of these so-called AGE structures, such as pentosidine (see Section 4.7.5.7), are coloured. Other, more complex chromophores include substances that are formed in systems containing furan-2-carbaldehyde. The lysyl residue in a peptide chain reacts in a neutral solution with furan-2-carbaldehyde to give chromophore (**4-110**). The chromophores formed between two lysyl residues (**4-111**) are also described. Reaction of the guanidyl group of an arginine residue with glyoxal and furan-2-carbaldehyde yields another chromophore (**4-112**). Amongst the structures that cause protein cross-linking is the radical 1,4-bis(5-amino-5-carboxy-1-pentyl)pyrazinum cation, abbreviated crosspy (**4-113**), bound to two lysyl residues of peptide chains. Its formation is shown in Figure 4.77. This structure is considered one of the key precursors of melanoidins in bread crust and in roasted coffee.

It is also possible that the chromophores of low molecular weight are only caught by the polymer via physical rather than covalent bonds. The colour is sometimes attributed to the presence of free radicals stabilised in the structure of Maillard polymers. Other kind of coloured structures may represent complexes with metal ions that create chromophores via chelation by colourless organic compounds, such as reductones.

The nature of melanoidins can differ according to the reactants present in processed food (Table 4.37). Melanoidins with higher molecular weight, or those derived from proteins (melanoproteins), which are cross-linked by α-dicarbonyl intermediates, are largely insoluble.

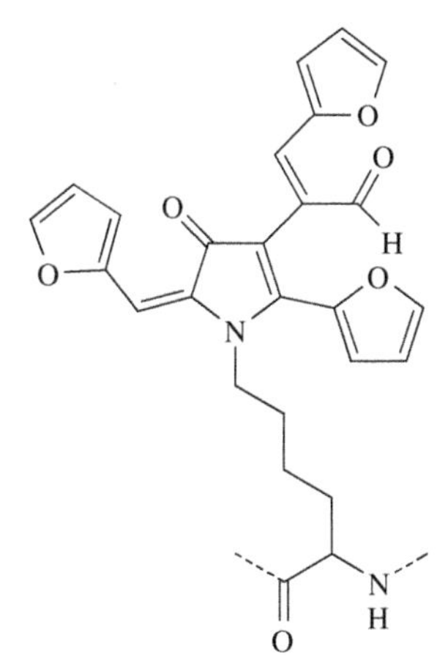

4-110, chromophore bound to protein through a lysine residue

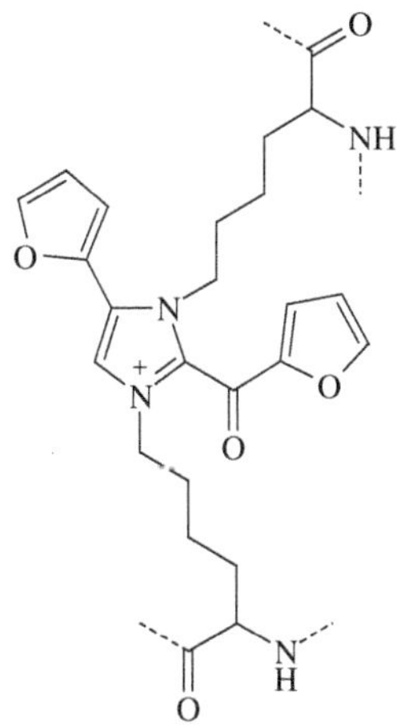

4-111, chromophore bound to a protein through two lysine residues

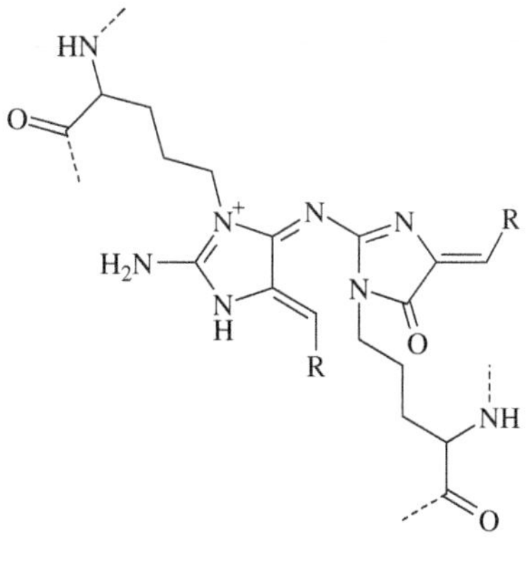

4-112, coloured products of glyoxal and furan-2-carbaldehyde
(R = furyl) with two arginine residues

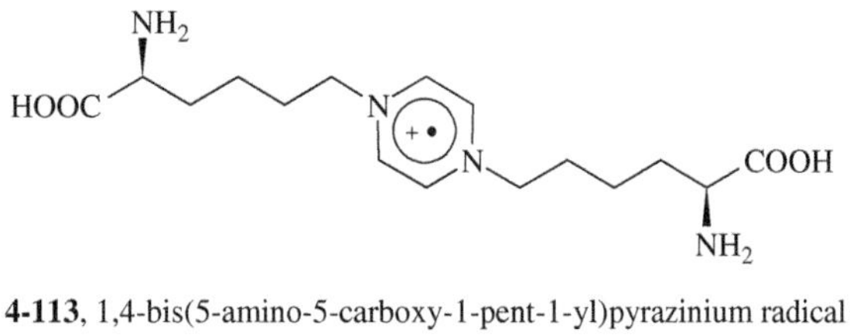

4-113, 1,4-bis(5-amino-5-carboxy-1-pent-1-yl)pyrazinium radical cation

Table 4.37 Properties of melanoidins isolated from different foods.

Food melanoidins	Water solubility	Other components
Coffee	High	Chlorogenic acid
Soy sauce	Moderate	–
Bread	Low	–
Nut and seeds	Low	Fats
Cocoa	Moderate	Catechins
Milk	Moderate	–
Dark beer	High	Phenolic compounds
Balsamic vinegar	Moderate	–
Sweet wine	Moderate	Phenolic compounds

Source: After F.J. Morales.

4.7.5.7 Melanoproteins and protein glycation

During processing and storage of foods, numerous reactions lead to protein structure modification. The changes comprise denaturation and a series of addition and other reactions. They lead to the formation of disulfide bridges and reversible oxidation of side chains of bound amino acids, formation of isopeptide bonds, and the artefacts of amino acids derived from dehydroalanine. The modifications of protein molecules are also caused by sugars and reactive dicarbonyl compounds formed during transformation of reducing sugars or oxidation of lipids. The bifunctional carbonyl compounds may react with one or two reactive amino acid residues in the peptide chains to form adducts causing inter- and intramolecular cross-linking of proteins. These reactions are of great importance for modifying functional properties of proteins and lowering their nutritional value. Proteins modified by reactions with carbonyl intermediates of the Maillard reaction or with secondary lipid oxidation products are often referred to as **melanoproteins**.

Significant in terms of health risks are spontaneous immoderate non-enzymatic reactions in living organisms that lead to the same or similar modifications of native protein as during food processing. This so-called protein **glycation** plays an important role both in physiological and in pathological processes in the organisms. The modified protein, or strictly the modified amino acids released from the protein, are known as advanced glycation/glycoxidation end-products (AGEs). Analogous structures called advanced lipoxidation end-products (ALEs) are produced by reactions of proteins with certain products of lipid oxidation (see Section 3.8.1.12.1). Excessive reactions are an indicator of tissue ageing and pathological manifestations of certain diseases, such as hyperglycaemia in patients with type 2 diabetes. AGEs accumulate at sites of microvascular injury in diabetes, including the kidney, the retina, and within the vasculature. One of the causes of oxidative stress in ageing, diabetes, and other diseases is considered to be the formation of oxygen radicals from glycated proteins. For example, even the Amadori compounds (products of the early glycation of proteins) can apparently generate the superoxide radical anion, which arises from hydrogen peroxide in the presence of oxygen. This can react with traces of transient metal ions to form hydroxyl radicals. The potential adverse effects on health of diet-derived AGEs is of current interest, due to their proposed involvement in the disease progression of diabetic and uraemic conditions.

The enhanced formation of AGEs also exists in various other diseases, such as atherosclerosis, Alzheimer's disease, end-stage renal disease (ESRD), rheumatoid arthritis, and liver cirrhosis. Whether dietary AGEs are also risk factors for these diseases is still unclear. Most modifications *in vivo* are found in proteins with long life, such as collagens, myelin, and α-crystalline. For example, a gradual change in the structure of collagen during ageing causes a decrease in the tissue elasticity. Ultimately, these changes lead to restrictions of joint mobility, decreased muscle performance, changes in the cardiovascular system, or changes in the ocular lens. Modification of proteins may similarly have a negative impact in cases where AGEs are located in areas of interaction with other protein chains, the substrate in interactions between the enzyme and the substrate, or DNA during transcription.

The most important Maillard reaction products of glucose metabolism and non-enzymatic degradation that react with proteins are reactive α-dicarbonyl intermediates: methylglyoxal, glyoxal, and (deoxy)glycosuloses. Other carbonyl precursors of AGEs are α-hydroxycarbonyl products of sugar fragmentation, transformation products of ascorbic acid, and some secondary decomposition products of lipid hydroperoxides. Some common intermediates of sugar metabolism also contribute to the formation of AGEs *in vivo*. For example, triose phosphates and fructose phosphates are effective modifiers of proteins.

The most frequently modified part of the protein is the ε-amino group of protein-bound lysine and the guanidyl group of protein-bound arginine, which are involved in the formation of cross-linked AGEs. These react with bifunctional reagents (such as dicarbonyl compounds), and the product may then react through the other functional group with another lysine or arginine residue of the same or another protein molecule. Tryptophan and histidine bound in proteins usually react with only one functional group of bifunctional reagents, yielding non-cross-linked AGEs. Little is known about the reactions of bound cysteine, whose thiol group is a better nucleophile than the amino groups of lysine or arginine. Cysteine conjugates can significantly influence the technological and physiological properties of glycated proteins, and also serve as carriers and reservoirs of reactive dicarbonyl compounds.

AGEs are released by controlled physiological or enzymatic hydrolysis of modified proteins to form modified low-molecular-weight peptides and modified amino acids. The number of AGEs identified to date may be just the tip of the iceberg, because numerous modifications, which usually occur only in trace quantities, are formed in tissue proteins, making it difficult to identify them. Adducts identified after hydrolysis are called by trivial names and acronyms. They can be classified according to the structure, the participating amino acids or carbonyl compounds, the origin of the precursors (carbohydrates or their degradation products, the oxidation products of lipids, or both groups), or by whether or not the AGEs are cross-linked. Many cross-linked products are fluorescent, whilst some of them are coloured or have reducing properties.

One of the non-cross-linked AGE N^ε-(carboxymethyl)lysine (CML) was the first stable AGE discovered. CML is present in a range of heat-treated foods, and as a major AGE structure is formed *in vivo* by oxidative cleavage of Amadori products between C-2 and C-3 of the carbohydrate chain (glyoxal) or by a gradual degradation of lactulosyllysine in dairy products. It is a marker of oxidative stress and long-term damage to protein in clinical studies (ageing, atherosclerosis, and diabetes). The protein modification has been demonstrated in collagen, eye lenses, serum proteins, and erythrocytes, amongst others. Together with some other AGEs, CML has also been demonstrated in lipofuscin granules, the intracellular proteolipid pigments of age-related spots. CML is also a marker of the Maillard reaction in foods. For example, cereal (26 mg/kg food) and fruit and vegetable (1.3 mg/kg food) categories have the highest and lowest level of CML.

The CML homologue N^ε-(carboxyethyl)lysine (CEL, **4-114**) is formed in analogous intracellular reactions with methylglyoxal, and is an important marker for age-dependent diseases such as cardiovascular disease in diabetic patients. The reaction of 3-deoxyhexosuloses with the ε-amino group of bound lysine yields pyrraline, bound 6-(2-formyl-5-hydroxymethylpyrrol-1-yl)-L-norleucine (**4-114**), identified in hydrolysates of milk proteins and in bread and found in relatively high levels in the serum of patients with diabetes, arteriosclerosis, or Alzheimer's disease (Figure 4.91).

The reaction of two lysyl residues with two molecules of glucose creates the so-called crossline (cross-linked AGE) (**4-114**), which is found in the serum and kidneys of diabetic patients. An analogous structure, called fluorolink (**4-114**), arises by dehydration and oxidation of crossline. Another cross-linked AGE is imidazolium salt of glyoxal-lysine dimer, trivially called GOLD (**4-114**), which is 2-ammonio-6-[1-(5-ammonio-6-oxido-6-oxohexyl) imidazolium-3-yl]hexanoate. The compound known as MOLD, (2-ammonio-6-[1-(5-ammonio-6-oxido-6-oxohexyl)-4-methylimidazolium-3-yl] hexanoate (**4-114**), results from the analogous reaction with methylglyoxal. These examples, however, do not cover all known AGEs that are involved in cross-linking of two lysyl residues. The non-cross-linked AGE referred to as argpyrimidine, N^δ-(5-hydroxy-4,6-dimethylpyrimidin-2-yl)-L-ornithine (**4-114**), is derived from the reaction of methylglyoxal with an arginine protein residue. Reactions of methylglyoxal (MG) or 3-deoxyglycosuloses (3-DG) with an arginine residue yield the corresponding imidazolones, MG- and 3-DG-imidazolone (**4-114**). Analogous imidazolones are formed in the reactions of methylglyoxal and 3-deoxyglycosuloses with protein-bound lysine. 3,4-Dideoxypentosulose formed exclusively from oligosaccharides with a (1 → 4) glycosidic bond is a dicarbonyl precursor in the modification of arginine residues and the formation of (N^δ-[5-(3′-hydroxypropyl)-4-oxoimidazolon-2-yl]-L-ornithine, PIO (**4-114**), in foods containing reducing (1 → 4)-disaccharides, such as lactose, maltose, and lactulose.

The result of cross-linking between the side chains of lysine and arginine is usually derivatives of imidazole. Pentosidine (**4-114**), 2-ammonio-6-{2-(4-ammonio-5-oxido-5-oxopentyl)-amino-3*H*-imidazo[4,5-b]pyridin-4-ium-4-yl}hexanoate, is a fluorescent compound that is formed on the reaction of pentoses (and possibly of ascorbic acid) with lysyl and arginyl residues in proteins. It was detected in plasma β₂ microglobulin from patients with haemodialysis-related amyloidosis (a disease that results from the abnormal deposition of the protein amyloid). Pentosidine is commonly present in acid hydrolysates of the tissues rich in collagen. Other cross-linkages, which include side chains of lysine and arginine, are formed by the reactions of glyoxal. An example of these compounds is 2-ammonio-6-({2-[(4-ammonio-5-oxido-5-oxopentyl)amino]-4,5-dihydro-1*H*-imidazol-5-ylidene}amino)hexanoate (GODIC, **4-114**).

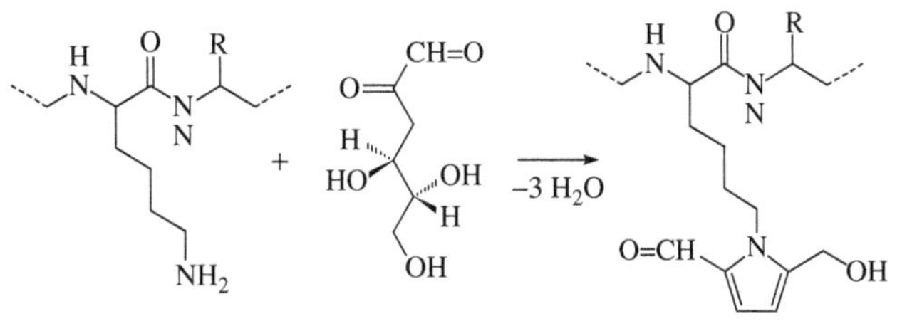

lysine bound in a protein   3-deoxy-D-*erythro*-hexos-2-ulose   modified lysine residue

Figure 4.91 Reaction of 3-deoxyhexos-2-uloses with bound lysine to form pyrralines.

By analogy, the reaction of methylglyoxal yields 2-ammonio-6-({2-[(4-ammonio-5-oxido-5-oxopentyl)amino]-4-methyl-4,5-dihydro-1*H*-imidazol-5-ylidene}amino)hexanoate, known as MODIC (**4-114**), whilst 3-deoxyhexosuloses give rise to 2-ammonio-6-({2-[(4-ammonio-5-oxido-5-oxopentyl)amino]-4-(2,3,4-trihydroxybutyl)-4,5-dihydro-1*H*-imidazol-5-ylidene}amino)hexanoate, or DOGDIG (**4-114**). The product of hexoses is 2-ammonio-6-({2-[(4-ammonio-5-oxido-5-oxopentyl)amino]-6,7-dihydroxy-4,5,6,7,8,8a-hexahydroimidazo[4,5-b]azepin-4-yl}hexanoate, known as glucosepane (**4-114**), the major protein cross-link of the senescent human serum albumin and lens protein.

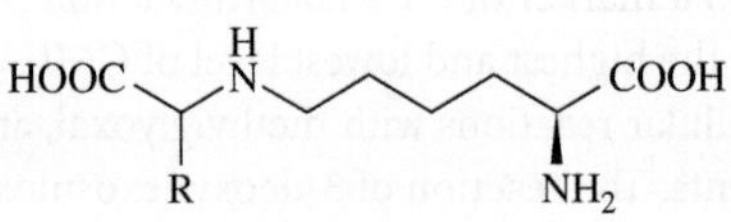

CML, R = H
CEL, R = CH$_3$

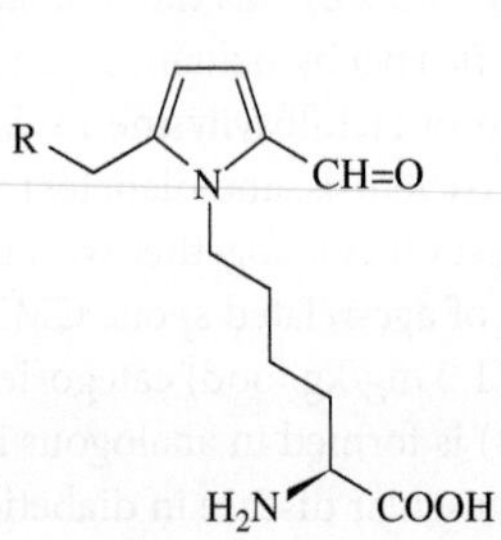

pyrraline, R = CH$_2$OH
formyline, R = H

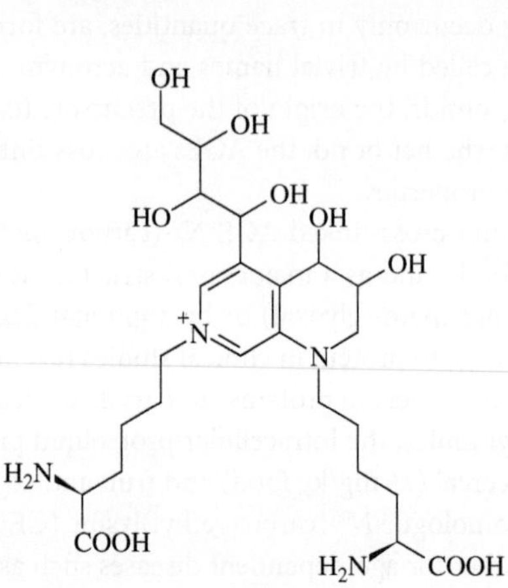

crossline

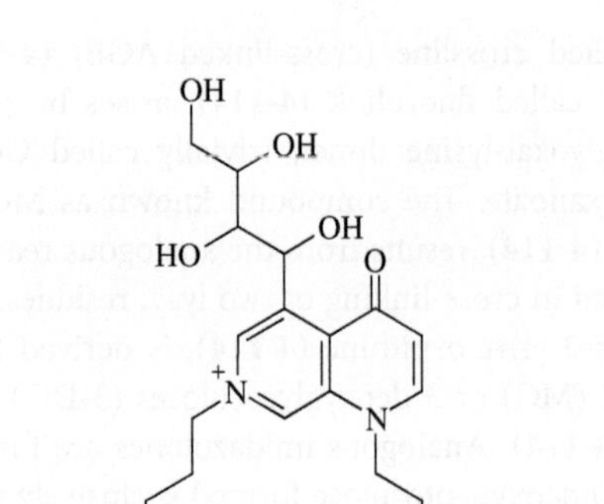

fluorolink

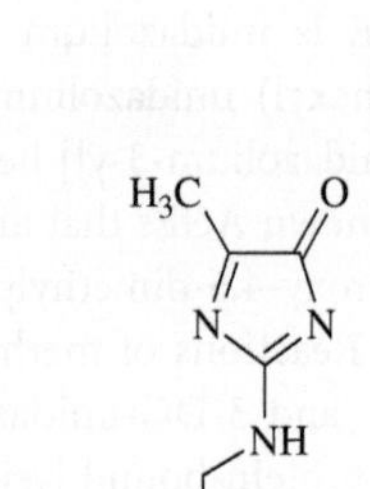

MG-imidazolone

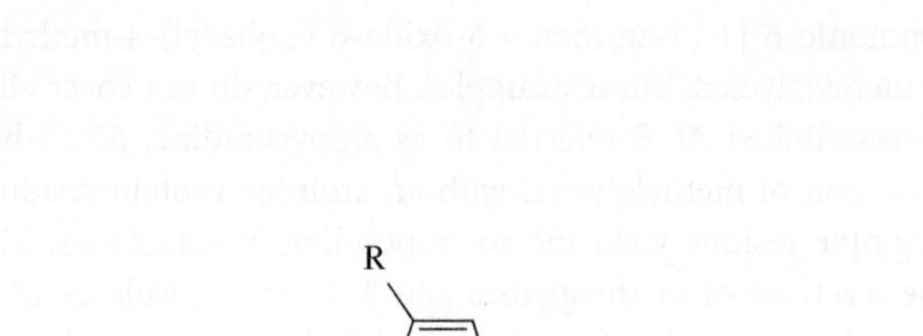

GOLD, R = H,
MOLD, R = CH$_3$

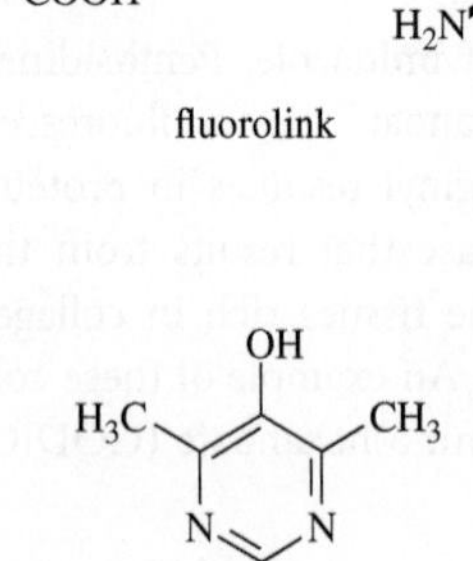

argpyrimidine

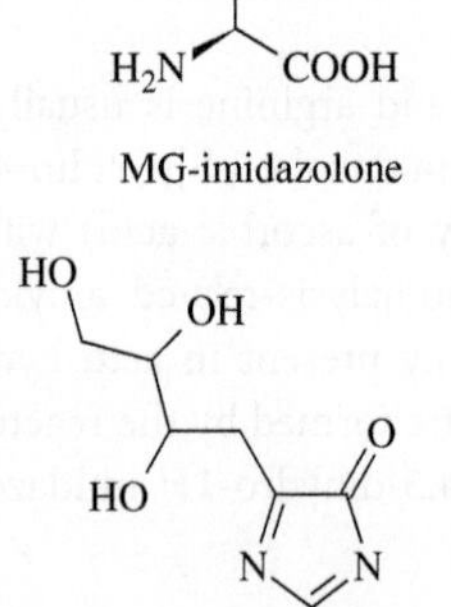

3DG-imidazolone

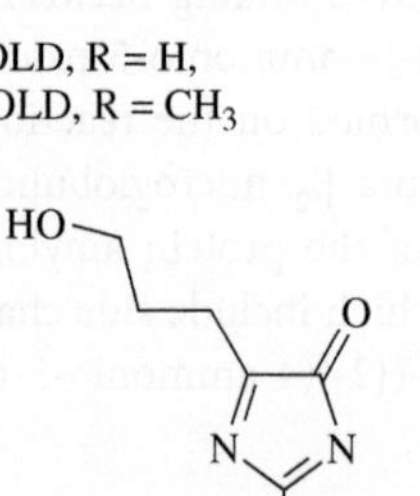

PIO

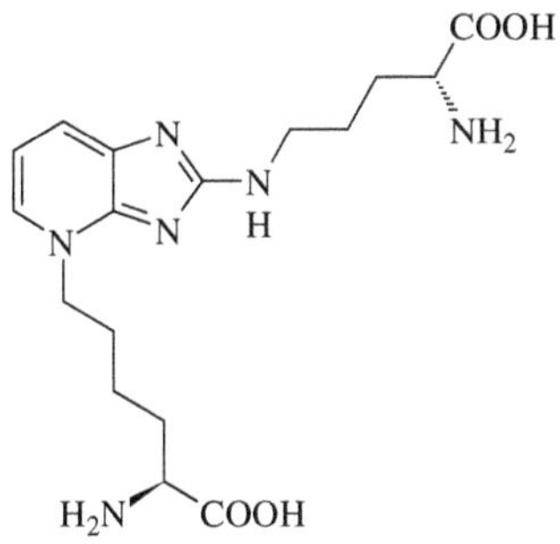

pentosidine

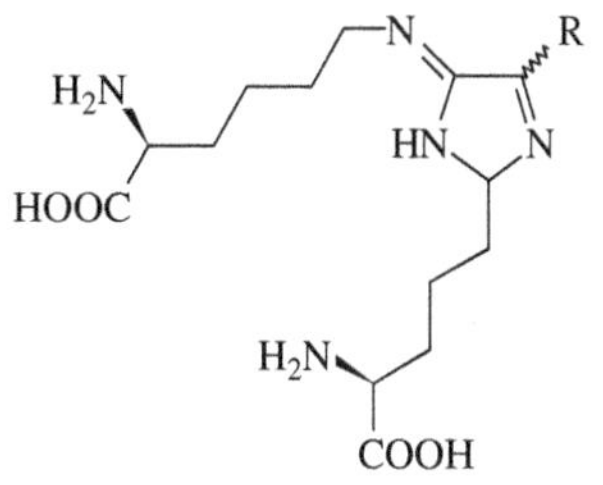

GODIC, R = H
MODIC, R = CH₃
DOGDIC, R = CH₂CH(OH)CH(OH)CH₂OH

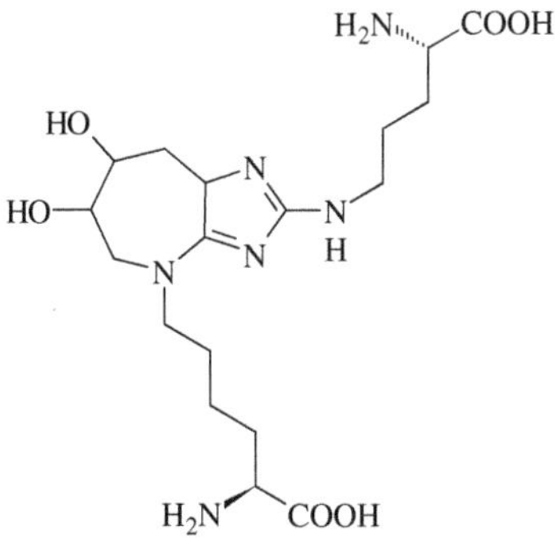

glucosepane

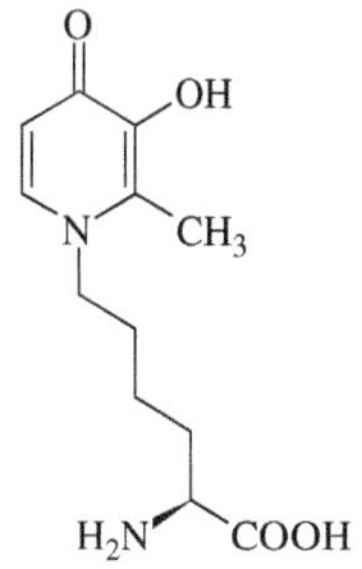

maltosine

4-114, structures of selected advanced glycation end products

The levels of these adducts in hydrolysed foods (biscuits, cereal bars, boiled egg white, milk products) are up to tens (exceptionally, hundreds) of mg/kg. In milk and heat-treated dairy products (including dried and UHT milk), the levels of CML range from 10 (raw milk) to 563 mg/kg. In roasted peanuts, the ε-Fru-Lys (Amadori product) can block up to 1.2% lysine, CML up to 0.2%, and pyrraline (**4-114**) up to 4% of protein-bound lysine. Considering that totally observed lysine modification was around 50%, most lysine modifications in roasted peanuts remain unknown. Low amounts of 6-(2-formyl-1-pyrrolyl)-L-norleucine (formyline, **4-114**), which results from the reaction of protein-bound lysine residues with 3-deoxypentosuloses, were detected in milk and whey products. In cereal products, formyline occurs in levels up to 34.8 mg/kg. In the outermost breadcrust, 0.6% of lysine residues can be modified to formyline. 6-(3-Hydroxy-4-oxo-2-methyl-4(1H)-pyridin-1-yl)-L-norleucine (maltosine, **4-114**) was found mainly in bread samples (0.1–4.2 mg/kg). The highest amounts (19 mg/kg) of this metal-chelating compound were found in the crust of wheat bread, representing a maximal lysine modification of 0.4%. Pyrraline, formyline, and maltosine were also found as consituents of beer proteins. After hydrolysis, these amino acid modifications are partly metabolised by yeasts (*S. cerevisiae*) via the Ehrlich pathway to the alcohols pyrralinol (up to 207 µg/l in beer), formylinol (up to 50 µg/l), and maltosinol (up to 6.9 µg/l), respectively.

Reactions between proteins and carbonyl compounds originating as the secondary products of lipid oxidation may also play an important role in the modification of protein-bound amino acids in certain foods. For example, *trans*-2-heptenal from heated oils reacts with bound lysine residues to provide *cis*- and *trans*-1-(5-amino-5-carboxypentyl)-4-butyl-3-(pent-1-en-1-yl)pyridin-1-ium, which can be good markers for a protein modification by the lipid peroxidation product.

The risk of physiological AGEs is also reduced by a number of substances occurring in foods. Antioxidants (flavanones and other phenolic compounds, vitamin E, ascorbic acid, and carotenoids) may inhibit oxidative stress in tissues and thus interfere with the formation of AGEs. Many of the antioxidants and many other compounds may also act as α-dicarbonyl scavengers, lowering the level of the ultimate glycation agents, protecting proteins from glycation, and thus delaying the formation of AGEs both in foods and *in vivo*. The scavenging or antiglycation effect of many particular compounds, as well as plant extracts, preparations, and waste materials, has been evaluated.

The trapping abilities can therefore be ascribed to flavanols and many other classes of flavonoids and other phenylpropanoids, pyridoxamine, creatine, low-molecular-mass thiols such as cysteine and glutathione, hydroxytyrosol, and so forth. For example, creatine reacts with methylglyoxal both under physiological conditions and in heat-processed meat to form methylglyoxal-derived hydroimidazolone of creatine. The scavengers are also naturally able to catch carbonyl compounds arising from lipid oxidation. For example, acrolein–hesperetin and acrolein–resveratrol adducts were isolated; acrolein reacts with resveratrol at the C-2 and C-3 positions through nucleophilic addition and forms an additional heterocyclic ring.

Similar protective antiglycation effects are found with the elements Se, Zn, Cu, and Mn, which are part of the redox enzyme systems. In addition to agents such as aminoguanidine, the reactive α-dicarbonyl precursors may also react with thiamine and its derivatives. Foods high in carbohydrates such as cereals (bakery products and pasta) are a principal source of protein-bound glycation compounds in the diet, followed by milk products and coffee. The important factor affecting the formation of AGEs is high temperature in processes such as baking or roasting. The dietary intake of AGEs substantially exceeds the amount generated physiologically. However, less than 10% of AGEs is usually absorbed from the dietary intake. Modified proteins are more resistant to proteolysis, and some AGEs even inhibit intestinal proteases. AGEs that are absorbed are mainly glycated amino acids (adducts released by protein hydrolysis), which are rapidly excreted in urine in healthy people. Peptides that contain AGEs are eliminated from the body in the same way as xenobiotics. The relationships with the increase in the physiological levels and other negative influences have not yet been demonstrated for dietary AGEs and their interactions and reactions in the colon have not yet been adequately documented.

A considerable but so far little explored technological potential of glycation reactions involves changes in the functional properties of proteins (formation of emulsions, foams, and gels, changes in solubility and stability, improvement of taste). For example, in whey powders, the denaturation of β-lactoglobulin is affected significantly by the extent of modification of lysine residues by lactose. Its solubility increased from 40% (4.6% lysine modification) to 82% (22.4% lysine modification). An increase in glycation leads to the increase of denaturation temperature (from 79.5 to 84 °C) and a slower denaturation-induced oligomerisation. Covalent attachment of lactose to whey proteins during preparation or storage significantly improves the heat stability of whey proteins. The degree of lysine modification may affect texture and other functional properties, thereby modifying the technological use of whey proteins.

The disadvantage of these reactions is the certain but usually negligible reduction of the nutritive value of proteins. Unlike acetylation, succinylation, and other chemical methods used to improve the functional properties of proteins, the Maillard reaction occurs spontaneously during thermal processes without the addition of foreign substances. For example, the increased solubility and emulsifying properties of proteins depend on the introduction of a limited number of hydrophilic sugar structures that are in contact with the aqueous phase whilst the hydrophobic protein areas are oriented to the lipid phase. The increased stability of proteins may be partly related to the blocking of lysine and arginine bindings that are attacked by proteolytic enzymes.

Modification of functional properties by glycosylation with polysaccharides is another effective means of improving the limited use of native proteins. Protein glycosylation can improve functionalities of native proteins, including solubility, rheological properties, emulsifying properties, foaming properties, gel property, film-forming properties, thermal stability, antioxidant activity, allergenicity, and antibacterial properties. Based on the improved functional properties, the conjugates have the potential to be used as an ingredient in certain new applications in food processing, such as printing food, gelled food, and colloid food. Protein–polysaccharide conjugates are extensively used as encapsulation or delivering materials in order to improve the bioaccessibility of bioactive compounds in food systems.

4.7.5.8 Antioxidant properties of reaction products

One of the important attributes of some melanoidins is their antioxidant activity. Antioxidant activity is highly dependent on the nature of the reactants from which melanoidins are formed. Some melanoidins also exhibit prooxidant properties.

The antioxidant properties of some reaction products of reducing sugars with amino acids and other amino compounds (ammonia, amines) have been known for several decades, and have been used in industrial processes. Amino acids are less active as antioxidants, but sugars and their transformation products do show these properties. For example, the addition of glucose and certain amino acids (particularly glycine, valine, and lysine) to the dough in the production of biscuits increased the stability of the fat towards autoxidation. The same effect is seen for the addition of reaction products of glucose with histidine in sausages stored at refrigerated temperatures. Heating milk with a mixture of glucose and histidine increases its oxidative stability after drying but reduces the available lysine content and can bring about discolouration during storage. Products of the Maillard reaction exhibit synergistic effects in the presence of some antioxidants. For example, products arising from D-xylose and ammonia can increase the antioxidant effects of tocopherols.

The chemical principle behind the antioxidant properties of Maillard reaction products is currently not well understood. It is assumed that these properties show both low-molecular-weight products and high-molecular-weight melanoidins. As the structure of melanoidins has

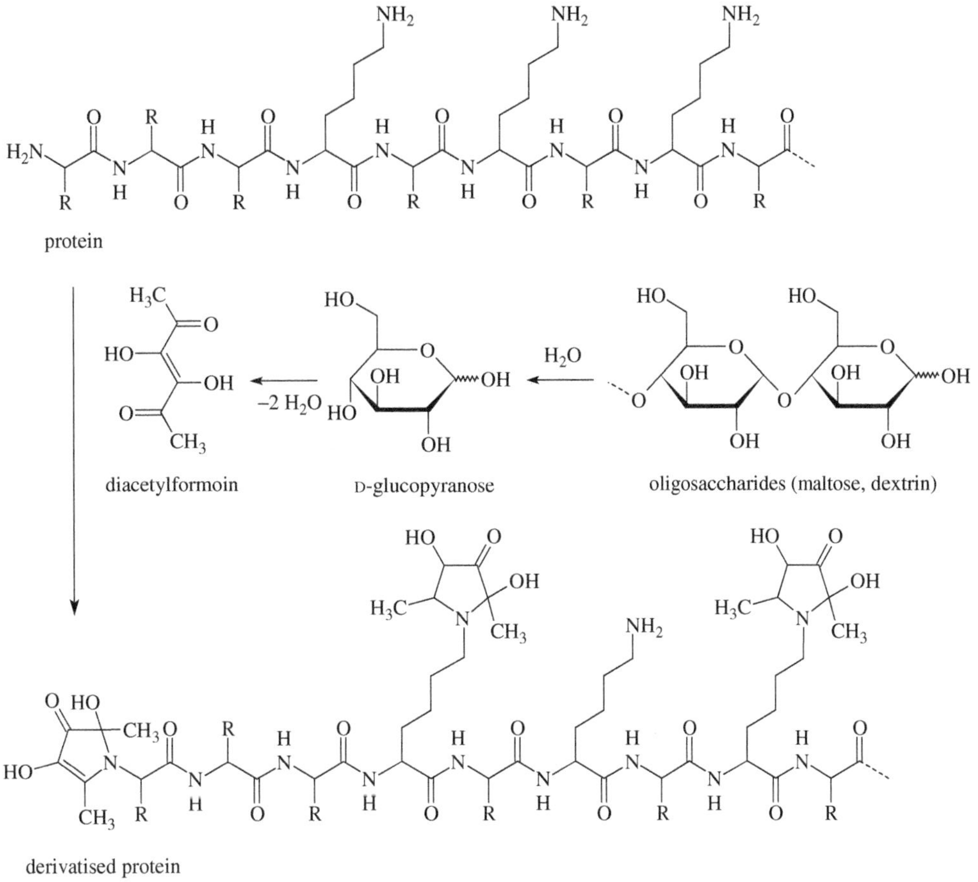

Figure 4.92 Formation of bound pronyl-L-lysine in bread crust.

not been clarified satisfactorily, it is difficult to explain the chemical nature of their antioxidant activity. The active structures are certainly reductones and aminoreductones bound in melanoidin molecules that reduce the products of autoxidation. One of a few identified reductone structures in real food melanoidins is 2,4-dihydroxy-2,5-dimethyl-1-(5-acetamino-5-methoxycarbonylpentyl)-3-oxo-2H-pyrrole bound by a peptide bond. This so-called pronyl-L-lysine (pyrrolinone reductonyl lysine) contributes significantly to the antioxidant properties of bread crust, where it is found in concentrations of about 60 mg/kg. Its amount depends on the type of bread dough, the pH during dough fermentation, and conditions during baking. Pronyl-lysine is formed in the reaction of diacetylformoin (dehydration product of D-glucose via 1-deoxy-D-*erythro*-hexo-2,4-diulose) with ε-amino groups of lysyl residues and also, by analogy, with the N-terminal lysyl residues of the peptide chain (Figure 4.92).

A few low-molecular Maillard-derived antioxidants have been identified to date. 2,3-Dihydro-3,5-dihydroxy-6-methyl-4H-pyran-4-one (Figure 4.33), one of the most common products of the Maillard reaction, is responsible for a significant part of the antioxidation capacity of the soluble portion of many cereal products (e.g. bread, coffee surrogates, and dark malts), dried fruits, and other foods. Other Maillard products possessing a free methylene group with acidic hydrogen are believed to have some reducing potential, but their levels in food are usually low. In aged garlic extract, derivatives of 1,2,3,4-tetrahydro-β-carboline-3-carboxylic acid (see Section 10.3.6.6.2) were found as significant antoxidants.

In addition to reductones, structures capable of binding metal ions that act as oxidation catalysts also have significant antioxidant activities. Although amino acids can bind metals to coordination compounds (see Section 6.2.2.1), metal chelates (M^{2+}) with Amadori compounds (**4-115**) have considerably higher stability and therefore higher antioxidant activity. The increased stability is due to additional metal ion binding by C-2 and C-3 hydroxyl groups. Significant amounts of N^{α}-(1-deoxy-D-fructos-1-yl)-L-histidine were found in dried tomatoes and some other dried products (**4-115**).

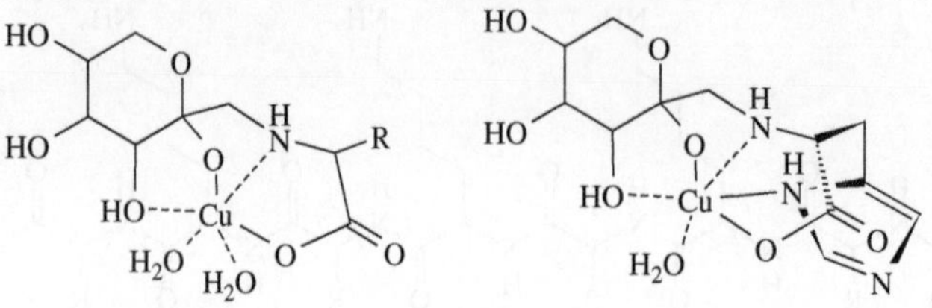

1-amino-1-deoxyketose N^{α}-(1-deoxy-D-fructos-1-yl)-L-histidine

4-115, typical complexes of Amadori compounds

The highest antioxidant activity is seen in reaction products formed from sugar degradation products (such as methylglyoxal and 1,3-dihydroxyacetone) and from dehydroascorbic acid. Less active as antioxidants are the reaction products of pentoses, and still less are those of hexoses. The products of sugars with a basic amino acid (arginine, lysine, and histidine) show higher antioxidant activity than products of other amino acids. Products of arginine and histidine with xylose are effective as antioxidants. Glutamic acid reaction products with glucose or fructose do not exhibit antioxidant properties. Reaction conditions play an important part. The antioxidant activity of the reaction products generally increases with growing pH and with increasing weight ratio of the amino compound to sugar.

4.7.5.9 Nutritional and toxicological aspects

The consequences of the Maillard reaction are desirable changes to the organoleptic properties of foods, such as the formation of the typical aroma, taste, and colour. However, some adverse changes can also appear, the most serious of which is the development of unusual taste and odour, undesirable colour (e.g. in dried foods), and reduced nutritional value. The decrease in nutritional value is caused by the actual loss of essential amino acids by irreversible reactions (e.g. in Strecker degradation), by binding of amino acids to non-utilisable complexes and covalent compounds, and by decreased protein digestibility due to resistant cross-links. The non-enzymatic browning reactions mainly affect the ε-amino group of lysine- and sulfur-containing amino acids.

The initial phase of the Maillard reaction, in which the Schiff base stabilisation by cyclisation occurs, does not negatively affect the nutritional value of foods, because these reaction intermediates are easily hydrolysed to the original components. The next stage, which is the rearrangement of N-glycosylamine to aminodeoxysugar, results in a reduction of nutritional value. For example, aldoses yield the Amadori compounds (1-amino-1-deoxyketoses), which are the main bound and non-utilisable form of lysine in foods. The reduction of nutritional value during the Maillard reaction occurs primarily in heat-stressed foodstuffs with low water content during drying, baking, frying, and roasting. Bread baking, for example, results in a 10–15% loss of total lysine, and up to a 70% loss of lysine in the crust. Higher losses occur during the drying of milk, where they can reach up to 30%, depending on the drying technology.

Melanoidins in foods high in saccharides, such as bread crust and coffee, can have particular biological functions of fibre, including the growth of desirable intestinal microflora (prebiotic effect). Melanoidins derived mainly from proteins do not show these properties.

As products of the Maillard reaction, numerous toxic substances can be formed that exhibit clastogenic, mutagenic, and carcinogenic effects. In this respect, various pyridoimidazoles, pyridoindoles, and tetraazafluoranthenes occupy significant positions (see Section 12.2.1). Some products of the Maillard reaction are secondary amines. Reaction with nitrous acid or with nitrogen oxides may transform these compounds into mutagenic and carcinogenic N-nitroso compounds. In addition, some other Maillard reaction products also show mutagenic activity, such as glyoxal, methylglyoxal, and 5-hydroxymethylfuran-2-carbaldehyde.

4.7.5.10 Factors influencing reactions

The practical application of the Maillard reaction during food processing requires it to be under control in order to suppress side reactions and highlight the desirable or beneficial reactions. The main factors that can be used for the control of the Maillard reaction during food processing are:

- temperature;

- reaction time;

- pH;

- water activity;

- type of reactants;

- availability of reactants.

Owing to the complexity of the Maillard reaction, its optimisation is not an easy task. Optimisation is also impeded by the fact that the individual factors, such as temperature, water activity, and pH, do not act in isolation, but can affect one another. The study of the influence of particular factors on the reaction is considerably difficult, and the interdependence of the various factors is often the cause of conflicting and contradictory results obtained in different situations.

The effect of temperature on chemical reactions is expressed as the activation energy of reaction (higher activation energy means that the reaction rate depends more on temperature). The values of activation energy of the Maillard reactions vary over a wide range from 10 to 160 kJ/mol. The activation energy is also highly dependent on water activity. For example, at low water activities, the activation energy of Amadori compound formation increases rapidly, whilst at medium and higher values, the activation energy is virtually independent of water activity. The activation energy (especially the activation energy of brown pigment formation) is also dependent on pH. For example, that of the formation of pigments in a lysine–glucose system increases with decreasing pH. Temperature affects not only the rate of the whole Maillard reaction, but also the particular reactions, in different ways. A typical example is the formation of flavour-active compounds, as the same reaction mixtures heated at different temperatures can produce very different aroma profiles. Higher temperatures generally lead to extensive fragmentation of reactants and a more diverse mixture of reaction products.

Water activity also significantly affects the formation of the brown colour. At high values, the reaction rate is very low due to the dilution of reactants. With decreasing activity, the reaction rate increases, especially at medium values (0.3–0.7), before reaching a maximum; further reduction in water activity leads to a decrease of reaction rates (see Section 7.9.2), and very low values can stop the reaction due to lack of mobility of the reactants. The value of water activity at which the Maillard reaction rate reaches a maximum can be influenced by the addition of glycerol or ethylene glycol. Whilst the addition of these agents in systems with higher water activity reduces the reaction rate (dilution of reactants), their addition to systems with low water activity increases the rate (higher mobility of reactants).

The Maillard reaction is also dependent on pH. Moreover, the reaction is usually accompanied by a decrease in pH as a result of formation of various carboxylic acids. The Maillard reaction rate generally increases with an increase in pH value. Brown colour formation in reactions of sugars with various amino acids increases with pH, reaching a maximum in the pH range 9–10. Glucose reacting with lysine produces about 500 times higher amounts of pyrazines in pH 9 in comparison with pH 5. As with temperature, pH affects not only the Maillard reaction rate (and, therefore, the quantitative composition of the reaction mixture), but also its qualitative composition. A typical example is the decomposition of ketosamines, where pH is a very important factor in determining whether the decomposition proceeds by 1,2- or 2,3-enolisation. Whilst 1,2-enolisation is favoured in sufficiently acidic media that allow protonisation of the Amadori compound nitrogen, during 2,3-enolisation, the Amadori compound must be partially deprotonised. Therefore, 2,3-enolisation dominates in alkaline solutions and under non-aqueous conditions.

The effect of reaction partners on the rate of the early Maillard reaction has already been discussed. The rate of the addition reaction is mainly related to the pK_a value of amino compound, which determines the concentration of reactive species at a certain pH of the reaction mixture. The most reactive free amino acid in a wide range of pH values is lysine; the least reactive are the aliphatic hydrophobic amino acids valine, leucine, and isoleucine. Under common conditions (pH 3–8), oligopeptides (mainly dipeptides) are more reactive than the corresponding amino acids, because their pK_a values are substantially lower. The reactivity of peptides, including the ability to chain cleavage of peptides, is highly dependent on their primary structure. For example, dipeptides with C-terminal glutamic acid are more reactive. It is reported that glutathione is easily degraded into glutamic acid and cysteinylglycine by heat. A high proportion of aliphatic α-amino acids increases the susceptibility of the peptide chain to hydrolysis. Peptides can directly contribute to some aroma and taste (especially meat-like and peptide-specific ones) and easily form specific products, such as 2,5-dioxopiperazines.

Another parameter that is compared in terms of reactivity is the ability of the reactants to form flavoured or coloured products. For example, lysine, glycine, and tryptophan produce a very intense colour with sugars, but the acidic, sulfur, and hydroxyamino acids contribute little to the browning reaction. Furan-2-carbaldehyde and glycolaldehyde belong to the most effective colour precursors in the presence of amino acids, and common carbonyl intermediates such as pyrrole-2-carbaldehyde, methylglyoxal, and glyoxal similarly have considerable potential. Amadori products provide fewer colours than the starting hexoses and pentoses. 5-Hydroxymethylfuran-2-carbaldehyde and acetoin are insignificant pigment precursors.

Due to the increased availability of free amino acids and sugars, some attention is paid to the Maillard reaction during the germination of plant seeds. Several intermediates of the Maillard reaction are apparently applied as stimulants of germination.

4.7.5.11 Inhibition of the Maillard reaction

Given that the Maillard reaction is not always desirable during the processing and storage of foods, a range of methods for its inhibition have been studied. These consist mainly of ways of creating conditions unfavourable to its progress. Given the factors that influence the course of the Maillard reaction, it is clear that a reduction of its rate or its complete inhibition can be achieved in many ways, but not all of them are applicable in practice. The choice of an inhibition method therefore greatly depends on the specific food and its processing technology. The most common methods of inhibition are:

- elimination of one of the reaction partners;

- adjustment of water content;

- decrease of temperature;

- reduction of heating period;

- adjustment of pH;

- addition of substances inhibiting or slowing down the reaction.

Removal of glucose has been used successfully in the production of powdered (dried) whole eggs and egg whites. Glucose is oxidised to inactive gluconic acid by the addition of a yeast preparation with glucose oxidase activity or by the addition of glucose oxidase enzyme. Simultaneous removal of oxygen slows down the autoxidation reaction. Fermentation of reducing sugars into non-reducing ones by the addition of starter cultures to cheese has been shown to prevent browning. Blocking or modification of amine or the thiol group by succinylation and through the use of oxidised phenolic compounds (quinones) is questionable due to the unavoidable modification of lysine and the risk of generating background bitterness, respectively.

Limiting the extent of the Maillard reaction in the production of dried foods is often achieved by reducing the drying temperature and time to a minimum. This can be done by frequently turning the food, or by drying it in the thinnest layer possible. It is important to lower the temperature at which the Maillard reaction proceedes quickly, especially when the food has a critical water content (water activity). For example, during spray drying of milk, it is necessary to avoid heating of the dried powder above 60 °C. Reducing the heating time is employed, for example, in the production of jams. The use of small production batches can shorten the time of heating by up to one-third, which has a significant positive effect on the sensory quality of the product. Whilst strawberry jams produced in large batches tend to have a reddish-brown colour and a caramel-like flavour, those prepared in smaller batches in the same way have a deep-red colour and a typical strawberry flavour. Reducing the heating time also indirectly affects the course of the Maillard reaction by lowering the amount of glucose produced by the inversion of saccharose.

Inhibition of the Maillard reaction in certain foods can also be achieved by the addition of sulfur dioxide or hydrogen sulfites (bisulfites). It is assumed that the inhibition occurs due to the addition of hydrogen sulfite to the carbonyl group of sugar, which is then blocked and cannot react with amino compounds. In a weakly acidic medium and at room temperature, glucose yields the corresponding 2-hydroxysulfonic acid, the systematic name of which is D-*glycero*-D-ido-1,2,3,4,5,6-hexahydroxyhexanesulfonic acid (**4-116**). In a neutral solution, 4-sulfohexos-2-ulose (3,4-dideoxy-4-sulfo-D-*glycero*-hexos-2-ulose) is produced (**4-116**). 4-Sulfohexos-2-ulose can be further transformed into isomeric acids by oxidation or benzilic acid rearrangement. Inhibition of the reactions using sulfur dioxide or hydrogen sulfites in the presence of ascorbic acid or pentose yields, analogously, 3,4-dideoxy-4-sulfopentos-2-uloses.

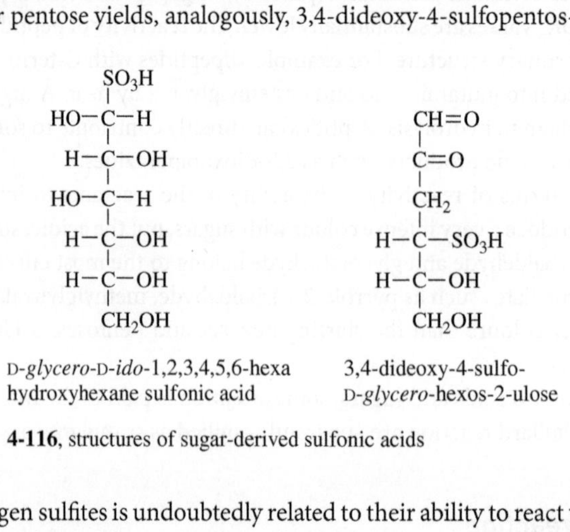

D-*glycero*-D-*ido*-1,2,3,4,5,6-hexa
hydroxyhexane sulfonic acid

3,4-dideoxy-4-sulfo-
D-*glycero*-hexos-2-ulose

4-116, structures of sugar-derived sulfonic acids

The effectiveness of inhibition by hydrogen sulfites is undoubtedly related to their ability to react with a wide range of products produced at all stages of the Maillard reaction. Enzymatic browning reactions are also inhibited. Carbon dioxide has a preservative function; alternatively, it may act as an antioxidant. Inhibition of the Maillard reaction can also be achieved by the addition of sulfur-containing amino acids such as cysteine; however, due to the possibility of degradation of sulfur amino acids and of the formation of unpleasant sulfur-containing flavour-active degradation products, this kind of inhibition is possible only exceptionally.

4.7.5.12 Significance for food technologies

Most food technologies have a long tradition, and control of the Maillard reaction is based on extensive empirical experience rather than scientific knowledge gained by systematic research. When new technologies have to be introduced, food processing technologists are faced with the task of optimising them in order to achieve the required product quality.

When a technology stands to benefit from improvements that enhance and complement its performance-in-use, the required product quality has to be at least comparable to that obtained by the traditional technology (e.g. microwave cooking versus traditional cooking). Apart from the lack of empirical experience, the optimisation of the Maillard reaction in new technological processes is more difficult, since technologists must often establish new parameters. For this reason, they are forced to use their scientific knowledge to achieve control of the reaction. Examples of relatively recent technologies in which the development of the Maillard reaction is not yet fully optimised are:

- extrusion;

- microwave cooking;

- infrared heating.

Other techniques such as ohmic heating, pulsed electrical field, high-pressure processing, and encapsulation of metal ions are promising tools for tailoring the extent the Maillard reaction in foods.

4.7.5.12.1 Roasting

One of the traditional processes in which the Maillard reaction is very intense is roasting (e.g. processing of cocoa and coffee, nut roasting). Roasting affects not only colour, but mainly the odour and taste of the processed raw materials.

During the roasting of cocoa beans, hundreds of volatile compounds arise, of which more than 400 have already been identified. The mechanisms of formation of many of these substances have been satisfactorily explained. It has also been found that the formation of volatiles is affected not only by the conditions of roasting, but to a large extent also by fermentation prior to roasting, which leads to the release of amino acids and reducing monosaccharides. Not all volatile products generated during roasting of cocoa beans, however, are important components of the cocoa aroma.

Roasting of green coffee beans also creates hundreds of volatile substances. Even more compounds have been identified in roasted coffee than in roasted cocoa. Depending on the roasting conditions (particularly temperature and roasting time), roasted coffee with different sensory properties can be obtained. During the roasting process, reducing monosaccharides are decomposed in the first instance. In later stages, non-reducing oligosaccharides and polysaccharides that have been cleaved to reducing sugars are also involved. At lower roasting temperatures, the rate of release of reducing sugars from polysaccharides and oligosaccharides is higher than sugar transformation. Coffee roasted in this way has a lighter colour and contains up to 1% of glucose and fructose. At high temperatures, rapid decomposition of reducing sugars occurs and higher amounts of polymeric brown pigments are formed. Reducing sugars are also partially decomposed to acids under these conditions, increasing the acid content. Thus, roasted coffee that normally contains only traces of reducing sugars has a dark colour and a slightly sour taste. In recent years, the so-called fast-roasting process has been introduced at temperatures higher than 230 °C, where the saccharose content in the coffee beans sharply decreases but the loss of polysaccharides is substantially lower than during conventional roasting, as reflected in reduced acidity.

The most famous coffee substitute (surrogate) is undoubtedly chicory root, which contains a considerable amount of inulin. Inulin is accompanied by free reducing sugars (glucose and fructose). Roasting chicory root gradually reduces the amount of inulin, and reducing sugars including fructose produced from inulin are subsequently degraded to dark caramelisation products (see Table 4.10).

The roasting processes produce a number of taste-active substances, particularly bitter compounds. Unlike the odour-active substances created by the Maillard reaction, knowledge of the bitter compounds is rather limited. Many of the identified substances were found in model reaction mixtures. It is known that pentoses and α-amino acids (such as xylose and alanine) produce intensely bitter substances identified as (E)-7-(2-furylmethyl)-2-(2-furylmethylidene)-3-(hydroxymethyl)-1-oxo-2,3-dihydro-1H-indolizinium-6-olate (quinizolate, **4-117**) and (E)-7-(2-furylmethyl)-2-(2-furylmethylidene)-3,8-bis(hydroxymethyl)-1-oxo-2,3-dihydro-1H-indolizinium-6-olate (homo-quinizolate, **4-117**). Diastereomers of 2H,7H,8aH-pyrano[2,3-b]pyran-3-one (**4-117**) have a taste reminiscent of capsaicin. Bitter-tasting compounds in reaction mixtures of hexoses with proline are a spirodiolone derivative (**4-117**), bispyrrolidinohexose reductones, and cyclopentaazepin-8(1H)-ones (**4-117**). Other products, such as 2,5-dimethyl-4-(1-pyrrolidinyl)-2H-furan-3-one (**4-117**) and 5-methyl-2-(1-pyrrolidinyl)-2-cyclopenten-2-one (**4-117**), cause afterimage cooling and a refreshing taste in the mouth. These substances have similarly been found in roasted malts. Mixtures of hexoses with alanine yield an inner salt of 1-(1-carboxyethyl)-5-hydroxy-2-(hydroxymethyl)pyridinium (**4-117**), which increases the intensity of the sweet taste perception.

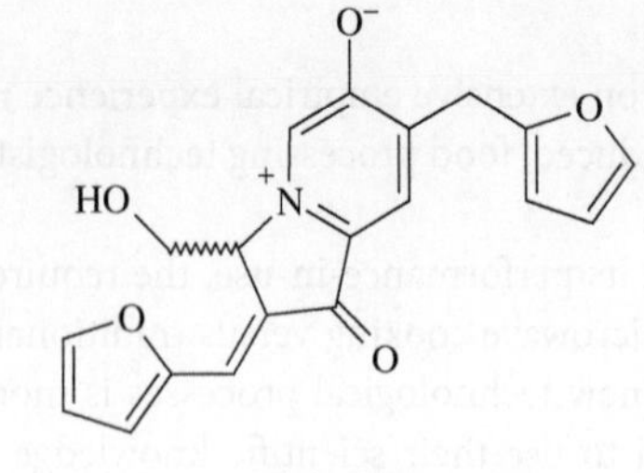

quinizolate

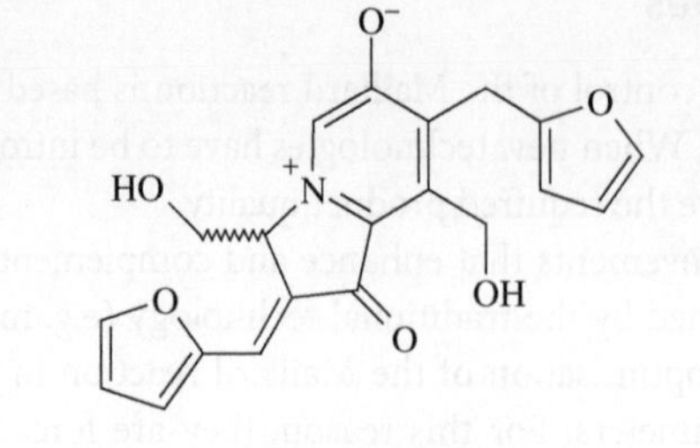

homoquinizolate

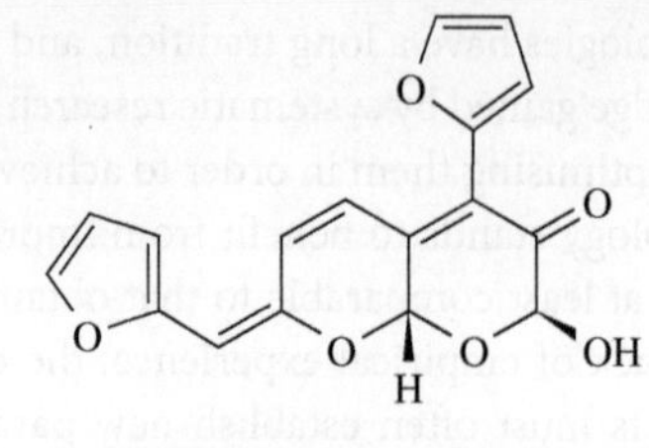

2*H*,7*H*,8a*H*-pyrano[2,3-*b*]-
pyran-3-one

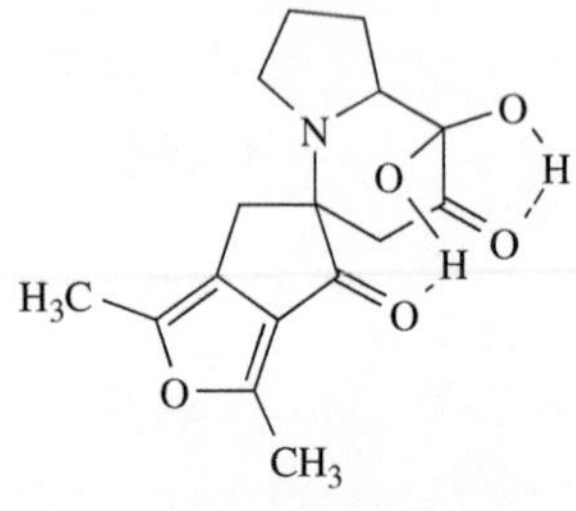

spirodiolone
derivative

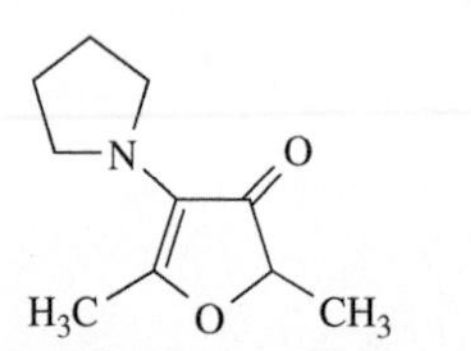

2,5-dimethyl-4-(1-pyrrolidinyl)-
2*H*-furan-3-one

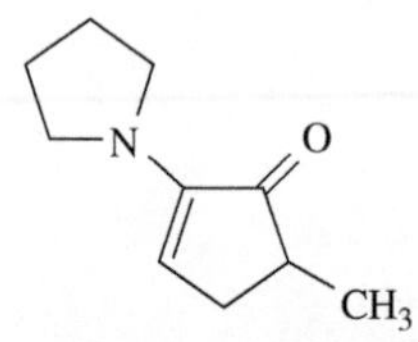

5-methyl-2-(1-pyrrolidinyl)-
2-cyclopenten-2-one

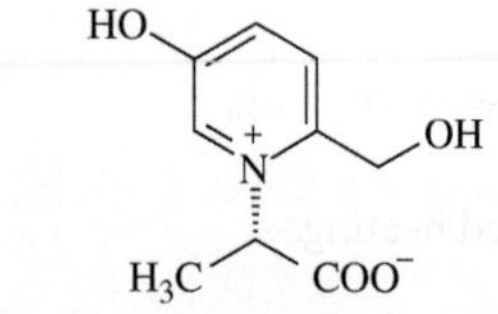

pyridinium
derivative

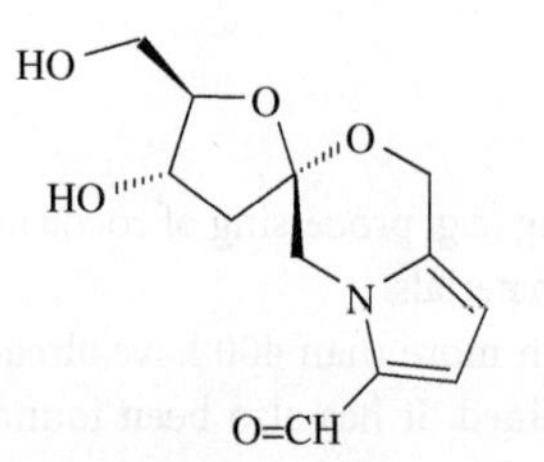

acortarin A

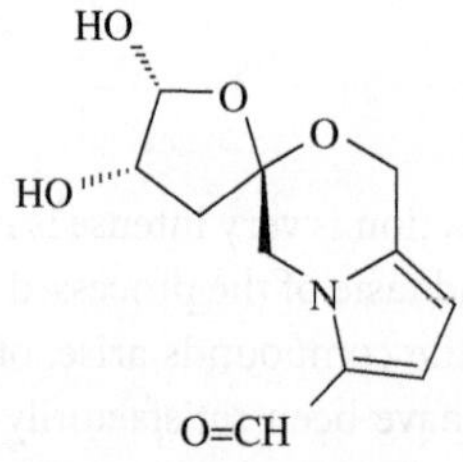

acortarin C

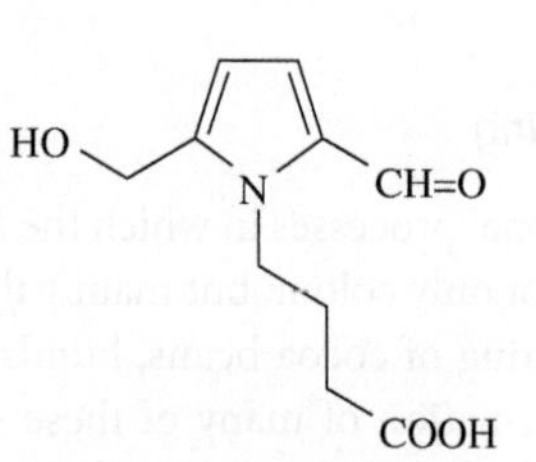

4-(2-formyl-5-(hydroxymethyl)-
1*H*-pyrrol-1-yl)butanoic acid

4-117, taste-active Maillard reaction substances

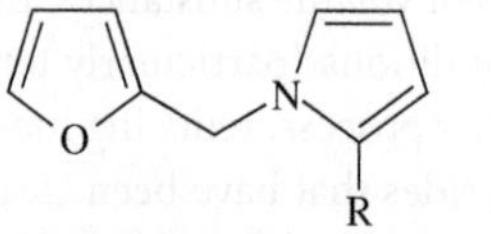

4-118, furfuryl pyrrole derivatives (R = H or CH=O)

4.7.5.12.2 Boiling, baking, and frying

Besides roasting, there are many other processes in which the Maillard reaction is desirable, including cooking, baking, and frying. The Maillard reaction can also have negative consequences, such as a certain reduction of the nutritional value of food or the production of mutagens, but the positive aspects (especially the typical colour, smell, and taste) significantly predominate (e.g. the formation of bread crust flavour and the flavour of cooked and roasted meat). For example, acortatarin A and acortatarin C (**4-117**; from the pyrrolomorpholine spiroketal family with furanose and pyranose isomeric forms, respectively, and with both epimeric configurations at the anomeric carbon), 5-hydroxymethylfuran-2-carbaldehyde, 2,3-dihydro-3,5-dihydroxy-6-methyl-4*H*-pyran-4-one, *N*-(1-deoxy-D-fructos-1-yl)-L-tryptophan (Amadori product), tryptophol, 2-(2-formyl-5-(hydroxymethyl-1*H*-pyrrol-1-yl)butanoic acid (**4-117**), and tryptophan are amongst the compounds with the highest bitterness intensities in wheat bread crust.

4.7.5.12.3 Drying

A typical example of a traditional technology in which the Maillard reaction manifests itself in a profoundly negative manner is the drying of milk, fruits, and vegetables.

In milk and dairy products, the intensity of the Maillard reaction is very high under certain conditions, due to the relatively high concentration of lactose and the presence of thermolabile proteins (especially whey proteins). During drying and heat treatment of milk, and also

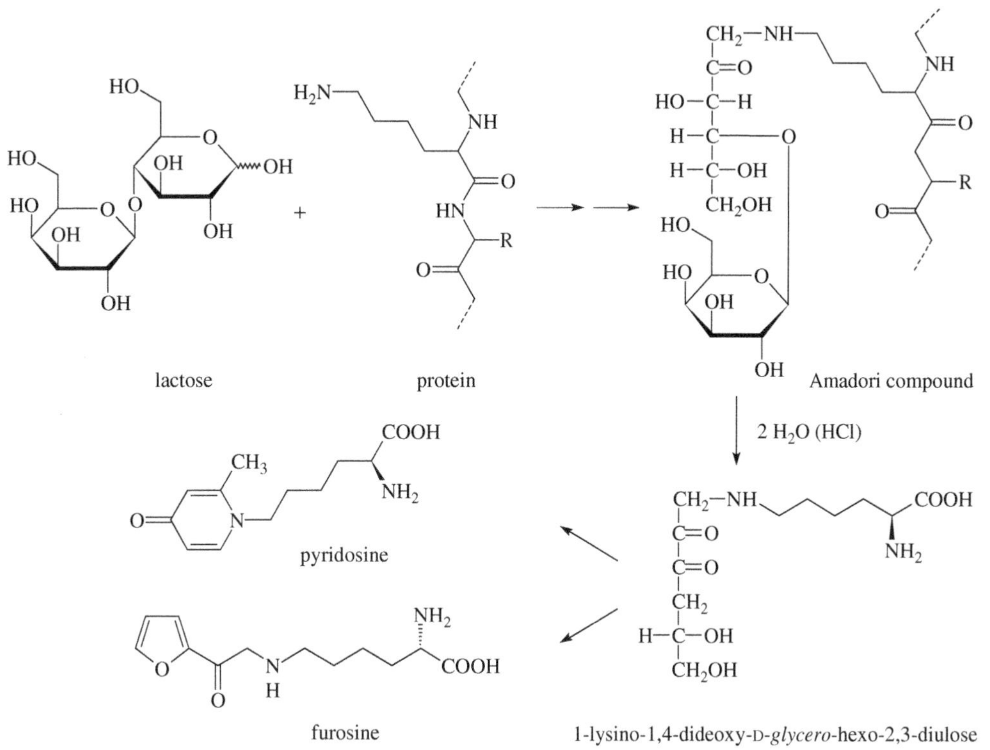

Figure 4.93 Formation of furosine and pyridosine from lactose and milk proteins.

during improper storage of milk powder, the non-enzymatic browning reactions may destroy lysine considerably. This amino acid becomes a limiting amino acid, even though it was present in sufficient amounts in the original raw material. Losses of lysine greatly depend on the technology of the milk processing, as well as on its appropriate practice. The total amount of lysine bound to lactose in raw milk is very small, but the bound lysine is non-utilisable. During the traditional drying process using steam-heated drying cylinders, lysine losses reached 10–30%, whilst spray drying losses reached only up to 3%. In freeze-dried milk, practically no loss of lysine occurs.

The Maillard reaction in dried milk affects the originally free ε-amino group of L-lysine bound in proteins that reacts with lactose, yielding the corresponding glycosylamine, which is fully utilisable as a source of lysine. Its rearrangement produces the non-utilisable protein bound Amadori compound (ε-N-deoxylactulosyllysine, also known as 1-lysino-1-deoxylactulose). Hydrolysis of these modified proteins initially gives the free Amadori compound, which is subsequently degraded to furosine, pyridosine, and lysine via the intermediate 1-lysine-1,4-dideoxy-D-*glycero*-hexo-2,3-diulose (Figure 4.93). Determination of pyridosine and furosine is used in practice for the tentative determination of the degree of damage in dried milk. The degree of damage to the milk powder (i.e. the amount of non-utilisable lysine) is very important information, especially for manufacturers of infant formulae.

The Maillard reaction is responsible for only about 20% of the losses of lactose in milk; the main cause (about 80%) is lactose isomerisation. Lactose isomerises even in the neutral medium of milk to produce lactulose (aldose–ketose isomerisation), which is accompanied by a small amount of epilactose (aldose–aldose isomerisation, known as epimerisation; Table 4.15). Lactulose in milk is a mixture of three isomers differing in the structure of bound fructose (β-furanose, α-furanose, β-pyranose in the ratio of about 75:10:15). It is commonly present in pasteurised, sterilised, or otherwise heat-treated milk and in milk products. UHT milk contains 50–750 mg/l of lactulose, whilst sterilised milk has a higher content. The lactulose content clearly correlates with the negative organoleptic properties (cooked flavour, darker colour); therefore, it is used as an indicator to differentiate pasteurised, sterilised, and UHT milk. The lactulose content in condensed milk is up to 1%, which corresponds to about 10% of isomerised lactose. Lactulose is partly hydrolysed to fructose and galactose, and these monosaccharides isomerise to produce small amounts of glucose and tagatose, respectively. Isomerisation of lactulose (2,3-enolisation) yields 4-deoxy-D-*glycero*-hexo-2,3-diulose as an intermediate product and 3-deoxy-D-*glycero*-pentos-2-ulose and formic acid as the main final products. Dehydration of 3-deoxy-D-*glycero*-pentos-2-ulose yields small amounts of furfuryl alcohol.

Amadori compounds have been identified in various processed fruit and vegetable products, including commercial carrot products (juices, baby foods, tinned and dehydrated carrots). Whilst these products showed fairly low rates of amino acid modification (up to 5%), dehydrated carrots contained Amadori products corresponding to a lysine derivatisation of up to 58% and a γ-aminobutyric acid derivatisation of nearly 100%. Besides the Amadori products formed during the early stage of the Maillard reaction, the AGEs $N^ε$-(carboxymethyl)lysine (CML) and 6-(2-formyl-5-hydroxymethylpyrrol-1-yl)-L-norleucine (pyrraline, **4-114**) also occurred in all carrot products (except raw carrot).

4.7.5.12.4 Extrusion

Extrusion technology is now widely used with a range of food applications, including the production of biscuits and breakfast cereals. It is a relatively complex process in which food is exposed to rather high temperature, high pressure, and shear forces for a short period of time. The quality of the final product depends on parameters such as the speed of extruder performance, the screw speed, and the configuration (shape). High temperature and low water activity of the processed material produce favourable conditions for the Maillard reaction. The negative aspect is the loss of lysine, but the positive aspect is the formation of a desirable flavour of the product. Under unfavourable conditions, losses of lysine during extrusion can reach 40–50%. These losses can be minimised in various ways, such as by lowering the temperature, increasing the water content, using non-reducing sugars (saccharose) instead of reducing sugars, and changing the geometry of the screw. For example, at a temperature of 210 °C, the increase of the moisture content from 13 to 18% results in a roughly seven times lower loss of lysine during the production of biscuits. To maintain lysine losses under 15% of the original lysine content, the recommended temperature is 180 °C and the water content in the material must be higher than 15%. The addition of reducing sugars is not recommended. Temperature and humidity are key parameters influencing the formation of flavour-active compounds and the processing of contaminants such as acrylamide during extrusion. The most important odour substances in the extruded materials are pyrazines. Their content increases as the temperature increases to 160 °C, but at higher temperatures the levels of pyrazines decrease.

4.7.5.12.5 Microwave cooking

In recent decades, microwave cooking has undergone considerable growth. The highest temperature is achieved not only on the surface of the food (as in classical heating), but also inside it. This can be a major disadvantage, because there is insufficient heating and drying of the surface, so the crust formation and colour and aroma development are insufficient. To reduce these shortcomings, the following principles have been proposed:

- modifications of recipes;

- changes of process parameters;

- alterations of packaging technology.

An example of recipe modification is the addition of artificial flavours or the coating of the food surface in pre-mixes containing reducing sugars and amino acids. The process can be modified by combination with another type of heating (such as grilling). Changes in packaging technology lie mainly in the use of packaging materials that can absorb the microwave radiation and heat the surface of the food with which they are in contact, such as a polyester film metallised with aluminum and glued onto paper. The temperature of the packaging material and the rate of heat transfer can be controlled by the thickness of the aluminum layer.

4.7.5.12.6 Infrared heating

Infrared heating has been successfully used in the cooking of meat and the production of biscuits and bread. The main advantage in comparison with conventional methods is the shorter cooking time and consequent energy savings. The sensory qualities of products obtained by conventional methods and by infrared heating are comparable; for example, bread forms a crust, albeit a slightly thinner one.

4.7.6 Caramelisation

Caramelisation of sugars is the process whereby brown to brown-black amorphous products of different compositions that are referred to as **caramel** are produced. A solution of caramel is called **couleur caramel** or **sugar couleur**. Caramel is produced by heating sugars or by adding substances that accelerate sugar caramelisation (catalysts). The temperatures applied are higher than about 120 °C: typically 150–190 °C, but not exceeding 240 °C.

The raw materials for the production of caramel usually include saccharose, glucose, fructose, glucose syrup, or starch. During the discontinuous (batch) mode of caramel production, these materials are heated at 120–180 °C for 5–10 hours in the presence of a catalyst. The choice of the catalyst depends on the intended use of the caramel. According to the technological process used (substances used in accelerating caramelisation), the following types of caramels can be distinguished:

- caramels with a positive electric charge (pH of isoelectric point is 4.0–7.0, preferably 6.0–7.0), produced by the addition of ammonia;

- caramels with a negative electric charge (pH of isoelectric point is <3.0, usually around 1.5), produced by the addition of ammonium sulfate or ammonium sulfite;

- caramels with no electric charge (produced in the presence of NaOH), known as spirit caramels.

Table 4.38 Classification of sugar couleurs and their use.

Class		Couleur name	Additives	Use
I	CP	Caustic (plain, spirit)	Na_2CO_3, K_2CO_3, NaOH, KOH, H_2SO_4, acetic acid, citric acid	Spirits with high alcohol content
II	CCS	Caustic sulfite	SO_2, H_2SO_4, Na_2SO_3, K_2SO_3, NaOH, KOH	Malt bread, vinegar, bear, spirits, flavoured wines, mead
III	AC	Ammonium	NH_3, $(NH_4)_2SO_4$, Na_2CO_3, H_2SO_4, NaOH, KOH	Beer and other alcoholic beverages, acidic foods
IV	SAC	Sulfite ammonium	NH_3, SO_2, $(NH_4)_2SO_3$, Na_2SO_3, K_2SO_3, Na_2CO_3, K_2CO_3, NaOH, KOH, H_2SO_4	Acidic foods, non-alcoholic beverages

Caramels are used to colour beer and other alcoholic beverages, soft drinks, vinegar, and confectionery, bakery, and meat products. Caramels with a positive electrical charge are particularly suitable for colouring beer, because the positively charged colloidal particles of tannin are not precipitated and do not form hazes in the presence of alcohol. Caramels with a negative electric charge show stability at low pH values and, therefore, are used in the manufacture of soft drinks. Spirit caramels are used for colouring alcoholic beverages such as rum, because they are soluble and stable in the presence of ethanol. Classification of sugar caramels and their uses is given in Table 4.38.

In addition to high-molecular-weight constituents, caramels contain a variety of substances of low molecular weight, which are formed in the Maillard reaction. These substances largely include unreacted sugars, acids (mainly pyruvic acid), sugar anhydrides, furan and pyran derivatives, and sugar fragments. Low-molecular-weight substances are the main precursors of reactions leading to the formation of brown polymeric melanoidins in sugar couleurs. The reactions taking place during the caramelisation of sugars are similar to those that occur during the Maillard reaction (such as dehydration, isomerisation, and retroaldolisation). Unlike the Maillard reaction, caramelisation, with one exception, does not produce nitrogenous compounds. The exception is caramelisation catalysed by ammonium salts or ammonia, in which nitrogen compounds are formed (e.g. pyrazines, imidazoles, and other heterocyclic compounds), but to a lesser extent than in the Maillard reaction.

By analogy with the Maillard reaction, some undesirable compounds can be formed in caramelisation. For example, 4(5)-methylimidazole, 2-methylimidazole, and 2-acetyl-4(5)-(*arabino*-1,2,3,4-tetrahydroxybutyl)imidazole (see Section 8.2.11.1.5) may occur in caramels of the class III or IV (Table 4.38). Caramels with a low level of imidazoles can also be obtained by other technologies, for example by rapid continuous production processes or by extrusion of pregelatinised starch or dextrin at a temperature of 150–220 °C. 4(5)-Methylimidazole and 2-methylimidazole are classified by the International Agency for Research on Cancer (IARC) as 2B group agents that are possibly

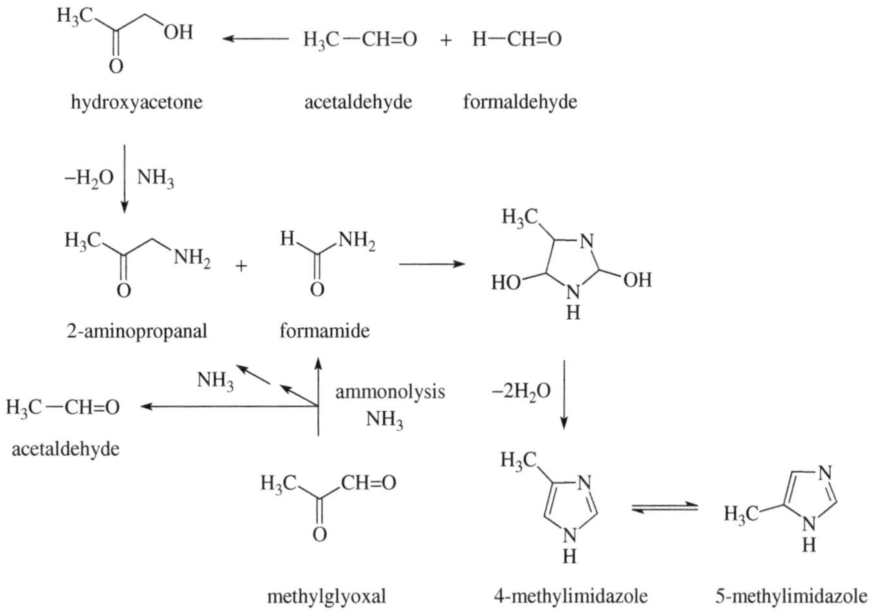

Figure 4.94 Formation of 4(5)-methylimidazole.

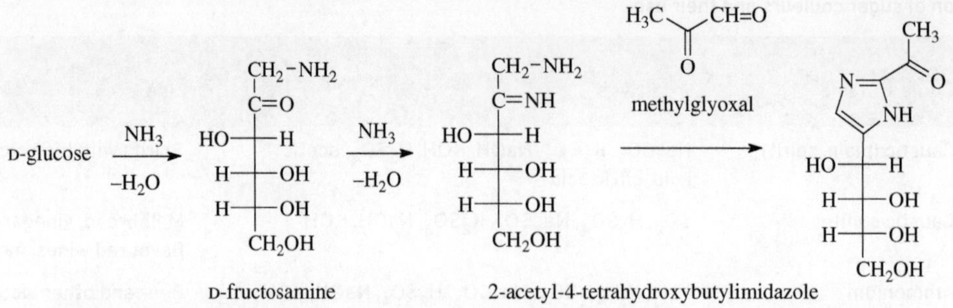

Figure 4.95 Formation of 2-acetyl-4(5)-(*arabino*-1,2,3,4-tetrahydroxybutyl)imidazole.

carcinogenic to humans. The Joint FAO/WHO Expert Committee on Food Additives (JECFA) has set the acceptable daily intake (ADI) of class II caramels to 160 mg/kg body weight per day and that of class III and class IV caramels to 200 mg/kg body weight per day. The European Commission has limited the 4(5)-methylimidazole content to 250 ppm (mg/kg). The ADI of class I caramels has not been specified. For example, the levels of 4(5)-methylimidazole found in commercial cola soft drinks range from 0.01 to 0.65 mg/l, whilst those of polyhydroxy-substituted imidazole are much lower (<0.007 mg/l). The 4(5)-methylimidazole contents in dark beers range from trace to 0.42 mg/l. A small quantity (0–8 μg/l) of 2-methylimidazole was found in dark beers only. The proposed formation mechanisms of 4(5)-methylimidazole from methylglyoxal and ammonia is given in Figure 4.94. Ammonolysis of methylglyoxal is proposed as the mechanism of the formation of acetaldehyde and formamide, which subsequently react with 2-aminopropanal arising as a product of hydroxyacetone and ammonia to yield 4(5)-methylimidazole.

Alkylimidazoles are also byproducts formed from natural constituents during the Maillard reaction in foods that are known not to contain caramel. The highest 4(5)-methylimidazole contents (up to 0.47 mg/kg) were observed in roasted barley, roasted malt, and cocoa powders, with the concomitant presence of 2-methylimidazole or 2-acetyl-4(5)-(*arabino*-1,2,3,4-tetrahydroxybutyl)imidazole (Figure 4.95) in some cases, albeit at significantly lower levels. Low amounts of 4(5)-methylimidazole (<0.06 mg/kg) were detected in cereal-based foods such as breakfast cereals and bread toasted to a brown color (medium toasted). The occurrence of alkylimidazoles are therefore not reliable markers of the addition of caramel to foods.

5

Vitamins

5.1 Introduction

Vitamins are organic substances that are essential for the human metabolism. They do not provide any building material, but mostly act as parts of enzymes catalysing biochemical reactions, although they have a number of other functions. Therefore, they are often referred to as **exogenous essential biocatalysts**.[1] Vitamins are synthesised almost exclusively by autotrophic organisms. Heterotrophic organisms synthesise vitamins only to a very limited extent (e.g. the liver can synthesise niacin from tryptophan), or they can be obtained as exogenous substances mainly in food and in some cases through the enteric (intestinal) microflora; insufficient amounts of vitamins in the diet may cause symptoms of deficiency.

Vitamins are substances with a variety of different chemical structures. In the past, before the structures of all the vitamins were disclosed, vitamin preparations represented mixtures of various substances; in that time, so-called biological units (such as mice or chicken units) were introduced for quantitative purposes. The biological unit of a vitamin was related to the amount of the vitamin that produced a physiological effect in the respective animal within a given time. Later, so-called international units (IUs) were derived, related to a particular vitamin weight. IUs are often used today for fat-soluble vitamins in pharmacy and medicine. In the past, vitamins were named for the diseases caused by their deficiency; for example, vitamin A was called antixerophthalmic vitamin or the vitamin against night blindness, whilst vitamin C was known as antiscorbutic vitamin or the vitamin against scurvy. Upper-case letters (vitamin A, vitamin C) were introduced later. Subsequently, it was discovered that several substances have the same physiological effects, and a numeric index on the upper-case letters began to be used (e.g. vitamins A_1 and A_2). Such designations are still commonly used, although some vitamins have a simple trivial name (e.g. retinol instead of vitamin A_1, ascorbic acid instead of vitamin C).

The most common method of classifying vitamins is according to their solubility in polar or non-polar solvents. Vitamins are thus divided into two groups:

- **fat-soluble vitamins**, **lipophilic vitamins** (four vitamins);

- **water-soluble vitamins**, **hydrophilic vitamins** (nine vitamins).

Fat-soluble vitamins are vitamin A, D, E, and K. Water-soluble vitamins include the B group vitamins or vitamin B complex and vitamin C. The B group vitamins include thiamine, riboflavin, niacin, pyridoxine derivatives, pantothenic acid, biotin, folacin, and the corrinoids.

Some substances, referred to as **provitamins**, do not show physiological effects themselves but can serve as precursors from which the body can synthesise vitamins. For example, a provitamin of vitamin A_1 (retinol) is β-carotene.

The fat-soluble vitamins have different functions. For example, vitamin A_1 (retinol) is required in the production of the visual pigment rhodopsin; its provitamin (β-carotene) likewise acts as a plant pigment and antioxidant. The function of hydrophilic vitamins is a catalytic effect, since in all organisms they generally occur as cofactors of various enzymes and play roles in the metabolism of nucleic acids,

---

[1] Another group of special biocatalysts is the hormones, which are synthesised within the body. Between vitamins and hormones, a sharp boundary exists; both groups were earlier called ergons. A hormone is a substance formed in one organ or part of the body and carried in the blood to another organ or part, depending on the specificity of their effects. Vitamin D_3 (cholecalciferol) is often considered to be one of the hormones.

The Chemistry of Food, Second Edition. Jan Velíšek, Richard Koplík and Karel Cejpek.
© 2020 John Wiley & Sons Ltd. Published 2020 by John Wiley & Sons Ltd.

proteins, carbohydrates, fats, and products of secondary metabolism. For example, vitamin B_1 (thiamine) is a cofactor of decarboxylases, dehydrogenases, and other enzymes.

The body's need for most vitamins is relatively low. The amounts needed to ensure the normal physiological function of humans are dependent on many factors, such as age, sex, health status, lifestyle, eating habits, and work-related activity. Many countries have recommendations for the daily intake of vitamins, which are continually revised in accordance with contemporary scientific knowledge and dietary guidelines. Presently, the Recommended Daily Allowances (RDAs), defined as the minimum daily amount required to avoid a deficiency, are set in the European Union (upper number) and United States (lower number) as follows: vitamin A (800/900 µg), vitamin D (5/15 µg), vitamin E (12/15 mg), vitamin K (75/120 µg), thiamine (1.1/1.2 mg), riboflavin (1.4/1.3 mg), niacin (16/16 mg), pantothenic acid (6/5 mg), vitamin B_6 (1.4/1.3 mg), biotin (50/30 µg), folacin (200/400 µg), vitamin B_{12} (2.5/2.4 µg), and vitamin C (80/90 mg).

Water-soluble vitamins are generally not stored in the body, or are stored only for a limited time, with the excess excreted in urine. Lipophilic vitamins are stored mainly in the liver. The **reserve capacity**, defined as the time during which the need for the vitamin is covered by the organism reserves, is longest for corrinoids (3–5 years) and vitamin A (1–2 years). The reserve capacity for folacin is 3–4 months, for vitamins C, D, E, and K, riboflavin, pyridoxine, and niacin is 2–6 weeks, and for thiamine, pantothenic acid, and biotin is only 4–10 days. Reserve capacity is affected by the history of vitamin intake, the metabolic need for the vitamin, and the health status of the individual.

The need for vitamins can also be affected by the presence of other food components that can interfere with vitamins in the diet. These substances are called **antivitamins** or **vitamin antagonists**. Antivitamins eliminate the biological effects of vitamins, which can lead to symptoms of deficiency. The activity of antivitamins is based on the following basic principles:

- Structural analogues of vitamins react with the apoenzymes (and act as competitive enzyme inhibitors) or with the proteins that transport vitamins (the antivitamin of thiamine is oxythiamine and the antivitamin of retinol is citral, also known as geranial).

- Certain enzymes convert some vitamins into inactive substances (e.g. lipoxygenase indirectly catalyses degradation of vitamin A and its provitamins, thiaminases decompose thiamine).

- Some substances (usually proteins, but also low-molecular-weight substances) form unusable complexes with vitamins (a typical example is the reaction of biotin with the egg protein avidin or the reaction of the amino acid linatin from flax seeds with pyridoxal).

Commonly used technological processes cannot usually remove antivitamins of the first group, the so-called true antivitamins. The remaining two groups can be largely eliminated by suitable processes or culinary practices (such as heat inactivation of enzymes or denaturation of proteins bound in the non-utilisable protein–vitamin complexes).

A disease resulting from a deficiency of one or more vitamins is **hypovitaminosis** (if a vitamin is supplied in insufficient quantity) or **avitaminosis** (a complete lack of a vitamin manifested by some biochemical processes disorder). The deficiency of vitamins was formerly one of the main causes of many diseases and deaths. Pellagra (deficiency of some B-complex vitamins), scurvy (vitamin C), beriberi (thiamine), rickets (vitamin D), pernicious anaemia associated with reduced ability to absorb vitamin B_{12} (corrinoids), and xerophthalmia (vitamin A) are now the best-known diseases caused by vitamin deficiency. Excessive intake of one or more vitamins (especially of lipophilic vitamins A and D) is called **hypervitaminosis**; it also represents an abnormal state (disease) resulting from disturbances in various biochemical processes.

In food, vitamins are found in different amounts, typically ranging from microgrammes per kilogramme to hundreds or thousands of milligrammes per kilogramme according to the particular vitamin, food type, and method of processing. Vitamins occur as free compounds and in various bound forms, usually bound to proteins or carbohydrates. Physiological activity generally encompasses more than one entity. For example, about 50 naturally occurring carotenoids have the activity of vitamin A, whilst two compounds, ascorbic acid and dehydroascorbic acid, have the activity of vitamin C.

The most important sources of vitamins are mainly the basic foods, such as meat and meat products, milk and dairy products, eggs (especially egg yolk), bread and other cereal products, fruits, and vegetables, which should adequately cover the body's vitamin requirements. Some foods have high or extremely high vitamin contents (e.g. vitamin C in rosehips), but these are eaten irregularly or rarely and therefore are not significant sources of vitamins for most of the population. Other vitamins are limited to a certain group of foods (e.g. vitamin B_{12} is only found in foods of animal origin, although it is also formed by some yeasts and bacteria).

The vitamin contents of foods are affected by a number of factors. In foods of animal origin, vitamin content depends mainly on the conditions during storage and processing of the raw materials. In foods of plant origin, the vitamin content depends particularly upon climatic conditions during growth, especially rainfall, fertilisation, the stage of ripeness, and post-harvest storage and processing. Vitamins, in general, are unstable compounds, and their loss can be induced by a number of factors. During food manufacturing, cooking, and storage of raw materials and foods, losses depend on the processing procedure, cooking method, cooking time, and temperature. Some vitamins are fairly heat-stable, whereas others are heat-labile. For this reason, vitamins are considered indicators of the use of good manufacturing practices or of the procedures that ensure good quality of food products for human consumption. Water-soluble vitamins are mainly leached out into the cooking water. The highest losses of fat-soluble vitamins are caused by oxidation. The stability of individual forms of vitamins varies, depending on external factors and on the specific food and technology used.

In the food industry, vitamin preparations are used to enrich many products, by the processes of restitution and fortification (enrichment). Restitution means the return of the vitamin content to the original level found in the raw material; fortification is an enrichment to a higher level needed for physiological or other reasons. Some vitamins are also used as natural dyes (riboflavin and provitamin A β-carotene) and as antioxidants (provitamins A, vitamin E, and vitamin C). The total intake of vitamins may be significantly influenced by the consumption of various concentrated sources of vitamins produced as dietary supplements and multivitamin preparations. In extreme cases, the use of such products can lead to hypervitaminosis.

This chapter describes the individual vitamins, their structure, nomenclature, activity in biochemical reactions occurring in the body, and importance in human nutrition. Their occurrence in foods of animal and vegetable origin and other sources, and their use in food technology, is reported in detail. The changes in vitamin contents and their reactions during storage of food materials and during culinary and industrial processing are also described.

5.2 Vitamin A

5.2.1 Structure and terminology

Vitamin A and its provitamins are classified as terpenoids or isoprenoids (see Section 8.2). Provitamins A are tetraterpenes (hydrocarbons) or tetraterpenoids (their oxygen derivatives) that contain 40 carbon atoms in a molecule. They originate, hypothetically, from eight molecules of isoprene (2-methylbuta-1,3-diene). The fission products, known as apocarotenoids, are widespread in living organisms, performing many key functions. In animals, apocarotenoids act as vitamins, visual pigments, and signalling molecules during cell division, growth, and differentiation of tissues and control of reproduction. In plants, they take on the role of hormones, pigments, and odorous compounds and perform a series of defensive functions.

The basic and most important biologically active apocarotenoid in animal tissues is **all-*trans*-retinol**, also known as axeroftol or vitamin A_1 (**5-1**). Retinol is an isoprenoid with 20 carbon atoms and five conjugated double bonds in the molecule; more precisely, it is a diterpenic alicyclic alcohol with the so-called β-ionone ring and a side chain of four conjugated double bonds attached to C-6. It is one of 16 possible stereoisomers.

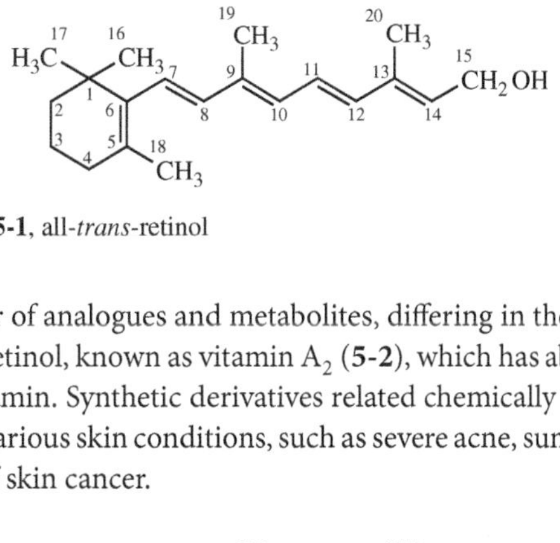

5-1, all-*trans*-retinol

In foods, retinol is accompanied by a number of analogues and metabolites, differing in the ionone cycle or side chain structures. Freshwater fish contain, for example, 3,4-didehydroretinol, known as vitamin A_2 (**5-2**), which has about 40% of the activity of retinol. Marine fish, birds, and mammals do not synthesise this vitamin. Synthetic derivatives related chemically to vitamin A are collectively called **retinoids**. These substances are used for the treatment of various skin conditions, such as severe acne, sun spots, wrinkles, and psoriasis. Some retinoids may even help treat or prevent certain forms of skin cancer.

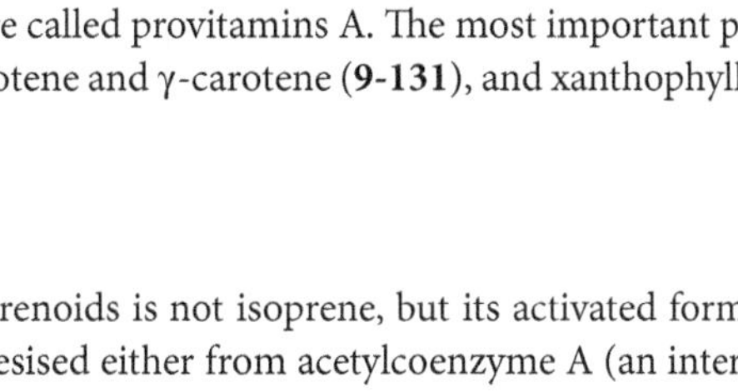

5-2, 3,4-didehydroretinol

In addition to vitamin A (antixerophthalmic vitamin), about 50 other naturally occurring compounds from the group of carotenoids also exert the same effect. These compounds are called provitamins A. The most important provitamin A is β-carotene. In foods, it is often accompanied by other carotenes, namely α-carotene and γ-carotene (**9-131**), and xanthophylls, such as β-cryptoxanthin and lutein (**9-132**).

5.2.2 Biochemistry

The building unit of vitamin A and other isoprenoids is not isoprene, but its activated forms, isopentenyl diphosphate and dimethylallyl diphosphate. Isopentenyl diphosphate is synthesised either from acetylcoenzyme A (an intermediate of its biosynthesis is mevalonic acid) or from 1-deoxy-D-xylulose 5-phosphate via 2-*C*-methyl- D-erythritol 4-phosphate and isopentenyl diphosphate. Gradual lengthening of the isoprenoid chain yields the immediate precursor of the carotenoids, geranylgeranyl diphosphate, with 20 carbon atoms in the molecule. The subsequent reactions of two molecules of geranylgeranyl diphosphate yield provitamins A (see Section 3.7.4.2).

In mammals, birds, and fish, provitamins A containing at least one β-ionone ring are transformed largely by symmetric (central) fission of the molecule catalysed by β-carotene-15,15′-dioxygenase to all-*trans*-retinal, also known as retinaldehyde (**5-3**). Two molecules of retinal are formed from one molecule of β-carotene, whilst other provitamins A provide only one molecule of retinal. Cleavage of β-carotene may also occur on other double bonds positioned between the two cyclohexen rings. This asymmetric (eccentric) cleavage provides two β-apocarotenals with different chain lengths. The subsequent cleavage of the β-apo-carotenal with the longer chain gives retinal, but only one molecule per molecule of β-carotene. Retinal is reversibly reduced to all-*trans*-retinol by retinol dehydrogenase. All-*trans*-retinal and other biologically active forms of vitamin A are stored mainly in the liver and transported, bound to specific proteins, by plasma. These other forms include all-*trans*-retinoic acid (**5-3**), which is the product of irreversible oxidation of retinal by retinal dehydrogenase, all-*trans*-retinol esters with higher fatty acids (**5-3**), all-*trans*-retinyl β-glucuronide (**5-3**), and other compounds. Retinoic acid occurs in a level of about 0.0001% in rosehip seed oil (the total oil content in seeds is approximately 9%).

In addition to β-carotene, freshwater fish can also convert the xanthophyl lutein, also known as 3,3′-dihydroxy-α-carotene or (3*R*,3′*S*,6′*R*)-β,ε-carotene-3,3′-diol, into vitamin A. Lutein eliminates one molecule of water (C-3′ hydroxyl group) and yields anhydrolutein (**5-4**), which cleaves to the corresponding aldehydes. These aldehydes are reduced to all-*trans*-3,4-didehydroretinol (vitamin A$_2$, **5-2**) and all-*trans*-3-hydroxyretinol, respectively. The latter compound can dehydrate to form another molecule of vitamin A$_2$. Trivial, specific, semi-systematic and systematic names of some provitamins A are given in Table 5.1.

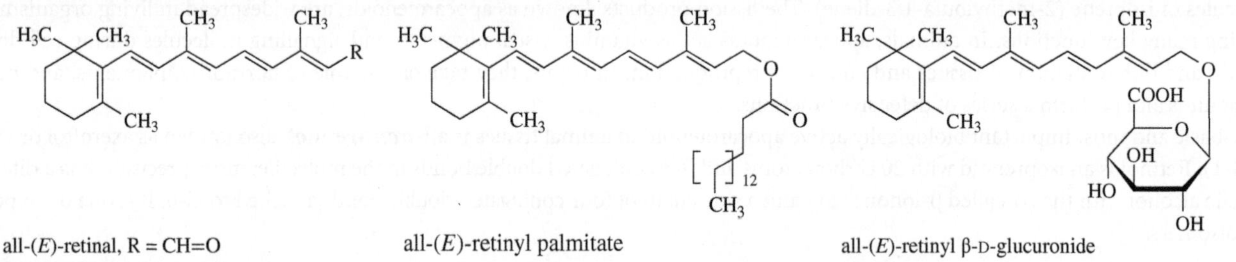

all-(*E*)-retinal, R = CH=O
all-(*E*)-retinoic acid, R = COOH

all-(*E*)-retinyl palmitate

all-(*E*)-retinyl β-D-glucuronide

5-3, retinoid forms of vitamin A

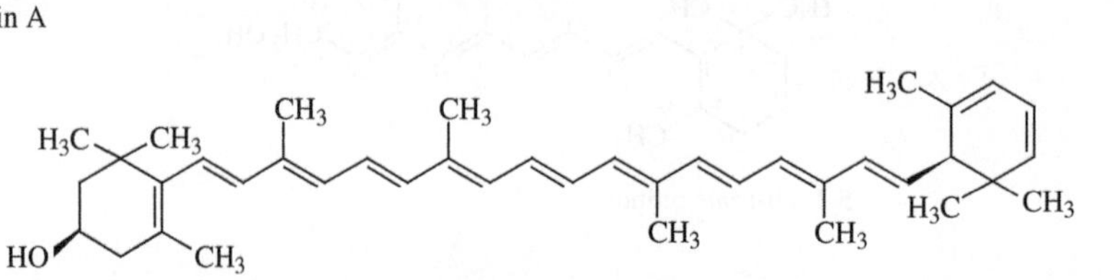

5-4, anhydrolutein

The biochemistry of vision is a very complex process, in which all-*trans*-retinol isomerises to 11-*cis*-retinol (**5-5**), which is then enzymatically oxidised to 11-*cis*-retinal (**5-5**). This compound associates with light-sensitive membrane-bound proteins (35–55 kDa) called opsins. Ciliary opsins (*c*-opsins) are typical of vertebrates, whilst invertebrates usually have rhabdomeric opsins (*r*-opsins). The resulting complexes are visual pigments (photoreceptors), which initiate a cascade that converts light (photons) falling on the retina into neural signals. Light destabilises opsins, which leads to protein conformation changes and isomerisation of 11-*cis*-retinal to all-*trans*-retinal. In order to function, each eye needs, in addition to opsins, the so-called shielding pigment that shields the retina from excess incoming light and exposes the photoreceptors to light coming from only a certain direction, thus ensuring the perception of directional light. Melanin (see Section 9.3.1.1) is a typical shielding pigment of vertebrates; invertebrates mainly have pteridines (see Section 9.3.4) and phenoxazines (see Section 9.3.6) at their disposal. During biological inactivation (catabolism), retinol is oxidised at carbon C-4 to a hydroxyl- or oxoderivative; alternatively, the side chain can be shortened or hydroxylated at the C-5 methyl group of the β-ionone ring.

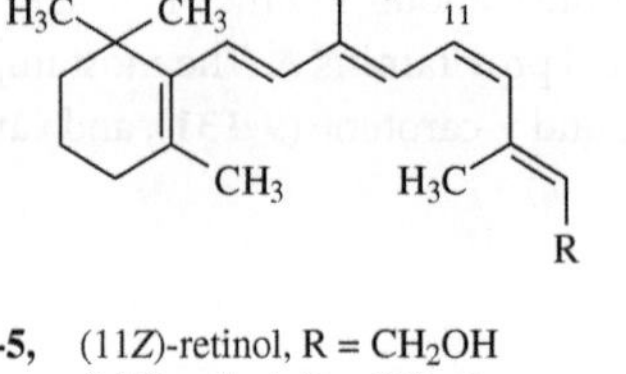

5-5,   (11*Z*)-retinol, R = CH$_2$OH
            (11*Z*)-retinal, R = CH=O

Table 5.1 Trivial, specific, semi-systematic, and systematic names of important provitamins A.

Compound
α-Carotene, β,ε-carotene
β-Carotene, β,β-carotene
γ-Carotene, β,ψ-carotene
β-Cryptoxanthin, 3-hydroxy-β,β-carotene
Echinenone, β,β-caroten-4-one
all-*trans*-Retinal, retinal, retinen, vitamin A₁ aldehyde, 15-apo-carotene-15-al, (2*E*,4*E*,6*E*,8*E*)-3,7-dimethyl-9-(2,6,6-trimethylcyclohex-1-en-1-yl)nona-2,4,6,8-tetraenal
all-*trans*-Retinol, retinol, vitamin A₁, (2*E*,4*E*,6*E*,8*E*)-3,7-dimethyl-9-(2,6,6-trimethylcyclohex-1-en-1-yl)nona-2,4,6,8-tetraen-1-ol
all-*trans*-Retinoic acid, retinoic acid, tretinoin, vitamin A₁ acid, (2*E*,4*E*,6*E*,8*E*)-3,7-dimethyl-9-(2,6,6-trimethylcyclohex-1-en-1-yl)nona-2,4,6,8-tetraen-1-carboxylic acid
all-*trans*-3,4-Didehydroretinol, 3,4-didehydroretinol, vitamin A₂, (2*E*,4*E*,6*E*,8*E*)-3,7-dimethyl-9-(2,6,6-trimethylcyclohexa-1,3-dien-1-yl)nona-2,4,6,8-tetraen-1-ol

5.2.3 Physiology and nutrition

The absorption of individual provitamins is not always quantitative. It depends on the food composition and the method of cooking, especially the presence of fats and their concentration. For example, the amount of β-carotene needed for the formation of 1 μg of retinol is 4 μg (if β-carotene is present in milk, margarine, vegetable oils, or animal fats), 8 μg (in cooked leafy vegetables or carrot cooked in fat), or even 12 μg (if present in carrot boiled in water). The provitamin in raw carrot is almost non-utilisable.

The recommended daily dose of retinol for children is 0.4–0.6 mg and for adults 0.8–1.0 mg. On average, it is 0.9 mg or 3000 IU of retinol, which is equivalent to 1.8 mg (3000 IU) of β-carotene in vitamin preparations or 10.8 mg (18 000 IU) in foods. For pregnant and lactating women, the recommended daily dose is 1.0–1.2 and 1.2–2.0 mg, respectively.[2] About 50% of the vitamin A requirement is covered (in some countries, for the most part) by provitamins in food of plant origin. About 40% of the required amount is provided by provitamins from vegetables, 20% by retinol and provitamins from meat and meat products, 15% by retinol and provitamins from milk and dairy products, 8% by fruit provitamins, 8% by retinol from fats (vegetable oils contain only provitamins), and 6% by retinol and provitamins from eggs. The biological activity of 9-*cis*-retinol is 21%, that of 11-*cis*-retinol is 24%, that of 13-*cis*-retinol is 75%, that of 9,13-di-*cis*-retinol is 24%, that of 11,13-di-*cis*-retinol is 15%, and that of all-*trans*-retinal is 90% of the activity of all-*trans*-retinol.

The avitaminosis manifests itself as disturbed vision (night blindness) and keratinisation of mucous membranes (which line the respiratory tract, intestines, urinary tract, and epithelium of the eye), inhibition of growth, and deformation of bones and reproductive organs. A reduced absorption of vitamin A can lead to hypovitaminosis, for example in the case of vegan diet.

High doses of vitamin A result in increased liver reserves of vitamin A and hypervitaminosis symptoms (acute or chronic intoxication with various symptoms including strumigenicity). Some individuals have genetically conditioned susceptibility to retinol, which is manifested by intolerance even at doses only slightly higher than normal doses. The excessive intake of β-carotene by vegetarians and children (hypercarotenosis or carotenaemia or xanthaemia) does not produce symptoms of hypervitaminosis (an excessive intake of provitamins from carrots or other vegetables worsens their absorption), but may manifest itself through the presence of β-carotene in plasma and by temporary yellow-orange discolouration of the skin, which dissipates when materials containing β-carotene are eliminated from or reduced in the diet.

Provitamins A exhibit anticarcinogenic effects because they are part of the control mechanisms that scavenge free radicals (toxic forms of oxygen). Their antioxidant potential is relatively low. Other carotenoids that do not act as provitamins A have significantly higher antioxidant activities, such as lycopene, zeaxanthin, and lutein. The enzymes that catalyse the oxidation of fatty acids (lipoxygenases or linoleate: O_2 oxidoreductases, formerly known as lipoxidases) are antivitamins A.

---

[2] The total content of vitamin A is expressed in international units (IU), or previously in retinol equivalents (RE). Thus, 1 IU is equal to 0.3 μg of retinol, 0.6 μg of β-carotene, or 1.2 μg of other provitamins A (such as α-carotene, γ-carotene, and β-cryptoxanthin) and 1 RE is equal to 1 μg of retinol, which is equivalent to 3.33 IU of vitamin activity of retinol or 10 IU of vitamin activity derived from β-carotene. Vitamin A in vitamin preparations is often in the form of the more stable retinyl acetate; 1 IU is equivalent to 0.33 μg of retinyl acetate. In animal materials, the amount of vitamin A (expressed in RE) equals the sum of microgrammes of retinol and microgrammes of β-carotene/6 or the IU of retinol/3.33 + IU of β-carotene/10. In plant materials, it is equal to the sum of microgrammes of β-carotene/6 and microgrammes of other retinoids/12.

5.2.4 Use

To enrich foods with vitamin A (e.g. table oils, margarine, butter, dairy products, and flour), synthetic and relatively stable retinyl acetate or retinyl palmitate are used. β-carotene is used as a lipophilic dye.

5.2.5 Occurrence

Retinol does not occur in foods of plant origin, higher fungi, or microorganisms (bacteria, yeasts, and moulds), but these materials often contain carotenes and xanthophylls that show the activity of provitamins A. Animals are unable to synthesise carotenoids *de novo;* they only convert plant pigments into substances of a different structure or store them as such (Table 5.2). These reactions generate retinol, 3,4-didehydroretinol and a retinol dimer kitol (**5-6**), which can be obtained from whale liver oil in particular. The latter substance has little or no biological activity, but is transformed into retinol upon heating to temperatures above 200 °C. Also, some rosy pink pigments of fish (such as salmon) and some crustaceans (such as shrimps), as well as the meat and feathers of some birds (such as flamingos), are transformed carotenoids (xanthophylls). The main pigment of fish and crustaceans is astaxanthin (see Section 9.9.2.4.3), whilst the feathers of flamingos contain canthaxanthin and astaxanthin as the main components, together with some minor pigments (echinenone, phoenicoxanthin, and phoenicopterone; see Section 9.9.3).

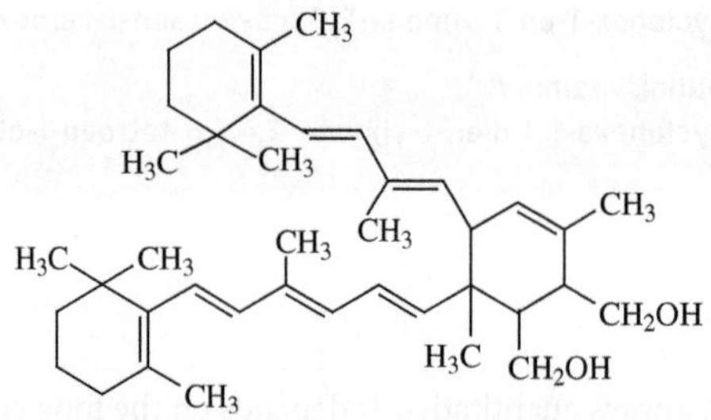

5-6, kitol

The main forms of vitamin A include retinol esterified with higher fatty acids and free retinol or retinal. Precursors, or provitamins A, occur in food of animal origin in relatively small quantities. The most common ester of retinol is palmitate, and esters with other fatty acids are also found in variable amounts. For example, the main component in milk is palmitate, followed by oleate and stearate. Other fatty acid esters (caprylate, caproate, linoleate, laurate, arachidonate, linolenate, myristate, palmitoleate, pentadecanoate, gadoleate, and heptadecanoate)

Table 5.2 Retinol and provitamin A contents of some foods.

Food	Content in edible portion (mg/kg or mg/l)		Food	Content in edible portion (mg/kg or mg/l)
	Vitamin A	Provitamin A		Provitamin A
Meat	0.1	0.4	Carrots	20–95
Liver	30–400	300	Parsley (root)	0.1
Milk	0.3–1.0	0.1–0.6	Parsley (curly)	30–260
Cheeses	1.6–3.2	0.3–8.0	Cabbage	3.0–74
Eggs	0.5–1.5	0.1–2.0	Savoy cabbage	50
Fish	0.5	0.7	Broccoli	25
Butter	5.0–10	4.0–8.0	Cauliflower	0.3
Apples	–	0.1–0.3	Lettuce	3.0–25
Apricots	–	6.0–20	Spinach	50–480
Bananas	–	0.3–2.3	Tomatoes	3.0–90
Oranges	–	0.5–4.0	Peppers	3.8–24
Melons	–	20	Peas	3.0
Mango	–	20	Beans	3.0–5.0

and free retinol are also present in smaller amounts. One particularly rich source of vitamin A is liver. For example, the retinol content in pork liver is about 30 mg/kg, and the liver of polar bears contains up to 60 g/kg; there is a clear correlation with their diet, which consists mainly of seals that feed on fish. Milk (as well as meat) contains relatively little vitamin A (the vitamin content is proportional to the fat content). Dairy products and butter are good sources of vitamin A because of their high fat content.

The most important provitamin A is β-carotene. Leafy vegetables such as spinach and cabbage are very rich sources of provitamins A as they contain 10–30 mg/kg retinol equivalents (RE), mainly in the form of β-carotene. Provitamins A form about 25% of total carotenoid pigments present in these vegetables. The classic source of β-carotene is carrots, containing about 20 mg/kg RE. The tomato has lycopene as the main pigment, which is not a provitamin A, and the amount of β-carotene is relatively low (about 6 mg/kg). Tomatoes also contain small amounts of γ-carotene (1 mg/kg). There is a high content of β-carotene in the orange varieties of tomatoes (Table 5.2). Some fruits (such as apricots and mango) are also good sources of vitamin A precursors. Margarine is usually fortified with synthetic retinyl acetate or palm oil (that contains mainly α-carotene), so that the vitamin content is the same as in butter.

Fish liver oil is a very abundant source of vitamin A. Cod liver oil contains 10–100 g/kg of retinol (or its esters). Liver oils of freshwater fish contain about 40 mg/kg of retinol and 110 mg/kg of 3,4-didehydroretinol (their ratio A is highly variable).

5.2.6 Reactions and changes in foods

Naturally occurring all-*trans* isomers of provitamin A and retinol are unstable compounds. They isomerise very easily during food storage, especially when exposed to light and higher temperatures (cooking, baking, and other thermal operations). They are also sensitive to oxidation (lipoxygenases, oxygen in the air, or chemical agents). These compounds likewise react with free radicals and thus inhibit unwanted radical oxidation reactions. Similarly to lipid oxidation, a range of products are formed due to the combined effect of various factors.

5.2.6.1 Retinol

Retinol and its derivatives isomerise to a mixture of products in which 13-*cis*- (**5-7**) and 9-*cis*-stereoisomers (**5-7**) dominate. These isomers of retinol generally have a less intense colour than all-*trans* isomers. At the same time, and especially in acidic media, there is a shift of the double bonds towards the β-ionone ring with the formation of a positional isomer, retrovitamin A, also known as α-retinol (**5-7**). Retrovitamin A partially dehydrates to all-*trans* anhydroretinol (anhydrovitamin A, **5-7**).

The oxidation of the β-ionone ring of retinol yields unstable hydroperoxides. Their subsequent reactions produce relatively stable epoxides, such as all-*trans*-5,6-epoxyretinol (**5-7**), and oxidation of the hydroxymethyl group produces all-*trans*-retinal. Other oxidation products include compounds with shorter side chains. These oxidation products of retinol or the corresponding free radicals yield polyene polymers. Similar reactions *in vivo* lead to the yellow-brown ageing pigment lipofuscin (see Section 3.6.1). Retinol esters are more stable to oxidation than free retinol. Tocopherols exhibit protective effects.

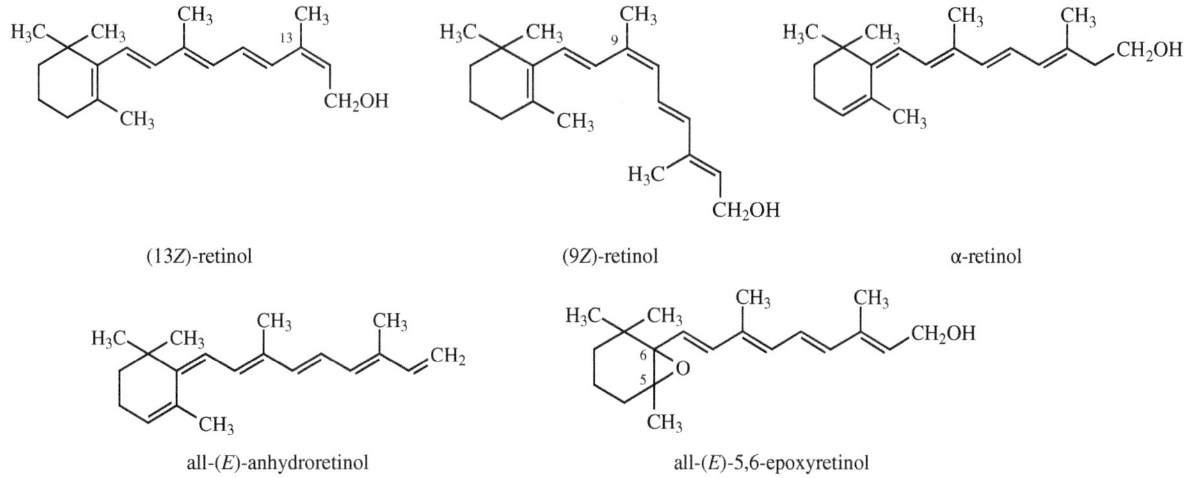

(13Z)-retinol (9Z)-retinol α-retinol

all-(*E*)-anhydroretinol all-(*E*)-5,6-epoxyretinol

5-7, retinol transformation products

5.2.6.2 β-Carotene

Similarly to retinol, isomerisation, oxidation, and degradation of β-carotene and other provitamins A may occur during food storage and processing through the combined effects of light, heat, oxygen, hydroxonium ions, and other factors. Some stereoisomers of β-carotene, such as 13-*cis*-β-carotene and 9-*cis*-β-carotene (**5-8**), appear as minor natural pigments in fruits and vegetables, especially in green species that contain chlorophyll pigments, because chlorophylls act as photosensitisers and catalyse photoisomerisation of β-carotene to 13-*cis*- and 9-*cis*-isomers. Stereoisomers of β-carotene are also formed during storage of flour under normal conditions and, in particular, at the higher temperatures common in cooking, baking, frying, and extrusion. Frying of foods or extrusion of cereals (at temperatures of 180–200 °C) yields 13-*cis*-β-carotene as the major product and other isomers, such as 9-*cis*-β-carotene, 15-*cis*-β-carotene, 13,13′-di-*cis*-β-carotene (**5-8**), 9,9′-di-*cis*-β-carotene, 9,13′-di-*cis*-β-carotene, and oxidation products such as 5,6-epoxide (**5-8**).

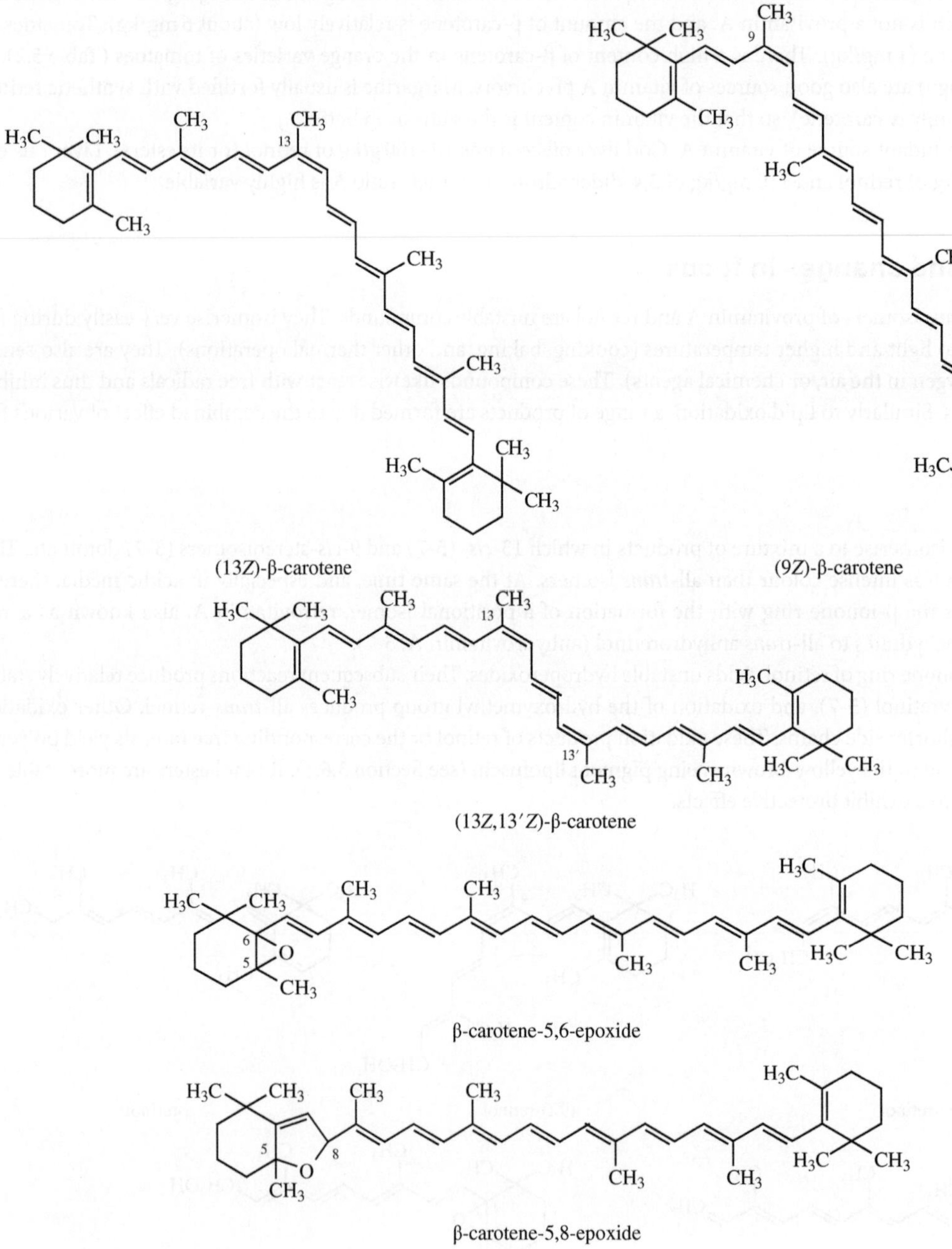

(13Z)-β-carotene

(9Z)-β-carotene

(13Z,13′Z)-β-carotene

β-carotene-5,6-epoxide

β-carotene-5,8-epoxide

5-8, selected isomerisation and oxidation products of β-carotene

In acidic media, 5,6-epoxides isomerise to dihydrofuran derivatives, also known as 5,8-epoxides (**5-8**); other products include diepoxides, β,β-carotene-3,3′-diol (zeaxanthin), β,β-carotene-4-one (echinenone), and their degradation products, such as β-apo-8′-carotenal (**5-9**), β-apo-10′-carotenal, β-apo-12′-carotenal, β-apo-14′-carotenal, and numerous other aldehydes. The main product is 5,8,5′,8′-diepoxide (also known as 5,8,5′,8′-diepoxy-5,8,5′,8′-tetrahydro-β,β-carotene, or aurochrome or ξ-carotene), and other products include 5,6,5′,6′-diepoxide (5,6,5′,6′-diepoxy-5,6,5′,6′-tetrahydro-β,β-carotene) and 5,6,5′,8′-diepoxide (5,6,5′,8′-diepoxy-5,6,5′,8′-tetrahydro-β,β-carotene or luteochrome).

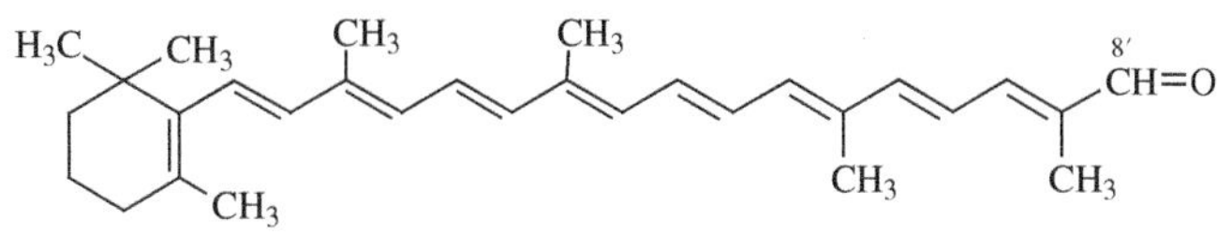

5-9, β-apo-8′-carotenal

Carotenoids, such as β-carotene, deactivate singlet oxygen ($^1O_2$; see Section 3.8.1.8.4) through physical quenching and chemical reactions and prevent lipid oxidation initiated by singlet oxygen. The products of physical quenching are oxygen in the triplet state ($^3O_2$) and β-carotene in the excited triplet state (^{3}β-carotene*). Oxidation of β-carotene by singlet oxygen yields endoperoxides, epoxides and apocarotenoids:

$$^1O_2 + \text{β-carotene} \rightarrow {}^3O_2 + {}^3\text{β-carotene}^*$$

Carotenoids also react with free radicals by numerous mechanisms. They deactivate them and are bleached (lose their colour). It is assumed that oxidising free radicals (e.g. superoxide radical anion, $O_2^{\bullet-}$) can abstract an electron from carotenoids and generate a carotenoid radical cation (β-carotene$^{\bullet+}$). The carotenoid radical cation may oxidise important molecules such as amino acids or may lose a proton under the formation of neutral radical. Under specific conditions, carotenoids can also create radical anions. The carotenoid radical anion can react with oxygen, producing superoxide radical anion. Carotenoids likewise react with sulfur-containing radicals such as thiyl (RS•) and sulfonyl ($RSO_2^{\bullet}$) radicals to form radical adducts. The reactions are illustrated in the example of β-carotene:

$$\text{β-carotene} + O_2^{\bullet-} + 2\,H^+ \rightarrow \text{β-carotene}^{\bullet+} + H_2O_2$$
$$\text{H-β-carotene}^{\bullet+} \rightarrow \text{β-carotene}^{\bullet} + H^+$$
$$\text{β-carotene}^{\bullet-} + O_2 \rightarrow \text{β-carotene} + O_2^{\bullet-}$$
$$\text{β-carotene} + RS^{\bullet} \rightarrow RS\text{-β-carotene}^{\bullet}$$

The carotenoid radical cation may be converted back to the parent carotenoid by water-soluble antioxidants such as ascorbic acid or phenolic compounds. In polar environments, α-tocopheroxyl radical cation (T—OH$^{\bullet+}$) is deprotonated and T—O$^\bullet$ does not react with carotenoids, whereas in non-polar environments, T—OH$^{\bullet+}$ is converted to the parent tocopherol (T—OH) by carotenoids:

$$\text{T-OH}^{\bullet+} \rightarrow T - O^{\bullet} + H^+$$
$$\text{β-carotene} + T - O^{\bullet} + H^+ \rightarrow \text{β-carotene}^{\bullet+} + T - OH$$

The strongly oxidising hydroxyl radicals (HO$^\bullet$) mainly add to carotenoids producing neutral radical adducts or they abstract a hydrogen atom (H$^\bullet$) from the carotenoid molecule:

$$\text{β-carotene} + HO^{\bullet} \rightarrow HO\text{-β-carotene}^{\bullet}$$
$$\text{β-carotene} + HO^{\bullet} \rightarrow \text{β-carotene}^{\bullet} + H_2O$$

The reaction of carotenoids with less strongly oxidising radicals such as hydroperoxyl radicals generated by autoxidation of lipids can lead to hydrogen atom transfer generating the neutral carotene radical. Lycopene is the easiest carotenoid to oxidise to its radical cation, whilst astaxanthin is the most difficult. Hydroperoxyl radicals ROO$^\bullet$ are not reduced to hydroperoxides as in the case of phenolic antioxidants, but are captured by the conjugated polyene system, represented by the formula **5-10**, with the formation of relatively stable neutral radicals that are stabilised by resonance:

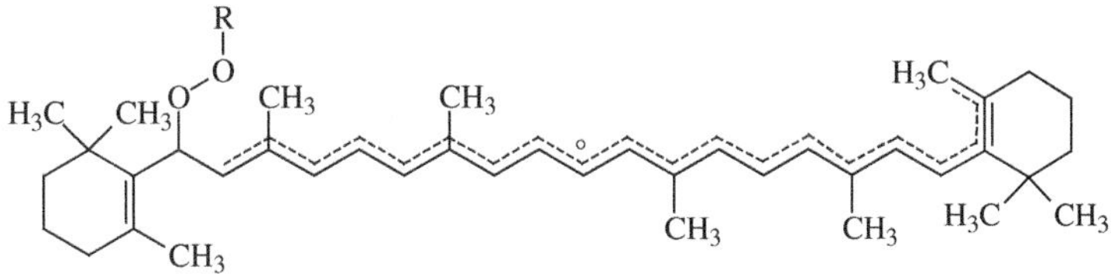

5-10, resonance-stabilised R—O—O—β-carotene radical

$$R - O - O^{\bullet} + \text{β-carotene} \rightarrow R - O - O - \text{β-carotene}^{\bullet}$$

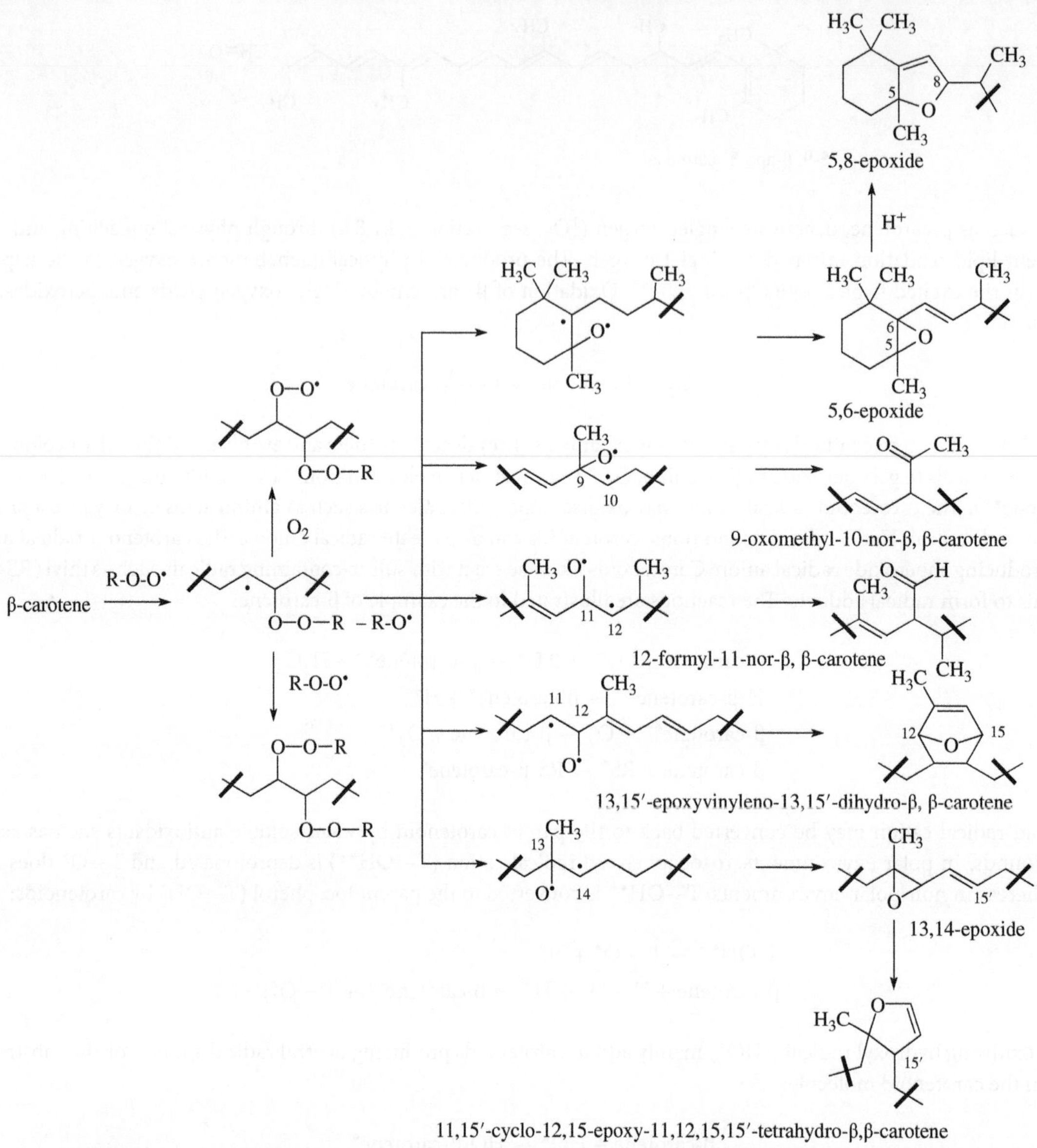

Figure 5.1 Reaction of β-carotene with lipid hydroperoxyl radicals (R = lipid residue).

These radicals break down to alkoxyl radicals (RO•) and stabilise through the formation of epoxides, carbonyl compounds, and a number of other products (Figure 5.1):

$$R—O—O—\beta\text{-carotene}^{\bullet} \rightarrow R—O^{\bullet} + \beta\text{-carotene epoxide}$$
$$\beta\text{-carotene epoxide} \rightarrow \text{polar products}$$

Under anaerobic conditions and in the presence of small amounts of oxygen, when carotenoids show higher antioxidant activity, β-carotene reacts with another hydroperoxyl radical with the formation of stable polar products:

$$R—O—O—\beta\text{-carotene}^{\bullet} + R—O—O^{\bullet} \rightarrow \text{polar products}$$

At higher partial pressure of oxygen (e.g. during storage of fat in contact with air), unstable peroxyl radicals and stable polar products are formed:

$$R—O—O—\beta\text{-carotene}^{\bullet} + O_2 \rightarrow R—O—O—\beta\text{-carotene}—O—O^{\bullet}$$
$$R—O—O—\beta\text{-carotene}—O—O^{\bullet} \rightarrow \text{polar products and radicals}$$

Natural provitamins A and vitamin A, β-carotene in foods of plant origin, and esters of retinol in foods of animal origin are relatively stable substances in the absence of air. At higher temperatures and in the light (e.g. during food preservation), they may isomerise to **neocarotenes**, which still exhibit vitamin A activity if they have preserved at least one β-ionone ring, but are less intensely coloured. Autoxidation (oxidation that occurs in the open air or in the presence of oxygen) of these substances is particularly rapid in dehydrated food. Retinoids also react with lipid oxidation products (fatty acid oxidation products), which similarly generate less coloured products.

During processing of meat and offal, vitamin A and provitamins A are very stable.

During pasteurisation, in ultra high-temperature (UHT) milk, and during milk drying, up to 6% of vitamin A is lost, and further losses occur during storage. In UHT milk, these losses are 3–7% over four weeks. In the presence of oxygen and in light (during storage in inappropriate containers), however, vitamin losses may reach up to 20–30% per hour. Vitamin A in dried milk powder is very stable, and losses, even during prolonged storage, do not exceed 10%. The vitamin content in cheeses is higher than in milk (often by 50% or more, according to the fat content). Storage of butter results in only small losses of retinol and carotenoids.

The changes to the provitamins A content of grains are negligible. During storage of flour, provitamins A react with fatty acid hydroperoxides produced by the action of lipoxygenases. During dough mixing and proofing, hydroperoxides oxidise carotenoids; the range of reactions depends on many factors, including water content. The result is a desirable lighter colour of the final product (e.g. of bread). On the other hand, the loss of carotenoids (up to 75%) is undesirable in the production of pasta. The addition of ascorbic acid, which inhibits lipoxygenase, has a colour-stabilising effect. Losses, often higher than 9%, occur in the production of extruded cereal products. The so-called bleaching of flour is also done by chemical oxidising agents such as halogen compounds and peroxides (see Section 11.4.2.2.1).

The extent of degradation reactions of carotenoid compounds in preserved fruits and vegetables after blanching and deaeration (removal of air) is usually small. For example, the retention of carotenoids in canned apricots, peaches, and plums ranges from 85 to 100% after 1 year of storage, depending on the type of fruit and the storage temperature. In the presence of oxygen, a number of isomerisation and fission products of carotenoids arise. Carotenoid substances during drying of fruits and vegetables are easily oxidised (the extent of oxidation depends, for example, on water activity, temperature, and oxygen content in the atmosphere). During storage of dried carrots in air, there may be up to 50% loss of carotenoids. There are also extensive losses in the production of fruit wines and spirits. Decomposition of carotenoids, however, also leads to a series of flavour-active compounds that can have an impact on the odour and taste of the product.

5.3 Vitamin D

5.3.1 Structure and terminology

Vitamin D is the common name for a group of closely related lipophilic 9,10-secosteroids, of which the most important are vitamin D_3, known as cholecalciferol (9,10-seco-$\Delta^{10(19),5,7}$-dehydrocholestatriene-3β-ol, **5-11**), and vitamin D_2, known as ergocalciferol (9,10-seco-$\Delta^{10(19),5,7,22}$-ergostatetraene-3β-ol, **5-11**).

Vitamins D are produced by ultraviolet (UV) irradiation of their precursors, provitamins D. Provitamins D are cyclopentaperhydrophenanthrenes with C-18 and C-19 methyl groups, a C-3 hydroxyl group, and a C-5(6), C-7(8) system of conjugated double bonds in ring B, which differ from one another in their length and the arrangement of the side chain at position C-17. Provitamin D_3 is **7-dehydrocholesterol**, also known as 7-procholesterol (**5-11**); provitamin D_2 is **ergosterol** (**5-11**). In the past, vitamin D_1 was the name given to a mixture of vitamin D_2 and lumisterol, a byproduct resulting from irradiation (photodegradation) of ergosterol (Table 5.3). Vitamin D_4 is 22-dihydroergocalciferol, which is produced by irradiation of 22-dihydroergosterol. Irradiation of 7-dehydrositosterol yields sitocalciferol, which was previously also known as vitamin D_5.

5.3.2 Biochemistry

Provitamins D, similarly to provitamins A, are terpenoids. Gradual lengthening of the isoprenoid chain gives farnesyl diphosphate; condensation of two molecules of farnesyl diphosphate and subsequent reactions yield provitamins D and other steroids. The precursor in the biosynthesis of cholesterol and of 7-dehydrocholesterol (provitamin D_3) is lanosterol (see Section 3.7.4.1). UV irradiation of 7-dehydrocholesterol present in the skin cells[3] at the wavelength range of 280–320 nm (with a maximum at 295–297 nm) brings about a photochemical reaction (due to energy absorption by the system of π-electrons) that yields as its first intermediate the so-called precholecalciferol (previtamin D_3) with an opened ring B, which spontaneously isomerises, with the migration of hydrogen, to cholecalciferol (Figure 5.2). Precholecalciferol shows about 35% of the activity of cholecalciferol.

---

[3]Electromagnetic radiation is classified into UV-A (315–400 nm), UV-B (280–315 nm), and UV-C (germicidal, 100–280 nm) radiation.

Table 5.3 Nomenclature of major vitamins D and their precursors.

Trivial name	Systematic name
7-Dehydrocholesterol	7-Dehydrocholest-5-en-3β-ol
(Z)-Tacalciol, precholecalciferol, previtamin D$_3$	(Z)-9,10-Secocholesta-5(10),6,8-trien-3β-ol
Cholecalciferol, calciol, vitamin D$_3$	(5Z,7E)-9,10-Secocholesta-5,7,10(19)-trien-3β-ol
25-Hydroxycholecalciferol, calcidiol	(5Z,7E)-9,10-Secocholesta-5,7,10(19)-trien-3β,25-diol
1α,25-Dihydroxycholecalciferol, calcitriol	(5Z,7E)-9,10-Secocholesta-5,7,10(19)-trien-1,3β,25-triol
(R)-1,24-Dihydroxycholecalciferol, (R)-24-hydroxycalcidiol	(5Z,7E)-9,10-Secocholesta-5,7,10(19)-trien-1,3β,24-triol
(R)-1,24,25-Trihydroxycholecalciferol, calcitetrol	(5Z,7E)-9,10-Secocholesta-5,7,10(19)-trien-1,3β,24,25-tetrol
Lumisterol	(E)-9β,10α-Ergosta-5,7,22-trien-3β-ol
Ergosterol	(E)-Ergosta-5,7,22-trien-3β-ol
Ergocalciferol, ercalciol, vitamin D$_2$	(5Z,7E,22E)-9,10-Secoergosta-5,7,10(19),22- tetraen-3β-ol

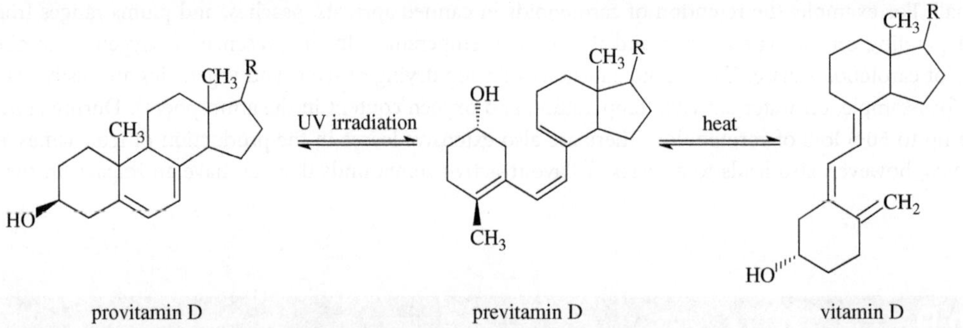

Figure 5.2 Formation of vitamins D from provitamins D.

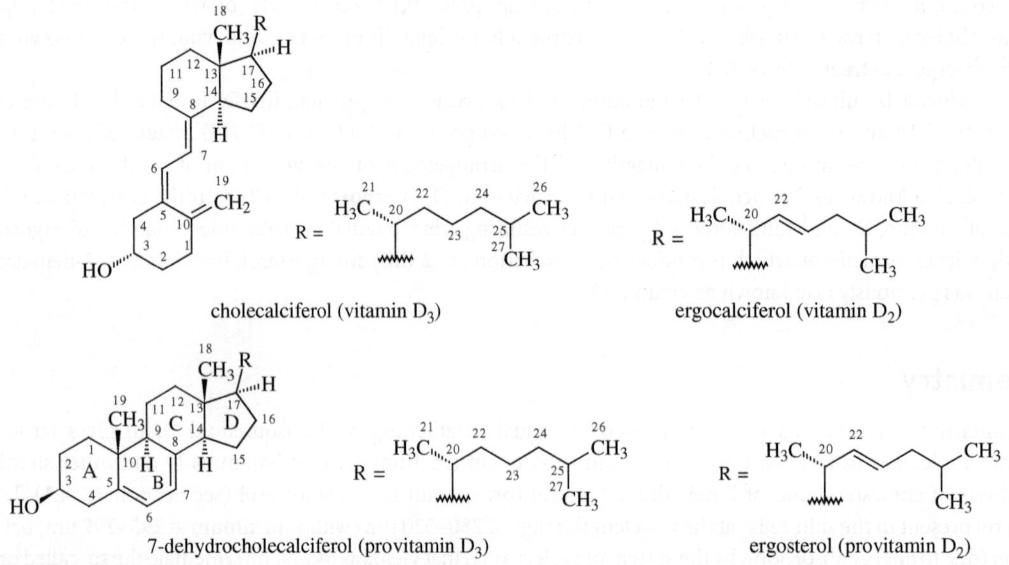

5-11, calciferols and their provitamins

Cholecalciferol binds to a specific globulin in blood plasma (vitamin D-binding protein, DBP) and is transported to the liver. Therefore, its concentration in blood plasma is relatively low (<20 nmol/l). In the liver, it is stored and, according to requirements, oxidised to 25-hydroxycholecalciferol, known as calcidiol (Figure 5.3), which is the main circulating metabolite of cholecalciferol. Its concentration in

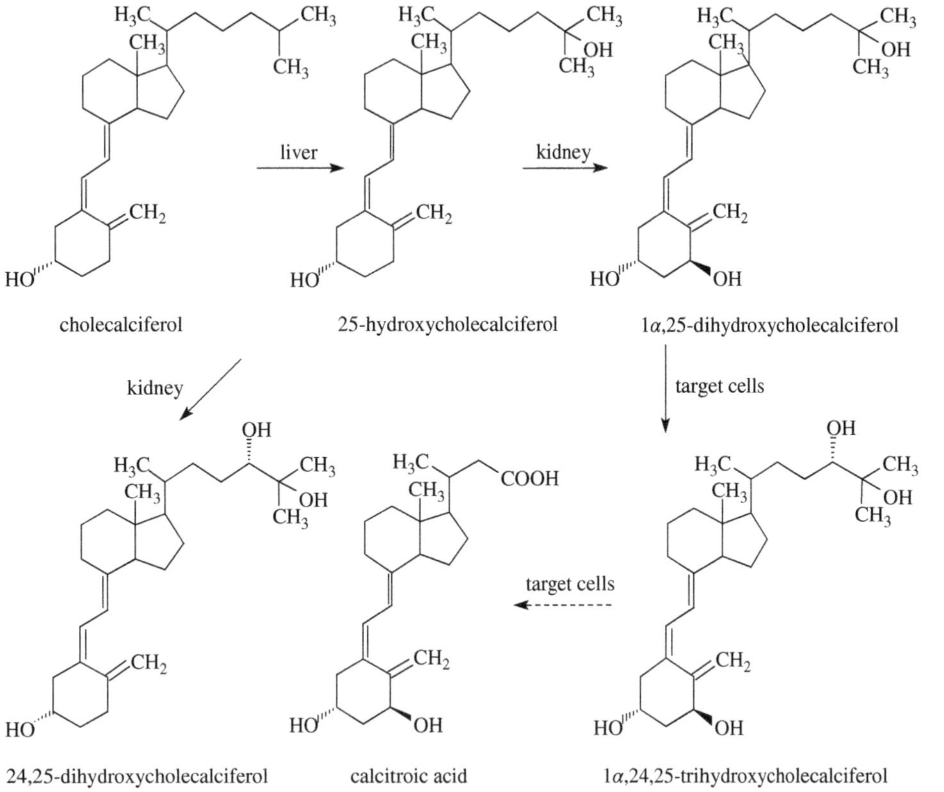

Figure 5.3 Metabolism of cholecalciferol.

plasma depends on many factors, such as the duration of exposure to sunlight and the season. Cholecalciferol is metabolised in the kidneys to a number of dihydroxysubstituted vitamins D_3. The major metabolites are the hormones 1α,25-dihydroxycholecalciferol (known as calcitriol) and (24R)-24,25-dihydroxycholecalciferol. The biological efficiency of 1α,25-dihydroxycholecalciferol is about 10 times higher than that of cholecalciferol. Along with two other hormones, calcitonin and parathyroid hormone (parathormone, parathyrin), these hormones act in the metabolism of calcium and phosphorus. The synthesis of 24,25-dihydroxycholecalciferol is a part of a detoxification mechanism; (24R)-1,24,25-trihydroxycholecalciferol is a starting compound for catabolism in the target cells, where calcitroic acid is formed as a product. Many other metabolites are known, such as (S)-1,25,26-trihydroxycholecalciferol and 1α,25-dihydroxycholecalciferol-26-lactone.

Because of the possibility of its biosynthesis in the body and due to its biological function (it is an antirachitic vitamin connected with the metabolism of calcium and phosphorus, which are necessary for growth, development, and maintenance of bone structure), some reports state that vitamin D_3 is more a hormone than a vitamin, but these do not reflect the definitions of either a vitamin or a hormone. Cholecalciferol is a vitamin in the true sense of the word, and its metabolite 25-hydroxycholecalciferol (or other deltanoids) is appropriately described as a prehormone. The hormonally active form of vitamin D_3 is 1α,25-dihydroxycholecalciferol. According to recent research, vitamin D acts not only in the metabolism of calcium and phosphorus, but also in cell differentiation, and plays an important part in the immune system.

Cholecalciferol, ergocalciferol, and their metabolites ingested via foods are resorbed in the intestines and (bound to DBP) transported to the liver by the lymphatic vascular system, where they are stored or metabolised.

Commonly used and recommended trivial and systematic names of key vitamins D and related steroids are listed in Table 5.3.

5.3.3 Physiology and nutrition

The vitamin D content of foods is still often given in IUs. One IU corresponds to 0.025 µg of vitamin D_3 or vitamin D_2. Both vitamins have the same biological activity. The most important form is cholecalciferol. The recommended daily dose of vitamin D, 2.5–10 µg (there is greater need for the upper limit in babies, children, and pregnant and lactating women), is accounted for by the vitamin D_3 obtained through biosynthesis from provitamin 7-dehydrocholesterol in conjunction with the vitamin D_3 and vitamin D_2 contained in foods.

Hypovitaminosis manifests itself as rickets (changes to the skeleton) in children and as softening and deformation of developed bones in adults, caused by defective mineralisation (osteomalacia). Doses significantly higher than the daily requirement result in a variety of symptoms. Long-term high doses of vitamin D may even cause hypercalcaemia, which is due to excessive skeletal calcium release and is characterised by elevated calcium level in blood; calcium is then deposited in various organs, such as the heart, blood vessels, and lungs.

5.3.4 Use

A common practice in some countries is the fortification of margarines by ergocalciferol or dehydrocholecalciferol. Milk and breakfast cereals are likewise often fortified. Ergocalciferol, produced industrially by photoisomerisation of ergosterol, is the main form of vitamin D found in fortified foods and pharmaceutical preparations. Some of the earlier procedures of feed enrichment by ergocalciferol (e.g. yeast for feed purposes) were based on irradiation of materials rich in ergosterol.

5.3.5 Occurrence

Cholecalciferol is also synthesised by mammals, birds, and fish, and is therefore found naturally in foods of animal origin, where it is also accompanied by its precursor (7-dehydrocholesterol) and related metabolites, such as calcidiol. Notable concentrations of cholecalciferol are found in the liver fats of marine fish (halibut, 30 mg/kg; mackerel, 15 mg/kg; cod, 2.5 mg/kg). Another valuable source is the meat of fatty fish (such as herring, mackerel, and salmon). Lower amounts of vitamin D_3 (cholecalciferol and metabolites) are present in the meat and offal of livestock and in other animal products, such as milk, dairy products, and eggs (Table 5.4). Vitamin levels in these foods are, of course, determined by many factors. The cholecalciferol content of milk in the winter, for example, is about four times lower than in the summer.

Ergosterol (provitamin D_2) is the main sterol of most fungi, and therefore is naturally present (together with small amounts of vitamin D_2, vitamin D_3, and its metabolites) in cheeses made with mould (such as Roquefort and Gorgonzola).

7-Dehydrocholesterol, cholecalciferol, calcitriol, and calcidiol can be found in some plant species of the *Solanaceae* family (also known as the nightshade or potato family) originating in South America. These plants (such as *Solanum glaucophyllum*, syn. *S. malacoxylon*) are calcinogenic plants responsible for producing the enzootic calcinosis of cattle and sheep in Argentina, Brazil, Paraguay, and Uruguay. The disease is characterised by the calcification of soft tissues, especially the aorta, heart, lungs, and kidneys.

The presence of ergosterol in oil seeds, grains, and cereal products (up to hundredths of microgrammes per kilogramme) is an indicator of microbial contamination. Similarly, ergosterol can be present for the same reason in some vegetables. The content in carrots is reportedly about 0.7 µg/kg, and in cabbage and spinach is about 0.1 µg/kg. Ergosterol is also the main sterol of the plant parasitic fungus ergot (*Claviceps purpurea*), commonly found on grains of rye or sometimes on other grasses and causing a devastating and sometimes deadly syndrome called ergotism in humans and other animals through consumption of contaminated foods and feeds (see Section 12.2.2.2.15).

The content of ergosterol in the species and strains of yeast *Saccharomyces cerevisiae* (a type of yeast used to make dough rise in baking) varies within wide limits from 600 to 1500 µg/kg (dry matter). An important source of ergocalciferol may be higher fungi (of the phylum *Basidiomycota*, which covers most of fungi often referred to as mushrooms), which also contain the provitamin D_2 ergosterol. Ergocalciferol is produced from ergosterol by direct solar radiation with the same mechanism that yields cholecalciferol from 7-dehydrocholesterol (via the corresponding previtamin) in humans. Therefore, mushrooms are the only non-animal-based food containing vitamin D, and hence the only natural vitamin D sources for vegetarians. Cultivated shiitake mushrooms (*Lentinula edodes*) and common button mushroom (*Agaricus bisporus*), however, have a significantly lower content of ergocalciferol (about 2–3 µg/kg) than wild mushrooms (chanterelle, *Cantharellus cibarius*, around 130 µg/kg, and penny bun, *Boletus edulis*, about 30 µg/kg f.w.). Generally, the content of ergocalciferol in wild mushrooms depends on many factors, such as climatic conditions and the colour of the stalk. Ergocalciferol represents about 90% of the vitamin content, with the remainder being previtamin D_2, 25-hydroxyergocalciferol, ergosta-5,7-dienol (22,23-dihydroergosterol), and ergosta-7-enol. The content of provitamin D_2 (ergosterol) in fungi varies considerably. For example, its content in white button mushrooms is 560 mg/kg, in shiitake mushrooms is 849 mg/kg, in oyster mushrooms (*Pleurotus ostreatus*) is 680 mg/kg, and in chanterelle mushrooms is 463 mg/kg (f.w.). Ergosterol usually represents 60–70% of the sterols present.

Table 5.4 Vitamin D content of some foods of animal origin.

Food	Edible portion (µg/kg or µg/l)	Food	Edible portion (µg/kg or µg/l)
Meat	3	Butter	10–20
Liver	2–11	Cheeses	8
Milk	1	Eggs	30–50
Cream	4	Sea fish	50–450

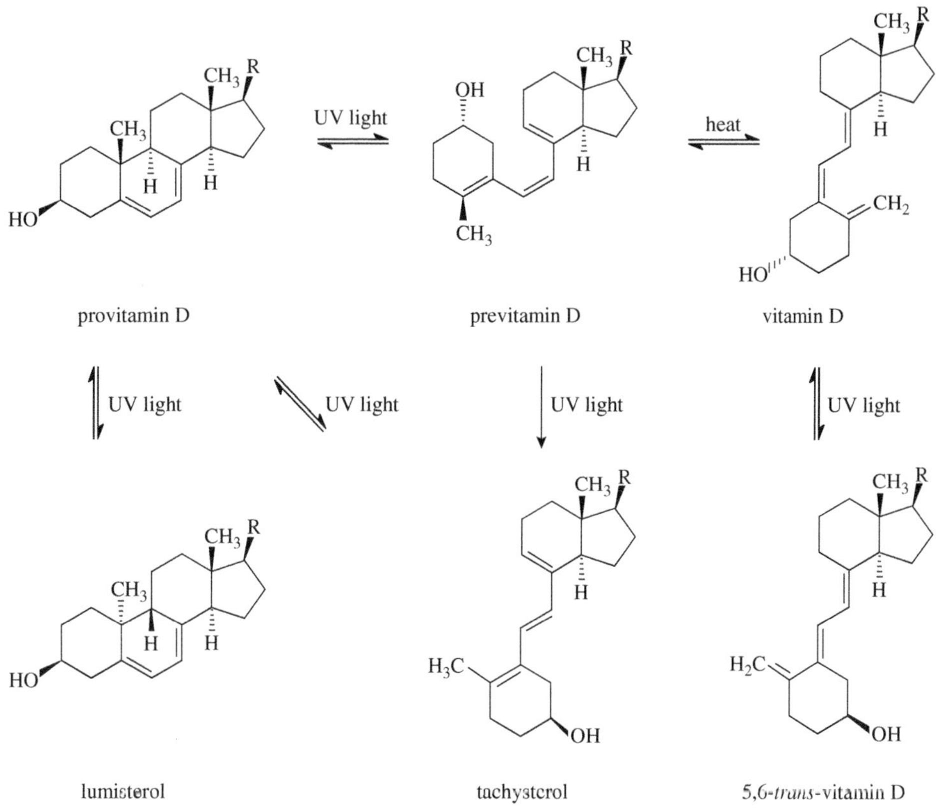

Figure 5.4 Photolytic products of vitamins D.

The concentration of vitamin D_2 in cultivated mushrooms increases if they are exposed to sunlight or artificial UV-B (315–280 nm) or UV-C (280–100 nm) light. After exposure to a UV-B light, the concentration of vitamin D_2 in shiitake mushrooms was increased to 36.7, 68.6, and 106 mg/kg for the pileus, middle, and gill parts, respectively. The concentration of vitamin D_2 produced in white button mushrooms after exposure to UV-C irradiation was 40.6–141 mg/kg, depending on irradiation dose and time of irradiation.

5.3.6 Reactions and changes in foods

Compounds belonging to the vitamin D group are oxylabile, just like other lipophilic vitamins. Some reactions, such as formation of oxidation products and isomers, can also be expected in foods. The irradiation of foods may produce photodegradation products that also occur during the industrial production of ergocalciferol from ergosterol. The most important product of ergosterol irradiation is previtamin D_2, but tachysterol, lumisterol, and other products are formed as byproducts (Figure 5.4). Thermal transformation of vitamins D (at temperatures around 200 °C) produces pyroisomers and isopyroisomers (Figure 5.5), whilst isomerisation in acidic media provides isovitamins D and isotachysterols (Figure 5.6).

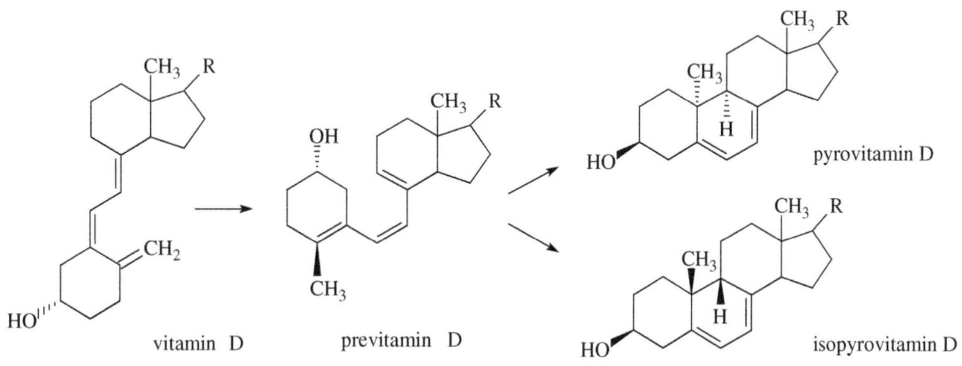

Figure 5.5 Thermal transformation of vitamins D.

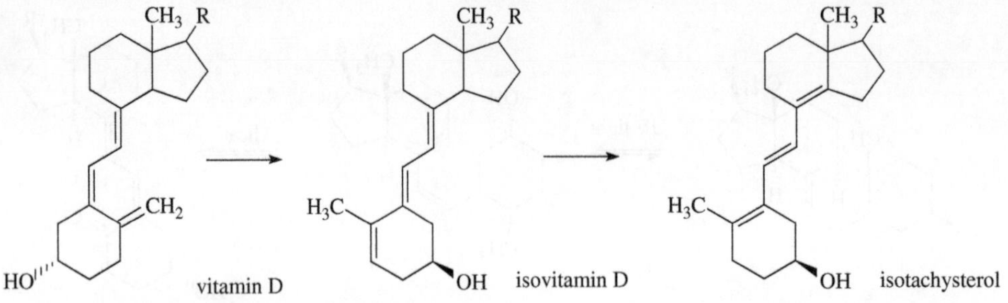

Figure 5.6 Isomerisation of vitamin D catalysed by acids.

5.4 Vitamin E

5.4.1 Structure and terminology

Vitamin E, formerly also known as antisterile vitamin, has eight basic structurally related derivatives of chroman-6-ol (chromane is benzodi-hydropyran or benzo[b]oxane). Structural bases common to all compounds with the reported activity of vitamin E (so-called vitagens E) are tocol (**5-12**) and tocotrienol (**5-13**), which contain a hydrophobic chromane ring with a saturated or unsaturated isoprenoid side chain of 16 carbon atoms. The systematic name of tocol is (2R,4′R,8′R)-3.4-dihydro-2-methyl-2-(4′,8′,12′-trimethyltridecyl)-2H-1-benzopyran-6-ol, whilst that of tocotrienol is (2R,3′E,7′E,4′R,8′R,11′E)-3,4-dihydro-2-methyl-2-(4′,8′,12′-trimethyltrideka-3′,7′,11′-trienyl)-2H-1-benzopyran-6-ol.

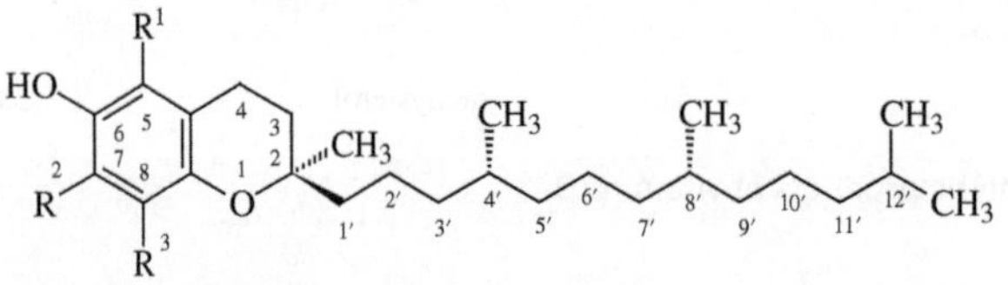

5-12, tocol derivatives

α-tocoferol (5,7,8-trimethyltocol), $R^1 = R^2 = R^3 = CH_3$
β-tocoferol (5,8-dimethyltocol), $R^1 = R^3 = CH_3$, $R^2 = H$
γ-tocoferol (7,8-dimethyltocol), $R^1 = H$, $R^2 = R^3 = CH_3$
δ-tocoferol (8-methyltocol), $R^1 = H$, $R^2 = H$, $R^3 = CH_3$
tocol, $R^1 = R^2 = R^3 = H$

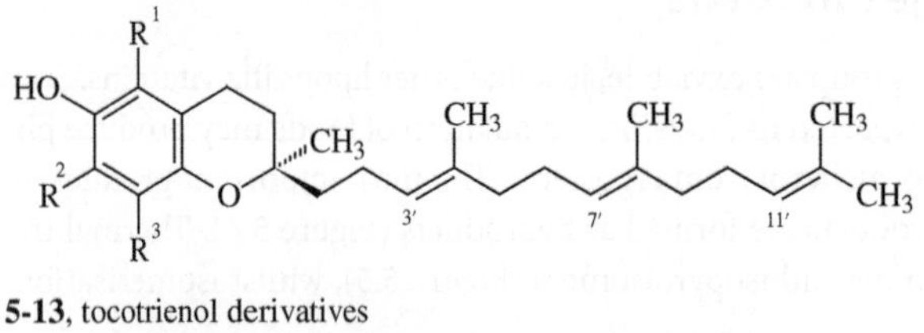

5-13, tocotrienol derivatives

α-tocotrienol (5,7,8-trimethyltocol), $R^1 = R^2 = R^3 = CH_3$
β-tocotrienol (5,8-dimethyltocol), $R^1 = R^3 = CH_3$, $R^2 = H$
γ-tocotrienol (7,8-dimethyltocol), $R^1 = H$, $R^2 = R^3 = CH_3$
δ-tocotrienol (8-methyltocol), $R^1 = H$, $R^2 = H$, $R^3 = CH_3$
tocotrienol, $R^1 = R^2 = R^3 = H$

The chromane ring is derived from the diterpenic alcohol phytol, (2E,7R,11R)-3,7,11,15-tetramethylhexadec-2-en-1-ol, but in addition to the hydroxyl groups in position C-6 it contains another methyl group at position C-2. The presence of these functional groups is essential for the biological activity of all vitamers. Four forms of vitamin E with a saturated terpenoid side chain derived from the tocol are called **tocopherols** (**5-12**); four forms with unsaturated side chains derived from tocotrienol are called **tocotrienols** (**5-13**). Tocopherols and tocotrienols are collectively known as **tocochromanols**. By virtue of the phenolic hydrogen on the 2H-1-benzopyran-6-ol nucleus, the individual compounds exhibit varying degrees of antioxidant activity, depending on the site and number of methyl groups and the type of isoprenoid side chain. The structurally related but biologically inactive 5,7-dimethyltocol was formerly called ζ-tocopherol, 5-methyltocol was ε-tocopherol, and 7-methyltocol was η-tocopherol.

Owing to the presence of three chiral centres (position C-2 of the chromane ring and position C-4′ and C-8′ in the side chain of tocol), each tocopherol can exist in eight diastereomeric forms. In nature, however, only (2R,4′R,8′R)-isomers of tocopherols exist. These are also known as (RRR)-isomers, 2D,4′D,8′D-tocopherols, d-tocopherols, or (+)-tocopherols.

(*RRR*)-α-Tocopherol is abbreviated to α-T. Tocotrienols containing three double bonds in the side chain may occur in eight different *Z* (*cis*) and E (*trans*) isomers and their combinations. In nature, only all-*trans* geometric isomers occur. α-Tocotrienol is abbreviated to α-T-3 or α-TT.

In photosynthetic organisms, including marine species, and as metabolic intermediates in animals, more than 200 chroman-6-ols and chromen-6-ols (**5-14**) can be found. Some vegetable oils also contain small amounts of 3,4-dehydrotocopherols (3,4-dehydrochromen-6-ols) and analogues of tocotrienols, tocodienols, and tocomonoenols (**5-14**) that have a partially saturated terpenoid side chain. α-Tocomonoenol occurs, for example, in crude palm oil (40 mg/kg), α-tocomonoenol (17.6 mg/kg) and γ-tocomonoenol (119 mg/kg) are found in roasted pumpkin seed oil, and δ-tocomonoenol is identified in the freeze-dried peel and pulp of kiwi fruits. α-Tocomonoenol, denoted α-T$_1$, is similar to α-tocopherol, but with a double bond at the side chain. There are only two known naturally occurring isomers of α-T$_1$, with double bonds in positions C-11′ and C-12′, respectively. Tocomonoenols also have three chiral centres at carbons C-2, C-4′, and C-8′, and all naturally occurring tocomonoenols have the *RRR* configuration. Rhizomes of plants from the genus *Dioscorea* (Dioscoreaceae), known as yams, are a common food in tropical areas and are widely used as a staple medicinal food. The oestrogenic activity of yams, *D. alata*, was attributed to α-tocopherol, γ-tocopherol-9 (**5-14**), hydro-Q$_9$ chromene (**5-14**), coenzyme Q$_9$, and 1-feruloylglycerol. Examples of bioactive 6-chromanols from lychee (*Litchi chinensis*, Sapindaceae) are litchtocotrienols A and E, cyclolitchtocotrienol A, and macrolitchtocotrienol A. (The names of recently described new compounds called litchtocotrienols are not properly created, as their molecules have only two double bonds in the side chain).

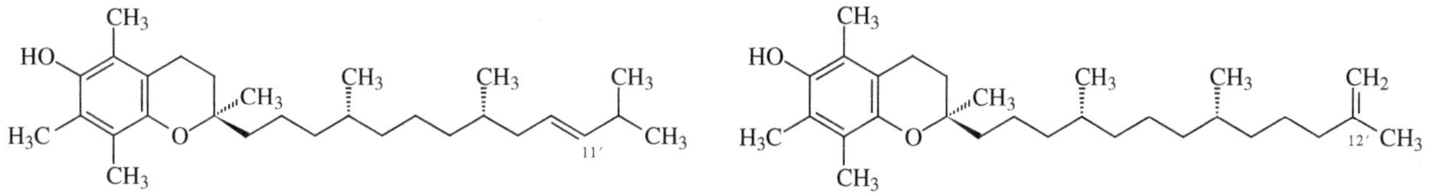

α-tocomonoenol isomers

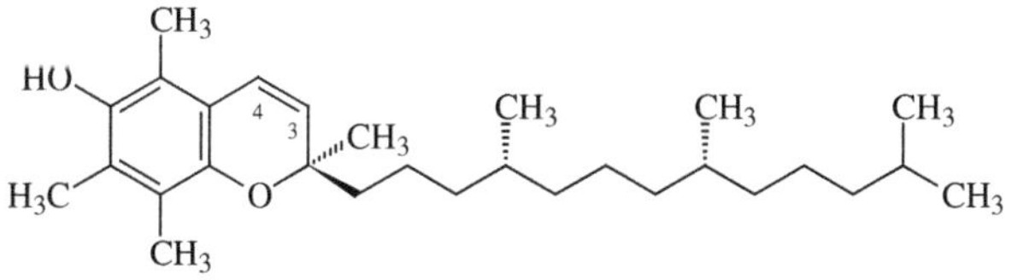

3,4-dehydro-α-tocopherol

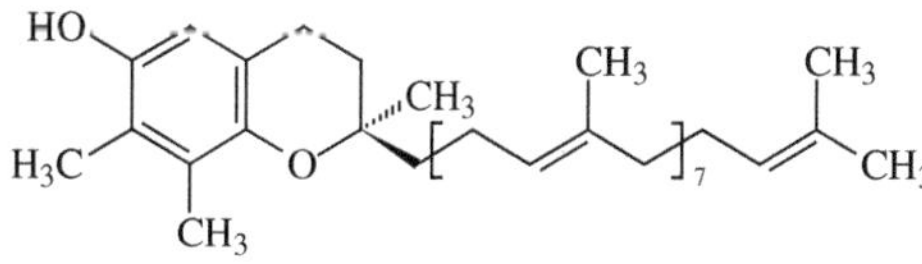

γ-tocopherol-9

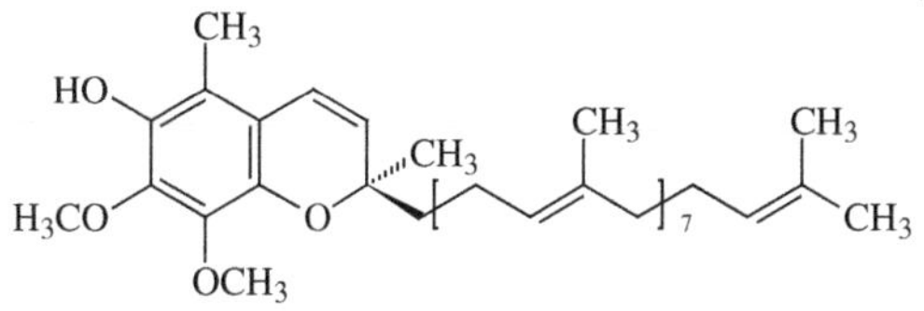

hydro-Q$_9$ chromene

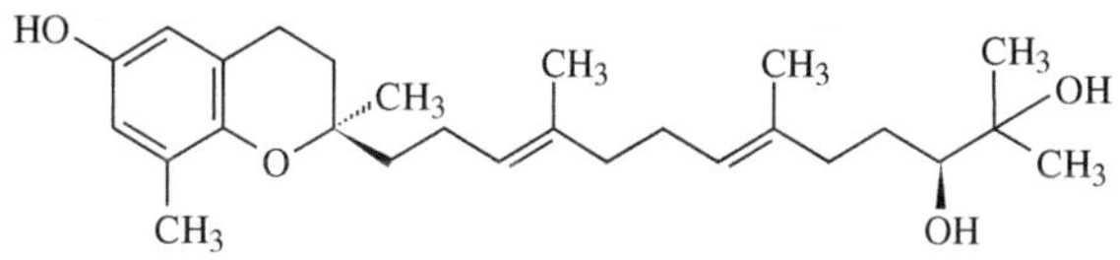

litchtocotrienol A [K1]

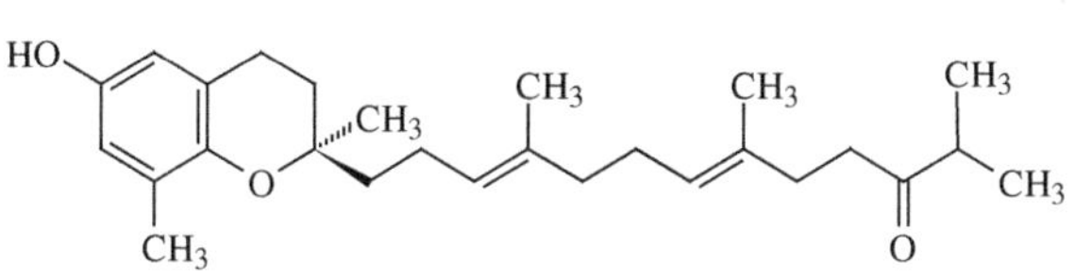

litchtocotrienol E

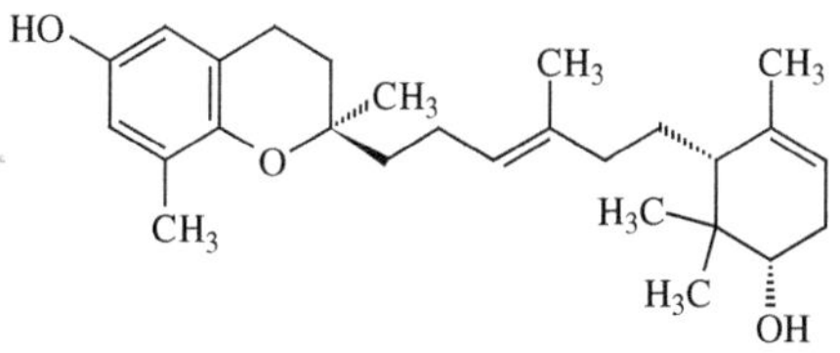

cyclolitchtocotrienol A

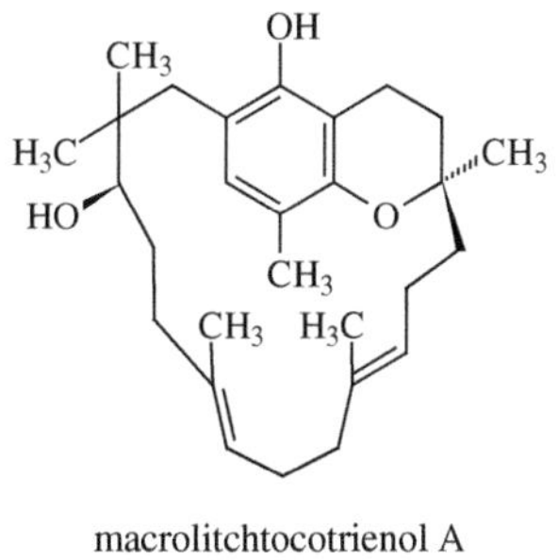

macrolitchtocotrienol A

5-14, examples of chromen-6-ols and chroman-6-ols related to tocopherols

5.4.2 Biochemistry

Vitamin E, as well as vitamin K, plastoquinones, and ubiquinones, is formed, in principle, from shikimic acid via 4-hydroxyphenylpyruvic and homogentisic acid. The terpenoid side chain is synthesised from geranylgeranyl diphosphate, which is gradually reduced to the less unsaturated phytyl diphosphate. The primary product of biosynthesis is δ-tocopherol; other tocopherols are the products of its methylation. The biosynthesis of tocotrienols lies in the condensation of homogentisic acid with geranylgeranyl diphosphate. Vitamin E is only synthesised by plants and some cyanobacteria.

Vitamin E (especially α-tocopherol) is the most important lipophilic antioxidant that acts in eucaryotic cells to protect unsaturated lipids against free radical damage. Along with β-carotene and coenzyme Q, it protects the structure and integrity of biomembranes, such as the cytoplasmic cell membrane and intracellular membranes and the lipoproteins present in plasma. It is transported in the bloodstream by association with low-density lipoprotein (LDL) particles (see Section 3.6.1). Each LDL particle contains six molecules of vitamin E.

Other functions of vitamin E are not yet fully elucidated. It probably directly contributes to the structure of biological membranes and modulation of their properties, because it specifically interacts with arachidonic acid and has some regulatory function in its metabolism (conversion of arachidonic acid into leucotrienes). Deficiency of vitamin E is manifested in females by placental disorders and in males by disorders of maturation of gametes.

5.4.3 Physiology and nutrition

An adequate intake of vitamin E is believed to prevent oxidation of lipids in biomembranes. Vitamin E is therefore a factor that slows the collection of changes known as the ageing process, and is applied in the prevention of cardiovascular diseases and cancer (oncogenesis).

The need for vitamin E is not yet precisely known. It depends greatly on the dietary intake of polyunsaturated fatty acids. For people whose average daily intake of these fatty acids is between 14 and 19 g, the recommended daily intake of vitamin E is 15 mg. More (RRR)-α-tocopherol (0.5–0.6 mg) is needed for each additional gramme of fatty acids taken in. For pregnant women, the recommended daily intake is higher by 2 mg, and for lactating women it is higher by 5 mg. The current increasing intake of unsaturated fats with low oxidation stability highlights the need for a revision of these recommendations. A real need is about 20–30 mg per day. The idea that there is a need for higher intake (of around 50 mg, but also up to 100 mg daily) has no basis in scientific knowledge or even in clinical trials.

The IU of activity is defined as an activity of 1 mg of synthetic all-rac-α-tocopheryl acetate (D,L-α-tocopheryl acetate) that is often used for food fortification and in multivitamin tablets. The biological activity of tocopherols and tocotrienols and their stereoisomers is difficult to determine, and is expressed in a relatively wide range. (RRR)-α-Tocopherol is generally considered the most effective substance, and its esters have similar effects. β-Tocopherol has about 50% the activity of α-tocopherol, γ-tocopherol about 10%, and δ-tocopherol approximately 3%.

Compared to natural α-tocopherol, synthetic α-tocopherol exhibits lower biological activity, because it consists of the C-2 enantiomers, (2R)-α-tocopherol and (2S)-α-tocopherol. In short, synthetic α-tocopherol is (2RS)-α-tocopherol (D,L-α-tocopherol or 2-ambo-α-tocopherol). It can also be a racemic mixture of all possible stereoisomers, in which case it consists of (2RS)-, (4'RS)-, and (8'RS)-α-tocopherol (or 2-ambo-, 4'-ambo, and 8'-ambo-α-tocopherol), which is called all-rac-α-tocopherol.

Commercial vitamin products often contain esters of α-tocopherol (acetate or hydrogen succinate), which are more stable to oxidation compared with free α-tocopherol. However, esterification of the C-6 hydroxyl group results in the formation of biologically inactive compounds, but these esters are rapidly hydrolysed to the biologically active α-tocopherol under the action of non-specific esterases.

The presence of double bonds in the molecule of the tocotrienols results in a decrease in their biological activity of about one-third compared with tocopherols. Biological activity of any significance is shown only by α-tocotrienol (about 30% of the activity of α-tocopherol) and β-tocotrienol (about 5% of the activity of α-tocopherol).

Vitamin requirements are mainly fulfilled by plant lipids, especially oils, and also by margarine enriched with tocopherols. Other foods of both plant and animal origin are important additional sources. Although they contain fewer vitamins, they are consumed regularly and in high quantities (e.g. bakery products, meat, eggs, and some vegetables).

Vitamin E deficiency is relatively rare, but occasionally occurs in newborns and adolescents. It manifests with similar symptoms as selenium deficiency, because the specific selenoproteins are also involved in the transport of tocopherols and protect the tocopherols against oxidation. Manifestation of vitamin E deficiency is mainly through degenerative nerve and muscle (neuromuscular) changes known as myopathy and encephalomalacia. Consumption of excessive amounts of tocopherols occurs only rarely.

5.4.4 Use

For fortification of foods and feeds and for pharmaceutical purposes, synthetic racemic α-tocopherol and its esters or a mixture of natural D-tocopherols obtained as a byproduct of refining (deodorisation) of oils from deodorisation condensates are used. The material produced should be protected from oxidation by addition of a phenolic antioxidant. Tocopherols β, γ, and δ can also be converted into α-tocopherol by methylation (tocotrienols by methylation and hydrogenation of the side chain).

5.4.5 Occurrence

Vitamin E is found primarily in foods of plant origin, and to a lesser extent (with a few exceptions) in foods of animal origin. Other sources do not have any practical significance (Table 5.5).

All eight biologically active tocopherols and tocotrienols occur in foods. Unsubstituted tocol does not occur in nature. Unsubstituted tocotrienol (also called desmethyl tocotrienol) was recently isolated from heated rice bran together with so-called didesmethyl tocotrienol, which lacks the C-2 methyl group. The main compounds are α-, β-, γ-, and δ-tocopherol and α-, β-, and γ-tocotrienol. Tocotrienols are also found in the form of esters.

In animal tissues, the composition of vitamers E is affected mainly by the feed. Vitamin content also varies according to the season. The main component (more than 90% of vitamin E) is always α-tocopherol. Animal fats contain much less vitamin E than vegetable oils. The total content of vitamin E in butter is up to 50 mg/kg (γ-tocopherol content is up to 10% of the total vitamin E content), in pork fat 6–30 mg/kg (about 5% as γ-tocopherol, 5% as α-tocotrienol), in beef tallow up to 20 mg/kg, and in chicken fat up to 25 mg/kg. Unlike other lipophilic vitamins, vitamin E does not occur in large quantities in fish oils. The content in cod liver oil is only 0.25 mg/kg.

In cereals, vitamin E is located mainly in the germ and bran, so white flours have lower vitamin content than whole-grain flours. The total vitamin E content is about 15–50 mg/kg. Details of the main forms of vitamin E in cereals and pseudocereals are shown in Table 5.6. The main form of vitamin E in the green parts of plants is α-tocopherol, located in the plastids; γ-tocopherol is the major form of vitamin E in cells that do not have chloroplasts (e.g. in seeds).

Table 5.5 Vitamin E content of selected food commodities.

Food	Content in edible portion (mg/kg or mg/l)	Food	Content in edible portion (mg/kg or mg/l)
Meet	2.5–7.7	Soybeans	2.7–13
Liver	4–14	Apples	1.8 7.4
Milk	0.2–1.2	Oranges	2.4–2.7
Butter	10–50	Cabbage	0.2–11
Cheeses	3.0–3.5	Spinach	16–25
Eggs	5–30	Tomatoes	3.6–4.9
Fish	4–80	Carrots	2.5–4.5
Flour (wheat)	15–50	Potatoes	0.6–0.9
Rice	0.4–4.5	Walnuts	35

Table 5.6 Tocopherols and tocotrienols content of cereals, pseudocereals, and plant oils (in mg/kg).

Vitamin	Wheat	Rye	Barley	Oat	Maize	Rice	Millet	Wheat bran oil	Rapeseed oil	Olive oil
α-T	9.7–14.0	8.0–16.0	3.7–11.6	4.3–8.9	0.2–22.1	0.6–7.5	0.8	1328	300	93–260
β-T	3.9–7.0	2.2–4.0	0.2–0.7	0.6–0.9	0.4	0.7	0.5	505		1.2–3.4
γ-T	–	6.0	0.2–12.9	0.9	17.5–56.0	0.8–4.0	23.3	112	481	2.6–9.8
δ-T	–	–	0.1–0.9	–	0.3–3.8	0.2–0.5	6.1	51	29	0.1–0.5
α-T$_3$	2.4–5.0	12.5–15.0	13.0–36.0	11.0–25.2	0.3–2.6	1.0–4.1	<0.2	40	<10	0.3–1.0
β-T$_3$	19.0–33.0	7.0–11.8	2.7–14.3	0.9–3.3	–	<0.2	–	140	–	–
γ-T$_3$	–	<0.2	3.6–8.4	0.2	3.3–9.9	2.0–9.5	0.7	–	–	0.4–0.9
δ-T$_3$	–	–	0.7–3.9	0.2	–	0.2–0.7	–	–	–	–
Total	35–59	32–44	31–80	19–38	22–94	4–27	31.6	2180	820	100–270

Oils from germs have even higher vitamin E contents than common vegetable oils. Virgin (crude) oils contain higher amounts of vitamin E than refined oils. The vitamin E content in crude rapeseed oil is 360–1000 mg/kg, whilst the refined oil vitamin content is 140–850 mg/kg. Crude and refined sunflower oils have vitamin E contents of 270–1240 and 270–900 mg/kg, respectively. Amongst the common vegetable oils, soybean oil contains the highest amount of vitamin E (530–2000 mg/kg); 11% of the vitamin is in the form of α-tocopherol, over 60% in the form of γ-tocopherol, and over 20% in the form of δ-tocopherol, with tocotrienols present in negligible quantities. The vitamin E content of olive oil is about 160 mg/kg. Germ oils contain large amounts of vitamin E. Wheat germ oil contains 1650–3000 mg/kg of vitamin E, which consists of about 61% of α-tocopherol, 23% of β-tocopherol, 5% of γ-tocopherol, 2% of δ-tocopherol, and approximately 8% of tocotrienols, of which β-tocotrienol is the dominating form (Table 5.6).

The content of vitamin E in fruits and vegetables usually does not exceed 10 mg/kg. The main vitamin form is α-tocopherol. Apples contain, on average, 1.8 mg/kg of α-tocopherol, but 6.1 mg/kg is found in the skin. The vitamin content in cabbage is about 0.9 mg/kg, of which only 0.1 mg/kg represents α-tocopherol; the rest is δ-tocopherol. Carrots contain about 2.5 mg/kg of α-tocopherol, 0.2 mg/kg of β- and γ-tocopherol, and 0.1 mg/kg of δ-tocopherol. Lettuces contain about 3.2 mg/kg of α-tocopherol, 1.5 mg/kg of β-tocopherol, and 2 mg/kg of γ-tocopherol. Potatoes contain very little vitamin E (about 0.1 mg/kg of α-tocopherol in the flesh and skin). Even tubers from transgenic plants (containing genes transferred from another species) accumulate approximately 10–100-fold less α-tocopherol than leaves or seeds. Relatively high amounts of vitamin E are found, of course, in potatoes fried in oil; for example, potato chips with reduced fat content (20.8%) contain about 55 mg/kg of α-tocopherol.

5.4.6 Reactions and changes in foods

Tocopherols and tocotrienols are cyclic monoethers of the respective hydroquinones and are therefore readily oxidised, for example by ferric ions, lipid hydroperoxides, and other oxidants. This creates the corresponding quinones (tocopheryl quinones or tocoquinones). Tocopheryl quinones can be reduced to tocopheryl hydroquinones (tocohydroquinones). The most important reactions are those with oxidised lipids (Figures 5.7 and 5.8).

The biological activity of vitamin E is related to its antioxidant effects; the most effective antioxidant (*in vivo*) is α-tocopherol. In food lipids, the situation is more complicated, because the antioxidant activity of tocopherols and tocotrienols depends on many factors. One important aspect is the amount and composition of unsaturated fatty acids. Under typical storage conditions, tocopherols are more effective antioxidants, for example, in animal fats (the main fatty acid is oleic acid) compared with their effect in vegetable oils that contain higher amounts of linoleic acid (e.g. soybean and sunflower oils). The antioxidant activity of vitamin E in emulsions depends on the structure of the emulsions and the presence of other antioxidants, such as 3,5-di-*tert*-butyl-4-hydroxytoluene (BHT) and ascorbyl palmitate. Temperature plays an important part, as does, particularly, the presence of oxygen and the stability of the radicals of tocopherols produced as intermediates in reactions with oxidised lipids. At 80 °C in the presence of air, δ-tocopherol, for example, is the only vitamin form which partially withstands heating for 6 h, when used as an antioxidant to protect linoleic acid against autoxidation. If lipids are in contact with an atmosphere containing only 10% oxygen (half of the amount of oxygen in air), β- and γ-tocopherols are also present, but α-tocopherol and all tocotrienols are absent. At 60 °C in the absence of oxygen, all tocopherols and tocotrienols are present.

For all these reasons, the following relative order of antioxidant efficiency of tocopherols in food is generally given, which is the opposite to the order of their biological activities (tocotrienols are less effective than the corresponding tocopherols):

$$\delta\text{-T} > \gamma\text{-T} > \alpha\text{-T} \ \text{ or } \ \delta\text{-T} > \gamma\text{-T} = \beta\text{-T} > \alpha\text{-T}$$

The mechanism of the antioxidant effect of vitamin E is similar to that of other lipophilic antioxidants. Tocopherols react with a number of free radicals, including active oxygen species. One tocopherol molecule can react with two hydroperoxyl radicals. Autoxidation of lipids is inhibited by reaction of tocopherols (abbreviated as T—OH) with hydroperoxyl lipid radicals (R—O—O•), with the formation of hydroperoxides (R—O—OH) and radicals of tocopherols (tocopheroxyl radicals, T—O•). This reaction interrupts the radical chain autoxidation reaction of lipids during the propagation phase:

$$R\text{—}O\text{—}O^{\bullet} + T\text{—}OH \rightarrow R\text{—}O\text{—}OH + T - O^{\bullet}$$

The resulting tocopheroxyl radical is not sufficiently reactive and therefore cannot split other lipid (fatty acid) molecules. In the termination phase of the chain autoxidation, the tocopheroxyl radical stabilises by irreversible reactions with other radicals, mostly with the second hydroperoxyl radical:

$$T\text{—}O^{\bullet} + R\text{—}O\text{—}O^{\bullet} \rightarrow \text{stable products}$$

Both reactions occur in lipids containing low concentrations of tocopherols and in the presence of sufficient amounts of oxygen. Alternatively, other reactions may take place. In the presence of large quantities of tocopherols, some tocopheroxyl radicals react with each other to form a dimer or a trimer:

$$T\text{—}O^{\bullet} + T\text{—}O^{\bullet} \rightarrow o\text{-quinone methide} \rightarrow \text{tocopherol dimer}$$

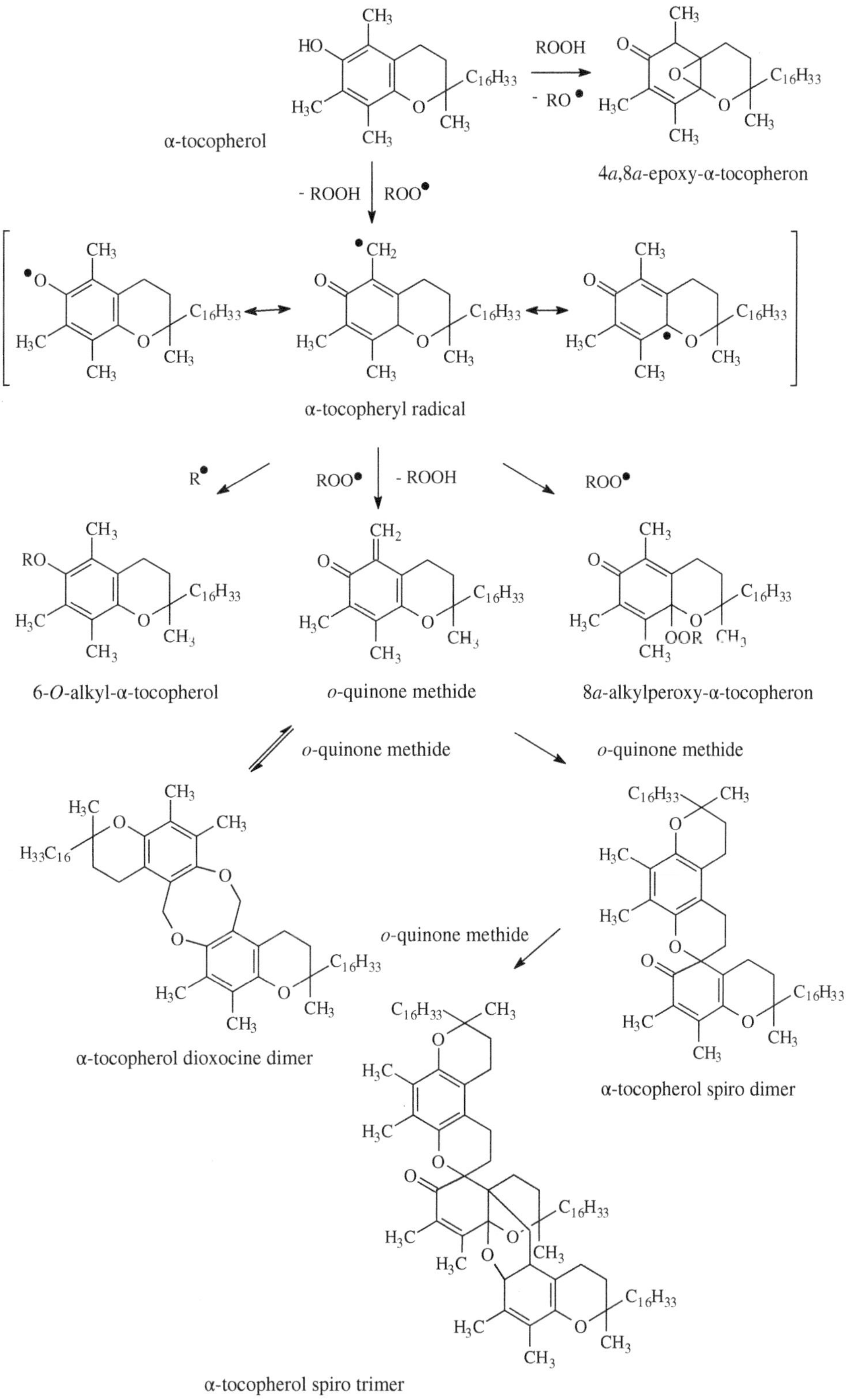

Figure 5.7 Reaction of α-tocopherol with oxidised lipids and the formation of oligomers.

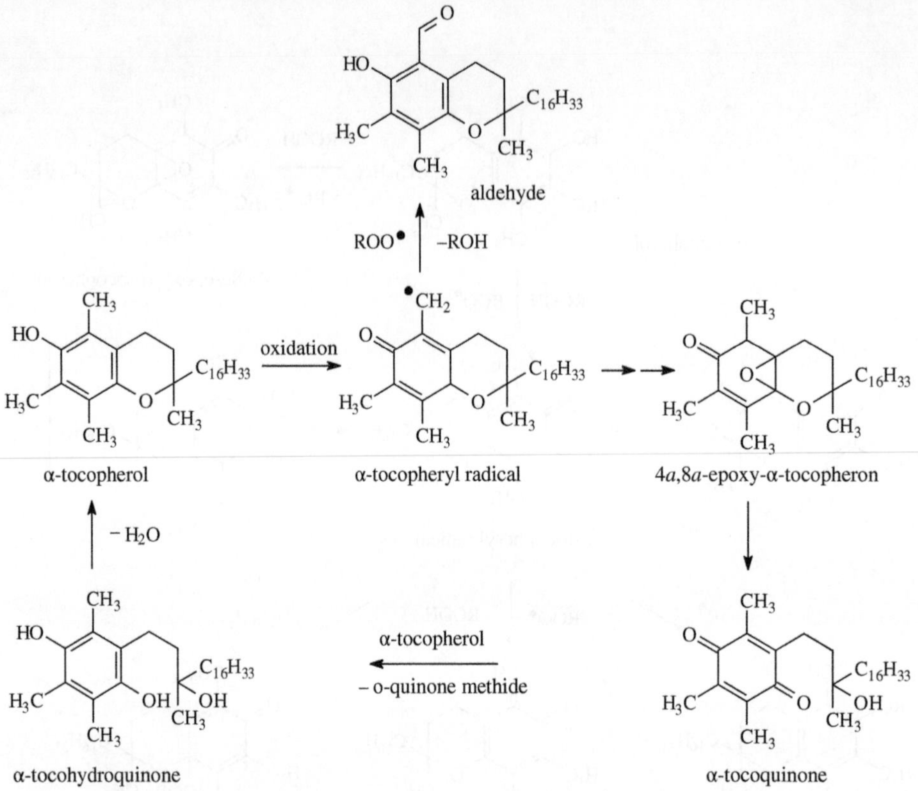

Figure 5.8 Some other reactions of α-tocopherol with oxidised lipids.

With an adequate supply of oxygen, the lipid radicals (R•), formed in the initial phase of the autoxidation chain reaction, preferentially yield hydroperoxyl radicals (R—O—O•). When there is a limited supply of oxygen (with a low partial pressure of oxygen) and in the absence of antioxidants, hydrocarbon radicals react with one another and form lipid dimers (R—R). In the presence of tocopherols, there is a competitive reaction of the lipid radicals with tocopheroxyl radicals that are created together with original lipids (R—H). The sequence of reactions then leads to stable products:

$$R^\bullet + R^\bullet \rightarrow R—R$$
$$R^\bullet + T—OH \rightarrow R—H + T—O^\bullet$$
$$R^\bullet + T—O^\bullet \rightarrow \text{stable products}$$

Figure 5.7 shows, as an example, the most important products of α-tocopherol in reactions with oxidised lipids. The main products, formed by reaction of tocopheroxyl radicals with hydroperoxyl radicals, are 8a-alkylperoxy-α-tocopherons.[4] To a lesser extent, the recombination of tocopheroxyl radicals with alkyl radicals (and alkoxyl radicals, R—O•) also occurs. The reaction products are the corresponding 6-O-alkyl-α-tocopherols. Dimers (which show reducing effects as well as tocopherols), trimers, and other products, such as epoxides, quinones, hydroquinones, and o-quinomethides, are likewise formed as minor products.

Other tocopherols and tocotrienols also react with oxidised lipids. Some reaction products of γ-tocopherol (**5-15**), such as tocopherol red (tocored, arising during bleaching of vegetable oils), C—C dimer, and C—O—C dimer and trimer, do not arise from α-tocopherol.

Tocopherols can also react with singlet oxygen as quenchers (analogously to β-carotene) with the formation of various oxidation products. The primary reaction is probably 1,4-cycloaddition of oxygen to form endoperoxides (Figure 5.9). If the phenolic group of the tocopherols is located at the lipid–water interface, tocopheroxyl radicals present in the lipid phase can be reduced back to tocopherols by water-soluble ascorbic acid, which thus also protects vitamin E against oxidation *in vivo*.

The refining of edible oils reduces the vitamin content to 10–50% of the original amount. The major losses occur during the deacidification step (removal of free fatty acids), if this step is performed by chemical methods (due to oxidation of vitamin E in the alkaline medium). High

---

[4]For example, oxidised linoleic acid esters (hydroperoxyl radicals of this acid) yield a mixture of (9R,8aR)-, (9R,8aS)-, (9S,8aR)-, and (9S,8aS)-9-(8a-peroxy-α-tocopheron)-(10E,12Z)-octadeca-10,12-dienoate, together with (13R,8aR)-, (13R,8aS)-, (13S,8aR)-, and (13S,8aS)-13-(8a-peroxy-α-tocopheron)-(9Z, 11E)-octadeca-9,11-dienoate.

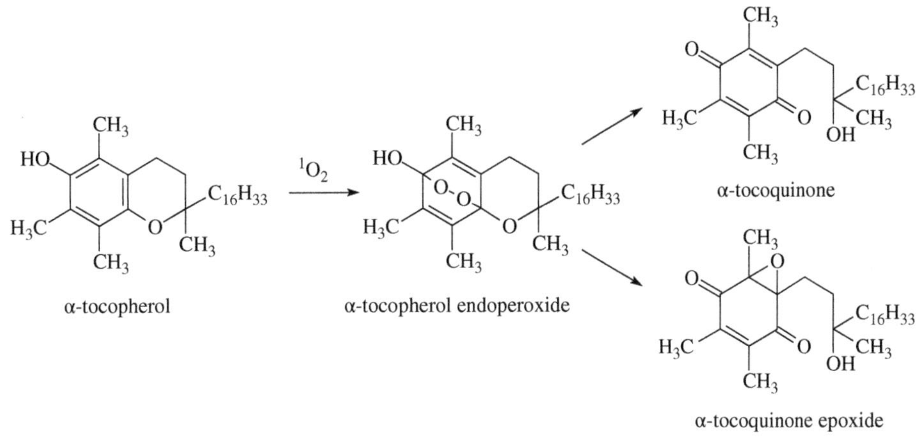

Figure 5.9 Oxidation of α-tocopherol by singlet oxygen.

losses of vitamin E also appear during oil bleaching (oxidation on the surface of bleaching earths catalysed mainly by ferric ions). Losses during the deodorisation step are mainly caused by volatilisation of vitamin E with water vapour under reduced pressure. Hydrogenation of fats using nickel catalysts (Raney nickel) leads to losses of 30–50%.

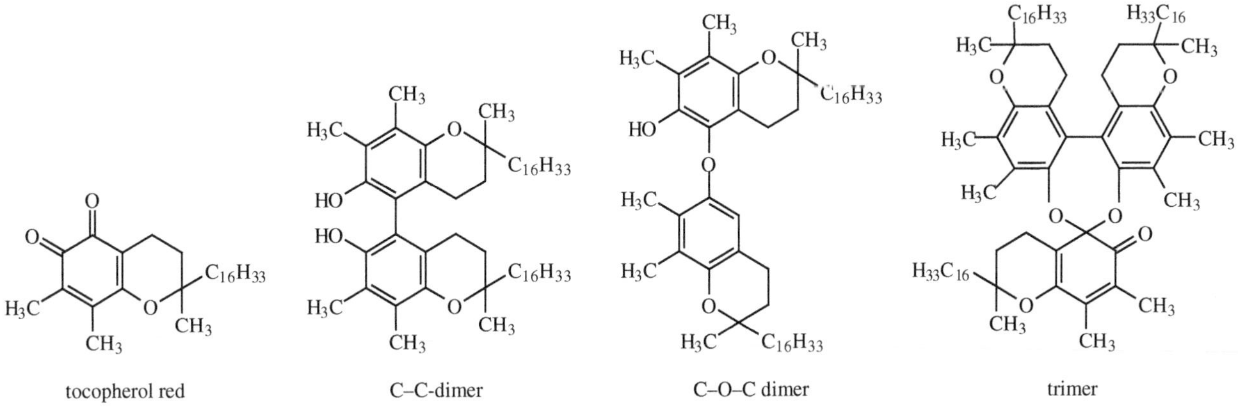

5-15, products of γ-tocopherol oxidation

In the absence of oxygen and oxidised lipids, vitamin E is relatively stable during normal culinary and industrial processing of foods. During processing and storage of meat, meat products, milk, dairy products, and cereals, the losses usually do not exceed 10% of the original amount. For example, the loss of vitamin E during pasteurisation of milk is about 5%. Stored grains lose about 10% of their vitamin content per month. The largest losses occur during frying and baking. In fats that are used repeatedly for frying food, tocopherols are virtually absent, because they are removed along with water vapour and decompose at higher temperatures. An analogous situation is in fried, frozen, and stored products, such as pre-fried potato chips (French fries). In general, the vitamin E content gradually decreases even under refrigerated storage of foods containing higher amounts of polyunsaturated fatty acids. Drying of fruits or vegetables results in a loss of 50–70% of vitamin E.

5.5 Vitamin K

5.5.1 Structure and terminology

All naturally occurring compounds that show vitamin K activity (the coagulation vitamin) are derivatives of menadione (2-methyl-1,4-naphthoquinone) with an isoprenoid unsaturated side chain at the C-3 position of the aromatic ring. Today, essentially two types of active substances are recognised. The other compounds with vitamin K activity are synthetic derivatives.

Vitamin K_1, also known as vitamin $K_{1(20)}$, phylloquinone, or phytylmenaquinone, contains a hexahydrotetraprenyl chain (2-methyl-3-phytyl-1,4-naphthoquinone, **5-16**). Phylloquinone occurs in foods of plant origin. The side chain derived from phytol has 20 carbon atoms (four isoprenoid units, of which three are reduced). The isomer with a *cis*-configuration in the side chain is not biologically active. The systematic name of vitamin K_1 is (2'E,7'R,11'R)-2-methyl-3-(3',7',11',15'-tetramethylhexadec-2-ene-1-yl)naphtho-1,4-quinone.

The second compound with the activity of vitamin K is vitamin K_2 (**5-16**), also known as vitamin $K_{2(n)}$, menaquinone, menaquinone-n, MK-n, or 2-methyl-3-multiprenyl-1,4-naphthoquinone (n = 0–13). The most common menaquinones contain 4–10 isoprene units, such as MK-7, also called pharnoquinone. Vitamin K_2 is produced by many bacteria and actinomycetes. A multiprenyl side chain with an all-*trans* configuration is the most common substituent. The bacteria usually produce menaquinone with different chain lengths. The main menaquinone of *Escherichia coli*, for example, is MK-8, whilst *Staphylococcus aureus* produces menaquinones MK-0 to MK-9. One or more isoprenoid groups in the side chain can be hydrogenated. Some bacteria also produce related 2-demethylmenaquinones. In bacteria and actinomycetes vitamin K_2 occurs at levels of about 600–1700 mg/kg dry matter. Yeasts do not produce vitamin K_2, but only related coenzymes Q.

phylloquinone   menaquinone ($n = 1$–13)

5-16, vitamin K

Vitamin K_3 (menadione, MK-0, 2-methyl-1,4-naphthoquinone) is a synthetic substance. The product of menadione reduction, known as menadiol (2-methylnaphthalene-1,4-diol), and derived compounds, such as the fat-soluble menadiol diacetate or menadiol dibutyrate and the water-soluble sodium salt of menadiol diphosphate, are referred to as vitamin K_4. Activity of vitamin K was also detected in the synthetic monoamino analogues and diamino analogues of menadiol, for example, in 1-amino-4-hydroxy-3-methylnaphthalene (vitamin K_5), 1,4-diamino-2-methylnaphthalene (vitamin K_6), and 1-amino-4-hydroxy-2-methylnaphthalene (vitamin K_7).

5.5.2 Biochemistry

Phylloquinone and menaquinone are derived from chorismic acid, which is formed from 3-phosphoenolpyruvic acid (a product of glycolysis) and D-erythrose 4-phosphate (a product of the pentose and Calvin cycles). It is transformed into isochorismic acid; other carbon atoms are derived from 2-oxoglutaric acid. The side chain is provided by phytyl diphosphate or by polyprenyl diphosphates, which are formed from geranylgeranyl diphosphate. The final reaction is a methylation at C-2.

Vitamin K in birds and mammals occurs in the reduced form (as a hydroquinone) and acts as an essential factor for the carboxylation of certain proteins (transformation of bound glutamic acid into γ-carboxyglutamic acid). Carboxyglutamic acid residues impart important properties to the relevant proteins, such as the ability to bind calcium ions and phospholipids, which are necessary for their activation and function in blood clotting. The best-known reaction is the conversion of inactive prothrombin into the active proteolytic enzyme thrombin. The main transport form in plasma is vitamin $K_{1(20)}$, which is bound to very-low-density lipoproteins (VLDLs). In the liver, about 90% of vitamin K occurs as menaquinones MK-7 to MK-12.

5.5.3 Physiology and nutrition

The daily requirement of vitamin K is estimated to be between 0.01 mg (for babies) and 0.14 mg (for adults). The daily intake by food is estimated at 0.3–0.5 mg, but only about 30–70% of dietary vitamin is absorbed in the intestines. According to some data, about 40–50% of the daily requirement of the vitamin is provided by foods, and under normal circumstances intestinal microflora produces the rest.

A deficiency in the production of vitamin K may occur during intestinal dysbiosis (an impaired balance of the intestinal microflora), in inflammatory intestinal diseases (such as Crohn's disease), in intestinal absorption disorders, and with an inadequate production of bile. Deficiency can be manifested by blood-clotting disorders (this is rare in humans). Chickens are particularly sensitive to the lack of vitamin K, mainly due to the use of antibiotics that inhibit the activity of intestinal microflora. Vitamin K is therefore added to feed mixtures to prevent bleeding into the muscles and skin.

Vitamin K antagonists (anticoagulants) are typically coumarins, and particularly dicoumarol (also known as melitoxin, **5-17**), which is produced in larger concentrations in the mouldy clover from 2-coumaryl-CoA under the action of fungi enzymes. Warfarin, that is 3-(3-oxo-1-phenylbutyl)-4-hydroxycoumarin (**5-17**), which also shows a significant anticoagulant effect, is used as a rodenticide (a chemical intended to kill rodents), but also in human medicine to prevent increased blood clotting, which can lead to embolisms. The mechanism of action of anticoagulants is that of competitive inhibition of the conversion of 2,3-epoxyvitamin K into vitamin K in the liver.

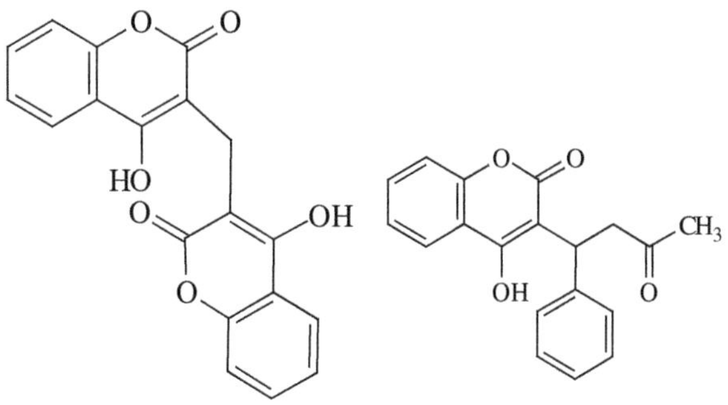

5-17, vitamin K antagonists

5.5.4 Use

Vitamin K_3 and its water-soluble forms, such as salts and complexes of its addition product with sodium hydrogen sulfite (trihydrate of sodium salt), as well as its complex with nicotinamide (5-18) or with other amino compounds, are used as additives in the fattening of chickens.

5.5.5 Occurrence

Meat and meat products have a moderately high vitamin K content. Liver is high in vitamin K (Table 5.7). In pig's liver, for instance, more than 10 active substances have been identified, amongst which vitamin $K_{1(20)}$ (0.012 mg/kg), MK-4 (0.011 mg/kg), MK-7 (0.016 mg/kg), MK-8 (0.025 mg/kg), MK-9 (0.006 mg/kg), and MK-10 (0.008 mg/kg) occur in significant quantities. Menaquinones MK-10 to MK 13, originally synthesised by bacteria in the rumen and subsequently absorbed, have been found in beef liver. The synthetic menadione and its derivatives are transformed by bacteria into MK-4 with a tetraprenyl side chain. This form of vitamin K also occurs in the tissue of animals whose feed was fortified with this vitamin.

Vitamin $K_{1(20)}$ is the only form of vitamin K that is found in foods of plant origin. It is a normal constituent of specialised cells for photosynthesis (chloroplasts). Rich sources of vitamin K_1 are mainly green leafy vegetables. For example, green cabbage leaves from the edge of the cone contain three to six times higher amounts of vitamin K_1 than the yellow leaves inside the cones. Vegetable oils are also good sources of vitamin K_1. Fruits, potatoes, and cereals contain lower amounts of vitamin K_1.

Table 5.7 Vitamin K content of some foods.

Food	Content in edible portion (mg/kg or mg/l)	Food	Content in edible portion (mg/kg or mg/l)
Pork liver	0.08	Cabbage	1.1–2.5
Beef liver	0.14	Broccoli	1.5–1.8
Pork meat	0.03	Spinach	2.0–14.4
Beef meat	0.04	Beans	0.1–0.5
Chicken meat	0.60	Peas	0.4
Fish (rainbow trout)	0.04	Tomatoes	0.02–0.06
Fish (Baltic herring)	0.01	Carrot	0.01–0.05
Eggs	0.02	Potatoes	0.01–0.02
Milk (fresh)	0.01–0.03	Strawberry	0.01
Yoghurt (plain)	0.01	Soybean oil	1.39–2.90
Cheese (Edam type)	0.49	Rapeseed oil	1.14–1.88
Cheese (Emmental type)	0.08–0.09	Olive oil	0.3–0.8
Bread	0.004	Sunflower oil	0.09

5.5.6 Reactions and changes in foods

Adverse reactions and losses of vitamin K activity occur under exposure to light (photodegradation to a number of products) or as a result of reactions with reducing agents or the transformations in alkaline media. The hydroquinone forms of the vitamin are oxidised to quinones by the air oxygen. Oxidation of phylloquinone (by hydrogen peroxide in an alkaline medium) yields 2,3-epoxide. During hydrogenation of vegetable oils, the side chain of phylloquinone is also partly hydrogenated. The hydrogenated fat contains the original phylloquinone, partly 2′,3′-dihydrophylloquinone (dihydrovitamin K$_1$, **5-18**), and decomposition products of phylloquinone, such as 2,3-dimethyl-1,4-naphthoquinone (**5-18**). The original vitamin content is decreased by 50% or more.

During storage and thermal processing of foods, vitamins K are relatively stable. Significant losses, however, occur when food is exposed to daylight. The vitamin content in vegetable oils decreases during frying (for 30 minutes at 190 °C), for example, to 85–90% of the original amount. If oil is exposed to daylight at ambient temperature, the vitamin content decreases to 50% in 1 day.

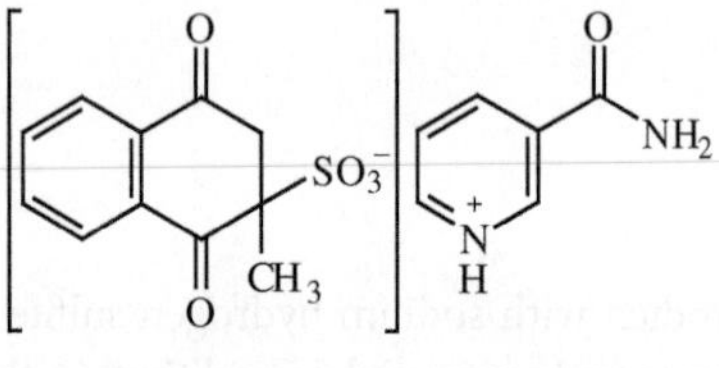

menadionbisulfite/nicotinamide complex

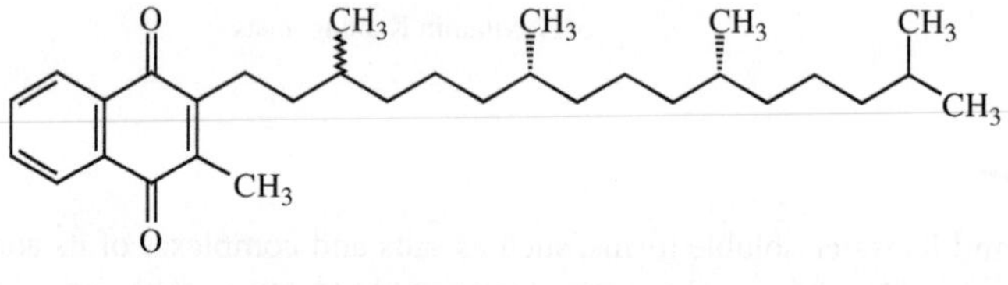

dihydrovitamin K$_1$

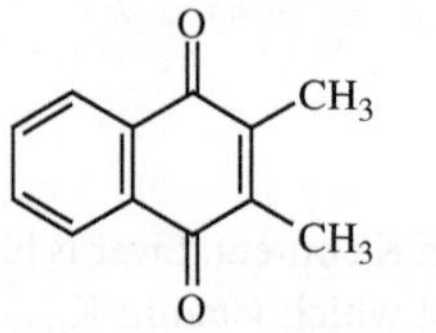

2,3-dimethyl-1,4-naphthoquinone

5-18, selected vitamin K derivatives

5.6 Thiamine

5.6.1 Structure and terminology

Thiamine (vitamin B$_1$, also formerly known as aneurine, **5-19**) contains a pyrimidine ring (4-amino-2-methylpyrimidine) attached by the methylene group at C-5 to the nitrogen of 5-(2-hydroxyethyl)-4-methylthiazole. It occurs primarily as a free compound, in the form of phosphate esters (**5-19**), as the monophosphate, diphosphate (pyrophosphate called cocarboxylase), and triphosphate, and as synthetic thiamine chloride hydrochloride, also known as thiamine hydrochloride (**5-19**).

thiamine (free base)

thiamine phosphates

thiamine phosphate, R = PO$_3$H$_2$
thiamine diphosphate, R = P$_2$O$_6$H$_3$
thiamine triphosphate, R = P$_3$O$_9$H$_4$

thiamine chloride hydrochloride

5-19, thiamine and thiamine derivatives

5.6.2 Biochemistry

Thiamine is synthesised *de novo* by all higher plants and many microorganisms. However, some microorganisms synthesise only the pyrimidine moiety (using 5-aminoimidazole ribonucleotide, an intermediate in the metabolism of purines) or only the thiazole part (from histidine and pyridoxol), and the remaining part of the thiamine molecule is obtained from the environment. Animals do not synthesise thiamine.

Free thiamine obtained from food (esters of thiamine after prior hydrolysis) is esterified to thiamine diphosphate, the active form of thiamine, in cells of various organs by pyrophosphokinase. Thiamine diphosphate is a cofactor of many important enzymes (decarboxylases, dehydrogenases, transketolases, and carboligases) associated primarily with the metabolism of carbohydrates and branched-chain aliphatic amino acids. Thiamine diphosphate bound to proteins is then esterified to thiamine triphosphate.

5.6.3 Physiology and nutrition

Thiamine is a cofactor of enzymes involved in energy metabolism; therefore, the required amount of the vitamin is mainly related to the amount of utilisable carbohydrates (glucose) received in the food. For every 4200 kJ (1000 kcal) of energy derived from carbohydrates, an intake of 0.4–0.6 mg of thiamine is recommended. For adults with a daily energy intake of 12 600 kJ (3000 kcal), the recommended intake of thiamine is 1.2 mg (for infants, typically 0.3 mg; for children, 0.7–1.2 mg; for men, 1.2–1.5 mg; for women, 1.0–1.1 mg; for pregnant women, 1.4 mg; and for lactating women, 1.5 mg).

Thiamine, like some other vitamins, is produced by intestinal microflora. The level of vitamin delivered in this way is too low, however, so in practice the required amount is only obtained through food. The most important source of thiamine is whole-grain cereal products, which supply about 40% of requirements (bread covers about 20%). Other important sources are meat and meat products (18–27%), milk and dairy products (8–14%), potatoes (10%), legumes (for reasons of low consumption, about 5%), vegetables (up to 12%), fruits (about 4%), and eggs (about 2%).

Thiamine deficiency is manifested by non-specific symptoms such as muscle fatigue, anorexia, weight loss, and irritability. The reason is an accumulation of lactic acid and pyruvic acid coming from glucose catabolism. Vitamin deficiency may appear as a result of its increased demand (e.g. during pregnancy, lactation, and sport activities), significantly inadequate or faulty nutrition, artificial (parenteral) nutrition over a long period of time, reduction in the diet, haemodialysis, treatment with antibiotics (which suppress the intestinal flora that synthesise B group vitamins), and chronic alcoholism. A deficient intake brings about alcoholic cardiomyopathy, the Wernicke–Korsakoff syndrome (encephalopathy and psychosis), or beriberi (a neurological and cardiovascular disease).

Thiamine has a high specificity, which means that even small changes of the molecule lead to a reduction of the vitamin effect, inefficiency, and in some cases even an antivitamin effect. A well-known thiamine antagonist is oxythiamine (5-20), which contains as a substituent of the pyrimidine ring a hydroxyl group instead of an amino group. Oxythiamine is formed from thiamine in strongly acidic media and therefore occurs, for example, in acid protein hydrolysates used as soup seasonings. Oxythiamine easily forms diphosphate, a competitive inhibitor of thiamine in enzymatic reactions.

Thiaminases – enzymes that decompose thiamine – also act as antivitamins. Higher activity of thiaminase I is seen in raw meat and some fresh raw sea fish, molluscs, the raw offal of farm animals, and some plants. The reaction catalysed by thiaminase I lies in the cleavage of the thiamine molecule by an exchange reaction with nitrogen bases (such as nicotinic acid) or thiols (such as cysteine). Thiaminase II is an enzyme present in many bacteria. This enzyme catalyses thiamine hydrolysis with the formation of the same products that form by non-enzymatic hydrolysis (Figure 5.10).

5.6.4 Use

In some countries, white wheat flour, breakfast cereals, and rice are fortified with thiamine. For further processing in the pharmaceuticals and food industries, the most commonly used compound is synthetic thiamine chloride hydrochloride.

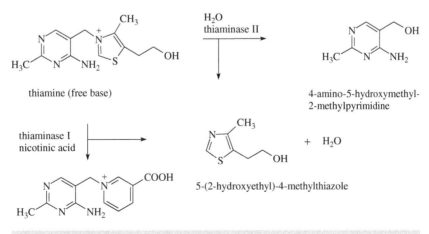

Figure 5.10 Degradation of thiamine by thiaminases.

Vitamin activity is also seen in some lipophilic thiamine derivatives with an opened thiazole ring (disulfides derived from the thiol form of thiamine). Oxidative cleavage of the thiazole ring in alkaline solution yields thiamine thiol, which can react with other thiols, forming thiamine alkyl disulfides. A number of disulfides known as allithiamines occur in plants, especially in members of the genus *Allium* (Amaryllidaceae). Thiamine allyl disulfide (**5-20**, R = CH_2—CH=CH_2) is a lipid-soluble form of thiamine that occurs naturally in garlic. In higher quantities, it forms via the heating of thiamine with garlic extract. Allithiamines are biologically active compounds as, on reductive cleavage of the disulfide bridge, they spontaneously dehydrate to yield thiamine. Some synthetic allithiamines (such as thiamine propyl disulfide, **5-20**, R = CH_2—CH_2—CH_3) have been used for the prevention and treatment of thiamine deficiency.

5.6.5 Occurrence

Free thiamine and its phosphoric acid esters occur in all foods, but only in some of them they are present in significant amounts (Table 5.8). In general, higher concentrations of thiamine (1–10 mg/kg) are found in foods that are rich in carbohydrates and undergo intensive metabolism of sugars (cereals and legumes, as well as pork meat and liver).

Animal tissues contain approximately 80–90% of the vitamin as thiamine diphosphate bound to proteins. Pork and chicken meat also contain higher concentrations of thiamine triphosphate (70–80% of the total vitamin content). A particularly rich source of thiamine is pork meat (which contains about 10 times more thiamine than other types of meat), ham, and other pork products, as well as other types of meat, milk, dairy products, and eggs.

Thiamine is present in milk mainly as a free compound, and the diphosphate is partly bound to proteins. Cow's milk contains about 50–75% of free thiamine and 18–45% of phosphorylated thiamine, which is more labile than free thiamine. About 5–17% of thiamine phosphates are bound to proteins. These forms contribute to about 90% of the activity of free thiamine.

Table 5.8 Group B vitamin content of common foods.

Food	Content in edible portion (mg/kg or mg/l)						
	Thiamine	Riboflavin	Niacin	Pyridoxine	Pantothenic acid	Biotin	Folacin
Pork meat	3.9–11	0.9–3.5	18–130	0.8–6.8	3.0–30	0.05	0.01–0.04
Beef meat	0.4–1.0	0.4–3.5	38–102	0.8–5.0	3.0–20	0.02–0.03	0.02–0.18
Chicken meat	1.0–1.5	0.7–2.8	93–122	2.6	5.3–9.6	0.11	0.10–0.12
Pork liver	2.7–7.6	29–44	164–223	1.7–5.9	4.0–200	0.90–1.00	1.36–2.21
Fish	0.6–1.7	1.0–3.3	22–84	4.5–9.7	1.2–25	0.02–0.26	0.12
Milk	0.3–0.7	0.2–3.0	0.8–5	0.2–2.0	0.4–4.0	0.01–0.09	0.03–0.28
Cheeses	0.2–0.6	3.3–5.7	0.3–16	0.4–0.8	2.9–4.0	0.02–0.05	0.08–0.82
Eggs	0.7–1.4	2.8–3.5	30	1.9–2.5	16–55	0.09–0.30	0.05–0.80
Flour (wheat)	0.6–5.5	0.2–1.2	9.0–57	1.2–6.0	8.0–13	0.01–0.06	0.60–1.46
Bread	0.6–3.0	0.6–1.5	8.0–34	0.3–3.0	4.0–5.0	0.01–0.02	0.26–0.54
Legumes	2.0–8.4	1.2–2.8	14–31	6.3	9.4–14	0.13–0.60	0.41–2.9
Cabbage	0.5–0.6	0.5	3.0	2.7	1.0–3.0	0.01–0.02	0.16–0.45
Spinach	0.5–1.5	0.6–3.4	6.0	2.2	1.8–27	0.03–0.07	0.50–1.92
Tomatoes	0.6	0.3–0.4	7.0	1.3–1.6	3.0–4.0	0.02–0.04	0.06–0.30
Carrot	0.3–1.4	0.5–2.6	5.0–15	1.0–7.0	3.0	0.03–0.04	0.4
Potatoes	0.5–1.8	0.3–2.0	10–20	1.4–2.3	3.0	0.01–0.02	0.08–0.20
Apples	0.4	0.1	1.0	0.3	1.0	0.01	0.06
Citrus fruits	0.4–1.0	0.2–0.4	1.0–4.0	0.2–1.7	2.0	0.01–0.03	0.05–0.40
Banana	0.5	0.4–0.6	7.0	2.6–3.1	2.0	0.04	0.28–0.36
Nuts	0.5–0.6	0.2–1.3	5.0–9.0	3.0	1.0	0.01–0.91	0.70
Yeast	7.1	17–44	112–200	11–55	50–200	0.80	15

In cereals, legumes, and generally all plant seeds, the most common form of thiamine is free thiamine. In cereals, which are the most important source of thiamine due to their large consumption, thiamine is found mostly in the outer layers (aleurone layer) of the grain and in the germ, which are removed during the milling process, and thus a large proportion of the thiamine is in the bran. White flours, therefore, contain about 10 times less thiamine than whole-grain flours. Whole-grain cereal products that are rich in thiamine, however, also contain relatively high concentrations of fibre and phytates, which inhibit intestinal absorption of thiamine. Brown rice (or hulled rice), which is unmilled or partly milled, contains a higher amount of thiamine than milled white rice (with the bran and germ removed). Parboiled rice contains a higher amount of thiamine than white rice, as it is absorbed during the process of parboiling. In this process, the harvested rice (rice with husk) is hydrated and then steamed, before drying. Once dried completely, the husk is removed. As a result, the majority of thiamine is absorbed in the inner parts of the grain. Other important sources of thiamine are potatoes and legumes.

Brewer's yeast is a rich source of thiamine, because the cells absorb thiamine present in the malt (containing about 160 mg/kg thiamine). Conversely, in beer (if it is filtered), the thiamine concentration is very low (0.01–0.06 mg/l).

5.6.6 Reactions and changes in foods

Thiamine is relatively stable only in acidic solutions (pH < 5). Thiamine diphosphate is unstable in weakly acidic and neutral solutions, and its hydrolysis yields thiamine monophosphate and thiamine. In neutral and alkaline solutions, thiamine exists as the free base, which is very unstable. It is hydrolysed to 4-amino-5-hydroxymethyl-2-methylpyrimidine and 5-(2-hydroxyethyl)-4-methylthiazole, and 4-amino-5-aminomethyl-2-methylpyrimidine arises as a minor product. Another non-volatile product, 2-methyl-4-amino-5-(2-methyl-3-furylthiomethyl)pyrimidine (**5-20**), is formed from the pyrimidine moiety of thiamine and a degradation product of thiamine 2-methyl-3-furanthiol. Similarly, the thiamine molecule splits into two parts by the action of sulfur dioxide or hydrogen sulfites (Figure 5.11). This cleavage produces 5-(4-amino-2-methyl-4-pyrimidinyl)methanesulfonic acid and 5-(2-hydroxyethyl)-4-methylthiazole. In aqueous solutions, the thiazole ring of thiamine opens to form thiamine thiol, which exists as a salt in alkaline solutions (both forms also arise in foods). Thiamine thiol is easily oxidised to thiamine disulfide, and thiochrome results as a minor product along with some other compounds (Figure 5.12).

A large number of other decomposition products of thiamine arise in aqueous solutions during boiling and by photodegradation. More than 70 degradation products have been identified. In addition to simple compounds, such as hydrogen sulfide, ammonia, formaldehyde, acetaldehyde, formic acid, and acetic acid, dozens of other minor sulfur compounds are formed: aliphatic sulfides, aliphatic carbonyl compounds with a thiol group, furans containing sulfur in the molecule (including the meaty aroma character imparting compound 2-methyl-3-furanthiol), thiophenes, thiazoles, and alicyclic and heterocyclic bicyclic sulfur compounds. An important precursor of many decomposition products is 5-hydroxy-3-mercaptopentan-2-one (Figure 5.13). Thiamine is therefore often used as a component in mixtures (traditionally mixtures of proteins or amino acids with sugars and other ingredients) called reaction flavours (also known as process flavours) that are powerful, highly concentrated foundations for flavour compounds or seasonings and can be used directly in processed foods. A particularly significant product of thiamine decomposition is bis(2-methyl-3-furyl)disulfide. Its threshold concentrations are extremely low. This disulfide also appears as a product of thiamine degradation during storage of multivitamin tablets or thiamine tablets, and is responsible for their distinctive odour.

Some reactions of thiamine, such as decarboxylation of α-oxocarboxylic acids, are the features of primary metabolism (decarboxylation of pyruvic acid to acetaldehyde in glycolysis and transformation of pyruvic acid to acetyl-CoA prior to its entry into the citric acid cycle) that depend on thiamine diphosphate as a coenzyme. The thiamine ring has an acidic hydrogen and is thus capable of producing the carbanion that acts as a nucleophile towards carbonyl groups. Analogous reactions proceed non-enzymatically and can thus be included as a part of the so-called Maillard reaction. In neutral solutions, for example, thiamine reacts with reducing sugars and other carbonyl compounds (Figure 5.14). Its reaction with glucose yields glucothiamine as a primary product. Reaction with furan-2-carbaldehyde, which is one of the main degradation products of ascorbic acid in acidic solutions, explains the decomposition of thiamine in the presence of vitamin C.

Amino acids catalyse the thiamine degradation in alkaline solutions. The products are hydrogen sulfide and dethiothiamine (**5-20**). With cysteine and other thiols (and cystine and other disulfides), thiamine reacts to form mixed thiamine disulfides. Such mixed disulfides are the previously mentioned allithiamines (**5-20**). Analogously, thiamine reacts with proteins containing cysteine thiol groups or disulfide bonds in the side chains. These reactions explain the protective effect of thiamine on protein degradation (Figure 5.15).

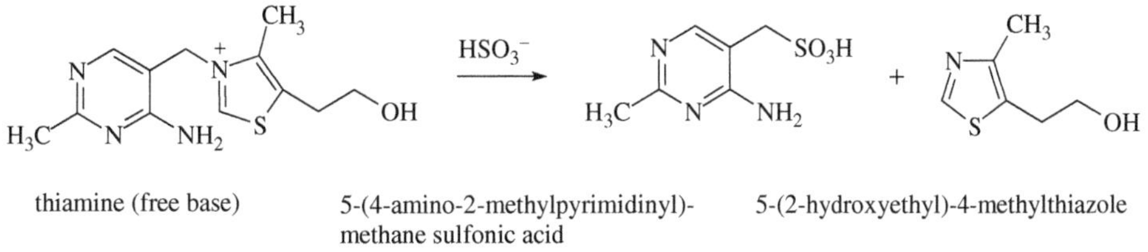

thiamine (free base)   5-(4-amino-2-methylpyrimidinyl)-   5-(2-hydroxyethyl)-4-methylthiazole
                                methane sulfonic acid

Figure 5.11 Degradation of thiamine by sulfur dioxide.

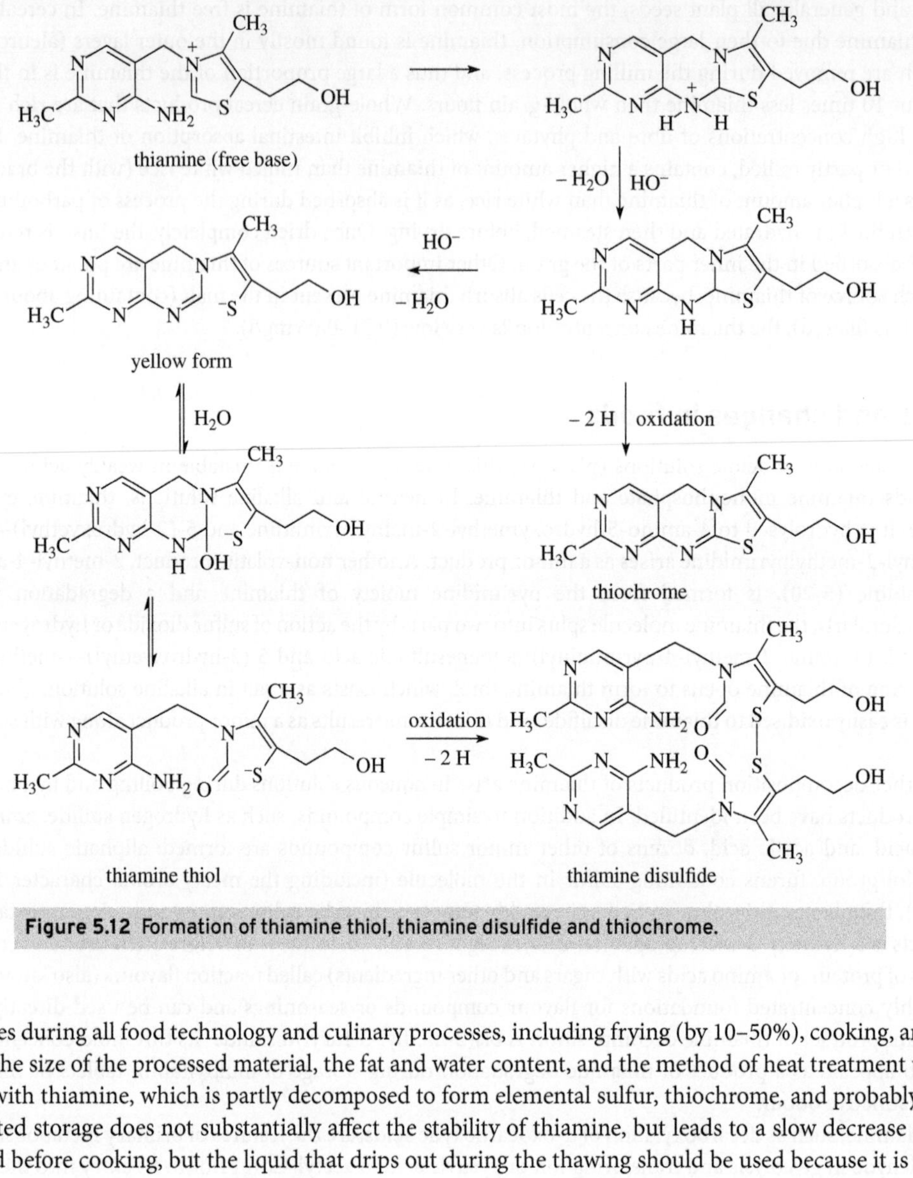

Figure 5.12 Formation of thiamine thiol, thiamine disulfide and thiochrome.

Thiamine decomposes during all food technology and culinary processes, including frying (by 10–50%), cooking, and stewing (50–70%). The losses depend on the size of the processed material, the fat and water content, and the method of heat treatment used. The nitrite used in meat curing reacts with thiamine, which is partly decomposed to form elemental sulfur, thiochrome, and probably oxythiamine (**5-20**). Freezing and refrigerated storage does not substantially affect the stability of thiamine, but leads to a slow decrease of its content. Frozen meat should be thawed before cooking, but the liquid that drips out during the thawing should be used because it is very rich in thiamine and other nutrients.

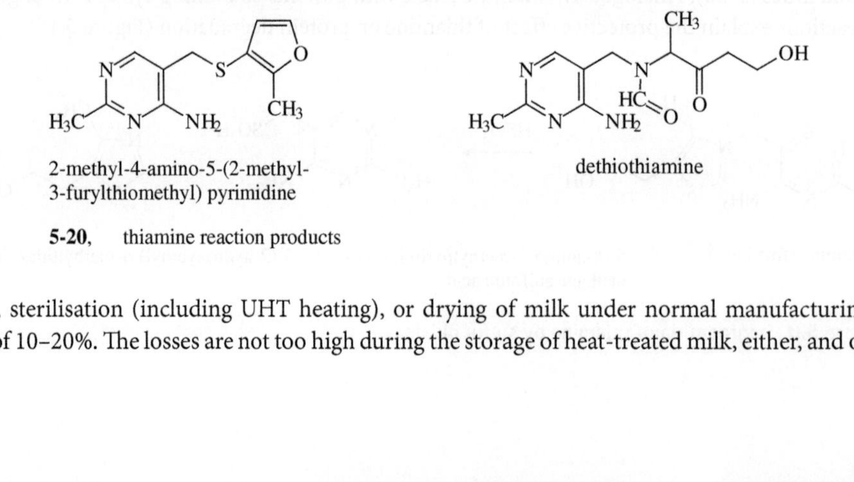

oxythiamine (free base)

allithiamines

2-methyl-4-amino-5-(2-methyl-3-furylthiomethyl) pyrimidine

dethiothiamine

5-20,   thiamine reaction products

During pasteurisation, sterilisation (including UHT heating), or drying of milk under normal manufacturing conditions, the loss of thiamine is in the range of 10–20%. The losses are not too high during the storage of heat-treated milk, either, and of course are proportional

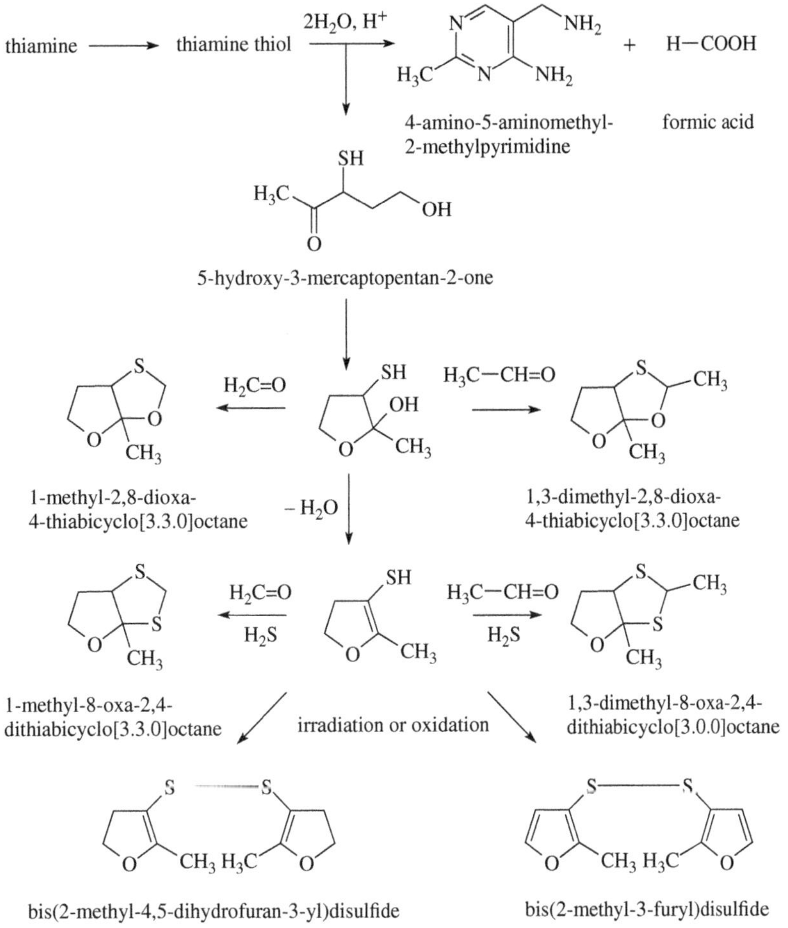

Figure 5.13 Formation of flavour-active products of thiamine.

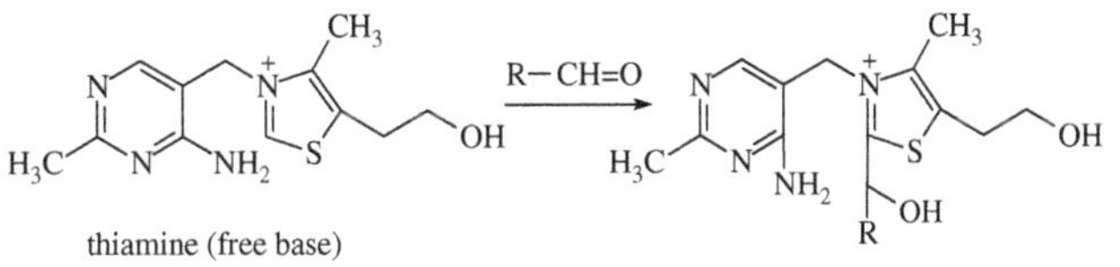

Figure 5.14 Reaction of thiamine with carbonyl compounds.

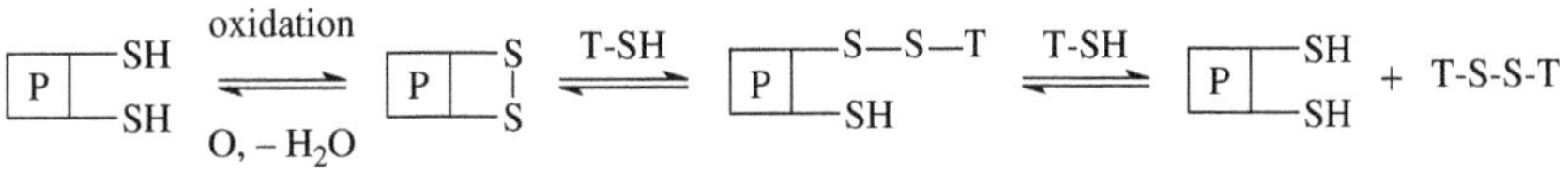

Figure 5.15 Reaction of thiamine with proteins. P, protein; T-SH, thiamine thiol; T-S-S-T, thiamine disulfide.

to storage time and temperature. The stability of thiamine in dried milk depends on the conditions during storage, mainly on temperature and the presence of oxygen. Losses usually do not exceed 20%.

Thiamine, like other vitamins, is not evenly distributed in cereal grains. A higher amount is found in the outer layers, and the concentration in flour depends on the degree of milling (known as flour extraction rate). For example, if the thiamine content of the original wheat is 3.87 mg/kg (expressed on a 13% moisture basis), the thiamine content in the 85% straight-run flour (yield of 85%, expressed as a percentage of the wheat represented) is 3.42 mg/kg, that in the 80% straight-run flour is 2.67 mg/kg, and that in the 70% straight-run flour is only

0.70 mg/kg, which is 18% of the original amount. Losses during storage of flour are relatively small (usually about 10% according to the conditions and length of storage).

During bread baking, the total losses are relatively small (about 20% of the amount in the flour), but in the crust they reach up to 90%. There are losses of up to 80% in the baking of biscuits. The higher losses are caused by the use of alkaline baking agents (such as carbonates of alkali metals). During grain extrusion, losses are relatively high (20–80%), depending on temperature, water content, and the presence of oxygen. Retention of thiamine generally decreases with increased temperature and decreased water content. Cooking of pasta leads to approximately 40% loss of thiamine, which is largely caused by vitamin leakage into the water.

Losses during the cooking of root vegetables are usually around 25% and in leafy vegetables are about 40%. The addition of sulfites, used to prevent enzymatic browning reactions in peeled potatoes or dried fruits and vegetables, results in extensive or total destruction of thiamine.

5.7 Riboflavin

5.7.1 Structure and terminology

The basis of the yellow-green riboflavin structure (vitamin B_2, formerly also known as vitamin G, lactoflavin, ovoflavin, or uroflavin) is an isoalloxazine skeleton, to which is bound the alditol ribitol at position N-10 (**5-21**). Riboflavin exists in an oxidised (flavoquinone) and a reduced form (flavohydroquinone).

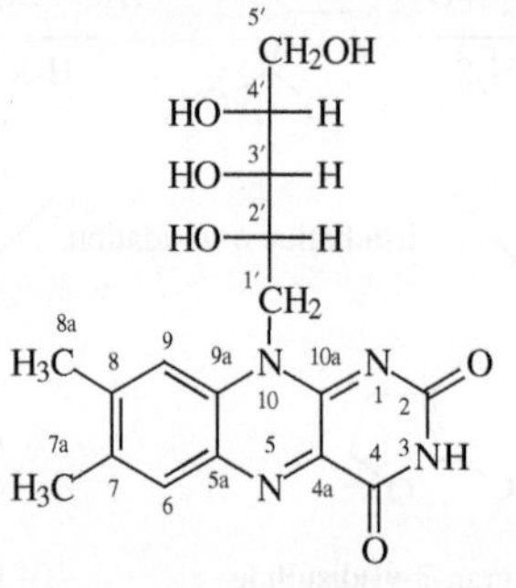

5-21, riboflavin (oxidised form)

Riboflavin, 7,8-dimethyl-10-(1′-D-ribityl)isoalloxazine, likewise occurs as a free compound, but predominantly exists in the form of riboflavin 5′-phosphate (flavin mononucleotide, FMN, **5-22**) and flavin adenine dinucleotide (FAD, **5-22**). These compounds, known as covalently bound riboflavin, are cofactors of enzymes known as flavoproteins, where carbons in positions C-6 or C-8a are involved in covalent bonding of flavins to apoproteins. The corresponding salts are present under physiological pH values. One-electron reduction of riboflavin generates two forms of riboflavin radicals, a red anion and a blue neutral species (called flavosemiquinone, Figure 5.16). A reduced, almost colourless form that acts in the enzyme-catalysed reduction–oxidation (redox) reactions is 1,5-dihydroriboflavin (another name is dihydroflavin or leucoflavin), which spontaneously oxidises by air oxygen to riboflavin. Another reduced form (generated by two-electron reduction in some enzymatic reactions) is 4a,5-dihydroriboflavin (**5-23**).

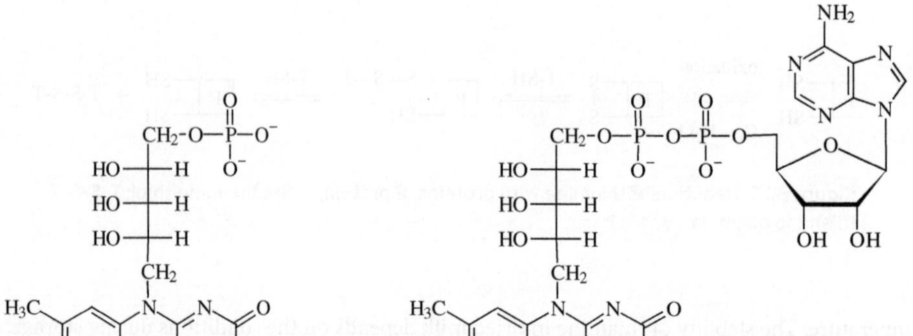

flavin mononucleotide   flavin adenin dinucleotide

5-22, phosphorylated forms of riboflavin

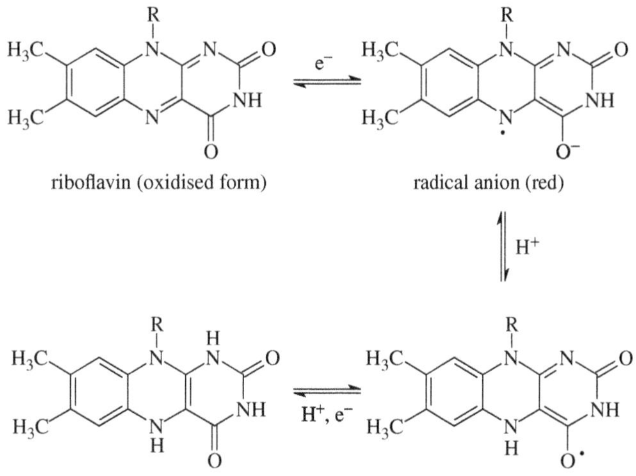

Figure 5.16 Oxidised and reduced forms of riboflavin (see **5-21** for R).

5.7.2 Biochemistry

Riboflavin is synthesised by a variety of microorganisms and plants. The starting substrate is a guanosine 5′-triphosphate (GTP). Animals only transform riboflavin received in food into FMN and FAD. The flavin enzymes are structurally and functionally similar to pteridine enzymes. Flavoproteins containing FMN and FAD cofactors are involved in both one-electron and two-electron redox reactions. One-electron acceptors are haem proteins or proteins containing sulfur and iron; a two-electron donor is, for example, the reduced form of nicotinamide adenine dinucleotide, NADH. Oxygen reacts with the flavins in several ways. For example, it acts as a one-electron acceptor, which yields the superoxide radical, or as a two-electron acceptor, which creates hydrogen peroxide. Most of the enzymes catalysing dehydrogenation, oxygen activation (hydroxylation, monooxygenation), and transmission of electrons are part of the respiratory chain located in the mitochondria. Some enzymes (such as glucose oxidase) are involved in other metabolic processes. In aerobic and some aerotolerant bacteria, riboflavin is involved in the biosynthesis of 5,6-dimethylbenzimidazole, which is a component of vitamin B_{12}.

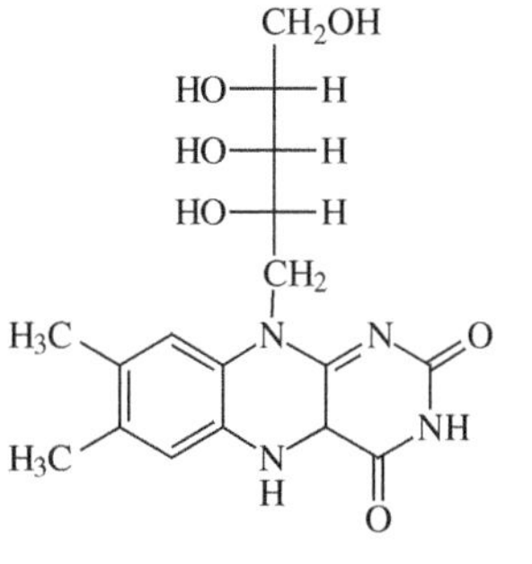

5-23, 4a,5-dihydroriboflavin

5.7.3 Physiology and nutrition

The daily requirement of this vitamin is between 0.4 mg (for infants) and 1.7 mg (for adolescents and adult men). In women, the daily requirement is somewhat lower (1.2–1.3 mg), whilst in pregnant and lactating women, it ranges from 1.6 to 1.8 mg or more. It is estimated that nearly 40% of the dietary vitamin comes from milk and dairy products, about 20% from meat and meat products, 15% from cereals, less than 10% from eggs, and less than 10% from vegetables. Riboflavin from foods of animal origin is more easily absorbed in the digestive tract than that from foods of plant origin, where covalently bound forms that are not easily hydrolysed by proteases dominate. Riboflavin deficiency, or ariboflavinose, is relatively rare. It manifests itself mainly in non-specific inflammatory changes (lesions) of the eye, mouth, skin (dermatitis), or mucous membranes. It is usually associated with other vitamin deficiencies such as pellagra.

5.7.4 Use

Like thiamine, riboflavin is also used for the fortification of certain foods, such as wheat flour and breakfast cereals. Because of its yellow-orange colour, it is also used as a colouring for cereal products, ice creams, and sugar-coated pills. Hydroxyl groups of ribitol can be easily esterified by carboxylic acids. For example, riboflavin-2′,3′,4′5′-tetrabutyrate has been used as an antioxidant.

5.7.5 Occurrence

Riboflavin is found in all foods, and its distribution is similar to the distribution of thiamine. In milk (and also in eggs), the prevailing form is riboflavin (about 82%), which is in part bound to α_S-casein and β-casein; about 14% is in the form of FAD and 4% in the form of FMN. Lower quantities of some other flavins, such as 10-(2-hydroxyethyl)flavin, 7a-hydroxyriboflavin (7-hydroxymethylriboflavin), and its 8a-isomer (8-hydroxymethylriboflavin) are also found in milk. A higher vitamin content is found particularly in meat and offal, cheeses, and some sea fish (Table 5.8).

Riboflavin is present in lower quantities than thiamine in bread and other cereal products, but whole-grain products have a higher content. In addition to riboflavin, FMN, and FAD, a large number of other riboflavin derivatives (esters and glycosides) that exhibit similar biological activity are found in higher plants and microorganisms. For example, the yellow ester of malonic acid (5′-malonylriboflavin) was found in oats growing in the dark.

Unlike thiamine, which remains in the yeast cells during fermentation, a significant amount of riboflavin is present in beer (approximately 0.5 mg/l). Brewer's and baker's yeast is, therefore, a rich source. Derivatives of riboflavin with the C-5′ hydroxyl group of ribitol oxidised to a carbonyl group (called riboflavinal) or a carboxyl group (riboflavinic acid) are present in some edible mushrooms (Basidiomycetes).

5.7.6 Reactions and changes in foods

Flavin coenzymes are very susceptible to enzymatic or chemical hydrolysis. FAD is hydrolysed in acidic solutions to FMN. In acidic solutions, the phosphate group of FMN migrates from the C-5′ position to C′-4, C′-3, and C′-2 positions, and subsequent hydrolysis of these monophosphates yields free riboflavin. Free riboflavin is a very stable vitamin in the absence of light, and in neutral and weakly acidic solutions is practically stable. Even during the production of acid protein hydrolysates (hydrolysis with 20% hydrochloric acid at temperatures above 100 °C), vitamin retention is around 20%. Urea and quinoxaline derivatives are riboflavin decomposition products in alkaline solutions.

In neutral and alkaline solutions by light, all flavins and especially free riboflavin and FMN show complex photochemical behaviour, act as photosensitisers of type I (free radical generation) and type II (oxidation by singlet oxygen) and are degraded to a variety of products. Depending on the pH of the solution, excited triplet riboflavin is degraded into non-volatile lumichrome, lumiflavin, and other products. The main product of riboflavin photodegradation, after cleavage of the ribitol side chain as 1-deoxy-D-*erythro*-pent-2-ulose (1-deoxy-D-ribulose) in acidic and neutral media, is lumichrome (7,8-dimethylalloxazine). In neutral and alkaline media, the reaction products are lumiflavin, 7,8,10-trimethylisoalloxazine, and D-erythrose (Figure 5.17). Both flavins formed by photodegradation of riboflavin are more effective oxidising agents than riboflavin itself. Flavins also transmit the absorbed light energy to air oxygen (triplet oxygen), which yields singlet oxygen. Singlet oxygen then oxidises flavins, which undergo photolytic fission (photodegradation) and similarly oxidises other organic compounds. For example, the reaction between riboflavin and singlet oxygen yields butane-2,3-dione (biacetyl) (Figure 5.18).

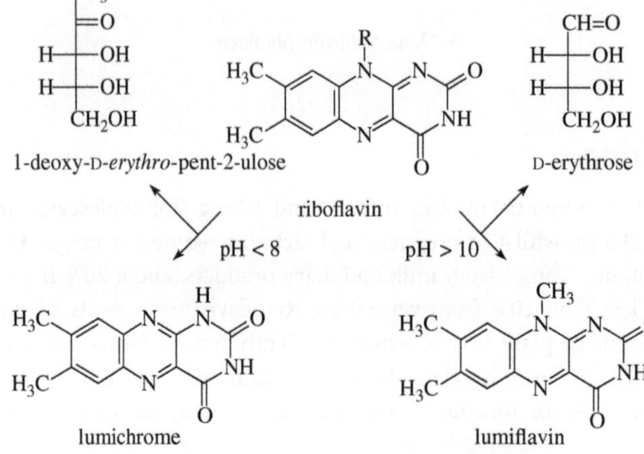

Figure 5.17 Photolysis of riboflavin (see **5-21** for R).

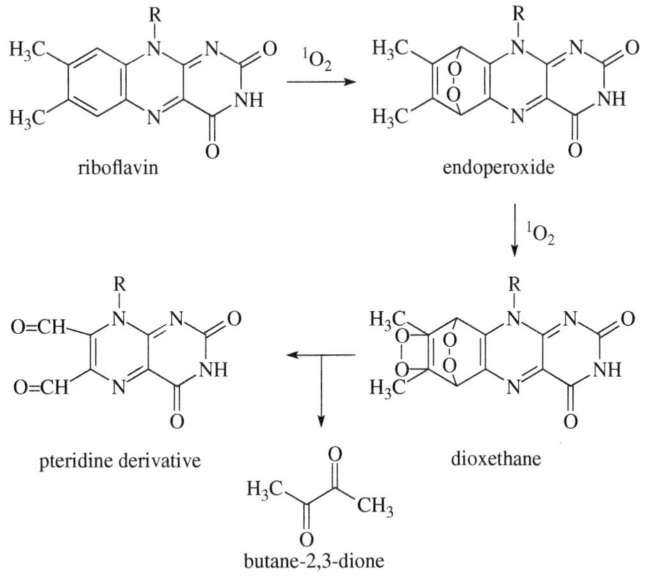

Figure 5.18 Oxidation of riboflavin by singlet oxygen (see **5-21** for R). Source: Jung et al. (2007), figure 3. Reproduced by permission of the American Chemical Society.

Riboflavin is slightly soluble in water (mainly in the reduced form), so losses during cooking meat are not very significant (they usually do not exceed 10%), and are caused mainly by extraction into water. During freezing and the refrigerated storage of meat, vitamin loss is also low. For example, the vitamin loss after 15 months of storage of meat is only about 15%.

Riboflavin in milk is very stable during normal technological operations (such as pasteurisation and sterilisation); losses do not exceed 5%. For example, retention in UHT milk is about 98%. Losses during storage of UHT milk are also small (about 10%). A significant loss of riboflavin in milk occurs when milk is exposed to direct sunlight. An hour's storage of milk in the sun results in a degradation of about 20–40% of the vitamin present. Diffused daylight also has a similar effect. Riboflavin photosensitisation in milk induces two distinctive off-flavours, which make it less acceptable to consumers. The first is the so-called sunlight flavour, giving a burnt and oxidised odour that is caused by methionine oxidation to volatile sulfur compounds, such as methional, methanthiol, and dimethyldisulfide. The second off-flavour is a cardboard or metallic flavour, which develops with a prolonged exposure to light and is attributable to some secondary fatty acid oxidation products that include aldehydes, especially (E)-alk-2-enals, ketones, alcohols, hydrocarbons, and other compounds. Fermented milk products contain, in most cases, higher concentrations of riboflavin than the original milk, because the vitamin is synthesised by the microorganisms used. During the drying of milk, about 2% of riboflavin is lost.

Cereals contain a considerable amount of riboflavin. The vitamin content in flour depends, as with other water-soluble vitamins, on the degree of milling (flour extraction rate). Its content is higher in dark flours than in white flours. Losses during cooking are also small (up to 10%); higher losses (up to 30%) are found in flours fortified with riboflavin. Losses in cooked pasta products (to a large extent caused by leaching) reach, according to the type of pasta, 35–55%.

Riboflavin losses in canned fruits and vegetables are in the range of 25–70%, according to the type of processed material. The main losses are caused by leaching. For example, cooked vegetables lose 30–40% of their riboflavin content to leaching into the cooking water.

Riboflavin in wine and beer acts as a photosensitiser when the beverages are exposed to direct sunlight in unsuitable containers.

5.8 Niacin

5.8.1 Structure and terminology

Niacin, formerly known as pellagra preventive factor (PP factor) or vitamin PP, is a common name for nicotinic acid (pyridine-3-carboxylic acid, **5-24**) and its amide, nicotinamide (niacin amide, formerly known as vitamin B_3, **5-24**). Both compounds have the same biological activity.

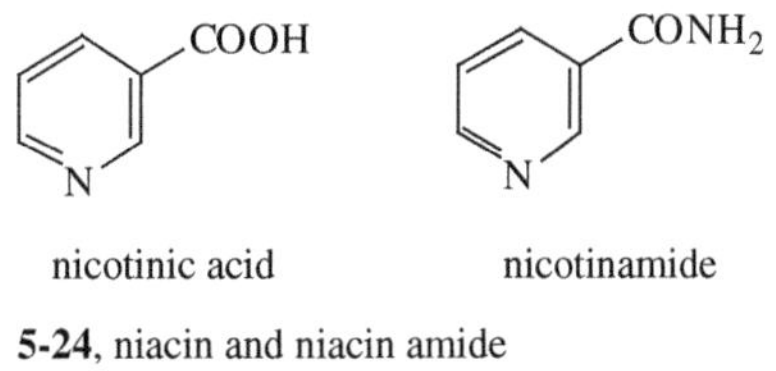

nicotinic acid   nicotinamide

5-24, niacin and niacin amide

5.8.2 Biochemistry

Nicotinamide is part of nicotinamide adenine dinucleotide (NAD; oxidised form NAD+, reduced form NADH) and its phosphoric acid ester, nicotinamide adenine dinucleotide phosphate (NADP; oxidised form NADP+, reduced form NADPH; 5-25), which are cofactors (coenzymes) of several hundred different enzymes. In the older literature, these substances are known as diphosphopyridine nucleotide (DPN+, DPNH) and triphosphopyridine nucleotide (TPN+, TPNH), respectively.

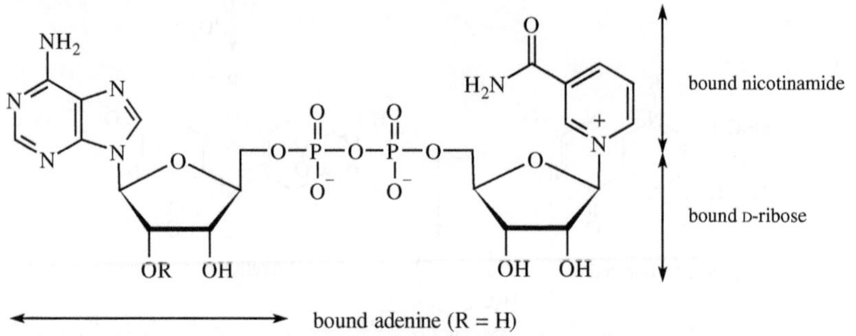

5-25, nicotinamide adenine dinucleotide (oxidised form), R = H
nicotinamide adenine dinucleotide phosphate (oxidised form), R = PO₃H₂

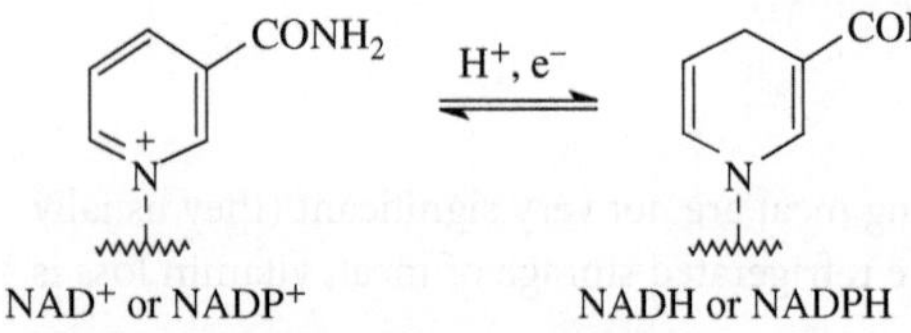

Figure 5.19 Oxidation and reduction of nicotinamide cofactors.

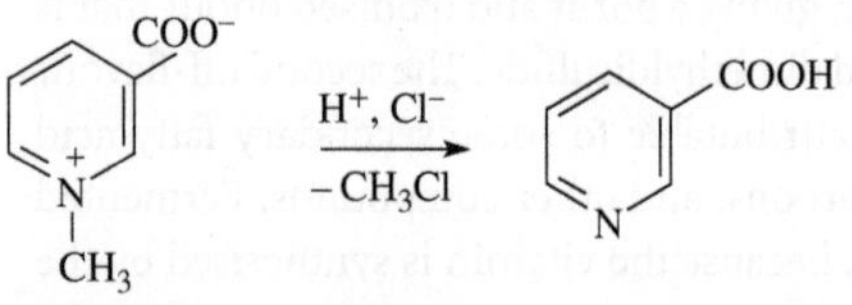

Figure 5.20 Pyrolysis of trigonelline.

Both cofactors are involved in respiratory electron transfer systems (Figure 5.19), for example in most redox reactions of the citric acid (Krebs) cycle. NAD is often involved in the degradation (catabolism) of sugars, fats, proteins, and ethanol, whilst NADP is involved mainly in biosynthetic (anabolic) reactions, such as synthesis of macromolecules, fatty acids, and cholesterol. Dinucleotides, in addition of their activities in redox reactions, participate in post-translational modifications of some proteins and other reactions.

NAD(P)+ from the diet is first hydrolysed to a mixture of nicotinic acid and nicotinamide. Nicotinic acid can be transformed into nicotinamide and then into NAD(P)+ in the body. These dinucleotides are *de novo* synthesised by bacteria and some plants from aspartic acid and 1,3-dihydroxyacetone phosphate. Quinolinic acid is an intermediate. It arises from tryptophan in some microorganisms and in animals.

Biosynthesis and catabolism of NAD(P)+ produce various derivatives of niacin, such as N^1-methylnicotinamide (a product of catabolism in humans and animals) and N-methylnicotinic acid (1-methylpyridinium-3-carboxylate), known as trigonelline (Figure 5.20), which is a product of catabolism in plants and fungi. Previously, trigonelline was known under the names caffearin, coffearin, and gynesine. It was first found in fenugreek seeds (*Trigonella foenum-graecum*, Fabaceae), and subsequently in many other plant species, such as legumes, cereals, and potatoes, and namely coffee plant. Trigonelline is a biologically active compound involved in the induction of leaf movement. It acts as an osmoprotectant and probably serves as a reserve source of nicotinic acid.

5.8.3 Physiology and nutrition

The need for niacin is not precisely known, since it depends on many factors, such as the amount of protein (tryptophan and leucine) consumed. Leucine is an inhibitor of NAD biosynthesis from tryptophan. The recommended daily dose is 2–12 mg for children, 14 mg for women, 16 mg for men, and 18 mg for pregnant and lactating women.

The niacin requirement is mainly provided by meat and meat products (about 40% or more), milk (about 10%), cereal products (about 20%), potatoes (about 10%), and other foods. Meat and meat products supply about 50% of the needs for tryptophan, milk and eggs about 25%. Lack of the vitamin, known as pellagra, is manifested mainly by skin damage, digestive tract disorders, and, later, mental disorders (dementia).

Niacin can be biosynthesised in a somewhat complicated way from tryptophan in the liver (using enzymes containing vitamin B₆ as a cofactor), but the amount is thought not to be sufficient to meet the requirement. It is reported that 34–86 mg (mean 60 mg) of tryptophan is required for biosynthesis of 1 mg of niacin. The possibility of biosynthesising niacin from tryptophan explains the beneficial effect of milk and eggs as a protection against pellagra. Both foods are good sources of tryptophan, although they contain low concentrations of niacin.

Some pyridine derivatives act as antagonists of niacin. Antagonists occurring in foods may include 3- and 4-acetylpyridine that arise as products of the Maillard reaction. *N*-Methylpyridinium cation generated by the thermal degradation of trigonelline inhibits gastric acid secretion and induces phase II detoxifying enzymes.

5.8.4 Use

In some countries, nicotinic acid is added to white wheat flour and to other cereal products. The complex with menadione bisulfite, which shows the activity of niacin and vitamin K, has been used in animal nutrition. Nicotinamide in combination with ascorbic acid is used as a meat colour stabiliser, but has not found wider application in the meat industry (under anaerobic conditions, metmyoglobin in raw meat is reduced back to myoglobin by enzymes that contain NAD as a cofactor).

5.8.5 Occurrence

Distribution of niacin in foods is similar to the distribution of other B group vitamins (Table 5.8). Niacin is present in small amounts in all foods, usually in a bound form.

Foods of animal origin mainly contain nicotinamide, mostly in the form of NAD and NADP. The richest source is offal, meat and meat products, and eggs, especially egg yolk (which contains approximately 60 mg/kg of niacin). Milk contains a surprisingly low amount of niacin; the niacin content of cheeses is higher.

The main form of this vitamin in foods of plant origin is nicotinic acid. Plants contain a number of nicotinic acid derivatives, such as *N*-(β-D-glucopyranosyl)nicotinic acid (**5-26**), the ester 1-*O*-nicotinoyl-β-D-glucopyranose, and other nicotinoyl glucose derivatives.

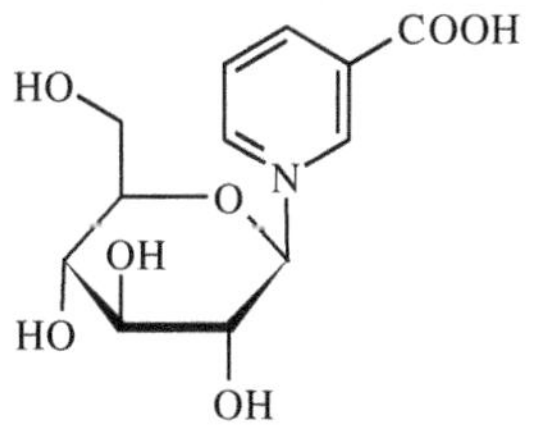

5-26, *N*-(β-D-glucopyranosyl)nicotinic acid

Cereals often have, at first sight, a considerable niacin content. Wheat contains about 30–70 mg/kg of niacin, and whole wheat flour and brown rice contain about 30–60 mg/kg. However, milling and peeling have a great influence on the vitamin content, because niacin is largely located in the germ and bran. For example, wheat grains containing niacin at a level of 57 mg/kg give 80 and 70% straight-run flours that have a vitamin content of 19 and 10 mg/kg, respectively. The same vitamin level (about 10 mg/kg) is found in white bread and white rice. Nicotinic acid in maize and sorghum grains is largely covalently bound to macromolecular glycopeptides and is generally not available to mammals. Partial hydrolysis of these glycopeptides gives 3-*O*-nicotinoyl-β-D-glucopyranose (**5-27**). The lack of bioavailability of nicotinic acid in maize, which leads to pellagra, is probably due to this type of binding. The biological availability of nicotinic acid in maize is significantly improved by pre-treatment of the maize flour with lime slurry, a procedure used in Mexico for the preparation of tortillas. For example, the total amount of the vitamin in raw maize grains is about 30 mg/kg, of which less than 2% is bioavailable. Boiling in water increases the vitamin utilisation to 16%, but when cooked in water with added lime, vitamin utilisation is 100%. Legumes, fruits, vegetables, and potatoes have average amounts of niacin.

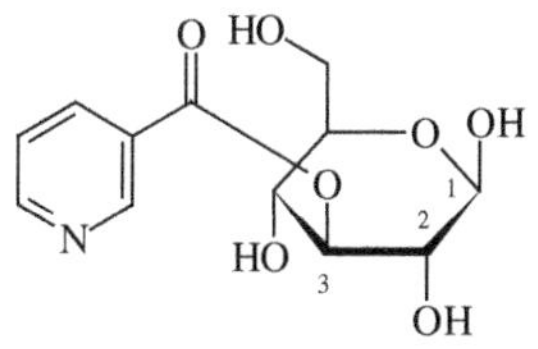

5-27, 3-*O*-nicotinoyl-β-D-glucopyranose

A surprisingly rich source of niacin is roasted coffee. Green coffee beans contain a large amount of the alkaloid trigonelline, which is demethylated on roasting, forming nicotinic acid (Figure 5.20). The content of niacin in roasted coffee is about 25 times higher than in green (unroasted) coffee, reaching up to 2% of dry matter. Together with other compounds, it also contributes to the bitter taste of roasted coffee. Decarboxylation of trigonelline gives a small amount of the *N*-methylpyridinium cation, transmethylation yields methyl nicotinate, and other reactions yield a number of volatile flavour-active pyrroles and pyridines. The caffeine content is virtually unchanged during roasting, so the ratio of caffeine to trigonelline serves as a criterion of the degree of coffee roasting.

5.8.6 Reactions and changes in foods

Nicotinic acid is very stable when heated in aqueous solutions, and is stable in acidic and alkaline media. Nicotinamide is very stable in neutral solutions, but in acidic and alkaline solutions is hydrolysed to nicotinic acid.

When processing raw meat and offal, the content of free nicotinamide increases due to the enzymatic hydrolysis of NAD and NADP. Vitamin losses during the thermal processing of meat usually do not exceed 10%, whilst losses due to draining from improper thawing can amount to 50%. Partial hydrolysis of nicotinamide to nicotinic acid may also occur.

Losses in milk are very small (up to 5%). In the manufacture of cheeses, the majority of the niacin goes into the whey (the liquid part of milk that remains after the manufacture of cheeses). Losses during maturation and storage of cheeses are negligible. The niacin content of yoghurts is slightly higher than in milk because it is produced by the employed microorganisms.

The loss of the vitamin does not usually exceed 10% in baked cereal products. The use of alkalising baking ingredients, such as baking powders based on ammonium bicarbonate, can increase the bioavailability of niacin due to its release from unavailable forms.

Typically, losses of niacin in preserved fruits and vegetables and in cooked potatoes do not exceed 30–40%; the main reason for any loss is leakage into water.

5.9 Pantothenic acid

5.9.1 Structure and terminology

Pantothenic acid (formerly also known as vitamin B$_5$, **5-28**) occurs in nature only as the (+)-D-form; that is, the (R)-enantiomer. (R)-Pantothenic acid is composed of (+)-D-pantoic acid, also known as (R)-pantoic acid, the systematic name of which is (R)-2,4-dihydroxy-3,3-dimethylbutanoic acid. Pantoic acid is linked by an amide bond to the 3-aminopropionic acid (β-alanine). The (S)-enantiomer of pantothenic acid, which is the (−)-L-isomer, is not biologically active and is an antimetabolite of (R)-pantothenic acid, as are some other structural analogues. The main biologically active natural forms of pantothenic acid are coenzyme A (CoA, HS-CoA or CoASH, **5-29**) and a protein called acyl-carrier protein (ACP, **5-30**). Coenzyme A is the active component of transacylases transporting the residues of carboxylic acids. The most common substance is acetyl-CoA, which carries acetyl groups. Other acyl coenzymes A include malonyl-CoA, succinyl-CoA, and other coenzymes that play a role in the metabolism of proteins, fats, and sugars. ACP plays a fundamental role in the biosynthesis of fatty acids and polyketides.

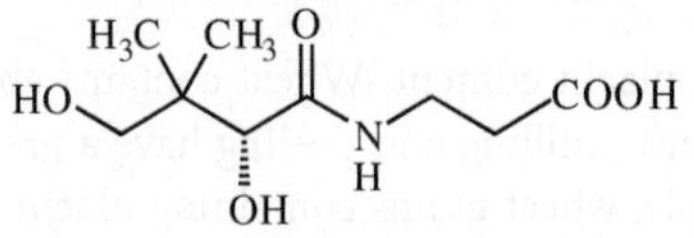

5-28, (R)-pantothenic acid

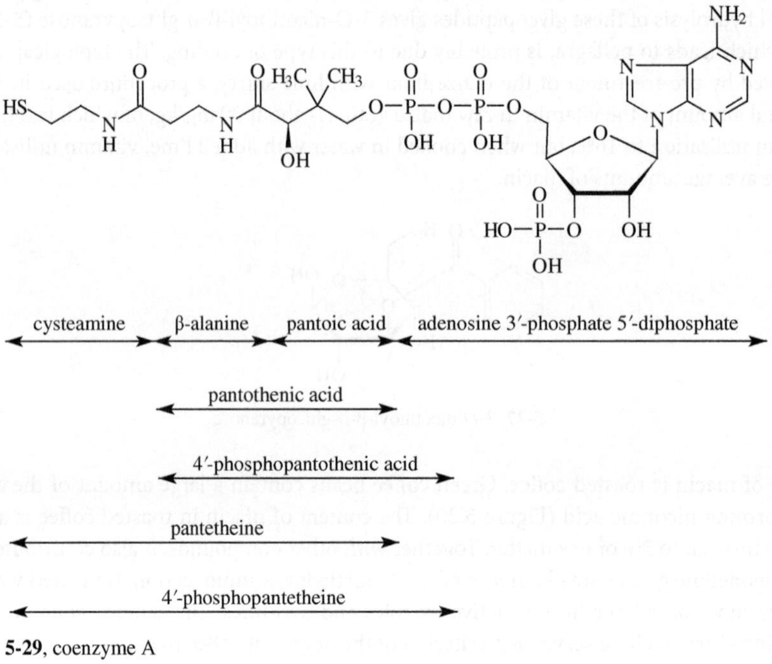

5-29, coenzyme A

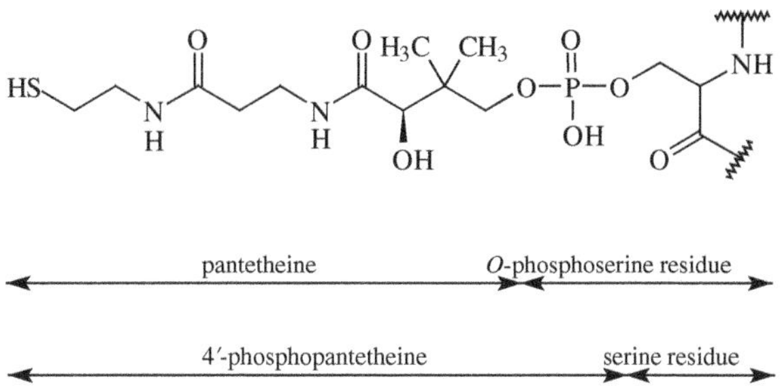

5-30, ACP

Pantothenic acid can be accompanied by its biologically active higher homologue, homopantothenic acid (**5-31**), which contains 4-aminobutanoic acid (γ-aminobutyric acid) instead of β-alanine and might act as an antagonist of pantothenic acid.

5.9.2 Biochemistry

Pantothenic acid is an essential nutritional factor for a range of yeasts, lactic acid and propionic acid bacteria, and other microorganisms. Some bacteria and plants synthesise this acid *de novo* from pantoic acid and β-alanine. The biosynthesis of pantoic acid uses 3-methyl-2-oxobutanoic (2-oxoisovaleric) acid, which is a precursor of valine; the donor of the hydroxymethyl group is 5,10-methylenetetrahydrofolic acid and decarboxylation of aspartic acid yields β-alanine.

Animals (and also some yeasts and bacteria) do not synthesise pantothenic acid and only convert exogenous vitamin obtained from food into the two metabolically active forms, coenzyme A and ACP. Exogenous coenzyme A is first hydrolysed to pantothenic acid and pantetheine (via 4′-phosphopantetheine); exogenous (R)-panthenol (5-31) is oxidised to pantothenic acid.

5.9.3 Physiology and nutrition

The recommended daily intake of pantothenic acid is 1.7 mg in infants under 6 months of age, 2 mg in children from 6 to 12 months, 3 mg in children from 4 to 8 years, 4 mg in children from 9 to 13.5 years, 5 mg in adolescents and men, and 6 mg in women (7 mg in lactating women). Normal dietary intake of this vitamin is 6–12 mg per day. Cases of deficiency, which is manifested mainly by skin irritation and paraesthesia (a tingling and itching sensation), are not common.

It is known that homopantothenic acid improves the metabolism of glucose in the brain and the higher functions of the brain, and it has been used in Japan to enhance mental functions, especially in Alzheimer's disease. A rare side-effect is an abnormal brain function resulting from the failure of the liver to eliminate toxins (hepatic encephalopathy). This condition was reversed by pantothenic acid supplementation, suggesting that it was due to pantothenic acid deficiency caused by homopantothenic acid.

5.9.4 Use

Pantothenic acid and its salts are added to foods only occasionally. Calcium and sodium pantothenate are more stable and less hygroscopic than the free acid. Salts of pantothenic acid and less polar (R)-pantothenyl alcohol, also known as (R)-panthenol (5-31), are used to fortify animal feed for farm animals (mainly poultry), and are also used in cosmetics. Proportionally, owing to increasing consumption of cooked foods and ready-to-eat meals in developed countries, more extensive fortification of foods can be expected.

5.9.5 Occurrence

Pantothenic acid is found in virtually all foods of plant and animal origin, usually in relatively small amounts (Table 5.8). Only a small proportion of the total vitamin content is free acid; bound forms such as coenzyme A, acyl-coenzymes A, and ACP prevail.

The pantothenic acid levels in individual foods are highly variable. Relatively large amounts of the vitamin occur in meat, and particularly in offal, eggs, milk, and some cheeses.

The pantothenic acid content in wheat flour is strongly influenced by the degree of milling. There is a higher vitamin level in whole-grain flour, cereal products, and legumes, and a roughly ten times lower level in fruits and vegetables. Homopantothenic acid occurs in cooked rice at concentrations of about 20 mg/kg.

Similarly to the other B group vitamins, relatively large amounts of pantothenic acid are found in yeasts. Homopantothenic acid occurs in dried yeasts in a concentration of 8500 mg/kg.

5.9.6 Reactions and changes in foods

The stability of pantothenic acid in aqueous solutions depends greatly on the pH value. The vitamin is most stable in weakly acidic (pH 4–5) solutions, but in more acidic and alkaline media the amide linkage is hydrolysed and pantothenic acid yields pantoic acid (or its salt) and β-alanine. The enzyme pantothenase of some bacteria specifically cleaves pantothenic acid into the same products. In acidic solutions, pantoic acid spontaneously dehydrates to form lactone, (R)-2-hydroxy-3,3-dimethylbutano-4-lactone, which is called pantoyllactone or pantolactone (5-31). Analogously, products of panthenol hydrolysis are pantoic acid and 3-aminopropane-1-ol (β-alanol). Solutions of coenzyme A and pantetheine are relatively stable in solutions of pH 2–6. Both compounds are easily oxidised to form disulfides, pantethine, and the disulfide form of coenzyme A.

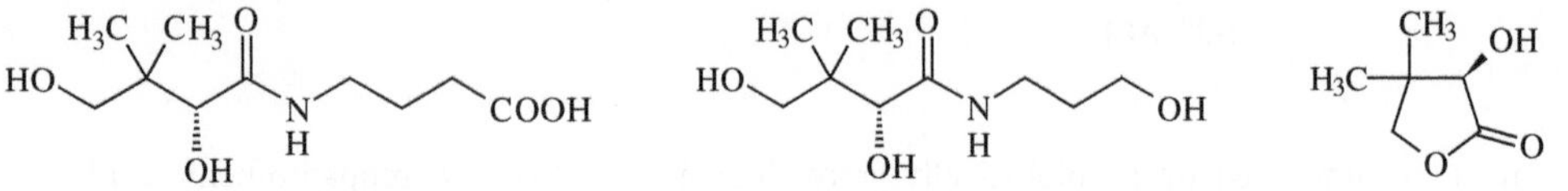

(R)-homopantothenic acid   (R)-panthenol   (R)-pantholactone

5-31, pantothenic acid derivatives

Pantothenic acid is relatively unstable during storage and especially during the thermal processing of foods. Losses caused by leaching into the water during operations such as washing, blanching, and cooking are often higher than those caused by hydrolysis.

Thermal processing of meat results in losses of pantothenic acid that reach about 12–50%, depending on the conditions. The extent of these losses is contingent on the type of heat treatment, the volume of water used, and other factors. Losses in canned meat and in meat products range from 20 to 35%.

Pasteurisation of milk does not significantly affect the vitamin content. In the case of UHT milk, the vitamin loss is about 5% of the original level, and the loss after 6 weeks of storage of UHT milk is 20–35%. The natural vitamin content in fermented dairy products is affected only slightly by fermentation. Milk powder manufacture and storage of milk powder result in small losses of the vitamin.

A small decrease in pantothenic acid concentration occurs during milling. The vitamin retention in bread is relatively high (up to 90%), whilst in cooked pasta, according to the type of product and method of preparation, it reaches 55–75%. Losses during the cooking of legumes are 25–56%, depending on soaking time. Higher amounts of pantothenic acid can be lost when exposed to alkali agents, such as baking soda.

In preserved fruits and fruit juices, the losses of pantothenic acid are on average about 50%. In pickles and brine-preserved vegetables, where the vitamin is exposed to vinegar, they can reach 45–80%.

5.10 Pyridoxal, pyridoxol, and pyridoxamine

5.10.1 Structure and terminology

The name vitamin B_6 (formerly known as adermine) refers to three structurally related, biologically active derivatives of 3-hydroxy-5-hydroxymethyl-2-methylpyridine and their 5′-phosphates. These three forms of vitamin B_6 differ in their substitution at position C-4 of the pyridine ring. The formyl derivative, 4-formyl-3-hydroxy-5-hydroxymethyl-2-methylpyridine, is called pyridoxal (5-32), the hydroxymethyl derivative, 3-hydroxy-4,5-bis(hydroxymethyl)-2-methylpyridine, is called pyridoxol or pyridoxine (5-32), and the aminomethyl derivative, 4-aminomethyl-3-hydroxy-5-hydroxymethyl-2-methylpyridine, is called pyridoxamine (5-32). The residue that results after splitting off the 4′-hydroxyl group is known as the pyridoxyl residue; the pyridoxylidene residue forms by splitting off the oxygen from the aldehyde group of pyridoxal.

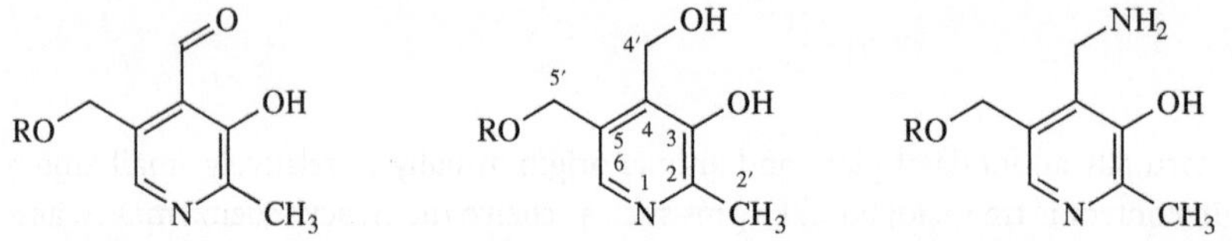

pyridoxal, R = H
(5′-phosphate, R = PO$_3$H$_2$)

pyridoxol, R = H
(5′-phosphate, R = PO$_3$H$_2$)

pyridoxamine, R = H
(5′-phosphate, R = PO$_3$H$_2$)

5-32, active forms of vitamin B_6

5.10.2 Biochemistry

Prokaryotic organisms synthesise pyridoxol 5′-phosphate from 1-deoxy-D-xylulose 5-phosphate (1-deoxy-D-threo-pent-2-ulose 5-phosphate) and the phosphate of 2-amino-2-deoxy-D-*threo*-tetronic (2-amino-2-deoxy-D-threonic) acid, known as 4-(phosphohydroxy)-L-threonine or 4-hydroxy-L-threonine 4-phosphate. Phosphohydroxythreonine arises from D-erythrose 4-phosphate, a product of decomposition of D-fructose 6-phosphate. Non-phosphorylated forms (pyridoxal, pyridoxol, and pyridoxamine) are produced by hydrolysis of the corresponding phosphates. Animals do not synthesise vitamin B_6 *de novo*, but only convert the non-phosphorylated forms in the liver, erythrocytes, and other tissues into the corresponding phosphates and one another's individual forms. Pyridoxal 5′-phosphate arises by oxidation of pyridoxol 5′-phosphate and its transamination yields pyridoxamine 5′-phosphate. Both these forms of vitamin B_6 are catalytically active. Pyridoxal 5′-phosphate (originally called codecarboxylase) is a cofactor for enzymes catalysing decarboxylation, deamination, racemisation, transamination, and transsulfuration of amino acids and also participates in many reactions related to metabolism of fats and sugars.

The dominant forms in animal tissues are pyridoxal 5′-phosphate and pyridoxamine 5′-phosphate. An important metabolite of pyridoxal excreted in the urine is biologically inactive 4-pyridoxic acid (**5-33**) and its lactone, which is called 4-pyridoxolactone (**5-33**). Minor metabolites generated by oxidation in position C-5′ are isopyridoxal (aldehyde), the corresponding carboxylic acid, and its lactone. Pyridoxol 5′-phosphate is the dominant form in plant tissues.

4-pyridoxic acid   4-pyridoxolactone   gingkotoxin   pyridoxol hydrochloride

5-33, examples of vitamin B_6 metabolites, antagonists and synthetic analogues

5.10.3 Physiology and nutrition

The recommended daily intake of vitamin B_6 is 0.3–2.6 mg (the lower limit is for babies and the upper limit for pregnant and lactating women). Vitamin B_6 deficiency manifests by dermatitis and neurological disorders (seizures in children). Long-term high-dose vitamin intake can cause neurological disorders, manifesting as loss of sensation in the feet and poor coordination. Such increased vitamin intake can be only provided by vitamin supplements. It is estimated that about 40% of the required vitamin is provided by meat and meat products, 22% by vegetables, 12% by milk and dairy products, 10% by cereals, 5% by legumes, 8% by fruits, and 2% by other sources.

Vitamin antagonists are compounds reacting with the carbonyl group of pyridoxal or structurally related substances. Natural antagonists may be tryptophan metabolites (hydrazines and hydroxylamines) that react to form the respective hydrazones and oximes, which are unavailable forms of vitamin B_6. Some reaction products of pyridoxal with amino acids are also unavailable or available to a small extent. Vitamin antagonists include linatin, an amino acid present in flax seeds and a neurotoxin gingkotoxin (4′-O-methylpyridoxol, **5-33**), which is a minor component of gingko leaves (*Ginkgo biloba*, Ginkgoaceae), known as the maidenhair tree.

5.10.4 Use

For the fortification of foods and for food supplements, synthetic pyridoxol hydrochloride (**5-33**) is used. Baby food, and in some countries white wheat flour, is the food most often fortified.

5.10.5 Occurrence

Meat, meat products, offal, and egg yolk are the rich sources of vitamin B_6 (Table 5.8). In foods of animal origin, the main compounds are pyridoxal and pyridoxamine, especially in the form of phosphate esters. For example, in meat, the main form is present as pyridoxal 5′-phosphate (about two-thirds of this vitamin is prosthetic groups of enzymes) bound to various proteins (such as imine); free pyridoxal 5′-phosphate appears to a lesser extent, followed by pyridoxamine 5′-phosphate. In contrast, milk contains only about 10% of vitamin in bound forms. The vitamin content in milk and cheeses is relatively low.

Foods of plant origin mainly contain pyridoxol and pyridoxal. Glycosides of pyridoxol are also common. The predominant form is 5′-*O*-(β-D-glucopyranosyl)pyridoxol (**5-34**), which in fruits and vegetables represents 5–80% of the total vitamin content. This form is less available than free pyridoxal. A minor component of pea (*Pisum sativum,* Viciaceae) is 5′-*O*-(β-D-glucopyranosyl)pyridoxol esterified by 3-hydroxy-3-methylglutaric acid at position C-6 of D-glucose. Cereals are good sources of vitamin B_6. A higher vitamin content can be found in whole-grain cereal products and cereal germs, as well as in some vegetables, potatoes, and legumes.

Vitamin B_6 is, like the other B vitamins, present in high concentrations in yeast.

5.10.6 Reactions and changes in foods

Vitamin B_6 is relatively stable in acidic solutions and less stable in neutral and alkaline solutions, particularly in the light. Pyridoxol is more stable than pyridoxal and pyridoxamine.

Amino acids, peptides, and proteins react with pyridoxal and its phosphate in neutral solutions to form imines (formerly known as Schiff bases). This reaction is analogous to reactions of reducing sugars in the Maillard reaction (Figure 5.21). Pyridoxal 5′-phosphate is more reactive than free pyridoxal. Reaction products with amino acids and proteins (imines) isomerise and are hydrolysed to pyridoxamine and the corresponding 2-oxocarboxylic acids (non-enzymatic transamination reaction). The addition of the second amino acid molecule (or thiol) to imines yields *N,N*′-substituted diamines. For example, reduction of imines by ascorbic acid or reductones arising in the Maillard reaction yields stable *N*-substituted pyridoxamines. Imines and diamines formed from imines and amino acids are physiologically fully available; pyridoxylamino acids are resistant to acid hydrolysis and therefore less physiologically active. Reaction of pyridoxal 5′-phosphate with proteins generates about 20% of bound forms (8% imines and diamines and 8% pyridoxylamino acids).

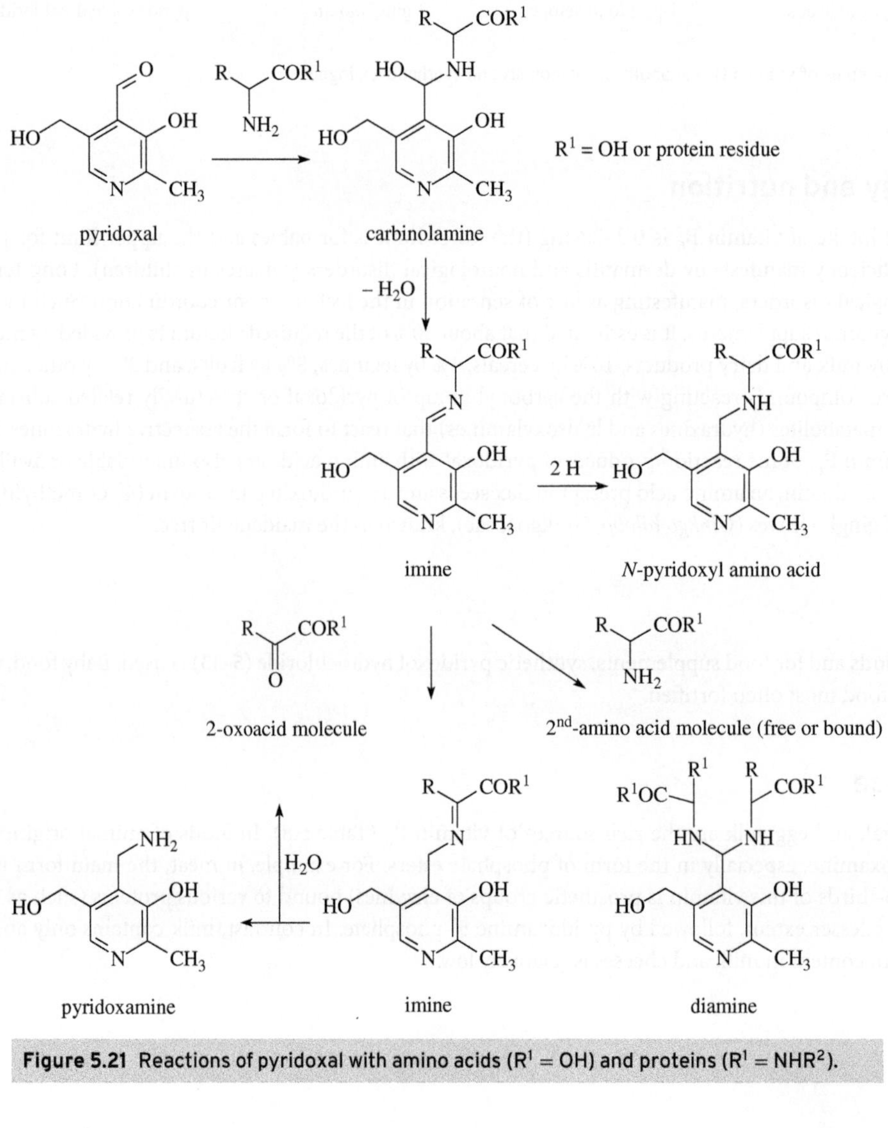

Figure 5.21 Reactions of pyridoxal with amino acids (R^1 = OH) and proteins (R^1 = NHR^2).

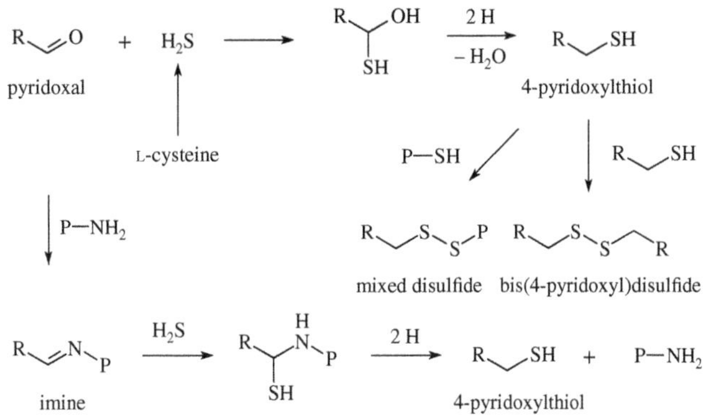

Figure 5.22 Reaction of pyridoxal with thiol group of cysteine and proteins (P = protein residue).

Amino acids containing other reactive functional groups react with pyridoxal to form heterocyclic compounds. Reaction with cysteine during sterilisation of milk gives a thiazolidine derivative (**5-34**). Reaction with tryptophan gives a β-carboline derivative (**5-34**). On a small scale, the Strecker degradation of amino acids yields the corresponding Strecker aldehydes. Hydrogen sulfide produced by the degradation of cysteine reacts with the carbonyl group of pyridoxal analogously to the amino group of amino acid or protein, and an unstable mercaptal forms as an intermediate. It is a precursor of pyridoxylthiol and bis(4-pyridoxyl)disulfide. Reaction with protein-bound cysteine yields mixed disulfides (Figure 5.22). Pyridoxylthiol and the corresponding disulfide show only low activity of vitamin B_6.

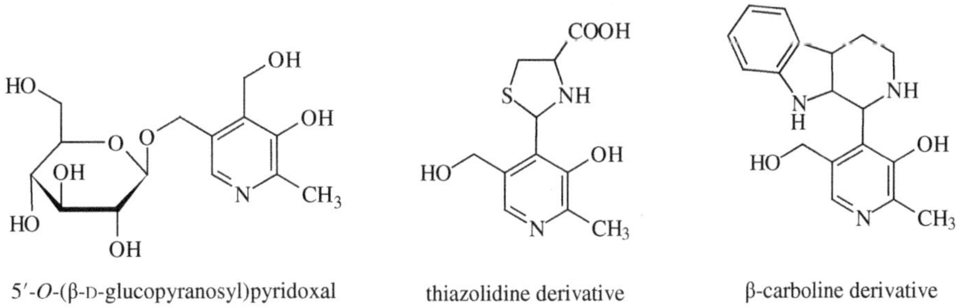

5'-O-(β-D-glucopyranosyl)pyridoxal   thiazolidine derivative   β-carboline derivative

5-34, examples of vitamin B_6 glycosides and reaction products

Vitamin losses during food storage and processing vary considerably in different commodities according to the predominant form of the vitamin. In foods of plant origin that contain pyridoxol, vitamin losses are usually small. In foods of animal origin containing reactive pyridoxal, losses are higher. Losses are also attributed to vitamin leaching and pyridoxal reaction with proteins.

Raw meat contains pyridoxal phosphate as the main vitamin form, whereas cooked meat predominantly contains a product of its transamination pyridoxamine phosphate. For example, the main form of vitamin B_6 in raw chicken meat is pyridoxal phosphate (56%), followed by pyridoxamine phosphate (42%) and pyridoxamine (2%). The main form in roasted chicken meat is pyridoxamine phosphate (70%), followed by pyridoxal phosphate (21%), pyridoxol (7%), and pyridoxamine (2%). Vitamin retention in roasted meat is about 45–65%.

During conventional methods of milk processing, vitamin losses are small. For example, during pasteurisation, the loss is up to 5%, whilst loss in UHT milk does not exceed 10%. Higher losses are caused by subsequent storage, so that the total losses in UHT milk are about 40–45%. During the heat treatment of milk, pyridoxal is partly transformed into pyridoxamine. Fresh milk mainly contains pyridoxic acid (38%), pyridoxal phosphate (32%), pyridoxal (24%), pyridoxamine (4%), and pyridoxamine phosphate (2%). In pasteurised milk, the main form is pyridoxal (41%), followed by pyridoxal phosphate (24%), pyridoxic acid (20%), pyridoxamine phosphate (9%), and pyridoxamine (6%).

During pasteurisation, and especially during drying, pyridoxal reacts with cysteine and lysine (free amino acids and amino acids bound in proteins). For example, milk powder contains 30–70% of the original amount of vitamin. The losses are higher when milk is exposed to sunlight.

The main forms of vitamin B_6 in cereals are free pyridoxol and pyridoxol bound to glucose. The vitamin content of flour depends on the degree of milling. In dough, and also in bread, the losses are small and do not exceed 15%. Vitamin retention in cooked pasta is 50–70%.

Relatively large losses occur during canning and cooking of vegetables (40–50%), whilst fruit losses are somewhat lower (approximately 40%).

5.11 Biotin

5.11.1 Structure and terminology

Biotin is composed of an ureido (tetrahydroimidizalone) ring fused with a tetrahydrothiophene ring substituted with the chain of pentanoic acid. The biotin molecule contains three asymmetric carbon atoms. Only one of eight possible isomers, (3aS,4S,6aR)-5-[2-oxohexahydro-1H-thieno[3,4-d]imidazol-4-yl] pentanoic acid, known as d-biotin or (+)-biotin (**5-35**), occurs in nature and is biologically active. Biotin has also been known as vitamin B$_7$ and vitamin H, and previously as bios II, factor X, and coenzyme R.

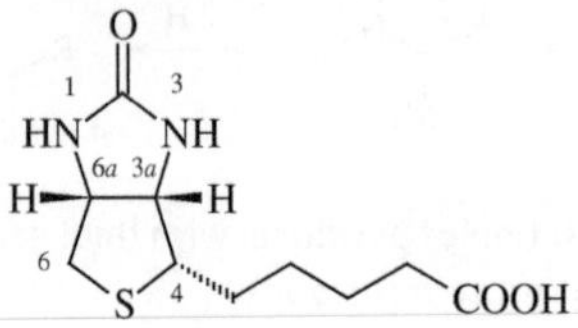

5-35, (+)-biotin

5.11.2 Biochemistry

Most microorganisms, fungi, and higher plants synthesise biotin from pimelic acid and alanine. Pimelic acid is a product of oxidative cleavage of higher fatty acids. The primary product of the biosynthesis is dethiobiotin (**5-36**), which reacts with methionine and gives biotin. Biotin is found as a prosthetic group of many enzymes that catalyse the transfer of carbon dioxide. An intermediate of these reactions is N-carboxybiotin (**5-36**). Biotin is bound in enzymes through an amide bond provided by the ε-amino group of lysine as the so-called N-ε-biotinyl-L-lysine (biocytin, **5-36**). Usually, three groups of enzymes with biotin as a cofactor are recognised: carboxylases, transcarboxylases, and decarboxylases. They act in the biosynthesis of fatty acids and the catabolism of branched-chain amino acids.

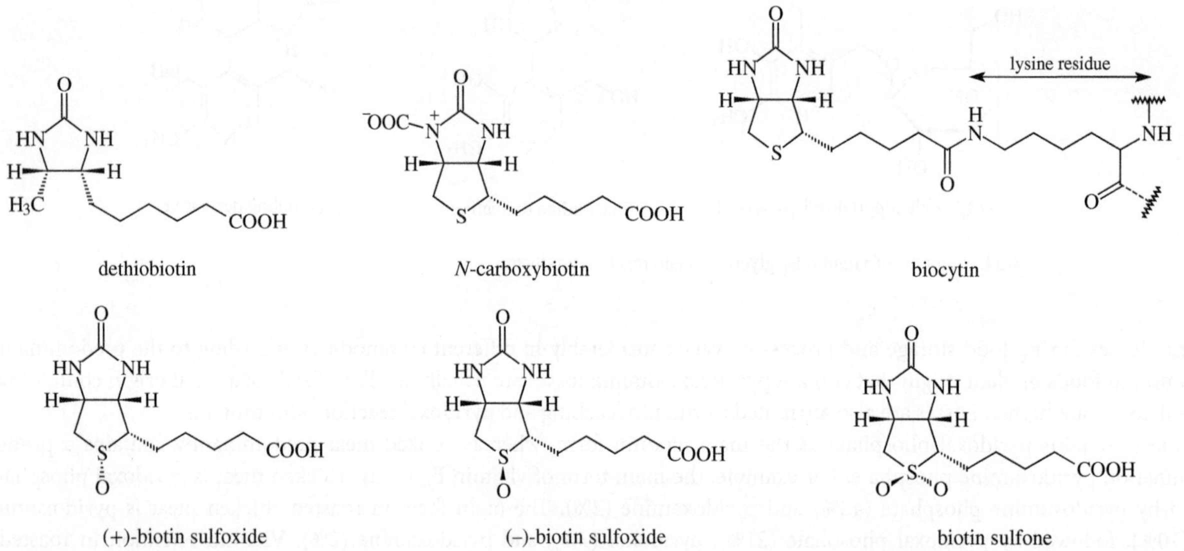

dethiobiotin

N-carboxybiotin

biocytin

(+)-biotin sulfoxide

(−)-biotin sulfoxide

biotin sulfone

5-36, precursors and metabolites of biotin

Only free biotin is absorbed from foods. Biotin bound to proteins must be hydrolysed by the enzyme biotinidase (biotinamide amidohydrolase). Excess free biotin is excreted in urine, along with its metabolites (such as the oxidation products biotin sulfoxides, **5-36**). Bound forms of biotin are excreted in the faeces.

5.11.3 Physiology and nutrition

The need for biotin is very low, and is usually covered by the vitamin from food and vitamin produced by intestinal microflora. The daily dietary intake is estimated at 50–100 μg, although some sources have reported an even higher intake (150–300 μg). The recommended daily intake is 5 μg for infants up to 6 months of age, 7 μg for infants aged 7–12 months, 8 μg for children aged 1–3 years, 12 μg for children aged 4–8 years, 20 μg for children aged 9–13 years, 25 μg for adolescents (14–18 years), 30 μg for elderly people and pregnant women, and 35 μg for lactating women. Spontaneous deficiency is rare and mild. The most common deficiency occurs when the diet is mainly based on the consumption of raw eggs, because raw egg white contains the basic, water-soluble glycoprotein avidin, which forms a very strong complex with biotin in which the vitamin is bound by non-covalent bonds. Biotin bound by avidin (avidin is a biotin antagonist) is not available. Denatured protein does not react with biotin. Vitamin deficiency and avitaminosis manifests by hair loss (alopecia), conjunctivitis (also called red eyes), and dermatitis (red rash around the eyes, nose, mouth, and genital area), and in adults by neurological symptoms (depression, lethargy, and hallucinations).

5.11.4 Use

The enrichment of foods with biotin is applied only rarely.

5.11.5 Occurrence

Biotin is found in a range of foods, but the concentration in most of them is usually low (Table 5.8). It is present partly as a free compound (milk, fruits, and vegetables) and partly bound to proteins (animal tissues, plant seeds, and yeast). Yeast autolysates, for example, contain free biotin and its precursors and analogues. These biotin vitamers, mainly dethiobiotin, biotin sulfone (5-36), and biocytin (5-36), are degradation products of enzymes containing biotin.

Good sources of biotin are egg yolk and organ meats (especially liver and kidney). Milk has a lower vitamin content.

Rich sources are some vegetables, such as peas and cauliflower (about 0.1 mg/kg), cereals, cereal products, and legumes. The biotin content in flour is strongly dependent (as is the content of all B group vitamins) on the flour extraction rate. In white flour, a little over 10% of the original vitamin present remains in the grain. However, only a small proportion of biotin in wheat is available; better availability comes from maize and soybeans. High levels of biotin are found in yeast and mushrooms (about 0.2 mg/kg).

5.11.6 Reactions and changes in foods

Biotin is stable in the light and when heated in neutral and acidic solutions. It is unstable in alkaline media. Biotin is easily oxidised by hydrogen peroxide and other oxidants to a mixture of isomeric (+)- and (−)-sulfoxides, and eventually to biotin sulfone (5-36). Sulfoxides also form as metabolic products of microorganisms. They are fully available, whereas sulfone is unavailable as a vitamin. Nitrogen of a ureido ring may be nitrosated to nitrosobiotin (5-37) in the presence of nitrites or nitrogen oxides.

5-37, nitrosobiotin

Biotin is very stable during food processing. Losses in hydrothermal processes are mainly caused by leaching into the water.

Biotin retention in cooked meat is relatively high. An average of 77% of the biotin was retained in the meat alone after cooking, and an average of 80% was retained in the meat plus drippings after cooking.

Pasteurisation of milk results in 10–15% losses of biotin. The losses during milk drying are higher, but subsequent storage does not influence the vitamin level substantially. Biotin concentration in cheeses is lower by about 20–35% compared with milk. The biotin content of yoghurts is influenced by the microflora (*Lactobacillus* spp.) used, and is lower than in milk by about 45–60%. Some bacteria (e.g. bacteria of the genus *Micrococcus*) produce biotin, and its content in yoghurt may then increase by 5–25%.

Biotin losses during the baking of bread and other cereal products are small. Cooking legumes results in loss, which reaches 5–15% depending on the time of soaking.

During preservation of fruit and vegetables, about 30% of biotin is lost. This relatively large proportion is caused by leaching and can therefore be regarded as a loss only when the brew is not consumed.

5.12 Folacin

5.12.1 Structure and terminology

Folacin is the name for the biologically active derivatives of folic (pteroylglutamic) acid (5-38), formerly also known as vitamin B_9, vitamin B_c, or vitamin M. The basis of its structure is pteroic acid, systematic name 4-[(pteridine-6-ylmethyl)amino]benzoic acid, which is hypothetically derived from 6-hydroxymethylpterin and 4-aminobenzoic acid. The carboxyl group of pteroic acid is conjugated, via an amide bond, with one or more L-glutamic acid units, typically with three to eight molecules (n = 3–8). These compounds are called glutamyl peptides or folate polyglutamates. The active form is (6S)-5,6,7,8-tetrahydrofolic or (6S)-5,6,7,8-tetrahydropteroylglutamic acid (**5-39**), $H_4PteGlu$ or FH_4 for short, with a reduced pteridine (more precisely pyrazine) ring. Heptaglutamate was previously known under the name chick growth factor or vitamin B_{11}.

5.12.2 Biochemistry

Many microorganisms and plants are capable of *de novo* synthesis of 7,8-dihydrofolic (7,8-dihydropteroylglutamic) acid (7,8-$H_2PteGlu$, **5-39**). Animals can only reduce this form of vitamin to tetrahydrofolic acid. They are also able to reduce the folic acid to dihydrofolic acid and can hydrolyse or add additional glutamic acid residues. The starting compound for the biosynthesis of 7,8-dihydrofolic acid is guanosine 5′-triphosphate (GTP). Reactions with 4-aminobenzoic acid (which results from chorismic acid via 4-amino-4-deoxychorismic acid) and with glutamic acid proceed in the later stages of biosynthesis.

The activity of folacin is similar to the activity of cobalamins. It is linked with the transfer of single-carbon functional groups, such as methyl (CH_3—), methylene (—CH_2—), formyl (—CH=O), and other groups whose donor is mainly choline, glyoxylic acid, serine, and other compounds. These functional groups are bound to N-5 or N-10 of tetrahydrofolic acid (**5-39**). The vitamin is a cofactor for enzymes that act mainly in the metabolism of amino acids (e.g. Ser/Gly or homocysteine/Met conversions and synthesis of creatine), purine, and pyrimidine nucleotides. Folacin, along with corrinoids, has an important role in homocysteine metabolism, which is one of the risk factors for the origination and development of cardiovascular diseases.

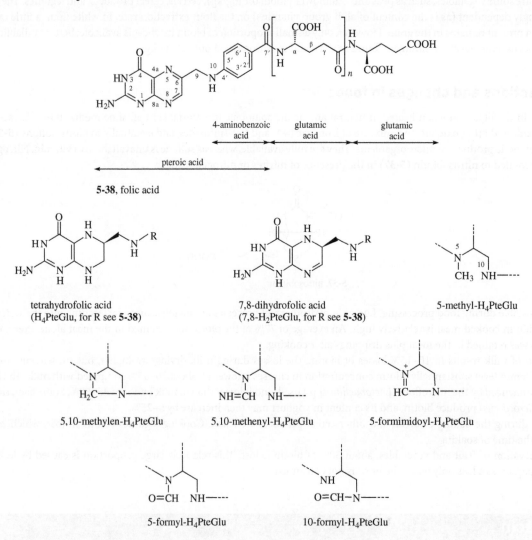

5-38, folic acid

tetrahydrofolic acid
($H_4PteGlu$, for R see **5-38**)

7,8-dihydrofolic acid
(7,8-$H_2PteGlu$, for R see **5-38**)

5-methyl-$H_4PteGlu$

5,10-methylen-$H_4PteGlu$

5,10-methenyl-$H_4PteGlu$

5-formimidoyl-$H_4PteGlu$

5-formyl-$H_4PteGlu$

10-formyl-$H_4PteGlu$

5-39, tetrahydro folic acid, dihydrofolic acid and active metabolites

5.12.3 Physiology and nutrition

The recommended daily intake of folic acid is 0.065–0.6 mg. Intake at the lower limit is recommended for infants (0.065 mg at 0–6 months, 0.08 mg at 7–12 months) and young children and teenagers (0.15 mg at 1–3 years, 0.2 mg at 4–8 years, 0.3 mg at 9–13 years), whilst intake at the upper income limit is recommended for older teenagers (14–19 years), older men, and women (0.4 mg) and pregnant and lactating women (0.6 mg).

Daily intake of 0.6 mg or higher can no longer be obtained from food due to our eating habits, so intake through food supplements is recommended. Because of the difference in bioavailability between folic acid in food supplements and the different forms of folacin found in food, the dietary folate equivalent (DFE) system was established. One DFE is defined as 1 μg of dietary folacin or 0.6 μg of supplemented folic acid. The tolerable upper intake level (UL) for folate is 1 mg for adult men and women and 800 μg for pregnant and lactating (breastfeeding) women under 18 years of age. Supplemental folic acid should not exceed the UL to prevent folic acid from masking symptoms of cobalamin deficiency.

Folacin deficiency may lead to glossitis, diarrhoea, depression, and confusion. Deficiency anaemia may develop especially in pregnancy and in elderly people. Symptoms of deficiency are similar to symptoms of cobalamin deficiency (known as macrocytic anaemia). Megaloblastic anaemia, the most common cause of macrocytic anaemia, is due to a deficiency of either cobalamin or folic acid (or both). Deficiency in the early stages of pregnancy can lead to developmental defects of the foetus (spinal cord defects and incomplete development of the brain). Women who are aware that they might be pregnant will need an increased daily intake of folic acid.

5.12.4 Use

For enrichment of foods, relatively stable synthetic forms of the vitamin are used, such as 5-formyltetrahydrofolic, 5-methyltetrahydrofolic, and tetrahydrofolic (**5-39**) acids.

5.12.5 Occurrence

Folacin exists primarily in the form of reduced tetrahydrofolates (**5-39**) containing a variable number of glutamic acid units. For example, oranges contain predominantly pentaglutamate (about 40% of the total vitamin content) and tetraglutamate (about 10%); the main form in lettuce is pentaglutamate and monoglutamate (both forms represent about 30% of the vitamin content), the main form in most leafy vegetables is heptaglutamyl conjugate, the main forms in fresh meat are penta-, hexa-, and heptaglutamate, and the main form in stored meat is triglutamate.[5] Fresh foods mostly contain 5-methyltetrahydrofolate and 10-formyltetrahydrofolate. The total vitamin content in common foods is given in Table 5.8.

Particularly important sources of folacin are eggs and offal. The main natural forms of folacin in animal materials are polyglutamylpeptides (the dominant form is the pentaglutamyl conjugate) derived from 5-methyltetrahydrofolic acid, abbreviated 5-methyl-$H_4PteGlu_n$ (about 50%), followed by 10-formyltetrahydrofolic acid, abbreviated 10-formyl-$H_4PteGlu_n$ (about 10%), and tetrahydrofolic acid or $H_4PteGlu_n$ (about 40%). In milk and dairy products, 25% of the total amount of vitamin is 5-methyl-$H_4PteGlu_n$, about 60% is 10-formyl-$H_4PteGlu_n$, and 15% is $H_4PteGlu_n$.

In intact plant tissue, folates are derived from 5-methyl-$H_4PteGlu_n$ and 10-formyl-$H_4PteGlu_n$. These vitamers exist to a small extent as monoglutamates but are mainly conjugated to a chain of 2–8 glutamate moieties. Hydrolysis of folate poly-γ-glutamates to shorter poly-γ-glutamates and monoglutamates is catalysed by endogenous γ-glutamyl hydrolase. For example, 5-methyltetrahydrofolate is the predominant vitamer in bell peppers, but a significant amount of 5-formyltetrahydrolfolate and some 10-formylfolate are also present. Folate content in red and green peppers differs markedly (0.70 and 0.21 mg/kg, respectively). A significant amount of folic acid ($PteGlu_n$) and its derivatives also occurs in cereals and legumes. In wheat and rye (also in bread and other cereal products and legumes), for example, 38–55% of the vitamin is in the form of 10-formyl-$H_4PteGlu_n$, 5–20% in the form of 5-methyl-$H_4PteGlu_n$, and 3–8% in the form of $H_4PteGlu_n$. In addition, 12–21% of folacin is present as 10-formyl-$PteGlu_n$ and 12–23% as $PteGlu_n$. A rich source of folate are rosehips. Their folate content is 4–6 mg/kg based on dry matter and 1.60–1.9 mg/kg based on the f.w. (edible part). Fresh citrus juices contain relatively high amounts of mono- and polyglutamyl forms of 5-methyl-$H_4PteGlu$ (0.01–0.13 mg/l) and folic acid (0.07–0.21 mg/l). The concentration of 5-methyl-$H_4PteGlu$ is generally greater than that of folic acid in orange juice; grapefruit juice contains the least amount of 5-methyl-$H_4PteGlu$ but relatively large quantities of folic acid. A majority of 5-methyl-$H_4PteGlu$ (60%) is present as the monoglutamyl

---

[5] Plant and animal organisms contain a large number of pteridine derivatives. They act as germination stimulants in potatoes and soybeans, for example, or as pigments in the wings of butterflies (see Section 9.3.4). Animal organisms contain about 100 different forms of folacin.

form. Good folate sources are also strawberries. Their folate content in different strawberry cultivars varied from 0.30 to 0.69 mg/kg (f.w.). Legumes also contain significant amounts of folate (lentil 2.2–2.9 mg/kg, chickpea 0.42–1.25 mg/kg, peas 0.41–2.0 mg/kg).

Rich sources of dietary vitamin are yeast and mushrooms. The dominant folate forms in commercial dry baker's yeast were found to be tetrahydrafolate and 5-methyltetrahydrofolate, with a total folate content of 28.9 mg/kg.

5.12.6 Reactions and changes in foods

Folacin is unstable in acid, neutral, and alkaline solutions, at higher temperatures, and especially in light in the presence of oxygen or in the presence of transition metals and riboflavin. Stability varies depending on the number of bound glutamate residues and groups at either N-5 or N-10, or at a bridge that connects the two. Folic and 5-formyltetrahydrofolic acids are relatively stable against oxidation; less stable is 5-methyltetrahydrofolic acid. Oxidation and other reactions commonly occurring during processing and storage of food are shown in Figure 5.23.

Folate degradation produces many different products, depending on pH and other factors. For example, tetrahydrofolic acid (H_4PteGlu) yields 4-aminobenzoyl-L-glutamic acid, 7,8-dihydrofolic acid, and a number of simple pterins (Figure 5.24). In the presence of reducing sugars, folic acid may react with their degradation products. For example, the non-enzymatic glycation of folic acid by 1,3-dihydroxyacetone yields the corresponding N^2-[1-(carboxyethyl)]folic acid (Figure 5.25).

Meat contains free folacin and folacin bound to polysaccharides. Losses during heat treatment can be as high as 95%. Most of the losses, as in other processed foods, are caused by leaching into water.

The stability of folacin in milk depends on the presence of oxygen. Typical losses induced by pasteurisation reach 5%, losses in UHT milk 10–20%, and in condensed milk about 25%. Folacin content in yoghurt depends on the type of microorganisms used. It may be lower, but it may also be higher than in the original milk. Hard cheeses contain 75–90% of folacin present in the raw material.

In fresh vegetables, 5-methyl-H_4PteGlu$_n$ is oxidised in acidic solutions, creating unstable but bioavailable vitamin 5-methyl-5,6-dihydrofolic acid (5-methyl-5,6-H_2PteGlu$_n$, 5-40). This compound isomerises in acidic solutions (during cooking or in the acidic environment of the stomach) to 5-methyl-5,8-dihydrofolic acid (5-methyl-5,8-H_2PteGlu$_n$, 5-40), which is not bioavailable. Both these forms can be reduced to 5-methyl-H_4PteGlu$_n$ by ascorbic acid. 10-Formyl-H_4PteGlu$_n$ is easily oxidised to 10-formyl-PteGlu$_n$. During heating and in an alkaline environment, this compound spontaneously isomerises to 5-formyl-H_4PteGlu$_n$.

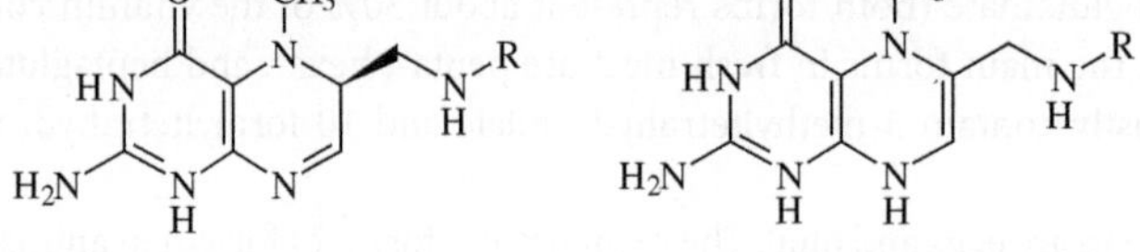

5-methyl-5,6-dihydrofolic acid   5-methyl-5,8-dihydrofolic acid

5-40, oxidised forms of 5-methyl-H_4PteGlu$_n$ (for R see **5-38**)

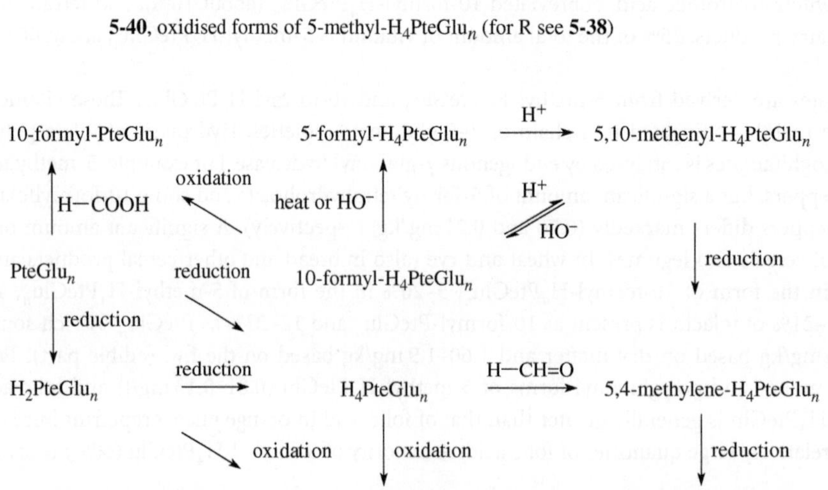

Figure 5.23 Non-enzymatic reactions of folylpolyglutamates.

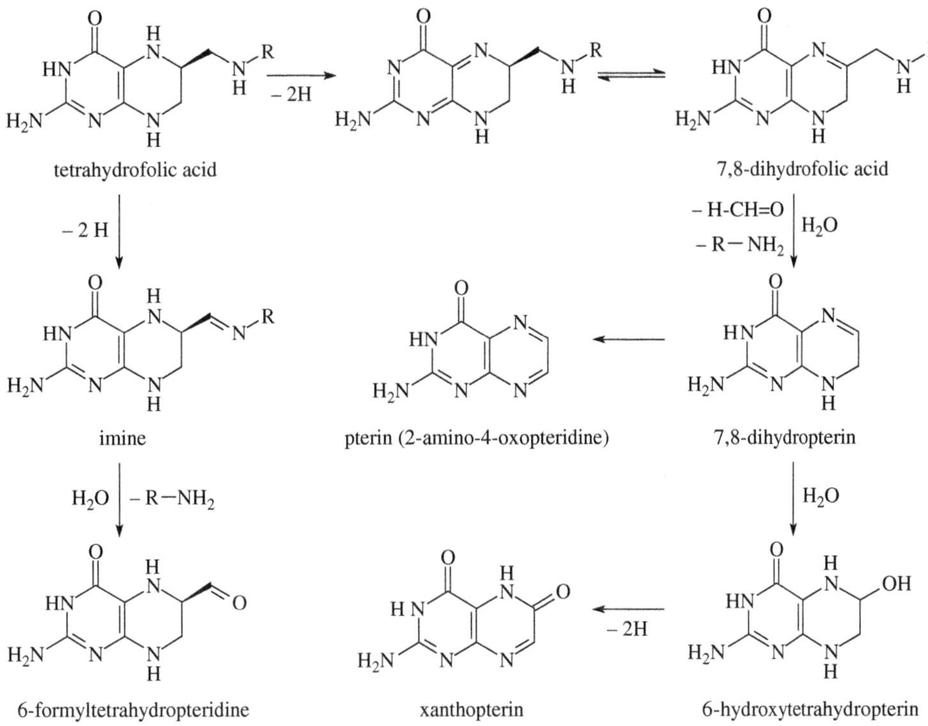

Figure 5.24 Degradation of tetrahydrofolic acid (see 5-38 for R).

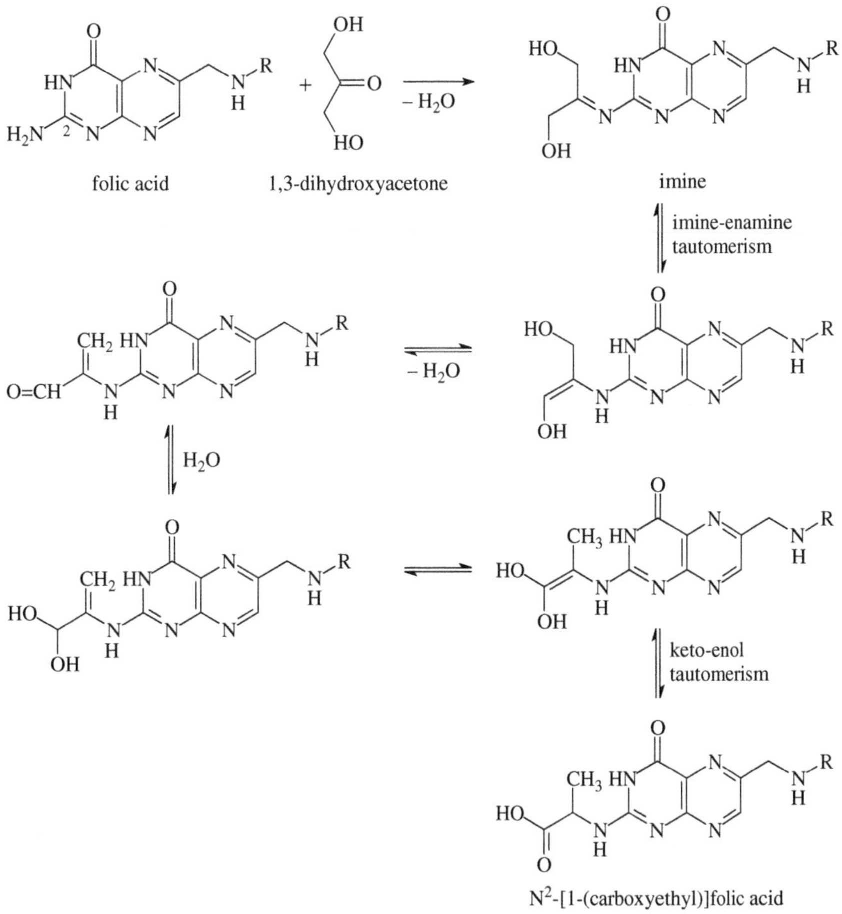

Figure 5.25 Reaction of folic acid with 1,3-dihydroxyacetone. Source: Schneider et al. (2002), figure 5. Reproduced by permission of the American Chemical Society.

Folacin content in cereals, like the content of all B group vitamins, is highest in the surface layers of the grain. The vitamin content of flour depends on the flour extraction rate. During the preparation of dough, this content does not change, but losses occur during bread baking (20% and more). Cooked pasta products lose about 20% of their folacin content.

When cooking and preserving vegetables, the vitamin loss is about 20–50%.

5.13 Corrinoids

5.13.1 Structure and terminology

Corrinoids are compounds with the activity of vitamin B_{12} that have the most complicated structure of all vitamins. The building block of corrinoids is the corrin ring (**5-41**), a partially hydrogenated, almost planar structure containing four reduced pyrrole rings joined into a macrocyclic ring by links between their α-positions, which resembles the porphyrine ring of haem and chlorophyll pigments. It differs from them, however, in that three of these links are formed by one-carbon units (methylidene bridges), as in a porphyrine ring, but the fourth is formed by a direct Cα—Cα bond; therefore, this structure does not contain the methylidene bridge between pyrrole rings A and D.

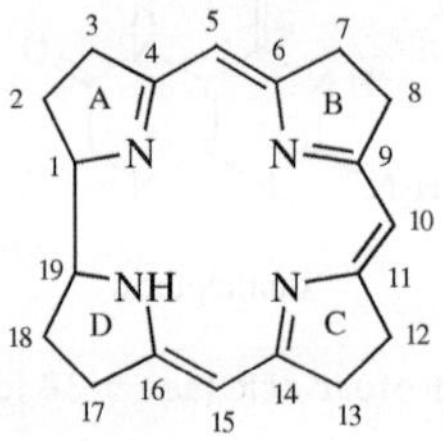

5-41, corrin

Substituents of the corrin ring are mostly methyl groups (six on pyrrole nuclei and two in positions 5 and 15) and amide residues, which include four propionamide residues R^2 (*bdef*) and three acetamide residues R^1 (*acg*) (**5-42**). The orientations of propionamide residues (R^2) are in the β-orientation, whilst the acetamide residues have α-orientation. Six of the seven carboxamide groups are primary amides, the rest of the propionamide in cycle D, which is R^2 (*f*), has a complicated structure, because it is an amide of propionic acid with (*R*)-1-aminopropan-2-ol that is esterified by α-D-ribofuranosyl-5,6-dimethylbenzimidazole 3′-phosphate.

The central cobalt atom can generate up to six coordination bonds with ligands. It is coordinated to four nitrogen atoms of the pyrrole nuclei; the fifth coordination bond binds the second nitrogen atom in 5,6-dimethylbenzimidazole in the α-position. In some corrinoids, this position is not occupied. The forms of vitamin B_{12} containing 5,6-dimethylbenzimidazole are called cobalamins. Cobalamins are the only important natural substances that contain a covalent bond, C—Co. Cobalt can have an oxidation degree 3+, 2+, or 1+ according to substitution and biochemical function. For example, cyanocobalamin, hydroxycobalamin, aquacobalamin, methylcobalamin, and some other compounds contain Co^{3+} and have a red colour. Enzymatic reduction (and the reaction with thiols) gives a brown product containing Co^{2+}, whilst further reduction yields a grey–green product containing Co^+. Substitution of cobalt by other metals leads to inactive products.

The sixth coordination bond in the β-position can occupy different groups or compounds; alternatively, this position may be not occupied at all. The molecule of naturally occurring cobalamin contains adenosine, and correspondingly a 5′-deoxy-5′-adenosyl residue, bound through the carbon C-5′. This biologically active form of vitamin B_{12} is called adenosylcobalamin, adenosylvitamin B_{12}, or coenzyme B_{12} (CoE-B_{12} for short). Another naturally occurring coenzyme is methylcobalamine, which has a methyl group as a ligand (methylvitamin B_{12}). Other ligands are a hydroxyl group (hydroxycobalamin, vitamin B_{12a}), water (aquacobalamine, vitamin B_{12b}), and a nitro group (nitritocobalamin, vitamin B_{12c}). Synthetic vitamin B_{12}, known as cyanocobalamin (metabolised to the active substance coenzyme B_{12}), is the form used in pharmaceutical preparations. It contains cyanide anion as a ligand. Depending on the reaction conditions, other ligands, such as sulfite anion in sulfitocobalamin, can also be bound to the carbon C-5′. All these structural analogues based on the corrin skeleton are collectively known as corrinoids (**5-43**).

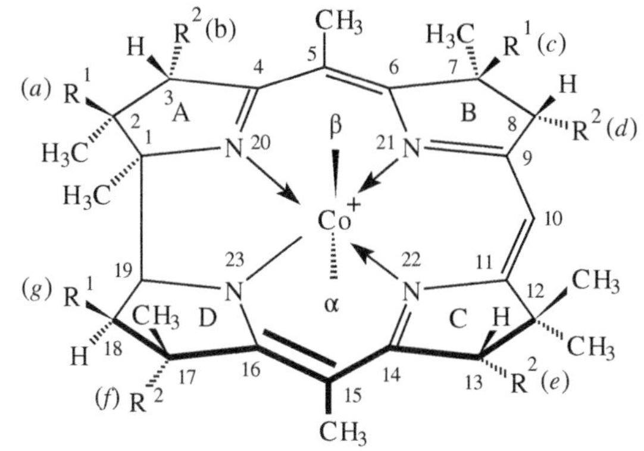

5-42, precursors of cobalamins

cobyrinic acid, $R^1 = CH_2COOH$, $R^2 = (CH_2)_2COOH$
cobyric acid, $R^1 = CH_2C(=O)NH_2$, R^2 (*bde*) = $(CH_2)_2C(=O)NH_2$, R^2 (*f*) = $(CH_2)_2COOH$
cobynic acid, $R^1 = CH_2COOH$, $R^2 = (CH_2)_2COOH$, R^2 (*f*) = $(CH_2)_2C(=O)NHCH_2CH(OH)CH_3$
cobinamide, $R^1 = CH_2C(=O)NH_2$, R^2 (*bde*) = $(CH_2)_2C(=O)NH_2$, R^2(*f*) = $(CH_2)_2C(=O)NHCH_2CH(OH)CH_3$

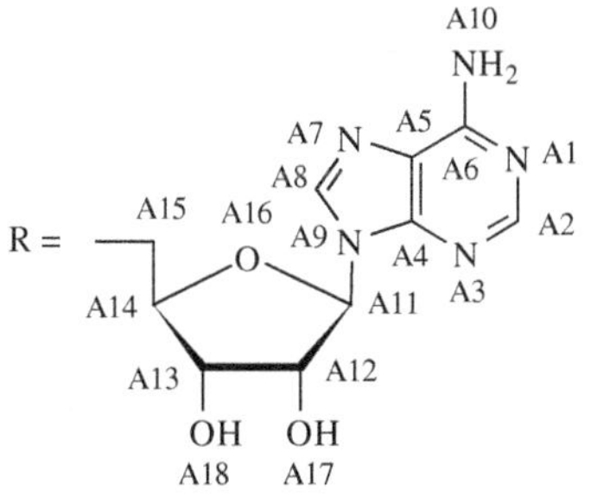

5′-deoxy-5′-adenosyl

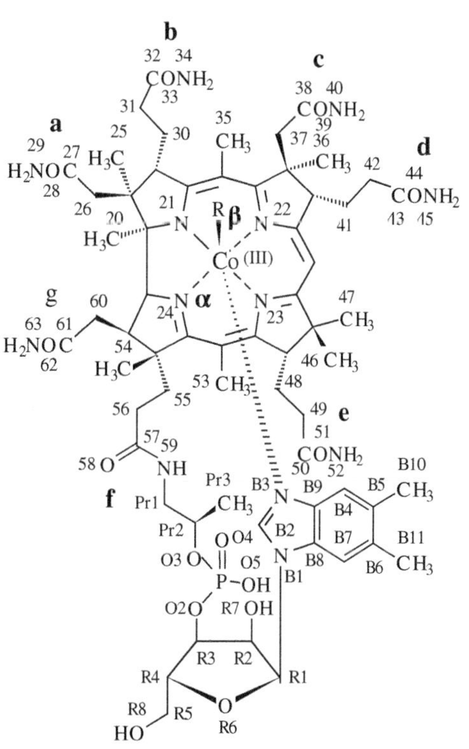

5′-deoxy-5′-adenosylcobalamine

5′- deoxy-5′-adenosylcobalamin, R = 5′-deoxy-5′-adenosyl
cyanocobalamin, R = CN
hydroxycobalamin, R = OH
aquacobalamin, R = H_2O
methylcobalamin, R = CH_3
nitritocobalamin, R = NO_2
sulfitocobalamin, R = SO_3

5-43, structures of important corrinoids

5.13.2 Biochemistry

Vitamin B$_{12}$ is synthesised by many bacteria, some cyanobacteria, and even some yeasts. The starting compound is 5-aminolaevulinic (5-amino-4-oxopentanoic) acid. The initial reaction sequences are the same as in the biosynthesis of porphyrins.

The vitamin produced by bacteria in the intestinal tract of herbivores is absorbed in the tissues and becomes the main vitamin source for other animals in the food chain. Cobalamins found in the diet are mostly bound to proteins and released from their protein complexes through the action of acids or pepsin in the stomach. Then, in a unique process, R-protein (also referred to as cobalophilin, haptocorrin, and transcobalamine I) secreted in saliva and in gastric juice picks up the vitamin and transports it through the stomach into the small intestine, where it is liberated by the action of pancreatic proteases. The free vitamin then attaches to a glycoprotein produced by the parietal cells of the stomach, known as intrinsic factor (IF) or gastric intrinsic factor (GIF), which carries it to the last section of the small intestine, the ileum, the cells of which contain receptors for the cobalamin–IF complex. In supplements, vitamin B$_{12}$ is not bound to protein and therefore does not need digestive enzymes or stomach acid to be detached. In addition to the IF mechanism, passive diffusion normally accounts for 1–3% of the vitamin B$_{12}$ absorbed when obtained through food. Another source of vitamin B$_{12}$ can be the food contaminated by bacteria producing vitamin B$_{12}$.

Coenzymes B$_{12}$ as cofactors catalyse two completely different types of reactions. Methylcobalamin (as with folacin) acts in some transmethylation reactions (such as biosynthesis of methionine from homocysteine) and collaborates with folic acid in the synthesis of DNA and red blood cells (biosynthesis of porphyrins) and in the fixation of carbon dioxide by some anaerobic acetogennic microorganisms. Enzymes using 5′-deoxy-5′-adenosylcobalamin catalyse a number of isomerisations that are otherwise only viable with difficulty (1,2-rearrangements, such as the formation of succinyl-CoA from methylmalonyl-CoA) and, in some organisms, reduce ribonucleotides to deoxyribonucleotides.

5.13.3 Physiology and nutrition

Vitamin B$_{12}$ is not absorbed very well, so relatively large amounts need to be supplied through the diet. The daily intake is estimated at 3–31 µg, and about 20–70% is absorbed. The amount of vitamin B$_{12}$ actually needed by the body of an adult is very small, probably only about 2–3 µg/day (the estimated average requirement is 2 µg/day; recommended dietary allowance (RDA) is 2.4 µg/day), whilst pregnant and lactating women need 4 µg/day. The need is met mainly by meat, meat products, and offal (about 70%; the richest dietary sources are the liver and kidney), followed by milk and dairy products (about 20%), eggs (about 9%), and cereal products (about 2%). Vegetables and fruits are very poor sources as they contain vitamin B$_{12}$ only if contaminated by faecal bacteria.

In contrast to other water-soluble vitamins, vitamin B$_{12}$ is not excreted quickly in the urine, but rather accumulates and is stored in the liver, kidney, and other body tissues. As a result, vitamin B$_{12}$ deficiency may not manifest itself until after five or six years of a diet supplying inadequate amounts. Since plants have no ability to synthesise vitamin B$_{12}$, strict vegetarians (vegans) have a greater risk of developing vitamin B$_{12}$ deficiency and hence must depend upon vitamin B$_{12}$-fortified foods or vitamin B$_{12}$-containing dietary supplements to meet their requirement. Cyanobacteria (blue-green bacteria) of the genus *Spirulina* (now *Arthrospira*; the older term remains in use for historical reasons) is a minimal source of vitamin B$_{12}$ and mostly contains the so-called pseudovitamin B$_{12}$, a derivative that is inactive in humans. Vitamin B$_{12}$ production in higher mushrooms is controversial. It has been suggested that the source is the microorganisms living on the surface of the mushrooms or in the compost containing horse manure–wheat straw that is used to cultivate them. For example, the highest vitamin B$_{12}$ content in cultivated white button mushrooms (*A. bisporus*) is found in the peel (0.7–3.5 µg/kg), as compared to the cap, stalk, or flesh (0.07–0.8 µg/kg). Lacto-ovo vegetarians usually get enough vitamin B$_{12}$ through consuming milk, dairy products, and eggs. Many elderly people (or people after stomach resection or peptic ulcer disease operations) are deficient because their production of the intrinsic factor needed to absorb the vitamin from the small intestine declines rapidly with age. People with intrinsic factor defects eventually develop a very serious pernicious (deadly) anaemia that manifests by a reduction of haem synthesis.

5.13.4 Use

Vitamin B$_{12}$ is used for the enrichment of some foods (such as breakfast cereals, soy products, energy bars, and yeast extract spreads), which may become the source of corrinoids for strict vegetarians and vegans. Along with other vitamins, vitamin B$_{12}$ is added to many multivitamin preparations and to food supplements. Cyanocobalamin, which is used in most supplements, is readily converted into the coenzyme forms of cobalamin (methylcobalamin and 5′-deoxyadenosylcobalamin) in the human body.

5.13.5 Occurrence

Corrinoids are present almost exclusively in foods of animal origin (Table 5.9). In milk, the main vitamins are adenosylcobalamin and methylcobalamin; cheeses and egg yolk contain mainly methylcobalamin.

Table 5.9 Corrinoid content of some foods.

Food	Content in edible portion (µg/kg or µg/l)	Food	Content in edible portion (µg/kg or µg/l)
Pork meat	6–10	Fish	13–28
Beef meat	112–20	Clams	11
Chicken meat	2–5	Milk	3–38
Pork liver	260–1220	Cheese	6–33
Beef liver	590–830	Eggs	7–9
Pork kidney	85–200	Yeast extract spread	5

Cobalamin does not occur in plants, although its presence has been acknowledged in legumes. Occasional findings in vegetables or fermented foods (such as beer and miso and soy sauce) may originate from contamination by organic fertilisers or by wild types of microorganisms. The vitamin probably comes from the biomass of these microorganisms. It is synthesised by many bacteria and some yeasts, such as *Candida utilis*.

5.13.6 Reactions and changes in foods

Despite their complex structure, cobalamins are relatively stable in solution at pH 4–7. In acidic solutions, cyanocobalamin gives hydroxycobalamin, which in acidic; neutral solutions exist as aquacobalamin. Acid hydrolysis of corrinoids produces mono-, di-, and tricarboxylic acids from the bound propionamides. Most susceptible to hydrolysis is the propionamide in position *e*. Hydrolysis of acetamide chains requires more drastic conditions, which simultaneously release the nucleotide and isopropylamine group in position *f*. Altogether, the acidic hydrolysis releases six ammonium ions. Hydrolysis in dilute solutions of alkali metal hydroxides yields biologically inactive dehydrovitamin B_{12} (the biologically inactive form with a lactam ring attached to ring B of the cobinamide ring). Alkaline hydrolysis further produces cyanocobinamide, 1-α-D-ribofuranosyl-5,6-dimethylbenzimidazole, and phosphate. Under alkaline conditions, a *c*-lactone ring in ring B also results.

Photodegradation of adenosylcobalamin in the absence of oxygen gives cobalamin and 5,8-cycloadenosine. In the presence of oxygen, aquacobalamin and adenosine-5′-carbaldehyde are produced. Photolysis of methylcobalamin in aqueous solutions and in the absence of oxygen produces aquacobalamin and methane as major products, or aquacobalamin and formaldehyde if in the presence of oxygen. In aqueous solutions in the presence of ascorbic acid, aquacobalamin is reduced, hydroxylated at C-5, and induced to form a lactone ring between the carboxyl *c* and C-6 of the corrin ring; yellow xanthocorrinoids are produced as the final products. The hydroxyl group in hydroxycobalamin can be easily substituted by another ligand, such as a cyanide anion and chlorine anion. The corrin ring can also be halogenated at C-10, for example, by chloramine-T (used as a disinfection and sanitation agent). The reaction of methylcobalamin with metal ions in aqueous solutions produces methyl derivatives of metals (transfer of methyl anion CH_3^-) and aquacobalamin.

During food processing and culinary operations, vitamin B_{12} seems to be very stable. The main reason for loss is leaching into water.

Vitamin losses during meat processing are dependent on the technology used. They can reach 55–70% of the original amount of the vitamin.

Under normal processing conditions, the vitamin content in milk does not change too much. The loss caused by pasteurisation is up to 10%, or about 10–20% in UHT milk. In the manufacture of hard cheeses, vitamin retention is about 60–90% of the original content. Dairy products fermented using *Propionibacterium shermanii* (such as Swiss-type cheeses) have increased vitamin content in comparison with the original milk (up to 30 times).

5.14 Vitamin C

5.14.1 Structure and terminology

The basic biologically active compound is ascorbic acid. Of the four possible stereoisomers (asymmetric carbons C-4 and C-5), the activity of vitamin C is seen only in L-ascorbic acid (L-*threo*-hex-2-enonic acid 1,4-lactone; formerly called ceritaminic acid, 2-keto- L-gulonic acid, L-xylo-hex-2-ulosonic acid, and, later, L-*xylo*-ascorbic acid). The D-isomer of ascorbic acid (D-*xylo*-ascorbic acid) and the second pair of enantiomers, known as L- and D-isoascorbic acids (**5-44**), do not show vitamin C activity. Previously, these acids, which are γ-lactones of L- and D-*erythro*-hex-2-enonic acids, were known as L- and D-erythorbic acids or L- and D-*arabino*-ascorbic acids.

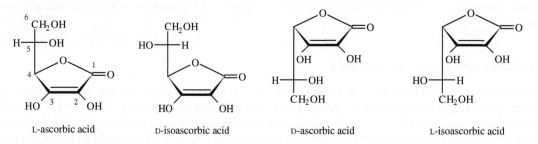

Figure 5.26 Biologically active forms of vitamin C.

The name vitamin C refers not only to L-ascorbic acid, but also to the whole reversible redox system, which includes the one-electron oxidation product of L-ascorbic acid known as L-ascorbyl radical (or L-monodehydroascorbic acid or semidehydroascorbic acid) and the two-electron oxidation product of L-ascorbic acid known as L-dehydroascorbic acid (Figure 5.26). Ascorbic acid and ascorbyl radical mainly occur as anions in solutions at physiological pH.

5-44, ascorbic acid isomers

5.14.2 Biochemistry

Ascorbic acid is synthesised by all green plants that provide their energy needs through photosynthesis (photoautotrophic plants). Almost all living organisms – insects and invertebrates, most fish, and many species of birds and mammals – have the ability to synthesise ascorbic acid. Notably, humans, but also gorillas, chimps, orangutans, and some monkeys, frutivorous (fruit eating) bats, guinea pigs, and capybaras, lack this ability.

Plants synthesise ascorbic acid from the active form of D-mannose (GDP-D-mannose). In mammals, the starting compound is UDP-D-glucose. Microorganisms and higher fungi synthesise 6-deoxy-L-ascorbic acid from L-fucose and D-erythroascorbic acid from D-arabinose.

The functions of ascorbic acid are related primarily to its redox properties. In plants, it has a role in photosynthesis; it regulates the amount of active oxygen species and acts in the growth and differentiation of cells. Cleavage of the molecule of ascorbic acid also creates specific plant metabolites, such as L-threonic, L-tartaric, L-glyceric, and oxalic acids.

In animals, ascorbic acid is involved primarily in hydroxylation reactions continuously occurring in the body. The most important oxidation reaction is hydroxylation of proline to 3-hydroxyproline and of lysine to 5-hydroxylysine in procollagen, which is related to protein biosynthesis of connective tissue. Another important reaction is hydroxylation of 3,4-dihydrophenylethylamine (dopamine) to norepinephrine (noradrenaline) and biosynthesis of betaine carnitine. Norepinephrine, along with other so-called catecholamines, acts as a carrier of chemical information (hormone and neurotransmitter); carnitine acts as a carrier of fatty acid residues through the membrane of the mitochondria. Vitamin C also participates in the biosynthesis of mucopolysaccharides, prostaglandins, and homogentisic acid (the precursor of tocopherols and plastoquinones), the absorption of ionic forms of iron, iron transport, and the transport of sodium, chloride, and possibly calcium ions; it further allows the transfer of sulfate in the form of ascorbyl-2-sulfate and acts in the metabolism of cholesterol and drugs and in many other reactions.

Very important reactions related to the antioxidant properties of vitamin C are those with active oxygen forms, free radicals, and oxidised forms of vitamin E, which protect vitamin E and lipid membranes from oxidation. Vitamin C also has a protective function for labile forms of folic acid. It inhibits the formation of nitrosamines and thus acts as a modulator of mutagenesis and carcinogenesis. Many other activities of vitamin C are still only partially understood or not understood at all.

5.14.3 Physiology and nutrition

A daily dose of 10 mg of L-ascorbic acid is sufficient to prevent scurvy. In the past, the recommended daily intake was 30 mg (50 mg for adolescents and 60 mg for pregnant women). Today, recommendations for vitamin C intake have been set by various national agencies and range from 40 to 95 mg/day (e.g. 40 mg in Great Britain, 45 mg by the WHO, and 60–95 mg in the United States). The US recommended daily dietary intake of vitamin C is 30 mg for infants, 45 mg for children, 60 mg for adult men and women, 80 mg for pregnant women (second and third trimesters), and 100 mg during lactation. For patients with respiratory diseases and during convalescence and in other cases (such as radiation therapy), the daily doses may reach 500–1000 mg or even more. The tolerable upper intake level for adult men and women is 2000 mg/day. The possible adverse effects of high doses of vitamin C must also be considered. These include so-called rebound scurvy (the vitamin C dependency state that occurs in the foetus of a woman taking megadoses of vitamin C during pregnancy), increased excretion of oxalic acid in the urine (which may lead to stones in the urinary tract), an increased absorption of iron in those susceptible to iron overload, increased absorption of toxic metals, and interference with certain medications (e.g. warfarin, aspirin, antidepressants, and contraceptive pills).

The total vitamin C requirement is met by foods, especially potatoes (about 20–30%), vegetables (about 30–40%), and fruits (30–35%). Milk covers less than 10% of the vitamin need. Deficiency or hypovitaminosis manifests in a number of non-specific symptoms, most commonly known as spring fatigue. The best-known syndrome of acute avitaminosis is scurvy. According to some new findings, dehydroascorbic acid has somewhat lower activity than ascorbic acid.

Oxidoreductases (such as ascorbate oxidase, ascorbate peroxidase, monodehydroascorbate reductase, dehydroascorbate reductase, superoxide dismutase, and ascorbate: cytochrome-*b* reductase) that act in the metabolism of vitamin C in animals and plants can be considered to be antivitamins C. Other oxidoreductases, such as the enzymes known trivially as polyphenoloxidases and some others, may indirectly cause loss of ascorbic acid.

5.14.4 Use

Ascorbic acid is widely used as a food additive due to its properties (it is a vitamin, antioxidant, and chelating agent), especially in canning and fermentation processes, but also in meat, fat, and cereal technologies. The water-soluble salt of L-ascorbic acid sodium L-ascorbate (**5-45**) and the lipophilic compound 6-palmitoyl-L-ascorbic acid (L-ascorbyl 6-palmitate, **5-45**), which inhibits the formation of nitrosamines in cured meat and meat products, are also used as antioxidants. Non-polar acetals of L-ascorbic acid derived from fatty aldehydes (dodecanal to octadecanal, **5-45**) were used as inhibitors of nitrosamine formation in the production of ham, as they were more stable than ascorbyl palmitate. D-Isoascorbic acid can in many cases replace L-ascorbic acid (e.g. as an antioxidant), but it cannot be used to improve the baking properties of flour (D- and L-isoascorbic acids are less active than L-ascorbic acid, and D-ascorbic acid is completely inactive), because the corresponding process in dough involves a stereospecific reaction.

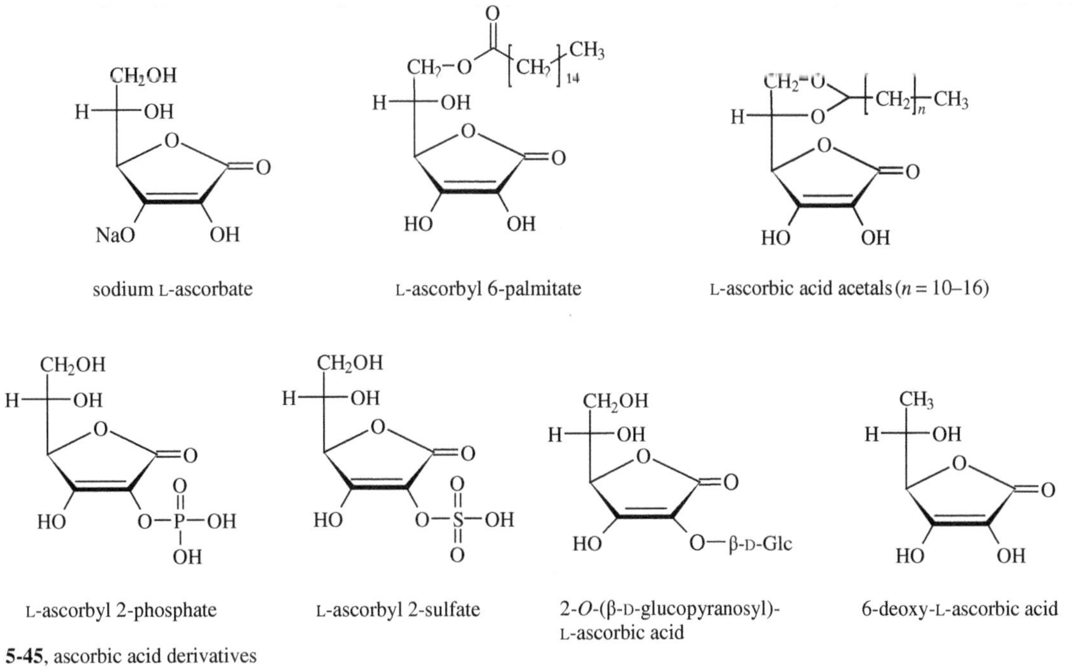

sodium L-ascorbate L-ascorbyl 6-palmitate L-ascorbic acid acetals ($n = 10–16$)

L-ascorbyl 2-phosphate L-ascorbyl 2-sulfate 2-*O*-(β-D-glucopyranosyl)-L-ascorbic acid 6-deoxy-L-ascorbic acid

5-45, ascorbic acid derivatives

Sodium ascorbate and ascorbic acid esters (**5-45**), such as ascorbyl 6-palmitate and ascorbyl 2-phosphate, are fully bioavailable, whilst ascorbyl 2-sulfate (**5-45**) is a completely inactive vitamin form. Phosphate and sulfate are about 20 times more stable to oxidation than the free acid. D-Isoascorbic acid (**5-44**) shows only 5–20% activity; 2-*O*-(β-D-glucopyranosyl)ascorbic acid (**5-45**) occurring in plants has the same biological activity as ascorbic acid and is also stable against oxidation. 6-Deoxy-L-ascorbic acid (**5-45**), found in fungi, has about 30% of the activity of ascorbic acid, and the bound ascorbic acid form ascorbigen about 15–20%.

Ascorbic acid is added to fruit juices, stewed fruit, and frozen fruit to prevent undesirable changes in flavour caused by oxidation during storage and processing. The removal of oxygen in airtight containers requires the addition of 3–7 mg of ascorbic acid per cubic centimetre of air present (depending on pH and temperature). Ascorbic acid at relatively low concentrations is often used as an inhibitor of enzymatic browning reactions during the peeling, slicing, and drying of fruits, vegetables, and potatoes. It is often used in combination with citric acid as it is more stable in acid solutions, and the optimum pH of enzymes (phenolases) lies in the range 6–7. A common practice is to soak fruits and vegetables for 3 minutes in a solution containing 1–3% ascorbic acid, 0.1–0.3% calcium chloride, and 0.015% hydrogen sulfites (they act as preservatives). In the absence of ascorbic acid, the amount of hydrogen sulfites in the bath has to be increased approximately 10-fold.

The addition of 20–30 mg/l ascorbic acid prevents the formation of colloidal turbidity (called chill haze) in beer, as well as adverse changes in flavour due to the oxidation that occurs during pasteurisation and storage. The use of ascorbic acid in winemaking can reduce the amount of sulfur dioxide used for fumigation.

The addition of ascorbic acid (or sodium ascorbate or ascorbyl palmitate) to meat and meat products, together with nitrites (at a level of 60–180 mg/kg), is of functional and economic importance as it enhances and speeds up production (e.g. of ham) considerably. The characteristic pigment of raw meat treated with nitrites, nitroxymyoglobin, is formed approximately three times faster. The addition of ascorbic acid allows the time of smoking to be shortened and stabilises the colour of the finished product. Ascorbic acid also enhances the inhibitory effects of nitrite on toxinogenic *Clostridium botulinum* bacteria. Additions of 300–1000 mg/kg of ascorbic acid (the optimal ratio of ascorbate and nitrite is 2 : 1; hydrophilic ascorbate is only partially effective and is often replaced by ascorbyl palmitate, which is soluble in fats) also inhibit the formation of nitrosamines. If nitrates are used instead of nitrites, they are reduced to nitrites by ascorbic acid.

The baking properties of flour improve after a certain period of storage, when the products of lipid oxidation react with thiol groups of flour proteins that are oxidised and form disulfide bridges between the protein molecules. Simultaneous degradation of carotenoid pigments results in a lighter colour of the flour. The addition of ascorbic acid (10–100 mg/kg) improves the baking properties of flour, especially during the Chorleywood industrial bread-making process, which reduces the time and labour input needed to make bread, thereby reducing manufacturing costs (see Section 5.14.6.1).

Ascorbic acid does not occur in fats and oils, but is used, as ascorbyl palmitate, as an antioxidant in amounts ranging from 0.006 to 0.040%.

5.14.5 Occurrence

Plasma and tissues of animals, as well as of foods of plant origin, usually contain 90–95% of vitamin C as ascorbic acid; the remainder is dehydroascorbic acid. Vitamin C is found in significant amounts only in the liver (Table 5.10). Other foods of animal origin (meat, eggs, and milk) have almost negligible importance as sources of vitamin C.

More than 90% of vitamin C in the human diet is supplied by fruits and vegetables (including potatoes). Fresh fruits and vegetables are the richest sources of vitamin C, but its amount varies considerably between species and cultivars. The vitamin content in horticultural crops is also strongly dependent on a number of climatic environments, preharvest and postharvest factors, such as preharvest climatic conditions, maturity, harvesting methods and postharvest handling procedures. The absolute highest concentration of ascorbic acid, reaching 17–46 g/kg edible portion, accumulates in the fruit of acerola (*Malpighia emarginata*, syn. *M. glabra*, Malpighiaceae), also known as Barbados cherry or West Indian cherry. A comparable content of vitamin C, 23–32 g/kg edible portion is in the Australian fruit of *Terminalia ferdinandiana* (Combretaceae), commonly known as kakadu plum or billygoat plum. Some other excellent sources of vitamin C (such as rosehips, blackcurrant, and curly parsley) are usually not too important in meeting the needs of vitamin C, because they are consumed only occasionally and in small quantities. Of much greater importance are the sources with an average or lower level of the vitamin, such as potatoes, which are known to be the most important source of vitamin C in the Western diet because of the large quantities consumed (although the vitamin content in potatoes significantly decreases during storage for the winter months). In winter and spring months, subtropical fruits (especially oranges) fulfil a considerable portion of the vitamin C requirements. In the past, good sources of vitamin C were common vegetables that could be eaten year-round, such as cabbage, but at present many other vegetables are regularly consumed (such as red and green peppers and tomatoes). Cereals contain only traces of vitamin C; somewhat higher levels are found only in germinating seeds.

A precursor of ascorbic acid, 2-*O*-(β-D-glucopyranosyl)ascorbic acid, was recently found in both the ripe fresh fruit and dried fruit of *Lycium barbarum* (Solanaceae), known as Chinese wolfberry or Chinese boxthorn, which is native to southeastern Europe and Asia. The dried fruit contains 2-*O*-(β-D-glucopyranosyl)ascorbic acid at a level of 5000 mg/kg, which is comparable to the ascorbic acid content of fresh lemons.

Table 5.10 Vitamin C content of some foods.

Food	Content in edible portion (mg/kg or mg/l)	Food	Content in edible portion (mg/kg or mg/l)
Meat	10–20	Carrot	50–100
Ham (pork, cured)[a]	300–500	Parsley (root)	230
Offal	50–340	Parsley (curly)	1500–2700
Milk	5–20	Chive	430
Apples	15–50	Leek	150–300
Pears	20–40	Onion	90–100
Plums	25–45	Garlic	150–160
Peaches	70–100	Horseradish	450–1200
Cherries, sweet cherries	60–300	Cabbage	170–700
Gooseberries	330–480	Cabbage(Savoy)	700–1400
Currant (red)	200–500	Brussel sprouts	1000–1030
Currant (black)	1100–3000	Broccoli	1100–1130
Grapes	20–50	Cauliflower	47–1610
Strawberries	400–700	Kohlrabi	280–700
Blackberries	90	Lettuce	60–300
Melons	130–590	Spinach	350–840
Oranges	300–600	Tomatoes	80–380
Lemons	300–640	Eggplant	80
Grapefruits	240–700	Pepper (various types)	620–3000
Pineapple	150–250	Cucumber	65–110
Bananas	90–320	Asparagus	150–400
Kiwi	700–1630	Peas	80–410
Mango	100–350	Beans	90–300
Papaya	620–980	Beetroot (red)	65
Rosehips	2500–10000	Potatoes	80–400

[a] Added ascorbic acid.

Higher fungi (commonly referred to as mushrooms) do not contain ascorbic acid, but do contain a number of structurally related compounds. The main compounds are 6-deoxy-L-ascorbic acid (**5-45**), the five-carbon ascorbic acid analogue D-erythroascorbic acid, which is (4R)-2,3,4,5-tetrahydroxypent-2-enonic acid 1,4-lactone, also known as D-*glycero*-pent-2-enonic acid 1,4-lactone or D-*erythro*-pent-2-ulosonic acid 1,4-lactone (**5-46**). Both acids occur as free compounds and as C-5 glycosides. 6-Deoxy-L-ascorbic acid, D-erythroascorbic acid, 6-deoxy-5-O-(α-D-xylopyranosyl)-L-ascorbic acid (**5-46**), 6-deoxy-5-O-(α-D-glucopyranosyl)-L-ascorbic acid (**5-46**), 5-O-(α-D-xylopyranosyl)-D-erythroascorbic acid (**5-46**), and 5-O-(α-D-glucopyranosyl)-D-erythroascorbic acid (**5-46**) were isolated from some edible fungi (of the phylum Basidiomycota), 5-O-(α-D-galactopyranosyl)-D-erythroascorbic acid (**5-46**) occurs in the pathogenic fungus *Sclerotinia sclerotiorum* (of the phylum Ascomycota), and 5-O-(α-D-glucopyranosyl)-D-erythroascorbic acid (**5-46**) was isolated from a filamentous fungus *Phycomyces blakesleeanus* (of the phylum Zygomycota). Yeast (*S. cerevisiae*) and red bread mould (*Neurospora crassa*, Ascomycota) contain only D-erythroascorbic acid. 6-Deoxyascorbic acid and its glycosides have lower activity than ascorbic acid (about 30%), and D-erythroascorbic acid does not show any vitamin C activity. All these compounds have similar functions in fungi to ascorbic acid in plants and animals, as they act as antioxidants.

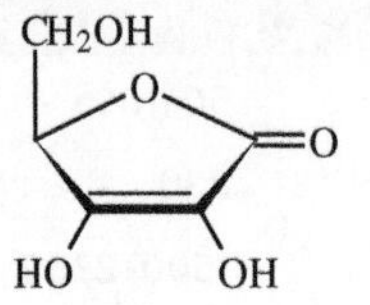

D-erythroascorbic acid

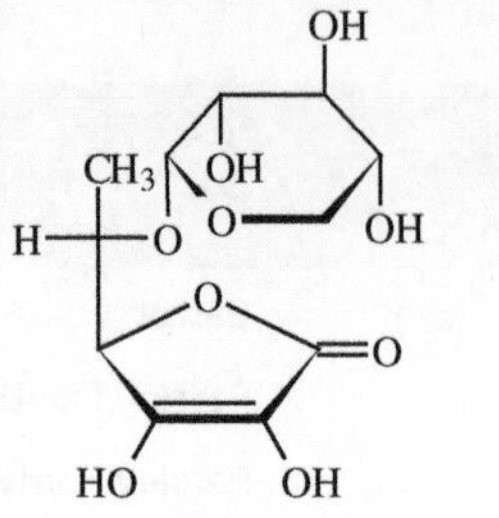

6-deoxy-5-*O*-(α-D-xylopyranosyl)-
L-ascorbic acid

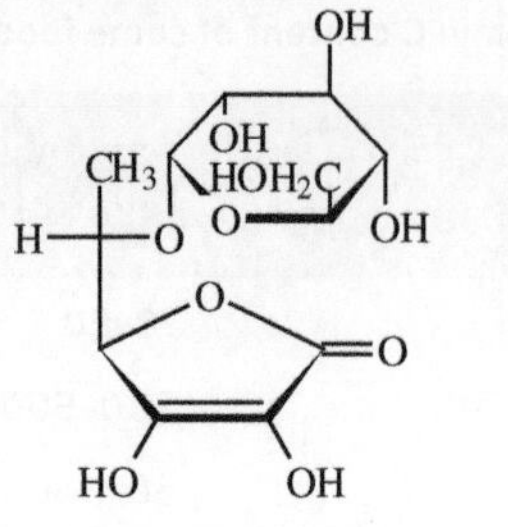

6-deoxy-5-*O*-(α-D-glucopyranosyl)-
L-ascorbic acid

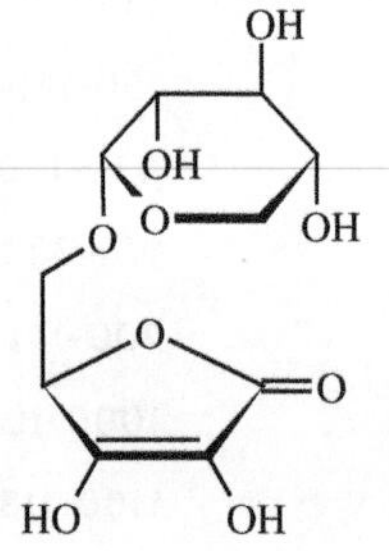

5-*O*-α-D-xylopyranosyl-
D-erythroascorbic acid

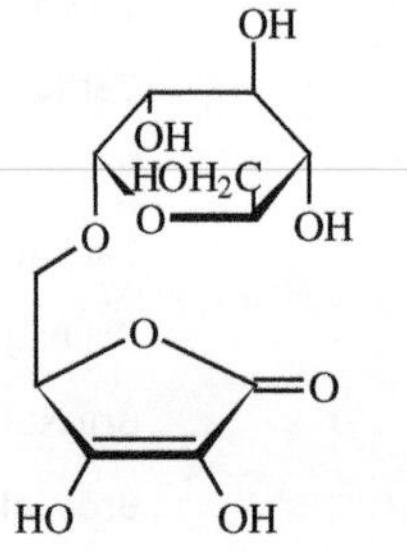

5-*O*-(α-D-glucopyranosyl)-
D-erythroascorbic acid

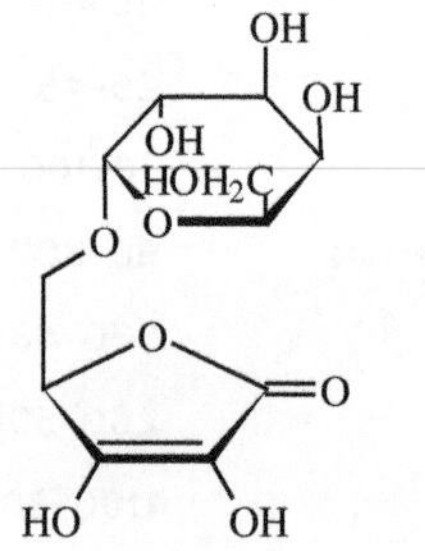

5-*O*-(α-D-galactopyranosyl)-
D-erythroascorbic acid

5-46, analogues of ascorbic acid in fungi

5.14.6 Reactions and changes in foods

Both enolic hydroxyl groups of ascorbic acid can dissociate, and ascorbic acid can be considered a diprotic acid ($pK_1 = 4.25$; $pK_2 = 11.8$). In solutions of physiological pH values, the prevailing ion is a monoanion. For example, in solutions of pH 7.4, 99.93% of ascorbic acid is present in the form of a monoanion; the rest is undissociated acid (0.06%) and dianions (0.01%). Only salts of the monovalent anion, such as sodium ascorbate, are known.

Ascorbyl radical that exists in solution as a resonance stabilised anion (as a cyclic compound, apparently with a double bond between carbons C-2 and C-3) and with the unpaired electron located in the C-4 region. Dehydroascorbic acid occurs as the bicyclic hydrated monomer (L-*xylo*-hexulono-l,4-lactone hydrate, Figure 5.27) in aqueous solutions.

5.14.6.1 Ascorbic acid

Oxidation of ascorbic acid to dehydroascorbic acid is catalysed by many oxidoreductases belonging to the category of ascorbic acid antivitamins. Ascorbic acid is also oxidised by atmospheric oxygen, hydrogen peroxide, and various other oxidising agents. Oxidation to dehydroascorbic acid is a reversible reaction that can be carried out by various mechanisms. The loss of one electron yields a radical of ascorbic acid as an intermediate, and the reaction is known as one-electron oxidation. Oxidation of ascorbic acid by the loss of two electrons yields dehydroascorbic acid, which is the first chemically stable product.

The oxidation of ascorbic acid in the enzymatically active and, especially, mechanically damaged plant tissues (e.g. by peeling and slicing) is mainly catalysed by ascorbate oxidase (L-ascorbate: O_2 oxidoreductase). The loss of vitamin activity in some plant tissues is associated with

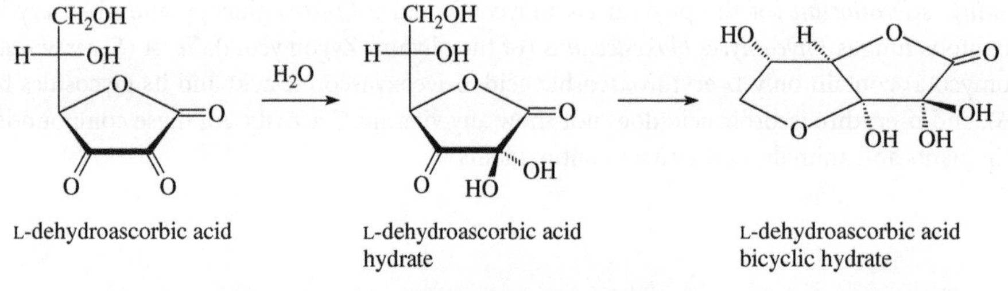

L-dehydroascorbic acid

L-dehydroascorbic acid
hydrate

L-dehydroascorbic acid
bicyclic hydrate

Figure 5.27 Formation of L-dehydroascorbic acid hydrate.

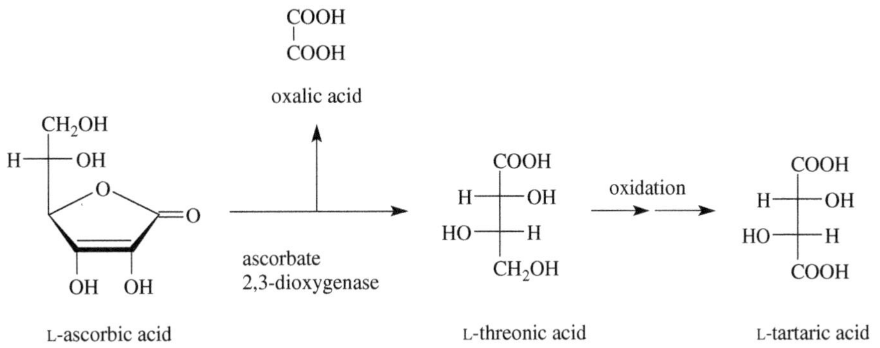

Figure 5.28 Formation of oxalic, threonic, and tartaric acids.

peroxidases and other enzymes. Ascorbate oxidase oxidises ascorbic acid in the presence of atmospheric oxygen. Generally, the reaction can be described by the following equation, where H_2A is ascorbic acid and A is dehydroascorbic acid:

$$2\,H_2A + O_2 \rightarrow 2\,A + 2\,H_2O$$

Ascorbate peroxidase employs hydrogen peroxide as a proton acceptor:

$$H_2A + H_2O_2 \rightarrow A + 2\,H_2O$$

Detailed reaction mechanisms are in fact more complex. In both cases, the primary product of oxidation of ascorbic acid is an ascorbyl radical ($HA^\bullet$) and its anion ($HA^{\bullet-}$), stabilised by resonance. It is relatively inert and does not react with oxygen, but comparatively quickly (the reaction half-life is about 0.2 seconds) provides an equimolar mixture of ascorbic acid and dehydroascorbic acid (or bicyclic dehydroascorbic acid hydrate, Figure 5.27) by disproportionation:

$$2\,H_2A \rightarrow 2\,HA^\bullet + 2\,H^+ + 2\,e^-$$
$$2\,HA^\bullet + 2\,e^- \rightarrow 2\,HA^{\bullet-}$$
$$2\,HA^{\bullet-} \rightarrow H_2A + A + 2\,e^-$$

The reaction is repeated until all the ascorbic acid is oxidised. It is reversible, and dehydroascorbic acid can be reduced back to ascorbic acid by reduced glutathione, cysteine and other thiols, hydroquinones, and other compounds. Dehydroascorbic acid is already fairly unstable and spontaneously hydrolyses with lactone ring opening. This reaction is not reversible. Losses of vitamin caused by enzymatic oxidation in fruits and vegetables during processing can be effectively reduced by blanching using steam or boiling water, which inactivates the enzymes oxidising ascorbic acid.

The enzyme ascorbate 2,3-dioxygenase (L-ascorbate: O_2 2,3-oxido-reductase) splits the molecule of ascorbic acid between carbons C-2 and C-3 (this requires triplet oxygen, water and Fe^{2+} ions) to oxalic acid and L-threonic acid, also known as (2R,3S)-trihydroxybutanoic acid. The oxidative cleavage of ascorbic acid plays an important role in the metabolism in most plants, because, in addition to oxalic acid that plays a role in calcium metabolism, the final product is (+)-L-tartaric acid (Figure 5.28), which is formed by oxidation of threonic acid. Threonic and oxalic acids are the main decomposition products of ascorbic acid also in weakly alkaline solutions and during oxidation by singlet oxygen (Figure 5.29). D-Erythroascorbic acid in fungi decomposes in an anlogous way, yielding oxalic acid and D-glyceric acid.

The most important reaction is the oxidation of ascorbic acid by air oxygen (autoxidation), which mainly causes losses in foods during processing. It takes place in both the presence and absence of transition metal ions. The most active ions are trivalent iron and divalent copper. The reaction depends on the pH; it is slow in an acidic solution, faster in a neutral solution, and fastest in an alkaline one. The catalytic efficiency of metal ions lies in the fact that ascorbic acid forms a very stable ternary complex (5-47) with a metal ion of higher valency and oxygen (in solutions of pH 2–8), in which it is present as an anion HA^-. Therefore, the reaction rate increases with pH. Other enediols react analogously.

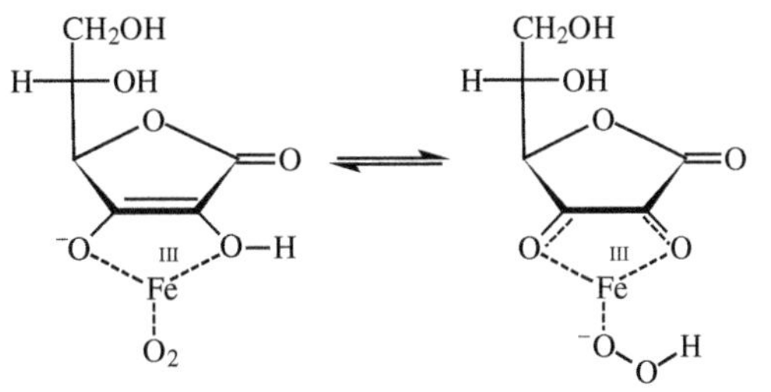

5-47, complex of L-ascorbic acid with ferric ion sand oxygen

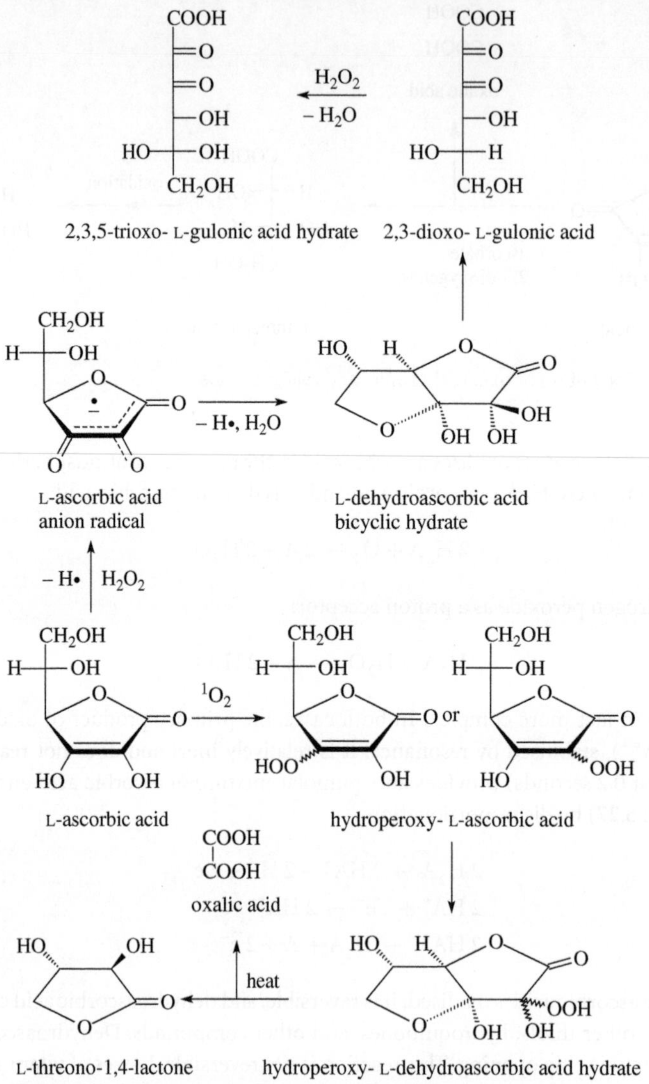

Figure 5.29 Oxidation of ascorbic acid by hydrogen peroxide and singlet oxygen.

Within the complex, two electrons are transferred from ascorbic acid to oxygen through the metal ion. The complex dissociates by the action of hydrogen ions (H^+), forming dehydroascorbic acid, hydrogen peroxide, and the metal ion. Reaction with Fe^{3+} (Fe^{2+} complex is less stable than the complex with Fe^{3+}) can be generally described by the following equations (similar reaction takes place with Cu^{2+}):

$$H_2A + O_2 \xrightarrow{Fe^{3+}} A + H_2O_2$$

The resulting hydrogen peroxide can oxidise another molecule of ascorbic acid to dehydroascorbic acid or other oxylabile compounds (e.g. anthocyanins), but can also react with Fe^{2+} ions to form hydroxyl ions and hydroxyl radicals (see the next page for Fenton reaction):

$$H_2A + H_2O_2 \rightarrow A + 2\,H_2O$$

The resulting reaction is the sum of the previous two.

Oxidation of ascorbic acid by hydrogen peroxide proceeds via ascorbyl radical and dehydroascorbic acid (apparently its hydrate), which yields 2,3-dioxo-L-gulonic acid. Hydrogen peroxide further oxidises 2,3-dioxo-L-gulonic acid to give unstable 2,3,5-trioxo-L-gulonic acid (Figure 5.29), which decomposes and produces other products. 2,3-Dioxo-L-gulonic acid and 2,3,5-trioxo-L-gulonic acid are probably hydrated in solutions.

Oxidation of ascorbic acid by singlet oxygen gives L-threonic acid 1,4-lactone and oxalic acid, analogously to the reaction catalysed by ascorbate 2,3-dioxygenase. Intermediates of this reaction are hydroperoxides of ascorbic acid and of dehydroascorbic acid hydrate (Figure 5.29).

Ascorbic acid reacts with metal ions to form complexes, but under certain conditions (especially at low pH values and if present in low concentrations) may also reduce metal ions. Reaction with Fe^{3+} ions (and by analogy with Cu^{2+} ions) is as follows:

$$H_2A + 2\,Fe^{3+} \rightarrow A + 2\,Fe^{2+} + 2\,H^+$$

The reducing effect of ascorbic acid in fact accelerates the oxidation reactions that lead to undesirable changes in the flavour and colour of foods. These changes are the result of a subsequent reaction of ions Fe^{2+} (and Cu^+) with oxygen, which yields a radical of superoxide anion ($O_2^{\bullet-}$ p$K_a = 4.8$), further oxidising Fe^{2+} or Cu^+ and producing hydrogen peroxide. Further oxidation of Fe^{2+} or Cu^+ ions with hydrogen peroxide (Fenton reaction) generates hydroxyl radicals ($HO^\bullet$), which are the most important oxidants in biological systems (resulting in many pathological processes) and in foods:

$$Fe^{2+} + O_2 \rightarrow Fe^{3+} + O_2^{\bullet-}$$
$$Fe^{2+} + O_2^{\bullet-} + H_2O + H^+ \rightarrow Fe^{3+} + HO^- + H_2O_2$$
$$Fe^{2+} + H_2O_2 \rightarrow Fe^{3+} + HO^- + HO^\bullet$$

Hydrogen peroxide can be replaced in these reactions by a fatty acid hydroperoxide (ROOH). Then, instead of hydroxyl radicals, alkoxyl radicals ($RO^\bullet$) are formed.

The prooxidant effects of ascorbic acid do not manifest when its concentration is sufficiently high, even though the increase of its concentration produces a higher amount of Fe^{2+} ions and hence $HO^\bullet$ radicals. In this case, a competitive reaction with free radicals takes place at the same time and ascorbic acid acts as an antioxidant.

Ascorbic acid, its isomers, and its derivatives can react with free radicals, causing oxidation of lipids and other oxylabile food components. It inhibits the radical chain autoxidation reaction and effectively acts as an antioxidant. The reaction of ascorbic acid with lipid (fatty acid) peroxyl radical ($ROO^\bullet$) or with alkoxyl radical ($RO^\bullet$) can be schematically represented by the following equation (ROOH is a fatty acid hydroperoxide):

$$H_2A + ROO^\bullet \rightarrow HA^\bullet + ROOH$$

The resulting ascorbyl radical ($HA^\bullet$) is no longer able to initiate a radical chain reaction and disproportionates into ascorbic acid and dehydroascorbic acid.

Ascorbic acid is generally a more effective antioxidant when used in combination with tocopherols. These preferentially react with lipid free radicals, and the resulting tocopheryl radicals, on the oil–water interface, are reduced back to tocopherols by ascorbic acid. Ascorbyl palmitate yields, via the corresponding radical, dehydroascorbyl palmitate, which, unlike dehydroascorbic acid, cannot form a cyclic hydrate. Ascorbic acid also reacts similarly with toxic forms of oxygen, such as a hydroxyl radical ($HO^\bullet$), an anion of superoxide radical ($O_2^{\bullet-}$), and singlet oxygen ($^1O_2$). Simultaneously, all of these reactions slow down the oxidation of lipids:

$$H_2A + HO^\bullet \rightarrow HA^\bullet + H_2O$$
$$H_2A + O_2^{\bullet-} + H^+ \rightarrow HA^\bullet + H_2O_2$$

In strongly acidic solutions, ascorbic acid can decarboxylate and, like other sugars, dehydrate. In model experiments, almost quantitative amounts of carbon dioxide and furan-2-carbaldehyde are produced. The reaction mechanism is shown in Figure 5.30. An important product is 3-deoxy-L-*threo*-pentos-2-ulose, also known by its trivial name, 3-deoxy-L-xylosone, which plays an important role in the Maillard reaction of pentoses.

In the absence of atmospheric oxygen, acid-catalysed degradation of ascorbic acid is considered a major cause of the loss of vitamin C in dried fruits, canned fruit compotes, and juices (pH value around 3.5), especially when stored at higher temperatures. For example, fruit juices lose 70–95% of ascorbic acid within 12 weeks of storage at 50 °C. The reaction rate is about 10 times lower than that of autoxidation catalysed by metal ions.

The loss of vitamin C may also occur during reactions of ascorbic acid with some of the reactive food components. In particular, reactions of ascorbic acid with quinones generated by enzymatic browning reactions and reactions with nitrites and haem pigments in meat and meat products are technologically significant.

The enzymatic browning reactions of fruits and vegetables occur especially in materials that have a low content of ascorbic acid and active enzymes belonging to the group of o-diphenol:O_2 oxidoreductases. In damaged plant tissues and in the presence of atmospheric oxygen, these enzymes catalyse the oxidation of phenolic substrates (so-called monophenols or o-diphenols) to o-quinones, which then form brown polymeric pigments. As long as ascorbic acid is present, quinones are reduced back to the parent diphenols and an equimolar amount of dehydroascorbic acid is formed.

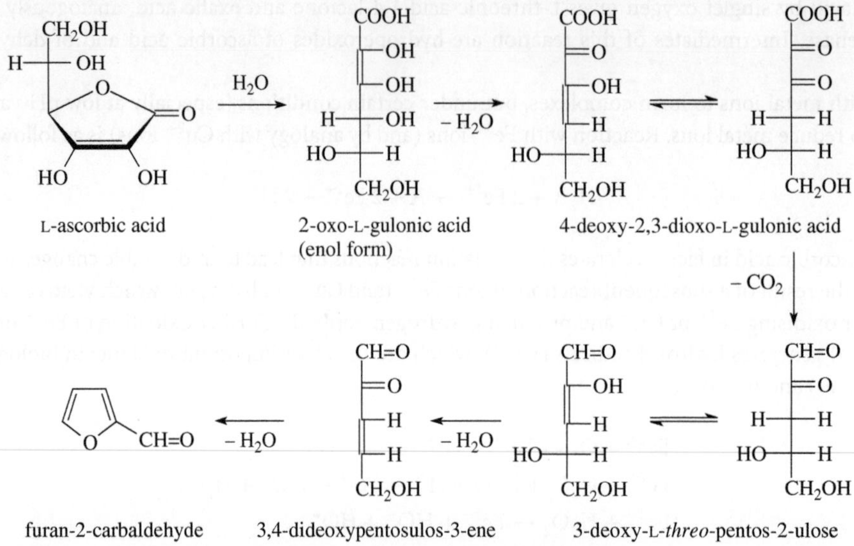

Figure 5.30 Degradation of L-ascorbic acid in acidic solutions.

The disadvantage of the use of nitrites in meat manufacturing processes is the formation of toxic N-nitrosamines by reaction of nitrous acid (nitrite) with naturally present secondary amines (see Section 12.3.7). Ascorbic acid added together with nitrites decomposes the excess of nitrous acid, which is a precursor of nitrosation agents (primarily dinitrogen trioxide, N_2O_3). It is assumed that the reaction of ascorbic acid with nitrous acid temporarily creates the corresponding 2-ester (**5-48**), which is decomposed to an ascorbyl radical and nitric oxide (nitrogen monoxide, NO), which reacts with haem pigments. Ascorbic acid 2,3-diester has also been proposed as an intermediate; one that decomposes to dehydroascorbic acid and nitric oxide. Minor products generated from nitrite (with myoglobin acts as a reducing agent) are nitrous oxide (N_2O) and nitrogen.

Even under physiological conditions, about 1% of haemoglobin in erythrocytes is present in the form of methaemoglobin, which is reduced to haemoglobin by methaemoglobin reductase. Myoglobin (Mb) is present in meat in the form of oxymyoglobin (MbO_2) and metmyoglobin (MetMb). Added ascorbic acid reduces both of these pigments to myoglobin (see Section 9.2.1.5.3). In the presence of ascorbic acid, a higher amount of myoglobin is available for the reaction, which creates the desired dye nitroxymyoglobin:

$$H_2A + MbO_2\ (Fe^{II}) + H^+ \rightarrow HA^{\bullet} + MetMb\ (Fe^{III}) + H_2O_2$$
$$H_2A + MetMb\ (Fe^{III}) \rightarrow HA^{\bullet} + Mb\ (Fe^{II}) + H^+$$

The aim of the Chorleywood bread process (see Section 5.14.4) is to use cheaper, lower-protein wheat and to reduce processing time. Flour, water, yeast, salt, fat, ascorbic acid, and minor ingredients (emulsifiers and enzymes) are mechanically mixed, which results in rapid oxidation of ascorbic acid to dehydroascorbic acid, caused mainly by ascorbase activity in the dough, which is a co-substrate of glutathione dehydrogenase. Glutathione dehydrogenase oxidises reduced glutathione (G-SH) to the corresponding disulfide (oxidised glutathione, G-S-S-G), which does not affect the rheological properties of dough:

$$A + 2\,G\text{-SH} \rightarrow H_2A + G\text{-S-S-G}$$

The content of reduced glutathione in wheat flour is 10–15 mg/kg. In the absence of ascorbic acid (dehydroascorbic acid, respectively), G-SH reacts with components of gluten (P) that are linked by disulfide bridges (P-S-S-P) to form mixed disulfides, which show negative effects on the baking properties of flour:

$$P\text{-S-S-P} + G\text{-SH} \rightarrow P\text{-S-S-G} + P\text{-SH}$$

Free cysteine (Cys-SH) formed by hydrolysis of glutathione can also depolymerise gluten proteins or react with mixed disulfides that also have negative effects on the baking properties of flour:

$$P\text{-S-S-P} + Cys\text{-SH} \rightarrow P\text{-S-S-Cys} + P\text{-SH}$$
$$P\text{-S-S-Cys} + G\text{-SH} \rightarrow P\text{-S-S-G} + Cys\text{-SH}$$
$$P\text{-S-S-G} + Cys\text{-SH} \rightarrow P\text{-S-S-Cys} + G\text{-SH}$$

Ascorbic acid carbanion (C-2) reacts with electrofilic agents (carbonium cations). A typical example of this reaction is the formation of ascorbigen in the reaction of ascorbic acid with the degradation products of indole glucosinolates in Brassicaceae vegetables. The glucosinolate glucobrassicin (and its analogues) is hydrolysed to glucose, hydrogensulfate anion, and unstable isothiocyanate by the enzyme myrosinase (thioglucosid glucohydrolase) in the damaged plant tissue (such as shredded cabbage). The anion of the tautomeric form of ascorbate then replaces thiocyanate (rhodanide) anion in this intermediate, which yields ascorbigen (see Section 10.3.10.2). Ascorbigen is more stable to oxidation than ascorbic acid and exhibits anticarcinogenic effects.

Another ascorbic acid derivative is a water-soluble tannin called elaeocarpusin, which was isolated from the leaves of the evergreen tropical tree *Elaeocarpus sylvestris* (*Elaeocarpaceae*). Elaeocarpusin is an ellagitannin derived by condensation of corilagin with dehydroascorbic acid. The reaction of ascorbic acid with aldehydes in aqueous solutions yields the corresponding hemiacetals. The reaction with methylglyoxal (a typical product of carbohydrate degradation) gives rise to bicyclic hemiacetals (**5-48**, R^1 = H or CH_3, R^2 = CH_3 or H). Similar products are produced with acrolein, the degradation product of glycerol. The reaction with (E)-hex-2-enal, the autoxidation product of linolenic acid, yields 6-propylbenzo[*b*]furan-7-ol (**5-48**).

| nitrous acid 2-ester | methylglyoxal acetals | 6-propylbenzo[*b*]furan-7-ol |

| hydroxysulfonate | 3,4-dideoxy-4-sulfopentos-2-ulose |

5-48, selected reaction and degradation products of L-ascorbic and L-dehydroascorbic acids

Sulfur dioxide reduces dehydroascorbic acid to ascorbic acid and is therefore frequently used in combination with ascorbic acid, particularly in fruits and vegetables. Under aerobic conditions, the bisulfite ion forms hydroxysulfonate of dehydroascorbic acid (**5-48**), and under anaerobic conditions, sulfite ion may catalyse degradation of ascorbic acid and add across the double bond of the formed 3,4-dideoxypentosulos-3-ene (Figure 5.30), yielding 3,4-dideoxy-4-sulfopentos-2-ulose (**5-48**).

5.14.6.2 Dehydroascorbic acid

Like other vicinal dicarbonyl derivatives of sugars, dehydroascorbic acid is involved in the Maillard reaction. Dehydroascorbic acid (or its bicyclic hydrate, Figure 5.27) is a γ-lactone that is readily hydrolysed under a base catalysis to its parent unstable compound, which undergoes a series of irreversible reactions. These reactions result in loss of vitamin C and the formation of coloured products, and the discoloration of fruit and vegetable products.

Dehydroascorbic acid hydrolysis yields biologically inactive 2,3-dioxogulonic (L-*threo*-hexo-2,3-diulosonic) acid. In aqueous solutions, this acid is present as a dihydrate (Figure 5.31). The reaction is generally acid and base catalysed. The activity of important ions and undissociated molecules decreases in the order: hydroxyl ions (HO^-) > hydronium ions (H_3O^+) > anions of carboxylic acids R—COO$^-$) > undissociated carboxylic acids (R—COOH) > water. The catalytic effect of hydroxyl ions is about 15.10^6 times higher than that of hydronium ions. This means that the reaction rate in the solution of pH 4 is 15.10^6 times lower than in solution of pH 10. Dehydroascorbic acid is most stable in solutions of pH 2.5–5.5, where the reaction is only catalysed by undissociated water molecules and is rapidly hydrolysed in neutral and alkaline solutions.

2.3-Dioxogulonic acid is an unstable compound and undergoes a series of reactions that are analogous to reactions of sugars. Its decomposition leads to many products, depending on the pH and the presence of atmospheric oxygen. In acidic solutions, the main transformation products are furan derivatives; decomposition in alkaline solutions produces organic acids (their salts) and reductones. The key reaction that occurs in a wide range of pH values (in acidic and alkaline solutions) is decarboxylation to (3R,4S)-3,4,5-trihydroxy-2-oxopentanal

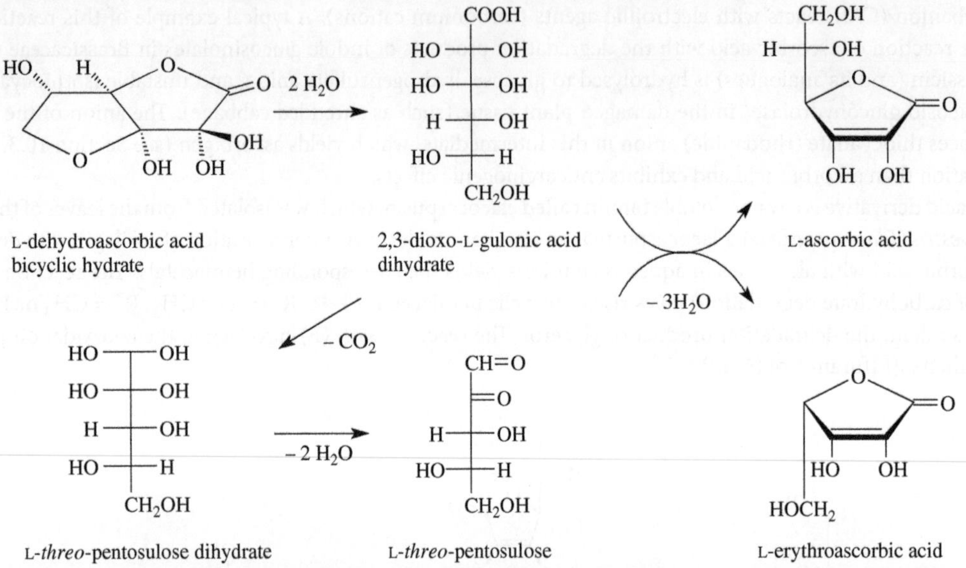

Figure 5.31 Degradation of L-dehydroascorbic acid.

(L-*threo*-pentosulose). In older literature, this substance was called L-xylosone and its enolform was known as reductone C. It is assumed that this enol form reacts with another molecule of 2,3-dioxo-L-gulonic acid, which is reduced to L-ascorbic acid, whilst L-*threo*-pentosulose is simultaneously oxidised to L-erythroascorbic acid, also known as (S)-2,3,4,5-tetrahydroxypent-2-enonic acid (Figure 5.31). Some other reactions of L-*threo*-pentosulose are shown in Figure 5.32. The main products formed by a sequence of isomerisation and (de)hydration reactions are 3-hydroxypyran-2-one and furan-2-carboxylic acid (also known as pyromucic or 2-furoic acid). The odour of 3-hydroxypyran-2-one is reminiscent of caramel (maltol and isomaltol have similar odours).

Dehydroascorbic acid reacts with amines, amino acids, and proteins, and participates in the non-enzymatic browning reactions that occur mainly in foods with low water activity and a relatively high content of vitamin C (e.g. dehydrated fruit juices). The primary product of the reaction with amino compounds is imine. Reduction of imine gives the corresponding amine. Isomerisation of imine and elimination of carbon dioxide and aldehyde yields scorbamic acid (Figure 5.33). This reaction is known as Strecker degradation of amino acids. A detailed reaction mechanism is given in Section 2.5.2.3. This 2-amino derivative reacts with another molecule of dehydroascorbic acid and yields the so-called red pigment (**5-49**). Formation of this pigment, which sometimes occurs during the preservation of cauliflower, leads to unwanted discoloration of the preserved product. The subsequent reaction of scorbamic acid with the red pigment creates the yellow pigment (**5-49**), which is probably an intermediate in reactions leading from dehydroascorbic acid to brown polymeric pigments called melanoidins. Other reaction products include cyclic nitrogenous products, and typically unstable free radicals and stable products, such as tris(2-deoxy-2-L-ascorbyl)amine (**5-49**).

red pigment   yellow pigment   tris(2-deoxy-2-L-ascorbyl)amine

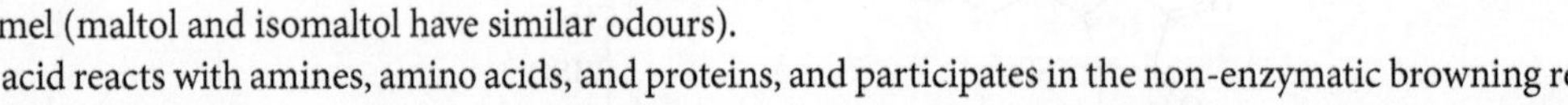

5-49, coloured reaction products of L-ascorbic acid

Ascorbic acid is one of the least stable vitamins. It is sensitive to heat, light, and oxygen. Storage of foods, cooking procedures, and industrial processing methods can result in significant loss of vitamin C. The most significant losses are caused by vitamin leaching and oxidation. In the absence of atmospheric oxygen, losses are mainly caused by acid catalysed degradation. Total losses generally range between 20 and 80%.

If fresh products are held at the appropriate temperature and consumed in a short period of time, they have more vitamin C than commercially canned products. Vitamin C degrades rapidly after harvest, and its amount depends on the commodity, temperature, pH, and

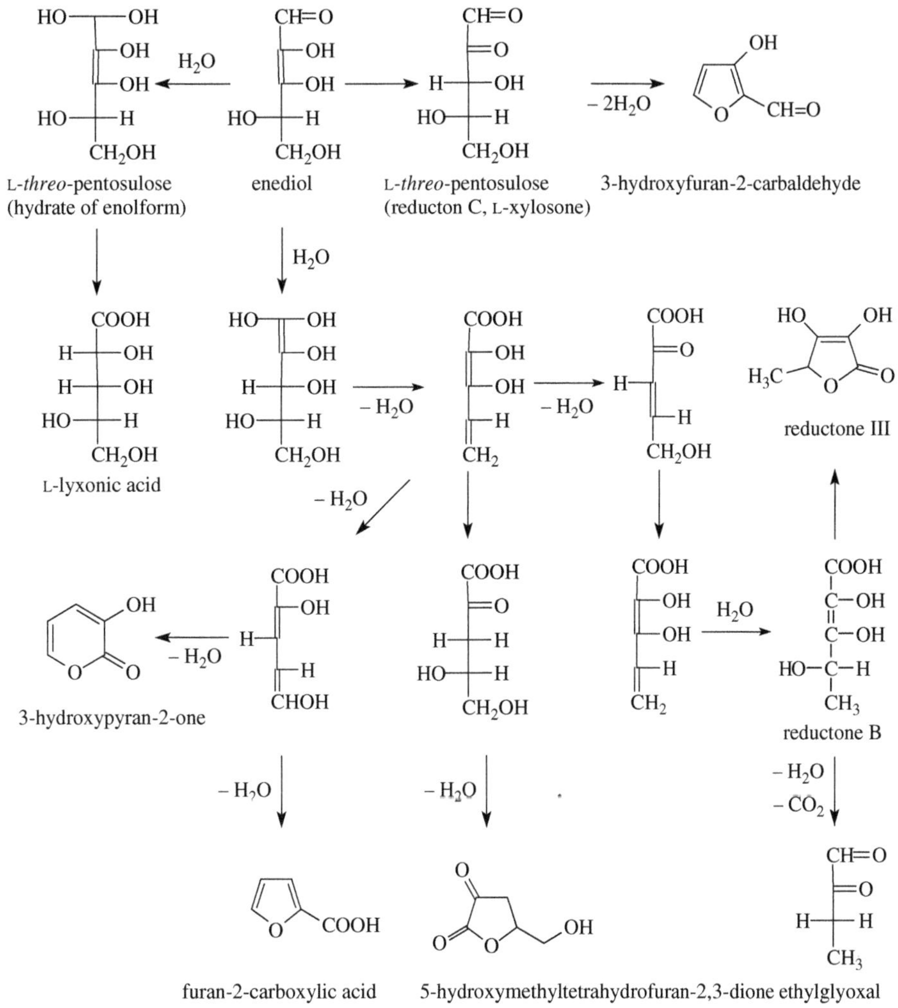

Figure 5.32 Degradation of L-*threo*-pentosulose.

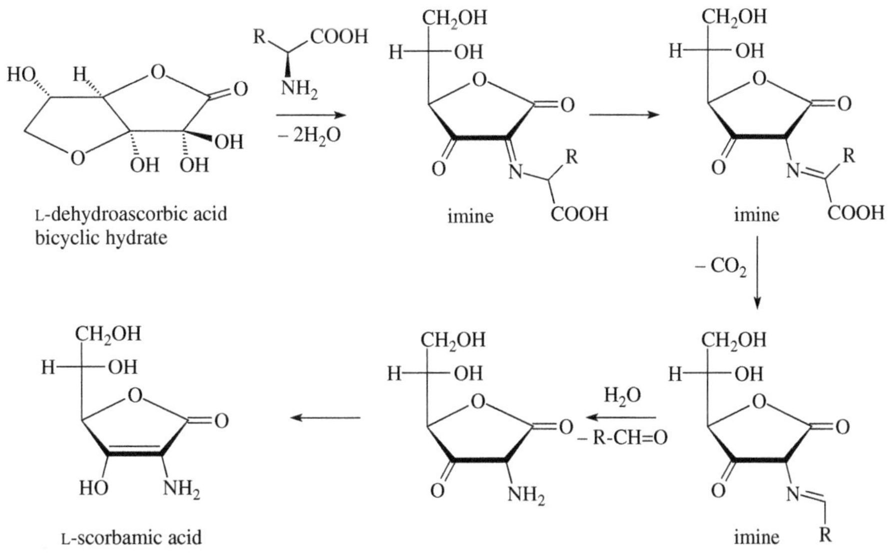

Figure 5.33 Formation of L-scorbamic acid from L-dehydroascorbic acid and amino acids.

other factors. During storage (and processing), the stability of ascorbic acid is higher in fruits, which have lower pH, than in vegetables. For example, the effect of storage on the vitamin C content in apples depends upon the time of picking, storage temperature and composition of the storage atmosphere. Apples (Golden Delicious variety) stored at 25 °C lose 41% of vitamin C in 2 weeks after harvesting, and almost 56% after 6 weeks of storage. Only small changes in the vitamin C content occur at 4 °C. Green beans, however, lose up to 77% of vitamin C in seven days of storage at 4 °C. The vitamin C level in fresh potatoes falls to 30–60% of the original level in the first 2 months after harvest, but then tends to stabilise at 25% of the original level.

Losses of ascorbic acid due to leaching occur during the washing, blanching, cooking, and preserving of fruits and vegetables in cases where the extract is not further processed. Losses during washing are lower than losses during blanching and cooking. The nature and extent of losses depends on pH, temperature, water quantity, surface area that is in contact with water, maturity, extent of contamination by metals, and presence of oxygen. A significant decrease in vitamin C content is also caused by peeling the fruit, when the surface layer rich in vitamin C is removed.

The average retention of fruit juices fortified with vitamin C is 60–80%. Between 10 and 90% of vitamin C is lost during canning of fruits, but changes are low during storage of canned products and little is lost during reheating, because the heating time is short. The lowest losses are achieved during short-term sterilisation at high temperature. In fruit treated with sulfur dioxide, the losses of ascorbic acid during technological processing are lower as sulfur dioxide reduces hydrogen peroxide produced by oxidation of ascorbic acid in the presence of heavy metals. Vitamin C is most stable during the freezing and refrigerated storage of fruits and vegetables. Storage at temperatures of −18 °C result in only minimal losses, whilst significant losses can occur during thawing (30–50%).

Losses are higher in green leafy vegetables with a large surface than in root vegetables. Degradation of vitamin C also occurs during processing of potatoes, with absolute losses on the order of 30%, although the concentration of vitamin C in potato crisps can be higher than in fresh potato due to a reduction of water content during frying. Vitamin loss also occurs during fermentation of vegetables. Sauerkraut, for example, contains about 50% of the vitamin as compared with fresh cabbage (90–190 mg/kg).

The best methods for small-scale processing are drying, chemical preservation, and heat processing; these and other methods may be applied in industrial processing. In order to retain the maximum amount of vitamin C in fruits and vegetables during storage and in processed products (based on reaction mechanisms already described), the following principles should be maintained:

- fruits and vegetables should be used when freshly harvested;

- they must not be subjected to long soaking, washing, or blanching;

- they must be processed immediately after preparation;

- the level of metal ions should be reduced by exclusion of any direct contact with copper, bronze, brass, and iron parts of processing equipment or copper, iron, or chipped pans; binding of metal ions to inactive complexes by chelating agents (such as ethylenediaminetetraacetic acid (EDTA), citrates, and phosphates); autoxidation of ascorbic acid is also slower in the presence of proteins, acidic polysaccharides, and flavonoids; unfavourable conditions for the formation of complexes of metal ions with ascorbic acid may be created by lowering water activity, pH value, using appropriate O-2 substituted derivatives of ascorbic acid, such as 2-phosphate or 2-O-α-D-glucoside;

- contact with air should be minimised by reducing the amount of oxygen (reduced pressure, inert atmosphere, addition of hydrogen bisulfites, fermentation, use of glucose oxidase and catalase); glucose oxidase catalyses the oxidation of D-glucose by oxygen to D-gluconic acid and hydrogen peroxide, which is then reduced by catalase to water:

$$2\,H_2O_2 \rightarrow 2\,H_2O + O_2$$

Losses of ascorbic acid during storage of raw milk are considerable. Cold storage causes about 50% loss of vitamin C, which increases with increased temperature. Heat treatment of milk decreases the content of vitamin C by 20–50%, depending on temperature and time of heating. The UHT treatment of milk causes about 10–30% loss. Ascorbic acid in dried, vitamin-enriched milk, packaged in an inert atmosphere, is relatively stable.

5.15 Other active substances

In the past, the category of vitamins included far more substances than today. The catalytic effect of some hydrophilic substances (such as 4-aminobenzoic acid, the building unit of folacin) was subsequently not reliably proven in the metabolism of humans. Some compounds can be synthesised in adequate amounts by the human body, such as adenine (vitamin B_4), 4-aminobenzoic acid (vitamin B_{10}), orotic acid (vitamin B_{13}), carnitine (vitamin B_{20}), *myo*-inositol (vitamin B_m), choline (vitamin B_p), 2-aminobenzoic (anthranilic) acid (previously known as vitamin L_1), 9′-(5-thiomethylribofuranosyl)adenine (a metabolite of RNA previously known as vitamin L_2), and salicylic acid (previously known as vitamin S). Furthermore, no signs of their essentiality and avitaminosis have been demonstrated.

The term 'vitamin' also included other essential nutrients, such as thioctic acid, taurine, coenzyme Q, and other biologically active compounds. Some substances are vitamins only for microorganisms (4-aminobenzoic acid, thioctic acid, and other growth factors of microorganisms). A range of biologically active compounds initially ranked amongst the vitamins are now categorised amongst other substances and are no longer considered true vitamins, such as essential fatty acids (previously known as vitamin or vitagen F; now classified as lipid constituents), bioflavonoids (previously vitamin P; now plant pigments called flavonoids), and vitamin J (a mixture of non-essential catechol and riboflavin). Coenzyme Q, carnitine, taurine, orotic acid, *myo*-inositol, choline, and other compounds are currently used as food supplements.[6] Some of the claimed benefits of food supplements have a more scientific basis than others, but all must be carefully evaluated and the supplemented compounds should not replace a varied and balanced diet.

The purine base adenine (6-aminopurine, previously also known as vitamin B_4, **5-50**) is generally widespread as a building unit of the ribofuranoside adenosine and its phosphates such as adenosine 5′-monophosphate (AMP, adenylic acid), adenosine 5′-diphosphate (ADP), and adenosine 5′-triphosphate (ATP), polynucleotides such as ribonucleic acid (RNA) and deoxyribonucleic acid (DNA), vitamins such as riboflavin and adenosylvitamin B_{12}, and cofactors such as $NAD(P)^+$ and FAD. Adenylic acid was once considered a member of the vitamin B complex (as vitamin B_8). The agent in shiitake (*L. edodes*) mushrooms that can reduce the level of cholesterol in the blood is a purine derivative, (2*R*,3*R*)-dihydroxy-4-(9-adenyl)butyric acid (**5-50**), called eritadenine (lentinacin, lentysine).

4-Aminobenzoic acid, also known as *p*-aminobenzoic acid (PABA), was previously ranked amongst the group B vitamins as vitamin B_{10}, vitamin B_x, or vitamin H_1 (**5-50**). It arises from chorismic acid and commonly occurs in foods of animal and plant origin. Good sources include organ meats, such as liver and kidney, eggs, whole-wheat flour, wheat germ, and brewer's yeast. Aminobenzoic acid is also supplied by intestinal microorganisms. Aminobenzoic acid is probably most important as an intermediate in the biosynthesis of folacin; therefore, it is a growth factor of some bacteria, but not for humans. Despite the lack of any recognised syndromes of aminobenzoic acid deficiency in humans, many claims of benefits are made by commercial suppliers of aminobenzoic acid as a nutritional supplement. Aminobenzoic acid supposedly participates in many metabolic processes in humans. It appears to function as a coenzyme in the conversion of certain chemical intermediates into purines. It has also been suggested that it has an antifibrosis activity and increases oxygen uptake at the tissues. These effects are, at present, still considered speculative. When it is in short supply, fatigue, irritability, nervousness, and depression may manifest, as well as constipation. Weeping eczema has been noted in people with 4-aminobenzoic acid deficiency, as well as patchy areas on the skin. Aminobenzoic acid physically blocks UV rays when it is applied to the skin (it is a UVB absorber) and is sometimes suggested for treatment of various skin diseases, such as weeping eczema (moist eczema), scleroderma (premature hardening of skin), depigmentation of skin (vitiligo), and premature grey hair, as well as for male infertility. The RDA is 0.05 mg for infants, 0.2–0.3 mg for children, 0.4 mg for adults, 0.8 mg for pregnant women, and 0.6 mg for lactating women.

Orotic acid (originally known as vitamin B_{13}, **5-50**) is an intermediate in the biosynthesis of pyrimidines required for DNA and RNA synthesis, but essentiality has not been demonstrated. It is synthesised from carbamoyl phosphate and aspartic acid through dihydroorotic acid and then transformed into orotidin 5′-phosphate and further to uridine 5′-monophosphate. When there is insufficient capacity to detoxify the ammonia load for urea synthesis, carbamoyl phosphate leaves the mitochondria and enters the pyrimidine pathway, where orotic acid biosynthesis is stimulated; orotic acid excretion in urine then increases. Orotic acid synthesis is abnormally high in hereditary deficiencies of urea-cycle enzymes or uridine monophosphate synthase. This acid occurs mainly in milk from ruminants. In cow's milk, the level is 20–100 mg/l; somewhat higher amounts are found in goat's and sheep's milk (200–400 mg/l). In human milk, concentrations are generally low (under 2 mg/l). Good sources of orotic acid are dairy products (at a level of about 150 mg/l in yoghurt), liver, and baker's yeast. Benefit claims include enhanced formation of albumen in the liver, especially in conditions of prolonged hypoxia that occur in some diseases, such as heart failure, and a positive effect on foetal development during pregnancy. The effectiveness of orotic acid was shown in children aged from 6 months to 10 years suffering from various skin diseases (eczema, atopic dermatitis, psoriasis, and ichthyosis). The daily dose suggested by commercial suppliers is 0.125–0.5 g for children 1–3 years old, 0.25–1 g for children 3–8 years old, and 0.5–1.5 g (sometimes up to 3 g) for adults.

Pangamic acid (formerly known as vitamin B_{15}, **5-50**) occurs mainly in cereals, legumes, seeds (sunflower and pumpkin seeds), and yeast. Its nutritional value is debatable, and no scientific evidence exists to substantiate any physiological function or biological effects. It probably acts (like choline) as a lipotropic factor, which stimulates the formation of VLDL in the liver. Although no evidence supports its use, pangamic acid has been used to treat cancer, schizophrenia, diabetes mellitus, heart diseases, alcoholism, hepatitis, and indigestion. It has also been asserted that pangamic acid detoxifies byproducts of human metabolism.

N,N-Dimethylglycine (originally called vitamin B_{16}) is an amino acid found in almost all animal and plant cells and as a building unit of pangamic acid. It can be formed by demethylation of *N,N,N*-trimethylglycine (glycine betaine), a very efficient cytoplasmic osmolyte of

---

[6]Some EU member states (e.g. the Czech Republic in Decree No. 225/2008 Coll.) have set a maximum permissible amount in daily intake for orotic acid (50 mg), acetylcarnitine (500 mg), and taurine (200 mg). The use of orotic acid salts as a source of minerals and choline is a safety concern.

some plants and bacteria or as an intermediate in the pathway leading from choline to glycine. Demethylation of dimethylglycine yields N-methylglycine (sarcosine). Dimethylglycine is contained in higher amounts in certain foods, including liver, beans, cereal grains, many seeds, and brewer's yeast. Deficiencies in dimethylglycine do not cause any adverse effects. Manufacturers of dimethylglycine supplements claim that as a supplement, vitamin B_{16} can improve athletic performance (it assists in oxygen utilisation within the body), enhance the immune system, stimulate neurological functions, help to manage autism (noticeable improvements in cognitive functions are exhibited in combination with vitamin B_6 supplementation) and epilepsy, and provide protection of cells against free radicals as an antioxidant.

The cyanogenic glycoside known as (R)-amygdalin, a β-gentiobioside of D-mandelic acid nitrile (see Section 10.3.10.2), is sometimes incorrectly referred to as vitamin B_{17}. Amygdalin is often confused with semi-synthetic laevomandelonitrile, laetrile for short, which is a β-glycoside of D-mandelic acid nitrile with D-glucuronic acid. Amygdalin occurs in plant seeds of the rose family (Rosaceae). Significant resources are bitter almonds and the pits of apricots, peaches, plums, and cherries. Small amounts of amydalin are also present in the cores of apples, pears, quinces, and other plant seeds. Amygdalin is capable of decomposing into a sugar molecule(s), benzaldehyde, and toxic hydrogen cyanide through the action of the hydrolase amygdalase (also called emulsin). The enzyme rhodanase (in bacteria, plants, and animals) acts in the catabolism of cyanides (R—C≡N), catalysing conversion of cyanides into thiocyanates (rhodanides, R—S—C≡N). Perhaps the most notable (yet controversial) benefit of amygdalin is its effectiveness in treating cancer, in particular prostate cancer, arthritic pain, and high blood pressure. The claimed anticarcinogenic activity is based upon the fact that cancer tissues are rich in hydrolase, which causes amygdalin to release cyanide, capable of destroying the cancer cells. According to this theory, non-cancerous tissues are protected from this fate by active rhodanase, which renders the cyanide harmless. However, no substantive benefit was observed in terms of cure, improvement, or stabilisation of cancer, improvement of symptoms related to cancer, or extension of life span. Furthermore, the hazards of amygdalin therapy were evidenced in several patients by symptoms of cyanide toxicity.

L-Carnitine, also known as (R)-carnitine or trimethylammonium-3-hydroxybutyrobetaine (formerly known as vitamin B_{20}, vitamin O, or vitamin B_t; see Section 2.2.1.2.1), is synthesised from lysine and methionine in many organisms, ranging from bacteria to mammals. The lysine becomes available in the form of ε-(N,N,N-trimethyl)lysine after lysosomal hydrolysis of proteins that contain this amino acid as a result of the post-translational methylation of lysine residues. Carnitine, in the activated form as acylcarnitine, serves as a carrier of fatty acid residues across the inner membrane of the mitochondria, which helps in the consumption and disposal of fat in the body. Then, fatty acids can be degraded by β-oxidation to acetyl-CoA in order to obtain energy via the citric acid cycle. The highest amount of carnitine is present in the muscles of animals; in red meat, levels range from 0.05 to 0.2% (f.w.). Good sources are also poultry and fish. Carnitine (free carnitine, acetylcarnitine, or propionylcarnitine) is often sold as a nutritional supplement for stimulation of energy metabolism of fatty acids, which is used in weight reduction and sports nutrition. Although carnitine has been marketed as a weight-loss supplement, there is no scientific evidence to show that it works. Some studies do show that oral carnitine reduces fat mass, increases muscle mass, and reduces fatigue, which may contribute to weight loss in some people. Some research suggests that carnitine may help prevent or reduce symptoms of an overactive thyroid gland, slow down the progression of Alzheimer's disease, relieve depression related to senility and other forms of dementia, improve memory in the elderly, improve male sexual function, and increase sperm count and mobility. Recommended doses of carnitine for adults vary depending on the health condition being treated. The usual dose is between 1 and 3 g/day.

The cyclic hexitol myo-inositol (incorrectly also known as $meso$-inositol; see Section 4.3.1.1.2) has previously been considered one of the group B vitamins (called i-inositol, phaseomannitol, nucitol, vitamin B_m, or bios I). Inositol is synthesised by conversion of D-glucose 6-phosphate and plays an important role as the structural basis for a number of secondary messengers in eukaryotic cells, including inositol phosphates and phosphatidylinositol in phospholipids. It is found in many foods, in particular legumes, cereals with high bran content, nuts, and fruits. Concentrations of inositol in legumes range from about 200 to 3000 mg/kg (d.w.). Phosphatidylinositol (and phosphatidylcholine) show lipotropic effects and prevent fatty liver disease, such as abnormal deposition of lipids within liver cells (steatosis). Inositol is a growth factor for various microorganisms. Deficiency in animals manifests by hair greying and falling out, but specific avitaminosis was not observed in humans. The manufactures of food supplements claim that inositol may be beneficial for treating certain psychiatric conditions, such as obsessive–compulsive disorder, panic attacks, trichotillomania (an impulse-control disorder), and bipolar disorder. In addition, inositol may be useful for ameliorating some of the symptoms of polycystic ovary syndrome (a female endocrine disorder). The daily intake of myo-inositol is estimated to be 1000 mg. The recommended daily dose from food supplements is 500 mg.

In plants (mainly cereals and legumes), inositol occurs in a bound form known as phytic acid (myo-inositol-1,2,3,4,5,6-hexakisphosphoric acid). Salts of this acid are called phytates. Calcium–magnesium salt of phytic acid, which also contains a number of other mineral elements, is called phytin (see Section 6.2.2.3). Phytic acid plays a role as an antioxidant (due to strong binding of transition metals), and anticarcinogenic effects have also been demonstrated. Partial esters of phytic acid are used to regulate the amount of calcium in cells, and exhibit anticoagulant, anti-inflammatory, and other effects. The other bound form of myo-inositol with sugar derivatives are pseudooligosaccharides (see Section 4.3.1.2.2) that occur in legumes.

Previous classification ranked choline, 2-hydroxy-N,N,N-trimethylethan-1-aminium (with chloride or hydroxyl group as a counter ion), amongst the vitamins of group B, when it was known as vitamin Bp (see Section 3.5.1.1.1). Plants and animals synthesise choline from glycine or serine, an intermediate of the biosynthesis is ethanolamine, which is methylated by folacin to yield choline. Choline is used for the biosynthesis of phospholipids (phosphatidylcholine and sphingomyelin) occurring in cell membranes and the biosynthesis of acetylcholine acting in the transmission of nerve impulses, participates in the transmethylation processes in the organism via its metabolite N,N,N-trimethylglycine (glycine betaine), and is a precursor of trimethylamine. Persons with the genetic disorder trimethylaminuria (the inability to metabolise trimethylamine) suffer from an unpleasant fishy body odour. Important food sources of choline include butter, egg yolk, pulses, oilseeds, nuts, and cereals. The daily choline requirement is around 600 mg. Mild deficiency has been linked to fatigue, insomnia, poor ability of the kidneys to concentrate urine, problems with memory, and nerve–muscle imbalances. Extreme dietary deficiency can result in liver dysfunction, cardiovascular disease, impaired growth, kidney failure, high blood pressure, and other symptoms. Soybean lecithin is the most common form of supplemental choline, but choline itself is also available in food supplements. It is used therapeutically (perorally and parenterally) in various liver diseases (e.g. steatosis and cirrhosis), for maintenance of cell-membrane integrity, in support of nervous system activity, and to lessen chronic inflammation. Adequate intake levels established by the US National Academy of Sciences are: 125 mg (0–6 months), 150 mg (6–12 months), 200 mg (1–3 years), 250 mg (4–8 years), 375 mg (males and females 9–13 years), 400 mg (females 14–18 years), 425 mg (females 19 years and older), 450 mg (pregnant females), and 550 mg (males 14 years and older and lactating females). Prevention of liver damage was the main criterion used in the establishment of these recommended levels.

Acetylcholine is found in foods of animal origin. In some plants, such as rapeseed, choline is found in the form of various phenolic acid esters, especially in sinapic (4-hydroxy-3,5-dimethoxycinnamic) acid esters, which are trivially called sinapine (see Section 8.2.7.1.1).

Bioflavonoids, also known as vitamin P (the permeability vitamin), are a group of biologically active flavonoids (flavonol and flavanone glycosides) that affect the permeability and elasticity of blood capillaries. These effects were first observed in a compound known as citrine obtained from citrus fruits. Later, it was found that citrine is a mixture of hesperidin and eriocitrin (also known as eriodyctin). Significant biological effects have been demonstrated mainly in quercetin-3-β-rutinoside, known as rutin (see Section 9.4.2.4). Bioflavonoids are found mainly in fruits and vegetables at a level of about 100–7000 mg/kg.

Vitamin U, also known as S-methyl-L-methionine, methylmethionine sulfonium chloride, cabigen, or antiulcer vitamin, was originally called vitamin U, because of its usefulness against ulceration of the digestive system. It is produced from methionine in an enzymatically catalysed methylation by S-adenosyl-L-methionine (SAM), but it can also arise in non-enzymatic reactions, for example by methylation of methionine with pectin. S-Methylmethionine serves as the storage form of labile methyl groups in plants and plays a role in preventing accumulation of the methylation agent SAM. It occurs mainly in *Brassica* vegetables, where it is a precursor of dimethyl sulfide during thermal processing. Its cabbage-like odour plays an important role in cooked vegetables. The content of S-methylmethionine is about 90 mg/kg in kohlrabi, 75–81 mg/kg in cabbage, 124 mg/kg in turnip, and 60 mg/kg in Savoy cabbage and Brussel sprouts. High concentrations of S-methylmethionine are also found in celery (60–176 mg/kg), leeks (60–94 mg/kg), and beetroot (89 mg/kg), but its concentration in tomatoes is only 2.8 mg/kg. In other vegetables, the levels are lower, and those found in fruits are about 1 mg/kg. S-Methylmethionine was recommended for the therapy of peptic ulcers, particularly of duodenal ulcers, and also in hyperlipidaemia. The traditional use of raw cabbage juice for the treatment of peptic ulcers would seem to support the use of vitamin U supplements as a healing aid for damaged and eroded intestinal mucous. There is currently no RDA for vitamin U. The recommended daily dose from food supplement manufacturers is 500–1000 mg.

Taurine (1-aminoethane-2-sulfonic acid, **5-50**) is the only known naturally occurring sulfonic acid that is essential for many biological processes. The major pathway of taurine metabolism is the synthesis of bile acids, which emulsify dietary fat and promote its processing in the intestines. Taurine is further involved in calcium metabolism modulation, neuroinhibition in the central nervous system, reproduction, osmoregulation, and the anti-inflammatory activity of leukocytes and platelet aggregation. It also plays a significant role as an antioxidant preventing the oxidative damage that occurs during the ageing process. Recent studies have provided evidence that taurine becomes a constituent of some biological macromolecules. For example, taurine-containing modified uridines have been found in the human mitochondria. Taurine combined with higher fatty acids such as 2-(octadecanoylamino)ethanesulfonic acid occurs in the cells of some protozoa (such as *Tetrahymena thermophila*). Taurine is biosynthesised from cysteine by two main distinct pathways: via cysteine sulfinic acid and hypotaurine and via cysteic acid. Hypotaurine can likewise be produced from L-cysteamine. In mammalian tissues, taurine is a ubiquitous semi-essential amino acid occurring as a free compound, although its concentrations in different tissues and fluids vary widely. It also occurs in insect tissues and it is particularly abundant in flight muscle and in eyes. On the other hand, taurine is completely absent in plants. The estimated mean daily intake of taurine in food is around 58 mg. Good sources of taurine include meat (0.02–0.1% f.w.) and some types of seafood, the latter being particularly rich. In mammalian tissues, taurine is the most abundant amino acid in the skeletal muscle, heart, retina, brain, and leukocytes. Taurine is used as an ingredient in many energy drinks and energy products, usually containing 4000 mg/l taurine. These drinks are marketed worldwide for the treatment of various physiological conditions, improvement of athletic performance, and general well-being. Taurine is also available in food supplements and is claimed to treat a wide range of conditions, from alcoholism and hepatitis to congestive heart failure. There is no RDA for this compound. The daily dose recommended by producers of dietary supplements varies between 2000 and 4000 mg.

The lipophilic sulfur-containing coenzyme (R)-lipoic (α-lipoic) or thiooctic acid has the systematic name (R)-5-(1,2-dithiolan-3-yl)pentanoic acid. Lipoic acid is synthesised as an offshoot of the fatty acid biosynthesis pathway in a wide number of organisms, including bacteria, fungi, plants, and animals. Lipoic acid is essential for the activity of a variety of enzyme complexes that catalyse oxidative decarboxylation. In the cell, very little lipoic acid exists as the free acid; almost all is bound to the ε-amino group of the lysine residue of target complexes. In plants, lipoic acid as the prosthetic group of pyruvate oxidase is a key substance in photosynthesis and acts as a growth factor of many bacteria. Besides its activity as a coenzyme, lipoic acid is considered as an efficient antioxidant since, in its reduced form, (R)-dihydrolipoic acid (6,8-disulfanyloctanoic acid or 6,8-dimercaptooctanoic acid), it constitutes a redox couple via modulation of the NADH/NAD$^+$ ratio (Figure 5.34). Dihydrolipoic acid can scavenge hydroxyl and peroxyl radicals, but can also chelate transition metals

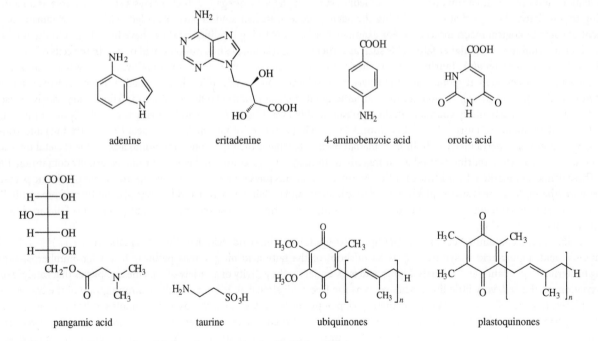

Figure 5.34 Reactions of lipoic acid.

(such as Fe and Cu). Lipoic acid occurs in almost all foods. Good sources are offal, yeast, and some vegetables (such as spinach and broccoli), but lipoic acid is not readily available as it occurs in bound forms. Lipoic acid is also available in food supplements, primarily as an antioxidant and weight-loss agent. It has also been suggested for treatment of cataracts, glaucoma, multiple sclerosis, burning mouth syndrome, Alzheimer's disease, and strokes. The daily dose suggested by commercial suppliers is 200–300 mg, but no RDA has been established.

A similar compound, 1,2-dithiacyclopentane-3-carboxylic acid (also known as tetranorlipoic or tetranorthioctic acid), was recently isolated from garlic.

Coenzymes Q_n (in short CoQ_n or ubiquinones-n) are 2,3-dimethoxy-5-methyl-1,4-benzoquinones with a long side chain attached to C-2, which is composed of varying number of isoprene units (**5-50**, $n = 1$–12) depending on the species. Most organisms synthesise a range of compounds, usually $n = 7$–10. Coenzyme Q_n functions as a mobile electron carrier within the mitochondrial inner membrane in all aerobic eukaryotic cells and many bacteria. Apart from this vital function, coenzymes Q_n act as effective antioxidants protecting cells from damage by free radicals, such as structurally similar vitamin E. In mammals, the most common form is coenzyme Q_{10} (also known as ubiquinone-10 or ubidecarenone), which contains 10 isoprene units. Coenzyme Q_{10} co-exists with its reduced form $CoQ_{10}H_2$, known as ubiquinol-10. Coenzyme Q_{10} is derived from 4-hydroxybenzoic acid (*p*-hydroxybenzoic acid), although the origin of this compound varies according to the organisms. Thus, bacteria are known to produce coenzyme Q_{10} from chorismic acid, whereas plants and animals utilise a route from L-phenylalanine or L-tyrosine via 4-hydroxycinnamic acid (4-coumaric acid). The adult human body pool of coenzyme Q_{10} has been found to be approximately 2 g and requires replacement of about 500 mg per day. This must be supplied either by endogenous synthesis or from exogenous sources. Reduced biosynthesis and increased utilisation by the body (increasingly with age) are the major factors that lead to deficiency of coenzyme Q_{10} in humans. Coenzyme Q_{10} is found in almost all foods of animal and vegetable origin. The average daily content of the western diet is 3–5 mg. It is met mostly by meat, poultry, and fish, and to some extent by food of vegetable origin. Higher amounts of coenzyme Q_{10} are found in meat (14–54 mg/kg), liver (26–50 mg/kg), fish (4–67 mg/kg), vegetable oils (4–280 mg/kg), and nuts (2–27 mg/kg). Supplementation appears to be the way for older people, and certainly the ill, to provide the major proportion of the 500 mg/day needed, but the optimal intake is not yet known, as the available clinical trials have provided conflicting results.

In plants, coenzymes Q_n are accompanied by other lipophilic compounds of similar structure, plastoquinones (**5-50**), vitamin E, and vitamin K_1. Similarly to tocopherols and tocotrienols, plastoquinones are produced from homogentisic acid by *C*-alkylation (*ortho* to the OH group) using polyisoprenyl diphosphate with $n = 3$–10, but most commonly with $n = 9$, which is called solanesyl diphosphate. Plastoquinones are involved in the photosynthetic electron transport chain in plants.

5-50, structure of compounds and their analogues previously classified as vitamins

6

Mineral Elements

6.1 Introduction

The chemical composition of food can be viewed either as detailed sets of information on the individual food components, or as the elemental composition – the content of the individual chemical elements. In anhydrous foods, the bulk of the materials are organic food substances. The main constitutional elements of the organic compounds are non-metals: carbon (C), oxygen (O), hydrogen (H), nitrogen (N), phosphorus (P), and sulfur (S). These are called **organogenic elements**. Other chemical elements contained in food are **mineral elements**. As both organic and inorganic compounds of phosphorus may occur in foodstuffs, phosphorus, and sometimes also sulfur, is included into both groups.

Mineral elements in food (also known as dietary minerals or mineral nutrients) are usually defined as the chemical elements (other than carbon and oxygen) contained in the ash of a food; that is, the elements that remain in the residue after the complete oxidation of the organic matter of food to carbon dioxide, water, and other oxidation products. The mineral fraction for most foods accounts for 0.5–3% by weight.

Mineral elements can be classified according to various criteria, for example with regard to concentration, biological and nutritional importance, dietetic effects, and origin. As far as the element concentration is concerned, two main groups of mineral elements are often distinguished: major elements and trace elements. These groups can be specified as follows:

- **major elements**, formerly referred to as **macroelements**, occur in food in larger amounts, usually in hundredths to units of weight per cent (or 10^2–10^4 mg/kg), and include the metals potassium (K), sodium (Na), magnesium (Mg), and calcium (Ca) and the non-metals chlorine (Cl), phosphorus (P), and sulfur (S);

- **trace elements** or **microelements** occur in food and biological materials in lower or even much lower concentrations (tens of mg/kg or less); important trace elements are the transition metals manganese (Mn), iron (Fe), nickel (Ni), copper (Cu), zinc (Zn), and molybdenum (Mo), the non-transition metal aluminum (Al), the semi-metals arsenic (As) and selenium (Se), and the non-metals boron (B), silicon (Si), and iodine (I); basically all elements except organogenic elements, major mineral elements, and noble gases can be considered as trace elements; some less common examples are lithium (Li), rubidium (Rb), caesium (Cs), strontium (Sr), and barium (Ba).

The border between the groups of major elements and trace elements is not unambiguously defined. Because the concentrations of some elements, especially iron and zinc (less commonly manganese), can lie in the range from tens to hundreds mg/kg, the term **minor elements** is sometimes used for elements like these. This designation of iron (or zinc or manganese) might be relevant in some food commodities with quite high content of these elements, such as tea leaves (Mn, Fe, Zn), pumpkin seeds (Zn), or legumes (Fe), whilst in many other foodstuffs, iron, zinc, and manganese are typical trace elements.

To emphasise that the contents of some trace elements is exceptionally low (10^{-3}–10^{-6} mg/kg), the corresponding elements are called **ultra trace elements**. Elements like beryllium (Be), vanadium (V), chromium (Cr), cobalt (Co), germanium (Ge), arsenic (As), selenium (Se), bromine (Br), zirconium (Zr), silver (Ag), cadmium (Cd), tin (Sn), antimony (Sb) tungsten (W), platinum group metals, rare-earth elements (from La to Lu), gold (Au), mercury (Hg), thallium (Tl), lead (Pb), and uranium (U) can be categorised as ultra trace elements. Nevertheless, on specific occasions, substantially higher levels of some elements (especially Cr, Co, As, Se, Br, Ag, Cd, Sn, Hg, and Pb) can be found in some food commodities.

The Chemistry of Food, Second Edition. Jan Velíšek, Richard Koplík and Karel Cejpek.
© 2020 John Wiley & Sons Ltd. Published 2020 by John Wiley & Sons Ltd.

Table 6.1 Mineral contents in the body of an adult (body weight 70 kg).

Element	Total amount	Unit	Element	Total amount	Unit
Ca	1000–1500	g	Si	1.4	g
Mg	25–40	g	Cu	100–180	mg
K	140–180	g	Mn	10–20	mg
Na	70–100	g	Mo	5–10	mg
P	420–840	g	Co	1–1.5	mg
S	c. 140	g	Ni	10	mg
Cl	70–110	g	Cr	5	mg
Fe	3–5	g	V	<1–20	mg
Zn	1.4–3	g	I	10–30	mg
F	0.8–2.5	g	Se	10–20	mg

The classification of minerals in foods to major and trace elements roughly corresponds to the occurrence of these elements in the human organism (Table 6.1).

The mineral elements contents in different individual foods can be substantially different, and are quite variable even within one commodity. This is due to differences in the metabolism of elements of diverse organisms, genetic factors, and, in particular, the conditions of production of the raw materials. Mineral elements present in food originate mostly from the soil. Plants absorb minerals from the soil and animals obtain their dietary minerals from the plants or other animals that they eat. Most of the mineral elements in the human diet come directly from foods of plant origin or indirectly from animal sources. Mineral elements from plant sources may vary from place to place, because the mineral content of the soil varies according to the location at which the plant was grown, as well as upon soil characteristics, mode and degree of fertilisation, climatic conditions, level of maturity of crops, and other factors. The mineral elements content of foods of animal origin depends on the nutrition, age, and state of health of the animals.

The general classification of elements as major, minor, trace, or ultra trace is therefore only approximate. For example, Al is present in various spices and tea in hundreds of milligrams per kilogram, so from this perspective it is one of the minor elements, whilst in milk it is classified as a trace element (because it is usually found in concentrations of <0.1 mg/kg). Another example of the difficulty in classification of elements is Mn. In animal products, such as beef and pork flesh, Mn is found only in traces (0.1–0.2 mg/kg), but about a 100-fold to 300-fold higher amount is found in cereals (wheat 35–49, rye 31–44, barley 16–25, oats 45–72 mg/kg) and approximately a 1000-fold higher level is present in tea leaves (300–1000 mg/kg). In some food materials that are highly purified (refined sucrose and oils), the total content of minerals is very low (10^0–10^1 mg/kg), so within these foods the elements generally classified as major elements (Ca, P, K) are actually trace elements. Another element of questionable inclusion into a group is sodium. Seen from the viewpoint of natural element content, Na is a trace element in most plant crops, but Na content in food products often achieves appreciable amounts (10^4 mg/kg) because NaCl and other sodium salts (hydrogen glutamate, nitrite, nitrate, phosphate) are often added to food.

According to their physiological significance, mineral elements in foods are divided into three groups:

- **essential elements**, also known as **obligatory** elements, are the elements that the body needs to receive from food in certain amounts to ensure important biological functions (construction of biological structures, catalytic, regulatory, and protective functions); these include all the major elements (Na, K, Mg, Ca, Cl, P, and S) and many trace elements (such as Fe, Zn, Mn, Cu, Ni, Co, Mo, Cr, Se, I, F, B, and Si);

- **non-essential elements** are physiologically indifferent elements or elements without any known biological function that are not significantly toxic; this group includes all other chemical elements in food that usually occur in trace amounts (Li, Rb, Cs, Ti, Au, Sn, Bi, Te, and Br); these elements sometimes regularly accompany essential elements (e.g. Li accompanies Na, Rb accompanies K);

- **toxic elements** are mostly elements without any biological role that are poisonous in the elemental form or in the form of soluble compounds; the toxic effects often consist in the inhibition of metabolically significant enzymes as a consequence of interaction between the element and the enzyme molecule; the most important toxic elements in food are particular metals (Pb, Cd, and Hg) and semi-metals (As).

The inclusion of an element into these groups is not definitive and depends on the specific biological species for which the element is essential. Moreover, some essential elements (e.g. Se and Ni) can be toxic at higher doses. On the other hand, As, whose toxic effects have been known for a long time, is a physiological stimulating factor for some animals.

A certain element may be included amongst the essential elements for a larger group of animals, if it meets the following conditions:

- it is present in all healthy tissues of the body;

- its concentrations in the same body tissues of different species are similar;

- its exclusion from the diet leads to repeated physiological abnormalities;

- re-introduction to the deficient diet results in normal physiological state;

- complete and long-term elimination of the element from the diet results in the death of the organism.

Elements that do not meet all these criteria, even if there is an evidence of positive effects of their presence in a nutritionally balanced diet, are not essential and are referred to as **functionally beneficial**. Despite demonstrations of their roles in experimental animals, their exact functions in human tissues and their importance for human health are uncertain. Some elements can be found in characteristic quantities in the bodies of all organisms (such as alkali elements, P, S, Cl, Fe, Zn, Cu, and Mn) and these are termed **invariable elements**. Other elements, known as **variable elements**, occur in higher concentrations only in some organisms. For example, V is present in blood cells of some tunicates (urochordates) or in fruiting bodies of some fungi (see Section 6.3.14.2), whilst in most other organisms it belongs to the trace or ultra trace elements.

Most of the elements discussed so far are natural food components. This means that they are present in a given food commodity in a characteristic amount, which is a consequence of a cycle of elements in nature and their natural distribution in different parts of the biosphere. The presence of an element in a foodstuff in larger quantities may be the result of contamination during the manufacturing process, or of the raw material during the agricultural production. This element is then considered a contaminant (e.g. toxic elements Pb, Hg, Cd, As, Tl, and Sb). Any essential element (e.g. Cu, Fe, Mn, Ni, Se, and Zn) may also become a contaminant if its content in food is significantly higher than the typical concentrations. Some anions also have toxic effects (see Section 6.6), as do some radionuclides (see Section 6.7).

For selected elements, the Recommended Daily Allowance (RDA) are set in the European Union (first number) and United States (second number) as follows: K (2000/4700 mg), Cl (800/2300 mg), Ca (800/1000 mg), P (700/700 mg), Fe (14/8 mg), Mg (375/400 mg), Zn (10/11 mg), Cu (1/0.9 mg), Mn (2/2.3 mg), F (3.5/4 mg), Se (55/55 µg), Cr (40/35 µg), Mo (50/45 µg), and I (150/150 µg). Insufficient intake of some essential elements through food can be solved by enrichment. For example, Ca can be added to dairy milk at a level of 30% and I at 20% of RDA. Generally, food enrichment with calcium is through different calcium salts (carbonate, chloride, citrate, gluconate, lactate, and phosphate), calcium oxide and hydroxide; iodine is added in the form of sodium or potassium iodide or iodate. If the element enrichment only compensates for the losses that occur during technological processing of raw materials, it is called **restitution**. If the level of the element is increased above its natural concentration in the food by enrichment, it is called **fortification**. In both cases, the element is considered to a food additive. Dietary supplements (also known as food supplements or nutritional supplements) may contain a single element, a mixture of different dietary minerals, or a combination of minerals with vitamins and other beneficial compounds.

This chapter is devoted to mineral substances occurring in foods, and deals first with their chemistry, binding options, and interactions with food components. All major elements and the most important trace elements are treated and their biochemical and physiological functions, metabolism, nutritional significance, and health consequences are discussed. The next part of the chapter deals with toxic elements and toxic anions, their occurrence in foods and the environment, dietary intake, metabolism, toxic effects, and toxicological evaluation. The final part, dealing with radionuclides and radioactivity, concentrates on radioactive nuclides occurring in the environment and foods and the health implications.

6.2 Chemistry of mineral elements

The chemical state of chemical elements in the food, their solubility, and the possibility of their absorption and bioavailability are determined by a number of factors, such as food composition and interaction with food components (e.g. water, amino acids, peptides, proteins, carbohydrates, lignin, and carboxylic acids), pH value, redox potential of the system, and the corresponding possibility of oxidation state changes.

6.2.1 Bonding possibilities

The chemical properties of the given element are crucial for the interactions of the element in the organic food matrix. These properties are determined by the position of the element in the Periodic Table and result from the electron configuration in the atom of the element.

Non-metals and semi-metals (P, As, S, and Se) with a medium value of electronegativity (the ability of a covalently bound atom of an element to attract a shared pair of electrons) form covalent compounds in biological systems, such as esters of phosphoric, diphosphoric, and triphosphoric acids, sulfur amino acids and their selenium analogues, arsenic analogues of amino compounds, and sulfur heterocycles.

The elements with very low electronegativity values (alkali metals, alkaline earth metals, such as Na, K, Ca, and Mg) and also the elements with high electronegativity values (halogens such as Cl, Br, and I) occur mainly as free ions (or hydrated ions) in biological materials, and are preferably involved in electrostatic interactions. However, even these elements can form less soluble compounds (e.g. calcium oxalate), covalent compounds (hormones thyroxine and triiodothyronine are iodinated aromatic amino acids, see Section 2.2.1.2.5), or complex compounds (chlorides as ligands and some metal ions as central atoms). A ligand is an entity (atom, ion, or molecule) that can act as an electron pair donor to create a coordinate covalent bond with the central ion. Cd and Hg also tend to form covalent compounds.

Transition metals and some post-transition metals (such as Al, Co, Cu, Fe, Mn, Ni, and Zn) have a strong tendency to form complex compounds. These elements have a large number of potential reaction partners (ligands) in biological materials. The stability of complexes depends on the type of ligand donor atoms, the metal and its oxidation state, together with steric factors.

Transition metal ions often exist in the form of relatively stable aqua complexes (hydrates) in aqueous solutions at the natural pH values of food materials. The binding of water is subject to electrostatic interactions between positive ions and coordinated water molecule dipoles. Aqua complexes of metal ions often accompany more complicated complexes of mineral compounds with various organic ligands. For example, the preferred ligands of Cu^{2+} ions are ligands containing sulfur and nitrogen or several nitrogen atoms as donors. The growth of the pH value leads to dissociation of hydrogen ions from aqua complexes and the formation of less soluble or insoluble hydroxides that have lower biologically availability or are fully unavailable. For example, cupric, ferrous, and ferric ions exist in acidic aqueous solutions as very soluble cations $Cu(H_2O)_4^{2+}$, $Fe(H_2O)_4^{2+}$, and $Fe(H_2O)_4^{3+}$, respectively, whilst in neutral and alkaline media these cations exist as the slightly soluble or insoluble hydroxides $Cu(OH)_2$, $Fe(OH)_2$, and $Fe(OH)_3$.

Conversely, Mn^{2+} ions form complexes fairly reluctantly, and in addition to ligands with sulfur and nitrogen, entities with oxygen donor atoms can also act as ligands. The complexes of metal ions in oxidation states III and IV are generally more stable than the corresponding cationic complexes in oxidation state II. In the neutral or weakly acidic media typical of foods, these metals form insoluble oxides (MnO_2) or oxocations (TiO^{2+}, VO_2^+, and BiO^+). Some metal ions, unless they are bound in complexes, are susceptible to disproportionation (e.g. $2\,Mn^{3+} + 2\,H_2O \rightarrow Mn^{2+} + MnO_2 + 4\,H^+$). With the exception of Mo, which is commonly in the very stable oxidation state VI, transition metals do not occur in high oxidation states in biological materials (e.g. oxoanions MnO_4^-, MnO_4^{2-}, and CrO_4^{2-}). Various forms of minerals in foods are summarised in Table 6.2.

Table 6.2 Overview of forms of minerals in foods.

Form	Example
Elemental form	Fe in some fortified foods
Free and hydrated ions of metals and non-metals in different oxidation states	Cu^{2+}, $Cu(H_2O)_4^{2+}$, $Fe(H_2O)_6^{2+}$, $Fe(H_2O)_6^{3+}$, AsO_3^{3-}, AsO_4^{3-}
Complex compounds of metals with inorganic ligands	$CuCl_4^{2-}$, $Cu(NH_3)_4^{2+}$
Complex compounds of metals with organic ligands	Compounds with amino acids, peptides, proteins, carbohydrates, phytic acid, hydroxycarboxylic acids, plant phenols (flavonoids), and porphyrins
Minerals bound to insoluble biopolymers	Binding to various components of fibre
Slightly soluble compounds	Sulfides, sulfates, phosphates, oxalates, hydroxides, phytates
Covalent compounds of non-metallic and semi-metallic elements	Phytic acid, sulfur amino acids and their selenium analogues, arsenic analogues of amino compounds
Organometallic compounds	Methylmercury, dimethylmercury, tetraethyllead

6.2.2 Interaction with organic food components

6.2.2.1 Aminocarboxylic acids

Aminocarboxylic acids can form **complexes** and **chelates** with metal ions. A complex consists of a central metal atom to which ligands are attached. Ligands are classified according to their dentecity, which is defined as number of lone pairs donated by a ligand to a metal atom. In a certain pH range, more functional groups can act as electron donors (polydentate ligands), which leads to the formation of cyclic complexes, *chelates*. Chelates are thermodynamically more stable than normal complexes with monofunctional ligands. The ability to form a chelate of suitable coordination geometry depends on whether the ligand molecule configuration allows the creation of five- to seven-membered rings.

Aminocarboxylic acids have amphoteric character and occur in aqueous media, depending on the pH, in various forms (see Section 2.2.3.1), and can bind metal ions as coordination compounds via dissociated carboxyl and amino groups. The donor atoms are the oxygen atom in the carboxyl group and the nitrogen atom in the amino group, which are able to provide an electron pair to the coordinate covalent bonds with the central metal ion. The stability of complexes of bivalent metal cations with amino acids decreases according to the bound metal in the order:

$$Cu^{2+} > Ni^{2+} > Zn^{2+} > Co^{2+} > Cd^{2+} > Fe^{2+} > Mn^{2+}$$

The effect of the acid–base equilibrium on the structure of amino acid–metal complexes can be demonstrated by glycine complexes with zinc. In neutral media, both functional groups of glycine (Gly) are involved in the binding of zinc, yielding the chelate $Zn(Gly)_2 \cdot 2H_2O$ (**6-1**). In acidic media, amino groups are protonised, so the metal is bound only by the carboxyl groups and the tetrahydrate of the zinc salt of glycine, $Zn(Gly)_2 \cdot 4H_2O$ (**6-1**), is formed. The nucleophilic character of nitrogen facilitates the coordination.

Other functional groups of amino acids may also contribute to the formation of complexes with metals and stabilise them through additional bonds. Metals and metalloids with a distinctive affinity for sulfur (such as Ag, As, Bi, Cd, Cu, Hg, and Sb) yield very stable compounds with thiols. The participation of the sulfhydryl (sulfanyl or thiol) group in cysteine in the bonds with metal ions leads to complexes (chelates) of higher stability (**6-1**). Stability constants of cysteine complexes with Mn^{2+}, Fe^{2+}, Co^{2+}, Ni^{2+}, Zn^{2+}, and Pb^{2+} are significantly higher than the corresponding glycine and histidine complexes. Unlike sulfhydryl groups, hydroxyl groups in serine, threonine, and tyrosine do not contribute significantly in the metal binding.

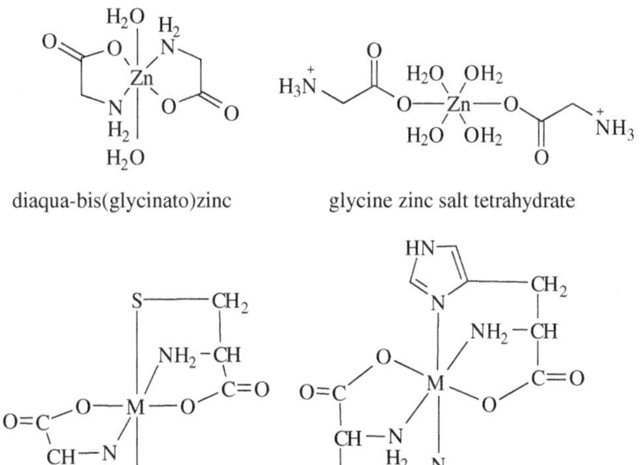

diaqua-bis(glycinato)zinc   glycine zinc salt tetrahydrate

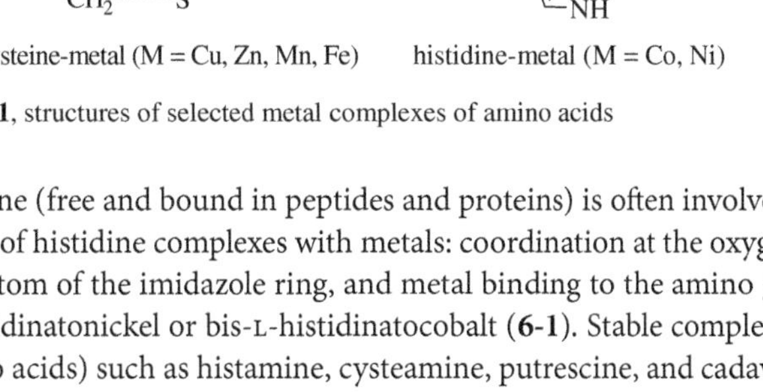

cysteine-metal (M = Cu, Zn, Mn, Fe)   histidine-metal (M = Co, Ni)

6-1, structures of selected metal complexes of amino acids

However, the imidazole ring of histidine (free and bound in peptides and proteins) is often involved in metal binding, which results in the following stages during the formation of histidine complexes with metals: coordination at the oxygen atom of the carboxyl, even at low pH value, metal binding to the nitrogen atom of the imidazole ring, and metal binding to the amino group nitrogen. All three types of bonds occur in complexes such as bis-L-histidinatonickel or bis-L-histidinatocobalt (**6-1**). Stable complexes also arise with amines and diamines (products of decarboxylation of amino acids) such as histamine, cysteamine, putrescine, and cadaverine.

Some less common amino acids also act as metal ligands. Examples are heterocyclic amino acids with an azetidine ring, especially nicotianamine and its hydroxylated derivative mugineic acid, 3-hydroxymugienic acid, and 2′-deoxymugienic acid. Related compounds without azetidine ring are avenic acid and distichonic acid (see Section 2.2.1.2.6). Amino acids of this type are, amongst other substances, excreted by plants in the vicinity of the roots, and via chelating action solubilise iron and some other elements in the soil. The metals are then more accessible to the root system of the plants. Mugienic acid and its derivatives occur especially in graminacous plants including some important crops as oat, rice, and maize. In relation to this function, these compounds are referred to as **phytosiderophores**. Nicotianamine was also found in some plants (called heavy metal hyperaccumulators), which have an extraordinary ability to accumulate particular metals from the soil, especially nickel and zinc. However, besides nicotianamine, further ligands, especially hydroxycarboxylic acids, are similarly involved in accumulating excessive amounts of metals in plant tissues.

6.2.2.2 Peptides and proteins

Peptides and proteins can bind metals through the N-terminal amino group and C-terminal carboxyl and functional groups in the side chains of amino acid residues. The involved amino acid residues are especially residues of lysine (Lys), ornithine (Orn), aspartic acid (Asp), glutamic acid (Glu), cysteine (Cys), and histidine (His). Another option is to bind the metal to the oxygen or nitrogen atom of the peptide bond. The oxygen atom of the carbonyl group of the peptide bond has only a slight nucleophilic character, so that metal binding is weak. The neighbouring amino group can provide chelate stabilisation. A metal ion binding on the nitrogen atom of the peptide bond requires a dissociation of the NH group.

In metal complexes of some oligopeptides, such as Gly–Gly, Gly–Gly–Gly–Gly, or Gly–Gly–His, a peptide chain surrounds the central metal ion and creates bonds with the participation of all sterically accessible donor groups. Alternatively, free coordination sites are occupied by water molecules (**6-2**). In some simple peptides, such as Gly–Gly, the structure of their complexes with metals may be rather complex and multinucleated complexes may arise, such as $Zn_2(Gly–Gly)_4 \cdot 2\ H_2O$.

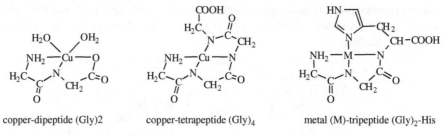

copper-dipeptide (Gly)2   copper-tetrapeptide (Gly)$_4$ metal (M)-tripeptide (Gly)$_2$-His

6-2, structures of selected metal complexes of peptides

Metallothioneins, a family of cysteine-rich metal-binding peptides and low-molecular-weight proteins (3.5–14 kDa), play a special role amongst naturally occurring metal-binding compounds, because they act as metal ligands (chelators) in metal detoxification. Metallothioneins are found in many microorganisms, plants, fungi, invertebrates, and vertebrates, including humans. They were originally found in the internal organs of mammals, such as the kidneys, liver, intestines, spleen, and pancreas of horses, cattle, and other livestock. They ensure detoxification of toxic metals (such as Cd) by formation of stable complexes. Stable complexes with essential metals (e.g. Zn and Cu) are employed for temporary storage of metals in tissues until they are required for the synthesis of metalloproteins or other substances. Metallothioneins also provide protection against oxidative stress. The first categorisation of metallothioneins established three classes: class I includes those that are homologous with horse metallothionein; class II includes the rest of the metallothioneins, with no homology with horse metallothionein; and class III includes **phytochelatins** and similar compounds. Four isoforms of the class I metallothioneins, designated MT1, MT2, MT3, and MT4, are expressed in the bodies of mammals. MT1 and MT2 occur in almost all tissues, whereas MT3 and MT4 are tissue-specific.

Metallothioneins of class I are thermostable and acid-resistant polypeptides (6–7 kDa) containing 60–68 amino acid residues, including 20 cysteine residues (cysteine residues represent about 30% of the amino acid content of metallothioneins). Histidine and aromatic amino acids are usually absent. The positions of the cysteine and lysine units in peptide chains are almost identical in various types of mammalian metallothioneins. The N-terminal amino acid is N-acetylmethionine (Figure 6.1). All cysteine residues occur in the reduced form and are coordinated to the metal ions through clusters of thiolate bonds (S⁻). Despite these common structural features, metallothioneins are organ-specific species. Their increased synthesis in the body is induced by a higher dietary intake of metals (Cd, Cu, Zn, Ni, Pb, Co, Bi, Hg, Au, and Ag), by an environmental exposure to metals (especially to Cd), or by other factors (reactive oxygen species and some hormones). One peptide molecule can bind up to seven atoms of divalent metals. Complexes isolated from animal tissues almost always contain Zn and Cd. These substances, however, still have some capacity, which allows the binding of other metal ions. Detailed studies of the metallothionein structure have found that the metal atoms are bound in two domains, α and β (**6-3**). The α-domain is located near the C-terminus of the peptide chain and, with the participation of 11 sulfur atoms, is capable of binding of up to four divalent metal ions (M), whilst the β-domain

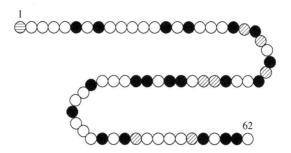

Figure 6.1 Structure of mammalian metallothionein. Full circles, cysteine; obliquely-shaded circles, lysine; horizontally-hatched circles, *N*-acetylmethionine.

(located near the *N*-terminus of the peptide chain) can bind up to three. Cu in the monovalent form may be bound in complexes with metallothionein in even larger quantities (12 Cu^I atoms in one molecule).

Metallothionein MT3 contains bound Cu and Zn and occurs in brain tissue, where it is involved in Cu and Zn homeostasis. MT3 molecules are composed of 68 amino acids. Similar binding peptides were also found in other animals exposed to toxic elements, for example in fish and invertebrates, but MTs of molluscs and crustaceans contain more amino acids (c. 77).

Metallothioneins of class II also contain a high proportion of cysteine units, and their positions in various polypeptides of this group are quite different. The corresponding molecules contain the characteristic repeating units Cys–Cys and Cys–X–Cys or Cys–X–X–Cys, where X is an amino acid other than cysteine. These polypeptides were found in several economically important crops such as wheat, maize, rice, buckwheat, and cotton, but often only in certain parts of the plant and at specific stages of their growth.

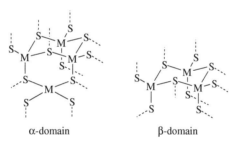

6-3, metallothionein binding domains (M = metal)

Class III metallothioneins represent plant and microbial peptides called **phytochelatins** (also called cadystines or EC-peptides) that are able to bind heavy metal ions into stable complexes, which are then stored in vacuoles. These peptides have a primary structure (γGlu–Cys)$_n$–Gly, where $n = 2 - 11$. A synthesis of phytochelatins starts from glutathione and is catalysed by phytochelatin synthase. Phytochelatins also include homophytochelatins, (γGlu–Cys)$_n$–β-Ala, that is, homoglutathione oligomers (in which glycine is replaced by β-alanine); hydroxymethylphytochelatins (hydroxyglutathione oligomers), (γGlu–Cys)$_n$–Ser, with serine as the *C*-terminal amino acid; and isophytochelatins that have their amino acid sequence terminated by glutamic acid or glutamine, (γGlu–Cys)$_n$–Glx. Other groups include desglycyl-phytochelatins, (γGlu–Cys)$_n$ and desglutamylphytochelatins with the structure Cys–(γGlu–Cys)$_n$–Gly. Phytochelatins can be found in yeasts, algae, mosses, and higher plants. They also occur in some agriculturally important plants, such as potatoes, beans, wheat, barley, and maize. Legume plants (family Fabaceae) primarily contain homophytochelatins, whilst grasses such as cereals (members of the monocotyledon family Poaceae), with the exception of maize, synthesise hydroxymethylphytochelatins.

Binding of metals in biological materials is largely mediated by proteins. The possibility of interactions of proteins with metal ions and the stability of the resulting complexes depends on temperature, pH, the type of metal ion, the presence of other components in the primary structure, and protein conformation. The binding of the metal ions by the protein molecule proceeds by similar mechanisms to that for the amino acids and peptides. The functional groups of proteins that are involved in metal binding are not necessarily the amino acid residues that are adjacent in the polypeptide chains. Due to the tertiary structure of the protein molecule, even the parts (electron-donating groups) fairly distant in amino acid sequence may come into a close contact with the metal ion, which results in a creation of coordination bonds. Another possibility is the binding of metals by interaction with phosphate groups of phosphoproteins. For example, glycophosphoprotein phosvitin of egg yolk forms complexes through the serine residues with Fe^{3+}, Ca^{2+}, or Mg^{2+}.

Metal complexes of proteins that occur in foods can be of two main types. The complexes of the first type, arising randomly, are relatively labile. Owing to the large number of reactive functional groups, practically every protein interacts with metal ions under appropriate conditions. The constitution of such a complex is not uniform and demonstrates the statistical distribution of the individual particles, which

Table 6.3 Examples of important metalloproteins.

Element	Metalloprotein	Occurrence	Element	Metalloprotein	Occurrence
Ca	Calmodulin	Widespread	Cu	Plantacyanin	Spinach
	Parvalbumin	Muscle		Ceruloplasmin	Blood plasma
	Troponin C	Muscle		Superoxide dismutases	Widespread
Fe	Myoglobin	Muscle		Haemocyanins	Molluscs, crustaceans, etc.
	Haemoglobin	Erythrocytes	Mo	Xanthinoxidase	Liver
	Cytochromes, catalases, peroxidases	Widespread	Fe (Cu, Zn, Mn)	Conalbumin	Egg white
	Transferrin	Blood, liver	Ni	Nickelplasmin	Blood plasma
	Ferritin	Spleen		Urease	Soy, rice
	Lactoferrin	Milk	Mn	Pyruvate decarboxylase	Widespread
	Ferredoxins	Widespread		Arginase	Liver

reflects the degree of dissociation of the functional groups of the protein. Higher concentrations of some metal ions can lead to protein denaturation. The metal complexes with proteins of the second group are **metalloproteins**, which have a regular structure, metal binding sites for the metal ion, and a characteristic way of binding. The metal can also be bound by substances other than amino acids in the protein macromolecule, for example prosthetic groups (such as Fe bound in the porphyrin structure of haem in haemoglobin, myoglobin, and haem enzymes). Some metalloproteins have very important biological functions; metalloproteins act as catalysts (metalloenzymes) and as transport and storage compounds (Table 6.3). A number of important metalloenzymes are mentioned in the relevant sections dealing with the individual elements.

6.2.2.3 Saccharides and their derivatives

Polyhydroxy compounds, cyclitols, and saccharides have in theory several coordinating sites for metal ions. The oxygen atoms of the hydroxyl groups are for most metals much weaker donors of electron pairs than atoms of nitrogen or sulfur. Therefore, only cyclitols and sugars with a favourable steric arrangement (an axial–equatorial–axial sequence of at least three hydroxyl groups in a six-membered ring, or a *cis–cis* sequence in a five-membered ring) form 1 : 1 complexes with metal cations in neutral aqueous solutions. An example of these weak complexes is the cationic *myo*-inositol-magnesium complex. Magnesium atom is coordinated to two *cis*-vicinal hydroxyl groups and four water molecules. Other examples are *epi*-inositol-calcium cation complex and anionic tridentate *epi*-inositol-borate ester (**6-4**). *cis*-Inositol forms complexes much more stable than those of any other polyol. An example of a sugar complex is β-D-apiose-borate ester in pectins (see Section 6.3.18).

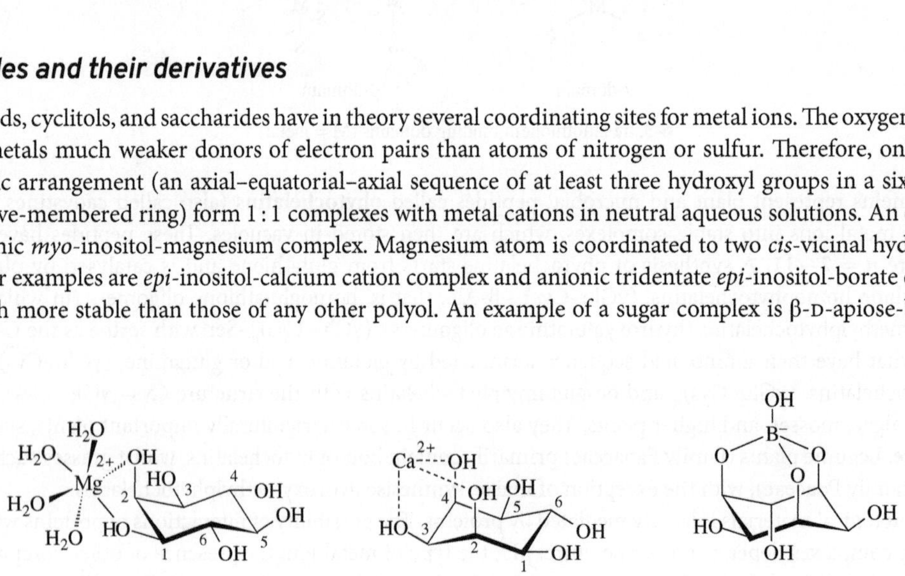

myo-inositol-Mg^{2+} *epi*-inositol-Ca^{2+} epi-inositol-B(OH)$_3$

6-4, structures of selected complexes of cyclitols

Phosphorylated organic substrates (e.g. phosphates of cyclitols and sugars) are acids and can form salts with various metals. An important binding agent for metallic elements is **phytic acid** (*myo*-inositol hexakis-(dihydrogen phosphate), Figure 6.2). The conformation of phytic acid in solutions of pH 0.5–9 has one phosphate group in the axial position and five phosphate groups in equatorial positions. In solutions of pH > 9.5, five phosphate groups are found in the axial positions and one in the equatorial position. At pH 9.5, both chair conformations are present in equimolar concentrations.

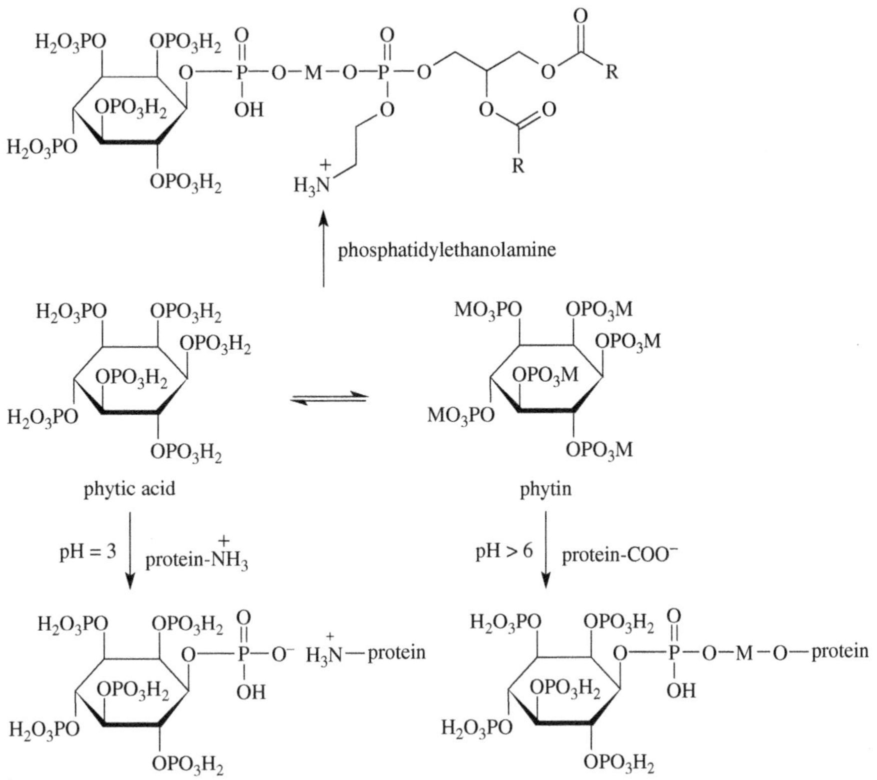

Figure 6.2 Reactions of phytic acid and phytin with metal ions (M = Ca, Mg, Zn, Fe, Cu, Mn), proteins, and phospholipids (R = fatty acid residue).

Phytic acid forms stable complexes with calcium, magnesium, iron, zinc, and other metal ions, so-called **phytates**, in ratios of $1:1–1:6$. The calcium–magnesium complex of phytic acid is known as **phytin** (or calcium–magnesium phytate). Binding of various metals on phytic acid at pH 7.4 drops in the order $Cu > Zn > Ni > Co > Mn > Fe > Ca$. The strong binding of the elements in these compounds and low solubility (iron phytate) result in decreased bioavailability of the bound elements in foods that contain higher amounts of phytic acid and phytin. On the other hand, some beneficial effects are ascribed to phytic acid, including protection against oxidative stress in the lumen of the intestinal tract and colon cancer prevention. Phytic acid and its salts with metals are found in foods of plant origin, especially cereals, legumes, oil seeds, and nuts (see Section 6.3.4.2.1).

Owing to the negative charges of the dissociated phosphate groups, phytic acid (even with bound metal) can react with ionised proteins or with phospholipids (Figure 6.2), and a complex of phytic acid–protein, metal–phytic acid–protein, or phytic acid–metal–phospholipid is formed. In an acid medium, the proteins are bound to phytic acid by positively charged amino groups of the basic amino acids (Lys, Arg, and His) and by the terminal amino groups. Around the protein isoelectric point, only weak interactions proceed. In neutral and alkaline media, phytate complexes with metals bind to proteins through ionised carboxyl groups. Crude soybean oil, for example, contains approximately 50–340 mg/kg phytate, which is largely removed during oil degumming and passes, together with the phospholipid fraction, into the lecithin fraction. Degummed oil contains phytates at a level of about 4–50 mg/kg.

In alkaline solutions, not commonly found in foods, hydroxyl groups of sugars, sugar alcohols, and cyclitols can lose protons and then form much stronger complexes. For example, aldoses and ketoses may form complexes with metals in acyclic (hydrated) or cyclic forms (furanoses and pyranoses). Complexes with ferric ions are yellow to dark brown, and the stoichiometric ratio of Fe and sugar is $1:1$. The study of these compounds was initiated because of the finding that the addition of some sugars in the diet (especially D-fructose and D-glucitol) increases the bioavailability of iron.

A typical example of practical importance is the formation of sucrose complexes with calcium ions. During sugar juice clarification (known as carbonation; see Section 4.4.2.1.2) using calcium hydroxide, saccharose dissociates in an alkaline medium in several steps (at 25 °C, $pK_1 = 12.60$, $pK_2 = 13.52$, $pK_3 = 13.72$, and $pK_4 = 13.77$). The most reactive atoms are hydrogens of the primary hydroxyl at C-6 of the glucose moiety and at C-1 and C-6 of the fructose moiety. In solutions of pH 11, about 10% of sucrose is present in the form of monovalent anions, at pH 12.2 about 50% of the molecules, and at pH 12.5 bivalent anions are also present. Depending on the amount of added slaked lime (affecting the pH of the juice) and the temperature, initially soluble monocalcium saccharate (molecular formula $C_{12}H_{22}O_{11}\cdot CaO$) and slightly soluble dicalcium saccharate ($C_{12}H_{22}O_{11}\cdot 2CaO$) form, followed at higher pH and temperatures by insoluble tricalcium saccharate ($C_{12}H_{22}O_{11}\cdot 3CaO$). In solutions, the ion $(C_{12}H_{21}O_{11}Ca)^+$ or $(C_{12}H_{21}O_{11}Ca_2)^{3+}$ is apparently present. After saturation of juice by carbon dioxide, which decreases the pH to 8.9–9.2, saccharates decompose. Formation of insoluble tricalcium saccharate is used to obtain saccharose from molasses. Similar complexes are formed with Sr^{2+} and Ba^{2+} ions.

Disaccharide lactose is especially capable of forming complexes. It complexes with metals (Ca, Ba, Sr, Mg, Mn, Zn, Na, and Li) in the ratio of 1 : 1 and in the range of pH 2–6.5. Calcium binding by lactose is connected with an increased absorption of calcium from milk.

Polysaccharides derived from glycuronic acids (such as pectins and alginates) form strong complexes with Ca^{2+} ions, which result in the formation of gels. Non-utilisable insoluble polysaccharides (e.g. cellulose and some hemicelluloses) and lignin that are classified as the insoluble fibre can also bind various minerals. These interactions can have a significant effect on the absorption of bound elements in the gastrointestinal tract. When studying metal binding to cellulose and hemicelluloses (such as arabinoxylans), it was found that the strength of interaction decreased in the order Cu > Zn > Ca. In slightly acidic media, cupric ions show a high affinity for cellulose, but in the presence of glycine, the binding of copper to the surface of cellulose molecules is substantially reduced. Lignin has two types of binding sites for binding metals. For sites with high affinity, the interaction strength decreases in the order Fe > Cu > Zn, whilst for those with low affinity, the order is Cu > Fe > Zn.

Various components of dietary fibre also have considerably different ion-exchange capacities. A long-term excessive intake of dietary fibre may bring about symptoms of calcium, iron, and zinc deficiency. The reduced absorption of minerals is especially pronounced for high doses of fibre and likewise of phytic acid.

6.2.2.4 Lipids

Food lipids usually contain only trace amounts of minerals. Non-polar triacylglycerols and waxes have virtually no possibility of binding mineral components. The exception is the opportunity for π-interaction of unsaturated fatty acids and unsaturated lipids that are able to complex with transition metals, such as silver. The complexes are of the charge-transfer type where the unsaturated compound acts as an electron donor and the Ag^+ cation as an electron acceptor. Free fatty acids yield salts with metals.

The situation is different in the polar lipids, as their molecules contain both hydrophobic chains of fatty acids and the hydrophilic residue. Phospholipids, especially phosphatidic and lysophosphatidic acids, form salts with various metal ions. The stability of salts of phosphatidic acid decreases as $UO_2^{2+} > Th^{4+} > Ce^{3+} > La^{3+} > Cd^{2+} > Pb^{2+} > Mn^{2+} > Cu^{2+} > Zn^{2+} > Co^{2+} > Ca^{2+} > Mg^{2+} > Ni^{2+} > Sr^{2+} > Ba^{2+} > Ag^+ > Li^+ > Na^+ > K^+$. Ampholytic phospholipids (e.g. phosphatidylserine and phosphatidylethanolamine) form complexes with metal ions. The phosphate group and probably the amino group are involved in the metal binding.

6.2.2.5 Carboxylic acids

Aliphatic mono- and dicarboxylic acids, oxocarboxylic acids, hydroxycarboxylic acids, and numerous aromatic acids are normal constituents of plant and animal foods. They form salts with metal ions, and their anions also have the properties of ligands. Carboxylic acids are the major binding partners of metals in fruits and some vegetables. They also contribute significantly to the binding of metal ions by humic substances in soils and sediments.

Calcium forms stable insoluble salt with oxalic acid (see Section 10.2.3.2). In plant cells with higher concentrations of oxalic acid, calcium oxalate can actually be present in the form of crystals. Some plants have been shown to bind metals in mixed complexes. For example, chromium can be bound in an oxalate–malate complex and nickel and zinc can form a citrate–malate complex. Citric acid has been proven to be a low-molecular-weight zinc ligand in human milk, and in casein micelles it binds calcium. It is also used as a food additive (acidulant, synergist to antioxidants, and sequestrant), so great attention has been paid to the formation of its complexes with metal ions. Its addition to cereal products leads to an increased solubility of naturally present iron, due to its release phytates.

Other acidic substances, such as ascorbic acid, also form metals complexes. Of particular interest is the ternary complex of ascorbic acid with oxygen and Fe^{3+} or Cu^{2+} ions, formed during autoxidation of vitamin C (see Section 5.14.6.1). Ascorbic acid can also reduce metal ions. During the reduction of Fe^{3+} ions to Fe^{2+} ions, hydroxyl radicals are generated as byproducts (see Section 5.14.6.1). In this reaction, ascorbic acid acts as a prooxidant. The stability of ferric complexes of aliphatic organic acids decrease in the series: citrate ≫ succinate ≈ ascorbate ≈ malate > lactate.

Examples of complexes of aliphatic and aromatic carboxylic acids with metal ions are represented by formulae (6-5). Aromatic 1,2-dicarboxylic acids form metal complexes with M^{2+} ions in a molar ratio of 1 : 1, but the metals may also be coordinated with a ligand belonging to two or more acid molecules.

6.2.2.6 Flavonoids and other plant phenols

The ions of transition elements form complexes with a number of aromatic compounds in which two adjacent carbon atoms of the aromatic ring bear hydroxyl groups or hydroxyl and carbonyl groups. This structural motif is common to a large number of plant phenolics.

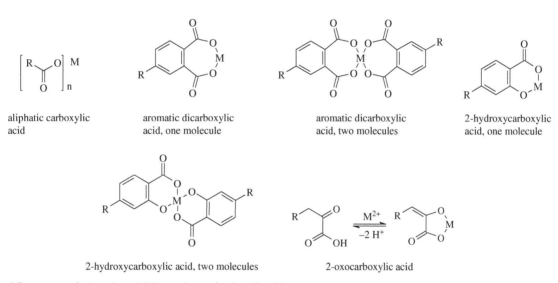

Figure 6.3 Complex of flavonols with cupric ions.

Reactions with metal ions are related to the antioxidant activity of phenols. For example, flavonoids with two hydroxyl groups in positions C-3′ and C-4′, a carbonyl group in position C-4, and a free hydroxyl at position C-3 or two hydroxyls in positions C-3 and C-5 have the highest antioxidant activity. As an illustration, the reaction of metal ions with flavonols is given in Figure 6.3, as flavonols belong to prominent yellow pigments of plants and act as very efficient antioxidants. Glycosides of flavonols, for example, have been identified as the main ligands of zinc in tea infusions. Aluminum in tea leaves is bound preferentially by catechins, and to a lesser extent by chlorogenic acid and other phenolic compounds.

aliphatic carboxylic acid

aromatic dicarboxylic acid, one molecule

aromatic dicarboxylic acid, two molecules

2-hydroxycarboxylic acid, one molecule

2-hydroxycarboxylic acid, two molecules

2-oxocarboxylic acid

6-5, structures of selected metal (M) complexes of carboxylic acids

6.2.2.7 Porphyrins and corrinoids

The macrocyclic tetrapyrrole skeletons of natural substances can act as ligands, binding metals to form complexes. In foods, the main important types of macrocyclic tetrapyrroles are porphin (see Section 9.2) and corrin (see Section 5.13.1) derivatives. Biologically important porphyrins (compounds with modified pyrrole subunits) are **haems** that contain iron as the central atom and function as prosthetic groups of important proteins, such as haemoglobin (the respiratory pigment in red blood cells of vertebrates) and myoglobin (the pigment in muscle fibres). Metalloproteins that have haem as their prosthetic group are known as haemoproteins. Of the various Fe-porphyrins, haem b (an iron chelate derived from protoporphyrin IX) is the most important compound contained in the molecules of globins, catalase, peroxidase, and proteins involved in the synthesis and transport of nitric oxide or in electron transport (P450 enzymes). Haems are also the prosthetic groups of a small number of enzymes, such as cytochrome c oxidase (haem a_3), cytochrome c 554 (haem c), nitrite reductase, sulfite reductase (sirohaem), cytochrome cd_1 sulfite reductase (haem d_1), quinol oxidase-cytochrome bo (haem o), and hydroxylamine oxidoreductase (haem P460). With the exceptions of haem c and P450 enzymes that are covalently bound to the protein via the cysteine sulfur atom, haems are bound to the corresponding proteins by non-covalent interactions, especially hydrophobic interactions. This allows the substitution of the porphyrin ring with hydrophobic vinyl (in the majority of haems) and farnesyl groups (in haems a and o). Iron bound to haemoproteins has the coordination number 5 or 6. In addition to the four nitrogen atoms of the pyrrole subunits, iron is coordinated in a direction perpendicular to the plane of the porphyrin cycle with the imidazole nitrogen atom of the histidine residue, with the sulfur atom of the cysteine residue, or with the phenolic oxygen atom of the tyrosine residue in the protein. The sixth coordination site of iron is usually occupied by a diatomic molecule, such as O_2 (in oxymyoglobin and haem oxidase).

Porphyrin structures known as **chlorophylls** (green pigments of many plants, algae, and cyanobacteria) contain magnesium as the central atom. Magnesium is not very tightly bound and in acidic media can be replaced by hydrogen ions. This substitution yields compounds called phaeophytins that can similarly rebind other metal ions, such as Cu^{2+}, Zn^{2+}, and Sn^{2+}. Some of these complexes are more stable than chlorophylls. Their spontaneous formation, however, proceeds only in metal-contaminated foods. The intentional addition of salts of these elements to foods in order to stabilise their colour is not allowed (see Section 9.2.2.3), but the corresponding copper–porphyrine complexes (namely so called sodium–copper chlorophylline) are used in specific cases as food additives (colourants).

A group of cobalt-containing organometallic compounds derived from corrin known as **cobalamins** have the activity of vitamin B_{12}. In addition to the four coordination bonds with nitrogens of the tetrapyrrole cycle, the central Co^{3+} atom is bound by two other coordination bonds. The fifth bond connects cobalt with the cyano group (in cyanocobalamin, the principal vitamin B_{12} form used in food and nutritional supplements), water (in aquacobalamin also known as vitamin B_{12a}), the methyl group (in methylcobalamin, methylvitamin B_{12}), or the 5′-adenosyl group (5′-deoxy-5′-adenosylcobalamin, often called 5′-deoxyadenosylcobalamin or adenosylvitamin B_{12}). The sixth ligand can contain different groups or compounds, or it may be not occupied at all (see Section 5.13.1).

6.2.2.8 Other complexing agents

Phosphorylated compounds containing heterocyclic nitrogen (such as nucleotides) have even stronger binding abilities than sugar phosphates mentioned above. For example, adenosine 5′-triphosphate (ATP) forms complexes with magnesium ions (**6-6**). Similarly, some vitamins (riboflavin and folic acid) and derived compounds (riboflavin 5′-phosphate and FAD) have donor heteroatoms in suitable positions and thus possess chelating properties. Isoalloxazine-derived compounds, such as riboflavin, yield chelates as shown in formulae **6-7**. Phosphorylated pyranopterins were identified as ligands of molybdenum and tungsten in cofactors of some bacterial metalloenzymes and metalloenzymes of mammals, such as cow's milk xanthinoxidase. Molybdopterin is a pyranopterin (pyran fused to a pterin molecule) with vicinal dithiol groups (pyranopterin dithiolate) that can bind Mo or W (**6-7**), where R represents hydrogen or adenosine phosphate and X is O (serine) or Se (selenocysteine). Free valences of metals in molybdenum cofactor are available for binding other pyranopterin molecules or other ligands. Chelates can also be formed with amino acids and nicotinic acid or with amino acids and imines (such as reaction products of pyridoxal and amino acids; **6-7**, R = amino acid residue).

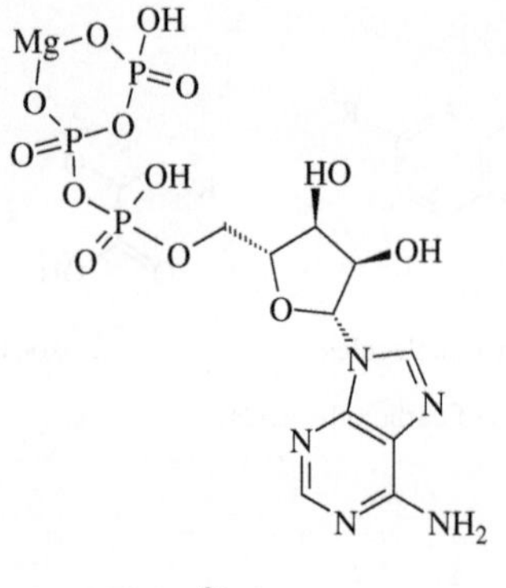

6-6, ATP-Mg^{2+} complex

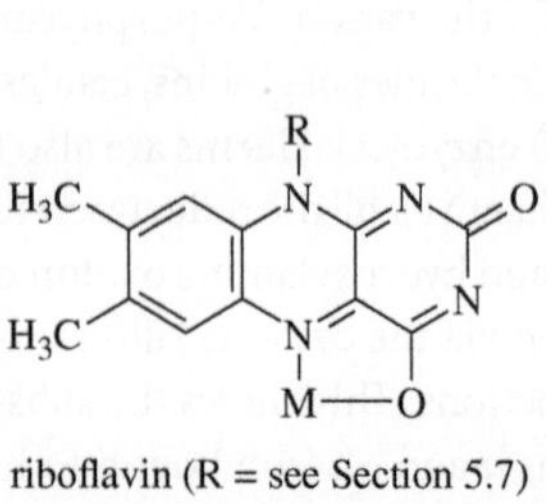

riboflavin (R = see Section 5.7)

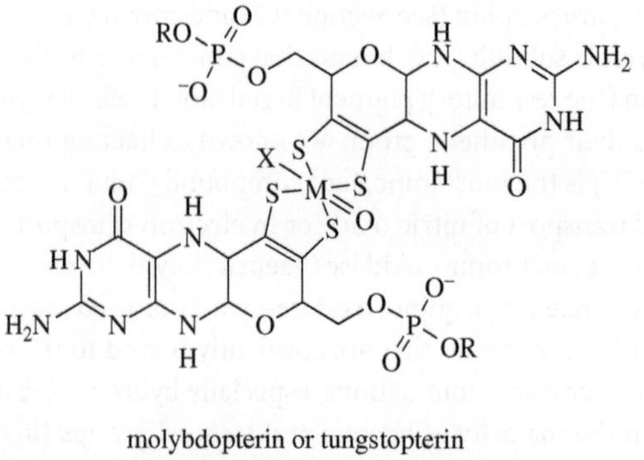

molybdopterin or tungstopterin (M = Mo or W, X = O or Se)

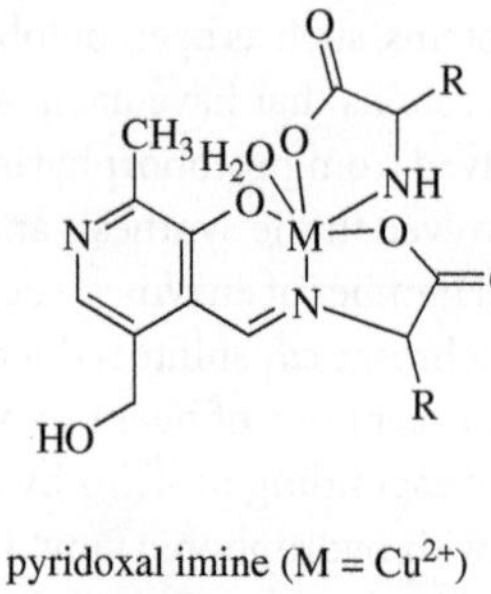

pyridoxal imine (M = Cu^{2+})

6-7, structures of metal (M) complexes of vitamins and related compounds

6.2.3 Bonding in covalent compounds

6.2.3.1 Compounds of non-metals and semi-metals

Covalent phosphorus compounds (sugar phosphates, nucleotides, and phytic acid) have already been mentioned several times. Foods also contain many sulfur-containing organic compounds. Their occurrence, properties, reactions, and importance are discussed elsewhere in this book.

Selenium is present in biological materials in a number of compounds that correspond in their structure to sulfur compounds. The amino acid L-selenocysteine (see Section 2.2.1.1.1) bound in proteins arises from selane (H_2Se) and O-acetyl-L-serine; its methylation yields Se-methyl-L-selenocysteine, whilst its reaction with O-succinyl-L-homoserine leads to L,L-selenocystathionine; cleavage of the last compound provides L-selenohomocysteine and methylation gives rise to L-selenomethionine (Figure 6.4). In plants growing in areas with a high level of selenium in the soil, the main compounds containing selenium are Se-methylselenocysteine, γ-L-glutamyl-Se-methylselenocysteine, selenocystathionine, and selenomethionine. In many foods of plant and animal origin, the main compound of selenium is selenocysteine bound in proteins.

In biological materials, arsenic occurs in both inorganic compounds and organic compounds. Methylarsonic acid, $CH_3AsO(OH)_2$, and dimethylarsinic (cacodylic) acid, $(CH_3)_2AsO(OH)$, are formed from inorganic arsenic compounds in aquatic organisms by biomethylation. Seafood and some other marine organisms accumulate arsenic in their bodies in the form of quaternary arsonium compounds, such as arsenobetaine and arsenocholine (see Section 6.5.3.2). These substances are also products of metabolic transformation of inorganic arsenic compounds that proceeds in mammals. In tissues of chondrichthyans (fish with a cartilaginous skeleton, such as sharks and rays), many species of marine fish (e.g. cod, flatfish, mackerel, herring, salmon, and others), crustaceans (e.g. lobsters, shrimps, and crabs), and molluscs (e.g. mussels and scallops), arsenobetaine is the main arsenic organic compound. In shrimps, the dominant form is arsenocholine, and arsenobetaine is present in lower amounts. In freshwater fish, arsenic is mostly found in other compounds that have not yet been satisfactorily described. Marine algae (such as *Ecklonia radiata*) contain **arsenosugars** derived from ribose, namely dimethylarsinoylribosides (**6-8**, R = OH, R^1 = OH or R = OH, R^1 = SO_3^- or R = OH, R^1 = OSO_3^- or R = NH_2, R^1 = OSO_3^- or R = OH, R^1 = $OPO_3^-OCH_2$ $CH(OH)CH_2OH$), the corresponding phosphatidylglycerol (R = $[CH_2]_{14}CH_3$), a glycolipid usually esterified by palmitic acid, and trimethylarsinoylriboside glycerol sulfate (**6-8**).

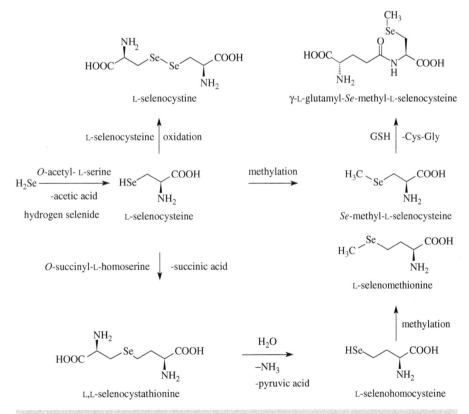

Figure 6.4 Biosynthesis of selenium amino acids.

dimethylarsinoylribosides

dimethylarsinoylriboside phosphatidylglycerol

trimethylarsinoylriboside glycerol sulfate

6-8, quaternary arsonium compounds and arsenosugars

Amongst the toxic organometallics in fish oil are the so-called arsenolipids, which include dimethylarsinoyl fatty acids and dimethylarsinoyl hydrocarbons. Their chain can be saturated or unsaturated. Arsenolipids have only recently been identified in different fish species, such as cod, capelin, and blue whiting fish. Examples of dimethylarsinoyl fatty acids are ω-(dimethylarsinoyl)alkanoic acids $(CH_3)_2As(O)(CH_2)_nCOOH$ (n = 12, 14, 16, and 18), ω-(dimethylarsinoyl)alkenoic acid with a chain length of 17 carbon atoms, and ω-(dimethylarsinoyl)alkapentaenoic acid with a chain length of 21 carbons (**6-9**). The most prevalent acids are 13-(dimethylarsinoyl) tridecanoic and 15-(dimethylarsinoyl)pentadecanoic acids. An example of dimethylarsinoyl hydrocarbons is 1-(dimethylarsinoyl) pentadecane $(CH_3)_2As(O)(CH_2)_{14}CH_3$. The long-chain fatty acids may also be conjugated to phosphatidylcholine in arsenic-containing phospholipids.

6-9, (7Z,10Z,13Z,16Z,19Z)-21-(dimethylarsinoyl)heneikosa-7,10,13,15,18-pentaenoic acid

6.2.3.2 Organometallic compounds

The bonds between a metallic element and a carbon atom in an organometallic compound are relatively polar. Depending on the metal valency, several bonds can be present in a molecule, such as in tetramethyllead, $(CH_3)_4Pb$ (the main use of this in the past was in antiknock additives for gasoline). Combustion of such fuel in car engines was an important source of environmental contamination. Some alkyl groups may be replaced by other groups or atoms, as is the case of dimethyllead dichloride $(CH_3)_2PbCl_2$. Organometallic compounds with the full number of alkyl substituents are non-polar substances with relatively low boiling point temperatures. Apart from tetramethyllead, traces of some other synthetic organometallic compounds may be found in food as contaminants. These include tetraethyl lead, $(CH_3CH_2)_4Pb$ (also used as an additive to gasoline), tributyl tin, a group of compounds containing the $(CH_3CH_2CH_2CH_2)_3Sn$ moiety, as in tributyltin hydride or tributyltin oxide (used as pesticides) and their decomposition products.

In bacteria, fungi, aquatic animals, and plants, organometallic and organometalloid compounds are produced by **biomethylation** of some elements. This applies especially to mercury and arsenic, but to a lesser extent also to antimony, bismuth, selenium, tellurium, lead, tin, and cadmium.

Biological methylation of mercury can be carried out under aerobic or anaerobic conditions. One donor of methyl groups (CH_3 anion) is methylcobalamin, which, in addition to mercury methylation, also participates in other methylation reactions (e.g. in the conversion of L-homocysteine into L-methionine). Other biomethylation agents are N^5-methyltetrahydrofolate and S-adenosyl-L-methionine, which operate mainly in the methylation of arsenic compounds and other elements that form anions. The microorganisms methylating mercury include bacterial species of the genera *Bifidobacterium*, *Chromobacterium*, *Clostridium*, *Enterobacter*, *Escherichia*, *Methanobacterium*, and *Pseudomonas* and the microscopic fungi *Aspergillus niger*, *Neurospora crassa*, and some other bacteria.

The product of biomethylation of the inorganic forms of mercury in fish and other aquatic organisms is methyl mercury. This is the name given to a compound CH_3HgX, where X can be halogen, hydroxyl anion, a sulfhydryl group, or a sulfide group. A minor product of mercury biomethylation and a breakdown product of methylmercury sulfide, $(CH_3Hg)_2S$, is dimethylmercury, $(CH_3)_2Hg$. Methylmercury is the predominant form of mercury in fish, crustaceans, and molluscs. Owing to the high affinity of mercury compounds for sulfur compounds, methylmercury CH_3Hg^+ cation bound through sulfhydryl groups of cysteine residues in proteins and peptides can thus be expected in fish tissue. It is also probable that this cation reacts with selenium compounds. In addition to fish, it is possible to find methylmercury in the bodies of piscivorous animals, such as aquatic birds, cetaceans, pinnipeds, and otters, and in the gastrointestinal tract of various other animals. In this case, enteric bacteria participate in the biomethylation of dietary mercury.

6.3 Essential elements

6.3.1 Sodium and potassium

6.3.1.1 Biochemistry and physiology

The total sodium and potassium contents in the human body are about 70–100 and 140–180 g, respectively. Sodium is found predominantly in the extracellular space, whilst potassium is located mainly inside the cells. Large amounts of sodium and potassium ions are found in the gastric juices.

The main function of sodium and potassium in the body, with chloride as the counter ion, is to maintain the osmotic pressure of fluids outside and inside cells and acid–base equilibrium. In addition, these elements are required for the activation of some enzymes, such as sodium for activation of α-amylase and potassium for activation of glycolytic and respiratory chain enzymes. Potassium significantly affects muscle activity, especially the activity of cardiac muscle.

Absorption of sodium and potassium in the gastrointestinal tract is rapid, and its effectiveness in a typical diet amounts to about 90%. The daily level of alkali metals obtained from food varies, with sodium ranging from 1.7 to 6.9 g and potassium from 2 to 5.9 g. Both elements are excreted from the body in urine, but a significant amount of sodium is lost in sweat as well. Excessive sweating in extreme physical exertion may lead to loss of 8 g of sodium per day (which corresponds to 20 g NaCl). Unless sodium is supplied in the diet at an increased level, in such cases muscle cramps, headaches, and diarrhoea develop. Although sodium is essential to bodily functions, too much can be harmful for people with kidney diseases, and long-term excessive sodium intake can cause hypertension. Other high-sodium-related complications include oedema (noticeable swelling in legs, hands, and face), heart failure, and shortness of breath. A deficiency of sodium is quite unlikely, but it represents a serious threat. It is often associated with a large loss of water (a case of enormous sweating or diarrhoea). Lack of potassium (caused, for example, by excessive loss of fluids in some diseases) can cause kidney failure, muscle weakness, and an irregular heartbeat.

6.3.1.2 Occurrence in foods

In foods, sodium and potassium are found mainly as free ions. The natural sodium content is highly variable. In many foods of plant origin, sodium is a trace element. In contrast, the potassium content in some plant materials is extremely high, reaching up to 2% (e.g. in tea and roasted coffee). In salted foods, sodium can achieve a content higher by several orders of magnitude than that in crude material. Table salt (sodium chloride) is added to preserve food and to increase flavour. The contents of sodium, potassium, and other major elements in selected foods are summarised in Table 6.4.

6.3.1.3 Nutrition

For an adult, the required daily dose is 500 mg sodium and 2000 mg potassium; for children up to the age of 1 year, it is 120–200 mg Na and 500–700 mg K; and for children aged 1–9 years, it is 225–400 mg Na and 1000–1600 mg K. Actual amounts of sodium intake are often considerably higher. Approximately 75% of dietary sodium comes from sodium chloride or sodium hydrogen glutamate (monosodium glutamate) used in food technology and culinary processing. Except in manual workers, the amount of dietary sodium should not be higher than 2.4 g/day (6 g NaCl).

6.3.2 Chlorine

6.3.2.1 Biochemistry and physiology

The amount of chlorine in the human body is about 80 g. In living organisms, chlorine mostly occurs in the form of anions, which are, together with sodium counter ions, located in the cytoplasm of cells and mainly in extracellular fluids (blood, gastric juice, and urine). Their main role is, similarly to sodium ions, to maintain osmotic pressure. In the gastric juices, chlorides act as counter ions of hydrogen in hydrochloric acid, which is excreted by the gastric wall. The concentration of hydrochloric acid in gastric juice is typically around 0.3%, but due to individual variability and the presence of other components, the corresponding pH value may range from 1.3 to 4. Chlorine is supplied in food mainly as sodium chloride at a level of 3–12 g/day. Chlorides are rapidly absorbed from the diet and excreted in urine.

Table 6.4 Major mineral element contents in some crops and foods.

Food	Content (mg/kg)						
	Na	K	Cl	Mg	Ca	P	S
Pork meat	450–600	2600–4000	480–490	80–220	50–90	1300–2200	1400–2600
Beef meat	580–690	3400	400–740	170–250	30–150	1200–2000	750–2100
Chicken meat	460	4100	610	130–290	60–130	1200–2500	2700
Pork liver	770	3500	1000	220–260	60–70	3600–4800	2300–2800
Fish	650–1200	2200–3600	570–1200	140–310	60–5200	1900–3900	1400–2300
Milk (whole)[a]	480–500	1550–1600	900–980	110–140	1100–1300	870–980	290–330
Curd	–	1000	–	90	960–990	2000	1500
Cheese	450–14100	1070–1100	12000–23000	170–550	1500–1512000	2900–8600	1900–2600
Yoghurt	660–770	1700–2200	–	140	1400	1100–1200	390–430
Egg (whole)	1350	1380	1600–1800	120–140	550–570	2100–2200	1700–2000
Egg (white)	1920	1480	1700	110	50–110	210–330	1800–2000
Egg (yolk)	500	1230	1400	140–150	1300–1400	5000–5900	1600–1700
Wheat	80	3500–5000	670	700–1500	230–500	3000–4100	1300–1500
Rye	20	5100	–	1100	240	3300	–
Flour (wheat)	20–30	1100–1300	360–480	210–1300	130–260	1000–3500	1300–1400
Bread (whole-wheat)	4000–6000	2300–2500	9100	230–550	140–650	1800–2000	800–1000
Rice (peeled)	60	1000	60–270	260–430	50–110	770–1200	690–860
Peas (mature seeds)	20–380	2900–900	390–600	1100–1300	440–780	3000–4300	1600–2000
Lentils	40–550	6700–8100	640	770	400–750	2400	1200
Beans	20–400	12000	20–250	230–1800	300–1800	3700–4300	1100–1700
Soybeans	60	16000	–	2400–2500	1300–1800	2900–7900	3500–3700
Cabbage	130	2300	220–450	120–230	300–750	280–680	440–900
Cauliflower	70–100	2100–4100	340	170	180–310	420–750	510–590
Spinach	600–1200	4900–7700	560–750	420–770	700–1250	250–550	270–400
Lettuce	30–100	2200	400	150–290	400–800	300–390	120–190
Tomatoes	30–60	900	500–600	110–180	60–140	210–260	110–140
Carrot	210	950	690	100–190	240–480	300–560	70–180
Peas (green)	20	3000	340–380	380–410	260–410	1000–1500	410–550
Onion	100–260	1300	190–270	70–160	200–440	300–480	360–530
Potatoes	30–280	4400–5700	450–790	200–320	30–130	320–580	240–350
Apples	16–30	900–400	<10–190	35–70	30–80	100–130	30–100
Oranges	14–30	1800–2000	32–40	110–140	400–730	230–240	90–130
Bananas	10	3500	790	310–420	50–120	230–310	80–130
Strawberries	15–30	1500	180	120–170	180–260	230–350	80–140
Walnuts	30	6900	230	1300	600	4300–5100	1000
Tea (black)	450	21600	5200	2500	4300	6300	1800
Coffee (roasted)	740	20200	240	2400	1300	1600	1100
Milk chocolate	2800	3500	1700	590–710	2200–3200	2200–3000	780–1100

[a]Breast milk contents of majority elements: Na 160, K 530, Cl 860, Mg 20, Ca 250–310, P 130–160, S 100–160 mg/kg.

6.3.2.2 Occurrence in foods

Chlorine is a major element in a range of foods. The chloride content in foods, however, depends mainly on whether table salt has been added to the material during food preparation and production. The concentrations of sodium and chlorides then correlate significantly. The chloride contents of some foods are summarised in Table 6.4. **Perchlorates** (ClO_4^-) and organic chlorine-containing compounds (see Chapter 12) are generally classified as contaminants. Perchlorates are important oxidisers for fireworks, airbags, and solid rocket propellants. They have also been used as total herbicides. Low levels of perchlorates can contaminate groundwater, drinking water, and even the milk of cows grazing on contaminated pastures. Perchlorates may interfere with the ability of the thyroid gland to utilise iodine in hormone production.

6.3.2.3 Nutrition

The minimum dose of chlorine required daily for an adult is 75 mg, for children up to the age of 1 year, 180–300 mg, and for children aged 1–9 years, 350–600 mg.

6.3.3 Magnesium and calcium

6.3.3.1 Biochemistry and physiology

The content of magnesium in the body of an adult is about 25–40 g, about 60% of which corresponds to magnesium bound in skeleton. In soft tissues, the highest magnesium concentrations are found in the pancreas, liver, and skeletal muscles. Blood and extracellular fluids contain only 1% of the total amount of magnesium in the body. Calcium is the most abundant mineral component in the human body; the total is about 1500 g, with 99% of this being in bones and teeth in the form of calcium phosphate.

Magnesium and calcium have many important biochemical functions in organisms. Magnesium is essential for all metabolic processes in which ATP arises or is hydrolysed, participates in the stabilisation of DNA, and is required to activate certain enzymes, such as phospho-transferases (kinases) and phosphatases. In this function, magnesium ions can sometimes be replaced by manganous ions (Mn^{2+}). Owing to the binding of magnesium in chlorophylls, this metal is essential for photosynthesising organisms. Together with calcium, magnesium affects the permeability of biological membranes and cell excitability. The concentration of magnesium ions in the extracellular fluids affects the function of nerve cells. Magnesium deficiency, especially with an excess of calcium, leads to increased irritation, and vice versa. A very high excess of magnesium causes suppression of nerve activity. In addition to structural functions together with proteins **osteocalcin** and **osteonectin**, the main biological function of calcium is participation in nerve and muscle activities. Calcium is also essential for blood clotting. A series of metabolic processes are regulated by calcium ions through the calcium-binding serum protein **calmodulin**, which affects the activity of certain enzymes (such as adenylate cyclase and, along with magnesium, the activity of ATPase).

The absorption of magnesium and calcium from food takes place in the small intestine. The efficiency of magnesium absorption from the diet at the normal dose of magnesium in healthy individuals is 40–50%. A higher portion is absorbed if the diet is low in magnesium. Excessive amounts are excreted from the body in urine. The efficiency of calcium absorption is low (about 5–15%), and is highly dependent on the chemical form of the calcium and the composition of the diet. For example, from spinach (containing calcium oxalate), wheat bread (containing phytin), and cabbage (where the main forms of calcium are soluble salts of organic acids), about 2–5, 40, and 40–70% of the total calcium, respectively, is absorbed. Phytic acid and some components of dietary fibre also reduce the absorption of magnesium, calcium, and some other elements from the diet (especially iron and zinc). Calcium absorption is also diminished by excessive dietary fluoride. A high protein content in the diet has an opposite effect on calcium absorption.

6.3.3.2 Occurrence in foods

The contents of magnesium and calcium in selected foods are summarised in Table 6.4.

6.3.3.3 Nutrition

The recommended daily dietary intake of magnesium is 50–70 mg for children under 1 year, 150–200 mg for children under 6 years, 350 mg for adult men, and 300 mg for adult women. During pregnancy and lactation, the daily dose of magnesium should be increased to 450 mg. The recommended daily intake is 400–500 mg of calcium for children under 1 year, 800–1200 mg for older children and adolescents, 800 mg for adults, and 1200 mg for pregnant and lactating women.

6.3.4 Phosphorus

6.3.4.1 Biochemistry and physiology

The adult human body contains about 420–840 g of phosphorus, 80–85% of which is found in the bones and teeth. The main elements of bones are calcium, phosphorus, and fluorine. The weight ratio of Ca/P in bones is approx. equal to 2. The phosphorus content in various tissues is as follows: in blood about 400 mg/l, in muscles 1700–2500 mg/kg, in nervous tissue 600 mg/kg, and in bones and teeth 22% by weight.

Phosphorus is the constitutive element of many compounds, and it has diverse biological functions in organisms, including structural functions, energy metabolism and activation functions, and regulatory and catalytic functions.

The phosphorus compounds occur in a number of important biological structures (inorganic phosphates in bones and teeth, phospholipids in biomembranes). The hydrolysis of 'high-energy' phosphates, such as ATP, GTP, phosphoenolpyruvate, and creatine phosphate, allows energy-demanding biosynthetic reactions to be accomplished. Conversely, in catabolic processes (oxidative phosphorylation, reactions of the citric acid cycle and glycolysis), the chemical energy gained from the substrate is stored in ATP. The transfer of the phosphate group (phosphorylation) activates common substrates (such as glucose). The participation of phosphate in the regulation of metabolism often consists in the conversion of inactive forms of some enzymes (such as glycogen phosphorylase or protein kinase) by phosphorylation. The allosteric activator of some enzymes (including protein kinase) is another phosphorus compound known as cyclic AMP. Phosphates also act as cofactors of enzymes (thiamine diphosphate, FAD, FMN, NADH, and pyridoxal phosphate). Phosphorus is also present in nucleic acids, which provide storage and expression of genetic information.

Phosphorus is absorbed in the small intestine, mainly in the form of the HPO_4^{2-} anion. Phosphorus absorption and excretion partly depends on the calcium content in the diet, and vice versa. This indicates that the optimum weight ratio of calcium to phosphorus in the diet is between 1:1 and 1:1.5. Moreover, several hormones (especially PTH, calcitriol, and calcitonin) participate in maintaining the homeostasis of phosphorus (and calcium) in the body.

The efficiency of phosphorus absorption depends on dietary composition (in particular, the levels and forms of phosphorus and calcium) and the age and health of the consumer. Newborns absorb 85–90% of phosphorus from breast milk and about 65–70% of phosphorus from cow's milk. In older children and adults, the rate of phosphorus absorption from the normal diet is about 50–70%, but this may increase to 90% in the case of low doses of phosphorus.

As regards the absorption of various phosphorus compounds, it is known that the best utilisable compounds are salts and esters of *ortho*-phosphoric (*o*-phosphoric) acid, whilst somewhat less utilisable are salts of *meta*-phosphoric (*m*-phosphoric) acid and polyphosphates. Phosphorus in the form of phytic acid is less effectively absorbed (20–50%), especially in the case of high dietary calcium doses. In some animals, the partial absorption of phytate phosphorus is probably caused by phytase activity of their intestinal alkaline phosphatase. In addition, a phytase of intestinal microorganisms can also participate in the absorption of phytate phosphorus.

6.3.4.2 Occurrence in foods

6.3.4.2.1 Total phosphorus content

Phosphorus is found in most foods in concentrations above 100 mg/kg. The exceptions are refined fats and refined sugar, which contain only traces of phosphorus. Nuts, cheeses, and other dairy products are rich sources of dietary phosphorus. For cow's milk, the ratio of Ca to P is about 1.2:1, whilst for breast milk, the ratio is about 2:1. The absolute amounts of both elements in breast milk compared with cow's milk are lower. The phosphorus contents in foods are summarised in Table 6.4.

6.3.4.2.2 Phytic acid

Phytic acid (*myo*-inositol-1,2,3,4,5,6-hexakis(dihydrogen phosphate)) occurs in a number of important crops, especially cereals, legumes, and oilseeds. The main form is a mixed calcium and magnesium salt, which is called phytin. Phytate phosphorus has reduced biological utilisation and phytic acid causes a lower utilisation than other minerals (Ca, Mg, Zn, and Fe in particular). The contents of phytic acid in some food materials and foods, and the ratio of phytate phosphorus to total phosphorus, are shown in Table 6.5.

In addition to cereals, legumes, and oilseeds, which are characterised by a high content of phytic acid, there are plant products that have a low amount of phytic acid (potatoes, artichokes, carrots, broccoli, strawberries, blackberries, and figs) and crops that do not contain phytic acid at all (lettuce, spinach, onion, celery, mushrooms, apples, bananas, pineapple, and citrus fruits).

Phytic acid is accompanied by a smaller amount of the partial esters of *myo*-inositol with phosphoric acid. Theoretically, there are 63 ($=2^6 - 1$) acyclic and cyclic esters. In lenses containing 0.49% of phytic acid, 0.07% of a mixture of pentakisphosphates and 0.01% of a mixture of tetrakisphosphates are present.

It is assumed that phytic acid and phytates in seeds of plants are used as a storage form of phosphorus and other minerals. Phytic acid has, however, numerous other biological effects in plant and animal organisms; for example, it acts as an antioxidant and anticarcinogen. Partial phosphoric acid esters, such as *myo*-inositol-1,4,5-trisdihydrogenphosphate and *myo*-inositol-1,3,4,5-tetrakisdihydrogenphosphate, act in the regulation of the intracellular calcium level. *Myo*-inositol-1,3,4,5,6-pentakisdihydrogenphosphate modulates the affinity of haemoglobin for oxygen, whilst *myo*-inositol-1,2,6-trisdihydrogenphosphate inhibits blood coagulation (platelet aggregation) and has anti-inflammatory effects. *Myo*-inositol itself was once considered a member of the B-group vitamins.

Table 6.5 Phytic acid and phytate phosphorus contents in some crops and foods.

Material	Phytic acid (g/kg)	Phytate phosphorus (%)	Material	Phytic acid (g/kg)	Phytate phosphorus (%)
Bread (whole- wheat)	4.3–8.2	38–66	Soy flour (defatted)	15.2–25.2	87
Wheat	3.9–13.5	60–80	Lentil	2.7–10.5	27–87
Rye	5.4–14.6	38–46	Peas	2.2–12.2	37
Barley	7.5–11.6	66–70	Almonds	12.9–14.6	82
Oat	7.0–11.6	49–71	Peanuts	17.6	57
Maize	8.3–22.2	71–88	Walnuts	6.5–7.7	42
Rice (unpeeled)	8.4–8.9	–	Cocoa	0.9	15
Rice (peeled)	3.4–5.0	61	Carrot	0.2–0.3	16
Soybeans	10.0–22.2	50–70	Potatoes	0.2–0.5	19–23

Table 6.6 *Myo*-inositol phosphate contents in rice cakes.

Phosphates of *myo*-inositol	Content (mg/kg)	Phosphates of *myo*-inositol	Content (mg/kg)
Hexakisphosphate	110	Trisphosphates	340
Pentakisphosphates	130	Bisphosphates	160
Tetrakisphosphates	460	Monophosphates	160

Under normal conditions, phytic acid is not hydrolysed chemically, but it may be hydrolysed enzymatically. The enzymatic hydrolysis proceeds in fermented cereal products such as bread and yields pentakis-, tetrakis-, tris-, bis-, and monodihydrogenphosphates, and possibly also free *myo*-inositol. In wheat and rye flour, the phytase of the grain is still partly active, and hydrolysis of phytic acid in dough is carried out by endogenous phytase (6-phytase) that hydrolyses the ester bond, predominantly in position C-6. Subsequent hydrolysis is achieved by the action of phytase of baker's yeasts and phytases of colon microorganisms. Microbial phytases preferentially hydrolyse the ester bond in position C-3. Partial hydrolytic cleavage of ester bonds of phytic acid occurs during baking. Bread production from wheat flour can result in a loss of up to 70–85% of the phytic acid originally present. Table 6.6 illustrates the overall extent of hydrolysis of phytic acid during the fermentation and baking of cereal products. Activity of phytase also increases during seeds germination. Particularly high activity is seen in phytase of germinating grains of wheat and rye and in phytase of pea seeds. Losses of phytic acid during the cooking of legumes are caused mainly by leaching.

Phytic acid is also used as an agent (food additive) for the clarification of wines, which ensures the removal of ferric ions in the form of insoluble ferric phytate.

6.3.4.2.3 Phosphates and phosphoric acid

The phosphorus content in some foods can be raised above the natural level using food additives based on polyphosphoric acid salts or phosphoric acid salts. Polyphosphates are either sodium or potassium salts of polyphosphoric acids with straight chains and different degrees of polymerisation, $M^I_{n+2}P_nO_{3n+1}$, or salts of cyclic polyphosphoric acids, $(M^IPO_3)_n$ (**6-10**), which are actually oligomers of hydrogenphosphoric acid (*m*-phosphoric acid, HPO_3). The most commonly used substances are sodium and potassium phosphates, especially disodium dihydrogendiphosphate, tetrasodium diphosphate, sodium triphosphate, and the so-called sodium hexametaphosphate.

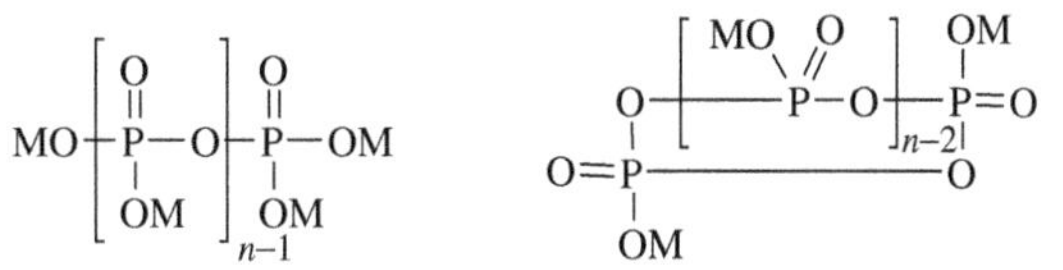

linear polyphosphate   cyclic polyphosphate

6-10, structures of straight-chain and cyclic polyphosphates (M = Na or K)

The addition of phosphates (polyphosphates) to food affects the hydration of proteins and polysaccharides and their colloidal properties. It is used to increase the water-holding capacity (WHC), sometimes referred to as water-binding capacity (WBC), when water is added to cured meat and certain meat products. Phosphates also provide the appropriate texture to processed cheeses made from traditional cheese and emulsifying salts, often with the addition of milk, salt, preservatives, and food colouring. Polyphosphates are also used as agents for the clarification and stabilisation of wine and beer, and additionally have an antimicrobial effect. Phosphoric acid is often used as an acidifying agent (acidulant) for soft drinks (e.g. in Coca-Cola). In drinks in cans, phosphates retard the corrosion of the packaging.

6.3.4.3 Nutrition

The recommended daily dietary intake of phosphorus is 300–500 mg for children under 1 year, 800 mg for children under 10 years, and 1200 mg for adults. These amounts are easily achieved in a normal diet. It is more important to maintain an appropriate ratio of calcium to phosphorus in the diet than to worry about the actual realised amount of phosphorus. Meat, poultry, and fish (without bones) contain about 15–20 times more phosphorus than calcium; eggs, cereals, and legumes contain about 2–4 times more. Only milk, cheese, leafy vegetables, and bones contain more calcium than phosphorus.

6.3.5 Sulfur

The inclusion of sulfur amongst the mineral elements is questionable, because the majority of sulfur in food is bound in organic compounds, such as sulfur-containing amino acids (cysteine and methionine), peptides (e.g. glutathione), proteins, S-glycosides, biocatalysts (thiamine, coenzyme A, biotin, lipoic acid), and heterocyclic compounds. Many sulfur compounds are important flavour-active compounds. On the other hand, inorganic sulfates are present in drinking water (especially some mineral waters) and sulfur dioxide, sulfites (Na_2SO_3, K_2SO_3), and disulfites ($K_2S_2O_5$) are used as food preservatives.

In contrast to other elements, the total content of sulfur in food is not commonly determined. Some data are summarised in Table 6.4. The recommended daily intake (approximately 0.1–0.6 g) is not officially established.

6.3.6 Iron

6.3.6.1 Biochemistry and physiology

The total amount of iron in the body of an adult is about 3–5 g. The highest concentrations are found in the blood (haemoglobin), liver, and spleen (ferritin and haemosiderin); lower concentrations are in the kidney, heart, and skeletal muscle (myoglobin). The concentration of iron in the pancreas and brain is about 2–10 times lower than the content of the liver or spleen. A list of the organic compounds of iron is given in Table 6.7.

In enzymes of the human body, only minute quantities of the total body iron are found. They can be divided into two groups:

- haem enzymes: cytochromes, oxygenases, and peroxidases;

- non-haem enzymes: succinate dehydrogenase, as well as liver xanthine oxidase, NADH-cytochrome c reductase (flavin enzymes, oxidoreductases), and aconitase (lyase).

The function of iron in the body depends on the compound in which it occurs. It mostly participates in oxygen transport through the bloodstream and oxygen storage in muscle tissue (iron bound in the haemoglobin and myoglobin), as well as in catalytic and oxidation–reduction reactions (iron in haem and flavin enzymes).

Table 6.7 Important iron compounds found in the human body.

Compound	Mass (g)	Mass of Fe (g)	Percentage of body Fe	Compound	Mass (g)	Mass of Fe (g)	Percentage of body Fe
Haemoglobin	900	3.0	60–70	Transferrin	10	0.004	0.1
Myoglobin	40	0.13	3–5	Catalase	5	0.004	0.1
Ferritin	2–4	0.4–0.8	7–15	Cytochrom c	0.8	0.004	0.1

The blood plasma contains non-haem glycometalloprotein **transferrin**, which serves as a transport form of iron. Transferrin is pink-coloured and contains two Fe^{3+} ions bound to the apoprotein (86 kDa), which belongs to the group of β_1-globulins. A transferrin molecule consists of two identical subunits, each of which contains one binding site for iron.

Ferritin, together with **haemosiderin**, is the principal iron storage protein in bacteria, plants, and animals. In animals, ferritins occur mainly in the spleen, liver, and bone marrow. Ferritins are metalloproteins that can contain up to 23% iron. The protein constituent of apoferritin (445 kDa) consists of 24 cubically arranged subunits. Apoferritin units form the protein shell of the mineral core, which contains hydrated ferric hydroxide. One ferritin molecule can bind from 2000 to 4500 iron atoms in trivalent state. The binding of ferritin to iron reduces the risk of formation of insoluble iron compounds. In animals, ferritin is found not only inside cells but also circulating in plasma, unless the body suffers from iron deficiency. Therefore, plasma levels of ferritin have sometimes been used as an index for the assessment of the iron body store. In some plants, structurally similar metalloproteins called **phytoferritins** are present. The amino acid composition of the phytoferritins shows some similarities to those of mammalian apoferritins. Although the quaternary structure is preserved, phytoferritins can store 1.2–1.4 times more iron than mammalian ferritins. Haemosiderin is an amorphous material, a complex of ferritin, denatured ferritin, and other material, containing 35% iron in the form of ferric hydroxide. It is most commonly found in macrophages and is especially abundant in situations following haemorrhage, suggesting that its formation may be related to phagocytosis of red blood cells and haemoglobin. Haemosiderin is accumulated in different organs during some diseases.

Low-molecular-weight iron chelates called **siderophores** are excreted by many aerobic and facultative anaerobic microorganisms growing in environments with low iron content. Bacteria, such as pathogenic bacteria of the genera *Escherichia*, *Aerobacteria*, and *Salmonella*, mainly contain catechol siderophores that are important for their acquisition of iron; aerobic and facultative anaerobic microorganisms (some yeasts and fungi) contain siderophores based on hydroxamic acids. A catecholate siderophore enterobactin is primarily found in Gram-negative bacteria, such as *Escherichia coli* and *Salmonella typhimurium*. Examples of the second group are ferrichrome and coprogen (**6-11**), which are produced by fungi of the genera *Aspergillus*, *Ustilago*, and *Penicillium*. Coprogen occurs in some microbial-processed cheeses, as *Penicillium roqueforti* and *Penicillium camemberti* are used to ripen blue and Camembert cheeses, respectively.

ferric enterobactin   ferrichrome   coprogen

6-11, structures of siderophores

Other types of biologically important compounds are proteins with iron and sulfur. In these compounds, iron is bound to sulfhydryl groups of cysteine residues or even to sulfide ions. This group of compounds includes the FeS-proteins **rubredoxins** and the Fe_2S_2-proteins and Fe_4S_4-proteins **ferredoxins**. These substances act as electron carriers via reversible change of the iron valency. Proteins with iron and sulfur are found in many organisms, including aerobic and anaerobic bacteria, algae, fungi, higher plants, and animals. Their main biological function is electron transfer. An important non-haem metalloenzyme with bound iron that occurs in prokaryotic organisms, higher plants, and animals is lipoxygenase, an enzyme catalysing the peroxidation of unsaturated fatty acids.

Iron absorption occurs predominantly in the duodenum and upper jejunum. The digestive tract absorbs 5–35% of iron present in the diet, depending on biological (health status, age, and sex) and chemical (forms of dietary iron and diet composition) factors, but the overall iron retention is lower (c. 5–15%), as a part of the absorbed iron is lost. The bioavailability of iron decreases in the series haem > Fe^{2+} > Fe^{3+}.

In a state of iron deficiency, the body is able to take up higher portions of dietary iron. Regulation mechanisms (e.g. enhanced or suppressed expression of ferroportin; see later) can increase or decrease the rate of iron absorption according to the current state of iron body stores. The mechanisms of haem iron absorption are completely different from those of inorganic iron salts; the process is more efficient (10–20% of iron intake) and occurs independently of duodenal pH. Dietary inorganic iron can be absorbed in the form of divalent ions.

During the digestion of food, a partial reduction of trivalent iron to divalent iron occurs. The reduction of ferric ions to ferrous ions is partly the result of some chemical components of partially digested food, but mainly of the ferric reductase enzyme (duodenal cytochrome *b*) immobilised on the outer membrane of the mucosal cells of the duodenum (enterocytes). The absorption of ferrous ions occurs in the duodenum and jejunum. Transport from the intestinal contents into enterocytes through the cell membrane is facilitated by a protein called the divalent metal transporter (DMT1), which transports some kinds of divalent metallic ions across the membrane into the cell. DMT1 works as a metal-proton symporter. The early phases of haem iron absorption are different. Haem iron is absorbed by the intestinal mucosa as a porphyrin complex in the presence of a specific carrier called haem carrier protein 1 (HCP1). Within the mucosal cells, haem is transformed into biliverdin and carbon monoxide by haem oxygenase, which results in liberation of Fe^{2+} ions. Enterocytes can store the iron so taken in the form of ferritin, or can move it into the body. Before binding to apoferritin, ferrous ions require oxidation to ferric ions. Inside the enterocyte, the oxidation is catalysed by the enzyme hephaestin, which is a metalloprotein containing copper. Membrane transfer of divalent iron from enterocytes into the blood circulation is facilitated by a membrane protein called ferroportin. The number of ferroportin channels in the membrane is controlled by hepcidin, a polypeptide hormone synthesised in the liver. The efficiency of the overall process is diminished by the life cycle of the enterocytes, which lasts only a few days (iron stored within the 'worn-out' enterocyte sloughs off intestinal wall, then passes back into the intestinal lumen).

Ferrous ions transferred into the blood are oxidised to ferric ions under the catalysis of ferro oxidase ceruloplasmin, the main plasma metalloprotein containing copper. Ferric ions are then taken by apotransferrin molecule to create transferrin, which ensures the transfer of iron to all tissues. Under normal conditions, about 30% of plasma transferrin is saturated with iron; the remainder is apotransferrin. The target tissue captures transferrin by specific receptors, and iron is immediately available for the synthesis of proteins and other haem metalloproteins or is temporarily stored in ferritin. A substantial portion of the transported iron from transferrin is taken away in the bone marrow for the production of erythrocytes. New erythrocytes absorb the whole transferrin molecules. The release of iron from transferrin occurs due to a lower pH compared with the pH in the extracellular space. Apotransferrin is then released from the erythrocytes and iron is built into the porphyrin skeleton under the catalysis of ferrochelatase. The cells of the reticuloendothelial system, which are located in the liver, spleen, and bone marrow, capture old and damaged erythrocytes. In these cells, the iron liberated from haem is stored in ferritin or haemosiderin, or is released into the plasma as Fe^{2+} ions through the ferroportin channel, where they are oxidised to Fe^{3+} and again captured in transferrin. Effective absorption is not only dependent on the iron valence, but can be enhanced or diminished in the presence of certain food components.

The most important substances that enhance iron absorption from the diet are:

- ascorbic acid, which acts as a reducing and chelating agent;

- organic acids (citric, lactic, malic, succinic, and tartaric acids);

- amino acids, especially histidine, lysine, and cysteine, which act as triple ligands, and peptides and proteins composed of these amino acids;

- carbohydrates, which positively influence the retention of iron (efficiency decreases in the order: lactose, sucrose, glucose, starch).

The mechanism of these effects probably lies in the formation of complexes with iron that prevent the formation of insoluble forms of trivalent iron $Fe(OH)_3$ and $FePO_4$ in the alkaline environment of the small intestine.

Most dietary factors negatively influencing iron absorption probably exert their action within the gastrointestinal lumen by making iron more or less bio-accessible for absorption. Substances that reduce iron absorption form either insoluble iron compounds (phytic acid) or stable soluble compounds that may release iron in the insoluble trivalent form (ferritin). The diminished absorption of iron is mainly caused by tannins and phenolic compounds, phytic acid, fibre, high doses of calcium and phosphorus, and extremely high doses of trace elements (cobalt, zinc, copper, and manganese).

Phenolic substances of tea are powerful factors that reduce iron absorption. Even in the presence of ascorbic acid, the absorption is diminished due to the formation of insoluble complexes with tannins. Non-haem iron absorption is reduced by up to 62 and 35% when food is administered simultaneously with tea and with coffee, respectively. In contrast, orange juice increases the absorption of iron by up to 85%.

There is no consistent information on the influence of fibre on the absorption of iron and other minerals. Many studies have focused on the effect of fibre added to food or of food with a high proportion of fibre. Cereal fibre contains fairly high amounts of phytic acid. Fibre itself does not affect iron absorption, but the combined effect of fibre and phytic acid is considerable. This illustrates, for example, a low utilisation of iron from foods with high contents of phytic acid, such as beans and lentils. Fruit fibre does not contain phytic acid; therefore, reduced iron retention was found only when the highly esterified pectin fraction was removed from the apple pomace. Iron absorption is also negatively affected by certain proteins, such as soy protein and phosphoprotein phosvitin, which is present in egg yolk.

6.3.6.2 Occurrence in foods

The main form of iron in animal tissues is haem (particularly myoglobin and haemoglobin). In egg white, iron is bound in conalbumin, and in the yolk, to phosphoprotein phosvitin. Milk contains the iron metalloprotein lactoferrin. Conalbumin and lactoferrin are structurally similar to serum transferrin. In plants, iron is bound in various complexes, especially with phytic acid, aliphatic hydroxycarboxylic acids, amino acids, thiols, phenolic substances, nucleotides, peptides, and proteins. The iron content in selected foods is shown in Table 6.8. Iron-rich foods include offal dishes, meat, eggs, pulses, tea, and cocoa. Moderate amounts of iron are found in fish, poultry, cereals, spinach, parsley, and nuts. Low levels of iron are present in milk, dairy products, fats and oils, potatoes, and most fruit.

For food fortification with iron, many compounds of iron and elemental iron are used. In the evaluation of the biological value of the individual compounds, the increased level of haemoglobin (which cannot be formed during iron deficiency) in experimental animals as a response to iron dose is used as a criterion. Compared with ferrous sulfate as a reference compound, the usable forms of iron can be divided into three groups:

- sources of iron with a relatively good biological value (over 70% compared with $FeSO_4$): ammonium ferric citrate, ferric chloride, ferric sulfate, ferric ammonium sulfate, ferrous fumarate, ferrous gluconate, ferrous sulfate, and ferrous tartrate;

- sources of iron with a mean biological value (20–70%): elemental iron;

- sources of iron with a low biological value (<20%): iron oxide, ferric phosphate, and ferrous carbonate.

In many countries, fortified milk products for infants contain ferrous fumarate. Other products that are often fortified with iron are cereal products.

Table 6.8 Iron and zinc contents in some crops and foods.

Food	Content (mg/kg)		Food	Content (mg/kg)	
	Fe	Zn		Fe	Zn
Pork meat	10–20	17–40	Beans	59–82	21–38
Beef meat	22–30	30–43	Soybeans	50–110	29–67
Chicken meat	4.3–8.4	8.1–12	Cabbage	3.1–9.0	1.5–2.9
Pork liver	130–370	56–112	Cauliflower	5.0–11	3.2–7.8
Fish	1.3–15	3.3–27	Spinach	10–40	4.3–13
Milk (whole)[a]	0.35–0.8	3.4–4.7	Lettuce	5.8–11	3.3–9.0
Curd	0.91–1.5	13–14	Tomatoes	2.2–5.0	1.2–4.8
Cheese	1.5–4.7	36–44	Carrot	3.4–7.4	2.5–5.9
Yoghurt	0.44–1.2	5.3–5.6	Peas (green)	18–22	11–15
Egg (whole)	21–26	13–15	Onion	3.0–6.1	3.1–5.2
Egg (white)	1.0–2.0	2.0	Potatoes	3.0–8.4	1.7–4.9
Egg (yolk)	61–72	38	Apples	2.3–4.8	0.2–4.9
Wheat	33–66	26–38	Oranges	1.3–5.0	0.9–1.2
Rye	25–28	22–40	Bananas	3.1–5.5	1.8–2.6
Flour (wheat)	12–25	8–36	Strawberries	3.6–9.6	1.1–1.9
Bread (whole-wheat)	24–33	13–29	Walnuts	21–24	24
Rice (peeled)	6.0–23	10–15	Tea (black)	110–310	23–38
Peas (mature seeds)	47–68	20–49	Coffee (roasted)	41	6.1–8.0
Lentils	69–130	28–32	Milk chocolate	11–19	18–19

[a]Breast milk concentrations: Fe 0.3–0.7, Zn 1.2 mg/kg.

6.3.6.3 Nutrition

The recommended daily intake of dietary iron is 6 mg for children under 6 months, 10 mg for children aged 6 months to 10 years, 12 mg for boys aged 11–18 years, 15–18 mg for girls and women aged from 11 to 50 years, and 10 mg for adult males and women aged over 50 years. The recommended dose for pregnant women is 30 mg, and that for lactating women is 15 mg. These values must be assessed with regard to the bioavailability of the iron. They were designed for healthy individuals, provided that the composition of the diet ensures good bioavailability of iron (the absorption efficiency of about 15%). This is only guaranteed by a diet containing a sufficient proportion of meat. The daily diet should provide an adult woman (weight 60 kg) with about 2.6 mg of absorbed iron, and an adult male (weight 80 kg) with about 1.4 mg. Taking into account the very low bioavailability of iron from the diet of vegetarians (usually 5%), which does not contain haem iron and is rich in fibre, phytates, and plant phenols, the total daily dose of iron for vegetarians should be about 50 mg for women and 30 mg for men.

Insufficient dietary intake of iron leads to a hypochromic microcytic anaemia and immune deterioration. In this type of anaemia, the amounts of haemoglobin and red blood cells are reduced, which results in reduction of transport of oxygen to the tissues and a decrease of body performance. Severe anaemia can lead to heart failure. Iron deficiency can be divided into three stages. In the first, iron stores are reduced and decreased plasma ferritin levels occur, but functional changes are not noticeable. In the second, the proportion of protoporphyrin in erythrocytes increases and the haemoglobin level only slightly decreases (the normal level of haemoglobin in blood ranges from 130 to 160 g/l); however, in plasma, a strong decrease in total iron, transferrin saturation, and total ferritin occurs, and simultaneously tissue iron content decreases. In the third stage, the symptoms of iron-deficiency anaemia appear: tissue iron stores are depleted, which leads to a significant decrease in haemoglobin levels (hypochromia, also called hypochromasia or hypochromatism) and to erythrocytes that are smaller than normal (microcytosis). Anaemia is more often encountered in women than in men and is often related to inadequate nutrition, especially in developing countries. There are several other types of anaemia, produced by a variety of underlying causes.

Excessive iron intake through food or dietary supplements (>1000 mg/day), particularly in the event of a disorder of iron absorption regulation, may cause an accumulation of haemosiderin in the liver (haemosiderosis), resulting in liver damage or other unwanted effects. The disorder of iron absorption known as hereditary haemochromatosis is a genetic disease. Iron overload can also appear as a result of repeated blood transfusion.

6.3.7 Zinc

6.3.7.1 Biochemistry and physiology

The adult human body contains from 1.4 to 3.0 g zinc. Roughly a half of the total body zinc is contained in muscles and about a third in bones. High concentrations of zinc are found mainly in skin, hair, nails, eye, liver, kidney, spleen, and male genital organs. In the cells of the liver, kidney, and some other organs, zinc is bound in metallothioneins.

Human blood contains 6–7 mg/l zinc, of which 75–88% is located in erythrocytes and 12–22% in blood plasma (plasma zinc concentration is about 1 mg/l). In blood plasma, the majority of zinc is bound to serum albumin and a smaller part to α_2-macroglobulin. The red blood cells contain zinc, mainly in the enzyme carbonate anhydratase.

Zinc is found in the bodies of all organisms. Hundreds of zinc-containing metalloenzymes are known. The presence of zinc in the molecules of some metalloenzymes is essential for their catalytic function. In other metalloproteins, zinc binding is involved in the fixation of the spatial structure of molecules (e.g. in protein transcription modulators, known as zinc-finger proteins, with a specific bond to certain sections of DNA). Well-known mammalian enzymes containing zinc are alcohol dehydrogenase, lactate dehydrogenase, superoxide dismutase (SOD), carboxypeptidases A and B, and dipeptidase or carbonate anhydratase. Important zinc-containing enzymes in the metabolism of yeast are phosphoglucomutase, phosphomannose isomerase, aldolase, alcohol dehydrogenase, and pyruvate carboxylase. Bacterial zinc-containing enzymes include alkaline phosphatase, RNA-polymerase, DNA-polymerase, and various proteinases. Zinc is therefore involved in the catalysis of reactions in many metabolic pathways. Zinc is also needed for the formation and activity of the pancreatic peptide hormone insulin. Human insulin is a peptide composed of 51 amino acids (5808 Da). The active form is a monomer, and six insulin molecules are assembled in a hexamer and stored in the body by means of zinc ions bound by histidine residues.

The absorption of zinc in the digestive tract occurs mainly in the duodenum, but also in other parts of the small intestine. Absorption efficiency is normally about 30%, but it varies in a very wide range (from <15 to 55%) according to the dietary composition. Divalent cations are involved in the transport of zinc into the intestinal cells (like in iron transport). In enterocytes, zinc is bound in metallothionein (at high doses in the diet) or in a complex with cysteine-rich intestinal protein (CRIP) (at lower doses). The absorption of zinc is higher in individuals with lower body weight and in cases of lower body zinc saturation. In contrast, high oral zinc doses decrease absorption efficiency.

The composition of diet strongly affects zinc absorption efficiency. A high content of proteins and amino acids increases the rate of absorption. Phytic acid and fibre operate in the opposite way. It was found that the molar ratio of phytate to zinc is a measure of the bioavailability of zinc from different foods. For example, the bioavailability of zinc is good (≥35%) from a diet low in fibre and phytic acid (ratio phytate: Zn ≤5), which is one with virtually no whole-grain cereal products and an appreciable portion of meat. Intermediate zinc utilisation (15–35%) is achieved in a mixed diet consisting of foods from plant and animal origins, where the ratio of phytate to Zn is

Table 6.9 Zinc and phytic acid contents of some wheat products and corresponding phytate/Zn ratios.

Product	Zn (mg/kg)	Phytate (%)	Phytate : Zn (mol/mol)
Wheat flour (white)	10.2–16	0.21–0.34	14–29
Wheat flour (whole)	29	0.68–0.92	23–31
Wheat germ	123	1.64–2.24	13–18
Wheat bran	73	1.98–2.69	27–37

approximately 5–10. Long-term intake of foods with a ratio of phytate to Zn greater than 20 leads to a zinc deficiency. Table 6.9 shows the ratios of phytate to Zn for some cereal products. Another factor reducing the bio-availability of zinc is an excess of some other mineral elements. High calcium content (mainly achieved by dietary supplements) decreases absorption of zinc by competing with it and, together with phytate, decreases zinc solubility in the intestinal contents. Zinc is excreted from the body via faeces. Faecal losses comprise both the zinc that was not absorbed from the diet and the zinc expelled into the small intestine in secreted bile and pancreatic juice.

6.3.7.2 Occurrence in foods

The zinc contents of selected foods are shown in Table 6.8.

6.3.7.3 Nutrition

The recommended daily dietary intake of zinc is 5 mg for children under 1 year, 10 mg for children from 1 to 10 years, 15 mg for boys and men, 12 mg for girls and women, and 10 mg for men and women over 50 years. The recommended dose for pregnant women is 15 mg, and that for lactating women is 16–19 mg. In diets containing a high proportion of substances that deteriorate the bioavailability of zinc, the recommended amounts are not sufficient.

Zinc deficiency can occur during long-term acceptance of a low-dose zinc diet or of dietary components that reduce the bioavailability of zinc. This can be especially dangerous in childhood, where zinc deficiency results in slow growth and poor development of sexual organs in males. These effects were observed in some Arab countries, in cases where the dominant component of the diet was whole-wheat bread prepared from unfermented dough. Other symptoms of deficiency can include loss of appetite and changes in the skin, hair, and nails. Many of these changes are reversible, and administration of higher doses of zinc may restore the normal state.

Zinc is toxic at higher doses. Oral administration of more than 2 g Zn (as $ZnSO_4$) causes irritation of the mucous membranes of the digestive tract and vomiting. Such a high dose of zinc is impossible to be received from a diet. Long-term intake of amounts 10–30 times higher than the recommended daily intake (100–300 mg) leads to some changes in the blood count, which are typical for copper deficiency, as zinc is an antagonist of copper.

6.3.8 Copper

6.3.8.1 Biochemistry and physiology

The adult human body contains about 100–180 mg copper, corresponding to about 1.7 mg/kg body weight. The concentration of copper in the bodies of babies is substantially higher, at about 4.7 mg/kg body weight. Average concentrations of copper in different tissues and organs of the human body are as follows: liver 15 mg/kg (newborns up to 230 mg/kg), kidney 2.1 mg/kg, muscles 0.7 mg/kg, brain 5.6 mg/kg, and lungs 2.2 mg/kg. In liver cells, most copper is bound in the enzyme superoxide dismutase (SOD, M_r = 32 kDa).

The average concentration of copper in blood is 1.10 mg/l in men and 1.23 mg/l in women. More than 90% of copper in the body is found in the blood plasma. The major copper-binding substance in blood plasma is monomeric glycometalloprotein **ceruloplasmin**, from a group of α_2-globulins, which is composed of 1046 amino acids and contains about 7–8% of sugar components (132 kDa). Ceruloplasmin has a blue colour, and in the normal state, one molecule contains six atoms of copper (at full saturation, additional binding sites may be occupied by an additional two copper atoms). Blood plasma contains about 300 mg/l of ceruloplasmin. In erythrocytes, copper occurs in another protein called erythrocuprein (31 kDa) and in the enzyme SOD.

Copper is an essential trace element for humans and other animals. Copper ions are included in the active centres of many enzymes, especially cytochrome *c* oxidase, SOD, various aminoxidases (such as lysyl oxidase), hydroxylases (e.g. dopamine β-hydroxylase and tyrosinase), galactose oxidase, and different phenoloxidases, such as laccase and other oxidoreductases. The so-called blue copper proteins, such as plastocyanin, azurin, and plantacyanin, occur in many prokaryotic organisms and plants. Through a change of copper valency, these proteins provide electron transfer in various redox processes.

In some invertebrates, oxygen is transported in the blood by the cupric metalloprotein **haemocyanin** (oxyhaemocyanin). Haemocyanin contains two copper atoms (Cu^+) linked by three bonds with histidyl residues. One molecule of oxygen in oxyhaemocyanin forms a bridge between two atoms of Cu^{2+}. **Cytochrome *c* oxidase** is an enzyme that requires haem and copper. It catalyses the final reaction of the respiratory chain (the electron transport from the cytochrome system to an oxygen molecule):

$$O_2 + 4 \text{ ferrocytochrome } c\,(Fe^{2+}) + 4\,H^+ \rightarrow 2\,H_2O + 4 \text{ ferricytochrome } c\,(Fe^{3+})$$

The enzyme SOD is essential for protecting subcellular structures from damage by oxidative reactions (free radicals), as it catalyses the disproportionation of superoxide anion radicals ($O_2^{\bullet-}$), which are toxic products of oxygen metabolism (see Section 3.8.1.13):

$$2\,O_2^{\bullet-} + 2\,H^+ \rightarrow H_2O_2 + O_2$$

Hydrogen peroxide is then decomposed by catalase or peroxidase. There are many different SODs, and all are metalloenzymes. The most common enzymes are SODs containing copper and zinc (CuZnSOD), but ones containing bound manganese (MnSOD), iron (FeSOD), and nickel (NiSOD) also exist. The course of SOD-catalysed reaction is accompanied by a cyclic change of the bound copper valency. First, the oxidised form (Cu^{2+}) of the enzyme produces oxygen with the first particle of superoxide. Second, the reduced (Cu^+) enzyme reacts with the second particle of superoxide to yield hydrogen peroxide, whilst the enzyme is regenerated:

$$Cu^{2+}ZnSOD + O_2^{\bullet-} \rightarrow Cu^+ZnSOD + O_2$$
$$Cu^+ZnSOD + O_2^{\bullet-} + 2\,H^+ \rightarrow Cu^{2+}ZnSOD + H_2O_2$$

Some aminoxidases provide oxidative deamination of biogenic amines or amino acids. Lysyl oxidase is essential for the integrity of connective tissue, as it catalyses the post-translational oxidative deamination of lysyl or hydroxylysyl residues in proteins, resulting in the loss of ε-amino group and the transformation of the terminal methylene group to an aldehyde group. Consequently, a cross-linking of peptide chains occurs, which makes the structure of some protein molecules (collagen, elastin) more rigid. This process leads to cross-linking of tropoelastin and continues to form desmosine and isodesmosine units (see Section 2.4.5.1.3) and to connect the individual chains to form extremely resistant protein elastin.

Tyrosinase is an enzyme found in melanocytes (melanin producing cells located in the bottom layer of the skin's epidermis) that participates in the pigmentation of skin and hair. This enzyme catalyses the hydroxylation of tyrosine to dopamine and two other reactions in the reactions sequence, leading to the main pigments known as phaeomelanins and eumelanins (see Section 9.3.1.1).

Another Cu-enzyme, α-amidating monooxygenase, is widespread in higher animals. It provides an oxidative elimination of the glycyl residue from the *C*-terminus of some peptide hormones (such as vasopressin, oxytocin, gastrin, or calcitonin) and amidation of the *C*-terminal amino acid. This reaction regulates the activity of the hormone.

Copper is also necessary for the efficient utilisation of iron and for the biosynthesis of some important compounds, such as the enzyme **ceruloplasmin**. Ceruloplasmin is produced in liver and acts in plasma as the main Cu compound, but it is not primarily involved in copper transport to the target organs. Plasma copper is transported mainly bound to albumin and partly in the form of complexes with low-molecular-weight ligands, such as histidine. Ceruloplasmin has the catalytic activity of ferrooxidase; it catalyses the oxidation of Fe^{2+} ions just absorbed to plasma to Fe^{3+} ions, allowing fixation of iron in the transferrin molecule. Copper deficiency, therefore, is similar to iron deficiency and leads to anaemia.

Copper is absorbed mainly in the duodenal part of the digestive tract. The efficiency of absorption is normally around 50%, but it depends on the current saturation of the organism (its deficit leads to a higher efficiency). As a binding agent, metallothionein may play a role in the intake of copper by intestinal epithelial cells. The absorption of copper is achieved by two mechanisms: active transport, which prevails in the copper deficit in the body, and simple diffusion. Both divalent metal transporter 1 (DMT1) and specific copper transporter 1 (Crt1) are involved in copper uptake by enterocytes (Crt 1 transports cuprous ions Cu^+). Copper homeostasis in the adult body is maintained by excretion of a part of the copper pool into intestinal lumen via bile and pancreatic juice. If the copper dietary intake is 1.6 mg/day, the amount absorbed into the blood circulation is about 0.8 mg and the amount excreted back into the intestine is about 0.7 mg. Only minor amounts are excreted via urine and lost in hair, nails, sweat, and desquamated skin.

6.3.8.2 Occurrence in foods

Most foods contain less than 10 mg/kg copper. Milk is especially poor in copper, but the bioavailability of copper from human milk is very high. Higher concentrations of copper are found in the liver (especially in calf liver), followed by legumes and some fungi. These concentrations are very variable. Copper content in selected foods is shown in Table 6.10. Copper naturally present in food is almost always

Table 6.10 Copper, manganese, nickel, cobalt, and chromium contents of some crops and foods.

Food	Content (mg/kg)					
	Cu	Mn	Ni	Co	Mo	Cr
Pork meat	<0.4–1.8	0.12–0.18	<0.01–0.03	<0.001–0.012	<0.1	<0.01–0.09
Beef meat	0.6–1.8	0.10–0.14	<0.01–0.04	0.001–0.02	<0.1	<0.01–0.05
Chicken meat	0.35–0.51	0.14–0.16	<0.02–0.04	<0.01	<0.1–0.14	0.01–0.08
Pork liver	10–23	3.4–4.4	<0.01–0.28	0.002–0.023	2.0	0.003–0.16
Fish	0.2–3.1	0.10–3.1	0.005–0.05	<0.001–0.012	<0.1	0.002–0.23
Milk (whole)[a]	0.05–0.2	0.03–0.09	<0.003–0.03	0.0004–0.0011	0.01–0.07	0.002–0.02
Curd	0.29–0.36	0.2–0.3	0.01–0.03	0.005	0.03–0.05	0.02
Cheese	0.3–19	0.4–0.8	0.02–0.2	0.01	0.05–0.1	0.01–0.13
Yoghurt	0.05–0.14	0.09–0.12	0.004–0.03	<0.005	<0.05	0.005–0.04
Egg (whole)	0.68–0.73	0.36–0.55	0.08	0.001–0.04	<0.05	0.005–0.02
Egg (white)	0.3	0.20	–	–	–	–
Egg (yolk)	1.6	1.0	–	–	–	–
Wheat	4.0–14	35–49	0.05–0.89	0.007–0.089	0.1–0.8	0.007–0.06
Rye	2.8–3.7	25.8	0.16	–	0.45	–
Flour (wheat)	2.0–6.5	7.3–36	<0.01–0.3	0.005–0.09	0.1–0.3	0.010–0.03
Bread (whole-wheat)	3.5	13–21	0.08–0.2	0.01–0.05	<0.2	0.01–0.13
Rice (peeled)	0.6–2.8	5.3–15	0.1	0.01–0.02	0.1–0.3	0.01–0.03
Peas (mature seeds)	4.9–8.5	8.1–15	0.4–3.0	0.013–0.2	0.1–2.6	0.02–0.09
Lentils	5.8–8.9	12–14	2.3–3.0	0.016–0.092	2.0–10	0.048–0.054
Beans	6.0–13	12–20	2.5–5.0	0.01–0.3	1.0–3.0	0.05–0.10
Soybeans	8.0–20	14–90	2.0–10	0.05–0.14	<0.06–10	0.05–0.08
Cabbage	0.3–1.0	1.1–3.6	0.01–0.3	<0.001–0.01	<0.1	0.001–0.03
Cauliflower	0.41–0.64	1.5–3.9	0.03–1.0	0.001–0.01	<0.1	0.001–0.01
Spinach	0.6–1.7	3.5–34	0.05–0.4	0.001–0.02	<0.006–0.10	0.01–0.12
Lettuce	0.4–1.5	1.3–12	0.01–0.3	<0.001–0.006	<0.1	0.005–0.08
Tomatoes	0.4–1.0	0.7–1.6	0.01–0.25	<0.005	<0.005–0.09	0.002–0.01
Carrot	0.37–0.8	1.5–6.9	<0.01–0.09	0.001–0.005	<0.006–0.06	0.001–0.13
Peas (green)	1.9–2.4	3.4–4.3	0.2–0.7	0.002–0.01	0.2	0.005–0.04
Onion	0.35–0.91	1.1–3.8	0.03–0.42	0.001–0.01	<0.006–0.06	0.005–0.02
Potatoes	0.3–1.6	0.9–4.4	0.01–0.26	0.002–0.02	0.01–0.09	0.002–0.035
Apples	0.24–0.63	0.3–4.1	0.004–0.03	<0.001–0.005	<0.1	0.003–0.03
Oranges	0.44–0.91	0.3–0.5	0.01–0.04	0.001–0.01	<0.1	<0.001–0.02
Bananas	0.7–1.6	1.5–3.1	0.01–0.05	<0.001–0.002	<0.1	0.02–0.05
Strawberries	0.54–0.74	1.4–7.5	0.02–0.13	<0.001–0.01	<0.1	<0.002–0.02
Walnuts	3.1	18	9.0	0.008–0.29	<0.2	0.08–0.29
Tea (black)	11–33	320–1040	1.9–12	0.60–1.0	0.13	0.62–2.6
Coffee (roasted)	8.2	15	0.6–1.0	0.34–0.88	<0.2	0.01–0.05
Milk chocolate	4.9	2.2–3.2	0.34	0.34	<0.2	0.04–0.1

[a]Breast milk concentrations: Cu 0.26–0.4, Mn 0.006–0.03, Ni 0.001–0.01, Co 0.0001, Mo 0.002–0.01, Cr 0.0003 mg/kg.

bound in coordination compounds with proteins and low-molecular-weight ligands. The level in foods can, in exceptional cases, be increased by contamination of crops (e.g. grapes or hops) treated with pesticides based on compounds containing copper. Another possibility of contamination by copper may occur when copper containers for food processing of food raw materials are used (e.g. in brewing technology). Copper ions act as a catalyst in the oxidation of labile components of food (L-ascorbic acid and lipids).

6.3.8.3 Nutrition

The recommended daily dietary intakes of copper are 0.4–0.7 mg for children under 1 year, 0.7–2.0 mg for children aged 1–10 years, 1.5–2.5 mg for adolescents, and 1.5–3.0 mg for adults. The absorption of copper depends on the chemical form in which this element is present in the diet. The high content of proteins, amino acids, and hydroxycarboxylic acids in the diet increases the availability of copper. In contrast, higher doses of ascorbic acid, fructose, molybdenum, sulfur compounds, and zinc significantly reduce copper absorption. The effect of phytate and dietary fibre on copper absorption is weaker than in the case of zinc.

Molybdenum and zinc are known as antagonists of copper, so that high dietary intakes of these elements can cause symptoms of copper deficiency. High doses of molybdenum cause a reduced absorption and increased copper urinary excretion. High dietary zinc induces an increased production of metallothionein in enterocytes, which leads to its retention in these cells, so that less copper is released into the bloodstream.

Copper deficiency in humans is very rare. The prolonged low intake of copper results in higher levels of cholesterol in the blood, changes in heart rhythm, and reduced glucose tolerance. Experiments on animals have shown that copper deficiency causes serious disorders of iron metabolism and subsequently hypochromic microcytic anaemia. Other symptoms are movement disorders and changes in skin, hair, nails (impaired pigmentation and formation of keratin), and bones (facile fragility and deformation).

Copper toxicity to mammals is relatively low (LD_{50} value for oral administration of cupric sulfate, $CuSO_4 \cdot 5H_2O$, in rats is 300 mg/kg b.w.). Copper ions are very toxic to fish.

6.3.9 Manganese

6.3.9.1 Biochemistry and physiology

The adult human body contains about 10–20 mg manganese. Higher concentrations are found in the bones (2.6 mg/kg), liver (1.4 mg/kg), and pancreas and kidney (1.2 mg/kg); lower concentrations are found in the brain, spleen, heart, and lungs (approximately 0.2–0.3 mg/kg); and even lower concentrations are found in skeletal muscle (around 0.06 mg/kg). With regard to the subcellular-level distribution, manganese is mostly present in the mitochondria. Biological structures containing keratin (such as bones and skin) are relatively rich in manganese. The manganese content in the blood may range from c. 10 to 70 µg/l, whilst in plasma it is only 0.6 to 4 µg/l. Manganese is mostly contained in blood erythrocytes, probably in the form of a porphyrin complex. In blood plasma, the majority of the trivalent manganese is bound to β_1-globulin. The binding of manganese by most biologically important ligands is weak.

Divalent manganese ions (Mn^{2+}) are necessary for the activity of diverse metallo-enzymes. Oxidoreductases catalyse oxidation and reduction (redox) reactions associated with changes in manganese valence (Mn^{II} vs Mn^{III} or Mn^{III} vs Mn^{IV}). SODs of some microorganisms and some animals contain Mn^{3+}/Mn^{2+} ions instead of copper and zinc ions as cofactor (see Section 3.8.1.13.2). Catalases of some microorganisms (e.g. the bacterium *Lactobacillus plantarum*) contain manganese instead of haem iron. In addition to participation of manganese in enzyme-catalysed oxidation–reduction processes, divalent manganese ions themselves can function as antioxidants because they scavenge peroxyl radicals formed in the course of lipid oxidation.

The most important manganese-containing enzymes belonging to other enzyme classes than oxidoreductases are pyruvate carboxylase and arginase. Both contain four Mn^{2+} ions per molecule. In the presence of biotin as a cofactor, pyruvate carboxylase catalyses the reaction pyruvic acid + CO_2 + ATP + H_2O → oxaloacetic acid + ADP + H_3PO_4. Oxaloacetic acid is an intermediate of the citric acid cycle and carbohydrate biosynthesis (gluconeogenesis).

Arginase is an enzyme that catalyses the hydrolysis of arginine to urea and ornithine. This reaction represents the terminal step of the urea (ornithine) cycle, which in mammals provides (in the form of urea) excretion of ammonia nitrogen released during decomposition of amino acids. The enzyme molecule is composed of two subunits, each containing two Mn^{2+} ions.

In addition to true metalloenzymes, which contain manganese, there are numerous enzymes activated by manganese ions. They include various hydrolases, kinases, decarboxylases, and glycosyltransferases. Manganese may also be bound, instead of magnesium, in enzyme molecules in which the catalytic function requires magnesium, whilst enzyme activity is maintained. In glutamine synthetase, the bound manganese probably has the function of fixing the molecular structure and is not involved in the catalytic function.

Manganese is also important in the light reactions of photosynthesis in plant chloroplasts. The oxygen evolving complex is a part of photosystem II, occurring in the thylakoid membranes of the chloroplasts, where the terminal photooxidation of water takes place. Photosystem II is the oxidoreductase complex containing a cluster of four atoms of Mn and one Ca atom. Chlorophyll *P680*, present in this photosystem, absorbs light and transfers an electron to a pheophytin molecule, from which it is further transferred by other electron carriers to plastoquinone. The resulting electron hole attracts electrons. Chlorophyll *P680* thus acts as a strong oxidant, ensuring the oxidation of O^{2-} to elemental oxygen through changes in the oxidation state of manganese. The reaction takes place as a cyclic process amongst five stages of the transelectronase, referred to as S_0 to S_4. In stage S_0, all four Mn atoms are in oxidation state III. Through four successive losses of electrons, stage S_4 corresponds to all four Mn atoms in oxidation state IV. In the final step, two water molecules yield one molecule of oxygen, and regenerated transelectronase is again in stage S_0.

Absorption of manganese from food takes place in all parts of the small intestine. The efficiency of manganese absorption in adult humans is about 3–4%. Experiments with animals showed that manganese absorption from food containing various manganese compounds of different solubilities (MnO, $MnCO_3$, $MnSO_4$, and $MnCl_2$) differed greatly. Although the mechanism of manganese absorption is not clear, there are some similarities with the metabolism of iron. The absorbed manganese is transported in blood plasma bound to transferrin, albumin, and α_2-macroglobulin. The binding of manganese to apotransferrin requires oxidation of Mn^{2+} to Mn^{3+}, which is probably provided by ceruloplasmin. The concentration of manganese in the cytoplasm of primary cells is approximately 1.10^{-7} mol/l. Significantly higher amounts are found in some organelles.

The absorption of manganese is increased in the presence of low-molecular-weight ligands (such as citric acid or histidine). High dietary intakes of iron, calcium, and phosphate can reduce the efficiency of manganese absorption. In contrast, high doses of manganese reduce the absorption of iron and consequently the haemoglobin level. There is also a competition in the absorption of manganese and cobalt.

The efficiency of manganese absorption is high if the diet contains low quantities of manganese and iron (e.g. in case of milk). The absorption of manganese from breast milk is higher than that from cow's milk, likely due to the different localisation of manganese in these fluids: in human milk, the majority of manganese is in the whey fraction, whereas in cow's milk, manganese is mostly contained in the casein fraction. Manganese from meat and fish is better absorbed than that from legumes.

The major route of excretion of manganese is through the bile excretion. The amount of manganese excreted in the urine is negligible.

6.3.9.2 Occurrence in foods

Food of animal origin contains low levels of manganese. Good sources are cereals and legumes. Some berries have high concentrations. For example, raspberries contain 6.7–18 mg/kg manganese and blueberries 23–48 mg/kg. Tea leaves and some spices have particularly high levels (e.g. cloves have about 600 mg/kg manganese, cardamom 320 mg/kg, and ginger 160 mg/kg). The manganese contents of selected foods are shown in Table 6.10. The major portion of manganese in food is in ionic form (especially as Mn^{2+}) or in the form of labile complexes.

6.3.9.3 Nutrition

Newborns fed on mother's milk need an extremely low intake of manganese (c. 3 µg daily). Adequate and safe daily dietary intakes are 0.3–1.0 mg for children under 1 year, 1.0–2.0 mg for the children aged from 1 to 10 years, c. 1.7 mg for adolescent girls and adult women, and c. 2.3 mg for adolescent boys and adult men. Higher intakes are recommended for pregnant and lactating women (2.0 and 2.6 mg, respectively).

Long-term deficiency in the diet can result in slow growth, abnormal bone development, and impaired reproductive function. In neonates, deficiency may result in movement disorders. Owing to the influence of manganese on carbohydrate metabolism, deficiency may result in a reduced ability to synthesise and utilise glucose. Manganese deficiency in experimental animals causes a decrease in pancreatic insulin secretion and reduced insulin activity in peripheral tissues, as well as changes in activity of pancreatic amylase.

Toxic effects of manganese (growth retardation, neurological symptoms, mental disorders, and anaemia) occur only at very high doses or with long-term inhalation exposure. Poisoning by manganese from food is almost impossible.

6.3.10 Nickel

6.3.10.1 Biochemistry and physiology

The body of an adult contains about 10 mg nickel. Data on the nickel content in various human tissues and organs are not entirely consistent. It can be influenced by many factors (such as age, gender, environment, and smoking). The concentration of nickel in lungs, kidneys, liver, heart, and bones is 85 ± 65 µg/kg, 10.5 ± 4.1 µg/kg, 8.2 ± 2.3 µg/kg, 6.4 ± 1.6 µg/kg, and 333 ± 147 µg/kg, respectively. The normal concentration of nickel in the blood is about 5 µg/l; in serum, it is about half of this amount. Serum nickel is bound to different proteins: albumin, histidine-rich glycoprotein, **nickelplasmin** (700 kDa), α_1-glycoprotein, and, in part, low-molecular-weight substances (mainly histidine).

To date, there is no known specific biochemical function for nickel in animal organisms. In plants and microorganisms, some metalloenzymes containing nickel have been found. One such enzyme is urease, which was first isolated from Jack bean (*Canavalia ensiformis*) seeds and has also been found in soybeans, other legumes, rice, and tobacco. The molecule of urease (580 kDa) contains 12 nickel atoms in six subunits. In the active enzyme centre, two nickel atoms are coordinated by four histidyl residues and one residue of ε-N-carbamoyl lysine; the carbamoyl group and one molecule of water form a bridge between them. Urease catalyses the hydrolysis of urea to ammonia and carbamate (carbamic acid), which is then hydrolysed spontaneously to hydrogen carbonate (bicarbonate) and ammonium ions.

Some bacteria of the genera *Acetobacterium*, *Alcaligenes*, *Clostridium*, *Desulfovibrio*, *Methanobacterium*, and *Vibrio* produce a variety of Ni-dependent oxidoreductases. Although no animal enzymes containing nickel are known, the element can act as an activator of some enzymes. For example, calcineurin, which has phosphatase activity against phosphoproteins, is activated by nickel ions. This enzyme contains bound iron and zinc, and its activity is regulated by conformational changes that depend on the binding of other metal ions (Ni^{2+}, Mn^{2+}, and Ca^{2+}) to some places in the molecule. A possible function of nickel is participation in the absorption of iron. Through as yet unknown mechanisms, it also apparently facilitates absorption of iron by conversion of Fe^{3+} into Fe^{2+}.

The effective absorption of nickel from food is usually lower than 10%. Increased absorption (20%) can occur in a state of iron deficiency. Absorbed nickel is excreted from the body mainly in urine.

6.3.10.2 Occurrence in foods

Fruits, cereals, and foods of animal origin, with the exception of some seafood (oysters), have very low nickel content (tens to hundreds of μg/kg). Higher concentrations (up to units of mg/kg) are found in legumes, nuts, tea leaves, cocoa beans, and chocolate. Refined vegetable oils and animal fats usually contain only trace amounts of nickel (units of μg/kg) and of other transition elements. Somewhat higher concentrations (tens to hundreds of μg/kg) are found in margarine, as the solid fats may be manufactured from oils by hydrogenation in the presence of a nickel catalyst. The concentration of nickel in food may increase during storage in metal containers, which has been shown in canned fruits. The nickel contents of selected foods are shown in Table 6.10. Extra nickel can appear in dishes as a result of contamination during culinary preparation of food by cooking in stainless-steel utensils.

6.3.10.3 Nutrition

The recommended daily dietary doses of nickel have not been determined. Actual dose depends on dietary habits, ranging between 150 and 700 μg/day. Animal experiments suggest that a deficit of nickel in the diet leads to slower growth.

Toxic effects of nickel occur only at levels ≥250 mg/kg food. Symptoms of nickel poisoning in animals are changes in fur and diarrhoea. High doses of dietary nickel are accompanied by an increased number of erythrocytes, increased haemoglobin level and serum proteins, increased urea, iron, zinc, copper, and nickel in the liver, increased hepatic activity of glutamate dehydrogenase, and decreased activity of glucose-6-phosphate dehydrogenase. At the same time, the amount of iodine in the thyroid gland decreases. Nickel compounds are also irritating to the skin. Allergy to nickel can appear when people wear cheap fancy jewellery or wristwatches made of nickel alloys. Highly toxic and carcinogenic volatile nickel tetracarbonyl, $Ni(CO)_4$, used as a reagent in organometallic chemistry, is readily absorbed through the lungs.

6.3.11 Cobalt

6.3.11.1 Biochemistry and physiology

The total amount of cobalt in the human body is <1.5 mg. The liver contains approximately 0.11 mg/kg of cobalt, skeletal muscle 0.2 mg/kg, bones 0.28 mg/kg, hairs 0.31 mg/kg, adipose tissue 0.36 mg/kg, and blood about 0.3 μg/l (mostly localised in plasma). The main biologically active compound is vitamin B_{12}.

Cobalt is an essential element for bacteria, algae, and ruminant mammals. For other organisms, including monogastric animals, cobalt is essential, in the form of vitamin B_{12} (cobalamin). Rumen microflora of ruminants synthesises cobalamin from cobalt in the diet. Derivatives of vitamin B_{12} cobamides are cofactors of some enzymes, including methylmalonyl coenzyme A mutase, glutamate mutase, and methionin synthetase.

In the gastrointestinal tract of humans, 20–97% of cobalt present in the diet is absorbed. For the absorption of vitamin B_{12}, a glycoprotein known as the intrinsic factor is essential, as it allows the transport of cobalamin into cells in the intestinal mucosa. The effectiveness of cobalt absorption increases with iron deficiency. Cobalt is excreted from the body mainly in the urine.

6.3.11.2 Occurrence in foods

The richest sources of cobalt in the diet are beans and offal. Poor sources are milk and dairy products and white flour products. Rich sources also include tea, coffee, and chocolate. Cobalt content in selected foods is shown in Table 6.10.

6.3.11.3 Nutrition

The recommended dose of cobalt has not been established. Actual daily dietary doses are very low, being estimated at 5–10 µg.

6.3.12 Molybdenum

6.3.12.1 Biochemistry and physiology

The total content of molybdenum in the human body is about 5–10 mg. The molybdenum content in the liver is c. 0.36–0.9 mg/kg, in kidney 0.4 mg/kg, and in brain and muscles 0.03 mg/kg. The molybdenum content in blood fluctuates over a very wide range (0.003–0.41 mg/l). Mean concentrations are 0.01–0.07 mg/l.

Molybdenum is an essential element for microorganisms, plants, and even animals, where some enzymes containing molybdenum have been found (e.g. xanthin oxidase, sulfite oxidase, and aldehyde oxidase). Other enzymes containing molybdenum include dimethyl sulfoxide oxidase, formate dehydrogenase from *E. coli*, and arsenite oxidase from *Alcaligenes faecalis*. These enzymes contain molybdenum cofactor **molybdopterin** (see Section 6.2.2.8). Xanthin oxidase also contains, besides flavin adenine dinucleotide cofactor (FAD), two clusters of iron and sulfur in each of the two subunits. Xanthin oxidase catalyses the oxidation of hypoxanthine to xanthine and the oxidation of xanthine to uric acid:

$$\text{hypoxanthine} + H_2O + O_2 \rightarrow \text{xanthine} + H_2O_2$$
$$\text{xanthine} + H_2O + O_2 \rightarrow \text{uric acid} + H_2O_2$$

Xanthine oxidase is a very important enzyme in the catabolism of purines. In the congenital metabolic disorder xanthinuria, it is missing. The enzyme sulfite oxidase catalyses the oxidation of sulfites to sulfates using oxygen. It occurs in animals, plants, and microorganisms. These oxidation–reduction processes are associated with a change in the oxidation state (from VI to IV and vice versa) of Mo bound in molybdopterin. Dimethyl sulfoxide oxidase is capable of reducing dimethyl sulfoxide to dimethyl sulfide. This enzyme serves as the terminal reductase under anaerobic conditions in some bacteria. Formate dehydrogenases catalyse the oxidation of formate ($H-COO^-$) to hydrogen carbonate (HCO_3^-).

Biologically active forms of molybdenum are also included in the bacteria-mediated process of nitrogen fixation. This consists in the conversion of molecular nitrogen into ammonia, which is then utilised by plants. In legumes and a few other plants, the specific nitrogenating bacteria (of genera *Rhizobium*, *Azotobacter*, *Clostridium*, and *Klebsiella*) live in small growths on the roots called nodules. Within these nodules, nitrogen fixation occurs. Closely related processes participating in the global nitrogen cycle are nitrification (conversion of ammonia to nitrate) and denitrification (back-conversion of nitrate to elemental nitrogen). The nitrification bacteria include those of the genera *Nitrosococcus*, *Nitrosomonas*, *Nitrobacter*, and *Nitrococcus*. The intermediates of nitrification are hydroxylamine and nitrite ions. Minor byproducts include nitrogen oxides, hydrazine, and molecular nitrogen. Some steps in these processes take place under the catalysis of various metalloenzymes.

Metalloenzymes of the just mentioned bacteria called nitrogenases ensure the conversion of elemental nitrogen into ammonia and nitrogen fixation in the soil. The overall reaction catalysed by a nitrogenase proceeds according to the following equation:

$$N_2 + 8\,H^+ + 8\,e^- + 16\,ATP + 16\,H_2O \rightarrow 2\,NH_3 + H_2 + 16\,ADP + 16\,H_3PO_4$$

Nitrogenases containing molybdenum are intermolecular complexes consisting of two parts: MoFe protein and Fe protein. The MoFe protein (230 kDa) has a heterotetrameric structure with an Fe—Mo cofactor and [Fe—S] cluster, whilst the Fe protein (64 kDa) has a homodimeric arrangement of subunits bound by an [4Fe—4S] cluster. In some nitrogenases, molybdenum can be replaced by vanadium or iron.

Nitrate reductase reduces nitrates (NO_3^-) to nitrites (NO_2^-). This enzyme occurs in microorganisms, algae, fungi, and plants, and its molecule is composed of two types of subunit and contains, in addition to molybdenum and haem, non-haem iron. The electron transfer ensuring substrate reduction by cyclic changes in Mo valency is coupled with the cytochrome system. The reduction of nitrate is the first step in the denitrification process, in which some microorganisms use nitrate ions as terminal electron acceptors instead of oxygen. The remaining steps lie in the gradual reduction to nitrogen (II) oxide, nitrous oxide, and elemental nitrogen. A part of the generated nitrous oxide is released into the atmosphere.

The absorption of molybdenum from food is quite effective (25–80%). With the exception of MoS_2, molybdenum is absorbed well from soluble compounds (e.g. sodium and ammonium molybdates) and less soluble compounds (e.g. molybdenum trioxide and calcium molybdate). Absorption of molybdenum is inhibited by sulfate anions. The excess of molybdenum is excreted in urine.

The absorption, metabolism, excretion, and physiological effects of molybdenum strongly depend on interactions with sulfur compounds, which govern the retention of molybdenum in various tissues and determine the tolerance to potentially toxic doses of molybdenum and its effects on the utilisation of copper. In the metabolic antagonism of molybdenum and copper, sulfur compounds mitigate the negative effects of higher molybdenum doses.

6.3.12.2 Occurrence in foods

Relatively high contents of molybdenum are found in pulses (units of mg/kg), moderate amounts are found in whole-grain cereals and offal (tenths to units of mg/kg), and low amounts are found in most vegetables, fruits, fish, meat, milk, dairy products, and fats (<0.1 mg/kg). The molybdenum contents of various foods are presented in Table 6.10.

6.3.12.3 Nutrition

The recommended dietary allowance (in the United States) of molybdenum for adults is 45 µg/day (50 µg/day during pregnancy and lactation). The symptoms of molybdenum deficiency are seen mainly in farm animals in areas with a low molybdenum content in the soil and vegetation.

6.3.13 Chromium

6.3.13.1 Biochemistry and physiology

The total chromium content in the human body is estimated to be 5 mg, which is fairly evenly distributed. The concentration of chromium in individual tissues and organs, except the lungs, decreases with age. The normal content of chromium in blood plasma is 0.1–0.4 µg/l.

Chromium is both beneficial and toxic, depending on its chemical species. Chromium in oxidation state III seems to be essential, whilst hexavalent chromium compounds (chromates and dichromates) are toxic and may be responsible for certain allergenic, mutagenic, and carcinogenic effects.

Trivalent chromium is important in carbohydrate metabolism. It probably facilitates the interaction between sulfhydryl groups in cell membranes and the disulfide groups of insulin, and thus allows insulin-stimulated utilisation of glucose. A complex compound of Cr^{III} with glycine, cysteine, glutamic acid, and nicotinic acid, known as the glucose tolerance factor, was previously considered to be a physiologically active substance containing chromium. This substance, isolated from the yeast biomass, was able to increase glucose tolerance in experimental animals, but later it was shown that it was an artefact produced by acid hydrolysis of the Cr^{III}-oligopeptide complex called **chromodulin**. Chromodulin occurs in the liver and kidney tissues of mammals. The second Cr^{III}-binding oligopeptide was isolated from cow colostrum. The oligopeptide apochromodulin (1.4 kDa), mainly containing glycine, cysteine, aspartic, and glutamic acids, strongly binds four Cr^{III} atoms, forming chromodulin. The peptide from colostrum has a similar amino acid composition. Both peptide complexes (but not apopeptides alone) stimulate glucose metabolism in adipocytes of rats. Chromodulin is incorporated into the amplification mechanism of insulin action. In the presence of insulin, chromodulin increases the activity of insulin receptors in adipocyte membranes. Apochromodulin is probably stored in cells sensitive to insulin.

Chromium compounds also interfere in the metabolism of lipids and proteins. Increasing doses of dietary chromium partially reduce the levels of cholesterol and triacylglycerols in plasma and simultaneously increase the proportion of high-density lipoproteins (HDLs). Chromium compounds are also involved in maintaining the structural integrity of nucleic acids, as chromium protects RNA against thermal denaturation. It accumulates in the cell nuclei.

Inorganic chromium compounds are absorbed very little in the gastrointestinal tract of humans and other mammals. Absorption efficiency ranges from 0.4 to 3% and decreases with increasing chromium dose (about 2% is absorbed at a dose of 10 µg/day and about 0.4% at a dose higher than 40 µg). The absorption of chromium is very fast. Even 15 minutes after administration, an increase in chromium concentration in the blood may be observed, and the same holds after 2 hours in urine. The absorption efficiency of natural chromium complexes is probably much higher. Oxalic acid significantly increases and phytic acid decreases the absorption of trivalent chromium.

The absorption of hexavalent chromium is three to five times more efficient than that of trivalent chromium. In blood, hexavalent chromium enters into blood cells and erythrocytes more easily than trivalent chromium and binds to haemoglobin. The absorbed trivalent chromium is bound in plasma to α-globulin and transferrin. Transferrin provides chromium transport to the tissues. In some target tissues, chromium from transferrin binds to apochromodulin to form chromodulin. In the cells, about 50% of chromium is contained in the nucleus and about 20% in the cytoplasm. Chromium is excreted from the body mainly in urine. Excessive intake of chromium by experimental animals or chromium intoxication also leads to excretion of chromodulin in the urine. Biosynthesis of this metallopeptide thus participates in the detoxification of chromium.

6.3.13.2 Occurrence in foods

Data on the content of chromium in biological materials derived from the years 1950 to 1970 place the chromium levels quite high, but these are now considered incorrect. Improved analytical techniques have enabled the real chromium contents of some materials (including blood plasma and milk) to be determined, which are found to lie in the ultra trace region. Despite this progress in analytical methods, the results of different studies are contradictory. A summary of the results suggests a very wide interval of chromium amounts in various materials. During manufacture and storage, foods come into contact with metallic materials (such as stainless steel) and the initially very low content of chromium (several µg/kg) can be increased significantly due to contamination. A rich source of biologically utilisable chromium is brewer's yeasts. The chromium contents of selected foods are shown in Table 6.10. Chromium contamination of food can occur during cooking in stainless-steel utensils.

Under the assumption that higher doses of chromium in the diet prevent diabetes and can to some extent increase weight loss, chromium compounds (usually complex of Cr^{3+} with picolinic acid) are often part of some food supplements. However, the effectiveness of such supplementation is still questionable.

6.3.13.3 Nutrition

An adequate daily dietary dose of chromium is considered to be 20–35 µg (30 and 45 µg for pregnant and lactating women, respectively). The actual dose of chromium found in several studies is in the range of 25–100 µg. In chromium deficiency, the following symptoms have been identified: impaired glucose tolerance, persistently elevated blood glucose level, elevated blood serum cholesterol and triacylglycerol levels, and the presence of carbohydrates in urine. There is thus a link between chromium deficiency and the development of diabetes and cardiovascular diseases. Nerve and brain disorders have also been recorded in connection with the lack of chromium.

Toxic effects of Cr^{3+} compounds are manifested only at high doses (much more than 1 mg/day). In contrast, hexavalent chromium is much more toxic (it causes liver and kidney damage), but the chromium in the hexavalent state does not occur in food. Nevertheless, some other means of exposure to hexavalent chromium can be dangerous. Chromate contact with the skin may cause dermatitis. Chronic exposure to dust containing chromates increases the risk of lung cancer development.

6.3.14 Vanadium

6.3.14.1 Biochemistry and physiology

The known levels of the transition metal vanadium in the human body are unreliable. Natural concentrations of vanadium in various tissues and organs are very low, but these may substantially increase on exposure to vanadium compounds. For example, the vanadium content in the kidney is in the range of 1–140 µg/kg, that in liver is 3–110 µg/kg, and that in muscles is 10–110 µg/kg. The vanadium concentration in blood plasma is 0.02–1.3 µg/l.

The biological functions of vanadium have not yet been defined clearly. However, vanadium has an effect on the activity of important enzymes, especially the inhibition of (Na, K)-ATPase, other ATPases, phosphatases, and phosphotransferases by metavanadate ions (VO_3^-). Vanadium also stimulates the synthesis of cyclic AMP by adenylate cyclase activation and may therefore interfere with the metabolism of carbohydrates and lipids.

In some species of algae, seaweed, lichen, and fungi, vanadium is an essential constituent haloperoxidase enzymes that catalyses the oxidation of Cl^- and Br^- by hydrogen peroxide.

In bacteria and plants, tetravalent vanadium interferes with iron transport mediated by siderophores, as vanadium has high affinity to the corresponding ligand groups.

In the human gastrointestinal tract, only 0.1–1% of vanadium contained in the diet is absorbed. For some laboratory animals, however, fairly high absorption of vanadium from the diet with a predominance of casein and carbohydrates has been observed. Vanadium absorbed in the form of metavanadate ions (VO_3^-) containing V^V is reduced by glutathione in the blood to VO^{2+} ions that contain V^{IV} and form 2 : 1 metal–protein complexes with ferritin and transferrin. Transferrin then obviously provides the distribution of vanadium in the tissues. Excessive intake of vanadium is excreted in the urine.

6.3.14.2 Occurrence in foods

The vanadium content of most foods is extremely low. Milk, fats, vegetables, and fruits contain <1–5 µg/kg of vanadium; whole-grain cereals, meat, fish, and liver contain 5–40 µg/kg. According to some sources, marine crustaceans and molluscs (e.g. oysters) have a higher content of vanadium, above 100 µg/kg. The marine animals called ascidians (class Ascidiacea, subphylum Tunicata, phyllum Chordata) concentrate considerable amounts of vanadium (c. 10 mg/kg) from seawater. In their bodies, vanadium is bound in metalloproteins called vanabins (or haemovanadins). High vanadium content is typical for fungi of the genus *Amanita* (Amanitaceae) (36–250 mg/kg d.w.), such as the toxic fly agaric (*A. muscaria*). These toxic toadstools contain a blue coloured complex (1 : 2) of V^{4+} cations with (2S,2'S)-N-hydroxyimino-2,2'-dipropionic acid, called **amavadin** (**6-12**). Amavadin is a stable compound that may act as a reversible redox system ($V^{IV} \rightarrow V^V$) coupled with peroxides.

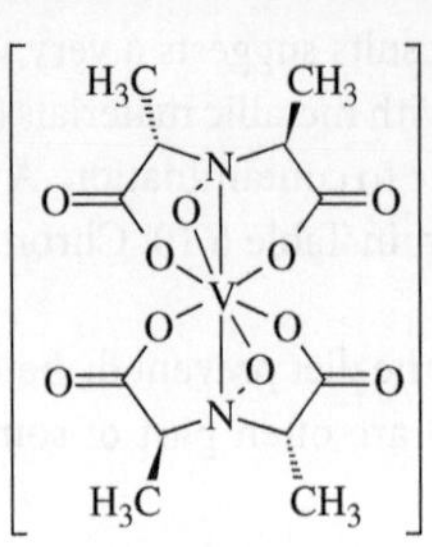

6-12, amavadin

6.3.14.3 Nutrition

The recommended daily intake of vanadium has not been determined. The actual food intake is estimated at 10–30 μg.

6.3.15 Selenium

6.3.15.1 Biochemistry and physiology

The body of an adult might contain c. 15 mg of selenium, but the actual content depends on the selenium dietary content, which may range in populations from different regions in very wide intervals. The highest concentrations are found in the kidney (0.2–1.5 mg/kg), liver (0.24–0.4 mg/kg), hairs (0.6–6 mg/kg), and bones (1–9 mg/kg), whilst muscles have a lower content (0.07–0.1 mg/kg). In blood, selenium ranges from 40 to 350 μg/l.

Selenium, as the amino acid selenocysteine, is an essential component of numerous enzymes and some other proteins called **seleno-proteins**. The best-known selenoenzyme is glutathione peroxidase. Glutathione peroxidase catalyses the reduction of hydrogen peroxide (H_2O_2) and hydroperoxides of fatty acids (R–O–OH) using a tripeptide glutathione (see Section 2.3.3.1.1) that exists in reduced (G–SH) and oxidised (G–S–S–G) states:

$$2\ G\text{–}SH + H_2O_2 \rightarrow G\text{–}S\text{–}S\text{–}G + 2H_2O$$
$$2\ G\text{–}SH + R\text{–}O\text{–}OH \rightarrow G\text{–}S\text{–}S\text{–}G + R\text{–}OH + H_2O$$

The first step involves oxidation of the selenol of a selenocysteine residue (R–SeH) by hydrogen peroxide, which yields a derivative of selenenic acid (R–Se–OH). Selenenic acid is then converted back into selenol by a two-step process:

$$R\text{–}SeH + H_2O_2 \rightarrow R\text{–}Se\text{–}OH + H_2O$$
$$R\text{–}Se\text{–}OH + G\text{–}SH \rightarrow G\text{–}S\text{–}Se\text{–}R + H_2O$$
$$G\text{–}S\text{–}Se\text{–}R + G\text{–}SH \rightarrow G\text{–}S\text{–}S\text{–}G + R\text{–}SeH$$

These reactions take place, for example, in erythrocytes, and ensure removal of lipid hydroperoxides from damaged biological membranes (lipoproteins). Glutathione peroxidase thus provides protection against oxidative damage of biological structures. After removing the hydroperoxyl group, the hydroxylated lipids may be metabolised normally by β-oxidation. Reduced glutathione is regenerated by NADPH under the catalysis of glutathione reductase:

$$G\text{–}S\text{–}S\text{–}G + NADPH + H^+ \rightarrow 2\ G\text{–}SH + 2\ NADP^+$$

In the human body, there are at least four different isoenzymes of glutathione peroxidase (GPx), referred as to cytosolic GPx (GPx1), gastrointestinal GPx (GPx2), plasma GPx (GPx3), and phospholipid hydroperoxide GPx (GPx4). The glutathione peroxidase isolated from erythrocytes (85–95 kDa) consists of four subunits, each of which contains one selenocysteine in its peptide chain. Selenium, as a part of glutathione peroxidase, also enhances the biological effects of vitamin E.

Another selenoenzyme is iodothyronine 5′-deiodinase (also known as thyroxine 5′-deiodinase), from a subfamily of deiodinase enzymes that are important in the activation and deactivation of the thyroid gland hormones. There are two types of deiodinase: type I occurs in the liver, kidney, muscle, and thyroid gland, whilst type II is present in the brain, pituitary, and adrenal glands. Iodothyronine 5′-deiodinase catalyses the conversion (deiodination) of the thyroid hormone thyroxine into the metabolically active triiodothyronine (see Section 2.2.1.2.5) and significantly enters into the metabolism of iodine and thyroid hormones. Other important selenoenzymes are thioredoxin reductase and selenophosphate synthetase.

Apart from selenoenzyme, a number of other selenoproteins have been discovered, but their functions are not yet sufficiently known. For example, selenoprotein P is a constituent of blood plasma and contains about half of the total plasma selenium. Selenoprotein W is contained in the muscles and brain. Prostate epithelial selenoprotein (15 kDa) probably has an antioxidative function. High selenium concentrations in male sexual organs are maintained in animals with selenium deficiency; therefore, selenium compounds seem to be important for reproductive function.

Some anticarcinogenic effects are also attributed to this element. Selenium reduces the toxic effects of arsenic, cadmium, mercury, thallium, and tellurium. With arsenic, the mechanism of this protective effect is the formation of bis(S-glutathionyl)arsineselenide anion, also called selenobis(S-glutathionyl)arsinium ion, which is formed in erythrocytes and excreted in bile (**6-13**).

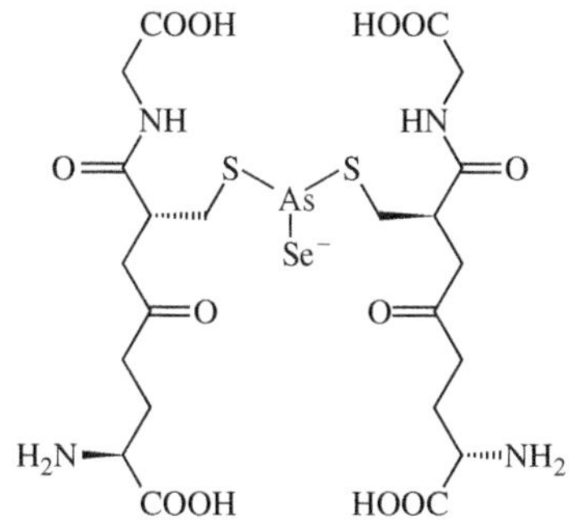

6-13, bis(S-glutathionyl)arsine selenide anion

The effectiveness of selenium absorption in the gastrointestinal tract of humans is quite high, but depends on the selenium form. Organic and inorganic dietary selenium are both typically well absorbed through the intestinal membrane (70–95%). Since selenium and sulfur have very similar properties, certain **selenocompounds** can be absorbed through the same pathways as their sulfur analogues. This is the case for selenate, which shares an active transport route with sulfate, and of selenomethionine, which passes the intestinal barrier using the same Na^+-dependent process as methionine. In contrast, selenite and selenocystine are taken up by a passive process (only in the direction of the concentration gradient) that is not affected by sulfite; however, selenocysteine does not seem to be influenced by cysteine. Selenomethionine absorption is 95–97%, whilst selenite (SeO_3^{2-}) absorption ranges from 44 to 76%. Following their absorption, selenium species in the blood are partially taken by blood cells and partially translocated towards different organs and tissues with the help of protein transporters (albumin and selenoprotein P) in the blood plasma. It is suggested that ingested selenium is initially bound to albumin, which transports the element to the liver, where it is released and serves in the synthesis of selenoproteins. The most common selenoprotein is selenoprotein P, which is released into the bloodstream to itself become a selenium transporter between the liver and various other organs and tissues. In the absorption and metabolism of selenium compounds, reduced glutathione (G–SH) and cysteine play an important role. Inorganic selenium compounds (e.g. selenites, SeO_3^{2-}) are metabolised in the body with the formation of selenotrisulfides (R–S–Se–S–R) and disulfides. Selenotrisulfides are relatively unstable, decomposing to disulfide and elemental selenium:

$$4\,R\text{–SH} + H_2SeO_3 \rightarrow R\text{–S–Se–S–R} + R\text{–S–S–R} + 3\,H_2O$$

$$R\text{–S–Se–S–R} \rightarrow R\text{–S–S–R} + Se$$

The reaction of G–SH with selenite yields selenotrisulfide, known as selenodiglutathione (G–S–Se–S–G). With an excess of glutathione, unstable selenodisulfide (G–S–Se–H) forms and decomposes to selane (hydrogen selenide, H_2Se) and oxidised glutathione (G–S–S–G). The amino acids containing selenium are metabolised in methylation reactions. For example, the methylation of selenocysteine gives Se-dimethyl-L-selenocysteine (**6-14**), which decomposes to L-alanine and volatile dimethylselenide $(CH_3)_2Se$. The methylation of dimethylselenide yields the trimethylselenonium cation $(CH_3)_3Se^+$.

Selenium is excreted from the body mainly in urine (approximately 60%). Major selenium compounds in urine are trimethylselenonium ions and **selenosugars**, such as 1,2-dideoxy-1-methylseleno-2-acetamido-D-galactopyranose (**6-14**). Part of the selenium is excreted by the lungs in volatile compounds such as dimethylselenide, dimethyldiselenide, and selane.

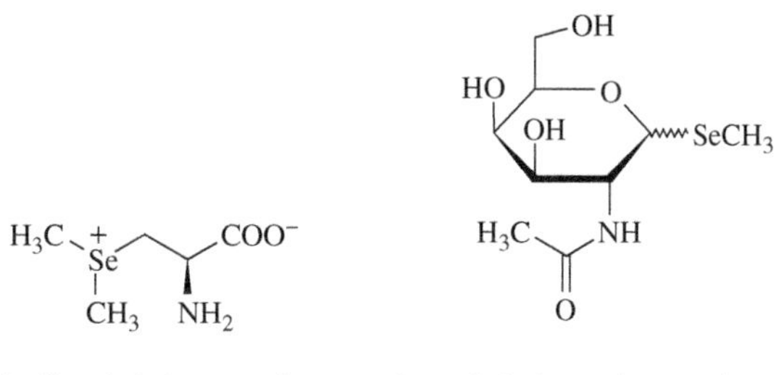

Se-dimethylselenocysteine Se-methylselenogalactosamine

6-14, examples of selenium metabolites

6.3.15.2 Occurrence in foods

The average natural content of selenium in the earth's crust is 0.09 mg/kg, but it is distributed very unevenly. Selenium accompanies sulfur in nature, so it is a minor component of sulfides of copper, silver, lead, and mercury. The highest concentrations of selenium are found in igneous and sedimentary rocks (in sulfide rocks, it is found at concentrations of more than 1000 mg/kg). Normal levels of selenium in the soil are 0.1–2.0 mg/kg dry matter. In natural waters, only trace amounts of selenium occur. Concentrations of selenium in lake and river waters range from about 0.02 to 10 µg/l. Seawater contains 0.03–0.25 µg/l of selenium.

In some territories, there are areas with high or low concentrations of selenium in the soil. For example, in China, there are areas with both selenium deficiency and extremely high concentrations of selenium. Mean concentrations of selenium are found in soils in large parts of the United States (in the states on the Atlantic and Pacific coasts). In Canada (excluding the province of Ontario), mean levels of selenium also occur. By contrast, in some European countries (such as Finland, Switzerland, and the Czech Republic) and New Zealand, concentrations of selenium in the soil are very low. If the land is used for growing crops or grazing, the concentration of selenium in crops and animal bodies is characteristic of the region. Therefore, there are very significant regional differences in concentrations of selenium in agricultural products and foods. An overview of the selenium content in foods is given in Table 6.11.

The foods that are rich in selenium are mainly marine fish, crustaceans, molluscs (e.g. oysters, prawns), freshwater fish, and offal (mainly kidneys). Eggs also have relatively high levels; most selenium is contained in the yolk. The content of selenium in milk, dairy products, and meat is lower, and is highly dependent on the animal's nutrition. Concentrations of selenium in fruits and vegetables (except garlic) are very low (<0.02 mg/kg). Mushrooms contain selenium at the level of 0.03–1.4 mg/kg. The selenium content in foods of plant origin is fundamentally influenced by the content of the soil or fertiliser and its availability to the plant. The concentration of selenium in foods of animal origin is determined by the selenium content in the animal feed. Some feeds are deliberately fortified with selenium compounds.

The level of selenium in the soil largely determines its concentration in plants. The ability of plants to absorb selenium depends on the pH, moisture, oxygen, and iron content of the soil. The highest availability of selenium for plants usually comes from slightly alkaline aerobic environments. Hexavalent selenium compounds are very soluble in the soil. Tetravalent selenium compounds are less soluble, and their solubility decreases in the presence of ferric oxide. The vegetation in areas with a high content of selenium in the soil can contain up to tens of mg/kg of selenium in dry matter. Such a concentration can poison grazing animals. In addition, some plant species growing on seleniferous soils are selenium-tolerant and accumulate very high concentrations of this element (selenium accumulators), but most plants are selenium non-accumulators and are selenium-sensitive. Concentrations of selenium in some species of the genus *Astragalus* (Fabaceae family), for example in the selenium-accumulating plant *A. bisulcatus*, commonly called two-grooved milkvetch or silver-leafed milkvetch (native to central and western North America), can reach thousands of mg/kg, and the plant can be toxic to cattle. Moderately selenium-accumulating plants belong to the Asteraceae family. In most plants, the selenium level does not usually exceed 1 mg/kg. In those that do strongly accumulate selenium, it is present mainly in the form of free selenium amino acids and peptides (especially those composed of *Se*-methylselenocysteine and glutamic acid).

Selenium in foods is mainly contained in the form of selenium amino acids bound in protein molecules. In most plant foods, the major form of selenium is **selenomethionine**. For example, total selenium concentrations for Canadian lentils ranged from 0.16 to 0.72 mg/kg, and almost all the selenium (86–95%) was present as selenomethionine, with a small part (5–14%) occurring as selenite. The main components of garlic are selenocysteine and organic selenides. In foods of animal origin, the appreciable part of selenium is bound in **selenocysteine** or selenocystine. Quite recently, a new low-molecular-selenium compound called **selenoneine** was identified in tuna fish, but it occurs also in some other fish species (e.g. Pacific sardine, horse mackerel, greeneye, skipjack, and swordfish). Selenoneine is a selenium-containing derivative of histidine (2-selenyl-N_α, N_α, N_α-trimethyl-L-histidine, **6-15**) that acts as an effective antioxidant, as the selenol form of the compound can be oxidised to the corresponding diselenide. Moreover, it is supposed that selenoneine has canceroprotective activity and moderates the toxic effects of methylmercury, which occurs in marine fish in significant amounts (see Section 6.5.2). The highest concentrations of selenoneine are found in the blood of fish, followed by the inner organs and muscles. Traces of selenoneine were also detected in pig liver and chicken liver.

reduced selenol form   reduced selenoxo form   oxidised diselenide form

6-15, selenoneine

Table 6.11 Selenium contents of major food materials and foods from different countries.

Foods	USA[a]	Canada[a]	Venezuela	Finland[b]	Germany	Czech and Slovak Republics
				Content in mg/kg		
Pork meat	0.04–0.24	0.31	0.83	0.01–0.09	0.19	0.02–0.07[c]
Beef meat	0.06–0.27	–	0.17	0.01–0.03	–	0.02
Chicken meat	0.10–0.12	–	–	0.08–0.14	–	0.07–0.11
Pork liver	0.64–0.70	0.36	0.36	0.34–0.51	0.17	0.09–0.34
Beef liver	0.43	0.50	0.69	0.03–0.13	0.09	0.02–0.14
Pork kidney	1.90–2.21	3.22	–	1.54–1.76	0.78	0.97–1.84
Beef kidney	1.45–1.70	2.31	–	0.62–0.78	0.95	0.20–1.02
Freshwater fish	0.34–0.37	0.59	–	0.12–0.53	0.38	0.05–0.38
Marine fish	0.12–1.41	0.75–1.48	–	0.11–0.80	–	–
Milk (whole)[d]	0.06	0.15	–	0.001–0.004	0.20	0.003
Curd	–	0.07	–	0.02–0.03	–	–
Cheese	0.09	–	0.43	0.01–0.06	–	0.02–0.04[c]
Yoghurt	0.05	–	–	0.003	–	0.004–0.008[c]
Egg (whole)	0.10	0.39	1.52	0.02–0.16	–	0.18–0.24
Egg (white)	0.03–0.05	0.12–0.15	–	–	–	0.06
Egg (yolk)	0.13–0.18	0.13–0.69	–	0.30	–	0.53
Wheat	0.20–0.61	0.58–1.09	0.25	0.004–0.025	0.34–0.88	–
Rye	0.36	–	–	0.01	–	–
Flour (wheat)	0.18–0.52	0.28–0.64	–	0.010–0.12	–	0.016
Bread (whole-wheat)	0.33–0.41	0.59–0.68	–	0.003–0.01	–	0.015–0.026[c]
Rice (peeled)	0.21–0.38	–	0.46	0.01–0.03	–	0.024–0.034[c]
Peas (mature seeds)	–	–	–	0.01	–	0.02
Lentils	–	0.61	–	–	0.10	0.03–0.08[c]
Beans	0.02–0.13	0.06	0.07	–	–	0.09
Soybeans	0.08–0.48	0.09	0.01	–	–	–
Cabbage	0.023	0.03	–	0.001–0.02	0.014	0.003
Cauliflower	0.007	0.004	0.01	<0.002	0.014	0.005
Spinach	0.012	–	–	<0.002	0.018	–
Lettuce	<0.001–0.011	0.008	–	<0.002	0.006	0.001[c]
Tomatoes	0.005	0.001	0.014	<0.002	0.007	< 0.001[c]
Carrot	0.022	0.006	–	<0.002	0.004	0.001–0.003[c]
Peas (green)	–	–	–	0.001–0.002	–	0.005
Onion	–	–	–	<0.002	–	0.003
Garlic	0.014–0.26	0.07	–	–	–	0.03–0.14
Potatoes	<0.002–0.055	0.023	0.016	0.001–0.002	0.017	0.003–0.018
Apples	0.005	0.004	0.006	0.001–0.003	0.01	0.001–0.003[c]

(continued overleaf)

Table 6.11 (continued)

Foods	Content in mg/kg					
	USA[a]	Canada[a]	Venezuela	Finland[b]	Germany	Czech and Slovak Republics
Oranges	0.013	0.015	0.008	<0.002	0.029	–
Bananas	0.01	–	0.005–0.06	0.001–0.01	–	–
Walnuts	0.08	–	–	–	–	–
Cocoa	0.21	–	–	–	–	–
Tea (black)	0.01–0.06	–	0.04	–	–	–
Coffee (roasted)	0.07–0.09	–	–	–	–	–

[a]Samples from areas with a medium content of selenium in the soil.
[b]Results from the 1970s; currently, the levels of selenium in most Finnish foods are significantly higher due to the use of fertilisers with added sodium selenate.
[c]Data from the Slovak Republic.
[d]Breast milk concentration of selenium: 0.006–0.028 mg/kg.

6.3.15.3 Nutrition

The recommended daily dietary dose of selenium is 15–20 μg for newborns and babies, 20–30 μg for children aged from 1 to 8 years, 40 μg for children aged from 9 to 13 years, and 55 μg for older teenagers and adults. For pregnant and lactating women, the daily dose should be increased to 60–70 μg.

The actual daily intake of selenium in the diet varies in different countries and depends on many factors. For example, the following amounts have been determined: 330 μg in Venezuela, 130–200 μg in Canada, 80–130 μg in the United States, about 60 μg in Great Britain, 30–40 μg in Finland (in the 1970s), and 23–33 μg in New Zealand. Owing to the low levels of selenium in the blood serum of the population of Finland, at the beginning of the 1980s it was decided to increase the selenium content in food crops by adding sodium selenate to the fertilisers used there. Within a few years (between 1984 and 1986), the content of selenium in important crops and livestock products had increased. For example, wheat selenium content increased from 0.01 to 0.23, potato from <0.002 to 0.02, milk from 0.008 to 0.03, and egg from 0.16 to 0.31 mg/kg. The average daily dose of dietary selenium increased to about 90 μg, and the selenium concentration in breast milk increased from 0.007 to 0.015 mg/kg.

Selenium deficiency in animals is manifested by hepatic necrosis and a set of symptoms called white-muscle disease. Very serious selenium deficiency in humans has been reported in China, in the Keshan region, where there is a very low selenium content in the soil. The heart disease known as Keshan's disease can be cured by administering doses of selenium.

Although selenium is an essential trace element, it is toxic if taken in excess. The symptoms of chronic selenium poisoning of animals (cattle, horses, and sheep) have been known since the nineteenth century and are called alkali disease, because it was assumed that they were caused by drinking water with a high salt content. Poisoning in its first stage involves loss of hair, hoof deformities, and hoof and motion disorders. The second stage causes blindness and often ends in the death of cattle.

Symptoms of poisoning in humans (selenosis) can occur even at doses 20 times higher than the recommended daily dose (1–2 mg). Chronic human exposure to higher doses of selenium is manifested by inflammation of the respiratory tract, pulmonary oedema, bleeding, skin changes, and depression. The characteristic garlic breath and metallic taste in the mouth are caused by the presence of dimethyldiselenide. In severe cases, jaundice, liver cirrhosis, hair and nails falling out, dental caries, and kidney failure can appear.

6.3.16 Iodine

6.3.16.1 Biochemistry and physiology

The body of an adult contains 10–30 mg of iodine. About 70–90% of this amount is contained in the thyroid gland, a vitally important hormonal gland that plays an essential role in metabolism, growth, and maturation of the human body.

Iodine is the constituent of the **thyroid hormones**, iodinated aromatic amino acids derived from tyrosine that are known by their trivial names as 3,5,3′,5′-tetraiodothyronine (thyroxine, often abbreviated as T4) and 3,5,3′-triiodothyronine (known as T3). Thyroid hormones regulate the rate of cellular oxidative processes that affect the consumption of oxygen in the liver, kidney, and heart tissue, breakdown of lipids (lipolysis), and hydrolysis of glycogen to glucose (glycogenolysis), enhance absorption of glucose and galactose, and affect thermoregulation. Activity of the thyroid gland is controlled by both the autonomic nervous system and the action of thyrotropin-releasing hormone (also known as TRH or thyroliberin), which is produced in the hypothalamus. Thyroliberin is a tripeptide composed of 5-oxopyrrolidin-2-carboxylic acid, histidine, and proline amide (362 Da). It releases thyrotropin-stimulating hormone (also known as thyrotropin, thyrostimulin, or TSH) from the anterior pituitary gland, which in the thyroid gland influences the formation of thyroglobulin (Tg), a dimeric glycoprotein (660 kDa) containing bound iodine.

In the form of iodide anions (I^-), which are the main form of iodine contained in foods, iodine is easily and completely absorbed in the gastrointestinal tract of humans. Other forms of iodine (salts of iodic acid iodates, IO_3^-) are first reduced to iodide ions. Iodine is rapidly transported by blood to the thyroid gland, which captures about 60 µg of iodine in the form of iodide a day. During lactation, a part of the iodine passes into the milk and the rest is eliminated from the body in urine.

The synthesis of thyroid hormones has several steps. In the thyroid gland, iodine ions are oxidised to the active form (cation I^+) by the action of a specific thyroid peroxidase (thyroperoxidase), which then reacts with tyrosyl residues of thyroglobulin to form 3-iodotyrosine. Subsequent iodisation of 3-iodotyrosine yields 3,5-diiodotyrosine. The condensation reaction of 3,5-diiodotyrosine with 3-iodotyrosine in the colloid of the thyroid follicle yields 3,5,3'-triiodothyronine. Two molecules of 3,5-diiodotyrosine combine to form thyroxine. These hormones, bound to thyroglobulin, are then released into the blood as a result of thyroglobulin proteolysis regulated by thyrotropin. In the blood, normal concentrations of 3,5,3'-triiodothyronine can vary by as much as 1–1.5 µg/l, and thyroxine concentrations range from 60 to 120 µg/l.

6.3.16.2 Occurrence in foods

The iodine content of most foods is in the tenths to hundredths of mg/kg. An overview of the iodine content in some common foods is given in Table 6.12. The content in foods of plant origin depends on the concentration of iodine in soil. The highest iodine concentrations are found in seafood and seaweed. The average concentration of iodine in seawater is about 60 µg/l, and in sea salt 82 mg/kg. The level of iodine in animal products depends on its level in feed, feed supplements, and any veterinary drugs. The natural iodine content in milk and dairy products can also be increased through contamination from disinfection preparations containing iodine compounds. These products are used, for example, to disinfect the udders of cows and production equipment in dairies.

Certain food additives contain iodine. For example, potassium iodate and calcium iodate are components of preparations for stabilising dough. The synthetic red food colouring erythrosine contains 58% iodine (four iodine atoms per molecule). Therefore, foods coloured using this pigment have a higher iodine content, but the bioavailability of erythrosine iodine is low (2–5%). The iodine content in foods and meals may also increase with the use of table salt fortified with iodine (as NaI or $NaIO_3$). The iodine concentration in salt is 20–50 mg/kg.

6.3.16.3 Nutrition

The required daily dose of iodine needed to prevent deficiency symptoms in adults is estimated at 50–75 µg. To ensure some reserves, however, higher doses are recommended. The official recommended daily intake of iodine is 40–50 µg for children aged under 1 year, 70 µg for children aged 1–3 years, 90–120 µg for children aged 4–10 years, and 150 µg for older children, adolescents, and adults. During pregnancy, the recommended daily dose is 175 µg, and during lactation 200 µg. In several European countries, and some others, iodine deficiency is a significant public health problem. For example, in the Czech Republic, the average daily intake of iodine is about 100 µg. In a study in the United Kingdom published in 2011, almost 70% of test subjects were found to be iodine deficient.

Decreased thyroid gland function, called **hypothyroidism**, may result from insufficient iodine intake or insufficient biosynthesis of hormones influenced by antithyroid agents also known as goitrogens. Iodine is necessary in the first 3 months of pregnancy in order for the foetal nervous system to develop. Congenital hypothyroidism then manifests by cretinism. Hypothyroidism in young organisms leads to failure of growth, known as nanism or dwarfism (dwarf growth in which intelligence is not violated), and later to excessive enlargement of the thyroid gland. In the past, thyroid gland enlargement occurred in inhabitants of the inland mountains, where the iodine content in food is low (e.g. in Austria and Switzerland). If the occurrence of such a disease is characteristic for certain areas, then we talk about endemic goitre.

Perchlorates (see Section 6.3.2.2) and some degradation products of glucosinolates, such as goitrin, isothiocyanates, nitriles, and thiocyanate (rhodanide) ions (SCN^-), have goitrogenic (antithyroid) activities. Thiocyanate ions also arise in the body as detoxification products of cyanides. The other compounds that show antithyroid activity are some congeners of polychlorinated biphenyls (PCBs), some pesticides, and a number of veterinary drugs containing thiourea residues (such as thiouracils, aminothiazoles, and mercaptoimidazoles) that inhibit the enzyme thyroid peroxidase. These substances may present in foods as contaminants (see Section 12.6.4).

Hyperfunction of the thyroid gland, called **hyperthyroidism**, manifests itself by production of an excessive amount of thyroid hormones. A disease where the thyroid gland is overactive is known as Graves' disease, the symptoms of which are increased metabolism, body temperature, and weight loss. Overproduction of the hormone in childhood leads to giant growth (gigantism), whilst overproduction in the later stages of growth leads to the growth of only the distal parts of the body (acromegaly).

Table 6.12 Iodine, fluorine, boron, and silicon contents of some crops and foods.

Food	Content (mg/kg)			
	I[a]	F	B	Si
Pork meat	0.009–0.016[b]	<0.2	0.1–0.2	–
Beef meat	0.015–0.019[b]	0.1–0.2	0.1–0.2	1
Chicken meat	<0.005	<0.2	0.1–0.4	1
Pork liver	–	<0.2	<0.2	–
Sea fish	0.28–1.75	0.3–2.2	<0.2	–
Milk (whole)[c]	0.016–0.75[b,d]	0.08–0.1	0.02–0.2	0.7
Curd	0.084–0.32[b]	0.2–0.4	0.1–0.2	–
Cheese	0.06–0.69	0.5–0.9	0.2–0.4	4–5
Yoghurt	0.022–0.26[b]	0.1	0.2	–
Eggs (whole)	0.029–0.73[b]	0.3	0.2–0.3	3
Wheat	0.024–0.043[b]	0.2–0.9	0.7–1.4	20–190
Flour (wheat)	0.017–0.025[b]	0.1–1.4	0.3–2.0	30–40
Bread (whole-wheat)	–	0.4–0.8	0.3–1.0	30–50
Rice (peeled)	–	0.3–0.6	0.7–0.8	30–90
Rye	–	0.3–2.0	0.7–1.5	30–290
Barley	–	0.4–1.6	0.7–1.4	1400–2900
Oats	–	0.4–1.5	0.5–1.4	3400–6300
Peas (mature seeds)	–	0.3–0.9	6.1–7.1	20–50
Beans	–	1.0–2.0	14–26	50–60
Soybeans	–	0.9–1.3	28	30
Cabbage	<0.01	0.02–0.2	1.7–2.2	<2
Cauliflower	<0.005	0.02–0.2	1.7–2.2	2
Spinach	0.022–0.028	0.3–0.4	2.4–2.9	20–50
Lettuce	<0.01–0.018	0.02–0.4	1.3–1.8	12
Tomatoes	<0.01	0.02–0.1	0.8–1.1	<2
Carrot	0.013	0.03–0.2	2.4–4.0	1
Peas (green)	0.047	<0.1	2.6–3.4	2–18
Onion	0.025	0.04–0.1	1.3–3.3	5
Potatoes	0.018–0.037	0.06–0.2	1.1–1.8	4–6
Mushrooms	0.013	0.2–0.3	0.2–0.3	10–30
Apples	0.002–0.007	<0.1–0.3	1.0–6.0	2–5
Oranges	0.008	0.04–0.1	2.7–3.0	<2
Bananas	<0.005	0.1	1.4–2.2	50–80
Strawberries	0.09	0.03–0.3	1.7–2.1	10–20
Peanuts	0.11	–	18	50
Tea (black)	–	115–450	–	–
Chocolate (milk)	0.33	1.0	1.7–2.9	10

[a]Unless otherwise specified, data on the iodine content come from the United States.

[b]Data from the Czech Republic.

[c]The breast milk concentration: I 0.06–0.18, F 0.016–0.04, B 0.06–0.08, Si 0.7 mg/kg.

[d]In the United States, typical iodine concentrations in milk are 0.12–0.29 mg/kg, but some extreme values were also found (0.08 and 1.9 mg/kg).

6.3.17 Fluorine

6.3.17.1 Biochemistry and physiology

The human body contains about 0.8–2.5 g of fluorine. Fluorides are structural components of bones and teeth, as they are bound in fluoroapatite $Ca_{10}(PO_4)_6F_2$, which is almost insoluble and makes the tooth surface much more resistant to acids. Therefore, fluorine has a protective effect against dental caries. For healthy tooth development, adequate doses of fluorine are important, especially in childhood. In some countries, a sufficient amount of fluoride is supplied in drinking water enriched with fluoride (NaF, H_2SiF_6, or Na_2SiF_6). The optimal concentration of fluoride in water is about 1 mg/l, but the original content of fluoride ions in freshwater sources normally ranges from 0.01 to 0.4 mg/l. During tooth development, concentrations of fluorine in water higher than 10 mg/l lead to dental fluorosis in children. Teeth impacted by fluorosis have visible discoloration, ranging from white spots to brown and black stains. Fluorides present in food and drinking water are rapidly and efficiently absorbed in the gastrointestinal tract. Absorption efficiency is 85–98%. The absorption of fluorides, however, is significantly reduced in patients taking antacids (substances neutralising stomach acidity) based on aluminum oxide or hydroxide. In the body, fluorine is incorporated into hydroxyapatite, an inorganic calcium-containing constituent of the bone matrix and teeth. Under normal conditions, 50–70% of the ingested amount is excreted from the body in urine.

6.3.17.2 Occurrence in foods

Fluorine occurs in foods exclusively in the form of fluoride. In most, its content ranges from hundredths to tenths of mg/kg (Table 6.12). Tea leaves have a very high fluorine content, so tea is the main source of fluorine in the diet. Some kinds of table salt are fortified with fluoride (as NaF or KF). Another regular source of fluoride in some countries (or specific regions) is drinking water intentionally enriched with fluoride (NaF, H_2SiF_6, or Na_2SiF_6).

6.3.17.3 Nutrition

The recommended daily dose of fluoride for babies under 6 months ranges from 0.1 to 0.5 mg, that for children aged from 6 months to 1 year is 0.2–1.0 mg, that for children aged 1–3 years is 0.5–1.5 mg, and that for children aged 4–10 years is 1.0–2.5 mg. For adults, a dose of 3–4 mg is recommended, but the total fluoride intake should not exceed 10 mg/day. Apart from dietary intake of fluoride, appreciable amounts of fluoride (units of mg) are taken by using toothpaste and other dental hygiene products that contain fluorine compounds (most often alkylammonium fluorides). Insufficient intake of fluorine in the diet and drinking water increases tooth decay and can lead to osteoporosis. When long-term elevated fluoride doses (20–80 mg/day) are taken, symptoms of poisoning (fluorosis) appear, which lead to damage of the teeth, bones, kidney, and nervous system, as fluoride ions act as inhibitors of some enzymes.

6.3.18 Boron

Boron is a non-metal that occurs in nature mainly as borate minerals, such as borax, $Na_2B_4O_5(OH)_4 \cdot 8\,H_2O$, and borosilicates. In biological material, compounds of boron (boric acid $B(OH)_3$ and borates) form stable complexes with polyhydroxy compounds, such as sugar alcohols, sugars, and derived substances (nucleotides and ascorbic acid). An example is the binding of borate ions to the *cis*-hydroxyls (at positions C-2 and C-3) of two β-D-apiose residues (**6-16**) located in different chains of a structural type of pectin called rhamnogalacturonan II, which are cross-linked in this way (see Section 4.5.6.6.1).

Boron is an essential element for plants. Whether it is essential for animals has not yet been unequivocally confirmed. But the biological effects of some compounds of boron are known. Boron has an impact on the effect of parathormone (PTH or parathyrin), a hormone secreted by parathyroid glands, which increases blood levels of calcium by its mobilisation from bones (its antagonist is calcitonin) and affects the metabolism of magnesium, calcium, phosphorus, and cholecalciferol. Boric acid also affects the activity of many enzymes, such as chymotrypsin, pyridine nucleotide–disulfide oxidoreductase, and flavin oxidoreductase.

6-16, β-D-apiofuranose-boric acid complex

The recommended dietary intake of boron has not been established, but doses up to 20 mg are tolerated as safe. Actual amounts depend on food composition and local conditions. The estimated dietary intake is 2–10 mg/day, but boron deficiency in humans has not yet been recorded. Boron contained in foods is easily absorbed in the gastrointestinal tract, but 30–92% of boron ingested is excreted in the urine. At higher doses of boric acid, boron accumulates in the nervous system.

An overview of the boron content in some foods is given in Table 6.12. Foods of animal origin are very poor sources of boron. The boron content of meat, fish, eggs, and dairy products does not usually exceed 0.3 mg/kg. The highest amounts of boron contain plant products such as legumes, nuts (tens of mg/kg), and fruits (units of mg/kg). The content of boron in vegetables and cereals is usually <2 mg/kg. Wines have a relatively high content (2–11 mg/l). The plant content of boron depends on its level in the soil. Under specific circumstances, much higher boron contents were found in some fruits (apples 468 mg/kg, tomatoes 1258 mg/kg dry matter). Boric acid and disodium tetraborate are allowed to be used for the preservation of certain foods in some countries (see Section 11.2.1.3.3).

6.3.19 Silicon

Silicon is a non-metal used as a structural material referred to as biogenic silica (polymerised silicic acid) by diatoms (*Bacillariophyceae*, a group of microscopic algae), by amoeboid protozoa (radiolarian), and by siliceous sponges. Some plants can deposit silicon within different intracellular and extracellular structures, and silica from decayed plants is re-deposited in the soil in the form of microscopic structures known as phytoliths. Some grasses and horsetails (plants of the genus *Equisetum*, Equisetaceae) contain up to 20% of silicon dioxide (SiO_2). The exact function of silicon in animals is still under discussion. It is assumed that it is essential for the synthesis of collagen and connective-tissue integrity. The silicon is probably in the form of *ortho*-silicic acid (H_4SiO_4) and disilicic acid ($H_2Si_2O_5$), bound by ester bonds in some mucopolysaccharides. The total silicon content of the human body is about 1.4 g.

The mechanism of silicon absorption in the digestive tract is not known. It is assumed that the available forms of silicon (contained in the diet) are *ortho*-silicic acid and soluble silicates. When given in the form of silicon aluminosilicates, only about 1% of silicon is absorbed. In contrast, some organosilicon compounds used in medicine (such as methylsilantriol salicylate) are absorbed at levels of up to 70%. The absorption efficiency of silicon contained in the diet varies over a wide interval of 5–100%. Silicon contained in beer has a high bioavailability (about 50%). The efficiencies of silicon absorption from meat, dairy, and soy products usually exceed 50%. From cereals and other foods containing high amounts of silicon, this element is absorbed poorly. Silicon is excreted mainly in the urine.

The silicon contents of certain foods are given in Table 6.12. Most foods of animal origin (milk, the meat of mammals and birds) contain only traces of silicon, but higher silicon content has been detected in marine molluscs. In contrast, cereals – mainly barley and oats – have high silicon contents. Wheat contains less silicon than barley. During brewing, most of the grain silicon remains in the husk, but significant quantities (bioavailable *ortho*-silicic acid) are extracted into the wort, and much of this gets into beer, which is relatively rich in silica (10–30 mg/l). A large quantity of silicon is also contained in the surface layers of rice. Brown rice contains 110–570 mg/kg, whilst peeled (polished) rice contains 30–90 mg/kg. Examples of fruits and vegetables with higher silicon contents are bananas and spinach.

The recommended daily dose of silicon has not yet been established. The total dietary intake is estimated to be 20–50 mg/day. Silicon intake (e.g. from beer) might reduce the bioavailability of aluminum, which has been implicated as one of the possible causal factors contributing to Alzheimer's disease. Silicone intake might also help to prevent bone thinning and osteoporosis. High doses of silicon may contribute to the formation of kidney and urinary stones.

6.4 Non-essential elements

6.4.1 Aluminum

Aluminum has long been considered a non-toxic and non-essential element. Today, however, some of its adverse health effects are known. The content of aluminum in the earth's crust is 8%, and it is the third most abundant element (after oxygen and silicon). The content of aluminum in biological materials is low to trace (e.g. 10–50 mg/kg dry matter in grass, 100–300 mg/kg dry matter in leaves of trees, 0.2–0.6 mg/kg in animal tissues), except in the lungs, which contain it at a level of 20–60 mg/kg.

Aluminum compounds contained in rocks, soil, and sediments are very slightly soluble, and in natural waters very low concentrations occur (mostly <20 µg/l). Aluminum ions are easily hydrolysed even in neutral media and form insoluble multinuclear complexes of aluminum ions or insoluble aluminum hydroxide $Al(OH)_3$. In addition, in a soil solution at neutral pH, the aluminum concentration is low and its uptake by plants is very limited. For example, soil acidification by acid rain increases the solubility of aluminum compounds, and aluminum can thus enter the hydrosphere and biosphere in higher amounts. In slightly acidic solutions (pH 5), the predominant cations are $[Al(OH)_2]^+ > [Al(H_2O)_6]^{3+} > [Al(OH)]^{2+}$. Under these conditions, aluminum is actually toxic to certain organisms. A higher intake of aluminum by birds leads to abnormalities of their eggs. Soluble compounds of aluminum in plants cause slower growth because aluminum is a phytotoxic substance.

6.4.1.1 Occurrence in foods and dietary intake

Older data on the content of aluminum in foods are often overestimated, since the determination of trace amounts of aluminum (<0.1 mg/kg) in biological materials is very difficult, and the published data should be assessed critically.

Aluminum is found in larger amounts in foods of plant origin than in those of animal origin. Tea leaves have very high concentrations of aluminum (850–1350 mg/kg; older leaves can have up to 5000 mg/kg dry matter), because the tea plants (*Camellia sinensis*, Theaceae) accumulate aluminum from the soil. Aluminum present in tea leaves is soluble in hot water (about one-third), so that the infusion prepared from 1 g of tea and 100 ml of water has an aluminum content of 2.7–4.9 mg/l. Aluminum in tea is mainly bound in the form of complexes with catechins, or less often with organic acids (such as chlorogenic, gallic, citric, malic, and oxalic acids) and fluoride ions. Fluoro complexes have the general formula $AlF_n^{(3-n)+}$; the plant obtains them from the soil mainly as AlF^{2+} and AlF_2^+ cations. Higher aluminum contents are also found in particular spices, such as basil (310 mg/kg), bay leaf (440 mg/kg), oregano (600 mg/kg), pepper (140 mg/kg), cinnamon (80 mg/kg), and thyme (750 mg/kg).

Cereals and legumes have moderate contents of aluminum (e.g. 4–14 mg/kg in wheat flour, 3–6 mg/kg in wheat bread, 5–6 mg/kg in peeled rice, 5–64 mg/kg in unpeeled rice, 2 mg/kg in maize, 15–22 mg/kg in lentils, 14–27 mg/kg in peas). Lower quantities are found in some root crops and vegetables (potatoes 1–6 mg/kg, carrots 1–6 mg/kg, spinach 5–25 mg/kg, cauliflower 0.2–12 mg/kg), meat (pork 2–5 mg/kg, beef 0.2–5 mg/kg), and fish (0.4–9 mg/kg). As already mentioned, recent results have quoted lower concentrations of aluminum in vegetables (0.1 mg/kg in cabbage, 0.05 mg/kg in carrots, 0.2 mg/kg in potatoes). Offal sometimes contains medium to high amounts (e.g. beef liver 4–46 mg/kg, pork kidney 2–18 mg/kg, pork brains 7–130 mg/kg). The concentrations in milk, dairy products, and eggs are very low (0.011–0.035 mg/kg in milk, 0.15–0.34 mg/kg in Edam cheese, <0.2 mg/kg in eggs). The aluminum concentration in breast milk generally ranges from 0.004 to 0.03 mg/kg, but can reach values of from 0.2 to 2.4 mg/kg.

Some aluminum compounds, such as sodium aluminum phosphates, $Na_3Al_2(HPO_4)(H_2PO_4)_7$ (acidic phosphate), and $Na_{15}Al_3(PO_4)_8$, are used in the United States and other countries as additives (e.g. as baking powder components). The aluminum content in foods can also be increased during their contact with metal aluminum. Some foods and beverages are commonly packed in aluminum foil, tubes, or cans. Until recently, aluminum barrels were used for beer and soft drinks. During the cooking of foods in aluminum cookware, the aluminum can be partially dissolved. This applies especially to acidic foods and drinks, such as fruits, fruit puree and juices, and vegetables in vinegar brine.

The average daily dietary dose of aluminum for adults in Western countries ranges from 4 (Switzerland 1985) to 27 mg (United States 1985). The tolerable daily intake (TDI) of aluminum for an adult (of the body weight of 70 kg) is 10 mg/day. Sources other than foods, such as some medicinal drugs, may sometimes be more important. For example, oral antacids neutralising gastric juices contain aluminum oxide or aluminum hydroxide. Daily dietary doses of aluminum compounds can then reach 2 g. Aluminum is also contained in various toothpastes and cosmetic preparations.

6.4.1.2 Metabolism and toxic effects

The effectiveness of aluminum absorption in the gastrointestinal tract of humans under normal conditions is about 0.1–0.3%. Lactic, citric, and ascorbic acids facilitate gastrointestinal absorption. Approximately 95% of orally administered aluminum binds to transferrin and albumin and is then excreted in urine. The normal concentration of aluminum in blood serum is 2–40 µg/l. That in various organs, except the lungs, is very low and does not increase with age. The total aluminum content in the human body is about 35 mg.

Aluminum can accumulate in the human body, especially when the gastrointestinal barrier is bypassed. If a significant load exceeds the excretory capacity of the body, the excess is deposited in various tissues, including bones, brain, liver, heart, spleen, and muscles. It is now commonly acknowledged that aluminum toxicity can be induced by oral exposure (as a result of aluminum-containing pharmaceutical products such as aluminum-based phosphate binders or antacid intake) and by infusion of aluminum-contaminated dialysis fluids. In the early 1970s, aluminum toxicity was implicated in patients with chronic renal failure who were undergoing dialysis. For example, with intravenously infused aluminum, 40% is retained in adults and up to 75% in neonates. The exact mechanism of aluminum toxicity is not completely understood. It is known that it combines with many proteins and cofactors that are required in intermediary steps of metabolism. There is probably a link between aluminum intake and increased incidence of Alzheimer's disease.

6.4.2 Tin

The average tin content of the earth's crust is 3 mg/kg. In nature, tin occurs mainly as the mineral cassiterite (stannic oxide, SnO_2) and as an accompanying metal in some sulfides. Metallic tin is an important component of conventional alloys (bronze) and of cans made of tinplate (tin-coated steel), and is used in the food industry. A large amount of tin is used for the production of organometallic compounds. Dibutyltin dilaurate and analogous octyltin compounds are used as stabilisers of plastics, such as poly(vinylchloride) (PVC). Tributyltin compounds are used for wood preservation and as ingredients in special paints for ships and other bodies exposed to the long-term effects of seawater. Triphenyltin acetate and triphenyltin hydroxide are used in agriculture as fungicides.

Natural freshwater contains only trace amounts of tin (0.01–1 µg/l), compared to concentrations of about 3 µg/l in seawater. Aquatic organisms can accumulate tin (bioconcentration factor is 10^2–10^3). Some microorganisms may methylate inorganic compounds and dealkylate synthetic organometallic ones. Tin concentrations in soils are in the range of 2–200 mg/kg dry matter.

The natural content of tin in foods is very low, and mean concentrations are <1 mg/kg in most foods (meat 0.007, meat products 0.18, offal 0.014, poultry 0.006, fish 0.032, milk 0.003, eggs 0.003, fresh fruit 0.019, green vegetables 0.003, other vegetables 0.05, potatoes 0.004, nuts 0.03 mg/kg f.w.). High levels of tin are often a result of migration from cans. Although tin is corrosion-resistant, acidic foods, such as fruits and vegetables, can cause corrosion of the tin layer. Hence, canned fruit juices and vegetables can contain a significantly higher level (e.g. 30–260 and c. 40 mg/kg, respectively) compared to uncanned products. A study in 2002 showed that 99.5% of 1200 tested food samples in cans contained less than the UK regulatory limit of 200 mg/kg of tin. In recent years, the use of tinplated cans for fruit juices, canned fruit, and other acidic foods has been restricted. In some foods, traces of tin organometallic compounds may be found; for example, wines transported in PVC containers contain dibutyl tin at concentrations of up to 160 µg/l.

The dietary intake of tin is estimated at 3 mg/day. The TDI for an adult man (weighing 70 kg) is 140 mg/day. The effectiveness of absorption of tin in the gastrointestinal tract depends on the metal valence (stannous compounds 3–8%, stannic compounds about 1%). The normal concentration of tin in blood plasma is 30–40 µg/l. The tin that is absorbed is excreted in urine and partly in bile. Nausea, vomiting, and diarrhoea have been reported after ingesting canned food containing more than 200 mg/kg of tin. Toxic effects of tin appear in cases of long-term consumption of foods with a high content of tin (1400 mg/kg). Inorganic tin in quantities higher than 150 mg/kg in beverages or 250 mg/kg in canned food may cause gastric irritation. Organometallic compounds of tin are highly toxic.

6.4.3　Bromine

Bromine is commonly found in the biosphere, mostly bound to metals in the form of inorganic bromides, but it is less abundant in the earth's crust (about 0.7–3 mg/kg) than other halogens (fluorine or chlorine). The concentrations of bromides in fresh water range from trace amounts to about 0.5 mg/l, whilst in seawater bromide occurs in concentrations 65–80 mg/l. In 2016, the worldwide annual production of bromine was estimated to be around 391 000 t. Many inorganic and organobromine compounds are used in flame retardants, drilling fluids, pesticides (mostly methyl bromide for fumigation of soils and post-harvest fumigation of grains), water-treatment chemicals, photographic chemicals, dyes, pharmaceuticals, and rubber additives.

Until 2014, when bromine (in the form of bromide anion) was found to be a necessary cofactor in the biosynthesis of collagen IV, bromine had not been shown to perform any essential function in plants, microorganisms, or animals.

Bromide ion has a low degree of toxicity and is not of toxicological concern in nutrition. However, bromide was once used as an anticonvulsant and sedative at doses as high as 6 g/day, and clinical symptoms (nausea, vomiting, abdominal pain, coma, and paralysis) of bromide intoxication (bromism) have been reported from its medical uses. The signs and symptoms relate to the nervous system, skin, glandular secretions, and gastrointestinal tract.

Organobromine compounds, also known as organobromides, are the most common organohalides in nature. Even though the concentration of bromide is only 0.3% of that of chloride in seawater, organobromine compounds are more prevalent in marine organisms than organochlorine derivatives. For example, ancient civilisations produced an expensive purple dye, Tyrian purple (also known as royal or imperial purple), from the secretions of *Murex* spp. sea snails. The main component of this dye is 6,6′-dibromoindigo (**6-17**).

6-17, 6,6′-dibromoindigo

The concentrations of bromide in selected food commodities are 3.7–14.4 mg/kg in bell peppers (capsicum), 4.5–9.3 mg/kg in potatoes, and 3.6–19.0 mg/kg in fungi. Crude fish oils contain between 2.6 and 9.6 mg bromide/kg.

Potassium bromate ($KBrO_3$) as the baking ingredient for bleaching flour (see Section 11.4.2.2.1) is banned in many countries. Brominated vegetable oil (used as a clouding agent to induce a turbid appearance for soft drinks in the United States; see Section 11.7.3) is banned from use in food in the European Union as bromine reacts with fatty acids, yielding brominated products. The main brominated fatty acids in brominated vegetable oil are dibromostearic acid (bromination product of oleic acid) and tetrabromostearic acid (bromination product of linoleic acid).

Bromine can be involved in the reaction with naturally occurring organic matter in drinking water, forming brominated and mixed chloro-bromo contaminants (such as trihalomethanes and halogenated acetic acids), and it can react with ozone to form bromate anion, BrO_3^-. Certain polybrominated compounds, similar in structure to persistent organic pollutants (POPs), are produced by sea biota.

Examples are simple natural brominated phenols produced by marine benthic animals. The production and use of brominated phenols as flame retardants or wood preservatives leads to their release into the environment. They are not readily biodegradable and therefore persist in the environment for a long time. Other examples are polybrominated diphenyl ethers, identified in algae, mussels, and fish, and polybrominated dibenzo-*p*-dioxins, found in mammals. These compounds are byproducts of chemical industry and of the combustion of brominated flame retardants.

6.4.4 Silver

Silver is a precious metal contained in minor quantities in the earth's crust. It occurs especially in sulfidic minerals (acanthite, Ag_2S; stephanite, Ag_5SbS_4) and as an admixture in ores of other metals. In some soils, the silver content can be up to tens of mg/kg. The natural content of silver in foodstuffs is low or even extremely low; for example, in human milk and infant formulas samples, silver contents range from <0.13 to 42 µg/kg. In common vegetables and legumes, silver contents are mostly below 2 µg/kg (<0.2–1.5 µg/kg), whilst in cereal grain, silver contents are up to 3 µg/kg. Amongst common crops, the highest content was found in buckwheat seeds (10–24 µg/kg). Some wild-growing edible mushrooms accumulate much higher quantities (e.g. the species of *Boletus* and related genera contained 0.7–11 mg/kg on d.w. basis). In contrast, cultivated common mushrooms (*Agaricus bisporus*) have a silver content comparable with that of cereal crops. Muscle tissue of oysters contains 0.5–1.0 mg/kg of silver (on d.w. basis). In the quantities contained in food, the silver in monovalent state does not represent a health risk. Metallic silver is allowed to be used as a food additive (E 174), namely for decoration of confectionery products. The compounds of monovalent silver are almost non-toxic. Food and drinking water from some sources may also contain trace amounts of elementary silver in the form of nanoparticles, as these are widely applied in various consumer products (cosmetics, clothing) for their antimicrobial effect and therefore get into the environment.

6.5 Toxic elements

The most important toxic elements include lead (Pb), cadmium (Cd), mercury (Hg), and arsenic (As). Some other trace metals (Cu, Zn, Cr, Sb, etc.) and semi-metals (Se) can be included amongst the toxic elements if they occur in sufficiently large quantities. The extent of an element's toxicity is given by its chemical species (e.g. Cr^{III} vs Cr^{VI}, organic As vs inorganic As). The term **heavy metals** is often used in place of toxic elements, but in this sense it is incorrect, because some toxic elements are not heavy (Be has very low atomic mass) and some are not strictly metals (As and Se are semi-metals).

Excessive levels of toxic elements can be damaging to an organism in case of bioaccumulation. Bioaccumulation is an increase in the concentration of a chemical in an organism over time, compared with its concentration in the environment. The presence of toxic elements in foods is mainly connected with environmental pollution. A number of anthropogenic and natural sources contribute to the toxic elements entering into the food chain. The main anthropogenic factors are burning fossil fuels, transport, industrial production of metals, the use of various elements in industry and technology and their related waste, the excessive use of mineral fertilisers and other agrochemicals (such as pesticides), and the application of sewage sludge to soil. Natural sources of toxic elements in the environment include the weathering of rocks, forest fires, and volcanic activity.

There are some sources that contribute substantially to the contamination of one particular element (e.g. lead contamination of the atmosphere from the exhausts of motor vehicles that use leaded gasoline), but most causes cannot be characterised in this way. For example, the environmental contamination in the vicinity of metallurgical plants is manifested by significantly elevated levels of many elements (Cd, Pb, Tl, In, As, Ga, Hg, Ni, Cr, Cu, Zn, and Ge), simultaneously in different sections of the environment. Emissions from thermal power plants and some other energy sources contain many elements simultaneously. It is therefore quite difficult to identify specific sources of contamination of individual elements. In addition, air and water flows act over large distances as transmission media of elements in the environment. In air, the elements can occur in aerosols of solid particles or even in the gas phase (mercury and organometallic compounds, arsines). Fallout of solid particles and wet deposition are the main routes of surface contamination of soil and plants. Wet deposition occurs when rain or snow removes metals from the atmosphere and delivers them to the ground, plants, or other surfaces.

The level of toxic elements in foods is amongst the main indicators of pollution, ecological risk, and food safety assessment. For lead, cadmium, and mercury, TDI levels have been established. It is supposed that when a person takes the TDI of an element (or compound) daily across his or her lifetime, this has no negative effect on his or her health. TDI is expressed in microgrammes (or milligrammes) of the element (or compound) per 1 kg of the body mass; alternatively, it can be calculated as the mass (in µg or mg) corresponding to average body mass (normally 60 or 70 kg).

The limits for maximum allowed contents of toxic elements in food are established by national or international authorities based on TDI values, toxicity data, and the consumption of individual food commodities. In addition, the 'common' contents of some toxic elements in specific foods (e.g. Hg in fish, Cd in vegetables, As in rice) are taken into considerations. Special care is taken with foodstuffs intended for infant and children nutrition. The maximum allowed contents of some toxic elements in the European Union are summarised in Table 6.13.

Table 6.13 Brief summary of maximum allowed contents of toxic elements in foods.

| Food | Maximum allowed contents (mg/kg f.w.) | | | | |
	Pb	Cd	Hg	Inorg. As	Inorg. Sn
Meat	0.1	0.05			
Fish	0.3	0.05–0.25	0.5–1.0		
Shellfish	0.5–1.5	0.5–1.0	0.5		
Liver, kidney	0.5	0.5–1.0			
Milk	0.02				
Cereals and legumes	0.2	0.1–0.2			
Rice and rice products				0.1–0.3	
Vegetables	0.05–0.3	0.05–0.2			
Fruit	0.1–0.2				
Mushrooms	0.3	0.2–1.0			
Fats and oils	0.1				
Fruit juices	0.03–0.05				
Wines	0.15–0.2				
Chocolate		0.1–0.8			
Honey	0.1				
Canned food					200
Canned beverages					100
Canned food for children					50
Infant formulas	0.01–0.05	0.005–0.04			

Commission Regulation (EC) No. 1881/2006 of 19 December 2006 setting maximum levels for certain contaminants in foodstuffs.

Poisoning with toxic elements (especially As, Hg, and Pb) is treated by taking chelating agents (or selenium compounds) that facilitate detoxification and excretion of the elements. Examples of chelating agents are 2,3-dimercaptopropan-1-ol, 2,3-dimercaptosuccinic acid, 2,3-dimercaptopropane-1-sulfonic acid, glutathione, and ethylenediaminetetraacetic acid (EDTA; see Section 11.3.3.1) and its salts.

6.5.1 Lead and cadmium

The average lead and cadmium contents in the earth's crust are 13 mg/kg and 0.1 mg/kg, respectively. In nature, these metals are found mainly in sulfide and carbonate ores (galena, PbS; greenockit, CdS; sphalerit, ZnS, and smithonite, $ZnCO_3$).

Metallic lead is used to make batteries, sheets, and pipes (including water pipes). Inorganic lead compounds serve as pigments (such as orange lead tetroxide, also called minium, Pb_3O_4, and yellow lead chromate, $PbCrO_4$) or are used in the production of lead glass (lead oxide, PbO). In the past, organometallic compounds of lead (especially tetraethyl lead) were used in gasoline as antiknock additives. The annual world production of lead amounts to about 6 million tons.

Cadmium is used in anticorrosive metal protection coatings and in battery production. Cadmium sulfide is used as a yellow pigment, and cadmium salts of fatty acids are used as stabilisers in PVC production. The annual world production of cadmium is about 23 000 tons.

6.5.1.1 Occurrence in the environment, transport, and distribution

The levels of lead and cadmium in air vary locally. In clean areas, air contains 5–200 ng/m^3 Pb and 0.3–2 ng/m^3 Cd, whilst in large cities it can contain 500–5000 ng/m^3 Pb and 2–50 ng/m^3 Cd. Natural waters contain only traces of lead and cadmium. Lead concentrations in uncontaminated river and lake waters range from 0.1 to 5 µg/l, and the cadmium content is even lower (0.07–0.8 µg/l). Significantly higher concentrations are found in sediments of uncontaminated rivers and reservoirs (10–30 mg/kg Pb, 0.04–0.8 mg/kg Cd). In sediments from contaminated localities, the contents can be up to several orders of magnitude higher. Numerous aquatic organisms (algae and other aquatic plants, zooplankton, crustaceans, and molluscs) accumulate cadmium and other elements (mercury, arsenic, and selenium) in their bodies to a significant extent. Even if they live in only slightly contaminated water (Cd concentration in units of µg/l), some animals (oysters, mussels, shrimps) can accumulate up to 100 mg/kg of cadmium.

The content of lead in uncontaminated soil can range from 5 to 40 mg/kg dry matter, and cadmium from 0.1 to 1 mg/kg dry matter (in polluted localities, the concentrations may be much higher).

Besides the total content of toxic elements, their availability to plants is another important factor. Plants take toxic elements either from the soil through their root system or from the atmosphere through foliar deposition. The root uptake of these elements depends not only on the type of plant and their content in soil, but also on their distribution in the soil horizon depths, their mobility in the soil, the organic matter content, and other factors. The bioavailability of lead and cadmium for a plant depends on the pH and redox potential of the soil solution. It increases in acidic and oxidising environments. Transition coefficients of elements for soil–plant transfer reach values ranging from 0.01 to 0.1 for lead and from 0.1 to 10 for cadmium. The uptake of elements by plants is quite different amongst different species and even cultivars. Some plants accumulate only certain elements. In these cases, the element does not exert a strong phytotoxic effects. Plants that accumulate cadmium and lead from the soil include spinach, lettuce, root vegetables, and some oilseed crops.

The distribution of metals in different parts of plants is not uniform. If foliar versus root uptake is negligible (less polluted localities) in comparison with root uptake, the concentrations of lead and cadmium (on d.w. basis) generally decrease in the order roots > leaves > stems > fruits > seeds. These factors should be taken into account when assessing the levels of toxic elements in plants, because only certain parts are consumed by herbivores, and only certain parts are processed for food or feed use.

Environmental pollution can increase the contents of lead and cadmium in animal bodies. Toxic elements enter animal bodies by the oral route via consumed feed. For human nutrition, the concentrations of these elements in muscles and viscera of animals (both farm and wild ones) are important. Lead and cadmium contents in the livers and kidneys of wild animals indicate the load in the habitat. Under normal conditions, the contents of lead and cadmium in meat are very low (units of µg/kg), in contrast to the concentrations in the kidney and liver, which are sometimes up to three orders of magnitude higher. In these organs, lead and cadmium become accumulated.

6.5.1.2 Occurrence in foods and dietary intake

The content of lead and cadmium in foods is low and very variable. A survey of toxic element levels in main foods and food raw materials is given in Table 6.14. In crops, the concentrations of lead and cadmium depend mainly on their content in soil. Quite high concentrations (in tens to hundreds of µg/kg) are typical for some vegetables (spinach, lettuce, and carrots), edible mushrooms, and oil seeds (e.g. poppy seeds contain 0.04–1.96 mg/kg of cadmium). In cereal grains (except rice), cadmium and lead concentrations rarely exceed 0.1 mg/kg. Relatively high concentrations of lead were found in wines (0.016–0.17 mg/l). Amongst foodstuffs of animal origin, the highest contents of lead and cadmium are in offal. Cadmium in kidney can reach especially high levels (up to units of mg/kg). The toxic metal content in the livers and kidneys of animals is related to their diet and age. In older animals, the concentrations are higher. Meat, eggs, milk, and dairy products contain only traces of lead and cadmium. Higher levels of lead might be found in foods packed in cans, due to contamination with lead contained in the alloy of tin, which is applied in the seam. Packing in glass rather than cans is recommended for foods that require more stringent standards (such as baby food). Exceptionally high amounts of lead are found in meat of game animals that were shot using lead bullets or pellets. In these cases, the metal particles are irregularly dispersed in the corresponding tissues, so some parts are not affected, whilst other are heavily contaminated.

Before 2010, a lead TDI value of 250 µg/day (for an adult of 70 kg body weight) was considered appropriate, but the EFSA re-evaluated this and established the so-called benchmark dose level (BMDL). The BMDL is related to the favourable value of some biomarker associated with exposure to lead (e.g. the BMDL equivalent to a lead blood level of 36 µg/l, which is associated with an acceptable cardiovascular risk, is 90 µg/day for an adult of 60 kg weight). The TDI of cadmium was recently adjusted to 25 µg/day for an adult. In total-diet studies carried out during last 40 years in Europe and the United States, a gradual decrease in the average daily dietary intakes of lead and cadmium was found. The recently estimated European average dietary intakes of lead and cadmium range from 32 to 41 µg/day and from 14 to 21 µg/day, respectively.

6.5.1.3 Metabolism and toxic effects

Lead and cadmium enter the body not only in food through the digestive tract, but also through the lungs. For smokers, inhalation exposure to cadmium intake is comparable to the intake from food. The level of cadmium in tobacco is about 1–2 mg/kg. The absorption of lead depends on age, dietary composition, and health. The effectiveness of absorption in adults is estimated at 10%. A child's body absorbs around 40–50% of lead from food. Lead absorption is more effective in foods containing higher amounts of protein, and less effective in the presence of large amounts of fibre, phytic acid, iron, and calcium. The normal levels of lead in the blood are in the range of 50–120 µg/l. The average efficiency of cadmium absorption is 6% or more. Blood levels of cadmium in non-smokers and smokers are 0.2–3 µg/l and 0.2–5 µg/l, respectively.

Table 6.14 Lead, cadmium, mercury, and arsenic contents of important food raw materials and foods.[a]

Food	Content (mg/kg)			
	Pb	Cd	Hg	As
Pork meat	0.005–0.05	0.001–0.01	0.002–0.006	0.003–0.03
Beef meat	0.004–0.07	<0.001–0.01	0.001–0.003	0.001–0.07
Chicken meat	0.008–0.04	0.001–0.005	0.001–0.002	0.001–0.03
Pork liver	0.014–0.04	0.025–0.10	0.007–0.014	0.005–0.02
Beef liver	0.01–0.42	0.03–0.17	0.001–0.005	0.005–0.07
Pork kidney	0.01–0.04	0.07–0.52	0.011–0.015	0.01
Beef kidney	0.06–0.22	0.06–2.0	0.003–0.014	0.02–0.13
Sea fish	0.01–0.14	0.001–0.07	0.03–0.85	0.50–1.4
Sweetwater fish	0.01–0.05	0.001–0.005	0.07–1.01	0.03–0.56
Milk (whole)[b]	0.001–0.002	<0.0001–0.001	<0.001	<0.001–0.003
Curd	0.02	<0.002	<0.001	0.01
Cheese	0.01–0.06	0.005–0.02	<0.002	<0.002–0.025
Yoghurt	0.01–0.03	0.001–0.003	<0.001	<0.005
Eggs	0.001–0.01	0.001–0.01	0.005–0.008	<0.002–0.01
Wheat	0.02–0.65	0.02–0.35	0.0001–0.006	0.005–0.29
Flour (wheat)	0.004–0.05	0.01–0.09	0.002–0.004	0.01–0.17
Bread (whole-wheat)	0.012–0.013	0.02–0.05	0.001–0.006	0.006–0.05
Rice (peeled)	0.003–0.08	0.004–0.14	0.002–0.008	0.04–0.31
Rye	0.01–0.17	0.004–0.04	0.002–0.007	0.03–0.10
Barley	0.03–0.27	0.004–0.04	0.001–0.006	0.005–0.38
Oats	0.03–0.30	0.004–0.07	0.0001–0.008	0.01–0.54
Peas (mature seeds)	0.01–0.43	0.01–0.03	0.002–0.02	0.01–0.05
Beans	0.02–0.10	0.003–0.02	0.004–0.02	<0.01
Soybeans	<0.002–0.32	0.04–0.09	<0.004	0.03–0.05
Cabbage	0.002–0.04	0.01–0.017	0.0003–0.001	<0.01
Cauliflower	0.002–0.02	0.002–0.02	0.0004–0.002	0.002–0.01
Spinach	0.01–0.29	0.01–0.35	<0.001–0.008	0.005–0.02
Lettuce	0.003–0.25	0.002–0.16	0.0005–0.01	0.002–0.14
Tomatoes	<0.001–0.04	0.002–0.05	0.0001–0.008	<0.001–0.002
Carrot	0.004–0.21	0.003–0.16	0.0006–0.005	0.003–0.11
Peas (green)	0.01–0.02	0.001–0.03	0.0005–0.002	0.01
Onion	<0.001–0.05	0.004–0.05	<0.001	0.01
Potatoes	0.006–0.04	0.002–0.06	0.0001–0.017	<0.001–0.04
Mushrooms	0.01–0.20	0.01–0.33	0.07–0.22	0.01
Apples	0.01–0.05	0.001–0.002	0.0003–0.002	0.001–0.22
Oranges	0.005–0.07	0.001–0.007	<0.001	0.004–0.02
Bananas	0.02–0.05	<0.002	0.001–0.002	0.04–0.09

Table 6.14 (*continued*)

Food	Content (mg/kg)			
	Pb	Cd	Hg	As
Strawberries	0.006–0.09	0.001–0.03	0.0002–0.001	<0.005
Grapes	0.012–0.024	0.001–0.002	0.0004–0.002	0.01–0.16
Peanuts	0.01–0.19	0.01–0.51	<0.004	–
Tea (black)	0.07–1.29	0.005–0.12	0.007–0.025	0.05–0.40
Coffee (roasted)	0.02–0.05	0.003–0.007	<0.004	0.05–0.22
Cocoa	0.03–0.07	0.095–0.17	<0.004	0.10
Chocolate (milk)	0.05	0.005–0.01	0.002–0.004	<0.05

[a]Data from Sweden, Finland, the Netherlands, Germany, Canada, and the United States.
[b]Breast milk concentrations: Pb 0.0004–0.002, Cd 0.0001–0.0006, Hg 0.0003–0.001, As <0.001 mg/kg.

The absorbed lead and cadmium are transported by blood to the liver and kidneys, where they are accumulated and bound in molecules of metallothioneines. Part of the lead in the liver is excreted via bile into the intestine. A small percentage is excreted in urine. Long-term exposure leads to lead accumulation in bones. The intoxication can damage the kidneys and liver. Lead also damages the nervous and cardiovascular systems. Cadmium has teratogenic and carcinogenic effects, damages reproductive organs, and affects blood pressure. Children are particularly susceptible to lead poisoning (due to the possibility of ingestion of soil or dust, higher absorbed amounts, and higher sensitivity of the organism). Even at a level of 150 µg/l in the blood, adverse effects may appear (such as slower mental and physical development, lower learning ability, lower intelligence, and decreased immunity).

Lead inhibits the synthesis of porphyrins, so that in chronic poisoning, the amount of haemoglobin in erythrocytes decreases and anaemia develops. Lead is an inhibitor of two enzymes essential for the synthesis of haem: aminolevulinic acid dehydratase and ferrochelatase. Lead can also damage the central and peripheral nervous systems. Lead poisoning may cause motor skills disorders, slowing of movement reactions, and similar symptoms. The neurotoxic effects of lead are particularly dangerous for children. Brain damage (encephalopathy) occurs during heavy exposure (blood levels of 800–1000 µg/l).

Cadmium poisoning can cause acute kidney failure, as the kidneys are the organ accumulating the most cadmium (in adults, they contain 4–10 mg Cd). The normal concentration in the renal cortex of humans is 10–30 mg/kg. The synthesis of metallothioneins prevents acute renal damage, if cadmium level in the renal cortex do not exceed c. 200 mg/kg. Intoxication with cadmium is manifested by the occurrence of proteins and sugars in urine, but only very small amounts are excreted in urine. Cadmium is also accumulated in the liver (normal amount 2–4 mg). At concentrations not exceeding 30 mg/kg of tissue, the binding of cadmium to liver metallothionein prevents liver damage. Chronic cadmium poisoning also results in decalcification and softening of bones. Itai itai disease was the first well-documented case of mass cadmium poisoning by rice (containing 1–3 mg/kg Cd) irrigated with polluted river water in Toyama, Japan (1968).

6.5.2 Mercury

The average mercury content in the earth's crust is about 0.05 mg/kg. In nature, mercury occurs in deposits mostly as mercuric sulfide, cinnabar (HgS). The annual global production is c. 7000 tons. Owing to its high toxicity, limits on the use of mercury have been introduced. It is used mainly:

- in electrical engineering for the production of batteries, switches, electrodes, and measuring instruments;

- for the electrochemical production of chlorine and caustic soda by electrolysis of sodium chloride solution (mercury is the cathode);

- for the manufacture of paints (a red pigment, vermilion, is mainly obtained by reduction from cinnabar), catalysts, and fungicides (phenylmercuric chloride for seed treatment);

- for the preparation of dental amalgam, used as a filling material in dentistry.

6.5.2.1 Occurrence in the environment, transport, and distribution

The input of mercury into the environment is mainly caused by volcanic activity, burning of coal, use in industry and agriculture, and handling of waste. The total amount of mercury entering the atmosphere is estimated at 150 000 tons/year. About two-thirds of this amount comes from natural resources.

The migration of mercury in the environment is related to the properties of mercury and its compounds, such as the volatility of elemental mercury, the solubility of mercury in fats (c. 5–50 mg/l), and the solubility of mercury in water (74 g/l for $HgCl_2$, 2 mg/l for Hg_2Cl_2, and just 2.10^{-15} ng/l for HgS).

Natural waters contain only trace amounts of mercury. In rivers, total mercury ranges from 0.01 to 0.2 µg/l (whilst the permissible limit for drinking water is 1 µg/l). The concentration of mercury in seawater is even lower (0.0004–0.002 µg/l). Nevertheless, the oceans and seas represent a large reservoir of mercury. For example, the Mediterranean Sea contains about 3700 tons of mercury, and each year through fallout and tributaries some 240 tons enter it, with about 80 tons/year going into the marine sediment and 150 tons evaporating into the atmosphere.

In the aquatic environment, important chemical transformations of mercury species take place, in particular the methylation reaction of bacteria and microscopic fungi, oxidation–reduction reactions, and precipitation reactions. The products of mercury biomethylation are the organometallic compounds methylmercury and dimethylmercury. In anaerobic conditions (as in the depths of the seas), mercuric salts can be reduced to elemental mercury or mercuric ions and to insoluble HgS, which deposits onto lake bottoms, streambeds, wetlands, and intertidal zones and is integrated into bottom substrates. The concentration of mercury in non-polluted sediments of rivers and lakes ranges from the tenths to units of mg/kg dry matter. About 1% of mercury in sediments is present in the form of methylated compounds. Aquatic organisms accumulate mercury from the water. The highest bioconcentration factors (see Section 12.1) were found in invertebrates (1.10^5), with lower factors in freshwater fish (6.10^4) and marine fish (1.10^4). The concentration of mercury in their bodies is several orders of magnitude higher than in the surrounding environment. Most mercury in fish is methylated.

The concentration of mercury in uncontaminated soils ranges from 0.02 to 0.2 mg/kg. The main forms of mercury in neutral and alkaline soils are the slightly soluble oxide (HgO) and carbonate ($HgCO_3$). Sulfur bacteria produce almost insoluble sulfide (HgS). Most of the remaining forms of mercury are adsorbed on organic matter, hydrated iron and manganese oxides, and clay minerals. Owing to the low mobility of mercury in the soil, only small amounts pass from the soil to plants.

The mercury content of plants is in the tenths to tens of µg/kg. Some mushrooms contain higher concentrations (tenths to units of mg/kg). The level in the bodies of animals is dependent on the composition of their food. High concentrations were found, for example, in the liver and kidneys of aquatic birds.

6.5.2.2 Occurrence in foods and dietary intake

The concentration of mercury in most foods ranges from units to tens of µg/kg (Table 6.14). High concentrations (tenths to units of mg/kg) are found in some edible mushrooms, fish, shellfish, and crustaceans. A survey of the mercury and methylmercury contents in marine and freshwater fish and shellfish is reported in Table 6.15.

The TDI of total mercury for an adult is 40 µg, and that of methylmercury is 13 µg (at 70 kg b.w.). The actual daily dietary dose of total mercury found in total-diet studies in European countries ranged from 3 to 16 µg/day, and that of methylmercury from 0.7 to 11 µg/day. The dietary intake of mercury is related to the eating of fish.

6.5.2.3 Metabolism and toxic effects

About 7% of mercury from food is absorbed in the small intestine. The absorbed mercury is taken up by the liver, kidneys, and brain. A part is eliminated from the liver via bile into the intestine. Mercury also accumulates in hair and nails. The toxic effects of mercury and its compounds are related to its high affinity for thiol groups in peptides and proteins. The binding of mercury to these functional groups occurs in enzymes and leads to their inhibition. The toxicity of various forms of mercury decreases as follows: alkylmercury compounds (mainly CH_3Hg^+) > elemental mercury vapours > mercuric salts (Hg^{II}) > arylmercury (diphenylmercury) and alkoxymercury compounds (RHgOR) > mercurous salts (Hg^I). In terms of food toxicology, methylmercury is the most important compound due to its presence in fish. In the 1950s and '60s, water in the Minamata Bay (Japan) was heavily polluted by wastewater containing mercury (units of µg/l) from a factory belonging to the Chisso Corporation. Methylmercury bioaccumulated in fish and shellfish consumed by local people resulted in the so-called Minamata disease, with more than 10 000 people affected. In 1970, Cargill Inc. (Minnesota, USA) sold seed grain treated with methylmercury to Basra, Iraq. Although this seed was not intended for human or animal consumption, but only for use in agriculture, a number of recipients consumed the surplus seed, which led to the deaths of hundreds of people.

Poisoning by inorganic mercury compounds and by elemental mercury occurs mainly through occupational exposure (workers in chemical plants and laboratories). The main organs that are damaged are the kidneys and the brain. The effects of the individual forms of mercury vary somewhat. In methylmercury poisoning, neurotoxic effects prevail, manifesting themselves by disturbances of sensory functions (vision, hearing, and balance), speech, swallowing disorders, morphological changes in the brain, and mental malfunction. Methylmercury also has teratogenic effects. There is a risk of foetal damage in expectant mothers where the concentration of mercury in the hair reaches

Table 6.15 Contents of total mercury (expressed in mg/kg) and methylmercury (expressed as mercury percentage of total mercury) in edible tissues (muscles) of fish, shellfish, and crustaceans.

Group	Marine species	Hg[a](MeHg)[a]	Freshwater species (except marked ones)	Hg[b](MeHg)[b]
Fish	Atlantic salmon (*Salmo salar*)	0.03 (82)	Carp (*Cyprinus carpio*)	0.006 ± 0.001
	Anchovy (*Engraulis encrasicolus*)	0.08 (48)	Chub (*Leuciscus cephalus*)	0.14 (82)
	Seabass (*Dicentrarchus labrax*)[c]	0.09 (49)	Rainbow trout (*Oncorhynchus mykiss*)	0.015 ± 0.001
	Pilchard (*Sardina pilchardus*)	0.09 (48)	Trout (*Salmo trutta*)	0.007 (90)
	Sole (*Solea solea*)	0.105 (50)	Bream (*Abramis brama*)	0.018 ± 0.008
	Seabream (*Sparus aurata*)	0.109 (50)	Flounder (*Platichthys flesus*)	0.06 ± 0.02
	Atlantic mackerel (*Scomber scombrus*)	0.114 (50)	Perch (*Perca fluviatilis*)	0.17 ± 0.12
	Hake (*Merluccius merluccius*)	0.132 (52)	Pike (*Esox lucius*)	0.08 (96)
	Red mullet (*Mullus barbatus*)	0.235 (56)	Pike-perch (zander) (*Sander lucioperca*)	0.34–1.9
	Dogfish (*Squalus acanthias*)	0.67 (64)	Eel (*Anguilla anguilla*)	0.13–2.4
	Bluefish tuna (*Thunnus thynnus*)[c]	1.54 (72)		
	Swordfish (*Xiphias gladius*)[c]	1.61 (73)		
Bivalves	European scallop (*Pecten jacobaeus*)	0.015 (40)	Duck mussel (*Anodonta anatina*)	0.04
	Mediterranean mussel (*Mytilus galloprovincialis*)[a]	0.03 (40)	Flat oyster (*Ostrea edulis*)[d]	0.04 (40)
Cephalopods	Squid (*Loligo vulgaris*)	0.10 (60)	Cuttlefish (*Sepia officinalis*)[d]	0.19 (60)
	Octopus (*Octopus vulgaris*)	0.10 (60)		
Crustaceans	Rose shrimp (*Parapenaeus longirostris*)	0.18 (60)	Lobster (*Homarus americanus*)[d]	0.14 (98)

[a]Median values.

[b]Mean ± standard deviation of total Hg or interval of total Hg or median values for both total Hg and methylmercury (in per cent total mercury).

[c]For these species, the ranges of Hg contents are quite wide; for swordfish, tuna, seabass, and mussel, ranges of 0.05–6.24, 0.32–2.00, 0.03–4.20, and 0.005–0.47 mg/kg, respectively, were reported.

[d]Marine species.

about 15–20 mg/kg. The symptoms of poisoning in adults appear when mercury concentrations in hair are above 30 mg/kg. Therefore, it is recommended that pregnant women significantly reduce their consumption of fish.

Poisoning by inorganic mercury compounds can lead to a reduced production of urine and even to kidney failure. Poisoning by elemental mercury can occur either by inhalation of mercury vapours or through ingestion of mercury. The ingestion of mercury manifests by increased salivation, a metallic taste in the mouth, swollen gums, loss of appetite, loss of teeth, vomiting, diarrhoea, fatigue, loss of self-control, and muscle weakness. The renal function deteriorates, and sometimes the thyroid gland is enlarged. Inhalation of mercury vapours leads to some further symptoms (bronchitis, chest pain, coughing, and difficult breathing).

6.5.3 Arsenic

The average content of arsenic in the earth's crust is 1.8 mg/kg. In nature, arsenic occurs in the form of sulfide minerals (e.g. arsenopyrite, $FeAsS$; realgar, As_4S_4; cobaltite $CoAsS$), and it is included as a minor component in sulfide ores of copper, lead, and other metals. The main industrially produced compound of arsenic is arsenic trioxide (As_2O_3). Annual world production of this oxide is about 60 000 tons. Elemental arsenic is used as an ingredient in alloys of lead and other metals. Some inorganic arsenic compounds, such as cupric hydrogen arsenite, $CuHAsO_3$ (Scheele's Green), and copper (II) acetoarsenite, $Cu(CH_3COO)_2 \cdot 3\,Cu(AsO_2)_2$ (Paris green), have been used as rodenticides and insecticides for fruit trees and vines and as pigments. Other substances, such as $Pb_3(AsO_4)_2$, $Ca_3\,(AsO_4)_2$, and Na_3AsO_3, have been used as pesticides for tobacco and cotton treatment. Synthetic organic arsenic compounds in the United States and some other countries are used as growth stimulators in pigs and poultry and as veterinary drugs. These compounds are mainly derivatives of phenylarsonic acid, $C_6H_5AsO(OH)_2$, such as 4-aminophenylarsonic acid (known as arsanilic acid) and 4-hydroxy-3-nitrobenzenearsonic acid (known as roxarsone). Arsanilic acid is used in the prevention and the treatment of swine dysentery in veterinary medicine. Roxarsone was widely used agriculturally as a chicken feed additive, in order to promote growth.

6.5.3.1 Occurrence in the environment, transport, and distribution

Arsenic enters the environment mainly as a result of smelting activities, the combustion of coal, and wood preserved by arsenic compounds. In the countries where arsenic compound-based pesticides are permitted, these agrochemicals may be the dominant source. Volcanic activity and weathering of rocks contribute to the entry of arsenic into the environment to a lesser extent. Annual world emissions of arsenic have been estimated at 120 000 tons.

For example, the annual mean concentrations of arsenic in air in the UK rural environment are in the range of 0.001–0.004 μg/m^3, whilst in urban areas they range from 0.005 to 0.007 μg/m^3. Typically, the highest concentrations are found at sites located in the immediate vicinity of industrial processes such as smelters, incinerators, and cement works, where they can rise temporarily to tens or even hundreds of μg/m^3. The concentrations of arsenic in the air are locally variable and are higher in winter. In the vicinity of thermal power plants, concentrations of arsenic are also higher.

Natural waters, with the exception of some mineral and thermal waters, contain only traces of arsenic. Concentrations in unpolluted river and lake waters are in the range of 0.15 to 0.45 μg/l, where the arsenic is found in inorganic compounds. Seawater contains 0.1–2 μg/l of arsenic in the form of arsenite (*ortho*-arsenite, AsO_3^{3-}), arsenate (AsO_4^{3-}), methylarsonic, and dimethylarsinic acids.

The level of arsenic in soil depends on the geological bedrock and the distance from the sources of contamination. The concentration of arsenic in uncontaminated soils ranges from 2 to 10 mg/kg dry matter. Land used for agricultural purposes should not contain more arsenic than 20 mg/kg dry matter. Arsenic in soil, unlike lead, cadmium, and mercury, is quite mobile under neutral or slightly alkaline and reducing conditions. This relates to better solubility of trivalent arsenic compounds in comparison with pentavalent compounds. The mobility of arsenic in the soil determines its availability to plants. The transition coefficient of arsenic for soil–plant transfer is about 0.02–0.1. In areas with high rates of deposition and areas where arsenic-containing pesticides are applied, the predominant intake by plants is foliar intake. Some crops (tobacco, cotton, and oats) have a high tolerance to arsenic and concentrate it more than other plants.

6.5.3.2 Occurrence in foods and dietary intake

The total arsenic contents of selected foods are shown in Table 6.14. Aquatic organisms, particularly marine organisms, contain arsenic at concentrations ranging from 1 to 100 mg/kg as a result of bioaccumulation and biotransformation processes. Marine organisms accumulate arsenic compounds from water and transform it particularly to arsenobetaine, arsenocholine (see Section 6.2.3.1) and some other organic compounds (**6-18**). Other arsenic compounds (arsenosugars and lipophilic compounds) are mentioned in Section 6.2.3.1. The organic compounds of arsenic are less toxic than the inorganic ones. According to Belgian research (2009), the average total As concentrations in fish and shellfish species from the North Sea were 12.8 and 21.6 mg/kg, respectively. An average As concentration of 0.132 mg/kg was found in fish and 0.198 mg/kg in shellfish (approximately 10% of total As exposure from foods is in the toxic forms). For instance, a marine bivalve mollusc variegated scallop (*Chlamys varia*) with a total arsenic content of 3.36 mg/kg contained 0.10 mg/kg of As as arsenite (As^{III}), 0.11 mg/kg of As as arsenate (As^V), 0.35 mg/kg of As as dimethylarsonic acid, 2.73 mg/kg of As as arsenobetaine, and 0.14 mg/kg of As as arsenocholine. Atlantic cod (*Gadus morhua*) with a total arsenic content of 34.9 mg/kg contained 0.05 mg/kg arsenite (As^{III}), 0.09 mg/kg arsenate (As^V), 0.13 mg/kg dimethylarsonic acid, 33.2 mg/kg arsenobetaine, and <0.03 mg/kg arsenocholine.

methylarsonic acid   dimethylarsinic acid   tetramethylarsonium   arsenocholine (2-hydroxyethyl)-trimethylarsonium

homoarsenocholine (3-hydroxypropyl)-trimethylarsonium   arsenobetaine carboxymethyltrimethylarsonium   homoarsenobetaine (2-carboxyethyl)-trimethylarsonium

trimethylarsine oxide   dimethylarsinoylpropionic acid

6-18, important organic arsenic compounds

Higher amounts of arsenic in foods of vegetable origin (tenths of mg/kg) occur in oats and rice, and some wines. The dietary exposition to arsenic from rice needs long-term attention, because an appreciable part (or even majority) of total arsenic in rice is toxic inorganic arsenic.

Higher amounts may be found in wine if arsenic pesticides are used in the vineyard. The normal levels of arsenic in wines range from 0.002 to 0.1 mg/l. The edible bolete (*Cyanoboletus pulverulentus*) found in Europe and eastern North America, commonly known as the ink-stain bolete, may hyperaccumulate arsenic up to 1300 mg/kg (d.w.) in fruit bodies. Besides occasional traces of methylarsonic acid, the arsenic occurs almost solely as dimethylarsonic acid.

The TDI of arsenic for an adult before 2009 was 150 μg (at 70 kg body weight). According to the EFSA, this value is no longer applicable, as carcinogenic effects may be manifested even in cases of low exposure. BMDL values for 1% extra cancer risk have thus been estimated for various kinds of cancer that might be induced by arsenic, with the value for a person of 70 kg weight calculated at 21–560 μg/day. The estimated dietary exposure to arsenic in EU countries for an adult (of body weight 70 kg) ranges from 5 to 46 μg/day (median value 11 μg/day). Total-diet studies showed a very wide range from 17 μg/day (Canada, 1987) to 130 μg/day (United Kingdom, 1985).

6.5.3.3 Metabolism and toxic effects

The absorption efficiency of inorganic arsenic compounds in the gastrointestinal tract is about 5–25%. Organic arsenic compounds are apparently completely absorbed. Inhalation is also a route of exposure, and a significant source is tobacco smoking, particularly when the tobacco has been treated with arsenic-based pesticides. Owing to the high affinity for keratin, arsenic accumulates in hair and nails. The normal level in hair is about 0.5 mg/kg. Inorganic arsenic compounds are metabolised in the body to methylarsonic and dimethylarsinic acids and are then excreted in urine. The half-life of arsenic in the human body is 10–30 days. The toxicity of arsenic compounds decreases in the order AsH_3 (arsine) > AsO_3^{3-} (*ortho*-arsenite) = As_2O_3 (arsenic trioxide) > AsO_4^{3-} (arsenate) > methylarsonic acid > dimethylarsinic acid > arsenobetaine ≈ arsenocholine.

Arsenic interferes with the activity of important enzymes by binding to their thiol groups. It inhibits glutathione peroxidase, alanine aminotransferase, aspartate aminotransferase, glucose 6-phosphate dehydrogenase, cholinesterase, various phosphotransferases, and enzymes with lipoic acid as the cofactor. In the glycolysis process, arsenate can replace inorganic phosphate, yielding 1-arseno-3-phosphoglyceric acid instead of 1,3-bisphosphoglyceric acid; as a result, one molecule of ATP gained by glycolysis is lost (see Section 8.2.6.1.3). Arsenate and other arsenic compounds also inhibit the citric acid cycle by blocking the conversion of pyruvate into acetyl-CoA, which results in further loss of ATP.

Chronic arsenic poisoning can occur even at a constant intake of 10 mg arsenic/day. Large-scale accidental arsenic poisonings have occurred in the past. In the early 1900s, more than 6000 British beer drinkers were apparently poisoned with arsenic in the well-known Staffordshire beer epidemic. In 1955, more than 12 000 Japanese infants were poisoned, causing 130 deaths, when the sodium phosphate used as a stabiliser in infant formula preparations was contaminated with arsenic. More recently, in 1973, 11 cases of arsenic poisoning in western Minnesota (United States) were attributed to the consumption of contaminated water. Chronic poisoning involves the loss of body weight, increased ptyalism, and deterioration of vision. Typical skin changes include swelling, eczema, and keratosis. Haematological and neurological changes can also occur (motor paralysis of the fingers, sleepiness, memory loss, confusion, and impaired hearing). Arsenic has carcinogenic, mutagenic, and teratogenic effects.

The symptoms of acute poisoning with arsenic trioxide (As_2O_3) are abdominal pain, vomiting, and diarrhoea. The skin is moist, and the pulse and breathing are weak and intermittent. A fatal dose for an adult is 70–180 mg, depending on their weight.

6.6 Toxic inorganic anions

Food, drinking water, and beverages contain numerous inorganic anions. Some anions are forms of organogenic elements (such as carbonates and hydrogen carbonates), essential elements (chlorides, phosphates, sulfates, iodides, fluorides, borates), and non-essential elements (bromides). These and many other anions are mostly beneficial or harmless, and exhibit toxic effects only if present in large quantities. Toxic effects can also result from an excessive accidental intake of these anions (fluorides, iodides, and bromides). Some anions exhibit toxic effects on the human organism even if ingested in low concentrations. Legislation only specifies the maximum levels for nitrates and nitrites. Other anions, however, may also have toxic effects, such as arsenates and arsenites (see Section 6.5.3.3), chromates and dichromates (see Section 6.3.13.1), perchlorates (see Section 6.3.2.2), and cyanides (see Section 10.3.9). Antinutritional effects are found in thiocyanate anions (rhodanides), which are metabolic products of cyanides (see Section 10.3.9.1.3) or of the degradation of some glucosinolates cruciferous (Brassicaceae) plants (see Section 10.3.10.2).

6.6.1 Nitrates and nitrites

Nitrates and nitrites are natural components of the environment and participate in the nitrogen cycle, a process involving the conversion of nitrogen into its various chemical forms. Important processes in the nitrogen cycle include fixation, mineralisation, nitrification, and denitrification. The conversion of gaseous nitrogen from the atmosphere into a form available to plants (and, hence, to animals and humans) is associated with leguminous plants through their symbiotic association with *Rhizobium* and other bacteria (e.g. of the genus *Azotobacter*) that produce ammonia (NH_3) or ammonium ions (NH_4^+) using the enzymes known as nitrogenases. The soil nitrogen also comes from decaying plant and animal residues (crop residues, green manures, and animal manure) containing proteins, nucleic acids, and other nitrogen compounds, which are decomposed to ammonia by various bacteria and fungi. Nitrification is the process by which ammonium ions and ammonia are oxidised into nitrites (NO_2^-) by ammonia-oxidising bacteria (e.g. of the genus *Nitrosomonas*) and the nitrites are further oxidised into nitrates (NO_3^-) by nitrite-oxidising bacteria (of the genus *Nitrobacter*). The two processes of nitrification are called nitritation and nitratation. Denitrifying bacteria release nitrogen from the nitrates, which returns to the atmosphere.

Soil nitrogen occurs mainly in the form of ammonium ions and nitrates available to plants. Ammonium ions are retained in the soil, since the cation is attracted to and held by the negatively charged soil clay and not leached to any great extent. The majority of the nitrogen utilised by plants is absorbed in the nitrate form. Ammonium salts and nitrates are used as commercial fertilisers. However, nitrate is highly leachable and readily moves with water through the soil profile. If there is excessive rainfall or over-irrigation, nitrate will be leached below the plant's root zone and can eventually contaminate the groundwater.

6.6.1.1 Occurrence

Nitrites and nitrates are an integral part of the nitrogen cycle in nature. Plants use nitrates from the soil to satisfy nutrient requirements and may accumulate them in their leaves and stems, and from there they can get into animal feed and human food. Nitrates and nitrites are likewise used as food additives to inhibit the growth of microorganisms in cured and processed meats. Very high concentrations of nitrates and nitrites have been implicated in the appearance of methaemoglobinaemia and possibly in gastric cancer. Actually, the vast majority of consumed nitrate and nitrite comes from natural vegetables and fruits rather than food additives (see Section 12.3.7). However, recently, these two ions have been considered essential nutrients which promote nitric oxide production and consequently help cardiovascular health.

6.6.1.1.1 Foods of plant origin

The nitrate content in plants is strongly influenced by genetic and environmental factors. A large variability of nitrate content is observed between plant species and even between cultivars of the same species. Nitrate distribution is unequal between plant parts. Leaf blades have a lower content than stems and petioles, and young leaves show a lower nitrate concentration than older leaves. Accumulation of nitrates occurs in situations where the plant does not reduce nitrates to more easily assimilable ammonium salts. These situations especially include adverse humidity, temperature, and insufficient light intensities that cause a lack of the carbon compounds necessary for conversion of accumulated nitrates into amino acids and proteins. A decrease of nitrate accumulation in the edible parts of plants during the day period and an increase during the night and under controlled irradiance in glasshouses, as well as a correlation between plant water and nitrate concentration, have been shown.

Vegetables (especially leafy vegetables, such as fresh lettuce and spinach) and potatoes are very important in human diet, but they unfortunately constitute a group of foods that contributes very much to dietary nitrate dose. Under excessive use of nitrogen fertilisers, these vegetables can accumulate high levels of nitrate which pose serious health hazards. Individual vegetables accumulate nitrates in varying quantities. According to their ability to accumulate nitrates, vegetables and root crops can be divided into three groups:

- vegetables with a high nitrate content (>1000 mg/kg), such as lettuce, spinach, endive, chard (mangold), Pekinensis and Chinese cabbages, radishes, small radishes, celery, rhubarb, and sweet maize;

- vegetables with a medium nitrate content (250–1000 mg/kg), such as cabbage, kale, cauliflower, eggplant, parsley, carrots, broccoli, garlic, and potatoes;

- vegetables with a low nitrate content (<250 mg/kg), such as Brussels sprouts, onions, tomatoes, peas, artichokes, asparagus, and cucumbers.

The common nitrate contents of some vegetables are listed in Table 6.16. In some crops, however, the nitrate content varies over a wide range (up to hundreds of per cent), due to genetic, climatic, and soil conditions (e.g. light intensity, rainfall intensity, and fertilisation). In fruits, nitrates are present in much lower concentrations. Only bananas and melons can have somewhat higher levels, of around 600–800 mg/kg.

6.6.1.1.2 Foods of animal origin

The natural content of nitrates in animal tissues is very low compared with plant tissues. The only exceptions are some meat products (such as ham and some sausages) and cheeses in which nitrates or nitrites are used as additives during manufacture.

Nitrates and nitrites are also used as food additives in order to produce a pink colour and a specific flavour in certain meats (e.g. cured meats). The nitrite used in meat curing is produced commercially as sodium nitrite, which possesses antimicrobial properties and prevents the growth of the bacterium *Clostridium botulinum*. This bacterium produces several toxins that cause a paralytic illness leading to

Table 6.16 Levels of nitrates in important vegetables and potatoes.

Vegetable	NO₃⁻ content (mg/kg) Minimum	NO₃⁻ content (mg/kg) Maximum	Vegetable	NO₃⁻ content (mg/kg) Minimum	NO₃⁻ content (mg/kg) Maximum
Celery	0	3640	Bell pepper	4	330
Onion	0	1435	Parsley	0	5400
Garlic	44	2400	Leek	30	2159
Beans	14	717	Tomatoes	0	136
Pea	10	58	Rhubarb	300	2525
Savoy cabbage	0	3192	Radish (daikon, mooli)	300	3770
Brussels sprouts	0	2500	Radish	390	5200
Kohlrabi	80	4380	Red beet	45	4700
Cauliflower	0	2685	Lettuce	60	6600
Aubergine (eggplant)	71	960	Spinach	20	4500
Carrots	0	3337	Cabbage	0	3230
Cucumbers	0	490	Potatoes	0	2795

respiratory failure, as seen in foodborne botulism (see Section 12.3.2.1). In some meat products (e.g. ham), sodium nitrate is used instead of sodium nitrite due to their long ageing period. Sodium nitrate is also used in the manufacture of some aged semi-hard and hard cheeses (such as Emmental, Gouda, and Edam) in order to prevent the late-blowing defect (characterised by eyes, slits, and cracks caused by the formation of gas bubbles) and an aberrant and unwanted aroma. These defects arise from the growth of *Clostridium tyrobutyricum*, a bacterium that produces hydrogen, butyric acid, and acetic acid as its major metabolites. The added sodium nitrate is reduced to nitrite, which inhibits the germination of bacterial spores.

6.6.1.1.3 Water

Nitrates and nitrites are present in all tap and bottled waters. They originate from natural deposits and are formed during the natural decay of vegetable material in soil. Rainfall washes them from the subsoil into groundwater. Nitrogenous fertilisers used on arable farmland and leaks from septic tanks can be significant sources of nitrates in ground and surface waters.

Many developed countries specify their own drinking water quality standards. For countries without a legislative or administrative framework for such standards, the World Health Organization (WHO) publishes guidelines (see Section 7.2.3.1). The following standards are included in the Drinking Water Directive (valid in the European Union) to ensure that drinking water will not cause methaemoglobinaemia: 50 mg/l nitrate (as NO_3^-) and 0.5 mg/l nitrite (as NO_2^-). The US Environmental Protection Agency sets values of 10 mg/l nitrate (as nitrogen) and 1 mg/l nitrite (as nitrogen).

6.6.1.2 Health and toxicological evaluation

The content of nitrates in food is regulated in many countries. Values for the European Union are shown in Table 6.17. At common concentrations, nitrates are relatively non-toxic for adults, because they are quite rapidly excreted in urine. The Acceptable Daily Intake (ADI), which quantifies the daily amount of nitrates in food that a person can ingest over a lifetime without a health risk, is 3.7 mg/kg body weight. Their potential toxicity results from the possibility of their reduction to nitrites (ADI = 0.07 mg/kg body weight).

Nitrates are partially reduced by the nitrate reductase of microorganisms present in raw materials during transport, storage, and processing of vegetable raw materials. Nitrites also form endogenously in the digestive tract by the action of microorganisms. The ingested nitrates are excreted in urine (about 80%, or about 50% in older people) in 4–12 hours, and the rest remains in the body. The nitrates remaining in the digestive tract are mainly transformed into ammonium salts. The toxic effect of nitrites, after their absorption into the blood, depends upon their possibility of inducing methaemoglobinaemia, a blood disorder in which an abnormal amount of methaemoglobin (containing Fe^{3+}) is produced by the oxidation of haemoglobin (containing Fe^{2+}). Methaemoglobin is unable to transport oxygen effectively to body tissues (see Section 9.2.1.5.1). Under normal physiological conditions, the amount of methaemoglobin is about 2%. The methaemoglobin reductase in the red blood cells of an adult converts methaemoglobin back into haemoglobin, but this enzyme may be produced in insufficient amounts or be absent in people who have an inherited mutation.

Table 6.17 Maximum levels for nitrates in foods.

Product		Maximum level (mg/kg)
Fresh lettuce (protected and open-grown), excluding iceberg-type lettuce	Harvested 1 October–31 March, grown under cover	4500
	Harvested 1 October–31 March, grown in the open air	4000
	Harvested 1 April–30 September, grown under cover	3500
Fresh spinach[a]	Harvested 1 October–31 March	3000
	Harvested 1 April–30 September	2500
Fresh lettuce (protected and open-grown), excluding iceberg-type lettuce	Harvested 1 April–30 September, grown in the open air	2500
Iceberg-type lettuce	Grown under cover	2500
	Grown in the open air	2000
Preserved, deep-frozen, or frozen spinach		2000
Processed cereal-based foods and baby foods for infants and young children[b]		200

Official Journal of the European Union, 20 December 2006, L 364/5. Commission Regulation (EC) No. 1881/2006 of 19 December 2006 setting maximum levels for certain contaminants in foodstuffs.

[a]Maximum levels do not apply for fresh spinach to be subjected to processing.

[b]Maximum level refers to products ready to use (marketed as such or after reconstitution as instructed by the manufacturer).

Nitrites causing methaemoglobinemia are especially dangerous in infants during the first two to four months (or even up to six months) of life (so-called blue-baby syndrome). During this period, the enzyme system is underdeveloped. The amount of foetal haemoglobin in infants is about 85% of total haemoglobin, and foetal haemoglobin is more easily oxidised than the haemoglobin of adults (haemoglobin A). In infants, the stomach has a lower concentration of acids (higher pH), so microorganisms may reduce nitrates to nitrites before their absorption. Nitrates can also partly move to the salivary glands, where they concentrate and re-enter the oral cavity. Endogenous reduction to nitrites takes place in the mouth, both in children and in adults. This forms up to 65% of the total content of nitrites. The manifestation of methaemoglobinaemia is a grey-blue to blue-violet colour of the mucous membranes, skin, and peripheral parts of the body (i.e. around the eyes and mouth). The first symptoms appear when the methaemoglobin concentration in the blood reaches 6–7%. After the age of six months, methaemoglobinaemia is no longer a threat, as the nitrate-converting bacteria are no longer present in the baby's stomach. Therefore, the nitrate and nitrite levels in water and food meant for infants are regulated.

6.7 Radionuclides

6.7.1 Radionuclides and radioactivity

Radionuclides are nuclides with an unstable nucleus that are subject to radioactive decay. There are three main types of radioactive decay:

- α-decay, a stream of particles composed of two protons and two neutrons (identical with the nucleus of helium ^{4}He);

- β-decay, a stream of electrons ($^-$β) or positrons ($^+$β) of high speed (about 10^8 m/s);

- γ-radiation (X-rays), an electromagnetic radiation of high frequency ($>10^{19}$ Hz) and short wavelength (<1 nm) that accompanies emissions of α- and β-particles.

Radionuclides can be roughly divided into α-emitters and β-emitters. The emission of α-particles is associated with a decrease in mass number by four units and in atomic number by two units. The loss of two protons results in an excess of two electrons, so α-decay is accompanied by the emission of electrons (radiation $^-$β). In the radioactive β-type conversion, the nucleon in the atom nucleus transforms and a neutron turns into a proton, or vice versa, which is associated with a change in particle charge and the emission of an electron or a positron. The capture of an electron (usually from sphere K of the electron cloud) by the atom nucleus is also possible, changing a proton into a neutron. In the electron cloud, an electron from the higher level jumps to fill the vacant place, accompanied by the emission of a photon (γ-radiation).

6.7.2 Sources of radioactivity

All the heavy elements, starting from the element with atomic number 84 (polonium, Po), have at least one radioactive nuclide (isotope). However, a number of light elements also form radioactive isotopes (such as tritium ^{3}H, carbon ^{14}C, and potassium ^{40}K). In nature, there are three types of radionuclides:

- heavy nuclides;

- their daughter products (daughter nuclides);

- light nuclides.

The first group includes very heavy elements such as uranium (U) and thorium (Th). Radionuclides of these elements have an extremely long half-life (e.g. ^{232}Th 1.41×10^{10} years, ^{238}U 4.51×10^9 years, ^{235}U 7.1×10^8 years). The second group includes daughter nuclides formed by the radioactive decay of the first group. For example, radium ^{226}Ra (half-life of 1600 years) arises from the decay of uranium ^{238}U. The decay of radium yields the radionuclides radon ^{222}Rn, polonium ^{210}Po, and lead ^{210}Pb. Natural heavy nuclides represent three decay chains. A parent isotope is one that undergoes α- or β-decay to form a daughter isotope. The daughter nuclide may be stable or it may decay to form a daughter nuclide of its own. Each decay chain thus has its initial mother nuclide and a final stable nuclide. By tracking the interrelationships of these nuclides, the radioactive nuclides are arranged into the following series:

- thorium series: begins with thorium ^{232}Th and ends with lead ^{208}Pb;

- uranium series: begins with uranium ^{238}U and ends with lead ^{206}Pb;

- actinium series: begins with so-called actinouranium ^{235}U and ends with lead ^{207}Pb.

Another group of radionuclides is generated by cosmic rays in the light elements in the atmosphere. For example, nitrogen forms radioactive carbon ^{14}C. The natural radioactivity in nature is mainly due to radionuclides of light elements. Approximately 90% of the radioactivity is attributable to ^{40}K (represents 0.012% of natural potassium and has a half-life of 1.3×10^9 years) and almost all the rest to carbon ^{14}C (half-life of 5730 years). Increased radioactivity in certain areas can be caused by an abnormally high incidence of radioactive elements (especially uranium) in rocks.

Anthropogenic sources of radioactivity in the environment include nuclear weapons testing and radioactive material handling, especially in nuclear power plants. In the explosion of atomic bombs and in nuclear reactors, a complex mixture of different radionuclides is produced, namely uranium ^{235}U, plutonium ^{239}Pu, caesium ^{137}Cs (half-life of 30 years), strontium ^{90}Sr (half-life of 28 years), cobalt ^{60}Co (half-life of 5.3 years), caesium ^{134}Cs (half-life of 2 years), ruthenium ^{106}Ru (half-life of 1 year), and iodine ^{131}I (half-life of 8 days). A number of other radionuclides result from an atomic explosion by collision of neutrons with the atoms of elements that are contained in the casing of the non-explosive parts of the atomic bomb. For example, these activation products include zinc ^{65}Zn (half-life of 245 days).

6.7.3 Content of radionuclides and radiation dose

The content of radioactive nuclides in environmental materials and foods is usually expressed as the specific activity, given most often in Bq/kg (for liquid materials, Bq/l). The dose of radiation absorbed is expressed as the mean energy absorbed per unit of weight or per unit of volume of material. The basic unit of the radioactive dose is the gray (Gy); 1 Gy (in J/kg) is defined as the absorption of 1 J of ionising radiation by 1 kg of material.

The individual types of ionising radiation resulting from nuclear processes have different biological effects. Therefore, in addition to the exposure dose (Gy), the so-called equivalent dose, absorbed by a given mass of biological tissue, was introduced. The equivalent dose (H) for tissue T and radiation type R is calculated by the formula $H_{T,R} = Q \cdot D_{T,R}$, where Q is a radiation (radiobiological) quality factor that depends on the type and energy of that radiation and $D_{T,R}$ is the total energy of radiation absorbed in a unit mass of any material. The quality factor Q is also known as the relative biological effectiveness of the radiation. For example, β- and γ-radiation have a value of $Q = 1$, whilst α-radiation has a value of $Q = 10$. The radiation quality factor of the neutron flux varies, depending on the neutron energy, from 2.5 to 10. Whilst the Gy measures the absorbed dose of radiation (D) by any material, the sievert (Sv, in J/kg) measures the equivalent dose of radiation (H) and evaluates the biological effects of ionising radiation. This means that a dose of α-radiation of 1 Gy corresponds to a dose-equivalent radiation of 10 Sv. The same dose-equivalent radiation has a dose of γ-radiation of 10 Gy. The load of the organism due to natural radioactivity is around 2 Sv per year, of which about 0.4 mSv is due to radionuclides occurring in food.

To assess the radiation hazard to humans, it is necessary to take into account several factors:

- radionuclide species, including the possibility of its conversion into other radionuclides and the related type of radiation;

- specific activity of the environment (food);

- type and length of exposure (external irradiation, internal irradiation by ingestion of radioactive food, inhalation of radioactive aerosols);

- chemical properties of radionuclides (in the case of internal exposure) affecting their behaviour in the body; their chemical analogy with essential elements and similar reactivity results in transportation to certain target organs, where they accumulate (e.g. accumulation of radioactive iodine ^{131}I in the thyroid gland, analogy of radioactive caesium ^{134}Cs with ^{137}Cs and of strontium ^{90}Sr with potassium and calcium).

According to their potential hazard, radionuclides are classified into four groups:

- very high radiotoxicity (^{90}Sr, ^{210}Pb, ^{211}At, ^{226}Ra, ^{227}Ac, ^{239}Pu, ^{241}Am, ^{242}Cm; deposited mainly in the bones);

- high radiotoxicity (^{45}Ca, ^{59}Fe, ^{89}Sr, ^{131}I, ^{140}Ba, ^{234}Th, ^{238}U);

- medium radiotoxicity (^{22}Na, ^{24}Na, ^{32}P, ^{35}S, ^{36}Cl, ^{42}K, ^{60}Co, ^{132}I, ^{137}Cs);

- low radiotoxicity (^{3}H, ^{7}Be, ^{14}C, ^{18}F).

6.7.4 Occurrence in the environment and in foods

A substantial part of natural radioactivity is produced by nuclides of light elements ^{40}K and ^{14}C that are dispersed in nature along with the major isotopes of potassium and carbon. Heavy radionuclides are not uniformly distributed in nature. For example, around 99.284% of natural uranium is ^{238}U that is strongly bound in some rocks (granite). Given the very long half-life (4.468×10^9 years), the effect of ^{238}U in the environment is permanent. The decay of this isotope creates a number of other radionuclides that are generated via ^{234}Th (half-life of 24 days) and ^{234}Pa (half-life of 1.2 minute) and themselves have long half-lives: ^{234}U (245 thousand years), ^{230}Th (75 000 years), and radium ^{226}Ra (1600 years). Radium gives, by α-decay, the radioactive noble gas radon ^{222}Rn, with a half-life of 90 hours. The activity of radon contributes significantly to the exposure of people in certain buildings, such as those that are thermally insulated from their surroundings by materials containing granite. Because it can be inhaled, a prolonged exposure to radon is quite dangerous. Other members of the decay series are mostly nuclides with short or very short half-lives. The only exceptions are lead ^{210}Pb (22 years) and polonium ^{210}Po (138 days). The stable end-product of a decay of uranium ^{238}U is lead ^{206}Pb.

Mining and processing of minerals, ores, and fossil fuels bring radioactive elements (and other toxic substances) into the environment. Handling of nuclear materials, nuclear accidents, and nuclear weapons tests also contribute to environmental radioactivity. The most famous incident was the atomic power plant disaster in Chernobyl, Ukraine, which occurred on 26 April 1986. From 1986 to 2000, over 350 000 people were evacuated and resettled from the most severely contaminated areas. Leakage of the radioactive material (with a predominance of ^{131}I, ^{137}Cs, and ^{132}Te) into the environment was so extensive that there was significant contamination not only of the area immediately around the atomic power plant but also, due to atmospheric transfer, of much of Europe and parts of Asia. The total radioactivity released as a result of this nuclear disaster is estimated at 3×10^{18} Bq. In order to minimise public exposure to radioactive elements (mainly ^{131}I), Poland and some Western European countries banned the distribution of milk originating from cows grazing freely outside, and children preventively received non-radioactive potassium iodide. This event significantly increased the interest in studying the behaviour of radionuclides in the environment. In the 1980s and '90s, as a consequence of the Chernobyl disaster, radioactivity was monitored in different compartments of the environment virtually worldwide. Significant attention was paid to long-range transport of radioactive aerosols, to radioactive fallout, to contamination of soil, and to the possibility of transport of radionuclides from soil to plants.

The mobility of radionuclides in environmental compartments is affected by the chemical properties of the source element, and is higher for halogens (e.g. ^{129}I, ^{131}I, and ^{36}Cl), alkali metals (^{40}K, ^{134}Cs, ^{137}Cs), non-metals (^{89}Se), and alkaline earth metals (^{90}Sr) than for transition metals and, especially, transuranium elements. Plants are partially contaminated by direct deposition on the leaves and partly by intake of radionuclides from the soil, which depends on the binding of the radionuclide in the soil components, the depth of the root system, and the developmental stage of the plant at the time of deposition. Study of the transfer of caesium radionuclides from soil to plants and fungi after the Chernobyl disaster shows the following results:

- alfalfa (*Medicago sativa*), which has deep roots, was unaffected by fallout by the second generation, unlike grasses and wild herbs, which are rooted largely in the surface soil layer;

- winter cereals were roughly twice as contaminated as spring cereals;

- summer fruits (such as currants) were more contaminated than late fruits (apples);

- needles of conifers (often used as bioindicators of pollution) showed a specific activity of caesium ranging from 1×10^2 to 1×10^3 Bq/kg; fungi and lichens showed values up to one order of magnitude higher ($1 \times 10^2 - 1 \times 10^4$ Bq/kg).

The Chernobyl disaster resulted in substantial radioactive fallout in parts of Norway. The deposition of radioactive dust caused contamination by caesium ranging from 5 to 200 kBq/m². In some areas of Czechoslovakia, the contamination was higher than 10 kBq/m²; in other areas, it reached only 2–3 kBq/m². In areas of Finland, the transfer of ^{137}Cs and ^{90}Sr from soil to some crops was studied for several years. Radioactive caesium follows the path of potassium and tends to accumulate in plant tissues, including fruits and vegetables. The ratio of specific activities of radionuclides in plants and soil ranged from 0.01 to 2.29 for ^{137}Cs and from 0.02 to 2.44 for ^{90}Sr (values in subsequent years gradually decreased). The value for ^{137}Cs decreased in the individual crops as follows: lettuce, cabbage > carrots, potatoes, and onions > cereals. The smallest transfer of ^{137}Cs to plants was recorded in crops grown on clay soils. The order for ^{90}Sr was: lettuce, cabbage > carrots, onions > cereals > potatoes.

Small amounts of ^{134}Cs and ^{137}Cs are released into the environment during nearly all nuclear weapon tests and some nuclear accidents, most notably the Chernobyl disaster. It is well documented that mushrooms, including edible species from contaminated forests, accumulate radionuclides (mainly ^{137}Cs) in their fungal sporocarps. For example, the edible mushroom commonly known as penny bun, porcino, or cep (*Boletus edulis*) from Poland (1987) showed about seven times higher activity of ^{137}Cs in comparison with the radioactive background, and about twice the activity of mushrooms from the United States. This activity (1100 Bq/kg) corresponded to a concentration of ^{137}Cs of around 0.0003 ng/kg, which is negligible in comparison with the content of stable caesium (^{133}Cs). To achieve radiotoxic doses, it would be necessary to consume 3000 kg of dry mushrooms. In addition, it was found that radioactive caesium could be effectively removed from the mushroom by washing with water or salt solution. High concentrations of radionuclides found in the fruiting bodies of the bay bolete *Boletus badius* after Chernobyl have been attributed to the complexation of ^{137}Cs by the so-called naphthalenoid pulvinic acids, which occur as their dipotassium salts in the cap skin of this mushroom. The main compounds are badione A and norbadione A (**6-19**), which are responsible for the chocolate-brown and golden-yellow colours of the cap skin of this and related bolete species. For example, complexation of ^{137}Cs with norbadione A involves a mixture of *Z/E* isomers and conformers with a broad diversity of binding modes. In the mycelia, ^{137}Cs is mostly trapped by polyphosphates in vacuoles and other organelles.

badione A   norbadione A

6-19, naphthalenoid pulvinic acids of bay bolete (*Boletus badius*)

The content of radioactive caesium in the milk and meat of livestock depends mainly on the contamination of the feed. For example, cows absorb about 30% of radioactive caesium in the feed. The specific activity of ^{137}Cs in beef in 1986 ranged from 14 to 129 Bq/kg. The maximum value observed was 240 Bq/kg in milk, and most samples did not exceed a level of 40 Bq/kg. In subsequent years, there was a significant decline (in 1988, only 0.7% of milk samples exceeded the level of 20 Bq/kg).

Significantly higher specific activity of caesium was detected in the meat of game (in 1986, depending on surface contamination of the territory, observed values for roe deer meat were 120–600 Bq/kg and for deer meat were 170–660 Bq/kg). Contamination of wildlife by radionuclides decreases in the order: deer > roe deer and fallow deer > mouflons > wild boars (wild pigs) > hares.

As regards the dietary intake of radionuclides, an average exposure of an adult amounting to c. 1.7 mGy/year was found in Germany in the 1970s. An amount of 0.2 mGy was accounted for by natural radionuclides (mainly ^{40}K and ^{14}C). The total intake of radionuclides ^{90}Sr and ^{137}Cs was in units of Bq per person per day. Following the Chernobyl disaster in 1986, the year-long dietary intake of radionuclides was 4600 Bq ^{131}I, 1760 Bq ^{134}Cs, and 3400 Bq ^{137}Cs, and the total dose equivalent increased by 0.04–0.26 mSv.

6.7.5 Fate in the organism

After radionuclides have entered into the body of an animal or human by ingestion or inhalation, distribution to individual tissues and organs and partial incorporation into these tissues takes place. For example, radioactive caesium as an analogue of potassium, after entering into the body, quickly appears in the kidney, liver, and spleen and accumulates in muscle tissue. As with other alkali metals, the excess of radioactive caesium is excreted from the body relatively quickly in sweat and urine. Radioactive iodine concentrates in the thyroid gland. Heavy radioactive metals accumulate in internal organs and bones. Subsequently, these metals are excreted and their concentration decreases due to radioactive decay. The time required for the radioactivity of material taken in by a living organism to be reduced to half its initial value by a combination of biological elimination processes and radioactive decay is called the effective half-life $T_{ef} = T_f T_b/(T_f + T_b)$, where T_f is the physical half-life of a radionuclide and T_b is the biological (elimination) half-life of the element in the given form.

6.7.6 Legislation and regulation

Radiocaesium is one of the most important artificial radionuclides produced by nuclear fission. Consumption of agricultural products contaminated with radiocaesium represents the principal route of human exposure to this radionuclide. Within the European Union, limits of activity of ^{134}Cs and ^{137}Cs (a strong emitter of γ-radiation) of 370 Bq/kg have been established for milk and foods for children. For other foods imported into the European Union, a limit of 600 Bq/kg has been set.

7
Water

7.1 Introduction

Water is one of the most widespread substances in the biosphere. In food chemistry, it is one of a group of nutrients (together with proteins, lipids, carbohydrates, vitamins, and minerals) that are necessary for the normal functioning of living organisms. Owing to its physico-chemical properties, it is particularly important:

- in temperature regulation of living organisms;
- as a transport medium for nutrients, metabolic products, and respiratory gases;
- as a solvent or dispersion medium;
- as a reactant involved in biochemical and chemical reactions.

The normal functioning of the human body inevitably leads to a continuous loss of water that must be compensated for by the water produced via the oxidation of primary nutrients: proteins, carbohydrates, and lipids. This water is sometimes referred to as **endogenous** or **metabolic water**. The amount of endogenous water is not sufficient to cover the lost water, which the organism has to compensate for with **exogenous water** contained in foods and especially in drinks. It is reported that the daily output of water by an adult is approximately 2500 g, of which about 1500 g (minimum 600 g) is excreted in the urine, about 550 g is excreted through the skin (sweating), 350 g is exhaled into the air, and 100 g is excreted in faeces. Replacement of the same amount is met by exogenous water in the form of various beverages (about 1300 g) and foods (about 900 g), and approximately 300 g of water is obtained by oxidation of nutrients. Oxidation of 1 g of proteins gives 0.37 g of water, oxidation of 1 g of fat gives 0.4 g, and oxidation of 1 g of carbohydrates (glucose) yields 0.6 g. Foods are very important sources of water, a fact that is often overlooked – we tend to concentrate only on the importance of drinking water and on water present in beverages.

This chapter deals with the production of drinking water and its quality requirements, the water content of foods, and the major changes to the water content that occur during culinary and technological operations. A substantial portion of the chapter is devoted to interactions of water with food components (inorganic compounds, proteins, carbohydrates, and lipids), various phenomena that occur on the phase interfaces (surface tension, adsorption, and capillary phenomena), food dispersal systems (sols, gels, emulsions, and foams), and water activity and its influence on the growth of microorganisms, biochemical and chemical reactions, texture, and the organoleptic properties of foods.

The Chemistry of Food, Second Edition. Jan Velíšek, Richard Koplík and Karel Cejpek.

7.2 Drinking water

Water is a renewable resource that exists in an endless hydrological cycle on the earth, moving between its gaseous, liquid, and solid forms. Drinking (potable) water comes from **surface water** and **ground water** resources. Surface water resources, including rivers, lakes, and reservoirs, are the largest and most reliable of all freshwater resources. They account for about 80% of all freshwater. Ground water is the water present beneath the earth's surface. Natural water is never chemically pure. According to its origin, various substances are dissolved and sometimes suspended in it. The quality of surface waters, as opposed to ground water, depends on many factors. In comparison with ground waters, surface waters usually have much higher concentrations of organic substances of various origins, higher amounts of dissolved oxygen, lower levels of carbon dioxide, and lower concentrations of iron and manganese ions. The level of microorganisms in surface waters is significantly higher than in ground waters.

7.2.1 Classification

Surface waters are categorised into several classes according to quality. Classification is based on the evaluation of mandatory water quality indicators that are guides to the oxygen regime (the amount of dissolved oxygen), basic chemical and physical parameters (pH, dissolved substances, conductivity, suspended solids, ammonia nitrogen, nitrate nitrogen, and total phosphorus), additional parameters (calcium, magnesium, chlorides, sulfates, anionic surfactants, hydrocarbons, and organically bound chlorine), heavy and toxic elements content (lead, cadmium, mercury, and arsenic), biological and microbiological parameters (especially coliform bacteria), and indicators of radioactivity. Local standards often differ by country.

Ground waters suitable for consumption are divided into two basic categories:

- **spring waters** (collected directly from an underground spring where it rises to the surface);

- **natural mineral waters** (also originate underground, but flow over and through rocks before being collected and bottled at source).

Natural mineral waters are evaluated according to different indicators, such as the total mineralisation, contents of dissolved gases and important components (carbonic, sulfuric, iodine, fluorine waters), actual pH, radioactivity of radon, spring temperature, and osmotic pressure. The presence of bicarbonates, sulfates, and chlorides of calcium and magnesium has practical significance in determining **water hardness**, and also the suitability and applicability of water for the production of foods and other purposes. Water hardness is assessed according to the level of bicarbonates (temporary hardness, also known as carbonate hardness) and of sulfates and chlorides (permanent hardness or non-carbonate hardness). Water hardness is often expressed in German degrees (°dH or dGH), American degrees (mg/l), English degrees (°e or °Clark), or French degrees (°f). For example, $1°dH$ corresponds to 10 mg of CaO or 7.17 mg of MgO in 1 l, $1°e$ corresponds to 64.8 mg of $CaCO_3$ per 4.55 l, and $1°f$ corresponds to 10 mg $CaCO_3$ per litre ($1°dH = 1.252°e = 0.5603°f$). Water with a hardness of 0–7 °dH is known as soft water, that with 7–14 °dH as moderately hard water, that with 14–21.3 °dH as hard water, and that with >21.3 °dH as very hard water. Temporary hardness can be removed by boiling or by adding calcium hydroxide. Calcium (or magnesium) bicarbonate is thus transformed into insoluble carbonate during cooking:

$$Ca(HCO_3)_2 \rightarrow CaCO_3 + H_2O + CO_2$$

$$Ca(HCO_3)_2 + Ca(OH)_2 \rightarrow 2CaCO_3 + 2H_2O$$

Removal of temporary and permanent hardness in practice can be achieved using ion exchangers or by the addition of sodium carbonate:

$$CaSO_4 + Na_2CO_3 \rightarrow CaCO_3 + Na_2SO_4$$

Bottled waters, strictly regulated under EU law, include three different categories:

- bottled drinking water;

- spring water;

- natural mineral water.

In the European Union, about 97% of all bottled water is either spring water or natural mineral water suitable for continuous use by the population. Bottled water may be called mineral water when it is bottled at the source. The US Food and Drug Administration classifies mineral water as water containing at least 250 ppm total dissolved solids, originating from a geologically and physically protected underground water source.

7.2.2 Production

High-quality, safe, and sufficient drinking water is essential for drinking, food preparation, and many other purposes, such as washing, cleaning, hygiene, and the watering of plants. Most water requires some type of treatment before use. Depending on the type (surface or ground) and quality of the water resource, the production of drinking water is achieved by various technological processes. Some water resources directly meet the requirements for drinking water, and others only require disinfection, aeration, or deacidification (removal of dissolved carbon dioxide and oxygen), but some are not suitable for drinking or for multiple uses in domestic, institution, and industrial application. Treatment processes typically consist either of chlorine disinfection only or of direct or conventional filtration and chlorination. Surface water sources from protected catchments are typically treated by chlorination only, whilst those from impacted catchments are treated by conventional coagulation, flocculation, sedimentation, filtration, and chlorination. Some resources require more complex treatment of the water, such as increasing the concentrations of Ca^{2+} and HCO_3^- ions (in soft water), reducing the concentration of Fe^{2+} and Mn^{2+} ions (using special manganese removal equipment), or removing NH_4^+ ions, heavy metals, and radioactive substances (radon and radionuclides).

Spring water and natural mineral water can undergo no or minimal treatment. The only treatments allowed prior to the bottling of waters are some physical processes (chemical processes are used only rarely) to remove unstable components, such as iron, sulfides, and arsenic through decantation, filtration, or treatment with ozone-enriched air. To protect the biological and microbiological quality, chlorine and its compounds should not be used. No additions are permitted except for carbon dioxide.

In some countries, bottled water for infants is not modified by any of the procedures described here, and hygienic quality assurance is possible only by ultraviolet radiation or ultrafiltration. It is also possible to stabilise this water using carbon dioxide.

7.2.3 Quality requirements

7.2.3.1 General requirements

Drinking water contains varying amounts of inorganic salts (cations and anions formed by dissociation of various salts), dissolved gases (carbon dioxide, oxygen, and other gases), and indicators of pollution, such as soluble portions of humus from the soil (called humins or humic substances), contaminants (e.g. phenols and petroleum fractions), and various bacteria. The presence of carbon dioxide and oxygen can increase the number of ions present; if the water is in contact with metals (copper, iron, and other metals), it can corrode kitchen and manufacturing equipment. Quality requirements for drinking water include microbiological, biological, physical, chemical, and radiological aspects. Some of these quality requirements are shown in Table 7.1 as an example, but local standards differ by country.

Bottled waters, other than those labelled 'natural mineral water', are expected to conform to essentially the same standards as the public water supply, and they are therefore suitable for giving to infants or for preparing infant feeds (as with tap water, bottled waters should be boiled and subsequently cooled before use). The requirements for bottled water are the highest for special water for infants, and the values of many parameters are several times lower. The limit is 50 mg/l of nitrate in the European Union and 44 mg/l in the United States (equivalent to 11.3 and 10 mg/l of nitrate nitrogen in 1 l, respectively) to ensure that drinking water will not cause methaemoglobinaemia, also known as blue-baby syndrome (see Section 6.6.1.2).

7.2.3.2 Food industry requirements

The food industry generally has high requirements for water quality, particularly in microbiological terms. For example, water hardness and its overall chemical composition are important factors determining the quality of beer produced in every brewery. Water with a certain degree of hardness is more suitable than soft water. For example, Dublin has hard water, which is suitable for producing Guinness, whilst Munich and Pilsen have softer water, which is particularly suited for producing lager beer. It is known that the calcium content is 14 mg/l in the Pilsner brewery waters, whilst in Munich brewing waters it reaches 109 mg/l. Both brewing waters also differ in their levels of magnesium (4 and 21 mg/l), bicarbonates (42 and 171 mg/l), sulfates (19 and 79 mg/l), and chlorides (9 and 53 mg/l). As a result of the water quality (amongst other factors), beers from these two cities have completely different characters (organoleptic properties). Beers of lighter colour require lower concentrations of bicarbonates and calcium and higher concentrations of sulfates, whilst in beers of darker colour the opposite is true. The presence of iron and manganese in large quantities is undesirable (iron concentration is limited to a value of 0.2 mg/l and manganese concentrations have to be even lower).

The dairy industry requires water with a low concentration of magnesium, otherwise the butter has a bitter taste. The iron concentration should not exceed 0.1 mg/l and manganese should be absent. Canning industries have similar requirements for the water they use. Higher nitrate concentrations are undesirable in the preservation of vegetables and meat.

The food industry also contributes significantly to the total volume of waste water (mainly from meat processing plants, sugar refineries, breweries, and dairies).

Table 7.1 Selected quality requirements for drinking water.

Parameter (units)	Value	Parameter (units)	Value
Indicator		Chromium (μg/l)	50
Colour, odour, taste, turbidity[a]	–	Copper (mg/l)	2
Conductivity (μS/cm at 20 °C)	2500	Lead (μg/l)	10
Hydrogen ion concentration (pH units)[b]	≥ 6.5 and ≤ 9.5	Mercury (μg/l)	1
Oxidisability (mg/l O_2)	5	Nickel (μg/l)	20
Aluminum (μg/l)	200	Selenium (μg/l)	10
Ammonium (mg/l)	0.5	Bromate (μg/l)	10
Chloride (mg/l)	250	Fluoride (mg/l)	1.5
Sulfate (mg/l)	250	Cyanide (μg/l)	50
Iron (μg/l)	200	Nitrate (mg/l)	50
Manganese (μg/l)	50	Nitrite (μg/l)	0.5
Sodium (mg/l)	200	Acrylamide (μg/l)	0.1
Total indicative dose (mS/year)	0.1	Benzene (μg/l)	1
Coliform bacteria (number/100 ml)	0	Benzo(a)pyrene (μg/l)	0.01
Microbiological		1,2-Dichlorethane (μg/l)	3
Escherichia coli (number/100 ml)[c]	0	Epichlorohydrin (μg/l)[c]	0.1
Enterococci (number/100 ml)[c]	0	Trichloroethene, tetrachloroethene (μg/l)	10
Chemical		Trihalomethanes -total (μg/l)	100
Antimony (μg/l)	5	Pesticides (μg/l)	0.1
Arsenic (μg/l)	10	Pesticides – total (μg/l)	0.5
Boron (mg/l)	1	Polycyclic aromatic hydrocarbons (μg/l)	0.1
Cadmium (μg/l)	5	Vinylchloride (μg/l)	0.5

[a]Acceptable to consumers and no abnormal change.
[b]May be reduced to pH 4.5 for still water in bottles or containers.
[c]The unit is number/250 ml for water in bottles or containers.

7.3 Water in foods

The water content in foods is highly variable. It is related to the chemical composition of the food raw materials (the original animal and plant tissues), the manner of their processing to give the final products, and the storage of these products. Water makes up 50–90% by weight of the raw materials of plant and animal origin and of many foods; the rest is called **dry matter**. According to their water content, foods are divided into those with high, medium, and low water contents. The amount of water significantly affects the organoleptic characteristics of a food (texture, smell, taste, and colour), its shelf life and resistance to microbial attack, and the enzymatic (biochemical) and non-enzymatic (chemical) reactions that occur during processing and storage.

7.3.1 Water content

The water contents of selected foods are given in Table 7.2.

Table 7.2 Water contents in selected foods.

Food	Water content (%)	Food	Water content (%)
Pork meat	30–72	Sugar (saccharose)	0–0.5
Beef meat	35–73	Fruits (juices)	81–94
Chicken	63–77	Vegetables	60–93
Fish	65–81	Potatoes	75–80
Milk (cows')	87–91	Legumes	10–12
Cheeses	30–78	Cereal grains	11–14
Eggs	74	Bread	35–45
Butter, margarines	15–18	Pasta products	9–12
Oils, pork lard	0–0.5	Nuts	3–6
Honey (syrups)	19 (40)	Beer	90–96

7.3.1.1 Foods of animal origin

The water content of meat depends on the species (origin) and, in particular, on the fat content. Owing to its relatively high fat content, pork meat tends to have lower water content (raw pork fat contains about 13% water), whilst beef has a higher water content. The liver contains 67–72% water. The water content in meat products is highly variable, typically ranging from 30 to 70%. Poultry meat contains more water than pork or beef and even more water has fish meat. For example, cod, carp, tuna, mackerel, trout, and eel contain 81, 78, 71, 68, 66, and 65% water, respectively.

The natural water content of milk varies within certain limits (87–91%) according to the fat content. Soft cottage cheese contains 78% water, whilst other cheeses can contain less (Camembert 52%, Cheddar, Emmental, and Roquefort 37–40%, and Parmesan 30%).

In butter and margarines, the water content varies over a relatively small range. Most butters and margarines contain 16% water. Some special products, such as butter and margarine spreads and products with reduced fat content, contain around 50% of water; rendered lard contains only traces.

The water content of eggs is relatively constant (on average 74%); egg white contains more (about 88%), whilst egg yolk contains less (about 49%).

The amount of water in honey depends upon the type and quality. Good-quality honey essentially has a low water content (generally about 17%). Honey is likely to ferment and lose its freshness if the water content is higher than 19%.

7.3.1.2 Foods of plant origin

The natural water content in the edible part of fruits and fruit juices mainly depends on the type. Bananas have a relatively low water content (about 76%), whilst pears (about 83%), apples (85%), peaches (89%), and strawberries (90%) have more. Oranges and lemons usually contain 86–87% water. Dried fruits normally contain 12–25%.

Fresh root vegetables (such as carrots and parsley) are usually about 90% water; cabbage contains about 92% water and lettuce and tomatoes about 95%. There are slightly lower water contents in some bulbous vegetables (89–93% in onion, 83–89% in leek, and 61–68% in garlic).

The water content of all cereals is virtually the same as that of flours and dry pasta products (9–14%). White bread usually contains 35–36% of water; there is more in rye bread (38–45%).

7.3.1.3 Beverages

The water content in beverages varies depending on the amount of added sugar and the concentrations of other substances. Fruit juices made from fresh fruit have approximately the same water content as the original fruit. Cola drinks usually contain about 90% water.

The water content in beer (90–96%) depends on the concentration of the original wort and the degree of fermentation. For the most common distilled drinks, such as whisky and vodka, the water content is around 60%. In liquors, the water content depends primarily on the amount of ethanol and on the added sugar.

7.3.2 Changes in water content

Changes to the water contents in foods and food raw materials take place in almost all modes of storage and during all methods of culinary and manufacturing processing. Storage of raw materials and foods in packaging materials that are permeable to water vapour leads to a decrease in water content through desiccation. In foods that easily become moist (dehydrated fruit juices and instant beverages, such as coffee and tea), the water content increases during storage under humid conditions.

In the manufacture of certain foods such as bread, but also of a wide range of meat, poultry, fish, and other products, water is intentionally added to the raw material in a specified amount. The water content also increases significantly during the steeping of pulses, whose chemical components (mainly polysaccharides) are able to bind water. Thermal processing of food by drying, cooking, baking, frying, grilling, and roasting usually decreases the water content. In cooked meat, myofibrillar proteins (such as actin and myosin) contribute significantly to water losses, as they release a certain amount of water during thermal denaturation, which leads to their aggregation. On the other hand, connective tissue proteins (such as collagen) can bind certain amount of water.

Losses of water and of water-soluble substances likewise occur during freezing and thawing of frozen foods, depending mainly on the speed of these two operations. If the processes are slow, losses are higher – due to cell damage by large ice crystals – than in the rapid freezing and thawing when the ice crystals are small.

A certain amount of water may also be formed by chemical reactions. An example is the thermal reactions during coffee roasting. The water naturally present in green coffee and water formed by the dehydration of carbohydrates during roasting escapes as steam, along with other volatile products of pyrolytic reactions.

Examples of changes in the water (dry matter) content of some common foods in different culinary procedures and during industrial processing are shown in Table 7.3.

Table 7.3 Changes of water content in foods during processing.

Food	Water content (%)	Food	Water content (%)
Pork meat		Sauce (shoyu)	63
Raw	68	Meat (dried)	9
Baked	55	Meat soaked in water for 1 h	65
Fried	53	Meat soaked in water for 10 h	72
Fish (cod)		Meat soaked (10 h) and boiled	79
Raw	81	*Potatoes*	
Preserved	79	Raw	80
Fried	65	Boiled (in skin)	80
Dried (salted)	52	Boiled (peeled)	83
Dehydrated (salted)	12	Flour	8
Milk (3.5% of fat)		French fries (pre-fried)	74
Raw	87	French fries (fried)	55
Pasteurised	87	Crisps (fried)	2
Evaporated (no sweetened)	74	*Apples*	
Condensed (sweetened)	27	After harvest	85
Dried	4	After storage	84
Soybeans		Boiled with sugar (puree)	80
Raw	10	Dried	24
Soaked in water for 1 h	35	Juice	88
Soaked in water for 10 h	60	*Onion*	
Soaked (10 h) and boiled	71	Raw	89
Flour	8	Boiled	92
Curd (tofu)	85	Fried	42
Milk	92	Dried	4

7.4 Structure

An oxygen atom in the ground state $(1s^2)$ $(2s^2)$ $(2p^4)$ has four valence p electrons, two of which are unpaired. The electronic structure of hydrogen $(1s)$ allows connection (hybridisation) of unpaired valence electron of two hydrogen atoms to two unpaired valence electrons p of the oxygen atom to four sp^3 orbitals. Two orbitals are bonding (molecular) orbitals, therefore two σ bonds or single covalent bonds O—H are formed and two orbitals are non-bonding orbitals (Figure 7.1). The water molecule H_2O, like other molecules, is geometrically and physically characterised by the bond (valence) angles of the three connected atoms H—O—H, bond length (distance between the centres of the atoms of oxygen and hydrogen), and energy of the covalent bonds O—H. The bond angle of the O—H covalent bond in a molecule of water is almost identical to the ideal tetrahedral angle 109°28′. The bond angle is 109°6′ in ice, 105° in liquid water, and 104°30′ in steam. The H—O bond length is 0.096 nm. The energy of covalent O—H bonds in a water molecule (dissociation energy) is about 461 kJ/mol. Intramolecular delocalisation of electrons in covalently bound hydrogen atoms towards the electronegative oxygen atom results in the formation of partial positive charges on both hydrogen atoms (indicated by δ+) and partial negative charge on the oxygen atom (δ−). The water molecule is therefore a polar molecule (it is polarised) and represents an electric (permanent) dipole. The measure of the polarity is the dipole moment, which is, compared with other compounds, relatively high: μ = 1.85 D or 6.17×10^{-30} C·m (D = debyes, C = coulomb).

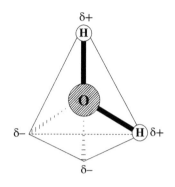

Figure 7.1 Hypothetical structure of a water molecule.

Water is a chemical substance that contains 18 different isotopic variants of the H_2O molecule due to the existence of isotopes of hydrogen and oxygen (^2H, ^3H, ^{17}O, and ^{18}O). In addition to undissociated water molecules, also present are hydrated hydrogen ions (protons), called oxonium (hydronium) ions (H_3O^+ and trimers $H_9O_3^+$, respectively) and hydroxyl ions (HO^-), resulting from dissociation of water and all the variants, which also contain the aforementioned isotopes of hydrogen and oxygen. The natural content of hydrogen and oxygen isotopes is very low (e.g. the amount of ^2H is only 0.000 165%); it is therefore sufficient to assume in many situations that the water molecule contains only ^1H and ^{16}O isotopes. Significant concentrations of ions in solution, however, exist at extremely low or extremely high pH values; that is, in strongly acidic or strongly alkaline solutions.

Biochemical reactions in living systems usually occur in neutral media (pH 6–8); only certain locations of the living organisms may have pH values that are lower or higher. One ion of H_3O^+ or one ion of HO^- in a solution of pH 7 and at 20 °C corresponds to 7.14×10^8 undissociated water molecules. The concentration of ions H_3O^+ or HO^- is 1×10^{-7} mol/l, whilst the concentration of undissociated water molecules is 55.6 mol/l. Food processing often takes place at the natural pH of the food materials, but the treatment of particular raw materials and the production of certain foods using acidifying or alkalising additives are also common, which strongly influences the course of many biochemical and chemical reactions.

7.5 Properties

Water, unlike related hydrides that contain elements of the sixth group of the Periodic Table (hydrogen sulfide, known as sulfan, H_2S, and hydrogen selenide or selan, H_2Se), has fairly unique physical and chemical properties due to mutual interactions of water molecules. The properties of water simultaneously determine many properties of foods. Under normal conditions, water occurs in all three states of matter, whose mutual relations, depending on temperature and pressure, summarise the phase diagram. Water has an anomalous freezing point (ice melting point) and boiling point, latent heat of fusion (melting or freezing), latent heat of vaporisation (boiling or condensing), specific heat capacity, relative permittivity (dielectric constant), surface tension, viscosity, and other physical properties (Table 7.4).

Table 7.4 Some important physical properties of water compared with ethanol.

Property	Water	Ethanol	Property	Water	Ethanol
Melting point (°C)[a]	0.0	−114.2	Specific heat (kJ/kg)	4.2	2.4
Boiling point (°C)[a]	100.0	78.3	Surface tension (mN/m)[b]	72.8	22.3
Latent heat of fusion (kJ/kg)	333.6	108.0	Dynamic viscosity (mPa/s)[b]	1.005	1.200
Latent heat of boiling (kJ/kg)	2258	841	Relative permittivity[b]	80.4	24.3

[a]Under normal pressure (101.325 kPa).
[b]At 20 °C.

Of note is the dependence of water density on temperature. Water has a maximum density of 0.999 97 kg/l at 3.98 °C; at all other temperatures, the density is lower. At 0 °C, the density of water is 0.9998 kg/l, but the density of ice at the same temperature is only 0.9168 kg/l. In other words, ice has about a 9% greater volume than water. This increase in volume on conversion of water into ice (volume expansion) has destructive effects, especially on plant tissues during freezing. The opposite process, melting of ice (volume contraction; e.g. during thawing of soft fruits, such as strawberries), has the same destructive effects. These effects are also manifested on even more resistant animal tissues. They cause meat to drip and lose juiciness. Slow thawing of frozen meat reduces the amount of lost moisture that contains nutritionally valuable substances.

The physical data for water, aqueous solutions of various substances, and dispersed systems where water is the dispersion medium (solvent) are commonly tabulated and are necessary for engineering calculations in all food technologies.

7.6 Interactions

Water molecules interact with one another, with inorganic substances, and with almost any compound or functional group of the organic food constituents, such as proteins, lipids, and carbohydrates. All these interactions significantly affect the properties of foods, including organoleptic properties, particularly texture, odour, and taste. These interactions are also known as **non-covalent interactions** or **non-covalent bonds**, or, generally, as **non-bonded interactions**. They are caused by various attractive and repulsive forces.

Electrostatic attractive and repulsive forces (between ions and dipoles) arising from **electrostatic interactions** and forces that are due to the **hydrophobic interactions** have the greatest significance. These are listed in Table 7.5.

Electrostatic forces between **ions** arising from Coulomb's law occur in the interactions of ions formed by dissociation of inorganic salts and in ionic interactions of ionised functional groups of organic substances, such as ionised protein functional groups, for example ionised carboxyl groups ($-COO^-$), ionised amino groups ($-NH_3^+$), and ionised functional groups of some polysaccharides ($-COO^-$, OSO_3^-).

An important type of electrostatic interaction is the interactions between ions and **dipoles** (permanent dipoles) that occur, for example, during interactions of salt ions with water molecules (the process is called hydration of ions) or during interactions of ionised functional groups of proteins and of some polysaccharides with water. The energy of interaction is several tens of kJ/mol (e.g. 21 kJ/mol in mutual interactions of ionised functional groups of proteins and 50–100 kJ/mol in interactions of monovalent salt ions with water).

Borderline cases of ion–ion and permanent dipole–permanent dipole interactions are the so-called **hydrogen bonds** or **H-bonds**. These occur in compounds containing hydrogen bound to one of the most electronegative elements (such as oxygen, nitrogen, or chlorine). This hydrogen has a relatively small but significant affinity for other electronegative atoms. Examples are mutual interactions of water molecules (interaction energy is 17 kJ/mol), interactions of water with polar non-ionised carboxyl groups ($-COOH$) and amino groups ($-NH_2$) of proteins or hydroxyl groups ($-OH$) of carbohydrates, and interactions of polar groups (e.g. carboxyl groups, amino groups, or other functional groups of proteins). The energy of these interactions is about 12 kJ/mol.

Electrostatic forces between dipoles, known as **dipole interactions (dipole–dipole interactions)** are other significant non-bonded interactions. The interactions occur between all types of dipoles, as well as between permanent dipoles, which are **permanent dipole–permanent dipole interactions**. Examples are interactions between polar functional groups, such as a carbonyl group: $-C=O$ (a special type of permanent dipole–permanent dipole type of interaction already mentioned is hydrogen bonds). Interactions of the dipole–dipole type are also important for non-polar molecules. At a certain time interval, the distribution of electric charge in non-polar molecules is not uniform.

Table 7.5 Overview of non-bonded interactions.

Interactions	Examples
Electrostatic interactions	
Ion–ion	Ions of salts, ionised functional groups of proteins and carbohydrates
Ion–permanent dipole	Ions of salts–water, ionised functional groups of proteins–water, functional groups of carbohydrates–water
Hydrogen bonds	Water–water, functional groups of proteins and carbohydrates
Permanent dipole–permanent dipole	Functional groups of proteins and carbohydrates
Induced dipole–induced dipole	Functional groups of proteins and carbohydrates
Hydrophobic interactions	Hydrophobic functional groups of proteins and lipids

At that given moment, however, some electrons accumulate on one side of the molecule, creating a temporary dipole, which induces dipoles in other molecules. These interactions between induced dipoles are referred to as **induced dipole–induced dipole interactions**. The third and final form of interaction of the dipole–dipole type is interactions between permanent and induced dipoles, known as **permanent dipole–induced dipole interactions**. The sum of the attractive or repulsive forces between molecules or between parts of the same molecule is known as the van der Waals forces (binding energy is about 1–4 kJ/mol). Special cases are forces between two permanent dipoles (known as Keesom forces), forces between two induced dipoles (London dispersion forces), and forces between permanent dipoles and induced dipoles (Debye forces).

The hydrophobic interactions occur in non-polar molecules or non-polar functional groups. They are important especially for lipids and proteins. A non-polar substance or non-polar functional group in an aqueous environment is situated in an area surrounded by water molecules that are associated with one another. In order to create this area, some hydrogen bonds between water molecules must be disrupted, which requires the release of energy. As a result, the free energy of the system decreases, and the decrease is compensated by a simultaneous decline in entropy (the system becomes less random, less disorderly, and more organised). If two molecules of a non-polar substance are in the same area, the number of broken hydrogen bonds is lower than when non-polar molecules exist separately and are separated by water molecules, because water molecules rearrange around this area so that they create the maximum number of hydrogen bonds. Transfer of non-polar molecules from a non-polar to a polar environment is therefore a spontaneous process, which is accompanied by a negative free energy change. For alkanes, this negative free energy change is −10 to −15 kJ/mol. It is higher for lipids. Water molecules, which otherwise create hypothetical six-membered structures, are partially transformed during the re-organisation into more compact five-membered structures.

7.6.1 Interactions with water molecules

Water in all states of matter is a highly organised dynamic system. The dipole character of the water molecule allows its association with other water molecules through hydrogen bonds (called hydrogen bridges), which are the interactions of a hydrogen atom covalently bound to the electronegative oxygen atom of one water molecule with the electronegative oxygen atom of another water molecule. In comparison with the length of the covalent bond O—H (0.096 nm), the length of a hydrogen bond (O … H distance) is almost double, at 0.177 nm; the distance between two adjacent oxygen atoms of water molecules (O … O) is 0.276 nm (in ice at 0 °C). The dissociation energy of hydrogen bonds between water molecules is relatively high, at around 5% of the covalent bond O—H dissociation energy (about 25 kJ/mol).

Figure 7.2 Interactions of water molecules. O, oxygen atom; •, hydrogen atom; –, covalent bond; ----, hydrogen bond.

Each water molecule in ordinary ice (called ice I_h, which is the only natural form of ice) associates with four other water molecules. The so-called coordination number (the number of neighbouring molecules located in the immediate vicinity of each water molecule) is therefore 4 (Figure 7.2). Taken together, water molecules in ice create a completely regular spatial association structure that is a hexagonal lattice (Figure 7.3). The crystal lattice structure is not static, since hydrogen bonds in the crystal lattice of ice and of water in liquid and gaseous states, as well as the covalent bond of the water molecules, are only temporary in character. The old links disappear and new links re-form, because hydrogen atoms oscillate, even at temperatures of −20 °C, between two neighbouring oxygen atoms: O—H … O↔O … H—O. The half-life of their duration is only about 1×10^{-11} seconds. Under normal freezing temperatures, only a small number of water molecules are not integrated into the crystal lattice of ice.

Liquid water at 0 °C has about 90% of the molecules associated by hydrogen bonds found in ice, whilst boiling water has about 80%. To a small extent, water molecules are associated even in the gas phase. In the liquid state, water molecules are associated in irregular clusters: aggregates of molecules, which are in equilibrium with randomly distributed free molecules. The number of water molecules in a cluster is estimated at 200–400 (at 20 °C). Water molecules in clusters are regularly oriented and arranged in a lattice similar to the ice lattice. The duration of clusters is about 1×10^{-8} seconds.

Figure 7.3 Crystal lattice of ice. O, oxygen atoms; hydrogen atoms not indicated. Fennema (1985, fig 1.1). Source: Reproduced by permission of Taylor & Francis – Marcel Dekker.

Liquid water, however, has a greater number of other structures. In addition to the hexagonal lattice found in ice, compact structures, such as five-membered rings, are also assumed. The consequence of the occurrence of these compact structures is an increase in the coordination number. The coordination number of water in ice at 0 °C is 4, the coordination number at 1.5 °C is 4.4, and this continues to rise with increasing temperature. The average distance between adjacent water molecules also increases with rising temperature. In ice at 0 °C, the distance between two neighbouring oxygen atoms of water molecules (O … O) is equal to 0.276 nm (Figure 7.2), whilst in water at 1.5 °C, this distance is 0.290 nm. The higher water density at 3.98 °C is explained by an increase in the value of the coordination number. At temperatures above 3.98 °C, the decrease in water density is due to the increased average distance between adjacent water molecules.

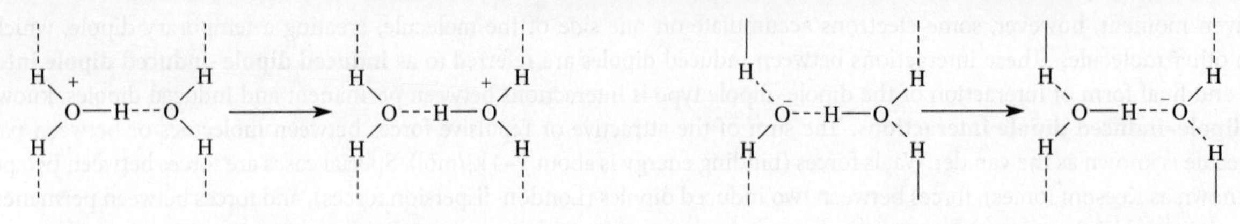

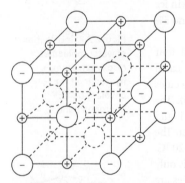

Figure 7.4 Interaction of oxonium and hydroxyl ions with water.

The disturbances in the lattice structure of ice and liquid water also occur in the presence of various non-electrolytes, such as sugars and ions resulting from dissociation of water or electrolytes, such as inorganic acids, hydroxides, and salts. Water molecules are associated with protons (oxonium ions) arising by dissociation of water or any acids present. Each molecule of an oxonium ion is associated with three water molecules through very strong hydrogen bonds (dissociation energy is about 100 kJ/mol), and the coordination number is 3. The associated structure is not static, since hydrogen bonds are involved in the transport of protons to other water molecules. The interaction is very fast (about 1×10^{12} per second). The equivalent conductivity and mobility of oxonium ions and of hydroxyl ions arethus very high in comparison with other ions (Figure 7.4). By analogy, hydrogen bonds are also involved in the transport of hydroxyl ions (electrons). Hydroxyl ions are associated with three molecules of water; therefore, the coordination number is 3, as in the case of oxonium ions.

7.6.2 Interactions with food components

Crucial for the properties of foods are processes taking place at the interface of water and air, water and other liquids and solids, and in the aqueous phase of aqueous solutions and dispersions.

7.6.2.1 Interactions with inorganic salts

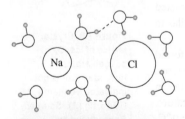

Figure 7.5 Crystal structure of sodium chloride. +, sodium cations; −, chlorine anions.

Inorganic salts are composed of ions, each bound by strong electrostatic attractive forces (interaction of the type ion–ion). This ion (electrovalent) bond generally occurs between two atoms of different electronegativities by transfer of one or more electrons from one atom to the other. For example, sodium (Na) is a typical metal (alkali metal) with low electronegativity (0.9 eV), whilst chlorine is a non-metal (halogen) with high electronegativity (2.9 eV). The positively charged sodium atom with 11 protons and 10 electrons (sodium cation Na^+) arises by transfer of one electron to the chlorine atom, giving rise to a negatively charged chlorine anion Cl^- (with 17 protons and 18 electrons). Both ions gain the electron structure of the rare gases neon and argon. The crystal lattice of sodium chloride results from the individual cations and anions; cations are surrounded by anions, and vice versa. Each sodium atom is surrounded by six equidistant chlorine atoms, and vice versa (sodium and chlorine have the coordination number 6). In the crystalline state, ions have a substantially lower energy than the atoms from which they were formed (Figure 7.5). The result of the strong electrostatic attractive forces between the ions is the high melting point (801 °C) and boiling point (1440 °C) of sodium chloride.

Figure 7.6 Water interactions with sodium and chloride ions. Fennema (1985, fig 10). Source: Reproduced by permission of Taylor & Francis – Marcel Dekker.

Crystals of inorganic salts dissolve well in dipolar water molecules. In order to produce the dissolution of crystals, water molecules break the electrostatic forces binding the ions in the crystal lattice. The energy of the hydrogen bonds between the water molecules (25 kJ/mol) is almost comparable to the energy of ionic bonds, which have electrostatic interactions between the ions (about 21 kJ/mol), and is high enough to overcome the electrostatic attractive forces. Cations and anions released from the crystal lattice are stabilised by interactions with water molecules (interaction type ion–dipole). The process is called hydration (generally known as solvation) of ions (Figure 7.6). This leads to association of dispersed ions with several molecules of the dispersion medium (water), and hydrated ions interfere with the regular arrangement of the water molecules (lattice) in their neighbourhood. The result is a somewhat lower water vapour pressure above the solution, lower water activity, higher boiling point, and lower freezing point of solution compared with pure water. Also, some properties of ions (such as mobility) are significantly affected by hydration. The degree of hydration of different ions varies depending on the concentration of the ions and other factors. In 0.5–1 mol/l solutions, the number of water molecules bound by Na^+ ion is eight or nine, whilst Cl^- ion binds only three water molecules.

Hydration of dissociated ions releases a certain amount of energy, called the **hydration energy**. The difference between the crystal lattice energy (energy of the ionic bonds) and hydration energy appears as a **heat of solution**, which is released (positive heat) into the environment or withdrawn (negative heat) from the surroundings. The heat of solution of sodium chloride taken from the surroundings is small (-86 kJ/kg, which corresponds to -5 kJ/mol), but a much higher negative heat of solution (-149 kJ/kg, -27 kJ/mol) arises by dissolving xylitol, which produces a perceived sensation of coolness in the mouth as it comes in contact with the saliva.

Aqua complexes (hydrates) of metal ions are described in detail in Section 6.2.1.

7.6.2.2 Interactions with proteins

Proteins are soluble in polar solvents such as water and in aqueous solutions of acids and bases. The majority of proteins form monodispersed systems (molecules in solution are separated by solvent), but some form oligodispersed systems.

The interaction of water with proteins is of paramount importance in maintaining the structure and functional properties of proteins *in vivo*, but also influences texture and other functional properties of foods. Proteins contain essentially three types of functional groups interacting with water:

- **ionised polar groups**, such as amino and carboxyl groups, for example in protein-bound lysine $-(CH_2)_4-NH_3^+$ and glutamic acid $-CH_2-CH_2-COO^-$;

- **electrically neutral polar groups**, such as the ε-amino group of lysine in alkaline media $-(CH_2)_4-NH_2$ and the carboxyl group of glutamic acid in acidic media $-CH_2-CH_2-COOH$ or the hydroxymethyl group of serine $-CH_2-OH$;

- **non-polar groups**, such as the alkyl groups of the branched aliphatic amino acids valine, leucine, and isoleucine, the methylthiomethylene group of methionine $-CH_2-S-CH_3$, or the aryl groups of aromatic amino acids such as phenylalanine.

The degree of hydration of protein molecules, which depends mainly on their structure and pH, is expressed as the ratio of bound water weight and protein weight (in grammes of water bound by 1 g of protein). The degree of hydration is different for each protein; for example, it is about 0.2 in ovalbumin, 0.3 in haemoglobin, and up to 0.8 in native β-lactoglobulin.

The pH has a great influence on the ionisation of polar functional groups in the protein-bound amino acids, and thus on the ability of proteins to interact with water (bind water). Ionised functional groups of proteins interact with water in a similar way to salt ions. The ionised basic side chains of lysine and histidine bind about four molecules of water via hydrogen bonds, the acidic side chains of glutamate and aspartate bind about six, whilst the neutral carboxyl groups of amino acids (interaction of dipole–dipole type) bind two, as well as the polar non-ionised side chains of serine and other amino acids.

A thermodynamically disadvantageous situation will occur if a hydrophobic substance or a non-polar functional group of proteins occurs in an aqueous environment, because non-polar functional groups associate and their hydrophobic interactions are especially important for maintaining the tertiary structure of proteins, but also play a role in other situations. Interactions of water with non-polar functional groups of proteins occur through the same mechanisms as interactions of water with lipids.

Only a small proportion of polar groups of native (undenatured) protein located in the hydrophobic area of the molecule are unable to bind water, for steric reasons. Thermal denaturation leads to changes in the higher structures of protein molecules (quaternary, tertiary, and secondary), whilst new, previously inaccessible functional groups can interact with water and the protein has an increased ability to bind water (by 30–45%) in the denatured state. If, however, denaturation leads to aggregation of denatured protein molecules, the functional groups of proteins react with themselves, thereby reducing the number of functional groups capable of entering into interactions with water, which decreases the ability of the proteins to bind water.

The behaviour of proteins is significantly affected by the presence of low-molecular-weight ions (cations and anions of inorganic salts). Many salts (including sodium chloride) increase protein solubility in concentrations up to about 1 mol/l. This phenomenon is explained by the fact that electrically charged dissociated polar groups on the surfaces of protein molecules bind salt ions via electrostatic forces tighter than water molecules. These ions can bring their own organised clusters of water molecules, stabilising the solvated (hydrated) proteins. This so-called **salting-in effect** (see Section 2.4.3.1), in which the solubility of the protein increases slightly, is of great importance in the production of many foods. Water binding of proteins is a particularly important phenomenon in the meat technology sector. For example, if the pH of meat decreases from 5.0 (which is approximately the value of the isoelectric point of muscle proteins) to about 3.5, the homogenate of muscle proteins may bind twice the amount of water. An example is some procedures in cured ham production, where meat salting increases the ability to bind water. The addition of salt solutions for curing, usually by injection or massage, guarantees the desired water and salt contents and makes the meat more tender. Sodium chloride and sodium chloride ions bind primarily to the myofibrillar proteins actin and myosin. In addition to sodium chloride, sodium and potassium phosphates, citrates, and tartrates added to meat and meat products affect muscle pH and increase the meat's water-holding capacity; they also act as emulsifying agents for processed cheeses (see Section 6.3.4.2.3). In the manufacture of processed cheeses, the added phosphates and polyphosphates act as emulsifiers. They release the casein molecules from micelles, and the caseins then bind the milk fat or added fat.

At higher salt concentrations (about 1–2 mol/l), the reverse process may occur, and many proteins become salted-out or precipitate. The concentration of available water molecules decreases as they are used to solvate (hydrate) the salt ions, and the amount of water needed to dissolve the protein decreases. Salts compete for water with the protein. This phenomenon, called the **salting-out effect**, occurs rarely in food processing, except in the surface layers of salted meat and fish and in some speciality meat and fish products with a high salt content.

7.6.2.3 Interactions with lipids

Lipid molecules (e.g. triacylglycerols) tend to aggregate in an aqueous medium due to hydrophobic interactions of the non-polar parts of the molecules. In living organisms, triacylglycerols are stored in the anhydrous form in special fat cells called adipocytes and act as reservoirs of energy. In hydrothermal processing of food, the fat is released from the adipocytes and often accumulates on the water surface as it is adsorbed at the interface with air. Depending on the amount and structure, lipids form a monomolecular film, individual droplets, or a continuous layer of fat on the water surface.

Polar lipids, such as glycerophospholipids, sphingolipids, and cholesterol, are the structural basis of animal- and plant-cell biomembranes and cell organelles, such as the nucleus, mitochondria, and chloroplasts. Polar lipids are amphiphilic molecules that adsorb on the surface of the water–air interface or water–oil dispersion and stabilise heterogeneous systems such as oil-in-water (o/w) emulsions and water-in-oil (w/o) emulsions. At concentrations higher than their solubility, polar lipids may form macromolecular aggregates called micelles and double films (double layers of molecules known as bimolecular lamellae).

7.6.2.4 Interactions with saccharides

Non-bonding interactions of water with monosaccharides, disaccharides, higher oligosaccharides, polysaccharides, and other biopolymers containing carbohydrates (such as glycoproteins and glycolipids) take place through hydrogen bonds with electrically neutral hydroxyl groups or possibly with other polar groups. These interactions are of great significance in living systems and influence the properties of many food dispersed systems.

7.6.2.4.1 Monosaccharides and oligosaccharides

Crystalline monosaccharides and disaccharides are generally very soluble in water, and during their dissolution the cohesive forces of crystals (holding the individual neutral molecules in the solid state) should be disrupted. These intermolecular attractive forces are the result of non-bonding interactions, usually attractive forces between permanent dipoles. However, other non-bonding interactions, such as van der Waals forces, may also be present.

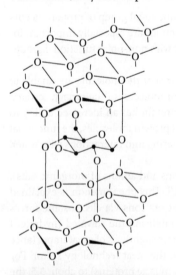

Figure 7.7 β-D-Glucopyranose in an ice lattice. O, oxygen atoms; •, carbon atoms; hydrogen atoms not marked. Fennema (1985, fig 1.1). Source: Reproduced by permission of Taylor & Francis – Marcel Dekker.

Water molecules located in the immediate vicinity of dispersed monosaccharide molecules are relatively tightly bound and contribute to the stability of some conformers or anomers. A glucose molecule, for example, is associated with 3.7 molecules of water. The $^4C_1$ glucose conformation (more than 99% of molecules) is thermodynamically preferred, in which the bulky substituents (secondary hydroxyls) on carbons C-2, C-3, and C-4 are in the equatorial position, and therefore are far apart. The primary hydroxyl group at C-6, which apparently does not interact with water, and the hemi-acetal hydroxyl group at C-1 of the β-anomer are also in equatorial positions. The conformer $^4C_1$ of β-D-glucopyranose thus has four equatorial hydroxyl groups interacting with water (the coordination number, in carbohydrates also called the hydration number, is 4), whilst α-D-glucopyranose has only three equatorial hydroxyl groups interacting with water (the hydration number is 3). The average number of equatorial hydroxyl groups of glucose in aqueous solution, and the resulting number indicating the number of water molecules located in the immediate vicinity of glucose molecule, is: $4 \times 0.67 + 3 \times 0.33 = 3.7$, because D-glucose solution contains approximately 67% β-D-glucopyranose and 33% α-D-glucopyranose in equilibrium.

The distance between the oxygen atoms of hydroxyl groups at C-1 and C-3 (or at C-2 and C-4) of the $^4C_1$ conformer of β-D-glucopyranose is 0.486 nm, which is almost the same as the distance between oxygen atoms in the lattice of liquid water (0.490 nm), and lies in the same plane (in the ice lattice, this distance is only 0.450 nm). By incorporation of glucose molecules into the water lattice, the relative positions of each layer of water lattice molecules virtually do not change, and, furthermore, the glucose molecules stabilise the surrounding clusters of water molecules (Figure 7.7). The structural similarity in the interaction with water explains the ability of glucose to slow the formation of ice crystals in frozen confectionery products (similar to the effect of glycerol in mixtures with water, which prevents ice from forming) and, particularly, its versatility in biochemical reactions and utility as a building block in living systems.

The range of interactions of other soluble carbohydrates with water also depends on their structure. The organised structure of water molecules is always relatively impaired. A ribose molecule, for example, is associated with 2.5 molecules of water, maltose and saccharose with 5.0–6.6 water molecules. In concentrated solutions of glucose and saccharose, the hydration number is lower: about 2 and 5, respectively. As with electrolyte solutions, solutions of carbohydrates likewise have a reduced water vapour pressure above the solution and a lower freezing point and water activity, but the boiling point of the solution is increased in comparison with pure water.

Many monosaccharides and disaccharides can form supersaturated solutions (syrups), usually by cooling saturated solutions prepared at a given temperature (e.g. during thickening). Syrup solutions exist in a higher energy state than saturated solutions. They are unstable, and the excess solids eventually crystallise to reach a lower energy state.

7.6.2.4.2 Polysaccharides

In the solid state, all polysaccharides contain about 10–20% of water at normal atmospheric humidity. Their molecules or parts of their molecules are either amorphous or involved in different, more or less ordered structures called crystalline regions. Components of starch (linear amylose with branched amylopectin) are arranged in starch granules, cellulose forms linear microfibrilles, and agar and carrageenan molecules form double helices. Water is tightly bound in the amorphous regions of molecules by hydrogen bonds through a number of free binding sites (hydroxyl, carboxyl, or other polar functional groups) that do not act in interactions with other binding sites of the same molecule (in intramolecular interactions) or with other polymer molecules (intermolecular interactions). In the crystalline regions, groups capable of interaction with water are not available.

Linear polysaccharides, such as homoglycans (with strong intra- and intermolecular non-bonding interactions), dissolve with difficulty or are completely insoluble. An example of a polysaccharide that is insoluble in cold water and soluble in hot water is amylose. Cellulose is insoluble even in hot water but is soluble in alkaline solvents. Linear polysaccharides usually form unstable solutions. During collisions of molecules, some bound water is lost as a result of the interactions of polysaccharide molecules that form larger aggregates of colloidal dimensions or even coarse dispersions, which may coagulate. These phenomena, illustrating the instability of linear polysaccharides, are well known in the retrogradation of amylose dispersions and starch gels. They also occur in dispersions of arabinoxylans.

Highly branched polysaccharides (such as glycogen, amylopectin, and gum arabic) are more soluble than linear polysaccharides, as their mutually interacting functional groups are more distant from one another and thus facilitate hydration. Their solutions have very low viscosity compared with linear polysaccharide solutions of the same molecular weight and the same concentration. This is attributed to the effective volume, the average hypothetical spheres that make up the rotating polysaccharide molecules during thermal motion (Figure 7.8). The tendency of branched polysaccharides to coagulation is very low. At high concentrations, they form sticky pastes, probably due to interactions of side chains. After drying, these substances can easily rehydrate.

Linear branched polysaccharides with a long main chain and a number of short side chains (such as guar and locust gums) combine the properties of linear and highly branched polysaccharides. Since the main chain is long, the viscosity of a solution is higher. Thanks to numerous short side chains, the intermolecular interactions are so strongly attenuated that the solubility and the ability to rehydrate are high, and concentrated solutions are stable.

Polysaccharides with carboxyl groups hydrate rapidly and form viscous solutions or gels. Examples of these polysaccharides are soluble pectins. Their solubility and ability to form gels are dependent on the degree of polymerisation and the number and distribution of methoxyl groups in the molecule. The solubility generally increases with decreased molecular weight and an increase in the number of methoxyl groups. The pH, temperature, and presence of other substances in solution (saccharose and calcium ions) have a large influence. Other polysaccharides occurring in the form of polyanions (e.g. as esters of sulfuric acid, such as agar and carrageenans) are readily soluble and form highly viscous solutions and gels, such as pectins. The formation of polysaccharide sols and gels is described in Section 7.8.3.

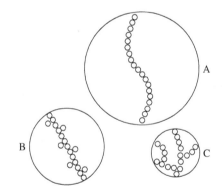

Figure 7.8 Schematic representation of effective volume of (A) linear, (B) substituted, and (C) branched polysaccharide molecules of the same relative molecular weight (number of monosaccharide units).

Modified polysaccharides significantly change their properties. The incorporation of acidic groups into molecules of neutral polysaccharides (such as starch phosphates or sodium carboxymethyl cellulose) increases the solubility and viscosity compared with the native polysaccharides. Substitution of neutral functional groups (e.g. in hydroxypropyl starch) increases the solubility, viscosity, and stability in solutions. Alkyl groups (in cellulose ethers) facilitate hydration. At a higher degree of substitution, the hydrophobic nature of the polymer increases, along with its solubility in organic solvents.

7.7 Phase interfaces

The outer boundary of liquid water (or any liquid phase in general), which is in contact with its vapour or the air, is called the **surface**. A surface forming a boundary between two or more separate phases (phase boundary), such as liquid–gas, liquid–solid, gas–solid, or, for immiscible materials, liquid–liquid or solid–solid, is called an **interface**. A surface or interface can be planar or curved. The thermodynamic properties of systems with planar and curved phase interfaces are different.

7.7.1 Surface tension

Water molecules present in the surface or in the interface have different properties in comparison with water molecules located deeper in the homogeneous liquid phase. The resulting attractive forces are directed into the liquid, because interactions with water molecules in the gas phase or with components in the air are insignificant. As a result of an imbalance of forces acting on molecules at the interface, a special type of force called **surface tension** is created in the surface. The existence of surface tension explains many phenomena taking place at planar interfaces, such as the behaviour of liquids on the surface of other liquids and on the surface of solid materials.

If the liquid is not exposed to external forces (e.g. in a vacuum), it occupies a spherical shape as it gains the smallest surface at the given volume. On the surface of another liquid, it either forms a drop that has a lenticular shape or it can be spread over its surface. The liquid's behaviour at a given temperature depends on the size of the liquid–air surface tension and on the interfacial liquid–liquid surface tension. On the solid surface, the liquid also either spreads out evenly or forms a drop. The shape and size of the drop depend on how the liquid wets the solid surface, which is related to the liquid–air surface tension, the surface energy of a solid, solid–liquid surface tension, and temperature.

When the surface of a solid substance is in contact with two immiscible liquids, one creates a film between the solid substance and the other. Which of the liquids forms the film depends on both the solid–liquid surface tensions and the liquid–liquid surface tension. The surface tension of pure liquids (such as water) and aqueous solutions (e.g. solutions of saccharose) depends on the temperature: with increasing temperature, surface tension generally decreases.

7.7.2 Adsorption

7.7.2.1 Liquid phase–gas phase systems

The surface tension of interfaces between phases where one phase is composed of two or more components (e.g. the interface of an aqueous solution of a solute–air; so-called moving interface) is affected by the solute. For example, the surface tension increases slightly with increasing concentration of polar substances (e.g. salts or sugars, Figure 7.9), and decreases with increasing concentration of substances containing polar and weakly non-polar functional groups (e.g. lower alcohols and lower carboxylic acids). Surface tension, however, quickly drops to a certain constant value after the addition of **surfactants**. At this value of surface tension, surfactant molecules aggregate to micelles, which remain in the aqueous phase and therefore do not affect the surface tension. Surfactants are substances having in the molecule both hydrophilic (polar) functional groups and strongly lipophilic (hydrophobic, non-polar) functional groups. Such substances are called **amphiphilic** or **amphipathic**. In practice, they are used mainly as detergents, emulsifiers, foam-forming substances, solubilisators, and wetting agents.

The reason that surface tension is influenced by dissolved substances is that the system (aqueous solution of a substance) can bring down its surface energy not only by reducing the phase boundary area as with pure water, but also by changing the concentration of solute at the interphase. Solutes dissolved in a liquid can generally migrate towards the surface or in the opposite direction. The tendency to remove the solute from the solution and concentrate it in the surface is manifested in the case where the mutual interactions of water molecules are stronger than the interactions between water and solute. The surface layer then contains an increased concentration of the solute that reduces the total surface energy (interfacial energy). The phenomenon is called **adsorption**. The substance that is adsorbed is the **adsorbate** and the substance on which another substance adsorbs is the **adsorbent**. Adsorption is a special type of surface aggregation.

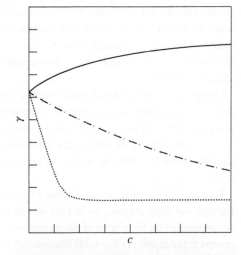

Figure 7.9 Schematic representation of water surface tension depending on the type and amount of dissolved substances. γ, surface tension (m.N/m); c, concentration; –, sodium chloride or saccharose; - - -, ethanol or acetic acid (water–air surface tension at 20 °C = 72.8 m.N/m); ......, surfactant.

Adsorption of liquids and gases can also occur in solid substances. If the liquid or gas does not stay on the surface layer of the solid substance, but penetrates into the interior, the phenomenon is known as **absorption** (formation of solid solutions). Resolution of both processes (adsorption and absorption) is usually not possible, because it is often a combination of the two. Therefore, the term often used is **sorption**.

The forces that cause adsorption at the liquid–gas phase are primarily non-bonding van der Waals forces and hydrophobic interactions. The adsorbed amount increases with the concentration of a substance in the liquid phase and decreases with temperature increase. The action is described by the Gibbs adsorption isotherm. Substances that migrate to the surface and lower the surface tension of liquids are surfactants. The layer in which a significant increase in the concentration of the adsorbing substance occurs usually has a thickness of only a few molecular diameters, and a film of the surfactant (Figure 7.10) is formed on the water surface.

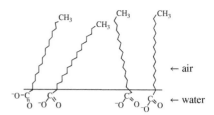

Figure 7.10 Sodium stearate molecules adsorbed on water–air interface at low concentration.

The formation of films of liquids at surfaces or interfaces is actually a special case of spreading a liquid on the surface of another liquid. If the strong adsorption leads to free spreading of the liquid, the adsorbed substance forms a film one molecule thick, which is called a **monomolecular film** (monomolecular layer). Monomolecular films also exist in gaseous and solid states. They represent a two-dimensional analogy of the three main three-dimensional states of matter (gaseous, liquid, and solid). Molecules of the substance in the surface film are oriented by their polar parts to the liquid (water) and by their non-polar parts to the less polar environment, which is air (Figure 7.11). Monomolecular films have a measurable effect on the mechanical, optical, and electrical properties of surfaces and interfaces. The external force, which the film causes during compression, is usually called **surface pressure**. If there is not enough space on the surface, the substance partly creates a monomolecular film and the excess remains in the form of drops, which are in equilibrium with the film. At higher concentrations of a substance, a multimolecular film or even a layer of a substance is formed (Figure 7.12).

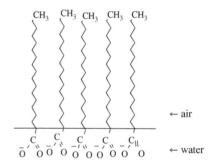

Figure 7.11 Sodium stearate molecules adsorbed in monomolecular film.

7.7.2.2 Liquid phase–liquid phase systems

Adsorption at the interface of two liquid phases has the same characteristics as the adsorption at the interface of a liquid phase–gas phase, since it is again a moving interface. In practice, the most important interfaces are those of the polar phase (water) with a less polar or non-polar phase and a water-immiscible liquid phase (oil). In these systems (e.g. in emulsions), surfactants are employed, whose polar functional groups interact with water molecules (solvation) and whose non-polar functional groups interact with the particles of the non-polar phase.

7.7.2.3 Solid phase–gas phase systems

Liquid–gas or liquid–liquid phase interfaces may be plain, concave, or convex, but the surfaces of solids are almost always non-homogeneous and contain a variety of edges, notches, pores, and capillaries. The forces that cause adsorption at the solid–gas interface are primarily non-bonding interactions of van der Waals forces and hydrophobic interactions. This so-called **physical adsorption** takes place on the entire surface. There is no monomolecular layer, but multiple layers of the adsorbing substance arise. Physical adsorption is very fast – almost instantaneous. The porous solids can also contain liquid phase in the pores. This phenomenon is called **capillary condensation**.

At the same time, in the active centres in solids, interactions between solid and liquid phases occur, having the character of chemical bonds. This type of adsorption is therefore called **chemisorption**. The chemisorption process is slower than physical adsorption, as it is preceded by diffusion of the substance to the outer surface of the adsorbent and diffusion through the adsorbent pores to the surface (to the active centres).

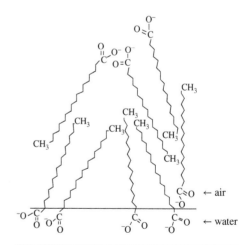

Figure 7.12 Sodium stearate molecules adsorbed on water–air interface at high concentration (multimolecular film).

7.7.2.4 Solid phase–liquid phase systems

Adsorption at the solid–liquid interface is generally similar to adsorption at the solid phase–gaseous phase interface. Theoretical modelling of the adsorption process is more difficult because, in addition to adsorbing dissolved substances, solvent is present (e.g. water), the molecules of which can also adsorb, and interactions of adsorbed molecules with molecules of the solvent may occur. Molecular adsorptions, when molecules of a substance are adsorbed, and ion adsorptions, in which ions of a substance are adsorbed, can similarly take place.

7.7.3 Capillary phenomena

The curvature of the surface of the phase boundary creates a number of important so-called **capillary phenomena**. The consequence is that the liquid substance in a capillary tube, with one end submerged in a container of liquid, rises. The free surface of a liquid (meniscus) can be flat or can have a convex or concave shape, depending on the solid and liquid surface. If the meniscus is concave upwards, the capillary liquid level rises above the liquid that surrounds it. This phenomenon is called **capillary elevation**. If the liquid substance does not wet the capillary wall, it creates a meniscus that is convex upwards, and its level is below the surface of the liquid surrounding it. This phenomenon is called **capillary depression**. Capillary rise of solutions in the pores is of paramount importance for the water supply of plant tissues, and also for the wetting and drying of foods. Capillary phenomena also explain the hysteresis that occurs during adsorption and desorption of water in food.

7.8 Food dispersed systems

Dispersed systems (dispersions) are mixtures of at least two substances, one of which (dispersed phase or internal phase) is often discontinuous and is distributed in the form of particles throughout another substance (continuous phase, also known as external phase or dispersion medium). Dispersions are systems that consist of at least two types of particles (molecules of dispersed phase and molecules of continuous phase) or two different phases (solid, liquid, or gaseous).

The dispersion medium of all living cells is water. If liquid water is in contact with its vapour, air, another liquid (with which it is not well mixed), or a solid, it forms a two-phase or polyphase system. Water is also a dispersion medium for low- and high-molecular-weight chemical components found in food. These components are water-dispersed in the form of ions, neutral molecules, or different particles that form the dispersed phase.

Dispersed systems are common aqueous solutions of low-molecular-weight compounds, salts, amino acids, and mono- and oligosaccharides (e.g. sodium chloride, glycine, and saccharose), as they contain at least two types of particles (molecules of the water-dispersion medium and the ions or molecules of the substance-dispersed phase). These systems are **monodispersed systems** because their components contain particles of about the same size with a certain relative molecular weight (in daltons, Da) or molar weight (in kg/mol or g/mol).

Dispersed systems are also solutions of macromolecular compounds, proteins, and polysaccharides, which are normally mixtures of polymer homologues containing particles of different sizes (often monomers, dimers, higher oligomers to polymers). They are therefore **polydispersed systems** that do not have an exact relative molecular weight. For their characterisation, the mean relative molecular weight or medium polymerisation degree is used (number of bound monomer units). An important point for the characterisation of polydispersed substances is the distribution curve, which indicates the frequency at which the individual particles occur in the dispersed system, depending on their molecular weight. In some polymers, the particles differ not only in size (such as fructans) but also in structure, as some of them can be branched in a different way (arabinoxylans) or can have different chemical compositions (alginates). In such cases, their distribution is known as chemical distribution.

7.8.1 Classification of dispersed systems

Dispersed systems are classified according to number of phases, shape of dispersed particles, particle size, state (gaseous, liquid, or solid) of dispersion medium, and state of dispersed phase. In practice, foods can contain all types of dispersed systems.

Depending on the number of phases, dispersed systems are divided into:

- **homogeneous dispersed systems**, where the dispersion medium and dispersed phase form one phase;

- **heterogeneous dispersed systems**, where the dispersion medium and dispersed phase are two different phases separated by a phase interface.

Homogeneous dispersed systems arise spontaneously and are stable (e.g. true solutions). Heterogeneous dispersed systems do not arise spontaneously, but are prepared artificially. They contain a dispersed phase not associated with molecules of solvent (usually water) and are non-equilibrium dispersed systems; therefore, they are unstable and spontaneously disappear.

According to the shape of the dispersed phase particles, dispersed systems are divided into:

- **globular dispersions** (all three dimensions are of the same order, they are isometric);

- **laminar dispersions** (one dimension is smaller by one order of magnitude);

- **fibrillar dispersions** (two dimensions are smaller by one order of magnitude).

Depending on the size of the dispersed phase particles (expressed as a radius, volume, or weight), which corresponds to the degree of dispersion of the dispersed phase in the dispersion medium, the dispersed systems are divided into three groups (real dispersed systems, however, form a continuous series without distinct boundaries):

- **molecular dispersions** are characterised by a high degree of dispersability; the dispersed phase particles (molecules or ions) have a size less than 1 nm, are not observable by electron microscopy, easily diffuse and penetrate through membranes (dialysis), show very intense thermal motion (Brownian motion), and lead to high osmotic pressures; systems are homogeneous and always form true solutions that arise spontaneously and are stable;

- **colloids** (colloidal dispersions or colloidal systems) are characterised by moderate dispersion; the size of colloidal particles is 1–1000 nm, and the particles can be observed by electron microscopy, diffuse slowly, penetrate some membranes, perform intense Brownian motion, and lead to a measurable osmotic pressure; colloids are often allocated into three categories: **associative colloids** or **micellar colloids** (which form a transition between molecular dispersions and colloids), **lyophilic colloids** (which are homogeneous and form colloidal solutions), and **lyophobic colloids** (which are heterogeneous and consist of two phases or form multiphase dispersed systems);

- **coarse dispersions** (suspensions) are heterogeneous dispersed systems in which the dispersed phase particles are larger than 1000 nm; systems with a particle size within 1000–50 000 nm are called **microdispersed systems**, whilst systems with a larger particle size (>50 000 nm) are called **macrodispersed systems**; particles of coarse dispersions are observable by optical microscope and are characterised by relatively fast sedimentation due to gravity or other forces, perform Brownian motion only to a maximum particle size of 4000 nm, do not cause osmotic pressure, and are always heterogeneous and unstable.

Dispersed systems are often classified into nine types according to the state of the dispersion medium and the state of the dispersed phase (gas, liquid, and solid) (Table 7.6). The most common dispersion systems are sols, gels, emulsions, and foams.

Table 7.6 Types of colloids.

Dispersed phase	Dispersion medium	Name of dispersion	Examples
Solid	Solid	Solid solution or solid sol or solid mixture	Iodised salt (mixture of NaCl and NaI)
Solid	Liquid	Sol (lyosol), true or colloidal solution, suspension	Whey
Solid	Gas	Aerosol (smoke)	Plant protection products (fumigants)
Liquid	Solid	Gel or solid emulsion	Cosmetic products
Liquid	Liquid	Emulsion or solution of liquid in liquid	Milk, mayonnaise, butter, margarine
Liquid	Gas	Aerosol (fog)	Plant protection products
Gas	Solid	Solid foam	Ice cream, foam sweets
Gas	Liquid	Foam or solution of gas in liquid	Whipped cream
Gas	Gas	Mixture of gases	Air

7.8.1.1 Molecular dispersions

Molecular dispersions are true solutions. They arise, for example, by dissolution of inorganic salts or low-molecular-weight organic compounds, such as monosaccharides and amino acids. The dispersed phase is dispersed into ions (e.g. sodium chloride) or molecules (such as glucose or air), which are of comparable size to the molecules of the dispersion medium (water). True solutions are usually kinetically stable (resistant to the effects of the gravitational field) and aggregative (still have the same degree of dispersion).

7.8.1.2 Colloidal dispersions

Colloidal dispersions (colloids) are basically formed in two ways. The first is spontaneous reversible association (aggregation) of some low-molecular-weight unstable organic and inorganic substances in true solutions. As a consequence of attractive forces, **associative colloids** or **micellar colloids** appear. An example of micellar colloids is casein micelles in milk. The second possibility is dissolution of high-molecular-weight substances, such as certain proteins or polysaccharides (including nucleic acids or synthetic polymers), which leads to homogeneous dispersion systems called **lyophilic colloids** that arise in various solvents. If the solvent is water, then **hydrophilic colloids** are formed. Colloids containing individual molecules of dispersed phase (substance) are sometimes called **molecular dispersions**. Heterogeneous dispersed systems called **lyophobic colloids** or **hydrophobic colloids** are prepared by physical or chemical methods, such as dispersion of non-polar substances in water. They resemble coarse dispersions and are therefore also called **phase colloids** or **dispersoids**. The common feature of all colloids is their large surface area, which is greater in smaller particles, and surface phenomena therefore often determine the properties of colloidal dispersions.

7.8.1.2.1 Micellar colloids

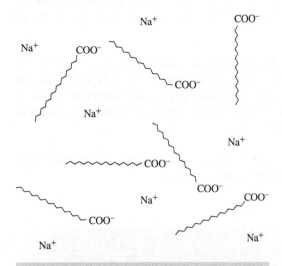

Figure 7.13 True solution of dissociated sodium stearate molecules.

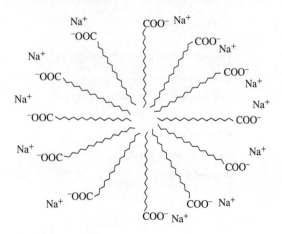

Figure 7.14 Micelle of dissociated sodium stearate molecules.

The formation of micellar colloids is mainly a result of hydrophobic interactions (the corresponding entropy change is about 13 kJ/mol) and van der Waals attractive forces between non-polar parts of the molecules. The resulting spherical associates (aggregates) of colloidal dimensions are called micelles. Micelles commonly contain tens of thousands of molecules, depending on the properties of the substances that form them, their concentration, and the temperature. Micellar colloids are usually globular or laminar dispersed systems. Micelle formation is typical for all amphiphilic substances. Spherical and ellipsoidal micelles typically consist of fatty acids, glycerophospholipids, and proteins that contain a hydrophobic and hydrophilic portion in the molecule (e.g. caseins in milk). Association of molecules occurs only at a concentration that is characteristic of the material, called the critical micelle concentration (dispersions of lower concentration provide true solutions or colloids). For amphiphilic constituents with one hydrophobic end, such as soaps (e.g. salts of stearic acid) and other surfactants (such as dodecyl sulfate), the critical micelle concentration is approximately 1 mmol/l; for amphiphilic substances with two hydrophobic ends (such as glycerophospholipids), the critical micelle concentration is generally 1 mol/l.

Classical examples of micellar colloids are dilute aqueous dispersions of fatty acids salts (soaps) and aqueous dispersions of other detergents. At a concentration below the critical micelle concentration, true solutions are formed (Figure 7.13), whilst at the critical micelle concentration and higher, micellar colloids are formed (Figure 7.14). The interiors of the micelles are filled with concentrically oriented hydrophobic groups, whilst on the surface hydrophilic ionised polar groups interact with counter-ions (e.g. Na⁺ cations) from the environment. These micelles are called ionic micelles. The arrangement into micelles is advantageous since it eliminates undesirable contact of water with the hydrophobic parts of amphion molecules and allows the polar groups to come into contact with water.

Concentrated solutions of soaps also yield laminar micelles that are formed by non-dissociated molecules. Their shape is similar to that of glycerophospholipid micelles. Glycerophospholipids (or sphingolipids) contain two non-polar chains in the molecule and, for steric reasons, tend to form large ellipsoidal micelles, which are actually a bilayer of molecules. They are called bimolecular sheets, bimolecular lamellae, or binary films (Figure 7.15). Milk is also a complex micellar colloid containing clusters of casein molecules called submicelles, which are associated to micelles by citric acid molecules and calcium ions (Figure 2.25).

7.8.1.3 Coarse dispersions

Coarse dispersions contain particles that are not individual chemicals (e.g. particles of food dressings containing solid particles of vegetables and spices), and are usually strongly polydispersed. The representation of a particle size in a polydispersed system is expressed by various distribution functions; for example, the mean radius of the particles and other particle characteristics can be determined.

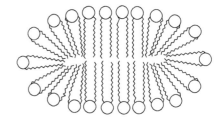

Figure 7.15 Phospholipid bilayers.

7.8.2 Stability of dispersed systems

Colloidal dispersed systems are characterised by high surface energy and high curvature of the surface particles. From a thermodynamic point of view, these systems have low aggregate and kinetic stability. Their aggregate instability manifests by two spontaneous processes:

- slow dissolution of small dispersed particles; the growth of larger particles is not usually significant (the speed of this process can be regulated in liquid dispersion media by changes in temperature or composition of the dispersion medium);

- **coagulation** (also called **flocculation**), which is clustering of dispersed particles into larger aggregates associated by adhesion forces of the particle surfaces; when aggregation leads to coarse dispersions, sedimentation of aggregates may occur.

7.8.3 Important dispersed systems

7.8.3.1 Sols

Dispersions of solids (dispersed phases) in liquids (dispersion media) are called **sols, or** in general **lyosols**. Heterogeneous dispersed systems with only weak interactions between the dispersed phase and dispersion medium are called **lyophobic sols**. Homogeneous dispersed systems with strong interactions between the dispersed phase and the dispersion medium are called **lyophilic sols**. The interaction of particles of lyophilic sols with the dispersion medium leads to **solvation**. If the dispersion medium is water, the preferred terms are **hydrosol**, **hydrophobic sol**, **hydrophilic sol**, and **hydration**.

Hydrophobic sols are, like all lyophobic colloids, heterogeneous. The dispersed phase and dispersion medium represent two different phases separated by the phase interface. Hydrophobic sols arise spontaneously, but are unstable. They are difficult to prepare (e.g. by mechanical agitation of solid particles, stirring, or condensation of small particles to micelles of colloidal dimensions) and require the presence of surfactants.

Hydrophilic sols are homogeneous colloids that give colloidal solutions. They can form spontaneously and are relatively stable. Typical hydrophilic sols are colloidal solutions of proteins and polysaccharides that are prepared by dissolving the solids in water. For example, proteins and polysaccharides are suspended in water, and the suspension usually requires heating. The sol, the dispersion of the solids (dispersed phase) in an aqueous dispersion medium, forms when there are only weak interactions between the macromolecules. In gelatine sol, individual protein molecules are compact, taking the shape of randomly coiled globules. A sol behaves as a liquid (Figure 7.16).

Dissolution of proteins and polysaccharides is not a simple process. The solids often absorb a certain proportion of the liquid (water) with which they are in contact, without losing their shape, which is their most important property. This process is called **imbibition**. Imbibition is the first step in seed germination and sprouting. It is also common in cooking rice, beans, and peas and in heating starch granules at temperatures lower than the gelatinisation temperature (see Section 4.5.6.1.3). Solids can take up additional fluid, which increases their volume and results in **swelling**. The particle volume during swelling does not increase proportionally to the volume of absorbed fluid (water), but the sum of volumes of the substance and absorbed water is always less than the volume of swollen product, which therefore has a higher density than the original material and subjects its surroundings to a certain pressure, the swelling pressure. Swelling is also accompanied by release of heat and by changes in mechanical properties, primarily involving increased strength and elasticity.

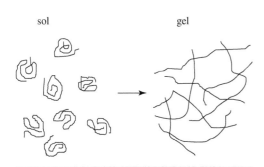

sol   gel

Figure 7.16 Scheme of gelatine gel formation.

Depending on the mechanism of swelling, the following categories are recognised:

- Capillary swelling, which occurs mainly with the fibrous structure of organic compounds, such as many polysaccharides; solvent (water) penetrates between the individual molecules of the constituents, which are solvated, and the particles move away from one another, which increases the degree of dispersal;

- Molecular swelling, during which the particles of the constituent material swell, their volume grows, and the degree of dispersal decreases.

According to the degree of swelling, the process can be:

- limited;

- unlimited.

Examples of limited swelling are gelatine and agar in warm water at room temperature. Examples of unlimited swelling are gum arabic or soluble starch, which swell in water at any temperature. Gelatine and agar show unlimited swelling in hot water (e.g. at 95 °C).

In gelatine and soluble polysaccharides (such as agar) in an aqueous environment, water molecules penetrate rapidly through the amorphous regions of macromolecules during swelling and interact with their free binding sites or interrupt the existing intermolecular bonds. Some parts of the macromolecular chains can easily be fully hydrated and, due to thermal motion, recede from one another. This reveals additional binding sites in the chain that are also solvated and solubilised. Swelling is limited only to this stage, as the hydrocolloid absorbs only a certain amount of liquid (usually through capillary swelling); gelatine and agar particles retain their original shape and do not dissolve further. Properties of the swollen material depend on the degree of hydration and the extent of the remaining intermolecular bonds. For example, all soluble polysaccharides go through this transition state during dissolution. Agar is not further hydrated, but other polysaccharides such as gum arabic continue to swell. During further hydration, a layer of water molecules surrounds the molecule of a polysaccharide, and the polymer is dispersed in an aqueous environment, creating a monodispersed system. With their change in shape, the solid polysaccharide molecules dissolve and form a solution (sol).

Attractive and repulsive forces between particles generally determine the stability of sols. If the particles remain dispersed, their mutual repulsive forces are larger than the attractive forces. The size of these forces is determined by various factors. Important factors in all hydrophobic and hydrophilic sols include the chemical structure of the substances, their particle size, the distance between particles, the presence of an electric charge (particles with the same charge are repelled, whilst those with opposite charges are attracted), other mutual non-bonding interactions of particles (mainly dipole-type interactions of all three types of dipoles), solute concentration, temperature, and so on. In hydrophilic sols, interactions of polarised or ionised functional groups with water through hydrogen bonds (hydration) are of great significance.

Water molecules of hydrophilic sols, in the vicinity of dispersed molecules of high-molecular-weight substances (proteins and polysaccharides), are relatively tightly associated by non-bonding interactions. They are usually in a state of thermodynamic equilibrium and are therefore relatively stable. Their viscosity and surface tension are significantly different (higher) than the viscosity and surface tension of the pure dispersion medium (water).

Coagulation of dispersed particles of sols can be attributed to different factors, such as:

- increased concentration of dispersed phase;

- temperature changes;

- mechanical agitation;

- addition of substances that remove the solvation (hydration) core of the macromolecules;

- addition of a sol whose particles have the opposite electrical charge;

- addition of electrolytes that affect the electric double layer of the particles and change the value of the ζ-potential (zeta potential in the interfacial double layer); all particles of the same substance in the same environment have the same charge sign, and in a collision between two particles, electrostatic repulsive forces act against the adhesion forces and no coagulation occurs.

The lowest concentration capable of causing coagulation of an electrolyte is called the **coagulation threshold**, and its reciprocal is the **electrolyte efficiency in precipitation** (the higher the precipitation value, the less efficient the electrolyte in the precipitation, and vice versa). The value of the ζ-potential at which coagulation occurs is the **critical potential**.

Adding small amounts of inorganic or organic salts to hydrophobic sols causes suppression of the electrical charge on the surface of the colloidal particles or in the vicinity of the particles, which leads to a decrease in the ζ-potential and to coagulation. Hydrophilic sols do not coagulate with the addition of small amounts of electrolytes, but their molecules are stabilised, which is manifested by the so-called **salting-in effect**. Precipitation occurs only if the electrolyte is present in excess. The electrolyte then acts as a dehydrating agent, which is manifested by the **salting-out effect**.

Salting-out abilities differ depending on the type of ion. Multivalent ions are predominantly effective (when they have opposite electric charge signs to the colloidal particles). The coagulation threshold ratios of monovalent, bivalent, and trivalent counter-ions are 1 : 0.016 : 0.0015. In accordance with the declining influence of the salting-out ability, the ions of inorganic and organic salts (the ions of some dyes, alkaloids, saponins, and neutral organic molecules such as sugars, which eliminate the hydration core of molecules, are effective) belong to the so-called **lyotropic series**, for example:

- anions: $citrates^{3-} > tartrates^{3-} > PO_4^{3-} > SO_4^{2-} > CO_3^{2-} > acetates > Cl^- > Br^- > NO_3^- > I^-$;

- cations: $Mg^{2+} > Ca^{2+} > Na^+ > K^+ > NH_4^+$;

- sugars: saccharose > glucose > xylose;

- sugar alcohols (alditols) and other polyols: glucitol > glycerol.

The electric charges of some hydrophilic sols (proteins or colloidal solutions of acid polysaccharides containing carboxyl and sulfate functional groups, such as pectin and agar, respectively) are significantly influenced by pH. At the isoelectric point, where the net charge of the molecule is zero, the stability of the colloidal solution is low. The stability of the sol can increase or decrease with a change of pH, as the dissociated functional groups bearing charges of the same sign repel one another and simultaneously interact with additional water molecules.

Electrostatic interactions between dispersed molecules are significantly affected by the dielectric constant of the medium. A decrease in the dielectric constant caused, for example, by the addition of some organic solvents (alcohols or ketones), makes the mutual electrostatic attractive forces stronger, and the dispersed particles can form clusters or coagulate. This phenomenon is employed for the isolation of some polysaccharides (such as agar, carrageenans, and pectin substances) from aqueous extracts by the addition of alcohol. The stability of hydrophilic sols is also affected by the concentration of the colloidal particles (they have a greater tendency towards coagulation at higher concentration) and temperature. The stability of sols can increase or decrease with increasing temperature and depends on the type of colloid.

The phenomenon known as **coacervation** frequently occurs during coagulation of lyophilic sols. The sol is stratified into two distinct liquid layers that consist of microscopic and macroscopic drops or of a continuous liquid phase.

A concentrated dispersion containing more than about 10% of solids in a liquid dispersion medium is known as a **paste**. Pastes have the characteristic properties of both sols and gels. For example, dough is a paste, and pastes are similarly formed by some polysaccharides (e.g. starch).

7.8.3.2 Gels

Gels found in foods are semi-solid materials with varying degrees of elasticity, brittleness, and stiffness (rigidity). They represent two-phase dispersed systems consisting of a dispersion medium, which is a solid phase (gelling agent), usually a protein or polysaccharide. The dispersion medium is water located in the spatial (three-dimensional) gel network and associated mainly by hydrogen bonds and weak physical bonds (capillary forces). Gel formation is usually accompanied by small changes in volume, Brownian motion ceases, and a certain amount of heat is released (with the exception of thixotropic gels). The mechanical properties of gels (viscosity, elasticity, and plasticity) are similar to those of solids.

7.8.3.2.1 Formation

Gels are formed from sufficiently concentrated colloidal sols (solutions), usually due to changes in temperature or by the addition of salts. Most gels are formed by the cooling of sols. In this way, for example, gels of gelatine, agar, and other polysaccharides are formed. In exceptional cases, the gels are formed by heating of a sol (e.g. the gel of methyl cellulose).

Protein gels Gels are formed when partially unfolded proteins form uncoiled polypeptide chains that interact at specific points to form a three-dimensional cross-linked network. For example, a gelatine gel forms by cooling a gelatine sol (at a concentration of about 1%) when the

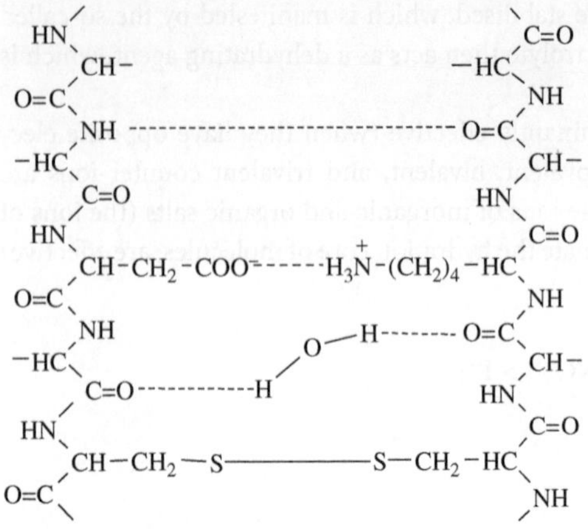

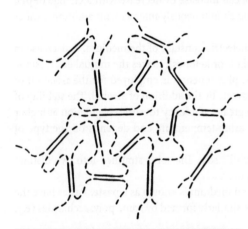

Figure 7.17 Bond types in gelatine gels.

protein globules uncoil into long fibrous molecules (Figure 7.16), which are mutually connected via non-bonding interactions. This creates an organised three-dimensional cross-linked structure (containing five or six crystalline domains per one polypeptide molecule) that binds a number of water molecules. The dispersion medium, unlike in a sol, now becomes the solid phase, and water becomes the dispersion medium. The sol thus creates a gel. In the formation of gels, ionic bonds mainly act between the amino groups of one polypeptide chain and the carboxyl groups of another. Secondary bonds are hydrogen bonds between the amide hydrogen atoms and the carbonyl groups of peptide bonds. Polypeptide chains can also be bound by covalent disulfide bonds (Figure 7.17).

Some globular proteins denature on heating, and the denatured molecules aggregate to form gels. In addition to intermolecular ionic and hydrophobic interactions, covalent disulfide bonds also play a part in the formation of these gels. The aggregation of molecules into gels occurs at higher protein concentrations (5–10%) in the solutions. An example is the thermal denaturation of egg white. In some cases, the aggregation of denatured protein molecules is caused by metal ions. An example is the formation of soybean curd (tofu) in the presence of calcium ions.

Polysaccharide gels The mechanism of gel formation in dispersions of polysaccharides is a more complex process. In the sol state (at temperatures above the gel melting point), the polymer chains occur (as in gelatine) in the shape of randomly coiled globules, and gel formation on cooling occurs only if an association with neighbouring macromolecules is possible. The ability to form gels depends mainly on the primary structure of the polysaccharide. In perfectly linear unbranched polysaccharides, such as amylose, interactions through hydrogen bonds occur along virtually the entire chain lengths. As a result of this association, dilute solutions of amylose precipitate, and concentrated solutions form rigid gels that are susceptible to retrogradation during further aggregation of molecules. At the other extreme are highly substituted and branched polysaccharides, such as gum arabic, which do not form gels, but only viscous solutions. Many other polysaccharide molecules have certain parts of the chain unsubstituted or unbranched, and in these so-called **binding zones** gelation occurs due to the formation of intermolecular junction zones between the binding zones of different chains, which leads to the formation of gels. For example, the binding zones of locust gum are the sequences of unsubstituted D-mannose units, whilst the pectin binding zones are regions composed of D-galacturonic acid units (Figure 7.18).

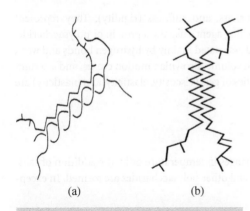

Figure 7.18 Polysaccharide gel with binding zones and randomly arranged chain residues.

Gels known as low-water-activity gels or sugar–acid–pectin gels, formed from highly methoxylated (esterified) pectins that contain more than 50% carboxyl groups esterified with methanol, are formed by this mechanism. A prerequisite of gel formation is the addition of sugar (saccharose), at a concentration of least 500 g/l, which acts as a dehydrating agent. Another condition is a sufficiently acidic medium maintained by adding fruit juice or acids (pH < 3.5). Such a gel is considered a three-dimensional network of pectin molecules with immobilised solvent (water), sugar, and acid. In this environment, free carboxyl groups are not dissociated, and therefore the otherwise charged chains are not repulsed.

The three-dimensional structure formation is based on chain associations stabilised by hydrogen bonding between undissociated carboxyl and secondary hydroxyl groups, and by hydrophobic interaction between methyl esters.

The association of chains and the formation of locust gum gel and of a mixed gel of locust gum with xanthan gum, where the molecules form double-helical structures, is shown schematically in Figure 7.19. Like locust gum galactomannan, glucomannan of xanthan gum also forms gels with glucomannan of konjak gum.

Agars and carrageenans (also furcellaran) form gels such that the molecules first associate to double helices, which are the gel-building structures. Double helices are transformed by further association to more complex helical aggregates known as superhelices (Figure 7.20). Agars form solid gels at a concentration of about 1% w/w. The average molecular weight of the basic building block of disaccharide agarose is about 150 Da; therefore, each polysaccharide molecule binds at least 550 water molecules. Association of basic double-helical structures to more complex helical aggregates also creates gellan gels (Figure 7.21).

(a)   (b)

Figure 7.19 Mutual association of (a) locust gum galactomannan chains and association of locust gum galactomannan with (b) xanthan gum glucomannan chains. Stephen (1995, fig 11.10). Source: Reproduced by permission of Taylor & Francis – Marcel Dekker.

Low-methoxylated (esterified) pectins can form gels in the presence of divalent cations, usually calcium. The interaction with calcium ions of fragments of two D-galacturonic acid adjacent units in two parallel chains of low methoxylated pectins is shown in Figure 7.22. Molecules of pectin with a low number of methoxyl groups (high content of free carboxyl groups) repel one another, which prevents the association of molecules via hydrogen bonds and hydrophobic interactions and the formation of gels. Therefore, gels are formed only in the presence of calcium ions. The initial strong association of two polymers into a dimer is followed by weak interdimer aggregation, mainly governed by electrostatic forces. Interactions along the entire length of the binding zones of low methoxylated pectin (also of sodium alginate salt) with calcium ions produce structures idealised in the so-called egg-box model (Figure 7.23). Alginates and gellan similarly form gels in the presence of metal ions, particularly calcium. The schematic structure of the alginate gel is shown in Figure 7.24. Binding zones comprising chain segments are mutually linked units of L-guluronic acid joined by calcium ions.

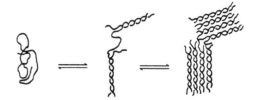

Figure 7.20 Formation of agar and carrageenan gels. Glicksman (1982, fig 12). Source: Reproduced by permission of Taylor & Francis – CRC Press.

7.8.3.2.2 Properties

The structure of the gel is influenced by temperature, mechanical stress, pH, and the presence of electrolytes (salts) and non-electrolytes (sugars). With the increase in temperature, gels become semi-solid materials, pastes, or even viscous liquids. In many cases, this process is reversible, and gelation (e.g. of gelatine or starch gels) occurs again on cooling. Gels of this type are called **thermo-reversible gels**. The process is related to the disappearance and appearance of hydrogen bonds between molecules of gelling agents. The opposite is **thermo-irreversible gels**, in which covalent bonds (e.g. disulfide bonds) necessary for gel formation are irreversibly interrupted on heating.

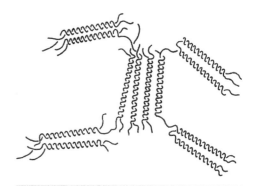

Figure 7.21 Formation of gellan gels.

The result of mechanical stress (when gels are shaken or agitated) is often the liquefaction of gels. If the gels that undergo liquefaction are left to stand, they rapidly regain their original consistency. This reversible property of some gels is called **thixotropy** (a time-dependent change in viscosity). Such gels are called **thixotropic gels**.

Gels usually retain their original volume for a long time, but ageing of some gels is manifested by **syneresis**, which is referred to as **retrogradation** if it takes place in starch gels. Ageing decreases the gel volume, meaning that the gel does not firmly bind all the water, which is partly released and excreted into the environment. Dried-out gels are called **xerogels**. The term xerogel is also used for dried compact macromolecular gels such as gelatine gels.

7.8.3.3 Emulsions

Emulsions are heterogeneous dispersed systems (hydrophobic colloids) similar to lyophobic sols. They are the result of dispersions of immiscible liquids. One of the liquids forms a dispersed phase, which is dispersed in the liquid dispersion medium of the second liquid into small particles (drops). Depending on the size of the dispersed particles and the associated stability, emulsions can be distinguished as:

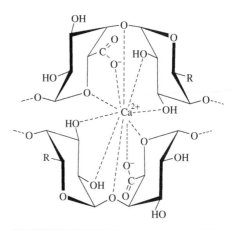

Figure 7.22 Interactions of low methoxylated pectin with calcium ions.

- **macro emulsions** with a particle size >100 nm, typically in the range 100–1000 nm (turbid, milky, and thermodynamically unstable dispersions of mutually immiscible liquids);

- **micro emulsions** with a particle size of 10–100 nm (clear and thermodynamically stable dispersions).

Emulsions in foods are macro emulsions, mainly mixtures of oil and water. The less polar phase of an emulsion (of lower relative permittivity) is referred to as the oil, whilst the other phase is water. Emulsions in food are of two types. An **oil-in-water** (o/w) emulsion (oil is the dispersed phase and water is the dispersion medium) contains small droplets of oil that are dispersed in water. A **water-in-oil** (w/o) emulsion has small

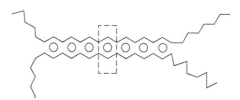

Figure 7.23 Schematic interactions of low methoxylated pectin or sodium alginate with calcium ions. O, calcium ions. Stephen (1995, fig 10.9). Source: Reproduced by permission of Taylor & Francis – Marcel Dekker.

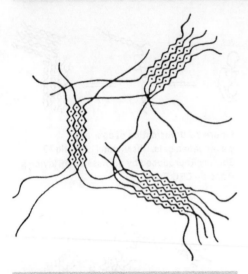

Figure 7.24 Formation of sodium alginate gels in the presence of calcium ions. Glicksman (1982, fig 14). Source: Reproduced by permission of Taylor & Francis – CRC Press.

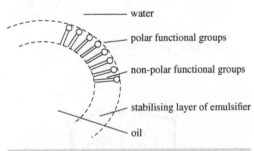

Figure 7.25 Schematic representation of an oil drop in an oil-in-water (o/w) emulsion.

droplets of water (dispersed phase) that are dispersed in oil (dispersion medium). Like all hydrophobic colloids, macro emulsions are unstable, because the dispersed phase tends to **coalescence** (small drops aggregate into larger drops and possibly even into a continuous phase, a layer), and the emulsions can then be divided into two phases, usually irreversibly. According to the density of the dispersion medium, the dispersed phase can either be concentrated on the surface or settle to the bottom of the container.

The stability of emulsions is influenced by many factors, such as the dispersed phase particle size (the smaller the particles, the more stable the emulsion), the amount of dispersed phase (more stable emulsions are formed with small amounts of dispersed phase), the densities of both phases (emulsions are stable when the differences between the densities of the two liquids are minimal), the viscosity of the dispersion medium (emulsions with a more viscous dispersion medium are more stable), temperature (extremely high or low temperatures are undesirable), dispersed phase electric charge (stability is influenced by the presence of identical electric charges on the dispersed particles preventing coalescence), the presence of electrolytes (stability is lower in the presence of electrolytes), interfacial tension (low interfacial tension increases emulsion stability), and other factors, including mechanical stress.

Emulsions, as well as other hydrophobic colloids (such as foams), can be stabilised mechanically or by the addition of natural or synthetic surfactants called emulsifiers, which reduce surface tension at the interface of both phases. For example, the surface tension at the interface of oil–water is 20–25 mN/m. In the presence of monoacylglycerol (0.3%), it is reduced to about 80%, and on adding a mixture of lecithin and monoacylglycerol, it drops to about 5% of the initial value, or less. In the interface, lipophilic groups of the emulsifier molecule interact with the lipophilic (non-polar) molecules of oil and hydrophilic (polar) groups of the emulsifier interact with water. The emulsifier (e.g. phospholipid or monoacylglycerol) forms a protective barrier that prevents coalescence of oil drops, connects the two liquids (oil and water), and stabilises the emulsion (Figure 7.25).

The stability of emulsions, as well as of other lyophobic colloids in the presence of emulsifiers, is limited. Therefore, emulsion stabilisers are often added. These can be low-molecular-weight hydrophilic substances (such as glycerol) or hydrophilic colloids (proteins or polysaccharides, e.g. gelatine, pectin, vegetable gums, or modified celluloses), which either increase the viscosity of the emulsions or interact with particles of hydrophobic colloids and thus allow their association with water.

Examples of o/w emulsions are milk and mayonnaise (where the dispersion media are water and vinegar and the dispersed phases are milk fat and vegetable oil, respectively). The emulsifier in milk is casein, but the emulsifiers in mayonnaise are egg yolk phospholipids. An example of a w/o emulsion is butter or margarine (where the dispersion medium is oil and the dispersed phase is water). Emulsifiers of butter are mainly natural phospholipids, whilst synthetic emulsifiers (monoacylglycerols and their derivatives and other compounds) are used to stabilise margarine.

In cases where the formation of emulsions is an undesirable phenomenon, their destruction can be achieved mechanically or chemically, for example by the addition of electrolytes.

7.8.3.4 Foams

Foam is a heterogeneous dispersed system strictly similar to an emulsion, but with **foaming** as its characteristic property. Foams consist of a gas (dispersed phase) dispersed into small bubbles in the liquid dispersion medium. The individual gas particles are separated by a liquid environment of different thicknesses (a few millimetres to a film thinner than 1 nm). Bubbles are of various shapes, according to the amount of gas located in the dispersion medium and where they occur (in the liquid phase or on the surface), but are typically round or hemispheric (Figure 7.26).

Foams exhibit characteristic elasticity and stiffness. Foam texture is related to the bubble size. A fine-textured foam consists of small bubbles, whereas a coarse foam contains larger bubbles. As heterogeneous dispersed systems, foams are thermodynamically unstable. When they collapse, which occurs in systems consisting of only the dispersion medium and dispersed phase (e.g. in air–water or water–carbon dioxide in soda water systems), the liquid phase flows from the liquid film that forms the bubble wall within a few seconds.

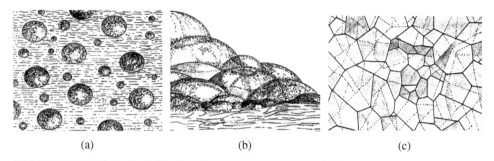

(a)   (b)   (c)

Figure 7.26 (a) Round gas bubbles in a diluted liquid phase of foam, (b) hemisphere bubbles on the liquid phase surface of a concentrated foam, and (c) the same bubbles viewed from below. Zapsalis and Beck (1985, fig 8.13). Source: Reproduced by permission of John Wiley and Sons.

The stability of foams is higher in the presence of surfactants, **foaming agents**, which adsorb onto the bubble walls and are arranged in the interphase so that their hydrophilic groups are situated in the water and their hydrophobic groups are directed into the air. They reduce the permeability of bubbles and increase the viscosity of the liquid phase, thus reducing the loss of liquid phase, which flows from the liquid film. In addition to the properties of the dispersed phase, dispersion medium, and the presence of surfactants, properties of foams are influenced by other factors, such as the method of foam preparation, the gravitational force, the forces associated with collisions between bubbles, evaporation of the liquid phase, and many others.

The dispersed phase of foams in foods is usually air or carbon dioxide. The gas is generally mechanically dispersed in a liquid, for example by whipping or bubbling. Foam texture affects dough and almost all bakery products, creams, and whipped creams. Foams form on the surface of non-alcoholic and alcoholic beverages. For example, beer foam is composed mainly of dispersed bubbles of carbon dioxide. The addition of nitrogen to a beer gives way to much smaller bubbles, creating a smoother and creamier product. A pivotal role in stabilising beer foam is attributed to barley proteins or polypeptides interacting with iso-α-bitter acids derived from hops. The bubbling properties of carbon dioxide dissolved in sparkling wines consist of the bubble nucleation on tiny particles stuck on the glass wall, bubble ascent, growth through the liquid, and bursting. A foam in which the dispersed phase is air is similarly produced when egg white is whipped. The stability of this foam is higher, as the protein is partially mechanically denatured. Hydrophobic groups of ovalbumin are then directed to the non-polar gaseous medium and hydrophilic functional groups interact with water. Thus oriented partially denatured albumin molecules aggregate and increase the viscosity of the foam. By heating, the whipped egg white proteins coagulate, producing a **solid foam**, which is found in some candies. Another example of a solid foam is ice cream, which also contains a considerable amount of air.

The naturally occurring foaming agents in foods are commonly heteroglycosides, known as saponins (which form stable foam at very low concentrations, around 0.005%), some polysaccharides, and proteins. Food additives used as foaming agents include polysaccharides (e.g. plant gums) and proteins with concentrations of about 0.5–2% by weight of liquid. Many other compounds are used as food additives, such as glycerol, its ethers, primary alcohols, and some saponins. Beer, for example, may contain propylene glycol alginate as a foam enhancer.

Foaming is sometimes unwelcomed in syrups, fruit concentrates, soft drinks, vegetable oils, tea and coffee extracts, and many other commodities. Reduction of foaming in these cases can be achieved by adding certain substances to cause the collapse of the foam. Their effect depends on their tendency to form a monomolecular film on the surface, which destabilises the foam. Often, these substances reduce the surface tension of the liquid phase to the threshold value at which the bubble walls are so thin that they burst. Commonly used additives are silicone oils in concentrations of 10–100 mg/kg and primary fatty alcohols, fatty amides, fatty acids, and their esters.

7.8.4 Properties of dispersed systems and foods

7.8.4.1 Mechanical properties

A branch of physics called **rheology** studies the behaviour of solids and liquids under mechanical stress, which manifests itself by deformation of solids and by flow of liquids under applied forces. The relationship between rheological and organoleptic properties of foods is the subject of **psychorheology**.

From the viewpoint of rheology, foods are very complex materials often composed of solid and liquid components. They are mixtures of various dispersion systems (homogeneous, heterogeneous, molecular, colloidal, and coarse dispersions). An example of such a complex material is milk (see Section 2.4.5.2). The behaviour of foods under mechanical stress (rheological behaviour) is closely related to their texture. The term texture includes properties of food producing the tactile (haptic) perception of tactile receptors, registered in the oral cavity and often applied by the touch of hands. Consumers assess, for example, the hardness of fruit according to the deformation resulting from finger pressure or the force that must be employed on chewing. The physical state and physico-chemical properties of foods are very important attributes, as they affect food behaviour during processing, storage, and consumption.

The term **consistency** is used to describe the mechanical aspects of the texture associated with the physical properties of food. Important rheological properties of solids are **elasticity** and **plasticity**, whilst an important feature of fluids is **viscosity**. These phenomena are responsible for the characteristic properties of many foods, since foods often simultaneously exhibit the properties of elastic solids and viscous liquids. Elasticity is the property of a material that enables it to resume its original shape or size when mechanical stress is removed. Plasticity is the ability of a material to be the subject of a continuous and permanent deformation when exposed to external, usually small forces that exceed its elastic limit (the stress point at which a material subjected to higher stress will no longer return to its original shape and will deform or break). The degree of plasticity depends on various factors, such as temperature. By increasing the temperature, an elastic gel, for example, is transferred from the elastic to the plastic (viscoplastic) state. The transition to the plastic state can be realised not only by heating, but also by the addition of water, which acts as a plasticiser. The plastic state of a material has certain characteristic and specific properties that are similar to those of both solids and liquids. Plasticity is to some extent related to viscosity, which is a characteristic property of liquids. Elasticity and viscosity are distinguishing properties of some dispersed systems (such as gels) and many other foods (e.g. dough and bread). Often, therefore, the mechanical properties of foods are described as **viscoelastic properties**. Geometric texture attributes, such as shape and size, are often called **appearance**. The appearance of foods is also closely related to their colour.

Various descriptive terms are used to characterise texture. For example, 'hardness' means resistance to deformation. 'Firmness' is essentially identical to 'hardness', and is sometimes used to express the ability to resist deformation caused by the material's own weight. 'Cohesiveness' relates to the strength of a material's internal ties. 'Flexibility' is the speed with which the deformed material regains its original shape when it is not subjected to the deforming force (it is associated with the relaxation time of the restoration to the original shape). Related terms are 'toughness', 'fragility', 'brittleness', 'friability', and 'crispness'. 'Stickiness' is the adhesion between the surface material and the adjacent surface. If both surfaces are of the same material, then the correct term is 'cohesion'. A material's mechanical characteristics are also described by such terms as 'viscosity', 'pastiness', and 'thinness'. The responsibility for the food properties represented by these terms generally belongs to macromolecular food components such as proteins, polysaccharides, and products of their mutual interactions and their interactions with other food constituents, first and foremost with water. The water content (amount of dry matter) and fat content in a food are related to the terms 'dry', 'watery', 'greasy', and 'tallow', amongst many others.

7.8.4.1.1 Elasticity

An important property of many materials is their ability to regain their original shape after the load is removed. These materials are called elastic. When the deforming force is weak (between zero and the elastic limit) and acts for a short time, it generally leads to elastic deformation, which is temporary, and the material will regain its initial dimensions. When the deforming force is weak but acts for a long time or exceeds the elastic limit of the material, the typical result is non-elastic (permanent) deformation. The ratio between elastic and non-elastic deformation is called the degree of elasticity. Deformation may be caused not only by pulling or compression, but also by more complex phenomena such as bending and torsion.

The deformation of a solid material by uniaxial stress caused by pressure (acts in one direction only) is in indirect proportion to the Young's modulus (E), also known as the modulus of elasticity, which characterises the material rigidity. A higher value of the modulus of elasticity indicates a rigid material (the E value is about 200 000 MPa for steel), a lower value indicates a more brittle material (such as glass, where E = 50 000–90 000 MPa), and a very low value represents a ductile material (such as rubber, E = 10–100 MPa). The E values of certain foods are, as an illustration, listed in Table 7.7. These values depend on many factors. For example, the modulus of elasticity in bread changes during storage and depends on the type of flour and addition of emulsifiers and a range of other variables.

A similar relationship as for the uniaxial stress caused by pressure can be expressed for the shear stress caused by stretching force.

7.8.4.1.2 Viscosity

The ideal fluid differs from the ideal elastic solid as the shear stress causes irreversible deformation manifested by fluid motion (flow). The liquid is trying to hinder the relative movement of one layer of liquid molecules relative to the other layers. It therefore has some internal

Table 7.7 Modulus of elasticity values of some foods.

Food	Modulus of elasticity (MPa)	Food	Modulus of elasticity (MPa)
Bread (fresh)	0.005	Bananas	0.8–3
Bread (almost fresh)	0.01	Apples, potatoes	6–14
Bread (old)	0.02	Carrots	20–40

friction (internal resistance), which is related to the mutual interaction of liquid molecules. This friction is called shear viscosity, often referred to simply as **viscosity**, which is a measure of a fluid's resistance to flow. An ideal fluid has constant viscosity when the value of the operating shear stresses changes.

Dynamic (absolute) viscosity, or the coefficient of absolute viscosity (η), *is* the measure of internal friction of the liquid (viscosity) at shear stress (τ) between the layers of non-turbulent fluid (Newtonian fluid) moving in straight parallel lines, which is given by the equation:

$$\eta = \tau / \dot{\gamma}$$

where $\dot{\gamma} = d\gamma/dt$ = shear strain rate. The dynamic viscosity of a Newtonian fluid (η) is a constant in Newton's law of viscosity: $\tau = F/A = \eta.dv/dx$, where τ = shear stress in the direction of the x-axis, caused by the force (F) between two planes (layers) of the liquid, A = area of planes, and dv/dx = shear stress or rate gradient between the planes perpendicular to both planes. Specific equations have also been proposed for non-Newtonian fluids.

The reciprocal of the dynamic viscosity, the fraction $1/\eta$, is called **fluidity**. The proportion of dynamic viscosity and fluid density (ρ) is called **kinematic viscosity** (v): $v = \eta/\rho$.

The viscosity of a fluid depends on pressure and temperature. With increasing pressure, viscosity increases, but the effect of pressure (with the exception of high pressures) on the viscosity of a fluid is usually negligible. With increasing temperature, the viscosity of a fluid generally decreases substantially. Most low-molecular-weight fluids obey Newton's equation and therefore are called **Newtonian fluids**. Their viscosity is constant at a given concentration of a substance in solution, and at the given temperature does not depend on the shear rate (dv/dx). Viscosity is the only parameter that characterises these liquids completely rheologically. The relationship between viscosity and concentration of dispersed phase for Newtonian fluids can be expressed by a number of equations. The viscosity of fluids decreases exponentially with increasing temperature due to the decrease of interactions of molecules and increasing distance. The dependence of the viscosity of aqueous solutions of saccharose on concentration and temperature is illustrated in Figure 7.27.

The viscosity of polymer dispersions depends, amongst other factors, on the structure of the macromolecules. Dispersions of soluble linear polysaccharides (such as amylose) are more viscous than dispersions of substituted polysaccharides (such as guar gum), and the less viscous dispersions are formed by branched polysaccharides (e.g. amylopectin and gum arabic). Viscosity is generally a function of radius of gyration, which is related to the volume that the molecules occupy in a dispersion medium under thermal motion. Linear branched molecules of about the same number of monomers have a much larger radius of gyration (Figure 7.8).

Newton's equation is not valid for **non-Newtonian** (non-ideal or real) **fluids** that contain asymmetrical particles, clusters of molecules, or macromolecules. Many common food systems, such as suspensions, emulsions, and gels, belong to non-Newtonian fluids. Their viscosity depends on the shear rate (dv/dx) or the tangential force. However, Newton's law also applies to certain non-Newtonian fluids at lower flow rates and to dilute polymer solutions (such as hydrophilic sols) at very low shear stresses. The rheological characterisation of non-Newtonian fluids requires knowledge of the dependence of shear stress on shear rate, which is called the **flow curve**. For example, the real shear strain rate ($\dot{\gamma}$) for spreading butter, margarine, or processed cheese on bread is 0–100 s^{-1}, that for pouring liquids from bottles is 50–200 s^{-1}, and that for pumping liquid chocolate in a pipe of a diameter of 100 mm at a flow rate of 50 l/min is 30 s^{-1}.

The non-Newtonian fluids are classified according to the behaviour that they show based on the rate gradient (dv/dx) and tangential force (Figure 7.28). The following five non-Newtonian fluids (systems) are recognised:

- **plastic** (Bingham plastic);

- **pseudoplastic** (shear thinning);

- **dilatant** (shear thickening);

- **thixotropic**;

- **rheopectic**.

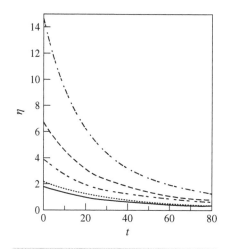

Figure 7.27 Dependence of dynamic viscosity of aqueous saccharose solutions on the concentration and the temperature. η, dynamic viscosity (mPa/s); t, temperature (°C); –, 0% (water); ......, 10%, – – –, 20%, (· – · –) 30%, (- - - -), 40% (· · · · · ·) saccharose solution.

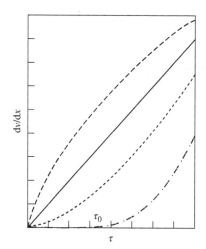

Figure 7.28 Schematic representation of shear rate (dv/dx) of Newtonian and non-Newtonian fluids on shear stress (τ). –, Newtonian; – – – – –, dilatant (shear thickening); ......, pseudoplastic (shear thinning); · – · – · plastic fluid; τ_0, yield stress. Zapsalis and Beck (1985, fig 8.24). Source: Reproduced by permission of John Wiley and Sons.

Plastic systems initially behave as solid (Bingham) bodies (for $\tau < \tau_0$) and resist deformation until a yield stress (τ_0) is reached. When that stress is exceeded, the shear rate grows. Further stress leads finally to linear (Newtonian) behaviour. Examples of plastic systems are chocolate, butter, cheese, various spreads, and ice cream. In pseudoplastic systems, the observed viscosity decreases with an increase in shear stress. An example of a pseudoplastic system is pudding. Dilatant systems resist deformation more than in proportion to the applied force. The shear rate grows much faster than that of Newtonian fluids and viscosity increases with an increase in shear stress. At low applied forces, the system behaves as a Newtonian fluid. Examples of dilatante systems are honey with added dextran and a slurry of wet beach sand. Thixotropic systems become more fluid (have lower viscosity) with increasing time of applied force. If the applied force ceases to operate, the original viscosity of the system is restored due to a reversible transformation of the sol–gel type. Examples of thixotropic systems are mayonnaise, ketchup, whipped and hardened fats, butter, and processed cheeses. Rheopectic systems exhibit behaviour opposite to that of thixotropic systems. Their viscosity increases with increasing time of applied force. An example is whipped egg white.

7.8.4.1.3 Viscoelasticity

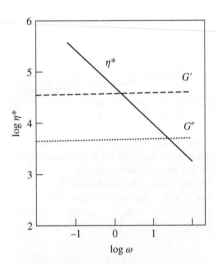

Figure 7.29 Typical mechanical spectra of polysaccharide gels. $\eta^\star$, complex dynamic viscosity (Pa.s); G', storage modulus (Pa.s); G'', loss modulus (Pa.s); ω, oscillation frequency (rad/sec). Stephen (1995, fig 16.1). Source: Reproduced by permission of Taylor & Francis – Marcel Dekker.

The majority of foods show viscoelastic properties. This means that these foods react, when exposed to applied force, in terms of both their elastic component (which behaves as in solids) and their viscous component (which behaves as in liquids). The rheological characterisation of these materials, in addition to flow curves, requires knowledge of other parameters, including complex (dynamic) modulus $G^\star$, which is related to the elastic component of the material (storage modulus G') and the viscous component of the material (loss modulus G''): $G^\star = (G'^2 + G''^2)^{1/2}$. With the so-called complex dynamic viscosity ($\eta^\star$), which is analogous to the dynamic viscosity resulting from Newton's law, the complex modulus ($G^\star$) is related by the equation $\eta^\star = G^\star/\omega$, where ω = oscillation frequency or strain rate under dynamic (sinusoidal) material loading.

An example of the mechanical spectra (dependences of G', G'', and $\eta^\star$ on ω; the SI physical unit is the Pascal second, Pa.s, equivalent to N.s/m^2 or kg/m.s) of a polysaccharide gel is shown in Figure 7.29. The gel has a much higher value of the storage modulus G than of the loss modulus G'' and therefore essentially behaves as a solid. Both modules are independent of strain rate (ω), and the gel is therefore highly elastic. The complex dynamic viscosity ($\eta^\star$) decreases with increasing strain rate (ω) values as the gel becomes more fluid, which means that it is thixotropic.

Important processes taking place during the formation of a gel, and the influences of the temperature and the concentration of the gelling agent, are shown in Figure 7.30. After an initial delay, during which the system behaves as a liquid, at time t_g the gel formation starts, which is rapid at first and then slows until a certain constant value of the storage modulus G' is achieved, which is a measure of elasticity of the material. The dependence of the storage modulus G' on the polysaccharide concentration in solution (for $t \gg t_g$) shows that the gel formation occurs only at concentrations that are higher than the critical concentration c_0. At concentrations higher than c_0, the gel rigidity grows rapidly, whilst at concentrations lower than c_0, the polysaccharide does not form a gel, and its solution remains just a viscous liquid.

7.8.4.2 Kinetic properties

Kinetic properties of dispersed systems (dispersions) determine the rate of processes that affect many aspects of food quality.

The dispersed phase particles of the colloidal and coarse dispersions, and also of foods, are in a permanent random motion called **Brownian motion**. The movement of molecules is caused by bumps of the dispersion medium molecules (mostly by water molecules) that perform the thermal motion. The speed of the particle movement increases with increased temperature and decreases with decreased particle size. At a certain particle size (approximately >4000 nm), their movement is not observable. The thermal motion of dispersed particles is the essence of some phenomena occurring in colloidal dispersion and coarse dispersed systems, and thus also in foods. These phenomena primarily include **diffusivity**, **sedimentation** due to gravitational or centrifugal forces, and **osmosis**.

7.8.4.2.1 Diffusivity

In dispersed systems with a dispersed phase or dispersion medium concentration gradient (chemical potential gradient), a spontaneous process occurs to balance the concentration. This kinetic behaviour, the ability of a substance to undergo diffusion, is called **diffusion**. The speed of diffusion is characterised by the diffusion coefficient (D). Its value increases with increasing temperature and decreases with increasing viscosity of the dispersion medium and radius of dispersed particles. The diffusion rate of the dispersed phase is therefore highest in molecular dispersed systems, lower in colloidal dispersion systems, and immeasurable in coarse dispersions.

Diffusivity determines the rate of drying, the rate of crystallisation of ice during freezing, and many other processes. In foods with low moisture content, diffusion of water depends mainly on the water content. The lower the water content, the lower the diffusion coefficient. At the critical water content, which corresponds roughly to the monomolecular layer surrounding the polymers, the diffusion coefficient is zero. In partially crystalline materials such as polysaccharides (starch granules or cellulose), diffusion occurs only in amorphous regions. Another important process is the diffusion of water through food packaging.

7.8.4.2.2 Sedimentation

Sedimentation is the process by which the dispersed phase particles (large molecules or macroscopic particles) are concentrated under the action of gravitational or centrifugal forces in the direction of the force. Friction acts against this movement, and diffusion can also be applied. The sedimentation rate is generally inversely proportional to the viscosity of the medium and directly proportional to the density difference between the dispersed phase and the dispersion medium, or to a square of particle diameter.

Diffusion of particles is immeasurable in coarse dispersions, where only the forces of gravity and friction act. Large particles settle faster than small particles. The movement of particles in the steady state is uniform. Molecular dispersions, on the other hand, do not show any measurable sedimentation, as the sedimentation rate is low; therefore, the molecular dispersions are kinetically stable and the rate of diffusion is high. In colloidal dispersions, both processes can be balanced and may establish sedimentation equilibrium in which the sedimentation rate is equal to the diffusion rate in the opposite direction. Such a dispersion system is then kinetically stable.

In practice, the sedimentation of food particles in coarse dispersions (such as some dressings) might be prevented by increasing the dispersion medium viscosity using suitable polysaccharides, such as carrageenan.

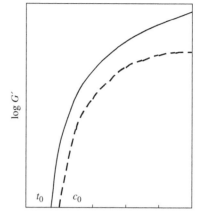

Figure 7.30 Influence of temperature and concentration on polysaccharide gels formation. G', storage modulus (Pa.s); —, temperature; - - - - -, concentration. Stephen (1995, fig 16.2). Source: Reproduced by permission of Taylor & Francis – Marcel Dekker.

7.8.4.2.3 Osmosis

Osmosis is a process in which dispersion medium (solvent) particles, such as water molecules, migrate from a region of higher concentration (higher chemical potential) to a region of lower concentration (lower chemical potential) through a semi-permeable membrane. This membrane is permeable only to dispersion medium particles and is impermeable to dispersed phase particles. The solvent migration tends to increase the solvent volume of the region of lower concentration and raises the so-called **osmotic pressure** of the solution. Osmotic pressure is the same as the pressure to which the solution would have to be exposed in order to prevent penetration of solvent through the membrane. It is directly proportional to the temperature and inversely proportional to the weight of dispersed particles (their molecular weight and size). Osmotic pressure, for example, decreases if the dispersed phase particles form aggregates or when they are present as a polymer.

Osmotic pressure is an extremely important phenomenon in living cells. Elastic semi-permeable cell membranes release water but are selective to solute transport (sugars, organic acids, and other substances). Very low concentrations of dissolved substances present in the cells produce high osmotic pressures that are necessary to maintain the internal pressure, called **turgor**, and a state of tension in animal tissues and plant tissues. Prevention of high osmotic pressure in plant tissues involves the synthesis of polysaccharides from monosaccharides (e.g. starch from glucose). Osmotic pressure is also an important phenomenon in animal cells post mortem and in the cells of plant tissues at the time of harvest. Higher osmotic pressures are deliberately used for growth control and inhibition of microorganisms in canned products (e.g. brines high in salt or sucrose).

7.8.4.3 Thermal properties

At higher temperatures, in frozen foods and in foods with low water content, macromolecular substances, such as proteins and polysaccharides, exist mainly in an amorphous state containing unorganised molecules. Amorphous or partially amorphous structures in foods commonly arise during various heat-treatment processes such as cooking, evaporation, drying, and extrusion, where water is removed relatively quickly or concentrated dispersions are rapidly cooled. Amorphous materials are found in a non-equilibrium state and therefore, in relation to time, undergo various changes of their physical and other properties. For example, amorphous lactose crystallises in powdered milk.

When a fluid or melt (such as a sucrose melt) cools, the kinetic energy of the molecules decreases until the motion of the molecules is so small, at a certain temperature $T = T_m$ (melting temperature), that the molecules can no longer escape from the secondary valence forces of neighbouring molecules, and the sucrose solution or melt crystallises. The volume of solution or melt changes in a step (usually decreases).

If the fluid or sucrose molecules become components of crystals, their spatial position is fixed, the molecules lose the freedom of movement, and Brownian motion ceases.

If water, other liquids, sucrose solutions, or melts are cooled rapidly, it may be that they do not crystallise for kinetic reasons. Generally, cooling the fluid to a temperature below its melting point ($T < T_m$, supercooling) changes its mechanical properties, and its viscosity increases according to the structure of the substance. The fluid passes the region of the iso phase transition temperature (T_f) to the **elastic state**, which is the state of fluid with a fixed structure.

Further cooling to temperatures close to T_g (glass transition temperature) results in a dramatic increase in viscosity, and at temperature $T = T_g$ the material passes from the fluid to the **glassy state**. This process is also called **nitrification** of fluids. At T_g and lower temperatures, the viscosity of vitreous material remains constant and very high (about 1×10^6 MPa/s). Other properties, such as permittivity, refractive index, diffusivity, and elastic modulus are also step changes. Between the elastic state and glassy state (also known as the supercooled or solid melt), there is no difference in the structure, and only the relaxation-time (the rate of changes of mechanical properties) characteristic for the elastic state is extended. The transition from the elastic to the glassy state is a reversible and kinetic process, since the value of T_g depends on the cooling rate.

The T_g values of food materials range from $-135\,^\circ$C (T_g of water) to tens of $^\circ$C or more above the melting point of ice (e.g. the T_g of sucrose is about 62 $^\circ$C). A decrease in T_g values is seen in the presence of water and other low-molecular-weight compounds that act as plasticisers (e.g. the T_g value of 50% aqueous solution of sucrose is about $-100\,^\circ$C). The T_g value of mixtures of two or more substances lies between the T_g values of pure substances. The changes in the physical state of sucrose solutions at various concentrations and temperatures are described in the state diagram given in Figure 7.31.

Similarly to sucrose, other low-molecular-weight substances and polymeric compounds (proteins and polysaccharides) present in foods may exist in various physical states, such as crystalline, glassy, elastic, or plastic, amongst which there are iso phase transitions given by temperatures T_m, T_f, and T_g, respectively. All polymers have similar properties, differing only in the temperatures where the iso phase transitions occur. Fully crystalline solids do not have the T_g temperature and amorphous substances do not have the T_f temperature. The T_g value has a great influence on the degree of crystallinity of polymers. Most biopolymers have similar T_g values. The T_g values of anhydrous substances lie in the range of $200 \pm 50\,^\circ$C (for starches, within 151–243 $^\circ$C). A decrease of the T_g value by $10 \pm 5\,^\circ$C is caused by the addition of 1% w/w of water. Biopolymers containing 20 ± 5% of water have the T_g value at room temperature, whilst biopolymers with 25–30% of water content have the T_g value at $-10 \pm 5\,^\circ$C. Generally, the T_g value increases with increasing relative molecular weight. Up to a relative molecular weight of about 104 Da, the T_g value increases linearly with its reciprocal value.

Knowledge of the T_g values and their changes according to the water content is important for predicting the shelf life of many foods and ongoing changes related to their organoleptic properties and other quality features. Foods in the T_g region, as well as solutions or melts of sugars, experience suddent changes in many properties, including the modulus of elasticity (which increases by several orders of magnitude) and the rate of relaxation of mechanical tension. Glassy material changes to the liquid (melt) or viscoelastic material when heated to a temperature above T_g.

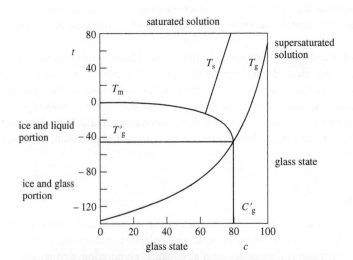

Figure 7.31 Saccharose-water state diagram. t, temperature ($^\circ$C); c, concentration; T_m, equilibrium: formation–melting of ice (equilibrium melting temperature of ice); T_s, equilibrium: saturated solution–oversaturated solution; T_g, glass transition temperature; T'_g, glass transition temperature of maximum concentrated liquid phase with saccharose content C'_g.

Viscoelastic behaviour can be seen in bread, which is actually a rigid foam. Typical bread containing 40% of water is glassy and brittle at temperatures below approximately $-15\,°C$, as it is at temperature $T < T_g$. Bread can be stored at such temperatures with no retrogradation of starch or any of the related consequences (ageing and hardening) that occur under storage at normal temperatures. With increasing temperature (or water content, which acts as a plasticiser), at $T = T_g$, the elasticity of bread increases rapidly and reaches a value that remains approximately constant during further temperature (water content) rise. Bread again becomes a viscoelastic material, but at $T > T_g$ the ageing process rapidly accelerates. Some bread types (e.g. flat and dry crispbreads, such as German Knäckebrot) are glassy and brittle, and durable even under normal temperatures, as they contain less than about 14% of water. Their glass transition temperature is higher than the ambient temperature, and therefore such breads can be stored for a long time without any changes. Wetting associated with the decrease of T_g value below room temperature results in the loss of crunchiness.

7.8.4.4 Optical properties

Light (electromagnetic waves) passing through homogeneous environments (e.g. true solutions) causes polarisation of particles (molecules). The resulting dipoles emit light of the same wavelength and, due to the interferences of secondary waves (Huygens' principle), spreads only in the direction of the primary light (incident) wave. In heterogeneous environments (e.g. in dispersions), this process leads to a different polarisation of dispersed phase particles and dispersion medium particles. The radiation is not compensated for according to Huygens' principle and is scattered in all directions. The light path through the dispersion system can be seen in a darkened room (the Tyndall effect). The intensity of light scattering in dilute dispersions on electrically non-conductive spherical particles, which is small in comparison with the wavelength (λ) of light ($< \lambda/20$), also depends on the wavelength of the incident light. Light scattering in the visible-spectrum region (approximately 400–700 nm) is about 10 times more intense for light of shorter wavelengths (violet and blue) than for light of longer ones (yellow and red). The dependence of the scattered light intensity on the wavelength is called **opalescence**. Dispersions composed of very small particles show weak opalescence, whilst dispersions of large particles do not show opalescence, because the light is reflected.

Opalescence often significantly affects the colour of dispersions. Dispersions exhibit different colours in incident and transmitted lights. These colours are complementary to one another; the blue part of the spectrum predominates in diffuse light and the red part in transmitted light. Therefore, diluted milk is pink, as it scatters light of shorter wavelengths whilst allowing light of longer wavelengths to pass through it.

7.8.4.5 Electric properties

Dispersed systems are externally electroneutral as a whole, but show some electric (electrokinetic) properties. Electrical properties of colloidal dispersion systems are associated with the ability to disperse particles that adsorb ions and molecules of the dispersion medium. The important food dispersed systems, lyophilic sols, preferentially adsorb ions resulting from dissociation of electrolytes. Electric charges of the same sign are then accumulated on the surface of colloidal particles as a result of adsorption of the ions that form an inner layer. The electric charges (ions) of opposite sign, which constitute the outer layer of ions, are in the surrounding part of the dispersion medium. Around the colloidal particles, an **electrical double layer** is thus formed, which stabilises the sols. Ions of the outer layer are arranged in two different sublayers. The first section of ions, partially neutralising the electric charge of the inner layer of ions, is bound by adsorption forces and forms a so-called Stern layer. The remainder of the ions of the same sign are bound by electrostatic forces and form a so-called diffusion layer (also known as Gouy–Chapman layer).

During the relative motion of dispersed particles with the electric double layer against the dispersion medium, the Stern layer and part of the diffusion layer move with the particle whilst the rest of the diffusion layer moves with the fluid. A potential thus arises in the interface with the liquid, which is called the **electrokinetic potential** or ζ-potential (zeta-potential). Its size depends on the type of electrolyte and the ability to adsorb ions. The existence of electrical charge on the dispersed phase particles (the existence of the electric double layer) significantly affects the stability of many dispersed systems. It is also associated with the phenomena that occur when one phase moves relative to another (in liquid–gas, liquid–liquid, and liquid–solid systems) and with the behaviour of dispersed systems under an external electric potential gradient.

The four possible types of electrokinetic phenomena are streaming (current) potential (electric potential generated by fluid movement relative to another phase), sedimentation potential (or the Dorn phenomenon or Dorn effect, due to the motion of dispersed particles relative to the fluid caused by sedimentation), electrophoresis, and electro-osmosis (movement of two phases caused by an external potential difference).

7.9 Water activity

The amount of water present in a food has only a very vague relationship with its resistance to microbial attack and biochemical and chemical reactions. A more important factor is the water's availability. Water availability is related to interactions of water with food components, with the bond strength of water bound by physical adsorption or chemisorption. Tightly bound water is less available than weakly bound water, which is again less available than free water. The measure of water availability is **water activity**. The water activity of a food is not the same as its moisture content.

All interactions of water in food (and also in solutions of electrolytes and non-electrolytes) result in a decrease in entropy, accompanied by a decrease of the vapour pressure. It can be inferred that the vapour pressure of water is related to water activity (a_w) by the following formula:

$$a_w = p_w/p_w^0$$

where p_w = partial pressure of water vapour above a solution of solids or liquids or above foods and p_w^0 = partial vapour pressure of pure water at the same temperature. Water activity values therefore range from 0 to 1.

When food is in equilibrium with the ambient air, its water activity is equal to the equilibrium relative humidity of the ambient air (φ; lowercase phi), which may range from 0 to 100%:

$$a_w = \varphi/100$$

The water activity of some selected foods is given in Table 7.8. The water content of many foods, and thus their activity, varies according to the humidity of the ambient air (and temperature), as there is a constant sorption or desorption of water. The term water activity should be used only for systems that are in thermodynamic equilibrium. Foods are often multicomponent multiphase systems, and only if there is a thermodynamic equilibrium between all phases is the water activity in the whole system equal; this does not happen often. It frequently requires hours, weeks, or even longer to achieve the equilibrium state. Many food systems may also exist in a non-equilibrium metastable state (e.g. sugar melts).

The water activity of a food increases with increasing temperature at constant moisture content. An increase in temperature of 10 °C causes an increase in water activity of about 0.03–0.2 and may, for example, have a negative effect on the stability of packaged foods, where the water content in the system is constant: undesirable microorganisms may grow or some adverse reactions may occur.

Knowledge of water activity in foods and of the relative humidity of the ambient air allow us to predict under what circumstances the food, in contact with air of a certain relative humidity, will dry up and when, instead, the food will become moist. However, water activity tells us nothing about how the water is bound in food. If the water activity in food is higher than the ambient humidity, the food loses water and dries up until equilibrium is established. In equilibrium, the water activity in the food is equal to the equilibrium relative humidity of the ambient air. On the other hand, if the water activity in the food is lower than the relative humidity of the ambient air, the food will take water from the ambient air and become moist. Prevention of adverse changes in water activity in food during production and storage (drying or wetting) and controlled removal of water by dehydration or drying (the purpose is to reduce weight and extend the food shelf life) is important in many food technologies.

7.9.1 Water activity and microorganisms

Knowledge of water activity is also important in assessing the potential susceptibility of foods to spoilage by microorganisms (Table 7.9). Most microorganisms grow well at a_w ranging from 0.91 to 0.99. Common bacteria need at least a water activity value of 0.94 for their growth, yeasts 0.90, and fungi around 0.75 (Figure 7.32). The bacterium *Clostridium botulinum* will grow and produce the deadly botulism toxin if water activity is above 0.85 and pH value is above 4.6. Most halophilic bacteria and mycotoxigenic aspergilli can reproduce in media where the water activity value is at least 0.75. Xerophilous fungi grow at a water activity of 0.60 and osmophilic yeasts are adapted to an environment of about the same water activity (0.65 or 0.60), which causes problems even in otherwise stable foods such as dried fruits and

Table 7.8 Water activity of selected foods.

Food	Water activity	Food	Water activity
Fresh meat and fish, milk, beer	0.99	Fruit jams, marmalade, jellies	0.75
Eggs, vegetables, and fruits	0.98–0.97	Honey, dried fruits, caramel	0.7–0.6
Bread	0.97–0.95	Pasta products (such as noodles)	0.5
Some cheeses (such as Swiss cheese)	0.91	Spices (10% of water)	0.4
Fermented sausages, margarine, aged cheese	0.87–0.85	Biscuits	0.3
Sausages	0.85–0.82	Dried milk and instant coffee	0.2
Condensed milk, legumes (15% of water), flour	0.8	Sugar, dehydrated soups	0.1

Table 7.9 Requirements of selected microorganisms to a minimum water activity.

Bacteria	Yeasts	Moulds	a_w	Bacteria	Yeasts	Moulds	a_w
Pseudomonas			0.96		Debaryomyces		0.87
Salmonela			0.95	Staphylococcus			0.86
Escherichia							
Bacillus							
Clostridium							
Salmonela			0.94			Penicillium	0.85
Escherichia							
Bacillus							
Clostridium		Rhizopus	0.93			Aspergillus	0.65
		Mucor					
	Rhodotorula		0.92		Zygosaccharomyces		0.62
	Pichia						
Micrococcus	Saccharomyces		0.90			Xeromyces	0.60
	Hansenula						
	Candida	Cladosporium	0.88				
	Torulopsis						

honey. This property of microorganisms is related to their chemical composition, the presence of polyols (glycerol and ribitol) in fungi, and the higher content of proline and glutamic acid in bacteria.

7.9.2 Water activity and reactions

In addition to microbial activity, water activity also affects many important enzymatic and chemical reactions that proceed in foods during processing and storage (Figure 7.32).

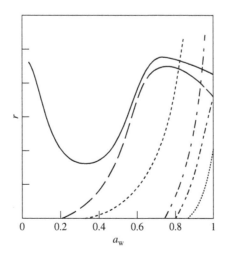

Figure 7.32 Influence of water activity on microorganisms and important reactions in foods. *r*, relative reaction rate; a_w, water activity; –·–·–·, moulds; - · · ·, yeasts; -- · -- ·, bacteria; . . . ., enzymatic activity; – – –, Maillard reaction; —, lipid autoxidation.

For example, the enzymatic hydrolysis of lipids in dried meat is a negligible process up to the point where water activity decreases below 0.4. Although lipid autoxidation rate decreases with decreasing water activity (up to about 0.3–0.5, because a certain amount of water inhibits the decomposition of lipid hydroperoxides), it then increases again with decreasing water activity, with a maximum reaction rate at water activity close to zero. This odd behaviour is explained by the removal of water from the hydrophilic sites of the material, meaning more lipid molecules are thus exposed to atmospheric oxygen.

The Maillard reaction rate increases with gradually decreasing water activity and the maximum reaction rate is at a water activity of about 0.7–0.8, which is attributed to the increase in the concentration of reactants (amino compounds and reducing sugars). With further decrease of water activity, the reaction rate decreases. The Maillard reaction does not proceed at all when the water activity decreases to less than 0.2–0.3, because the mobility (diffusivity) of reactants is too low.

7.9.3 Water activity and food organoleptic properties

Water activity is also related to the organoleptic properties of foods, especially foods with low or medium water content. At water activity values 0.35–0.50, some foods show desirable organoleptic properties such as crispness and crunchiness (e.g. potato crisps and extruded cereal products), which are mostly lost when the food becomes moist. In this range of water activities, instant beverages also become moist and, as a result, undesirable changes may occur, such as crystallisation of amorphous sugars, wetting and dissolution of their crystals, stickiness, and colour changes.

7.9.4 Free and bound water

Water contained in foods was previously classified as **bound water** and **free water**, or as **movable water** and **immobilised water**. Such a classification takes into account the strength of water bonds in food and is related to how easily water can be removed from food by mechanical (e.g. by pressing) or physical (e.g. drying) processes. These terms are very vague, however, often poorly understood, and mostly incorrectly used. According to some opinions, all the water contained in foods can be classified as bound water. In common foods with high water content (about 90%), bound water exists at water activities ranging from 0.0 to 0.7, and is often divided into the following categories:

- Very tightly bound water is present as **constitutional water**, which is an integral part of hydrates (<0.03% of total water). About 0–1% of the total water is found in the immediate vicinity of molecules of organic substances in foods, and has the following attributes: it has a lower mobility in comparison with the bulk water in the same food; it is bound mainly by chemisorption, particularly by the hydrophilic polar groups of proteins and polysaccharides (by water–ion and water–dipole associations); it can form a continuous layer surrounding hydrophilic molecules (proteins and polysaccharides) or their hydrophilic sites; it does not have the function of a solvent; and it does not freeze even at −40 °C. This proportion of water (generally 0.5 ± 0.4% of the total water) is referred to as **vicinal water**. At this water content, which roughly corresponds to the monomolecular layer, no chemical reactions proceed. For example, the water content corresponding to the monomolecular layer is reportedly 11% in gelatine and starches, 6% in dried potatoes, 3% in whey powder, and 0.4% in crystalline saccharose.

- Another proportion of bound water (3 ± 2% of the total water content) exists at water activities ranging from 0.2 to 0.7. This water occupies the remaining first-layer sites and forms several layers around the monomolecular layer. In these layers, mutual hydrogen bonds between water molecules already dominate, but there are also interactions between water molecules and ions or dipoles. Some water molecules penetrate into the capillary pores in the food structures by physical sorption. This water has the limited function of a solvent, and the main proportion does not freeze at −40 °C. The boundary between the first and second categories of water is a value of water activity of about 0.25. This water is known as **multilayer water**.

- Water activities ranging from 0.7 to 1.0, amounting to approximately 90–96% of the total water content, occurs more or less as **free water**, which has similar properties to water in diluted solutions of salts, or all the attributes of pure water. It mainly acts as a solvent of inorganic and organic substances. This water does not come out of foods freely, as it is bound exclusively by physical sorption (capillary forces), but it can be released from foods using suitably small forces (such as the gravitation force). The boundary between the waters of the second and third categories is at a water activity value of around 0.7 and higher. Sometimes, this category of water is known as **condensed water**; two types are recognised, these being **trapped water** and **free water**.

Classification of water is also possible on the basis of thermodynamic properties. The binding enthalpy of water of the first category (vicinal water) is about −4 to −6 kJ/mol, that of the second category (multilayer water) is approximately 1–3 kJ/mol, and that of the third category (condensed water) is around −0.3 kJ/mol.

7.9.5 Sorption isotherms

The relationship between water content in food (or equilibrium relative humidity of ambient air) and water activity is not simple. It is best described by the **sorption isotherm** of a particular food, which is the dependence of its water content on water activity. The sorption isotherm of a hypothetical food with high water content is shown in Figure 7.33, whilst the same isotherm over a narrow range of water contents and indicating the existence of vicinal water, multilayer water, and condensed water is shown in Figure 7.34.

Various foods have different sorption isotherm shapes (Figure 7.35) depending on their physical microstructure and macrostructure, qualitative and quantitative chemical composition, and distribution of chemical components. Attempts have been made to explain the shape of isotherms of food of different chemical compositions, but their importance lies more in practical application than in theoretical interpretation. From the sorption isotherms of dried acid protein hydrolysate, which is strongly hygroscopic, it can be seen that even a very small change in water activity (or the humidity of the ambient air) can cause a significant increase in water content in the hydrolysate. At the point in the figure marked *, the visible changes in the hydrolysate water content are beginning, as originally powdery particles stick together and change their colour from light brown to dark brown. At the point marked **, the hydrolysates become a viscous sticky mass. Agglomeration of particles of initially powdery material, between which the liquid forms a film that agglutinates them, depends on the surface tension and viscosity of the film. An example of a moderately hygroscopic material is saccharose, whilst that of a non-hygroscopic material is ground coffee.

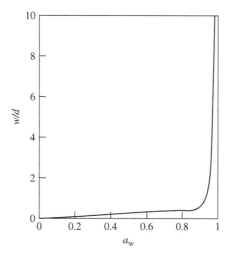

Figure 7.33 General sorption isotherm of food in a wide range of water contents. w/d, grammes of water per gramme of dry matter; a_w, water activity.

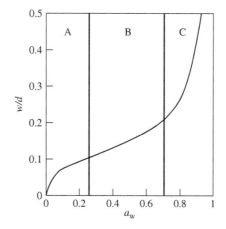

Figure 7.34 Detail of general sorption isotherm. w/d, grammes of water per gramme of dry matter; a_w, water activity; A, vicinal water; B, multilayer water; C, condensed water.

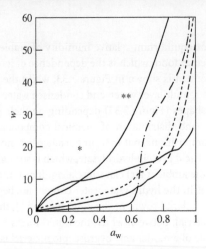

Figure 7.35 Various types of sorption isotherms. *w*, water amount (%); *a*ₘ, water activity; –, dry hydrolysed vegetable protein; ...., dry whey; - · · ·, ground roasted coffee; ----, wheat flour; - ·· - ··, saccharose.

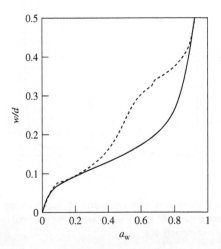

Figure 7.36 Hysteresis during water absorption and desorption. *w/d*, grammes of water per gramme of dry matter; *a*ₘ, water activity; –, absorption; - - -, desorption.

The sorption isotherms of foods are only valid at the particular temperature at which the results were obtained. Many materials (such as fibrous proteins, cellulose, and most foods) have different water activities at the same water content depending on whether the system absorbs or desorbs water. The corresponding isotherms then have a different shape, and therefore the absorption and desorption isotherms are recognised. This phenomenon is called a **hysteresis** (Figure 7.36). In general, the water activity is higher when a food absorbs water. If it loses water, then the water activity is lower. The physical nature of the hysteresis appears to be related to capillary phenomena in foods and their eventual collapse. The practical significance of the existence of a hysteresis is that, when adjusting the water content in food to a water activity value at which a certain reaction does not take place, it is better to reach this value by absorption, because the material then binds water more tightly than material of the same water activity that has been achieved by desorption.

The importance of sorption isotherms is generally in the evaluation of the water content in food at which the adverse effects on the food quality can be minimised. This is usually the moment when other layers are formed around the monomolecular layer of water (vicinal water), which is when the multilayer water arises. Most adverse events in the storage of foods with medium and low water content, such as crystallisation of amorphous sugars (e.g. lactose in powdered or condensed milk), agglomeration of powder materials, stickiness, and formation of large crystals in frozen products (e.g. in ice cream), relate to water content, water activity, and storage temperature, and are therefore closely related to the glass transition temperature of the material.

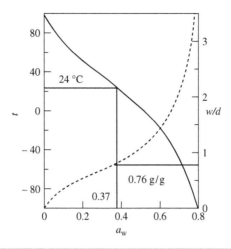

Figure 7.37 Sorption isotherm of powdered milk in association with glass transition temperature. t, temperature (°C); w/d, grammes of water per gramme of dry matter; a_w, water activity; – – –, sorption isotherm; –, glass transition temperature (T_g in °C).

Figure 7.37 illustrates, together with the sorption isotherm, the glass transition temperature of dried milk (virtually identical to the glass transition temperature of lactose), according to the activity of water. This dependence allows the prediction of the influence of water activity (a_w) or equilibrium relative humidity (φ) on glass transition temperature (T_g) and the evaluation of product stability or localisation of critical values of water activity. The critical value at 24 °C is the value $a_w = 0.37$ at which the T_g value is the same as the storage temperature (24 °C). Under such conditions, the stored dried milk contains 0.76 g of water in 1 g of dry matter, which leads to agglomeration of particles of milk and to the crystallisation of amorphous lactose, which causes a sandy consistency of the product. The rate of these changes is determined by the difference between storage and glass transition temperatures ($T - T_g$). Minor changes occur during storage at lower temperatures (at $T < T_g$).

8

Flavour-Active Compounds

8.1 Introduction

The most important psychological factor in human nutrition is the sensory quality of food, which significantly affects the type and quantity of food consumed and its utilization. Sensory quality of food depends on the presence of **sensory-active substances** that are perceived by chemical sensors and which determine the **sensory value** of foods by inducing olfactory, gustative, and other sensations (such as visual, haptic, or tactile, and auditory; see Chapter 1). The **organoleptic properties** of foods are usually more important to the consumer than other attributes, such as the vitamin content, because this information is observed first, and thus contributes significantly to the overall impression of the food raw materials and foods. Many foods and dishes are therefore aromatised, flavoured, coloured, or have their texture improved, whilst many others are consumed solely for their sensory value, such as various delicatessen items (delicacies, gourmet or fine foods).

Olfactory sensation results when olfactory receptors in the nose are stimulated by a particular substance in gaseous form called an **odorant**, which is capable of being translated into the subjective responses of neural brain stimulation that we term **odour, smell, aroma, flavour**, or **scent**. Gustatory sensation that determines the **taste** is a sensation elicited by substances acting on taste receptors in taste cells in the mouth.

Depending on their origin, the odour- and taste-active substances can be divided into two basic groups:

- Those that are already present in food raw materials or foods, as products of the secondary metabolism. These secondary metabolites are produced by intracellular processes and their quality and quantity depend mainly on the genetic predisposition of the organism (plant or animal species). In plants, animals, and other organisms, there is a variability within certain limits that is caused by some external factors. In plants, it may be location, age, degree of maturity (vegetative stage), harvest time, various environmental factors (the amount of moisture, nutrients, temperature, and light), and conditions during post-harvest storage. These flavour-active substances are also known as **primary flavour-active substances**.

- Many odour- and taste-active substances occur in foods in a bound, sensorially inactive form, mainly as non-volatile glycosides or esters. These compounds release the sensorially active substances by the action of enzymes, glycosidases, and hydrolases of carboxylic acid esters during tissue disintegration. Sensorially active substances also result during storage and processing of food as products of enzymatic, and non-enzymatic reactions from proteins, carbohydrates, and lipids (the primary products of the metabolism), or from other chemical substances, such as vitamins and pigments. Fermentation processes and thermal processing of foods (during cooking, baking, frying, smoking, and drying) are the main processes in which these substances are formed from their precursors. Autoxidation, the Maillard reaction (non-enzymatic browning reactions) and enzymatic browning reactions are the major reactions that lead to the formation of these flavour-active substances during storage, industrial, and culinary processing of foods. Sensorially active substances generated during these processes are often referred to as **secondary flavour-active substances**.

The resulting sensations of smell and taste, but also the sensations of colour and texture, are due to substances that are usually complex mixtures of several compounds. Odoriferous (odour-active) substances (odorants) are compounds that act on the olfactory receptors and give the impression of odour. They can also affect the taste receptors, in which case they are also taste-active substances. An example of such a compound is acetic acid, which has a penetrating odour resembling vinegar, and a sharp sour taste. Odoriferous substances are predominantly low-polar or non-polar (water insoluble to slightly soluble) compounds, producing a wide variety of sensory perceptions. Taste substances are those that act on taste receptors and give the impression of taste in the mouth. Gustatory sensation includes five basic tastes: **sweet, salty, sour, bitter**, and **umami**. Common sweet substances are monosaccharides and disaccharides, such as saccharose (sucrose, table sugar), whilst the best-known salty substance is sodium chloride (table salt). A sour taste is found in the already mentioned

The Chemistry of Food, Second Edition. Jan Velíšek, Richard Koplík and Karel Cejpek.
© 2020 John Wiley & Sons Ltd. Published 2020 by John Wiley & Sons Ltd.

acetic acid found in vinegar and pickled vegetables, and bitter-tasting compounds are caffeine in coffee and quinine in tonic water. In the 1980s, scientists recognised a fifth taste, umami, or savoriness, translated from Japanese as savoury or yummy. Less acknowledged is a sixth taste candidate, kokumi, widely accepted in Japan since 2010. Kokumi is translated as heartiness or mouthfulness and describes compounds in food that do not have their own flavour, but enhance the flavours with which they are combined.

The oral cavity also responds to other sensations (somatosensations), such as those described as astringent, pungent (hot, burning), and harsh tastes. There is an increasingly popular belief that taste is more complicated than was originally believed as it is related to a combination of sensations such as smell, colour, texture, and the sound of foods when chewing, and even to the emotional circumstances of the consumer when eating.

Taste-active substances are usually water-soluble polar non-volatile substances. Some taste-active substances may additionally be odour-active substances, but not necessarily. The complex (uniform) sensory perception of taste and smell is called **flavour**; a strong pleasant smell, usually from food or drink, is called **aroma**, and aromatic food or drink is that which has a pleasant smell. However, the term 'aromatic substance' has a completely different meaning in organic chemistry. An aromatic substance, put simply, is a benzene derivative. Unwanted, altered, modified, unnatural, and unpleasant odours, tastes, and flavours are called **off-odours**, **off-tastes**, and **off-flavours**, respectively.

Substances influencing **colour** (such as dyes and pigments) are also important sensory-active substances present in foods. They determine not only the characteristic colour of the food, but also the taste threshold concentrations of substances and the ability to identify odours. Food that has a satisfactory nutritional and hygienic-toxicological quality or excellent odour, taste, and texture will still not be accepted by consumers unless the characteristic colour corresponds to the standard product.

The properties included under the heading **texture** are found mainly in macromolecular food components, especially proteins and polysaccharides, as well as products of their interactions and associations with other food ingredients, first and foremost with water. Texture implies those characteristics of foods that cause tactile or haptic sensations registered by receptors in the oral cavity. Touch by hands is also very important. Auditorial perceptions such as crispness are related to a range of textural characteristics. Geometric attributes of texture that simultaneously cause visual and haptic sensations, often referred to as shape and appearance (such as particle size or size of the whole food), are closely related to food colour. The term **consistency** describes the texture aspects related to physical (mechanical) properties of food, which are also called rheological properties (see Section 7.8.4.1).

This relatively extensive chapter is devoted to components that affect food odour and taste. In the first section, the aromatic compounds are classified into their common groups: hydrocarbons, their oxidation products (including alcohols, aldehydes, ketones, carboxylic acids, and functional derivatives of carboxylic acids: esters, lactones, and nitriles), phenols, and sulfur- and nitrogen-containing compounds, including heterocyclic compounds. Each section presents the structures of individual compounds and their properties, occurrence, and important reactions. The next part is devoted to the aromas of individual food commodities (meat and meat products, poultry, fish, milk and dairy products, eggs, cereal products, fruits, vegetables, alcoholic beverages, tea, coffee, cacao, and chocolate, nuts, honey, mushrooms, and spices). Next, the importance of aromatic compounds in the diet, and their organoleptic properties, biological effects, production, and use, is dealt with. The last part is devoted to substances that affect the taste of foods. These include the five universally accepted basic tastes: sweet, salty, sour, bitter, and umami, as well as other taste attributes: astringent, pungent (sharp, burning), and cooling, and commodities in which they are found. The structures of these compounds and their occurrence, properties, reactions, significance in nutrition, physiology, technology, and practical use are all described.

8.2 Odour-Active substances

The aroma of food is often a very complex phenomenon caused by a large number of odour-active substances. The total number of odorants identified in foods is estimated to be about 10 000, and each food commonly contains several hundred different aromatic compounds. Some compounds are not involved in the characteristic odour of a particular food at all, some contribute very little, but others are of fundamental importance for various reasons, such as their odour character, low threshold concentration of perception, or high concentration. The resulting odor impression is often made up of only a small number of compounds. The intensity and quality of odour, however, depends not only on the odoriferous substances present, but also on other food components, especially proteins, carbohydrates, and lipids, with which the odorants interact. These non-bonding interactions influence the concentration of aromatic substances in the gaseous phase. For example, the aroma of oranges and roasted coffee is a mixture of about two dozen different compounds. In only a limited number of cases are the typical odours associated with a single substance or a few compounds, which are called **key components**. An example is vanillin, which is practically the only substance responsible for the typical aroma of dry vanilla beans. Some other examples are listed in Table 8.1.

Volatile flavour-active components of foods include virtually all groups of organic compounds, hydrocarbons, and their oxygen derivatives (such as alcohols, aldehydes, acetals, ketones, carboxylic acids, their esters, lactones, and other derivatives), nitrogen derivatives (amines, nitrogen heterocyclic compounds), sulfur derivatives (such as thiols, sulfides, and sulfur heterocyclic compounds), and many more. Particularly important primary flavour-active substances are **terpenes**. Terpenes are naturally occurring hydrocarbons derived biosynthetically from units of isoprene (2-methylbuta-1,3-diene) linked together head to tail (or, more rarely, head to head) to form linear chains, or which may be arranged to form rings. Terpenes that are modified chemically, such as by oxidation or rearrangement of the

Table 8.1 Characteristic (key) odorous components of foods.

Compound[a]	Descriptor	Occurrence
(−)-(R)-Oct-1-en-3-ol	Mushroom-like	Mushrooms, moulds
(−)-Geosmin	Earthy	Red (garden, table) beetroot
Anethole	Anise-like	Anise seeds
Cinnamaldehyde	Cinnamon-like	Cinnamon bark
Vanillin	Vanilla-like	Dry vanilla beans
Eugenol	Clove-like	Clove plant fruits
(E)-Citral (geranial, citral a) and (Z)-citral (neral, citral b)	Lemon-like	Lemons
(2E,6Z)-Nona-2,6-dienal	Cucumber-like	Fresh cucumbers
Benzaldehyde	Bitter almond	Bitter almonds, sour cherries
(+)-(2E,5S)- and (−)-(2E,5R)-5-methylhept-2-en-4-one[b]	Hazelnut-like	Roasted hazelnuts
4-(4-Hydroxyphenyl)butan-2-one (raspberry ketone)	Raspberry-like	Raspberries
(+)-(S)-Carvone	Caraway-like	Caraway and dill seeds
(2E,4Z)-Ethyl deca-2,4-dienoate	Pear-like	Pears
5-Ethyl-3-hydroxy-4-methyl-5H-furan-2-one (abhexon)	Hydrolysate-like	Acid protein hydrolysates
(+)-(R)-p-Menth-1-ene-8-thiol	Grapefruit-like	Grapefruits
Diallyl disulfide	Garlic-like	Garlic
Maltol and isomaltol	Caramel-like	Caramel
2-Acetyl-1-pyrroline	Crust-like	Bread crust, aromatic rice
2-Isobutylthiazole	Tomato-like	Tomato leaves, fresh fruits

[a] The structures of compounds are given further in the text.
[b] Trade name of the mixture of (E)-isomers is filbertone.

carbon skeleton, are called **terpenoids**, or sometimes **isoprenoids**. Terpenoids are the largest class of plant secondary metabolites with over 20 000 known compounds, produced primarily by a wide variety of plants, where they serve a range of different functions in basic and specialised metabolism. Depending on the number of isoprene units, isoprenoids acting as odour-active substances are divided into:

- **hemiterpenes** and **hemiterpenoids** (C_5 compounds containing one isoprene unit);

- **monoterpenes** and **monoterpenoids** (C_{10} compounds containing two isoprene units);

- **sesquiterpenes** and **sesquiterpenoids** (C_{15} compounds containing three isoprene units).

The main volatile components in essential oils are monoterpenes and monoterpenoids. Historically they have been used in the food, perfume, and pharmaceutical industries because of their culinary, fragrant, and antimicrobial properties. In fruits and vegetables, herbs and spices, and wines, they express a wide spectrum of odours, most of which are perceived as very pleasant. Sesquiterpenes and sesquiterpenoids are amongst the most widely occurring odorants. Numerous sesquiterpenic hydrocarbons, alcohols, and derived metabolites found in plant essential oils are highly valued for their desirable flavour characteristics. They also display a broad range of physiological properties, including antibiotic, antiviral, antifungal, antitumour, and hormonal activities. Synthetic variations and derivatives of natural terpenes and terpenoids are additionally used to greatly expand the variety of aromas used in perfumery and of flavours used in food additives.

Secondary flavour-active substances arise in particular:

- as metabolic products of microorganisms in fermentation processes;

- by oxidation and degradation of labile constituents (such as lipids and carotenoids);

- in thermal processes, especially from proteins and carbohydrates in the Maillard reaction.

For completeness, this chapter also includes some non-volatile substances that do not have the potency of odour-active compounds. These substances may frequently be precursors of volatile compounds or have other important properties.

8.2.1 Hydrocarbons

Hydrocarbons are common components of many foods often occurring as constituents of essential oils and food lipids. They are either natural components of food raw materials and foods (primary substances) or form during food processing and storage by enzymatic and chemical reactions as secondary substances (e.g. as lipid oxidation products and degradation products of carotenoids). Some hydrocarbons may be present in food as endogenous or exogenous contaminants (e.g. polycyclic aromatic hydrocarbons, PAHs).

For food flavouring, hydrocarbons are used relatively rarely. Some aliphatic hydrocarbons (such as hexane) and their mixtures (e.g. petroleum ether) are also used as solvents in the flavour and fragrance industry and in oleochemistry. Their residues may therefore be present in essential oils, oilseed meals, and other materials obtained by extraction with hydrocarbons.

8.2.1.1 Classification, structure, terminology, and occurrence

Depending on their structure, hydrocarbons occurring in foods can be divided into three basic groups: aliphatic, alicyclic, and aromatic hydrocarbons. The most important flavour-active substances are terpenic hydrocarbons.

8.2.1.1.1 Terpenic hydrocarbons

Compounds with the formula $(C_5H_8)_n$, which includes monoterpenes $(C_{10}H_{16})$ and sesquiterpenes $(C_{15}H_{24})$, can be active as odorous substances. Higher terpenic hydrocarbons, starting from diterpenes $(C_{20}H_{32})$, are indifferent as flavouring materials, but may act as taste-active substances or precursors of flavour-active substances (see Figure 3.6). Terpenic hydrocarbons occur in almost all fruits, vegetables, and spices. For example, about 90–99% of orange fruit volatiles and about 80% of carrot root volatiles are terpenic hydrocarbons. About 70–80% of black pepper volatiles are monoterpenic hydrocarbons and 20–30% are sesquiterpenic hydrocarbons. However, terpenic hydrocarbons are usually not very important compounds in defining the typical aroma of food commodities. More important are various monoterpenoids and sesquiterpenoids, alcohols, aldehydes, ketones, esters, and other oxygen-containing compounds. About 1000 monoterpenoids, more than 300 distinct sesquiterpene carbon skeletons, and more than 7000 oxidised or otherwise modified sesquiterpenic derivatives have been identified in nature. Most of these terpenoids are optically active compounds. Individual enantiomers and diastereomers may occur in different organisms or just in a single organism and often as mixtures.

Monoterpenes Monoterpenic hydrocarbons found in foods are linear (acyclic), monocyclic, bicyclic, and tricyclic compounds. Linear monoterpenes are mainly present in fruits and essential oils. Examples of common hydrocarbons are myrcene (β-myrcene, 7-methyl-3-methyleneocta-1,6-diene) and ocimene (β-ocimene, 3,7-dimethylocta-1,3,6-triene, **8-1**), which occurs as *trans*-isomer (*E*)-ocimene and *cis*-isomer (*Z*)-ocimene. α-Ocimene is (*Z*)-3,7-dimethylocta-1,3,7-triene. The ocimenes are found within a variety of plants and fruits.

Monocyclic monoterpenic hydrocarbons are derived predominantly from the optically active hydrocarbon 1-methyl-4-(propan-2-yl)cyclohexane, known as *p*-menthane. An exception is 1-methyl-4-(propan-2-yl)benzene also known as *p*-cymene or cymene (**8-1**), which is an aromatic hydrocarbon. Cymene is a common component of many essential oils, especially the essential oils of cumin (the seed of the herb *Cuminum cyminum* of the parsley family Apiaceae) and common thyme (*Thymus vulgaris*, from the mint family Lamiaceae) listed in Table 8.32.

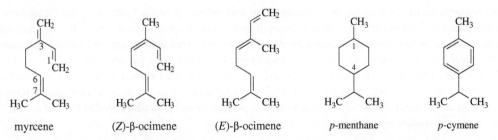

myrcene   (*Z*)-β-ocimene   (*E*)-β-ocimene   *p*-menthane   *p*-cymene

8-1, linear aliphatic, alicyclic, and aromatic monoterpenes

Bicyclic hydrocarbons can be divided into seven structural types that are combinations of three- and five-membered rings (thujane, also known as sabinane), three- and six-membered rings (carane), six- and four-membered rings (pinane), or two- and five-membered rings (fenchane, camphane (also known as bornylane), isocamphane, and isobornylane) (**8-2**).

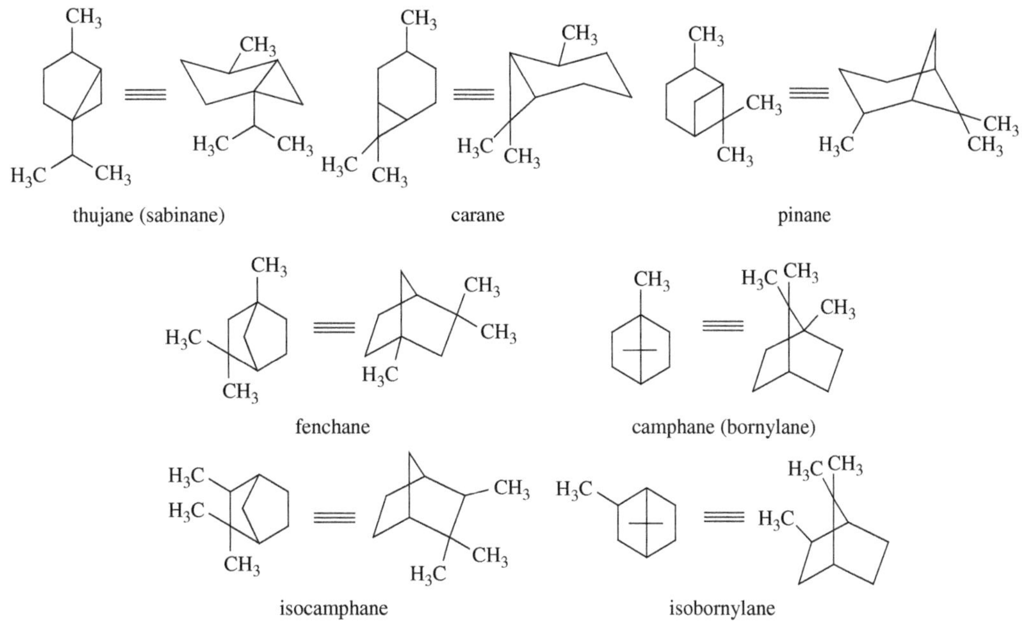

thujane (sabinane)   carane   pinane

fenchane   camphane (bornylane)

isocamphane   isobornylane

8-2, bicyclic monoterpenic hydrocarbons

The most important compounds are menthadienes, derived from *p*-menthane, which differ in the positions of the double bonds (**8-3**). A common hydrocarbon is limonene (*p*-mentha-1,8-diene), which typically occurs in many essential oils (such as essential oils of citruses, mints, and conifers) and turpentine. For example, the (+)-limonene isomer is the major component (>90%) of essential oils of citrus peels, whilst (−)-limonene is a component of essential oils of different types of mint (*Mentha* spp., Lamiaceae) and conifers. Racemate, which is trivially known as dipentene, occurs in many essential oils. Other important menthadienes include α-terpinene (*p*-mentha-1,3-diene), γ-terpinene (*p*-mentha-1,4-diene), α-phellandrene (*p*-mentha-1,5-diene), and β-phellandrene, also known as *p*-mentha-1(7),2-diene.

A number of bicyclic hydrocarbons are present in turpentine, and also as components of fruits, vegetables, and spices (**8-4**). A frequently occurring compound is sabinene, also known as thujene or 4(10)-thujene (which is found in higher concentrations in the essential oil of black pepper), car-3-ene (Δ³-carene), α-pinene (2-pinene), β-pinene (also known as 2(10)-pinene, nopinene, or pseudopinene), and camphene. Trivial and systematic names of the main compounds are listed in Table 8.2.

Table 8.2 Trivial and systematic names of some monoterpenic hydrocarbons.

Trivial name	Systematic name (IUPAC)
(+)-α-Phellandrene	(S)-2-Methyl-5-propan-2-ylcyclohexa-1,3-diene
(+)-β-Phellandrene	(S)-3-Methylene-6-propan-2-ylcyclohex-1-ene
(+)-Camphene	(1R,4S)-6,6-Dimethyl-5-methylidenebicyclo[2.2.1]heptanes
(+)-Car-3-ene	(1S,6R)-3,7,7-Trimethylbicyclo[4.1.0]hept-3-ene
(+)-Limonene	(R)-1-Methyl-4-prop-1-en-2-ylcyclohex-1-ene
Myrcene	7-Methyl-3-methyleneocta-1,6-diene
(Z)-β-Ocimene	(Z)-3,7-Dimethylocta-1,3,6-triene
(E)-β-Ocimene	(E)-3,7-Dimethylocta-1,3,6-triene
(+)-α-Pinene	(1R,5R)-2,6,6-Trimethylbicyclo[3.1.1]hept-2-ene
(+)-β-Pinene	(1R,5R)-6,6-Dimethyl-2-methylenebicyclo[3.1.1]heptane
(+)-Sabinene	(R)-4-Methylene-1-propan-2-ylbicyclo[3.1.0]hexane
α-Terpinene	1-Methyl-4-propan-2-ylcyclohexa-1,3-diene
γ-Terpinene	1-Methyl-4-propan-2-ylcyclohexa-1,4-diene

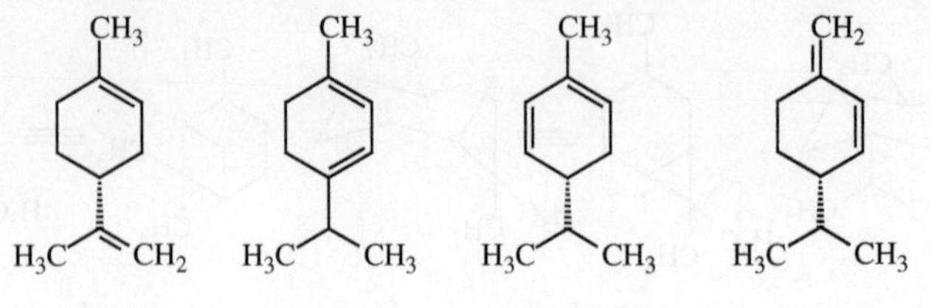

(+)-limonene   α-terpinene   (+)-α-phellandrene   (+)-β-phellandrene

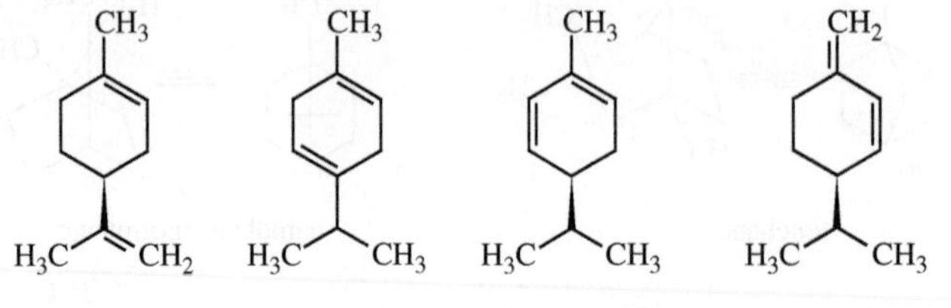

(–)-limonene   γ-terpinene   (–)-α-phellandrene   (–)-β-phellandrene

8-3, cyclic monoterpenic hydrocarbons menthadienes

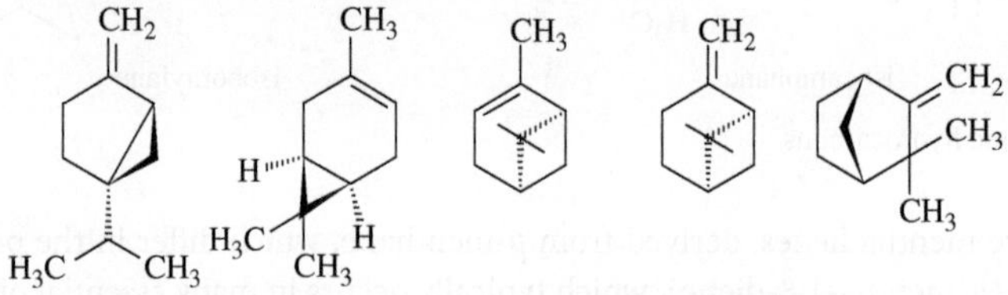

(+)-sabinene   (+)-car-3-ene   (+)-α-pinene   (+)-β-pinene   (+)-camphene

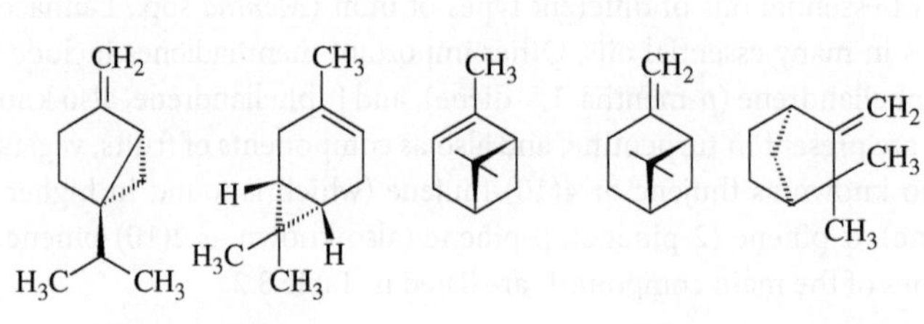

(–)-sabinene   (–)-car-3-ene   (–)-α-pinene   (–)-β-pinene   (–)-camphene

8-4, bicyclic monoterpenic hydrocarbons

Sesquiterpenes About 300 different basic structures are found in nature from which sesquiterpenes and sesquiterpenoids are derived. Often present in foods are stereoisomeric acyclic sesquiterpenic hydrocarbons called α-farnesene and β-farnesene. α-Farnesene exists as four stereoisomers that differ in the geometry of two (C-3 and C-6) of their three internal double bonds. The most common isomers are (3E,6E)-α-farnesene, known as *trans,trans*-α-farnesene, and (3Z,6E)-α-farnesene, known as *cis,trans-α-farnesene* (**8-5**). (3E,6E)-α-Farnesene, which represents about 90% of the α-farnesene isomers in some apple and pear cultivars, accumulates in the surface wax of fruits during low-temperature storage. Conjugated trienes, such as (7E,9E)-2,6,10-trimethyldodeca-2,7,9,11-tetraen-6-ol, resulting from its oxidation, have been linked with the development of a serious physiological storage disorder known as superficial scald, which manifests as brown or black patches on the fruit skin and in a bitter taste in the fruit. (3Z,6E)-α-Farnesene has been isolated from the perilla oil (*Perilla frutescens*, Lamiaceae), which dries faster than linseed oil; therefore, it is used in the production of varnishes. The herb is used in Chinese medicine, either alone or in combination with other herbs, especially as a remedy for coughs and asthma. Both isomers act as insect pheromones. β-Farnesene can exist as two stereoisomers about the geometry of its central (C-6) double bond. The (6E)-isomer, known as *trans*-β-farnesene (**8-5**), is a constituent of various essentials oils, such as ginger oil.

Cyclic sesquiterpenes (**8-6**) are found in large numbers in food volatiles and essential oils. The representatives of monocyclic hydrocarbons with six-membered cycle are α-bisabolene and β-bisabolene, α-zingiberene, and *ar*-curcumene. Representatives of macrocyclic hydrocarbons are germacrene A, germacrene D, and α-humulene. Bicyclic hydrocarbons with two six-membered rings are α-selinene, valencene, and α-, β-, γ-, and δ-cadinene. Muurolenes are isomeric with cadinenes,[1] whilst examples of other structures are β-caryophyllene

---

[1] The name cadinene is used to refer to all sesquiterpenes having the structure of cadalane (4-isopropyl-1,6-dimethyldecahydronaphthalene). However, a large number of isomers exist with different positions and stereochemistry of the double bonds. These compounds are divided into four subgroups according to the stereochemistry of the isopropyl group at C-1 and the hydrogen atoms on C-4a and C-8a: cadinenes (1S,4aR,8aR), muurolenes (1S,4aS,8aR), amorphenes (1S,4aR,8aS), and bulgarenes (1S,4aS,8aS).

and α-bergamotene. Some sesquiterpene hydrocarbons are precursors for the biosynthesis of other important compounds. For example, germacrene D is the precursor of two major odour-active components of grapefruit peel, valencene and nootkatone. Valencene is the main volatile emitted by flowers of the vine (*Vitis vinifera*, Vitaceae) and a precursor of ketone nootkatone in grapefruits. (−)-δ-Cadinene is a precursor of the dimeric sesquiterpenoid pigment of cotton plants (*Gossypium hirsutum* and other species, Malvaceae) called gossypol.

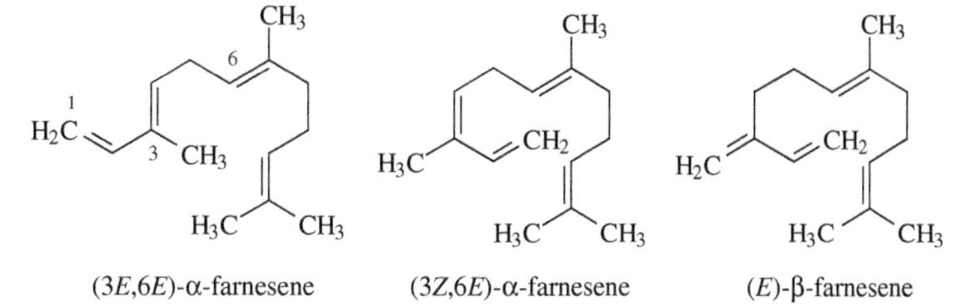

(3E,6E)-α-farnesene   (3Z,6E)-α-farnesene   (E)-β-farnesene

8-5, linear sesquiterpenic farnesenes

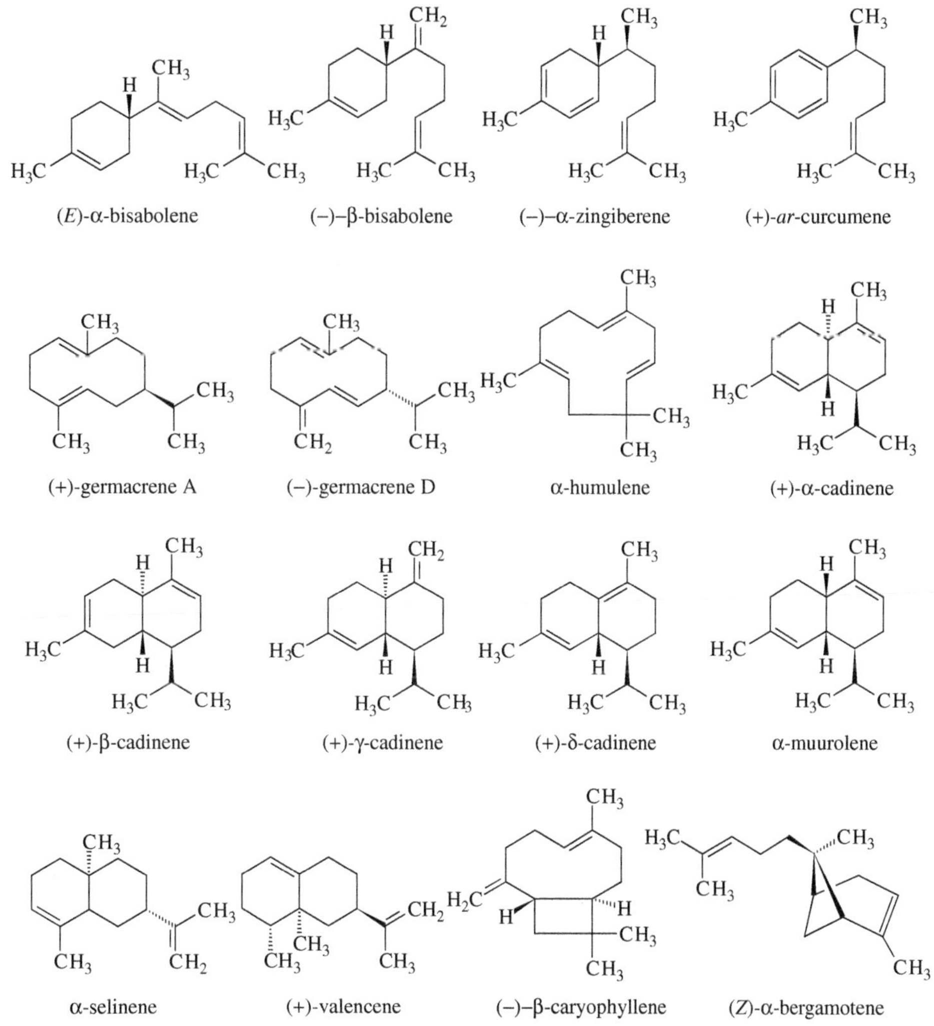

(E)-α-bisabolene   (−)-β-bisabolene   (−)-α-zingiberene   (+)-*ar*-curcumene

(+)-germacrene A   (−)-germacrene D   α-humulene   (+)-α-cadinene

(+)-β-cadinene   (+)-γ-cadinene   (+)-δ-cadinene   α-muurolene

α-selinene   (+)-valencene   (−)-β-caryophyllene   (Z)-α-bergamotene

8-6, cyclic sesquiterpenic hydrocarbons

β-Caryophyllene is a common component of many essential oils. For example, the essential oil of black pepper (see Table 8.32) contains about 20% of β-caryophyllene. Its α-isomer, humulene (also known as α-caryophyllene), occurs in the hops essential oil (*Humulus lupulus*, Cannabaceae); both compounds are often present in the mixture. Common components of essential oils are also bisabolenes. Higher quantities of β-bisabolene (10–15%), and also of α-zingiberene (22–30%), *ar*-curcumene (20%), α-selinene, and β-farnesene, are found in ginger essential oil. The essential oil of white sandalwood (*Santalum album*, Santalaceae) contains, in addition to sesquiterpenic alcohols as major components, many sesquiterpenic hydrocarbons, including α-bergamotene, *ar*-curcumene, β-bisabolene, and many others. Trivial and systematic names of the main sesquiterpenic hydrocarbons are listed in Table 8.3.

Table 8.3 Trivial and systematic names of some sesquiterpenic hydrocarbons.

Trivial name	Systematic name (IUPAC)
(−)-(E)-α-Bergamotene	(1S,5S,6R)-2,6-Dimethyl-6-(4-methylpent-3-en-1-yl)bicyclo[3.1.1]hept-2-ene
(E)-α-Bisabolene	(E)-1-Methyl-4-(6-methylhepta-2,5-dien-2-yl)cyclohex-1-ene
(−)-β-Bisabolene	(S)-1-Methyl-4-(6-methylhepta-1,5-dien-2-yl)cyclohex-1-ene
(Z,E)-α-Farnesene	(3Z,6E)-3,7,11-Trimethyldodeca-1,3,6,10-tetraene
(E,E)-α-Farnesene	(3E,6E)-3,7,11-Trimethyldodeca-1,3,6,10-tetraene
(E)-β-Farnesene	(E)-7,11-Dimethyl-3-methylidenedodeca-1,6,10-triene
(+)-Germacrene A	(1E,5E,8R)-1,5-Dimethyl-8-prop-1-en-2-ylcyclodeca-1,5-diene
(−)-Germacrene D	(1E,6E,8S)-1-Methyl-5-methylidene-8-propan-2-ylcyclodeca-1,6-diene
α-Humulene	(1E,4E,8E)-2,6,6,9-Tetramethylcycloundeca-1,4,8-triene
(+)-α-Cadinene	(1S,4aR,8aR)-1-Propan-2-yl-4,7-trimethyl-1,2,4a,5,6,8a-hexahydronaphthalene
(+)-β-Cadinene	(1S,4aR,8aR)-4,7-Dimethyl-1-propan-2-yl-1,2,4a,5,8,8a-hexahydronaphthalene
(+)-γ-Cadinene	(1S,4aR,8aR)-7-methyl-4-methylidene-1-propan-2-yl-1,2,3,4a,5,6,8a-hexahydronaftalene
(+)-δ-Cadinene	(1S,8aR)-4,7-Dimethyl-1-propan-2-yl-1,2,3,5,6,8a-hexahydronaphthalene
(−)-β-Caryofyllene	(1R,4E,9S)-8-Methylen-4,11,11-trimethylbicyclo[7.2.0]undec-4-ene
(+)-ar-Curcumene	(S)-1-(1,5-Dimethylhex-4-enyl)-4-methylbenzene
(+)-α-Muurolene	(1S,4aS,8aR)-4,7-Dimethyl-1-propan-2-yl-1,2,4a,5,6,8a-hexahydronaftalen
(+)-α-Selinene	(4aR,7R)-1,4a-Dimethyl-7-prop-1-en-2-yl-3,4,4a,5,6,7,8,8a-octahydronaphthalene
(+)-Valencene	(1R,7R,8aS)-1,8a-Dimethyl-7-prop-1-en-2-yl-1,2,3,5,6,7,8,8a-octahydronaphthalene
(−)-α-Zingiberene	(2S,5R)-2-Methyl-5-(6-methylhept-5-en-2-yl)cyclohexa-1,3-diene

Diterpenes Diterpenes are biosynthesised by plants, animals, and fungi; more than 3000 different structures have been defined in nature. Diterpenic hydrocarbons, although not directly involved in aroma of foods, are precursors of important biologically active compounds, such as retinol and phytol. Like monoterpenes and sesquiterpenes, diterpenes are mostly cyclic compounds. An example of an important tricyclic hydrocarbon is (−)-ent-kaur-16-ene (**8-7**), a precursor of the plant hormones gibberellins, the flavour-active diterpenoid alcohols cafestol and kahweol (see **8-18**) found in unroasted coffee beans, the arabica-specific bitter glycoside mozambioside (see Section 8.3.5), the diterpenoid antioxidants carnosic acid and bitter carnosol (see Section 10.3.11.2) occurring in the medical herb rosemary (*Rosmarinus officinalis*, Lamiaceae), and sweet stevioside (see Section 11.3.2.1.3) found in the leaves of South American stevia (*Stevia rebaudiana*, Asteraceae), known as sweet leaf.

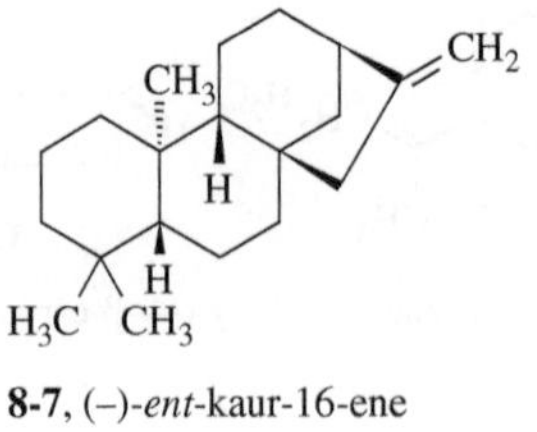

8-7, (−)-ent-kaur-16-ene

8.2.1.1.2 Other hydrocarbons

Aliphatic and alicyclic hydrocarbons Saturated and unsaturated hydrocarbons with odd and even numbers of carbon atoms in the molecule (about C-11 to C-35) are present as the primary substances in all vegetable oils and animal fats. Alkanes, alkenes, alkadienes, and alkatrienes also arise as oxidation products of unsaturated fatty acids, catalysed by lipoxygenases or by autoxidation of fatty acids during food storage and processing. Only the lower hydrocarbons can play a role as odour-active substances. The main hydrocarbons resulting from oxidation of unsaturated fatty acids are ethane from linolenic acid, pentane and butane from linoleic acid, and hexane and octane from oleic acid.

Table 8.4 Some hydrocarbons formed by oxidation of unsaturated fatty acids.

Hydrocarbon	Hydroperoxy acid	Fatty acid
Dec-1-ene	(Z)-8-Hydroperoxyoctadec-9-enoic	Oleic
Non-1-ene	(E)-9-Hydroperoxyoctadec-10-enoic	Oleic
Octane	(E)-10-Hydroperoxyoctadec-8-enoic	Oleic
Hexane	(Z)-11-Hydroperoxyoctadec-9-enoic	Oleic
(Z)-Deca-1,4-diene	(9Z,12Z)-8-Hydroperoxyoctadeca-9,12-dienoic	Linoleic
(Z)-Nona-1,3-diene	(10E,12Z)-9-Hydroperoxyoctadeca-10,12-dienoic	Linoleic
(Z)-Oct-2-ene	(8E,12Z)-10-Hydroperoxyoctadeca-8,12-dienoic	Linoleic
Hept-1-ene	(9Z,12Z)-11-Hydroperoxyoctadeca-9,12-dienoic	Linoleic
Hex-1-ene	(9Z,13E)-12-Hydroperoxyoctadeca-9,13-dienoic	Linoleic
Pentane	(9Z,11E)-13-Hydroperoxyoctadeca-9,11-dienoic	Linoleic
Butane	(9Z,12Z)-14-Hydroperoxyoctadeca-9,12-dienoic	Linoleic
(3Z,6Z)-Nona-1,3,6-triene	(10E,12Z,15Z)-9-Hydroperoxyoctadeca-10,12,15-trienoic	Linolenic
(2Z,5Z)-Octa-2,5-diene	(8E,12Z,15Z)-10-Hydroperoxyoctadeca-8,12,15-trienoic	Linolenic
(Z)-Hepta-1,4-diene	(9Z,12Z,15Z)-11-Hydroperoxyoctadeca-9,12,15-trienoic	Linolenic
(Z)-Hexa-1,3-diene	(9Z,13E,15Z)-12-Hydroperoxyoctadeca-9,13,15-trienoic	Linolenic
(Z)-Pent-2-ene	(9Z,11E,15Z)-13-Hydroperoxyoctadeca-9,11,15-trienoic	Linolenic
But-1-ene	(9Z,12Z,15Z)-14-Hydroperoxyoctadeca-9,12,15-trienoic	Linolenic
Ethane	(9Z,12Z,14E)-16-Hydroperoxyoctadeca-9,12,14-trienoic	Linolenic

The immediate precursors of hydrocarbons are the fatty acid hydroperoxides (Table 8.4). The unsaturated hydrocarbons are predominantly (Z)-isomers. Numerous other hydrocarbons, including alicyclic hydrocarbons, appear as secondary lipid oxidation products.

Some aliphatic, alicyclic, and aromatic hydrocarbons may be the products of oxidation and of the degradation of substances other than lipids; for example, penta-1,3-diene arises by decarboxylation of sorbic acid (see Section 11.2.1.1.2), which is used as a preservative.

Aromatic hydrocarbons Aromatic hydrocarbons are relatively rare natural components of foods. An important natural component of essential oils of many spices and vegetables is *p*-cymene (1-isopropyl-4-methylbenzene, **8-1**). Together with the related hydrocarbon α,*p*-dimethylstyrene, *p*-cymene is also formed by degradation of citral. Various alkyl benzenes and alkyl xylenes are found in small amounts in olive oil. The degradation products of carotenoids are, for example, 1,2-dihydro-1,1,6-trimethylnaphthalene, an odorant of tomatoes (see Section 9.9.5.2.1), and (*E*)-1-(2,3,6-trimethylphenyl)buta-1,3-diene (**8-8**), an odorant in wines.

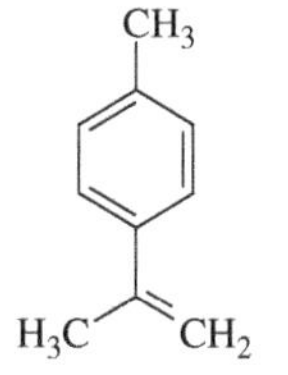

α,*p*-dimethylstyrene

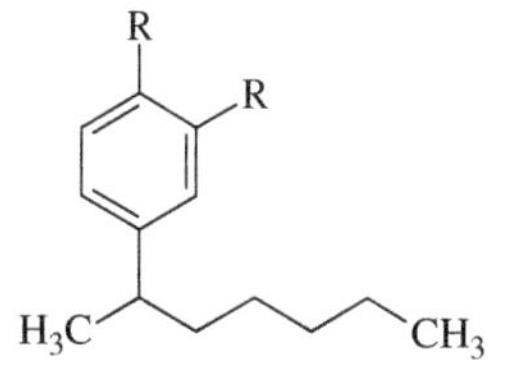

olive oil alkylbenzenes (R = H)
olive oil alkylxylenes (R = CH₃)

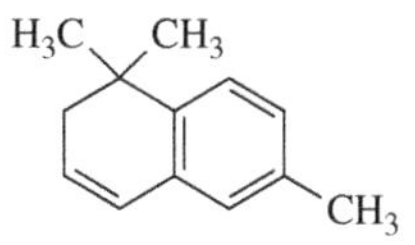

1,2-dihydro-1,1,6-trimethylnaphthalene

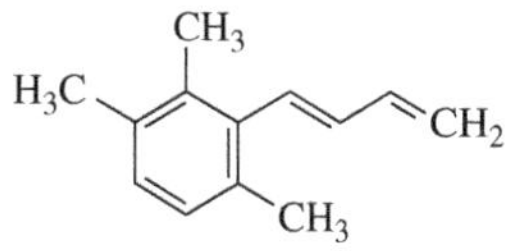

(*E*)-1-(2,3,6-trimethylphenyl)-
buta-1,3-diene

8-8, selected aromatic hydrocarbons

Undesirable contaminants include monocyclic aromatic hydrocarbons (MAHs), toluene, xylenes, and ethyl benzene, which may be present in small quantities in foods as exogenous contaminants, along with PAHs (see Section 12.3.5). Together with benzene and styrene (vinyl benzene), these hydrocarbons are also formed as processing contaminants (see Section 12.3.6).

8.2.1.2 Properties and reactions

Autoxidation of saturated hydrocarbons is, like autoxidation of saturated fatty acids, important at higher temperatures (around 150 °C). The final odorous products formed are mainly fatty acids, their lactones, alcohols, and ketones with relatively few carbon atoms in the molecule.

Terpenic hydrocarbons are stable in the absence of air, but are easily oxidised. The autoxidation reaction proceeds by similar mechanisms as autoxidation of unsaturated fatty acids and depends greatly on their structure. The primary autoxidation products are hydroperoxides. In branched hydrocarbons, the hydroperoxyl group mainly occurs in the secondary or tertiary carbon adjacent to the quaternary carbon of the double bond. The final autoxidation products are usually epoxides, alcohols, and ketones. The primary site of oxidation is the carbon adjacent to the double bond, as in monounsaturated fatty acids. Hydroperoxides can also be formed by the addition of singlet oxygen to the olefin double bond.

An example is the oxidation of limonene, the (+)-(R)-isomer of which is the major component of essential oils from the peel of citrus fruits. The oxidation products are the alcohols (2R,4S)-carveol, known as (+)-(E)-carveol, and (1S,4S)-carveol, known as (+)-(Z)-carveol, as well as ketone (+)-(S)-carvone (see **8-36**), an important component of caraway essential oil, and isomeric limonene-1,2-epoxides, also known as limonene oxides. The reaction mechanism, including the hydroperoxide intermediates, is shown in Figure 8.1.

Grapefruit peel oil aroma, a strange, extraneous type of aroma normally not present in oranges, may arise in orange juices as an off-flavour through metal-catalysed oxidation or photooxidation of the hydrocarbon (+)-valencene (**8-6**) to ketone (+)-nootkatone via hydroxylation to the corresponding alcohol (+)-β-nootkatol (see **8-17**) and its oxidation. Cyclisation of β-ionone to 1,1,6-trimethyl-1,2-dihydronaphthalene (**8-8**) in fruits containing carotenoid pigments causes off-flavour in fruit juices and wines (see Section 9.9.5.2.1).

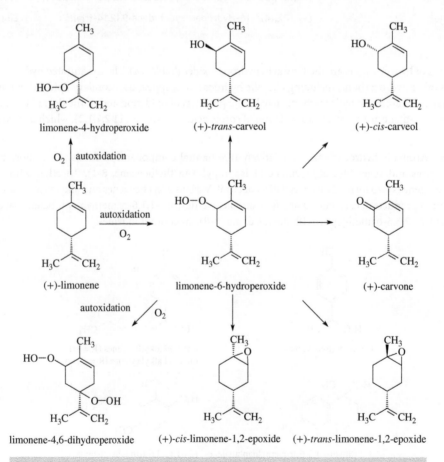

Figure 8.1 Main products of (+)-limonene oxidation.

8.2.2 Alcohols

Alcohols and phenols can be formally considered as the primary oxidation products of hydrocarbons. Alcohols are primary and secondary odour- and taste-active substances in foods. Typical compounds are primary, secondary, and tertiary aliphatic, alicyclic, aromatic, and heterocyclic alcohols. Lower primary alcohols and their esters mainly act as odour-active compounds, especially in fruits and alcoholic beverages. Slight or no odour-activity is shown by higher aliphatic alcohols (such as the fatty alcohols, which are components of waxes), polar aliphatic and alicyclic diols, triols, other polyols, diterpenic, triterpenic alcohols and sterols, amino alcohols, hydroxycarboxylic acids, and other polar hydroxy derivatives. Some of these substances exhibit a sweet (e.g. glycerol, sugar alcohols) or a sour (e.g. hydroxycarboxylic acids) taste.

Alcohols (especially terpenic alcohols) that have fewer than 15 carbon atoms in the molecule are used for food flavouring. The use of higher alcohols is an exception. Lower alcohols are used for the production of esters, acetals, and other flavour-active compounds or as food additives (e.g. solvents).

8.2.2.1 Classification, structure, terminology, and occurrence

8.2.2.1.1 Aliphatic and alicyclic alcohols

Methanol Methanol (**8-9**) is the first member of the homologous series of saturated aliphatic alcohols found in plants in the form of various esters, most often in pectins and esters of aromatic carboxylic acids (benzoic, salicylic, cinnamic, and others). Free methanol results mainly from hydrolysis of pectin catalysed by pectin esterases; therefore, it occurs regularly in small amounts as a component of all natural fruit juices and is present in larger amounts in wines, ciders, and fruit distillates.

The methanol content in citrus juices ranges from 24 to 47 mg/l, whilst that in blackcurrant juices reaches 70–176 mg/l. The methanol content of ciders is 36–88 mg/l and that of fruit wines is usually 20–240 mg/l, but exceptionally it may reach more than 600 mg/l, when a further portion of the pectin is hydrolysed during the fermentation process. The amount of methanol formed in grape wines depends on many factors. Red wines have about twice as much as white wines. Vodka has a very small amount (79–158 mg/l). Spirits produced by distilling wine (such as brandy and cognac) have methanol contents similar to those of wines (320–400 mg/l). Fruit brandies (distilled from fermented stone fruits) tend to have much higher contents. The content of methanol in cherry brandy is usually 0.48–0.95% (4800–9500 mg/l), and that in plum brandy is around 1.2% (12 000 mg/l).

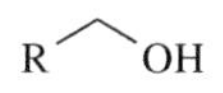

$$R \diagup OH$$

8-9, aliphatic alcohols
methanol, R = H
ethanol, R = CH_3
propan-1-ol, R = CH_2CH_3
butan-1-ol, R = $CH_2CH_2CH_3$
2-methylpropan-1-ol, R = $CH(CH_3)_2$
2-methylbutan-1-ol, R = $CH(CH_3)(CH_2CH_3)$
3-methylbutan-1-ol, R = $CH_2CH(CH_3)_2$

Ethanol Ethanol (ethyl alcohol, **8-9**) is a common alcohol that is present in food as the free substance or bound in various esters. A small amount of ethyl alcohol (0.003–0.015% in blood) is necessarily present in our bodies even if we do not drink alcohol. Free ethanol, together with carbon dioxide and many minority substances, is formed as the main product of anaerobic degradation of sugars by yeast: the so-called alcoholic fermentation or alcoholic glycolysis. Therefore, ethanol is present in variable amounts in alcoholic beverages and in batter, bread, and all fermented milk products. Ethanol is generally not considered an important flavouring, but it still has a significant impact on the flavour and energy value of alcoholic beverages. The energy value of ethanol is 29 kJ/g (7 kcal/g).

Alcoholic fermentation is a sequence of reactions that generates pyruvic acid from sugars as the key intermediate (Figure 8.2). The decarboxylation of pyruvic acid by pyruvate decarboxylase yields acetaldehyde (ethanal), which is reduced to ethanol by alcohol dehydrogenase.

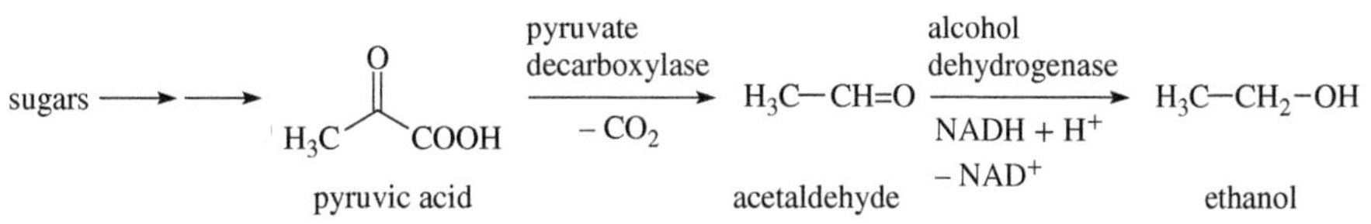

Figure 8.2 Formation of ethanol in alcoholic fermentation of sugars.

The amount of ethanol depends on the amount of fermentable sugars in the raw material, the type and strain of yeast used, the fermentation temperature, the nutrient content in the medium, and other factors. The same reaction also takes place in plants when there is a reduced oxygen level. For the industrial production of ethanol, certain strains of *Saccharomyces cerevisiae* yeast are employed, which are able to ferment glucose, mannose, fructose, saccharose, xylulose, and raffinose. The ability to ferment galactose, maltose, and melibiose is variable; trehalose, lactose, and xylose are not fermented at all. Glucose, fructose, and saccharose can enter yeast cells directly (after cleavage of saccharose by invertase); maltose requires specific permeases to enter the cells. To produce ethanol from starchy materials such as cereal grains, the starch must first be converted into fermentable sugars, which are formed by starch degradation by amylases and maltases (see Section 4.5.6.1.3).

The ethanol content of beers varies over a relatively wide range. For example, beers produced in the Czech Republic are classified according to their ethanol content in per cent by volume (alcoholic strength or alcohol by volume, ABV) and to the original concentration of the extract in the wort before fermentation. Recognized are non-alcoholic beers with 0.5% ABV and low alcohol beers with 0.5–1.2% ABV. Table beers have up to 6 % extract m/m, lager beers have 7–10% extract m/m, lager beers have 11–12% extract m/m, full beers with 11–12% extract m/m (only top-fermented beers), and the extract of strong beers is 13% and more. For example, the pilsner type of beer known as Pilsner Urquell, produced in the Czech Republic, has an original wort concentration of 11–12% and ethanol content of about 3.6% w/w. Beers for diabetics are low-carbohydrate beers with a higher alcohol content. Dark beers have an original wort content that is generally higher (13–20% w/w). The dark British beers stout, porter, and an Irish dry stout Guinness, have high ethanol contents (up to 6.8% w/w). Designation according to the ethanol content is mandatory for beers with reduced alcohol content (from 0.5 to 1.2% ABV) and non-alcoholic beers, which contain up to 0.5% ABV. Similar classification of beers is also used in other EU countries.

Cider alcohol content varies from 1.2 to 8.5% ABV or more. Ethanol content in wines varies from 8–9 to 18–18.5% ABV, depending on the content of sugars in the must (see Section 4.2.2.6). Normal table wines contain 10–14% ethanol by volume, whilst dessert wines of the sherry type contain 17–24% ABV.

The ethanol content in grape brandy and other spirits is around 40% ABV. A number of special products have lower alcohol content, but some spirits have also higher alcohol content. In the United States, the proof of an alcoholic beverage is twice its alcohol content, expressed as percentage by volume at 60 °F (15.6 °C). So, an 80-proof whisky is 40% ethanol ABV.

Milk contains a small amount of ethanol, about 0.003% on average; yoghurts and other dairy products obtained by fermentation contain ethanol at levels not exceeding 0.04–0.05%. Of the dairy products, only fermented mare's milk (called kumiss or kumys) has a higher alcohol content (1–3%).

Higher alcohols In addition to ethanol, alcoholic fermentation produces a series of higher aliphatic alcohols (**8-9**) with a strong aroma, which are collectively known as the **fusel oil alcohols**. These alcohols accompany ethanol in beer, wine, and other alcoholic beverages, but also in dough and fermented dairy products (Table 8.5). Fusel oil alcohols are produced either by catabolic processes (some aminocarboxylic acids are their precursors) or anabolic processes (they are formed from sugars in the biosynthesis of aminocarboxylic acids). The immediate precursors of fusel oil alcohols are aldehydes generated as byproducts of metabolism. Alcohol dehydrogenases reduce these aldehydes to the corresponding alcohols. Formation of fusel oil alcohols is described in Section 2.5.1.3.2, which deals with transamination and oxidative deamination of aminocarboxylic acids. The level of fusel oil alcohols in alcoholic beverages depends on the raw materials processed. In wines, for example, important variables are the grape variety (red wines contain somewhat higher amounts of fusel oil alcohols than white wines), conditions during fermentation, and the yeast strains used.

2-Methylpropan-1-ol, known as isobutyl alcohol or isobutanol (resulting from valine via 2-oxoisovalerate), and 3-methylbutan-1-ol, known as isoamyl alcohol (created from leucine via 2-oxoisocaproate), are the main alcohols found in fusel oils in relatively large amounts. Both have considerable influence on the aroma of alcoholic beverages. Fusel oils contain some other alcohols in smaller quantities, such as (−)-(S)-2-methylbutan-1-ol, known as optically active amyl alcohol (produced from isoleucine via 2-oxo-3-methylvaleric acid),

Table 8.5 Fusel oil and other higher alcohols in some alcoholic beverages (mg/l).

Alcohol	Beer	Wine	Whisky	Alcohol	Beer	Wine	Whisky
Fusel oil alcohols				2-Phenylethanol	4–102	5–138	0.6–131
Propan-1-ol	4–60	11–93	20–187	Tyrosol	0.6–29	5–45	–
Butan-1-ol	–	3–9	–	Tryptophol	0.2–12	0–1.6	–
2-Methylpropan-1-ol	2–98	15–184	110–670	*Other alcohols*			
(S)-2-Methylbutan-1-ol	3–41	12–311	60–1390	Butan-2,3-diol[a]	40–250	165–1615	–
3-Methylbutan-1-ol	19–160	40–523	150–1465	Glycerol	1100–3200	1400–26700	–

[a]Butan-2,3-diol is a mixture of isomers, amongst which (2R,3R)-butan-2,3-diol, also known as (−)-D-butan-2,3-diol, prevails.

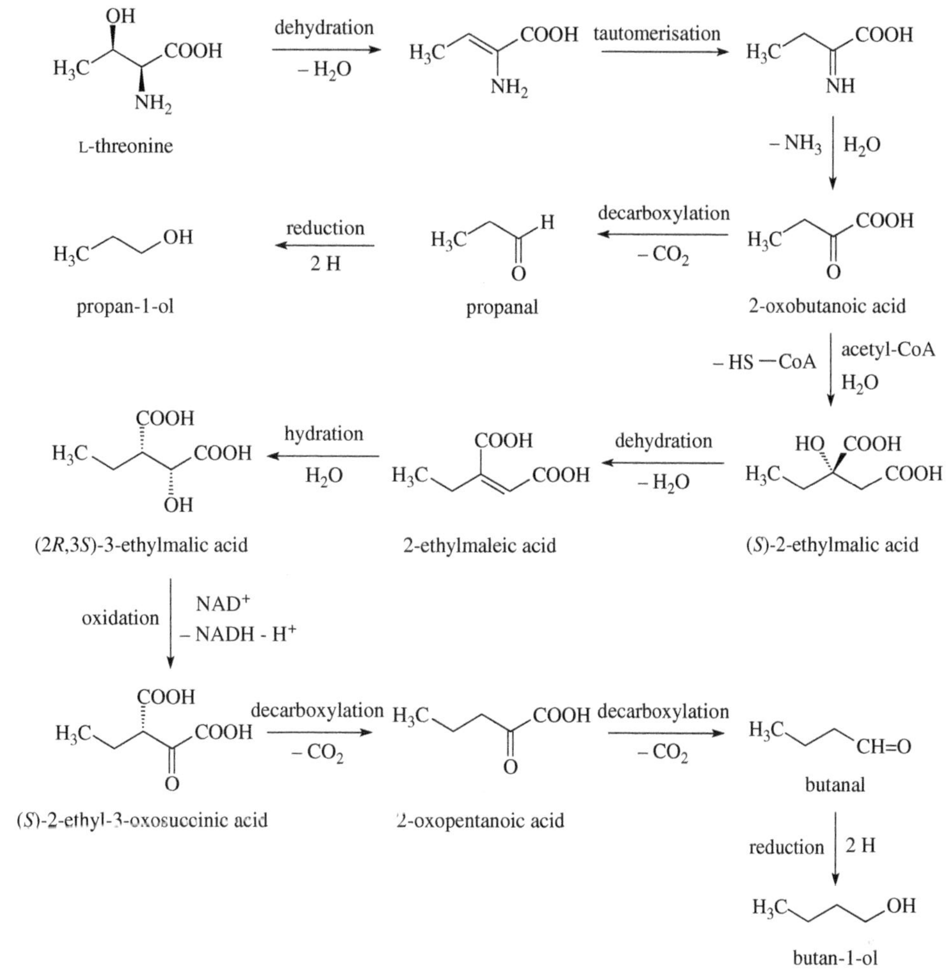

Figure 8.3 Formation of propan-1-ol and butan-1-ol from threonine.

propane-1-ol (which is produced from threonine via 2-oxobutyric acid), and butane-1-ol (created as a byproduct of isoleucine biosynthesis from threonine, by decarboxylation of 2-oxovaleric acid and reduction of butanal that is formed as the decarboxylation product; Figure 8.3). In the butanol fermentation method employed in industrial processes (also known as acetone–butanol fermentation) with bacteria of the genus *Amylobacter*, butan-1-ol is the main product, followed by acetone, propan-2-ol (isopropyl alcohol), and ethanol. During fermentation processes, the aromatic aminocarboxylic acid phenylalanine forms 2-phenylethanol, also known as 2-phenylethyl alcohol, tyrosine yields the non-volatile alcohol tyrosol, 3,4-dihydroxyphenylalanine (DOPA) yields 3-hydroxytyrosol, and tryptophol is produced from tryptophan. Some mycobacteria, such as *Mycobacterium diernhoferi*, *M. fortuitum*, and *M. chelonei*, as well as some yeast strains, such as *Saccharomyces rouxii*, are known to oxidise histamine to give histaminol (**8-10**). Histaminol has also been found in red and white wines at concentrations of 0.3–1.1 mg/l.

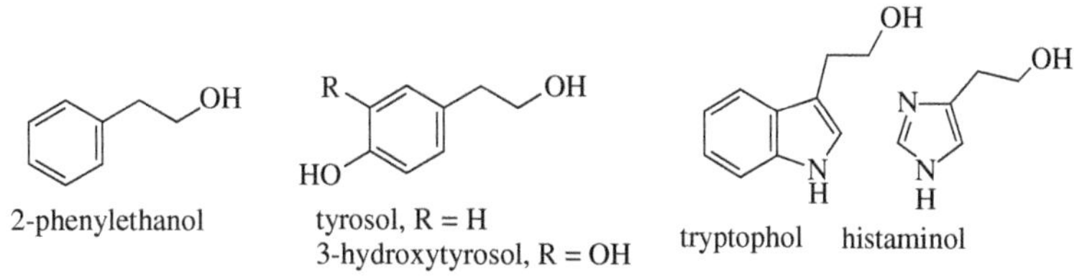

2-phenylethanol

tyrosol, R = H
3-hydroxytyrosol, R = OH

tryptophol   histaminol

8-10, aromatic and heterocyclic alcohols derived from amino acids

During the fermentation process, sulfur aminocarboxylic acids yield sulfur-containing alcohols. Methional is formed from methionine and reduced to the corresponding alcohol methionol (**8-11**) in beers and wines. Important sulfur-containing alcohols formed in wines, including 3-mercapto-3-methyl-1-ol, 4-mercapto-4-methylpentan-2-ol, and 3-mercaptohexan-1-ol, are formed as degradation products of cysteine conjugates (**8-11**) with sulfur-containing alcohols under the action of CS lyases. These sulfur-containing alcohols are characteristic components of wine aroma. For example, (+)-(S)-3-mercaptohexan-1-ol (**8-11**) has on dilution an interesting tropical fruit aroma, reminiscent of passion fruit. The (−)-(R)-isomer has a fruitier aroma, reminiscent of grapefruit.

Some higher alcohols present in alcoholic beverages at low concentrations exist even in the raw materials. For example, pentan-1-ol, hexan-1-ol, heptan-1-ol, octan-1-ol, and other higher alcohols are common constituents in grape musts. They are formed by oxidation of essential fatty acids by lipoxygenases and through cleavage of fatty acid hydroperoxides by lyases, followed by reduction of the saturated or unsaturated aldehydes through the action of alcohol dehydrogenases found in the raw material (Figure 8.4) and of yeast alcohol dehydrogenases during fermentation. These aliphatic alcohols having a chain of at least six carbon atoms are known as fatty alcohols. Detailed mechanisms of these reactions are given in Section 3.8.1.8.5.

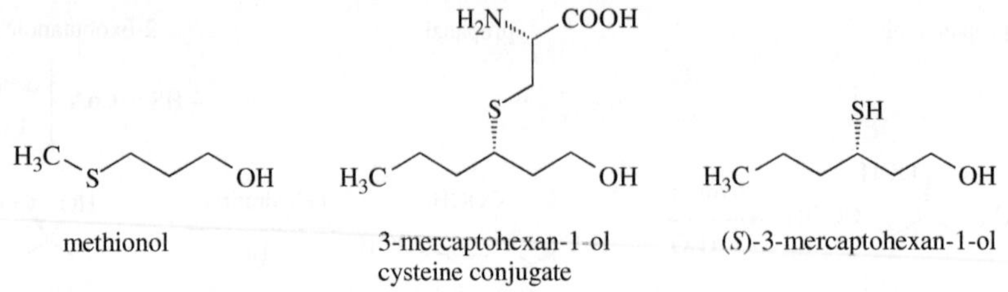

methionol

3-mercaptohexan-1-ol cysteine conjugate

(S)-3-mercaptohexan-1-ol

8-11, sulfur-containing alcohols and their precursors

As an example, reduction of propanal yields propan-1-ol, butan-1-ol is produced from but-2-enal or butanal, and pentanal gives rise to pentan-1-ol, which is present in wines at levels of about 0.1 mg/l. Hexan-1-ol (usually 0.3–12 mg/l in wines) arises mainly by reduction of hexanal (Figure 8.5) and (E)-hex-2-enal (Figure 8.6) by the action of alcohol dehydrogenases. It has a faint green and tallowy odour (see later). Hexanal forms by decomposition of linoleic acid (13S)-hydroperoxide with hydroperoxide lyase, whilst (E)-hex-2-enal forms by enzymatic cleavage of linolenic acid (13S)-hydroperoxide and isomerisation of the thus formed (Z)-hex-3-enal, catalysed by enal isomerase. Reduction of methylketones produced by oxidation of fatty acids generates alkan-2-ols.

Alcohol dehydrogenases are quite specific to the substrate and reduce higher aldehydes (>C_5) very slowly. For this reason, the content of higher (fatty) alcohols in alcoholic beverages, as well as in fruits and vegetables, is relatively low, and the main products of fatty acid oxidation are aldehydes and not alcohols.

In addition to aminocarboxylic and fatty acids, also some other food components can be the precursors of higher alcohols. For example, minor degradation products of β-carotene include butan-1-ol, 2-methylpropan-1-ol and pentan-1-ol.

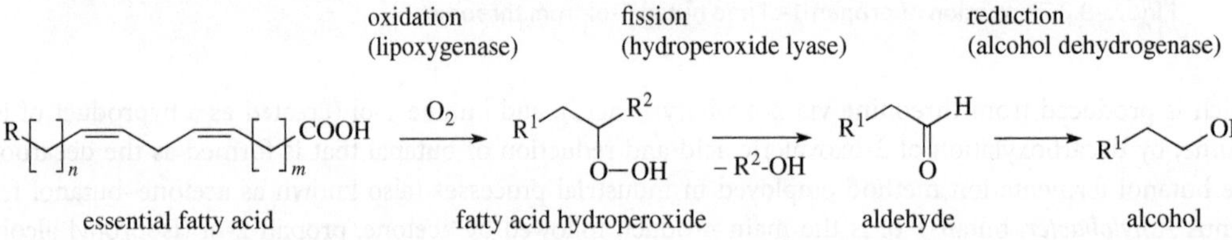

Figure 8.4 Formation of alcohols from fatty acids.

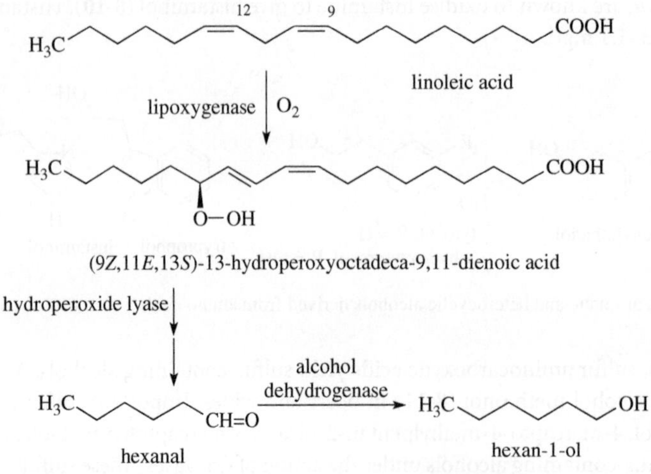

Figure 8.5 Formation of hexan-1-ol from linoleic acid.

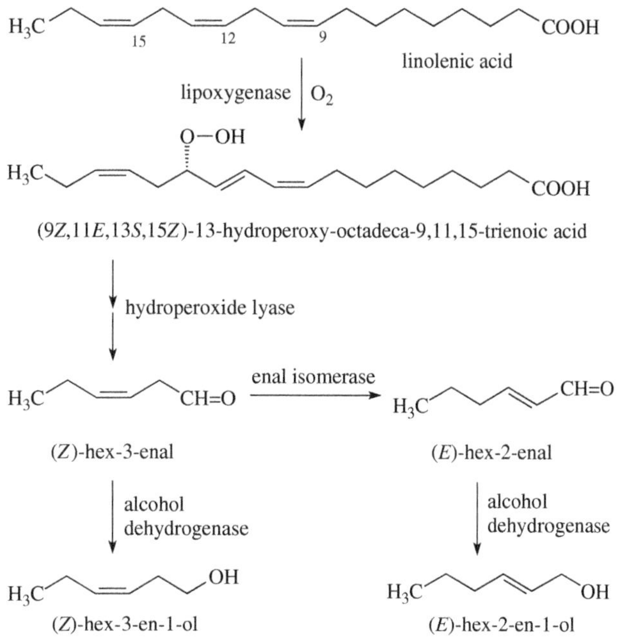

Figure 8.6 Formation of hex-2-en-1-ol and hex-3-en-1-ol from linolenic acid.

Unsaturated alcohols The simplest aliphatic unsaturated alcohol, allyl alcohol (prop-2-en-1-ol), is formed from alliin during high-temperature processing of garlic and is one of the major degradation products of alliin heated at 80–200 °C in the presence of variable amounts of water. Its amount increased with the amount of added water and was the highest at 140 °C (37 mg/g amino acid). Allyl alcohol at a level of 1.1 mg/kg was also found in Boursin Ail cheese flavoured with garlic. It is proposed that allyl alcohol is formed either through [2,3]-sigmatropic rearrangement of alliin to the corresponding sulfenate, which decomposes to allyl alcohol and cysteine (Figure 8.7), or by transformation of alliin via thiosulfonium ion. Cysteine then decomposes further into acetaldehyde, hydrogen sulfide, and ammonia.

Some unsaturated aliphatic alcohols arising from essential fatty acids are important flavour-active components of fresh fruits, vegetables, and mushrooms. Hydroperoxides of fatty acids resulting from regioselective and stereospecific oxidation by lipoxygenases are broken down in different ways. In animal tissues, hydroperoxides are reduced to non-volatile hydroxycarboxylic acids by the enzyme glutathione peroxidase. In plants and fungi, hydroperoxides decompose by lyases, dehydrases, epoxygenases, and hydroperoxide hydrolases. Differences in the sensory quality of active products are related to the substrate and reaction specificity of these enzymes. The main products formed in plants are aldehydes, whilst mushrooms produce allyl alcohols.

(Z)-Alkenals produced by the cleavage of peroxides of linoleic and α-linolenic acids can be subsequently transformed to (E)-alkenals by enal isomerases and reduced to (Z)- or (E)-alkanols by alcohol dehydrogenases (Figure 8.6). These products often have a characteristic aroma of fruits and vegetables. For example, isomerisation of (Z)-hex-3-enal, known as leaf aldehyde, yields (E)-hex-2-enal; on reduction, this gives (Z)-hex-3-en-1-ol, which is trivially called leaf alcohol. The aroma of alk-3-en-1-ols resembles that of alk-3-en-1-als.

The most important compound that has the smell of fresh mushrooms is (−)-(R)-oct-1-en-3-ol, which arises in a similar way from linoleic acid (Figure 8.8). The formation of (10S)-hydroperoxide is catalysed by a specific lipoxygenase. Oct-1-en-3-ol is accompanied by oct-1-en-3-one, octan-1-ol and (R)-octan-3-ol, and other minority compounds with the mushroom-like smell, such as (3R,5Z)-octa-1,5-dien-3-ol (**8-12**).

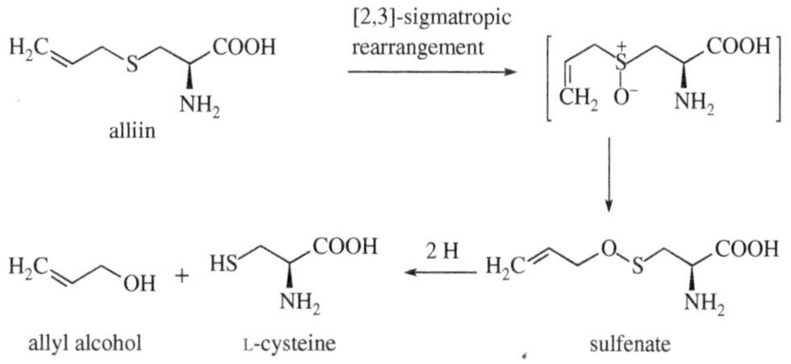

Figure 8.7 Formation of allyl alcohol from alliin.

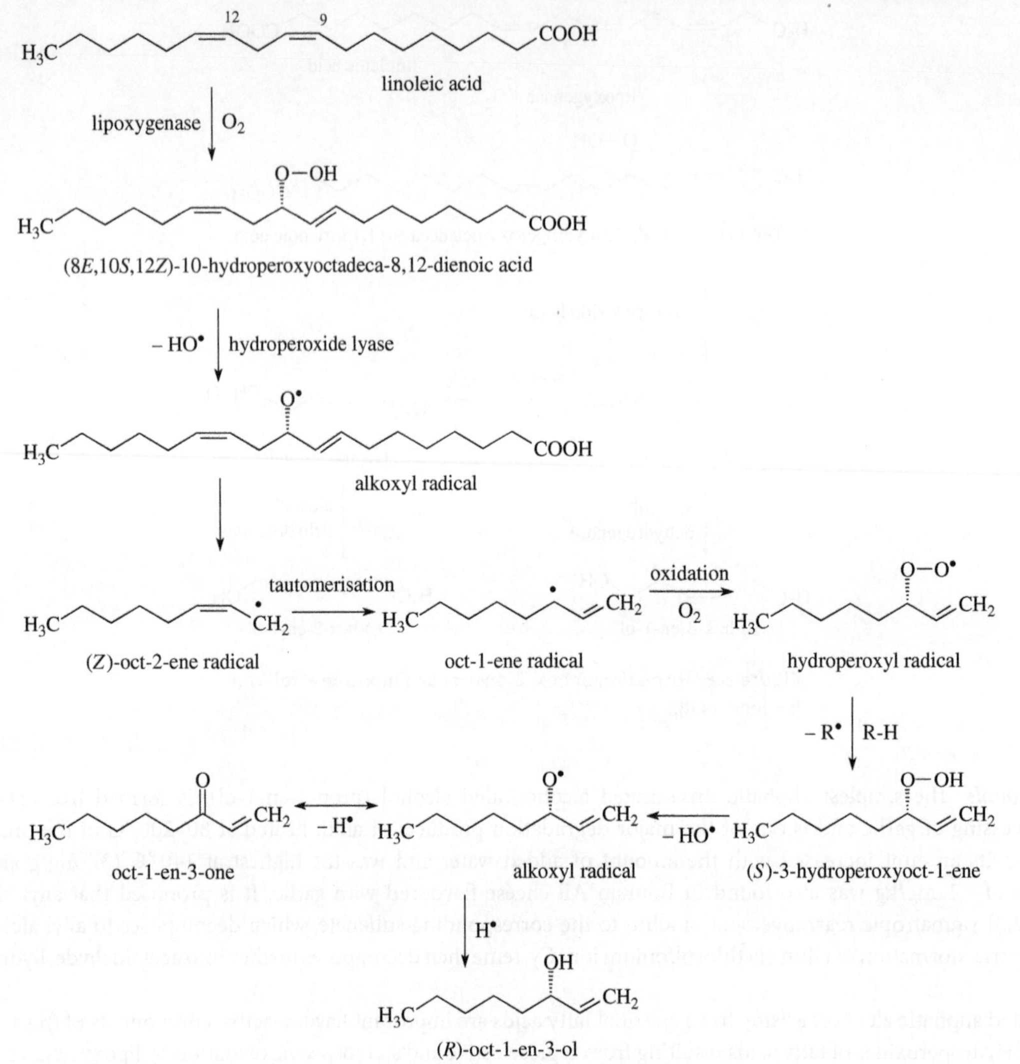

linoleic acid

lipoxygenase | O_2

(8E,10S,12Z)-10-hydroperoxyoctadeca-8,12-dienoic acid

– HO• | hydroperoxide lyase

alkoxyl radical

tautomerisation

(Z)-oct-2-ene radical   oct-1-ene radical   hydroperoxyl radical

oxidation O_2

– R• | R-H

– HO•

oct-1-en-3-one   alkoxyl radical   (S)-3-hydroperoxyoct-1-ene

H•

(R)-oct-1-en-3-ol

Figure 8.8 Formation of oct-1-en-3-ol and oct-1-en-3-one from linoleic acid.

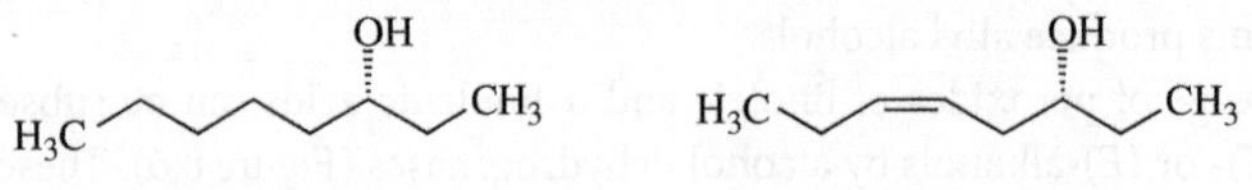

(R)-octan-3-ol   (3R,5Z)-octa-1,5-dien-3-ol

8-12, minority alcohols with mushroom-like smell

In meat, oct-1-en-3-ol and oct-1-en-3-one are off-flavour compounds produced analogously by decomposition of 12-hydroperoxy-5,8,10,14-eicosatetraenoic acid, which is one of the arachidonic acid autoxidation products. Oct-1-en-3-ol in soybean products is in general an off-flavour compound formed not *de novo* from unsaturated fatty acids but instead from the corresponding β-primeveroside hydrolysed by β-glycosidase.

Terpenic alcohols

Monoterpenic alcohols   Monoterpenic alcohols are widespread odour-active components of spices, fruits, vegetables, and various essential oils from flowers and other plant parts that are often used in perfumery. They exhibit frequently sweet, heavy floral odours in different shades.

The most important monoterpenic acyclic alcohol with one double bond is citronellol. Alcohols with two double bonds include linalool, geraniol, and nerol. An example of an alcohol with three double bonds is hotrienol (**8-13**). (+)-Citronellol (β-citronellol) is a major component of essential oils of *Zieria citriodora* (syn. *Boronia citriodora*, Rutaceae, about 80%), a plant native to Australia and known as lemon-scented zieria. It also occurs in *Corymbia citriodora* (syn. *Eucalyptus citriodora*, Myrtaceae; 15–20%), an Australian tree known as lemon-scented gum or lemon eucalyptus. (−)-Citronellol isolated from natural sources is often called rhodinol. It is the predominant enantiomer in geranium (*Pelargonium graveolens* and other species, Geraniaceae) and Bulgarian Damask rose essential oils (*Rosa damascena* var. *bulgaria*, Rosaceae), which contain up to 50% citronellol. A mixture of both enantiomers is present in many essential oils, such as that of aromatic citronella grass *Cymbopogon nardus* and *C. winterianus* (Poaceae), native to tropical Asia. (+)-Linalool (coriandrol) is found at a level of 60–80% in coriander essential oil (see Table 8.32), whilst (−)-linalool (licareol) is the major component (80–85%) of cinnamon essential oil. Geraniol is, at about 8%, a component of geranium oil; its *cis*-isomer, known as nerol, occurs in Bulgarian rose oil together with citronellol. Hotrienol is the product of linalool oxidation and further transformation of linalool oxidation products. It occurs in many essential oils; for example, it is an odorous component of some aromatic wines and of the essential oil of elderflower (*Sambucus nigra*, Adoxaceae). In foods and beverages, elderflower is used as a flavouring component.

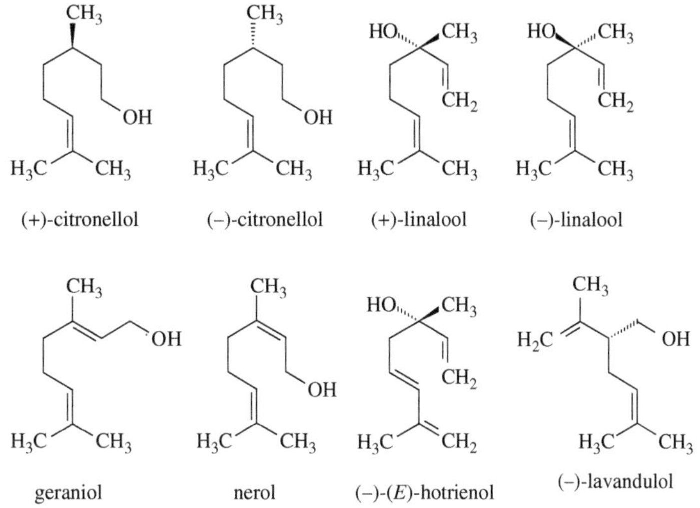

(+)-citronellol   (−)-citronellol   (+)-linalool   (−)-linalool

geraniol   nerol   (−)-(*E*)-hotrienol   (−)-lavandulol

8-13, acyclic monoterpenic alcohols

Other common components of essential oils are the monoterpenic monocyclic alcohols α-terpineol (found in lilac, marjoram, cardamom, star anise oil and other oils) and terpinen-4-ol (4-terpineol). Also known as 4-carvomenthenol, terpinen-4-ol is a component of the essential oils of pine (*Pinus* spp., Pinaceae), eucalyptus (*Eucalyptus* spp. Myrtaceae), marjoram, and thyme. It is also the main component of the antiseptic essential oil of the Australian tree *Melaleuca alternifolia* (Myrtaceae), known as tea tree oil. It often occurs as a racemate. (+)-Terpinen-4-ol occurs at a level of about 10% in the essential oil of lavender (*Lavandula* spp., Lamiaceae), (−)-terpinen-4-ol is a component of orange essential oil, (+)-(*E*)-carveol is an intermediate in the biosynthesis of the characteristic caraway oil component (+)-carvone (see **8-36**) from (+)-limonene, and the isomer (−)-(*E*)-carveol (**8-14**) is an intermediate of the biosynthesis of (−)-carvone from (−)-limonene. (−)-Carvone (see **8-36**) is an important component of spearmint essential oil (see Table 8.32).

(+)-α-terpineol   (−)-α-terpineol   (+)-terpinen-4-ol   (−)-terpinen-4-ol

(+)-*trans*-carveol   (−)-*trans*-carveol   (−)-menthol   (+)-isomenthol

(+)-neomenthol   (+)-neoisomenthol   (−)-*trans*-isopiperitenol   (−)-perillyl alcohol

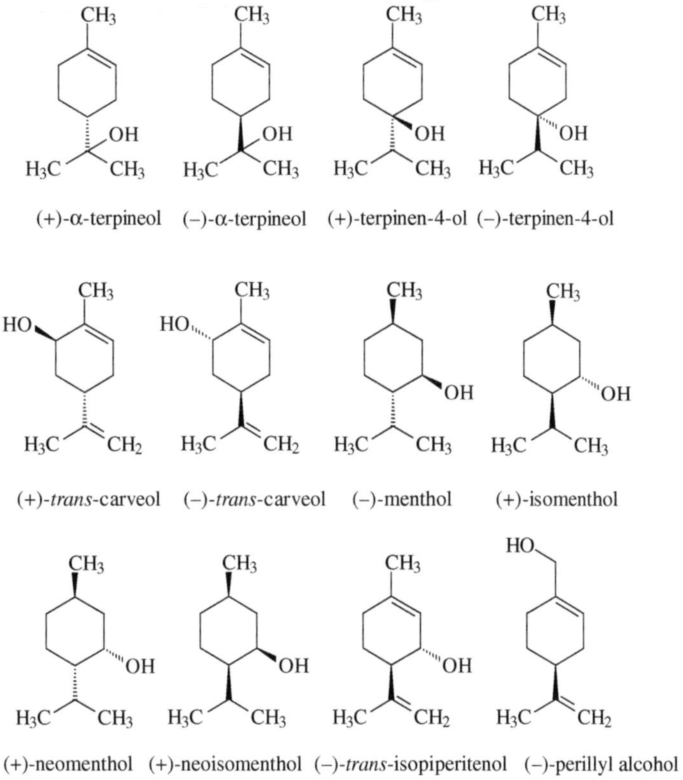

8-14, monocyclic monoterpenic alcohols

(−)-Menthol, which evokes coolness through stimulation of the somatosensory system, is, along with its isomers (+)-isomenthol, (+)-neomenthol, and (+)-neoisomenthol (**8-14**), a key component of peppermint essential oil (see Table 8.32). Good-quality peppermint essential oil has a high menthol content, a moderately high content of menthone, and a low content of (+)-(R)-pulegone (see **8-36**) and (+)-(R)-menthofuran (see **8-21**). Pulegone and menthofuran are undesirable because of their hepatotoxicity, and therefore appear in the list of substances, which should not be added to foods as flavouring. Their natural amount in foods is restricted to 2000 and 3000 mg/kg in micro-breath-freshening confectionery, to 250 and 500 mg/kg in other peppermint-containing confectionery, to 350 and 1000 mg/kg in chewing gums, and to 100 and 200 mg/kg in peppermint-containing alcoholic beverages, respectively. The level of pulegone is restricted to 20 mg/kg in peppermint-containing non-alcoholic beverages. Pulegone and menthofuran were found in low amounts as natural components in red Bordeaux wine.

(−)-Perillyl alcohol is a precursor of (−)-perillyl aldehyde (perillal). Fenchol (also known as α-fenchyl alcohol or fenchan-2-ol, **8-15**) is a bicyclic monoterpenic alcohol occurring as a minor component in citrus, fennel, and sage essential oils. Another common compound is the bicyclic alcohol thujan-4-ol (sabinene hydrate), which occurs at a high level in marjoram essential oil (see Table 8.32). Borneol (bornan-2-ol) is a component of camphor oils (*Cinnamomum camphora*, Lauraceae). Trivial and systematic names of selected monoterpenic alcohols are listed in Table 8.6.

Terpenic alcohols are found in flowers, fruits, and other plant materials mainly as glycosides. The predominant glycosides are β-D-glucopyranosides substituted with L-rhamnose, L-arabinose, and D-apiose and acylated with malonic acid. In apricots and grapes, for example, glycosides are primarily localised in the skin. Linalool, nerol, and α-terpineol are mainly present as β-D-glucosides, geraniol occurs in the form of β-rutinoside, and linalool and α-terpineol occur as 6-O-α-L-arabinofuranosyl-(1 → 6)-β-D-glucopyranoside. Glycosides of diols and other polyols have also been identified.

Table 8.6 Trivial and systematic names of monoterpenic alcohols.[a]

Trivial name	Systematic name (IUPAC)
(+)-Borneol	(1R,2S,4R)-1,7,7-Trimethylbicyclo[2.2.1]heptan-2-ol
(+)-Citronellol	(R)-3,7-Dimethyloct-6-en-1-ol
(+)-Fenchol	(R)-1,5,5-Trimethylbicyclo[2.2.1]heptan-6-ol
Geraniol	(E)-3,7-Dimethylocta-2,6-dien-1-ol
(−)-(E)-Hotrienol	(3R,5E)-3,7-Dimethylocta-1,5,7-trien-3-ol
(+)-Isomenthol	(1S,2R,5R)-5-Methyl-2-propan-2-ylcyclohexan-1-ol
(+)-(E)-Carveol	(1R,5S)-2-Methyl-5-prop-1-en-2-ylcyclohex-2-en-1-ol
(−)-(E)-Carveol	(1S,5R)-2-Methyl-5-prop-1-en-2-ylcyclohex-2-en-1-ol
(−)-Lavandulol	(R)-5-Methyl-2-prop-1-en-2-ylhex-4-en-1-ol
(+)-Linalool	(S)-3,7-Dimethylocta-1,6-dien-3-ol
(−)-Menthol	(1R,2S,5R)-5-Methyl-2-propan-2-ylcyclohexan-1-ol
(−)-2-Methylisoborneol	(1R-exo)-1,2,7,7-Tetramethylbicyclo[2.2.1]heptan-2-ol
(+)-Neomenthol	(1S,2S,5R)-5-Methyl-2-propan-2-ylcyclohexan-1-ol
Nerol	(Z)-3,7-Dimethylocta-2,6-dien-1-ol
(−)-Perillyl alcohol	[(S)-4-Prop-1-en-2-ylcyclohex-1-en-1-yl]methanol
(+)-(Z)-Sabinene hydrate	(1β,2β,5α)-4-Methyl-1-propan-2-ylbicyclo[3.1.0]hexan-4-ol
(+)-Terpinen-4-ol	(S)-4-Methyl-1-propan-2-ylcyclohex-3-en-1-ol
(+)-α-Terpineol	(R)-2-(4-Methyl-1-cyclohex-3-en-1-yl)propan-2-ol

[a]Numbers in the trivial names derived from *p*-menthane (4-isopropyl-1-methylcyclohexane) differ from the numbers of the names recommended by IUPAC. For example, (+)-(E)-dihydrocarvone is (1S,4S)-dihydrocarvone or (2S,5S)-2-methyl-5-prop-1-en-2-ylcyclohexan-1-one; (+)-(E)-carveol is (2R,4S)-carveol or (1R,5S)-2-methyl-5-prop-1-en-2-ylcyclohex-2-en-1-ol.

8-15, bicyclic monoterpenic alcohols

Complex enzymatic and non-enzymatic transformations of glycosides of monoterpenic alcohols and free alcohols during the development of flowers, fruit ripening, and fermentation provide a range of new oxygenated monoterpenoids, which are often characteristic constituents of flowers, fruits, and alcoholic beverages (wine and spirits). For example, a glycoside of terpenic alcohol can be enzymatically oxidised, or the free alcohol formed by hydrolysis of glycoside can be oxidised, and then transformed into other products. These transformations are shown in Figure 8.9, using the example of (+)-linalool and its glucoside. Linalool in lilac flowers (*Syringa vulgaris*, Oleaceae) is oxidised

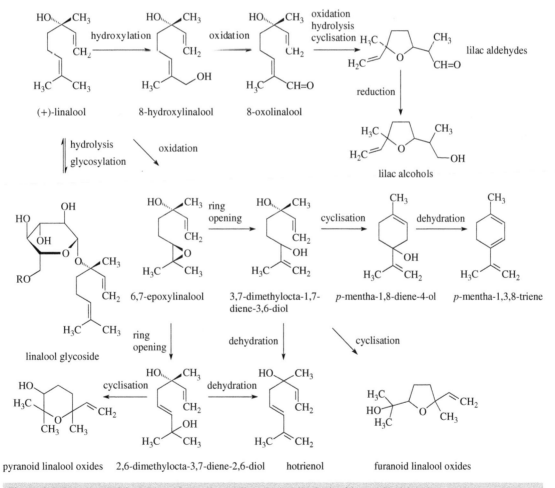

Figure 8.9 Reactions of linalool and formation of lilac alcohol, linalool oxides, and hotrienol.

to 8-hydroxylinalool, which is the precursor of four (5'S)-isomers of lilac aldehydes, reduction of which yields lilac alcohols, which are typical components of the odour of lilac flowers and of the honey of citrus flowers. Linalool glucoside can be oxidised analogously to 8-hydroxylinalool glucoside, hydrolysis of which provides free 8-hydroxylinalool. Under the acidic conditions in musts, especially during heating, glycosides are easily hydrolysed non-enzymatically to monoterpenic alcohols or their oxidation products, such as 6,7-epoxylinalool. Enzymatic ring opening of this epoxide yields either 3,7-dimethyl-1,7-octadien-3,6-diol or 2,6-dimethyl-3,7-octadien-2,6-diol. The former compound then becomes the precursor (after cyclisation and dehydration) of p-mentha-1,3,8-triene, which is the main flavour-active component of parsley essential oil. Its dehydration yields (−)-(E)-hotrienol, which is a component of many fruit aromas. In acidic solutions, spontaneous cyclisation of the mentioned diols yields furanoid and pyranoid linalool oxides. Linalool oxides are also present as glycosides.

The monoterpenic alcohol biosynthesised from two molecules of dimethylallyl diphosphate linked together head to head is (−)-(R)-lavandulol (8-13). Along with its acetate, lavandulol occurs in the essential oil of common lavender (Lavandula angustifolia, syn. L. officinalis, Lamiaceae).

The characteristic odour of raw, wet land, called earthy off-odour, is caused by the presence of two terpenoids, monoterpenic alcohol (−)-2-methylisoborneol and (−)-geosmin (8-16), which are produced by soil streptomycetes, myxobacteria, and cyanobacteria. Geosmin has only 12 carbon atoms as its biosynthesis starts from the sesquiterpenic hydrocarbon germacrene D, but during this biosynthesis three carbon atoms are eliminated as acetone. Both compounds cause an unpleasant earthy odour in drinking water and some foods, such as beetroot during cooking, canned mushrooms, wheat, coffee, aged wine, and the meat of freshwater fish (such as carp) that depend on the nutrients from benthos, animal (zoobenthos), and vegetable (phytobenthos) organisms inhabiting the bottom waters. For example, 2-methylisoborneol has been reported to occur in soil in quantities of about 4 µg/kg, which is more than 100-fold above its threshold value. Its reported occurrence in green coffee beans (80–420 ng/kg in arabica and 740–1280 ng/kg in robusta) is most likely of microbial origin.

2-methylisoborneol   geosmin

8-16, terpenic alcohols causing earthy off-odour

Sesquiterpenic alcohols   Examples of sesquiterpenic alicyclic alcohols (**8-17**) are farnesol and nerolidol (also known as peruviol). Both alcohols, smelling of flowers, are components of many essential oils used in perfumery. Of the four possible geometric isomers, the (2E,6E)-isomer of farnesol is the most common in nature, occurring, for example, in basil oil and ambrette (*Abelmoschus moschatus*; Malvaceae) seed oil. The (2Z,6E)-isomer occurs in the petit grain oil bigarade, which is derived from the bitter orange tree leaves (*Citrus aurantium* var. *amara*, Rutaceae). Farnesol is a natural pesticide for mites and a pheromone for several species of insects. Nerolidol with a double bond at C-6 occurs in the form of (Z)- and (E)-isomers, each of which can exist as an enantiomeric pair (chiral carbon C-3). The individual enantiomers and their mixtures are found in many essential oils.

Table 8.7 Trivial and systematic names of selected sesquiterpenic alcohols.

Trivial name	Systematic name (IUPAC)
(Z)-α-*trans*-Bergamotol	(1S,3E,5S)-4-Hydroxymethyl-7-methyl-7-(4-methylpent-3-enyl)bicyclo[3.1.1]hept-3-ene
(−)-α-Bisabolol	(S)-6-Methyl-2-(4-methylcyclohex-3-en-1-yl)hept-5-en-2-ol
(E,E)-Farnesol	(2E,6E)-3,7,11-Trimethyldodeca-2,6,10-trien-1-ol
(−)-Geosmin	(4S,4aS,8aR)-4,8a-Dimethyl-1,2,3,4,5,6,7,8-octahydronaphthalen-4a-ol
(+)-Hernandulcin	(2S,6S)-6-(2-Hydroxy-6-methylhept-5-en-2-yl)-3-methylcyclohex-2-en-1-one
Capsidiol	(1R,3R,4S,4aR,6R)-4,4a-Dimethyl-6-prop-1-en-2-yl-1,2,3,4,5,6,7-heptahydronaphthalen-1,3-diol
(Z)-Lanceol	(Z)-6-Methyl-2-(4-methylcyclohex-3-en-1-yl)hepta-1,5-dien-7-ol
(+)-(Z)-Nerolidol	(3S,6Z)-3,7,11-Trimethyldodeca-1,6,10-trien-3-ol
(+)-(E)-Nerolidol	(3R,6E)-3,7,11-Trimethyldodeca-1,6,10-trien-3-ol
(+)-Nootkatol	(2R,4S,4aS,6R)-4,4a-Dimethyl-6-prop-1-en-2-yl-2,3,4,5,6,7,8-heptahydronaphthalen-2-ol
(−)-(Z)-β-Santalol	(1S,2Z,4R,6R)-2-Methyl-5-(6-methyl-5-methylidene-6-bicyclo[2.2.1]heptanyl)pent-2-en-1-ol

An example of monocyclic sesquiterpenic alcohols is (+)-α-bisabolol (**8-17**), which occurs in citrus essential oils along with other sesquiterpenic alcohols and represents the major component of the essential oil of chamomile flowers (*Matricaria chamomilla*, Asteraceae). α-Bisabolol (0.26%) and other sesquiterpenoids, such as monocyclic sesquiterpenic alcohol (Z)-lanceol (1.7%), bicyclic sesquiterpenic alcohol *trans*-α-(Z)-bergamotol (3.7%), (−)-(Z)-β-santalol (about 21%), and its isomers are the major and essential components of white sandalwood (*S. album*, Santalaceae) oil used in perfumery, cosmetics, and aromatherapy. (Z)-Lanceol is also a component of sage species essential oil. Sesquiterpenic bicyclic alcohol (+)-β-nootkatol is the precursor of ketone (+)-nootkatone, an important component of grapefruit essential oil. Trivial and systematic names of the main sesquiterpenic alcohols are listed in Table 8.7.

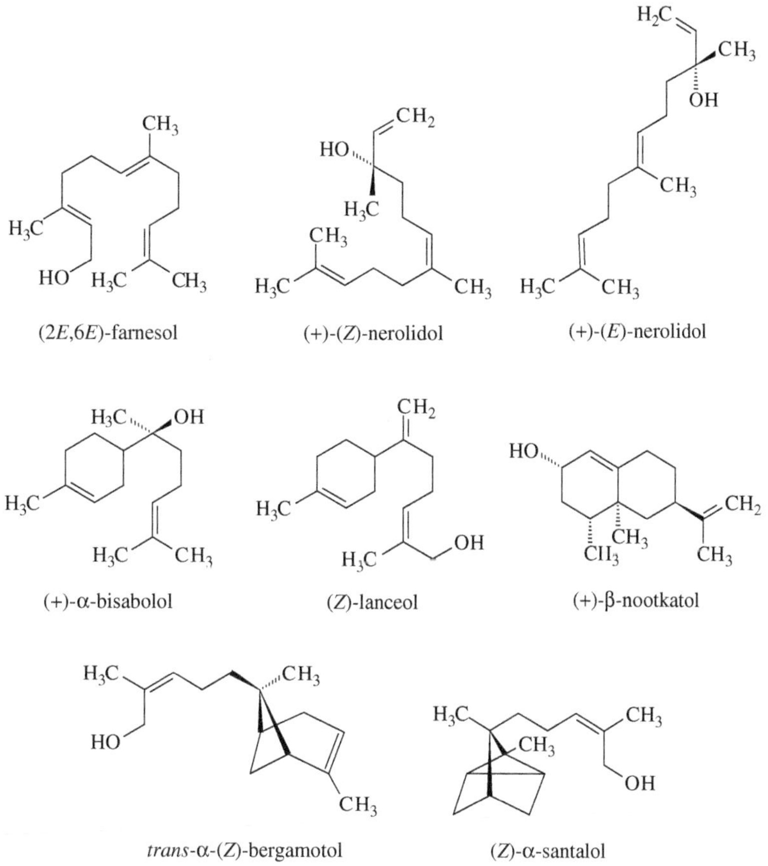

(2E,6E)-farnesol (+)-(Z)-nerolidol (+)-(E)-nerolidol

(+)-α-bisabolol (Z)-lanceol (+)-β-nootkatol

trans-α-(Z)-bergamotol (Z)-α-santalol

8-17, sesquiterpenic alcohols

Many sesquiterpenoids have antimicrobial and insecticidal properties. For example, germacrene A is the precursor of capsidiol (see Section 10.3.11.2), which is the main phytoalexin in bell peppers and tobacco plants, when attacked by pathogenic fungi.

Diterpenic alcohols Diterpenic alcohols are present in nature as free compounds or bound in fatty acid esters and glycosides. They have no significance as flavourings, but may be precursors of flavour- and biologically active products. Important representatives of acyclic diterpenoid alcohols are retinol (vitamin A₁) and phytol (2E,3,7R,11R,15)-3,7,11,15-tetramethylhexadec-2-en-1-ol, a constituent of chlorophylls, tocopherols, and K group vitamins. Tetracyclic diterpenoid acids gibberellins act as plant hormones universally in plant tissues.

Alicyclic diterpenoid alcohols occur as toxic components of latex of many plant species from the spurge family (Euphorbiaceae). Their representatives are, for example, esters of phorbol, such as 13-acetyl-12-myristoylphorbol (**8-18**). Cafestol, kahweol, and related alicyclic diterpenoids (such as 16-O-methylcafestol and 16-O-methylkahweol) are found exclusively in coffee beans (mainly esterified to fatty acids at the C-16 or C-17 position) and unfiltered coffee (Scandinavian-style boiled coffee and Turkish-style coffee). In filtered coffee drinks, they are present in only negligible amounts. Kahweol is specific to *Coffea arabica* coffee, where it occurs in concentrations of about 5890 and 5200 mg/kg (f.w.) in endosperm and perisperm, respectively. The amount of cafestol in *C. arabica* is about 3000 mg/kg in the endosperm and 1300 mg/kg in the perisperm; in *C. canephora*, (syn. *C. robusta*), known as robusta coffee, the amounts are about 940 mg/kg and 1100 mg/kg, respectively. 16-O-Methylcafestol and 16-O-methylkahweol occur as minor compounds in *C. arabica* and in somewhat greater amounts in *C. robusta*. During roasting of green coffee seeds, these alcohols partially dehydrate to dehydroalcohols (dehydrocafestol and dehydrokahweol) and are oxidised to the corresponding aldehydes (cafestal and kahweal). Cafestol, and partially also kahweol, increases the cholesterol level in the blood serum but also exhibits some beneficial effects, including anticarcinogenic activities.

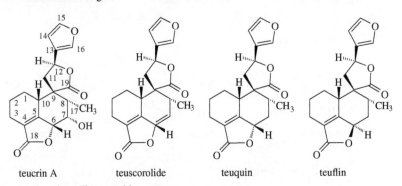

phorbol

cafestol, R = H
16-*O*-methylcafestol, R = CH₃

kahweol, R = H
16-*O*-methylkahweol, R = CH₃

8-18, diterpenic alcohols

Teucrin A is the major component of the so-called *neo*-clerodane diterpenoids (**8-19**) from the plant known as wall germander (*Teucrium chamaedrys*, Lamiaceae), native to Europe and the Near East. The alcoholic extracts of the plant are used for the aromatisation (bittering) of wines and aperitifs (as it contains a number of odour-active mono- and sesquiterpenoids), and in folk medicine for their antiseptic (anti-inflammatory) and choleretic (increased bile excretion) properties. Teucrin A is accompanied by other diterpenoids, such as teuscorolide, teuquin, teuflin (**8-19**), and their glycosides. Because of its hepatotoxicity, teukrin A is in the list of substances of Regulation (EC) No. 1334/2008 that must not be added to foods as flavourings. Its natural content is restricted to 5 mg/kg in bitter-tasting spirit drinks or bitters and to 2 mg/kg in other alcoholic beverages.

teucrin A   teuscorolide   teuquin   teuflin

8-19, *neo*-clerodane diterpenoids

8.2.2.1.2 Aromatic and heterocyclic alcohols

Aromatic alcohols, unlike phenols, have a hydroxyl group attached to the side chains but not directly to the aromatic carbon atom. They are the natural constituents of many plant materials, including food, where they arise primarily by enzymatic and thermal processes. The simplest alcohol in this group is benzyl alcohol (**8-20**), a compound with sweet, floral, and fruity odour. It is assumed that benzyl alcohol arises by a gradual reduction of benzoic acid by dehydrogenase enzymes via benzaldehyde. In alcoholic beverages, it arises by reduction of benzaldehyde, which is a product of cyanogenic glycosides decomposition and of Strecker degradation of phenylalanine. In fruit brandies produced from fruit fermented with stones (such as cherry, apricot, and plum brandies), quantities amounting to 20–70 mg (per litre of pure ethanol) are found. Benzyl alcohol also occurs in the form of esters in some essential oils. The most common compounds are benzyl acetate and benzyl benzoate.

benzyl alcohol   2-phenylethanol   (*R*)-1-phenylethanol   (*E*)-cinnamyl alcohol

8-20, examples of aromatic alcohols

Common non-volatile components of plants are hydroxy derivatives of benzyl alcohol, which mainly occur in nature in the bound form as glycosides, esters of aromatic carboxylic acids (caffeic, protocatechuic, 4-hydroxybenzoic, and vanillic acids), and glycosides of these esters. For example, 4-hydroxybenzyl alcohol was first isolated from muskmelon seedlings (*Cucurbita moschata*, Cucurbitaceae). The compound acts as a cofactor for indoleacetic acid oxidase. The glycoside of its ester with protocatechuic acid, 4-(3,4-dihydroxybenzoyloxymethyl)phenyl-*O*-β-ᴅ-glucopyranoside, occurs in oregano (*Origanum vulgare*, Lamiaceae), which has been reported to possess antithrombin, anti-*Helicobacter pylori*, antibiotic, antihyperglycaemic, and antioxidation effects.

2-Phenylethanol (phenethyl alcohol, **8-20**), the higher homologue of benzyl alcohol with floral (lilac, rose) odour, is formed during fermentation processes from phenylalanine via reduction of phenylacetaldehyde. It is one of the main components of rose oil and occurs in small quantities in many other essential oils. 1-Phenylethanol (**8-20**) is a chiral positional isomer of phenethyl alcohol with a floral odour resembling honey. It is found in nature as a glucoside and primeveroside of racemate in tea flowers, black tea, some fruits, vegetables, and alcoholic beverages.

Hydroxy alcohols derived from other aromatic amino acids are non-volatile compounds. 4-Hydroxyphenylethanol (tyrosol) arises from tyrosine, 3,4-dihydroxyphenylethanol (hydroxytyrosol) from 3,4-dihydroxyphenylalanine (DOPA), and tryptophol from tryptophan. Tyrosol and hydroxytyrosol belong to the main phenolic compounds in olives. One component of cinnamon oil (see Table 8.32), fruit brandies, and other materials is the *trans*-isomer of cinnamyl alcohol, (E)-cinnamyl alcohol (**8-20**), which is produced by the reduction of cinnamic acid via cinnamic acid aldehyde (cinnamaldehyde). In the form of various esters, cinnamyl alcohol is found in the leaves and bark of evergreen aromatic trees and shrubs of the genus *Cinnamomum* (Lauraceae), growing in tropical and subtropical regions of America, Asia, and Australia, and in Peru balsam, the secretion of trees of *Myroxylon balsamum* (Fabaceae), native to Central and South America, which is used in perfumery. A number of related substituted alcohols derived from substituted cinnamic acids, such as 4-coumaryl, caffeoyl, coniferyl (feruloyl), and 5-hydroxyconiferyl (sinapoyl) alcohol, and the corresponding aldehydes, are building units of lignin (Figure 8.10).

The most common oxygen-containing heterocyclic alcohol is furfuryl alcohol, which is a degradation product of sugars. Furfuryl alcohol forms primarily from furan-2-carbaldehyde by hydrogenation or in a Cannizzaro reaction. A large number of other alcohols derived from furan, pyran, and other heterocyclic compounds are products of the Maillard reaction.

8.2.2.1.3 Glycols and polyols

Glycols and polyols are non-volatile compounds that are formed in foods as secondary products of fermentation processes and in the Maillard reaction, or which may be used as food additives. Frequently occurring polyols present in foods are sugar alcohols that are used as sweeteners.

The lowest member of the homologous series of glycols is ethylene glycol. It may be released to the environment as a contaminant, with the major source being the disposal of used antifreeze and deicing solutions. It is also used in hydraulic brake fluids, inks in stamp pads, and ballpoint pens. Another two-carbon compound is the oxidation product of ethylene glycol, glycolaldehyde, which is produced as a degradation product of sugars in the Maillard reaction.

The three-carbon compound methylglyoxal is the byproduct of fermentation processes and of the Maillard reaction. Products of its reduction, namely acetol (hydroxyacetone), (R)-propane-1,2-diol (D-propane-1,2-diol), and (R)-lactaldehyde (D-lactic acid aldehyde or D-lactaldehyde), are relatively common compounds present in small amounts in foods (Figure 8.11). Under anaerobic conditions, some microorganisms (such as *Bacillus amaracrylus*, *Citrobacter freundii*, and *Klebsiella pneumoniae*) can ferment glycerol as the sole substrate and transform it into 1,3-dihydroxyacetone (which is then involved in metabolic pathways) or 3-hydroxypropionaldehyde (by a coenzyme B_{12}-dependent glycerol dehydratase), which was first discovered in wine spoiled by *B. amaracrylus*. It is further reduced to propane-1,3-diol. 3-Hydroxypropionaldehyde dimer (see Section 4.7.1.2.3) was patented under the name reuterin and is thought to be responsible for the probiotic effects of lactic acid bacteria *Lactobacillus reuteri*. The bacterium *L. brevis* can produce propane-1,3-diol from glycerol as a product of sugar and lactic acid co-fermentation.

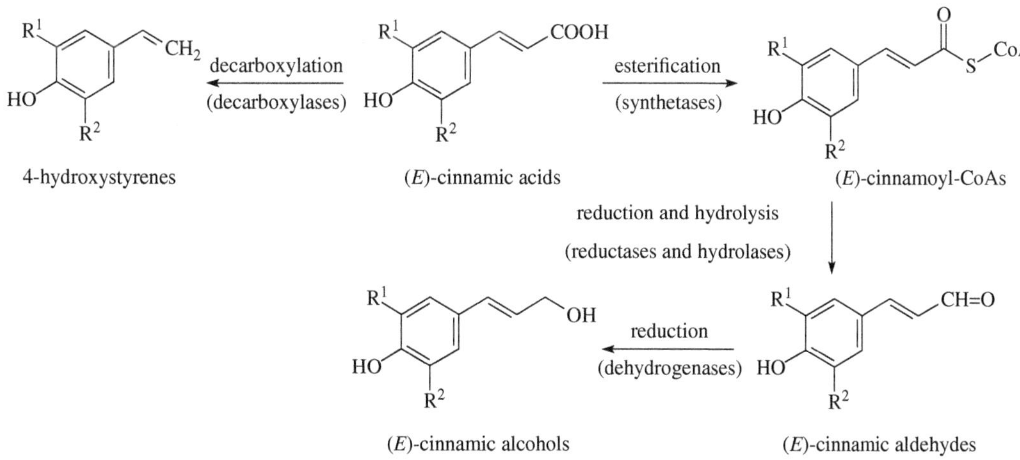

Figure 8.10 Biosynthesis of cinnamic alcohols from cinnamic acids.

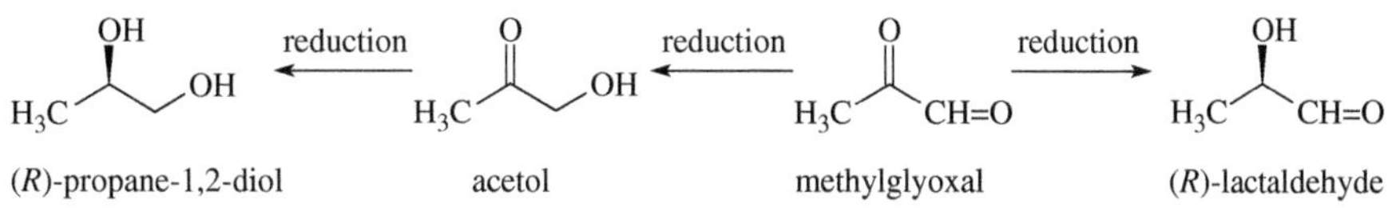

Figure 8.11 Formation of acetol, lactaldehyde, and propane-1,2-diol.

Commercial propane-1,2-diol (propylene glycol) is a racemic mixture of both stereoisomers. It is used to absorb excess water, if used as a humectant, to maintain moisture in some medicines, cosmetics, and foods, and as a solvent for food colourings and stabilisers of carbonyl compounds in food flavours. It is also used in antifreeze and de-icing solutions and as a solvent in the paint and plastic industries.

The most important diols occurring in alcoholic beverages and fermented dairy products are butane-2,3-diol and pentane-2,3-diol, which are not, however, odour-active substances. These diols are produced as byproducts of the activity of certain microorganisms, along with the corresponding acyloins (acetoin and 3-hydroxypentane-2-one) and sensory-active α-diketones (biacetyl and pentane-2,3-dione).

In fermented dairy products, such as yoghurt, and butter, all of these four-carbon compounds are produced from pyruvic acid, which is formed from citric acid. In the so-called malolactic fermentation of wine, pyruvic acid arises mainly by decarboxylation of (S)-malic (L-malic) acid. The conversion of pyruvic acid into lactic acid (catalysed by L-lactate dehydrogenase and D-lactate dehydrogenase) requires nicotinamide adenine dinucleotide (NADH). If NADH is not available, some lactic acid bacteria (such as *Streptococcus, Leuconostoc, Lactobacillus, Pediococcus,* and *Oenococcus*; in fermented dairy products, the lactic acid bacterium *Streptococcus diacetylactis* is important, and in wines, the bacterium *Oenococcus oeni*) transform most of the pyruvic acid and acetaldehyde, which is generated by decarboxylation, to an unstable intermediate (S)-2-hydroxy-2-methyl-3-oxobutanoic acid, also known as (S)-2-acetolactic acid. Decarboxylation and oxidation of 2-acetolactic acid provide butane-2,3-dione (biacetyl), whilst decarboxylation yields (R)-3-hydroxybutan-2-one (acetoin) and reduction of acetoin gives (2R,3R)-butane-2,3-diol, also known as (−)-D-*threo*-butane-2,3-diol (Figure 8.12). In small quantities, it is accompanied by *meso*-butane-2,3-diol and (2S,3S)-butane-2,3-diol, also known as (+)-L-*threo*-butane-2,3-diol. The amount of the *meso* isomer is 20–38% of the total amount of butane-2,3-diol. Not all lactic acid bacteria produce all of these compounds, and in some cases the reaction stops early and some intermediates accumulate. Acetic acid is also a fermentation product.

Five-carbon compounds are formed from (S)-aceto-2-hydroxybutyric acid, also known as (S)-2-ethyl-2-hydroxy-3-oxobutanoic acid, which is the intermediate of isoleucine biosynthesis from threonine. This acid is formed in the reaction of 2-oxobutanoic acid with

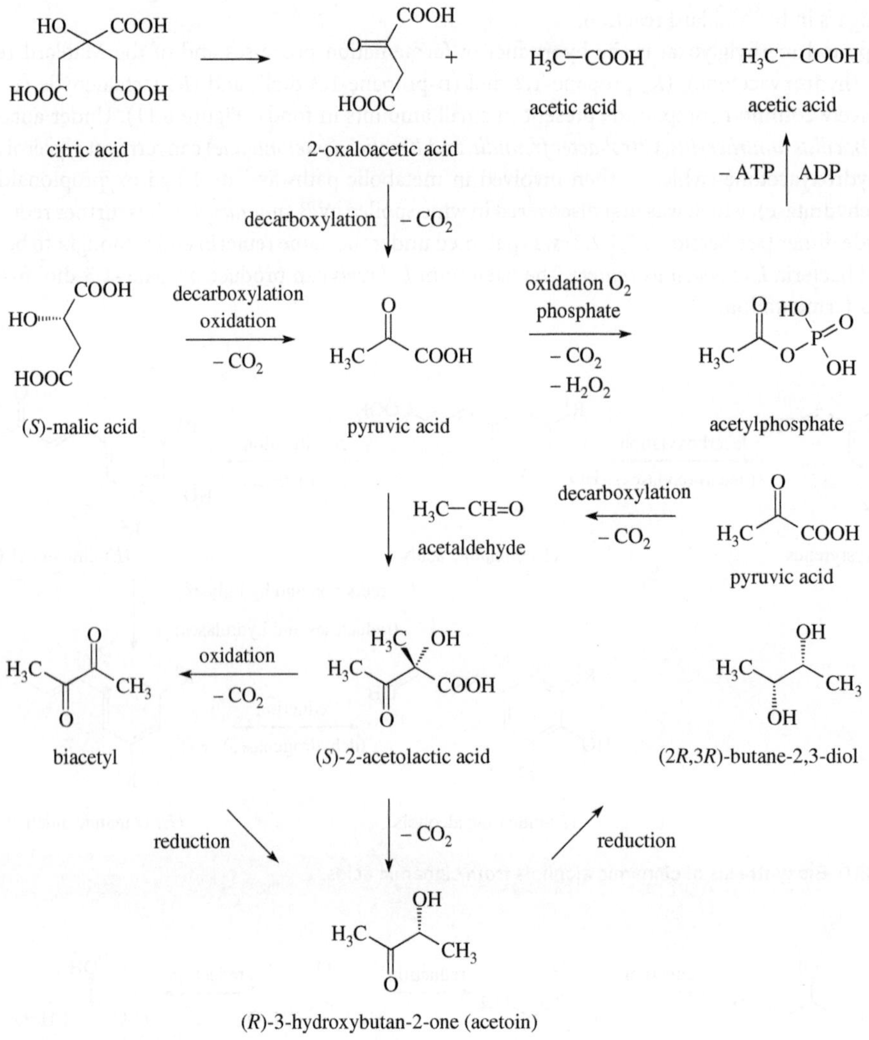

Figure 8.12 Formation of butane-2,3-diol, acetoin, and biacetyl.

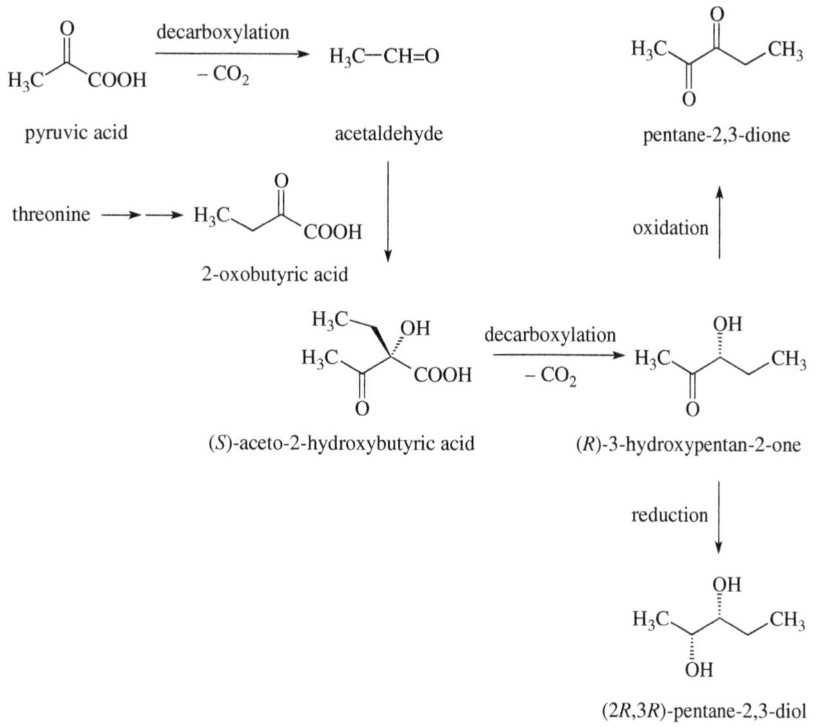

Figure 8.13 Formation of pentane-2,3-diol, 3-hydroxypentan-2-one, and pentane-2,3-dione.

acetaldehyde, and its decarboxylation provides (R)-3-hydroxypentan-2-one, which is oxidised to pentane-2,3-dione and reduced to (2R,3R)-pentane-2,3-diol (Figure 8.13).

Glycerol (propane-1,2,3-triol) mainly occurs in foods in the form of triacylglycerols and many other lipids known as glycerolipids. Glycerol is also produced as a byproduct of alcoholic fermentation (also in alkaline media together with acetic acid salts) by reduction of the intermediate 1,3-dihydroxyacetone phosphate via glycerol 3-phosphate. It is often found in beer, wine, and other alcoholic beverages, in an amount depending mainly on the temperature during fermentation, yeast strain, and, in wine, the presence of sulfites. Glycerol concentrations in pilsner-type beers are 1.5–2.9 g/l. In wine, levels between 1 and 15 g/l are frequently encountered, with average values of approximately 7 g/l. Higher amounts of glycerol result from higher temperatures, so its content in red wines is usually higher (by about 20–30%) than in whites. In the presence of higher amounts of sulfites, the yeast *Saccharomyces cerevisiae* produces virtually no alcohol, because acetaldehyde is blocked in the form of the α-hydroxysulfonic acid and cannot be reduced to ethanol. However, 1,3-dihydroxyacetone phosphate is still reduced to glycerol 3-phosphate, which is hydrolysed to glycerol. A higher glycerol content is found in wines produced from grapes infected with the *Botrytis cinerea* fungus. Glycerol affects the organoleptic properties of beer and wine; higher levels are thought to contribute to the viscosity, sweetness, and smoothness of a beverage.

8.2.2.2 Properties and reactions

Low-molecular-weight alcohols and glycols are toxic. Methanol and ethanol are both biotransformed by alcohol dehydrogenase; however, ethanol has the greater affinity for the enzyme. The toxicity resulting from higher doses of methanol is very well documented in both humans and animals and is attributed to its toxic metabolite formic acid, which is known to be toxic to the optic nerve. Formic acid requires folic acid as a cofactor for its elimination. Animal studies have shown that when folate levels are low, the elimination of formic acid is slower. Therefore, folate deficient chronic drinkers may be at higher risk of organ damage, blindness, and even death. The lethal oral dose of methanol is 340 mg/kg body weight. Levels of methanol in fruit brandies are usually around 4 g/l, but can be much higher (e.g. 12 g/l in plum, apple, pear, raspberry, and blackberry brandies and 2 g/l in wine brandies). The current EU general limit for naturally occurring methanol is 10 g of methanol per litre of ethanol (which equates to 0.4% v/v of methanol in a 40% alcohol beverage) – this provides a safety margin.

Ethanol acts on the central nervous system, causing euphoria at first, but eventually – if consumed in sufficient quantities – leading to death. Ethanol is an addictive poison that causes alcoholism (ethylism). Ethylene glycol is a sweet toxic compound, metabolised to glycolic acid and oxalic acid, that can damage the kidneys, nervous system, heart, and lungs.

As in water, hydrogen bonds also play the major role in the molecular attractive forces in alcohols (especially in lower alcohols and polyols). This is related to the relatively good solubility of alcohols and glycols in water, and to their relatively high boiling points. The large number of hydroxyl groups allows polyols to create complex molecular associates through hydrogen bonding in solutions.

The most important reactions taking place on the hydroxy groups of alcohols are O—H and C—O bond cleavage. With O—H bond cleavage, reactions with strong acids, oxidations of primary alcohols to aldehydes, oxidations of secondary alcohols to ketones, and reactions with organic acids (formation of esters) all take place. In foods, the latter three are particularly important, and are usually enzymatically catalysed. Other important reactions are dehydration and its opposite, hydration, which yield unsaturated hydrocarbons from alcohols and isomeric alcohols from unsaturated hydrocarbons, respectively. They are particularly important in terpenic alcohols. In oleochemistry, oxidation and esterification reactions are used for the production of various lipid derivatives.

8.2.3 Ethers

Symmetric and asymmetric aliphatic, alicyclic, and aromatic ethers and ethers with the oxygen atom bound in the ring can all be found in foods. Volatile dialkyl ethers are virtually absent, but some are synthesised and used as flavourings. Odour- and taste-active ethers are mainly terpenoid ethers (derived from monoterpenes and sesquiterpenes) and aromatic ethers.

8.2.3.1 Classification, structure, terminology, and occurrence

8.2.3.1.1 Terpenoid ethers

Ethers whose oxygen atom is bound to vicinal carbons or is part of a larger alicyclic ring include 1,2-epoxides (oxiranes), 1,4-epoxides (furans), and 1,5-epoxides (pyrans). These and other terpenic epoxides are the primary components responsible for the odours of many foods. Terpenoid ethers also form as secondary products of the oxidation and dehydration of carotenoid pigments, steroids, fatty acids, PAHs, and many other compounds.

An example of a terpenic 1,2-epoxide is β-caryophyllene oxide, also known as (−)-epoxycaryophyllene (8-21), which occurs in many essential oils. An example of a terpenic 1,4-epoxide is the so-called (+)-dill ether, (3R,4S,8S)-3,9-epoxy-p-menth-l-ene (8-21), which is a typical component of the essential oil of caraway (30%) and dill. An example of an unsaturated 1,4-epoxide is (+)-menthofuran (8-21), the metabolite of ketone (+)-pulegone. Both compounds are components of peppermin oil (see Table 8.32), and are hepatotoxic. The monoterpenoid compound (+)-1,8-cineole (also known as limonene oxide, eucalyptol, or 1,8-epoxy-p-menthane; 8-21) is an example of a more complex structure. It is present in essential oils of many types of spice, and in higher quantities in the essential oil of trees of the genus *Eucalyptus* (Myrtaceae). Trivial and systematic names of selected ethers are given in Table 8.8.

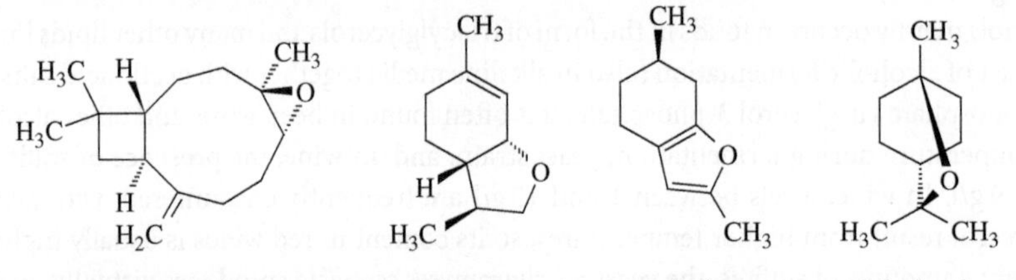

epoxycaryophyllene   dill ether   menthofuran   1,8-cineole

8-21, terpenoid epoxides

Table 8.8 Trivial and systematic names of some terpenoid ethers.

Trivial name	Systematic name (IUPAC)
(−)-Epoxycaryophyllene	(1R,4R,6R,10S)-9-Methylene-4,12,12-trimethyl-5-oxatricyclo[8.2.0.0⁴,⁶]dodecane
(+)-Dill ether	(3S,3aS,7aR)-3,6-Dimethyl-2,3,3a,4,5,7,7a-hexahydro-1-benzofuran
(+)-Menthofuran	(R)-3,6-Dimethyl-4,5,6,7-tetrahydro-1-benzofuran
(+)-1,8-Cineol	(1S,4S)-1,3,3-Trimethyl-2-oxabicyclo[2.2.2]octane
(−)-Nerol oxide	(S)-3,6-Dihydro-4-methyl-2-(2-methylprop-1-enyl)-2H-pyrane
(−)-(Z)-Rose oxide	(2S,4R)-4-Methyl-2-(2-methylprop-1-enyl)oxane
(2R,5S)-(Z)-Linalool oxide (furanoid)	(2R,5S)-2-methyl-5-prop-1-en-2-yl-2-vinyltetrahydrofuran
(2R,5R)-(E)-Linalool oxide (furanoid)	(2R,5R)-2-methyl-5-prop-1-en-2-yl-2-vinyltetrahydrofuran
(3R,6R)-(Z)-Linalool oxide (pyranoid)	(3R,6R)-2,2,6-Trimethyl-6-vinyltetrahydro-2H-pyran-3-ol
(3S,6S)-(Z)-Linalool oxide (pyranoid)	(3S,6S)-2,2,6-Trimethyl-6-vinyltetrahydro-2H-pyran-3-ol

Numerous other ethers containing pyran or furan rings are formed by the dehydration of aliphatic diols (e.g. linalool oxides, rose oxide, or nerol oxide) and are components of many essential oils. For example, the furanoid (2R,5R)-*trans*- and (2R,5S)-*cis*- and pyranoid (3R,6R)-*cis*- and (3S,6S)-*cis*-linalool oxides (**8-22**) are odorants of jasmine tea, aromatic wines, elderberry bush flowers, and linden honey. In grapes, (−)-*cis*-rose oxide from (−)-citronellol is analogously produced. Rose oxide is a component of rose and geranium essential oils (**8-22**). Nerol oxide in rose oil is a racemate.

8.2.3.1.2 Aromatic ethers

Foods may contain a number of alkyl aryl ethers that are components of essential oils of different spices. These ethers are most often derived from anisole (methoxybenzene, **8-23**) or veratrole (1,2-dimethoxybenzene, **8-23**), which are substituted by a prop-1-en-yl or 1-prop-2-en-1-yl (allyl) group at the C-4 position of the benzene ring. An important ether is estragole (also known as 4-allylanisole or methyl chavicol, **8-23**), which is the main component of basil essential oil (over 80%) and of tarragon (dragon's wort) essential oil (over 60%) (see Table 8.32). Estragole is a genotoxic carcinogen in experimental animals after chronic exposure, and therefore appears in the list of substances of Regulation (EC) No. 1334/2008 that shoud not be added to foods as flavourings. Its natural content in foods is restricted to 50 mg/kg in dairy products, processed fruits, vegetables including mushrooms (roots, tubers and legumes), nuts and seeds, and fish products, and to 10 mg/kg in non-alcoholic beverages.

Isomers of anethole (isoestragole, **8-23**) with a typical anise aroma are the main components of anise (>95%), fennel (>80%), and star anise (>95%) essential oils. In natural essential oils, the (E)-isomer of anethole dominates, which is used as a flavour and fragrance agent as it has a liquorice-type odour and an anise-type taste. The (Z)-isomer of anethole has an anise-type odour. It was associated with liver cancer in rats, but is now regarded as safe.

Components of clove oil (see Table 8.32) and many other oils are β-caryophyllene (4–21%), eugenol (49–87%), eugenyl acetate (0.5–21%), methyl eugenol, and elemicin (**8-23**). Another compound derived from veratrole is methyl isoeugenol, which can occur as both (E)- and (Z)-isomers (**8-23**). (E)-Methyl isoeugenol occurs in high concentrations in the seed oil and oil obtained from aerial parts of carrot (*Daucus carota*), which is reported to be an antimicrobial against the human enteropathogen *Campylobacter jejuni*. Methyl isoeugenol is found in lower amounts in many other essential oils (such as citronella, calamus, nutmeg, and laurel leaf oils). Elemicin is also a component of carrot essential oil and banana aroma. A component of the essential oil obtained from the root of calamus (sweet flag, *Acorus calamus*, Acoraceae) is asarone, which is present as a mixture of two isomers, with α-asarone (**8-23**) as the (E)-isomer and β-asarone (**8-23**), also known as *cis*-isoasarone, as the (Z)-isomer. Calamus root is used as a medicinal plant for wound healing, as an antipyretic drug, for its effect against dyspepsia (indigestion, upset stomach), and as a bittering agent for alcoholic beverages (including vermouth and beer) and in cosmetics (such as flavourings for toothpastes). β-Asarone has toxic effects in mammals (it acts as a chemosterilant) and therefore appears in the list of substances of Regulation (EC) No. 1334/2008 that should not be added to foods as flavourings. Its natural content in alcoholic beverages is limited to a maximum level of 1 mg/kg, and tetraploid forms of the plant should not be used for the production of flavourings or of food ingredients with flavouring properties. A related North American species, *Anaxyrus americanus*, today described as a variety of calamus (*A. c.* var. *americanus*), does not contain β-asarone at all. β-Asarone and the related 2,4,5-trimethoxybenzaldehyde (also known as gazarin) have also been found in carrot seeds.

(2R,5S)-*cis*-furanoid linalool oxide   (2R,5R)-*trans*-furanoid linalool oxide   (−)-*cis*-rose oxide

(3S,6S)-*cis*-pyranoid linalool oxide   (3R,6R)-*cis*-pyranoid linalool oxide   (−)-nerol oxide

8-22, monoterpenoid oxides

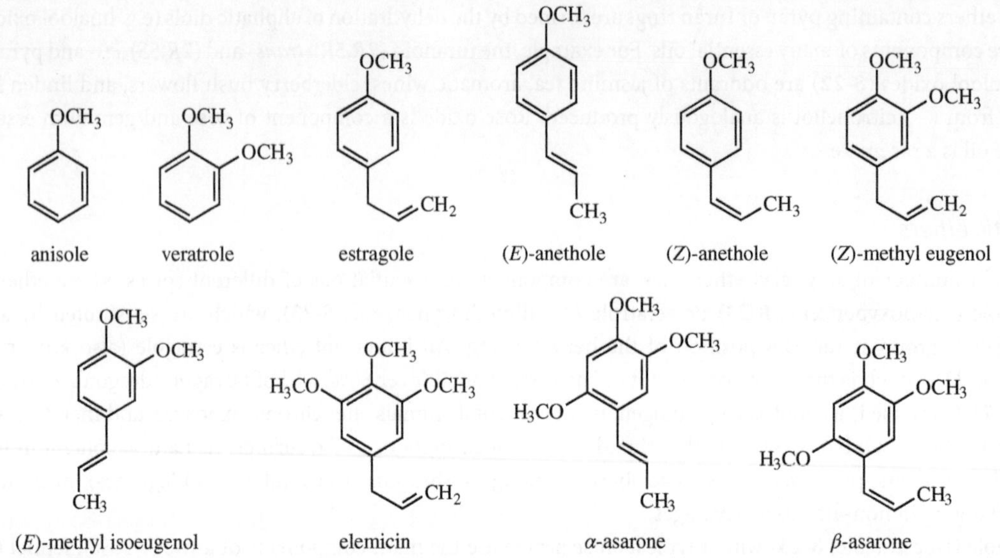

anisole   veratrole   estragole   (*E*)-anethole   (*Z*)-anethole   (*Z*)-methyl eugenol

(*E*)-methyl isoeugenol   elemicin   α-asarone   β-asarone

8-23, aromatic ethers

Methyl eugenol (allylveratrol) (**8-23**), a component of several essential oils of, for example, fennel, citronella, basil, bay, and tea tree (*Melaleuca* spp.), is related, along with eugenol and asarons, to compounds with an attached 1,3-dioxol ring (methylenedioxy group). Most substances are simple prop-2-en-1-yl (allyl) benzenes, substituted by methoxyl groups (**8-24**). Important compounds are 5-(prop-2-en-1-yl)-2*H*-1,3-benzodioxole (safrole or shikimole), 5-(prop-1-en-1-yl)-2*H*-1,3-benzodioxole (isosafrole), and 4-methoxy-6-(prop-2-en-1-yl)-2*H*-1,3-benzodioxole, also known as myristicin or 5-methoxysafrole. Myristicin is a characteristic component of essential oils from the seeds of some root vegetables (carrots, parsley, and celery) and herbs, and particularly from nutmeg and mace (see Table 8.32). Safrole is the main component (representing more than 80%) of sassafras oil (shikimol), derived from the root, bark, or fruit of the sassafras tree (*Sassafras albidum*, Lauraceae; grows in North America), which was used in the United States to flavour beer and soft drinks until 1978. Locally, sassafras herbal tea is still in use in the treatment of rheumatism and skin diseases. Safrole is also a component of essential oils of nutmeg and mace (the content is about 0.1%) and of anise, cinnamon, and some other essential oils. Isosafrole is also a common component of many essential oils (such as sassafras, laurel, and clove essential oils). Apiole, 4,7-dimethoxy-5-(prop-2-en-1-yl)-2*H*-1,3-benzodioxole, known as parsley apiole or parsley camphor, is a flavour-active compound occurring in all parts of parsley and in celery leaves. Its isomer, 4,5-dimethoxy-6-(prop-2-en-1-yl)-2*H*-1,3-benzodioxole, known as dill apiole, occurs in dill leaves.

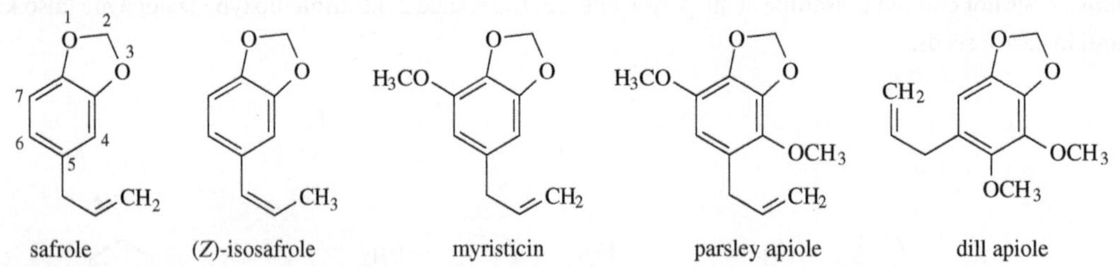

safrole   (*Z*)-isosafrole   myristicin   parsley apiole   dill apiole

8-24, aromatic ethers with a methylenedioxyphenyl moiety

Methyl eugenol and safrole are weak carcinogens. Myristicin is a psychomimetic compound tending to induce narcotic effects, hallucinations, and other symptoms of a psychosis. Methyl eugenol and safrole are in the list of substances of Regulation (EC) No. 1334/2008 that should not be added to foods as flavourings. The natural content of methyl eugenol in foods (the maximum level of safrole is in parenthesis) is restricted to 60 (25) mg/kg in soups and sauces, to 20 mg/kg in dairy products and ready-to-eat savouries, to 15 (15) mg/kg in meat preparations and meat products, including poultry and game, to 10 (15) mg/kg in fish products, and to 1 (1) mg/kg in non-alcoholic beverages. Sassafras leaves must be safrole-free to be used as food additives, and sassafras oil is not used at all.

8.2.3.1.3 Other ethers

Simple oxiranes (e.g. oxirane, known as ethylene oxide, and methyloxirane, known as propylene-1,2-oxide) are used as food additives (such as preservatives). Some ethers can also be regarded as phenols, aromatic aldehydes (such as vanillin), or acetals.

Particular non-volatile ethers are classified as lipids (e.g. plasmalogens, esters of 1-alkoxypropan-2,3-diols such as chimyl alcohol). Some also occur in small amounts as substances accompanying lipids, such as dialicyclic ethers derived from sterols that occur in refined vegetable oils. Of great importance in oleochemistry are non-volatile ethers of glycerol and other polyols (dimers and higher oligomers) that are used, for example, as emulsifiers. The so-called vinyl ethers are formed as secondary oxidation products of essential fatty acids, representatives of which are pentylfuran and two isomeric 2-(pent-2-en-1-yl)furans.

Dehydration of carbohydrates in acidic solution yields derivatives of furan-2-carbaldehyde. An example of a heterocyclic ether is the ether derived from 5-hydroxymethylfuran-2-carbaldehyde that accompanies the parent aldehyde in sugar hydrolysates. Furfuryl ethyl ether is an important flavour compound indicative of beer storage and ageing conditions. It is most likely formed by protonation of furfuryl alcohol or furfuryl acetate followed by substitution of the leaving group by the nucleophilic ethanol. Another example of a heterocyclic ether is that formed by dehydration of 3-hydroxymethylindole in cruciferous vegetables, where 3-hydroxymethylindole arises by degradation of glucosinolate glucobrassicin.

8.2.3.2 Properties and reactions

Ethers are relatively stable in acidic and alkaline media. In the presence of oxygen, they slowly autoxidise to form the corresponding hydroperoxides (α-hydroperoxyoxaalkanes) and dialkyl peroxides, which are often transformed into thermolabile polymers.

8.2.4 Carbonyl compounds

8.2.4.1 Classification, structure, terminology, and occurrence

The carbonyl compounds molecules contain either an aldehyde group —CH=O or a keto group (oxo group) —C(=O)—, and can thus be divided into:

- aldehydes;

- ketones.

Volatile aldehydes and ketones are the most important odour- and taste-active substances. They occur in foods as primary substances and as components of various essential oils, and they also result from enzymatic and chemical reactions of various precursors as secondary substances. They are often desirable flavour-active components of foods, but in some cases they may also carry undesirable odours and tastes and serve as indicators of unwanted changes in sensory or nutritional value (e.g. autoxidation of lipids).

Carbonyl compounds also include a range of non-volatile polar compounds, such as reducing sugars and some products of their transformation (degradation), which are often taste-active substances, usually with a sweet taste. A special group of carbonyl compounds is the oxocarboxylic acids and, in a broader sense, all carboxylic acids, which often carry a sour taste. A special group of unsaturated diketones derived from aromatic systems is the quinones, which are often significant natural pigments in foods.

8.2.4.1.1 Aldehydes

Aliphatic saturated and unsaturated aldehydes Almost all saturated aliphatic aldehydes, starting with formaldehyde (methanal) and ending with dodecanal, are important odour-active compounds. The precursors of aldehydes are common aminocarboxylic acids, unsaturated fatty acids, sugars, and some other food components. Particularly important odour-active compounds are terpenic aldehydes.

Amino acids produce aldehydes as secondary products of alcoholic or lactic acid fermentations and via Strecker degradation during thermal processes. Formaldehyde (methanal) is formed from glycine, acetaldehyde (ethanal) from alanine, propanal and butanal from threonine (Figure 8.3), 2-methylpropanal from valine, 3-methylbutanal from leucine, 2-methylbutanal from isoleucine (**8-25**), 2-mercaptoethanal (mercaptoacetaldehyde) from cysteine (see Figure 8.75), and 3-(methylthio)propanal (methional, **8-26**) from methionine.

Autoxidation and oxidation of fatty acids catalysed by lipoxygenases also results in the formation of a number of aldehydes. Particularly important are straight-chain saturated aldehydes and various unsaturated aldehydes, such as alk-2-enals, alk-3-enals, alka-2,4-dienals, and others with two, three, or four double bonds (Tables 8.9 and 8.10). In some vegetables (e.g. cucumbers), aldehydes also result from α-oxidation of fatty acids (Figure 8.14). The primary oxidation products of essential fatty acids are hydroperoxides, which break down to aldehydes and other products under the action of lyases and can be isomerised by isomerases (Figures 8.5 and 8.6). Autoxidation of saturated fatty acids proceeds at higher temperatures. An overview of the main aldehydes produced from oleic, linoleic, and linolenic acids is given in Table 8.10. The amount of autoxidation products of unsaturated fatty acids commonly ranges from units to thousands of mg/kg. An important indicator of the rancidity of fats is malondialdehyde.

Table 8.9 Organoleptic properties and precursors of aldehydes arising from amino acids and fatty acids.

Aldehyde	Odour	Precursor
Methanal (formaldehyde)	Pungent, sharp	Glycine
Ethanal (acetaldehyde)	Pungent, fruity, fresh	Alanine
Propanal	Pungent	Threonine, linolenic acid
2-Methylpropanal	Pungent, green	Valine
3-Methylbutanal	Green, bitter almond	Leucine
2-Methylbutanal	Green, bitter almond	Isoleucine
Methional	Boiled potatoes	Methionine
2-Phenylethanal (phenylacetaldehyde)	Floral, honey	Phenylalanine
Pentanal	Pungent	Linoleic acid
Hexanal	Tallowy, green	Linoleic acid
Heptanal	Oily, greasy	Oleic acid
Nonanal	Tallowy	Linolenic acid
(*E*)-Pent-2-enal	Oily, greasy, green	Linolenic acid
(*Z*)-Hex-3-enal	Green	Linolenic acid
(*E*)-Hex-2-enal	Oily, greasy, green	Linolenic acid
(*E*)-Hept-2-enal	Oily, greasy	Linoleic acid
(*Z*)-Oct-2-enal	Walnuts-like	Linoleic acid
(*E*)-Oct-2-enal	Oily, greasy	Linoleic acid
(*E*)-Non-2-enal	Oily, greasy	Linoleic acid
(2*E*,4*Z*)-Hepta-2,4-dienal	Oily, greasy, frying fats	Linolenic acid
(2*E*,4*E*)-Hepta-2,4-dienal	Oily, greasy	Linolenic acid
(3*Z*,6*Z*)-Nona-3,6-dienal	Cucumber-like	Linolenic acid
(2*E*,6*Z*)-Nona-2,6-dienal	Cucumber-like	Linolenic acid
(2*E*,4*Z*)-Deca-2,4-dienal	Frying fats	Linoleic acid, arachidonic acid
(2*E*,4*E*)-Deca-2,4-dienal	Frying fats	Linoleic acid
(2*E*,4*Z*,7*Z*)-Deca-2,4,7-trienal	Fish oil	Linolenic acid

Volatile carbonyl compounds are also formed by degradation of sugars (e.g. formaldehyde, acetaldehyde, biacetyl, and furan-2-carbaldehyde).

$$R-CH=O$$

8-25, saturated aliphatic aldehydes
methanal (formaldehyde), R = H
ethanal (acetaldehyde), R = CH_3
propanal (propionaldehyde), R = CH_2CH_3
butanal (butyraldehyde), R = $CH_2CH_2CH_3$
2-methylpropanal (isobutyraldehyde), R = $CH(CH_3)_2$
pentanal (valeraldehyde), R = $CH_2CH_2CH_2CH_3$
2-methylbutanal, R = $CH(CH_3)CH_2CH_3$
3-methylbutanal (isovaleraldehyde), R = $CH_2CH(CH_3)_2$

Formaldehyde is present in various foods (such as milk, cheeses, and alcoholic beverages), but only in very small amounts, as it is highly reactive and enters into reaction with many food components. Acetaldehyde is a precursor of ethanol (Figure 8.2), formed in large quantities during fermentation processes in alcoholic beverages as a degradation product of sugars caused by microorganisms. Small amounts of acetaldehyde can likewise be found in fermented dairy products, such as yoghurt.

Table 8.10 Aldehydes arising by oxidation of unsaturated fatty acids.

Primarily formed aldehyde	Aldehyde after isomerisation	Hydroperoxy acid
(Z)-Undec-2-enal	(E)-Undec-2-enal	(Z)-8-Hydroperoxyoctadec-9-enoic
(E)-Dec-2-enal	–	(E)-9-Hydroperoxyoctadec-10-enoic
Nonanal	–	(E)-10-Hydroperoxyoctadec-8-enoic
Octanal	–	(Z)-11-Hydroperoxyoctadec-9-enoic
(2Z,5Z)-Undeca-2,5-dienal	(2E,4E)-Undeca-2,4-dienal	(9Z,12Z)-8-Hydroperoxyoctadeca-9,12-dienoic
(2E,4Z)-Deca-2,4-dienal	(2E,4E)-Deca-2,4-dienal	(10E,12Z)-9-Hydroperoxyoctadeca-10,12-dienoic
(Z)-Non-3-enal	(2E)-Non-2-enal	(10E,12Z)-9-Hydroperoxyoctadeca-10,12-dienoic
(Z)-Oct-2-enal	(2E)-Oct-2-enal	(9Z,12Z)-11-Hydroperoxyoctadeca-9,12-dienoic
(E)-Hept-2-enal	–	(9Z,13E)-12-Hydroperoxyoctadeca-9,13-dienoic
Hexanal	–	(9Z,11E)-13-Hydroperoxyoctadeca-9,11-dienoic
Pentanal	–	(9Z,12Z)-14-Hydroperoxyoctadeca-9,12-dienoic
(2E,4Z,7Z)-Deca-2,4,7-trienal	–	(9E,12Z,15Z)-9-Hydroperoxyoctadeca-10,12,15-trienoic
(3Z,6Z)-Nona-3,6-dienal	(2E,6Z)-Nona-2,6-dienal	(8E,12Z,15Z)-10-Hydroperoxyoctadeca-8,12,15-trienoic
(2Z,5Z)-Octa-2,5-dienal	(2E,4E)-Octa-2,4-dienal	(9Z,12Z,15Z)-11-Hydroperoxyoctadeca-9,12,15-trienoic
(2E,4Z)-2,4-Heptadienal	(2E,4E)-Hepta-2,4-dienal	(9Z,13E,15Z)-12-Hydroperoxyoctadeca-9,13,15-trienoic
(Z)-Hex-3-enal	(E)-Hex-2-enal	(9Z,11E,15Z)-13-Hydroperoxyoctadeca-9,11,15-trienoic
(Z)-Pent-2-enal	(E)-Pent-2-enal	(9Z,12Z,15Z)-14-Hydroperoxyoctadeca-9,12,15-trienoic
Propanal	–	(9Z,12Z,14E)-16-Hydroperoxyoctadeca-9,12,14-trienoic

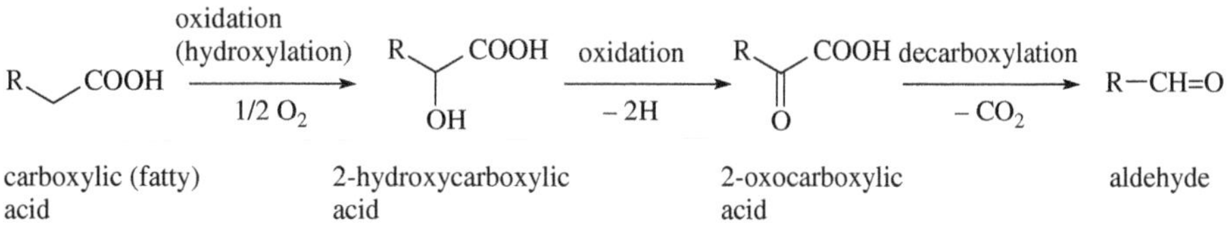

Figure 8.14 Formation of aldehydes by α-oxidation of fatty acids.

The simplest representative of alk-2-enals is propenal (acrolein). It is formed from heat-processed fats and oils or directly by dehydration of glycerol (Figure 8.15); for example, 23.4 mg of acrolein/kg arise in rapeseed oil after 24 hours of heating at 180 °C. Small amounts of acrolein are also present in alcoholic beverages, such as beer, wine, and spirits. Acrolein is an irritating aldehyde with a disagreeable odour and toxic effects. According to the International Agency for Research on Cancer (IARC), it is not classifiable with respect to its carcinogenicity to humans.

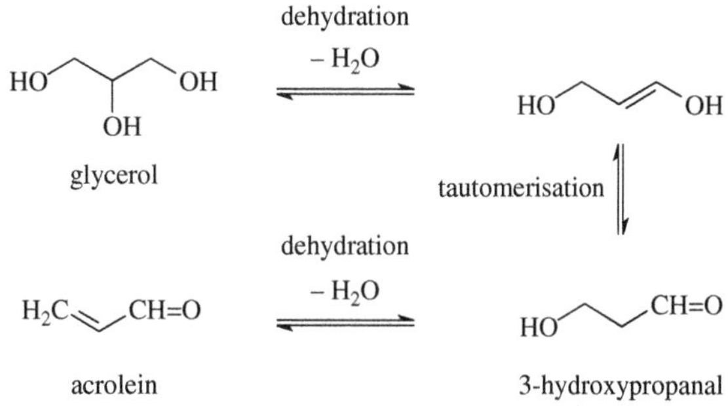

Figure 8.15 Formation of acrolein from glycerol.

The higher homologue of acrolein is (E)-but-2-enal (crotonaldehyde). It is formed by aldolisation of acetaldehyde and dehydration of the aldolisation product, which is facilitated by the presence of conjugated double bonds (see Figure 8.31). In heat-processed fats and oils, the main source of crotonaldehyde seems to be linolenic acid (see Figure 12.6). The amount of crotonaldehyde formed depends on the type of oil, applied temperature, and time. For example, about 30 mg/kg of crotonaldehyde is produced from coconut or linseed oils after heat processing for 24 hours at 180 °C. Crotonaldehyde is also present in fried food, occuring in potato chips and doughnuts in concentrations of 12–25 and 8–19 µg/kg, respectively.

Some aldehydes that are formed from oxidised oils and fats are desirable flavour-active compounds in low concentrations. The odour qualities in the series of alk-3-enals (and related alcohols) change from green and grassy to an overall citrus-like, fresh, soapy, and coriander-like odour with increasing chain length. An example is (Z)-hex-3-enal, also known as leaf aldehyde, which contributes to the pleasant fresh, herbal scent of fresh vegetables and fruits, called green odour, and acts as an attractant to many insects (Figure 8.6). It is produced in small amounts by most plants from (13S)-hydroperoxide of α-linolenic acid, which is the main fatty acid of green plants, by hydroperoxide lyase. Other compounds responsible for the green odour are hexanal (Figure 8.5), (E)-hex-2-enal (Figure 8.6) and (E)-pent-2-enal (8-27), along with some alcohols. The cleavage of (9S)-hydroperoxide of α-linolenic acid provides (3Z,6Z)-nona-3,6-dienal with an oily and green odour (Table 8.9). Its isomerisation provides (2E,6Z)-nona-2,6-dienal (8-27) with a typical aroma of fresh cucumbers. The mixture of arachidonic acid oxidation products, (Z)-non-3-enal, (Z)-dec-4-enal, (2Z,5Z)-undeca-2,5-dienal, and (2E,4Z,7Z)-2,4,7-tridecatrienal (8-27) resembles the aroma of cooked chicken. Both isomers of deca-2,4-dienal (8-27) that arise from linoleic and arachidonic acids contribute to the pleasant smell of fried foods. The key odour-active component of oatmeal is (2E,4E,6Z)-nona-2,4,6-trienal (8-39), formed by oxidation of linolenic acid. It has a sweet and cereal-like smell. Higher aliphatic aldehydes, such as 12-methyltridecanal, have an odour resembling beef. They are formed during the thermal processing of meat by degradation of plasmalogens.

Aldehydes produced from fatty acids are also principles of a rancid odour and taste. The organoleptic properties of some aldehydes produced by oxidation of fatty acids and organoleptic properties of the so-called Strecker aldehydes are listed in Table 8.9, together with their precursors. Sometimes, they are formed even from minor unsaturated fatty acids. In beef and mutton tallow and butter, for example, small amounts of (11Z,15Z)-octadeca-11,15-dienoic acid occur, autoxidation of which yields (Z)-hept-4-enal (8-27), an important bearer of the rancid smell of these fats. Important carriers of the so-called hydrogenation flavour of hydrogenated soybean oils are (E)-non-6-enal (8-27), which arises from (9Z,15E)-octadeca-9,15-dienoic acid; this does not occur in natural fats, but is a product of the partial reduction of linolenic acid. An important contribution to the hydrogenation flavour of soybean oil comes from (2E,6E)-octa-2,6-dienal, formed from (12E,16E)-11-hydroperoxyoctadeca-12,16-dienoic acid.

A component of the aroma of some vegetables (e.g. tomatoes) is the sulfur-containing aldehyde 2-methylthioethanal (8-26), which arises by Strecker degradation of S-methylcysteine and its sulfoxide. The higher homologue of 2-methylthioethanal is methional (3-methylthiopropanal, 8-26), which is an important aroma component of cooked potatoes and many other foods. However, the presence of methional in some foods has a negative effect. For example, it is the key compound responsible for the defect of alcohol-free beers described as worty flavour. Methional (formed by riboflavin-induced photooxidation of methionine) and its degradation product dimethyl disulfide give a burnt and oxidised odour to milk, as well as wine and beer. The other off-flavour in milk is cardboard-like or metallic flavour, which develops on prolonged exposure to light. Compounds responsible for this off-flavour are secondary lipid oxidation products, such as hexanal, pentanal, and some other substances formed by riboflavin-catalysed photooxidation of fatty acids.

8-26, methylthioalkanals

2-methylthioethanal, n = 0
3-methylthiopropanal (methional), n = 1

(E)-pent-2-enal   (Z)-hept-4-enal   (2E,4Z)-hepta-2,4-dienal   (Z)-non-3-enal

(E)-non-6-enal   (3Z,6Z)-nona-3,6-dienal   (2E,6Z)-nona-2,6-dienal

(2E,4E,6Z)-nona-2,4,6-trienal   (Z)-dec-4-enal   (2E,4Z)-deca-2,4-dienal

(2E,4E)-deca-2,4-dienal   (2Z,5Z)-undeca-2,5-dienal   (2E,4Z,7Z)-trideca-2,4,7-trienal

8-27, selected unsaturated aldehydes arising from polyenoic fatty acids

Terpenic aldehydes Monoterpenic and some sesquiterpenic aldehydes are important flavour-active compounds with practical significance. The aliphatic aldehyde derived from (+)-citronellol is (+)-citronellal (**8-28**), which represents about 32–45% of citronella oil from the leaves and stems of grasses of the genus *Cymbopogon* (such as *C. citratus*, Poaceae), better known as lemongrass, which is widely used as a culinary herb in Asian cuisines. (−)-Citronellal (**8-28**) occurs in amounts of up to 80% in the essential oil found in the subtropical Australian bush, *Backhousia citriodora* (lemon myrtle, Myrtaceae). Its crushed leaves have a culinary use.

The aliphatic unsaturated aldehyde citral is one of the most frequently occurring compounds. Natural citral is always a mixture of geranial (**8-28**), also known as (*E*)-citral or citral a, and neral (**8-28**), known as (*Z*)-citral or citral b. Citral occurs in the already mentioned citronella oil in concentrations of 11–13%. In much smaller amounts, it is a component of many essential oils, especially oils of citrus fruits, as well as of ginger and pepper essential oils. It also occurs in wine and other products.

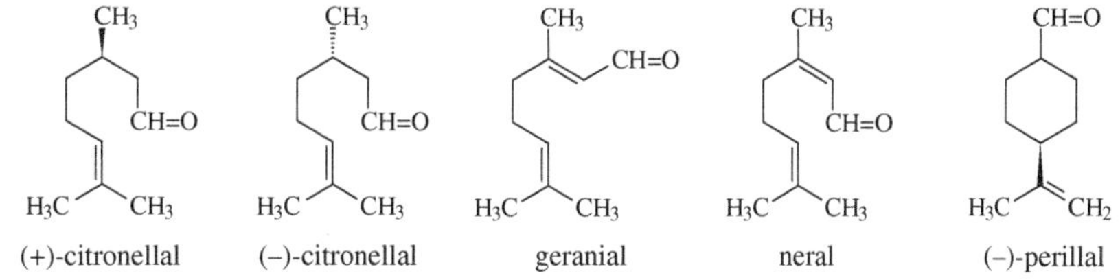

8-28, monoterpenic aldehydes

The essential oil from the leaves of the edible African plant *Perilla frutescens* (Lamiaceae), which has a strong minty smell, contains as a main component (50–60%) monocyclic (−)-perillyl aldehyde (perillal, **8-28**), whose biosynthesis starts with (−)-limonene and proceeds via (−)-perillyl alcohol as intermediate. The leaves of various perilla varieties (such as *P. f.* var. *crispa*) are used as a vegetable, and the seeds as a source of cooking oil.

A significant acyclic sesquiterpenic aldehyde occurring in orange essential oil is α-sinensal (**8-29**), which is formed by oxidation of (3*E*,6*E*)-α-farnesene (**8-5**). The mandarin essential oil contains β-sinensal (**8-29**), which is produced by oxidation of (*E*)-β-farnesene (**8-5**). Some other sesquiterpenic aldehydes are minor but still significant components of citrus oils, such as monocyclic aldehyde lanceal (**8-29**) and bicyclic aldehyde bergamotenal (**8-29**).

The alicyclic monoterpenic aldehyde safranal has a different biochemical origin, and is the main characteristic odorous component of saffron. Safranal is classified as a degraded carotenoid (apocarotenoid) as it is produced from zeaxanthin via hydrolysis of the bitter intermediate picrocrocin. Degradation of carotenoids produces a number of other aromatic compounds (see Section 9.9.5.2). A list of the names of major terpenoid aldehydes is given in Table 8.11.

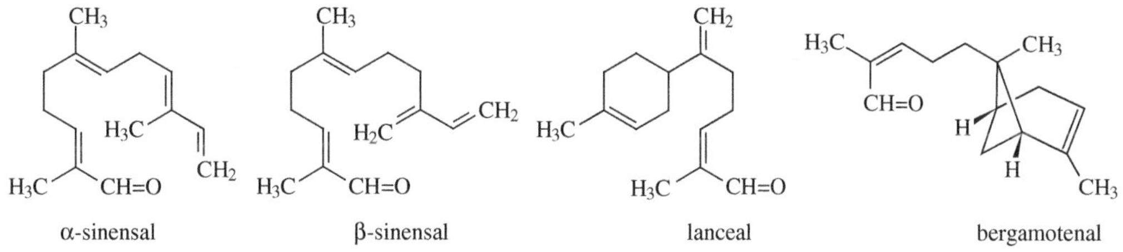

8-29, sesquiterpenic aldehydes

Table 8.11 Trivial and systematic names of selected terpenoid aldehydes.

Trivial name	Systematic name (IUPAC)
(+)-Citronellal	(*R*)-3,7-Dimethyloct-6-enal
(−)-Citronellal	(*S*)-3,7-Dimethyloct-6-enal
Geranial	(*E*)-3,7-Dimethylocta-2,6-dienal
Neral	(*Z*)-3,7-Dimethylocta-2,6-dienal
(−)-Perillal	(*S*)-4-Prop-1-en-2-ylcyclohex-1-ene-1-carbaldehyde
α-Sinensal	(2*E*,6*E*,9*E*)-2,6,10-Trimethyldodeca-2,6,9,11-tetraenal
β-Sinensal	(2*E*,6*E*)-2,6-Dimethyl-10-methylidenedodeca-2,6,11-trienal
Lanceal	(*Z*)-6-Methyl-2-(4-methylcyclohex-3-en-1-yl)hepta-1,5-dienal
Bergamotenal	(1*S*,3*E*,5*S*)-4-Formyl-7-methyl-7-(4-methylpent-3-enyl)bicyclo[3.1.1]hept-3-ene

Aromatic aldehydes Aromatic aldehydes very often occur in essential oils and as odour-active components of different foods. Their biosynthesis is based, in principle, on transformations of phenylalanine, the key decomposition product of which is cinnamic acid (Figure 8.10).

A widespread aromatic aldehyde is benzaldehyde (**8-30**), which is produced by the reduction of benzoic acid, the precursor of which is cinnamic acid. Benzaldehyde can also be formed thermally from phenylalanine, phenylacetaldehyde, and some other compounds. Benzaldehyde is present in bound form in some cyanogenic glycosides and is released by their hydrolysis (see Section 10.3.9.1.2). It is therefore a key component of bitter almond oil and is also present in the essential oil of cinnamon, and it occurs as an important odorous component of all alcoholic beverages obtained by the fermentation of stone fruits (such as plum brandy).

Alkyl derivatives of benzaldehyde include cuminaldehyde (4-isopropylbenzaldehyde, **8-30**), which is a component of cumin, cinnamon, and basil essential oils (see Table 8.32). Anisaldehyde (4-methoxybenzaldehyde, **8-30**) is a fragrance ingredient of anise, star anise, vanilla, and some other essential oils. A higher homologue of benzaldehyde phenylethanal (phenylacetaldehyde, **8-30**) is a common component of many essential oils. It has a floral aroma resembling hyacinth flowers and has therefore been used in perfumery. The cinnamon essential oil also contains its higher homologue, 3-phenylpropanal (dihydrocinnamaldehyde, **8-30**), which accompanies (together with cuminaldehyde) the key odorant of cinnamon, cinnamaldehyde (**8-30**). Cinnamon essential oil (representing about 0.5–1% by weight of cinnamon bark) contains around 90% cinnamaldehyde. The dominant isomer is the (*E*)-isomer. However, in fresh cinnamon bark, cinnamyl acetate is predominant. Cinnamaldehyde arises from this ester by enzymatic hydrolysis during fermentation and reduction of the resulting cinnamic acid by aldehyde–alcohol oxidoreductase. Without the participation of enzymes, cinnamaldehyde is produced by Strecker degradation of phenylalanine.

benzaldehyde   cuminaldehyde   anisaldehyde   phenylethanal   3-phenylpropanal   (*E*)-cinnamaldehyde

8-30, selected aromatic aldehydes

Other cinnamic acids also produce aldehydes, but they have no significance as odour-active substances. By analogy, 4-hydroxycinnamaldehyde (coumaraldehyde) is produced from 4-coumaric acid, caffeoyl aldehyde from caffeic acid, coniferyl aldehyde (also known as ferulyl aldehyde) from ferulic acid, 5-hydroxyconiferyl aldehyde from 5-hydroxyferulic acid, and sinapyl aldehyde from sinapic acid. These aldehydes can be reduced to the corresponding alcohols, with which they play a role as the building units of lignin biosynthesis.

Some hydroxyaldehydes derived from benzaldehyde are also important: these can be simultaneously classified as phenols. The main characteristic component of vanilla beans (*Vanilla planifolia*, Orchidaceae) is vanillin (**8-31**; see Table 8.32). Vanillin is present in the green pods of orchids exclusively in the conjugated form as β-D-glucopyranoside, called glucovanillin (also known as avenin, **8-31**). The characteristic aroma of vanilla appears after fermentation (hydrolysis by β-D-glucosidase). The vanillin content in fermented pods is usually 2–2.5%. Low-quality species of vanilla also contain the aromatic aldehyde piperonal (heliotropin, **8-130**). Ethylvanillin (bourbonal, **8-130**) has an odour resembling vanillin, but slightly rougher. It does not occur in nature, but synthetic ethylvanillin is used as food flavouring, especially in the production of vanilla sugar.

vanillin   glucovanillin

8-31, vanillin and its glucoside

Heterocyclic aldehydes Carbohydrate-containing foods often contain furan-2-carbaldehyde arising from pentoses and ascorbic acid as a dehydration product. Other common furan-derived aldehydes are 5-hydroxymethylfuran-2-carbaldehyde (resulting from hexoses) and 5-methylfuran-2-carbaldehyde (from 6-deoxyhexoses). Many other heterocyclic aldehydes, derived from pyrrole, thiophene, pyridine, pyrazine, and other heterocyclic compounds, are the Maillard reaction products.

8.2.4.1.2 Ketones

As with aldehydes, various ketones can occur as primary constituents of food raw materials and foods, or may arise as secondary compounds in various reactions. Many ketones are categorised by a characteristic odour and therefore may be either desirable or undesirable substances. An important group originating from the degradation of sugars (e.g. abhexon and sotolon) is described in Section 8.2.7.1.2, dealing with lactones. A number of odour-active ketones arise by the degradation of fatty acids and carotenoids.

Aliphatic and alicyclic ketones Foods frequently contain saturated and unsaturated aliphatic ketones with between 3 and 17 carbon atoms in the molecule. Frequently occurring aliphatic ketones are methylketones, the most common of which is acetone (propanone, **8-32**, $n = 0$). Acetone is present, usually in small quantities, in all biological substrates, where it arises by decarboxylation of acetoacetic (3-oxobutanoic) acid. Acetoacetic acid is formed as an intermediate during degradation of fatty acids by β-oxidation. Acetone in the skins of apples is produced from pyruvic acid via citramalic acid (Figure 8.16). The relatively large amount is generated by acetone–butanol fermentation. Many other methylketones occur as odour-active components of essential oils. For example, a component of cinnamon and star anise essential oils is heptan-2-one, also known as methyl pentyl ketone (**8-32**, $n = 4$).

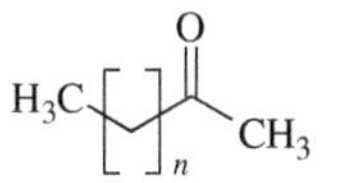

8-32, methylketones

Many methylketones occur in foods as products of lipid degradation during so-called ketonic (perfume) rancidity. These methylketones have higher odour detection threshold concentrations than the corresponding aldehydes, but nevertheless play important roles as flavour-active components of some foods. They appear as desirable flavour components in blue cheeses (such as Roquefort, Stilton, Cabrales, Danish blue, and Gorgonzola), the manufacture of which involves the spores of the moulds *Penicillium roqueforti* and *Penicillium glaucum*. Significant methylketones (**8-32**) in blue cheeses are mainly acetone ($n = 0$), pentan-2-one ($n = 2$), heptan-2-one ($n = 4$) nonan-2-one ($n = 6$), and undecan-2-one ($n = 8$). The amount of individual compounds is between 5 and 180 mg/kg of d.w. The latter three methylketones are also used for aromatisation of cheeses.

The presence of methylketones is undesirable in some fats, such as butter, coconut oil, and palm oil. Also undesirable are unsaturated ketones produced by oxidation of fatty acids, which are called alkyl vinyl ketones. For example, pent-1-en-3-one (**8-33**), an impact aroma compound of orange juice, grapefruits, black tea, olive oil, and tomatoes, smells of fish oil; oct-1-en-3-one (**8-33**) has a strong metallic mushroom-like odour; and (*Z*)-octa-1,5-dien-3-one (**8-33**) has an oily, greasy, and geranium-like metallic odour.

A related ketone formed by oxidation of fatty acids is (*E*)-5-methylhept-2-en-4-one (**8-33**), known as filbertone (Table 8.1). Filbertone is a key component of the odour of hazelnuts and some chocolate products, such as nougat. Its content in raw nuts is low (about 1.4 μg/kg), but increases to 660 μg/kg in nuts roasted at 180 °C for 9 minutes and to 1150 μg/kg in nuts roasted for 15 minutes. Hazelnuts contain a mixture of (+)-(2*E*,5*S*)- and (−)-(2*E*,5*R*)-isomers (**8-33**). The enantiomeric composition of raw hazelnuts falls in the ratio of 80–85% (+)-(*E*,*S*) and 15–20% (−)-(*E*,*R*), whereas for roasted hazelnuts the ratio is about 71.5–72.5% (+)-(*E*,*S*) and 27.5–28.5% (−)-(*E*,*R*). Filbertone is also present in hazelnut oils. Oils from raw nuts contain filbertone at a level of <10 μg/kg, but its content in oils from roasted nuts is higher (around 315 μg/kg). Identification of filbertone as a characteristic marker has been used to detect the presence of cheaper hazelnut oil in

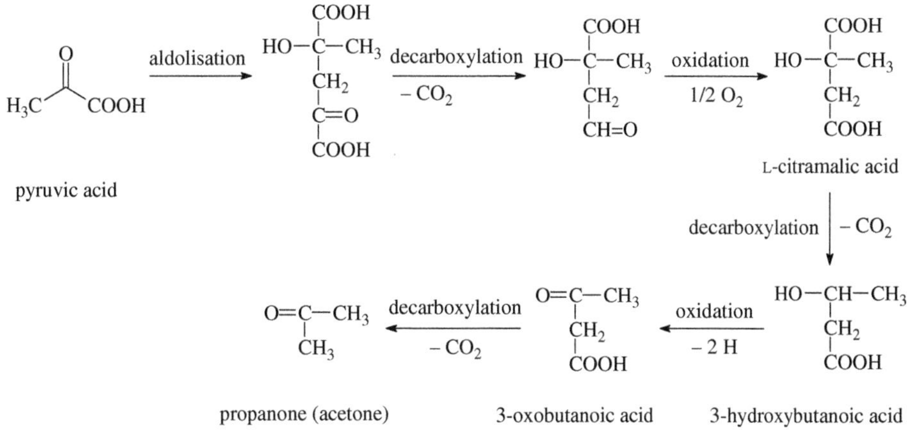

Figure 8.16 Formation of acetone from pyruvic acid.

olive oil, hazelnuts in cocoa spreads, and hazelnut traces in products during monitoring for hazelnut allergies (known to be one of the most common types of food allergy).

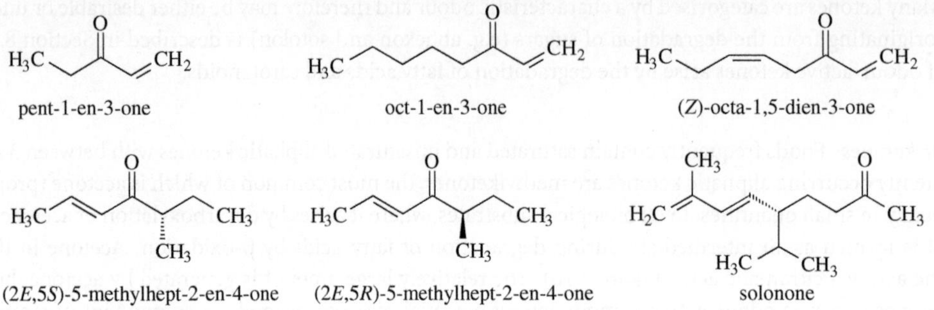

pent-1-en-3-one　　　　　oct-1-en-3-one　　　　　(Z)-octa-1,5-dien-3-one

(2E,5S)-5-methylhept-2-en-4-one　　　(2E,5R)-5-methylhept-2-en-4-one　　　solonone

8-33, aliphatic unsaturated ketones

A number of unsaturated aliphatic and alicyclic ketones are formed by degradation of carotenoid pigments. For example, components of tomato flavour are unsaturated methylketones (E)-6-methylhept-5-en-2-one, (3E,5E)-6-methylhepta-3,5-dien-2-one, and other methylketones, and the flavour-active component of tobacco is (5R,6E)-5-isopropyl-8-methylnona-6,8-dien-2-one, known as solonone (**8-33**). Alicyclic ketones, which are the result of degradation of carotenoid pigments, include flavour-active ionones and damascones (see Fig. 9.29).

An important alicyclic ketone obtained from the essential oil of jasmine flowers (*Jasminum grandiflorum*, Oleaceae) is (Z)-3-methyl-2-(pent-2-en-1-yl)-2-cyclopenten-1-one, called *cis*-jasmone (**8-34**), which belongs to the group of structurally related compounds called jasmonoids (see Section 3.8.1.8.5) that are used in perfumery and cosmetics. Jasmine flowers are used to scent tea in China.

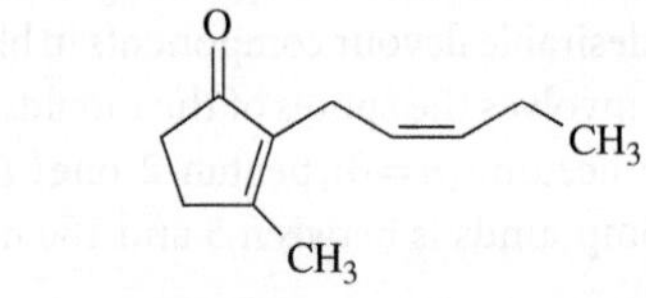

8-34, (Z)-jasmone

Cyclotene (2-hydroxy-3-methylcyclopent-2-en-1-one, Figure 4.54) is a secondary transformation product of sugars, the odour of which resembles caramel. It is a characteristic aroma component of maple syrup obtained from the sugar maple tree (*Acer saccharum*, Aceraceae).

Unusual genotoxic ketones are 2-alk(en)ylcyclobutanones (**8-35**), which develop in irradiated meat and fish. The list of foods and food ingredients that may be treated with ionising radiation in the EU (Directives 1999/2/EC and 1999/3/EC) does not include meat and meat products, except for chicken meat (the Netherlands), mechanically recovered chicken meat (France), and poultry (France and the United Kingdom). 2-Alkylcyclobutanones are formed by the loss of an electron from the oxygen on the carbonyl of a free fatty acid or fatty acid bound in triacylglycerol, followed by a rearrangement to produce products having the same number of carbon atoms as the parent fatty acid. For example, 2-dodecylcyclobutanone, 2-tetradecylcyclobutanone, and (Z)-2-(tetradec-5′-en-1-yl)cyclobutanone have been identified in irradiated cured pork products. The amount of 2-dodecylcyclobutanone in salami irradiated at a dose of 5 kGy was 0.026 mg/kg immediately after the irradiation and 0.068 mg/kg after 60 days of storage.

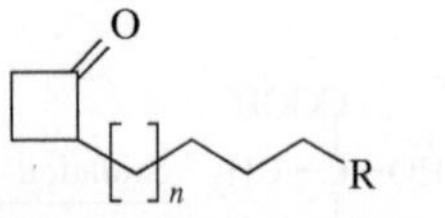

8-35, 2-alk(en)ylcyclobutanones

2-dodecylcyclobutanone, R = CH$_3$, n = 8
2-tetradecylcyclobutanone, R = CH$_3$, n = 10
(Z)-2-(tetradec-5′-en-1-yl)cyclobutanone, R = CH=CH [CH$_2$]$_7$CH$_3$, n = 1

Terpenic ketones Monoterpenic ketones (**8-36**) are frequently very important aromatic substances of many food raw materials, spices, and medicinal herbs. Monocyclic ketone (+)-carvone is a key component of caraway and dill oils, where it is accompanied by (+)-(E)-dihydrocarvone (i.e. (1S,4S)-dihydrocarvone), which also possesses caraway odour. (−)-Carvone is a typical component of spearmint essential oil, where it is accompanied by (−)-(E)-dihydrocarvone, (1R,4R)-dihydrocarvone, which has the same odour (see Table 8.32). Isomeric (−)-menthone and its precursor (+)-pulegone occur in the essential oils of the European pennyroyal, a traditional culinary herb and folk remedy, and other *Mentha* species, as well as in marjoram essential oil. In the essential oil of mint, (−)-menthone is accompanied by (+)-isomenthone.

An example of bicyclic monoterpenic ketones is thujone, which occurs as four stereoisomers, (−)-α-thujone, (+)-α-thujone, (−)-β-thujone, and (+)-β-thujone (**8-37**), in the essential oil of common (garden) sage (see Table 8.32). (+)-α-Thujone and (−)-β-thujone do not occur in other common thujone-containing essential oils, such as species of the genus *Artemisia* (Asteraceae), including mugwort, wormwood, and sagebrush. Thujones show neurotoxic effects and therefore appear in the list of substances of Regulation (EC) No. 1334/2008 that should not be added to foods as flavourings. Their natural content is restricted to 35 mg/kg in alcoholic beverages and to 0.5 mg/kg in non-alcoholic beverages produced from *Artemisia* species.

Bicyclic ketone camphor (**8-37**), formed by oxidation of borneol, is a component of cinnamon, sage, and rosemary essential oils. In nature, camphor is formed by the oxidation of borneol, and usually it occurs as a mixture of two isomers, (+)-camphor, which is more common, and (−)-camphor. Camphor is obtained from the camphor laurel tree wood (*Camphor officinalis*, syn. *C. camphora*, Lauraceae) originating in Southeast Asia. A further sesquiterpenic ketone is (−)-fenchone (see **8-51**), which is found in many essential oils, such as fennel essential oil.

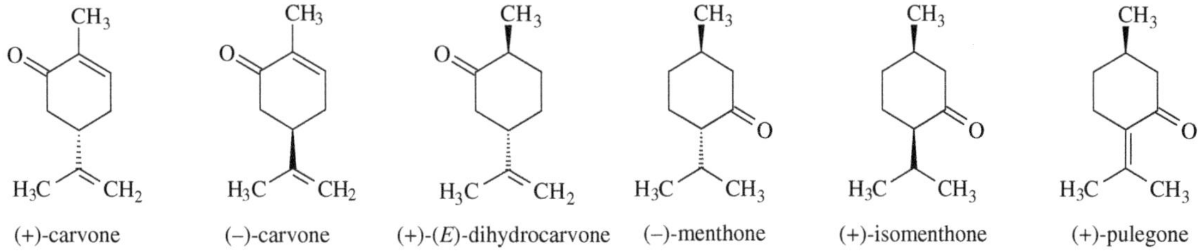

| (+)-carvone | (−)-carvone | (+)-(*E*)-dihydrocarvone | (−)-menthone | (+)-isomenthone | (+)-pulegone |

8-36, cyclic monoterpenic ketones

| (−)-α-thujone | (+)-β-thujone | (+)-α-thujone | (−)-β-thujone |

| (+)-camphor | (−)-camphor | (−)-fenchone |

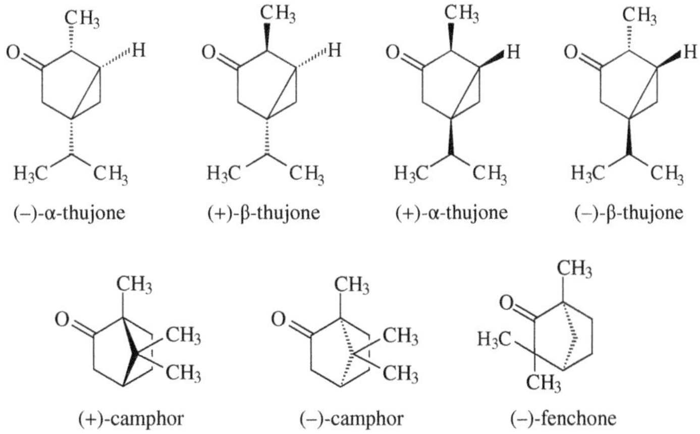

8-37, bicyclic monoterpenic ketones

The monocyclic sesquiterpenic ketone turmerone is the main component (25–50%) of turmeric essential oil (see Table 8.32). Turmerone is a mixture of α-turmerone, β-turmerone (curlone), and (+)-*ar*-turmerone (**8-38**). The ratio between these three components is approximately 3.5 : 1 : 1.5. *ar*-Turmerone is formed by oxidation of hydrocarbon (+)-*ar*-curcumene, which is present in turmeric rhizomes at a level of about 6%. The cyclobisabolane ketol bicycloturmeronol (**8-38**) is a recently identified turmeric constituent. Bicyclic sesquiterpenic ketone is (+)-nootkatone (**8-38**), an important odorous compound of grapefruits, in which it is formed by oxidation of hydrocarbon (+)-valencene via alcohol (+)-β-nootkatol. Another oxidation product of valencene is (+)-8,9-didehydronootkatone (**8-38**). The trivial and systematic names of important terpenic ketones are listed in Table 8.12.

Aromatic ketones The basic substance is acetophenone, also known as phenyl methyl ketone (**8-39**), which occurs in a small amount in some essential oils. A component of fennel (see Table 8.32) and star anise essential oils is anise ketone, also known as 4-methoxyphenylacetone (**8-39**). 1-(4-Hydroxyphenyl)butan-3-one, called raspberry ketone (**8-39**), is a natural component of raspberry aroma.

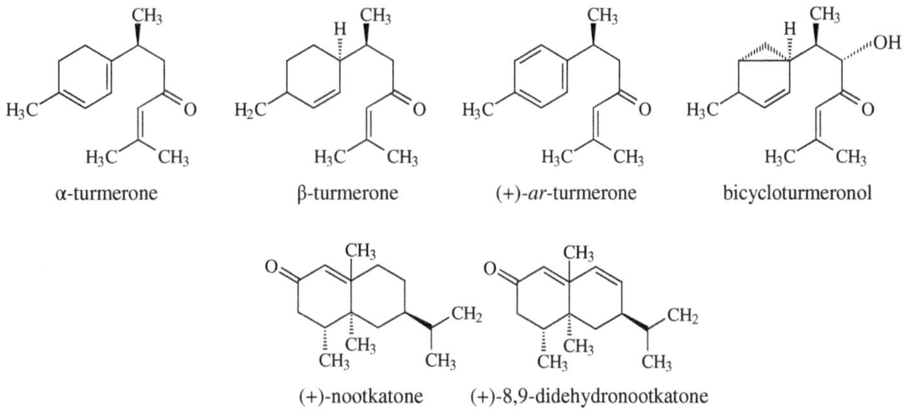

| α-turmerone | β-turmerone | (+)-*ar*-turmerone | bicycloturmeronol |

| (+)-nootkatone | (+)-8,9-didehydronootkatone |

8-38, sesquiterpenic ketones

Table 8.12 Trivial and systematic names of selected terpenic ketones.

Trivial name	Systematic name (IUPAC)
(−)-Fenchone	(1R,4S)-1,3,3-Trimethylbicyclo[2.2.1]heptan-2-one
(+)-Isomenthone	(2R,5R)-5-Methyl-2-propan-2-ylcyclohexan-1-one
(+)-Camphor	(1R,4S)-1,7,7-Trimethylbicyclo[2.2.1]heptan-2-one
(+)-Carvone	(S)-2-Methyl-5-prop-1-en-2-ylcyclohex-2-en-1-one
(−)-Carvone	(R)-2-Methyl-5-prop-1-en-2-ylcyclohex-2-en-1-one
(−)-Menthone	(2S,5R)-5-Methyl-2-propan-2-ylcyclohexan-1-one
(+)-Nootkatone	(4R,4aS,6R)-4a,5-Dimethyl-1,2,3,4,4a,5,6,7-octahydro-7-oxo-3-prop-1-en-2-ylnaphthalene
(+)-Pulegone	(R)-5-Methyl-2-propan-2-ylidencyclohexan-1-one
(−)-α-Thujone	(1S,4R,5R)-Thujan-3-one; [1S-(1α,4α,5α,)]-4-methyl-1-propan-2-ylbicyclo[3.1.0]hexan-3-one
(+)-β-Thujone	(1S,4S,5R)-Thujan-3-one; [1S-(1α,4β,5α)]-4-methyl-1-propan-2-ylbicyclo[3.1.0]hexan-3-one
(+)-ar-Turmerone	(S)-2-Methyl-6-(4-methylphenyl)hept-2-en-4-one
(+)-(E)-Dihydrocarvone	(2S,5S)-2-methyl-5-prop-1-en-2-ylcyclohexanone
(−)-(E)-Dihydrocarvone	(2R,5R)-2-methyl-5-prop-1-en-2-ylcyclohexanone

acetophenone   anise ketone   raspberry ketone

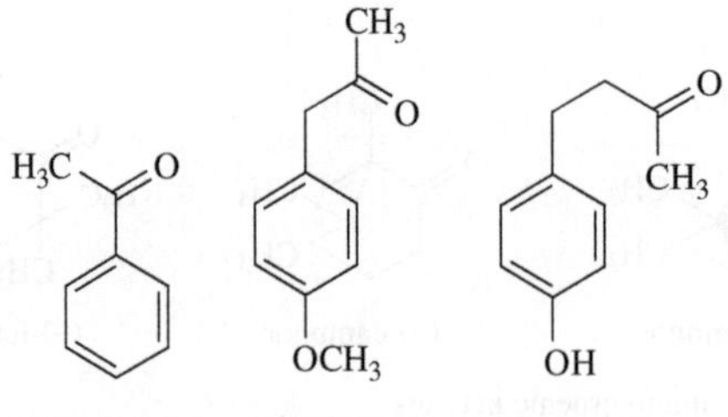

8-39, aryl-containing ketones

Hydroxycarbonyl and dicarbonyl compounds Common α-hydroxycarbonyl compounds in foods include various acyloins (**8-40**), also referred to as ketols or α-hydroxyketones, which together with related α-dicarbonyl compounds (**8-40**) can be considered oxidation products of ethylene glycol (ethane-1,2-diol) and other polyols, but which are produced from other precursors. Some important hydroxycarbonyl and dicarbonyl compounds are formed as physiological secondary fermentation metabolites that can be either desirable or undesirable substances in foods.

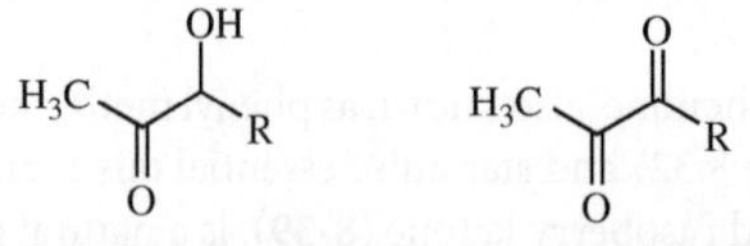

α-hydroxyketones (acyloins) α-dicarbonyls

8-40, hydroxycarbonyl and dicarbonyl compounds

Important three-carbon compounds are methylglyoxal and products of its reduction, acetol and D-lactic acid aldehyde (D-lactaldehyde) (Figure 8.11). The four-carbon compounds comprise acetoin (2-hydroxybutanone) and its oxidation product biacetyl (Figure 8.12). Important compounds are also their higher homologues, 3-hydroxypentan-2-one and pentane-2,3-dione. Acetoin and biacetyl are flavour-active compounds that have a pleasant butter-like odour and creamy taste in diluted solutions and are actually components of butter volatiles. For example, the biacetyl content in butter made from sour cream is around 4 mg/kg. In alcoholic beverages, such as beer and wine, biacetyl is an undesirable substance as it imparts a characteristic buttery aroma to beer when present above its odour-detection threshold (around 0.1–0.2 ppm in lager and 0.1–0.4 ppm in ales). Biacetyl at detectable concentrations is acceptable in some beer styles, such as Bohemian pilsner and some English ales. Biacetyl is an indicator of unwanted bacterial contamination. The contents of the main hydroxycarbonyl and dicarbonyl compounds in wine are given in Table 8.13. Red wines tend to have slightly higher levels of acetoin and biacetyl than white wines. Pentane-2,3-dione has a similar flavour to biacetyl, although it is often described as more toffee-like, but with a higher flavour threshold.

Table 8.13 Contents of main hydroxycarbonyl and dicarbonyl compounds in wines.

Compound	Content (mg/l)
Propane-1,2-dione (methylglyoxal)	0.06–0.42
3-Hydroxybutan-2-one (acetoin)	0.7–53
Butane-2,3-dione (biacetyl)	0.1–7.5
3-Hydroxypentan-2-one	0.5–2.8
Pentane-2,3-dione	7.0–18

In addition to the already mentioned acyloins, food products contain numerous other odorous α-hydroxyketones. For example, 3-hydroxypentan-2-one has ben identified in cheeses, durian, wines, sherry, asparagus, honey, tea, butter, and soy sauce, 2-hydroxypentan-3-one in cheeses, durian, coffee, wine, sherry honey, butter, and soy sauce, 2-hydroxyhexan-3-one in wine, 4-hydroxyhexan-3-one in durian and tea, 3-hydroxy-5-methylhexan-2-one in cheeses, 3-hydroxyoctan-2-one in beef and heated mutton fat, 5-hydroxyoctan-4-one in cocoa, and 3-hydroxy-4-phenylbutan-2-one in wine, sherry, and honey.

Some α-hydroxycarbonyl and α-dicarbonyl compounds, such as glycolaldehyde, 1-hydroxypropanone (hydroxyacetone or acetol), glyoxal (oxalic acid dialdehyde), methylglyoxal (pyruvic acid aldehyde), and 3-hydroxybutan-2-one (acetoin) are important products of carbohydrate degradation. These compounds are generally not characterised by distinct organoleptic properties, but are precursors to many important compounds, especially heterocyclic volatiles formed in the Maillard reaction. The 1,2-dicarbonyl compound glyoxal, the 1,3-dicarbonyl compound malondialdehyde, and various ketols are similarly formed as products of fatty acid oxidation.

8.2.4.2 Properties and reactions

Aliphatic aldehydes with 1–7 carbon atoms in the molecule generally have a sharp, pungent, and sometimes rancid aroma, whilst aldehydes with 8–14 carbon atoms (including monoterpenic aldehydes) are generally characterised by a pleasant aroma. Higher aldehydes are usually odourless.

Symmetric and asymmetric aliphatic ketones of low molecular weight have a pleasant smell. Higher ketones, especially methyl ketones, have a characteristic smell, which is sometimes desirable and sometimes not. Flavour-active substances in the essential oils of different spices are various monoterpenic ketones. Aliphatic ketones of medium chain length and terpenic ketones are used as flavouring substances in foods and cosmetic products. In the food flavouring industry, some α-hydroxycarbonyl compounds (e.g. acetoin) and some α-dicarbonyl compounds (e.g. biacetyl) are usually mixed with other ingredients to produce a butter flavour or other flavours in a variety of food products, including butter made from sweet cream, margarine, and popcorn. Pentane-2,3-dione does not have a significant impact on the flavour of foods.

The most important reactions of carbonyl compounds are addition, oxidation, and reduction taking place on the carbonyl group. Owing to the presence of free electron pairs on the carbonyl-group oxygen, carbonyl compounds are weak bases, and in acidic media they may exist in protonated forms (as conjugated acids; Figure 8.17). Bases split the acidic hydrogen on the carbons adjacent to the carbonyl group to form reactive carbanions (Figure 8.18). Cleavage of hydrogen from the carbon in position α to the carbonyl group and its addition to the carbanion oxygen leads to an equilibrium between the oxo form and the enol form. This isomerisation reaction is called enolisation (Figure 8.19). Enolisation is catalysed by both acids and bases, and thus is an acid–base-catalysed reaction. Under normal conditions, the equilibrium in simple aldehydes and ketones is almost completely shifted in favour of carbonyl compounds. However, if the methylene-group hydrogens are activated, as is the case with malondialdehyde, the proton splits easily, and the dominating form of such a carbonyl compound is then the thermodynamically stable enol form.

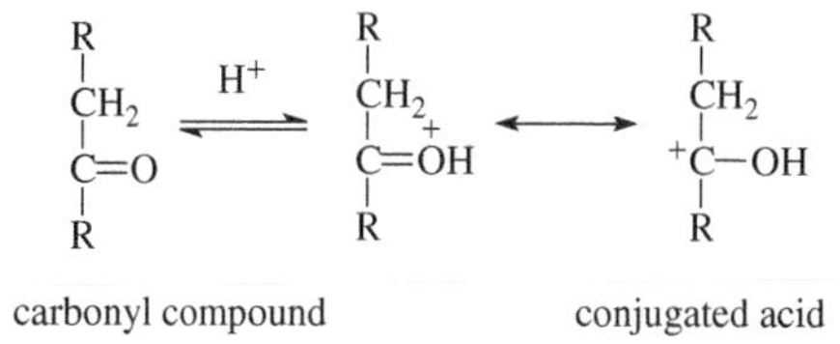

Figure 8.17 Formation of conjugated acids from carbonyl compounds in acidic media.

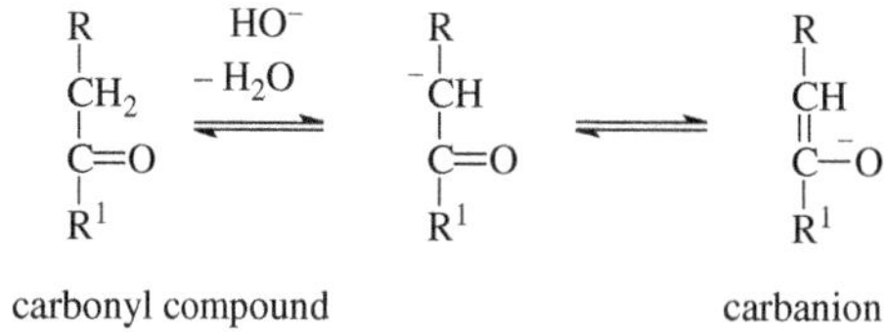

Figure 8.18 Formation of carbanions from carbonyl compounds in basic media.

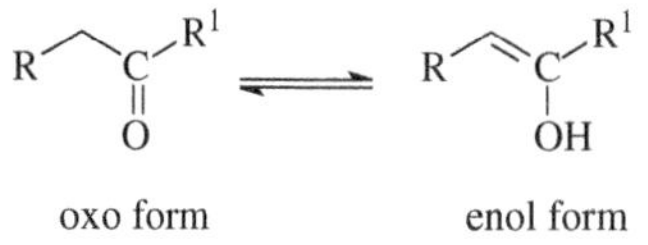

Figure 8.19 Enolisation of carbonyl compounds.

8.2.4.2.1 Addition reactions

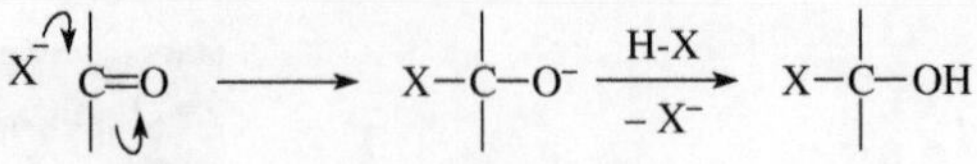

Carbon of the carbonyl group reacts readily with nucleophilic reagents. The first step is an addition of a nucleophile on the carbon atom of the polarised carbonyl-group double bond, which is followed by reaction with an electrophile, usually a proton (Figure 8.20). The most important addition reactions of carbonyl compounds are reactions with water, alcohols, sulfan, thiols, ammonia, amines, and aminocarboxylic acids. Reactions are generally acid–base-catalysed, so are influenced to a great extent by pH. Addition of acid leads to the formation of a more reactive conjugated acid,

Figure 8.20 Nucleophilic addition to carbonyl compounds (X = OH, SH, NH$_2$, NHR).

but also reduces the concentration of the nucleophile. An addition of hydroxides has the opposite effect. The ease of addition primarily affects the inductive effect of substituents. Alkyl and aryl substituents have a positive inductive effect (+I), and the reactivity of the carbonyl compounds therefore decreases in the following order: formaldehyde > aromatic aldehydes > higher aliphatic aldehydes, unsaturated aldehydes > methyl ketones > alkyl aryl ketones > dialkyl ketones. The presence of electronegative functional groups with negative inductive effect (−I) in the aliphatic substituent in the position adjacent to the carbonyl group (such as another carbonyl group or hydroxy group) results in correspondingly higher reactivity of the carbonyl group. Reactivity then declines in the order: α-dicarbonyl compounds (such as glycosuloses) and α-hydroxycarbonyl compounds (such as common aldoses and ketoses). If the carbonyl group is conjugated with a double bond (e.g. in alk-2-enals), the 1,4-addition of nucleophilic reagents may occur (Figure 8.21).

Reactions with water and alcohols The addition of water to the carbonyl group in aqueous solutions can lead to the formation of hydrates. The reactions and stability of hydrates depend primarily on the inductive effect of substituents. Hydration of formaldehyde readily yields methylene glycol (methanediol), which polymerises to linear oligomers and to polymers known as paraformaldehyde. Hydrates of α-dicarbonyl and α-hydroxycarbonyl compounds spontaneously dimerise into various cyclic 1,4-dioxanes. These hydrates are intermediates of other α-dicarbonyl and α-hydroxycarbonyl compounds in the oxidation–reduction reactions and precursors of carboxylic acids. Dialkyl ketones do not form hydrates.

Adducts resulting from the reaction of carbonyl compounds with alcohols (also with sugars) are called hemiacetals. This reaction (1,2-addition) is catalysed by acids, and primary alcohols are more reactive than secondary alcohols. Splitting a water molecule in acidic media yields a conjugate acid that reacts with another molecule of alcohol with the formation of **acetals**. Products of ketones with alcohols were formerly known as ketals (Figure 8.22). The formation of acetals has considerable importance in the development of flavour-active compounds in alcoholic beverages. In the case of alk-2-enals (such as acrolein), the reaction proceeds in a manner identical with alkene electrophilic addition and, in addition to the 1,2-addition, 1,4-addition is also possible (Figure 8.23). Reactions of glycols with carbonyl compounds yield cyclic acetals, known as that 1,3-dioxolanes (Figure 8.24). The condensation reaction between glycerol and acetaldehyde under acid conditions (at the pH of wine) leads analogously to the formation of four isomers: (Z)- and (E)-5-hydroxy-2-methyl-1,3-dioxane and (Z)- and (E)-4-hydroxymethyl-2-methyl-1,3-dioxolane (**8-41**). It is expected that these acetals contribute to the aroma of old Sherry and Port wines.

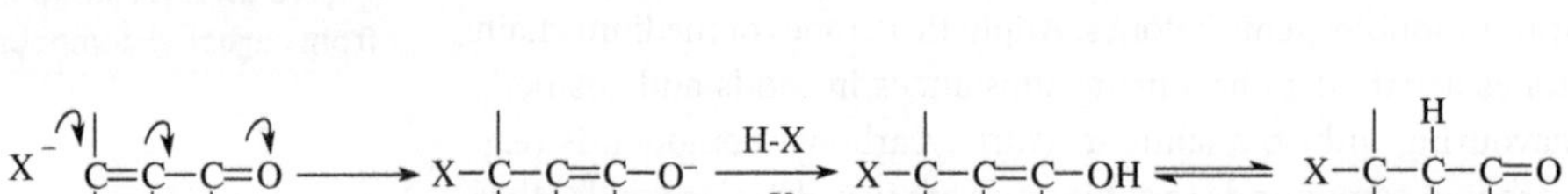

Figure 8.21 1,4-Addition to carbonyl compound double bond.

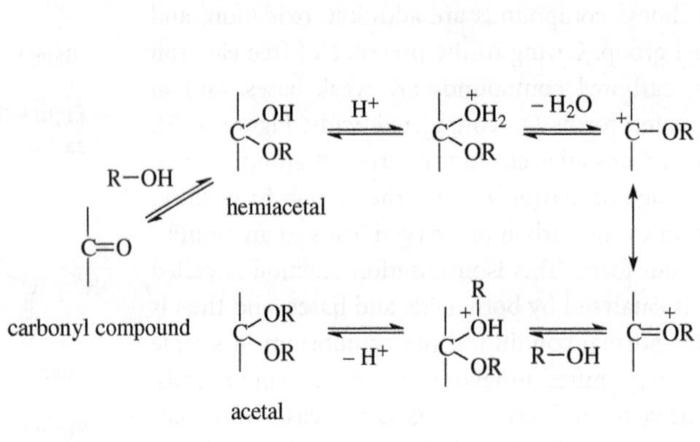

Figure 8.22 Addition of alcohols and formation of acetals.

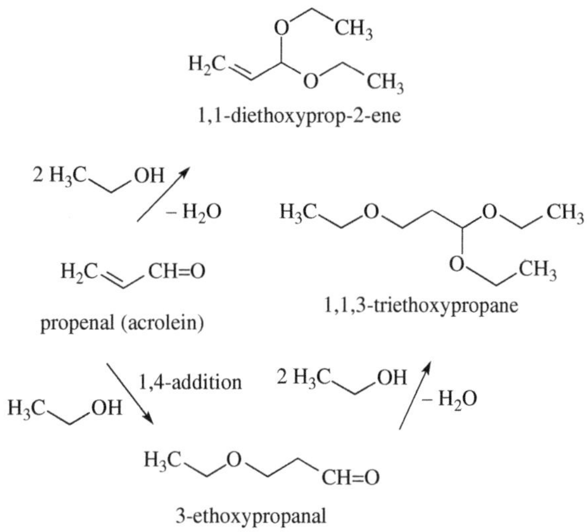

Figure 8.23 Reactions of acrolein with ethanol.

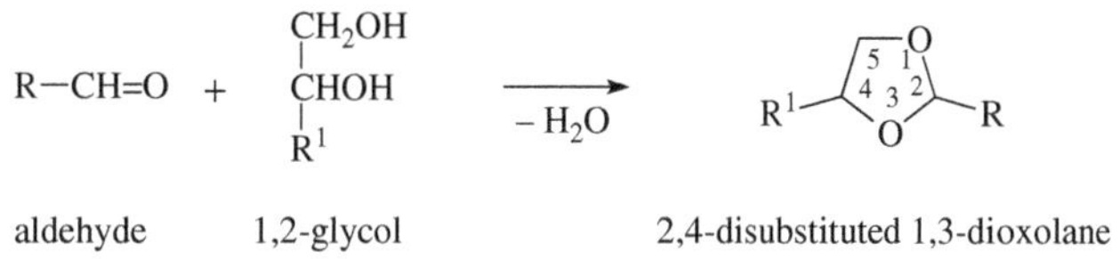

Figure 8.24 Formation of cyclic acetals from aldehydes and 1,2-glycols.

Acetals are stable in neutral and alkaline solutions. Those derived from aldehydes are more stable than those derived from ketones, whilst cyclic acetals are more stable than aliphatic ones. In acidic solutions, they are hydrolysed to the parent compounds.

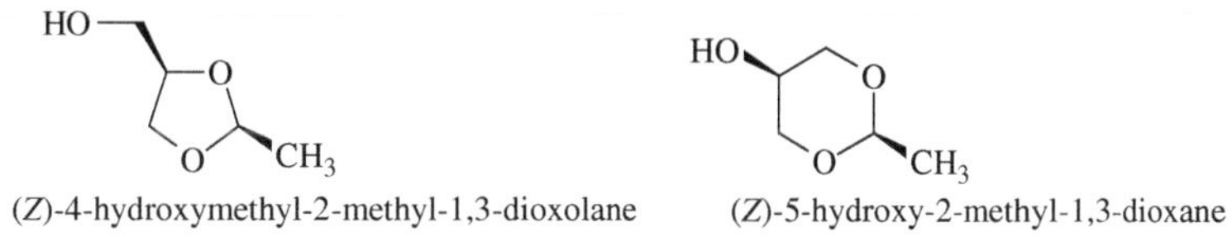

(Z)-4-hydroxymethyl-2-methyl-1,3-dioxolane   (Z)-5-hydroxy-2-methyl-1,3-dioxane

8-41, reaction products of glycerol with acetaldehyde

Reactions with sulfan and thiols Similarly to water and alcohols, carbonyl compounds react with sulfan (hydrogen sulfide) and thiols. Addition of sulfan gives unstable α-hydroxythiols (mercapto alcohols) as intermediates (Figure 8.25). Mercapto alcohols easily dehydrate to thioaldehydes or thioketones, forming cyclic compounds (mostly trimers, which are known as trithioaldehydes and trithioketones) or linear polymers. This reaction is of great importance, for example, in the formation of many aromatic compounds arising during the thermal processing of meat. An example is the reaction of acetaldehyde with sulfan (Figure 8.26).

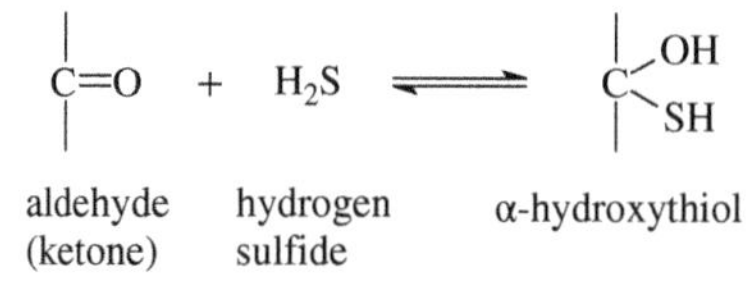

Thiols react with aldehydes and ketones to form thioacetals (mercaptals). In comparison with the reaction of alcohols, the reaction equilibrium is shifted more in favour of products that are mainly involved as flavour-active components of meat, vegetables, mushrooms, and other foods (Figure 8.27). The reaction intermediates (semithioacetals) give rise to a number of other products.

Figure 8.25 Reaction of carbonyl compounds with hydrogen sulfide.

Reactions with ammonia and amines Addition products of aldehydes and ketones with ammonia and amines (hemiaminals) are unstable compounds that are stabilised by a number of subsequent reactions (Figure 8.28). Dehydration of addition products yields imines. Other possible products are enamines, amines containing a double bond linkage —CH=CH—NH— (Figure 8.29), cyclic trimers (symmetrical 1,3,5-triazines), and various coloured polymers, generally of unknown structure. Reaction of the addition products with another amine

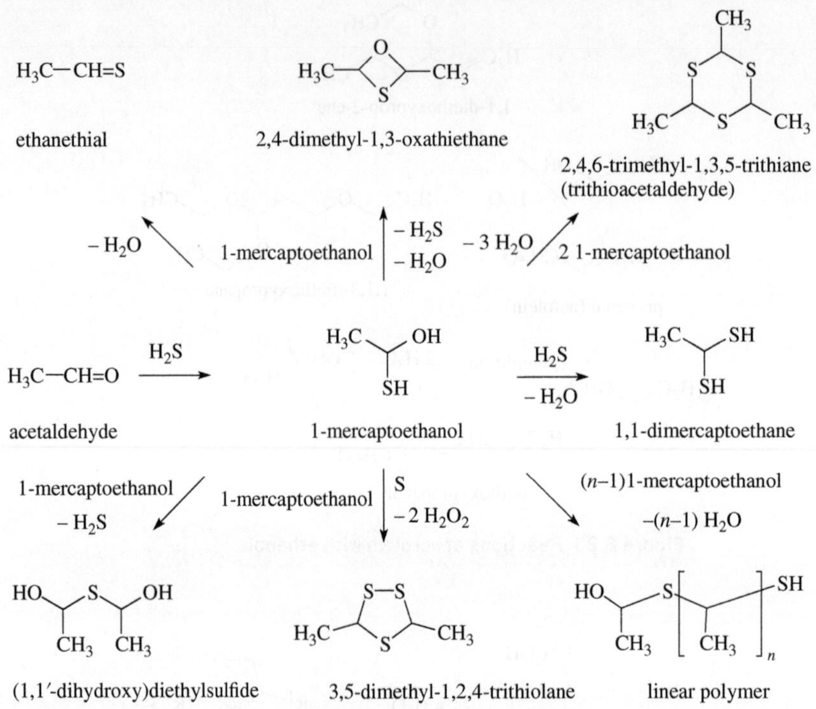

Figure 8.26 Reactions of acetaldehyde with hydrogen sulfide.

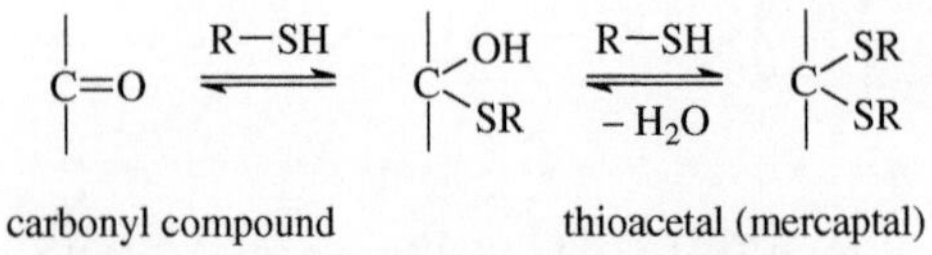

Figure 8.27 Reactions of carbonyl compounds with thiols.

molecule gives rise to aminals, whilst reaction with another molecule of the carbonyl compounds yields α,α′-dihydroxyamines. For example, in addition to 1,3,5-triazines, it is mostly polymers that are formed from aliphatic aldehydes and ammonia or amines. The reaction of formaldehyde with ammonia yields hexamethylenetetramine (**8-42**; see Section 11.2.1.2.3). If cyclisation of the intermediate is possible, stable heterocyclic compounds may be formed, such as those produced in reactions of aldehydes with some aminocarboxylic acids. However, the main reaction products are polymeric brown pigments called melanoidins (see Section 4.7.5.7).

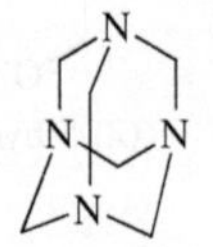

8-42, hexamethylenetetramine

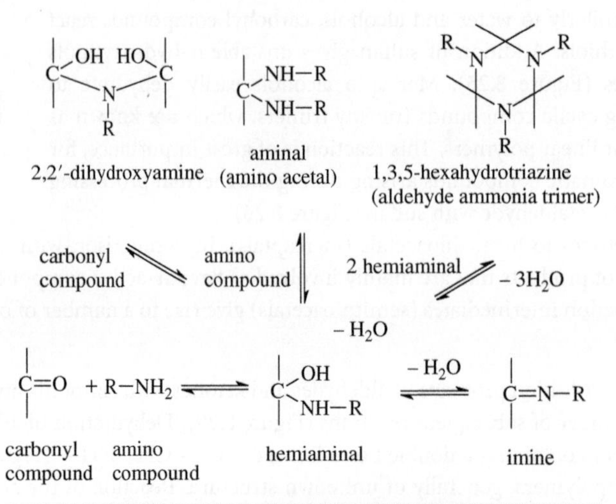

Figure 8.28 Reactions of carbonyl compounds with ammonia (R = H) and amines.

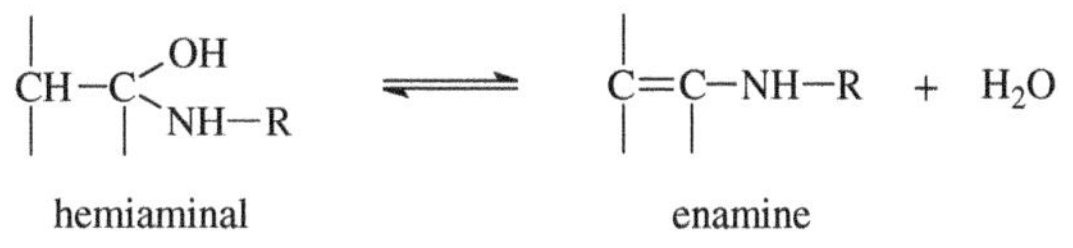

Figure 8.29 Formation of enamines.

Aldol condensation Aldol condensation or aldolisation is a reaction of aldehydes and ketones that have at least one hydrogen on the α-carbon (in position C-2 to the carbonyl group). The nucleophile in aldolisation is an anion of the carbonyl compound. It reacts with the carbonyl group of the second aldehyde or ketone molecule to form α-hydroxycarbonyl compounds, called aldols. Up to this point, the reaction is an aldol reaction. The following reaction step is an elimination reaction of water. The aldol that is produced dehydrates at elevated temperatures, in the presence of a strong base or in acidic media, yielding an α,β-unsaturated aldehyde with conjugated double bonds. Saturated aldehydes produce unsaturated aldehydes, alk-2-enals (Figure 8.30). For example, aldolisation of acetaldehyde and dehydration of the product provides both enantiomers of but-2-enal, which is also known as crotonic acid aldehyde or crotonaldehyde (Figure 8.31). Formaldehyde can react with other aldehydes or ketones having at least one hydrogen atom on the α-carbon to the carbonyl group. High reactivity of formaldehyde causes aldolisation of other molecules of formaldehyde, the number of which corresponds to the number of α-hydrogen atoms available in the reaction partner. For example, acetaldehyde reacts with formaldehyde in the formation of mono-, di-, and trimethylolacetaldehyde, which yields pentaerythritol by a Cannizzaro reaction (Figure 8.32).

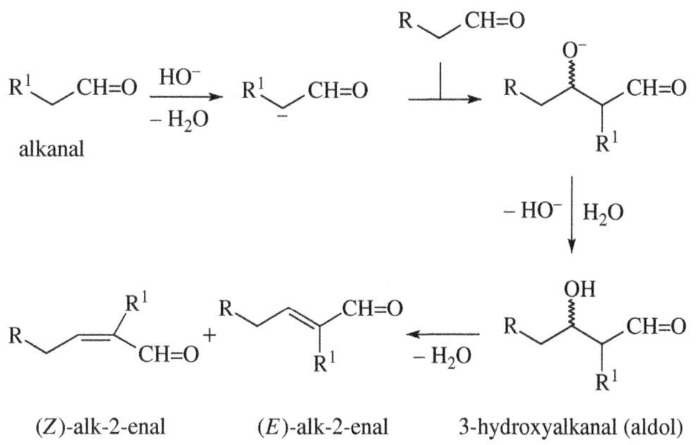

Figure 8.30 Aldolisation of carbonyl compounds.

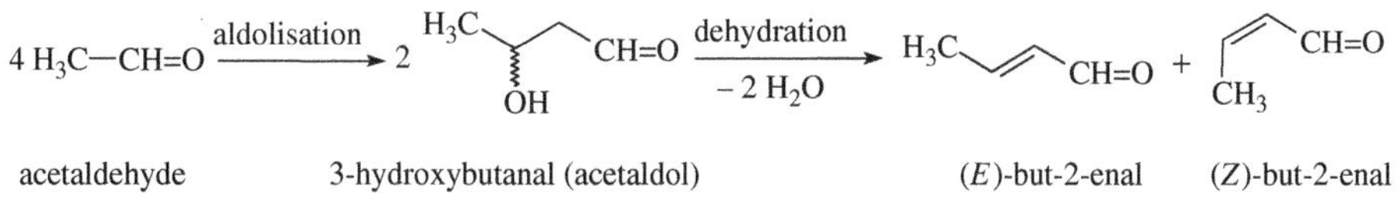

Figure 8.31 Aldolisation of acetaldehyde.

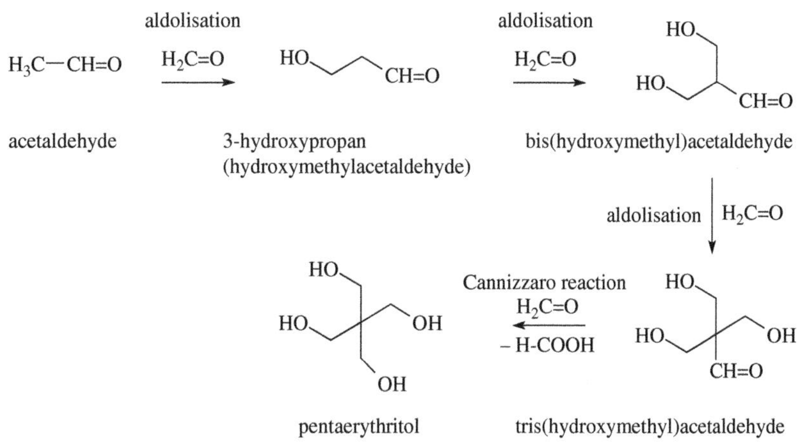

Figure 8.32 Reactions of formaldehyde with acetaldehyde.

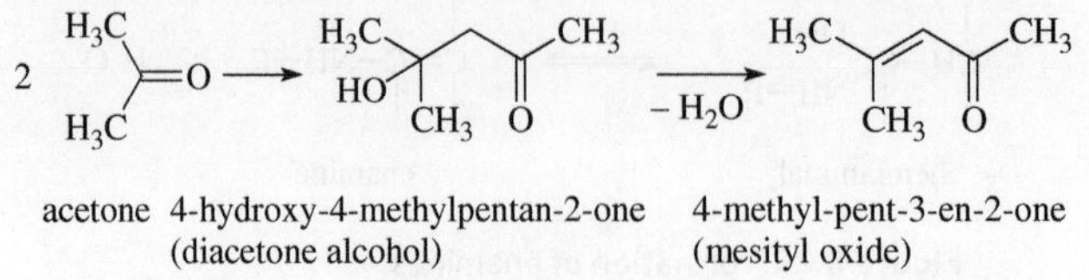

acetone   4-hydroxy-4-methylpentan-2-one   4-methyl-pent-3-en-2-one
          (diacetone alcohol)   (mesityl oxide)

Figure 8.33 Aldolisation of acetone.

The aldolisation of acetone in alkaline media yields diacetone alcohol, which, on heating in the presence of traces of acids, produces mesityl oxide (4-methylpent-3-en-2-one) with an aroma resembling peppermint (Figure 8.33). Mesityl oxide is a precursor of other aromatic compounds.

8.2.4.2.2 Oxidations and reductions

The carbonyl group of aldehydes, in contrast to that of ketones, is readily oxidised by oxygen to a carboxyl group. The primary product of autoxidation is a peroxyacid that oxidises another molecule of aldehyde to form two molecules of carboxylic acids (Figure 8.34). The autoxidation of formaldehyde gives rise to formic acid, acetaldehyde yields acetic acid, benzaldehyde gives benzoic acid, and furan-2-carbaldehyde gives 2-furancarboxylic acid, also known as furoic or pyromucic acid. Oxidation of a ketone carbonyl group in foods does not occur. The hydrocarbon chains of aldehydes and ketones are also subject to autoxidation, as are the hydrocarbon chains of fatty acids in lipids.

Reduction of aldehydes generates primary alcohols and reduction of ketones yields secondary alcohols. This is usually an enzymatic reaction. For example, benzyl alcohol arises from benzaldehyde.

Cannizzaro reaction   A specific case of an oxidation–reduction reaction is the Cannizzaro reaction, in which one molecule of aldehyde is oxidised with reduction of the second molecule. A Cannizzaro reaction occurs in aldehydes that do not have a hydrogen atom in position α to the carbonyl group. The reaction products are mixtures of primary alcohols and carboxylic acids. A Cannizzaro reaction is catalysed by hydroxyl ions; the key step is the transfer of the hydride from the hydride anion to the second aldehyde molecule (Figure 8.35).

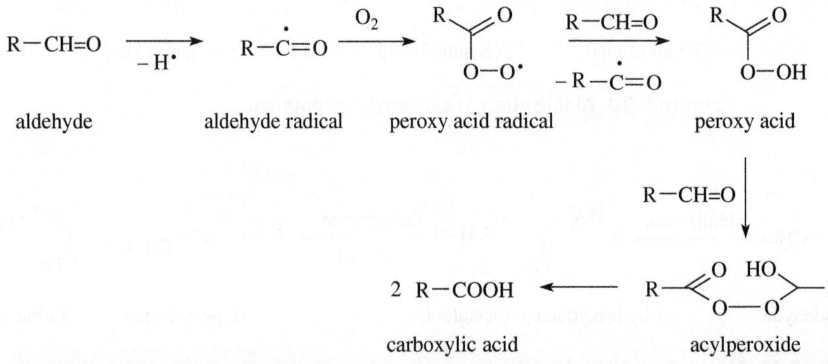

aldehyde   aldehyde radical   peroxy acid radical   peroxy acid

2 R—COOH   acylperoxide
carboxylic acid

Figure 8.34 Oxidation of aldehydes to carboxylic acids.

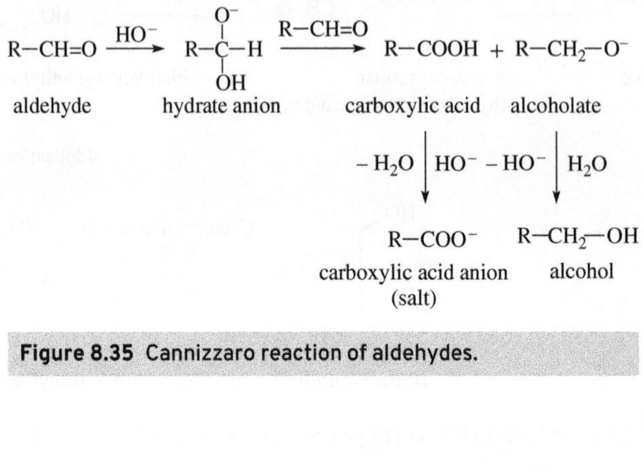

aldehyde   hydrate anion   carboxylic acid   alcoholate

R—COO⁻   R—CH₂—OH
carboxylic acid anion   alcohol
(salt)

Figure 8.35 Cannizzaro reaction of aldehydes.

$$R-CH=O \; + \; H_2C=O \; \xrightarrow{HO^-} \; R-CH_2-OH \; + \; H-COOH$$

aldehyde   formaldehyde   alcohol   formic acid

Figure 8.36 Cannizaro reaction of aldehydes and formaldehyde.

For example, Cannizzaro reaction of formaldehyde gives methanol and formic acid. Methanol can react with another molecule of formaldehyde (its hydrate) to form acetal 1,1-dimethoxymethane, also known as 2,4-dioxapentane. The reaction of methanol with formaldehyde hydrate oligomers yields, by analogy, higher dioxaalkanes. The Cannizzaro reaction of benzaldehyde also proceeds easily (especially in the presence of heavy metal traces), yielding benzyl alcohol and benzoic acid. Benzoic acid is otherwise also formed by autoxidation of benzaldehyde. The Cannizzaro reaction of furan-2-carbaldehyde yields furfuryl alcohol and 2-furancarboxylic acid.

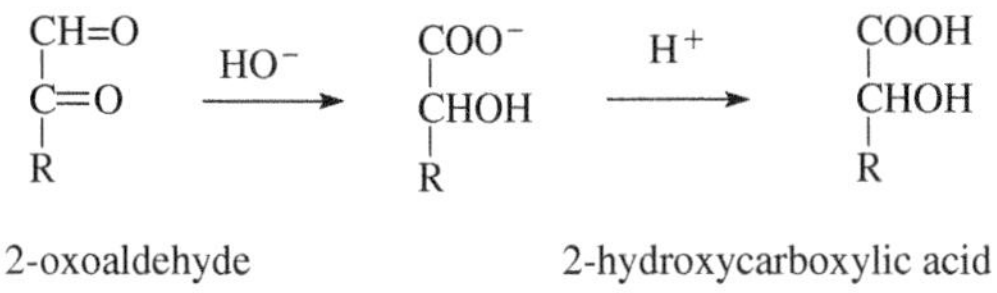

In the presence of formaldehyde, a Cannizzaro reaction occurs with aldehydes, which do not have hydrogen atoms on the carbon in position α to the carbonyl group. Formaldehyde is then oxidised to formic acid, and the aldehyde is reduced to the corresponding alcohol (Figure 8.36). Analogously, the reaction of formaldehyde with aldoses produces small amounts of formic acid and the corresponding alditol.

In the case of 2-oxoaldehydes, such as glyoxal and methylglyoxal, the Cannizzaro reaction proceeds as the intramolecular reaction and the product is 2-hydroxycarboxylic acid. The aldehyde group of oxo aldehyde is oxidised to a carboxylic group with the simultaneous reduction of the oxo group to a hydroxyl group (Figure 8.37). For example, glyoxal yields glycolic acid and methylglyoxal yields lactic acid (racemate).

Figure 8.37 Cannizaro reaction of α-dicarbonyl compounds.

2-oxoaldehyde   2-hydroxycarboxylic acid

8.2.5 Acetals

8.2.5.1 Classification, structure, terminology, and occurrence

Acetals occur wherever aldehydes and ketones are present with an excess of alcohol. Relatively large amounts of acetals are therefore found in alcoholic beverages, particularly spirits with higher ethanol content. However, even there the proportion of aldehydes bound as acetals does not usually exceed 15–30% of the total aldehyde content.

Acetaldehyde, present in alcoholic beverages in a relatively high concentration, is more reactive than higher alkanals, alkenals, and aromatic aldehydes. Therefore, the most common substance found in alcoholic beverages is diethylacetal (1,1-diethoxyethane) resulting from the reaction of acetaldehyde with ethanol (**8-43**). It is characterised by a sharp fruity aroma and a flavour reminiscent of nuts. The content of 1,1-diethoxyethane in whiskies and brandies, depending on the time of ripening, varies from about 4 to 60 mg/l. Acetals derived from formaldehyde and ethanol and acetals derived from higher aldehydes are present in lower amounts, as well as acetals arising from fusel oil alcohols.

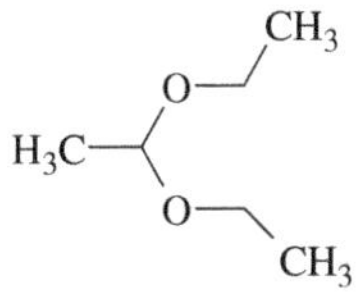

8-43, 1,1-diethoxyethane

Reaction of the unsaturated aldehyde acrolein with ethanol leads to either 1,1-diethoxyprop-2-ene or 3-ethoxypropanal. Both compounds can react with another ethanol molecule to give the final product, 1,1,3-triethoxypropane (Figure 8.23). All three are found in alcoholic beverages.

1,3-Dioxolanes, cyclic acetals resulting from the condensation of carbonyl compounds with glycols in acidic media, are common in alcoholic beverages. Brandies and other spirits contain 4-hydroxymethyl-2-methyl-1,3-dioxolane ($R^1 = CH_2OH$, $R^2 = H$), which forms from acetaldehyde and glycerol, 2,4-dimethyl-1,3-dioxolane ($R^1 = CH_3$, $R^2 = H$), which forms from acetaldehyde and propane-1,2-diol (propylene glycol, a product of methylglyoxal reduction), and 2,4,5-trimethyl-1,3-dioxolane ($R^1 = R^2 = CH_3$), which forms from acetaldehyde and butane-2,3-diol.

The most commonly occurring cyclic acetals of commercial aromas are 1,3-dioxolane derivatives that are derived from propane-1,2-diol used as a solvent. In vanilla flavours, for example, a considerable amount of 2-(4-hydroxy-2-methoxyphenyl)-4-methyl-1,3-dioxolane (**8-44**) occurs, which is formed in the reaction of propane-1,2-diol with vanillin.

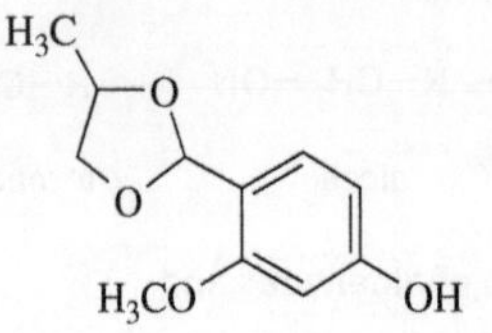

8-44, 2-(4-hydroxy-2-methoxyphenyl)-4-methyl-1,3-dioxolane

8.2.5.2 Properties and reactions

Acetals derived from lower or higher alcohols generally have a pleasant smell similar to the original carbonyl compounds, but weaker and softer. Acetals are more stable than the corresponding aldehydes and ketones, especially in alkaline media and during oxidation. Therefore, they are often used as flavourings for soaps. Under acidic conditions, acetals are easily hydrolysed to the parent compounds.

8.2.6 Carboxylic acids

Carboxylic acids are particularly important components of foods of plant origin. They influence the course of enzymatic and chemical reactions, the microbiological stability of foods during storage and processing, and the organoleptic and technological properties of the foods. In foods, the predominant compounds that can be found are aliphatic, alicyclic, aromatic, and heterocyclic carboxylic acids that contain one (monocarboxylic acids) or more (polycarboxylic acids) carboxyl groups in the molecule. Many carboxylic acids also contain other functional groups such as hydroxyl groups (hydroxycarboxylic acids), carbonyl groups (oxocarboxylic acids), amino groups (aminocarboxylic acids), and mercapto groups (mercaptocarboxylic acids). Certain halogen-containing carboxylic acids, the molecule of which contains one or more chlorine atoms, for example, are food contaminants.

It is mainly the lower carboxylic acids and some aromatic carboxylic acids that act as odour- or taste-active compounds. Taste-active substances are predominantly polyhydric carboxylic acids, such as citric and malic acids, and some aliphatic carboxylic acids, such as acetic and lactic acids, which are major carriers of the sour taste in food raw materials and foods. Short-chain fatty acids also have some importance as odour- and taste-active substances (C_4 and C_6; as well as medium-chain acids, C_8–C_{12}). A number of carboxylic acids can become precursors of important flavour-active compounds, such as esters and lactones.

8.2.6.1 Classification, structure, terminology, and occurrence

8.2.6.1.1 Aliphatic monocarboxylic acids

Important odour-active acids are mainly saturated monocarboxylic acids containing 1–10 carbon atoms in the molecule (**8-45**), which, apart from in the carboxyl group, do not contain other functional groups. Acids with four or more carbon atoms in the molecule are classified as fatty acids.

$$R-COOH$$

8-45, lower carboxylic acids

formic (methanoic), R = H
acetic (ethanoic), R = CH$_3$
propionic (propanoic), R = CH$_2$CH$_3$
butyric (butanoic), R = [CH$_2$]$_2$CH$_3$
isobutyric (2-methylpropanoic), R = CH(CH$_3$)$_2$
valeric (pentanoic), R = [CH$_2$]$_3$CH$_3$
isovaleric (3-methylbutanoic), R = CH$_2$CH(CH$_3$)$_2$
caproic (hexanoic), R = [CH$_2$]$_4$CH$_3$
caprylic (octanoic), R = [CH$_2$]$_6$CH$_3$
capric (decanoic), R = [CH$_2$]$_8$CH$_3$

The basic member of the homologous series of these acids is formic acid. It occurs as a free compound and as esters mainly in vegetables, fruits, and alcoholic beverages, where it arises (in addition to ethanol and acetic acid) as a byproduct of the anaerobic fermentation of sugars by certain microorganisms. The precursor of formic acid is pyruvic acid, which reacts with HS-CoA in the formation of acetyl-CoA and formic acid, which is then decomposed into carbon dioxide and hydrogen. This reaction is catalysed by pyruvate formate-lyase. Formic acid is also produced in the seeds and leaves of plants by α-oxidation of fatty acids (see Figure 8.14). The product of oxidation is 2-hydroxycarboxylic acid, which is transformed into formic acid and carbon dioxide. The final product is a fatty acid containing one less carbon atom.

Sugars (hexoses) yield formic acid as a product of the Cannizzaro reaction of formaldehyde and as one of the final products of the dehydration of sugars in acidic media and their degradation in neutral and alkaline media. At high concentrations (in quantities of up to 2%), formic acid therefore occurs in acid protein hydrolysates, where it is formed by 5-hydroxymethyl-2-furancarbaldehyde degradation. In rancid fats, it arises by oxidative decomposition of aldehydes. It is also used as a preservative.

Acetic acid (**8-45**) is the most common monocarboxylic acid occurring in foods. It is a typical component of fruits (as the free acid and its esters) and foods produced by fermentation processes. Acetic acid is also produced as a degradation product of sugars in the Maillard reaction and in thermal processes. For food purposes, acetic acid is produced from ethanol or fruit wines as vinegar by microbial oxidation of ethanol via acetaldehyde using aerobic bacteria of the genus *Acetobacter* (Figure 8.38). In the manufacture of vinegar, the bacterium *A. aceti* is traditionally used. The fermented ethanol (including wines) has a resulting acetic acid concentration of about 5%, with a pH of about 2.4. Any type of vinegar can be distilled to produce a colourless solution that contains about 8% acetic acid.

The higher homologues of aliphatic monocarboxylic acid, such as propionic, butyric, isobutyric, valeric, isovaleric, and caproic acids (**8-45**), are formed by different mechanisms. Butyric, caproic, caprylic, and capric acids occur, along with other acids (such as isoacids, anteisoacids, and acids with an odd number of carbon atoms), in relatively high quantities in milk fat, in the form of triacylglycerols. Their biosynthesis is primarily based on acyl-CoA. Propionic acid is biosynthesised if the parent compound is propionyl-CoA instead of acetyl-CoA. Numerous other acyl-CoAs are intermediates in the biosynthesis of amino acids formed by elongation of 2-oxoacids under the catalytic activity of acyl-CoA synthetases. Examples of these acyl-CoAs are butanoyl, 2-methylpropanoyl-, pentanoyl-, 3-methylbutanoyl-, 4-methylpentanoyl-, and hexanoyl-CoA. The corresponding 2-oxoacids are formed as amino acid catabolic products by transamination catalysed by transaminases or by oxidative deamination catalysed by oxidases. Their decarboxylation catalysed by decarboxylases yields aldehydes, which are finally oxidised to carboxylic acids by aldehyde dehydrogenases. For example, valine is the precursor of isobutyric acid, isovaleric acid arises from leucine, 2-methylbutyric acid arises from isoleucine, and propionic acid arises from threonine. These acids are the byproducts of fermentation processes (Table 8.14).

Carbohydrates are also precursors of carboxylic acids. Bacteria of the genera *Clostridium*, *Butyribacterium*, and *Butyrivibrio* ferment sugars primarily to acetic and butyric acids. Bacteria of the genus *Propionibacterium* (*P. freudenreichii* subsp. *shermanii*) ferment lactose to lactic acid, which is reduced to propionic acid, an important acid of the famous Emmental cheese (Figure 8.39).

Alk-2-enoic acids are frequent components of pyrrolizidine alkaloids known as senecio alkaloids (see Section 10.3.6.1.1). An important member of these acids is (*E*)-but-2-enoic (crotonic) acid (**8-46**). This also occurs in small quantities, together with other unsaturated acids with 5–10 carbon atoms in the molecule, in beer, other alcoholic beverages, and fermented milks. Examples of acids with five carbon atoms in the molecule (**8-46**) are (*Z*)-2-methylbut-2-enoic (angelic) acid and (*E*)-2-methylbut-2-enoic acid (tiglic) acid, the precursor of which is threonine. Valine is converted into (*Z*)-3-methylbut-2-enoic (senecioic) acid with five carbon atoms.

Alk-3-enoic acids formed from fatty acids are commonly found in foods and other natural sources, playing a vital role not only in the attractiveness of foods but also in chemo-communication in insects. Their odour quality changes from sweaty via plastic-like to sweaty and waxy.

An important unsaturated dienoic acid is (2*E*,4*E*)-hexa-2,4-dienoic acid, known as sorbic acid (**8-46**). It was first isolated from fruits of the rowan tree (*Sorbus aucuparia*, Rosaceae), where it results from dehydration and ring opening of (*S*)-2-methyl-2,3-dihydropyran-6-one, which is called parasorbic acid (**8-79**). Sorbic acid has antimicrobial effects and is therefore used as a preservative.

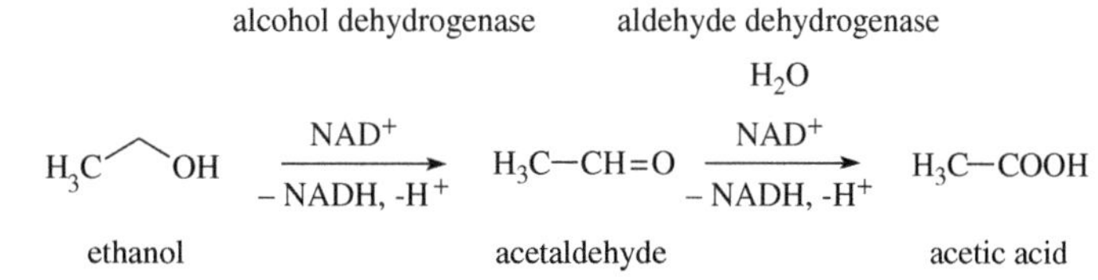

Figure 8.38 Oxidation of ethanol and formation of acetic acid.

Table 8.14 Contents of lower carboxylic acids in beer.

Acid	Content (mg/l)	Acid	Content (mg/l)
Acetic	12–155	Isovaleric	0.4–3.4
Butyric	0.6–2.6	Caproic	0.9–22.8
Isobutyric	0.3–3.3	Caprylic	1.8–15.4
Valeric	0.1–0.2	Capric	0.1–5.2

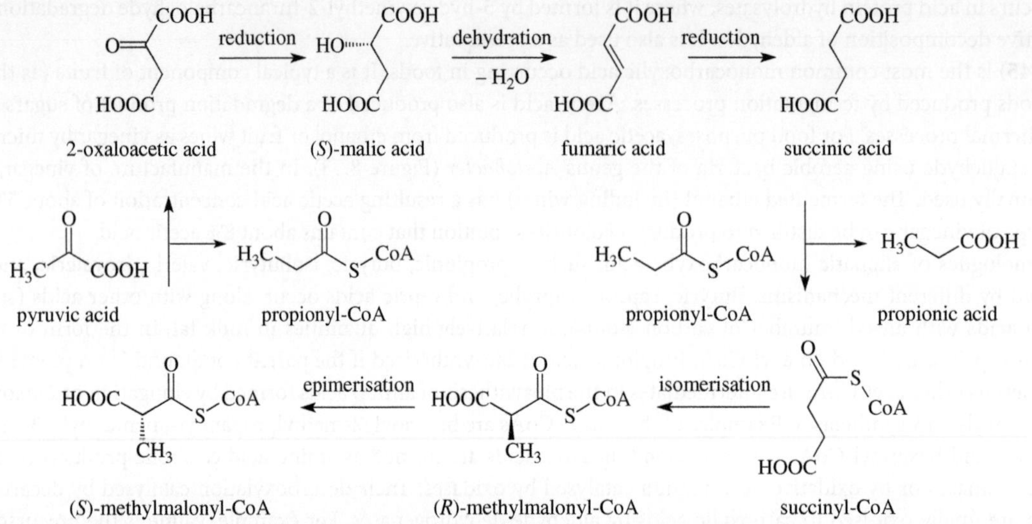

Figure 8.39 Mechanism of propionic acid fermentation pathway.

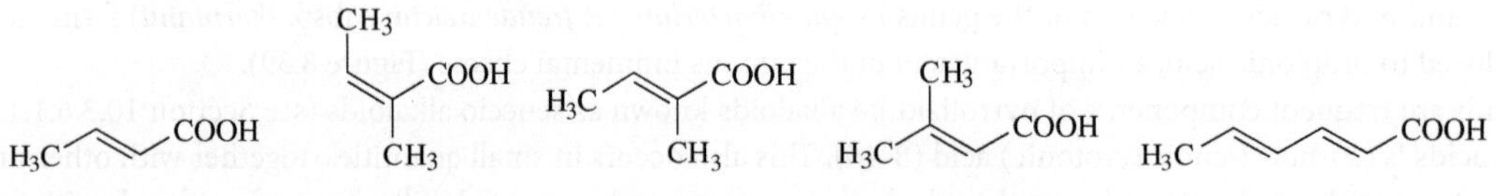

8-46, unsaturated aliphatic carboxylic acids

8.2.6.1.2 *Aliphatic dicarboxylic and tricarboxylic acids*

Aliphatic dicarboxylic and tricarboxylic acids are non-volatile compounds and may therefore have a role just as taste-active (acidic) food components.

The basic member of the homologous series of aliphatic dicarboxylic acids (**8-47**) is oxalic (ethanedioic) acid. The main precursor of oxalic acid is L-ascorbic acid; therefore, oxalic acid is a normal constituent of fruits and vegetables (Figure 8.40). It occurs as a soluble potassium salt (e.g. in bananas) or as a calcium salt (especially in fruits and vegetables) that is soluble in acidic and insoluble in neutral and alkaline media. Oxalic acid is categorised amongst the antinutritional compounds, as it interferes with the metabolism of calcium. Regulation of the calcium content in plants is based on the same principle. At higher concentrations, oxalic acid occurs in rhubarb and spinach, where its content can be somewhat decreased by blanching. A similar amount of oxalic acid is found in ripe carambola fruits (*Averrhoa carambola*, Oxalidaceae), tea, and cocoa (Table 8.15). Tea is considered the major source of oxalic acid in countries where it is consumed in large quantities (such as the United Kingdom). Oxalic acid also occurs in the so-called beer stone (composed of calcium oxalate and organic substances) that forms on the inside surfaces of brewing apparatus.

Approximately 75% of all human kidney stones are composed primarily of calcium oxalate, and hyperoxaluria, also known as oxalosis (urinary oxalate excretion that exceeds 40 mg/day), is a primary risk factor for this disorder. Urinary oxalate originates from a combination of absorbed dietary oxalate and endogenously synthesised oxalate; nevertheless, boiling of vegetables markedly reduces dietary soluble oxalate content by 30–87% and is more effective than steaming (5–53%). The loss of insoluble oxalate during cooking varies greatly, ranging from 0 to 74%.

8-47, dicarboxylic acids

oxalic (ethanedioic), $n = 0$
malonic (propanedioic, 1,1-methanedicarboxylic), $n = 1$
succinic (butanedioic, 1,2-ethane dicarboxylic), $n = 2$
glutaric (pentanedioic, 1,3-propane dicarboxylic), $n = 3$
adipic (hexanedioic, 1,4-butane dicarboxylic), $n = 4$
pimelic (heptanedioic, 1,5-pentane dicarboxylic), $n = 5$
suberic (octanedioic, 1,6-hexane dicarboxylic), $n = 6$
azelaic (nonanedioic, 1,7-heptane dicarboxylic), $n = 7$
sebacic (decanedioic, 1,8-octane dicarboxylic), $n = 8$

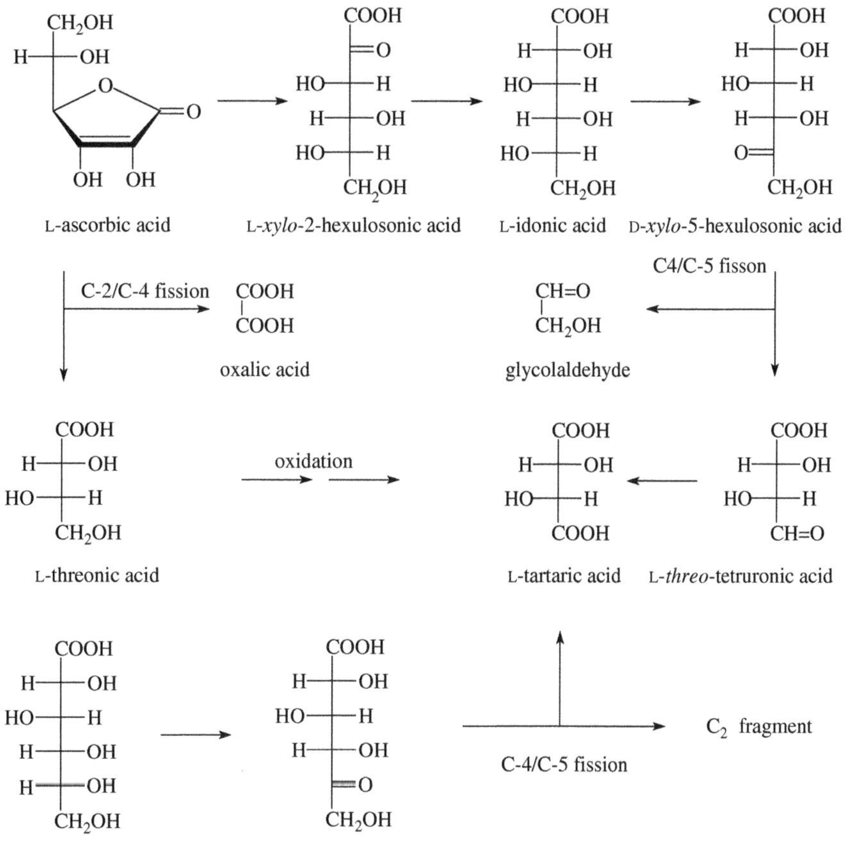

Figure 8.40 Biosynthesis of oxalic and L-tartaric acids.

Higher saturated aliphatic dicarboxylic acids, such as malonic, succinic, and glutaric acids and other higher homologues, are found in many food raw materials and foods, but usually in smaller quantities. They are mainly intermediates of fatty acids biosynthesis, the citric acid cycle, and other metabolic processes, but only succinic acid occurs in a somewhat larger amount in some fruits (e.g. in currants and strawberries). All the just mentioned dicarboxylic acids, including oxalic, adipic, pimelic, suberic acids, and their higher homologues, are present in small amounts in wine and beer.

Amongst the unsaturated dicarboxylic acids, (E)-but-2-enedioic acid, known as fumaric acid (**8-48**), has a certain importance. It is formed in the ornithine cycle, and occurs in small quantities in virtually all products of animal and vegetable origin and in greater quantities in some fungi. Fumaric acid arises as a non-enzymatic deamination product of aspartic acid. Along with isomeric (Z)-but-2-enedioic acid, known as maleinic (maleic) acid (**8-48**), fumaric acid forms by pyrolysis of malic acid during thermal processes, such as coffee roasting.

A higher homologue of fumaric acid is (E)-2-methylbut-2-enedioic acid, (E)-prop-1-ene-1,2-dicarboxylic acid, which is known trivially as mesaconic or methylfumaric acid (**8-48**). It occurs in significant quantities in sugar beets. It is used as a fire retardant. The cis-isomer of mesaconic acid is (Z)-2-methylbut-2-enedioic acid, also called (Z)-prop-1-ene-1,2-dicarboxylic acid, methylmaleic acid, or citraconic acid (**8-48**). It is found in higher amounts in blueberries. 2-Methylidenebutanedioic acid, also known as prop-1-ene-2,3-dicarboxylic acid or itaconic acid (**8-48**), is produced, together with mesaconic and citraconic acids, by thermal degradation of citric acid. Citric acid degradation occurs, for example, during the roasting of coffee and the refining (deodorisation) of vegetable oils, where citric acid is used as an additive agent (about 100 mg of 20% citric acid per kg oil) to chelate traces of metals, resulting in increased oil stability during storage. The common, older name for these three acids is thus pyrocitric acid.

Tricarboxylic acid, known as aconitic acid (prop-1-ene-1,2,3-tricarboxylic acid or achilleic or citridinic acid), occurs in two geometric isomers (**8-48**). (Z)-Aconitate is widespread as a product of the isomerisation of citrate to D-isocitrate (catalysed by aconitase) in the citric acid cycle. About 5% aconitic acid is found in molasses from cane sugar production, where the (E)-isomer prevails, as it is formed by isomerisation of (Z)-aconitic acid at elevated temperatures and low pH. The amount of (Z)-aconitic acid in the growing cane is low, because it is used in the citric acid cycle and not stored in the plant. Decarboxylation of aconitic acid at elevated temperatures yields itaconic acid.

Table 8.15 Oxalic acid contents in some foods.

Food	Content (mg/kg)	Food	Content (mg/kg)
Vegetables		Kiwi	180–450
Bamboo shoots	1600–4600	Oranges	62
Bell peppers	400	Orange juice	44
Broccoli	<100–500	Raspberries	22
Brussel sprouts	3600	Strawberries	19
Cabbage	130–1250	*Cereals*	
Carrots	100–5000	Wheat (*Triticum aestivum*)	533
Celery (stems)	175	Rye	712
Cucumbers	20	Oats	163
Garlic	360	Basmati rice	172
Green beans	200–450	Maize	386
Onion	30–50	*Other foods*	
Potatoes	23	Wheat wholemeal flour	700
Parsley (curled)	1660	Rye wholemeal flour	289
Pea	500	Bread (white)	49
Radishes	3	Cornflakes	56
Rhubarb	2300–9600	Oatmeal porridge	10
Salad beet	300–1380	Lentils	1180
Spinach	5400–9800	Tea (dry leaves)	3750–14500
Tomatoes	50–100	Coffee (Nescafé powder)	570
Fruits		Cacao powder	6230
Apples	15	Peanut butter	7050
Banana	7	Beef (roasted)	4
Black currants	43	Milk	5
Carambola (star fruit)	500–9600	Mushrooms (*Boletus edulis*)	136–536

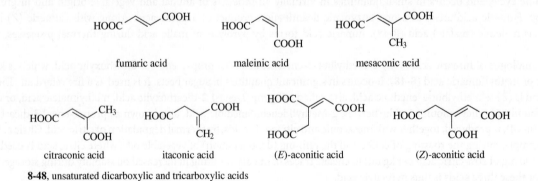

8-48, unsaturated dicarboxylic and tricarboxylic acids

8.2.6.1.3 Aliphatic hydroxycarboxylic acids

Hydroxycarboxylic acids are mostly polar non-volatile compounds influencing the sour taste of fruits, vegetables, and other foods. They are found predominantly as free substances and often in the form of various derivatives. Some of them are important odour-active compounds, especially as esters and lactones. Rarely hydroxycarboxylic acids occur as glycosides.

The simplest hydroxycarboxylic acid is glycolic (hydroxyethanoic) acid (**8-49**), which occurs in small amounts in most plant materials as a natural component. It also arises by the Cannizzaro reaction of glyoxal.

8-49, selected hydroxycarboxylic acids

The most important representative of this group of carboxylic acids is lactic (2-hydroxypropanoic) acid. (+)-L-Lactic acid, (S)-2-hydroxypropanoic acid (**8-49**), is present in meat. For example, fresh beef usually contains lactic acid at a level of 0.2–0.8%. It is formed in anaerobic glycolysis from glycogen.

Homofermentative lactic acid bacteria (such as *Lactococcus lactis* and *Streptococcus lactis*) produce (+)-L-lactic acid (e.g. in sour cream). Both isomers, (+)-L-lactic acid and (−)-D-lactic acid (**8-49**), are formed during milk fermentation by heterofermentative bacteria (lactic acid bacteria are mostly heterofermentative bacteria), and lactic acid thus also occurs as a racemate in sauerkraut, pickled cucumbers, olives, and silage. For example, bacteria of the genus *Leuconostoc* produce D-lactic acid, whilst *Pediococcus acidilactici* and other bacteria produce racemic lactic acid. The content of lactic acid in dairy products is 0.5–1.0%. L-Lactic acid in yoghurt represents about 54%, and in sour cream 96%, of the total lactic acid content. The total lactic acid content in sauerkraut is 1.5–2.5%, in fermented cucumbers it is 0.5–1.5%, and in fermented green olives it is 0.8–1.2%.

Fermentation with homofermentative lactic acid bacteria involves phosphorylation of α-lactose to α-lactose 6′-phosphate and hydrolysis to α-D-galactose 6-phosphate and β-D-glucose. α-D-Galactose 6-phosphate is then degraded via D-tagatofuranose 6-phosphate and D-tagatofuranose 1,6-bisphosphate to D-glyceraldehyde 3-phosphate. β-D-Glucose is phosphorylated to β-D-glucose 6-phosphate. All sugars are then activated by different mechanisms and converted into D-fructose 1,6-bisphosphate, which is degraded, via D-glyceraldehyde 3-phosphate and other metabolites, to pyruvic acid, which is then further reduced to L-lactic acid, virtually the only product of homofermentative lactic acid fermentation (Figure 8.41). The reaction sequences involve phosphoglycerate kinase, phosphoglyceromutase, enolase, pyruvate kinase, L-lactate dehydrogenase, galactokinase, UDP-galactose pyrophosphorylase, UDP-galactose 4-epimerase, UDP-glucose pyrophosphorylase, phosphoglucomutase, glucokinase, fructokinase, phosphoglucose isomerase, fructose bisphosphate aldolase, 1-phosphofructokinase, phospho-β-galactosidase, D-galactose 6-phosphate isomerase, D-tagatose 6-phosphate kinase, and tagatose 1,6-bisphosphate aldolase.

Heterofermentative lactic acid bacteria produce L- and D-lactic acids, ethanol, carbon dioxide, and a small amount of acetic acid from glucose and fructose. Sugars are phosphorylated in the bacterial cells: glucose to β-D-glucose 6-phosphate by glucokinase and fructose usually to D-fructose 1-phosphate by fructokinase. D-Fructose 1-phosphate, with the catalysis of glucose-6-phosphate isomerase, isomerises to β-D-glucose 6-phosphate (sometimes produced directly), which is then further metabolised (Figure 8.42; structures of compounds are shown in Figure 8.41). Saccharose is first hydrolysed to glucose and fructose by invertase, and maltose is hydrolysed to glucose by maltase.

Polycarboxylic hydroxy acids are important carriers of acidic taste of foods. Dicarboxylic hydroxy acids primarily include malic (hydroxysuccinic) acid (**8-50**), occurring only as (−)-L-isomer, the (S)-isomer, which arises as an intermediate in the citric acid cycle. It is found in abundance primarily in fruits and vegetables as a free compound. (S)-malic acid 1′-O-β-gentiobioside was identified in lettuce. Malic acid is synthesised from fumaric acid using *Lactobacillus brevis* or yeasts of the genus *Candida* and is used as a food additive (acidulant).

8-50, hydroxydicarboxylic acids

In the skins of apples, pyruvic acid is the precursor of (+)-L-citramalic acid, which is also known as (S)-2-hydroxy-2-methylsuccinic acid or (S)-2-methylmalic acid (**8-50**). During fruit ripening, citramalic acid is oxidised to 2-oxobutyric acid, and its decarboxylation yields acetone, which is a component of apple aroma. The same isomer of citramalic acid is produced by some enterobacteria during the degradation of glutamic acid to pyruvic acid. The (R)-isomer of citramalic acid is an intermediate in the biosynthesis of isoleucine in spirochetes.

Tartaric (2,3-dihydroxysuccinic) acid is an important representative of dihydroxydicarboxylic (aldaric) acids and the most prominent acid in wine. In nature, it occurs almost exclusively as (+)-L-tartaric acid (**8-50**), also known as dextrotartaric acid or L-threaric acid, which is the (2R,3R)-isomer. Occasionally, tartaric acid occurs as (−)-tartaric, laevotartaric acid, which is the (2S,3S)- isomer, also known as D-threaric acid (**8-50**). An optically inactive racemic mixture of both isomers, called racemic (uvic) acid, has been demonstrated in grape juice. The optically active symmetric (2R,3S)-tartaric acid (**8-50**), called meso-tartaric or erythraric acid, does not occur in nature. Synthetic racemic and meso-tartaric acids are used as acidulants.

L-Tartaric acid is found in wines as the poorly soluble potassium hydrogen tartrate salt (potassium bitartrate), which often crystallises in young wines and on cooling as cream-of-tartar, known in winemakers' jargon as tartrates. The crystals are harmless, but their presence is generally undesirable to consumers. The cream-of-tartar formation can be prevented by cold stabilisation of wines.

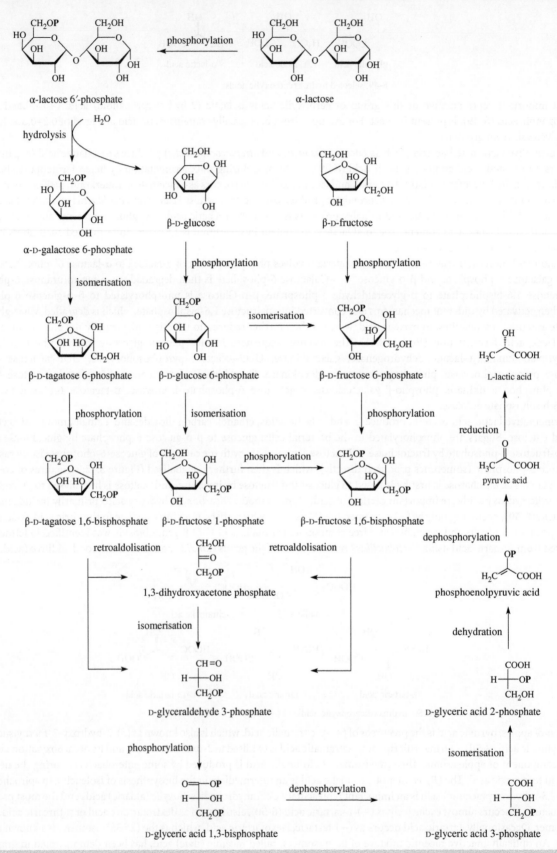

Figure 8.41 Mechanism of homofermentative lactic acid bacteria fermentation (P = phosphate residue).

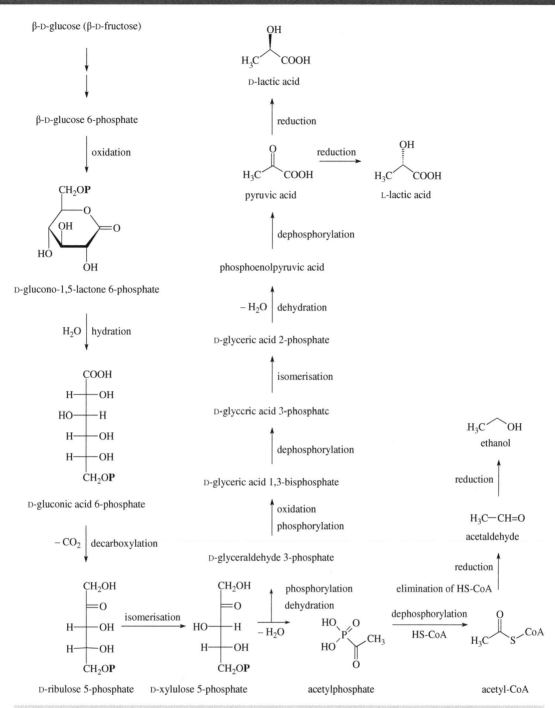

Figure 8.42 Mechanism of heterofermentative lactic acid bacteria fermentation (P = phosphate residue).

L-Tartaric acid biosynthesis starts from L-ascorbic acid or D-gluconic acid (Figure 8.40). In grapes and other plants of the *Vitaceae* family, tartaric acid arises from the ascorbic acid intermediate D-xylo-5-hexulosonic acid via the cleavage between carbons C-4 and C-5. Another option, and one which is apparently used by most plants (the reaction is catalysed by ascorbate 2,3-dioxygenase), is the cleavage of ascorbic acid between carbons C-2 and C-3, which simultaneously yields oxalic acid. A third possibility is seen mainly in plants of the *Fabaceae* family.

The optically inactive citric acid (2-hydroxypropane-1,2,3-tricarboxylic acid, **8-51**) is the most important representative of tricarboxylic hydroxy acids. Citric acid, like malic acid, occurs in many fruits (especially lemons). Industrially, it is obtained from lemon juice or produced by fermentation of molasses using the fungus *Aspergillus niger*. Along with malic acid, citric acid is used as an acidulant in widely different canning products and soft drinks, but it is also employed in the refination (degumming) of vegetable oils and for other purposes.

The malic and citric acids contents in fruits and vegetables are summarised in Table 8.16. In addition to these major acids, other carboxylic acids are also present in smaller amounts, such as tartaric acid, which occurs in higher amounts in grapes. Generally, fruits and vegetables contain a number of other acids, amongst which succinic, fumaric, quinic, pyruvic, and oxalic acids (together with ascorbic acid)

Table 8.16 Malic acid and citric acid contents in selected fruits and vegetables.

Fruits	Content (mg/kg)		Vegetables	Content (mg/kg)	
	Malic acid	Citric acid		Malic acid	Citric acid
Apples	2000–13 000	75–100	Asparagus	600–670	430–2000
Apricots	8000–14 100	7000	Broccoli	1200	2100
Bananas	2770	680	Cabbage	500–2000	700–4510
Black currant	4000–4400	25 000–31 600	Carrots	800–5200	40–930
Gooseberry	7000–000	6000–8000	Cauliflower	3900	2100
Grapefruits	400–600	11 900–21 000	Cucumbers	2400	100–3260
Grapes[a]	1680–15 360	305–1160	Garlic	2400	190
Lemons	1700–3000	40 000–43 800	Green beans	1120–1300	300–340
Oranges	600–2000	5600–9800	Green peas	750–1730	1100–1790
Peaches	3000–7700	<2000	Lettuce	920–2430	120–270
Pears	1000–4150	1800	Onion	1550–1960	230–1100
Pineapple	1000–5000	4000–12 000	Rhubarb[b]	9100–10 250	1350–1370
Red currant	1000–3000	12 000–20 200	Spinach	360–640	82–101
Strawberries	900–2000	6000–11 000	Tomatoes	1350–4700	8800–26 300

[a] Tartaric acid in the amount of 4190–13 510 mg/kg.
[b] Oxalic acid in the amount of 2300–9600 mg/kg.

predominate. For example, depending on variety, growing stage, and many other factors, the total amount of organic acids in carrots is about 2000–3000 mg/kg. After harvest, isocitric and malic acids represent about 90–95% of the total acids. The contents of isocitric and malic acids are approximately 1000 and 800 mg/kg, respectively. Tartaric (18–55 mg/kg), succinic (22–130 mg/kg), fumaric (5–8 mg/kg), quinic (42–60 mg/kg), oxalic (100 mg/kg), and some other acids are found in smaller amounts.

A D-*threo* isomer of citric acid, (1*R*,2*S*)-1-hydroxypropane-1,2,3-tricarboxylic acid (**8-51**), also known as D-isocitric acid, is another intermediate of the citric acid cycle. This is the dominant acid in blackberries (Table 8.17), but in other fruits it occurs in small or insignificant amounts (e.g. apples). The isocitric acid content is one of the most important markers in the estimation of the proportion of blackberries in fruit mixtures.

Table 8.17 Isocitric acid contents in selected fruits.

Fruits	Isocitric acid content (mg/kg)
Apples	<5
Apricots	75–200
Black currant	160–500
Blackberries	8000–10 000
Grapefruits	140–350
Oranges	65–200
Peaches	30–160
Pears	<40
Raspberries	60–220
Strawberries	30–90

The hydroxy derivative of citric acid, 1,2-dihydroxypropane-1,2,3-tricarboxylic acid, known as hydroxycitric acid, has four possible isomers: (+)- and (−)-hydroxycitric acids and (+)- and (−)-*allo*-hydroxycitric acids. The (1*S*,2*S*)-diastereomer is known as (−)-hydroxycitric or garcinic acid (**8-51**). Together with the corresponding γ-lactone, (2*S*,3*R*)-3-hydroxy-5-oxooxolane-2,3-dicarboxylic acid (hibiscic acid lactone, **8-52**), it occurs in the skin of the Southeast Asian fruit of the *Garcinia gummi-gutta* (syn. *G. cambogia*, Clusiaceae) tree, known as the Malabar tamarind, in the amount of 16–18% of dry matter. This fruit is related to the more common mangosteen (*G. mangostana*). Extracts from the fruit rind, called citrin, are used as dietary supplements for weight loss because garcinic acid inhibits the activity of ATP citrate lyase, an enzyme involved in the conversion of glucose to fat, thereby reducing the extent of lipogenesis by increasing the formation of glycogen in liver and suppression of appetite.

The garcinic acid isomer, (1*S*,2*R*)-1,2-dihydroxypropane-1,2,3-tricarboxylic acid or (+)-*allo*-hydroxycitric acid, also known as hibiscus or hibiscic acid (**8-51**), together with its methylester, γ-lactone, and glucoside, is located in the spicy leaves of a species of hibiscus called roselle (*Hibiscus sabdariffa*, Malvaceae), which are prepared like spinach in the tropics. Its fleshy flower calyces, containing intense staining deep red anthocyanin pigments, are increasingly used as food colourings and are often added to fruit and herbal teas.

8-51, hydroxytricarboxylic acids

citric acid   D-isocitric acid   garcinic acid   hibiscic acid

8-52, hibiscic acid lactone

An unusual citric acid derivative is toxic agaricinic (agaric) acid (**10-87**), which occurs in some fungi. Foods also contain numerous hydroxy acids derived from sugars.

8.2.6.1.4 Aliphatic oxocarboxylic acids

2-Oxocarboxylic acids are major products of the metabolism of proteins, lipids, and carbohydrates and arise as intermediates in fermentation processes, as products of other enzymatic reactions not linked with the metabolism of major nutrients, and as products of some non-enzymatic reactions. Oxoacids usually occur in low concentrations in all foods of animal and vegetable origin. They are polar, non-volatile compounds, but some of their reaction products are important odour-active substances in foods.

The simplest 2-oxocarboxylic acid is glyoxylic acid (oxoacetic acid, **8-53**), which occurs in the immature fruits of many plants. Glyoxylic acid is produced as the main intermediate of the glyoxylic acid cycle (ongoing in plants and microorganisms) and by other biochemical reactions, such as degradation of purine bases, 4-hydroxyproline, and other compounds.

The key compound of metabolism of carbohydrates (such as glycolysis and gluconeogenesis), lipids, and proteins in animals and plants is pyruvic (2-oxopropanoic) acid (**8-53**). It is similarly formed in fermentation processes and by enzymatic cleavage of alliin and other *S*-alk(en)yl-L-cysteine derivatives in garlic, onion, and other plants. Some higher homologues of pyruvic acid, such as 2-oxobutanoic, 2-oxopentanoic (2-oxovaleric), 2-oxo-3-methylbutanoic (2-oxoisovaleric), 2-oxo-3-methylpentanoic (2-oxo-3-methylvaleric), and 2-oxo-4methylpentanoic (2-oxoisocapronic) acids, are precursors of fusel oil alcohols in alcoholic beverages.

8-53, 2-oxoacids

glyoxylic, R = H
2-oxopropanoic (pyruvic), R = CH$_3$
2-oxobutanoic, R = CH$_2$CH$_3$
2-oxosuccinic (oxaloacetic), R = CH$_2$COOH

Foods also contain various oxodicarboxylic and oxotricarboxylic acids that arise predominantly in the citric or glyoxylic acid cycle. Common oxodicarboxylic acid is oxaloacetic (oxosuccinic, oxobutanedioic) acid (**8-53**), a metabolic intermediate in the citric acid cycle, glyoxylate cycle, gluconeogenesis, and other processes. 2-Oxoglutaric (2-oxopentanedioic) acid, an intermediate in the citric acid cycle, is a product of glutamate deamination. The simplest 3-oxocarboxylic acid is acetoacetic (diacetic or 3-oxobutanoic) acid (**8-54**). 4-Oxopentanoic (laevulinic) acid (**8-54**) occurs in foods containing carbohydrates, as it is one of the final dehydration products of hexoses via 5-hydroxymethylfuran-2-carboxaldehyde in acidic media. It is present in very high concentrations (about 1–2% w/w) in acid protein hydrolysates.

Table 8.18 Contents of dicarboxylic acids, oxoacids, and alicyclic acids in selected fruits and vegetables (mg/kg).

Acid	Orange	Lemon	Peach	Grape	Melon	Bell pepper	Tomato	Lettuce
Malonic	2.4	1.4	0.4	1.5	1.9	4.8	0.7	6.0
Succinic	2.1	0.5	23	7.2	8.1	7.4	1.1	30
Fumaric	0.0	0.0	7.7	7.5	7.1	1.0	0.8	90
Oxoglutaric	9.8	23	16	6.0	0.0	1.0	4.7	13
Quinic	140	440	34	35	0.8	320	27	20
Shikimic	2.0	3.0	4.4	60	1.0	7.9	0.8	5.3

The contents of major non-volatile dicarboxylic acids, oxoacids, and alicyclic acids in selected fruits and vegetables are given in Table 8.18.

8-54, acetoacetic acid, $n = 1$
laevulinic acid, $n = 2$

8.2.6.1.5 Alicyclic carboxylic acids

One of the most important substances derived from tetrahydroxycyclohexane is (1S,3R,4S,5R)-1,3,4,5-tetrahydroxycyclohexane-1-carboxylic acid, known as (−)-L-quinic acid (**8-55**). The trihydroxycyclohexene derivative is related (3R,4S,5R)-3,4,5- trihydroxycyclohex-1-ene-1-carboxylic acid, known as (−)-L-shikimic acid (**8-55**), which is a key intermediate in the biosynthesis of phenolic compounds. As free acids, but mainly as esters (depsides) with (E)-cinnamic acids (caffeic, ferulic, and 2-coumaric acids), both compounds are the major components of coffee beans. In smaller amounts, they are found in tea leaves, cocoa beans, many fruits and vegetables, potatoes, and other plant materials. In the juices of berries (e.g. gooseberries, currants, and blueberries) and mango, these acids occur in hundreds of thousands of mg/l, but in other fruits their content is about 10 times lower. As an example, quinic acid content can be used to differentiate North American blueberry products (juices or jams) made from northern high bush blueberries (*Vaccinium corymbosum*, Ericaceae), which contain from 0.37 to 2.53 g/kg quinic acid, from European blueberry (bilberry) products prepared from *V. myrtillus* berries, which contain 3.74–7.84 g/kg quinic acid.

L-quinic acid   L-shikimic acid

8-55, alicyclic carboxylic acids

8.2.6.1.6 Aromatic carboxylic acids

The simplest aromatic carboxylic acid is benzoic acid (**8-56**), which is relatively widespread in plant materials, mainly as 1-O-benzoyl-β-D-glucopyranose; both compounds are non-volatile polar substances that do not have any impact on food flavour. In essential oils, benzoic acid is found in the form of flavour-active esters. It arises from cinnamic acid or cinnamoyl-CoA through side-chain shortening by a C_2 unit. Its content in fruits and vegetables is generally very low, at around 0.05%. In cranberries, it is present predominantly as the glycoside 6-O-benzoyl-β-D-glucopyranose, known as vaccinin (see **4-92**), in an amount of around 0.2%. Free benzoic acid is also present in a very small amount (about 0.0015%) in yoghurts, where it results from hippuric acid hydrolysis. It is often also added to foods as a preservative. A typical example of a preserved product is table mustard, where concentrations typically reach 1000 mg/kg.

Benzoic acids substituted by hydroxyl and methoxyl groups (**8-56**) are produced through different biochemical mechanisms. Their representatives are 2-hydroxybenzoic (salicylic), 3-hydroxybenzoic (*m*-salicylic), 4-hydroxybenzoic (*p*-hydroxybenzoic), 2,3-dihydroxybenzoic (2-pyrocatechuic or hypogallic), 2,4-dihydroxybenzoic (β-resorcylic, β-resorcynolic or *p*-hydroxysalicylic), 2,5-dihydroxybenzoic (gentisic), 3,4-dihydroxybenzoic (protocatechuic), 3,5-dihydroxybenzoic (α-resorcylic), and 3,4,5-trihydroxycarboxylic (gallic) acids. Salicylic acid occurs in small quantities (0.3–3 mg/kg) in different types of fruit, mainly in the form of glycosides and esters, and participates in thermogenesis, metabolism of minerals, plant flowering, ethylene biosynthesis, resistance to pathogens, and other processes. 3-Hydroxybenzoic acid was found in pineapple fruit. Owing to its antimicrobial effects, 4-hydroxybenzoic acid has found use as a preservative mainly in the form of esters: so-called parabens (see Section 11.2.1.1.3). 2,3-Dihydroxybenzoic acid occurs in the fruit of

Phyllanthus acidus (Phyllanthaceae, known as Malay gooseberry), is a part of various bacterial siderophores (6-11), and is a product of human aspirin metabolism. 2,4-Dihydroxybenzoic acid is a common phenolic acid in wheat and rye and a degradation product of cyanidin glycosides. 2,5-Dihydroxybenzoic acid occurs in cocoa and wine and is used as an antioxidant in some pharmaceutical products. 3,4-Dihydroxybenzoic acid is an effective antioxidant in fruits, vegetables, and *in vivo*. 3,5-Dihydroxybenzoic acid is a metabolite of alkylresorcinols (10-1). An example of a less common derivative is a glucoside of protocatechuic acid ester with 4-hydroxybenzyl alcohol, 4-(3,4-dihydroxybenzoyloxymethyl)phenyl-*O*-β-D-glucopyranoside, which is isolated from oregano (see Section 8.2.2.1.2). Gallic acid is present in foods of plant origin in small quantities as a free acid, but frequently it is seen as a component of hydrolysable tannins (see Section 8.2.5.1).

4-Methoxybenzoic (*p*-anisic) acid is found in anise, 4-hydroxy-3-methoxybenzoic (vanillic) acid accompanies vanillin in vanilla pods, and other acids such as 3-hydroxy-4-methoxybenzoic (isovanillic), 3,4-dimethoxybenzoová (veratric) acid, 3,4-dihydroxy-5-methoxybenzoic (3-*O*-methoxygallic) acid, and 4-hydroxy-3,5-dimethoxybenzoic (syringic) acid occur in small amounts in all foods of plant origin.

8-56, benzoic acid and substituted benzoic acids

A mixture of 6-alkatrienyl, 6-alkadienyl, 6-alkenyl, and 5-alkyl substituted salicylic acids, collectively known as anacardic acid, occur in the cashew nut shell liquid at a level of 60–65%, together with decarboxylated products known as cardol and the corresponding phenols, called cardanol. The main components are (8*Z*,11*Z*)-2-hydroxy-6-(pentadeca-8,11,14-trien-1-yl)benzoic acid, (8*Z*,11*Z*)-2-hydroxy-6-(pentadeca-8,11-dien-1-yl)benzoic acid, (8*Z*)-2-hydroxy-6-(pentadeca-8-en-1-yl)benzoic acid, and 2-hydroxy-6-pentadecylbenzoic acid. The anacardic acid content of cashew nuts ranges from 6.2 to 82.6 mg/g and is attributed to the incorporation of cashew nut shell liquid during processing. Anacardic acid possesses antimicrobial, antiacne, and other medicinal properties. Related to anacardic acid from cashew nut are ginkgolic acids occurring in ginkgo tree (*Ginkgo biloba*, Ginkgoaceae). These 2-hydroxy-6-alk(en)ylbenzoic acids are a mixture of 13:0 and 15:0 saturated and 8 *t* and 10 *t*-15:1, 8 *t*, 10 *t*, and 12 *t*-17:1, and 17:2 unsaturated acids. Extracts of *G. biloba* leaves have long been used as dietary supplements in traditional medicine and have multiple potential therapeutic applications, including in ameliorating dementia.

Higher homologues of benzoic acid are phenylacetic and 3-phenylpropionic (3-phenypropanoic) acids (8-57), which arise by the microbial degradation of organic matter (such as lignin) and by oxidation of the corresponding aldehydes. The flavour of phenylacetic acid is reminiscent of honey. In the form of esters, it is often present in essential oils, and it is also used in perfumery. 3-Phenylpropionic acid is found in low concentrations in cheeses produced in the presence of moulds. 2-Oxo-3-phenylpropanoic acid (phenylpyruvic acid) is a product of the oxidative deamination of phenylalanine existing in equilibrium with its *E*- and *Z*-enol tautomers. The rare enolic phenylpyruvic acid-2-*O*-glucoside, (*Z*)-2-(β-D-glucopyranosyloxy)-3-phenylprop-2-enoic acid, is one of the major constituents of fermented rooibos (*Aspalathus linearis*, Fabaceae) infusions. It has been shown to enhance insulin release and glucose uptake in muscle cells.

The most common unsaturated aromatic carboxylic acid is cinnamic acid (8-57). In cinnamon and other spices, the (*E*)-isomer predominates. Cinnamic acid oxidation products are 4-hydroxycinnamic acid (also known as 4-coumaric or *p*-coumaric acid) and 3,4-dihydroxycinnamic acids, known as caffeic acid. The 3-methoxy derivative of caffeic acid is ferulic acid and the 3,5-dimethoxy derivative is sinapic acid. These so-called cinnamic acids are present in plant tissues as free substances, but in larger quantities they are found as esters, amides, or glycosides.

8-57, higher homologues of benzoic acid and common (*E*)-cinnamic acids

Cinnamic acids containing prenyl side chains represent a rare class of natural products together with prenylated coumarins, flavonoids, and some other compounds. The prenyl skeleton is attached to the aromatic compound either directly or through another atom such as oxygen or, less frequently, nitrogen or sulfur. Four types of prenyl side chain have been identified (Figure 3.10), based on the size of the carbon: C_5 (isopentenyl), C_{10} (geranyl), C_{15} (farnesyl), and C_{20} (geranylgeranyl). Compounds having the isopentenyl and geranyl side chains are more abundant in nature, whilst those with farnesyl and geranylgeranyl are rare. Examples of prenylated ferulic acids are boropinic and 4'-geranyloxyferulic acids (**8-58**) identified in grapefruit skin.

boropinic acid   4'-geranyloxyferulic acid

8-58, structures of prenylated cinnamic acids

The contents of the major phenolic acids in different fruits and vegetables differ qualitatively and quantitatively, and change significantly during fruit growth and ripening. For example, apples at harvest time contain in their f.w. about 12 mg/kg *p*-coumaric, 85 mg/kg caffeic, and 4 mg/kg ferulic acid. At maturity, the contents of these acids are 1.9, 10.4, and 0.4 mg/kg, respectively. Ripe strawberries contain as the main acid *p*-coumaric acid (175 mg/kg). Present in larger quantities are gallic (121 mg/kg), 4-hydroxybenzoic (108 mg/kg), caffeic (39 mg/kg), and vanillic (34 mg/kg) acids.

In cereal grains, phenolic acids are concentrated in the aleurone cells and other outer layers, which form the bran on milling. They are present mainly in the bound forms, linked to cell-wall structural components such as cellulose, lignin, and proteins. The majority of the feruloyl groups are attached to the cell-wall arabinoxylan via the carboxyl groups acylating the primary hydroxyl groups at the C-5 positions of the α-L-arabinofuranosyl residues. It has been shown that structural function in plant cell walls involves ferulic acid as it can form dimers (diferulic acids, also known as dehydrodiferulic acids) through oxidative cross-linking between esterified feruloyl groups. These dimers serve to cross-link cell-wall polysaccharide chains. The main ferulic acid dehydrodimer is 8-*O*-4'-diferulic acid. Other dehydrodimers include 5,5'-, 8,5'-, and 8,8'-diferulic acid, the 8,5'-diferulic acid benzofuran form, and the 8,8'-diferulic acid aryl form (**8-59**). Ferulic acid can also form trimers and tetramers. Sinapic acid dehydrodimers and sinapate–ferulate heterodimers have also been identified in cereal grains. In wheat, millet sprouts, Chinese cabbage, and parsley, dimers of 8,3-dicoumaric, 8,5-diferulic, 8-*O*-4-diferulic, and 8,8-disinapic acids were found in concentrations 0.05–2.8 mg/kg (d.w.). The monomer-to-dimer ratio ranged from 2 to 850. The phenolic acid contents in wheat and rye are given in Table 8.19. Contents in beer are illustrated in Table 8.20.

Table 8.19 Free, conjugated, and bound phenolic acid contents in wheat and rye.

Phenolic acid	Wheat (mg/kg dry matter)			Rye (mg/kg dry matter)		
	Free	Conjugated	Bound	Free	Conjugated	Bound
Benzoic acids						
4-Hydroxybenzoic	0–0.10	2.3–16.2	0–8.3	–	5.5–13.0	1.2–14.0
2,4-Dihydroxybenzoic	0–4.6	5.2–147	0–215	0.5–5.0	44.6–107	0–109
Vanillic	0–4.4	7.0–24.5	1.7–9.0	1.1–1.7	4.6–7.8	1.4–10.4
Syringic	0–5.1	1.2–22.0	0.2–13.4	1.1–2.3	0.2–3.0	0.3–12.4
Cinnamic acids						
2-Hydroxycinnamic	0–1.1	1.3–3.6	2.4–7.4	0.1–0.5	1.6–2.4	3.3–25.7
4-Coumaric	0–2.3	1.7–14.6	2.4–19.1	0.7–1.7	7.4–25.8	2.7–59.5
Caffeic	0–4.3	–	–	3.1–3.7	–	–
Sinapic	0–12.3	21.5–137	12.9–40.0	8.8–9.6	51.6–91.4	26.1–60.1
Ferulic	1.2–6.2	9.4–87.8	162–721	3.0–6.4	34.8–123	146–521
Total acids	300–>1000			491–1082		
Total free acids	3–30			11–29		
Total conjugated acids	76–416			153–349		
Total bound acids	208–964			216–711		

Table 8.20 Phenolic acid contents in beer.

Acid	Content (mg/kg)	Acid	Content (mg/kg)
Benzoic	0.4	Phenylpropionic	<0.1
Salicylic	0.02–3.1	Phenylpyruvic	0.8–1.0
4-Hydroxybenzoic	0.1–15.9	Cinnamic[a]	0.6–26.1
Gentisic	0.3–4.6	o-Coumaric	0.2–0.3
Protocatechuic	1.0–19.0	m-Coumaric	<0.1
Vanillic	0.3–1.1	p-Coumaric	1.2–7.0
Syringic	1.3–7.0	Caffeic	2.0–8.0
Gallic	1.1–29.2	Ferulic	2.2–20.8
Phenylacetic	0.5–0.9	Sinapic	0.7–3.9

[a]The (Z)-isomer content is <0.01 mg/l.

Phenolic acids and their derivatives can act as primary antioxidants. Their activity depends on the number of hydroxyl groups in the molecule. Generally, cinnamic acids and o-diphenols are more active as antioxidants (e.g. caffeic acid and its ester chlorogenic acid or the ester of caffeic acid and 3,4-dihydroxyphenyllactic acid, known as rosmarinic acid). In addition to these esters (depsides), many other phenolic acid derivatives, such as amides and glycosides, show antioxidant activity.

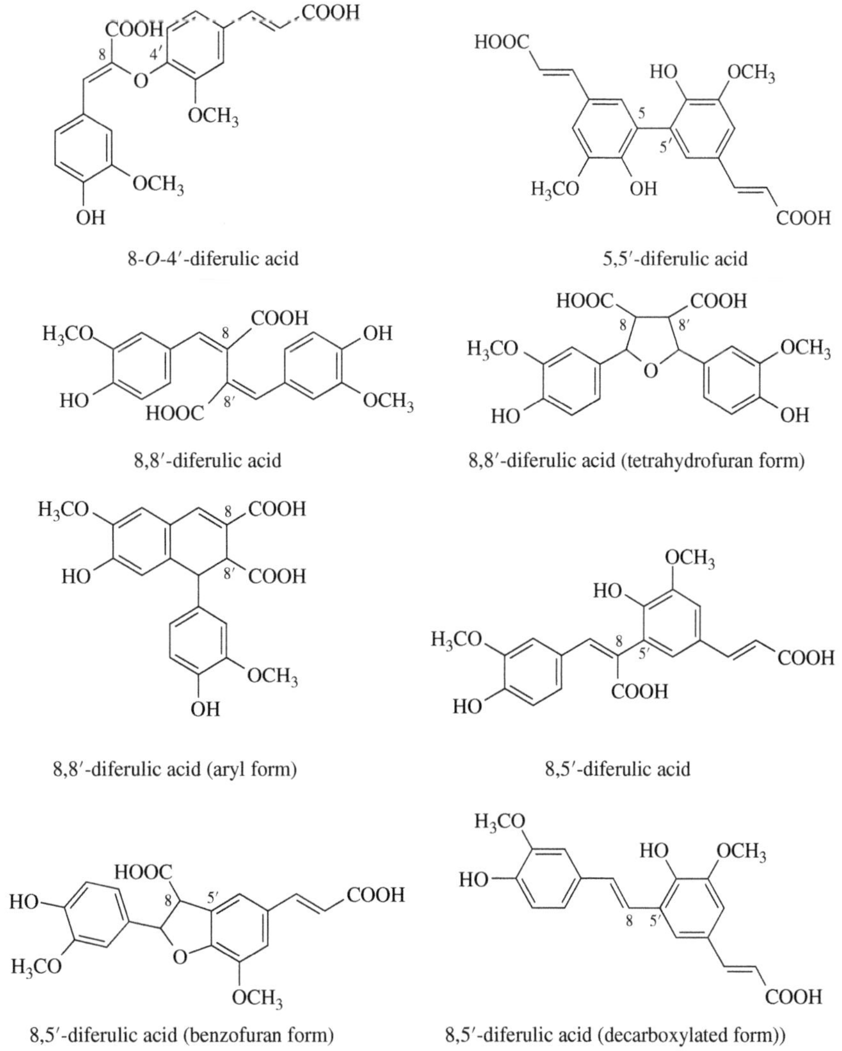

8-O-4′-diferulic acid

5,5′-diferulic acid

8,8′-diferulic acid

8,8′-diferulic acid (tetrahydrofuran form)

8,8′-diferulic acid (aryl form)

8,5′-diferulic acid

8,5′-diferulic acid (benzofuran form)

8,5′-diferulic acid (decarboxylated form))

8-59, selected ferulic acid dimers

8.2.6.1.7 Heterocyclic carboxylic acids

Plants contain a number of heterocyclic carboxylic acids, the most common of which are carboxylic acids with substituted furan and 4H-pyran (γ-pyran) skeletons. An example is furan-2-carboxylic (furoic) acid, which arises by oxidation of furan-2-carboxaldehyde and by degradation of ascorbic (dehydroascorbic) acid. Analogously, oxidation of the sugar-degradation product 5-hydroxymethylfuran-2-carbaldehyde yields 5-hydroxymethylfuran-2-carboxylic acid. A number of biologically active C_7 dicarboxylic acids are synthesised by plants across a range of plant families, such as carrots, wheat, sunflower, sugar beet, and tobacco, as secondary metabolites. Examples are vasorelaxation-active (+)-osbeckic acid, found, for example, in tartary buckwheat, (−)-daucic acid with antioxidative activity, found in carrots, chelidonic acid with acitivity in allergic reactions, first identified in celandine (*Chelidonium majus*, Papaveraceae) sap, and antiinflammatory and antibacterial meconic acid, found in plants of the Papaveraceae family, such as opium poppy (*Papaver somniferum*). These acids occur either free or esterified with various phenolic acids. An example is 4,5-di-O-caffeoyldaucic acid (**8-60**), occurring in tubers of sweet potato (*Ipomoea batatas*, Convolvulaceae).

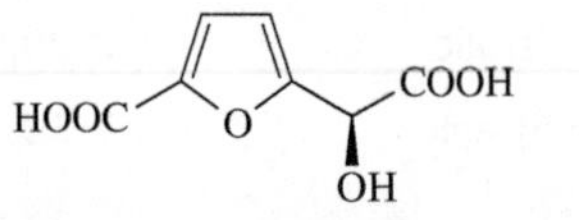

(+)-osbeckic acid

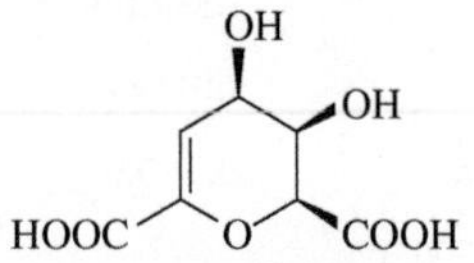

(−)-daucic acid

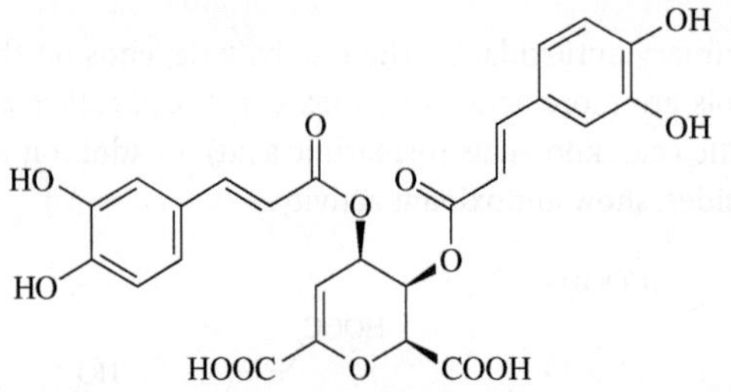

4,5-di-O-caffeoyldaucic acid

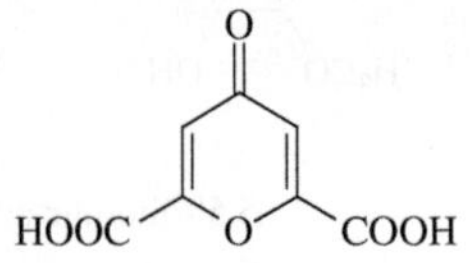

chelidonic acid

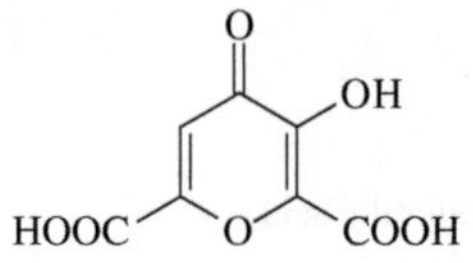

meconic acid

8-60, O-heterocyclic carboxylic acids

Carboxylic acids are also derived from S- and N-heterocyclic compounds. Many of them are natural biologically active compounds, whilst others are formed during food processing. Examples are some vitamins, such as biotin, lipoic acid, nicotinic acid, 4-pyridoxic acid, and orotic acid. Yeast autolysates, for example, contain carboxylic acids derived from thiophene.

8.2.6.2 Properties and reactions

8.2.6.2.1 Properties

In solutions and to some extent even in the gaseous state carboxylic acids are molecules associated by hydrogen bonds. The acidity of carboxylic acids is conditioned, according to electron theory, by the presence of carbonyl groups, which facilitate the dissociation of the hydroxyl to the carboxylate anion.

$$R - COOH + H_2O \rightarrow R - COO^- + H_3O^+$$

The strength of acids (acidity) is expressed by their dissociation constants K_a, which depend on temperature and the type of solvent (its relative permittivity). The acidity of carboxylic acids lies between those of the strong mineral acids and carbonic acid. An overview of the pK_a values of some acids is given in Table 8.21.

The dissociation of acids is influenced considerably by other functional groups in the vicinity of the carboxyl group. If, in the neighbourhood, there is a group repelling electrons (+I effect), such as alkyl, the acid is a weak acid (the K_a value is lower). Electronegative substituents in the vicinity of the carboxyl group, such as a hydroxyl or carbonyl group, facilitate dissociation (−I effect). The inductive effects are less observable (or unobservable) with increasing distance from the carboxyl group. The strongest aliphatic monocarboxylic acid is therefore formic acid, and hydroxycarboxylic and oxocarboxylic acids are stronger than the corresponding unsubstituted carboxylic acids.

Table 8.21 Dissociation constants of acids.

Acid	pK_a value at 25 °C			Acid	pK_a value at 25 °C		
------	pK_{a1}	pK_{a2}	pK_{a3}	------	pK_{a1}	pK_{a2}	pK_{a3}
Formic	3.75	–	–	Glyoxylic	2.98	–	–
Acetic	4.53	–	–	Quinic	3.58	–	–
Propionic	4.87	–	–	Shikimic	4.76	–	–
Butyric	4.83	–	–	Oxalic	1.20	3.67	–
Isobutyric	4.84	–	–	Succinic	4.22	5.70	–
Valeric	4.81	–	–	Fumaric	3.09	4.60	–
Sorbic	4.76	–	–	Malic	3.46	5.21	–
Benzoic	4.19	–	–	Tartaric	2.98	4.34	–
Glycolic	3.70	–	–	Citric	2.79	4.30	5.65
Lactic	3.83	–	–	Isocitric	3.28	4.71	6.39
Pyruvic	2.39	–	–	Phosphoric	2.15	7.10	12.40

8.2.6.2.2 Reactions

The most common chemical transformations of carboxylic acids in food are reactions caused by the cleavage of the O—H bond of the carboxylic group (salt formation), esterification, decarboxylation, and reactions ongoing on the hydrocarbon residue.

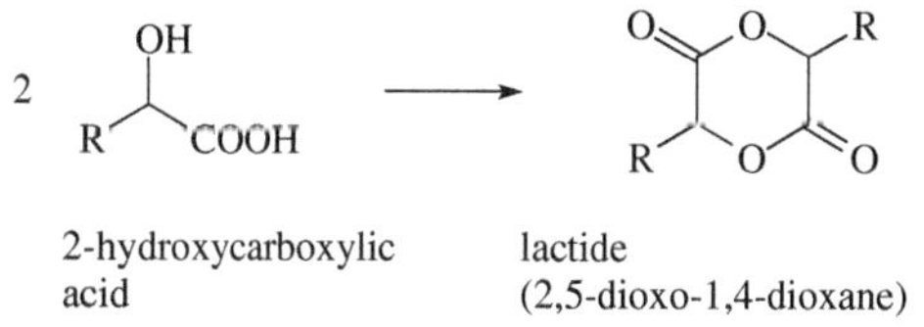

Reactions of aliphatic unsubstituted carboxylic acids The most important reaction is the formation of salts and esters. Acids usually eliminate carbon dioxide with difficulty. The prominent reaction of higher fatty acids is oxidation of the hydrocarbon chain.

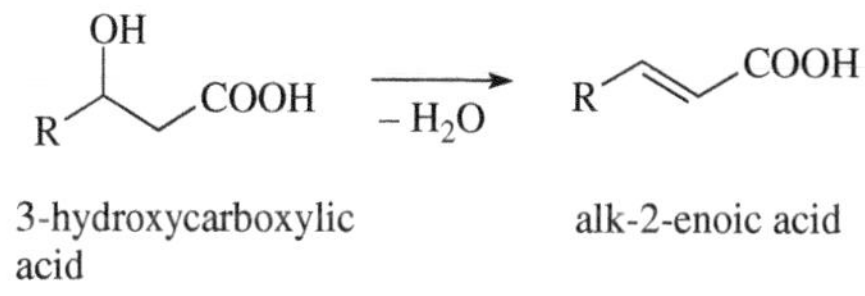

Reactions of hydroxycarboxylic acids Heating of aliphatic α-hydroxycarboxylic acids leads to their dehydration and the formation of six-membered cyclic esters, derivatives of 1,4-dioxane, called lactides. Lactic acid is an α-hydroxy acid that forms a cyclic diester 3,6-dimethyl-2,5-dioxo-1,4-dioxane (Figure 8.43, R = CH_3). Heating of 3-hydroxy acids (β-hydroxy acids) leads to elimination of water to form α,β-unsaturated (alk-2-enoic) acids (Figure 8.44). Heating of γ- and δ-hydroxy acids leads to intramolecular esterification to form lactones (γ- and δ-lactones, Figures 8.45 and 8.46). Higher lactones are not formed from hydroxy acids.

Hydroxycarboxylic acid with the hydroxyl groups at distant carbons in the molecule, such as ricinoleic acid and related hydroxy fatty acids, yields internal esters called **estolides** upon heating. Estolides have also been found in oxidised lipids (see Section 3.8.1.14).

Malic acid dehydrates to form maleic acid when heated. During wine ageing, tartaric acid yields 2,3-dihydroxymaleinic, (S)-2-hydroxy-3-oxobutanedioic, 2,3-dioxobutanedioic, formylglyoxylic (the precursor of glyoxal), and 2-hydroxymalonic (tartronic) acids (Figure 8.47). Heating of tartaric acid to temperatures of about 150 °C yields metatartaric acid lactide (see **11-39**). At higher temperatures, dehydration, and isomerisation (as with β-hydroxy acids) leads to the formation of an intermediate oxosuccinic acid (Figure 8.48), decarboxylation of which (it behaves as α-hydroxy acid) yields pyruvic acid. Citric acid at temperatures around 150 °C (e.g. during coffee roasting or when it is used as a synergist in oil deodorisation) decomposes to form anhydrides, which react with water to form aconitic, itaconic, and citraconic acids (Figure 8.49). Isomerisation of itaconic acid gives rise to citraconic acid.

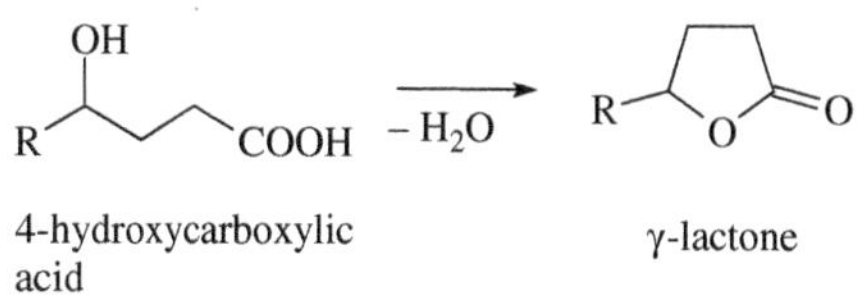

2-hydroxycarboxylic acid → lactide (2,5-dioxo-1,4-dioxane)

Figure 8.43 Dehydration of α-hydroxycarboxylic acids.

3-hydroxycarboxylic acid → alk-2-enoic acid

Figure 8.44 Dehydration of β-hydroxycarboxylic acids.

4-hydroxycarboxylic acid → γ-lactone

Figure 8.45 Dehydration of γ-hydroxycarboxylic acids.

5-hydroxycarboxylic acid → δ-lactone

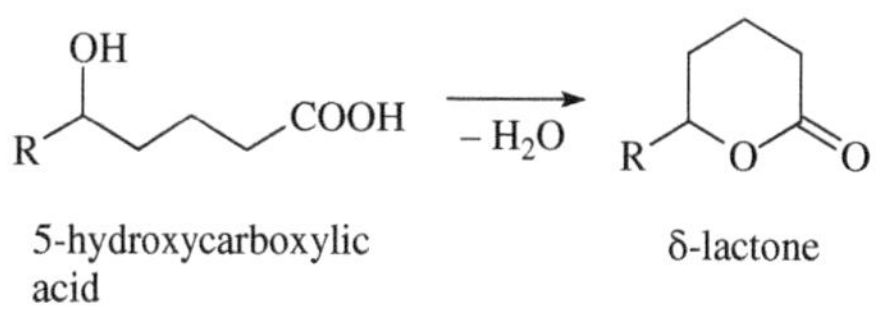

Figure 8.46 Dehydration of δ-hydroxycarboxylic acids.

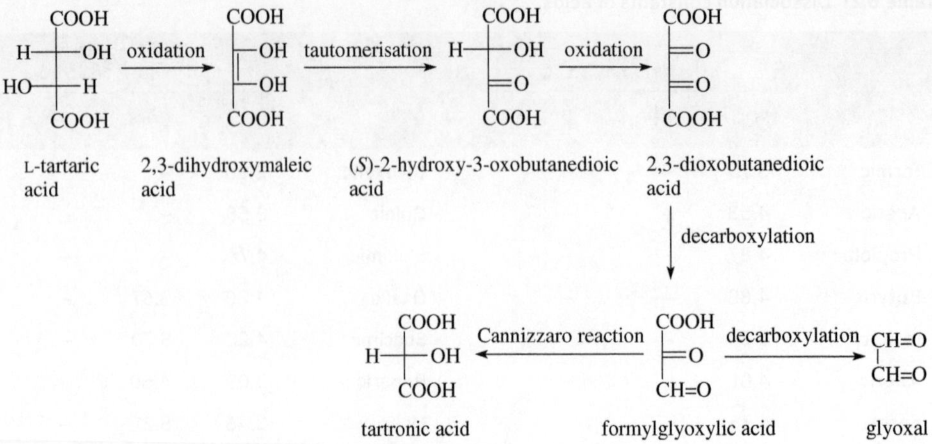

Figure 8.47 Reactions of tartaric acid during wine ageing.

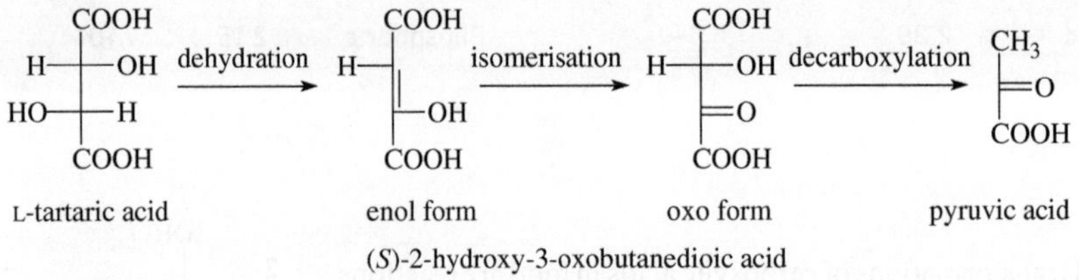

Figure 8.48 Thermal degradation of tartaric acid.

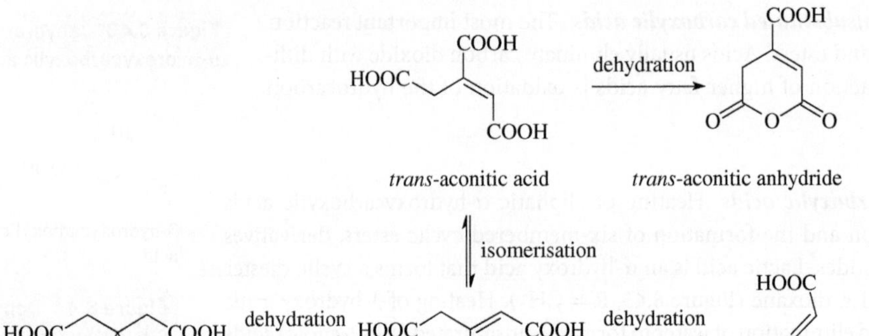

Figure 8.49 Thermal degradation of citric acid.

Reactions of oxocarboxylic acids Thermal degradation of 2-oxocarboxylic acids (α-ketocarboxylic acids) proceeds with the elimination of carbon monoxide (decarbonylation, Figure 8.50) and the formation of the corresponding carboxylic acids. Heating of 3-oxoacids (β-keto acids) leads to decarboxylation and the formation of alkyl methyl ketones (Figure 8.51). Higher keto acids, such as γ-keto acids, partly isomerise by heating to the corresponding unsaturated hydroxy acids, which form lactones by dehydration. For example, laevulinic acid yields two isomeric lactones, α-angelica lactone and β-angelica lactone (see **8-75**).

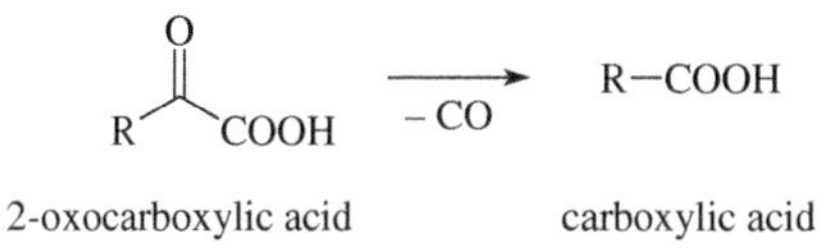

Figure 8.50 Thermal degradation of α-oxocarboxylic acids.

Reactions of aromatic carboxylic acids Phenolic acids are precursors of a number of simple phenols, which result from the activities of microorganisms or during thermal processes. The main products of thermal degradation (decarboxylation) of cinnamic acids are 4-vinyl phenols. The subsequent reactions yield the corresponding 4-formyl phenols, 4-ethyl phenols, 4-(prop-2-en-1-yl) phenols, 4-acetyl phenols, and many other compounds. The oxidation of phenolic acids and the formation of condensation products are described in Section 9.12.4.

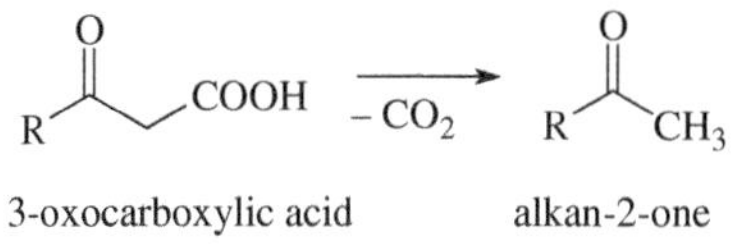

Figure 8.51 Thermal degradation of β-oxocarboxylic acids.

8.2.7 Functional derivatives of carboxylic acids

Numerous functional derivatives of carboxylic and substituted carboxylic acids have been found to be natural food components. In particular, important compounds are esters and lactones that act as odour-active substances. Other important functional derivatives are anhydrides, amides, and nitriles.

8.2.7.1 Classification, structure, nomenclature, and occurrence

8.2.7.1.1 Esters

Volatile esters of aromatic carboxylic acids are often important components of plant floral scents and the aromas of fruits, vegetables, and spices. In the flowers of plants, esters serve as signalling molecules, attractants, and repellents and act in various defence mechanisms. More than 1000 different compounds have been identified.

Flowers, medicinal herbs, spices, fruits, and vegetables also contain numerous non-volatile esters. For example, common esters in grapes are diethyl malate, diethyl tartrate, and depsides of tartaric acid with phenolic acids. An important group of non-volatile esters occurring in many plant materials is depsides of phenolic acid with quinic, shikimic, malic, and tartaric acids. For example, in coffee, esters of (E)-cinnamic acids with L-quinic acid, known as chlorogenic acid, occur. Most have virtually no effect on the odour of foods. However, they can influence their taste and colour, either by themselves or through some of their reaction products. For example, chlorogenic acid contributes to the bitter taste of coffee and has antioxidant, hypoglycaemic, antiviral, hepatoprotective, and immunoprotective effects.

Esters of higher fatty acids are classified either as lipids or as lipid-accompanying substances. Some of these (esters of higher fatty acids with lower aliphatic alcohols, such as ethanol) are odour-active constituents found mainly in alcoholic beverages, but they usually only affect the taste or are flavour-indifferent. Certain esters of higher fatty acids (e.g. esters of xanthophylls) are important colouring substances in foods of plant origin.

Volatile esters Esters are reaction products of carboxylic acids with alcohols of general formula R^1-CO-OR^2. They belong to the most widespread compounds found in foods, where they are often accompanied by the corresponding carboxylic acids and alcohols. The odour-active esters are mostly common esters of monocarboxylic acids (R^1-COOH), whilst esters of polycarboxylic acids are found less frequently. Alcohols bound in esters are either monohydric alcohols (R^2-OH containing only one hydroxyl group in each molecule) or polyhydric alcohols. Esters of lower aliphatic acids with lower aliphatic and aromatic alcohols are odour-active compounds. They are especially important components of the primary flavour of fruits, vegetables, beverages, and various spices, where they arise by alcoholysis of acyl coenzymes A:

$$R^1 - CO - SCoA + R^2 - OH \rightarrow R^1 - CO - O - R^2 + HS - CoA$$

For example, ethanolysis of acetyl-CoA produces ethyl acetate. In disrupted plant tissues, such as during production of juices, esters are rapidly broken down by various hydrolases, resulting in a change of the flavour character. During heating and long storage of foods, esters may also be formed in small quantities as secondary odour-active compounds. For example, esters are formed during ageing of wines and spirits by esterification of carboxylic acids with alcohols (ethanol or fusel oil alcohols, although non-enzymatic esterification is a very slow reaction), acidolysis (reaction of acids with esters), alcoholysis (reaction of alcohols with esters), or ester exchange (reaction of esters with other esters).

The lower fatty acid most frequently bound in esters is acetic acid, whilst formic, propionic, butyric, and isobutyric acids occur less often. The common alcohol bound in esters is ethanol. However, esters of methanol, allyl alcohol, and butan-1-ol, higher alcohols, and very often esters of monoterpenic and aromatic alcohols also occur in foods, and esters of sulfur-containing alcohols are common. Esters of low-molecular-weight acids and alcohols usually have a fruity odours; those of terpenic alcohols with low-molecular-weight acids tend to have fragrant odours resembling flowers. Esters of aromatic acids and aromatic alcohols generally have heavy balsamic odours.

The most common ester of alcoholic beverages is ethyl acetate (8-61). In beer, for example, ethyl acetate concentrations range from 25 to 33 mg/l, and in wines from 11 to 261 mg/l. The threshold concentration is 150–170 mg/l; therefore, ethyl acetate in wines very often acts as an odour-active component. Ethyl acetate in spirits (such as whisky) is present at levels of 45–460 mg/l. Other esters present in high concentrations in alcoholic beverages include isopentyl (isoamyl) acetate and isobutyl acetate (8-61), which arise, along with ethyl acetate, by alcoholysis of acetyl-CoA through the respective alcohols. Isobutyric and isovaleric acids yield virtually no esters in beer and wine, as the non-enzymatic esterification is too slow and the equilibrium is shifted to the detriment of esters. In spirits, these esters occur at levels up to units of mg/l.

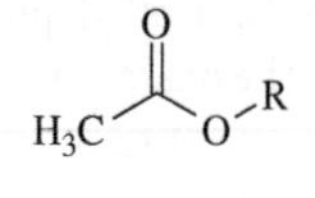

8-61, selected acetic acid esters (acetates)

ethyl acetate, $R = CH_2CH_3$
isobutyl acetate, $R = CH_2CH(CH_3)_2$
isopentyl acetate, $R = CH_2CH_2CH(CH_3)_2$

Fatty acids bound in esters tend to have an even number of carbon atoms. These esters belong to the aromatic substances influencing the bouquet of wines. For example, they are important carriers of the fine flavour of Riesling wines and occur in larger quantities in heavy aromatic varieties such as Traminer, Muscat Ottonel, and Sauvignon. The content of the main esters in wines is presented in Table 8.22. The ester content of musts is about 10 times lower. For example, the aroma of hexyl acetate is described as green, fruity, sweet, fatty, fresh, apple, and pear, whilst the aroma of ethyl hexanoate is sweet, fruity, pineapple, waxy, fatty, and estery with a green banana nuance. Some sulfur-containing esters are also characteristic aromatic constituents of wines. For example, 3-mercaptohexyl acetate, produced by the fermentation of 3-mercaptohexan-1-ol (8-11), may occur in some wines (such as Traminer and Riesling) at a level of 0.5 μg/l, which is perceived as a fruity flavour.

Terpenoid esters occur in various fruits and many spices. Examples are the fairly widespread geranyl and neryl acetates (8-62), which are components of citrus oils and odours of some spices (such as coriander).

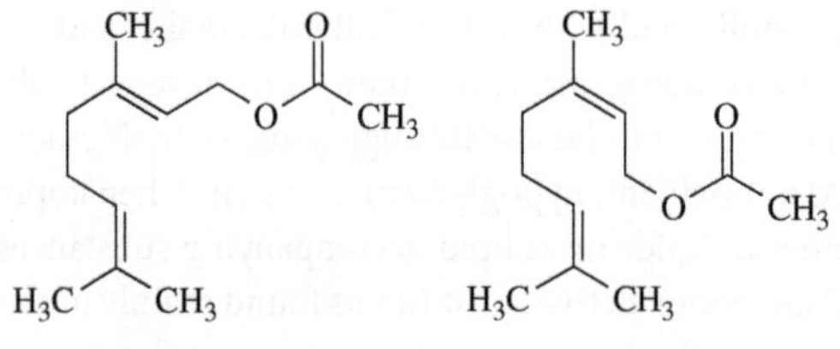

geranyl acetate   neryl acetate

8-62, selected acetataes of monoterpenic alcohols

Table 8.22 Contents of main esters in wines.

Ester[a]	Content(mg/kg)	Ester[a]	Content(mg/kg)
Ethyl acetate	11–261	Ethyl butyrate	0.2–4
Isobutyl acetate	0.1–11	Ethyl capronate	0.2–9
Isoamyl acetate	0.9–12	Ethyl caprylate	0.3–9
Hexyl acetate	0–9	Ethyl caprinate	0.1–9
Phenylethyl acetate	0–9	Ethyl lactate	12–378

[a]Composition refers to wines from the common grape vine *Vitis vinifera* (Vitiaceae). The natural component of grapes and wines from fox grape (*V. labrusca*) and some of its hybrids, originating in North America, is methyl anthranilate and ethyl anthranilate. In wines, these esters occur at levels up to 3.1 mg/l. Along with 2-aminoacetophenone, formed by degradation of tryptophan, they may cause the characteristic unpleasant odour.

Non-volatile esters An example of a simple non-volatile phenolic acid ester is caffeic acid phenylethyl ester occurring in propolis that possesses many biological activities, including antibacterial, antiviral, antioxidant, anti-inflammatory, and anticancer effects. The main aromatic carboxylic acids of coffee are **chlorogenic acids**, esters (depsides) of L-quinic acid with predominantly (*E*)-cinnamic, caffeic, ferulic, and 4-coumaric acids. In smaller amounts, chlorogenic acids are found in tea, cocoa, apples, pears, other fruits, vegetables, potatoes, and other products. Green and roasted coffees, for example, contain more than 50 cinnamic acids esters. Their level in green coffee beans ranges from 4 to 14%. The main groups are caffeoylquinic, dicaffeoylquinic, feruloylquinic, *p*-coumaroylquinic, and caffeoylferuloylquinic acids. Moreover, all esters of quinic acid are found in three types of positional isomers that include 3-*O*-esters, 4-*O*-esters, and 5-*O*-esters. The main component is chlorogenic acid (i.e. 3-*O*-caffeoyl-L-quinic acid), but this name also refers to all the natural quinic acid esters. The structures of the basic esters of caffeoylquinic acids are given in formulae **8-63**.

Numerous minor chlorogenic acids have recently been identified in coffee, including diferuloylquinic, di-*p*-coumaroylquinic, dimethoxycinnamoylquinic, and other chlorogenic acids, such as 3,4-di-*p*-coumaroylquinic, 3,5-di-*p*-coumaroylquinic, and 4,5-di-*p*-coumaroylquinic acids; 3-*p*-coumaroyl-4-caffeoylquinic, 3-*p*-coumaroyl-5-caffeoylquinic, 4-*p*-coumaroyl-5-caffeoylquinic, 3-caffeoyl-4-*p*-coumaroylquinic, 3-caffeoyl-5-*p*-coumaroylquinic, and 4-caffeoyl-5-*p*-coumaroylquinic acids; 3-*p*-coumaroyl-4-feruloylquinic, 3-*p*-coumaroyl-5-feruloylquinic, and 4-*p*-coumaroyl-5-feruloylquinic acid; and 4-dimethoxycinnamoyl-5-*p*-coumaroylquinic acid. The minor chlorogenic acids amount to less than 1% of all chlorogenic acids.

The content of chlorogenic acids in roasted coffee and coffee beverages depends on the type and method of roasting. Of about 80 varieties of three coffee plants, namely Arabic coffee (*Coffea arabica*, Rubiaceae), robusta coffee (*C. canephora*), and Liberian coffee (*C. liberica*), only the first two are of commercial importance. *C. arabica* (the most common varieties are *typica*, *bourbon*, *mocha*, and *maragogips*) represents about 75%, *C. canephora* (varieties *robusta*, *typical*, and others) about 25%, and *C. liberica* about 1% of world production. All varieties of *C. canephora* are commercially known by the name robusta. Their seeds have a higher content of chlorogenic acids than those of *C. arabica*. Dicaffeoylquinic acids and mixed diesters (caffeoyl-feruloyl quinic acids) and derivatives of some amino acids

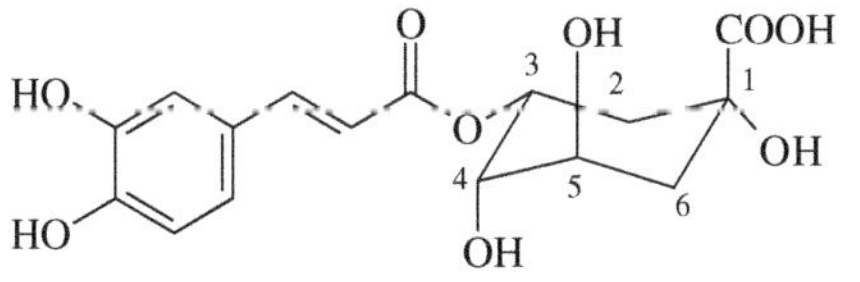

3-*O*-caffeoyl-L-quinic(chlorogenic) acid

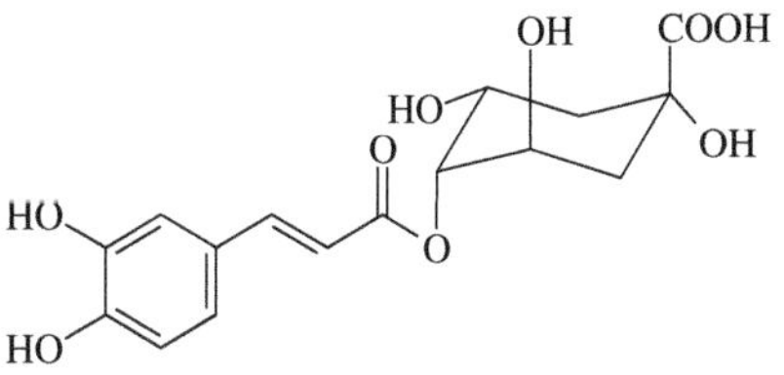

4-*O*-caffeoyl-L-quinic (cryptochlorogenic) acid

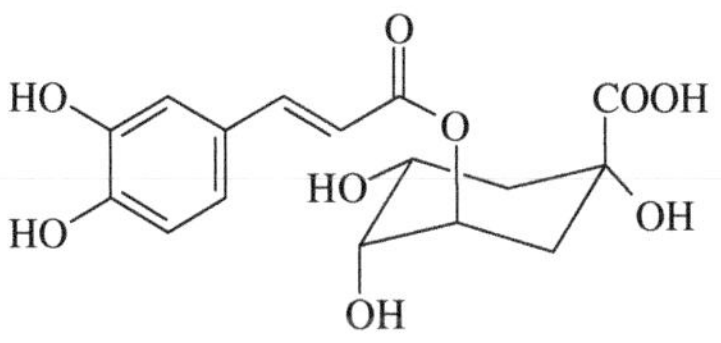

5-*O*-caffeoyl-L-quinic (neochlorogenic) acid

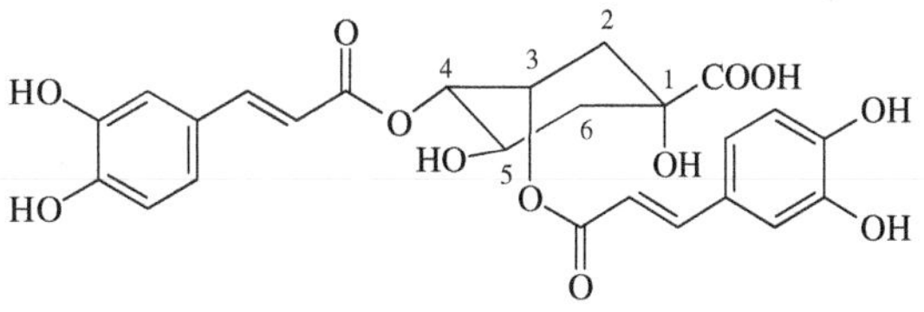

3,4-di-*O*-caffeoyl-L-quinic (isochlorogenic a) acid

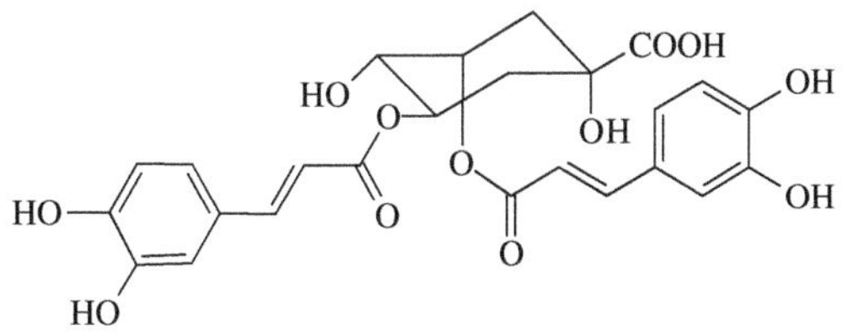

3,5-di-*O*-caffeoyl-L-quinic (isochlorogenic b) acid

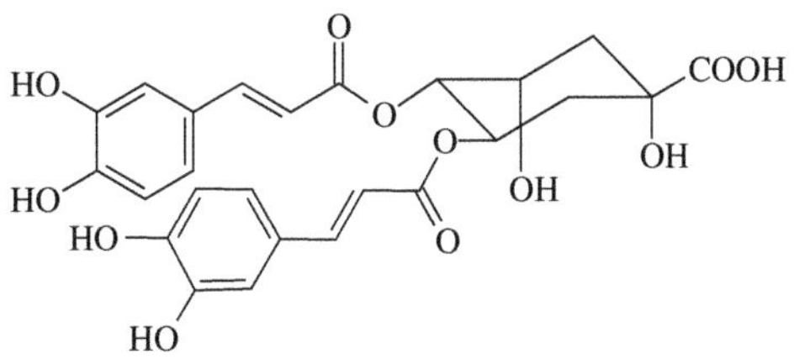

4,5-di-*O*-caffeoyl-L-quinic (isochlorogenic c) acid

8-63, major chlorogenic acids

Table 8.23 Contents of major hydroxycinnamic acids in roasted coffee (*Coffea arabica*).

Acid	Content (mg/kg)	Acid	Content (mg/kg)
3-Caffeoylquinic (chlorogenic)	20	3,4-Dicaffeoylquinic (isochlorogenic a)	0.1
4-Caffeoylquinic (cryptochlorogenic)	2	3,5-Dicaffeoylquinic (isochlorogenic b)	0.9
5-Caffeoylquinic (neochlorogenic)	1	4,5-Dicaffeoylquinic (isochlorogenic c)	0.1

(such as *N*-caffeoyl-L-tryptophan and *N*-caffeoyl-L-tyrosine) are found only in robusta coffee. The chlorogenic acid content in coffee surrogates, such as roasted chicory root, is lower than in roasted coffee seeds.

During roasting of coffee seeds, about 30–70% chlorogenic acids are transformed into γ-lactones of 3-*O*- and 4-*O*-cinnamoylquinic acids (cinnamoyl-1,5-quinolactones) and other products. These lactones are the most important bitter substances of roasted coffee. The main component is 3-*O*-caffeoyl-L-quino-1,5-lactone, also known as 3-*O*-caffeoyl-γ-quinide (see Section 8.3.5.1.4). The chlorogenic acid contents in roasted coffee are given in Table 8.23.

The brewing process leads to the migration of bound acyl groups, so, for example, 4,5-dicaffeoylquinic acid produces 3,4- and 3,5-dicaffeoylquinic acids. Hydroxylation of the chlorogenic acid olefinic cinnamoyl moieties by the addition of water results in the formation of 3-hydroxydihydrocaffeic acid derivatives. This reaction is reversible, and elimination of water from 3-hydroxydihydrocaffeic acids yields (*Z*)-isomers of caffeoylquinic acids (*trans/cis*-isomerisation).

One of the best sources of chlorogenic acids in nature is yerba mate (*Ilex paraguariensis*, Aquifoliaceae). Toasted leaves of this shrub are used to make the beverage known as mate in South America and can also be found in various energy drinks on the market today (see Section 10.3.6.10). Total chlorogenic acids contents in green mate varies from 87 to 132 g/kg (d.w.). Content in toasted mate varies from 15 to 46 g/kg. Overall, caffeoylquinic acid isomers were the most abundant chlorogenic acids in both green and toasted mate, followed by dicaffeoylquinic acids and feruloylquinic acids. These classes accounted for 58.5, 40.0, and 1.5 of chlorogenic acids, respectively, in green mate and 76.3, 20.7, and 3.0%, respectively, in toasted mate. When leaves are toasted, some isomers are partly transformed into 1,5-quinolactones. Average contents of 3-*O*-caffeoyl-L-quino-1,5-lactone, and 4-*O*-caffeoyl-L-quino-1,5-lactone in commercial toasted samples were 1015 and 618 mg/kg (d.w.), respectively.

Raw potatoes contain 100–200 mg/kg of chlorogenic acids, boiled potatoes contain about 35% of this amount, and baked potatoes contain almost none.

Some caffeic acid esters found in aromatic and medicinal plants have significant biological effects. An example is caffeic acid ester with (+)-(*R*)-3,4-dihydroxyphenyllactic acid, known as rosmarinic acid (**8-64**), which is an important antioxidant distributed mainly in plants of the mint family (Lamiaceae), such as rosemary, oregano, sage, and thyme. Another example is 1,5-di-*O*-caffeoyl-L-quinic acid, known as cynarin (**8-64**), which occurs in the globe artichoke (*Cynara cardunculus* var. *scolymus*), a thistle cultivated as a food. Extracts containing cynarin have positive effects on hepatobiliary diseases (liver and biliary diseases), hyperlipidemia, and cholesterol metabolism. A representative of shikimic acid depsides is 3-*O*-caffeoyl-L-shikimic acid, known as dactyliferic acid (**8-64**), which occurs in dates. This and other related depsides are substrates of oxidoreductases in enzymatic browning reactions in dates.

The seeds and shoots of plants of the genus *Brassica* of the cabbage family (Brassicaceae) contain various esters of cinnamic acids with malic acid, such as 2-*O*-(4-coumaroyl)-L-malic, 2-*O*-caffeoyl-L-malic (**8-64**), 2-*O*-feruloyl-L-malic, and 2-*O*-sinapoyl-L-malic acids. Grape musts and wines contain depsides of some phenolic acids with L-tartaric acid that are prone to enzymatic oxidation reactions during the winemaking process, forming *o*-quinones and leading to colour darkening. Common compounds are depsides of caffeic (**8-64**), 4-coumaric, ferulic, and vanillic acids. For example, the vanilloyltartaric acid content in musts and wines ranges from 1.4 to 11.7 mg/l, the 4-coumaroyltartaric (cutaric) acid content is is 0.6–5.5 mg/l, and the caffeoyltartaric (caftaric) acid content is 10.2–26.9 mg/l, reaching up to 100 mg/l in air-protected fresh juices. A few mg/l of fertaric acid (the ferulic analogue) are usually found. In all these depsides, (*E*)-isomers dominate. 2,3-Di-*O*-caffeoyl-L-tartaric acid, known as cichoric acid (**8-64**), is an example of a diester of L-tartaric acid, occurring in plants of the Asteraceae family. Cichoric acid is situated in the root and leaves of chicory (*Cichorium intybus*), endive (*Cichorium endivia*), and lettuce (*Lactuca sativa*).

Examples of phenolic acid derivatives with less common skeletons are origanine A (**8-65**), B, and C, polyphenolic compounds found in oregano. Their skeletons of cyclohexenetetracarboxylic acid are attached with bioactive moieties including 3,4-dihydroxyphenyl, 4-*O*-β-D-glucoside of 4-hydroxybenzyl alcohol (known as gastrodine), and 3-(3,4-dihydroxyphenyl)lactic acid residues.

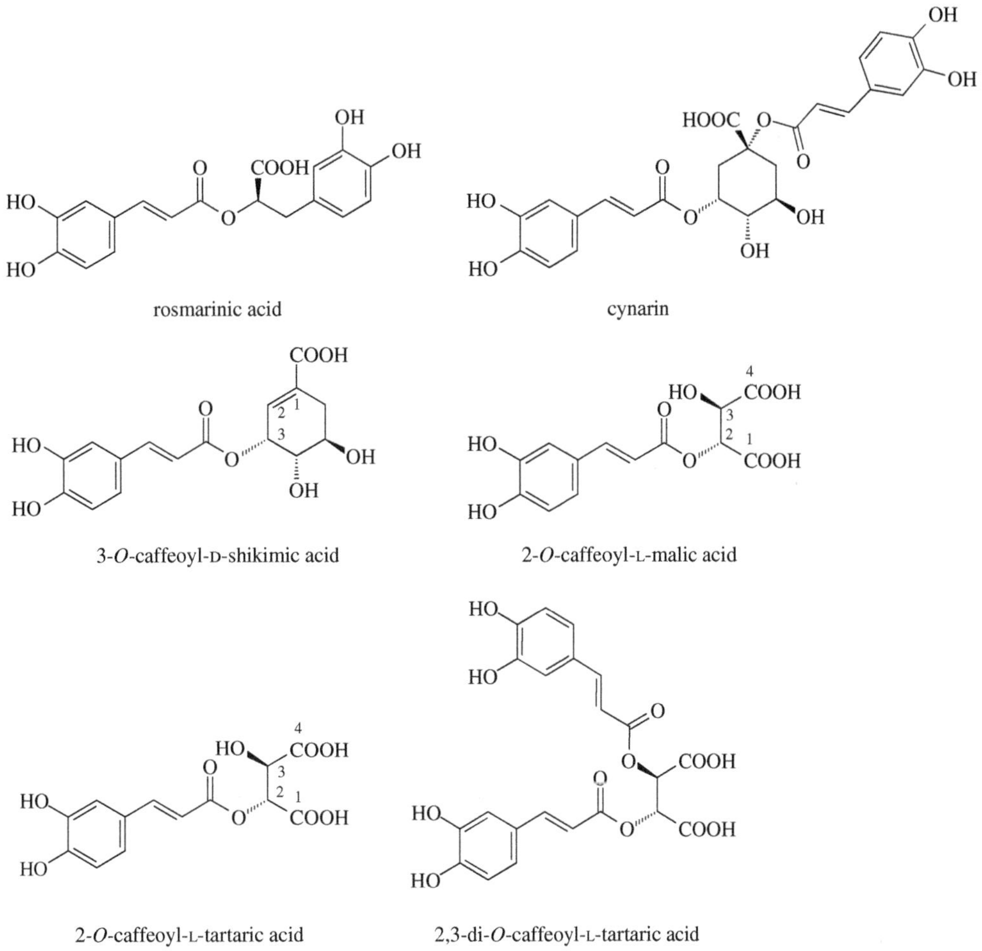

rosmarinic acid

cynarin

3-*O*-caffeoyl-D-shikimic acid

2-*O*-caffeoyl-L-malic acid

2-*O*-caffeoyl-L-tartaric acid

2,3-di-*O*-caffeoyl-L-tartaric acid

8-64, selected caffeic acid esters

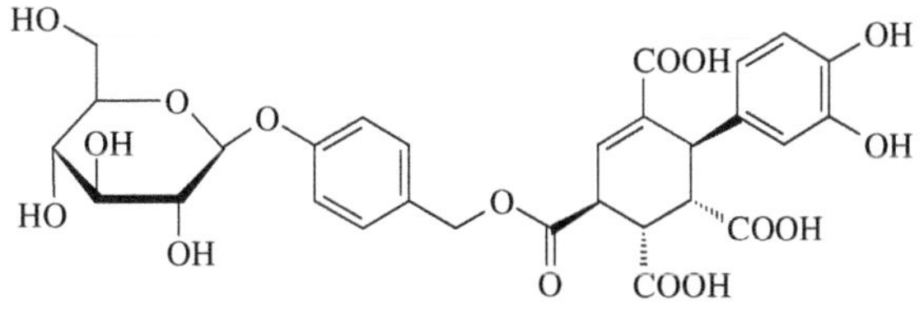

8-65, origanine A

Fruit juice concentrate of Japanese apricot of the Armeniaca section of the genus *Prunus* (*Prunus mume*), which is effective against influenza A infection, contains, as the active component, esters of 5-hydroxymethyl-2-furancarbaldehyde with malic (**8-66**) and citric acids.

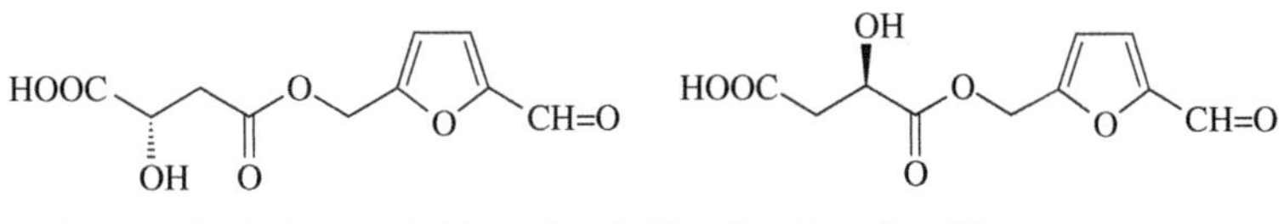

8-66, esters of 5-hydroxymethylfuran-2-carbaldehyde with malic acid

Esters of sinapic acid, such as sinapoyl choline and sinapoyl malate (2-O-sinapoyl-L-malic acid) (**8-67**), are found primarily in cruciferous plants (Brassicaceae), where they belong to the main phenylpropanoids accumulated. Their counter-ions are normally thiocyanate ions (S=C=N⁻ ↔ ⁻S—C≡N) and, in the intact seeds, glucosinolates. Sinapine-glucoraphanin salt has recently been isolated from broccoli seeds.

Table 8.24 Phenolic acid contents in rapeseed flour.

Acid	Content (mg/kg)		Acid	Content (mg/kg)	
	Free acids	Esters		Free acids	Esters
Salicylic	1–31	7–10	o-Coumaric	3–11	–
Gentisic	trace–8	trace–9	p-Coumaric	trace–30	trace–8
4-Hydroxybenzoic	0–22	trace–27	(E)-Caffeic	trace–18	0–trace
Protocatechuic	4–14	trace–18	(Z)-Caffeic	–	trace–21
Vanillic	trace–9	trace–12	(E)-Ferulic	9–47	8–79
Syringic	1–24	trace–23	(E)-Sinapic	35–516	1713–5971
Cinnamic	trace–10	–	(Z)-Sinapic	32–101	445–989

Sinapoyl choline is found exclusively in plant seeds, which means that it occurs also in the seeds of oilseed rape (rapeseed, *Brassica napus*) and turnip rape (*B. rapa*, syn. *B. campestris*), used as a source of rapeseed oil. The seeds contain up to 1% sinapoyl choline, known as sinapine. In sinapine, both sinapic acid isomers occur, but the (E)-isomer always prevails. The term 'sinapines' is used generally for all choline esters with phenolic acids. The composition of the phenolic acids in rapeseed flour is shown in Table 8.24. About 85% are bound in sinapines, whilst the rest are free acids or their glycosides. 4-Hydroxybenzoic acid, rather than sinapic acid, prevails in white mustard seeds (*Sinapis alba*), which are used raw or roasted in dishes and, mixed with other ingredients, produce a mustard paste or sauce that is used as a condiment. Synthetic lipophilic sinapates such as octyl sinapate significantly improve the oxidative stability of rapeseed oil.

In germinating seeds, the choline esters are transformed into the corresponding malates. In young plants, 2-O-sinapoyl-L-malic acid predominates, whilst the main depsides in older plants are 2-O-(4-coumaroyl)-L-malic acid and 2-O-caffeoyl-L-malic acid. Sinapoyl malate protects plants against UV-B radiation (280–320 nm); the function of sinapoyl choline is still mainly unknown.

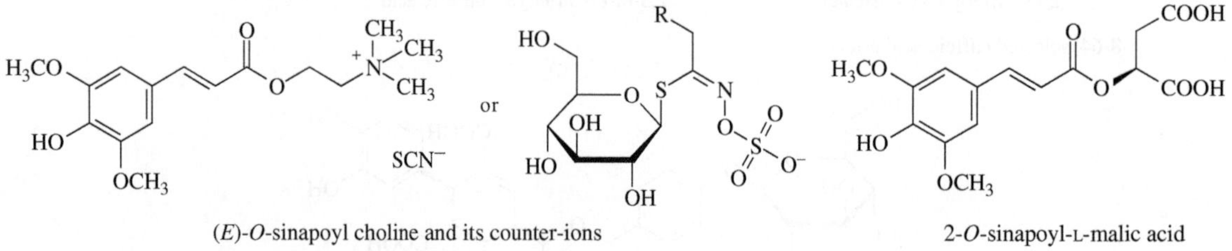

(E)-O-sinapoyl choline and its counter-ions   2-O-sinapoyl-L-malic acid

8-67, sinapoyl choline, its counter-ions, and sinapoyl malic acid

Sinapoyl esters are considered antinutritional compounds because they cause a dark colour and bitter and astringent taste in rapeseed meal and extracted protein products and have negative effects on the digestibility of rapeseed meal. The intensity of the bitter taste is comparable to that of the bitter taste of caffeine. In rapeseed oil, sinapines form complexes with proteins that are removed in the refining process. Sinapines (or fish meal) present in the diet of some breeds of laying hens can cause an unusual fishy off-flavour from freshly laid eggs. The eggs with fishy flavour have higher trimethylamine concentration in the yolk. *In vivo*, trimethylamine is produced in the gut by bacteria and metabolised to the odourless N-oxide (**8-95**).

8.2.7.1.2 Lactones

Lactones are internal esters of hydroxycarboxylic acids. They are usually characterised by the number of carbon atoms (aceto = 2 carbon atoms, propio = 3, butyro = 4, valero = 5, capro = 6), to which is added the suffix 'lactone' and a Greek letter prefix that indicates the size of the lactone ring (**8-68**). The nomenclature of lactones is based on the nomenclature of hydroxycarboxylic acids. The simplest α-lactone is hypothetically derived from α-hydroxycarboxylic acid and is ethano-2-lactone. β-Lactone derived from β-hydroxycarboxylic acids is propano-3-lactone (formerly also known as 3-propanolide), which can be considered an internal ester of 3-hydroxypropanoic acid. The simplest γ-lactone is butano-4-lactone, also known as 4-butanolide, which is derived from 4-hydroxybutanoic acid. The basic member of the homologous series of δ-lactones is pentano-5-lactone, also known as 5-pentanolide, which is derived from 5-hydroxypentanoic acid, whilst ε-lactone is hexano-6-lactone or 6-hexanolide. Lactones can also be considered as oxygen heterocyclic compounds. α-Lactones are actually oxiranes or 1,2-ethyleneoxides, β-lactones are oxetanes or 1,3-propyleneoxides, γ-lactones are oxolanes or tetrahydrofuranes, δ-lactones are oxanes or tetrahydropyranes, and ε-lactones are oxepanes, hexahydrooxepines, or hexamethylene oxides.

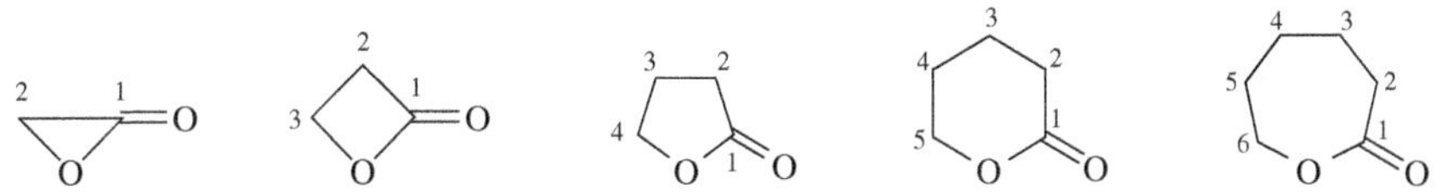

α-acetolactone β-propiolactone γ-butyrolactone δ-valerolactone ε-caprolactone

8-68, lactones

Naturally occurring lactones are mainly saturated and unsaturated γ- and δ-lactones, and to a lesser extent macrocyclic lactones. γ-Lactones and δ-lactones are intramolecular esters of the corresponding hydroxycarboxylic acids and contribute to the aroma of many foods. Other lactones arise from other reactions that are not common in foods. For example, dehydration of β-hydroxy acids does not produce β-lactones, but gives rise to α,β-unsaturated carboxylic acids.

Lactones occur as natural odorants in all major food commodities, including meat and meat products, milk, dairy products, cereals, fruits, vegetables, and various beverages, such as tea, wine, and spirits. Odour-active compounds in foods, mainly γ- and δ-lactones, are formed from aliphatic saturated and unsaturated γ-hydroxycarboxylic and δ-hydroxycarboxylic acids derived from fatty acids or sugars, but some lactones also arise from other precursors (e.g. mint lactone is a terpenoid compound and pantolactone is produced by hydrolysis of pantothenic acid via pantoic acid).

Some furanones and pyranones may also be seen as a subclass of lactones The most important representatives of these compounds are phthalides that are 2-benzofuran-1(3H)-ones or 3,4-dihydroisocoumarins (3,4-dihydroisochromen-1-ones) and, especially, coumarins (**8-69**) that are 2H-1-benzopyran-2-ones or 2H-chromen-2-ones (δ-lactones of 2-hydroxycinnamic acids).

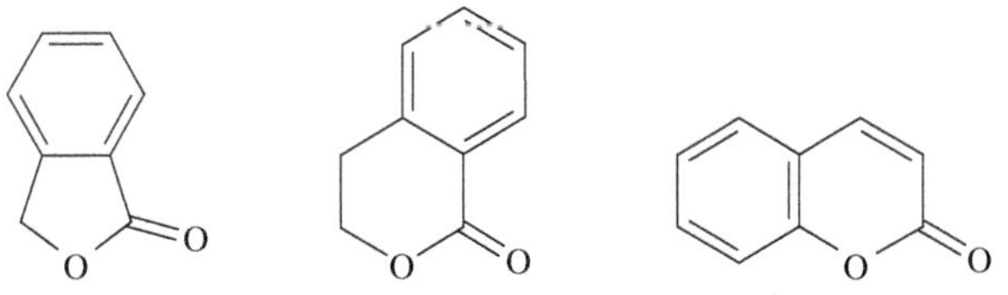

phthalide   3,4-dihydroisocoumarin   coumarin

8-69, common benzolactones

Lactones of aliphatic hydroxycarboxylic acids The simplest lactone is γ-butyrolactone (butane-4-lactone, **8-68**), which is found as common constituent of many foods in small amounts. γ-Butyrolactone has a faintly sweet odour reminiscent of rancid butter. γ-Lactones (**8-70**) and δ-lactones (**8-71**) of higher aliphatic straight-chain hydroxycarboxylic acids are odour-significant components of fruits and dairy products. The odour of γ-nonalactone (nonano-4-lactone) is reminiscent of coconut. γ-Decalactone (decano-4-lactone) has a fruity odour reminiscent of peaches, and is a component of apricots and strawberries. δ-Decalactone (decano-5-lactone), the odour of which resembles peaches, is a component of raspberries and coconuts. γ-Undecalactone (undecano-4-lactone) has an intense peach-like odour, and the odour of γ-dodecalactone (dodecano-4-lactone) is reminiscent of peaches with butter notes. δ-Dodecalactone (dodecano-5-lactone) and γ-decalactone are the major carriers of the flavour of condensed milk and heated milk fat (butter). Dishes prepared with butter then have a typical odour reminiscent of clarified butter. Some lactones, such as γ-undecalactone and δ-dodecalactone, are used for flavouring dairy products and margarines. All these lactones exist as pairs of enantiomers, the abundance of which varies greatly, but the dominating forms are (+)-(R)-stereoisomers. For example, coconut oil contains 75% of the (R)-isomer of δ-decalactone and 25% of the desired (S)-isomer (**8-72**), the content of (R)-γ-dodecalactone in raspberries is about 50% of the total γ-dodecalactone concentration, whilst strawberries contain this enantiomer exclusively.

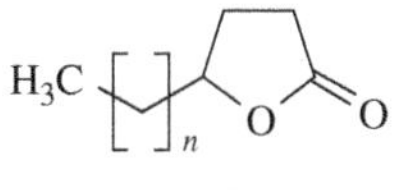

8-70, γ-lactones of saturated hydroxycarboxylic acid

nonano-4-lactone, $n = 4$
decano-4-lactone, $n = 5$
undecano-4-lactone, $n = 6$
dodecano-4-lactone, $n = 7$

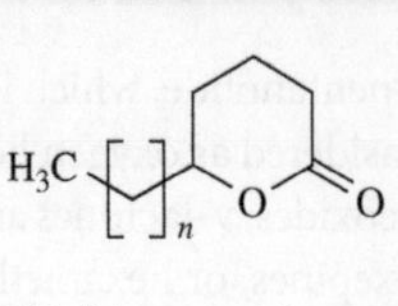

8-71, δ-lactones of saturated hydroxycarboxylic acids

nonano-5-lactone, $n = 3$
decano-5-lactone, $n = 4$
undecano-5-lactone, $n = 5$
dodecano-5-lactone, $n = 6$

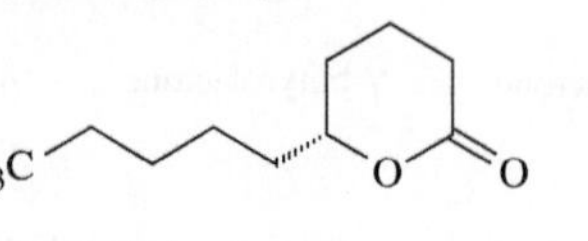

8-72, (R)-δ-decalactone

Other important odorous components of foods are fragrant lactones derived from branched aliphatic hydroxycarboxylic acids. The trivial names are normally preferred. An important component of the aroma of wines and spirits aged in oak barrels (such as whisky) is the so-called whisky lactone, 5-butyl-4-methyl-4,5-dihydro-3H-furan-2-one (3-methyloctano-4-lactone, 3-methyl-4-octanolide, or 4-butyl-3-methylbutyrolactone), also known as cognac or oak lactone, which can exist in four stereoisomers, amongst which the (3S,4S)-isomer (**8-73**) predominates. In brandy, for example, the content of the (Z)-isomers, (3R,4R)- and (3S,4S)-isomers, that have a desirable flavour ranges between 30 and 247 µg/l. The (E)-isomers, (3S,4R)- and (3R,4S)-isomers, are found at a concentration of 115–736 µg/l. The precursor of the (3S,4S)-isomer in oak wood (*Quercus petraea*, Fagaceae) is 6'-O-gallate of (3S,4S)-4-β-D-glucopyranosyloxy-3-methyloctanoic acid (**8-74**).

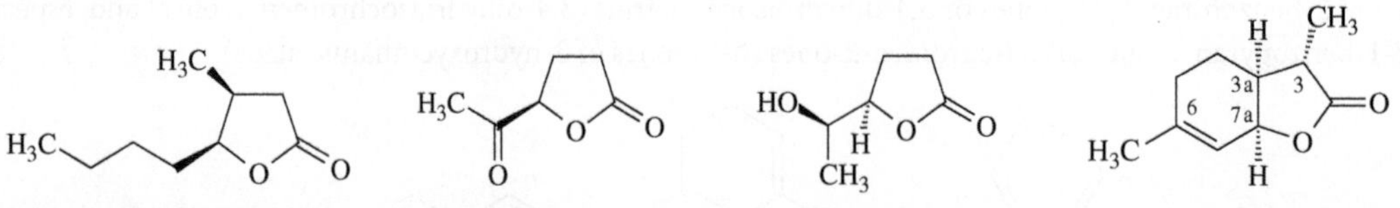

(3S,4S)-whisky lactone   (R)-solerone   (4R,5R)-solerole   (3S,3aS,7aR)-wine lactone

8-73, examples of saturated γ-lactones

An important γ-lactone in wines is 5-oxohexano-4-lactone, known as solerone, which recalls the smell of wine. It occurs as a mixture of (R)- (**8-73**) and (S)-enantiomers. The reduced form of solerone with a fruity odour, 4,5-dihydroxyhexano-4-lactone, is known under the trivial name of solerole or sherry lactone (**8-73**) as it occurs in sherry-type wines, as a mixture of (4R,5R)- and (4S,5S)-diastereomers. The almost racemic composition of solerone and solerole is attributed to racemisation during wine storage. An important lactone related to phthalides is 3a,4,5,7a-tetrahydro-3,6-dimethyl-3H-benzofuran-2-one, also known as 3H-benzofuran-2-one or wine lactone (**8-73**). Of the eight possible stereomers, only the (3S,3aS,7aR)-isomer with a lactonic, herbal, and sweet odour is found in white wines, in amounts of 0.07–0.2 µg/l.

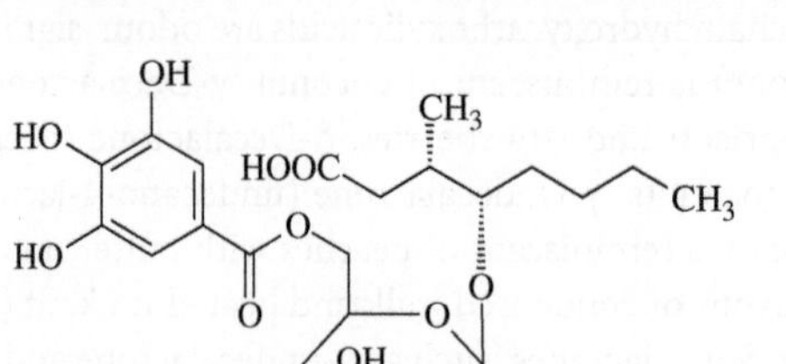

8-74, (3S,4S)-whisky lactone precursor

The odour of unsaturated γ-lactone furan-2-one (γ-crotonolactone, **8-75**) resembles rancid fat. The methyl derivatives pent-3-eno-4-lactone (5-methyl-3H-furan-2-one or α-angelica lactone) and more stable pent-2-eno-4-lactone (5-methyl-2H-furan-2-one or β-angelica lactone, **8-75**) are produced as degradation products of hexoses in acid solutions via laevulinic acid. They have a sweet, herbal odour. Non-2-en-4-olide, also known as 5-pentyl-5H-furan-2-one (**8-75**), contributes to a specific aroma of dessert wines.

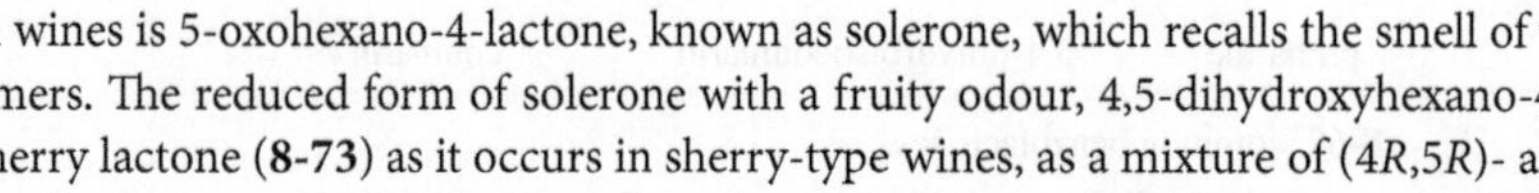

furan-2-one   (R)-5-methyl-2H-furan-2-one, R = CH₃   5-methyl-3H-furan-2-one
                     (R)-5-pentyl-2H-furan-2-one, R = (CH₂)₄CH₃

8-75, unsaturated γ-lactones

Unsaturated 2-hydroxy-substituted γ-lactones (3-hydroxyfuran-2-ones) derived from γ-crotonolactone have a specific odour. 2-Hydroxy-3-methylpent-2-eno-4-lactone (3-hydroxy-4,5-dimethyl-5*H*-furan-2-one), also known as sotolon (by trade name), caramel furanone, sugar lactone, or fenugreek lactone (**8-76**), is a key odorant of lovage leaves (*Levisticum officinale*, Apiaceae) and fenugreek seeds (*Trigonella foenum-graecum*, Fabaceae) and occurs as a flavour-active component of meat, bread, roasted coffee, wine, saké, mushrooms, and other food products as a mixture of (−)-(*R*)- and (+)-(*S*)-enantiomers. Typically, sotolon has a seasoning-like odour resembling caramel, maple syrup, or burnt sugar at lower concentrations and resembles fenugreek or curry at higher ones. The main contributor to the characteristic aroma of wines is (*S*)-sotolon.

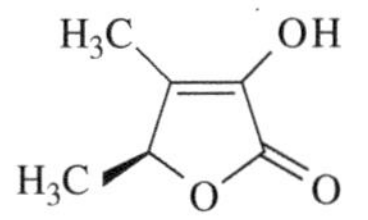

8-76, (*S*)-sotolon

Sotolon is formed from various precursors. In fenugreek seeds, its precursor is the amino acid 4-hydroxyisoleucine (see Section 2.2.1.2.1) and the supposed intermediate is 3-amino-4,5-dimethyl-5*H*-furan-2-one (Figure 8.52). In model wine solutions, sotolon is probably formed from biacetyl and glycolaldehyde (Figure 8.53). The product of pyruvic acid aldolisation and decarboxylation reacts with formaldehyde and an amino acid, and sotolon is formed via the intermediate 3-amino-4,5-dimethyl-5*H*-furan-2-one (Figure 8.54). Precursors of 3-amino-4,5-dimethyl-5*H*-furan-2-one can also be glyoxylic acid and acetaldehyde. In protein hydrolysates, sotolon probably forms by aldolisation of pyruvic and 2-oxobutanoic acids ((like in Figure 8.55). The sotolon precursor in wine, saké, and sugar molasses may be the serine that produces 2-oxopropanoic (pyruvic) acid by oxidative deamination. 2-Oxobutanoic (2-oxobutyric) acid, a product of oxidative transamination of threonine (Figure 2.46), is also produced by decarboxylation of 2-oxoglutaric acid.

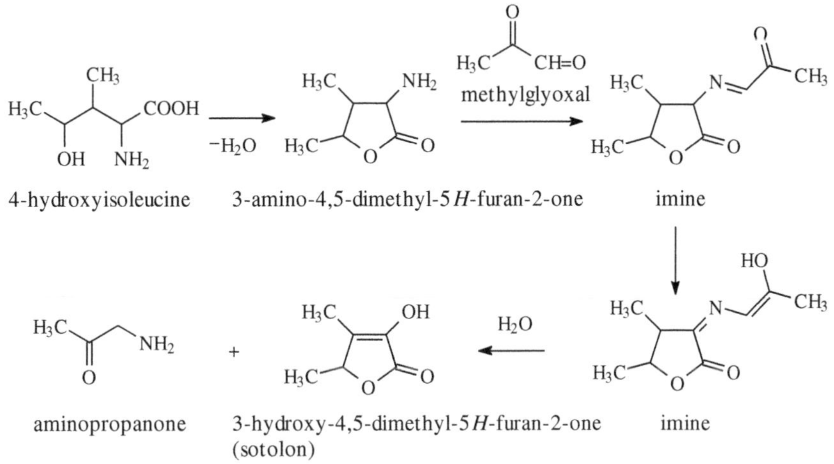

Figure 8.52 Formation of sotolon from 4-hydroxyisoleucine in fenugreek seeds.

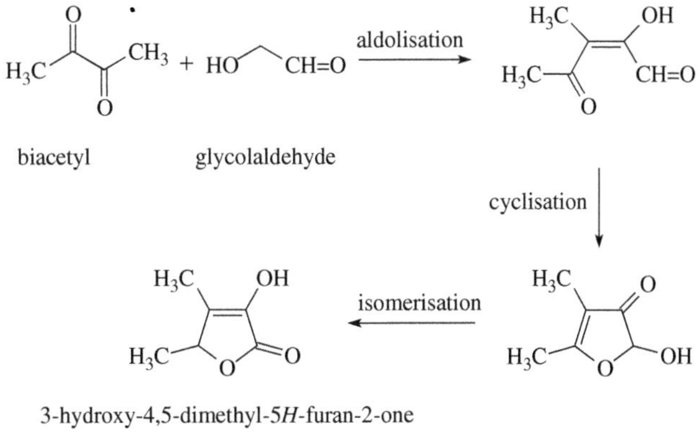

Figure 8.53 Formation of sotolon from biacetyl and glycolaldehyde.

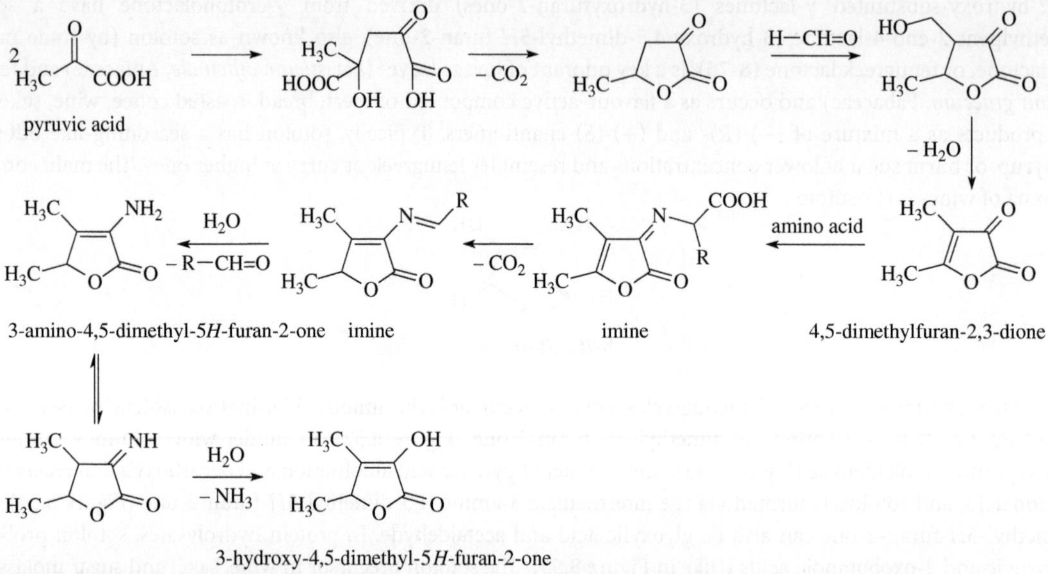

Figure 8.54 Formation of sotolon from pyruvic acid, formaldehyde, and amino acids.

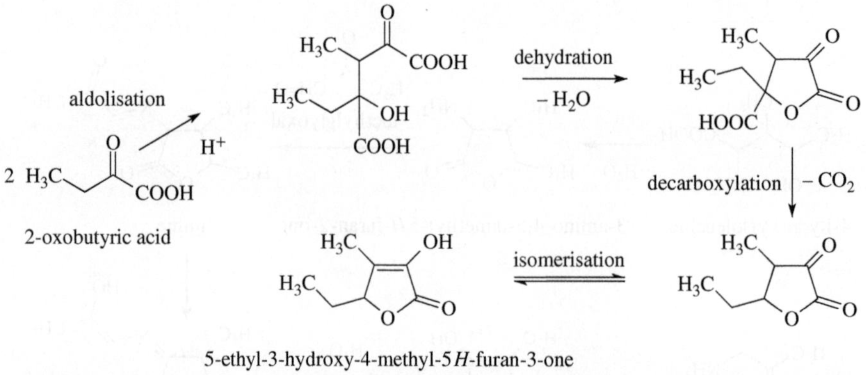

Figure 8.55 Formation of lactones from 2-oxocarboxylic acids.

Two molecules of 2-oxobutanoic acid produce analogously the higher homologue of sotolon, known under the trade name abhexon (see Figure 8.55). Abhexon, 5-ethyl-3-hydroxy-4-methyl-5H-furan-2-one, also known as maple furanone or maggi lactone, is the key aroma component of acid protein hydrolysates. The (+)-(R)-abhexon (8-77) has an intense typical hydrolysate and maple syrup-like odour, whilst the (−)-(S)-isomer has a less intense odour. Its content in soup seasonings based on acid protein hydrolysates is 1–4 mg/l.

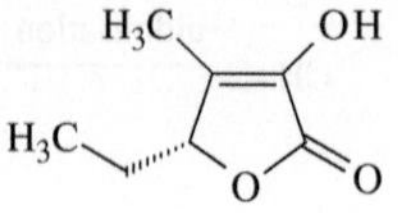

8-77, (R)-abhexon

An example of a bicyclic p-menthane γ-lactone occuring naturally in the essential oil of peppermint (*Mentha x piperita*, Lamiaceae), pennyroyal (*M. pulegium*), and other mints is 3,6-dimethyl-5,6,7,7a-tetrahydro-4H-benzofuran-2-one, known as mint lactone or mint furanone (8-78). Two mint lactone diastereomers with a mint, herbaceous, and coumarinic odour occur in red Bordeaux wine and contribute to its mint notes. Mint lactones are accompanied by diastereomers of the related terpenoid lactone 3,6-dimethyl-4,5,6,7-tetrahydrobenzofuran-2(3H)-one, known as menthofurolactone (8-78).

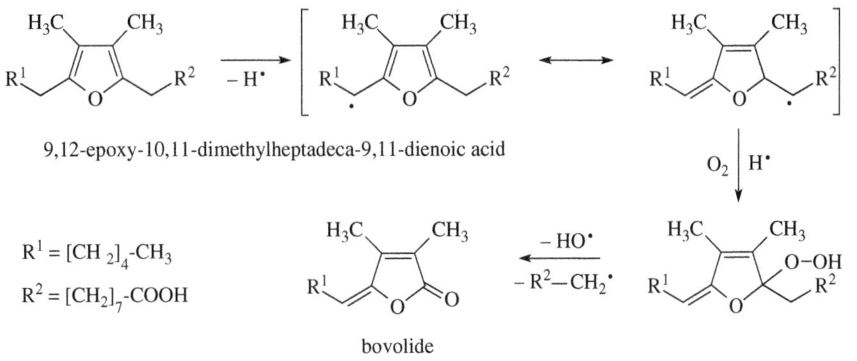

Figure 8.56 Formation of bovolides.

Dimethylsubstituted unsaturated γ-lactones that result from the autoxidation or photooxidation of furan fatty acids are called bovolides. Bovolides have been found in some plants, butter, cooked meats, and seafood. The most important odour-active compounds are 3,4-dimethyl-5-pentylidene-5H-furan-2-one, called bovolide (Figure 8.56), and 3,4-dimethyl-5-pentyl-5H-furan-2-one, known trivially as dihydrobovolide. Both compounds have a characteristic odour, which resembles celery. The mechanism of autoxidation of fatty acids is described in detail in Section 3.8.1.8.2.

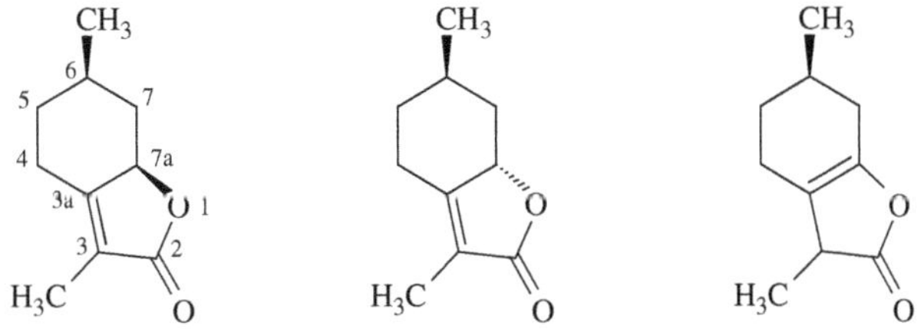

(−)-(6R,7aR)-mint lactone   (+)-(6R,7aS)-isomint lactone   menthofuro lactone

8-78, *p*-menthane lactones

Important taste- and odour-active substances include some unsaturated δ-lactones. 3-Hydroxypyran-2-one (3-hydroxy-2H-pyran-2-one), also known as isopyromucic acid (**8-79**), is the main degradation product of ascorbic and dehydroascorbic acids. Its odour, which is reminiscent of liquorice and caramel, can be detected, for example, in dehydrated orange juices. The so-called parasorbic acid (**8-79**), (S)-6-methyl-5,6-dihydro-2H-pyran-2-one or (S)-hex-2-en-5-olide, has mutagenic and carcinogenic properties. By heating in solutions, parasorbic acid yields sorbic acid via hydrolysis of 5-hydroxyhex-2-enoic acid, which arises on opening the lactone ring. Parasorbic acid is found in rowan berries (*Sorbus aucuparia*, Rosaceae). (−)-(R)-5,6-dihydro-6-pentyl-2H-pyran-2-one, known as massoia lactone (**8-79**), was identified in Merlot and Cabernet Sauvignon musts and wines. It is reduced to (R)-δ-decalactone during alcoholic fermentation.

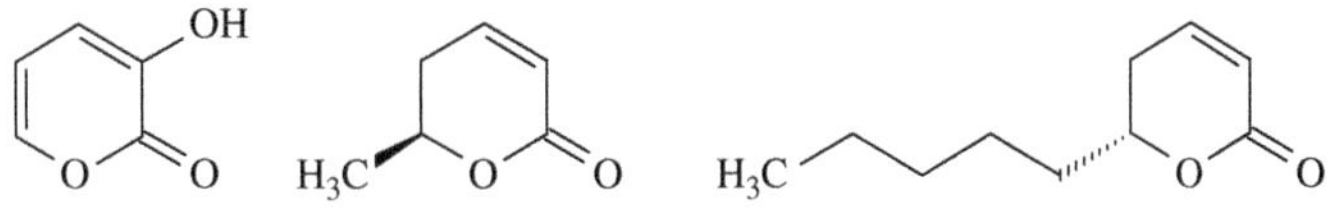

3-hydroxy-2H-pyran-2-one   parasorbic acid   (R)-massoia lactone

8-79, examples of unsaturated δ-lactones

Styrylpyrones Typical lactones, polyketides related to lignans and flavonoids, are styrylpyrones derived from cinnamic acids (the corresponding acyl-CoAs). An example of a styrylpyrone is (2R,1'E)-4-methoxy-2-(2'-phenyleth-1'-en-1'-yl)-2,3-dihydropyran-6-one, better known as kavain or kawain, and formerly as genosan (**8-80**). Kavain, along with more than 18 related compounds called kava pyrones (kava lactones), is present in kava-kava (*Piper methysticum*, Piperaceae), which grows on the islands of Micronesia and Polynesia. The other major constituents are methysticin, demethoxyyangonin, yangonin, dihydrokavain, dihydromethysticin, and 5,6-dehydromethysticin (**8-80**), which belong to three different systems A–C. The A system is characterised by the absence of double bonds in both positions 5,6 and 7,8, B is a completely unsaturated system, and C has a double bond in the 7,8 position.

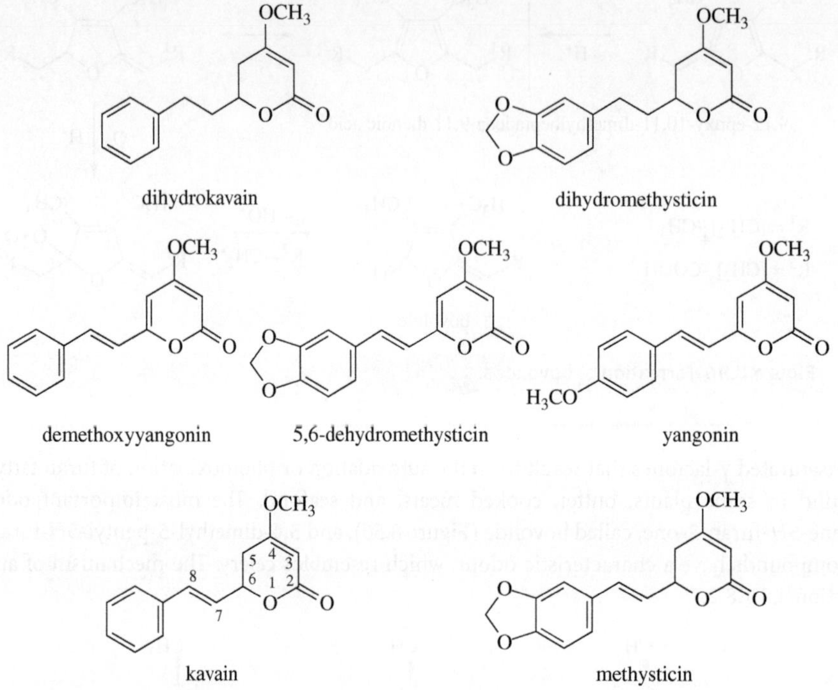

dihydrokavain

dihydromethysticin

demethoxyyangonin

5,6-dehydromethysticin

yangonin

kavain

methysticin

8-80, kava lactones

The second group of biologically active compounds occurring in kava is the chalcones, known as flavokavains (see Section 9.4.2.5). A drink called kava or kava-kava is prepared from the rhizomes, which contain in the dry state about 12% kava pyrones. The traditional kava preparation has been part of the Pacific Islanders' culture for thousands of years, serving as a beverage and medication and used during socio-religious functions. Phytochemicals based on kava are sold worldwide for the treatment of anxiety, nervous tension, agitation, and insomnia.

Phthalides Phthalides are unique components of aromatic vegetables of the carrot family Apiaceae. They are a group of related γ-lactones, (3*H*)-isobenzofuran-1-ones. The benzene ring of phthalides may be partially or fully unsaturated (in hexahydrophthalides, tetrahydro-hydrophthalides, and dihydrohydrophthalides), and there can be alkyl or alkylidene substituents in position C-3 of the heterocyclic ring.

The biosynthesis of phthalides is associated with the biosynthesis of styrylpyrones, as both types of compounds are ketides, but phthalides are derived from aliphatic carboxylic acids (the corresponding acyl-CoAs). Phthalides in plants have various regulatory functions at the cellular level. They exhibit spasmolytic, hypotensive, sedative, and diuretic effects.

In particular, phthalides are found in celeriac (*Apium graveolens* var. *rapaceum*), celery (*A. g.* var. *dulce*), and lovage (*Levisticum officinale*). In smaller quantities, they are present as aromatic components of garden parsley (*Petroselinum hortense*), dill (*Anethum graveolens*), coriander (*Coriandrum sativum*), fennel (*Phoeniculum vulgare*), and other highly aromatic and flavourful plants of the Apiaceae family that have culinary and medicinal uses. The main compounds present in celery are 3-butyl-4,5-dihydrophthalide (also known as sedanenolide or senkyunolide, **8-81**), 3-butyl-3a,4,5,6-tetrahydrophthalide (also known as sedanolide or neocnidilide), 3-butylphthalide, and 3-butylidene-4,5-dihydrophthalide (called (*Z*)-ligustilide). Other dihydro-, tetrahydro-, and hexahydrophthalides are present in small quantities. The main fragrant component of lovage is (*Z*)-ligustilide, and another important compound is sedanenolide. The main phthalides of the leaves and roots of parsley cultivars are sedanenolide, (*E*)-ligustilide, and 3-butylphthalide. Sedanolide is the main phthalide of dill and coriander.

Isomeric phthalides have an odour reminiscent of celery or acid protein hydrolysates. Phthalides occur as numerous isomers in varying amounts in various plants. For example, the content of 3-butylphthalide (with the celery stem-like odour typical of celery and celeriac) in tubers of celery is about 4 wt%; about 88% of this amount is the (*S*)-isomer and 12% the (*R*)-isomer, which has about a 1000-times less intense flavour. The amount of 3-butylphthalide in bulb-like stem bases of fennel is only about 0.01%, but the main isomer (64%) is (*R*)-3-butylphthalide. The odour of (−)-(*S*)-sedanenolide resembles celery stem, whilst that of (+)-(*R*)-sedanenolide is more like celery leaf. The (−)-(3*S*,3a*R*)-sedanolide has a natural celery-leaf odour, whilst the (+)-(3*E*,3a*S*)-sedanolide has a more celery-seed odour with a weak hint of celery.

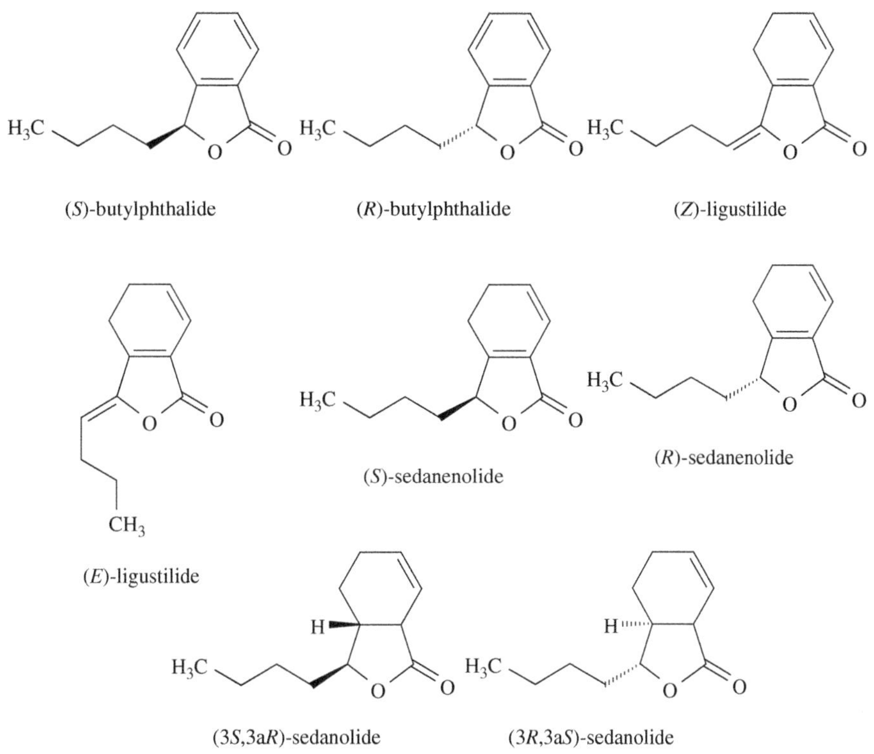

(S)-butylphthalide   (R)-butylphthalide   (Z)-ligustilide

(E)-ligustilide

(S)-sedanenolide

(R)-sedanenolide

(3S,3aR)-sedanolide   (3R,3aS)-sedanolide

8-81, phthalides

Coumarins A special group of δ-lactones formally derived from the hydroxycinnamic acids are coumarins with a skeleton of 2*H*-benzopyran-2-one, which is also called chromen-2-one or 5,6-benzo-2-pyrone (**8-69**). More than 1000 coumarins are found in nature, but only the basic member of the homologous series performs as an odorous compound; this is called coumarin (**8-69**). Plant materials also contain a number of non-volatile coumarins substituted with hydroxyl and methoxyl groups and their glycosides. These and other coumarins, generally taking on the role of phytoalexins, such as isocoumarins, furanocoumarins, and pyranocoumarins, are described in Section 10.3.12.1. Some isocoumarins are intensely sweet (e.g. phyllodulcin) or bitter (e.g. 6-methoxymellein) substances, whilst 3-phenylcoumarins and 4-phenylcoumarins are coloured compounds that are classified as isoflavones and neoflavonoids, respectively.

Coumarin is found in many fresh plant tissues. Natural coumarin is obtained from the seeds of the tree *Dipteryx odorata* (Fabaceae), native to northern South America, which are known as tonka beans. High quantities of coumarin also contain the aromatic bark of an evergreen tree native to South Asia called Chinese cinnamon or cassia (*Cinnamomum aromaticum*, Lauraceae), which is used as a spice. European health agencies have warned against consuming large amounts of cassia, as the tolerable daily intake may be exceeded in consumers with a high intake of this spice containing high levels of coumarin (on average, 2100–4400 mg/kg, but reaching up to 10 000 mg/kg). Coumarin is also found in *Hierochloe odorata* (commonly known as sweet grass, vanilla grass, holy grass in the UK, and bison grass in Poland), an aromatic herb native to northern Eurasia and North America. Small amounts of coumarin occur in strawberries, apricots, liquorice rhizome (*Glycyrrhiza glabra*, Fabaceae), and other fruits and plant materials.

Plants only contain the β-ᴅ-glucoside of (*E*)-2-hydroxycinnamic (*o*-coumaric) acid, called melilotin or melilotoside (Figure 8.57), which is the precursor of coumarin accumulating in the vacuoles along with the glucoside of (*Z*)-2-hydroxycinnamic (coumarinic) acid. Coumarin is only produced by plants during tissue injury when glucosides of coumaric and coumarinic acids come into contact with β-glucosidase. The glucoside of coumaric acid is isomerised to the glucoside of coumarinic acid by 2-coumarate β-ᴅ-glucoside isomerase, but the *trans–cis* isomerisation also occurs spontaneously by means of UV light. The last step is the spontaneous cyclisation (lactonisation) of coumarinic acid to coumarin, which can be further transformed into dihydrocoumarin and melilotic acid.

The coumarin scent vaguely recalls fresh clover (trefoil) and vanilla, and is thus added to tobacco products, artificial vanilla substitutes in perfumes, cosmetic products, and alcoholic beverages, (such as Polish vodka Żubrówka Bison, known as Grass Vodka, which contains a bison grass blade in every bottle). Coumarin was banned as a food additive in the 1950s because of its moderate toxicity (LD$_{50}$ of 275 mg/kg), manifested by liver (hepatotoxicity) and kidney (nephrotoxicity) damage; it is on the list of substances of Regulation (EC) No. 1334/2008 that should not be added as such to foods as flavourings. The maximum level of naturally occurring coumarin in traditional and seasonal bakery products is restricted to 50 mg/kg, in fine bakery products (containing a reference to coumarin in the labelling) to 15 mg/kg, in breakfast cereals to 20 mg/kg, and in desserts to 5 mg/kg. According to the German Federal Institute for Risk Assessment, its amount in food must not exceed the tolerable daily intake of 0.1 mg per kg body weight.

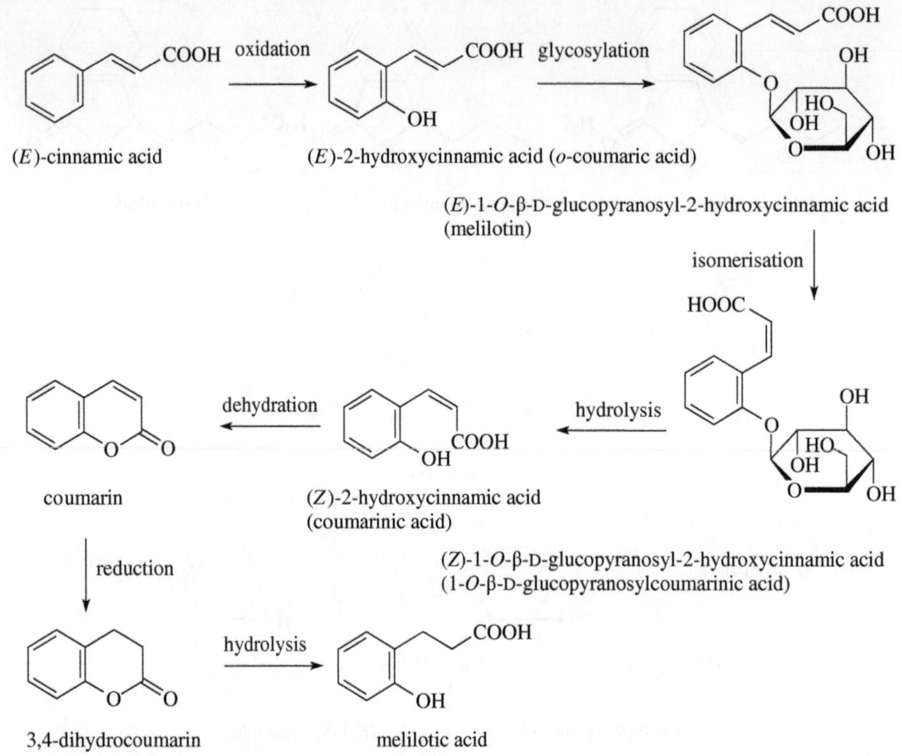

Figure 8.57 Biosynthesis of coumarin.

8.2.7.1.3 Nitriles

Only nitriles (**8-82**) are found in food; isomeric isonitriles (isocyanides) are not. Precursors of volatile nitriles are almost exclusively glucosinolates that occur in cruciferous plants of the family Brassicaceae. Nitriles (together with a number of other products) are formed by their enzymatic and thermal degradation (see Section 10.3.10.2).

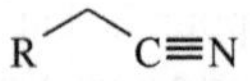

8-82, nitriles

A common nitrile produced during processing of the vast majority of *Brassica* vegetables is but-3-enenitrile (allyl cyanide, **8-82**, R = CH = CH$_2$), which arises by degradation of glucosinolate sinigrin. For example, but-3-enenitrile acts (along with other compounds) as an odour-active compound in cooked cabbage and sauerkraut. Other volatile nitriles include pent-4-enenitrile (but-3-en-1-yl cyanide, **8-82**, R = CH$_2$−CH = CH$_2$), which contributes considerably to the typical aroma of cabbage and some types of mustard pastes made from black and brown mustard seeds (*Brassica nigra* and *Brassica juncea*, respectively), and its higher homologue, hex-5-enenitrile (pent-4-en-1-yl cyanide, **8-82**, R = CH$_2$CH$_2$CH = CH$_2$). These nitriles are formed from glucosinolates known as gluconapin and glucobrassicanapin, respectively. Different cyanoepithioalkanes – glucosinolates with hydroxyalkenyl substituents – are formed as minor glucosinolate hydrolysis products. For example, glucosinolate progoitrin yields hydroxyalkenylnitriles and cyanohydroxyepithioalkanes.

An example of an important non-volatile nitrile is 4-hydroxybenzyl cyanide, which is found in mustard pastes made from white mustard seeds (*Sinapis alba*) containing glucosinolate sinalbin. Non-volatile nitriles also include cyanogenic glycosides, cyanogenic lipids, and amino acids that contain the cyano group.

8.2.7.2 Properties and reactions

8.2.7.2.1 Esters

The organoleptic properties of some esters are given in Table 8.25. One of the most important reactions of esters is hydrolysis (usually enzymatic hydrolysis), which can have a negative effect on the flavour of many foods, especially fruits, where esters are important components of the primary flavour. Water is present in large excess in most foods, so heating or storage usually leads to the hydrolysis of esters, even in the absence of hydrolytic enzymes. Chemical hydrolysis of esters is faster in alkaline solutions, where salts of carboxylic acid are formed. Hydrolytic reactions of esters in foods are of great importance in materials rich in lipids and in many other cases. For example, hydrolysis of pectins alters their ability to form gels and produces methanol.

Table 8.25 Organoleptic properties of important esters.

Ester	Odour character	Ester	Odour character
Allyl capronate	Pineapple	Ethyl lactate	Mandarin
Amyl acetate	Pear	Ethyl 2-methylbutyrate	Strawberries
Amyl cinnamate	Cacao	Geranyl acetate	Rose
Amyl formate	Black currents, plum	(Z)-Hex-3-en-1-yl isovalerate	Rum
Amyl caprylate	Brandy, whisky	(Z)-Hex-3-en-1-yl butyrate	Apple
Butyl acetate	Pineapple	Isopulegyl acetate	Green apple
Butyl isobutyrate	Pineapple	Methylphenyl acetate	Mint
Butyl valerate	Apple	Methyl isovalerate	Honey
Cyclohexyl formate	Sour cherry	Methyl capronate	Apple
Diethyl malonate	Apple	Methyl N-methylanthranilate	Pineapple

In enzymatically inactive foods, continual reactions proceed between acids, alcohols, and esters. Owing to low reaction rates, these never reach the equilibrium state. During heating or prolonged storage, less volatile acids bound in esters may be displaced by more volatile ones, through a process called acidolysis. A common example is the reaction of an ester with an alcohol, which is called alcoholysis. Alcoholysis helps improve the flavour of alcoholic beverages (especially fruit spirits) during long-term storage, known as ageing. One reason is the reaction of ethanol (which is present in large excess) with esters, the constituents of which are fusel oil alcohols.

8.2.7.2.2 Lactones

Generally, lactones with five- and six-membered rings are not too reactive. They are relatively stable in acidic solutions, where they typically arise by spontaneous dehydration of hydroxycarboxylic acids. In aqueous solutions, particularly alkaline ones, the lactone ring opens with the formation of a salt of the hydroxycarboxylic acid. The original lactone may form again after acidification, especially on heating. Lactones react with alcohols to form esters, and similarly, the lactone ring opens by the action of amino compounds to form amides.

8.2.7.2.3 Nitriles

Addition reactions to the nitrile group and hydrolysis of nitriles to the corresponding carboxylic acids may, to a certain extent, take place in foods containing nitriles. Nitriles occurring in trace amounts in rapeseed oils can be reduced to the corresponding amines during oil hydrogenation.

8.2.8 Phenols

8.2.8.1 Classification, structure, terminology, and occurrence

Phenols are components of virtually all foods. They are a very heterogeneous group of compounds, some of which may act also as odour-active compounds, including certain simple phenols formed as the degradation products of phenolic acids and the products of their reduction, such as aldehydes and alcohols. Phenols are also important compounds influencing the taste of foods (such as simple phenols and condensed tannins, responsible for astringent taste). Phenolic compounds include many natural pigments (some quinones, lignans, flavonoids, and related stilbenes, xanthones, and other compounds). Many phenols exhibit significant biological effects, and therefore act as defensive substances in plants (phytoalexins) and as natural antioxidants. An overview of the basic groups of phenolic compounds occurring in foods is given in Table 8.26.

Phenols that act as odorous substances in foods are either primary food components of some essential oils or are produced as secondary substances in food processing. The primary components are phenols structurally related to alkyl aryl ethers. Secondarily formed phenols are produced mainly from phenolic acids and lignin during thermal processes and by the action of microorganisms. Particularly important compounds are derivatives of phenol, guaiacol (2-methoxyphenol), and syringol (2,6-dimethoxyphenol) (**8-83**).

Simple alkyl phenols (**8-83**) found as constituents of essential oils include two major monoterpenes with a similar odour, carvacrol (5-isopropyl-2-methylphenol) and thymol (2-isopropyl-5-methylphenol). These phenols occur, for example, in the essential oil of thyme. Chavicol (4-prop-2-en-1-ylphenol or 4-allylphenol) is a component of the essential oil of basil.

Table 8.26 Main types of food phenols.

Number of C atoms	Basic skeleton type	Groups of phenolic compounds
6	C_6	Simple phenols, benzoquinones
7	$C_6–C_1$	Benzoic acids
8	$C_6–C_2$	Acetophenones, phenylacetic acids
9	$C_6–C_3$	Phenylpropanoids, phenylpropenes, chromenes, isochromenes, cinnamic acids, coumarins
10	$C_6–C_4$	Naphthoquinones
13	$C_6–C_1–C_6$	Xanthones
14	$C_6–C_2–C_6$	Anthraquinones, stilbenes, diarylethanoids
14	$C_6–C_1–C_6–C_1,\ C_6–C_1–C_1–C_6,\ C_1–C_6–C_6–C_1$	Hydrolysable tannins (dimers)
15	$C_6–C_3–C_6$	Flavonoids, pterocarpans, diarylpropanoids
16	$C_6–C_4–C_6$	Diarylbutanoids
16	$C_6–C_{10}$	Gingerols
17	$C_6–C_5–C_6$	Diarylpentanoids
18	$C_6–C_3–C_3–C_6,\ C_6–C_3–C_6–C_3$	Lignans, neolignans
18	$C_6–C_6–C_6$	Terphenylquinones
19	$C_6–C_7–C_6$	Diarylheptanoids, curcuminoids
28	$(C_6–C_2–C_6)_2$	Bianthrones
30	$(C_6–C_3–C_6)_2$	Biflavonoids
n	$(C_6–C_3)_n$	Lignin
n	$(C_6–C_3–C_6)_n$	Condensed tannins, flavolans
n	$(C_6–C_1–C_6–C_1)_n,\ (C_6–C_1–C_1–C_6)_n,\ (C_1–C_6–C_6–C_1)_n$	Hydrolysable tannins (oligomers)

A number of other phenols are derived from guaiacol (**8-83**). Isoeugenol, 2-methoxy-4-(prop-1-en-1-yl)phenol occurs in the essential oil of basil as the (*E*)-isomer. Clove essential oil contains as a key aroma compound 2-methoxy-4-(prop-2-en-1-yl)phenol, known as eugenol. Eugenol and chavicol (along with myrcene) are major components of the essential oil of allspice (pimento). 2-Methoxy-5-(prop-2-en-1-yl)phenol, known as chavibetol (5-allylguaiacol or 5-allyl-2-methoxyphenol), is an isomer of eugenol with a spicy odour and is one of the primary constituents of the essential oil from the leaves of the betel plant (*Piper betle*, Piperaceae, see Section 10.3.6.3). It is used as an Ayurvedic remedy and in cooking, for its peppery taste.

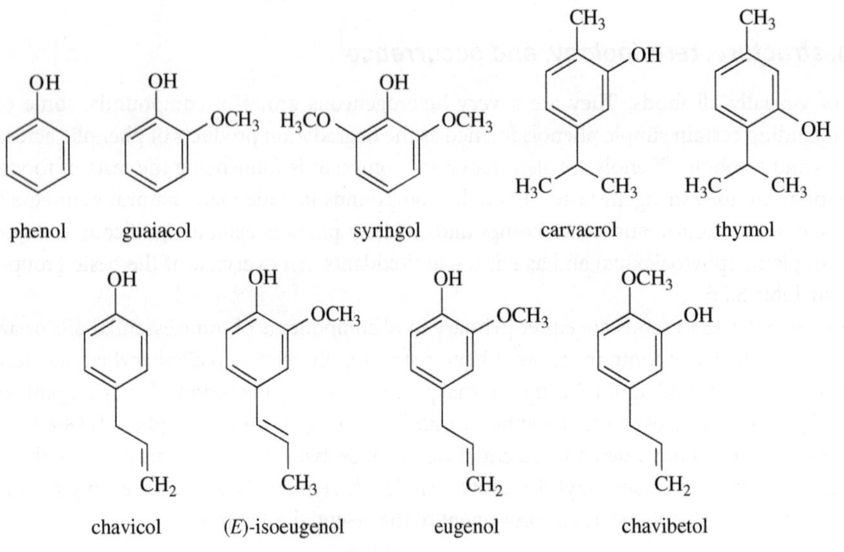

8-83, phenol and its methoxy/alk(en)yl derivatives

3-Alkatrienyl, 3-alkadienyl, 3-alkenyl, and 3-alkyl phenols, collectively known as cardanol, occur in the cashew nutshell liquid at a level of 10%. The main components of this mixture are (8Z,11Z)-3-(pentadeca-8,11,14-trien-1-yl)phenol, (8Z,11Z)-3-(pentadeca-8,11-dien-1-yl)phenol, (8Z)-3-(pentadeca-8-en-1-yl)phenol, and 3-pentadecyl phenol. Derivatives of cardanol find applications as polymeric resins, dyestuffs, plasticisers, surface-active agents, and pesticides.

Phenols resulting from pyrolysis of lignin or occurring in liquid smoke (used for smoking foods) are typically found in smoked foods. Pyrolysis of softwood gives rise mainly to guaiacols (2-methoxyphenols), whilst pyrolysis of hardwood yields a mixture of guaiacols and syringols (2,6-dimethoxyphenols). Phenols not only affect the flavour of smoked products, but also stabilise their colour and act as preservatives. The content of phenols in smoked meat is around 40 mg/kg, depending on the smoking process. Phenolic substances in alcoholic beverages (such as whisky) are extracted from lignin from oak barrels by ethanol.

Phenolic compounds produced by degradation of phenolic acids or lignin by the activity of microorganisms occur as byproducts of lactic and alcoholic fermentations. The content of phenols in alcoholic beverages is highly variable. Concentrations of phenols are relatively low, but they also have low odour-detection thresholds. Phenols only rarely carry an unusual after-taste or off-taste. In wines, for example, phenols are mainly formed from free hydroxycinnamic acids and their ethyl esters by the effect of esterase, cinnamate decarboxylase, and vinylphenol reductase of Dekkera/Brettanomyces yeasts. The presence of 4-ethylphenol, 4-ethylguaiacol, and 4-ethylcatechol (4-ethylbenzene-1,2-diol) in red wines negatively affects their aroma, conferring horsy, barnyard, smoky, and medicinal aromatic notes. Higher amounts of phenols are found in various speciality beers (such as those made from malt directly dried by flue gas), sherry-type wines (Table 8.27), and spirits aged in oak barrels. The main phenols in whisky are eugenol, vanillin, guaiacol, phenol, cresols, and some other compounds.

Phenol also occurs in sour cream and butter (in butter, in an amount of 9–16 µg/kg), alongside other phenols in lower concentrations, especially cresols and guaiacol.

Roasting of coffee, nuts, or almonds, drying of malt, bread baking, production of vegetable oils, and other thermal processes generate sensory active volatile phenols that mainly arise from phenolic acids. As an example, the mechanism of phenol formation from ferulic acid is shown in Figure 8.58. At temperatures around 200 °C, the main products formed are fragments resulting from the gradual degradation of the three-carbon side chain at position C-4, its oxidation, and, partly, recombination of radicals. The compositions of these phenols are shown in Table 8.28. The majority of the pyrolytic fragments are produced in the temperature range 360–410 °C. This includes again the cleavage of C—C bonds in the side chain (as at lower temperatures, but to a greater extent than at 200 °C), but the methoxyl group also splits off as a methoxyl radical. The results are three basic structures that represent, in addition to compounds derived from guaiacol, phenol and catechol derivatives. The main guaiacols are 4-ethylguaiacol and 4-vinylguaiacol. The main phenols are 2-methyl-4-vinylphenol, 2-methyl-4-ethylphenol, 4-vinylphenol, 4-ethylphenol, 2,4-dimethylphenol, and 4-methyl-2-methylphenol. The main catechols (benzene-1,2-diols) are 4-vinyl-, 4-ethyl, 4-methylcatechols, and catechol (pyrocatechol). Recombinations and disproportionations of radicals give rise to other radicals and other products. Some of these reactions are illustrated by the following equations:

$$2\,CH_3{}^\bullet \rightarrow CH_3 - CH_3$$
$$CH_3 - CH_3 \rightarrow CH_3 - CH_2{}^\bullet + H^\bullet$$
$$CH_3 - CH_2{}^\bullet \rightarrow CH_2 = CH_2 + H^\bullet$$
$$H_3C^\bullet + H^\bullet \rightarrow CH_4$$
$$H_3CO^\bullet + H \rightarrow CH_3OH$$

Table 8.27 Main phenol contents in selected alcoholic beverages (µg/l).

Compound	Beer	Sherry	Whisky
Phenol	trace	trace–100	trace–12
o-Cresol (2-methylphenol)	trace	trace	0–75
m-Cresol (3-methylphenol)	–	10 000	0–35
p-Cresol (4-methylphenol)	–	10 000	9–50
2-Ethylphenol	–	50 000	2
4-Ethylphenol	90	350 000	35–39
2-Methoxyphenol (guaiacol)	10–50	–	0–12
2-Methoxy-4-methylphenol (4-methylguaiacol)	trace–90	trace	trace–5
4-Ethyl-2-methoxyphenol (4-ethylguaiacol)	0–300	80 000	2–36
4-Vinylphenol	5–170	20	–
2-Methoxy-4-vinylphenol (4-vinylguaiacol)	7–100	50	–
4-Allyl-2-methoxyphenol (eugenol)	80	10 000	11–195
4-Ethyl-2,6-dimethoxyphenol	30–1200	40 000	–

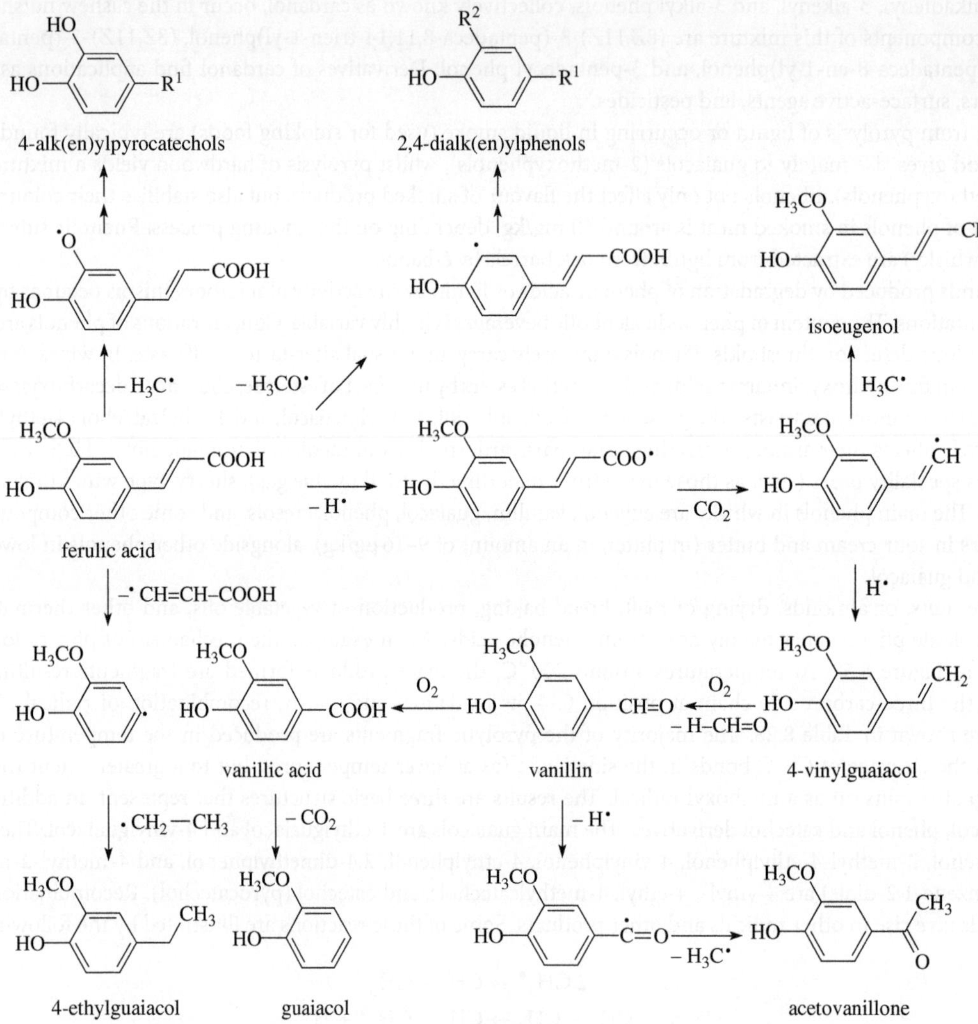

Figure 8.58 Ferulic acid pyrolysis.

Table 8.28 Pyrolysis products of ferulic acid in air at 200 °C.

Product	Composition (%)	Product	Composition (%)
4-Vinylguaiacol	79.9	4-Ethylguaiacol	5.5
Vanillin	6.4	Acetovanillon	2.6
Guaiacol	3.1	Isoeugenol	2.5

Analogous products are also formed from substituted cinnamic acids, such as 4-coumaric, caffeic, and sinapic acids. For example, the main phenol with antioxidative and antimutagenic activities that occurs in roasted high-erucic mustard-seed oil and crude canola oil is 2,6-dimethoxy-4-vinylphenol, called canolol, which is produced from sinapic acid. The major products of caffeic acid pyrolysis at 225 °C are catechol (pyrocatechol), 4-vinylcatechol, and 4-ethylcatechol. A more complex mixture of products arises by pyrolysis of lignin. In the smoke condensates used in the meat industry, more than 150 different phenols and dozens of aromatic alcohols, phenolic acids, and hydroxylated heterocyclic compounds have been identified.

8.2.8.2 Properties and reactions

At higher concentrations, simple phenols are toxic substances that block oxidative phosphorylation. They also exhibit nephrotoxic effects and can act as co-carcinogens. Some alkyl and alkenyl derivatives of phenol, catechol (pyrocatechol), and resorcinol may cause allergic dermatitis.

The most important reaction of natural phenols in foods is oxidation. In many plant foods, there is an enzyme-catalysed oxidation of monophenols to *o*-diphenols (1,2-dihydroxybenzenes), and a subsequent oxidation of the *o*-diphenols thus formed to *o*-quinones. These and subsequent reactions belong to the set of enzymatic browning reactions described in Section 9.12.

8.2.9 Sulfur compounds

8.2.9.1 Classification, structure, terminology, and occurrence

Sulfur compounds are an important group of aroma components of foods that determine or significantly affect their flavour. However, they may also cause various undesirable off-flavours. Sulfur compounds are formed either from the precursor by enzymatic reactions in damaged plant tissues as primary aromatic substances or during the heat treatment of foods of plant and animal origin by non-enzymatic reactions as secondarily produced aromatic substances (e.g. diallyl disulfide is the characteristic substance of garlic-like aroma, allyl isothiocyanate gives the typical aroma of horseradish, 2-isobutylthiazole is a major characteristic component of fresh tomato aroma). A number of sulfur compounds have beneficial effects in human health.

Amongst the most important volatile sulfur compounds are hydrogen sulfide (sulfan), thiols, sulfides, isothiocyanates, and heterocyclic sulfur compounds. Precursors of volatile sulfur compounds are usually non-volatile, sensory-indifferent sulfur compounds, especially sulfur-containing amino acids cysteine, cystine, methionine, and their derivatives, such as *S*-alk(en)ylcysteine sulfoxides, glucosinolates, and thiamine. Another important sulfur compound is sulfur dioxide, which is used as a preservative and an inhibitor of enzymatic browning reactions or the Maillard reaction. It also occurs in a small amount as a metabolite in fermentation processes. Sulfur compounds may be accompanied by their selenium analogues at very low concentrations.

Very intense-smelling sulfur compounds with low thresholds of perception often act as key aromatic compounds. They frequently have very similar structures and contain an important odorophore (a group responsible for the odour character), which is sulfur and oxygen in positions 1 and 3 of the molecule (**8-84**). This structure occurs in mercaptans, thioacetals, mercapto ethers, mercapto ketones, alkylthio esters, and other compounds.

8-84, structure of *S*-compounds with an intense odour

8.2.9.1.1 Sulfane

Sulfane (hydrogen sulfide) is formed in all foods containing proteins (bound cysteine or cystine) during prolonged storage or heat treatment through the reaction known as desulfuration. It is also formed as a byproduct of the Strecker degradation of sulfur-containing amino acids, by degradation of isothiocyanates via carbonyl sulfide, and by microbial reduction of sulfates and sulfites. It is found in some mineral waters. The human body produces sulfane in trace amounts and uses it as a signalling molecule. It has a very low odour threshold (below 1 µg/l in air). The odour increases as it becomes more concentrated, and its smell resembles rotten eggs up to a concentration of about 30 µg/l. Higher concentrations in foods that are rich in protein are associated with putrefactive processes. Sulfane is a very reactive compound that produces a number of volatile odour-active products with various food constituents, especially carbonyl compounds.

8.2.9.1.2 Thiols

Compounds containing a carbon-bonded mercapto group (sulfhydryl group, —SH), called mercaptans or thiols (**8-85**), are the sulfur analogues of hydroxy compounds.

At low concentrations, thiols are often important components of the aroma of many foods. The basic member of the homologous series of thiols is methanethiol (methyl mercaptan), which is produced during the thermal processing of foods, mainly by Strecker degradation of methionine via the unstable primary reaction product methional. Propane-1-thiol is an important component of leek and onion flavours, whilst prop-2-ene-1-thiol (allyl mercaptan) is generated during the processing of garlic by dismutation of diallyl disulfide. At lower concentrations, it shows a typical garlic and meat-like aroma. It is also the main active component of 'garlic breath' after eating garlic. 3-Methylbut-2-ene-1-thiol is an important component of roasted coffee aroma. It arises from hydrogen sulfide and prenyl alcohol (3-methylbut-2-ene-1-ol) and occurs in green coffee at levels of about 0.5 mg/kg.

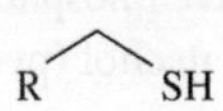

8-85, thiols

methanethiol, R = H
propane-1-thiol, R = CH$_2$CH$_3$
prop-2-ene-1-thiol, R = CH=CH$_2$
3-methylbut-2-ene-1-thiol, R = CH=C(CH$_3$)$_2$

The tropical aroma of the guava fruit (*Psidium guava*, Myrtaceae) is credited to (S)-pentane-2-thiol (**8-86**). The higher homologue of (S)-pentane-2-thiol, (S)-heptane-2-thiol, is found in red and green bell peppers. 2-Mercaptoheptan-4-one (the prefix mercapto- can be replaced by the prefix sulfanyl-), 4-mercaptoheptan-2-one, and the corresponding alcohols, 2-mercaptoheptan-4-ol and 4-mercaptoheptan-2-ol, occur as a mixture of stereoisomers in cooked bell peppers. Their aroma properties are described as sulfury, onion-like, and vegetable-like. The characteristic sulfur note of blackcurrants is caused by the presence of 4-methoxy-2-methylbutane-2-thiol (2-mercapto-4-methoxy-2-methylbutane), called blackcurrant mercaptan (**8-86**), which occurs in juices at concentrations of 0.2 μg/l and higher. It seems that it can be used to express a fruity aroma to tropical fruit juices, coffee, and other products. The same compound is also found in virgin olive oils at levels of about 2 μg/l. Another component resembling blackcurrant aroma is 4-mercapto-4-methylpentan-2-one, the so-called cat ketone (**8-86**), which also occurs in small amounts in grapefruit, Japanese green tea, basil, and aromatic wines (such as Sauvignon Blanc) and contributes to their musky aroma. 4-Mercapto-4-methylpentan-2-one, 3-mercapto-4-methylpentan-1-ol, and almost 50 other simple and polyfunctional thiols (mainly β-sulfanylalkyl esters, β-sulfanylalkyl alcohols, and β-sulfanylalkyl carbonyls) were evidenced in hop cultivars.

In passion fruit (*Passiflora edulis*, Passifloraceae), contributions to the aroma come from 3-mercaptohexan-1-ol and a number of other sulfur compounds, such as 3-mercaptohexyl acetate, 3-mercapto-3-methylbutan-1-ol, and 3-mercapto-3-methylbutyl acetate. Both enantiomers of 3-mercaptohexan-1-ol also occur in wines. The (S)-enantiomer (**8-86**) smells more of passion fruit, whilst the (R)-enantiomer is fruitier and more reminiscent of grapefruit. A component of raw and cooked onion with a very intense aroma is 3-mercapto-2-methylpentan-1-ol (2-methyl-1-hydroxypentan-3-thiol).

An example of an important 1-(alkylsulfanyl)alkanethiol is 1-(methylthio)ethanethiol (**8-86**), which is a component of meat aroma. Its higher homologue 1-(ethylthio)ethane-1-thiol, resembling roasted onion, is a key aroma compound of durian.

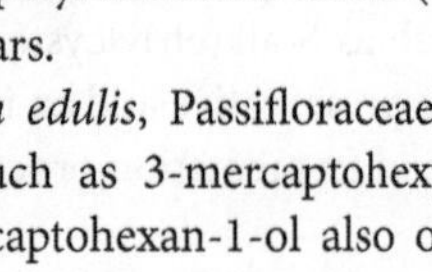

(S)-pentane-2-thiol

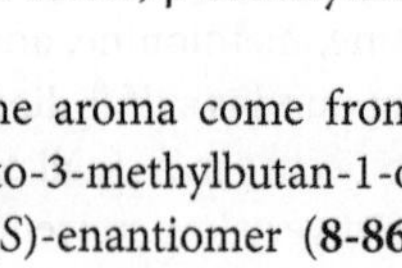

3-mercapto-3-methylbutan-1-ol, R = H
2-mercapto-4-methoxy-2-methylbutane, R = CH$_3$

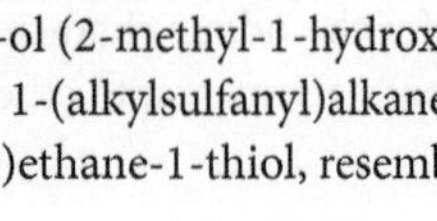

4-mercapto-4-methylpentan-2-one

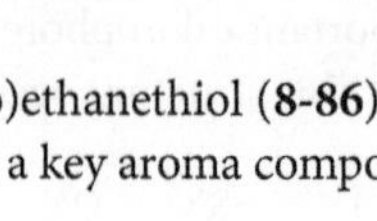

1-(methylthio)ethane-1-thiol, R = H
1-(ethylthio)ethane-1-thiol, R = CH$_3$

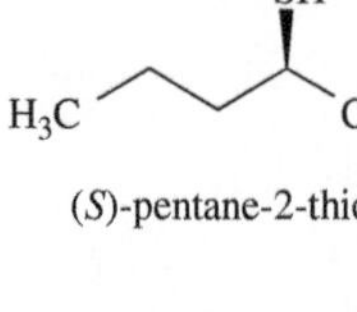

(R)-3-mercaptobutan-2-one 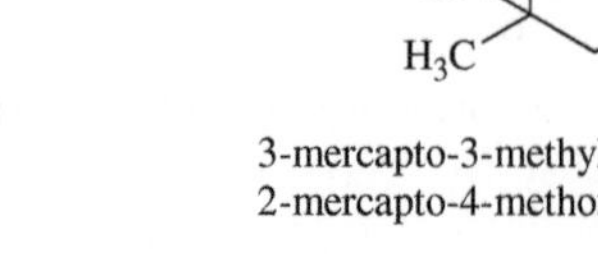   (S)-3-mercaptohexan-1-ol 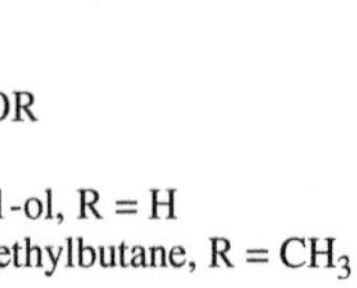   (R)-1-phenylethane-1-thiol

8-86, structures of selected aliphatic and aromatic thiols

Alicyclic thiols are also significant aromatic compounds. The characteristic flavour component of grapefruit is monoterpene *p*-mentha-1-ene-8-thiol. Another related thiol, (1S,4E)-*p*-methane-8-thiol-2-one, is a key aroma component of blackcurrant.

A unique sulfury and burnt odour is exhibited by 1-phenylethane-1-thiol, which constitutes the character impact compound of the tropical to subtropical curry tree (*Murraya koenigii*, syn. *Bergera koenigii*, Rutaceae), the leaves of which (containing (R)- and (S)-enanthiomers in a ratio of about 2:1) are used in many dishes in India and Sri Lanka.

Many lower aliphatic thiols and polyfunctional thiols exhibit an intense, often penetrating, unpleasant and repulsive smell. For example, 3-methyl-2-sulfanylbutan-1-ol (2-mercapto-3-methylbutan-1-ol), arising from 2,3-epoxy-3-methylbutanal in hops, imparts an onion-like

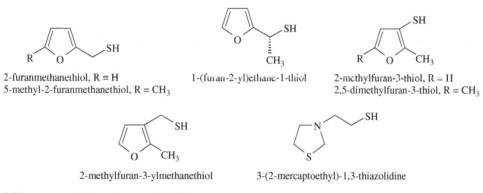

Figure 8.59 Formation of thiols by degradation of cysteine conjugates.

off-flavour to beer. 3-Methylbut-2-ene-1-thiol (1-mercapto-3-methylbut-2-ene, also known as prenyl thiol, **8-85**) is an important component of roasted coffee aroma, but is also a carrier of an off-flavour in beer known as light-struck flavour that appears on exposure to light. In well-stored beer, it occurs at a level of 1–5 ng/l, but in beer stored in the light its content increases up to 0.01–1.05 μg/l. Its precursors are iso-α-acids (isohumulones) derived from hops. 1-Mercapto-3-methylbut-2-ene results from exposure of beer to UV light or from a photosensitised reaction (on exposure to sunlight or visible light) where, in the first instance, riboflavin acts as photosensitiser. Currently, modified (hydrogenated) iso-α-acids are used to enable cold-hopping and provide increased resistance against the formation of this compound after irradiation of beer.

Thiols are often present in fruits and vegetables in combined forms, as S-cysteine and S-glutathione conjugates, and are released from these precursors by the action of L-cysteine-S-conjugate thiol-lyase (deaminating; pyruvate-forming) during fermentation and during eating by β-lyases that originate from oral microflora (Figure 8.59). For example, the precursors of 3-mercaptohexan-1-ol in passion fruit and wines are the individual diastereomers of 3-S-(hexan-1-ol)-L-cysteine and (in wines) of 3-S-glutathionylhexan-1-ol. The precursor of 4-mercapto-4-methylpentan-2-one in wines is 4-S-glutathionyl-4-methylpentan-2-one.

A representative of heterocyclic thiols formed in the Maillard reaction is 2-furanmethanethiol (furan-2-ylmethanethiol, furfuryl mercaptan, **8-87**), a key odorant in roasted and ground coffee. This compound was also identified in a wide range of thermally treated foods, such as meat, bread, roasted sesame seeds, and wine. Important coffee volatiles are 5-methyl-2-furanmethanethiol (also found in cooked beef), 2-methyl-3-furanthiol, and 2,5-dimethylfuran-3-thiol. On the other hand, 2-methyl-3-furanthiol can contribute to stored orange juice off-flavour. 2-Methylfuran-3-thiol (fish thiol) is responsible for a boiled note in milk and also occurs in cooked beef, chicken broth, and yeast extract. 1-(Furan-2-yl)ethane-1-thiol and 2-methylfuran-3-ylmethanethiol were identified in coffee and wine, whilst synthesised 1-(furan-2-yl)- and 1-(thiophene-2-yl)alkane-1-thiols have interesting passion fruit-like aroma. 3-(2-Mercaptoethyl)-1,3-thiazolidine has a very strong odour resembling popcorn.

2-furanmethanethiol, R = H
5-methyl-2-furanmethanethiol, R = CH₃

1-(furan-2-yl)ethane-1-thiol

2-methylfuran-3-thiol, R = H
2,5-dimethylfuran-3-thiol, R = CH₃

2-methylfuran-3-ylmethanethiol

3-(2-mercaptoethyl)-1,3-thiazolidine

8-87, structures of selected heterocyclic thiols

8.2.9.1.3 Sulfides and oligosulfides

In foods, a range of aliphatic sulfides, disulfides, trisulfides (**8-88**), and higher oligosulfides, but also some cyclic sulfides, are important compounds. Sulfides and oligosulfides have an unpleasant odour at higher concentrations, but in low concentrations they contribute to the characteristic full flavour of many foods, such as onion, garlic, cheese, meat, vegetables, and chocolate, and are used in various flavour compositions.

8-88, sulfides

sulfides, $n = 1$
disulfides, $n = 2$
trisulfides, $n = 3$

Dialkyl sulfides are sulfur analogues of ethers. They are formed in foods (along with dialkyl trisulfides) mainly by disproportionation of disulfides. The homolytic S—R bond cleavage of disulfides proceeds by photolysis or heating:

$$R - S - S - R \rightarrow R - S - S^{\bullet} + R^{\bullet}$$
$$R - S - S - R + R^{\bullet} \rightarrow R - S - R + R - S$$
$$R - S - S - R + R - S^{\bullet} \rightarrow R - S - S - S - R + R^{\bullet}$$

The basic member of the homologous series of dialkyl sulfides is dimethyl sulfide (**8-88**, n = 1, R = R^1 = CH$_3$), which has a characteristic sulfurous, cabbage-like smell at low concentrations, similar to that of dimethyl disulfide (**8-88**, n = 2, R = R^1 = CH$_3$), which also has garlic-like notes. Dimethyl trisulfide (**8-88**, n = 3, R = R^1 = CH$_3$) odour is described as sulfurous, alliaceous, cooked, savoury, meaty, eggy, and vegetative, with a fresh, green, and onion top note.

Dimethyl sulfide is an important odorous component of tea, coffee, cocoa, and other foods. Dimethyl sulfide arises by disproportionation of dimethyl disulfide (e.g. in garlic and cruciferous vegetables). In many vegetables (such as cooked cabbage, celery, and some other vegetables), dimethyl sulfide forms by decomposition of S-methylmethionine during food storage and thermal treatment (Figure 8.60). S-Methylmethionine, known as vitamin U, is a natural component of *Brassica* and other vegetables. In cruciferous vegetables, dimethyl disulfide and trisulfide result from degradation of S-methylcysteine sulfoxide. Dimethyl disulfide and trisulfide are, along with the degradation products of glucosinolates, important components of a cruciferous vegetable aroma. Dimethyl trisulfide contributes significantly to the flavour of cabbage, garlic, and also chicken.

Allyl-, propyl-, methyl-, and prop-1-en-1-yl sulfides and di-, tri-, and tetrasulfides belong to the most important sensory-active components of bulb vegetables, such as garlic, chives, leeks, onions, shallots, and other varieties of onions, where they arise by enzymatic (and also thermal) decomposition of S-alk(en)ylcysteine sulfoxides, mainly via the corresponding thiosulfinates. Ethyl methyl disulfide (**8-88**, n = 1, R = CH$_3$, R^1 = CH$_2$CH$_3$) is a flavour-active component of the highly prized fruit, durian.

In beer and milk, dimethyl sulfide causes an unpleasant off-flavour. Sulfides also contribute to an off-flavour of aged beer resembling cooked vegetables, especially creamed maize, cabbage, and tomato, which becomes shellfish/oyster-like in high concentrations. The main undesirable compounds responsible for this off-flavour include dimethyl disulfide (first perceived at concentration of about 30 µg/l), diethyl sulfide, diisopropyl sulfide, and dimethyl sulfide, plus other compounds at considerably lower concentrations.

An important sulfide is methional (**8-26**). Methional in beer and wine is formed from methionine by the activity of microorganisms. It is partly reduced to the corresponding alcohol methionol (**8-11**) and reaction with acetyl-CoA yields 3-methylthiopropyl acetate (**8-89**), which is an important component of various fermented foods. Another acetic acid ester, 3-(methylthio)hexyl acetate, is a compound that possesses attractive tropical fruity notes on dilution. The less odoriferous (−)-(R)-enantiomer (**8-89**) is reminiscent of passion fruit, whilst the (+)-(S)-form has a more herbaceous odour. Both isomers have been found in passion fruit (*Passiflora edulis*, Passifloraceae), guava (*Psidium guajava*, Myrtaceae), and aromatic white wines. Ethyl 3-(methylthio)propionate, known as pineapple mercaptan (**8-89**), has a flavour reminiscent of pineapple. S-(Methylthio)hexanoate (S-methylhexanethioate, **8-89**) is a component of the durian fruit. Condensation of methional with ethanol yields (Z)-2-(methylthio)methylbut-2-enal, also known as 2-ethylidenemethional (**8-89**), which is an important component of potato chips aroma. It also occurs in coffee, cocoa, and other foods. Methyl (2-methyl-3-furyl) disulfide (**8-89**) is a component of meat volatiles, and also of chocolate aroma. Bis-(2-methyl-3-furyl)disulfide, found in cooked beef and black and green tea, has an overall roasty, meat-like, and sulfur-like odour, and is also formed as a decomposition product of thiamine. Many other sulfides are, similarly, important components of foods.

Cyclic disulfides, such as dithiines, are components of garlic aroma. The mushroom shiitake (*Lentinula edodes*), native to East Asia, which has a sulfurous aroma, contains 4,4,7,7-tetramethyl-1,2,3,5,6-pentathiepan, known as lanthionine, as a characteristic component. A representative of the thienofurans is diterpene kahweofuran, known by the systematic name 2-methyl-3-oxa-8-thiabicyclo[3.3.0]octa-1,4-diene or 2,3-dihydro-6-methylthieno[2,3-c]furan (**8-89**), which is only found in roasted coffee.

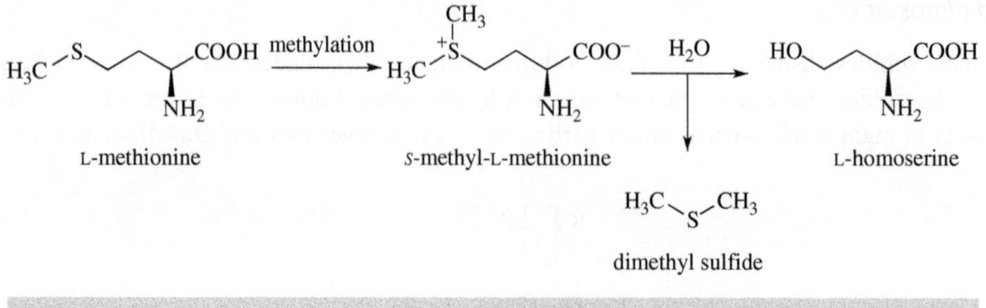

Figure 8.60 Formation of dimethyl sulfide from S-methyl methionine.

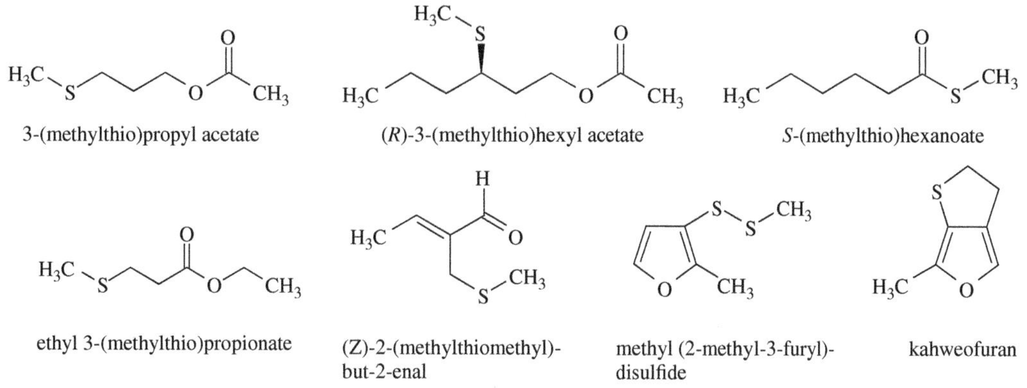

8-89, structures of selected sulfides

8.2.9.1.4 Sulfenic acids and their reaction products

Following tissue damage, S-alk(en)ylcysteine sulfoxides (see Section 2.2.1.2.2), which typically occur in garlic, onion, and other plants of the Amarillidaceae and Brassicaceae families, decompose enzymatically to alk(en)ylsulfenic acids under the catalysis of C—S lyase(cysteine sulfoxide lyase known as alliinase). Alk(en)ylsulfenic acids possess enol and oxo tautomers: R-S-O-H ⇌ R-S(=O)-H, with the tautomeric equilibrium shifted almost exclusively to the enol form. Alk(en)ylsulfenic acids are very unstable compounds that undergo a large number of reactions.

The self-condensation of alk(en)ylsulfenic acids yields dialk(en)ylthiosulfinates, also known as alkanethiosulfinic acid esters, of a general structure R—S(=O)—S—R. Dimethylthiosulfinate is produced in *Brassica* and *Allium* vegetables from S-methylcysteine sulfoxide via methylsulfenic acid. A well-known thiosulfinate allicin is diallylthiosulfinate, also known as di(prop-2-en-1-yl)thiosulfinate, formed from S-allylcysteine sulfoxide (alliin), which is found mainly in garlic via (prop-2-en-1-yl)sulfenic acid and is thought to be responsible for its potent antioxidant activity. Its lifetime is less than one second. Allicin is a chiral compound, but it occurs naturally only as a relatively unstable racemate. Its half-life in water is 30–40 days at 23 °C. Analogously, but to a lesser extent, other thiosulfinates are also found in garlic, including dimethylthiosulfinate, allylmethylthiosulfinate, and methylallylthiosulfinate. Another very important thiosulfinate is di(prop-1-en-1-yl)thiosulfinate, which is formed mainly in onion via (E)-(prop-1-en-1-yl)sulfenic acid, a degradation product of (E)-S-(prop-1-en-1-yl)cysteine sulfoxide (isoalliin). Spontaneous cyclisation of isoalliin in raw and cooked onion gives (1S,3R,5S)-3-carboxy-5-methyl-1,4-thiazan-S-oxide, also known as cycloalliin (see Section 2.2.1.2.7). The general mechanism for the formation of dialk(en)ylthiosulfinates is given in Figure 8.61.

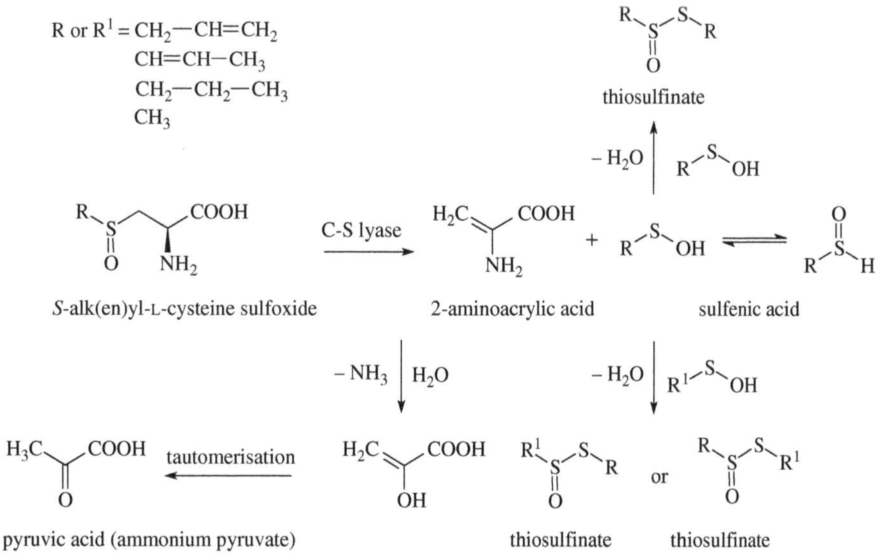

Figure 8.61 Formation of dialk(en)ylthiosulfinates by decomposition of S-alk(en)ylcysteine sulfoxides.

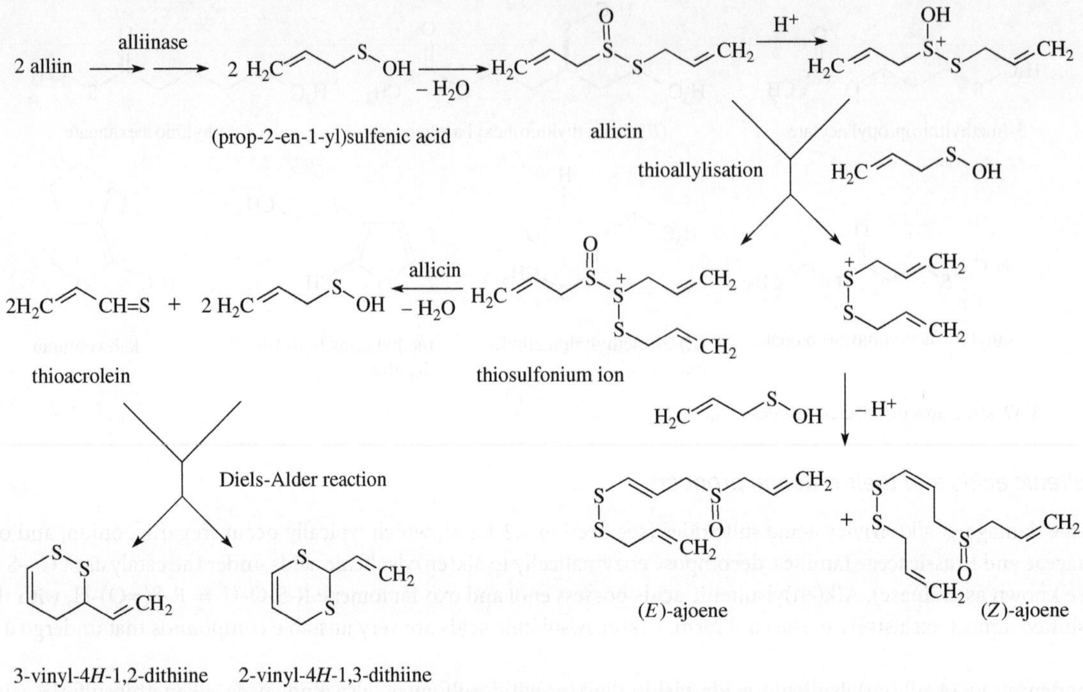

Figure 8.62 Transformation of allicin to dithiines and ajoenes in non-polar media. **Source:** Modified according to Block et al. (2010), Figure 8.1. Reproduced by permission of the American Chemical Society.

In non-polar media (e.g. vegetable oils), allicin is transformed into dithiines (about 70%), diallyl disulfide and diallyl oligosulfides (about 18%), 4,5,9-trithiadodeca-1,6,11-triene 9-oxides known as ajoenes (*ajo* is Spanish for garlic), and some other products (Figure 8.62). Vinyldithiines occur as regioisomers, 2-vinyl-4*H*-1,3-dithiine (the main product) and 3-vinyl-4*H*-1,2-dithiine, with (*E*)-ajoene usually present at twice the amount of (*Z*)-ajoene. The main products of alliin transformation in polar media are sulfides. At room temperature or on heating, allicin is converted into diallyl disulfide, diallyl trisulfide, and diallyl polysulfides, which are the principal components of garlic essential oil (Figure 8.63). Other reaction products are allyl alcohol, sulfur dioxide, and propene. Disproportionation of thiosulfinates in aqueous (polar) solutions also leads to disulfides and thiosulfonates of general formula R–SO$_2$–S–R. For example, disproportionation of allicin yields diallylthiosulfonate (pseudoallicin) and diallyl disulfide, which subsequently give diallyl sulfide and diallyl trisulfide. In addition to alliin, garlic contains lesser amounts of methiin and isoalliin, which decompose by alliinase to the corresponding alk(en)ylsulfenic acids. These acids can self-condense to the corresponding dialk(en)ylthiosulfinates or react with allicin to yield mixed alk(en)ylthiosulfinates.

The (*E*)-(prop-1-en-1-yl)sulfenic acid derived from isoalliin in onion is a unique compound. It is rapidly rearranged by a second enzyme, LF-lyase (lachrymatory factor-lyase), absent in garlic, to give (*Z*)-thiopropanal *S*-oxide (thiopropanal sulfoxide), which is a lachrymatory factor that causes tears when cutting onions (Figure 8.64). Leek and chives contain the same sulfur-containing amino acids as onion; only the propiin content is somewhat higher, at the expense of isoalliin, which is the major cause of their much lower lachrymatory properties. (*E*)-(Prop-1-en-1-yl)sulfenic acid or (*Z*)-thiopropanal sulfoxide yields the corresponding dimer and (*Z*,*Z*)-2,3-dimethyl-1,4-butanedithial-1,4-dioxide, also known as bis(sulfine). Reaction of (*Z*)-thiopropanal sulfoxide with sulfenic acids derived from isoalliin or methiin yields cepaenes with antithrombotic properties. Self-condensation of (prop-1-en-1-yl)thiosulfenic acid gives (prop-1-en-1-yl)thiosulfinate, which is the precursor of 2,3-dimethyl-5,6-dithiabicyclo[2.2.1]hexane 5-oxides, known as zwiebelanes (*Zwiebel* is German for onion). The main product is (*E*)-zwiebelane. Spontaneous rearrangement of thiosulfinate yields 3,4-dimethylbutanedithial *S*-oxide and cyclic sultene, the precursor of cepathiolanes. 3,4-Dimethylbutanedithial *S*-oxide can react with thiosulfenic acids or thiosulfinates to produce bitter allithiolanes. Their main representatives are described in Section 8.3.5.1.2.

Technological processing of garlic, onions, and leeks (e.g. cutting, preservation in salt, or drying) often leads to the formation of undesirable intensely coloured compounds, which are purple/blue or yellow in garlic and pink to deep red in onions and leeks. These pigments are not stable and slowly change to yellow-brown or brown within a few days of storage. The pinking of onions and leeks and the greening of garlic are chemically similar reactions consisting of enzymatic and non-enzymatic processes. The primary precursor is the sulfur amino acid isoalliin (see Figure 8.64). In disrupted tissues, isoalliin is degraded by the enzyme alliinase to form (prop-1-en-1-yl)thiosulfinate or mixed thiosulfinates that react via 2,3-dimethylbutanedithial *S*-oxide with free amino acids to form *N*-substituted 3,4-dimethylpyrroles, the pigment precursors. Molecules containing a 3,4-dimethyl pyrrole ring then react with acrolein/thioacrolein and with formaldehyde/thioformaldehyde (derived from allicin and other thiosulfinates) to form conjugated purple/blue and yellow pigments in garlic (Figure 8.65).

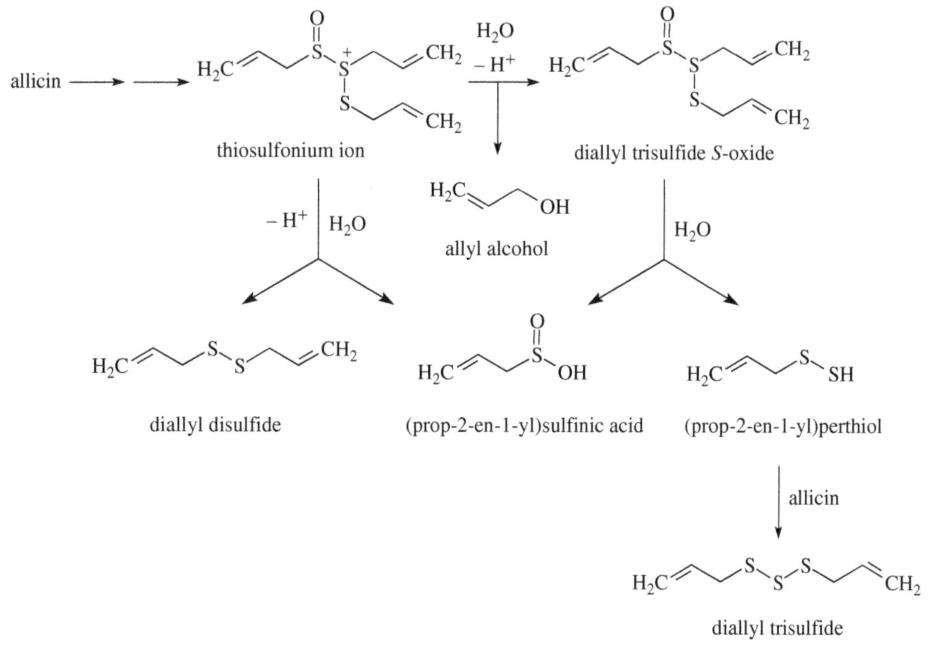

Figure 8.63 Transformation of allicin to oligosulfides and allyl alcohol in polar media. Source: Modified according to Block et al. (2010), Figure 8.1. Reproduced by permission of the American Chemical Society.

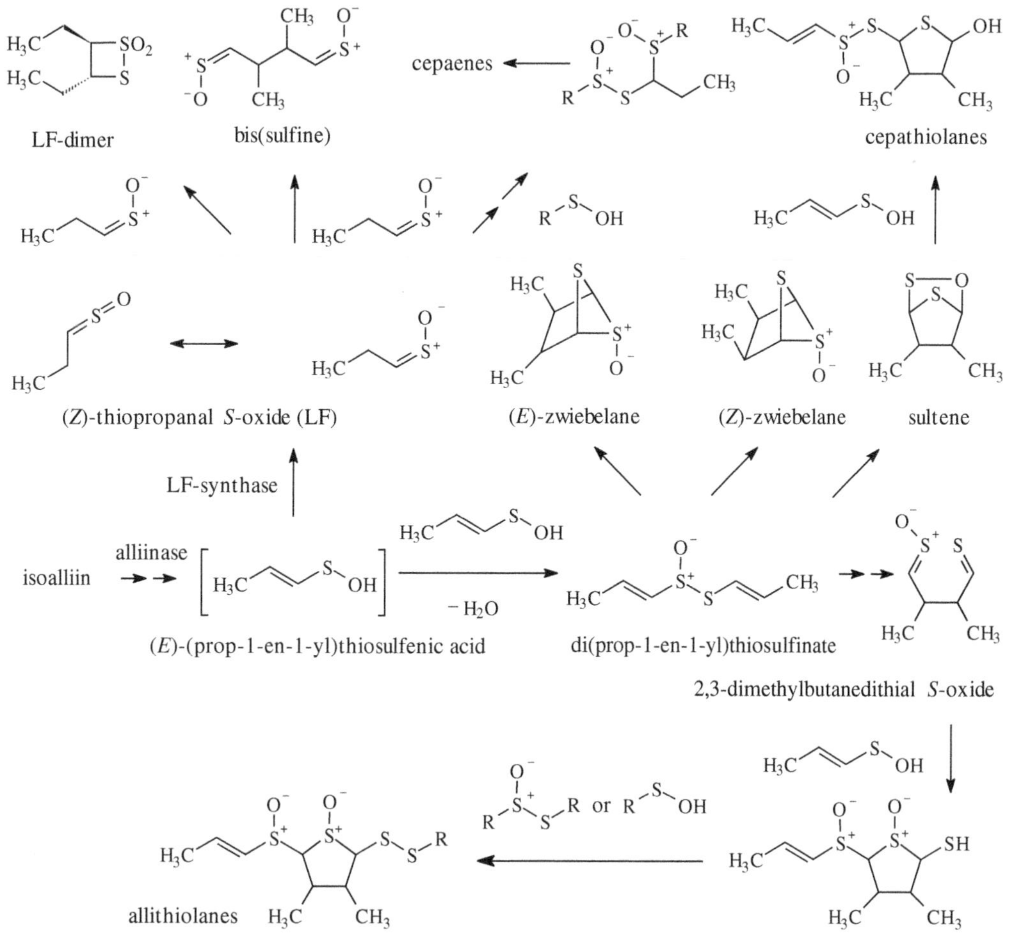

Figure 8.64 Proposed formation pathways of thiopropanal S-oxide, allithiolanes cepaenes, cepathiolanes, and zwiebelanes in onion (R = CH$_3$ or CH$_2$ = CHCH$_3$).

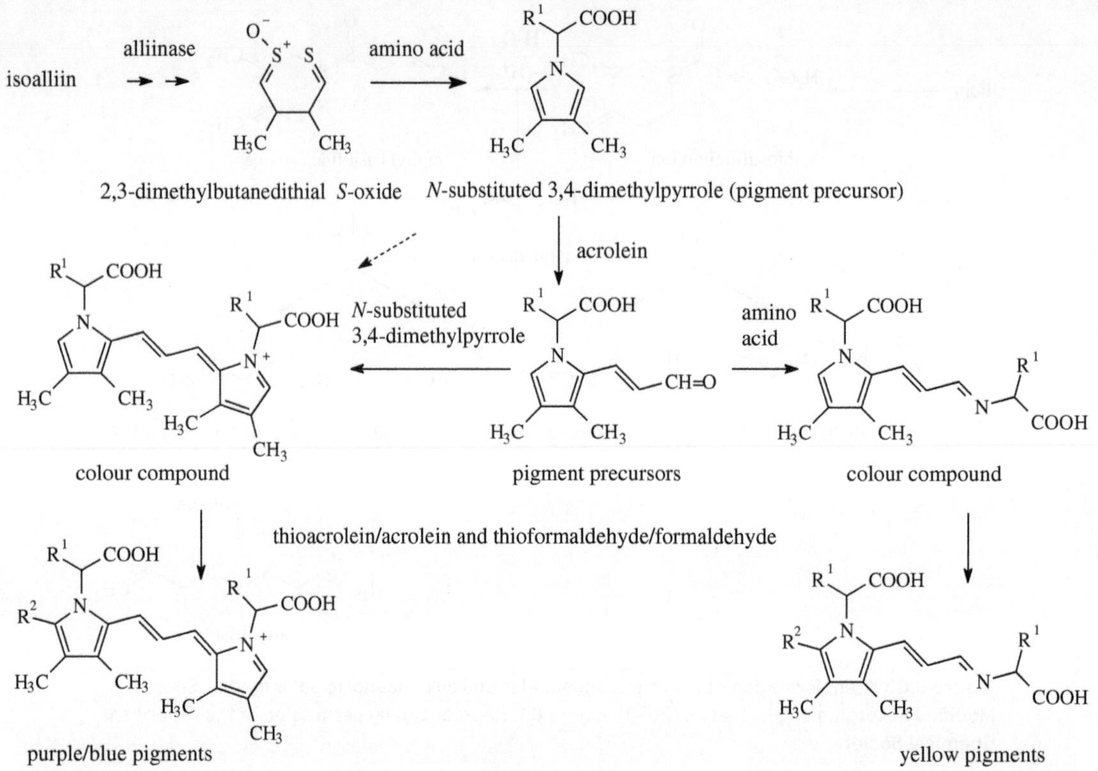

Figure 8.65 Proposed formation of compounds responsible for garlic greening. $R^1 = CH_3$ in alanine, $R^2 = CH_3$ or $CH = CHCH_3$.

The compounds responsible for pink discoloration of onion and leek represent a complex mixture of two N-substituted 3,4-dimethylpyrrole rings linked by either a methine or a propenylidene bridge and modified by various C_1 (derived from methiin) and C_3 (derived from isoalliin) side chains (**8-90**).

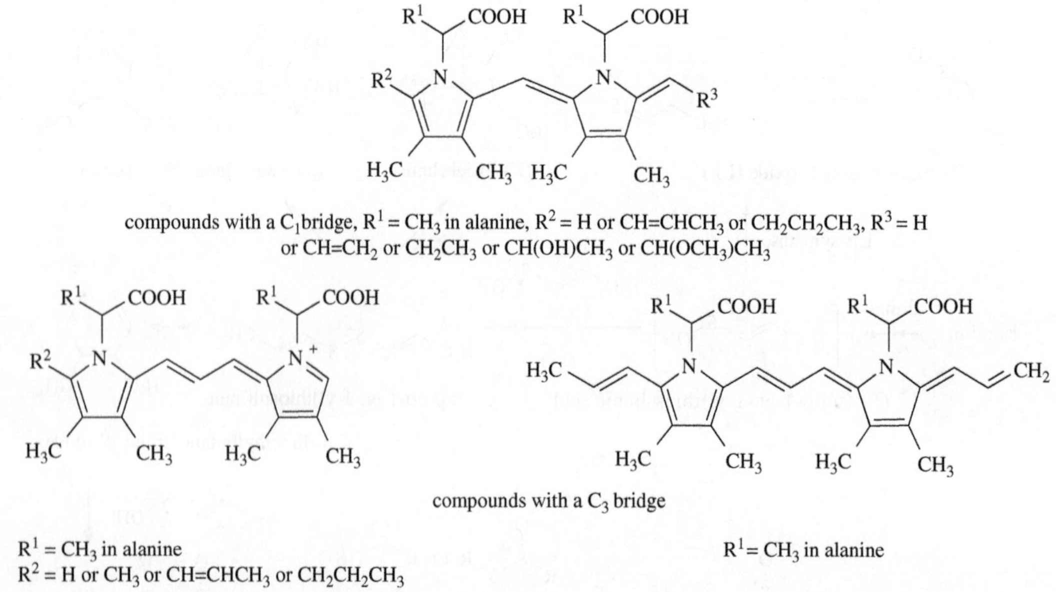

compounds with a C_1 bridge, $R^1 = CH_3$ in alanine, $R^2 = H$ or $CH=CHCH_3$ or $CH_2CH_2CH_3$, $R^3 = H$ or $CH=CH_2$ or CH_2CH_3 or $CH(OH)CH_3$ or $CH(OCH_3)CH_3$

compounds with a C_3 bridge

$R^1 = CH_3$ in alanine
$R^2 = H$ or CH_3 or $CH=CHCH_3$ or $CH_2CH_2CH_3$

$R^1 = CH_3$ in alanine

8-90, tentative structures of compounds responsible for pinking of onion and leeks

Most of the species of the subgenus *Melanocrommyum* of genus *Allium* are characterised by a deep orange to red ichor (fluid discharge) occurring after damage of the cells. This red pigment, 3,3′-dithio-2,2′-dipyrrole (**8-91**), forms spontaneously from (R_C,S_S)-S-(pyrrol-3-yl)cysteine sulfoxide via 2-lactyl-3′-pyrrolyl sulfoxide.

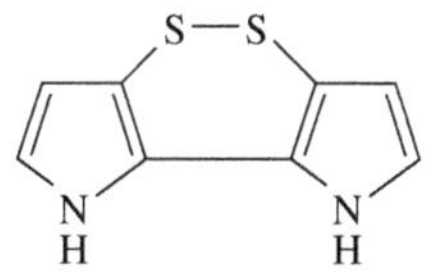

8-91, 3,3′-dithio-2,2′-dipyrrole

8.2.9.1.5 Isothiocyanates

Isothiocyanates (formerly known as mustard oils, **8-92**) are compounds with a cumulated system of double bonds. They belong to the heterocumulenes, compounds that are formally derived from the hydrocarbon allene ($H_2C = C = CH_2$). In small quantities, isothiocyanates are often accompanied by isomeric thiocyanates. Related cyanates and isocyanates do not occur in foods.

$$R - N = C = S$$

8-92, isothiocyanates

Precursors of isothiocyanates in food are the sulfur-containing glycosides known as glucosinolates (see Section 10.3.10) that are found particularly in the cruciferous plant family (Brassicaceae). When the plant is damaged, for example by cutting or masticating, the enzyme β-glucosidase is released and catalyses the glucosinolate degradation to isothiocyanates and a number of other substances. Isothiocyanates have a very intense and sharp odour and taste. One of the major representatives of this group of substances is allyl isothiocyanate (prop-2-en-1-yl isothiocyanate, **8-92**, $R = CH_2-CH=CH_2$), which is an active component of hot horseradish, wasabi, pastes made from brown and black mustard seeds, and a number of *Brassica* vegetables.

8.2.9.2 Properties and reactions

Sulfur compounds are reactive and enter into reactions with many other food components. Most sulfur compounds are substances with characteristic intense flavours that contribute to the odour and taste of many foods.

8.2.9.2.1 Sulfane, thiols, and sulfides

Sulfane and thiols react readily with the carbonyl groups of carbonyl compounds (Figures 8.25 and 8.27) and often undergo addition reactions to double bonds of unsaturated compounds. Unlike alcohols, thiols are easily oxidised by atmospheric oxygen to disulfides. The reaction mechanism is given in Section 2.5.1.1.1. Like methionine, cysteine, and cystine, simple sulfides and disulfides are also oxidised to sulfoxides, sulfones, and other products. Important oxidation reagents are fatty acid hydroperoxides generated in the autoxidation of unsaturated fatty acids. An important reaction of oligosulfides is the interchange reaction that also proceeds in peptides and proteins containing bound cystine:

$$R - S - S - R + R^1 - S - S - R^1 \rightarrow 2\,R - S - S - R^1$$

In veterinary science, it is known that onion and garlic are oxidatively toxic to erythrocytes resulting in haemolytic anaemia in domestic animals, such as dogs, cats, horses, sheep, and cattle. The causative agents have been identified as di(prop-2-en-1-yl)trisulfide ($R-S-S-S-R$, R = prop-2-en-1-yl, **8-88**), tetrasulfide and pentasulfide, di(prop-2-en-1-yl)thiosulfonate ($R-SO_2-S-R$, R = prop-2-en-1-yl), and other transformation products of isoalliin.

8.2.9.2.2 Thiosulfinates

Aliphatic thiosulfinates show antimicrobial and antiaggregatory activity with human blood platelets, but they are rather unstable. They are mainly oxidised to less active thiosulfonates and undergo other reactions. Thiosulfinates derived from vegetables such as garlic and onions can freely permeate cell membranes and rapidly react with reduced glutathione (G–SH) to form intracellular mixed disulfide conjugates of the type G–S–S–R, where R = allyl or propyl. Analogous conjugates with cystine are found in extracellular fluids. Conjugation reactions of thiosulfinates in the presence of free thiols, including those in proteins, also occur in foods. Some conjugates of thiosulfinates with cysteine and degradation products of thiosulfinates have demonstrable biological activities. Decomposition products of thiosulfinates similarly have various physiological effects, such as lowering blood pressure, cholesterol, and blood sugar levels, as well as antimutagenic and anticarcinogenic activities.

$$R-N=C=S \rightleftharpoons R^+ + [^-S=C=N \longleftrightarrow ^-S-C\equiv N] \rightleftharpoons R-S-C\equiv N$$

Figure 8.66 Heterolytic cleavage of isothiocyanates and thiocyanates.

8.2.9.2.3 Isothiocyanates

Isothiocyanates are not only flavour-active compounds, but also have some antibacterial and fungicidal effects and a certain toxicity, such as mild strumigenic and cytotoxic activities. These properties are found especially in methyl isothiocyanate and allyl isothiocyanate, which are used as pesticides, as their insecticidal, herbicidal, fungicidal, and nematocidal effects can control soil-borne plant pathogens and parasitic nematodes. Allyl isothiocyanate have found use in the canning industry in modified atmospheres, extending the shelf-life of packaged food (e.g. of meat products). An ability of some isothiocyanates of cruciferous vegetables to inhibit chemically induced cancer has also been demonstrated. In this respect, particular attention has been paid to sulforaphane (4-methylsulfinylbutyl isothiocyanate), which arises from its precursor, glucoraphanin, in broccoli and radishes.

In aqueous solutions, isothiocyanates rearrange into thiocyanates. The rearrangement takes place via various mechanisms, but usually heterolytic or homolytic cleavage of the molecule (Figures 8.66 and 8.67). Allyl isothiocyanate preferably isomerises via the six-membered cyclic allyl thiocyanate intermediate (**8-93**). Isothiocyanates are extremely reactive substances, due to the electrophilic nature of their functional group, and they enter into reactions with a number of nucleophilic reagents. Particularly important reactions in foods are those of isothiocyanates with mercapto, amino, and hydroxyl groups of amino acids and proteins. Some reactions of isothiocyanates with amino acids also produce coloured products (see Section 2.5.2.4).

$$R-S-C\equiv N \rightleftharpoons R^\bullet + SCN^\bullet$$
$$R\cdot + R-N=C=S \rightleftharpoons R^\bullet + R-S-C\equiv N$$

Figure 8.67 Homolytic cleavage of isothiocyanates.

$$R-N=C=S \ + \ H-X \rightleftharpoons R-\underset{H}{N}-\overset{S}{\underset{}{C}}-X$$

isothiocyanate

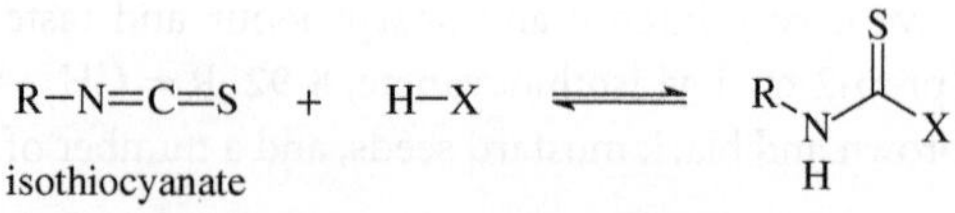

Figure 8.68 Reaction of isothiocyanates with nucleophilic reagents. HX = water (H_2O) – thiocarbamoic acid, alcohol (R^1–OH) – thiocarbamate, sulfan (H_2S) – dithiocarbamoic acid, ammonia (NH_3) – N-alkylthiourea, amine (R^1–NH_2) – N,N'-dialkylthiourea.

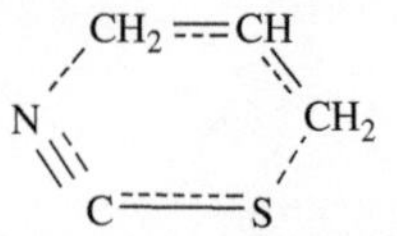

8-93, cyclic allyl thiocyanate intermediate (transition state)

Additions of various nucleophilic reagents to the isothiocyanate functional group are shown in Figure 8.68. The nucleophilic reagent can be water (hydroxyl ions), which yields thiocarbamoic acid salts, alcohols, which give thiocarbamic acid esters, sulfane, which gives dithiocarbamoic acid salts, thiols, which give dithiocarbamoic acid esters, ammonia, which gives N-substituted thioureas, or amines, which give N,N'-disubstituted thioureas. Reactions are usually complex and depend on pH and other factors. The main reactions of allyl isothiocyanate in aqueous solutions are shown in Figure 8.69. The main products in weakly acidic solutions (pH 4) are carbon disulfide and allylamine, whilst those in slightly alkaline solutions (pH 8) are allylamine, allyldithiocarbamate, diallylthiourea, and carbon disulfide. Diallyl disulfide and other products are also produced in small amounts. For example, diallyl disulfide imparts a faint odour resembling garlic to older mustard pastes. In the presence of hydrogen sulfites (HSO_3^- ions), sometimes used as preservatives, allyl isothiocyanate yields allylaminothiocarbonylsulfonic acid (**8-94**).

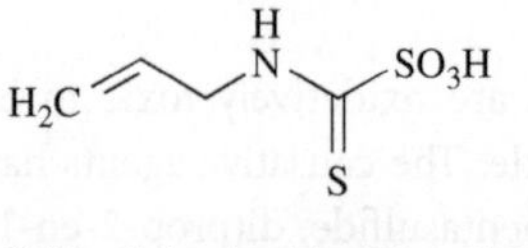

8-94, allylaminothiocarbonyl sulfonic acid

Aromatic and heterocyclic isothiocyanates are unstable and easily hydrolyse to the corresponding alcohols. For example, 4-hydroxybenzyl isothiocyanate (the degradation product of sinalbin, which occurs in the seeds of white mustard, *Sinapis alba*, Brassicaceae) decomposes in normal table mustard to 4-hydroxybenzyl alcohol and thiocyanate ions (Figure 8.70). Isothiocyanates derived from indolyl glucosinolates behave similarly.

8.2.10 Nitrogen compounds

8.2.10.1 Classification, structure, terminology, and occurrence

Ammonia, volatile amines, imines, amides, and, in particular, heterocyclic compounds containing nitrogen, are important flavour-active substances in non-acidic foods.

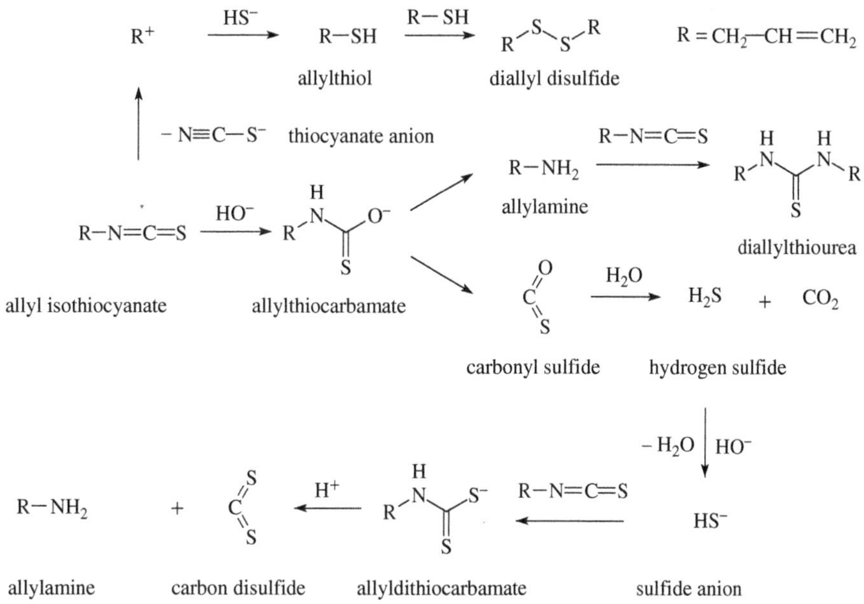

Figure 8.69 Decomposition of allyl isothiocyanate in aqueous solutions.

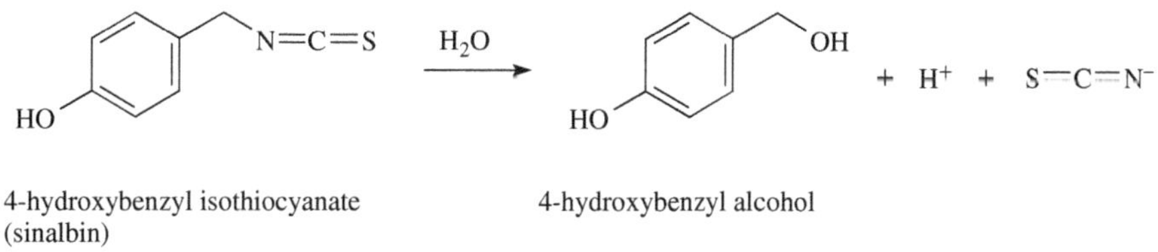

Figure 8.70 Hydrolysis of 4-hydroxybenzyl isothiocyanate.

8.2.10.1.1 Ammonia

Ammonia (NH_3) found in foods is mainly a product of deamination of free nucleotides (e.g. adenosine $5'$-monophosphate, AMP, deaminates to inosine $5'$-monophosphate, IMP), or of the amino acid amides asparagine and glutamine. In acidic foods, ammonia is present almost exclusively in the form of ammonium salts.

8.2.10.1.2 Amines

Amines are structurally derived from ammonia by substitution. The basic amines are primary amines (RNH_2), secondary amines (RR^1NH), tertiary amines (RR^1R^2N), and quaternary amines with the general structure $RR^1R^2R^3N^+(OH^-)$. Tertiary amines can form N-oxides of the type $RR^1R^2N^+(O^-)$.

Primary amines in foods most often arise as products of enzymatic reactions that include decarboxylation of amino acids catalysed by non-specific decarboxylases and amination (reductive amination) or transamination of aldehydes by transaminases (Figure 8.71).

The decarboxylation of amino acids takes place mainly in materials of animal origin, whilst the formation of aldehydes and amines occurs largely in plant materials. Common products are aliphatic amines. In aromatic compounds, such as benzylamine (**8-95**), the amino group is bound in the side chain. Amines are present in virtually all foods, and therefore can also be found in fruits and alcoholic beverages, but they act as odour-active substances only in some non-acidic foods of animal origin, especially cheeses, fish, and meat. In acidic foods, amines are present in the form of non-volatile salts. In plants and fungi, they often function as insect attractants. An overview of major amine precursors is given in Table 8.29. The common volatile amines contents in selected foods are shown in Table 8.30. Enzymatic decarboxylation also occurs in amino acids other than those listed in Table 8.29. Sulfur amino acids and hydroxyamino acids are decomposed with the formation of non-volatile amines. For example, cysteamine (2-mercaptoethylamine, **8-95**) is produced from cysteine, homocysteamine (3-mercaptopropylamine, **8-95**) from homocysteine, 3-(methylthio)propylamine (**8-95**) from methionine, ethanolamine (2-aminoethanol, **8-95**) from serine, and $(-)$-(R)-1-aminopropan-2-ol (**8-95**, R = CH_3) from threonine. Ethanolamine also results from the hydrolysis of some glycerophospholipidis, for example from (3-sn-phosphatidyl)ethanolamine.

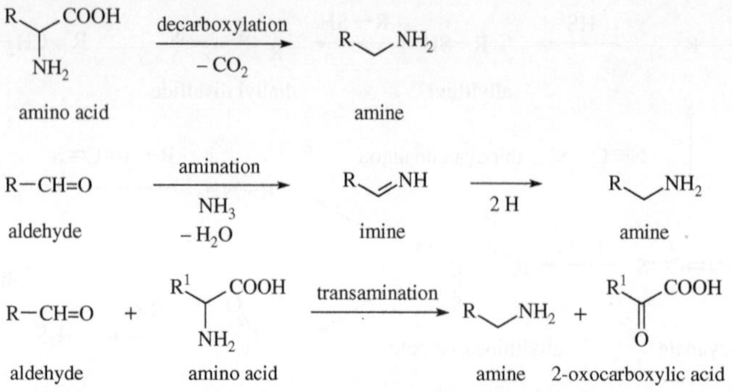

Figure 8.71 Main reactions of amine formation.

Table 8.29 Precursors of volatile amines in foods.

Amine	Precursor		Amine	Precursor	
	Amino acid	Aldehyde		Amino acid	Aldehyde
Methylamine	Glycine	Formaldehyde	Butylamine	Norvaline	Butanal
Ethylamine	Alanine	Acetaldehyde	Pentylamine	Norleucine	Pentanal
Propylamine	2-Aminobutyric acid	Propanal	Isopentylamine	Leucine	3-Methylbutanal
Isobutylamine	Valine	2-Methylpropanal	Hexylamine	1-Aminoheptanoic acid	Hexanal
2-Methylbutylamine	Isoleucine	2-Methylbutanal	Benzylamine	Phenylglycine	Benzaldehyde

Table 8.30 Ammonia and volatile amine contents in selected foods.

Compound	Content (mg/kg f.w.)				Compound	Content (mg/kg f.w.)			
	Cabbage	Carrot	Cheeses	Beer		Cabbage	Carrot	Cheeses	Beer
Ammonia	15 260	3970	16 440	10–68	Isobutylamine	–	–	0.2	<0.22
Methylamine	16.6	3.8	3–12	0.02–0.32	Pentylamine	0.4	–	1.2	–
Ethylamine	1.3	1	1–4	0.31–2.12	Isopentylamine	0.5	–	<0.2	<0.14
Propylamine	–	–	2–8.7	<0.17	Dimethylamine	2.8	–	–	0.6
Butylamine	–	–	3.7	<0.07	Benzylamine	3.8	2.8	–	–

Some aliphatic, aromatic, and heterocyclic amines and diamines formed by decarboxylation of basic, aromatic, and heterocyclic amino acids are biologically active substances called biogenic amines. Subsequent enzymatic transformations of biogenic amines yield other biologically active products that play important roles in living organisms, such as adrenal hormones called catecholamines (see Section 10.3.4).

Decarboxylation of amino acids can also proceed as a non-enzymatic reaction. Analogously to the enzyme-catalysed decarboxylations, amines are formed as byproducts of the Strecker degradation of amino acids and arise by thermal decarboxylation of amino acids, especially sulfur amino acids, hydroxyamino acids, and aromatic amino acids. For example, thermal decarboxylation and subsequent reactions of cysteine and cystine produce ammonia and ethylamine, whilst those of methionine produce ethylamine, pentylamine, and crotylamine (but-2-en-1-ylamine). Furfuryl amine arises from 1-amino-1-deoxyketopentoses in the Maillard reaction (**8-95**).

A number of aromatic and heterocyclic amines that have the amino group attached to the aromatic (heterocyclic) ring may be formed by pyrolysis of amino acids and proteins in the Maillard reaction. Some of these amines (e.g. aminoimidazoazaarenes; see Section 12.2.1.1) are mutagenic and probably carcinogenic compounds.

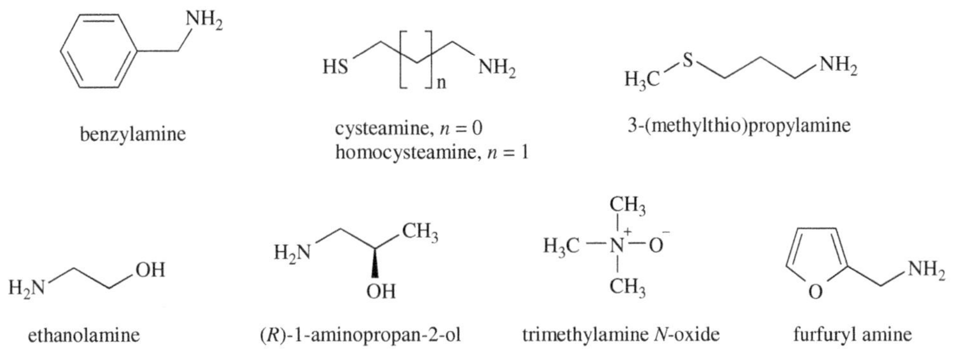

benzylamine

cysteamine, $n = 0$
homocysteamine, $n = 1$

3-(methylthio)propylamine

ethanolamine

(R)-1-aminopropan-2-ol

trimethylamine N-oxide

furfuryl amine

8-95, structures of selected amines and amino alcohols

Secondary and tertiary amines are formed from precursors other than amino acids. Dimethylamine results from degradation of choline (which is present in some phospholipids) and some alkaloids (e.g. in beer, it is produced from gramine present in germinating barley grains). It is also formed in non-enzymatic browning reactions from methylamine and formaldehyde or by decarboxylation of sarcosine. Trimethylamine, together with dimethylamine, methylamine, and ammonia, is an odorous compound of fish and other aquatic animals. It is formed by reduction of sensory indifferent trimethylamine oxide (trimethylaminoxide, 8-95) in tissues post mortem. Trimethylamine oxide is an osmolyte found in many marine species in quantities of 1–5% (d.w.) of the muscle tissue. Under the catalysis of trimethylamine oxide demethylase, it is decomposed to dimethylamine and formaldehyde.

8.2.10.1.3 Amides

Carboxylic acid amides are organic compounds which are formed by replacing the OH group of the carboxyl group with the NH_2 amide group. Unsubstituted amides and the corresponding N-substituted and N,N-disubstituted amides (8-96) are polar compounds of low volatility, yet some may participate in the aroma of non-acidic foods. Commonly occurring compounds are mainly amides derived from formic and acetic acids. For example, beer contains hundredths to tenths mg/kg of N-methylformamide (8-96, R = H, $R^1 = CH_3$, $R^2 = H$), N-methylacetamide (8-96, R = $R^1 = CH_3$, $R^2 = H$), N,N'-dimethylacetamide (8-96, R = $R^1 = R^2 = CH_3$), N-(2-methylbutyl)acetamide (8-96, R = CH_3, $R^1 = CH_2CH(CH_3)CH_2CH_3$, $R^2 = H$), and its isomer N-(3-methylbutyl)acetamide (8-96, $R^1 = CH_2CH_2CH(CH_3)_2$, $R^2 = H$), which are produced by condensation of carboxylic acids with amino acids followed by decarboxylation (Figure 8.72). Many non-volatile N-substituted formamides and acetamides are also produced in the Maillard reaction. Decarboxylation of asparagine catalysed by decarboxylases yields 3-aminopropionamide (8-96, R = $CH_2CH_2NH_2$, $R^1 = R^2 = H$), which may become a precursor of acrylamide (see Section 12.2.2).

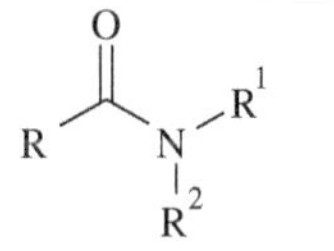

8-96, carboxylic acid amides

unsubstituted amides, $R^1 = R^2 = H$
N-substituted amides, $R^1 = H$
N,N-disubstituted amides

Lactams are cyclic amides hypothetically formed by dehydration of the corresponding amino acids (8-97). The simplest structures include aziridin-2-one (α-lactam, α-acetolactam, or ethane-2-lactam), which has been postulated as an intermediate in numerous processes. An azetidin-2-one (β-lactam, β-propiolactam, or propane-3-lactam) ring is part of the structure of β-lactam antibiotics. Important components of the aroma of shrimps, crabs, other marine crustaceans, molluscs, and other marine animals include a number of cyclic amides (lactams), such as pyrrolidin-2-one (2-pyrrolidone, γ-lactam, γ-butyrolactam, or butane-4-lactam), piperidin-2-one (2-piperidone, δ-lactam, δ-valerolactam, or pentane-5-lactam), 2-perhydroazepinone (ε-lactam, ε-caprolactam, or hexane-6-lactam), and their N-methyl-substituted analogues.

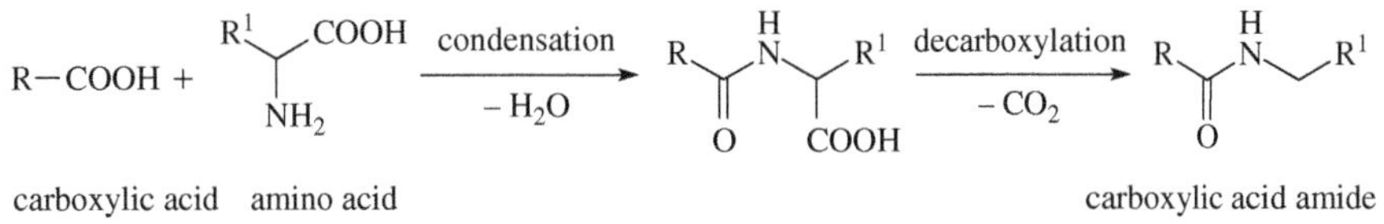

carboxylic acid   amino acid

carboxylic acid amide

Figure 8.72 Formation of amides from carboxylic acids and amino acids.

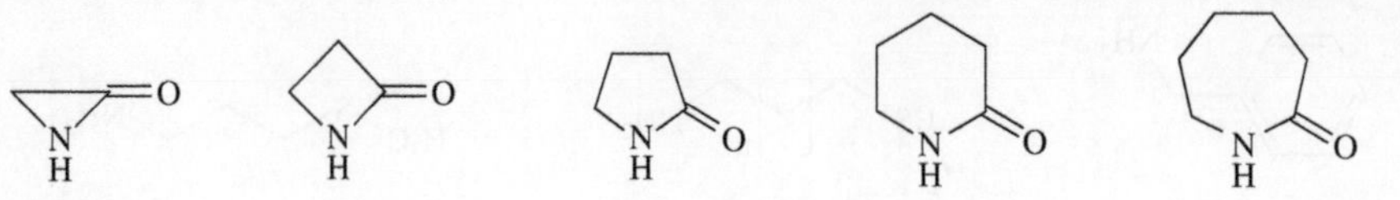

aziridin-2-one   azetidin-2-one   pyrrolidin-2-one   piperidin-2-one   perhydroazepin-2-one

8-97, cyclic amides (lactams)

A group of about 40 amides and their glycosides derived from 5-hydroxyanthranilic, 5-hydroxy-4-methoxyanhranilic, or 4-hydroxyanthranilic acids and cinnamic acids (cinnamic, *p*-coumaric, caffeic, ferrulic, and sinapic acids), trivially named avenanthramides, are important phytoalexins unique to oat grains (*Avena sativa*, Poaceae). They were found to exert very high antioxidant activities both *in vitro* and *in vivo*. Although the number of identified anthramides is high, the most common and dominant are the following derivatives of 5-hydroxyanthranilic acid (**8-98**): avenanthramide 2p, derived from *p*-coumaric acid (also called avenanthramide A, AF-1, or Bp), avenanthramide 2f, derived from ferulic acid (also called B, AF-2, or Bf), and avenanthramide 2c, derived from caffeic acid (also called C, AF-6, or Bc). Avenanthramides $2p_d$ (also called O), $2f_d$ (also called P), and $2c_d$ (also called Q) differ from 2p, 2f, and 2c in their additional double bond in the cinnamic acid moiety. Avenanthramides are localised in subaleurone and aleurone parts and in the pericarp of the oat grain. Avenanthramides 2p, 2f, and 2c are the major representatives, with 2p being the most abundant. The most abundant avenanthramide glycoside is 2c-3′-*O*-β-D-glucoside. After removal of the hulls, the main compound (40–132 mg/kg) in the oat groats used for human consumption is avenanthramide 2f. Germination increases the levels of avenanthramides several times, and heat treatments cause only a slight decrease in their concentration.

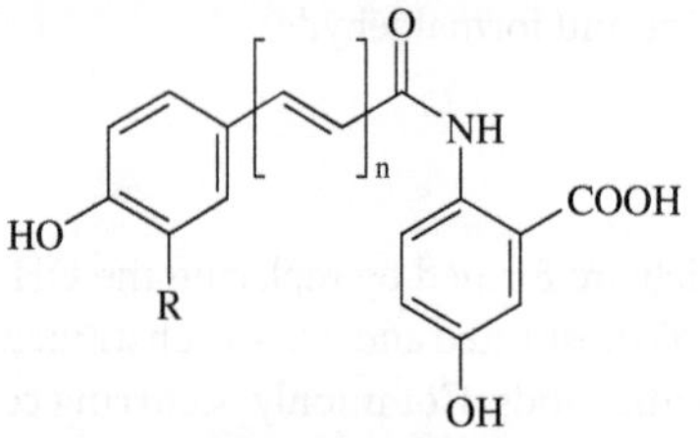

8-98, avenanthramide 2p, R = H, n = 1
avenanthramide 2c, R = OH, n = 1
avenanthramide 2f, R = OCH₃, n = 1
avenanthramide $2p_d$, R = H, n = 2
avenanthramide $2c_d$, R = OH, n = 2
avenanthramide $2f_d$, R = OCH₃, n = 2

Apart from aspartic and glutamic acid amides (glutamine and asparagine), foods contain a number of other non-volatile amides derived almost exclusively from glutamine or glutamic acid. An example is the amino acid L-theanine (see Section 2.2.1.2.3) in tea leaves. Other examples of important amides are vanillylamides (see Section 10.3.6.12) of bell peppers, which are derived from vanillylamine and fatty acids, and piperic acid amide piperine, occurring in black pepper (see Section 10.3.6.4). Numerous amides derived from biogenic amines and polyamines and cinnamic acids (cinnamic acid amides or phenolamides) are described in Section 10.3.4.1.2.

Properties and reactions Aldehydes and ketones react with ammonia and primary amines to form imines. Reaction with aldehydes yields aldimines, whilst reactions with ketones yield ketimines (Figure 8.28). Imines derived from aliphatic carbonyl compounds are generally unstable and are transformed to more stable products, such as amines and diamines. Secondary amines react with aldehydes and ketones with the formation of enamines (Figure 8.29). Flavour-active compounds are imines derived from furan-2-carbaldehyde. For example, the aromas of *N*-furfuryl(isobutylidene)amine and *N*-(furfurylidene)isobutylamine (**8-155**) reportedly resemble chocolate.

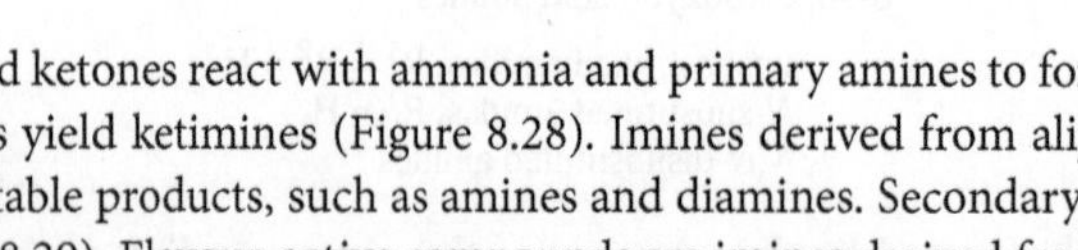

N-(isobutylidene)furfuryl-
methylamine, R = CH(CH₃)₂

N-(furfurylidene)-
isobutylamine, R = CH(CH₃)₂

8-99, amines and imines derived from furan

8.2.11 Heterocyclic compounds

Heterocyclic compounds are those in which one or more carbon atoms in the ring are replaced by another element, called a heteroatom. Most common in foods are five- and six-membered heterocyclic compounds containing as the heteroatom oxygen, sulfur, or nitrogen. Also common are heterocycles containing more of the same or different heteroatoms. Heterocyclic compounds are important odour- and taste-active components of many foods. The basic structures of most oxygen, sulfur, and nitrogen heterocycles are shown in formulae **8-100**.

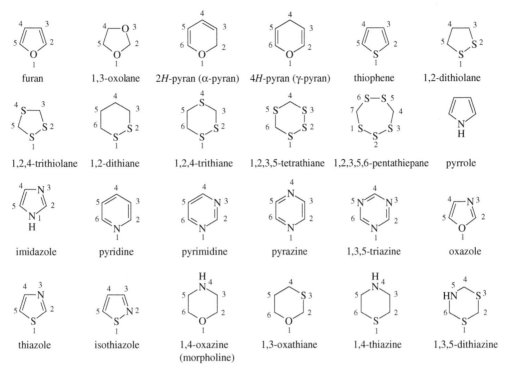

8-100, structures of selected heterocyclic compounds

More important than basic heterocycles, however, are their derivatives, particularly alkyl, acyl, hydroxy, oxo, and certain others. A number of important aromatic compounds are derived from dihydro, tetrahydro, or hexahydro derivatives of these heterocyclic compounds. Heterocycles containing polar functional groups are non-volatile, including virtually all imidazole and pyrimidine derivatives, but they often have other functions in living organisms. Heterocyclic compounds are formed mainly in non-enzymatic browning reactions, especially in food surface layers during the thermal processing of foods by baking, frying, and roasting. They also arise by pyrolysis of proteins, carbohydrates, lipids, and other food components, and some result from reactions catalysed by enzymes.

8.2.11.1 Classification, structure, terminology, and occurrence

8.2.11.1.1 Furans

A large number of furan derivatives are produced in non-enzymatic browning reactions as dehydration products of carbohydrates and related compounds, such as ascorbic acid (γ-lactones can also be considered furan derivatives). Most furans are common to many thermally processed foods. Higher levels of these compounds are also present in acid protein hydrolysates. Furan-2-carbaldehyde has a characteristic pleasant woody aroma resembling nuts, which arises from pentoses, ascorbic acid, and 5-methylfuran-2-carbaldehyde, which is a product of 6-deoxyhexoses. A fatty herbal-type flavour is provided by 5-hydroxymethylfuran-2-carbaldehyde, arising from hexoses. 2-Acetylfuran arising from hexoses and pentoses is an important component of a number of foods (e.g. fruits, wine, and beer) with a sweet, balsamic flavour. 2-Acetyl-3-hydroxyfuran, known as isomaltol (**8-101**), has a caramel-like flavour (Figure 8.73). Isomaltol is formed from hexoses and 4-*O*-substituted glucoses such as maltose.

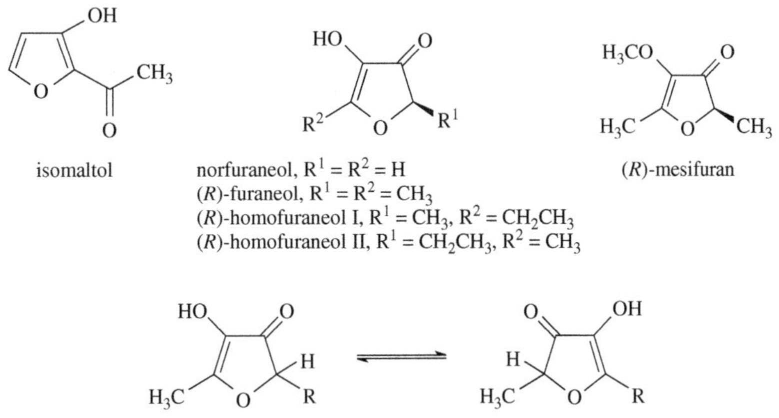

isomaltol

norfuraneol, R^1 = R^2 = H
(*R*)-furaneol, R^1 = R^2 = CH$_3$
(*R*)-homofuraneol I, R^1 = CH$_3$, R^2 = CH$_2$CH$_3$
(*R*)-homofuraneol II, R^1 = CH$_2$CH$_3$, R^2 = CH$_3$

(*R*)-mesifuran

8-101, structures of selected furans and general structure of compounds with caramel flavour

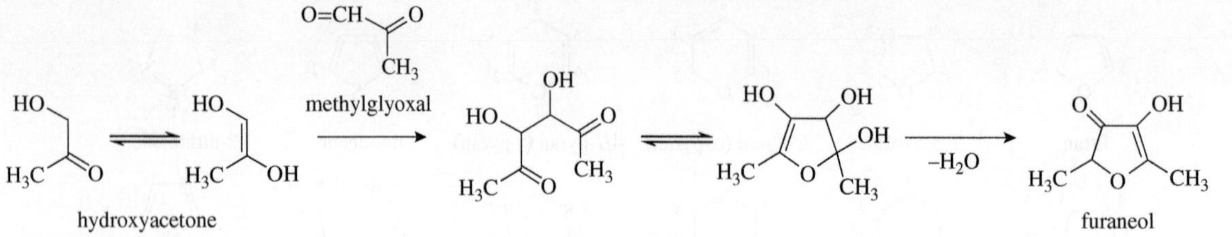

Figure 8.73 Formation of furaneol from methylglyoxal and hydroxyacetone.

The most important furans in major food commodities are the 4-hydroxy-2*H*-furan-3-ones or 4-hydroxy-3(2*H*)-furanones (**8-101**) norfuraneol, furaneol, and homofuraneol, which are characterised by a planar arrangement of molecules and the same configuration of the enol, hydroxyl, and oxo groups. Furanones occur as racemic mixtures because of their keto-enol tautomerism (**8-101**). Similar organoleptic properties to those of isomaltol has 2,4-dihydroxy-2,5-dimethyl-2*H*-furan-3-one, the cyclic form of so-called diacetylformoin (see Section 4.7.1.1. Some other compounds have analogous structures and properties to furanones, including maltol (3-hydroxy-2-methyl-4*H*-pyran-4-one), dihydromaltol (2,3-dihydro-5-hydroxy-6-methyl-4*H*-pyran-4-one), and alicyclic cyclopentenolones. All compounds with this structure more or less have a caramel flavour with sugary and sweet notes; this flavour differs, however, if the hydroxyl group is methylated or missing. 4-Hydroxy-5-methyl-2*H*-furan-3-one (norfuraneol), also known as toffee furanone or chicory furanone, arises in caramel, roasted chicory root, and meat broth from pentoses and hexoses in the Maillard reaction (see Section 4.7.1.1.3). 4-Hydroxy-2,5-dimethyl-2*H*-furan-3-one, also known as furaneol, strawberry furanone, or pineapple furanone, arises in the Maillard reaction from L-rhamnose (hexoses in general). An alternative reaction pathway to furaneol is that of methylglyoxal with hydroxyacetone (1-hydroxypropan-2-one) (Figure 8.73). Furaneol occurs in strawberries, pineapple, roasted almonds, popcorn, meat broth, and a number of other foods as a racemic mixture. The structure of (+)-(*R*)-furaneol responsible for the characteristic odour is given in formula **8-101**. The odour of furaneol is sugary, jammy, and reminiscent of strawberries and, at higher concentrations, caramel (threshold concentration is 1–4 µg/l in air). Racemic homofuraneol, also known as shoyu furanone, exists in two tautomeric forms through keto-enol isomerisation, as 5-ethyl-4-hydroxy-2-methyl-2*H*-furan-3-one (homofuraneol I) or 2-ethyl-4-hydroxy-5-methyl-2*H*-furan-3-one (homofuraneol II). Homofuraneol I and homofuraneol II occur in a ratio of 1 : 3 to 1 : 2, putting them in equilibrium with each other. Structures of (+)-(*R*)-isomers are given in formulae **8-101**. Only one isomer, (*R*)-homofuraneol I, was found to have a strong roasted sweet scent. The others have a sweet but less roasted odour. Homofuraneol was found in soy sauce, among others places.

An example of a furanone with methylated hydroxyl in position C-4 is 2,5-dimethyl-4-methoxy-2*H*-furan-3-one, also known as mesifuran, which is an enzymatic methylation product of furaneol. (+)-(*R*)-mesifuran (**8-101**) has a burnt and caramel-like odour, which is more subtle and mellow than the odour of furaneol. Mesifuran, identified for the first time in a northern berry known as arctic bramble (*Rubus arcticus*, *Rosaceae*), accompanies furaneol in strawberries, pineapple, and other fruits. Unsubstituted 2,5-dimethyl-2*H*-furan-3-one is a constituent of bread and coffee flavours; its odour resembles bread.

Autoxidation of linoleic and linolenic acids yields 2-pentylfuran and isomeric 2-(pent-2-en-1-yl)furans. These compounds, along with other oxidation products possess a green odour resembling beans. Avocados (*Persea americana*, Lauraceae) and other *Persea* species contain a group of furans commonly referred to as avocadofurans that are substituted at C-2 with long saturated or unsaturated side chains. The first two avocadofurans identified from avocado fruit and seeds were 2-(tridec-12-yn-1-yl)furan (avocadynenofuran) and 2-(tridec-12-en-1-yl)furan (avocadienofuran) (**8-102**). Since then, many other avocadofurans with growth-inhibitory and insecticidal activities have been identified.

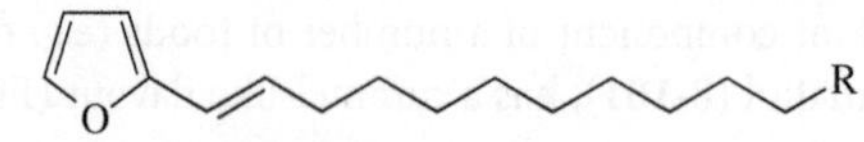

8-102, avocadienofuran, R = CH=CH₂
avocadynenofuran, R = C≡CH

The leaves of the small perennial shrubby herb *Helichrysum italicum* (Asteraceae), growing in the Mediterranean and known as curry plant, release a distinct curry aroma when crushed. Their key aroma compound is 5-acetyl-2,3-dihydro-3-hydroxy-2-[1-(hydroxymethyl)ethenyl]benzofuran, known as (2*R*,3*R*)-10-acetoxytoxol (**8-119**). Numerous sulfur-containing furan derivatives significantly contribute to the aroma of many foods and beverages. Examples are given in Sections 8.2.9.1.2 and 8.2.9.1.3.

8.2.11.1.2 Pyrans

Pyrans occurring in foods are hypothetically derived from α-pyran or γ-pyran, or from α- or γ-pyrones. δ-Lactones can also be considered derivatives of α-pyran. An important α-pyrone is 3-hydroxypyran-2-one, also known as 3-hydroxy-2*H*-pyran-2-one, which is a δ-lactone of 2,5-dihydroxypenta-2,4-dienoic acid with a caramel-like flavour that arises from ascorbic acid. Undoubtedly, the most important γ-pyrone is 3-hydroxy-2-methyl-4*H*-pyran-4-one, known as maltol (**8-103**), which features a caramel (or malt) smell and taste. About 40-times more potent than maltol is dihydromaltol (2,3-dihydro-5-hydroxy-6-methyl-4H-pyran-4-one) identified as a novel potent aroma compound in a Russian style kefir made from cooked milk, known as Ryazhenka kefir. 3-Hydroxy-6-hydroxymethyl-4H-pyran-4-one, known as kojic acid (**8-103**), arises from glucose during the fermentation of rice by the fungus *Aspergillus oryzae*. It has antimicrobial and antiviral activities and is widely used as a food additive preventing the enzymatic browning of shrimps. Allixin (3-hydroxy-5-methoxy-6-methyl-2-pentyl-4H-pyran-4-one) is a phytoalexin produced by garlic. A large number of other γ-pyrones are formed in non-enzymatic browning reactions.

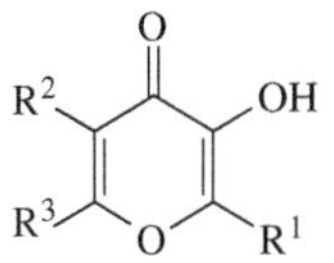

8-103, 3-hydroxy-4H-pyran-4-ones

maltol, $R^1 = CH_3$, $R^2 = R^3 = H$
kojic acid, $R^1 = R^2 = H$, $R^3 = CH_2OH$
allixin, $R^1 = (CH_2)_4CH_3$, $R^2 = OCH_3$, $R^3 = CH_3$

Additionally, aroma components of some foods are six-membered heterocycles containing an oxygen atom and sulfur in the molecule. An example of such a compound is (2R,4S)-2-methyl-4-propyl-1,3-oxathiane (**8-104**), also known as (Z)-tropathiane, which occurs in the yellow passion fruit (*P. edulis*, f. *flavicarpa*, Passifloraceae), pineapple (*Ananas comosus*, Bromeliaceae), white wines, and whisky. It arises by condensation of 3-mercaptohexan-1-ol (**8-11**) with acetaldehyde.

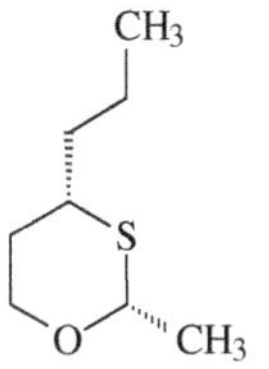

8-104, (Z)-tropathiane

8.2.11.1.3 Thiophenes

Thiophenes are found in many foods and often contribute significantly to their organoleptic properties. They are important as odorous components of meat, roasted coffee, roasted nuts, onion, and other foods. For example, whilst unsubstituted thiophene has an odour reminiscent of benzene, 2-methyl-substituted thiophenes positively affect the flavour of canned meat. 2-Methylthiophene's odour has been described as green, heated onion, sulfurous, and sweet. Dimethylthiophenes are important components of fried onion aroma; 3-butylthiophene-2-carbaldehyde and 3-(3-methylbut-2-en-1-yl)thiophene-2-carbaldehyde exhibit strong citrus and citral-like notes in fried chicken, 2-acetyl-3-methylthiophene is reminiscent of honey and roasted nuts, and 3-acetyl-2,5-dimethylthiophene has a sulfurous odour.

Thiophenes are formed by different reactions. The general one is that of furan derivatives with hydrogen sulfide (Figure 8.74). This mechanism is assumed, for example, in the formation of some thiophene derivatives from 4-hydroxy-2H-furan-3-ones. Alkylthiophenes substituted in positions C-2 and C-3 may arise in reactions of 2-mercaptoethanal with unsaturated aldehydes, such as acrolein and but-2-enal (Figure 8.75). 2-Mercaptoethanal is a product of Strecker degradation of cysteine.

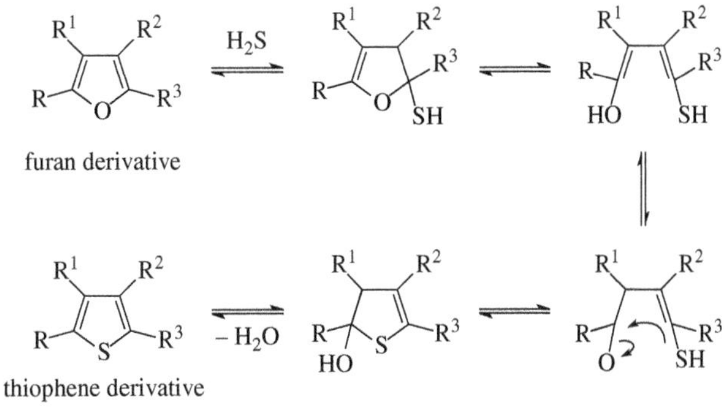

Figure 8.74 Formation of thiophenes from furans.

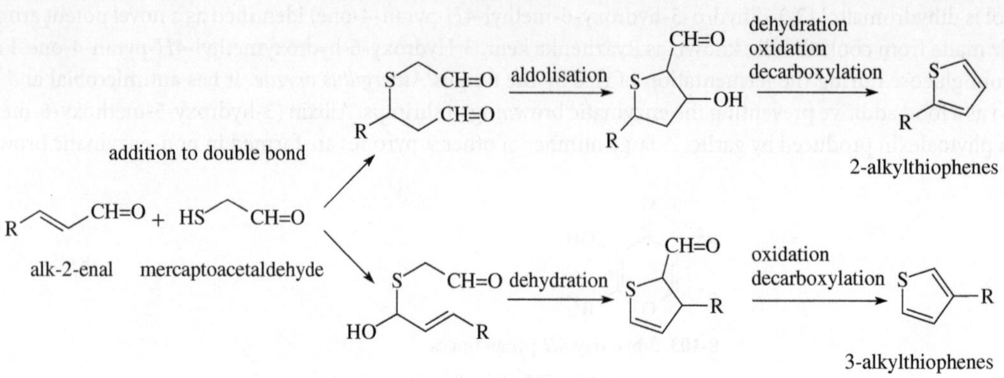

Figure 8.75 Formation of 2-alkylthiophenes and 3-alkylthiophenes.

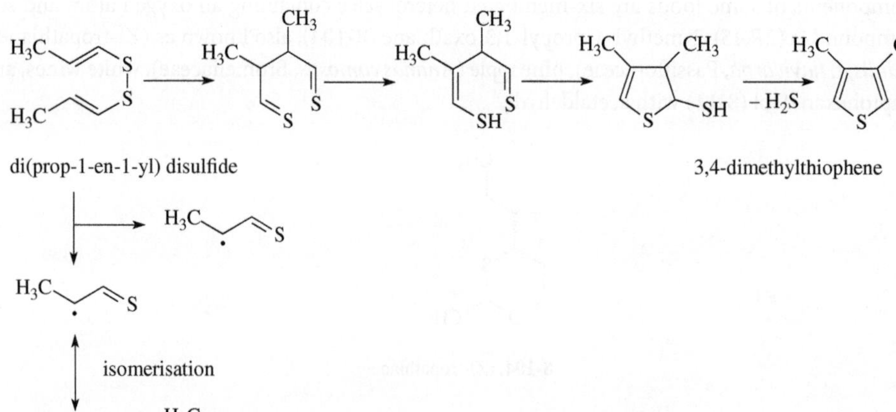

Figure 8.76 Formation of 2,4-dimethylthiophene and 3,4-dimethylthiophene in roasted onion.

The characteristic components of fried onion aroma are 2,4-dimethylthiophene and 3,4-dimethylthiophene. These are produced from di(prop-1-en-1-yl)disulfide (Figure 8.76), which is formed from isoalliin via the corresponding thiosulfinate in the same way that diallyl disulfide gives alliin via allicin. Thiophenes found in beef represent the structures shown in formula **8-110** (see later).

Common heterocyclic compounds generated by thermal processes are various thienothiophenes (**8-105**), components of roasted coffee volatiles.

thieno[2,3-*b*]-thiophene   thieno[3,2-*b*]-thiophene

8-105, thienothiophenes

8.2.11.1.4 Pyrroles

Pyrroles generally have an intense and unpleasant odour, but some (and their derived compounds) resemble caramel, nuts, bread, or chocolate upon dilution. A non-volatile pyrrole derivative is 5-oxopyrrolidine-2-carboxylic acid, produced by heating glutamic acid or glutamine.

Many pyrroles arise as products of the Maillard reaction from carbohydrates and proline. Unsubstituted pyrrole and some alkylsubstituted pyrroles also result from pyrolysis of other amino acids. For example, pyrrole appears in the pyrolysate of glycine, serine, threonine, and proline, 1-methylpyrrole arises from hydroxyproline, and 2-methylpyrrole and 3-ethyl-4-methylpyrrole are products of serine and threonine degradation. In general, pyrroles may arise in reactions of furans with ammonia (Figure 8.77), analogous to the reaction of furans with

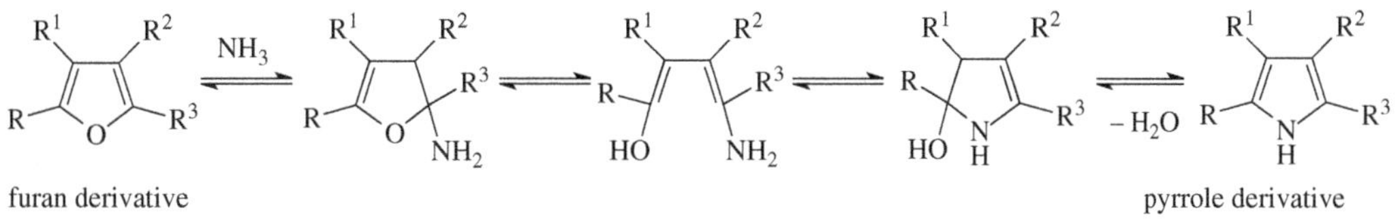

Figure 8.77 Formation of pyrroles from furans.

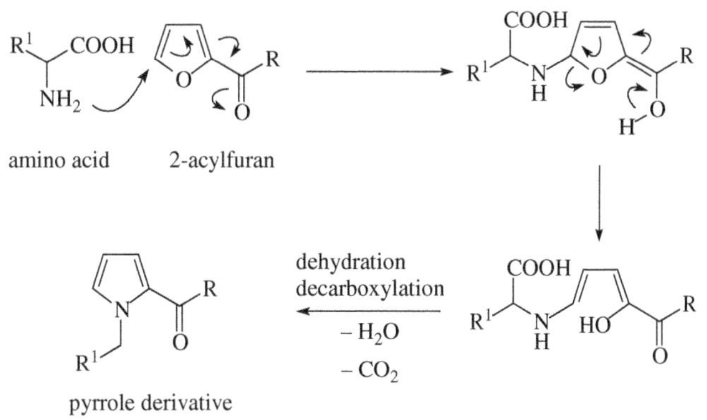

Figure 8.78 Formation of 2-acylpyrroles.

sulfane, which yields thiophenes. Pyrroles acylated in position C 2 of the nuclei are formed in reactions of amino acids with 2 acylfurans (Figure 8.78), and the corresponding pyridines are similarly produced. A particularly important compound is 2-acetyl-1-pyrroline (see Figure 8.89), which is a characteristic aroma component of white bread crust, fragrant rice varieties (such as basmati and jasmine rice), and popcorn.

Indoles (4,5-benzopyrroles) are components of a generally undesirable odour of certain foods. Typical compounds are indole and 3-methylindole (**8-106**), trivially called skatole. At higher concentrations, these compounds have a faecal smell (they are also found in faeces), but in low concentrations, they have a flowery one. Both indoles are the result of degradation of tryptophan by microorganisms (also in the digestive tract of animals) or of pyrolysis of tryptophan. They contribute to a specific boar odour in pork meat and to the typical odour of cooked beef tripe. For this reason, the traditional method for cooking tripe soup includes heating and removing the first broth. The acceptable concentration in pork fat is 0.15–0.25 mg/kg.

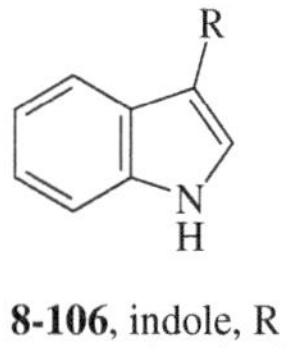

8-106, indole, R = H
skatole, R = CH$_3$

8.2.11.1.5 Imidazoles

Imidazoles are virtually non-volatile compounds and therefore do not act as odour-active substances. The main reaction leading to the formation of simple imidazoles is that of α-dicarbonyl compounds with aldehydes and ammonia (Figure 8.79). The basic member of the homologous series, unsubstituted imidazole (Figure 8.79, $R^1 = R^2 = R^3 = H$), thus arises from glyoxal, formaldehyde, and ammonia, whilst 4-methylimidazole (Figure 8.79, $R^1 = CH_3$, $R^2 = R^3 = H$) is analogously formed from methylglyoxal, formaldehyde, and ammonia and the precursors of 2-methylimidazole (Figure 8.79, $R^1 = R^2 = H$, $R^3 = CH_3$) are glyoxal, acetaldehyde, and ammonia. 4-Methylimidazole is found in large quantities (50–700 mg/kg) in caramel (see Section 4.7.6) when ammonia or ammonium salts are used as catalysts. Common dark beers and cola drinks may contain more than 250 µg/kg. 4-Methylimidazole acts as a convulsant (an agent that causes convulsions) for some animals at very high doses and is a probable carcinogen; therefore, in the European Union, the legal limit was established at 250 mg/kg of caramel.

Similar effects were observed for 2-methylimidazole, also found in caramel, and for polyhydroxy-substituted imidazoles produced from glycosuloses, 3-deoxyglycosuloses, and ammonia. These imidazoles are unstable, and at higher temperatures they decompose to imidazole, its methyl derivatives, and the corresponding glycols. An example is 2-acetyl-4-(arabino-1,2,3,4-tetrahydroxybutyl)imidazole.

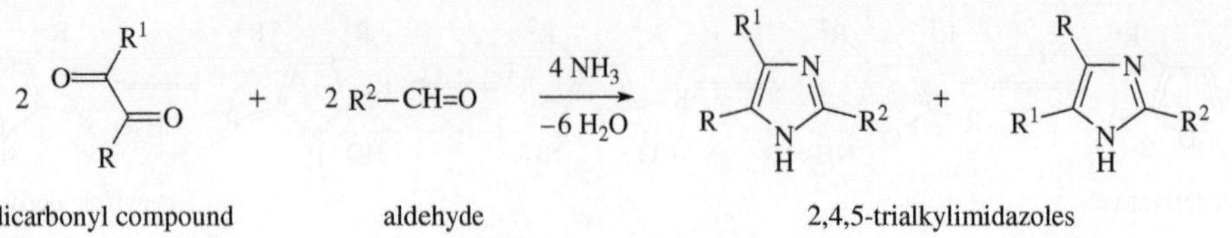

Figure 8.79 Formation of imidazoles from α-dicarbonyl compounds, aldehydes, and ammonia.

Non-volatile imidazoles produced from amino acids in reactions with α-dicarbonyl compounds and aldehydes are betaines. Aldehydes are formed from amino acids by Strecker degradation. The basic member of the homologous series of these imidazoles is 3-carboxymethyl-1-imidazolium ethanoate, which arises as the main product in the reaction of glycine with glyoxal (**8-107**).

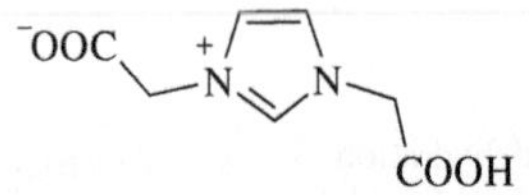

8-107, 3-carboxymethyl-1-imidazolium ethanoate

8.2.11.1.6 Pyridines

Descriptive terminologies for the odours of pyridines use terms such as green, bitter, astringent, roasted, burnt, pungent, solvent, and fishy, none of which could be considered desirable. Their presence in some food commodities, such as beer and whisky, is disagreeable and associated with a cardboard, oxidised, and harsh flavour. In roasted coffee, pyridines may contribute to a pleasant smell that is, however, less pleasant than that of pyrazines.

Some pyridines may arise, like thiophenes and pyrroles, in reactions of amino compounds with 2-acylfurans (Figure 8.80). Many are formed by thermal degradation of amino acids, proteins, and other nitrogenous compounds. For example, a large number of pyridines present in roasted coffee result from the pyrolysis of alkaloid trigonelline. During the roasting process, trigonelline breaks down in two ways (Figure 8.81). The first is the migration of the N-methyl group, which gives rise to methyl nicotinate (about 7%). Its content in roasted coffee is about 6 g/kg. In the presence of water, methyl nicotinate is hydrolysed to nicotinic acid. The second pathway involves decarboxylation of about 88% of trigonelline to a number of pyridines. An important volatile product is unsubstituted pyridine, produced at a concentration of about 200 mg/kg (5.3% of degraded trigonelline). Higher amounts in coffee cause an unpleasant aroma. Another major product is 3-methylpyridine (0.25%). An important intermediate is considered to be 1-methylpyridinium hydroxide, which decomposes to give methylamine, acetaldehyde, and malondialdehyde. The N-methylpyridinium ion shows chemopreventive effects. Reactions of these and certain carbonyl compounds and other fragments yield large quantities of different products, in particular various alkylpyridines and arylpyridines, amongst which 4-phenylpyridine, formed at a level of about 5 g/kg, is particularly significant.

A significant alkyl pyridine is 2-pentylpyridine, which is found, for example, in meat. It may be formed by ring closure of deca-2,4-dienals by ammonia. Important components of cocoa, bread, aromatic rice, popcorn, and other cereal products are 2-acetylpyridine and 6-acetyl-1,2,3,4-tetrahydropyridine (see Figure 8.90), which occurs in tautomeric equilibrium with 6-acetyl-2,3,4,5-tetrahydropyridine. Tetrahydropyridines develop together with 2-acetyl-1-pyrroline during the baking and roasting process (see Section 8.2.12.4.1).

The aroma components of coffee (and other foods) include different quinolines and isoquinolines, which result from trigonelline. Some pyridines (aminoimidazopyridines) and quinolines (aminoimidazoquinolines) produced in the Maillard reaction are classified as processing contaminants (see Section 12.2.1).

Pyridines and pyrazines, such as 2-ethylpyridine, 2,3,5-trimethyl pyridine, 2-ethyl-3,5-dimethyl pyridine, 2,3-dimethylpyrazine, and 2,5-dimethylpyrazine, are the most prominent classes of odorous compounds identified as being responsible for the odour of a cigar smoker's breath. They may be generated during cigar pyrrolysis by cleavage of nicotine or by the Maillard reaction.

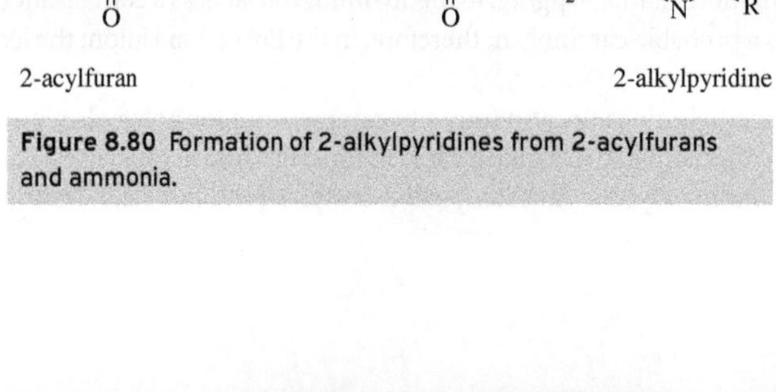

Figure 8.80 Formation of 2-alkylpyridines from 2-acylfurans and ammonia.

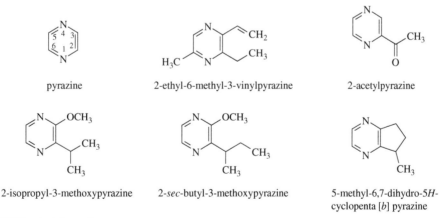

Figure 8.81 Main products of trigonelline pyrolysis.

8.2.11.1.7 Pyrazines

Pyrazine derivatives (e.g. alkylpyrazines, acylpyrazines, and alkoxypyrazines, **8-108**) are present in virtually all heat-processed foods (such as meat, bread, cocoa, roasted coffee, and nuts), where they are the major carriers of the characteristic burnt, roasted, and nutty odour. Pyrazines arise mostly as products of the Maillard reaction and by pyrolysis of some amino acids. 2-Alkyl-3-methoxypyrazines, which are odour components of various vegetables, are formed as primary odorous compounds through enzymatic reactions.

The general scheme of formation of pyrazines as byproducts of Strecker degradation of amino acids with α-dicarbonyl compounds is shown in Figure 8.82. For example, the main product of glyoxal reaction with amino acids is unsubstituted pyrazine, the reaction of amino acids with a mixture of glyoxal and methylglyoxal yields methylpyrazine, and the reaction of methylglyoxal itself gives 2,5-dimethylpyrazine as its main product. 2,5-Dimethylpyrazine is likewise produced by pyrolysis of threonine via the corresponding 2,5-dioxopyrazine. From simple alkylpyrazines, the methyl- and ethyl-substituted derivatives (Figure 8.83) are produced on reaction with formaldehyde and acetaldehyde, respectively. Recently, an alternate route to pyrazine formation was proposed, based on dimerisation of azomethine ylides formed in the reaction of amino acids with 2-oxo acids. Simple alkylpyrazines may also form as pyrolysis products of non-volatile polyhydroxyalkyl-substituted pyrazines, which are the reaction products of reducing sugars and amino acids. 2,3-Dimethylpyrazine and 2,5-dimethylpyrazine have also been identified as important odorous constituents responsible for the odour of a cigar smoker's breath.

pyrazine

2-ethyl-6-methyl-3-vinylpyrazine

2-acetylpyrazine

2-isopropyl-3-methoxypyrazine

2-sec-butyl-3-methoxypyrazine

5-methyl-6,7-dihydro-5*H*-cyclopenta [*b*] pyrazine

8-108, selected pyrazines

Many pyrazines are synthesised and used to flavour instant coffee, ice creams, potato products, and many other foods. For example, the odour of 2,5-dimethylpyrazine resembles roasted hazelnuts, and 2,6-dimethylpyrazine and trimethylpyrazine have a chocolate-like odour. Important components of coffee aroma include a number of pyrazines, such as 2,3-diethyl-3-methylpyrazine, 2-ethyl-3,5-dimethylpyrazine, 2,6-dimethyl-3-vinylpyrazine, and 2-ethyl-6-methyl-3-vinylpyrazine. The flavour-active components of popcorn and white bread are acetylpyrazine, 2-ethyl-3-methylpyrazine, 3-ethyl-2,5-dimethylpyrazine, and 5-methyl-6,7-dihydro-5*H*-cyclopenta[*b*]pyrazine. Cyclopentapyrazines, specifically 6,7-dihydro-5*H*-cyclopenta[*b*]pyrazines, are formed in reactions of cyclopentenolones with aminoketones and ammonia.

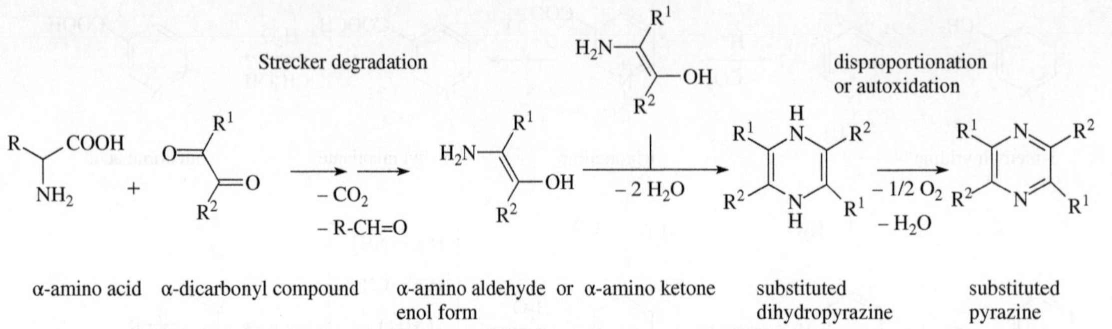

Figure 8.82 Formation of alkylpyrazines from α-dicarbonyl compounds and amino acids.

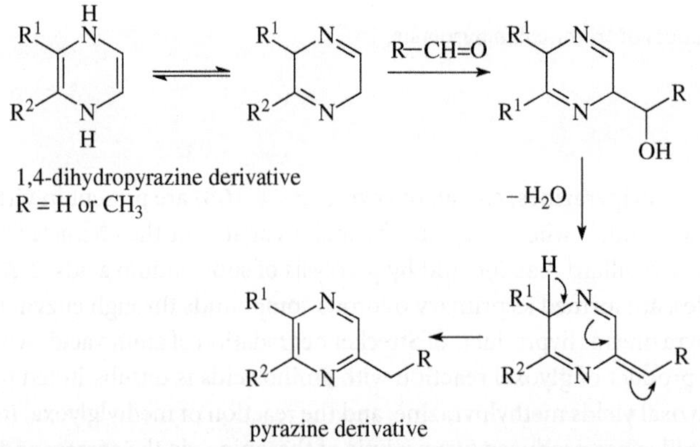

Figure 8.83 Formation of methylsubstituted and ethylsubstituted pyrazines.

In some plants and microorganisms, tetramethylpyrazine occurs as a product of metabolism, and 2-alkyl-3-methoxypyrazines are important odorous components of many types of vegetables. 2-Isopropyl-3-methoxypyrazine is important in the aroma of parsley, peas, peanuts, and potatoes, 2-isobutyl-3-methoxypyrazine in the aroma of pepper, and 2-sec-butyl-3-methoxypyrazine in the aroma of parsley, peas, and salad beet. 2-Alkyl-3-methoxypyrazines are also significant in the varietal character of wines, such as Cabernet Sauvignon and Sauvignon Blanc. The predominant compound is 2-isobutyl-3-methoxypyrazine, which occurs in these wines in concentrations of 3–56 ng/l; its odour threshold concentration is 1 ng/l. The 2-isobutyl-3-methoxypyrazine content in wines depends on the degree of maturation of the grapes (decreasing with advancing maturity) and the climate in which they are grown (increasing in cool climates and with low levels of sunlight). On the other hand, 2-isopropyl-3-methoxypyrazine has been suggested to be one of the key compounds responsible for the off-flavour produced by Asian lady beetle (*Harmonia axyridis*) in Frontenac and Leon Millot wines produced in Minnesota, and 2-methoxy-3,5-dimethylpyrazine was found to be responsible for the unpleasant, musty, and mouldy aroma of wine corks in Australian wines, which is similar to that produced by 2,4,6-trichloroanisole.

Thermal reactions such as coffee roasting produce various benzopyrazine (quinoxaline) derivatives. Quinoxalines (aminoimidazoquinoxalines) that arise in the Maillard reaction are classified as processing contaminants (see Section 12.2.1).

8.2.11.1.8 Oxazoles

From the olfactive point of view, an unsubstituted oxazole odour is reminiscent of pyridine; alkylsubstituted oxazoles resemble melons with very ripe kiwi notes and with other more or less aggressive nuances and pungent notes. Oxazolidines, oxazolines, and oxazoles (like other heterocycles) arise as reaction products of amino acids and reducing sugars, especially in the Maillard reaction, and occur as aroma components of roasted coffee, cacao, meat, and other commodities. An important reaction is the Strecker degradation of cysteine. Formation of alkyloxazoles can be explained by condensation of aminoketones (products of the Strecker degradation of amino acids with α-dicarbonyl compounds) or hydroxyketones and ammonia with aldehydes via oxazolines that are oxidised to oxazoles with the elimination of water. For example, 4,5-dimethyloxazole (Figure 8.84, $R^1 = R^2 = CH_3$, $R^3 = H$) results from biacetyl and glycine (precursor of formaldehyde) and, in an even greater amount, from formaldehyde and 2-aminobutan-3-one, which is a product of the Strecker degradation of amino acids with biacetyl. 2,4,5-Trimethyloxazole (Figure 8.84, $R^1 = R^2 = R^3 = CH_3$) arises from 2-aminobutan-3-one and acetaldehyde, 2,5-dimethyl-4-ethyloxazole (Figure 8.84, $R^1 = CH_2CH_3$, $R^2 = R^3 = CH_3$) and 2,4-dimethyl-5-ethyloxazole (Figure 8.84, $R^1 = R^3 = CH_3$,

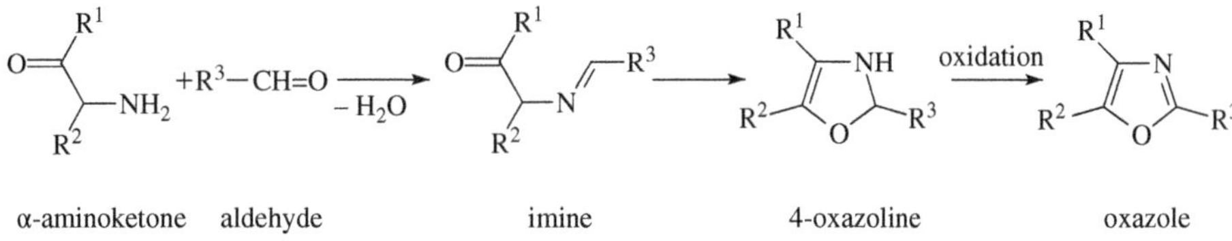

Figure 8.84 Formation of alkyloxazoles.

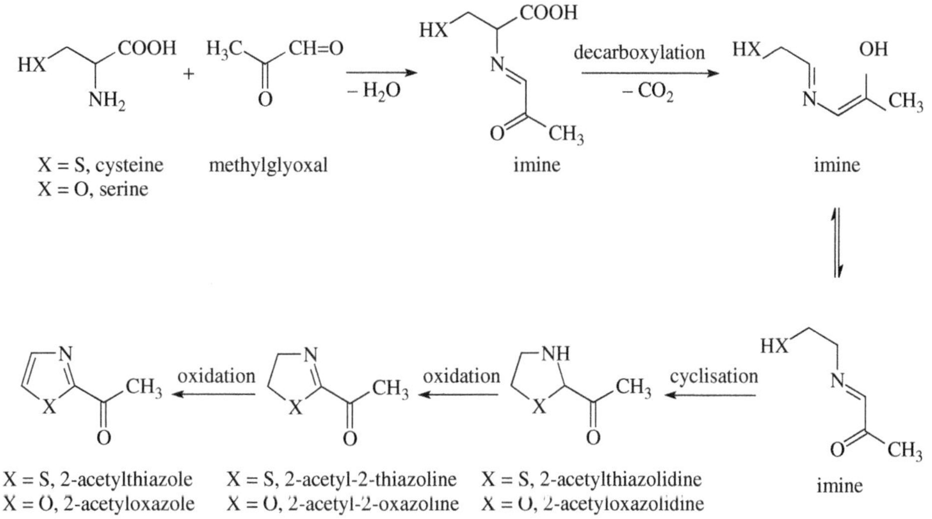

X = S, cysteine   methylglyoxal
X = O, serine

X = S, 2-acetylthiazole   X = S, 2-acetyl-2-thiazoline   X = S, 2-acetylthiazolidine
X = O, 2-acetyloxazole   X = O, 2-acetyl-2-oxazoline   X = O, 2-acetyloxazolidine

Figure 8.85 Formation of 2-acetyloxazole and 2-acetylthiazole.

$R^2 = CH_2CH_3$) are products of pentane-2,3-dione and acetaldehyde, and 2,4,5-trimethyl-3-oxazoline is formed as a reaction product of acetoin, acetaldehyde, and ammonia (Figure 8.85). Analogously, 2-acetyloxazole is produced from serine and methylglyoxal.

8.2.11.1.9 Thiazoles and other sulfur heterocycles

Thiazoles are substances found in many fresh and thermally processed foods. For example, they are present in fresh fruits and vegetables, boiled potatoes, roasted coffee, cocoa, and roasted meat. A number of thiazoles can be synthesised and used to flavour various foods.

Thiazole itself smells like pyridine, but substituted thiazoles, such as 2,4-dimethyl-5-vinylthiazole and 2-acetylthiazole, generally have a desirable odour, frequently described as nutty, green, roasted, and vegetable-like. 2-Methylthiazole and particularly 2-isobutylthiazole (Figure 8.86) have an odour resembling vegetables, with the latter typically having the smell of fresh tomato leaves and tomatoes. Related 2-isopropyl-4-methylthiazole, enzymatically synthesised from cysteine and valine, is an odorous component of durian. Some thiazoles arise from degradation of other sulfur-containing compounds, for example non-volatile 4-methyl-5-(2-hydroxyethyl)thiazole, which is formed by degradation of thiamine either through the action of thiaminases or by non-enzymatic reactions. Its dehydration yields 4-methyl-5-vinylthiazole, which is a constituent of cocoa aroma.

The Strecker degradation of cysteine by methylglyoxal yields 2-acetylthiazole in thermally processed foods (Figure 8.85). A possible intermediate of this reaction is 2-acetyl-2-thiazoline with an intense aroma of fresh bread crust, which is similar to the odour of 2-acetyl-1-pyrroline. 2-Acetyl-2-thiazoline has been identified amongst the volatiles of beef broth and was reported as being an important component of roasted beef (14–28 μg/kg) and other foods. The reaction pathways leading to 2-acetyl-2-thiazoline involve the cysteine decarboxylation product cysteamine and the participation of sugar degradation products, such as methylglyoxal. The reaction mechanism (Figure 8.87) involves the condensation of cysteamine with methylglyoxal, isomerisation of the condensation product, its cyclisation to 2-(1-hydroxyethyl)-4,5-dihydrothiazole, and its isomerisation to 2-(1-hydroxyethylene)thiazolidine, which is oxidised by atmospheric oxygen in the presence of transition metals via hydroperoxide (with elimination of hydrogen peroxide) to 2-acetyl-2-thiazoline. Alternative mechanisms of oxidation of 2-(1-hydroxyethylene) thiazolidine (or 2-acetylthiazolidine) have also been suggested. In addition to the pathway involving methylglyoxal, another suggested pathway starts with cysteamine and 4-deoxy-D-*glycero*-hexo-2,3-diulose, which provides, on elimination of formaldehyde and glycolaldehyde, the necessary three carbons for the formation of 2-acetyl-2-thiazoline.

2-Ethyl-4-methyl-3-thiazoline and 2-isopropyl-4-methyl-3-thiazoline were identified as aroma constituents of the toasted sesame seed oil. In stored orange juices, hexanal, octanal, nonanal, and decanal react with cysteine and yield the corresponding 2-alkylsubstituted thiazolidine-4-carboxylic acids (see Section 2.5.2.3.1).

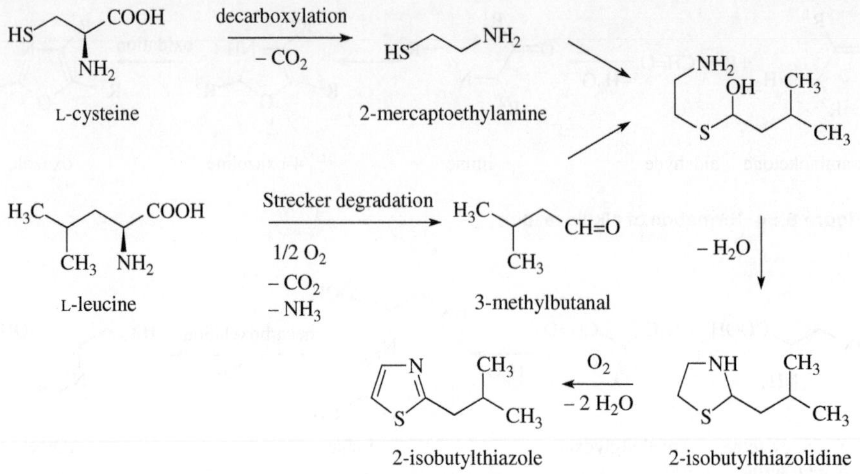

Figure 8.86 Biosynthesis of 2-isobutylthiazole.

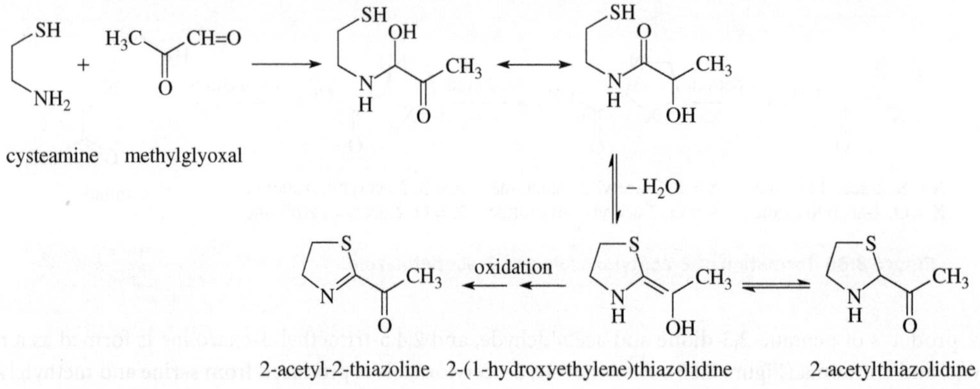

Figure 8.87 Formation of 2-acetyl-2-thiazoline.

Thiazoles are also carriers of undesirable odours and tastes, such as 2-acetylthiazole (Figure 8.85) in beer and 2-acetyl-2-thiazoline (Figure 8.87) or benzothiazole (**8-109**) in condensed and powdered milk.

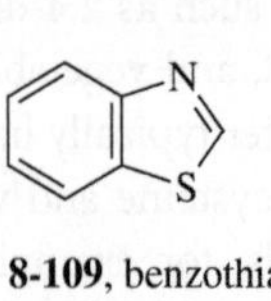

8-109, benzothiazole

8.2.11.2 Properties and reactions

Reactions of heterocyclic compounds are very complex. Significant reactions are discussed in the chapters dealing with the individual heterocyclic compounds; other important reactions are related to the Maillard reaction.

8.2.12 Aromatic substances of foods

Although more than 10 000 volatile compounds have been classified in foods, only about 200 compounds have been assigned as key food odorants contributing to the overall aromas of foods, such as beverages, meat products, cheeses, or baked goods.

8.2.12.1 Meat and meat products

Hundreds of compounds contribute to the taste and odour of meat. These include non-volatile substances, which mainly affect the taste, and volatile substances, which particularly affect the odour. Raw meat has only a weak, faint odour, and aromatic substances arise mainly during thermal processing from various precursors found in raw meat. Many of these precursors are formed during ageing of meat due to important chemical, structural, and functional changes in tissues. The characteristic flavour of cooked meat derives from thermally induced reactions occurring during heating, principally the Maillard reaction and degradation of lipids. Aroma character depends on the type of meat (pork, beef, mutton, chicken, and turkey, each having its own different, distinctive odour) and the method of heat treatment (e.g. cooking, baking, frying, and grilling), which is related to the water and fat contents in the material and the temperature during the treatment. Parameters such as diet, sex, age, differences between muscles, and storage conditions also play a significant role in determining the overall flavour sensation. For example, because of differences in the digestive systems between ruminants and non-ruminants, pork meat typically has more linoleic acid than beef or lamb, and pork and poultry meat contain higher amounts of polyunsaturated fatty acids than beef and lamb.

Lipid oxidation products and their reaction products with amino acids (proteins) have a considerable influence on the typical odour and taste of meat. Particularly significant aminocarboxylic acids include glutamic acid, alanine, threonine, lysine, guanidine compounds (creatine and creatinine), quaternary ammonium compounds (choline and carnitine), peptides (β-alanylhistidine peptides and some products of proteolysis), free nucleotides, nucleosides and their bases (especially inosine 5′-monophosphate, IMP), proteins, carboxylic acids (especially lactic acid), sugars (mainly glucose, fructose, and their phosphates and ribose formed by hydrolysis of free nucleotides), and some vitamins (especially thiamine).

The most important reaction that leads to flavour-active compounds in meat is the Maillard reaction, and the Strecker degradation of amino acids in particular. Many secondary compounds are formed by reactions of primary degradation products (hydrogen sulfide, methanethiol, and ammonia), with Strecker aldehydes and carbonyl compounds being formed from lipid oxidation. Active aromatic substances include numerous alcohols, aldehydes, ketones, carboxylic acids, esters, lactones, various aliphatic sulfur compounds, and oxygen, sulfur, and nitrogen heterocycles. Not all volatile compounds are equally important to the meat aroma, but the resulting flavour is always determined by a relatively large number of different compounds.

Important substances for the basic flavour of cooked and baked meats are aliphatic thiols, sulfides, aldehydes, furans, pyridines, and thiophenes. The list of sensory-active compounds is far from complete, and many components of meat flavour are yet unknown.

Examples of important aliphatic thiols are 3-mercaptobutan-2-one and 3-mercaptopentan-2-one, found in cooked beef. A mixture of 3-mercapto-2-methylpentan-1-ol diastereomers (Figure 8.88) has a broth-like, sweaty, and leek-like flavour. Very low odour threshold concentrations and an odour reminiscent of roasted meat are found in 2-methylfuran-3-thiol (fish thiol) and 2,5-dimethylfuran-3-thiol, their corresponding disulfides, and 2-furanmethanethiol (furfuryl mercaptan). Important sulfur compounds in beef are thiophenes, such as thiophene-2-thiol, 2-methylthiophene-3-thiol, 4,5-dihydro-2-methylthiophene-3-thiol, along with 2-pentylpyridine, and 5-acetyl-2,3-dihydro-1,4-thiazine (**8-110**). Odorous components of heat-treated beef also include 2-acetyl-1-pyrroline (Figure 8.89), which is otherwise an important component of bread crust aroma, 4-hydroxy-2,5-dimethyl-2*H*-furan-3-one (furaneol), its lower homologue 4-hydroxy-5-methyl-2*H*-furan-3-one (norfuraneol), and lactone 3-hydroxy-4,5-dimethyl-5*H*-furan-2-one (sotolon).

A key role is played by compounds formed from oxidised meat lipids, including carbonyl compounds such as hexanal, octanal, nonanal, oct-1-en-3-one, (*Z*)-octa-1,5-dien-3-one, (*Z*)- and (*E*)-non-2-enal, (2*E*,6*Z*)-nona-2,6-dienal, (2*E*,4*Z*)-deca-2,4-dienal, and (*E*)-4,5-epoxydec-2-enal. Plasmalogenes produce 12-methyltridecanal (also used as a food additive) and other long-chain aldehydes with the aroma of beef stew.

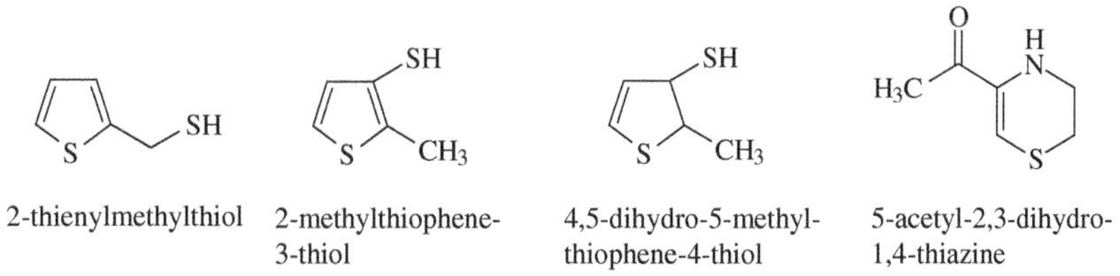

2-thienylmethylthiol 2-methylthiophene-3-thiol 4,5-dihydro-5-methyl-thiophene-4-thiol 5-acetyl-2,3-dihydro-1,4-thiazine

8-110, selected *S*-heterocyclic substances in meat

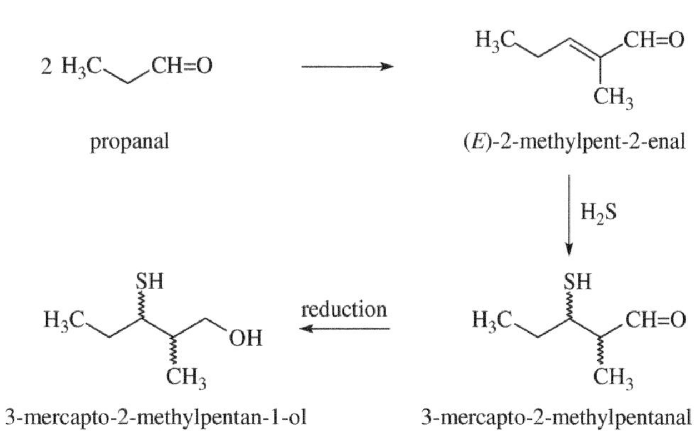

Figure 8.88 Formation of 3-mercapto-2-methylpentan-1-ol from propanal and hydrogen sulfide.

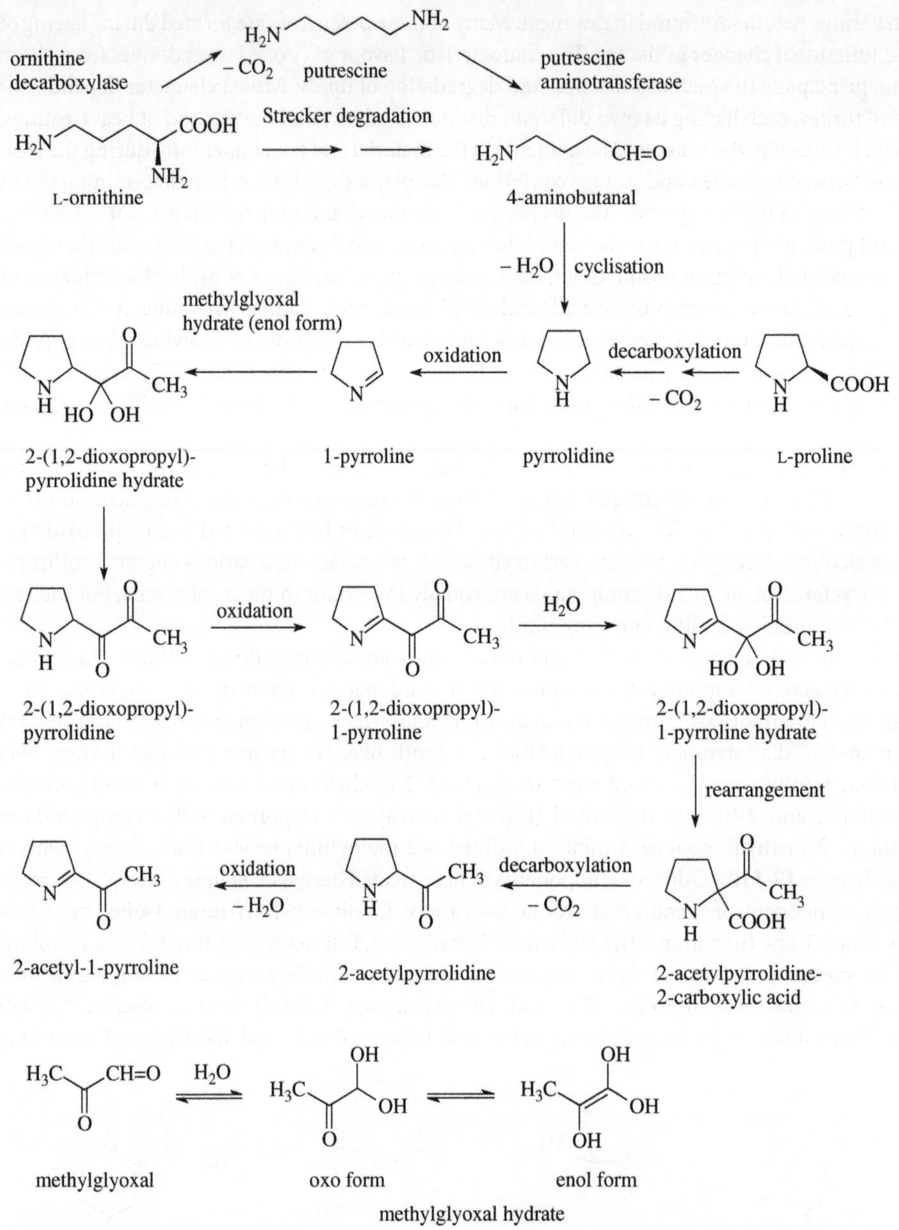

Figure 8.89 Formation of 2-acetyl-1-pyrroline from ornithine and proline.

For example, lamb/mutton has a distinctive aroma, which to some extent determines its lesser popularity. The main aromatic component of heat-treated meat is an unusual branched fatty acid 4-ethyloctanoic acid (**8-111**) that has a waxy, fatty, creamy, mouldy, sour, sweaty, cheesy odour with animal-like nuances and occurs as an almost racemic mixture of the enantiomers. 4-Ethyloctanoic acid (also identified in stewed beef gravy and goat and sheep cheeses) is accompanied by 4-methyloctanoic acid. The 4-ethyloctanoic acid content is approximately the same in raw meat but higher in fat.

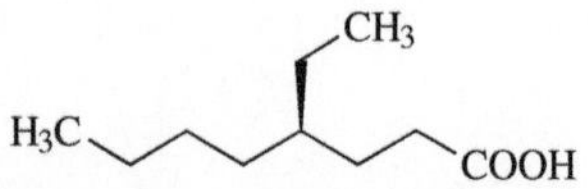

8-111, (*R*)-4-ethyloctanoic acid

Aromatic substances in meat products depend on production technology and the recipe. In dry-fermented and mould-ripened sausages, such as Hungarian-type salami, the key aroma compounds are 2-methoxyphenol, eugenol, (*E*)-isoeugenol, 5-methyl-2-methoxyphenol, 4-propyl-2-methoxyphenol, 4-ethyl-2-methoxyphenol and 3-ethylphenol, phenylacetaldehyde, methional, acetic acid, and 3-methylbutanoic acid.

The characteristic unpleasant off-flavour that can be evident during the cooking or eating of pork or pork products is boar taint, which is derived from non-castrated boars as they reach puberty. Rarely, this defect may also occur in meat of castrates and sows. The major boar taint compounds accumulating in the fat of animals are the boar pheromone (+)-5α-androst-16-en-3-one (androstenone), produced in the testes, and its metabolic products 3α,5α-androst-16-en-3-ol (3α-androstenol) and 3β-androstenol (**8-112**). Androstenone has a musk aroma, and the corresponding alcohols have a smell resembling urine. In wild boars, these compounds were found in extraordinarily high concentrations: 3.3 ng/g androstenone, 1.3 µg/g 3α-androstenol, and 0.5 µg/g 3β-androstenol. Other compounds contributing to the boar taint are skatole (also linked with faecal contamination of the skin) and 4-phenylbut-3-en-2-one (related to cinnamic acid metabolism and phenylalanine degradation).

5α-androst-16-en-3-one   3α,5α-androst-16-en-3-ol

8-112, androst-16-ene derivatives

8.2.12.1.1 Poultry and fish

The volatile components of raw chicken breast muscle include mainly carbonyls, thiols, sulfides, and alcohols. The major volatile components of fried chicken are similar to volatiles of the meat of farm animals (e.g. common aldehydes, ketones, sulfur-containing compounds etc.). Feed plays an important role in imparting certain flavour characteristics to poultry meat.

The characteristic essential flavour-active components of fish and other aquatic animals are amines and other nitrogenous compounds. Trimethylamine arises by reduction of sensorially indifferent trimethylamine oxide (acting in the regulation of osmotic pressure in cells) in the tissue post mortem. The amount of trimethylamine oxide, as well as of a number of simultaneously produced biogenic amines, depends primarily on the species, type, and time of storage. Its content in freshwater fish is about 5 mg/kg, and in seafood is 40–120 mg/kg. Other important compounds are dimethylamine and ammonia.

Flavour-active components of fresh fish are further oxidation products of unsaturated fatty acids formed by the action of lipoxygenases. Important compounds are mainly alcohols and carbonyl compounds. For example, oxidation of eicosapentaenoic acid via 12-hydroperoxide yields (3Z,6Z)-nona-3,6-dienal, (2E,6Z)-nona-2,6-dienal, (3Z,6Z)-nona-3,6-dien-1-ol, (Z)-octa-1,5-dien-3-ol, and (Z)-octa-1,5-dien-3-one, whilst the products of 15-hydroperoxide decomposition include (Z)-hex-3-enal, (E)-hex-2-enal, (Z)-hex-3-en-1-ol, and pent-1-en-3-ol.

The fishy and greasy off-flavour appearing in long-term refrigerated storage of fish is almost entirely caused by lipid oxidation products that are present in larger quantities than in fresh fish. The important components are mainly (Z)-hex-3-enal, (Z)-hept-4-enal, (Z)-octa-1,5-dien-3-one, and (3Z,6Z)-nona-3,6-dienal. An important component is methional, the product of the Strecker degradation of methionine.

8.2.12.2 Milk and dairy products

8.2.12.2.1 Milk

Raw or gently pasteurised milk (e.g. for 10 seconds at 73 °C) has a fine characteristic odour and sweet taste. Typical components present in low concentrations are dimethylsulfide, biacetyl, 2-methylbutan-1-ol, (Z)-hept-4-enal, and (E)-non-2-enal. Milk pasteurised at higher temperatures and ultra high-temperature (UHT) milk present the so-called cooked flavour, the appearance of which is the first measurable manifestation of the chemical changes that occur in heated milk. The cooked off-flavour of UHT milk is associated with the presence of a variety of sulfur-containing compounds, such as sulfane, dimethylsulfide, dimethyldisulfide, and dimethyltrisulfide, that are produced from proteins contained in the membranes of fat particles and from thiamine. The so-called stale flavour of UHT milk is characterised by the presence of alkan-2-ones (methylketones), lactones, and aliphatic aldehydes. Methylketones are generated by thermal decarboxylation of β-oxocarboxylic acids (mainly hexan-2-one, heptan-2-one, and nonan-2-one), whilst γ-lactones and δ-lactones are produced by dehydration of γ- and δ-hydroxycarboxylic acids (mainly δ-decalactone and γ- and δ-dodecalactones). Important carbonyl compounds include hexanal, 3-methylbutanal, (Z)-hept-4-enal and (E)-non-2-enal, and biacetyl. In the more intensive thermal treatment of milk (sterilisation), products of the Maillard reaction play a role, such as maltol and isomaltol, 5-hydroxymethylfuran-2-carbaldehyde, 4-hydroxy-2,5-dimethyl-2H-furan-3-one (furaneol), and 2,5-dimethylpyrazine.

Evaporation and drying lead to more extensive reactions in condensed and powdered milk that include degradation of proteins and thiamine, Maillard reaction between proteins (preferably of bound lysine) and lactose, degradation of oxo acids (to alkan-2-ones) and hydroxycarboxylic acids (to lactones), and oxidation of fatty acids. Particularly susceptible to oxidation are phospholipids of fat particles.

In addition to the compounds found in UHT milk, other odorous compounds are benzaldehyde (formed by reaction between phenylalanine and lactose), furfuryl alcohol (product of lactose degradation), and acetophenone, 2-aminoacetophenone, and benzothiazole (produced by decomposition of tryptophan).

Flavour defects of milk and dairy products may have a number of different causes. Compounds causing an off-flavour can pass into milk from unsuitable feed or milk may absorb flavours from the surroundings during storage and distribution. Other defects can occur in both raw and pasteurised milk as endogenous odorous substances are formed by enzymatic and chemical reactions. Oxidation of lipids is the major problem during storage of dried milk, leading to unpleasant stale off-flavour. For example, (Z)-hept-4-enal, oct-1-en-3-one, and hexanal are important carriers of a full, creamy flavour, but at higher concentrations they confer cardboard, metal, and green flavours. Long-term storage of condensed milk is reflected as old flavour, due to higher concentrations of normal compounds such as benzothiazole and 2-aminoacetophenone (>1 µg/kg). Often confused with oxidised flavour is the so-called sunlight flavour (characterised as burnt protein or medicinal-like), which develops in unprotected milk stored in the light. Photooxidation activates riboflavin, which is responsible for catalysing the conversion of methionine into sulfur compounds via methional.

Bacterial off-flavours result from the growth of psychrophilic bacteria that are present in milk due to poor sanitation or handling practices. Fruit odour is caused by carboxylic acid esters produced by the bacterium *Pseudomonas fragii*, a malt smell (caused by methylbutanal, 2-methylbutanal, and 2-methylpropanal) is produced by the bacterium *Streptococcus lactis* var. *maltigenes*, and a phenolic odour is caused by the bacterium *Bacillus circulans*. Rancid smell, caused by lower fatty acids from butyric to lauric acids, is manifested as a result of lipolysis by milk lipase or bacterial lipases.

8.2.12.2.2 Cream and butter

For cream flavour, the most important components are lactones of hydroxycarboxylic acids, in particular δ-decalactone, δ-dodecalactone, and (Z)-dodec-6-eno-γ-lactone. Less important are other lactones, such as δ-tetradeca-, δ-hexadeca-, γ-tetradeca-, γ-hexadeca-, γ- and δ-octadeca-, and γ- and δ-eicosalactones, and other compounds, such as methanethiol and skatole.

The aroma of butter made from sweet cream is affected primarily by free fatty acids (especially capric and lauric acids), δ- and γ-lactones, dimethylsulfide, (Z)-hept-4-enal, and the degradation products of tryptophan (indole and skatole). The butter obtained from sour cream contains mainly metabolic products of microorganisms (so-called starter cultures). Especially important compounds are biacetyl, lactic, and acetic acids.

Lipases in stored butter gradually release fatty acids from triacylglycerols, and their presence can be detected as a rancid and soapy flavour when they reach 30–40% of their threshold concentrations, which is a result of their synergism. The reaction is called hydrolytic rancidity. Mainly responsibe for the rancid flavour is butyric acid, followed by caproic acid. Caprylic acid has a rancid soap-like flavour, whilst capric and lauric acids have only soapy flavours.

In long-term stored butter, active oxidative rancidity products are (E)-non-2-enal and, in particular, (Z)-non-2-enal, whilst less active products are (Z)-hept-4-enal, oct-1-en-3-one, and others. The rancid and soapy odour in butter can also be caused by contamination with anion active detergents, such as natriumdodecyl sulfate.

8.2.12.2.3 Fermented dairy products

Characteristic flavour-active substances of fermented dairy products are metabolites of lactic acid bacteria, especially biacetyl, acetaldehyde, dimethylsulfide, lactic and acetic acids, various aldehydes, ketones, and esters. An important product is carbon dioxide. The acetaldehyde content in good quality yoghurts is 13–16 µg/kg, whilst the biacetyl content is about four times higher.

8.2.12.2.4 Cheeses

Cheeses contain large amounts of aromatic compounds that differ qualitatively and quantitatively according to the type. Important components of hard cheeses (Gouda type) include some carboxylic acid esters (ethyl butanoate, ethyl hexanoate), as well as carboxylic acids (acetic, butyric, isobutyric, valeric, isovaleric, 2-methylbutyric, and caproic acids). Cheeses manufactured using bacteria of the genus *Propionibacterium* (such as Emmental and Gruyère) contain propionic acid and other lower fatty acids, methyl thioacetate, some oxocarboxylic acids, various alcohols, esters (such as ethyl butanoate), lactones (such as δ-decalactone), amines and other basic compounds (also skatole, in addition to aliphatic amines), alkylpyrazines (e.g. 2-*sec*-butyl-3-methoxypyrazine), 4-hydroxy-2,5-dimethyl-2*H*-furan-3-one (furaneol), 2-ethyl-4- hydroxy-5-methyl-2*H*-furan-3-one (homofuraneol), and a range of other compounds.

Cheeses with a very pungent aroma caused by bacteria that live in the rind and on the surface (such as Pont l'Eveque, Limburger, and Romadur) contain phenol, cresol, acetophenone, and methylthioesters of lower fatty acids (**8-113**) such as methyl thioacetate (R = H), methyl thiopropionate (R = CH$_3$), and methyl thiobutyrate (R = CH$_2$CH$_3$) as flavourings. Other important flavouring substances include C$_4$–C$_{10}$ fatty acids, especially capric acid, alkan-2-ones (methylketones), and alkan-2-ols.

8-113, fatty acid methylthioesters

Camembert-type cheeses with a powdery rind of white mould (*Penicillium camemberti*) smell pleasantly of mushrooms, earth, and garlic. The characteristic component of the mushroom-like odour is oct-1-en-3-ol, the floral odour components are primarily 2-phenylethanol and 2-phenylethyl acetate, and a nutty odour is caused by 1,3-dimethoxybenzene and methyl cinnamate. The garlic note of matured cheese is caused by the presence of sulfur compounds such as 2-bis(methylthio)methane (also known as bis(dimethylsulfanyl)methane or 2,4-dithiapentane), tris(methylthio)methane (also known as tris(methylsulfanyl)methane or 3-methylthio-2,4-dithiapentane), methyl(methylthio)methyl disulfide (also known as (methyldisulfanyl)methylsulfanylmethane or 2,3,5-trithiahexane), and bis(methylthiomethyl) sulfide (**8-114**).

bis(methylthio)methane   tris(methylthio)methane   methyl(methylthiomethyl)disulfide   bis(methylthiomethyl)sulfide

8-114, methylthio and methylthiomethyl compounds

8.2.12.3 Eggs

Fresh eggs contain more than 100 volatile compounds in low concentrations. An important group is the sulfur compounds, mainly dimethylsulfide and dimethyldisulfide, which are characteristic components of egg aroma. Higher levels of these compounds are found in older eggs. Therefore, the long-term storage of eggs produces defects in smell and taste. The main flavour-active compounds are the degradation products of amino acids (indole and lower alkylbenzenes, especially toluene) and aldehydes and ketones (especially acetaldehyde, propionaldehyde, acetone, and butanone). A significant proportion of volatiles represent C_7–C_{17} hydrocarbons, with the main hydrocarbon being heptadec-5-ene; pyrroles and pyrazines are also present in very small amounts.

Heat treatment of eggs by cooking and frying creates a variety of other volatile compounds, more from egg yolk than from egg white. The main volatile components are aldehydes, alcohols, free fatty acids, esters, and other compounds. Those occurring in the highest quantities are (2Z,4Z)- and (2Z,4E)-deca-2,4-dienals, (Z)-oct-2-enal, nonanal, hexanal, phenylacetaldehyde, and hexyl butyrate. Egg white odour resembles (2E,4Z,7Z)-2,4,7-tridecatrienal, which arises as a product of arachidonic acid autoxidation (see Section 3.8.1.12.1). Sulfane and ammonia are also important protein degradation products.

Some eggs may have an unusual or unacceptable odour or taste, although their appearance is normal. The cause might be age (they are 'past their best'), high storage temperatures, or poor storage conditions, resulting in fishy or other undesirable flavours. The fishy off-flavour, more common in brown-shelled eggs, is caused by the presence of trimethylamine, produced by microbial decomposition of choline when hens are fed excessive amounts of fishmeals or fish oils and rapeseed and mustard meals. In the case of oilseed meals, trimethylamine arises from sinapine, an ester of choline with sinapic acid. In most breeds of hen, trimethylamine is enzymatically oxidised to odourless trimethylamine oxide, which is excreted from the body.

8.2.12.4 Cereals

8.2.12.4.1 Bread and cereal products

Two processes produce aroma substances from bread. Activities of yeast in the dough give volatiles of breadcrumbs, whilst baking yields volatiles of bread crust.

The main aroma substances produced by yeast in breadcrumbs are aldehydes, alcohols, acetals, and sulfur compounds. The most important components are 2-phenylethanol and 3-methylbutan-1-ol, with acetaldehyde, 2-methylpropanal, 3-methylbutanal, ethanol, 2-methylpropan-1-ol, 1,1-diethoxyethane, dimethylsulfide, and dimethyldisulfide also present in significant quantities. Other important volatiles are products of oxidation of fatty acids, (E)-non-2-enal, and (2E,4E)-deca-2,4-dienal.

Volatiles of the bread crust arise mainly in the Maillard reaction and through caramelisation of sugars. Important precursors of these compounds are the amino acids ornithine, proline, arginine, and lysine. Their partners are reducing sugars and their degradation products (such as methylglyoxal and hydroxyacetone). The basic odorous substances of wheat and rye breads are pyrazines, especially acetylpyrazine and 2-ethyl-3-methylpyrazine. Two key compounds have a major impact on the aroma of bread crust, namely 2-acetyl-1-pyrroline and 6-acetyl-1,2,3,4-tetrahydropyridine, which is in equilibrium with its tautomer 6-acetyl-2,3,4,5-tetrahydropyridine. In wheat bread crust, concentrations of 2-acetyl-1-pyrroline and 6-acetyl-1,2,3,4-tetrahydropyridine were found to be 19–78 and 53 µg/kg, respectively. The precursors of all three compounds are the amino acids proline and ornithine. Ornithine is produced by yeast during fermentation of the dough from glutamic acid, and an alternative route is the formation from proline in a reaction catalysed by ornithine cyclodeamidase. In cases where ornithine is not present, the only precursor is proline. Decarboxylation and oxidation of proline to 1-pyrroline is facilitated by reaction with reactive α-dicarbonyl compounds, such as 1-deoxy-D-*erythro*-hexo-2,3-diulose arising from D-fructose, whilst 1-deoxy-D-*erythro*-hexo-2,3-diulose is transformed into 1,6-dideoxy-D-*glycero*-hexo-2,4,5-triulose (acyclic form of diacetylformoin) by the formation of iminium ion, followed by its decarboxylation and dehydration. 1-Pyrroline then condenses with hydrated methylglyoxal to yield 2-(1,2-dioxopropyl)pyrrolidone, which spontaneously oxidises with air oxygen to the corresponding 1-pyrroline, which further undergoes an addition of water with the formation of a hydrate and subsequent semibenzilic rearrangement to a β-ketoacid. Its decarboxylation generates 2-acetylpyrrolidine, which is spontaneously oxidised to 2-acetyl-1-pyrroline. The immediate precursor of 2-acetyl-1-pyrroline from ornithine is 4-aminobutanal, which is produced in the Strecker degradation of ornithine, but also as a product of ornithine catabolism by yeast. Ornithine is first decarboxylated to putrescine by ornithine decarboxylase, and putrescine transamination yields 4-aminobutanal with catalysis of aminotransferase. A hypothetical mechanism of formation of 2-acetyl-1-pyrroline is outlined in Figure 8.89.

The formation of tetrahydropyridines starts with the reaction of 1-pyrroline with hydroxyacetone (acetol, 1-hydroxypropan-2-one), which gives rise to 2-(1-hydroxy-2-oxopropyl)pyrrolidine. The ring enlargement requires opening of the pyrrolidine ring with the formation of 7-aminoheptane-2,3-dione, which finally cyclises to 6-acetyl-1,2,3,4-tetrahydropyridine (Figure 8.90). Hydroxyacetone is produced in the baking process by thermal decomposition of methylglyoxal.

Additional odorous components, such as 3-ethyl-2,5-dimethylpyrazine, biacetyl, 3-methylbutanal, and 2-phenylethanal, are found in dark rye bread. Toasted bread contains the same compounds that carry the aroma of bread crust and crumb. A key component is again 2-acetyl-1-pyrroline, with other major ones being methional, 4-hydroxy-2,5-dimethyl-2*H*-furan-3-one (furaneol), (*E*)-non-2-enal, and biacetyl, the odour of which resembles butter.

Important flavour components of other cereal products are, as in bread, 2-acetyl-1-pyrroline, 6-acetyl-1,2,3,4-tetrahydropyridine, acetylpyrazine, and similar compounds.

Cereal flours and legume products, such as soybean products, often assume rancid flavours on prolonged storage. The carriers of this off-flavour are carbonyl compounds (such as hexanal and other aldehydes) produced by oxidation catalysed by lipoxygenases.

8.2.12.4.2 Rice

The aroma of cooked rice, especially of the basmati and jasmine aromatic rice varieties, is mainly influenced by the formation of 2-acetyl-1-pyrroline, which has the odour of bread crust and popcorn. Its concentration can reach approximately 0.6 mg/kg. Significant compounds are aldehydes formed by oxidation of fatty acids, especially (2*E*,4*E*)-deca-2,4-dienal, nonanal, hexanal, octanal, and (*E*)-non-2-enal, and the phenols guaiacol and 4-vinylphenol, resulting from the degradation of phenolic acids. Other important compounds are 3-hydroxy-4,5-dimethyl-5*H*-furan-2-one (sotolon) with protein hydrolysate aroma and 2-aminoacetophenone with a phenolic odour, formed by the degradation of tryptophan.

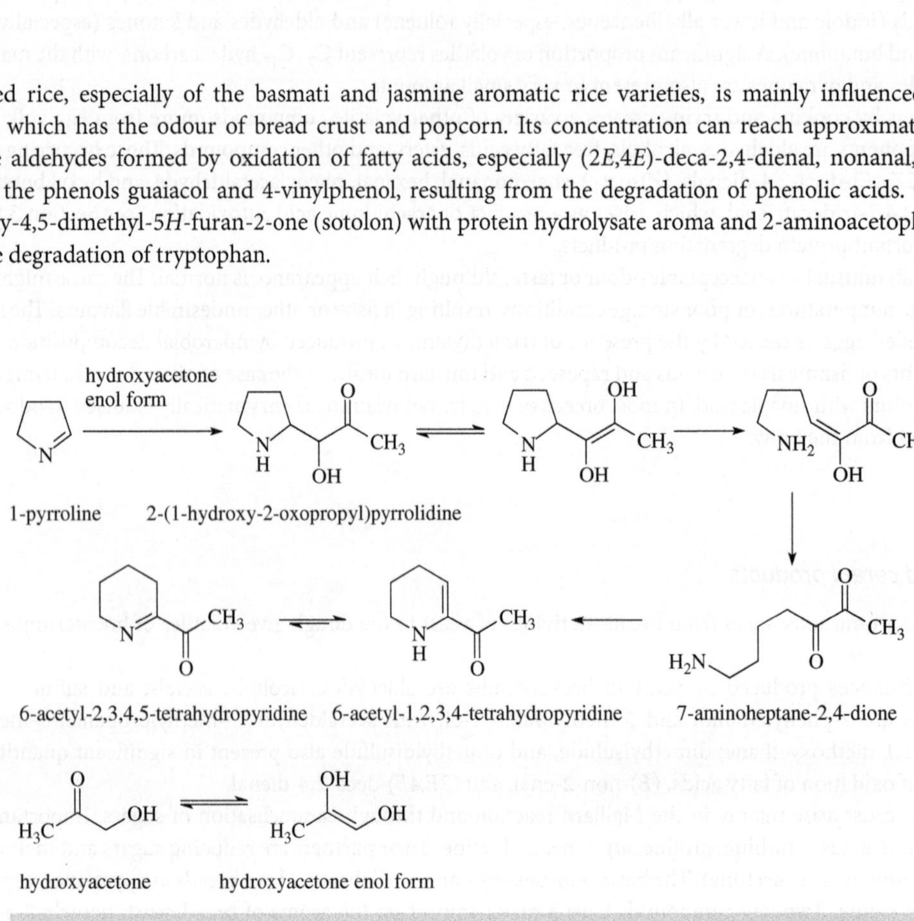

Figure 8.90 Formation of 6-acetyltetrahydropyridines from ornithine and proline.

8.2.12.5 Fruits

8.2.12.5.1 Apples

The aroma of apples (*Malus pumila*, syn. *Malus domestica*, Rosaceae) consists of more than 300 different compounds, of which the most important are C_5 carboxylic acids, alcohols, and esters. Important acids are 2-methylbutyric and 3-methylbutyric (isovaleric) acids, which are found in the fruits in a ratio of about 80 : 20 and in the juices in the ratio 99 : 1 due to the activities of microorganisms. In some apple varieties, esters predominate, whilst in others, alcohols are in the majority. The key odorous ester is ethyl 2-methylbutanoate, but there are other butanoates and acetates present, such as butyl acetate, 2-methylbutyl acetate and 3-methylbutyl acetate, ethyl butanoate, ethyl 2-methylbutanoate, methyl 2-methylbutanoate, hexyl 2-methylbutanoate, and ethylesters of 5-hydroxyoctanoic and 5-hydroxydecanoic acids. Compounds responsible for the green apple aroma are hexanal, (*E*)-hex-2-enal, (*Z*)-hex-3-en-1-ol (leaf alcohol), (*E*)-hex-2-en-1-ol, and related esters such as (*Z*)-hex-3-en-1-yl butyrate. Other alcohols include butane-1-ol, hexan-1-ol, linalool, and 2-phenylethanol. An important flavour-active compound is (*E*)-β-damascenone, resulting from the decomposition of carotenoids, followed by oct-1-en-3-ol with a mushroom odour, methional with the odour of boiled potatoes, and dimethyldisulfide with a sulfurous smell. Cooking of apples causes partial hydrolysis of esters and the formation of lactones from hydroxycarboxylic acids.

8.2.12.5.2 Cherries and plums

A typical component of the flavours in cherries (*Cerasus vulgaris*, syn. *Prunus cerasus*, Rosaceae) is benzaldehyde. Important compounds for cherry flavour are β-ionone, hexanal, (*E*)-hex-2-enal, (*E*)-hex-2-en-1-ol, (2*E*,6*Z*)-nona-2,6-dienal, phenylacetaldehyde, benzylalcohol, 2-phenylethanol, 4-vinylphenol, eugenol, and others. During heating of cherry juice and in the preparation of jams and confectionery products, yet more benzaldehyde is produced by hydrolysis of cyanogenic glycosides. Similarly, hydrolysis of linalool, geraniol, and other glycosides yields free alcohols. Loss of volatile C_6 aldehydes and nona-2,6-dienal changes the formerly green aroma to more floral notes.

The characteristic aroma component of plums (*Prunus domestica*), in addition to benzaldehyde, is provided by linalool, methyl cinnamate, δ-decalactone and, aldehydes with green odour. Plum compotes also contain nonanal, benzyl acetate, and degradation products of carotenoids.

8.2.12.5.3 Apricots

The aroma of apricots (*Armeniaca vulgaris*, syn. *Prunus armeniaca*, Rosaceae) is made up of a large number of different substances. Important components are monoterpenic hydrocarbons, alcohols, and aldehydes (myrcene, limonene, *p*-cymene, terpinolene, α-terpineol, geranial, geraniol, and linalool in particular) and aldehydes with green flavour such as (*Z*)-hex-3-enal and acetaldehyde. Other volatile components include products of oxidation of fatty acids, such as (2*E*,6*Z*)-nona-2,6-dienal and (*Z*)-octa-1,5-dien-3-one, lactones (γ-hexalactone, γ-octalactone, γ-decalactone, γ-dodecalactone, δ-decalactone, and δ-dodecalactone), carboxylic acids (especially 2-methylbutanoic and acetic acids), and degradation products of carotenoids, such as β-ionone.

8.2.12.5.4 Peaches

The basic flavour of peaches (*Persica vulgaris*, syn. *Prunus persica*, Rosaceae) is typified by the presence of γ-lactones (C_6–C_{12}) and δ-lactones (C_{10} and C_{12}). Individual varieties differ mainly in their content of esters and monoterpenoids. As with all stone fruits, an important component is benzaldehyde; others include benzyl alcohol, ethyl cinnamate, isopentyl acetate, linalool, α-terpineol, hexanal, (*Z*)-hex-3-enal, (*E*)-hex-2-enal, and decomposition products of carotenoids.

8.2.12.5.5 Strawberries

The large-fruited strawberry and its hybrids (*Fragaria x ananassa*, Rosaceae), with its unique aroma, is one of the most popular fruits worldwide. Almost 1000 volatile organic compounds have been identified in its volatilome during the past several decades, but its composition still remains largely controversial despite considerable progress made during the past several decades. Besides a number of esters (mainly butanoates) and aldehydes, such as (*Z*)-hex-3-enal, which generally have green and fruity flavours, 4-hydroxy-2,5-dimethyl-2*H*-furan-3-one (furaneol, also known as strawberry or pineapple furanone) and its methyl ether 2,5-dimethyl-4-methoxy-2*H*-furan-3-one (mesifuran) have special importance for the aroma of ripe fruit. These compounds are present in the order of units of mg/kg. The content of furaneol decreases during ripening and storage of strawberries, but the content of its ether mesifuran, whose smell is reminiscent of sherry-type wines, increases. The aroma of strawberry juice reportedly represents a mixture of only furaneol, its methyl ether, (*Z*)-hex-3-enal, methyl and ethyl butanoate, ethyl 2-methylbutanoate, ethyl 3-methylbutanoate (ethyl isobutyrate), biacetyl, and acetic and butanoic acids. Sulfur volatiles have also been reported in strawberries, and some may have an impact on strawberry aroma as the concentrations of most increase with increasing maturity. In addition to hydrogen sulfide, methanethiol, dimethyl sulfide, dimethyl disulfide, and dimethyl trisulfide, sulfur volatiles of straberries include methyl thioacetate, methyl thiopropionate, methyl thiobutanoate, ethyl thiobutanoate, methyl thiohexanoate, methyl (methylthio)acetate, ethyl (methylthio)acetate, methyl 2-(methylthio)butanoate, methyl 3-(methylthio)propionate, ethyl 3-(methylthio)propionate (pineapple mercaptan)), and methyl thiooctanoate, which can be used to distinguish overripe from fully ripe and commercially ripened berries.

8.2.12.5.6 Raspberries

The characteristic flavour component of raspberries (*Rubus idaeus* ssp. *vulgatus*, Rosaceae), 4-(4-hydroxyphenyl)butan-2-one, is known as raspberry ketone. Other important compounds include methyl cinnamate, non-1-en-3-one, (Z)-hex-3-en-1-ol with a green flavour, and degraded carotenoids such as α-ionone, β-ionone, and α-irone.

8.2.12.5.7 Blackcurrants

The characteristic aroma of blackcurrants (varieties of *Ribes nigrum*, Grossulariaceae) is mainly a consequence of aliphatic and alicyclic thiols. The bearer of the characteristic cat odour is 4-methoxy-2-methyl-2-thiol, called blackcurrant mercaptan, which also occurs in olive oil and green tea. Other important components are 4-mercapto-4-methylpentan-2-one, called cat ketone, and (1S,4R)-p-methane-8-thiol-2-one. Cat ketone is also found in grapefruits, some hop cultivars, basil, and some aromatic wines. Other volatiles showing the highest aroma activity values are (Z)-hex-3-enal, ethyl butanoate, 1,8-cineole, oct-1-en-3-one, and alkyl-substituted 3-methoxypyrazines. The leaves contain about 0.7% essential oil, whose main component is *p*-cymene. *p*-Methane-8-thiol-2-one contributes to the characteristic strong smell of the leaves.

8.2.12.5.8 Citrus fruits

The main odour-active components of citrus fruits (fruits of plants of the genus *Citrus* of the rut family Rutaceae) are listed in Table 8.31. The aroma of citrus fruits normally consists of between several tens and several hundreds of compounds. For example, it is estimated that lemon oil contains about 300 different substances. Some of these are of fundamental importance for the typical aroma character; others have little or none. The main component of all citrus fruits is always (+)-limonene, which determines the basic sensory character of essential oils, but its presence is not necessary. Removal of the limonene, whilst leaving other important components (such as citral) in place, produces a more soluble, stronger, and more stable oil, which, however, lacks the characteristic freshness associated with the complete oil. Important components of citrus fruit odours are saturated straight-chain C_8–C_{12} aldehydes (fatty aldehydes), as well as citral, which is always a mixture of two stereoisomers, known as geranial and neral, and the unsaturated aldehyde (E)-dec-2-enal. For the aroma of fresh fruit, aldehydes with green odour are important. Limonene readily undergoes autoxidation and is therefore a source of instability in juices and essential oils obtained from citrus fruits (Figure 8.1).

Oranges The presence of (+)-valencene distinguishes orange and grapefruit aroma from the aroma of other citrus fruits. Acetaldehyde and (Z)-hex-3-enal are the main contributors to the smell of fresh fruit. Some aldehydes also influence the aroma of oranges, such as octanal, nonanal and decanal, and deca-2,4-dienals are important components in Spanish oranges. β-Sinensal has a pleasant orange smell (α-sinensal is typical for the aroma of mandarin); other important compounds in the fresh juice are citral (mixture of isomers) and vanillin and some ketones, such as pent-1-en-3-one and β-ionone. The aroma of fresh oranges can be substituted by about 15 components, which include some ethyl esters, such as ethyl propionate, ethyl butanoate, ethyl 2-methylpropanoate, (S)-ethyl 2-methylbutanoate, ethyl 3-hydroxyhexanoate, neryl acetate, and wine lactone. Important alcohols include 3-methylbutan-1-ol, (+)-α-terpineol, (−)-terpinen-4-ol, and (+)-linalool. Other significant components are some hydrocarbons, especially β-caryophyllene with a spice-like odour, α-pinene, and myrcene.

The compositions of volatiles of juices and beverages prepared from the oils extracted from citrus fruit peels are slightly different. For example, juices contain ethyl vinyl ketone, (E)-pent-2-enal, and ethyl butanoate as major aromatic substances.

During storage of juices, the impact aroma compound pent-1-en-3-one is transformed into 1-hydroxypentan-3-one, 4-hydroxydecane-3,8-dione, and S-(3-oxopentyl)-L-cysteine (a reaction product with cysteine). Hexanal, octanal, nonanal, and decanal are oxidised to the corresponding acids and react with cysteine to 2-alkylsubstituted thiazolidine-4-carboxylic acids. The main products of limonene oxidation, carvone and carveol, cause a terpenic off-flavour in juices and essential oils. Valencene oxidation products, such as (+)-nootkatone, cause a grapefruit-like off-flavour. Large amounts of (+)-α-terpineol, which arises during storage of juices by acid-catalysed hydration or microbial transformation of limonene, is also perceived as an off-flavour.

Bergamot oranges The essential oil of bergamot oranges is rich in (+)-limonene, (+)-linalool, and (+)-linalyl acetate. The dominant sesquiterpenes are (E)-α-bergamotene (0.2–0.4%), whilst other major sesquiterpenes are β-caryophyllene, germacrene D, α-humulene, β-farnesene, and β-bisabolene. The typical aroma of bergamot oranges is attributed to oxygen compounds, such as (−)-guaienol, (+)-spathulenol, nerolidol, farnesol, α-bisabolol, and β-bisabolol. Other important trace compounds are the aldehydes β-sinensal, lanceal, and bergamotenal. Further compounds that make a contribution are sesquiterpenic ketones (+)-nootkatone, (+)-8,9-didehydronootkatone, camphorenone, and other substances.

Table 8.31 Basic composition of citrus essential oils (in %).

Compound	Orange (*C. aurantium* var. *dulcis*)	Bergamot orange (*C. aurantium* var. *bergamia*)	Mandarin orange (*C. reticulata*)	Lemon (*C. limonum*)	Grapefruit (*C. paradisi*)
Hydrocarbons					
α-Bergamotene	0.06	0.3	–	0.4	–
β-Bisabolene	–	0.6	–	0.2–0.8	–
Camphene	–	–	0.02	0.2–0.5	0.01–0.4
Caryophyllene	0.1	0.2	0.03	0.3	0.02–0.1
p-Cymene	0.2	0.5–3.5	0.2–8	0.6–1	0.4
Farnesenes	0.07	0.04	0.1	–	0.1
Limonene	88–97	26	80–94	60–80	86–95
Myrcene	1–2	0.6–1.1	2–7	1–12	2–3.7
α-Phelandrene	–	–	0.05–0.3	0.2	–
β-Phelandrene	0.1	0.1	0.5	0.8	–
α-Pinene	0.2–0.6	1–2.2	1–2.5	1.5–5	0.4
β-Pinene	–	4–10	1–2	6–14	0.05
Sabinene	0.2–0.6	1	1–2	0.8–1.9	1.1
α-Terpinene	–	–	0.1–0.4	0.7	–
β-Terpinene	–	3.2	–	–	–
γ-Terpinene	0.1	5–11	3–17	6–12	0.1–0.8
Terpinolene	0.1	0.3–1	–	0.6–0.9	–
Valencene	0.2	–	–	–	0.4
Alcohols					
Citronellol	–	–	0.02	0.5	–
Fenchol	–	0.01			–
Geraniol	–	0.6	0.04	0.1	–
Linalool	0.55	16	1–6	0.2	0.4
Nerol	–	0.08	0.05	0.1	–
Octan-1-ol	–	0.02	0.1	–	0.8
p-Menth-3-en-1-ol	–	–	0.2	–	–
α-Terpineol	0.1–0.5	0.3	0.05–1	0.3	0.2
β-Terpineol	–	–	0.4	–	–
Terpinen-4-ol	0.06–0.2	0.06	0.06–0.3	0.01–0.4	0.08
Aldehydes					
Citral	0.1–0.2	0.6	0.2	2–13.2	–
Citronellal	0.1	0.02	0.15	0.03–0.2	0.1
Decanal	0.1–0.7	–	0.2–0.9	0.15	0.3–0.6
Dodecanal	0.05–0.2	–	0.02–0.15	0.1	0.15
Nonanal	0.06–0.2	0.08	0.03	0.1–0.3	0.04–0.1
Octanal	0.2–2.8	0.07	0.3	0.15	0.3–0.8

(*continued overleaf*)

Table 8.31 (continued)

Compound	Orange (*C. aurantium* var. *dulcis*)	Bergamot orange (*C. aurantium* var. *bergamia*)	Mandarin orange (*C. reticulata*)	Lemon (*C. limonum*)	Grapefruit (*C. paradisi*)
Perillaldehyde	0.02	–	0.02–0.1	–	0.2
α-Sinensal	0.03	–	0.2	–	–
β-Sinensal	0.1	–	0.2	–	–
Ketones					
Carvone	0.1	0.09	0.005–0.03	0.04	0.02
Nootkatone	0.01	–	0.01	0.06	0.5–1.8
Esters					
Citronellyl acetate	–	–	0.1	0.2	0.04
Geranyl acetate	–	–	0.1	0.1–1	0.2
Neryl acetate	0.1	–	0.1	0.7	0.2
Octyl acetate	0.1	–	0.3	0.04	0.1

Mandarin oranges and tangerines The aroma-active compounds of mandarin oranges (*Citrus reticulata*) include esters, alcohols, aldehydes, ketones, and monoterpenes. Nonanal, hexanal, decanal, linalool, (*R*)-limonene, γ-terpinene, β-ionone, and methyl butyrate are the key odorants. The essential oil of mandarin orange hybrids, called tangerines (*C. tangerina*), contains about 0.85% of methyl ester of *N*-methylanthranilic acid (**8-115**), which is the key component. Other important substances are terpenic hydrocarbons such as γ-terpinene and β-pinene, α-sinensal, and other terpenoids.

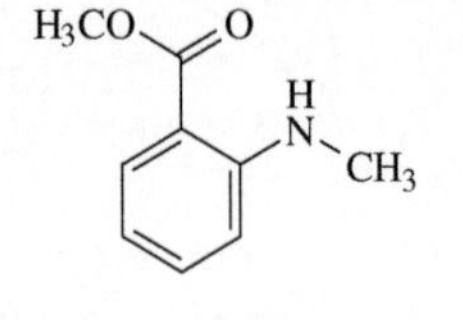

8-115, methyl *N*-methylanthranilate

Grapefruits The key components of fresh grapefruit juices with a typical grapefruit odour are both enantiomers of *p*-mentha-1-en-8-thiol (**8-116**). The (+)-(*R*)-enantiomer is present in minute concentrations (less than 1 μg/kg) but has a very low odour threshold concentration. The (−)-(*S*)-*p*-mentha-1-ene-8-thiol has a weak and non-specific smell. Of the other sulfur compounds, 4-mercapto-4-methylpentan-2-one is significant; it also occurs in blackcurrants, some hop cultivars, aromatic wines, and basil. A relatively high content of sesquiterpenoids is also typical. The smell and bitter taste of grapefruits arise from (+)-nootkatone and (+)-8,9-didehydronootkatone. Important odour-active compounds are numerous cyclic ethers, which are likewise found in other essential oils. For example, the essential oil contains about 13% of linalool oxides, which arise from linalool via 5,6-epoxide, and another important epoxide is (*E*)-4,5-epoxydec-2-enal. The fresh odour of juices is mainly influenced by aliphatic aldehydes, such as acetaldehyde, (*Z*)-hex-3-enal and decanal, as well as by some esters, such as ethyl 2-methylpropionate, ethyl (*S*)-2-methylbutanoate, and wine lactone.

(*R*)-*p*-menth-1-ene-8-thiol   (*S*)-*p*-menth-1-ene-8-thiol

8-116, *p*-menth-1-ene-8-thiol enantiomers

Lemons Principal carriers of the aroma of lemons are β-pinene, (−)-terpinen-4-ol, and citronellol, whose odour is reminiscent of lemon peel, whilst α-bergamotene contributes significantly to the basic lemon odour. A high content of citral (isomeric aldehydes geranial and neral) and the presence of some alkanals, esters, and alcohols, especially geranyl acetate, neryl acetate, and α-bisabolol, are very important.

The off-flavour of lemon juices may be caused by degradation products of citral (*p*-cymene and α,*p*-dimethylstyrene) that result from reactions catalysed by acids.

8.2.12.5.9 Bananas

The characteristic aroma components of bananas (*Musa* x *paradisiaca*, Musaceae) are esters. Important components are largely acetic acid esters, and the most significant compound is isopentyl acetate. The typical banana odour comes from esters of pentan-1-ol with acetic, propionic, and butyric acids, whilst esters of butanols and hexanols with acetic and butyric acids show a general fruity aroma. Other compounds also contribute to the full fine aroma, such as eugenol and its derivatives (methyl eugenol and elemicin).

8.2.12.5.10 Pineapple

Pineapple (*A. comosus*, Bromeliaceae) aroma consists of about 200 alcohols, esters, lactones, aldehydes, ketones, monoterpenes, sesquiterpenes, and other volatiles. About 80% of the total volatile substances are esters. The main components in the green fruit are ethyl acetate, ethyl 3-(methylthio)propionate (known as pineapple mercaptan) with a distinctive pineapple aroma, and ethyl 3-(acetoxy)hexanoate (**8-116**). The ripe fruit contains, as the main esters, ethyl acetate, (2*R*,3*R*)-butane-2,3-diol diacetate (**8-116**), and ketone 3-hydroxybutan-2-one. An important compound for the typical character of pineapple aroma, as in strawberry aroma, is 2,5-dimethyl-4-hydroxy-2*H*-furan-3-one (furaneol), present as a glycoside, as well as 2,5-dimethyl-4-methoxy-2*H*-furan-3-one (mesifuran).

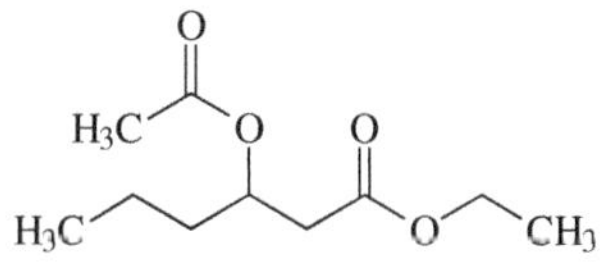

ethyl 3-(acetoxy)hexanoate 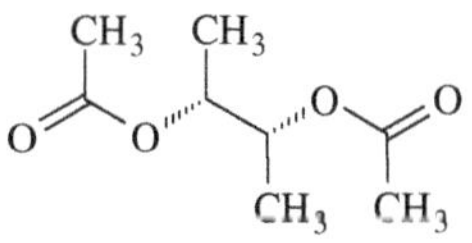   (2*R*,3*R*)-butane-2,3-diol diacetate

8-116, acetoxy esters in pineapple

8.2.12.5.11 Durian

Durian (*Durio zibethinus*, Malvaceae) aroma is special in that it consists of a number of different sulfur compounds. The overall odour of durian pulp can be mimicked by only two compounds, ethyl (*S*)-2-methylbutanoate (fruity, **8-117**) and 1-(ethylsulfanyl)ethane-1-thiol (**8-86**, roasted onion). Other important compounds include ethanethiol (rotten onion), methanethiol (rotten cabbage), ethane-1,1-dithiol (sulfury, durian), and ethyl 2-methylpropanoate (fruity).

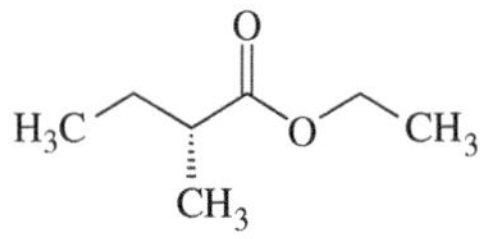

8-117, ethyl (*S*)-2-methylbutanoate

8.2.12.6 Vegetables

8.2.12.6.1 Cabbage, kale, and kohlrabi

The aroma of cabbage (*Brassica oleracea* var. *capitata*), kale (*B. oleracea* var. *sabauda*), kohlrabi (*B. oleracea* var. *gongylodes*), and other vegetables of the genus *Brassica* (Brassicaceae) is characterised by the presence of isothiocyanates, which result from the enzymatic and non-enzymatic decomposition of glucosinolates. Important compounds are aldehydes with green odour and 2-alkyl-3-methoxypyrazines. Cooking cabbage produces, as the main products, allyl isothiocyanate (prop-2-en-1-yl isothiocyanate) arising from sinigrin, but-3-en-1-yl isothiocyanate (from gluconapin), and 2-phenylethyl isothiokyanate (from gluconasturtiin). The latter is especially important, as it has a low odour threshold concentration. The resulting aroma, in addition to isothiocyanates, is also influenced by a large number of other sulfur compounds, such as dimethylsulfide and dimethyltrisulfide, that are formed by decomposition of *S*-methylcysteine sulfoxide, cysteine, and methionine (via methional). Nitriles produced as byproducts of degradation of glucosinolates also have some relevance, and to some extent recall the aroma of garlic. The main component is allyl cyanide (but-3-ene nitrile).

8.2.12.6.2 Cauliflower and broccoli

The components of cooked cauliflower (*B. oleracea* var. botrytis) and broccoli (*B. oleracea* var. *italica*) aroma are similar to the aroma components of cabbage. Another important compound is the aldehyde nonanal. The typical odour of cauliflower comes from 3-(methylthio)propyl isothiocyanate (**8-118**), which is produced from glucosinolate glucoibervirin. The typical odour of broccoli is also due to the presence of 3-(methylsulfinyl)propyl isothiocyanate (**8-118**), which develops from glucosinolate glucoiberin.

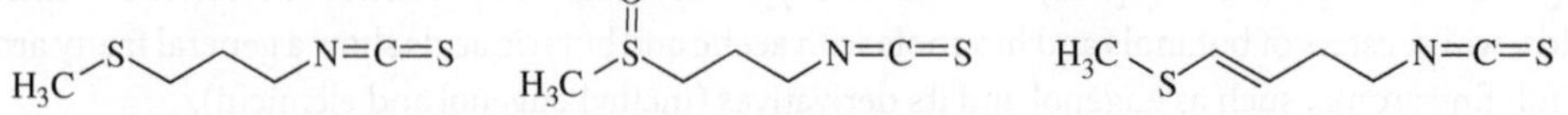

| 3-(methylthio)propyl isothiocyanate | 3-(methylsulfinyl)propyl isothiocyanate | (*E*)-4-(methylthio)but-3-en-1-yl isothiocyanate |

8-118, selected isothiocyanates of *Brassica* vegetables

8.2.12.6.3 Radish and horseradish

The main component of numerous varieties of radish (*Raphanus sativus*, Brassicaceae) is (*E*)-4-(methylthio)but-3-en-1-yl isothiocyanate (**8-118**), which is produced from glucosinolate glucoraphasatin and is responsible for pungent taste of many radishes. The odour and pungent taste of horseradish (*Armoracia rusticana*) is caused by the presence of allyl isothiocyanate, which is a degradation product of glucosinolate sinigrin.

8.2.12.6.4 Carrots

The aroma of carrots root (*Daucus carota*, Apiaceae) consists of various aldehydes, ketones, mono- and sesquiterpenic hydrocarbons, and other compounds. The important hydrocarbons are myrcene, sabinene, terpinolene, β-caryophyllene, γ-bisabolene, and α-pinene, which are present in the largest quantities. The significant carbonyl compounds are acetaldehyde and (2*E*,6*Z*)-nona-2,6-dienal. One of the typical aromatic substances determining the basic odour of carrots is 2-*sec*-butyl-3-methoxypyrazine. During the cooking, the contents of methanal, ethanal, propanal, octanal, (*Z*)-dec-2-enal, and some sulfur compounds, such as dimethylsulfide and ethanethiol, increase and the contents of monoterpenes and β-caryophyllene decrease. The compounds showing the highest odour activity in cooked carrots are linden ether (3,6-dimethyl-2,4,5,7a-tetrahydro-1-benzofuran, **8-119**), a constituent of linden honey, β-damascenone, (*E*)-non-2-enal, (*E*,*E*)-deca-2,4-dienal, β-ionone, octanal, (*E*)-dec-2-enal, eugenol, and *p*-vinylguaiacol.

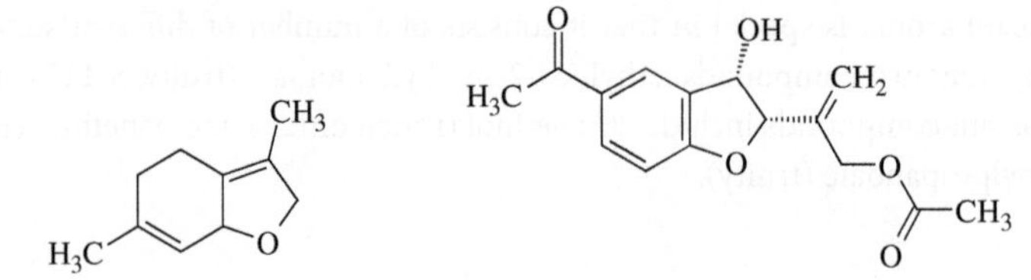

| linden ether | (2*R*,3*R*)-10-acetoxytoxol |

8-119, selected benzofurans

8.2.12.6.5 Parsley

The compound responsible for the typical smell of fresh parsley (*Petroselinum hortense*) and the main aroma constituent of parsley leaf oil are *p*-mentha-1,3,8-triene (Figure 8.9) and myrcene. As in other vegetables of the Apiaceae family, important substances of root parsley and leaf parsley are phthalides. The main phthalides are sedanenolide, (*E*)-ligustilide, and butylphthalide, whilst (*Z*)- and (*E*)-butylidenephthalide, (*Z*)-ligustilide, (*Z*)-sedanolide, and 3-butyl-5,6-dihydrophthalide are present in smaller amounts. Other important components of leaf parsley are linalool, β-citronellol, methyl 2-methylbutanoate, oct-1-en-3-one, (*Z*)-octa-1,5-dien-3-one, 2-isopropyl-3-methoxypyrazine, 2-*sec*-butyl-3-methoxypyrazine, (*Z*)-hex-3-enal, (*E*)-dec-6-enal, (2*E*,4*E*)-deca-2,4-dienal, and β-ionone, resulting from the degradation of carotenoids. Other components are the monoterpenes myrcene, α-pinene, β-pinene, α-thujene, camphene, sabinene, car-3-ene (Δ³-carene), α- and β-phellandrene, (*S*)-limonene, γ-terpinene, *p*-cymene, and terpinolene.

8.2.12.6.6 Celery and celeriac

The typical components of the aroma of celery (*Apium graveolens* var. *rapaceum*, Apiaceae) and celeriac (*A. graveolens* var. *dulce*) are numerous dihydrophthalides, tetrahydrophthalides, hexahydrophthalides, and phthalides. The main odour-active compound is 3-butyl-4,5-dihydrophthalide (sedanenolide), whilst 3-butyl-3,4,5,6-tetrahydrophthalide (sedanolide, neoknidilide), 3-butylphthalide, and 3-butylidene-4,5-dihydrophthalide, known as (Z)-ligustilide, are present in significant quantities. Also present in smaller amounts are other dihydrophthalides, tetrahydrophthalides, and hexahydrophthalides and other compounds such as (3E,5Z)-undeca-1,3,5-triene (galbanolene).

8.2.12.6.7 Beetroot

The characteristic earthy flavour of taproot portion of table beet, also known as red beet, garden beet or simply as beet (*Beta vulgaris ssp. vulgaris*, Amaranthaceae, is due to the presence of tertiary sesquiterpenic alcohol (−)-geosmin, which is commonly produced by many soil microorganisms. Other important compounds include pyridine, 4-methylpyridine, 2-*sec*-butyl-3-methoxypyrazine, carbonyl compounds such as 3-methylbutanal, biacetyl, hexanal, furan-2-carboxaldehyde, benzaldehyde, and phenylacetaldehyde, and alcohols such as 2-methylpropan-1-ol and 3-methylbutan-1-ol.

8.2.12.6.8 Tomatoes

The aroma of tomatoes (*Solanum lycopersicum*, Solanaceae) represents about 400 different compounds. Important components of the aroma of fresh ripe tomatoes are 2-isobutylthiazole, which is also the bearer of the typical aroma of the whole plant. The precursors of 2-isobutylthiazole are 3-methylbutanal, formed by deamination and decarboxylation of leucine, and cysteamine (2-mercaptoethylamine), formed by decarboxylation of cysteine (Figure 8.86). Other important components of tomato aroma are C_6 aldehydes hexanal, (E)-hex-2-enal and (Z)-hex-3-enal (green, grassy), and (E)-4,5-epoxy-dec-2-enal (metallic) (the products of enzymatic oxidation of fatty acids), as well as methional (potato-like), 3-methylbutanal, 2-methylbutanal, and their corresponding alcohols. Oxidation products of carotenoids also play a role, such as unsaturated methylketones (E)-6-methylhept-5-en-2-one and (E)-6-methylhepta-3,5-dien-2-one, as do some other products, such as β-ionone, β-cyclocitral, 5,6-epoxy-β-ionone, and (E)-β-damascenone. Significant odorous substances are wine lactone (coconut and dill) and 2,5-dimethyl-4-hydroxy-2H-furan-3-one (furaneol, caramel-like), which is released from the corresponding β-D-glucoside. In addition, a recently identified fruity, almond-like odorant is 2-methyl-2-ethoxytetrahydropyran.

The aroma of thermally processed tomatoes, such as tomato paste, is mainly determined by the presence of acetic and isovaleric (3-methylbutanoic) acids, 3-methylbutanal, methional, eugenol, 4-vinylguaiacol, dimethylsulfide, β-ionone, (E)-β-damascenone, and other compounds such as linalool, 4-hydroxy-2,5-dimethyl-2H-furan-3-one (furaneol), and its lower homologue 4-hydroxy-5-methyl-2H-furan-3-one (norfuraneol). The aroma of tomato concentrates is influenced by the presence of some Maillard-reaction products, such as 2-formyl-, 2-acetylpyrrole, and some pyrazines.

8.2.12.6.9 Bell peppers

Bell pepper (*Capsicum annuum*, Solanaceae), also known as sweet pepper, pepper, or capsicum, and chilli pepper (*C. frutescens*) contain as their key compound 2-isobutyl-3-methoxypyrazine, which carries the typical sharp spice-like odour of fresh vegetables. Other major components identified include terpenic hydrocarbons (E)-β-ocimene and limonene, alcohols (Z)-pent-2-en-1-ol, (E)-hex-2-en-1-ol, and (Z)-hex-3-en-1-ol with their fruity and green aroma, and linalool, methyl salicylate, and the aldehydes *p*-menth-1-en-9-al, (2E,6Z)-nona-2,4-dienal (as in cucumbers) and (2E,4E)-deca-2,4-dienal.

8.2.12.6.10 Cucumbers

The key odour-active components of fresh cucumbers (*Cucumis sativus*, Cucurbitaceae) are aldehydes generated by enzymatic oxidation of unsaturated fatty acids. The most important compounds are (3Z,6Z)-nona-3,6-dienal, (2E,6Z)-nona-2,6-dienal, and (Z)-non-3-enal, which recall the smell of fresh cucumbers. Additional important components are some other aldehydes, such as (Z)-hex-3-enal, (E)-hex-2-enal, (E)-non-2-enal, nonanal, (Z)-hex-6-enal, certain alcohols, and 2-alkyl-3-methoxypyrazines.

The odour of fermented cucumbers and cucumbers pickled in sour brine is almost exclusively influenced by the odours of the fermentation products used, vinegar and spices.

8.2.12.6.11 Garlic

The characteristic aroma of fresh garlic (*Allium sativum*, Amaryllidaceae) is almost exclusively represented by diallyl thiosulfinate (allicin), and to a lesser extent by allylmethyl thiosulfinate.

The flavour of roasted garlic is associated with heterocyclic Maillard-reaction products derived from reducing sugars, S-allyl-L-cysteine, and thiosulfinates. In pan-fried garlic slices, various pyrrolomorpholine spiroketals, allylthio substituted formylpyrroles, and pyrrolopyrazines were identified (see Section 4.7). At higher temperatures (during frying or baking), a number of dithiolanes and trithiolanes, dithianes, trithiepanes, and tetrathiepanes are produced.

The aged garlic extract prepared by soaking sliced garlic in water or aqueous ethanol and storing it for several months exhibits strong antioxidant activity. The main changes during the ageing process are complete hydrolysis of the γ-glutamylcysteines (γ-glutamyl-*S*-allylcysteine to *S*-allylcysteine and γ-glutamyl-*S*-prop-1-en-1-ylcysteine to *S*-prop-1-en-1-ylcysteine, see Section 2.2.1.2.2), catalysed by γ-glutamyltransferase, an increase in cystine concentration due to protein hydrolysis and in *S*-allylmercaptocysteine probably due to the reaction of allicin with protein-derived cysteine, and the formation of thiosulfinates and their gradual transformation to other products (volatile allyl sulfides are evaporated almost completely). The key compounds in aged garlic extract are *S*-allylcysteine (Cy-S-allyl) and *S*-allylmercaptocysteine (Cy-S-S-allyl), which is believed to be a major bioactive component with antioxidant, antiproliferative, apoptosis-inducing, and antimetastatic activities. Another active compound is *S*-allylmercaptoglutathione (G-S-S-allyl). The major antioxidants identified in aged garlic extracts include the Maillard-reaction product *N*α-(1-deoxy-D-fructos-1-yl)-L-arginine, various β-carbolines such as (3*S*)-1,2,3,4-tetrahydro-β-carboline-3-carboxylic acid, (1*R*,3*S*)- and (1*S*,3*S*)-1-methyl-1,2,3,4-tetrahydro-β-carboline-3-carboxylic acid, and (1*R*,3*S*)- and (1*S*,3*S*)-1-methyl-1,2,3,4-tetrahydro-β-carboline-1,3-dicarboxylic acids (see Section 10.3.3.1), coniferyl alcohol and derived dilignols, (−)-(2*R*,3*S*)-dihydrodehydrodiconiferyl alcohol, (+)-(2*S*,3*R*)-dehydroconiferyl alcohol, and *erythro*- and *threo*-guaiacylglycerol-β-*O*-4′-coniferylethers (**8-120**).

(2*S*,3*R*)-dehydroconiferyl alcohol   *threo*-guaiacylglycerol-β-*O*-4′-coniferylether

8-120, coniferyl alcohol-derived compounds in aged garlic extract

8.2.12.6.12 Onions, leeks, and chives

The most important aromatic compounds of freshly sliced onions (*Allium cepa*, Amaryllidaceae), leeks (*A. ampeloprasum*), and chives (*A. schoenoprasum*) are dialk(en)yl thiosulfinates (prop-1-en-1-yl-, methyl, and propyl derivatives), which give rise to a diverse mixture of odour-active degradation products. The tear factor of onion is (*Z*)-thiopropanal-*S*-oxide (Figure 8.64). The odorous constituent of raw and heat-treated onion is the intensively smelling 3-mercapto-2-methylpentan-1-ol (Figure 8.88), which occurs as a mixture of diastereomers. Its content in freshly cut raw onion ranges from 8 to 32 μg/kg, rising in raw stored and then cooked onion to 34–246 μg/kg. Important odorants are only (2*R*,3*S*)- and (2*S*,3*R*)-diastereomers (**8-121**). Spontaneous degradation of dialk(en)yl thiosulfinates yields the corresponding disulfides and trisulfides, which, together with 2-methylbut-2-enal and 2-methylpent-2-enal, represent the main aromatic components of cooked onions. The characteristic aroma components of fried onions are dimethylthiophenes, especially 2,4-dimethylthiophene and 3,4-dimethylthiophene (Figure 8.76).

(2*R*,3*S*)-diastereomer   (2*S*,3*R*)-diastereomer

8-121, 3-mercapto-2-methylpentan-1-ol

8.2.12.6.13 Peas

Essential to the odour of green peas (*Pisum sativum*, Viciaceae) are aldehydes and some pyrazines. The typical smell is of 2-isopropyl-3-methoxypyrazine, whilst an important component is 2-*sec*-butyl-3-methoxypyrazine, which resembles the odour of bell peppers.

8.2.12.6.14 Asparagus

The characteristic aroma component of asparagus (*Asparagus officinalis*, Asparagaceae) is 3*H*-1,2-dithiol, arising during cooking from asparagusic (1,2-dithiolane-4-carboxylic) acid (Figure 8.91), which forms a redox system together with dihydroasparagusic acid. Asparagusic acid is present as free acid, as esters of 6-*O*-α- and 6-*O*-β-D-glucopyranose, exhibiting an interesting buttery mouth-coating effect, and as methyl and ethyl esters together with methyl esters of the epimeric sulfoxides of asparagusic acid, in which the *S*-oxide takes up either a *syn* or an *anti* configuration relative to the methyl ester (**8-122**). A minor constituent of asparagus is 1,2,3-trithiane-4-carboxylic acid, which is a contact allergen from asparagus. Other odorous compounds of cooked asparagus are dimethyl sulfide, different thiophenes, thiazoles (e.g. 2-acetylthiazole), pyrroles, pyrazines, aldehydes, ketones, and phenols (e.g. vanillin). The key contributors to the typical bitter taste of white asparagus spears are steroidal saponins (see Section 10.3.3.2).

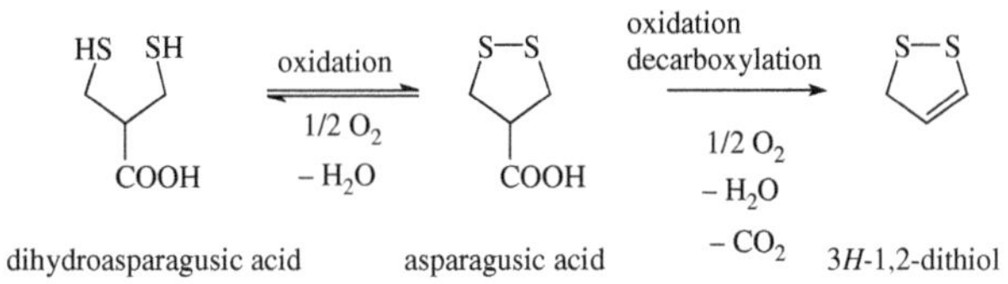

Figure 8.91 Decomposition of asparagusic acid during boiling of asparagus.

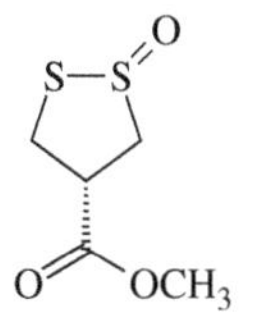

8-122, asparagusic acid *anti-S*-oxide methyl ester

Phytoalexins occurring in asparagus are the polyacetylenes asparenyol, asparenyn, and 2-hydroxyasparenyn (**10-123**), which have cancer chemopreventive properties.

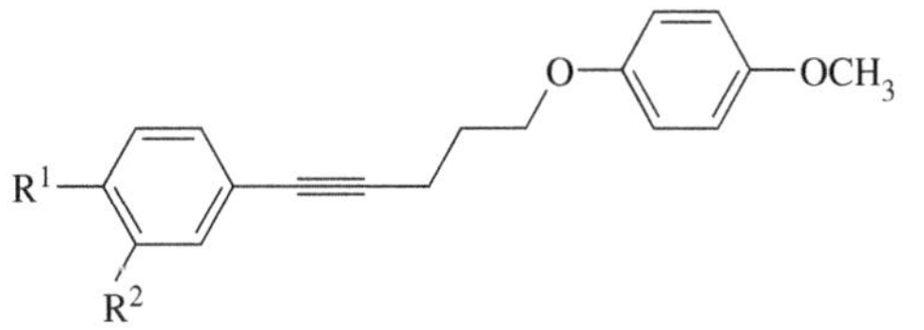

8-123, asparenyol, $R^1 = OH$, $R^2 = H$
asparenyn, $R^1 = H_3CO$, $R^2 = H$
2-hydroxyasparenyn, $R^1 = H_3CO$, $R^2 = OH$

8.2.12.7 Potatoes

2-Isopropyl-3-methoxypyrazine and 2,5-dimethylpyrazine, with their typical earthy flavour, have great importance for the aroma of raw potatoes (*Solanum tuberosum*, Solanaceae). In addition, raw potatoes contain a range of carbonyl compounds and alcohols. A very important component of the aroma of cooked potatoes is methional.

The aroma of potato chips (French fries) is a mixture of more than 500 compounds. The important components are 2,3-diethyl-5-methylpyrazine, 2-ethenyl-3-ethyl-5-methylpyrazine, (2*E*,4*Z*)-deca-2,4-dienal, (*E*)-4,5-epoxydec-2-enal, oct-1-en-3-one, (*Z*)- and (*E*)-non-2-enal, 2-methylpropanal, 2-methylbutanal and 3-methylbutanal, biacetyl, and 4-hydroxy-2,5-dimethyl-2*H*-furan-3-one (furaneol). Another principal component is (*Z*)-2-(methylthiomethyl)but-2-enal (also known as 2-ethylidene methional, **8-124**), which arises in the reaction of methional with acetaldehyde. Both isomers of 2-(methylthiomethyl)but-2-enal are used as flavour enhancers. Less significant compounds are methanethiol and methional.

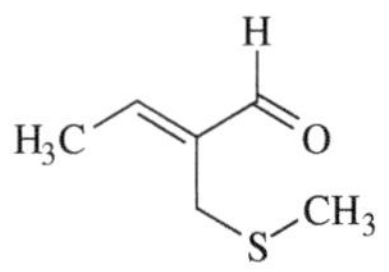

8-124, (*Z*)-2-(methylthiomethyl)but-2-enal

8.2.12.8 Alcoholic beverages

The characteristic aroma of beer, wine, spirits, and other alcoholic beverages is produced by a large number of volatile compounds that are formed due to activities of yeasts or which come from the raw materials used. In addition to fermentable sugars, higher fatty acids, organic nitrogen and sulfur compounds, and many other substances pass into yeast cells from the media and participate in biochemical reactions in which aromatic components are formed as byproducts. Their number depends on the conditions during fermentation, and also, in spirits, on the method of distillation. A number of components arise from chemical reactions during the ripening and storage of beverages.

8.2.12.8.1 Hops and beer

Worldwide, several hundred brands of beer are produced. Types of beer vary, being based on the basic classification into lagers and ales, which roughly corresponds to the division between bottom-fermented and top-fermented beers. Beers contain about 1000 different compounds, many of which can affect the organoleptic properties, which depend on the raw materials used and their quantity, the brewing technology, the yeast used, possible contamination by other microorganisms, and other factors. According to the current taxonomy, the yeasts used to produce alcoholic beverages are classified under one common species named *Saccharomyces cerevisiae*. Previously used names for bottom-fermented yeast were *S. carlsbergensis* and, later, *S. uvarum*, whilst *S. cerevisiae* was the name for top-fermenting yeast. Bottom-fermented yeast was also classified as a subspecies of *S. cerevisiae* yeast.

The primary aromatic substances in beer are derived from raw materials (barley and hops) that confer the beer's typical taste and odour. Bitter acids of hops have a bitter taste, whilst polyphenols (condensed tannins) have an astrigent taste and antioxidant effects. Hop cones also contain 0.3–1% m/m of terpenoids (60–80% of hop essential oil), which have a considerable influence on the smell of beer. The main components of aromatic hop oils are sesquiterpenic hydrocarbons, in which α-humulene, β-caryophyllene, and farnesene dominate. The major monoterpenic hydrocarbon is myrcene. For example, the essential oil content of fine aromatic varieties such as Saaz is 0.8% m/m, of which 23% is myrcene, 20.5% α-humulene, 14% farnesene, and 6% β-caryophyllene. Significant components of the hop aroma in beer are mainly terpenoid epoxides and diepoxides resulting from autoxidation of α-humulene and β-caryophyllene, but also other terpenoids. Important components of hops odour are various alcohols (such as geraniol and linalool), esters (ethyl 2-methylpropanoate, methyl 2-methylbutanoate, propyl 2-methylbutanoate, and esters of terpenic alcohols such as geranyl isobutanoate), hydrocarbons, aldehydes, and ketones formed by oxidation of fatty acids such as (3*E*,5*Z*)-undeca-1,3,5-triene, (galbanolene), (*Z*)-hex-3-enal, nonanal, (*Z*)-octa-1,5-dien-3-one, their epoxides such as (*E*)-4,5-epoxydec-2-enal, and sulfur compounds.

Other aromatic substances are formed by enzymatic and chemical reactions during malting, and many compounds arise during technological processes such as mashing and fermentation. Fermentation changes the smell and taste of a young beer through reactions related to the metabolism of yeast. Top-fermented beers are sometimes very aromatic. Some beers, especially wheat beers, may contain a higher amount of 4-vinylguaiacol, which grants them a typical flavour. Other flavours can come from the addition of various spices, such as coriander, which is added to some Belgian beers.

Generally, some of the odour-active components of beer are considered desirable or desirable in small quantities, whilst others are indifferent or undesirable. For example, for the pilsner type of beer, important odourous components include some alcohols (ethanol and linalool), aldehydes (acetaldehyde), acids (2- and 3-methylbutanoic acids), ethyl esters (butanoate, 2-methylpropanoate, 4-methylpentanoate, and caprylate), sugar degradation products such as 4-hydroxy-2,5-dimethyl-2*H*-furan-3-one (furaneol), and degradation products of carotenoids such as (*E*)-β-damascenone. Interestingly, the feeling of an empty taste is associated with lower ethanol content. The bearers of caramel aroma in dark beers are maltol and isomaltol, which can come from caramel malts used in the production of dark beers or from caramel used for colouring. Various esters of organic acids, such as 3-methylbutyl acetate and ethyl esters of fatty acids, have a flowery and ester aroma. An important descriptor of the taste of beer is fullness, which may be caused by increased extract in unfermented beer, higher protein and carbohydrate content, or the composition of the brewing water, which may for example have increased chloride content. Carbon dioxide content also significantly affects the physiological and sensory perception and beer foaming.

A range of different reasons may lead to the formation of undesirable sensory-active substances, such as microbial contamination, pasteurisation, and long-term storage. Various secondary substances formed during these processes may affect the smell/taste profile of beer, usually negatively. The carriers of sulfur odour are sulfane and various thiols, dimethyl sulfide, and dimethyltrisulfide (arising either from hop *S*-methylcysteine sulfoxide or from methional). Their smell resembles cooked vegetables. Biacetyl's smell resembles butter, acetic acid's resembles vinegar, volatile phenols have a phenolic aroma, and some lactones (especially δ-decalactone) give rise to a fruity odour. The development of a solvent-like stale flavour is associated with the formation of furfuryl ethyl ether, which is formed in an acid-catalysed substitution reaction of ethanol and furfuryl alcohol.

In general, there exist flavours that are considered 'off' or faulty when found at a certain level (or any level) in beer. For example, during transport and storage, beer may be exposed to visible light (in the presence of photosensitisers, such as riboflavin), which leads to sun-struck (light-struck) flavour (skunky or catty) caused by 3-methylbut-2-en-1-thiol (prenylthiol). The oxidation of unsaturated fatty acids to alkenals, such as (*E*)-non-2-enal, leads to an off-flavour, known as cardboard (papery or sherry-like) flavour, associated with the ageing process of beer. Concentrations of (*E*)-non-2-enal ranging from 0.2 to 0.5 g/l are usually found after three to five months at 20 °C, but then decrease below the threshold concentration after one year of ageing, probably due to hydration to 3-hydroxynonanal or oxidation to non-2-enoic acid.

8.2.12.8.2 Grapes and wine

The mainstream grape varieties for winemaking belong to *V. vinifera* (Vitaceae). The organoleptic properties and quality of grapes and wine depend on the vine variety, the ripeness of the grapes, conditions during must fermentation (a dominant species of microorganisms is the yeast *S. cerevisiae*), conditions during ageing, and many other factors.

In wine terminology, **aroma** is the smell of grape berries and young wines. Aromas from grapes determine the varietal typicality, whereas fermentation aromas can supplement the varietal aroma. It is estimated that wines contain 400–600 flavour-active compounds in a total amount of 0.8–1.2 g/l. They exist as free volatiles and as bound aroma components in grapes and wine. The latter group of compounds is more than 10 times as abundant as the former. The important bound aroma substance classes are glycosides and cysteinylated conjugates, from which free volatiles are released during fermentation. The transformation of aroma during ageing by chemical reactions leads to a wine's **bouquet**.

Compared with other alcoholic beverages, wines are very acidic (their pH value is within 2.8–3.8; tartaric acid is a major carrier of sour taste), and they have a relatively high ethanol content. The ethanol content is related to the sugar content in the must, and to the degree of grape ripeness and the extent of fermentation. For red wine taste, tannins and other flavonoid molecules are important compounds. Ethanol corrects their bitter and astringent flavour.

Wines contain several hundreds of odorous compounds, so the following overview gives only typical examples. Terpenoids, together with thiols, esters, and higher alcohols, are key wine odorants. According to the grape varieties, wines can have a neutral (e.g. Müller Thurgau) or strong (e.g. Muscat, Sauvignon, Gewürztraminer, and Scheurebe) aroma derived from terpenoids. The primary aroma of Muscat grapes is determined by the presence of terpenic alcohols, especially linalool (the content is about 0.4 mg/l), geraniol (0.3 mg/l), nerol, α-terpineol, and linalooloxides (0.1 mg/l). The characteristic impact component of Gewürztraminer and Scheurebe wines is *cis*-rose oxide. Some of thse terpenes are also involved in the odour of other grape varieties (such as Riesling), where their concentration only reaches 0.1–0.3 mg/l. In varieties lacking the characteristic aroma, these terpenoids are present only in traces, or not at all. Eucalyptol (1,8-cineole) may contribute to the hay, dried herbs, and blackcurrant aromas reported in Australian Cabernet Sauvignon wines and may be a potential marker of regional typicality of these wines.

The peppery character of grapes and wines is attributed to the presence of the sesquiterpenoid (−)-rotundone, (3*S*,5*R*,8*S*)-5-(isoprop-2-en-1-yl)-3,8-dimethyl-3,4,5,6,7,8-hexahydro-2*H*-azulen-1-one (**8-125**), the only known impact odorant of wines with a peppery aroma. Rotundone in wines arises by aerial oxidation of α-guaiene, which is derived from the sesquiterpene guaiane, (1*S*,3a*S*,4*S*,7*R*,8a*S*)-7-isopropyl-1,4-dimethyldecahydroazulene (see **8-138**). This compound was first found in Syrah (Siraz), Mourvèdre, and Durif wines produced in Australia, at concentrations of up to 145 ng/l, and later in higher amounts (up to 561 ng/l) in Schioppettino and Vespolina red wines and in Grüner Veltliner white wines produced in Europe. As a potent flavour-active compound, rotundone was considered to be the most potent odorant in roasted chicory. It was also found in several kinds of fruits, such as grapefruit, orange, apple, and mango, and in some common herbs and spices, such as black pepper, marjoram, oregano, rosemary, basil, and thyme. Its levels were 2180 and 1920 ng/kg in white and pink grapefruit peels and 29.6 and 49.8 ng/kg in white and pink grapefruit juices, respectively. In black and white peppercorns, it is present at about 10 000 times higher levels in comparison with very peppery wines.

The secondary aromatic compounds are mainly higher alcohols (Table 8.5) and esters (Table 8.25). Higher alcohols are present in somewhat larger quantities than those that correspond to their odour thresholds. At low concentrations, they have a positive effect on the wine aroma, but in concentrations higher than about 400 mg/l they have a negative effect. An important role is played by ethyl acetate. In sub-threshold concentrations of 50–80 mg/l (the odour threshold concentration in wine is 160 mg/l), together with other substances, ethyl acetate is a desired aroma component, but in concentrations higher than 160 mg/l it has a harsh taste and smell. In red and white wines, (*S*)-2-methylbutyl acetate is associated with blackberry–fruit and banana notes. Its levels are generally higher in red than in white wines of the same age. Ethyl (*S*)-2-hydroxy-4-methylpentanoate (ethyl leucate, **8-126**) was identified in Bordeaux red and white table wines as a compound associated with a fresh blackberry aroma. Ethyl 2-methylbutanoate has a similar fruity (blackberry) aroma. Ethyl esters of fatty acids (such as caproic, caprylic, capric, and lauric acids) occur in relatively low concentrations (5–10 mg/l), which are nevertheless about 10 times higher than the threshold concentrations. The smell of ethyl caproate is fruity, but that of ethyl laurate is soap-like.

8-125, (−)-rotundone

8-126, ethyl leucate

Thiols released from non-volatile precursors during alcoholic fermentation, in particular enantiomers of 4-mercapto-4-methylpentan-2-one, 4-mercapto-4-methylpentan-2-ol, and 3-mercaptohexan-1-ol and 3-mercaptohexyl acetate, are responsible for the grapefruit and passion fruit scents of young white wines, especially of the variety Sauvignon Blanc and some others (e.g. Gewürztraminer, Scheurebe, Riesling, Muscat, Sylvaner, Cabernet Sauvignon, and Merlot). 3-Mercaptohexan-1-ol is produced from (*E*)-hex-2-enal and (*E*)-hex-2-en-1-ol precursors during fermentation. Various other thiols have likewise been identified in wines, such as 2-furanmethanethiol and benzylmercaptan.

Certain heterocyclic compounds are also important aroma substances in wines, such as pyrazines in Cabernet Sauvignon and Sauvignon Blanc and both enantiomers of 3-hydroxy-4,5-dimethyl-5*H*-furan-2-one (sotolon), which occur in white wines and sherries and are key components of the typical aroma of aged port wines. The terpenoid ether menthofuran and three derived *p*-menthane lactones with mint-like odours have been identified in red wines. A monoterpenic ketone piperitone is a contributor to the positive mint aroma of aged red Bordeaux wines. The musts and red wines marked by dried fruit and cooked fruit aromas contain (*R*)-5,6-dihydro-6-pentyl-2*H*-pyran-2-one, known as massoia lactone; in Merlot and Cabernet Sauvignon musts marked by dried fruit flavours from overripe grapes, its concentration reached 68 μg/l. (*R*)-Massoia lactone is reduced during alcoholic fermentation to (*R*)-δ-decalactone. (*Z*)-Dodec-6-eno-γ-lactone (detection threshold 700 ng/l) is a potent aroma compound found in Australian Riesling and Viognier white wines.

The precise chemical reactions leading to the formation of bouquet substances during ageing are not yet widely known. There are two types of reactions that produce bouquet constituents: oxidation, which is characterised by the presence of aldehydes and acetals (e.g. in Madeira-type wines), and reduction (such as in quality table wines after a period of bottle maturation). During wine ageing, glycosides of terpenic alcohols and some esters of carboxylic acids hydrolyse. Aldehydes are oxidised to carboxylic acids and partially esterified, the content of degradation products of carotenoids called norisoprenoids (e.g. vitispirane) increases, and tannins polymerise.

Markers of oxidation in wines can be some oxidation products of fats and oils. (*Z*)-Octa-1,5-dien-3-one (green and geranium-like odour) contributes to the flavour of Merlot and Cabernet Sauvignon musts. Its concentration in musts marked by dried fruit flavours reaches 90 ng/l. Small amounts of 3-methylnonan-2,4-dione (2.9 ng/l), the major contributor to reversion in soybean oil, were found in non-oxidised white wines, but the highest levels were found in oxidised botrytised red wines (293.8 ng/l). The impact odorants contributing to mushroom off-flavour in wines are oct-1-en-3-one and non-1-en-3-one. Compounds contributing to a specific aroma (typical overripe orange notes) of great noble rot dessert wines associated with *Botrytis cinerea* development are γ-nonalactone and non-2-en-4-olide, also known as 5-pentyl-5*H*-furan-2-one. The amount of this lactone correlates with wine ageing, whilst the *S*-form is more dominant in young dessert wines. The N-heterocycles 1-methylpyrrole-2-methanethiol and 1-ethylpyrrole-2-methanethiol are responsible for the typical aromatic nuances of some Chardonnay wines, reminiscent of hazelnuts.

Compounds associated with oak wood ageing are eugenol and 3-methyloctan-4-olide, also known as whisky lactone. Vanillylthiol (reminiscent of clove and smoke) has been identified in the highest concentrations (more than 8.3 μg/l) in red wines aged in new oak barrels.

Wines infected with either lactic acid bacteria (particularly heterofermentative strains) or *Dekkera/Brettanomyces* yeast can potentially produce mousy off-flavour, for which 2-ethyltetrahydropyridine, 2-acetyltetrahydopyridine, and 2-acetylpyrroline are responsible.

8.2.12.8.3 Spirits

In addition to ethanol, distillates contain a large number of volatile compounds, which either come from raw materials or are produced during fermentation and ageing. Their content varies according to the quality of raw material, the technology used, and a number of other factors. Many aromatic compounds are fundamental to the final product; additional components have little importance or no importance at all. The main components are ethanol, methanol, and higher alcohols. Other compounds include carbonyl compounds and acetals, of which acetaldehyde and 1,1-diethoxyethane are the major components. Fatty acid esters of lower aliphatic alcohols are particularly important, as they carry a fruity flavour (especially in fruit spirits). The main ester is always ethyl acetate. Ethyl esters of higher fatty acids (up to ethyl palmitate) carry the soapy flavour. Lactones, phenols, terpenoids, nitrogen heterocycles, and many other compounds also play important roles.

For example, the total content of methanol and higher alcohols in whisky is usually about 1 g/l. The total content of carboxylic acids ranges from 100 (in Scotch whisky) to 400 (in Bourbon) mg/l, of which the main component (amounting to 40–95% of the total acidity) is acetic acid. The main odour-active components of Bourbon are aldehydes, the most important of which are 3-methylbutanal, 2-methylbutanal, 2-methylpropanal, and (2*E*,4*E*)-deca-2,4-dienal. Other important components are ethyl esters, such as butanoate, 2-methylpropanoate, (*S*)-2-methylbutanoate, 3-methylbutanoate, hexanoate, and octanoate, lactones, especially (3*S*,4*S*)-whisky lactone, δ-nonalactone, and γ-decalactone, some phenols (vanillin and eugenol), and (*E*)-β-damascenone, which is a degradation product of carotenoids.

The key aroma compounds in rum are (*E*)-β-damascenone, (3*S*,4*S*)-whisky lactone, alcohols (2- and 3-methylbutanol), carbonyl compounds (3-methylbutanal, butane-2,3-dione), and ethyl esters (ethyl butanoate, ethyl (*S*)-2-methylbutanoate, and esters of higher fatty acids) that arise mainly during fermentation. Another important group of aroma-active compounds is the phenols (vanillin, 4-ethylphenol, 2-methoxyphenol, 4-ethyl-2-methoxyphenol, and 2-methoxy-4-propylphenol), which are mainly extracted from the wood barrels used for ageing.

8.2.12.9 Additional commodities

8.2.12.9.1 Tea

The content of tea (*Camellia sinensis* and *C. assamica*, Theaceae) volatiles is about 0.01–0.02% of dry matter. Black tea contains about four to five times the number of aromatic compounds as green tea. Of the more than 300 known substances in tea, only some are important for the aroma. The basic aroma consists mainly of regular compounds having green odour, such as (Z)-hex-3-en-1-ol, (E)-hex-2-enal, and hexanal, and usually 3-methylbutan-1-ol, 2-phenylethanol, methyl salicylate, phenol, and guaiacol. In Darjeeling tea, other important components are (2E,4E,6Z)-nona-2,4,6-trienal, 4-hydroxy-2,5-dimethyl-2H-furan-3-one (furaneol), and vanillin. Active terpenoids and degradation products of carotenoids include α-terpineol, linalool, nerol, linalooloxides, β-damascone, (E)-β-damascenone, β-ionone, dihydroactindiolide, and theaspiran and its derivatives, such as theaspiron and hydroxytheaspiran. The latter is a very important compound with low odour threshold concentration. It constitutes about 1% of volatile substances. Compared with black tea, green tea contains lower amounts of linalool, linalool oxides, and some other compounds, but higher amounts of nerolidol and β-ionone. A significant compound is 3-methylnonan-2,4-dione.

8.2.12.9.2 Coffee

Volatile components constitute about 0.1% of roasted coffee by weight (*Coffea* species, Rubiaceae), and more than 200 substances have been shown in green coffee. More than 800 compounds are known to make up the aroma of roasted coffee, although only about 60 have a significant role. Especially typical are a large number of heterocyclic compounds, mainly furans, pyrroles, indoles, pyridines, quinolines, pyrazines, quinoxalines, thiophenes, thiazoles, and oxazoles, which arise in caramelisation and the Maillard reaction during coffee roasting. In addition to heterocyclic products, other important volatiles are some aliphatic compounds (hydrocarbons, alcohols, carbonyl compounds, carboxylic acids, esters, and aliphatic sulfur and nitrogen compounds), alicyclic compounds (especially ketones), and aromatic compounds (hydrocarbons, alcohols, phenols, carbonyl compounds, and esters).

The aroma of coffee is very complex, and virtually none of the just mentioned constituents provides the typical aroma by itself. Exceptions are 2-furanmethanethiol and 5-methyl-2-furanmethanethiol, which have odours reminiscent of coffee. In freshly roasted coffee, 2-furanmethanethiol is present in an amount ranging from 0.01 to 0.5 μg/kg. A highly efficient sulfur-containing substance is 2-methyl-3-furanthiol, arising from thiamine. Important heterocyclic compounds are 2-acetyl-4-methylthiazole and diterpene kahweofuran.

Significant aliphatic sulfur compounds are methional, 3-methyl-but-2-en-1-thiol, 3-mercapto-3-methylbutan-1-ol, its ester 3-mercapto-3-methylbutyl formate, methanethiol, and dimethyltrisulfide. 3-Mercapto-3-methyl 1 ol also occurs in passion fruit and blackcurrant, and as a putative cat pheromone in cat urine, where it is formed as a degradation product of amino acid L-felinine (see Section 2.2.1.2.2). Of more than 70 known pyrazines, the most important in roasted coffee are isopropylpyrazine, 2-isobutyl-3-methoxypyrazine, 2-ethyl-3,5-dimethylpyrazine, 2,3-diethyl-5-methylpyrazine, 2,6-dimethyl-3-vinylpyrazine, and 2-ethyl-6-methyl-3-vinylpyrazine. Pyridine and its alkyl derivatives and bicyclic pyridines have a negative impact on the quality of coffee aroma. Important aromatic compounds include some furans and pyrans, especially γ-lactones, such as 3-hydroxy-4,5-dimethyl-5H-furan-2-one (sotolon), 5-ethyl-3-hydroxy-4-methyl-5H-furan-2-one (abhexon), and furan-2-carbaldehyde and maltol. Important phenols are vanillin, 4-vinylguaiacol, and 4-ethylguaiacol, whilst important carbonyl compounds include (E)-β-damascenone, biacetyl, and pentane-2,3-dione, as well as some Strecker aldehydes such as acetaldehyde, 2-methylbutanal, 3-methylbutanal, propanal, (E)-non-2-enal, and 2-hydroxy-3-methylcyclopent-2-en-1-one, known as cyclotene.

During storage of raw coffee beans, atypical odours may develop, which are suggested to influence the aroma of coffee beverages. The compounds responsible for this atypical odour include (E)-β-damascenone with cooked apple-like odour, 2-methoxy-4-vinylphenol with clove-like odour, and methyl 2-methylbutanoate and methyl 3-methylbutanoate with fruity odour.

Components of roasted coffee aroma, especially sulfur compounds, are easily oxidised, and ground roasted coffee kept in storage soon loses its typical aroma due to exposure to air. In order to retain the aroma, it is necessary to store the coffee in impermeable containers under an inert atmosphere. Reactions of the odour-active thiols, such as 2-furanmethanethiol, with phenols generated from chlorogenic acid degradation products are responsible for the rapid aroma staling of coffee beverages. Examples of such reaction products are conjugates of catechol (pyrocatechol) with one, two, and three molecules of 2-furanmethanethiol (**8-127**). Thiols also react with cysteine bound in proteins through thioether bond formation.

8.2.12.9.3 Cocoa and chocolate

Cocoa originates from beans of the cocoa tree (*Theobroma cacao*, Sterculiaceae) and is the main ingredient in chocolate manufacture. Fresh cocoa beans have a sour odour and a sour, bitter, and astringent taste. The main contributors to the bitter taste are purine alkaloids. The astringent taste is mainly influenced by oligomeric procyanidins, flavan-3-ol-C-glycopyranosides, and N-phenylpropenoyl amino acids (see Section 2.2.1.2.1). The formation of the characteristic aroma of cocoa depends on a number of processes, including the fermentation, drying, and roasting of cocoa beans.

About 600 compounds have been identified as odour-active components of cocoa. These include carboxylic acids, esters, aldehydes, ketones, alcohols, pyrazines, quinoxalines, furans, pyrones, lactones, and pyrroles. Important aldehydes are 2-methylpropanal and 2-methylbutanal, which recalls the smell of cocoa and malt, as well as (E)-2-phenyl-5-methylhex-2-enal (**8-128**), which resembles chocolate. The latter is formed by aldol condensation of 3-methylbutanal with phenylacetaldehyde and dehydration of the aldolisation product. Linalool and 2-phenylethanol carry sweet and floral scents, and 2-acetylpyridine has a roasted smell. Also important as aroma constituents are sulfides, especially dimethylsulfide, dimethyltrisulfide, and benzyl methyl sulfide (**8-128**). Other important compounds include 2- and 3-methylbutyric acids, methyl anthranilate (**8-128**), and the Maillard-reaction products 4-hydroxy-2,5-dimethyl-2H-furan-3-one (furaneol) and maltol, trimethylpyrazine, 2-ethyl-3,5-dimethylpyrazine, 2,4,5-trimethyloxazole, and 4-methyl-5-vinylthiazole. A mixture of 24 compounds can reportedly simulate cocoa aroma.

The key components of chocolate aroma are C$_5$ carboxylic acids with a sweet smell (2-methylbutyric), lactones, aldehydes (such as isovaleraldehyde with a sharp odour resembling malt), (*E*)-non-2-enal (with green and tallowy smell), (2*E*,4*E*)-nona-2,4-dienal and (2*E*,4*E*)-deca-2,4-dienal (with an oily smell resembling fried foods), ketones such as oct-1-en-3-one (with a mushroom-like smell), (2*E*,5*S*)- and (2*E*,5*E*)-5-methyl-hept-2-en-4-one (known as filbertone, with the odour of hazelnuts, **8-128**), 2-ethyl-3,5-dimethylpyrazine (with a smell reminiscent of fried potato chips), 2-ethyl-3,6-dimethylpyrazine (with the smell of nuts with earthy tones), and methyl 2-methyl-3-furyl disulfide (with a meaty and sulfurous smell).

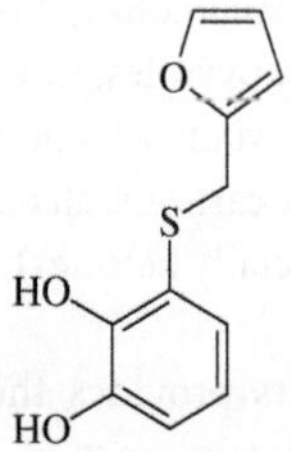

3-[(2-furylmethyl)sulfanyl]catechol

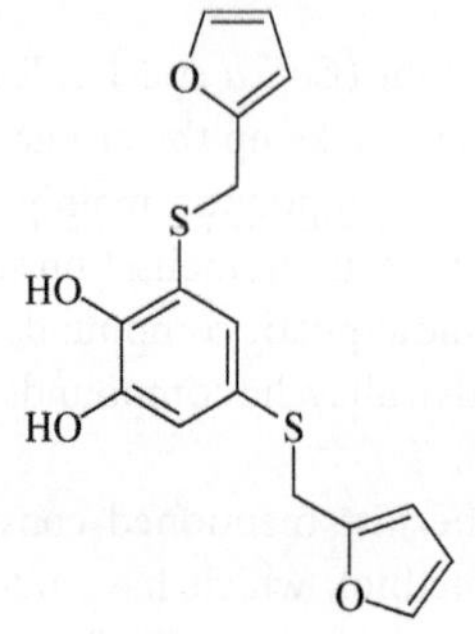

3,5-bis[(2-furylmethyl)sulfanyl]catecho

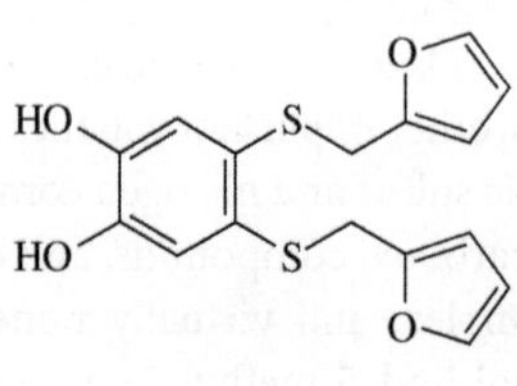

4,5-bis[(2-furylmethyl)sulfanyl]catechol

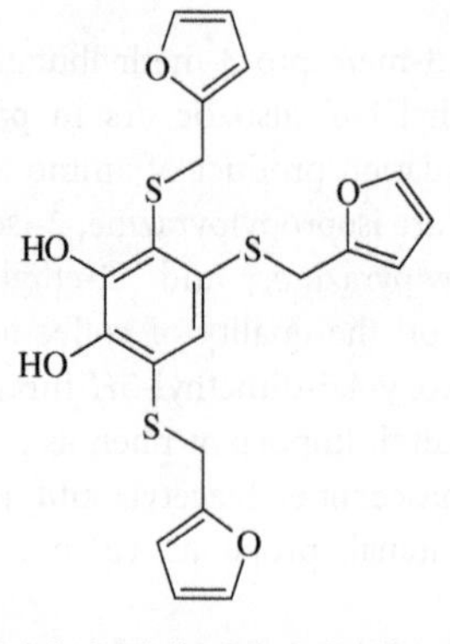

3,4,6-tris[(2-furylmethyl)sulfanyl]catechol

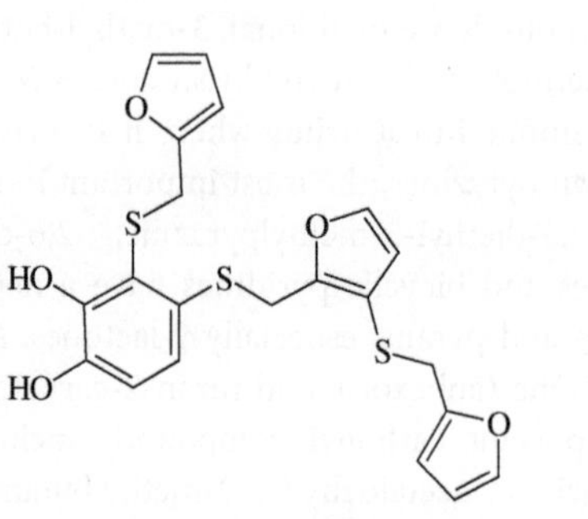

3-[(2-furylmethyl)sulfanyl]-4-{2-[3-(2-furylmethyl)sulfanyl]furylmethyl)sulfanyl}catechol

8-127, conjugates of catechol and 2-furylmethanethiol

8.2.12.9.4 Peanuts

Unroasted peanut seeds (*Arachis hypogaea*, Fabaceae) contain about 75 volatiles, whilst roasted seeds contain around 320. The compounds of the greatest importance for the aroma of raw seeds are 2-isopropyl-3-methoxypyrazine (earthy and pea-like), 2-isobutyl-3-methoxypyrazine (bell pepper-like and earthy), and (*E*)-4,5-epoxydec-2-enal (metallic).

Aroma components of roasted seeds include 2-acetyl-1-pyrroline, 2-propionyl-1-pyrroline, 6-acetyl-1,2,3,4-tetrahydropyridine, its tautomer, 2-acetylpyridine (roasted aroma), 4-hydroxy-2,5-dimethyl-2H-furan-3-one (furaneol, with a somewhat caramel odour), oct-1-en-3-one (with a mushroom-like odour), 2-isopropyl-3-methoxypyrazine, 2-isobutyl-3-methoxypyrazine, methional (resembling cooked potato), and phenylacetaldehyde and phenylacetic acid (resembling honey). Important aldehydes formed by the oxidation of fatty acids include (2*E*,4*E*)-deca-2,4-dienal (fried foods odour), (*E*)-4,5-epoxydec-2-enal (metallic odour), (*E*)-non-2-enal, and (2*E*,4*E*)-nona-2,4-dienal (oily odour).

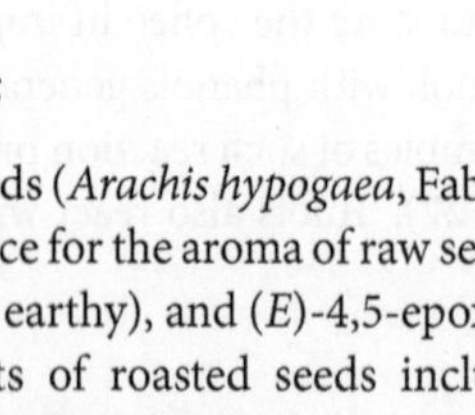

(*E*)-2-phenyl-5-methyl-hex-2-enal

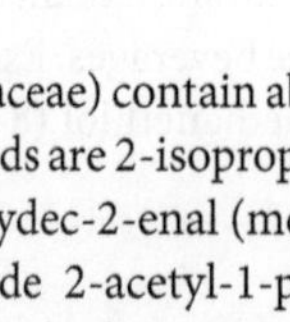

benzyl methyl sulfide

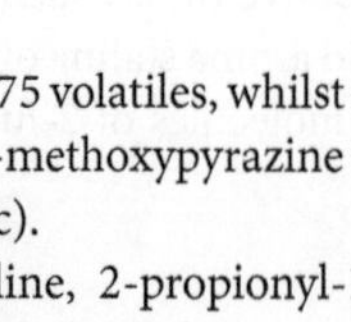

methyl anthranilate

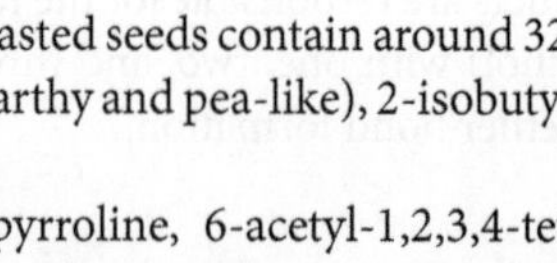

(2*E*,5*S*)-5-methyl-hept-2-en-4-one

8-128, selected cocoa volatiles

8.2.12.9.5 Honey

Typical honeys contain from 50 to more than 100 different aromatic compounds. The most important factors are the plants used by the bees for the production of the honey and the geographic area in which it originated. For example, the key aroma compounds in monofloral rape honey were (E)-β-damascenone (cooked apple-like), phenylacetic acid (honey-like), phenylacetaldehyde (honey-like, sweet), 4-methoxybenzaldehyde (anise seed-like), 3-phenylpropanoic acid (flowery, waxy), and 2-methoxy-4-vinylphenol (clove-like), as well as dimethyl trisulfide (cooked, savoury with a fresh, green, onion topnote).

8.2.12.9.6 Vinegar

Vinegar made from ethanol contains almost exclusively acetic acid, ethanol, and ethyl acetate, which is the result of esterification of acetic acid with ethanol during storage. The aroma of vinegars made from grapes and other fruits depends on the constituents formed during the alcoholic and acetic fermentations and in ageing. In addition to the characteristic substances of the plant material, these vinegars (including the balsamic vinegar) contain fermentation products such as 3-methylbutyric, 2-methylbutyric and butyric acids, 2-phenylethanol, acetoin, biacetyl, fusel oil alcohols (mixture of 2- and 3-methylbutan-1-ol), esters, and a number of non-volatile taste modulators.

8.2.12.9.7 Mushrooms

Most types of wild and cultivated edible mushroom of the phyla Basidiomycota (club fungi) and Ascomycota (sac fungi) contain as their key compound (R)-oct-1-en-3-ol (mushroom-like), which is produced by enzymatic oxidation of linoleic acid. Oct-1-en-3-ol is accompanied by a number of other compounds. The components of common mushroom (*Agaricus bisporus*) aroma are the following alcohols and carbonyl compounds: 3-methylbutan-1-ol, (R)-octan-3-ol, (3R,5Z)-octa-1,5-dien-3-ol, (Z)-oct-2-en-1-ol (weak mushroom odour), octan-3-one, benzyl alcohol, benzaldehyde, and furan-2-carbaldehyde.

The aromatic components of dried mushrooms are numerous carboxylic acids, their lactones, aliphatic sulfur compounds, and heterocyclic compounds (especially pyrazines and pyrroles) formed in the Maillard reaction. Some other odour-active substances are formed during cooking. A particularly significant compound is oct-1-en-3-one, which has a metallic and faintly mushroom odour that is important in the transformation of raw mushroom smell to the cooked mushroom smell. Important aroma compounds in pan-fried common mushrooms are 3-methylbutanal (malty), methional (cooked potato), oct-1-en-3-one, phenylacetaldehyde, 2-acetyl-1-pyrroline, 2-propionyl-1-pyrroline (popcorn-like), 4-hydroxy-2,5-dimethyl-2H-furan-3-one (furaneol, caramel-like), 3-hydroxy-4,5-dimethyl-5H-furan-2-one (sotolon, seasoning-like), and 2,3-diethyl-5-methylpyrazine.

The typical odorous components of truffle, the fruiting body of a subterranean fungus of the genus *Tuber* (*T. brumale*, brumale truffle or winter truffle, and *T. melanosporum*, black truffle) are various aliphatic sulfur compounds, such as methyl (methylthio)methyl disulfide, bis(methylthio)methane (garlic-like), tris(methylthio)methane, and dimethyl sulfide. The musky odour of these mushrooms is derived from (3α,5α)-androst-16-en-3-ol, which is present at concentrations of about 50 mg/kg. The key contributor to the aroma of white truffle (*T. magnatum*) is 3-(methylthio)propanal (potato-like), followed by 2- and 3-methylbutanal (malty), butane-2,3-dione (buttery), bis(methylthio)methane, and 1-pyrroline. The key components of burgundy truffle (*T. uncinatum*) are butane-2,3-dione (buttery), phenylacetic acid (honey-like), and vanillin (vanilla-like).

The widely used shiitake mushroom (*Lentinula edodes*), with a distinctive sulfurous aroma, contains a characteristic component, 4,4,7,7-tetramethyl-1,2,3,5,6-pentathiepane, known as lenthionine. Lenthionine arises from lentinic acid along with other cyclic polysulfides by enzymatic reaction (Figure 8.92). The first step is the hydrolysis of lentinic acid with γ-glutamyltransferase to deglutamyllentinic acid and the decomposition of this acid into pyruvic acid, ammonia, and 1,2-dithiran. 1,2-Dithiran then polymerises to both cyclic and linear polysulfides. Other important compounds are oct-1-en-3-ol, 1,3-dithiethane, and dimethyl disulfide.

The major constituents of the unpleasant smell of common stinkhorn (*Phallus impudicus*) are dimethyl disulfide and dimethyl trisulfide. Minor components are linalool, *trans*-β-ocimene, and phenylacetaldehyde.

8.2.12.9.8 Spices

Fresh, dried, or otherwise prepared parts of some plants have an intense, pleasant, and characteristic odour and taste, and are therefore used as spices for food flavouring. The main constituents of the aroma of spices are various volatile mono- and sesquiterpenoids, in particular hydrocarbons, alcohols, aldehydes, ketones, esters, and ethers (Table 8.32). The total content of essential oils is in the tenths to units of per cent. Some spices also contain components with a characteristic spicy, sharp, and burning taste and important pigments.

8.2.13 Physiology and nutrition

8.2.13.1 Organoleptic properties

8.2.13.1.1 Structure of odour-active compounds

Characteristic organoleptic properties of odoriferous substances are related to their structure and stereochemistry. Enantioselectivity and diastereoselectivity of substances are of clear importance for their biological activity, and play a role in chemical communication systems in bacteria, algae, insects, fish, and higher animals. It is not surprising, therefore, that the enantiomeric forms of compounds often show

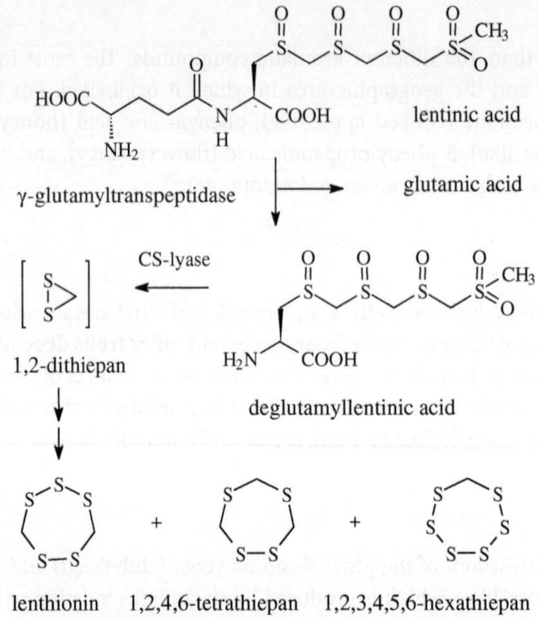

Figure 8.92 Formation of cyclic polysulfides from lentinic acid.

differences in smell (and taste). Small changes in the structures of molecules often lead to drastic changes in their quality and quantity of sensory perception. Structural analogues of compounds or structural isomers (positional as well as functional isomers) therefore often exhibit different organoleptic properties. Odour-active substances are mostly chiral molecules, and the individual enantiomers or diastereomers have different organoleptic properties. Smell is a phenomenon associated with the elementary composition of substances, their spatial arrangement, and, in unsaturated compounds, their molecular geometry. Biochemical reactions in which these compounds arise are enantioselective and diastereoselective.

For example, the pleasant smell of α-terpineol (**8-14**) resembles lilac flowers, but the substitution of oxygen for sulfur turns this into a very aggressive smell of *p*-menth-1-ene-8-thiol (**8-116**), which at very low concentrations (<1 µg/kg) recalls the scent of grapefruits, for which (*R*)-*p*-menth-1-en-8-thiol is a key component. The natural isomer (+)-(*R*)-1-*p*-menth-1-ene-8-thiol has the aroma of grapefruit, but the (*S*)-enantiomer has a weak and non-specific flavour.

Blackcurrant aroma is reminiscent of only one diastereomer of *p*-menthan-8-ol-3-one, (1*S*,4*R*)-*p*-menthan-8-ol-3-one (**8-129**). This compound is also a characteristic component of blackcurrant leaves and fruits. In contrast, (1*R*,4*S*)-*p*-menthan-8-ol-3-one has an unpleasant smell resembling burnt rubber and mercaptans, (1*R*,4*R*)-*p*-menthan-8-ol-3-one smells like onion and has a weak fruity odour, and (1*S*,4*S*)-*p*-menthan-8-ol-3-one has a similar but stronger smell of tropical fruits and a faint sulfur smell.

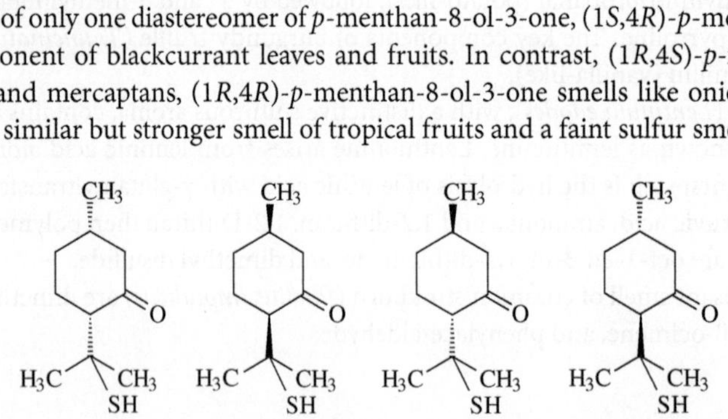

8-129, *p*-menthan-8-ol-3-one isomers

Vanillin (**8-31**) has a typical smell of vanilla, isovanillin (**8-130**) does not resemble vanilla at all, and heliotropin (**8-130**) has a floral and spicy smell reminiscent of both vanillin and isovanillin.

Table 8.32 Basic compositions of common spices.

Latin and English names	Main components of essential oil[a]
Apiaceae	
Anethum graveolens Dill (seeds, leaves)	Seeds: **(+)-carvone, limonene, (+)-dihydrocarvone, α-phellandrene**, (−)-carveol
	Leaves: **(+)-dill ether**, methyl 2-methylbutanoate, (+)-α-phellandrene, myristicin
Carum carvi Caraway (seeds)	**(+)-Carvone, (+)-limonene**, α-pinene, α-phellandrene, dihydrocarvyl acetate, 1,8-cineole, linalool
Coriandrum sativum Coriander (seeds, leaves)	Seeds: **(+)-linalool**, γ-terpinene, α-pinene, *p*-cymene, camphor, geranyl acetate
	Leaves: **decanal, (*E*)-dec-2-enal, dec-2-en-1-ol, cyclodecane, (*Z*)-dodec-2-enal**, dodecanal, dodecan-1-ol
Cuminum cyminum Cumin (seeds)	**Cuminaldehyde**, α-pinene, **β-pinene**, *p*-cymene, 2-ethoxy-3-isopropypyrazine, 2-methoxy-3-*sec*-butylpyrazine, 2-methoxy-3-methylpyrazine
Foeniculum vulgare Fennel (seeds)	**(*E*)-Anethol, fenchol**, α-phellandrene, (+)-limonene, estragol, camphene
Levisticum officinale Lovage (leaves)	3-Butylidene-4,5-dihydrophthalide (ligustilide), other phthalides, α-terpinyl acetate, α-terpineol, carveol, carvacrol, coumarin
Petroselinum crispum Parsley *(leaves)*	β-Phellandrene, apiole, *p*-mentha-1,3,8-triene, myristicin, α-terpinolene, β-myrcene, β-sesquiphellandrene, caryophyllene, γ-cadinene, α-phellandrene
Pimpinella anisum Anise (seeds)	**(*E*)-Anethol**, estragol, 4-methoxyacetofenone, acetaldehyde, 4-isopropylbenzaldehyde, 1,8-cineole
Asteraceae	
Artemisia dracunculus Tarragon (leaves)	**Estragol**, 4-methoxycinnamaldehyde, 5,7-dimethoxycoumarin, anethol, α-pinene, myrcene
Helichrysum italicum Curry plant (leaves)	**10-Acetoxytoxol, neryl acetate, α-curcumene, α-pinene**, nerol, neryl propionate, β-caryophyllene, γ-elemene, limonene, β-selinene
Cupressaceae	
Juniperus communis Common juniper (seeds)	**α-Pinene, myrcene, β-pinene**, (+)-limonene, *p*-cymene, camphene, terpin-1-en-4-ol, α-terpineol, borneol, β-cadinene, sabinene
Fabaceae	
Trigonella foenum-graecum Fenugreek	**Sotolon**, hexanol, heptanoic acid, dihydroactindiolide, α-muurolene, β-elemene, tetradecane, pentadecane
Illiciaceae	
Illicium verum Star anise (seeds)	**(*E*)-Anethol**, α-phellandrene, α-pinene, β-cadinene, Δ^3-carene, *p*-cymene, dipentene, α-terpineol, estragol
Lamiaceae	
Mentha piperita Peppermint (leaves)	**(−)-Menthol, (−)-menthone, (−)-menthyl acetate, (+)-menthofuran**, α-pinene, α-phellandrene, β-caryophyllene
Mentha spicata Spearmint (leaves)	**(−)-Carvone, limonene**, dihydrocarvone, sabinene hydrate, (−)-menthone
Ocimum basilicum Basil (leaves)	**Estragol, linalool**, 1,8-cineole, eugenol, methyl cinnamate, nerol, linalyl acetate
Origanum majorana Marjoran (leaves)	**Terpin-1-en-4-ol, (*Z*)-sabinene hydrate**, (*E*)-sabinene hydrate, **1,8-cineole, estragol, α-terpineol, eugenol**
Origanum vulgare Oregano (leaves)	Carvacrol, thymol, *p*-cymene, carvacryl methyl ether, linalool, α-pinene, bornyl acetate, camphor
Rosmarinus officinalis Rosemary (leaves)	**1,8-cineole**, α-pinene, **camphor**, camphene, borneol, bornyl acetate
Salvia officinalis Common sage (leaves)	α-Thujone, β-thujone, camphor, 1,8-cineole, α-pinene, β-pinene, bornyl acetate, myrcene, borneol, linalyl acetate, β-ocimene

(continued overleaf)

Table 8.32 (*continued*)

Latin name	Main components of essential oil[a]
Satureja hortensis Summer savoury (leaves)	**Carvacrol, γ-terpinene**
Thymus vulgaris Common thyme (leaves)	**Thymol, carvacrol**, *p*-cymene, linalool, limonene, α-pinene, camphene, γ-terpinene, β-caryophyllene, geraniol, carvacrol
Lauraceae	
Cinnamomum verum True cinnamon (barks)	**Cinnamaldehyde, eugenol, safrol, linalool**, camphor
Laurus nobilis Bay laurel (leaves)	**1,8-Cineole, linalool**, eugenol, α-pinene, α-phellandrene, geraniol, γ-terpinene, *p*-cymene, β-caryophyllene, eugenyl acetate, camphene
Myristicaceae	
Myristica fragrans Nutmeg (seeds)	**α-Pinene, β-pinene, sabinene, limonene**, 1,8-cineole, safrol, myristicin, γ-terpinene, terpinen-4-ol, eugenol, isoeugenol
Myrtaceae	
Syzygium aromaticum Clove (flowers)	Eugenol, β-caryophyllene, eugenyl acetate, α-pinene, methyl benzoate, methyl salicylate, heptanol, nonanol, furfuryl alcohol
Pimenta dioica Allspice (seeds)	**Eugenol**, 1,8-cineole, β-caryophyllene, methyl eugenol, α-phellandrene
Piperaceae	
Piper nigrum Black pepper (seeds)	**α-Pinene, β-pinene, sabinene, β-caryophyllene, Δ³-carene, limonene**
Piper longum Long pepper (fruits)	**Pentadecane**, β-caryophyllene, β-caryophyllene oxide, β-bisabolene, sabinene, α-zingiberene, *p*-cymene, *p*-methoxy acetophenone
Ranunculaceae	
Nigella sativa Black caraway (seeds)	**(*E*)-anethole, *p*-cymene**, limonene, carvone
Rutaceae	
Murraya koenigii Curry tree (leaves)	1-Phenylethanethiol, linalool, α-pinene, 1,8-cineole, (3*Z*)-hex-3-enal, 3-(methylsulfanyl)propanal, myrcene, (*Z*)-hex-3-en-1-ol, (2*E*,6*Z*)-nona-2,6-dienal
Zingiberaceae	
Curcuma longa Turmeric (rhizomes)	**(+)-*ar*-Turmerone, α-turmerone, β-turmerone (curlone), *ar*-curcumene, zingiberene**, 1,8-cineole, borneol, α-phellandrene
Elettaria cardamomum Cardamom (seeds)	**1,8-Cineole, α-terpinyl acetate, (+)-limonene**, sabinene, borneol
Zingiber officinale Ginger (rhizomes)	**(−)-Zingiberene, β-sesquiphellandrene, β-bisabolene, citral**, geranial, eucalyptol, α-pinene, β-linalool, **citronellyl acetate**, bornyl acetate

[a]Components occurring in quantities higher than 10% are highlighted in bold.

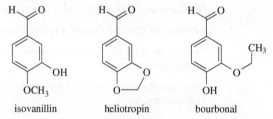

8-130, structural analogues of vanillin

isovanillin　　　heliotropin　　　bourbonal

Similarly, different organoleptic properties are also found in *cis/trans* or (*E,Z*)-isomeric substances. For example, hydrocarbon galbanolene, which is (3*E*,5*Z*)-undeca-1,3,5-triene (**8-131**) occurring in celery root, celery and parsley leaves, and mandarin oil, has a green smell, but its all-*trans* isomer (3*E*,5*E*)-undeca-1,3,5-triene has an oily one (**8-131**). (*Z*)-Isomers are generally considered natural and are therefore more acceptable, whilst (*E*)-isomers are often associated with synthetic compounds and various off-flavours.

(3*E*,5*Z*)-undeca-1,3,5-triene   (3*E*,5*E*)-undeca-1,3,5-triene

8-131, undeca-1,3,5-triene isomers

Analysis of individual forms is also important in determining the authenticity of foods and essential oils. The presence of certain forms, such as the geometrical (*E*)-isomers, is often evaluated as an off-flavour. For example, (+)-(*R*)-limonene (**8-3**) is a natural component of citrus essential oils and has a citrus smell like oranges. The isomer (−)-(*S*)-limonene (**8-3**), which is the main component of essential oils of various types of mints (*Mentha* spp.), has a smell reminiscent of turpentine. The alcohol (−)-(*R*)-linalool, also known as licareol (**8-13**), has a strong woody smell like lavender, but (+)-(*S*)-isomer, also known as coriandrol (**8-13**), has a sweet smell with lavender tones. Another alcohol, (−)-(*R*)-oct-1-en-3-ol (Figure 8.8), has an intense smell of fresh mushrooms, but (+)-(*S*)-isomer (**8-16**) has a smell resembling vegetables with faint mushroom notes. The ketone (−)-(*R*)-carvone (**8-36**) has a mint-like smell and is the main component of spearmint essential oil, whilst its isomer (+)-(*S*)-carvone (**8-36**) smells like caraway and is a major component of caraway essential oil.

Menthol has three asymmetric carbon atoms in the molecule, so there are four pairs of optical isomers, but only (−)-(1*R*,3*R*,4*S*)-menthol (**8-14**) has any practical significance. It has a sweet, fresh, minty smell and a cool, refreshing taste. In the other isomers, the mint smell and cool taste are weaker, and the dominating characteristic is vegetable-like.

The only isomer of nootkatone that resembles the odour of grapefruits is (+)-nootkatone (**8-38**); the remaining isomers have a more woody and spicy smell.

8.2.13.1.2 Sensitivity to perception of odours

Sensitivity to odour perception is different for every person and depends on the physiological state of the organism, psychological conditions, pathological changes, and other factors. Generally, women are more sensitive to odours than men. Both sexes show a maximum sensitivity at puberty, and sensitivity to the perception of odours is reduced after the 70th year of life, and drastically after the 80th year. Sensitivity is also lower during pregnancy, colds, and some diseases. Smokers perceive some odours as less intense (e.g. odour of cloves and thiols) and some as more so (such as banana aroma).

Part of the population suffers from so-called specific **anosmia** (specific olfactory blindness), which is manifested by an inability to perceive certain odours. About 100 substances are known that are not perceived by a limited number of human beings. For example, just 70.5% of women and 62.8% of men reportedly perceive 5-androst-16-en-3-one, whilst the scents of cloves and bananas are perceived by nearly 100% of the population. About 1.2% of the total population suffer from total anosmia, which is a condition in which people have no sense of smell. **Hyperosmia** is an extremely sensitive sense of smell, whilst **parosmia** causes people to misinterpret smells, detecting something unpleasant where an odour is neutral or pleasant. In **phantosmia** (a sort of olfactory hallucination), people detect smells where there are none. Anosmia differs from temporary olfactory adaptation (fatigue), which is adaptation to the smell of an odoriferous substance for a very short time after the smell ceases. People with anosmia often also experience **ageusia**, the inability to taste, because smell plays an important role in taste perception.

The ability to perceive the odour of a particular substance also depends on its **odour threshold** (aroma, flavour) **value**, which is the concentration of the substance detectable by the sense of smell. The **odour detection threshold value** is the lowest concentration of a stimulus (odoriferous substance) that can be detected in comparison with an environment which does not contain this substance. The **odour recognition threshold value** is the concentration at which a substance can be not only detected, but also recognised. It corresponds to a concentration that allows identification of the odour quality of a substance, which is usually higher than the odour detection threshold. Both values are measures of odour intensity, but they depend considerably on the environment, solubility, partition coefficients between air and water (oil), and some other factors. For example, values measured in air are typically several orders of magnitude lower than those measured above aqueous solutions. A substance with a high odour threshold value must be present in foods in higher concentrations than a substance with a low odour threshold value, otherwise its smell is imperceptible. **Orthonasal** and **retronasal** detection thresholds represent two distinct brain networks for smell. The orthonasal system deals with odours (traditional smell) 'sniffed in' to the nose. The retronasal system is involved in sensing odours when we eat food (mouth smell). It is best described as a combination of traditional smell (orthonasal smell) and taste. The measure of whether a substance acts as an odour-active substance is its **odour unit**: its actual concentration divided by its odour threshold concentration.

The presence of lipids, proteins, carbohydrates, and other substances significantly influences the retention of aromatic compounds in foods and has an effect on their odour intensity and quality (e.g. the odour threshold values of non-polar substances measured in water are usually lower than values measured in emulsions or fats). The effects known as **synergism** and **antagonism** play important roles in mixtures of odoriferous substances, causing a higher (synergism) or lower (antagonism) response to the mixture of compounds than that expected from simple additivity amongst the components. For example, the odour of butyric acid is more intense in the presence of isovaleric acid and less intense in the presence of acetic acid. Antagonistic odorants are, for example, eugenol and methyl isoeugenol. Table 8.33 displays the odour detection threshold values of selected odoriferous substances in water as examples.

8.2.13.2 Physiological effects

Odoriferous substances and their mixtures (e.g. essential oils) show a number of beneficial effects, for which they have found use as pharmaceuticals, pharmaceutical ingredients, and food additives. Bactericidal and anti-inflammatory effects are seen in borneol, eugenol, pinenes, camphor, thymol, and menthol, whilst cholinolytic (spasmolytic) effects preventing a drop in blood pressure and suppressing the secretory activity of various organs are found in camphor, pinenes, and camphene, and analeptic effects (stimulating activity of circulation and respiratory system) are seen in camphene. The essential oils of many spices (such as marjoram, sage, and thyme, amongst many others) exhibit antioxidant effects, and therefore find use as natural antioxidants of fats.

Some terpenoids and other aromatic substances also show a variety of toxic effects, and their content in foods is therefore restricted by legislation in many countries.[2] The monoterpenes α-thujone, known as (−)-thujone, and β-thujone, known as (+)-isothujone (**8-37**), are the dominant components of various essential oils of plants of the genus *Artemisia* (Asteraceae), referred to under various names, such as mugwort, sagebrush, sagewort, wormwood, and tarragon. The content of essential oils ranges from 0.25 to 1.32%, in which thujones are present in amounts of 3–12%. Thujones also occur in the essential oil of sage (*Salvia officinalis*, Lamiaceae) and tansy (*Tanacetum vulgare*, Asteraceae), and are found in lower levels in yarrow (*Achillea millefolium*, Asteraceae) and other plants. All of these are used as medicinal herbs or as bittering agents in alcoholic beverages. For example, tansy is used in bittering Chartreuse liquor and wormwood in absinthe. α-Thujone is perhaps best known as the active ingredient of the green liquor absinthe, which was a very popular French drink in the 1800s and gained considerable notoriety as the preferred liqueur of artists and writers. By 1910, the French were drinking 36 million litres per year. Although absinthe became an epidemic health problem, leading to a ban in France in 1915, its use continues on a small scale today, either legally or illicitly. The acute toxicity of α-thujone can be attributed to its blocking of the γ-aminobutyric acid (GABA)-gated chloride channel. Both thujones show chronic neurotoxicity manifested by hallucinations, hyperexcitability, sleeplessness, tremors, convulsions, and, later, damage of the cerebral cortex. Manifestations of chronic toxicity are called absinthism, the disorder associated with the habitual abuse of absinthe. The thujone content is therefore restricted.

Toxic effects have are also been found for (+)-(*R*)-*p*-mentha-4(8)-en-3-one, known as (+)-pulegone (**8-36**), which is a component of essential oils of different mint species, especially squaw mint, also know as pennyroyal (*M. pulegium*, Lamiaceae). Mints are mostly used as carminative agents (antiflatulence agents), spasmolytic agents (release spasms), stimulants of bile and gastric juice production, and food flavourings. Pulegone in peppermint oil, for example, exhibits neurotoxicity and relatively high hepatotoxicity leading to necrosis of liver cells. Its content in food is therefore restricted.

The bitter sesquiterpene santonin occurs mainly in the flowering tops of plants of the genus *Artemisia* and in their essential oils. Previously, these drugs were used as intestinal anthelmintic drugs. Santonin promotes chromatopsia (an abnormal condition in which all objects appear in purple colour). Its level in food is legally restricted.

Carcinogenic effects are shown by some substituted alkenylbenzenes (e.g. prop-1-en-1-ylbenzenes and prop-2-en-1-ylbenzenes) such as safrole (**8-24**) and methyl eugenol (**8-23**) that occur in several essential oils. Myristicin, found in seed essential oils of some root vegetables (such as carrots), is a psychomimetic substance having narcotic and hallucinogenic effects similar to the effects of ethanol. It also displays antimicrobial effects. β-Asarone (**8-23**), occurring mainly in wild ginger (*Asarum europaeum*), has chemosterilation effects. Another aromatic substance the use of which is restricted is coumarin (**8-69**), which occurs, for example, in tonka beans (*Dipteryx odorata*, Fabaceae) and many other plants. Coumarin has hepatotoxic effects, but certain anticarcinogenic properties and some positive effects in the treatment of varicose vein syndrome (such as rutin and hypericin) have also been demonstrated.

8.2.14 Production and use

Flavourings are products that are intended to be added to food in order to impart or modify odour or taste. They are made or consist of the following categories: flavouring substances, flavouring preparations, thermal process flavourings, smoke flavourings, flavour precursors,

---

[2] The influence of odoriferous substances on human health is the concern of the International Fragrance Association (IFRA) and various other organisations. In the Generally Recognized as Safe (GRAS) system published by the Flavour and Extract Manufacturer's Association (FEMA) and the Code published by the Council of Europe (CoE), non-toxic flavour-active substances are allocated numbers (FEMA numbers or CoE numbers). This is in addition to Chemical Abstracts Service (CAS) numbers.

Table 8.33 Odour detection threshold values of powerful odoriferous substances in µg/l (ppb) in water at 20 °C.

Compound	Occurrence	Odour detection threshold
Acetaldehyde	Widespread (alcoholic beverages)	9
Acetic acid	Vinegar	25 000
2-Acetyl-1-pyrroline	Bread, cereal products	0.1
6-Acetyl-1,2,3,4-tetrahydropyridine	Bread, cereal products	1
Benzaldehyde	Essential oils (bitter almonds)	400
Biacetyl	Fermentation product (butter)	2000
2-sec-Butyl-3-methoxypyrazine	Parsley, peas	0.001
Butylphthalide, (S)-	Celery, celeriac	5
Butyric acid	Rancid butter	200
Cyclotene	Roasted products containing sugar	300
Deca-2,4-dienal, (2E,4E)-	Rancid fats, some essential oils	0.1
Decalactone, δ-	Butter, fruits	400
Decanal	Rancid fats, essential oils	0.1
Dimethyl disulfide	Widespread (vegetables and meat)	100
Dimethyl sulfide	Widespread (vegetables and meat)	0.1
3,4-Dimethylthiophene	Roasted onion	1
Dimethyl trisulfide	Widespread (meat, chocolate)	0.01
Ethanol	Alcoholic beverages	10 000
Ethyl acetate	Alcoholic beverages	5000
Ethyl butyrate	Fruits	1
Ethyl palmitate	Alcoholic beverages	2000
Furfuryl mercaptan	Roasted coffee	0.005
Geraniol	Essential oils	100
Hex-3-en-1-ol (leaf alcohol), (Z)-	Widespread in plant materials	1550
Hexan-1-ol	Essential oils, alcoholic beverages	1000
Hexanal	Widespread (alcoholic beverages)	5
2-Isobutyl-3-methoxypyrazine	Bell pepper	0.002
2-Isobutylthiazole	Fresh tomatoes	30
Limonene, (R)-	Citrus fruits	200
Maltol	Caramel, bread	35 000
Maple furanone	Acid protein hydrolysates	0.01
Mesifuran	Sherry wines, fruits	0.3
Methanethiol	Widespread (vegetables and meat)	0.02
Methional	Widespread (potatoes, vegetables, meat)	0.2
Oct-1-en-3-ol, (R)-	Mushrooms	1
Oct-1-en-3-one	Mushrooms, moulds	1
Octanal	Rancid fats, essential oils	0.7
Phenylacetaldehyde	Widespread (phenylalanine degradation)	40

(continued overleaf)

Table 8.33 (*continued*)

Compound	Occurrence	Odour detection threshold
p-Menth-1-en-8-thiol, (*R*)-	Grapefruit juice	0.000 02
Rotundone	Wines, fruits	0.008
Skatole	Pork meat (off-flavour)	0.003
Sotolon	Fenugreek, caramel, maple syrup	0.001
Sulfane	Widespread (vegetables and meat)	10
2,4,6-Trichloroanisole	Cork bottle stoppers (off-flavour)	0.01
Trimethylamine	Fish, eggs (off-flavour)	2000
Vanillin	Essential oils (vanilla)	20

and other flavourings or mixtures thereof. Food flavourings are derived from natural materials or produced synthetically. It is estimated that of the food flavourings used in developed countries, about 75% are natural compounds and the rest are synthetic. About 99% of synthetic substances are substances occurring in nature (nature-identical substances) and 1% are artificial, synthetic flavourings.

The most important materials for the production of natural food flavourings (but also perfume compositions for cosmetic purposes) are **essential oils**, **oleoresins**, extracts, juices, pulps, and distillates. They are obtained almost exclusively from plant materials, fresh or dried plants, or parts of plants. Materials called **resinoids** are primarily used for the fixation of fragrance compositions in cosmetics, and only occasionally in foods. Drugs of animal origin are not used for the production of food flavourings, in contrast to cosmetic products.

8.2.14.1 Essential oils

Essential oils are complex mixtures of volatile substances contained in natural plant materials. They are obtained from different parts of plants, their flowers (such as jasmine oil), flowering stalks or straw (mint and thyme oils), fruits or seeds (cumin, pepper, and juniper oils), fruit pericarp (citrus oils), wood (sandal oil) leaves (bay oil), bulbs (garlic oil), rhizomes (sweet flag and turmeric oils), and roots (gentian oil).

Essential oils are obtained mostly by distillation with water vapour, by extraction with non-polar solvents, and by pressing and separation of the oil layer (oils of citrus fruit peels). These processes produce an essential oil known as a **concrete**, which is a mixture of essential oil, various lipophilic substances, and waxes. The accompanying substances are separated by dissolution in hot ethanol, freezing, and filtration, and the essential oil thus obtained is **absolute**. Monoterpene-free or sesquiterpene-free essential oils (which do not contain even monoterpenic hydrocarbons) are collectively called **deterpenic essential oils**. They are more concentrated than the starting oils are more stable against autoxidation.

8.2.14.2 Oleoresins

Oleoresins are the flavour extracts obtained by the solvent extraction of ground spices or vegetables. Their characteristics are given in Table 8.34. They are mixtures of essential oils, resins, and other components that are extracted with various solvents or by supercritical carbon dioxide. Oleoresins have the aroma of the original spice or vegetable, as they contain essential oils in varying proportions, and possess attributes which contribute to the taste, such as pungency (black pepper and ginger oleoresins) and colour (turmeric and hot bell pepper oleoresins). For example, the essential oils of black pepper (Table 8.32) obtained by steam distillation contain as major components (about 20%) the terpenes α-pinene, sabinene, and β-caryophyllene, and in smaller amounts, car-3-ene, limonene, β-pinene, and other compounds. Oleoresin of black pepper obtained by extraction contains about 13–14% essentials oil and 50% of hot piperine and related non-volatile substances.

8.2.14.3 Resinoids

Resinoids are products obtained by the extraction of natural materials, mostly resins, gum resins, and balsams. Resinoids have a characteristic smell as they contain essential oils.

Resins are the products of various shrubs and trees, especially conifers. In addition to essential oils, resins contain resin acids such as tricyclic diterpene abietic acid (see Section 3.8.2.2.2), resin alcohols (resinols), resin esters, and other substances. Particularly well known

Table 8.34 Characteristics of some common oleoresins.

Oleoresin	Essential oil (ml/kg)	Spice equivalent[a]	Oleoresin	Essential oil (ml/kg)	Spice equivalent[a]
Bay leaf	300	3.5	Allspice	600	4.5
Clove	700	6	Bell pepper	trace[b c]	10
Caraway	600	5	Black pepper	150[c d]	4.5
Coriander	330	10	Cinnamon	160	5.5
Turmeric	trace[b e]	15	Sage	250	2
Marjoram	400	3	Vanilla	50[f]	5
Mace	500	7.5	Thyme	500	4
Nutmeg	800	6	Ginger	250[c g]	4

[a]Number of weight units that have the same flavour as 100 units of natural spices.
[b]A minimum content of dyes is required.
[c]Sharpness (heat) of hot oleoresins, expressed in Scoville heat units (SHU).
[d]50% of piperine.
[e]At least 12.5% of curcumine.
[f]10% of vanillin.
[g]12% of hot ketones.

is Chios mastic gum. Real mastic is only produced in the south part of Chios Island, and comes from the mastic tree (*Pistacia lentiscus* var. *chia*, Anacardiaceae). It is used to flavour alcoholic beverages (especially liqueurs and ouzo), wine, baked goods, chewing gum, and some cosmetic products. The essential oil, with a balsam-like odour, has antiseptic properties as it contains α- and β-pinene as essential ingredients.

8.2.14.4 Other materials

Extraction is used to obtain not only essential oils and oleoresins, but also other materials for food flavouring, such as **extracts** and **tinctures**. Extraction with a solvent (mostly ethanol) by standing at room temperature is called maceration, and the product is the **macerate**. Extraction with the flow of a solvent that passes through a material is percolation, and the product thus obtained is the **percolate**. Products obtained by cold **maceration** or **percolation** usually have better organoleptic properties. Extraction at higher temperatures is often called **digestion**. An alcoholic (or aqueous) extract (or extract obtained using other solvents) is usually called a **tincture**. In practice, the basic extraction procedures are variously combined and modified. For example, hop extract is produced by two-stage extraction. The first stage is percolation of a volatile solvent (such as dichloromethane) to obtain the oil, and this is followed by hot-water extraction (extraction of tannins and so-called soft resins). Both extracts, after evaporation of solvents, are combined and homogenised. The name **essence** is correctly used for alcoholic extracts of materials, but today it is used more for special composed products employed as food flavourings, which are mainly known as **aroma compositions**. Other raw materials for the production of food flavourings include **musts**, **juices**, and **pulps** from fruits and vegetables.

8.2.14.5 Synthetic compounds

Hundreds of different essential oils are used in current industrial practice. Their scarcity, price, and differences in quality have led to the production of cheaper **reconstituted essential oils**, which contain the same fragrances as natural essential oils but are synthetic and usually lack more subtle characteristics.

Examples of synthetic substances that do not occur in nature and yet find a use in food flavouring are ethyl vanillin (bourbonal, **8-31**) and (*E*)-2-ethoxy-5-(prop-1-en-1-yl)phenol (propenyl guaethol, **8-132**), which has a vanilla-like smell. The advantage is that the smell of ethyl vanillin is two to four times more intense than the smell of vanillin. Another commonly used compound is ethylmaltol (**8-132**), which has a caramel odour four to six times more intense than maltol. Allyl phenoxy acetate (**8-132**) has a sweet smell of honey and pineapple. One of the first synthetic substances with a strawberry flavour, used for confectionery, ice cream, and other products, was (*E*)-isomer of ethyl 3-methyl-3-phenylglycidylic acid (ethyl 3-methyl-3-phenyloxiran-2-carboxylate, also known as fraseol or strawberry aldehyde, **8-132**). Only the (2*R*,3*R*)-isomer, however, has a desirable strawberry flavour. Ethyl 3-phenylglycidylic acid (ethyl 3-phenyloxiran-2-carboxylate, also known as raspberry aldehyde found use as a flavouring agent in similar products that require the aroma of raspberries.

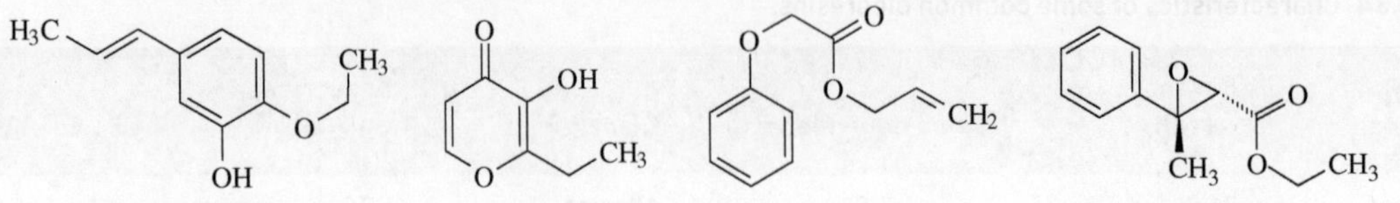

propenyl guaethol   ethylmaltol   allyl phenoxyacetate   ethyl 3-methyl-3-phenyl-
                                                                   oxirane-2-carboxylate

8-132, examples of selected synthetic odorants

8.3 Taste-Active substances

Oral ingestion of food in humans and higher animals is inseparable from subjective taste sensations. The stimuli for irritation of taste receptors localised in the mouth, especially on the tongue, are **taste-active substances**. They are usually polar, water-soluble, and non-volatile compounds. The resulting sensations are usually combinations of fundamental and other tastes:

- **sweet**;

- **salty**;

- **acid**;

- **bitter**;

- **umami** (now considered a fifth basic taste).

These basic taste sensations arise at relatively specialised receptors located in different parts of the mouth. Sweet substances are perceived primarily at the tip of the tongue, salty substances in defined areas of the upper surface of the tongue, sour substances to the sides, and bitter substances in the root of the tongue and soft palate. Taste receptors also respond to other stimuli (tastes) that are incorporated virtually in the whole oral cavity:

- **astringent**;

- **pungent** (burning or hot);

- **cooling** and others.

The interpretation of taste in terms of molecular interactions of taste-active compounds with biopolymers in taste receptors is probably analogous to the general scheme of pairs of interactions of the type enzyme–substrate, hormone–receptor, and antigen–antibody. These interactions and relationships between the chemical structure of a substance and its taste are still largely incomplete.

The measure of taste intensity is the lowest detectable concentration of a substance in the solution causing the sensation, called the **threshold value**. As in the case of odour-active substances, both **taste detection threshold values** and **taste recognition threshold values** are recognised. Both are measures of taste intensity.

8.3.1 Sweet substances

Sweet taste is commonly associated with sugars, especially sucrose (saccharose). Sweet substances are, with few exceptions, monosaccharides, oligosaccharides, and sugar alcohols, but most are less sweet than sucrose. Many sugars are not sweet at all, and some, such as β-D-mannose and some oligosaccharides, are even bitter. In contrast, there are other compounds that have a completely different structure than sugars and yet are much sweeter than sucrose (e.g. synthetic sweeteners).

8.3.1.1 Sweet taste quality and intensity

Sugar and all sweet substances differ in their taste quality and intensity. Sucrose has a particularly full taste, which is acceptable even at high concentrations. It is therefore used as a standard for sweet taste in the sensory evaluation of sweet substances. The threshold values of some sugars in aqueous solutions are listed in Table 8.35. For practical reasons, the sweetness of a substance is expressed as a multiple of the sweetness of the sucrose solution (mainly 10% solution). The relative sweetnesses of certain sugars and sugar alcohols are given in Table 8.36. These values are only approximate, as they depend on sugar concentration, type and amount of anomers, temperature, presence of other substances, and other factors.

For example, D-fructose solutions are sweeter than sucrose solutions, but in pastries and hot coffee both show the same sweetness. The sweetest form is β-D-fructopyranose, which has about 180% of sucrose sweetness, but as a result of mutarotation, the sweetness of solutions decreases to about 150% of sucrose sweetness because the individual anomers each have a different sweetness. The sweetness of fructose syrups depends on the fructose content. Syrups containing 42% fructose are about as sweet as sucrose solutions, those with 55% fructose show 100–110% of sucrose sweetness, and those with 90% fructose have 120–160% of sucrose sweetness.

In 10% solutions, the relative sweetness of D-glucose is about 50–60% of sucrose sweetness, whilst in solutions containing 50–60% glucose, the sweetness is about 90–100% of sucrose sweetness. β-D-Glucopyranose has about 66% of the sweetness of α-D-glucopyranose. The sweetness of glucose syrups depends on the glucose content and its lower oligomers. A syrup with a DE (dextrose equivalents) value of 30 has about 30–35%, one with a DE value of 36 has 35–40%, one with a DE value of 42 has 45–50%, one with a DE value of 54 has 50–55%, and one with a DE value of 62 has 60–70% of sucrose sweetness.

Table 8.35 Taste threshold values of selected sugars.

Sugar	Detection threshold		Recognition threshold	
	mol/l	%	mol/l	%
D-Glucose	0.065	1.17	0.090	1.63
D-Fructose	0.020	0.24	0.052	0.94
Maltose	0.038	1.36	0.080	2.89
Saccharose	0.011	0.36	0.024	0.81
Lactose	0.072	2.60	0.116	4.19

Table 8.36 Relative sweetnesses of selected sugars and sugar alcohols (10% saccharose solution = 1).

Compound	Sweetness	Compound	Sweetness
Monosaccharides		Trisaccharides	
D-Xylose	0.70	Raffinose	0.15–0.20
D-Glucose	0.40–0.80	Alditols	
D-Mannose	0.30–0.60	L-Arabinitol	1.00
D-Galactose	0.30–0.60	Xylitol	0.90–1.20
D-Fruktose	0.90–1.80	D-Glucitol	0.40–0.60
Disaccharides		D-Mannitol	0.50–0.70
Invert sugar	0.95–1.80	Galactitol	0.40
L-Rhamnose	0.30	Maltitol	0.70–0.90
α,α-Trehalose	0.60	Isomaltitol	0.40–0.50
Maltose	0.30–0.60	Palatinitol	0.45
Lactose	0.20–0.60	Lactitol	0.30–0.40
Lactulose	0.60	*Various*	
Saccharose	1.00	Glycerol	0.50

Besides the sweet taste, some sweet substances show other side-tastes. For example, maltose and D-glucitol have the flavour of syrups, and the taste of D-fructose solutions is fruity and slightly sour. Sucrose taste in threshold concentrations is even described as slightly bitter. Xylitol shows a cooling effect on dissolution.

Sweet substances are classified according to various criteria:

- origin, as natural, synthetic identical to the natural, modified natural, or synthetic (absent in nature);

- nutritionally, as substances that are a source of energy and substances that have no nutritional value;

- medically, e.g. as contraindicated substances in diabetics (people with *diabetes mellitus*) and substances that do not increase blood glucose levels; also recognised are cariogenic (causing tooth decay) and non-cariogenic sweet substances.

The relative sweetnesses of some non-sugar sweet natural, modified natural, and synthetic compounds and other aspects of sweet substances, which include food additives (sweeteners), are described in Section 11.3.2.

8.3.1.2 Physiology, nutrition, and use

Sucrose has major significance as a natural sweet substance, as do starch syrups (mixtures of D-glucose, maltose, and maltooligosaccharides), D-glucose, invert sugar (equimolar mixture of D-glucose and D-fructose), D-fructose, and lactose. Sugar alcohols, D-glucitol (also known as D-sorbitol), D-mannitol, and xylitol have found widespread use as sweeteners for diabetics. Other sugars, sugar alcohols, and non-sugar natural sweeteners are, with a few exceptions, primarily the subject of academic rather than industrial interest. All these sweet substances have some nutritional value, whilst alcoholic sugars are non-cariogenic or slightly cariogenic substances. Modified natural and synthetic sweet substances used as sweeteners have, with some exceptions, no nutritional value, and are also non-cariogenic.

8.3.2 Salty substances

Salty tastes are exhibited almost exclusively by some inorganic salts (especially halides, sulfates, phosphates, nitrates, and carbonates of alkali metals, alkaline earth metals, and ammonium salts). Salty tastes combined with other basic tastes, such as sour, bitter, and umami, are also shown by salts of some salts of carboxylic acids (salts of formic, acetic, succinic, adipic, fumaric, lactic, tartaric, and citric acids), amino acids (such as salts of glutamic acid), some oligopeptides, and choline. Some compounds increase the sensation of salt taste. Typical representatives of these compounds are phthalides (**8-69**), occurring in some vegetables of the Apiaceae family. That is why celery leaves taste more salty than other vegetables.

8.3.2.1 Salty taste quality and intensity

The quality of salty taste varies in different substances depending on the type of compound, its concentration, and the presence of ingredients. In the case of inorganic salts, both types of ions, cations and anions, are involved in the salty taste perception. For example, sodium chloride (NaCl) and sodium sulfate (Na_2SO_4) differ in the intensity of salty taste (effect of anion), whilst sodium chloride (NaCl) and potassium chloride (KCl) do not have the same salty taste quality (the effect of cation). The intensity of the bitter taste usually increases with increased relative molecular weight (apparent diameters of the two ions) of salt. Only sodium chloride has a pure salty taste (the sum of the apparent diameters of the two ions is 0.556 nm), corresponding to the hydrated sodium cation in combination with the hydrated chloride anion released from the crystal lattice of NaCl. Other salty substances also exhibit more or less intense bitter taste or salty taste when combined with some other tastes, such as metallic taste. For example, potassium bromide (KBr, 0.658 nm) tastes salty and bitter, whilst potassium iodide (KI, 0.706 nm) is bitter, as is magnesium chloride ($MgCl_2$, 0.850 nm). The character of salinity in many cases varies with salt concentration. Sodium chloride, which is considered the standard of salty taste, has a sweet taste at very low concentrations. Potassium chloride at low concentrations has a sweet taste that is enhanced with increasing concentration, whilst the intensity of the bitter taste also increases. At higher concentrations, it tastes bitter and salty with a slightly sour taste.

The quality of the salty taste of foods depends on the ratio of sodium cations (Na^+) and chloride anions (Cl^-). Foods with natural levels of these ions, however, do not always taste salty, because they may not be present in the required stoichiometric ratio. The quality of salty taste of a mixture of salty substances depends on their type and mutual ratio, which is employed to compose the salt substitutes used in various diets. The intensity of the salty taste depends on the concentration of salty substances and the presence of other components in the mixture. The salinity of various salts is an additive property, but some mixtures exhibit synergism, which means that the intensity of a mixture of salty compounds is higher than the sum of salty tastes of its components. The threshold concentrations of the most common salty substances are listed in Table 8.37.

Table 8.37 Detection threshold concentrations of selected salty inorganic salts.

Salt		Threshold concentration (mg/l)
Sodium chloride	NaCl	160–240, 584, 296–839,[a]1750
Potassium chloride	KCl	1267, 345–1270[b]
Ammonium chloride	NH_4Cl	214
Magnesium chloride	$MgCl_2$	1430

[a]In men, higher by about 50% compared with women.
[b]In men, higher by about 40% compared with women.

Table 8.38 Sodium chloride contents in edible portions of selected foods.

Food	NaCl (g/kg)	Food	NaCl (g/kg)
Meat (beef, pork, poultry)	1.82–2.02	Rice	0.15
Meat products	12–50	Lens	0.15
Fish	0.98–5.50	Fresh fruits	0.04–0.08
Milk	0.88–2.25	Fresh vegetables	0.05–1.55
Dairy products	0.83–80	Canned vegetables	5.0–24.5
Butter	1.30	Potato	0.48
Eggs	1.50	Beer	0.13
Flour	0.050	Wine	0.05–0.08
Bread and pastries	0.55–10.4	Nuts	0.05–0.75

In practice, the most important salty compound is sodium chloride, which occurs in lower or higher quantities in virtually all foods as a natural constituent or an intentionally added additive. The content of NaCl in selected foods is given in Table 8.38. According to the sodium content (salt contains 38.4% sodium), which roughly corresponds to the NaCl content, foods can be divided into four categories:

- foods with very low Na content (usually containing a thousandths of a gramme, but always less than 0.4 g of sodium per 1 kg food), such as fruits, fresh vegetables, most fats, sugar, confectionery, and some dairy products;

- foods with low Na content (0.4–1.2 g/kg), such as fresh meat, fish, poultry, milk and dairy products (except hard and processed cheeses), and some edible fats;

- foods with high Na content (1.2–4.0 g/kg), such as some types of bread and pickles;

- foods with very high Na content (more than 4.0 g/kg), such as smoked meat products, hard and processed cheeses, some bakery products, dried soups, olives and vegetables in brine, salty snacks, and similar products.

Poultry products usually have a lower NaCl content than pork meat products; the highest content of NaCl is found in uncooked and raw smoked meat, dry or hard salami, some cheeses, and fish. For example, the NaCl content of dry pork salami may be 58 g/kg, whilst beef jerky contains 50 g/kg NaCl. Salted fish has an extremely high salt content (dry and salted cod contains 180 g/kg Na), so before eating it is necessary to remove NaCl by dipping in water. With regard to dairy products, the highest content of NaCl is found in cheeses (Parmesan 40 g/kg, blue cheese 36 g/kg, feta 28 g/kg, Camembert 21 g/kg, Cheddar 16 g/kg). Egg white and yolk contain NaCl as a natural component at levels of 0.25 and 12.5 g/kg, respectively. Examples of fruit products with high salt content are olives pickled in salty brine. Fermented olives have an NaCl content of 60–90 g/kg, black unfermented olives only 10–30 g/kg. Various paste-type soup products have even higher sodium chloride contents, such as broths (about 650 g/kg), soy sauces (about 180 g/kg), and acid protein hydrolysates used as soup seasonings (200 g/kg).

Sodium added to food is not always in the form of NaCl. Common food additives, such as baking soda, some preservatives, and monosodium glutamate also contribute to the total amount of sodium consumed (see Section 6.3.1).

8.3.2.2 Physiology, nutrition, and use

Salty substances exhibit a variety of pharmacological effects, whose character depends on the type of cation and anion. Some substances are toxic at higher concentrations. The compound consumed in the largest amount is sodium chloride. The daily intake of salt in developed countries is estimated at 8–15 g. Sodium chloride supports the perception of taste of foods at the required intensity and fullness, stimulates not only receptors for salty taste but also (and significantly) the perception of the sweet taste of sucrose and some other sweet substances, as well as sour taste perception, suppresses the sensation of metallic taste and some other off-tastes and the feel of diluted (watery) taste, optimises the resulting taste sensation, and promotes balance between various basic tastes.

Sodium chloride is essential for the human body. Excessive intake of NaCl, however, causes fluid retention and swelling, burdens the kidneys, heart, and blood circulation, and contributes to the emergence of hypertension. In industrialised countries, the consumption of NaCl is excessive in the majority of the population. It is recommended that it should be reduced below 10 g per person per day (see Section 6.3.1). Some diseases (hypertension, renal insufficiency, and oedema) require a diet with a limited supply of salt or that is completely salt-free. In these cases, the palatability of food is achieved by recipe modification, especially through the addition of spices or salt substitutes, in which the Na^+ cation is replaced by other cations, in particular the K^+ cation.

Sodium chloride is added to food for the following reasons:

- to achieve the desired organoleptic properties of food products and dishes;

- to improve the processing conditions (e.g. salt in the manufacture of bread strengthens gluten in the dough and thus contributes to the stability of dough during mechanical processing; in the manufacture of processed cheeses, NaCl, which is a part of the cheese-melting salts, displaces calcium from milk proteins; in the meat industry, NaCl increases the meat binding capacity – in sausages, it increases protein solubility and the emulsification of coagulated proteins with fat and water to the desired structure);

- to achieve conservation, which lies in the ability of NaCl to reduce water activity below the level required by the growth of undesirable microorganisms, which is known as the bacteriostatic effect;

- to control desirable fermentation processes by suppression of growth of undesirable microflora (e.g. during dough rising, cheese ripening, and lactic acid fermentation of olives, cucumbers, and cabbage).

Other saline substances have found application almost exclusively in salt substitutes, which are used in diets with reduced salt content (low sodium diets) for patients who have to radically reduce sodium intake. A salt-free diet is suitable in cases of failure of the excretion of sodium ions and water retention in the body, which results in massive swelling, especially of the feet (occurring during certain heart and kidney diseases and in pregnancy). The main salt substitute is potassium chloride, the properties of which are similar to those of sodium chloride. This suits quite well in terms of technological and preservative properties, but less so in terms of taste, because it has a strong bitter taste in addition to its salty one. Potassium chloride is therefore used in mixtures with other salty compounds to correct its off-taste. The cations Mg^{2+}, Ca^{2+}, and NH_4^+ are also used in salt substitutes in various combinations with inorganic and organic anions (chlorides, sulfates, phosphates, citrates, succinates, lactates, acetates, and formates). Some dipeptides also have a salty taste (see Section 2.3.3.2), and monosodium glutamate is often used as a salty taste enhancer. Compounds used as substitutes for table salt are often very complex and are subject to a number of patents.

8.3.3 Sour substances

The acidity of a food is related to the amount of undissociated and dissociated forms of carboxylic acids and oxonium ions, respectively. The major substances that give a sour taste in foods are undissociated hydroxycarboxylic acids, mainly citric and malic acids. Often, however, other carboxylic acids occur, such as ascorbic acid in most types of fruits, tartaric acid in grapes, isocitric acid in blackberries, oxalic acid in rhubarb, and lactic acid in some dairy products (such as yoghurt), fermented cucumbers, cabbage, and olives. Acetic acid occurs in vinegar and propionic acid in Emmental-type cheeses. The acidity of cola drinks is provided by phosphoric acid, sometimes accompanied by citric or other acids.

8.3.3.1 Sour taste quality and intensity

Individual acids differ in the nature of their sour taste, and often in the quality. Threshold concentrations of sour taste perception of different acids also differ to a certain extent. They vary over a rather wide range depending on a number of subjective and objective factors, such as sensory analysis methodology, quality of respondents, and quality of water (Table 8.39).

Table 8.39 Detection thresholds of selected acids in water according to various sources.

Acid	Detection threshold (mg/l)	Acid	Detection threshold (mg/l)
Acetic	19.9,[a] 54.1,[b] 110, 175[c]	Succinic	556
Lactic	15.1,[a] 133,[b] 200	Malic	110
Pyruvic	22.6,[a] 47.2[b]	Tartaric	80
Oxalic	18.6[a]	Citric	26.4,[a] 100–130, 150, 350[b]

[a] In water, pH 4.3.
[b] In water, pH 6–7.
[c] In beer.

Less important for the perception of sour taste are hydrogen ions (H^+) and, correspondingly, oxonium cations (H_3O^+) resulting from the dissociation of acids:

$$R-COOH + H_2O \rightarrow R-COO^- + H_3O^+$$

The number of oxonium ions, related to the pH of the biological system, is an important criterion as it affects the redox potential of the system, ongoing enzymatic and chemical reactions, growth of microorganisms, and odour, taste, and colour of the food. The pH value depends on the concentration of the acids, their dissociation constants, and the degree of neutralisation of the acids by basic components. For various reasons (microbiological, technological, and other), it is useful to divide foods into:

- very acidic (pH < 4.0);

- less acidic (pH 4.0–6.5);

- non-acidic (pH > 6.5).

Fruit is, with few exceptions, always very acidic, and the maximum acid levels in fruits occur in the period before full maturity. The pH of fruit juices is generally lower than 4.0, but in overripe pears, cherries, and peaches it may increase up to pH 4.5, and in overripe elderberries it may reach a value of 4.7. The level of acids in a fruit depends on its type, but is usually 10–30 g/kg, with lower amounts found in pears (about 1 g/kg) and higher amounts in citrus fruits (about 80 g/kg). A sour taste is often modified by the presence of carbohydrates, tannins, ethanol, or various cations and other substances. Carbohydrates weaken the sour taste of acids, whilst tannins and alcohol emphasise it.

Fresh vegetables are, in comparison with fruits, relatively less acidic. The pH values of common vegetables vary, usually from 5.0 to 6.6. An exception is rhubarb, where the pH is around 3.2; tomatoes are also quite acidic, with a pH of about 4.3. The content of acids in fresh vegetables usually ranges between 2 and 4 g/kg.

Meat and other animal products are even less acidic (pH 6.6–7.2). Immediately after slaughter, glycogen in muscles is broken down and the acidity of the meat increases somewhat due to a temporary increase in lactic acid content (pH of about 5.8), although it never reaches the acidity of very acidic food (pH < 4.0).

Fresh milk is a non-acidic food (pH varies in the narrow range of 6.50–6.75), whilst the pH of yoghurt and other fermented dairy products ranges from 4.0 to 4.2. The pH value of a typical hard cheeses is about 5.1.

Chicken egg white is completely non-acidic, but during storage the pH increases from its original value of 7.5 to 9.0 or higher. Egg yolk is more acidic, with a pH around 6.2.

8.3.3.2 Physiology, nutrition, and use

Organic acids, hydrogen salts, and some mineral acids and hydroxycarboxylic acid lactones have found use as acidifying agents and acidulants for the modification of pH, but they are also used as antimicrobial agents. The addition of some salts and alkaline agents, on the other hand, reduces acidity (e.g. the acidity of cocoa beans during roasting). The use and medical evaluation of these compounds are described in Section 11.3.3.

8.3.4 Umami and kokumi tastes

The term **umami** is relatively new in the West, but it is not new to the Japanese, who have used it since the early 1900s to describe a delicious, full-bodied, savoury, or meaty taste that brings a sense of satisfaction and increased viscosity in the mouth. It is translated from Japanese as 'savoury' or 'yummy'. Being able to distinguish the umami taste takes some practice, because it is not as obvious as the other basic tastes, such as sweet and bitter. Umami is caused mainly by glutamic acid and its salt, sodium hydrogen glutamate, which is the predominant form of glutamic acid at pH values ranging from 4.3 to 9.5. The taste intensity of glutamate is quite strong. The detection threshold is 120–170 mg/l, with a recognition threshold of 300 mg/l. The word corresponding to umami in Chinese is *xian wei*.

Umami, spicy, and meaty tastes are also found in a number of substances derived from glutamic acid (**8-133**). For example, umami taste in potato chips and fermented corn sauce displayed stereoisomers of pyroglutamic acid. Glycoconjugates of glutamic acid produced in the Maillard reaction, such as *N*-glycoside and *N*-(D-glucos-1-yl)-L-glutamic acid, and the corresponding Amadori compounds *N*-(1-deoxy-D-fructos-1-yl)-L-glutamic acid and *N*-(1-carboxyethyl)-6-hydroxymethylpyridinium-3-ol inner salt (alapyridaine), were identified in heated sugar/amino acid mixtures and in beef bouillon. (*S*)-Malic acid 1-*O*-D-glucopyranoside, known as (*S*)-morelid, contributes to the umami taste of morel mushrooms (*Morchella deliciosa*, Morchellaceae), whilst (*R*)-2-(carboxymethylamino)propanoic acid, known as (*R*)-strombine, is an umami compound isolated from dried scallops.

Umami compounds also include many small peptides such as Glu-Glu-Leu (and lactoyl-Glu), which have been found in Parmesan cheese, β-alanyl dipeptides found in bouillons (see Section 2.3.3.1.3), octapeptide Lys-Gly-Asp-Glu-Glu-Ser-Leu-Ala isolated from beef broth, and peptides separated from soy sauce (such as Leu-Pro-Glu-Glu-Val, Ala-Gln-Ala-Leu-Gln-Ala-Gln-Ala, and Glu-Gln-Gln-Gln-Gln). By the end of 2016, 52 peptides had been reported to show umami taste, including 24 dipeptides, 16 tripeptides, 5 octapeptides, 2 pentapeptides, 2 hexapeptides, 1 tetrapeptide, 1 heptapeptide, and 1 undecapeptide.

In the past, sodium hydrogen glutamate was linked with the so-called Chinese-restaurant syndrome, which is manifested by fairly typical and unspecific sensations such as headache, flushing, sweating, facial pressure or tightness, numbness, tingling or burning in the face, neck, and other areas, chest pain, and nausea. Latterly, sodium hydrogen glutamate has been classified as a food ingredient that is generally recognised as safe, but its use as a flavour enhancer remains controversial. For this reason, when this compound is added to foods, legislation requires that it be listed on the label.

Some other amino acids also show umami taste but have not found practical applications. An example is ibotenic acid (occurring in fly agaric, *Amanita muscaria*, and panther cap, *Amanita pantherina*, Amanitaceae), which has umami taste 5–30 times more intense than glutamate.

Kokumi is another food attribute identified by Japanese researchers. The literal translation from Japanese is 'rich' (*koku*) 'taste' (*mi*). It is sometimes translated as 'heartiness' or 'mouthfulness', and it describes compounds in foods that do not have their own flavour or a distinct flavour, but which enhance the flavours with which they are combined by triggering calcium-sensing receptors in the tongue. In foods system, there are three types of flavour sensation attributed to kokumi:

- mouthfulness and continuity (long-lasting taste development);

- punch (initial taste and impact);

- mildness (roundness and balance).

Kokumi sensation appears to be related to a number of γ-glutamyl peptides, for example glutathione, γ-Glu-Val-Gly in soy sauces, and γ-glutamyl peptides in matured cheeses. Examples of these peptides are γ-Glu-Gly, γ-Glu-Leu, γ-Glu-Glu, γ-Glu-Gln, γ-Glu-Met, and γ-Glu-His in Gouda chesse and γ-Glu-Gly, γ-Glu-Ala, γ-Glu-Thr, γ-Glu-Asp, γ-Glu-Lys, γ-Glu-Glu, γ-Glu-Trp, γ-Glu-Gln, and γ-Glu-His in Parmesan cheese. Another example of a mouthfulness (kokumi)-enhancing molecule is the bitter hydrophobic compound (12*Z*,15*Z*)-1-acetoxy-2-hydroxy-4-oxoheneicosa-12,15-diene (see **8-136**), identified in thermally processed avocado fruits.

A special group of taste-active compounds is the umami and kokumi enhancers used as food additives. Examples of these substances are 5′-nucleotides, such as inosine 5′-monophosphate (5′-IMP, inosinic acid disodium salt) and guanosine 5′-monophosphate (5′-GMP, about twice as active in the enhancement of the umami taste of glutamate solutions as 5′-IMP), as well as the less active xanthosine 5′-monophosphate (5′-XMP) and adenosine 5′-monophosphate (5′-AMP), which show only about 0.5 and 0.1 times the impact of 5′-IMP, respectively (see Section 11.3.5). Umami-enhancing molecules are also particular derivatives of 5′-nucleotides arising in the Maillard reaction of 5′-GMP. The umami-enhancing nucleotide diastereomers (*R*)- and (*S*)-N^2-(1-carboxyethyl)guanosine 5′-monophosphate (**8-133**) is derived from the reaction of guanosine 5′-monophosphate (5′-GMP) with reactive sugar breakdown products such as glyceraldehyde and dihydroxyacetone in yeast extract. Another compound is lactamide N^2-lactoylguanosine 5′-monophosphate, found in dried bonito. It is formed in the reaction of 5′-GMP with lactic acid.

Another group of taste enhancers arises from bitter-tasting creatinine and reducing hexoses in the Maillard reaction. Identified compounds include *N*-(1-methyl-4-oxoimidazolidin-2-ylidene)aminoacetic acid derived from glycine, *N*-(1-methyl-4-oxoimidazolidin-2-ylidene)aminopropionic acid derived from alanine, and *N*-(1-methyl-4-oxoimidazolidin-2-ylidene)amino-4,5,6-trihydroxyhexanoic acid (**8-133**) derived from glucometasaccharinic acids, such as 3-deoxy-D-gluconic acid. At subthreshold concentrations, these substances and some other synthesised *N*-(1-methyl-4-oxoimidazolidin-2-ylidene)-α-amino acids enhanced the typical thick-sour and mouth-drying sensation and the mouthfulness imparted by stewed beef juice.

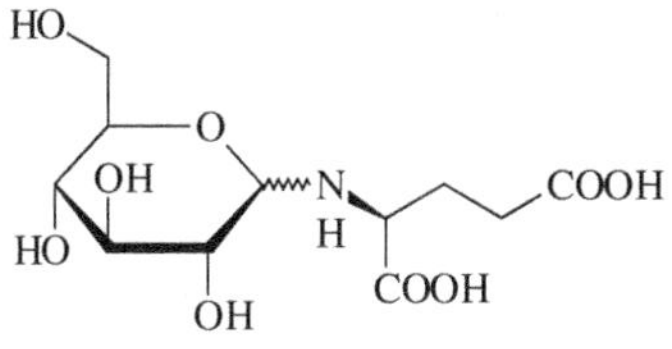

N-(D-glucos-1-yl)-L-glutamic acid

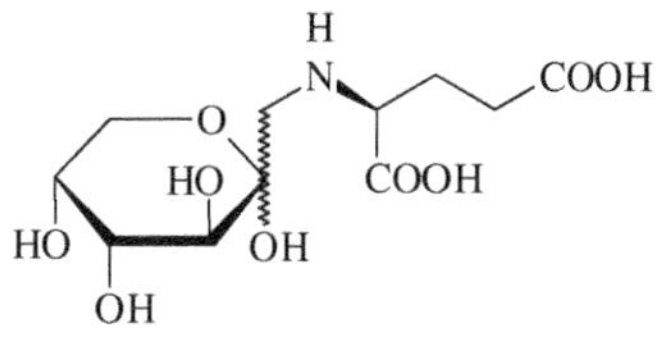

N-(1-deoxy-D-fructos-1-yl)-L-glutamic acid

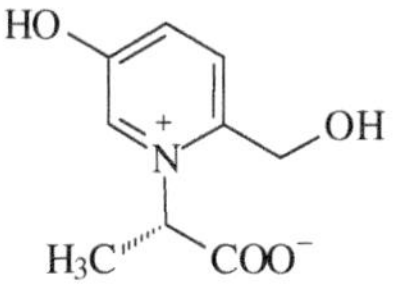

(+)-(S)-alapyridaine

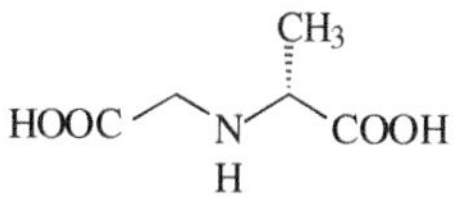

(R)-strombine

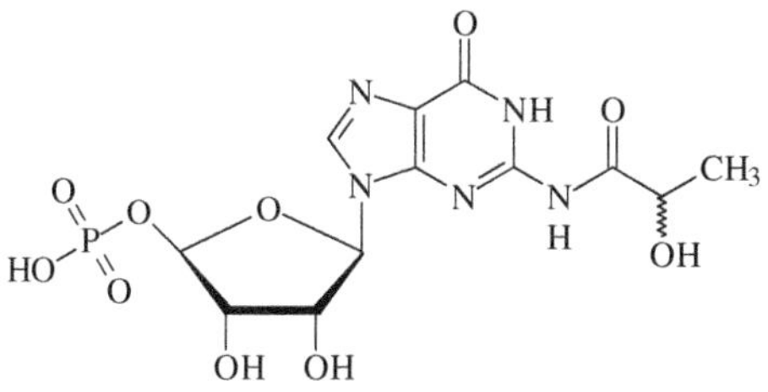

N^2-lactoylguanosine 5′-monophosphate

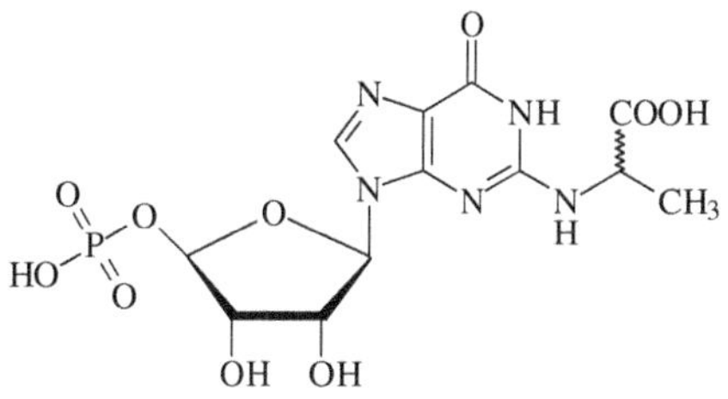

N^2-(1-carboxyethyl)guanosine 5′-monophosphate

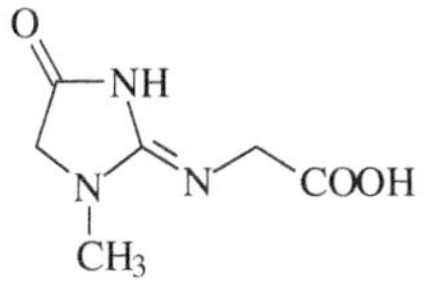

N-(1-methyl-4-oxoimidazolidin-2-ylidene)
aminoacetic acid

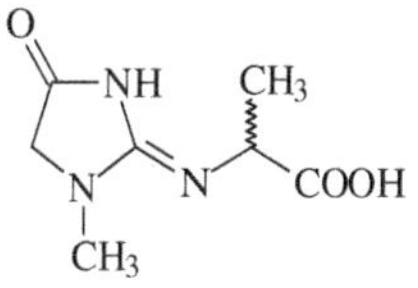

N-(1-methyl-4-oxoimidazolidin-2-ylidene)-
aminopropionic acid

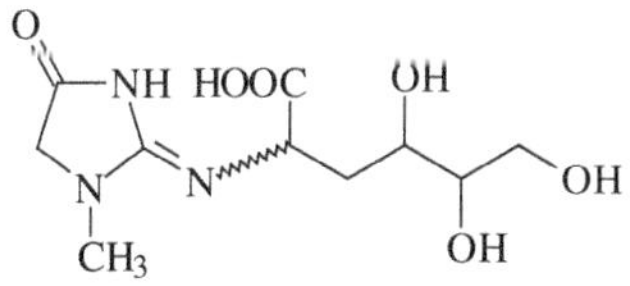

N-(1-methyl-4-oxoimidazolidin-2-ylidene)amino-
4,5,6-trihydroxyhexanoic acid

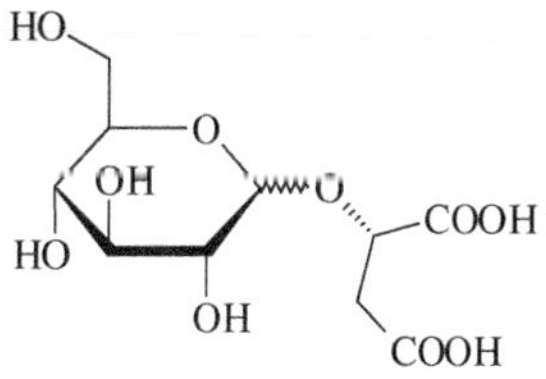

(S)-morelid

8-133, compounds with umami taste and taste enhancers

8.3.5 Bitter substances

Bitter substances of foods are usually classified according to origin into compounds that:

- are characteristic natural components of certain foods and whose occurrence is genetically determined;

- are formed during processing and storage of foods by chemical reactions or activities of their own enzyme systems;

- result from contamination by certain microorganisms growing on food materials;

- have been intentionally added to foods as bitter additives.

A number of organic compounds commonly present in foods have a bitter taste, such as certain fatty acids, amino acids, peptides, amines, amides, ketones, nitrogen-containing heterocyclic compounds (including alkaloids), and many others. Certain inorganic salts are also bitter.

A bitter taste is desirable for some foods, where it is a typical taste. Examples include grapefruits, chicory buds, cocoa, coffee, beer, and tonic drinks. However, sometimes a bitter taste is considered undesirable (off-flavour); affected foods may then have an unacceptable taste and can even be inedible (e.g. oranges, carrots, cottage cheese).

Bitterness is related to the hydrophobicity of the molecules of the bitter compound, the size of the non-polar part, and the molecular configuration; additionally, the presence of at least one polar functional group is required. Data on bitter compounds can be obtained using sensory analysis under defined conditions. They are often referred to a bitter taste standard, which is usually quinine or caffeine (which is about 60 times less bitter than quinine). Compounds with threshold concentrations lower than 0.1 mol/l are generally considered to be very bitter. Table 8.40 gives some examples of threshold concentrations of selected bitter substances.

8.3.5.1 Bitter substances naturally present and formed during food processing and storage

8.3.5.1.1 Fruits

Bitter taste may be a natural feature of some types of fruit, but most are not bitter in the time of full maturity. Unripe fruit may be hard, sharp, bitter, astringent, and not tasty, whilst ripe fruit can be juicy, sweet, and delicious.

The typical natural bitter taste of certain citrus fruits and juices is caused by the presence of flavanone-7-glycosides, which have as a sugar component disaccharide neohesperidose and are therefore known as **neohesperidosides**. Aglycones are not bitter. The bitter component of grapefruits is naringin (naringenin-7-O-neohesperidoside). Some other bitter odour-active compounds also contribute to the bitter taste of grapefruits, such as the sesquiterpenic ketone nootkatone. The bitter constituent of bitter oranges, also known as bigarade oranges (*Citrus aurantium* subsp. *amara*), is neohesperidine (hesperetin-7-O-neohesperidoside).

A bitter taste also often occurs in otherwise sweet oranges during juice processing and storage, due to the presence of metabolically altered (degraded) triterpenoids (tetranortriterpenoids) known as **limonoids**. The basic structure consists of four six-membered rings and one furan ring. These secondary metabolites occur as aglycones, glucosides, or A-ring lactones. More than 300 limonoids are known, which are found in the rue (citrus) family Rutaceae and in some other plant families. The most abundant aglycone and glucoside for most citrus species are limonin and limonin 17β-D-glucopyranoside with the sugar moiety bound to the C-17 hydroxyl group (the aglycone is called limonoic acid lactone or limonate A-ring lactone). These compounds are biosynthesised by transformation (oxidation, isomerisation, rearrangement, and elimination of four carbon atoms of the side chain) of 4,4-dimethylsterol euphol via butyrospermol and other metabolites, such as non-bitter

Table 8.40 Detection threshold concentrations of selected bitter compounds according to various sources.

Compound type	Bitter compound	Detection threshold (mmol/l)	Compound type	Bitter compound	Detection threshold (mmol/l)
Inorganic salts	Magnesium sulfate	5	Flavonoids	(+)-Catechin	2
Amides	Propionamide	52		Naringin	0.2 (0.04)
Amines	Propylamine	20		Procyanidin B₃	0.07
	Butylamine	6	Diarylheptanoids	Asadanin	0.013
Amino acids	L-Tryptofane	4	N-Heterocycles	Pyridine	2
	L-Leucine	15		Pyrazine	13
Fatty acids	Oleic acid	9–12		Methylpyrazine	7
	Elaidic acid	22		Acetylpyrazine	1
	Linoleic acid	4–6		Imidazole	6
	α-Linolenic acid	0,6–1,2	Limonoids	Limonin	0.08 (0.003)
	γ-Linolenic acid	3–6	Phloroglucinols	Iso-α-acids	0.02
	Arachidonic acid	6–8		Tetrahydroiso-α-acids	0.007
Alkaloids	Quinine	0.03 (0.025, 0.01, 0.008)	Quassinoids	Quassin	0.0002
	Caffeine	2 (0.7, 0.02)	Secoiridoids	Amarogentin	0.000 03
Cucurbitacins	Cucurbitacin C	0.000 2	Diterpenoids	Mozambioside	0.06

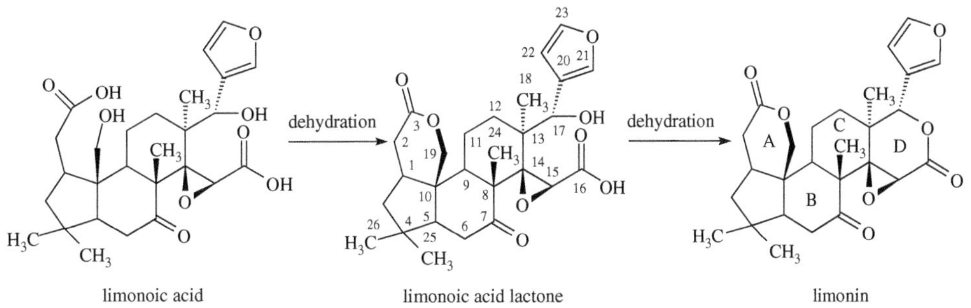

Figure 8.93 Formation of limonin from non-bitter precursor.

limonoic acid and its lactone (ring A). The biosynthesis takes place in the leaves, and limonoids are then transported to the fruits and seeds in particular. Limonin glucoside is stable in neutral media, but hydrolysis of the sugar moiety in acidic juices under the action of β-glucosidase and dehydration (to form another δ-lactone in ring D) yield intensely bitter limonin (Figure 8.93) and its epimer, which is formed through the inversion of the C-17 substituent. The formation of limonin is accelerated by juice pasteurisation. The bitter taste is reflected at a limonin content higher than 6 mg/l. Limonin can also be found in lemons and grapefruits.

Several dozen limonin-related compounds (**8-134**) have been found in citrus fruits. All citrus limonoids contain a furan ring attached to ring D at C-17 and oxygen-containing functional groups in positions C-3, C-4, C-7, C-16, and C-17, as well as C-14,15 of the epoxy ring. In addition to limonin, only three other limonoids – nomilin, ichangin, and nomilinoic acid – are bitter compounds. Analogously to limonin, nomilin, nomilinoic acid, and other limonin-related compounds occur in citrus fruits as 17β-D-glucopyranosides. Of these, only nomilin (which is about twice as bitter as limonin, but occurs in smaller quantities) may be involved in the bitter taste in citrus juices, together with limonin. Limonoids are also reported to possess multiple health-promoting properties.

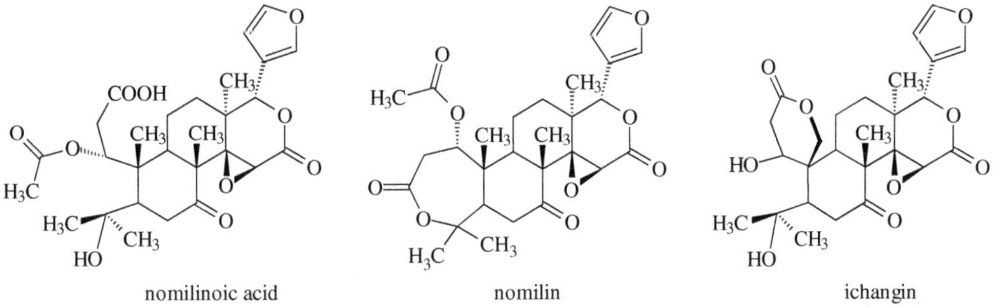

nomilinoic acid   nomilin   ichangin

8-134, minor bitter limonoids of oranges

Olives – fruits of the olive tree (*Olea europea*, Oleaceae) – have a markedly bitter taste, because they contain a group of **secoiridoids** (see Section 9.10) and their β-D-glucosides, known as **oleosides** (**8-135**), which protect against attack by microorganisms and insects. Small green olives contain the so-called methyl oleoside (also known as oleoside-11-methyl ester or elenolic acid glucoside, hydrolysis of which gives rise to the oleoside demethylelenolic acid glucoside). Small green olives also contain ligstroside, which is a derivative of tyrosol and methyl oleoside. During fruit growth and maturation, ligstroside is oxidised (hydroxylated) to oleuropein, a compound derived from 3,4-dihydroxyphenylethanol (3-hydroxytyrosol); in olives that reach normal size, ligstroside does not occur. Oleuropein is found in green and yellow fruits, where its concentration can reach up to 140 mg/kg (d.w.). Its content decreases during fruit ripening and is almost zero in the fully mature blue fruits. Ripe fruits contain only demethyloleuropein formed by hydrolysis of oleuropein. Hydrolysis by esterase yields hydroxytyrosol and elenolic acid glucoside, which is then hydrolysed by β-glucosidase to elenolic acid and glucose. Hydrolysis of oleuropein by β-glucosidase yields glucose and the corresponding aglycone, which is hydrolysed by esterase to 3-hydroxytyrosol and elenolic acid. Olive leaves and fruits contain structurally related (*E*)-elenolide recognized as an antihypertensive agent. Elenolide is present in olive oils in quantities up to about 2800 mg/kg. Tyrosol and hydroxytyrosol also contribute to the bitter taste of olives. Hydroxytyrosol is likewise present in olives as 4-β-D-glucopyranoside. Its amount increases with increasing ripeness. A related tyrosol ester, a phenylethanoid oleocanthal, is found in extra virgin olive oil. Concentrations of oleacanthal (which shows anti-inflamatory properties) in extra virgin olive oil range from 22 to 190 mg/kg. Oleocanthal is accompanied by oleacein, a 3-hydroxytyrosol derivative that is also found in green vegetables, tea, herbs, and spices. Oleuropein aglycon, ligstroside aglycon, decarboxymethyl oleuropein aglycon, decarboxymethyl ligstroside aglycon, oleacein, and elenolic acid show bitterness and pungency (a slightly tangy, peppery bite in the back of the throat). Oleacein is considered the most powerful antioxidant in olive oil.

Debittering of immature green olives is achieved either by lactic acid fermentation or by hydrolysis of bitter glycosides through dipping in solutions of sodium hydroxide. In alkaline media, bitter oleuropein is hydrolysed to glucose and the corresponding aglycone, which yields 3-hydroxytyrosol (3,4-dihydroxyphenylethanol), methanol, and demethylelenolic acid.

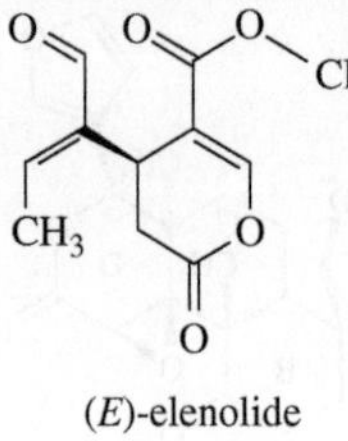

(E)-elenolide

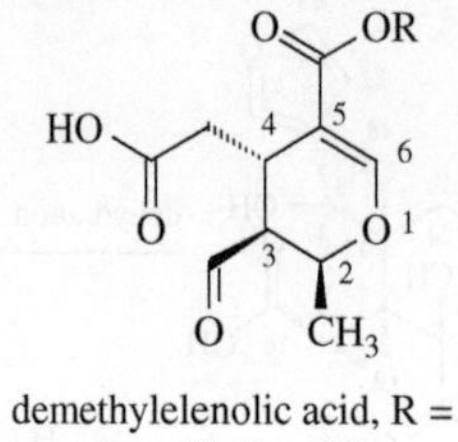

demethylelenolic acid, R = H
elenolic acid, R = CH$_3$

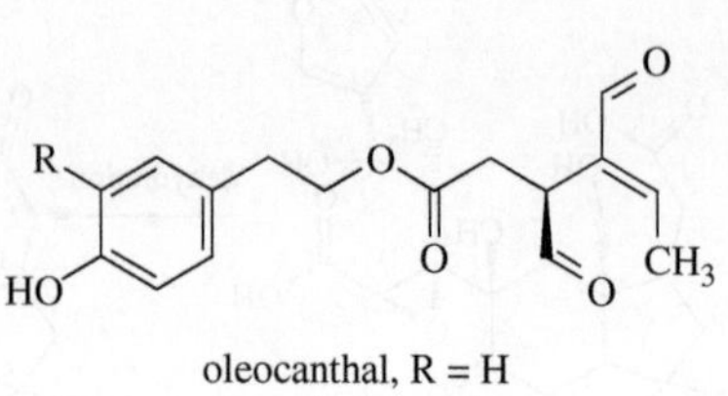

oleocanthal, R = H
oleacein, R = OH

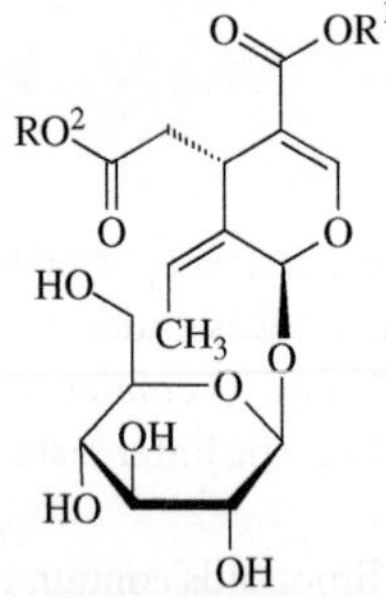

oleoside, R^1 = R^2 = H
methyl oleoside, R^1 = CH$_3$, R^2 = H

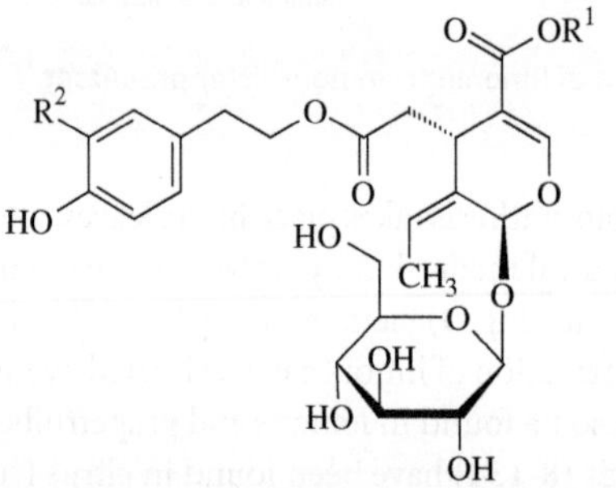

ligstroside, R^1 = CH$_3$, R^2 = H
oleuropein, R^1 = CH$_3$, R^2 = OH
demethyloleuropein, R^1 = H, R^2 = OH

8-135, oleosides and related compounds of olives

Raw avocado (*Persea americana*, Lauraceae) contains only trace levels of bitter compounds, but heat treatment (canning and preserving fruit) or air-drying can induce the development of an unpleasant bitter after-taste. The substances responsible for the bitter taste are various saturated and unsaturated C$_{17}$–C$_{21}$ fatty acid derivatives with 1,2,4-trihydroxy-(triols), 1-acetoxy-2,4-dihydroxy-(acetates of triols), and 1-acetoxy-2-hydroxy-4-oxo groups (acetates of oxodiols). The most bitter compound is fungicidal toxin, (12Z,15Z)-1-acetoxy-2-hydroxy-4-oxoheneicosa-12,15-diene, known as persin, closely followed by other structurally related compounds (**8-136**). The mixture of avocadyne and avocadene, two 17-carbon polyhydroxylated fatty alcohols (known as avocatin B), was recently demonstrated to possess potent anticancer activity. Natural constituents of avocado fruits of analogous structures are acetates of long-chain unsaturated fatty acids and furans with long-chain saturated and unsaturated substituents.

The bitter taste sometimes occurring in stewed sour cherries, plums, and other stone fruits of the Rosaceae family is caused by the presence of cyanogenic glycosides, especially prunasin (see Section 10.3.9.1). Rowan berries, the fruit of the rowan (*Sorbus aucuparia*), which are used to make jam or jelly with a distinctive bitter taste, contain 3-hydroxy-5-hexanolide β-D-glucoside (**8-137**). Enzymatic hydrolysis of this glycoside and dehydration of aglycone yields parasorbic acid; ring opening and further dehydration of parasorbic acid produces sorbic acid. Some selected rowan cultivars have less bitter or non-bitter fruits.

The ethyl ester of glucose, ethyl β-D-glucopyranoside, contributes to the bitterness of sea buckthorn (*Hippophae rhamnoides*, Elaeagnaceae) juice. Its content in the juice may vary between 0.6 and 19.8 g/l.

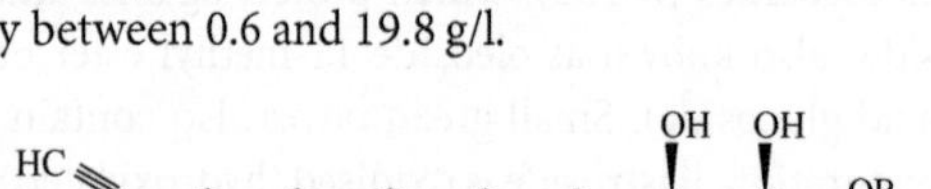

1,2,4-trihydroxyheptadeca-16-yne (avocadyne), R = H
1-acetoxy-2,4-dihydroxyheptadeca-16-yne, R = COCH$_3$

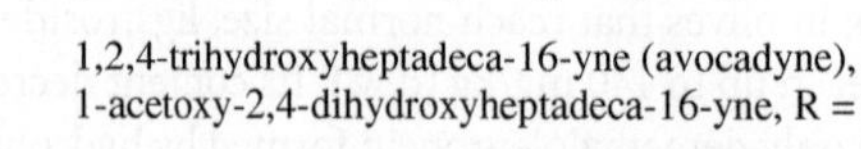

1,2,4-trihydroxyheptadeca-16-ene (avocadene), R = H
1-acetoxy-2,4-dihydroxyheptadeca-16-ene, R = COCH$_3$

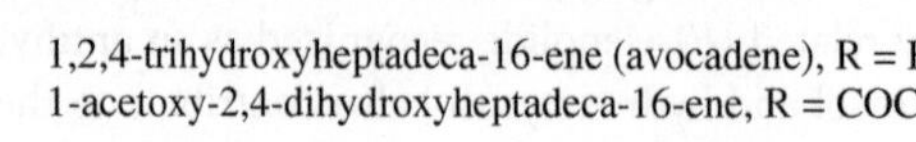

1-acetoxy-2-hydroxy-4-oxoheptadeca-16-ene

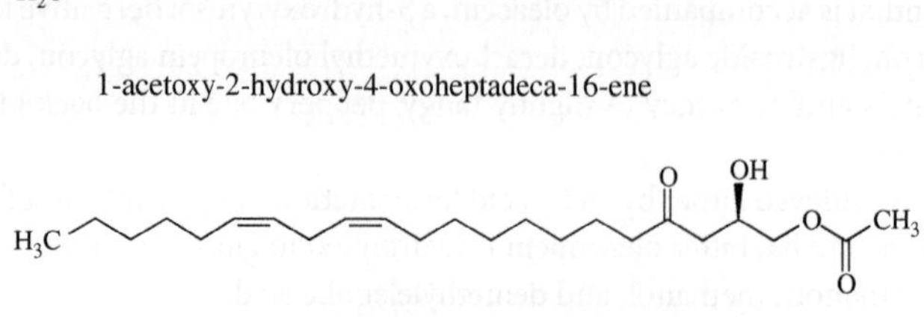

(12Z,15Z)-1-acetoxy-2-hydroxy-4-oxoheneicosa-12,15-diene

8-136, bitter compounds in avocado

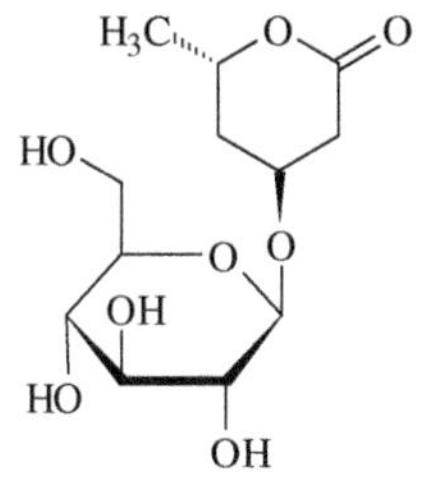

8-137, (3*S*,5*S*)-3-(β-D-glucopyranosyloxy)-
hexano-5-lactone

8.3.5.1.2 Vegetables

A number of vegetables from the family Asteraceae have a bitter taste, such as lettuce (*Lactuca sativa*), particularly the stalk and the white milky juice (latex), and endive (*Cichorium endivia*), the aerial part (leaf rosette) of which is eaten as a salad. Chicory (*C. intybus* var. *foliosum*) is slightly bitter, and is cultivated for salad leaves (etiolated buds) called chicons, which grow from the root vertex in the dark. Roasted chicory root (*C. intybus* var. *sativum*) is used in the manufacture of coffee surrogates.

Bitter substances of these vegetables are primarily various sesquiterpenic lactones derived from sesquiterpenic hydrocarbon guaiane, known under the IUPAC name of (1*S*,3a*S*,4*S*,7*R*,8a*S*)-7-isopropyl-1,4-dimethyldecahydroazulene (**8-138**). Their presence within salad lettuce and chicory has considerable economic impact. The principal bitter sesquiterpene lactones found in species and cultivars of lettuce have been reported to be lactucin, 8- and 15-deoxylactucin, lactucopicrin, and their derivatives, such as 11,13-dihydroanalogues, esters, and glycosides (**8-138**). Major latex components are 15-oxalate and 8-sulfate conjugates of lactucin. The oxalates are unstable, reverting to the parent sesquiterpene lactones on hydrolysis, whilst the sulfates are stable, and 15-deoxylactucin-8-sulfate contributes most strongly to bitterness perception. Glycosides of sesquiterpenic lactones may also be important components in latex of various lettuca species. Example are picriside A (lactucin-15-glucoside) and crepidiaside A (8-deoxylactucin-15-glucoside), identified in aerial parts of *L. sativa*. Examples of 11,13-dihydroanalogues of sesquiterpene lactones are 11β,13-dihydrolactucin-8-*O*-sulfate (jaquinelin-8-*O*-sulfate) and cichorioside B, identified in *L. sativa* var. *capitata* (iceberg lettuce). One of the most toxic phytoalexins, lettucenin A, contributes significantly to the browning of lettuce. It was found in *Lactuca* spp. but not in chicory.

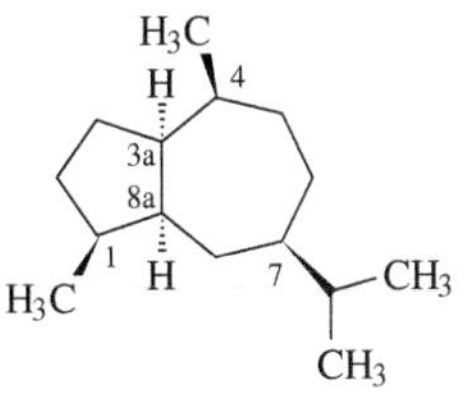

guaiane

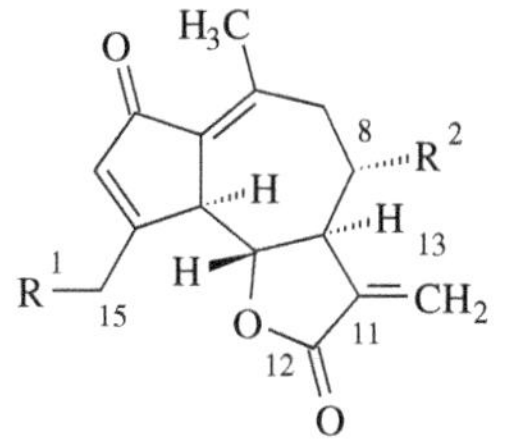

lactucin, R¹ = R² = OH
8-deoxylactucin, R¹ = OH, R² = H
15-deoxylactucin, R¹ = H, R² = OH

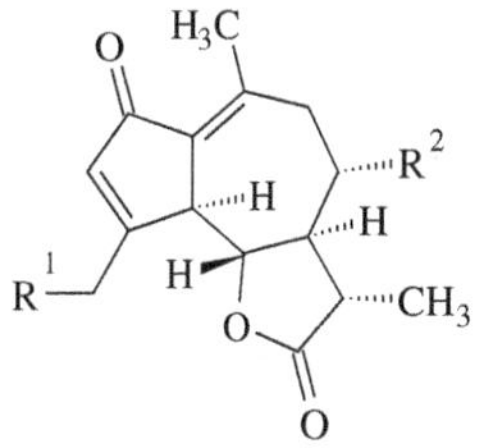

jaquinelin, R¹ = OH, R² = H
jacquinelin glycoside, R¹ = OGlc*p*, R² = H
jacquinelin-8-*O*-sulfate, R¹ = OH, R² = OSO₃H
cichorioside B, R¹ = OGlc*p*, R² = OH

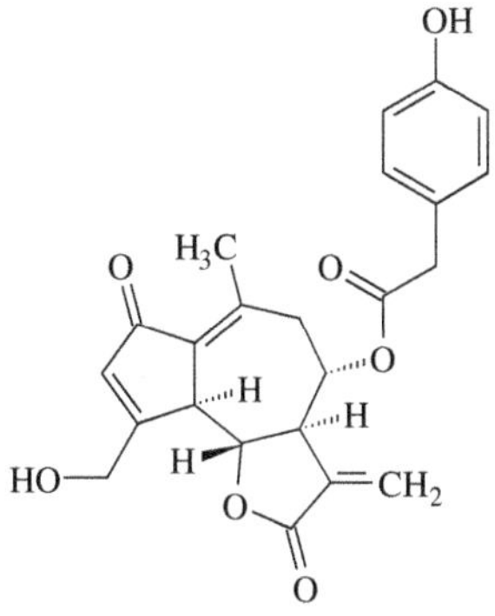

lactucopicrin

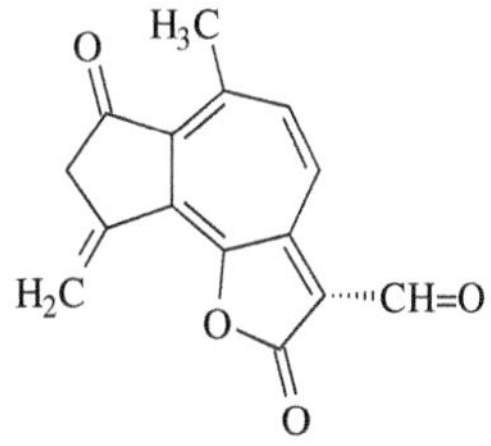

lettucenin A

8-138, guaiane and guaiane-derived bitter compounds in lettuce

In fresh chicory buds, the highest concentration of bitter substances (about 100–250 mg/kg) is present in the central axis, of which approximately 50% is lactucin, 30% 8-deoxylactucin, and 5% lactucopicrin. The 11,13-dihydro derivatives are present in small amounts. The washing procedure significantly reduces the sesquiterpene lactone content and bitterness in fresh-cut chicory. Profiles of sesquiterpene lactones change substantially during storage, revealing lactucopicrin to be the major one directly after washing, whilst 11β,13-dihydrolactucin prevails at the end of storage. The bitter taste of chicory buds also comes from coumarins and their glycosides (see Section 10.3.12.1), especially scopoletin, umbelliferone, aesculin (aesculetin 6-*O*-β-D-glucoside), and cichoriin (aesculetin 7-*O*-β-D-glucoside).

The main bitter substances of the leaves of the artichoke thistle, known as cardoon (*Cynara cardunculus*), and of cultured varieties of globe artichoke (*C. cardunculus* var. *scolymus*) of the aster (sunflower) family Asteraceae, are guaiane-derived sesquiterpene lactones, of which grosheimin (up to 19.5 g/kg f.w.) and 8-deoxy-11,13-dihydroxygrosheimin are the most abundant, followed by cynaropicrin, 11β,13-dihydrocynaropicrin, and cynaratriol (**8-139**). Cynaropicrin is about twice as bitter as quinine and is sometimes used as a bittering substance in citrus drinks.

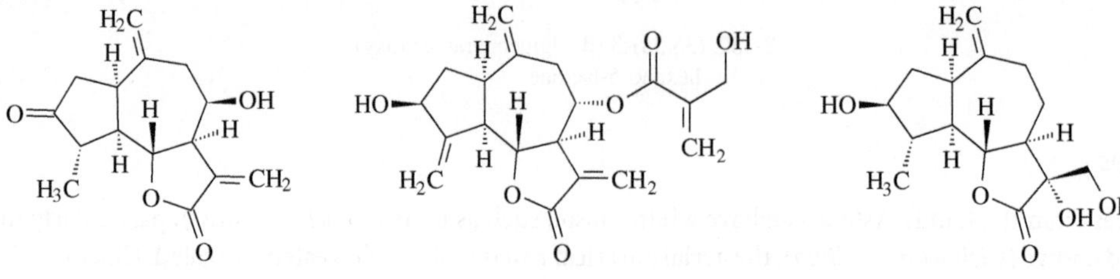

grosheimin   cynaropicrin   cynaratriol

8-139, bitter compounds in cardoon

Vegetables of the cucurbit family (Cucurbitaceae), which includes squashes, melons, pumpkins, cucumbers, and gourds of the genera *Benincasa, Citrullus, Cucumic, Cucurbita, Lagenaria,* and *Luffa,* occasionally show bitter taste in response to various stress factors. The bearers of bitter taste are tetracyclic triterpenoids, called **cucurbitacins**, which are derived from the triterpenic hydrocarbon 5α-cucurbitane (**8-140**). Cucurbitacins are found in vegetables in levels of <0.01% as free compounds (aglycones) or as glycosides. Examples are cucurbitacins A-I, the most common of which are cucurbitacins C and D. The main bitter substance of cucumber is cucurbitacin C, pumpkins contain cucurbitacins E and B (and sometimes D and I), and watermelons contain cucurbitacin E. Cucurbitacins A–I are found in cucumber seeds. Cucurbitacins and their glycosides are also found in plants belonging to several other families, such as Brassicaceae, Begoniaceae, Scrophulariaceae, Primulaceae, and Rosaceae. For example, under normal conditions, cucumbers contain cucurbitacins throughout the plant (in the stem, leaves, and cotyledonary leaves) but not in the fruit. The highest concentrations of cucurbitacin C in the bitter fruit of cucumbers are found in parts adjacent to the stem (up to 10 mg/kg) and just under the skin. Various stress factors, such as significant differences between night and day temperatures, waterlogging, and overdrying, however, manifest themselves with bitter fruits. Modern hybrid varieties of cucumbers do not synthesise cucurbitacins. Insecticidal, antitumour, and anti-inflammatory effects of cucurbitacins are currently being studied.

A special group of cucurbitacins is found in bitter melon (*Momordica charantia*, Cucurbitaceae), originated in India and now widely cultivated in Asia and Africa. Bitter melon is also known as bitter gourd or bitter squash and belongs to the most bitter of the edible fruits. Its taste is caused by a number of cucurbitane-type triterpenic glycosides called momordicosides, such as momordicosides K and L (**8-141**). Bitter melon is associated with a wide range of beneficial effects on health: anticancer, antiviral, anti-inflammatory, analgestic, hypolipidemic, and hypocholesterolemic effects.

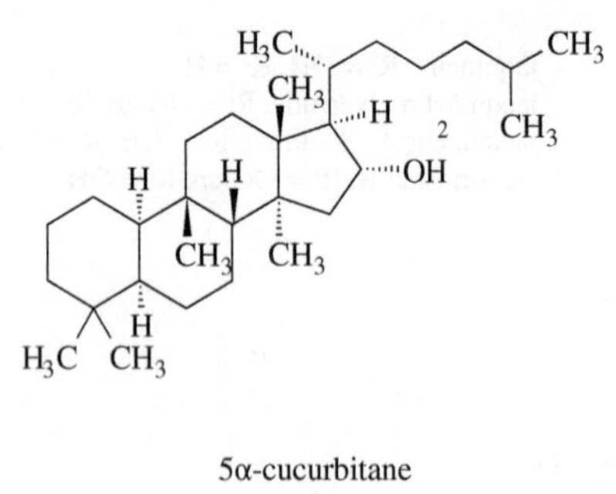

5α-cucurbitane

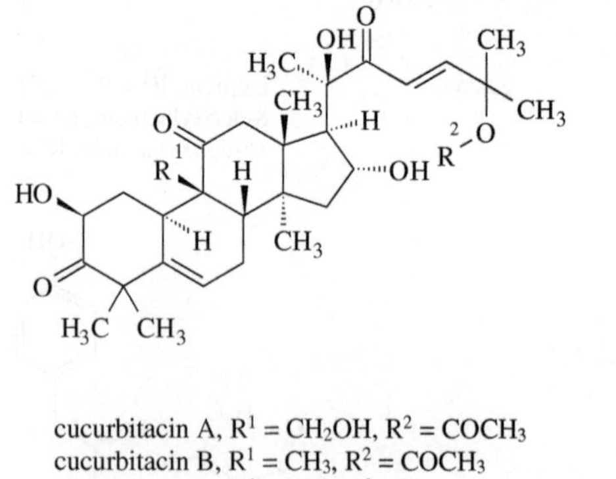

cucurbitacin A, R^1 = CH_2OH, R^2 = COCH_3
cucurbitacin B, R^1 = CH_3, R^2 = COCH_3
cucurbitacin D, R^1 = CH_3, R^2 = H

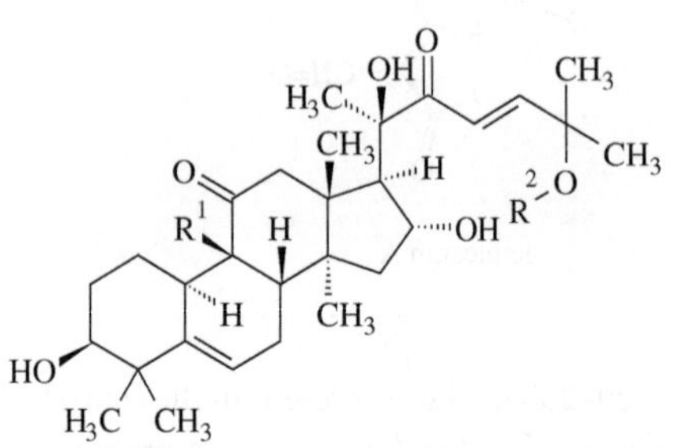

cucurbitacin C, R^1 = CH_2OH, R^2 = COCH_3

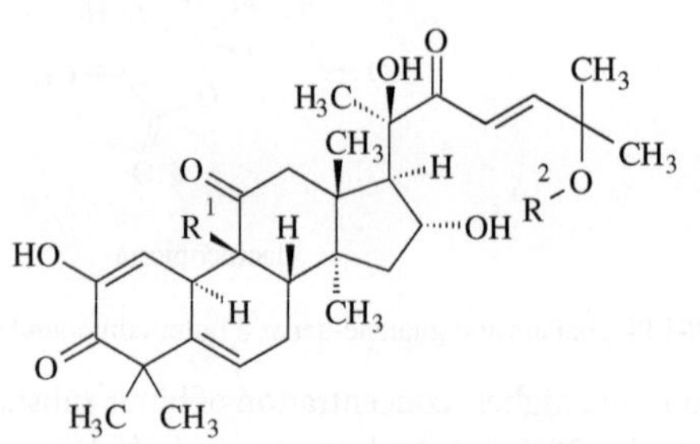

cucurbitacin E, R^1 = CH_3, R^2 = COCH_3
cucurbitacin I, R^1 = CH_3, R^2 = H

8-140, 5α-cucurbitane and derived cucurbitacins

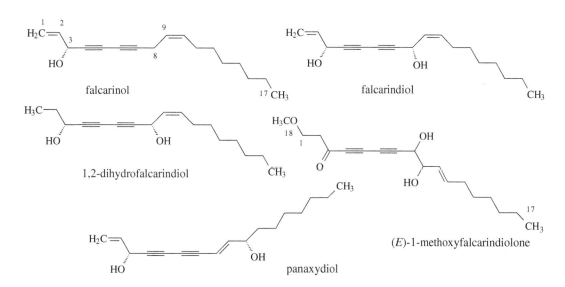

8-141, momordicoside K, R = CH₃
momordicoside L, R = OH

The bitter component of carrot seeds is 2,4,5-trihydroxybenzaldehyde. The main contributors to the bitter off-taste of carrot roots are polyacetylenic oxylipins. In the higher plants, more than 1400 polyacetylenes have been identified, 12 of which are found in carrots. The bitter C_{17} phytoalexin found in carrots, parsley, parsnip, fennel, celery, celeriac, and other umbelliferous (Apiaceae) vegetables, and also in ginseng (*Panax ginseng*, Araliaceae), is (3R,9Z)-1,9-heptadecadiene-4,6-diyn-3-ol, known as falcarinol, carotatoxin, or panaxynol (**8-142**). In carrots, falcarinol may occur in a concentration of about 290 mg/kg dry matter. The corresponding 3-ketone is a constituent of ginseng. Another polyacetylene of carrots is a product of falcarinol oxidation, (3R,8S,9Z)-heptadeca-1,9-dien-4,6-diyne-3,8-diol, known as falcarindiol. This diol occurs in about the same concentrations as falcarinol (240 mg/kg) and is accompanied by the corresponding 3- and 8-acetate, 1,2-dihydrofalcarindiol, its 3-acetate, 8- and 9-acetates of (E)-falcarindiolone and (E)-1-methoxyfalcarindiolone, and (3R,8E,10S)-panaxydiol (**8-142**). Fungal infection of carrot roots produces the bitter phytoalexin (R)-6-methoxymellein (8-hydroxy-6-methoxy-3-methyl-3,4-dihydroisocoumarin, **10-150**).

Falcarinol and falcarindiol were also found in parsnip in concentrations of 1600 and 5770 mg/kg dry matter, respectively, and in fennel in concentrations of 40 and 240 mg/kg, respectively. Falcarindiol, (3R,8S,9Z)-8-O-methylfalcarinol and panaxydiol occur in parsley at concentrations of 2320, 350, and 120 mg/kg, respectively. Celeriac contains up to 1620 mg/kg falcarinol, 2070 mg/kg falcarindiol, 170 mg/kg 8-O-methylfalcarinol, and 60 mg/kg falcarindiol.

falcarinol

falcarindiol

1,2-dihydrofalcarindiol

(E)-1-methoxyfalcarindiolone

panaxydiol

8-142, aliphatic polyacetylenes

Aside from their antibacterial and antifungal activities, aliphatic polyacetylenes of the falcarinol type have a number of other interesting bioactivities, including anti-inflammatory, anti-platelet-aggregatory, neuritogenic, serotonergic, and potential anticancer effects. A medicinal usage of pure polyacetylenes is not feasible, because of their pronounced chemical instability and their ability to induce allergic reactions. However, consumption of food containing polyacetylenes may have a chemopreventive benefit. Adverse effects due to an excessive intake of polyacetylenes with the human diet are not to be expected, because polyacetylenes have a bitter off-taste in high concentrations.

The bitter taste of some vegetables of the genus *Brassica* (Brassicaceae), such as kohlrabi and Brussels sprouts, which can occur in the cooked vegetables, is caused by the presence of goitrin (Figure 10.19) formed by non-enzymatic decomposition of glucosinolate progoitrin.

Bitter taste often occurs during storage of sliced onion (*A. cepa*). Carriers are probably decomposition products of the lachrymatory factor (Z)-thiopropanal sulfoxide, which arises from S-(prop-1-en-1-yl)cysteine sulfoxide (isoalliin). Chopped onions processed for cooking can sometimes develop an unpleasant bitter taste due to the presence of structurally related derivatives of 3,4-dimethylthiolane S-oxide, called trivially allithiolanes A-I (**8-143**), which form spontaneously when the onion is damaged. Allithiolanes are also present in leeks, and one of the groups was found in garlic.

8.3.5.1.3 Spices and herbs

A large number of bitter plants or plant parts (such as roots, rhizomes, bark, leaves, flowers, and others) are used as drugs in folk and official medicine, for the manufacture of infusions, tinctures, and bitter non-alcoholic and alcoholic beverages, or as spices for food seasoning.

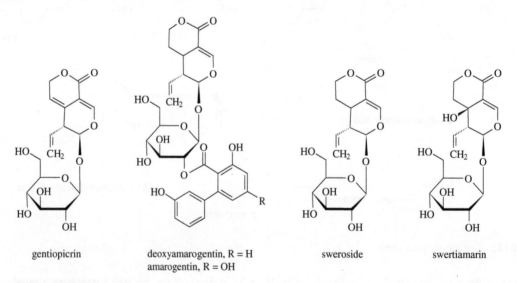

8-143,   allithiolane A, R = CH₃   allithiolane D   allithiolane E
       allithiolane B, R = CH=CHCH₃
       allithiolane C, R = CH₂CH₂CH₃

An important bitter substance used for making bitter drinks (such as tonic waters) is the alkaloid quinine, which is obtained from the bark of different species of the genus *Cinchona* (Rubiaceae), especially the bark of *C. officinalis*, native to Amazon rainforest (see Section 10.3.6.8).

Many plants used for their bitter taste contain strongly bitter glycosides of **iridoids** and **secoiridoids** (see Section 9.10), which they synthesise as protection against invading microorganisms and grazing herbivores. Iridoids show a number of biological activities (such as cardiovascular, antihepatotoxic, hypoglycaemic, antiviral, anticarcinogenic, and immunomodulatory effects). Secoiridoid glycosides occur in the root of yellow gentian (*Gentiana lutea*), the bitter herb centaury (*Centaurium erythraea*), and almost all plants of the Gentianaceae family. The root of yellow gentian contains primarily gentiopicrin (gentiopicroside, **8-144**), amarogentin (amarogentoside, **8-144**), and sweroside (**8-144**). Also present are additional related compounds, such as swertiamarin (swertiamaroside, **8-144**) and deoxyamarogentin (deoxyamarogentoside, **8-245**). Amarogentin is considered the most bitter substance occurring in nature, as its bitter taste is noticeable even at a dilution of 1 : 58 000 000 (which corresponds to a solution prepared by dissolving 17 µg of amarogentin in 1000 ml of water). The gentiopicrin content in fresh root is 1.5–3.5%, whilst the content of amarogentin is only 0.05%, but due to its low threshold value amarogentin is a significant bitter substance. In acidic media, swertiamarin is transformed into erythrocentaurin (5-formyl-3,4-dihydroisocoumarin) almost quantitatively.

      gentiopicrin   deoxyamarogentin, R = H   sweroside   swertiamarin
                      amarogentin, R = OH

8-144, secoiridoid glycosides

Bitter substances of centaury are similar to bitter substances of yellow gentian. The main component is usually gentiopicin (originally known under the name erytaurin), along with swertiamarin, amarogentin, and other minor bitter substances such as dimeric centauroside (**8-145**). Some centaury species contain sweroside in almost the same amount as gentiopicrin.

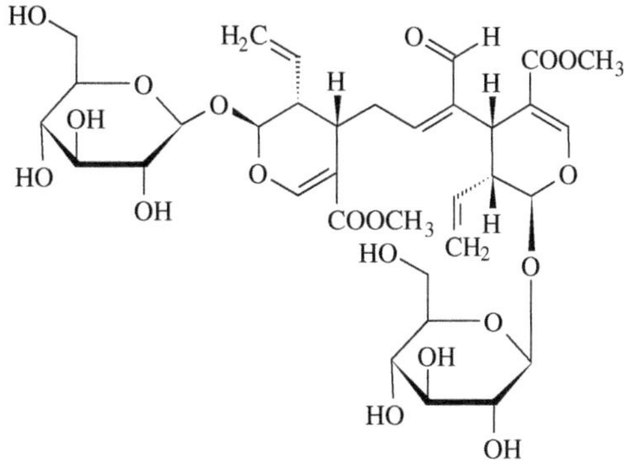

8-145, centauroside

Plants of the family Asteraceae, which are used as spices and for the production of bitter liqueurs and vermouths (aromatised fortified wine flavoured with various botanicals), contain sesquiterpenic lactones often derived from guaiane (**8-138**), eudesmane, and germacrane (**8-146**). Eudesmane (also known as selinane) has the systematic name (1*S*,4a*R*,7*R*,8a*S*)-7-isopropyl-1,4a-dimethyldecahydronaphthalene, whilst germacrane is (1*R*,4*S*,7*S*)-4-isopropyl-1,7-dimethylcyclodecane.

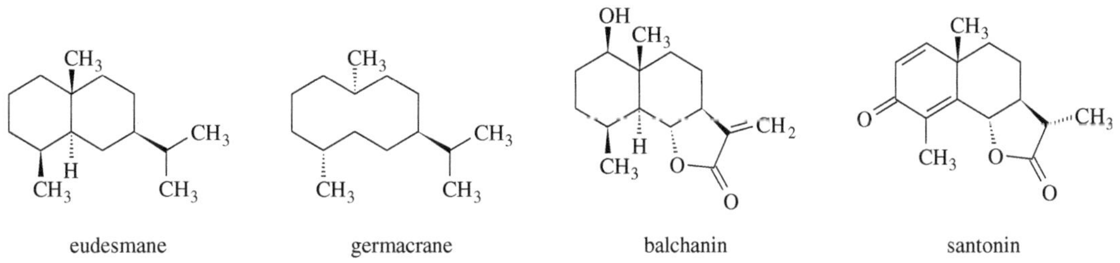

eudesmane   germacrane   balchanin   santonin

8-146, eudesmane, germacrane, and derived bitter compounds

The main bitter component of wormwood (*Artemisia absinthium*) is a dimeric sesquiterpenic lactone, absinthin (**8-147**), which is produced by intramolecular cycloaddition (Diels–Alder reaction) probably of two molecules of astabisin (**8-147**), derived from guaiane. In wormwood extracts, absinthin is transformed into isomeric anabsinthin (**8-147**). The bitter taste of mugwort (*A. vulgaris*) is derived primarily from the related compounds estafiatin (**8-147**) and balchanin (**8-146**), which are derived in turn from eudesmane. Another bitter substance occurring in different plants of the genus *Artemisia*, such as *A. absinthium*, *A. vulgaris*, *A. pontica* (Roman wormwood), and *A. maritima* (see wormwood), is santonin (**8-146**), which is also derived from eudesmane. It is toxic, so safe maximum levels have been set for its presence in alcoholic beverages and foods. Toxic components of wormwood and some other essential oils are α- and β-thujones.

Common sage and rosemary (see Table 8.32), plants of the Lamiaceae family, contain the diterpenes carnosic acid, also known as rosmaricin (**10-135**), derived from *ent*-kaur-16-ene (**8-7**), and bitter carnosol (picrosalvin, **10-135**), which are potent antioxidants. Carnosic acid is a major component of fresh rosemary tops (1–2%), but it is unstable and is enzymatically transformed to carnosol. These two diterpenoids represent about 15% w/w of plants haulm extracts and exhibit about 90% of extract antioxidant activity. Other transformation products of carnosic acid are rosmanol (7α-hydroxy derivative, **10-136**), epirosmanol (7β-isomer), and similar compounds.

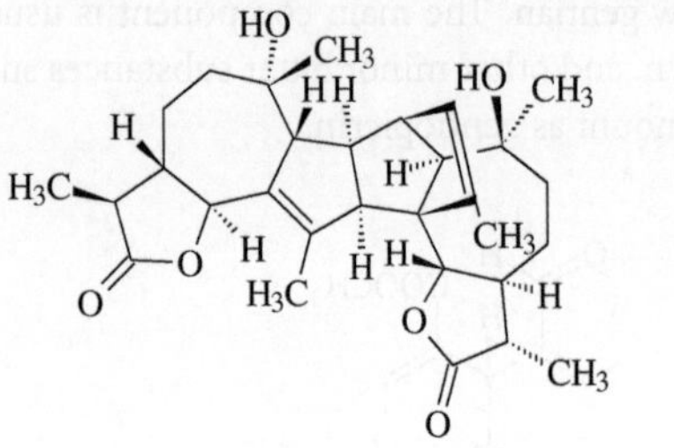

absinthin

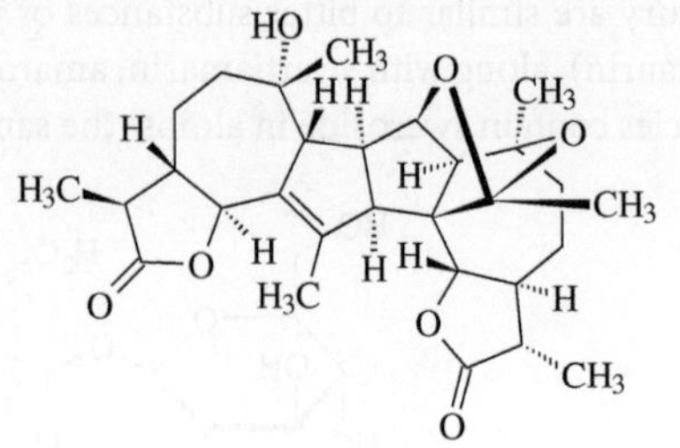

anabsinthin

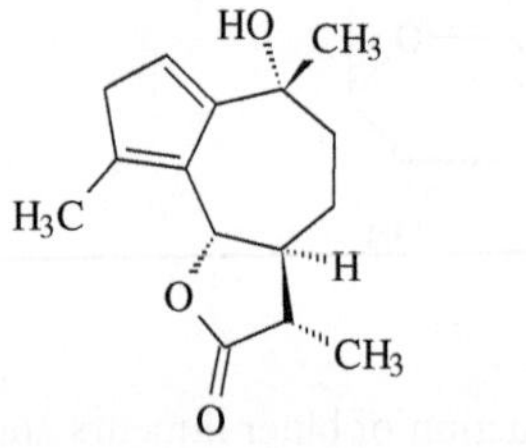

astabisin

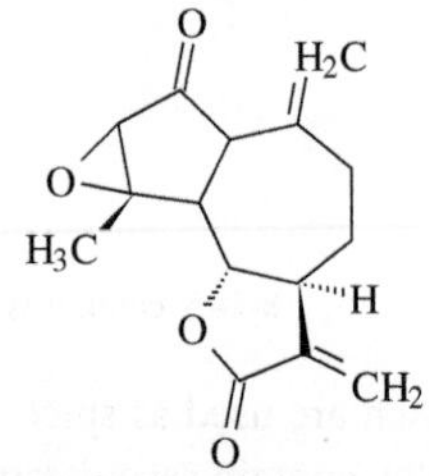

estafiatin

8-147, bitter sesquiterpenic lactones derived from guaiane

Another member of the Asteraceae family is St. Benedict's thistle, also known as the blessed or holy thistle (*Cnicus benedictus*), which contains bitter cnicin (**8-148**) derived from germacrane.

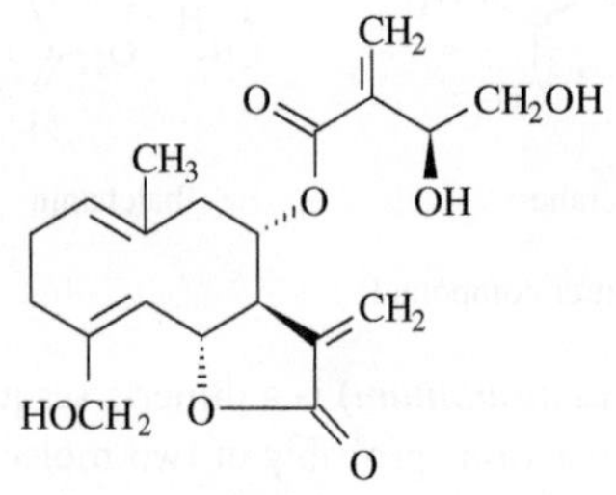

8-148, cnicin

The principal bitter constituent of the sweet flag (calamus) rhizome (*Acorus calamus*, Araceae) is a glycoside of acorine (2-acetyl-14-hydroxyhetisine, **8-149**), which is derived from the diterpenoid alkaloid (2α,11α,13R)-hetisan-2,11,13-triol, known as hetisine.

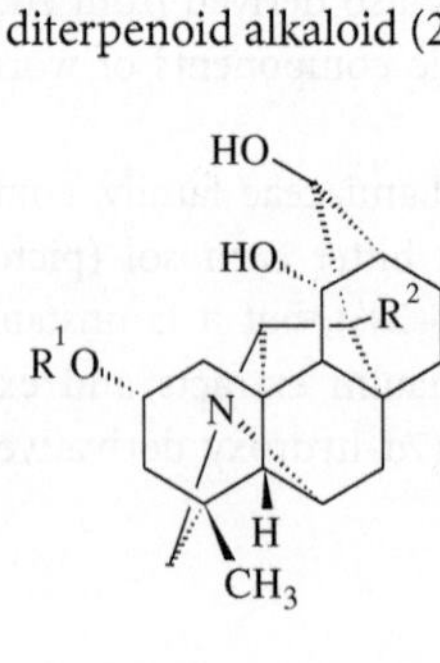

8-149, hetisine, R¹ = R² = H
acorine, R¹ = COCH₃, R² = OH

Typical bitter substances of shrubs of the family Simaroubaceae are degraded triterpenoids known as **decanortriterpenoids**, related to limonoids known as **quassinoids**. The so-called bitter quassia wood from the shrub *Quassia amara*, native to tropical America, contains as its main bitter substance quassin (**8-150**), one of the bitterest natural substances known. For toxicological reasons (negative effect on fertility of tested animals), flavourings and food ingredients with flavouring properties produced from the source material (and also from the West Indian tree *Picrasma excelsa*, of the same family) may only be used for the production of beverages and baked products. Examples of other bitter substances of *Q. amara* are neoquassin (with a reduced oxo group in position C-16), 14,15-dehydroquassin, and 18-hydroxyquassin (with a methyl group in position C-18 replaced by a hydroxymethyl group).

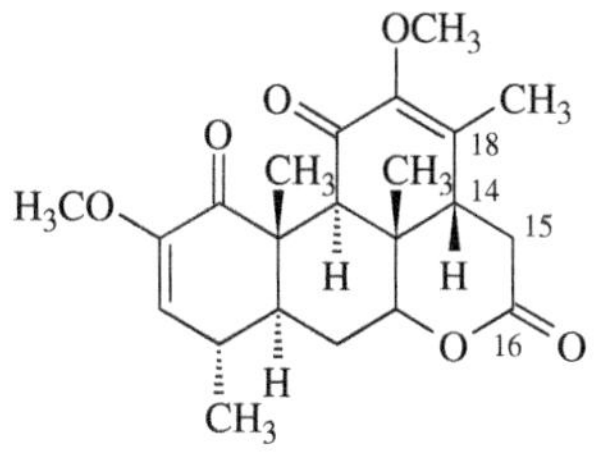

8-50, quassin

The most precious and expensive spice in the world is saffron (dried stigmas of the saffron flower, *Crocus sativus*, Iridacaeae), which is used for its bright orange-yellow colour and intense flavour. The bitter substance of saffron picrocrocin is β-D-glucoside of (*R*)-4-hydroxy-2,6,6-trimethylcyclohex-1-ene-1-carbaldehyde, known as (*R*)-4-hydroxy-β-cyclocitral. Such fragments of carotenoids are called apocarotenoids. Picrocrocin arises by oxidative cleavage of the carotenoid pigment zeaxanthin via enzyme dioxygenase and glycosylation of (*R*)-4-hydroxy-β-cyclocitral (Figure 8.94). Hydrolysis of picrocrocin to (*R*)-4-hydroxy-β-cyclocitral via dehydration of this aglycone yields safranal (see Section 9.9.5.2.3), whose flavour is reminiscent of saffron.

8.3.5.1.4 Other foods and beverages

Dairy products and protein hydrolysates Precursors of bitter substances in milk and dairy products are quite often proteins and mineral salts. Bitter substances include peptides produced by enzyme proteolysis (see Section 2.3.3.2) that contain varying numbers of bound amino acids, usually between 2 and 23. Peptides with a molecular weight higher than 3000 Da are not bitter because they cannot interact with the taste receptors. Bitterness is especially typical for certain dairy products, such as cheeses, yoghurt, and casein hydrolysates (see Section 2.3.3.2). For example, the bitterness of matured Gouda cheese was found to be primarily induced by calcium ($CaCl_2$) and magnesium ($MgCl_2$) chlorides, as well as various bitter-tasting free amino acids, whereas bitter peptides, which arise from the hydrolysis of β-casein and $α_{S1}$-casein, were found to influence its bitterness quality rather than its bitter intensity. Bitter peptides isolated from Cheddar cheese formed from $α_{S1}$-casein have the following sequences of amino acids: Arg-Pro-Lys-His-Pro-Ile-Lys-His-Gln-Gly-Leu-Pro-Gln, Leu-Pro-Gln-Glu, Arg-Pro-Lys-His-Pro-Ile-Lys, and Lys-Pro-Trp-Ile-Gln-Pro-Lys. The bitter peptide of Cheddar cheese originating from β-casein is the linear nonapeptide Val-Pro-Gly-Glu-Ile-Val-Glu-Ser-Leu. Another example of a bitter peptide occurring in β-casein hydrolysate is the linear pentadecapeptide Tyr-Gln-Gln-Pro-Val-Leu-Gly-Pro-Val-Arg-Gly-Pro-Phe-Pro-Ile. Bitter substances are also formed during pyrolysis of proteins and in reactions of proteins with sugars (the Maillard reaction).

Cereals and bread The bitter compounds in whole-wheat bread crumb were reported to be flavonoid 8-*C*-glycosides apigenin-6-*C*-galactoside-8-*C*-arabinoside and its isomer apigenin-6-*C*-arabinoside-8-*C*-galactoside, found in wheat. Contributors to bitter taste are tryptophan and (9*S*,10*E*,12*S*,13*S*)-trihydroxyoctadec-10-enoic acid, known as (−)-pinellic acid (see Section 3.3.1.5), a metabolite of linoleic acid generated during bread-making (hydration of flour). The pyrrolomorpholine spiroketals acortatarin A and C (see Section 4.7.5.12.1) were isolated as bitter components of whole-wheat bread crust as Maillard reaction products.

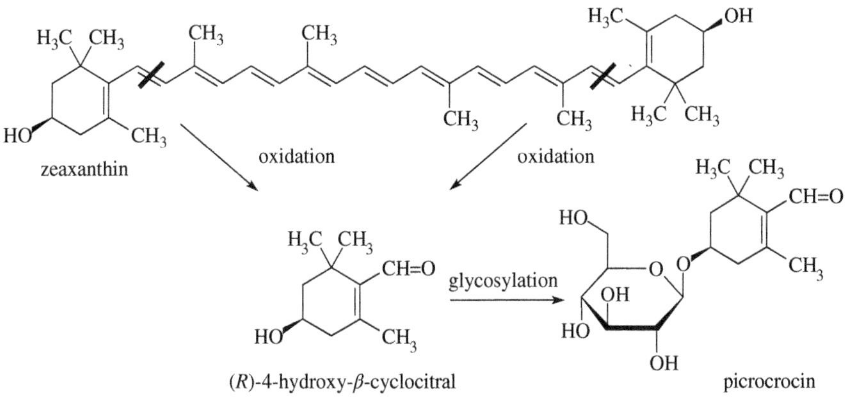

Figure 8.94 Formation of picrocrocin from zeaxanthin.

The key phytochemicals contributing to the typical astringent and bitter off-taste of oat are avenanthramides and avenacosides (see Section 10.3.3.2).

Legumes and oilseeds The most important carriers of bitter and astringent taste in the seeds of plants of the family Fabaceae, which includes certain varieties of pea (*Pisum sativum*) and peanuts (*Arachis hypogaea*), are the saponins (see Section 10.3.2.2), chiefly sojasaponin B_b (also known as sojasaponin I). The content of saponins in sweet varieties typically ranges between 0.1 and 0.2%, with a content higher than this perceived as a bitter taste. The content of saponins in peanuts may range from 0.01 to 1.6%. The characteristic bitter and astringent tastes of soy flour (*Glycine max*) are mainly due to isoflavones and their glycosides (daidzin, genistin, and glycitein-7-*O*-β-D-glucoside), but it may also contain saponins in the range from 0.2 to 5.6%.

The cyclic diarylheptanoid asadanin (**8-151**) is the major inducer of the bitter off-taste of hazelnut kernels (*Corylus avellana*, Betulaceae), accompanied by some other bitter-tasting diarylheptanoids (Table 8.40). The major inductor of asadanin biosynthesis is infection of hazelnuts by the fungus *Eremothecium coryli* (originally *Nematospora coryli*).

8-151, asadanin

The bitter taste of the extraction meal and flour from rapeseeds (*Brassica napus*, Brassicaceae) is caused by sinapine (present as sinapine chloride) and its components, choline chloride and sinapic acid, which show about 80% of the sinapine bitterness. Goitrin may also act as a bitter substance, as in some *Brassica* vegetables (e.g. Brussels sprouts).

Hops and beer Hops are the cone-shaped female inflorescences of the vine-like plant *Humulus lupulus* (Cannabaceae). They are harvested, dried, and used as whole hops (also called whole leaf or whole cone hops), hop pellets, and hop extracts. Hops are one of four essential ingredients in beer (alongside barley, yeast, and water) and are also used in other beverages and herbal medicine.

The lupulin glands of cones hops contain substances that impart bitterness, flavour, and stability to the finished product. Today, up to about 50 individual bitter components can be found in beer. These are often categorised as:

- hop resins;

- hop essential oil (monoterpenes and sesquiterpenes with characteristic aroma;

- hop polyphenols (a varied group of about 100 compounds, comprising up to 30% of the polyphenols in the wort, with the rest coming from the barley malt; the main groups include prenyl flavonoids, proanthocyanidins, cinnamic acids, and many other polyphenols; polyphenols are often involved in haze formation in finished beer).

The **hop resins** (compounds extractable from hops with diethyl ether and soluble in cold methanol) are divided into two main groups: hard resins (insoluble in hexane) and soft resins (soluble in hexane). The hard resins mainly contain prenylflavonoids (see Section 10.3.12.4) and oxidation products of α- and β-acids. The soft resins are mainly composed of the hop bitter acids humulones and lupulones and their isomers:

- humulones (humulone homologues), called **α-acids** or α-bitter acids;

- lupulones (lupulone homologues), called **β-acids** or β-bitter acids.

The bitter acids in fresh hops are bitter only in name, and in fact have an indifferent taste. According to their biochemical origin, α- and β-acids are ketides, derivatives of 1,3,5-benzenetriol (phloroglucinol). They contain a residue of carboxylic acid (acyl) and two (in humulones) to three (in lupulones) 3-methylbut-2-en-1-yl (prenyl) side chains as substituents of the aromatic ring. Their level usually amounts to about 5–20% of the cones weight (9–10% of α-acids and up to 10% of β-acids; water content is about 11%) and about 25% of the dry weight of the hop cones. The main α-acids are *n*-humulone, *co*-humulone, and *ad*-humulone, whilst minority components are *post*-humulone, *pre*-humulone, and *adpre*-humulone (**8-152**). Analogously, β-acids include *n*-lupulone, *co*-lupulone, *ad*-lupulone, *post*-lupulone, *pre*-lupulone, and *adpre*-lupulone (**8-153**). The approximate compositions of these two groups of bitter acids in different hop varieties are shown in Table 8.41.

Table 8.41 Contents of α- and β-bitter acid contents in acids hops.

Compound	Content (% of α-bitter acids)	Compound	Content (% of β-bitter acids)
α-Bitter acids		*β-Bitter acids*	
n-humulone	35–70	*n*-lupulone	30–55
co-humulone	20–65	*co*-lupulone	37–68
ad-humulone	10–15	*ad*-lupulone	11–12
pre-humulone	1–10	*pre*-lupulone	1–5

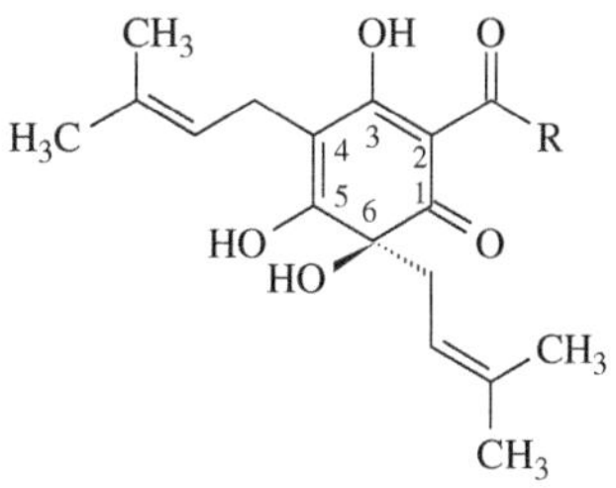

8-152, α-acids

n-humulone, R = CH$_2$CH(CH$_3$)$_2$
co-humulone, R = CH(CH$_3$)$_2$
ad-humulone, R = CH(CH$_3$)CH$_2$CH$_3$
post-humulone, R = CH$_2$CH$_3$
pre-humulone, R = CH$_2$CH$_2$CH(CH$_3$)$_2$
adpre-humulone, R = (CH$_2$)$_4$CH$_3$

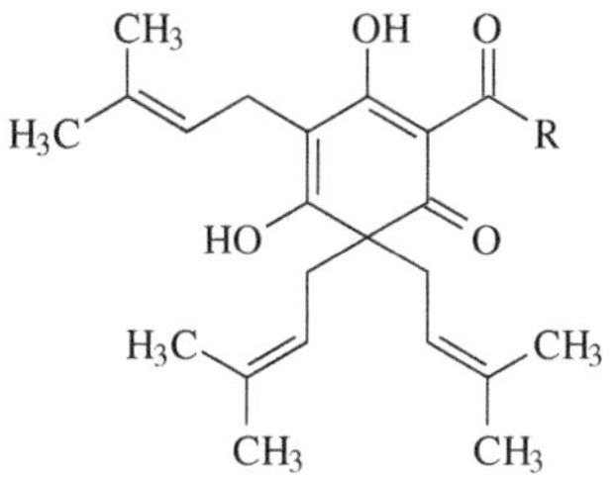

8-153, β-acids

n-lupulone, R = CH$_2$CH(CH$_3$)$_2$
co-lupulone, R = CH(CH$_3$)$_2$
ad-lupulone, R = CH(CH$_3$)CH$_2$CH$_3$
post-lupulone, R = CH$_2$CH$_3$
pre-lupulone, R = CH$_2$CH$_2$CH(CH$_3$)$_2$
adpre-lupulone, R = (CH$_2$)$_4$CH$_3$

Both types of hop acids (α- and β-acids) are very reactive compounds, particularly in the light. During the drying and storage of hops and during the brewing process, they are transformed into a large number of different products.

α-Acids are stable under acidic conditions and isomerise under basic conditions or during the wort-boiling step of the brewing process into desired compounds with bittering, aromatising, and preservative properties. Boiling the wort transforms about 50–70% of non-bitter humulones into bitter isohumulones, also known as iso-α-acids or iso-α-bitter acids, which are the main contributors to the bitter taste of beer, beer foam stability, and inhibition of Gram-positive bacterial growth. The isomerisation reaction is reversible and proceeds as a stereoselective tautomerisation of enolate into ketone followed by ring contraction (at C-5 and C-6) of acyloins (Figure 8.95).

Iso-α-acids have two chiral centres and therefore exist as (4*R*,*Z*)-isomers (*cis*-isomers) and (4*S*,*E*)-isomers (*trans*-isomers), depending on the spatial arrangement of the tertiary alcohol function at C-4 and the prenyl side chain at C-5. The ratio of these two isomers in wort is about 2 : 1 in favour of the more stable *cis*-isomers. The amount of iso-α-acids in beer is rather low (10–40% of α-acids), for various reasons. Six major intensely bitter iso-α-acids, namely *cis*-iso-*n*-humulone, *cis*-iso-*ad*-humulone, *cis*-iso-*co*-humulone, *trans*-iso-*n*-humulone,

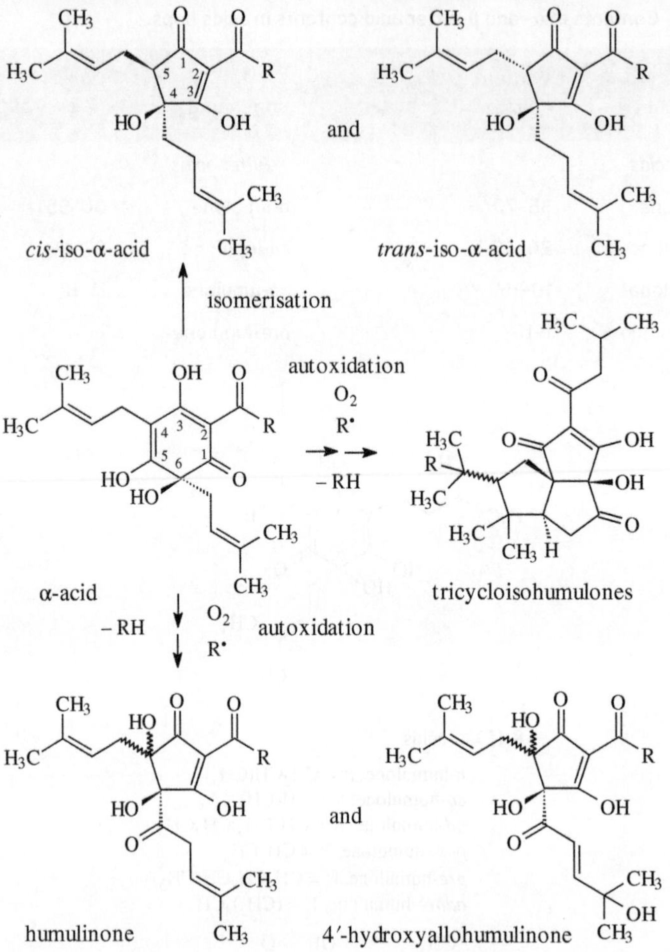

Figure 8.95 Isomerisation and oxidation of hop α-acids.

trans-iso-ad-humulone, and *trans-iso-co*-humulone generally occur in beers at concentrations of 15–80 mg/l, or even 100 mg/l in very bitter English ales.

The α-acids are also prone to oxidative degradation, generally accompanied by ring contraction. The main autoxidation products of α-acids in stored hops are humulinones and 4′-hydroxyallohumulinones. Minor products are various tricycloisohumulones, such as tricycloperoxyisohumulones (R = OOH) and tricyclooxyisohumulones (R = OH), products without the prenyl side chain (e.g. deisopropyltricycloisohumulones), and similar compounds.

During beer storage, the iso-α-acids undergo acid- and base-catalysed degradation and oxidative reactions (Figure 8.96). It has been demonstrated that the ageing of beer in brown bottles induces a preferential acid-catalysed degradation of the *trans*-isomers of iso-α-acids, whereas the corresponding *cis*-isomers are more stable. Acid-catalysed degradation leads to a range of tricyclic and tetracyclic products, exemplified by tricyclohumol and tricyclodehydrohumol. Their bitterness is about 70% of the bitterness of isohumulones. Base-catalysed degradation proceeds via isomerisation of the side chain in iso-α-acid, yielding *allo*-isohumulones, which are transformed by retro-aldolisation into deacylated (4*R*,*Z*)- and (4*S*,*E*)-humulinic acids, also known as *cis*- and *trans*-humulinic acids, respectively, and to other products. In direct sunlight (UV or visible light, 350–500 nm/photosensitiser), the side chain of iso-α-acids is cleaved, which yields dehydrohumulinic acid. Decarbonylation of the formed acyl radical and recombination with a thiol radical, originating from sulfur compounds in beer, yields but-2-en-1-thiol, causing the so-called light-struck off-flavour. The storage of beer in polyethylene terephthalate bottles, permeable to oxygen, induces autoxidation of both isomers of iso-α-acids to peroxides (*cis*- and *trans*-hydroperoxy-*allo*-iso-humulones) via the corresponding hydroperoxyl free radicals. Decomposition of peroxides yields, via the corresponding alkoxy radicals to allylic hydroxyl derivatives.

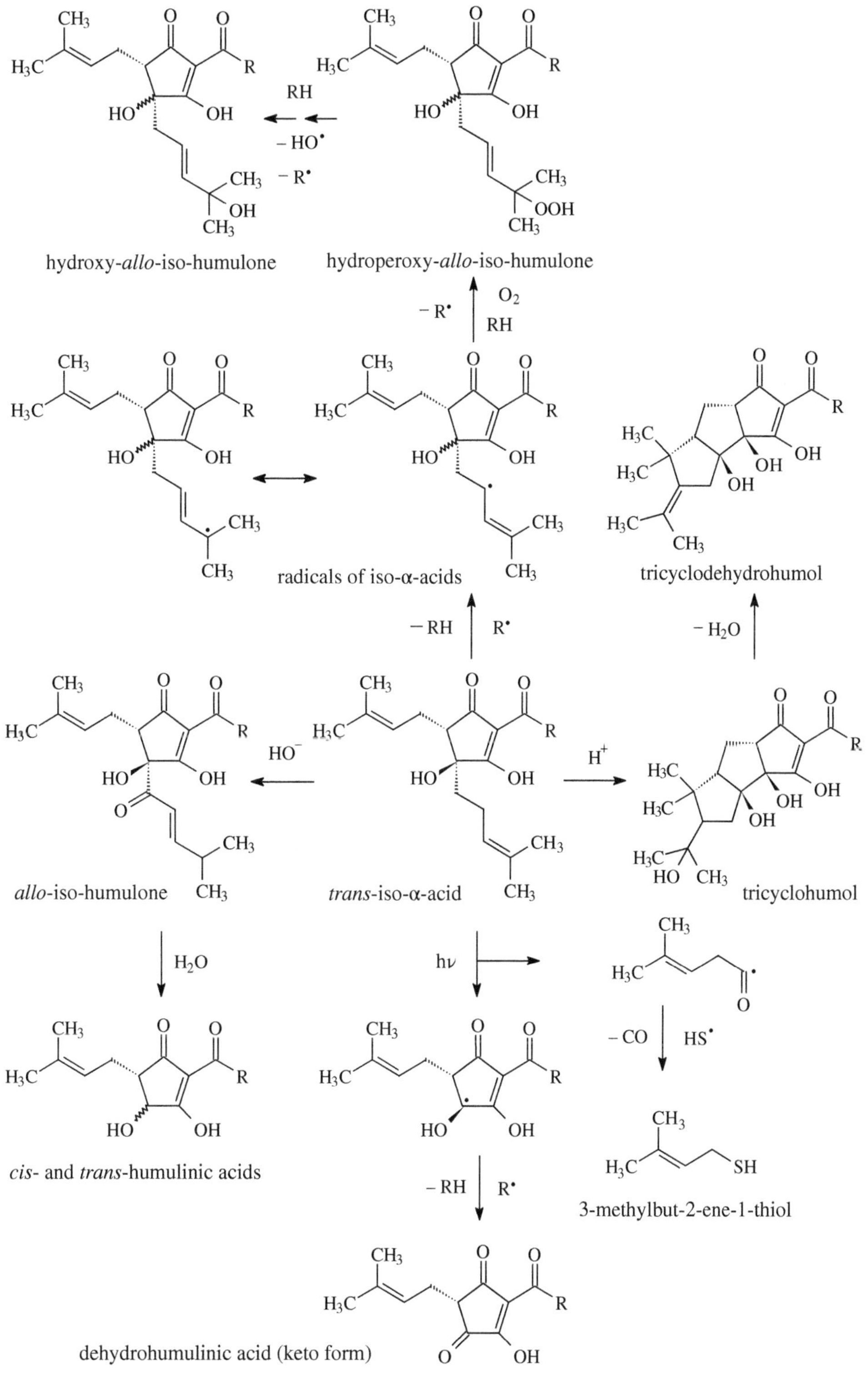

Figure 8.96 Degradation and oxidation of iso-α-acids. Source: Intelmann and Hofmann (2010), Figure 8.7. Reproduced by permission of the American Chemical Society.

Easy oxidation of humulones and lupulones in hops is the reason for the use of various hop products, including hydrogenated (reduced) iso-α-acids. Examples of hydrogenated substances are dihydro-iso-α-acids (dihydroisohumulones, also known as *rho*-isohumulones) and dihydro-iso-α-acids (tetrahydroisohumulones, **8-154**), the taste threshold of which is 0.007 mg/l in water. Dihydro- and tetrahydro-iso-α-acids are stable even in the light and do not undergo photofragmentation. Beers containing these compounds can be stored in white glass bottles. The bitter taste of roasted malt used in the production of dark beers is caused by the presence of 2,5-dioxopiperazines and other products of the Maillard reaction. In comparison with bitter substances of hops, the contribution of these compounds to the bitterness of beer is rather questionable.

cis-dihydro-iso-α-acids   *cis*-tetrahydro-iso-α-acids

8-154, reduced iso-α-acids

β-Bitter acids form some oxidation products during the wort boiling process that contribute to the bitter taste of beer to a minor extent, but many possess unpleasant organoleptic characteristics. Therefore, β-bitter acids are considered a negative factor in brewing, and hop varieties with low content of lupulones are preferred. The main bitter transformation products of lupulones during storage of hops are the hulupones, which belong to traditionally recognised non-specific soft resins called resupones. Upon wort boiling, the abstraction of a hydrogen atom from lupulones gives a corresponding alkoxy radical, which upon mesomerisation reacts with oxygen and another lupulone molecule to form a hydroperoxide. This hydroperoxide is degraded to hulupones, and the hulupones in turn are degraded to hulupinic acids. The radical may undergo a series of intramolecular cyclisations to produce the tricyclic isopropyl radical, which is a precursor of a range of epimeric tricyclolupones (Figure 8.97).

Grapes and wine The bitterness and astringency of red wines is mainly induced by bitter and astringent phenolic acid ethyl esters (for example, ethyl esters of gallic, *p*-coumaric, and syringic acids) and flavan-3-ols (catechin, epicatechin, procyanidins B$_1$, B$_2$, B$_3$, and C$_1$), astringent flavonol glucosides (quercetin, isorhamnetin, and syringetin 3-*O*-β-D-glucopyranosides), flavanonol rhamnosides (aromadendrin and taxifolin 3-*O*-α-L-rhamnopyranosides), and polymers of molecular weight >5 kDa (see Section 9.4.1.5.9) and is amplified by carboxylic acids.

Hydrolysable tannins and lignans can contribute to the bitterness of oaked wines and other alcoholic beverages. The most bitter compounds released from oak wood (*Q. petraea*, Fagaceae) are derivatives of gallotannins with phenols, such as 3-methoxy-4-hydroxyphenyl-1-*O*-β-D-(6'-*O*-galloyl)glucopyranoside and quercoresinosides A (**8-170**) and B.

For more information see Section 8.3.6.2.2 dealing with tannins.

Tea The bitterness and astringency of fresh tea leaves is related primarily to the presence of catechins and procyanidins (see Section 9.12.5.4), which constitute 10–30% of dry tea leaves. The key taste-active compounds responsible for the typical bitter and astrigent taste of tea infusions are bitter caffeine, astringent and bitter epigallocatechin-3-*O*-gallate, and astringent catechin and flavonol 3-glycosides, namely kaempherol- and quercetin-derived 3-*O*-β-D-glucopyranosides, 3-*O*-β-D-galactopyranosides, 3-*O*-[α-L-rhamnopyranosyl-(1 → 6)-β-D-glucopyranosides], quercetin 3-*O*-[β-D-glucopyranosyl-(1 → 3)-α-L-rhamnopyranoside], and 3-*O*-β-D-glucopyranoside and 3-*O*-β-D-galactopyranoside of myricetin. Less important tea constituents are bitter amino acids (valine, leucine, isoleucine, phenylalanine, and tyrosine), sweet and umami-like theanine, and bitter saponins with characteristic aglycones (sapogenols), such as theasapogenols, assamsapogenols, and related compounds (see Section 10.3.2.2). During the fermentation of tea leaves, procyanidins react with proteins, which reduce the original astringent taste of tea leaves, but enzymatic browning reactions create new astringent products.

Coffee The bitterness of roasted coffee is affected by roasting of the beans (the degree of bitterness increases with prolonged roasting), preparation of brine, water hardness, temperature, and the use of sugar and milk. It is the result of many compounds, but the majority of the bitter taste is shown by caffeine and the transformation products of chlorogenic acids formed during the roasting of the coffee seeds by dehydration, decarboxylation, and subsequent reactions.

Chlorogenic acids partially dehydrate to form lactones. For steric reasons, the main lactones of roasted coffee are γ-lactones derived from 3-*O*-cinnamoyl-L-quinic (chlorogenic) and 4-*O*-cinnamoyl-L-quinic (cryptochlorogenic) acids. The main products are 3-*O*-L-caffeoyl-quino-1,5-lactone (also known as 3-*O*-caffeoyl-γ-quinide, **8-155**), produced from chlorogenic acid, and 4-*O*-caffeoyl-L-quino-1,5-lactone

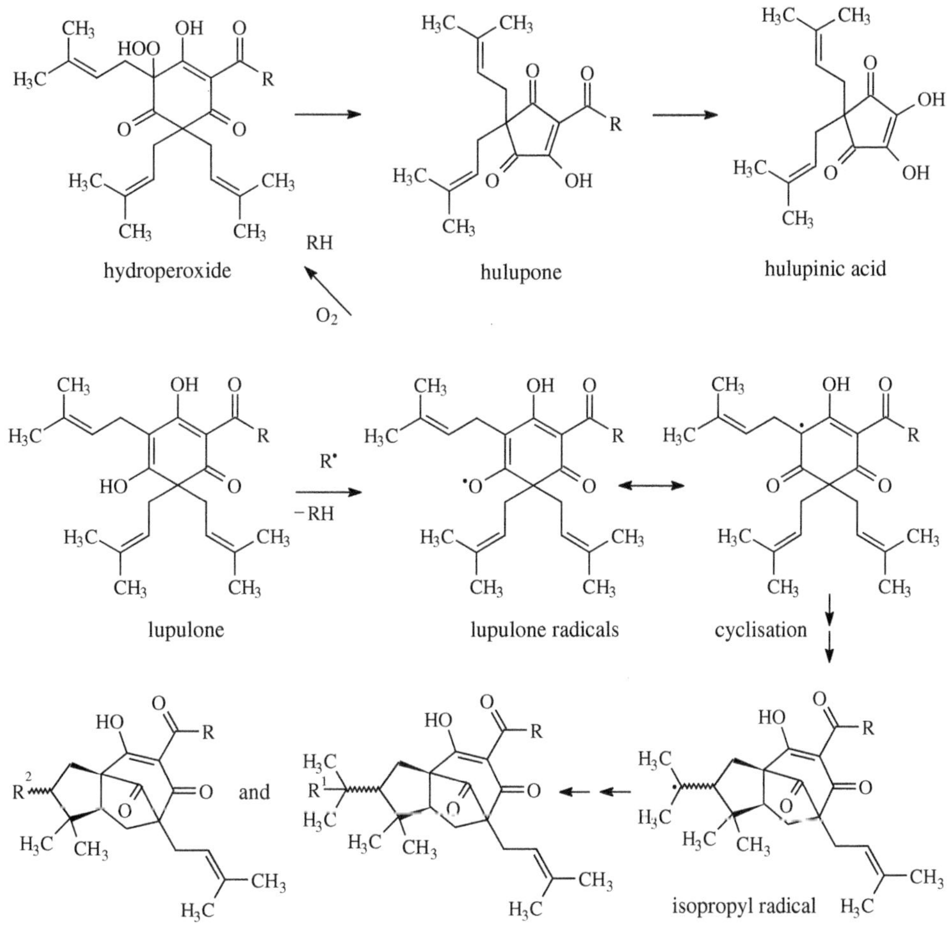

nortricyclolupone, $R^2 = H$
dehydrotricyclolupone, $R^2 = CH_3C(=CH_2)$

hydroperoxytricyclolupone, $R^1 = OOH$
hydroxytricyclolupone, $R^1 = OH$
tricyclolupone, $R^1 = H$

Figure 8.97 Oxidation of hop β-acids.

(4-*O*-caffeoyl-γ-quinide, **8-156**), produced from cryptochlorogenic acid. These are followed by γ-lactones derived from 3-*O*- and 4-*O*-feruloyl-L-quinic acid and other depsides.

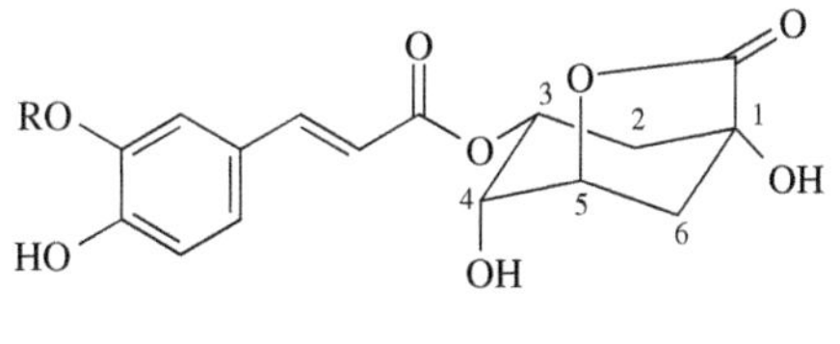

8-155, 3-*O*-caffeoyl-L-quino-1,5-lactone, R = H
3-*O*-feruloyl-L-quino-1,5-lactone, $R = CH_3$

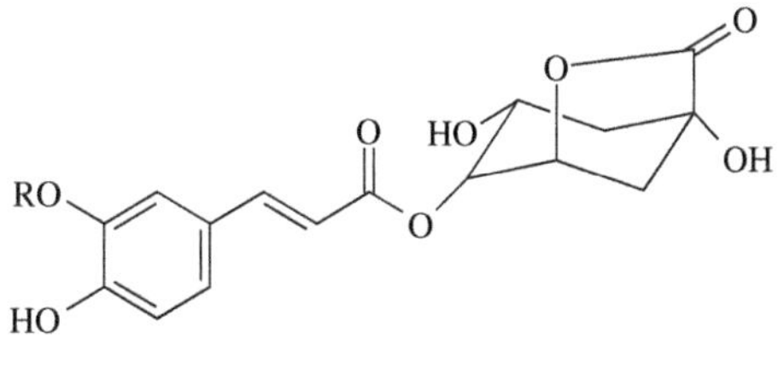

8-156, 4-*O*-caffeoyl-L-quino-1,5-lactone, R = H
4-*O*-feruloyl-L-quino-1,5-lactone, $R = CH_3$

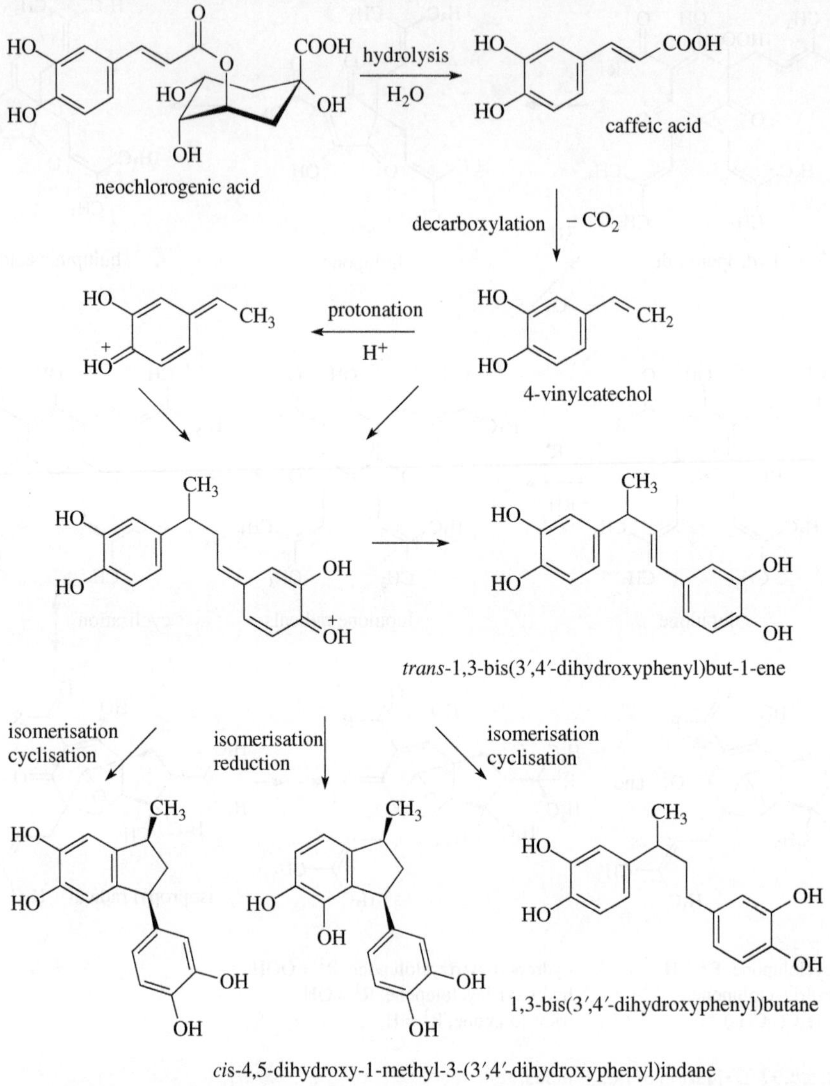

Figure 8.98 Formation of bitter substances from neochlorogenic and caffeic acids.

Reactions of chlorogenic acids during coffee roasting are very complex. Along with dehydration, chlorogenic acids are decarboxylated and hydrolysed. For example, 5-*O*-caffeoyl-L-quinic (neochlorogenic) acid yields, by hydrolysis, caffeic acid, which on decarboxylation or *syn*-elimination of quinic acid gives 4-vinylcatechol. Protonation of 4-vinylcatechol gives the reactive electrophilic cation with a quinone methide structure, which is condensed with 4-vinylcatechol, and through subsequent reactions produces a group of polyhydroxylated phenylindans such as 1,3-bis(3′,4′-dihydroxyphenyl)butane, *trans*-1,3-bis(3′,4′-dihydroxyphenyl)but-1-ene, *cis*- and *trans*-5,6-dihydroxy-1-methyl-3-(3′,4′-dihydroxyphenyl)indane, and *cis*- and *trans*-4,5-dihydroxy-1-methyl-3-(3′,4′-dihydroxyphenyl)indane, which exhibit a harsh and lingering bitter taste profile. Subsequent reactions with other 4-vinylcatechol cations yield higher oligomers. Only *cis*-isomers of indanes are shown in Figure 8.98. Dihydroxybenzene and trihydroxybenzene derivatives arising by decarboxylation of phenolic acids, such as catechol (pyrocatechol), 3-methyl- and 4-methylcatechol, and pyrogallol, condense with furan derivatives resulting from the Maillard reaction of sugars (such as furfuryl alcohol) to form bitter (furan-2-yl)methylated benzene diols and triols. Compounds of this group identified in coffee included 4-(furan-2-ylmethyl)benzene-1,2-diol (**8-157**, $R^1 = R^2 = H$, $R^3 = OH$), formed from catechol, 3-(furan-2-ylmethyl)-6-methylbenzene-1,2-diol (**8-157**, $R^1 = CH_3$, $R^2 = OH$, $R^3 = H$), formed from 3-methylcatechol, 4-(furan-2-ylmethyl)-5-methylbenzene-1,2-diol (**8-157**, $R^1 = CH_3$, $R^2 = H$, $R^3 = OH$), formed from 4-methylcatechol, and 4-(furan-2-ylmethyl)benzene-1,2,3-triol, formed from pyrogallol (**8-157**, $R^1 = R^2 = OH$, $R^3 = H$).

Quinic acid resulting from chlorogenic acids also has a bitter taste, but the contribution of 2,5-dioxopiperazines (see Section 2.3.3.2) and other heterocyclic compounds generated during roasting by the Maillard reaction is not very significant.

An arabica-specific bitter-tasting glucoside of the (−)-ent-kaure-16-ene (**8-7**) type, with a β-D-glucose unit attached to C-17, is mozambioside (**8-157**; Table 8.40). Green coffees contain, aside from the related diterpenoid alcohols cafestol and kahweol, approximately

0.4–1.2 mmol/kg mozambioside, whereas roasted coffees contain a reduced concentration (0.2 mmol/kg), indicating partial degradation of mozambioside during roasting. Traces of mozambioside (<0.005 mmol/kg) were detected in robusta coffees. Mozambioside is nearly quantitatively extracted into the aqueous brew during coffee-making.

Cocoa The bitterness of cocoa, like the bitterness of coffee, increases with the degree of roasting. The main bitter components are the bitter purine alkaloids theobromine and caffeine, as well as cyclic dipeptides (2,5-dioxopiperazines) formed during thermal fragmentation of proteins and as products of the Maillard reaction.

(furan-2-ylmethyl)benzene
diols and triols

mozambioside, R = β-D-Glc*p*

8-157, minor bitter compounds in roasted coffee

8.3.5.2 *Methods of debittering and masking the bitter taste*

In certain foods, a bitter taste is definitely not desirable, so different debittering methods have been developed. Methods for removing the bitter taste of enzymatic protein hydrolysates (such as casein hydrolysates) are described in Section 2.3.3.2. These are mainly based on controlled proteolysis, plastein reaction, extraction with azeotropic mixtures of alcohols, and masking of bitter substances.

Debittering of citrus (grapefruit) juices that contain bitter flavanone-7-glucosides is based on enzymatic hydrolysis of the sugar residue in bitter glycosides. The necessary enzymes exhibiting the activities of α-rhamnosidase and β-glucosidase are derived from the microorganisms *Phomopsis citri* and *Cochliobolus miyabeanus*, amongst others. Debittering of orange juice (removal of bitter limonin) is done using special enzymes or immobilised cells (such as the bacterium *Arthrobacter globiformis*).

Debittering of olives is based on hydrolysis of the bitter oleuropein by fermentation or in alkaline solutions. Hydrolysis of oleuropein in immature green or yellow fruits is done by dipping in a 1.3–2.6% sodium hydroxide solution for 6–10 hours. To obtain black unfermented olives from ripe blue olives that are no longer bitter, the fruits are repeatedly soaked in 1–2% NaOH solution with aeration. Under these conditions, intense non-enzymatic browning reactions occur, and the dark colour of the fruits is stabilised by the addition of ferrous gluconate that forms black complexes with oxidised polyphenols.

8.3.5.3 *Physiology, nutrition, and use*

Some bitter substances exhibit different physiological effects and may even be highly toxic (e.g. some alkaloids). The bitter taste of food is often unconsciously associated with the presence of toxic substances, for example in mushrooms, potatoes, and lupine seeds. The physiological effects of bitter substances such as alkaloids, cyanogenic glycosides, saponins, degradation products of glucosinolates (such as goitrin), and plant phenols are discussed in Chapter 10, along with health assessments and appropriate legislation. A number of other bitter compounds have beneficial physiological effects as they act as antioxidants (e.g. many phenolic compounds) and, in some cases, as anticarcinogens (e.g. cucurbitacins and many phenolics). Bitter substances generally support the appetite, which lies in the increased secretion of gastric juices and improved food digestion.

Bitter substances harmless to health have found use mainly in the manufacture of bitter soft drinks, such as bitter lemon (a carbonated drink flavoured with quinine and lemon) and tonic water (a carbonated water flavoured with quinine), and alcoholic beverages, such as beer, specialty wines (vermouths), liqueurs, and appetisers.

8.3.6 Astringent substances

Astringency (sometimes called adstringency) is a feeling of dryness in the mouth. It is often accompanied by bitter or sour taste properties, or both. The different subqualities perceived can be caused by different astringent compounds and the influences of other food constituents, such as acids and sugars. The primary cause of astringency is the interaction of salivary proteins with astringent compounds present in foods of plant origin. These interactions lead to denaturation of saliva proteins and loss of their protective effects, and astringent compounds may consequently interact with the oral-cavity proteins, which is perceived as astringency. In addition to phenolic compounds, other compounds can likewise contribute to the astringent taste of food commodities.

Phenolic compounds interacting with proteins are collectively called **tannins**. The common feature of tannins is their ability to react with proteins and precipitate them from aqueous solutions. This reaction is applied *in vivo* in the regulation of enzymatic functions of plant proteins in damaged tissues (structurally similar lignins are missing this feature). The affinity of tannins for proteins depends on the number of hydroxyl groups and their arrangement, the degree of polymerisation of phenols, the primary, secondary, and tertiary structure of proteins, and other factors. In the formation of complexes with proteins, tannins primarily interact via hydrogen bonds and hydrophobic interactions. Particularly strong hydrogen bonds exist between hydroxyl groups of phenols and the secondary amino groups of bound proline (saliva proteins are proline rich) and amide groups of peptide bonds. Ionised hydroxyl groups (in solutions of pH > 10) do not react with proteins. Low-molecular-weight phenolic compounds, such as phenolic acids and simple flavonoids, although they react with proteins, do not form cross-links between polypeptide chains, meaning there is no precipitation of proteins.

Tannins are divided into two large groups:

- **hydrolysable tannins;**

- **condensed tannins.**

Hydrolysable tannins are polymers of gallic acid esters known as polygalloyl esters. Condensed tannins, formerly also known as flavolans, are polymers of some flavonoid compounds with the structure of 3-hydroxyflavan (flavan-3-ol). However, virtually any number of combinations of condensed tannins and hydrolysable tannins, called complex tannins, similarly exist.

Owing to their properties, tannins have been used for millennia to tan animal hides into leather. As natural food components, tannins are of great importance, as they often substantially affect both desirable and undesirable taste characteristics. A desirable property is a reasonable astringency of tea, coffee, cocoa (added milk or cream removes the astringency as a result of interaction of tannins with milk proteins), red wine, and beer. Undesirable astringency is shown by unripe fruits, such as bananas, persimmons, and unripe walnuts. In some drinks, such as fruit juices, beer, and wine, hazes or sediments are formed, in which proteins and tannins participate. Beer and brewing technology distinguishes between a chill haze and a permanent haze. The polyphenols combine slowly with proteins to form a chill haze when beer is cooled, but the haze redissolves when it is warmed up. As the polyphenols polymerise into larger units, they become insoluble at room temperature, forming an irreversible permanent haze. A common practice is, therefore, to eliminate tannins in beer and wine by the use of additives such as gelatine, polyvinylpolypyrrolidone, or polyamide.

8.3.6.1 Hydrolysable tannins

Hydrolysable tannins are usually divided into:

- **gallotannins;**

- **ellagitannins.**

Both groups are derivatives of O-galloyl-β-D-glucopyranose in which gallic acid and products of it biochemical transformations are bound by ester (depside) bonds to glucose. Gallic acid can bind other molecules of gallic acid via hydroxyl groups to form chains containing two to five gallic acid molecules. The depside formed from two molecules of gallic acid is called m-digallic acid (**8-158**). Mono- to pentasubstituted glucoses are often called simple galloyl glucoses, unlike the complex molecules, which are called gallotannins. Less common components of gallotannins are some other saccharides, quinic acid, and other compounds. Hydrolysable tannins have a relative molecular weight up to about 5 kDa. Hydrolysis of gallotannins by acids, alkalis, or esterases (tannase, also known as tannin acyl hydrolase) yields D-glucose and gallic acid as the main products.

Ellagitannins are formed during biosynthesis of hydrolysable tannins by oxidation of two adjacent gallic acid residues and their combining through covalent C—C bonds. Ellagitannins can therefore contain hexahydroxybiphenyl residues derived from the 3,4,5,3′,4′,5′-hexahydroxybiphenylic acid (**8-158**), which arises through their acid hydrolysis, and which in the hydrolysates spontaneously dehydrates to form ellagic acid (**8-158**).

m-digallic acid   hexahydroxybiphenylic acid   ellagic acid

8-158, gallic acid derivatives

8.3.6.1.1 Gallotannins

Biosynthesis of gallotannins is based on gallic acid and an activated form of glucose (UDP-glucose). The first intermediate is 1-*O*-galloyl-β-D-glucopyranose (β-glucogallin, **8-159**), which is the donor of the galloyl residue in the biosynthesis of di- to 1,2,3,4,6-penta-*O*-galloyl-β-D-glucopyranoses (**8-159**). Typical gallotannins are products known commercially as tannin or tannic acid (containing residues of gallic and *m*-digallic acids) that are extracted from oak apples (*Quercus infectoria*, Fagaceae), tara pods (*Caesalpinia spinosa*, Fabaceae), gallnuts from *Rhus semialata* (Anacardiaceae), or Sicilian sumac leaves (*R. coriaria*). Oak apple is the common name for a large, round (20–50 mm in size) gall commonly found on many species of oak and caused by metabolites injected by the larvae of certain kinds of gall wasps (e.g. *Cynips quercusfolii*), which lay their eggs on oak leaves.

Tannic acid is a mixture of esters containing between 3 and 12 molecules of gallic acid, but also *m*-digallic acid and higher depsides. The composition is highly variable depending on the origin of the acid. Tannic acid has found use as a food additive and as an agent preventing the formation of a chill haze in beer and turbidity in wine and vinegar.

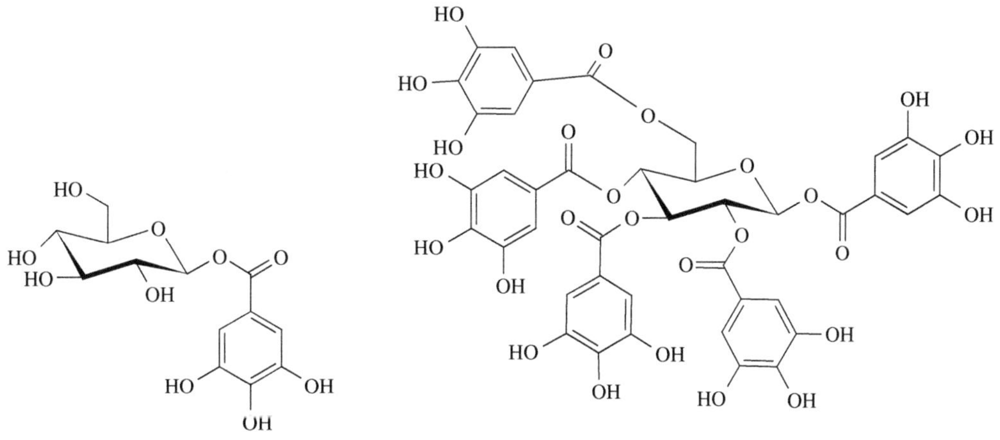

1-*O*-galloyl-β-D-glucopyranose   1,2,3,4,6-penta-*O*-galloyl-β-D-glucopyranose

8-159, simple gallotannins

8.3.6.1.2 Ellagitannins

Ellagitannins occur as minor components of commercial tannic acid and as various extracts and infusions (e.g. teas from medicinal herbs and the bark of trees). They are also natural taste-active constituents of some alcoholic beverages matured in oak barrels, such as high-quality wines and spirits (e.g. cognac, brandy, whisky, bourbon, and rum). All components extracted from the wood are degraded to some extent (Table 8.42).

Table 8.42 Contents of lignin, ellagitannins, and important phenols in cognac during ageing in an oak barrel.

Compound	Content (mg/l)				
	1 year	2 years	10 years	20 years	30 years
Lignin	12	29	127	201	219
Ellagitannins	10	25	31	17	4
Ellagic acid	7	12	32	44	55
Gallic acid	3	3	22	23	26
Vanillin	1	2	6	7	7
Syringaldehyde	1	3	11	13	14
Vanillic acid	1	2	3	4	5
Syringic acid	1	1	4	6	6

Monomers The structures of several hundred simple ellagitannins occurring in nature are known. An example of a relatively simple compound is ellagitannin strictinin (**8-160**), which occurs in green and black teas in quantities of 0.2–1.5 g/kg. Another is ellagitannin corilagin (**8-160**), from the leaves of cranberries. Tellimagrandin I (**8-160**) is an ellagitannin found in many plants (such as tea, rose hips, and walnuts) as monomers or as oligomeric ellagitannins. Punicalagins (**8-160**) are more complex ellagitannins that are found naturally as α- and β-anomers in pomegranate juice at concentrations of about 2 g/l or even higher, depending on the cultivar. Pomegranate ellagitannins of similar structure are punicalins with non-esterified C-2 and C-3 glucose hydroxyl groups.

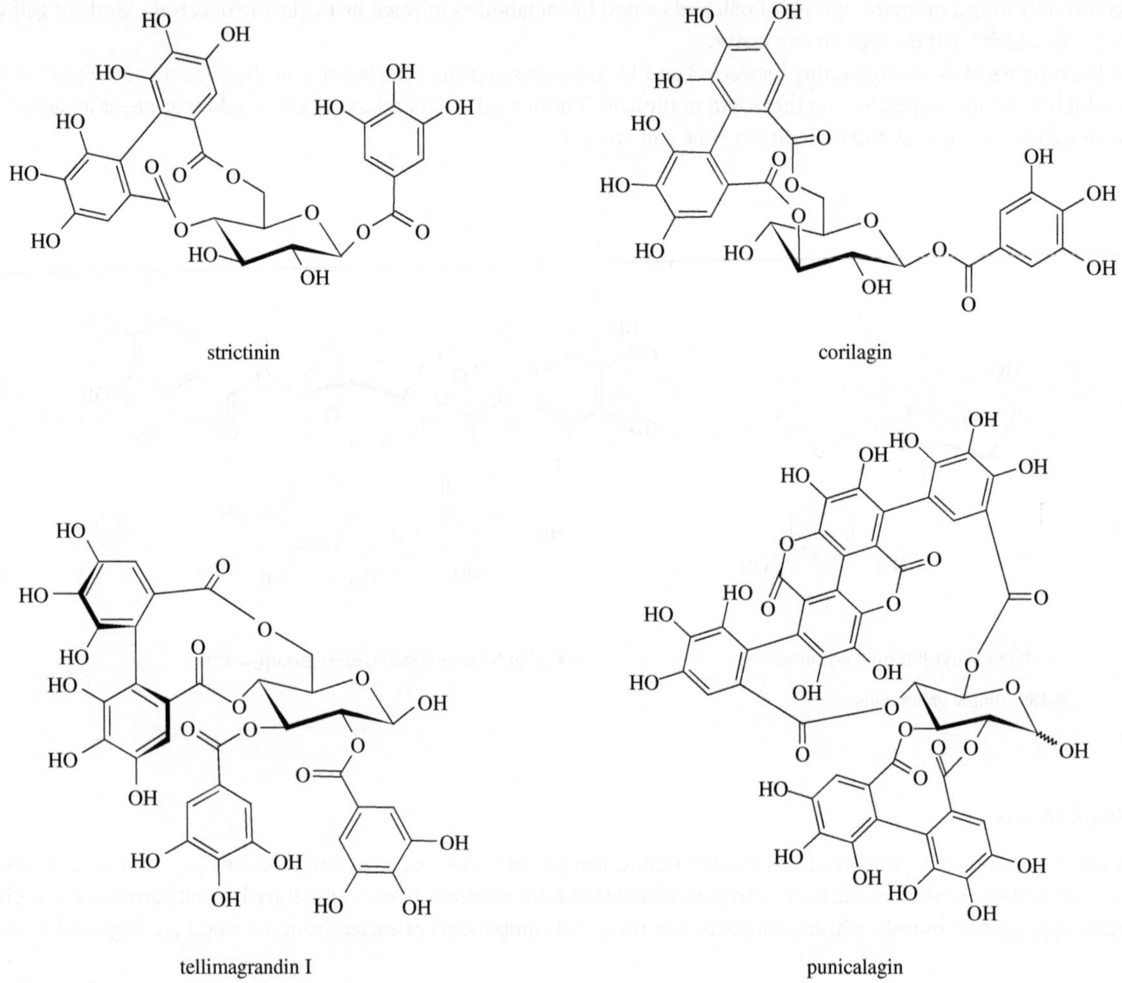

strictinin

corilagin

tellimagrandin I

punicalagin

8-160, common monomeric ellagitannins

Examples of the so-called *C*-glucosidic monomeric ellagitannins (formed from 1,2,3,4,6-pentagalloylglucose and further modified via oxidative dehydrogenation and opening of the glucopyranose ring) are the oak-derived astringent diastereomeric monomers vescalagin and castalagin (castalagin is the 33β-isomer of vescalagin) and the related *C*-glycosides grandinin and roburin E (**8-161**). For the production of oak barrels, wood of the English oak (*Quercus robur*, Fagaceae) or Sessile oak (*Q. petraea*) (in Europe) or of the white oak (*Q. alba*) (in the United States) is used. In addition to insoluble polymers (cellulose, hemicelluloses, and lignin), oak wood contains about 10% of phenolic compounds in dry matter, the predominant part of which is ellagitannins.

Oligomers and other products Only about 20 monomeric ellagitannins are components of oligomeric ellagitannins that arise from monomers by biochemical reactions (mainly oxidation). The most common oligomers are dimers, trimers, and tetramers.

Oligomeric ellagotannins are separated according to their structure into type GOG (where two galloyl units G are connected by an ether bond as in *m*-digalloyl acid; in short, *m*-GOG), type DOG (formed by joining hexahydroxybiphenyloyl and galloyl units), type D(OG)$_2$, type *C*-glucosidic (composed of monomers joined by C—C bonds), and so on. An example of a GOG oligomer is sanguiin H-6 (**8-162**), which occurs in raspberries and blackberries. An example of a *C*-glycosidic dimeric ellagitannin is roburin D (vescalagin-castalagin dimer, **8-163**), found in oak wood and in oak-matured red and white wines and bourbon whisky, where it occurs alongside other oligomeric ellagitannins.

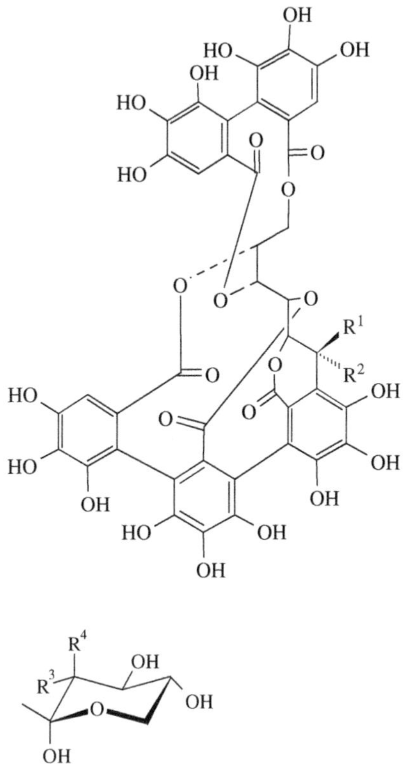

β-D-Lyx*p*-(1⟶, R³ = H, R⁴ = OH
β-D-Xyl*p*-(1⟶, R³ = OH, R⁴ = H

8-161, oak-wood ellagitannins

vescalagin, R¹ = H, R² = OH
castalagin, R¹ = OH, R² = H
grandinin, R¹ = H, R² = D-lyxose
roburin E, R¹ = H, R² = D-xylose

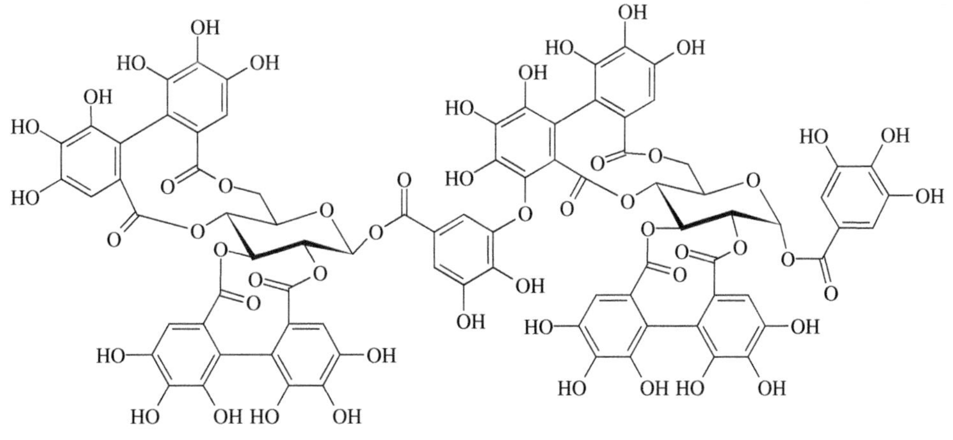

8-162, sanguiin H-6

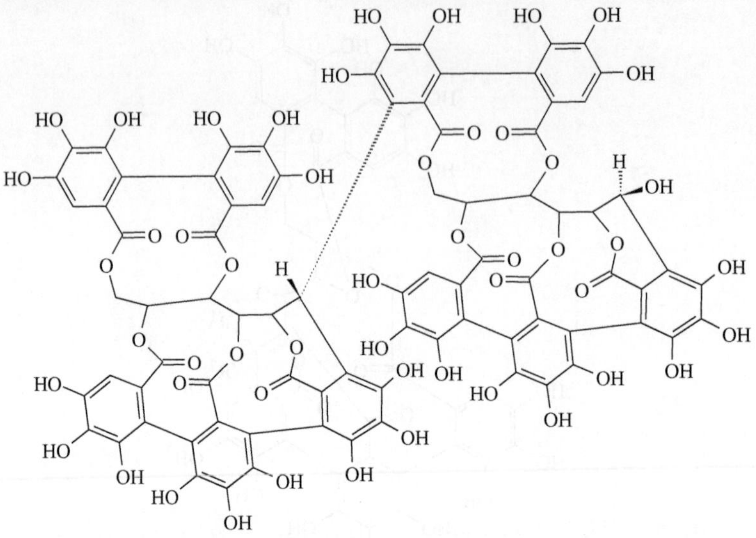

8-163, roburin D

During ageing of alcoholic beverages in oak barrels, wood tannins can be extracted from the wood and subsequently hydrolysed and transformed into bitter products by reactions with wood phenols and other flavonoids (see **8-170**).

8.3.6.2 Condensed tannins

Condensed tannins, called **proanthocyanidins** and sometimes **tannoids**, are structurally diverse oligomers and polymers of flavonoid compounds with the structure of flavan-3-ol. Obsolete names are leucoanthocyanidins, anthocyanogens, flavolanes, flavylans, and flavylogens. The term **leucoanthocyanidins** is reserved for monomeric flavan-3,4-diols. Proanthocyanidin oligomers confer an astringent and bitter taste to fruits, fruit juices, tea, wine, beer, and many other foods and drinks. Oligomers resulting from the condensation of 2–10 basic units (the number of hydroxyl groups is also important), whose molecular weight ranges up to about 5 kDa, show this taste. Higher polymers with relative molecular weights up to 400 kDa do not have the taste of tannins. These polymers, however, have a more important role as the pigments of red wines, and are also involved in the formation of hazes and sediments of wines, beers, and fruit juices.

8.3.6.2.1 Monomers

Monomeric units of condensed tannins, which lack their astringent and bitter taste, are colourless catechins, also known as 3-hydroxyflavans or flavan-3-ols (**8-164**). They are intermediates in the biosynthesis of other flavonoids. Flavan-3-ols are found in virtually all fruits, vegetables, and other plant materials. Their structure depends on the stereochemistry of the flavan-3-ol units, the number of hydroxyl groups, the stereochemistry of mutual bonds of units forming oligomers, the degree of polymerisation, and possible modification of the C-3 hydroxyl group. Monomeric flavan-3-ols with one hydroxyl group at carbon C-4′ in ring C are called afzelechins (ring B is derived from 4-hydroxybenzoic acid), monomeric flavan-3-ols with two hydroxyl at carbons C-3′ and C-4′ in ring B are called catechins (ring B is derived from protocatechuic acid), and, finally, the monomeric flavan-3-ols bearing three hydroxy groups in ring C (at C-3′, C-4′ and C-5′) are known as gallocatechins (ring B is derived from gallic acid). Other common compounds include esters of afzelechins, catechins, and gallocatechins with gallic acid bound at the C-3 position of ring C, known as afzelechin-3-O-gallates, catechin-3-O-gallates, and gallocatechin-3-O-gallates, respectively (**8-165**).

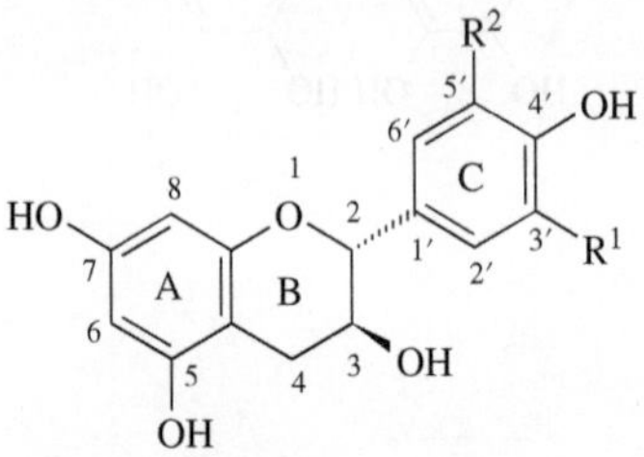

(+)-afzelechins, $R^1 = R^2 = H$
(+)-catechins, $R^1 = OH$, $R^2 = H$
(+)-gallocatechins, $R^1 = R^2 = OH$

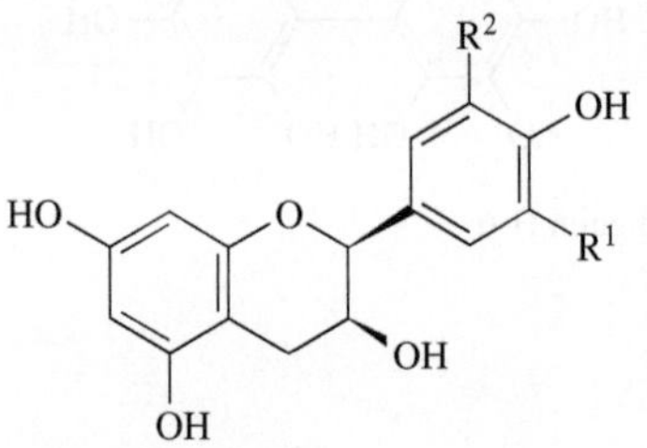

(−)-epiafzelechins, $R^1 = R^2 = H$
(−)-epicatechins, $R^1 = OH$, $R^2 = H$
(−)-epigallocatechins, $R^1 = R^2 = OH$

8-164, flavan-3-ols

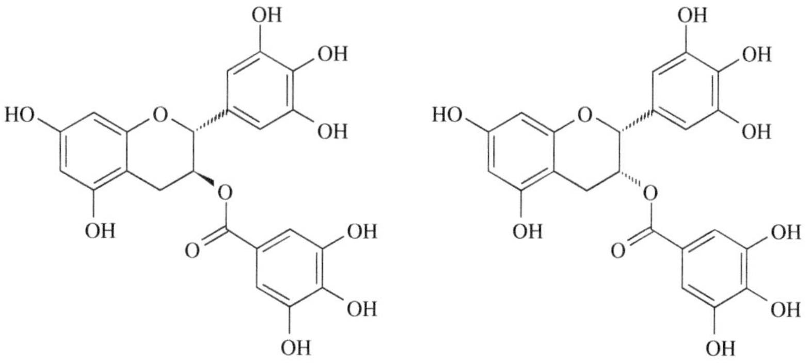

(+)-gallocatechin-3-*O*-gallate   (−)-epigallocatechin-3-*O*-gallate

8-165, gallic acid esters of gallocatechins and epigallocatechins

Flavan-3-ols and their gallates have two chiral carbon atoms (C-2 and C-3) in the molecule and may therefore occur in four isomers. The so-called (+)-afzelechins, (+)-catechins, (+)-gallocatechins, (−)-afzelechins, (−)-catechins, and (−)-gallocatechins have the C-2 and C-3 hydrogens in the (*E*)-configuration, whilst the corresponding epicatechins and epigallocatechins have the C-2 and C-3 hydrogen atoms in the (*Z*)-configuration. In nature, only (+)-afzelechins, (+)-catechins, and (+)-gallocatechins that are (2*R*,3*S*)-isomers and (−)-epiafzelechins, (−)-epicatechins, and (−)-epigallocatechins that are (2*R*,3*R*)-isomers, are found.

(+)-Afzelechins occur abundantly in lichens and in some families of higher plants, such as the plant families Ericaceae, Aesculaceae, Lauraceae, Rhizophoraceae, and Rosaceae.

The amount of catechins in fruits commonly ranges from units up to hundreds of units of mg/kg. For example, apples contain the major flavan-3-ols (+)-catechin (4–16 mg/kg) and (−)-epicatechin (72–103 mg/kg). Red wines, in addition to (+)-catechin (16–53 mg/l) and (−)-epicatechin (9–42 mg/kg), also contain (−)-epigallocatechin, whilst white wines contain (+)-catechin and (−)-epicatechin in significantly lower levels (2–6 mg/l and about 1 mg/l, respectively). Cocoa contains (−)-epicatechin, (+)-catechin, (+)-gallocatechin, and (−)-epigallocatechin. In green tea leaves, the major component is (−)-epigallocatechin-3-*O*-gallate (about 60% of all catechines), followed by (−)-epicatechin-3-*O*-gallate (10%) and, in smaller amounts (as in wine and other materials), virtually all the other catechins.

Relatively rare compounds are flavan-3-ol glycosides. For example, (+)-catechin-7-*O*-β-D-glucoside has been isolated from several plant species, such as barley, buckwheat, vigna, and rhubarb. Usually, 7-*O*-β-D-glucosides coexists in plants with other isomeric glucosides, such as 5-*O*-, 4′-*O*- and 3′-*O*-β-D-glucosides.

Monomeric flavan-3,4-diols, also known as leucoanthocyanidins (**8-166**), have two hydroxy groups in ring B at carbons C-3 and C-4, (+)-leucoanthocyanidins have these two hydroxyls in position β (above the ring plane). On heating in acidic solutions, these are transformed into colourless catechins and coloured anthocyanidins (aglycones of anthocyanins), from which are derived the trivial names of the leucoanthocyanidins. Leucoanthocyanidin with one hydroxyl group in ring C at C-4′ is called leucopelargonidin, that with two hydroxyls (at C-3′ and C-4′) is called leucocyanidin, and that with three (at C-3′, C-4′, and C-5′) is called leucodelphinidin. According to IUPAC nomenclature, (+)-leucopelargonidin is (2*R*,3*S*,4*S*)-2-(3-hydroxyphenyl)-3,4-dihydro-2*H*-chromene-3,4,5,7-tetrol.

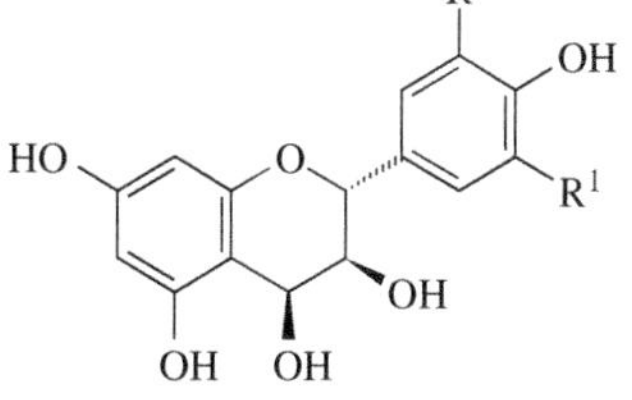

(+)-leucopelargonidin, R^1 = R^2 = H
(+)-leucocyanidin, R^1 = OH, R^2 = H
(+)-leucodelphinidin, R^1 = R^2 = OH

8-166, flavan-3,4-diols

8.3.6.2.2 Proanthocyanidins

Common types of proanthocyanidins are procyanidins that are composed of the flavan-3-ol monomers: (+)-catechin and (−)-epicatechin units. Procyanidins with 2–10 units are oligomeric procyanidins, whilst those with >10 units are defined as polymeric procyanidins. Monomeric flavanols yield oligomeric and polymeric procyanidins of several types (**8-167**).

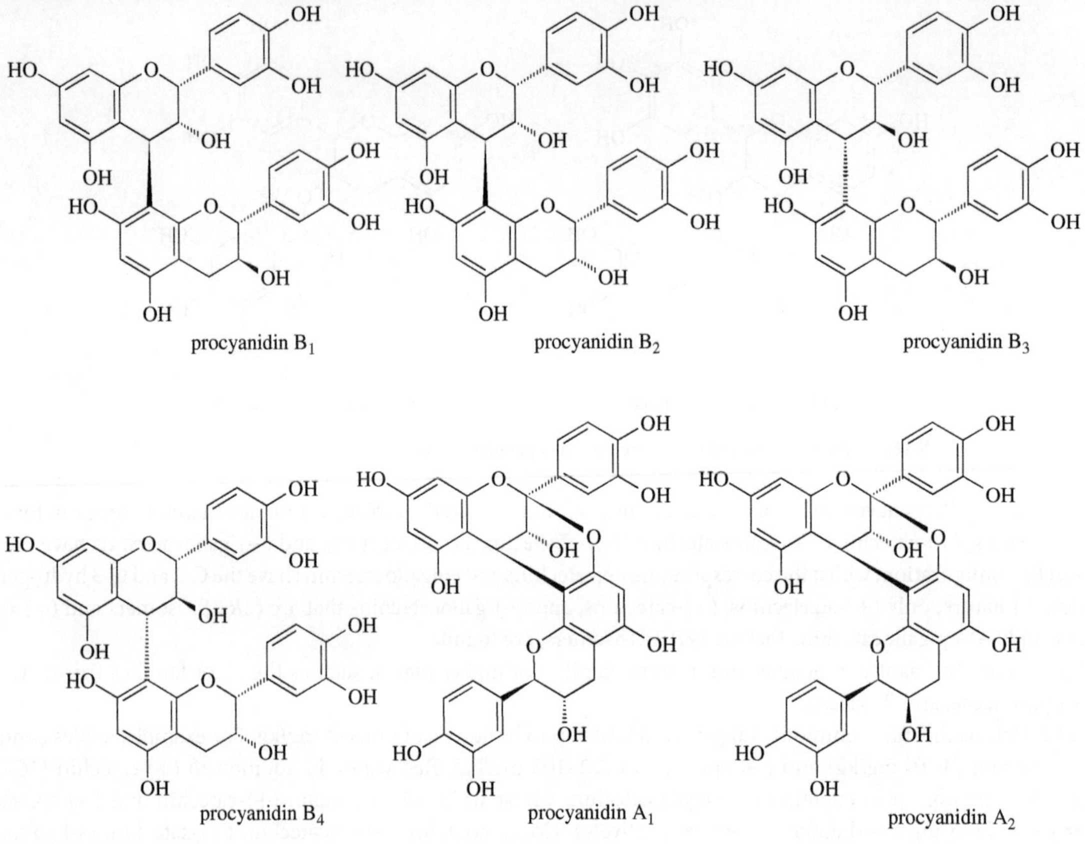

procyanidin B₁   procyanidin B₂   procyanidin B₃

procyanidin B₄   procyanidin A₁   procyanidin A₂

8-167, common dimeric procyanidins

In procyanidins of type B, the link between the monomeric units is usually in position C-4 of the upper unit and in position C-8 of the lower unit (bond $C_4 \rightarrow C_8$), in dimers B_1 to B_4, and can be either α- (below the ring plane) or β- (above the ring plane). The link between the monomeric units in dimers B_5 to B_8 is in position C-6 of the upper unit and in position C-8 of the lower unit (bond $C_6 \rightarrow C_8$). Normally, the C-3 hydroxyl group of flavan-3-ol units is esterified, mostly by gallic acid.

B-type procyanidin dimers can be converted into A-type dimers by oxidation that probably involves a free radical-driven process or an enzyme-catalysed free radical reaction via a quinone methide mechanism. Type A procyanidins have an additional ether bond between the C-7 hydroxyl of ring A of the lower unit and the C-2 hydroxyl of the upper unit (bond $C_2 \rightarrow O \rightarrow C_7$). Procyanidin A_1 is epicatechin-(4β → 8, 2β → O → 7)-catechin and procyanidin A_2 is epicatechin-(4β → 8, 2β → O → 7)-epicatechin. For example, oxidation of proanthocyanidin B_2 is converted only to proanthocyanidin A_2 ($C_4 \rightarrow C_8$ and C_2-O-C_7), whereas proanthocyanidin B_5 may result in two distinct A-type structures ($C_6 \rightarrow C_8$ and C_2-O-C_5 or C_2-O-C_7). Type C procyanidins are trimeric procyanidins.

Procyanidin B_1 is epicatechin-(4β → 8)-catechin, procyanidin B_2 is epicatechin-(4β → 8)-epicatechin, procyanidin B_3 is catechin-(4α → 8)-catechin, and procyanidin B_4 is catechin-(4α → 8)-epicatechin. The less common procyanidin B_5 is epicatechin-(4β → 6)-catechin, procyanidin B_6 is catechin-(4α → 6)-catechin, procyanidin B_7 is epicatechin-(4β → 6)-epicatechin, and procyanidin B_8 is catechin-(4α → 6)-epicatechin. Procyanidin A_1 is epicatechin-(4β → 8, 2β → O → 7)-catechin and procyanidin A_2 is epicatechin-(4β → 8, 2β → O → 7)-epicatechin. Examples of trimeric procyanidins are procyanidin C_1, epicatechin-(4β → 8)-epicatechin-(4β → 8)-epicatechin, and procyanidin C_2, catechin-(4α → 8)-catechin-(4α → 8)-epicatechin. Tetrameric procyanidins are exemplified by cinnamtannin A_2, epicatechin-(4β → 8)-epicatechin-(4β → 8)-epicatechin-(4β → 8)-epicatechin, and pentameric procyanidins by cinnamtannin A_3, epicatechin-(4β → 8)-epicatechin-(4β → 8)-epicatechin-(4β → 8)-epicatechin(4β → 8)-epicatechin (**8-168**).

It is still not known whether the condensation of monomers proceeds by enzymatic or non-enzymatic mechanisms, or both. The main building units are *trans*-2,3-flavan-3-ols; that is, (2R,3S)-isomers and *cis*-2,3-flavan-3-ols that are (2R,3R)-isomers. The majority of procyanidins arise from (+)-catechin, which is the lower unit, and (−)-epicatechin, which is the upper unit. It is assumed that the formation of the upper units involves quinones or carbocations arising from leucoanthocyanidins, anthocyanidins, and flavan-3-ols and perhaps some enzymes, such as leucoanthocyanidin reductases and polyphenol oxidases. The expected mechanism of formation of these procyanidin units is outlined in Figure 8.99.

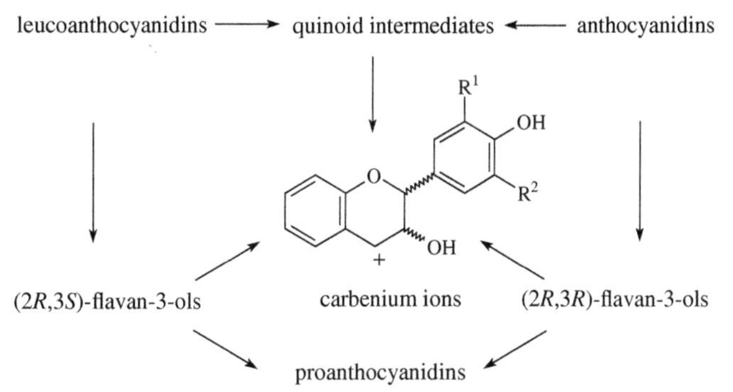

Figure 8.99 Formation of proanthocyanidins.

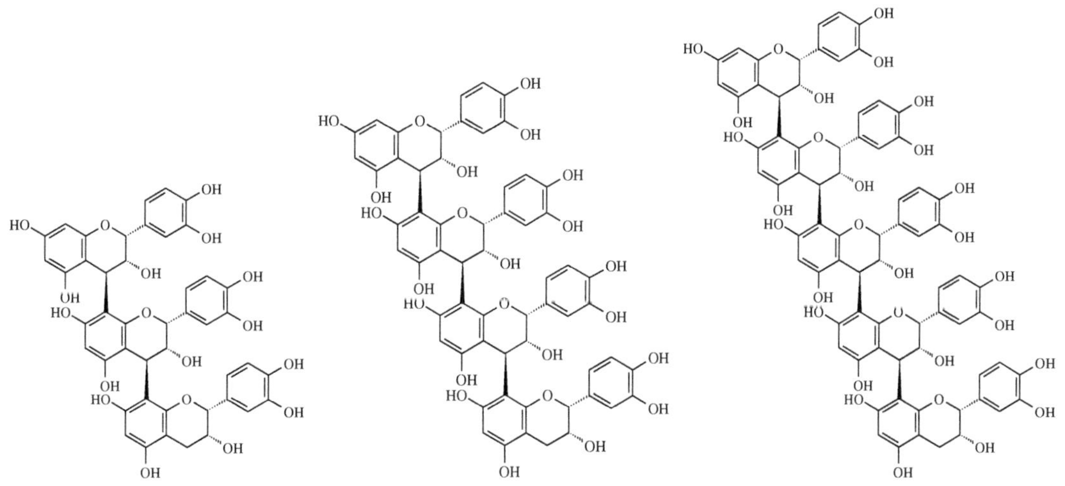

procyanidin C$_1$ procyanidin cinnamtannin A$_2$ procyanidin cinnamtannin A$_3$

8-168, common trimeric, tetrameric, and pentameric procyanidins

In food raw materials and foods, procyanidins B are the most common type. Plants usually contain different procyanidins. For example, procyanidin B$_1$ prevails in grapes, cranberries, and sorghum, procyanidin B$_2$ in apples, cherries, and cocoa beans, procyanidin B$_3$ in strawberries, and procyanidin B$_4$ in raspberries and blueberries. In blackcurrants, the dimers gallocatechin-(4α → 8)-gallocatechin and gallocatechin-(4α → 8)-epigallocatechin are present. Procyanidins from hops consist mainly of oligomeric catechins ranging from dimers to octamers, with minor amounts of catechin oligomers containing one or two gallocatechin units. Typical compounds are procyanidin dimers (B$_1$, B$_2$, B$_3$, and B$_4$) and one trimer, epicatechin-(4β → 8)-catechin-(4α → 8)-catechin. Procyanidins A$_1$ and A$_2$ are found, for example, in the shell of peanuts (*Arachis hypogaea*, Fabaceae), in the textured inedible rind of lychee fruit (*Litchi sinensis*, Sapindaceae), and in cocoa beans.

Procyanidins and other polyphenols are highly reactive compounds and suitable substrates for numerous enzymatic and chemical reactions. Frequently, these compounds are also important components of many plant foods and often have complex structures. For example, polymeric condensed tannins derived from procyanidin and prodelphinidin units form fibrous tissue filling the shells of hazelnuts (*Corylus avellana*, Betulaceae). The red filling of pecan shells (*Carya illinoensis*, Juglandaceae) is formed exclusively from prodelphinidin polymers.

Flavan-3-ol-*C*-D-glycopyranosides, such as (−)-catechin-6-*C*-D-glucopyranoside, (−)-catechin-8-*C*-D-glucopyranoside, (−)-catechin-6-*C*,8-*C*-D-diglucopyranoside (**8-169**), and related (−)-epicatechin-derived compounds, have been shown to exhibit astringent taste and modify the bitter taste intensity of cocoa beverages. These conjugates probably arise through non-enzymatic C-glycosylation of flavan-3-ols by oligosaccharides and polysaccharides.

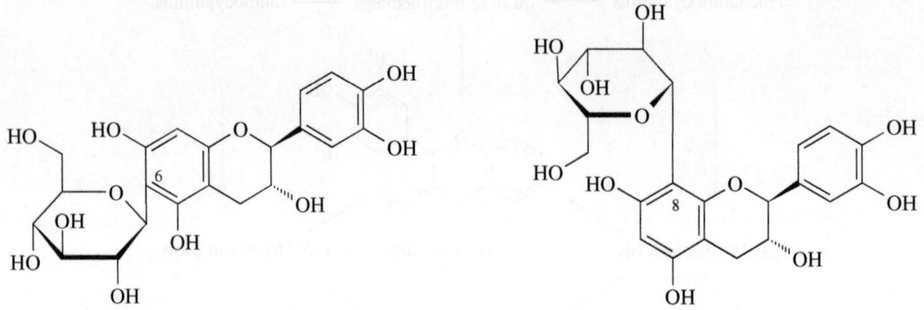

(−)-catechin-6-C-D-glucopyranoside

(−)-catechin-8-C-D-glucopyranoside

8-169, examples of catechin glucosides

Grape and wine tannins The content of procyanidins of the procyanidin type in the seeds of red grapes is two to five times higher than in the skin. In the seeds, it is mainly oligomers with two to six flavanol units that are found, whilst in the skin higher oligomers are also present. The main dimer in the skin of grapes is procyanidin B_1, which is epicatechin-$(4\beta \rightarrow 8)$-catechin, and the major trimer is epicatechin-$(4\beta \rightarrow 8)$-epicatechin-$(4\beta \rightarrow 8)$-catechin. The seeds contain mainly procyanidin B_2, which is epicatechin-$(4\beta \rightarrow 8)$-epicatechin, and the trimer epicatechin-$(4\beta \rightarrow 8)$-epicatechin-$(4\beta \rightarrow 8)$-epicatechin (procyanidin C_1). In addition to these compounds, other dimers are found in smaller quantities, such as procyanidin B_2-$3'$-O-gallate, which is epicatechin-$(4\beta \rightarrow 8)$-epicatechin-$3'$-O-gallate, as well as trimers and higher oligomers. In addition to procyanidins, prodelphinidins are also present in small amounts. Tannins of red wines have similar composition. Tannins pass to the wine mainly from the grape seeds. Their content in red wines (including monomeric catechins) is about 80–270 mg/l, whilst in white wines it is approximately 4–13 mg/l.

Catechins and lower oligomers show low astringency but have a rather bitter taste. Oligomers containing more than four molecules of monomers are slightly bitter and very astringent. In addition to colourless tannins, red wines contain tannin complexes with polysaccharides and minerals and coloured reaction products of tannins with anthocyanins.

The degradation of flavan-3-ol oligomers (procyanidins) to flavan-3-ols and flavan-3-ol 4-sulfonates (see Section 9.4.1.5.7) during wine ageing may provide an explanation for the reduction in astringency of aged red wines.

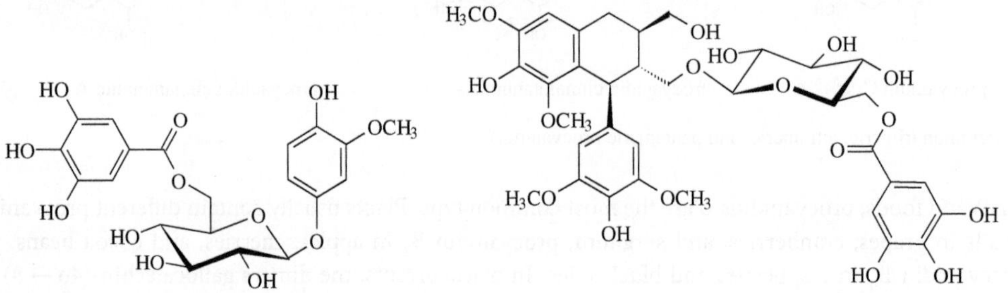

3-methoxy-4-hydroxyphenyl-1-O-
β-D-$(6'$-O-galloyl)glucopyranoside

quercoresinoside A

8-170, bitter glucosides from oak wood

Cider tannins The bitter and astringent taste of cider from specific varieties of bitter apples is caused by the presence of oligomeric procyanidins, mainly tetramers and higher oligomers.

Cocoa tannins Flavan-3-ols and procyanidins are the main tannins in cocoa and cocoa products. Their content in various cocoa products varies between 2 and 500 g/kg. The main flavan-3-ol in unroasted (raw) cocoa beans is (−)-epicatechin. The main dimeric procyanidins are B_2 (340–650 mg/kg), B_5, and B_1, followed by dimeric procyanidin A_2 (9–13 mg/kg), trimeric procyanidin C_1, tetrameric procyanidin cinnamtannin A_2, and pentameric procyanidin cinnamtannin A_3. Procyanidins can be partially epimerised through cocoa bean fermentation and partially degraded during roasting.

Tea tannins Like other plant materials, fresh tea leaves contain flavan-3-ols and bitter and adstringent proanthocyanidins, which are also found in green tea. Condensed with other polyphenols, they occur in oolong and black tea (see Section 9.12.5.4).

8.3.6.3 Phlorotannins

The phlorotannins are a group of more than 150 tannins found in brown algae (Phaeophyceae) and in a lower amount in some red algae (Rhodophyta). The basic building units of phlorotannins (also known as seaweed polyphenols), rather than gallic acid, are dehydrooligomers or dehydropolymers of phloroglucinols (1,3,5-trihydroxybenzenes) bound in the form of glycosides. Examples of phlorotannins from the brown alga *Fucus vesiculosus* (Fucaceae), known by the common name bladder wrack, are tetrafucol with a phenyl linkage between phloroglucinol molecules and phlorofucofuroeckol (**8-171**) with two ether linkages and the dibenzo-1,4-dioxin and dibenzofuran structural elements. Phlorotannins are integral structural components of cell walls in algae, but they also play a role in protection against UV radiation, in osmoprotection, and in defence against grazing. They may inhibit the formation of advanced glycation end products (AGEs) by scavenging reactive carbonyls.

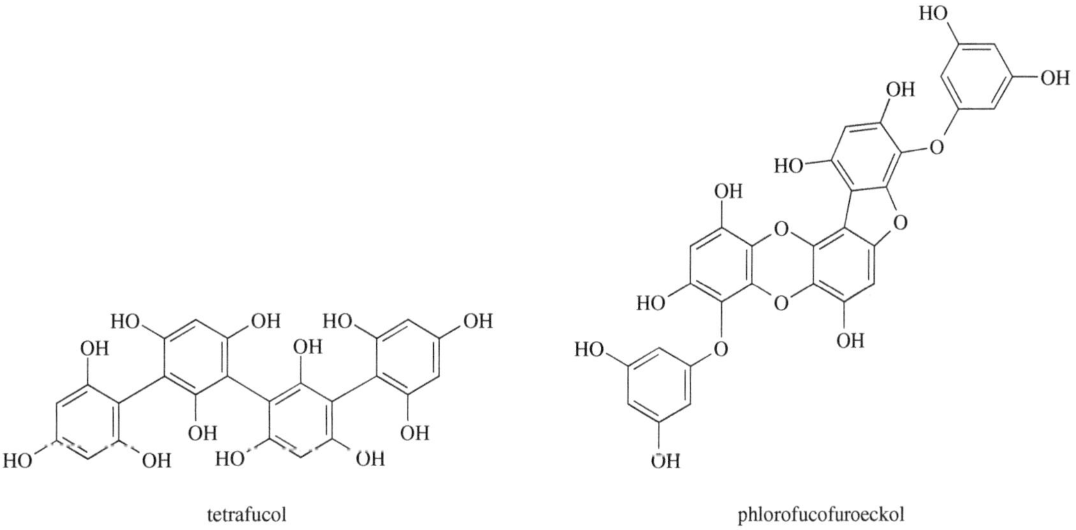

tetrafucol

phlorofucofuroeckol

8-171, phlorotannins

8.3.6.4 Miscellaneous astringent compounds

Astringency is traditionally thought to be induced by plant tannins in foods through tannin–protein interactions. However, a wide range of other phenolic compounds and some non-phenolic compounds in various foods can elicit astringency. Many ellagitannins that contribute to astringent properties of foods do not interact with salivary proteins and may be directly perceived through some receptors. The astringency of organic acids, such as (*Z*)- and (*E*)-isomers of aconitic acid in red currants, may be directly linked to the perception of sourness. In addition to flavan-3-ol glycosides and aconitic acids, the key astringent and mouth-drying compounds in red currants are astringent indoles and nitriles (**8-172**). Examples of these compounds are 3-carboxymethylindole-1-*N*-β-D-glucopyranoside (R = H), 3-methylcarboxymethylindole-1-*N*-β-D-glucopyranoside (R = CH$_3$), (*E*)-2-(4-hydroxybenzoyloxymethyl)-4-β-D-glucopyranosyloxybut-2-enenitrile (R = H) derived from 4-hydroxybenzoic acid, and (*E*)-2-(4-hydroxy-3-methoxybenzoyloxymethyl)-4-β-D-glucopyranosyloxybut-2-enenitrile derived from vanillic acid (R = CH$_3$). It is mainly isoflavones (see Section 9.4.2.8) and various saponins (see Section 10.3.2.2) that contribute to the astringent taste of soya beans. The key contributors to the astringent taste of rapeseeds and rape meals are some esters of sinapic acid called sinapines (see Section 8.2.7.1.1), whilst those in non-fermented cocoa beans and cocoa products are *N*-phenylpropenoyl amino acids (see Section 2.2.1.2.1).

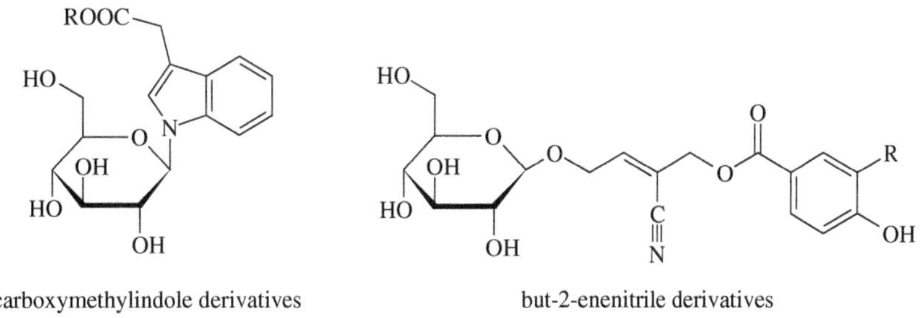

carboxymethylindole derivatives

but-2-enenitrile derivatives

8-172, astringent indoles and nitriles

8.3.7 Pungent substances

Burning, stinging, sharp, hot, and pungent flavour is a characteristic phenomenon accompanying the consumption of some spices, such as hot chillies, pepper, ginger, and cloves, spice mixtures that contain these spices, special sauces, some vegetables of the Brassicaceae family such as mustard, horseradish, and radish, and vegetables of the Amaryllidaceae family such as garlic and onion.

8.3.7.1 Hot pepper

The substances responsible for the hot taste of different domesticated pepper genera (*Capsicum annuum*, *C. frutescens*, *C. chinense*, *C. pubescens*, and *C. baccatum*) and their varieties are protoalkaloids **capsaicinoids**, which include a group of vanillylamides derived from C_8–C_{11} (*E*)-monoenic branched-chain fatty acids (isoacids and anteisoacids) and saturated fatty acids with branched or straight chains. (*Z*)-Isomers do not occur in peppers. About 90% of the capsaicinoids are represented, in approximately equal proportions, by capsaicin and dihydrocapsaicin; the rest are other, related compounds (see Section 10.3.6.12). The capsaicinoid content of fresh bell (sweet) pepper cultivars (*C. annuum*) is often negligible (0.001% or less), but in some varieties of *C. annuum* (such as cayenne and jalapeño peppers) the level is in the range of 0.2–1%, and in others, as well as in some varieties of *C. frutescens* (such as tabasco and chilli peppers), it may be even higher. The hottest peppers are some varieties of *C. chinense* (such as habanero).

The taste-recognition threshold of capsaicin is about 0.1 mg/kg, and a concentration of 10 mg/kg causes a strong burning sensation. The burning sensation of dihydrocapsaicin is almost the same. The sharpness of both capsaicinoids is about 150–300 times higher than that of the hot components of black pepper and ginger. The pungency of hot peppers is evaluated using Scoville heat units (SHU). Typical relative pungencies in SHU for the main hot pepper substances capsaicin, dihydrocapsaicin, nordihydrocapsaicin, homodihydrocapsaicin, and homocapsaicin are 16 000 000, 16 000 000, 9 000 000, 8 600 000, and 8 600 000, respectively. Red habanero pepper has 150 000, tabasco pepper 120 000, jalapeño pepper 25 000, and sweet bell pepper 0 SHU.

8.3.7.2 Black pepper

Black peppers are fruits of *Piper nigrum* (Piperaceae), a plant native to south India, which is highly appreciated by consumers for its aroma and taste. The pungent, burning, and tingling oral impression imparted by black pepper is mainly produced by protoalkaloid piperine, along with over 20 other related compounds (see Section 10.3.6.4.1).

8.3.7.3 Ginger and galangal

Ginger is the rhizome of the *Zingiber officinale* (Zingiberaceae) plant, native to Southeast Asia. The pungent spicy components of ginger are phenolic alkanones known as gingerols and shogaols (**8-173**). Their total content is 0.7–0.9% in the fresh rhizome and 1.1–1.6% in the dry rhizome.

The main and most important component is [6]-gingerol, (5*S*)-5-hydroxy-1-(4-hydroxy-3-methoxyphenyl)decan-3-one, which represents about 75% of hot substances in oleoresins. The end of its side chain is derived from C_6 aldehyde (hexanal). [6]-Gingerol is accompanied by its higher homologues [8]-gingerol (11%) and [10]-gingerol (15%), derived from octanal and decanal, respectively, which occur in smaller amounts in ginger rhizome. Minor pungent components are [6]-shogaol (0.5%), [8]-shogaol, and [10]-shogaol. The hottest component of ginger is [6]-gingerol; its homologues [8]-gingerol and [10]-gingerol are about 4–15 times less spicy, whilst [6]-shogaol is 2–60 times less pungent.

In addition to these compounds, several tens of structurally related minor compounds occur in *Z. officinale* and other ginger species, such as a homologue of [6]-gingerol called [12]-gingerol, methoxy-[6]-gingerol, deoxy-[6]-gingerol, 1-dehydro-[6]-gingerdione, diketone [10]-gingerdione, (3*R*,5*S*)-diol known as [6]-gingerdiol (**8-174**), and the curcuminoids tetrahydrocurcumin and hexahydrocurcumin (see Section 9.6). [6]-Gingerdiol occurs in the form of 4′-*O*- and 5-*O*-β-D-glucopyranosides.

The compositions of ginger rhizome and of ginger oleoresin components vary within certain limits, depending primarily on the variety of plant and the post-harvest operations. Thermally processed and stored materials contain higher quantities of shogaols, created by dehydration of gingerols. Retroaldolisation of gingerols yields pungent zingerone and the corresponding alkanals, which cause an off-flavour. [6]-Gingerol yields hexanal (Figure 8.100), [8]-gingerol octanal, and [10]-gingerol decanal on degradation.

Ginger and its components are also known to possess such physiological features as antimicrobial, antioxidative, antitumor, and antiplatelet aggregation activities.

Galangal, also known as Thai or Siamese ginger, is the aromatic and pungent rhizome of *Alpinia galangal*, another plant in the ginger family. Galangal rhizome is widely used as a spice for flavouring food in East Asia. It is to Thai cooking what common ginger is to Chinese cooking. The pungent principle of galangal was identified as 1′-acetoxychavicol acetate (galangal acetate, **8-175**). Galangal acetate is about 10 times less pugent than capsaicin and lacks its lingering effect. In aqueous solutions, it undergoes hydrolysis and isomerisation reactions.

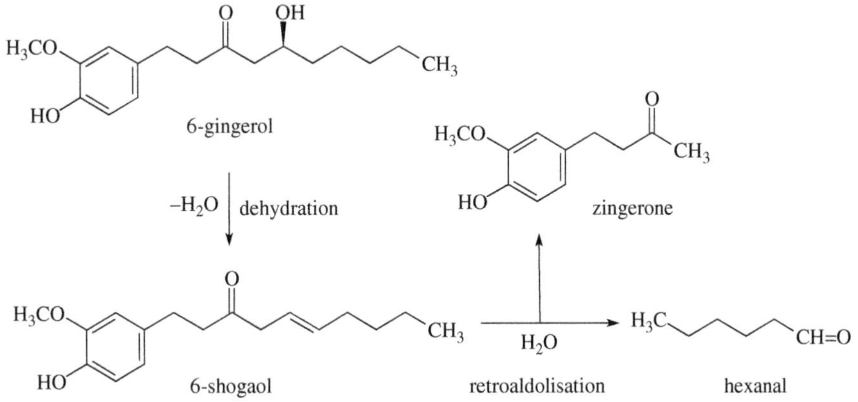

Figure 8.100 Degradation of gingerols.

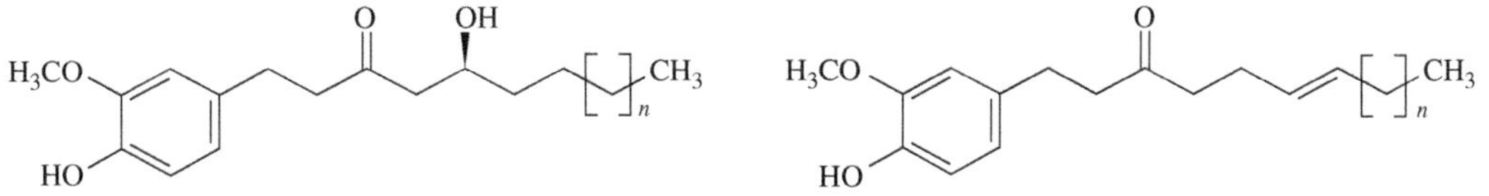

8-173, gingerols and shogaols

[6]-gingerol, $n = 2$
[8]-gingerol, $n = 4$
[10]-gingerol, $n = 6$

[6]-shogaol, $n = 2$
[8]-shogaol, $n = 4$
[10]-shogaol, $n = 6$

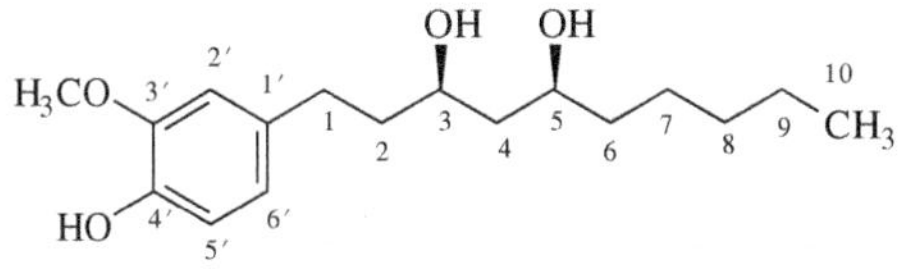

[6]-gingerdiol

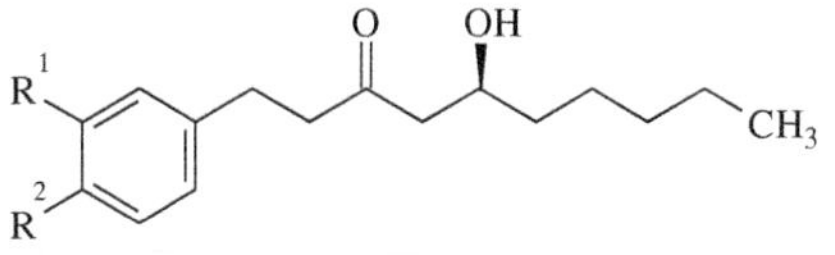

methoxy-[6]-gingerol, $R^1 = OCH_3$, $R^1 = OCH_3$
deoxy-[6]-gingerol, $R^1 = H$, $R^2 = OH$

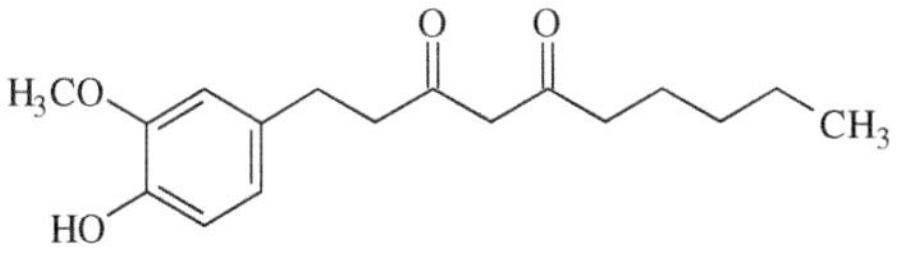

[6]-gingerdione

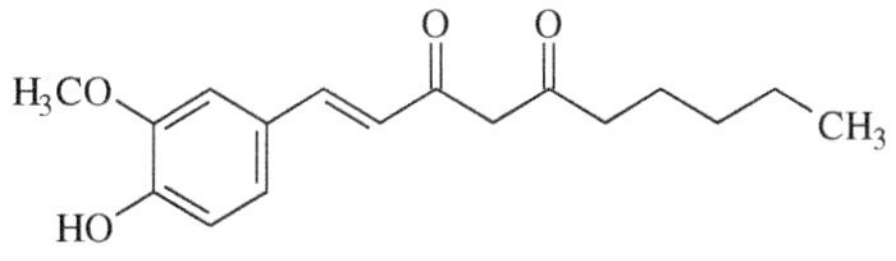

1-dehydro-[6]-gingerdione

8-174, selected minor constituents of ginger rhizome

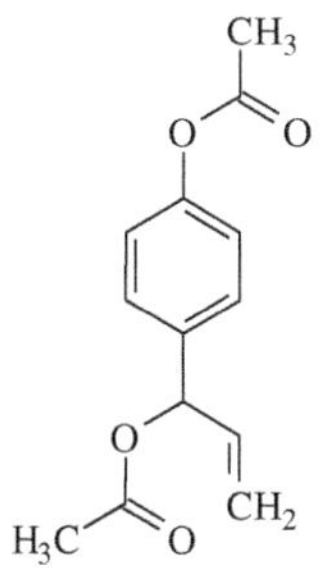

8-175, galangal acetate

8.3.7.4 Clove

Cloves, dry flower buds of the *Eugenia caryophyllata* (Myrtaceae) tree, native to Zanzibar and a number of South and Southeast Asian countries, contain 15–20% of the essential oil, the main part (80–90%) of which is eugenol (4-allyl-2-methoxyphenol, **8-83**). Eugenol has the characteristic clove smell and a slightly pungent taste.

8.3.7.5 Vegetables and condiments

The family of the cruciferous plants (Brassicaceae) includes a number of vegetables that have a pungent and spicy taste if eaten raw. An example of a vegetable with extreme burning taste is grated horseradish (*Armoracia rusticana*) root. A less pungent taste is found in some raw vegetables of the genus *Brassica*, such as cabbage, Savoy cabbage, kale, radishes, and black (*B. nigra*), brown (Indian or Chinese mustard, *B. juncea*), and white (*B. alba*) mustard seeds, which are used to produce mustard pastes. For example, the French-style Dijon mustard is produced from black mustard seeds and some varieties of brown mustard, English mustard is made using white, black, and brown mustard seeds (flour), and Krems mustard is made from a mixture of white and black mustard seeds.

The characteristic aroma and pungent taste of Brassica vegetables and mustard seeds is related to the presence of isothiocyanates, which are produced as a result of the enzyme myrosinase (thioglucoside glucohydrolase) activity in damaged plant tissues on sensory-indifferent glucosinolates (see Section 10.3.2.4). One of the most common isothiocyanates is allyl isothiocyanate. Allyl isothiocyanate is produced from glucosinolate sinigrin, the content of which in horseradish is 27–29 g/kg (which roughly corresponds to 8–9 g/kg of allyl isothiocyanate). Other isothiocyanates also contribute to the spicy flavour of horseradish, such as 2-phenylethyl isothiocyanate, which arises from glucosinolate gluconasturtiin (the content of which is within 4.2–7.2 g/kg). In the cones of cabbage and other cruciferous vegetables, the level of sinigrin and gluconasturtiin is much lower (4–146 and 0.7–6.1 mg/kg f.w., respectively). In the case of cabbage, the amount of sinigrin corresponds to 1–45 mg/kg of allyl isothiocyanate. Some radish varieties, such as the grey-black radish with round roots, contain as the main pungent substance (*E*)-4-(methylthio)-but-3-en-1-yl isothiocyanate, which is produced from the glucosinolate glucoraphasatin. Its concentration in radishes is about 1–3 g/kg, resulting in an isothiocyanate content of 0.5–1 g/kg.

Allyl isothiocyanate is also the main ingredient of table mustard pastes made from black mustard seeds. The level of sinigrin in seeds is 89–100% of the total glucosinolate content, which corresponds to 18–45 g/kg. Its content in table mustards made from black mustard seeds is 10–18 g/kg. The glucosinolates gluconapin (0–11%) and glucobrassicanapin (0–0.3%) are found in lower amounts. The brown mustard seed content of allyl isothiocyanate is just 7 g/kg. Another major isothiocyanate of brown mustard seeds is but-3-en-1-yl isothiocyanate, which is present at a concentration of 3–4 g/kg. White mustard seeds contain 4-hydroxybenzyl isothiocyanate as the main substance, which arises from glucosinolate sinalbin. 4-Hydroxybenzyl isothiocyanate is usually only mildly spicy. The sinalbin content in seeds ranges between 20 and 50 g/kg. The content of 4-hydroxybenzyl isothiocyanate in autolysed seeds is 15–37 g/kg.

Allyl isothiocyanate is also the main pungent and spicy component of Japanese horseradish, called wasabi (*Eutrema japonicum*), of the crucifer family (Brassicaceae). Its content in wasabi is 7–10 g/kg. The unique flavour of wasabi is a result of complex chemical mixtures from the broken cells of the plant, allyl isothiocyanate, and methylthioalkyl isothiocyanates: 6-(methylsulfinyl)hexyl isothiocyanate, 7-methylthioheptyl isothiocyanate, and 8-methylthiooctyl isothiocyanate.

8.3.7.5.1 Alliaceous vegetables

The key flavour precursors of garlic (*Allium sativum*, Amarillidaceae), onion (*A. cepa*), shallot (*A. cepa* var. *aggregatum*), leek (*A. ampeloprasum*), chives (*A. schoenoprasum*), and other alliaceous vegetables are the sulfur-containing amino acids *S*-alk(en)ylcysteine sulfoxides, also known as *S*-alk(en)ylcysteine-*S*-oxides (see Section 2.2.1.2.2). The typical aroma and taste of these vegetables develops as a result of the decomposition of *S*-alk(en)ylcysteine sulfoxides by the C-S lyase called alliinase to pungent dialk(en)ylthiosulfinates. Dialk(en)ylthiosulfinates, however, are extremely unstable substances that quickly decompose on prolonged standing, depending on the temperature and polarity of the environment, to form a variety of secondary products (such as thiosulfonates, disulfides, and trisulfides), which are responsible for the aroma and taste of processed alliaceous vegetables.

8.3.8 Cooling substances

(−)-Menthol (**8-14**) is a common cooling substance used in food products. The effect of coolness evoked through stimulation of the somatosensory system can also be produced by several other naturally occurring molecules, mainly derived from terpenes and sesquiterpenes (**8-176**). One such molecule is (−)-isopulegol, occurring, for example, in the essential oil of Australian lemon-scented gum (*Corymbia citriodora*, Myrtaceae) trees, which has 20% of the cooling power of (−)-menthol. Similar cooling agents are (+)-(*Z*)- and (−)-(*E*)-*p*-menthane-3,8-diols. Alternative natural cooling molecules are (−)-menthone in peppermint oil and (+)- and (−)-piperitone in mint oils, which are weakly cooling. Several other terpenes, such as (+)-1,8-cineole (eucalyptol, **8-21**), which occurs in essential oils of various spices and trees of the genus *Eucalyptus*, and verbenol, from marjoram and sage oils, are slightly cooling substances. The only cooling sesquiterpene is cubebol, which is found in cubeb (tailed pepper, *Piper cubeba*, Piperaceae) essential oil. A considerable number of compounds showing a cooling effect have been synthesised and evaluated for the physiological sensation of cooling. Many artificial cooling compounds are menthol derivatives (esters, ethers, and acetals) or have very different chemical structures.

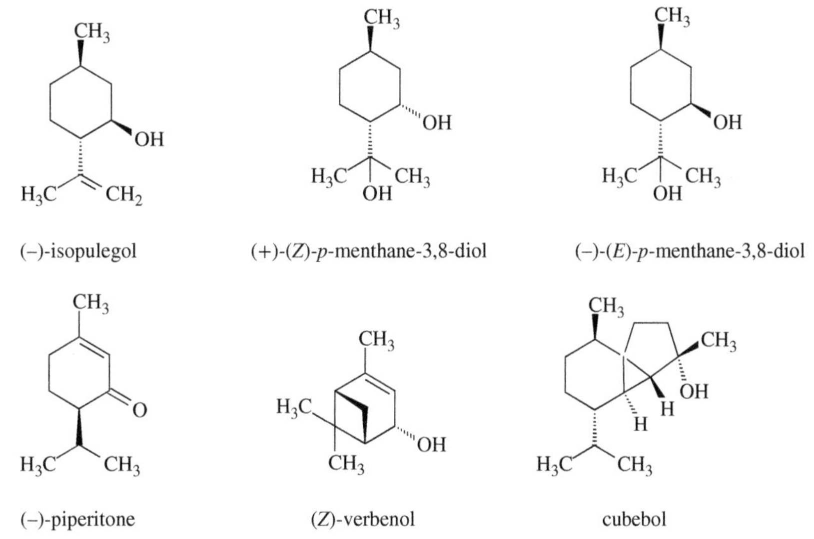

(–)-isopulegol   (+)-(Z)-p-menthane-3,8-diol   (–)-(E)-p-menthane-3,8-diol

(–)-piperitone   (Z)-verbenol   cubebol

8-176, cooling-active terpenoids

Some cooling compounds are also present in raw food materials and foods. For example, 3-methyl- ($R^1 = CH_3$, $R^2 = H$) and 5-methyl-2-(1-pyrrolidinyl)cyclopent-2-en-1-one ($R^1 = H$, $R^2 = CH_3$, **8-177**) were recently identified as intense cooling compound in roasted dark malt. It was shown that these 2-aminocyclopent-2-en-1-ones were produced by the Maillard reaction from hexose-derived cyclotene (2-hydroxy-3-methylcyclopent-2-en-1-one) and 1-pyrrolidine formed by the Strecker degradation of proline.

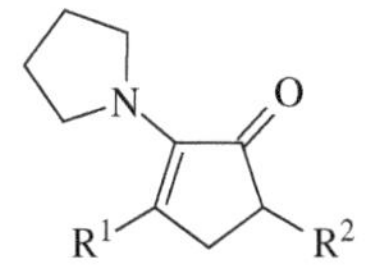

8-177, 1-pyrrolidinylcyclopent-2-en-1-ones

8.3.9 Physiology, nutrition, and use

The main sources of flavan-3-ols and proanthocyanidins are tea, fruits, cocoa products, and wine. In the European Union, the mean daily intake of flavan-3-ols is 77 mg, whilst that of proanthocyanidins is 123 mg. Proanthocyanidins have different physiological effects. Recently acknowledged are their anti-inflammatory and anti-allergic effects and their beneficial effects in the development of atherosclerosis (related to their reaction with free radicals). Condensed tannins have also found use as food additives. A diet that is high in tannins may also, however, show negative effects, such as lower utilisation of proteins.

The physiological effects of hot piperidine alkaloids (see Section 10.3.6.4) and of capsaicinoids (see Section 10.3.6.12) are described elsewhere. Gingerols have only low acute toxicities, but show carminative (prevent the formation of gas in the gastrointestinal tract), antiemetic (prevent or alleviate nausea and vomiting), and antioxidative effects. Allyl isothiocyanate is used as a food flavouring agent (e.g. in mayonnaise, horseradish, mustards, and various special sauces), as a preservative in animal feed, as a rubefacient (counterirritant), as a fungicide, in ointments, in mustard plasters, as an adjuvant (enhancing the action of medical treatment), and as a repellent for cats and dogs. The physiological properties of garlic constituents were described in Section 8.2.9.2.2. Eugenol is potentially hepatotoxic; an overdose may cause a wide range of symptoms. It is subject to restrictions on its use in perfumery as it may cause an allergic reaction in sensitised people.

9

Pigments and Other Colourants

9.1 Introduction

Colour is the property possessed by an object of producing different sensations on the eye as a result of the way it reflects or emits light. Colour plays a crucial role in food choice and acceptance, and alongside flavour and texture is considered a major quality factor of food. Substances whose presence in living cells determines a characteristic colour are called biological pigments.[1] In food raw materials and in many foods, these substances are known as **food pigments**. In general contexts, a pigment is any material from which a dye may be prepared. The term 'dye' refers to a colourant used to colour or stain various materials using dyestuffs. The general term 'colourant' or 'colour additive' means any substance, such as a pigment or dye, that imparts colour. The harmless substance used as a food additive to impart colour to food or drink is a **food dye** or **food colourant**. Food colourants are also used in a variety of non-food applications, including cosmetics and pharmaceuticals.

Food pigments are usually divided into three main types:

- natural pigments (approximately 1500 natural pigments have been isolated from foods);

- synthetic pigments identical with natural pigments (nature identical pigments);

- synthetic pigments not found in nature.

Natural pigments are coloured substances that are synthesised, accumulated, or excreted to the environment by living cells. These are:

- natural parts of foods of animal or vegetable origin due to the genetic dispositions of the given organism;

- parts of other natural materials (pigments of algae, fungi, lichens, or microorganisms) that are obtained in the original state as such or are structurally modified and used for food colouring as colour additives.

Coloured products derived from natural raw materials through various technological processes, such as caramel and malt extracts, which contain melanoidins, are also considered natural colourants. Natural colourants additionally include copper complexes of chlorophylls and chlorophyllins, which do not occur in nature (or which may be present in foods in negligible amounts), inorganic pigments such as calcium carbonate, iron(III) oxide (iron trioxide) and titanium(V) oxide (titanium dioxide), and nature identical synthetic dyes.

Natural pigments are classified according to their structure, occurrence in biological materials, or important properties (such as their solubility in water and fat).

This chapter describes pigments occurring in individual food raw materials and food commodities, their properties and stability, their importance in human nutrition, and their potential use as food colourings in food technology. Attention is also paid to the desirable and undesirable enzymatic and chemical colour changes that proceed during the processing and storage of food raw materials and foods. Substances used as colour additives are described in Chapter 11.

---

[1] Pigment colour differs from structural colour in that it is the same for all viewing angles, whereas structural colour is the result of selective reflection or iridescence (change of colour as the angle of view/illumination changes).

The Chemistry of Food, Second Edition. Jan Velíšek, Richard Koplík and Karel Cejpek.
© 2020 John Wiley & Sons Ltd. Published 2020 by John Wiley & Sons Ltd.

9.2 Tetrapyrroles

Tetrapyrrole pigments are a numerically small but very widespread and significant group of differently coloured pigments. Their structure, a resonance-stabilised planar ring system, consists of four pyrrole rings A, B, C, and D connected in a tetrapyrrole (porphyrin) circle with a conjugated double bond system through methine bridges in the α-positions of pyrrole structures (**9-1**) or arranged linearly (**9-2**). Accordingly, two basic groups of tetrapyrrole pigments are recognised:

- cyclic **porphyrins** (porphyrin pigments);

- linear (open chain) **bilins** (bilin pigments).

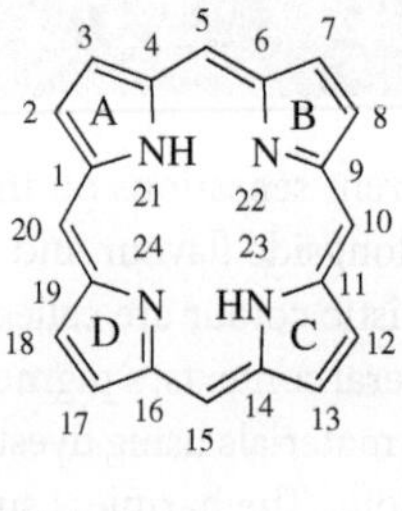

9-1, basic structure of porphyrin pigments

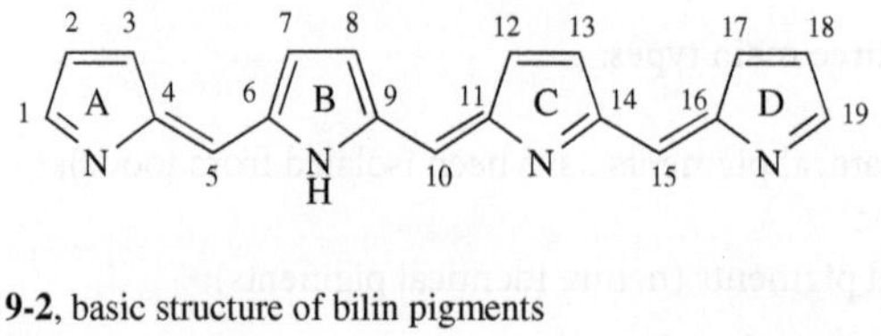

9-2, basic structure of bilin pigments

It has been reported that 28 porphyrin pigments and their precursors occur in nature. Porphyrins are chromophores of two basic groups of metalloproteins:

- pigments of animal tissues, called haem pigments or **haems**;

- pigments of plant tissues, some algae, and some microorganisms, called chlorophyll pigments or **chlorophylls**.

Bilins are usually classified according to their origin as:

- phycobilins, which are pigments of some algae;

- bile pigments resulting from degradation of haem pigments (pigments in urine and faeces of mammals).

9.2.1 Haem pigments

The most important haem pigments are metalloproteins:

- **myoglobin**, the red pigment of muscle tissue;

- **haemoglobin**, the pigment of red blood cells (erythrocytes).

Interest in haem pigments is focused primarily on their biochemical properties in relation to the transfer of oxygen in tissues, and also to the metabolism of bile pigments. However, food chemists and technologists are mainly interested in the stability and changes of haem pigments during the processing and storage of meat and meat products.

9.2.1.1 Structure and nomenclature

The basis of the structure of myoglobin, haemoglobin, and derived pigments is a substituted cyclic tetrapyrrole protoporphyrin IX (**9-3**) with a central atom of divalent iron, which is called **haem** or protohaem (**9-4**). The haem molecule is almost planar. Haem is a conjugated system and although there are two coordination-covalent Fe—N bonds (the bonding electron pair apparently belongs to only one of the atoms); in fact, all four bonds with pyrrole rings are equivalent. The fifth and sixth coordination bonds of iron (above and below the plane of the planar molecule) can be occupied by different ligands. **Haemin** is protoporphyrin IX with Fe(II), where the counter-ion is a chloride ion. Protoporphyrin IX with trivalent iron, an oxidation product of haem, is called **haematin**, in which a hydroxyl group is bound to the central Fe(III) atom as a counter-ion.

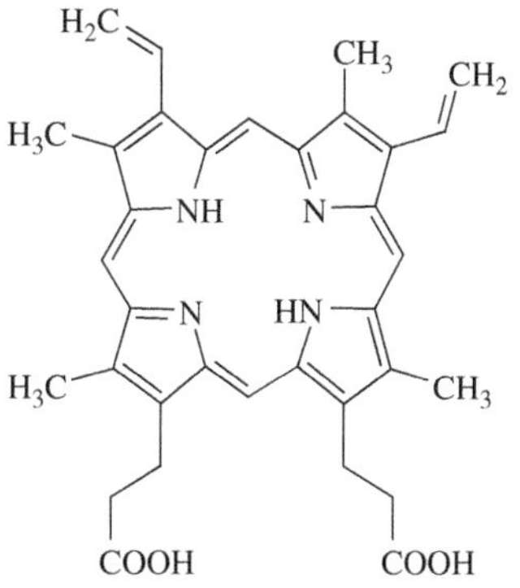

9-3, protoporphyrin IX

In muscle tissue, other haem pigments also occur. One example is cytochromes, which carry out electron transport and contain iron in a similar porphyrin protein complex to that in myoglobin. Another is zinc protoporphyrin IX, which is a normal minor metabolite. During iron insufficiency or impaired utilisation, its concentration in muscles increases. Haem pigments can similarly be found in the plant kingdom in the form of cytochromes and some oxidoreductases, such as catalase and peroxidase.

In myoglobin (**9-5**) and its derivatives, the fifth ligand is bound to the divalent iron atom via the imidazole group of histidine (His^{93}). The fifth ligand is a protein (globin) with a relative molecular weight of 16.8 kDa. The sixth ligand is water, but this may be absent. The molecule of haemoglobin (referred to as haemoglobin A) includes four polypeptide chains (two pairs of identical subunits α and β; $\alpha_2\beta_2$ for short), each containing a single haem molecule (of relative molecular weight 64.5 kDa).

When myoglobin comes into contact with oxygen, it forms a reversible bright red pigment **oxymyoglobin** (**9-6**), where the sixth ligand donating both electrons to iron to form a coordinate bond is oxygen. This reaction, known as oxygenation, proceeds at oxygen partial pressures greater than 70–80 mmHg, and its partial pressure in air is always higher, at approximately 160 mmHg.[2] At the very low oxygen partial pressure (<1.4 mmHg) typical of vacuum-packed meat, the electron deficiency of the iron allows it to interact ionically with water in the absence of stronger electron pair-donating ligands, which can form covalent linkages, making water the sixth ligand. The relevant purple–red pigment is called **deoxymyoglobin** (**9-6**). Analogously, haemoglobin gives rise to **oxyhaemoglobin** and **deoxyhaemoglobin**.

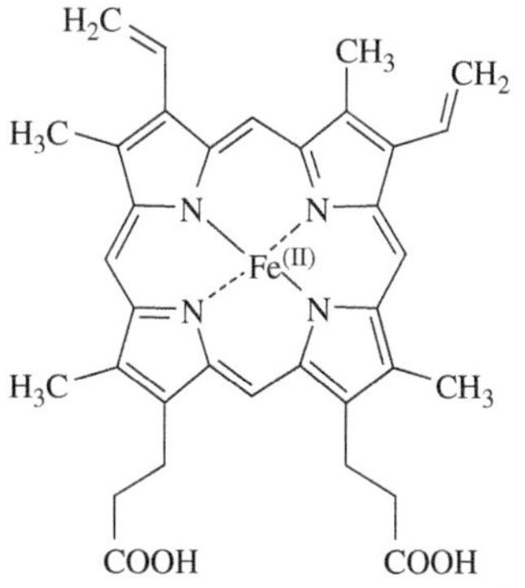

9-4, haem

---

[2] The sea-level standard atmospheric pressure is 101.325 kPa, which is equal to 1 atm or 760 mm Hg (torr) or 14.696 psi. So, if the air is 20.9% oxygen, the oxygen tension (partial pressure of oxygen) is 21.2 kPa, which corresponds to 159 mm Hg (1 mm Hg = 133.3 Pa).

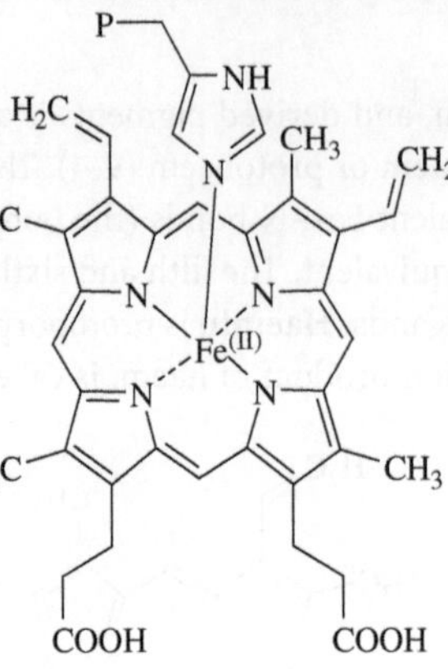

9-5, myoglobin, P = protein residue

An oxymyoglobin oxidation product containing trivalent iron is **metmyoglobin** (**9-6**), which is brown or grey-brown. Analogously, oxidation of haemoglobin (a loss of one electron) yields **methaemoglobin**.

Haem and its derivatives often have several different trivial names that describe the degree of oxidation of iron and bound ligands. An overview of these names is given in Table 9.1.

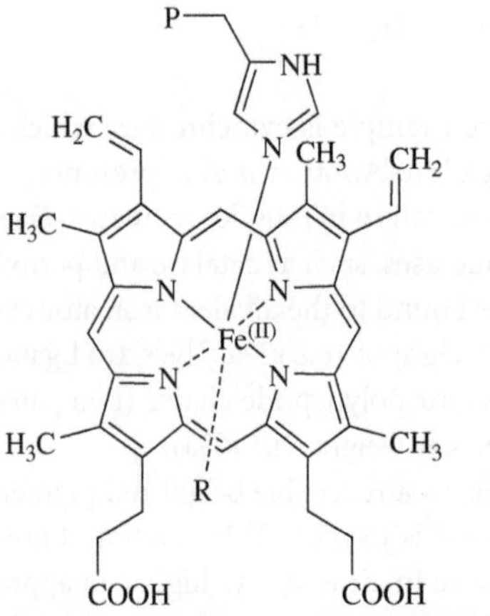

9-6, oxymyoglobin, P = protein residue, R = O$_2$
deoxymyoglobin, P = protein residue, R = H$_2$O
metmyoglobin, P = protein residue, R = OH
nitrosomyoglobin, P = protein residue, R = NO

Table 9.1 Nomenclature of haem pigments.

Name	Synonym	Oxidation degree Fe/other ligand	Name	Synonym	Oxidation degree Fe/other ligand
Haem	Ferroprotohaem IX	Fe(II)/H$_2$O	Haematin	Ferriprotohaem IX	Fe(III)/OH
	Protohaem			Hydroxyferriprotoporphyrin	
	Ferroprotoporphyrin IX			Alcaline haematin	
	Ferrous protoporphyrin IX		Haemin	Protohaemin IX	Fe(III)/Cl
	Reduced haematin			Chlorohaemin	
				Chloroferriprotoporphyrin	

Table 9.2 Haem pigment contents in beef and pork.

Meat	Myoglobin (mg/kg)	Haemoglobin (mg/kg)	Proportion of myoglobin (%)
Beef	3140–7020	340–520	90–94
Pork	790–2320	360–1200	50–75

9.2.1.2 Occurrence

The main respiratory pigment found in vertebrates is haemoglobin, whilst myoglobin represents only about 10% of the total iron present in the body. After slaughter and bleeding of animals, the predominant muscle pigment is myoglobin (Table 9.2). The content of myoglobin depends not only on the type of animal, but also on its age, the type of muscle (Table 9.3), and other factors. The total average content of haem pigments in meat of various animals is given in Table 9.4. Dark meat (e.g. horse meat) contains more myoglobin than light meat (e.g. pork or veal). For example, turkey legs are dark meat, whilst the breast and wings are light. The dark red muscle fibres contain a lot of myoglobin in the capillaries, require oxygen, contract slowly, are rich in mitochondria, and split ATP for energy at a relatively low rate via aerobic respiration. White muscle fibers, on the other hand, relax/contract quickly, split ATP at a fast rate (anaerobic respiration), and contain a large amount of glycogen, which can be used as a fast energy source. Meat colour is also linked to defects. The content of haem pigments in dark and light fish muscles is given in Table 9.5.

9.2.1.3 Use

The natural content of myoglobin is particularly important for the colour of meat and meat products. The colouring properties of pork blood haemoglobin are employed throughout the world in some speciality products, such as blood soup, blutwurst, blood pudding, and black pudding. The blood has to be aseptically collected from healthy pigs and processed swiftly. In China, Thailand, and Vietnam, pig,

Table 9.3 Myoglobin contents in beef and pork.

Meat	Animal age	Muscle[a]	Myoglobin (mg/kg)
Veal	12 days	*Longissimus dorsi*	700
Beef	3 years	*Longissimus dorsi*	4600
Pork	5 months	*Longissimus dorsi*	300
Pork	7 months	*Longissimus dorsi*	440
Pork	7 months	*Rectus femoris*	860

[a] *Longissimus dorsi* is a muscle of pork loin (cutlet), *Rectus femoris* is the main muscle of pork leg (oyster piece).

Table 9.4 Haem pigment contents in various animals.

Meat	Haem pigments (mg/kg)	Meat	Haem pigments (mg/kg)
Beef	1700–7500	Boar	5500
Veal	438–1490	Deer	6000–7000
Pork	254–3500	Whales	9100
Horse	3620–8000	Chicken	126–158
Lamb	2500	Turkey	125–456
Goat	6350	Goose	1586
Rabbit	200	Duck	1168

Table 9.5 Haem pigment contents in dark and light fish muscles.

Fish	Haemoglobin (mg/kg)		Myoglobin (mg/kg)	
	Dark	Light	Dark	Light
Chub mackerel (*Scomber japonicus*)	5700	100	3900	<10
Pacific saury (*Cololabis saira*)	4800	350	270	10
Pacific bluefin tuna (*Thunnus orientalis*)	4900	1700	3200	700

chicken, duck, and goose blood is used in soups and other dishes. Drinking blood from cattle is a part of the traditional diet of Maasai people in Kenya and Tanzania.

9.2.1.4 Biochemistry, physiology, and nutrition

Haemoproteins are a large group of proteins performing different functions in animal and plant organisms. In animals, haemoglobin plays a fundamental role as a carrier of oxygen and carbon dioxide, whilst myoglobin is a reservoir of oxygen. Haem pigments also play a key role in energy conversion in cytochromes, and are responsible for procuring the energy that animals produce by oxidation of nutrients in the respiratory chain and which plants produce by photosynthesis.

As well as oxygen, myoglobin and haemoglobin can also bind other ligands. For example, poisoning by carbon monoxide (CO) transforms myoglobin into carbonylmyoglobin and haemoglobin into carbonylhaemoglobin, as the affinity of these pigments for carbon monoxide is about 130 times higher than that for oxygen, meaning oxygen is displaced by the new ligand carbon monoxide. The same pigments are formed in meat packaged in a carbon monoxide atmosphere. Poisoning by cyanides leads analogously to the formation of cyanomyoglobin and cyanohaemoglobin.

From the nutritional point of view, haem pigments are of considerable importance in the supply of iron to the human organism. The so-called haem iron, which includes iron bound in myoglobin and haemoglobin and that bound in muscle respiratory enzymes, is absorbed in the body at a level of 10–30%, whilst non-haem iron is absorbed a level of only 1–5%.

Endogenous haem is degraded by oxidative cleavage of the porphyrin ring via the enzyme haem oxygenase inside spleen macrophages to blue-green biliverdin IX α (releasing carbon monoxide and Fe), which is reduced by biliverdin reductase to orange bilirubin IX α. Bilirubin forms a conjugate with D-glucuronic acid (diglucuronide, the major bile pigment) binding to propionic acid residues. The action of bacterial enzymes in the colon creates free bilirubin, which is further converted into other compounds. Of these, the most important is colourless urobilinogen. Part of urobilinogen reaches the bloodstream and the kidneys, where it is oxidised to yellow urobilin IX α, the characteristic pigment of urine. Most urobilinogen is oxidised in the intestine (via colourless stercobilinogen) to brown-red stercobilin IX α, which stains faeces.

9.2.1.5 Properties and reactions

The colour of meat and meat products depends on many factors, especially the oxidation level of the central iron atom, the ligands that surround the central atom, and the structure of the protein moiety.

9.2.1.5.1 Pigments of raw meat

Consumers usually consider the bright red colour of meat as a sign of freshness. The colour of freshly cut or minced meat comes from two types of pigments, myoglobin/oxymyoglobin and metmyoglobin. After several hours or days of exposure, oxymyoglobin can convert into metmyoglobin, which has a brown-grey colour. The colour of vacuum-packed meat containing deoxymyoglobin is purplish-pink. The colour of meat packaged in an atmosphere of carbon monoxide is clearly cherry-red (typical carbon monoxide concentrations range from 0.4 to 0.5%).

Atmospheric oxygen diffuses into the surface layer of meat (with a thickness of up to 10 mm); myoglobin is transformed by oxygenation into bright-red oxymyoglobin, and so meat acquires an attractive bright-red colour (Figure 9.1). The binding of oxygen is enabled by the second histidine residue of the myoglobin molecule (His[64], the so-called distal histidine) through hydrogen or O—N bonds (Figure 9.2). The reaction is reversible, since

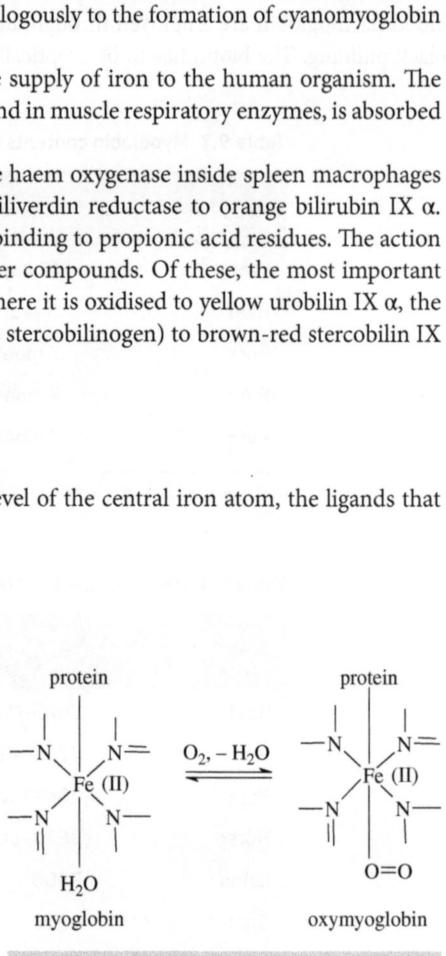

Figure 9.1 Formation of oxymyoglobin from myoglobin.

oxygen dissociates from oxymyoglobin continuously. At a higher oxygen partial pressure, oxymyoglobin is fairly stable, since the ligand oxygen is stabilised by a three-dimensional structure of apoprotein surrounding the haem. At low oxygen pressure, however, which occurs at the interface between the oxymyoglobin and myoglobin layers during meat storage in the air, both meat pigments are slowly oxidised by air oxygen to the brown-grey, unattractive metmyoglobin, as the binding of oxygen in oxymyoglobin involves a charge migration from the haem to oxygen, probably via superoxoferrihaem $[Fe^{3+} O_2^-]^{2+}$ intermediate, which dissociates to metmyoglobin and superoxide radical (Figure 9.3).

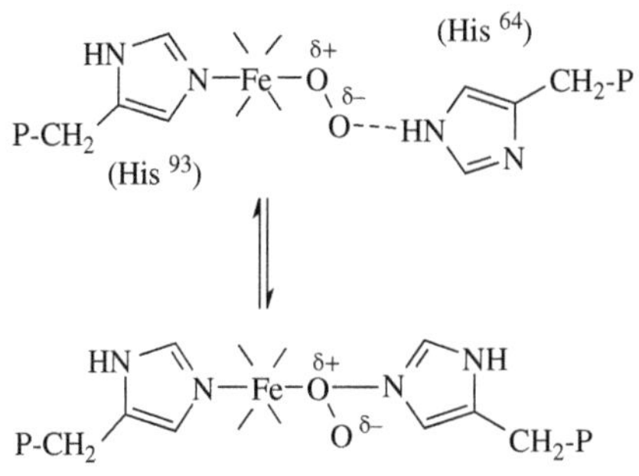

Fresh meat contains reducing agents (e.g. thiol groups of proteins and oxidoreductases with NADH as cofactors), which continuously reduce metmyoglobin to myoglobin. After oxidation of these reducing substances, a brown layer of metmyoglobin is gradually formed below the surface of the meat, and eventually the entire surface turns brown. Browning is an indication that the meat is not particularly fresh. The autoxidation of haemoglobin to methaemoglobin occurs in the same way. The colour of vacuum-packed meat, and of meat packaged in a carbon monoxide atmosphere, is more stable.

The oxidation of Fe^{2+} to Fe^{3+} is accelerated by light and at higher temperatures and lower pH. The oxidation is faster after rigor mortis dissipates (when the pH of the meat is 5–6) than immediately post mortem (when the pH is almost neutral).

In addition to the change in meat colour, oxygen bound in oxymyoglobin dissociates after protonation and autoxidation of oxymyoglobin to metmyoglobin produces free superoxide radicals (Figure 9.3). The distal histidine acts as a proton donor, as do certain protein functional groups of the environment. Superoxide free radicals rapidly dismutates to oxygen and hydrogen peroxide, but the eventually formed hydroxyl radicals can initiate autoxidation of fatty acids, which can lead to rancidity in fatty parts of the meat.[3]

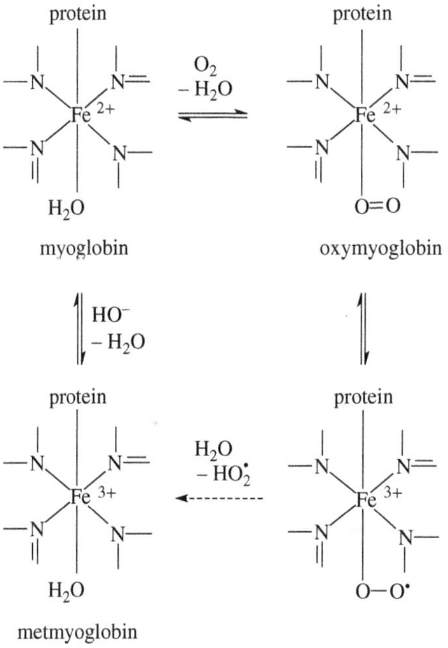

Figure 9.2 Oxygen binding and its stabilisation in myoglobin molecule (P = protein residue).

9.2.1.5.2 Meat colour stabilisation

The prevention of changes in colour during heat treatment and simultaneous inhibition of microbial growth in cured meat, sausages, and other products can be achieved by curing. Curing is one of various food-preservation processes achieved by addition of curing salts. Curing salts are mixtures of table salt (sodium chloride) with nitrites or nitrates and other ingredients, the addition of which is necessary to maintain the colour of products and to prevent the growth and toxin production by the bacteria *Clostridium botulinum*. The use of nitrates and nitrites in cured meat and meat products must comply with EU legislation. Currently authorised as food additives are sodium and potassium nitrites and sodium and potassium nitrates, which may be sold only in a mixture with salt or a salt substitute. Typically, the product is cured by injection with a curing mixture containing nitrites/nitrates followed by immersion in brine or by immersion only. Dry curing involves the dry application of the curing salt to the surface of the meat, followed by a period of stabilisation/maturation. Nitrates have no direct activity against *C. botulinum*, but they act as reservoirs of nitrites, generated by microbial activity.

Figure 9.3 Formation of metmyoglobin in meat.

The reaction of meat pigments with nitrites, which provides the cured meat colour and flavour, is very complex. Nitrites are reduced to nitrogen monoxide (nitric oxide NO) by myoglobin and other reducing agents, whilst myoglobin is oxidised to metmyoglobin:

$$\text{myoglobin} + NO_2^- \rightarrow \text{metmyoglobin} + NO$$

The resulting nitrogen monoxide reacts with another molecule of myoglobin to form the red pigment **nitrosomyoglobin** (formerly also called nitroxymyoglobin or nitrosylmyoglobin, **9-6**). The reaction of nitrogen monoxide with metmyoglobin yields nitrosometmyoglobin, which can also be formed by reaction of nitrogen monoxide with metmyoglobin nitrite (instead of N=O, the NO_2^- anion is present). Nitrosomyoglobin also occurs in meat smoked in the home, which does not contain nitrites, as flue gas contains various nitrogen oxides. Metmyoglobin formed by oxidation of myoglobin is reduced back to myoglobin either chemically (the electron donors are thiol groups of proteins and other thiols) or by the system of dehydrogenases containing NAD as a cofactor. In the same way, nitrosometmyoglobin is also

---

[3] In all organisms *in vivo* (except bacteria), the superoxide radical is removed by the action of superoxide dismutase, which catalyses the dismutation of superoxide radical into oxygen and hydrogen peroxide. Hydrogen peroxide is then decomposed to water and oxygen by catalase.

reduced to form nitrosomyoglobin. The reaction of nitrites with oxymyoglobin in the surface layers of meat leads to oxidation of nitrite, and even of haem iron (the oxidising agent is oxygen bound in oxymyoglobin). This reaction explains the grey-brown colour of the surface layers of cured meat, which remains pink inside.

Nitrosomyoglobin is a very stable pigment at normal temperatures, as the bound ligand nitrogen monoxide is stabilised by the N—N bond with the distal histidine (His[64]). During heat treatment, nitroxymyoglobin is denatured and splits off the globin, and the vacant coordination bond of iron is occupied by another nitric oxide molecule, which yields the most important pigment of heat-treated cured meat and meat products with NO ligands in both axial positions, **nitrosohaemochrome** (also called nitroxyhaemochrome and nitrosylmyochromogen, **9-7**). At the same time, other meat proteins denature, and their accessible thiol groups then reduce the metmyoglobin back to myoglobin, which can be used for the formation of a further portion of nitrosohaemochrome.

9-7, nitrosohaemochrome

Better colour of smoked meat products is achieved by adding reducing agents together with nitrites. The most commonly used agents are L-ascorbic acid, sodium L-ascorbate, L-ascorbyl palmitate, and sodium L-isoascorbate (erythorbate). The addition of ascorbate is preferred because it does not impair the pH of the meat. Reducing agents reduce metmyoglobin back to myoglobin, produce nitrogen monoxide from nitrites without the participation of myoglobin, and reduce nitrates to nitrites (H_2A = ascorbic acid, A = dehydroascorbic acid):

$$2 \text{ metmyoglobin} + H_2A \rightarrow 2 \text{ myoglobin} + A$$
$$2 NO_2^- + H_2A + 2 H^+ \rightarrow 2 NO + A + 2 H_2O$$
$$2 NO_3^- + H_2A + 2 H^+ \rightarrow 2 NO_2^- + A + 2 H_2O$$

If protected from light, nitrosohaemochrome is stable to oxidation even at elevated temperatures, but in light the molecule dissociates to nitrogen monoxide, which is oxidised to nitrogen dioxide, and the haem produced is rapidly oxidised to haematin or degraded in the presence of oxygen or other oxidising agents. Meat products then become greyish and fade.

Myoglobin can degrade through a number of pathways producing dark-brown, yellow, and even green pigments that may cause colour defects, for example in minced meat and some meat products. The colour change can also be caused by air oxygen, fatty acid hydroperoxides, or hydrogen peroxide generated, for example, from added ascorbic acid or by lactobacilli in fermented salami. The oxidation occurs even when nitrites or ascorbic acid are over-dosed and in the absence of reducing substances in meat, which leads to grey and green pigments.

Myoglobin readily reacts with hydrogen peroxide and fatty acid hydroperoxides. The oxidation of myoglobin by hydrogen peroxide starts with the formation of a radical in the globin portion of the molecule and conversion of the haem Fe^{3+} iron into the Fe^{4+} state. The reaction proceeds via a red radical intermediate **perferrylmyoglobin** (containing tetravalent iron), which rapidly breaks down to green **ferrylmyoglobin** (containing tetravalent iron, native globin, and oxygen atom as the ligand):

$$\text{myoglobin} (Fe^{2+} - O_2) + H_2O_2 \rightarrow \text{perferrylmyoglobin}^{\bullet} \rightarrow \text{ferrylmyoglobin} (Fe^{4+} - O_2) + H_2O + O_2^-$$

Hydrogen peroxide can also irreversibly oxidise myoglobin, yielding green **choleglobin** (containing native globin, Fe^{3+} or Fe^{2+}, and a covalently or ionically bound hydroperoxide O—OH group as the ligand). Choleglobin is rapidly degraded to yield globin, iron, and a tetrapyrrole. In the presence of sulfhydryl substances, myoglobin can be reversibly reduced at one double bond by reaction with sulfane and oxygen (via ferrylmyoglobin) to green **sulfmyoglobin** containing native globin, Fe^{2+}, and a covalently bound SH group as the ligand:

$$\text{ferrylmyoglobin} (Fe^{4+} - O_2) + HS^- \rightarrow \text{sulfmyoglobin}(Fe^{2+} - SH) + HO^-$$

Oxidation of sulfmyoglobin (loss of one electron) yields red metsulfmyoglobin containing Fe^{3+} and an ionically bound SH group as the ligand. During heat treatment, the globin of green pigments is denatured and split off with the formation of green **verdochrome** (containing Fe^{3+} and ionically bound water) with porphyrin ring opening.

9.2.1.5.3 Pigments of heat-processed meat

When meat is cooked, globin denatures at temperatures above about 65 °C (along with other proteins) to a degree that depends on pH and the intensity and time of thermal processing, but the haem portion remains intact. The colour of the haem pigments containing denatured globin is determined by the oxidation state of haem iron and the sixth ligand molecule, which is attached to haem. The denatured globin cannot actually maintain iron in the reduced state; therefore, haem is oxidised to haematin (metmyoglobin with denatured globin) and the red meat colour is changed to brown and then to grey-brown.

9.2.2 Chlorophylls

Chlorophyll pigments or **chlorophylls** are a group of green pigments found in the tissues of green plants and some other organisms that allow them to carry out photosynthesis – the process of transforming light energy into chemical energy. Several different types of chlorophylls have been described.

9.2.2.1 Structure and nomenclature

The structural base of most chlorophylls is the cyclic tetrapyrrole **17,18-dihydroporphyrin** (**9-8**), derived from protoporphyrin IX. Unlike haem pigments, chlorophylls have a partially reduced ring D (they are 17,18-dihydroporphyrins), another ring E generated by cyclisation of the propionic acid residue in position C-13, and chelated magnesium (magnesium ion of oxidation number +2) as the central atom. The pigments of some bacteria known as **bacteriochlorophylls** are **7,8,17,18-tetrahydroporphyrins** (**9-9**).

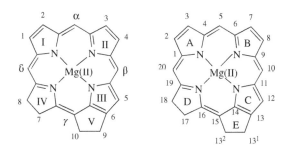

9-8, basic structure of dihydroporphyrin pigments by the older Fischer system (left) and the IUPAC recommended system (right)

From the food perspective, the most important chlorophylls are chlorophyll *a* and chlorophyll *b* and their degradation products the **phaeophytins**: 17,18-dihydrophaeophytins (phaeophytin *a* and phaeophytin *b*). Some other products of chlorophyll biosynthesis and degradation are also components of foods, such as **chlorophylides**, 17,18-dihydrophaeophorbideato-Mg(II) (chlorophylide *a* and chlorophylide b), and **phaeophorbides**, 17,18-dihydrophaeophorbides (phaeophorbide *a* and phaeophorbide *b*).

Chlorophyll *a*, 17,18-dihydrophaeophytinato-Mg(II), is 17,18-dihydroporphyrin substituted in positions C-2, C-7, C-12, and C-18 by methyl groups, in position C-3 by a vinyl group, and in position C-8 by an ethyl group. In position C-17, there is a residue of propionic acid esterified by diterpenic alcohol *trans*-phytol, (2E,7R,11R)-3,7,11,15-tetramethylhexadec-2-en-1-ol (see Section 3.7.2), which constitutes a lipophilic side chain of chlorophylls. In position C-13^1 of ring E, there is an oxo group, and in position C-13^2, there is a carboxymethyl group (**9-10**). Chlorophylls *a* and *b* differ only by the substituent at C-7: a methyl group in chlorophyll *a* and a formyl group in chlorophyll *b*. Chlorophylls *a'* and *b'*, which differ in their configuration of C-13^2 substituents, arise by epimerisation of chlorophyls *a* and *b* in thermal processes and in alkaline media. The structures of currently known chlorophylls and bacteriochlorophylls are listed in Table 9.6.

9-9, basic structure of tetrahydroporphyrin pigments

Table 9.6 Overview of chlorophyll and bacteriochlorophyll structures.

Pigment	Substituent at carbon[a]					
	C-3	C-7	C-8	C-17	C-18	C-20
Dihydroporphyrins						
Chlorophyll *a*	CH=CH$_2$	CH$_3$	CH$_2$CH$_3$	Phytyl and H	CH$_3$ and H	
Chlorophyll *b*	CH=CH$_2$	CH=O	CH$_2$CH$_3$	Phytyl and H	CH$_3$ and H	
Chlorophyll *d*	CH=O	CH$_3$	CH$_2$CH$_3$	Phytyl and H	CH$_3$ and H	
Tetrahydroporphyrins						
Bacteriochlorophyll *a*	(C=O)CH$_3$	CH$_3$ a H	CH$_2$CH$_3$ a H	Phytyl and H	CH$_3$ and H	
Bacteriochlorophyll *b*	(C=O)CH$_3$	CH$_3$ a H	CH$_2$CH$_3$ a H	Farnesyl and H	CH$_3$ and H	
Bacteriochlorophyll *c*	CH(OH)CH$_3$	CH$_3$ a H	CH$_2$CH$_3$ a H	Farnesyl and H	CH$_3$ and H	CH$_3$ or CH$_2$CH$_3$
Bacteriochlorophyll *d*	CH(OH)CH$_3$	CH$_3$ a H	CH$_2$CH$_3$ a H	Farnesyl and H	CH$_3$ and H	
Bacteriochlorophyll *e*	CH(OH)CH$_3$	CH=O a H	CH$_2$CH$_3$ a H		CH$_3$ and H	CH$_3$ or CH$_2$CH$_3$
Porphyrins						
Chlorophyll c$_1$	CH=CH$_2$	CH$_3$	CH$_2$CH$_3$	CH=CHCOOH	CH$_3$	
Chlorophyll c$_2$	CH=CH$_2$	CH$_3$	CH=CH$_2$	CH=CHCOOH	CH$_3$	

[a]The substituent in positions C-2, C-12, and C-18 is always a CH$_3$ group (with the exception of bacteriochlorophyll e, which in position C-12 is a CH$_2$CH$_3$ group), that in positions C-13 and C-15 is the rest of the cyclopentanone. Other positions are not substituted.

Phaeophytins are derived from chlorophylls, with hydrogen replacing magnesium (**9-11**). Hydrolysis of phytol in chlorophylls yields chlorophylides (**9-12**). Replacing the magnesium ion by two hydrogen ions in chlorophylides gives rise to phaeophorbides (**9-13**). Replacing phytol (phytyl residue) in phaeophorbides with methanol (methyl group) gives methylphaeophorbides (**9-13**). Substitution of the carboxymethyl group on C-13^2 with hydrogen produces pyrophaeophytins (**9-11**) from phaeophytins and phaeophorbides form pyrophaeophorbides (**9-13**).

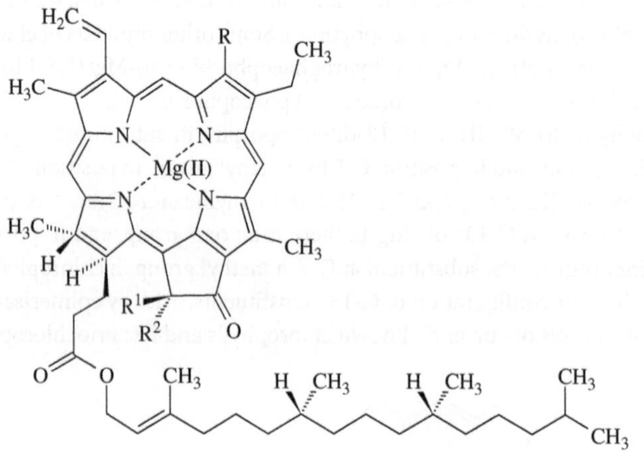

9-10, chlorophylls

chlorophyll *a*, R = CH$_3$, R^1 = H, R^2 = COOCH$_3$
chlorophyll *b*, R = CH=O, R^1 = H, R^2 = COOCH$_3$
chlorophyll *a*′, R = CH$_3$, R^1 = COOCH$_3$, R^2 = H
chlorophyll *b*′, R = CH=O, R^1 = COOCH$_3$, R^2 = H

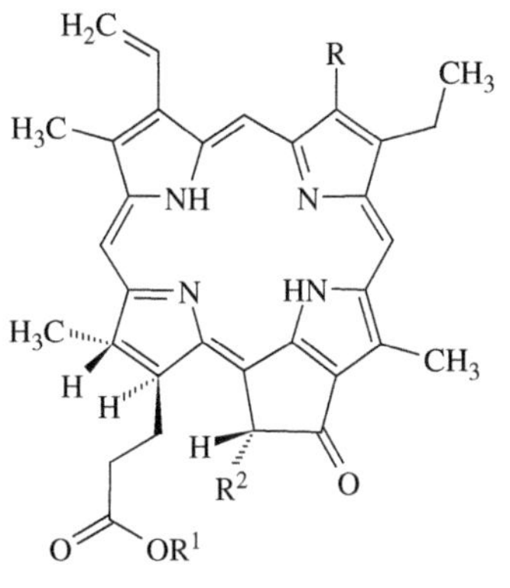

9-11, phaeophytins and pyrophaeophytins

phaeophytin *a*, R = CH$_3$, R^1 = fytyl, R^2 = COOCH$_3$
phaeophytin *b*, R = CH=O, R^1 = fytyl, R^2 = COOCH$_3$
pyrophaeophytin *a*, R = CH$_3$, R^1 = H, R^2 = H
pyrophaeophytin *b*, R = CH=O, R^1 = H, R^2 = H

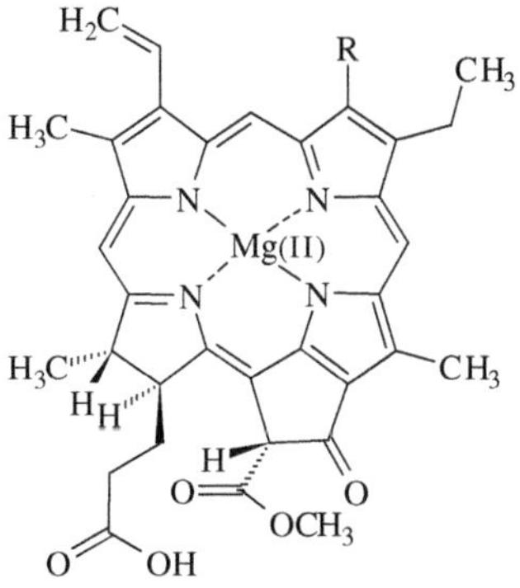

9-12, chlorophylides

chlorophylide *a*, R = CH$_3$
chlorophylide *b*, R = CH=O

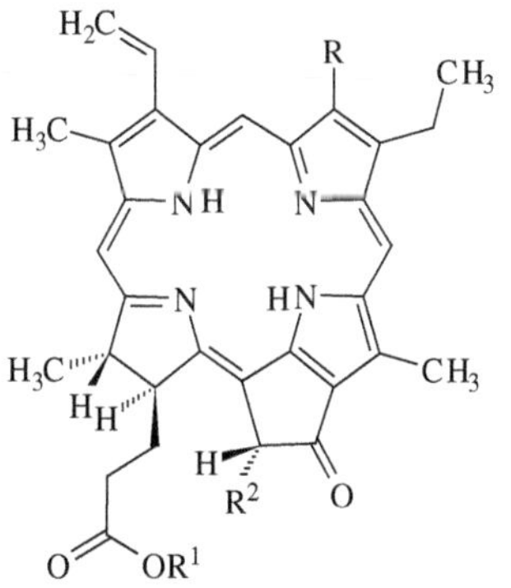

9-13, phaeophorbides, methylphaeophorbides and pyrophaeophorbides

phaeophorbide *a*, R = CH$_3$, R^1 = H, R^2 = COOCH$_3$
phaeophorbide *b*, R = CH=O, R^1 = H, R^2 = COOCH$_3$
methylphaeophorbide *a*, R = CH$_3$, R^1= CH$_3$, R^2 = COOCH$_3$
methylphaeophorbide *b*, R = CH = O, R^1 = CH$_3$, R^2 = COOCH$_3$
pyrophaeophorbide *a*, R = CH$_3$, R^1 = H, R^2 = H
pyrophaeophorbide *b*, R = CH = O, R^1 = H, R^2 = H

9.2.2.2 Occurrence

Chlorophylls are found in all green plants, mosses, some algae, and some microorganisms (Table 9.7). Germ cells of flowering plants (angiosperms, Angiospermae) that grow in the dark lack chlorophyll, because the enzyme protochlorophylide reductase (which reduces the

Table 9.7 Occurrence of chlorophylls and bacteriochlorophylls.

Pigment	Organism
Chlorophyll *a*	All photosynthetic organisms, higher plants, algae (*Cyanophyta, Prochlorophyta*), bacteria
Chlorophyll *b*	Higher plants, algae (*Chlorophyta, Euglenophyta, Prochlorophyta*), bacteria
Chlorophyll *c*	Algae (*Phaeophyta, Pyrrophyta, Bacillariophyta, Chrysophyta, Prasinophyta, Cryptophyta, Xanthophyta*)
Chlorophyll *d*	Algae (*Rhodophyta, Chrysophyta*)
Bacteriochlorophyll *a, b*	Bacteria (*Chromatiaceae, Rhodospirillaceae*)
Bacteriochlorophyll *c, d, e*	Bacteria (*Chlorobiaceae, Chloroflexaceae*)

colourless intermediate protochlorophylide to chlorophylide) requires light. These so-called etiolated plants will start to synthesise chlorophyll after exposure to light. Anoxygenic photosynthetic bacteria do not need light to catalyse this reaction, as they contain other enzymes. Cyanobacteria, algae, lower plants, and gymnosperms (Gymnospermae), such as conifers, synthesise chlorophyll even in the dark.

Food contains mainly chlorophyll *a* and chlorophyll *b* originating from higher plants. These two forms are present in the ratio of about 3 : 1. The total chlorophyll contents in the green parts of plants are very different. Grapes, for example, only contain chlorophyll at a level of about 6 mg/kg, whereas spinach contains up to 790 mg/kg.

Chlorophylls in living cells are located in chloroplasts associated by non-bonding interactions with proteins (26–28 kDa). These complexes are further associated along the phytol chain with carotenes, xanthophylls, and tocopherols. Ten molecules of chlorophylls are associated with about one lipophilic molecule.

9.2.2.3 Use

The most common forms of green pigments used as food additives are fat-soluble mixtures commercially known as **chlorophylls** and **chlorophyllin copper complexes**.

Chlorophylls are mixtures of products consisting mainly of pheophytins *a* and *b* and their epimers (known as pheophytins *a'* and *b'*), which differ in the arrangement of position C-13^2. Most of the chlorophylls for food use are obtained from edible higher plants (mainly nettles, alfalfa, parsley leaves), edible single-cell green algae (*Chlorella*), green algae known as sea lettuces (*Ulva*), cyanobacteria (blue-green algae *Spirulina*), and some other sources. In Japan, chlorophylls are extracted from green silkworm (*Bombyx mori*) excrements.

The water-soluble greyish-green semi-synthetic mixture of pigments called chlorophyllins (or phytochlorins; the recommended name is rhodochlorins) is prepared by alkaline hydrolysis of chlorophylls, which results in hydrolysis of methyl and phytyl esters. Chlorophyllins contain a central magnesium atom and sodium or potassium salts of various products resulting from chlorophylls after alkaline hydrolysis, such as salts of phaeophorbides *a* and *b*, salts of phaeophytins *a* and *b* and their *allo*-forms (diastereomers), pigments with a carboxyl group at position C-13 and a carboxyethyl group at position C-17, and products with an opened cyclopentanone E ring that contains a newly formed carboxymethyl group in position C-15. Other products may include compounds with an oxidised ring D not connected to ring E and possibly a reduced vinyl group in position C-3. These products were previously known, for example, as chlorins, purpurins, and rhodins. Typically, chlorin e$_6$ is a phaeophorbide-derived product (it does not contain a magnesium ion or bound phytol) with a carboxyl group at position C-13, a carboxyethyl group at position C-17, and a carboxymethyl group at position C-15.

Stable bright-green pigments with the trade name chlorophyllin copper complex are obtained by acidification of chlorophyllins in the presence of copper salts. Instead of magnesium, these products contain Cu(II) as the central atom. An example of a trisodium salt of chlorophyllin copper complex is given in formula **9-14**.

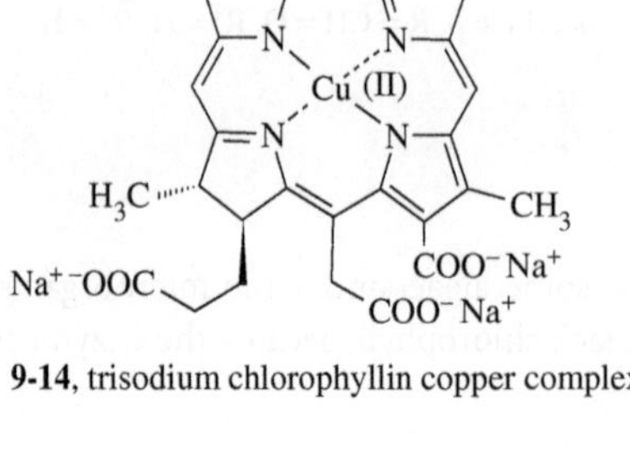

9-14, trisodium chlorophyllin copper complex

Chlorophylls are used as food pigments (e.g. for pasta products, beverages, sweets, soups, frozen yoghurts, and creams) and, most heavily, in cosmetic products. The use of chlorophyllin copper complexes (under the name of natural green 3, E number E141) is usually limited to certain products, such as chewing gum. These pigments have also found use in alternative medicine, for example in odour control of wounds, injuries, and radiation burns and treatment of calcium oxalate kidney stones.

9.2.2.4 Biochemistry, physiology, and nutrition

Chlorophylls are normal constituents of foods of plant origin, and are ingested daily over a lifetime by both animals and humans, without any appreciable health risks. In people with certain genetic abnormalities such as albinism, also called achromia (a disorder characterised by the complete or partial absence of pigment in the skin, hair, and eyes, due to the absence or defect of tyrosinase producing pigments called melanins), inflammation of the skin (dermatitis) is observed after exposure to sunlight. Photosensitivity is associated with the presence of chlorophylides, particularly phaeophorbides, methyl-phaeophorbides, and hydroxyphaeophorbides (which have a hydroxyl group on C-13^2 instead of H atom) in certain foods and food supplements containing green algae.

9.2.2.5 Properties and reactions

Chlorophylls and phaeophytins are fat-soluble pigments. Chlorophylides and phaeophorbides are hydrophilic pigments soluble in water due to the absence of phytol. The colours of chlorophylls and derived products are given in Table 9.8.

An overview of the main reactions of chlorophylls and their degradation products is given in Figure 9.4. The most important reaction occurring during food processing is the replacement of the central magnesium atom by hydrogen and the formation of phaeophytins. This reaction, called **phaeophytinisation**, proceeds even in weakly acidic media, which are normally found in fruits and vegetables, such as sweet pickled greengages and sour pickled cucumbers. The action of the enzyme chlorophyllase and of strong acids results in phytol hydrolysis in phaeophytins and the formation of phaeophorbides. In enzymatically active plant tissues, the hydrolysis of phytol in chlorophylls by chlorophyllase yields chlorophylides. Replacing magnesium with hydrogen in chlorophylides in weakly acidic media leads to phaeophorbides, which are coloured like phaeophytins. This is one of reasons why green vegetables (such as beans, peas, and broccoli) have to be blanched in order to destroy the enzyme activity before preservation by freezing.

The culinary and manufacturing processes and storage of green fruits and vegetable result in a greater or lesser degree of chlorophylls degradation, which is accelerated by temperature, light, and ionising radiation, the presence of acids, and some enzymes. Degradation of chlorophylls by enzymes occurs even during cold storage of vegetables. For example, changes in chlorophyll pigments during storage of green beans and peas are attributed to degradation of chlorophyll pigments by lipoxygenase, which leads to the formation of free radicals and chlorophyll degradation. Changes of chlorophylls occurring during blanching of green beans and sterilisation of spinach are evident from the data presented in Table 9.9. Deeper changes take place when the material is exposed to heat in an acidic environment, which results in the formation of phaeophytins and other degradation products. For example, the main pigments in fermented cucumbers are phaeophorbides, followed by phaeophytins, chlorophylls, and chlorophylides. Phaeophorbides and phaeophytins are also the main pigments in fermented olives. The negative effect of light is reflected especially during drying, as fruits and vegetables pre-treated with sulfur dioxide to prevent enzymatic browning reactions are more sensitive.

Phaeophytinisation and other reactions described in this section negatively affect the sensory quality of foods containing chlorophyll pigments. The main principles of colour stabilisation are based on the positive effect of added alkaline agents (such as carbonates), high-temperature short-time (HTST) sterilization, and enzymatic conversion of chlorophylls to chlorophylides, or a combination of these methods. It is necessary to combine these procedures with storage at low temperatures.

Table 9.8 Colours of chlorophylls and some of their degradation products.

Pigment	Colour	Pigment	Colour
Chlorophyl a	Blue-green	Bacteriochlorophyll c	Green
Chlorophyl b	Green	Bacteriochlorophyll d	Green
Chlorophyl c_1	Yellow-green	Bacteriochlorophyll e	Green
Chlorophyl c_2	Yellow-green	Phaeophitin a and b	Olive brown
Chlorophyl d	Blue-green	Chlorophylide a and b	Green
Bacteriochlorophyll a	Grey-pink	Phaeophorbide a and b	Olive brown
Bacteriochlorophyll b	Brown-pink	Open ring products	Colourless

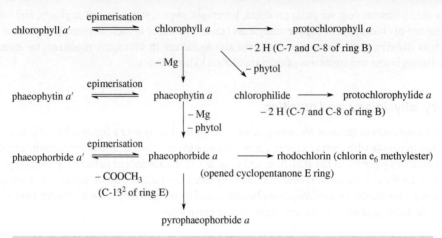

Figure 9.4 Main reactions of chlorophylls during food processing.

Table 9.9 Changes in chlorophylls and phaeophitins in vegetable processing (% of total pigment content in the original material).

Pigment	Green beans		Spinach	
	Raw	Blanched (100 °C/240 s)	Raw	Sterilised (127 °C/96 s)
Chlorophyll a	49	37	48	27
Chlorophyll b	25	24	20	16
Phaeophytin a	18	29	26	47
Phaeophytin b	8	10	6	10

Chlorophyll pigments are readily extracted from oil seeds and impart a greenish colour to crude oils. They are prominent pigments in unrefined virgin and extra virgin olive oils. The chlorophyll content in olive oil depends on genetic factors (olive variety), geographical origin, stage of fruit ripeness (olives picked green produce greener oil), environmental conditions, the extraction process, and storage conditions. The olive oil concentrations of chlorophylls vary over a wide range, from 2 to 76 mg/kg.

In oils processed by conventional refining techniques, chlorophylls are converted into phaeophytins, which gives the oil a dark, dull-brown colour and contributes to an off-flavour. In the light, chlorophyll pigments may promote formation of oxygen radicals and speed up oil oxidation and reduce its storage stability. In oils protected from light, however, chlorophylls act as primary antioxidants. Chlorophyll pigments also act as catalyst poisons in oil hardening. Together with other undesirable substances, chlorophylls are therefore removed during refining (bleaching).

The main pigments of rapeseeds are chlorophyll a and b, whilst minority pigments are phaeophytin a, phaeophorbide a, and methylphaeophorbide a. Crude rapeseed oils contain as the main pigment phaeophytin a, with smaller amounts of pyrophaeophytin a, whilst the minority pigments are phaeophorbide a, methylphaeophorbide a, and phaeophytin b. The total content of chlorophyll pigments in crude oils is 5–50 mg/kg. Refined (bleached) rapeseed oils may typically contain only traces (<1 mg/kg) of chlorophyll pigments. Soya oil typically contains less than 1 mg/kg of chlorophylls.

Chlorophylls form complexes with certain metal ions (e.g. Zn^{2+}, Cu^{2+}, Sn^{2+}; Figure 9.5), in which these metals, unlike Mg^{2+}, are bound tightly and are not released during manufacturing. The complexes are not cleaved even in the human digestive tract, and the metals are therefore not utilised. In earlier technological practice, the transformation of chlorophyll to phaeophytin and the related undesirable colour changes were prevented by the exchange of Mg^{2+} ions for Cu^{2+} ions in a solution of cupric sulfate ($CuSO_4$). Today, this procedure is not used for health reasons, but the use of a chlorophyllin copper complex is permitted in some countries for food colouring. Grey-brown Sn^{2+} complexes sometimes occur in canned fruits and vegetables in containers made of tinplate. Phaeophytins, chlorophylides, and phaeophorbides may also react with metal ions.

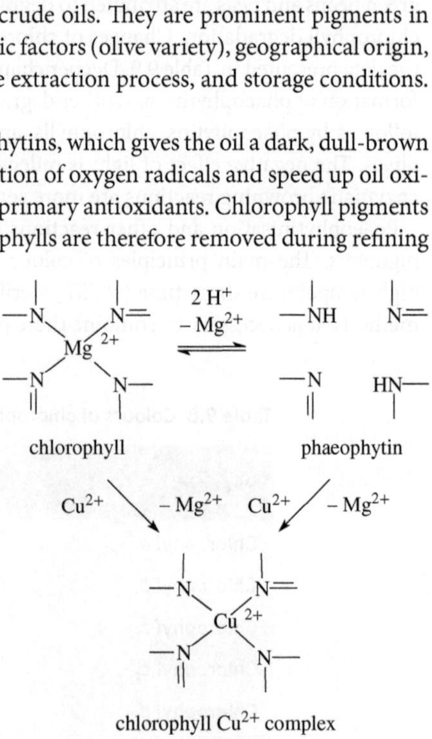

Figure 9.5 Formation of Cu^{2+} chlorophyll complex.

9.2.3 Phycobilins

Bilins (phycobilins or phycobiliproteins) are fluorescent, water-soluble complexes of linear tetrapyrroles with proteins. Phycobilins are light-capturing pigments (chromophores) found in cyanobacteria. The range of colours is extremely wide and depends on the source and other factors. Phycobilins are divided into three main groups:

- red **phycoerythrins;**

- blue **phycocyanins;**

- blue **allophycocyanins.**

The structure of phycobilins is very similar to the structure of mammalian bile pigments biliverdin and bilirubin, which arise as products of the catabolism of haem pigments. Chromophores of phycoerythrins are called phycoerythrobilins (**9-15**) and chromophores of blue phycocyanins and allophycocyanins are called phycocyanobilins (**9-15**). Phycoerythrins and phycocyanins occur in red algae (rhodophyta) and cryptomonads (phylum Cryptophyta, superclass Cryptomonada). Allophycocyanin is the pigment found in rhodophyta and bacteria of the phylum Cyanobacteria, also known as Cyanophyta or blue-green algae. Pigments obtained from cyanobacteria were originally called C-phycobilins, and pigments of rhodophyta were R-phycobilins. The chloroplasts of red algae contain phycobilins and chlorophylls a and c.

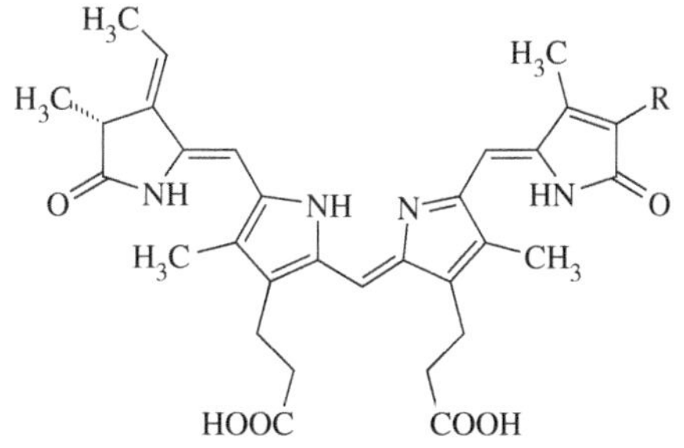

9-15, phycoerythrobilin, R = CH=CH$_2$
phycocyanobilin, R = CH$_2$CH$_3$

Examples of edible red algae of the genera *Porphyra* and *Pyropia* are the seaweed products *nori* (in Japan) and *gim* (in Korea), used chiefly as a wrap of sushi and to make seaweed rice rolls. Cyanobacteria of the genus *Arthrospira* (*A. platensis* and *A. maxima*) are used for the preparation of a dietary supplement known as spirulina. Phycocyanins are thought to have antiviral, anticancer, and some other activities and act as immune system stimulants.

9.3 Other nitrogen pigments

In addition to tetrapyrrole pigments, foods contain many other nitrogenous pigments theoretically derived from indole, isoquinoline, and pyrimidine. The indole pigments include widespread pigments melanins. Other important indole (dihydroindole or derived dihydropyridine) pigments are water-soluble red and yellow plant pigments, which are collectively called betalains. Substituted pyrimidines include three main groups of pigments with four nitrogen atoms in the molecule purines, pterines, and isoalloxazines. Compounds structurally related to pterines are fenazine and phenoxazine derivatives.

9.3.1 Indoles and related compounds

9.3.1.1 Melanins

Melanins are an important group of pigments found in all living organisms. They are divided into three basic groups (combined melanins are products of co-polymerisation of nitrogen-containing and sulfur-containing melanogens):

- **eumelanins** (polymerisation products of nitrogen melanogens);

- **phaeomelanins** (polymerisation products of sulfur melanogens);

- **allomelanins** (polymerisation products of polyphenols).

Melanins are formed by radical reactions *in vivo* in a process called melanogenesis, probably as a protection against harmful environmental effects upon the organism. For example, mammalian eumelanins efficiently absorb UV radiation and protect cells against reactive forms of oxygen and peroxyl radicals arising from lipids. Melanogenesis in pathogenic fungi and bacteria is associated with increased virulence, and melanins in bacteria are involved in fixation of atmospheric nitrogen.

Brown to black eumelanins (**9-16**) and yellow to red phaeomelanins (**9-16**) in various stages of oxidation and polymerisation are common pigments of skin, hair, eyes, feathers, and scales of mammals, birds, reptiles, amphibians, fish, and other organisms. Eumelanins and phaeomelanins in human skin arise in the basal layer of epidermis in the specialised cells melanocytes. In melanocytes, melanins occur in the cytoplasmic vacuoles, melanosomes, often in complexes with metals (Fe, Cu, and Zn) and proteins. Melanins are often accompanied by other pigments, resulting in countless shades of black, brown, yellow, and red colours. Phaeomelanins and eumelanins also contribute to the anomalous colour effects, such as iridiscence (interference) of colours often seen in the butterflies, reptiles and fish.

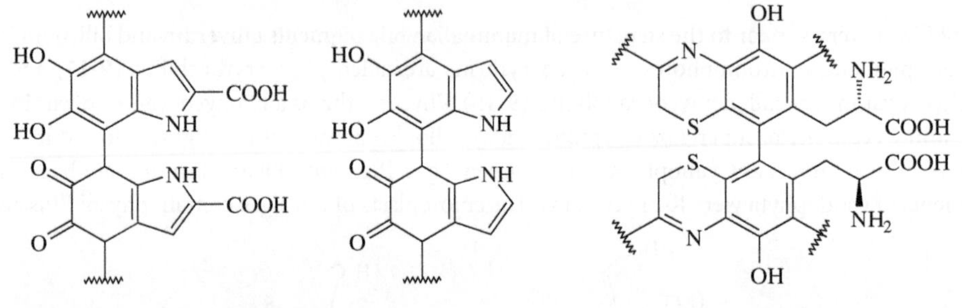

eumelanins   phaeomelanins

9-16, basic structures of melanins

Brown to black allomelanins are a heterogeneous group of melanin pigments, which are common in many fungi, higher plants, and even some animals. They include nitrogen-free polymeric products formed from phenols and quinones that are located, for example, in nut shells, leaves of plants in the final stage of evolution (senescence), and the bark of trees. Some allomelanins formed by condensation of quinones with amino acids (proteins) may also contain nitrogen and sulfur. Many plant materials form allomelanins during storage (such as bananas), tissue injury (such as apples and potatoes), or processing (e.g. during green tea fermentation) in enzymatic browning reactions. Quinones formed by oxidation of phenols and by activity of soil microorganisms are precursors of high-molecular-weight humic acids occurring in humus and related compounds. Structurally similar but simpler brown melanins are, for example, components of the dark-brown pigments derived from the ink sac of some cephalopods, such as the common cuttlefish (*Sepia officinalis*) and squids (of the order Teuthida), found in special sauces consumed in the Mediterranean region.

9.3.1.2 Betalains

The betalains are a group of about 100 water-soluble purple, red, orange, and yellow pigments of higher plants and some higher fungi. Two groups include:

- **betacyanins** (formerly called nitrogen anthocyanins), resembling anthocyanins with their purple and red colour;

- yellow and orange **betaxanthins**.

9.3.1.2.1 Structure and nomenclature

All betalains have the same basic structure. The chromophore system of conjugated double bonds is derived from dihydropyridine (Figure 9.6). The individual compounds differ from one another by the structure of substituents R and R^1, which are aliphatic or part of the nitrogen heterocycles.

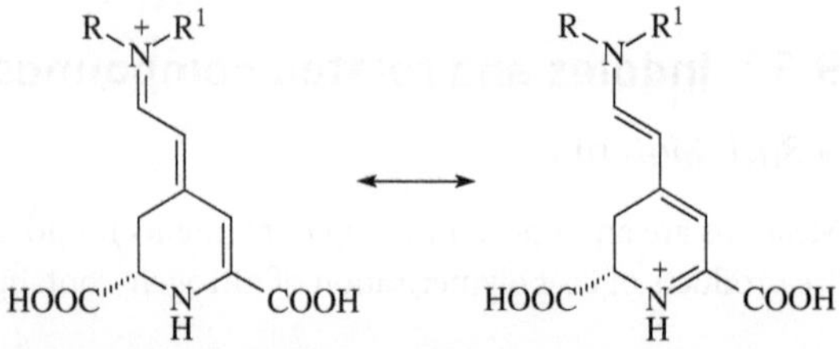

Figure 9.6 Basic structure of betalains.

Betacyanins   All known betacyanins (about 50 compounds) are found exclusively as glycosides and acylated glycosides. The main aglycone of the important betacyanins is betanidin (**9-17**). In small quantities, betanidin is accompanied by its C-15 epimer isobetanidin, whilst neobetanidin (14,15-dehydrobetanidin) and 2-decarboxybetanidin (**9-18**) occur sporadically. Recent publications demonstrated the possibility of the generation of neoderivatives and 2-decarboxylated derivatives during heating of red beet and purple pitaya. Various other decarboxylated pigments may also be produced.

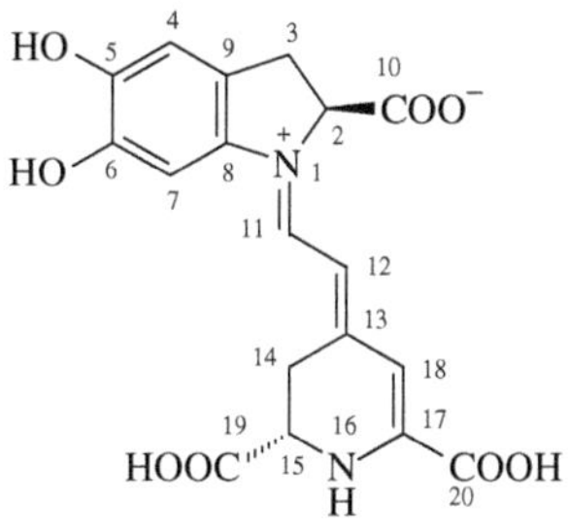

9-17, betanidin

The dominant betacyanin is the 5-*O*-β-D-glucoside of betanidin, (15*S*)-betanidin-5-*O*-β-D-glucopyranoside, which is called betanin (**9-19**). It is usually accompanied by isobetanin, which is the (15*R*)-isomer of betanin. Minor pigments are betanin 6-sulfate and isobetanin 6-sulfate, known as prebetanin (**9-19**) and isoprebetanin, respectively. A less common pigment is betanidin 6-*O*-β-D-glucoside, known as gomphrenin I (**9-20**).

Less common sugars bound in betacyanins include L-rhamnose, D-glucuronic acid, and β-sophorose. Amaranthin is an example of a pigment in which the disaccharide β-D-Glc*p*A-(1 → 2)-β-D-Glc*p* is bound at position C-5; this is related to sophorose, β-D-Glc*p*-(1 → 2)-β-D-Glc*p*. The sugar component of betalains can be esterified with malonic, ferulic, *p*-coumaric, sinapic, caffeic, citric, and 3-hydroxy-3-methylglutaric acids. For example, phyllocactin is betanin with glucose at position C-6, esterified with malonic acid; the same position in hylocerenin is esterified with 3-hydroxy-3-methylglutaric acid. Gomphrenin II is gomphrenin I esterified with *p*-coumaric acid at position C-6 of glucose (**9-20**).

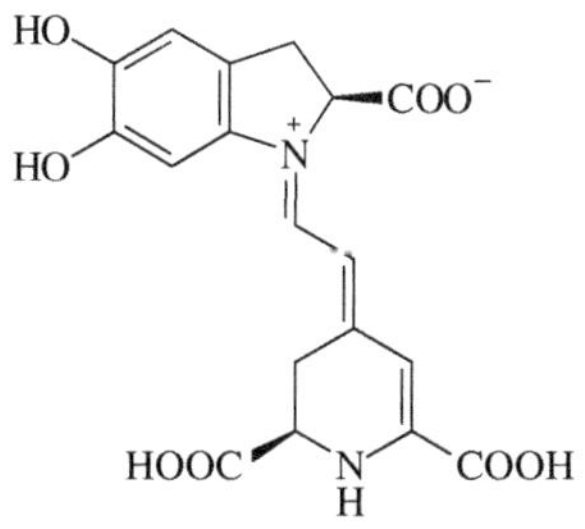

isobetanidin 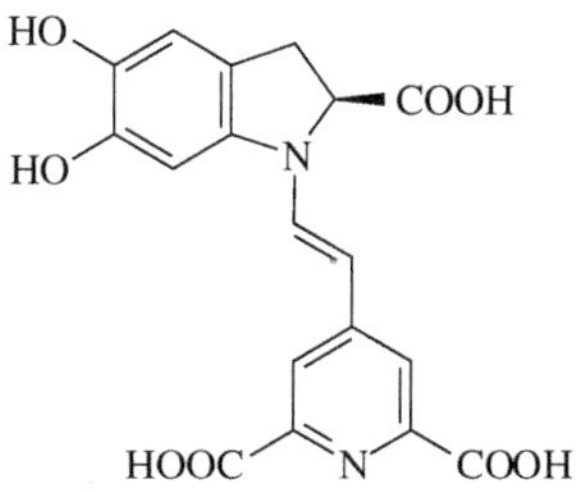   neobetanidin 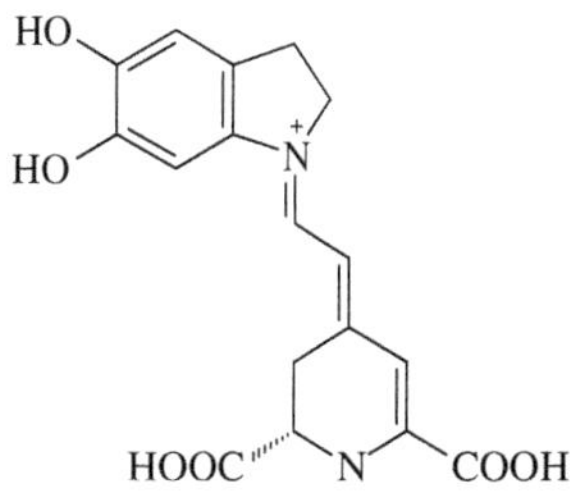   2-decarboxybetanidin

9-18, isobetanidin and related aglycones

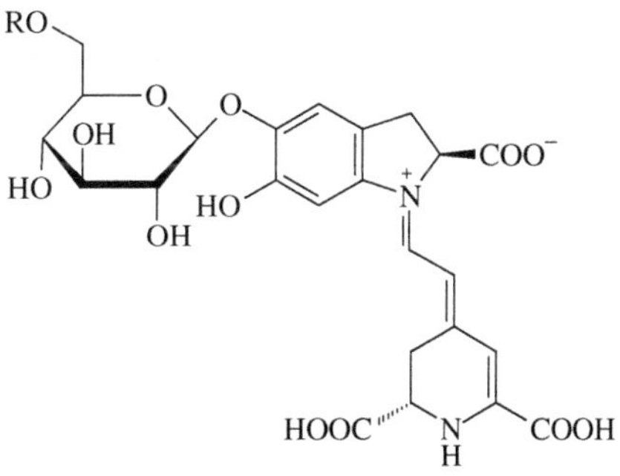

9-19, betanin, R = H
prebetanin, R = SO$_3$H

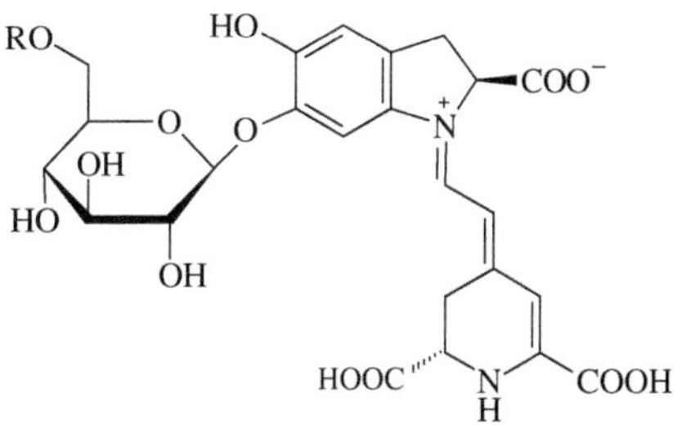

9-20, gomphrenin I, R = H
phyllocactin, R = COCH$_2$COOH
hylocerenin, R = COCH$_2$C(OH)(CH$_3$)CH$_2$COOH

Betaxanthins Betaxanthins are dihydropyridine derivatives (Figure 9.6) that arise as condensation products of betalamic acid (Figure 9.7) with amino acids or biogenic amines. Examples of such – mainly yellow – pigments are vulgaxanthin I ((*S*)-glutamine-betaxanthin; R = H, R^1 = glutamine residue), vulgaxanthin II (R = H, R^1 = glutamic acid residue, **9-21**), vulgaxanthin III (R = H, R^1 = asparagine residue), and vulgaxanthin IV (R = H, R^1 = leucine residue). The R^1 substituents of other pigments are derived from methionine sulfoxide (miraxanthin I), aspartic acid (miraxanthin II), tyramine (miraxanthin III), tyrosine (portulaxanthin II), or dopa (dopaxanthin). Less widespread are indicaxanthin, which is (*S*)-proline-betaxanthin (**9-22**), portulaxanthin I (**9-22**), which contains 4-hydroxyproline residue, and muscaurin I and muscaurin II (**9-23**), which are derived from ibotenic and stizolobic acids, respectively.

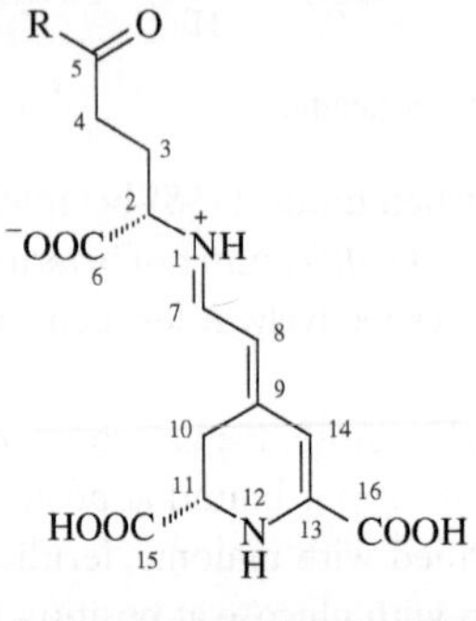

9-21, vulgaxanthin I, R = NH$_2$
vulgaxanthin II, R = OH

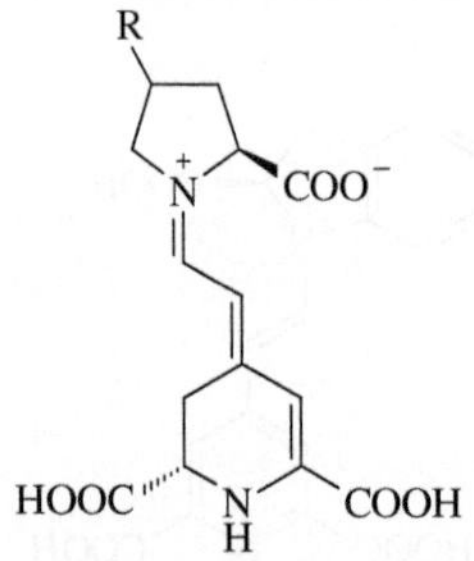

9-22, indicaxanthin, R = H
portulaxanthin I, R = OH

9.3.1.2.2 Occurrence

Betalains occur in nature in a relatively small number of plant genera belonging to the order of flowering plants Caryophyllales. The only exceptions are plants of the pink (carnation) family (Caryophyllaceae) and of the family Molluginaceae that accumulate anthocyanins instead of betacyanins. These two types of plant pigments never occur together in one plant. The most important sources of betacyanins are the cultivated red varieties of beetroot (cultivars of *Beta vulgaris*, Amaranthaceae, formerly Chenopodiaceae), according to which these pigments were named, and a shrub with red berries known as American pokeweed (*Phytolacca americana*) of the family *Phytolaccaceae*, which is native to eastern North America.

Beetroot has an average betalain content of 0.1%. Some varieties, however, contain up to twice as much, which is about 2 g/kg (f.w.). Betanin represents about 75–95% of all betacyanins of beetroot and dominates significantly over yellow betaxanthins. The main yellow pigment is vulgaxanthin I, representing about 95% of the total; the rest are vulgaxanthin II and betalamic acid, a key intermediate of the biosynthesis and degradation of betalains (Figure 9.7). The pigment of American pokeweed was originally called phytolaccanin, but was later identified as betanin.

The main pigment of red flowers, leaves, and stems of the ornamental plant *Amaranthus tricolor* and of other species of the family Amaranthaceae, some of which are consumed for their seeds, leaves, and stems, is amaranthin derived from betanidin, which is accompanied by isoamaranthin derived from isobetanidin. The main pigments of red spinach (*A. dubius*), also known as Chinese spinach, are betanin, amaranthin, isoamaranthin, and decarboxyamaranthin. Amaranthin, isoamaranthin, betanin, and isobetanin are the main pigments of quinoa grains (*Chenopodium quinoa*) of the same plant family. Gomphrenin I and II (accompanied by their decarboxylation products, 2-, 17-, and 2,17-decarboxygomphrenins, their diastereomers and dehydrogenated derivatives) are the main pigments of fruits of Malabar (Ceylon) spinach (*Basella alba*, Basellaceae), an edible perennial vine growing in tropical Asia and Africa, whose leaves are eaten as a vegetable, even though it is not related to true spinach (*Spinacia oleracea*, Amaranthaceae).

Betalains also occur in edible fruits of some cacti (nopales) of the genus *Opuntia* (Cactaceae). Fruits of *O. ficus-indica*, referred to as prickly pear or Indian fig, contain yellow indicaxanthin (yellow fruits) and red betanin (red fruits). Betanin, isobetanin, and the acylated betacyanins, phyllocactin and hylocerenin, were found in the red-fleshed pitahaya, better known as dragon fruit (*Hylocereus polyrhizus*). The red- and orange-fleshed pitaya fruits (*Stenocereus pruinosus* and *S. stellatus*) contain indicaxanthin, gomphrenin I, phyllocactin, and their isomers.

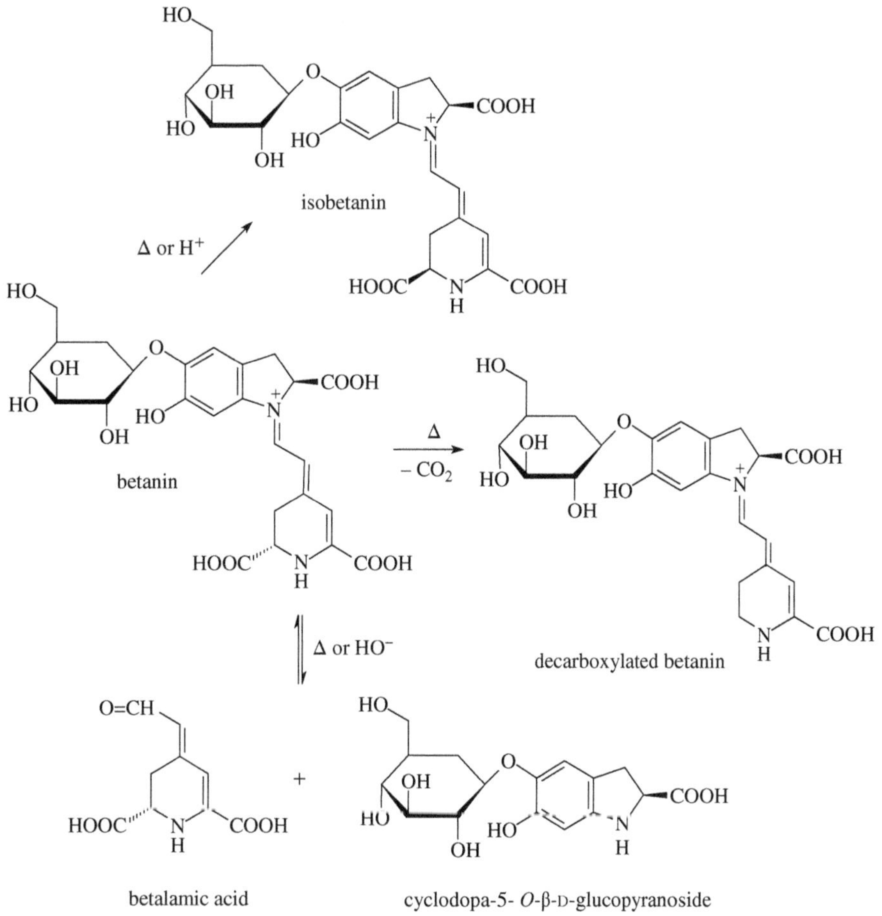

Figure 9.7 Mechanism of betanin degradation.

Betalains are also pigments of some mushrooms. For example, the common mushroom (*Agaricus bisporus*, Agaricaceae) supposedly contains small quantities of the same pigments as beetroot. Following tissue damage, the light pink colour of the fungus changes to grey-black, since L-dopa is oxidised to melanin pigments. The striking orange-red pigment of the cap of fly agaric *Amanita muscaria* (Amanitaceae) is a mixture of the purple betacyanin muscapurpurin (**9-23**), orange betaxanthins muscaurins I–VII, and yellow muscaflavin (**9-23**). Muscaurin I and muscaurin II (**9-23**) are derived from unusual mushroom non-protein amino acids ibotenic and stizolobic acids, respectively, which are the major agaric pigments. Pigments termed muscaurins III–VII were later identified as mixtures of pigments derived from common protenogenous amino acids. Muscaurin III is a mixture of vulgaxanthin I (**9-21**), known as (*S*)-glutamine-betaxanthin, miraxanthin III, known as (*S*)-aspartic acid-betaxanthin, and betaxanthin, derived from 2-aminoadipic acid. Muscaurin IV is a mixture of vulgaxanthin I (**9-21**) and miraxanthin III. Muscaurin V is a mixture of vulgaxanthin II (**9-21**), known as (*S*)-glutamic acid-betaxanthin, indicaxanthine (**9-22**), known as (*S*)-proline-betaxanthin, and betaxanthins, derived from valine and leucine (vulgaxanthin IV). Muscaurin VI is a mixture of vulgaxanthin II (**9-21**) and indicaxanthin (**9-22**). One of the dominant pigments, muscaurin VII, is derived from histidine.

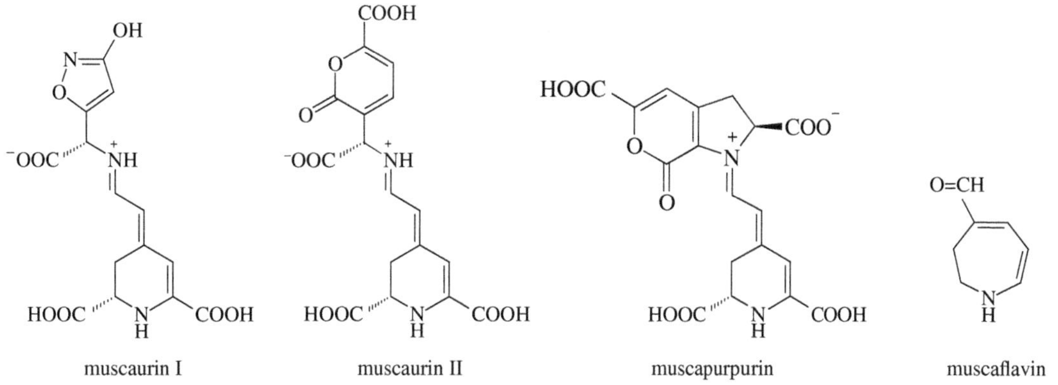

9-23, pigments of *Amanita muscaria*

Fly agaric and mushrooms of the genus Hygrocybe (*Hygrophoraceae*), such as *H. conica*, commonly known as the witch's hat, synthesise a yellow isomer of betalamic acid called muscaflavin (**9-23**), which is derived from dihydroazepine, and store it in the cup skin. In the same way that betalamic acid is involved in the formation of various betalains, muscaflavin can spontaneously condense with amino acids to form aldimine bonds, yielding yellow hygroaurins (**9-24**).

9.3.1.2.3 Use

Only pigments from beetroot have found significant practical use; these are commercially called betanin or beetroot red. Owing to its low stability, beetroot red is used for colouring foods with a short shelf-life, such as dairy and meat products (e.g. sausages of poultry meat with a light colour), acidic beverages (e.g. soft drinks), and some sweets. The pigment is supplied as a concentrated syrup or powder. The juice of the American pokeweed was used in the past to colour wine; today, this is considered falsification and is illegal.

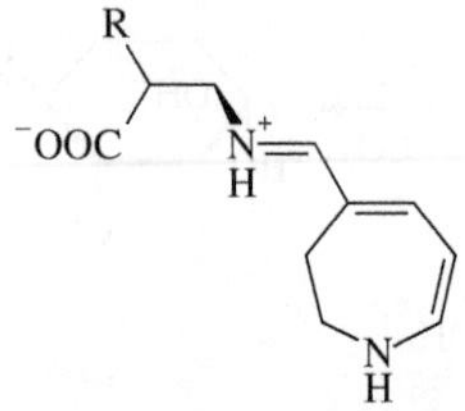

9-24, hygroaurins, R = amino acid residue

9.3.1.2.4 Biochemistry, physiology, and nutrition

The precursors of betacyanins are the amino acid L-tyrosine and, correspondingly, 3,4-dihydroxy-L-phenylalanine (L-dopa). By a sequence of oxidation reactions via 4,5-seco-L-dopa, betalamic acid (Figure 9.7) and cyclo-3-(3,4-dihydroxyphenyl)-L-alanine (L-cyclodopa) are produced. Condensation of betalamic acid with cyclodopa yields betacyanins, whilst its condensation with amino acids produces betaxanthins. Muscaflavin in fungi also arises from dopa, but via 2,3-seco-L-dopa spontaneous cyclisation. Reaction of muscaflavin with amino acids yields hygroaurins (Figure 9.8).

Betalains exhibit antioxidant activity and react with free radicals. Their adverse effects have not been detected; therefore, beetroot pigments are generally accepted as safe food colourants.

9.3.1.2.5 Properties and reactions

Beetroot betacyanins have a very intense colour but are sensitive to oxidation (on light and especially in the presence of divalent and trivalent metals) and the presence of sulfur dioxide (which decolourises betacyanins). In the absence of these agents, the rate and extent of betacyanin

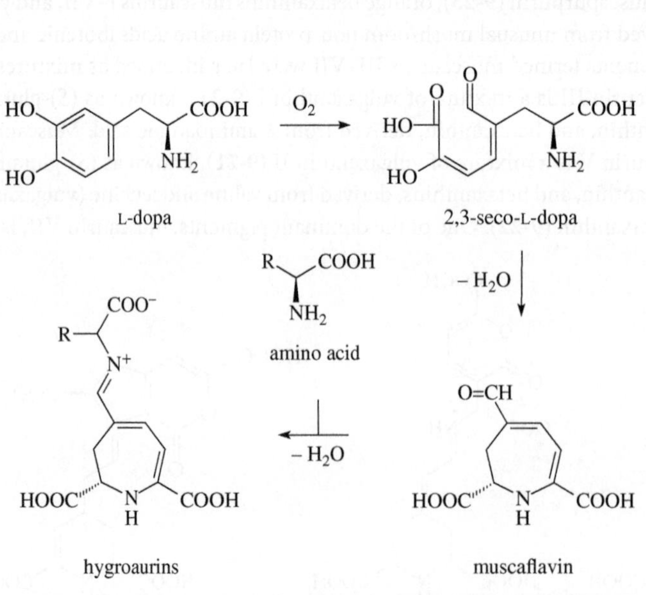

Figure 9.8 Biosynthesis of muscaflavin and hygroaurins.

degradation depends mainly on water activity and pH (Figure 9.7). They exhibit the highest stability in solutions of pH 4–5, but at pH > 7 are rapidly degraded. In acidic solutions, betacyanins hydrolyse to the corresponding aglycones and glucose. The degradation products in alkaline solutions are 4-methylpyridine-2,6-dicarboxylic acid, formic acid, and (S)-5,6-dihydroxy-2,3-dihydroindole-2-carboxylic acid (9-25). The latter compound also arises, together with the corresponding amino acids or amines, as a degradation product of betaxanthins in aqueous solutions.

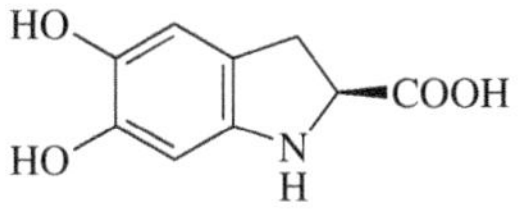

9-25, 5,6-dihydroxy-2,3-dihydroindole-2-carboxylic acid

9.3.2 Isoquinolines

The isoquinoline (9-30) or benzyltetrahydroisoquinoline skeleton can be found in the tetracyclic alkaloid berberine, also called umbellatin (9-30). Berberine occurs as the main component in all parts of the European barberry shrub (*Berberis vulgaris*, Berberidaceae), in particular the cortex, but also in leaves and immature fruits. Barberry alkaloids are carriers of the characteristic yellow colour of barberry wood and bark, but do not practically influence the colour of edible berries. The mildly poisonous cortex of the roots, containing 12–15% of alkaloids, was used in the past as a drug for medical purposes (the bark of the stem contains 5.5–8% and the bark of the branches about 3.5% of the mixture of alkaloids). The main components of the root bark are the quaternary bases berberine, jatrorrhizine, columbamine, and palmatine (9-30) and related tertiary bases. Various Asian species of barberry, such as *B. amurensis* and *B. asiatica*, are used in medical preparations in India and the Far East. Berberine also occurs in a widespread evergreen shrub called Oregon grape (*Mahonia aquifolium*, Berberidaceae), related to the barberry, and in goldenseal, also known as orangeroot (*Hydrastis canadensis*, Ranunculaceae).

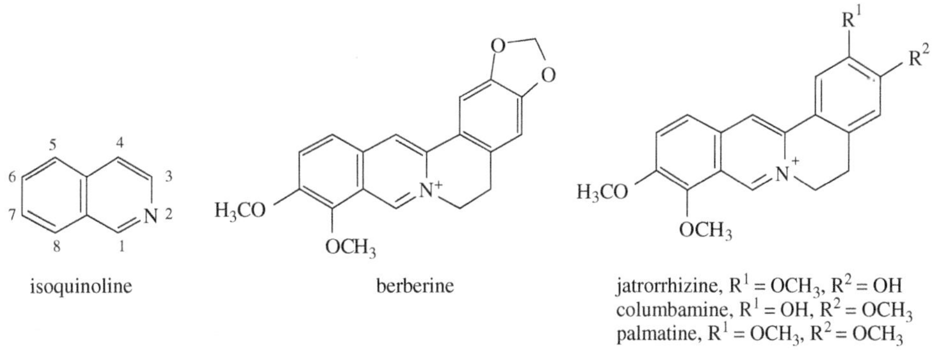

isoquinoline   berberine

jatrorrhizine, $R^1 = OCH_3$, $R^2 = OH$
columbamine, $R^1 = OH$, $R^2 = OCH_3$
palmatine, $R^1 = OCH_3$, $R^2 = OCH_3$

9-30, isoquinoline and derived pigments

Overdose or overuse of drugs containing berberine leads to transient irritation of the central nervous system, stupor, and diarrhoea and can cause kidney damage. Therefore, their use is restricted in many countries. The use of berberine as a yellow food pigment and bittering agent for liqueurs and spirits was reported by a Japanese company, Kakko Honsha, in 1980. In the European Union, the berberine content in aromatised foods is restricted by law. The tolerated content (maximum level), originating exclusively from natural sources, is 10 mg/l for alcoholic beverages and 0.1 mg/kg for other foods and beverages.

9.3.3 Purines

Pigments derived from purine (9-27), such as guanine, xanthine, and uric acid (9-27), are very important in the animal kingdom. They occur in two tautomeric forms: the major oxo form and the rare enol form. The actual compounds are colourless, but in the form of granules or microcrystals they are the bases of white, cream, and silvery semi-transparent pigments that are found, for example, in fish scales.

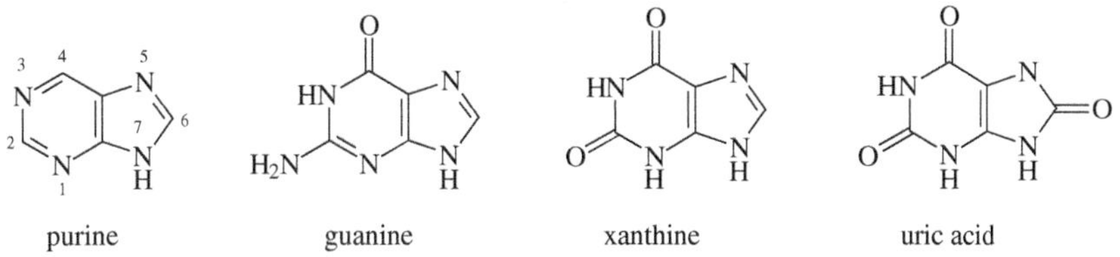

purine   guanine   xanthine   uric acid

9-27, purine and oxoforms of derived pigments

9.3.4 Pterines

The basic skeleton of pterine pigments is pteridine (**9-28**). Pteridine substituted in position C-2 with an amino group and in position C-4 with a hydroxyl group (tautomeric oxo group) is called pterine (**9-29**). Natural pterines are biologically active forms of pteroylglutamic (folic) acid (see Section 5.12), but due to the quantities in which they occur in foods, they have no importance as yellow food pigments.

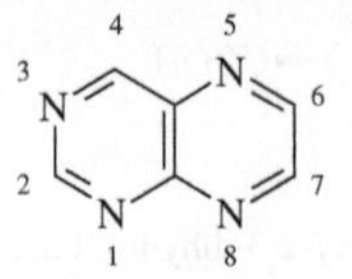

9-28, pteridine

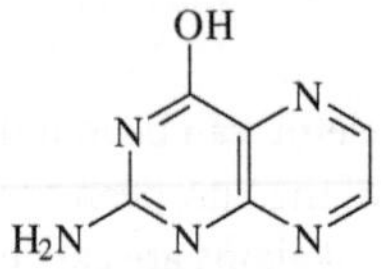

9-29, pterine

White, yellow, orange, and red pterines are pigments of butterfly wings. Pterines also occur in some shellfish, fish, amphibians, and reptiles and act as shielding pigments of the eye in some invertebrates.

9.3.5 Isoalloxazines

Flavins derived from isoalloxazine (**9-30**) are a group of pale-yellow, greenly fluorescent biological pigments (biochromes) widely distributed in small quantities in foods of plant and animal origin. Flavins are synthesised only by bacteria, yeasts, and green plants; for this reason, animals are dependent on plant sources. An important representative of flavins is riboflavin (vitamin B_2), the parent precursor to the coenzymes FMN and FAD (see Section 5.7). Milk whey (serum phase) is a rich source of riboflavin that gives whey its yellow-green colour. Riboflavin (the E number is E101) is also employed to colour convenient foods, soft drinks, cheese and cheese products, dairy products, bakery goods, fish products, canned fruits and vegetables, confectionery, desert powder, jams and jellies, soups, mayonnaise and salad dressing, mustard, and flavorings.

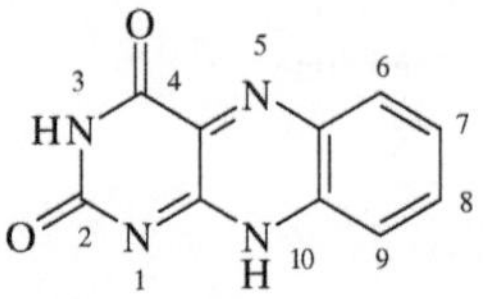

9-30, isoalloxazine

9.3.6 Phenoxazines

Phenoxazine (**9-31**) pigments occur mainly in lichens (symbiotic organisms of fungi belonging to the phylum Ascomycota, algae, or cyanobacteria), mostly in the form of uncoloured precursors (different depsides). To a lesser extent, phenoxazines occur in higher fungi of the Polyporaceae family. Pigments extracted from lichens with urine (a source of ammonia) or ammonia solutions and oxidised by air oxygen have been used as textile dyes from time immemorial.

An important group of phenoxazine pigments known as orcein or orchil, is a purple-to-red pigment extracted from lichens of the genera *Roccella* and *Orchella* (orchella weeds). Orcein is a mixture of different hydroxyphenoxazones and aminophenoxazones. Examples of these components are α-hydroxy and α-amino orceins (**9-31**) and β-hydroxy and β-amino orceins (**9-31**). Their precursors in lichens are depsides of some aromatic acid, such as lecanoric (orsellic) acid and erythrin (**9-32**), that produce as a degradation product orcinol (5-methylbenzene-1,3-diol) and analogous products, which on condensation yield orcein pigments. The dye is very slightly soluble in water, but soluble in ethanol. Orcein can be used to dye wool and silk, and was once approved as a food colourant (with E number E121), but is now banned throughout the European Union. Sulfonation of orcein produces a water-soluble pigment, which was used for colouring soft drinks and confectionery.

An example of a phenoxazine pigment occurring in higher fungi is the cinnabar red pigment of the wood-rotting fungus *Pycnoporus cinnabarinus*, 2-amino-3*H*-phenoxazin-3-one-1,9-dicarboxylic acid, named cinnabaric acid (**9-33**).

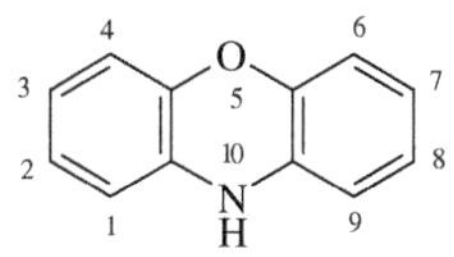

phenoxazine

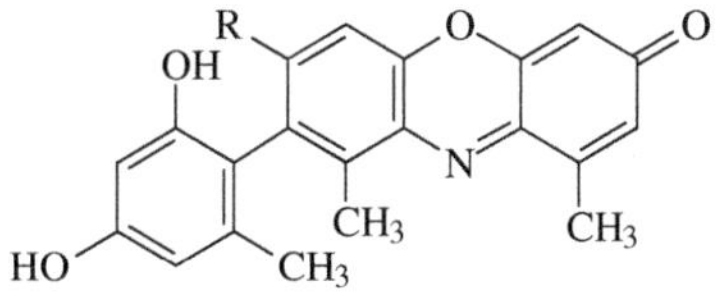

α-hydroxy orcein (R = OH)
α-amino orcein (R = NH$_2$)

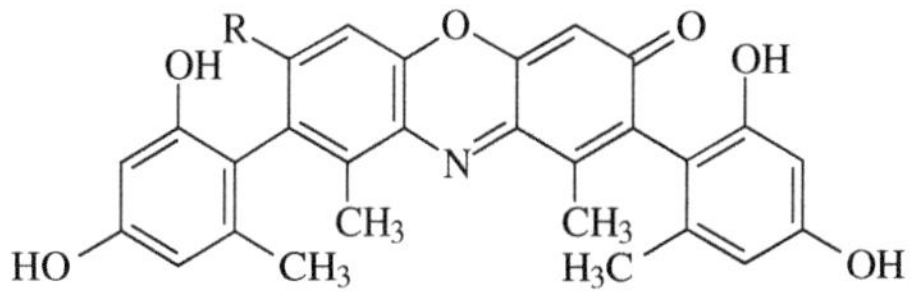

β-hydroxy orcein (R = OH)
β-amino orcein (R = NH$_2$)

9-31, phenoxazine and derived pigments

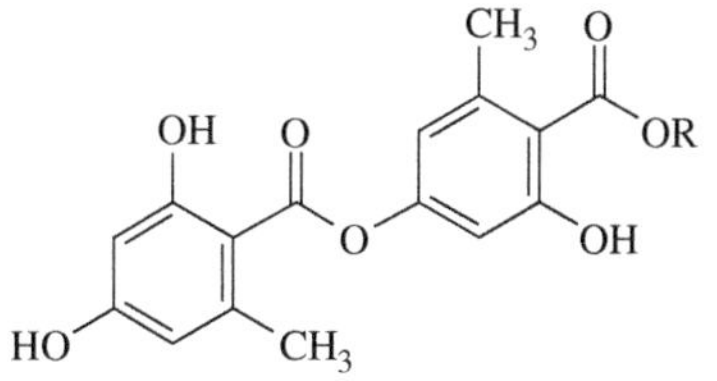

9-32, erythrin, R = H
lecanoric acid, R = CH$_2$CH(OH)CH(OH)CH$_2$OH

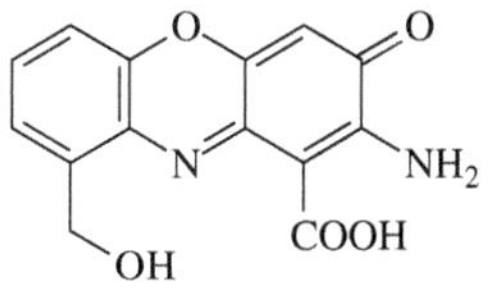

9-33, cinnabaric acid

9.4 Flavonoids

The flavonoids are a very large group of plant phenols containing two benzene rings (ring A and C) in the molecule, associated by a three-carbon chain. The number of flavonoid compounds is now estimated at about 5000, and new compounds are still being found in various plant sources. Most flavonoid compounds have the C$_3$ chain as a part of a heterocyclic ring derived from 2*H*-pyran (ring C). Flavonoids have an arrangement of C$_6$—C$_3$—C$_6$ and are therefore derived from 2*H*-chromene (**9-34**) substituted in position C-2 by a phenyl group, which is called flavan (**9-35**). Compounds of this type are often called 1,3-diarylpropanoids. Typically, all three rings of flavonoids are substituted with hydroxyl or methoxyl groups, and individual derivatives differ only in the degree of substitution and oxidation. Flavonoids occur as free compounds, but more frequently as glycosides, acylated glycosides, and polymers.

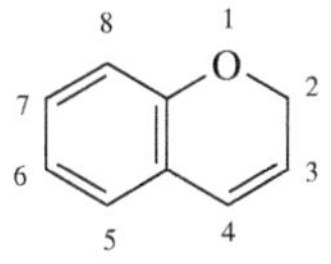

9-34, 2*H*-chromene

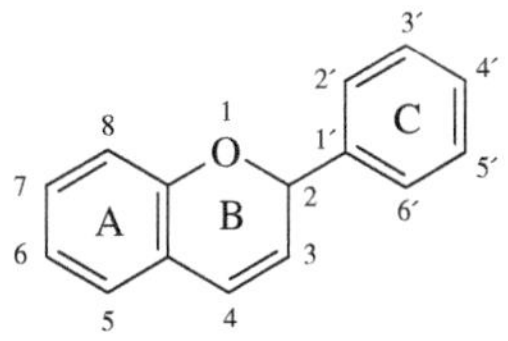

9-35, flavan

Depending on the degree of oxidation and C-3 chain substitution, the following basic structures of flavonoids can be identified:

- **catechins** (flavan-3-ols);

- **leucoanthocyanidins** (flavan-3,4-diols);

- **flavanones**;

- **flavanonoles**;

- **flavones**;

- **flavonols** (dihydroflavones);

- **anthocyanidins**.

The basic structure of these flavonoid compounds is shown in formulae **9-36**. Their degree of oxidation increases along the row from left to right, along with the colour intensity. The compounds listed in the columns have the same degree of oxidation. A special case is anthocyanidins, which contain a system of conjugated double bonds; therefore, they are flavylium (2-phenylbenzopyrylium or 2-phenylchromenylium) cations.

In several cases, the six-membered heterocyclic ring B exists in the isomeric open form; alternatively, it can be substituted by a five-membered heterocyclic ring. The structurally related compounds, in which an aliphatic C-3 chain (or a chain that is partially included into the furan ring) connects rings A and C, are divided into:

- **chalcones** and **dihydrochalcones**;

- **aurones**.

Structures of these flavonoids are given in formulae **9-37**. Less common compounds with ring B associated with pyran ring C in position C-3 (derived from isoflavan) are called **isoflavonoids** (also known as 3-phenylchromen-4-ones or 1,2-diarylpropanes, **9-37**). If the connection between these two rings is in position C-4, the corresponding compounds, derived from neoflavan, are called **neoflavonoids** or 4-phenylcoumarins, 5,6-benzo-2-pyrones, or 1,1-diarylpropanes (**9-37**).

Only certain flavonoids are important as natural plant pigments; others act as taste-active compounds, as they are precursors of astringent and bitter substances or have various important biological effects. All coloured flavonoids were previously logically divided into two large groups according to their colour: the red and blue **anthocyanins** and the yellow **anthoxanthins**. Their names were derived from the Greek word *anthos* (flower), *cyaneos* (blue), and *xanthos* (yellow). The names of flavones and other yellow flavonoids are derived from the Latin word *flavus* (yellow). Chalcones and aurones were previously called anthochlor pigments or **anthochlors** (from Greek *chloros*, green).

Catechins and leucoanthocyanidins are colourless compounds, but the brown pigments that arise from them in enzymatic browning reactions are important pigments of many foods. Colourless leucoanthocyanidins also give rise to the corresponding coloured anthocyanidins during processing of fruits and vegetables in acidic solutions. Oligomers of catechins and leucoanthocyanidins with a bitter taste are classified as condensed tannins. Flavanones and flavanonoles are colourless or pale-yellow compounds, and do not have great significance as plant pigments. Some flavanones, however, are important bitter components of grapefruits. The most important flavonoid pigments are yellow flavones and flavonols and red (also yellow or orange), purple, and blue anthocyanins. Chalcones and dihydrochalcones are yellow pigments and aurones are golden-yellow pigments mostly of plant flowers, but as food pigments they are not particularly important. Some other pale-yellow pigments, the isoflavones, similarly have negligible importance as pigments, but they show important biological properties (oestrogenic activity, see Section 10.3.12.3). They are found in certain food commodities (notably soya beans and derived products).

Flavonoids are primary antioxidants. Flavonols and 5-hydroxysubstituted flavones also bind metals into inactive complexes. Important for the antioxidant activity of flavonoids is the number of hydroxyl groups in the molecule and their positions. Active compounds are all dihydroxy derivatives with hydroxyl groups at positions C-3′ and C-4′. The presence of other hydroxy groups in ring B further increases the antioxidant activity (e.g. robinetin and myricetin with an additional hydroxyl group at C-5′ are more active as antioxidants than quercetin). Low antioxidative activity is exhibited by flavonoids with one hydroxy group in ring B (such as flavanones naringenin and hesperetin). Other important functional groups include the carbonyl group at C-4 and free hydroxyl group at C-3 (or C-5). Very efficient antioxidants have two hydroxyl groups in the *o*-position in one ring and two hydroxy groups in the *p*-position in other rings, such as 3,5,8,3′,4′- and 3,7,8,2′,5′-pentahydroxy derivatives.

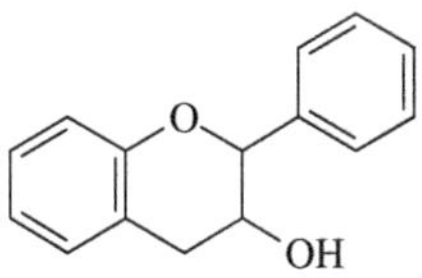

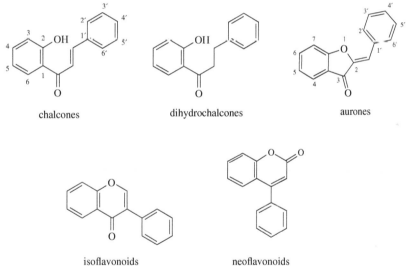

9-36, common flavonoid compounds

flavan-3-ols (catechins)

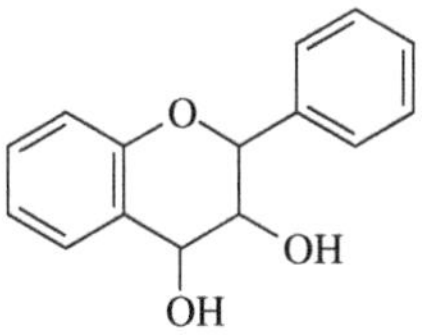

flavan-3,4-diols (leucoanthocyanidins)

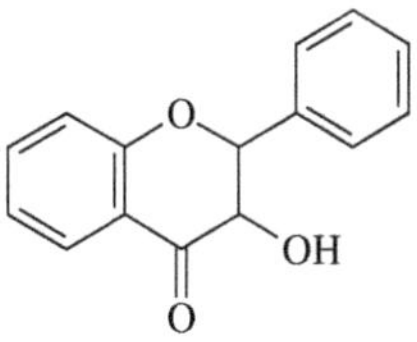

flavanonoles (dihydroflavonols)

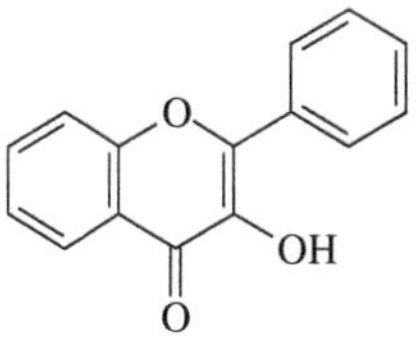

flavonols

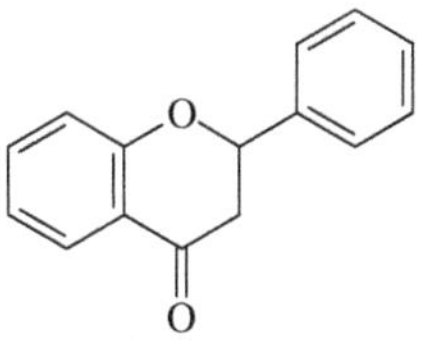

flavanones (dihydroflavones)

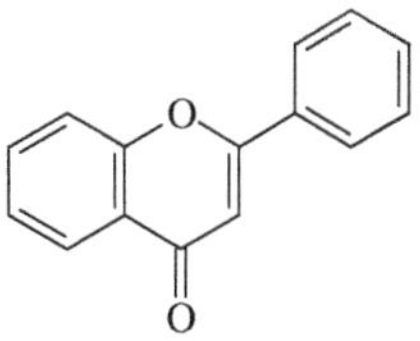

flavones

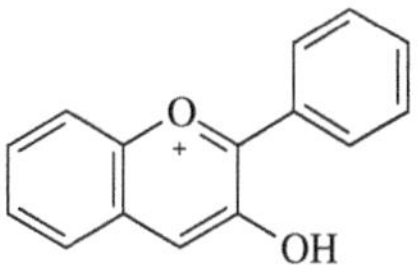

anthocyanidins

chalcones

dihydrochalcones

aurones

isoflavonoids

neoflavonoids

9-37, less common flavonoid compounds

9.4.1 Anthocyanins

Anthocyanins are the most widespread (and numerically a very large group of) hydrophilic plant pigments. So far, about 300 different anthocyanins have been identified in nature. Many fruits, vegetables, flowers, and other plant materials owe their attractive orange, red, purple, and blue colour to this group of water-soluble pigments.

9.4.1.1 Structure and nomenclature

Anthocyanins are glycosides of different aglycones that are called anthocyanidins. All anthocyanidins are derived from one basic structure, which is flavylium cation (**9-36**). It is reported that 17 different anthocyanidins exist in nature. All compounds are substituted in position C-4 by a hydroxyl group, and they mutually differ by substitutions at positions C-3, C-5, C-6, C-7, C-3′, and C-5′. In positions C-5, C-7, C-3′, and C-5′, methoxyl groups may occur. Table 9.10 lists the 15 most common anthocyanidins, the last two of which are 6-hydroxyanthocyanidin (abbreviated 6-OHCy) and 5-methylcyanidin (5-MCy). The most important compounds occurring in foods are six anthocyanidins (**9-38**)

Table 9.10 Overview of anthocyanidin structures.

Name[a]	Abbreviation	Substituents					
		C-3	C-5	C-6	C-7	C-3′	C-5′
Apigeninidin	Ap	H	OH	H	OH	H	H
Luteolinidin	Lt	H	OH	H	OH	OH	H
Tricitinidin	Tr	H	OH	H	OH	OH	OH
Pelargonidin	Pg	OH	OH	H	OH	H	H
Aurantinidin	Au	OH	OH	OH	OH	H	H
Cyanidin	Cy	OH	OH	H	OH	OH	H
Peonidin	Pn	OH	OH	H	OH	OCH$_3$	H
Rosinidin	Rs	OH	OH	H	OCH$_3$	OCH$_3$	H
Delfinidin	Dp	OH	OH	H	OH	OH	OH
Petunidin	Pt	OH	OH	H	OH	OCH$_3$	OH
Pulchellidin	Pl	OH	OCH$_3$	H	OH	OH	OH
Europinidin	Eu	OH	OCH$_3$	H	OH	OCH$_3$	OH
Malvidin	Mv	OH	OH	H	OH	OCH$_3$	OCH$_3$
Hirsutidin	Hs	OH	OH	H	OCH$_3$	OCH$_3$	OCH$_3$
Capensinidin	Cp	OH	OCH$_3$	H	OH	OCH$_3$	OCH$_3$

[a]Most important anthocyanidins are indicated in bold.

with a hydroxyl group at position C-3. In descending frequency of occurrence, these anthocyanidins are: cyanidin (from the Latin name for the cornflower, *Centaurea cyanus*, Asteraceae), pelargonidin (found in geraniums, *Pelargonium* spp., Geraniaceae), peonidin (found in peonies, *Paeonia* spp., Paeoniaceae), delphinidin (found in delphiniums, *Delphinium* spp., Ranunculaceae), petunidin (found in petunias, *Petunia* spp., Solanaceae), and malvidin (found in mallow, *Malva* spp., Malvaceae), formerly also known as oenidin.

Materials that contain anthocyanins generally have the colour of their aglycones. Free aglycones of anthocyanins occur in plant tissues rarely and only in trace amounts as anthocyanin hydrolysis products and products derived from leucoanthocyanidins. In all plant materials, the main pigments are glycosides and acylated glycosides of anthocyanidins. Present in some fruits and vegetables are anthocyanins derived from a single anthocyanidine (e.g. in apples, red cabbage varieties, and elderberries, these are cyanidin glycosides), but sometimes pigments are derived from several different anthocyanidins. For example, blackcurrant anthocyanins are delphinidin and cyanidin derivatives, whilst strawberry anthocyanins are derived from pelargonidin and cyanidin.

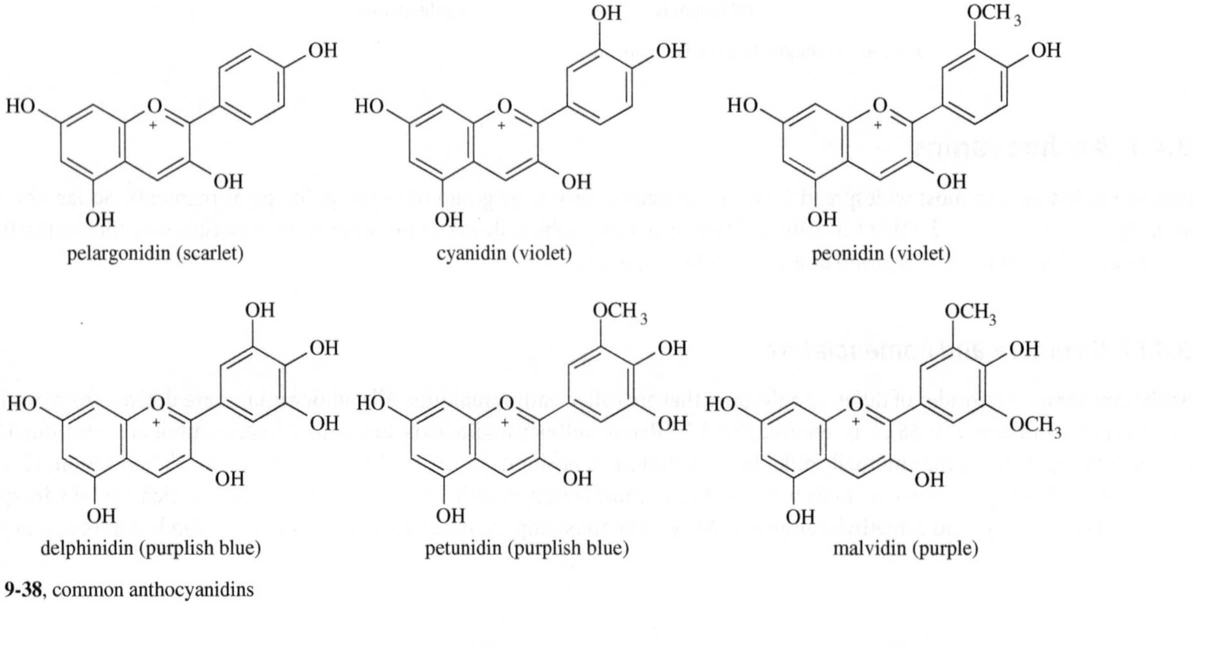

pelargonidin (scarlet)   cyanidin (violet)   peonidin (violet)

delphinidin (purplish blue)   petunidin (purplish blue)   malvidin (purple)

9-38, common anthocyanidins

The anthocyanin pigments contain five monosaccharides: D-glucose, L-rhamnose, D-galactose, D-xylose, and L-arabinose (in descending order of frequency). Sugars are always bound in position C-3, and if another hydroxyl group is glycosylated, it is the hydroxyl at C-5 (the bound sugar is glucose and less often rhamnose). Free hydroxyl at C-3 destabilises the anthocyanidin chromophore, so the hydrolysis of sugar bound as *O*-glycoside at C-3 results in rapid and irreversible decomposition of anthocyanidin and leads to discolouration. In exceptional cases, sugars are also bound in positions C-7, C-3′, C-5′, and C-4′ as mono- (mainly glucose), di-, or trisaccharides. The most common disaccharides are rutinose, sambubiose, sophorose, laminaribiose, and gentiobiose.

Depending on the number of bound sugar molecules, anthocyanins are divided into 18 groups, of which the most important are:

- monosides with glucose, galactose, rhamnose, or arabinose in position C-3 (3-monosides);

- biosides with disaccharides (such as rutinose, sambubiose, sophorose, gentiobiose, neohesperidose, laminaribiose, robinobiose, and other disaccharides) bound in position C-3 (3-biosides);

- triosides with linear or branched trisaccharides bound in position C-3 (3-triosides);

- 3,5-diglycosides with monosaccharides at positions C-3 and C-5;

- 3,7-diglycosides with monosaccharides at positions C-3 and C-7;

- 3-biosides-5-monosides with disaccharide in position C-3 and monosaccharide in position C-5.

The most frequently occurring pigments are cyanidin-3-*O*-glycosides, which occur approximately three times more often than 3,5-di-*O*-glycosides. The type of sugar has little effect on the chemical properties of a pigment; much more significant is the position in which it is bound. An example of 3-*O*-glycosides is cyanidin-3-*O*-β-D-glucopyranoside (**9-39**).

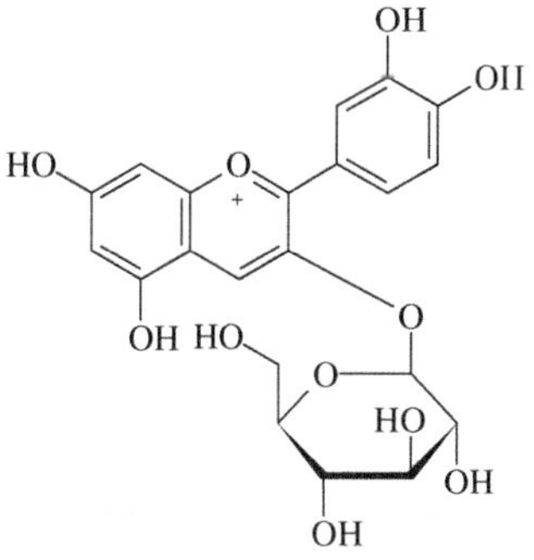

9-39, cyanidin-3-*O*-β-D-glucopyranoside

Besides the common anthocyanins, some unusual compounds differing from the normal structures in the substituent position are found in nature. Examples of rare **3-deoxyanthocyanins (9-40)** lacking C-3 hydroxyl group are apigeninidin, luteolinidin, related 5- and 7-methoxylated derivatives, and glycosides. These 3-deoxy forms have a yellow to orange colour in acidic media and, unlike the red 3-hydroxy derivatives, are stable even as aglycones. Red tricetinidin is found in black tea infusions as a product of epigallocatechin gallate degradation.

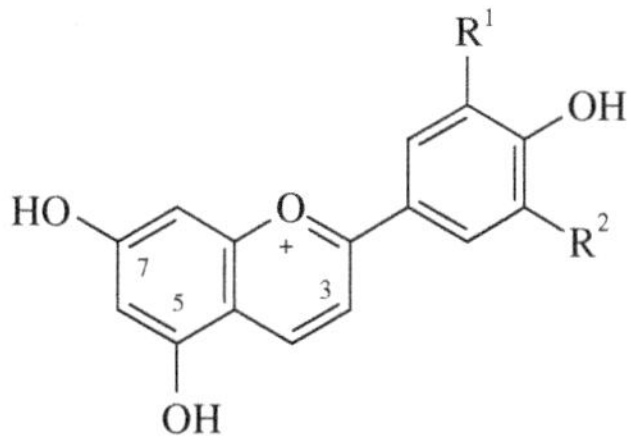

9-40, apigeninidin, $R^1 = R^2 = H$
luteolinidin, $R^1 = OH$, $R^2 = H$
tricetinidin, $R^1 = R^2 = OH$

Sugars bound in anthocyanins are often acylated by phenolic acids (*p*-coumaric, caffeic, ferulic, sinapic, and, less frequently, *p*-hydroxybenzoic acids), malonic acid, and acetic acid. Acids are usually linked to the C-6 hydroxyl group of glucose and the C-4 hydroxyl group of rhamnose. 3-Glycosides are only rarely acylated. An exception is delphinidin-3-*O*-rutinoside-5,3′,5′-tri-*O*-glucoside, which contains four molecules of phenolic acids, some of which are bound to glucose occurring in other positions. An example of acylated pigments is (*E*)-petunidin-3-*O*-[6-*O*-(4-*O*-*p*-coumaroyl-α-L-rhamnopyranosyl)-β-D-glucopyranoside]-5-*O*-β-D-glucopyranoside, trivially called petanin (**9-41**), which is the characteristic pigment in the skin of red potato varieties. Other examples of acylated anthocyanins are pigments of black or purple carrots (*Daucus carota* ssp. *sativus* var. *atrorubens*). In recent years, many new varieties with an extremely high content of pigments have been cultivated. The anthocyanin content of several of these black carrot cultivars was reported to range up to 1750 mg/kg (f.w.). The dominating anthocyanins are (*E*)-cyanidin-3-*O*-[2-*O*-β-D-xylopyranosyl-6-*O*-(4-*O*-*p*-coumaroyl)-β-D-glucopyranosyl]-β-D-galactopyranoside and related pigments derived from caffeic and sinapic acids (**9-42**).

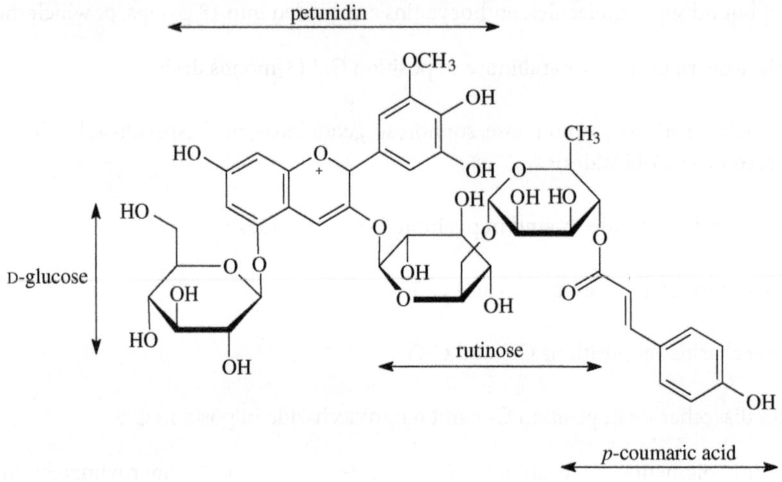

9-41, petanin

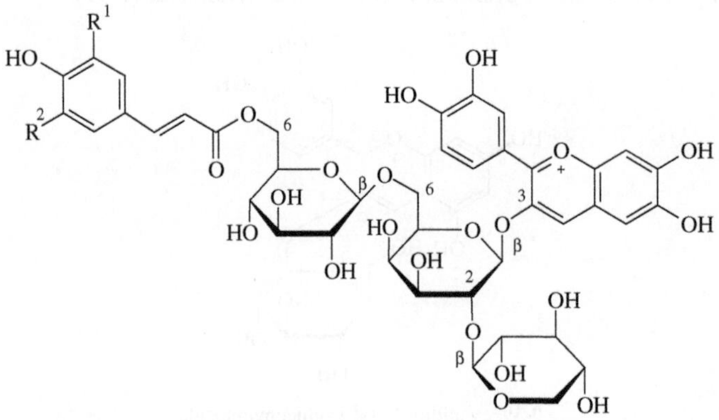

9-42, pigment of black carrot cultivars, R^1= R^2= H

Many other anthocyanins were known by trivial names in the past. Examples are cyanidin-3-*O*-glucoside (chrysanthemin), cyanidin-3-*O*-galactoside (idaein), cyanidin-3-*O*-rutinoside (ceracyanin), cyanidin-3-*O*-sambubioside (sambycyanin), cyanidin-3-*O*-(2-glucosyl) rutinoside (mecocyanin), delphinidin-3-*O*-glucoside (myrtillin), malvidin-3-*O*-glucoside (oenin), and malvidin-3,5-di-*O*-glucoside (malvin).

9.4.1.2 Occurrence

Anthocyanins are found in many plant species located in cell vacuoles and stabilised by interactions of the type ion–ion with organic acids (malonic, malic, and citric acids). The main sources of anthocyanin pigments in food are fruits of plants of the grapevine family (Vitaceae), which includes grapes, and of the rose family (Rosaceae), which includes cherries, plums, raspberries, strawberries, blackberries, apples, pears, and many others. Other significant plants containing anthocyanine pigments belong to other plant families, such as eggplants (aubergines) and potato varieties with red skin (Solanaceae), black- and redcurrants, red varieties of gooseberries (Grossulariaceae), blueberries, and cranberries (Ericaceae), olives (Oleaceae), red cabbage, radishes, and red varieties of other cruciferous vegetables (Brassicaceae). Lists of the main sources of anthocyanins and the individual main pigments are given in Tables 9.11 and 9.12. The number of anthocyanins present in individual fruits and vegetables is variable, ranging from a few (e.g. in strawberries, blackberries, and red cabbage) to more than 10 different pigments (in blueberries and red varieties of grapes and maize). In addition to anthocyanins, many fruits and vegetables also contain yellow-to-orange carotenoids and anthoxanthins, green chlorophylls, and other pigments (but never betalains), the presence of which often affects the resulting colour.

Table 9.11 Main anthocyanins of selected fruits.

Fruit	Latin name	Main anthocyanins[a, b]
Apples (skin)[c]	*Malus pumila*	**Cy-3-gal**, 3-glc, 3-xyl, 3- and 7-ara, and acyl derivatives
Apricots (skin)	*Armeniaca vulgaris*	**Cy-3-glc**
Açai	*Euterpe oleracea*	**Cy-3-glc, Cy-3-rut**, Df-3-galactoside, Df-3-glucoside, Df-3-rutinoside, Pn-3-glucoside
Acerola	*Malpighia emarginata*	**Cy-3-rha, Pg-3-rha**
Bilberries	*Vaccinium myrtillus*	**Df-3-gal, Df-3-glc, Cy-3-ara, Df-3-ara, Pt-3-glc, Mv-3-glc**
Blackberries	*Rubus* sp.	**Cy-3-glc**, Cy-3-rut, Cy-3-xyl, Cy-3-(6-malonyl-glc), 3-[6-(3-hydroxy-3-methylglutaroyl]glc, Pg-3-glc
Blueberries (lowbush)	*Vaccinium myrtilloides*	**Mv-3-glc, Mv-3-gal**, Mv-3-ara, **Pt-3-glc**, Pt-3-ara, Pt-3-(malonoyl)glc, Cy-3-(malonoyl)glc, Dp-3-(malonoyl)glc
Blueberries (highbush)	*Vaccinium corymbosum*	**Dp-3-gal, Mv-3-ara, Mv-3-gal, Dp-3-ara, Pt-3-gal, Pt-3-ara**
Cherries (sour)	*Cerasus vulgaris*	**Cy-3-(2-Glc)rut, Cy-3-glc**, Cy-3-rut, Cy-3-sop, Cy-3-sam, Pn-3-glc
Cherries (sweet)[c]	*Cerasus avium*	**Cy-3-rut, Cy-3-glc**, Cy-3-sop (some varieties), Pn-3-glc, Pn-3-rut, and their acyl derivatives
Chokeberries	*Aronia melanocarpa*	**Cy-3-gal, Cy-3-ara**, Cy-3-glc, Cy-3-xyl
Cornel berries	*Cornus mas*	**Cy-3-gal, Pg-3-gal**, Pg-3-rha-gal, Dp-3-gal
Lingonberries	*Vaccinium vitis-idaea*	**Cy-3-gal, Cy-3-ara**, Cy-3-glc, Cy-3-xyl-Glc, Df-3-glc, Mv-3,5-di-glc
Currant (black)	*Ribes nigrum*	**Dp-3-rut, Dp-3-glc, Cy-3-glc**, Cy-3-rut, Pt-3-rut, Pn-3-rut
Currant (red)	*Ribes silvestre*, syn. *R. rubrum*	**Cy-3-xyl-rut, Cy-3-sam, Cy-3-rut**, Cy-3-glc, Cy-3-sop, Cy-3-glc-rut
Elderberries	*Sambucus nigra*	**Cy-3-glc, Cy-3-sam**, Cy-3-sam-5-glc, Cy-3,5-di-glc
Figs (skin)	*Ficus carica*	**Cy-3-rut**, Cy-3-glc, Cy-3,5-di-glc, Pg-3-rut
Gooseberries[c]	*Grossularia uva-crispa*	**Cy-3-rut, Cy-3-glc**, Cy-3-(6-*p*-coumaroyl)glc, Pe-3-glc, Pn-3-glc
Grapes[c]	*Vitis vinifera*	**Mv-3-glc, Mv-3-*p*-coumaroyl-glc** (in some varieties), **Mv-3-acetyl-glc**, Pt-, Pe-, Df- and Cy-Glc, and their acyl derivatives
	Vitis labrusca	**Dp**, Cy, Pt, Pn and **Mv-3-(*p*-coumaroyl)-glcs-5-glcs**, 3,5-di-glc, 3-*p*-coumaroyl-glc, **3-glc**
Lichee (skin)	*Litchi chinensis*	**Cy-3-rut**, Cy-3-gal, Cy-3-glc, Pg-3,5-di-glc, Pg-3-glc
Mango	*Mangifera indica*	**Pn-3-gal**
Mangosteen (rinds)	*Mangifera mangostana*	**Cy-3-sop**, Cy-3-glc
Mulberries	*Morus nigra*	**Cy-3-glc, Cy-3-rut**, Cy-3-sop, Pg-3-glc, Pg-3-rut
Olives	*Olea europea*	**Cy-3-glc, Cy-3-rut**, and their acyl derivatives
Oranges[c]	*Citrus sinensis*	**Cy-3-glc, Cy-3-(6-malonyl)glc**, Dp-3-glc, Pn-3-glc, pyranoanthocyanins
Passion fruit[c]	*Passiflora edulis*	**Pg-3-di-glc, Dp-3-glc**
Peaches[c]	*Persica vulgaris*	**Cy-3-glc, Cy-3-rut**
Pears (skin)[c]	*Prunus persica*	**Cy-3-gal, Cy-3-ara**, Cy-3-glc, Cy-3-xyl, Pe-3-gal, Pe-3-rut, and acyl derivatives
Plums	*Prunus domestica*	**Cy-3-xyl, Cy-3-rut**, Pn-3-glc, Pn-3-rut
Pomegranate	*Punica granatum*	**Dp-3,5-di-glc, Cy-3,5-di-glc**, Pg-3,5-di-glc, **Cy-3-glc**, Dp-3-glc, Pg-3-glc

(continued overleaf)

Table 9.11 (continued)

Fruit	Latin name	Main anthocyanins[a] [b]
Raspberries	*Rubus idaeus*	**Cy-3-rut, Cy-3-sop, Cy-3-glc-rut, Cy-3-glc,** Pg-3-glc-rut, Pg-3-sop, Pg-3-glc, Mv-3-glc
Saskatoon berries	*Amelanchier alnifolia*	**Cy-3-gal, Cy-3-glc, Cy-3-ara,** Cy-3-xyl, Dp-3-glc, Dp-3-rut, Mv-3-glc
Sloeberries	*Prunus spinosa*	**Cy-3-rut, Cy-3-glc, Pn-3-rut,** Pn-3-glc
Strawberries	*Fragaria* spp.	**Pg-3-glc,** Cy-3-glc, Pg-3,5-di-glc, Cy-3,5-di-glc and their **succinates**
Wolfberries (goji)[d]	*Lycium ruthenicum*	**Pt-3-rut-5-glc, Pt-3,5-di-glc**

[a]Cy = cyanidin, Pg = pelargonidin, Pn = peonidin, Dp = delphinidin, Pt = petunidin, Mv = malvidin, ara = arabinoside, gal = galactoside, glc = glucoside, rut = rutinoside, sam = sambubioside, sop = sophoroside, xyl = xyloside.

[b]Components occurring in quantities higher than about 10% are highlighted in bold.

[c]Red variety.

[d]Black variety.

3-Deoxyanthocyanins are not commonly found in higher plants. They occur in high quantities primarily in ferns and mosses and in sorghum cereal crops (such as *Sorghum bicolor*, Poaceae). Apigeninidin and luteolinidin represent on average 36–50% of total black and brown sorghum anthocyanins (2.8–4.3 g/kg).

9.4.1.2.1 Grapes and wine

European grapevine (*Vitis vinifera*, Vitaceae) varieties contain only 3-*O*-monoglucosides of anthocyanidins. North American and other species, such as river bank grape (*V. riparia*), rock (mountain) grape (*V. rupestris*), fox grape (*V. labrusca*), and their hybrids with *V. vinifera* also contain the corresponding 3,5-di-*O*-glucosides and their acyl derivatives. The total amount of anthocyanins can range from about 300 to more than 7000 mg/kg (f.w.).

Distribution of anthocyanins in red grapes (*V. vinifera*) is highly variable and differs according to species, variety, and a number of other variables. The most abundant pigments are the 3-*O*-glucosides of malvidin and peonidin, as well as their 6″-acetyl and *p*-coumaroyl derivatives. The 3-*O*-glycosides of petunidin, delphinidin, and cyanidin are also widespread. For example, Cabernet Sauvignon grapes contains 42.6% malvidin-3-glucoside, 20.5% malvidin-3-acetylglucoside, 10.0% delphinidin-3-glucoside, 6.4% malvidin-3-*p*-coumaroylglucoside, 6.1% petunidin-3-glucoside, 5.3% peonidin-3-glucoside, 2.5% delphinidin-3-acetylglucoside, 2.2% petunidin-3-acetylglucoside, 1.3% cyanidin-3-glucoside, 0.9% peonidin-3-acetylglucoside, 0.6% peonidin-3-*p*-coumaroylglucoside, 0.5% delphinidin-3-*p*-coumaroylglucoside, 0.4% petunidin-3-*p*-coumaroylglucoside, 0.1% malvidin-3-caffeoylglucoside, and smaller amounts of cyanidin-3-acetylglucoside and cyanidin-3-*p*-coumaroylglucoside. The most important pigment in non-European grape species is malvidin-3,5-diglucoside.

Essentially the same pigments that are found in the grape skins from which they were extracted during fermentation are responsible for the young red wine colour. During wine maturation and ageing, the colour of wine significantly changes. The amount of the original anthocyanins decreases, as they react with colourless flavan-3-ols (such as catechin, epicatechin, and condensed tannins to form oligomeric pigments) and undergo multiple other reactions to form a heterogeneous mixture of typical darker and more stable red pigments. These pigments are less sensitive to changes in pH (nucleophilic attacks of water, which results in the formation of carbinol pseudo base and subsequent loss of colour) or discolouration by sulfur dioxide. This is why mature wines are darker than young wines. The reaction mechanisms of formation of pigments in aged red wines are described in Section 9.4.1.5.9.

Polymerisation reactions in red wines lead to the gradual formation of water-soluble astringent polymers and insoluble, brownish-red, high-molecular-weight condensation products called **phlobaphens**, which form sediments in wine. Certain other wine components, such as proteins, ascorbic acid, reducing sugars, and metal ions, can apparently be involved in their formation.

9.4.1.2.2 Apples and pears

Anthocyanin pigments responsible for the red colour of apple and pear peels are derived from cyanidin. The main pigment is cyanidin-3-galactoside (idaein), which represents about 94% of pigments in the Jonathan and 85% of all anthocyanins in the Red Delicious cultivars. Other pigments of these apple cultivars are cyanidin-3-arabinoside (10 and 4%) and cyanidin-3-glucoside (5 and 3%). Cyanidin-3-xyloside and acyl derivatives of the previously listed anthocyanins are found in traces. The total amount of anthocyanins is about 100–200 mg/kg (f.w.).

Table 9.12 Main anthocyanins of selected vegetables, cereals, and other crops.

Material	Latin name	Main anthocyanins[a, b]
Vegetables		
Cabbage (leaves)[c]	*Brassica oleracea var. capitata*	**Cy-3-sop-5-glc** (malonyl, ***p*-coumaroyl**, di-*p*-coumaroyl, feruloyl, diferuloyl, **sinapoyl**, and disinapoyl esters)
Carrot (black)		**Cy-3-xyl(sinapoyl-glc)gal, 3-xyl(feruloyl-glc)gal**, other Cy-3-xyl-gal, Pn-3-xyl-gal
Chicory[c]	*Cichorium intybus*	**Cy-3-(6-malonyl)glc, Dp-3-(6-malonyl)glc**, Dp-3-(6-malonyl)glc-5-(malonyl)glc
Eggplant (skin)	*Solanum melongena*	**Dp-3-[4-(*p*-coumaroyl)rut]-5-glc**[d](nasunin), **Dp-3-glc**, Dp-3-rut-5-glc, Dp-3-rut, Dp-3-*p*-coumaroyl-rut, Dp-3,5-di-glc
Lettuce (leaves)[c]	*Lactuca sativa*	**Cy-3-(3-malonoyl)glc, Cy-3-(3-acetyl)glc**, Cy-3-(6-malonoyl)glc, Cy-3-glc
Onion[c]	*Allium cepa*	**Cy-3-(malonoyl)glc-5-glc, Cy-3-(3-acetyl)glc**, Cy-3-glc, Cy-3-gal, Cy-3-lam, Pn-3-glc
Radishes[c](skin)	*Raphanus sativus*	**Pg-3-(*p*-coumaroyl)di-glc-5-(malonyl)glc, Pg-3-(feruloyl)di-glc-5-(malonyl)glc**, Pg-3-(caffeoyl)di-glc-5-(malonyl)glc
Cereals		
Maize[e]	*Zea mays*	**Cy-3-glc, Pg-3-glc, Pn-3-gc** (free and acylated by malonic acid)
Blue barley[e]	*Hordeum vulgare*	**Pt-3-glc, Cy-3-glc**
Black-purple rice[e]	*Oryza sativa*	**Cy-3-glc, Pn-3-glc**, Cy-3,5-di-glc, Cy-3-ara, Pg-3-glc, Cy and Pn-3-(6-*p*-coumaroyl)glc
Sorghum	*Sorghum bicolor*	**Ap-glc, Lt-glc**, Mv-glc
Other crops		
Red kidney beans[e]	*Phaseolus vulgaris*	**Pg-3-sam, Pg-3-glc**, Cy-3-glc
Black beans	*Phaseolus vulgaris*	**Dp-3-glc, Pt-3-glc, Mv-3-glc**
Potatoes[c]	*Solanum tuberosum*	Pg, Cy, Pn, Dp, Pt and Mv-5-(*p*-coumaroyl)glc-3-ruts, Pg, Cy, Dp and Pt-3-rha-glcs, Pg and Pn-3-(*p*-coumaroyl)rut-5-glcs, 3-(feruloyl)rut-5-glc
Roselle	*Hibiscus sabdariffa*	**Cy-3-sam, Dp-3-sam**, Cy-3-glc, Dp-3-glc
Sweet potato[f]	*Ipomoea batatas*	**Cy-3-caffeoyl-sop-5-glc, Pn-3-caffeoyl-sop-5-glc**, Cy-3-*p*-hydroxybenzoyl-sop-5-glc, Pn-3-*p*-hydroxybenzoyl-sop-5-glc, Pn-caffeoyl-feruloyl-sop-5-glc

[a]Cy = cyanidin, Pg = pelargonidin, Pn = peonidin, Dp = delphinidin, Pt = petunidin, Mv = malvidin, Ap = apigeninidin, Lt = luteolinidin, ara = arabinoside, gal = galactoside, glc = glucoside, lam = laminaribioside, rut = rutinoside, sam = sambubioside, sop = sophoroside, xyl = xyloside.
[b]Components occurring in quantities higher than about 10% are highlighted in bold.
[c]Red variety.
[d](*E*)- and (*Z*)-isomers.
[e]Black, blue, and purple grains produced in order to make specialty foods or for use in ornamentation.
[f]Purple-fleshed.

Compared with other red fruits, the concentration of anthocyanins in pear peels is distinctly lower, ranging from 50 to 100 mg/kg (f.w.). Pigments of red varieties of pears are derived mainly from cyanidin (unlike in apples, anthocyanidins of pears are not acylated by phenolic acids). The main pigments are cyanidin-3-galactoside, cyanidin-3-arabinoside, and 3-rutinoside; in some cultivars, peonidin-3-galactoside is also found.

9.4.1.2.3 Sour cherries

The main pigments of sour cherries are anthocyanins derived from cyanidin. Their mean total content ranges from 350 to 820 mg/kg (f.w.), but may be even higher. For example, in the popular Morello variety, the levels are 69–77% of the anthocyanin pigment cyanidin-3-(2-glucosyl)rutinoside (mecocyanin), 11–16% of cyanidin-3-rutinoside, 3–15% of cyanidin-3-glucoside, and 1–3% of cyanidin-3-sophoroside. The sour cherry variety Montmorency contains as the main pigment cyanidin-3-rutinoside (27%), followed by cyanidin-3-(2-glucosyl) rutinoside (25%), cyanidin-3-glucoside (19%), and cyanidin-3-sophoroside (16%). Minor pigments include cyanidin-3-sambubioside and 3-glucoside and peonidin-3-rutinoside (3%). Glycosides are accompanied by free aglycones cyanidin (7%) and peonidin (3%).

9.4.1.2.4 Olives

The immature olive is grey-green, and later green to yellow-green; the fruit darkens on ripening and becomes violet and almost black. The main pigment of immature green olives is chlorophyll. Ripe purple to almost black olives contain anthocyanin pigments at a concentration of about 5000 mg/kg (f.w.) in practically all cultivars.

The main anthocyanins of ripe olives, which are concentrated mainly in the skin, are cyanidin-3-glucoside, cyanidin-3-rutinoside, and their acylated derivatives. Contents of monoglycosides and diglycosides are roughly in the ratio 1 : 1 to 1 : 4. For example, the Spanish olive variety Manzanilla contains about 15% cyanidin-3-glucoside and other cyanidin glycosides, traces of 3-rutinoside, but 60% 3-rutinoside acylated by caffeic acid, traces of 3-glucosylrutinoside, 25% 3-(2-glucosyl)rutinoside, and traces of this anthocyanine acylated by caffeic acid. In numerous other varieties, further pigments were also found, such as cyanidin-3-(2-xylosyl) rutinoside, 3-rhamnosyldiglucoside acylated by caffeic acid, and 3-glucoside acylated by p-coumaric acid. Particular anthocyanins derived from peonidin may also be present. In Greek olives, anthocyanins derived from delphinidin are also present. Pigments of pickled black olives are formed in the enzymatic browning reactions.

9.4.1.2.5 Potatoes

There are more than 5000 varieties of potato in the world. Some are distinguished only by a deeply coloured skin caused by anthocyanins; their flesh is the usual white or cream, which is due to the presence of carotenoid pigments. Others have coloured flesh. In South America, potatoes with a diversity of colours are found: purple, pink, orange, and yellow, often with a contrasting one around the eyes. The colour of the flesh is often similar to that of the skins. The so-called purple potatoes (or blue potatoes), even if not commonly found in our habitual diet, are now available on the European market. They have purple skin and flesh, which becomes blue once cooked.

The concentration of anthocyanins in potatoes varies over a wide range. The level of anthocyanins in skin tissue is quite high. However, the skin is such a small volume of the whole tuber that, generally, red-skinned white-fleshed potatoes contain about 15 mg anthocyanins per f.w. However, the amount of anthocyanins in potatoes with pigments in the flesh range from 150 to about 400–500 mg (f.w.). The red-flesh potatoes contain predominantly acylated glucosides of pelargonidin. Purple-fleshed potatoes have a more complex content of acylated glucosides of pelargonidin, petunidin, cyanidin, and malvidin. For example, the purple-flesh variety Vitelotte Noire, with a total anthocyanin content of 486 mg/kg (f.w.), contains as its main anthocyanin pigments (E)-malvidin-3-O-[6-O-(4-O-p-coumaroyl-α-L-rhamnopyranosyl)-β-D-glucopyranoside]-5-O-β-D-glucopyranoside (294 mg/kg) and (E)-petunidin-3-O-[6-O-(4-O-p-coumaroyl-α-L-rhamnopyranosyl)-β-D-glucopyranoside]-5-O-β-D-glucopyranoside (125 mg/kg), known as petanin (**9-41**).

9.4.1.3 Use

Anthocyanins isolated from natural sources have been used as food colourings for more than 100 years, and for much longer in the form of various fruit juice concentrates. The disadvantage is that they only have the intense colour in solutions of pH < 3.5, so they are only suitable for acidic foods. In recent years, their importance as food colourings has increased in relation to growing consumer interest in natural substances. Potential sources of these pigments are limited by the availability of the plant materials and overall economic conditions of their production, so only a few plant species may be used industrially.

The anthocyanin pigments most commonly used for food colouring are derived from grapes (skin or juice sediment), where the anthocyanin content ranges from 0.3 to 7.5 mg/kg. Historically, the oldest product of this type, produced in Italy since 1879, is called enocyanin (or enocianina). A rich source of anthocyanins is elderberries containing pigments at a concentration of of 2–10 g/kg (f.w.), or chokeberries (10 g/kg) that have similar composition of pigments. A very high content of anthocyanins is also found in blackberries (0.8–3.3 g/kg) that sometimes precipitate in juices. Other sources include red cabbage (containing 0.7–0.9 g/kg of anthocyanins), roselle calyces (15 g/kg d.w.), and sometimes sweet potatoes, red pulp oranges, leaves and seeds of red varieties of maize, and different local products.

9.4.1.4 Biochemistry, physiology, and nutrition

Anthocyanin pigments are generally allowed for food colouring, and in most countries there are no limits on their use. Very low toxicity was observed even in their metal complexes with Al(III) and Sn(II) ions. Apart from imparting colour to food raw materials and foods, anthocyanins also have an array of health-promoting benefits, as they can protect against a variety of oxidants through a number of mechanisms.

9.4.1.5 Properties and reactions

From the technological perspective, the most important property of anthocyanins is colour and its stability, which is usually relatively low. The main factors affecting the colour stability are the structure of the molecule, presence of certain enzymes, pH, temperature, presence

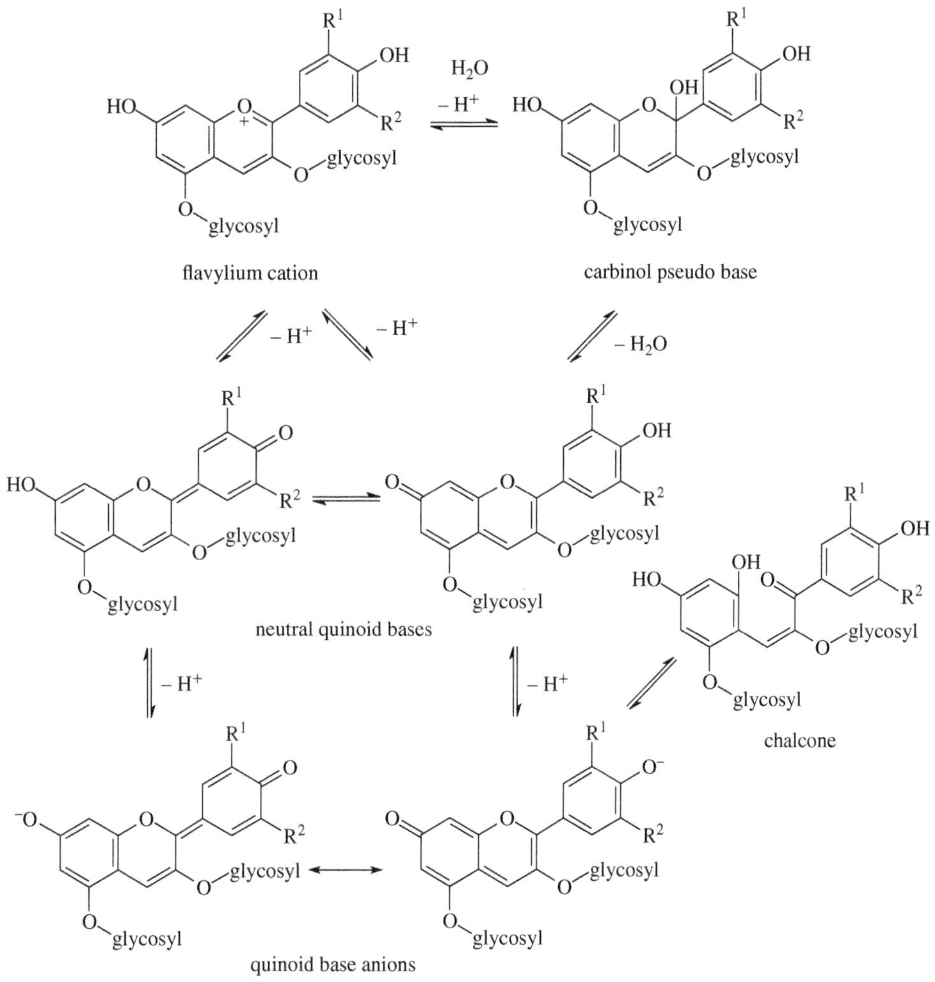

Figure 9.9 Transformations of anthocyanins according to pH (R^1 and R^2 = H, OH, OCH_3, or O-glycosyl).

of oxygen, and exposure to radiation. Anthocyanins may produce various coloured or colourless products in reactions with other food components, such as ascorbic acid, sulfur dioxide, other phenols, and metal ions.

9.4.1.5.1 Effect of pH

In aqueous solutions, anthocyanins are transformed into various products of different colour. Depending on pH, the following five different aglycone structures are in equilibrium (Figure 9.9):

- red flavylium cation;

- colourless carbinol pseudo base;

- purplish-red neutral quinoid base;

- blue quinoid base;

- yellow chalcones.

In solutions of pH 1.0 and lower, anthocyanins exist solely as red-coloured flavylium salts. When increasing pH, the equilibrium shifts in favour of colourless carbinol pseudo base and the red colour fades. Around the range of pH values of 4.0–4.5, anthocyanins are completely colourless. Another increase in pH is manifested by the purplish-red colour, which is caused by formation of a neutral quinoid base that requires the presence of free hydroxyl groups on one of C-5, C-7, or C-4′ carbons. In solutions of pH 7, a blue-coloured quinoid base is formed. After some time, or following an increase in pH value, a gradual decrease of blue colour intensity occurs as a result of yellow chalcone formation. If the solution is acidified to around pH 1.0, the blue quinoid and colourless carbinol bases are converted back into red flavylium cations. The transformation of chalcones is slower and not quantitative.

In processed fruit and vegetables, the situation is more complex. Anthocyanins in plants (pH 2.5–7.5) occur as a purplish-red neutral quinoid base, but in food products they may be in media of different pH. However, they are mostly stabilised by inter copigmentation (interactions with other flavonoids) or intra copigmentation (acylated forms) or by interactions with other food components. Many products therefore retain their original colour or possess some discolouration. In products stored long-term (such as strawberry jams, olives) oligomers are then formed with colour similar to the original colour of the anthocyanins, whilst the original pigments may be absent.

9.4.1.5.2 Effect of structure

The colour of non-acylated and monoacylated anthocyanins in acidic media is mainly dependent on the number and type of aglycone (anthocyanidin) substituents. Anthocyanidins with a higher number of hydroxyl groups tend to have a blue tint, and methoxyl derivatives have a red tint. Anthocyanidins with a higher number of hydroxyl groups are less stable, and their stability increases with a growing number of methoxyl groups. Glycosides and their acyl derivatives generally have a red-blue colour and are more stable than the corresponding aglycones (anthocyanidins). During storage, heat treatment, and exposure to light, diglycosides are more stable than monoglycosides. The type of bound sugar has a minor influence on the stability of the pigments.

The presence of one or more acyl groups stabilises anthocyanins (due to the so-called intramolecular copigmentation), and their reaction with water in neutral or weakly acidic environments does not lead to the formation of a colourless quinoid base, but preferably to a blue quinoid base. These pigments are less sensitive to changes in pH, and colour is stable in weakly acidic and neutral media.

9.4.1.5.3 Effect of temperature

Like most chemical reactions, the stability of anthocyanins and the rate of their degradation are affected by temperature, and also depend on the pH value, their structure, the presence of oxygen, and the possibility of entering into reactions with other components of the system.

Degradation of anthocyanins in the absence of oxygen, at pH ranging from 2.0 to 4.5, is virtually independent of pH and usually takes place under aerobic and anaerobic conditions as a first-order reaction. In the presence of oxygen, 3-glycosides of anthocyanidins show the highest stability at elevated temperatures in the pH range of 1.8–2.0, and 3,5-diglycosides are stable at pH 4.0–5.0. Most anthocyanins somewhat paradoxically exhibit increased stability at the elevated temperatures used in the processing of fruits and vegetables. This phenomenon is explained by the protective effect of various components of the system, and by condensation of monomers that leads to the formation of more stable oligomeric pigments, whose content increases with temperature and storage time. Oligomeric pigments are important colour carriers, especially of stored fruit juices and red wines.

The mechanism of anthocyanin degradation reactions depends not only on temperature, but also on the structure of substances. Decomposition reactions of 3-glycosides of anthocyanidins are shown in Figure 9.10. The major products are glycosides of the corresponding chalcones. Hydrolysis of glycosidic bonds in chalcones yields 1,2-diketones (α-diketones). Subsequent reactions of the primary degradation products (most of them are colourless compounds) produce brown polymers. Somewhat different is the decomposition mechanism of 3,5-diglycosides of anthocyanidins, where the main reaction products are diglycosides of coumarins (Figure 9.11).

9.4.1.5.4 Enzymes

Loss of colour of anthocyanins may also be caused by enzymatic reactions catalysed by two groups of enzymes:

- glycosidases, which hydrolyse the glycosidic bond to form sugar and anthocyanin aglycone (anthocyanidin), which is unstable and spontaneously transformes into colourless products;

- polyphenol oxidases that also act in the enzymatic browning reactions.

9.4.1.5.5 Oxygen and peroxides

Atmospheric oxygen oxidises anthocyanins to colourless or brown-coloured products directly or through other labile compounds that are preferentially oxidised by oxygen (such as ascorbic acid). Degradation of anthocyanins that is induced by ascorbic acid proceeds indirectly

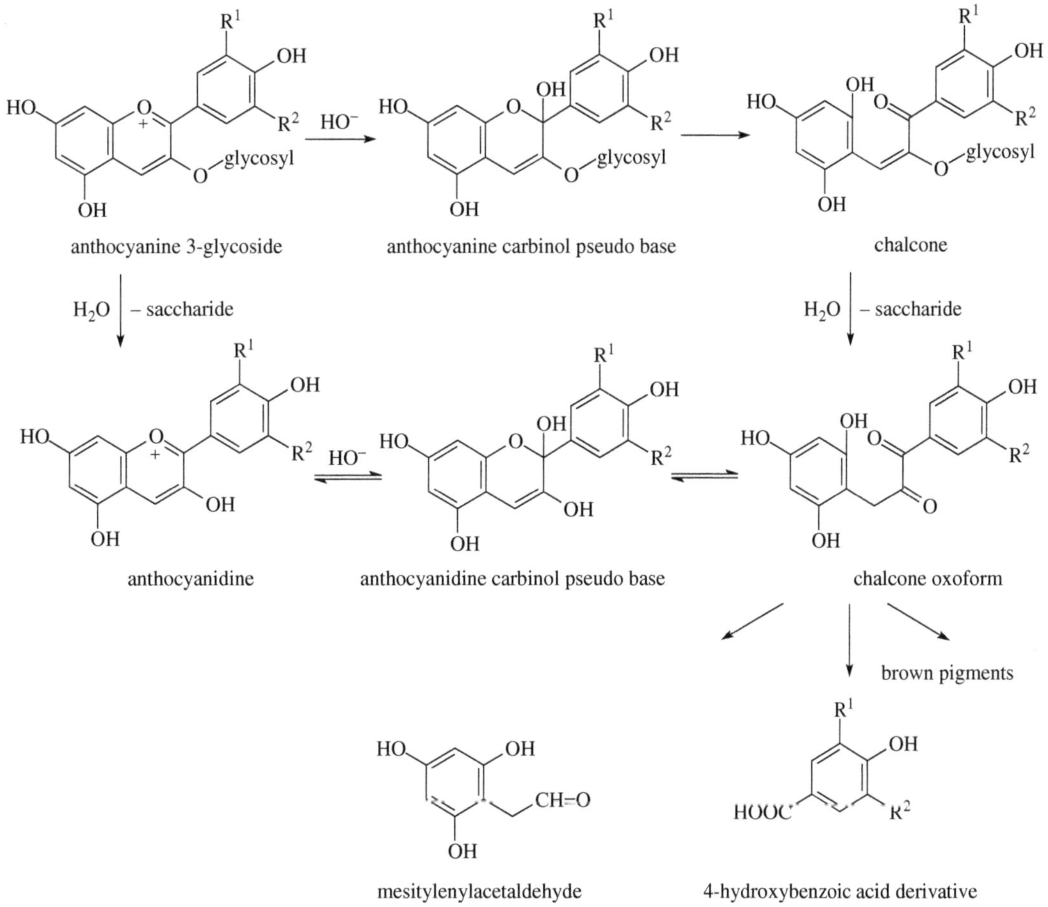

Figure 9.10 Degradation of 3-glycosides of anthocyanidins (R^1 and R^2 = H, OH, OCH_3, or O-glycosyl).

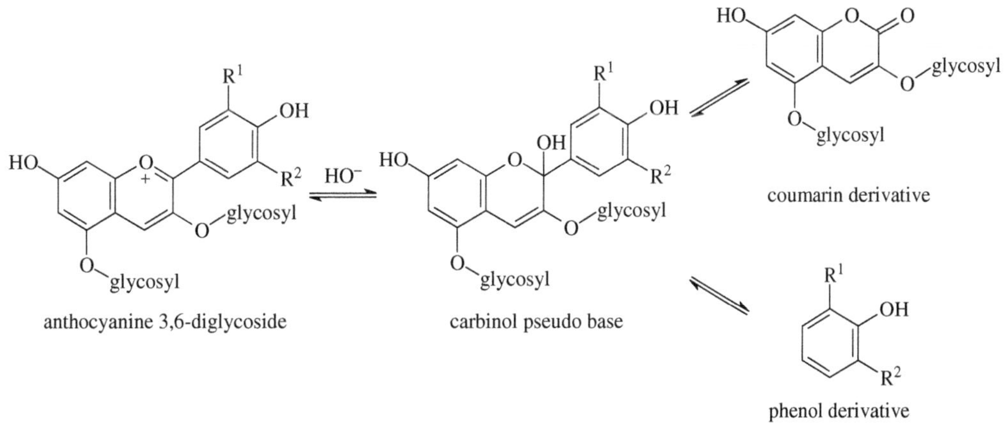

Figure 9.11 Degradation of 3,5-diglycosides of anthocyanidins (R^1 and R^2 = H, OH, OCH_3, or O-glycosyl).

via hydrogen peroxide, which arises by ascorbic acid oxidation. For example, the anthocyanin pigment malvin (malvidin-3,5-diglucoside), found in a variety of fruits and flowers, is oxidised by hydrogen peroxide under the opening of the heterocyclic ring between C-2 and C-3, which yields a colourless substance malvone (**9-43**). A hydrogen atom may replace the hydroxyl group at C-4. On alkaline hydrolysis, malvone affords syringic acid. A similar degradation of anthocyanin hirsutin (hirsutidin-3,5-diglucoside) affords colourless hirsulone.

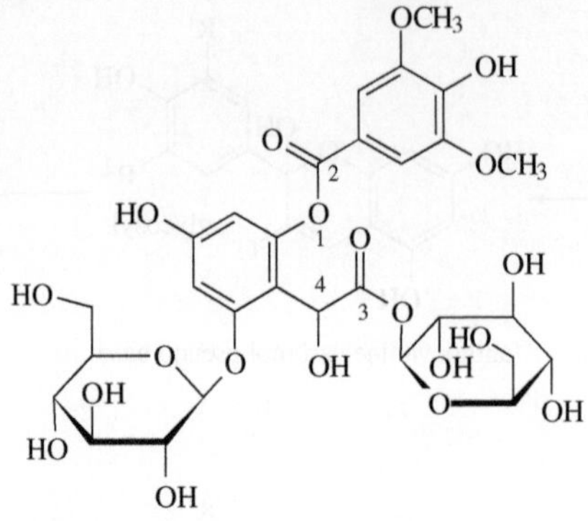

9-43, malvone

9.4.1.5.6 Radiation

Anthocyanins are unstable when exposed to visible, ultraviolet, or ionising radiation. Decomposition takes place mainly as photooxidation. Anthocyanins substituted at C-5 hydroxyl group, which are fluorescent compounds, are more sensitive to photochemical degradation in comparison with C-5 unsubstituted anthocyanins.

9.4.1.5.7 Sulfur dioxide

The natural anthocyanin pigments form adducts with bisulfite ion at position C-2 or C-4 of the flavylium nucleus, which simultaneously stabilises the glycosidic bond at position C-3 (Figure 9.12). The adducts are colourless and are stable at pH about 3, but acidification to pH < 1 and heating lead to reversal of the reaction with quantitative recovery of the pigment. The same reversible reaction is responsible for the pinking in white wines after bottling. The responsible compound is mainly adducts of malvidin-3-O-glucoside (oenin). Such adducts are probably unstable also under gastric conditions. Adducts with bisulfite ions are also produced with flavan-3-ols, their oligomers and other flavonoids.

9.4.1.5.8 Sugars and their degradation products

Sugar concentrations higher than 20% (found in jams and similar products) have a stabilising effect on the colour of anthocyanins, mainly due to decreased water activity. Degradation of anthocyanins is accelerated in the presence of sugar degradation products, especially by furan-2-carbaldehyde and 5-hydroxymethylfuran-2-carbaldehyde, which produce complex brown-coloured condensation products with anthocyanins.

9.4.1.5.9 Transformed pigments in red wines

Anthocyanins are relatively unstable and undergo numerous chemical reactions in red wines during fermentation processes, maturation, and ageing. The monomeric anthocyanins extracted from the grape skin are of a crucial contribution to the colour of young red wine. Their intramolecular interactions and reactions with other wine constituents, especially phenolic compounds, can further enhance the colour of

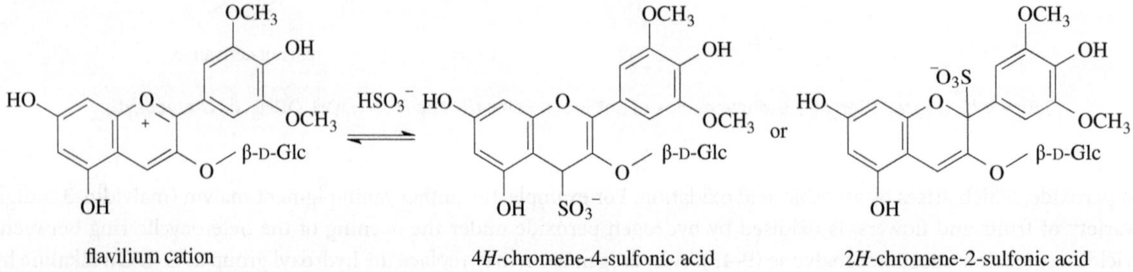

Figure 9.12 Reaction of malvidin-3-O-glucoside with bisulfites.

wine during maturation and ageing. One of these reactions is based on the reaction (called copigmentation) of anthocyanin with flavan-3-ols (procyanidins), called copigments, through which a complex (non-covalent interaction) of these two compounds initially arises, followed by the final pigments, which are dimers linked by covalent bonds. Copigmentation of anthocyanins accounts for over 30% of fresh red wine colour, whilst during storage, the colour of polymeric pigments formed between anthocyanins and flavan-3-ols predominates. An example is a dimer of anthocyanin and (+)-catechin linked by a C4 → C8 bond, which is formed in the nucleophilic attack of the phloroglucinol ring of catechin on the electron deficient C-4 position of anthocyanin (**9-44**). This dimer has a red colour similar to the colour of anthocyanin, even though the copigment is a colourless flavan-3-ol. Another possibility for the formation of more stable pigments in red wines is the reaction of anthocyanins with colourless flavan-3-ols through their nucleophilic C-8 position (less likely through the C-6 position) mediated by aldehydes, such as the fermentation product acetaldehyde, which leads to coloured anthocyanin-alkylene-(epi)catechin conjugates (Figure 9.13). Acetaldehyde may also mediate analogous self condensation of anthocyanins. Other aldehydes arising in wines as fermentation products and intermediates (such as propionaldehyde and isobutyraldehyde) or aldehydes extracted from oak wood during wine ageing (such as benzaldehyde and cinnamaldehyde derivatives) may be involved in similar reactions leading to various anthocyanin-alkylene-(epi)catechin dimers. Some aldehydes, such as methylglyoxal, furan-2-carbaldehyde, and 5-hydroxymethylfuran-2-carbaldehyde, may react directly with flavan-3-ols, yielding alkylene-linked dimers. Their oxidation and dehydration yields xanthylium-type pigments (Figure 9.14). Cinnamic aldehydes, such as coniferyl aldehyde ($R^1 = H$, $R^2 = OCH_3$) and sinapyl aldehyde ($R^1 = R^2 = OCH_3$), produce, via alkylene-linked dimers, orange pigments called **oaklins** (Figure 9.14). In addition to flavan-3-ols, other compounds, such as cinnamic acids, can act as copigments.

9-44, anthocyanin-catechin dimer

One of the most interesting groups of anthocyanin reaction products in red wines is **pyranoanthocyanins** (Figure 9.15), possessing an additional pyran ring structure between C-4 and the hydroxyl group at C-5 of anthocyanin, which seems to be responsible for relatively higher stability of pyranoanthocyanins (e.g. their resistance to pH changes and sulfur dioxide bleaching). The newly formed pyran ring is either unsubstituted or substituted at C-5 by COOH (in carboxypyranoanthocyanins), CH_3 (in methylpyranoanthocyanins), phenols (in pyranoanthocyanin-phenols), or flavanols (in pyranoanthocyanin-flavanols).

Among the pyranoanthocyanins bearing different moieties of fermentation products, the major and most important group occurring naturally in red wines is **vitisins**. The substituted carboxypyranoanthocyanins (A-type vitisins with carboxyl group at C-5) arise from the cycloaddition reaction of pyruvic acid to an anthocyanin moiety (Figure 9.16), whilst unsubstituted pyranoanthocyanins (B-type vitisins) are formed by reaction of anthocyanins with acetaldehyde as minor pigments. Vitisin A and vitisin B are derived from the main pigment of red wines, malvidin-3-O-glucoside (oenin). Also identified in aged wines are yellow methylpyranoanthocyanins (with a methyl group at C-5) derived from malvidin-3-O-glucoside, which form by cycloaddition reaction of anthocyanins with acetoacetic acid followed by decarboxylation and isomerisation. For example, vitisin A derived from malvidin-3-O-glucoside and its acylated forms (acetyl and p-coumaroyl glucosides) have been identified in aged Port red wine.

Recently, a group of neutral non-oxonium pyranoanthocyanins occurring in aged Port red wine, called **oxovitisins A**, has been identified. Oxovitisins A arise by hydration of vitisins A, decarboxylation of the formed intermediate, and isomerisation of the decarboxylation product (Figure 9.17).

In red wines, 4-hydroxycinnamic acids (such as p-coumaric, caffeic, ferulic, and syringic acids) and their decarboxylation products, 4-vinylphenols (such as 4-vinylphenol, 4-vinylcatechol, 4-vinylguaiacol, and 4-vinylsyringol), can react with free anthocyanins to yield pyranoanthocyanin-phenols (also known as hydroxyphenyl-pyranoanthocyanins), some of which are named **pinotins**, since they were first isolated from *V. vinifera* cv. Pinotage wines. The formation of the best-known representative of this family of pigments, pinotin A (**9-45**), is analogous to that of A-type vitisins (Figure 9.16). The nucleophilic C-2-position of caffeic acid attacks the electrophilic C-4 position of malvidin-3-O-glucoside to form an intermediate carbonium ion, in which the reaction of C-5 hydroxyl of the anthocyanin moiety forms a pyran ring.

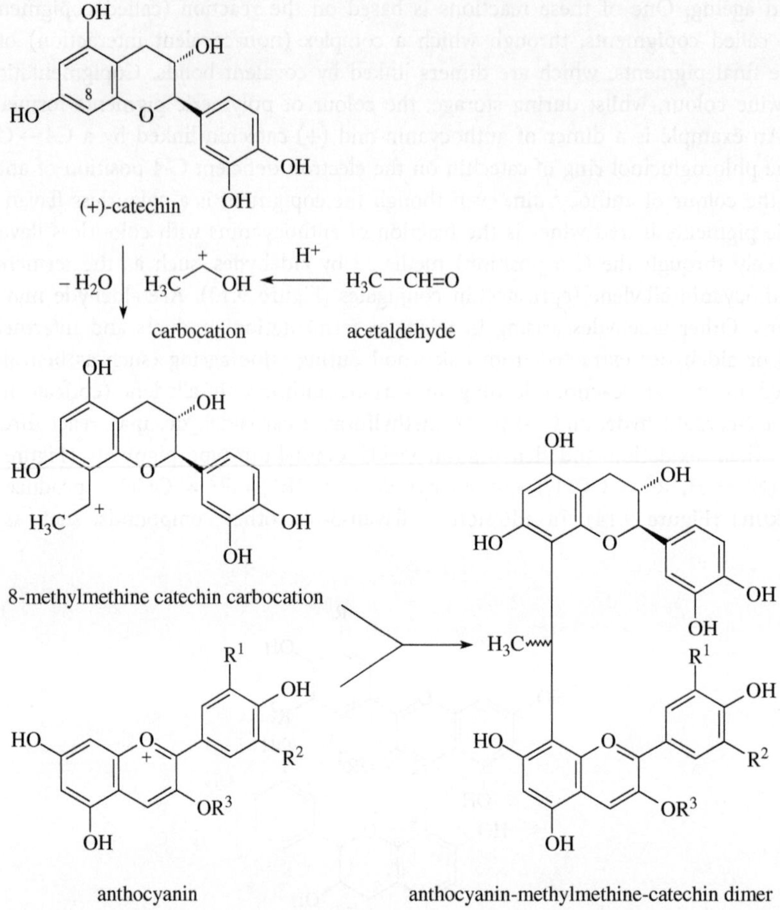

Figure 9.13 Reactions of anthocyanins with flavan-3-ols and aldehydes.

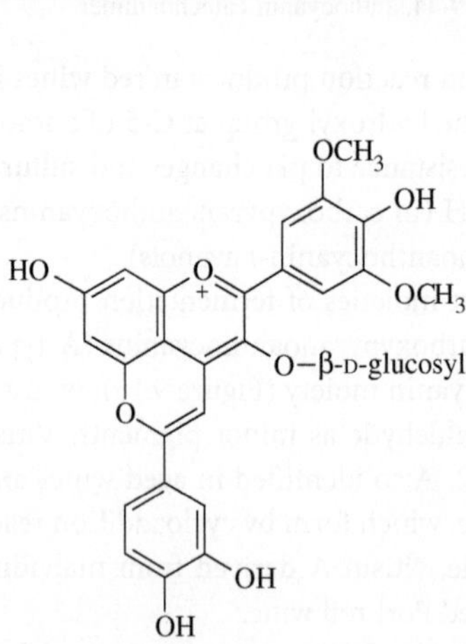

9-45, pinotin A

Orange pyranoanthocyanin-flavanols (also known as flavanylpyranoanthocyanins or vinylflavanol-pyranoanthocyanins) isolated from Port wine contain a pyroanthocyanin moiety linked directly to flavan-3-ols or procyanidins. These pigments arise in wines from the reaction of anthocyanins with 8-vinylflavan-3-ols (such as 8-vinylcatechin) or 8-vinylprocyanidins (such as 8-vinylprocyanidin B₃) mediated by acetaldehyde, as in the case of anthocyanin-alkylene-(epi)catechin conjugates (Figure 9.13). 8-Vinylflavanols are not present in grapes, but they have been proposed to result from the dehydration of the flavanol–ethanol adducts formed by reaction of flavanols with acetaldehyde. Examples of pyranoanthocyanin-flavanols generated from malvidin-3-*O*-glucoside and (+)-catechin and from malvidin-3-*O*-glucoside and

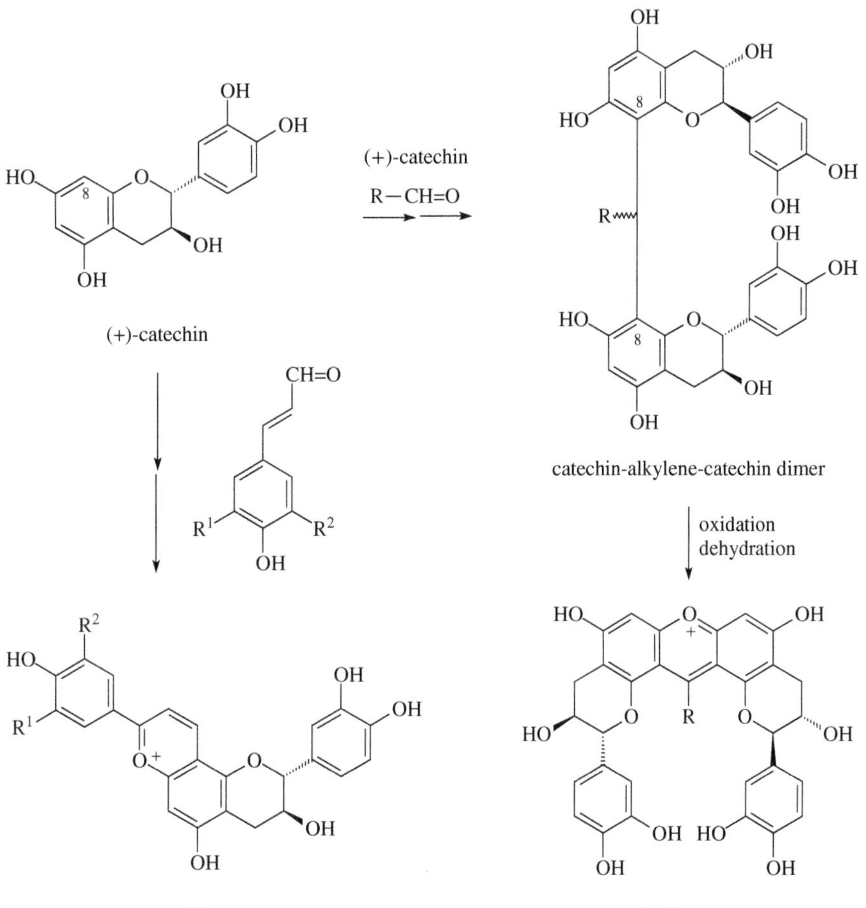

Figure 9.14 Reactions of aldehydes with flavan-3-ols.

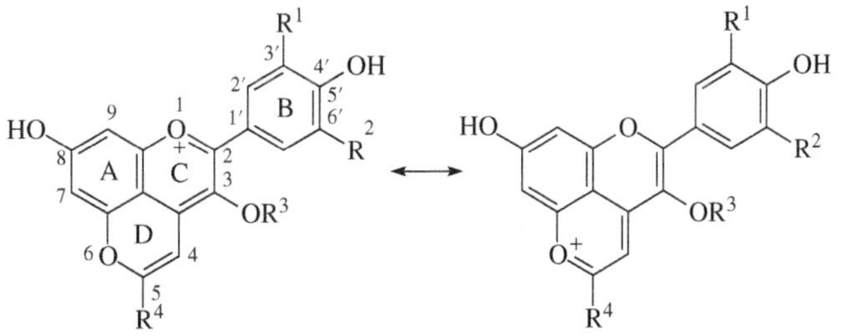

Figure 9.15 Basic structure of pyranoanthocyanins (R^1 and R^2 = H, OH, or CH_3, R^3 = sugar residue).

procyanidin B_3 are given in formulae **9-46**. In addition to pyranoanthocyanin-flavan-3-ol monomers and pyranoanthocyanin–procyanidin dimers, more polymerised pyranoanthocyanin-flavanols (up to tetramers) have been found in some red table wines and port.

Related blue-violet vinylpyranoanthocyanins (also known as flavanyl/phenyl-vinylpyranoanthocyanins) named **portisins** are produced in Port wine during ageing. A-type portisins arise from the reaction between 8-vinylflavanols and carboxypyrananthocyanins (vitisins A), followed by the elimination of formic acid, which gives rise to the vinyl bridge (Figure 9.18). B-type portisins arise from vitisins A and 4-hydroxycinnamic acids. The corresponding reaction mechanism is analogous to that of portisins A, but it involves further decarboxylation of the intermediate.

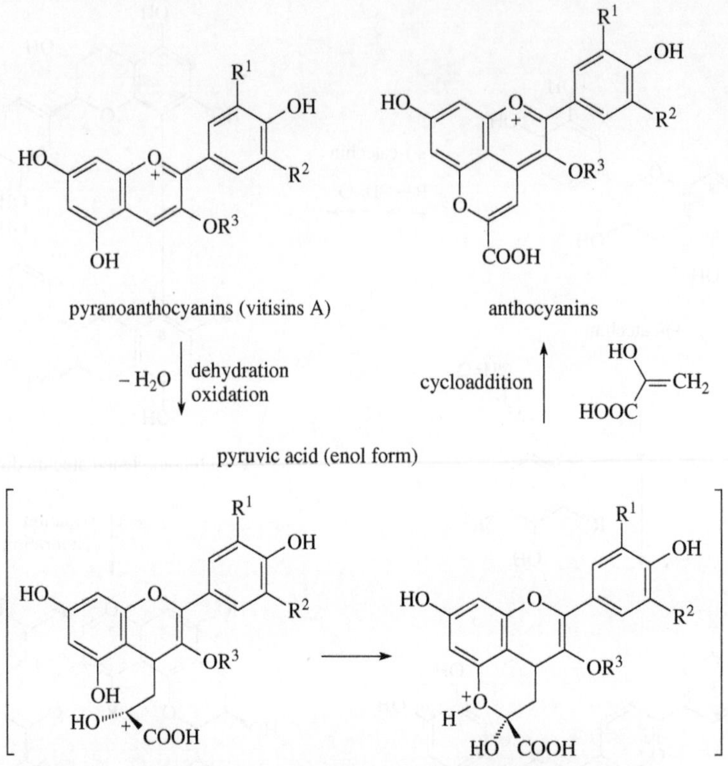

Figure 9.16 Reaction of anthocyanins with pyruvic acid and formation of A-type vitisins (R^1 and R^2 = H, OH, or CH_3, R^3 = sugar residue).

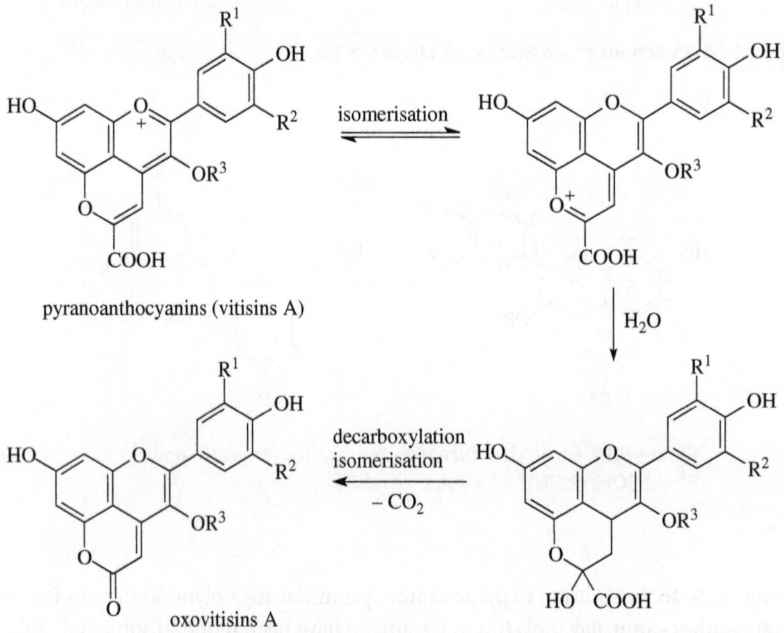

Figure 9.17 Formation of A-type oxovitisins from A-type vitisins (R^1 and R^2 = H, OH, or CH_3, R^3 = sugar residue).

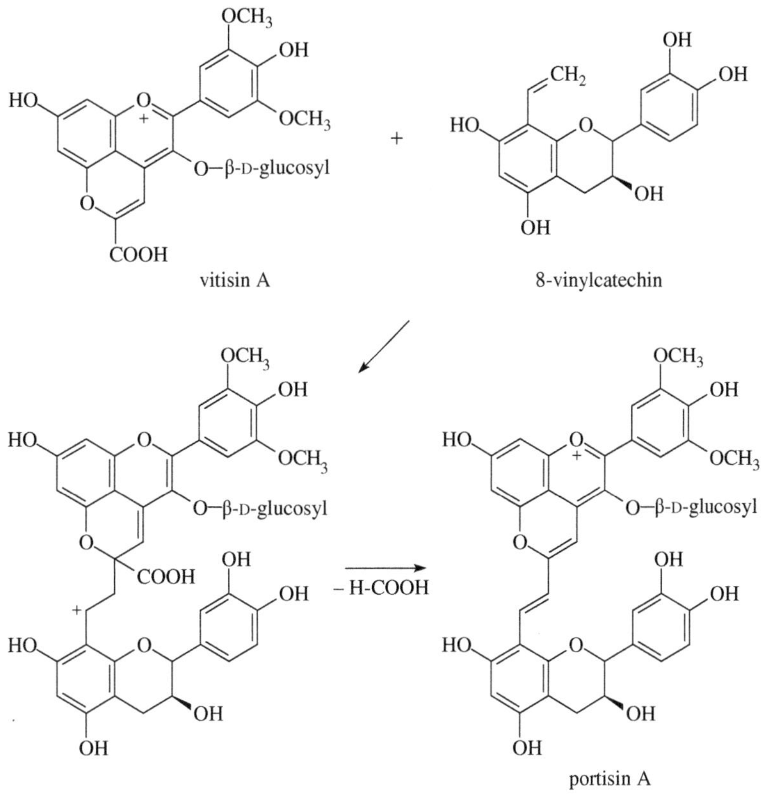

Figure 9.18 Formation of portisin A.

A relatively new class of turquoise coloured pyranoanthocyanin dimers derived from delphinidine glycosides was identified in aged Port wine and shown to be formed by reactions of carboxypyranoanthocyanin with methyl-pyranoanthocyanin. The structure of the identified pyranoanthocyanin methine dimer is given in formula **9-47**.

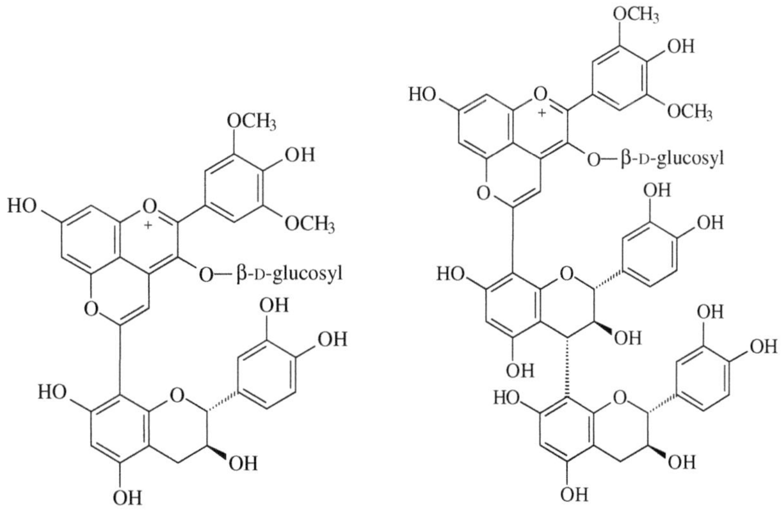

malvidin-3-*O*-glucoside-(+)-catechin
9-46, pyranoanthocyanin-flavanols

malvidin-3-*O*-glucoside-procyanidin B_3

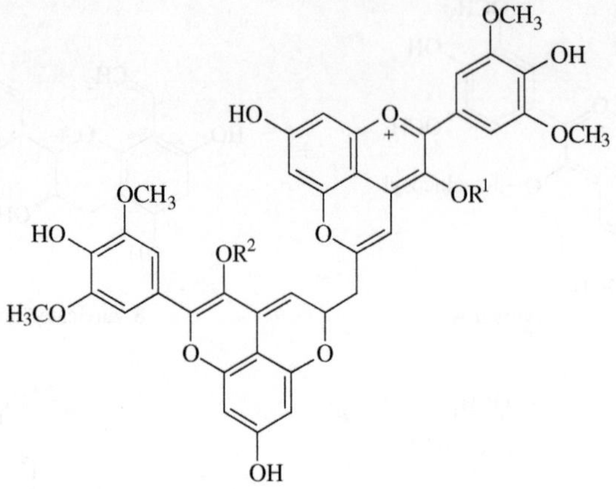

9-47, pyranoanthocyanin dimer

The structural backbone of astringent polymers (molecular weights >5 kDa) is composed of procyanidin chain with (−)-epicatechin, (+)-catechin, and (−)-epicatechin-3-O-gallate units as extension and terminal units and with (−)-epigallocatechin as extension units. Some procyanidins are linked at the A-ring with methylmethine bridges produced by acetaldehyde, and some are attached to anthocyanins and pyranoanthocyanins through a C—C linkage at position C6 or C8. The polymer is esterified with various carboxylic acids (Figure 9.19).

9.4.1.5.10 Transformed pigments in fruits and vegetables

Some of the pigments occurring in aged wines also arise in anthocyanin-rich fruits and vegetables. Three pyranoanthocyanins, the 3-glucosides of 5-carboxypyranodelphinidin, 5-carboxypyranopetunidin, and 5-carboxypyranomalvidin, were formed in extract of black beans (*Phaseolus vulgaris*, Fabaceae) fortified with pyruvic acid. The purple pigment of adzuki beans (*Vigna angularis*) is a condensation product of cyanidin and (+)-catechin called vignacyanidin (**9-48**), in which 5-hydroxy and C-4 positions of the cyanidin moiety are substituted by the addition of 5-hydroxy and C-6 position of the (+)-catechin moiety, respectively. Pyranoanthocyanin-phenols derived from cyanidin-3-O-glucoside and 4-vinylphenol, 4-vinylcatechol, 4-vinylguaiacol, and 4-vinylsyringol were identified in blood orange juice. Anthocyanin-flavanol reaction products (derived from cyanidin and delphinidin) were also found in blackcurrant and strawberries. Vinylcatechol adducts of cyanidin (derived from 4-vinylphenol, 4-vinylcatechol, and 4-vinylguaiacol) were isolated from a black carrot variety.

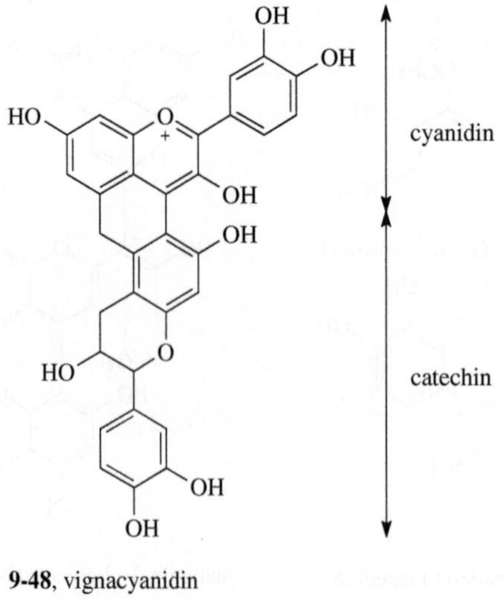

9-48, vignacyanidin

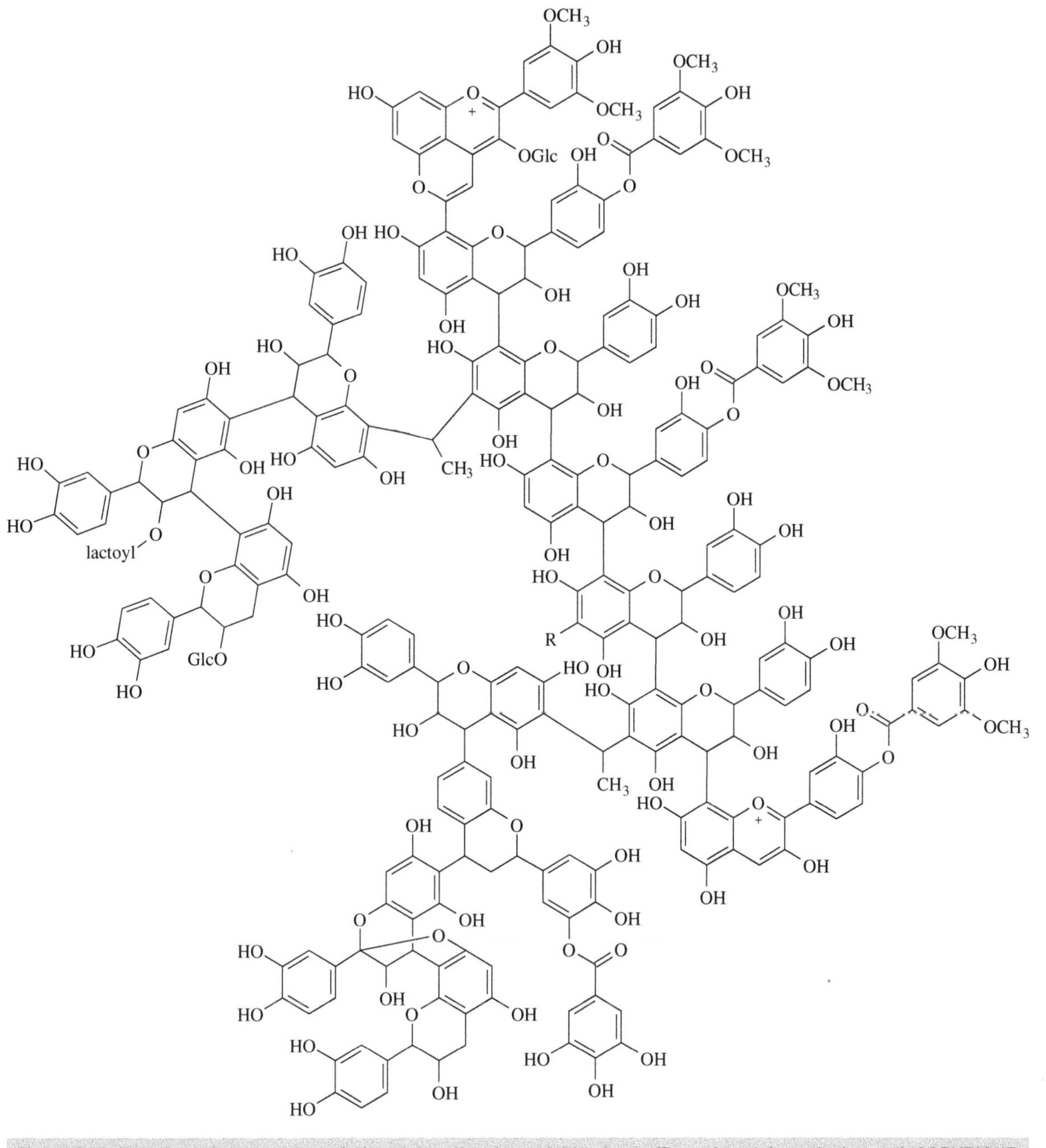

Figure 9.19 Proposed structure of a red wine high-molecular-weight polymer (R = poly-α-1,4-ᴅ-galacturonyl). COPYRIGHT Wollmann and Hofmann (2013), Figure 10.

9.4.1.5.11 Other reactions

Anthocyanin with the structure of *o*-diphenols forms complexes with metals (such as Al, K, Fe, Cu, Ca, and Sn) that can stabilise the colour of products but may cause unwanted discolourations. For example, complexes with tin formed in cans may change the red colour of fruits (such as strawberries) to purple.

9.4.2 Other flavonoids

Flavonoids that are usually pale-yellow-to-dark-yellow pigments include flavanones, flavanonoles, flavones, flavonols, chalcones, aurones, and isoflavones. The most important food pigments are flavones and flavonols. However, with few exceptions, these flavonoid substances are not used as food colourings. Previously, some of these compounds were ranked among the so-called bioflavonoids (see Section 5.15) for their biological effects.

Flavonoids usually exist in the form of *O*-glycosides, which contain either free or acylated D-glucose, L-rhamnose, D-galactose, L-arabinose, D-xylose, D-apiose, or D-glucuronic acid. Sugars are bound at positions C-7, C-5, C-4′, and C-3′, but mostly through the hydroxyl group attached to the C-7 carbon. In addition to *O*-glycosides, flavonoids are quite often found as *C*-glycosides (mainly derived from flavones and flavonols), in which glucose is bound by the C—C bond in positions C-6 or C-8 of flavonoid molecules.

Hydrolysis of glycosides in the manufacturing of fruits and vegetables may in some cases (in acidic media and, in particular, at higher temperatures) lead to increased concentrations of aglycones. Most flavonoids in foods are involved in enzymatic browning reactions. The ability to bind heavy metals, along with the ability to terminate the radical oxidation reactions, gives flavonoids antioxidant activities. However, metal complexes of flavonoids in foods sometimes cause unwanted discolourations.

9.4.2.1 Flavanones

Colourless to pale-yellow flavanones are widespread in foods but are not important as food pigments. At higher concentrations, they are found only in citrus fruits. The main components are glycosides derived from (2*S*)-5,7-dihydroxyflavanones, which differ in substituents of the ring C (**9-49** and **9-50**). The most important flavanone aglycones are hesperetin (5,7,3′-trihydroxy-4′-methoxyflavanone) and naringenin (5,7,4′-trihydroxyflavanone). Hesperetin is the major aglycone occurring in oranges and lemons. Its contents in juices is about 1000 mg/l. Grapefruits contain naringenin as the main glycoside component.

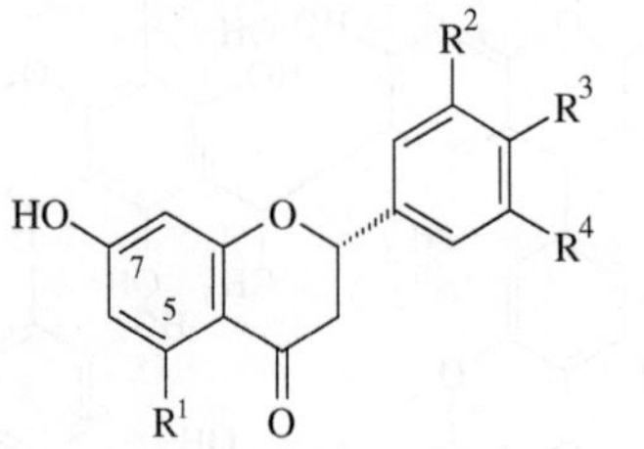

9-49, flavanones

liquiritigenin, R^1 = H, R^2 = H, R^3 = OH, R^4 = H
butin, R^1 = H, R^2 = OH, R^3 = OH, R^4 = H
pinocembrin, R^1 = OH, R^2 = H, R^3 = H, R^4 = H
naringenin, R^1 = OH, R^2 = H, R^3 = OH, R^4 = H
eriodictyol, R^1 = OH, R^2 = OH, R^3 = OH, R^4 = H
dihydrotricetin, R^1 = OH, R^2 = OH, R^3 = OH, R^4 = OH

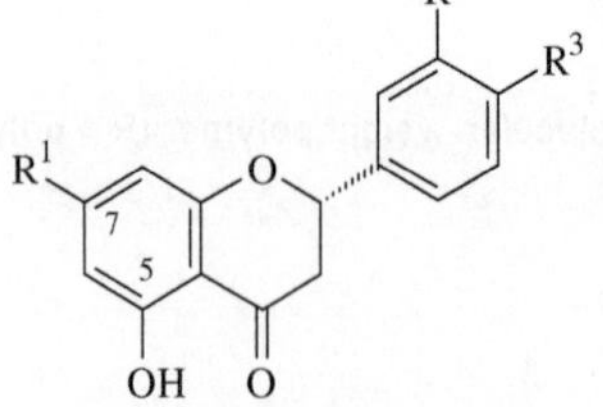

9-50, flavanone methoxyderivatives

sakuranetin, R^1 = OCH_3, R^2 = H, R^3 = OH
sterubin, R^1 = OCH_3, R^2 = OH, R^3 = OH
isosakuranetin, R^1 = OH, R^2 = H, R^3 = OCH_3
hesperetin, R^1 = OH, R^2 = OH, R^3 = OCH_3
homoeriodyctiol, R^1 = OH, R^2 = OCH_3, R^3 = OH

Common glycosides of citrus fruits and other plant materials are sakuranin (sakuranetin-5-glucoside), narirutin (naringenin-7-rutinoside, also known as neoponcirin), naringin (naringenin-7-neohesperidoside), eriodictin (eriodictyol-7-rhamnoside), ericitrin (eriodictyol-7-rutinoside), didymin (isosakuranetin-7-rutinoside), poncirin (isosakuranetin-7-neohesperidoside), hesperidin (hesperetin-7-rutinoside or cirantin), and neohesperidin (hesperetin-7-neohesperidoside). In citrus fruits, flavanone glycosides are located mainly in the albedo. Oranges contain only rutinosides (such as hesperidin and narirutin), the main glycoside is hesperidin. Its content in growing fruits increases till full ripeness and then remain constant at the level of 1–6 g per fruit. Grapefruits contain rutinosides and neohesperidosides; the dominating glycoside (up to 90% of total glycosides) in grapefruits is naringin. Grapefruit juice processed by blending has significantly higher levels of flavonoids (didymin, hesperidin, naringin, narirutin, neohesperidin, and poncirin) and limonin (see Section 8.3.5.1.1) compared to juicing and hand squeezing. Naringin (naringenin-7-neohesperidoside) occurring in grapefruits, neohesperidin (hesperitin-7-neohesperidoside) found in bitter bigarade oranges (*Citrus aurantium* var. *amara*, Rutaceae), and all other flavanone neohesperidosides are intensely bitter substance, unlike the corresponding rutinosides and glucosides.

Liquiritin (liquiritigenin-4′-glucoside) occurs in licorice (*Glycyrrhiza glabra*, Fabaceae), sakuranin and prunin (naringenin-7-glucoside) in plums (*Prunus* spp., Rosaceae), pinocembrin glycosides in legumes, and pyracanthoside (eriodictyol-7-glucoside) in firethorn fruits (*Pyracantha coccinea*, Rosaceae). Butin and its 7,3′-diglucoside butrin are components of the tree *Butea monosperma*, native to tropical and sub-tropical parts of the Indian Subcontinent and Southeast Asia, which has orange-red flowers that are used in colouring food and in medical products. Prenylated flavanones (see Section 10.3.12.3) occur in hop cones.

9.4.2.2 Flavanonoles

Flavanonoles and their glycosides are not very important flavonoids, because they do not occur at higher concentrations in food materials. An example of a flavonol is taxifolin (dihydroquercetin, **9-51**), which occurs in larger quantities in peanuts and as a component of pollen, and which along with other flavanonoles is quite a common component of other plants.

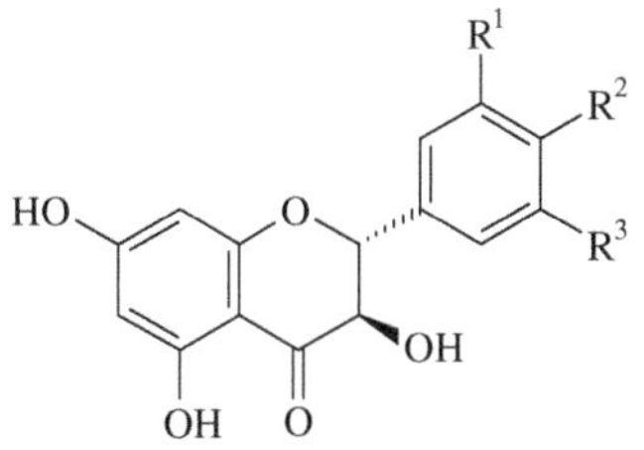

9-51, flavanonoles

pinobanksin, $R^1 = R^2 = R^3 = H$
aromadendrin (dihydrokaempherol), $R^1 = R^3 = H, R^2 = OH$
taxifolin (dihydroquercetin), $R^1 = R^2 = OH, R^3 = H$
ampelopsin (dihydromyricetin), $R^1 = R^2 = R^3 = OH$

9.4.2.3 Flavones

Flavones are, together with flavonols, the most widespread yellow pigments of plants. Typical compounds in foods are *O*-glycosides of flavones substituted at C-5 and C-7, or less often in position C-6 of ring A and at C-4′ of ring B. If the substituent occurs in position C-4′ then carbons C-3′ and C-5′ and rarely carbon C-2′ are also often substituted. Common substituents are hydroxyl and methoxyl groups. Particularly frequent flavones are apigenin and luteolin (**9-52**); less often, tricetin and other flavones occur.

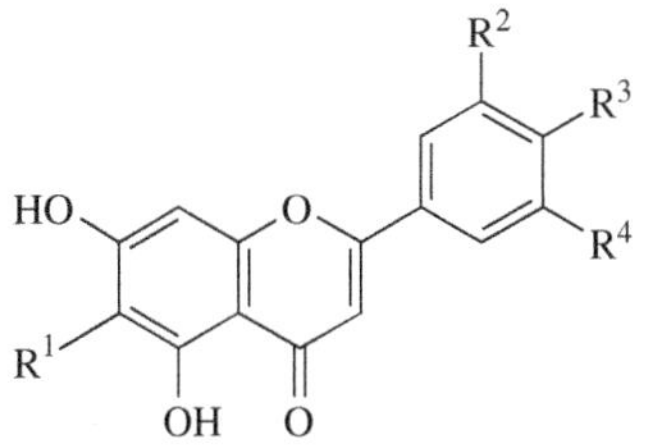

9-52, flavones

chrysin, $R^1 = H, R^2 = H, R^3 = H, R^4 = H$
apigenin, $R^1 = H, R^2 = H, R^3 = OH, R^4 = H$
luteolin, $R^1 = H, R^2 = OH, R^3 = OH, R^4 = H$
tricetin, $R^1 = H, R^2 = OH, R^3 = OH, R^4 = OH$
baikalein, $R^1 = OH, R^2 = H, R^3 = H, R^4 = H$
scutellarein, $R^1 = OH, R^2 = H, R^3 = OH, R^4 = H$

Examples of flavones with methylated hydroxyl groups in ring A are hispidulin, nepetin, and cirisiliol, whilst acacetin, diosmetin, chrysoeriol, and tricin have methoxyl groups in ring C. Methoxyl groups in both rings A and C have limocitrin, tangeretin, nobiletin and sinensetin (**9-53**), which occur in citrus fruits, especially in peel, together with a number of other hydroxylated polymethoxyflavones and other flavonoids. Tricin is found mainly in cereal grains, such as wheat, rice, barley, oat, and maize.

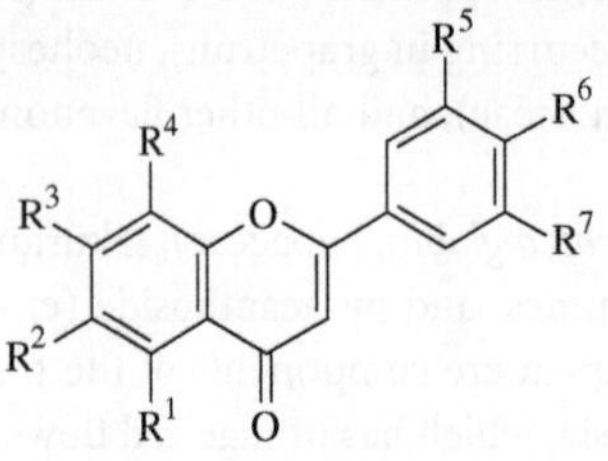

9-53, flavone methoxy derivatives

acacetin, $R^2 = R^4 = R^5 = R^7 = H$, $R^1 = R^3 = OH$, $R^6 = OCH_3$
chrysoeriol, $R^2 = R^4 = R^7 = H$, $R^1 = R^3 = R^6 = OH$, $R^5 = OCH_3$
cirsiliol, $R^4 = R^7 = H$, $R^1 = R^5 = R^6 = OH$, $R^2 = R^3 = OCH_3$
diosmetin, $R^2 = R^4 = R^7 = H$, $R^1 = R^3 = R^5 = OH$, $R^6 = OCH_3$
hispidulin, $R^4 = R^5 = R^7 = H$, $R^1 = R^3 = R^6 = OH$, $R^2 = OCH_3$
limocitrin, $R^2 = R^7 = H$, $R^1 = R^3 = R^6 = OH$, $R^4 = R^5 = OCH_3$
nepetin, $R^4 = R^7 = H$, $R^1 = R^3 = R^5 = R^6 = OH$, $R^2 = OCH_3$
nobiletin, $R^7 = H$, $R^1 = R^2 = R^3 = R^4 = R^5 = R^6 = OCH_3$
sinensetin, $R^4 = R^7 = H$, $R^1 = R^2 = R^3 = R^5 = R^6 = OCH_3$
tangeretin, $R^5 = R^7 = H$, $R^1 = R^2 = R^3 = R^4 = R^6 = OCH_3$
tricin, $R^2 = R^4 = H$, $R^1 = R^3 = R^6 = OH$, $R^5 = R^7 = OCH_3$

Sugars (mainly D-glucose, D-galactose, and L-rhamnose) in O-glycosides of flavones are preferentially bound to the hydroxyl group in position C-7. For example, curly mint (*Mentha crispa*, Lamiaceae) contains the glycoside diosmin (diosmetin-7-rutinoside). Other O-glycosides and C-glycosides are rare. Unique tetraglycosides teaghrelins (**9-54**), derived from apigenin (teaghrelin-1) and luteolin (teaghrelin-2), occur in oolong tea.

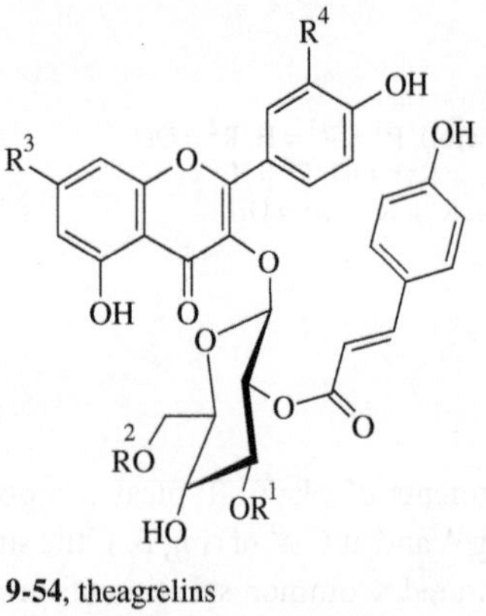

9-54, theagrelins

theagrelin 1, R^1 = β-D-Glc*p*, R^2 = β-D-Glc*p*-(1→3)-α-L-Rha*p*, R^3 = H, R^4 = H
theagrelin 2, R^1 = β-D-Glc*p*, R^2 = β-D-Glc*p*-(1→3)-α-L-Rha*p*, R^3 = H, R^4 = OH

The most common C-glycosides are vitexin (8-C-glucosylapigenin) and orientin (8-C-glucosylluteolin, **9-55**). These occur, for example, in rice bran and many fruits. Also common are isovitexin (6-C-glucosylapigenin) and isoorientin (6-C-glucosylluteolin). Orientin and isoorientin, the main components of fermented roiboos tea (*Aspalathus linearis*, Fabaceae) and acai fruits (*Euterpe oleraceae*, Arecaceae), show antioxidant properties. Vitexin, isovitexin, flavone C-glycoside chafuroside A (with a potent anti-inflammatory activity), and its regioisomer chafuroside B (**9-56**) were isolated from oolong tea. C-Glycoside, called schaftoside (apigenin-6-C-β-D-glucopyranosyl-8-C-α-L-arabinopyranoside), is found in figs, and its presence may be used to detect fig juice in other fruit juices (e.g. grape juice).

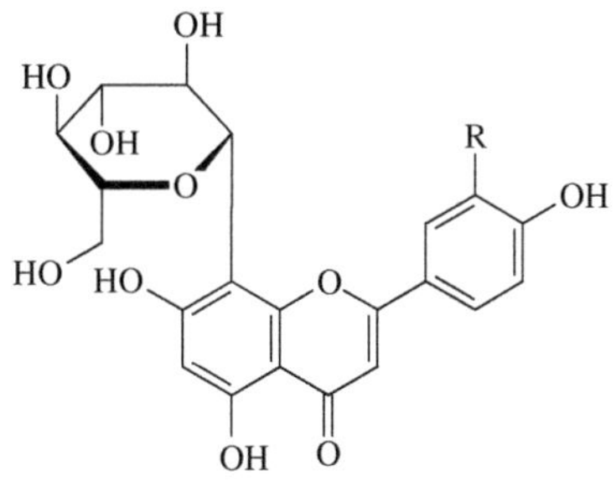

9-55, flavone *C*-glucosides

vitexin, R = H
orientin, R = OH

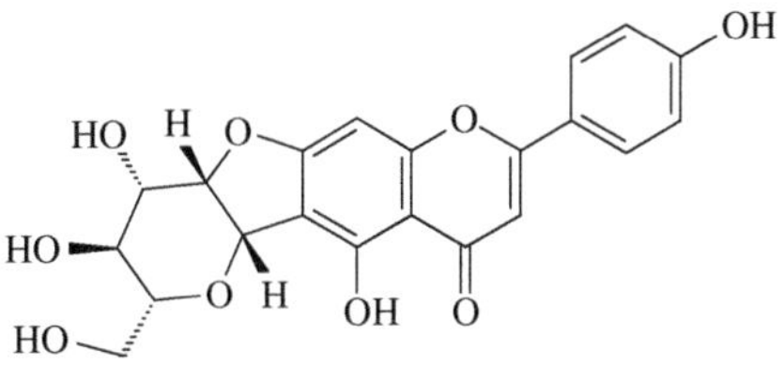

chafuroside A 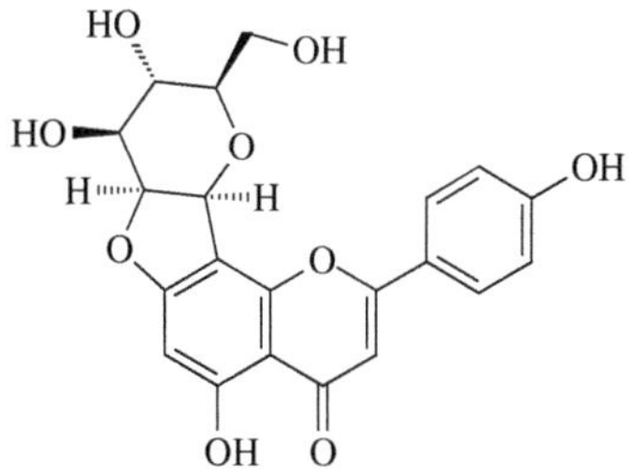   chafuroside B

9-56, chafurosides

9.4.2.4 Flavonols

Flavonols are, together with flavones, an important group of yellow plant pigments that also have multiple biological activities. All major flavonols occurring in foods have hydroxyl groups in positions C-3, C-5, C-7, and C-4 of the flavan skeleton (**9-57**) and mutually differ by substitutions at positions C-3′, C-4′, and C-5′. Examples of flavonol methyl ethers are rhamnetin, isorhamnetin, and other substances (**9-58**). Free aglycones are found in relatively small quantities, and the main forms are flavonol *O*-glycosides. Flavonol glycosides are mainly 3-*O*-glycosides and, less often, 7-*O*-glycosides.

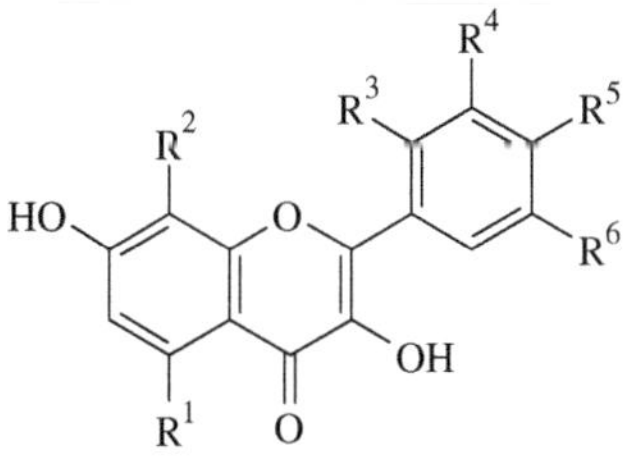

9-57, flavonols

galangin, $R^2 = R^3 = R^4 = R^5 = R^6 = H$, $R^1 = OH$
datiscetin, $R^2 = R^4 = R^5 = R^6 = H$, $R^1 = R^3 = OH$
kaempherol, $R^2 = R^3 = R^4 = R^6 = H$, $R^1 = R^5 = OH$
quercetin, $R^2 = R^3 = R^6 = H$, $R^1 = R^4 = R^5 = OH$
myricetin, $R^2 = H$, $R^1 = R^4 = R^5 = R^6 = OH$
physetin, $R^1 = R^2 = R^3 = R^6 = H$, $R^4 = R^5 = OH$
robinetin, $R^1 = R^2 = R^3 = H$, $R^4 = R^5 = R^6 = OH$
morin, $R^1 = R^2 = R^4 = R^6 = H$, $R^3 = R^5 = OH$
herbacetin, $R^3 = R^4 = R^6 = H$, $R^1 = R^2 = R^5 = OH$
gosypetin, $R^3 = R^6 = H$, $R^1 = R^2 = R^4 = R^5 = OH$

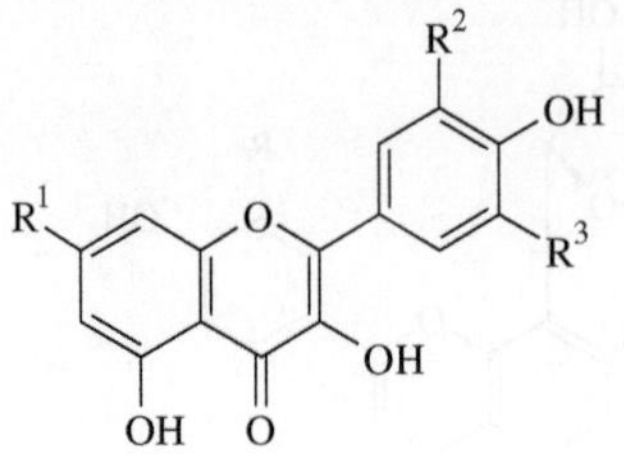

9-58, flavonol methoxy derivatives

rhamnetin, $R^1 = OCH_3$, $R^2 = OH$, $R^3 = H$
isorhamnetin, $R^1 = OH$, $R^2 = OCH_3$, $R^3 = H$
laricitrin, $R^1 = OH$, $R^2 = OCH_3$, $R^3 = OH$
syringetin, $R^1 = OH$, $R^2 = OCH_3$, $R^3 = OCH_3$

The most widespread compounds are glycosides derived from quercetin, kaempherol, and myricetin; glycosides of isorhamnetin, laricitrin, and syringetin are less common. Trivial and systematic names of common quercetin glycosides are given in Table 9.13. Their contents in selected fruits are given in Table 9.14.

Table 9.13 Trivial and systematic names of selected quercetin glycosides.

Trivial name	Systematic name	Trivial name	Systematic name
Avicularin	quercetin-3-O-α-L-arabinofuranoside	Spiraein (spiraeoside)	quercetin-4'-β-D-glucopyranoside
Reinoutrin	quercetin-3-O-β-D-xylopyranoside	Hyperin (hyperoside)	quercetin-3-β-D-galactopyranoside
Isoquercitrin	quercetin-3-O-β-D-glucopyranoside	Quercitrin	quercetin-3-α-L-rhamnopyranoside
Quercimeritrin	quercetin-7-O-β-D-glucopyranoside	Rutin	quercetin-3-β-rutinoside

Table 9.14 Contents of flavonol glycosides in selected berries (in mg/kg).

Berry	Latin name	Kaempferol	Quercetin	Myricetin	Isorhamnetin	Laricitrin	Syringetin
Bilberry	*Vaccinium myrtillus*	1.51	79.4	22.3	1.61	4.90	4.50
Blackberry	*Rubus fruticosus*	6.70	42.6		20.1		
Blueberry (highbush)	*Vaccinium corymbosum*	2.53	136	22.0	9.51	4.00	11.7
Currant (black)	*Ribes nigrum*	134	35.7	31.6			
Currant (red)	*Ribes rubrum*	25.0	23.6	3.40			
Elderberry	*Sambucus nigra*	4.40	549	3.70	10.8		
Gooseberry (white)	*Ribes grossularia*	0.36	28.6	0.23	49.9	0.79	0.59
Chokeberry	*Aronia melanocarpa*	0.54	263	0.75	2.7		
Kiwifruit	*Actinidia arguta*	2.04	25.9				0.36
Mulberry (black)	*Morus nigra*	1.63	105				
Raspberry	*Rubus idaeus*	1.20	26.3		2.5		
Rowanberry	*Sorbus aucuparia*	9.22	184	0.52			
Strawberry	*Fragaria x ananassa*	1.96	7.34		1.21		

A generally widespread glycoside of many plants is rutin. Particularly high content of its aglycone quercetin (2.5–6.5%) is found in the dry outer skin of red onion cultivars (onions are a primary source of dietary quercetin glycosides in the Western diet). Apples contain, as major glycosides, quercetin-3-glucoside and 3-galactoside (depending on the variety and other factors, they are found in levels of about 0.01–0.95 g/kg), as well as 3-rhamnoside (0.01–0.37 g/kg), 3-arabinoside (0.02–0.78 g/kg), 3-xyloside (0.03–0.25 g/kg), and 3-rutinoside (0.01–0.12 g/kg). In smaller amounts, apples contain 3-glucosides of kaempferol and myricetin. Elderberries contain quercetin-3-rutinoside, which is missing in strawberries. Quercetin-3-glycoside (trisaccharide containing two molecules of L-rhamnose and one molecule of D-glucose) occurs in blackcurrants but not in redcurrants.

In tea leaves, flavonols and their mono-, di-, and tri-glycosides are found in larger quantities than they are in fruits. The main flavonol glycosides are 3-glucosides, 3-galactosides (the major component is myricetin-3-galactoside), and 3-rutinosides; 3-rhamnoside of quercetin and 3-rhamnodiglucosides and 3-glucorhamnogalactosides of quercetin and kaempferol significantly contribute to the bitter taste of tea infusions (see Section 8.3.5.1.4). Black teas usually contain 0.4–1.7%, green tea 1.5–1.7%, and instant teas 2.6–3.1% flavonol glycosides (in d.w.)

For example, avicularin and isoquercitrin occur in bilberries, whilst spiraein, together with rutin, is a constituent of onion and chestnuts (seeds of horse chestnut, *Aesculus hippocastanum*, Sapindaceae). Isorhamnetin occurs as 3-rutinoside (narcissin) in citrus and other fruits. Galangin, a flavonol with an unsubstituted pyran ring C, is found in rhizomes of greater galanga (*Alpinia galanga*, Zingiberaceae), commonly known as galangal, which are widely used in Chinese and Thai traditional medicine and as a condiment for foods for their characteristic fragrance and pungency (see Section 8.3.7.3).

Rutin and several other flavonoid glycosides exhibit antioxidant properties and affect the flexibility and permeability of blood capillaries. Rutin (formerly vitamin P) is thus used in pharmaceutical preparations and food supplements. Together with other substances called bioflavonoids (see Section 5.15), rutin increases levels of ascorbic acid in various animal organs, either by providing protection against oxidation catalysed by metal ions or through increased ascorbic acid utilisation in the body. Natural sources of ascorbic acid containing flavonoids (such as rosehips with a considerable amount of rutin) are thus more effective than synthetic vitamin C.

The first hydrothermal degradation product of rutin at higher temperatures is isoquercitrin (and L-rhamnose), which is further degraded to quercetin (and D-glucose). Quercetin is degraded via 3,4-dihydroxybenzoic acid to catechol or via 2-(2-hydroxy-1-oxoethyl)phloroglucinol to 2H-benzofuran-3-one (coumaran-3-one), the precursor of aurones (Figure 9.20). The degradation products of sugar moieties are the corresponding furan-2-carbaldehydes. A complex of rutin with iron causes dark discolouration of asparagus in tins, and the stannous complex of rutin is yellow.

9.4.2.5 Chalcones

Various chalcones, dihydrochalcones (chalconoids), and aurones (auronoids) are not particularly important components of plant food materials, but they occur as notable pigments of flowers of many ornamental plants, such as common snapdragon (*Antirrhinum* spp., Scrophulariaceae), cosmos (*Cosmos* spp.), and dahlia (*Dahlia* spp.) from the Asteraceae family. They absorb UV radiation (unlike the carotenoid pigments in the other part of flowers) and act as attractants for bees and other insects that pollinate flowers.

Chalcones are also pigments of legume seeds (**9-59**) and woods. Isoliquiritigenin occurs in soybeans as 4′-O-β-D-glucopyranoside. The most common chalcone is butein. Its 4-glucoside is called coreopsin and 4,3′-diglucoside is isobutrin. These glycosides are found as pigments of flowers of the tropical legume plant *Butea monosperma* (Fabaceae), which are used locally for food colouring. Chalconaringenin is the predominant compound in tomatoes. It occurs in amounts of 9–182 mg/kg, which is 35–71% of the total flavonoid content.

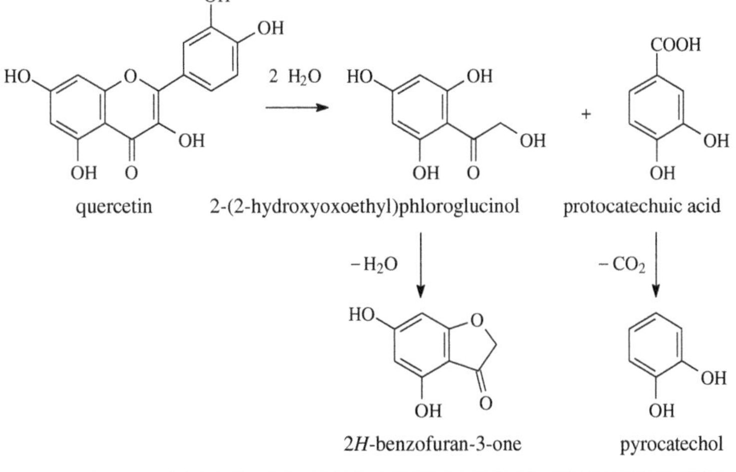

Figure 9.20 Hydrothermal degradation of quercetin.

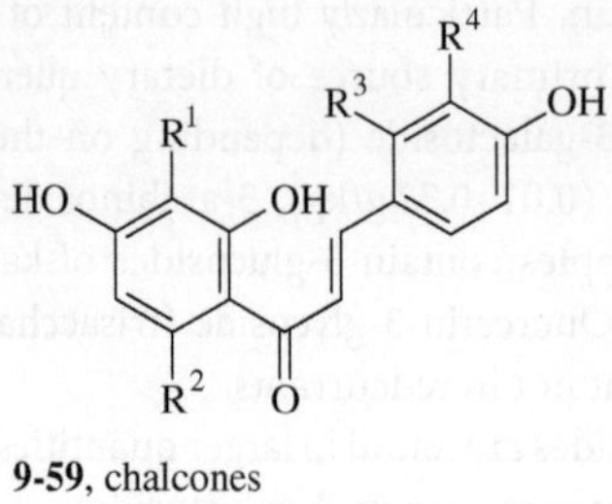

9-59, chalcones

isoliquiritigenin, $R^1 = R^2 = R^3 = R^4 = H$
butein, $R^1 = R^2 = R^3 = H, R^4 = OH$
marein, $R^1 = R^4 = OH, R^2 = R^3 = H$
okanin, $R^1 = R^3 = R^4 = OH, R^2 = H$
chalconaringenin, $R^1 = R^3 = R^4 = H, R^2 = OH$

Chalcones with methoxy groups in positions 4 and 6 of ring A represent a class of biologically active chalconoids known as flavokavains (also called flavokawains or flavokawins, **9-60**) found in the root of kava plant (*Piper methysticum*, Piperaceae; see Section 8.2.7.1.2). Currently identified types include flavokavain A (2-hydroxy-4,4',6-trimethoxychalcone), flavokavain B (2-hydroxy-4',6-dimethoxychalcone), and flavokavain C (2,4'-dihydroxy-4,6-dimethoxychalcone). The inhibitory effect of flavokavain A may be related to low bladder tumour incidences in the Pacific Island nations. Other biologically active components of kava root are styrylpyrones (see Section 8.2.7.1.2). Prenylated chalcones are found in hop cones (see Section 10.3.12.4).

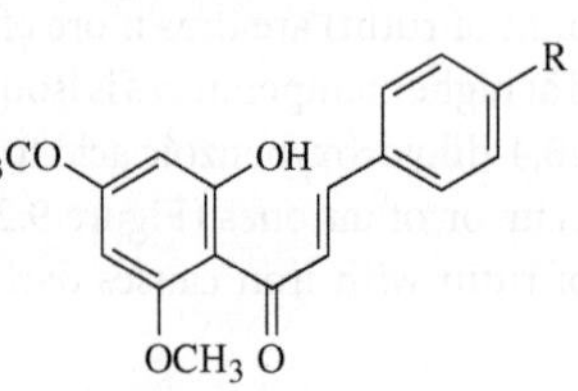

9-60, methoxychalcones

flavokavain A, R = OCH₃
flavokavain B, R = H
flavokavain C, R = OH

Chalcones always contain a hydroxyl group at C-2; this comes from pyrane ring C of flavanones, which produce chalcones in alkaline medium (Figure 9.21). Under acidic conditions, especially with heating, the opposite reaction proceeds: conversion of chalcones into flavanones. For example, naringenin chalcone gives rise to naringenin, hesperidin arises from hesperidin chalcone, and isobutrin arises from butrin. These reactions are related to the enzymatic reactions of chalcone biosynthesis, which catalyses chalcone-flavanone isomerase.

9.4.2.5.1 Quinochalcones

Carthamin (carthemon, **9-61**) is the main water-soluble red pigment contained in the flower petals of a thistle-like herbaceous plant known as safflower (*Carthamus tinctorius*, Asteraceae), which is utilised in producing herbal medicines and natural red dye used as a food colourant. Safflower seeds are utilised in making edible oil for use in cooking. The main safflower pigment is red carthamin, which is composed of two *C*-glucosylquinochalcone moieties. A minor red pigment is hydroxyethylcarthamin (R = glucose with CH_2CH_2OH substituent in position C-6). The major yellow pigments with the *C*-glucosylquinochalcone moieties are hydroxysafflor yellow A (**9-62**) and safflor yellow B (**9-63**). Besides these, about 15 structurally similar *C*-glucosylquinochalcones have been identified, including yellow compounds safflomin A, safflomin B, safflomin C, isosafflomin C, methylsafflomin C, methylisosafflomin C, safflor yellow A, anhydrosafflor yellow B, and precarthamin (**9-64**), decarboxylation and oxidation of which yields carthamin (**9-61**).

Isomerisation of carthamin in an acidic environment yields yellow isocarthamin, whilst its hydrolysis yields glucose and two aglycones: the flavanones carthamidin (6-hydroxynaringenin) and isocarthamidin (8-hydroxynaringenin). Recently, three novel quinochalcone

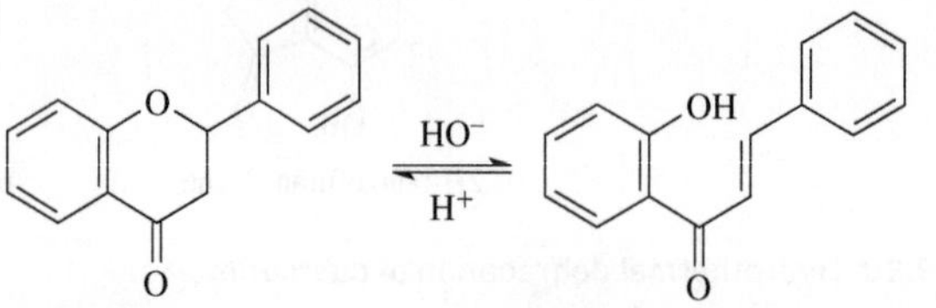

Figure 9.21 Interconversion of flavanones and chalcones.

C-glycosides saffloquinosides and two novel carthorquinosides were isolated from the safflower florets. Saffloquinoside A (**9-65**) and saffloquinoside C have uncommon five- or six-membered dioxaspirocycles, whilst saffloquinoside B has a cyclohexatrione skeleton with a benzyl group and two *C*-glycosyl units. Carthorquinoside A is a quinochalcone-flavonol structure linked via a methylene bridge, whilst carthorquinoside B contains two glucopyranosylquinochalcone moieties linked via the formyl carbon of an acyclic glucosyl unit. In addition, the *N*-containing yellow pigments, tinctormin (**9-66**) and cartormin (**9-67**), have recently been isolated.

Carthamin was used as a wool dye in ancient times and is now the only chalcone-type pigment recommended in some countries for food colouring. As a food additive, it is known as Natural Red 26 (also known as safflower red and carthamic acid). It may become a promising food colouring, for example for yoghurt and other dairy products. Hydroxysafflor yellow A has attracted attention because of its remarkable cardiovascular activities. Carthorquinosides exhibit anti-inflammatory activities.

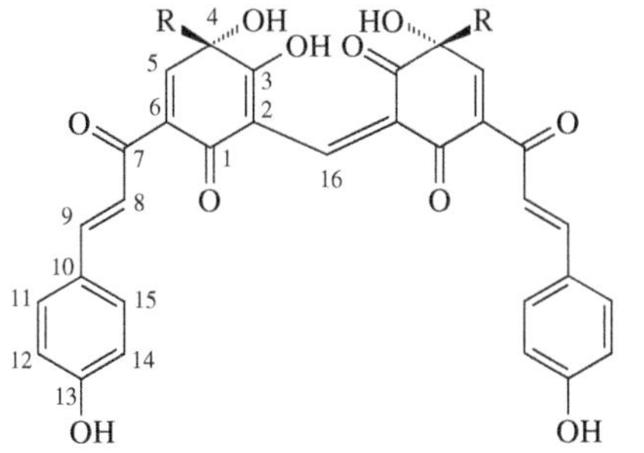

9-61, carthamin, R = β-D-Glc*p*

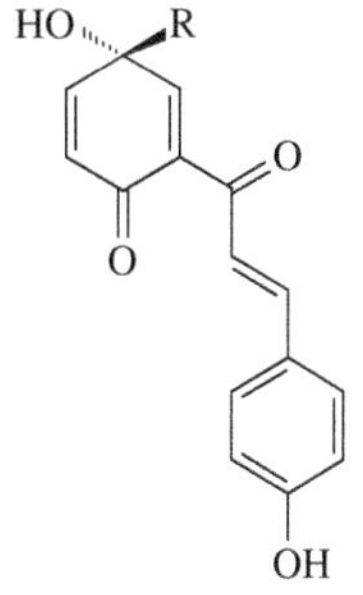

9-62, hydroxysafflor yellow A, R = β-D-Glc*p*

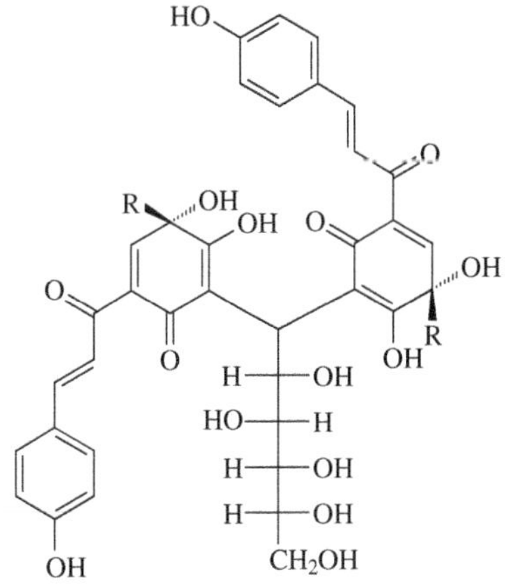

9-63, safflor yellow B, R = β-D-Glc*p*

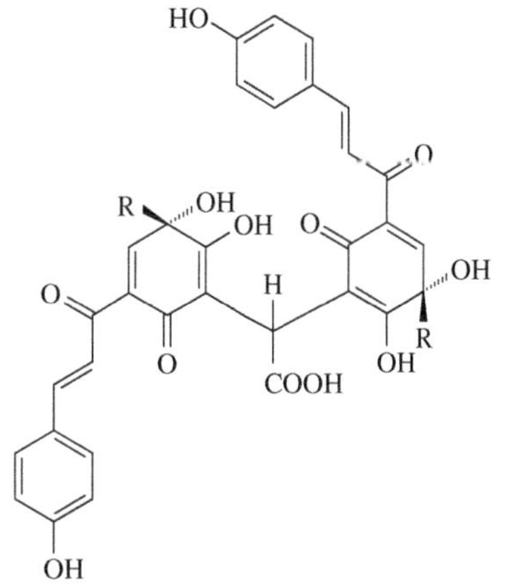

9-64, precarthamin, R = β-D-Glc*p*

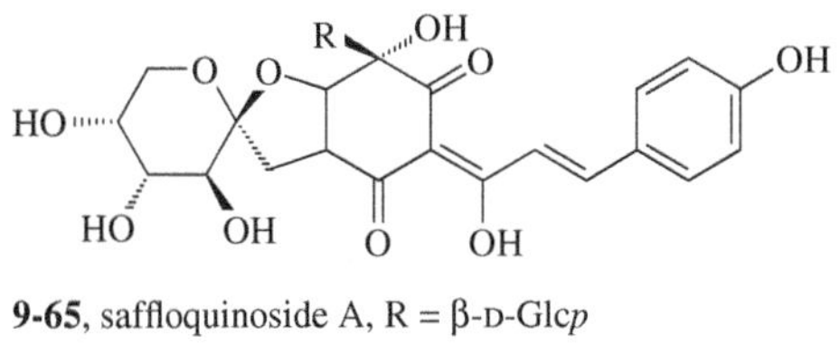

9-65, saffloquinoside A, R = β-D-Glc*p*

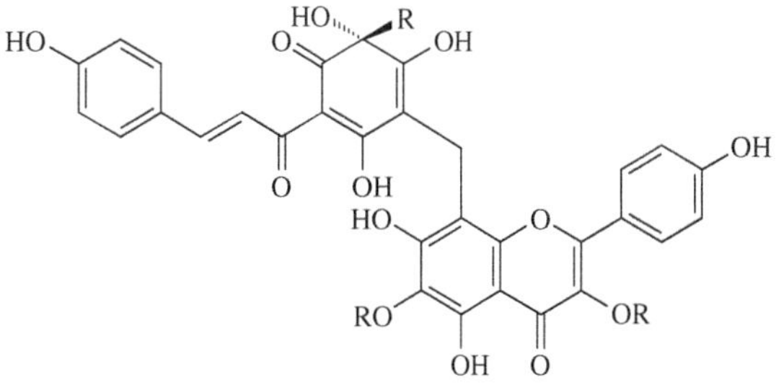

9-66, carthorquinoside A, R = β-D-Glc*p*

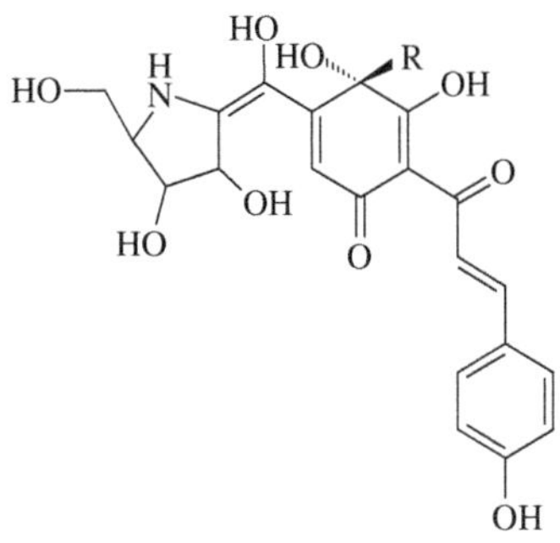

9-67, tinctormin, R = β-D-Glc*p*

9.4.2.6 Dihydrochalcones

An example of a natural dihydrochalcone is phloretin (**9-68**). Its glycoside, phloretin-6-O-β-D-glucoside, is called phlorizin (also referred to as phloridzin). Its occurrence is practically limited to unripe apples (the botanical genus *Malus*, Rosaceae), where phlorizin functions as an inhibitor of seed germination. In small quantities, phlorizin is also found in ripe apples and apple products (at a level of 0.1–22 mg/kg d.w. in the skin, 0.03–0.3 mg/kg in the flesh, and 0.01–0.4 mg/kg in the juices), where it is accompanied by a trace amount of 6-O-xyloglucoside. The presence of these characteristic compounds may be used to detect the addition of apple juice to other fruit juices, although small amounts of phlorizin have also been found in strawberries. Phlorizin has a bitter taste, like naringin. In mammals, it causes glycosuria (glucose excretion in the urine). Human intestinal flora transforms phlorizin to phloretin, 3-(4-hydroxyphenyl)propanoic acid (also known as phloretic acid or desaminotyrosine), phloretic acid methyl ester, and phloroglucinol (1,3,5-benzenetriol).

Enzymatic and acid hydrolysis of phlorizin releases the aglycone phloretin, whilst alkaline hydrolysis yields 4-hydroxyphenylpropionic acid and β-D-glucopyranoside of phloroglucinol, which is called phlorin. It has been shown that phlorin is present in peel of citrus fruits. In species and varieties of oranges, phlorin was found in juices and peel extracts with a mean of 22 and 492 mg/l, respectively, whilst in grapefruits, means were 108 mg/l in juices and 982 mg/l in peel extracts. In contrast, phlorin was not found in mandarin and clementine juices, except in a few mandarin varieties (30–33 mg/l).

Phloretin-3′,5′-di-C-β-D-glucopyranoside is the first C-glycoside identified in tomatoes and the first dihydrochalcone from *Solanum* species. In tomatoes, its amount ranges from 2 to 15 mg/kg.

Dihydrochalcones prepared semi-synthetically by reduction of the flavanones neohesperidin and naringin have intensely sweet taste. Neohesperidin dihydrochalcone is used as a sweetener (see Section 11.3.2.1.2).

The dihydrochalcone C-glucopyranosides aspalathin and nothofagin (**9-69**) are unique antioxidants to the rooibos plant (*A. linearis*, Fabaceae), the leaves and stem of which are fermented in order to develop the characteristic rooibos flavour and red colour. The key red compounds arising via oxidation of aspalathin and subsequent reactions are two structures with dibenzofuran skeleton, (*S*)- and (*R*)-3-(7,9-dihydroxy-2,3-dioxo-6-β-D-glucopyranosyl-3,4-dihydrodibenzo[*b,d*]furan-4*a*(2*H*)-yl) propionic acid (**9-70**), which are ultimately degraded to more stable tannin-like structures. The aqueous infusion is consumed as a caffeine-free rooibos tea (also known as red tea) or in iced teas and yoghurt. Aqueous spray-dried extracts of fermented rooibos usually contain up to 0.5% aspalathin, which is roughly half the quantity in the original green material. Nothofagin content is about 10 times lower.

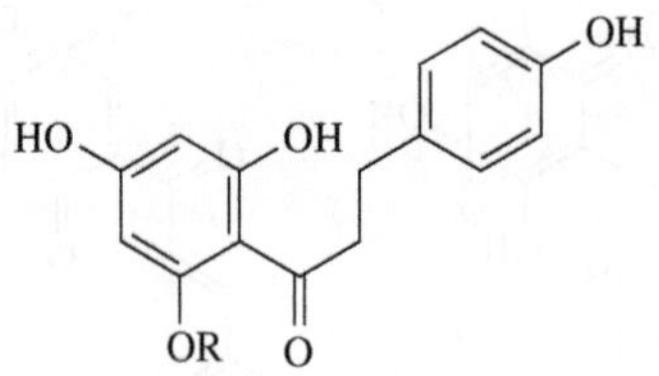

9-68, dihydrochalcones

phloretin, R = H
phlorizin, R = β-D-glucosyl

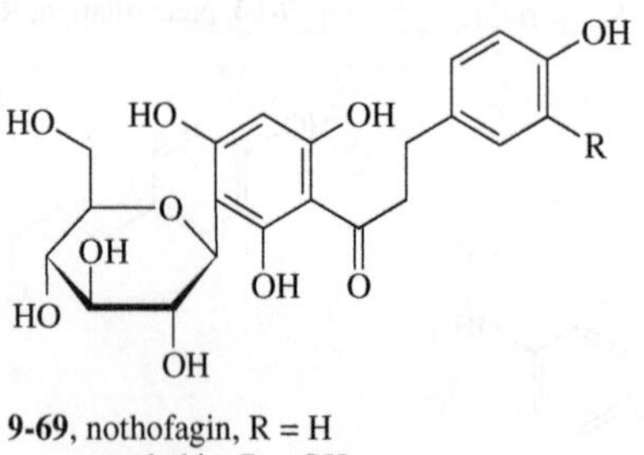

9-69, nothofagin, R = H
aspalathin, R = OH

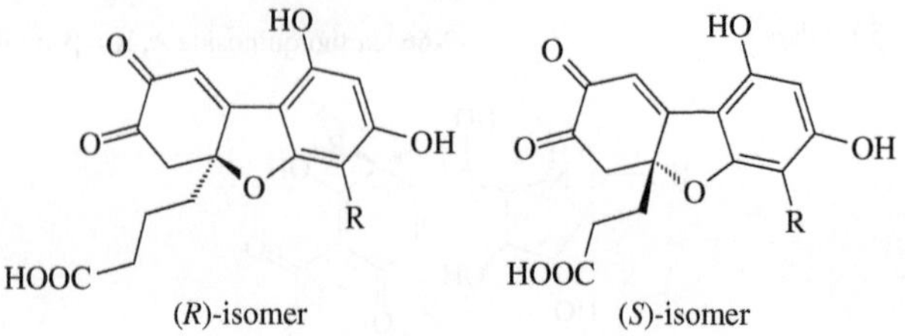

(*R*)-isomer   (*S*)-isomer

9-70, red dibenzofurans in roiboos tea, R = β-D-Glc*p*

9.4.2.7 Aurones

Aurones, 2-benzylidenebenzo-2*H*-furan-3-ones, are widespread in various flowers, where they play a significant role in their pigmentation. Sulfuretin (**9-71**) is a pigment of dahlia (*Dahlia* spp., Asteraceae) flowers, whilst aureusidin occurs in flowers of the common snapdragon (*Antirrhinum majus*, Plantaginaceae), where it occurs as 6-*O*-β-D-glucoside (called aureusin) together with 4-*O*-β-D-glucoside of aureusidin (cernuoside) and 6-*O*-β-D-glucoside of bracteatin. Hispidol occurs in soybeans as 6-*O*-β-D-glucoside.

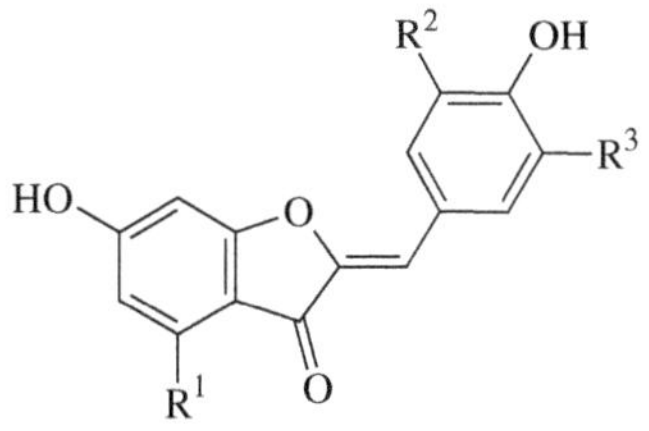

9-71, aurones

hispidol, R^1 = H, R^2 = H, R^3 = H
sulfuretin, R^1 = H, R^2 = OH, R^3 = H
aureusidin, R^1 = OH, R^2 = OH, R^3 = H
bracteatin, R^1 = OH, R^2 = OH, R^3 = OH

9.4.2.8 Isoflavones

Isoflavones are mainly found in the subfamily Faboideae of the legume family Fabaceae, commonly known as the legume, pea, or bean family. These pale-yellow compounds have also been found in a number of other families outside of the Fabaceae, but as plant pigments are of little importance. The principal plant that produces isoflavones (1–2 g/kg) is soy (*Glycine max*), and generally only small amounts are found in most other plants belonging to the same family (see Section 10.3.12.4). An example of a less common isoflavones is cajanin (2′,4′,5-trihydroxy-7-methoxyisoflavone, **9-72**), found in pigeon pea (*Cajanus cajan*) and jack bean (*Canavalia ensiformis*). Another example is pratensin (3′,5,7-trihydroxy-4′-methoxyisoflavone), identified in red clover (*Trifolium pratense*) and later as a constituent of chickpea (*Cicer arietinum*). Prunetin (4′,5-dihydroxy-7-methoxyisoflavone) occurs in *Glycyrrhiza glabra* (licorice), but also in several *Prunus* species (Rosaceae). Santal (3′,4′,5-trihydroxy-7-methoxyisoflavone) is present as one of the pigments in white sandalwood (*Santalum album*, Santalaceae).

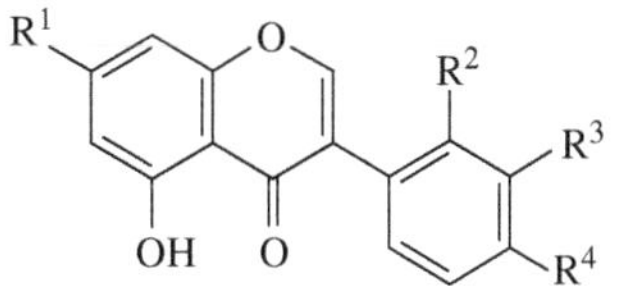

9-72, isoflavones

cajanin, R^1 = OCH$_3$, R^2 = OH, R^3 = H, R^4 = OH
pratensin, R^1 = OH, R^2 = H, R^3 = OH, R^4 = OCH$_3$
prunetin, R^1 = OCH$_3$, R^2 = H, R^3 = H, R^4 = OH
santal, R^1 = OCH$_3$, R^2 = H, R^3 = OH, R^4 = OH

9.4.2.9 Santalins

The lipophilic red pigments santalin A and santalin B (**9-73**) are (along with related compounds such as the coumarin derivative santalin AC, **9-73**) components of fragrant Indian (white) sandalwood (*Santalum album*, Santalaceae) and extracts obtained from red sandalwood (*Pterocarpus santalinus*, Fabaceae). Their main use was previously in dyeing wool and in folk medicine for antiseptic properties. Currently, red sandalwood is used for colouring tea infusions, acid-pickled vegetables, and some sausages. The so-called white sandalwood, unlike red sandalwood, is used to flavour foods and other materials, because of its high essential oil content (about 2.5%).

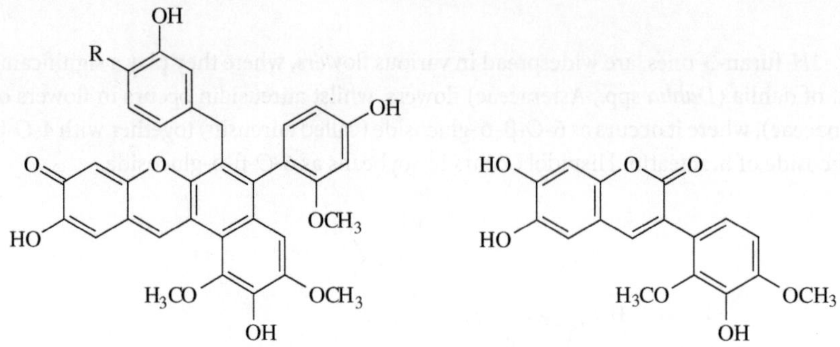

santalin A, R = OH
santalin B, R = OCH₃

santalin AC

9-73, santalins

9.5 Xanthones

Xanthones are a group of about 200 yellow pigments of basic structure C_6—C_1—C_6. They are derived from xanthone (xanthen-9-one, **9-74**) and are biogenetically related to flavonoids. Their occurrence is limited to a few families of higher plants, Clusiaceae (syn. Guttiferae), Gentianaceae, Anacardiaceae, Moraceae, and Polygalaceae, as well as some fungi and lichens. From time immemorial, some xanthones have been used as textile dyes and food colourants. Xanthones also exhibit various pharmacological effects. Various parts of plants containing xanthones have been used for ages in tradicional folk medicine in different part of world.

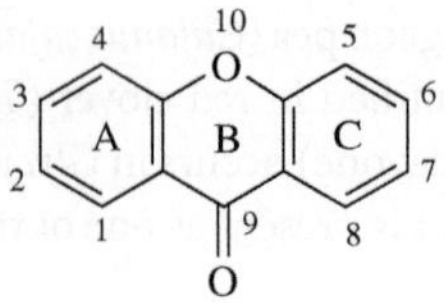

9-74, xanthone

Important xanthones are yellow pigments gentisin (1,3,7-trihydroxyxanthone) and gentisein (1,3,6-trihydroxyxanthone, **9-75**), occurring along with other xanthones as glycosides in the yellow gentian root (*Gentiana lutea*, Gentianaceae), which is used for its bitter taste and specific aroma in the production of bitter liqueurs. Gentisein also occurs in centaury (*Centaurium erythraea*), belonging to the same plant family.

A representative of other important xanthones is norathyriol (1,3,6,7-tetrahydroxyxanthone), which occurs as 2-*C*-glucoside called mangiferin (**9-76**) in a variety of plants, for example in the leaves, peals, and immature fruits of the mango (*Mangifera indica*, Anacardiaceae), accompanied by a minor 4-*C*-glucoside known as isomangiferin. Mangiferin has anti-inflammatory, antiviral, and antioxidant properties. The occurrence of mangiferin-6′-*O*-gallate has been described in mango leaves. In India, mango tree leaves were fed in a limited amount with cattle, whose urine was used as a yellow dye for fabrics and carpets.

The major secondary metabolites of the tropical evergreen tree *Garcinia mangostana* (Clusiaceae), originating in Sunda Islands and the Moluccas, known as mangosteen, are prenylated xanthone derivatives, some of which possess important pharmacological activities. The mangosteen fruits have an inedible dark purple to red-purple rind containing anthocyanins and an edible soft and juicy white pulp (arils) with a sweet, slightly acid taste and a pleasant aroma. All parts of the fruit and the mangosteen drink contain identical main xanthones, which are α-mangostin, γ-mangostin, gartanin, 8-deoxygartanin, garcinone E, 1,7-dihydroxy-3-methoxy-2-(3-methylbut-2-enyl)xanthan, and 1,3,7-trihydroxy-2,8-di-(3-methylbut-2-enyl)xanthan (**9-77**). The total amounts of these xanthones in pericarp, aryl segments, and beverage are 17 g/kg (f.w.), 1.1 g/kg, and 0.19 g/l, respectively. The main component is α-mangostin, which is present in amounts of 12 g/kg, 0.21 g/kg, and 0.12 g/l, respectively. Another closely related yellow pigment in the mangosteen family is prenylated benzophenone derivative garcinol (**9-78**). It occurs, for example, in the amount of 2–3% in culinary fruits of *G. indica* (and other species), commonly known as kokum. The dried rind of fruit is used as a garnish for curry and in traditional medicine in India.

The European mushrooms of the genus *Cortinarius* (Cortinariaceae) characteristically contain the xanthone dermoxanthone (**9-79**) and its methyl ester, found in the stem of the surprise webcap *C. semisanguineus*. These xanthones are responsible for the bright yellow fluorescence of the mushroom under UV light.

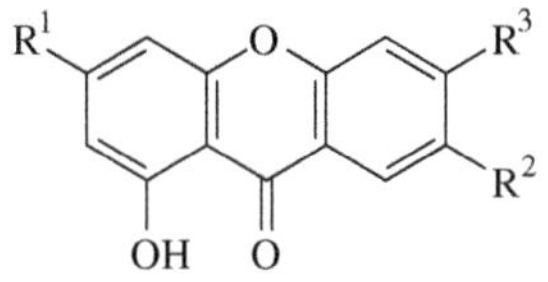

9-75, simple xanthones

gentisein, R^1 = OH, R^2 = H, R^3 = OH
gentisin, R^1 = OH, R^2 = OH, R^3 = H
norathyriol, R^1 = OH, R^2 = OH, R^3 = OH

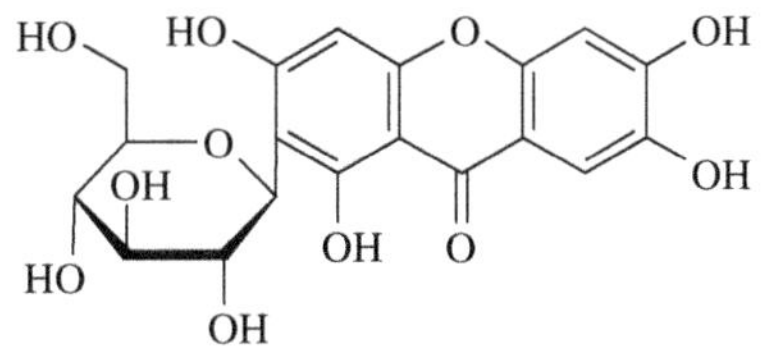

9-76, mangiferin

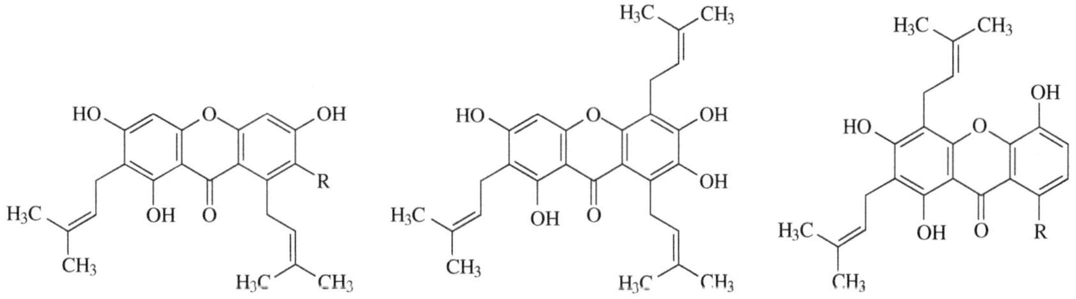

γ-mangostin, R = OH
α-mangostin, R = OCH$_3$

garcinone E

8-deoxygartanin, R = H
gartanin, R = OH

9-77, prenylated xanthones in mangosteen fruits

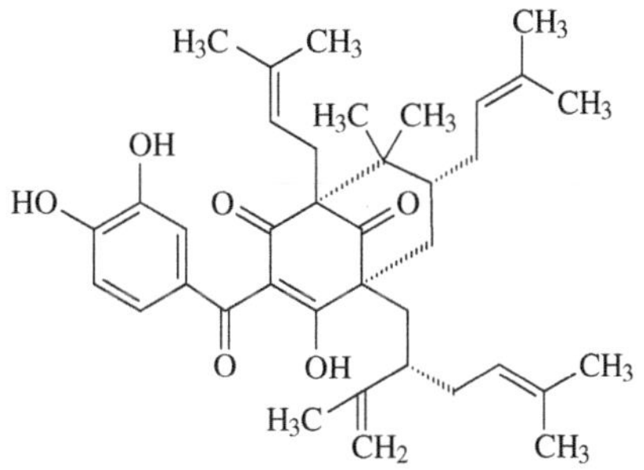

9-78, garcinol

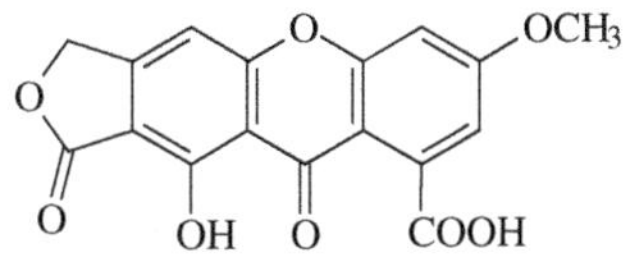

9-79, dermoxanthone

9.6 Curcuminoids

Curcuminoids are a group of phenolic compounds related to lignans and belonging to diarylheptanoids (C_6—C_7—C_6 compounds) that occur in about 30 species of plants of the genus *Curcuma*. The best known curcuminoids are yellow pigments of a rhizomatous plant known as turmeric (*C. longa*, Zingiberaceae), also known as yellow ginger, which is native to India and tropical South Asia. The dried and grind rhizome is used as a spice in curries and to impart colour to various dishes and condiments. Commercial turmeric contains the diketone curcumin (**9-80**), (1*E*,6*E*)-1,7-bis(4-hydroxy-3-methoxyphenyl)hepta-1,6-dien-3,5-dione (71.5%), which is accompanied by demethoxy-curcumin (19.4%), bisdemethoxycurcumin (9.1%), and other compounds of similar structures. Representation of the pigments may vary according to species, origin, and maturity of rhizomes. For example, rhizomes of *C. aromatica* contain only traces of bisdemethoxycurcumin, and the main pigment in some species may be demethoxycurcumin. Minor structurally related curcuminoids in turmeric are curcumin metabolites tetrahydrocurcumin, hexahydrocurcumin, cyclocurcumin, and lower homologues diarylpentanoids (**9-81**). Diarylpentanoids are the main curcuminoids of *C. xanthorrhiza*.

Autoxidation of curcumin results in a variety of products, characterised mainly by a modified heptadienedione chain. The products carry various types of oxygen substitutions at C-1 and C-7 of the C7 linker between the two guaiacol rings. A second characteristic feature is the cyclisation involving C-2 and C-6 of the heptadienedione unit to a cyclopentadione in most of the degradation products, such as bicyclopentadione and spiroepoxide. Curcumin autoxidation also yields 7-norcyclopentadione (**9-81**), a degradation product exhibiting a loss of one carbon removed via a peroxide-linked curcumin dimer in conjunction with radical-mediated 1,2-aryl migration of a guaiacol moiety.

A series of unusual curcuminoids, cassumunins A, B, and C, and the complex curcuminoids cassumunarins A, B, and C (**9-82**) have been isolated from the rhizomes of *Zingiber montanum* (syn. *Z. cassumar*). Another unusual diarylheptanoid (4*E*,6*E*)-1,7-bis(4-hydroxyphenyl)hepta-4,6-dien-3-one (**9-83**) was isolated from banana fruits (*Musa x paradisiaca*, Musaceae).

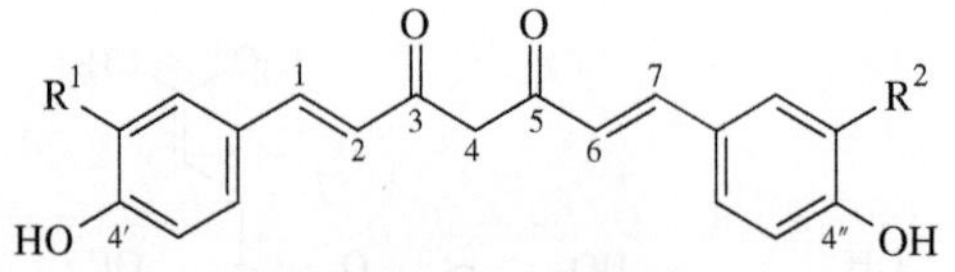

9-80, main turmeric pigments (keto forms)

curcumin, $R^1 = R^2 = OCH_3$
demethoxycurcumin, $R^1 = OCH_3$, $R^2 = H$
bisdemethoxycurcumin, $R^1 = R^2 = H$

The diketones form stable enols and are readily deprotonated to form enolates. Recombination of curcumin radicals formed in foods yields various dimers as radical termination products (Figure 9.22), in addition to the coupling products of curcumin with lipid hydroperoxides. In alkaline solutions, curcumin is degraded to ferulic acid, feruloylaldehyde (coniferyl aldehyde), feruloyl methane, and vanillin.

Curcumin, cassumunins, and cassumunarins exhibit various pharmacological effects and show potent anti-inflammatory, antioxidant, and anti-tumour-promoting activities. Curcumin has been shown to inhibit formation of advanced glycation end products (AGEs) by trapping reactive dicarbonyl compounds, such as methylglyoxal. The most favoured antioxidant is the enol form. Biotransformation of curcuminoids by human gut microbiota leads to a mixture of demethylation products. Curcumin is transformed to demethylcurcumin and subsequently to bisdemethylcurcumin.

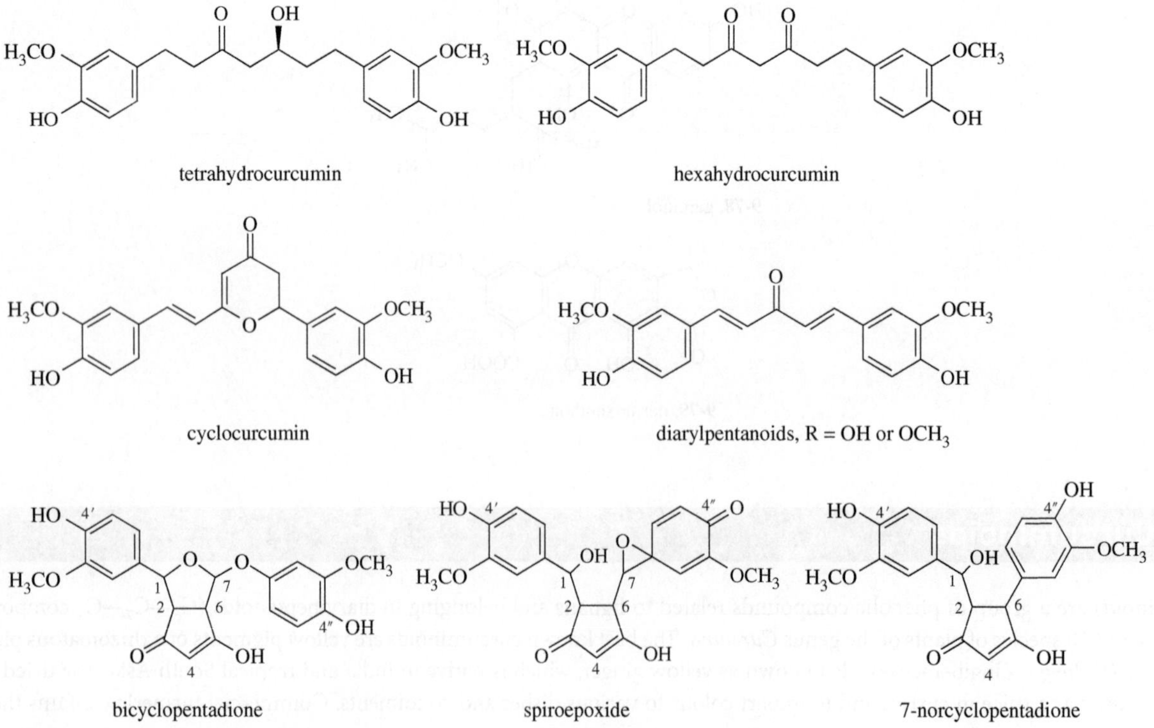

tetrahydrocurcumin

hexahydrocurcumin

cyclocurcumin

diarylpentanoids, R = OH or OCH_3

bicyclopentadione

spiroepoxide

7-norcyclopentadione

9-81, minority curcuminoids in turmeric

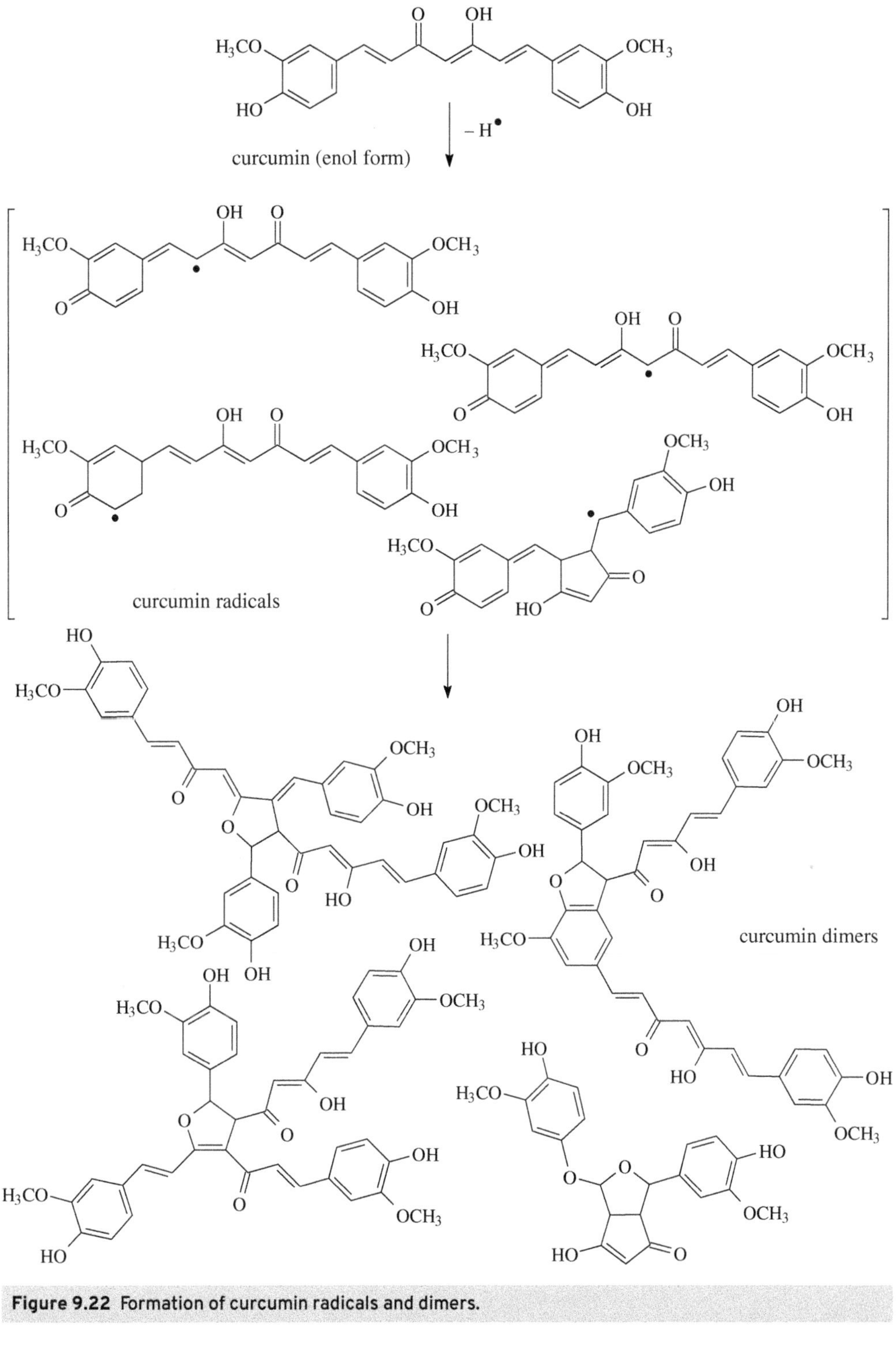

Figure 9.22 Formation of curcumin radicals and dimers.

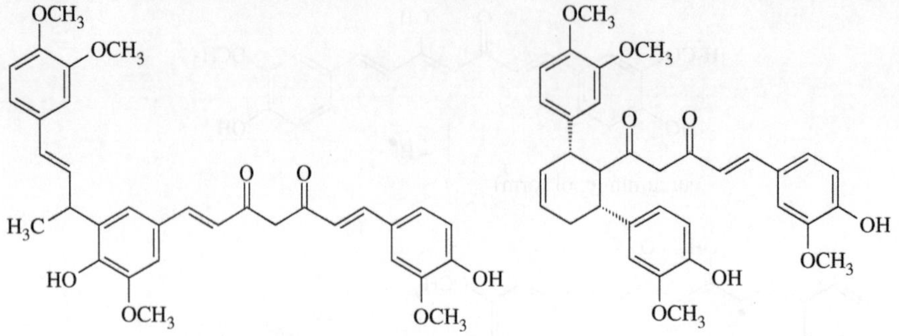

cassumunin A

cassumunarin A

9-82, cassumunins and cassumunarins

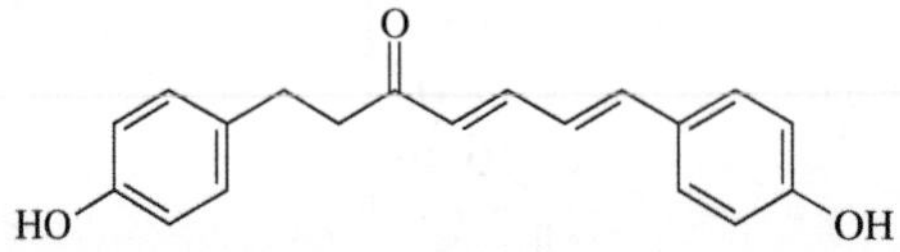

9-83, 1,7-bis(4-hydroxyphenyl)hepta-4,6-dien-3-one

9.7 Isochromenes

Red yeast rice (in China referred to as *ankak*, *anka*, or *ang khak* and in Japan as *red rice koji*) is the product of yeasts of the genus *Monascus*, especially *M. purpureus*, grown on white rice. The dried and powdered yeast-rice has been used in traditional Chinese medicine. Its colour is acquired from a mixture of yellow, orange, and red pigments derived from 1*H*-isochromene (**9-92**). In this mixture of lipophilic substances, the yellow pigments monascin and ancaflavin and their orange oxidised forms rubropunctatin and monascorubin (**9-92**) predominate. Some preparations of red yeast rice (monascus or monascus red) are used in food products in Chinese cuisine to impart colour to meat (including Peking duck), fish, soybean curd (tofu), rice wine (sake), beans, candy, and other foods.

Some red yeast rice products may contain mycotoxin citrinin, which is suspected of being a renal carcinogen. Yellow citrinin was originally mistakenly regarded as a non-toxic pigment called monascidin A (see Section 12.3.1). Some products also contain monacolins, dehydromonacolins, and structurally similar decalins. Monacolin K is identical to the active ingredient in the cholesterol-lowering drug lovastatin (**9-85**), which is one of the drugs in the category known as statins. These drugs lower blood cholesterol levels by reducing the production of cholesterol by the liver.

Native pigments react easily with amino compounds (amino acids, amino alcohols, amino sugars, and proteins) to form extracellular, water-soluble nitrogen analogues of native pigments, which are collectively called **azaphilones**. Representatives of azaphilones are the red to purple pigments rubropunctamine and monascorubramine (**9-86**).

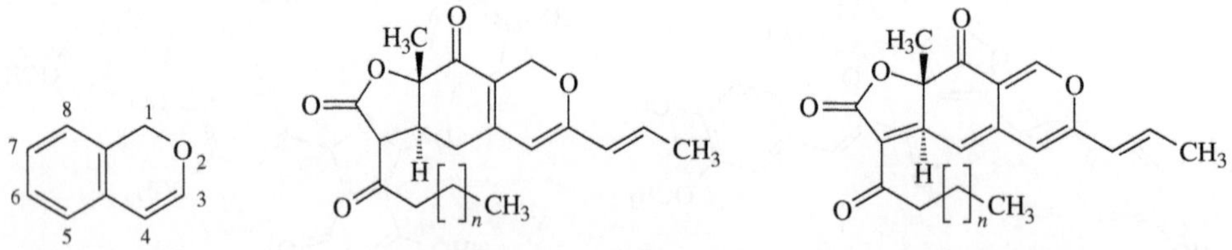

1*H*-isochromene

monascin, *n* = 3
ancaflavin, *n* = 5

rubropunctatin, *n* = 3
monascorubin, *n* = 5

9-84, isochromene and derived pigments

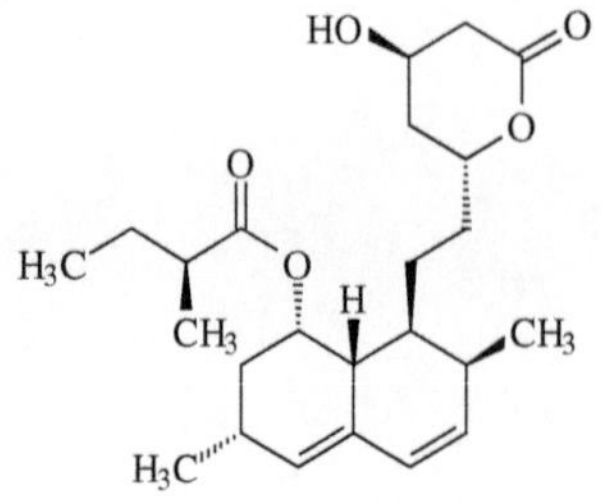

9-85, lovastatin

9-86, rubropunctamine, *n* = 3
monascorubramine, *n* = 5

9.8 Quinoid pigments

Food raw materials and foods of plant origin contain a large number of various phenolic substances that are transformed by biochemical and chemical reactions to coloured chinones and other pigments.

9.8.1 Quinones

Quinones represent a group of about 200 yellow, red, brown, and almost black pigments with variable structure (**9-87**). They include simple quinones, dimers, trimers, and condensation products that mutually differ in the number of hydroxyl groups and other substituents. The naturally occurring quinoid pigments are mostly derived from:

- benzo-1,4-quinone (also known as *p*-benzoquinone or 1,4-dioxobenzene);

- naphtho-1,4-quinone (1,4-dioxonaphthalene);

- anthra-9,10-quinone (9,10-dioxoanthracene).

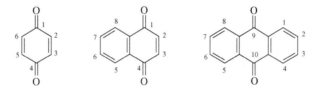

benzo-1,4-quinone   naphtho-1,4-quinone   anthra-9,10-quinone

9-87, basic skeletons of natural quinones

Many quinones have more complex structures. For example, a group of terpenoid quinones has been identified that includes biologically active plastoquinones, ubiquinones (coenzymes Q), tocopherols, and tocotrienols with vitamin E activity. Phylloquinone and its derivatives possess vitamin K activity (see Chapter 5).

In the past, some quinoid pigments were used as textile and leather dyes (they were later replaced by synthetic dyes), and many plants were specially grown for this function. Some quinoid pigments are now used for cosmetic and pharmaceutical purposes and also as food additives. Compared with other pigments, natural quinoid pigments are of less importance.

9.8.1.1 Benzoquinones

Benzo-1,4-quinoid structures occur in nature as the final oxidation products of various mono- and polycyclic compounds. Most simple benzo-1,4-quinones occur in microorganisms (moulds), higher fungi, and lichens, and less frequently in higher plants and some insects. Common substances are glycosides occurring in colourless reduced forms (leucoforms), commonly hydroxy derivatives and their glycosides and esters. Coloured quinones are formed from these precursors by hydrolysis catalysed by saccharases and by enzymatic oxidation or autoxidation of aglycones.

The basic compound is a pale-yellow benzo-1,4-quinone, which occurs in plants as a colourless hydroquinone β-D-glucoside called arbutin (**9-88**). This compound is an active component of leaves of the lingonberry (cowberry, *Vaccinium vitis-idaea*) and the related bearberry (*Arctostaphylos uva-ursi*, Ericaceae), native to the northern areas of the Northern Hemisphere. The drug is mainly used as a disinfectant and urinary antiseptic agent. Arbutin is found in the leaves of cowberry in quantities of 3.3–5.4% and in bearberry leaves in quantities of 4.2–7.7%. The aglycone hydroquinone, which arises via hydrolysis of arbutin by arbutase, is present at a low level (0.1–0.2% and 0.3%, respectively). Other related glycosides are methylarbutin (1.3%) and pyroside (hydroquinone 6′-acetyl-β-D-glucoside, **9-88**), hydroquinone β-gentiobioside and salidroside (tyrosol β-D-glucoside).

Arbutin is also found in small quantities in cereals (e.g. wheat and rice) and some fruits (especially in pears); its incidence in fruit juices is an indicator of the presence of pear juice. The arbutin derivative 6′-*O*-(4-hydroxybenzyl)arbutin occurs in marjoram (*Origanum majorana*, Lamiaceae). Also widespread in plants (e.g. wheat grains) are 2-methoxy and 2,6-dimethoxy arbutin derivatives, which may cause a pink discolouration in flour. In olives (*Olea europaea*, Oleaceae) a different type of quinoid glycoside occurs, called cornoside (**9-89**). Cornoside is related to oleuropein, which is a bitter substance in olives (see Section 8.3.5.1.1). Thymohydroquinone glycosides are precursors of thymoquinone (2-isopropyl-5-methyl-1,4-benzoquinone) in black caraway (cumin) seeds (*Nigella sativa*, Ranunculaceae).

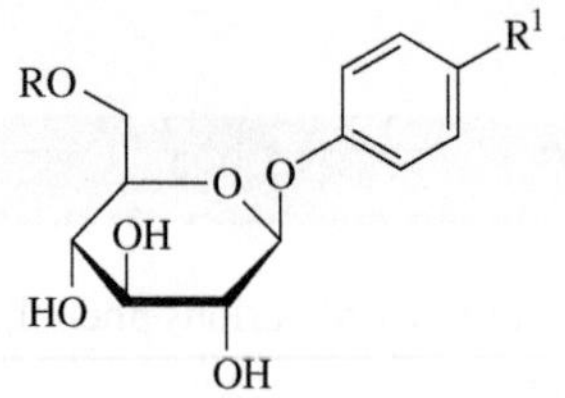

9-88, hydroquinone glucosides

arbutin, R = H, R^1= OH
methylarbutin, R = H, R^1 = OCH$_3$
pyroside, R = COCH$_3$, R^1 = OH

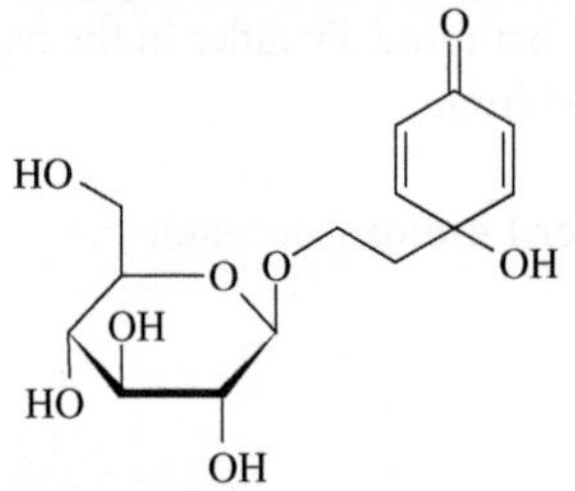

9-89, cornoside

Prenylated benzoquinones also occur, albeit rarely, in Boletales mushrooms. Suillin (**9-90**) occurring in weeping (granulated) bolete (*Suillus granulatus*, Suillaceae) is an example of acetylated and prenylated 1,2,4-trihydroxybenzenes. Its oxidation products are responsible for the brown colour of mushroom caps. Prenylated benzoquinones mostly appear as meroterpenoids named boviquinones. Boviquinone-4 (2,5-dihydroxy-3-geranylgeranyl-1,4-benzoquinone, **9-91**) is an example of a prenylated benzoquinone found in Jersey cow mushroom (*S. bovines*). The analogue of agaritine (see Section 10.3.12.3), found in mushrooms of the genus *Agaricus* and derived from 4-aminophenol, is γ-glutamyl-4-hydroxybenzene (**9-92**). It is readily oxidised, via γ-glutamyl-3,4-dihydroxybenzene, to the corresponding quinone. The quinone decomposes to 2-hydroxy-4-iminocyclohexa-2,5-dienone (**9-92**), which imparts a pink-red colour to some agarics, such as common mushroom (*A. bisporus*, Agaricaceae). The yellow nitrogen-containing pigment characteristic of the yellow staining of *A. xanthodermus* and some other species is agaricone (**9-92**), a metabolite of glutamic acid derived hydrazine agaritine. Agaricone forms by oxidation of the corresponding leucophenol in the damaged tissue.

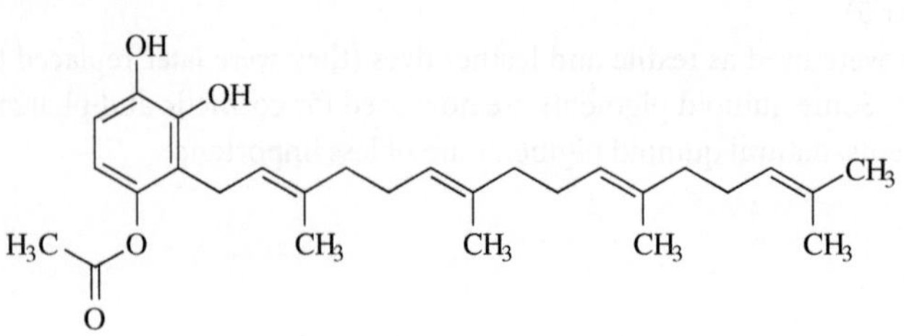

9-90, suillin

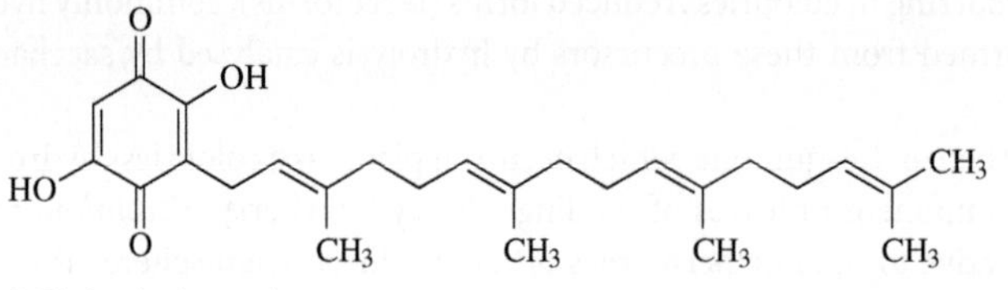

9-91, boviquinone-4

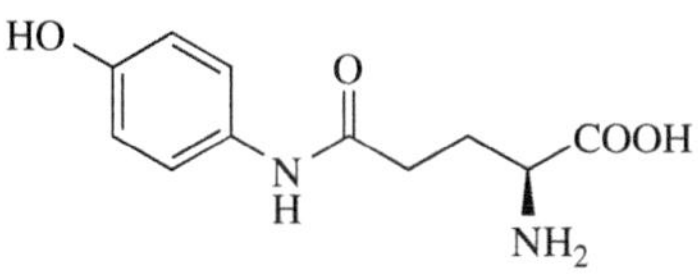

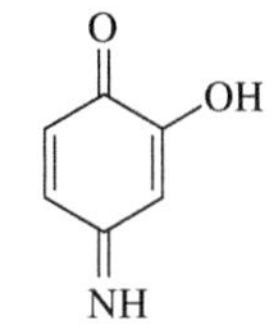

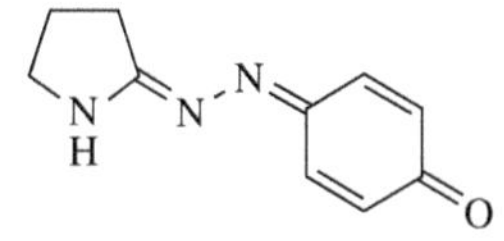

γ-glutamyl-4-hydroxybenzene   2-hydroxy-4-iminocyclohexa-2,5-dienone   agaricone

9-92, pigments and their precursors in mushrooms of the genus *Agaricus*

9.8.1.1.1 Terphenylquinones

Terphenylquinones (benzo-1,4-quinones substituted at positions C-2 and C-5 by phenyl groups) and their alkyl derivatives are red, violet to brown pigments of many species of lichens, moulds, and higher fungi. They often occur in leuco forms (such as dihydroxy derivatives or their acetates). Terphenylquinones, exemplified by the dark-red polyporic acid and bronze-brown atromentin (**9-93**), are mainly produced by wood-rotting higher fungi of the order Polyporales growing on various deciduous trees, but in other higher fungi, such as the Boletales fungi, they appear only sporadically. For example, polyporic acid, the parent compound of numerous terphenylquinones and related compounds, is the major component of *Hapalopilus nidulans*, amounting to up to 43% of its dry weight. The orange tooth (*Hydnellum aurantiacum*) colour is derived from atromentin and 3,6-dibenzoylatromentin, known as aurantiacin. Atromentin occurs in the intact fruit bodies in the form of colourless precursors (such as dihydroaurantiacin in *Hydnellum aurantiacum*). Atromentin is the key intermediate for many conversions leading to more hydroxylated terphenylquinones, such as variegatin and similar pigments. The yellow pigment muscarufin (**9-94**), a putative derivative of terphenylquinones, is one of the main pigments of the fly agaric (*Amanita muscaria*) hat. In dimeric forms, terphenylquinones contribute to the yellow colour of some lichens and the chocolate-brown hats of mushrooms of the order Boletales, which are characterised by a diversity of colours.

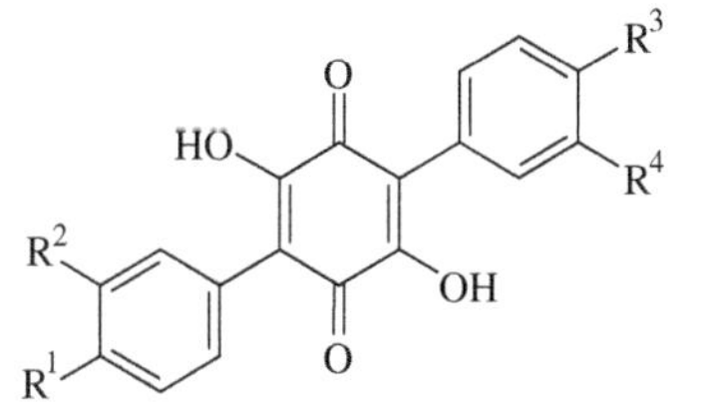

9-93, polyporic acid, $R^1 = R^2 = R^3 = R^4 = H$
atromentin, $R^1 = R^3 = OH$, $R^2 = R^4 = H$
variegatin, $R^1 = R^2 = R^3 = R^4 = OH$

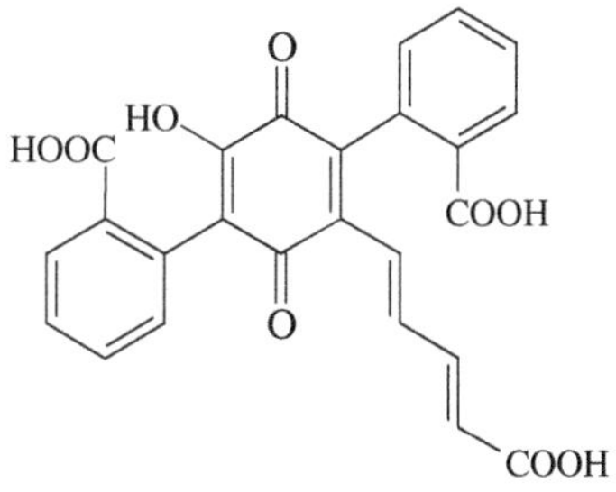

9-94, muscarufin

9.8.1.1.2 Pulvinic acids

Yellow pigments of lichens, moulds, and higher fungi called pulvinic (pulvic) acids (**9-95**) are formed, along with other products, from the corresponding terphenylquinones through lactone formation, after the terphenylquinone ring has been oxidised and opened. The unsubstituted parent compound, called pulvinic acid, only occurs in the form of its methyl ester, named vulpinic acid. Xerocomic and variegatic acids play the most important role, being responsible for the blue colours acquired in many boletes after their fruiting bodies are injured, which results in the oxidation of these acids to the corresponding blue quinonmethide anions (**9-96**). Pulvinic acids are especially widespread in mushrooms belonging to the genera *Gomphidius* (Gomphidiaceae) and *Suillus* (Suillaceae). The yellow pigment gomphidic acid was found for the first time in the slimy spike-cap (*G. glutinosus*). The yellow-brown cap and stem of the larch bolete (*S. grevillei*) contain at least 11 yellow, orange, and red pigments derived from decarboxylated pulvinic acids, of which 3′,4′,4-trihydroxypulvinone (**9-97**), derived from variegatic acid (**9-95**), is the major one. Simple oxidation products of terphenylquinones under the preservation of the central quinone ring are cycloleucomelone and cyclovariegatin (**9-98**). Cycloleucomelone occurs in the fruiting bodies of *Boletopsis leucomelaena* (Suillaceae), accompanied by a series of colourless analogues containing five, four, and three acetyl residues. Cyclovariegatin is a minor pigment of larch bolete and a precursor of violet thelephoric acid (**9-99**). Variegatorubin and xerocomorubin (**9-100**) exemplify red pigments formed from pulvinic acids by the second lactone ring formation. Derived from pulvinic acids are many other more complex structures, such as badione A, which is responsible, together with other polyphenols, for the chocolate-brown and golden-yellow colours of the cap skin of bay bolete (*Boletus badius*, Boletacaee; see Section 6.7.4).

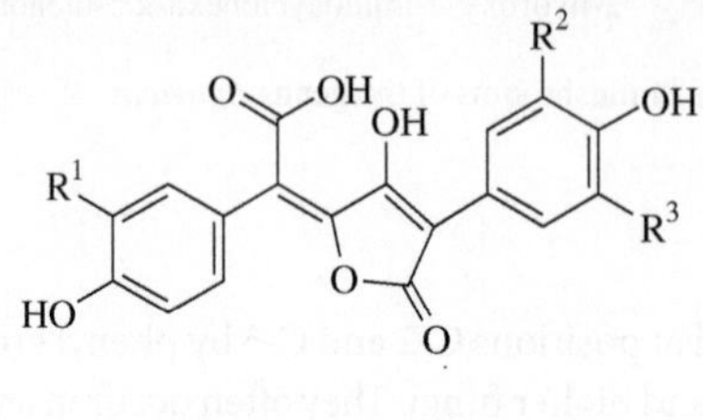

9-95, atromentic acid, R^1 = R^2 = R^3 = H
xerocomic acid, R^1 = H, R^2 = OH, R^3 = H
gomphidic acid, R^1 = H, R^2 = R^3 = OH
isoxerocomic acid, R^1 = OH, R^2 = R^3 = H
variegatic acid, R^1 = OH, R^2 = OH, R^3 = H

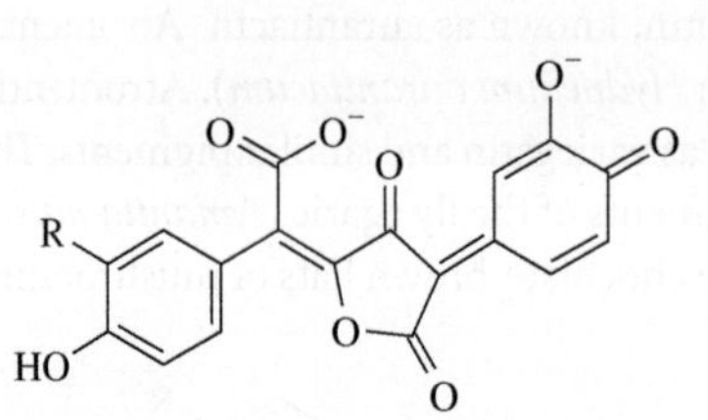

9-96, xerocomic acid quinonemethide anion, R = H
variegatic acid quinonemethide anion, R = OH

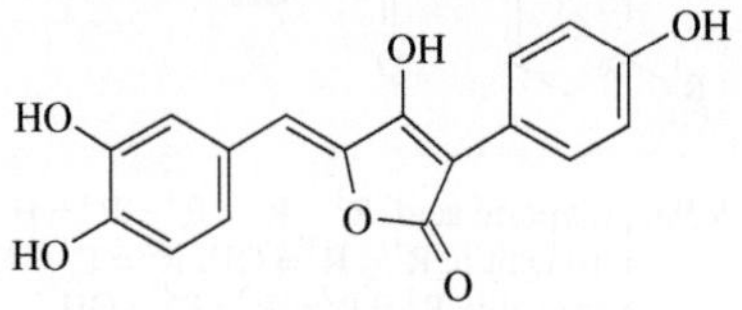

9-97, 3′,4′,4-trihydroxypulvinone

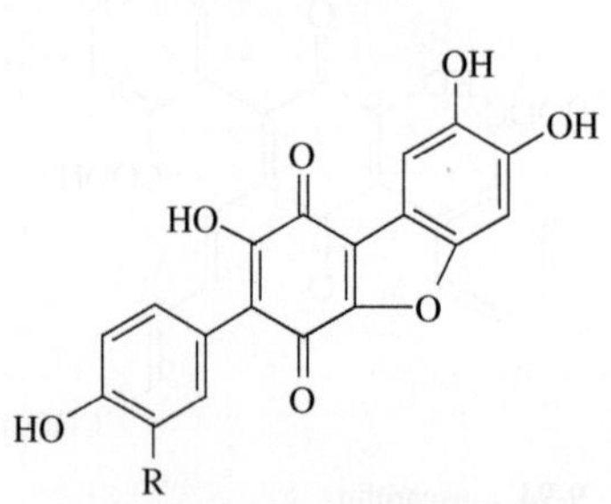

9-98, cycloleucomelone, R = H
cyclovariegatin, R = OH

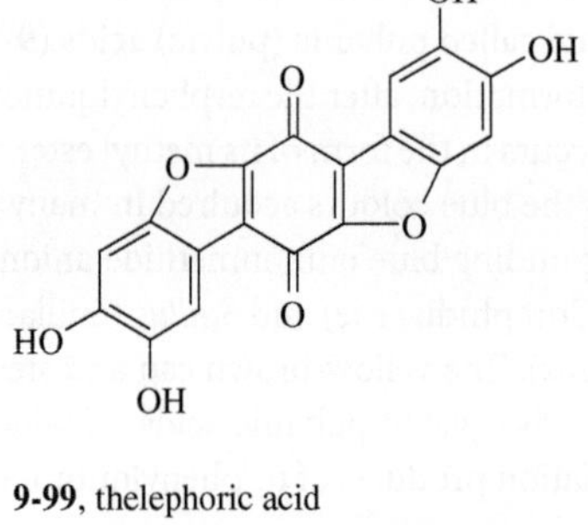

9-99, thelephoric acid

9-100, xerocomorubin, R = H
variegatorubin, R = OH

9.8.1.1.3 Troponoids

Troponoids are compounds derived from tropone (cyclohepta-2,4,6-trien-1-one, **9-101**), which contains as a basic skeleton a seven-membered ring with three conjugated double bonds and a carbonyl group. Troponoids arise by transformation of phenolic compounds and are synthesised mainly by fungi and higher plants – but they can also arise in non-enzymatic browning reactions. The basic compound with the enol hydroxyl group is 2-hydroxytropone (2-hydroxycyclohepta-2,4,6-trien-1-one), called tropolone (**9-101**). Possible pathways leading from quinone methide to the tropolone skeleton are outlined in Figure 9.23. Chinone methide arises either from a 2-methylphenyl radical and oxygen or from a benzyl radical and hydroperoxyl (perhydroxyl) radical.

tropone   tropolone

9-101, basic structures of troponoid pigments

Natural tropone analogues are represented by the brownish-red pigment purpurogallin (**9-102**), which occurs, for example, in oak apples, where it arises as a product of gallic acid (gallotannines) oxidation and decarboxylation. Enzymatic oxidation of gallic acid and elimination of carbon dioxide produces its dimer, 8-carboxypurpurogallin (**9-102**). A purpurogallin derivative, fomentariol (**9-102**), is produced by the plant pathogen known as the tinder fungus (*Fomes fomentarius*, Polyporaceae). Most troponoids are antibacterial and antifungal agents. Pigments of black tea with a tropone ring (such as theaflavins) are produced from catechins in enzymatic browning reactions.

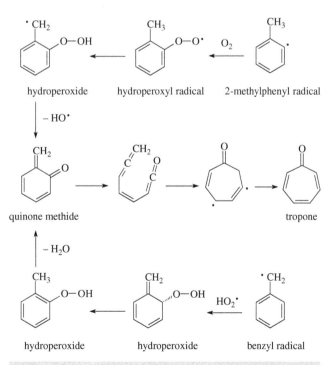

Figure 9.23 Formation of tropone ring from quinone methide.

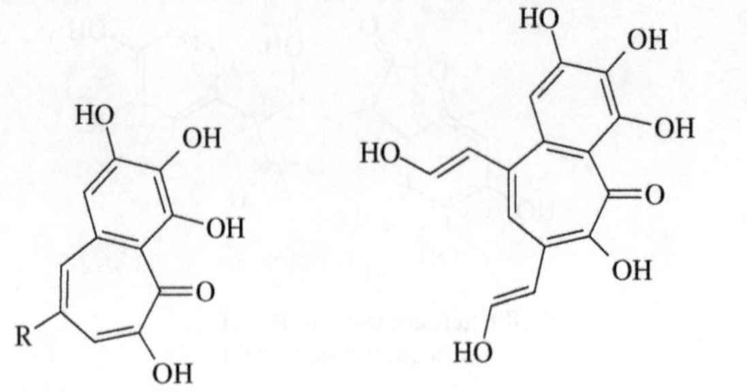

purpurogallin, R = H
8-carboxypurpurogallin, R = COOH

9-102, troponoid pigments

fomentariol

9.8.1.2 Naphthoquinones

Most naphthoquinones occur in higher plants in the form of glycosides of non-coloured reduced forms. Coloured naphthoquinones are released by glycosidases and enzymatic oxidation or autoxidation of aglycones.

A representative of simple naphthoquinones is yellow-orange lawsone (2-hydroxynaphtho-1,4-quinone, **9-103**), which occurs as a mixture of several naphthoquinones in henna (*Lawsonia inermis*, Lythraceae), growing in the Middle East and India, but also located in many other plants. The pigment, which exhibits antimicrobial effects, is obtained from the leaves of henna bushes, used for centuries in decorating hands, feet, nails, and skin and as a hair dye.

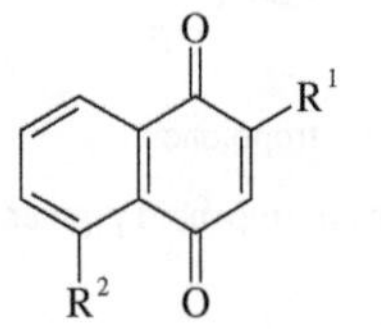

9-103, lawsone, R^1 = OH, R^2 = H
juglone, R^1 = H, R^2 = OH
plumbagine, R^1 = CH$_3$, R^2 = OH

Red-brown juglone (5-hydroxynaphtho-1,4-quinone, **9-103**) is present in the leaves and immature fruits of walnuts (*Juglans regia*, Juglandaceae) and other species as a colourless 4-β-D-glucopyranoside of 1,4,5-trihydroxynaphthalene (1,5-dihydroxy-4-naphthalenyl-β-D-glucopyranoside), known as hydrojuglone. Juglone arises either by hydrolysis of the glycoside and oxidation of aglycone or by glycoside oxidation and its subsequent hydrolysis. Juglone is an allelopatic substance that inhibits the growth of many other plants and is responsible for the change in skin colour to yellow-brown on handling of unripe walnuts. Its homologue is a yellow pigment, plumbagine (5-hydroxy-2-methylnaphtho-1,4-quinone, **9-103**), which occurs in different parts of the European common leadwort (*Plumbago europaea*, Plumbaginaceae). Plumbagine is also found in the common sundew (*Drosera rotundifolia*, Droseraceae) and other medical plants. Juglone and plumbagine may produce dermatitis in susceptible people, and juglone has a laxative effect in the same way as some anthraquinones.

The roots of alkanet, also known as common bugloss (*Alkanna tinctoria*, syn. *Anchusa tinctoria*, Boraginaceae), which grows in the south of Europe, contain the reddish pigment alkannin (alkannet, **9-104**), once used as a textile dye. The red-brown (S)-isomer of alkannin is today used as a food colouring E103, imparting colour to ice creams, sweets, and other products. The red (R)-isomer called shikonin was isolated from the roots of the medicinal plant *Lithospermum erythrorhizon* of the same family, which is native to China.

Naphthalene and naphthoquinone structures are also widespread as red, purple, brown, and black pigments in higher fungi of the phylum Ascomycota. For example, xylariaceous wood fungi of the genus *Daldinia* (Xylariaceae) are dominated by metabolites produced by oxidative coupling of colourless naphthalene-1,8-diol. Examples of these structures are the 4,4′-binaphthyl derivatives 1,1′,8,8′-tetrahydroxybinaphthyl and its ether 1,8-dihydroxy-1′,8′-dimethoxybinaphthyl (**9-105**), as well as more condensed brown-to-black pigments called perylenequinones. A simpler perilenequinone known as hypocrellin (**9-106**), isolated from the bamboo fungus *Hypocrella bambusae* (Hypocreaceae), is a photodynamic pigment.

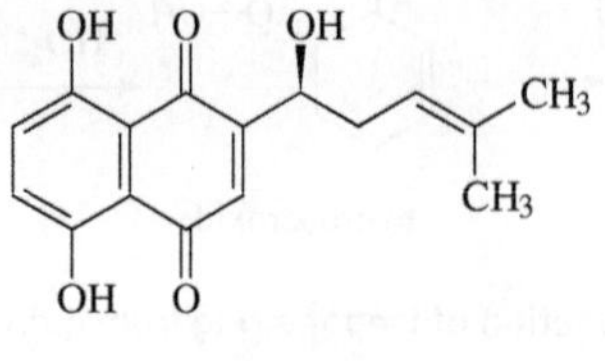

9-104, alkannin

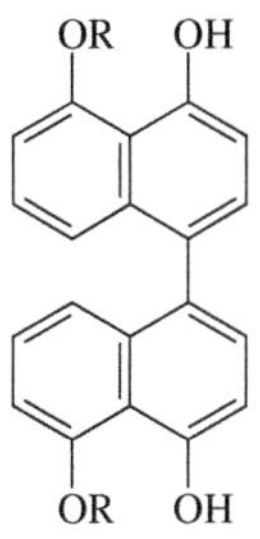

9-105, 1,1′,8,8′-tetrahydroxy-4,4′-binaphthyl, R = H
1,8-dihydroxy-1′,8′-dimethoxy-4,4′-binaphthyl, R = CH$_3$

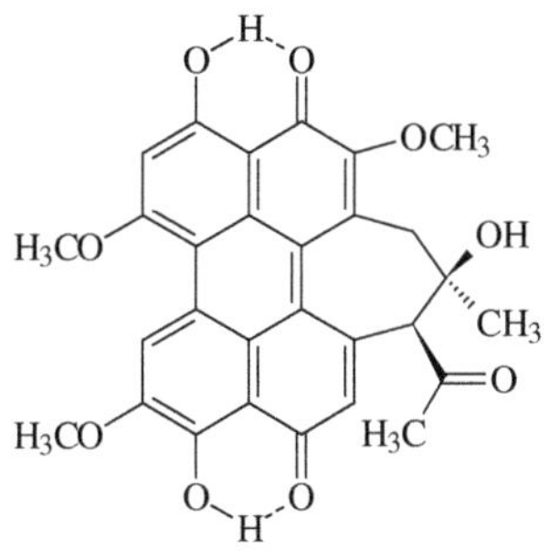

9-106, hypocrellin

9.8.1.3 Anthraquinones

Anthraquinones are the most widespread group of natural quinones. They occur in higher plants, fungi, lichens, and some insects. They are accompanied by substituted anthranols and anthrahydroquinones (**9-107**) and their oxo forms anthrones and oxanthrones (**9-108**), which are, as glycosides, precursors of anthraquinone pigments.

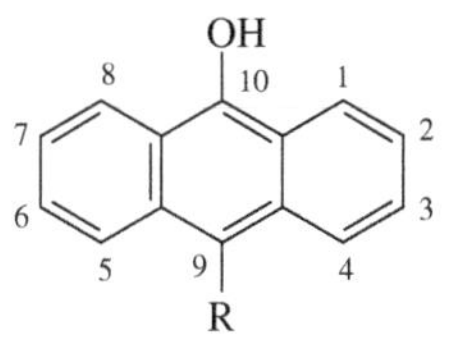

9-107, anthranol, R = H
anthrahydroquinone, R = OH

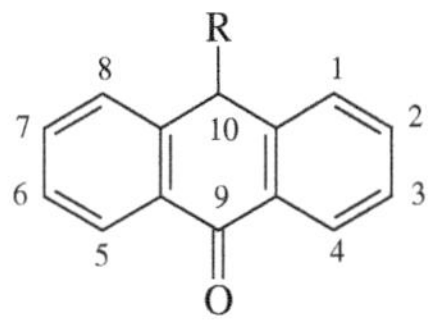

9-108, anthrone, R = H
oxanthrone, R = OH

The most familiar single anthraquinone, substituted with hydroxyl groups on only one benzene ring, is yellow-orange 1,2-dihydroxyanthraquinone, called alizarin (or Turkish Red or Mordant Red 11). Also known is red pigment 1,2,4-trihydroxyalizarin, called purpurin (**9-109**), which occurs in roots of the common madder (*Rubia tinctorum*, Rubiaceae). Other anthraquinone aglycones in madder are lucidin and quinizarin (**9-109**). The corresponding alizarin glycoside is ruberythric acid (**9-110**), the sugar moiety of which is primeverose (6-*O*-β-D-xylopyranosyl-β-D-glucopyranose), bound to the C-2 alizarin hydroxyl group. Another main glycoside is lucidin-3-*O*-primeveroside. The pigments arise from glycosides by fermentation or acid hydrolysis. The related anthraquinone anthragallol (**9-109**) occurs with its hydroxy derivatives, methyl ethers, glycosides, and other derivatives in a number of plants.

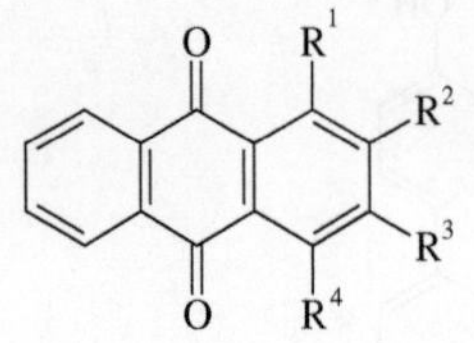

9-109, alizarin, R^1 = OH, R^2 = OH, R^3 = H, R^4 = H
purpurin, R^1 = OH, R^2 = OH, R^3 = H, R^4 = OH
lucidin, R^1 = OH, R^2 = CH$_2$OH, R^3 = OH, R^4 = H
quinizarin, R^1 = OH, R^2 = H, R^3 = H, R^4 = OH
anthragallol, R^1 = OH, R^2 = OH, R^3 = OH, R^4 = OH

Alizarin is used as a dye mostly for textile purposes, but also for food colouring. Purpurin is normally not coloured, but is red when dissolved in alkaline solutions. Mixed with clay and treated with alum and ammonia, it gives a brilliant red colourant, madder lake (**9-111**). Alum is a hydrated double sulfate salt of aluminum with the general formula $XAl(SO_4)_2 \cdot 12\ H_2O$, where X is a monovalent cation such as K^+ or NH_4^+.

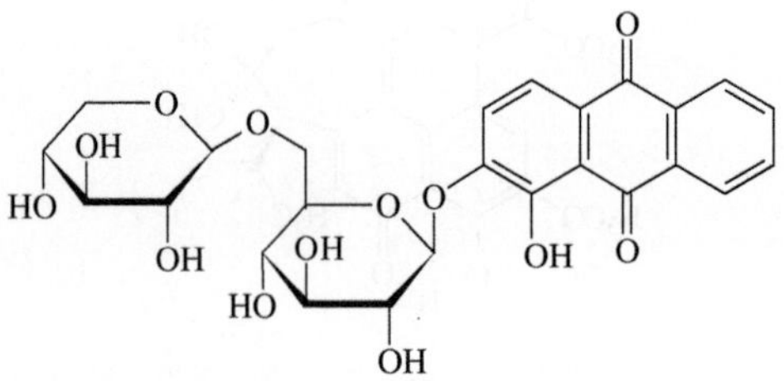

9-110, ruberythric acid

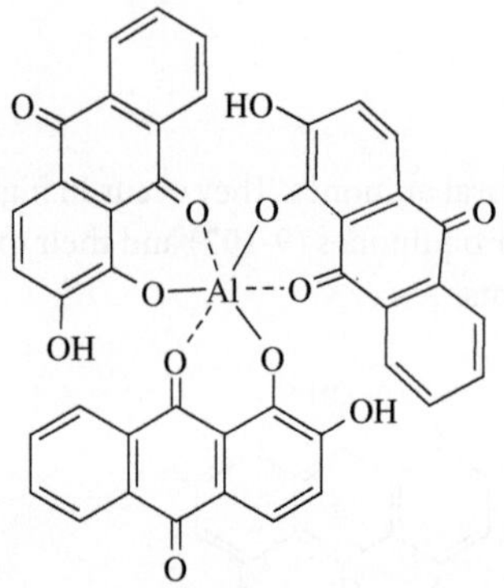

9-111, madder lake structure

9.8.1.3.1 Emodins

A very important group of polyhydroxysubstituted anthraquinones, substituted at both aromatic rings, is the emodins (**9-112**). A common feature of these anthraquinones is the presence of at least two OH groups (at C-1 and C-8) and a methyl group (at C-3) or its oxidised forms (hydroxymethyl or carboxyl group).

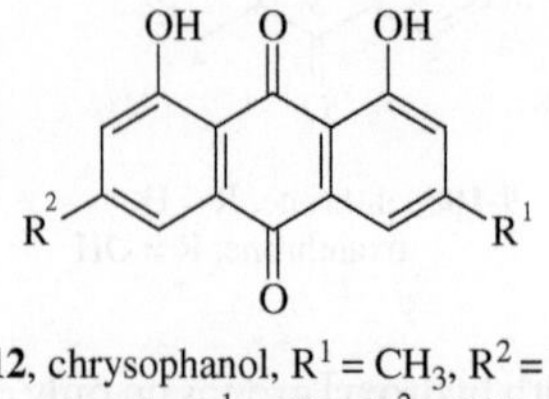

9-112, chrysophanol, R^1 = CH$_3$, R^2 = H
emodin, R^1 = CH$_3$, R^2 = OH
aloe-emodin, R^1 = CH$_2$OH, R^2 = H
rhein, R^1 = COOH, R^2 = H
parietin, R^1 = CH$_3$, R^2 = OCH$_3$

Emodin (formerly also known as frangula-emodin) is widespread in fungi, lichens, and higher plants, some of which are used in medicine. It occurs, for example, in the bark and fruits of common shrubs of the genus *Frangula* (such as alder buckthorn, *F. alnus*) and *Rhamnus* (such as buckthorn, *R. cathartica*) of the Rhamnaceae family and in the leaves and pods of Egyptian senna (*Senna alexandrina*, syn. *Cassia angustifolia*, Fabaceae). The aqueous extract of Egyptian senna is allowed as a dye for cigarette paper. Chrysophanol, emodin, rhein, parietin (also known as physcion), and chrysarone (**9-113**) are found in the red crisp stalks and especially the roots of various species of rhubarb (*Rheum* spp., Polygonaceae). Aloe-emodin is found in some aloe species (*Aloe* spp., Xanthorrhoeaceae) and is synthesised by the mold *Aspergillus wentti*, which is used for the production of pectolytic enzymes. Parietin occurs in lichens of the genus *Xanthoria*.

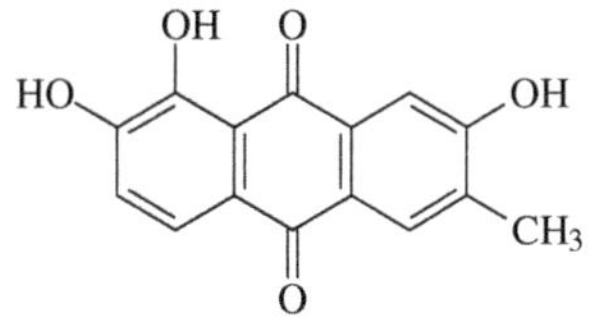

9-113, chrysarone

Like other quinones, emodins occur in nature as glycosides. For example, in rhubarb (*R. rhabarbarum*, syn. *R. undulatum*, and false rhubarb *R. rhaponticum*), chrysophanol is present as 1-*O*-β-D-glucopyranoside, called chrysophanein or chrysophaniin, and parietin occurs as 1-*O*-glucoside, known as physcionin. The same glycosides and chrysophanol-8-*O*-glucoside, called pulmatin, occur in *R. pulmatum*. In young leaves, glycosides derived from anthranols are prevalent, whilst in older leaves, anthraquinone glycosides prevail.

Plants of the genus *Aloe* contain a reduced form of aloe-emodin as the *C*-glucoside of the corresponding anthrone, which is called aloin. Lemon-yellow aloin is a mixture of two isomers. The major product, (*S*)-aloin or (*S*)-β-D-glucopyranoside-aloe-emodin, is referred to as aloin A (or barbaloin, **9-114**), whilst the (*R*)-isomer is aloin B (isobarbaloin). Related glycosides, called aloinosides *a* and *b*, contain α-L-rhamnopyranose, bound by an *O*-glycosidic bond to the hydroxymethyl group at C-3.

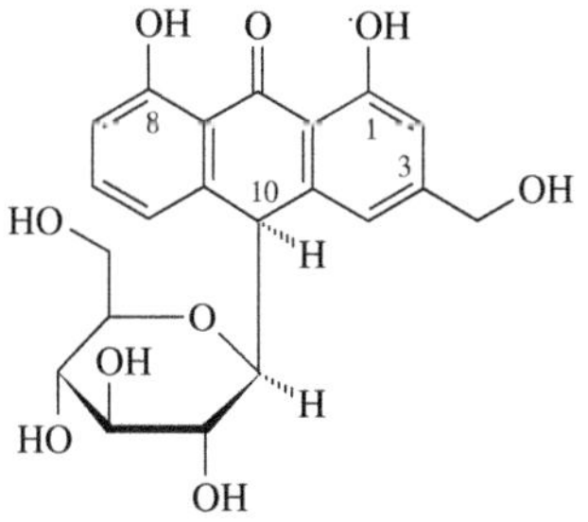

9-114, aloin A

Some emodins have a laxative effect, and related purgative drugs (containing about 10% of aloin) were used for centuries in medicine. *Aloe* spp. plants are largely used in foods and beverages as flavoring agents. Because of the averse pharmacological effects of aloe constituents on consumers, the EEC lists aloin as a marker of *Aloe* occurrence in food and limits its content to 0.1 mg/kg in foods, 0.1 mg/l in beverages, and 50 mg/l in alcoholic beverages. The choice of aloin as a marker of aloe in beverages is inadequate because aloin is unstable in solutions. The degradation products include aloin dimers and trimers and aloe-emodin, so the more stable 5-methylchromones aloesin and aloeresin A (**9-115**) have been suggested as alternative markers.

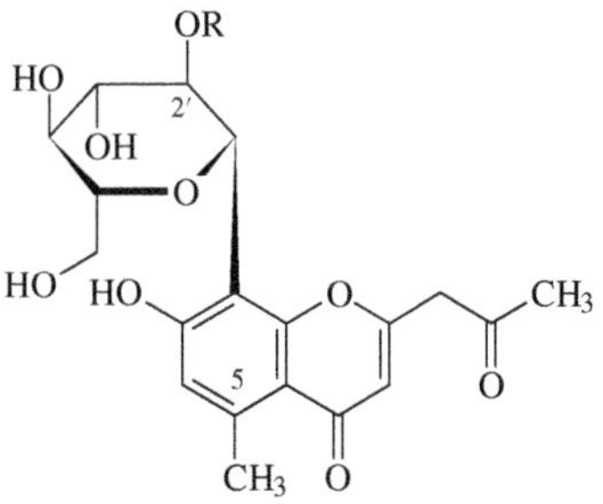

aloesin, R = H
aloeresin A, R = *p*-coumaroyl

9-115, 5-methylchromones of *Aloe* spp.

Fungi contain a range of anthraquinones with both rings substituted. In many cases, anthraquinones are found in fungi as their colourless reduced forms, which may occur as glycosides. Many natural anthraquinones are oligomers formed by the coupling of two or more anthraquinone molecules. These oligomers further differ in the points through which monomers are attached and may have more than one polymorphic form. For example, the major dark-orange pigment of the European toadstool *Cortinarius cinnabarinus* (Cortinariaceae) is fallacinol (teloschistin, 6-*O*-methoxycitreorosein, **9-116**). Several species of the genera *Cortinarius*, *Dermocybe*, and *Tricholoma*, such as *C. cinnamomeoluteus*, trivially known as man on horseback, produce a bright-yellow dimeric anthraquinone flavomannin-6,6′-di-*O*-methyl ether (**9-117**). This pigment arises by 7,7′-coupling of the corresponding green dihydroanthracenone (*R*)-torosachrysone (**9-118**). In its homochiral form, it occurs in the European *C. citrinus* and *C. croceus*, whilst in Australian fungi, it forms a mixture of (3*R*,3*R*′,*M*)- (**9-117**) and (3*R*,3*R*′,*P*)-atropoisomers. The green (*R*)-atrochrysone (**9-118**) occurs in *C. atrovirens* and *C. odoratus*, whilst (*S*)-torosachrysone-8-*O*-methyl ether is a constituent of *C. fulmineus*, *C. citrinus*, *C. splendens*, *T. equestre*, and the sulfur knight *T. sulfureum*.

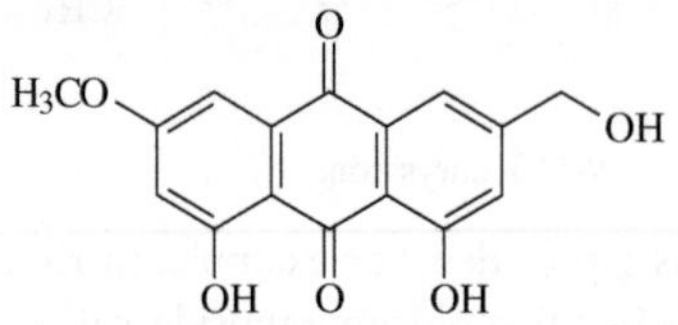

9-116, fallacinol

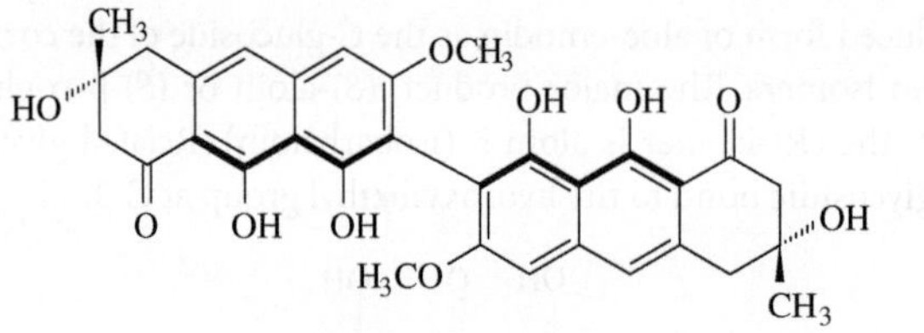

9-117, (3*R*,3*R*′,*M*)-flavomannin-6,6′-di-*O*-methyl ether

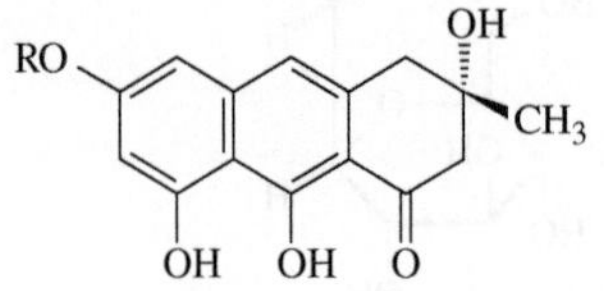

9-118, (*R*)-atrochrysone, R = H
(*R*)-torosachrysone, R = CH₃

9.8.1.3.2 Bianthrones and related compounds

In addition to emodins, rhubarb root also contains dimeric reduced forms of emodins derived from 10,10′-bianthronyl (**9-119**), which are called bianthrones, dianthrones, or dianthraquinones. These can occur as homobianthrones, such as emodin bianthrone and parietin bianthrone, or as heterobianthrones (mixed dimers), such as palmidin A (**9-120**), which is bianthrone of emodin and aloe-emodin.

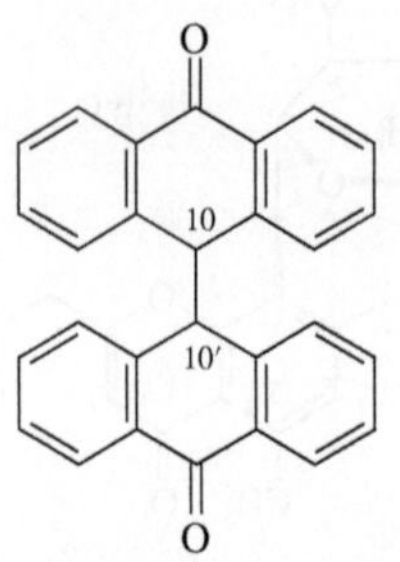

9-119, 10,10′-bianthronyl

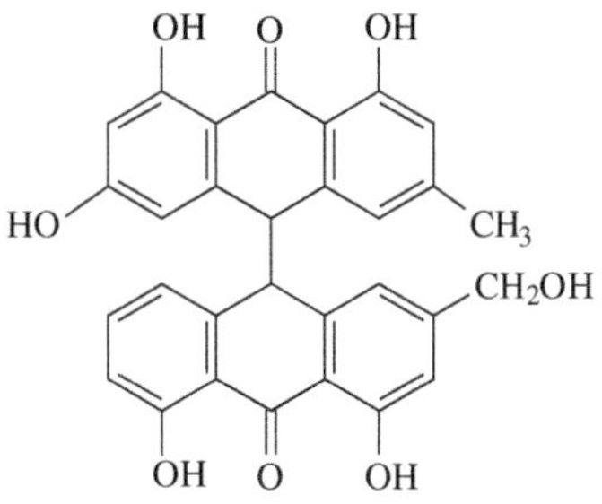

9-120, palmidin A

The related sennosides A, B, C, and D (**9-121**) occur in amounts of 1.5–3% in the leaves of Egyptian senna shrubs (*Senna alexandrina*, Fabaceae), cultivated primarily in India. Senna leaves were used by Arab physicians as a laxative drug and digestive stimulant. The heterobianthrones sennoside A (*meso*-derivative) and B (*trans*-derivative) are glycosides derived from aloe-emodin and rhein, whilsts the homobianthrones sennoside C (*meso*-derivative) and D (*trans*-derivative) are derived from rhein.

An even more condensed derivative of emodin bianthrone is the purple photodynamic pigment hypericin, 1,3,4,6,8,13-hexahydroxy-10,11-dimethylphenanthro[1,10,9,8-*opqra*]perylen-7,14-dione (**9-122**), which occurs in the flowers of St John's wort (*Hypericum perforatum*, Hypericaceae), a yellow-flowering perennial herb indigenous to Europe, and other *Hypericum* species. Hypericin is accompanied by the related compounds pseudohypericin and isohypericin (**9-122**) and their precursors protohypericin and protopseudohypericin (**9-123**), phloroglucinols (hyperforin and adhyperforin, **9-124**), flavonol glycosides, and biflavones. It is a phototoxic substance that can act as a photosensitiser upon excitation with visible light. Photosensitivity is often seen in animals that have been allowed to graze on St John's wort. Hyperflorin and its derivative adhyperflorin act as inhibitors of catecholamines (such as serotonin) and have antidepressant effects; St John's wort is widely known as a herbal treatment for depression. The presence of hypericin in food is consequently regulated by legislation. According to Regulation (EC) No 1334/2008, flavourings and food ingredients with flavouring properties produced from *H. perforatum* may only be used for the production of alcoholic beverages.

The fagopyrins (**9-125**) are a mixture of red fluorescent naphthodianthrones structurally related to hypericin that occur in common buckwheat (*Fagopyrum esculentum*, Polygonaceae), mainly in the leaves (445–636 mg/kg d.w) and stems (143–264 mg/kg d.w). The individual phagopyrins differ in that they contain, in positions 2 and 5, either pyrrolidine or piperidine moieties, or a combination of the two; the same can be seen in fagopyrin B ($R^1 = R^2$ = 2-piperidinyl, $R^3 = R^4$ = H), fagopyrin C (R^1 = 2-piperidinyl, R^2 = 2-pyrrolidinyl, $R^3 = CH_3$, R^4 = H), fagopyrin D ($R^1 = R^2$ = 2-piperidinyl, $R^3 = CH_3$, R^4 = H), and fagopyrin E (R^1 = 2-pyrrolidinyl, R^2 = 2-piperidinyl, $R^3 = R^4 = CH_3$). They are accompanied by their precursors, such as protofagopyrin, which is related to protohypericin. Like hypericin, the fagopyrins are phototoxic agents with anticarcinogenic properties. Diets extensively composed of buckwheat sprouts, herbs, and particularly flowers or of fagopyrin-rich buckwheat extracts may cause fagopyrism (buckwheat poisoning). Fagopyrism has primarily been observed in white and white-spotted domestic animals (sheep, pigs, cattle, and horses), but rare human cases have also been reported. Symptoms of fagopyrism are manifested by a skin irritation.

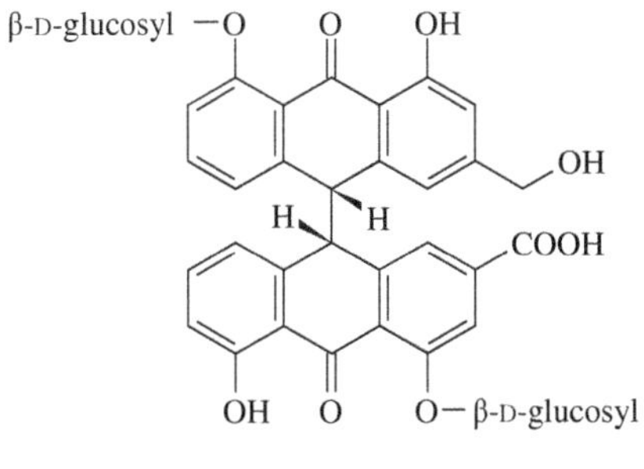

sennoside A, R = COOH
sennoside C, R = CH₂COOH

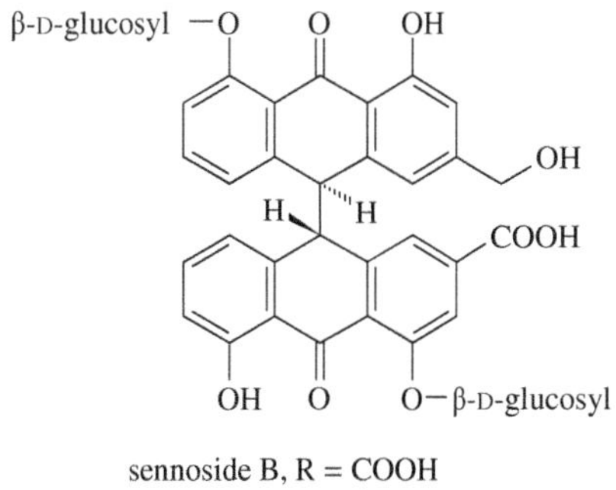

sennoside B, R = COOH
sennoside D, R = CH₂COOH

9-121, sennosides

9-122, hypericin, R¹ = OH, R² = CH₃, R³ = CH₃
pseudohypericin, R¹ = OH, R² = CH₃, R³ = CH₂OH
isohypericin, R¹ = CH₃, R² = OH, R³ = CH₃

9-123, protohypericin, R¹ = OH, R² = CH₃, R³ = CH₃
protopseudohypericin, R¹ = OH, R² = CH₃, R³ = CH₂OH

9-124, hyperflorin, R = CH₃
adhyperflorin, R = CH₂CH₃

general structure
9-125, fagopyrins

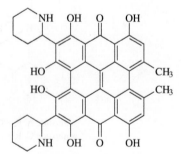

fagopyrin A, R¹ = R²= 2-pyrrolidinyl   fagopyrin F, R¹ = R² = 2-piperidinyl

9.8.1.3.3 Carmine pigments

Carmine is a natural red food colour (also known as Natural Red 4) isolated from the coccid insect *Dactylopius coccus* (*Coccus cacti*, Dactilopiidae), which lives as a parasite on cacti of genera *Opuntia* and *Nopalea*, especially *N. coccinellifera* (Cactaceae), originally grown in Mexico. The E number E120(i) is the pure colour, whilst E120(ii) is the crude extract. The term **cochineal** is used to describe both the fertilised females of the coccid insect and the pigment obtained from the same.

The main component of cochineal is carminic acid (**9-126**), a water-soluble and stable substance that is the *C*-glucoside of kermesic acid (**9-127**). The shade of colour in solution depends on pH, being orange at pH 3, red at pH 5.5, and purple at pH 7. The colour intensity is relatively low, so carminic acid is used mainly as the red aluminum salt called carmine lake or crimson lake (containing about 50% of the acid). It is prepared from powdered insect bodies hydrolysed in diluted ammonia (or sodium carbonate) and by the addition of alum, and is used in the colouring of some aperitifs (vermouths), meat products, specialty bakery and confectionery products, jams, and dairy products.

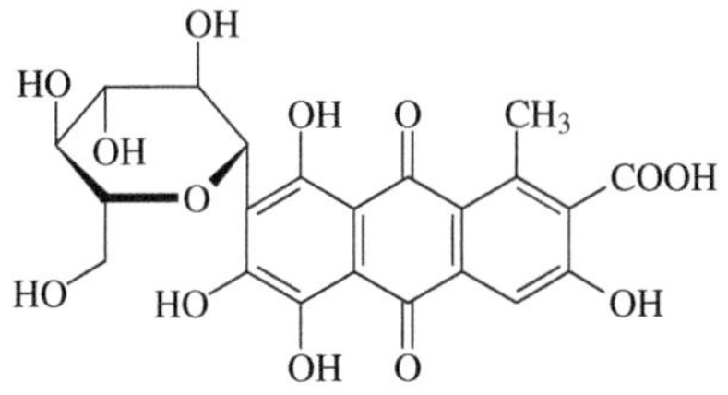

9-126, carminic acid

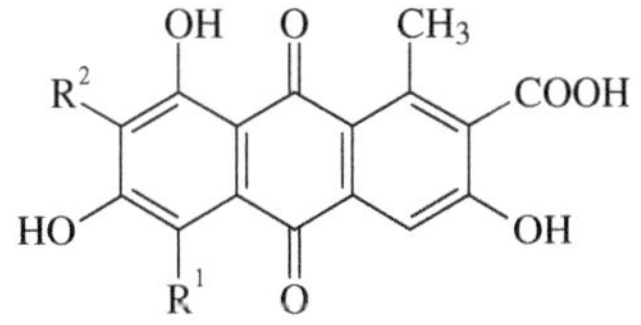

9-127, kermesic acid, $R^1 = OH$, $R^2 = H$
ceroalbolinic acid, $R^1 = H$, $R^2 = OH$

A similar pigment to carmine is **kermes**, which was mainly used in the Middle Ages, in the Mediterranean region. Kermes is derived from European species of insects, mainly *Kermes vermilio and K. illicis* (Kermesidae), living on the kermes oak (*Quercus coccifera*, Fabaceae). The aglycones of kermes are kermesic acid (the same aglycone as in carminic acid) and its isomer, ceroalbolinic acid (**9-127**).

Lac is the scarlet secretion of a number of species of insects of several genera, the most commonly cultivated of which is *Kerria lacca* (syn. *Laccifer lacca*, Kerriidae). The main producers of lac are India and Malaysia. Lac pigment is a complex mixture of laccaic acids (**9-128** and **9-129**), erythrolaccin, and deoxyerythrolaccin (**9-129**).

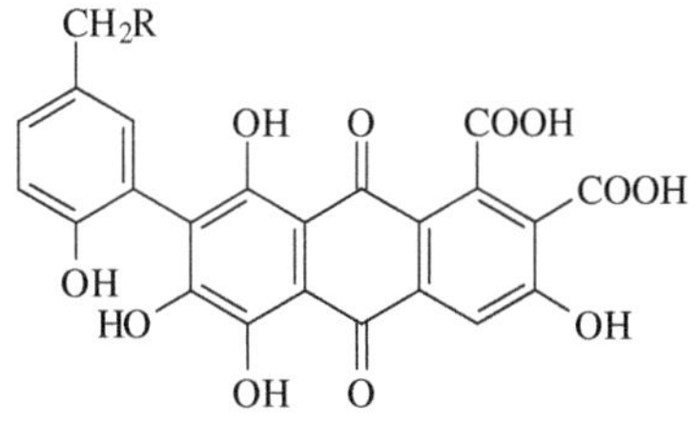

9-128, laccaic acid A, $R = CH_2NHCOCH_3$
laccaic acid B, $R = CH_2OH$
laccaic acid C, $R = CH(NH_2)COOH$
laccaic acid E, $R = CH_2NH_2$

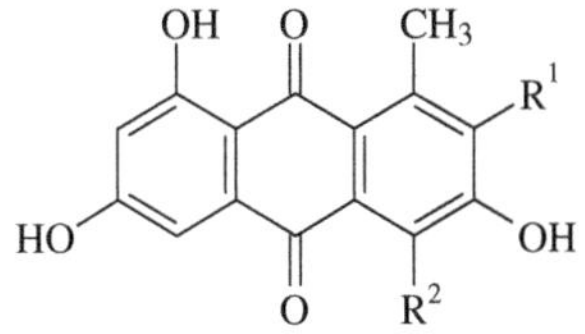

9-129, laccaic acid D, $R^1 = COOH$, $R^2 = H$
erythrolaccin, $R^1 = H$, $R^2 = OH$
deoxyerythrolaccin, $R^1 = R^2 = H$

9.9 Carotenoids

Carotenoids are widespread yellow and orange (rarely yellow-green and red) lipophilic pigments of plants, fungi, algae, microorganisms, and some animals (crustaceans, fish, birds, and mammals). Their annual production in nature is estimated at 1×10^8 tons. In plants, carotenoids occur in plastids called chromoplasts. Today, about 700 naturally occurring carotenoid pigments are known. Of this number, about 50 may act as provitamins A.

9.9.1 Structure and nomenclature

Most carotenoid compounds are tetraterpenoids formally containing eight isoprene units. They owe their colour to a chain of conjugated double bonds, which occurs in several basic structures and their combinations. Carotenoids are divided into two main groups:

- hydrocarbons called **carotenes**;

- compounds carrying one or more oxygen atoms (such as alcohols, aldehydes, ketones, and epoxides), which are known as **oxocarotenoids** or **xanthophylls**.

9.9.1.1 Carotenes

The simplest prototype of acyclic carotenes (**9-130**), called ψ-**carotenes**, is the polyunsaturated acyclic hydrocarbon (15Z)-7,8,11,12,7′,8′,11′, 12′-octahydro-ψ,ψ-carotene, known as phytoene. Isomerisation of phytoene yields the *trans*-isomer phytofluene (7,8,11,12,7′,8′-hexahydro-ψ,ψ-carotene). Oxidation of phytofluene gradually gives ζ-carotene (7,8,7′,8′-tetrahydro-ψ,ψ-carotene), neurosporene (7,8-dihydro-ψ,ψ-carotene), and lycopene (ψ,ψ-carotene), which is the main pigment of tomatoes (30–200 mg/kg), watermelons (33–121 mg/kg), and rose hips (101–834 mg/kg f.w.). Acyclic carotenes, with the exception of lycopene, are found in small quantities in foods. They accompany alicyclic carotenes and xanthophylls, which are the main carotenoids.

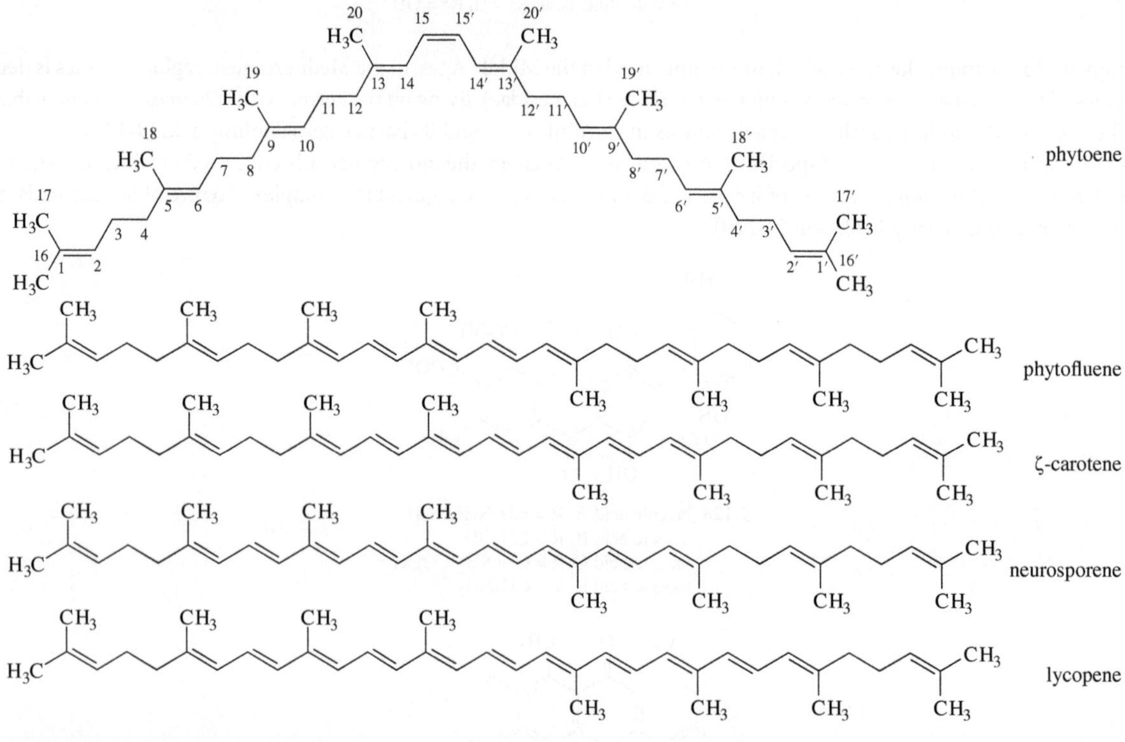

9-130, acyclic carotenes

Alicyclic carotenes are formed by enzymatically catalysed cyclisation at one or both ends of the acyclic ψ-carotenes, which leads to formation of a β-ionone structure in β-**carotene**s or of an α-ionone structure in ε-**carotene**s (Figure 9.24). Examples of hydrocarbons with a β-ionone ring at only one end of the molecule are β-zeacarotene and γ-carotene (β,ψ-carotene). Cyclisation at both ends produces structures that are present, for example, in β-carotene, α-carotene, or α-zeacarotene. β-Carotene with two β-ionone rings is therefore systematically called β,β-carotene, whilst α-carotene is called β,ε-carotene, because it has only one β-ionone ring and one α-ionone ring. Carotene with

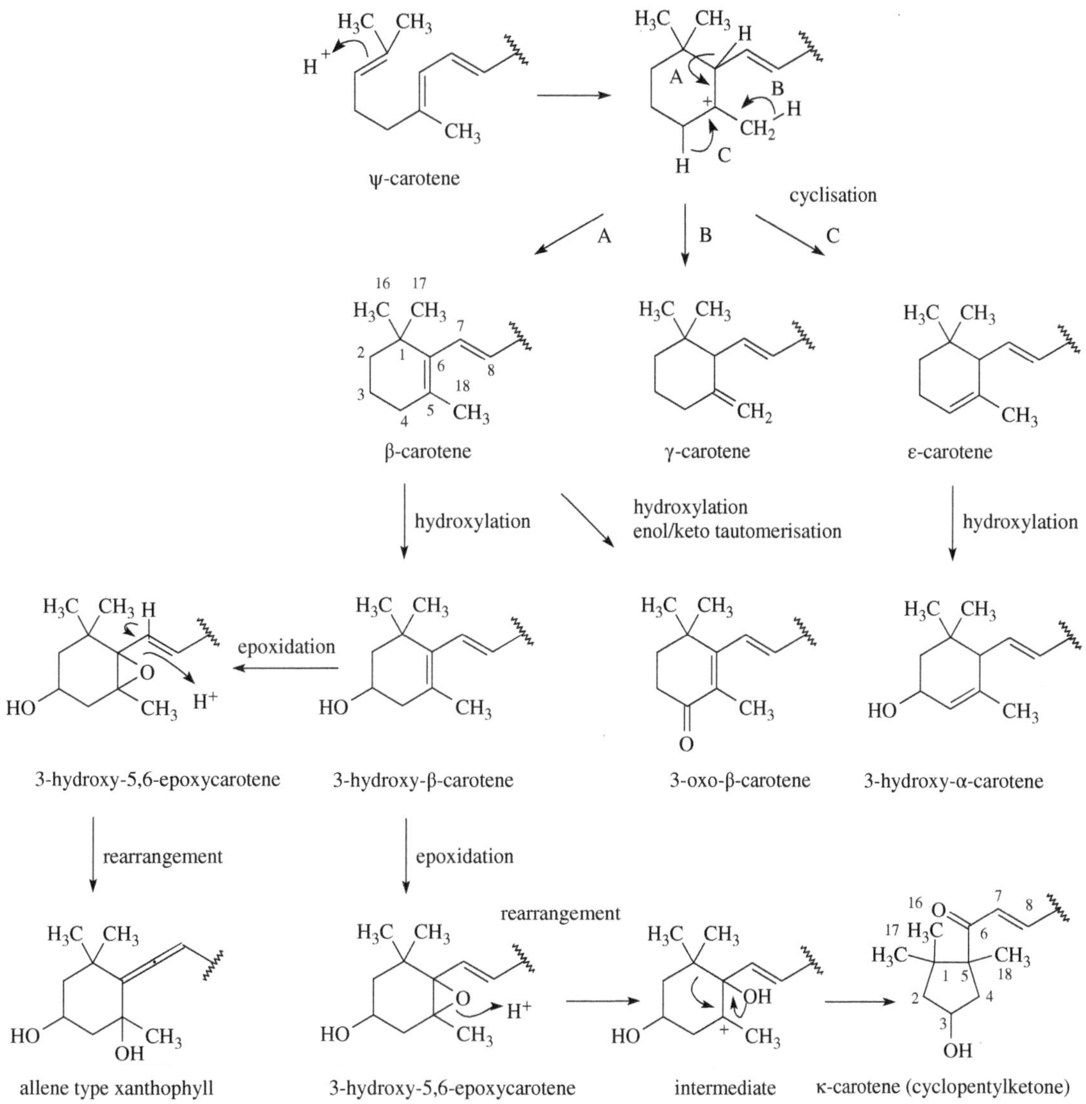

Figure 9.24 Formation of alicyclic carotenes and xanthofylls.

two α-ionone rings (ε-rings) is ε-carotene or ε,ε-carotene. Carotenes with a β-ionone ring, such as α-carotene, β-carotene, and γ-carotene, are precursors of retinol (a provitamin A). β-Zeacarotene is not a provitamin A, because it has a partially reduced side chain (**9-131**).

9.9.1.2 Xanthophylls

Xanthophylls are the main carotenoids of plants. They primarily arise as products of biochemical oxidation (hydroxylation and epoxidation) of carotenes. Xanthophylls derived from acyclic carotenes occur in foods in small quantities. For example, tomatoes contain as minor pigments 1,2-epoxylycopene, 5,6-epoxylycopene, 1,2-epoxyphytoene, and some other compounds. Much more common are monohydroxysubstituted alicyclic derivatives of carotenes called **cryptoxanthins**. Rubixanthin is a xanthophyll pigment derived from β-zeacarotene (**9-132**). Most plant materials contain small amounts of α-cryptoxanthin (also called zeinoxanthin), derived from α-carotene, and β-cryptoxanthin, derived from β-carotene, which are precursors of xanthophylls containing two hydroxyl groups in the molecule. The xanthophyll β-cryptoxanthin is provitamin A. Examples of dihydroxysubstituted pigments are zeaxanthin (arising from β-cryptoxanthin and β-carotene, respectively) and lutein (the precursor of which is α-cryptoxanthin or α-carotene, respectively). The precursor of lactucaxanthin is ε-carotene. The roles that lutein and zeaxanthin play in photosynthesis, excess light-energy dissipation, and general antioxidant functions cause them to be ubiquitous in all plant tissues.

The systematic name of rubixanthin is (3R)-β,ψ-carotene-3-ol, that of zeaxanthin is (3R,3'R)-7,8-didehydro-β,β-carotene-3,3'-diol or 3,3'-dihydroxy-β-carotene, that of lutein is (3R,3'S,6'R)-β,ε-carotene-3,3'-diol or 3,3'-α-carotene, that of antheraxanthin is 5,6-epoxy-5,6-dihydro-β,β-carotene-3,3'-diol, that of taraxanthin is (3S,5R,6S,3'S,6'R)-5,6-epoxy-5,6-dihydro-β,ε-carotene-3,3'-diol, that of violaxanthin is (3S,5R,6S,3'S,5'R,6'S)-5,6,5',6'-diepoxy-5,6,5',6'-tetrahydro-β,β-carotene-3,3'-diol or zeaxanthin diepoxide, and that of neoxanthin is (3S,5R,6R,3'S,5'R,6'S)-5',6'-diepoxy-6,7-didehydro-5,6,5',6'-tetrahydro-β,β-carotene-3,5,3'-triol.

9-131, alicyclic carotenes

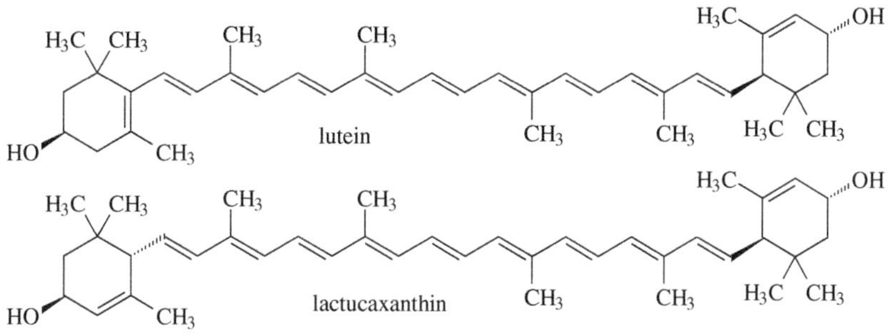

9-132, hydroxy and dihydroxy carotenes

Many carotenoids possess an epoxy group, usually in position 5,6 or 5,8 (often referred to as furanoids or furanoid oxides), although some 1,2-epoxides, 3,6-epoxides, and 4,5-epoxides have also been reported. Examples of 5,6-epoxides are antheraxanthin derived from β-carotene and taraxanthin (lutein epoxide) derived from α-carotene. Oxidation at both ends of the molecule gives 5,6,5′,6′-diepoxides such as violaxanthin. Neoxanthin, formerly also called foliaxanthin, which is found in the chloroplasts of higher plants, and fucoxanthin (**9-133**), occurring in brown algae kelps and most other heterokonts, are examples of rarely occurring compounds called **allenes** (dienes with cumulated double bonds).

5,6-Epoxides readily rearrange to 5,8-epoxides (epoxide-furanoid rearrangement); the configuration at C-5 remains unchanged, whilst the corresponding C-8 epimers arise and co-exist in plants. Examples of 3S,5R,8R-epimers of 5,8-epoxides (**9-134**) are muta-tochrome, 5,8-epoxy-5,8-dihydro-β,β-carotene (also known as citroxanthin) derived from β-carotene 5,6-epoxide, flavoxanthin derived from lutein 5,6-epoxide, mutatoxanthin (5,8-epoxy-5,8-dihydro-β,β-carotene-3,3′-diol) derived from antheraxanthin, luteoxanthin (5,6,5′,8′-diepoxy-5,6,5′,8′-tetrahydro-β,β-carotene-3,3′-diol), and auroxanthin (5,8,5′,8′-diepoxy-5,8,5′,8′-tetrahydro-β,β-carotene-3,3′-diol, 3S,5R,8R,3′S,5′R,8′R-epimer. Two luteoxanthin epimers and three auroxanthin epimers are both 5,6- and 5,8-diepoxides derived from violaxanthin. Chrysanthemaxanthin is 3S,5R,8S-epimer.

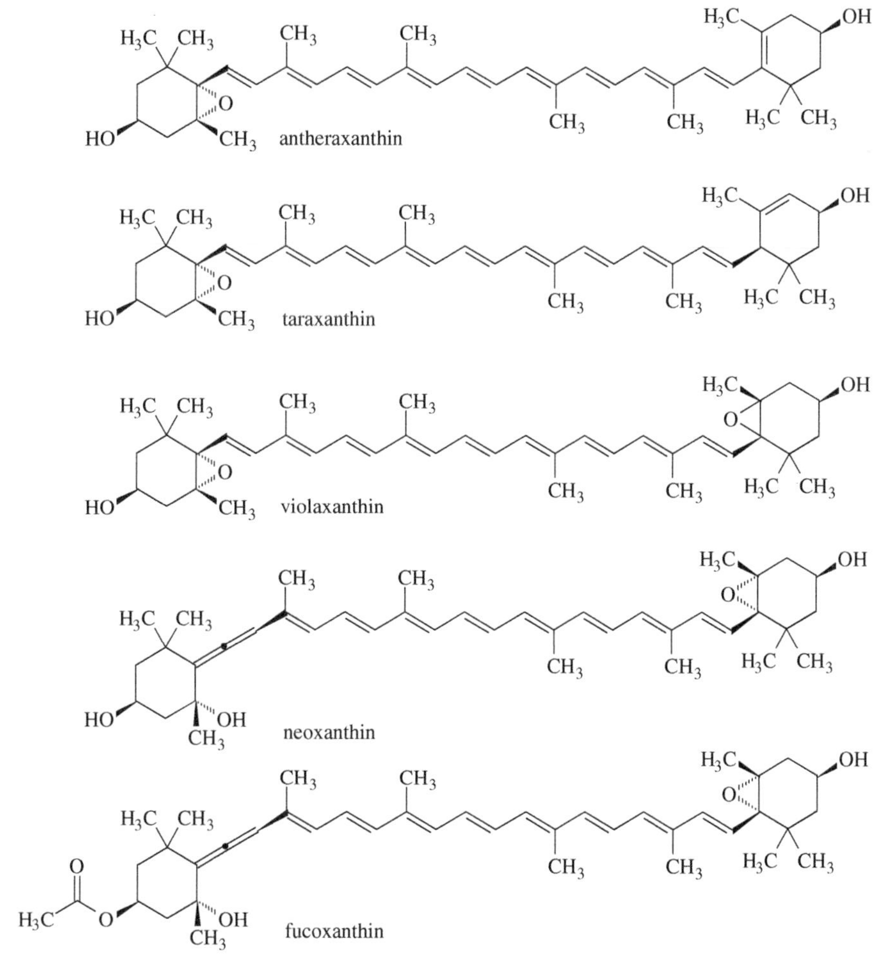

9-133, 5,6-epoxides and allenes

By rearrangement of the 5,6-epoxy group, an additional group of xanthophylls called cyclopentylketones or κ-carotenes (**9-135**) can be created, the most important of which are capsanthin, (3*R*,3′*S*,5′*R*)-3,3′-dihydroxy-β,κ-caroten-6′-one, and capsorubin, (3*S*,5*R*,3′*S*,5′*R*)-3,3′-dihydroxy-κ,κ-caroten-6,6′-dione.

The reaction mechanisms of biosynthesis of most xanthophylls are shown schematically in Figure 9.24.

9.9.1.3 Apocarotenoids

A relatively small but very important group of xanthophylls is compounds containing fewer than 40 carbon atoms in the molecule. These compounds, resulting from the breakdown of carotenoids, are called **degraded carotenoids** or **apocarotenoids**. Apocarotenoids often exhibit different biological functions and many are important flavorings (see Section 9.9.5.2).

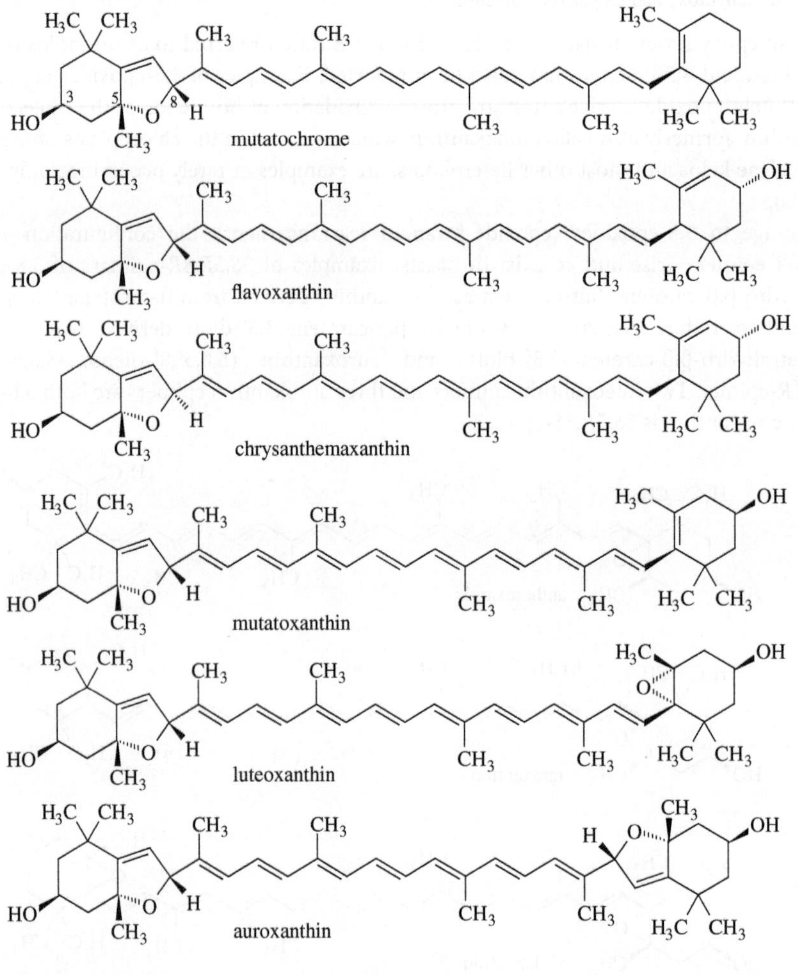

9-134, 5,8-epoxides

The most important apocarotenoid is vitamin A₁ (all-*trans*-retinol), with 20 carbon atoms in the molecule. Other important apocarotenals are crocetin with 20 carbon atoms and bixin (*cis*-bixin) with 22. A quite widespread product of the catabolism of carotenoids is C_{30} apocarotenal β-citraurin, (*R*)-3-hydroxy-8′-apo-β-carotene-8′-al (**9-136**). Citranaxanthin (**9-137**) is a yellow C_{33} pigment found in citrus fruits.

9.9.2 Occurrence

Carotenoids are important, and are the most widespread lipophilic pigments of many fruits and vegetables. They occur in all photosynthetic plant tissues, where they are found as photochemically active components of chromoplasts. They are often accompanied by other pigments, such as anthocyanins in peaches and apricots. The presence of carotenoids in the green parts of plants is often masked by chlorophyll pigments.

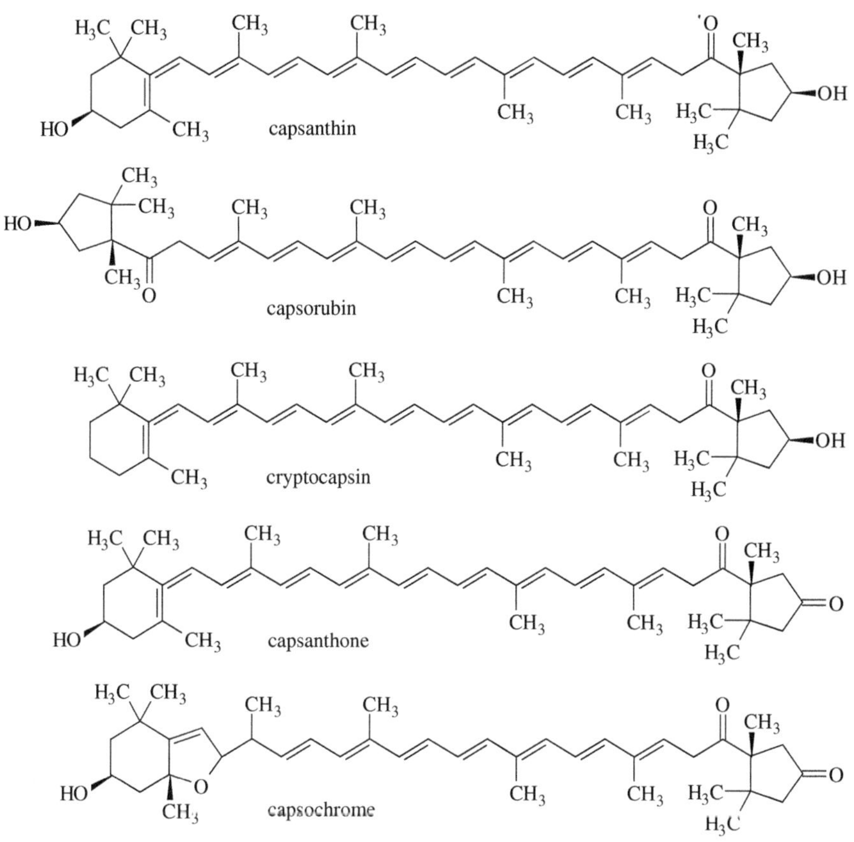

9-135, cyclopentanols and cyclopentanones

Qualitative and quantitative composition of carotenoids depends on many factors, such as the species and variety of the plant, season, maturity stage, and method of processing. In some fruits and in potatoes, carotenoids occur in units of mg/kg, in most fruits and vegetables they are present in dozens of mg/kg, and in carrots, tomatoes, and peppers they are found at the level of hundreds of mg/kg.

A certain proportion of carotenoids in plants is associated with proteins, as is the case with chlorophyll pigments. These are generally known as **carotenoproteins**.

Xanthophylls are also present as fatty acid esters and glycosides. For example, the main yellow pigment of the flowering head of sunflower (*Helianthus annuus*, Asteraceae) is lutein dipalmitate.

All-*trans*-isomers of carotenoids in fresh and thermally processed materials are accompanied by small amounts of *cis*-isomers, called **neo-carotenoids**. β-Carotene is accompanied mainly by the geometric isomers 9-*cis*, 13-*cis*-, and 15,15′-*cis*-β-carotene. Lutein is accompanied mainly by 9-*cis*, 9′-*cis*, 13-*cis*-, and 13′-*cis* isomers; less common are 15-*cis* and 15′-*cis* isomers. Neoxanthin is accompanied by 9-*cis*-, 9′-*cis*-, 13-*cis*-, and 13′-*cis* isomers. Thermal processing can induce *trans* to *cis* isomerisation.

9.9.2.1 Fruits

Fruits usually contain several dozens of carotenoids, some of which are present as dominant pigments. The compositions of carotenoids and their contents in selected fruits, where they are the dominant pigments, are presented in Table 9.15. Rarely, the main pigments of fruits are acyclic carotenoids such as lycopene. For example, in the pericarp and pulp of pink guava cultivars, which contain about 20 carotenoids, the main pigments in all the ripening stages are β-carotene, lycopene (63–92% of total carotenoids), and 15-*cis*-lycopene. β-Carotene is the main pigment in apricot (97%), durian (80%), mango (70%), and goldenberry (55%), with other carotenes and xanthophylls present in smaller amounts. More commonly, the main pigments in most fruits are various xanthophylls, which are partly present in the form of fatty acid monoesters and diesters. The principal fatty acids bound to xanthophylls are palmitic and myristic acid. For example, peaches contain higher amounts of xanthophylls than apricots and oranges, which contain a relatively small amount of carotenes but variable amounts of cryptoxanthin, lutein, antheraxanthin, and violaxanthin as their major pigments.

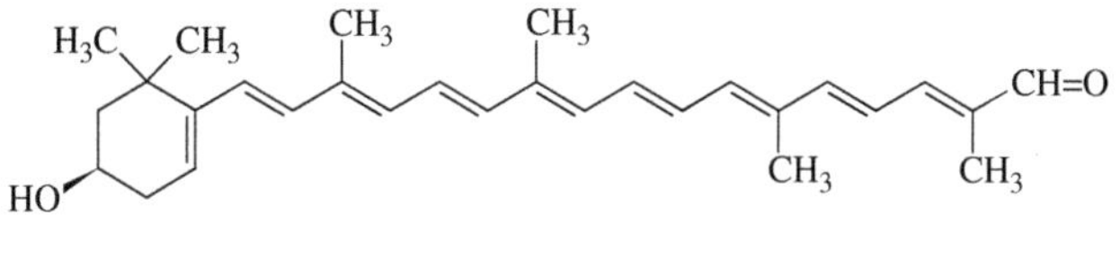

9-136, β-citraurin

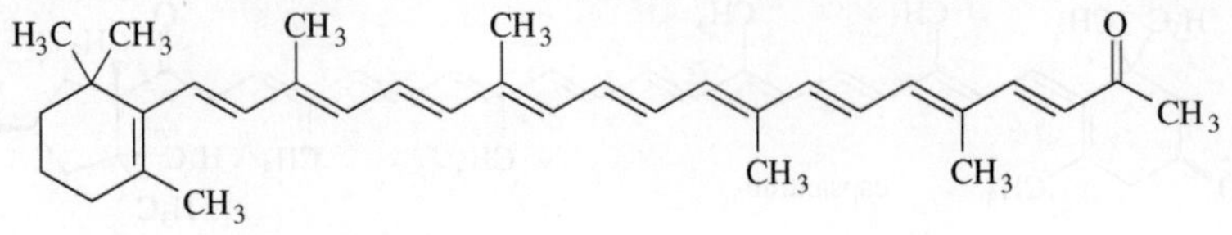

9-137, citranaxanthin

9.9.2.2 Vegetables

Carrot has some of the highest levels of β-carotene, along with cruciferous leafy vegetables and some bell pepper and pumpkin varieties. Pumpkin varieties with a high content of β- and α-carotene have an orange colour, whilst varieties with a high lutein content and low carotene content show a bright-yellow colour. The highest concentrations of lutein can be found in dark-green leafy vegetables such as spinach. Zeaxanthin accompanies lutein in leaf vegetables, but in most vegetables it is usually found in low amounts. A special case is tomato, whose red colour is caused by its major carotene, lycopene, which accumulates as the fruit ripens. Other major sources of lycopene are watermelon and rose hips.

Table 9.15 Compositions and contents of the main carotenoids of fruits and vegetables.

Carotenoid	Content (mg/kg f.w.)							
	Apricot	Mango	Orange	Persimmon	Carrots	Spinach	Tomato	Bell pepper
Carotenes								
Lycopene	0.1	–	–	trace–1.1	–	–	16–750	–
Neurosporene	–	–	–	–	–	–	3.0	–
ζ-Carotene	0.4	trace–0.1	0.5	–	–	–	8.4	–
Phytofluene	0.3	–	1.3	–	–	–	5.1	–
Phytoene	0.6	–	0.4	–	–	–	6.0	–
β-Carotene	64	4.9–27	0.1–0.4	0.1–1.2	46–103	33–89	2.8–5.8	51–275
α-Carotene	–	–	0.1–0.2	trace–0.14	22–49	trace	–	–
γ-Carotene	0.2	–	–	–	6.3–27	–	0.4–1.6	–
Xanthophylls								
5,6-Epoxylycopene	–	–	–	–	–	–	5.3	–
β-Cryptoxanthin	–	–	0.1–7.1	0.6–9.4	–	–	–	36–79
α-Cryptoxanthin	–	0.3–1.1	–	–	–	–	–	–
Zeaxanthin	–	–	0.5	0.5–9.4	–	–	–	40–125
Lutein	–	–	0.3	trace–0.6	1.1–5.6	42–81	0.4–1.3	
Antheraxanthin	–	–	0.6	–	–	–	–	33–44
Violaxanthin	–	0.7–3.0	0.7	0.1–0.9	–	74	–	53–98
Neoxanthin	–	–	–	0.1–0.7	–	24	–	174
Mutatochrome	–	trace–0.1	–	–	–	–	–	–
Mutatoxanthin	–	trace–2.0	0.6	–	–	5.0	–	164
Luteoxanthin	–	0.8–5.5	1.7	–	–	–	–	85
Auroxanthin	–	0.1–0.4	1.2	–	–	–	–	–
Capsanthin	–	–	–	–	–	–	–	523–1207
5,6-Epoxycapsaxanthin	–	–	–	–	–	–	40–216	–
Capsorubin	–	–	–	–	–	–	53–179	–
Cucurbitaxanthin A	–	–	–	–	–	–	69–213	–
Cryptocapsin	–	–	–	–	–	–	814	–

9.9.2.2.1 Carrots

The predominant pigment of carrots is β-carotene (Table 9.15). Its content is generally 60–120 mg/kg, but may reach 300 mg/kg in some varieties. Other pigments are various carotenes and xanthophylls partly associated with proteins.

9.9.2.2.2 Leafy vegetables

In leafy vegetables, β-carotene usually amounts to 10–20% of the total carotenoid content. Other carotenoid pigments are various xanthophylls. Lutein, violaxanthin, and neoxanthin usually occur in large amounts, whilst cryptoxanthin, zeaxanthin (the main carotenoid of maize), and antheraxanthin are found in smaller ones (Table 9.15). The presence of carotenoids is masked by chlorophylls. Lettuce is an example of a vegetable that accumulates a higher amount of lactucaxanthin.

9.9.2.2.3 Tomatoes

The main pigment in red tomato, lycopene, typically represents 85% of total carotenoids. The content of β-carotene is relatively low, at about 6 mg/kg, whilst that of γ-carotene is about 1 mg/kg (see Table 9.15). The lycopene content increases during ripening, reaching a final amount in the range of 30–200 mg/kg (about 95% of which represents the all-*trans* isomer). Some orange hybrid tomatoes contain lower amounts of lycopene and higher amounts of β-carotene (up to about 80 mg/kg) and γ-carotene (about 7 mg/kg). The yellow tomatoes have a minimum of lycopene and a higher amount of β-carotene. Tomato products are mostly prepared by evaporation of tomato juice, so they contain higher concentrations of lycopene (e.g. concentrated juice, 62 mg/kg; puree, 133 mg/kg; ketchup, 102 mg/kg) than the starting material (30 mg/kg), but also less stable *cis*-isomers, the content of which increases with temperature and heating time. Up to 20% of the *trans*-isomer initially present may isomerise.

9.9.2.2.4 Bell peppers and paprika

Red bell peppers (*Capsicum* spp., Solanaceae) are an example of a plant species that contain unusual carotenoid pigments known as cyclopentylketones (Table 9.15). The main pigments in green peppers, in addition to chlorophylls, are lutein (8–14 mg/kg), violaxanthin (8–10 mg/kg), neoxanthin (8–9 mg/kg), and β-carotene (6–8 mg/kg). During maturation, the minority yellow xanthophylls present in green fruits (β-cryptoxanthin and zeaxanthin) are enzymatically oxidised via 5,6-epoxides (5,6-epoxycryptoxanthin, antheraxanthin, and violaxanthin) to red κ-carotenes (Figure 9.24). The principal red pigment is always capsanthin, which constitutes 32–38% of the carotenoid pigments of peppers. Present in amounts of 1–4% are capsanthin 5,6-epoxide, capsorubin (6–10%), and a small number of other substances, such as cucurbitaxanthin A (capsolutein) and cryptocapsin. Cucurbitaxanthin A is 3,6-epoxy-5,6-dihydro-β,β-carotene-3′,5-diol. The relative contents of major carotenoids formed during ripening of pepper pods are given in Table 9.16. Biosynthesis of κ-carotenes proceeds simultaneously with esterification of pigments by fatty acids. The majority of pigments of mature peppers (about 80%) are totally or partially esterified with fatty acids. The main fatty acids of yellow xanthophylls are linoleic, myristic, and palmitic acids, whilst those of red xanthophylls are lauric, myristic, and palmitic acids.

9.9.2.2.5 Rose hips

The carotenoid content in fruits of the genus *Rosa* (Rosaceae) differs between species and is influenced by degree of ripening, climate, variations within and between years, and growing and storage conditions. For example, the mean total amount of carotenoids from different species (*R. dumalis*, *R. rubiginosa*, and *R. spinosissima*) harvested at full maturity was 1024 mg/kg (d.w.), of which 503 mg/kg represented carotenes, 39 mg/kg xanthophylls, and 483 mg/kg esterified xanthophylls. The main carotene was lycopene (292 mg/kg), followed by β-carotene (196 mg/kg), (7Z,9Z,7′Z,9′Z)-lycopene, also known as (9Z,9′Z)-ζ-carotene or prolycopene (17 mg/kg), ζ-carotene (16 mg/kg), and γ-carotene (12 mg/kg). The main xanthophyl was rubixanthin (27 mg/kg), followed by lutein and zeaxanthin (together 11 mg/kg), neochrome (11 mg/kg), neoxanthin (9 mg/kg), and violaxanthin (8 mg/kg). The same main carotenoids were found in the fruit of dog rose (*R. canina*). The total amount of carotenoids from fruits of multiflora rose (*R. multiflora*) was 612 mg/kg (d.w), of which 33% represented lycopene and its isomers, 22% β-carotene, 20% xanthophyll esters, and 6% free xanthophylls. The dominant xanthophyll esters were lutein dilaurate, zeaxanthin and violaxanthin dilaurates, whilst in other species the dominant esters were rubixanthin laurate and myristate.

9.9.2.3 Other plant materials

9.9.2.3.1 Cereals

The common wheat varieties (*Triticum aestivum*, Poaceae) contain as the main carotenoid lutein (about 2 mg/kg), accompanied by small amounts of β-carotene (about 1% of total carotenoids) and traces of zeaxanthin. The total carotenoid content in *T. durum* is about three times higher than that in common wheat, with lutein (1.5–4 mg/kg) the main component. Lutein is similarly the main carotenoid component in oats and barley.

Table 9.16 Changes of main carotenoids during pepper pod ripening (% of total carotenoids).

Carotenoid	Degree of ripeness					
	Green	Light yellow	Yellow	Orange	Red	Dark red
α-Carotene	0.3	0.7	0.4	0.5	0.2	0.2
β-Carotene	11.3	12.2	7.1	5.9	9.2	8.9
α-Cryptoxanthin	1.0	1.8	0.6	0.3	0.3	0.3
β-Cryptoxanthin	0.7	0.8	3.5	6.5	5.8	5.1
Zeaxanthin	6.7	6.8	10.5	19.5	16.9	15.3
Lutein	31.9	28.8	1.5	0.1	0.0	0.0
Antheraxanthin	2.3	0.3	1.4	3.6	1.4	1.6
Violaxanthin	4.8	1.4	6.3	4.2	1.3	2.0
Neoxanthin	3.7	5.5	4.0	0.0	0.0	0.0
Mutatoxanthins[a]	1.2	1.4	1.5	2.1	3.1	3.0
Luteoxanthins[a]	5.6	5.1	3.9	1.6	1.5	1.5
Auroxanthins[a]	0.1	0.8	0.2	0.0	0.0	0.0
Capsanthin	0.9	1.3	24.6	28.2	29.3	28.3
Capsorubin	0.0	0.0	2.8	2.7	2.0	2.6
Cryptocapsin	0.3	0.3	0.3	0.8	0.5	0.8
Capsanthon	0.0	0.0	0.1	0.4	0.4	0.4
Total content (mg/kg)	115	168	448	1327	6107	9947

[a]Mixture of epimers.

Maize is exceptionally high in lutein, at a concentration of about 20 mg/kg. It also has high concentrations of zeaxanthin (6–10 mg/kg), β-cryptoxanthin (2 mg/kg), and β-carotene (1 mg/kg) and smaller concentrations of 15-*cis*-lutein, 13-*cis*-lutein, 13′-*cis*-lutein, 9-*cis*-lutein, 9′-*cis*-lutein, and 9-*cis*-zeaxanthin.

The major brown-rice carotenoids are β-carotene and lutein (both about 0.1 mg/kg), whilst zeaxanthin levels are lower (about 0.030 mg/kg).

9.9.2.3.2 Annatto

Annatto (also known as achiote) is the pigment from fruits (red seeds and pulp) of the achiote tree (*Bixa orellana*, Bixaceae), indigenous to Central and South America. The main component of annatto is apocarotenoid bixin, also known as (9′*Z*)-bixin (9′-*cis*-bixin, **9-138**), which arises as a product of lutein catabolism. The amount of bixin in seeds is about 2%.

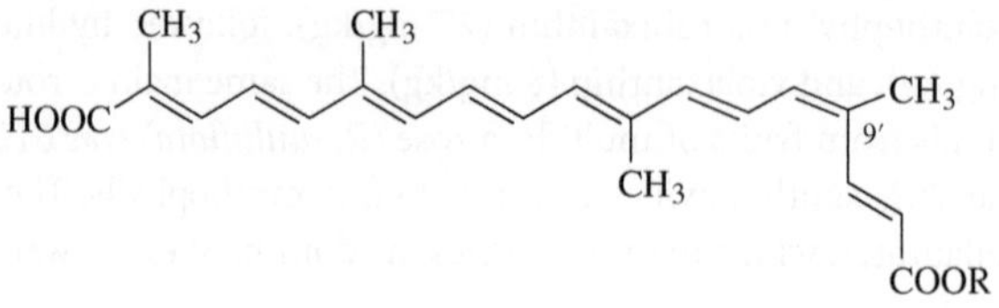

9-138, (*Z*)-bixin, R = CH₃
(*Z*)-norbixin, R = H

Commercially, the seeds and flesh are processed by extraction with vegetable oils at reduced pressure and at temperatures lower than 130 °C. Under these conditions, 9′-*cis*-bixin undergoes partial isomerisation to the all-*trans* isomer called (*E*)-bixin (*trans*-bixin, **9-139**). Extracts contain about 0.2–0.5% of a mixture of these two pigments in different proportions according to the conditions of extraction. The resulting product, called *trans*-bixin, is a red, relatively stable, and fairly fat-soluble pigment (it dissolves to give about 5% solutions). During extraction, 9'-*cis*-bixin is partly decomposed to orange and pale-yellow products. The main product is 14-methyl-hydrogen-4,8-dimethyltetradeca-2,4,6,8,10,12-hexaene-1,14-dioic acid (**9-140**), which exists in various stereoisomers, such as the all-*trans* isomer and 15-*cis*, 13-*cis*, and 9-*cis* isomers. Related compounds with 18 and 13 carbon atoms per molecule, *m*-xylene, toluene, and other products also arise. Coloured degradation products of 9'-*cis*-bixin may constitute up to 40% of pigments in commercial products.

The extraction of annatto seeds with aqueous alkaline solutions (at temperatures up to 70 °C) produces orange dicarboxylic acid called norbixin or 9'-*cis*-norbixin (**9-138**), which is soluble in polar media. Isomerisation of norbixin yields red all-*trans*-norbixin (**9-139**), which is slightly soluble in fats.

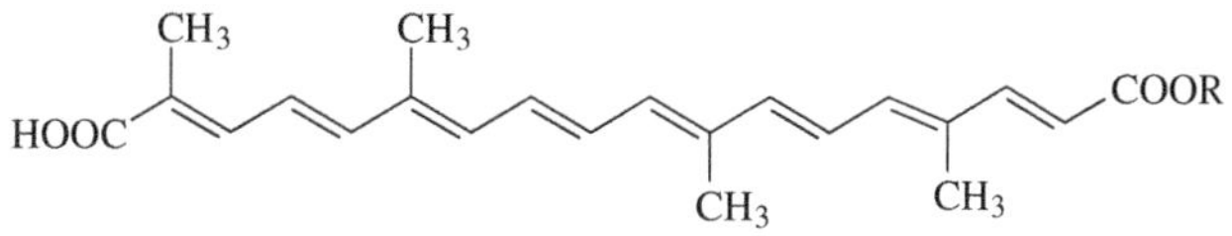

9-139, (*E*)-bixin, R = CH₃
(*E*)-norbixin, R = H

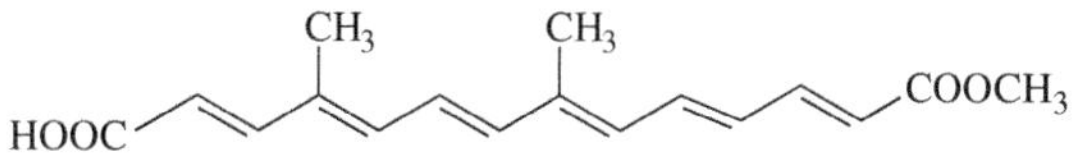

9-140, degradation product of (9′Z)-bixin

9.9.2.3.3 Saffron

Each flower of the saffron crocus (*Crocus sativus*, Iridaceae) has three filiform (20–30 mm long) purplish-brown stigmas, which are dried and used as a culinary spice for their pungent, characteristic spice odour, bitter taste (see Section 8.3.5.1.3), and colour. Saffron contains water-soluble yellow-orange apocarotenoid pigments, the chromophore of which is water-insoluble brick-red aglycone crocetin, also known as α-crocetin (8,8′-diapocarotene-8,8′-dicarboxylic acid, **9-141**), formed by cleavage of zeaxanthin. Crocetin in saffron occurs as a yellow-orange ester of crocetin with disaccharide gentiobiose (di-β-gentiobiosylcrocetin), which is called α-crocin or just crocin (**9-142**). Crocin is accompanied by some minor pigments: β-gentiobiosyl-β-D-glucosyl-, di-β-D-glucosyl-, β-gentiobiosyl-, and β-D-glucosylesters, and also by esters derived from trisaccharide neapolitanose, such as β-gentiobiosyl-β-neapolitanosylcrocetin and di-β-neapolitanosylcrocetin.

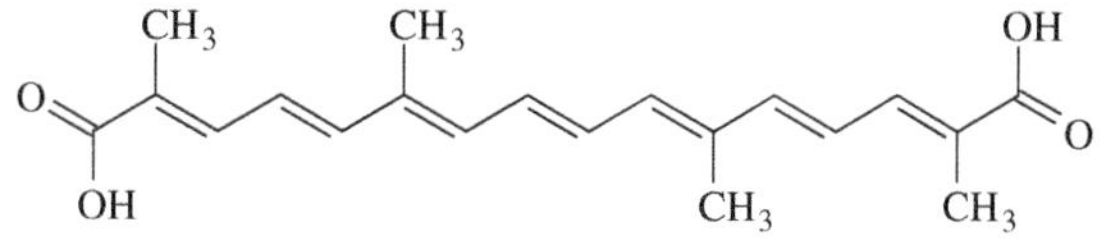

9-141, crocetin

The disadvantage of saffron is its high price, so it is often falsified. Falsification is usually achieved using dry marigold petals (*Calendula officinalis*, Asteraceae), or alternatively the petals of safflower (*Carthamus tinctorius*, Asteraceae). The pigment of yellow marigold petals is flavoxanthin, whilst the pigment of orange petals is lycopene. The pigment of yellow-to-red safflower petals is carthamin. Crocin is also found in fruits of the gardenia (*Gardenia augusta*, syn. *G. jasminoides*, Rubiaceae).

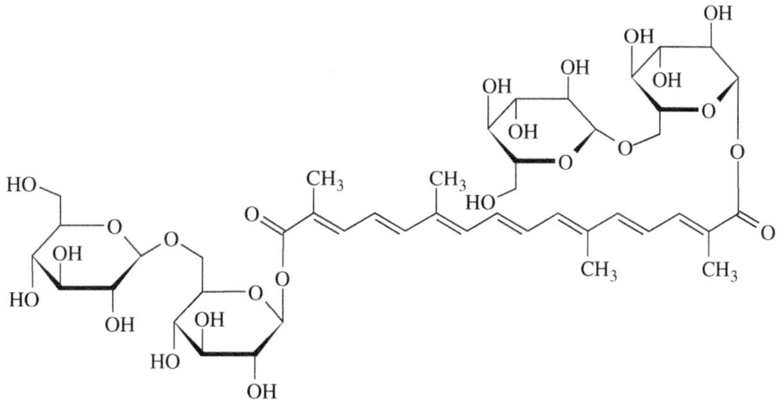

9-142, crocin

9.9.2.3.4 Vegetable oils

The concentration of carotenoids in crude vegetable oils is about 0.03–0.25%. Their content in refined oils is lower, and depends on the conditions during refining. Pigments that are found in refined oils differ in their structure from natural pigments as they contain products of the isomerisation and degradation of natural pigments. A relatively high content of carotenoids (0.05–0.2%) is found in palm oil obtained from mesocarp of oil palm seeds (*Elaeis guineensis*, Arecaceae), which has a light orange colour as the oil contains α- and β-carotene, occurring in a ratio of about 2 : 3, as its main components.

9.9.2.4 Foods of animal origin

Animals are unable to synthesise carotenoids *de novo*. They can only convert plant pigments occurring in food into substances of different structures or else store them as such.

9.9.2.4.1 Depot fats of mammals and birds

The main pigments of depot fats of birds (poultry) and mammals are the xanthophylls lutein and zeaxanthin. Also present are small amounts of β-carotene and other pigments.

The feathers in many birds are also coloured by carotenoid pigments. For example, canthaxanthin occurs as a major pigment of pink flamingo feathers, along with astaxanthin and other minor oxocarotenoids, such as echinenone (myxoxanthin, β,β-caroten-4-one) and phoenicoxanthin (adonirubin), (3S)-3-hydroxy-β,β-carotene-4,4'-dione.

9.9.2.4.2 Eggs

Egg yolk and shell colour are important aspects of egg quality in many countries. Egg yolk pigmentation is a complex process influenced by many factors, such as the hen's diet, carotenoid source, and so on. Egg yolk contains as its main pigments the same carotenoids that occur in the depot fat of hens. The most abundant is lutein/zeaxanthin, followed by β-cryptoxanthin and β-carotene, and then by the other pigments occurring in the feed (such as canthaxanthin and β-apo-8'-carotenal). The amount of β-carotene is low (0.3–1% of total carotenoids), as it is quickly metabolised.

The colour of the shell is largerly determined by genetics, so the breed of the hen is the main determining factor. Most often, white hens lay white eggs and brown/reddish hens lay brown eggs. There are some breeds that will lay green or blue eggs. The pigments of brown eggs are mixtures of protoporphyrin IX, biliverdin IXα, and its zinc chelate, with traces of other haemoglobin-derived pigments in various proportions. In green and blue eggs, biliverdin IXα dominates. The shell or yolk colour bears no relation to the flavour or nutritional quality of the egg.

9.9.2.4.3 Fish and crustaceans

The carotenoids in salmonids (salmon and trout) and many crustaceans (shrimps, crabs, and others) represent a group of red oxocarotenoids: astacene, astaxanthin, and canthaxanthin (**9-143**). The main pigment, astaxanthin, (3S,3S')-3,3'-dihydroxy-β,β-carotene-4,4'-dione, is bound to proteins in dark blue-red and green-red water-soluble carotenoprotein complexes and partly occurs as esters of higher fatty acids. During cooking, proteins denature and liberate the typical red carotenoid astaxanthin. For example, the light blue crab *Homarus gammarus* (Nephropidae), known as European lobster or common lobster, contains a blue carotenoprotein called α-crustacyanin. This protein (320 kDa) is composed of 16 apoprotein units, each of which contains one molecule of astaxanthin. Astaxanthin is also abundant in some mushrooms, algae, yeasts, and bird feathers. Astacene (3,3'-dihydroxy-2,3,2',3'-tetradehydro-β,β-carotene-4,4'-dione, found in the enol form in neutral solutions) and semiastacene (3,3'-dihydroxy-2,3-didehydro-β,β-carotene-4,4'-dione) are oxidative artifacts. Astacene has been also reported in small quantities in some green algae, higher plants, and animals. Canthaxanthin (β,β-carotene-4,4'-dione) has been found in some bacteria, green algae, crustaceans, and beetle species, and accumulates in some fish such as Pacific salmon and carp.

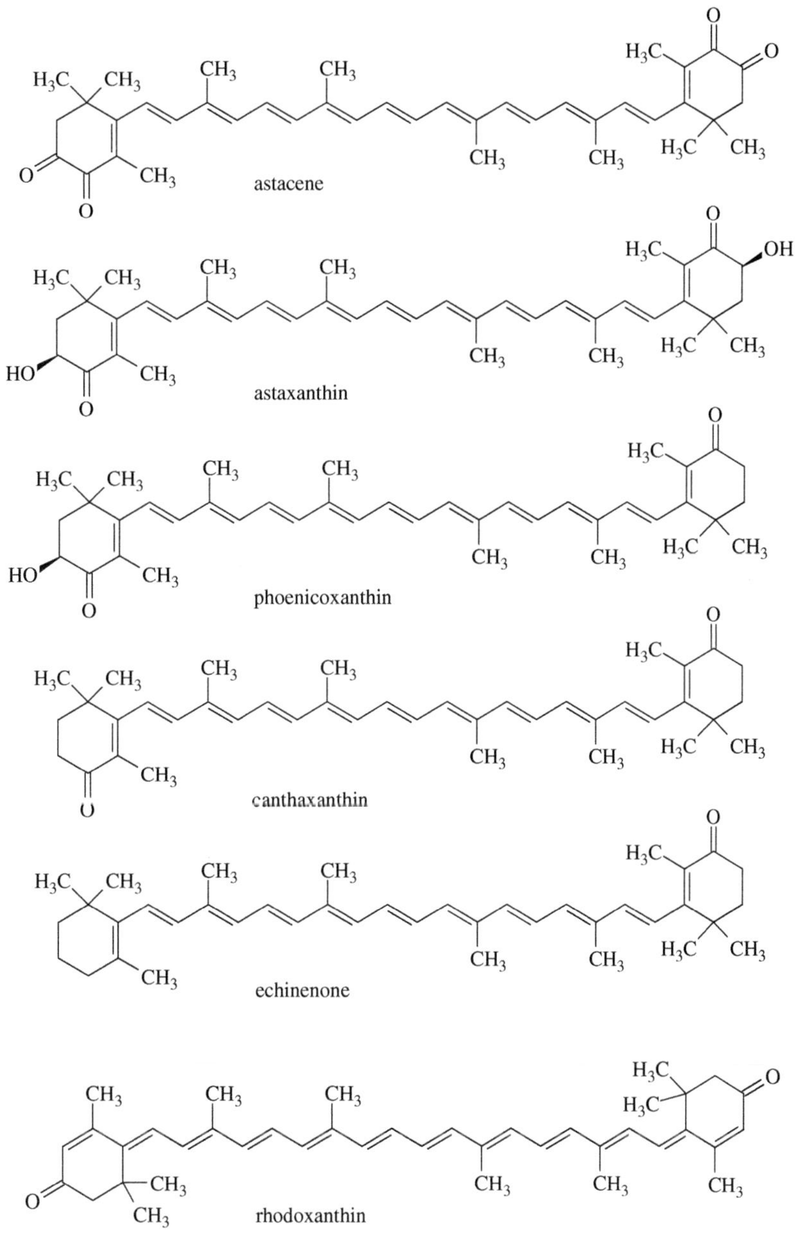

astacene

astaxanthin

phoenicoxanthin

canthaxanthin

echinenone

rhodoxanthin

9-143, major oxocarotenoids

9.9.3 Use

Natural and synthetic carotenoids, fresh or dry parts or extracts (so-called oleoresins) of certain plants containing carotenoids (e.g. carrots, orange peels, tomatoes, annatto, and paprika, which is made from the dried and ground fruits of various *Capsicum* species and cultivars), and crude red palm oil have found use as food dyes and, relatively recently, as antioxidants.

In amounts of 1–10 mg/kg, various carotenoid pigments are approved for use in different countries in imparting colour to many foods, such as margarines, cheeses, yoghurt, ice creams, fruit juices, dressings, flour, pasta products, and cakes. Common colourants are β-, α-, and γ-carotene (E160a), annatto (E160b), paprika oleoresin (E160c), lycopene (E160d), β-apo-8′-carotenal (E160), ethyl ester of β-apo-8′-carotenic acid (Food Orange 7, E160f), flavoxanthin (E161a), lutein (E161b), cryptoxanthin (E161c), rubixanthin (E161d), violaxanthin (E161e), rhodoxanthin (E161f), canthaxanthin (E161g), zeaxanthin (E161h), citranaxanthin (E161i), and astaxanthin (E161j).

Some of these colourants are also added to the feed of dairy cows to ensure desirable pigmentation of milk and dairy products. β-Carotene is stored in adipose tissue and also passes partly into milk. The most important colourants in modern poultry production include carotenoids obtained from different plant materials, lutein, and canthaxanthin (egg yolk and meat pigmentation). Less common are citranaxanthin, ethyl β-apo-8′-carotenoate, and paprika extract. β-Carotene is not used, as it is quickly converted to retinol, which has almost no effect on yolk and meat colours. Canthaxanthin is also used for skin tanning in cosmetics and in combination with astaxanthin for salmon feeds.

9.9.4 Biochemistry, physiology, and nutrition

Some carotenoids are precursors (provitamins) of vitamin A. Others, such as cryptoxanthin, zeaxanthin, and lutein, exhibit, by contrast, about half of the activity of provitamins A. Some carotenoids, such as lycopene, astaxanthin, and canthaxanthin, are more effective in quenching singlet oxygen than β-carotene. Carotenoids such as β-carotene also react with free radicals. Because of their antioxidative properties, they are used in the prevention of degenerative processes and as anticancer agents.

9.9.5 Reactions and changes

The combined effect of oxidoreductases, light, heat, oxygen, hydronium ions, and other factors can lead to isomerisation, oxidation, and degradation of carotenoids and xanthophylls (see Section 5.2.6.2). Carotenoids present in the form of carotenoproteins are more stable than free substances. Xanthophylls, especially epoxides of carotenoids, are more susceptible to changes under the conditions used during food processing. Apocarotenoids, which contain a carboxyl group, form soluble salts in alkaline solutions.

9.9.5.1 Carotenoids and colour changes

Food products that undergo undesirable colour changes in the presence of enzymes and oxygen are sterilised in order to achieve the inactivation of enzymes and stored in an inert atmosphere or in the presence of antioxidants. In cases where it is impossible to prevent the degradation of carotenoids, the material can be coloured using synthetic carotenoids. For example, during the storage of flour or the manufacture of pasta, losses of carotenoids can reach 30–60%.

In acidic citrus juices, spontaneous conversion of 5,6- and 5′,6′-epoxides to 5,8- and 5′,8′-epoxides occurs. Relatively rapid enzymatic degradation of carotenoids (about 50% of pigments are degraded within 20 minutes) occurs in stored raw spinach puree. Lipoxygenases are indirectly involved in the degradation of carotenoids, as a result of their reaction with singlet oxygen and free radicals (see Section 3.8.1.8.5). Therefore, it is necessary to inhibit oxidoreductases of spinach leaves by blanching them in hot water before freezing. During storage of frozen blanched spinach puree, changes in the content of carotenes and xanthophylls are minimal.

Thermal processing of vegetables using mild processes such as steaming results in relatively low losses of carotenes (up to 10%) and somewhat higher losses of xanthophylls. Epoxy xanthophylls, such as violaxanthin and taraxanthin, however, undergo significant losses. When cooking spinach, broccoli, or green beans by steaming or boiling in water for 3–10 minutes, the loss of taraxanthin can reach 35–100% and the loss of violaxanthin 35–65%, depending on the type and method of processing. Cooking beans in water for 10 minutes results in a loss of about 60% and cooking for one hour leads to complete degradation of both epoxy xanthophylls.

The pigment content during drying and grinding of bell pepper pods is reduced to about half. Most losses are caused by dilution of the product by non-coloured parts of fruits, such as seeds, but some degradation of carotenoids also occurs. The least stable are the yellow pigments, especially β-carotene (see Section 3.8.1.8.5), whilst the red pigments (capsanthin and its derivatives) only decompose to a small extent. With singlet oxygen, capsanthin yields 5,6- and 5,8-endoperoxides, and with reactive oxygen species (superoxide anion radical and hydroxyl radical), 5,6- and 5,8-epoxides (**9-144**). More stable capsorubin produces 7,8-endoperoxide with singlet oxygen and 7,8-epoxide with superoxide anion radical. In paprika stored in the presence of oxygen, the hydroxyl groups of the capsanthin cyclopentane ring are oxidised with the formation of the corresponding diketone capsanthone, also known as capsanthinone, which is (3R,5′R)-3-hydroxy-β,κ-carotene-3′,6′-dione. Capsanthone decomposes further to β-citraurin, which yields 3-oxo-β-apo-8′-carotene, which is then degraded to low-molecular-weight products. Products containing a carbonyl group in the molecule enter the non-enzymatic browning reactions, which results in a change of the red colour of paprika to brown.

Similar reactions also occur in dried and concentrated tomato products (such as dried soups, juice concentrates, pickles, and ketchup). From 5,6-epoxylycopene (5,6-epoxy-5,6-dihydrolycopene) in fresh fruits arises, through epoxide ring opening under acidic conditions, 5,6-dihydroxy-5,6-dihydrolycopene.

| 5,6-epoxide | 5,8-epoxide | 5,6-endoperoxide | 5,8-endoperoxide |

9-144, endoperoxides and epoxides of capsaicin

In stored foods containing natural bixin (9′-cis isomer), in the presence of light, all-trans-bixin may be formed and hydrolysed to all-trans-norbixin. In cheeses containing bixin, these substances bind to phosphoproteins (e.g. some caseins), causing pink discolouration of products. In addition to trans-isomers, some degradation products of bixin may also arise, such as methyl ester of all-trans-4,8-dimethyltetradeca-2,4,6,8,10,12-hexaenoic acid and aromatic hydrocarbons (e.g. m-xylene).

9.9.5.2 Carotenoids and aroma

Carotenoids are precursors of many important compounds, which are formed as products of their catabolism or oxidation (e.g. during fruit ripening or processing). Oxidation of carotenoids into fragments called **apocarotenoids** or **diapocarotenoids** is provided by a group of enzymes known as dioxygenases, which catalyse the transfer of molecular oxygen to the substrate. Cleavage of carotenoids in the central double bond (so-called symmetric fission) results in the formation of two C_{20} molecules of apocarotenoids. In this way, for example, the visual signalling molecules of retinal and retinoic acid is produced (see Section 5.2.2). Asymmetric (eccentric) cleavage of carotenoids yields apocarotenoid molecules with different carbon-chain lengths. The most important degradation products are C_{15}, C_{13}, C_{11}, C_{10}, and C_9 compounds. The biologically important C_{15} compound is phytohormone abscisic acid, which is regarded as the major player in mediating the adaptation of plants to stress and preparing plants for a period of dormancy (loss of leaves and other phenomena). Degradation products of carotenoids responsible for the smell and taste of food (mainly fruits and vegetables) and the smell of many flowers are mainly C_{13}, C_{11}, C_{10}, and C_9 compounds, but C_{14}, C_8, and other apocarotenoids may also arise. The most important apocarotenoids are C_{13} compounds. Degraded carotenoids with more original carbons (C_{39}, C_{38}, or C_{37}) are called **norcarotenoids**. One to three carbon atoms are formally removed from carotenoids by other reactions than those occuring with apocarotenoids (such as oxidation and decarboxylation).

9.9.5.2.1 Apocarotenoids C_{13}

Apocarotenoids C_{13} result from the cleavage of C-9/C-10 and C-9′/C-10′ bonds. In the latter case, another fission product is a C_{14} diapocarotenoid. The main pigment of red varieties of tomatoes is lycopene (ψ,ψ-carotene). Fission of the C-9/C-10 (C-9′/C-10′) bond in lycopene yields the aromatic compound of tomatoes (3E,5E)-6,10-dimethylundeca-3,5,9-trien-2-one (pseudoionone) and, obviously, C_{14} diapocarotene 4,9-dimethyldodeca-2,4,6,8,10-pentaenedial. Acyclic carotene phytoene gives rise to C_{13} methylketone (E)-6,10-dimethylundeca-5,9-dien-2-one (known as geranylacetone or dihydropseudoionone) and diapocarotenoid 4,9-dimethyldodeca-4,6,8-trienedial (Figure 9.25). One molecule of geranylacetone may arise from ζ-carotene.

The C-9/C-10 (or C-9′/C-10′) double bond cleavage in alicyclic carotenes and xanthophylls yields products with a structure derived from the hydrocarbon megastigmane (**9-145**). Common carotenes may produce eight C_{13} methylketones (**9-146**). For example, the cleavage of β-carotene, α-carotene, and β-cryptoxanthin produces β-ionone, which is a fragrant component of raspberries, blueberries, passion fruit, apricot, carambola, cherry, mango, plums, black tea, tomatoes, carrots, bell peppers, and tobacco. α-Carotene is a precursor of (R)-ionone, also known as (+)-α-ionone. This compound, with a fruity and floral aroma reminiscent of violets and raspberries, is a fragrant component of blackcurrants, blueberries, raspberries, bananas, cherries, plums, peaches, vanilla, tomatoes, carrots, celery, black tea, tobacco, and other products.

Hydroxy ionones and epoxy ionones arise by fission of various xanthophylls. For example, (S)-3-hydroxy-β-ionone arises from β-cryptoxanthin, zeaxanthin, and lutein whilst 5,6-epoxy-β-ionone arises from antheraxanthin and violaxanthin. (3R,6R)-3-Hydroxy-α-ionone may arise from lutein, (3S,6S)-3-hydroxy-α-ionone from lactucaxanthin, and (3S,6R)-3-hydroxy-α-ionone from 3′-epilutein.

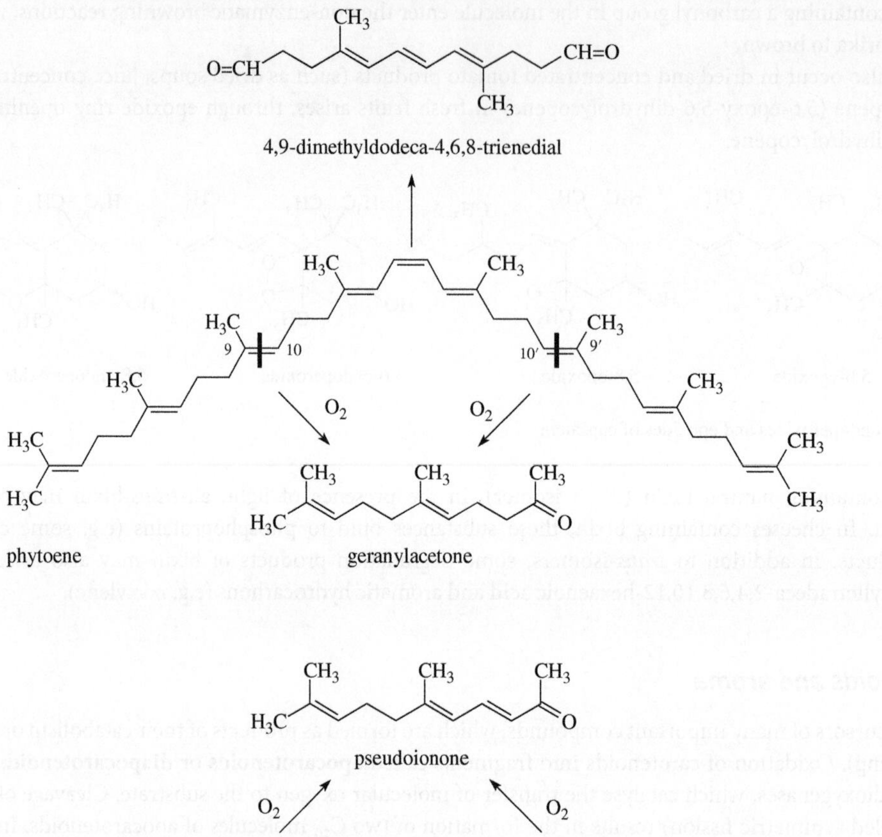

Figure 9.25 Fission of acyclic carotenes.

The special ketone known as grasshopper ketone arises from neoxanthin. Whilst maintaining the stereochemistry at carbon C-3, hydroxy ionones and grasshopper ketone can be glycosylated and become precursors of many other compounds. About 50 products formed by oxidation, reduction, dehydration, and cyclisation have been identified, some of which are formed by more complex mechanisms. Some of these reactions are shown in Figure 9.26, which illustrates the conversion of (S)-3-hydroxy-β-ionone into 1,2-dihydro-1,1,6-trimethylnaphthalene. In small amounts, this compound is a fragrant component of peaches, strawberries, tomatoes, and wine. In passion fruit juices, it causes an off-flavour described as aroma flattening during pasteurisation, whilst in older vintages of Riesling wines it may be responsible for the kerosene-like smell.

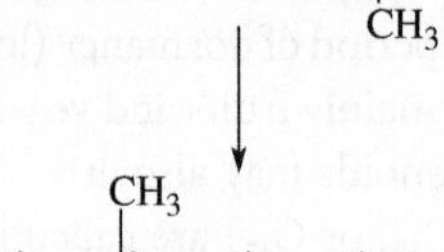

9-145, megastigmane

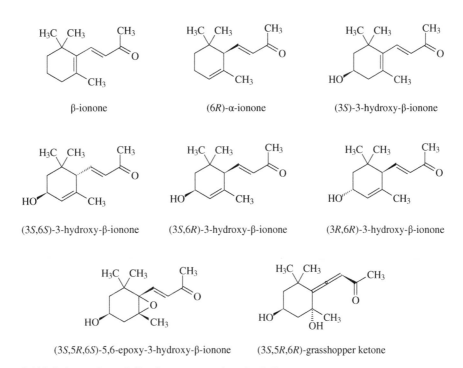

Figure 9.26 Formation of 1,2-dihydro-1,1,6-trimethylnaphthalene from (S)-3-hydroxy-β-ionone.

β-ionone

(6R)-α-ionone

(3S)-3-hydroxy-β-ionone

(3S,6S)-3-hydroxy-β-ionone

(3S,6R)-3-hydroxy-β-ionone

(3R,6R)-3-hydroxy-β-ionone

(3S,5R,6S)-5,6-epoxy-3-hydroxy-β-ionone

(3S,5R,6R)-grasshopper ketone

9-146, fission products of alicyclic carotenes and xanthophylls

Figure 9.27 illustrates the mechanism of transformation of (3S,6R)-3-hydroxy-α-ionone into very important odour-active compounds, 3-oxo-α-ionone and 3,4-dihydro-3-oxoedulanes (components of wine aroma), a decalin derivative (a component of passion fruit aroma), and megastigmatrienones found in tobacco.

Several important C_{13} compounds are derived from epoxy apocarotenoids. These epoxides yield, by epoxy group protonation and C-4 deprotonation, C-6 alcohols (epoxy apocarotenoids with hydroxyl group in the side chain give rise to diols), which may occur as glycosides. Their hydrolysis during fruit ripening yields the corresponding alcohols. An example is oxidation of 5,6-epoxy-3-hydroxy-β-ionone to dehydrovomifoliol, which is reduced to blumenol A, from which arises blumenol B and subsequently the important fragrance compound theaspirone. Theaspirone occurs, for example, in passion fruit, black tea, and tobacco (Figure 9.28).

Other spiroethers (8,9-dehydrotheaspirones) may arise analogously from 7,8-dehydrovomifoliol via 7,8-dihydrovomifoliol by reduction of two double bonds. 7,8-Dihydrovomifoliol is found in fruits and vegetables in the form of glycosides. The (S)-7,8-dehydrovomifoliol has a fruity and flowery odour, but the (R)-isomer has an odour resembling wood. Dehydrovomifoliol and blumenol A may also be formed by singlet oxygen oxidation of 3-oxo-β-ionone. Structurally similar theaspiranes are important odorous components of many fruits (strawberries, blackberries, grapes, passion fruit, guava, and others) and black tea. (2S,5S)-Isomers and (2R,5R)-isomers (**9-147**) have a slight smell resembling camphor, the (2S,5R)-isomer has the smell of camphor and naphthalene, but the (2R,5S)-isomer has a very attractive fruity

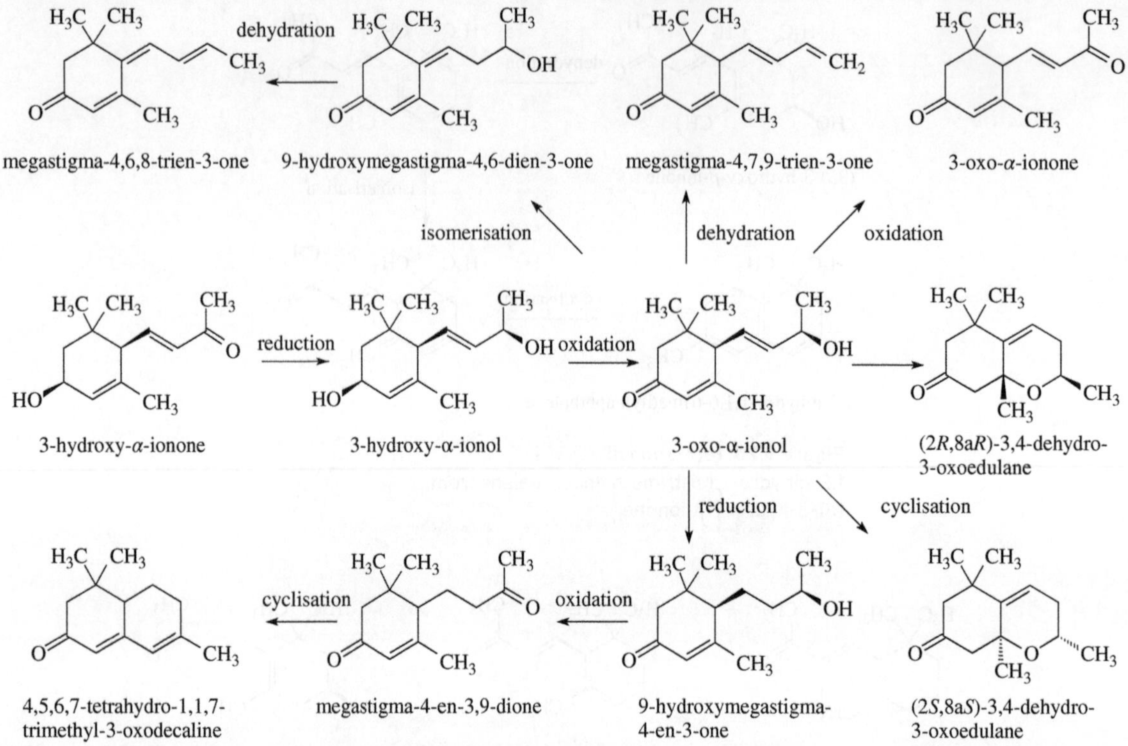

Figure 9.27 Formation of odorous compounds from (3S,6R)-3-hydroxy-α-ionone.

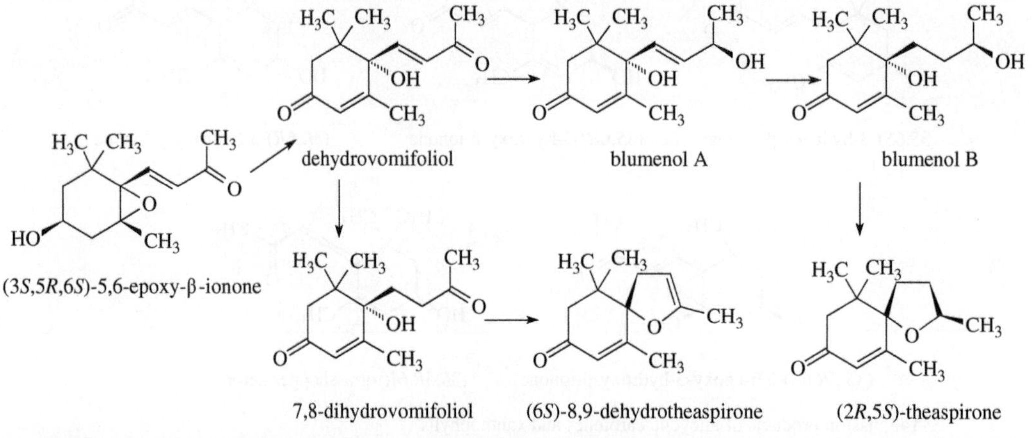

Figure 9.28 Formation of odorous compounds from epoxycarotenoids.

odour reminiscent of blackcurrants. Thanks to the enantioselectivity of biochemical mechanisms, some chiral compounds predominate in certain plant materials. For example, green tea contains about 85% (2S)-theaspiranes and 15% (2R)-theaspiranes. Related vitispiranes (**9-148**) are important components of the aroma of grapes and wines and also occur in black tea and tobacco. 4-Oxo-β-ionone (**9-149**) is formed by photooxidation of α-ionone.

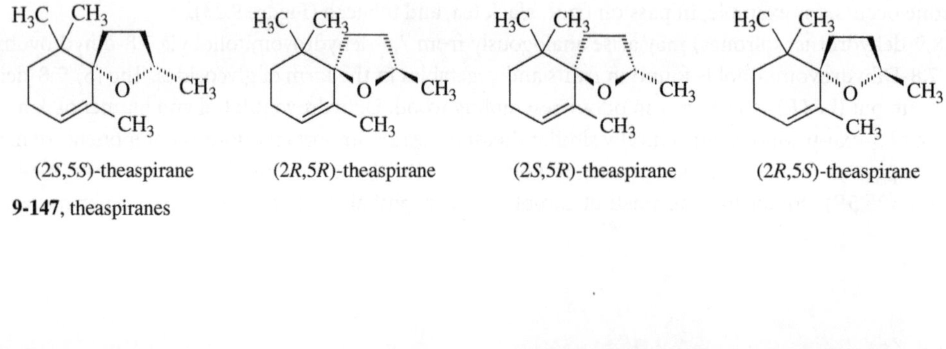

(2S,5S)-theaspirane   (2R,5R)-theaspirane   (2S,5R)-theaspirane   (2R,5S)-theaspirane

9-147, theaspiranes

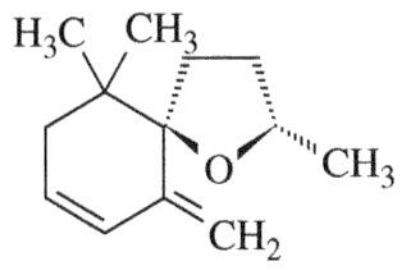

9-148, (2S,5R)-vitispirane

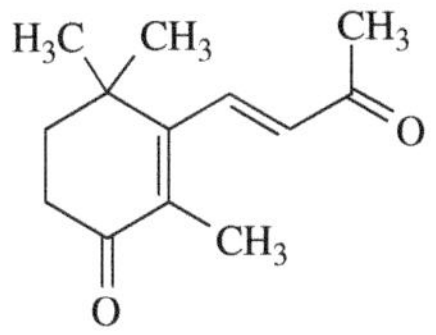

9-149, 4-oxo-β-ionone

Very important fragrant compounds are the C_{13} damascones oxidised on carbon C-7. β-Damascol, as a racemate, is an odorous component of papaya, black tea, rum, and tobacco. It arises by oxidation of β-ionol with singlet oxygen (Figure 9.29) via β-damascone, which is reduced to β-damascol. α-Ionol may analogously be produced by oxidation of α-damascone. Its aroma resembles roses and fruits. The precursor of the related fragrant substance (E)-β-damascenone is grasshopper ketone, formed by the cleavage of neoxanthin. Grasshopper ketone (Figure 9.30) is found in many fruits and vegetables (apricots, star fruit, grapes, kiwi, mango, apples, raspberries, and tomatoes), coffee, black tea, honey, and alcoholic beverages (beer, wine, rum, and brandy). Hydroxylated damascones (such as grasshopper ketone and 3-hydroxy-β-damascenone) and other C_{13} alcohols are commonly found in plants as glycosides, the aglycones of which are released by enzymatic or acid hydrolysis. For example, glycosides present in grapes are hydrolysed during must production and wine fermentation.

9.9.5.2.2 Apocarotenoids C_{11}

An important compound of this group of apocarotenoids is the C_{11} apocarotenoid dihydroactindiolide, presenting a cooling effect in the oral cavity. Dihydroactindiolide typically occurs as a constituent of tomato, black tea, and tobacco aroma. It is formed by oxidation of β-carotene, β-ionone, and β-ionol with singlet oxygen (Figure 9.31).

9.9.5.2.3 Apocarotenoids C_{10}

Cleavage of carotenoids in the C-7/C-8 (C-7′C-8′) double bond gives products that are formally monoterpenoids. Cleavage of β-carotene yields 2,6,6-trimethylcyclohex-1-ene-1-carbaldehyde, which is known as β-cyclocitral (**9-150**). Cyclocitral has been found in apricots, sugar melons (*Cucumis melo*), watermelon (*Citrullus lanatus*, Cucurbitacaeae), tomatoes, bell peppers, peas, broccoli, black tea, rum, and other materials. The best-known aromatic substance of this group of apocarotenoids is safranal (4,5-dehydro-β-cyclocitral, **9-150**), which has the smell and taste of saffron (see Section 9.9.2.3.2). Safranal arises by hydrolysis of a bitter-tasting glycoside, picrocrocin, and by dehydration of aglycone (R)-4-hydroxy-β-cyclocitral (see Section 8.3.5.1.3). It is also a minor component of grapefruit juice, black tea, and bell pepper aroma.

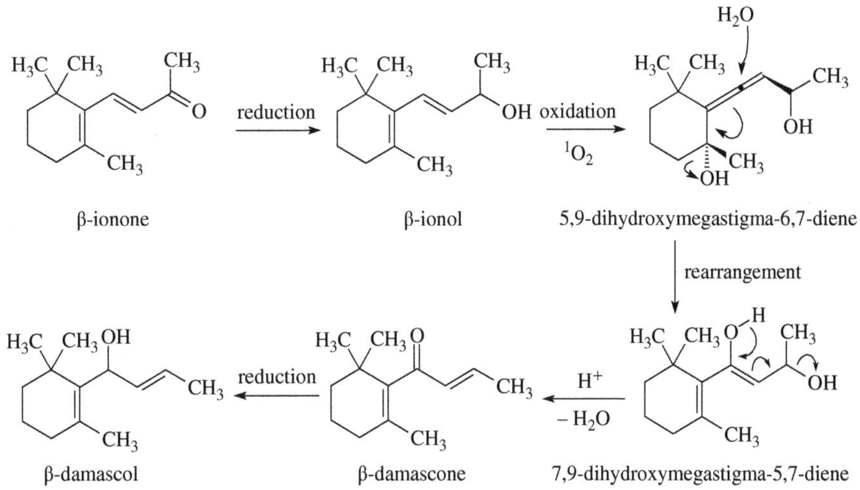

Figure 9.29 Formation of β-damascone and β-damascol from β-ionone.

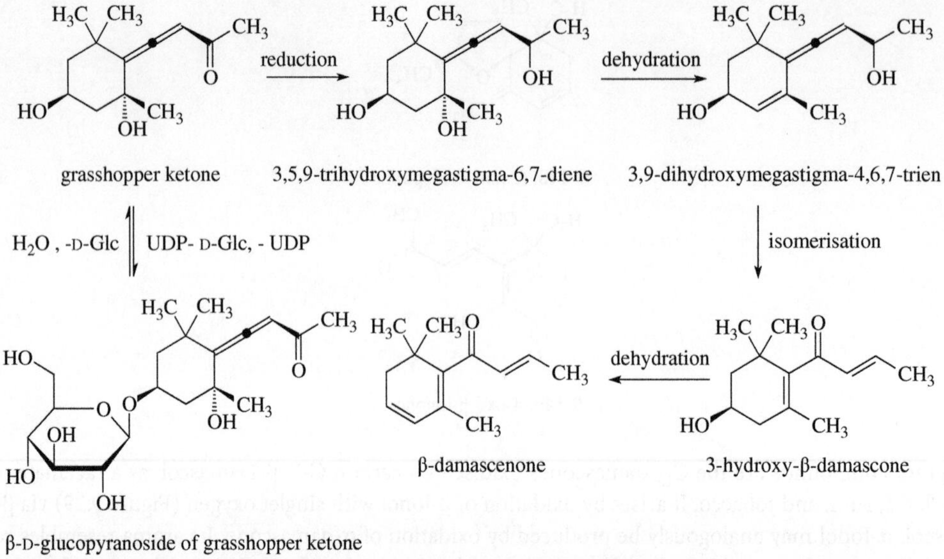

Figure 9.30 Formation of β-damascenone from grasshopper ketone.

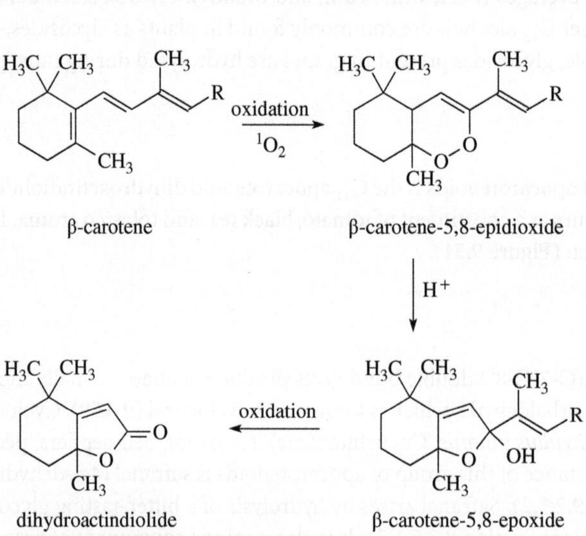

Figure 9.31 Formation of dihydroactindiolide from β-carotene.

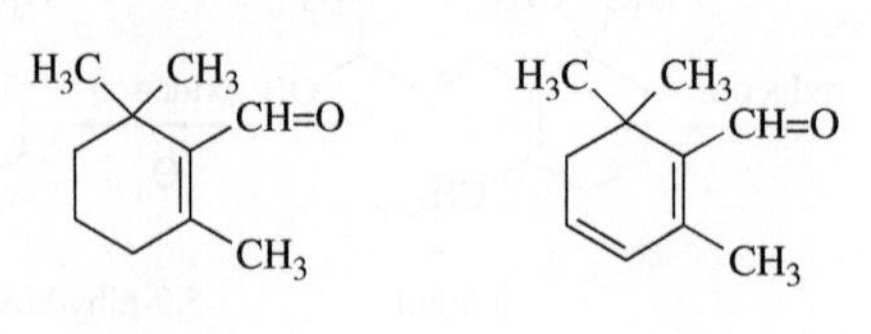

β-cyclocitral   safranal

9-150, C9 apocarotenoids

9.9.5.2.4 Apocarotenoids C₉ and C₈

The main formation pathway of C_9 apocarotenoids with cyclohexanone and cyclohexenone structures is conversion of hydroperoxides derived from β-damascol (Figure 9.32). Hydroperoxides generated by autoxidation of carotenoids are further oxidised, reduced, and hydrated to form a variety of different structures. The most important compound of this apocarotenoid group is probably 2,6,6-trimethylcyclohex-2-en-1,4-dione (4-oxoisophorone, 9-151), whose scent recalls dry straw. It occurs in black tea, saffron, and

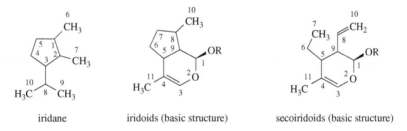

Figure 9.32 Formation of cyclohexanones and cyclohexenones.

1,1-dimethyl-3-methylene-cyclohexan-2-one

β-damascol

1,1,3-trimethylcyclo-hex-3-en-2-one

tobacco. Acyclic carotene dehydrolycopene gives rise to methylketone (3*E*,5*E*)-6-methylhepta-3,5-dien-2-one (**9-151**), occurring in tomatoes. Another example of C_8 apocarotenoids in tomatoes is methylketone 6-methylhept-5-en-2-one (**9-151**), which is produced by C-5/C-6 (C-5′/C-6′) double-bond cleavage of acyclic carotene lycopene. It may also arise from phytoene.

4-oxoisophorone

6-methylhepta-3,5-dien-2-one

6-methylhept-5-en-2-one

9-151, C_9 and C_8 apocarotenoids

9.10 Iridoids

Iridoid monoterpenes called **iridoids** are a group of compounds derived from the skeleton of the hydrocarbon iridane (**9-152**). Their usual structure is represented by formulae **9-152**. The cleavage of the cyclopentane ring and introduction of various functional groups (mainly by oxidation) to the fission products yields **secoiridoids** (**9-152**), which include many bitter plant metabolites.

iridane

iridoids (basic structure)

secoiridoids (basic structure)

9-152, structures of iridane, iridoids and secoiridoids

Pigments derived from yellow-to-red fruits of common gardenia, also known as cape jasmine (*Gardenia jasminoides*, Rubiaceae), an evergreen shrub growing in India, China, and Japan, are iridoid pigments, important for their use as dyes in the textile industry and for imparting colour to foods. Fruits located in pods contain three main groups of yellow and orange pigments: iridoid pigments, apocarotenoids (crocin and crocetin such as saffron), and flavonoids. Interest in the gardenia pigment was initiated in Japan through the pursuit of substitution of annatto and saffron with cheaper pigments in cases where their specific flavour was not required. The main iridoid pigments of gardenia are the glucosides gardenoside, geniposide, and gardoside (**9-153**). They are accompanied by a number of related minor pigments.

gardenoside

geniposide

gardoside

9-153, examples of main gardenia iridoids

Preparation of pigments is based on their extraction with water, enzymatic hydrolysis of glycosides, and reaction of aglycones with amines. By adjusting the process conditions (temperature, pH, oxygen concentration, type of reacting amino compounds, and presence of metal ions), a number of pigments can be obtained whose colour is yellow-to-green, red, purple, or blue: a combination of the colours of the pigments produced. Gardenia pigments can be used to impart colour to ice creams, confectionery, pasta products, and other foods.

Relatively rare iridoid β-D-glucosides, known as cornusfurosides A–D, containing a furan ring, and their dimers, called cornusides A–O (**9-154**), were isolated from the fruits of *Cornus mas* and related *C. officinalis* (Cornaceae), species of dogwood also known as European cornel and Japanese cornel, respectively.

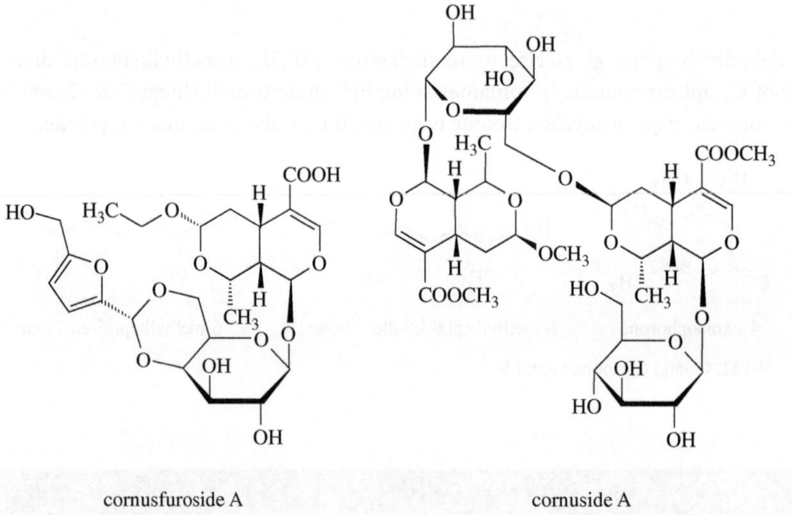

cornusfuroside A   cornuside A

9-154, examples of cornusfurosides and cornusides

9.11 Other terpenoid pigments

The toxic yellow pigment gossypol, 2,2'-bis(8-formyl-1,6,7-trihydroxy-5-isopropyl-3-methyl)naphthalene, resembles phenolic pigments, but its biosynthetic precursor is a sesquiterpene hydrocarbon (+)-δ-cadinene with the chemical name (1*S*,8a*R*)-1-isopropyl-4,7-dimethyl-1, 2,3,5,6,8a-hexahydronaphthalene (see Section 8.2.1.1.1). In a sequence of oxidation reactions, δ-cadinene is transformed into the gossypol intermediate 4-formyl-1-isopropyl-2,3,5,7-tetrahydroxynaphthalene, which is known as hemigossypol. Further one-electron oxidation of hemigossypol gives rise to a C-5 radical, the resonance form (C-6) of which dimerises to gossypol.

Gossypol occurs in buds and seeds of upland (Mexican) cotton (*Gossypium hirsutum*, Malvaceae) and in other *Gossypium* species. Cotton seeds usually contain 0.6–1.5% of gossypol localised in special pigment cells. Gossypol occurs in two enantiomers, as (+)-gossypol and (−)-gossypol (**9-155**), in the ratio of 3 : 2. In some varieties of cotton, 6-methoxygossypol and 6,6'-dimethoxygossypol are found as minor products. During cottonseed oil extraction, gossypol passes from the seeds to the oil. It is present in flour, used in some countries for human nutrition, and in meal, used as feed for poultry and livestock. Currently cultivated cotton varieties have a low content of gossypol.

Only (−)-gossypol is toxic, and only for omnivores. Chronic intoxication caused by low, long-term income is reflected in reduced appetite, liver damage, and reduced blood clotting, as well as a reversible sterility in males at a dose of about 10 mg/day. In addition to these effects, gossypol has some activity as an antimalarial drug and is studied for its anticarcinogenic properties.

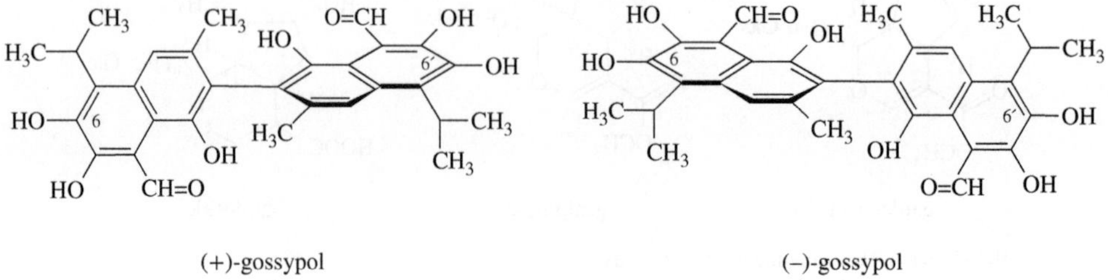

(+)-gossypol   (−)-gossypol

9-155, gossypol

Gossypol is a very reactive compound, forming salts with metal cations, which involves hydroxyl groups at C-1 and C-1′ (Figure 9.33). By oxidation with oxygen or with ferric ions, the corresponding *o*- and *p*-naphthoquinones arise, which mutually polymerise and react with proteins and phospholipids with the formation of dark pigments. In scavenging free radicals, gossypol shows greater activity than both methylated derivatives.

The reaction of gossypol in milled seeds with globulins, to form hydrophilic and hydrophobic pigments, has been used to detoxify cottonseed oil. The detoxifying effect is increased by seed steaming with the addition of ferrous sulfate, which catalyses decarbonylation of gossypol to apogossypol, which is not stable and is less toxic than gossypol. The detoxified material contained 0.6–0.9% bound gossypol and 0.01–0.02% free gossypol. Small quantities of gossypol in oil can react (either directly or after oxidation to quinone) with amino groups of phospholipids to form imines, which polymerise to fat-soluble macromolecular dark pigments; these cannot be effectively removed during subsequent refining.

The colours of some higher fungi of the genus *Lactarius*, collectively known as milk caps, and of the closely related genus *Russula* (Russulaceae) are based on sesquiterpenoid pigments. The colours of the injured flesh and the latex of several mushroom species of the genus *Lactarius* undergo changes as a result of the presence of sesquiterpenoids. An example is *L. uvidus*, which has white latex that rapidly turns violet on exposure to the air. The compounds responsible for this colour change are mostly derivatives of sesquiterpene hydrocarbon (4a*R*,5*S*,6*S*,8a*S*)-1,1,4a,5,6-pentamethyltetrahydronaphthalene (called drimane), such as drimenol (**9-156**), the corresponding 6,7-epoxide known as uvidin A, and their transformation products.

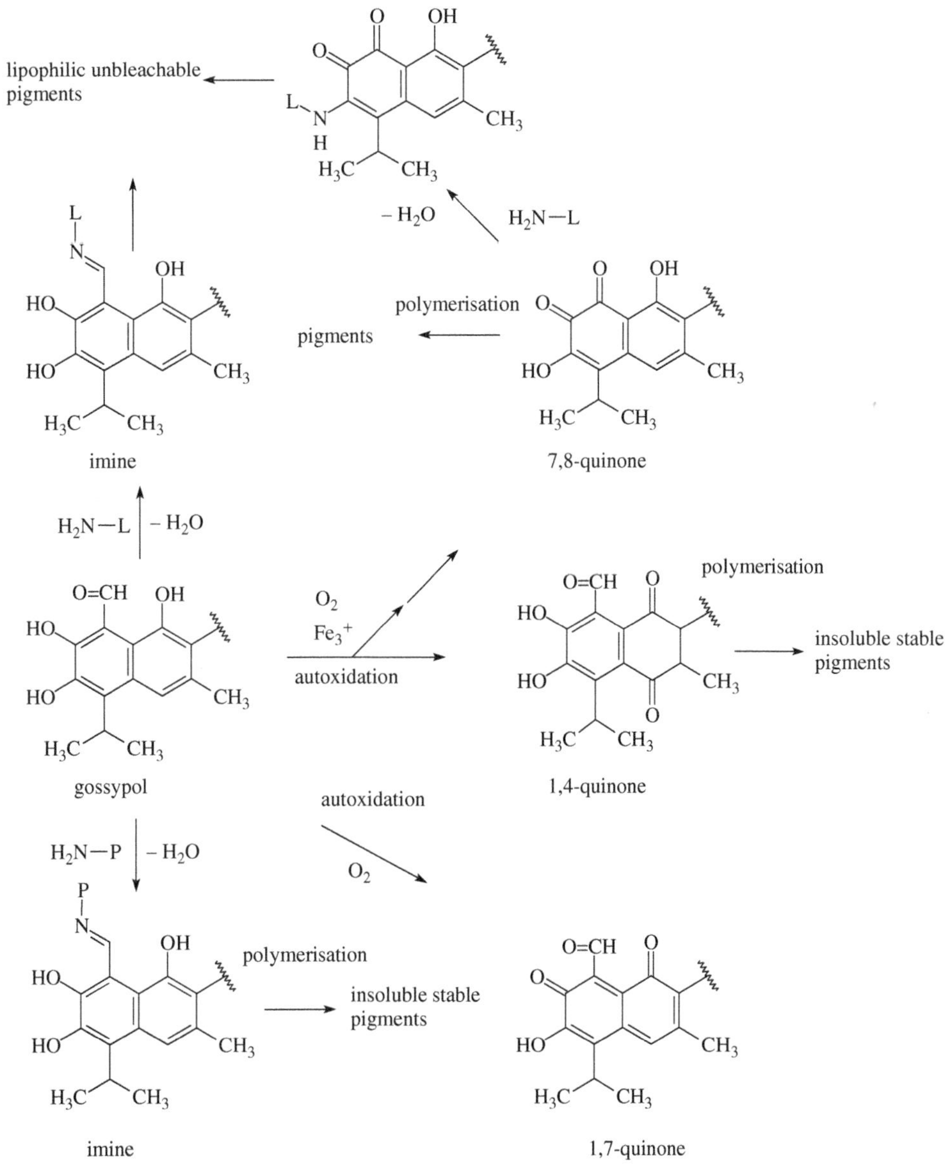

Figure 9.33 Reactions of gossypol (H_2N-P = protein with bound lysine, H_2N-L = phospholipid with bound ethanolamine).

Blue pigments containing seven-membered rings are derived from sesquiterpene bicyclo[5.3.0]deca-1,3,4,6,8-pentaene, known as azulene (**9-157**). The pigments vetivazulene (4,8-dimethyl-2-isopropylazulene) and guaiazulene (1,4-dimethyl-7-isopropylazulene) are found in mushrooms and some marine invertebrates. For example, the young fruiting body of *Lactarius deliciosus* starts out carrot-coloured but slowly turns green on ageing. The compounds responsible for these colour changes are the blue pigment 1,4-dimethyl-7-(1-methylethenyl)azulene, known as lactarazulene, lipophilic red-violet 4-methyl-7-(1-methylethenyl)azulene-1-carbaldehyde, called lactaroviolin (**9-157**), and related pigments, such as the red 7-acetyl-4-methylazulene-1-carbaldehyde.

drimane   drimenol

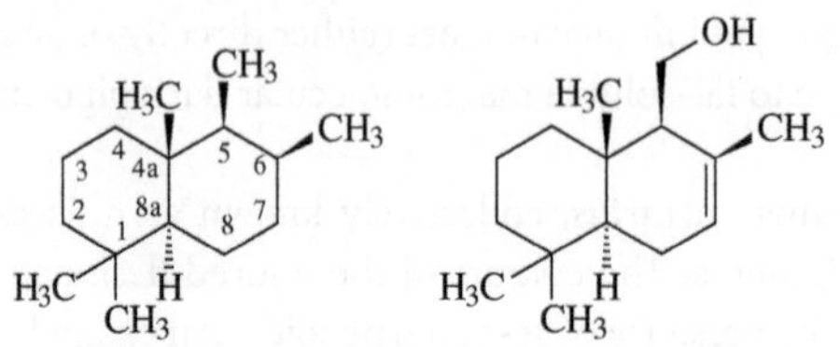

9-156, drimane and derived pigment

The lipophilic pigment 1-hydroxymethyl-4-methyl-7-(1-methylethenyl)azulene stearate (R = $CH_2OCO(CH_2)_{16}CH_3$, **9-157**) is responsible for the brilliant blue colour of the indigo milk cap (*Lactarius indigo*), native to America and Asia, which has also been reported from southern France. Azulene also gives the characteristic blue colour to chamomile oil (*Matricaria chamomilla*, Asteraceae).

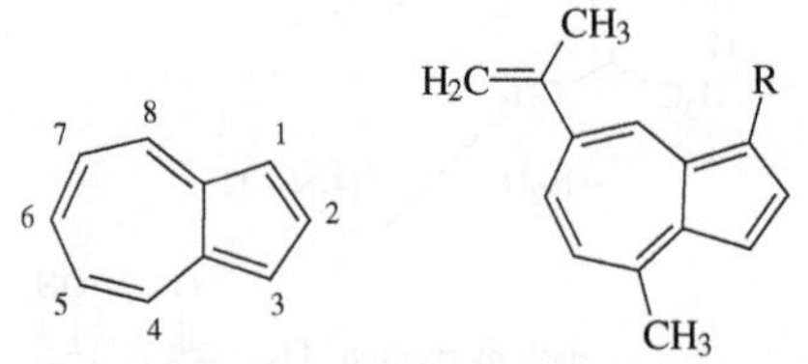

azulene   lactarazulene, R = CH_3
                 lactaroviolin, R = CH=O

9-157, azulene and derived pigments

9.12 Enzymatic browning

Enzymatic browning is a complex enzyme-catalysed and spontaneous (non-enzymatic) process that creates melanin-like brown pigments and other coloured products. Substrates in enzymatic browning are polyphenols oxidised by polyphenol oxidases and other enzymes to quinones, which are subsequently transformed into coloured pigments.

In many cases, enzymatic browning is to some extent desirable, because it leads to the formation of the characteristic colour and flavour of some food products, such as in the fermentation of tea and of cocoa beans, ripening of dates, and production of black olives, raisins, sherry-type wines, and red wines.

In general, enzymatic browning is an undesirable phenomenon. It is responsible for up to 50% of all losses during fruit and vegetable production, leads to browning in some fresh-cut fruits, vegetables, and other products (e.g. apples, potatoes, and mushrooms), and has negative effects on colour, taste, and nutritional value, particularly in post-harvest storage of fresh fruits. Adverse enzymatic browning also occurs in the processing of some seafood (e.g. shrimp and crabs). All of these colour changes are usually adverse reactions often classified as discolourations and defects.

In foods, many adverse reactions also occur that cannot be strictly classified as enzymatic browning. For example, procyanidins may cause the pink discolouration of canned pears (see Section 8.3.6.2.2), whilst adducts of anthocyanins with sulfur dioxide can be responsible for the pinking of white wines (see Section 9.4.1.5.7). A red discolouration originating from ascorbic acid and amino acids may be produced in pickled cauliflower (**5-49**), whilst a pink and yellow discolouration caused by reactions of isothiocyanates with amino acids can occur in *Brassica* vegetables (see Section 2.5.2.4). The precursors of the green-to-blue discolouration of garlic and the pink-to-red discolouration of onion and leek are products of amino acids with sulfenic acid resulting from isoalliin and other *S*-alk(en)ylcysteine *S*-oxides (see Section 8.2.9.1.4). The green-to-blue discolouration of sweet potatoes results from the reaction of the amino acids with chlorogenic acid (see Section 9.12.4).

9.12.1 Enzymes

Enzymes from the class of oxidoreductases, catalysing enzymatic browning reactions, are most often known trivially as polyphenol oxidases. Two groups can be identified:

- **catechol oxidases;**

- **laccases**.

Catechol oxidases, formerly known as monophenol oxidases, catalyse two different reactions (Figure 9.34). The first is the oxidation (hydroxylation) of monophenols to o-diphenols (o-hydroquinones, 1,2-dihydroxybenzenes), which relates to the enzyme activity known as cresolase activity. These enzymes are now called monophenol oxygenases. The second reaction is the oxidation of o-diphenols to o-quinones (benzo-1,2-quinones also called 1,2-dioxobenzenes), which concerns the enzyme activity known as catecholase activity. Enzymes showing this activity were previously called diphenol oxidases or o-diphenolases but are now known under the systematic name o-diphenol:O_2 oxidoreductases. An example is tyrosinase, a copper-containing enzyme present in plant and animal tissues that produces dopa from tyrosine and catalyses the two-electron oxidation of o-diphenols (not p-diphenols) to their corresponding quinones. Some catechol oxidases lack this type of activity. Laccases oxidise not only o-diphenols to o-quinones, but also p-diphenols to p-quinones (benzo-1,4-quinones also called 1,4-dioxobenzenes) (Figure 9.34). Other enzymes that are able to oxidise phenols to quinones are peroxidases. Their direct participation in enzymatic browning reactions is not yet fully understood.

Polyphenol oxidases are widespread in most plant species. They occur mainly in plastids of plant cells. For example, polyphenol oxidases in apples are located in chloroplasts and mitochondria, usually tightly bound to the membrane. The cytoplasm contains soluble polyphenoloxidases, whose activity increases with maturation.

The activity of polyphenol oxidases depends mainly on the type, variety, and age of plant material (e.g. unripe fruits turn brown faster), cultivation conditions, and method of processing, and increases with mechanical injury of plant tissue. The optimal pH for the enzyme activity is 4.5–5.5, but the enzymes are active even in more acidic media. For example, they show approximately 40% of optimal activity at pH 3. The highest activity of polyphenol oxidases is in the outer layers of fruits, especially in the skin. In animals, the enzyme tyrosinase is located mainly in the skin, hair, and eyes.

9.12.2 Substrates

The tendency of plant materials towards browning varies within plant species, but the enzyme activity also depends on the type and amount of substrates available. The substrates of polyphenol oxidases may be a large number of different phenolics, but usually only a few are of practical importance (Table 9.17). Exceptionally, other compounds may also contribute to the browning, such as sesquiterpenoids in lettuce.

Phenolic compounds are present separately from the substrates, predominantly located in the cell vacuoles. Particularly prominent substrates are caffeic acid and its esters, as well as some flavonoid substances, of which monomeric flavan-3-ols (catechins) are the most important. Other groups of phenolic compounds, such as condensed forms of flavan-3-ols and flavan-3,4-diols (tannins), flavonols, flavones, flavans, chalcones, dihydrochalcones, and anthocyanins, are only partly oxidised. One of the reasons for this is probably the steric hindrance caused by the corresponding glycosides and substrate specificity of polyphenol oxidases. The content of phenolic compounds depends on genetic factors (on plant species and varieties), the degree of maturity, and external (environmental) factors (light, temperature, nutrients, and so on). The only substrate in animal tissues is the amino acid tyrosine.

Quinones themselves are mostly coloured compounds. Simple quinones are usually red when they contain an o-quinoid structure in the molecule (such as o-benzoquinone), whilst compounds with the p-chinoid arrangement are yellow (e.g. p-benzoquinone). The catechin oxidation product is bright yellow, quinone formed of chlorogenic acid has a yellow-orange colour, and quinone derived from amino acid dopa (dopaquinone) is pink.

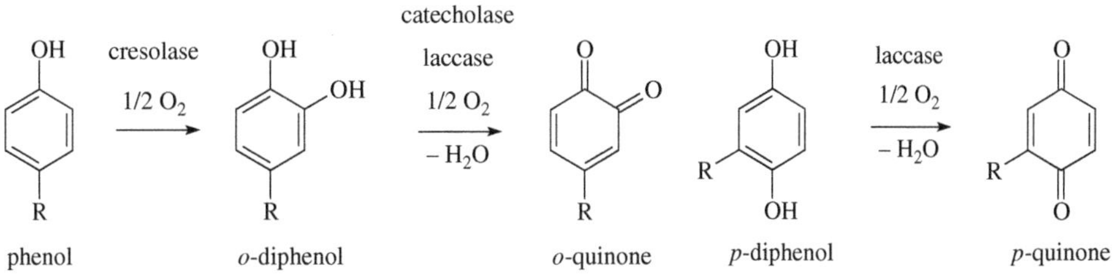

Figure 9.34 Reactions catalysed by polyphenol oxidases with creasolase, catecholase, and laccase activities.

Table 9.17 Important substrates of polyphenol oxidases in food materials.

Food	Substrate	Food	Substrate
Fruits		*Vegetables and potatoes*	
Apples	Chlorogenic acid, catechins, tannins	Lettuce	Tyrosine, lettucenins
Pears	Chlorogenic acid	Beans (green)	3,4-Dihydroxyphenylalanine (dopa)
Grapes	Caftaric acid, catechins	Onion	Protocatechuic acid
Peches	Catechins	Potatoes	Tyrosine, chlorogenic acid, catechins
Olives	3,4-Dihydroxyphenylethanol and its derivatives	*Others*	
Bananasy	3,4-Dihydroxyphenylethylamine	Mushrooms	Tyrosine, pulvinic acids, and other phenols
Dates	Dactyliferic acid	Tea leaves	Catechins
Mango	Gallic acid, gallotannins	Coffee beans	Chlorogenic and caffeic acids
Avocado	Phenolic acids	Cocoa beans	Catechins

9.12.3 Non-enzymatic oxidation and subsequent reactions

Plant phenols can be oxidised to the corresponding *o*-diphenols and subsequently to *o*-quinones not only by enzymes but also by hydrogen peroxide (Figure 9.35), which is a product of ascorbic acid oxidation catalysed by transition metal ions (see Section 5.14.6.1), autoxidation of *o*-diphenols, or autoxidation of cuprous ions in an acidic solution:

$$o\text{-diphenol} + O_2 \rightarrow o\text{-quinone} + H_2O_2$$

$$O_2 + 2\,Cu^+ + 2\,H^+ \rightarrow 2\,H_2O_2 + 2\,Cu^{2+}$$

In acidic fruit juices, hydrogen peroxide preferably oxidises other substances. For example, in wine, it oxidises ethanol to acetaldehyde, and in strawberries, it oxidises anthocyanins to colourless products.

Quinones are highly redox-active molecules that can redox cycle with their semiquinone radicals ($SQ^{\bullet-}$), formed by one-electron reduction, leading to the formation of reactive oxygen species, including superoxide radical ($O_2^{\bullet-}$), hydrogen peroxide, and, ultimately, hydroxyl radical ($HO^{\bullet}$):

$$o\text{-diphenol} + o\text{-quinone} \rightleftharpoons 2\;SQ^{\bullet-} + 2\;H^+$$

$$o\text{-quinone} + O_2^{\bullet-} \rightleftharpoons SQ^{\bullet-} + O_2$$

$$O_2^{\bullet-} + 2\;H^+ \rightleftharpoons H_2O_2$$

Production of reactive oxygen species can cause severe oxidative stress within cells through the formation of oxidised cellular macromolecules, including lipids, proteins, and DNA.

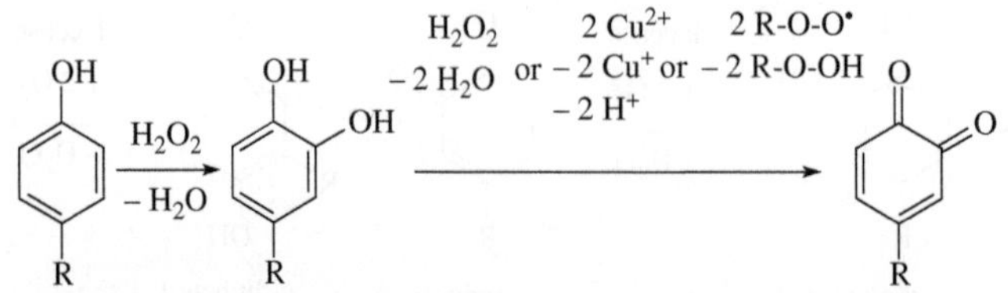

Figure 9.35 Non-enzymatic oxidation of phenols to *o*-diphenols and *o*-quinones.

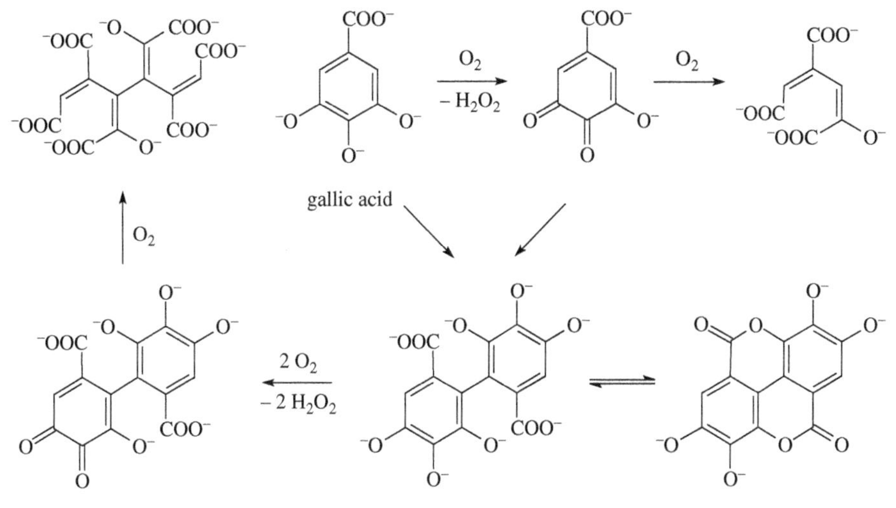

Figure 9.36 Gallic acid autoxidation in alkaline solutions.

In a neutral and alkaline medium, when the phenolic OH groups (and carboxylic groups of phenolic acids) ionise, the anions spontaneously yield products of higher molecular weights. Some of these reactions are shown by the example of gallic acid in Figure 9.36. The main reaction product is hexahydroxybiphenylic acid, which is a biochemical precursor of ellagic acid. Hexahydroxybiphenylic acid and, in part, gallic acid also produce acyclic oxidation products. A large number of autoxidation products are similarly produced from cinnamic acids. The structures of certain dimers produced by autoxidation of caffeic acid in alkaline media at elevated temperatures are given in formulae **9-158**. These dimers are lignans, trivially called **caffearins**. Products of similar structures arise from other cinnamic acids; for example, the autoxidation product of sinapic acid is thomasidioic acid (**9-158**). In neutral media of pH 7, thomasidioic acid is produced from about 30% of the initially present sinapic acid. In alkaline media of pH 8.5, virtually all sinapic acid is transformed into thomasidioic acid.

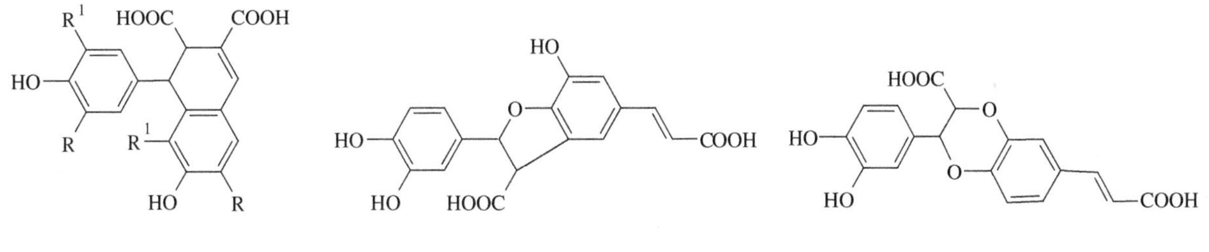

cyclohexene type, R = OH, R^1 = H
thomasidioic acid, R = R^1 = OCH_3

furan type

dioxolane type

9-158, caffeic acid dimers

9.12.4 Reactions of quinones

o-Quinones, formed by the oxidation of phenolic compounds, are highly reactive compounds. The majority of the reactions of quinones involve nucleophilic Michael additions. The addition and substitution reactions of quinones with nucleophiles, particularly amino and thiol groups, have been extensively documented. Some of these are shown in simplified form in Figure 9.37. Quinones form adducts with the regenerated structure of the original diphenol, which have a lower redox potential than the original. These adducts can be easily oxidised to the corresponding substituted *o*-quinones. The reactivity of *o*-quinones depends on the structure of the starting phenol and is determined by the redox potential of the reacting compounds. For example, *o*-quinones resulting from the oxidation of chlorogenic acid may oxidise catechins to *o*-quinones, regenerating chlorogenic acid, but the opposite reaction is not possible. Reactivity also depends on pH, temperature, and the presence of other compounds. Reaction with water yields derivatives of 1,2,4-trihydroxybenzene as intermediates, which on oxidation produce hydroxy derivatives of *o*- and *p*-quinones. Reactions of this type with tyrosine residues in the molecules of oxidoreductases lead to the inactivation of enzymes.

Reactions leading to brown pigments are based on condensation of *o*-quinones with the original *o*-diphenols to dimeric products and on condensation of *o*-quinones with amino acids, peptides, and proteins. A similar reaction also occurs, for example, in the formation of

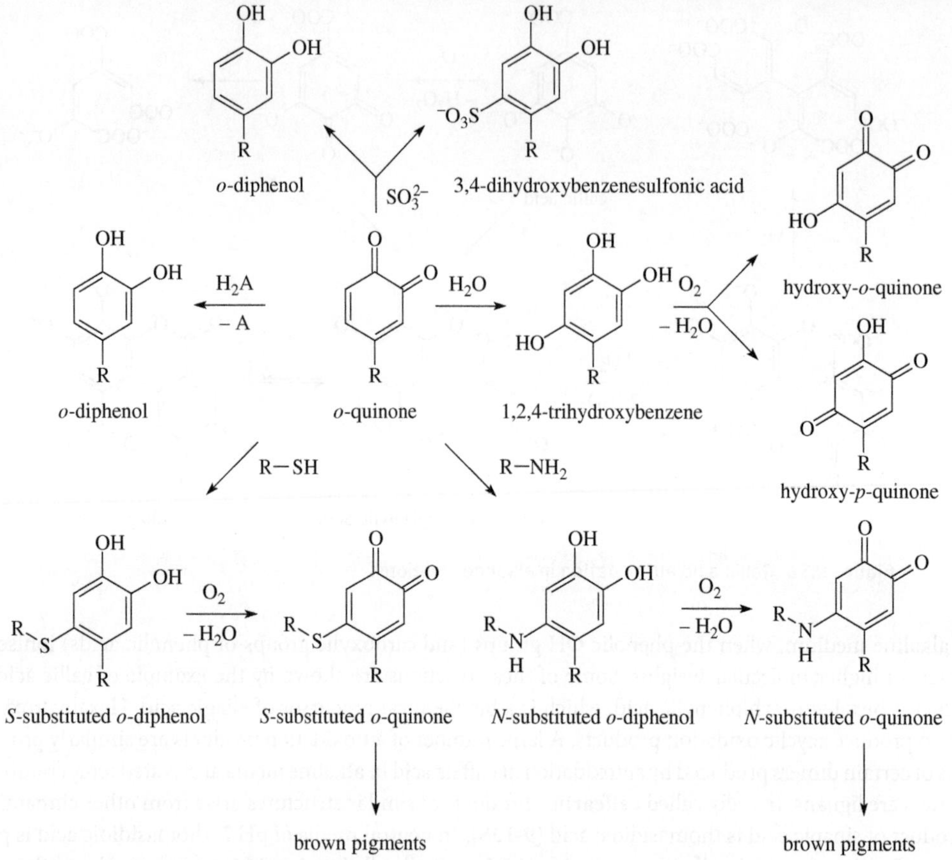

Figure 9.37 Reaction of quinones with food components (H$_2$A = ascorbic acid).

astringent tannins and pigments formed during red wine maturation. Dimers of phenolic compounds with the structure of *o*-diphenols are further oxidised by the action of polyphenol oxidases or by other quinones, and oxidation products condense again with phenols to give brown coloured higher oligomers and polymers (Figure 9.38). Quinones can even react with such *o*-diphenols, which are not otherwise oxidised to *o*-quinones. This reaction regenerates the original phenol, from which quinone was derived (Figure 9.37). Seven-membered tropone derivatives can also arise as reaction products. Their formation is analogous to the formation of theaflavin, which is produced during fermentation of green tea leaves (see Section 9.12.5.4).

Reaction of amino acids, peptides, and proteins with amino groups in neutral solutions leads to mono- and disubstituted *o*-quinones. In alkaline solutions, an amino group is added to the carbonyl group of quinones, which results in the replacement of one or both quinone oxygen atoms =O by an =N-R group and formation of quinone imines (Figure 9.39). Strecker degradation of amino acids also takes place as a side reaction. The reaction of proteins with plant phenols during the alkaline extraction of oil meals (e.g. in sunflower meal, which contains a considerable amount of polyphenols) leads, therefore, to a range of discolourations of the products, and reduces the protein nutritional value due to the loss of lysine (bound in the non-available products) and methionine (by oxidation) and a lower availability of tryptophan.

In the presence of certain amino acids, quinones may yield stable low-molecular-weight coloured products, in which case the browning reaction cannot proceed. Some plant materials, such as apples, sweet potatoes (*Ipomoea batatas*, Convolvulaceae), and other raw materials, especially when treated with alkaline agents, such as baking powder, turn green during culinary processing. Greening can also be seen during extraction of sunflower meal protein with alkaline reagents. Greening cause caffeic acid esters, particularly chlorogenic acid, to be oxidised

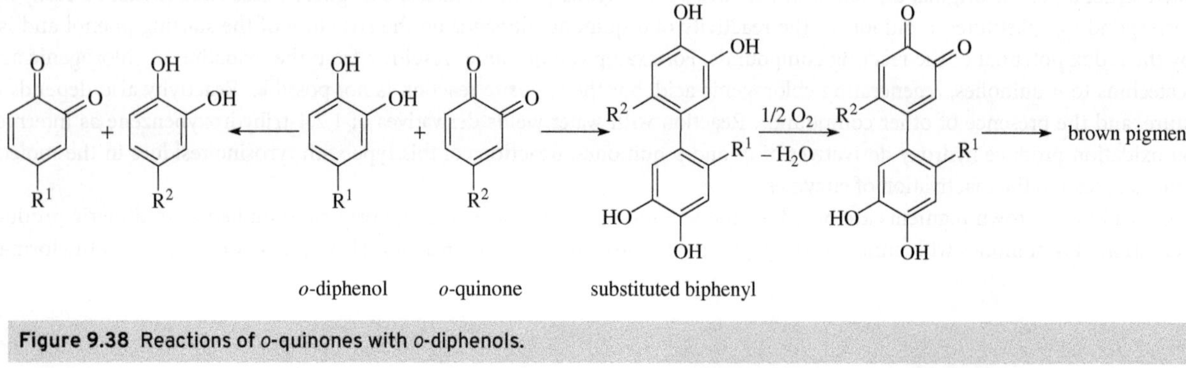

Figure 9.38 Reactions of *o*-quinones with *o*-diphenols.

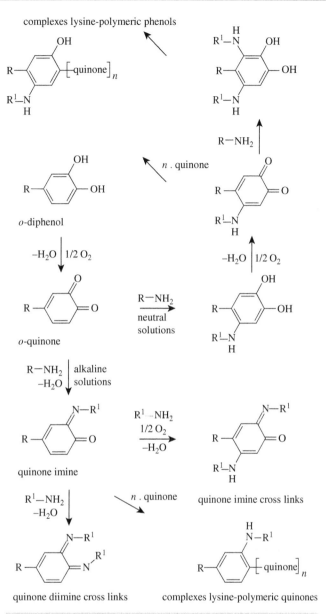

Figure 9.39 Reaction of lysine with o-quinones
(R^1-NHg = ε-amino group of lysine). Stephen (1995), Figure 10.
Source: Reproduced by permission of Taylor & Francis – Marcel Dekker.

to the corresponding o-quinones, which react with the primary amino group of amino acids or with primary amines. The mechanism of this green pigment formation is shown in Figure 9.40. The immediate precursor of the green pigment is yellow substituted trihydroxybenzacridine, the oxidation of which yields the green pigment. Reduction of the green pigment (e.g. with ascorbic acid) temporarily yields the yellow pigment. Oxidation of the green pigment in an alkaline solution produces a blue pigment. Amino acids with additional functional groups, such as serine and threonine, as well as proline and cysteine, do not form the green pigment.

The reaction of amino acids, peptides, and proteins with thiol groups occurs in weakly acidic and neutral solutions and proceeds analogously to the reaction with amino groups. An example might be the reaction of hydroxycinnamic acids with glutathione in white musts and wines. The main compounds involved in oxidation reactions are caftaric acid (**8-64**) and structurally similar depsides of hydroxycinnamic acids, which are prone to enzymatic oxidation reactions to o-quinones and other products during the winemaking process, leading to colour darkening. Glutathione is capable of trapping these o-quinones and thus limiting juice browning. It has been shown that the favoured reaction is the substitution of the sulfanyl group of glutathione at C-2 of the aromatic ring of hydroxycinnamic acids, resulting in the formation of 2-S adducts. The main product is (E)-2-(S-glutathionyl)caftaric acid (**9-159**). Several minor isomers, namely the cis-isomer of the 2-S adduct, trans-isomers of the 5-S and 6-S adducts, and the 2,5-di(S-glutathionyl)caftaric acid, were also identified. Aged bottled wines had an increased content of caffeic acid from hydrolysis of caftaric acid and of 2-(S-glutathionyl)caffeic acid from hydrolysis of glutathionylcaftaric acid.

In materials containing ascorbic acid (abbreviated as H_2A), o-quinones are reduced to the original 1,2-diphenols of lighter colour and ascorbic acid is oxidised to dehydroascorbic acid (abbreviated A). This reaction can be monitored at home by the addition of lemon juice

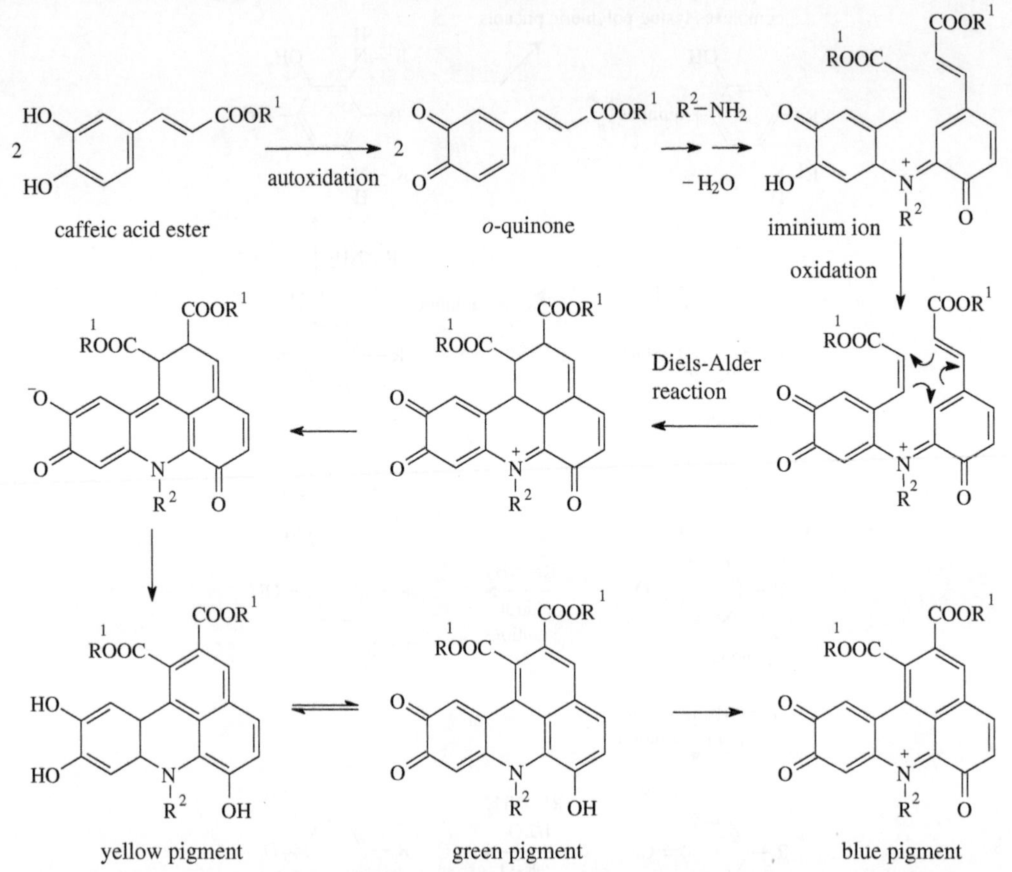

Figure 9.40 Formation of pigments from chlorogenic acid and amino acids.

to tea, which results in a lighter tea colour. Quinones can also be reduced by hydrogen sulfites (bisulfites) or sulfites, with the formation of the corresponding colourless 2′-sulfo derivatives (2′-sulfonic acids). The addition of hydrogen sulfites or sulfites to peeled potatoes is employed, for example, to prevent potato browning. The main 2′-sulfo derivatives identified in hydrogen sulfite-treated potatoes were derived from 5-*O*-caffeoyl-L-quinic acid (neochlorogenic acid) and 4-*O*-caffeoyl-L-quinic acid (cryptochlorogenic acid). An example is 2′-sulfo-5-*O*-caffeoyl-L-quinic acid (**9-160**). The reaction with hydrogen sulfite includes the caffeic acid quinone, produced by oxidation with polyphenol oxidase. The reaction is analogous to that of quinones with amino acids (see Section 2.5.2.5). It proceeds via the nucleophilic attack of hydrogen sulfite anions to the C-2′ of quinones derived from chlorogenic acids, which yields 4′-anions of chlorogenic acids as intermediates. Other caffeoyl derivatives and derivatives of quercetin glycosides are produced as minor sulfo-adducts. Feruloyl and sinapoyl derivatives do not form adducts with hydrogen sulfites.

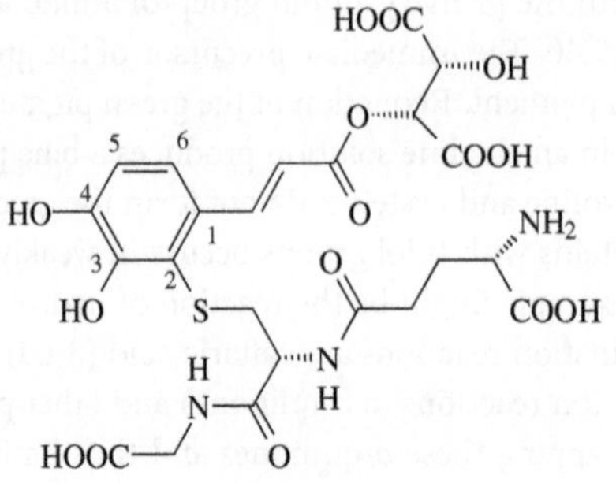

9-159, (*E*)-2-(S-glutathionyl)caftaric acid

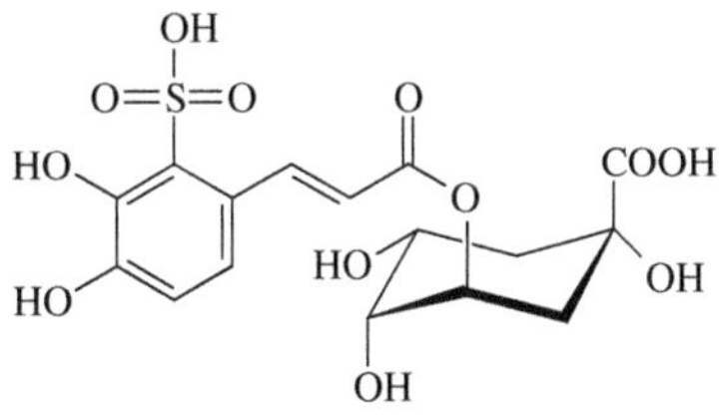

9-160, 2'-sulfo-5-*O*-caffeoyl-L-quinic acid

9.12.5 Reactions in foods

9.12.5.1 Apples

The main substrates of polyphenol oxidases in apples are hydroxycinnamic acids, especially caffeic acid depsides, such as chlorogenic acid, and some flavonoids, such as flavan-3-ols, flavonols, dihydrochalcones, and anthocyanins (in red-skinned apple varieties). In the flesh of apples, the main phenolics are chlorogenic (3-caffeoylquinic) acid, flavan-3-ol epicatechin, and procyanidin B_2, representing condensed tannins. These three compounds represent more than 90% of the phenolic compounds present. The main substrates in apple peel are flavan-3-ols and flavonols (quercetin glycosides), and to a lesser extent chlorogenic acid and some other depsides.

The concentration of phenolic compounds is highest in young fruit and decreases rapidly with the fruit's development. After harvesting, the total content of phenolic compounds is roughly constant or slightly lowered. The total content of phenolic compounds in ripe apples ranges between 0.5 and 11 g/kg in the flesh and between 8.7 and 19.2 g/kg in the skin. The phenolic compounds concentration in apple juice is around 1.5 g/kg.

9.12.5.2 Potatoes

Common phenolics of potatoes are caffeic acid, its depsides (chlorogenic acid), other hydroxycinnamic acids and their conjugates, and the flavonol rutin. The content of phenolic compounds varies over a wide range depending on several factors, such as variety. Chlorogenic acid and its isomers have been found in potato tubers in a range from 0.3 to 13.7 g/kg (d.w.), and rutin in a range from 0 to 190 mg/kg (d.w.).

9.12.5.3 Lettuce

Caffeic acid derivatives are the main phenolics in green lettuce varieties (*Lactuca sativa*, Asteraceae), whilst flavonols were detected in higher quantities in red varieties and endive and anthocyanins were only present in red-leafed varieties. Wound-induced formation of yellow-brown pigments in processed lettuce (cultivar Iceberg) have been typically attributed to the enzymatic oxidation of polyphenols. Besides the enzymatic polyphenol browning, reactions of yellow sesquiterpenes known as lettucenins (structurally similar to bitter lactucin, see Section 8.3.5.1.2) also play a role in the formation of coloured products. The main compounds involved in these colour changes are yellow lettucenins A and B (**9-161**), which accumulate especially near the lettuce butt after cutting. Lettucenin A is also present at trace levels in common chicory (*Cichorium intybus*, Asteraceae).

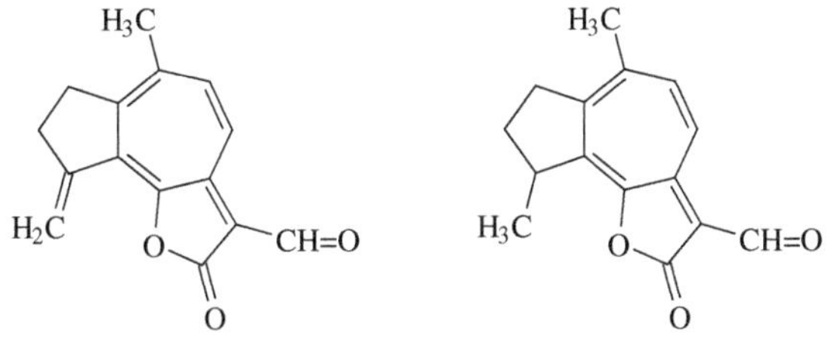

lettucenin A   lettucenin B

9-161, lettucenins

9.12.5.4 Tea

The commercial types of beverage commonly prepared from the leaves and buds of two major tea varieties grown today, *Camellia sinensis* var. *sinensis* and *C. sinensis* var. *assamica* (Theaceae), are green tea, oolong tea, and black tea (usually called red tea in China). The processing methods used for the leaves are similar, except in the manner of fermentation and firing. The fermentation level increases from unfermented green tea to partially fermented oolong tea to well-fermented black tea. Firing involves heating by means of a hot plate, hot air, or flame after drying. Some species of green tea and oolong tea leaves are subjected to firing at >160 °C in order to enhance stability, taste, and aroma. Owing to differences in manufacturing, the types of polyphenols in green tea, oolong tea, and black tea are very different.

Tea is one of the richest sources of flavonoids and phenolic acids. Green tea flavonoids are mainly flavan-3-ols (catechins) and their gallates, which constitute 80–90% of the phenolic compounds of leaves and 25–35% of their dry weight. For example, the most prominent catechins (**9-162**) in mature leaves from the Assam variety are (in d.w. %) (−)-epigallocatechin-3-*O*-gallate (4.7), (−)-epicatechin-3-*O*-gallate (1.2), (−)-epigallocatechin (2.4), (−)-epicatechin (1.0), (−)-epigallocatechin-3,5-di-*O*-gallate (0.2), (−)-epicatechin-3,5-di-*O*-gallate (0.04), (+)-gallocatechin (0.4), and (+)-catechin (0.3). Catechins in green tea can be partly esterified with higher fatty acids. An example of a catechin fatty acid ester is 4′-*O*-palmitoyl epigallocatechin-3-*O*-gallate.

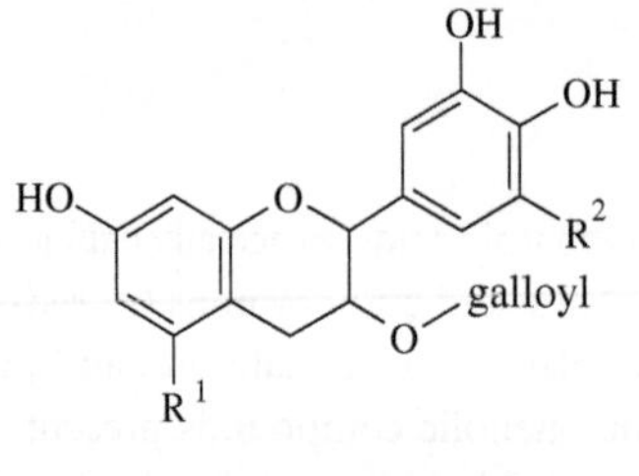

9-162, main tea catechin gallates

(-)-epigallocatechin-3-*O*-gallate, R^1 = R^2 = OH
(-)-epicatechin-3-*O*-gallate, R^1 = OH, R^2 = H
(-)-epigallocatechin-3,5-di-*O*-gallate, R^1 = gallate, R^2 = OH
(-)-epicatechin-3,5-di-*O*-gallate, R^1 = gallate, R^2 = H

Green tea also contains colourless dimeric bisflavanols, such as bisflavanol A (**9-163**), originally considered an intermediate product in the formation of theaflavins. It arises from the paired condensation of two molecules of epigallocatechins. The bisflavanols were later rediscovered and reclassified under the wider name of theasinensins (**9-164**). Theasinensins were found in green tea leaves and in oolong tea. For example, theasinensin F is a catechin–gallocatechin dimer. Similar compounds are dehydrotheasinensins (**9-165**).

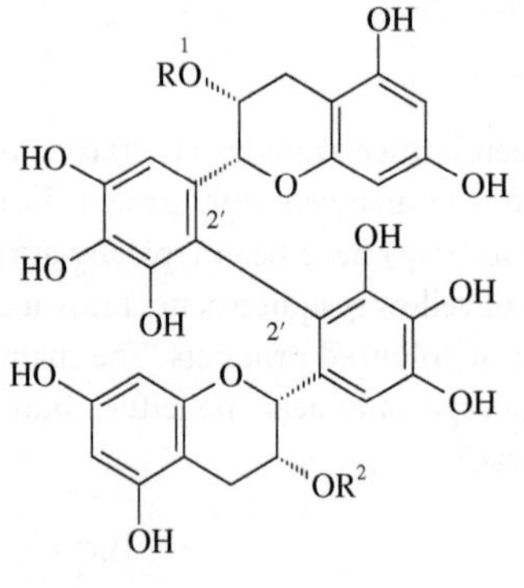

9-163, bisflavanol A (R^1 = R^2 = galloyl)

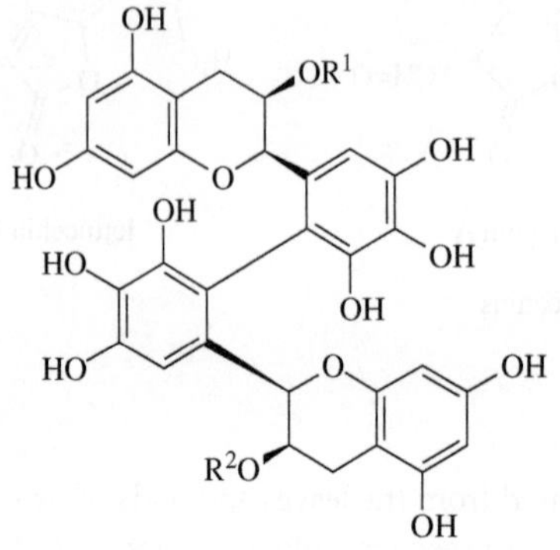

9-164, theasinensin A (R^1 = R^2 = galloyl)

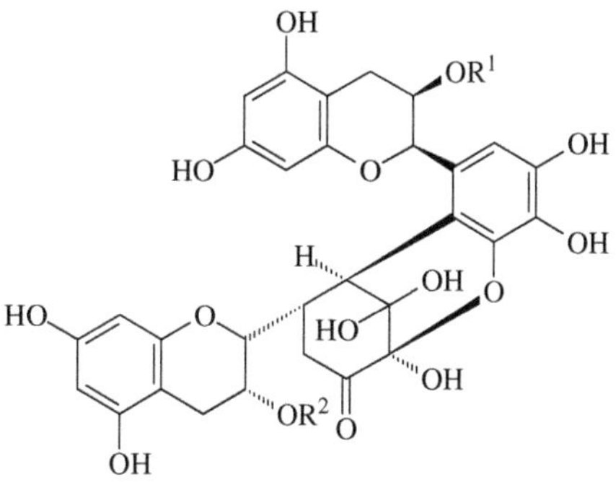

9-165, dehydrotheasinensin A ($R^1 = R^2 = $ galloyl)

Procyanidins in tea leaves are oligomers of epicatechin and catechin, such as epicatechin-($4\beta \to 8$)-epicatechin (procyanidin B_2) and catechin-($4\alpha \to 8$)-catechin (procyanidin B_3), but their amount is lower than that of flavan-3-ols. The average content of procyanidins (sum of 16 different compounds) in green tea was 0.84%, whilst that in black tea was considerably lower (0.5%). Important polyphenols are also hydrolysable tannins, such as ellagitannin strictinin (see Section 8.3.6.1.2), free gallic acid, and its depsides with quinic acid (e.g. 1-*O*-galloylquinic acid, 3-*O*-galloylquinic acid (the umami-enhancing compound theogallin), 4-*O*-galloylquinic acid, and 5-*O*-galloylquinic acid).

Catechins remain unoxidised in green tea, but during production of oolong tea and black tea oxidoreductases and peroxidases oxidise catechins to *o*-quinones, which condense with parent catechins to form a heterogenous mixture of novel dimeric, oligomeric, and polymeric pigments. The dark colour of tea is also significantly influenced by phaeophytins arising from chlorophylls during fermentation. Oxidised catechins and procyanidins also react with amino and thiol groups of proteins and with other compounds, reducing the original astringent taste of tea leaves, but enzymatic browning reactions create new astringent products. Black tea is characterised by two main pigments (collectively referred to as oxytheotannins):

- **theaflavins**, soluble dimeric flavonoids containing a seven-membered benzotropolone ring, which therefore give a bright-orange to red colour in black tea infusions (about 10% of d.w);

- **thearubigins**, the most abundant heterogeneous mixtures of red-and-yellow to orange-brown soluble to insoluble pigments, with a relative molecular weight of 700–400 000 Da (low-molecular-weight compounds exhibit an astringent taste), which account for about 75% of the solids in a typical black tea infusion.

Black tea contains six main theaflavins: theaflavin (product of epicatechin and epigallocatechin, **9-166**), theaflavin-3-*O*-gallate ($R^1 = $ H, $R^2 = $ galloyl, product of epicatechin and epigallocatechin-3-*O*-gallate), theaflavin-3'-*O*-gallate (product of epicatechin-3-*O*-gallate and epigallocatechin), theaflavin-3,3'-di-*O*-gallate (product of epicatechin-3-*O*-gallate and epigallocatechin-3-*O*-gallate), theaflavin-3,5,3'-tri-*O*-gallate (product of epicatechin-3-*O*-gallate and epigallocatechin-3,5-di-*O*-gallate), theaflavin-3,3'5'-tri-*O*-gallate (product of epicatechin-3, 5-di-*O*-gallate and epigallocatechin-3-*O*-gallate), and theaflavin-3,3',5,5'-tetra-*O*-gallate (product of epicatechin-3,5-di-*O*-gallate and epigallocatechin-3,5-di-*O*-gallate). It has been reported that the relative proportions of theaflavins in black tea are 18% theaflavin, 18% theaflavin-3-*O*-gallate, 20% theaflavin-3'-*O*-gallate, and 40% theaflavin-3,3'-di-*O*-gallate.

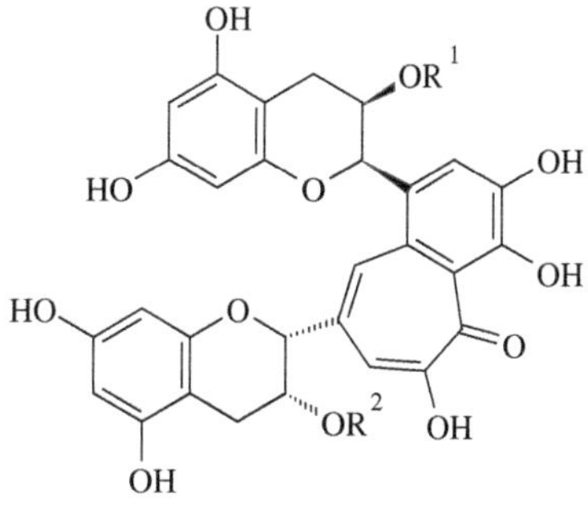

9-166, theaflavin, $R^1 = R^2 = $ H

Structurally related minority pigments are **isotheaflavins**, **neotheaflavins**, **theaflavic acids**, **isotheaflavic acids**, **theaflagallins**, and **epitheaflagallins**, which also possess a similar benzotropolone unit. Isotheaflavins (**9-167**) and neotheaflavins (**9-168**) arise in part from the non-*epi*-forms of the catechin pairs. Theaflavic acids (**9-169**) and epitheaflavic acids (**9-170**) are formed by oxidative condensation of gallic acid with the non-*epi*- and *epi*-forms of catechins, respectively. Similarly, the minor polyphenols in black tea theaflagallins (**9-171**) are enzymatically synthesised from gallic acid and gallocatechins, whilst epitheaflagallins (**9-172**) arise from gallic acid and epigallocatechins.

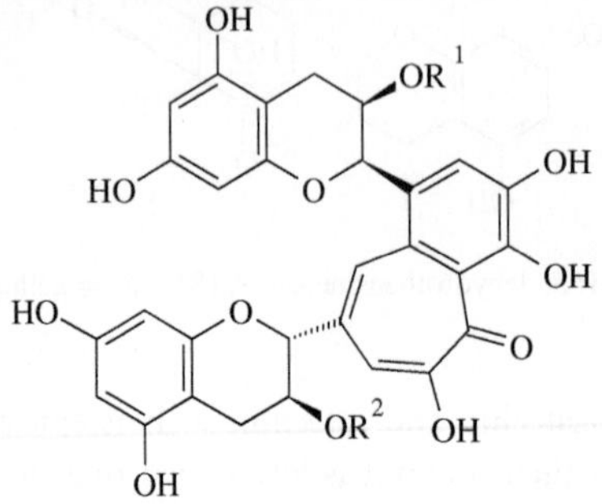

9-167, isotheaflavin, $R^1 = R^2 = H$

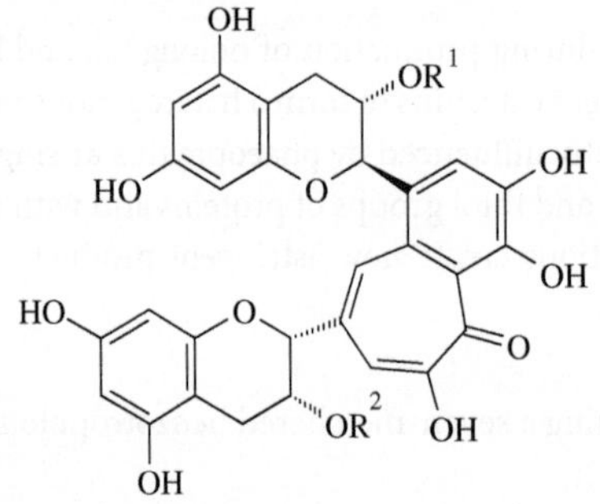

9-168, neotheaflavin, $R^1 = R^2 = H$

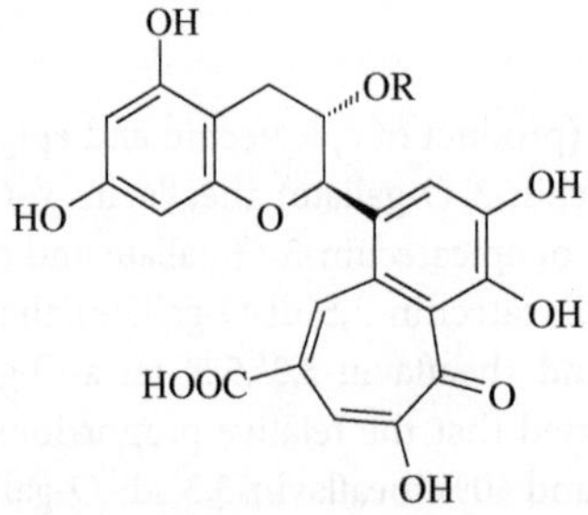

9-169, theaflavic acid, R = H

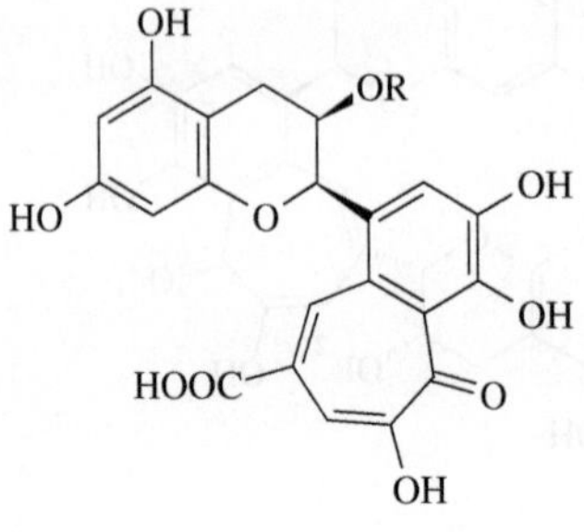

9-170, epitheaflavic acid, R = H
3-*O*-galloylepitheaflavic acid, R = galloyl

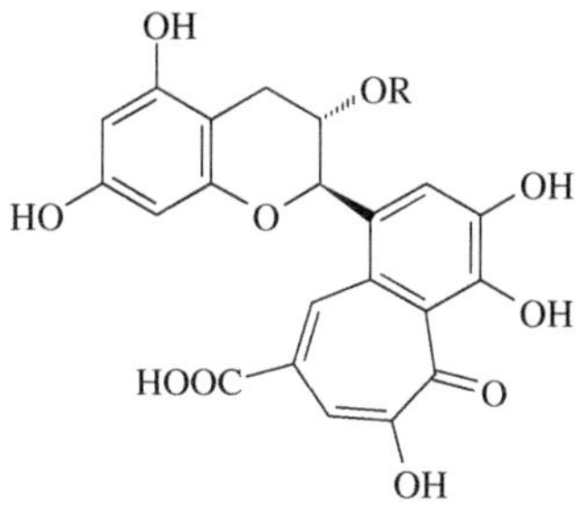

9-171, theaflagallin, R = H

Tea leaf fermentation also yields oxidation products not arising solely from catechins and gallocatechins. For example, condensation of catechins with the umami-enhancing compound theogallin (3-O-galloylquinic acid) yields theogallinin (**9-173**). Theaflavonin (**9-174**) represents compounds formed by condensation of catechins with myricitrin (3-O-α-L-rhamnopyranoside of myricetin). Reaction with chlorogenic acid generates an unusual theaflavin-type compound with a benzotropolone unit (**9-175**).

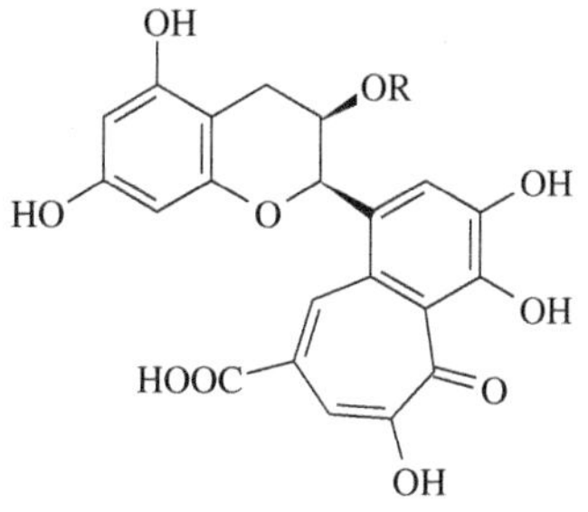

9-172, epitheaflagallin, R = H
3-O-galloylepitheaflagallin, R = galloyl

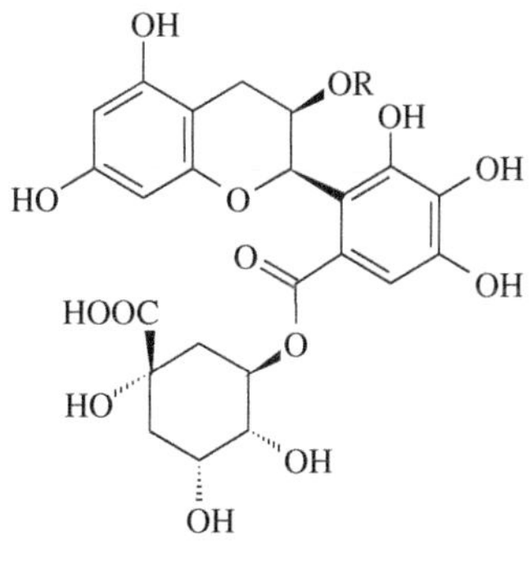

9-173, theogallinin, R = galloyl

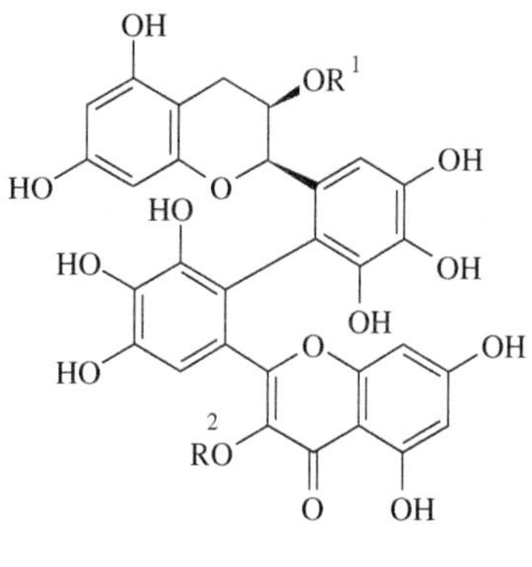

9-174, theaflavonin, R^1 = galloyl, R^2 = β-D-glucosyl

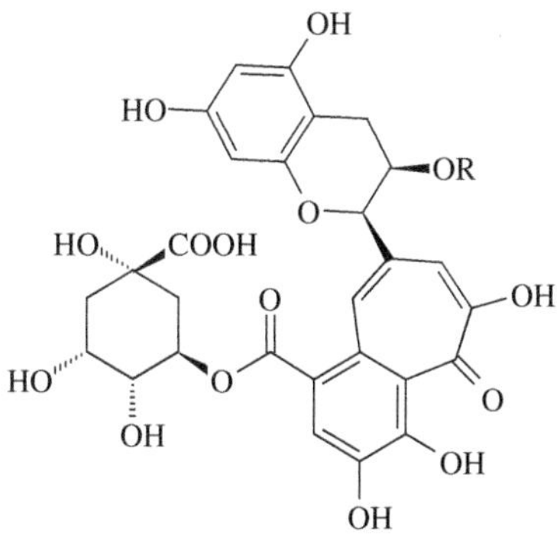

9-175, theaflavin-chlorogenic acid adduct, R = H or gallate

Rare compounds identified in the leaves of *C. sinensis* var. *pubilimba*, a special green tea that undergoes fermentation by *Aspergillus* molds, known as *pu-erh* tea, are diastereomers of 8-*C* *N*-ethylpyrrolidin-2-one flavan-3-ols, known as **puerins** (**9-176**). The substituted part of the *N*-ethylpyrrolidin-2-one is derived from theanine.

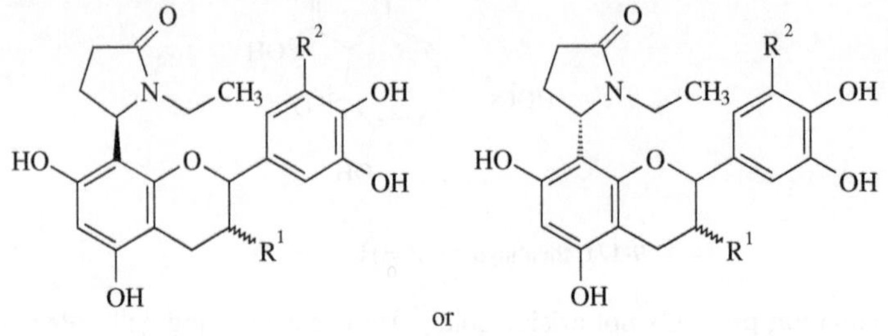

9-176, puerins, $R^1 = \alpha$-*O*-galloyl, α-OH or β-OH, $R^2 = H$ or OH

Thearubigins are a very diverse mixture of many thousands of compounds with poorly characterised structure. However, some lower-molecular-weight compounds have already been identified. Examples of these polyhydroxylated products are yellow **theacitrins** (**9-177**), their degradation products **theacitrinins** (**9-178**), **theanaphthoquinone** (**9-179**), and various peroxo−/epoxy-compounds in the series of theasinensin A (**9-164**). Theacitrins are produced by oxidative coupling of gallocatechins; that is, coupling of flavan-3-ols with pyrogallol-type B-rings. These are covalently bound to proteins via thiol groups and remain in the remnants of tea leaves in the preparation of brine. About 20% of tea leaf flavonoids are transformed into insoluble polymeric brown products known as theafulvins (whose structure is not yet fully known) or other polymers.

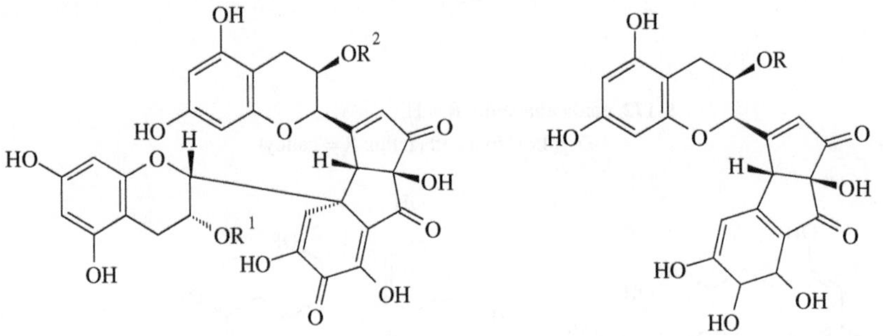

9-177, theacitrin A, R^1 = galloyl, $R^2 = H$
theacitrin C, $R^1 = R^2$ = galloyl

9-178, theacitrinin A, R^1 = galloyl, $R^2 = H$
theacitrinin C, $R^1 = R^2$ = galloyl

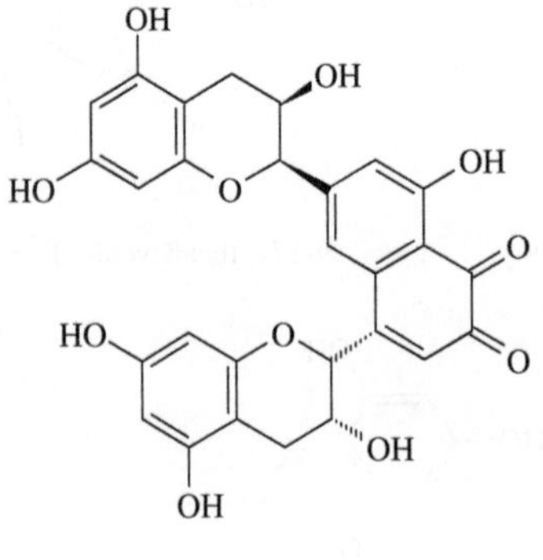

9-179, theanaphthoquinone

9.12.6 Inhibition

The enzymatic browning reaction is often an important technological problem, especially in the case of subsequent cold or freezing storage of fruit and vegetable products. Therefore, in addition to its mechanisms, considerable attention is paid to its possible prevention. Prevention should begin with the selection of materials suitable for the given method of processing.

Fruits and vegetables with white flesh are prone to enzymatic browning, especially apples and pears, but also apricots, peaches, plums, potatoes, and other materials. In most berries (raspberries, currants, and blueberries), the enzymatic browning process has practically no impact due to the high content of natural pigments. The tendency to browning is due to activity of polyphenol oxidases and the content and composition of phenols. Generally, it is possible to use the total content of phenols, the tyrosine content, or the activity of polyphenol oxidase as a criterion for selecting varieties for a particular type of treatment. One promising route may be the breeding of varieties with a lower tendency to enzymatic browning, or the use of gene manipulation. Technological inhibitory interventions can be either physical methods or chemical methods.

9.12.6.1 Physical methods

This includes gentle material handling before and during the process, where the intensity of browning, induced by mechanical means, may be influenced by the style of some technological operations, such as clean cutting of the fruit.

The most important physical method of polyphenol oxidase inhibition is exposure of the material to high temperatures. Polyphenol oxidases are inhibited by temperatures above 70 °C and are most stable in media of pH around 6.0; in both directions from this value, their resistance against heating decreases quite sharply. Blanching, consisting in a rapid immersion of the material in hot water (salt or sugar solution) or exposure to water vapour, is used in the processing of vegetables that are eaten cooked and sometimes for the treatment of fruit pickles. Blanching is not very suitable for the inactivation of enzymes of fruits, as it causes changes in consistency and organoleptic properties; nor is it suitable for the treatment of aromatic herbs used as spices.

Another physical method of browning prevention is restriction of oxygen access. The production processes for frozen sensitive materials make use of packaging in hermetically sealed containers under either normal or reduced pressure, packaging in an inert atmosphere, and addition of sucrose. The first three methods differ only in the input treatment, and their common feature is a temporary effect only for freezer storage. After the package is opened at normal temperature, the browning reaction usually proceeds even faster due to tissue disruption. Dipping in sucrose solution before freezing or filling the entire package with sugar has, in addition to the suppression of browning, even more positive effects, especially in terms of increased viscosity of the liquid phase within the tissue. Significant inhibition of enzymatic reactions, however, occurs only at concentrations of sucrose higher than 20%, as a result of increased osmotic pressure. Sugar also enhances the taste and aroma of fruit, but limits the possibility of further use of such processed raw material, which is only suitable for the production of liqueurs, syrups, and fruit purees.

9.12.6.2 Chemical methods

Another option for reducing or inhibiting enzymatic browning is the use of chemical reagents. There are many substances capable of inhibiting polyphenol oxidases, but the mechanism of their action is often unknown. Complexing agents can bind copper ions, which the enzymes require. Interaction with copper is also the cause of the inhibitory effects of some inorganic ions, especially halides. Substances structurally similar to substrate (such as benzoic and cinnamic acids and their derivatives) may competitively inhibit polyphenol oxidases. Also considered is the possibility of direct influence on the active sites of enzymes by free radicals, which can be produced, for example, from added ascorbic acid. The inhibitory effect of hydrogen peroxide on mushroom tyrosinase is explained by the same mechanism. Acidification below the polyphenol oxidase optimum pH (pH 4–4.5) may also slow or stop the browning.

Another group of inhibitors of polyphenol oxidases are reducing agents capable of reducing quinones back to diphenols (ascorbic acid, thiols, and some other agents). The antibrowning effect of ascorbic acid has been associated with its ability to reduce quinones to their precursor phenolics and with lowering the pH through a concomitant inhibition of polyphenol oxidase activity. However, if the effect of the inhibitor is a mere reduction of quinones, browning is inhibited only until exhaustion of the reagents. The sulfur-containing agents (such as bisulfites) seem to control the browning reaction by irreversible inactivation of polyphenol oxidase, as well as by reacting with quinones to produce colourless compounds. An overview of the main inhibitors is given in Table 9.18.

The practical use of chemical inhibitors is limited by several factors. Most substances with a strong inhibitory effect, when tested in model systems, are not effective in real materials, where other factors also play a role. It is easier to use an inhibitor in a homogeneous material, such as juice or puree. In the case of cut fruits, leaves, and spices, there are two possible alternatives: soaking for some time in the inhibitor solution and spraying the inhibitor over the material surface. Individual substances differ in their ability to penetrate the plant tissue. Bisulfites in particular are very effective. Another limitation may be material properties. For example, the use of acid is avoided in green plants, where browning will be suppressed but the acid will accelerate the conversion of chlorophylls to phaeophytins (phaeophytinisation).

Health regulations normally allow only the use of ascorbic acid and some of its derivatives, citric acid, sodium chloride, and hydrogen bisulfites. Hydrogen bisulfites and sulfur dioxide are very effective in protecting against browning and are widely applied (e.g. in packed peeled potatoes). In practical terms, mixtures of citric and ascorbic acids are used. As a very effective means to stabilise the colour of apple slices and other fruits, a mixture of phosphates, citric acid, and glucose is recommended. The use of an inhibitor is often part of a complex procedure also involving other methods, especially the limitation of air access. No agent (perhaps with the exception of hydrogen sulfites) and no treatment is universally applicable.

Table 9.18 Inhibitors of enzymatic browning reactions.

Compound	Mechanism of action
Tropolone (2-hydroxycyclohepta-2,4,6-trien-1-one)	Copper binding
Kojic acid	Copper binding
Citric acid	Copper binding, acidification
Phosphoric acid, polyphosphates	Copper binding, acidification
Ethylenediamintetraacetic acid (EDTA)	Copper binding, slight acidification
D-Gluconic acid	Copper binding, slight acidification
Sodium chloride, sodium fluoride	Copper binding
Benzoic, p-coumaric, cinnamic, and feruic acids	Competitive enzyme inhibition
4-Hexylresorcinol and other 4-substituted resorcinols	Competitive enzyme inhibition
Proteases (ficin from figs, actinidin from kiwi, papain from papaya, bromelain from pineapple)	Enzyme inhibition
Sodium hypochlorite, calcium hypochlorite	Enzyme inhibition
Modified substrates, such as (+)-catechin 3'-O-α-D-glucoside	Enzyme inhibition
Cyclodextrins	Formation of complexes with phenols
Ascorbic acid and its derivatives	Reduction of quinones, slight acidification
Hydrogensulfites, sulfur dioxide	Reduction of quinones and addition products, enzyme inhibition
Thiols (cysteine, N-acetylcysteine, glutathione)	Reduction of quinones and addition products, enzyme inhibition

10
Natural Antinutritional, Toxic, and Other Bioactive Compounds

10.1 Introduction

In addition to beneficial and neutral substances, foods can contain substances that are harmful to health. There are two basic types of compounds with potential adverse health effects. The first is **antinutritional compounds**, which may interfere with the body's ability to digest and utilise nutrients. The second is **toxic compounds** or **toxicants**, which can produce adverse biological effects of any nature and may, depending on the extent of exposure (dose), cause health problems, disease, or damage to the body. It is estimated that about 99% of all toxic food components are **natural toxic compounds**, also known as **natural toxins** produced by living organisms as a result of genetic predisposition. The ability of an organism to produce toxins is called toxinogenicity. The most common toxins found in food are plant toxins, which may be produced as a natural defense mechanism against predators, insects, or microorganisms, as a consequence of infestation with microorganisms, or in response to various stress factors. Toxins are not harmful to the organism itself, but they may be toxic to other creatures, including humans, when consumed in excess. Toxins have diverse structures and differ in biological function and toxicity.

Compounds that are a permanent part of the plant tissues, synthesised by the plant at a constant rate, and therefore always present in the tissues are called **phytoanticipins**. They are produced by healthy plants from the beginning of growth and serve only as a passive defence against potential harmful factors. **Phytoalexins** are antimicrobial substances synthesised *de novo* by plants from phytoanticipins to combat infection by pathogens. Phytoalexins are also called **phytoncides**, **plant antibiotics**, or **plant pesticides**, as they are toxic to pathogenic ilicitors (viruses and microorganisms) and animal pests (insects and higher animals). The distinction between phytoalexins and phytoanticipins is not always clear, as some compounds may be phytoalexins in one species and phytoanticipins in another. In addition, the same substance may often serve as both phytoalexin and phytoanticipin. In general, the distinction between the two compounds depends on when they are produced, before or after infection.

A specific category of compounds is substances added intentionally to improve food quality during processing, storage, and other operations (extending the shelf-life or improving sensory, nutritional, and technological quality), which are therefore classified as additive substances or **food additives** (see Chapter 11). Some foods may contain compounds which are not their natural components or compounds which may under certain circumstances be harmful. Compounds that accidentally get into food from the environment during production, storage, transportation, marketing, and other manipulations or arising in physical and chemical processes of food production are classified as **pollutants** or **food contaminants** (see Chapter 12). Products of anthropogenic activities that do not occur in nature are called **xenobiotics** (from the Greek *xenos*, meaning foreign). According to some classifications, food additives are not regarded as foreign substances but natural toxic substances in food are regarded as food contaminants.

Natural toxic substances and some food contaminants may exhibit different biological effects upon living organisms, largely depending on where and how they are taken in orally (through the mouth), by inhalation (through breathing), or by absorption through the mucous membranes and the skin (dermally). Adverse biological effects, bringing the risk of possible poisoning, are called **toxic effects**. Toxic effects may manifest as:

- **acute**, after a single dose of a toxin;[1]

- **chronic**, after repeated or continuous administration of toxic substances (over a period of time).

[1] The acute toxicity of a compound is generally measured by experiments on animals (often rats). The most common measure of acute toxicity is lethal dose value (LD_{50}), usually given in weight of compound per kilogramme of body weight, which causes death in 50% of test animals.

The Chemistry of Food, Second Edition. Jan Velíšek, Richard Koplík and Karel Cejpek.
© 2020 John Wiley & Sons Ltd. Published 2020 by John Wiley & Sons Ltd.

On the basis of their biological action, toxic substances may act as:

- **asphyxiants** (exerting their effects through a depletion of oxygen to the tissues – asphyxia, such as carbon monoxide and hydrogen cyanide);

- **carcinogenic agents** or carcinogens[2] (inducing cancer, such as some mushroom toxins and food contaminants);

- **hepatotoxic agents** or hepatotoxins (producing liver damage, such as pyrrolizidine alkaloids, amatoxins, and phallotoxins);

- **irritants** and **sensitising agents** (causing irritation on contact with the skin, such as ammonia, allyl isothiocyanate, and some essential oils of spices; sensitisation is a reaction after exposure to sensitising agents that may cause allergic responses, such as hypericin, fagopyrin, and some coumarins);

- **mutagenic agents** or mutagens (causing cancer or undesirable mutations, such as reactive oxygen species, some carbonyl compounds, and aromatic amines);

- **nephrotoxic agents** or nephrotoxins (producing kidney damage, such as amatoxins and cortinarins);

- **neurotoxic agents** or neurotoxins (damaging nervous tissue, such as 3-(N-oxalyl)-L-2,3-diaminopropionic and ibotenic acids);

- **teratogenic agents** or teratogens (causing defects in the foetus in pregnant women, such as alkaloids chaconine, tomatine, anabasine, and anatabine).

According to current knowledge, a number of natural compounds formerly classified as toxins or indifferent substances are desirable bioactive and even beneficial food components for human well-being and health at low concentrations, and many find use in the pharmaceutical industry and medicine. The term food component refers to non-essential molecules that are present in foods and exhibit the capacity to modulate one or more metabolic processes, resulting in the promotion of better health. In general, it is thought that bioactive food components are predominantly found in plant foods, such as cereals, nuts, fruits, and vegetables. For example, besides providing energy as food, cereal grains can also affect human health either positively or negatively through many minor components (vitamins, steroids, flavonoids, lignans, and phenolic acids, as well as avenanthramides in oats, benzoxazinoids in wheat and rye, hordatines in barley, and alkylresorcinols in wheat, rye, and barley). Isothiocyanates and other degradation products of glucosinolates, occurring in broccoli, cauliflower, Brussels sprouts, and other *Brassica* vegetables, were known as potential antithyroid compounds, but it has been shown that many of them induce detoxifying enzyme systems and have antimicrobial, immunomodulatory, and anticancer properties. Phytooestrogens of soybeans, soy-based products, and flaxseeds have antioestrogenic, antiosteoporotic, and antiproliferative properties. Whilst a few plant saponins are toxic in large doses, most saponins found in food (cereals, legumes, potatoes, spinach, tomatoes, and other products) are safe and may even have many beneficial health effects. Studies have illustrated beneficial effects of some saponins on blood cholesterol levels (by forming complexes with cholesterol and bile acids), a hypolipidaemic effect or antioxidant, antimutagenic, and anticarcinogenic properties, and stimulation of the immune system.

This chapter looks at natural antinutritional, toxic, and also beneficial substances occurring in food and food raw materials that are used in the diet either directly or following technological and culinary processing. Attention is paid to natural substances that enter the food chain indirectly through livestock feed, but not all toxic compounds are included; many others are described in the chapters devoted to amino acids, peptides, proteins, lipids, carbohydrates, minerals, flavourings, pigments, food additives, and food contaminants. No attempt has been made to cover toxic compounds that are primarily of pharmaceutical and medicinal interest, nor toxic compounds accidentally ingested or taken as illicit drugs. Each section concentrates on the structure and occurrence of antinutritional, toxic, or beneficial compounds and their biological effects, changes, and reactions in raw materials and foods.

10.2 Antinutritional compounds

Antinutritional substances include:

- enzyme inhibitors;

- antivitamins;

---

[2]The International Agency for Research on Cancer (IARC), a part of the World Health Organization (WHO), recognises the following groups of carcinogenic substances: Group 1, carcinogenic to humans; Group 2A, probably carcinogenic to humans; Group 2B, possibly carcinogenic to humans; Group 3, unclassifiable as to carcinogenicity in humans; and Group 4, probably not carcinogenic to humans.

- compounds interfering with the metabolism of minerals;

- antinutritional phenolic compounds;

- antinutritional oligosaccharides.

10.2.1 Enzyme inhibitors

Enzyme inhibitors, also known as antienzymes, are various low-and high-molecular-weight natural food components and foreign substances that affect the activity of various enzymes. In nutrition terms, the most important antienzymes are inhibitors of digestive enzymes, particularly **inhibitors of proteases**, also known as antiproteases. A less important group is **inhibitors of saccharases**.

10.2.1.1 Inhibitors of proteases

Protease inhibitors, which are important from the nutritional point of view, are proteins or polypeptides capable of inhibiting the digestive proteolytic enzymes.

10.2.1.1.1 Classification

Protease inhibitors are classified according to the type of protease they inhibit. Recognised inhibitors are:

- serine protease inhibitors (inhibitors of trypsin, chymotrypsin, elastase, and plasmin);

- sulfhydryl protease inhibitors (inhibitors of pepsin and thrombin);

- acidic protease inhibitors (inhibitors of cathepsin D);

- metaloprotease inhibitors (inhibitors of pancreatic carboxypeptidase).

The most important inhibitors are inhibitors of serine proteases, which include two main groups:

- Kunitz-type inhibitors (KIs)

- Bowman–Birk-type inhibitors (BBIs).

10.2.1.1.2 Occurrence and properties

Protease inhibitors are naturally present in plant foods, especially legume seeds, but they also occur in cereals and some other plant materials (such as potatoes and tomatoes). Protease inhibitors of microbial and animal origins also exist. In plants, they fulfill several functions. They probably serve as a cytosol protection against endogenous proteases released in disrupted cellular structures. They also act as storage proteins during seed germination and are involved in the protection of plant tissues against elicitors (viruses, bacteria, and fungi) and predators (animals).

Kunitz-type inhibitors Kunitz-type inhibitors have a relative molecular weight of about 20 kDa and two disulfide bridges in the molecule. They primarily exhibit specificity to trypsin. Examples are Kunitz-type inhibitors of soya beans, called soybean trypsin inhibitors (STIs), which are a group of isoinhibitors with relative molecular weights of 18–24 kDa. The main component is a protein containing 181 amino acid residues and two disulfide bridges (20 kDa). The binding sites, where the inhibitor interacts with trypsin, are the amino acid residues Arg[63] and Ile[64]. A complex with trypsin arises in a stoichiometric ratio (one inhibitor molecule interacts with one molecule of trypsin). The complex is analogous to the enzyme–substrate complex, but does not in practice dissociate into the original protein (trypsin) and enzyme inhibitor.

Bowman–Birk-type inhibitors Inhibitors of the Bowman–Birk type have a relative molecular weight of about 6–10 kDa and a higher number of disulfide bridges compared to Kunitz-type inhibitors. They exhibit specificity against trypsin and chymotrypsin, as they contain two independent binding sites in the molecule. Bowman–Birk-type inhibitors belong to the most common inhibitors. They occur in legumes, cereals, pseudocereals, potatoes, and some other plant materials.

Table 10.1 Protease inhibitor contents in selected legumes.

Legume seed	Content of protease inhibitors (g/kg)	
	Kunitz-type inhibitors	Bowman–Birk-type inhibitors
Soya bean (*Glycine max*)	20	2–3
Peas (*Pisum sativum*)	–	0.05–3
Cowpea (*Cicer arietinum*)	–	2.3

In soya beans (*Glycine max*, Fabaceae), five isoinhibitors of this type occur, which belong to the group of potato inhibitors (PIs) and are labelled PI-I to PI-V. The inhibitors PI-I and PI-II are also found in potatoes and barley. Inhibitor PI-I is a polypeptide of molecular weight 8 kDa, which contains a polypeptide chain consisting of 71 amino acids and 7 disulfide bridges. The inhibitor PI-I has dual binding specificity. The binding sites of trypsin are amino acids Lys^{16}-Ser^{17}, whilst interaction with chymotrypsin occurs via amino acids Leu^{44}-Ser^{45}. Similar structures and properties have inhibitors present in other legumes and oilseeds. Common bean (*Phaseolus vulgaris*) contains garden bean inhibitor (GBI), lima (butter) bean (*Phaseolus lunatus*) contains lima bean inhibitor (LBI), chickpeas (*Cicer arietinum*) contain cow pea inhibitor (CPI), and peanuts (*Arachis hypogaea*) contain groundnut inhibitor (GI). Buckwheat (*Fagopyrum esculentum*, Polygonaceae) contains three isoinhibitors, called buckwheat trypsin inhibitors (BTI)-1 to BTI-3, which consist of a single polypeptide chain containing 69 amino acids. The contents of protease inhibitors in selected legumes are shown in Table 10.1.

Sulfhydryl protease inhibitors are found, for example, in the seeds of legumes of the genus *Vigna*, inhibitors of acid proteases occur in potatoes, and inhibitors of metaloproteases occur in potatoes, tomatoes, and other crops.

10.2.1.1.3 Mechanism of action

Knowledge of the antinutritive effects of protease inhibitors in humans is still limited and mainly derived from the knowledge gained in animal nutrition. When feeding raw or undercooked legumes to farm animals, inhibitors of proteases (trypsin inhibitors) cause disturbances that are manifested by slow growth of the animals. In chronic cases, this leads to enlargement of the pancreas, which is histologically described as hypertrophy (enlargement of the pancreas or its parts due to enlargement of individual cells) and hyperplasia (enlargement of pancreas by pancreatic cell multiplication). This is due to increased secretion of digestive enzymes, including trypsin, chymotrypsin, and elastase, by the hyperactive pancreas. It is thought that the growth retardation of animals is actually the result of endogenous loss of amino acids, particularly sulfur amino acids, which are used for the synthesis of proteases that subsequently pass into the faeces and cannot be used for the synthesis of muscle proteins.

Trypsin inhibitors are not the only antinutritive factors in soya beans that may slow the growth of farm animals. For example, the slowdown of animal growth, for which trypsin inhibitors are responsible, is about 40% of the total antinutritive activity of raw soya beans. About 25% of the antinutritive activity can be attributed to lectins and the rest to other antinutritional substances.

10.2.1.1.4 Inactivation

Adverse effects of protease inhibitors are relatively easy to eliminate by thermal processing of plant materials. Reduction of inhibitor activity depends on temperature, time of heating, material particle size, and water content. For soya beans, the most common method of thermal inactivation of protease inhibitors is the so-called toasting process (effect of water vapour), but other methods are also effective, such as boiling in water, dry roasting, microwave heating, and extrusion. For example, trypsin inhibitors of most commercial soya products for human consumption (tofu, soya beverage (incorrectly called soya milk), soya isolates and concentrates, and textured soya meat substitutes) are sufficiently inhibited, showing only about 20% of the activity of raw soya beans. Inactivation of inhibitors to 50–60% is necessary to eliminate their adverse effects (animal growth retardation and impaired pancreatic function). Germination of mature soya beans also gradually decreases the activity of Kunitz-type inhibitors, because modified forms of these proteins arise by proteolysis and *de novo* synthesis.

10.2.1.2 Inhibitors of saccharases

Many cereals and cereal products (such as wheat, rye, breakfast cereals, and bread) contain proteins inhibiting animal, but not plant, amylases. Invertase inhibitors are found in potato tubers. The significance of these enzyme inhibitors is negligible and the consequences of their presence in food are not yet well known. It is believed that amylase inhibitors could potentially be used in preparations for diets promoting weight loss.

10.2.2 Antivitamins

Antivitamins (vitamin antagonists) are substances that eliminate the biological effects of vitamins, which can lead to deficiency symptoms (see Chapter 5).

10.2.3 Compounds binding minerals

In foods of plant origin, certain compounds are found as natural components that interfere with the metabolism of minerals. The most important groups include:

- **phytic acid** and **phytin;**

- **oxalic acid;**

- **glucosinolates** and their degradation products.

10.2.3.1 Phytic acid and phytin

Phytic acid, *myo*-inositol-1,2,3,4,5,6-hexakisdihydrogenphosphate, is the main storage form of phosphorus used during germination of seeds of cereals, pulses, and oilseeds. Phytic acid in seeds occurs primarily as a mixed calcium and magnesium salt, called phytin. Insoluble salts and salts that are utilised only slightly are also produced with other di- and trivalent ions, in particular Fe and Zn (see Section 6.3.4.2.1).

10.2.3.2 Oxalic acid

Oxalic acid is a common component in many vegetables and other plant foods. It yields insoluble calcium oxalate with calcium ions, which may under certain circumstances (low intake of calcium and vitamin D) seriously interfere with the metabolism of calcium (see Section 8.2.6.1.2).

10.2.3.3 Glucosinolates

Glucosinolates, especially the glucosinolate progoitrin (whose decomposition yields the antithyroid substance goitrin), but also some others that produce antithyroid isothiocyanates (such as sinigrin, which produces allyl isothiocyanate) or antithyroid thiocyanates (such as indole glucosinolates), are often classified as compounds that interfere with the metabolism of iodine (see Section 10.3.2.4).

10.2.4 Carbohydrates

A special group of antinutrients is the α-galactosides, also called α-D-galactosides of saccharose. These are mainly found in legumes, where they have a role as storage carbohydrates. An important representative of these sugars is trisaccharide raffinose and its higher homologues, whilst less common are trisaccharide manninotriose and α-galactosides of cyclitols, which are classified as pseudooligosaccharides (see Section 4.3.1.2.2). The presence of these substances in legumes to some extent limits the use of legumes in human consumption and in the nutrition of monogastric animals, because their flatulent activity results in gastrointestinal discomfort. In ruminants, these problems do not occur, because α-galactosides are hydrolysed by α-galactosidase.

10.2.5 Phenolic compounds

Plants contain many phenolic secondary metabolites involved in plant defense, which are collectively known as antiherbivory compounds and have the characteristics of antinutritional substances. These include some alkaloids and cyanogenic glycosides, as well as some phenolic acids, flavonoids, tannins, and lignans of cereals, benzoxazinoids of wheat and rye, avenanthramides of oat, hordatines of barley, and other compounds, which have been described in other contexts.

10.2.5.1 Tannins

Tannins are found in relatively large quantities in the seeds of leguminous plants (in amounts up to 0.45 g/kg in soya beans and 20 g/kg in common beans) and in oilseed extraction meals. Their complexes and reaction products with proteins are resistant to enzymatic hydrolysis, which results is lower digestibility and subsequently in reduced weight gain in livestock. Excessive consumption of tannins may also lead to decreased absorption of some minerals and can cause damage to the intestinal mucosa.

10.2.5.2 Alkylresorcinols

Alkylresorcinols, also known as resorcinolic lipids, are present in several families of higher plants, algae, mosses, fungi, and bacteria. Alkyresorcinols are mostly straight-chain 5-alkyl derivatives of resorcinol (benzene-1,3-diol, R=H, **10-1**) with an odd chain length in the range of 17–25 carbon atoms. Different derivatives including 5-alkenyl, 5-alkadienyl, 5-(2-oxoalkyl), 5-(2-oxoalkenyl), 5-(2-hydroxyalkyl), and even-alkyl-chain and branched-chain resorcinols are also present to a minor extent.

basic structure   5-heneicosylresorcinol

10-1, structures of substituted resorcinols

Amongst the plants used for human nutrition, alkylresorcinols are present in relatively high amounts in some cereals (rye, wheat, and triticale) and pseudocereals (quinoa), in smaller amounts in barley, and in very small amounts in oats, maize, and legumes (such as garden peas). They are located mainly in the embryo (germ) and the outer layer of the kernel; therefore, they remain in wholegrain cereals.

The content of alkylresorcinols vary widely, depending on environmental, genetic, and other factors. Their amount in rye kernels (*Secale cereale*) is 720–2000 mg/kg (in the germ about 3000 mg/kg) and in wheat (*Triticum aestivum*) 200–1429 mg/kg (around 2000 mg/kg in the germ). In durum wheat (*T. durum*), the content of alkylresorcinols is on average 455 mg/kg, and in triticale it is between the values for rye and wheat. The most common alkylresorcinols are compounds with alkyl chains C15:0—C27:0. The main compound in rye, wheat, and triticale is 5-heneicosylresorcinol (**10-1**). Alkenylresorcinols represent about 20–30% of the content of alkylresorcinols in rye and about 10% in wheat. For comparison, the main compound in barley is 5-pentacosylresorcinol and the mean content of alkylresorcinols is 8–210 mg/kg. The mean content of straight-chain alkylresorcinols in quinoa is 58 mg/kg, whilst the content of branched-chain alkylresorcinols is 182 mg/kg.

The content of alkylresorcinols is significantly reduced during dough fermentation and baking. The baking process has been reported to reduce their content in flour by 42–100%. In extruded cereal products, the original amount of alkylresorcinols was reduced to about 50–80%.

Alkylresorcinols in cereals negatively influence the growth of animals. Their toxicity is based on their effect on the hydrophobic properties (permeability) of membranes for potassium ions and some organic compounds (such as glycerol). It decreases with increasing length of the substituent and grows with the number of double bonds in the chain. The most toxic alkylresorcinols are homologues with unsaturated and shorter (13 or fewer carbon atoms) chains. Alkenylresorcinols also exhibit haemolytic activity. Some synthetic resorcinols (e.g. 4-hexylresorcinol, approved as a food additive in the European Union, E586) proved to be inhibitors of enzymatic browning reactions that cause blotches on canned shrimps.

5-Alkyl, 5-alkenyl, and 5-alkadienyl resorcinols also occur as allergenic compounds in mango (*Mangifera indica*, Anacardiaceae). Depending on the cultivar, their content ranges from about 80 to 1850 mg/kg of d.w in mango peels and from about 5 to 190 mg/kg of d.w in the pulp. The main compound is the C17 : 1 homologue (52.5%).

Toxic 5-alkatrienyl, 5-alkadienyl, 5-alkenyl, and 5-alkyl resorcinols, collectively known as cardol, occur in the cashew nutshell liquid, also known as cashew shell oil, a natural resin found in the honeycomb structure of the cashew nutshell. It is a byproduct of cashew nuts processing. A typical cashew shell oil contains 15–20% of cardol. The main components are (8Z,11Z)-pentadeca-8,11,14-trien-1-yl, (8Z,11Z)-pentadeca-8,11-dien-1-yl, (8Z)-pentadeca-8-en-1-yl, and pentadecyl derivatives.

10.3 Toxic compounds

Adverse reactions to food can be classified into toxic (caused by toxins that are toxic to all individuals) and non-toxic reactions (occurring only in certain individuals). The non-toxic types may be divided further into immune- and non immune-mediated reactions. The term **intolerance** is used for non immune-mediated reactions, and the term **hypersensitivity** is used for immune-mediated reactions. Food intolerance may be enzymatic, pharmacologic or undefined. Immune-mediated reactions may be IgE (Immunoglobulin E)-mediated (i.e., **allergy** or type I hypersensitivity) or non-IgE-mediated. Allergens are registered in the database of the International Union of Immunological Societies (IUIS).

10.3.1 Food intolerance-inducing compounds

Food intolerance, also known as non-immunoglobulin E (IgE)-mediated food hypersensitivity or non-allergic food hypersensitivity, is an abnormal non-immunological reaction of the body to a particular food or drink that involves difficulty in digestion. The symptoms for food intolerance or mild to moderate food allergy may be similar.

Food intolerance does not include psychological reactions to food and eating disorders, such as **anorexia nervosa** and **bulimia nervosa**. Anorexia nervosa, commonly referred to simply as anorexia, is an eating disorder that consists in a person's complete opposition to eating. The person feels sick or constantly hunger-free, which can lead to malnutrition and other problems associated with starvation. Bulimia nervosa, also called bulimia, is an eating disorder that consists in seizure eating and attempts to vomit (purge) food by inducing diarrhoea and the use of appetite suppressants or other substances to control weight loss.

Conditions resulting in food intolerance include metabolic disorders, sensitivity to certain foods (*anaphylaxis*), and aversion to certain foods (*idiosyncrasy*). It is estimated that about 0.3–7% of the population suffers from food-intolerance symptoms, which include nausea, bloating, abdominal pain, and diarrhoea, beginning hours or days after ingestion; these symptoms are not usually life-threatening. The cause of these metabolic disorders of the organism is usually insufficient activity of some enzymes, which is hereditary, so food intolerance of this type is a congenital metabolic disorder (present at birth). A common example is alcohol intolerance caused by deficiency of an enzyme alcohol dehydrogenase, which is common amongst Asian people, where about 50% are affected. Other common non-immunological reactions caused by insufficient activity of enzymes include **lactose intolerance**, **phenylketonuria**, and **favism**. A well-known type of hypersensitivity to food (anaphylaxis) is a reaction to strawberries, which is manifested by urticaria (hives). Amongst the manifestations known as aversion to food (*idiosyncrasy*) are the so-called Chinese restaurant syndrome (associated with a higher intake of sodium-hydrogen glutamate; its mechanism is not yet well known) and migraine occurring after eating certain foods.

10.3.1.1 Lactose intolerance

Lactose intolerance, often confused with milk allergy, is one of the most common metabolic disorders. Under normal conditions, lactose in the digestive tract is hydrolysed by β-galactosidase (lactase) to glucose and galactose, which are then used as an energy source. In the case of lactose intolerance, this key enzyme in the body is missing or not very active. Disaccharides are generally not absorbed by the wall of the small intestine, so in the absence of lactase, lactose passes from food into the colon and is fermented there by intestinal bacteria, similarly to other non-utilisable carbohydrates, forming gases (hydrogen, carbon dioxide, and methane). Three types of lactose intolerance are recognised:

- primary lactose intolerance, which is common amongst non-breastfed children in many Asian and African countries;

- secondary lactose intolerance, in which lactase is not biosynthesised as a result of various gastrointestinal diseases;

- congenital lactose intolerance, a genetic disorder in which lactase is not biosynthesised from birth (since it is a metabolic disorder, the disease is lifelong, but it may show a slight improvement with age).

Lactose intolerance generally increases with age. It is estimated that globally, about 75% of the adult population shows a decrease in lactase activity during adolescence and adulthood (in Northern Europe, about 5% of the population; in southern Europe, about 71%; and in some African and Asian countries, up to 90%). Elimination of milk or a control of the diet containing milk is the best way to prevent this metabolic disorder. It can also be avoided through the use of appropriate technological procedures, such as enrichment of milk by β-galactosidase and milk fermentation. Fermented milk products largely contribute to reducing adverse patient reactions to milk, as much of the lactose present in milk is converted into lactic acid. Low-lactose foods intended for human consumption must contain up to 10 g/kg (or 10 g/l) lactose; lactose-free foods can contain just 100 mg/kg (or 100 mg/l) lactose and no galactose.

10.3.1.2 Phenylketonuria

Phenylketonuria is a congenital metabolic disorder consisting in the metabolism of the amino acid phenylalanine, which is a component of all proteins. Very low activity or absence of phenylalanine hydroxylase, which breaks down phenylalanine, is caused by mutations (damage) of the gene for this enzyme. If this condition is not treated, phenylalanine metabolites, such as phenylpyruvic, phenyllactic, and phenylacetic acids (urine of patients has a characteristic mice-like odour), accumulate in the body and cause total disruption of child development. The result is damage to the development of the nervous system, leading to mental disability (*microcephaly*) manifested by termination of the growth of the brain and delayed psychomotoric development. The prevalence of phenylketonuria shows considerable geographic variation. It is estimated to be present in 1/10 000 live births in Europe and the United States, with a higher rate in some countries (Ireland and Italy). Prevalence is particularly high in Turkey, where it is 1/4000 live births, and is far lower in the Finnish, African, and Japanese populations.

Phenylketonuria of children cannot be cured, and so they need to follow a controlled diet based on nutrition with a reduced content of natural proteins. The phenylalanine content in foods cannot exceed 200 mg/kg (200 mg/l). The aim is to keep the level of phenylalanine in the blood at the lowest possible level. According to a person's age and nutritional needs, the natural protein can be replaced by an artificial mixture of individual amino acids without phenylalanine or a mixture of phenylalanine-free higher peptides. Higher peptides can be prepared by plastein synthesis (see Section 2.3.3.2), which consists of partial protein hydrolysis, treatment with protease under suitable conditions leading to release of hydrophobic amino acids, and plastein reaction in the presence of added tryptophan and tyrosine, which yields practically phenylalanine-free plastein.

10.3.1.3 Favism

Favism is caused by the toxic pyrimidines divicine (2,4-diamino-5,6-dihydroxypyrimidine) and isouramil (4-amino-2,5,6-tri-hydroxypyrimidine, **10-2**), which occur in broad (fava) beans (*Vicia faba*, Fabaceae) and horse beans (*V. faba* var. *equina*), and in small quantities in some other legumes as 5-*O*-β-D-glucopyranosides known as vicine and convicine (**10-3**), respectively. These glycosides probably offer some protection against elicitors (bacteria and fungi), as they exhibit bacteriostatic and fungistatic effects. Sweet pea (vetchlings) (*Lathyrus* spp.) and pea (*Pisum* spp.) species do not contain pyrimidine glycosides. The vicine content of fava beans generally ranges from 4.2 to 10.8 g/kg, whilst the convicine content tends to be lower, within 0.3–0.5 g/kg. Divicine and isouramil arise from the respective glucosides, mainly in the large intestine, by hydrolysis with microbial β-glucosidases. The glycoside content in the beans can be reduced by extraction with water. In this way, it is possible to remove 80–90% of the original amount of glycosides in beans.

10-2, divicine, R = NH$_2$
isouramil, R = OH

10-3, vicine, R = NH$_2$
convicine, R = OH

The main favism symptom is acute haemolytic anaemia, accompanied by high fever, jaundice, and swelling of the liver and spleen, as toxic pyrimidines oxidise the reduced form of glutathione in erythrocytes. Favism is manifested especially in individuals with low (usually hereditary) activity of the enzyme glucose 6-phosphate dehydrogenase in erythrocytes, which reduces the oxidised form of glutathione. This results in a lower concentration of reduced glutathione, which has a protective effect on the erythrocyte membrane. In addition to active substances of fava beans, medical drugs that oxidise glutathione can also invoke favism.

The presence of fava beans in feed mixtures for laying hens negatively affects the quality of the eggs, which have a lower weight and increased yolk fragility.

10.3.2 Food allergens

Food allergy is an abnormal response of the immune system to even a tiny amount of a particular food, which is manifested by symptoms such as sneezing, itching, nasal obstruction (hay fever), redness of the eyes (conjunctivitis), nausea, diarrhoea, vomiting, eczema, coughing, and wheezing (asthma). This type of allergic reaction is often called a hypersensitivity reaction. Rarely, allergic reactions to food can cause serious illness and death, such as a life-threatening set of symptoms called anaphylaxis or anaphylactic shock. It is estimated that about 2.5% of adults and about 6–8% of children, mainly younger than six years, have true food allergies.

A food allergy is an allergic reaction of the organism in which the **allergen** (in immunological terms, an antigen) is a component of food. Allergens, the agents causing the allergy, may be proteins, polysaccharides, or low-molecular-weight compounds called **haptens**. Allergens acquire immunological properties through interactions with serum proteins of the organism. Immunogenicity of proteins that are hosts for foreign substances is bestwoed by the amino acid sequences and their constitution and conformation. The mechanism of the immune-mediated reactions takes place in two stages. The first is a specific immunological reaction, in which the allergen reacts with IgE antibodies to release mediators such as prostaglandins and biogenic amines, one of which is histamine. The second, non-specific stage sees the hypersensitive organism respond to mediator stimulation by pathological manifestations. The most common manifestations of an allergy are skin reactions (such as atopic eczema, hives, and swelling), respiratory-system reactions (rhinitis), and digestive-system reactions. Non-IgE mediated food allergies are caused by a reaction involving other components of the immune system apart from IgE antibodies.

In theory, any food can be an allergen, but all foods do not have the same ability to allergise the organism. Amongst those that tend to cause the most allergic problems are milk, eggs (egg white), some fish, some fruits (apples, pears, apricots, peaches, and strawberries), some vegetables (tomatoes, celery, spinach, and parsley), cereals (hypersensitivity to cereals usually persists to a certain stage of life and improves with age), legumes (such as soya beans), and seeds and nuts (sesame seeds and peanuts; reactions are usually very severe).

Allergens are substances that are very thermostable and stable in acidic media, so resist the digestive processes. A variety of approaches to decreasing the allergenicity of foods have been attempted. The overall conclusion is that processing does not completely abolish the allergenic potential of allergens. Currently, only fermentation and hydrolysis have potential to reduce allergenicity. For example, allergenicity of milk proteins can be reduced about 1000 times by combining enzymatic hydrolysis, thermal denaturation, and ultrafiltration. Reduction of soybean allergenicity can be achieved by thermal and enzymatic treatment, non-thermal procesing methods (such as irradiation and hydrostatic pressure), and genetic modification. For the treatment of this allergy, avoidance is at present the only solution.

10.3.2.1 Milk

Non-toxic adverse reactions to milk are primarily caused by either lactose intolerance or milk allergy. The cow milk proteins prevalently implicated in allergic responses in children are the serum (whey) proteins α-lactalbumin (Bos d 4) and β-lactoglobulin (Bos d 5), in addition to casein fraction (Bos d 8). In adults, the predominant allergen is casein, whereas sensitization to whey proteins is rare. The casein fraction is composed of four proteins α_{s1}- (a major allergen), α_{s2}-, β-, and κ-casein, in approximate proportions of 40%, 10%, 40%, and 10%, respectively.

10.3.2.2 Eggs

Egg allergy is one of the most prevalent food allergies in industrialised countries. The estimated prevalence of egg allergy varies between 1.6 and 3.2%, making it the second most common cause of food allergy in children. The majority of the relevant egg allergens (Gal d 1 to Gal d 6) have been identified in the white: Gal d 1 (ovomucoid, 28 kDa), Gal d 2 (ovalbumin, 44 kDa), Gal d 3 (ovotransferrin, 78 kDa), and Gal d 4 (lysozyme C, 14 kDa). The allergens Gal d 5 (serum albumin, 14 kDa) and Gal d-6 (YGP42, 35 kDa) are found in the yolk.

10.3.2.3 Cereals

Cereal proteins are not significant as allergens. For example, buckwheat (*Fagopyrum esculentum*, Polygonaceae) contains the thermostable allergens Fag e 2 (2S albumin, 16 kDa) and Fag e 3 (vicilin fragment, 19 kDa), but some allergenicity has been observed even for rice (containing allergens Ori s 1 and Ori s 12), barley (Hor v 1 identical to Hor v 15, Hor v 12, Hor v 16, Hor v 17, and Hor v 20), and wheat (Tri a 12, Tri a 14, Tri a 15, Tri a 18, Tri a 19, Tri a 21, and Tri a 25 to Tri a 37). The disease caused by wheat allergens is called **Duhring's disease** or *dermatitis herpetiformis*. Wheat flour may cause **baker's asthma**, one of the most common forms of occupational asthma. This allergic disease is caused by inhalation of flour dust containing proteins of relative molecular weight around 15 kDa, which act as α-amylase inhibitors in the grain. It can affect workers in bakeries, flour mills, and kitchens.

A malabsorption disorder of the small intestine caused by sensitivity to gluten, a protein found in wheat, and similar proteins in rye, barley, and oats is not a food allergy or intolerance but an autoimmune disease known by the Latin name ***coeliac sprue*** (coeliac disease, non-tropical sprue, or gluten enteropathy). The disease is caused by gliadine (prolamine) and gluteline fractions of gluten or gluteline fractions of barley (hordeine) and rye (secaline) having two amino acid sequences: Pro-Ser-Gln-Gln and Gln-Gln-Gln-Pro.

Coeliac disease may occur in genetically predisposed people of all ages from middle infancy onward. In children, it usually occurs soon after they are first given a diet containing gluten (semolina porridge, biscuits, or soup thickened with flour). Infants, toddlers, and young children may often exhibit growth failure, vomiting, bloated abdomen, behavioural changes, and failure to thrive. In adults, coeliac disease can be triggered for the first time at between 30 and 50 years of age following surgery, viral infection, severe emotional stress, pregnancy, or childbirth. It can manifest similarly as in children, but often there are cases with less developed symptoms.

At present, the only sure way to treat coeliac disease is to follow a gluten-free diet, avoiding wheat, rye, and barley. As alternatives, products containing millet, maize, rice, amaranth, buckwheat, soybeans, and edible chestnuts may be consumed. Vegetables including potatoes, fruits, meat, fish, eggs, milk, and milk products are also allowed.

The European Union adopts common rules concerning the composition and labelling of foods for people who are intolerant to gluten. Very low-gluten foods containing ingredients made from wheat, rye, barley, oats, or their crossbred varieties that have been especially processed to reduce gluten must not contain a level of gluten exceeding 100 mg/kg. The content of gluten in products labelled 'gluten-free' must be lower than 20 mg/kg.

10.3.2.4 Legumes

Soya is one of the most important sources of the food allergens Gly m 1–Gly m 7, which are responsible for about 90% of allergic reactions in the child population. At least 16 such allergens have been identified. Gly m 1 is a hydrophobic protein, Gly m 2 is a small cysteine-rich cationic protein (defensin), Gly m 3 is an actin-binding protein (profiling), Gly m 4 is a member of the PR-10 protein family, Gly m 5 is β-conglycinin (vicilin, 7S globulin), Gly m 6 is glycinin (legumin, 11S globulin), and Gly m 7 is a seed biotinylated protein.

10.3.2.5 Nuts

Peanuts contain several allergens (Ara h 1–Ara h 13) that are highly resistant to heat. The three major ones, Ara h 1, Ara h 3, and Ara h 2 (members of the cupin superfamily of proteins) are 7S and 11S globulins and 2S albumins, respectively. Peanut allergy (affecting 1.3% of the general population) is the most common cause of severe allergy attacks, especially in children, and appears to be on the rise. Peanut allergy symptoms can range from a minor irritation to a life-threatening reaction (anaphylaxis). For some people, even tiny amounts of peanuts can cause a serious reaction.

10.3.2.6 Other foods

Allergenicity has been demonstrated in a number of other plant materials, exemplified by green peas, rapeseed, mustard, sesame seed, and other materials. The common characteristics of these allergens are a low molecular weight and a high content of cysteine/cystine.

10.3.3 Lectins

One of the protective mechanisms of plants invovles the accumulation of certain proteins in seeds and vegetative parts related to reproduction. These organs are unacceptable, unpalatable, or toxic to parasites and predators. The proteins so accumulated are inhibitors of proteases, saccharases, and lectins.

10.3.3.1 Structure, nomenclature, and occurrence

Lectins, formerly called phytohaemagglutinins, were originally defined as proteins capable of red blood cell (erythrocytes) agglutination in animal organisms. Their activity is due to their interaction with sugars, which are components of glycoproteins or glycolipids in cell membranes. The current definition refers to lectins as all proteins with at least one centre other than the active (catalytic) centre (by which proteins can reversibly bind to specific monosaccharides and oligosaccharides). Many lectins can bind to monosaccharides and their derivatives (such as D-glucose, D-mannose, D-galactose, N-acetyl-D-galactosamine, N-acetyl-D-glucosamine, and L-fucose), but they have a higher affinity to oligosaccharides that do not occur in plants but are typical components of animal glycoproteins. Such compounds include N-acetylneuraminic acid and N-acetyl-D-galactosamine, located as building units in the chains of complex carbohydrates.

Lectins are a highly heterogeneous group, comprising several hundred plant proteins, which are divided into three categories:

- **merolectins**;

- **hololectins**;

- **chimerolectins**.

Merolectins are simple proteins, incapable of agglutinating cells, that contain only one centre binding sugars and no catalytic centre. Merolectin, for example, is the protein of amaranth (*Amaranthus caudatus*, Amaranthaceae), which binds polysaccharide chitin.

Hololectins are proteins containing at least two centres that bind sugars and, like merolectins, have no catalytic functions. They behave as true agglutinins. The majority of plant lectins are hololectins. An example is soya bean lectin (soya bean agglutinin, SBA), a tetrameric glycoprotein (or metalloprotein, as it contains one binding site for Mn^{2+} and four sites for binding transition metals) with a relative molecular weight of 120 kDa. It is composed of two structurally slightly different subunits (30 kDa) and contains four centres binding N-acetyl-D-galactosamine.

Chimerolectins are complex proteins containing one or two centres that bind sugars and an independent centre with catalytic or other biological activity. Depending on the number of sugar-binding centres, they behave either as merolectins or as hololectins. For example, lectin inactivating ribosomes type 2, such as toxic protein ricin occurring in castor seeds (*Ricinus communis*, Euphorbiaceae), has a catalytically active chain A covalently linked to two chains B by disulfide bridges. Each of these polypeptide chains B contains one centre, which binds sugars; therefore, ricin behaves as a hololectin.

The majority of known lectins belong to one of four major subgroups:

- legume lectins;

- mannose-binding lectins (monocot plant lectins);

- chitin-binding lectins;

- ribosome type 2-inactivating lectins.

The most important mannose-binding lectins occur in different families of plants, mainly Amarillidaceae, Araceae, Liliaceae, and Orchidaceae. Chitin-binding lectins are located in five mutually taxonomically unrelated families of plants, Poaceae, Solanaceae, Urticaceae, Papaveraceae, and Amaranthaceae. Some lectins cannot be classified into any of these subgroups, because they appear to belong to other, not yet specified subgroups.

Lectins are widely distributed in both the plant and the animal kingdoms, and are even present in some microorganisms. The largest amounts are found in the seeds of plants. The most frequently occurring lectins, their contents, and their important properties are shown in Table 10.2. Some lectins traditionally have trivial names; for example, the lectin (albumin) of castor seeds is ricin and the lectin of jack beans (*Canavalia ensiformis*, Fabaceae) is concavalin A.

10.3.3.2 Reactions and changes

Lectins can be denatured by heat, and a decrease of their biological activity occurs during proteolysis. Detoxification procedures of lectins in food materials commonly include soaking and cooking. The efficiency of these operations depends on the time of soaking, temperature, and time of heat treatment. Soaking itself is not sufficient. For example, detoxification of beans by soaking for 16 hours at 22–25 °C decreases the activity of lectins by only 4–5%. Cooking itself is also not appropriate for food detoxification. For example, detoxification of beans by cooking takes about 90 minutes, whilst the same material soaked for 16 hours is detoxified during 4–10 minutes of cooking. Lectins are largely dissolved into the soaking water and cooking water. It is not recommended to consume this water. Autoclaving, especially at higher pressures, is also a suitable detoxification procedure, but during extrusion, only small losses of lectin activity occur (12–18%).

Cereal products containing soya beans generally show higher residual activity of lectins than other materials. Seed germination leads to a significant decrease in the activity of lectins. The germination time required to substantially reduce the activity of lectins is four to six days. Detoxification of plant products, such as potatoes, can be achieved by common manufacturing and culinary practices.

10.3.3.3 Biological effects

Lectins applied intravenously are highly toxic, and some exhibit toxicity even when applied orally. Acute toxicity is usually low, but long-term exposure to even small amounts of lectins can be harmful. Consuming raw or inadequately cooked beans may cause stomach upset, vomiting, and diarrhoea. In comparison with the lectins of common beans, soya bean lectins are relatively well digested in the small intestine, but they are still antinutritional or toxic compounds. Approximately 60% of ingested lectins passes through the digestive tract unchanged and bind to the carbohydrate receptors of the small-intestine epithelium, which is manifested by reduced viability of epithelial cells and a possible increase of small-intestine weight as a result of cell hyperplasia. At concentrations of 0.5–0.6%, the growth of experimental animals is slow; at higher concentrations, they experience weight loss and, in some cases, death.

Non-toxic lectins are lectins of garlic, onions, leeks, tomatoes, and amaranth. Of the common lectins, those of peanuts, lentils, peas, common beans, and soya beans are slightly toxic, those of wheat are moderately toxic, those of some beans (such as Jack beans) are highly toxic, and those of castor seed are lethal (Table 10.2). Some lectins (such as those of garlic) have prebiotic effects and inhibit undesirable intestinal microflora. Often, lectins are not the only toxic substances of materials. For example, those of soybean seeds participate in their antinutritional and toxic effects at a level of about 25%, whilst trypsin inhibitors act at about 40%; the rest is covered by saponins and other substances.

10.3.4 Biogenic amines and polyamines

Biogenic amines are a group of aliphatic, aromatic, or heterocyclic bases derived from amino acids that exhibit a variety of biological effects, as they perform different functions in animal and plant tissues. Some biogenic amines are building materials for the biosynthesis of phytohormones of the auxin group, plant protoalkaloids (such as hordenine and gramine), true alkaloids, and other secondary plant metabolites. In animal tissues, they have the function of tissue hormones (e.g. histamine) and are precursors of adrenal hormones (catecholamines).

Table 10.2 Main lectins in economically important plants.

Plant	Latin name	Occurrence	Content (g/kg)	Thermal stability[a]	Toxicity in foods	
					Raw	Processed
Legume lectins						
Peanuts	*Arachis hypogaea*	Seeds	0.2–2	Unstable	Yes	Yes
Peas	*Pisum sativum*	Seeds	0.2–2	Unstable	Possibly	No
Common beans	*Phaseolus vulgaris*	Seeds	1–10	Medium	Yes	Possibly
Jack beans	*Phaseolus lunatus*	Seeds	1–10	Medium	Yes	Possibly
Tepary beans	*Phaseolus acutifolius*	Seeds	1–10	Medium	Yes	Possibly
Runner bean	*Phaseolus coccineus*	Seeds	1–10	Medium	Yes	Possibly
Lens	*Lens culinaris*	Seeds	0.1–1	Unstable	Yes	No
Soya beans	*Glycine max*	Seeds	0.2–2	Low	Yes	No
Broad beans	*Vicia faba*	Seeds	0.1–1	Unstable	Possibly	No
Mannose-binding lectins						
Onion	*Allium cepa*	Bulb	<0.01	Medium	No	No
Garlic	*Allium sativum*	Bulb	0.5–2	Medium	No	No
Leek	*Allium ampeloprasum*	Leaf sheaths	<0.01	Medium	No	No
Chitin-binding lectins						
Wheat	*Triticum aestivum*	Seeds	<0.01	High	Yes	Yes
Wheat	*Triticum aestivum*	Germ	0.1–0.5	High	Yes	Yes
Rye	*Secale cereale*	Seeds	<0.01	High	Yes	Possibly
Barley	*Hordeum vulgare*	Seeds	<0.01	High	Yes	Possibly
Rice	*Oryza sativa*	Seeds	<0.01	High	Yes	Possibly
Potatoes	*Solanum tuberosum*	Tubers	0.01–0.05	High	–	Possibly
Tomatoes	*Lycopersicon esculentum*	Fruits	<0.01	High	Possibly	Possibly
Ribosome type 2-inactivating lectins						
Castor oil plant[b]	*Ricinus communis*	Seeds	1–5	Unstable	–	–
Elderberries	*Sambucus nigra*	Fruit	0.01	Medium	Yes	Possibly
Other plant lectins						
Amaranth	*Amaranthus caudatus*	Seeds	0.1–0.5	Unstable	Possibly	Unknown
Banana	*Musa x paradisiaca*	Fruit	<0.01	Unknown	Unknown	Unknown

[a]Thermal stability: high = lectin does not lose activity on heating at 80 °C; medium = lectin does not lose activity on heating at 70 °C; low = lectin does not lose activity on heating at 60 °C; unstable lectin is denatured at 60 °C.
[b]Highly toxic.

Biogenic amines are formed from amino acids by the action of carboxylyases (decarboxylases containing as a cofactor pyridoxal 5′-phosphate) or arise from amino acids and carbonyl compounds by the action of transaminases (see Section 8.2.10.1.2). The so-called endogenous biogenic amines are the products of metabolism and at low concentrations are natural components of almost all foods. Exogenous biogenic amines are formed in foods as a result of microbial contamination and fermentation processes.

10.3.4.1 Structure, nomenclature, and occurrence

Decarboxylation of arginine (by arginine decarboxylase) yields (4-aminobutyl)guanidine known as agmatine, histidine gives rise to 2-(1H-imidazol-4-yl)ethanamine known as histamine (histidine decarboxylase), phenylalanine gives phenylethylamine known as phenethylamine (phenylalanine decarboxylase), tyrosine gives 4-(2-aminoethyl)phenol known as tyramine (tyrosine decarboxylase), and 3,4-dihydroxyphenylalanine (dopa) and tryptophan (aromatic amino acid decarboxylase, also known as dopa decarboxylase or tryptophan decarboxylase) give 4-(2-aminoethyl)benzene-1,2-diol (dopamine) and 2-(1H-indol-3-yl)ethanamine (tryptamine), respectively. Decarboxylation of lysine (lysine decarboxylase) provides 1,5-diaminopentane (cadaverine), whilst ornithine (formed from arginine by the action of arginase) produces 1,4-diaminobutane (putrescine) through ornithine decarboxylase (**10-4**).

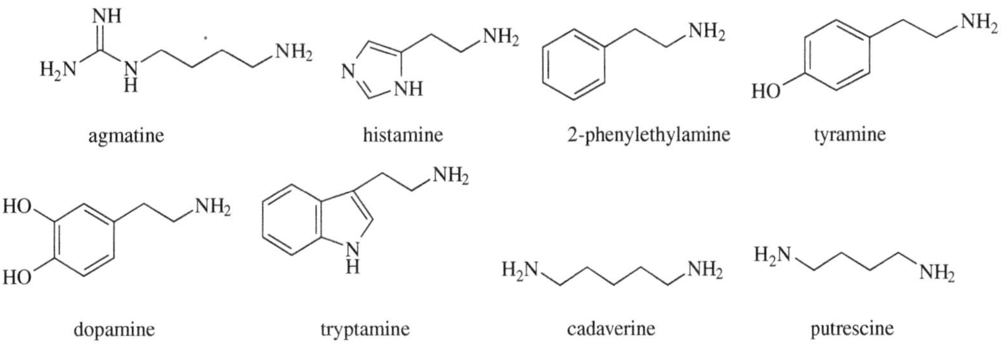

10-4, structures of biogenic amines

Putrescine, produced by decarboxylation of ornithine and also arised from agmatine with catalysis by agmatinase, becomes the starting compound for the biosynthesis of spermidine and spermine. These reactions are catalysed by spermidine synthase and spermine synthase, respectively, involving S-adenosyl-L-methionine (AdoMct or SAM), and occur between bacteria and mammals (Figure 10.1). S-Adenosyl-L-methionine amide (dSAM), formed by decarboxylation of SAM, provides trimethylene amine residue for this biosynthesis, which yields S-methyl-5′-thioadenosine (MTA).

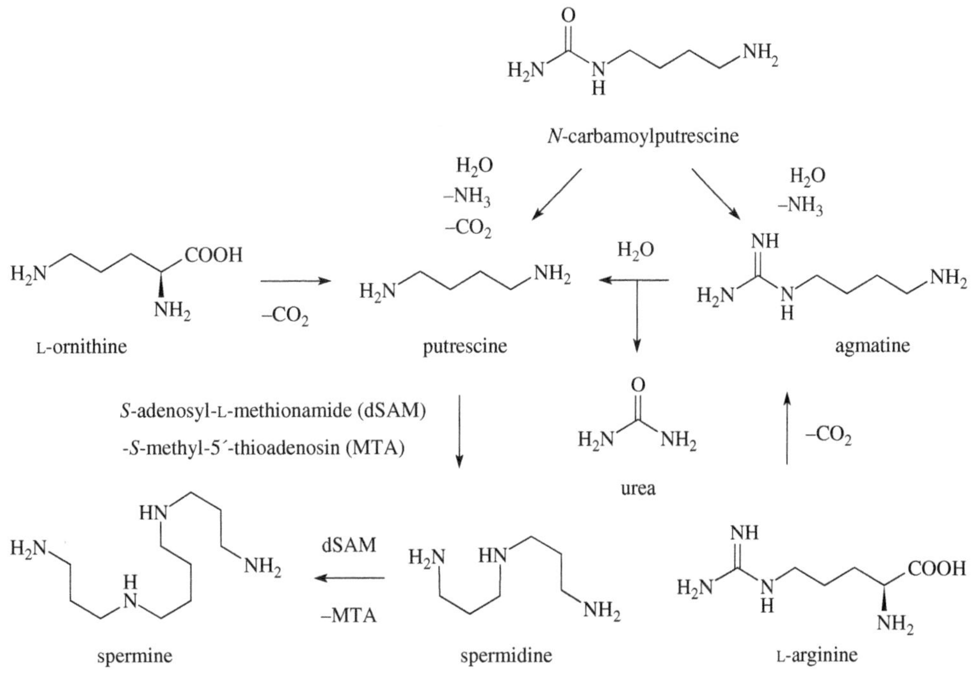

Figure 10.1 Biosynthesis of spermidine and spermine.

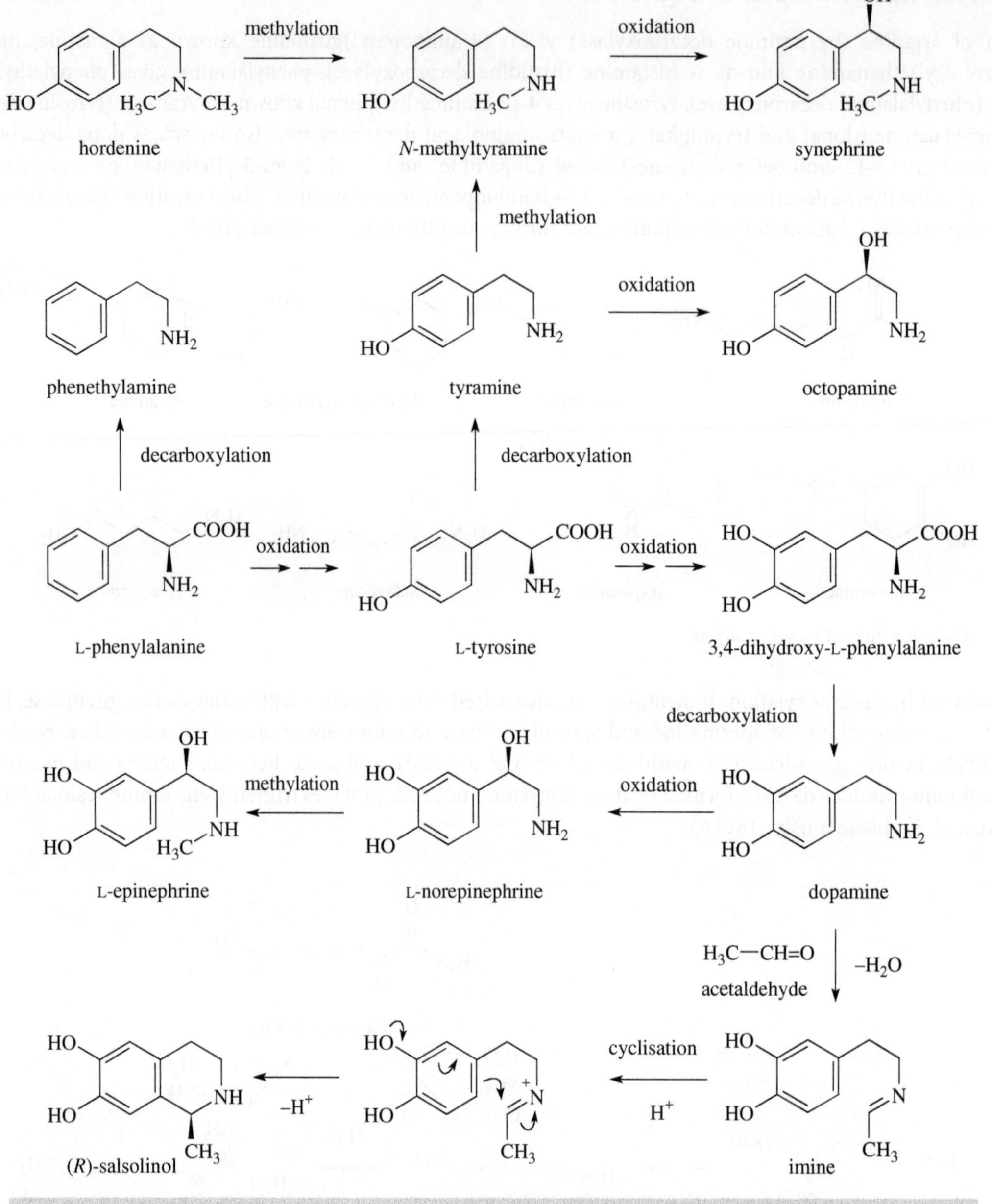

Figure 10.2 Formation of catecholamines and other metabolites of biogenic amines.

Biologically active polyamines putrescine, spermidine, and spermine differ from the classical group of biogenic amines. Spermidine and spermine arise under specific conditions and have different biological effects in comparison with the 'classical' biogenic amines, which are mainly histamine, tyramine, 2-phenylethylamine, putrescine, cadaverine, tryptamine, and agmatine.

In the transformation of biogenic amines to other biologically active products, a number of different oxygenases, methyltransferases, and other enzymes are involved (Figure 10.2). Oxidation of tyramine yields 4-(2-amino-1-hydroxyethyl)phenol, known as octopamine, first demonstrated in common octopus (*Octopus vulgaris*), which acts as a neurohormone and neuromodulator in invertebrates. Methylation of tyramine gives rise to *N*-methyltyramine (a precursor of synephrine), *N,N*-dimethyltyramine (hordenine), and *N,N,N*,-trimethyltyramine (candicine or maltoxin), which acts in animals as a neurotoxin.

Oxidation of dopamine provides catecholamines, such as adrenal hormone (*R*)-norepinephrine (also known as L-norepinephrine or noradrenaline), in animals, methylation of which yields another adrenal hormone (*R*)-epinephrine (L-epinephrine or adrenaline). The formal metabolite of norepinephrine is 3,4-dihydroxyphenylglycol, which belongs to the major C_6—C_2 phenolic compounds found in olive fruits. Oxidation of dopamine via dopachrom leads to melanin pigments (see Section 9.3.1.1), whilst reaction with acetaldehyde yields salsolinol (6,7-dihydroxy-1,2,3,4-isoquinoline), which is, together with dopamine and epinephrine, the major metabolite in bananas. This isoquinoline alkaloid is biosynthesised by non-enzymatic Pictet–Spengler condensation from dopamine and acetaldehyde (enzymatically

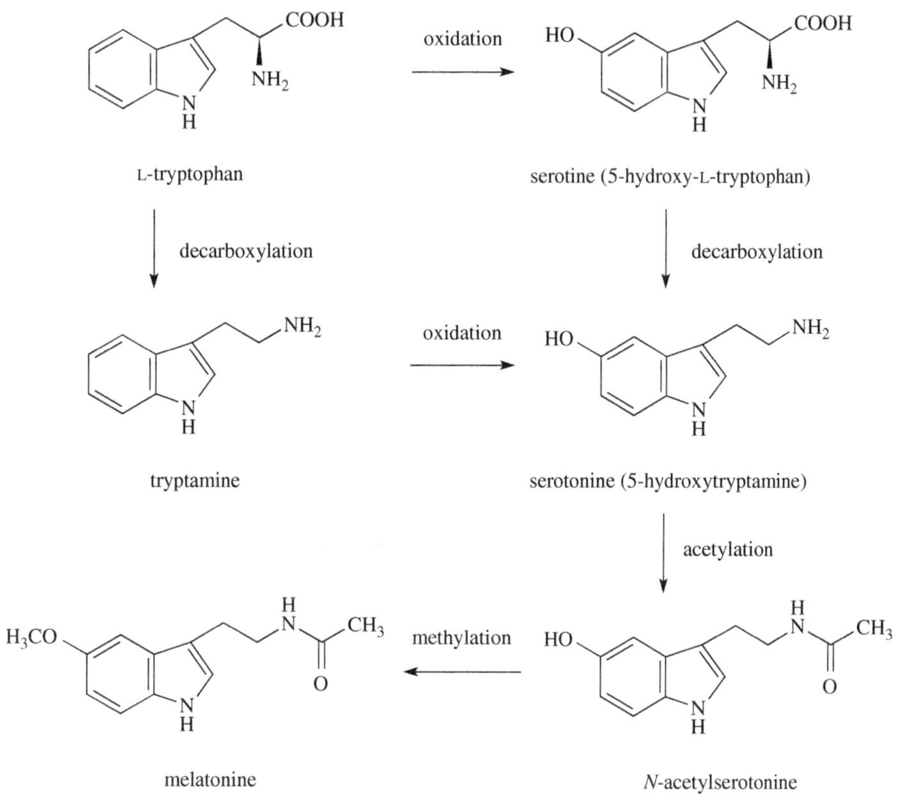

Figure 10.3 Decarboxylation and other reactions of tryptophan.

generated from ethanol) during the post-climacteric phase in banana (Figure 10.2). In dried banana chips, both $(+)$-(R)-salsolinol and $(-)$-(S)-salsolinol occur at the 40 mg/kg level as a racemic mixture. In plants, catecholamines have a role in protection against predators and act in plant growth regulation. They inhibit oxidation of phytohormone 3-indolylacetic acid (see **10-114**), an auxin that controlles cell division, germination, phototropism (the tendency of growing plant organs to move or curve under the influence of light), and other vital plant processes.

The oxidation of tryptophan yields serotine, whose decarboxylation gives tryptamine, 2-(1H-indol-3-yl)ethanamine, the oxidation of which leads to the hormone serotonine, from which arises N-acetylserotonine, the precursor of melatonine (N-acetyl-5-methoxytryptamine) (Figure 10.3). Melatonine is a neurohormone produced in animals that regulates sleep and wakefulness. It is also present in many taxonomically distant groups of organisms, including bacteria, fungi, and plants. In plants, it is produced as a defense against oxidative stress. The levels in foods were found to range from none to ppt levels. Higher concentrations are reportedly found in beverages such as coffee, tea, wine, and beer, cereals (including wheat and rise), and bread.

Biogenic amines are present in almost all foods as normal products of metabolism. They are found in higher levels in fermented products (such as cheeses, durable sausages, beer, wine, and sauerkraut), where they are formed by microbial activities. Contaminating microflora primarily causes their presence in fish and meat during storage. High concentrations of biogenic amines also occur in fruits, vegetables, and mushrooms in advanced stages of spoilage under improper storage conditions. An overview of biogenic amines occurring in foods, their precursors, their biological activities, and their main transformation products is given in Table 10.3.

Biogenic amines with more significant adverse effects on human health are histamine, tyramine, and 2-phenylethylamine. The occurrence of tryptamine and agmatine is generally low and is not associated with adverse effects on human health. The increased content of diamines putrescine and cadaverine serves primarily as an indicator of deficiencies in processing technology and storage of foods and food raw materials. An overview of microorganisms involved in the production of biogenic amines in various foods is given in Table 10.4. The contents of the main biogenic amines in some common foods are shown in Table 10.5 (considerable variation in their contents is characteristic, as they depend on a number of factors).

10.3.4.1.1 Foods of animal origin

The main biogenic amines of meat, fish, and cheese are histamine, cadaverine, putrescine, and tyramine. During meat storage, the content of biogenic amines increases due to enzyme activity of the microflora, therefore the content of some biogenic amines can be used as an indicator

Table 10.3 Biogenic amines and their precursors, transformation products, and biological activities.

Biogenic amine	Original amino acid	Other products	Biological ativity
Histamine	Histidine		Local tissue hormone, reduces blood pressure, effect on gastric juice secretion, participation in anaphylactic shock and allergic reactions
Cadaverine	Lysine		Stabilisation of macromolecules (nucleic acids), subcellular structures (ribosomes), stimulation of cell differentiation, vegetable hormone
Putrescine	Arginine via ornithine or citrulline	N-Methylputrescine, spermidine, spermine	Stabilisation of macromolecules (nucleic acids), subcellular structures (ribosomes), stimulation of cell differentiation, vegetable hormone
Agmatine	Arginine	Putrescine, N-methylputrescine, spermidine, spermine	Stabilisation of macromolecules (nucleic acids), subcellular structures (ribosomes), stimulation of cell differentiation, vegetable hormone
Phenethylamine	Phenylalanine	Tyramine, dopamine, epinephrine, norepinephrine	Neuromodulator, neurotransmitter in mammalian central nervous system
Tyramine	Tyrosine	Dopamine, epinephrine, norepinephrine, synephrine, hordenine	Precursor of dopamine, local tissue hormone, increases blood pressure, effect on smooth muscle contraction
Dopamine	Dopa	Norepinefrine, epinefrine	Mediators of sympathetic nerves
Tryptamine	Tryptophan	Serotonine, melatonine	Locan tissue animal and plant hormones (catecholamines), effect on blood pressure, intestinal peristalsis, mental functions

Table 10.4 Important microorganisms producing biogenic amines.

Food	Microorganisms	Amines produced
Fish	*Morganella morganii, Klebsiella pneumoniae, Hafnia alvei, Proteus mirabilis, Proteus vulgaris, Clostridium perfringens, Enterobacter aerogenes, Bacillus* spp., *Staphylococcus xylosus*	Histamine, tyramine, cadaverine, putrescine, agmatine
Cheeses	*Lactobacillus buchneri, L. bulgaricus, L. plantarum, L. casei, L. acidophilus, L. arabinosae, Streptococcus faecium, S. mitis, Bacillus macerans, Propionibacterium* spp.	Histamine, cadaverine, putrescine, tyramine, tryptamine
Meat and meat products	*Pediococcus* spp., *Lactobacillus* spp., *Pseudomonas* spp., *Streptococcus* spp., *Micrococcus* spp., *Enterobacteriaceae*	Histamine, cadaverine, putrescine, tyramine, phenethylamine, tryptamine
Fermented vegetables	*Lactobacillus plantarum, Leuconostoc mesenteroides, Pediococcus* spp.	Histamine, cadaverine, putrescine, tyramine, phenethylamine, tryptamine
Fermented soya products	*Rhizopus oligosporus, Trichosporon beigliii, Lactobacillus plantarum*	Histamine, cadaverine, putrescine, tyramine, tryptamine

Table 10.5 Main biogenic amine and polyamine contents in foods.

	Content (mg/kg or mg/l)									
Foods	Histamine	Cadaverine	Putrescine	Spermidine	Spermine	Agmatine	Phenethylamine	Tyramine	Tryptamine	Serotonine
Meat										
Pork	0–45	0–171	trace–702	trace–5	5–40	–	–	1–35	1–48	–
Beef	0–217	0–27	trace–26	trace–5	5–40	2–112	–	trace–61	–	–
Chicken	1	9	trace–10	5–10	20–60	–	trace	23	–	–
Meat products										
Ham	1–271	trace–97	trace–20	trace–8	20–60	–	trace–215	trace–618	8–67	–
Bacon	15	trace–1	trace–8	2–42	1–212	–	–	1–3	4	–
Sausages	trace–550	trace–787	1–396	trace–10	10–60	–	0–696	0–1 240	0–29	
Fish										
Tuna	trace–8 000	trace–447	trace–200	1–10	2–35	–	trace–45	trace–1 060	–	–
Mackerel	trace–3 000	trace–226	trace–40	2–4	trace–8	–	trace–126	trace–75	–	–
Cheeses										
Soft cheese	0	0–1.5	0–3.1	0–0.8	0–1.1	0	0	0–0.6	0	–
Hard cheese[a]	0–301	0–710	0–612	0–43	0–19	0–22	0–32	0–301	0–45	–
Hard cheese[b]	0–609	0–389	0–670	0–40	0–22	0–27	0–30	0–609	0–34	–
Cheddar	0–1300	0	1–996	–	–	–	0–303	0–1500	0–300	–
Emmental	trace–2000	0–460	1–130	–	–	–	0–490	1–1000	0–210	–
Gouda	0–850	1–140	1–200	–	–	–	0–46	0–670	10–200	–
Edam	0–88	trace	trace	–	–	–		trace–320		–
Roquefort	0–4100	42–905	44–830	–	–	–	10–25	trace–1350	10–1100	–
Fruits										
Bananas			5	10	trace	–	–	7–95	–	12–78
Pineapples	2–65	–	–	–	–	–	–	0–4	–	–
Oranges	–	–	95–150	trace–10	trace	–	–	1–10	trace	–
Grapefruits	–	–	20–90	2–15	trace	–	–	0–1400	–	–
Vegetables										
Spinach	60	–	trace–120	1–15	trace–4	–	–	0–680	–	–
Tomatoes	trace–1	–	10	trace	trace	–	–	0–1200	4	12
Other foods										
Souerkraut	1–200	1–311	6–550	trace–45	trace	–	0–9	2–310	–	–
Soya sauce	0–274	–	trace–500		–	–	trace	trace–882	trace–100	–
Beer	0–22	0–40	2–15	0–7	0–4	1–41	0–8	1–68	0–5	–
Malt	1–4	–	4–10	–	–	23–117	–	9–28	–	–
Wine (red)	0–30	0–47	2–20	trace	trace	–	trace	0–90	–	–
Wine (white)	0–20	3–108	1–11	trace	trace	–	–	trace–212	–	–
Sherry	0–31	1	3–25	–	–	–	1	1–17	–	–
Chocolate	0–10	0–8	0	1–2	trace–11	–	0–27	0–2	trace–1	–

[a]Produced from pasteurised cow's milk.
[b]Produced from raw sheep's and cow's milk.

Table 10.6 Changes in biogenic amine amounts during fermentation of durable sausages.

Starting culture	Days	Content (mg/kg)				
		Histamine	Cadaverine	Putrescine	Phenethylamine	Tyramine
High activity of decarboxylases	0	3	1	1	2	2
	1	4	7	21	2	69
	2	5	35	18	9	95
	9	8	64	15	11	142
	21	6	84	13	11	120
Low activity of decarboxylases	0	3	1	1	2	2
	1	3	1	1	2	2
	2	3	1	1	2	5
	9	4	1	2	2	20
	21	3	1	3	2	21

of meat freshness. Fresh pork meat contains, for example, up to 7 mg/kg of cadaverine and putrescine, whilst their content in rotten meat is 60 mg/kg or more. Cooking has relatively little influence on the content of biogenic amines, which only decreases with partial degradation and leaching (which is higher in pork meat). The level of the polyamines spermidine and spermine during thermal processing of meat and offal decreases by several tens of per cent, whilst higher decreases occur during baking and frying than during boiling or stewing.

The content of biogenic amines increases during the production of fermented sausages and ripening of cheeses. This increase is especially noticeable in the early stages of fermentation and is dependent on the types of microorganisms. The formation of biogenic amines in durable salami and cheeses is caused by microorganisms of the processed raw material (Table 10.6). In rare cases, products may contain 100–1000 mg/kg of histamine, up to 580 mg/kg of putrescine, up to 90 mg/kg of spermidine, and up to 100 mg/kg of spermine. The currently used starter cultures should have very low or negligible activity of amino acid decarboxylases.

Biogenic amines occur in fresh fish meat in small amounts. For example, tuna meat contains 0–10 mg/kg of histamine and 0–2 mg/kg of tyramine, but the content of biogenic amines increases with improper storage. During storage of fish at temperatures around 0 °C and lower, biogenic amines are formed in almost negligible quantities. At higher temperatures, the microflora in fish produces mainly histamine, and tissues of Scrombroidae fish (tuna and mackerel) can contain up to 3000 mg/kg (mackerel) or even 8000 mg/kg (tuna). The optimum temperature for histamine formation is 5–38 °C, depending mainly on the type of contaminating microflora. This is due to high levels of free, easily accessible histidine in the muscles of these fish species. Other biogenic amines also arise in relatively high amounts, including tyramine, cadaverine, and putrescine. Agmatine is usually found in meat and fish in amounts of 0–3 mg/kg. High concentrations of agmatine, however, are found in some species of shellfish and dried fish. For example, fresh Japanese abalone (*Haliotis sieboldii*) contains 40–200 mg/kg of agmatine. Its level in dried fish is up to 650 mg/kg.

Ripening of cheeses leads to a significant formation of biogenic amines only in plants with poor hygiene levels, where it is caused by contaminating microflora. With good technology and following good hygienic principles, even long-term ripened cheeses contain only small amounts of biogenic amines and polyamines, of which spermidine is present in the highest amount. The situation is different in cheeses made from unpasteurised milk (mainly milk of sheep and goats), which are often produced in small factories with hygiene deficiencies, even in developed countries.

The removal of already formed biogenic amines from foods is very difficult. A decrease in their concentration can be achieved by diaminooxidase, but in practice it is impossible to use this enzymatic method of decontamination. Partial reduction of amines also occurs in heat-treated products, where biogenic amines can react with reducing sugars and their decomposition products in the Maillard reaction. The best way of producing foods containing small amounts of biogenic amines is through the observance of processes and hygienic conditions that prevent their formation.

10.3.4.1.2 Foods of plant origin

The main biogenic amine in fruits and vegetables is tyramine, with many others present in smaller quantities. For example, tyramine is the main biogenic amine of bananas, followed by phenylethylamine, histamine, dopamine, serotonine, and norepinephrine. Octopamine,

N-methylated derivatives of tyramine, and some of their glycosides (tyramine-*O*-β-D-glucoside, *N*-methyltyramine-*O*-β-D-glucoside, and hordenine-*O*-β-D-glucoside) were reported to occur in juice of *Citrus* fruits, such as sweet orange, bitter orange, bergamot, citron, lemon, mandarin, and pomelo. The total amount of biogenic amines in bitter orange (*C. aurantium*), for example, ranged from about 10 to 80 g/kg. The main component was synephrine (8.8–71.5 g/kg), which is used in slimming diets. Synephrine also occurs at low concentrations (about 13–35 mg/l in juice) in sweet oranges, which are, however, one of its main sources in a normal human diet. Putrescine occurs in higher amounts in citrus fruits. In the United States, for example, oranges and grapefruits (including juices) represent the largest dietary item in the intake of polyamines, due to the presence of putrescine. Histamine (at a concentration of approximately 60 mg/kg), *N*-methylhistamine, *N*-acetylhistamine, and histamine amides with short-chain fatty acids occur in spinach leaves. Histamine amide with capronic acid was found in tomato products together with amides derived from histidinol (see Section 8.2.2.1.2). Hordenine occurs naturally in a variety of plants, taking its name from barley. In germinating barley, hordenine levels reach a maximum within 5–11 days of germination, being about 20–30 mg/kg in malt and 12–24 mg/l in beer. In barley, the hordenine is accompanied by the corresponding quaternary base candicine (**10-6**). Biogenic amines, including polyamines, also occur in beer, where they are produced by contaminating lactic acid bacteria. Their concentrations are as follows: histamine <0.2–22 mg/kg, cadaverine <0.2–49 mg/kg, putrescine <0.3–31 mg/kg, spermidine <0.2–7 mg/kg, spermine <0.2–15 mg/kg, agmatine 0.5–47 mg/kg, phenethylamine <0.2–8 mg/kg, tyramine <0.3–68 mg/kg, and tryptamine 0–10 mg/kg.

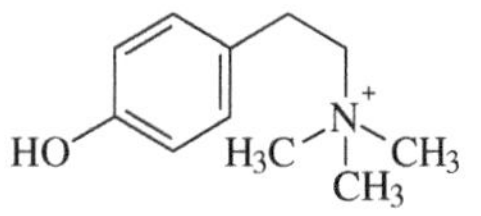

10-6, candicine

Amides derived from cinnamic acids and biogenic amines (e.g. tyramine and octopamine) or polyamines (known as cinnamic acid amides or **phenolamides**) are frequent constituents of plant cell walls. It is known that these phytoalexins exhibit fungicidal activity and act as a barrier against the penetration of pathogens, because they increase tissue rigidity and reduce its digestibility. The number of cinnamic acid residues in the polyamine backbone is usually one to three. Examples are tyramine amides (**10-7**), namely *N*-feruloyltyramine (moupinamide) and *N*-sinapoyltyramine. Antioxidants occurring in black pepper, and *N*-*p*-coumaroyloctopamine and *N*-feruloyloctopamine, identified as major antioxidants in garlic skin. Dimers of phenolic amides, named lyciumamides A (**10-8**), B, and C, together with two monomers, *N*-*p*-coumaroyl tyramine and *N*-feruloyl tyramine, were isolated from the fruits of *Lycium barbarum* (Solanaceae), known as wolfberry or goji berry.

Polyamine conjugates with cinnamic acids are potent antioxidants. Example are phenolamides derived from ferulic acid, which were found in rice (**10-9**). *N,N'*-dicoumaroylputrescine, *N*-*p*-coumaroyl-*N'*-feruloylputrescine, and *N,N'*-diferuloylputrescine were isolated from maize bran and *N*-caffeoylputrescine from tobacco. Wheat contains ferulic acid amides and *p*-coumaric acid amides of 2-hydroxyputrescine.

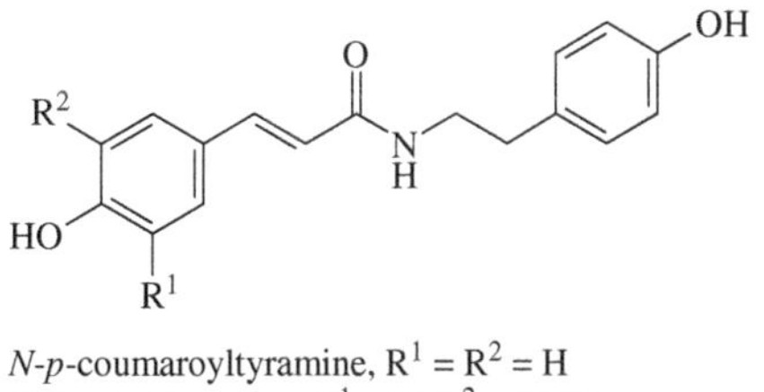

N-*p*-coumaroyltyramine, R¹ = R² = H
N-feruloyltyramine, R¹ = H, R² = OCH₃
N-caffeoyltyramine, R¹ = R² = OH
N-sinapoyltyramine, R¹ = OH, R² = OCH₃

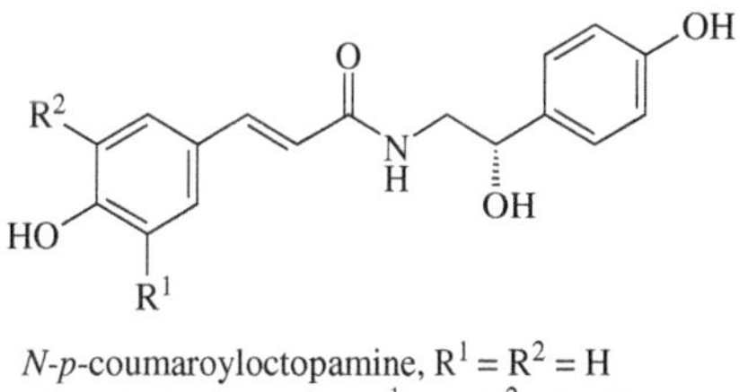

10-7, amides of (*E*)-cinnamic acids

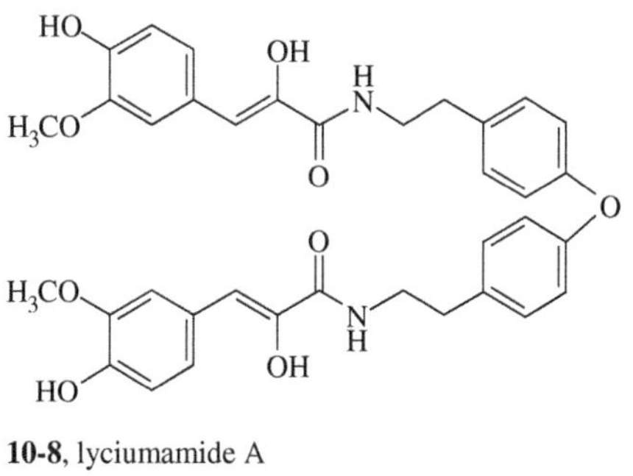

N-*p*-coumaroyloctopamine, R¹ = R² = H
N-feruloyloctopamine, R¹ = H, R² = OCH₃

10-8, lyciumamide A

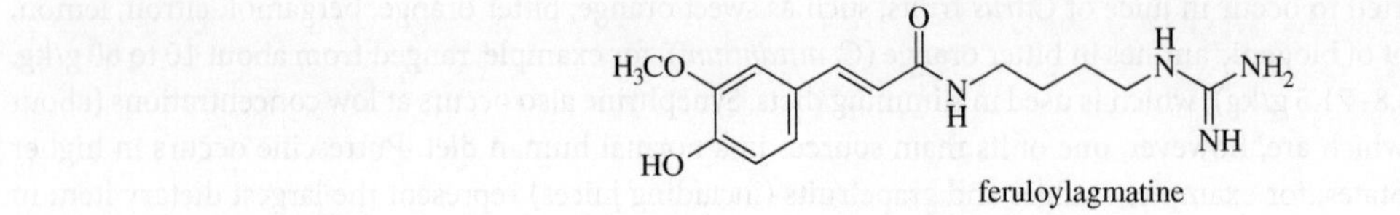

feruloylagmatine

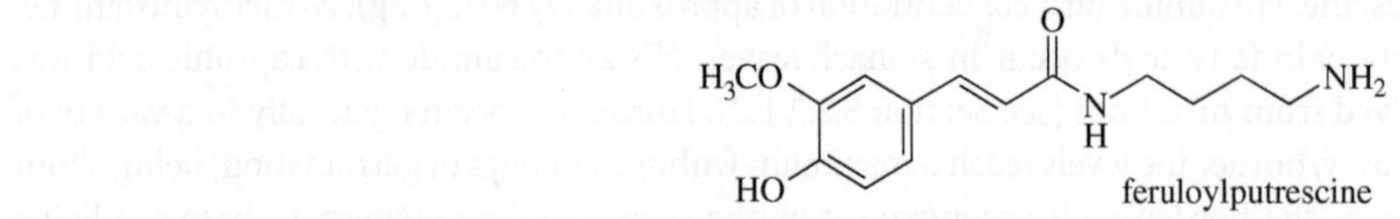

feruloylputrescine

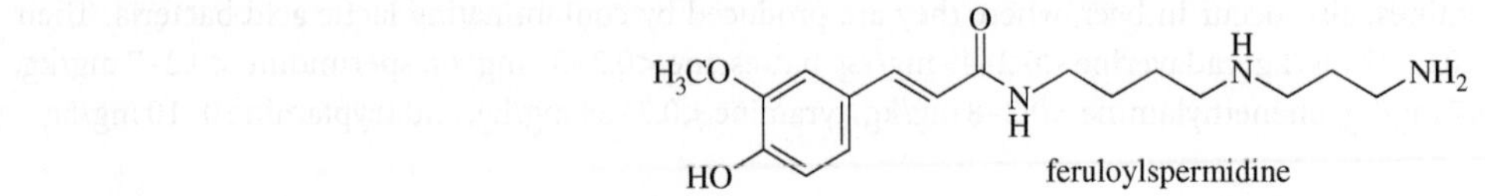

feruloylspermidine

10-9, examples of ferulic acid-derived phenolamides

Hydroxycinnamoylagmatines are precursors of hordatines, phenolamides typical to ungerminated barley grains, although small amounts can also be found in wheat. Hordatine A is a dimer of *p*-coumaroylagmatine, hordatine B is a dimer of *p*-coumaroylagmatine and feruloylagmatine, hordatine C (**10-10**) is a dimer of feruloylagmatine, and hordatine D is a dimer of feruloylagmatine and sinapoylagmatine. The glycosides of hordatines A and B (β-D-glucopyranosides) have collectively been known as hordatine M; glycosides and diglycosides of hordatines A–D were also reported. Hydroxy derivatives are known as hordatines A1, B1, and C1 (R^3 = OH, R^4 = H) and dihydroxy derivatives as hordatines A2, B2, and C2 (R^3 = R^4 = OH). For example, the content of hordatine A ranged from 103 to 254 nmol/g (d.w.). Hordatines withstand barley processing, and thus are also found in malts and beer.

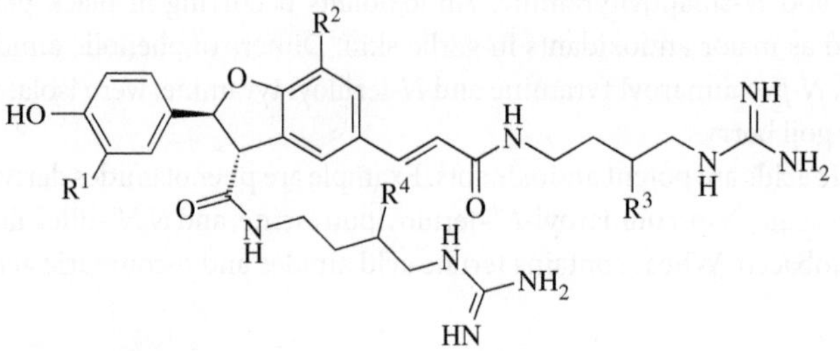

10-10, hordatine A, $R^1 = R^2 = R^3 = R^4 = H$
hordatine B, $R^1 = R^3 = R^4 = H$, $R^2 = OCH_3$
hordatine C, $R^1 = R^2 = OCH_3$, $R^3 = R^4 = H$

Dicaffeoylspermidine derivatives, a rare kind of plant secondary metabolite, are primarily distributed in the family Solanaceae. Examples of dicaffeoyl spermidine derivatives are lycibarbarspermidines A and N (**10-11**), isolated from wolfberries.

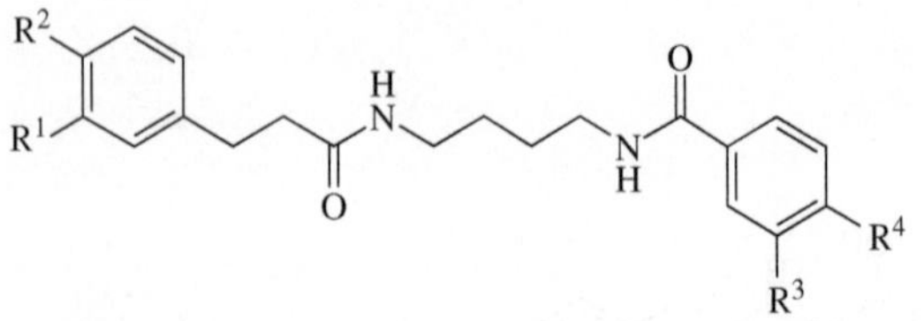

lycibarbarspermidine A, $R^1 = R^3 = R^4 = OH$, $R^2 = O$-β-D-Glc
10-11, lycibarbarspermidines

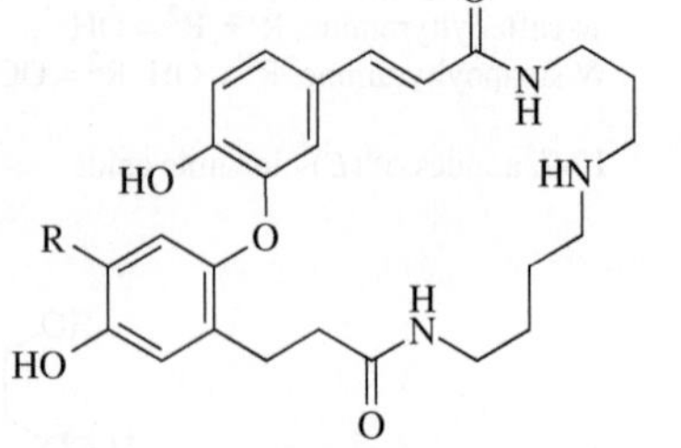

lycibarbarspermidine N, R = O-β-D-Glc

Citrus genus fruits contain indole metabolites derived from the *N*-methylation of tryptamine and 5-hydroxytryptamine (serotonine), which are probably involved in plant defense mechanisms. The identified compounds include serotonie 5-*O*-β-D-glucoside, *N*-methylserotonine 5-*O*-β-D-glucoside, *N,N*-dimethylserotonine (bufotenine) 5-*O*-β-D-glucoside, and *N,N,N*-trimethylserotonine (bufotenidine) 5-*O*-β-D-glucoside.

Derived from tryptamine are also phytoalexins of the family Brassicaceae, called brassinins (**10-12**). Their structures are illustrated in four commonly occurring indole phytoalexins: brassinin, cyclobrassinin, spirobrassinin, and brassilexin.

Tryptamine is a precursor of protoalkaloids gramine (also known as donazine) and *N*-methylgramine (**10-13**) in germinating seeds of monocotyledonous plants (grasses of the Poaceae family), which accumulate mainly in the leaves. They have the function of protecting against sucking insects and plant pathogens and act as inhibitors of the growth of certain other plants. For example, the maximum gramine concentration in barley leaves of around 900 mg/kg (f.w.) is reached within the first 2 weeks of growth. Partial decomposition of gramine in malt yields dimethylamine (see Section 8.2.10.1.2). Tryptamine is also a precursor of a range of β-carbolines, indole alkaloids, and pseudo alkaloids, such as psilocin and psilocybin in hallucinogenic mushrooms. Tryptamine is a precursor of a plant hormone (phytohormone), 3-indolylacetic acid (see **10-114**).

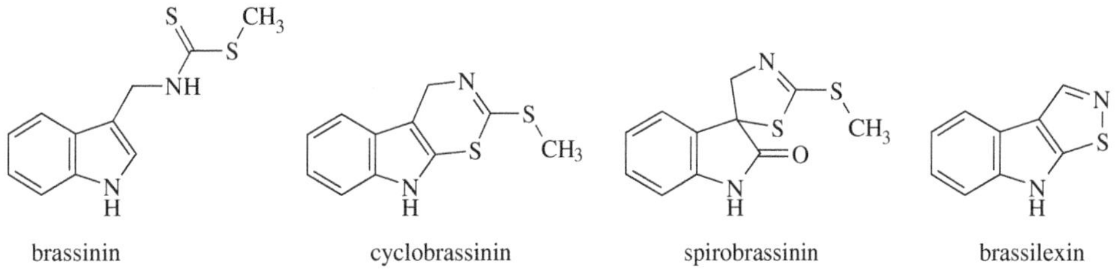

brassinin   cyclobrassinin   spirobrassinin   brassilexin

10-12, brassinins

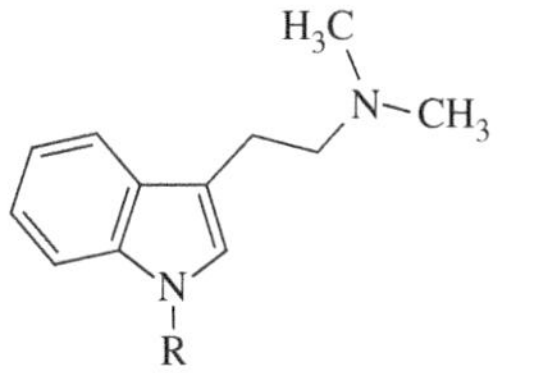

10-13, gramine, R = H
N-methylgramine, R = CH₃

Cocoa shells contain 330–395 mg/kg of *N*-alkanoyltryptamides (**10-14**, R^1 = H), mainly docosanoyl-2-(3-indolyl)ethylamide (behenic acid tryptamide, n = 17) and tetracosanoyl-2-(3-indolyl)ethylamide (lignoceric acid tryptamide, n = 19), but the cotyledons contain only 7–10 mg/kg. Their content can be employed as an indicator of the shell content in cocoa products, which is an important quality parameter for products that are made from roasted cocoa beans. *N*-Alkanoyl-5-hydroxytryptamides (**10-14**) occurring in the wax layer of coffee beens are derived from serotonine, where R^1 = OH and R^2 = CH₃ (n = 12, 14, 16, 17, 18, 19, or 20) or CH₂OH (n = 18 or 20). It has been shown that all these amides are responsible for the stomach irritation caused by stimulation of gastric juice secretion after ingestion of coffee brews.

Melatonine (*N*-acetyl-5-methoxytryptamine, produced mainly by the pineal gland in vertebrates) and its isomer (*N*-acetyl-6-methoxytryptamine) are also formed in foods of plant origin, for example during bread dough fermentation.

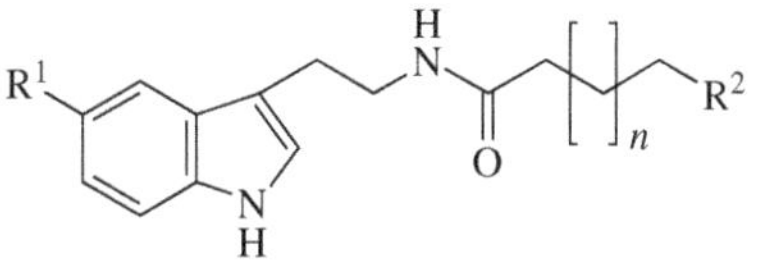

10-14, *N*-alkanoyltryptamides, R^1 = H
N-alkanoyl-5-hydroxytryptamides, R^1 = OH

10.3.4.2 Reactions and changes

Biogenic amines are reactive substances (Figure 10.4). In addition to enzymatic reactions leading to biogenic amine derivatives and other compounds, biogenic amines can give rise to aldehydes by oxidative deamination. Under long-term storage or at elevated temperatures, they react with triacylglycerols to form fatty acid amides. Analogously with other amino compounds, biogenic amines react with reducing sugars in the Maillard reaction to yield corresponding imines and other products. Imines also arise by oxidation of biogenic amines with hydrogen peroxide or lipid hydroperoxides. Secondary amines can react with nitrogen oxides to form carcinogenic nitrosamines. Proteins react with biogenic amines such as phenylethylamine, putrescine, histamine, tyramine, and spermidine to form β-*N*-substituted derivatives of diaminopropanoic acid. The likely mechanism of formation of these amino acid derivatives is β-elimination of cysteine residues and subsequent addition of an amine to the double bond of dehydroalanine, which arises analogously as in the case of lysinoalanine formation (see **2-121**). A condensation reaction of tryptamine (and tryptophan) with aldehydes in many foods leads to 1,2,3,4-tetrahydro-β-carboline and β-carboline

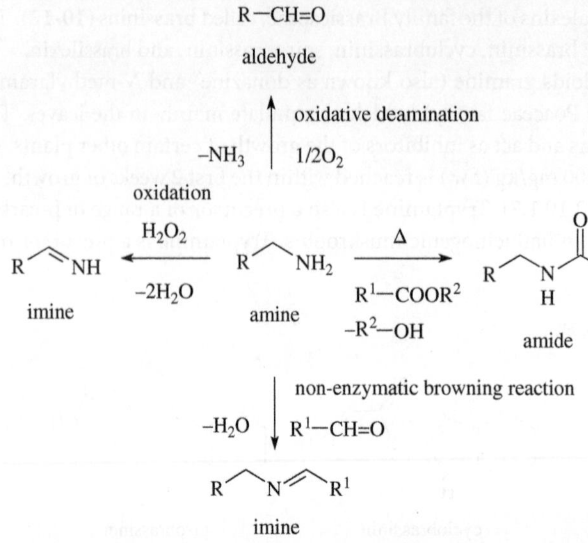

Figure 10.4 Main reactions of biogenic amines.

alkaloids (see Section *10.3.6.6.1*). Reaction of histamine with acetaldehyde yields 4-methyl-4,5,6,7-tetrahydro-1*H*-imidazo[4,5-*c*]pyridine), known as 4-methylspinaceamine (**11-1**), which is present in fermented foods, such as oriental sauces and some cheeses (**10-15**).

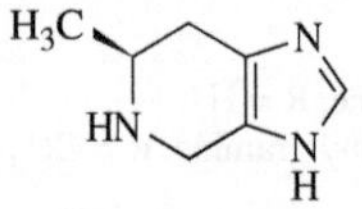

10-15, (*S*)-4-methylspinaceamine

10.3.4.3 *Biological effects*

Biogenic amines (in the narrower sense, biogenic amines do not include the polyamines spermidine and spermine) are indispensable for the body, but at high concentrations they may act as:

- psychoactive amines;

- vasoactive amines.

Psychoactive amines act as transmitters in the central nervous system, whilst vasoactive amines act directly or indirectly on the vascular system. According to the action, vasoactive amines can be divided into:

- vasocontractibile amines (e.g. tyramine);

- vasodilating amines (e.g. histamine).

Symptoms of consuming high doses of biogenic amines are vomiting, difficulty breathing, sweating, palpitations, hypotension (histamine) or hypertension (tyramine), and migraine (phenylethylamine and tyramine). Monoamine oxidase and diamine oxidase are the main enzymes that decompose biogenic amines in the gut. Spermine and spermidine are decomposed by polyamine oxidase, but polyamine oxidase-mediated catabolism of spermine and spermidine produces putrescine and toxic acrolein (Figure 10.5). The toxic effect of biogenic amines is strongly influenced by the activity of these enzymes, which may be different for different individuals and depends on many factors, such as the presence of inhibitors (certain medicines, particularly from the group of psychotropic drugs, and to a lesser extent alcohol) or potentiators. High concentrations of biogenic amines cannot be eliminated by this enzyme system. Part of its capacity is needed particularly for the detoxification of histamine and tyramine depletes putrescine and cadaverine; whilst they do not themselves possess a health risk, their content may be quite high in a variety of foods (Table 10.5).

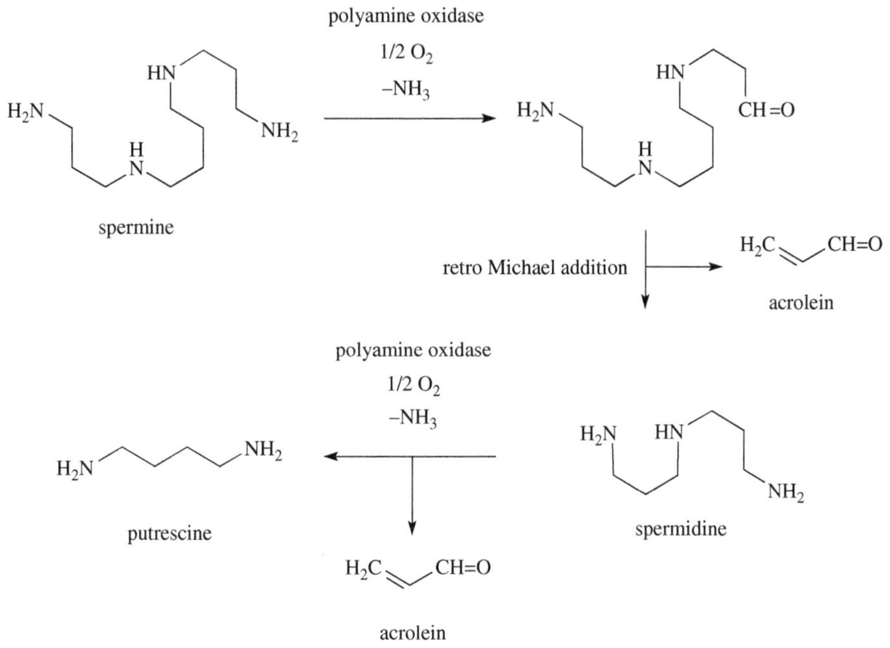

Figure 10.5 Catabolism of spermine and spermidine.

When evaluating the toxic effect of biogenic amines, it is necessary to consider not only the presence of a particular amine, but also other factors, such as the amount of food consumed and the presence of other toxic substances. For this reason, it is very difficult to set the levels that have adverse effects in food. No adverse health effects were observed after exposure to the following biogenic amine levels in food (per person per meal): 50 mg histamine for healthy individuals, or below detectable limits for those with histamine intolerance; 600 mg tyramine for healthy individuals not taking monoamino oxidase inhibitor (MAOI) drugs, but 50 mg for those taking third-generation MAOI drugs and 6 mg for those taking classical MAOI drugs and for putrescine and cadaverine. Concentrations of histamine higher than 500–1000 mg/kg are considered dangerous to humans. Increased amounts of histamine can cause anaphylactic shock (a very serious allergic reaction manifested by dizziness, loss of consciousness, laboured breathing, swelling of the tongue and breathing tubes, blueness of the skin, low blood pressure, heart failure, and death). Histamine poisoning (histaminosis) from marine fish and fish products in coastal countries is the most common cause of foodborne poisonings. In many countries, the maximum amount of histamine and tyramine is set, but information on the toxicity of other biogenic amines is inadequate. Commission regulation (EC) No. 2073/2005 on microbiological criteria for foodstuffs sets the limits for fishery products from fish species associated with a high amount of histidine (particularly fish species of the families Scombridae, Clupeidae, Engaulidae, Coryfenidae, Pomatomidae, and Scombresosidae) to 100–200 mg/kg, or 200–400 mg/kg for those that have undergone enzyme maturation treatment in brine. For example, in the Czech Republic, histamine limits are set to 200 mg/kg for fish and fish products and 20 mg/l for beer and wine.

As multifunctional cations polyamines, spermidine and spermine perform many important biological roles, the most important of which is participation in the protein biosynthesis and biosynthesis of nucleic acids. Biological processes in animals involve polyamines from three sources: endogenous (produced in cells), produced by the gastrointestinal-tract microflora, and foodborne. Increased intake of dietary polyamines is desirable for accelerated healing of wounds, burns, development and recovery of the intestinal mucosa, and other processes, but the dietary intake should be low in people with cancer. There is an increased level of polyamines in young, fast-growing, and metabolically active tissues (such as the liver and kidney). Plant products contain more spermidine than spermine, whilst animal products contain more spermine than spermidine. The physiological need for polyamines is not known.

The protoalkaloid gramine, found in higher amount in grasses of the genus *Phalaris* (Poaceae), may cause sudden collapse and death of the grazing cattle. In sheep, the chronic poisoning is manifested by degenerative changes in the central nervous system.

10.3.5 Toxic amino acids

Common proteinogenic amino acids may show some toxic effects in animals, including L-tryptophan and its transformation products and the transformation products of L-tyrosine, such as L-3,4-dihydroxyphenylalanine (dopa). In plants, there are about 700 aminocarboxylic and iminocarboxylic acids (see Section 2.2.1.2) that do not occur in proteins and are classified as secondary metabolites. Many of these compounds are structurally similar to the basic (proteinogenic) amino acids, and their toxicity may be manifested in a variety of symptoms. The mechanism of toxicity is usually due to one of the following effects:

- competitive inhibition of enzymes because of resemblance to normal substrates;

- interference in the activation and transfer of normal amino acid to transfer RNA;

- interference in the assembly of amino acids into protein and interruption of protein synthesis;

- incorporation into proteins resulting in the formation of non-functional proteins.

With the exception of the first effect, these are fairly subtle metabolic disorders that lead to diseases in humans and animals rather than to poisoning. Toxicologically important amino acids that are part of human food or animal feed are usually divided into several groups:

- lathyrogenic amino acids (lathyrogens);

- arginine analogues;

- analogues of sulfur and selenium amino acids;

- other toxic amino acids.

10.3.5.1 Structure, nomenclature, and occurrence

10.3.5.1.1 Lathyrogens

Lathyrogens are toxic amino acids and their derivatives (peptides, nitriles, and other substances) occurring in the seeds of certain vetchlings (*Lathyrus* spp.) and vetches (*Vicia* spp.) belonging to the legume family (Fabaceae). Lathyrism is manifested by deformations of the lower limbs (**osteolathyrism**), damage to the blood vessels (**angiolathyrism**), and disorders of the nervous system (**neurolathyrism**).

Osteolathyrism appears in livestock fed with seeds of some vetchling species, namely sweet pea (*L. odoratus*), caley pea (*L. hirsutus*), tiny pea (*L. pussilus*), and pink vetchling. (*L. roseus*). Forms of osteolathyrism in poultry fed with sweet pea seeds are sometimes called **odoratism**. Osteolathyrism also occurs in people suffering from chronic neurolathyrism. The toxic substance is β-aminopropionitrile (see Section 2.2.1.2.4), occurring as such or as the γ-glutamyl derivative. β-Aminopropionitrile arises from 2-cyanoethylisoxazolin-5-one, the precursor of which is 2-(3-amino-3-carboxypropyl)isoxazolin-5-one (see **2-74**). The content of β-aminopropionitrile in seeds of sweet pea ranges up to 0.8% of dry matter. In vetches, β-aminopropionitrile results probably by decarboxylation of cyanoalanine (see Figure 2.2).

Neurolathyrism accompanies consumption of seeds of some vetchlings and vetches. Neuronal activity is shown by L-2,4-diaminobutanoic (L-α,γ-diaminobutyric) acid, which is a lower homologue of L-ornithine. It is present in particular in the seeds of flat pea (*L. sylvestris*) and perennial pea (*L. latifolius*), along with one of the degradation products of 2-(3-amino-3-carboxypropyl)isoxazolin-5-one, but it may also arise from L-3-cyanoalanine. 2,4-Diaminobutanoic acid in seeds occurs in quantities up to 1.5% (d.w.) and is accompanied by 4-(*N*-oxalyl)-L-2,4-diaminobutanoic acid. The main neurolathyrogen is 3-(*N*-oxalyl)-L-2,3-diaminopropanoic acid, also known as β-(*N*-oxalylamino)-L-alanine, which is produced by the reaction of L-2,3-diaminopropanoic acid with oxalyl-CoA and from L-3-cyanoalanine. This neurotoxin is found primarily in seeds of grass pea (*L. sativus*), red pea (*L. cicera*), winged vetchling (*L. ochrus*), and crimson pea (*L. clymenum*). Its content in seeds of grass pea ranges from 0.1 to 1.5% and in seeds of red pea from 0.1 to 0.3%. Diaminopropanoic acid similarly occurs in many rattle pods (*Crotalaria* spp.) and acacias (*Acacia* spp.), plants of the Fabaceae family. At a level of about 5% w/w, diaminopropanoic acid is accompanied by the non-toxic α-isomer, which arises from the β-isomer by intramolecular rearrangement via 2,3-dioxopiperazine-5-carboxylic acid (Figure 10.6). The reaction sequences of the formation of neurotoxic amino acids and other toxic compounds in vetchlings seeds are illustrated in Figure 10.7.

10.3.5.1.2 Arginine analogues

A number of legume plants, especially seeds of tropical legumes, but also seeds of common vetch species, contain structural analogues of arginine, which may interfere with ornithine metabolism in the urea cycle. An example is the unique amino acid L-canavanine (see **2-44**), which was first isolated from Jack beans (*Canavalia ensiformis*, Fabaceae), probably native to South America, and is cultivated for its edible pods and seeds or as a reclamation plant and for green manure. The concentration in seeds is around 50 g/kg (d.w.). It was later identified in a number of other legumes.

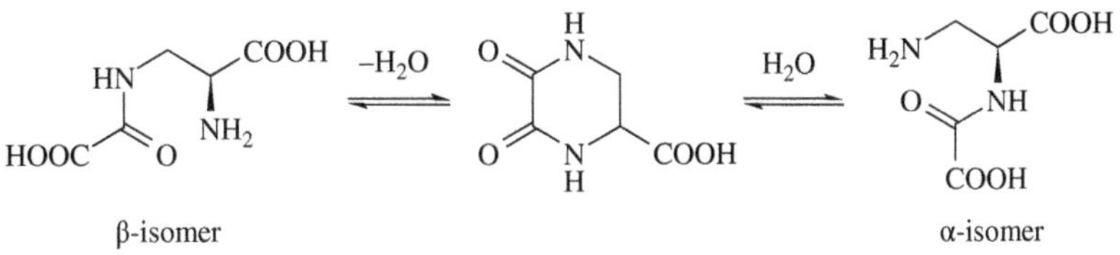

Figure 10.6 Isomerisation of 3-(*N*-oxalyl)-2,3-diaminopropanoic acid.

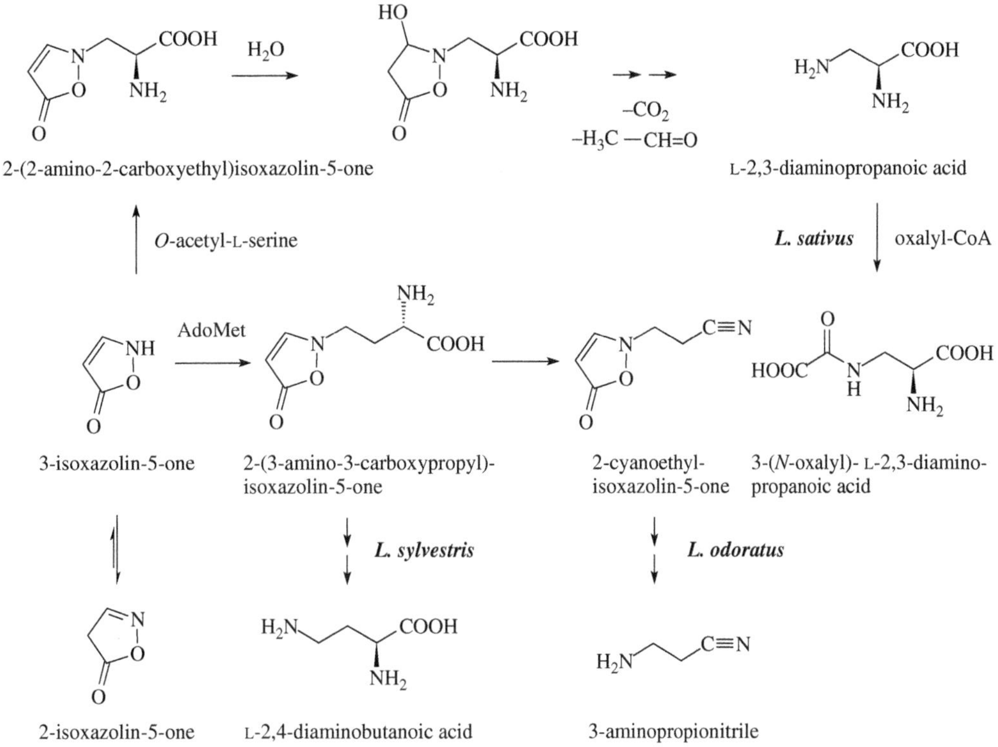

Figure 10.7 Biosynthesis of lathyrogens in vetchings (AdoMet is *S*-adenosyl-L-methionine).

An unusual toxic amino acid is L-indospicine (see **2-45**), occurring in seeds of indigo shrubs (*Indigofera hendecaphylla*, syn. *I. spicata*, Fabaceae) at about 20 g/kg (d.w.). Indospicine is accompanied by canavanine (9 g/kg). Another analogue of arginine is homoarginine, which occurs in the seeds of vetchlings and vetches together with L-4-hydroxyarginine and homocitrulline (see **2-42**). Seeds of red pea contain homoarginine at a level of about 12 g/kg (d.w.).

10.3.5.1.3 Sulfur and selenium amino acid analogues

Some sulfur amino acids or their selenium analogues may also exhibit toxic effects in livestock. An example of a sulfur-containing amino acid that is toxic to cattle, sheep, goats, and other ruminants is a derivative of L-cysteine, *S*-methyl-L-cysteine sulfoxide, known as methiin (see **2-23**). Methiin commonly occurs in forage brassicas, such as forage rape (*Brassica napus*), leaf turnips or forage brassica hybrids (*B. campestris*), kale (*B. oleracea*), turnips (*B. rapa*), and swedes (*B. napobrassica*) (see Section 2.2.1.2.2). Its content in forage brassicas varies over a wide range of about 3–14 g/kg (d.w.), depending on the variety, the time of harvest, and other factors. For example, its content in forage rape is commonly around 0.1 g/kg (d.w.), but may reach 5–7 g/kg. Methiin also occurs in cabbage and broccoli, where its content is about 0.2 g/kg, in Brussels sprouts at 0.7 g/kg, and in onion and garlic at about 1.6 and 3 g/kg (f.w.), respectively.

An analogue of L-cystine is 3,3'-(methylendithio)di-L-alanine, known as L,L-djenkolic acid (see **2-32**), which may cause poisoning if consumed by humans or livestock. It occurs, for example, in mildly toxic jengkol (djenkol) beans (*Archidendron pauciflorum*, syn. *Pithecellobium lobatum*, Fabaceae). Despite their strong smell, the beans are a popular food in Southeast Asia. Djenkolic acid also occurs in *Leucaena* forage legumes, such as *L. esculenta* (Mimosaceae), at about 2 g/kg.

Another cause of livestock poisoning is selenium analogues of sulfur amino acids, present in greater amounts in plants that grow on seleniferous soils (see Section 2.2.1.2.2).

10.3.5.1.4 Other toxic amino acids

Toxic effects, reflected particularly in farm animals, are exhibited by many other amino acids present in higher plants. Less frequently, these amino acids may occur as toxic substances in human nutrition. Toxic amino acids also contain some higher fungi.

Also toxic is the common amino acid L-3,4-dihydroxyphenylalanine (dopa), which occurs in relatively large quantities (about 50–60 g/kg f.w.) in the immature seeds of faba beans (*Vicia faba*, Fabaceae), which are also involved in the formation of favism. The mature seeds contain about 10 times lower levels of this amino acid. Similar concentrations of dopa also occur in the velvet bean (*Mucuna pruriens*) native to Africa and tropical Asia. Toxic for ruminants in higher amounts are the proteinogenous amino acid L-tryptophan and its degradation product, 3-methylindole (skatole, see **2-112**). Skatole is produced from tryptophan in the rumen as a product of decomposition by bacteria of the genus *Lactobacillus*.

An important amino acid that is toxic to livestock is L-mimosine, β-(3-hydroxy-4-pyridone-1-yl)-L-alanine (see **2-62**). It occurs in plants of genera *Leucaena* and *Mimosa* of the family Mimosaceae. Its quantity in the fodder plant *L. leucocephala* can reach up to 30–50 g/kg (d.w.).

An unusual toxic amino acid containing a cyclopropane ring in the molecule is hypoglycin A, L-3-(methylencyclopropyl)alanine. Hypoglycin occurs as a mixture of (2*S*,4*R*)- and (2*S*,4*S*)-diastereomers in immature fruits (pulp and seeds) of ackee (akee) fruit (*Blighia sapida*) from the family Sapindaceae, originating from the island of Jamaica. It has a hypoglycaemic effect but also exhibits other toxic effects. In immature seeds, the content of hypoglycin A is 1000–1110 mg/kg, but at maturation it decreases to less than 100 mg/kg. Hypoglycin B is a less toxic γ-glutamyl derivative of hypoglycin A, which is only present in the seeds of the fruit. The US Food and Drug Administration and Health Canada have set the maximum permissible level for hypoglycin A to 100 mg/kg.

10.3.5.2 Reactions and changes

Toxic amino acids undergo similar reactions and changes as the other amino acids. Compounds with a lathyrogenic effect also occur in protein isolates obtained from seeds of vetchlings and vetches, but their content is reduced by 50–85% during protein isolation. Reduction of lathyrogens can also be achieved by various culinary practices. The amount of β-*N*-oxalylaminoalanine may be reduced by up to about 50% by cooking and by up to 95% by seed soaking, cooking under pressure, and subsequent fermentation. The detoxification partly consists in the conversion of the toxic β-form to the non-toxic α-form (Figure 10.6).

The content of canavanine in Jack bean seeds can be reduced from 50 to 8 g/kg by soaking, depending on the ratio of water to extracted material and temperature. The procedure for the preparation of immature seeds of ackee fruit is based on the removal of the pericarp and cooking, which reduces the hypoglycin concentration to an acceptable level of about 1 g/kg.

10.3.5.3 Biological effects

Lathyrisms, and correspondingly neurolathyrism, occurs in the poorer sections of the population, particularly in certain parts of India, Bangladesh, Ethiopia, and Nepal, after approximately three to six months' consumption of food that contains more than two-thirds of the components originating from seeds of vetchlings. Cases of neurolathyrism have also been reported in other countries, including some in Europe. Symptoms of this disease are a neurodegenerative muscular rigidity and weakness of the lower limbs, which may extend to their paralysis. This disease affects more young men than women, and in extreme cases ends in death.

The lower homologue of lysine and ornithine, 2,4-diaminobutanoic acid, interferes (acts as the arginine analogue) with reactions of the ornithine cycle, which leads to increased ammonia content in the blood and brain. 2,3-Diaminopropanoic acid acts as an analogue of glutamate in the nervous system.

Canavanine and other arginine analogues (indospicine and homoarginine) are not very important in human nutrition, but the crops in which they are present are important in animal nutrition. Poisoning is primarily a manifestation of growth retardation. Most sensitive to this are poultry (and other birds), which do not have the ornithine cycle. Toxicity is based upon degradation to canaline (by arginase) and reaction with aldehydes, for instance with pyridoxal 5'-phosphate in molecules of decarboxylases and aminotransferases, which yields stable oximes. Indospicine additionally has hepatotoxic and teratogenic effects.

Kale poisoning, or a severe haemolytic anaemia in ruminants, broke out in cattle in Europe in the 1930s, but its link to the degradation product of methiin – dimethylsulfide – was only discovered about 35 years later. The toxic agent dimethyldisulfide arises as a product of methiin degradation by intestinal microflora. Dimethyldisulfide oxidises glutathione in red blood cells, the function of which is to protect haem from oxidation to haematin. In humans and other monogastric animals, such symptoms were not demonstrated.

Djencolic acid affects kidney function due to precipitation of amino acids in body fluids. Symptoms of djenkol bean poisoning (djenkolism) are nausea, vomiting, bilateral loin pain, and urinary obstructions such as haematuria (presence of blood in the urine) and oliguria (low urinary output; less than 500 ml in every 24 hours).

The side effects of higher amounts of dopa in humans have been associated with dizziness, staggering, increased heart rate, vomiting, and psychiatric disturbances, which are consistent with its role of a precursor of the neurotransmitter dopamine. In pigs, dopa in feeds results in reduced intake and depressed weight gains. In ruminants, however, the seeds of *Vicia* and *Mucuna* have been used without apparent ill effects.

Higher amounts of tryptophan (its degradation product skatole) in feeds of ruminants may cause acute pulmonary oedema (a buildup of fluid in the spaces outside the blood vessels of the lungs) and emphysema (a chronic respiratory disease causing a decrease in lung function and often breathlessness).

Feeding crops with a higher content of mimosine (a thyreotoxic amino acid) cause reduction of weight gain, hair loss (mimosine acts as an epilator), and the possible appearance of ophthalmic catarrh and goitre. The strumigene is 3,4-dihydroxypyridine formed from mimosine degradation in the rumen. Mimosine poisoning can lead to the death of affected animals. Mimosine also has teratogenic effects. It is excreted from the body in the urine, either as a free compound or after decarboxylation as mimosinamine.

Acute toxic effects of hypoglycin are manifested mainly in malnourished subjects and by strong vomiting in children. The toxic substances are hypoglycin metabolites, which are (3*R*)- and (3*S*)-2-methylenecyclopropylacetic acids (see Section 2.2.1.2.1). Symptoms of poisoning, known as Jamaican vomiting sickness or toxic hypoglycaemic syndrome, occur 6–48 hours after ingestion of seminal immature follicles. The poisoning is manifested by vomiting, drowsiness, fatigue, and hypoglycaemia (a blood glucose level below the level necessary to properly support the body's need for energy and stability throughout its cells). The toxin (the active metabolite is methylencyclopropylacetyl-CoA) interferes with the metabolism of branched-chain amino acids, irreversibly binds to FAD, and inhibits acyldehydrogenases acting in β-oxidation of fatty acids.

10.3.6 Alkaloids

Alkaloids are a heterogeneous group comprising more than 10 000 compounds of different structures. Alkaloids are basic nitrogenous compounds produced as secondary metabolites that exhibit a variety of biological effects, depending on the amount consumed. They often occur as a mixture of compounds of related structures, either as free compounds, *N*-oxides, salts of carboxylic acids, their esters or amides, or glycosides. Some products are considered as detoxification plant products, growth regulators, and spare forms of nitrogen that may have an important role in protecting plants against elicitors and pathogens and in the evolution of plant species.

Alkaloids are found in about 15–20% of vascular plants, in seeds, leaves, roots, bark, and other parts, but they similarly occur in certain species of mosses, fungi, bacteria, and some invertebrates (centipedes, beetles, butterflies, and crustaceans) and vertebrates (e.g. in frogs and salamanders). Alkaloids of certain species of insects and other animals are often of plant origin, but may also be synthesised *de novo*.

No evaluation criterion exists that would allow the characterisation of alkaloids as a single group of natural substances. Some alkaloids are also classified as herbal antibiotics (natural pesticides), because they are a part of the defence mechanisms of plants against elicitors and predators. Some alkaloids are classified as natural toxic amino acids (e.g. ibotenic acid and agaritine), biogenic amines (e.g. histamine, hordenine, or psilocin), and natural dyes (betalains or berberine). Many alkaloids and related compounds also arise during thermal processing of foods from essential nutrients in the Maillard reaction (e.g. β-carbolines).

Alkaloids are normally classified into three main basic groups:

- **true alkaloids**;

- **pseudo alkaloids**;

- **protoalkaloids**.

True alkaloids are typically heterocyclic nitrogen bases derived from amino acids. They exhibit a wide range of physiological effects and are often highly toxic to humans and other animals (such as nicotine in tobacco). Pseudo alkaloids are heterocyclic nitrogen bases, yet their precursors are not amino acids, but terpenoids or purines. They are generally less toxic than true alkaloids. Examples of pseudo alkaloids are caffeine in coffee and solanine in potatoes. Protoalkaloids are basic amines derived from amino acids, but the nitrogen is not a part of the aromatic (heterocyclic) system. An example is capsaicin present in hot peppers.

Common, but not very accurate, is the classification of the individual alkaloid groups according to the heterocyclic compounds from which the basic skeletons are derived. True alkaloids of medicinal plants called necines are derived from pyrrolizidine, whilst alkaloids with a pyridine ring are nicotine and other tobacco alkaloids. Besides the pyridine ring, these alkaloids also contain other *N*-heterocycles, such as 1,2,5,6-tetrahydropyridine and piperidine. Tetrahydropyridine and piperidine rings occur too in areca nut alkaloids, and the black pepper alkaloid piperine is also derived from the piperidine ring. Lupine alkaloids are derived from chinolizidine, quinine occurring in the bark of the cinchona tree is derived from quinoline, and opium alkaloids are derivatives of isoquinoline. Pseudo alkaloids, such as caffeine in coffee

Table 10.7 Overview of important alkaloids occurring in foods.

Structural types of basic skeleton	Precursors	Occurrence	Examples
True alkaloids			
Pyridine, piperidine, and pyrrolidone alkaloids[a]	Arg, Lys, Orn, nicotinic acid	Tobacco leaves	Nicotine, nornicotine, anatabine, anabasine
	Nicotinic acid	Areca nut	Arecoline, arecaidine
	Lys, Phe	Black pepper seeds	Piperine
Pyrrolizidine alkaloids[b]	Arg, Ile, Leu, Orn, Val, Thr	Ragworts, groundsels	Senecionine
Quinolizidine alkaloids	Lys	Lupin seeds	Lupanine, lupinine, sparteine
Quinoline alkaloids[c]	Trp, mevalonic acid	Cinchona bark	Quinine, quinidine
Isoquinoline and phenanthrene alkaloids	Phe, Tyr, acetyl-CoA	Poppy straw and latex	Morphine, codeine, papaverine
(Nor)tropane alkaloids	Putrescine	Potatoes, tomatoes	Calystegines
Protoalkaloids			
Capsaicinoids (vanillylamides)[d]	Leu, Phe, Val, malonyl-CoA	Chili pepper	Capsaicine, nordihydrocapsaicine
Pseudo alkaloids			
Purine alkaloids	Purines	Coffee, tea, cocoa	Coffeine, theobromine, theophylline
Steroid (terpenoid) glycoalkaloids	Mevalonic acid	Potatoes, tomatoes	Solanine, tomatine

[a]The pyridine ring in nicotine and tetrahydropyridine ring in arecoline are derived from nicotinic acid, which arises from aspartic acid and glyceraldehyde 3-phosphate via quinolinic acid; the pyrrolidine ring in nicotine and nornicotinu arises from Orn; the piperidine ring in anatabine and anabasine arises from Lys; the dihydropyridine ring of the natural pigments betalaines arises from Tyr via DOPA and betalamic acid.

[b]The pyrrolizidine skeleton is derived from Orn. Necic acids, by which some alkaloids are esterified, such as senecic acid, arise from Ile.

[c]Precursors are Trp and monoterpenic glycoside loganin. Not listed in this table are numerous other alkaloids derived from Trp (except β-carbolins), such as the indole protoalkaloids tryptamines, which are covered under biogenic amines (e.g. gramine), toxic components of higher fungi (e.g. psilocin), and hordenine arising from Tyr.

[d]Phe is a precursor of vanillylamine. The fatty acid in capsaicine is derived from Val.

and tea, are derived from purine, whilst glycoalkaloids containing nitrogen in the molecule, such as solanine in potatoes, are derived from steroids. Protoalkaloids of hot peppers, called capsaicinoids, are vanillic acid amides.

Another aspect is also frequently applied, which is the alkaloid origin. Nicotine, for example, is included amongst the tobacco alkaloids, arecoline belongs to areca nut alkaloids, and quinolizidine alkaloids occurring in lupine seeds are lupine alkaloids. For practical reasons, classification based on these principles is also used in the following sections. Today, classification according to the precursors of alkaloid biosynthesis is often promoted. This, however, requires a detailed knowledge of the biogenesis mechanisms (Table 10.7).

10.3.6.1 Pyrrolizidine alkaloids

Pyrrolizidine alkaloids are a large group of about 250 alkaloids produced by approximately 6000 plant species belonging to 13 families, such as Boraginaceae, Asteraceae, Heliotropiaceae, Fabaceae, and Rhamnaceae. Pyrrolizidine alkaloids occur in relatively small amounts (0.1–1% of dry matter) as a complex mixture of more than 10 different compounds. Important sources of alkaloids are some herbs and fodder plants.

The skeletons of pyrrolizidine alkaloids are derived from 2,3,5,6,7,8-hexahydro-1*H*-pyrrolizine, known as pyrrolizidine, and from pyrrolizidine-*N*-oxide (**10-16**). The bases of the pyrrolizidine alkaloids are necines (necine bases) derived from bicyclic amino alcohols, which have their origin in 1-hydroxymethylpyrrolizidine. Necines may be saturated or may have a double bond at C-1 of ring B, and may also have an additional one or two hydroxyl groups at C-2, C-6, or C-7. Necines are esterified with carboxylic acids, which are called necic acids.

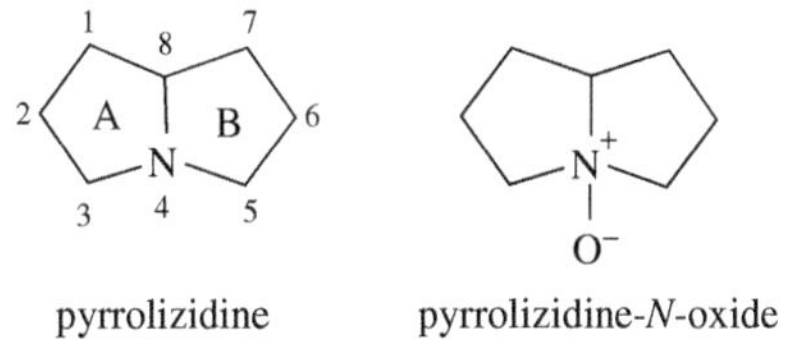

pyrrolizidine   pyrrolizidine-*N*-oxide

10-16, skeletons of pyrrolizidine alkaloids

10.3.6.1.1 Structure, occurrence, and terminology

The most frequently occurring necines in medicinal plants are supinidine, isomeric retronecine and heliotridine, platynecine, hastanecine, and otonecine (**10-17**).

Necine bases are esterified by different carboxylic acids. With the exception of acetic acid, these acids have 5–10 carbon atoms and include mono- and dicarboxylic acids with branched chains substituted with hydroxyl, methoxyl (other alkoxyl), epoxy, carboxy, and acetoxy groups. Necines with two hydroxyls, such as 7,9-necinediol, may be esterified with carboxylic acids in positions C-7 and C-9, or both. Esterification with dicarboxylic acids produces macrocyclic alkaloids with 11- to 14-membered rings. Depending on the ester type, the following alkaloids can be recognised:

- monoesters;

- diesters;

- macrocyclic diesters.

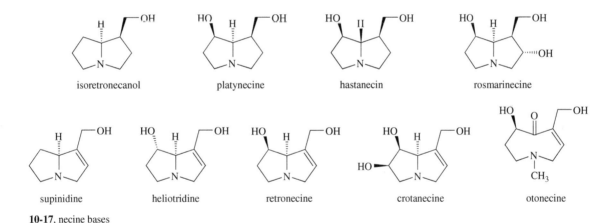

10-17, necine bases

The most frequently occurring carboxylic acids with five carbon atoms per molecule are (*Z*)-2-methylbut-2-enoic acid, also known as (*Z*)-2,3-dimethylacrylic or angelic acid (**8-62**), its isomer (*E*)-2-methylbut-2-enoic acid, known as (*E*)-2,3-dimethylacrylic (tiglic) acid, (*Z*)-3-methylbut-2-enoic acid (senecioic) acid (**8-62**), and (*Z*)-2-hydroxymethylbut-2-enoic (sarracinic) acid (**10-18**).

Frequent carboxylic acids with seven carbon atoms per molecule are (+)-trachelanthic acid, known as (2*S*,3*R*)-2,3-dihydroxy-2-isopropylbutanoic acid, (−)-trachelanthic acid, known as (2*R*,3*S*)-2,3-dihydroxy-2-isopropylbutanoic acid, (+)-viridifloric acid, known as (2*R*,3*R*)-2,3-dihydroxy-2-isopropylbutanoic acid, and (−)-viridifloric acid, known as (2*S*,3*S*)-2,3-dihydroxy-2-isopropylbutanoic acid (**10-18**). Methyl derivatives of these acids have eight carbon atoms in the molecule. For example, a methyl derivative of (+)-trachelanthic acid is (2*S*,3*R*)-2-hydroxy-3-isopropyl-3-methoxybutanoic (heliotric) acid, with three hydroxyl groups in the molecule it is (2*R*,3*R*)-2,3-dihydroxy-2-(1-hydroxyethyl)-3-methylbutanoic (echimidinic) acid, and its methyl derivative is (2*R*,3*R*)-2,3-dihydroxy-2-(1-methoxyethyl)-3-methylbutanoic (lasiocarpic) acid (**10-18**). An example of necic acid bound in the macrocyclic diesters (10 carbon atoms) is (2*S*,3*R*,5*Z*)-5-ethylidene-2-hydroxy-2,3-dimethylhexane-1,5-dioic acid (senecinic) acid (**10-18**).

Pyrrolizidine alkaloids are found in some medicinal herbs that are used as a traditional medicine in many countries. These plants grow mainly in temperate climates, but some require tropical or subtropical climates. The Boraginaceae family is represented by common comfrey (*Symphytum officinale*), borage (*Borago officinalis*), common bugloss (*Anchusa officinalis*), viper's bugloss or blueweed (*Echium vulgare*), and houndstongue (*Cinoglossum officinale*). The Asteraceae family includes coltsfoot (*Tussilago farfara*) and butterbur (*Petasites hybridus*) and the Rhamnaceae family includes buckthorn (*Rhamnus cathartica*). Butterbur has a total alkaloid content in the root of about 0.01%, but the alkaloid content in other parts of the plant is negligible. The main alkaloid is senecionine.

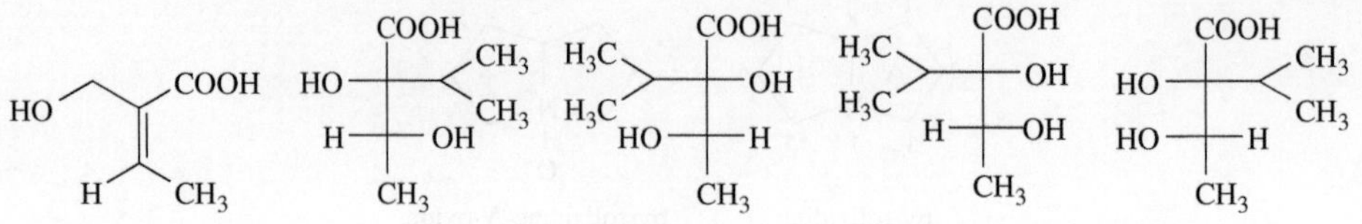

sarracinic acid

(+)-isomer   (−)-isomer
trachelanthic acid

(+)-isomer   (−)-isomer
viridifloric acid

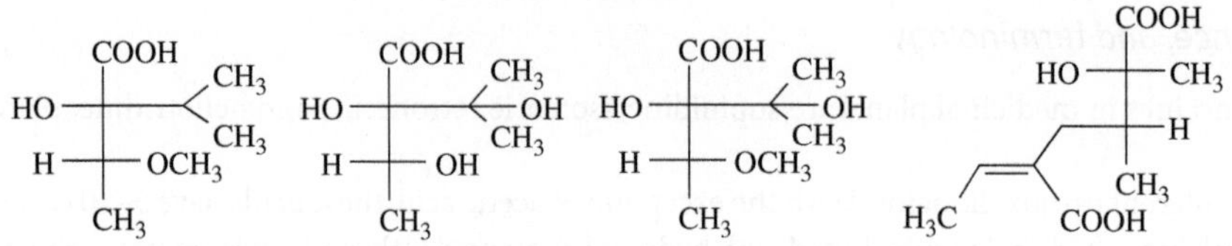

heliotric acid   echimidinic acid   lasiocarpic acid   senecinic acid

10-18, necic acids

Pyrrolizidine alkaloids also occur in a number of plants that can become part of the feed, and therefore may cause poisoning of livestock. Examples are the legume forage plants of the genus *Crotalaria* (Fabaceae), commonly known as rattle pods, and especially *C. spectabilis*, used for green manure and as a cover crop in subtropical and tropical regions of Africa, Asia, and the United States. Also poisonous to cattle are some plants of the Asteraceae family, such as ragwort (*Senecio jacobaea*), hemp-agrimony (*Eupatorium cannabinum*), summer ragwort (*Ligularia debtata*), and bill goat weed (*Ageratum conyzoides*), as well as European heliotrope (*Heliotropium europium*, Boraginaceae).

The main comfrey alkaloids (echinatine, echimidine, heliosupine, and lasiocarpine) and European heliotrope alkaloids (heliotrine, lasiocarpine, and symphytine, **10-19**) are examples of monoesters and diesters of various necines. Echimidine is derived from retronecine; other alkaloids are derived from heliotridine. Examples of macrocyclic diesters derived from retronecine are alkaloids of ragwort senecionine and seneciphylline and alkaloids of the *Crotalaria* legumes fulvine and monocrotaline (**10-20**).

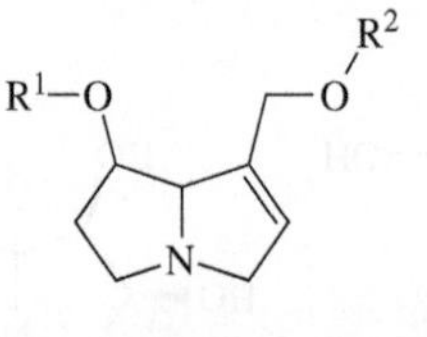

10-19, monoesters and diesters
echinatine, R^1 = H, R^2 = viridifloroyl
heliotrine, R^1 = H, R^2 = heliotroyl
echimidine, R^1 = angeloyl, R^2 = echimidinoyl
heliosupine, R^1 = angeloyl, R^2 = echimidinoyl
lasiocarpine, R^1 = angeloyl, R^2 = lasiocarpoyl
symphytine, R^1 = tigloyl, R^2 = viridifloroyl

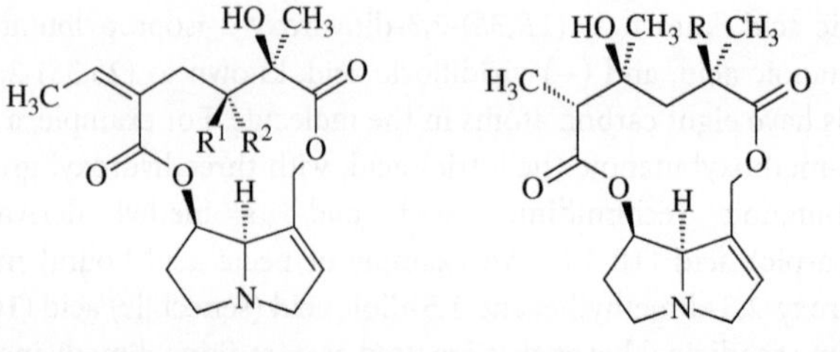

10-20, macrocyclic diesters
senecionine, R^1 = H, R^2 = CH_3 fulvine, R = H
seneciphylline, R^1 and R^2 = CH_2 monocrotaline, R = OH

10.3.6.1.2 Reactions and changes

The concentration of alkaloids in plants is highly variable, even between different parts of the same plant, and depends on many other factors. When preparing medicinal herbal teas, most alkaloids leach into water. Stability of individual compounds is not known, but it can be supposed that they are partly hydrolysed to necine bases and carboxylic acids.

10.3.6.1.3 Biological effects

Most plants containing pyrrolizidine alkaloids are toxic to humans and domestic animals. Poisoning occurs most frequently after eating different parts of plants as food (such as salads from young leaves of comfrey and butterbur), extracts from medicinal herbs, or contaminated cereals. Cattle, horses, farmed deer, and pigs are most susceptible; sheep and goats require about 20 times more plant material than cattle. Acute poisoning is rare and is characterised by sudden death from haemorrhagic liver necrosis and visceral haemorrhages. Chronic exposure is more typical. The signs in horses and cattle include loss of condition, anorexia, dullness, and constipation or diarrhoea. Residues of alkaloids may appear in milk and other animal products, but these residues are not risky. Another example may be contamination of honey, to which the pyrrolizidine alkaloids are transferred by bees. For example, honey from viper's bugloss flowers may contain 0.3–1 mg/kg of alkaloids with echimidine as the main component.

Pyrrolizidine alkaloids are hepatotoxins (substances that deplete the liver) and carcinogens. Characteristic symptoms of toxicose are magnification of liver cells (hepatocytes) called megalocytosis, hyperplasia (enlargement of the liver due to multiplication of cells), liver fibrosis (extensive scarring of the liver), and subsequent symptoms of liver dysfunction, such as hyperbilirubinaemia (an elevated blood level of the pigment bilirubin), hypoalbuminaemia (an abnormally low blood level of albumin), oedema, jaundice, and cirrhosis. Alkaloids are not responsible for the toxic effects as such; rather, it is the pyrrole structures (**10-21**) arising from alkaloids in the liver by enzymatic transformation of 1,2-unsaturated compounds (supinidine, retronecine, heliotridine, and otonecine derivatives) that lead to toxicity. Toxic substances are also aldehydes formed as breakdown product of alkaloids, such as (*E*)-4-hydroxyhex-2-enal.

10-21, toxic metabolites

10.3.6.2 Pyridine alkaloids

The most important alkaloids, which contain the pyridine nucleus in the molecule, are **tobacco alkaloids**. Besides the pyridine nucleus, tobacco alkaloids also contain hexahydropyridine (piperidine), 1,2,5,6-tetrahydropyridine, pyrrole, 3,4-dihydro-2*H*-pyrrole (1-pyrroline), or tetrahydropyrrole (pyrrolidine) units (**10-22**).

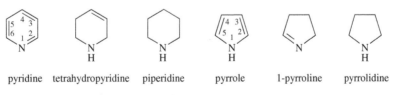

pyridine   tetrahydropyridine   piperidine   pyrrole   1-pyrroline   pyrrolidine

10-22, *N*-heterocycles in tobacco alkaloids

10.3.6.2.1 Structure, nomenclature, and occurrence

The main alkaloids of different cultivars of commercial tobacco species (*Nicotiana tabacum* and *N. rustica*, Solanaceae) are L-nicotine or (−)-nicotine and (*S*)-1-methyl-2-(pyrid-3-yl)pyrrolidine or (*S*)-3-(1-methylpyrrolidin-2-yl)pyridine, according to the IUPAC nomenclature (**10-23**). Nicotine is also present in small quantities in other plants (about 24 species of 12 plant families), but especially in plants of the nightshade family, which also includes potatoes, tomatoes, and eggplants (aubergines).

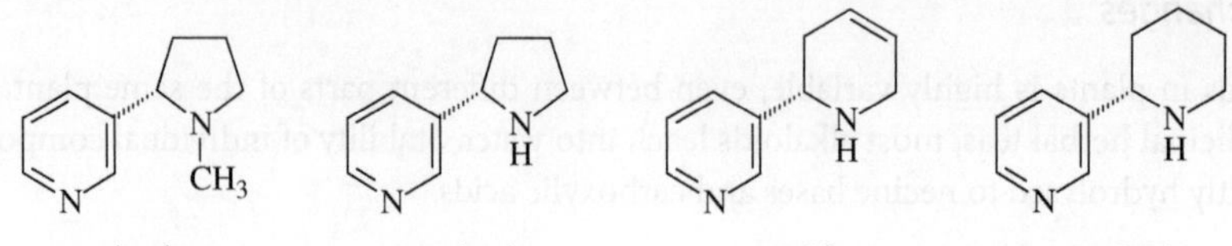

nicotine   nornicotine   anatabine   anabasine

10-23, main tobacco alkaloids

Nicotine in tobacco is always accompanied by three other prominent alkaloids: nornicotine, (*S*)-2-(3-pyridyl)pyrrolidine, anatabine, (*S*)-2-(pyrid-3-yl)-1,2,5,6-tertrahydropyridine, and anabasine, (*S*)-2-(pyrid-3-yl)piperidine (**10-23**). Apart from these, more than 20 other minor tobacco alkaloids have been identified.

The total content of alkaloids in tobacco ranges from 0.3 to 3% of d.w, depending on the species and variety of plant, climatic and soil conditions, and other factors. Approximately 95% of this amount is represented by nicotine. For example, leaves of the so-called bright-leaf tobacco, commonly known as Virginia tobacco (*N. tabacum*), with a total alkaloid content of 1.93%, containing 1.85% of nicotine, has 0.04% of anatabine, 0.03% of nornicotine, and 0.01% of anabasine. The content of other (minority) alkaloids is less than 0.01%.

10.3.6.2.2 Reactions and changes

Anabasine and anatabine are formed as products of nicotine catabolism. The content of nicotine and other alkaloids in fresh tobacco leaves decreases slightly during post-harvest treatment, drying, and fermentation. During tobacco drying and fermentation, nicotine is oxidised to nicotine N'-oxide, known as oxynicotine, then via nicotine $\Delta^{1'(5')}$ iminium ion to cotinine, and through transmethylation to nornicotine, which partially decomposes to nicotinic acid via myosmine (Figure 10.8). In tobacco products, the content of alkaloids does not change. One cigarette (about 0.8 g of tobacco) contains 1 mg of absorbable nicotine, and new products with reduced nicotine content (light, superlight, and ultra-light cigarettes) usually have a nicotine content of 0.1–0.8 mg.

The nicotine remains in the body for about six hours after cigarette smoking and is metabolised by liver enzymes to cotinine and other metabolites. Cotinine is then excreted in the urine but is present in the body for about 16 hours after smoking. During tobacco burning, nicotine is partially oxidised to cotinine and oxynicotine that rapidly decomposes at higher temperatures. Pyrolysis of nicotine also produces hydrogen cyanide (0.004–0.270 mg per one cigarette) and other products. The resulting nitrogen oxides react with the decomposition products of nicotine to form carcinogenic nitrosamines.

10.3.6.2.3 Biological effects

Nicotine is a potent neurotoxin, particularly for insects. In the past, a tobacco extract was used as an insecticide. In mammals, nicotine in low concentrations acts as a stimulant. Moderate nicotine doses increase respiratory and motor activity and lead to vomiting; high doses result in tremors and end in coma. The sensitivity of animals to nicotine ranges from 6 to 30 mg/kg. Somewhat lower toxicities are exhibited by nornicotine and anabasine. Anabasine (and probably also anatabine) shows teratogenic effects. Chronic pathological changes in the respiratory system and carcinogenicity of tobacco smoke have been clearly demonstrated; therefore, Directive 90/239/EEC establishes maximum limits for the carcinogenic tar, nicotine, and carbon monoxide yields of cigarettes marketed in the EU member states. Since 1 January 2004, the amount per cigarette has been set at not greater than 10 mg for tar, 1 mg for nicotine, and 10 mg for carbon monoxide.

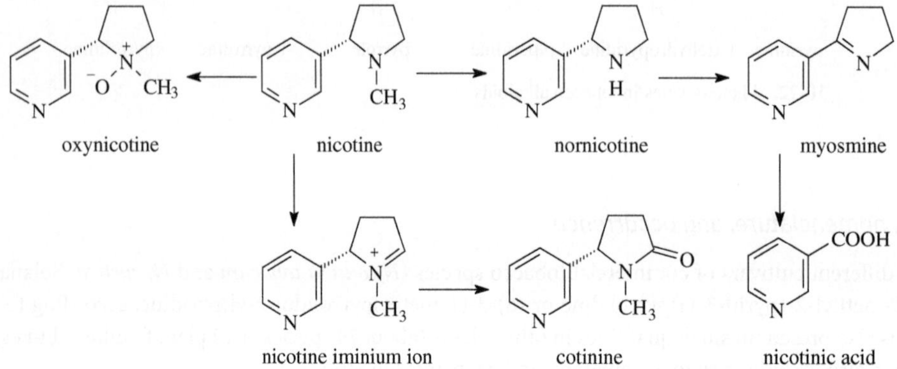

oxynicotine   nicotine   nornicotine   myosmine

nicotine iminium ion   cotinine   nicotinic acid

Figure 10.8 Products arising during tobacco fermentation.

10.3.6.3 Tetrahydropyridine alkaloids

The areca nut is the fruit of the areca palm (*Areca catechu*, Arecaceae), which grows in the tropical Pacific (Melanesia and Micronesia), Southeast and South Asia, and parts of east Africa. This seed is commonly referred to as betel nut, so it is easily confused with betel (*Piper betle*), leaves of a vine belonging to the Piperaceae family (see Section 8.2.8.1). In many Asian cultures, areca nuts are ground and mixed with lime (calcium hydroxide), and may have clove, cardamom, or other spices added for extra flavouring. The mixture is wraped in betel leaves to form a little ball called betel quid, which is chewed or kept in the mouth to allow the slow release of areca nut alkaloids. Tobacco leaf is often added to the mixture, thereby adding the effect of nicotine. Areca nut chewing is the fourth most popular habit worldwide, after the use of tobacco, alcohol, and caffeine.

10.3.6.3.1 Structure, nomenclature, and occurrence

The areca nut contains four main 1,2,5,6-tetrahydropyridine alkaloids: arecoline (1-methyl-1,2,5,6-tetrahydropyridine-3-carboxylic acid methyl ester), arecaidine (1-methyl-1,2,5,6-tetrahydropyridine-3-carboxylic acid, also known as arecaine or *N*-methylguvacine), guvacoline or norarecoline (1,2,5,6-tetrahydropyridine-5-carboxylic acid methyl ester), and guvacine (1,2,5,6-tetrahydropyridine-5-carboxylic acid), accompanied by nipecotic acid optical isomers (piperidine-3-carboxylic acid) (**10-24**). In nuts obtained from India, China, and the United States, guvacine was the most abundant alkaloid (1.39–8.16 mg/g), followed by arecoline (0.64–2.22 mg/g), arecaidine (0.14–1.70 mg/g), and guvacoline (0.17–0.99 mg/g). The total alkaloid content per nut was between 2 and 10 mg/g, with an average amount of about 0.45 mg/g.

arecaidine, R = H
arecoline, R = CH$_3$

guvacine, R = H
guvacoline, R = CH$_3$

(*R*)-(-)-nipecotic acid

10-24, areca nut alkaloids

10.3.6.3.2 Reactions and changes

Arecoline is an oily liquid that is soluble in water and alcohols. It is hydrolysed (especially in alkaline media) to relatively stable arecaidine. Hydrolysis of guvacoline yields guvacine.

10.3.6.3.3 Biological effects

About 200 million people commonly chew betel quid from the West Pacific to South Asia in order to obtain a stimulating effect. It has been estimated that about 600 million people consume areca nut worldwide. The primary active ingredient responsible for the central nervous system effects of the areca nut is arecoline. It is absorbed into the blood via the mucous membranes of the mouth and intestine, and affects both the peripheral and the central nervous system. Human studies have shown that the systemic effects of betel quid chewing are evident with an increase in heart rate, blood pressure, and body temperature within 2 minutes after chewing, reaching a peak within 6 minutes, and lasting from 15 minutes to 3 hours. Arecoline (with arecaidine) is about 15 times more potent than guvacoline and guvacine.

Current science is confident that areca nut chewing is associated with oral and oesophageal cancers. Research suggests this is probably at least partly because of arecoline itself, although it could also be due to the other constituents of the nut, some of which are precursors to nitrosamines that form in the mouth during chewing. There are not specific legislative restrictive provisions regarding areca nut or its active principles in the European Community or the United States.

10.3.6.4 Piperidine alkaloids

The most important alkaloids derived from piperidine (**10-22**) are alkaloids of black pepper (*Piper nigrum*, Piperaceae).

10.3.6.4.1 Structure, nomenclature, and occurrence

Black pepper has a unique attractive aroma and typical pungent and tingling taste. The predominating pungent compound identified in peppercorns is piperine, also known as piperic acid amide or (2*E*,4*E*)-piperoyl-1-piperidine or (2*E*,4*E*)-5-(1,3-benzodioxol-5-yl)-*N*-piperidinylpenta-2,4-dienamide (**10-26**). Black pepper also contains a number of related minor amides of two types: those with the 3,4-methylenedioxy moiety derived from ferulic acid and those with an unsaturated long chain derived from fatty acids (**10-27**). Both types of compounds are further derived from piperidine (piperanine, piperettine, and piperoleines), pyrrolidine (piperyline), or isobutylamine (pellitorine and piperlonguminine).

piperine

piperanine

piperettine

piperoleine

piperoleine B

piperyline

piperlonguminine

pellitorine

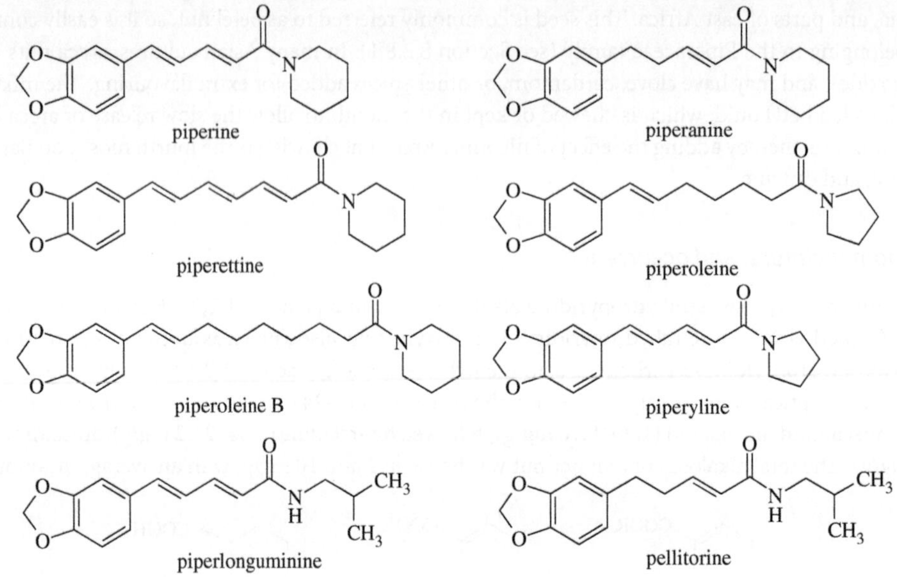

10-26, piperine and related alkaloids of black pepper

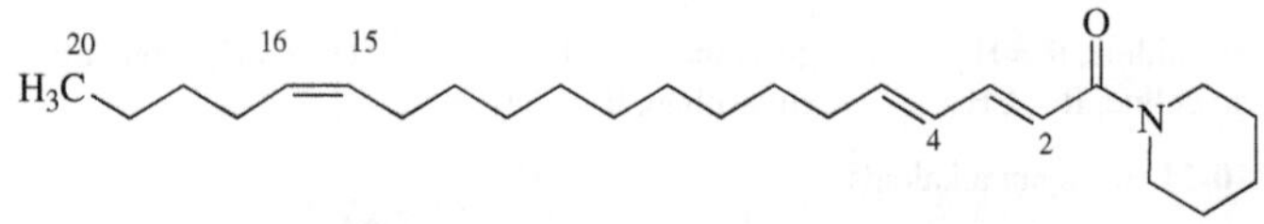

10-27, (2E,4E,15Z)-1-eicosa-2,4,15-trieny-1-yl piperidine

Piperine and its analogues with the 3,4-methylenedioxy moiety induce only pungent sensation, whilst the conjugated 2,4-dienoic acid amides with an additional Z-double bond, derived from C_{18} and C_{20} fatty acids and piperidine, pyrrolidine, or isobutylamine, exhibit both burning and tingling sensations. For example, the threshold concentrations for pungency of piperine, piperyline, piperettine, and piperlonguminine are 3.0, 5.1, 5.2, and 10.4 nmol/cm^2, respectively. The threshold concentration for the tingling sensation of (2E,4E,15Z)-1-eicosa-2,4,15-trienylpiperidine is 260 nmol/cm^2, and 260 nmol/cm^2 for pungency.

Piperine content is typically 3–8% (and sometimes more), but it depends on the origin and other variables. The highest amount of piperine in pepper fruits occurs just before full maturity. That is why black pepper is obtained from immature but already fully developed green fruits that are fermented during sun drying. Green pepper (quickly sun-dried unripe berries) contains a considerable amount of piperine, as does black pepper. White pepper (ripe fruit seeds) is produced from fully ripe red peppercorns by removal of the skin. The content of piperine in white pepper is somewhat lower. The levels of the minor alkaloids piperyline and piperettine are only 0.2–0.3% and 0.2–1.6%, respectively. Ripe red peppercorns can also be dried or preserved in brine. The piperine content in oleoresins is normally 35–40%, which corresponds to minced spice in a ratio of about 1 : 25.

Many other unsaturated amides have been identified in *Piper* species from all tropical regions of the world. Much hotter and more pungent than black pepper is long pepper (*P. longum*). The plant is native to India, Indonesia, and other Southeast Asian countries and finds its application in culinary practice and a number of common health benefits. The hot and pungent taste of fruits and seeds is mainly due to the presence of piperine (about 5% d.w.). Besides piperine, the other alkaloids are piperettine, piperanine, piperlonguminine, and pellitorine (4,5-dihydropiperlonguminine) (**10-26**).

10.3.6.4.2 Reactions and changes

Piperine is very stable in intact seeds and in ground seeds. If protected from light, it is stable for at least 6 months if stored in paper, or for 10 months if kept in aluminum packaging, but it is easily oxidised in air. Piperine in solutions, in ground spice, and in oleoresins is accompanied by its diastereomers, (2Z,4E)-piperoyl-1-piperidine (isopiperine), (2Z,4Z)-piperoyl-1-piperidine (chavicine), and (2E,4Z)-piperoyl-1-piperidine (isochavicine, **10-28**), which arise by enzyme-catalysed or light-induced *cis/trans*-isomerisations of the double bonds of the parent piperine molecule. However, they show little or no (isochavicine) pungency. Alkaline hydrolysis of piperine yields the salt of piperic acid, (2E,4E)-5-(3,4-methylenedioxyphenyl)-2,4-pentadienoic acid, and piperidine. In acidic solutions, piperine may react with nitrites to form nitrosamines.

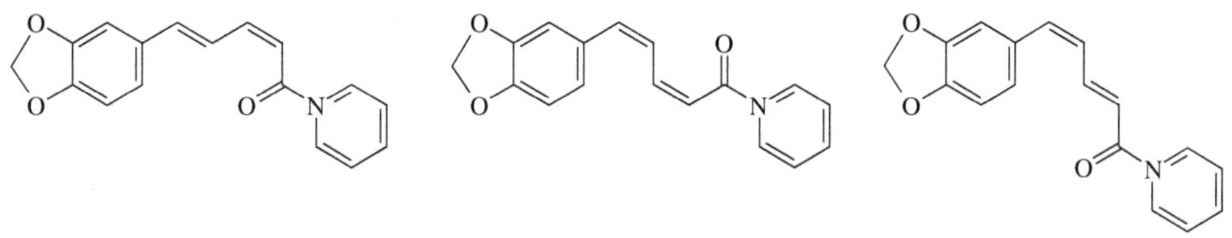

isopiperine   chavicine   isochavicine

10-28, piperine isomers

10.3.6.4.3 Biological effects

Piperine acts as a stimulant of the central nervous system and has weak antipyretic and mutagenic effects. Reportedly, it is beneficial in increasing thermogenesis (the process of generating energy in cells), which may be helpful in reducing inflammation, improving digestion, and relieving pain and asthma. At higher concentrations, piperine damages tongue tissue and lowers blood pressure and breathing rate. Piperine derivatives of pepper show antioxidative, antimicrobial, and insecticidal effects.

10.3.6.5 Nortropane and tropane alkaloids

The most important group of alkaloids with a nortropane skeleton (**10-29**) found in foods is the bicyclic polyhydroxy alkaloids, called **calystegines**. Calystegines have been isolated from the roots of the larger bindweed *Calystegia sepium* (Convolvulaceae) and later identified in the plant families Solanaceae, Moraceae, and Brassicaceae. They are biosynthesised together with tropane alkaloids from putrescine via *N*-methylputrescine and other metabolites. The pathway to tropane alkaloids branches and the loss of the *N*-methyl group provides the nortropane skeleton of calystegines.

10.3.6.5.1 Structure, nomenclature, and occurrence

Tropane alkaloids are a group of more than 200 compounds derived from tropane skeleton (**10-29**) best known for their occurrence in the Solanaceae family.

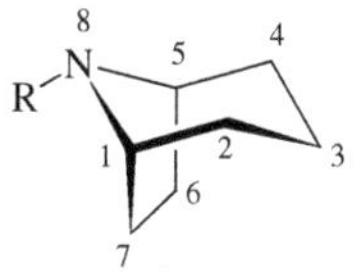

10-29, nortropane skeleton, R = H
tropane skeleton, R = CH$_3$

The structure of calystegines falls into three main classes, calystegines A, B, and C (**10-30**), having three, four, and five hydroxyl groups, respectively. Diversity within these three classes arises from the positional and stereochemical configuration of these hydroxyl groups. A few calystegines exist that deviate in structure. For example, calystegine N$_1$ carries an amino group instead of a hydroxyl group on C-1 of the nortropane structure, in addition to three hydroxyl groups on C-2, C-3, and C-4. This calystegine is very labile and undergoes conversion to calystegine B$_2$. Conversely, calystegine N$_1$ may be an artefact formed from calystegine B$_2$ and ammonia.

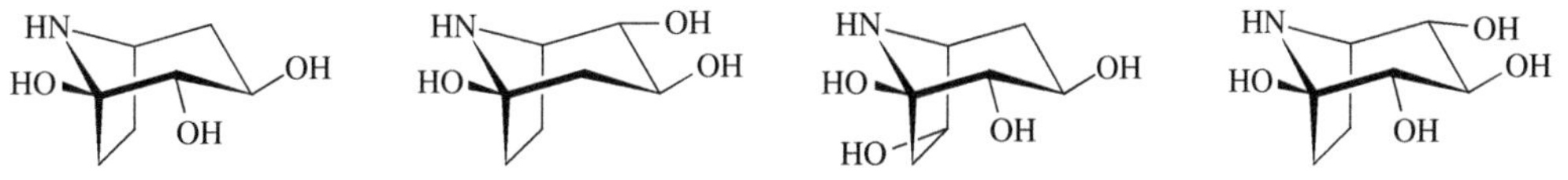

calystegine A$_3$ calystegine A$_5$ calystegine B$_1$ calystegine B$_2$

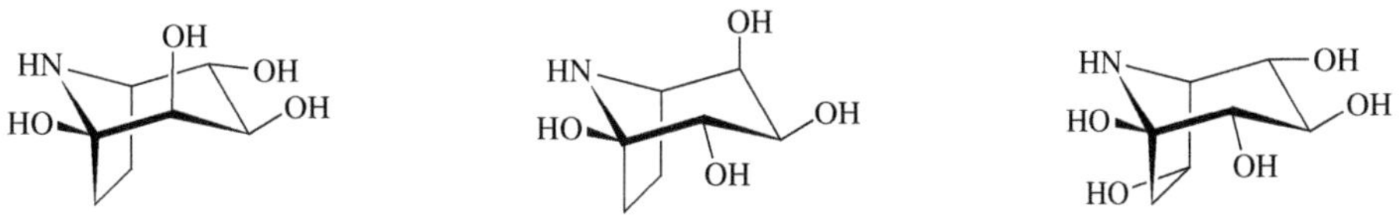

calystegine B$_3$ calystegine B$_4$ calystegine C$_1$

10-30, calystegines

Calystegines have been found in numerous edible vegetables and fruits, especially members of the Solanaceae family, such as potatoes, tomatoes, eggplants, and sweet and chili peppers, but also in mulberries (Moraceae) and some *Brassica* vegetables and spices, such as kohlrabi, Brussel sprouts, and black mustard leaves.

The major calystegines of potatoes are calystegine A_3 and calystegine B_2. In potato cultivars grown in the United States, calystegine A_3 and calystegine B_2 concentrations were 6–316 mg/kg in dry flesh, 218–2581 mg/kg in dry peel, 34–326 mg/kg in dry whole potatoes, 1–68 mg/kg in wet flesh, 35–467 mg/kg in wet peel, and 5–68 mg/kg in wet whole potatoes. Concentrations in sproutings are on average 100 times higher than in the tuber flesh and 8 times higher than in the peel. The calystegines in sprouts include, in addition to the more abundant calystegines A_3 (on average 3999 mg/kg d.w.) and B_2 (7425 mg/kg d.w.), small amounts of four additional types, B_3 (86 mg/kg d.w.), B_4 (319 mg/kg d.w.) and N_1 (15 mg/kg d.w.).

White mulberries (*Morus alba*) contain 4-*O*-β-D-galactopyranosyl-calystegine B_2 (10 mg/kg d.w.), calystegines A_3 and B_2 occur in kohlrabi leaves in the amount of 26 and 7 mg/kg, respectively, calystegine A_3 occurs in the leaves of Brussel sprouts (5 mg/kg d.w.), and calystegines A_3, A_5, and B_2 have been identified in black mustard leaves in concentrations of 13, 3, and 2 mg/kg (d.w.), respectively.

Tropane alkaloids comprise more than 200 compounds, which mainly occur in plants of the families Erythroxylaceae (the coca family) and Solanaceae. Tropane alkaloids can be exemplified by (−)-hyoscyamine (also known as daturine, the levorotary isomer of atropine) and (−)-scopolamine (also known as hyoscine, **10-31**), which are found in plants of the genus *Datura* (Solanaceae), such as *D. stramonium* (also known as thorn apple or Jimson weed). When growing in direct vicinity of crops, tropane alkaloid containing plants are potential contaminants of various food and feed materials.

hyoscyamine   scopolamine

10-31, tropane alkaloids

10.3.6.5.2 Reactions and changes

Removal of the peel reduces the calystegine content by an average of over 50% in common potato varieties. Calystegines are water-soluble compounds and their leaching into water can be expected during potato processing.

10.3.6.5.3 Biological effects

Calystegines exhibit potent specific inhibition of glycosidases that are universally required for normal cell function. Although no human toxicity data for calystegines have been reported, consumption of calystegines has been associated with gastrointestinal upset and it is reasonable to suppose that the calystegines may also inhibit digestive and hepatic enzymes.

Tropane alkaloids can act as anticholinergic drugs, and as a result have effects on heart rate, respiration, and functions in the central nervous system. Since 2016, maximum limits of 1 µg/kg for atropine and scopolamine in infant foods have been applicable in the European Union.

10.3.6.6 Indole alkaloids

The indole alkaloids are a large class of more than 4100 compounds containing a structural moiety of indole (see **2-112**), which is derived from the amino acid tryptophan. Many of them possess biological activities and some are used in medicine. Certain indole-derived compounds are not universally assigned to the alkaloids, and they are therefore mentioned in other sections, such as the pigments melanin, betalain (Section 9.3.1), and phenoxazine (Section 9.3.6), glucosinolates (Section 10.2.3.3), phytoalexins, and biogenic and aromatic amines (Section 12.2.1) occurring in tryptophan pyrolysates.

10.3.6.6.1 Structure, nomenclature, and occurrence

Depending on their biosynthesis, two types of indole alkaloids are distinguished: isoprenoids and non-isoprenoids. More than 3000 terpenoid indole alkaloids occur in nature. The most prominent groups in foods are ergot alkaloids produced by the rye ergot fungus *Claviceps purpurea* (see Section 12.2.2.2.15). Non-isoprenoid alkaloids include several groups of indole-derived compounds. The most important compounds arising in foods are **simple indoles** and **simple β-carbolines** (Figure 10.9).

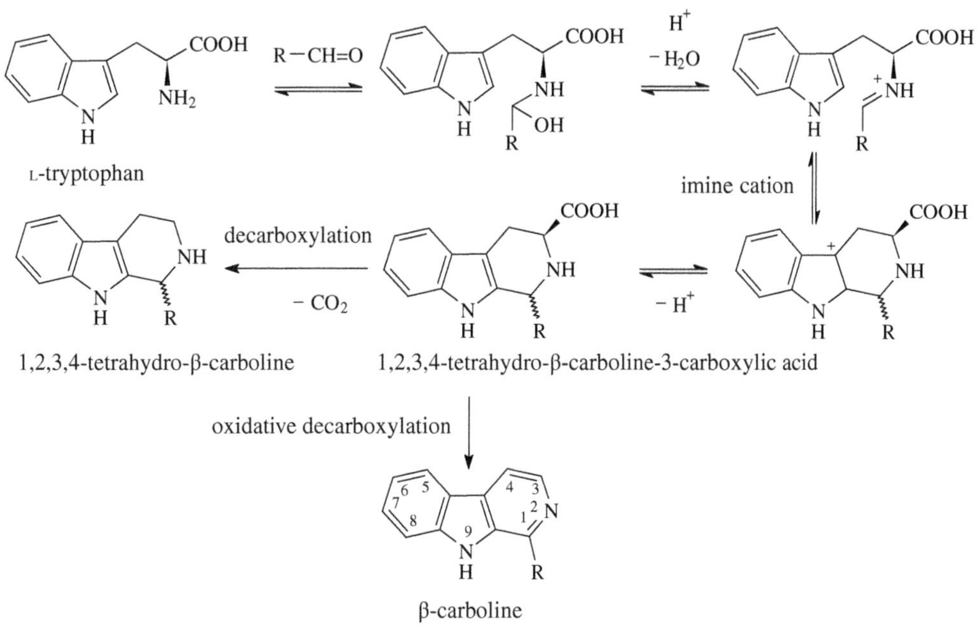

Figure 10.9 Reaction of tryptophan with aldehydes.

10.3.6.6.2 Reactions and changes

The prevalence of non-isoprenoid β-carboline alkaloids in foods is associated with the ease of forming the β-carboline core from indoleamines such as tryptophan and tryptamine, various aldehydes, and even aldoses through a Pictet–Spengler condensation (a special case of the Mannich reaction). The mechanisms of this reaction and of subsequent reactions leading to 1,2,3,4-tetrahydro-β-carboline-3-carboxylic acids, 1,2,3,4-tetrahydro-β-carbolines, and β-carbolines are given in Figure 10.9.

1,2,3,4-Tetrahydro-β-carboline-3-carboxylic acid derived from formaldehyde (Figure 10.9, R = H) and acetaldehyde (Figure 10.9, R = CH$_3$) and their decarboxylation products, 1,2,3,4-tetrahydro-β-carbolines, have been identified in various products, such as cashew nuts, walnuts, pineapple, banana, wine, vinigar, beer, green vegetable soybeans, soy sauce, raisins, fruit juices, syrups, purees, and jams. For example, (3S)-1,2,3,4-tetrahydro-β-carboline-3-carboxylic acid and (1S,3S)- and (1R,3S)-1-methyl-1,2,3,4-tetrahydro-β-carboline-3-carboxylic acid diastereomers (**10-32**) have been found in commercially prepared fruit juices in amounts of 0.01–1.45, 0.02–9.1, and 0.01–2.48 μg/g, respectively. The content was higher in citrus juices (orange and grapefruit) than in other juices (grape, apple, pineapple, peach, banana, pear, and tomato). The major antioxidants identified in aged garlic extracts (see Section 8.2.12.6.11) include (3S)-1,2,3,4-tetrahydro-β-carboline-3-carboxylic acid, (1R,3S)- and (1S,3S)-1-methyl-1,2,3,4-tetrahydro-β-carboline-3-carboxylic acid, and (1R,3S)- and (1S,3S)-1-methyl-1,2,3,4-tetrahydro-β-carboline-1,3-dicarboxylic acids (derived from glyoxylic acid). In green vegetable soybeans were identified, (1R,3S)- and (1S,3S)-1-methyl-1,2,3,4-tetrahydro-β-carboline-3-carboxylic acids and unusual alkaloids.

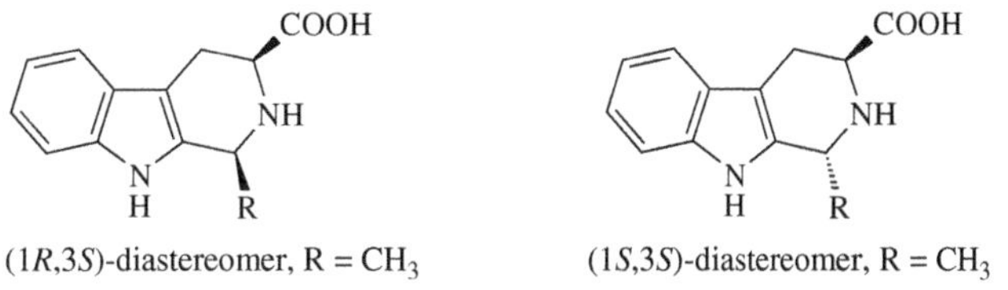

(1R,3S)-diastereomer, R = CH$_3$ (1S,3S)-diastereomer, R = CH$_3$

10-32, 1-methyl-1,2,3,4-tetrahydro-β-carboline-3-carboxylic acids

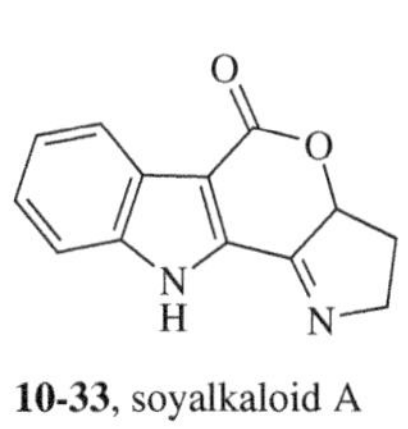

10-33, soyalkaloid A 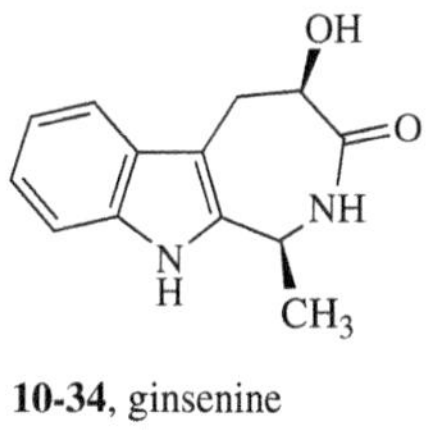   **10-34**, ginsenine

Decarboxylation and oxidation products of 1,2,3,4-tetrahydro-β-carboline-3-carboxylic acids derived from formaldehyde include 9H-pyrido[3,4-b]indole, also known as β-carboline or norharmane (Figure 10.9, R = H). The corresponding product of acetaldehyde is 1-methyl-9H-pyrido[3,4-b]indole, known as harmane (Figure 10.9, R = CH$_3$). These β-carbolines were identified in a number of foods at levels up to 700 mg/kg, although more typically their concentrations in smoked, cooked, and fermented foods range from a few mg/kg to 1–2 orders of magnitude less. Typical norharmane and harmane findings in pan-fried minced beef patties and ground beef prepared at a temperature of 175–230 °C and with a cooking time of 2–10 minutes ranged from 0.8 to 11.3 µg/kg and from 1.4 to 10.3 µg/kg, respectively. In raisins, norharmane concentrations ranged from 2 to 120 µg/kg and concentrations of harmane were 6–644 µg/kg (the maximum concentrations of 1,2,3,4-tetrahydro-β-carboline-3-carboxylic acids reached 50 mg/kg). Dark-brown raisins (sun-dried) contained higher levels of these alkaloids than golden raisins. Arabica coffees contained norharmane and harmane in the amount of about 136 and 51 µg/l, respectively. Commercial blends (usually with a maximum of 30% robusta) ranged from the cited arabica values to 345 µg/l of norharmane and 145 µg/l of harmane. Both alkaloids were also found in tobacco smoke, with 11.2 µg of norharmane and 3.6 µg of harmane per cigarette.

Reactions of tryptophan with aldohexoses (at temperatures above 50 °C and in acidic solutions) are a prerequisite for the formation of glycotetrahydro-β-carbolines, which arise together with N-glycosides and C-glycosyl conjugates of tryptophan. Pictet–Spengler-type condensation of tryptophan and glucose gives (1R,3S)- (10-35) and (1S,3S)-1-(D-gluco-1,2,3,4,5-pentahydroxypent-1-yl)-1,2,3,4-tetrahydro-β-carboline-3-carboxylic acid diastereomers. These compounds were also found in fruit juices. Grape juices and mixed fruit juices containing grape juice exhibited the highest level of pentahydroxypentyl-tetrahydro-β-carboline-3-carboxylic acids (up to 3.8 mg/l). Relatively high concentrations of these β-carbolines were also found in tomato, pineapple, multifruit, and tropical fruit juices. In contrast, they were not detected in apple juices or banana and peach nectars, although a low amount was detected in orange juices. In the presence of oxygen or mild oxidants such as L-dehydroascorbic acid, 1-(D-gluco-1,2,3,4,5-pentahydroxypent-1-yl)-1,2,3,4-tetrahydro-β-carboline-3-carboxylic acid readily eliminates carbon dioxide and is oxidised to 1-(D-gluco-1,2,3,4,5-pentahydroxypent-1-yl)-β-carboline (10-35). Diastereomeric 1-(1,3,4,5-tetrahydroxypent-1-yl)-β-carbolines, 1-(1,4,5-trihydroxypent-1-yl)-β-carbolines, and E/Z-isomers of 1-(1,5-dihydroxypent-3-en-1-yl)-β-carbolines, arising by oxidative decarboxylation and dehydration, isomerisation, and reduction of the carbohydrate-derived side chain, were found in various food products. The highest concentrations of these β-carbolines were found in ketchups, soy, and fish sauces. The main components were diastereomers of 1-(1,3,4,5-tetrahydroxypent-1-yl)-β-carbolines (1.4–3.5 mg/kg).

The β-carbolines flazine and perlolyrine (10-36), derived from a degradation product of the hexose 5-hydroxymethylfuran-2-carbaldehyde, were found in soy sauce. Common concentrations of these alkaloids are 24 and 2.8 mg/kg, respectively.

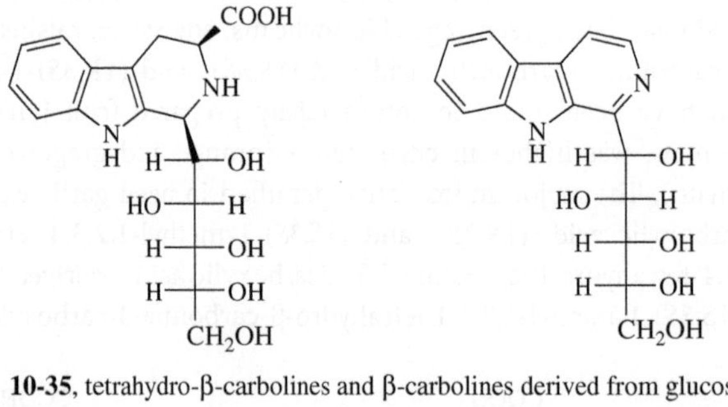

10-35, tetrahydro-β-carbolines and β-carbolines derived from glucose

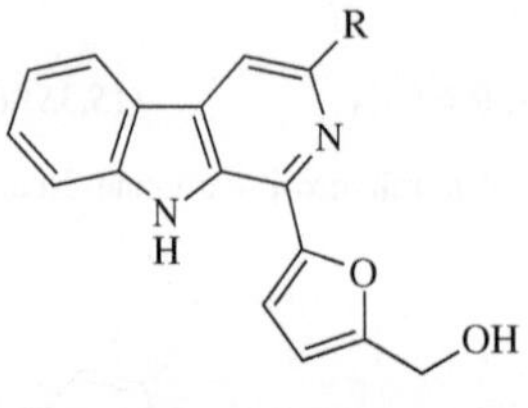

10-36, perlolyrine, R = H
flazine, R = COOH

10.3.6.6.3 Biological effects

β-Carbolines may act as neuromodulators, and some may have an endocrinological function, including inhibition of monoamine oxidase, which catalyses the oxidative deamination of monoamines, competitive inhibition of monoamine neurotransmitter serotonin uptake, inhibition of Na^+-dependent transports, and binding to opiate receptors.

10.3.6.7 Quinolizidine alkaloids

Quinolizidine alkaloids, hypothetically derived from quinolizidine (**10-37**), are a special group of bicyclic, tricyclic, and tetracyclic secondary metabolites of some legumes, especially of the genera *Lupinus*, *Baptisia*, *Thermopsis*, *Genista*, *Cytisus*, *Chamaecytisus*, *Laburnum*, and *Sophora* (Fabaceae), which occur as a complex mixture of several compounds. Quinolizidine alkaloids are also secondary metabolites of some plants of the Chenopodiaceae, Ranunculaceae, Berberidaceae, and Solanaceae families.

10-37, quinolizidine

10.3.6.7.1 Structure, nomenclature, and occurrence

Important quinolizidine alkaloids occur in seeds of certain species of lupines (*Lupinus* spp., Fabaceae), including domesticated varieties. In Europe, Africa, and America, lupines have been used as valuable pulses for human nutrition and the feeding of livestock since ancient times. Native species growing in southwestern Europe are white lupine (*L. albus*), blue (narrow-leaf) lupine (*L. angustifolius*), and yellow lupine (*L. luteus*). Yellow lupine is currently grown in Western Australia, particularly for feed, and many Asian countries, for human nutrition. Beans of pearl (Andean) lupine (*L. mutabilis*), called tarhui or tarwi, are eaten by South American Indians of the Andes region and used as an oil crop (mainly in Chile).

Seeds of lupines contain more than 100 bicyclic, tricyclic, and tetracyclic quinolizidines. With respect to economically important lupine species, the major alkaloids are lupinine, angustifoline, sparteine (also known as lupinidine), lupanine, α-isolupanine, 3β-hydroxylupanine, 13α-hydroxylupanine, albine, and multiflorine (**10-38**). These alkaloids are accompanied by a series of related alkaloids and esters of hydroxylated alkaloids with benzoic, *p*-coumaric, sinapic, and other organic acids. The hydroxyl group of *p*-coumaric acid is often glycosylated by α-L-rhamnose. The content of alkaloids in plant seeds is highly dependent on the plant species and climatic conditions. The content in seeds of original bitter varieties of *L. angustifolius* normally ranges from 1 to 3 and can be up to 5%. The alkaloid content of sweet varieties obtained by breeding is only 0.01–0.03%, but can be also 0.001%. For example, the maximum alkaloid content of seeds of sweet lupine varieties in Australia is 0.002% and in Chile 0.05%. The compositions of the major alkaloids of different lupine seed species are shown in Table 10.8. The content of other toxic and antinutritional substances (lectins, trypsin inhibitors, and phytates) in lupine seeds is similar to that in other legume seeds.

Table 10.8 Main alkaloid contents in lupine seeds.

Alkaloid	Content (% of total alkaloid content)			
	L. albus	L. angustifolius	L. luteus[a]	L. mutabilis
Lupinine	–	–	60	–
Sparteine	<1	<1	30	16
Albine	15	–	–	–
Angustifoline	<1	10–16	–	1
Lupanine	70	70	<1	46
3-Hydroxylupanine	–	–	–	12
13-Hydroxylupanine	8	12–38	–	7
Multiflorine	3	–	–	–

[a]The seeds of some varieties contain up to 1200 mg/kg of protoalkaloid gramine.

lupinine, angustifoline, sparteine, lupanine, α-isolupanine, 3-hydroxylupanine, 13-hydroxylupanine, albine, multiflorine

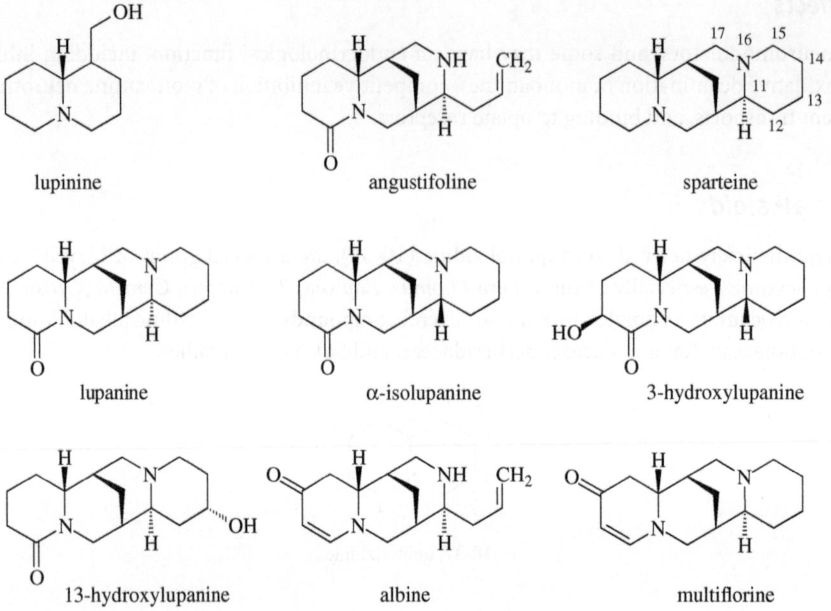

10-38, quinolizidine alkaloids of lupin seeds

10.3.6.7.2 Reactions and changes

Lupine alkaloids are very stable compounds. During processing and storage of seeds, their content does not change. The most common method of debittering (removal of alkaloids) is extraction of ground seeds with water (by soaking or boiling). These procedures can reduce the alkaloid content by about 100 times.

The fat content of lupine is relatively small, except in the South American species *L. mutabilis*, which contains about 22–24% of oil in dry matter. In the technological production of oil from this lupine, about 60% of alkaloids present in bitter varieties are removed during extraction with hexane. Subsequent refining decreases their concentration in the oil down to 0.0005%. Meals are debittered by extraction with ethanol containing hydrochloric acid, which decreases the alkaloid content from the original amount of 3.2% to 0.1–0.2%.

10.3.6.7.3 Biological effects

Individual alkaloids exhibit different toxicities as they are hepatotoxic and teratogenic and may represent a hazard for humans and livestock. Lupanine and sparteine are the most toxic substances. Ingestion of 11–25 mg of lupine alkaloids per kilogram of body weight may cause serious health disorders, manifested by nervousness, vomiting, breathing difficulties, impaired vision, sweating, progressive weakness, and, in extreme cases, coma. Alkaloids obtained during lupine seed debittering are used as pesticides and in human medicine.

10.3.6.8 Quinoline alkaloids

Alkaloids hypothetically derived from quinoline (**10-39**) are a large group of secondary plant metabolites. These alkaloids are found in many plant families, such as Apocynaceae, Loganiaceae, Rubiaceae, and Nyssaceae. The most important quinoline alkaloids are those derived from the basic skeleton of rubane (**10-40**), which are found in the bark of cinchona or quina trees (*Cinchona* spp., Rubiaceae), native to the tropical forests of western South America, but today also grown in other countries.

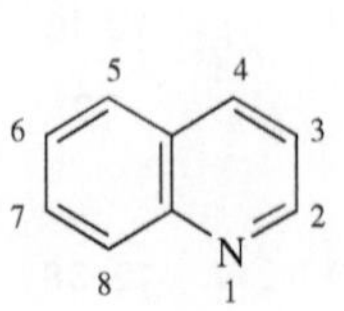

10-39, quinoline

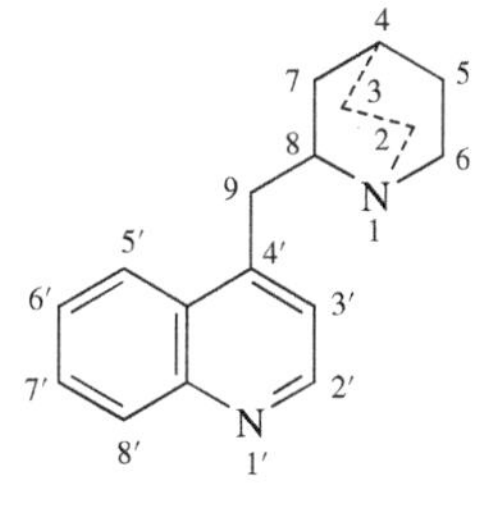

10-40, rubane

10.3.6.8.1 Structure, nomenclature, and occurrence

The quinine molecule consists of a quinoline ring with a methoxyl group in position C-6′ and a quinuclidine bicyclic structure with a vinyl group (at C-5) in position C-4′, which is bound through carbon C-9 bearing a hydroxyl group. Quinine has asymmetric centres at carbons C-4, C-5, C-8, and C-9. The steric configurations at C-4 and C-5 in the major cinchona bark alkaloids quinine, quinidine, cinchonine, and cinchonidine (**10-41**) are the same. Quinidine, the optical isomer of quinine, has a different spatial arrangement of C-8 and C-9, whilst cinchonidine, which lacks the C-6′ methoxyl group, has the same one as quinine. Its isomer is cinchonine, which corresponds in spatial arrangement to quinidine.

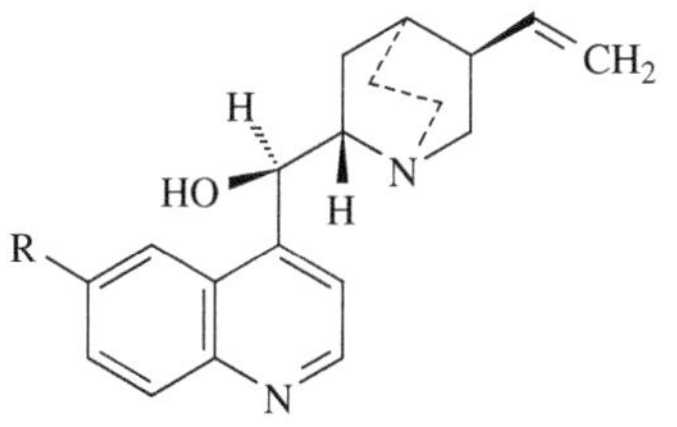

(−)-cinchonidine, R = H
(−)-quinine, R = OCH₃

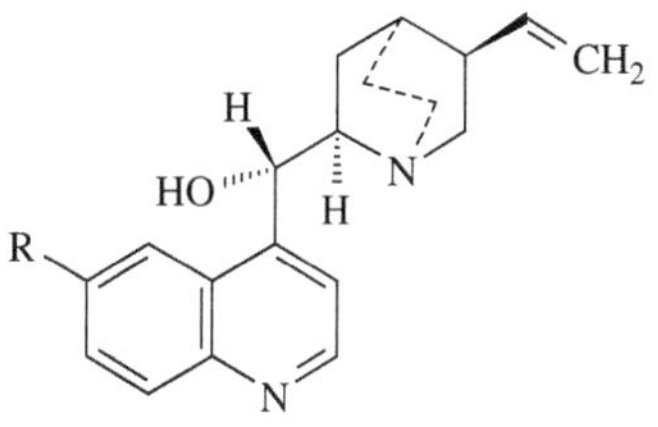

(+)-cinchonine, R = H
(+)-quinidine, R = OCH₃

10-41, cinchona bark alkaloids

The genus *Cinchona* includes about 40 tree species, of which about 12 are of commercial importance. Alkaloids in the bark of *C. officinalis* occur in an amount of 5–8%, of which 2–7.5% is quinine (**10-41**). The total alkaloid content of *C. succiruba* (syn. *C. pubescens*) bark is 6–16% (quinine content is 4–14%), that of *C. ledgeriana* bark is 5–14% (quinine content is 3–13%), and that of *C. calisaya* bark is 3–7% (quinine content is 0–4%).

Quinine, isolated from cinchona bark in the form of hydrochloride or sulfate, is mainly used in medicine as an antimalarial and antipyretic medication. In sensory analysis, it is used as a standard of bitterness, and in the food industry it is used for the production of bitter soft drinks (e.g. bitter lemon, Indian tonic waters) and some alcoholic beverages (e.g. the flavoured wine Barolo Chinato in Italy), due to its distinctive bitter taste. The taste threshold concentration of bitter taste perception is about 10 mg/l.

10.3.6.8.2 Reactions and changes

In light, quinine is degraded by photochemical reactions. The main product in acidic media is 9-deoxyquinine (**10-42**), whilst those in slightly acidic and neutral solutions (e.g. beverages) are 6-methoxy-4-methylquinoline, 6-methoxyquinoline, and 2-formyl-5-vinylquinuclidine (**10-43**). In beverages exposed to direct sunlight, quinine is totally degraded within several hours. The degradation is accompanied by a light opalescence and a loss of bitter taste. Dark bottles substantially reduce quinine degradation.

10.3.6.8.3 Biological effects

Quinine is a teratogenic substance and should never be used by pregnant women, as birth defects and miscarriages may occur. Side effects can include ringing in the ears, nausea, blurred vision, chest pain, upset stomachs, and breathing problems. The use of quinine is therefore regulated. In the European Union, for non-alcoholic beverages, quinine (sulfate or hydrochloride) may be used in the highest allowable amount of 75 mg/l, whilst for spirits (bitter alcoholic beverages), the highest allowable amount is 300 mg/l (calculated as free base).

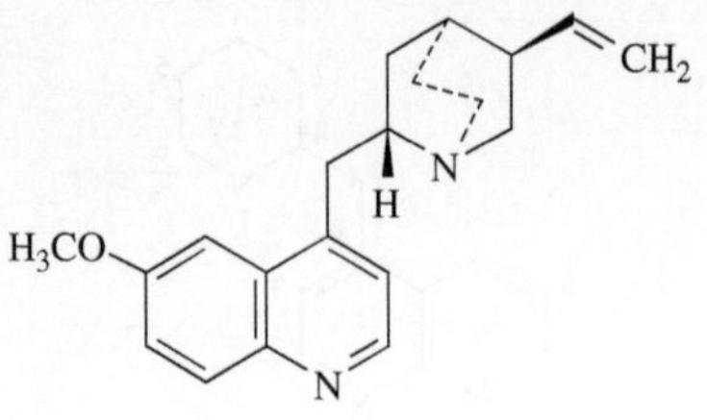

10-42, 9-deoxyquinine

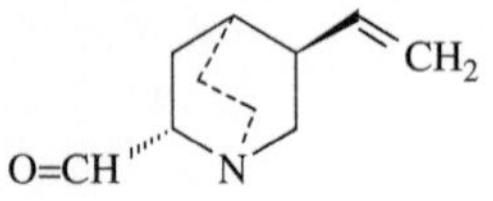

10-43, 2-formyl-5-vinyl-quinuclidine

10.3.6.9 Isoquinoline and phenanthrene alkaloids

The most important alkaloids with phenanthrene (**12-44**) and isoquinoline (**8-45**) skeletons occur in the opium poppy (*Papaver somniferum*, Papaveraceae). Some other alkaloids (e.g. salsolinol, Figure 10.2) are classified as biogenic amines.

10.3.6.9.1 Structure, nomenclature, and occurrence

Opium poppy is a plant cultivated in a number of varieties as a source of oilseed, straw, and opium. The garden forms are sometimes grown for decoration. Whole or crushed poppy seeds are used in various bakery products, on top of dishes, and in cake fillings in both Europe and the United States. Poppy seed oil has been used as a culinary salad oil, cooking oil, and semi-drying oil for use in art (Table 3.14). Poppy straw (the upper stalk of the plant including the crushed capsules) and opium (the air-dried milky exudate with a bitter taste, obtained by incising the unripe seed capsules) contain pharmaceutically significant alkaloids conventionally divided into two distinct classes, phenanthrene derivatives (**10-46**) and isoquinoline derivatives (**10-47**). The phenanthrene alkaloids are derived biosynthetically from benzylisoquinoline intermediates.

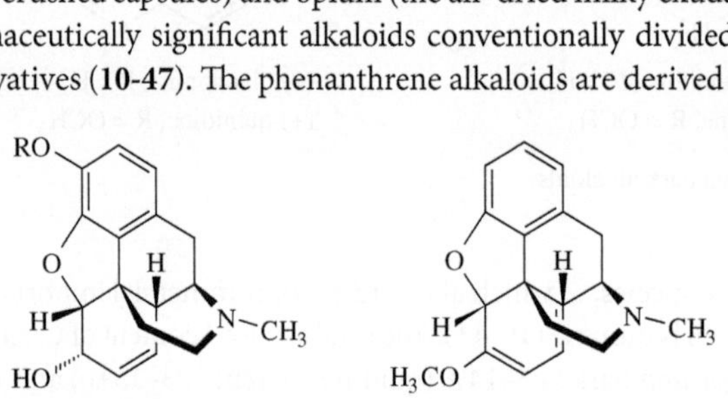

morphine, R = H
codeine, R = CH$_3$

thebaine

10-46, phenanthrene alkaloids

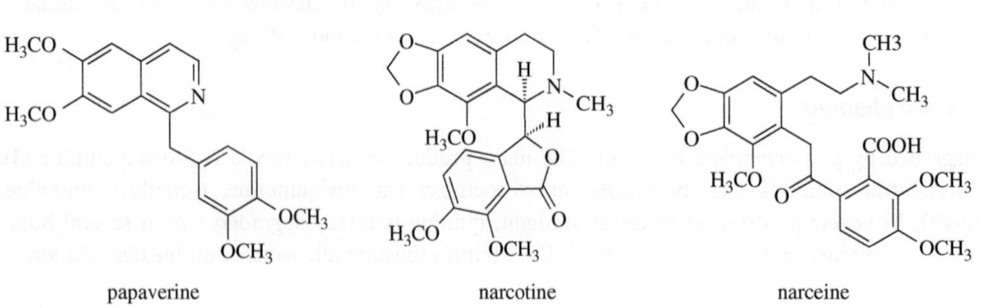

papaverine

narcotine

narceine

10-47, isoquinoline alkaloids

10.3.6.9.2 Reactions and changes

The amount of alkaloids in plants depends primarily on poppy varieties, but also on climatic and soil conditions. In capsules, the opium content is highest in the 10–12 days prior to full maturation of the seeds. The six opium alkaloids that occur naturally in the largest amounts are morphine (8–17%), narcotine (also known as noscapine, 1–10%), codeine (0.7–5%), thebaine (0.1–2.5%), papaverine (0.5–1.5%), and narceine (0.3%). Of these, morphine, codeine (3-methylmorphine), and thebaine (paramorphine) are phenanthrene alkaloids, usually called morphinans. Amongst the other three, the phenanthrene alkaloid narceine is a derivative o 1-(2-aminoethyl)phenanthrene, and although it does not contain a nitrogen heterocycle, it is considered alkaloid, whilst papaverine is a benzylisoquinoline derivative and narcotine is a phthalid-isoquinoline derivative.

Opium contains approximately 5–10% water, 20% various sugars, and simple carboxylic acids, including dibasic meconic acid (3–5%, **8-77**), which is a chemotaxonomic marker of the genus *Papaver* and other closely related genera. The alkaloid content is approximately 10–20%, with more than 40 individual alkaloids occurring in the plant as meconates and other simple carboxylic acid salts. The four economically significant alkaloids of opium are morphine, codeine, thebaine, and papaverine. A number of other opium alkaloids used medically are commercial products obtained by conversion of morphine, codeine, and thebaine. Some of them, such as heroin (diacetylmorphine, diamorphine), remain widely available on the illicit market within Europe and United States.

The alkaloids of the poppy straw are the same as those of opium. Those occurring in the largest amounts are morphine (0.04–1.47%), codeine (0.01–0.35%), thebaine (0.00–0.06%), papaverine (0.00–0.10%), and narcotine (0.00–0.04%).

Trace amounts of alkaloids found in the seeds are caused by contamination by poppy straw particles at harvest or by latex staining on harvesting insufficiently mature vegetation. Consumption of poppy seeds contaminated with opium alkaloids can lead to adverse health effects, especially in babies, infants, and people with severe health issues. Analysis of 41 samples collected in 2015 in the Netherlands and Germany showed that opium alkaloids were found in 35 samples in the range of 0.2–240 mg/kg morphine, 0.1–340 mg/kg codeine, 0.1–100 mg/kg thebaine, 0.1–5.2 mg/kg narcotine, 0.1–3.4 mg/kg papaverine, and 0.1–1.7 mg/kg narceine.

The alkaloid content of poppy seed-containing foods can be reduced by several forms of pre-treatment and processing. Food processing may decrease the alkaloid content by up to 90%. The most effective methods include washing, soaking, and heat treatments, as well as grinding and combinations of all four.

10.3.6.9.3 Biological effects

Morphine is an exogenous opioid (xenobiotic opioid) that produces analgesia as well as other pharmacological actions as a result of its affinity to bind to receptors normally acted upon by endogenous opioids: peptides called endorphins (**endo**genous mo**rphins**). Many other benzylisoquinoline alkaloids possess potent pharmacological activities, including codeine with the same pharmacological activity as morphine, the less potent vasodilator and antispasmodic drug papaverine, and the potential anticancer drug narcotine. Morphine and some other opioids are under international control.

Legislation on opium alkaloids in poppy seeds for food uses exists only in some EU countries. The Czech Republic, the leader in the cultivation and consumption of poppy seeds for food purposes, limits the content of morphine alkaloids to a maximum of 25 mg/kg, whilst Hungary has national maximum levels of 40 mg/kg for morphine and narcotine, 30 mg/kg for morphine, and 20 mg/kg for narcotine, thebaine, and codeine.

10.3.6.10 Purine alkaloids

Alkaloids derived from purine (**10-48**) and the purine oxidation product xanthine (**10-49**) are called purine alkaloids. Purine alkaloids are the most widespread alkaloids in foods.

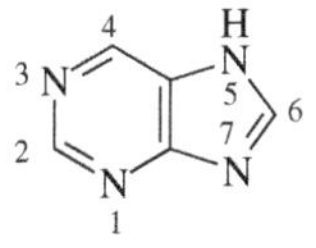

10-48, purine

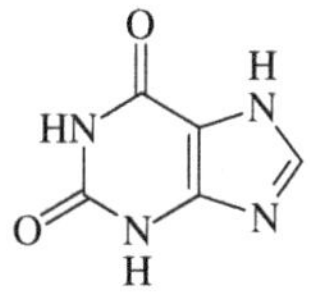

10-49, xanthine

10.3.6.10.1 Structure, nomenclature, and occurrence

Purine alkaloids are methyl derivatives of xanthine. The most common is 1,3,5-trimethylxanthine, 1,3,7-trimethyl-1*H*-purine-2,6(3*H*,7*H*)-dione, known trivially as caffeine (**10-50**). It is accompanied by dimethylxanthines theobromine, theophylline, and paraxanthine, and also by methylxanthine heteroxanthine and methyluric acids, which are minority alkaloids, with the exception of cocoa and chocolate. Theacrine,

Table 10.9 Alkaloid contents in green coffee beans.

Alkaloid	Content in dry matter (%)		
	Coffea arabica	*Coffea canephora*	*Coffea liberica*
Coffeine	0.53–1.45	2.11–2.72	1.28–1.35
Theobromine	<0.005	<0.005–0.01	<0.005
Theophylline	<0.005	<0.005–0.01	0.01
Trigonelline	0.97–1.31	0.57–0.88	0.25–0.29

1,3,7,9-tetramethyluric acid (**10-51**), is a purine alkaloid obtained, for example, from Assam tea, which is manufactured specifically from the plant *Camellia sinensis* var. *assamica* in Assam in India, the world's largest tea-growing region. Caffeine is found in the seeds, leaves, and fruits of more than 60 species of plants. It is assumed that the reason for the accumulation of purine alkaloids by plants is to protect young leaves, flowers, and fruits against damage caused by pests.

The contents of caffeine, theobromine, and theophylline and alkaloid trigonelline in common species of green coffee beans (*Coffea* spp., Rubiaceae) are given in Table 10.9. The amount of caffeine is dependent on the method of brine preparation (water temperature, time of extraction, and other factors). The average content of caffeine in one cup of coffee (100 ml) is about 80 mg, or 1–6 mg in decaffeinated coffee, 29–91 mg in instant, and 93–127 mg in filter coffee. Decaffeinated coffee is produced by extraction of caffeine with organic solvents, in particular dichloromethane or supercritical carbon dioxide. The caffeine obtained is used in the pharmaceutical industry and for the enrichment of soft drinks.

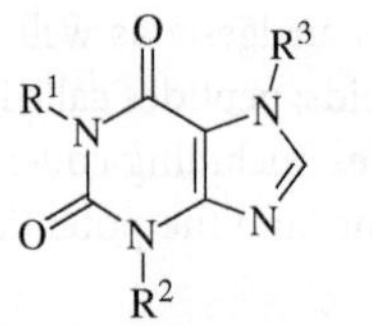

10-50, purine alkaloids

caffeine, $R^1 = R^2 = R^3 = CH_3$
theobromine, $R^1 = H$, $R^2 = R^3 = CH_3$
theofylline, $R^1 = R^2 = CH_3$, $R^3 = H$
paraxanthine, $R^1 = R^3 = CH_3$, $R^2 = H$
heteroxanthine, $R^1 = R^2 = H$, $R^3 = CH_3$

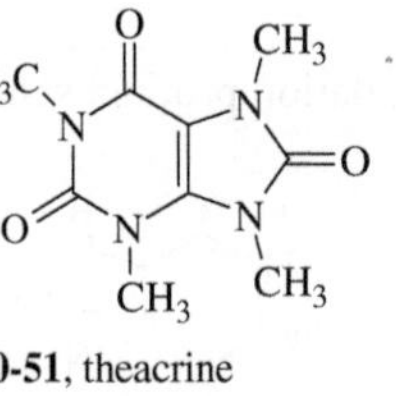

10-51, theacrine

Tea leaves (varieties of *C. sinensis*, Theaceae) contain about 2% caffeine and less than 0.2% theobromine per d.w. The mixture of tea alkaloids was previously called **theine**. A normal cup of tea contains about one-half to one-third of the caffeine found in a cup of coffee of the same size.

Leaves of the South American shrub yerba mate or mate plant (*Ilex paraguariensis*, Aquifoliaceae) contain 0.7–2.7% of caffeine per d.w, between 0.3 and 0.9% of theobromine, and traces of theophylline. Mate aqueous extracts have a caffeine content of 61–245 mg/l and theobromine content of 5.2–15.6 mg/l. The mixture of mate alkaloids was previously called **mateine**.

The total alkaloid content of cocoa beans (*Theobroma cacao*, Sterculiaceae) is in the range of 0.7–3.2% of d.w. The main alkaloid is theobromine, the content of which is 0.6–3.1%, whilst the content of caffeine ranges from 0.02 to 0.5%. Dark chocolate contains 0.3–0.7% of theobromine and 0.02–0.03% of caffeine, whilst milk chocolate contains 0.1–0.4% of theobromine and 0.01–0.02% of caffeine. Chocolate drinks contain theobromine at a level of 260–440 mg/l and caffeine at 10–12.5 mg/l.

In cola drinks (such as Coca-Cola and Pepsi), part of the caffeine usually comes from the nuts of some cola species (*Cola acuminata* and *C. nitida*, Malvaceae), whose caffeine content is 1.5–2.5%. The rest is caffeine from other sources (mainly the production of decaffeinated coffee). The caffeine content in soft drinks is generally in the range of 50–250 mg/l.

Caffeine occurs at levels of 2.5–7.5% in guarana nuts, seeds of *Paullinia cupana* (Sapindaceae) native to the tropics of South America. These nuts are roasted to a mass that tastes like chocolate, which serves in the preparation of a refreshing drink. The mixture of guarana alkaloids was previously called **guaranine**.

10.3.6.10.2 Reactions and changes

Methylxanthines are very stable compounds that, except in the case of non-enzymatic browning reactions during the fermentation of tea leaves (see Section 9.12.5.4) and cocoa beans, are virtually stable during the storage and technological processing of raw materials. In the manufacture of green and black tea, dimethylxanthines (including theophylline) and other purines result as products of caffeine catabolism.

When roasting coffee, the caffeine content remains virtually unchanged. Trigonelline (see Section 5.8.2) that accompanies coffee alkaloids decomposes to nicotinic acid and volatile pyridines, so the trigonelline and caffeine content ratio is used as an indicator of coffee roasting intensity.

10.3.6.10.3 Biological effects

Caffeine is classified as a gustatory and stimulating substance and is used as a food additive. Formerly, its concentration in soft drinks was limited to a highest allowable amount of 250 mg/l, or 320 mg/l in energy drinks. The European legislation states that if a beverage intended for consumption contains caffeine in a proportion in excess of 150 mg/l, the product must feature the message, 'High caffeine content' on the label. In a small daily dose (<3 mg/kg), caffeine acts as a stimulant of the central nervous system and as a diuretic agent. High doses, however, have different neuroendocrine effects, and very high doses are reportedly teratogenic. Theobromine and theophylline exhibit weaker stimulatory effects than caffeine and may cause abnormalities of spermatogenic cells. Theobromine is toxic to some animals, such as dogs, which metabolise it more slowly than humans. Theophylline is used therapeutically in respiratory diseases, such as chronic inflammatory disorder of the bronchial tube (bronchial asthma). In sensory studies, some flavanones and amides of aromatic amines with hydroxylated benzoic acids could significantly decrease the bitter taste of caffeine. Theacrine shows anti-inflammatory and analgesic effects.

10.3.6.11 Steroid glycoalkaloids

Steroid glycoalkaloids (also called steroid alkamines) occur in dicotyledonous plants of the families Solanaceae and Liliaceae and of the subfamily Asclepiadoideae in the Apocynaceae. The fully saturated core of steroid alkaloids is derived from C_{27} hydrocarbon 5α-cholestane, which is composed of three cyclohexane rings (designated as rings A, B, and C) and one cyclopentane ring (ring D) with a C_8 side chain attached to C-17 in (20R)-configuration (see Section 3.7.4.1). Only plants of the genus *Solanum* of the nightshade family, which includes potatoes (*S. tuberosum*), eggplants (*S. melongena*), and tomatoes (*S. lycopersicum*), have practical importance in foods.

10.3.6.11.1 Structure, nomenclature, and occurrence

In more than 300 species of plants of the nightshade family, there are at least 90 different steroid glycoalkaloids, divided into five structural types according to the structure of the aglycone (most occur as glycosides). Only glycoalkaloids, the aglycones of which are 3β-hydroxy derivatives of solanidane (solanidanes **10-52**) or spirosolane (spirosolanes or spiroaminoacetals, **10-53**), are found in plant foods used for human nutrition.

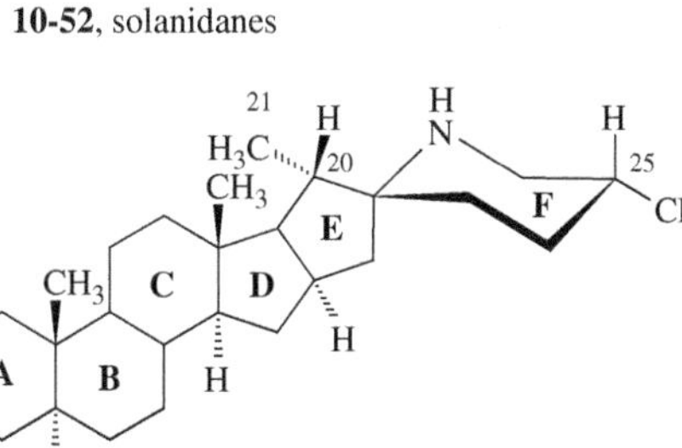

10-52, solanidanes

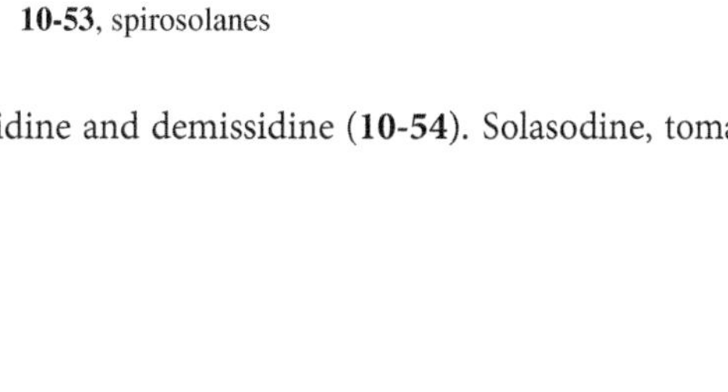

10-53, spirosolanes

Aglycones Solanidane derivatives are solanidine and demissidine (**10-54**). Solasodine, tomatidenol, and tomatidine (**10-55**) are derived from spirosolane.

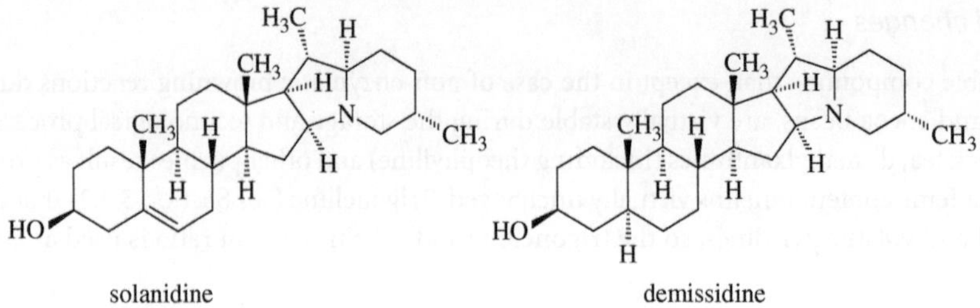

solanidine

demissidine

10-53, solanidane derivatives

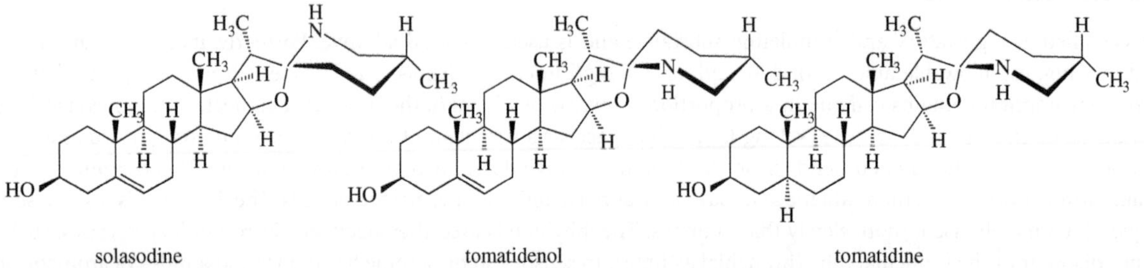

solasodine

tomatidenol

tomatidine

10-55, spirosolane derivatives

Glycosides The main compounds in plants are glycosides of sterols, which are accompanied by small amounts of free aglycones. Sugars (linear and branched tetra-, tri-, di-, and monosaccharides) are bound to aglycones via the hydroxyl group at C-3.

Solanidine (solanid-5-en-3β-ol) is the aglycone of two major potato glycoalkaloids known as α-solanine (**10-56**) and α-chaconine (**10-57**). Their mixture was previously called solanine. These glycoalkaloids represent about 95% of the potato glycoalkaloids. Sugar bound in α-solanine is β-solatriose; that bound in α-chaconine is β-chacotriose. Besides α-solanine and α-chaconine, cultural potato varieties contain some minority alkaloids, such as β-solanines and β-chaconines, γ-solanine, γ-chaconine, α-solamarine, and β-solamarine (Table 10.10).

Table 10.10 Potato glycoalkaloids.

Glycoalkaloid	Aglycone	Saccharide	Abbreviated notation
α-Solanine	Solanidine	β-Solatriose	β-D-Glcp-(1 → 3)-[α-L-Rhap-(1 → 2)]-β-D-Galp
β₁-Solanine	Solanidine	β-Neohesperidose	α-L-Rhap-(1 → 2)-β-D-Galp
β₂-Solanine	Solanidine	β-Solabiose	β-D-Glcp-(1 → 3)-β-D-Galp
γ-Solanine	Solanidine	β-D-Galactose	β-D-Galp
α-Chaconine	Solanidine	β-Chacotriose	α-L-Rhap-(1 → 4)-[α-L-Rhap-(1 → 2)]-β-D-Glcp
β₁-Chaconine	Solanidine	β-Neohesperidose	α-L-Rhap-(1 → 2)-β-D-Glcp
β₂-Chaconine	Solanidine	β-Chacobiose	α-L-Rhap-(1 → 4)-β-D-Glcp
γ-Chaconine	Solanidine	β-D-Glucose	β-D-Glcp
α-Solasonine	Solasodine	β-Solatriose	β-D-Glcp-(1 → 3)-[α-L-Rhap-(1 → 2)]-β-D-Galp
α-Solamargine	Solasodine	β-Chacotriose	α-L-Rhap-(1 → 4)-[α-L-Rhap-(1 → 2)]-β-D-Glcp
α-Solamarine	Tomatidenol	β-Solatriose	β-D-Glcp-(1 → 3)-[α-L-Rhap-(1 → 2)]-β-D-Galp
β-Solamarine	Tomatidenol	β-Chacotriose	α-L-Rhap-(1 → 4)-[α-L-Rhap-(1 → 2)]-β-D-Glcp
α-Tomatine	Tomatidine	β-Lycotetraose	β-D-Xylp-(1 → 3)-[β-D-Glcp-(1 → 2)]-β-D-Glcp-(1 → 4)-β-D-Galp
Demissine	Demissidine	β-Lycotetraose	β-D-Xylp-(1 → 3)-[β-D-Glcp-(1 → 2)]-β-D-Glcp-(1 → 4)-β-D-Galp
Commersonine	Demissidine	β-Commertetraose	β-D-Glcp-(1 → 3)-[β-D-Glcp-(1 → 2)]-β-D-Glcp-(1 → 4)-β-D-Galp
Leptinine I	Leptinidine	β-Chacotriose	α-L-Rhap-(1 → 4)-[α-L-Rhap-(1 → 2)]-β-D-Glcp
Leptinine II	Leptinidine	β-Solatriose	β-D-Glcp-(1 → 3)-[α-L-Rhap-(1 → 2)]-β-D-Galp
Leptine I	23-Acetylleptinidine	β-Chacotriose	α-L-Rhap-(1 → 4)-[α-L-Rhap-(1 → 2)]-β-D-Glcp
Leptine II	23-Acetylleptinidine	β-Solatriose	β-D-Glcp-(1 → 3)-[α-L-Rhap-(1 → 2)]-β-D-Galp

In the wild potato species growing from the United States to southern Chile, there occur a number of other glycosides with different aglycones (e.g. leptines, leptinines, commersonine, and demissine).

Spirosolanes, namely solasodine, the (25R)-stereoisomer, and tomatidenol, the (25S)-stereoisomer, are found in wild potato species and varieties obtained by cross-breeding them, such as *S. berthaultii* and *S. vernei*. They mutually differ only in the position of the heterocyclic nitrogen in ring F. Solasodine occurs as glycoside α-solasonine containing trisaccharide β-solatriose, which is also bound in α-solanine. The same aglycone is in α-solamargine containing β-chacotriose, which is also bound in α-chaconine. Both glycosides, α-solasonine and α-solamargine, are found, for example, in wild potato species (*S. berthaultii* and *S. vernei*) and eggplants (*S. melanogena*).

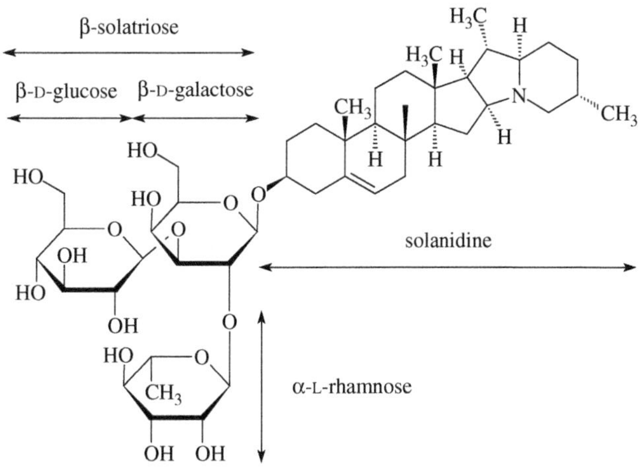

10-56, α-solanine

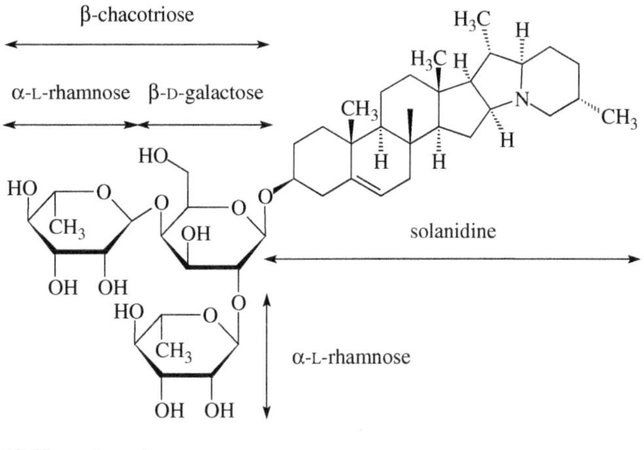

10-57, α-chaconine

The (25S)-stereoisomer tomatidenol (**10-55**) is present in potatoes in a steroid glycoside α-solamarine containing bound solatriose, or as β-solamarine, which is coupled with β-chacotriose. Most potato varieties containing solamarines come from *S. demissum* species, which also contain (only in the leaves, not in tubers) alkaloids known as leptines and leptinines. The aglycone of leptinines I and II is 23-hydroxysolanidine, called leptinidine, whilst the aglycone of leptines I and II is leptidine (23-acetylleptinidine, **10-58**). Leptinine I and leptine I contain β-chacotriose, whilst leptinine II and leptine II contain β-solatriose (Table 10.10).

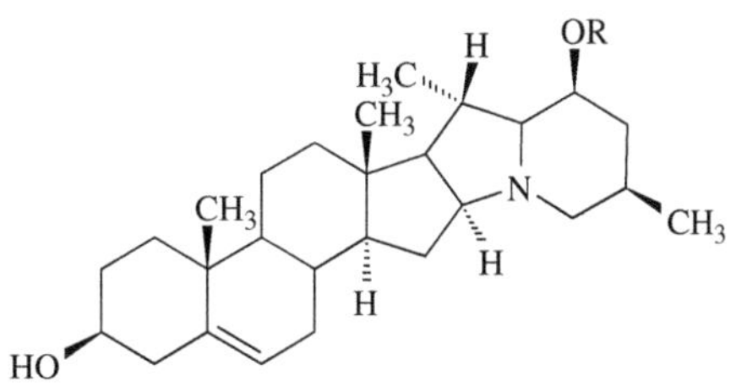

10-58, leptinidine (23-hydroxysolanidine), R = H
leptidine (23-acetylleptinidine), R = COCH₃

Table 10.11 Glycoalkaloid distribution and content in potatoes.

Part of tuber	Content (mg/kg f.w.)
Unpeeled	75
Peel (2–3% of tuber weight)	300–600
Peel (10–15% of tuber weight)	150–300
Peel with germ (3 mm)	300–500
Peeled	12–50

The β-glycoside of tomatidenol, called dehydrotomatine, accompanies the main glycoside α-tomatine (lycopersicin) in tomatoes. Both contain bound lycotetraose. The aglycone of the main tomato glycoalkaloid α-tomatine (**10-59**) is the (25*S*)-stereoisomer tomatidine (**10-55**), also known as 5,6-dihydrotomatidenol. α-Tomatine also occurs in small amounts in some potato species and other species of the genus *Solanum*. In wild potatoes (*S. acaule*), its content ranges from to 74 to 497 mg/kg (f.w.).

The distribution of glycoalkaloids differs in the different parts of a plant. α-Solanine and α-chaconine occur in roughly equal amounts. The largest amount of solanine in potato is present in the flowers (5000 mg/kg) and germ (1950–4360 mg/kg). The highest concentrations in tubers are in the surface layers, decreasing towards the centre (Table 10.11). Higher amounts are found in smaller tubers. Usually, the amount of solanine does not exceed 200 mg/kg in potato tubers, except in the wild species. Potatoes containing more than 140 mg/kg of glycoalkaloids exhibit bitter taste, whilst those containing more than 220 mg/kg have a hot taste. The amount of alkaloids is strongly dependent on soil and climatic conditions. In tubers exposed to light and in injured tubers, alkaloid biosynthesis increases by up to 400%. The stimulating effect mainly comes from light of shorter wavelengths; that of longer wavelengths primarily stimulates the biosynthesis of chlorophyll.

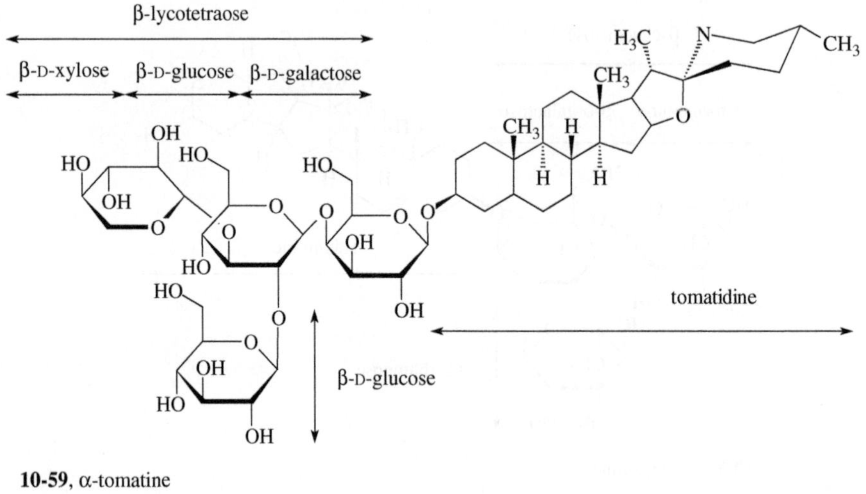

10-59, α-tomatine

Tomatine is practically the only glycoalkaloid of tomatoes. It is accompanied by glycoalkaloids containing as sugar components the products of partial hydrolysis of lycotetraose (β$_1$-tomatine with trisaccharide produced by hydrolysis of xylose from lycotetraose, β$_2$-tomatine with trisaccharide without glucose, and γ-tomatine with disaccharide without xylose and glucose). Tomatine is present in all parts of the plant, with the highest amount located in the leaves, flowers, and small green fruits. The tomatine content decreases during fruit maturation and is very small in mature red fruits (Table 10.12).

The amount of α-solasonine and α-solamargine in various species, hybrids, and varieties of eggplant ranges from 36 to 691 mg/kg dry matter and from 140 to 7700 mg/kg dry matter, respectively, with current eggplant genotypes in the range of 92–125 and 229–299 mg/kg dry matter, respectively.

10.3.6.11.2 Reactions and changes

Alkaloids in potatoes are mainly removed by peeling (60–90%), but they are relatively stable during technological and culinary processing. The degradation of alkaloids does not occur even during the production of potato crisps and French fries, but their relative content increases due to loss of water during frying. When cooking potatoes, some loss of alkaloids occurs due to leaching into water. The addition of 0.3% of acetic acid reduces the solanine content in cooked potatoes by about 84%.

Table 10.12 Tomatine content in tomatoes.

State of ripeness	Days after flowering	Content (mg/kg f.w.)	
		Tomatine	Dehydrotomatine
Small and green	10	795	60
Large and green	20	49	2.6
Pink	30	20	trace
Light red	40	1.5	0
Red and ripe	50	3.7	0

Source: Kozukue et al. (2004), table 1. Reproduced by permission of the American Chemical Society.

Opinions on changes of alkaloids during potato storage vary. Some studies have shown an increase in alkaloid content at lower temperatures. Higher temperature and lower relative humidity are the optimal conditions for preventing the formation of glycoalkaloids during potato storage.

The activity of hydrolases or hydrolysis in acidic media leads to the hydrolysis of glycoalkaloids, but the released aglycones are practically stable. For example, α-chaconine in germinating potatoes is gradually hydrolysed to β$_2$-chaconine, but the α-solanine content is unchanged. Partial acid hydrolysis of α-chaconine and α-solanine leads gradually to approximately equimolar quantities of β$_1$-glycosides, β$_2$-glycosides, and γ-glycosides, and the final product is aglycone solanidine. The same products may also arise in the digestive tract. Upon heating in mineral acids, solanidine partially dehydrates to solanthrene (**10-60**) and other minor products, such as the oxidation product solanid-4-en-3-one (**10-60**). These reactions can be expected to a lesser extent during thermal processing of potatoes.

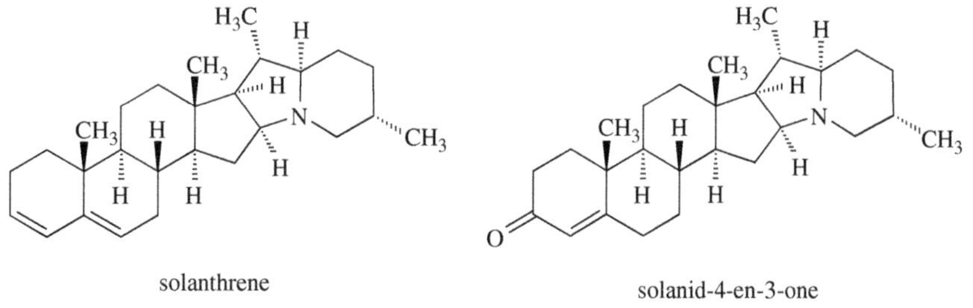

solanthrene   solanid-4-en-3-one

10-60, solanidine degradation products

Tomatine is relatively stable under normal storage and processing conditions. In acid media, partial hydrolysis of lycotetraose can be expected. Freezing of unripe tomatoes (unpeeled, green-yellow tomatoes) results in approximately 8% loss of tomatine, whilst freezing of peeled tomatoes leads to an 18% loss of alkaloids, which may be attributed to tissue damage and the action of hydrolytic enzymes.

10.3.6.11.3 Biological effects

The levels of the cholinesterase inhibitors α-solanine and α-chaconine normally found in potatoes (20–100 mg/kg) are not of toxicological concern. Typical manifestations of higher levels (so-called solanine poisoning) are vomiting, diarrhoea, stomach cramps, headaches, and dizziness. α-Solanine toxicity is attributable either to the inhibition of cholinesterase or to damage to the digestive-tract membranes and some organs. High doses of α-chaconine have teratogenic effects. In some countries, and formerly also in the European Union, permissible concentrations of α-solanine and α-chaconine are set to 200 mg/kg.

Tomatine has antifungal and teratogenic effects. Toxic effects are often associated with the ability of 3β-hydroxysterols to bind in membranes, which destabilises the lipid bilayer. For tomato alkaloids (α-solanine, α-chaconine, and α-tomatine), the same maximum amount has been set as for potato alkaloids.

10.3.6.12 Capsaicinoids

Protoalkaloids capsaicinoids are amides derived from vanillylamine (**10-61**), whose occurrence is restricted to species of hot pepper (*Capsicum* spp.) of the Solanaceae family (see Section 8.3.6.1). The main representatives of capsaicinoids are vanillylamides (*N*-vanillylacylamides) derived from C$_8$–C$_{11}$ branched- and straight-chain *trans*-monoenic and saturated fatty acids (**10-62**).

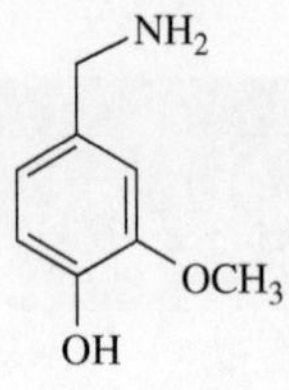

10-61, vanillylamine

10.3.6.12.1 Structure, nomenclature, and occurrence

The main components causing a burning sensation when in contact with mucous membranes are capsaicin, (*E*)-8-methyl-*N*-vanillylnon-6-enamide, and dihydrocapsaicin, 8-methyl-*N*-vanillylnonanamide, which constitute about 90% of total capsaicinoids. These are accompanied by minority alkaloids such as nordihydrocapsaicin, nordihydrocapsaicin II, homodihydrocapsaicin, homodihydrocapsaicin II, homocapsaicin, homocapsaicin II, and some related compounds, including the pungent (−)-capsaicinol (**10-63**), which also acts as an effective antioxidant.

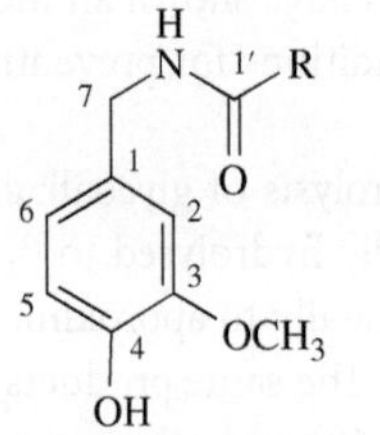

10-62, capsaicinoids

capsaicin, R = $[CH_2]_4CH=CHCH(CH_3)_2$
dihydrocapsaicin, R = $[CH_2]_6CH(CH_3)_2$
nordihydrocapsaicin, R = $[CH_2]_5CH(CH_3)_2$
nordihydrocapsaicin II, R = $[CH_2]_4CH(CH_3)CH_2CH_3$
homodihydrocapsaicin, R = $[CH_2]_7CH(CH_3)_2$
homodihydrocapsaicin II, R = $[CH_2]_6CH(CH_3)CH_2CH_3$
homocapsaicin, R = $[CH_2]_4CH=CHCH_2CH(CH_3)_2$
homocapsaicin II, R = $[CH_2]_4CH=CHCH(CH_3)CH_2CH_3$

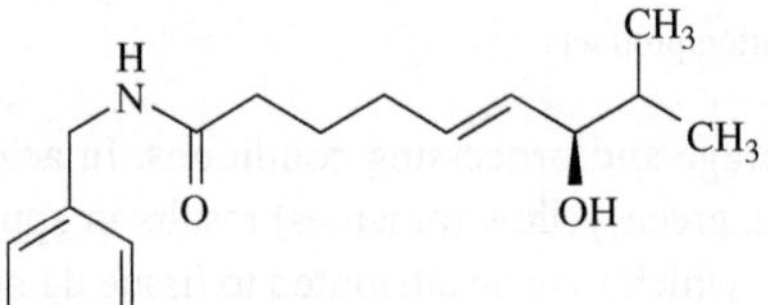

10-63, capsaicinol

The content of capsaicinoids in fruits of *Capsicum* species depends on the variety, age, maturity, season, and agronomic conditions. Their amount in large bell peppers (*C. annuum*) is usually very low, becoming higher in medium-sized fruits (such as Tabasco, *C. frutescens*) and highest in small chilli peppers (*C. frutescens*). Most alkaloids are found in the flesh, with lower concentrations in the seeds and skin. Lower amounts of capsaicinoids occur in the young green fruits. The alkaloid content increases during the growing and ripening of fruits, reaches a maximum shortly before harvest, and then decreases slightly. The amount of capsaicinoids in sweeter pepper varieties cultivated in Europe varies in the range of 0.001–0.01%. Chilli peppers typically contain about 0.2–1.5% of alkaloids. For example, of the total amount of capsaicinoids in chilli peppers (0.4%), about 49% are represented by capsaicin, 44% by dihydrocapsaicin, 6% by nordihydrocapsaicin, 1% by homodihydrocapsaicin, and 0.3% by homocapsaicin. Capsaicin and dihydrocapsaicin can also be found in the form of β-D-glucopyranosides (**10-64**). For example, the concentration of capsaicin β-D-glucopyranoside in the fruits of hot *C. annuum* ranged from 0.11 to 2.65 mg/kg, whilst that of capsaicin was 0.03–0.48% (f.w.). Capsaicinoids were also found to be present in vegetative organs of *Capsicum* plants, such as stems and leaves. In this case, the proportion of individual capsaicinoids was different to that in fruits, and dihydrocapsaicin was the more abundant compound.

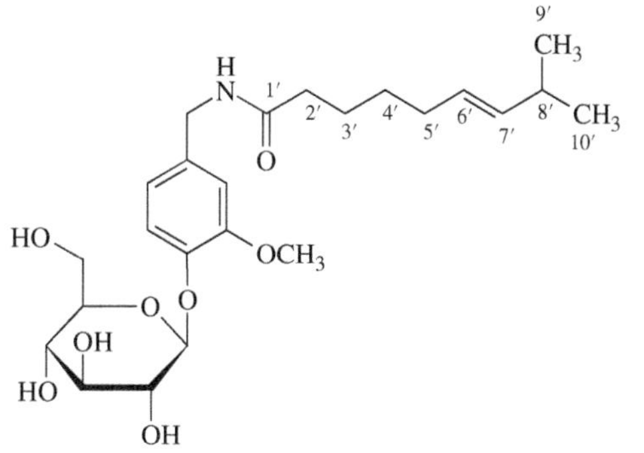

10-64, capsaicin β-D-glucopyranoside

The burning sensation caused by capsaicinoids in the mouth and throat is their characteristic property. Vanillylamides with a longer or shorter acyl chain than capsaicin have less pungency. There are not particularly large differences between individual capsaicinoids (see Section 8.3.6.1). At a concentration of about 10 mg/kg, capsaicin causes burning and pungency, which is perceptible even at a concentration of 0.1 mg/kg. The burning effect is amplified by sucrose and reduced by sodium chloride and solutions with higher viscosity (such as yoghurt).

Capsicum fruits also contain capsaicinoids with a long-chain acyl moiety. In *Capsicum* oleoresins, vanillylamides derived from palmitic acid (palvanil), stearic acid (stevanil), elaidic acid (olvanil), and (9*E*,12*E*)-octadeca-9,12-dienoic acid (livanil) were identified, and the existence of myristic acid (myrvanil) and (9*E*,12*E*,15*E*)-octadeca-9,12,15-trienoic acid (linvanil) derivatives was suggested. The content ratios of the total long-chain vanillylamides (except myrvanil) versus capsaicin in the oleoresins were significantly larger (0.1–41%) than in fresh fruits (<0.01%).

Recent studies have revealed that many kinds of pungent and non-pungent *Capsicum* cultivars contain a novel group of almost non-pungent capsaicinoid-like substances named **capsinoids**. The fundamental structure of capsinoids is a fatty acid ester with vanillyl alcohol, whereas in capsaicinoids, a fatty acid amide is linked to vanillylamine. A new class of fatty acid ester with coniferyl alcohol, namely the **capsiconinoids**, has recently been found in several *Capsicum* cultivars. Examples of identified compounds (**10-65**) are ester of vanillyl alcohol with (*E*)-8-methylnon-6-enoic acid (capsiate) and ester of coniferyl alcohol with (*E*)-8-methylnon-6-enoic acid (capsiconiate). The highest content of capsiconinoids was found in *C. baccatum* var. *praetermissum* (3314 mg/kg d.w.), whilst that in *C. annuum* and *C. frutescens* varieties ranged from 0 to 239 and from 0 to 39 mg/kg (d.w.), respectively.

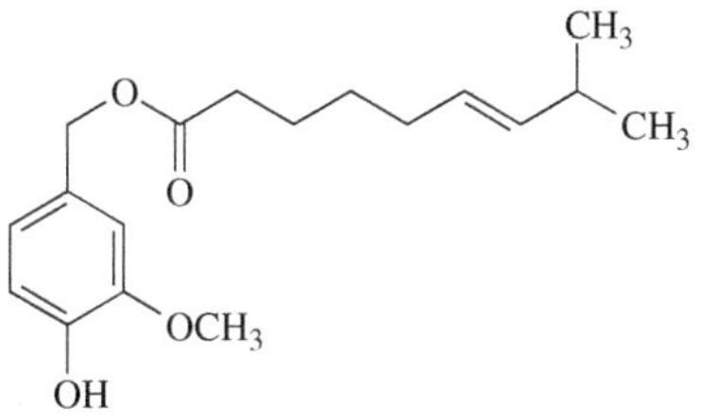

capsiate

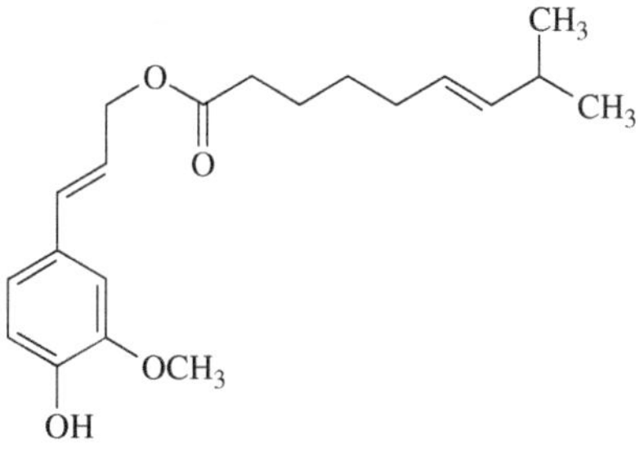

capsiconiate

10-65, capsiconinoids

10.3.6.12.2 Reactions and changes

Boiling of hot peppers results in a partial leaching of capsaicinoids into water. Their concentration also decreases during other cooking procedures, but in some cases it may actually increase, probably due to hydrolysis of capsaicinoid glycosides (Table 10.13). Peppers dried in the sun likewise contain lower amount of capsaicinoids. Hydrolysis in acidic and alkaline media yields vanillylamine and corresponding fatty acids (their salts). During culinary procedures, which result in tissue disruption, the released enzymes partly hydrolyse capsaicinoids, which are transformed into the higher-molecular-weight compounds. The action of endogenous peroxidase results, for example, in the formation of 5,5′-dicapsaicin, dimeric 4-O-5′-dicapsaicin ether (**10-66**), higher oligomers, and copolymers with proteins. In special beverages flavoured with pepper, 5,5′-dicapsaicin may arise by photooxidation in the presence of riboflavin and ascorbic acid (where redox system riboflavin/1,5-dihydroriboflavin acts as a photocatalyst).

Table 10.13 Influence of technological processing on the content of capsaicinoids in jalapeño peppers.

Pepper	Content (mg/kg)			
	Capsaicin	Dihydrocapsaicin	Nordihydrocapsaicin	Total
Raw	7300	6300	1200	14 800
Boiled (100 °C, 10 minutes)	8600	7700	1300	17 000
Blanched and sterilised (100 °C, 50 minutes)	4700	4500	700	9900
Blanched (100 °C, 3 minutes) and frozen (−18 °C)	4000	3700	600	8300

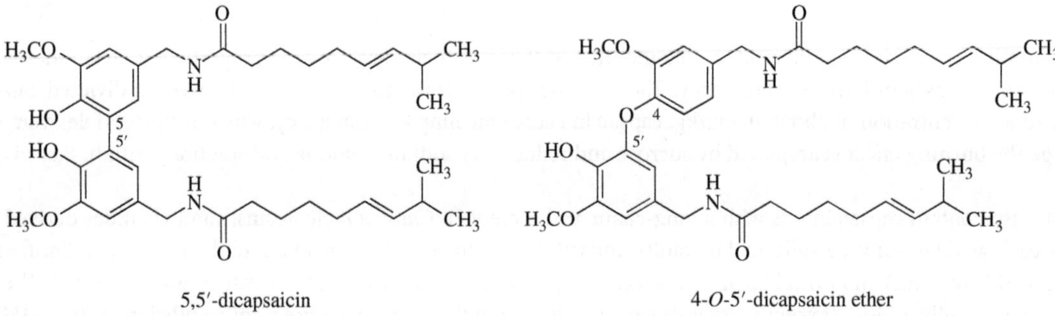

5,5′-dicapsaicin

4-*O*-5′-dicapsaicin ether

10-66, products of capsaicin oxidation

Thermal decomposition of capsaicin yields capsaicin dimer (*E,E*)-*N*-vanillyl-di(8-methylnon-6-en)imide (**10-67**), various amides, acids, hydrocarbons, and phenols. Roasting of peppers in oil may produce fatty acid amides. Reaction with oleic acid yields (*Z*)-*N*-vanillyloctadec-9-enamide, whilst reaction of oleic acid with ammonia (formed from capsaicin) gives rise to (*Z*)-octadec-9-enamide (oleic acid amide, oleamide).

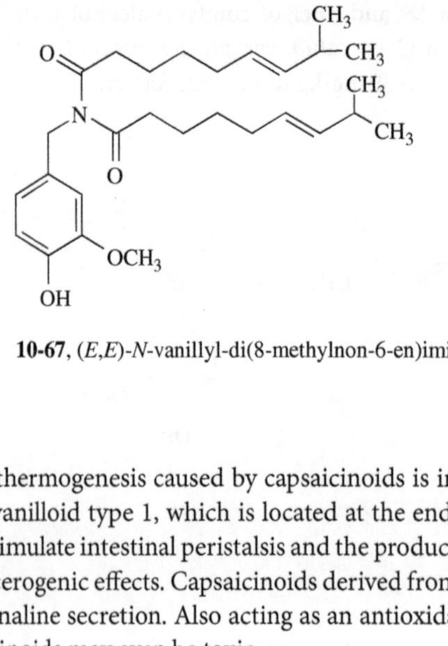

10-67, (*E,E*)-*N*-vanillyl-di(8-methylnon-6-en)imide

10.3.6.12.3 Biological effects

The burning sensation and enhancement of thermogenesis caused by capsaicinoids is induced by the direct activation of a non-selective cation channel, transient receptor potential vanilloid type 1, which is located at the ends of sensory nerves. Capsiconinoids have agonist activity for this receptor. Capsaicinoids also stimulate intestinal peristalsis and the production of bile, suppress fat accumulation, and exhibit weak antimicrobial, antioxidant, and anticancerogenic effects. Capsaicinoids derived from long-chain fatty acids are anti-inflammatory and antinociceptive substances and enhance adrenaline secretion. Also acting as an antioxidant is (−)-capsaicinol, which has similar effects to α-tocopherol. High concentrations of capsaicinoids may even be toxic.

Capsaicin reduces the sensitivity of nerve cells (nocireceptors) for pain and is therefore used in medicine (in creams and sprays) against infectious diseases of the skin and mucous membranes, such as the viral infections *herpes simplex* and *psoriasis*, and seems to be active in the treatment of lung and prostate cancer.

10.3.7 Mushroom toxins

Mushrooms have inspired the cuisines of many cultures (notably Chinese, Japanese, and European) for centuries, and many species have been used in folk medicine for thousands of years. Roughly 700–1000 of the 14 000 described species of mushrooms are believed to have immuno-pharmacological properties. Only about 50 of these have been used medicinally. Medicinal mushrooms are reported to exhibit a wide variety of beneficial effects. The main attention in recent years has focused on immunological and anticancer properties, but mushrooms also show antioxidative, antihypertensive, anti-inflammatory, antidiabetic, antiviral, and cholesterol-lowering activities. The most popular and most studied medicinal mushrooms are reishi or lingzhi (*Ganoderma lucidum*, Ganodermataceae), hen of the woods (*Grifola frondosa*, Meripilaceae), and turkey tail (*Trametes versicolor*, Polyporaceae), mushrooms of the order Polyporales. The range of medically active compounds that have been identified include various triterpenic compounds similar to steroid hormones, polysaccharides (such as ß-glucans), and alkaloids.

It is also true that approximately 70% of intoxications caused by natural substances can be attributed to mushrooms. Mushroom poisoning refers to the harmful effects of ingestion of toxic substances present in some higher fungi, which are most commonly called mushroom toxins. The term 'mycotoxin' is usually reserved for the toxic products produced by moulds that readily colonise crops (see Section 12.3.1).[3] From the culinary point of view, mushrooms are conventionally divided into:

- edible;

- conditionally edible;

- inedible;

- poisonous.

Just 200–300 species have been clearly established to be safe to eat. Conditionally edible mushrooms are suitable for human consumption if they can be eaten after soaking, cooking, drying, or some other culinary pretratment. Inedible mushrooms cannot be consumed because they have, for example, a bitter taste, a disgusting aroma, or inappropriate consistency. Approximately 50–100 mushroom species are known to be poisonous, of which only about 10 are fatal. However, many people are still hospitalised after eating mushrooms every year, a number of them fatally. Mushroom poisoning is more common in Europe than in America. The only way to tell if a particular fungus is edible is to correctly identify its species. There are no short cuts.

Poisonous mushrooms are mostly divided according to their toxic effects, which are related to chemical composition. Poisonous mushrooms can cause several types of poisoning:

- **hepatonephrotoxic syndrome** (poisoning belongs to the most difficult poisonings with the greatest mortality and leads to liver and kidney damage);

- **nephrotoxic syndrome** (poisoning leads to kidney damage);

- **muscarinic syndrome** (poisoning acts on the neurotransmitter acetylcholine-type receptors, known as muscarinic receptors, and is manifested by various symptoms);

- **gastroenterodyspeptic syndrome** (poisoning irritates the gastrointestinal tract mucosa and causes dyspepsia – impaired digestion);

- **disulfirame (antabus) syndrome** (toxicity depends primarily on the amount of alcohol consumed);

- **hallucinogenic syndrome** (toxins act selectively on the central nervous system and cause hallucinations);

- **pseudo poisoning** (according to individual sensitivity, health problems can occur due to inappropriate culinary treatment of mushrooms, inappropriate storage, or food combination).

Toxic substances occurring in mushrooms can be chemically divided into toxic proteins, toxic cyclopeptides, toxic amino acids, toxic amines, hydrazines, alkaloids, and other nitrogenous compounds, and toxic terpenoids. According to the severity, the most dangerous form of mushroom poisoning, accounting for the most mushroom-related deaths, is that caused by cyclopeptides of toxic species of toadstools (*Amanita* spp.), called **phalloidin poisoning**, followed by poisoning due to cortinars (webcaps, *Cortinarius* spp.), called **orellanine poisoning**. The most common form of mushroom poisoning is caused by a wide variety of gastrointestinal irritants. Toxins of many fungi that cause gastrointestinal syndromes are not yet well known. An example may be a poisonous mushroom commonly known as livid entomoma or livid agaric (*Entoloma sinatum*, Entolomataceae), which is responsible for about 10% of all mushroom poisonings in Europe. Poisoning is mainly gastrointestinal in nature, but the identity of toxic substances is unknown, the toxins are likely to be alkaloids.

10.3.7.1 Proteins

10.3.7.1.1 Structure, nomenclature, and occurrence

Toxic protein of false parasol, also known as green-spored parasol (*Chlorophyllum molybdites*, Agaricaceae), is the most common cause of mushroom poisoning in the United States. It has a relative molecular weight of 400 kDa and is composed of several subunits of molecular

---

[3]Mycotoxins are toxic secondary metabolites produced by eukaryotic organisms classified into the domain Eukaryota and kingdom Fungi. Fungi include microorganisms such as yeasts and moulds, as well as the more familiar mushrooms. They can be divided into micromycetes (microfungi) and macromycetes (macrofungi), also known as higher fungi. The term 'mushroom' is most often applied to higher fungi (also known as club fungi) belonging to divisions Basidiomycota and Ascomycota that have a stem (stipe) and a cap (pileus) and produce a large fleshy fruiting body.

weights of 40–60 kDa. Bolaffinin is a toxic protein isolated from spotted bolete (*Xanthoconium affine*, syn. *Boletus affinis*, Boletaceae) containing 234 amino acid residues with one disulfide bridge (molecular weight of 22 kDa). Bolesatin is toxic glycoprotein with a molecular weight of 63 kDa isolated from devil's bolete (Satan's mushroom, *Boletus satanas*, Boletacae). Wood pink gill (*Entoloma rhodopolium*, syn. *Rhodophyllus rhodopolius*, Entoolomataceae) contains a toxic protein with a molecular weight of 40 kDa. Toxic proteins are also present in other mushrooms.

10.3.7.1.2 Reactions and changes

The toxic protein of false parasol is labile when heated and is cleaved by proteolytic enzymes. Bolaffinin loses toxicity after being boiled for 15 minutes. Cooked mushrooms are likely to be less toxic than raw mushrooms. Bolesatin is relatively thermostable and is resistant to hydrolysis by proteolytic enzymes.

10.3.7.1.3 Biological effects

Toxic protein of false parasol is a gastrointestinal irritant that is capable of producing severe symptoms when ingested (**gastroenterodyspeptic syndrome**). Symptoms manifest one to two hours after consumtion and can include nausea, vomitting, diarrhoea, loss of fluids, and hypersensitivity to light and noise, but only rarely do they lead to death. Bolaffinin inhibits protein synthesis and damages the liver. Bolesatin manifests similar effects to lectins (e.g. cytotoxicity, inhibition of protein synthesis). The symptom-free period lasts between half an hour and four hours. The first symptoms are nausea and vomiting with pain, and later diarrhoea. Poisoning disappears within 24 hours (**gastroenterodyspeptic syndrome**).

10.3.7.2 Peptides

10.3.7.2.1 Structure, nomenclature, and occurrence

The most important compounds of this group of mushroom toxins are amatoxins (amanita toxins, **10-68**) and phallotoxins (**10-69**), bicyclic oligopeptides whose main cycle is bridged by cysteine with a sulfur-linked indole, which cause **phalloidin poisoning**. They occur in the death cap (*Amanita phalloides*, Amanitaceae). The main toxin is phalloidine (about 100 mg/kg of fresh mushroom), whilst α-amanitine and β-amanitine occur in comparable amounts (about 80 and 50 mg/kg, respectively); other toxins are present in smaller amounts. Amanita toxins are similarly found in several other Amanita species (*A. bisporigera, A. hygroscopia, A. ocreata, A. suballiacea, A. tenuifolia,* and *A. virosa*) and some members of the genera *Galerina* (*G. autumnalis, G. marginata,* and *G. venenata*) and *Lepiota* (*L. brunneoincarnata, L. chlorophyllum, L. helveola,* and *L. josserandii*). A number of other related peptides known as cycloamanides have no biological activity. Similar structures and toxicities to those of phallotoxins are seen in virotoxins (**10-70**) present in the European destroying angel (*A. virosa*).

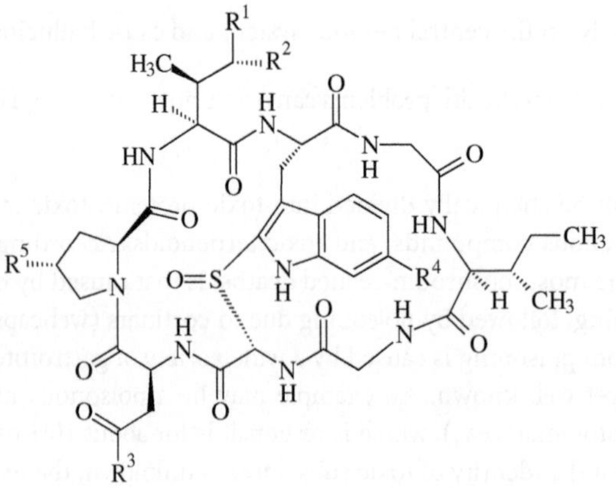

10-68, α-amanitin, $R^1 = CH_2OH$, $R^2 = OH$, $R^3 = NH_2$, $R^4 = OH$, $R^5 = OH$
β-amanitin, $R^1 = CH_2OH$, $R^2 = OH$, $R^3 = OH$, $R^4 = OH$, $R^5 = OH$
γ-amanitin, $R^1 = CH_3$, $R^2 = OH$, $R^3 = NH_2$, $R^4 = OH$, $R^5 = OH$
ε-amanitin, $R^1 = CH_3$, $R^2 = OH$, $R^3 = OH$, $R^4 = OH$, $R^5 = OH$
amanine, $R^1 = CH_2OH$, $R^2 = OH$, $R^3 = OH$, $R^4 = H$, $R^5 = OH$
amaninamide, $R^1 = CH_2OH$, $R^2 = OH$, $R^3 = NH_2$, $R^4 = H$, $R^5 = OH$
amanulline, $R^1 = CH_3$, $R^2 = H$, $R^3 = NH_2$, $R^4 = OH$, $R^5 = OH$
amanullic acid, $R^1 = CH_3$, $R^2 = H$, $R^3 = OH$, $R^4 = OH$, $R^5 = OH$
proamanulline, $R^1 = CH_3$, $R^2 = H$, $R^3 = NH_2$, $R^4 = OH$, $R^5 = H$

Less toxic cyclic peptides are also found in other fungi. Examples are cortinarins A, B, and C (**10-71**), fluorescent decapeptides found in fool's web cap (*Cortinarius orellanus*, Cortinariaceae) and in more that 40 species of the genus *Cortinarius*. However, the main toxins of *C. orellanus* are not peptides but the bipyridyl derivative orellanine.

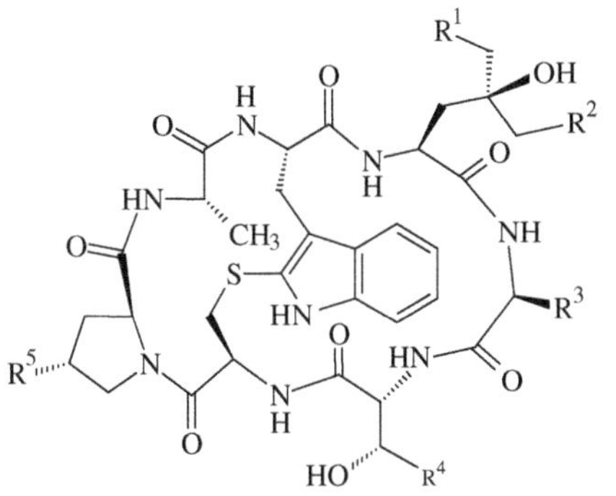

10-69, phalloidin, R^1 = OH, R^2 = H, R^3 = CH, R^4 = CH, R^5 = OH
phalloin, R^1 = H, R^2 = H, R^3 = CH$_3$, R^4 = CH$_3$, R^5 = OH
prophalloin, R^1 = H, R^2 = H, R^3 = CH$_3$, R^4 = CH$_3$, R^5 = H
phallisin, R^1 = OH, R^2 = OH, R^3 = CH$_3$, R^4 = CH$_3$, R^5 = OH
phallacin, R^1 = H, R^2 = H, R^3 = CH(CH$_3$)$_2$, R^4 = COOH, R^5 = OH
phallacidin, R^1 = OH, R^2 = H, R^3 = CH(CH$_3$)$_2$, R^4 = COOH, R^5 = OH
phallasacin, R^1 = OH, R^2 = OH, R^3 = CH(CH$_3$)$_2$, R^4 = COOH, R^5 = OH

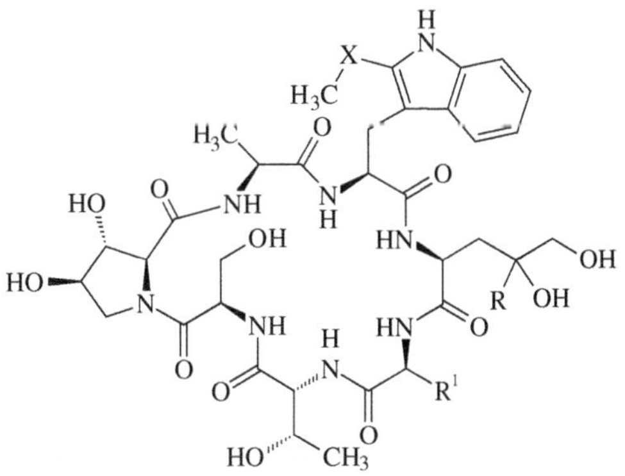

10-70, viroidin, X = SO$_2$, R = CH$_3$, R^1 = CH(CH$_3$)$_2$
deoxyviroidin, X = SO, R = CH$_3$, R^1 = CH(CH$_3$)$_2$
[Ala^1]viroidin, X = SO$_2$, R = CH$_3$, R^1 = CH$_3$ [Ala^1]
deoxyviroidin, X = SO, R = CH$_3$, R^1 = CH$_3$
viroisin, X = SO$_2$, R^1 = CH$_2$OH, R = CH(CH$_3$)$_2$
deoxyviroisin, X = SO, R = CH$_2$OH, R^1 = CH(CH$_3$)$_2$

10.3.7.2.2 Reactions and changes

Amatoxins, phallotoxins, and virotoxins are stable under all normal conditions of storage and processing, including heat treatment, and technological processes do not reduce their high toxicity.

10.3.7.2.3 Biological effects

Amatoxins, of which α-amanitine is present in the highest concentration in the death cap, being responsible for its toxicity (**hepatonephrotoxic syndrome**), are more toxic than the related phallotoxins, which exhibit toxicity only after intravenous administration. They are amongst the most toxic natural substances, as they are inhibitors of important enzymes responsible for the metabolism of the major nutrients and cause damage to the liver and kidneys. The toxicity of amatoxins is caused by the inhibition of RNA polymerase II in eukaryotic cells; that of phallotoxins is caused by selective binding of toxins to the membrane of F-actin and stabilisation of this complex. Death cap poisoning is manifested by effects similar to those of cholera: gagging, vomiting, and diarrhoea about 8–48 hours after ingestion, with other symptoms occurring in the following days. Death occurs in 4–12 days. Poisoning is fatal in 50% of cases.

Cyclic peptides cortinarins A, B, and C are nephrotoxic (**nephrotoxic syndrome**); intoxication manifests by acute or chronic damage to the kidneys 3–14 days after ingestion of mushrooms.

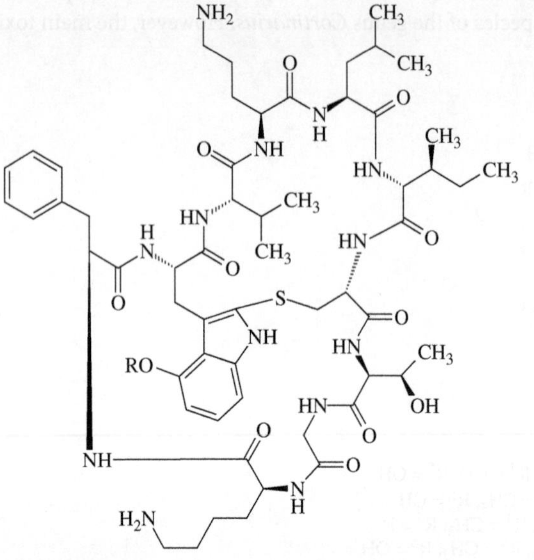

10-71, cortinarin A, R = CH₃
cortinarin B, R = H

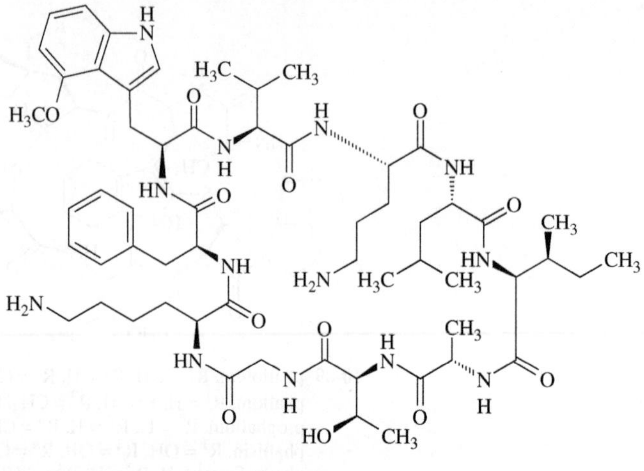

cortinarin C

10.3.7.3 Amino acids

10.3.7.3.1 Structure, nomenclature, and occurrence

Mushrooms contain a number of free amino acids, which cause poisoning manifested by symptoms similar to those of the death cap. For example, (S)-2-amino-pent-4-ynoic acid, also known as L-2-aminopent-4-ynoic acid or L-propargylglycine, (S)-2-aminohexa-4,5-dienoic acid, also called L-2-aminohexa-4,5-dienoic acid, (S)-allylglycine, or L-allylglycine (**10-72**), and related amino acids are found in some other toadstool species, such as the abrupt-bulbed lepidella (*Amanita abrupta*) and European solitary lepidella (*A. echinocephala*, syn. *A. solitaria*), where they occur together with (2S,4Z)-2-amino-5-chloro-6-hydroxyhex-4-enoic acid. In addition to 2-amino-5-chloro-6-hydroxyhex-4-enoic acid, four other chlorinated allylic amino acids have been isolated from various A. species. A number of other C_6 and C_7 unbranched and branched amino acids have been isolated in recent years, reflecting great variability in the metabolism of branched-chain amino acids (valine, leucine, and isoleucine) in different mushroom species.

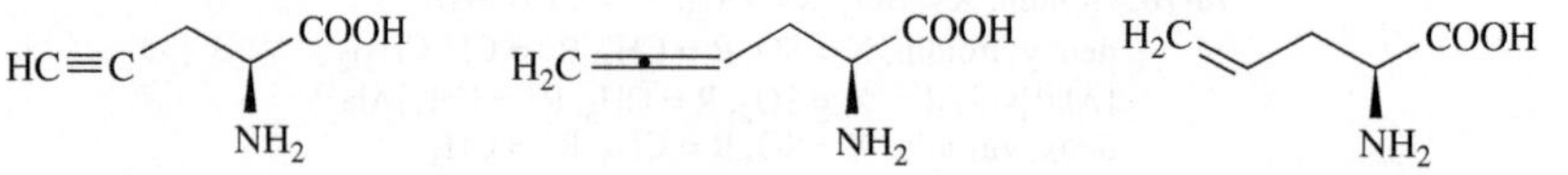

L-propargylglycine   L-2-aminohexa-4,5-dienoic acid   L-allylglycine

10-72, toxic amino acids in mushrooms

A number of other amino acids structurally related to L-glutamic acid exhibit neurotoxic (psychotropic) and other toxic effects. Glutamic acid acts as an excitatory neurotransmitter in the central nervous system of mammals. Glutamate receptors are involved in higher neural functions (such as memory and learning) and disorders (such as epilepsy). For example, the main toxic agent of the fly agaric (*Amanita muscaria*), which derives its name from the ability of its juice to stun and sometimes kill house flies (*Musca domestica*), as well as of panther cap (*A. pantherina*) and European pine cone lepidella (*A. strobiliformis*), is (S)-2-amino-(3-hydroxy-5-isoxazolyl)acetic acid, known as ibotenic acid and formerly as pantherin, agarin, α-toxin, and premuscimol (see **2-52**); enol and keto tautomers of ibotenic acid occur in the ratio of 96 : 4. Also present is a product of ibotenic acid decarboxylation, 5-aminomethyl-3-hydroxyisoxazol, known as muscimol (or pyroibotenic acid or β-toxin). The ibotenic acid content in hats of toadstools is approximately 0.02%. Muskazone (2-amino-2,3-dihydro-2-oxo-5-isooxazoleacetic acid), which shows antibacterial and antifungal activities, is probably formed by rearrangement of ibotenic acid. *A. strobiliformis* and *Tricholoma muscarium* (Tricholomataceae) contain another important neurotoxic amino acid, a dihydro analogue of ibotenic acid (see **2-53**), 2-amino-2-(3-oxoisoxazolidin-5-yl)acetic acid, known as tricholomic acid (Figure 10.10).

Other neurotoxins are acromelic acids (**10-73**). Acromelic acids A and B occur in *Clitocybe acromelalga* (Tricholomataceae), growing in Japan, together with the minority congeners acromelic acids C, D, and E and other related biologically active compounds.

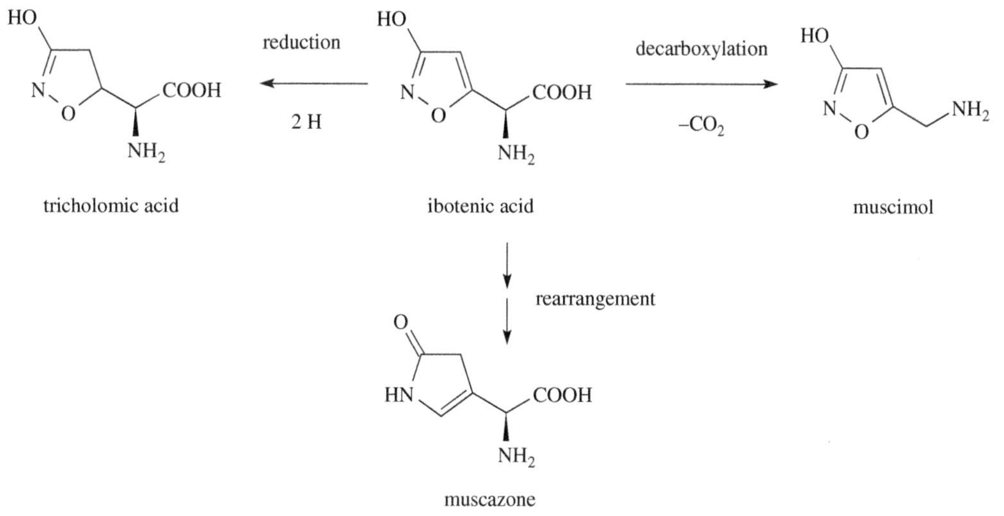

Figure 10.10 Reactions of ibotenic acid.

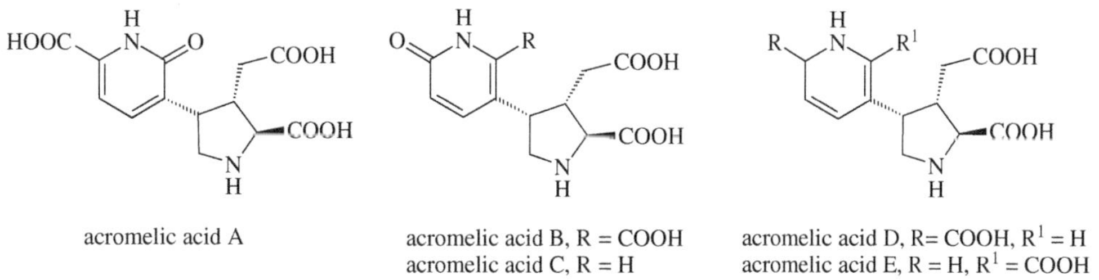

acromelic acid A

acromelic acid B, R = COOH
acromelic acid C, R = H

acromelic acid D, R= COOH, R^1 = H
acromelic acid E, R = H, R^1 = COOH

10-73, acromelic acids

Present in the common ink cap (*Coprinopsis atramentarius*, Psathyrellaceae) and some other mushrooms, such as fat-footed *Clitocybe* (*C. claviceps*, Tricholomataceae), sharp-scaly pholiota (*Pholiota squarrosa*, Strophariaceae), and lurid bolete (*Boletus luridus*, Boletaceae), is the toxic glutamine derivative, N^5-(1-hydroxycyclopropyl)-L-glutamine, known as coprine (**10-74**).

10.3.7.3.2 Reactions and changes

Most toxic amino acids of fungi are stable substances that do not undergo reactions leading to reduction of their toxicity. Decomposition of the original amino acid in some cases yields degradation products that are also toxic, such as the decarboxylation product of ibotenic acid called muscimol (Figure 10.10).

Coprine is a protoxin without intrinsic toxicity but which hydrolyses to the active toxic principle 1-aminocyclopropanol and cyclopropanone hydrate (cyclopropane-1,1-diol, **10-75**).

10.3.7.3.3 Biological effects

Poisoning by propargylglycine and 2-aminohexa-4,5-dienoic acid has similar symptoms to phalloidin poisoning (**hepatonephrotoxic syndrome**). Although not as toxic as the death cap or the destroying angel, the ingestion of the abrupt-bulbed lepidella may cause changes in liver function similar to these species. The poisoning begins with violent vomiting, diarrhoea, and dehydration after a delay of 10–20 hours. The toxic effects of allylglycine are not yet precisely known.

Ibotenic acid is a likely carcinogen. In humans, poisoning is manifested by hallucinations. Ibotenic and tricholomic acids exhibit taste properties similar to those of monosodium glutamate (umami taste, see Section 8.3), but much more intense. Muscimol has similar effects as γ-aminobutyric acid (GABA or 4-aminobutanoic acid), which acts as an inhibitory neurotransmitter of the central nervous system (**hallucinogenic syndrome**). Acromelic acids A and B behave as neurotoxins (**hallucinogenic syndrome**), acting as agonists (activators) of the neurotransmitter L-glutamic acid. Their activity is manifested by special symptoms as they evoke powerful burning and reddish swellings on the legs, which last a month or longer. Deaths caused by poisoning are rare.

Many mushrooms fall into the category of causing only gastrointestinal-specific irritation, distress, abdominal cramping, diarrhoea, nausea, and vomiting. Most of these symptoms are poorly understood. To this category belongs the common ink cap containing coprine, a substance that is toxic only when ingested in combination with alcohol. The degradation products of coprine (**10.74**), 1-aminocyclopropanol and cyclopropane-1,1-diol (**10.75**), inhibit the enzyme aldehyde dehydrogenase, which catalyses conversion of acetaldehyde to acetic acid, resulting in accumulation of acetaldehyde in the blood. Inhibition of aldehyde dehydrogenase produces a clinical syndrome similar to disulfiram (Antabuse) alcohol reaction, manifested by flushing, throbbing in the temples, and usually headache starting 5–30 minutes after ingestion of alcohol (**disulfiram syndrome**).

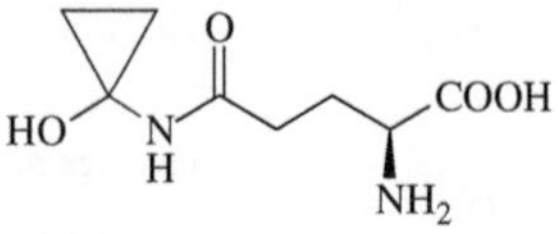

10-74, coprine

10-75, 1-aminocyclopropanol, R = NH_2
cyclopropane-1,1-diol, R = OH

10.3.7.4 Amines, alkaloids, and other nitrogenous compounds

10.3.7.4.1 Structure, nomenclature, and occurrence

Many mycorrhizal fungi of the genera *Clytocybe* (for example in the ivory funnel *C. dealbata* and the fool's funnel *C. rivulosa*, Tricholomataceae) and *Inocybe* (for example in the deadly firecap *I. erubescens*, Inocybaceae) contain, in concentrations up to 1.6%, the toxic amine L-(+)-muscarine, which causes **muscarinic poisoning**. The muscarine molecule has three asymmetric centres, and thus can exist in eight stereoisomers – but only three muscarine isomers, epimuscarine, allomuscarine, and epiallomuscarine (**10-75**), have been found in nature. About 500 times lower quantities of muscarine occur in the fly agaric (*Amanita muscaria*), in addition to the main neurotoxin ibotenic acid.

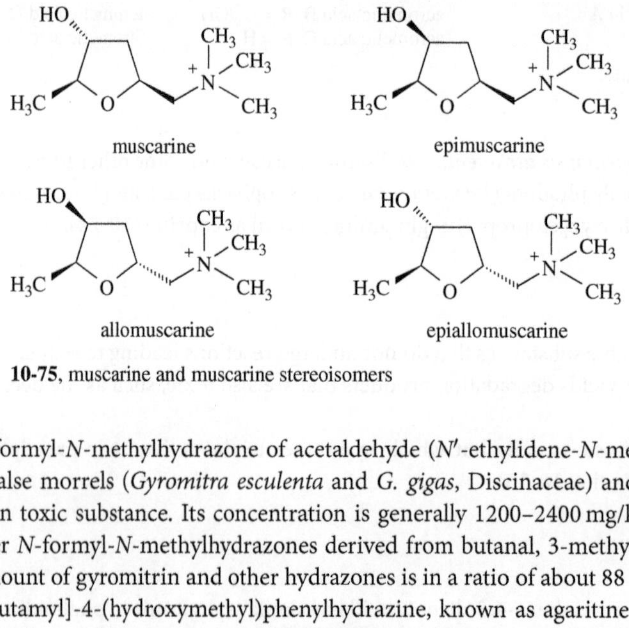

muscarine   epimuscarine

allomuscarine   epiallomuscarine

10-75, muscarine and muscarine stereoisomers

Gyromitrin (**10-76**) is a toxic *N*-formyl-*N*-methylhydrazone of acetaldehyde (*N′*-ethylidene-*N*-methylformohydrazide) that occurs in the fruiting body of the so-called false morrels (*Gyromitra esculenta* and *G. gigas*, Discinaceae) and some other species of the phylum Ascomycota. Gyromitrin is the main toxic substance. Its concentration is generally 1200–2400 mg/kg, but can reach up to 3600 mg/kg. Gyromitrin is accompanied by other *N*-formyl-*N*-methylhydrazones derived from butanal, 3-methylbutanal, pentanal, hexanal, octanal, and (*E*)- and (*Z*)-oct-2-enal. The amount of gyromitrin and other hydrazones is in a ratio of about 88 : 12.

Another hydrazine is β-*N*-[γ-L-glutamyl]-4-(hydroxymethyl)phenylhydrazine, known as agaritine (**10-77**). Agaritine is found only in fungi of the genera *Agaricus* and *Macrolepiota* (Agaricaceae), to which belong the dominant cultivated button mushrooms (*A. bisporus*) and field mushrooms (*A. campestris*). The best-known member of the genus *Macrolepiota* is the parasol mushroom (*M. procera*). The concentration of agaritine in fresh common button mushrooms varies over a wide range (100–1700 mg/kg). The agaritine content is also dependent on the growth cycle of the mushroom. Harvesting periods are repeated over 3–5 day cycles, with breaks when the harvest is very small. The agaritine content is higher in fruiting bodies harvested in the latter days of the cycle and lower in mushrooms grown on natural substrates as compared to synthetic or mixed ones. The horse mushroom (*A. arvensis*) has a higher agaritine content when growing in nature: sometimes 2000 mg/kg or more.

In mushrooms, agaritine is enzymatically oxidised at the C-4 hydroxymethyl group to the corresponding formyl and carboxyl derivatives, β-*N*-[γ-L-glutamyl]-4-(formyl)phenylhydrazine and β-*N*-[γ-L-glutamyl]-4-(carboxy)phenylhydrazine, repectively. The presence of β-*N*-[γ-L-glutamyl]-4-(carboxy)phenylhydrazine at a level of 40 mg/kg was demonstrated only in field mushroom (*A. campestris*). Agaritine is hydrolysed by γ-glutamyl transferase to L-glutamic acid and 4-(hydroxymethyl)phenylhydrazine (**10-77**) during storage of fresh, chilled, or frozen mushrooms; enzymatic hydrolysis of β-*N*-[γ-L-glutamyl]-4-(carboxy)phenylhydrazine yields 4-(carboxy)phenylhydrazine (**10-77**). 4-(Hydroxymethyl)phenylhydrazine is subsequently oxidised to the corresponding 4-(hydroxymethyl)benzenediazonium ion (**10-78**), the amount of which in the basal–stalk sections of the cultivated mushroom is about 1 mg/kg. 4-(Carboxy)phenylhydrazine concentrations in mushrooms are about 10 mg/kg. In addition to these compounds, the common mushrooms contain some other precursors, metabolites, and degradation products of agaritine, such as 4-carboxymethylbenzoic acid and agaritine analogues derived from 4-aminophenol, γ-glutamyl-4-hydroxybenzene, the characteristic yellow pigment agaricone (**9-102**) of the yellow-staining mushroom *A. xanthodermus*.

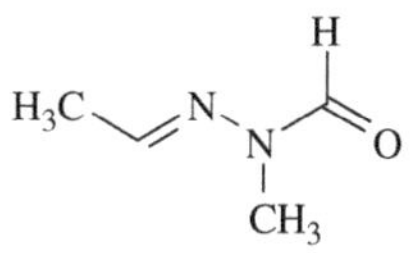

10-76, gyromitrin

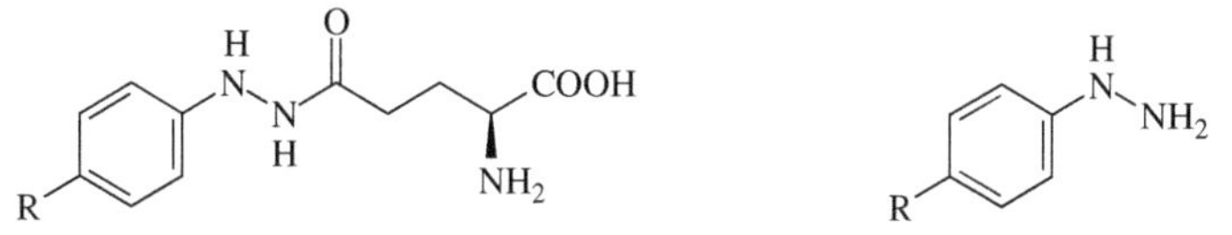

agaritine, R = CH$_2$OH

10-77, agaritine and related hydrazines

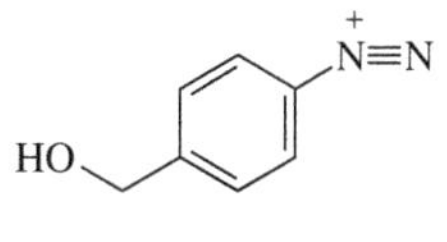

4-(hydroxymethyl)phenylhydrazine, R = CH$_2$OH
4-(carboxy)phenylhydrazine, R = COOH

10-78, 4-(hydroxymethyl)benzenediazonium ion

The fungi fool's web cap (*Cortinarius orellanus*, Cortinariaceae), deadly web cap (*C. rubellus*), and other species, which cause so-called **orellanine poisoning**, contain a toxic pyridine alkaloid orellanine and its degradation products, orellinine and yellow orelline (**10-79**), which are produced by heating or through exposure to UV radiation. Orellanine is a polar compound, due to tautomerism between the neutral and internal ionic form. In the mushroom, orellanine is present in an insoluble protonated form or as a soluble sodium salt. In addition to orellanine, these fungi contain toxic cyclic peptides cortinarins.

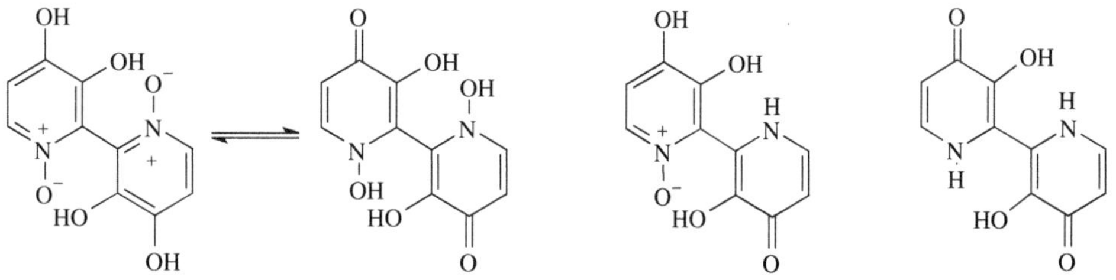

orellanine

orellinine

orelline

10-79, orellanine and orellanine degradation products

The psychedelic mushrooms of the genus *Psilocybe*, such as *P. mexicana* (Strophariaceae), and various fungi belonging to the genera *Paneolus*, *Pluteus*, and *Conocybe*, amongst others, contain as the active component psilocin and its phosphoric acid ester psilocybin (**10-80**), which are carriers of hallucinogenic effects. *P. mexicana*, called *teonáncatl*, was used by the Mayans for religious ceremonies over 2000 years ago. Other related hallucinogenic substances, such as baeocystine and norbaeocystine (**10-80**), occur in *P. baeocystis*. Another hallucinogenic amine of similar structure is *N,N*-dimethylserotonin (bufotenine), which is found in false death cap, also known as citron amanita (*Amanita citrina*). The same amine also occurs in skin secretions of some toads, amphibians, and higher plants. For example, bufotenine is present in a concentration of 25 mg/l in the red latex (called *takini*) of the tree *Brosimum acutifolium* (Moraceae), used in shamanic practices in the Amazon rainforest.

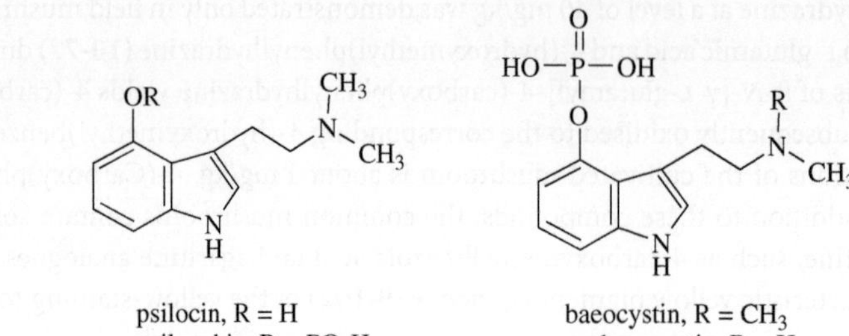

psilocin, R = H
psilocybin, R = PO₃H₂

baeocystin, R = CH₃
norbaeocystin, R = H

10-80, hallucinogenic indole derivatives

10.3.7.4.2 Reactions and changes

Muscarine is stable during technological and culinary operations, and remains so even after prolonged heating and on exposure to digestive juices.

Gyromitrin poisoning is sporadic, as it only occurs when eating raw mushrooms. Cooked and dried mushrooms are edible. On cooking or drying, gyromitrin decomposes to form acetaldehyde and N-formyl-N-methylhydrazine, which is more stable than its precursor, although in acidic media and in the milieu of the human stomach it decomposes to volatile N-methylhydrazine (Figure 10.11). The concentrations of N-formyl-N-methylhydrazine and N-methylhydrazine in mushrooms can reach 500 and 40–350 mg/kg, respectively. Similar reactions proceed *in vivo*, and 25–30% of gyromitrin may be transformed into N-methylhydrazine. Prolonged drying in the air decreases the N-methylhydrazine content by about 30–70%. A significant decrease of hydrazine content (to 10–15% of the original value) occurs through leaching and volatilisation of products during cooking. The content of toxins in canned mushrooms is also much lower (6–65 mg/kg).

Only consumption of raw button mushrooms is potentially harmful, because agaritine partly decomposes under technological and culinary treatments. During storage at temperatures of 2–12 °C, the agaritine concentration decreases by about 32%. Frozen mushrooms stored for 1 month and then defrosted contained about 25% of the agaritine present in fresh mushrooms. Drying does not have substantial influence on agaritine decomposition, so the content of agaritine in dried mushrooms is relatively high (up to 4600 mg/kg). A decrease of agaritine content of up to about 75% occurs during cooking (about 50% of the loss is due to leaching into water). Approximately 20–50% of agaritine remained after 120 minutes of heating mashroom extracts at 120 °C. Canned mushrooms contain 15–18 mg/kg of agaritine, which corresponds to about 5–8% of the original amount.

Orellanine is relatively thermostable and therefore also occurs in the dried musrooms. Its degradation products orellinine and orelline (**10-79**) are produced by intense heating or through exposure to UV radiation.

Figure 10.11 Metabolism and non-enzymatic decomposition of gyromitrin.

Under normal temperature, both psilocin and psilocybin are stable in aqueous solutions for several days. Psilocybin is dephosphorylated into psilocin under acidic conditions and also enzymatically by phosphatases when ingested. In dried mushrooms, the content of toxic substances is lower than in fresh ones.

10.3.7.4.3 Biological effects

Muscarine poisoning (**muscarinic syndrome**) is manifested about 15–30 minutes after ingestion of mushrooms and can last up to two hours. Nausea, blurred vision, sweating (cold sweat), paleness, salivation, tearing, heart rhythm disturbances, abdominal pain, diarrhoea, vomiting, and severe asthmatic breathing may occur. Lethal poisonings are described primarily by ingestion *Inocybe erubescens*, which contains a large amount of toxin (at least 100 times more than *Amantia muscaria*). Mortality is 8–10%.

It is believed that the main cause of the toxicity of *Gyromitra* genus mushrooms (**hepatotoxic syndrome**) is the hepatotoxic and carcinogenic substances *N*-formyl-*N*-methylhydrazine and *N*-methylhydrazine (Figure 10.22). Poisoning from consumption of false morel is widespread in Europe, although it can only occur on consumption of the raw mushrooms. Poisoning with raw mushrooms is relatively common, and symptoms of gyromitrin poisoning are similar to those of toadstool poisoning. They occur after 6–12 hours of latency, in two phases. The first, which lasts for one or two days, is manifested by fatigue, headaches, dizziness, stomach upset, and later vomiting and stomach and liver pain. In the second stage, hepatitis appears due to liver damage. Poisoning is fatal in 50% of cases. Frequent consumption of false morel can generally cause jaundice and neurological disorders.

Symptoms of poisoning with button mushrooms are nausea, vomiting, and diarrhoea (**gastroenterodyspeptic syndrome**). The metabolic fate of agaritine has been linked with the carcinogenity of the mushroom due to hydrazines arising from agaritine and 4-(hydroxymethyl)benzenediazonium salts. The risk arising from the consumption of mushrooms, with regard to the amount consumed and content of agaritine and its derivatives, requires more detailed evaluation.

Orellanine and orellinine are nephrotoxins, but the related orelline is a non-toxic compound. Manifestations of poisoning (**nephrotoxic syndrome**) consist of renal impairment after latency of 2–17 days. The most important symptoms are fatigue, loss of appetite, nausea or even vomiting, headache, severe thirst, urge to urinate, diarrhoea, fever, chills without fever, and pain in the limbs and the muscles generally. Progressive kidney damage later leads, conversely, to limited production of urine and constipation. Poisoning often results in death or permanent kidney damage, requiring regular dialysis or kidney transplantation.

Most of the approximately 200 species of mushrooms containing psilocybin fall in the genus *Psilocybe*. Psilocybin and psilocin act as hallucinogens (**hallucinogenic syndrome**), with activity similar to that of the known hallucinogen lysergic acid diethylamide (LSD), as they also affect receptors of serotonin (5-hydroxytryptamine) in the central nervous system. For this reason, many countries have some level of regulation or prohibition of these so-called magic mushrooms.

10.3.7.5 Terpenoids and other compounds

10.3.7.5.1 Structure, nomenclature, and occurrence

In addition to toxic nitrogen compounds, fungi also contain toxic terpenoids. The jack-o'-lantern mushroom (*Omphalotus olearius*, syn. *Clitocybe illudens*, Marasmiaceae) and some other fungi contain the toxic sesquiterpenoids illudins M and S, accompanied by the non-toxic dihydroilludins M and S. The similarly poisonous Japanese mushroom tsukiyotake (*Omphalotus japonicus*, syn. *Pleurotus japonicus*) contains 6-deoxyilludins M and S (**10-81**). A compound of similar structure is leaianafulvene (**10-82**), which is found in *Mycena laevigata* (Mycenaceae).

illudin M, R = H
illudin S (lampterol), R = OH

dihydroilludin M, R = H
dihydroilludin S, R = OH

6-deoxyilludin M, R = H
6-deoxyilludin S, R = OH

10-81, illudins, dihydroilludins, and desoxyilludins

10-82, leaianafulvene

Lanostane triterpenoid alcohols fasciculols A to F (**10-83**) are examples of biologically active components of the mushroom called sulfur tuft (*Naematoloma fasciculare*, Strophariaceae) and related species. Fasciculols B, C, and F and three fasciculol esters, fasciculic acids A, B, and C (**10-83**), have calmodulin inhibitory activity. Calmodulin (**cal**cium-**modul**ated prot**ein**) is a calcium-binding messenger protein of eukaryotic cells that mediates many crucial processes, such as inflammation, metabolism, smooth-muscle contraction, intracellular movement, short- and long-term memory, and immune response. Toxic substances related to fasciculols and referred to as HS-A, HS-B, and HS-C (**10-84**) are toxins of *Hebeloma spoliatum*, a species of mushroom of the Hymenogastraceae family.

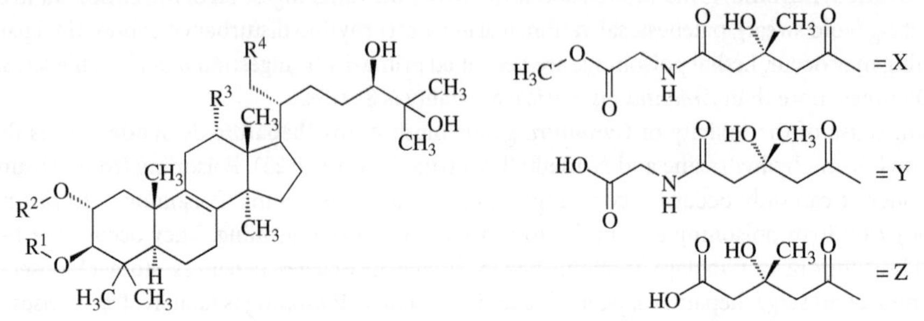

10-83, fasciculols and fasciculic acids

fasciculol A, $R^1 = H$, $R^2 = H$, $R^3 = H$, $R^4 = H$
fasciculol B, $R^1 = H$, $R^2 = H$, $R^3 = OH$, $R^4 = H$
fasciculol C, $R^1 = H$, $R^2 = H$, $R^3 = OH$, $R^4 = OH$
fasciculol D, $R^1 = H$, $R^2 = X$, $R^3 = OH$, $R^4 = H$
fasciculol E, $R^1 = X$, $R^2 = H$, $R^3 = OH$, $R^4 = OH$
fasciculol F, $R^1 = H$, $R^2 = X$, $R^3 = OH$, $R^4 = OH$

fasciculic acid A, $R^1 = H$, $R^2 = Z$, $R^3 = H$, $R^4 = H$
fasciculic acid B, $R^1 = H$, $R^2 = Z$, $R^3 = OH$, $R^4 = H$
fasciculic acid C, $R^1 = Y$, $R^2 = H$, $R^3 = OH$, $R^4 = OH$

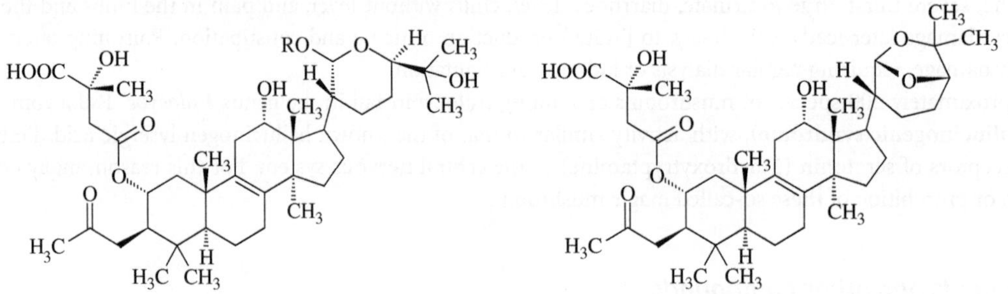

HS-A, R = H
HS-B, R = OCOCH$_3$

HS-C

10-84, HS toxins

Hebevinosides are toxic triterpenoid diglycosides and monoglycosides. The aglycone structure is similar to that of related bitter cucurbitacins occurring in cucurbit plants (Cucurbitaceae, **8-140**). Hebevinosides were found in the mushroom *Hebeloma vinosophylum* (Cortinariaceae) growing on high-nitrogen media. Approximately 14 different compounds with different substitutions (R^1 = H or CH$_3$ and R_2–R_5 = H or C(=O)CH$_3$) have been identified. Only glycosides containing D-glucose linked at C-16 are toxic. Examples of hebevinosides are toxic C-3/C-16 diglycosides and C-16 glycosides (**10-85**).

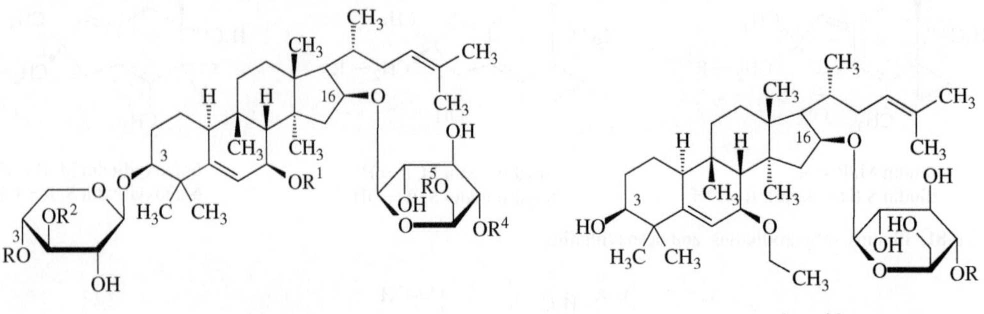

diglycosides

monoglycosides

10-85, basic structures of hebevinosides

The linear terpenoid gymnopilin (**10-86**) is found in the hallucinogenic mushroom *Gymnopilus junonius* (syn. *G. spectabilis*, Cortinariaceae), called big laughing gym, as a mixture of congeners with different numbers of isoprene units (m = 1–3, n = 5–7). Compounds with more than one double bond (m > 1) are toxic. Toxicity also depends on the number of segments (n) with hydroxyl groups. Gymnopilin is further responsible for the mushroom's bitter taste.

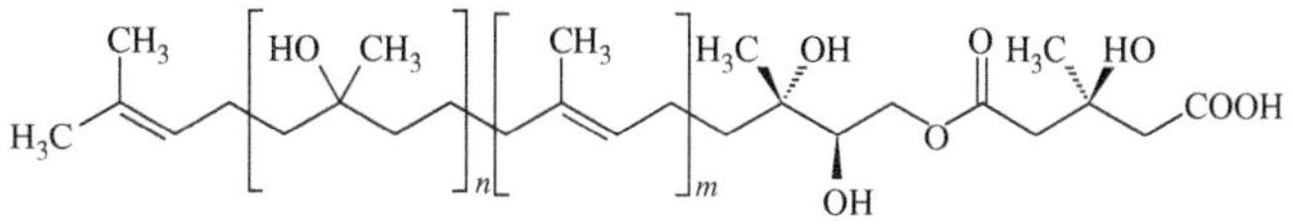

10-86, gymnopilin

The wood-decay fungus *Laricifomes officinalis* (Fomitopsidaceae), known as agarikon, quinine conk, or white agaric mushroom, contains high amounts (14–16% of dry matter) of bitter-tasting citric acid derivative with a lipophilic C-2 substituent derived from palmitic acid. This substance is known as 2-hydroxynonadecane-1,2,3-tricarboxylic, α-hexadecylcitric, α-cetylcitric, agaricinic, agaric, or laricic acid (**10-87**). In the past, the mushroom was used as a bitter additive for beverages and as an antiperspirant for medical purposes (treatment of sweats in wasting conditions such as phthisis); a dry sponge impregnated with potassium nitrate was used as tinder.

10.3.7.5.2 Reactions and changes

Toxic terpenoids are stable and fungal foods do not lose their toxicity even during cooking. Insignificant losses of these lipophilic toxins occur by leaching.

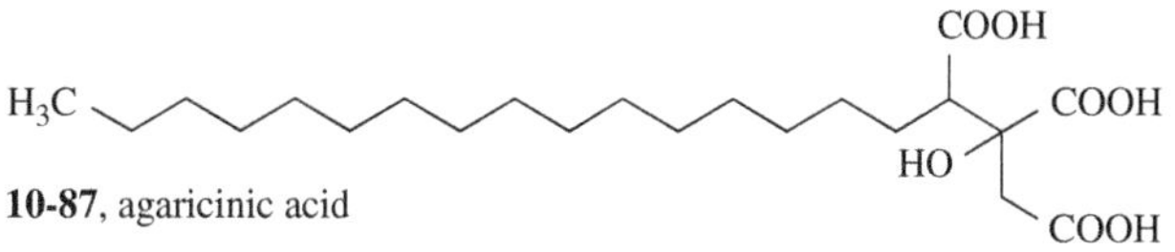

10-87, agaricinic acid

10.3.7.5.3 Biological effects

Poisonings caused by illudin M and S, not regarded as deadly, are manifested by typical digestive problems, vomiting, and diarrhoea. Illudin S and 6-deoxyilludins M and S show some anticarcinogenic activity. Fasciculols E and F cause vomiting, diarrhoea, and, in some cases, convulsions and paralysis. Poisoning may lead to death. Toxins HS-A, HS-B, and HS-C may cause paralysis of limbs at higher concentrations (around 100 mg/kg) and diarrhoea at lower ones (about 45 mg/kg). These substances affect not only the central nervous system, but also the autonomic one. Hebevinosides act as paralytic toxins. Gymnopilin is a neurotoxic hallucinogenic substance. Due to its adverse biological effects (it acts as an antiperspirant, a substance preventing cutaneous respiration), EU legislation allows agaric acid as a bitter-tasting flavouring only for alcoholic beverages (see Section 11.3.1.2).

10.3.8 Marine toxins

The meat of some marine fish, shellfish, and many other animals can be toxic. In most cases, the toxic substance is present throughout the organism, as it is usually caused by the consumption of phytoplankton, microscopic algae (Dinoflagellates), blue-green algae (Cyanobacteria), and diatoms (Diatomeae), which contain various hepatotoxins, neurotoxins, dermatotoxins, or intestinal toxins. The primary sources of toxins are bacteria that live with these algae in symbiosis or as epiphytes. A number of toxins are also produced by some fish and other marine animals. The most important seafood toxins include:

- saxitoxin and its derivatives;

- okadaic acid and its analogues (dinophysistoxins);

- azaspiracids;

- domoic acid and its analogues;

- brevetoxins;

- tetrodotoxin and its analogues;

- ciguatoxins;

- palytoxin.

10.3.8.1 Bivalve mollusc toxins

The class Bivalvia (bivalves) of the phyllum Mollusca (molluscs) includes molluscs known as Pacific oysters (*Crassostrea gigas*, Ostreidae), three closely related taxa of blue mussels (*Mytilus edulis*, Mytilidae), hard (round) clams (*Mercenaria mercenaria*, Veneridae), common whelks (*Buccinum undatum*, Buccinidae), and many others. The number of cases of serious poisoning by oysters, mussels, clams, and other bivalves has been increasing in recent years. Shellfish (a culinary term for exoskeleton-bearing aquatic invertebrates used as food, including molluscs and crustaceans) take in algae containing toxins from the water as food. With the growing proliferation of toxic algae in the oceans, due to eutrophication and climate change, the degree and extent of contamination of shellfish has been increasing. Global aquaculture development, along with the growing popularity of seafood and other factors, is contributing to the increase in health problems following consumption of these marine animals.

Toxins of shellfish are characterised by strong, sometimes fatal, acute human toxicity. They are thermostable and show no sensory warning signals. Occurrence of toxins in various areas is sporadic, and probably dependent on meteorological and climatic conditions (such as wind direction and speed, water temperature, depth and currents, and microlocation) that control the spread of toxic algae. This leads to the utter unpredictability of the degree of contamination and expansion of shellfish, with a consequent high demand for continuous monitoring and control of production sites.

The most important groups of biologically active compounds in shellfish are distinguished by the type of their toxic effects (Table 10.14). Some marine toxins are found in the South Seas, whilst others predominate in colder waters. In Europe, the two most common syndromes of shellfish poisoning are paralytic shellfish poisoning (PSP) and diarrhoeic shellfish poisoning (DSP). Other types of shellfish poisoning are azaspiracid poisoning (AZP), amnaesic shellfish poisoning (ASP), and neurotoxic shellfish poisoning (NSP).

10.3.8.1.1 Paralytic neurotoxins

Structure, nomenclature, and occurrence   The cause of PSP is the consumption of shellfish contaminated by hydrophilic heterocyclic guanidines, a group of more than 20 related carbamate alkaloid neurotoxins that are either non-sulfated (saxitoxins, STX), singly sulfated (gonyautoxins, GTX), or doubly sulfated (N-sulfocarbamoyl-11-hydroxysulfate toxins, C-toxins). Saxitoxin (**10-88**) is accompanied by the structurally related toxins neosaxitoxin and gonyautoxins I–IV and the corresponding N-sulfocarbamoyl derivatives ($R^4 = O\text{-}CO\text{-}NH\text{-}SO_3^-$) and decarbamoyl derivatives ($R^4 = OH$). These toxins accumulate in shellfish through ingestion of the dinoflagellates during blooms. The most important sources are dinoflagellates of genera *Gymnodium* (in Europe, Japan, and North America), *Alexandrium* (such as *A. tamarense* and *A. catenella* in South America), and *Pyrodinium* (in the Indo-Pacific region). Saxitoxins and related analogues are known to inhibit nerve–muscle transmission by blocking the sodium channels in the excitable membrane and cause a lethal toxicity through breathing muscle paralysis.

10-88 saxitoxin and its derivatives

saxitoxin, $R^1 = R^2 = R^3 = H$, $R^4 = OCONH_2$
neosaxitoxin, $R^1 = OH$, $R^2 = R^3 = H$, $R^4 = OCONH_2$
decarbamoylsaxitoxin, $R^1 = R^2 = R^3 = H$, $R^4 = OH$
gonyautoxin-1, $R^1 = OH$, $R^2 = H$, $R^3 = OSO_3^-$, $R^4 = OCONH_2$
gonyautoxin-2, $R^1 = R^2 = H$, $R^3 = OSO_3^-$, $R^4 = OCONH_2$
gonyautoxin-3, $R^1 = H$, $R^2 = OSO_3^-$, $R^3 = H$, $R^4 = OCONH_2$
gonyautoxin-4, $R^1 = H$, $R^2 = OSO_3^-$, $R^3 = OH$, $R^4 = OCONH_2$
gonyautoxin-5, $R^1 = R^2 = R^3 = H$, $R^4 = OCONHSO_3^-$
gonyautoxin-6, $R^1 = oH$, $R^2 = R^3 = H$, $R^4 = OCONHSO_3^-$
C1, $R^1 = R^2 = H$, $R^3 = OSO_3^-$, $R^4 = OCONHSO_3^-$
C2, $R^1 = H$, $R^2 = OSO_3^-$, $R^3 = H$, $R^4 = OCONHSO_3^-$
C3, $R^1 = OH$, $R^2 = H$, $R^3 = OSO_3^-$, $R^4 = OCONHSO_3^-$
C4, $R^1 = OH$, $R^2 = OSO_3^-$, $R^3 = H$, $R^4 = OCONHSO_3^-$

Saxitoxin may be found in oysters, common whelks, and other species of marine molluscs, but they also occur in some microscopic freshwater blue-green algae (*Aphanizomenon flos-aquae*) and red macroscopic algae of the genus *Jania*, which become the source of saxitoxin in crab meat.

Reactions and changes   The toxins are extremely thermostable and cannot be removed by cooking.

Biological effects   Saxitoxin and its derivatives are neurotoxins. The mechanism of action is based on the blockade of conductance in sodium tubules of nerve membrane with subsequent blockade of transmission of nerve impulses. Clinical symptoms may appear within one hour and are characterised by numbness in the extremities, which spreads to the neck and face, along with symptoms of severe nausea. In severe cases, paralysis of respiratory muscles may occur – at high doses, within two hours after exposure. PSP toxins are quickly eliminated from the body, and survival for 24 hours means a good chance of recovery. Treatment must be implemented as soon as possible and is only symptomatic. On a global scale, about 2000 cases of poisoning occur each year, with approximately 15% mortality. Total toxicity of saxitonins is calculated using toxicity equivalents (TEQ, see Section 12.4.1.4) based on the relative toxicities of individual congeners

Table 10.14 Overview of toxins in molluscs.

Syndrome	Toxin	Symptoms	Occurrece	Limit (EU)
PSP	Saxitoxins	Neurological (mortality)	Europe, Japan, America, Pacific Ocean	800 µg/kg
DSP	Okadaic acid and dinophysis- toxins	Gastrointestinal	Europe (northwest), South America	160 µg/kg
AZP	Azaspiracids and analogues	Gastrointestinal	Ireland (Europe)	160 µg/kg
ASP	Domoic acid	Gastrointestinal/short-term memory loss (mortality)	Global (Canada)	20 µg/kg
NSP	Brevetoxins	Neurological (gastrointestinal)	Florida, Caribean, New Zealand	

(toxicity equivalency factors, TEF) in the edible portion of molluscs. The effective dose in humans is estimated at 2–30 µg of TEQ per kg body weight; severe poisonings were observed at doses of 10–300 mg TEQ per kg body weight. Poisoning periodically appears in various localities.

The hygienic limit in the European Union is currently 80 µg of paralytic toxins in 100 g of edible shellfish portion. The European Union and other countries have drawn up detailed legislative regulations for the control of saxitoxin complexes.

10.3.8.1.2 Diarrhoetic toxins

Structure, nomenclature, and occurrence Diarrhoetic shellfish poisoning is associated with the consumption of marine shellfish (scallops, oysters, and other shellfish), accumulating toxins from dinoflagellates such as *Dinophysis* and *Prorocentrum* spp. Poisonings of this type are widespread throughout the world, mainly in Japan, Scandinavia, northwestern Europe, and South America. The cause is the okadaic acid group of lipophilic toxins, which include okadaic acid (**10-89**) and its analogues (dinophysistoxins), such as dinophysistoxin-1, known as (*R*)-35-methylokadaic acid, and dinophysistoxin-2, distinguished from okadaic acid by the position of the methyl group, which are the main toxic components. Also important are a number of substituted congeners occurring as sulfated diol esters in algae and as fatty acid esters in shellfish, such as dinophysistoxin-3, also known as 31-demethyl-35-methylokadaic acid.

10-89, okadaic acid, R^1 = H, R^2 = H
dinofysistoxin-1, R^1 = H, R^2 = CH$_3$
dinofysistoxin-3, R^1 = fatty acid residue, R^2 = CH$_3$

A poisoning with similar symptoms as those caused by okadaic acid can be the result of other related toxins, such as pectenotoxins (PTX, **10-90**) from marine dinoflagellates of the genus *Dinophysis* and disulfates of cyclic polyethers yessotoxins (YTX, **10-91**), which are produced by planktonic algae *Protoceratium reticulatum* and possibly also by dinoflagellates *Lingulodinium polyedrum*. These toxins have a different mechanism of action and are not included in the hygienic limits for dinophysistoxins.

10-90, pectenotoxin-1, R = CH$_2$OH
pectenotoxin-2, R = CH$_3$
pectenotoxin-3, R = CH=O
pectenotoxin-6, R = COOH

The most commonly found pectenotoxin in algae is pectenotoxin-2. Pectenotoxin-2 is enzymatically hydrolysed into a pectenotoxin-2 seco acid formed by many shellfish species, a reaction that constitutes a detoxification mechanism. Spontaneous epimerisation at C-7 of pectenotoxin-2 seco acid produces 7-epi-pectenotoxin-2 seco acid. Both compounds occur in blue mussels (*Mytilus edulis*) from Ireland as fatty acid esters. Other enzyme-mediated conversions of pectenotoxin-2 include oxidation at C-43 in the scallop *Patinopecten yessoensis*, Pectinidae, yielding pectenotoxin-1, pectenotoxin-3, and pectenotoxin-6 (**10-90**).

The structures of the yessotoxins differ from those of toxins in the diarrhoetic shellfish poisoning toxins group and the pectenotoxins, but are similar to the brevetoxins and ciguatoxins in that they have a ladder-shaped polycyclic ether skeleton. Yessotoxin (**10-91**) was evidenced in a marine bivalve mollusc called yesso scallop (*P. yessoensis*), and has since been found in a wide range of shellfish from around the world. More recently, 45-hydroxyyessotoxin and 45,46,47-trinoryessotoxin were also isolated from scallops. 45-Hydroxyyessotoxin was isolated from mussels, as were homoyessotoxin, 45-hydroxyhomoyessotoxin, 1-desulfoyessotoxin, adriatoxin, carboxyyessotoxin, carboxyhomoyessotoxin, and numerous minor related compounds.

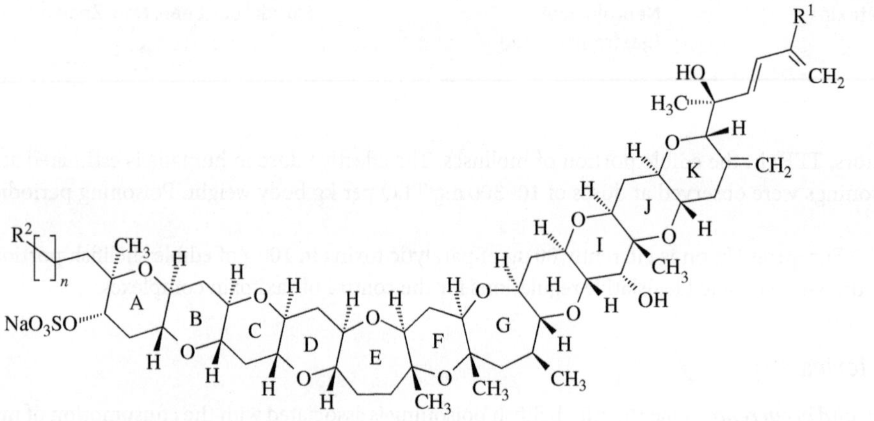

yessotoxin, R^1 = CH₂CH=CH₂, R^2 = NaO₃SO, n = 1
45-hydroxyyessotoxin, R^1 = CH(OH)CH=CH₂, R^2 = NaO₃SO, n = 1
45,46,47-trinoryessotoxin, R^1 = H, R^2 = NaO₃SO, n = 1
homoyessotoxin, R^1 = CH=CH₂, R^2 = NaO₃SO, n = 2
45-hydroxyhomoyessotoxin, R^1 = CH(OH)CH=CH₂, R^2 = NaO₃SO, n = 2
1-desulfoyessotoxin, R^1 = CH(OH)CH=CH₂, R^2 = H, n = 1

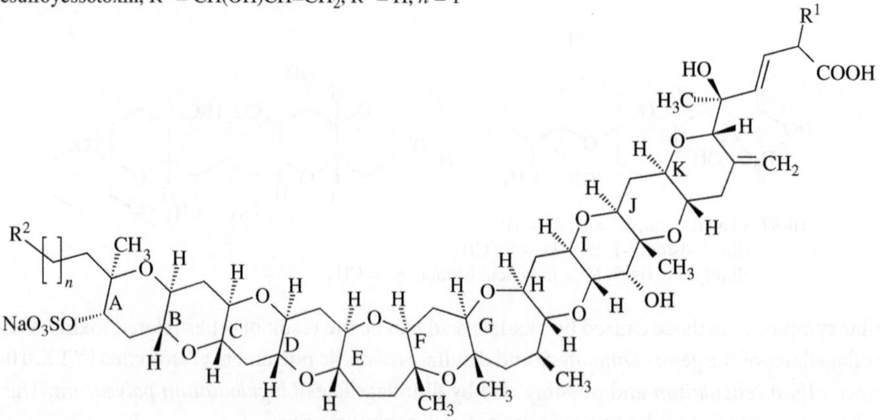

carboxyyessotoxin, R^1 = CH₂CH=CH₂, R^2 = NaO₃SO, n = 1
carboxyhomoyessotoxin, R^1 = CH₂CH=CH₂, R^2 = NaO₃SO, n = 2

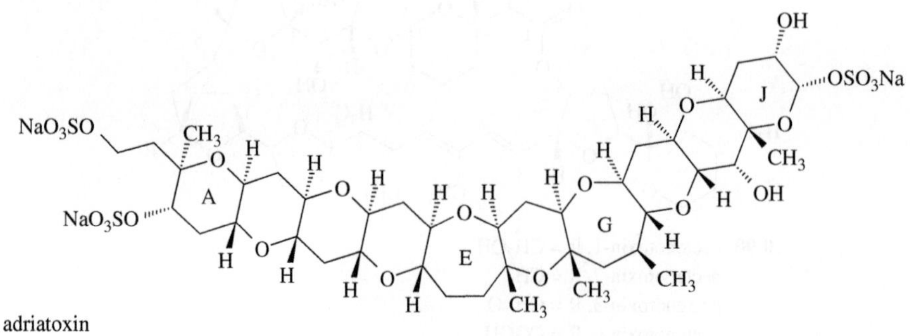

adriatoxin

10-91, yessotoxin and related toxins

Reactions and changes Toxins are thermostable (as are other toxins in shellfish and structurally similar ciguatera toxins of fish) and cannot be removed by heat treatments.

Biological effects Okadaic acid is an inhibitor of phosphatases, which leads to phosphorylation of proteins controlling excretion of fluids from the intestine. The main symptoms of poisoning are indigestion exhibited by stomach pain, nausea, vomiting, and diarrhoea. These symptoms occur quickly (30 minutes after ingestion of shellfish) and disappear spontaneously after three or four days, without the need for hospitalisation. Data on the long-term toxicity of dinophysistoxins in humans are lacking. In laboratory rodents, high doses of these toxins act as carcinogenicity promoters, but their genotoxicity has not been confirmed.

The existing hygienic limits for diarrhoetic toxins in the European Union are based on acute toxicity in the form of an acute reference dose (ARfD). Toxicity of the whole group of toxins is translated into okadaic acid using TEF (toxic equivalency factor): for okadaic acid and dinophysistoxin-1, TEF = 1, whilst for dinophysistoxin-3, TEF = 2; TEF for dinophysistoxin-3 is equal to that of the corresponding non-esterified compounds (okadaic acid, dinophysistoxin-1, and dinophysistoxin-2). Cases of human poisoning have been observed in areas having 0.8 mg okadaic acid TEQ (toxic equivalent) per kg body weight. In the European Union, the currently applied hygienic limit is 160 µg of okadaic acid TEQ per kg for edible molluscs, but it is under discussion whether this provides adequate protection when eating large portions of this food.

10.3.8.1.3 Azaspiracids

Structure, nomenclature, and occurrence The group of marine toxins known as azaspiracids (AZA) was just recently discovered to be the cause of shellfish poisoning in humans. Their source is the algae of the genus *Protoperidinium*. Azaspiracids are a group of polyethers (such as okadaic acid), but they contain a nitrogen heterocyclic amine piperidine (**10-92**). About 20 different derivatives have been identified, amongst which azaspiracid-1 (AZA1) and azaspiracid-3 (AZA3) are relevant to the toxicity.

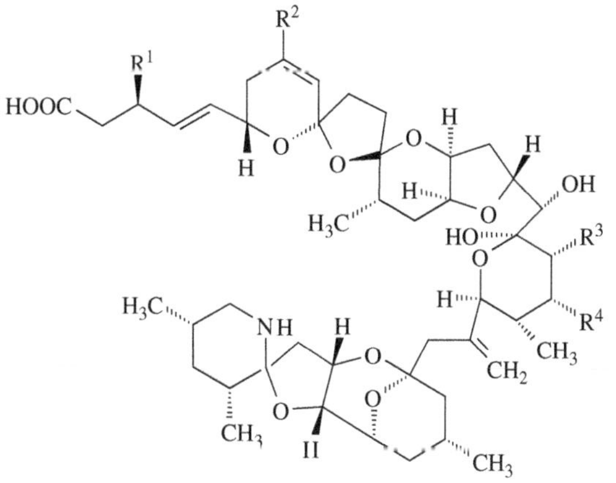

10-92, azaspiracids

AZA1, $R^1 = R^2 = H$, $R^3 = CH_3$, $R^4 = H$
AZA2, $R^1 = H$, $R^2 = CH_3$, $R^3 = CH_3$, $R^4 = H$
AZA3, $R^1 = R^2 = R^3 = R^4 = H$
AZA4, $R^1 = OH$, $R^2 = R^3 = R^4 = H$
AZA5, $R^1 = R^2 = R^3 = H$, $R^4 = OH$

Reactions and changes Azaspiracids are thermostable and cannot be removed by heat treatments.

Biological effects AZP syndrome was first described in 1995 after a poisoning case involving consumption of blue mussels (*Mytilus edulis*) from Ireland. In subsequent years, poisoning cases have been reported from other European countries, but the mussels were always of Irish origin. Poisoning is characterised by digestive problems: nausea, vomiting, diarrhoea, and stomach pain. In all cases, there is spontaneous recovery within a few days.

As in other toxins, the hygienic limits are based on the ARfD, and the total exposure to individual azaspiracid analogues is converted into AZA1 using TEF equivalents. The lowest dose that causes toxic effects is estimated to be approximately 2 mg of AZA1 equivalents per adult. The EU hygienic limit is (as with the case of DSP) 160 µg of AZA1 TEQ per kg of edible shellfish portion.

10.3.8.1.4 Amnestic toxins

Structure, nomenclature, and occurrence ASP is a less common type of poisoning. It occurs after ingestion of shellfish contaminated by a specific toxic amino acid referred to as domoic acid, a naturally occurring neuroexcitatory toxin produced primarily by the marine diatom *Pseudonitzschia multiseries*, which is widespread in warm and colder seas. Other compounds of similar structure include domoic acid (**10-93**), its diastereomer 5′-domoic acid, and 10 domoic acid isomers, exemplified by isodomoic acids A, D, and H (**10-94**). The first human poisoning case was reported in 1987 and occurred through ingestion of domoic acid-contaminated blue mussels (*Mytilus edulis*). Contamination was later found in various molluscs. Domoic acid was originally isolated in Japan from the red alga *Chondria armata*, called *doumoi*.

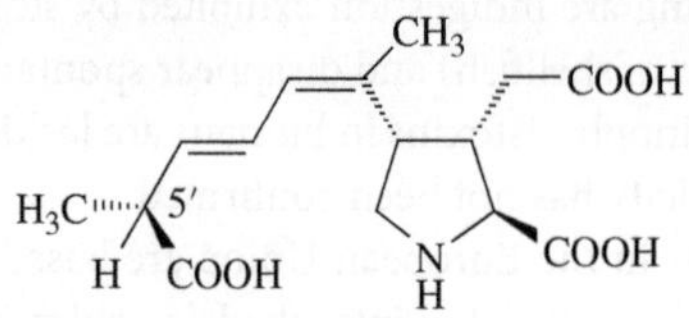

10-93, domoic acid

Reactions and changes Domoic acid is thermostable and cannot be removed by heat treatments. It has been shown that domoic acid may degrade in acidic media, but the practical importance of such treatment (e.g. pickling in vinegar) is not known.

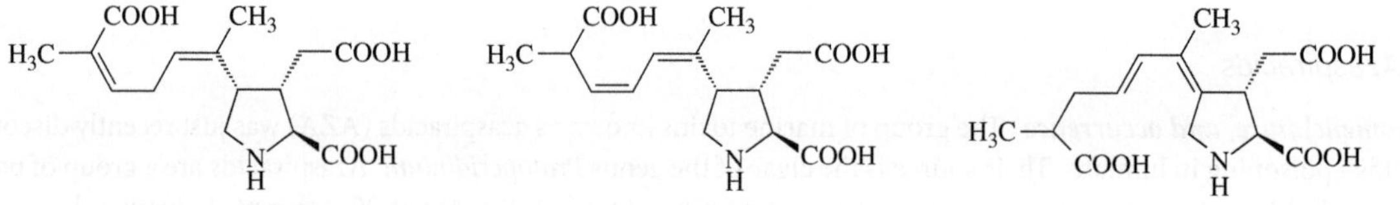

isodomoic acid A   isodomoic acid D   isodomoic acid H

10-94, isodomoic acids

Biological effects Amnaesic shellfish poisoning is characterised by relatively mild gastrointestinal symptoms (vomiting, stomach pain, and diarrhoea). In serious cases, especially in the elderly and in impaired and sick individuals, the neurotoxic symptoms accompanied by persistent amnaesia may outweigh the gastrointestinal ones. A case of public poisoning was reported in Canada in 1987 (over 100 people, with four deaths) after ingestion of blue mussels (*M. edulis*). The mechanism of the toxic effect is the interaction of toxins with specific glutamate receptors in the brain. Continuous stimulation of receptors leads to damage of the nervous tissue. Mild gastrointestinal symptoms of poisoning in humans were found after the intake of domoic acid at a concentration of 0.9–1.9 mg per kg body weight. Severe neurological changes start at domoic acid levels ranging from 1.9 to 4.2 mg per kg of body weight.

In the Canadian study, the lowest concentration of domoic acid in shellfish leading to mild gastrointestinal problems was estimated at 200 mg/kg of edible portion, and the hygienic limit was set at 20 mg/kg of edible portion. This limit is also applied in other countries, including EU countries. Contamination of shellfish is occasionally detected in various countries, but since the introduction of the limit, no cases of poisoning have occurred.

10.3.8.1.5 Neurotoxins

Structure, nomenclature, and occurrence Neurotoxic brevetoxins are metabolites that are found in the microscopic alga *Gymnodinium breve* (syn. *Ptychodiscus brevis*), which occurs in the Gulf of Mexico, the Caribbean, and New Zealand. The extreme algal blooms, for example along the coast of Florida, create a so-called red tide, which is a frequent cause of mass fish poisoning. Toxins present in the air as an aerosol may cause temporary inhalation problems when inhaled.

In humans, these neurotoxins cause irritation of the eyes and occasionally poisoning when consumed along with bivalves that have ingested these algae. The most effective toxin produced by algae *G. breve* is brevetoxin A (**10-95**), which occurs together with brevetoxin B (**10-96**).

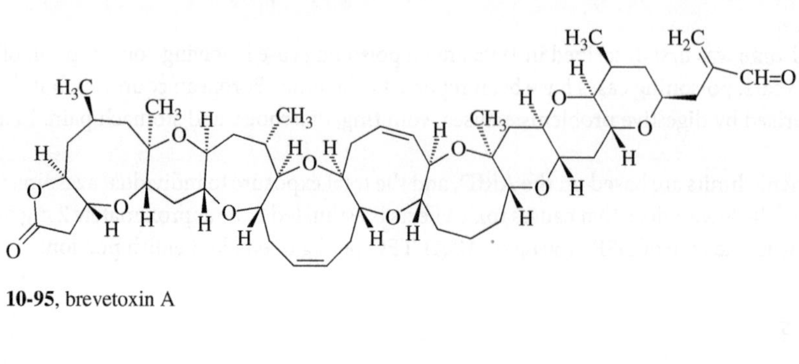

10-95, brevetoxin A

Reactions and changes Brevetoxins, as well as structurally similar ciguatoxins, are thermostable and cannot be removed by heat treatments.

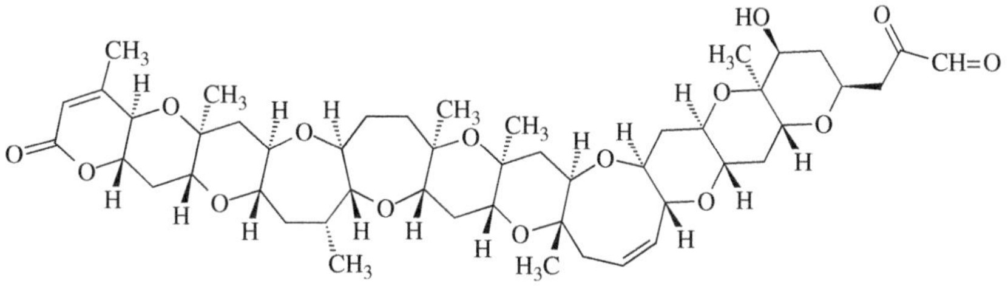

10-96, brevetoxin B

Biological effects Symptoms of poisoning by brevetoxins are similar to those of ciguatera fish poisoning (CFP). These neurological symptoms, accompanied by digestive difficulties, disappear spontaneously within 48 hours.

10.3.8.2 Fish toxins

10.3.8.2.1 Tetrodotoxin

Structure, nomenclature, and occurrence The most effective fish toxin is tetrodotoxin (**10-97**), O-methyl-O',O''-isopropylidenetetrodotoxin hydrochloride, which is accompanied by the related compounds 6-epitetrodotoxin, 11-deoxytetrodotoxin, 11-oxotetrodotoxin, (R)-11-nortetrodotoxin-6-ol, (S)-11-nortetrodotoxin-6-ol, and chiriquitoxin. Both epimers of 11-nortetrodotoxin are probably decarboxylation products of hypothetical tetrodonic acid. Tautomerism is typical for tetrodotoxin and its derivatives; these substances exhibit some properties of ketones and some of alcohols (Figure 10.12).

Tetrodotoxin is found in fish known by various names, including puffer fish, puffers, balloon fish, blowfish, and puffu, of the family Tetraodontidae, genera *Tetraodon* and *Fugu*, which are consumed as a delicacy in Japan. The toxin is not present in the meat, but is in the female roe, liver, intestines, and skin. The source is algae consumed by the fish; accordingly, the bacteria were originally classified in the genus *Pseudomonas*, and later in the genus *Alteromonas*. The current name is *Shewanella alga*. Tetrodotoxin and chiriquitoxin are also found in some Central American salamanders and frogs (*Atelops chiriquiensis*).

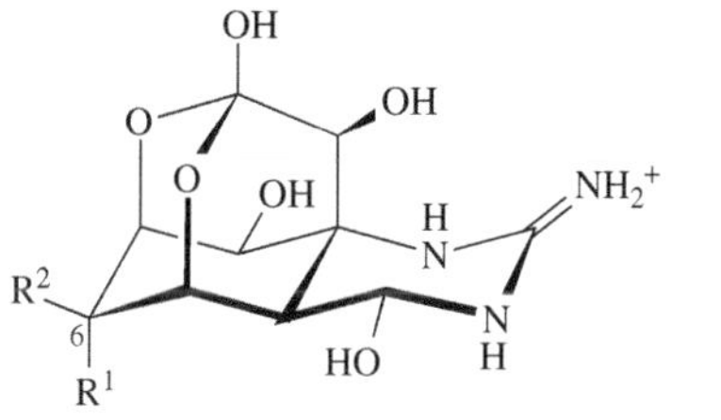

10-97, tetrodotoxin, (cation), $R^1 = OH$, $R^2 = CH_2OH$
6-epitetrodotoxin, $R^1 = CH_2OH$, $R^2 = OH$
11-deoxytetrodotoxin, $R^1 = OH$, $R^2 = CH_3$
11-oxotetrodotoxin, $R^1 = OH$, $R^2 = CHO$
(R)-11-nortetrodotoxin-6-ol, $R^1 = H$, $R^2 = OH$
(S)-11-nortetrodotoxin-6-ol, $R^1 = OH$, $R^2 = H$
chiriquitoxin, $R^1 = OH$, $R^2 = CH(OH)CH(NH_2)$
COOH, (1R, 2S)-isomer

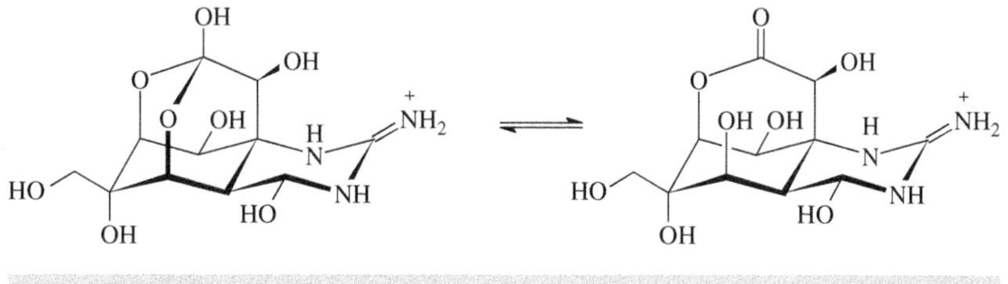

Figure 10.12 Tetrodotoxin salt tautomers.

Reactions and changes   Toxicity of the fish can be reduced by extraction with water or water acidified with acetic acid. Toxicity of the roe can likewise be reduced by the action of alkaline reagents. During heating, the toxic compounds are partially degraded to form less toxic derivatives. Sometimes, even the meat may be toxic due to migration of the toxin from the skin.

Biological effects   Tetrodotoxin is a powerful neurotoxin (paralytic poison, preventing the function of neuronal channels for Na^+ ions). More than 60% of poisonings are fatal. About as toxic as tetrodotoxin ($LD_{50} = 8$–$10\,\mu g/kg$ in mice) is chiriquitoxin; other compounds are less so.

10.3.8.2.2 Ciguatoxins

Structure, nomenclature, and occurrence   Poisoning caused by species of moray eels, such as the giant moray (*Gymnothorax javanicus*, Muraenidae), found at coral reefs in tropical seas in the Indo-Pacific, is known as CFP. Mainly responsible for the poisoning are ciguatoxin and maitotoxin and its congeners, which are members of the polycyclic ether family of marine toxins, found in fish flesh. Ciguatoxin (**10-98**) and its polar congeners arise by the action of fish oxidases from polar congeners derived from dinoflagellates (organisms of the phylum Dinoflagellata). Dinoflagellates adhere to coral, algae, and seaweed, where they are eaten by herbivorous fish, which in turn are eaten by carnivorous fish. For example, ciguatoxin-3C (formerly known as gambiertoxin 4b, **10-99**) is detected in the epiphytic dinoflagellate species *Gambierdiscus toxicus* (Goniodomataceae) along with related compounds, such as gambierol (**10-100**). The congener 54-deoxyciguatoxin (ciguatera-4B, **10-99**), which is found only in fish, is the immediate precursor of ciguatoxin. A related polycyclic ether is maitotoxin (**10-101**).

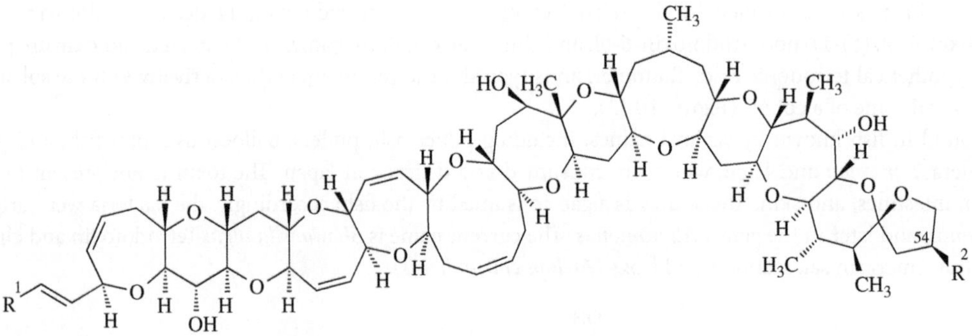

10-98, ciguatoxin, $R^1 = CH(OH)CH_2OH$, $R^2 = OH$
ciguatoxin-4B, $R^1 = CH=CH_2$, $R^2 = H$

Reactions and changes   Ciguatoxins do not affect the organoleptic characteristics of meat. They are thermostable and cannot be removed by thermal procedures. Long-term water extraction may partially reduce their content.

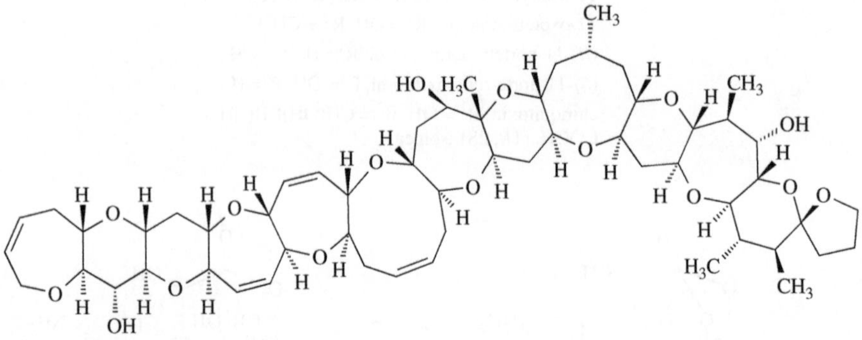

10-99, ciguatoxin-3C

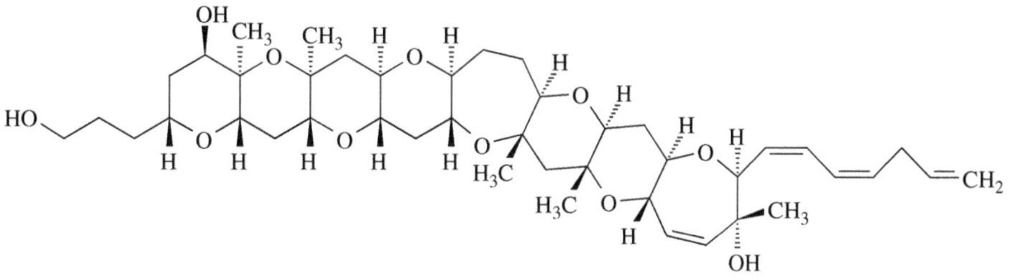

10-100, gambierol

Biological effects CFP is the most common form of poisoning on consumption of fish. Symptoms appear a few hours after eating as digestive problems and subsequent neurological symptoms, muscle pain, weakness, low blood pressure, and changes in heart rate. The variety of symptoms is often associated with the presence of maitotoxin and other bioactive compounds. Mortality caused by this poisoning is low. Ciguatoxin is about 10 times more toxic than less polar congeners (such as ciguatoxin-4B). Maitotoxin belongs to the most toxic non-protein compounds (LD_{50} = 50 ng/kg in mice).

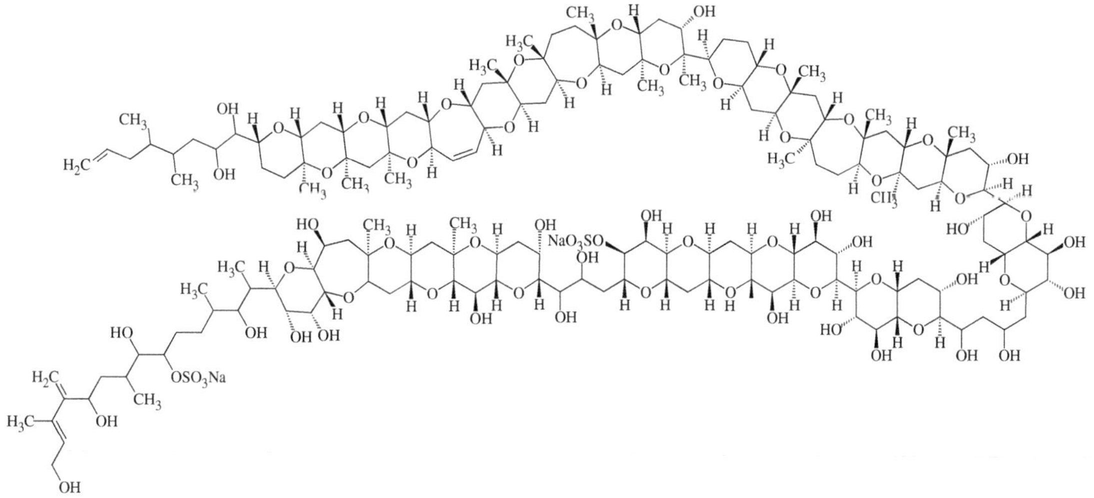

10-101, maitotoxin

10.3.8.2.3 Palytoxin

Structure, nomenclature, and occurrence Palytoxin is a group of related lipophilic toxins, the structure of which may vary in different species of zoanthids (animals within the order Zoantharia), commonly found in coral reefs. Palytoxin (**10-102**) was first isolated from the Hawaiian seaweed-like coral zoanthid *Palythoa toxica*. Later, it was found in the red alga *Chondria armata* (Rhodomelaceae), crabs of the genera *Demania* and *Lophozozymus* (Xanthidae), pink tail triggerfish (*Melichthys vidua*, Balistidae), and scrawled filefish (*Aluterus scriptus*, Monacanthidae), as well as in the liver of parrot fish (*Ypsiscarus ovifrons*). Even though the appearance of palytoxin was initially restricted to tropical areas, its recent occurrence in microalga, dinoflagellate of the genus *Ostreopsis* in the Mediterranean Sea, points to a worldwide dissemination probably related to climate change.

Reactions and changes Palytoxin group toxins are relatively stable. No data on their changes during cooking and industrial processing are known.

Biological effects Palytoxin targets the sodium–potassium pump protein by binding to the molecule analogously as tetrodotoxin. It probably also has carcinogenic effects. It is considered one of the most toxic non-peptide substances known, second only to maitotoxin

($LD_{50} = 0.1$–$0.3\,\mu g$ in mice) of the same class as botulotoxin. It is responsible for lethal poisonings caused by eating the meat of the crab *Demania reynaudii* in the Philippines. Poisoning causes similar symptoms to tetrodotoxin poisoning, including angina-like chest pains, asthma-like breathing difficulties, tachycardia (racing pulse), and unstable blood pressure with episodes of low blood pressure and haemolysis. Onset is rapid, with death occurring within minutes.

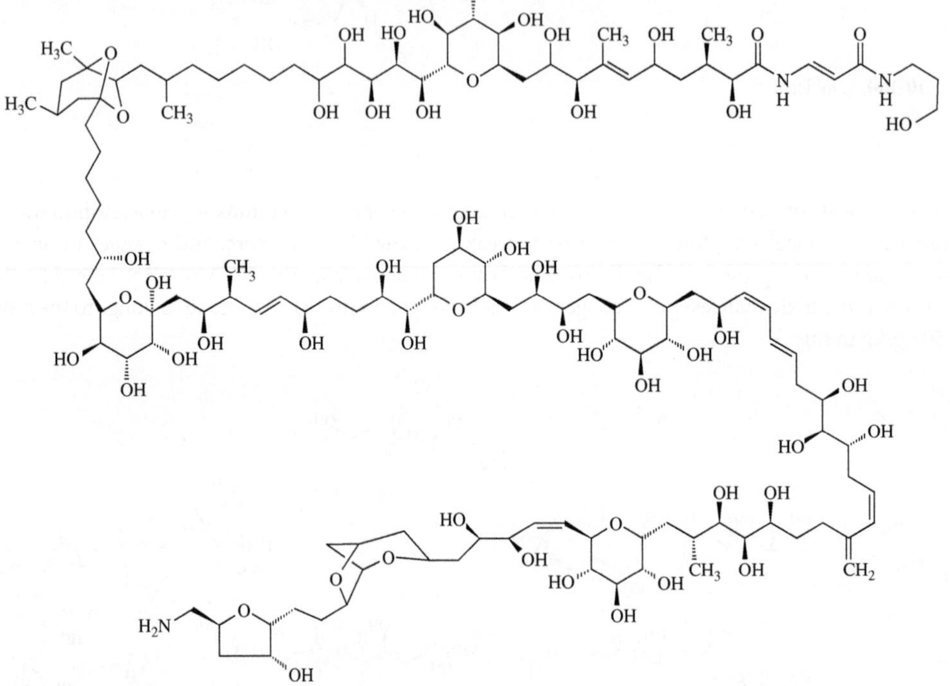

10-102, palytoxin

10.3.8.2.4 Other toxins

In saltwater fish, a number of other biologically active metabolites of algae occur that cause different types of poisoning. In addition to these secondary metabolites, some fish also produce their own toxins. For example, *Stichaeus grigorjewi* (Stichaeidae), common in the Northwest Pacific, contains toxic lipoproteins, which are known as α-, β-, and γ-lipostichaerins, in the roe. In addition to neutral lipids, lipostichaerins contain the unusual toxic phospholipid dinogunellin, which is composed of adenosine and 2-aminosuccinamide (**10-103**, R = fatty acid residue). The same ichthyotoxin was found in the eggs of other fish, which are toxic, especially at the time of spawning. Lipoproteins easily denature by heat, but the toxic component dinogunellin is stable. The toxin affects the central nervous system, but poisoning is not usually fatal.

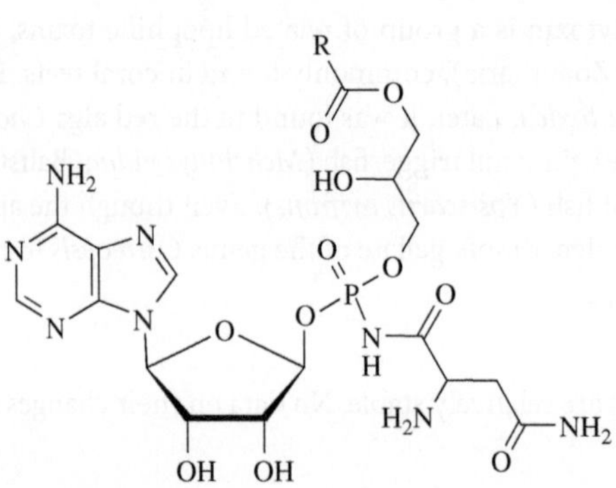

10-103, dinogunellin

The blood of some fish species of the garden eel family Congridae and the freshwater eel family Anguillidae, such as the European eel (*Anguilla anguilla*), contains ichthyohaematoxin. This protein decomposes with acids and bases and by the action of proteolytic enzymes, and is denatured by heating to a temperature of 70 °C or higher or by UV irradiation. Different individuals are sensitive to the eel's blood in different ways. Poisoning occurs rarely because the blood is toxic only when consumed in large quantities. Toxicity depends on the season and is highest at the time of spawning.

Ichthyocrinotoxins occur in toxic skin secretions of many fish species of the families Soleidae, Ariidae, Ostraciidae, Tetraodontidae, Gobiidae, Serranidae, Batrachoididae, and Muraenidae. They show strong haemolytic effects and serve as protective agents against microorganisms and parasites. The structure of the toxins is dependent on their origin. Toxic secretion components are peptides, steroids, and various fatty acid derivatives that have a bitter taste, leaving the food unacceptable; poisoning of humans is unknown.

For example, the skin secretion of the fish peacock sole *Pardachirus pavoninus* (Soleidae), occurring in the Indo-Pacific region, contains linear peptides produced by a series of toxic glands along the bases of the dorsal- and anal-fin rays. These toxic peptides, composed of 33 amino (2800 Da), are called pardaxin 1, pardaxin 2, and pardaxin 3. The primary structure of pardaxin 1 is formed by the following sequence of amino acids: Gly-Phe-Phe-Ala-Leu-Ile-Pro-Lys-Ile-Ile-Ser-Ser-Pro-Leu-Phe-Lys-Thr-Leu-Leu-Ser-Ala-Val-Gly-Ser-Ala-Leu-Ser-Ser-Ser-Gly-Glu-Gln-Glu. As with other polypeptides, paradaxins are labile compounds that are relatively easily hydrolysed by proteolytic enzymes, and partly also by cooking.

Other compounds present in the toxic secretion of peacock sole are glycosides of steroid toxins called pavoninins. The skin secretion of a related fish species known as finless sole (*Pachygrapsus marmoratus*) is structurally related lipophilic toxins called mosesins. Examples of these metabolites are pavonins 1 and 2 and mosesin 1 (**10-104**).

The skin secretion of *Ostracion lentiginosus* and of various other members of the Ostraciidae family contains toxic pahutoxin, (S)-2-(3-acetyloxyhexadecanoyl)oxyethyltri methylazanium chloride (choline chloride ester of 3-acetylhexadecanoic acid, **10-105**). Pahutoxin is a cationic surfactant. It can be released to the surrounding area, which negatively affects other fish, as the toxic effect is similar to the effect of saponins. The secretion of Hawaiian boxfish *Opisthothylax immaculatus* contains pahutoxin and the related homopahutoxin (choline chloride ester of 3-propionylhexadecanoic acid), which has the systematic name (S)-2-(3-propanoyloxyhexadecanoyl)oxyethylazanium chloride (**10-105**).

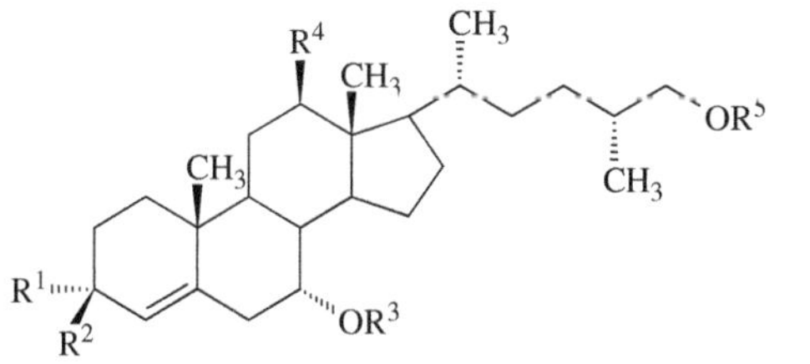

10-104, pavoninin 1, $R^1 = OH$, $R^2 = OH$, $R^3 = \beta$-D-Glc$p$NAc-(1→, $R^4 = H$, $R^5 = COCH_3$
pavoninin 2, $R^1 = OH$, $R^2 = OH$, $R^3 = \beta$-D-Glc$p$NAc-(1→, $R^4 = H$, $R^5 = H$
mosesin 1, $R^1 = OH$, $R^2 = H$, $R^3 = \beta$-D-Gal$p$6Ac-(1→, $R^4 = OH$, $R^5 = COCH_3$

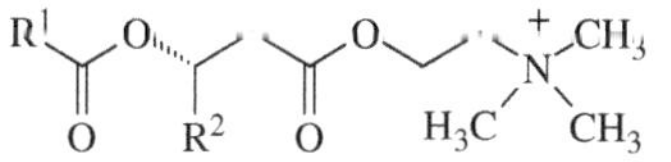

10-105, pahutoxin, $R^1 = CH_3$, $R^2 = (CH_2)_{12}CH_3$
homopahutoxin, $R^1 = CH_2CH_3$, $R^2 = (CH_2)_{12}CH_3$

10.3.9 Cyanogens

Cyanogenesis is the ability of plants and some other organisms to produce hydrogen cyanide by decomposition of cyanogenic compounds. Cyanogenic compounds or cyanogens have been detected in various parts of the roughly 3000 species of plants indexed to 110 families. It is assumed that cyanogens in plants repel predators by their bitter taste and odour and through the toxicity of their breakdown products. They may also participate in nitrogen metabolism as nitrogen storage forms during seed germination and early stages of plant development. Cyanogens are divided into three basic groups:

- cyanogenic glycosides (β-glycosides of 2-hydroxynitriles);

- pseudocyanogenic glycosides (glycosides or methylazoxymethanol, azoxyglycosides);

- cyanogenic lipids (cyanolipids or fatty acid esters of cyanohydrins).

In addition, some higher plants also accumulate different nitriles during the assimilation of hydrogen cyanide. Examples are β-cyano-ʟ-alanine and β-aminopropionitrile in vetches (*Vicia* spp.) and sweet peas or vetchlings (*Lathyrus* spp.) of the legume family (Fabaceae). Many plants also produce cyanides as byproducts of the biosynthesis of plant hormone ethylene (Figure 2.1).

10.3.9.1 Cyanogenic glycosides

10.3.9.1.1 Structure, nomenclature, and occurrence

The approximately 75 documented cyanogenic glycosides are all *O*-β-glycosides of 2-hydroxynitriles (formerly called cyanohydrins) that are glycosides of nitriles of 2-hydroxycarboxylic acids (**10-106**). Cyanogenic glycosides are the most important and most widely occurring cyanogens of many plants consumed as human food or used as livestock feed. They are located mainly in dicotyledonous plants belonging to the families Fabaceae, Asteraceae, Euphorbiaceae, and Passifloraceae and in monocotyledonous plants of the families Poaceae and Araceae. A certain cyanogenic glycoside occurs in only one or two families of plants, and one or two cyanogenic glycosides occur in only one plant.

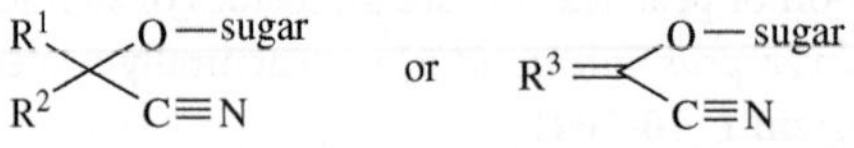

10-106, general structure of cyanogenic glycosides

| acacipetalin | amygdalin, lucumin passiedulin, prunasin vicianin | deidaclin | dihydrogynocardin |

| dhurrin | gynocardin | heterodendrin | holocalin |

| linamarin linustatin | lotaustralin neolinustatin | passicoriacin | sambunigrin |

| suberin A | taractophyllin | taxiphyllin | tetraphyllin A |

| tetraphyllin B | triglochinin | volkenin | zierin |

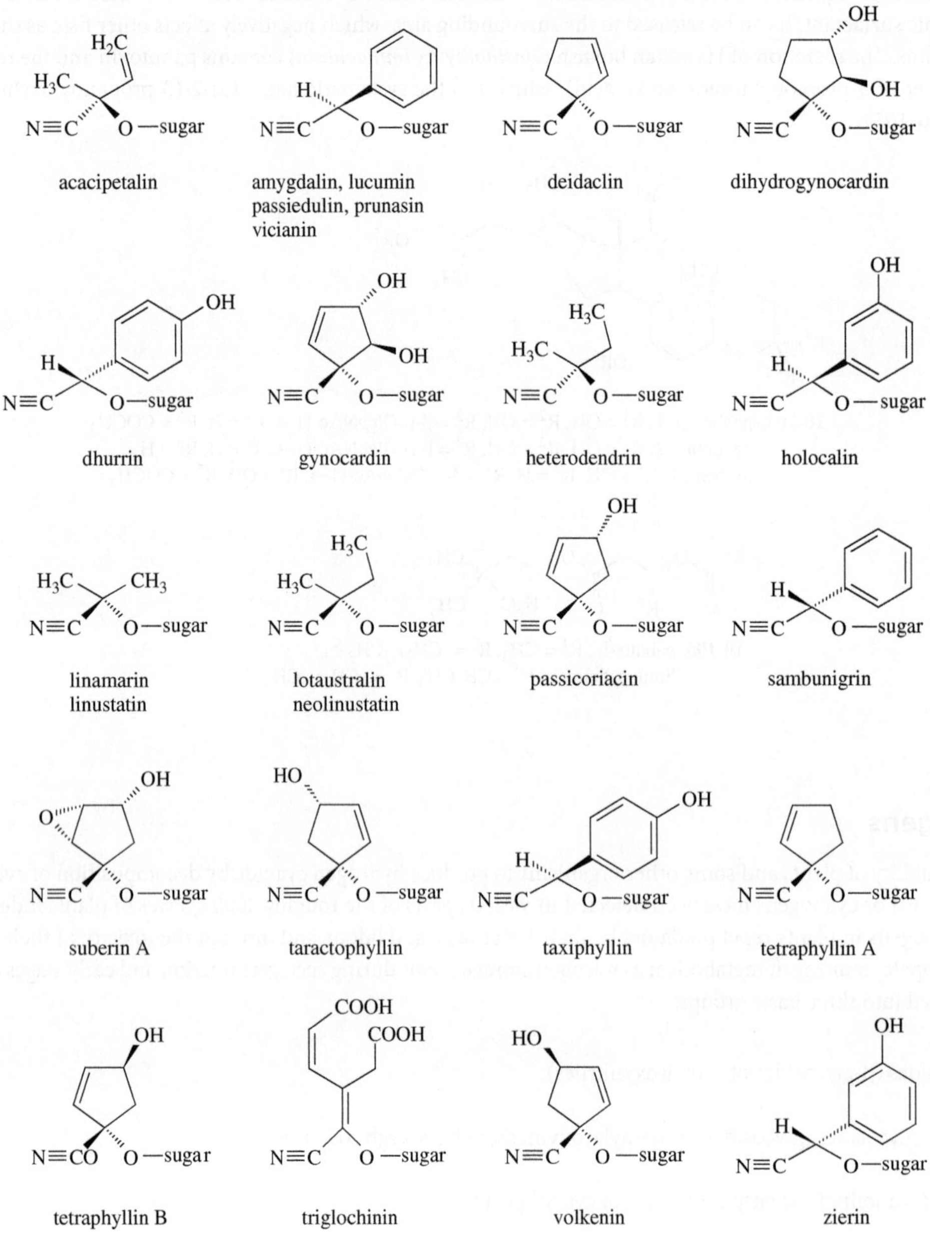

10-107, cyanogenic glycosides

Table 10.15 Structures and occurrences of cyanogenic glycosides.

Trivial name	Sugar	Isomer	Occurrence (Latin names of plant species)
Acacipetalin (proacacipetalin)	D-Glucose	S	Acacia
Amygdalin	Gentiobiose	R	Prunus
Deidaclin	D-Glucose	R	Deidamia, Passiflora
Dhurrin	D-Glucose	S	Sorghum
Dihydrogynocardin	D-Glucose	1S,4S,5R	Passiflora
Gynocardin	D-Glucose	1S,4S,5R	Gynocardia, Pangium, Taractogenes, Rawsonia
Heterodendrin (dihydroacipetalin)	D-Glucose	S	Acacia
Holocalin	D-Glucose	R	Sambucus
Linamarin (phaseolunatin)	D-Glucose	−	Trifolium, Lotus, Phaseolus, Manihot
Linustatin	Gentiobiose	−	Linum
Lotaustralin (methyllinamarin)	D-Glucose	R	Lotus, Manihot
Lucumin	Primeverose	R	Lucuma
Neolinustatin	Gentiobiose	R	Linum
Passicoriacin	D-Glucose	S	Passiflora
Passiedulin	D-Allose	R	Passiflora
Prunasin	D-Glucose	R	Prunus, Malus, Pyrus, Sorbus, Cydonia, Carica
Sambunigrin	D-Glucose	S	Prunus, Sambucus
Taractophyllin	D-Glucose	1R,4S	Passiflora
Suberin A	D-Glucose	1R,2R,3R,4R	Passiflora
Taxiphyllin	D-Glucose	R	Taxus, Triglochin, Bambusa
Tetraphyllin A	D-Glucose	S	Tetrapathaea, Passiflora
Tetraphyllin B	D-Glucose	1S,4S	Tetrapathaea, Passiflora, Mathurina, Carica
Triglochinin	D-Glucose	−	Triglochin, Glyceria, Melica
Vicianin	Vicianose	R	Vicia
Volkenin	D-Glucose	1R,4R	Passiflora, Mathurina
Zierin	D-Glucose	S	Sambucus

Cyanogenic glycosides differ in the type of bound sugar, substituents R^1, R^2 (or R^3), and chirality of carbinol carbon atom. With some exceptions, the sugar commonly bound in cyanogenic glycosides is monosaccharide β-D-glucose. The exceptions are disaccharides β-vicianose, α-L-Ara$p$-(1 → 6)β-D-Glc$p$, β-primeverose, β-D-Xyl$p$-(1 → 6)-β-D-Glc$p$, β-gentiobiose, β-D-Glc$p$-(1 → 6)-β-D-Glc$p$, and some others. In most cyanogenic glycosides, the substituent R^1 is an aliphatic or aromatic substituent and R^2 is hydrogen. In this case, the carbinol carbon atom (C-2 carbon of aglycone) is chiral and epimeric pairs of compounds therefore exist. The structures of the major cyanogenic glycosides are shown in formulae **10-107** and Table 10.15, which also lists their occurrence.

Cyanogenic glycosides are usually subdivided according to the amino acids from which they arise. Most cyanogenic glycosides are derived from valine (linamarin), isoleucine (lotaustralin), leucine (heterodendrin and epiheterodendrin), phenylalanine (amygdalin and prunasin), and tyrosine (dhurrin, taxifyllin and triglochinin). Cyclopentanoid cyanogens (such as gynocardin) arise from cyclopentenylglycine (**2-16**).

The simplest cyanogenic glycoside derived from valine is linamarin (formerly also called phaseolunatin). Its higher homologue is (R)-lotaustralin, which is derived from isoleucine. Both cyanogens occur together, because the corresponding enzymes are not specific and transform both amino acids (valine and isoleucine) into cyanogenic glycosides. Of the products containing these cyanogens, the one that is the most important for human nutrition is cassava (*Manihot esculenta*, Euphorbiaceae), sometimes also called yucca or tapioca, which is a major staple food in the developing world, providing a basic diet for around 500 million people. Cassava is native to South America and is now extensively cultivated in tropical and subtropical regions, especially in sub-Saharan Africa, but also in Indonesia and other countries. Cassava contains linamarin as the main cyanogenic glycoside, with (R)-lotaustralin also present in an amount about 20 times smaller. Both glycosides occur in the leaves and starchy tubers, especially in the cortex. Linamarin is similarly the cyanogenic glycoside of butter (lima) beans (*Phaseolus lunatus*, Fabaceae) and other plants from the same family, which are often a part of livestock forage. The main cyanogen in butter beans is (*R*)-lotaustralin. Linustatin has the same aglycone as linamarin, and the same aglycone as in (*R*)-lotaustralin is found in (*R*)-neolinustatin. These glycosides, containing disaccharide gentiobiose, occur as the major cyanogenic glycosides in the linseed (*Linum usitatissimum*, Linaceae). From leucine, the β-glucosides (*S*)-heterodendrin (dihydroacacipetalin) and (*S*)-acacipetalin (proacacipetalin) are derived, occurring in many acacias (*Acacia* spp.), shrubs, and trees belonging to the subfamily Mimosoideae of the family Fabaceae.

A number of cyanogenic glycosides are derived from the aromatic amino acids phenylalanine and tyrosine. From phenylalanine is derived the epimeric pair of β-glucosides known as (*R*)-prunasin and (*S*)-sambunigrin. (*R*)-Passiedulin of purple passion fruit, also known as purple granadilla (*Passiflora edulis*, Passifloraceae), originating in tropical South America, contains the rare hexose D-allose instead of D-glucose. Prunasin is a cyanogenic β-glucoside occurring in seeds of many plants of the rose family (Rosaceae), such as plums, apricots, peaches, apples, pears, quinces, cranberries, and many acacia species; it is also present in passion fruits, accompanied by other cyanogens. Sambunigrin, accompanied by prunasin and other glycosides, is the main cyanogenic β-glucoside of elderberries (*Sambucus nigra*, Adoxaceae). It is found in all parts of the plant, especially the leaves and immature fruits. During maturation, its content decreases, and by the time of maturity, it is no longer present in the fruits. Juice made from unripe elderberries does not contain sambunigrin at concentrations that represent a health risk, but can be unacceptably bitter. (*R*)-Epiheterodendrin occurs in some varieties of malting barley (*Hordeum* spp., Poaceae) and in malt.

The same aglycone as prunasin (D-mandelic acid nitrile) contains (*R*)-amygdalin, which is β-gentiobioside, (*R*)-vicianin, containing β-vicianose, and (*R*)-lucumin, which contains β-primeverose. The most important cyanogenic glycoside of this group of cyanogens is amygdalin. Relatively high amounts of amygdalin (together with prunasin and sambunigrin) are found in seeds from Rosaceae species, as compared with seeds from non-Rosaceae species (0.01–0.2 g/kg). Significant resources include bitter almonds and apricot, peach, plum, and cherry pits. In small amounts, amygdalin also occurs in apple, pear, and quince seeds. Amygdalin is responsible for the bitterness of semi-bitter and bitter almond phenotypes. Its levels range from 2 to 157 mg/kg in non-bitter, from 524 to 1773 mg/kg in semi-bitter, and from 33 007 to 53 998 mg/kg in bitter almonds. These amounts are not life-threatening. Poisoning usually occurs only accidentally, for example when children consume large quantities of bitter almonds. The amygdalin content in apple seeds ranges from 1 to 4 g/kg. That in commercially available apple juice is low, ranging from 0.01 to 0.04 g/l. The precursor of amygdalin is prunasin, which is also present in the flesh of immature fruits.

From phenylalanine is derived the cyanogenic β-glucoside vicianin, found in vetches, such as the common vetch (*Vicia sativa*, Fabaceae). Lucumin is present in tropical plants of the genus *Lucuma* of the same plant family. Other cyanogenic β-glucosides are (*R*)-holocalin and its epimer (*S*)-zierin, which occur in small amounts in elder leaves and unripe berries.

Tyrosine yields the β-glucoside (*R*)-taxiphyllin and epimeric (*S*)-dhurrin. Taxiphyllin is present in bamboo shoots (*Bambusa* spp., Poaceae) and needles of juniper (*Juniperus* spp., Cupressaceae). Dhurrin occurs in grasses (Gramineae) such as sorghum (*Sorghum bicolor*), especially in the young green parts of the plant. Tyrosine is also the precursor of the β-glucoside triglochinin, which occurs in grasses such as melic grasses (*Melica* spp.), reed manna grass (*Glyceria maxima*), and wetland plants such as marsh arrow grass (*Triglochin palustris*, Junkaginaceae), which may become a part of forage for livestock.

An unusual cyanogenic glycoside is (1*S*,4*S*,5*R*)-gynocardin. It occurs along with other β-glucosides and fatty acids with a cyclopentene ring (see Section 3.3.1.3.3) in the seeds of chaulmoogra trees of the genus *Hydnocarpos* (syn. *Taraktogenos*, Achariaceae, formerly Flacourtiaceae), native to Indonesia, Malaysia, and the Philippines.

The cyanogenic glucoside contents of selected foods, expressed in terms of the amount of bound hydrogen cyanide (cyanogenic potential), are given in Table 10.16.

10.3.9.1.2 Reactions and changes

Cleavage of cyanogenic glycosides may occur either enzymatically or chemically. Glycosides containing β-D-glucose are more or less hydrolysed by specific β-glucosidases, which are often referred to by their common names according to origin. For example, the bitter almond enzyme is called amygdalase (amygdalinase, formerly also emulsin) and the cassava enzyme is known as linamarase. The enzyme hydrolyses the glycosidic bond to produce sugar and cyanohydrin. Glycosides containing disaccharides are broken down in two stages. In the first, the corresponding glycoside (containing bound monosaccharide) is produced, and in the second, the monosaccharide and aglycone (cyanohydrin) are released. For example, prunasin arises as an intermediate from amygdalin and is further hydrolysed to glucose and 2-hydroxynitrile. Enzymes involved in the decomposition of 2-hydroxynitriles are aldehyde lyases, which catalyse the cleavage of a cyanohydrins to aldehydes or ketones and hydrogen cyanide (Figure 10.13). Examples are hydroxynitrilase, which catalyses the cleavage of acetone cyanohydrins, mandelonitril lyase, also known as (*R*)-oxynitrilase, which cleaves a number of aromatic and aliphatic cyanohydrins, and hydroxymandelonitrile lyase, which cleaves (*S*)-4-hydroxymandelonitriles. The final product of triglochinin degradation is not the corresponding carbonyl compound, but a product of its hydration, (*E*)-but-2-ene-1,2,4-tricarboxylic acid.

The hydrolysis products of major cyanogenic glycosides are summarised in Table 10.17. In all cases, other products are sugars and hydrogen cyanide. The enzymatic degradation of amygdalin is shown in Figure 10.14 as an example. Hydrogen cyanide released from stone fruit cyanogens is a precursor of toxic ethyl carbamate, which occurs mainly in stone fruit distillates (see Section 12.3.8).

Table 10.16 HCN contents in some plant cyanogenic glycosides.

Origin	HCN (mg/kg fresh material)	Origin	HCN (mg/kg fresh material)
Cassava		*Stone fruits*	
Leaves	650–1040	Bitter (sweet) almonds	2800–4110 (0–100)
Whole tubers	550	Apricot seeds	3200
Tuber bark	840–2450	Sour cherry seeds	3540
Tuber flesh	100–330	Sour cherry flesh	10
Bamboo		*Sorghum*	
Unripe shoots	3000	Germinating plants	2400
Tops of unripe shoots	8000	Young leaves	600
Purple passion fruit		*Legumes*	
Unripe fruit	700	Butter beans	100–4000
Ripe fruit	100	Common beans	20
Linseeds	200–380	Peas	23

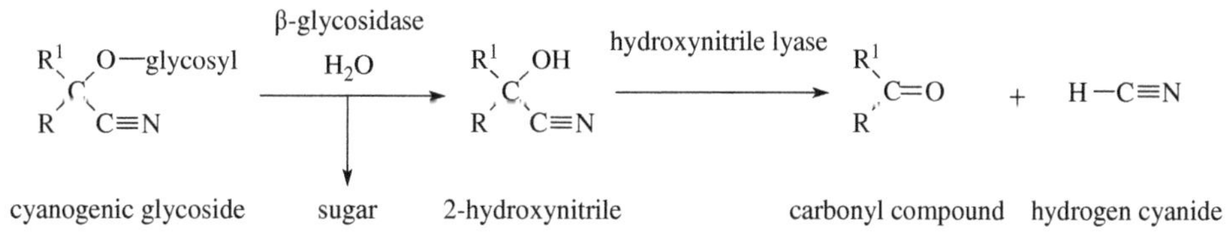

Figure 10.13 Enzymatic hydrolysis of cyanogenic glycosides.

Table 10.17 Hydrolysis products of cyanogenic glycosides.

Glycoside	Carbonyl compound	Glycoside	Carbonyl compound
Linamarin, linustatin	Acetone	Vicianin	Benzaldehyde
Lotaustralin, neolinustatin	Butan-2-one	Holocalin	3-Hydroxybenzaldehyde
Prunasin	Benzaldehyde	Zierin	3-Hydroxybenzaldehyde
Sambunigrin	Benzaldehyde	Taxiphyllin	4-Hydroxybenzaldehyde
Amygdalin	Benzaldehyde	Dhurrin	4-Hydroxybenzaldehyde

Cyanogenic glycosides are also hydrolysed by diluted acids at elevated temperatures, producing sugars and 2-hydroxynitriles. For example, amygdalin produces D-mandelic acid nitrile. In concentrated acids, the hydrolysis of nitriles yields 2-hydroxycarboxylic acids and ammonium salts. Amygdalin thus yields D-mandelic acid.

In aqueous solutions, and rapidly in alkaline solutions, cyanogenic glycosides epimerise. (R)-amygdalin yields a mixture of epimers known as isoamygdalin. The epimer of amygdalin derived from L-mandelic acid, called (S)-neoamygdalin, does not occur in nature. The epimeric pair of prunasin and sambunigrin was formerly called prulaurasin. In strongly alkaline media, both epimers yield glycosides of 2-hydroxycarboxylic acids. The product of amygdalin alkaline hydrolysis is an epimeric mixture of mandelic acid gentiobioside, called amygdalinic acid (Figure 10.15). (S)-epilucumin, which arises from (R)-lucumin, likewise does not occur in nature.

Detoxication of cassava in culinary practice is achieved by sun drying, crushing and grinding, soaking in water, boiling, and fermentation. During fermentation, about 90% of cyanogens are removed; simultaneously formed substances carry the favourable organoleptic properties of the products. After 30 minutes of cooking, the content of cyanogens decreases by about 8–30% of the original amount. The main cyanogen of bamboo shoots, taxiphyllin, is labile and decomposes during cooking.

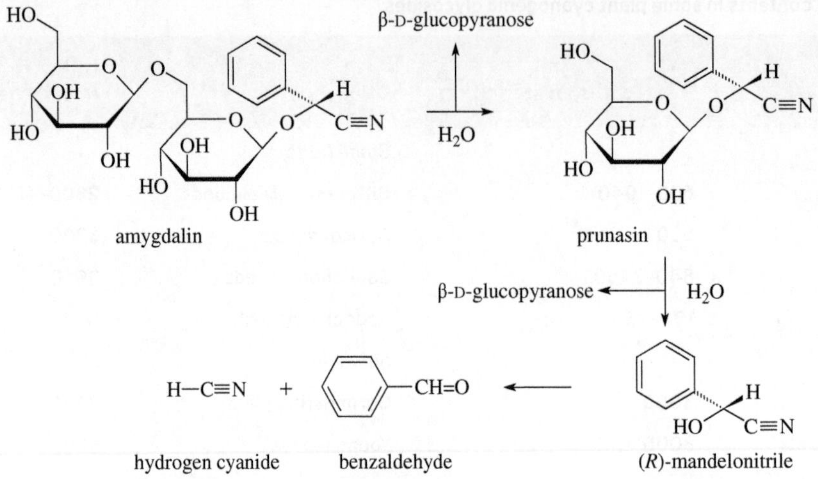

Figure 10.14 Enzymatic hydrolysis of amygdalin.

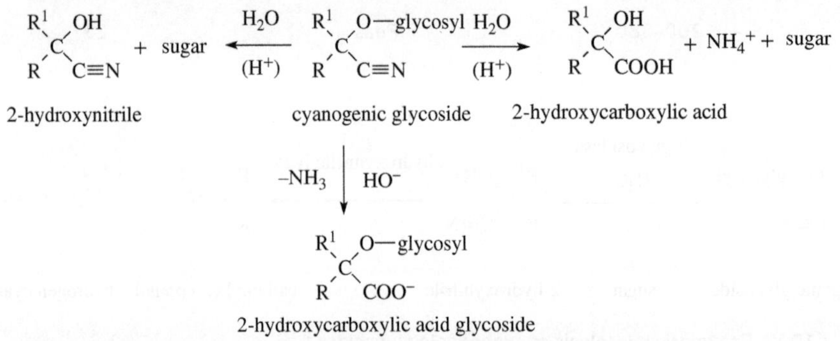

Figure 10.15 Degradation of cyanogenic glycoside in acidic and alkaline media.

The most common source of cyanogens in the European diet is stone fruits. Cyanogens are mainly present in the inedible part of the fruit, the stone. When whole fruit is processed, cyanogens may gradually pass from the stone into the edible part. For example, fruit juices obtained from unstoned fruit may contain up to 15 mg/kg of hydrogen cyanide. Higher concentrations also occur in compotes made of unstoned fruit. For example, apricot compotes may contain up to 33 mg/kg of hydrogen cyanide. The amount of released hydrogen cyanide depends on the technological conditions, especially the temperature. The enzymatic degradation can occur only if there is thermal damage of cell membranes and if the enzymes have not been inhibited. The higher the inactivation effect, the lower the concentration of hydrogen cyanide in the finished product. Benzaldehyde arising from the enzymatic degradation of stone fruit cyanogens is often the bearer of flavour in the stone fruit products. Benzoic acid formed by partial oxidation of benzaldehyde has antimicrobial properties.

Kernels of some stone fruits (such as bitter almonds) are used to make marzipan, almond jelly, and other confectionery products. Again, even in this case, enzymatic hydrolysis of cyanogenic glycosides releases hydrogen cyanide, which is, however, removed in the subsequent technological operations. Similar changes of cyanogens occur in the production of fruit spirits, as the major quantity of hydrogen cyanide formed in the fermentation process is vaporised during fermentation and subsequent distillation. Stone fruit distillates usually contain 0.3–3 mg/kg of hydrogen cyanide (cyanides).

10.3.9.1.3 Biological effects

Cyanogenic glycosides are not toxic; only their decomposition product, hydrogen cyanide, is. Acute toxicity is due to inhibition of cytochrome oxidase in the respiratory chain by cyanide (reaction with copper ions) and reaction with haemoglobin (formation of cyanohaemoglobin). The lethal dose of cyanide for humans is from 0.5 to 3.5 mg/kg body weight, which corresponds to 35–245 mg for an adult. Symptoms of ingestion of a lethal dose are manifested by stiffness of limbs, dazed consciousness, cyanosis (bluish discoloration of the skin), convulsions, and coma. Ingestion of lower cyanide concentrations leads to headaches, anxiety, uneasiness in the throat and chest, and heartbeat and muscle weakness. Chronic symptoms of poisoning (manifested by disease of the peripheral nerves) are referred to as degenerative tropical neuropathy.

In cases where fresh cyanogenic plants containing intact glycosides are eaten, only a small amount of hydrogen cyanide is released, since glycosidases hydrolysing cyanogenic glycosides are inhibited in the stomachs of humans and other monogastric animals. In ruminants (polygastric animals), the plant enzymes are not inhibited and cyanogenic glycosides are partly decomposed by the rumen microflora, which can lead to poisoning.

Detoxification products of cyanides in the human body are thiocyanate (rhodanide) ions. The conversion of cyanides to thiocyanates in the presence of thiosulfates as sulfur donors is catalysed by mitochondrial rhodanese (thiosulfate: cyanide sulfurtranferase). Another enzyme that provides detoxification of cyanides is 3-mercaptopyruvate sulfurtransferase. Certain anticarcinogenic effects of cyanogenic glycosides (vitamin B_{17}) described in the past have not been demonstrated (see Section 5.15).

Plants have different detoxification mechanisms from animals. Cysteine (or serine) reacts with hydrogen cyanide in a reaction catalysed by β-cyanoalanine synthase to form β-cyanoalanine, which is transformed by β-cyanoalanine hydratase into asparagine, an amino acid important for nitrogen storage (see Figure 2.2). Asparagine can be further metabolised to aspartic acid and ammonia by the action of asparaginase. Pathogenic fungi that attack plants, such as *Gloecercospora sorghi* and *Colletotrichum graminicola*, infecting sorghum, contain the enzyme cyanide hydratase, which catalyses the addition of water to hydrogen cyanide, yielding formamide ($H-CONH_2$).

10.3.9.2 Pseudocyanogenic glycosides

Toxic pseudocyanogenic glycosides (azoxyglycosides) are present in many plants of the cycad family Cycadaceae, found across the subtropical and tropical parts of the world. The pseudocyanogenic glycoside cycasin occurs in seeds of some cycad species used as a source of starch. Depending on the type of plant, starchy dishes made from cycad seeds may contain up to 0.22 mg/kg of azoxyglycosides.

For example, sago cycad (*Cycas revoluta*), the only native cycad in Japan, has been used as a starch source on Amami Oshima, an island in the Amami Islands, for centuries. The stems are mashed into a pulp and then fermented, slowly removing the toxins. Likewise, seeds have been used to make a cycad cake called *sotetsu mochi*. Moreover, crushed seeds are made into *sotetsu miso*, a paste much like the normal miso made from soya beans. All of these practices are considered dangerous, but the traditions persist in local areas to this day. If eaten, cycad products containing azoxyglycosides may lead to gastrointestinal distress and liver failure. Cycasin and other azoxyglycosides are not mutagenic or carcinogenic, but the main metabolite methylazoxymethanol is. Incidence of neurological disorders resembling Parkinson's disease has been reported in various areas of the Pacific (e.g. in the US island territory of Guam), but these may also be caused by the presence of lathyrogenic 2,4-diaminobutyric acid (see Section 2.2.1.2.4).

Pseudocyanogenic glycosides are decomposed by pathways other than cyanogenic glycosides. In a weakly alkaline medium, they release cyanides, formates, and elemental nitrogen; in an acidic medium, they yield formaldehyde, methanol, and elemental nitrogen. Their enzymatic degradation produces methylazoxymethanol (Figure 10.16).

10.3.9.3 Cyanogenic lipids

Cyanogenic lipids (cyanolipids) occur together with conventional triacylglycerols in the seeds of many plants of the soapberry family (Sapindaceae); in certain cases, they represent more than 50% of the seed oil. Members of this family have been widely studied for their pharmacological (antioxidant, anti-inflammatory, and antidiabetic properties) and anti-insect activities, and many species have economically valuable tropical fruits and wood. Cyanogenic lipids also occur in plants of the Hippocastaneaceae and Boraginaceae families, which are not used for human consumption.

The five-carbon skeletons of all cyanolipids, derived from leucine, have the cyano group and one or two hydroxyl groups esterified by fatty acids. Type I cyanolipids are diesters of 1-cyano-2-hydroxymethylprop-2-en-1-ol, type II are diesters of 1-cyano-2-hydroxy-methylprop-1-en-3-ol, type III are esters of 1-cyano-3-hydroxy-prop-1-ene (**10-108**), and type IV are esters of 1-cyano-2-methylprop-2-en-1-ol (Figure 10.17).

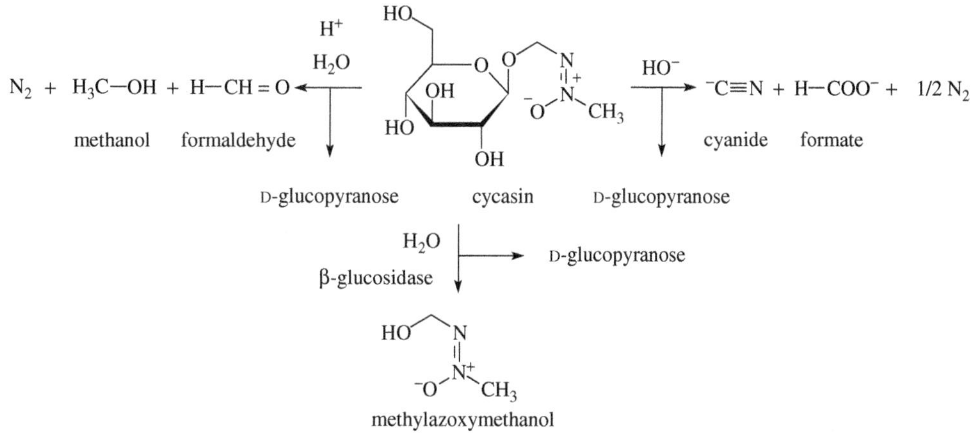

Figure 10.16 Degradation of pseudocyanogenic glycosides.

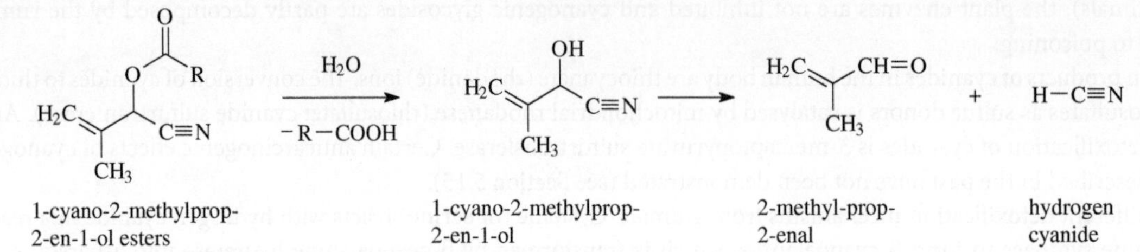

Figure 10.17 Hydrolysis of cyanogenic lipids.

In seeds of some plants, these basic skeletons are not esterified with fatty acids, but exist as glucosides. Most common are the type I cyanolipids, often accompanied by type II. Compounds of type III are only found together with type II. Cyanolipids of type IV are relatively rare. Compounds of types I and IV are cyanohydrins with a chiral centre, whilst compounds of types II and III are α,β-unsaturated nitriles that are not optically active. The cyanohydrins released by acid or base hydrolysis of type I and IV cyanolipids spontaneously decompose to form hydrogen cyanide (Figure 10.17), parent carbonyl compounds, and fatty acids.

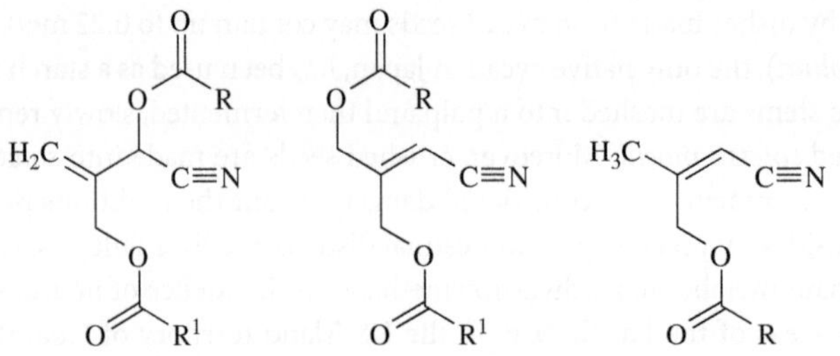

10-108, general structures of cyanogenic lipids

10.3.10 Glucosinolates

Glucosinolates, also known as thioglucosides, and formerly as mustard oil glycosides, are an important group of more than 150 secondary metabolites found exclusively in flowering dicotyledonous plants belonging to the order Brassicales. In terms of their importance, the dominant position is occupied by the crucifers (mustard or cabbage) family (Brassicaceae), which includes economically important crops (oilseeds, vegetables, and condiments). For example, glucosinolates are responsible for the typical pungent taste and odour of rapeseed, mustard, horseradish, wasabi, other vegetables, and some spices.

Considerable attention was given to glucosinolates mainly in connection with the strumigenic (goitrogenic) effects of rapeseed meal used as feed for livestock. Glucosinolates were therefore classified as antinutritional factors that interfere with iodine metabolism. Current research is focused primarily on the biological properties of glucosinolate degradation products and their positive effects, but they also have some negative effects on human health.

10.3.10.1 Structure, nomenclature, and occurrence

The glucosinolate molecule (**10-109**) is formed by the sugar component (in most cases β-D-glucose or β-D-thioglucose, which may be esterified by sinapic or some other carboxylic acid) and the aglycone, which is a sulfonated oxime anion in the (E)-position to the side chain R and in the (Z)-position to thioglucose residue. Generally, glucosinolates are found in nature as potassium salts. The vast number of glucosinolates occurring in nature results from the chemical diversity of the side chain, which is derived from a few amino acids (methionine and its higher homologues, phenylalanine, tyrosine, and tryptophan). An overview of the main glucosinolates is given in Table 10.18.

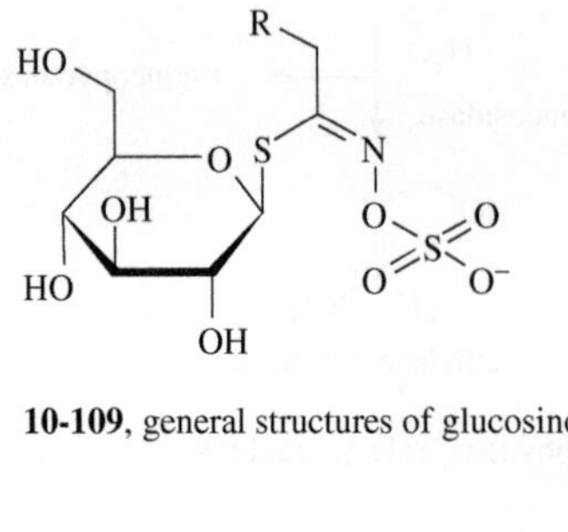

10-109, general structures of glucosinolates

Table 10.18 Names and structures of widespread glucosinolates[a].

No.	Trivial name	Substituent R	No.	Trivial name	Substituent R
1	Glucocapparin	Methyl	46	Glucoarabishirsuin	8-Methylthiooctyl
2	Glucoslepidiin	Ethyl-	47	Glucoarabispurpureain	9-Methylthiononyl
3	Sinigrin	Prop-2-en-1-yl (allyl)	48	Glucoiberin	(R)-3-Methylsulfinylpropyl
4	Glucoputranjivin	Isopropyl	49	Glucoraphenin	(R)-4-Methylsulfinylbut-3-en-1-yl
5	Gluconapin	But-3-en-1-yl	50	Glucoraphanin	(R)-Methylsulfinylbutyl
6	Glucocapparisflexuosain	Butyl	51	Glucoalyssin	(R)-5-Methylsulfinylpentyl
7	Glucoconringianin	Isobutyl	52	Glucohesperalin	(R)-6-Methylsulfinylhexyl
8	Glucochlearin	(S)-Methylpropyl	53	Glucoarabidopsithalianin (Glucoibarin)	(R)-7-Methylsulfinylheptyl
9	Glucobrassicanapin	Pent-4-en-1-yl	54	Glucohirsutin	(R)-8-Methylsulfinyloctyl
10	Glucocapparilinearisin	3-Methylbut-3-en-1-yl	55	Glucoarabin	(R)-9-Methylsulfinylnonyl
11	Glucokohlrabiin	Pentyl (Amyl)	56	Glucocamelein	(R)-10-Methylsulfinyldecyl
12	Glucoarmoracialapathin	3-Methylbutyl (Isoamyl)	57	Gluconesliapaniculatin	(R)-11-Methylsulfinylundecyl
13	Glucojiaputin	(S)-2-Methylbutyl	58	Glucocamelinain	(R)-11-Methylsulfinyldodecyl
14	Glucowasabiamin	Hex-5-en-1-yl	59	Glucoerysimumhieracifolinin	(R)-3-Hydroxy-5-methylsulfinylpentyl
15	Glucoraphasativusain	Hexyl	60	Glucoarabishirsutain	(R)-3-Oxo-8-methylsulfinyloctyl
16	Glucoraphanusativasin	4-Methylpentyl	61	Glucocheirolin	3-Methylsulfonylpropyl
17	Glucowasajaponicain	Hept-6-en-1-yl	62	Glucoerysolin	4-Methylsulfonylbutyl
18	Glucocapparimasiakain	2-Hydroxyethyl	63	Glucoerysihieracifoliumin	5-Methylsulfonylpentyl
19	Glucosisymbrin	(R)-2-Hydroxy-1-methylethyl	64	Glucobenzosisymbrin	(R)-2-Benzoyloxy-1-methylethyl
20	Glucoarabidopsithalin	2-Hydroxypropyl	65	Glucomalcolmiin	3-Benzoyloxypropyl
21	Glucoerysimumhieracifolium	3-Hydroxypropyl	66	Glucoaustriacuin	(R)-Benzoyloxy-1-ethylethyl
22	Progoitrin	(R)-2-hydroxybut-3-en-1-yl	67	Glucoarabidopsithain	5-Hydroxybenzoyloxypentyl
23	Epiprogoitrin	(S)-2-hydroxybut-3-en-1-yl	68	Glucotropaeolin	Benzyl
24	Glucoresedalbain	2-Hydroxy-2-methylpropyl	69	Gluconasturtiin	2-Phenethyl
25	Glucosisaustriacin	(R)-2-Hydroxy-1-ethylethyl	70	Glucoarmoracialapicin	3-Phenylpropyl
26	Glucocappariflexin	3-Hydroxybutyl	71	Glucoarmoracialafolicin	4-Phenylbutyl
27	Glucoarabidopsithalianain	4-Hydroxybutyl	72	Glucoreluteolain	2-Hydroxybenzyl
28	Glucoerypestrin	4-Methoxy-4-oxobutyl	73	Glucolepigrmin	3-Hydroxybenzyl
29	GlucoNapoleiferin	(S)-2-Hydroxypent-4-en-1-yl	74	Sinalbin	4-Hydroxybenzyl
30	Epinapoleiferin	(R)-2-Hydroxypent-4-en-1-yl	75	Glucolimnantin	3-Methoxybenzyl
31	Glucoarmoracialapathin	2-Hydroxypentyl	76	Glucoaubrietin	4-Methoxybenzyl
32	Glucoarabidopsin	5-Hydroxypentyl	77	Glucohesmatrolin	3,4-Dihydroxybenzyl
33	Glucocleomin	(R)-2-Hydroxy-2-methylbutyl	78	Glucoelongatin	4-Hydroxy-3-methoxybenzyl
34	Glucocapparisgrandisin	4,5,6,7-Tetrahydroxydecyl	79	Glucolongifoliain	3,4-Dimethoxybenzyl
35	Glucocapparisangulatain	5-Oxoheptyl	80	Glucosibarin (epiglucobarbarin)	(R)-2-Hydroxy-2-phenylethyl
36	Gluconorcappasalin	6-Oxoheptyl	81	Glucobarbarin	(S)-2-Hydroxy-2-phenylethyl
37	Glucocappasalin	5-Oxooctyl	82	Glucoarabihirsuin	(S)-2-Hydroxy-2-(4-hydroxyphenyl)ethyl
38	Glucosativin	4-Sulfanylbutyl	83	Glucoarabihirin	(R)-2-Hydroxy-2-(4-hydroxyphenyl)ethyl
39	Glucoamoracialapathifolin	2-Methylthioethyl	84	Glucobrassicin	3-Indol-3-ylmethyl
40	Glucoiberverin	3-Methylthiopropyl	85	4-Hydroxyglucobrassicin	4-Hydroxyindol-3-ylmethyl
41	Glucoraphasatin	4-Methylthiobut-3-en-1-yl	86	Sulfoglucobrassicin	1-Sulfoindol-3-ylmethyl
42	Glucoerucin	4-Methylthiobutyl	87	N-acetylglucobrassicin	1-Acetylindol-3-ylmethyl
43	Glucoberteroin	5-Methylthiopentyl	88	Neoglucobrassicin	1-Methoxyindol-3-ylmethyl
44	Glucoarabidopthalianin	6-Methylthiohexyl	89	4-Methoxyglucobrassicin	4-Methoxyindol-3-ylmethyl
45	Glucoarabishirsutain	7-Methylthiohexyl	90	1,4-Dimethoxyglucobrassicin	1,4-Dimethoxyindol-3-ylmethyl

[a]In addition to the mentioned glucosinolates, in plants occur glycosides of glucosinolates with sugars bound to the hydroxyl groups (apiofuranosyl, arabinopyranosyl, and rhamnopyranosyl glucosinolates) and esters with phenolic acids bound to glucose. (isoferuloyl, diisoferuloyl, sinapoyl, and disinapoyl glucosinolates).

Semisystematic names of glucosinolates are formed from the chemical names of the variable part of the molecule (side chain R) and the extension glucosinolate (e.g. 2-hydroxybut-3-en-1-ylglucosinolate). However, trivial names are more often used, derived mostly from the Latin names of the plants from which the glucosinolates were first isolated. For example, 4-hydroxybenzylglucosinolate was first isolated from white mustard seeds (*Sinapis alba*) and was therefore termed sinalbin. Prop-2-en-1-ylglucosinolate, isolated from seeds of black mustard (*Brassica nigra*, syn. *Sinapis nigra*), was called sinigrin. 2-Hydroxybut-3-en-1-ylglucosinolate, which is a common constituent of cruciferous vegetables and rapeseeds, is called progoitrin as it is the precursor of goitrin, showing significant goitrogenic (antithyroid, strumigenic) effects.

The side chain R determines the chemical, physical, and biological properties of the individual glucosinolates and the type of their degradation products. According to the structure of the side chain, glucosinolates can be divided into several groups:

- aliphatic or alk(en)ylglucosinolates (Nos. 1–6 in Table 10.18) or aliphatic hydroxy-substituted glucosinolates (Nos. 7–9);

- sulfur-containing glucosinolates with a methylthio group in the side chain (Nos. 10–13), or their oxidised forms (Nos. 14–19);

- aromatic glucosinolates with unsubstituted or substituted benzene rings (Nos. 20–26);

- indole glucosinolates (with a substituted indole skeleton, Nos. 26–30).

An overview of significant plants of the Brassicaceae family that contain glucosinolates is given in Table 10.19. Some plants belonging to other families are also occasionally consumed. An example is the caper bush (*Capparis spinosa*, Capparidaceae), with edible flower buds known as capers, containing glucosinolate glucocapparin, which are used as a seasoning. The seeds of papaya (*Carica papaya*, Caricaceae), containing glucosinolate glucotropaeolin, are used for their spicy and pungent flavour as a black pepper substitute. Glucotropaeolin is also the main glucosinolate in the roots of *Moringa oleifera* (Moringaceeae), commonly known as the drumstick tree. But the most predominant glucosinolate in leaves, flowers, pods, and seeds is glucomoringin, 4-*O*-(α-L-rhamnopyranosyloxy)benzylglucosinolate. The immature pods, flowers, and foliage of this tree are used for culinary purposes in different parts of the world, and the seed oil (yield 30–40% w/w), also known as 'ben oil', is used for the production of biodiesel. The main glucosinolate in the roots of wasabi (*Eutrema japonicum*, Brassicaceae), used widely as a pungent spice for sushi and sashimi, is sinigrin, although unusual thioglucosides known as wasulfisides A–D (**10-110**) have

Table 10.19 Overview of dominant glucosinolates of selected vegetables and oilseeds.

Vegetable	Latin name	Typical glucosinolates[a]
Cabbage	*Brassica oleracea* convar. *capitata*	3, 5, 7, 10, 22, 42, 48, 50, 69, 74, 80, 84, 85, 88
Brussels sprouts	*Brassica oleracea* convar. *oleracea* var. *gemmifera*	3, 5, 10, 14, 22, 48, 80, 85, 88
Cauliflower	*Brassica oleracea* var. *botrytis*	3, 5, 7, 14, 22, 42, 48, 50, 61, 80, 84, 88
Kale	*Brassica oleracea* var. *sabellica*	3, 5, 7, 9, 22, 42, 48, 74, 80, 81, 84
Kohlrabi	*Brassica oleracea* var. *gongylodes*	3, 48, 80, 84, 85, 88
Broccoli	*Brassica oleracea* var. *italica*	5, 7, 10, 14, 22, 40, 42, 48, 50, 51, 69, 74, 84, 85, 88, 89
Chinese cabbage	*Brassica chinensis* var. *chinensis*	76, 80
Turnip rape	*Brassica rapa* subsp. *rapa*	3, 5, 10, 14, 22, 59, 69, 80, 84
Rapeseed[b]	*Brassica napus* var. *napus*	5, 9, 22, 29, 84
Brown mustard	*Brassica juncea*	3, 5
Black mustard	*Brassica nigra*	3
White mustard	*Sinapis alba*	74
Radish	*Raphanus sativus*	7, 42, 48, 49, 74, 84
Horseradish	*Armoracia rusticana*	3, 5, 8, 9, 38, 48, 50, 68, 69, 74, 84, 85, 88, 89
Garden cress	*Lepidium sativum*	3, 68, 69
Watercress	*Nasturtium officinale*	69, 80, 81, 84, 85, 88, 89

[a] Names of substances are given in Table 10.18.

[b] Crambe (*Crambe abyssinica*), an oilseed crop native to the Mediterranean area, contains epiprogoitrin as the main glucosinolate instead of progoitrin. Rapeseeds contain a number of minor glucosinolates. For example, the 00 varieties (with low erucic acid and glucosinolate contents) contain, glucosinolates 1, 23, 24, 48–51, 61, 69, 72, 74, 76, 80, 81, 85, and 88 (Table 10.18) and glycosides of glucosinolates with sugar bound to the hydroxyl group of the aromatic ring (4-α-L-rhamnopyranolsyloxybenzylglucosinolate, known as glucomoringain, and glucosinolate esters with phenolic acids bound to glucose (e.g. 6′-sinapoylglucoraphenin, and (Z)-6′-feruloylglucosibarin).

recently been identified. Wasulfisides A (exhibiting anti-inflammatory activity) and B possess a disulfide bond connecting the sugar moiety with the aglycone. The disulfide bond in wasulfiside A is a part of an unusual 1,4,5-oxadithiocane ring. Wasulfisides B (3R enantiomer) and C (3S enantiomer) are accompanied by the corresponding 3-epiwasulfisides B and C, respectively.

wasulfiside A

wasulfiside B, R = SCH(CH₃)CH₂C(=O)OCH₃
wasulfiside C, R = CH(CH₃)CH₂C(=O)OCH₃
wasulfiside D, R = CH₂CH₂CH₂CN

10-110, wasulfisides

The intake of glucosinolates by farm animals is associated with *Brassica* fodder crops (such as fodder kale, fodder rape, swede, and turnip) and rapeseed meal, a byproduct of crushing, expelling, and extracting oil from oilseed rape. In human food, the income of glucosinolates is almost exclusively associated with the consumption of cruciferous vegetables. The most important source of glucosinolates are cabbage, cauliflower, kale, broccoli, and Brussels sprouts, followed by occasionally consumed vegetables such as radishes, garden cress, watercress, and horseradish. Vegetables of the cruciferous family commonly contain approximately 20–30 glucosinolates. On any one plant, there are usually only a few in significant quantities (Table 10.19). Whilst the composition of glucosinolates in each plant (or variety) is typical and is determined mainly by genetic factors, the total glucosinolate content is influenced by a number of external factors during cultivation (such as climate conditions, use of fertilisers, and pest attack). The average glucosinolate content in the vegetative parts of plants (f.w.) ranges from 100 to 2500 mg/kg, but in seeds it may reach values of up to 60 000 mg/kg.

10.3.10.2 Reactions and changes

In plant tissue, glucosinolates are accompanied by the enzyme myrosinase (thioglucoside glucohydrolase), which catalyses their decomposition. In intact tissue, myrosinase is located separately from glucosinolates, but in mechanically damaged cells (damaged by cutting, biting, or freezing), it comes into contact with glucosinolates, which are then hydrolysed relatively rapidly. The rate of glucosinolate hydrolysis is determined by enzyme activity, which is influenced by many factors (temperature, pH value, type and part of the plant, presence of substances that act as activators or inhibitors).

The actual enzymatic hydrolysis is initiated by myrosinase-catalysed cleavage of the thioglucoside bond of glucosinolate. Hydrolysis yields, in addition to D-glucose, an unstable intermediate (aglycone) thiohydroxamate-O-sulfonate, which is spontaneously degraded with the cleavage of bisulfate ion (HSO_4^-) and stabilisation of the remaining part of the molecule to yield one or more stable products (Figure 10.18). The most common glucosinolate decomposition products are isothiocyanates and nitriles; however, depending on the glucosinolate structure and some external factors, a number of other products can arise. Isothiocyanates (and partly also nitriles) formed by hydrolysis of aliphatic glucosinolates contribute to the typical spicy flavour of cruciferous vegetables (see Section 8.3.7.5).

For example, degradation of the most common glucosinolate, sinigrin, yields, in addition to allyl isothiocyanate (see Section 8.2.9.1.5) and allyl cyanide (but-3-enenitrile, see Section 8.2.7.1.3), allyl thiocyanate. It is not clear whether allyl thiocyanate arises directly from the unstable aglycone or by isomerisation of isothiocyanate. The final product is 2-(cyanomethyl)thiirane, also known as 2-(cyanomethyl)episulfide or 1-cyano-2,3-epithiopropane (**10-111**), which arises at the expense of allyl cyanide in the presence of a special protein cofactor, called epithiospecifier protein (ESP), whose existence has been demonstrated in many plants. This protein is essential for myrosinase-induced formation of (R)- and (S)-enantiomeric epithionitriles of all glucosinolates with unsaturated side chains. Gluconapin yields both enantiomers of 2-(thiiran-2-yl)acetonitrile (1-cyano-3,4-epithiobutane), glucobrassicanapin yields 1-cyano-4,5-epithiopentanes, progoitrin (except (R)-1-cyano-2-hydroxybut-3-ene) yields diastereomeric (2R)-1-cyano-2-hydroxy-3,4-epithiobutanes (**10-112**), and gluconapoleiferin yields 1-cyano-2-hydroxy-4,5-epithiopentanes.

Isothiocyanates resulting from 2-hydroxyalkenylglucosinolates are unstable and spontaneously cyclise (by intramolecular addition of hydroxyl group) to substituted 5-alkenyl-1,3-oxazolidine-2-thiones. Progoitrin, the dominant glucosinolate of rapeseeds, yields (R)-5-vinyl-1,3-oxazolidine-2-thione called goitrin (Figure 10.19), napoleiferin yields 5-allyl-1,3-oxazolidine-2-thione, and glucobarbarin yields 5-phenyl-1,3-oxazolidine-2-thione.

(R)-enantiomer   (S)-enantiomer

10-111, 1-cyano-2,3-epithiopropane

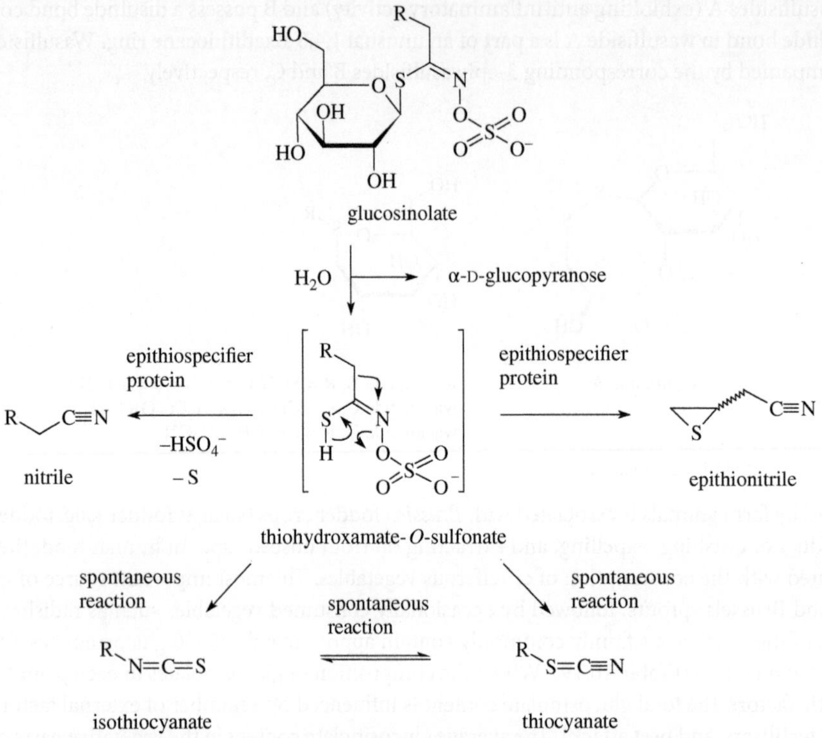

Figure 10.18 General mechanism of glucosinolate degradation.

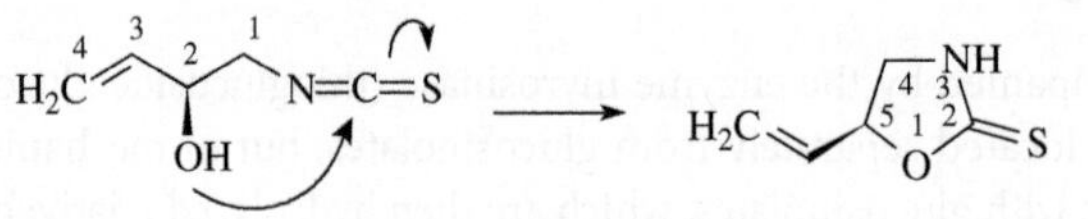

(R)-2-hydroxybut-3-en-1-ylisothiocyanate   (R)-5-vinyloxazolidine-2-thione

Figure 10.19 Formation of goitrin from 2-hydroxybut-2-en-1-yl isothiocyanate.

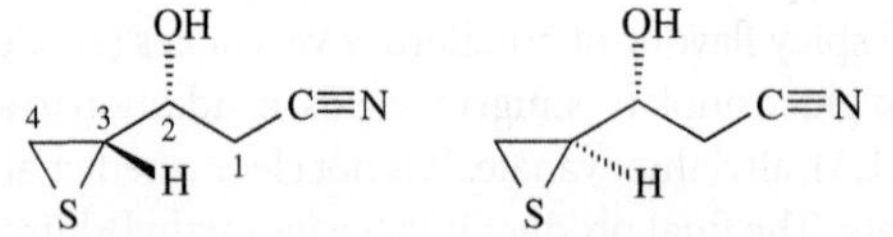

(2R,3R)-diastereomer   .   (2R,3S)-diastereomer

10-112, 1-cyano-2-hydroxy-3,4-epithiobutane

The situation is somewhat more complicated in the case of the so-called indole glucosinolates (indol-3-ylmethylglucosinolates), known as glucobrassicins. In seeds of oilseed rape and other *Brassica* species, the total pool of glucosinolates is most often dominated by 4-hydroxyindol-3-ylmethyl glucosinolate (4-hydroxyglucobrassicin) and methionine-derived glucosinolates. In contrast, vegetative parts of *Brassica* species, including various types of vegetables, have a relatively high content of the indol-3-ylmethyl glucosinolates: glucobrassicin, 4-methoxyglucobrassicin, and 1-methoxyglucobrassicin (neoglucobrassicin). Under the catalysis of myrosinase, the indol-3-ylmethylglucosinolates (as well as some other arylmethylglucosinolates) are degraded to indol-3-ylacetonitriles and unstable isothiocyanates, with release of the thiocyanate (rhodanide) ion and a reactive carbonium ion, which gives a complex mixture of reaction products, such as 3-hydroxymethylindoles (indol-3-ylmethanols) and ascorbigens, depending on the nucleophiles available and the reaction conditions. Their subsequent reactions give various dimers and oligomers. Self-condensation of 3-hydroxymethylindole, after cleavage of formaldehyde, yields 3,3′-diindolylmethane and even polyindolylmethanes (Figure 10.20).

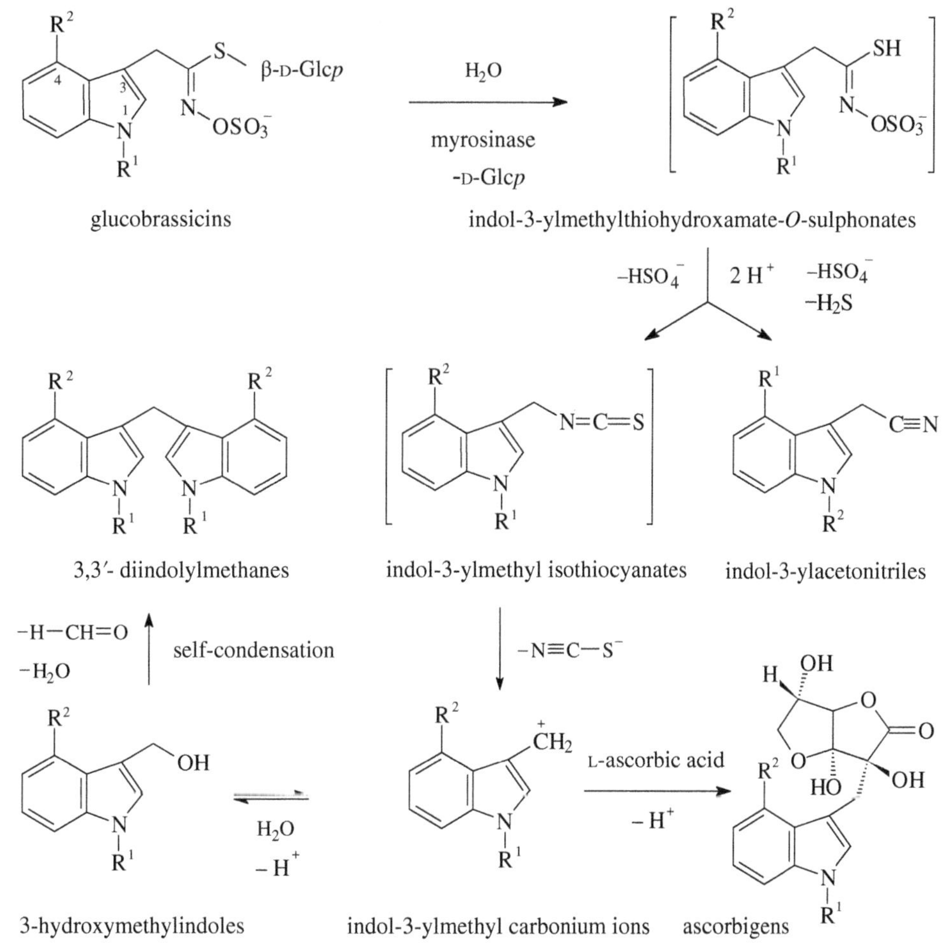

Figure 10.20 Degradation of indol-3-ylmethylglucosinolates. Glucobrassicins: glucobrassicin ($R^1 = R^2 = H$), 4-hydroxyglucobrassicin ($R^1 = H$, $R^2 = OH$), 4-methoxyglucobrassicin ($R^1 = H$, $R^2 = OCH_3$), neoglucobrassicin ($R^1 = OCH_3$, $R^2 = H$). Ascorbigens: ascorbigen ($R^1 = R^2 = H$), 4-hydroxyascorbigen ($R^1 = H$, $R^2 = OH$), 4-methoxyascorbigen ($R^1 = H$, $R^2 = OCH_3$), neoascorbigen ($R^1 = OCH_3$, $R^2 = H$).

Ascorbigen derived from glucobrassicin was formerly called ascorbigen A. In vegetables rich in ascorbic acid, ascorbigen is the main transformation product of glucobrassicin. Its content in cruciferous vegetables can reach 5–60 mg/kg. Depending on various factors, about 20–50% of glucobrassicin can be converted into ascorbigen, which has 15–20% the activity of ascorbic acid. Stabilities of 3-hydroxymethylindole and ascorbigen are limited, particularly at higher temperatures, and both compounds provide a wide range of transformation products. At higher temperatures, ascorbigen splits off ascorbic acid and the remaining residue can bind as a cation to another molecule of ascorbigen to form a dimer and a trimer (**10-113**). Binding to other nucleophilic compounds is also possible. A mixture of the dimer and trimer was previously called ascorbigen B.

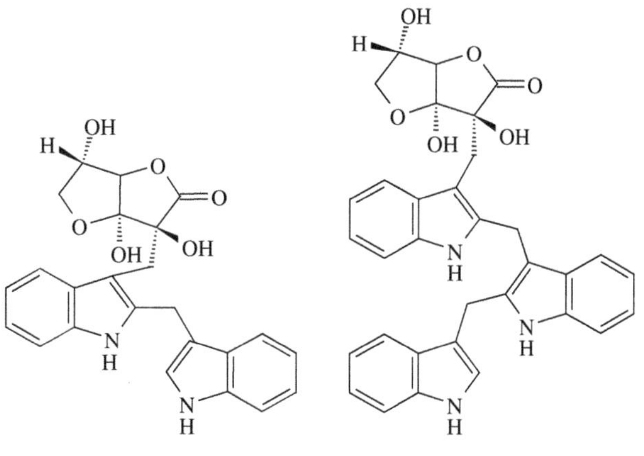

dimer   trimer

10-113, ascorbigen oligomers

Thermal degradation of glucobrassicin produces 3-indolylacetonitrile as the main product (Figure 10.20), along with 3-indolylacetic acid, 3-indolylacetamide, 3-formylindole (**10-114**), 3-methylindòle (skatole, see Section 2.5.1.1.3), and other compounds. 3-Indolylacetic acid, a common plant hormone synthesised by several independent pathways, is found in many foods of plant origin. In wines, for example, it is transformed to 3-indolylethanol, 3-indolyl-L-lactic acid (**10-114**), its ethyl ester, and O-β-D-glucoside.

10-114, 3-indolylacetic acid, R = COOH
3-indolylacetamide, R = CONH$_2$
3-indolylacetaldehyde, R = CH=O
3-indolylethanol, R = CH$_2$OH
3-indolyllactic acid, R = CH(OH)COOH

The pH value of the reaction medium affects the rate of enzymatic hydrolysis, as myrosinases of various plants usually have several pH optima, which lie in the interval between pH 5 and 8, due to the presence of several isoenzymes and various degradation mechanisms of unstable aglycones (Figure 10.18). Generally, under acidic conditions (pH < 4), nitriles predominantly arise, whilst at higher pH values, larger amounts of isothiocyanates are formed. Another factor affecting the decomposition of glucosinolates is temperature. Thermal processing of vegetables (such as cooking, blanching, and sterilisation) leads to inactivation of enzymatic systems, including myrosinase and ESP. Depending on the temperature and duration of heat treatment, glucosinolates are also partially decomposed along with other thermolabile products (such as 3-hydroxymethylindole and ascorbigen). In addition, a number of volatile substances are evaporated, in particular isothiocyanates and nitriles, which significantly changes the flavour characteristics of processed vegetables. The disruption of plant tissue, preceded by a sufficiently effective thermal intervention, no longer results in enzymatic degradation of glucosinolates, and the remaining glucosinolate content stays in the food. Glucosinolate hydrolysis may be affected by the presence of various chemicals, the most significant of which is ascorbic acid, which occurs in cruciferous vegetables in amounts of about 100–1200 mg/kg. In addition to the formation of ascorbigen, ascorbic acid also participates in myrosinase activation, thus speeding up the whole process of enzymatic degradation of glucosinolates. Changes in the content of glucosinolates in selected cruciferous vegetables during cooking are summarised in Table 10.20.

Generally, blanching and cooking of vegetables results in a considerable leakage of glucosinolates into water: on average, 30–40% of glucosinolates are extracted, and in white cabbage 28% are extracted during 10 minutes of boiling. Losses of glucosinolates by leakage and degradation are lowest for sinigrin (about 30%) and highest for neoglucobrassicin (about 60%). Approximately one-third to one-half of the original content of glucosinolates remains unchanged during cooking. During storage of boiled cabbage, further losses of glucosinolates occur. The most significant were observed for sinigrin (20–45%), and the least significant for glucobrassicin (12–32%). In frozen vegetables, due to myrosinase activity in damaged tissues, about 50% of the originally present glucosinolates decomposes. Virtually complete degradation of glucosinolates occurs during the first week of cabbage fermentation.

Table 10.20 Glucosinolate contents in fresh and cooked cruciferous vegetables.

Vegetables	Content (mg/kg f.w.)		Content of individual glucosinolates (mg/kg f.w.)[a]										
	Range	Average	2	4	6	7	9	12	13	14	24	26	28
Cabbage													
Raw	360–2754	1089	263	18	–	38	–	–	–	450	–	295	25
Cooked	315–1651	786	202	13	–	27	–	–	–	300		174	11
Cauliflower													
Raw	138–2083	620	142	7	–	23	–	–	–	173	–	227	48
Cooked	94–1111	420	100	3	–	14	–	–	–	122	–	151	30
Brussel sprouts													
Raw	1455–3939	2260	445	252	–	478	–	–	–	353	–	624	110
Cooked	597–2452	1237	264	148	–	299	–	–	–	195	–	298	33
Turnip													
Raw	392–1657	560	–	42	37	371	42	46	71	–	97	48	96
Cooked	205–944	291	–	24	26	206	23	23	39	–	58	23	38

[a]Compound numbers are given in Table 10.18.

10.3.10.3 Biological effects

The mean daily intake of glucosinolates is difficult to evaluate as there are large variations between countries, income groups (e.g. it is likely that certain individuals, such as vegetarians, will consume more than 300 mg total glucosinolates per day), and times of year (the amount consumed is approximately doubled in the winter months). The nutritional and toxicological consequences of such an intake are largely unknown at present. As an example, the mean total glucosinolate intake in Germany in 2009 was 14.2 mg/day for men and 14.8 mg/day for women, increasing with age and education. Quantitatively, the most important individual glucosinolates were glucobrassicin and sinigrin, with mean daily intakes of 3.5 and 1.7 mg/day for men and 4.2 and 2.5 mg/day for women, respectively. Broccoli, Brussels sprouts, and cauliflower contributed most to the total glucosinolate intake.

Glucosinolates as such are in practice indifferent compounds and thus are probably neither harmful nor beneficial. Only products of their degradation show biological effects. The situation is aggravated by the fact that one substance often exhibits several different effects.

10.3.10.3.1 Toxic effects

Decomposition of aliphatic glucosinolate leads to slightly strumigenic isothiocyanates, some of which exhibit strong antimicrobial and insecticidal effects (such as allyl isothiocyanate formed from sinigrin). Other products include nitriles and cyanoepithioalkanes, which may be hepatotoxic and nephrotoxic. Strumigenic (R)-5-vinyl-1,3-oxazolidine-2-thione (goitrin), formed by progoitrin hydrolysis, inhibits the synthesis of thyroid hormones (thyroxine and triiodothyronine) and transmission of iodine in the thyroid gland. Thiocyanate anions, arising from unstable indole and aromatic isothiocyanates, exhibit moderate strumigenic activities, but unlike goitrin and isothiocyanates, they act competitively with respect to iodine. Both mechanisms can lead to an enlarged thyroid gland (goitre) and disturbances of its function. In humans, the strumigenous effects of these compounds have not been demonstrated.

The presence of glucosinolates may be a problem in fodder production. Major deleterious effects of glucosinolates ingestion in animals are reduced palatability and decreased growth and reproduction. Progoitrin and epiprogoitrin impair palatability at a level of between 2.3 and 4.7 mol/g diet, whilst at higher levels feed intake decreases. Nitriles are known to affect liver and kidney functions. The thiocyanates interfere with iodine availability, whereas goitrin is responsible for the morphological and physiological changes of thyroid glands. Rapeseed meal feeding (containing 36–40% of protein) does not impair carcass quality, but it does increase erucic and elaidic acid contents in the carcass and milk fat. However, the newly introduced rapeseed cultivars of canola quality (commonly known as double-low or 00) have, in addition to a low content of erucic acid (0.2%, which corresponds to about 0.5% in the oil), a reduced glucosinolate content (20 mmol/kg). Ruminants are less sensitive to dietary glucosinolates. Pigs are more severely affected by dietary glucosinolate compared with rabbits, poultry, and fish. The tolerance levels of glucosinolates in ruminants, pig, rabbits, poultry, and fish are 1.5–4.2, 0.78, 7.0, 5.4, and 3.6 mol/g of diet, respectively. Results using techniques based on detoxification of oilseed meals based on removal of glucosinolates and their breakdown products by alkali extraction or exposure to water vapour at higher temperatures (toasting) are not satisfactory, especially with regard to their economic demands. Iodine supplementation in the diet of pigs and ruminants seems to be promising.

10.3.10.3.2 Beneficial effects

Experimental studies with laboratory animals and the results of epidemiological studies have confirmed that increased consumption of cruciferous vegetables reduces the risk of chemically induced cancer, which is cancer induced by intake of carcinogens. The active principles are some glucosinolate degradation products, although the presence of ascorbic acid, α-tocopherol, β-carotene, dimethylthiosulfinate, phenolic antioxidants (e.g. flavonoids), and fibre cannot be ignored.

Benzyl isothiocyanate and 2-(phenylethyl) isothiocyanate (**10-115**), which arise by hydrolysis of gluconasturtiin and glucotropaeolin, respectively, have the ability to activate some important enzymatic system deactivating carcinogens. Also worth mentioning is 4-(methylsulfinyl)butyl isothiocyanate, known as sulforaphane (**10-116**), whose beneficial effects have been appreciated. Its precursor is glucosinolate glucoraphanin, occurring, for example, in broccoli and radishes. Another product of glucoraphanin degradation is the corresponding 5-methylsulfinylpentanenitrile which is called sulforaphane nitrile. In broccoli, glucoraphanin is the main glucosinolate. It is present at a level of about 1600 mg/kg (f.w.) (approximately 55% of the total glucosinolate content). The other broccoli glucosinolates (in descending order of quantity) are glucoiberin, glucobrassicin, glucoerucin, progoitrin, 4-hydroxyglucobrassicin, glucoibervirin, 4-methoxyglucobrassicin, gluconapin, glucoalyssin, and 1-methoxyglucobrassisin. The major glucosinolates found in broccoli seeds were glucoraphanin and glucoerucin, and progoitrin in a minority of samples. The main glucosinolate of radishes is glucoraphasatin (178–2230 mg/kg f.w.), followed by glucobrassicin, glucoraphenin, and glucoraphanin (0–24 mg/kg f.w., 0–7% of the total glucosinolate content). Sulforaphene (4-methylsulfinylbut-3-enyl isothiocyanate), produced by myrosinase hydrolysis of glucoraphenin, shows significant chemopreventive activity, but is unstable and is quickly converted to 6-[(methylsulfinyl)methyl]-1,3-thiazinan-2-thione (Figure 10.21).

10-115, benzyl isothiocyanate, $n = 1$
2-(phenylethyl) isothiocyanate, $n = 2$

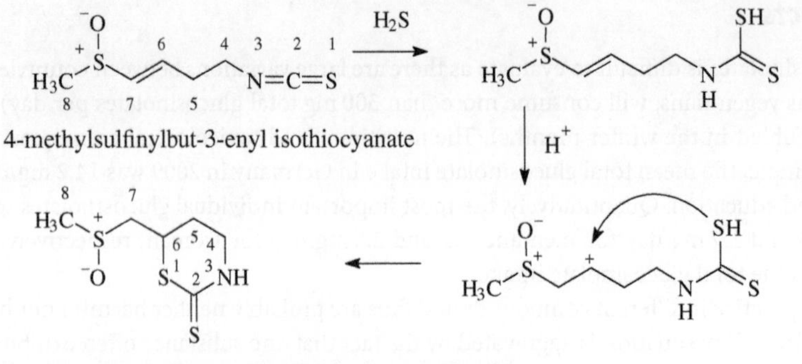

4-methylsulfinylbut-3-enyl isothiocyanate

6-[(methylsulfinylmethyl)]-1,3-thiazinan-2-thione

Figure 10.21 Sulforaphene degradation pathway.

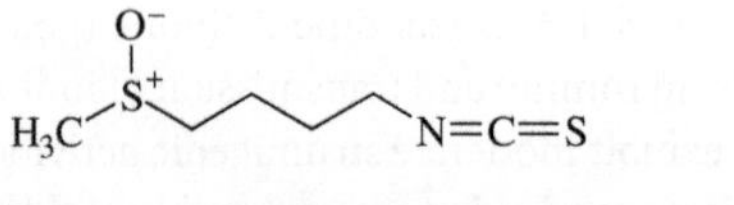

10-116, 4-(methylsulfinyl)butyl isothiocyanate

Much attention has been devoted to degradation products of indole glucosinolates, as they are found in virtually all commonly consumed vegetables and sometimes form a significant proportion of total glucosinolates (e.g. 95% in Chinese cabbage). One of the major degradation products of glucobrassicin is 3-hydroxymethylindole, a compound possessing an unusual ability to induce enzymes of the first and second phases of detoxification. Ascorbigen resulting from the reaction of 3-hydroxymethylindole with ascorbic acid is partially degraded in the stomach, ascorbic acid splits off, and the molecule residue is transformed into products with antimutagenic and anticarcinogenic properties, such as 5,11-dihydroindolo[3,2-*b*]carbazole, which can inhibit certain DNA-damaging carcinogens. In neutral media, ascorbigen is converted into indole sugars (**10-117**), the toxicological evaluations of which (along with the evaluation of the effects of other ascorbigen metabolites and 3-hydroxymethylindole) are not yet sufficiently developed.

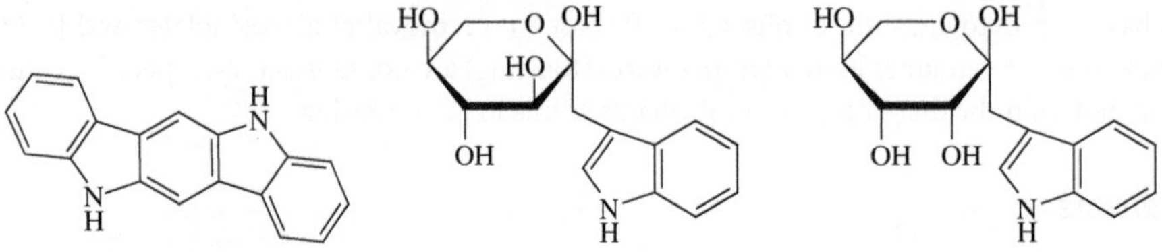

5,11-dihydroindolo[3,2-*b*]-
carbazole

10-117, metabolites of ascorbigen

1-deoxy-1-(3-indolyl)-
α-L-sorbopyranose

1-deoxy-1-(3-indolyl)-
α-L-tagatopyranose

10.3.11 Saponins and other terpenoids

10.3.11.1 Saponins

10.3.11.1.1 Structure, nomenclature, and occurrence

Saponins are a diverse group of surface-active heteroglycosides occurring mainly in plants. Hydrophobic aglycones of saponins, which are called sapogenols (formerly sapogenins), are compounds derived from:

- C_{30} triterpenoids, also known as triterpenic alcohols;

- C_{27} steroids.

The aglycone is bound to one or more sugar residues. In **monodesmosides**, one sugar residue (mono- or oligosaccharide) is bound to the aglycone (sugar is normally bound to the C-3 hydroxyl), in **bisdesmosides**, two sugar residues are bound in different positions, and in **trisdesmosides**, three sugar residues are found. Common sugars are usually L-arabinose, D-glucose, D-mannose, D-galactose, L-rhamnose,

Table 10.21 Saponin contents in legumes and other plants.

Plant	Latin name	Content (%)	Plant	Latin name	Content (%)
Fabaceae			Lentil	*Lens culinaris*	0.11–0.51
Soya bean	*Glycine max*	0.22–5.6	Peanut	*Arachis hypogaea*	0.01–1.6
Common bean	*Phaseolus vulgaris*	0.35–1.6	Liquorice	*Glycyrrhiza glabra*	2.2–15.0
Lima bean[a]	*Phaseolus lunatus*	0.10	*Amaranthaceae*[b]		
Mung bean[c]	*Vigna radiata*	0.34	Spinach	*Spinacia oleracea*	4.7
Chickpea	*Cicer arietinum*	0.23–6.0	Beet	*Beta vulgaris*	5.8
Pea	*Pisum sativum*	0.11–0.18	Quinoa	*Chenopodium quinoa*	0.14–2.3

[a]Also known as butter bean.
[b]Formerly Chenopodiaceae.
[c]Also known as green gram.

D-glucuronic acid, and D-galacturonic acid, and less frequently D-xylose and D-apiose, but some others may also be present. Sugars can also be acylated with organic acids, such as acetic acid. The amount of saponins depends mainly on the plant species and climatic conditions. The largest concentration of saponins is located in the roots, bark, and fast-growing parts (Table 10.21). At low concentrations, saponins are also present in certain marine organisms and bacteria.

The physiological role of saponins in plants is not yet fully understood; nevertheless, it is assumed that they represent part of the defence system of protective molecules termed phytoprotectants – chemical agents called phytoanticipins or phytoalexins that interact with pathogens on the plant surface to prevent infection.

Triterpenoid saponins Triterpenoid saponins are common components of many plants. Their structure is mainly derived from pentacyclic triterpenoids, such as lupeol, α-amyrine, and β-amyrine (see Section 3.7.4.1.1). In positions C-4 (C-23 methyl), C-17 (C-28 methyl), and C-20 (C-30 methyl) of the aglycone system, C_6—C_6—C_6—C_6—C_6, a carboxyl group formed by oxidation of a methyl group, may occur. The methyl groups of some saponins are only partially oxidised to hydroxymethyl and formyl groups, and some aglycones contain other hydroxyl groups at C-21 and C-22. Positions C-11 and C-16 may also be oxidised. Sugar moieties (1–6) are bound to the C-3 hydroxyl. Some hydroxyl groups may be esterified with carboxylic acids, for example acetic acid (C-22 hydroxyl) or angelic acid (C-21 hydroxyl).

The best-known saponins are soyasaponins, aglycones of which are soyasapogenol A (oleane-12-en-3β,22β,24-triol), soyasapogenol B (olean-12-en-3β,22β,24-triol, **10-118**), soyasapogenol E (oleane-12-en-3β,24-diol-22-one, **10-119**), and hederagenin (**10-120**). The content of saponins in seeds of the soybean ranges from 0.6 to as much as 6.5% dry matter, depending on the variety, cultivation year, growth location, and degree of maturity.

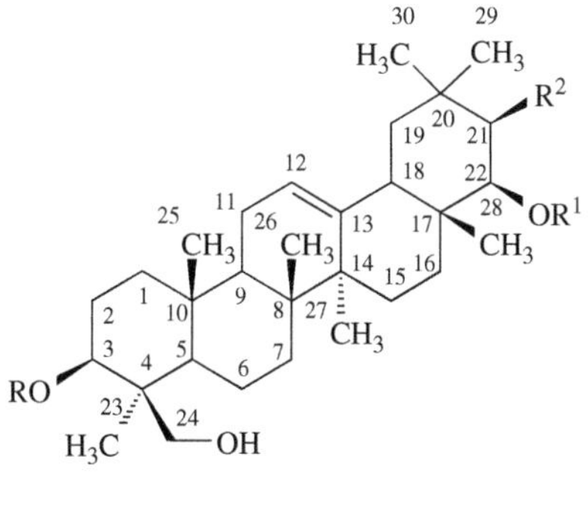

10-118, soyasapogenol A, R = H, R¹ = H, R² = OH
   soyasapogenol B, R = H, R¹ = H, R² = H
   soyasaponin A_a, R=β-D-Glcp-(1→2)-β-D-Galp-(1→2)-β-D-GlcpA-(1→, R¹= β-D-Xylp2,3,4Ac₃-
      (1→3)-α-L-Araf-(1→, R²= OH
   soyasaponin A_b, R=β-D-Glcp-(1→2)-β-D-Galp-(1→2)-β-D-GlcpA-(1→, R¹=β-D-Glcp2,3,4,6Ac₄-
      (1→3)- α-L-Araf-(1→, R²= OH
   soyasaponin B_a, R=β-D-Glcp-(1→2)-β-D-Galp-(1→2)-β-D-GlcpA-(1→, R¹ = H, R² = H
   soyasaponin B_b, R=α-L-Rhap-(1→2)-β-D-Galp-(1→2)-β-D-GlcpA-(1→, R¹ = H, R² = H

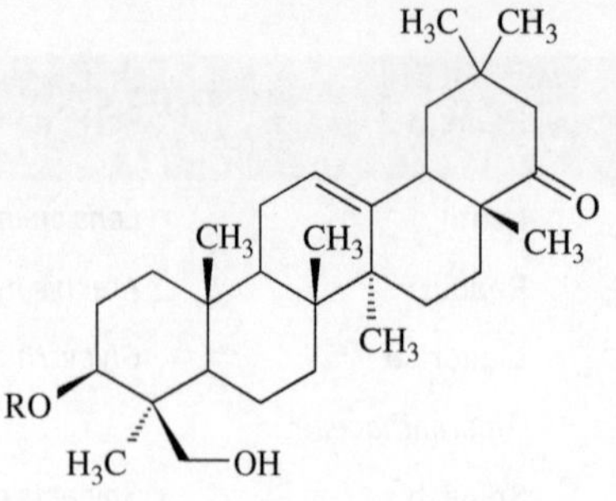

10-119, soyasapogenol E, R = H
soyasaponin B$_d$, R = β-D-Glc$p$-(1→2)-β-D-Gal$p$-(1→2)-β-D-Glc$p$A-(1→
soyasaponin B$_e$, R = α-L-Rha$p$-(1→2)-β-D-Gal$p$-(1→2)-β-D-Glc$p$A-(1→

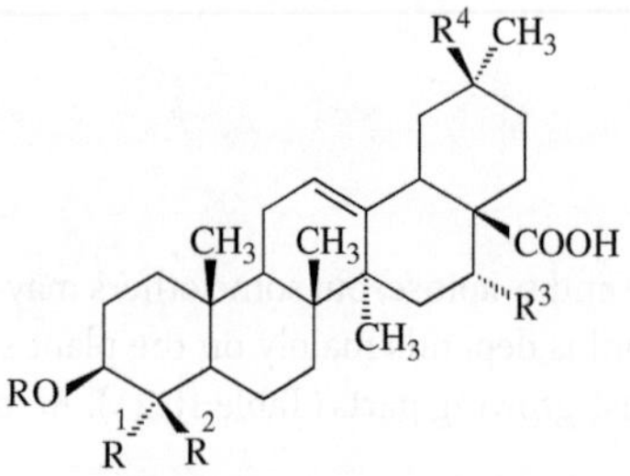

10-120, oleanolic acid, R = H, R^1 = R^2 = CH$_3$, R^3 = H, R^4 = CH$_3$
hederagenin, R = H, R^1 = CH$_2$OH, R^2 = CH$_3$, R^3 = H, R^4 = CH$_3$
phytolaccagenic acid, R = H, R^1 = CH$_2$OH, R^2 = CH$_3$, R^3 = H, R^4 = COOCH$_3$
quillajic acid, R = H, R^1 = CH = O, R^2 = CH$_3$, R^3 = OH, R^4 = CH$_3$
gypsogenic acid, R = H, R^1 = COOH, R^2 = CH$_3$, R^3 = H, R^4 = CH$_3$

Saponins derived from soyasapogenol A are bisdesmosides containing sugars bound to C-3 and C-22 hydroxyls. Some of the sugars can be acetylated. Acetylated derivatives have a bitter and astringent taste. Most varieties of soya contain soyasaponin A$_a$ and A$_b$ (**10-118**) as the main representative of the acetylated compounds. Glycosides of soyasapogenol B, such as soyasaponin B$_a$ and soyasaponin B$_b$, still often referred to as soyasaponin I (**10-118**), lack the C-22 hydroxyl group; they belong to monodesmosides, and the sugar moiety is attached to C-3 hydroxyl. Additional soyasaponins are derived from soyasapogenol E, including saponin B$_d$ and saponin B$_e$ (**10-119**). A recently identified group of soyasaponins B$_a$A and B$_b$A (also known as soyasaponin βg, soyasaponin VI, or chromosaponin I) are conjugates of soyasaponin B$_b$ with γ-pyrone 2,3-dihydro-2,5-dihydroxy-6-methyl-4H-pyran-4-one, which is bound to carbon C-22 (**10-121**). These saponins are thus actually bisdesmosides. Soyasaponin B$_b$ and its conjugates with γ-pyrane occur as the main saponins in some other legumes, such as common beans and lima beans, peas, lentils, and chickpeas.

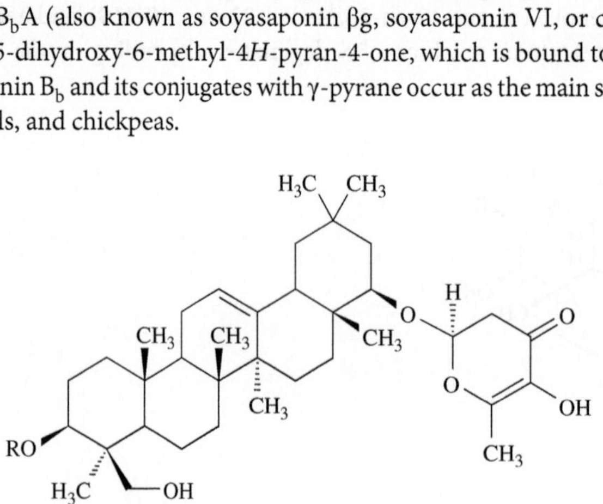

10-121, soyasaponin B$_a$A, R = β-D-Glc$p$-(1→2)-β-D-Gal$p$-(1→2)-β-D-Glc$p$A-(1→,
soyasaponin B$_b$A, R = α-L-Rha$p$-(1→2)-β-D-Gal$p$-(1→2)-β-D-Glc$p$A-(1→

Plants of the Amaranthaceae (formerly Chenopodiaceae) family, such as sugar beet (*Beta vulgaris*), contain about 5.8% saponins. The main sapogenin is oleanolic acid. Spinach contains about 4.7% saponins, derived from oleanolic acid and hederagenin (**10-120**). The edible grain quinoa of the same plant family (*Chenopodium quinoa*), originating in South America, contains 2–5% saponins in the form of oleanane-type triterpenoid glycosides or sapogenins that confer an undesirable bitter taste. The predominant sapogenin is phytolaccagenic acid at 16.72 mg/g, followed by hederagenin (**10-120**) at 4.22 mg/g, representing ~70 and 30% of the total sapogenin content, respectively. Aglycones of soap tree bark (*Quillaja saponaria*, Quillajaceae) saponins, which are used industrially for pharmaceutical and cosmetic products, are quillajic and gypsogenic acids (**10-120**). Their content in soap tree bark is around 10%.

The intensely sweet rhizome of liquorice (*Glycyrrhiza glabra*, Fabaceae), with typical flavour, used in the production of various pharmaceutical and confectionery products, contains 2.2–15.0% of saponins. The predominant sweet saponin is glycyrrhizin (**10-122**), a mixture of potassium and calcium salts of a monodesmoside, with two β-D-glucuronic acid units bound to the C-3 hydroxyl, known as glycyrrhizic acid (glycyrrhizinic acid). Hydrolysis to the aglycone known as glycyrrhetic (glycyrrhetinic) acid or enoxolone proceeds via the corresdponding mono-β-D-glucuronide. The sweetness of glycyrrhizin is about 50–150 times higher than that of sucrose. In addition to glycyrrhizin, the liquorice rhizome contains a large number of similar monodesmosidic saponins.

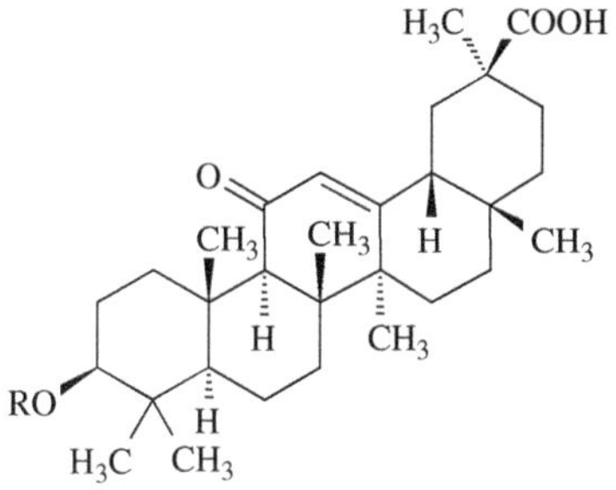

10-122, glycyrrhetic acid, R = H

glycyrrhizic acid, R = β-D-Glc*p*A-(1→2)-β-D-Glc*p*A-(1→.

Saponins of tea leaves (*C. sinensis*, Theaceae) contain as the main components theasapogenol A and theasapogenol B (**10-123**). Related compounds include theasapogenols C, D, and E and assamsapogenols A, B, C, and D.

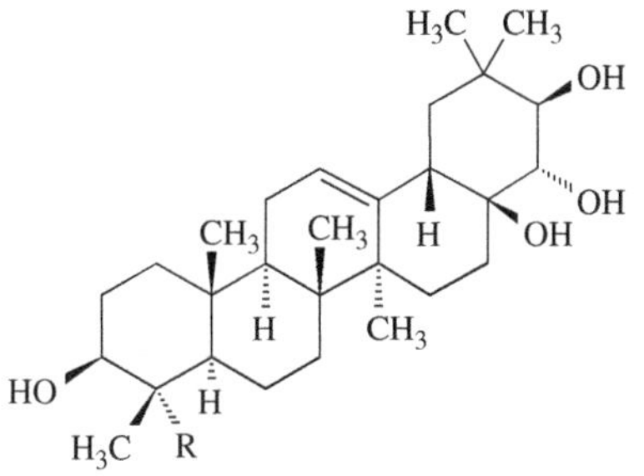

10-123, theasapogenol A, R = CH₂OH

theasapogenol B, R = CH=O

Steroid saponins Saponins that have in the molecule steroids as aglycones (steroid saponins) contain aglycones based on the C_{27} structures of spirostanol or furostanol formed by modifications of the cholesterol side chain. Sugars (two to five residues of hexoses or pentoses) are attached to the C-3 hydroxyl group of aglycones. Other conjugates of this type include other triterpenes with sugars attached at different positions.

Steroidal saponins are less widespread in comparison with triterpenoid saponins. They typically occur in monocotyledonous plants of the families Agavaceae, Amaryllidaceae, Asparagaceae, Dioscoreaceae, and Poaceae, but are also present in some dicotyledonous plants, such as legumes of the family Fabaceae. Both types of saponins are often present in the same plant material.

Plants of the Allioideae subfamily (the old Alliaceae family), such as bulb onion and shallot (*Allium cepa*), contain saponins derived from oleanolic acid, β-amyrin (see **3-99**), gitogenin (**10-124**), diosgenin (**10-125**), β-chlorogenin (**10-126**), and cepagenin (**10-127**). Recently identified were spirostane-type saponins with the trivial names alliospirosides A–D (**10-128**), tropeosides A1, A2, B1, and B2, and ascalonicosides A1, A2, and B. For example, the sugar bound in allispiroside A is a disaccharide with R = α-L-Ara*f*-(1 → 2)-6-deoxy-α-L-Man*p*-, tropeoside A1 (**10-129**) contains two monosaccharides, R^1 = β-D-Gal*p*, R^2 = α-L-Rha*p*, and ascalonicoside A1 (**10-129**) contains one monosaccharide (R^1 = β-D-Gal*p*) and one disaccharide (R^2 = α-L-Rha*p*-(1 → 2)-β-D-Glc*p*-). Saponins of garlic (*A. sativum*) are derived from phytosterol sitosterol (see **3-119**), whilst saponins of leek (*A. ampeloprasum*) contain oleanolic acid and gitogenin (**10-120**) as aglycones.

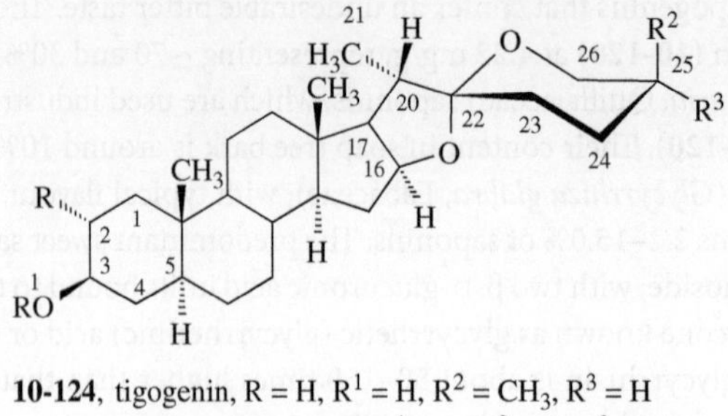

10-124, tigogenin, R = H, R¹ = H, R² = CH₃, R³ = H

neotigogenin, R = H, R¹ = H, R² = H, R³ = CH₃

gitogenin, R = H, R¹ = H, R² = H, R³ = CH₃

neogitogenin, R = OH, R¹ = H, R² = H, R³ = CH₃

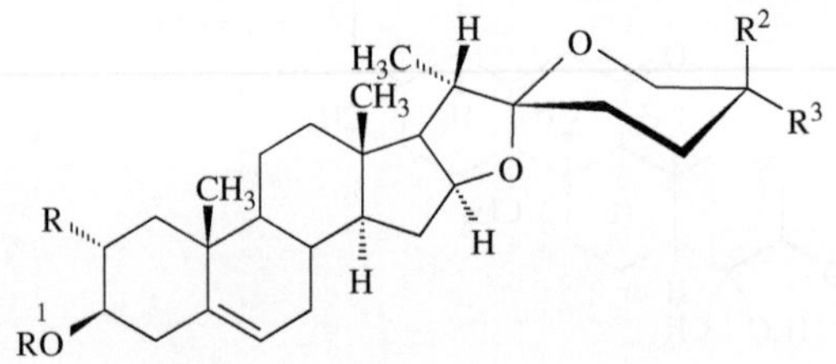

10-125, diosgenin, R = H, R¹ = H, R² = CH₃, R³ = H

yamogenin, R = H, R¹ = H, R² = H, R³ = CH₃

yuccagenin, R = OH, R¹ = H, R² = CH₃, R³ = H

lilagenin, R = OH, R¹ = H, R² = H, R³ = CH₃

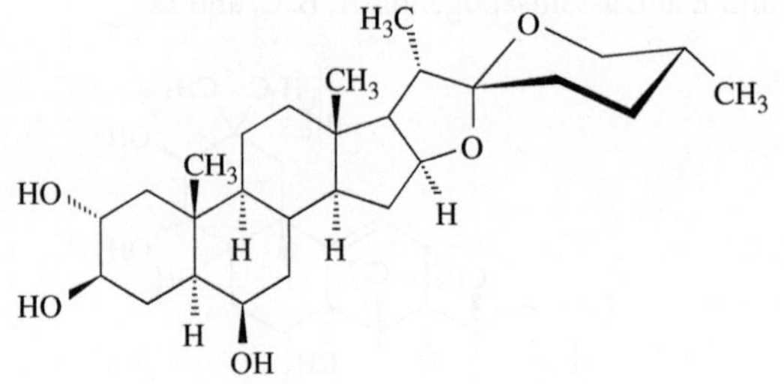

10-126, β-chlorogenin

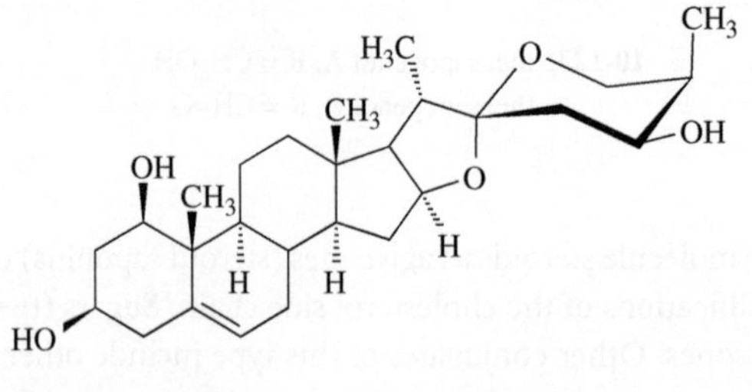

10-127, cepagenin

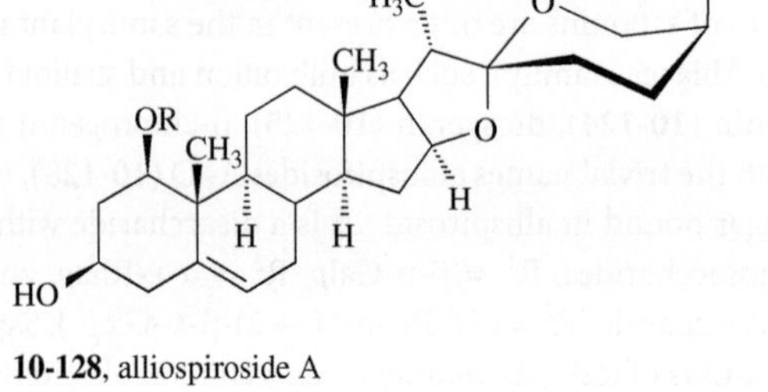

10-128, alliospiroside A

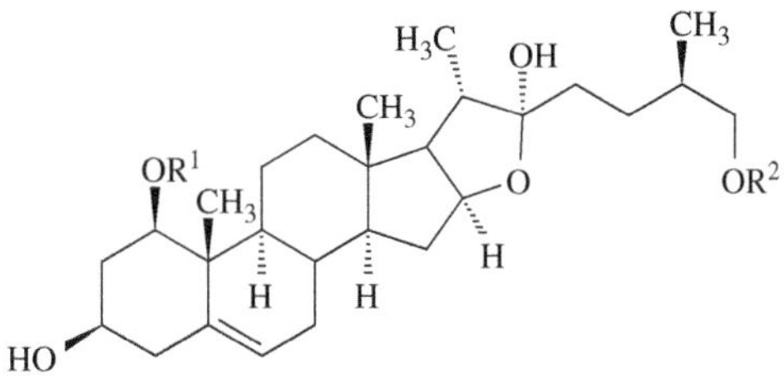

10-129, tropeoside A1/ascalonicoside A1

Saponins of asparagus (*Asparagus officinalis*, Asparagaceae) were once trivially called officinalisins and asparagosins. Officinalisins contain 5β-furostan-3β,22α,26-triol as their aglycone, whilst the aglycone of asparagosins is (25S)-spirost-5-en-3β-ol, trivially called yamogenin (**10-125**). Present as a minor aglycone is sarsapogenin (**10-130**) derived from spirostane. At least nine compounds, known as asparagosides A–I, have been characterised. Saponins are located in larger quantities at the bottoms of the stalks, to which they give a bitter taste. Recently identified was a series of bitter steroidal saponins, for which the lowest threshold concentration was determined for the epimeric mixture of bisdesmosidic diosgenyl saponin (25R/S)-furost-5-en-3β,22α,26-triol-3-O-[α-L-Rha*p*-(1 → 4)-β-D-Glc*p*]-26-O-β-D-Glc*p* (10.9 µmol/l).

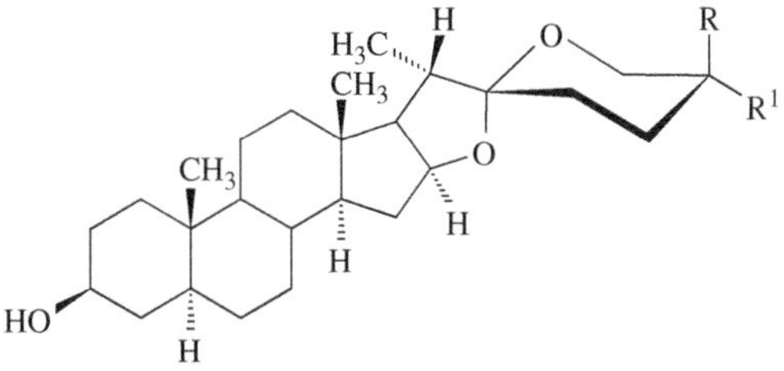

10-130, smilagenin, R = CH₃, R¹ – H
sarsapogenin, R = H, R¹ = CH₃

Yamogenin and its isomer diosgenin, (25R)-spirost-5-en-3β-ol (**10-125**), are the aglycones of saponins in yams (*Dioscorea* spp., Dioscoraceae), the starchy tubers of which are eaten in Africa, Southeast Asia, and the Pacific region (*D. alata* and *D. esculenta* in Southeast Asia, *D. trifida* in South America, *D. rotundata* and *D. cayenensis* in West Africa). Yams typically contain 4–8% saponins, which are also a source of steroids for the pharmaceutical industry.

A number of steroid saponins called fenugrins and grecunins are found in the legume plant fenugreek (*Trigonella foenum-graecum*, Fabaceae), which is used as a spice and fodder plant. In India, for example, fenugreek leaves are eaten as a salad, and in European countries the seeds are used as a spice. The main aglycone of the seeds is diosgenin; yamogenin (**10-125**), gitogenin, neotigogenin (**10-124**), smilagenin, sarsapogenin (**10-130**), and their 3α-epimers, known as epismilagenin and episarsapogenin, are found in smaller quantities. Minor fenugreek components are the dihydroxy derivatives yuccagenin, lilagenin (**10-125**), gitogenin, and neogitogenin (**10-124**).

A similar structure is seen in steroidal bisdesmosides of oats (*Avena sativa*, Poaceae), which are called avenacosides A and B (**10-131**). Their total level in commercial oat products varies from 49.6 to 443.0 mg/kg. Specific β-glucosidase splits off glucose bound at C-26, and these bisdesmosides are transformed to monodesmosides called deglucoavenacosides, which are less bitter but exhibit higher antimicrobial and haemolytic effects. Besides avenacosides A and B, two other bitter-tasting bisdesmosidic saponins were recently identified, namely 3-(O-α-L-rhamnopyranosyl-(1 → 2)-[β-D-glucopyranosyl-(1 → 3)-β-D-glucopyranosyl-(1 → 4)]-β-D-glucopyranosid)-26-O-β-D-glucopyranosyl-(25R)-furost-5-ene-3β,22,26-triol and 3-(O-α-L-rhamnopyranosyl-(1 → 2)-[β-D-glucopyranosyl-(1 → 4)]-β-D-glucopyranosid)-26-O-β-D-glucopyranosyl-(25R)-furost-5-ene-3β,22,26-triol.

10.3.11.1.2 Reactions and changes

Saponins are relatively stable during common technological and culinary operations. Their amount can be reduced by adequate washing, maceration, or removal of the surface layer (peeling).

Sugar beet saponins are removed during sugar refining. Soya bean saponins are removed by debittering via various technological processes, such as peeling, acid hydrolysis, and fermentation processes (using suitable microorganisms such as the moulds *Aspergillus oryzae* and *A. niger*), where bitter saponins are enzymatically hydrolysed to non-bitter aglycones. On cooking of legumes, bitter glycosides are partly dissolved and partly hydrolysed to aglycones and sugars. For example, losses of saponins during cooking of lentils may reach 6–14% of the original amount, which is about 700–1100 mg/kg (d.w.). The highest reduction of saponins in legumes can be achieved by soaking, followed by germination for 40 hours and boiling. Under these circumstances, the content of saponins may be reduced by up to 75%.

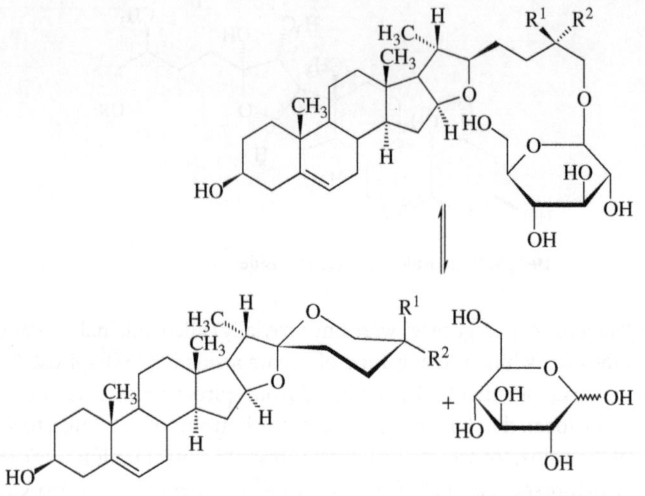

Figure 10.22 Interconversion of furostane and spirostane structures.

Triterpenic saponins derived from soyasapogenol E (B_d and B_e saponins) are highly unstable, and during processing of legumes, they are reduced to B_a and B_b saponins, respectively. During the thermal processing of legumes, soyasaponin B_eA is partially degraded to soyasaponins B_b and B_e, yielding the respective aglycone soyasapogenol E. Glucose at position C-26 of steroid furostane-type saponins can be split off by enzymatic hydrolysis or the activity of microorganisms, and furostane derivatives are transformed into spirostane derivatives (Figure 10.22). Saponin aglycones may partially eliminate the hydroxyl group at C-3 as water, with the formation of corresponding unsaturated hydrocarbons.

10.3.11.1.3 Biological effects

In the past, practically all saponins were considered antinutritional or toxic substances, as they negatively influence the organoleptic properties of foods, causing unwanted bitterness and astringency in soybeans and other legumes. Animal nutritionists have generally considered saponins to be deleterious compounds that negatively affect growth, feed intake, and reproduction. Saponins are also highly toxic to cold-blooded organisms, such as insects and fish. Toxic saponins are often called **sapotoxins**. Their toxic effect is manifested by haemolysis of erythrocyte cells and damage of intestinal mucosa. The main reason for this is the interaction of the saponins with cholesterol in the cell walls. High doses of toxic saponins may damage the liver, which can lead to respiratory failure. In retrospect, however, only some saponins are really toxic.

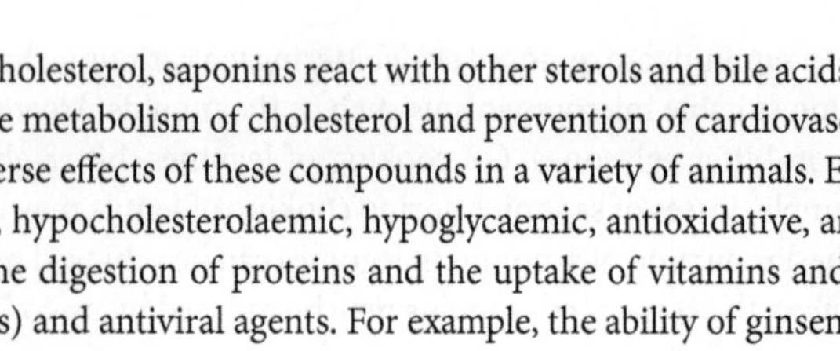

10-131, avenacoside A, R =β-D-Glc*p*-(1→2)-[α-L-Rha*p*-(1→4)]-β-D-Glc*p*-(1→, R^1=β-D-Glc*p*-(1→
avenacoside B, R =β-D-Glc*p*-(1→3)-β-D-Glc*p*-(1→2)-[α-L-Rha*p*-(1→4)]-β-D-Glc*p*-(1→, R^1=β-D-Glc*p*-(1→

Analogously to cholesterol, saponins react with other sterols and bile acids (under micelle formation) and thereby inhibit their absorption; this is related to the metabolism of cholesterol and prevention of cardiovascular diseases. Recently, a number of studies have reported both beneficial and adverse effects of these compounds in a variety of animals. Extensive research carried out on the membrane-permeabilising, immunostimulant, hypocholesterolaemic, hypoglycaemic, antioxidative, and anticarcinogenic properties of saponins has shown that they may also impair the digestion of proteins and the uptake of vitamins and minerals in the gut, as well as acting as antifungal (e.g. some asparagus saponins) and antiviral agents. For example, the ability of ginseng (*Panax ginseng*, Araliaceae) to slow the ageing of organisms is attributed to the antioxidant activity of ginsenosides (panaxosides), which represent a class of steroid glycosides and triterpene saponins.

In the form of concentrates, some saponins are used as foaming agents, emulsifiers, and antioxidants; glycyrrhizin from liquorice is used as a sweetener in the manufacture of confectionery and tobacco. In the manufacture of some soft drinks, as well as the production of the famous English ginger beer, saponins from the soap tree bark (*Quillaja saponaria*, Quillajaceae) are used as foaming agents. Saponins are also commonly used in cosmetic products (shampoos and other hair preparations).

10.3.11.2 Other terpenoids

Sweet potato (*Ipomoea batatas*, Convolvulaceae) root tissue produces in response to various exogenous stimuli structurally related phytoalexins known as furanoterpenoids and coumarins (such as umbelliferone and scopoletin). The first phytoalexin isolated from the crop affected by the mould *Fusarium solani* was hepatotoxic ipomeamarone. Also of toxicological concern are four other related derivatives: ipomeanine, a mixture of enantiomers of ipomean-4-ol and ipomean-1-ol (**10-132**), and a mixture of diastereomers of ipomea-1,4-diol. The pneumotoxin (a respiratory tract toxicant) ipomean-4-ol is mainly responsible for the toxic effects of sweet potatoes. The use of mechanically damaged and microbially infected sweet potato as feed for cattle represents a significant risk, as in some instances it has led to the death of livestock as a result of pulmonary oedema. It seems very likely that these substances significantly contribute to the chronic respiratory diseases in the Pacific regions, where sweet potatoes are an important staple food.

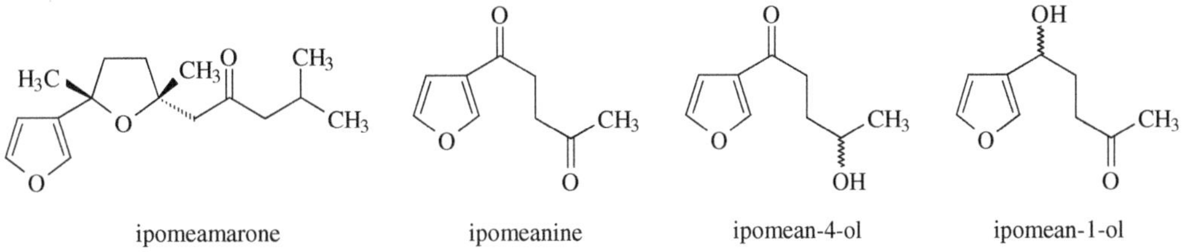

ipomeamarone   ipomeanine   ipomean-4-ol   ipomean-1-ol

10-132, furanoterpenoids of sweet potatoes

Representatives of sesquiterpenic phytoalexins found in bacterially rotted potato tubers (*Solanum tuberosum*) and other economically important solanaceous plants (Solanaceae) are phytuberin and lubimin. Lubimin is also a major phytoalexin of eggplants (*S. melongena*). Another terpenoid is rishitin, occurring in potatoes and tomatoes (*S. lycopersicum*). Capsidiol (**10-133**) is a phytoalexin of bell peppers (*Capsicum* spp.) and tobacco (*Nicotiana* spp.). Several diterpenoid glycosides named capsianosides from bell peppers exhibit antihypertensive effects and have been found to be related to the improvement and prevention of hypertension. An example is capsianoside I (**10-134**).

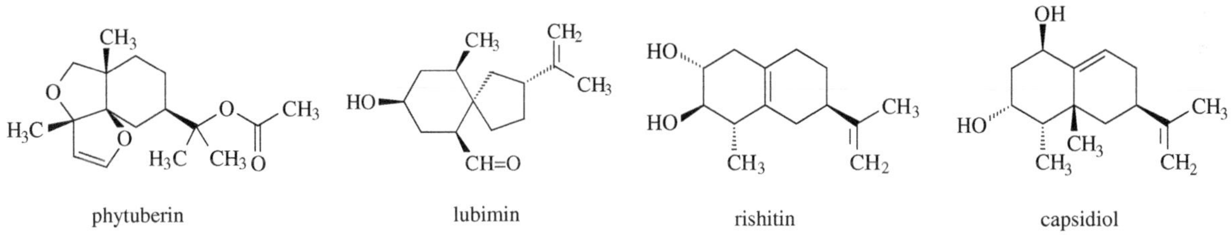

phytuberin   lubimin   rishitin   capsidiol

10-133, sesquiterpenic phytoalexins of solanaceous plants

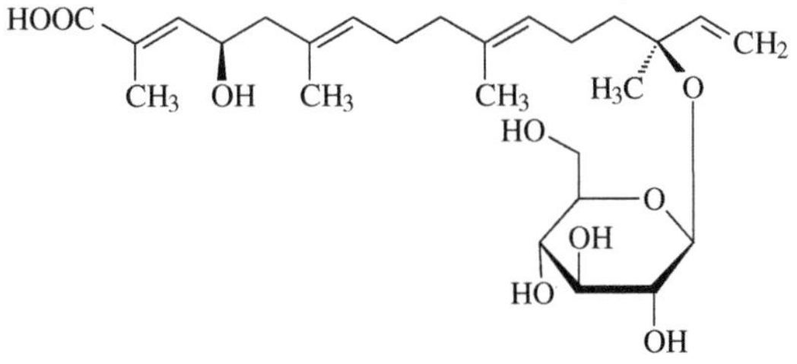

10-134, capsianoside I

Mint plants (Lamiaceae), such as common thyme (*Thymus vulgaris*), contain substituted diterpenoid quinones derived from biphenyl (**10-135**) and related to monoterpenic alcohol thymol, which exhibit marked antioxidant activities. Amongst the most active natural diterpenoid antioxidants with anti-inflammatory effects is carnosic acid, also known as rosmaricine, which is accompanied by carnosol, also known as picrosalvin (**10-135**). These two antioxidants represent about 15% by weight of commercial extracts of rosemary (*Rosmarinus officinalis*) and more than 90% of their antioxidative activity. The carnosic acid content in fresh spices is about 1–2%.

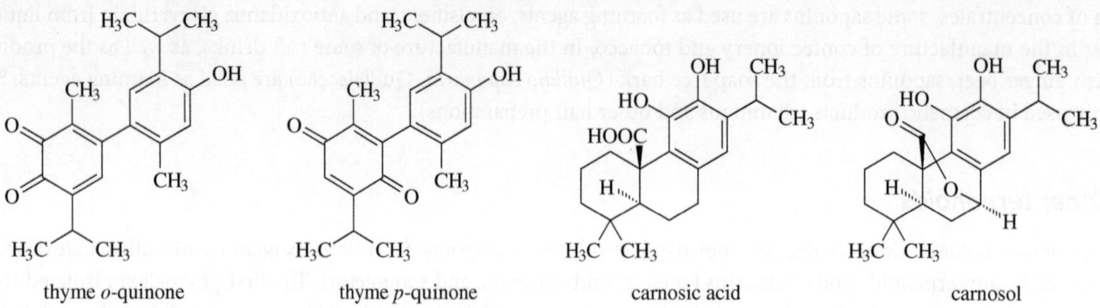

thyme *o*-quinone   thyme *p*-quinone   carnosic acid   carnosol

10-135, important diterpenic phytoalexins and antioxidants

Carnosic acid is unstable and is transformed into carnosol. Other active products are rosmanol (7α-isomer), epirosmanol (7β-isomer), and 7-methylepirosmanol. Other minority compounds of rosemary extracts are isorosmanol, rosmariquinone, and rosmaridiphenol (**10-136**), which is related to tropones. The same compounds are also found in common (garden) sage (*Salvia officinalis*).

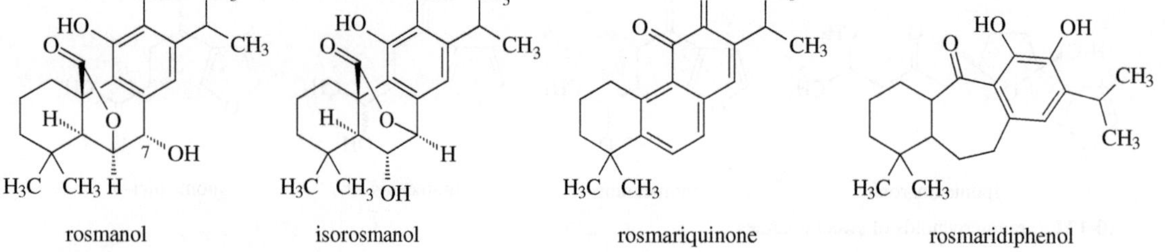

rosmanol   isorosmanol   rosmariquinone   rosmaridiphenol

10-136, diterpenes related to carnosic acid

Carnosic acid has a typical *o*-diphenol structure and is easily oxidised. The antioxidation mechanism is based on a coupling reaction with the peroxyl radical at the 12- or 14-position of carnosic acid and the subsequent transformation reactions of intermediates to an *o*-quinone and a hydroxy *p*-quinone (Figure 10.23).

Extract of rosemary is currently approved in the European Union as a natural antioxidant (E392). It is used in both food and non-food products (e.g. toothpaste, mouthwash, and chewing gum).

The antioxidant potential of plants (such as rosemary) also comes from some triterpenic acids (such as betulinic, oleanolic, and ursolic acids), triterpenic alcohols (β-amyrin), other compounds (see Section 3.7.4.1.1), and derived saponins. Usually, low antioxidant activity (up to temperatures around 180 °C) is exhibited by phytosterols, the most active compound of which is Δ^5-avenasterol (see Section 3.7.4.1.3).

10.3.12 Plant phenols

10.3.12.1 Coumarins

Phototoxicity and other toxic effects are exhibited by some representatives of a group of phenolic compounds called coumarins, plant secondary metabolites with the structure of C_6—C_3. Furanocoumarins and some plant pigments (see Section 9.8.1.3.1), such as hypericin of St John's wort and fagopyrin of common buckwheat, act as the primary photosensitisers.

More than 1000 coumarins occur in nature, of which about 300 are simple coumarins. They occur most often as a mixture of 10–20 related compounds in about 100 families of plants, but especially Rutaceae, which includes various citruses, Moraceae, which includes mulberries (*Morus* spp.) and figs (*Ficus* spp.), Apiaceae, which includes many species of aromatic vegetables (carrots, parsley, parsnips, and celery) and spices (caraway, cumin, coriander, dill, anise, and fennel), and Asteraceae, which includes some medicinal herbs. Simple coumarins (including 2*H*-1-benzopyran-2-ones and 5,6-benzo-2-pyrones) are derived from γ-lactones of 2-hydroxycinnamic acids. The basic member of the homologous series of coumarins (**10-137**) is 2*H*-chromen-2-one, called coumarin. Almost all other simple coumarins are compounds hydroxylated at C-7. Other skeletal positions may also be hydroxylated, and some hydroxyl groups are protected by methylation or isoprenylation. Coumarins also often occur as glycosides, which are accompanied by their respective aglycones. Some coumarins act as blastocolins (inhibiting seed germination). Relatively widespread are **furanocoumarins**, which are linear (6,7-furanocoumarins, also known as psoralen-type furanocoumarins or psoralens) or angular (7,8-furanocoumarins, also known as angelicin-type furanocoumarins or angelicins) compounds. Linear and angular **pyranocoumarins** are relatively rare; they are found mainly in plants of the Rutaceae family. **Isocoumarins** (3,4-benzo-2-pyrones) are also rare, and 3-phenylcoumarins (**isoflavonoids**) and 4-phenylcoumarins (**neoflavonoids**) are rarer still (**10-137**).

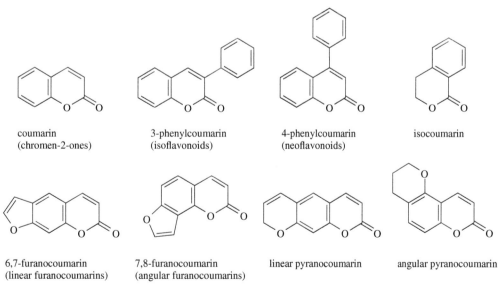

Figure 10.23 Antioxidant mechanism of carnosic acid.

coumarin
(chromen-2-ones)

3-phenylcoumarin
(isoflavonoids)

4-phenylcoumarin
(neoflavonoids)

isocoumarin

6,7-furanocoumarin
(linear furanocoumarins)

7,8-furanocoumarin
(angular furanocoumarins)

linear pyranocoumarin

angular pyranocoumarin

10-137, types of coumarins

10.3.12.1.1 Simple coumarins

Typical representatives of simple coumarins substituted at position C-6 or at positions C-6 and C-7 are umbelliferon (7-hydroxycoumarin), derived from 4-coumaric acid, and its 7-*O*-β-D-glucopyranoside skimmin; aesculetin (6,7-dihydroxycoumarin, formerly also known as cichorigenin), its 6-*O*-β-D-glucopyranoside aesculin (or aesculoside), and its 7-*O*-β-D-glucopyranoside cichoriin; scopoletin (7-hydroxy-6-methoxycoumarin) and its 7-*O*-β-D-glucopyranoside scopolin; and scoparon (6,7-dimethoxycoumarin, **10-138**). These coumarins are present in small amounts in some fruits and vegetables as phytoalexins. Aesculetin and scopoletin were found at a level of about 1 mg/kg in carrots, celery, and other root vegetables and some fruits (such as apricots). The same compounds, aesculetin 6-*O*-β-D-glucoside (aesculin) and aesculetin 7-*O*-β-D-glucoside (cichoriin), also occur in the roots and buds of chicory, and together with other substances contribute to the bitter taste of this vegetable. The bark of horse chestnut (*Aesculus* spp., Hippocastanaceae) contains a large amount of aesculin. Scoparone occurs in citrus fruits exposed to UV radiation or significant temperature fluctuations. Hemiarin (7-methoxycoumarin) and ajapin (6,7-methylenedioxycoumarin, **10-139**) are found in tubers of Jerusalem artichokes (*Helianthus tuberosus*, Asteraceae).

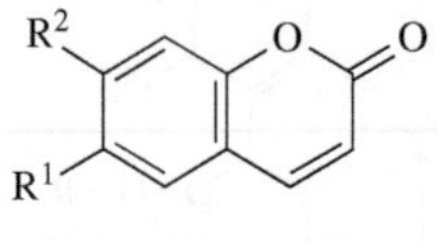

10-138, important coumarins

    umbelliferon, R¹ = H, R² = OH
    aeculetin, R = R² = OH
    hemiarin, R¹ = H, R² = OCH₃
    scopoletin, R¹ = OCH₃, R² = OH
    scoparon, R¹ = R² = OCH₃

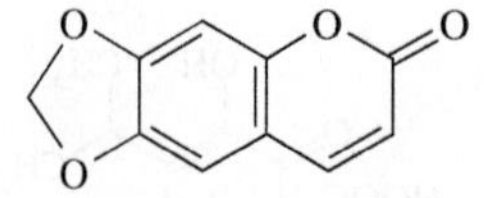

10-139, ajapin

Examples of coumarins substituted at other positions of the benzene ring are daphnetin (7,8-dihydroxycoumarin), limettin (5,7-dimethoxycoumarin), and fraxidin (6,7-dimethoxy-8-hydroxycoumarin, **10-140**). Limettin occurs in citrus fruits, daphnetin-8-*O*-β-D-glucoside, known as daphnin, is present in higher quantities in the bark of daphne (*Daphne mezereum*, Thymelaeaceae), commonly known as mezereon, and fraxidin-8-*O*-β-D-glucoside or fraxin occurs together with related coumarins in the bark of common ash (*Fraxinus excelsior*, Oleaceae).

Examples of coumarins with isoprenoid substituents are osthenol, 7-hydroxy-8-(2-hydroxy-3-methyl-3-en-1-yl)coumarin (**10-141**), osthol, 7-methoxy-8-(3-methyl-2-en-1-yl)coumarin, and auraptene, 7-geranyloxycoumarin, which occur in citrus fruits.

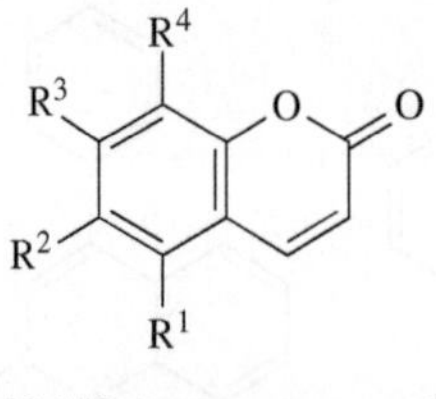

10-140, important coumarins

    daphnetin, R¹ = R² = H, R³ = R⁴ = OH
    limettin, R¹ = R³ = OCH₃, R² = R⁴ = H
    fraxidin, R¹ = H, R² = R³ = OCH₃, R⁴ = OH

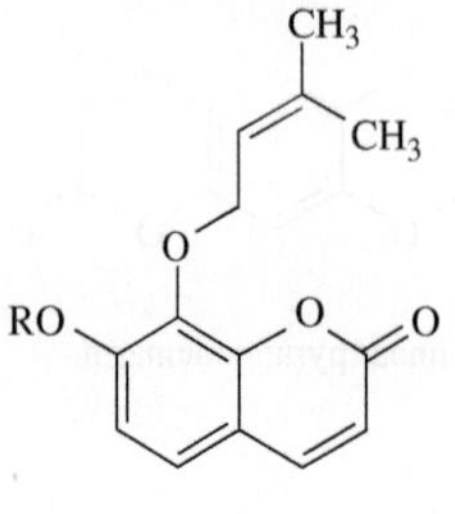

10-141, isoprenoid coumarins

    osthenol, R = OH
    osthol, R = OCH₃

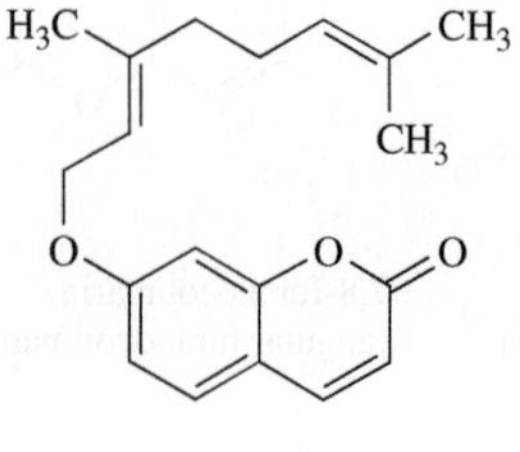

auraptene

10.3.12.1.2 Furanocoumarins

Furanocoumarins, synthesised from umbelliferon, are frequently found in many plants. There are two basic types:

- **linear** coumarins or 6,7-furanocoumarins, or psoralen-type coumarins or psoralens;

- **angular** coumarins or 7,8-furanocoumarins, or angelicin-type coumarins or angelicins.

Common linear hydroxy- and methoxy-substituted furanocoumarins (**10-142**) include psoralen, bergaptol (5-hydroxypsoralen), bergapten (5-methoxypsoralen), xanthotoxin (8-methoxypsoralen), and isopimpinellin (5,8-dimethoxypsoralen). Other linear furanocoumarins, such 8-(3-methyl-2-en-1-yloxy)psoralen (imperatorin), 5-(3-methyl-2-en-1-yloxy)psoralen (isoimperatorin) (**10-143**), and 5-(3,7-dimethyl-2,6-dien-1-yloxy)psoralen (5-geranoxypsoralen or bergamottin), and other compounds have an alkenoxyl isoprenoid chain, as do the simple coumarins osthol and auraptene. These compounds are found mainly in citrus essential oils and root vegetables. Psoralen and bergapten are also the main coumarins of fig tree leaves. A more complex structure is oxypeucedanin, 5-(3,3-dimethyloxiranylmethoxy)psoralen (**10-144**), which occurs as a mixture of (*R*)- and (*S*)-isomers in parsley.

10-142, hydroxy- and methoxy-substituted psoralens

psoralen, $R^1 = R^2 = R^3 = H$
bergaptol, $R^1 = OH$, $R^2 = R^3 = H$
bergapten, $R^1 = OCH_3$, $R^2 = R^3 = H$
xanthotoxin, $R^1 = R^3 = H$, $R^2 = OCH_3$
isopimpinellin, $R^1 = R^2 = OCH_3$, $R^3 = H$

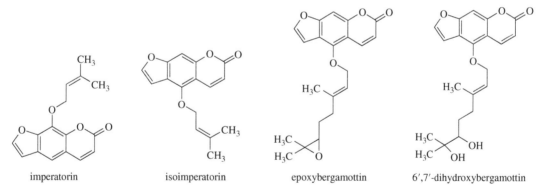

imperatorin   isoimperatorin   epoxybergamottin   6′,7′-dihydroxybergamottin

10-143, psoralen-type furanocoumarins with isoprenoid chains

10-144, oxypeucedanin

Table 10.22 Furanocoumarin contents in selected fruits, vegetables, and herbs (in µg/kg).

Food	Bergaptol	Psoralen	Xanthotoxin	Bergapten	Bergamottin	Total furanocoumarins
Fruits						
Clementine	–	–	–	–	0.2	0.2
Grapefruit	135	–	–	0.7	8357	21 858
Lemon	117	<0.1	0.3	–	212	330
Lime	313	0.1	2.5	136	8694	9151
Orange	–	–	–	–	0.5	0.5
Fig (fresh)	–	0.1	–	–	0.2	0.5
Fig (dried)	–	14	–	1.8	0.2	16
Vegetables						
Carrots	–	0.1	0.9	1.1	44	68
Celeriac	–	1.7	161	234	–	396
Celery	–	14	145	86	4	252
Parsnip	10	4	208	113	0.1	335
Herbs						
Coriander	–	–	–	1.5	–	2
Cumin	–	3.9	0.5	3.2	0.6	8
Dill	–	9.5	0.6	2.4	0.5	13

Angular-type furanocoumarins (**10-145**) include angelicin, isobergapten (5-methoxyangelicin), sphondin (6-methoxyangelicin), and pimpinellin (5,6-dimethoxyangelicin). In small amounts, these compounds accompany linear furanocoumarins.

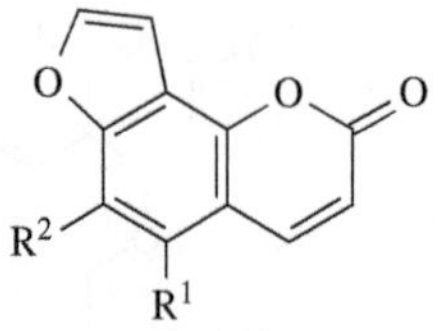

10-145, angelicin-type furanocoumarins
angelicin, $R^1 = H$, $R^2 = H$
isobergapten, $R^1 = OCH_3$, $R^2 = H$
sphondin, $R^1 = H$, $R^2 = OCH_3$
pimpinellin, $R^1 = OCH_3$, $R^2 = OCH_3$

The major sources of psoralens are citrus fruits, root vegetables, herbs, and spices (Table 10.22). Relatively high concentrations of furanocoumarins (9.5 mg/kg) are found in grapefruit juices. The main ingredients are bergamottin (4.3 mg/kg), bergapetol (1.9 mg/kg), and bergapten (0.03 mg/kg). Grapefruit peel contains other major components: 6′,7′-dihydroxybergamottin (158 mg/kg), epoxybergamottin (96 mg/kg), and auraptene (69 mg/kg). Juice from the leaves of the common fig tree (*Ficus carica*, Moraceae) contains psoralen (2000 mg/l) and bergapten (500 mg/l).

The concentration of coumarins in plants is dependent on a number of factors. In healthy plants, their content varies in the different parts and greatly depends on climatic conditions. For example, in celery (*Apium graveolens* var. *dulce*, Apiaceae), whose leaves are eaten raw as a salad and cooked as a vegetable, the content of furanocoumarins in older outer leaves is 50 mg/kg, that in the inner leaves is 8 mg/kg, and that in the small central leaves is 4 mg/kg fresh matter. The content in leaf petioles is about 0.2 mg/kg fresh matter. The amount of furanocoumarins is also low in the root, reaching just 1.1 mg/kg. A similar amount of furanocoumarins is found in the tuberous thick base of stems of celeriac (A. *graviolens*, var. *rapaceum*), which is cooked and eaten as a salad vegetable. High levels of furanocoumarins, however, appear as a result of different stress factors, such as attacks of microorganisms, exposure to UV radiation, storage at low temperatures, and mechanical damage. For example, attack of the fungus *Sclerotinia sclerotiorum* on celery root results in a significant increase in the concentrations of these compounds (up to 44 mg/kg from about 2 mg/kg).

10.3.12.1.3 Pyranocoumarins

Also found in plants, but in small amounts, are pyranocoumarins, of both linear (**10-146**) and angular (**10-147**) types. Common linear pyranocoumarins are present mainly in plants of the Rutaceae family, and thus also in citrus fruits and, especially, citrus essential oils extracted from the peel. A similar pyranocoumarin is graveolon (**10-148**), which is present in small amounts in the leaves and roots of parsley.

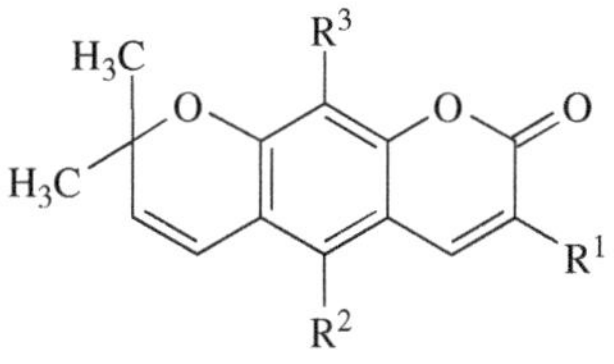

10-146, important linear pyranocoumarins

xanthiletin, R^1 = H, R^2 = H, R^3 = H
xanthoxylethin, R^1 = OCH$_3$, R^2 = H, R^3 = H
lugangetin, R^1 = OCH$_3$, R^2 = H, R^3 = H
trachyphyllin, R^1 = OH, R^2 = CH=CHCH(CH$_3$)$_2$, R^3 = H

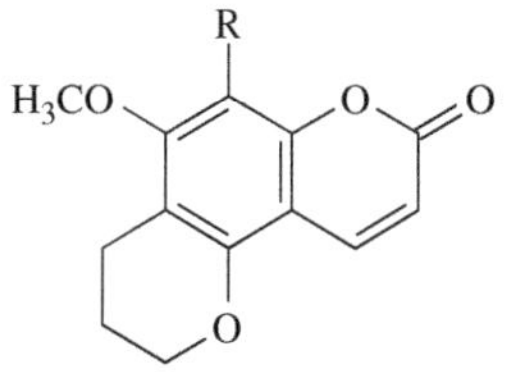

10-147, important angular pyranocoumarins

alloxanthoxyletin, R = H
dipetalin, R = CH=CHCH(CH$_3$)$_2$
aviceunin, R = CH=CHC(CH$_3$)=CH$_2$

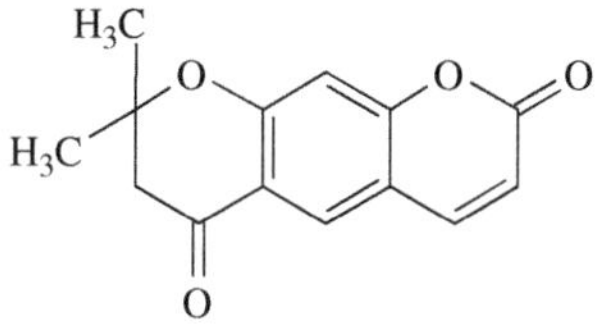

10-148, graveolon

10.3.12.1.4 Phenylcoumarins and isocoumarins

A coumarin substituted at the C-3 phenyl group (having an anisoflavone skeleton), known as santalin AC (6,7,3′-trihydroxy-2′,4′-dimethoxy-3-phenylcoumarin), is a minor component of red sandalwood colour (see Section 9.4.2.9). 4-Phenylcoumarins (also known as neoflavonoids) are found in trees of the genus *Dalbergia* (Fabaceae), grown in tropical regions of Asia and America. A typical representative is dalbergin (**10-149**).

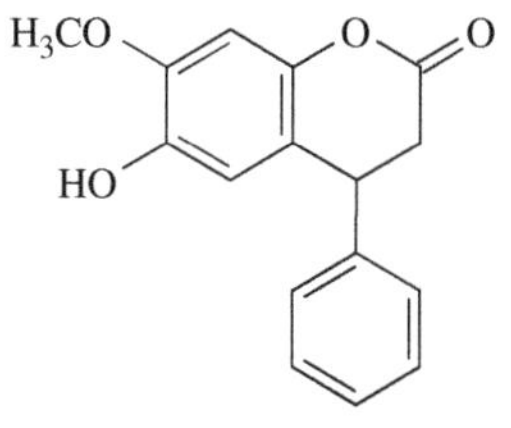

10-149, dalbergin

Rarely found in plants are isocoumarins (3,4-benzo-2-pyrones or 1*H*-2-benzopyran-1-ones). Isocoumarin phyllodulcin is used locally as a sweetener (see Section 11.3.2.1.3). Infection of carrot roots (*Daucus carota*, Daucaceae) by the fungi *Ceratocystis fimbriata*, *Chetomium globosum*, *Botrytis cinerea*, and *Thielaviopsis basicola* produces the bitter phytoalexin (*R*)-6-methoxymellein

(8-hydroxy-6-methoxy-3-methyl-3,4-dihydroisocoumarin, **10-150**), which likewise has oestrogenic activity. In Europe, levels of 6-methoxymellein ranged from 0.02 to 76.00 mg/kg in fresh carrots and from 0.04 to 15.64 mg/kg in processed carrot products. The recognition threshold concentration of 6-methoxymellein is 20 mg/l. Levels are reduced by 69 and 33% in blanched carrots processed by boiling in water and steam treatment, respectively.

10-150, 6-methoxymellein

Coumarins exhibit a wide variety of biological effects. They mainly act as vasodilatant agents and anticoagulants. For example, the synthetic dicoumarol warfarin is used in human medicine and as a rhodenticide (see Section 5.5.3). Furanocoumarins exhibit spasmolytic and vasodilatory effects, and coumarins and furanocoumarins have antibacterial and antifungal properties. For example, umbelliferon, scopoletin, and furanocoumarins act as molluscocides (bergapten and isopimpinellin) and may have oestrogenic effects (some isocoumarins), and furanocoumarins can act as photosensitisers and thus exhibit phototoxic effects.

Photosensitisation caused by furanocoumarins is manifested by an increased sensitivity of unpigmented skin to sunlight (near-UV radiation of wavelength 400–300 nm, 3.10–4.13 eV, visible to birds, insects, and fish) and is associated with a higher incidence of skin cancer. The primary cause is binding of furanocoumarins to DNA (cycloaddition reactions with pyrimidine bases) and other macromolecular components of cells (such as fatty acids in lipoproteins), with the formation of free radicals activated by UV radiation. This disease is different from photo-dermatitis or sunburn, which is damage to the skin without a photosensitiser. Photosensitising diseases are categorised according to the origin of the photosensitiser and the method by which it gets into the peripheral circulation as primary photosensitisation (the photosensitiser is an exogenous substance) and several other forms. Furanocoumarins and the toxic pigments hypericin and phagopyrin (see Section 9.8.1.3.2) cause primary photosensitisation.

The toxicity of furanocoumarins occurring in foods may manifest in humans if consumed in amounts of 0.14–0.38 mg/kg body weight, which is out of the question in normal consumption of root vegetables. Total daily intake is estimated at 1.3 mg. Photosensitisation caused by furanocoumarins may be a problem for workers handling celery or parsley leaves or coming into contact with the juice of fresh figs. It is reported that this so-called acute contact dermatitis appears after contact with celery leaves containing at least 18 mg/kg furanocoumarins. Chronic dermatitis may appear on repeated contact with celery leaves even if the material contains about 7 mg/kg of furanocoumarins in f.w. In this respect, particularly infamous is the giant hogweed (*Heracleum mantegazzianum*, Apiaceae), native to Central Asia and introduced as an ornamental plant to some European countries, the United States, and Canada, the sap of which causes severe dermatitis resulting in blisters, long-lasting scars, and blindness (on contact with the eyes).

Properties of furanocoumarins are also used therapeutically. Bergapten and xanthotoxin are used to treat psoriasis and the idiopathic depigmentation called vitiligo, which is caused by the formation of melanin pigment disorders associated with the activity of tyrosinase. After irradiation, psoralens act on the surrounding undamaged melanocytes, which are bigger and produce more melanin, and thus give the skin its normal colour.

10.3.12.2 Stilbenes

Stilbenes (also called diarylethanoids) are a group of plant secondary metabolites with the structure C_6—C_2—C_6. They are hypothetically derived from the hydrocarbon (*E*)-1,2-diphenylethylene, known as (*E*)-stilbene or *trans*-stilbene (**10-151**). Ring A usually carries two hydroxy groups in the *m*-position, whereas ring B is replaced by hydroxy and methoxy groups in the *o*-, *m*-, or *p*-position. Free stilbenes accompanied by the corresponding glycosides are found in small quantities in several plant species. The basic member of a number of common stilbenes is (*E*)-pinosylvin (stilbene-3,5-diol). It occurs along with other stilbenes (**10-152**) especially in conifers, such as pines (*Pinus* spp., Pinaceae).

10-151, (*E*)-stilbene

A representative of stilbenes with antimicrobial and antioxidant effects is (*E*)-resveratrol (3,4′,5-trihydroxystilbene, **10-152**), an antifungal substance produced by a relatively limited number of plant species in response to biotic and abiotic stress and UV light irradiation. Resveratrol is synthesised by common grape vine (*Vitis vinifera*, Vitaceae), some legumes such as peanuts (*Arachis hypogea*, Fabaceae), conifers, and some other plants.

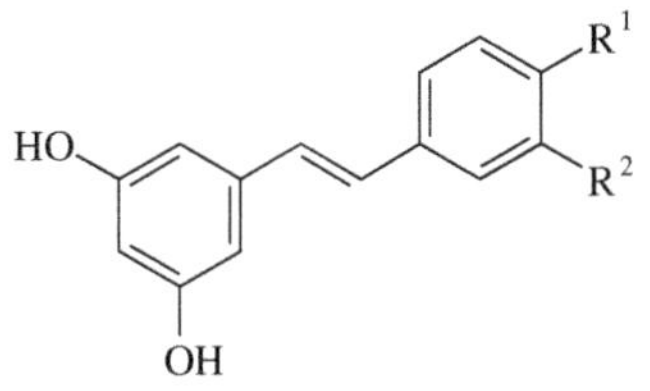

10-152, (*E*)-pinosylvin, R^1 = R^2 = H
(*E*)-resveratrol, R^1 = OH, R^2 = H
(*E*)-piceatannol, R^1 = R^2 = OH
(*E*)-rhapontigenin, R^1 = OCH$_3$, R^2 = OH
(*E*)-isorhapontigenin, R^1 = OH, R^2 = OCH$_3$

Resveratrol in grapes is mainly found in the skins of red grape varieties. From there, it passes into the wine in amounts of about 1–5 mg/l. It is accompanied by (*Z*)-resveratrol and some oxidation products, such as (*E*)- and (*Z*)-piceatannol, (*E*)- and (*Z*)-3,3′,4′,5,-tetrahydroxystilbene. Larger amounts of resveratrol and piceatannol are bound in the form of β-D-glucopyranosides. The glucopyranoside of (*E*)-resveratrol is called (*E*)-piceid, whilst the glucopyranoside of (*E*)-piceatannol is (*E*)-astringin (**10-153**). In recent years, the anticarcinogenic and cardioprotective effects of resveratrol have been intensively studied.

ε-Viniferins, the dehydrodimers δ-viniferins (also known as viniferifurans), gnetin C, scirpusin A, and pallidol (**10-154**) are examples of resveratrol dimers found in wine grapes. For example, pallidol is a constituent of Riesling wine as 3-*O*-glucopyranoside and 3,3′-di-*O*-glucopyranoside. In addition, grapevine canes contain higher resveratrol oligomers, examples of which are the trimers α-viniferin and miyabenol C, the tetramers vitisin B and hopeaphenol, and the hexamer viniphenol A (**10-155**). Resveratrol oligomers show even higher antioxidant and fungicidal activities than resveratrol. Plants of the genus *Vitis* (Vitaceae) contain many other stilbenoids that are of chemotaxonomic interest.

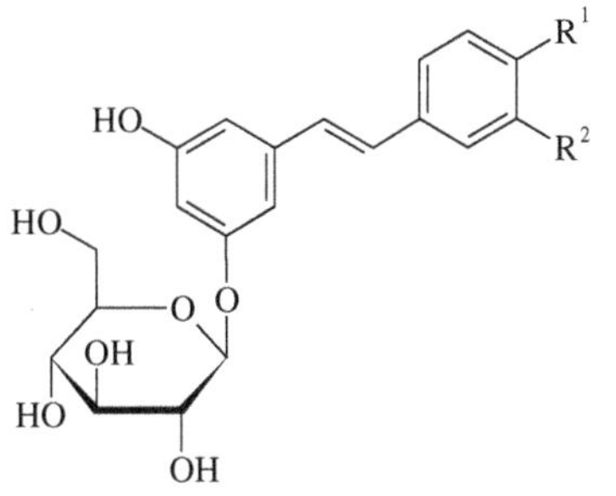

10-153, (*E*)-piceid, R^1 = OH, R^2 = H
(*E*)-astringin, R^1 = R^2 = OH

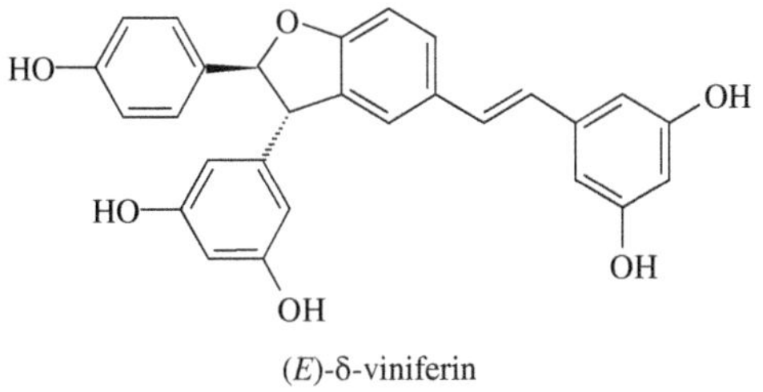

(*E*)-δ-viniferin

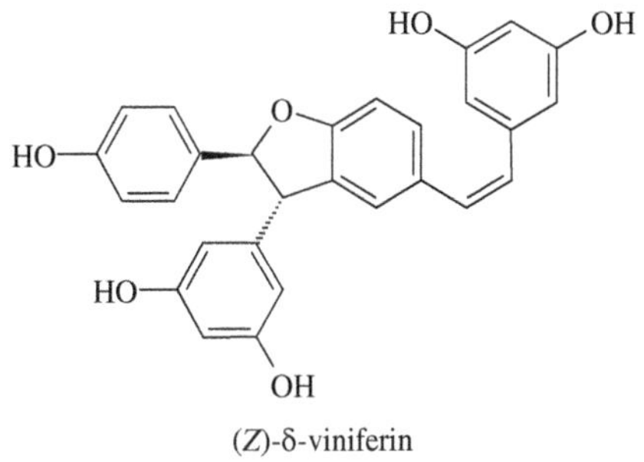

(*Z*)-δ-viniferin

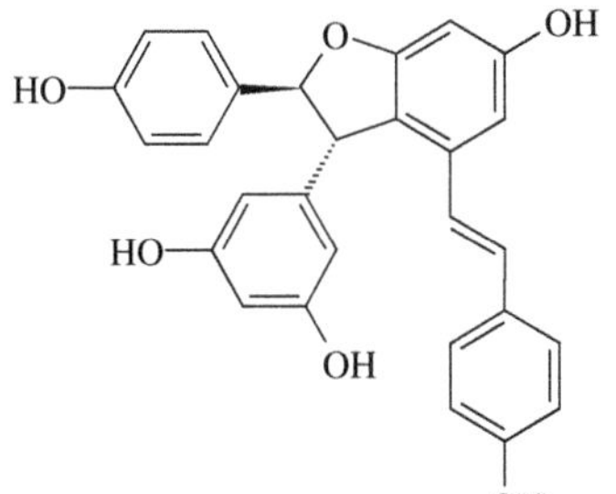

(*E*)-ε-viniferin

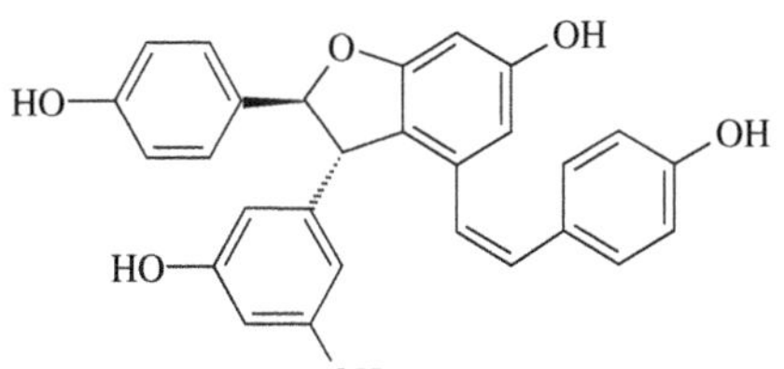

(*Z*)-ε-viniferin

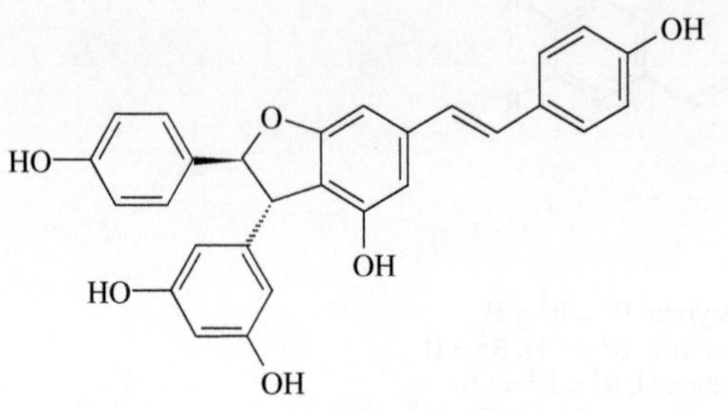

(E)-gnetin C

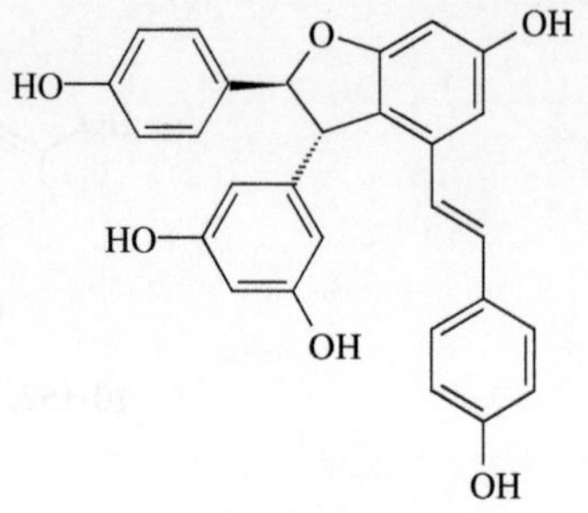

(E)-scirpusin A

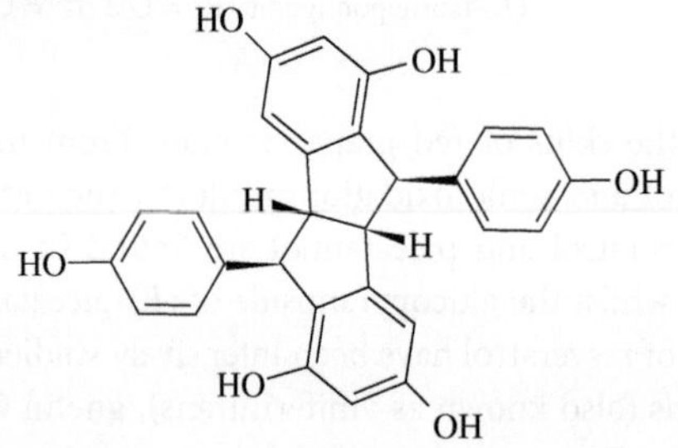

pallidol

10-154, resveratrol dimers and analogues

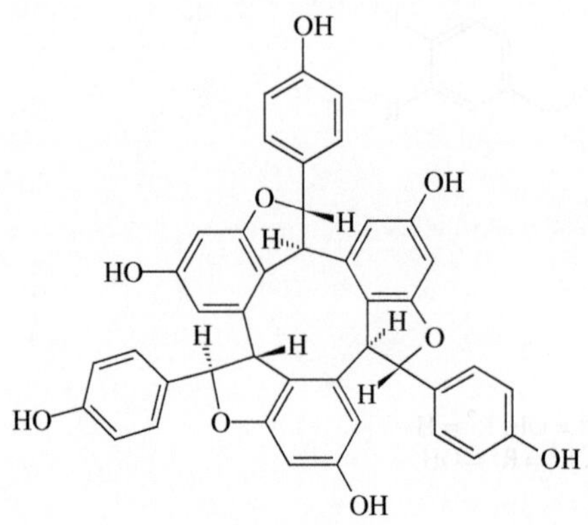

α-viniferin

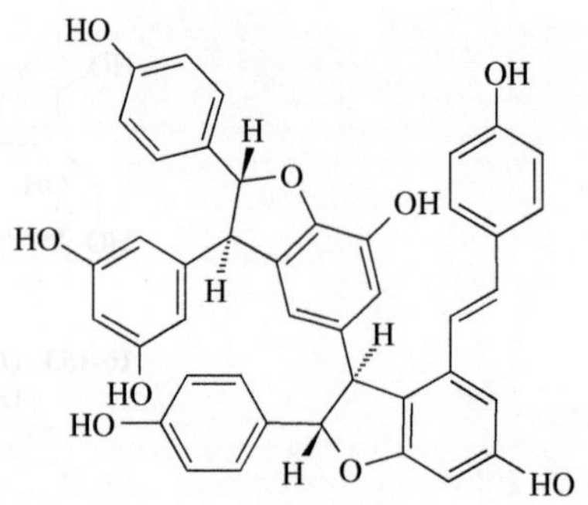

miyabenol C

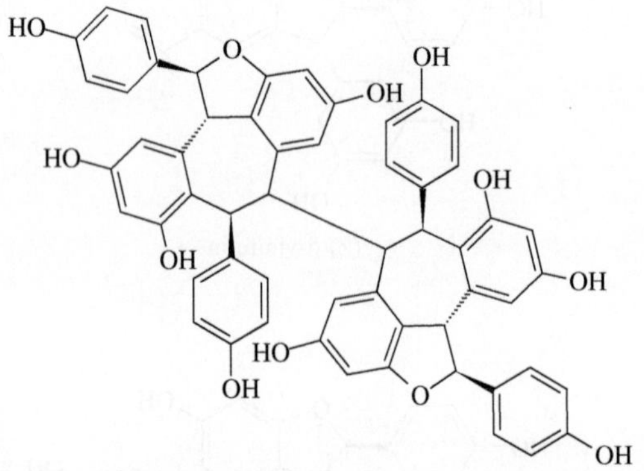

hopeaphenol

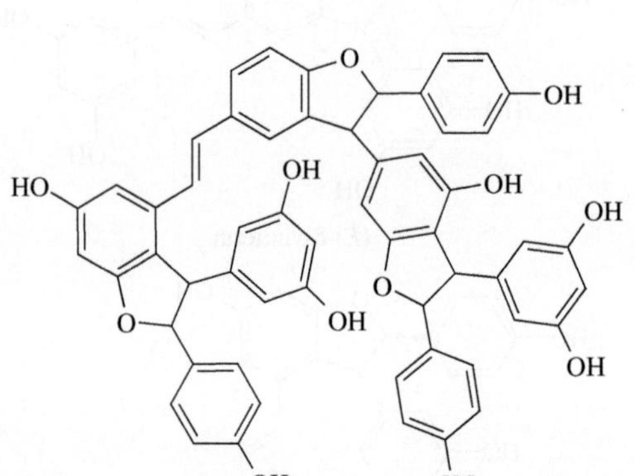

vitisin B

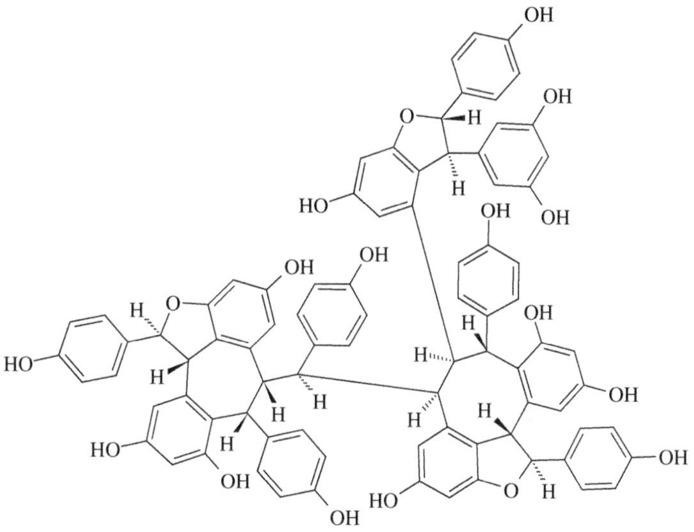

viniphenol A

10-155, examples of resveratrol trimers, tetramers, and hexamers

The stilbenes reported for several varieties of peanuts (*Arachis hypogaea*), found in the leaves, roots, and seeds, seem to be derived mainly from (*E*)-resveratrol. Examples of peanut stilbenes are (*E*)-piceatannol (**10-152**), (*E*)-piceid (**10-153**), and the prenylated stilbenes arachidins, (*E*)-4′-deoxyarachidin-2 (chiricanine A) and (*E*)-4′-deoxyarachidin-3 (arahypin-1) (**10-156**).

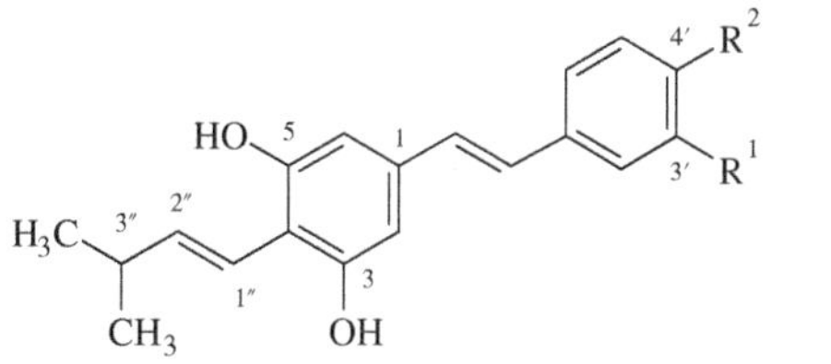

arahypin-1, $R^1 = R^2 = H$
arachidin-1, $R^1 = R^2 = OH$
arachidin-3, $R^1 = H$, $R^2 = OH$

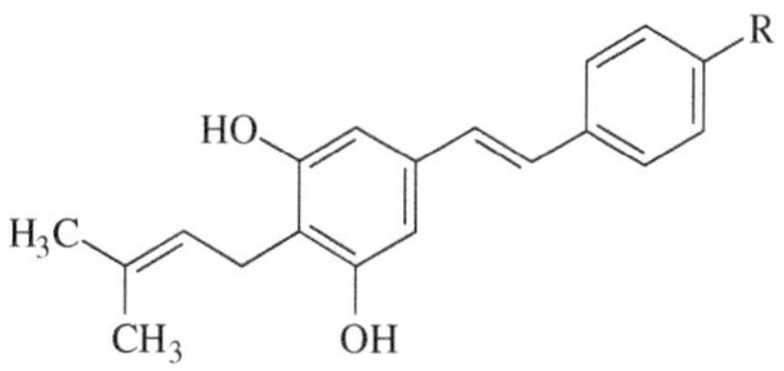

chiricanine A, R = H
arachidin-2, R = OH

10-156, arachidins and related compounds

Hydroxysubstituted stilbenes are often methylated. Examples are rhapontigenin (3,3′,5-trihydroxy-4′-methoxystilbene) and isorhapontigenin (3,4′,5-trihydroxy-3′-methoxystilbene, **10-152**). Rhapontigenin occurs in berries of crimson glory vine (*Vitis coignetiae*), a plant used to produce wines in Korea and Japan. Rhapontigenin 3-*O*-β-ᴅ-glucopyranoside, called rhaponticin (**10-152**), is a constituent of rhubarb (*Rheum undulatum*, Polygonaceae); it is used for its haemostatic, antioxidant, and antimicrobial effects in East Asia. Isorhapontigenin (**10-152**) occurs in wine grapes; its 3-*O*-β-ᴅ-glucopyranoside is called isorhaponticin.

10.3.12.3 Isoflavones

The occurrence of about 200 known isoflavonoids is practically limited to legumes (plants in the family Fabaceae). Isoflavonoids are also found in smaller quantities in some other plant families, such as Amaranthaceae, Iridaceae, Moraceae, and Rosaceae. The main representatives of isoflavonoid substances are isoflavones, compounds isomeric with flavones, plant secondary metabolites with the structure of C_6—C_3—C_6.

Isoflavones exhibit various biological effects, the most serious of which are oestrogenic effects. Their steric structure is similar to that of steroidal oestrogens (female sex hormones), so these compounds exhibit different oestrogenic and antioestrogenic effects in animals. The most active women's oestrogen derived from hydrocarbon 5β-oestrane (**10-157**) is 3,17β-oestradiol (abbreviated E2, **10-158**), which is supplied to the tissues by blood circulation, where it binds to specific receptors (proteins). In the body, it is transformed into oestrone (**10-159**) and oestriol (**10-158**). Oestrone was also found in small quantities in palm oil, palm kernel oil (obtained from oil palm, *Elaeis guineensis*, Arecaceae), and pomegranate seeds (*Punica granatum*, Punicaceae). Male hormones are androgens, but very little is known about the plant androgens (such as testosterone, **10-158**) found, for example, in the pollen of pines (*Pinus silvestris*, Pinaceae). Androgenic activity is also seen in the sweetener stevioside.

Phytooestrogens are non-steroidal substances found in about 300 plant species. The oestrogenic activity is also found in many flavones, flavonols, flavanones, chalcones, pterocarpans, lignans, and some other polyphenols. The most common phytooestrogens are isoflavones and lignans.

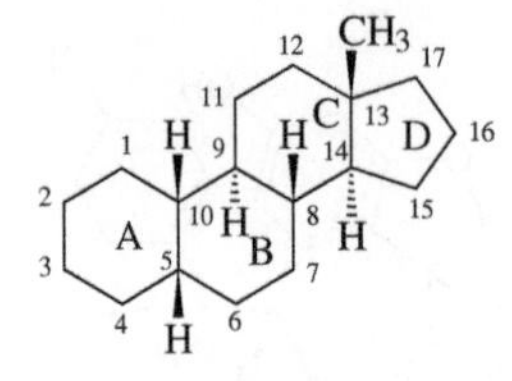

10-157, 5β-oestrane

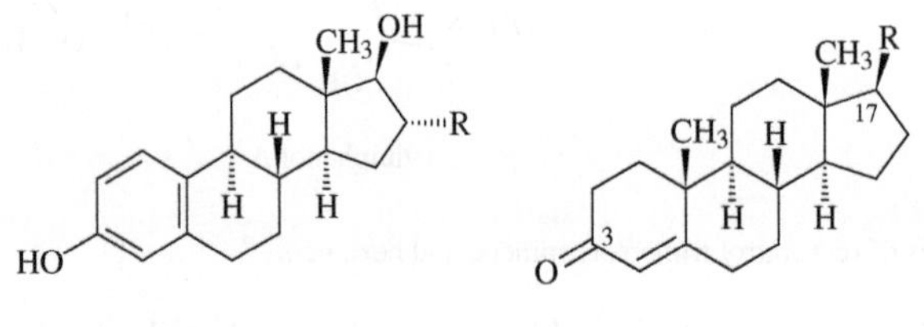

10-158, examples of female and male sex hormones

oestradiol, R = H   testosterone, R = OH
oestriol, R = OH   progesterone, R = COCH₃

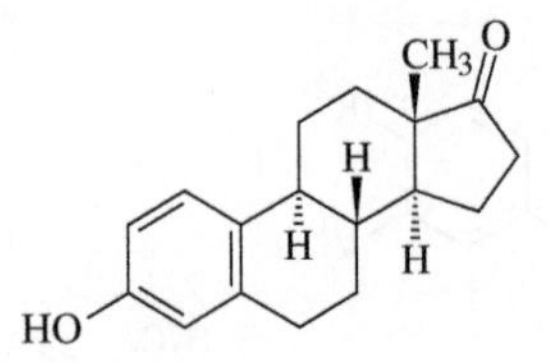

10-159, oestrone

Oestrogenic activity is not limited to mammalian steroid hormones and plant oestrogens. A number of exogenous compounds present in foods of plant origin as natural components or substances that get into the food as contaminants (mould metabolites, pesticides, and some other substances) also possess oestrogenic activity. These substances mimic the effect of sex hormones and induce *oestrus*, as seen in experimental animals. According to the origin, exogenous oestrogenic compounds include:

- **phytooestrogens;**

- **mycooestrogens;**

- **xenooestrogens**, also known as anthropogenic oestrogens.

The most effective mycooestrogens are lactones of resorcylic acid produced by fungi of the genus *Fusarium* and synthetic analogues (see Section 12.2.2.2.11). Comparable (or even greater) oestrogenic activity to that of oestradiol is exhibited by synthetic xenooestrogen stilbene (*E*)-stilboestrol (*trans*-diethylstilboestrol, **10-123**), which was originally used in the fattening of cattle and other livestock because it accelerated their growth. In human medicine, *trans*-diethylstilbestrol was used to prevent miscarriages, but because of its carcinogenic effects it is no longer employed.

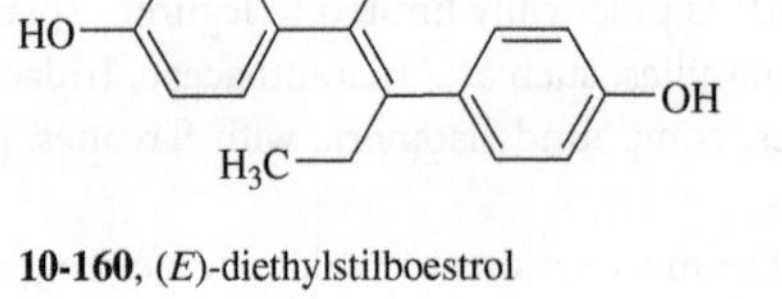

10-160, (*E*)-diethylstilboestrol

Oestrogenic activity has been demonstrated in a number of other anthropogenic substances. An example is bisphenol A, a monomer of polycarbonate and epoxy resins, which can be leached from lacquered plates for cans or dental compositions and seals. Other examples are some halogenated pesticides, polychlorinated biphenyls, their metabolites, phthalates (used as plastic softeners), veterinary pharmaceuticals, and some other xenobiotics (see Chapter 12).

10.3.12.3.1 Structure, nomenclature, and occurrence

Soya beans contain the isoflavone daidzein (7,4′-dihydroxyisoflavone), which is the most active oestrogenic isoflavone, accompanied by genistein (5,7,4′-trihydroxyisoflavone), formononetin (7-hydroxy-4′-methoxyisoflavone), glycitein (7,4′-dihydroxy-6-methoxyisoflavone), and biochanin A (5,7-dihydroxy-4′-methoxyisoflavone, **10-161**). The biochemical precursor of daidzein, formononetin, and glycitein is flavanone liquiritigenin (7,4′-dihydroxyflavanone), whilst the precursor of genistein and biochanin A is naringenin (5,7,4′-trihydroxy-flavanone).

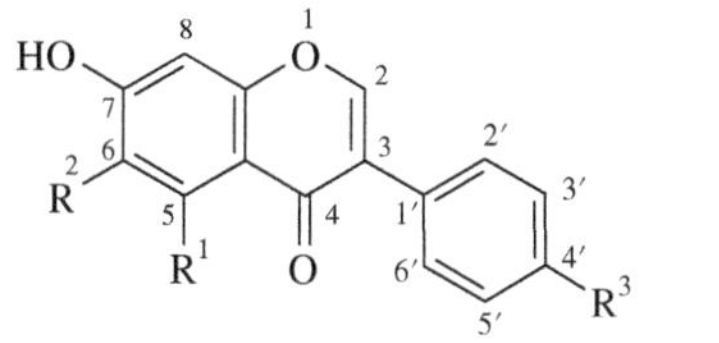

10-161, isoflavones

daidzein, R^1 = H, R^2 = H, R^3 = OH
genistein, R^1 = OH, R^2 = H, R^3 = OH
formononetin, R^1 = H, R^2 = H, R^3 = OCH_3
glycitein, R^1 = H, R^2 = OCH_3, R^3 = OH
biochanin A, R^1 = OH, R^2 = H, R^3 = OCH_3

Isoflavones occur predominantly as 7-β-D-glucosides. The glucoside of daidzein is daidzin, that of genistein is genistin, that of formononetin is ononin, and that of glycitein is glycitin (**10-162**). The main components of soya beans are the glycosides genistin, daidzin, and glycitin and their malonic acid esters. Free isoflavones and acetyl derivatives of glycosides (decarboxylation products of malonyl esters) also occur in small amounts (**10-162**). Contents of the individual compounds are given in Table 10.23. The concentrations of isoflavones in soya beans vary over a wide range, from about 0.13 to 0.42%, whilst their content in soya bean flour is about 0.2%, that in soya isolates is 0.06–0.10%, and that in soya concentrates is 0.07%. Free aglycones even occur in acid hydrolysate of soybean meal, which is used as a food seasoning. Sprouting beans contain formononetin as one of their major isoflavones.

The main isoflavone in chickpea is biochanin A. Other major isoflavones are daidzein, formononetin, biochanin A glucoside, and malonylated biochanin A glucoside. The isoflavone content in chickpea sprouts germinated under different conditions is increased compared to untreated chickpea seeds, with the maximum amount found in sprouts germinated for eight days.

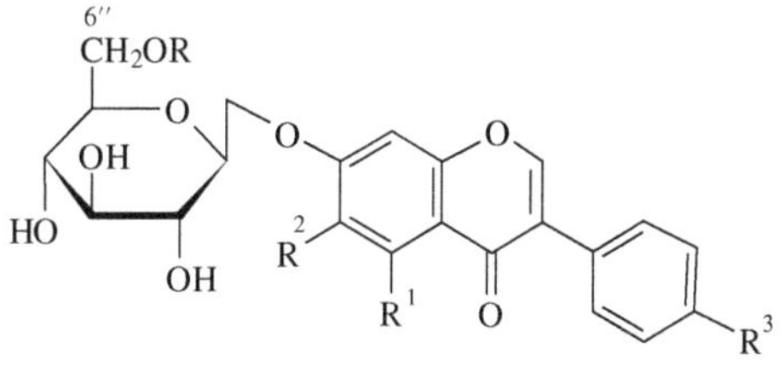

10-162, glucosides of isoflavones and their esters

daidzin, R = H, R^1= H, R^2= H, R^3= OH
genistin, R = H, R^1= OH, R^2= H, R^3= OH
ononin, R = H, R^1= H, R^2= H, R^3= OCH_3
glycitin, R = H, R^1= H, R^2= OCH_3, R^3= OH
6″-acetylgenistin, R = $COCH_3$, R^1= OH, R^2= H, R^3= OH
6″-malonylgenistin, R = $COCH_2COOH$, R^1= OH, R^2= H, R^3= OH

Table 10.23 Isoflavone and isoflavone derivative contents in soya beans and soya bean products (mg/kg).

| Product | Glucoside | | | Aglykon | | | Malonate | | | Acetate | | |
	Daidzin	Genistin	Glycitin	Daidzein	Genistein	Gycitein	Daidzin	Genistin	Glycitin	Daidzin	Genistin	Glycitin
Beans	234–637	326–888	60–66	10–28	11–30	19–22	121–690	290–1756	58–72	trace	2–5	25–33
Flour	147	407	41	4	22	19	261	1023	57	trace	1	32
Isolate	trace–88	137–301	34–49	11–63	36–136	25–53	18–20	88–100	36–39	6–74	0–215	33–46
Concentrate	trace	18	31	0	0	23	0	trace	0	trace	1	0
Tofu[a]	25	84	8	46	52	12	159	108	0	8	1	29
Tempeh[b]	2	65	14	137	193	24	255	164	0	11	0	0
Miso[c]	0–72	96–123	18–21	34–271	93–183	15–54	0	0	19–22	1	2–11	0

[a]Tofu (also known as bean curd), originating in ancient China, is made by coagulating soy milk with a coagulant, such as calcium sulfate.
[b]Tempeh is an Indonesian cake-like food prepared by fermentation of soya beans.
[c]Miso is a traditional Japanese seasoning produced by fermenting rice and soybeans.

Oestrogenic isoflavones also occur in other plants, but in much smaller quantities. For example, peanuts contain about 0.50 mg/kg of daidzein and 0.83 mg/kg of genistein (free and bound as glycosides and their derivatives), sunflower seeds contain 0.08 and 0.14 mg/kg, and poppy seeds contain 0.18 and 0.07 mg/kg, respectively. Examples of more complex structures are isoflavones of *Apios americana* (Fabaceae), sometimes called American groundnut potato bean or Indian potato, the ancient North American Indian crops providing sweet tubers that contain 7-O- and 4′,7-di-O-glycosides of 2′-hydroxy-, 5-methoxy-, 2′-hydroxy-5-methoxy-, and 3′,5-dimethoxygenistein. Formononetin is the main isoflavone of some forage plants. For example, red clover (*Trifolium pratense*), also used for the manufacture of supplements for human use, contains formononetin (1–30 g/kg in dry matter), biochanin, daidzein, and genistein. Minority forage isoflavones are genistein, daidzein, and biochanin.

In addition to oestrogenic activity, isoflavones exhibit relatively high antioxidant activity, which requires the presence of two hydroxyl groups (at C-7 and C-4′) in the molecule, as found in daidzein. The activity of aglycones is higher than in glycosides, but generally isoflavones have lower antioxidant activity than the corresponding flavones. Isoflavones are also contributors to the astringent and bitter taste of soya beans.

Dihydroisoflavones (isoflavanones) occur together with isoflavones in legumes, predominantly as 6″-malonylglucosides. Examples of common isoflavanones are homoferreirin and cicerin (**10-163**), which occur in chickpea (*Cicer arietinum*).

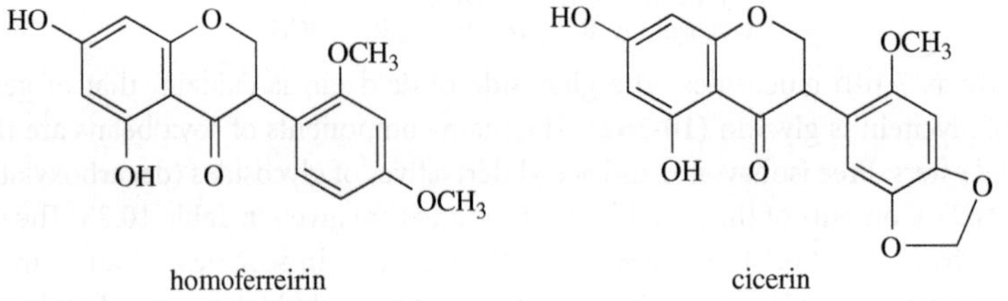

homoferreirin   cicerin

10-163, examples of isoflavanones

10.3.12.3.2 Reactions and changes

Isoflavones are transferred virtually unchanged from soya beans into flour. The loss caused by leaching during soaking of soya beans is about 11%. Cooking results in about a 50% loss of the original content. About 40% of isoflavones are lost during extraction with alkaline agents during production of protein isolates. Acid hydrolysates of soybean meal contain only isoflavone aglycones formed by hydrolysis of glycosides.

10.3.12.3.3 Biological effects

Oestrogens are both beneficial and harmful for the organism. Oestradiol, the endogenous oestrogen of mammals, is essential for normal development and reproduction of the organism, and also affects other important processes related to the immune system and central nervous system. At the same time, however, oestradiol is associated with cancers of the mammary glands, rectum, and probably prostate. Despite continuing research, the role of phytoestrogens in health is still under discussion and remains controversial.

Phytooestrogens show several orders of magnitude lower oestrogenic activity than oestradiol (Table 10.24), but given the quantities in which they are consumed, they may cause male cattle infertility and abortions in female cattle.

Table 10.24 Relative activities of selected oestrogens.

Oestrogen	Relative activity (%)
(*E*)-Diethylstiboestrol	100
Oestrone	6.9
Coumoestrol	0.035
Genistein	0.001
Daidzein	0.00075
Biochanin A	0.00046
Formononetin	0.00026

In women consuming food rich in phytooestrogens, irregularities of the menstrual cycle were observed. On the other hand, in a population with a high intake of soy isoflavones (such as in some Asian countries), lower incidence of breast cancer in women and of prostate cancer in men was reported. The activity of phytooestrogens is generally an additive property. If the level of endogenous (steroid) oestrogens is low (typically in animals), phytooestrogens bind together with them to the receptors. If the level of steroid oestrogens is high (usually in women), phytooestrogens act antagonistically (as antioestrogens).

Some isoflavones do not exhibit oestrogenic activity themselves. The isoflavone formononetin is not an oestrogenic substance, but in the rumen of ruminant animals (as well as in silage feed and the digestive tract of humans, as it also occurs in the urine of women after consumption of soya products) it is transformed by bacteria following the hydrolysis of sugar to aglycone and, further, via daidzein, to the oestrogenic equol (7,4′-dihydroxyisoflavan, **10-164**), which is the main product of this transformation. Byproducts of formononetin transformations are angolensin and 4′-O-methylequol, from which equol arises by a side-path reaction. Equol is partially transformed into O-demethylangolensin (**10-164**). Equol represents about 70% of formononetin degradation products, O-demethylangolensin 5–20%, and 4′-O-methylequol and angolensin minor amounts. For example, the infertility syndrome called clover disease of sheep, occurring in Western Australia in the 1940s, was caused by equol. The disease was manifested by a decrease of sheep gestation, frequent abortions and post-natal complications. Similar syndromes have been found in other animals.

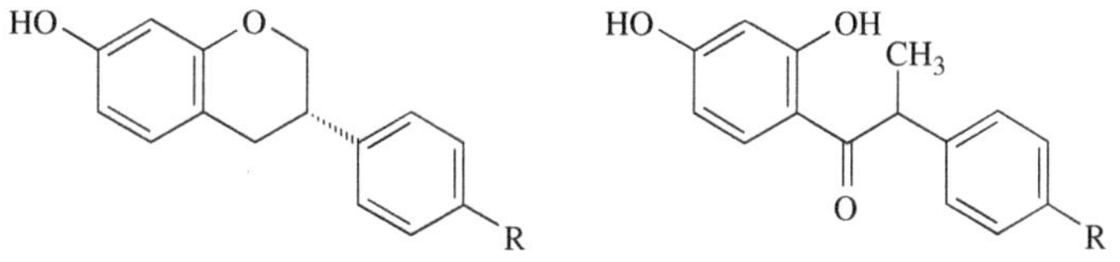

equol, R = OH
4′-O-methylequol, R = OCH₃

O-demethylangolensin, R = OH
angolensin, R = OCH₃

10-164, transformation products of formononetin

Biochanin A is degraded to genistein and further to dihydrogenistein, from which 4-ethylphenol and other simple products arise. The bacterial degradation products of matairesinol in the digestive tract are enterolactone (**10-165**) and lariciresinol. Secoisolariciresinol gives rise to enterodiol (**10-166**), which can be oxidised to the enterolactone. These compounds are excreted in the urine.

The result of exposure to anthropogenic oestrogens in men is believed to be lower sperm quality, and in women there is a higher incidence of lung cancer. Oestrogens may also cause anomalies in wildlife (such as feminisation of male fish). These findings are still mainly speculative due to the small amounts of consumed xenooestrogens compared with the much higher amount of oestrogens produced endogenously and the amount of consumed phytooestrogens.

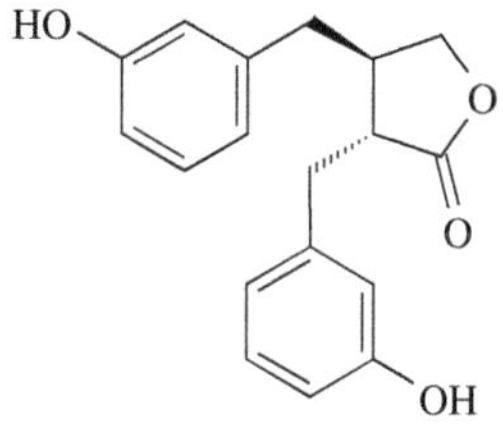

10-165, enterolactone

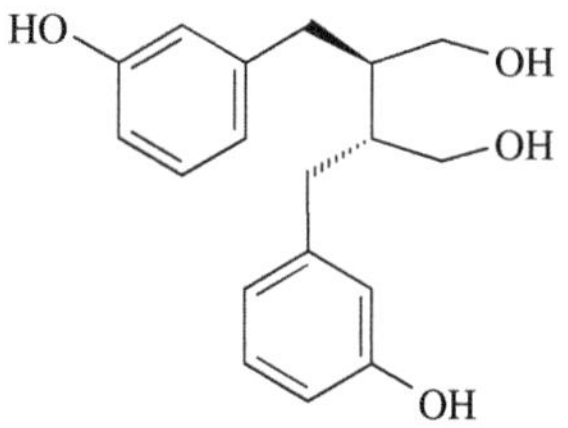

10-166, enterodiol

10.3.12.4 Prenylflavonoids

In addition to isoflavones, legume plants (Fabaceae) also synthesise lipophilic prenylated isoflavones, which are more efficient defence substances (phytoalexins) against pathogenic ilicitors (microorganisms) and animal pests than the original isoflavones. They arise by prenylation of isoflavone precursors and may possess oestrogenic activity. A representative of many structurally similar prenylated isoflavones is wighteone (**10-167**), found in liquorice and white lupin.

Prenylated isoflavones related to prenylated flavonoids (chalcones and flavanones or dihydroflavones) are located in mature female cones of hops (*Humulus lupulus*, Cannabaceae). Thanks to their remarkable bioactive (oestrogenic and anticarcinogenic) effects, they have recently become the subject of medical and pharmaceutical research. The most common compound is the chalcone xanthohumol (**10-167**), a promising anticancer agent that does not exhibit oestrogenic activity. Its content in hop cones is around 1%. Xanthohumol is accompanied by the flavanone isoxanthohumol, with potent oestrogenic activity, which arises from xanthohumol by conversion during heating. Isoxanthohumol is therefore the main prenylated flavonoid in beer. At concentrations about 10–100 times lower, prenylated flavonoids are accompanied by chalcone demethylxanthohumol, which is a precursor of most hop prenylflavonoids. The main oestrogen is flavanone 8-prenylnaringenin (**10-167**), accompanied by 6-prenylnaringenin and other compounds. Concentrations of 8-prenylnaringenin in various beer brands range from <1.6 μg/l in non-alcoholic beers to 138.5 μg/l in stouts. 8-Prenylnaringenin also results from isoxanthohumol by the action of bacteria in the colon or cytochrome P450 enzymes in liver. Chalcone xanthogalenol (**10-167**) has only been detected in US and East Asian hops varieties.

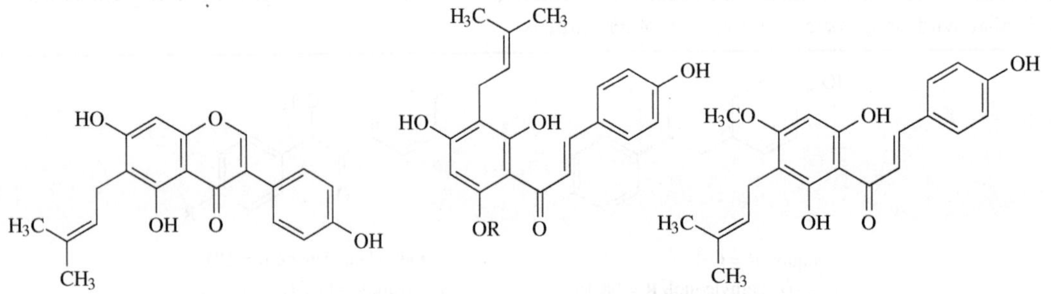

wighteone

demethylxanthohumol, R = H
xanthohumol, R = CH₃

xanthogalenol

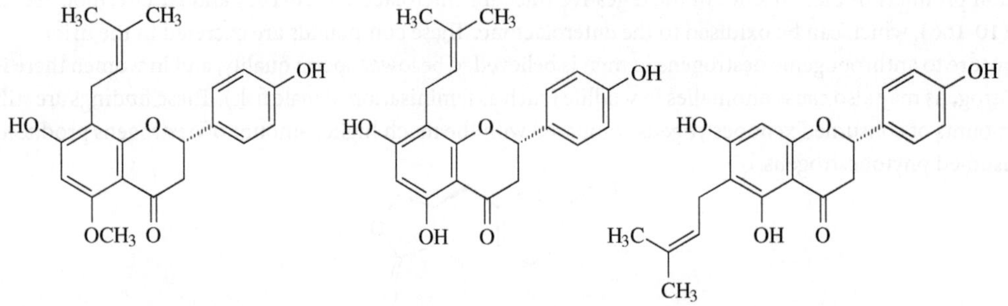

isoxanthohumol

8-prenylnaringenin

6-prenylnaringenin

10-167, examples of prenylated flavonoids

10.3.12.5 Pterocarpans

Pterocarpans are a group of modified isoflavonoids consisting of a 2*H*-chromene moiety fused to a benzo[*b*]furan (1-benzofuran or coumarone) moiety. They are formed by coupling of the B ring to the 4-one position in isoflavones. Pterocarpans occur in small amounts in legume plants (Fabaceae) and in several other families, and typically act as inducible phytoalexins synthesised in response to biotic and abiotic stress factors, such as fungi or harmful insects and physical injury. They may possess oestrogenic activity.

In germinated soybeans and other legumes, the major oestrogen is coumestrol (**10-168**), present in the amount of 0.05–0.2 mg/kg. During germination, its concentration increases about 70–150 times. The highest amount of coumestrol occurs in hulls of beans. The oestrogenic activity of coumestrol is 30–40 times higher than the activities of isoflavones. Small quantities of other related compounds are also present. Different tissues of the soybean plant produce a similar prenylated phytoalexin with antioestrogenic activity, called glyceollin. Glyceollin is derived from pterocarpan (−)-glycinol, formed by the cyclisation of isoflavone daidzein. It is a collective name for soybean metabolites possessing pterocarpanoid skeletons and a cyclic ether moiety originating from a C5 prenyl moiety (**10-168**). Four isomers of glyceollins (I, II, III, and IV) were described in infected or stressed soybean plants.

Pterocarpan (−)-maackiain occurs in chickpea (*C. arietinum*), (−)-pisatin in garden pea (*Pisum sativum*), and (−)-phaseolin (**10-168**) in French beans (*P. vulgaris*).

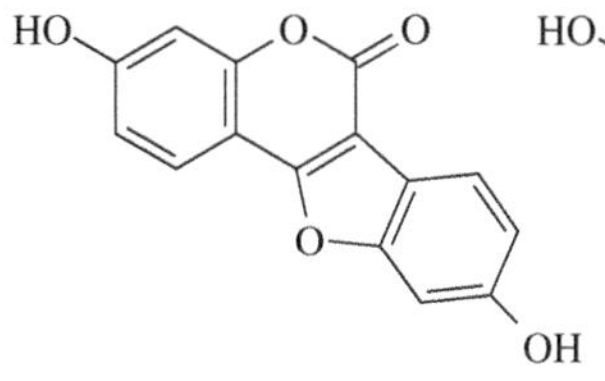

coumestrol 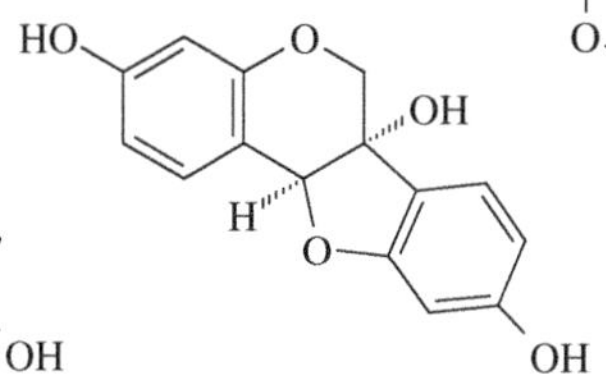   glycinol 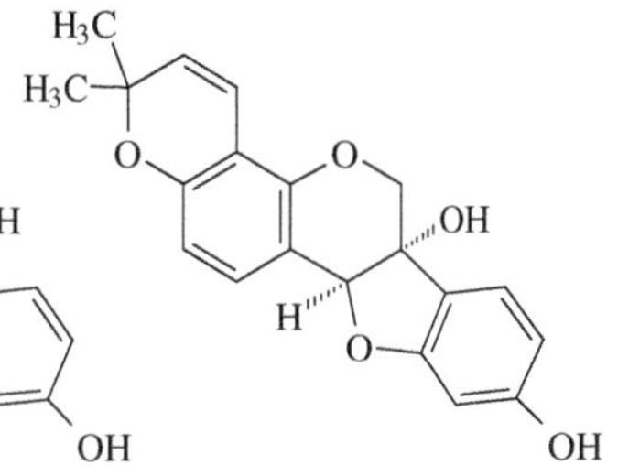   glyceollin I

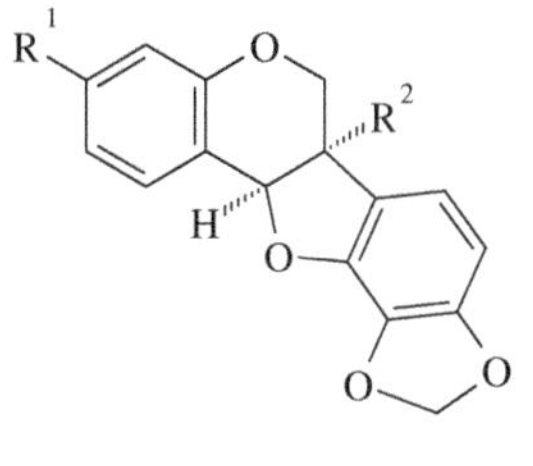

maackiain, R^1 = OH, R^2 = H
pisatin, R^1 = OCH$_3$, R^2 = OH

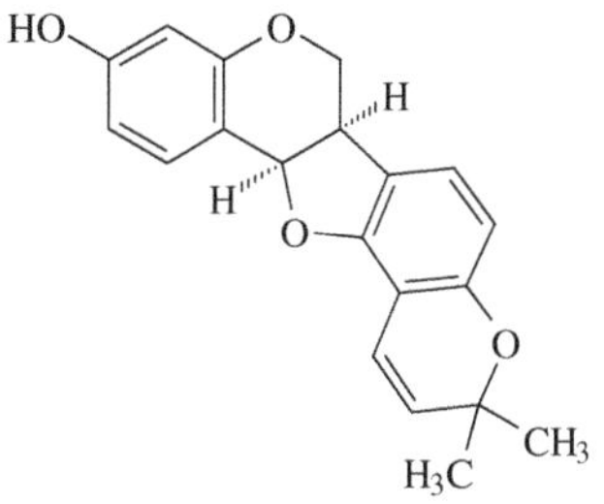

phaseolin

10-168, common pterocarpans in legume plants

10.3.12.6 Lignans

Lignans are a group of plant phenolic compounds of high structural diversity classified as phenylpropanoids. They are dimers of type $(C_6-C_3)_2$ formed by association of two C_3 phenylpropanoid units (monolignol units) on the central carbon atom of the side chain via a C-8/C-8' (8,8' or β-β'link) bond, accompanied by the formation of a $C_6-C_3-C_3-C_6$ skeleton (**10-169**). Lignans can occur as free compounds or attached to sugars and carboxylic acids. Less common dimers of type $C_6-C_3-C_6-C_3$ are called **neolignans**. The polymer of C_3 units known as lignin (see Section 4.5.6.7.1) has an analogous structure. Together with lignin, lignans are the most widespread plant components; they possess a range of biological activities, including oestrogenic activity.

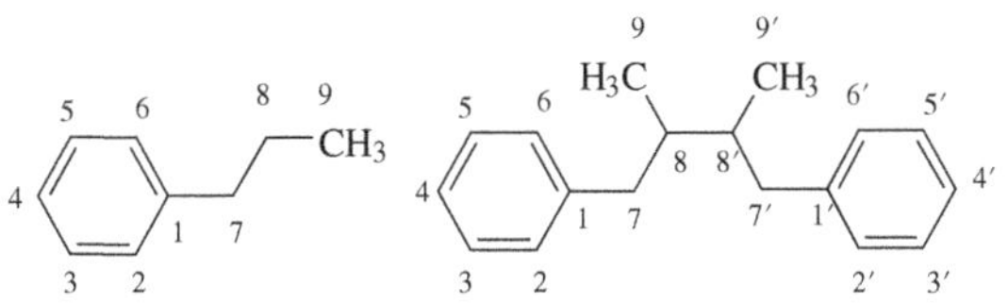

10-169, structure and carbon numbering of phenylpropane units and C-8/C-8' dimers

Several hundred lignans have been identified, amongst which the linear and cyclic (cyclolignan) types are the most important in food materials. Depending on the degree of skeleton oxidation, several types of lignans are distinguished (**10-170**):

- dibenzylbutane(diol)s (known as diarylbutanoids);

- dibenzylbutyrolactones (derivatives of butanolide known as lignanolides);

- tetrahydrofurans (known as monoepoxylignans);

- furofurans (derivatives of 3,7-dioxabicyclo[3.3.0]octane known as bisepoxylignans);

- tetralins (known as tetrahydronaphthalenes);

- naphthalenes;

- dibenzocyclooctadienes.

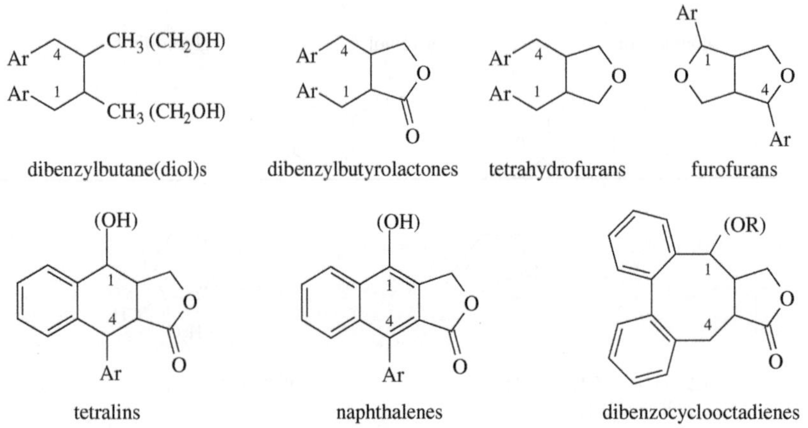

10-170, basic structures of lignans and cyclolignans

Lignans have a function in protecting the plant against pathogens, as well as a variety of other biological effects; they additionally act as phytooestrogens, anticarcinogens, antioxidants, and antiviral, antibacterial, and insecticidal agents. Secoisolariciresinol, matairesinol, lariciresinol, and pinoresinol act as mammalian oestrogens as they can be converted into enterolignans by the intestinal microflora.

In foods, lignans mainly occur as components of whole-grain cereal products, in rice, legumes, nuts, and various other seeds, and in vegetables and fruits (Table 10.25). The most common lignans in cereals are secoisolariciresinol (dibenzylbutane type) and matairesinol (dibenzylbutyrolactone type), making up about 25% of total lignan intake. Other important lignans are lariciresinol (tetrahydrofuran type) and syringaresinol, pinoresinol, and medioresinol (furofuran types). Most are linked to lignin structures, although free and esterified lignans also occur.

Table 10.25 Lignan contents in selected foods.

Product	Water (%)	Content (µg/kg)				
		Pinoresinol	Lariciresinol	Secoisolariciresinol	Matairesinol	Total
Flexseed	7.0	332	304	29 420	55.3	30 110
Sesame seed	4.7	2933	947	6.6	48.1	3935
Peanuts	6.5	0	4.1	5.3	0	9.4
Broccoli	89	56.8	21.2	0.8	0	78.7
Carrots	92	1.9	8.0	9.3	0	17.1
Onion	90	0	1.9	1.8	0	3.6
Spinach	94	1.2	6.8	0.2	0	8.2
Tomatoes	94	1.4	4.2	0.2	0	5.8
Potatoes	80	0	1.7	0.2	0	2.0
Oranges	86	2.4	4.7	0.5	0.2	7.8
Peaches	80	18.6	8.0	2.7	0	29.3
Olive oil	0.0	24.3	0.4	0	0	24.8

The longest known lignan of the dibenzylbutane type is nordihydroguaiaretic acid, (2S,3R)-4-[4-(3,4-dihydroxyphenyl)-2,3-dimethyl-butyl]benzene-1,2-diol, abbreviated NDGA (**10-171**). In the 1950s and '60s NDGA was used as an antioxidant, initially in the form of a resinous product obtained from the North American evergreen creosote bush (*Larrea tridentata*, Zygophyllaceae) and later as a synthetic substance. The antioxidant properties of NDGA are similar to those of gallates, but because of its adverse toxicological effects it has no practical significance.

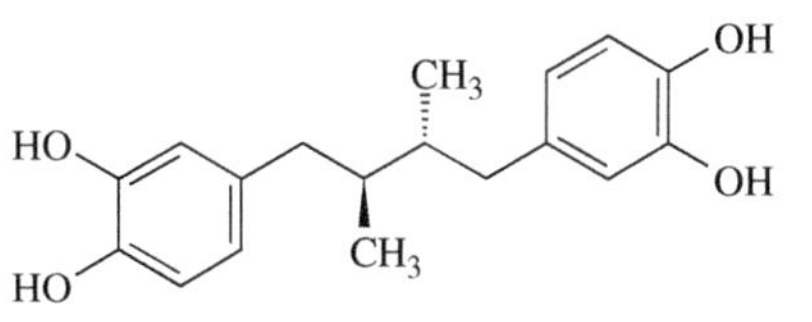

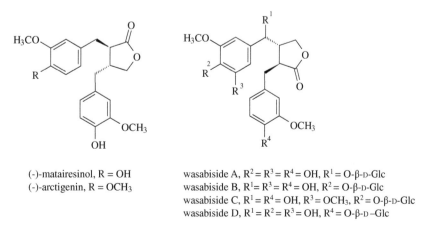

nordihydroguaiaretic acid

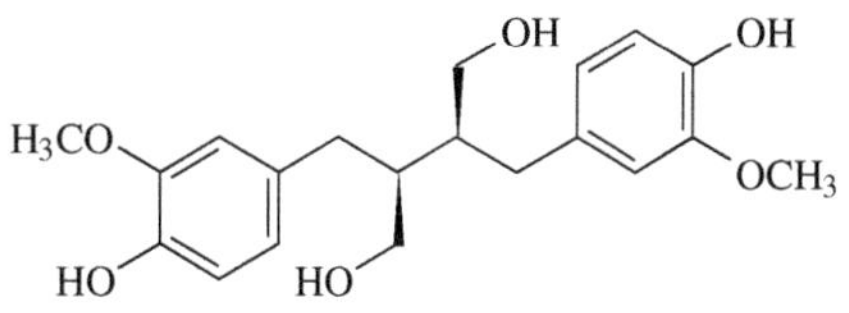

(−)-secoisolariciresinol

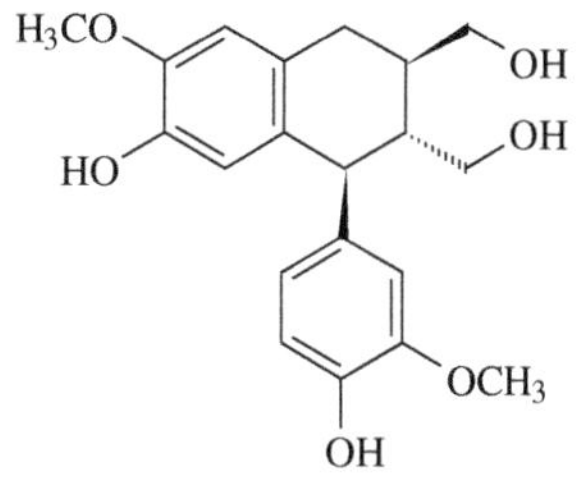

(+)-isolariciresinol

10-171, dibenzylbutane(diol)-type lignans

Common dibenzylbutan-type lignans derived from coniferyl alcohol are (−)-secoisolariciresinol, (2R,3R)-2,3-bis[(4-hydroxy-3-methoxyphenyl)methyl]butane-1,4-diol (**10-171**), and (+)-isolariciresinol, (6R,7R,8S)-8-(4-hydroxy-3-methoxyphenyl)-6,7-bis(hydroxymethyl)-3-methoxy-5,6,7,8-tetrahydronaphthalen-2-ol (**10-171**). Unlike in cereals, secoisolariciresinol is the main lignan (82–6907 mg/kg d.w) in flexseed (linseed, *Linum usitatissimum*, Linaceae), where it predominantly occurs as 2,3-di-*O*-β-D-glucoside esterified with 3-hydroxy-3-methylglutaric acid. Isolariciresinol is particularly abundant (5 mg/kg dry matter) in pomegranate fruits (*Punica granatum*, Lythraceae).

The representative of relatively simple lignans of the dibenzylbutyrolactone type derived from coniferyl alcohol is (−)-matairesinol, (3R,4R)-3,4-bis[(4-hydroxy-3-methoxyphenyl)methyl]oxolan-2-one (**10-172**), which occurs together with secoisolariciresinol in numerous foods. In flexseed, it is found in concentrations of 0.4–91 mg/kg (d.w.), predominantly as 4-*O*-β-D-glucoside or 4,4′-di-*O*-β-D-glucoside. Arctigenin, (3R,4R)-4-[(3,4-dimethoxyphenyl)methyl]-3-[(4-hydroxy-3-methoxyphenyl)methyl]-2-tetrahydrofuranone (**10-172**), was isolated from wolfberry (goji berry), the fruit of *Licium barbarum* (Solanaceae). The root of wasabi, also called Japanese horseradish (*Eutrema japonicum*, Brassicaceae), contains uncommon lignan glycosides of the dibenzylbutyrolactone type known as wasabisides (**10-172**), which have neuroprotective and anticarcinogenic effects.

(-)-matairesinol, R = OH
(-)-arctigenin, R = OCH₃

wasabiside A, R² = R³ = R⁴ = OH, R¹ = O-β-D-Glc
wasabiside B, R¹ = R³ = R⁴ = OH, R² = O-β-D-Glc
wasabiside C, R¹ = R⁴ = OH, R³ = OCH₃, R² = O-β-D-Glc
wasabiside D, R¹ = R² = R³ = OH, R⁴ = O-β-D–Glc

10-172, dibenzylbutyrolactone-type lignans

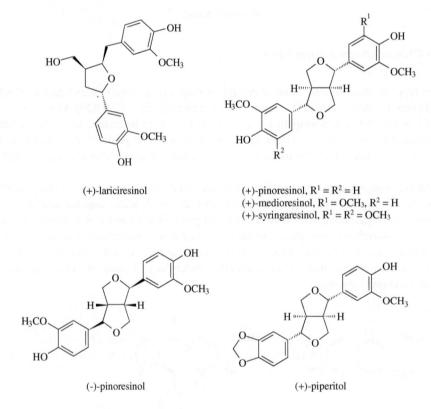

Figure 10.24 Formation of samin and sesamol from sesamolin.

The most common furan-type lignan derived from coniferyl alcohol is (+)-lariciresinol, (2S,3R,4R)-4-[4-(4-hydroxy-3-methoxybenzyl)-3-(hydroxymethyl)tetrahydrofuran-2-yl]-2-methoxyphenol (**10-173**). Its concentration in flexseeds is within the range of 3.4–42 mg/kg of dry matter.

The most common representative of lignans of the furofuran type is (+)-pinoresinol (**10-173**). Based on coniferyl alcohol, the furofuran (+)-pinoresinol is a precursor of many other lignans, such as (+)-lariciresinol (furan, **10-173**), (−)-secoisolariciresinol (dibenzylbutane, **10-171**), and (−)-matairesinol (dibenzylbutyrolactone, **10-172**), which are intermediates in the biosynthesis of cyclolignans derived from aryltetrahydronaphthalene. Isomerisation of (+)-pinoresinol gives (−)-pinoresinol (**10-173**), which is analogously the precursor of (−)-lariciresinol and (+)-secoisolariciresinol. (+)-Pinoresinol is further a precursor of the furofurans (+)-piperitol (**10-173**), (+)-sesamolinol (**10-174**), (+)-sesamin (**10-174**) also known as (+)-episesamin or (+)-asarinin, and (+)-sesamolin (Figure 10.24). The related furofuran (+)-medioresinol (**10-173**) is derived from coniferyl alcohol and sinapoyl (5-hydroxyconiferyl) alcohol, whilst (+)-syringaresinol (**10-173**) is derived exclusively from sinapoyl alcohol.

(+)-lariciresinol

(+)-pinoresinol, $R^1 = R^2 = H$
(+)-medioresinol, $R^1 = OCH_3$, $R^2 = H$
(+)-syringaresinol, $R^1 = R^2 = OCH_3$

(-)-pinoresinol

(+)-piperitol

10-173, furan- and furofuran-type lignans

Sesamin is a fairly widespread furofuran in higher plants, found in large quantities in sesame oil obtained from the seeds of Indian sesame (*Sesamum indicum*, Pedaliaceae), which is highly resistant to oxidation. The main components of sesame seed are sesamin (0.1–10 g/kg) and sesamolin (0–10 g/kg). Several minor furofuran lignans are also found, such as (+)-pinoresinol, (+)-piperitol, (+)-sesamolinol, (−)-sesamin (**10-174**), along with other substances. Oestrogenic (−)-sesamin (a precursor of enterolactone) occurs in some medicinal plants as well, including *Asiasarum sieboldi* of the birthwort family Aristolochiaceae (syn. *Asarum sieboldi*), known as Chinese wild ginger, which is used in cosmetic skin-conditioning preparations.

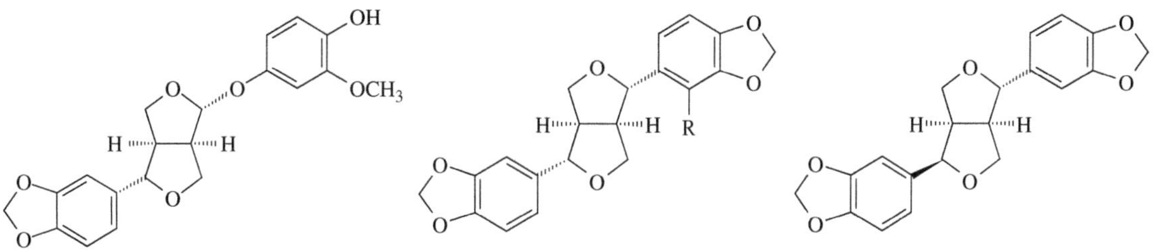

(+)-sesamolinol

(+)-sesamin, R = H
(+)-sesaminol, R = OH

(-)-sesamin

10-174, lignans in sesame seeds

Hydroxysubstituted lignans such as sesaminol (**10-174**) and compounds with an opened dioxol ring are present as β-glycosides (gentiobiosides, sophorosides, and glucosides). Water-soluble lignan glucosides are, for example, sesaminol-2′-*O*-β-D-glucoside (sesaminol glucoside), sesaminol-2′-*O*-β-D-glucopyranosyl-(1 → 6)-β-D-glucopyranoside (sesaminol diglucoside), sesaminol-2′-*O*-β-D-glucopyranosyl-(1 → 2)-*O*-[β-D-glucopyranosyl-(1 → 6)]-β-D-glucopyranoside (sesaminol triglucoside, **10-175**), and (+)-sesamolinol-4′-*O*-β-D-glucoside.

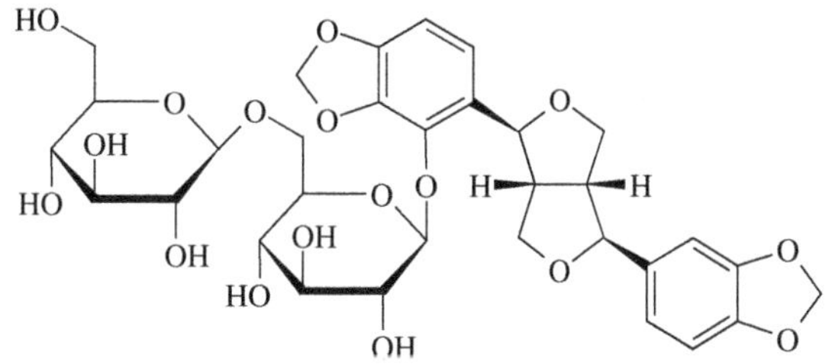

sesaminol diglucoside

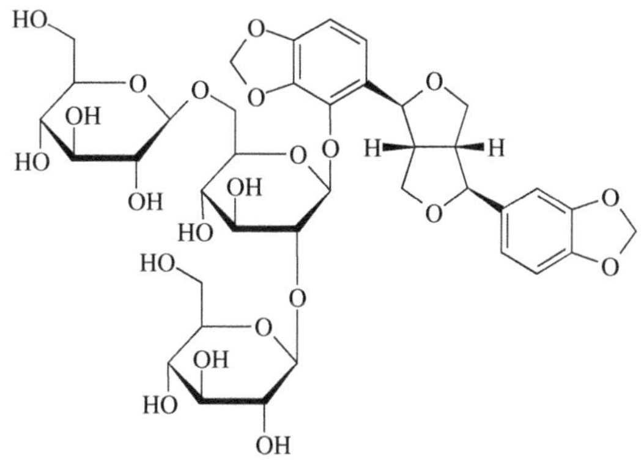

sesaminol triglucoside

10-175, sesaminol glucosides

Sesamin, in addition to its significant antioxidant properties, also shows an insecticidal effect. The actual antioxidant (having an effect comparable to BHA and BHT) is sesamol (Figure 10.24), which is present in small amounts in seeds but arises in larger quantities from hydrolysis of sesamolin during sesame oil refining (especially in bleaching) and during thermal operations (such as frying). Sesamol is easily oxidised in the presence of Fe(III) ions and produces toxic oxidation products (**10-176**), which are structurally oligomers of sesamol. Extramolecular dehydratation of samin gives disaminyl ether.

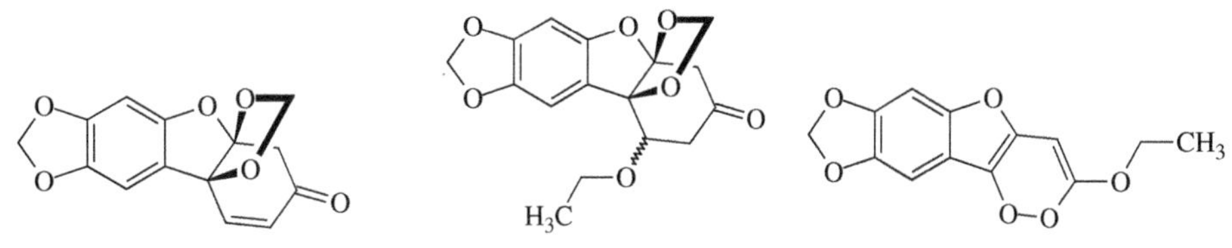

10-176, sesamol oxidation products

A representative of the tetralins is (−)-α-conidendrin, also known as tsuga lactone, (3aR,9S,9aR)-7-hydroxy-9-(4-hydroxy-3-methoxyphenyl)-6-methoxy-3a,4,9,9a-tetrahydro-1H-benzo[f][2]benzofuran-3-one (10-177), which occurs in lignin hydrolysates.

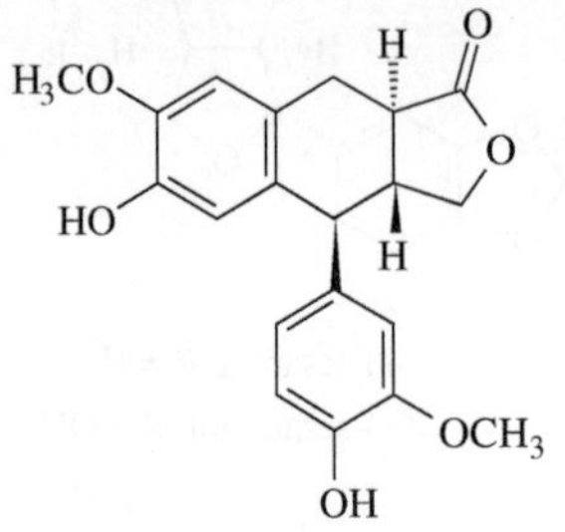

10-177, α-conidendrin

10.3.12.7 Flavonolignans

Related to lignans are the flavonolignans, which do not exhibit oestrogenic activity. Their biosynthesis is based on flavonoids (such as dihydroflavonol taxifolin or dihydroquercetin) and phenylpropanoids, usually coniferyl alcohol. Examples are components of the pericarp of milk thistle (*Silybum marianum*, Asteraceae) called silymarin, 70–80% of which are structurally related flavonolignans. The main component of the silymarin complex is silybin (also known as silibinin), which is a mixture of diastereomers A and B (10-178) occurring in the ratio of 1 : 1. Milk thistle is recommended for indigestion, as a supportive agent in chronic hepatitis therapy and liver cirrhosis, and for gallbladder problems and loss of appetite.

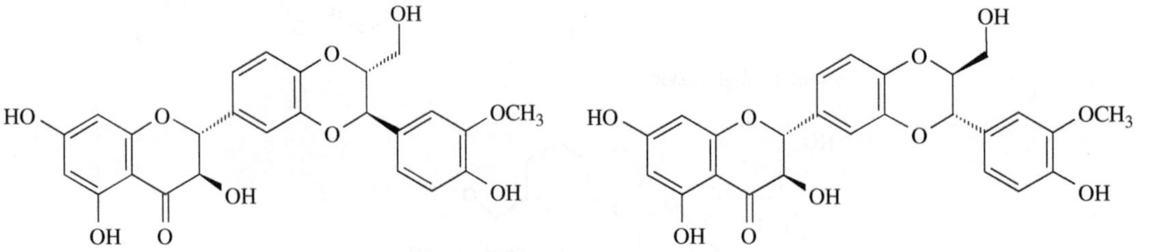

silybin A

silybin B

10-178, sylibins

10.3.12.8 Benzoxazinoids

Benzoxazinoids, cyclic hydroxamic acids, are a class of tryptophan-derived alleochemicals found in cereal grains and consequently in cereal foods. They are not typical phenols but have some common properties with them. Plants use benzoxazinoids as a chemical barrier against other organisms (herbivores, pests, pathogenic fungi, and bacteria). The term 'benzoxazinoid' is used to refer to **benzoxazinones**, their glucosides, and the corresponding aglycones containing a 2-hydroxy-2H-1,4-benzoxazin-3(4H)-one skeleton, as well as their degradation products the **benzoxazolinones**. The health benefits of benzoxazinoid intake (antimicrobial, anticancer, reproductive- and central nervous system-stimulatory, immunoregulatory, and appetite- and weight-reducing effects) may be associated with the solitary or overlapping biological effects of fibres, lignans, phenolic acids, alkylresorcinols, and other bioactive compounds.

The biosynthetic pathway to benzoxazinoids diverges from tryptophan biosynthesis at indole-3-glycerol phosphate, which is converted to indolin-2-one. Subsequently, four oxygen atoms are enzymatically introduced into the indolin-2-one moiety via 3-hydroxyindolin-2-one, (R)-2-hydroxy-2H-1,4-benzoxazin-3(4H)-one (HBOA), and (R)-2,4-dihydroxy-2H-1,4-benzoxazin-3(4H)-one (DIBOA) (Figure 10.25). After glucosylation, the DIBOA glucoside is modified by hydroxylation to (R)-2,4,7-trihydroxy-(2H)-1,4-benzoxazin-3(4H)-one (TRIBOA glucoside) and by O-methylation at C-7 to form (R)-2,4-dihydroxy-7-methoxy-2H-1,4-benzoxazin-3(4H)-one (DIMBOA glucoside). The glucosides are stored in vacuoles. On experiencing tissue damage, specific glucosidases present in the plastids spontaneously produce bioactive aglycones benzoxazolinones (also known as 3H-benzoxazol-2-ones or 2,3-dihydro-1,3-benzoxazol-2-ones), such as 2-benzoxazolinone (BOA) and 6-methoxy-2-benzoxazolinone (MBOA).

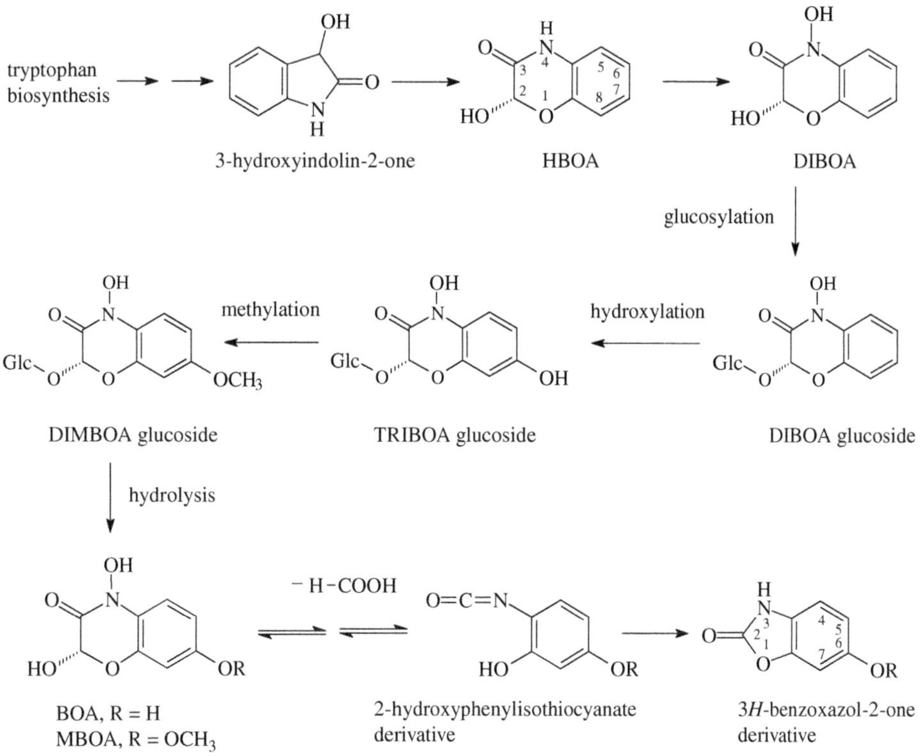

Figure 10.25 Biosynthesis and degradation of benzoxazinoids.

Young plants of wheat, corn, rice, and other cereals contain mainly 3H-1,3-benzoxazol-2-one (BOA) and 7-methoxy-3H-benzoxazol-2-one (MBOA), which do not have oestrogenic activities but do stimulate the secretion of oestradiol from ovarian follicle, showing signs of oestrogen activities. Rye breads have larger amounts of benzoxazinoids (143–3560 mg/kg dry matter) than breads containing wheat (11–449 mg/kg dry matter). More benzoxazinoids were found in whole-grain (57–449 mg/kg dry matter) than in refined-wheat (11–92 mg/kg dry matter) breads. The main components are DIBOA and MBOA.

Table 11.1 Antimicrobial spectrum of benzoic acid.

Microorganism	pH	Minimum inhibitory amount (mg/l)	Microorganism	pH	Minimum inhibitory amount (mg/l)
Bacteria			Yeasts		
Bacillus cereus	6.3	500	Hansenula spp.	4.0	180
Escherichia coli	5.2–5.6	50–120	Rhodotorula spp.	–	100–200
Lactobacillus spp.	4.3–6.0	300–1800	Saccharomyces bayanus	4.0	330
Lysteria monocytogenes	5.6	2000–3000	Zygosaccharomyces spp.	4.8	1000–4800
Micrococcus spp.	5.5–5.6	50–100	Moulds		
Pseudomonas spp.	6.0	200–480	Aspergillus spp.	3.0–5.5	200–4000
Streptococcus spp.	5.2–5.6	200–400	Bysochlamys nivea	3.3	500
Yeasts			Cladosporium herbarum	5.1	100
Candida crusei	–	300–700	Mucor racemosus	5.0	30–120
Debaromyces hansenii	4.8	500	Penicillium spp.	2.6–5.0	30–2000
Pichia membranefaciens	–	700	Rhizopus nigricans	5.0	30–120

In industrial and culinary practice, preservation methods based on physical principles are widely used in addition to chemical preservatives. Of these, food preservation by heat treatment (pasteurisation and sterilisation), cold (chilling and freezing), drying (dehydration), irradiation, and more recently by high pressure attract the most attention.

11.2.1.1 Acids and their derivatives

11.2.1.1.1 Benzoic acid

Benzoic acid (E210) and benzoates (E211–E213) are primarily used as antifungal agents. Most fungi (yeasts and moulds) are inhibited at concentrations ranging from 500 to 1000 mg/kg. Some bacteria are inhibited at a concentration of 100–200 mg/kg, but many are inhibited by much higher concentrations (Table 11.1). According to the type of food, the concentration used ranges from 150 to 2000 mg/kg.

The active form is undissociated acid (pK = 4.19 at 25 °C), which is about 100 times more effective than its anion. The antimicrobial effect is probably due to inhibition of amino acid utility by microorganisms, inhibition of transport of substrates, and inhibition of enzymes involved in acetic acid metabolism, oxidative phosphorylation, and the citric acid cycle.

In small amounts, benzoic acid (free or bound in esters) occurs as a natural component of foods. It is mainly present in fruits and some fermented dairy products (see Section 8.2.6.1.6).

11.2.1.1.2 Sorbic acid

(2E,4E)-Hexa-2,4-dienoic acid (Figure 11.1), known as sorbic acid (E200), and its salts (sorbates, E201–E203) are potent inhibitors of a number of fungi, yeasts, and some bacteria. Sorbic acid is used (according to the type of food) in quantities of 200–2000 mg/kg. The active form is the undissociated acid (pK = 4.76 at 25 °C), which is roughly 10–600 times more effective than the anion. The effect is related partly to inhibition of dehydrogenases involved in the oxidation of fatty acids, sulfhydryl enzymes (it is added to the cysteine thiol group), and partly to interference with the transport of substances through the cytoplasmic membrane.

Sorbic acid has a conjugated system of double bonds, which makes it susceptible to autoxidation and nucleophilic attack, sometimes giving it mutagenic products. Autoxidation of sorbic acid yields unstable hydroperoxides, which decompose to acetaldehyde and fumaric acid monoaldehyde (β-carboxyacrolein) as final products. This aldehyde reacts with amino acids and proteins and is responsible for the browning of certain foods preserved with sorbic acid. Other products of oxidation are acrolein and crotonaldehyde. In the presence of sulfur dioxide, α-angelica lactone and 2-acetyl-5-methylfuran are produced. Nucleophiles attack the sorbic acid molecule in position C-5. Reaction with amines under conditions typical of food processing (50–80 °C) yields cyclic N-alkyl-6-methyl-3,6-dihydropyrid-2-ones, resulting from a double addition reaction. The reaction with sulfur dioxide (bisulfites) and thiols yields mainly 5-substituted hex-2-enoic acids. Because sorbic acid inhibits the growth of *Clostridium botulinum* bacteria, it has been proposed as a partial replacement for nitrite in meat curing. However, this practice may lead to other toxicological problems, since sorbic acid reacts with nitrite to yield mutagenic products.

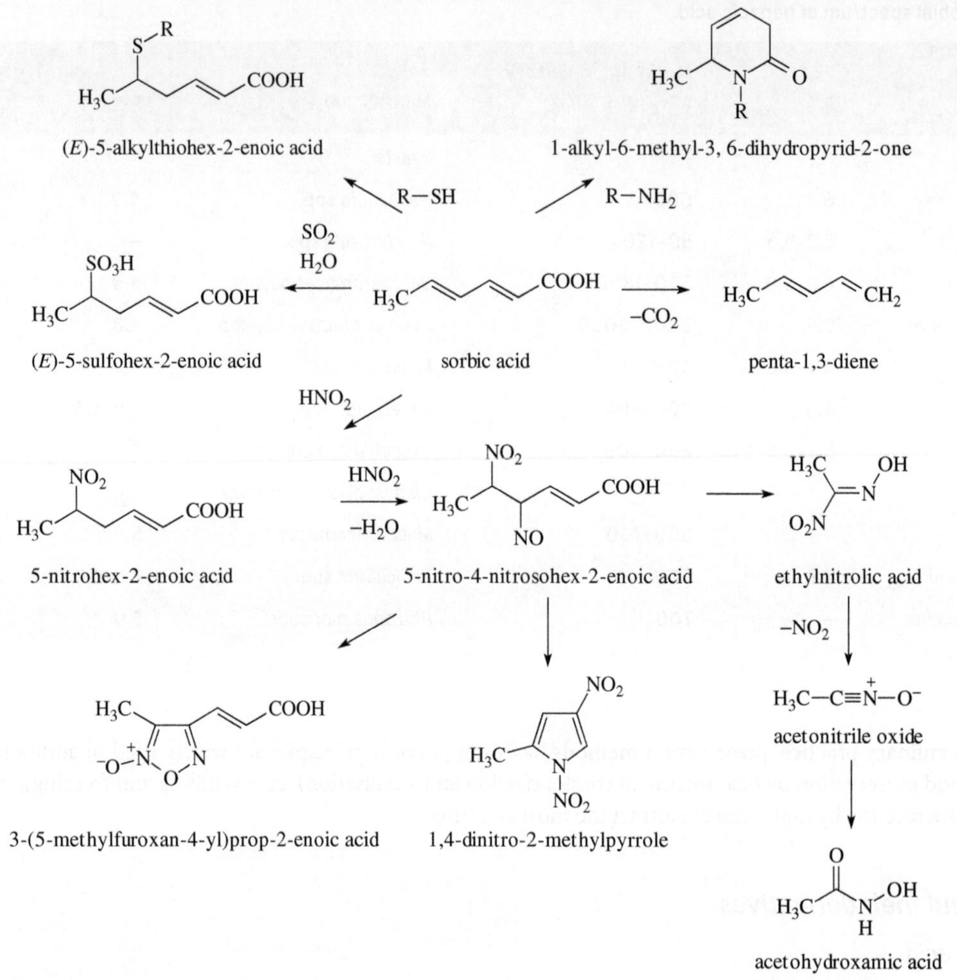

Figure 11.1 Reactions of sorbic acid in foods.

At pH 3.5–4.2, the main mutagens are unstable 1,4-dinitro-2-methylpyrrole, ethylnitrolic acid, and a derivative of furoxan (1,2,5-oxadiazole 2-oxide) 3-(5-methyl-4-furoxanyl)prop-2-enoic acid, which undergo decomposition to other products. Ascorbic acid completely eliminates the mutagenicity of 1,4-dinitro-2-methylpyrrole by reduction of the C-4 nitro group to a C-amino group in 1-nitro-2-methyl-4-amino pyrrole (Figure 11.1).

Some microorganisms (such as *Penicillium roqueforti*) decarboxylate sorbic acid to penta-1,3-diene, which may cause an off-flavour (resembling kerosene) in cheeses and fermented dairy products (Figure 11.1). Also reported are cases of contaminated margarines, fruit drinks, jams, and marzipan.

11.2.1.1.3 Parabens

Alkyl esters (methyl, ethyl, propyl, and heptyl esters, previously also benzyl ester) of *p*-hydroxybenzoic (4-hydroxybenzoic) acid (see **8–56**) are collectively called parabens. Formerly, these esters were also known as abegins or under various trade names. Unlike free acids (benzoic and sorbic acid), parabens and their salts (E209, E214–E219) are effective in less acidic and slightly alkaline environments. Their antimicrobial effectiveness increases with the length of the alkyl residue, which, on the other hand, decreases their solubility in water and thus the possibility of their practical use. Salts of parabens are particularly soluble. Their effect on the taste and smell of protected foods is very small. Parabens are mainly active against fungi and yeasts, and less active against Gram-positive bacteria (Table 11.2). Their preservative effect is based on influencing the permeability of cell membranes. Parabens are used in amounts ranging from 500 to 1000 mg/kg.

11.2.1.1.4 Other acids and their derivatives

Formic acid (pK = 3.75 at 25 °C, see Section 8.2.6.1.1) is the simplest carboxylic acid. It shows the highest antimicrobial activity of all lower carboxylic acids (E236). Formic acid and formates are especially effective against bacteria and yeasts. Lactic acid bacteria and yeasts

Table 11.2 Antimicrobial spectrum of parabens.

Microorganism	Concentration (mg/l) causing inhibition			Microorganism	Concentration (mg/l) causing inhibition		
	Methyl	Propyl	Heptyl		Methyl	Propyl	Heptyl
Bacteria				*Fungi (yeasts and moulds)*			
Bacillus cereus	2000	125–400	12	*Candida albicans*	1000	125–250	–
Escherichia coli	2000	400–1000	–	*Saccharomyces bayanus*	930	220	–
Lysteria monocytogenes	> 512	512	–	*Saccharomyces cerevisiae*	1000	125–200	100
Pseudomonas aeruginosa	4000	8000	–	*Aspergillus flavus*	–	200	–
Salmonella typhosa	2000	1000	–	*Aspergillus niger*	1000	200–250	–
Staphylococcus aureus	4000	350–500	12	*Rhizopus nigricans*	500	125	–

are relatively resistant. Formic acid is used to preserve acidic fruit juices or purées whose colour is derived from anthocyanins (such as strawberries) and which therefore cannot be preserved by sulfur dioxide. During thickening, formic acid is volatilised with water vapour and removed from the product. Because of its smell, its use is very limited. The other disadvantage is that formic acid partially hydrolyses pectin. In the European Union, formic acid is not approved.

Acetic acid ($pK_a = 4.53$ at 25 °C, see Section 8.2.6.1.1) is used as a preservative and also as an acidifying agent (acidulant, E260). It is effective against yeasts and bacteria, but less so against moulds. Tolerant bacteria are acetic acid and lactic acid bacteria and Gram-positive bacteria producing butyric acid under anaerobic conditions. Acetic acid is mainly used for the preparation of pickles for preserved vegetables and in the production of ketchups, dressings, mayonnaises, fish, and other products. The mechanism of action is the inhibition of the transport of substances through cytoplasmic membranes and inhibition of electron transport. Sodium acetate and sodium diacetate (E262, a 1 : 1 mixture of sodium acetate and acetic acid, also known as sodium hydrogen acetate) are effective preservatives against bacteria and fungi.

Dehydroacetic acid, δ-lactone of 2-acetyl-5-hydroxy-3-oxo-hex-4-enoic acid (E265, **11–1**) and its sodium salt are mainly active against fungi at higher pH values, where other acids are ineffective. It is not approved in the European Union, but in the United States it is allowed for the preservation of cut or peeled squash. It has similarly been used as a fungistatic agent for surface treatment of cheeses and for the suppression of the microbial defects of bread and related products, called ropiness or ropy bread, caused by the thermophilic spore-forming bacterium *Bacillus mesentericus* (a variant of *B. subtilis*), whose spores survive cooking temperatures that do not exceed 100 °C inside the bread and can germinate under favourable conditions. Bread ropiness is a result of the production of slime casings by these bacteria and enzymatic hydrolysis of starch and gluten. On the second or third day, the bread crumb becomes damp and sticky as a result of the formation of gummy products, which stretch into strands when the bread is pulled apart. In addition to ropiness, the spoiled bread has an off-flavour, sometimes characterised as fruity or pineapple-like, and turns yellow. Formerly, when ropiness occurred, bakers acidified doughs with vinegar as a protective measure. Dehydroacetic acid has also been shown to be an effective antifungal agent against black bread moulds, particularly *Aspergillus niger* and *Rhizopus nigrificans*.

Propionic acid arises in Emmental cheeses by propionic acid fermentation (see Section 8.2.6.1.1). Propionic acid (E280) and its salts (E281–E283) are active as preservatives in weakly acidic media (to pH 5, pK = 4.87 at 25 °C), especially against moulds but less so against Gram-negative bacteria; activity against yeasts is almost completely absent. Propionic acid is mainly used as a preservative in stored animal feed and grain. It is also used for extending the shelf-life of bread, tortillas, and other cereal products, as it prevents the growth of moulds (as calcium propionate and sodium propionate). The inhibition of bacteria that cause bread ropiness occurs at concentrations of 2000–3000 mg/kg. The mechanism of action is mainly based on accumulation of propionic acid in the cells of microorganisms and inhibition of important enzymes.

Lactic acid (pK = 3.83 at 25 °C, see Section 8.2.6.1.3) occurs as a natural compound in many fermented products (e.g. yoghurt, fermented sauerkraut, and olives). As an additive, lactic acid (E270) is mainly used as an acidulant. The undissociated form diffuses through the cell membrane of many bacteria and lowers the pH within the cell. It has a bacteriostatic effect on the pathogenic bacterium *Mycobacterium bovis*, a member of the *M. tuberculosis* bacteria, the causative agents of bovine tuberculosis in cattle. Lactic acid is also effective against other bacteria, so it is used, for example, for surface decontamination of meat and in the production of delicacies. Usually, it is applied in combination with sodium lactate.

Fumaric acid ($pK_1 = 3.09$, $pK_2 = 4.60$ at 25 °C, see Section 8.2.6.1.2) is used as an acidulant and preservative (E297) for the inhibition of lactic acid fermentation in wines. Fumaric acid and its esters (especially monomethyl and monoethyl esters) may slow the formation of botulinum toxin in canned meat and prevent the growth of moulds in bread.

Other organic acids are important mainly as acidulants. Approved substances in the European Union are citric, malic, tartaric (see Section 8.2.6.1.3), and ascorbic (see Section 5.14.1) acids and their sodium, potassium, and calcium salts.

11.2.1.2 Other organic substances

In addition to carboxylic acids and their salts and esters, some other organic compounds that have antimicrobial effects can similarly be used as food preservatives. These include some antibiotics, the enzyme lysozyme, biphenyl, *o*-phenylphenol, thiabendazole, dialkyl dicarbonates, and alkylene oxides. Other substances with antimicrobial effects belong to other categories of food additives. For example, sucrose esters with palmitic and stearic acid, used as emulsifiers, are active against fungi of the genera *Aspergillus*, *Penicillium*, and *Cladosporium*. Some other organic compounds used for food preservation are not considered food additives, such as sucrose, as well as different natural substances with antimicrobial effects occurring in some vegetables (e.g. allicin in garlic and allyl isothiocyanate in horseradish) and spices (e.g. piperine in black pepper and capsaicin in hot peppers). Many other natural substances show antimicrobial effects some of which are discussed in Chapter 10.

During the last few decades, some previously used preservatives have been dropped for hygienic-toxicological reasons (in some cases, it was found that they were potential carcinogens) and are prohibited in the European Union and other countries. Such substances include, for example, dehydroacetic acid (**11–1**) and its sodium salt (previously used to preserve jams and margarines), 8-hydroxyquinoline (tobacco), thiourea (fresh fruits), salicylic acid (fruits and vegetables), and 3-(5-nitrofuran-1-yl)acrylic acid (wines, **11–1**).

11.2.1.2.1 Biphenyl and its derivatives

Biphenyl (E230) and *o*-phenylphenol (E231), also known as 2-biphenylol (**11–1**) or its sodium salt sodium *o*-phenylphenol (E232), have been used for the surface treatment of citrus fruits. They are active against fungi.

11.2.1.2.2 Thiabendazole

The fungicide 2-(4-thiazolyl)benzimidazole, known as thiabendazole (E233, see **12–49**) has been used for the same purpose as biphenyl and *o*-phenylphenol and for the surface treatment of bananas.

11.2.1.2.3 Hexamethylenetetramine

Hexamethylenetetramine (E239), also known as urotropine, hexamine, methenamine, or 1,3,5,7-tetraazaadamantane (see **8–42**), arises as a reaction product of formaldehyde with ammonia. Its preservation effect appears to be due to the gradual liberation of formaldehyde (E240, which is not approved in the European Union) and its oxidation product formic acid under acid conditions or in the presence of proteins. The most abundant end-product of formaldehyde in cheeses preserved with hexamethylenetetramine is spinacine, 4,5,5,7-tetrahydro-3*H*-imidazo[4,5-*d*]pyridine-6-carboxylic acid (**11–1**), derived from the *N*-terminal histidine residue in γ-casein. (see **10–15**).

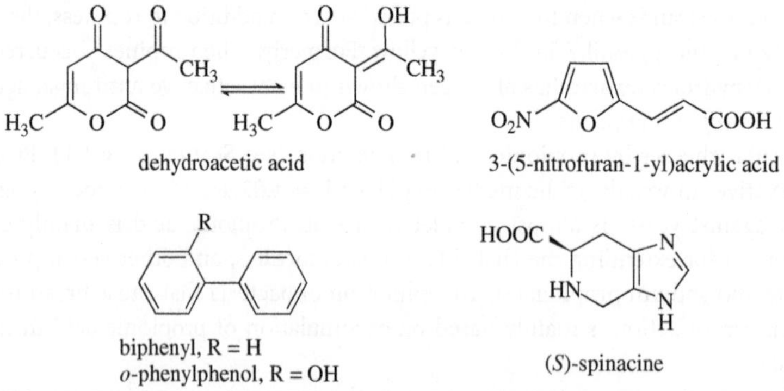

11-1, structures of selected preservatives and derived products

11.2.1.2.4 Dialkyl dicarbonates

Diethyl dicarbonate (diethyl pyrocarbonate) was used in the past as a preservative for soft drinks and some alcoholic beverages (including wine). In aqueous solution, however, it is rapidly hydrolysed to ethanol and carbon dioxide, and in alcoholic beverages it is transformed into diethyl carbonate by reaction with ethanol. In the presence of ammonium salts, toxic ethyl carbamate then results, also known as urethane (Figure 11.2). For these reasons, diethyl dicarbonate is not approved in the European Union; the approved compound is dimethyl dicarbonate (E242), which can be used for the preservation of soft drinks, teas, and herbal teas and in some countries to stabilise wine. The methyl carbamate produced from dimethyl dicarbonate is a non-toxic substance. In treated wines, it occurs in amounts up to $10 \, \mu g/l$.

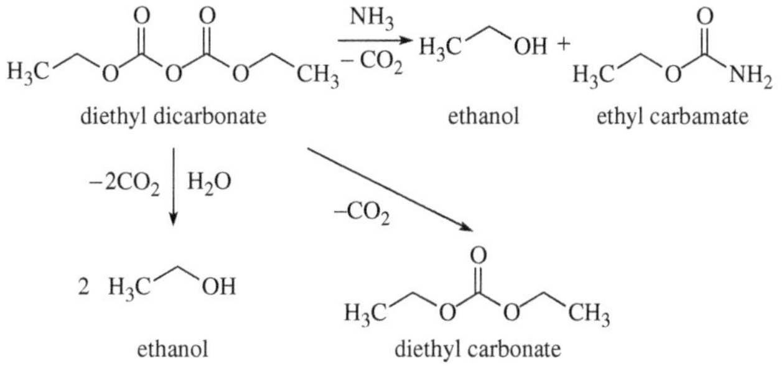

Figure 11.2 Degradation of diethyl dicarbonate in foods.

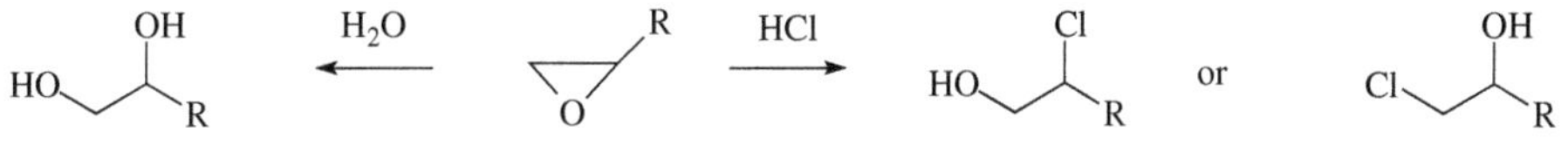

ethane-1,2-diol, R = H   oxirane, R = H   2-chloroethanol, R = H   2-chloroethanol, R = H
propane-1,2-diol, R = CH₃   methyloxirane, R = CH₃   2-chloropropan-1-ol, R = CH₃   1-chloropropan-2-ol, R = CH₃

Figure 11.3 Reactions of oxiranes with water and hydrogen chloride.

11.2.1.2.5 Alkylene oxides

Effective against all microorganisms (vegetative forms and spores) and possessing insecticidal properties are ethylene oxide (oxirane) and propylene oxide (methyloxirane). In some countries, these compounds are approved for fumigation of foods with low water contents, where other methods of preservation are not applicable (such as sterilisation of spices, nuts, starch, and flour). The modern alternative is irradiation.

Oxiranes are highly toxic alkylating agents; the hydrolysis product of ethylene oxide, ethylene glycol, and the reaction products of oxiranes with chloride ions are likewise toxic. The latter reaction yields 2-chloroethanol, which arises from oxirane, and isomeric vicinal chloropropanols (chlorohydrins), resulting from methyloxirane (Figure 11.3). Oxiranes react with a number of other food components, such as vitamins (riboflavin, pyridoxine, niacin, and folic acid) and amino acids (methionine and histidine), to form biologically inactive products.

11.2.1.2.6 Antibiotics

The use of antibiotics in food is problematic, because the substances used in human and veterinary medicine are not allowed. With regard to this requirement, only polypeptide antibiotics produced by lactic acid bacteria from certain strains of the genera *Lactococcus*, *Lactobacillus*, *Leuconostoc*, and *Pediococcus* are approved in EU and other countries. They are known under the general common name **bacteriocins**.

Nisin The thermoresistant polypeptide nisin (E234) is produced by some strains of *Lactococcus lactis* subsp. *lactis* (syn. *Streptococcus lactis*) and used as a preservative in dairy technology to stabilise fermented products, such as aged and processed cheeses, puddings, and creams. Nisin is effective against Gram-positive bacteria (vegetative forms and spores). Gram-negative bacteria are resistant to it, because their cell walls are far less permeable than those of Gram-positive ones. The molecule of nisin (**11–2**) contains 34 amino acid units arranged in five cycles. It possesses some unusual amino acid residues: Abu, 2-aminobutanoic acid; Dha, dehydroalanine (2-aminoacrylic acid); and Dhb, dehydrobutyrine (2-aminocrotonic acid, which occurs as the L- or D-enantiomer), as well as thioether cross-linkages: Lan, *meso*-lanthionine; MeLan, 3-methyllanthionine residues, which are indicated as Ala-S-Ala and Abu-S-Ala (where the amino-terminal moieties have the D-configuration), respectively.

Natamycin Natamycin (E235), also known as pimaricin or tennecetin (**11–3**), is a polyene macrolide substituted by the sugar residue derived from 3-amino-3,6-dideoxy-β-D-mannose. It is produced by certain strains of the bacteria *Streptomyces natalensis* and *S. lactis* and is allowed for surface treatment of cheeses and durable meat products. Natamycin is active against most fungi and yeasts at concentrations of 5–10 mg/kg, but is ineffective against bacteria. Like other polyene antibiotics, it inhibits fungal growth by binding to ergosterol in the plasma membrane, preventing ergosterol-dependent fusion of vacuoles, as well as membrane fusion and fission.

11.2.1.2.7 Enzymes

The active substance against Gram-positive bacteria is the enzyme lysozyme (neuramidinase), from the group of hydrolases. Lysozyme is allowed in many countries for the treatment of some dairy products and wine.

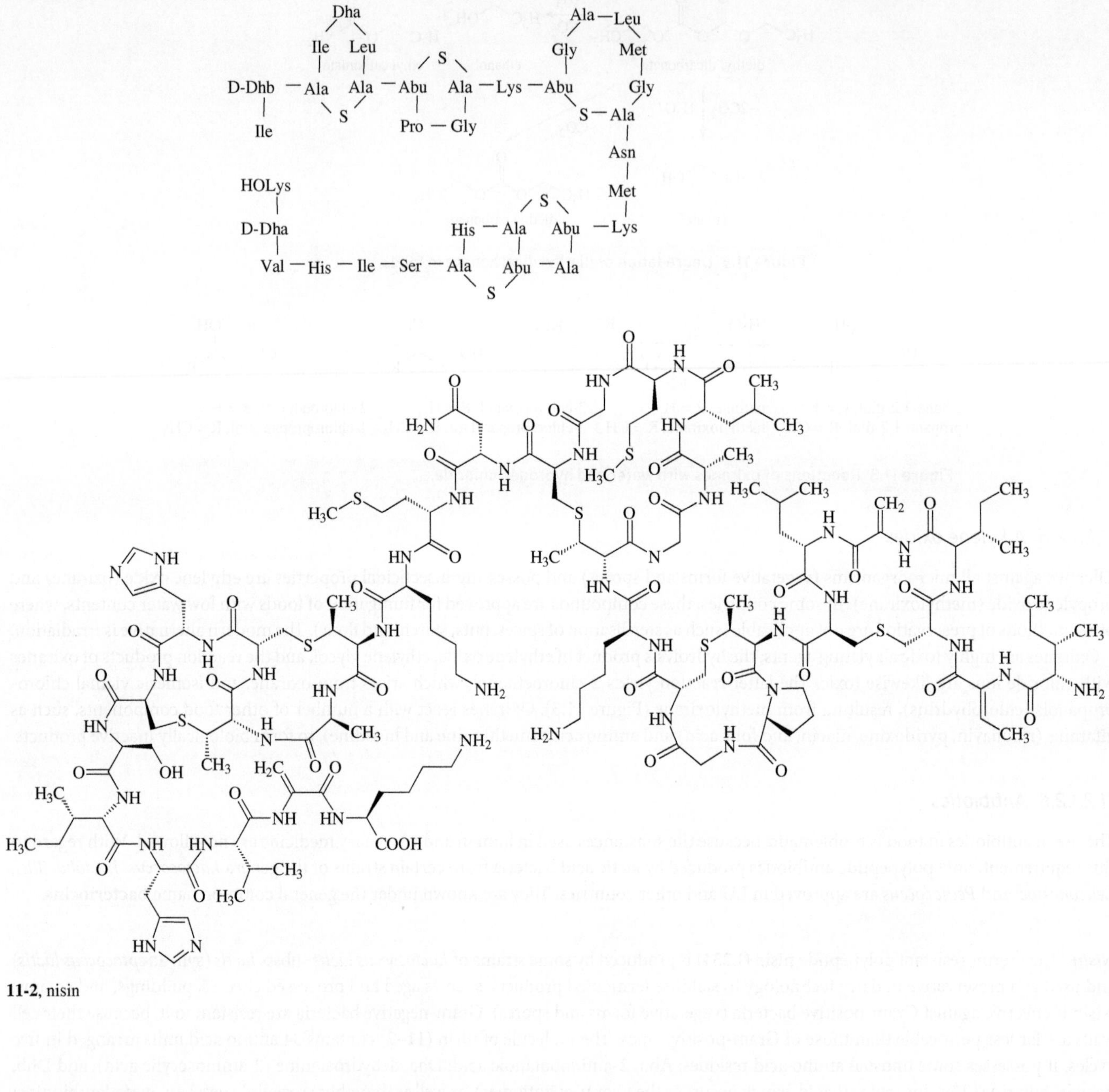

11-2, nisin

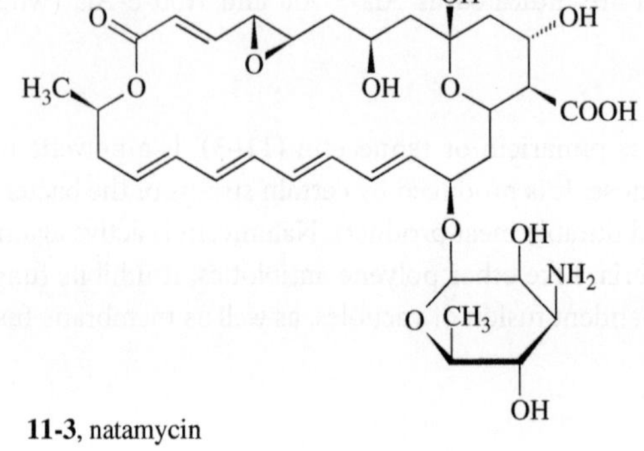

11-3, natamycin

11.2.1.3 Inorganic compounds

11.2.1.3.1 Sulfur dioxide and sulfites

Sulfur dioxide (SO_2, E220) and some of its derivatives, such as sulfites (SO_3^{2-}; E221, E225, E226), sodium hydrogen sulfite (also known as bisulfite, $NaHSO_3$, E222) and disulfites (commonly known as pyrosulfates or metabisulfites, $S_2O_5^{2-}$; E223, E224), can be used as preservatives, but also as inhibitors of enzymatic and non-enzymatic browning reactions, bleaching agents and antioxidants. Calcium hydrogen sulfite, $Ca(HSO_3)_2$ (E227) is used as a firming agent.

Aqueous solutions of sulfur dioxide (SO_2 dissolves up to 9.5% solution at 20 °C) gives rise to sulfurous acid (H_2SO_3). Sulfurous acid dissociates into two stages ($pK_1 = 1.76$, $pK_2 = 7.20$). Depending on the pH of the medium, solutions of sulfurous acid contain, in addition to SO_2 and undissociated acid, hydrogen sulfite (HSO_3^-) and sulfite (SO_3^{2-}) ions. In acidic foods of pH 3–4, hydrogen sulfites dominate:

$$SO_2 + H_2O \rightleftharpoons H_2SO_3 \xrightarrow{\text{pK} = 1.81} H^+ + HSO_3^- \xrightarrow{\text{pK} = 6.91} 2\,H^+ + SO_3^-$$

In acidic aqueous solution, disulfite decomposes to hydrogen sulfite and sulfur dioxide:

$$S_2O_5^{2-} + H^+ \rightarrow HSO_3^- + SO_2$$

The active form is undissociated acid, which is the only form effective against yeasts, therefore sulfur dioxide and sulfites are only effective in acidic foods (pH < 4). In some countries, sulfur dioxide solutions can be used to inhibit the growth of bacteria on the surface of meat and meat products. The main applications of sulfur dioxide and sulfites concern the growth inhibition of acetic acid and lactic acid bacteria and wild yeasts in wine. The levels of $Na_2S_2O_5$ or $K_2S_2O_5$ (sodium or potassium disulfites) and of $KHSO_3$ (potassium hydrogen sulfite) allowed by the European Union for some types of wine vary from 160 to 260 mg/l. Other applications are related to protecting fruit from the growth of moulds. Sulfur dioxide acts against certain bacteria in concentrations of around 1–2 mg/l, which is a bacteriostatic effect, but at higher concentrations its effect is bactericidal. Concentrations of 1–10 mg/l (at pH 3.5), for example, inhibit the lactic acid fermentation of fruit products. Concentrations inhibiting yeasts (*Saccharomyces cerevisiae*) and moulds vary from 0.1 to 20 mg/l. The amounts of sulfur dioxide used may range, according to the type of food, from 10 to 2000 mg/kg.

Sulfur dioxide in foods yields a number of addition products with aldehydes and ketones, and also with sugars (e.g. in wine), allyl isothiocyanate (e.g. in mustard), and a number of other food constituents in which it is reversibly bound. Aliphatic aldehydes form adducts (also known as bisulfites), whereas ketones are found predominantly in their free form. Analogous products are formed with some sugar degradation products in non-enzymatic browning reactions (see Section 4.7.1), with oxidised phenols (such as chlorogenic acids, see Section 9.12.4), and with proteins and peptides containing disulfide bonds, which are cleaved with the formation of thiosulfates (*S*-sulfonates, $R\text{–}S\text{–}SO_3^-$) and thiols:

$$R - S - S - R + SO_3^{2-} \rightarrow R - S - SO_3^- + RS^-$$

Sulfur dioxide (hydrogen sulfites) also reacts with thiamine to yield inactive compounds (see Section 5.6.6), forms adducts with riboflavin, nicotinamide, and vitamin K, inhibits ascorbic acid oxidation (see Section 5.14.6.1), reacts with ascorbic acid degradation products, reduces *o*-quinones produced in enzymatic browning reactions back to 1,2-diphenols (see Section 9.12.6.2), causes decolourisation of fruit anthocyanins (see Section 9.4.1.5.7), and reacts with synthetic azo dyes to form coloured or colourless products (see Section 11.4.1.3.2). Sulfur dioxide also reacts with pyrimidine bases *in vitro*, specifically with cytosine and 5-methylcytosine. Important reactions are shown in Figure 11.4.

Sulfur dioxide and hydrogen sulfites are rarely used in combination with nitrites, because their reactions in acidic media yield sulfonates of either hydroxylamine or ammonia, which destroys the preservative activity of the individual additives. The reaction of alkali metal nitrites with bisulfite at lower temperatures leads to the formation of hydroxylamine *N,N*-disulfonate ($HON(SO_3)_2^{2-}$), which is hydrolysed to hydroxylamine *N*-sulfonate ($HONHSO_3^-$) and further to hydroxylamine (H_2NOH). At elevated temperatures, complete substitution proceeds with the formation of ammonia *N,N,N*-trisulfonate, $N(SO_3)_3^{3-}$, which is hydrolysed to sulfamate ($H_2NSO_3^-$). Subsequent reaction of sulfamate with nitrous acid leads to production of bisulfate ion and nitrogen gas:

$$NO_2^- + 2HSO_3^- \rightarrow HON(SO_3)_2^{2-} + HO^-$$
$$HON(SO_3)_2^{2-} + H_2O \rightarrow HONHSO_3^- + HSO_4^-$$
$$HONHSO_3^- + H_2O \rightarrow H_2NOH + HSO_4^-$$
$$HON(SO_3)_2^{2-} + HSO_3^- \rightarrow N(SO_3)_3^{3-} + H_2O$$
$$N(SO_3)_3^{3-} + 2H_3O^+ \rightarrow H_2NSO_3^- + 2H_2SO_4$$
$$H_2NSO_3^- + HNO_2 \rightarrow N_2 + HSO_4^- + H_2O$$

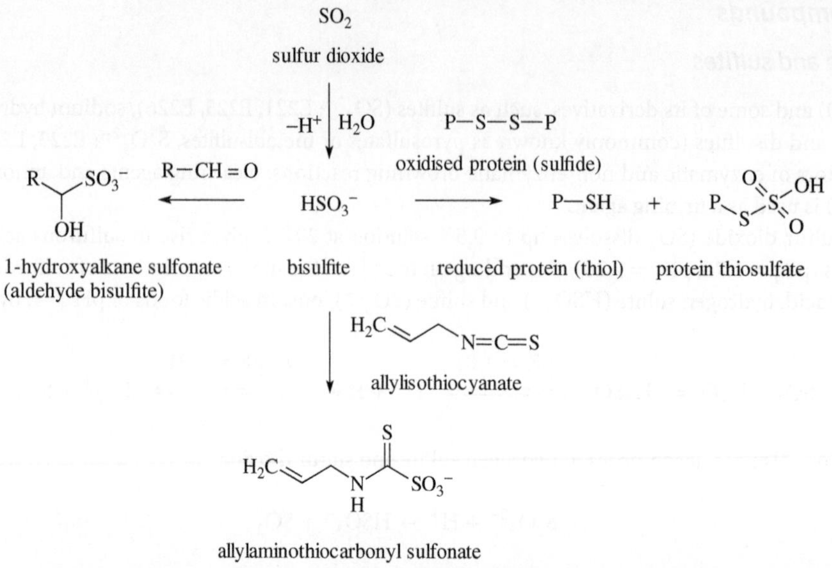

Figure 11.4 Important reactions of sulfur dioxide in foods.

11.2.1.3.2 Nitrites and nitrates

In addition to their use in meat colour stabilisation (see Section 9.2.1.5.2), nitrites (NO_2^-, E249 and E250), together with salt (sodium chloride), can also be used as antimicrobial agents. This use is especially important in non-sterile meat products, as they inhibit the growth of *Clostridium botulinum* bacteria. The efficiency of the inhibition depends on the pH, because it is proportional to the concentration of nitrous acid (HNO_2) formed from nitrites. Nitrates (NO_3^-) may also be used as preservatives, namely sodium nitrate (E251, known as Chile saltpeter) and potassium nitrate (E252, saltpeter), which are reduced to nitrites by microorganisms.

11.2.1.3.3 Boric acid and its salts

Boric acid (H_3BO_3, E284) and disodium tetraborate ($Na_2B_4O_7$, known as borax, E285) are permitted for use as meat preservatives in some countries, although in the European Union are only approved to preserve caviar. Their preservative effect is based on inhibition of decarboxylases of amino acids and of phosphate metabolism.

11.2.1.3.4 Sodium chloride

Sodium chloride (NaCl) is commonly used in combination with other preservatives and preservation methods, but is not classified as a food additive (see Section 8.3.2.1). Its antimicrobial activity is related to its ability to reduce water activity, thus creating unfavourable conditions for microbial growth (sucrose is similarly active). The sensitivity of microorganisms varies considerably. Intolerant bacteria can be inhibited by a level of 10 g/kg, mesophilic bacteria and psychrotropic Gram-negative rods tolerate concentrations 6–10 times higher, lactic bacteria survive in an environment with concentrations ranging from 60 to 150 g/kg, and spore-forming bacteria can even tolerate concentrations of 160 g/kg.

11.2.1.3.5 Other inorganic substances

Inorganic compounds exhibiting antimicrobial activity also include hydrogen peroxide, phosphates, and carbon dioxide, which is effective against fungi and Gram-negative psychrotropic bacteria, but less so against lactic acid bacteria and anaerobes.

11.2.1.4 Legislation

Compounds that are conditionally permitted for food preservation in the European Union are listed in Table 11.3. All preservatives may only be used for the preservation of listed foods and to the maximum amount allowed.

Benzoic acid and its salts can be used for jams, jellies, and marmalades, either alone or in mixtures with sorbic acid. Benzoic acid in quantities of up to 30 mg/kg in fermented dairy products is not considered a food additive, as it arises from the hippuric acid (see **2-17**). For

Table 11.3 List of current EU-approved preservatives and their E numbers.

E number	Name[a]	E number	Name[a]
E200	Sorbic acid	E228	Potassium hydrogen sulfite
E202	Potassium sorbate	E230	Biphenyl (diphenyl)
E203	Calcium sorbate	E234	Nisin
E210	Benzoic acid	E235	Natamycin
E211	Sodium benzoate	E239	Hexamethylenetetramine
E212	Potassium benzoate	E242	Dimethyl dicarbonate
E213	Calcium benzoate	E249	Potassium nitrite
E214	Ethyl p-hydroxybenzoate	E250	Sodium nitrite
E215	Sodium ethyl p-hydroxybenzoate	E251	Sodium nitrate
E218	Methyl p-hydroxybenzoate	E252	Potassium nitrate
E219	Sodium methyl p-hydroxybenzoate	E280	Propionic acid
E220	Sulfur dioxide	E281	Sodium propionate
E221	Sodium sulfite	E282	Calcium propionate
E222	Sodium hydrogen sulfite	E283	Potassium propionate
E223	Sodium metabisulfite	E284	Boric acid
E224	Potassium metabisulfite	E285	Sodium tetraborate (borax)
E226	Calcium sulfite	E1105	Lysozyme
E227	Calcium hydrogen sulfite		

[a]Directive No. 95/2/EC of 20 February 1995 on food additives other than colours and sweeteners.

example, the maximum permissible level of benzoic acid for cooked red beet is 2000 mg/kg, and that for cooked crustaceans and molluscs is 1000 mg/kg. Sorbic acid alone is allowed for potato dough, pre-fried potato slices, and processed cheeses (2000 mg/kg), dried fruit and gnocchi (1000 mg/kg), and wines (200 mg/l). Spirits with less than 15% alcohol by volume can contain a mixture of benzoic acid (200 mg/l) and sorbic acid (200 mg/l), whilst olives and olive-based preparations may contain benzoic acid (500 mg/kg) and sorbic acid (1000 mg/kg). Sorbic acid with p-hydroxybenzoates or their salts may be used in preserved jelly coatings of meat products and cereal- or potato-based snacks and coated nuts (1000 mg/kg).

The use of propionic acid and its salts is allowed for pre-packed sliced bread and rye bread (3000 mg/kg) and for energy-reduced bread, partially baked bread, pre-packed bread, and pre-packed fine bakery wares (including flour confectionery) with a water activity of more than 0.65 (2000 mg/kg). Fermented milk products (Emmental-type cheeses) contain natural propionic acid arising in the fermentation process, which is not regarded as a food additive.

Nisin can be used for semolina and tapioca puddings and similar products (3 mg/kg), ripened and processed cheeses 12.5 mg/kg), and Mascarpone cheese (10 mg/kg). Natamycin may be used for surface treatment of hard, semi-hard, and semi-soft cheeses and dried, cured sausages (1 mg/dm^2 surface up to a depth of 5 mm). Lysozyme is used in *quantum satis* (a Latin term meaning 'the amount that is needed'), for example, to preserve ripened cheeses.

Biphenyl is only used for the surface treatment of citrus fruits. Hexamethylenetetramine is only approved for preservation of Provolone cheese that originates in Casilli near Napoli. Variants of Provolone cheese are produced in North America and Japan. In the past, hexamethylenetetramine was also used as a preservative for collagen casings and fish products. Dimethyl dicarbonate is allowed for non-alcoholic flavoured drinks, alcohol-free wine, and liquid-tea concentrates at the amount of 250 mg/l.

The maximum level of sulfur dioxide and sulfites (expressed as SO_2) differs significantly in different foodstuffs. For example, SO_2 content can be 50 mg/kg in peeled potatoes, 100 mg/kg in processed potatoes (including frozen and deep-frozen), potato dough, dried mushrooms, and jams, jellies, and marmalades made with sulfited fruit, 200 mg/kg in dried tomato, 400 mg/kg in dried white vegetables and dehydrated potatoes, 450 mg/kg in burger meat with a minimum vegetable or cereal content of 4% and breakfast sausages, 600 mg/kg in dried apples and pears, 1000 mg/kg in dried bananas, and 2000 mg/kg in dried apricots, peaches, grapes, prunes, and figs. Under EU law, any wine containing more than 10 mg/l of SO_2 must be labelled as 'containing sulfites'. The maximum permitted level of SO_2 is 160 mg/l in red wines, 200 mg/l in rose and white wines, and 400 mg/l in sweet wines. In beer, including low-alcohol and alcohol-free beer, where SO_2 arises through yeast activity, its content may be up to 20 mg/l, or even 50 mg/l where there is a second fermentation in the cask.

The maximum amount of nitrites that may be added during the manufacture of meat products and sterilised meat products is 150 and 100 mg/kg (expressed as $NaNO_2$), respectively. The maximum residual levels are also set. Potassium and sodium nitrates can be used for non-heat-treated meats at a level of 150 mg/kg, and for traditional immersion cured meat speciality products (such as British Wiltshire bacon and Wiltshire ham, German *Rohschinken*, and Spanish *jamón curado*) at 300 mg/kg.

Boric acid and borax are allowed for caviar at a level of 4 g/kg (expressed as boric acid).

11.2.1.5 Health assessment

Benzoic acid and benzoates have low toxicity (average daily intake, ADI = 5 mg/kg b.w.) because there is an effective detoxification mechanism against them, consisting of conjugation of benzoates with glycine to hippuric acid, which is excreted in the urine. Around 66–95% of benzoates can be removed in this way, with the rest taking the form of glucuronate. Some individuals, however, show increased sensitivity to benzoic acid. Sorbic acid (also sorbates) is regarded as one of the least toxic preservatives (ADI = 25 mg/kg b.w.), but in cosmetic and pharmaceutical products it may irritate the skin in susceptible individuals. Parabens and the corresponding free acid are even less toxic than sorbic acid (ADI = 70 mg/kg b.w.), but they are rapidly hydrolysed and the free acid is excreted as a conjugate with glycine and glucuronic acid. Parabens exhibit local anaesthetic effects and may cause dermatitis in susceptible individuals. The intake of propionic acid is not limited. The use of formic acid (ADI = 3 mg/kg b.w.) is prohibited in food preservation in some countries (including EU countries) and formic acid (as well as methanol, from which formic acid arises through the action of alcohol dehydrogenase via formaldehyde) is known to be toxic to the optic nerve (see Section 8.2.2.2).

An ADI equal to 0.13 mg/kg (b.w.) has been set for nisin, and one equal to 0.3 mg/kg (b.w.) for natamycin. In 2003, the European Parliament announced that nisin should not be used because of possible worsening effect of antibiotics on humans, but the Scientific Panel on Food Additives, Flavourings, Processing Aids and Materials in Contact with Foods (AFC Panel) approved its use as a food additive in 2006. This substance is also permitted in the United States.

For the antifungal phenols, ADI = 0.05 mg/kg (b.w.) for biphenyl and 0.2 mg/kg (b.w.) for *o*-phenylphenol and its sodium salt. Hexamethylenetetramine may liberate free formaldehyde in the stomach. Both hexamethylenetetramine (ADI = 0–0.15 mg/kg b.w.) and formaldehyde (ADI = 0.15 mg/kg b.w.) have been shown to act as mutagens. Acute and short-term toxicological studies on spinacine found an ADI of 3 mg/kg (b.w.). It was concluded that there is no appreciable health risk from consumption of cheese made using formaldehyde (Grana Padano cheese) or hexamethylenetetramine (Provolone cheese). The ADI value for dimethyl dicarbonate was not determined, as this ester decomposes when dissolved in the product.

Bisulfites are oxidised in the body by sulfite oxidase (sulfite: ferricytochrom c oxidoreductase) to sulfates, which are excreted in the urine. The enzyme activity is individual, and the toxic effects of sulfur dioxide and sulfites are therefore variable. Some individuals tolerate amounts of up to 50 mg/kg, whilst in sensitive individuals such concentrations will cause headaches, nausea, and diarrhoea. In asthmatics receiving steroids, sulfites may cause allergic reactions (ADI = 0.7 mg/kg b.w.).

A detailed health and toxicological evaluation of nitrates and nitrites is given in Section 6.6.1.2. ADI values of 3.7 mg/kg (b.w.) for nitrates and 0.07 mg/kg (b.w.) for nitrites have been set. Nitrites can induce tumours either directly through effects on immune functions or indirectly via carcinogenic *N*-nitrosamines produced from nitrites and secondary amines. Repeated consumption of higher amounts of nitrites may cause methaemoglobinaemia, especially in children.

Boric acid and borax (ADI = 0.1 mg/kg b.w.) resorb rapidly in the body and are slowly excreted. In comparison with other acids, which act as preservatives, these compounds are more toxic, but no side effects are known from foods. Chronic administration causes bleeding, dermatitis, and anaemia. Concentrations of boric acid (borates) in pharmaceutical preparations are much higher and may cause several side effects.

11.2.2 Antioxidants

Antioxidants are substances that prolong the shelf-life of foods by protecting them against deterioration caused by oxidation, which is reflected in rancidity in fats and other easily oxidising food components. Lipid oxidation causes other chemical changes in foods that negatively affect their nutritional value (oxidation of vitamins), sensory value (oxidation of flavour-active components and pigments), and hygienic-toxicological quality (some oxidation products may be toxic). On the other hand, the oxidation of essential fatty acids also generates desirable aromas in certain foods (fruits, vegetables, and mushrooms; see Section 3.8.1.8.5).

Table 11.4 List of current EU-approved antioxidants and their E numbers.

E number	Name[a]	E number	Name[a]
E300	Ascorbic acid[b]	E311	Octyl gallate
E301	Sodium ascorbate	E312	Dodecyl gallate
E302	Calcium ascorbate	E315	Erythorbic (isoascorbic) acid
E304	Ascorbic acid fatty acid esters	E316	Sodium erythorbate (isoascorbate)
E306	Tocopherols (natural mixture)	E319	*tert*-Butylhydroquinone (TBHQ)
E307	α-Tocoferol	E320	*tert*-Butylhydroxyanisol (BHA)
E308	γ-Tocoferol	E321	*tert*-Butylhydroxytoluene (BHT)
E309	δ-Tocoferol	E392	Rosemary extracts
E310	Propyl gallate	E586	4-Hexylresorcinol

[a]Directive No. 95/2/EC of 20 February 1995 on food additives other than colours and sweeteners
[b]Ascorbic acid (E300) is classified as a food supplement.

11.2.2.1 Classification

Antioxidants interfere with the process of oxidation of lipids and other oxylabile compounds so that they:

- react with free radicals (**primary antioxidants**) or reduce the resulting hydroperoxides (**secondary antioxidants**);

- bind catalytically active metals to ineffective complexes;

- eliminate the oxygen present.

The reaction mechanism is detailed in Section 3.8.1.8 on the oxidation of lipids. The primary antioxidants include all authorised substances: ascorbic and erythorbic (isoascorbic) acid and their derivatives, tocopherols, and phenolic antioxidants (Table 11.4). The secondary antioxidants include cysteine, cysteine-containing peptides (such as glutathione), methionine, lipoic acid, and other naturally occurring compounds, which are not however used as antioxidants. A synthetic secondary antioxidant used in the past was 3,3′-thiodipropionic acid dilaurate (**11–4**).

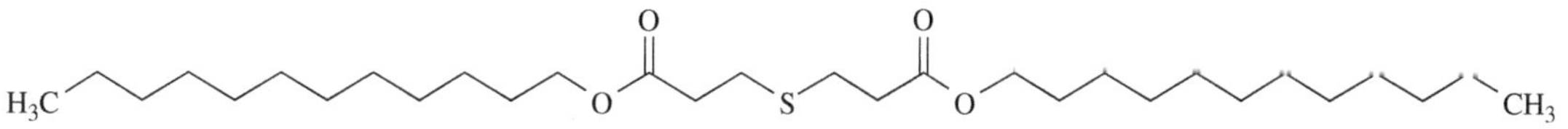

11-4, didodecyl-3,3′-thiodipropionate

Equally important is the classification of antioxidants according to their origin into:

- **natural antioxidants**;

- **synthetic antioxidants**.

The authorised natural antioxidants include tocopherols, which today are usually synthesised. A variety of other natural antioxidants are present in a number of essential oils (especially in spices). However, they do not usually have a constant composition and tend to be less effective and more expensive compared to synthetic antioxidants.

According to their structure, the synthetic antioxidants are divided into:

- **phenolic** antioxidants (of approved compounds: tocopherols and phenolic antioxidants, also a number of other phenolics present in foods, spices, and other natural materials);

- **enediols** (of approved compounds; enediols include ascorbic and erythorbic acids, their salts and other derivatives);

- **other substances**.

11.2.2.2 Mechanism of action

The mechanism of action of ascorbic acid and its derivatives is given in Section 5.14.6.1, and the mechanism of action of tocopherols is described in Section 5.4.6. Phenolic compounds (Ar–OH) as primary antioxidants may interfere with the oxidation of lipids (R–H) in competition with the propagation phase reaction of autoxidation. They react with hydroperoxyl radicals (ROO$^\bullet$) produced by lipid oxidation or with alkoxyl radicals (RO$^\bullet$) arising by decomposition of lipid hydroperoxides. They provide these radicals with hydrogen, thereby interrupting the radical chain autoxidation reaction. The resulting products are phenoxyl (aryloxyl) radicals of antioxidants:

$$Ar - OH + R - O - O^\bullet \rightarrow R - O - OH + Ar - O^\bullet \text{ or}$$
$$Ar - OH + R - O^\bullet \rightarrow R - OH + Ar - O^\bullet$$

These free radicals react with hydroperoxyl and alkoxyl radicals of oxidised fatty acids in the termination phase of the reaction:

$$R - O - O^\bullet + Ar - O^\bullet \rightarrow R - O - Ar + O_2$$
$$R - O^\bullet + Ar - O^\bullet \rightarrow R - O - O - Ar$$

The reaction mechanism is shown in Figure 11.5. Phenoxyl radicals, however, cannot enter the chain radical reaction and cannot initiate the fission of other lipid molecules. However, at high concentrations of an antioxidant, these reactions can take place and the antioxidant then acts as a prooxidant, and the antioxidant then acts as a prooxidant:

$$Ar - O^\bullet + R - O - OH \rightarrow Ar - OH + R - O - O^\bullet$$
$$Ar - O^\bullet + R - H \rightarrow Ar - OH + R^\bullet$$

The relative stability and low reactivity of phenoxyl radicals is associated with the unpaired electron delocalisation in the aromatic system. The attack of atmospheric oxygen can be therefore very difficult. This situation is apparent, for example, in unsubstituted phenol, which is ineffective as an antioxidant. Recombination of phenoxyl radicals produces a number of dimeric and oligomeric products (Figure 11.6). Reported products from oxidation of phenol with molecular oxygen and ozone include a range of aromatic compounds, such as o- and p-benzochinone, catechol, and hydroquinone and aliphatic compounds, such as maleic anhydride, maleic, fumaric, propionic, oxalic, glyoxylic, acetic and formic acids, carbon dioxide, and carbon monoxide.

Substitution of phenol by an alkyl groups in the *ortho* or *para* position increases (by the conjugation effect) the density of electrons on the OH group, which increases the ability of phenol to react with free radicals. Stability of the resulting phenoxyl radical also increases the

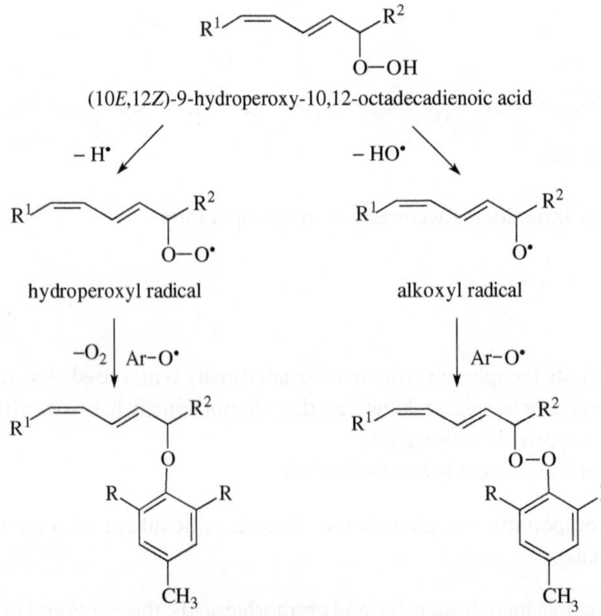

Figure 11.5 Reaction of a free antioxidant radical (BHT) with a free radical derived from oxidised linolenic acid (R = *tert*-butyl, R^1 = [CH$_2$]$_4$-CH$_3$, R^2 = [CH$_2$]$_7$-COOH).

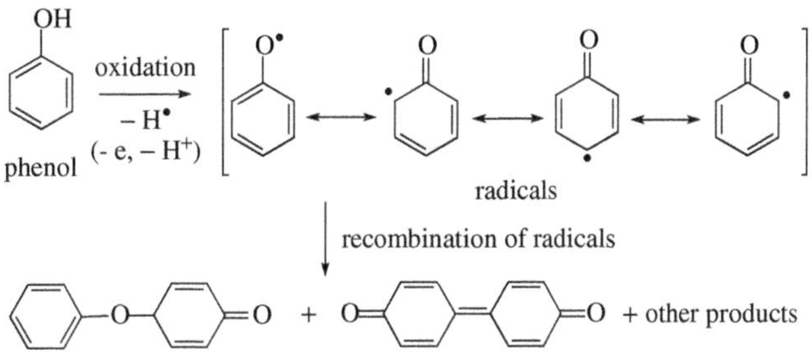

Figure 11.6 Mechanism of phenol oxidation.

presence of bulky substituents in the *ortho* position. Such antioxidants are the synthetic phenolic antioxidants butylhydroxyanisole (BHA, **11–5**) and butylhydroxytoluene (BHT, **11–5**).

The antioxidant activity of phenol is also increased by the presence of additional hydroxyl group in the *ortho* or *para* position. An example of such an antioxidant is *tert*-butylhydroquinone (TBHQ, **11–5**).

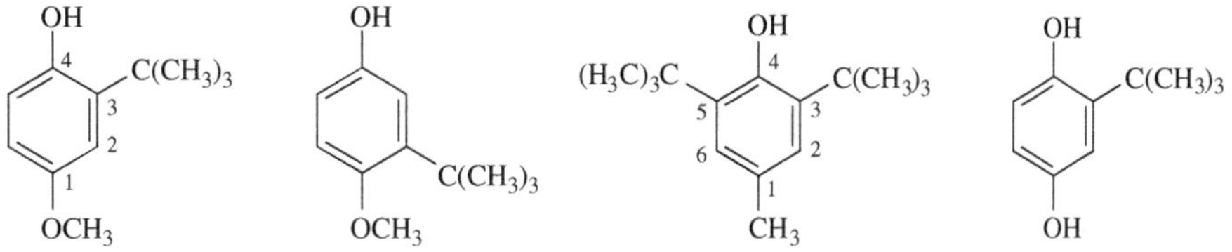

11–5, synthetic phenolic antioxidants

The effectiveness of 1,2-dihydroxybenzene derivatives is attributed to a phenoxyl radical stabilised by an intramolecular hydrogen bond (Figure 11.7). The activity of 2-methoxyphenol is lower, because the generated radical cannot be stabilised by a hydrogen bond. The antioxidant activity of 1,2- and 1,4-dihydroxybenzene is partly caused by the fact that the semiquinone radical can be further oxidised to the corresponding *o*-quinone or *p*-quinone, respectively, by reaction with another lipid radical; alternatively, it may disproportionate to the corresponding quinone and hydroquinone (Figure 11.8). Primary antioxidants also include phenolic acids, in particular the substituted cinnamic acids and their esters (depsides), glycosides, and amides. The phenoxyl radical of very effective 4-hydroxysubstituted cinnamic acids is stabilised by resonance (Figure 11.7). Many flavonoid substances are also primary antioxidants. Particularly effective compounds are chalcones, which provide resonance stabilised radicals (Figure 11.7).

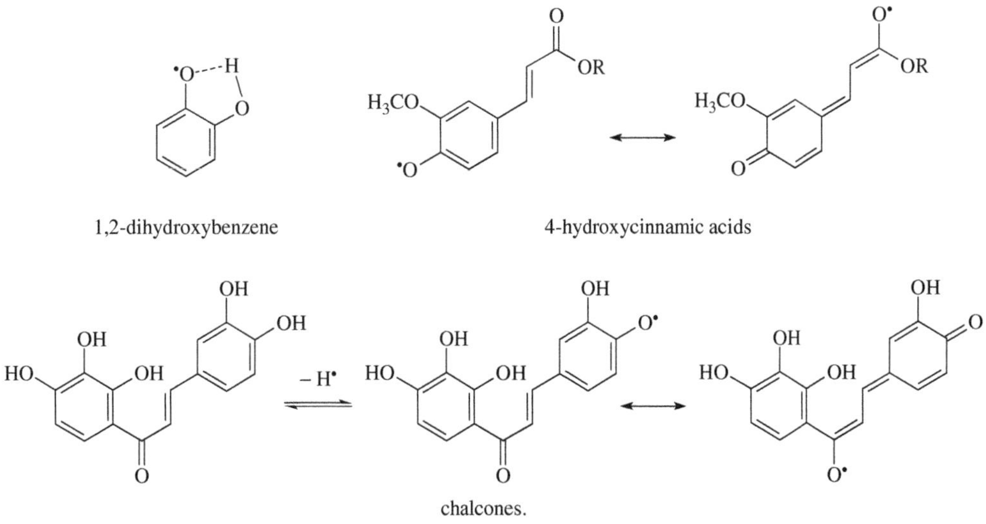

1,2-dihydroxybenzene   4-hydroxycinnamic acids

chalcones.

Figure 11.7 Stabilised radicals of selected phenolics.

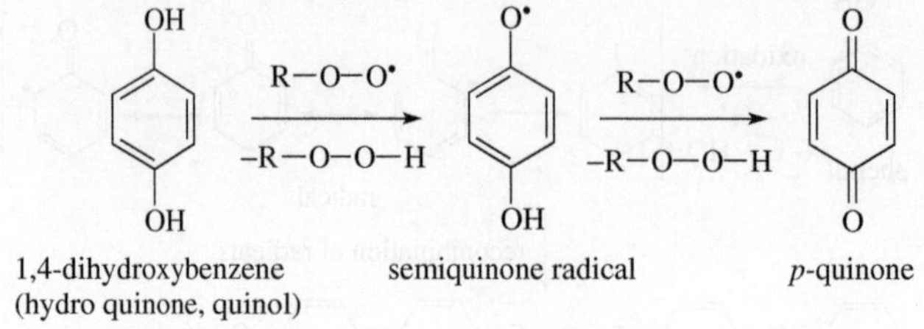

Figure 11.8 Oxidation of 1,4-dihydroxybenzene.

11.2.2.3 Synthetic antioxidants

Modern antioxidant technology is only about 100 years old. The only synthetic antioxidants widely used in foods are the phenolic antioxidants BHA, BHT, and diphenol TBHQ (**11–5**), all of which have an intermediate polarity; more polar antioxidants are esters of gallic acid (gallates) and some esters of ascorbic acid. Polar antioxidants are ascorbic and erythorbic acids and their salts. Ascorbic acid, its derivatives, and its analogue erythorbic (isoascorbic) acid are described in Section 5.13.1.

11.2.2.3.1 BHA

Commercial butylated hydroxyanisole (BHA, E320) is a mixture of two isomers: approximately 90% (at least 88%) 3-*tert*-butyl-4-hydroxyanisole (3-BHA) and about 10% 2-*tert*-butyl-4-hydroxyanisole (2-BHA, **11–5**). BHA is particularly effective for the protection of lipids containing fatty acids with shorter chains (such as coconut and palm kernel oils) and in the aroma and colour of essential oils. Like BHT, BHA is often used in packaging materials, where it can migrate into food. Both compounds may (in contrast to the low-volatile gallates) show slight odour reminiscent of phenols. BHA exhibits synergism with BHT and gallates, and in comparison with BHT it displays a somewhat higher so-called **carry-through effect**, which means that it is also effective as an antioxidant in the final heat treated product.

As with all other antioxidants, BHA undergoes transformation reactions during oxidation of lipids. The most common products are dimers, biphenyls (2,2′-dihydroxy-5,5′-dimethoxy-3,3′-di-*tert*-butylbiphenyl, R = *tert*-butyl, **11–6**), and ethers (2′,3-di-*tert*-butyl-2-hydroxy-4′, 5-dimethoxydiphenylether, R = *tert*-butyl, **11–6**). Most primary oxidation products still preserve the antioxidant activity.

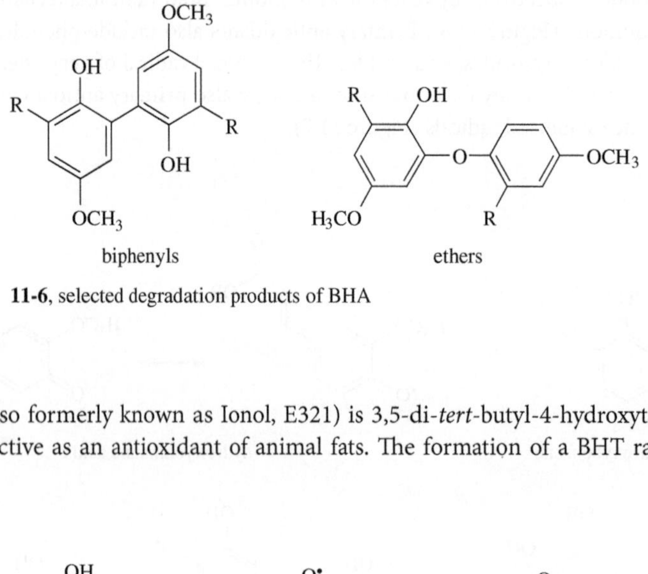

biphenyls   ethers

11-6, selected degradation products of BHA

11.2.2.3.2 BHT

Butylated hydroxytoluene (BHT, also formerly known as Ionol, E321) is 3,5-di-*tert*-butyl-4-hydroxytoluene (**11–5**). In comparison with BHA, BHT is somewhat more effective as an antioxidant of animal fats. The formation of a BHT radical and its reactions is shown in Figure 11.9.

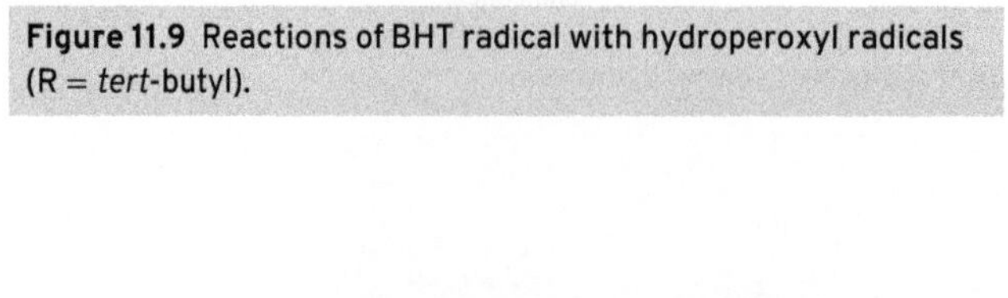

Figure 11.9 Reactions of BHT radical with hydroperoxyl radicals (R = *tert*-butyl).

Important degradation products of BHT, which are also active as antioxidants, are 3,5-di-*tert*-butyl-4-hydroxybenzaldehyde (R = *tert*-butyl, R^1 = CH=O), the related primary alcohol (R^1 = CH$_2$OH) and 1,2-bis(3,5-di-*tert*-butyl-4-hydroxyphenyl)ethane (R = *tert*-butyl) (**11-7**). In addition to these products, lower amounts of other phenols, quinones (**11-7**, R = *tert*-butyl), and stilbenes (**11-7**, R = *tert*-butyl) are formed. Further, mixed products such as 3,3',5'-tri-*tert*-butyl-5-methoxy-2,4'-dihydroxydiphenylmethane (**11-7**, R = *tert*-butyl) arise in mixtures of BHT and BHA.

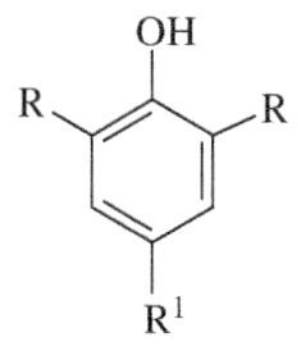

3,5-di-*tert*-butyl-4-hydroxy-benzaldehyde

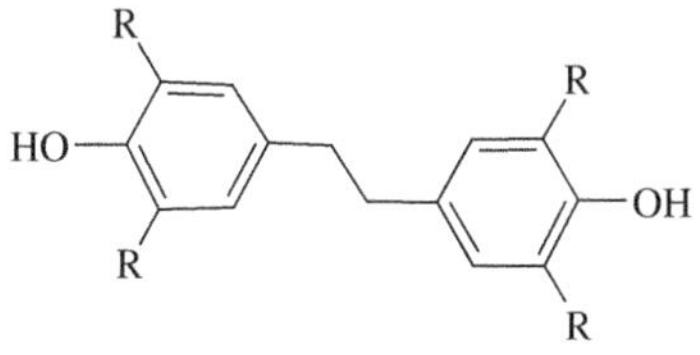

1,2-bis(3,5-di-*tert*-butyl-4-hydroxyphenyl)ethane

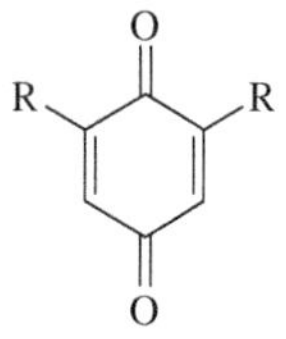

2,6-di-*tert*-butylbenzoquinone

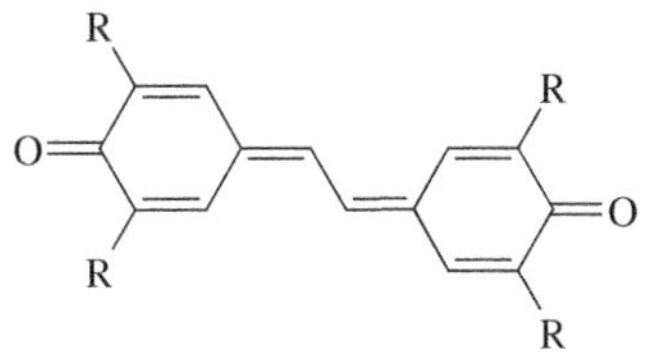

3,5,3',5'-tetra-*tert*-butylstilbenequinone

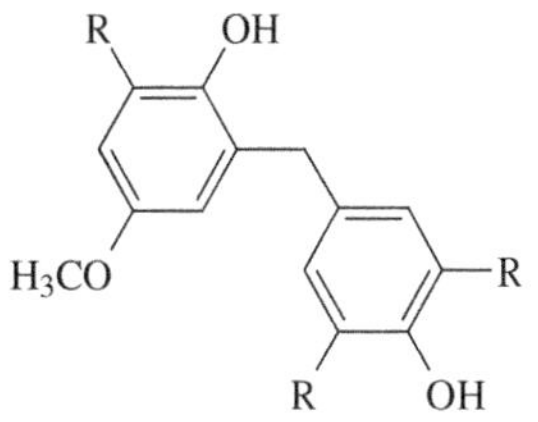

3,3',5'-tri-*tert*-butyl-5-methoxy-2,4'-dihydroxydiphenylmethane

11-7, selected degradation products of BHT

11.2.2.3.3 TBHQ

TBHQ (E319), 2-*tert*-butylhydroquinone (**11-5**), is the only diphenol used as an antioxidant. It is more effective in vegetable oils than BHA and BHT and is one of the best antioxidants designated for fats used for frying. Its carry-through effect is comparable with that of BHA. A further increase of TBHQ antioxidant activity, especially for the protection of vegetable oils, is possible in combination with chelating agents (such as citric acid). For example, a ternary mixture of TBHQ, monoacylglycerol citrate, and ascorbyl palmitate exhibits high thermal stability and provides optimum protection in oils during high-temperature processing. In European countries and Japan, the addition of TBHQ to foods is not allowed.

As the diphenol antioxidant, TBHQ reacts with hydroperoxyl radicals to form semiquinone radicals stabilised by resonance. These intermediates yield dimers, and the original hydroquinone regenerates by dismutation. The intermediates can also react with other lipid radicals (Figure 11.10). All degradation products of TBHQ exhibit antioxidant activity; 2,3-dihydro-2,2-dimethylbenzofuran-5-ol and 2-(2-hydroxy-2-methyl-1-propyl)hydroquinone (**11-8**) have even higher activity than TBHQ.

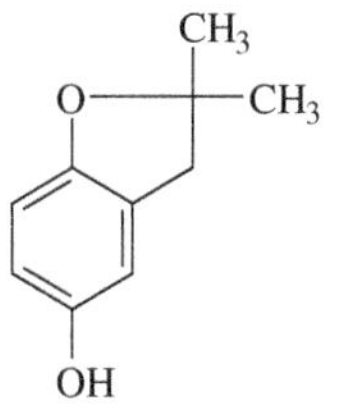

2,3-dihydro-2,2-dimethyl-benzofuran-5-ol

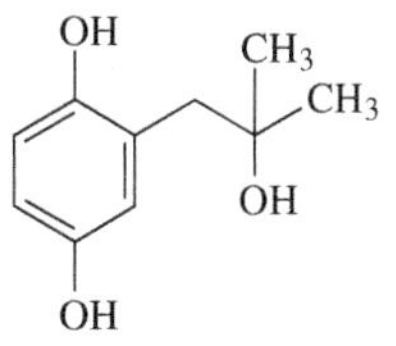

2-(2-hydroxy-2-methyl-1-propyl)hydroquinone

11-8, selected degradation products of TBHQ

11.2.2.3.4 Gallates

Esters of gallic acid, gallates (**11-9**), are found in small amounts in foods of plant origin as natural constituents. Three synthetic esters can be used as antioxidants in the European Union: propyl gallate (E310), octyl gallate (E311), and dodecyl gallate (E312). Propyl gallate is the only gallic acid ester permitted in foods in the United States and Canada. The antioxidant activity of gallates, which are more polar compounds than BHA, BHT, and TBHQ, is higher in anhydrous fats or fats with minimum moisture, which is related to their solubility in the two phases, the oil and water. Propyl gallate is therefore suitable, for example, for the stabilisation of animal fats (e.g. lard and tallow). Gallates are more soluble in emulsions but less active than BHA and BHT.

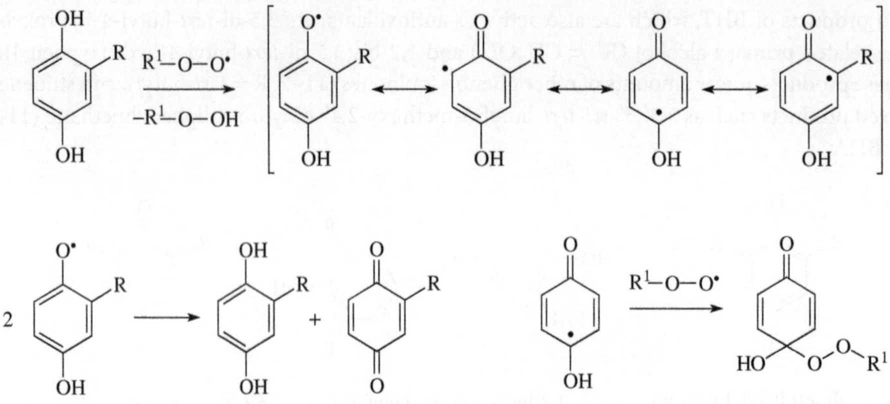

Figure 11.10 Reactions of TBHQ radicals with hydroperoxyl radicals (R = *tert*-butyl).

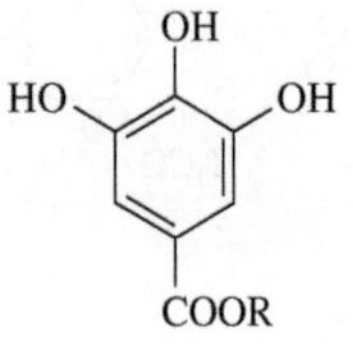

11-9, propyl gallate, R = CH$_2$CH$_2$CH$_3$
octyl gallate, R = CH$_2$[CH$_2$]$_6$CH$_3$
dodecyl gallate, R = CH$_2$[CH$_2$]$_{10}$CH$_3$

Propyl gallate is a relatively unstable compound, and is therefore not suitable for fats used for frying (where the temperature exceeds 190 °C). For the same reason, propyl gallate has a weak carry-through effect (unlike octyl gallate and dodecyl gallate). Propyl gallate forms blue-black complexes with iron and copper ions; therefore, it is always used in combination with chelating agents (such as citric acid). Gallates exhibit synergism with BHA and BHT, but using them together with TBHQ is illegal in the United States. One of the major degradation products arising from propyl gallate is ellagic acid (see Section 8.3.5.1), which also has antioxidant properties. In addition to degradation products of propyl gallate and BHA, in mixtures of propyl gallate with BHA various mixed dimers arise, such as the propyl-3,5-dihydroxy-4-(2-hydroxy-5-methoxy-3-*tert*-butylphenoxy) benzoate (**11-10**, R^1 = COOCH$_2$CH$_2$CH$_3$, R^2 = R^4 = H, R^3 = OH). Their antioxidant activity is comparable to that of propyl gallate.

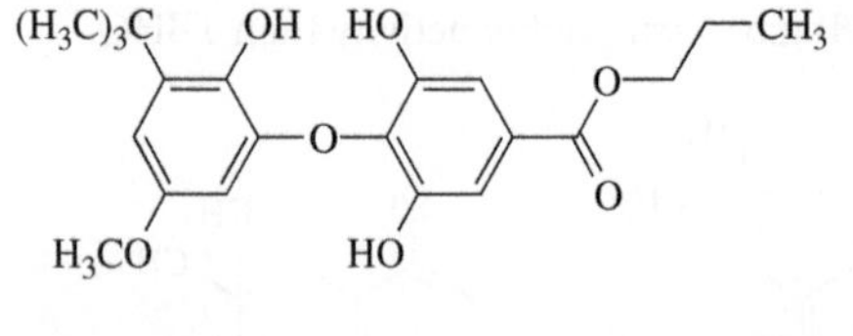

11-10, degradation product in BHA-propyl gallate mixtures

11.2.2.3.5 Other antioxidants

Antioxidants with a nitrogen heterocycle (such as dihydropyridine or dihydroquinoline derived compounds) are used only uncommonly, because of their relatively high toxicity. An example is ethoxyquin, also known as santokin (6-ethoxy-2,2,4-trimethyl-1,2-dihydroquinoline), which is used by the feed industry. The active form in fats is a free radical (**11-11**). The water-soluble analogue of α-tocopherol, 6-hydroxy-2,5,7,8-
tetramethylchromane-2-carboxylic acid, with the trade name Trolox (**11-11**), has found use in biological, biochemical, and food research. It is mainly used as a standard for measuring the antioxidant capacities of other compounds. Many other antioxidants are used in cosmetics, the petrochemical industry, and elsewhere.

Synthetic antioxidants (4-hexylresorcinol, see Section 10.2.5.2) and natural antioxidants approved as food additives in the European Union, such as ascorbic acid and its salts, fatty acid esters, and analogues (erythorbic acid) and tocopherols are discussed in Chapter 5.

Natural compounds with antioxidative activity occurring in herbs and spices, including the extracts of rosemary (E392) are addressed in Chapter 10.

11-11, less common synthetic antioxidants

11.2.2.4 Legislation

Substances with antioxidant effects (Table 11.4) that may be used in food production in the necessary quantities include fatty acids esters of ascorbic acid (palmitate and stearate), natural products containing tocopherols and synthetic α-, β-, γ-, and δ-tocopherol (which are identical to natural tocopherols), and rosemary extracts.

Erythorbic acid and sodium erythorbate may be used for preserved and semi-preserved fish products and frozen and deep-frozen fish at a concentration of 1500 mg/kg. The amount of these antioxidants allowed for the preservation of cured and preserved meat products is only 500 mg/kg (expressed as erythorbic acid). The use of 4-hexylresorcinol (4-hexyl-1,3-benzenediol) as a colour-retention agent for treatment of crustacean meat (fresh and frozen) is permitted in amounts up to 2 mg/kg, in order to prevent black spots arising by enzymatic browning reactions, known as shrimp melanosis.

To a limited extent, synthetic esters of gallic acid (propyl gallate, octyl gallate, and dodecyl gallate) and the phenolic antioxidants BHA and BHT may be used. When they are used in combination, the individual levels must be reduced proportionally, so that the sum of the individual antioxidants does not exceed the maximum amount of any one allowed for that food. For example, the maximum permissible amount of gallates and BHA is 200 mg/kg and for BHT 100 mg/kg in fats and oils used in the professional manufacture of heat-treated foodstuffs, frying oil, and frying fat, excluding olive pomace oil, fish oil, lard, beef, poultry, and sheep fat. Dehydrated potatoes may only contain 25 mg/kg of gallates and BHA, individually or in combination, whilst chewing gums may contain 400 mg/kg of these antioxidants or BHT.

11.2.2.5 Health assessment

Vitamin E can cause several side effects in high concentrations that are associated with the use of tocopherols not as additives, but as vitamin supplements. The ADI value for tocopherols is 0.15–2 mg/kg (b.w.). A higher single dose of α-tocopheryl acetate (400–500 mg) could cause negative effects, manifested by a higher intake of iodine by the thyroid gland. No side effects are known to be caused by erythorbic acid or sodium erythorbate in the concentrations that are commonly used. The ADI value has not been established for 6-hexylresorcinol. There is negative evidence of carcinogenesis, and no significant untoward effects were observed in humans when it was used as an anthelmintic agent.

The toxicological evaluation of phenolic antioxidants is not yet entirely definitive. The only problem with the use of BHA (ADI = 1.0 mg/kg b.w., classified as a Group 2B agent, possibly carcinogenic to humans, by the International Agency for Research on Cancer, IARC) seems to be the formation of lesions observed in experimental animals, although some pseudoallergic reactions have also been reported. BHA in combination with high concentrations of vitamin C can produce free radicals, which can cause damage to the components of cells, including DNA. BHA is excreted in the urine after conjugation as D-glucuronide and sulfate; its oxidation is not significant. At high doses, BHT (ADI = 0.25 mg/kg b.w., IARC Group 3, not classifiable as to its carcinogenicity to humans) can cause internal bleeding, which is related to its ability to reduce vitamin K. BHT can cause migraine in some people and liver damage in high concentrations; (pseudo)allergic symptoms have also been reported. BHT is metabolised primarily by oxidation of *tert*-butyl groups. TBHQ (ADI = 0.7 mg/kg b.w.) is not allowed in some countries due to the lack of information on its toxicity (possible mutagenic activity). It is metabolised through oxidation and conjugation (sulfate and glucuronide). Gallates are metabolised after hydrolysis to the corresponding alcohols and gallic acid and methylation of gallic acid to 4-O-methylgallic acid. Gallic acid can cause eczema, stomach problems, and hyperactivity. Contact dermatitis has been observed in bakers and in people handling gallates (ADI of propyl gallate = 0.5 mg/kg b.w.).

11.3 Substances regulating odour and taste

Substances used for food flavouring are the most comprehensive group of additives. The following main groups are recognised:

- flavourings;

- sweeteners;

- acidulants and acidity regulators;

- bitter substances and stimulants;

- flavour enhancers.

11.3.1 Flavourings

Flavourings are substances that affect olfactory and gustatory receptors and induce human perception of smell or taste. They are used to give food a taste or odour that would otherwise be absent or that is not present in a characteristic intensity. Individual compounds responsible for the odour and taste of food raw materials and foods are described in detail in Sections 8.2 and 8.3, respectively.

11.3.1.1 Legislation

EU legislation defines different types of flavourings:

- natural flavourings, which are derived from natural materials using physical, biotechnological (biochemical and microbiological), and other procedures;

- natural-identical, which are obtained by synthesis, but are identical to substances present in natural materials;

- artificial flavouring substances, which are obtained by synthesis, and are not identical to natural flavouring substances.

In flavourings that are not chemically defined compounds but mixtures of many substances, the following categories can be distinguished:

- flavouring preparations obtained by physical, enzymatic, or microbiological processes from material of vegetable or animal origin or from other foodstuffs;

- process flavourings that evolve flavour after heating and are obtained by heating a mixture of mostly amino acids and other nitrogen compounds with sugars or with other compounds for a maximum of 15 minutes;

- smoke flavourings, which are extracted from the products of the pyrolysis of materials used traditionally for smoked foods.

EU legislation also sets out general rules for the use of flavourings, requirements for labelling, and maximum levels for substances that raise concerns for human health. Flavourings are not substances that exclusively have sweet, sour, bitter, or salty tastes. Legislative regulations list more than 100 plant materials classified as foods or spices that, either in themselves or in the form of different products, can be used for food flavouring. Other flavouring substances are allowed in the production of tobacco products.

11.3.1.2 Health assessment

Substances that should not be added as such to food include (in alphabetical order) agaric acid, aloin, capsaicin, coumarin, hypericine, β-asarone, estragole, hydrogen cyanide, menthofuran, methyl eugenol, pulegone, safrole, quassin, safrole, teucrin A, α-thujone, and β-thujone. Table 11.5 lists the plant materials that contain these natural toxic substances. Examples of maximum levels of these substances naturally present in flavourings and food ingredients with flavouring properties, as well as their occurrence, health assessment and properties are given in Chapters 9 and 10.

Table 11.5 Main natural toxic compounds in plant materials used for aromatisation of foods and beverages.

Toxic compound	Plant (English and Latin names)
Agaric acid	Wood-decay fungus, known as agarikon (*Laricifomes officinalis*)
Aloin	Aloe (various species, *Aloe* spp.)
β-Asarone	Calamus (sweet flag, *Acorus calamus*)
Coumarin	Woodruff (*Asperula odorata*), yellow sweet clover (yellow melilot, *Melilotus officinalis*), cumaru (*Dipterix odorata*), sweet grass (*Hierochloe odorata*)
Estragole	Tarragon (*Artemisia dracunculus*)
Hydrogen cyanide	Elder (*Sambucus nigra*), plum (*Prunus domestica*)
Hypericin	St John's wort (*Hypericum perforatum*)
Menthofuran	Peppermint (*Mentha piperita*)
Methyl eugenol	Clove (*Syzygium aromaticum*)
Pulegone	Wrinkled-leaf mint (*Mentha crispa*), peppermint (*M. piperita*), pennyroyal (*M. pulegium*)
Quassin	Quassia (bitterwood, *Quassia amara*)
Quinine	Cinchona trees (various species, *Cinchona* spp.)
Safrole and isosafrole	Nutmeg (*Myristica fragrans*)
Teucrin A	Wall germander (*Teucrium chamaedrys*)
α- and β-Thujone	Mugwort (*Artemisia vulgaris*), wormwood (*A. absinthium*), santonica (*A. cina*), Roman wormwood (*A. pontica*), tansy (*Tanacetum vulgare*), common yarrow (*Achillea millefolium*), musk yarrow (*A. moschata*), common sage (*Salvia officinalis*)

11.3.2 Sweeteners

Naturally occurring sweet substances are mainly monosaccharides, disaccharides, and sugar alcohols (see Section 8.3.1). In addition to sugars and sugar alcohols, many other much sweeter substances are found in nature and a number of sweet compounds have been synthesised. Health, nutritional, and economic aspects have led to the introduction of some natural, natural-identical, and synthetic substances into human nutrition.

11.3.2.1 Classification

Sweeteners are classified according to their origin into the following groups:

- natural (e.g. protein thaumatin);

- natural-identical, which includes synthetic substances identical to natural (e.g. sugar alcohols) or modified natural (e.g. neohesperidin dihydrochalcone) substances;

- synthetic (e.g. saccharin).

From the nutritional point of view, two categories of sweeteners are recognised:

- nutritional (e.g. sugar alcohols)

- non-nutritional (virtually all other natural, modified natural, and synthetic substances).

Closely related to sweeteners are sweet taste-modifying compounds, which include:

- sweetness-inducing substances (the proteins curculin, miraculin, and strogin)

- antisweet substances (the triterpenic glycosides gymnemic acid, hodulcin, and ziziphin).

11.3.2.1.1 Nutritional sweeteners

Monosaccharides, disaccharides, sugar alcohols, and other nutritional sweeteners derived from sugars are not considered food additives. These sweeteners are described in detail, including their relative sweetness, in Chapter 4, which deals with carbohydrates.

Sweet peptides (such as aspartame), proteins (e.g. thaumatin), and glycosides, (stevioside) are a source of energy, but in the quantities in which they are used their contribution to the total energy intake is insignificant. These sweeteners are therefore ranked amongst non-nutritional sweeteners in Section 11.3.2.1.2.

11.3.2.1.2 Synthetic non-nutritional sweeteners

The relative sweetnesses of chemically modified natural and synthetic sweeteners are given in Table 11.6.

Acesulfame K   Acesulfame K (E950, K being the symbol for potassium) is a potassium salt of 6-methyl-1-oxa-2-thia-3H-azin-4-one-2,2-dioxide (**11–12**). Trade names are Sunett and Sweet One. This substance (approved in the European Union in 1983 and the United States in 2003 for general use, but not in meat or poultry) is often blended with other sweeteners (usually sucralose or aspartame) to mask its slightly bitter aftertaste. Its stability makes it suitable for products that require a long shelf-life (it decomposes at temperatures exceeding 235 °C). It is used in many applications, and in particular in baked goods and beverages. In carbonated drinks, it is almost always used in conjunction with another sweetener, such as aspartame or sucralose.

Advantame   Advantame, N-[3-(3-hydroxy-4-methoxyphenyl)propyl]-α-L-aspartyl-L-phenylalanine 1-methyl ester (**11–12**), is the most potent sweetener by far. Structurally, it is rather similar to neotame, being based on Aspartame. At ultra-low levels, it acts as flavour enhancer. It is more stable than Aspartame, it does not break down under high temperatures, and so it can be used in all processed foods and in cooking.

Table 11.6 Relative sweetnesses of chemically modified and synthetic sweet substances (5% sucrose equivalent = 1).

Compound	Sweetness
Modified substances	
Naringin dihydrochalcone	100–350
Neohesperidin dihydrochalcone	500–2000
(*E*)-Perillaldehyde oxime	350
Synthetic substances	
Acesulfame K	80–250
Advantame	20 000
Alitame	2000
2-Amino-4-nitro-1-propoxybenzene	4000–5000
Aspartame	100–200
Cyclamates	30–60
Dulcin	70–350
Lugundame	220 000–300 000
Neotame	7000–13 000
Saccharin	200–700
Sucralose	600

The US FDA has approved advantame for general use as a food additive in all foods and beverages except meat and poultry, and has issued a guideline for acceptable daily intake.

Alitame Like aspartame and neotame, alitame (in some countries known as aclame, **11–12**), L-α-aspartyl-N-(2,2,4,4-tetramethyl-3-thietanyl)-D-alanine amide, is an aspartic acid-based dipeptide that is about 10 times sweeter than aspartame and has no aftertaste. As with aspartame, it is hydrolysed in acidic media but does not contain bound phenylalanine and can therefore be used by people with phenylketonuria. Alitame is not approved for use in the European Union or United States, but is approved in Australia, China, and some other countries.

Aspartame Aspartame (E951) is a methyl ester of linear dipeptide L-aspartyl-L-phenylalanine (**11–12**) that was used by several EU countries in the 1980s, gaining EU-wide approval in 1994 (and approval for general purpose in the United States in 1996). It does not show any aftertaste. In non-aqueous media (such as powdered drinks, chewing gum, and instant coffee), it is stable. In acidic aqueous solution, depending on the pH and temperature, its ester bond can be hydrolysed, creating the corresponding dipeptide (L-aspartyl-L-phenylalanine) and methanol. A product of the dehydration of this linear dipeptide is cyclic dipeptide cyclo-(L-aspartyl-L-phenylalanine, **11–13**). On hydrolysis, it provides a linear dipeptide, which is further hydrolysed to free aspartic acid and phenylalanine. Aspartame reactions are associated with a decrease in sweetness. Aspartame is therefore not suitable for all foods (especially acidic foods), or for all types of food processing. In non-alcoholic beverages, for example, about 20% or more of aspartame can be degraded at room temperature during four months of storage.

Cyclamate Cyclamate is the generic name for cyclamic (cyclohexylsulfamic) acid and its salts, sodium cyclamate (**11–12**) and calcium cyclamate (E952), which became popular in the 1950s. In 1958, cyclamate was classified as Generally Recognized as Safe (GRAS) in the United States, but it was subsequently banned in 1970 (see Section 11.3.2.3). Cyclamates are stable to heat, but exhibit minor aftertaste; therefore, they are used as sweeteners in baked goods, confectionaries, desserts, soft drinks, preserves, and salad dressings, often in a mixture with saccharin (10 : 1) to produce a synergistic sweetening effect.

Dihydrochalcones Some dihydrochalcones exhibit a sweet taste (see Section 9.4.2.6). They are obtained by the catalytic reduction of the flavanone glycosides neohesperidin and naringin, which contain bound β-neohesperidose, α-L-Rha$p$-(1 → 2)-β-D-Glc$p$, in the C-7 position. Catalytic reduction of neohesperidin (hesperetin-7-neohesperidoside) obtained from bitter oranges yields neohesperidin dihydrochalcone (**11–12**, R^1 = H, R^2 = OH, with β-neohesperidose bound to C-4), which has been approved in the European Union as a sweetener since 1994 (E959). Naringin (naringenin-7-neohesperidoside) gives rise to naringin dihydrochalcone (**11–12**, R^1 = OH, R^2 = CH_3). Neohesperidin dihydrochalcone has some properties that limit its use as a sweetener: it has an intense cooling effect on the tongue, liquorice-like and bitter off-tastes that make it decidedly not sucrose-like, and is slow in sweet taste onset. Because of these properties, it is particularly suitable for use in the pharmaceutical industry and in animal nutrition. With respect to its properties as a sweetener and flavouring, neohesperidin dihydrochalcone has applications in sweets, canned vegetables, desserts, sauces, and food supplements. Other important fields of application are the livestock industry, cosmetics (such as toothpastes), and pharmacy.

Dulcin Dulcin (4-ethoxyphenylurea, **11–12**), also known as sucrol or valzin, was an important sweetener of the early twentieth century. It had an advantage over saccharin in that in the past it had been used in blends. It has a more pleasant sweet taste and does not possess a bitter aftertaste. However, in 1954, animal testing revealed unspecified carcinogenic properties, which led to a ban on its use.

Lugduname Lugduname (**11–12**) is one of the sweetest compounds known. It is a member of the group of guanidine sweeteners, which also include bernardame, carrelame, and sucrononate. The guanidine sweeteners are not expected to be used in foods. They exhibit different levels of toxicity, aftertaste, bitterness, and other properties.

Neotame Neotame, N-(3,3-dimethylbutyl)-L-α-aspartyl)-L-phenylalanine 1-methyl ester (**11–12**), is another sweetener from the family of dipeptides, approved in the European Union since 2011 as a flavour enhancer (E961) and in the United States since 2002. Neotame is a derivative of the dipeptide composed of aspartic acid and phenylalanine methylester, but unlike aspartame, aspartic acid occurs in the form of 3,3-dimethylbutylamide. Neotame functions effectively in a wide range of products, such as beverages, desserts, candy, ice creams, and bakery goods. It can be used alone or as part of a blend system with other non-nutritive or nutritive sweeteners.

Saccharin Saccharin (E954) is a common name for the corresponding acid, 1,1-dioxo-1,2-benzothiazol-3-one, and its sodium (**11–12**), potassium, and calcium salts. Substances in which the group –SO_2–NH–is a part of the rings are called sultames. It is the oldest sweetener, and became widespread during World War I (due to the shortage of sugar), increasing in popularity amongst dieters in the 1960s and '70s.

In the United States, it was classified as GRAS. The onset of sweetness from saccharin is rapid. The disadvantage is that it shows faint metallic and bitter aftertastes, but they can be masked by lactose or when used in combination with aspartame and other sweeteners. A 10 : 1 cyclamate : saccharin blend is common in countries where both are legal. Saccharin is very stable in foods even during heat treatments. The food industry uses saccharin salts in a variety of foods and drinks, such as baked goods, breads and cookies, sweetened diet drinks, sweetened and fruit-flavoured yoghurt, jams, jellies, and ice creams.

Sucralose Sucralose (E955) or chlorogalactosaccharose is a non-reducing disaccharide, known as 4,1′,6′-trichloro-4,1′,6′-trideoxygalactosaccharose or by the systematic name 1,6-dichloro-1,6-dideoxy-β-D-fructofuranosyl-4-chloro-4-deoxy-α-D-galactopyranoside (**11–12**). In fact, it is a sucrose molecule in which three of the hydroxyl groups have been replaced by chlorine atoms. It is a relatively new sweetener, and has not yet been introduced in many countries. Sucralose was first approved for use in Canada in 1991, in the United States as a general-purpose sweetener in 1999, and in the European Union in 2005. It appears to be only partially absorbed, followed by its excretion in the kidneys. The sweet flavour profile is similar to that of sucrose, with a slightly bitter off-taste. Sucralose is resistant to acidic and enzymatic hydrolysis and is stable during thermal operations. At high temperature in aqueous solution, it is partly hydrolysed, producing 4-chloro-4-deoxy-D-galactose and 1,6-dichloro-1,6-dideoxy-D-fructose. Sucralose is widely used by the food, beverage, and pharmaceutical industries and can be found in products such as diet drinks, yoghurts, and breakfast cereals.

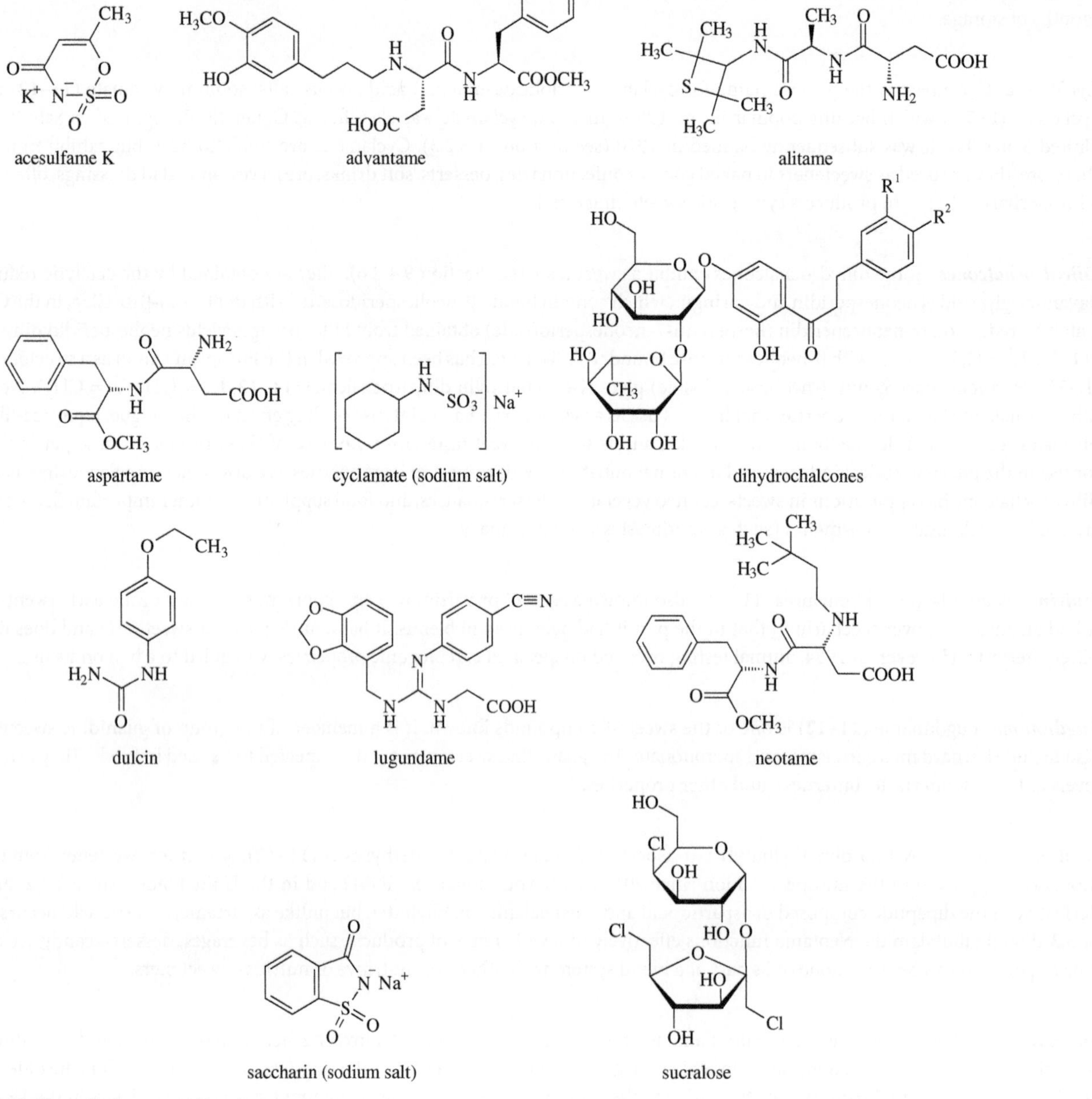

acesulfame K   advantame   alitame

aspartame   cyclamate (sodium salt)   dihydrochalcones

dulcin   lugundame   neotame

saccharin (sodium salt)   sucralose

11-12, synthetic non-nutritional sweeteners

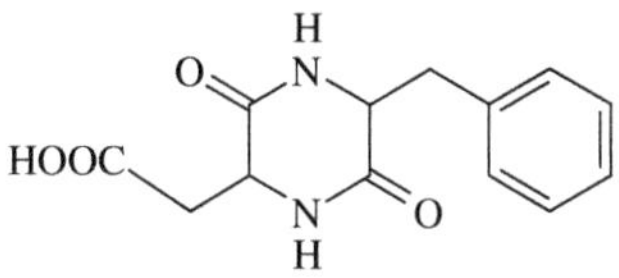

11-13, 3-benzyl-6-carboxymethyl-2,5-dioxopiperazine

Other sweet compounds In the past, many other synthetic sweet substances were temporarily used as sweeteners, such as 2-amino-1-propoxy-4-nitrobenzene or 6-chlorotryptofan (about 100 times sweeter than sucrose). They have not found wider application, particularly for toxicological reasons.

11.3.2.1.3 Natural non-nutritional sweeteners

Natural sweeteners have received much interest due to increasing health concerns over the consumption of sugar as well as problems related to the safety of some non-nutritional synthetic (artificial) sweeteners. Natural sweeteners belong to different chemical families. They include various terpenoids and other natural compounds. Although many natural compounds are sweet in taste, none of them has actually replaced sucrose.

Glycyrrhizin The rhizome of liquorice (*Glycyrrhiza glabra*, Fabaceae), native to southern Europe and parts of Asia, contains triterpenoid saponins, the main sweet component of which, with a characteristic liquorice cooling taste, is a glycoside known as glycyrrhizin (**10-122**). Glycyrrhizin is a mixture of potassium and calcium salts of glycyrrhizic acid. The aglycone is glycyrrhetic (glycyrrhetinic) acid. The sugar bound at C-3 is a disaccharide composed of two β-D-glucuronic acid units: β-d-Glc$p$A-(1 → 2)-β-D-Glc$p$A. The dried rhizome has been used as a laxative and in confectionery. Glycyrrhizin is used to sweeten and flavour some candies, pharmaceutical preparations, and tobacco products. It is reported to be relatively heat-stable. It is classified as a flavouring agent, but not as a sweetener. The relative sweetnesses of glycyrrhizin and some other natural sweet substances are shown in Table 11.7.

Hernandulcin The sweet sesquiterpenoids (+)-hernandulcin, 6-(1,5-dimethyl-1-hydroxyhex-4-en-1-yl)-3- methylcyclohexen-2-one, and (+)-4β-hydroxyhernandulcin (**11-14**) occur at a level of about 0.04% in the sweet herb *Phyla dulcis* (syn. *Lippia dulcis*, Verbenaceae), native to tropical Central and South America. Leaves with the sweet taste (hernandulcin is about 1250 times sweeter than saccharose) are used in traditional medicine. Because of its slightly bitter taste and minty aftertaste, hernandulcin has only limited use as a sweetener.

Table 11.7 Relative sweetnesses of natural sweet compounds (5% sucrose equivalent = 1).

Compound	Sweetness
Brazzein	1000–2000
Curculin	500 (430–2070)
Dulcoside A	30
Glycyrrhizin	50
(+)-Hernandulcin	1250
Mabinlins	100–400
Mogrosides-4 and 5	400
Monatin	3000
Monellin	1500–3000
Osladin	3000
Perillaldehyde	12
Phyllodulcin	200–800
Rebaudioside A	130
Stevioside	100–300
Strogin	150
Thaumatin	2000–3000

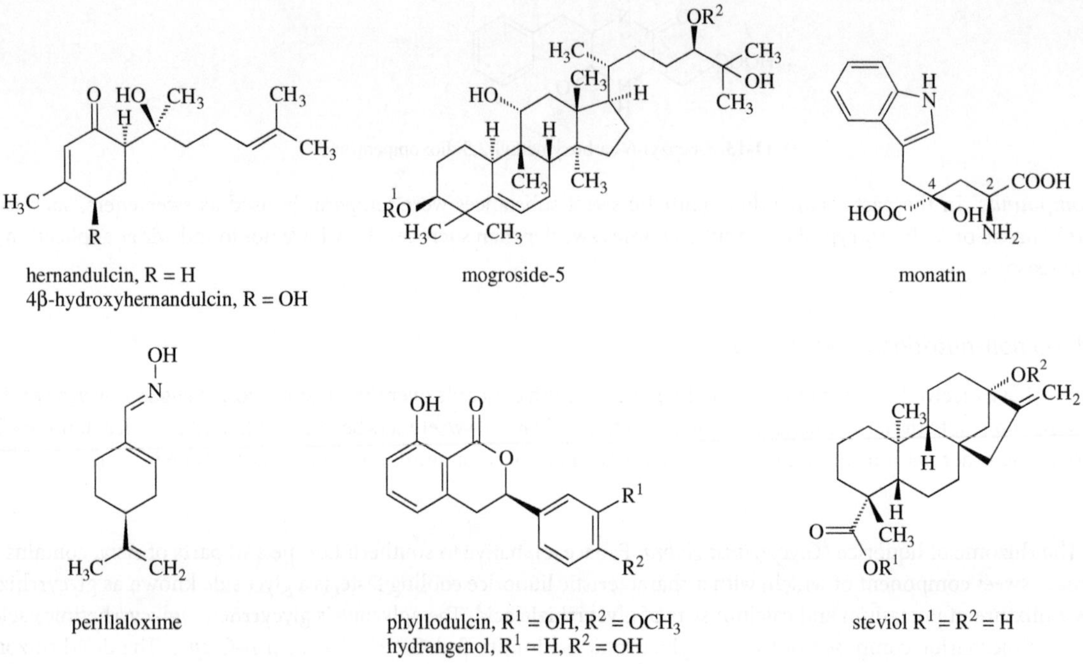

hernandulcin, R = H
4β-hydroxyhernandulcin, R = OH

mogroside-5

monatin

perillaloxime

phyllodulcin, R¹ = OH, R² = OCH₃
hydrangenol, R¹ = H, R² = OH

steviol R¹ = R² = H

11-14, natural non-nutritional sweeteners

Mogrosides The mogrosides are a group of sweet cucurbitane glycosides present at the level of about 1% in the fleshy part of the fruit of the gourd vine *Siraitia grosvenorii* of the Cucurbitaceae family, native to southern China and northern Thailand. The mogrosides have been numbered I–V, and the main component is called mogroside-5 (**11-14**, R¹ = β-gentiobiosyl, R² = β-neapolitanosyl), previously known as esgoside. Other structurally similar compounds are siamenoside and neomogroside. The plant's fruit has become known as monk fruit. Both the fresh and dried fruits are extracted to yield a powder that is 80% or more mogrosides. These extracts have been used in China for hundreds of years as low-calorie sweeteners for cooling drinks and in traditional medicine. Mogrosides have a good clean taste better than that of stevioside, and without the bitter aftertaste.

Monatin The *R,R*-isomer of monatin (**11-14**), known by the common and usual name arruva, is a naturally occurring sweetener isolated from the shrub *Sclerochiton ilicifolius* (Acanthaceae), found in the Transvaal region of South Africa. The sweet compound represents about 0.01% of the bark of the root by weight. It has a superior taste to stevioside, with a profile much closer to sugar and without an aftertaste. Monatin is an indole derivative, so its degradation products in the presence of light (hydroxylated and peroxide species formed on the indole ring as well as multiple ring and side-chain oxidation and scission products) smell unpleasantly. Also intensely sweet are the (2*R*,4*S*)-isomer and the (2*S*,4*S*)-isomer.

Perillaldehyde The monoterpenic aldehyde (-)-perillaldehyde, also known as perillal (see Section 8.2.4.1.1), which occurs in small quantities as a component of various essential oils (such as oils of citrus fruits), has a sweet taste. In the past, perillaldehyde was used as a starting compound for the synthesis of much sweeter (*E*)-perillaldehyde oxime (perillaloxime), called perillartin (**11-14**).

Phyllodulcin The dihydroisocoumarin phyllodulcin (**11-14**) occurs in the leaves of the large-leaf hydrangea (*Hydrangea macrophylla*, Hydrangeaceae), native to Japan, and the leaves of the mountain hydrangea (*H. serrata*), native to Japan and Korea, along with related compounds such as hydrangenol. The sweetness of phyllodulcin is characterised by a slow onset and long-lasting sensation of sweet taste and by an off-taste resembling liquorice. Phyllodulcin has found use in sweetening confectionery and chewing gums and in making regionally popular herbal teas, such as *amacha* tea in Japan, used in the celebration of Buddha's birth.

Stevioside The sweet substance called stevioside is an *ent*-kaurene diterpene glycoside containing steviol (**11-14**) as the aglycone and β-D-glucose and disaccharide β-sophorose as sugar components. Stevioside is accompanied by other related glycosides, which include rebaudioside A, rebaudioside B, rebaudioside C, rebaudioside D, rebaudioside F, dulcoside A, rubusoside, and steviolbioside (Table 11.8). It occurs in amounts of up to 6% in the leaves of stevia (*Stevia rebaudiana*, Asteraceae), a shrub native to tropical South America (Paraguay), which can be used directly for sweetening, although the taste quality is not as good as that of isolated glycosides. Today, stevia is cultivated in East

Table 11.8 Structures of stevioside and related substances.

Trivial name	Substituents in steviol (see 11-14)	
	R^1	R^2
Stevioside	β-D-Glc*p*	β-D-Glc*p*-(1 → 2)-β-D-Glc*p*
Rebaudioside A	β-D-Glc*p*	β-D-Glc*p*-(1 → 3)-β-D-Glc*p*-(1 → 2)-β-D-Glc*p*
Rebaudioside B	H	β-D-Glc*p*-(1 → 3)-β-D-Glc*p*-(1 → 2)-β-D-Glc*p*
Rebaudioside C (dulcoside B)	β-D-Glc*p*	β-D-Glc*p*-(1 → 3)-β-D-Glc*p*-(1 → 2)-α-L-Rha*p*
Rebaudioside D	β-D-Glc*p*-(1 → 2)-β-D-Glc*p*	β-D-Glc*p*-(1 → 3)-β-D-Glc*p*-(1 → 2)-p-D-Glc*p*
Rebaudioside E	β-D-Glc*p*-(1 → 2)-β-D-Glc	β-D-Glc*p*-(1 → 2)-β-D-Glc*p*
Rebaudioside F	β-D-Glc*p*	β-D-Glc*p*-(1 → 3)-β-D-Glc*p*-(1 → 2)-β-D-Xyl*p*
Dulcoside A	β-D-Glc*p*	β-D-Glc*p*-(1 → 2)-α-L-Rha*p*
Rubusoside	β-D-Glc*p*	β-D-Glc*p*
Steviolbioside	H	β-D-Glc*p*-(1 → 2)-β-D-Glc*p* (β-soforosa)

Asia (China, Japan, Korea, and Malaysia). In Japan, it has been available to sweeten soft drinks, candy, and chewing gums for decades. In some countries, health concerns have limited its availability; for example, the United States banned stevioside in the early 1990s (because of possible mutagenicity of steviol), but in 2008 it approved rebaudioside A as a food additive. Relatively recently (in 2011), stevioside was approved for use in the European Union as steviol glycoside (E960), and maximum content levels for different types of food and beverage were established.

11.3.2.1.4 Natural nutritional sweeteners

Natural nutritional sweeteners are proteins, but their potency is so high that they are not significant sources of energy. All protein sweeteners have a slow onset of sweet taste, and the sweetness lingers for some time. They are hydrolysed in acidic media, which results in the loss of the taste. It is expected that sweet proteins are digested just like any other dietary proteins.

Brazzein Brazzein is a sweet protein composed of 54 amino acids (6.5 kDa). It occurs in very sweet edible fruits of the climbing plant *Pentadiplandra brazzeana* (Pentadiplandraceae), native to West Africa. Brazzein has an excellent taste similar to sugar, with a slight liquorice-like cooling effect in the mouth. It is water-soluble and remarkably heat-stable and so is suitable for cooking and for processed foods. It has a synergistic effect if mixed with stevioside.

Mabinlins Mabinlins-1, 2, 3, and 4 are sweet-tasting proteins extracted from the seed of a caper known as mabinlang (*Capparis masaikai*, Capparaceae) and used in traditional Chinese medicine. The main component is mabinlin-2, an exceptionally thermostable protein, which consists of two subunits containing 33 and 72 amino acids (10.4 kDa) linked by a pair of disulfide bridges. The other two disulfide bridges stabilise the amino acid chain.

Monellin Monellin is extracted from the fruit *Dioscoreophyllum volkensii* (syn. *D. cumminsii*, Menispermaceae), native to tropical African rainforests. It is a sweet protein with a liquorice-like flavour that consists of two peptide chains, A and B, composed of a sequence of 45 and 50 amino acids, respectively (11.5 kDa). Under food processing conditions, it is unstable and has no practical significance as a sweet substance.

Thaumatin Thaumatin (also called thalin) is a mixture of sweet proteins extracted from the fruit of the tropical West African flowering plant *Thaumatococcus danielli* (Marantaceae), known as miracle fruit or miracle berry. It is the only protein sweetener approved in the European Union (E957). The main components, with sweet notes resembling liquorice, are the proteins thaumatin I and thaumatin II, with relative molecular weights about 22 kDa (thaumatin I consists of 207 amino acids and contains one disulfide bond and no histidine). The commercially available preparations additionally contain several other sweet proteins (e.g. thaumatins a, b, and c) and small amounts of polysaccharides (arabinogalactans and arabinoglucuronoxylans). Thaumatin is unstable with heat treatment. It is mainly used in food and drinks in combination with other sweeteners, for its flavour-modifying properties and not as a sweetener, because it acts synergistically in combination with acesulfame K, saccharin, stevioside, and other sweeteners.

11.3.2.1.5 Sweet taste-modifying compounds

Sweetness-inducing substances (the proteins curculin, miraculin, and strogin) have the ability to modify a sour taste into a sweet taste. Antisweet substances or sweetness inhibitors (gymnemic acid, hodulcin, and ziziphin) suppress the sweet taste sensation.

Curculin Curculin is a sweet protein from the fruit of *Curculigo latifolia* (Hypoxidaceae), a plant growing in Malaysia. Curculin is a dimer consisting of curculin 1 (12.5 kDa) and curculin 2 (12.7 kDa) units. Each unit contains 114 amino acids, bound through two disulfide bridges. Like miraculin, curculin exhibits taste-modifying activity, making water, sour solutions, and sour foods (such as lemon and vinegar) taste sweet. This effect lasts about 10 minutes. Unlike miraculin, it also exhibits a sweet taste on its own. Curculin has been approved in Japan as a harmless food additive.

Gymnemic acids Gymnemic acids belong to a class of triterpenic glycosides isolated from the leaves of *Gymnema sylvestre* (Asclepiadaceae). The aglycone gymnemagenin is esterified with various organic acids (acetic acid, tigloic acid, and 2-methylbutanoic acid) and bound to glucuronic acid. More than 20 homologues of gymnemic acid are known, of which gymnemic acid I (**11–15**, R^1 = tigloyl, R^2 = acetyl, R^3 = glucuronic acid residue) has the highest antisweet activity. This activity is reversible, but sweetness recovery on the tongue can take more than 10 minutes.

Hodulcine Hodulcines (or hodulosides) are dammarane-type triterpenic glycosides isolated from the leaves of *Hovenia dulcis* (Rhamnaceae), also known as Japanese raisin tree. The highest antisweet activity is shown by hoduloside I (**11–15**, R^1 = β-neohesperidosyl, R^2 = β-D-glucopyranosyl), but this is still lower than that of gymnemic acid I.

Miraculin The glycoprotein miraculin occurs in small red berries of the West African shrub *Synsepalum dulcificum* (syn. *Richadella dulcifica*, Sapotaceae). Although it is tasteless itself, it adjusts the sour taste of acids to a sweet taste (e.g. the taste of lemon juice), a perception lasting about one hour. Miraculin occurs as a tetramer (98.4 kDa), where a combination of two monomers forms two dimers linked by disulfide bridges. The relative molecular weight of the monomer is 24.6 kDa. Sugars bound in the glycoprotein (4 kDa, 13.9% of molar weight) are D-glucosamine (31%), D-mannose (30%), L-fucose (22%), D-xylose (10%), and D-galactose (7%). When heated over 100 °C, miraculin loses its taste-modifying property. Miraculin has no importance as a food additive (taste-modifying agent).

Strogin Strogin is a mixture of strogins I–V, oleanane-type triterpenic glycosides isolated from the leaves of *Staurogyne merguensis* (Acanthaceae), native to Malaysia. Strogins themselves have a sweet taste, which diminishes within a few minutes. Subsequent application of cold water to the mouth then elicits a further sweet taste (sweetness-inducing activity), as with circulin. The structure of strogin resembles that of gymnemic acid, which has antisweet activity.

Ziziphin Ziziphin (**11–15**, R^1 = 4-O-α-L-arabinopyranosyl-α-L-rhamnosyl, R^2 = 2,3-diacetyl-α-L-rhamnopyranosyl) is a triterpenic glycoside isolated from the leaves of *Ziziphus jujuba* (Rhamnaceae) together with several damarane-type saponins designated jujubasaponins. The antisweet activities of zizaphin and jujubasaponins II and III are less effective than that of gymnemic acid I.

hoduloside I   gymnemic acid I   ziziphin

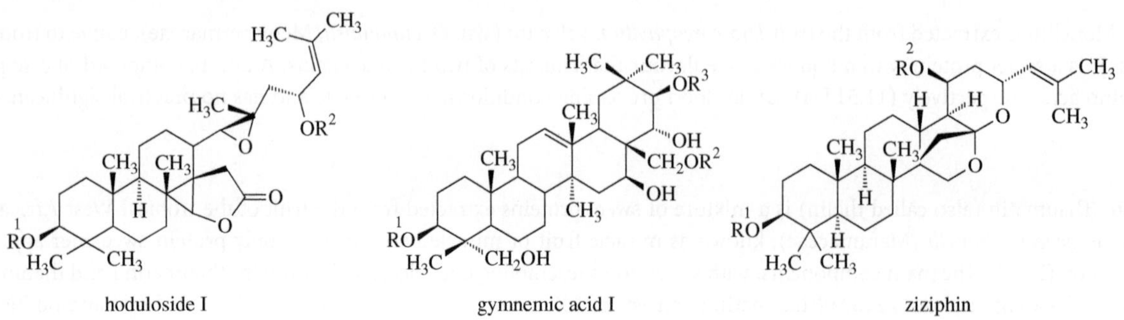

11-15, compounds with antisweet activities

Table 11.9 List of current EU-approved sweeteners and their E numbers.

E number	Name
E420	Sorbitol and sorbitol syrup
E421	Mannitol
E950	Acesulfame K
E951	Aspartame
E952	Cyclamic acid and its Na and Ca salts
E953	Isomalt
E954	Saccharin and its Na, K, and Ca salts
E955	Sucralose
E957	Thaumatin
E959	Neohesperidin dihydrochalcone
E960	Steviol glycoside
E961	Neotame (as a flavour enhancer)
E962	Salt of aspartame-acesulfame
E965	Maltitol and maltitol syrup
E966	Lactitol
E967	Xylitol
E968	Erythritol

11.3.2.2 Legislation

In the European Union, most sweeteners are only permitted to be used in a limited number of foods and are subject to specific quantitative limits (Table 11.9). Sorbitol (the correct name is glucitol, E420), mannitol (E421), maltitol (E965), lactitol (E966), xylitol (E967), and erythritol (E968) are approved as food sweeteners, and are described together with other sugar alcohols in Section 4.3.1.3.1. They are permitted to be used in *quantum satis* for confectionery, desserts, and similar products. Common carbohydrates occurring in foods (glucose, fructose, sucrose, and lactose) and honey are not considered sweeteners.

Sweeteners may be used to sweeten foods and for the preparation of table-top sweeteners, but are not intended to be processed for baby food. Foods and table-top sweeteners containing more than 10% sugar alcohols must carry a warning on the packaging that excessive consumption may have a laxative effect. Table-top sweeteners and foods containing aspartame must be labelled as containing a source of phenylalanine (see Section 10.3.1.2). Table 11.10 provides the maximum usable doses of approved sweeteners in some selected foodstuffs.

11.3.2.3 Health assessment

The ADI value is not specified for sugar alcohols derived from monosaccharides (xylitol, glucitol, and mannitol) and disaccharides (maltitol, isomaltitol, and lactitol).

Acesulfame K is rapidly absorbed and excreted mainly in the urine as an unchanged compound. In certain countries (including EU countries), acesulfame K has many uses, but in others it is only used for toothpastes and mouthwashes. The ADI value is 15 mg/kg (b.w.) for children and adults. Acute toxicity in humans may manifest itself by headaches.

Aspartame has been the most controversial artificial sweetener by far because of its potential toxicity. It is hydrolysed to phenylalanine, aspartic acid, and methanol and partially metabolised to glutamate. The daily intake of phenylalanine from food, which is estimated at 3.6 g, remains virtually unchanged using aspartame, but in people with impaired phenylalanine metabolism (phenylketonuria), some problems can arise. In some foods, where up to 5% of aspartame may be transformed into 2,5-dioxopiperazine, there has been some doubt about its possible carcinogenic properties. The toxicological data suggest that normal consumption of aspartame is not risky (ADI = 40 mg/kg b.w. for aspartame and 7.5 mg/kg b.w. for the corresponding 2,5-dioxopiperazine).

Table 11.10 Quantitative limits and applications for synthetic non-nutritional sweeteners.

Maximum usable dose (mg/kg or mg/l)	Sweetener[a]						
	Ac	As	Cy	Nc	Sa	Su	Th
Non-alcoholic beverages	350	600	250	30–50	80–100	200–300	–
Desserts and similar products	350	500–1000	250	50	100	400	–
Confectionery with no added sugar	500	1000	500	100	500	1000	50
Energy-reduced jams, jellies, and marmalades	1000	1000	1000	50	200	400	–
Chewing gums with no added sugar	2000	5500	1500	400	1200	3000	50
Cider and perry	350	600	–	20	80	50	–
Beer (various types)	350	600	–	10	80	250	–

[a] Ac = acesulfame K, As = aspartame, Cy = cyclamates, Nc = neohesperidine chalcone, Sa = saccharin, Su = sucralose, Th = thaumatin.

Owing to the presence of the 3,3-dimethylbutyl group in neotame (ADI = 2 mg/kg b.w.), peptidases are effectively blocked and the availability of phenylalanine is thus reduced, which is important for people suffering from phenylketonuria. At high doses of neotame, acute toxicity may be manifested by headache and hepatotoxicity. In contrast to aspartame, advantame (ADI = 5 mg/kg b.w.) does not form dioxopiperazine derivatives. Advantame slowly degrades under acidic conditions and at high temperatures under baking conditions and the main product is advantame acid. Unlike advantame, alitame (ADI = 5 mg/kg b.w.) does not contain bound phenylalanine. It is relatively stable at high temperatures and in a broad pH range. In the body, it is partly hydrolysed to aspartic acid and alanine amide.

Cyclamates (ADI = 11 mg/kg b.w.) are partially absorbed in the digestive tract (individually within 1–60%), and the rest is then transformed by intestinal bacteria into cyclohexylamine. Cyclohexanol and cyclohexane are produced in small amounts. Problems arose in 1969 when a single scientific study showed bladder carcinogenicity in rats fed with cyclamates due to the formation of cyclohexylamine, which led to an immediate ban on their use in many countries. More than two dozen other studies on their safety reportedly failed to show the same results, which has led to re-approval of the use of cyclamates in many countries, including in the European Union, although they remain banned in the United States.

Ingested neohesperidin dihydrochalcone (ADI = >5000 mg/kg b.w.) is hydrolysed in the same way as other related, naturally occurring flavonoids, yielding sugars and the corresponding aglycone. The aglycone is partly absorbed, metabolised, conjugated, and excreted via bile and urine, partly undergoes bacterial ring cleavage (of the C-ring), and subsequently experiences cleavage of the three-carbon bridge. It has been shown that neohesperidin dihydrochalcone can produce some side effects, such as nausea and migraine, at concentrations of around 20 mg/kg and above, but this effect is not well documented.

Saccharin (ADI = 5 mg/kg b.w.) is slowly but almost completely absorbed in the digestive tract and is rapidly excreted unchanged in the urine. The unabsorbed portion is excreted in faeces. Some doubts persist about its potential carcinogenicity or co-carcinogenicity, as saccharin has been shown to produce urinary bladder tumours in rats. This finding has led to a ban in some countries and a proposed ban in the United States, but the phenomenon has not been observed in mice or any other species, including humans. Acute toxicity may be manifested by nausea, vomiting, and diarrhoea.

Sucralose (ADI = 15 mg/kg b.w.) consumption does not have adverse health effects at doses up to 10 mg/kg (b.w.) or with repeated doses of up to 5 mg/kg per day for several weeks. The acute toxicity of sucralose may be manifested by migraine in sensitive individuals and by diarrhoea.

Stevioside (ADI = 4 mg/kg b.w.) is not degraded in the stomach juice, and its uptake by the gastrointestinal tract is very low. All of the stevioside reaching the colon is degraded by microorganisms into steviol, the only metabolite found in faeces. In urine, no stevioside or free steviol is present, except small amounts of steviol glucuronide. Studies of the toxicity of stevioside found no clinical signs of toxicity or morphological or histopathological changes in test animals.

Thaumatin (there is currently no listed ADI) is completely hydrolysed in the digestive tract to amino acids (about as rapidly as egg albumin). Its consumption does not pose any health risk.

In the United States, glycyrrhizin (ADI = 0.2 mg/kg b.w.) is generally recognised as safe. The European Union suggests that an intake of 100 mg/day (equivalent to approximately 50 g of liquorice sweets) would be unlikely to cause adverse effects (pseudohyperaldosteronism leading to hypocalcaemia alkalosis associated with a low serum potassium level and lower aldosterone secretion) in the majority of adults. People suffering from hypertension should avoid excessive consumption of liquorice products.

11.3.3 Acidulants and acidity regulators

Acidulants used as food additives are inorganic and organic acids generally identical to those that occur naturally in foods. They are most commonly used for their sour taste (see Section 8.3.3) but often have other beneficial properties. For example, some acids:

- exhibit antimicrobial effects and are therefore used simultaneously as preservatives, such as vinegar – acetic acid (E260), propionic acid (E280), and other acids;

- have significant organoleptic properties (taste and smell) and are used as flavouring agents, such as acetic acid (E260), succinic acid (E363), fumaric acid (E297), adipic acid (E355), lactic acid (E270), citric acid (E330), and malic acid (E296);

- act as colour stabilisers, such as ascorbic acid (E300) in meat products and citric acid (E330) in fruit products;

- act as sequestrants and synergists of antioxidants, such as calcium disodium salt of ethylenediaminetetraacetic acid (E385) and citric (E330), tartaric (E334), malic (E296), ascorbic (E300), and phosphoric acids (E338);

- modify texture; for example, citric acid (E330) allows the formation of some pectin gels or milk clotting by chymosin and inhibits the formation of crystals in confectionery;

- suppress the formation of hazes, such as lactic acid (E270) in brine of fermented olives;

- hydrolyse proteins, such as hydrochloric acid (E507) in the production of acidic protein hydrolysates.

Acidifying agents also include substances that produce acids by hydrolysis or during heating. This group of additives includes salts used as raising agents, such as sodium (E500), potassium (501), and ammonium carbonates (E503), which release carbon dioxide in the dough. They are also used in the production of sparkling beverages. δ-Lactone of D-gluconic acid (D-glucono-1,5-lactone, E575) is used in durable fermented salami, dairy, and other products, producing D-gluconic acid by hydrolysis (see Section 4.3.2.1).

Acidity and pH regulators maintain the acidity or alkalinity of foods. They mostly include salts of various acids with buffering effects and alkaline agents. For example, dispersion stabilisers for dairy products (melting salts) and meat products (substances that increase meat water holding capacity) are mostly various di-, tri-, and polyphosphates (E339, E400, E411, E450–E452). Sodium hydrogen carbonate (E500) is used for pH adjustment in the manufacture of Dutch-processed cocoa powder, neutralising its natural acidity. Sodium carbonate (E500) or sodium hydroxide (E524) is used for the neutralisation of acidic protein hydrolysates, whilst sodium hydroxide (E524) is employed in olive debittering (see Section 8.3.5.2) and in peeling fruits and vegetables.

11.3.3.1 Legislation

Many substances may be used in food production only in the amount necessary to achieve a desired effect. Phosphates, or mixtures thereof, may be used only for listed foods, and their amount is limited.

For example, phosphoric acid (E338) is allowed in flavoured soft drinks (such as cola drinks) in quantities up to 700 mg/l, whilst phosphates (as P_2O_5) are allowed in soft fresh cheeses (2000 mg/kg), melted cheese (melting salt, 20 000 mg/kg), meat products (to increase water binding capacity, 5000 mg/kg), and powder whiteners for beverages (30 000 mg/kg). The lactide of tartaric acid, called metatartaric acid (E353, **11–16**), prepared by heating tartaric acid, is allowed only for wines (100 mg/l), whilst the calcium disodium salt of ethylenediaminetetraacetic acid (E385, **11–17**) is allowed at a concentration of 75 mg/kg for emulsified sauces, mayonnaises, pickled products from crustaceans, molluscs, and fish and up to 250 mg/kg for canned legumes, artichokes, and mushrooms. To highlight the colour of black olives, ferrous salts of some organic acids are allowed, such as lactate (E579) or gluconate (E579) at a level of 150 mg/kg (calculated as iron), which form black complexes with oxidised fruit polyphenols. Urea (E927) may only be used in food supplements (200 mg/kg), and triethyl citrate in egg whites (and only in the specified quantity).

11-16, metatartaric acid

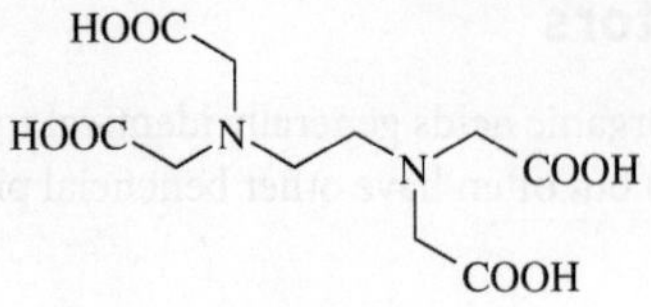

11-17, ethylenediaminetetraacetic acid

11.3.3.2 Health assessment

Acids and their salts are generally considered to be natural food constituents. Citric, fumaric, and succinic acids are intermediates of the citric acid cycle, whilst propionic acid is metabolised like other fatty acids (ADI values are not given). There are, however, reservations over the use of certain acids. Acetic acid induces epidermal reactions and other allergic-type symptoms in susceptible individuals. Adipic acid (ADI = 5 mg/kg b.w.) may affect the growth of animals. Restrictions also relate to fumaric acid (ADI = 6 mg/kg b.w.), racemic malic acid (ADI = 100 mg/kg b.w., but L-isomer is a normal food constituent), and L-tartaric acid (ADI = 30 mg/kg b.w.). Racemic and D-lactic acid may cause acidosis in infants, vomiting, and dehydration. The ADI value is not set, but it is not recommended to use these acids in infant nutrition.

11.3.4 Bitter substances and stimulants

A large number of organic and inorganic compounds have a bitter taste, as discussed in Section 8.3.5. Bitter-tasting substances derived from plant materials (such as hops, wormwood, and other herbs), which are used for aromatisation, are classified as flavourings or as bitter-tasting additive substances and stimulants, which include the alkaloids caffeine (see **10-50**) and quinine (see **10-41**). None of these compounds have E numbers. Another bitter additive substance is octaacetylsaccharose, better known as sucrose octaacetate (**11-18**). Pesticide products containing sucrose octaacetate as an inert ingredient are used as insect repellents, herbicides, flea and tick sprays, and other insecticides. Other commercial uses of sucrose octaacetate include impregnating and insulating papers, as well as uses in lacquers and plastics.

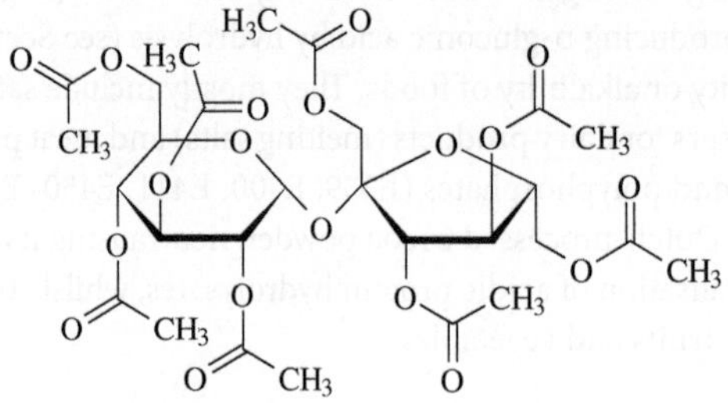

11-18, octaacetylsaccharose

11.3.4.1 Legislation

In the production of food, caffeine and quinine may be used either directly or as components of flavourings. Their amount was limited in the past, but according to new legislation only beverages that have a caffeine concentration higher than 150 mg/l must be labelled as having 'High caffeine content'. The maximum content of quinine in bitter drinks is not limited, either. Sucrose octaacetate has been approved by the FDA as a food additive in the United States (but not in the European Union), where it may be added to foods only in the specified quantities.

11.3.4.2 Health assessment

In the gastrointestinal tract, octaacetylsaccharose is hydrolysed to acetic acid and sucrose which are metabolised in the normal way. The health evaluation of caffeine and quinine is described in Chapter 10.

11.3.5 Flavour enhancers

Flavour enhancers are substances that intensify or modify the original flavour of certain foods, even though they do not have flavours of their own. Additives of this category include:

- L-glutamic acid (E620) and its salts, sodium hydrogen glutamate (E621), potassium hydrogen glutamate (E622), ammonium hydrogen glutamate (E624), calcium glutamate (E623), and magnesium glutamate (E625);

Table 11.11 Natural content of glutamic, inosinic, and guanylic acids in selected foods.

Food	Content (mg/kg)		
	Free Glu	IMP	GMP
Pork meat	230	1860	37
Chicken meat	440	1150	22
Peas	750	0	0
Tomatoes	2460	0	0

- purine 5′-nucleotides, which include inosine 5′-phosphate (inosinic acid, IMP, E630), disodium (E631), dipotassium (E632), and calcium inosinate (E633), guanosine 5′-phosphate (guanidylic or guanylic acid, GMP, E626), disodium (E627), dipotassium (E628), and calcium guanylate (E629).

Of particular importance are glutamic acid and sodium hydrogen glutamate. This salt shows the taste referred to as umami (see Section 8.3.4). At the concentrations in which it is added as an additive (0.05–0.8%), it amplifies and enhances the flavour of meat and vegetable products, such as soups and sauces, meat and vegetable preserves, tomato juice, ketchup, mayonnaise, and other products.

Synergistic effects to the taste of glutamate are shown by disodium 5′-ribonucleotides: IMP and GMP (**11–19**), which are added to foods in amounts ranging from 0.001 to 0.2%. When used, the amount of glutamate can be reduced about 10 times whilst maintaining the same intense umami taste. Commonly used is a mixture containing 95% w/w sodium hydrogen glutamate, 2.5% w/w IMP, and 2.5% w/w GMP. Detection thresholds of IMP and GMP are 250 and 125 mg/l, respectively. The detection threshold of a mixture of these substances (in weight ratio 1 : 1) is 63 mg/l; in combination with glutamate (at a concentration of 8 g/l; its detection threshold is 120 mg/l), it is as low as 0.31 mg/l. All three compounds are common food ingredients (Table 11.11). In some products, however, the glutamate content is very high (e.g. about 40 g/kg in acid protein hydrolysates, which corresponds to 100 g/kg of dry matter). In meat and fish, IMP prevails, as it arises from ATP via hydrolysis post mortem to ADP and AMP and deamination of AMP. IMP is then degraded via inosine, hypoxanthine, and xanthine to uric acid. The amount of IMP in meat extracts (highly concentrated meat stocks) used in cooking reaches about 10 g/kg. In the meat of crustaceans, the main nucleotide is adenosine 5′-phosphate (AMP). Some mushrooms and yeast extracts used as food additives (flavourings) and as nutrients for bacterial culture media in microbiology have a higher content of GMP. Xanthosine 5′-phosphate (xanthylic acid, XMP, **11–19**) has similar properties as IMP and GMP.

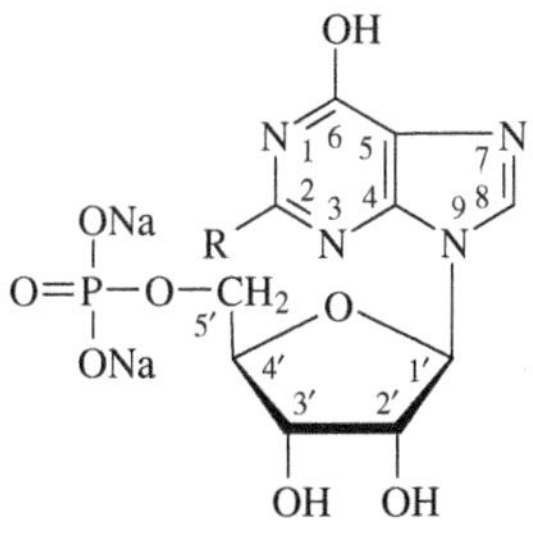

11-19, 5′-nucleotides

5′-IMP, R = H
5′-GMP, R = NH$_2$
5′-XMP, R = OH

11.3.5.1 Legislation

Glutamic, inosinic, and guanylic acids and their sodium and potassium salts can be used as flavour enhancers, individually or in combination, up to the maximum allowable amount. The permissible amount of glutamic acid in foodstuffs in general (excluding soft drinks) is 10 g/kg. Specified amounts of this amino acid and its nucleotides are prescribed for condiment preparations.

11.3.5.2 Health assessment

In the past, there were some reservations over excessive intake of glutamic acid and its salts because it was associated with the so-called Chinese restaurant syndrome, manifested in sensitive individuals by headaches, anxiety, digestive problems, and burning in the upper parts of the body, but these side effects have not been scientifically proven. Nevertheless, when glutamic acid or glutamates are added to foods, legislation requires that it be listed on the label (see Section 8.3.4). Many countries have restricted the amount of glutamic acid, glutamate, and 5′-nucleotides added to foods legislatively, but in some (e.g. Japan), these substances are not regulated at all. The ADI for sodium hydrogen glutamate is 120 mg/kg (b.w.); those for 5′-nucleotides are not specified, but salts of IMP (inosinates) may not be used in products intended for children under 12 weeks. As 5′-nucleotides are metabolised to uric acid, they should be avoided by people suffering from the arthritis form known as gout and by asthmatic people.

11.4 Substances modifying colour

Stabilisation of natural food colours and colouring of food has been carried out since time immemorial for aesthetic reasons, to make food more visually appealing, but the physiological reasons are also significant. Colours are used to restore the original appearance of food whose pigments have been affected by processing, storage, packaging, and distribution and to give colour to food that would otherwise be colourless, which might impair its visual acceptability. Another reason is the standardisation of colour, for example to compensate for seasonal fluctuations. A food's colour is related to its likeability, increases the secretion of gastric juices in the consumer, and leads to better utility. In some cases the natural colour may be unwanted and can be removed using bleaching agents.

11.4.1 Colours

11.4.1.1 Classification

Pigments found in foods are divided according to their origin into:

- natural pigments;
- natural-identical synthetic pigments;
- synthetic dyes.

11.4.1.2 Natural and natural-identical pigments

Natural and natural-identical pigments approved for the colouring of food in the European Union are listed in Table 11.12 and described in detail in Chapter 9. Caramel colours, one of the oldest and most widely used food colourings, are described in Section 4.7.6. As can be seen, some E numbers in the list are missing. For example, under E161a, E161c, and E161d are the natural xanthophyll pigments flavoxanthin, cryptoxanthin, and rubixanthin (also known as Natural Yellow 27), respectively (see Section 9.9.1.2). As food additives, they are not approved for use in the European Union or United States, but they are approved in Australia and New Zealand.

Table 11.12 List of current EU-approved natural and natural-identical colours and their E numbers.

E number	Name	E number	Name
E100	Curcumin	E160a	Carotenes
E101	Riboflavin and riboflavin 5′-phosphate	E160b	Annatto, bixin, and norbixin
E120	Cochineal, carminic acid, and carmines	E160c	Paprika extract, capsanthin, capsorubin
E140	Chlorophylls and chlorophyllins	E160d	Lycopene
E141	Copper complexes of chlorophyll, chlorophyllins	E160e	β-Apo-8′-carotenal
E150a	Plain caramel	E160f	Ethyl β-apo-8′-carotenoate
E150b	Caustic sulfite caramel	E161b	Lutein
E150c	Ammonia caramel	E161g	Canthaxanthin
E150d	Sulfite ammonia caramel	E162	Beetroot Red, betanin
E153	Vegetable carbon	E163	Anthocyanins

11.4.1.3 Synthetic dyes

11.4.1.3.1 Structure and nomenclature

Synthetic dyes generally have more intense colour than natural dyes, are more stable, and do not introduce any characteristic odours or tastes into coloured food. Therefore, they are widely used in food practice, mainly for practical and economic reasons. Food colours are contained in many foods, including snack foods, margarine, cheese, jams and jellies, desserts, drinks, and other products.

The following dye classes are recognised, according to structure:

- azo dyes (monoazo, bisazo-, trisazo-to polyazo dyes);

- diphenylmethane and triphenylmethane dyes;

- pyrazolone dyes;

- nitro dyes;

- xanthene dyes;

- anthraquinone dyes;

- quinoline dyes;

- indigo dyes.

Other dyes, such as acridine, diazonium, phthalocyanin, tetrazolium, and thiazole dyes, are not approved for foods. According to their physico-chemical properties, synthetic dyes may be classified into sour, alkaline, and neutral dyes. According to their solubility, they are divided into lyophylic dyes (soluble in water) and lipophilic dyes (soluble in fats).

The list of synthetic dyes approved in the European Union is given in Table 11.13. All synthetic dyes approved as food dyes are water-soluble compounds. The most represented are acidic dyes containing sulfonic, carboxyl, and hydroxyl groups. Most of them belong to the azo dyes; some are di- and triphenylmethane dyes, nitro dyes, or xanthene dyes. Basic dyes contain one or more free or substituted amino groups. These include the majority of the di- and triphenylmethane dyes and certain azo dyes. All dyes are used in the form of salts (usually sodium, potassium, or calcium salts).

Specific properties of dyes depend on the functional groups present. A characteristic is the presence of two types of functional groups, **chromophores** and **auxochromes**. Chromophore groups are related to the class of dye (e.g. azo dyes, nitro dyes) and are responsible for its behaviour in oxidation and reduction reactions. Auxochrome groups are responsible for the staining properties and behaviour in relation to acids, alkalis, light, and heat. The structures of synthetic dyes approved in the European Union are given in formulae **11–20**.

11.4.1.3.2 Properties and applications

In addition to toxicological criteria, it is required that synthetic dyes be chemically pure substances that do not influence any other organoleptic properties of food (with the exception of colour). They must be stable to changes of pH and on exposure to light. Generally, there is no one dye that is suitable for all applications and situations, so those used are usually made up of several components and represent a mixture of dyes.

The majority of synthetic dyes have sufficient stability, particularly in dry foods and foods protected from light. Their stability is also sufficient under normal conditions of production, processing, and storage. Azo dyes can relatively easily reduce metal ions and certain reducing agents (such as sulfur dioxide and ascorbic acid present in beverages), yielding colourless products. Triphenylmethane, indigo, and xanthene dyes are more stable, but due to UV radiation indigotine (also known as indigo carmine) and erythrosine may be decolourised.

Synthetic dyes are available:

- in the form of dispersions, pastes, aqueous or non-aqueous solutions (mainly in propylene glycol or glycerol), or in the solid state (such as water-soluble granules or powders);

- in the form of water-soluble lake pigments (commonly known as lakes).

Table 11.13 Synthetic EU-approved food dyes, their E numbers, and some attributes.

E number	Name	Other names[a]	Class	Colour
E102	Tartrazine	Acid yellow T, Hydrazine yellow, FD&C Yellow No. 5; CI Acid Yellow 23	Monoazo	Yellow
E104	Quinoline yellow	Food Yellow-13, D&C Yellow No. 10, CI Acid Yellow-3	Quinoline	Yellow
E110	Sunset Yellow FCF, Orange Yellow S	FD&C Yellow No. 6, CI Food Yellow 3	Monoazo	Orange
E122	Azorubine, Carmoisine		Monoazo	Blue-red
E123	Amaranth	FD&C Red No. 2, CI Acid Red 27, CI Food Red 9	Monoazo	Blue-red
E124	Ponceau 4R, Cochineal Red A		Monoazo	Red
E127	Erythrosine	FD&C Red No. 3	Xanthene	Red
E129	Allura Red AC	FD&C Red No. 40	Monoazo	Blue-red
E131	Patent Blue V		Monoazo	Red
E132	Indigotine (indigo carmine)[b]	FD&C Blue No. 2	Triphenylmethane	Green-blue
E133	Brilliant Blue FCF	FD&C Blue Dye No. 1, CI Acid blue 9, CI Food Blue 2, CI Pigment Blue 24	Indigo	Dark blue
E142	Green S	CI Acid Green 50, CI Food Green 4	Triphenylmethane	Green-blue
E151	Brilliant Black BN, Black PN	CI Food Black 1	Triphenylmethane	Green
E154	Brown FK		Bisazo	Black
	Brown HT		Monoazo, bisazo, and trisazo	Brown
E180	Litholrubine BK		Bisazo	Brown

[a]FD&C numbers indicate that the US FDA has approved the colorant for use in foods, drugs, and cosmetics. Colorants without FD&C numbers are banned in the United States. Colour additives FD&C Green No. 3 (for general use), Orange B (for casings or surfaces of frankfurters and sausages), and Citrus Red No. 2 (for skins of oranges not intended or used for processing) are approved in the United State but are not approved in the European Union. CI = Colour Index number.

[b]Indigotine occurs as a natural pigment (in the shrub *Indigofera tinctoria*), although commercially it is produced synthetically.

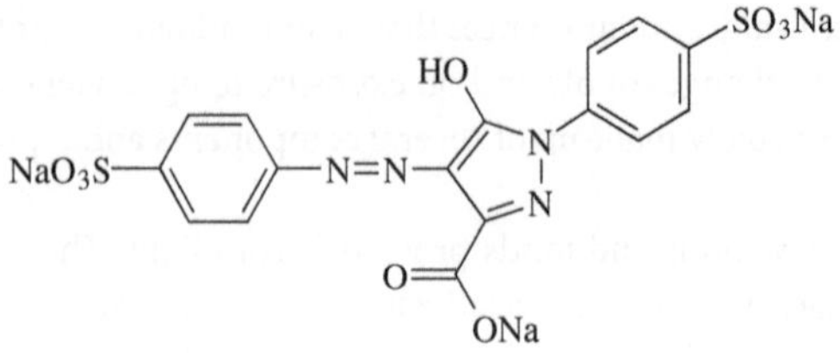

Tartrazine

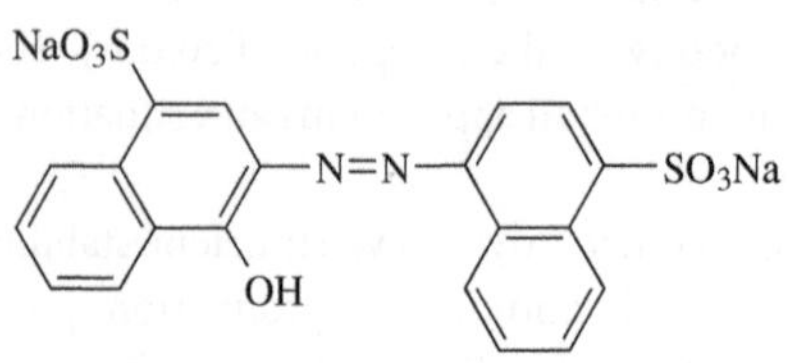

Azorubine

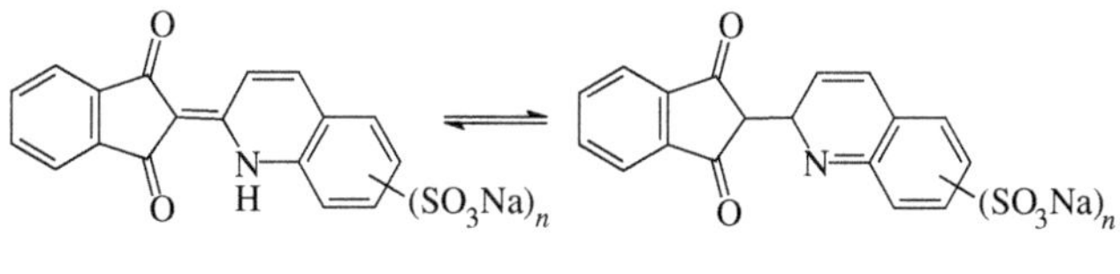

Quinoline Yellow (n = 1–3)

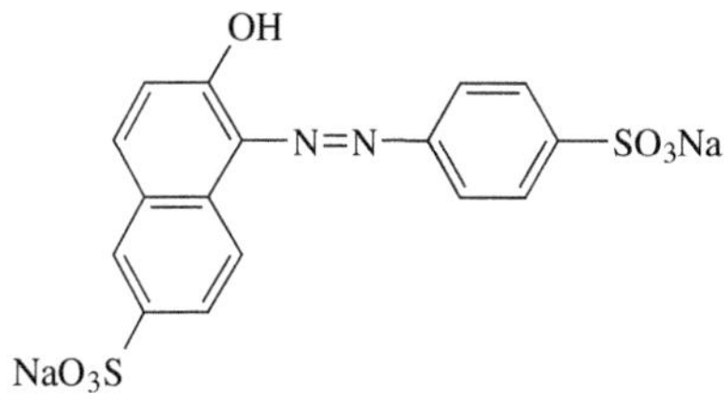

Sunset Yellow FCF

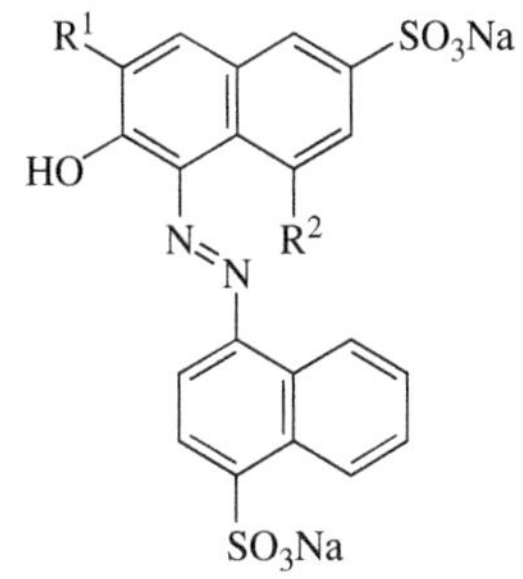

Amaranth, R^1 = SO$_3$Na, R^2 = H
Ponceau 4R, R^1 = H, R^2 = SO$_3$Na

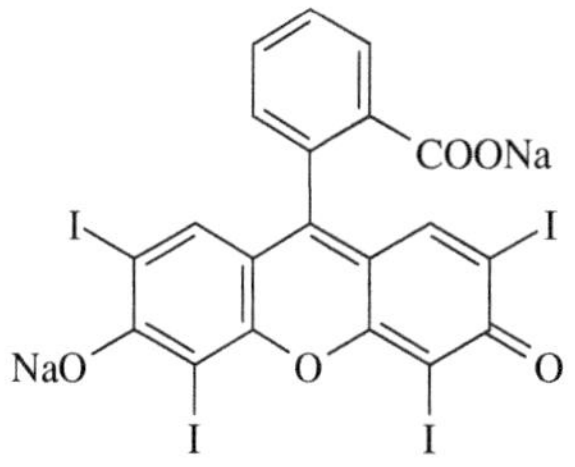

Erythrosine

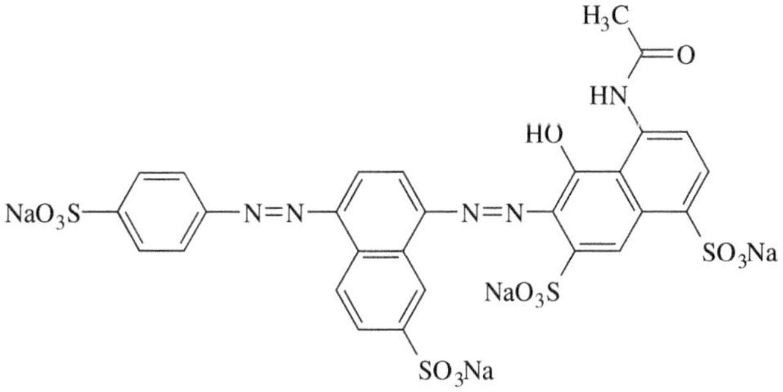

Green S

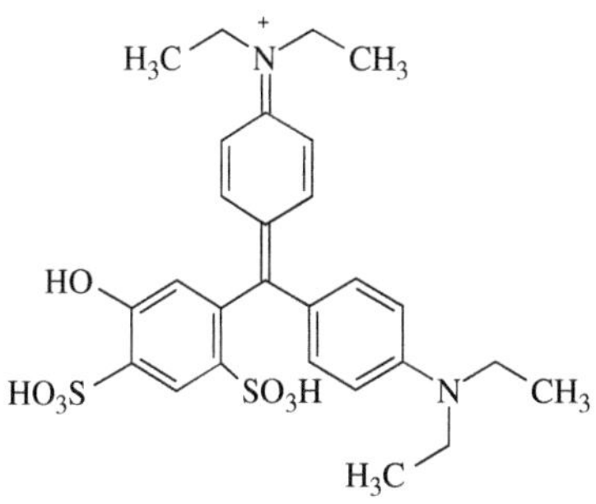

Patent Blue V

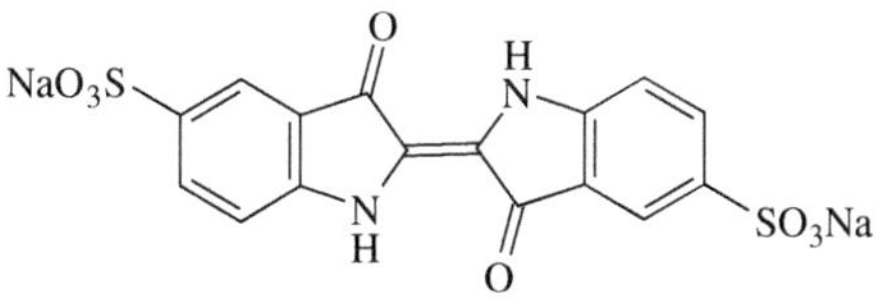

Indigotine

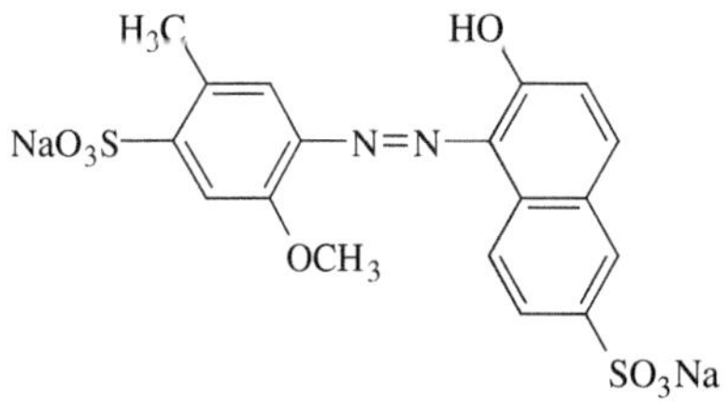

Allura Red AC

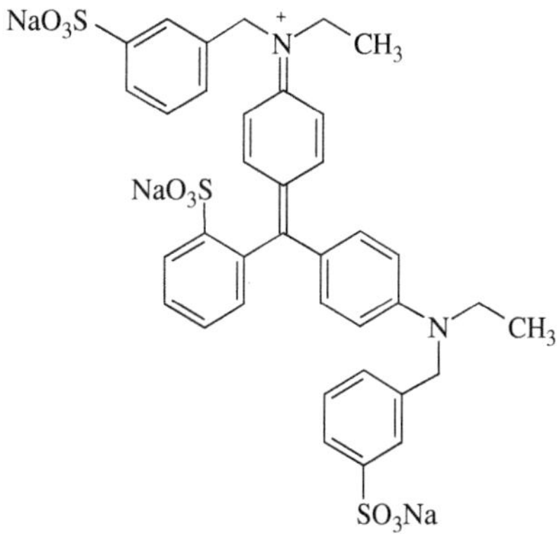

Brilliant Blue FCF

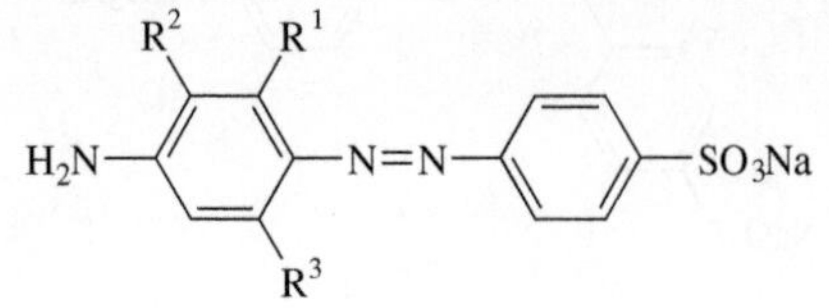

Brown FK (monoazo components,
R^1 = H or NH_2, R^2 = H or CH_3, R^3 = H or NH_2)

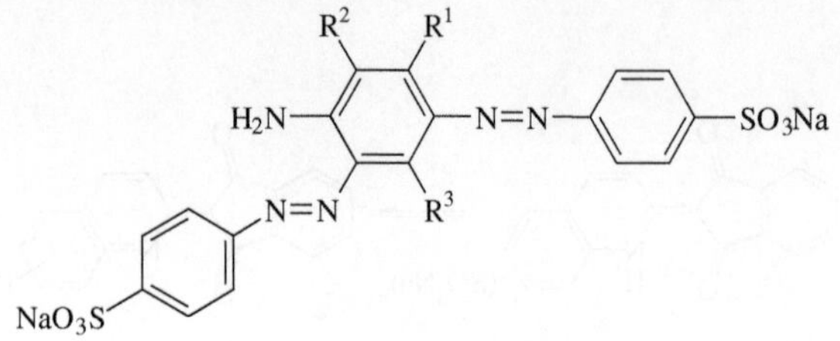

Brown FK (didiazo components,
(substituents: see monoazo components)

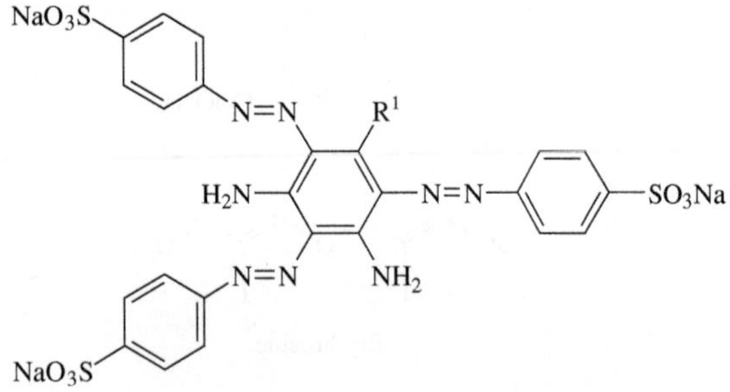

Brown FK (triazo components, substituents:
see monoazo components)

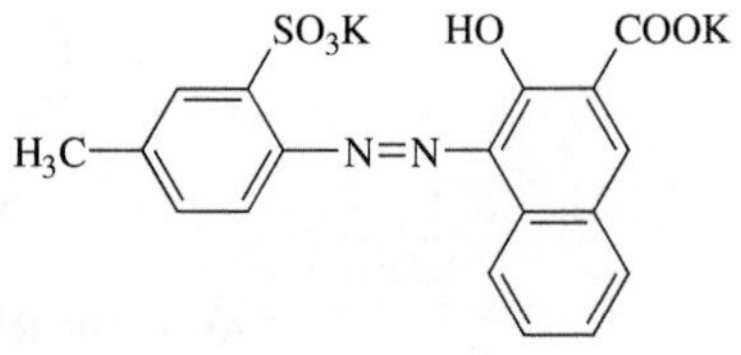

Litholrubine BK

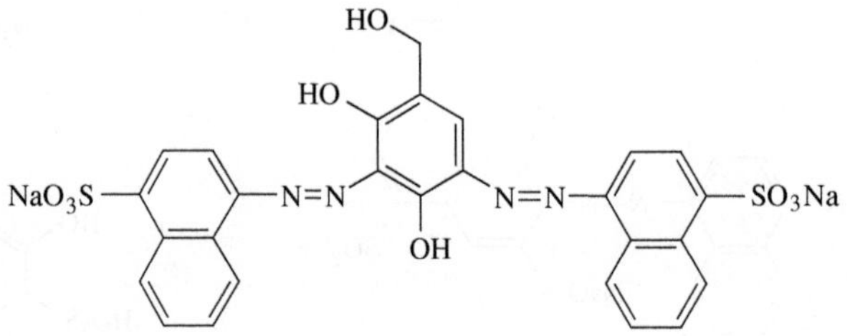

Brown HT

11-20, structures of water-soluble food dyes

Solid products are especially suitable for colouring beverages and emulsions and pastes for colouring confectionery and bakery products, whilst liquid paints are used for colouring dairy products. Lake pigments are obtained by adsorption of dyes on hydrated alumina. They have different dye contents, normally 10–40%. Lakes are not oil-soluble, but are oil-dispersible. Lakes that contain individual dyes or mixtures thereof are delivered in solid form or as dispersions in hydrogenated vegetable oils, propylene glycol, and glycerol or sugar syrup. These forms are more stable than dyes and are ideal for colouring products containing fats and oils or items lacking sufficient moisture to dissolve dyes.

Some dyes, such as the azo dyes amaranth and Sunset Yellow FCF, can react in slightly acidic media with bisulfites (sulfur dioxide) to produce coloured products (e.g. a yellow product arises from Sunset Yellow FCF), but some, such as tartrazine and azorubine, yield colourless products. The former group of azo dyes has hydroxyl groups in the *ortho*-position to the azo bond as well as unsubstituted *para*-positions in the naphthalene nucleus. Such dyes exist predominantly as hydrazone tautomers rather than strictly azo compounds, which facilitates the addition of bisulfite ion to the *para*-position (Figure 11.11). With azorubine, the addition of bisulfite ion at the *para*-position appears not

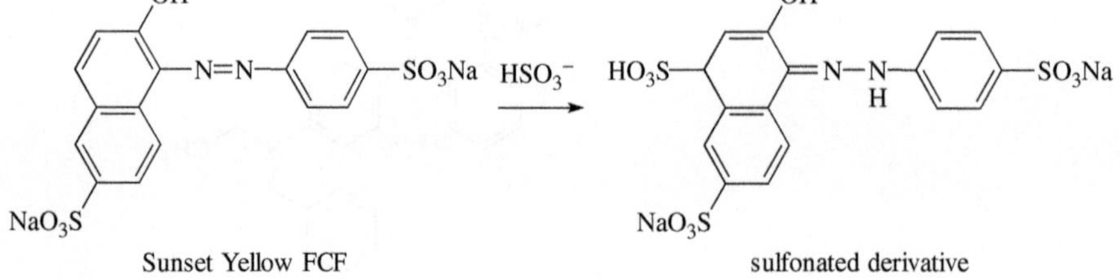

Figure 11.11 Reaction of bisulfite with Sunset Yellow FCF

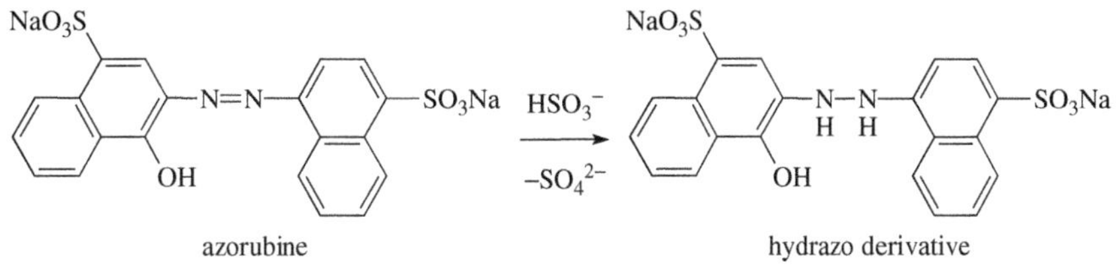

Figure 11.12 Reaction of bisulfite with azorubine.

to take place, probably as a result of charge delocalisation via the fused aromatic ring of the naphthalene system. It is probable that the dye reacts with bisulfite to form an unstable complex that hydrolyses to a colourless hydrazo product (Figure 11.12).

The reaction of ascorbic acid with azo dyes (frequently used in combination, for example in fruit drinks) leads to dye degradation. The stability of the most frequently used dyes (at pH 3.0 and 4.0) in the presence of ascorbic acid decreases as follows: tartrazine > Sunset Yellow FCF = amaranth, Ponceau 4R > azorubine > Brilliant Black BN.

11.4.1.4 Inorganic pigments

Inorganic compounds in food are only used in special cases. For example, for the surface treatment and decoration of dragees, candies, and similar products, the white pigments calcium carbonate (E170) and titanium dioxide (E171), red, yellow, and black iron oxides and hydroxides (E172), and aluminum (173), silver (E174), and gold (E175) pigments are used. Silver and gold pigments are similarly used for the decoration of specialty liqueurs.

11.4.1.5 Legislation

Flavour-active materials with simultaneous colouring effects (such as paprika, saffron, and turmeric) are not considered colorants, nor are colorants intended to colour the inedible external parts of foods (such as paraffin wax coatings of cheeses and sausage casings).

Some dyes may only be used for certain purposes, and to the maximum amount allowed. For example, amaranth is allowed for aperitif wines and spirits (30 mg/kg), erythrosine for cocktail and candied cherries (200 mg/kg), and Litholrubine BK for edible coatings of cheeses (in specified quantities). For meat and meat products, only a single dye may be used: Allura Red AC, Brilliant Blue FCF, or Brown HT or a mixture of Allura Red AC and Brilliant Blue FCF. For the purpose of food colourings used in the home, the pigments and dyes listed in Tables 11.12 and 11.13 are available, except for annatto, canthaxanthin, amaranth, erythrosine, Brown FK, Litholrubine BK, and aluminum pigment. The rules also specify foods that can be coloured only by certain dyes. For example, beer, vinegar, whisky, brandy, and rum-type alcoholic beverages may be coloured with caramel (in a specified quantity), butter may be coloured with carotene (10 mg/kg), and margarines may be coloured with carotene and curcumin (in specified quantities) or annatto (10 mg/kg). Foods that cannot be coloured are also identified, mostly including unprocessed foods such as milk, vegetable oils, animal fats, mineral and table water, egg contents, flour, bread, pasta products, sugar, meat, fish, wine, honey, fruit and vegetable juices, tomato paste and sauces, coffee, cocoa and chocolate products, and infant and child nutrition.

11.4.1.6 Health assessment

Food pigments and dyes should not represent any health risk if used up to the maximum amount allowed, as each substance authorised for use in the European Union is subject to a rigorous scientific safety assessment. Some dyes can, however, cause health problems. Side effects are known for azo dyes, such as tartrazine, Sunset Yellow FCF, and azorubine, in people intolerant to salicylates (aspirin and some fruits). In combination with benzoates, these dyes are implicated in a large percentage of cases of hyperactivity in children, as is carmine pigment. Asthmatics may also experience symptoms following consumption, as azo dyes are histamine-liberating agents. Erythrosine may cause hyperactivity and increased photosensitivity in people with sensitivity to sunlight, and possible mutagenicity has also been reported. In high concentrations, erythrosine may interfere with iodine metabolism, but such concentrations cannot be reached through the consumption of food. Patent Blue V and indigotine can function as histamine-liberating agents and may cause allergic reactions due to coupling of the colour to body proteins, as can Brilliant Blue FCF and Green S.

A promising solution is the use of natural pigments extracted from tissue cultures. Other encouraging prospects are new dyes such as high-molecular-weight pigments that are non-absorbable in the digestive tract, with chromophores fixed to the polymer so that they pass through the digestive tract in an unaltered state.

11.4.2 Bleaching agents

Bleaching agents include compounds that unwanted dyes:

- reduce;

- oxidise to colourless or less intensely coloured products.

11.4.2.1 Reducing agents

Substances with reducing effects are sulfur dioxide and sulfites, which are also used as preservatives, inhibitors of enzymatic browning reactions, and inhibitors of non-enzymatic browning reactions. The bleaching activity is mainly based on their ability to reduce the primary products of enzymatic browning reactions: quinones, whose subsequent reactions would otherwise lead to undesirable discoloration in dried fruits, vegetables, potatoes, and other products. Furthermore, these compounds are used for the bleaching of hops, lecithin concentrates, mushrooms, nuts, and fish products.

11.4.2.2 Oxidising agents

Substances with oxidising effects (not allowed under EU legislation) include compounds with active oxygen and compounds with active chlorine.

11.4.2.2.1 Compounds with active oxygen

Halides and peroxides have some importance as bleaching agents with active oxygen. The most widely used agent for bleaching flour is potassium bromate ($KBrO_3$), which has been in use for about 100 years. Potassium bromate bleaches carotenoid pigments in flour and simultaneously oxidises gluten and glutathione, thereby improving the flour's baking properties (see Section 5.14.6.1). It is reduced to bromide. The problem is its toxicity and carcinogenicity, and it has never been allowed as a flour improver in a number of countries, but it is commonly used in much of Latin America and East Asia. The suggested quantity of potassium bromate for improving bread flour is 50 mg/kg. Certain other compounds have also been used for the same purpose, such as potassium and calcium iodates (IO_3^-), cupric sulfate, ammonium and potassium peroxodisulfates (also known as persulfates, $S_2O_8^{2-}$), hydrogen peroxide, and ozone.

Organic compounds used as bleaching agents and dough conditioners include dibenzoyl peroxide (**11–21**) and azodicarbonamide (azoformamide). Azodicarbonamide-releasing nitrogen gas was previously used as a blowing agent in rubber and plastic products that were permitted in food packaging applications. In certain countries, azodicarbonamide is approved as a food additive in amounts up to a maximum of 45 mg/kg in flour, such as the United States, Canada, and much of Asia, but is banned in Singapore, Australia, and Europe. In Singapore, it is considered so dangerous that using it can get you 15 years in prison and a hefty fine. Azodicarbonamide is stable in dry flour, but it reacts in moist flour and yields biurea as the main reaction product, which is relatively stable during baking but partly decomposes to urazole, semicarbazide (an IARC Group 3 substance not classifiable as to its carcinogenicity to humans), and unstable carbamic acid (Figure 11.13), which yields ethyl carbamate on reaction with the ethanol produced by yeast. Semicarbazide was not detected after dough maturation at room temperature or elevated temperature, but only in bread. Commercial breads showed a wide range of semicarbazide concentrations (10–1200 µg/kg).

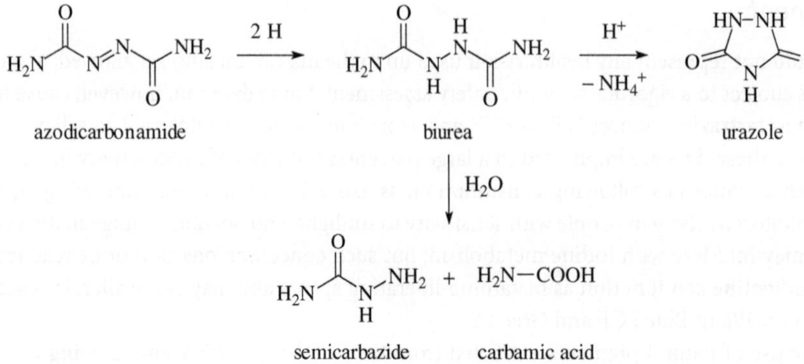

Figure 11.13 Decomposition of azodicarbonamide.

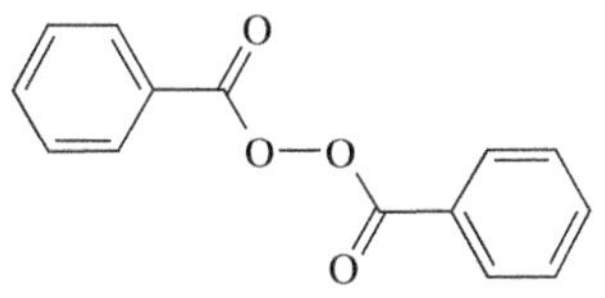

11-21, dibenzoylperoxide

11.4.2.2.2 Compounds with active chlorine

Gaseous chlorine, chlorine dioxide (ClO_2), and sodium and potassium hypochlorites (OCl^-) are used as bleaching agents and improvers of flour baking properties, and also for chemical disinfection of water. The most frequently used agent is chlorine dioxide, in amounts up to about 30 mg/kg flour in some countries.

The use of compounds with active chlorine is problematic from the hygienic and toxicological perspectives, because their reactions with food components may produce a number of potentially toxic chlorinated products. Chlorine dissolved in water reacts to form hypochlorous acid and hydrochloric acid (at pH 4–6 and higher):

$$Cl_2 + 2\,H_2O \rightleftharpoons HOCl + H_3O^+ + Cl^-$$

Under acidic conditions, the reaction equilibrium is shifted away from the formation of hypochlorous acid, which explains the formation of hypochlorite and chlorine, and also the chlorinating action of chloramines, which are used as disinfectants and sanitising agents in the food industry. In acidic media, hypochloric acid cation and chloride arise in the reaction with hydronium ions:

$$HOCl + H_3O^+ \rightleftharpoons H_2OCl^+ + H_2O \rightleftharpoons Cl^+ + 2\,H_2O$$

Hypochlorous acid and its anion also arise in alkaline media (pH 10.5–12.5):

$$Cl_2 + HO^- \rightleftharpoons HOCl + Cl^-$$

$$HOCl + HO^- \rightleftharpoons ClO^- + H_2O$$

In addition, irreversible reactions also proceed, in which oxygen is formed as the reaction product, which explains the oxidative properties of chlorine, hypochlorites, and chloramines:

$$2\,Cl_2 + 2\,HO^- + 2\,H_2O \rightarrow 4\,Cl^- + 2\,H_3O + O_2$$
$$\text{or } 2\,ClO^- \rightarrow 2\,Cl^- + O_2$$
$$2\,Cl^+ + 4\,HO^- \rightarrow 2\,Cl^- + 2\,H_2O + O_2$$

In the reaction of chlorine dioxide with water, hypochlorous, hydrochloric, and chloric acids are formed temporarily, and in alkaline solutions, chlorites (ClO_2^-), chlorates (ClO_3^-), and other products arise. The cation H_2OCl^+, formed in aqueous solutions of chlorine,

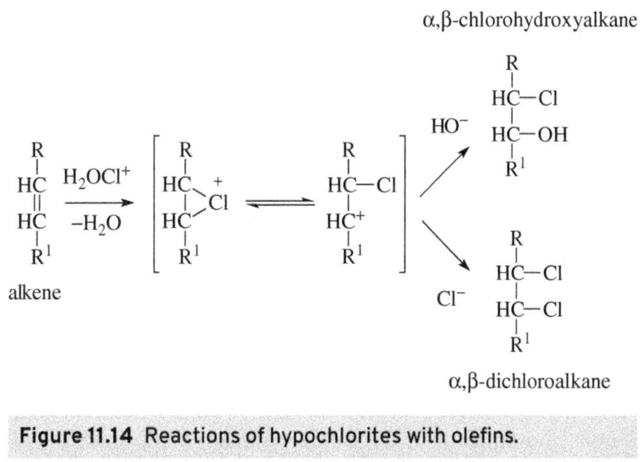

Figure 11.14 Reactions of hypochlorites with olefins.

Table 11.14 Chlorinated fatty acids in flour bleached with chlorine.

Chlorine in flour	Chlorinated fatty acids (mg/kg of flour)					
(mg/kg)	$C_{18:1}$(Cl,Cl)	C18:0(Cl,Cl)	$C_{18:0}$(Cl,Cl,Cl,Cl)	$C_{18:1}$(OH,Cl)	$C_{18:0}$(OH,Cl)	$C_{18:0}$(OH,Cl,Cl,Cl)
0	0	0	0	2	<1	0
100	7	0	0	37	6	0
500	130	12	2	170	28	22
1000	330	75	39	160	65	130
2000	100	160	210	26	71	540
3000	50	160	240	41	160	1010

chlorine dioxide, and hypochlorites, may react with alkenes and other unsaturated compounds. The electrophilic addition of HOCl to alkenes is an established reaction mechanism for α,β-chloroalcohol (chlorohydrin) and α,β-dichloro derivative formation (Figure 11.14). The reaction yielding chlorohydrins follows the Markovnikov rule, with the hydroxyl group adding to the more substituted carbon. Oxidation of chlorohydrins by hypochlorites in acidic media may yield α-oxochloroalkanes, whilst epoxides arise in neutral (or slightly alkaline) media by elimination of hydrogen chloride and are hydrolysed by acids and bases to the corresponding α, β-dihydroxyalkanes.

The reaction of chlorine in flour with unsaturated fatty acids, which are also alkenes, yields a number of fatty acid chlorinated derivatives (dichloroacids and chlorohydroxyacids). Chlorohydroxyacids are formed preferentially with chlorine dioxide. For example, oleic acid produces 9,10-dichlorostearic, 9-chloro-10-hydroxystearic, and 10-chloro-9-hydroxystearic acids. From linoleic acid, the corresponding disubstituted derivatives of oleic acid or tetrasubstituted derivatives of stearic acid are formed. Their content in flour, depending on the amount of chlorine used for bleaching, is given in Table 11.14. Chlorinated fatty acids are also formed in the fat of chicken carcasses that are cooled in chlorinated water after slaughter.

The use of nitrogen trichloride, also known as trichloroamine (NCl_3), to bleach flour was found to evoke hysteria in experimental animals (dogs) due to the presence of methionine sulfoximine (**11–22**). Methionine sulfoximine, a methionine antagonist, arises in the reaction of methionine with nitrogen trichloride.

Unsaturated terpenes also react with electrophilic reagents, such as hypochlorous acid (HOCl). Reactions of HOCl with a variety of terpenes, including limonene monoxide, α-pinene, and α-terpineol, have been reported. For example, limonene, the major component of orange essential oil, reacts with the hypochlorous acid found in chlorinated water to form the diequatorially substituted (1R,2R,4R)-2-chloro-p-menth-8-en-1-ol, the diaxial (E)-stereoisomer, (1S,2S,4R)-2-chloro-p-menth-8-en-1-ol as the major chlorohydrin, and the dichlorohydrin, (1R,2R,4R)-2,9-dichloro-p-menth-8-en-1-ol (**11–23**).

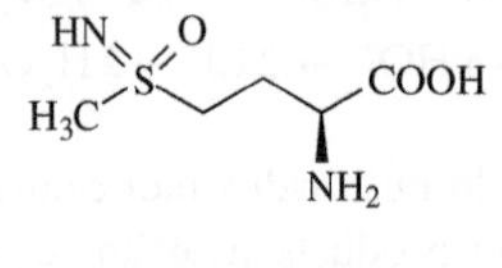

11-22, methionine sulfoximine

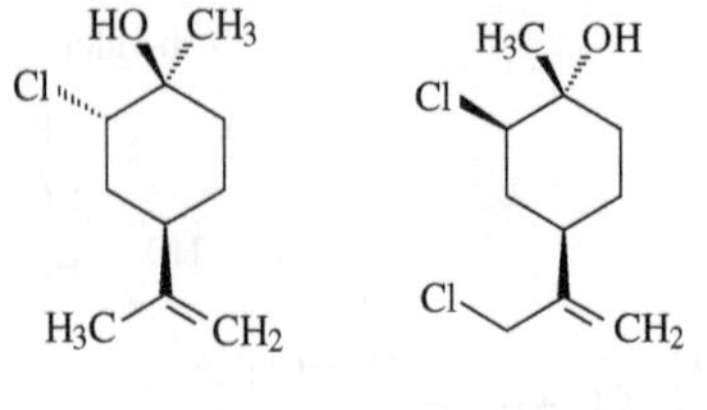

(1S,2S,4R)-2-chloro-
p-menth-8-en-1-ol

(1R,2R,4R)-2,9-dichloro-
p-menth-8-en-1-ol

11-23, products of limonene and hypochlorous acid

One of the major problems associated with the disinfection of water supplies by chlorination is that the chlorinated water may produce the so-called chlorophenolic taste, caused by a reaction between the added chlorine and phenol and some of its homologues, which are present in trace amounts. The chlorination of phenol proceeds by the stepwise substitution of the 2-, 4-, and 6-positions of the aromatic ring. Initially, phenol is chlorinated to form either 2- or 4-chlorophenol. Then, 2-chlorophenol is chlorinated to form either 2,4- or 2,6-dichlorophenol, whilst 4-chlorophenol forms 2,4-dichlorophenol. Both 2,4- and 2,6-dichlorophenol are chlorinated to form 2,4,6-trichlorophenol, which reacts with aqueous chlorine to form a mixture of non-phenolic oxidation products. 2,4,6-Trichloroanisole was the first compound identified as the source of taints in wines, perceived as musty or mouldy flavour. 2,4,6-Trichloroanisole in wine originates from chlorination of lignin-related substances during bleaching of the cork with chlorine, the product(s) subsequently being leached into the wine during storage. This taint is known as corkiness and is widely attributed to the interaction of bacteria and fungi with constituents of the cork that methylate 2,4,6-trichlorophenol. 2,4,6-Trichloroanisole can be similarly formed during microbial contamination of packaging materials and in soil contaminated with phenols. 2,4,6-Trichlorophenol is also an industrial agent used to decontaminate wooden objects, including the floors, beams, and barrels in wine cellars. In addition, 2,4,6-trichloroanisole is produced together with 2,4,6-trichlorophenol by degradation of some pesticides (such as fungicide pentachlorophenol). Pentachlorophenol and the lower chlorinated phenols, tetra- and trichlorophenols, have been used increasingly as fungicides, herbicides, and insecticides and in the synthesis of other pesticides (**12-48**).

11.5 Substances modifying texture

Substances modifying and regulating food texture and other physical properties of foods are the main additive substances used in terms of quantity. The most important groups are:

- thickeners and gelling agents;

- emulsifiers.

Many of these substances can be simultaneously classified into several categories of food additives, because they have a number of different properties and are used for a number of different purposes.

11.5.1 Thickeners and gelling agents

The reason for the use of thickeners and gelling agents is to create and maintain a desirable texture in food. Thickeners are substances that increase the viscosity of food and gelling agents produce gels. These additives include natural plant polysaccharides, such as starch (not considered a food additive), cellulose (E460) and pectins (E440), gum arabic (E414), seaweed polysaccharides agar (E406) and carrageenan (E407), extracellular bacterial polysaccharides, gellan (E418) and xanthan gums (E415), modified starches (E1404, E1410, E412–E1414, E1420, E1422, E1440–1442), and modified celluloses (E461–E469). Their structures, occurrence, properties and applications are described in Chapter 4, dealing with polysaccharides. Gels and other dispersion systems are described in Section 7.8.3.2.2.

11.5.1.1 Legislation

Thickeners and stabilisers of dispersions and emulsions that can be used to the maximum amount allowed include karaya gum (E416), konjak gum (E425), sucrose acetate isobutyrate (E444), and glycerol esters of wood rosins (E445). For example, karaya gum (E416) is allowed at levels of 5000 mg/kg for chewing gums, confectionery fillings, toppings for pastries, and biscuits, at 6000 mg/kg for desserts, and at up to 10 000 mg/kg for cold emulsified sauces and egg liqueurs.

11.5.1.2 Health assessment

Some thickeners and gelling agents are considered as foods (e.g. starches), and for many others ADI values are not specified, such as carrageenan (E407), locust bean gum (E410), guar gum (E412), tragacanth (E413), powdered and microcrystalline cellulose (E460), and gum arabic (E414). ADI values are set for karaya gum (20 mg/kg b.w.), tara gum (E417, 12.5 mg/kg b.w.), methyl, ethyl, hydroxymethyl, hydroxypropyl, and sodium carboxymethyl cellulose (25 mg/kg b.w.), and dextrins (E1400, 70 mg/kg b.w.). The role of thickeners and gelling agents in nutrition is described elsewhere (see Section 7.8.3.2.2). No side effects of natural polysaccharides, modified starches, and celluloses are known in the concentrations used, although high concentrations may bring about flatulence and bloating due to fermentation by intestinal microflora and high concentrations of cellulose (E460), and modified celluloses can cause intestinal problems, such as bloating, constipation, and diarrhoea. Tara gum (E417) and gellan gum (E418) may have laxative properties. Short-chain carrageenans can cause intestinal leakage and are not permitted for use in foods. High concentrations of alginic acid and alginates (E400) could lead to impairment of iron uptake, as iron is efficiently bound by these polysaccharides.

11.5.2 Emulsifiers

Emulsifiers (also known as emulgents) are surfactants enabling the formation of emulsions (especially dispersions of fat in various products). In addition to their ability to form an emulsion, they also have the ability to interact with other food ingredients. An emulsifier may be an aerating agent, starch complexing agent, or crystallisation inhibitor. In flours, they act as conditioners, softening the crust of the pastry, and in confectionery they act as modifiers of crystallisation of fats and have other beneficial properties. Formation of emulsions, their properties, and the factors affecting their formation are described in Section 7.8.3.3.

11.5.2.1 Classification

Food emulsifiers are classified according to several criteria. Any emulsifier consists of a hydrophilic head derived from a variety of polar compounds (e.g. glycols and sugar alcohols) and a hydrophobic tail, which is a residue of a fatty acid. The hydrophilic head is directed to the aqueous phase and the hydrophobic tail to the oil. According to the origin, emulsifiers are identified as:

- natural (such as lecithin and partial esters of glycerol);

- synthetic (other emulsifiers).

According to the structure of the polar part of the molecule (hydrophilic head), the following groups of emulsifiers are recognised:

- esters of glycols (e.g. esters of propane-1,2-diol);

- glycerol esters and their derivatives (e.g. partial esters of glycerol);

- esters of sorbitans (esters of glucitol dehydration products);

- sucrose esters (partial esters);

- esters of hydroxycarboxylic acids (such as lactic and tartaric acids);

- lecithin and its derivatives.

According to the properties of hydrophilic and lipophilic moieties (expressed by the so-called hydrophilic lipophilic balance (HLB) value), emulsifiers are recognised as:

- hydrophilic;

- lipophilic.

According to their ability to form or not to form ions, emulsifiers are classified into:

- ionogenic, also known as ionic (the hydrophilic moiety can be an anion or a cation, or it may have an amphoteric character);

- non-ionogenic, also known as non-ionic (the hydrophilic part of the molecule is not ionised).

Salts of fatty acids are ionogenic emulsifiers, and lecithins have an amphoteric character. The most common food emulsifiers (e.g. fatty acid esters of glycerol, sorbitans, sucrose, and hydroxycarboxylic acids) are non-ionogenic emulsifiers.

The applicability of food non-ionogenic emulsifiers greatly depends on their HLB value, which is a measure of the degree to which they are hydrophilic or lipophilic. Substances with a low HLB value tend to dissolve in oil (o), but substances with a high HLB value dissolve better in water (w). Numerical HLB values range from 0.0 to 20.0, with a value of 0 corresponding to lipophilic substances and a value of 20 to hydrophilic substances (see Table 11.15).

11.5.2.1.1 Lecithin and its derivatives

The term 'lecithin' is used for (3-*sn*-phosphatidyl)choline (1,2-diacyl-*sn*-glycero-3-phosphocholine, or phosphatidylcholine for short) and the mixture of natural phospholipids used as emulsifiers (see Section 3.5.1). Crude food-grade lecithins (E322) are mainly obtained from

Table 11.15 HLB values and E numbers of selected emulsifiers.

Substance	HLB	Substance	HLB
Oleic acid	1.0	Polyoxyethylene sorbitan tristearate (Tween 65, E436)	10.5
Sorbitan tristearate (Span 65, E492)	2.1	Polyoxyethylene sorbitan monostearate (Tween 60, E435)	14.9
Monostearoylglycerol (mixture of isomers)	3.4	Polyoxyethylene sorbitan monooleate (Tween 80, E433)	15.0
Sorbitan monooleate (Span 80, E494)	4.3	Polyoxyethylene sorbitan monopalmitate (Tween 40, E434)	15.6
Sorbitan monostearate (Span 60, E491)	4,7	Polyoxyethylene sorbitan monolaurate (Tween 20, E432)	16.7
Sorbitan monopalmitate (Span 40, E495)	6.7	Sodium oleate	18.0
Sorbitan monolaurate (Span 20, E493)	8.3	Potassium oleate	20.0

the processing (degumming, also known as hydration) of vegetable oils. Almost all commercially available lecithin is derived from soybean oil, which contains 1–3% of lecithin. Less important sources include maize and safflower oils and eggs. The main component of lecithin is phosphatidylcholine, where the hydrophilic part of the molecule $X = (CH_2)_2N^+(CH_3)_3$. Commercial lecithin may contain up to 35% triacylglycerols.

Commercial lecithin is an important product because of its dietary significance and multifaceted industrial applications. Lecithin products fall into three categories: natural lecithin (such as unbleached and bleached lecithin), refined lecithin, and chemically modified lecithin. Bleaching of dark-brown crude lecithin is carried out using hydrogen peroxide or benzoyl peroxide to obtain a product of lighter colour, but it may result in oxidation of unsaturated fatty acids. Chemically modified lecithin is more hydrophilic. For example, hydroxylated lecithin is prepared by treating lecithin with hydrogen peroxide and acetic or lactic acid. Two of the earliest applications of lecithin were in lowering the viscosity in chocolate and confectionery products and providing emulsification/antispatter properties in margarine. Other uses are in the production of bakery goods, pasta products, textiles, insecticides, and paints.

11.5.2.1.2 Fatty acid salts

Salts of fatty acids (sodium, potassium, and calcium salts, E470a, and magnesium salts, E470b) are produced mainly from vegetable oils, but can also be produced from animal fats. The acids are mixtures of oleic, palmitic, stearic, and myristic acids. Salts of fatty acids are used as emulsifiers and as anticaking agents in powdered foods to prevent clumping.

11.5.2.1.3 Esters of glycols

Fatty acid esters of propane-1,2-diol (E477), also known as propylene glycol, and fatty acid esters of polyethylene glycol 8000 (11-24, E1521; the HLB value depends on the degree of ethoxylation) are obtained by direct esterification or by enzymatic reactions. These substances are used for oil-in-water (o/w) emulsions.

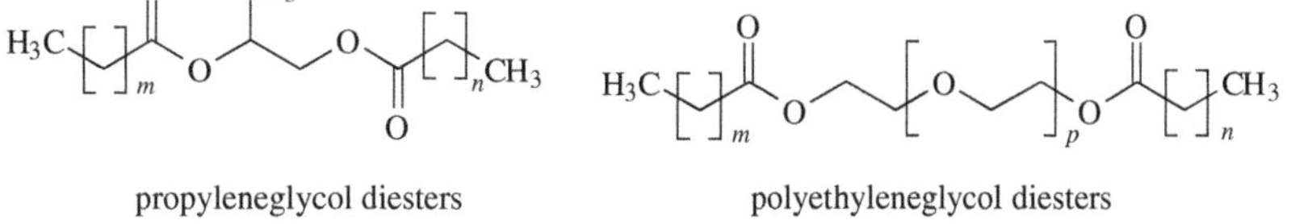

propyleneglycol diesters   polyethyleneglycol diesters

11-24, esters of glycols

11.5.2.1.4 Esters of glycerol and their derivatives

Monoacylglycerols and diacylglycerols Partial esters of glycerol, mono- and diacylglycerols (E471), are obtained by glycerolysis of fats (usually hydrogenated oils) or by direct esterification of glycerol. The predominant components of monoacylglycerols are 1-acyl-*sn*-glycerols, whilst diacylglycerols are dominated by 1,3-diacyl-*sn*-glycerols. Monoacylglycerols obtained by molecular distillation (containing about 90% of monoacylglycerols) are suitable for most purposes.

Esters of monoacylglycerols and diacylglycerols The emulsifying efficiency of partial glycerol esters is higher when these esters are esterified with carboxylic acids, such as acetic (E472a), lactic (E472b), citric (E472c), fumaric, succinic, tartaric (E472d), monoacetyl, and diacetyl tartaric acids (E472e), or mixtures thereof, such as mixed acetic and tartaric acid esters of mono- and diacylglycerols (E472f). For example, the reaction of 1-acyl-*sn*-glycerols with lactic acid yields a mixture of products containing 1-acyl-3-lactate (**11–25**), its isomer 1-acyl-2-lactate, and the fully esterified product 1-acyl-2,3-dilactate. Corresponding products are produced with diacetyl tartaric acid (**11–25**). Analogous to these derivatives containing one fatty acid in the molecule are esters of glycerol with two fatty acids and one hydroxycarboxylic acid, produced for special purposes.

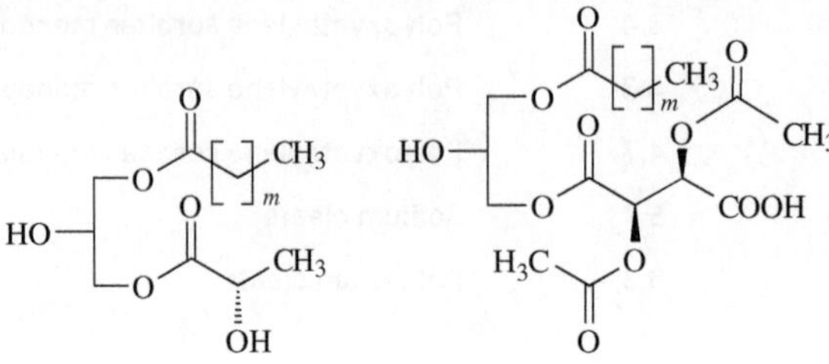

(*S*)-1-acyl-3-lactoyl-   (2*R*,3*R*)-1-acyl-3-(2,3-diacetyl-
sn-glycerol   tartaroyl)-*sn*-glycerol

11-25, esters of glycerol with hydroxycarboxylic acids

Esters of mono- and diacylglycerols with acetic acid (E472a) are used as substances preventing crystallisation of fats, esters of succinic acid have found use as flour conditioners (enhancers), and esters of citric acid are used as emulsifiers, solvents for antioxidants, and fat substitutes in some foods. They also enhance the baking properties of flour.

Ethers of monoacylglycerols and diacylglycerols   Another way of modifying the properties of partial esters of glycerol is by their reaction with ethylene oxide in alkaline media (ethoxylation), which yields products with different lengths of side chain (containing up to 40 residues of ethylene oxide), sometimes branched and therefore of different polarities. The structure of 1-monoacyl-*sn*-glycerol ethers is represented by formula **11–26**. Instead of ethylene oxide, propylene oxide may similarly be used, which gives rise to analogous products. Mixed copolymers may also be obtained. Ethoxylated monoacylglycerols have good dough-strengthening characteristics but very little crumb-softening. Dosage of these viscous liquids is rather critical as excess amounts cause excessive expansion, which can lead to collapse of the bread when in the oven. Ethoxylated monoacylglycerols are usually combined with monoacylglycerols, but they are not approved for use in the European Union.

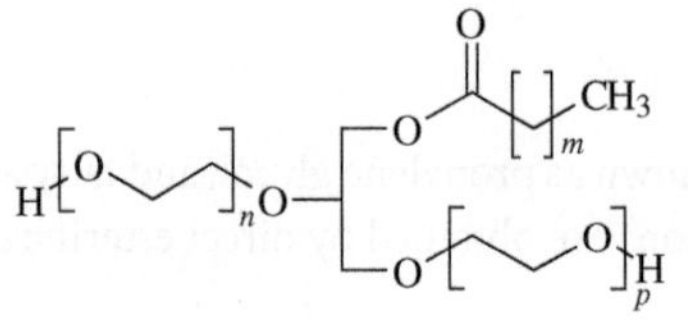

11-26, polyethyleneglycol ether
of 1-monoacyl-*sn*-glycerol

Esters of polyglycerols   Fatty acids can also be esterified by polyglycerols (E475), which are formed from glycerol in alkaline media via 2,3-epoxypropan-1-ol (glycidol, Figure 11.15). Glycidol can isomerise to 3-hydroxypropanal or 1-hydroxypropanone, which give various coloured products in the Maillard reaction. Glycidol additionally condenses with another molecule of glycerol to 1,1-diglycerol (4-oxa-heptane-1,2,6,7-tetrol), and to a lesser extent to 1,2-diglycerol (3-oxa-2-hydroxymethylhexan-1,5,6-triol) or, via epoxide, to triacylglycerols (such as 4,8-dioxaundecan-1,2,6,10,11-pentol) and higher oligomers. The dominant products under alkaline conditions are diglycerols, but also common are products that contain three to six or more glycerol molecules. In addition to linear products, branched products are formed to a lesser extent (the reactivity of secondary hydroxyl groups is lower). In acidic media, various cyclic products are produced, the main one being 2,5-bis(hydroxymethyl)-1,4-dioxane, formed as a product of diglycerol dehydration. The same products result from glycerol in acid protein hydrolysates.

Phosphatidic acids   Monoacylglycerols and diacylglycerols can also be esterified using phosphorus pentoxide. For example, esterification of 1,2-diacyl-*sn*-glycerol yields phosphatidic acid (1,2-diacyl-*sn*-glycerol 3-phosphate) and the corresponding bisphosphatidic acid and their positional isomers. Both acids are natural-identical products, which serve as substitutes for lecithin. Ammonium phosphatides (E442) are approved in the European Union as emulsifiers.

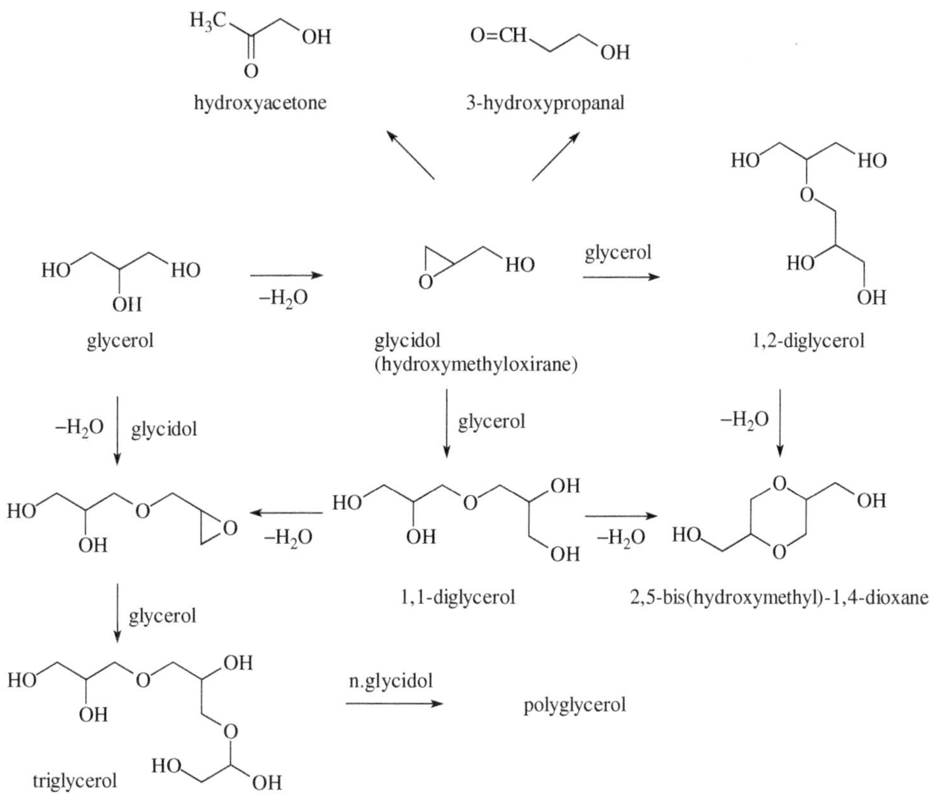

Figure 11.15 Reactions of glycerol with glycidol.

11.5.2.1.5 Derivatives of sorbitans

Sorbitan esters An important group of non-ionogenic emulsifiers is esters of sorbitol (correctly D-glucitol) with higher fatty acids. During esterification, sorbitol is simultaneously dehydrated, creating a mixture of free anhydro derivatives called sorbitans and their monoesters, diesters, and triesters. The most important products are esters derived from 1,4-sorbitan, 1,5-sorbitan, and 2,5-sorbitan (**11–27**). The reaction can be continued to yield a dianhydride ester, which is called isosorbide (**11–27**). The resulting mixtures of various mono-, di-, and triesters of optically active anhydrides (e.g. the corresponding 1,4-sorbitan and the isosorbide structures) are known as **Spans**. Their trade names, E numbers, and HLB values are listed in Table 11.15. Spans act as lipophilic emulsifiers with properties similar to those of monoacylglycerols (sorbitan monoesters) and diacylglycerols (sorbitan triesters) but they have a higher emulsifying ability.

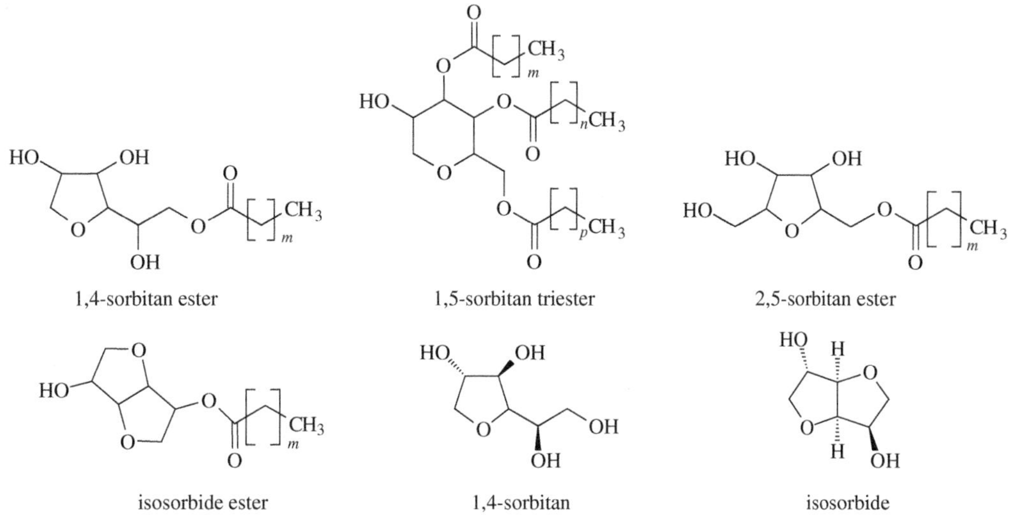

11-27, structures of sorbitan esters

Acylsorbitan ethers Partial esters of sorbitans with fatty acids can be modified by reaction with ethylene oxide (under pressure, at elevated temperatures, and in the presence of sodium ethoxide or other catalysts), which yields a mixture of products substituted to varying degrees in free hydroxyl groups, the side chains of which may be branched. These polar emulsifiers are also known as polysorbates or **Tweens** (**11–28**). They are very effective in slowing the ageing of pastry products. Their trade names, E numbers, and HLB values are listed in Table 11.15.

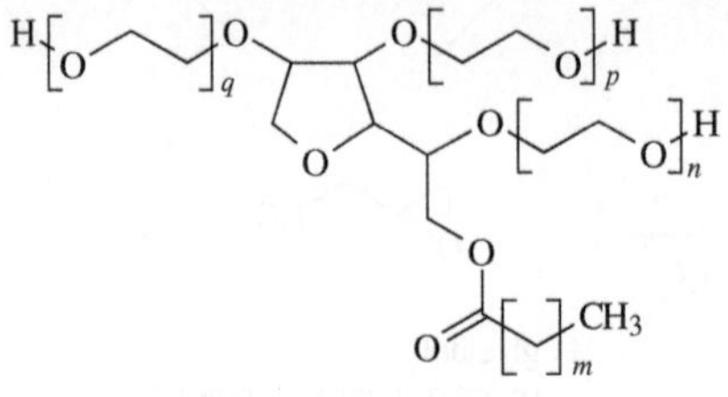

11-28, Tween derived from 1,4-sorbitan

11.5.2.1.6 Sucrose derivatives

Esters A special group of emulsifiers is sucrose esters of higher fatty acids (E473), which are synthesised from fatty acid methyl esters re-esterified by sucrose (in dimethylformamide or dimethyl sulfoxide under catalysis of alkaline reagents). The most common compounds are strongly polar, water-soluble monoesters (**11–29**) with HLB values >16, which are used as stabilisers of emulsions, and diesters, which have a wide range of HLB values (7–13) and are used to stabilise emulsions of the type o/w. Adjusting the reaction conditions can also provide triesters (HLB < 1) and tetraesters. Like sucrose, sucrose esters are easily decomposed on heating to give coloured products.

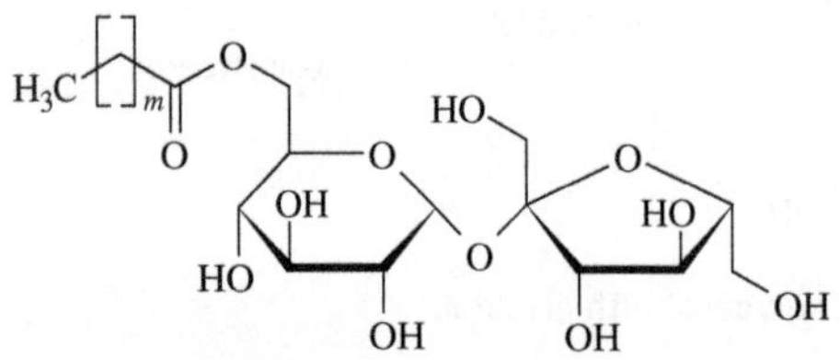

11-29, sucrose-6-ester

Sucrose esters containing more than five fatty acid residues do not have emulsifying ability, but can be used as low-energy fat substitutes, because they are not absorbed in the digestive tract. Such a product is Olestra, also known by its brand name Olean, which was approved by the US FDA as a food additive in 1996. In the late 1990s, Olestra lost its popularity due to the discovery of side effects (including inhibition of absorption of some fat-soluble vitamins), and it is not approved for sale in many countries.

Sugarglycerides Esterification of glycerol esters by sucrose yields the so-called sugarglycerides (E474), which represent a mixture of various sucrose esters and glycerol esters. Their polarity roughly corresponds to the polarity of monoacylglycerols.

11.5.2.1.7 Hydroxycarboxylic acid esters

Fatty acids may also be directly esterified by hydroxycarboxylic acids, the most common of which are lactic acid and (2S,3S)-tartaric (D-tartaric) acid. Lactic acid gives rise to esters called lactylates: first a monoester (**11–30**), which reacts with another molecule of lactic acid to yield the ester of a dimeric acid (**11–30**); this latter ester may also arise by reaction of a fatty acid with lactides. Other reactions can produce emulsifiers in which one molecule of a fatty acid accounts for a greater number of molecules of lactic acid (**11–30**). Stearoyl tartrate, also known as stearoylpalmitoyl tartrate (E483), is approved in the European Union as an emulsifier. The main components of this product are distearoyl tartrate, dipalmitoyl tartrate, and stearoylpalmitoyl tartrate. Esters of lactic acid (lactylates) can be converted into salts and thereby become highly effective polar, anionic emulsifiers. Sodium (E481) and calcium (E482) stearoyl 2-lactylates are approved in the European Union as emulsifiers and flour improvers.

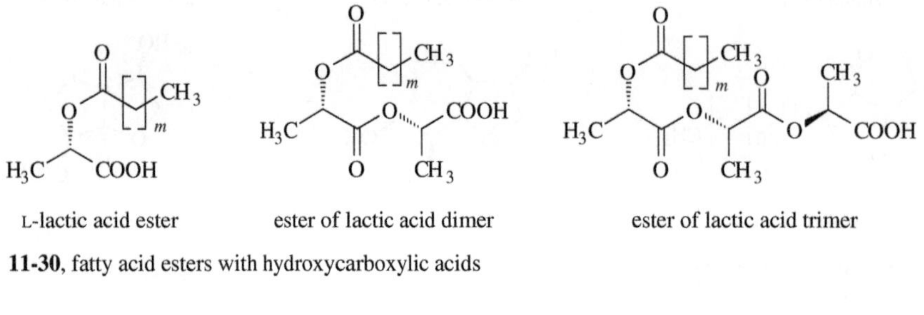

L-lactic acid ester   ester of lactic acid dimer   ester of lactic acid trimer

11-30, fatty acid esters with hydroxycarboxylic acids

11.5.2.1.8 Other emulsifiers

Fatty acids can be esterified by a number of sugar alcohols (such as D-mannitol, maltitol, and lactitol) or sugars (D-glucose, D-fructose, maltose, and lactose).

11.5.2.2 Properties

For application in food systems and preparation of w/o emulsions, emulsifiers with an HLB value of 3–6 are used, whilst for the preparation of o/w emulsions, substances with an HLB value of 15–18 are used. Emulsifiers with an HLB value of 7–9 are commonly used as moisturisers (emollients). Approximate HLB values of selected emulsifiers and other surfactants are given in Table 11.15. For example, an emulsifier of HLB value 17 is needed for the preparation of an o/w emulsion of oleic acid, an emulsifier of HLB value 9 for beeswax emulsion, an emulsifier of HLB value 6 for rapeseed oil and cocoa butter emulsions, and an emulsifier of HLB value 5 for pork lard emulsion. In practice, mixtures of two compatible emulsifiers in calculated proportions (one with low HLB value, the other with high HLB value) are preferred.

11.5.2.3 Legislation

Emulsifiers that may be used in the necessary quantities are E322, E471, and E472a–f. Polysorbates (E432–E436), sugar esters (E473, E474), stearoyl lactates (E481, E482), and emulsifiers E475, E476, E477, E479b, and E483 can only be used up to the maximum amount permitted for certain foods. For example, sugar esters (E474) are allowed for fine pastry, biscuits, and confectionery products in quantities up to 10 000 mg/kg, for ice creams in quantities up to 5000 mg/kg, and for soups and broths in quantities up to 2000 mg/kg. The maximum amount allowed varies according to the type of emulsifier and type of food. For example, for polyoxyethylene sorbitans, it ranges from 1000 (dietary foods for weight control) to 10 000 mg/kg (emulsified fats for bakery purposes).

11.5.2.4 Health assessment

There is no evidence in the available information on lecithin and lecithin bleached with hydrogen peroxide that demonstrates or suggests reasonable grounds to suspect a hazard to the public when they are used at levels that are now current or that might reasonably be expected in the future. For fatty acid salts, monoacylglycerols and diacylglycerols, and esters of mono- and diacylglycerols with acetic, lactic, and citric acids, no acute or chronic effects have been observed and ADI values are not specified.

Certain toxic effects may be observed at higher doses of other emulsifiers, but side effects are not known in the concentrations used. ADI values have been set for the remaining emulsifiers: 25 mg/kg (b.w.) for propane-1,2-diol esters and polyglycerol esters of fatty acids (high concentrations of propyleneglycol can cause eczema in sensitive persons, but not normally from use in foods), none for esters of mono- and diacylglycerols with carboxylic acids, except 30 mg/kg (b.w.) for tartaric acid esters (E472d–f), 30 mg/kg (b.w.) for ammonium phosphatides, 25 mg/kg (b.w.) for sorbitan esters (Spans) and the entire group of polyoxyethylene sorbitans (Tweens) in the E432–E436 range (people intolerant of propylene glycol should also avoid this group of emulsifiers), 16 mg/kg (b.w.) for sucrose esters and sucroglycerides, and 20 mg/kg (b.w.) for stearoyl tartrate and sodium and calcium stearoyl 2-lactylates.

11.6 Substances increasing biological value

Important nutritional factors include vitamins, minerals, amino acids, some fatty acids, fibre, and other substances with important biological effects. Some of these substances may be used as food supplements, increasing the nutritional value of the food or showing other beneficial effects, and may have a function as pigments (riboflavin) or antioxidants (such as ascorbic acid). The use of substances enhancing the biological value of food closely follows the development of knowledge in nutrition and is focused on the intake of certain essential exogenous substances that may prevent various, previously endemic or just regional diseases. Two basic reasons exist for the use of dietary supplements:

- they preserve the nutritional quality of food consumed at levels consistent with modern knowledge (such as adding vitamin D to margarines);

- they correct a deficiency of some nutritionally valuable substances in the diet (such as iodination of table salt).

11.6.1 Legislation

The Recommended Daily Allowances (RDAs) for vitamins and mineral elements set in the European Union and United States are given in Chapters 5 and 7 of their respective legislations. According to European Regulation (EC) No 1925/2006, the enrichment of foods with vitamins and mineral elements is only allowed with approved (listed) vitamins, vitamin formulations, and mineral substances and to a maximum per cent proportion of the reference daily intake. For the majority of vitamins and minerals, Directive 90/496/EEC on nutrition labelling of foodstuffs applies, which defines a significant amount as 15% of RDA (Recommended Dietary Allowances).

11.6.2 Health assessment

Medical evaluation and other aspects of dietary supplements are listed in the chapters dealing with main nutrients (amino acids, essential fatty acids), vitamins, and minerals.

11.7 Other food additives

In food production, many other additives are used, which have different properties and effects. Most often, this category of substances is classified into: firming agents, processing aids, synergists and potentiators, propellants, and solvents. Some technologies have identified other groups. For example, for tobacco products, combustion modifiers (activated carbon and ammonium chloride), substances for direct printing on cigarette paper, additives for chewing and snuff tobacco, and other additives are used.

11.7.1 Firming agents

Firming agents are those that restore or maintain the texture of food. Generally, they are soluble compounds that penetrate well into the material and do not have their own aroma or colour. These substances are mainly used in canned fruits and vegetables, jams, and other products of plant origin, but also for animal products (such as cheeses). Typical firming agents are calcium and magnesium salts, such as calcium carbonate (E170), calcium hydrogen sulfite (E227), calcium citrates (E333), calcium phosphates (E341), calcium sulfate (E516), calcium chloride (E509), calcium gluconate (E578), magnesium chloride (E511), magnesium sulfate (E518), and magnesium gluconate (E580). For example, the effect of calcium chloride in fruits is based on the interaction of calcium ions with pectin, but it also binds metals and acts as an acidity regulator.

11.7.2 Processing aids

This category of additives includes aroma carriers, filling agents, adhesives, finishing substances, humectants, and plasticisers.

The use of solid aroma carriers facilitates the application of delicate or water-insoluble additives (particularly essential oils and various aroma compositions) in products. Aroma carriers also enhance the retention of aroma, since foods can be flavoured after heat or other treatments, which minimises the loss of aromatic compounds. Examples of aromatic carriers are starch, dextrins (E1400), cellulose (E460), silica (silicon dioxide, E551), and in particular β-cyclodextrin (E459), in quantities up to 1000 mg/kg. Aromatic substances are also applied after dissolution in appropriate solvents.

Filling agents increase the volume or weight of a food and generally do not significantly affect its energy yield. Fillers do not have their own flavour and do not change the colour of products. Some oligosaccharides and polysaccharides have found use in the manufacture of confectionery, chewing gum, vitamin preparations, and cereal mixtures, and especially in various dietary and low-energy products. Fillers can also include some flour conditioners (others than emulsifiers), which increase the volume of bakery products, such as calcium stearoyl 2-lactylate (E482).

Adhesives bind food particles (e.g. reconstituted poultry and fish meat and the so-called soy meat). They have also found use in extruded foods and in the production of chewing gum, sweets, and tablets. The most commonly used additives are starch, dextrins (E1400), various plant gums, and oils and some salts (such as phosphates, which increase the solubility of proteins that denature during subsequent heat treatments and strengthen the material). Various adhesive agents are also used for food packaging.

Films on the food surface provide protection against oxidation and slow other reactions occurring in foods, prevent evaporation of water and wetting, and facilitate dissolution of products. Glossy coatings provide an attractive appearance to the food. In some cases, these films and coatings are a barrier against invasion by microorganisms. Edible coatings and coatings that are easily removable are not considered coatings as such. Some additives are used as glazing agents for fresh fruits (such as carnauba wax, E903), chocolates (such as a synthetic mixture of hydrocarbons: microcrystalline wax, E905), and eggs (to prevent access of air through pores of the shell; for example, mineral oil, E905a). In the production of milk substitutes (coffee creamers), sodium caseinate is used for fat encapsulation.

Plasticisers are substances that affect the mechanical properties of a food. Monoacylglycerols, oils, waxes and resins, and, in particular, various wetting agents (humectants) are used as plasticisers. Humectants retain water in the food, prevent it from drying out, reduce volatilisation of odorous substances, or promote the dissolution of some substances in aqueous media. Polyols are mainly used as humectants. For example, glycerol (E422) is used in grated coconut, propane-1,2-diol (E477), glycerol, butane-1,3-diol, triethylene glycol

(HOCH$_2$CH$_2$OCH$_2$CH$_2$OCH$_2$CH$_2$OH), and triacetin (triacetyl-*sn*-glycerol, E1518) in tobacco products, monoacylglycerols in caramel and margarine, and waxes in chewing gums. Related groups are substances that increase the water-binding capacity in foods containing proteins, such as phosphates in dairy and meat products. Softening agents are used for printing inks and varnishes for packaging materials (such as triacetin in paper for cigarette filters).

11.7.3 Auxiliary agents

Auxiliary food additives include anticaking agents, clarifying agents, haze-forming agents, foaming agents, dispersion stabilisers, antifoaming agents, lubricants and release agents, sequestrants, synergists, and potentiators, packaging gases, and catalysts.

Anticaking agents form coatings on the surface of food particles and reduce their tendency towards mutual adhesion. Potassium (E340) and magnesium phosphates (E343), silicon dioxide (silica, E551), various silicates (E551–E556, E559), and other substances may be used as anticaking agents.

Clarifying agents stabilise beverages by removing hazes (e.g. in beer, wine, and fruit juices). Proteins (such as gelatine), polyphenols (such as tannin), polyvinylpolypyrrolidone (E1202), phytic acid, and bentonite (E558) can act as clarifying agents. Some enzymes that hydrolyse polysaccharides (such as pectins in fruit juices) are also clarifying agents.

To induce a turbid appearance to non-alcoholic drinks and beverages, especially citrus fruit beverages, ice creams, and other products, vegetable gums or, in the past, brominated vegetable oils (E443) are used. These are no longer approved in the European Union. For fruit beverages other than citrus fruit beverages, the pulp and peels of citrus fruits are used.

Foaming agents are surfactants that allow the creation of dispersions of gaseous substances in liquid or solid food. Foaming gases are carbon monoxide and carbon dioxide (E290), and in some countries natural saponins.

Dispersion stabilisers help maintain the desirable physical properties of emulsions and other disperse systems. Various polysaccharides, such as gum arabic (E414), are used.

Antifoaming agents are food additives that prevent the formation of foam or reduce foaming, such as fatty acid esters of polyoxyethylene sorbitans (E432–E436) and silicone oils (dimethylpolysiloxane, E900, **11–31**).

Lubricants and releasing agents are applied on the surface of products or of production equipment. The purpose is to reduce the mutual attractive forces between the individual parts of the products and the tack on packaging, production equipment, and teeth during chewing, and to allow easier handling of products and simplify processing. They are used mainly for dehydrated and frozen vegetables (magnesium silicate, E553a), sweets and chewing gum (starch), milk powder, cheeses, pasta (mono- and diacylglycerols, E471), and roasted nuts (starch). To reduce the stickiness of the product towards the manufacturing equipment, silicone oils and lecithin may be used. Lubrication of sheets for baking pastry is also done with silicone oil.

Sequestrants (chelating agents) form complexes with metal ions, preventing oxidation, undesirable discoloration, and turbidity (calcium disodium ethylenediaminetetraacetate, E386, and its salts and phosphates, E450–E452).

Synergists and potentiators are substances that increase the effects of other additives. They are used to increase the activity of antioxidants (polyphenols, E452, and citric acid, E330), emulsifiers (phosphates, E450), aromatic compounds (sodium hydrogen glutamate, E621, and 5′-nucleotides such as inosine 5′-phosphate, E630).

Packaging gases (other than air), which are introduced into the containers before, during, or after filling with foods, have the role of an inert or modified atmosphere (e.g. nitrogen in ground coffee packaging, E941).

Propellants are substances that expel a food from a container or facilitate the formation of foam. For whipped creams and other dairy products, nitrous oxide (E942) is used. In cases where acidic agents can be applied, it is possible to use carbon dioxide (E290); if foaming is desired, nitrogen (E941) can be used.

Catalysts speed up chemical reactions in which they do not actually enter. They are typically used in small quantities. Raney nickel is used as a catalyst in the hydrogenation of oils, and sodium methoxide (sodium methanoate) in the transesterification of fats. Acids are used as catalysts in the production of certain modified starches. Catalysts also include enzymes used as food additives, such as amylase (E1100), proteases (E1101, for example papain, bromelain, and ficin), glucose oxidase (E1102), invertase (E1103), and lipases (E1104).

Solvents are additives that allow the extraction of desirable compounds and their dissolution and dilution. They also serve as carriers of aromatic compounds. For the extraction of hops, coffee, tea, and spices, hexane, dichloromethane, acetone, trichloroethylene, and supercritical carbon dioxide (for the extraction of caffeine from coffee or tea) are most commonly used. Ethanol is used as a solvent for confectionery, monoacylglycerols for antioxidants, and polyols for flavour potentiators.

11.7.4 Legislation

Auxiliary agents include a vast array of food additives. In many cases, it is permissible to use these substances in the amount of *quantum satis* or under good manufacturing practice, but for some commodities maximum levels of additives are set.

Anticaking agents may be used up to the maximum amounts and only for specified foods. For example, for normal table salt, cocoa powder, and potato flakes, silicon dioxide (E551, 20 g/kg) is used as the anticaking agent. Other agents may be used for coffee and tea whiteners, such as silicon dioxide, calcium phosphate (E341), and magnesium carbonate (E504), at levels of 10 g/kg. Talc (E553b) may be added to rice in specified quantities. Dimethylpolysiloxane (E900, **11–31**) may be used as an antifoaming and anticaking agent at an amount that does not exceed the specified maximum limits (100 mg/kg in chewing gum and 10 mg/kg in other foods). For the polishing and surface treatment of confectionery, chocolate, walnut kernels, and fresh fruits (apples and pears), some waxes may be used, such as beeswax (E901), candelila wax (E902), carnauba wax (E903), and shellac (E904). For tableting and coating of tablets, polyvinylpyrrolidone (E1201, **11–31**) and polyvinylpolypyrrolidone (E1202) may be used. For dissolution, dilution, and other preparations of colourings, emulsifiers, antioxidants, and flavourings, only propane-1,2-diol (propylene glycol, E1520) may be used, in a maximum amount of 1000 mg/kg (except in baby food), or else polyethylene glycol 6000, $HO–CH_2–(CH_2–O–CH_2–)_n–CH_2–OH$, (E1521). The numbers that are often included in the names of polyethylene glycols indicate their average molecular weights. For antioxidants and fat-soluble pigments, calcium carbonate (E170), calcium acetate (E263), sodium citrates (E331) and numerous others, lecithins (E322), a number of polysaccharides, sugar alcohols, and emulsifiers are allowed as carriers. For food dyes, sorbitan esters (E491–E495) and silicon dioxide (silica, E551) are permitted as emulsifiers. For emulsifiers, colours, and flavourings, some silicates can be used as anticaking agents, such as calcium silicate (E552). Beeswax (E901) is only allowed for food dyes. Polyvinylpyrrolidone (E1201, **11–31**) and polyvinylpolypyrrolidone (E1202) and various modified starches and esters (triethyl citrate, E1505, and triacetin, E1518) are allowed as stabilisers for sweeteners.

dimethylpolysiloxane   polyvinylpyrrolidone

11-31, structures of auxiliary food additives

11.7.5 Health assessment

Some food additives are considered as food (such as gelatin, non-fat milk powder, honey, sucrose, and starch), and for a number of other food additives ADI values are not specified, including various oxides, hydroxides, salts (such as magnesium oxide and calcium chloride), polysaccharides (such as carrageenans), glycerol, polyvinylpolypyrrolidone, ethanol, and waxes (beeswax, candilla wax, and canauba wax). Other substances are considered to be inert (such as dimethyl polysiloxane and polyvinylpolypyrrolidone). Carbon dioxide in concentrations >10% by volume is a smothering agent, as is nitrous oxide in high concentrations. Calcium oxide and sodium methoxide are corrosive chemicals.

ADI values are determined for phosphoric acid and phosphates (the maximum total intake of phosphorus from phosphates should be 70 mg/kg b.w.), calcium D-gluconate and polyvinylpyrrolidone (**11–31**, 50 mg/kg b.w.), propane-1,2-diol and ethyl acetate (25 mg/kg b.w.), and tannin (0.6 mg/kg b.w.). Raney nickel is classified as a possible carcinogen.

Some chlorinated solvents previously used for the extraction of oils react with proteins to form toxic products. For example, soybean meal extracted with trichloroethylene, when fed to calves, caused aplastic anaemia (a condition that occurs when the body does not produce enough new blood cells) due to toxic (Z)- and (E)-isomers of S-(1,2-dichloroethenyl)-L-cysteine, also known as S-(1,2-dichlorovinyl)-L-cysteine (**11–32**), which arises as a reaction product of protein-bound cysteine with trichloroethylene and subsequent proteolysis. The reaction of 1,2-dichloroethane with proteins in cod fillets did not lead to chlorinated reaction products, but the nutritional value was reduced due to the formation of non-utilisable cysteine, histidine, and methionine. Protein-bound cysteine yields, for example, S,S′-ethylenebiscysteine (**11–32**), which is resistant to proteolysis.

(E)-S-(1,2-dichloro-   S,S′-ethylenebis-L-cysteine
vinyl)-L-cysteine

11-32, products of cysteine with chlorinated solvents

12

Food Contaminants

12.1 Introduction

Contamination is any unintentional biological, chemical, or physical process that reduces the quality or safety of a food. Harmful substances that can get into food unintentionally or accidentally from microorganisms and environmental contamination and during agricultural production, technological or culinary processing, storage, transportation, and marketing are called food contaminants.

Biological contamination includes microorganisms such as viruses, bacteria, yeasts, moulds, and parasites. Some of these are pathogens or may produce toxins, which can result in food poisoning. Sources of environmental food contamination include air, water, soil, processing equipment, and packaging materials. Contaminants from these sources may be certain radionuclides and toxic metals, persistent organic pollutants (POPs), polycyclic aromatic hydrocarbons (PAHs), persistent organohalogen compounds such as polychlorinated dibenzo-p-dioxins (PCDDs), polychlorinated dibenzofurans (PCDFs), polychlorinated biphenyls (PCBs), and the DDT group of insecticides, plasticisers and additives from plastics (phthalates, bisphenol A), flame retardants, and various industrial chemicals. Food contaminants from agricultural activities and technological operations may include pesticides, plant growth regulators, fertilisers, and veterinary drugs. Some technological and culinary procedures lead to toxic heterocyclic amines, PAHs, and nitrosamines. A number of food contaminants have been discovered relatively recently (so-called emerging food contaminants), including acrylamide, furan, benzene, 3-chloropropane-1,2-diol, and its fatty acid esters, malondialdehyde, and 4-hydroxynon-2-enal.

Food contaminants can be classified according to the source of contamination and the mechanism by which they enter the food product. Important groups of contaminants include:

- **primary** or **exogenous** contaminants originating from external sources;

- **secondary** or **endogenous** contaminants arising from natural components due to various physical and chemical factors during food processing, better known as **foodborne, process-induced, processing,** or **technological** contaminants.

This chapter is divided into seven main sections. The first is focused on microbial toxins (bacterial toxins and mycotoxins) and the second deals with processing (technological) contaminants. Others deal with environmental food contaminants: various persistent organohalogen pollutants, pesticides (persistent chlorinated hydrocarbons and modern pesticides), veterinary drugs, contaminants from packaging materials, and other toxic compounds. In each case, the structure, properties, occurrence, main sources of dietary intake, reaction mechanisms, prevention and mitigation, and toxicological evaluations are presented.

Radionuclides and toxic metals are dealt with in Chapter 6, natural antinutritional and toxic food constituents in Chapter 10, and food additives in Chapter 11.

12.2 Microbial toxins

Food, by its nature, begins to spoil the moment it is harvested. Since ancient times, various preservation processes, such as cooking, drying, freezing, smoking, and fermenting, have been used to increase shelf-life and safeguard desired organoleptic properties (smell, taste, colour, and texture). The primary objective of food preservation is to prevent or slow the growth of microorganisms, as many microorganisms invading food raw materials and foods cause spoilage and a number of health problems. According to the nature of the agent and

The Chemistry of Food, Second Edition. Jan Velíšek, Richard Koplík and Karel Cejpek.
© 2020 John Wiley & Sons Ltd. Published 2020 by John Wiley & Sons Ltd.

the mechanism of its action, foodborne diseases are classified as either infection (toxo infections) or poisoning (intoxications). Studies of intestinal infectious diseases indicate that between 50 and 60% of all causative agents are unidentified.

The following sections deal only with toxic metabolites of bacterial toxins called **bacteriotoxins** and with metabolites of microscopic filamentous fungi known as **mycotoxins**. The inherent vegetative forms of microorganisms and their spores do not constitute a health hazard.

12.2.1 Bacteriotoxins

Many bacterial pathogens that cause diseases have the ability to produce toxins (toxigenesis). Two main groups of bacteriotoxins are recognised, lipopolysaccharides (lipooligosaccharides) and proteins (polypeptides). Lipopolysaccharides (lipooligosaccharides) are **endotoxins** associated with the cell walls of Gram-negative bacteria. Proteins (polypeptides) are **exotoxins**, which are usually secreted into the environment or released by lysis of bacterial cells and act at a site removed from bacterial growth. Exotoxins can be further classified by their toxic effect at the site of damage:

- **enterotoxins** attack the intestinal mucosa and cause diarrhoea, and are generated in the intestines;

- **cytotoxins** are toxic to cells of exposed organisms;

- **neurotoxins** interfere with the transmission of nerve impulses.

Bacterial protein toxins are the most powerful human poisons. Their production is generally species-specific. Most cases of food poisoning are infections caused by bacteria, such as *Salmonella* and *Campylobacter*; less common, but fatal, are intoxications caused by the bacterium *Clostridum botulinum*. Fortunately, bacterial toxins occur relatively rarely in the human diet; therefore, generally accepted hygienic limits have not been established.

The most effective way to minimise foodborne poisoning caused by bacterial toxins is to respect the principles of HACCP (Hazard Analysis Critical Control Points) in food production and handling. Uniform principles for food producers and their responsibility for food safety are governed by Regulation (EC) No. 852/2004 concerning hygiene in the production, distribution, and sale of food and the introduction of critical control points (HACCP) as preventive tools for ensuring food safety. For foods of animal origin, these requirements are further specified in Regulation (EC) No. 853/2004.

12.2.1.1 Botulotoxins

Botulinum toxins (botulotoxins) produced by the bacterium *C. botulinum* are extremely toxic proteins consisting of a sequence of about 1300 amino acids, although not all strains of *C. botulinum* produce them. Seven toxigenic types of bacteria exist, each producing a distinct form of botulinum toxin that may cause botulism. These are designated A–G, and most have several subtypes. Particularly toxic are types A, B, E, and F. In Europe, the most frequently occurring toxin is type B, whilst in the United States, type A is the most significant cause of botulism. Botulinum toxins may enter the human body by ingestion of toxin from foods (foodborne botulism), but also by certain other pathways. Botulinum toxins have been known since the eighteenth century, when they were described as a 'sausage poison' because they arose from improperly handled or prepared meat products. Even today, they can be found in very rare cases in non-acidic canned products, such as sausages and some other meat products.

Production of botulinum toxins proceeds under anaerobic conditions, and the optimum conditions for their reaction in foods are pH 4.8–8.5 and temperature about 30 °C, although they are also produced at lower temperatures. Owing to the extreme danger of botulinum toxins (intoxication might be lethal), the possibility of preventing their occurrence in foods has been studied extensively. The most important aspect of botulism prevention is proper food handling and preparation. The spores of *C. botulinum* can survive boiling (100 °C at normal pressure) for more than 1 hour, although they are killed by autoclaving. Because the toxin is heat-labile, boiling or intense heating of contaminated food (e.g. heating at 80 °C for 10 minutes or cooking at 100 °C for several seconds) will inactivate it. Food containers that bulge may contain gas produced by *C. botulinum* and should not be opened or tasted. Nitrites have inhibitory effects (see Section 11.2.1.3.2).

12.2.1.2 Other bacterial toxins

Toxic products may, under certain circumstances, produce a range of bacteria. Their toxins may already be present in the food at the time of consumption or they may be produced in the digestive tract. The pathogens present in the food may also include some viruses, such as the hepatitis A virus, which causes infectious hepatitis in humans.

Staphyllococcal food poisoning is a gastrointestinal illness caused by eating foods contaminated with toxins produced by the *Staphylococcus aureus* bacteria. *S. aureus* produces seven neurotoxins that are not destroyed by cooking. The most common cause of poisoning is toxins of types A and D, individually or in combination. *Clostridium perfringens* (of types A–E) produces four major toxins, α, β, ε, and ι. The cause of food poisoning (gastroenteritis) is mainly *C. perfringens* of type A producing toxin α, which shows the activity of phospholipase C. Another cause of food poisoning (gastroenteritis) is often the bacterium *Bacillus cereus*.

In non-invasive infections, bacteria reproduce and produce toxins in the digestive tract. Most attention in this regard is given to toxins produced by the bacteria *Vibrio cholerae* (the causative agent of cholera), strain 0157:H7 of *Escherichia coli* (a cause of poisoning called colitis), and substances produced by bacteria of the genus *Salmonella*, especially *S. enteritidis* PT4 (*S. enterica* ssp. *enterica* serovar *enteritidis*), which causes diarrhoea (the poisoning is called salmonellosis, which in people at risk, such as infants, small children, and the elderly, can become very serious). Often a cause of diarrhoea known as schigellose are toxins of the bacterium *Shigella sonnei*. Previously rare, but more often encountered today, is campylobacteriose, caused by the bacterium *Campylobacter jejuni* and other infections, such as jersiniose (*Yersinia enterocolitica*), vibriose (*Vibrio parahaemolyticus*), and listeriose (*Listeria monocytogenes*).

12.2.2 Mycotoxins

Microscopic filamentous fungi (multicellular, eukaryotic, heterotrophic, saprophytic, or parasitic microorganisms) can cause adverse human health effects through three specific mechanisms: generation of a harmful immune response (e.g. allergy or hypersensitivity pneumonitis), direct infection by the organism, and toxic-irritant effects from toxic secondary metabolites, **mycotoxins**, produced by toxigenic species. In medical and veterinary mycology, about 150 species of microscopic pathogenic fungi (moulds) are known, and in foods, 114 fungal and 12 yeast species have been characterised. Of the total number of 114 species that are of importance in foods, 65 are toxigenic. Mycotoxins are produced by vegetative parts of fungi known as mycelia and secreted into the substrate, but can also occur in the reproductive structure – spores that contaminate the human environment. Poisoning associated with exposure to mycotoxins in food and feed is demonstrated by various symptoms, known collectively as mycotoxicoses. The symptoms of a mycotoxicosis depend on the type of mycotoxin, its concentration and length of exposure, and the age, health, and sex of the exposed individual. Mycotoxicoses can be categorised as acute or chronic. Acute toxicity generally has a rapid onset and an obvious toxic response. Chronic toxicity is characterised by low-dose exposure over a long time period, resulting in cancers and other generally irreversible effects. Mycotoxins, in view of their ubiquity, can occur at virtually all levels of the food chain of humans and livestock.

Of nearly 1000 known mycotoxins, more than 300 have been at least partially characterised, and approximately 20 can occur in food or feed in toxicologically relevant concentrations. To ensure the protection of consumer and livestock health, the maximum concentrations for these mycotoxins have been set in selected food commodities. According to US Food and Agriculture Organization (FAO) estimates, contamination by mycotoxins can be demonstrated in up to 25% of food consumed, with more severe levels found particularly in developing countries.

The negative health effects resulting from dietary intake of mouldy food have occurred for as long as the human race has existed, especially since humans left the nomadic way of life and began to grow different crops and store surpluses. The oldest described mycotoxicosis is ergotism, induced by metabolites (alkaloids) of ascomycetes known as ergot fungi, the most prominent member of which is the fungus *Claviceps purpurea*, which grows on rye and related plants. Another mycotoxicosis is alimentary toxic aleukia (ATA), which is induced by T-2 toxin, and related trichothecenes, produced by fungi of the genus *Fusarium* and certain other fungi. The so-called yellow rice disease was caused by fungi of the genus *Penicillium* (e.g. *P. citreonigrum*), which producesthe causative agent yellow mycotoxin citreoviridin and another mycotoxin citrinin. A systematic study of mycotoxins was initiated by increased signs of dangerous mouldy food and feed in the 1960s. The major impetus for the research was an event that took place in Britain in 1960, when there was a series of mass deaths of about 120 000 turkeys and other poultry. The causative agents of the intoxication were identified as then-unknown fungal toxins from peanut meal that were called aflatoxins (from the Latin name of the fungus, *Aspergillus flavus*).

12.2.2.1 Classification

Mycotoxins are an extremely diverse group of compounds. Their relative molecular weight usually does not exceed 1000 Da. Just as in the case of other secondary metabolites, mycotoxins cannot be simply classified into groups of compounds on the basis of their chemical structure, without consideration of their occurrence, production, or toxic effects. Production of mycotoxins by toxigenic fungi (moulds) is subject to a number of factors. Under certain conditions, mycotoxins may not be produced at all, as not all species and strains of potentially toxic fungi are toxigenic. A specific mycotoxin can also be produced by representatives of several genera of fungi, and two or more mycotoxins can be produced by one fungal species.

The criterion for classification of mycotoxins is the mechanism for the biosynthesis of their basic skeleton from primary metabolites (their precursors). A number of important mycotoxins are biosynthesised by the so-called polyketide pathway, specifically patulin, penicillic acid (tetraketides), ochratoxin A and citrinin (pentaketides), zearalenone (nonaketide), and sterigmatocystin and aflatoxins (decaketides). The intermediate directly involved in the biosynthesis of the polyketide is acetyl coenzyme A. The second major pathway of biosynthesis starts with mevalonic acid, from which are formed trichothecene mycotoxins containing sesquiterpenic skeletons in their molecules. For completeness, other metabolic pathways also exist, which play a role in the biosynthesis of some of the less common fungal metabolites. Amino acids are biochemical precursors of cyclic polypeptides such as ergot alkaloids. So-called tremorgenic mycotoxins, a typical example of which is roquefortine C, are formed by the reaction of amino acids with mevalonate. Reactions similar to those of the Krebs cycle yield the so-called nonadrides, a small group of fungal metabolites characterised by the presence of a nine-membered alicyclic ring and two five-ring anhydride functions. Examples of nonadrides are rubratoxins.

The most important producers of toxicologically relevant mycotoxins are microscopic filamentous fungi of the genera *Aspergillus*, *Penicillium*, and *Fusarium* (Table 12.1). Some currently monitored mycotoxins are produced by representatives of the genera *Claviceps*, *Alternaria*, *Chaetomium*, and *Sordaria*.

12.2.2.2 Structure, occurrence, and properties

Contamination can occur at all stages of food production. Figure 12.1 shows the dissemination of mycotoxins in human and animal food chains and some possibilities for their further fate. A comprehensive list of dietary exposure sources is as follows:

- foods of plant origin, in the case of pre-harvest primary infection of crops and contamination of final products (secondary infection);

- foods of animal origin, where the animals are fed contaminated feed (milk and dairy products, meat and meat products);

- foods whose manufacture involves cultural microscopic filamentous fungi (blue cheese, fermented meat products, fermented oriental soy sauce, and other plant products);

- biotechnology products consumed as food additives, aids, or supplements (microbial proteins, technical enzymes, amino acids concentrates, vitamins, and other substances).

Factors affecting the extent of potential primary mycotoxin contamination of agricultural crops under field conditions in the pre-harvest period include cultivar resistance to attack by fungi, degree of physiological stress to which the plants are exposed (lack of minerals, small, or excessive amounts of water, soil salinity, pollution, mechanical damage caused by the attack of insects or other pests), virulence of pathogenic fungi, type of mycotoxin produced (mechanism of its biosynthesis), phase of plant growth cycle, in which the plant was infected by fungi and the interval between the onset of a period of mycotoxin production and harvest, ability of plants (enzyme systems) to transform mycotoxins into non-toxic products, and climatic conditions in the pre-harvest period (especially by high relative humidity in the later stages of crop ripening). In addition to fungi of the genera *Fusarium* and *Claviceps* requiring relatively high humidity, cultured plants can also be infected in the period before the harvest by toxigenic representatives of the genera *Aspergillus* and *Alternaria*, and possibly by fungi of the genera *Chaetomium* and *Sordaria*.

Contamination of plant products is also possible during transport, processing, and storage. Producers of mycotoxins in this case can be, in particular, fungi of genera *Aspergillus* and *Penicillium*, which can grow on materials of lower water activity. A critical factor limiting the growth of fungi during storage of agricultural crops is the content of available water. The outer layers of grains are a natural barrier to fungi. If the layers are damaged during harvest, transport, or processing, mould spores can easily access the necessary nutrients that are found in the inner parts of the grains. Generally, smaller cereal grains (e.g. wheat, barley, and rice) are attacked by fungi less often than larger grains (e.g. maize).

12.2.2.2.1 Aflatoxins

Aflatoxins belong to the most often monitored mycotoxins, due to their high toxicity. The most important producers of aflatoxins are two closely related fungi of the genus *Aspergillus*: *A. flavus* and *A. parasiticus*. A rarer one is *A. nomius*. The mycelia of aflatoxins grow, under favourable conditions (temperature between 13 and 37 °C, water activity < 0.82), on virtually any substrate; however, they have been observed in the highest amount (in some cases, hundreds of thousands of mg/kg) in maize, peanuts, pistachios, Brazil nuts, cottonseed, and copra. Lower levels of aflatoxins are found in almonds, pecans, walnuts, raisins, figs, and various spices. As evidenced by the Rapid Alert System for Food and Feed (RASFF), aflatoxins represent the contaminants most often found in the control of food imported into the European Union (most often nuts and nut products).

Currently, 16 natural aflatoxins are known. Formulae **12-1** show the four major naturally occurring representatives of the B and G groups of aflatoxins, known as aflatoxins B_1, B_2, G_1, and G_2. The designation B (blue) or G (green) is associated with fluorescence under UV light. Whilst *A. flavus* produces mainly aflatoxins B_1 and B_2, contamination by *A. parasiticus* leads to aflatoxins G_1 and G_2 in the contaminated material. The basis of the skeleton of all aflatoxins is coumarin condensed with bisdihydrofurofuran (dihydrobisfuran), plus cyclopentanone in the case of aflatoxins B_1 and B_2 or 5,6-dihydropyran-2-one in the case of aflatoxins G_1 and G_2.

Table 12.1 Examples of the most important mycotoxins produced by fungi of genera *Aspergillus*, *Penicillium*, and *Fusarium*.

Fungus	Dominant mycotoxins
Aspergillus spp.	
A. carneus	Citrinin
A. clavanus	Patulin
A. flavus	Aflatoxins B_1, B_2, cyclopiazonic acid
A. ochraceus	Ochratoxins, penicillic acid
A. oryzae	Cyclopiazonic acid
A. parasiticus	Aflatoxins B_1, B_2, G_1, G_2
A. tereus	Citreoviridin, citrinin, patulin
A. tamarii	Cyclopiazonic acid
A. versicolor	Sterigmatocystin, cyclopiazonic acid
Penicillium spp.	
P. aurantiogriseum	Cyclopiazonic acid, penicillic acid
P. camemberti	Cyclopiazonic acid
P. chrysogenum	Cyclopiazonic acid
P. citreonigrum	Citreoviridin
P. citrinum	Citrinin
P. commune	Cyclopiazonic acid
P. aethiopicum	Citrinin, patulin
P. griseofulvum	Cyclopiazonic acid, patulin
P. purpurescens	Ochratoxin A
P. roqueforti	Patulin
P. crateriforme	Rubratoxins
P. simplissimum	Penicillic acid
P. brasilianum	Citrinin, ochratoxin A, cyclopiazonic acid
P. crustosum	Cyclopiazonic acid, ochratoxin A
Fusarium spp.	
F. acuminatum	Diacetoxyscirpenol, monoacetoxyscirpenol, HT-2 toxin, T-2 toxin, moniliformin
F. anthopilum	Moniliformin
F. avenaceum	Moniliformin
F. chlamydosporium	Moniliformin
F. crookwellense	Deoxynivalenol, nivalenol, zearalenone
F. culmorum	Fusarin C, deoxynivalenol, nivalenol, zearalenone
F. graminearum	Deoxynivalenol, diacetoxyscirpenol, zearalenone
F. verticillioides	Fumonisins, fusarin C, moniliformin
F. oxysporum	Moniliformin, T-2 toxin
F. poae	Fusarin C, diacetoxyscirpenol, monoacetoxyscirpenol, HT-2 toxin, T-2 toxin, zearalenone
F. sambucinum	Fusarin C, diacetoxyscirpenol, monoacetoxyscirpenol, HT-2 toxin, T-2 toxin
F. semitectum	Moniliformin, zearalenone, fusarenone-X
F. sporotrichioides	Diacetoxyscirpenol, HT-2 toxin, T-2 toxin, zearalenone
F. tricinctum	Fusarin C

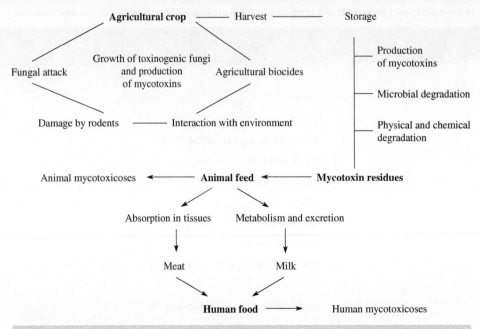

Figure 12.1 Factors influencing the occurrence of mycotoxins in foods and feeds.

Aflatoxin B_1 is the main representative of this group in contaminated foods. The total content (the sum of the four major aflatoxins) is on average 24% or higher, but the variability in different types of food is large (2–70%). Distribution of aflatoxins in contaminated commodities is very uneven, with the highest concentration often found in foci of high humidity or in mechanically damaged areas.

aflatoxin B_1 aflatoxin B_2 aflatoxin G_1 aflatoxin G_2

12-1, aflatoxins B and G

Like other xenobiotics, in animal organisms biotransformation of aflatoxins proceeds in three main stages (mainly in the liver): bioactivation, conjugation, and deconjugation. Phase I reactions lead, apart from the formation of relatively non-toxic products, to toxic intermediates. The toxic intermediate of aflatoxin B_1 is *exo*-8,9-epoxide. Phase II represents detoxification, such as hydroxylation of aflatoxin B_1 to aflatoxins B_{2a}, M_1, and Q_1, demethylation to aflatoxin P_1, and oxidation to aflatoxicol, which is accompanied by significantly reduced toxicity (Figure 12.2). These and other aflatoxins were identified as metabolites of aflatoxins of the B and G groups in the bodies of humans and animals and in tissue cultures. Phase III proceeds under the activity of intestinal microflora and may lead to the reabsorption of the released aflatoxin.

Exo-8,9-epoxide of aflatoxin B_1 reacts with DNA to form a covalent product with N^7-guanine (**12-2**), forms an adduct with glutathione (which is excreted in the urine), reacts with amino acids in proteins with the formation of imines, and is hydrolysed by water to 8,9-dihydroxyaflatoxin B_2, which forms adducts with proteins (as well as aflatoxicol) and is partly excreted in the urine.

When feed contaminated with aflatoxins B_1 and B_2 is given to cows, aflatoxins M_1 and M_2 resulting from hydroxylation of the parent compounds may be found in milk after roughly 12 hours. The transition factor, which is the ratio between the amount of ingested precursor (aflatoxin B_1) and the amount of excreted aflatoxin M_1, is in the range of 100 : 1–300 : 1. The major part of this metabolite is excreted in the urine. Transient factors of aflatoxins to muscles are low in cattle; for aflatoxin B_1, they are in the range 1000 : 1–14 000 : 1. Somewhat lower values were found in the livers of pigs and poultry, but foods of animal origin (except milk and dairy products) do not generally represent a significant dietary source of aflatoxins.

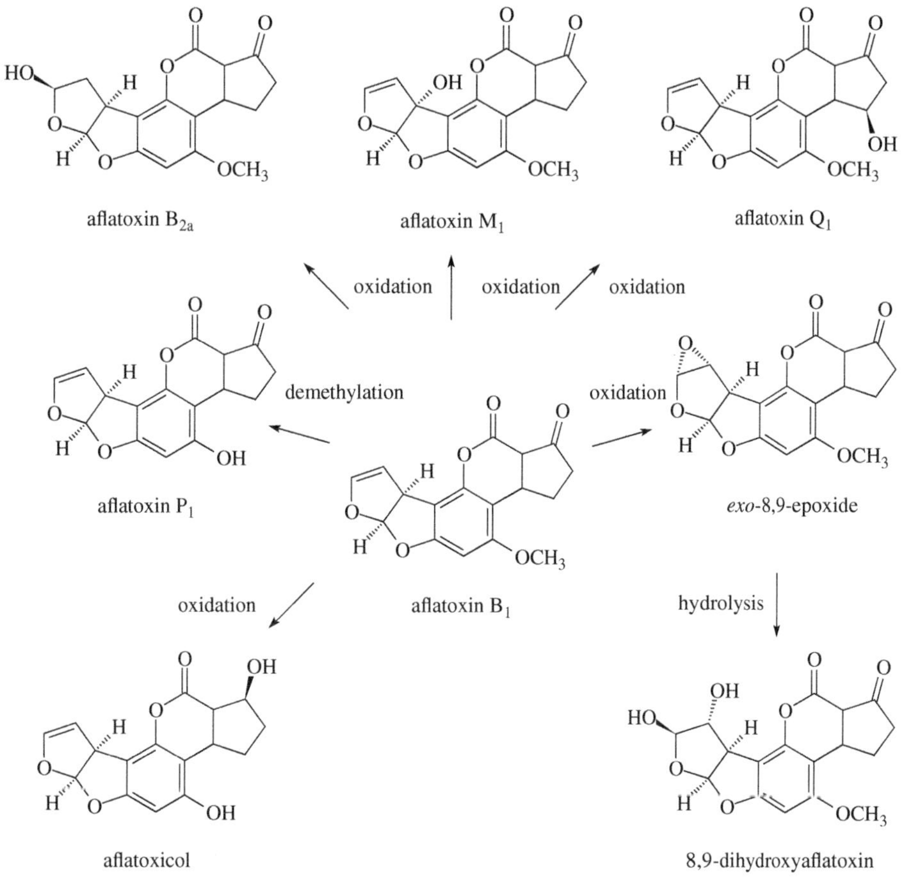

Figure 12.2 Aflatoxin B₁ metabolites.

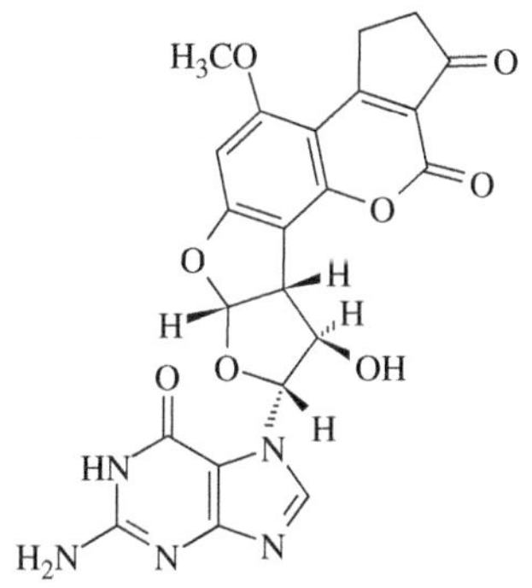

12-2, adduct of aflatoxin B₁ with guanine

During milling of wheat grains, aflatoxins, found mostly in the outer (aleurone) layer of the grain and in the germ, are redistributed amongst individual fractions according to the degree of milling, known as the flour extraction rate. For example, the decrease of contamination in high-extraction wheat flour (type 85) by about 15% is accompanied by higher aflatoxin concentrations in the bran.

Aflatoxins are relatively hydrophilic compounds, and in the production of vegetable oils they are concentrated in oilseed meals, which may become a source of contamination of the food chain when used for livestock feeding.

Similarly, separation of fat during milk processing increases the content of aflatoxin M₁ in some products, such as low-fat milk, cottage cheese, whey, buttermilk, and other products.

Fermentation processes lead to a small decrease in aflatoxin content, but on average, for example, 20–25% of all aflatoxins contained in barley are transferred to beer. The highest losses occur in the final stage of the brewing process, when, after boiling with hops, proteins, the remaining hops, and other insoluble particles are separated by filtration. In the production of spirits, aflatoxins from contaminated material do not pass to the distillate.

Table 12.2 Changes in aflatoxin B_1 and M_1 levels during processing of contaminated materials.

Product	Processing conditions	Losses (%)	Product	Processing conditions	Losses (%)
Peanuts	Roasting (150 °C, 30 minutes)	20	Rice	Steaming	0
	Dry roasting	31		Boiling under pressure (120 °C)	27
	Roasting in oil	35		Regular cooking	51
Peanut products	Roasting (204 °C)	50–60	Milk	Pasteuristion (72 °C, 45 seconds)	35
Peanut oil (raw)	Heating (120 °C, 10 minutes)	0		Sterilisation (115 °C, 20 minutes)	19

Aflatoxins are relatively stable compounds, so thermal processes do not lead to their complete elimination in materials with low moisture content. The extent of their degradation depends on the content of water, fat, and other components. Roasting of coffee at 180 °C for 10 minutes results in only a 50% decrease in aflatoxin B_1 concentration. Other examples are shown in Table 12.2.

Aflatoxin B_1 is considered the most potent known natural hepatocarcinogen. One of the most tragic cases of acute human aflatoxin poisoning was reported in India in 1974, when the consumption of mouldy maize afflicted about 1000 people, some 100 of whom died. Based on the evaluation of aflatoxin concentrations and estimates of aflatoxin exposure dose, it was considered that the LD_{50} value for humans was somewhat higher than that for the most sensitive animal species, such as dogs, where the LD_{50} value for aflatoxin B_1 is in the range 0.35–0.50 mg/kg b.w. Numerous independent epidemiological studies have pointed to the high incidence of liver cancer in the populations of some regions of East Asia and Africa, which correlates with a high dietary intake of aflatoxins. The most obvious correlation was found in areas with a frequent occurrence of chronic hepatitis B. Mutagenicity and carcinogenicity of aflatoxins generally follow the order $B_1 > G_1 > B_2 > G_2$ and are related to the interaction of reactive metabolites with DNA and inhibition of its replication and transcription, as well as to inhibition of other processes, such as RNA and protein biosynthesis. In addition, aflatoxins induce a number of other adverse biological responses in humans, such as so-called Reye's syndrome (a potentially fatal disease that causes detrimental effects to many organs), respiratory diseases, chronic gastritis, mental disability of children, and immunosuppression. According to the International Agency for Research on Cancer (IARC),[1] aflatoxins are classified as Group 1 carcinogens to humans.

In livestock, the consumption of aflatoxins in feed leads to a number of problems. Chronic exposure to aflatoxins leads to reduced growth and greater susceptibility to infections. This condition is often referred to as total failure to thrive. An extreme consequence of giving animals contaminated feed is lethal acute poisoning.

Contamination of plant materials with mycotoxins can become a significant economic problem, if it concerns a large volume of biomass. Because of the high carcinogenic potential of aflatoxins, special attention is paid to the possibilities of detoxification of contaminated materials, which would at least partially facilitate their use. The most common method is washing with ammonium hydroxide, often at elevated temperature and pressure. The efficiency achieved under optimal conditions can reach as high as 99%, but the principle of such treatment has not yet been satisfactorily explained. Amongst the degradation products of aflatoxin B_1, in model experiments, a decarboxylation product aflatoxin D_1 (12-3) has been identified. However, decontamination may cause a number of negative changes in the composition of the treated material, which results in a decrease of its nutritional or sensory quality. The use of decontaminated products for human nutrition is prohibited. Alternatively, the contaminated feedstock may be used for the production of ethanol or other fermentation products.

12-3, aflatoxin D₁

Owing to the high toxicity of aflatoxins, maximum amounts have been set for aflatoxin B_1 and for the sum of aflatoxins B_1, B_2, G_1, and G_2, as well as for aflatoxin M_1 in milk and certain dairy products. The situation is quite complicated, because the limits in Europen Commission

---

[1] The IARC, a part of the World Health Organization (WHO), recognises the following groups of carcinogenic substances: Group 1, carcinogenic to humans; Group 2A, probably carcinogenic to humans; Group 2B, possibly carcinogenic to humans; Group 3, unclassifiable as to carcinogenicity in humans; and Group 4, probably not carcinogenic to humans.

Table 12.3 Examples of typical limits for mycotoxins in food.

Mycotoxin	Product	Maximum levels (μg/kg)[a]
Sum of aflatoxins B_1, B_2, G_1, and G_2	Nuts, dried fruits, cereals, spices	4.0–15.0
Aflatoxin B_1	Nuts, dried fruits, cereals, spices, foods for infants and young children, dietary foods for special medical purposes	0.1–8.0
Aflatoxin M_1	Infant formulae and follow-on formulae, dietary foods for special medical purposes	0.025–0.05
Ochratoxin A	Cereals, dried vine fruit (currants, raisins and sultanas), coffee, wine, foods for infants and young children, beer, cacao, meat products, spices	0.5–10
Patulin	Apple juice and drinks from apples, cider, compotes, purees, baby food	10.0–50.0
Deoxynivalenol	Cereals, flour, pastries, biscuits, cereal snacks and breakfast cereals	200–1750
Zearalenone	Cereals, flour, pastries, biscuits, cereal snacks, breakfast cereals, baby food	20–350
Fumonisins	Maize and maize products	200–4000
T-2 toxin and HT-2 toxin	Not yet defined	–

[a]Maximum level can vary in the given range depending on the exact specification of food and on whether it is consumed directly or further processed. For example, the EU limit for patulin is set to 50 μg/kg in both apple juices and ciders, to 25 μg/kg in solid apple products, and to 10 μg/kg in products for infants and young children.

Regulation (EC) No. 1881/2006 have been subjected to rapid change, as seen for example in Commission Regulation (EC) No. 1126/2007, since for many countries strict limits can bring about serious economic difficulties. This concerns aflatoxins and *Fusarium* mycotoxins (the problematic crop is maize). To illustrate the maximum amounts set for mycotoxins, Table 12.3 summarises typical values of different EC documents. In the regulation of mycotoxins in foods, further development can be expected based on the results of toxicological evaluation and monitoring of real levels of mycotoxins in foods, depending on the geographical origin and climatic conditions. Last but not least, the influences of processing technologies and transformation products of mycotoxins will also be taken into consideration, such as their conjugates with saccharides.

12.2.2.2.2 Sterigmatocystin

Sterigmatocystin (**12-4**) is a toxic secondary metabolite of fungi of the genus *Aspergillus*. One of its more prominent producers is the fungus *A. versicolor*, but sterigmatocystin is also produced by the fungi *A. flavus* and *A. parasiticus*. The presence of sterigmatocystin (along with aflatoxins) has been demonstrated in mouldy cereals, coffee beans, and some animal products such as ham, sausages, and hard cheeses, where it is located mainly on the surface.

Sterigmatocystin, a xanthen-7-one derivative, which contains the bisdihydrofurofuran skeleton, is related to aflatoxins and even regarded as their precursor. Sterigmatocystin is very acutely toxic to humans and animals, and like aflatoxins acts as a hepatocarcinogen. According to the IARC, it is classified as a Group 2B carcinogen to humans. As one of the priorities in the dietary risk assessment, the European Food Safety Authority (EFSA) identified the need to gather a more complete set of data on the occurrence of this mycotoxin in foods. Maximum levels in foods are not currently set due to a lack of data from which to assess the associated health risks.

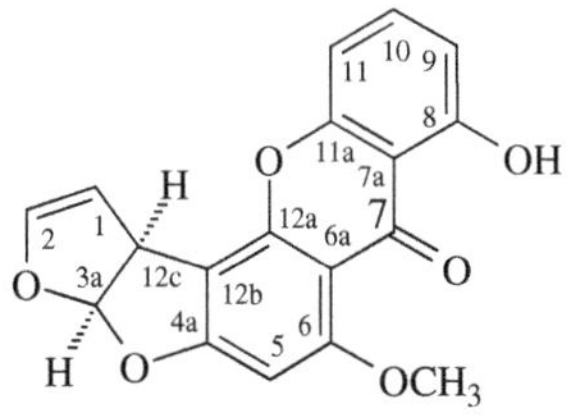

12-4, sterigmatocystin

12.2.2.2.3 Ochratoxins

Ochratoxins are a group of mycotoxins discovered in South Africa in 1965 during investigations of toxic filamentous fungi isolated from agricultural crops. The most important producers of these mycotoxins are fungi of the genus *Aspergillus*, in particular *A. ochraceus*, and also some industrial strains of *Aspergillus niger*. In the colder climates of Europe (especially in Scandinavia), important producers of ochratoxins are fungi of the genus *Penicillium*, mainly *P. verrucosum*, *P. nordicum*, and *P. carbonarius*.

From a toxicological point of view, the most important representative of this group of mycotoxins is ochratoxin A (**12-5**), the molecule of which contains L-phenylalanine *N*-substituted with a derivative of (3*R*)-3,4-dihydro-3-methylisocoumarine that contains at C-5 a chlorine atom to which are attributed the toxic effects of ochratoxin A. The incorporation of the chlorine atom into the skeleton of ochratoxin A is achieved by the action of chloroperoxidase; the chlorine donor is inorganic chloride. Ochratoxin B (**12-5**) differs from ochratoxin A only in the absence of the chlorine atom, whilst ochratoxin C (**12-5**) is an ethyl ester of ochratoxin A. (4*R*)- and (4*S*)-Hydroxyochratoxin A, 4-hydroxyochratoxin B, and ochratoxin A analogue, containing a (3*R*)-hydroxymethyl group, are naturally produced by certain fungi. Ochratoxins B and α, β, and γ, resulting from parent compounds through the loss of phenylalanine caused by amide bond hydrolysis, are virtually non-toxic.

12-5, ochratoxin A, R^1 = Cl, R^2 = H
ochratoxin B, R^1 = H, R^2 = H
ochratoxin C, R^1 = Cl, R^2 = CH$_2$CH$_3$

Concentrations of ochratoxin A in cereals rarely exceed 50 µg/kg, but improper storage may cause a massive increase in contamination. In addition to cereals and cereal products, ochratoxin A has also been identified in some legumes, coffee, cocoa, grapefruit juices, raisins, nuts, spices, and wine – especially red wine, which acquires ochratoxin A from the grape skins.

Ochratoxin A is a relatively stable compound, and losses under normal processing conditions are not too significant (Table 12.4). The occurrence and concentration of ochratoxin A in wines is likely to increase in the southern wine regions of Europe, which is connected with the infection of grapes by the toxigenic fungi *Aspergillus carbonarius* and other species (*A. niger*), which are almost absent in the northern wine regions. Attention must also be paid to the strains of *A. niger*, which are commonly used in biotechnology for the production of a variety of hydrolytic enzymes and organic acids, particularly citric acid. Their ability to produce ochratoxin A was discovered in the 1990s. Conjugated forms were found in cell cultures of wheat and maize: methyl ester of ochratoxin A and methyl ester and glucosides of 4-hydroxyochratoxins A.

Ochratoxin A is probably the only mycotoxin whose dietary intake is mainly associated with foods of animal origin. If contaminated feed is fed to animals, ochratoxin A is correspondingly found in the bodies of farm animals; it accumulates mainly in the kidney. In pigs, concentrations of ochratoxin A in the kidney may reach hundreds of mg/kg. The transient factor is in the range of 20–60 in the kidney and between 400 and 660 in the liver. Trace concentrations of ochratoxin A were also detected in the flesh, and occasionally in cheeses with moulds on the surface and also in cured salami with casings treated with edible mould cultures.

In humans, exposure to ochratoxin A is associated with Balkan endemic nephropathy (BEN), also known as Danubian endemic familial nephropathy, a chronic kidney disease that affects the kidney and leads to fibrosis and decreased kidney function. It is also characterised by a high frequency of urothelial cancer (cancer of the urinary tract). The most serious biological effect of ochratoxin A for exposed animals is nephrotoxicity, but hepatotoxicity, genotoxicity, and immunotoxicity have also been demonstrated, and its carcinogenic potential is under discussion. When contaminated feed is fed to livestock, their growth is slow and relates, *inter alia*, to lower utilisation of nutrients.

Hygienic limits for ochratoxin A residues exist in many countries, examples of which are given in Table 12.3. Maximum levels are exceeded only in exceptional cases.

Table 12.4 Changes in the content of ochratoxin A during processing of contaminated crops.

Product	Processing conditions	Losses (%)	Product	Processing conditions	Losses (%)
Coffee beans	Roasting, unspecified	10–20	Brewing mash	Cooking	27–28
Coffee beans	Roasting (200 °C, 5 min)	0	Cereal products	Autoclaving (120 °C, 3 h)	30

12.2.2.2.4 Patulin

Initially, patulin was evaluated as a pharmacological agent due to its antimicrobial properties. However, the gastrointestinal and dermal irritation observed in human trials prevented its use as such. The most important producers of patulin are microscopic fungi of the genus *Penicillium*, especially *P. patulinum* and *P. expansum*, which are common pathogens of many fruits and vegetables. The production of this mycotoxin was also detected in some fungi of the genera *Aspergillus* and *Byssochlamys*. Within the food industry, patulin contamination is considered of greatest concern in apples and apple products, which are the main sources of human patulin intake. Nevertheless, it has also been found at significant concentrations in other fruits, such as pears, peaches, strawberries, blueberries, cherries, apricots, and grapes, and in cheeses and animal feed, especially in silages.

Patulin, 4-hydroxy-4*H*-furo(3,2-*c*)pyran-2(6*H*)-one (**12-6**) is a furopyrone mycotoxin that occurs in nature as a racemate. It is a relatively common contaminant of apple concentrates and juices, especially when overripe or damaged fruit has been used. Its concentrations do not normally exceed 0.1 mg/kg, but can reach about 2500 µg/kg if the juice is obtained from rotten apples, which may contain up to about 45 000 µg/kg of patulin.

Patulin is highly soluble in water and relatively stable in acidic media (pH 3.0–6.5); therefore, special attention is devoted to the reduction of its content in foods. The recommendations can be summarised in the following points:

- removing mycelium from the product, even with the adjacent biomass, is not sufficient for decontamination, because patulin easily diffuses into the whole material;

- patulin present in apple pomace, a byproduct of juice production, may pose a risk to livestock if used as feed;

- storage of products containing patulin (especially at higher temperatures) leads to a gradual decrease of its content;

- thickening of juices by vacuum distillation reduces patulin content by an average of 25% and pasteurisation at 80 °C leads to a negligible reduction in the patulin concentration, but sterilisation may reduce patulin levels in apple juice by about 20% or more;

- decrease of patulin content in foods (to about 40–95% of the original concentration) probably occurs during microwave heating.

Patulin is virtually absent from wines, because fermentation of contaminated juices by the yeast *Saccharomyces cerevisiae* (and other microorganisms, such as *Gluconobacter oxydans*) leads to relatively rapid transformation of patulin into ascladiols, 5-(2-hydroxyethylidene)-4-hydroxymethyl-5*H*-furan-2-ones (**12-6**). However, practically no information exists about the toxicity of ascladiols. Sulfites and compounds containing sulfhydryl functional groups (cysteine and glutathione) can contribute to an acceleration of patulin degradation. Owing to the electrofilicity of patulin, reactions with other nucleophilic substances, such as amino acids (especially lysine and histidine) and proteins, can be expected. Increased patulin degradation induced by hydroxyl and other free radicals was observed in fruit juices fortified with ascorbic acid to the level of 500 mg/l.

In addition to patulin's antibiotic effects against Gram-negative and Gram-positive bacteria, antifungal and antiviral properties (inhibition of replication of mycoviruses) have also been reported. Its carcinogenic and mutagenic effects have subsequently been proven, in association with the inhibition of RNA transcription and selective DNA damage. Also demonstrated were negative effects on the gastrointestinal tract and neurotoxic and immunotoxic effects; therefore, the content of patulin (a Group 3 carcinogen – a compound for which there are not enough data to allow its classification) in food should be reduced to a minimum technically achievable level. The limits set in the European Union for apples and apple products are summarised in Table 12.3.

patulin   (*Z*)-ascladiol   (*E*)-ascladiol

12-6, patulin and its transformation products

12.2.2.2.5 Cyclopiazonic acid

Cyclopiazonic acid or α-cyclopiazonic acid (**12-7**) is a toxic metabolite that occasionally accompanies aflatoxins in contaminated materials. Its major producers are some strains of *Aspergilus flavus* and *A. versicolor*, but it is also produced by some other fungi, especially *Penicillium commune*, *P. griseofulvum*, and occasionally *P. camembertii*. Its presence has been demonstrated, for example, in maize (highest levels close to 10 mg/kg), sunflower seeds, and peanuts, but also in meat products with edible moulds on their casings and cheeses that have moulds on the rind or throughout, which may indicate that the strict criteria for selecting non-toxigenic strains of moulds were not met. Some findings document frequent occurrence of cyclopiazonic acid in animal feed (hay and silage), but information about the deposition of this mycotoxin in animal tissues is not available. Generally, the data on the occurrence of cyclopiazonic acid are very limited, as is information about its stability in contaminated materials. Japanese studies have reported a significant decrease of the level of this mycotoxin in the production of soy sauce.

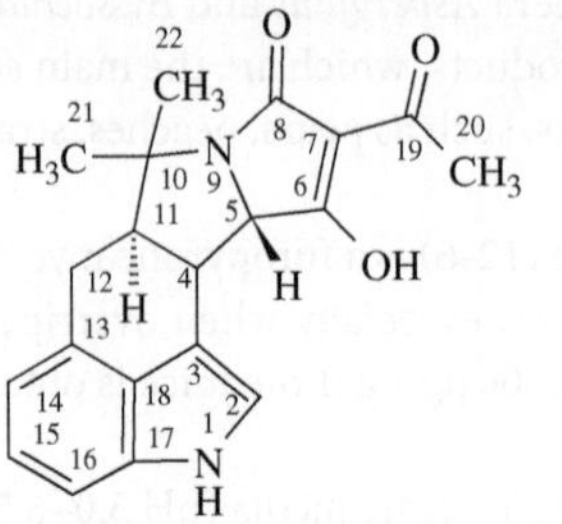

12-7, cyclopiazonic acid

The adverse health effects of cyclopiazonic acid manifest at higher concentrations. It is a specific inhibitor of calcium-ATPase, which transfers calcium after a muscle has contracted. In animals, it causes necrosis of the liver and necrotic changes in the gastrointestinal tract and muscles. It is considered a potential carcinogen, but exposure limits are not yet defined.

12.2.2.2.6 Roquefortine C

Roquefortine C (**12-8**) is a basic compound that can be categorised amongst the alkaloids. The basis of its structure is a dihydroindole skeleton. It can be produced by some strains of the fungus *Penicillium roquefortii*, used to make Roquefort cheeses (blue cheeses). These fungi are commonly found in nature and were discovered by cheese makers when ageing cheeses in damp, cool caves. *P. roquefortii* comprises three accepted species: *P. carneum*, associated with meat, cheese, and bread; *P. paneum*, associated primarily with bread and silage; and *P. roqueforti*, associated with various processed foods and silage.

At the time of its discovery, concentrations of roquefortine C in cheeses reached as much as units of mg/kg, but today the cultural strains used either do not produce this mycotoxin at all or produce it only in trace amounts. The toxicity of roquefortine is generally low, but detailed data on its effect on the human organism are not available.

Roquefortine C was often accompanied by a structurally related mycotoxin isofumigaclavine A (**12-8**). *P. roqueforti* moulds may occasionally produce patulin, citrinin, penicillic acid, the so-called PR toxin (**12-8**), and certain other toxins that have been implicated in incidents of mycotoxicoses. However, PR toxin is not stable in cheese and breaks down to the less toxic PR imine (**12-8**).

Other secondary metabolites of *P. roqueforti* found in blue cheeses are andrastins A–D, with skeletons of *ent*-5α,14β-androstane. In European blue cheeses, the content of andrastin A (**12-8**) ranged from 0.1 to 3.7 mg/kg, whilst the contents of andrastins B, C, and D were on average five times lower. The most significant biological activity of adrastins is the ability to inhibit the enzyme farnesyltransferase, which catalyses the transfer of farnesyl residue from farnesyl diphosphate to proteins. It is a part of the apparatus carrying post-translational modification of proteins in the cell. Attachment of farnesyl residue to a protein changes its physico-chemical and biological effects. Such modified proteins are able to cross the lipophilic membranes, and many play a role as cell-signalling molecules. They have also anticarcinogenic properties.

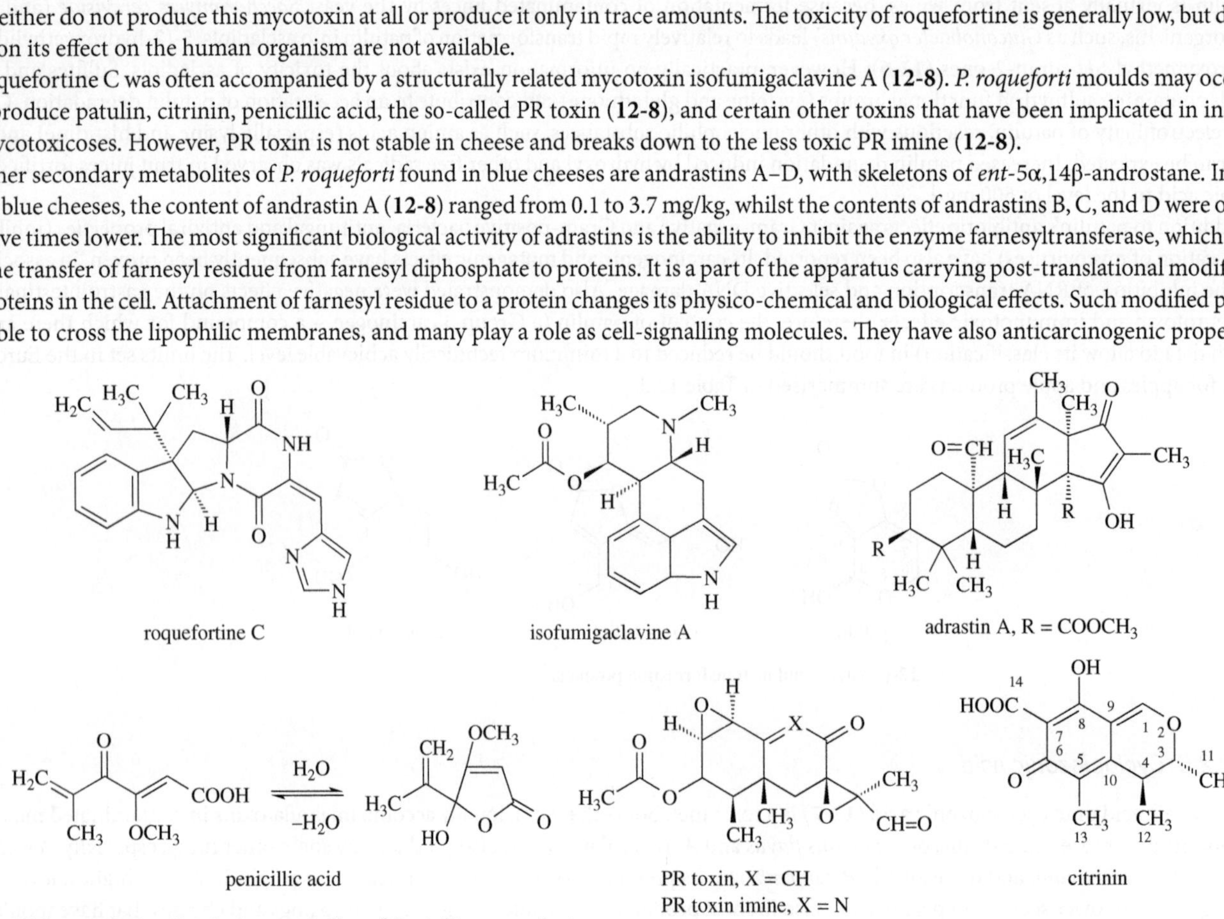

roquefortine C   isofumigaclavine A   adrastin A, R = COOCH₃

penicillic acid   PR toxin, X = CH   citrinin
                                   PR toxin imine, X = N

12-8, roquefortine C and other metabolites of *Penicillium roqueforti*

12.2.2.2.7 Citrinin

Citrinin, also known as antimycin (**12-8**), can exist in two tautomeric forms. The quinoid form is normally present in neutral media, whilst the tautomeric phenol form occurs in alkaline media. Citrinin is a secondary metabolite of fungi of the genera *Aspergillus* and *Penicillium*, especially the fungi *P. citrinum* and *P. verrucosum*. In temperate climates, it is found mainly in cereals, but data concerning its occurrence are limited. Citrinine has also been shown to be a contaminant in some red yeast rice products obtained using the Monascus yeast (see Section 9.7).

Shortly after its discovery in the early 1930s, citrinin was considered an antibiotic, but later its strong nephrotoxicity was demonstrated, as well as its possible teratogenicity and its interference with metabolic processes leading to damage in the liver. The intoxication of livestock (especially pigs, but also other monogastric animals) is manifested by diarrhoea and reduced weight gain. Toxic effects in humans have not been clearly documented.

12.2.2.2.8 Citreoviridin

Citreoviridin (**12-9**) is a typical secondary metabolite of the fungus *Penicillium citreoviride*. It was first identified in crops grown in Asia; its occurrence has been reported even in pecans and maize that had not been harvested at the time of full maturity. Information on its presence in the diet of the European population is not available, but some findings – particularly in maize – have been reported.

12-9, citreoviridin

Citreoviridin is a potent inhibitor of soluble mitochondrial ATPase. In experimental animals, it caused a disease called acute cardiac beriberi, whose symptoms are similar to those of classic beriberi (convulsions and paralysis), but which is not curable with thiamine.

12.2.2.2.9 Penicillic acid

Penicillic acid, (*Z*)-3-methoxy-5-methyl-4-oxohexa-2,5-dienoic acid, and the lactone of its hydrate, 5-hydroxyisoprop-5-en-1-yl-4-methoxyfuran-2-one (**12-8**), are produced by a wide range of fungi. An example is *Penicillium aurantiogrisseum*, found in maize. The presence of this mycotoxin has also been documented in some sausages and on the surface of hard cheeses. Occurrence of moulds in the deeper layers below the surface is limited by the availability of oxygen.

Penicillic acid is a cytotoxic agent with antibacterial and antiviral effects. During repeated administration to experimental animals, hepatotoxicity and nephrotoxicity have been demonstrated. It is a relatively unstable compound, reacting readily with sulfhydryl substances under opening of the lactone ring (as patulin) and undergoing loss of toxicity.

12.2.2.2.10 Trichothecenes

Trichothecenes are one of the most important groups of mycotoxins. Their main producers are fungi of the genus *Fusarium*. Unlike other toxigenic fungi, in which saprophytic species prevail, *Fusarium* species are primarily parasitic fungi, plant pathogens, and in some cases opportunistic infectious agents of humans and animals. The production of trichothecenes has been demonstrated even in some strains of fungi of the genera *Myrothecium* (e.g. *M. verrucaria* and *M. roridum*), *Trichoderma*, *Cephalosporium*, *Verticimonosporium*, and *Stachybotrys*. *Stachybotrys atra* moulds were known in ancient times and are now attracting attention again. Representatives of this genus grow preferentially on substrates rich in cellulose, such as wood, paper, and cotton. An unusual feature, compared with other moulds, is the high content of trichothecenes, especially T-2 toxin, in the spores. Inhalation of spores can cause pulmonary haemorrhage, and some fatalities have been reported in cases involving children.

Trichothecenes constitute a diverse group of compounds derived from tricyclic sesquiterpenes. The most important structural features causing their biological activities are six-membered rings containing a double bond between carbons C-9 and C-10, an epoxy ring between C-12 and C-13, and hydroxyl or acetyl groups at appropriate positions on the trichothecene nucleus, as well as the structure and position of the side chain. All trichothecenes have the same stereochemistry: α at C-3, C-7, and C-8 and β at C-4 and C-5 in A and B trichothecene types. According to the characteristic structures related to their properties, trichothecenes can be divided into four principal types:

- type A trichothecenes (**12-10**), which do not have the C-8 oxo group (e.g. the T-2 toxin, T-2 tetraol, HT-2 toxin, diacetoxyscirpenol, neosolaniol, and some other trichothecenes produced by *Fusarium acuminatum*, *F. sporotrichiodes* var. *tricinctum*, and *F. poae*);

- type B trichothecenes (**12-11**), which have the C-8 carbonyl group (e.g. nivalenol, deoxynivalenol (also called vomitoxin), 3-acetyldeoxynivalenol, 15-acetyldeoxynivalenol, trichothecin, and fusarenone-X produced by *F. nivale* and *F. episphaeria*);

- type C trichothecenes, which have a second epoxide group at C-7/C-8 (or at C-9/C-10) (e.g. crotocin, **12-12**, produced by *Cephalosporium crotocingigenum*);

- type D trichothecenes, which contain a macrocyclic ring between C-4 and C-15 with two ester linkages (e.g. verrucarin A, **12-13**).

Two additional trichothecene types are recognised: type E includes mycotoxins with an opened macrocyclic ring, whilst type F does not have the 12,13-epoxide ring, as it is replaced by a vinyl linkage with the oxygen atom in the epoxide ring removed.

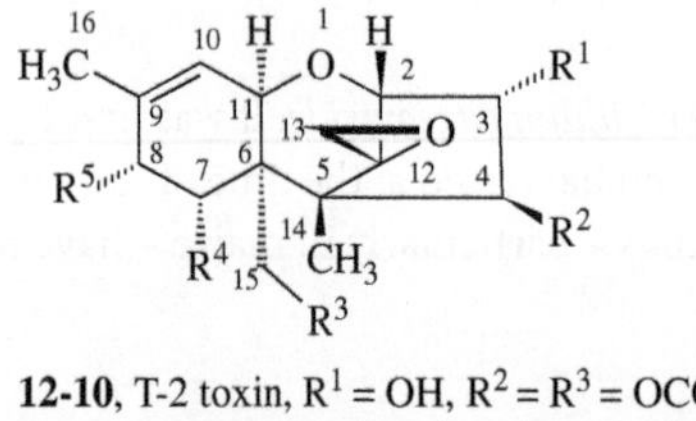

12-10, T-2 toxin, R^1 = OH, R^2 = R^3 = $OCOCH_3$, R^4 = H, R^5 = $COCH_2CH(CH_3)_2$
T-2 tetraol, R^1 = R^2 = R^3 = R^5 = OH, R^4 = H
HT-2 toxin, R^1 = R^2 = OH, R^3 = $OCOCH_3$, R^4 = H, R^5 = $COCH_2CH(CH_3)_2$
diacetoxyscirpenol, R^1 = OH, R^2 = R^3 = $OCOCH_3$, R^4 = R^5 = H
neosolaniol, R^1 = OH, R^2 = R^3 = $OCOCH_3$, R^4 = H, R^5 = OH

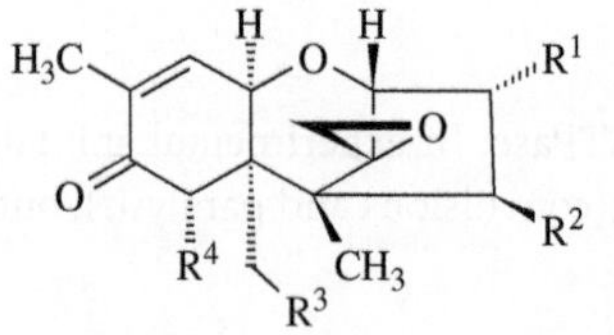

12-11, deoxynivalenol, R^1 = R^3 = R^4 = OH, R^2 = H
nivalenol, R^1 = R^2 = R^3 = R^4 = OH
trichothecin, R^1 = R^3 = R^4 = H, R^2 = $OCOCH=CHCH_3$
fusarenone-X, R^1 = R^2 = R^3 = OH, R^4 = $OCOCH_3$

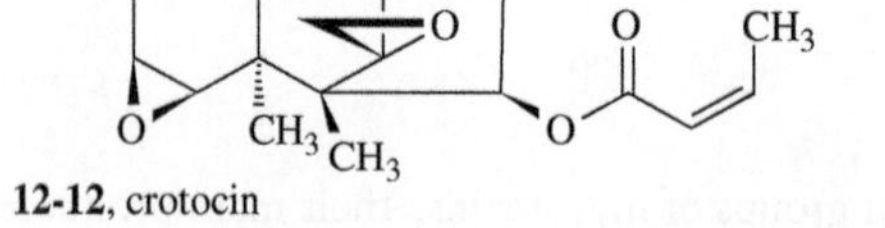

12-12, crotocin

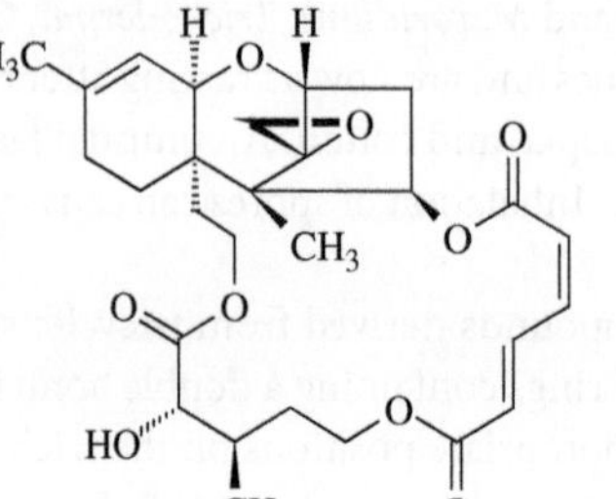

12-13, verrucarin A

Nearly 180 trichothecenes and their derivatives are currently known. Whilst types C and D do not commonly occur in crops, types A and B are widespread. The presence of one and sometimes more trichothecenes has been demonstrated mainly in cereals (wheat, barley, oats, and maize) grown in the temperate zones of Europe, America, and Asia. Trichothecenes A and B have also been reported in soybeans, other oil seeds, banana, and mango. Grains attacked by *Fusarium* moulds are characterised by a typical reddish colour.

Type A mycotoxins T-2 and HT-2 (a product of T-2 deacetylation) are found mainly in oats, in amounts up to 1000 μg/kg, but the annual variability of contamination is considerable.

One of the most trailed trichothecenes is a type B trichothecene deoxynivalenol, which is mainly produced by the fungus *Gibberella zeae*, an anamorph (the asexual reproductive form in the life cycle) of *Fusarium graminearum* found mainly in warmer geographic regions (its optimum growth temperature is 25 °C). Another producer is *F. culmorum*, which requires a lower temperature (optimum 21 °C). These microscopic fungi cause a disease in wheat, barley, maize, and other cereals called Fusarium head blight (FHB) that produces losses in grain yield and quality. Under conditions favourable for the growth of fungi, the deoxynivalenol concentration can reach up to 10 mg/kg; however, typical findings are significantly lower. The incidence of deoxynivalenol (as well as other mycotoxins) in cereals is highly variable, depending on the climatic conditions in the locality, the type of crop, and, of course, the resistance of crop varieties.

The physico-chemical properties of trichothecenes are dependent on the presence of polar substituents; therefore, type B trichothecenes are generally more polar than type A. Under normal conditions of technological and culinary processing of cereals, trichothecenes are relatively stable toxins and pass from contaminated raw materials into the final products, such as bakery products, breakfast cereals, and beer.

The only way to reduce the level of type A trichothecenes in oats is by the removal of bran from grains, as bran may contain 75–90% of mycotoxins. Similarly, during the processing of wheat in mills, a significant portion of type B mycotoxins (up to 50%) is removed, but their contamination in the bran is roughly two to three times higher than the original contamination in the grain. Whole-wheat flour derived by grinding of the whole grain generally contains higher amounts of trichothecenes than white flour, as its content of mycotoxins depends on the degree of milling (flour extraction rate) and is generally 20–70% of the original amount in the grain. For example, flour produced from wheat with a deoxynivalenol content of 0.51 mg/kg contained 0.35 mg/kg of this mycotoxin, whilst its concentration in the bran increased to 1.12 mg/kg.

The effect of baking on the deoxynivalenol content in non-yeast products is variable, ranging from no effect to 35% reduction. Higher decreases can be observed in fermented products, where the loss of deoxynivalenol ranges from 15 to 56%. A decrease of up to 40% in concentrations of deoxynivalenol and some other mycotoxins (nivalenol and acetyloxynivalenols) occurs with the addition of sulfites to the dough, as the sulfites react with trichothecene mycotoxins with the addition to the C-9/C-10 double bond yielding the corresponding hydroxysulfonates. A significant reduction in the content of trichothecenes, especially group B mycotoxins, is also achieved by washing the contaminated grains, as mycotoxins mainly occur on the surface. For example, in barley containing deoxynivalenol at a level of 4.1 mg/kg, the content decreased to 0.78 mg/kg in the grain after steeping but increased to 7.4 mg/kg in green malt and to 10.0 mg/kg in malt after germ separation. Concentrations in beer are in the range of about 0.13–0.5 mg/l.

In addition to free trichothecenes, plant materials contain various conjugates of mycotoxins with mono- and oligosaccharides, which form during the detoxification process in contaminated plants. These conjugates are either water-soluble (so-called masked mycotoxins) or insoluble (so-called bound mycotoxins). The best known conjugated trichothecene is deoxynivalenol-3-β-D-glucoside (**12-14**), which may occur, for example in wheat, in amounts corresponding to up to 30% of the molar concentration of free deoxynivalenol. Barley with a content of deoxynivalenol-3-glucoside 0.58 mg/kg had a content of this conjugate in grains after steeping of 1.1 mg/kg, in green malt of 17 mg/kg, and in malt after germ separation of 19 mg/kg. The increase over the initial level is due to deoxynivalenol-glycosyl-transferase enzymatic activity. Deoxynivalenol-3-glucoside can be hydrolysed to deoxynivalenol in human and animal gastrointestinal tracts.

Foods of animal origin do not contribute significantly to the exposure of consumers, as the transfer of trichothecenes into meat, milk, and eggs is insignificant because trichothecenes undergo rapid detoxification (epoxide ring opening and formation of glucuronides) after absorption in the digestive tract. In livestock (including cattle and pigs), acetylated trichothecenes are easily hydrolysed by enzymes of the intestinal microflora.

12-14, deoxynivalenol-3-β-D-glucopyranoside

Trichothecenes exhibit a wide range of biological effects, such as antibacterial, antiviral, cytostatic, and fungistatic effects; some are also phytotoxic. Toxicity to vertebrates varies over a considerable range. Based on experiments with animals, some (especially T-2 toxin as a representative of the type A trichothecenes) are mutagenic and immunotoxic (immunosuppressive) compounds that lower the body's normal immune response. In humans, exposure to T-2 toxin is associated with ATA, which is manifested by a marked reduction or the complete absence of leukocytes or platelets in blood serum. The largest ATA epidemic was recorded in the former Soviet Union in the 1940s, when, as a result of the war, unharvested grains (mainly millet) in fields were covered with snow and attacked by *Fusarium* moulds. The grain harvested the following spring was extremely contaminated and caused the deaths of more than 17 000 people. To a lesser extent, ATA also occurred in Hungary and France in the 1950s and '60s. Typical symptoms of intoxication are gastrointestinal inflammation, vomiting, and diarrhoea, often with headaches. In later stages, the intoxication leads to a reduction in the number of platelets and white blood cells, and in the third stage, to various infections caused (in healthy people) by banal harmless microflora. Another type A trichothecene, diacetoxyscirpenol, is significantly less acutely toxic than T-2 toxin, but is characterised by significant teratogenic potential. Deoxynivalenol, a representative of the type B trichothecenes, is the least toxic mycotoxin of this group, but is often the main one in contaminated foods. Signs of acute intoxication caused by deoxynivalenol are vomiting, abdominal pain, diarrhoea, and headaches associated with dizziness. Extremely high toxicity is shown by macrocyclic trichothecenes of type D. For example, the acute toxicity of verrucarin is still about 10 times higher than that of T-2 toxin. In comparison with aflatoxins, trichothecenes exhibit direct biological effects without metabolic activation. It is expected that they react with some cellular components, such as ribosomes, which results in inhibition of protein synthesis. The IARC, however, has not included trichothecenes amongst human carcinogens. Currently, maximum limits are set only for deoxynivalenol (Table 12.3).

12.2.2.2.11 Zearalenone

Zearalenone, also known as mycotoxin F_2 (**12-15**), is another toxicologically significant metabolite of fungi of the genus *Fusarium*. The major producers are the fungi *F. culmorum*, *F. graminearum*, *F. cerealis*, *F. equiseti*, and *F. semitectum*. Zearalenone is found mostly in cereals, especially maize but often also wheat, barley, and oats, and sometimes in spices. Producers of zearalenone are found in virtually all climatic zones, and almost without exception the positive samples (typical concentrations are in the range of tens to hundreds of µg/kg) are also contaminated by fusariotoxins (deoxynivalenol and nivalenol) and fumonisins. Some *Fusarium* strains also produce other zearalenone derivatives, whose concentrations can be 10–20% of those of zearalenone. For example, (*E*)-zearalenone-related metabolites of *F. graminearum* are 13-formylzearalenone, 5,6-dehydrozearalenone, epimers of 5-hydroxyzearalenone and 10-hydroxyzearalenone, α- and β-epimers of zearalenol, α-zearalanol (zeranol), β-zearalanol (taleranol or teranol), (*Z*)-zearalenone, and (*Z*)-zearalenol. α-Zearalanol is produced semisynthetically for veterinary use; such use is prohibited in the European Union.

12-15, (*E*)-zearalenone

The dominant metabolite found in maize (but also in wheat) is zearalenone-14-β-D-glucoside, accompanied by the corresponding malonate and diglycosides (diglucoside and xylosylglucoside). The fungus *Thamnidium elegans* also produces zearalenone-14,16-diglucoside. Another metabolite identified only in maize is zearalenone-14-sulfate. The formation of zearalenone conjugates was studied intensively in the model plant thale cress (*Arabidopsis thaliana*), which led to identification of glucoside, malonylglucoside, diglucoside, triglucoside, xylosylglucoside, and α-zearalenol and β-zearalenol. It is assumed that these metabolites may also be present in cereals (Figure 12.3).

Zearalenone is relatively thermostable and does not decompose even when heated to 120 °C for 4 hours; therefore, its loss is mainly attributed to the formation of complexes with components of the material. Irradiation by UV light causes isomerisation of the natural (*E*)-isomer to (*Z*)-isomer (**12-16**).

Washing of maize, which is effective in decreasing the concentration of deoxynivalenol and other trichothecenes, does not lead to a significant decrease in the amount of zearalenone, probably due to its hydrophobicity. As a relatively stable compound, zearalenone is transferred to cereal products, including bread, malt, and beer. Zearalenone is relatively lipophilic, and therefore is also found in vegetable oils, especially germ oils. As in the case of fusariotoxins, the content of zearalenone in flour is lower than that in the original grains and depends on the flour extraction rate. For example, white flours made from contaminated wheat contain only 30–50% of the initial amount of zearalenone, whilst its concentrations in bread (compared with flour) decrease by 34–40%. Other experiments found that the residual content of zearalenone was 60% of the initial amount in bread and 80% in biscuits. In the production of pasta in the presence of 1% potassium carbonate, the loss of zearalenone was 48–62%.

Zearalenone is rapidly metabolised in the liver. Primarily, the C-7 oxo group is reduced to yield α- and β-zearalenol and the C-11 double bond is reduced to yield the corresponding zearalanols. These products are then secreted as soluble glucuronides, but in small amounts they can be found in milk and eggs, although this route of human dietary exposure is marginal. Acute toxicity of zearalenone is low; however,

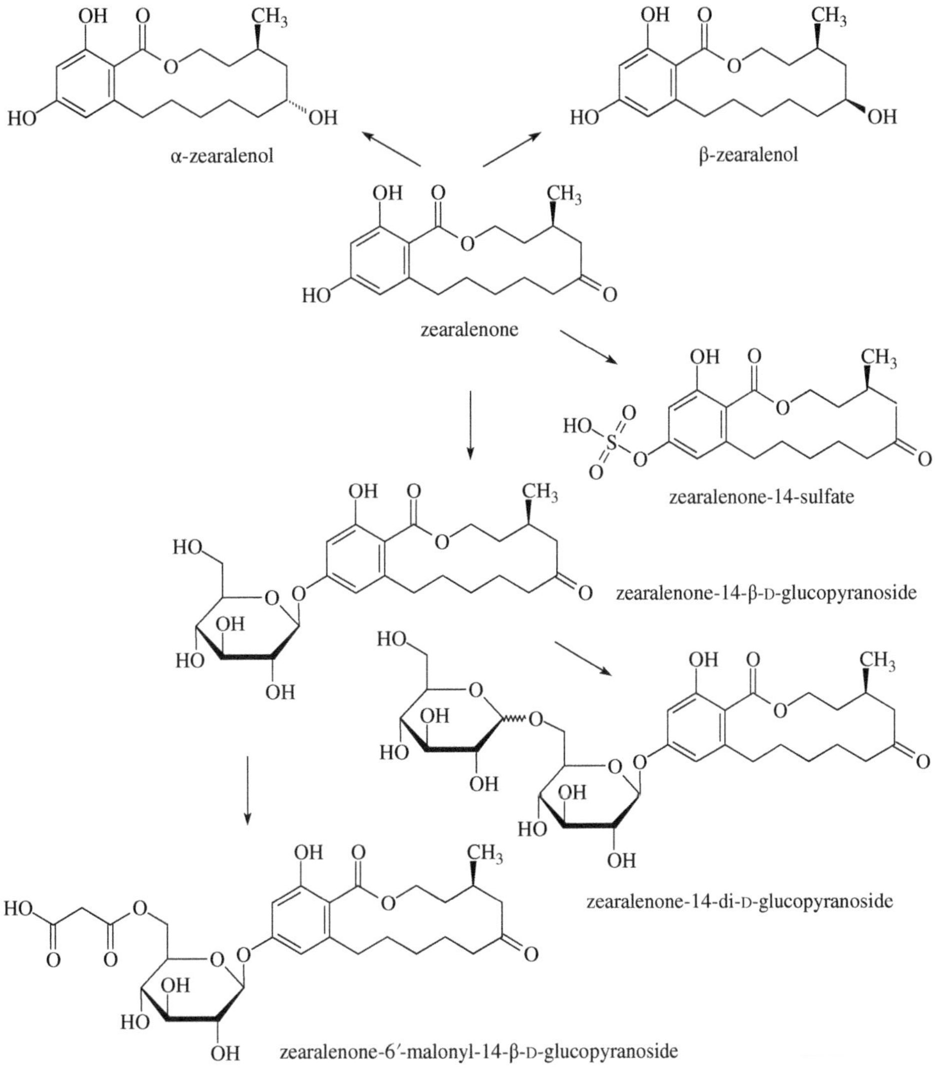

Figure 12.3 Zearalenone derivatives in plants.

zearalenone and its derivatives exhibit significant oestrogenic and anabolic effects. Dietary intake may cause (thanks to structural similarity with steroid hormones oestrogens) hyper oestrogenic syndrome. Oestrogenic potential is approximately one-tenth of the potential of 3,17β-estradiol. Generally, the binding affinity to the oestrogenic receptors decreases as follows: α-zearalanol > α-zearalenol > β-zearalanol > β-zearalenol. The IARC ranks zearalenone in Group 3 (compounds unclassifiable as to carcinogenicity in humans), but maximum limits are set for selected cereal commodities (Table 12.3).

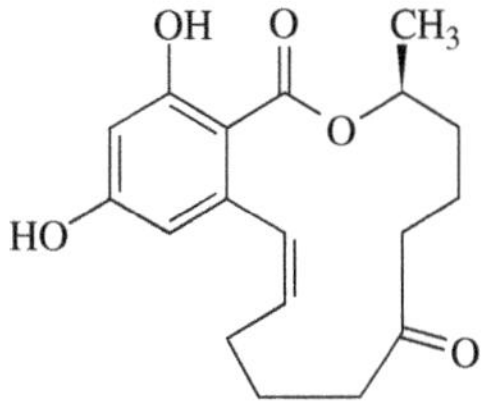

12-16, (Z)-zearalenone

12.2.2.2.12 Fumonisins

The fumonisins were only discovered in the late 1980s, in South Africa. Twenty fungi of the genus *Fusarium* have been idenitified that biosynthesise these secondary metabolites, but the main producers are the fungi *F. moniliformis* and *F. proliferatum*. Fumonisins are found in cereals, most notably maize (where their content can reach tens of mg/kg) and maize products, such as maize flakes and polenta. Fumonisins have also been identified in rice, millet, and other cereals. Findings in animal feed, especially in silage, are also common.

Dozens of secondary metabolites from the fumonisin group are currently known. These are divided into four groups (A, B, C, and P) according to their structure, which mainly include relatively polar diesters of propane-1,2,3-tricarboxylic acid with 2-amino- or 2-acetylamino-12,16-dimethyl-3,5,10,14,15-pentahydroxyeicosane (**12-17**). In terms of toxicity and food safety, monitored mycotoxins are group B fumonisins. The most common compounds occurring in the highest concentration are fumonisins B_1 and B_2, which often occur in a ratio of about $3:1$. Fumonisin B_3 can often be found in materials with higher levels of contamination. Fumonisin B_1 (macrofusine), the predominant metabolite, is known to cause a range of species-specific toxic responses, including leucoencephalomalacia in horses, pulmonary oedema in swine, and hepatosis and nephrotoxicity in rodents.

12-17, fumonisin B_1, $R^1 = R^2 = OH$
fumonisin B_2, $R^1 = OH$, $R^2 = H$
fumonisin B_3, $R^1 = H$, $R^2 = OH$

Like many other mycotoxins, fumonisins are relatively stable even at higher temperatures. For example, a significant decrease in the concentration of fumonisins in maize meal occurs during heating to 200 °C for 60 minutes. During baking of muffins made from contaminated maize flour at 220 °C, a noticeable reduction of the amount of fumonisins B_1 and B_2 occurred after about 30 minutes. With normal thermosterilation of sweet maize, losses of fumonisins were around 15%, rarely more. Higher pH values may cause hydrolysis of one or both propane-1,2,3-tricarboxylic acids. This reaction proceeds, for example, during the production of some speciality maize products, such as authentic Mexican tortillas and nachos prepared from maize flour with the addition of lime milk (see Section 5.8.5), and results in a reduction of fumonisin B_1 content of up to 60%. Hydrolysed fumonisins may occur in products at a level of up to 20% of the parent mycotoxin amount. Their toxicity has not yet been evaluated, but it is suggested that they may be more toxic than the parent compounds, because they are less polar and therefore more easily absorbed by the intestinal mucosa.

Losses of fumonisins during technological processing may be due to reactions with some components of the material, which yield bound forms of mycotoxins. Reaction depends mainly on the composition of the raw materials and the conditions employed during processing. Figure 12.4 illustrates the formation of fumonisin B_1 conjugates with starch and proteins, which takes place at high temperatures. The first reaction step is dehydration to fumonisin bisanhydride, which subsequently reacts with either protein (via the lysine ε-amino group and sulfhydryl group of bound cysteine) or starch. In the reaction of fumonisin B_1 with reducing sugars, the amino group of the side chain enters the reaction, yielding analogous products that also arise in the Maillard reaction. In model experiments and real foods, the corresponding glycosylamines, Amadori products, and other products arise. The identified adducts from the reaction of fumonisin B_1 with glucose include *N*-(1-deoxy-D-fructos-1-yl)fumonisin B_1, *N*-methylfumonisin B_1, *N*-(carboxymethyl)fumonisin B_1, *N*-(3-hydroxyacetonyl)-fumonisin B_1, *N*-(2-hydroxy-2-carboxyethyl)fumonisin B_1, and *N*-methylfumonisin B_1.

Toxicity of fumonisins is caused by inhibition of the enzyme sphingosine *N*-acetyltransferase (ceramide synthase) due to the structural similarity of fumonisins with sphingosine (see Section 3.5.1.1.3). The toxic effects are closely related to the presence of the amino group in the molecule; the reaction products with reducing sugars are virtually non-toxic as the amino group is blocked. Fumonisin B_2 is more cytotoxic than fumonisin B_1. The assessment of dietary risk of exposure to fumonisins is relatively complicated with regard to a variety of individual mycotoxins.

Fumonisins also cause a number of livestock diseases, such as a mycotoxicosis known as equine leucoencephalomalasia (ELEM) in horses and porcine pulmonary oedema (PPE) in pigs. Intoxication by higher doses of mycotoxins can be fatal in both cases.

In connection with the possible occurrence of fumonisins in foods, their hepatotoxicity and nephrotoxicity to humans and carcinogenicity to animals are often discussed. Studies from South Africa and China document a possible contribution of high doses of fumonisins from maize in the aetiology of human oesophageal cancer. According to the IARC, fumonisins are classified as Group 2B carcinogens and characterised as promoters of carcinogenic processes. For fumonisins B_1 and B_2, maximum limits are set for the sum of two major mycotoxins (Table 12.3) in selected cereal commodities. The tolerable daily intake (TDI) for fumonisins B_1, B_2, and B_3, alone or in combination, is 2 μg/kg b.w.

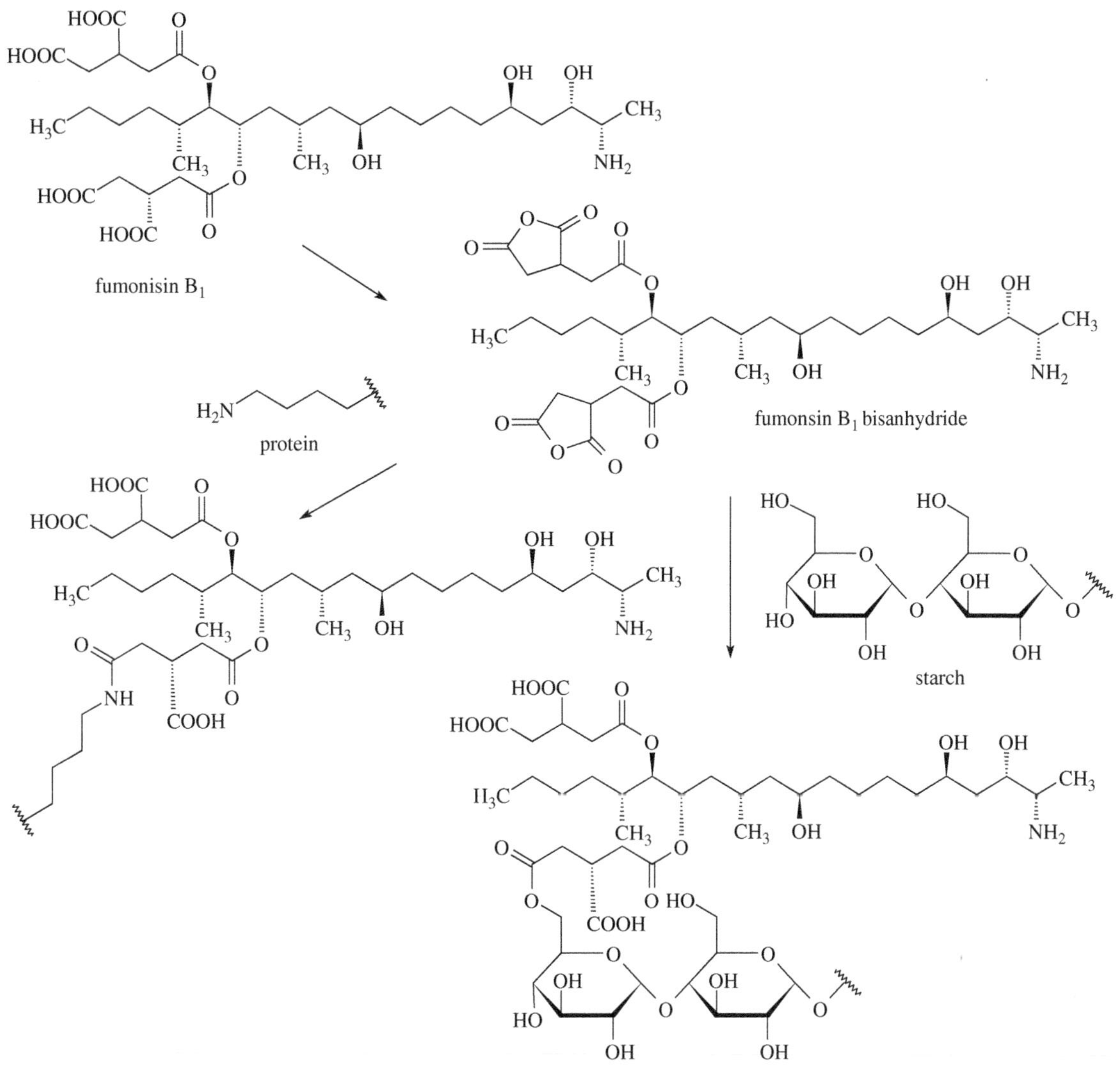

Figure 12.4 Reaction of fumonisin B₁ with proteins and starch.

12.2.2.2.13 Moniliformin

Moniliformin sodium salt (sodium 3-hydroxycyclobut-3-ene-1,2-dione, **12-18**) was originally isolated from substrates attacked by fungus *Fusarium moniliformis* (syn. *F. verticillioides*). Other producers of this genus include *F. avenaceum*, *F. subglutinans*, and *F. proliferatum*. Moniliformin is most often found in maize, but is also phytotoxic to wheat and tobacco. Poultry are very sensitive to this mycotoxin, especially chickens and ducklings. Its toxicological evaluation is not yet complete, and hygienic limits have not been established.

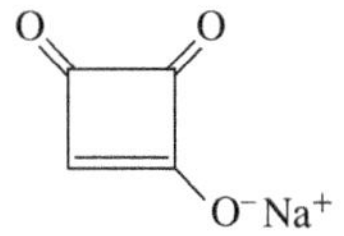

12-18, moniliformin

12.2.2.2.14 Fusarin C

Fusarin C (**12-19**) is a co-metabolite of the parasitic fungus *Fusarium moniliformis* (syn. *F. verticillioides*), occurring primarily on maize but also found, along with moniliformin, in other cereals. Of the nine known producers of fusarin C (including *F. graminearum*, teleomorph of *Gibberella moniliformis*, *G. fujikuroi*, and *F. venenatum*), seven have been isolated from European crops and soils. Fusarin C is a potent mutagen and a potential human carcinogen. From this perspective, its toxicity is similar to that of aflatoxin B_1 and sterigmatocystin.

12.2.2.2.15 Ergot mycotoxins

Ergot fungi are a group of fungi of the genus *Claviceps*, which includes about 50 species occurring mostly in the tropical regions. The most prominent member is the common fungus *C. purpurea*, which parasitises on certain cereals, especially rye and triticale, but also wheat. The infested grain is transformed by the fungus into black-to-dark purple sclerotium, known as ergot. Cereal heads may contain one or several ergot bodies, which contain an interesting group of ergot toxins: alkaloids with a wide range of biological effects. Ergot alkaloids are used in medicine today, especially in neuroendocrinology and the treatment of diseases associated with impaired neural transmission in both the central and the peripheral nervous systems (e.g. migraine, Parkinson's disease, senile dementia).

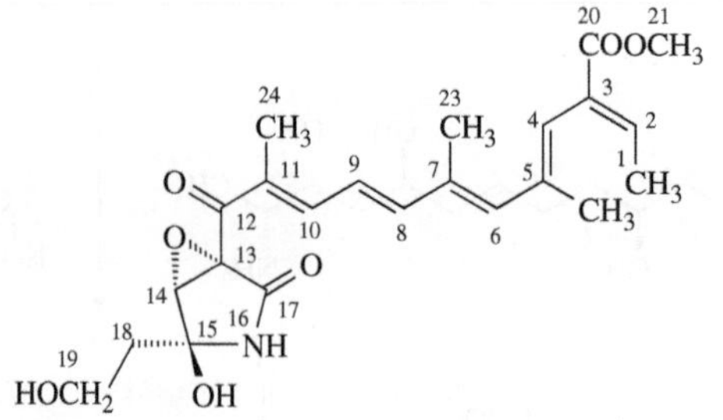

12-19, fusarin C

Historically, these alkaloids sometimes caused a very serious disease called ergotism, the first mycotoxicoses described in humans. Epidemics of ergotism were documented in ancient Greece and in the Middle Ages across Europe, particularly in populations that relied on bread prepared from rye flour contaminated with ergot alkaloids. *Ergotism gangraenosus* was a type of poisoning that affected the circulatory system (symptoms included swelling and inflammation of the limbs and, in the last stage, gangrene), whilst *ergotism convulsivus* was manifested primarily by an impaired nervous system (symptoms were hallucinations and ecstasy). The apparent randomness of these epidemics and the helplessness of people against their effects were still fertile ground for the spread of religious beliefs in the early modern period, as evidenced by the name given to the diseased state: Holy Fire (*ignis sacer* in Latin) or St Anthony's fire (*ignis sanati Antonii*). The last case of mass poisoning in Europe was described in the French village of Pont-Saint-Esprit in 1951. Outside Europe, a milder convulsive form of ergotism with no fatalities occurred in India in 1975. An outbreak of gangrenous ergotism occurred in Ethiopia in 1977–78, when 140 people showed signs of intoxication with high mortality (34%). In 2002, a severe outbreak was again reported in Ethiopia.

The ergot sclerotia contain from 0.15 to 0.5% alkaloids synthesised by a combination of fungal and plant metabolisms; more than 50 have been characterised. Ergot alkaloids are 3,4-disubstituted indole derivatives derived from the tetracyclic ring of ergoline, (6a*R*)-4,6,6a,7,8,9,10,10a-octahydroindolo[4,3-*fg*]quinoline (**12-20**), which has a C-9 double bond and a methyl group at N-6 and is substituted at C-8; therefore, ergot alkaloids are in fact derivatives of (+)-D-lysergic acid (**12-20**), known by the chemical name 7-methyl-4,6,6a,7,8,9-hexahydroindolo[4,3-*fg*]quinoline-9-carboxylic acid or 6-methylergol-9-ene-8-carboxylic acid, which is typically bound as an amide with an amino alcohol as in ergometrine (**12-20**) or with a small polypeptide structure as in ergotamine (**12-20**). The building blocks for lysergic acid are tryptamine and an isoprene unit. The relative representation of individual alkaloids differs in different producer strains and depending on the host plant and environmental factors. Ergot alkaloids are accompanied by simpler bases, biogenic amines (such as histamine and tyramine) and betaines (2-mercaptohistidine, known as ergothioneine). Ergot alkaloids can be divided according to the structure into three main groups:

- clavines (hydroxy- or dehydroderivatives of 6,8-dimethylergoline);

- simple lysergic acid derivatives (typically amides);

- peptide ergot alkaloids, known as ergopeptines (peptides derived from lysergic acid).

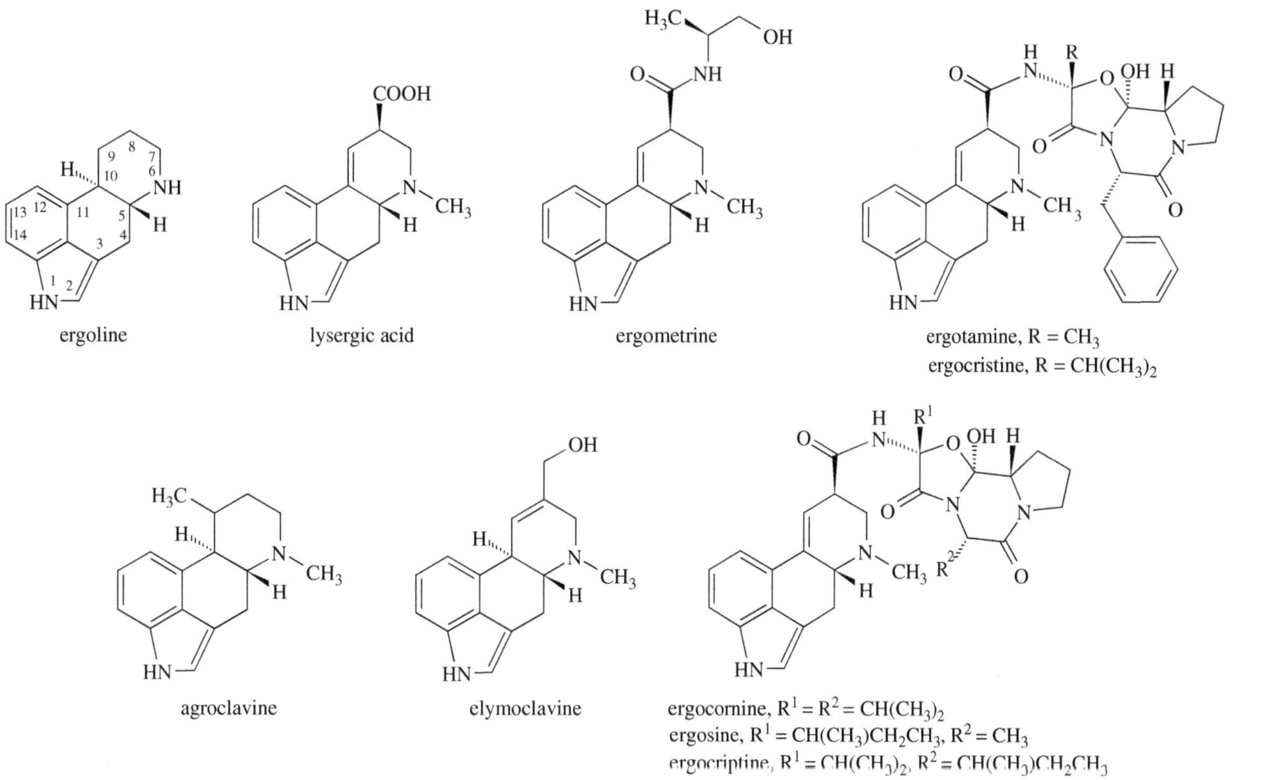

12-20, ergot alkaloids and parent compounds

Clavines (clavine alkaloids), such as agroclavine and elymoclavine (**12-20**), are considered precursors to other ergot alkaloids. At the end of the 1930s, a hallucinogenic drug, lysergic acid diethylamide (LSD), was synthesised by modification of natural derivatives of lysergic acid. In terms of chemical food safety, the EFSA has identified ergometrine, ergotamine, ergocristine, ergocornine, ergosine, and ergocryptine (α-ergocryptine) (**12-20**) as the most important mycotoxins.

During processing of contaminated grains, alkaloids from the sclerotium may pass into the flour and then into the final products, but concentrations of ergot alkaloids in contaminated flours rarely exceed 0.1 mg/kg. Rye flours may commonly contain higher amounts of alkaloids than wheat flours. For example, monitoring of Swiss flours in 1985 found that the total alkaloid content in rye flours was 0.015–0.397 mg/kg, that in white flours was 0.004 mg/kg, and that in brown wheat flours was 0.103 mg/kg. Monitoring in Sweden in 1993 found the highest concentration in wheat and rye products to be 0.024 mg/kg, with ergotamine as the most frequently identified toxin, whilst in Canada in 1980 ergocristine was the major alkaloid detected in wheat and rye flour, at levels up to 0.062 mg/kg.

With the exception of ergometrine, ergot alkaloids are relatively non-polar compounds and therefore are not very soluble in water. Lysergic acid is a chiral compound with two stereo centres. The isomer with inverted configuration at C-8 close to the carboxy group is called D-isolysergic acid. Inversion at C-5 close to the nitrogen atom leads to L-lysergic acid and L-isolysergic acid. Alkaloids with a double bond between carbons C-9 and C-10 (ergolenes) are susceptible to epimerisation, especially in non-acidic media (Figure 12.5). The laevo-rotating ergopeptins with configuration (*R*) at C-8 thus give rise to dextro-rotating (*S*)-isomers (D-isolysergic acid derivatives) called ergopeptinins. The individual isomers differ in both biological and physico-chemical properties. Their amount increases with unsuitable conditions of grain storage.

In addition to ergot, ergot alkaloids occur in some higher plants. For example, derivatives of D-lysergic and D-isolysergic acids were isolated from the seeds of beach moonflower (*Ipomoea violacea*) and the related Christmas vine *Turbina corymbosa* (Convolvulaceae). These seeds were used traditionally amongst Mexico's Zapotec Indians for ceremonial and curative purposes.

Currently, no country regulates ergot alkaloids in food, and only in Canada and Uruguay are they regulated in feeds. The Canadian guideline limit for ergots in feed for swine is 6000 μg/kg, that for feed for dairy cattle, sheep, and horses is 3000 μg/kg, and that for feed for chicks is 9000 μg/kg. In the European Union, the issue of ergot alkaloids is currently under discussion, and maximum limits of 400–500 μg/kg in grain intended for human consumption have been proposed. In India, regulatory measures exist for the occurrence of ergot bodies in cereals, and the designed limit is 0.01% (1 ergot body per 10 000 grains). At present, the maximum permissible level in the United States and Canada is 300 mg ergot per kg grain. Feed materials exceeding this limit are labelled 'ergoty'.

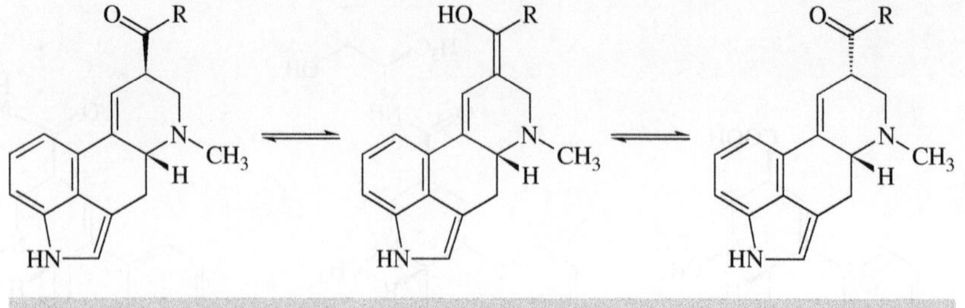

Figure 12.5 Isomerisation of ergot alkaloids.

12.2.2.2.16 Alternaria mycotoxins

Many species of the genus *Alternaria* can invade crops in the pre- and post-harvest periods and cause considerable losses due to rotting of various fruits and vegetables. Owing to their ability to grow even at low temperatures, they are also responsible for spoilage of these commodities during refrigerated transport and storage. Several *Alternaria* species are known producers of toxic secondary metabolites called *Alternaria* mycotoxins or altertoxins. The best-known representative of these fungi is *A. alternate*, found on cereals, sunflower seeds, oilseed rape, olives, and various other fruits and vegetables, especially tomatoes.

Alternaria mycotoxins can be divided into three groups based on their structure. The first is derivatives of dibenzopyrone, which include alternariol (isolated in 1953 as the first alternaria toxin), alternariol monomethyl ether, and altenuene (**12-21**). The second group includes derivatives of perylene, represented by altertoxin I and derived epoxide altertoxin II (**12-21**). The third group contains derivatives of tetramic acid (1,5-dihydro-4-hydroxy-2*H*-pyrrole-2-one), the most studied of which is tenuazonic acid, (5*S*,6*S*)-3-acetyl-5-*sec*-butyl-4-hydroxy-3-pyrrolin-2-one (**12-21**).

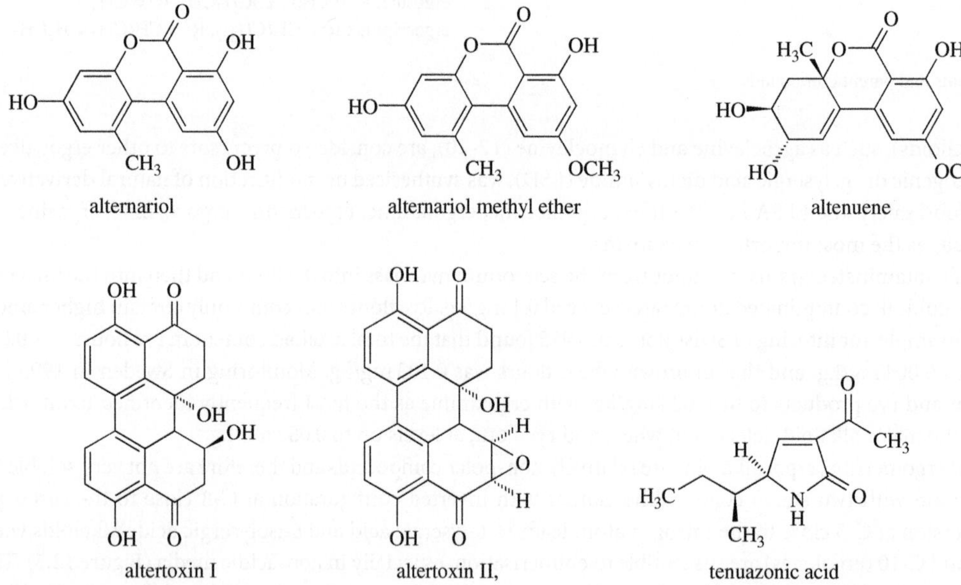

alternariol   alternariol methyl ether   altenuene

altertoxin I   altertoxin II,   tenuazonic acid

12-21, alternaria mycotoxins

Information on the occurrence of alternaria mycotoxins in food and feed is relatively limited. For example, alternariol, its methyl ether, and tenuazonic acid have frequently been detected in apples, apple juice concentrates, and other apple products, as well as mandarins, olives, red peppers, tomatoes, tomato products, rapeseed meal, sunflower seeds, sorghum, wheat, and edible oils (olive oil, rapeseed oil, sesame oil, and sunflower oil). The maximum levels reported in foods are in the range of about 1–1000 µg/kg; higher levels are commonly found in visibly infected rotted products that are not suitable for consumption.

Alternaria mycotoxins are relatively stable under conditions of technological and culinary food processing.

Exposure to altertoxins has been linked to a variety of adverse health effects. They exhibit only low acute toxicity (the most toxic substance is tenuazonic acid), and their chronic toxic effects are not yet fully described. Mycotoxicoses, however, have not only been described in animals, but have been associated with ATA, a frequently fatal mycotoxicosis that followed the ingestion of grain or grain byproducts infested with fungi. Toxicological studies have demonstrated the ability of tenuazonic acid to inhibit protein synthesis; alternariol and its methyl ether were shown to be teratogenic. Alternaria toxins are generally cytotoxic for mammalian cells, and from a toxicological point of view are of the highest concern, being probably mutagenic, although their mutagenicity is at least one order of magnitude lower in comparison with that of aflatoxin B_1. Exposure limits as TDI or a provisional maximum tolerable daily intake (PMTDI) for alternaria toxins have not been derived.

12.2.2.2.17 Tremorgens

Tremorgens include a rather heterogeneous group of more than 30 alkaloids (indole diterpenoids). Producers of tremorgens are fungi of the genus *Aspergillus* and some other genera, such as *Penicillium* and *Claviceps*. The first tremorgen, aflatrem (12-22), was isolated from the mycelium of *A. flavus* and is also produced by other *Aspergillus* species, such as *A. minisclerotigenes*. Another example of this group of mycotoxins is paxilline (or paxiline, 12-22), which is produced by the fungus *Penicillium paxilli*. The fungus *Claviceps paspali* produces ergot alkaloids in sclerotia, as well as tremorgens called paspalitrems, such as paspalitrem A (12-22), B, and C. Paspalitrem B contains a 3-hydroxy-3-methylbut-1-(en-1-yl) unit instead of a 3-methylbut-2-en-1-yl unit. Paspalitrem C differs from paspalitrem A only by the position of attachment of the 3-methylbut-1-en-1-yl substituent to the indole ring.

Tremorgens have been found in a number of commodities, such as maize, wheat, nuts, and silage and other feeds. Tremorgenic mycotoxins induce neurologic symptoms ranging from mental confusion to tremors, seizures, and death, and are apparently the only class of mycotoxins with significant central nervous system activity. Tremorgens have been implicated in a number of neurologic diseases of cattle, collectively known as staggers syndromes. They pose significant agricultural and health problems for both cattle and humans. Toxicological evaluation of tremorgens is not complete, and hygienic limits have not been established.

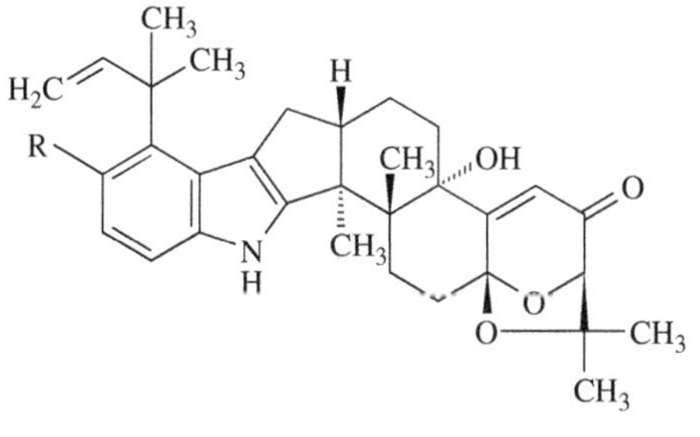

aflatrem, R = H
paspalitrem A, R = CH₂CH=CH(CH₃)₂

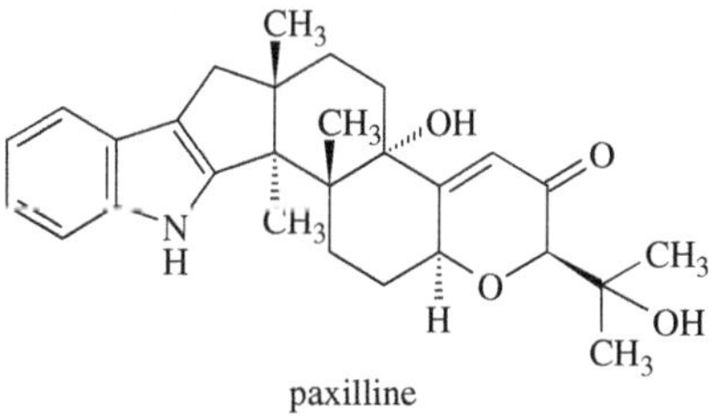

paxilline

12-22, structures of selected tremorgens

12.2.2.3 Health and toxicological evaluation

Unequivocally, the most important route of human exposure to mycotoxins is contaminated foods, although in certain circumstances a significant risk may be associated with the inhalation of mould spores or dust particles containing mycotoxins. Contamination of pharmaceutical preparations created by biotechnological processes, in which the filamentous fungi are used, is also possible. The health limits for individual mycotoxins in foods are exceeded in the European Union only in extremely rare cases. The health risks of life-long (chronic) exposure even to very low concentrations of mycotoxins (often represented by several compounds) are essentially unknown.

12.2.2.4 Mitigation

The optimal way of reducing the occurrence of mycotoxins in human food represents a complex of three basic preventive measures:

- reduction of infection by toxigenic fungi in the growth period of crops;

- fast and efficient drying of harvested crops and their proper storage;

- use of fungicides to inhibit the growth of fungi and fungal spores in the growth period of crops and during crop storage.

Measures to minimise contamination of agricultural crops in the pre-harvest period can significantly reduce the incidence of toxins produced by fungi of the genus *Fusarium* (trichothecenes and zearalenone), ergot alkaloids, tremorgens, and even aflatoxins. Historically, the oldest successful method is a reduction in the amounts of ergot alkaloids through effective application of preventive measures, such as crop varieties resistant to mould infestation, crop rotation, and application of fungicides. Besides conventional techniques, innovative approaches are now increasingly applied, such as the outputs of genetic engineering or the use of so-called biocompetitive factors. For example, cereals deliberately infected with spores of competitive non-toxigenic strains of *A. flavus* contain only traces of aflatoxins.

The successful protection of crops against fungal attack continues at the moment of harvest. Crops must be harvested at full maturity, when their humidity is low, and they must be dry before storage. With regard to the manufacturing industry, all types of physical methods can extend food shelf-life, including pasteurisation, sterilisation, cooling, freezing, vacuum packaging, and irradiation; these can all be seen as precautionary measures. In practice, chemical methods are also occasionally used, including the use of various preservatives, such as sorbic, benzoic, and propionic acids.

Although the previously preventive measures can help reduce the occurrence of mycotoxins in the human food chain, exclusion of toxigenic microflora is technically impossible, and in many cases it is necessary to detoxify and decontaminate large amounts of crop or feed in order to protect the health of consumers or livestock and prevent economic losses. In practice, detoxification or decontamination procedures include degradation of mycotoxins to inactive products. Generally, there are three possible solutions, based on physical, chemical, or biological principles.

A decrease in the content of mycotoxins may be achieved by mechanical removal of attacked particles and by washing or milling of the infected grains. Heat treatment of contaminated materials decreases mycotoxin concentrations in many cases. Effective removal of aflatoxins from oilseed meals can be achieved using different solvents (such as acetone or ethanol), but this process is cost-intensive, leads to a loss of nutrients, and may affect the organoleptic properties of the meals. Any aflatoxin residues that may be present in crude vegetable oils are removed during refining. Aflatoxins can be removed from milk, cream, or peanut oil using hydrated calcium aluminosilicate clay or bentonite. These mineral materials are sometimes added directly to feed in order to immobilise mycotoxins in the digestive tract of livestock (reducing the transfer into the blood stream). For example, the addition of hydrated calcium aluminosilicate to the feed of dairy cows decreases excretion of aflatoxin M_1 to milk by 24–44%. However, the question remains as to whether the use of chemisorbents reduces the intake of some essential minerals (such as copper, zinc, and iron) and water-soluble vitamins.

The greatest attention has been paid to the possibility of chemical degradation of aflatoxins. The most important method is decontamination with ammonia. Other agents tested include sodium hypochlorite and hydrogen peroxide, the practical application of which is problematic due to the unwanted oxidation of a number of treated material components. Aflatoxins B_1 and G_1 and patulin are partially degraded by sulfur dioxide.

Biological detoxification in practice means biotransformation or biodegradation of mycotoxins by the action of enzymes, which generate metabolites that are non-toxic or less toxic than the starting toxin and can be easily eliminated from the body. Bacteria studied in this respect are *Flavobacterium aurantiacum* and some fungi of the genus *Rhizopus*. Biotechnological methods have many advantages over chemical methods, as they do not use aggressive agents and are often beneficial in improving the digestibility and usability of proteins. Promising results in the detoxification of trichothecenes, for example, have been brought about by biotechnological methods utilising microflora of the gastrointestinal tract of monogastric animals.

12.3 Technological contaminants

Processes operating at elevated temperatures during various industrial operations and culinary processing induce, depending on the specific physical conditions and chemical composition, a variety of enzymatic (biochemical) and non-enzymatic (chemical) reactions leading to the formation of new compounds. Many compounds produced in this way have antioxidant and antimutagenic properties, but industrial operations and culinary processing may also lead to the formation of compounds with proven toxic potential, which in some cases may show mutagenicity and carcinogenicity. This group of contaminants is mostly considered processing or technological contaminants.

Precursors of undesired and toxic compounds may be common amino acids or proteins, reducing sugars (including ascorbic acid), fatty acids, triacylglycerols, other lipids, and some other food components. Often, there are multiple precursors. Some additives may also participate in the formation of contaminants, such as hydrochloric acid (known as protein hydrolysing agent in the production of acid protein hydrolysates, and in technological practice considered an ancillary material), chlorine (chlorine dioxide) from drinking water, and nitrites (used as antimicrobial agents and stabilisers of meat colour).

Examples of processing contaminants arising at higher temperatures are primary aromatic amines produced from creatinine, PAHs formed by pyrolysis of a number of compounds, acrylamide, and furan arising in the Maillard reaction. Under normal food processing conditions, chloropropanols and their esters with higher fatty acids arise from lipids, and nitrosamines from amines and nitrites (and nitrogen oxides). Some processing contaminants may be simultaneously classified as exogenous contaminants, but exogenous sources are minimal in comparison with endogenous ones. For example, the exogenous source of PAHs in the food chain may be a contaminated environment or tobacco smoke. Tobacco smoke can also be a source of nitrosamines. Some contaminants classified as exogenous may also arise in smaller amounts during food processing. Typical examples are monocyclic aromatic hydrocarbons (MAHs), such as benzene, which is produced by decarboxylation of benzoic acid. Styrene is a product of cinnamic acid decarboxylation. A number of other compounds that can be classified as technological contaminants are covered elsewhere in this book. For example, thermal processes in alkaline media give rise to lysinoalanine (see Section 2.5.1.3.3) and D-isomers of amino acids (see Section 2.5.1.2), whilst the Maillard reaction gives rise to potentially toxic (mutagenic) compounds such as methylglyoxal (see Section 4.7.1.2.2), and 5-hydroxymethylfuran-2-carbaldehyde (see Section 4.7.1.1.3). Other toxic products may be formed from lipids, such as acrolein (see Section 3.8.2.2.3), malondialdehyde, and 4-hydroxynon-2-enal (see Section 3.8.1.12.1). Also discussed is the health hazard of transformed glycosylated proteins, known as AGEs, which are produced in the advanced stages of the Maillard reaction, or of the products of the subsequent oxidation of lipids (lipoxidation), known as ALEs (see Section 3.8.1.12.1).

The following sections deal in detail only with processing contaminants whose occurrence (and amount) in foods is either regulated by EU legislation (for which maximum residue limits, MRLs are set) or subject to various monitoring studies.

12.3.1 Heterocyclic amines

12.3.1.1 Classification, structure, and properties

The study of toxic products of the Maillard reaction was initiated in 1977 with the finding that the mutagenicity of charred grilled meat and fish is produced, in addition to neutral PAHs such as benzo[*a*]pyrene, by certain basic compounds. Later research has shown that the basic compounds are mainly primary heterocyclic amines. These substances were divided into two groups:

- non-IQ mutagens;

- IQ mutagens.

Non-IQ (also known as aminocarbolines or pyrolytic heterocyclic aromatic amines) and IQ mutagens are typical technological contaminants. More than 20 mutagenic heterocyclic aromatic amines have been isolated from various cooked meats, fish, and poultry. Mutagens of the first group are mainly **pyridoimidazoles** and **pyridoindoles**. These substances were first isolated from pyrolysates of amino acids and proteins obtained at temperatures of 300–800 °C. Particularly high mutagenicity is shown by pyrolysates of tryptophan, glutamic acid, lysine, ornithine, and phenylalanine and pyrolysates of some proteins (casein, wheat gluten, and soy globulin). It was subsequently found that these substances likewise arise during thermal processing of foods, such as roasting, pan frying, grilling, and grilling directly over an open flame. The first group of mutagens (**12-23**) includes 3-amino-1,4-dimethyl-5*H*-pyrido[4,3-*b*]indole, also known as Trp-P-1, which was found in tryptophan pyrolysates together with 3-amino-1-methyl-5*H*-pyrido[4,3-*b*]indole (Trp-P-2). To this group also belong 2-amino-6-methyldipyrido[1,2-*a* : 3′,2′-*d*]imidazole (Glu-P-1), 2-aminodipyrido [1,2-*a* : 3′,2′-*d*]imidazole (Glu-P-2), found in glutamic acid pyrolysates, 3,4-cyclopentenopyrido[3,2-*a*]carbazole (Lys-P-1) from lysine pyrolysates, 4-amino-6-methyl-1*H*-2, 5,10,10*b*-tetraazafluoranthene (Orn-P-1) from ornithine pyrolysates, 2-amino-5-phenylpyridine (Phe-P-1) arising from phenylalanine, and 2-amino-9*H*-pyrido[2,3-*b*]indole (AαC) and 2-amino-3-methyl-9*H*-pyrido[2,3-*b*]indole (MeAαC).

Pyrolysates of aliphatic amino acids, such as alanine, valine, leucine, and isoleucine (obtained at 230–250 °C), contain, in addition to the corresponding 2,5-dioxopiperazines (see Section 2.5.1.3.3), compounds of similar structure called **imidazopyrazinediones**, such as 7,8-dihydro-2*H*-imidazo[1,2-*a*]pyrazine-3,6-dione (**12-24**), which are formed in the reaction of 2,5-dioxopiperazines with another molecule of amino acid and subsequent cyclisation. **Diimidazopyrazinediones**, such as 2,5,7,10-tetrahydrodiimidazo[1,2-*a* : 1′,2′-*d*] pyrazine-3,6-dione (**12-24**), are formed analogously. These compounds do not show toxic effects as they are not primary aromatic amines.

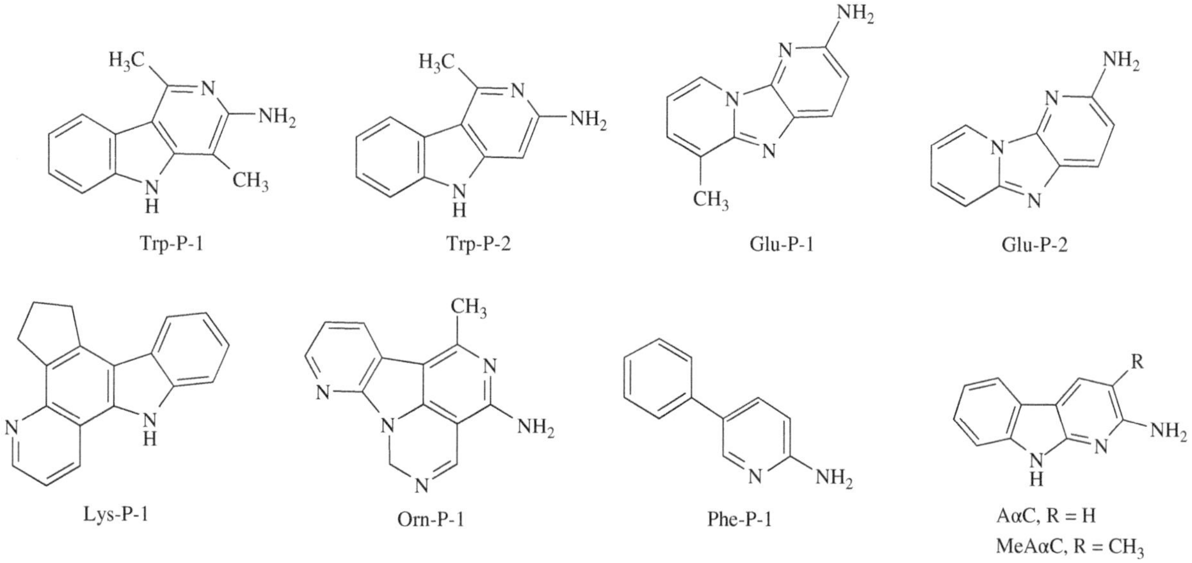

Trp-P-1   Trp-P-2   Glu-P-1   Glu-P-2

Lys-P-1   Orn-P-1   Phe-P-1   AαC, R = H
                                                   MeAαC, R = CH₃

12-23, products in pyrolysates of amino acids

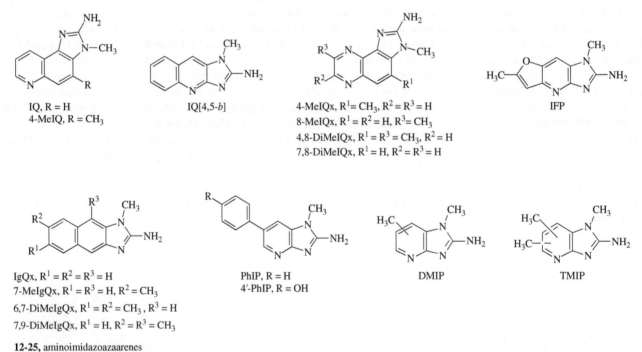

12-24, imidazopyrazinediones and diimidazopyrazinediones

Mutagens of the second and most important group include **aminoimidazoquinolines**, **aminoimidazoquinoxalines**, and **aminoimidazopyridine** (**12-25**), namely 2-amino-3-methylimidazo [4,5-*f*] quinoline (IQ), 2-amino-3,4-dimethylimidazo [4,5-*f*]quinoline (MeIQ or 4-MeIQ), 2-amino-3,4-dimethylimidazo[4,5-*f*] quinoline known as IQ [4,5-*b*], 2-amino-3-methylimidazo[4,5-*f*]quinoxaline (IQx), 2-amino-3,4-dimethylimidazo [4,5-*f*] quinoxaline (4-MeIQx), 2-amino-3,8-dimethylimidazo [4,5-*f*]quinoxaline (MeIQx or 8-MeIQx), 2-amino-3,4,8-trimethylimidazo [4,5-*f*]quinoxaline (4,8-DiMeIQx), 2-amino-3,7,8-trimethylimidazo-[4,5-*f*]quinoxaline (7,8-DiMeIQx), 2-amino-1-methyl-imidazo[4,5-*g*]quinoxaline (IgQx), 2-amino-1,7-dimethyl imidazo[4,5-*g*]quinoxaline (7-MeIgQx), 2-amino-1,6,7-tri methylimidazo[4,5-*g*]quinoxaline (6,7-DiMeIgQx), 2-amino-1,7,9-trimethylimidazo[4,5-*g*]quinoxaline (7,9-DiMeIgQx), 2-amino-1-methyl-6-phenylimidazo[4,5-*b*]pyridine (PhIP), 2-amino-1-methyl-6-(4-hydroxyphenyl)imidazo-[4,5-*b*]pyridine (4′-OH-PhIP), IFP, 2-amino-(1,6-dimethylfuro[3,2-e]imidazo[4,5-b])pyridine, 2-amino-dimethylimidazopyridine (DMIP), and 2-aminotrimethylimidazopyridine (TMIP). All these IQ mutagens containing an imidazole cycle derived from creatine/creatinine and an amino group in the C-2 position of the cycle are collectively called **aminoimidazoazaarenes**. The remaining parts of their molecules are formed from Maillard reaction products. Aminoimidazoquinoline (IQ) and aminomethylimidazoquinoline (MeIQ) were first isolated from broiled sun-dried sardines by Japanese scientists in 1980, and one year later aminomethylimidazoquinoxaline (MeIQx) was isolated from fried beef.

IQ, R = H
4-MeIQ, R = CH₃

IQ[4,5-*b*]

4-MeIQx, R¹= CH₃, R² = R³ = H
8-MeIQx, R¹ = R² = H, R³= CH₃
4,8-DiMeIQx, R¹ = R³ = CH₃, R² = H
7,8-DiMeIQx, R¹ = H, R² = R³ = H

IFP

IgQx, R¹ = R² = R³ = H
7-MeIgQx, R¹ = R³ = H, R² = CH₃
6,7-DiMeIgQx, R¹ = R² = CH₃, R³ = H
7,9-DiMeIgQx, R¹ = H, R² = R³ = CH₃

PhIP, R = H
4′-PhIP, R = OH

DMIP

TMIP

12-25, aminoimidazoazaarenes

12.3.1.2 Occurrence, main sources, and dietary intake

Exposure to aminoimidazoazaarenes primarily occurs through meat and fish, but small amounts may also be present in flavourings, beer, wine, and cigarette smoke. Exposure can similarly occur through inhalation of aerosol particles generated during cooking or that are present in the combustion gas, especially in diesel engine exhaust gas. In foods, PhIP is found in the highest levels, followed by MeIQx, MeIQ, and IQ. The total amount in thermally processed meat varies by meat type, cooking method, and 'doneness' level (rare, medium, or well done), but usually ranges from <1 to about 500 μg/kg; the current average concentrations are less than 100 μg/kg. Total daily intake of aminoimidazoazaarenes is within the limits of 160–1800 ng, and the dietary exposure is estimated to be 1–17 ng/kg b.w per day.

PhIP is the main aromatic amine formed in very-well-done barbequed chicken (up to 305 μg/kg), oven-broiled bacon (16 μg/kg), pan-fried bacon (4.9 μg/kg), oven-broiled ground beef (0.04–2.9 μg/kg), and pan-fried beef steak (0.04–12.46 μg/kg). The amount of PhIP in meat depends greatly on its type and the method of heat treatment. Available data vary widely. For example, salmon cooked in a pan, in a convection oven, or grilled at 200 °C may contain PhIP at the level of 0.02–2.2 μg/kg, and in extreme cases even 73 μg/kg. PhIP has also been identified in flavourings, beer, and wine (0.01–480 μg/kg), in cigarette smoke, and in the air and surface waters. Daily intake varies from 286 to 458 ng, and its dietary exposure is estimated at 17 ng/kg b.w. Cooked meats also contain the 4′-hydroxy analogue of PhIP, 4′-OH-PhIP. The highest content of 4′-OH-PhIP was found in fried and griddled chicken breast, at concentrations of 43.7 and 13.4 μg/kg, respectively; the corresponding PhIP concentrations were 19.2 and 5.8 μg/kg. MeIQx was found in beef, pork, and chicken meat and in fish. The highest amount in beef (steaks) was 8.2 μg/kg and that in hamburgers 4.6 μg/kg, but in pork the concentrations were relatively low (except bacon, at 0.9–27 μg/kg). Concentrations in grilled chicken and in chicken roasted in the oven or in a pan were 9 μg/kg, and in fish about 1.2 μg/kg. MeIQx was also found in thickened meat broth, wine, surface water, and air. Its daily intake is estimated at 33–44.8 ng, and its dietary exposure at 2.61 ng/kg b.w.

MeIQ occurs less frequently and at lower concentrations as compared with other aminoimidazoazaarenes. The highest concentrations (0.03–72 μg/kg) have been found in fish, at the upper limit in grilled sardines and at the lower limit in baked fish. It is found in low or undetectable quantities in beef, pork (0.02 μg/kg in roasted meat, 1.7 μg/kg in bacon), and chicken. MeIQ has also been identified in thickened meat broth, roasted coffee beans, and cigarette smoke. Dietary exposure is estimated at 0.6 ng/kg b.w.

IQ was first found in grilled fish, fried minced beef, and beef extracts, and later in many other foods (steak, roasted chicken, fried minced pork, and fried eggs) and cigarette smoke. Its concentration ranges from <0.1 to >150 μg/kg, but usually does not exceed 1 μg/kg. For example, in oven-broiled ground beef its concentration ranges from 0.01 to 0.1 μg/kg and in barbecued chicken from 0.02 to 0.91 μg/kg. Dietary exposure to IQ is estimated to be 0.28 ng/kg b.w per day.

The other aromatic amines are generally found in variable amounts. For example, well-done pan-fried beef steak may contain 23.7 μg/kg 7-MeIgQx, 6.31 μg/kg 7,9-DiMeIgQx, 4.03 μg/kg IgQx and IFP (well-done meats may contain IFP at levels of 1.4–46 μg/kg), 2.06 μg/kg 4,8-DiMeIQx, 0.68 μg/kg 6,7-DiMeIQx, and 0.16 μg/kg IQ[4,5-*b*]; AαC and MeAαC are not present.

12.3.1.3 Reactions

Heterocyclic amines known as aminoimidazoazaarenes are formed during thermal processing of meat of warm-blooded animals, poultry, and fish by reactions of creatine/creatinine (**2-118**) with certain products of the Maillard reaction, but detailed mechanisms of their formation are not yet fully understood. Creatine was postulated to form the 2-amino-3-methylimidazo (2-aminoimidazo) moiety of aminoimidazoazaarenes by cyclisation and water elimination, a reaction that takes place spontaneously at temperatures above 100 °C. This part of the molecule (especially its 2-amino group) is a common moiety of all aminoimidazoazaarenes and is responsible for their mutagenicity. The remaining parts are assumed to arise from the Maillard reaction between sugars and amino acids, which produces various reactive intermediates for the formation of pyridines and pyrazines, such as glycolaldehyde and glyoxal imines, pyridinium cation, pyrazinium cation radical, and pyrazinium dication. Aldol condensation of Strecker aldehydes is believed to link the two parts of aminoimidazoazaarenes together (Figure 12.6).

During cooking, heterocyclic amines can react with proteins by condensation of the amino groups of heterocyclic amines (such as PhIP) and carboxyl groups of proteins, yielding the corresponding amides. The presence of such adducts may be an important factor in evaluating the carcinogenic risk of heterocyclic amines, because a portion of these mutagens formed by cooking changes into non-mutagenic adducts. However, the parent heterocyclic amines can be released from high-molecular-weight compounds by acid hydrolysis or proteolytic digestion in the gastrointestinal tract and may influence human exposure to heterocyclic amines and carcinogenic risk.

12.3.1.4 Health and toxicological evaluation

Four compounds show high specific mutagenicity and carcinogenicity, namely 2-amino-3-methylimidazo[4,5-*f*]quinoline (IQ), 2-amino-3,4-dimethylimidazo[4,5-*f*]quinoline (4-MeIQ), 2-amino-3,8-dimethylimidazo[4,5-*f*]quinoxaline (4-MeIQx), and 2-amino-1-methyl-6-phenylimidazo[4,5-*b*]pyridine (PhIP). In comparison with known mutagens and carcinogens, such as benzo[*a*]pyrene, these compounds cause serious DNA damage and induce chromosome aberrations (changes in structure and number). They are easily absorbed and distributed in the tissues. IQ mutagens are metabolised by the phase I enzymes (activation by *N*-hydroxylation) and the phase II enzymes (further activation by the formation of arylnitrenium ions and conjugation with DNA). DNA adducts have been detected in various tissues, such as the colon and prostate. The risk of carcinogenesis from the intake of individual aminoimidazoazaarenes is difficult to estimate, because there are always mixtures of several compounds in meat, along with other contaminants (PAHs, nitrosamines, and other carcinogens).

12.3.1.5 Mitigation

The content of aminoimidazoazaarenes in foods is dependent on the particular temperature and time of heat treatment and also on the water content, pH, and concentration of precursors. Higher temperatures and longer processing times generally result in higher quantities of aminoimidazoazaarenes. The content of these contaminants also depends on the type of heat treatment. Direct heating operations and operations with efficient heat transfer, such as grilling, produce higher amounts of aminoimidazoazaarenes than indirect methods, such as cooking in water or steaming. The highest amount of aminoimidazoazaarenes is therefore formed in the surface layers of food exposed to

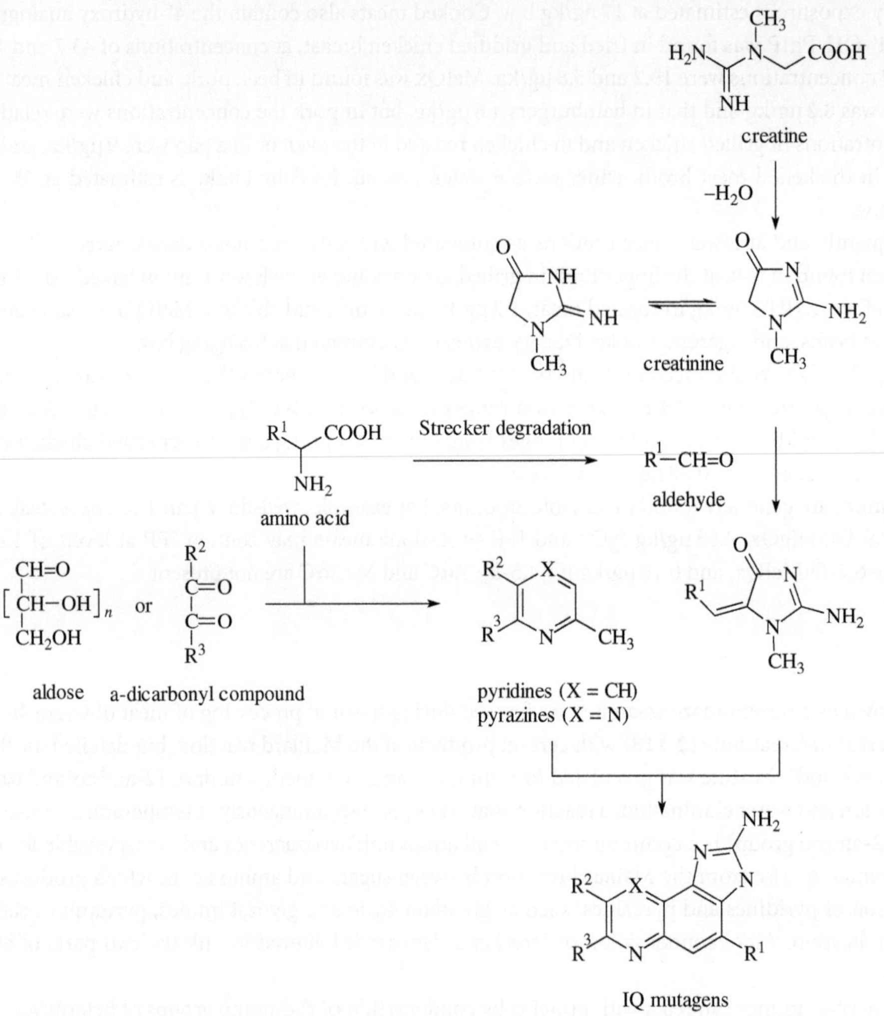

Figure 12.6 Formation of aminoimidazoazaarenes.

higher temperatures (especially during grilling and roasting) that have lower water content. During cooking in water, these mutagens are either not formed or arise in low quantities. Their concentrations in commercially prepared foods are therefore very low.

The formation of heterocyclic amines of the IQ type is effectively inhibited by antioxidants. The presence of phenolic compounds such as flavanones and flavan-3-ols significantly inhibits their formation. The mechanism of the inhibitory effect is partly based on the elimination of Strecker aldehydes, precursors of IQ mutagens, by condensation with phenols.

12.3.2 Acrylamide

12.3.2.1 Classification, structure, and properties

Acrylamide, also known as acrylic acid amide or prop-2-enamide (Figure 12.3), occurs in a wide range of fried, roasted, baked, and toasted foods, especially foods rich in starch. The discovery of acrylamide in food is primarily associated with an accident that occurred in Sweden during the construction of a railway tunnel. Dissemination of synthetic acrylamide into the environment due to a technical error induced acute neurotoxic symptoms in construction workers. Livestock and other animals in the area were also affected, and surface-water contamination led to the deaths of huge numbers of fish. In the extensive investigations that followed, concentrations of the exposure biomarker, acrylamide–haemoglobin adduct, were monitored. The surprising finding was that there were relatively high levels in unexposed persons from the reference remote locations, which led to identification of thermally processed foods with high starch content as the acrylamide source. Until then, the primary known potential sources of acrylamide had been drinking water and tobacco smoke.

12.3.2.2 Occurrence, main sources, and dietary intake

The dietary exposure of Western populations to acrylamide mainly comes from fried potato chips (16–30%), potato crisps (6–46%), coffee (13–39%), pastry and sweet biscuits (10–20%), bread and crisp bread (10–30%), and to a lesser extent other foods (<10%). Significant differences in contributions to the total intake of acrylamide depend on the composition of the food basket in different countries. For example, in the United States, 35% of exposure is attributed to fried potato products, but only 7% to coffee.

According to the Joint FAO/WHO Expert Committee on Food Additives (JECFA), the daily intake of acrylamide for the general population is estimated to be between 0.3 and 2.0 µg/kg b.w. It is expected that the daily intake of acrylamide in children may be two to three times higher than in adults. The average daily intake for the general population is estimated at 1 µg/kg b.w. For populations with a high intake of acrylamide and for children, the average daily intake is estimated at 4 µg/kg b.w.

Table 12.6 presents summary data collected in the monitoring implemented in 2002–2008 in EU member states. Concentrations of acrylamide fluctuate over a wide range, depending on the raw materials, recipe, and method of culinary or industrial processing. Raw or cooked foods (if temperatures do not exceed 100 °C) usually do not contain acrylamide at all, or contain only very small quantities. Surprisingly high concentrations (>1000 µg/kg) were found in some types of dried fruits, such as pears and plums, although the processing temperatures did not exceed 80 °C.

12.3.2.3 Reactions

Acrylamide is formed primarily in the Maillard reaction. The key precursors are the amino acid asparagine, reducing sugars, and various carbonyl compounds resulting from sugar degradation and lipid oxidation that enable the decarboxylation of asparagine (Figure 12.7). In the absence of reducing sugars, α-hydroxycarbonyl, α-dicarbonyl, and other active compounds, deamination of asparagine yields fumaric acid monoamide, known as fumaramic acid. A primary product of the reaction is N-glycosylasparagine, which dehydrates to form the corresponding imine. In media with low water activity, N-glycosylasparagine and imine are relatively stable, but in aqueous media imine is hydrolysed or may be rearranged (Amadori rearrangement) to ketosamine (oxoform of 1-amino-1-deoxysugar). Ketosamine is not significant as a precursor of acrylamide as it is degraded to products involved in the formation of coloured and flavour-active products.

Decarboxylation of imines derived from α-hydroxycarbonyl compounds, having a free hydroxyl group in the β-position relative to the nitrogen atom, yields the corresponding azomethine ylid. The azomethine ylid can exist in two forms stabilised by resonance, which differ in the position of the C=N bond. Another alternative to imine conversion is the formation of imine betaine and intramolecular cyclisation of the product to an oxazolidin-5-one derivative. An imine with a double bond between nitrogen and carbon derived from the sugar may be hydrolysed to the parent sugar and 3-aminopropionamide, but it may also isomerise to produce a decarboxylated Amadori compound. Acrylamide is then formed by cleavage of the covalent bond between carbon and nitrogen. The imine double bond between nitrogen and carbon from the asparagine can only be hydrolysed to the Strecker aldehyde of asparagine and 1-amino-1-deoxyalditol, but the decarboxylated Amadori compound does not arise in this case. Another portion of acrylamide may be produced by enzymatic deamination of 3-aminopropionamide.

The amount of acrylamide formed from α-hydroxycarbonyl compounds, such as glucose, fructose, and hydroxyacetone (acetol), is generally much higher than for α-dicarbonyl compounds – represented, for example, by butane-2,3-dione (biacetyl), 2-oxopropanal (methylglyoxal), and other substances. Trace amounts of acrylamide may result from precursors other than asparagine, such as acrolein and acrylic acid. The thermal degradation of gluten in the production of wheat bread may also give rise to marginal amounts of acrylamide.

12.3.2.4 Health and toxicological evaluation

Acrylamide is a neurotoxic compound with genotoxic potential, and according to the IARC is classified as a potential human carcinogen (Group 2A). In the organism, acrylamide is rapidly absorbed and evenly distributed. The primary metabolites are glutathione conjugates and the oxidation product glycidamide. Acrylamide and glycidamide can react (form adducts) with macromolecules such as haemoglobin and DNA (Figure 12.8). It is mainly glycidamide that is responsible for the carcinogenic and genotoxic effects. Glycidamide may also be formed by oxidation of acrylamide by fatty acid hydroperoxides.

The European Commission has not yet set limits for acrylamide, but in 2017 it issued a food safety regulation that aimed to reduce the presence of the chemical acrylamide in food products (Commission Regulation 2017/2158).

12.3.2.5 Mitigation

Acrylamide levels can be lowered by mitigation approaches based on the selection of a suitable raw material (with low contents of reducing sugars and asparagine), modification of the processing conditions, or use of less risky processing/cooking procedures (suitable temperature, pH, and water content profiles). Knowledge of the formation mechanisms of acrylamide has led to a number of recommendations for food producers and consumers.

For example, the fermentation processes in the production of bread and pastry, leading to a decrease of dough pH, reduce the formation of acrylamide during baking (the effect is based on protonation of asparagine amino group, which inhibits the reaction with carbonyl groups of sugars). The use of rising agents based on ammonium carbonate in the production of certain bakery products (such as gingerbread) should be avoided.

When frying potato chips, the acrylamide formation correlates with temperature, duration, and the colour of the products (Table 12.5). Mitigation measures to reduce the presence of acrylamide in French fries and other cut deep-fried or oven-fried potato products include

asparagine   aldose   N-glycosylasparagine

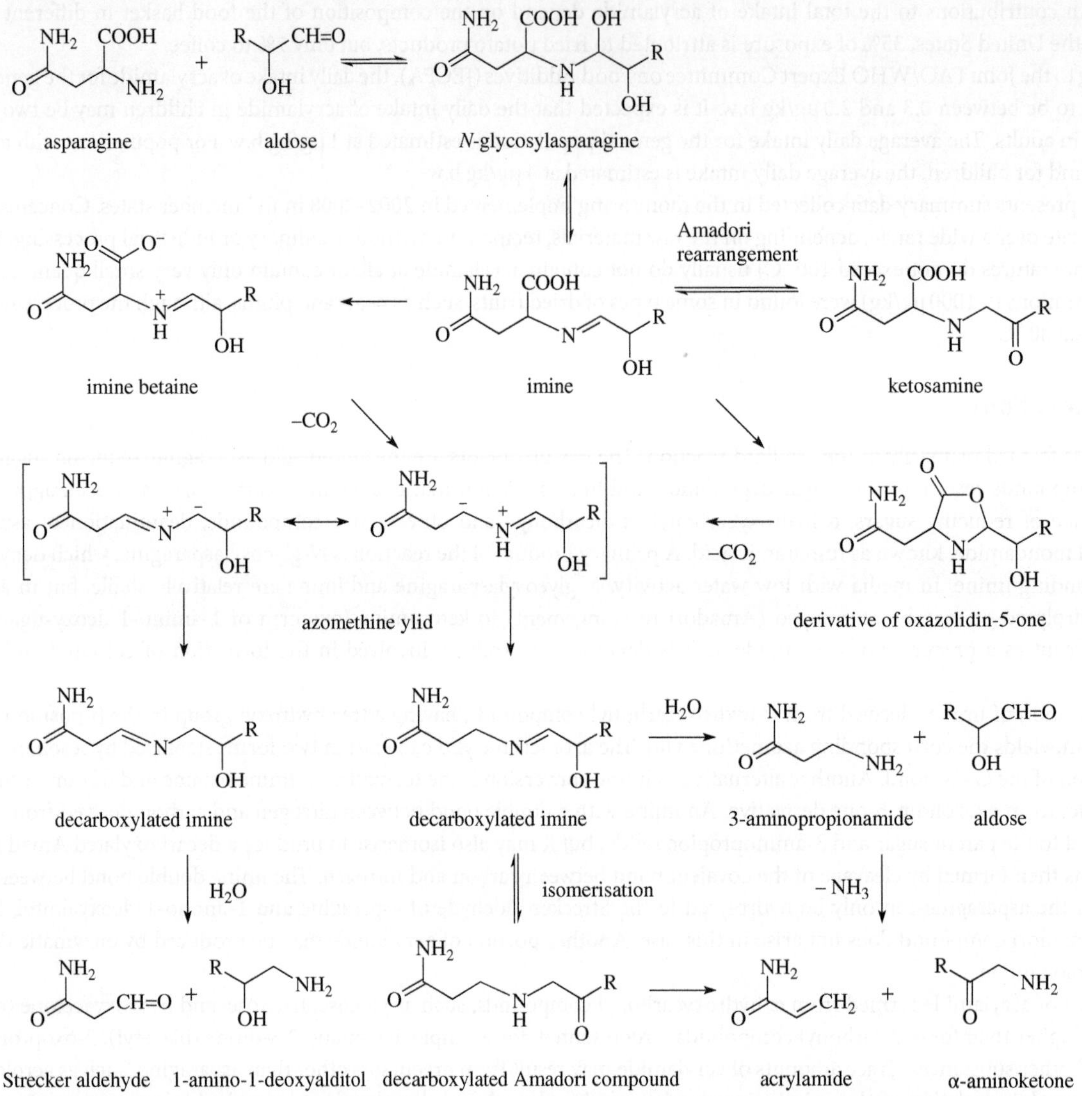

Amadori rearrangement

imine betaine   imine   ketosamine

$-CO_2$

$-CO_2$

azomethine ylid   derivative of oxazolidin-5-one

decarboxylated imine   decarboxylated imine H_2O 3-aminopropionamide   aldose

H_2O isomerisation $-NH_3$

Strecker aldehyde   1-amino-1-deoxyalditol   decarboxylated Amadori compound   acrylamide   α-aminoketone

Figure 12.7 Mechanism of acrylamide formation in the presence of α-hydroxycarbonyl compounds.

tests for reducing sugars prior to use and the removal of immature tubers having high reducing sugar levels (storage of potatoes below about 4 °C increases the level of reducing sugars). The use of reducing sugars as a browning agent should be avoided. For end users and consumers, recommended cooking instructions should be specified. For example, French potatoes should be cooked until a golden-yellow colour and the frying temperature should be kept between 160 and 175 °C.

The production of acrylamide during the the roasting of coffee is different. A sharp increase in content to about 12 500 µg/kg after 4–5 minutes of roasting at 230 °C is followed by a gradual decrease. After 25 minutes, the acrylamide content may be 6000 µg/kg, and after 50 minutes, it is just 2000 µg/kg. This could be partly due to losses caused by volatilisation, but is mainly attributable to reactions with coffee bean components. The decrease of acrylamide level continues during storage and in ground roasted coffee, with a loss of 40% of acrylamide over six months of storage. In considering coffee blend composition, products based on robusta beans tend to have higher acrylamide levels than those based on arabica beans. In coffee substitutes made from chicory, addition of other ingredients, such as chicory fibres or roasted cereals, is effective in reducing the acrylamide content in the final product.

Under certain circumstances, acrylamide may infiltrate into drinking water treated using polymers based on it, which are used as flocculants. With regard to the toxicity of acrylamide and health risks associated with its intake, the hygienic limit for water intended for human consumption is 0.1 µg/l (European Council Directive 98/83/EC).

12.3.3 Furan

12.3.3.1 Classification, structure, and properties

Furan is a simple heterocyclic compound (see Section 8.2.11) that typically arises by thermal decomposition of natural materials containing cellulose and pentoses. Its presence in foods was demonstrated in the 1970s in studies of volatile compounds produced during cooking. With

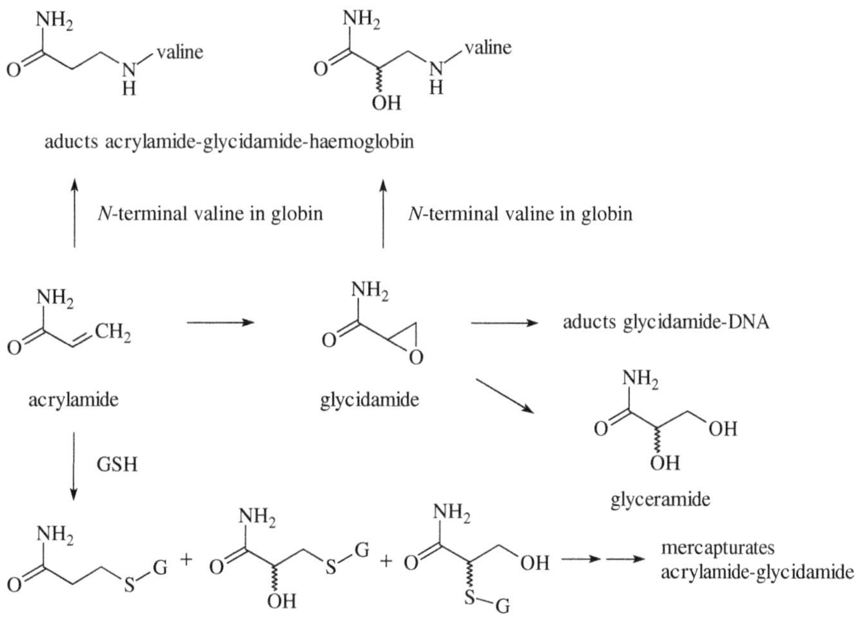

Figure 12.8 Metabolism of acrylamide in organism (GSH = reduced glutathione).

Table 12.5 Changes of acrylamide concentrations in potato chips, depending on the temperature and time of frying.

Temperature (°C)	Time (minutes)	Acrylamide (μg/kg)	Temperature (°C)	Time (minutes)	Acrylamide (μg/kg)
130	7.0	12 000	170	2.3	26 000
140	6.0	14 000	180	2.1	32 000
150	4.7	18 000	190	1.6	38 000
160	3.0	22 000	200	1.3	47 000

Table 12.6 Typical concentrations of acrylamide in various foods in the European market.[a]

Food	Mean value (μg/kg)	Maximum (μg/kg)	Food	Mean value (μg/kg)	Maximum (μg/kg)
Cookies	313–317	4200	Potato chips	348–350	2668
Bread	126–136	2430	Potato crisps	626–628	4180
Cereal breakfast	135–156	1600	Homemade potato products	310–319	2175
Baby cereal nutrition	52–74	353	Canned baby food	23–44	162
Coffee	249–253	1158	Other products[b]	305–313	4700

[a]Data collected according to the recommendation of the European Commission 2007/331/EC.
[b]Pizza, pancakes, waffles, fish fingers, meat balls, chicken nuggets, fried fish, vegetarian steak, and roasted cauliflower.

regard to its low boiling point (34.1 °C), its concentrations were not often quantified, as attention was mainly focused on the flavour and the biologically active derivatives of furan formed mainly in the Maillard reaction. In this context, the results presented by the US Food and Drug Administration (FDA) in 2004, which documented relatively high levels of furan in various heat-processed food products, especially roasted, pasteurised, and sterilised foods, were very surprising.

12.3.3.2 Occurrence, main sources, and dietary intake

The highest furan concentrations were found in roasted coffee, caramel, canned baby foods (mainly vegetable foods), and many other canned foods. Typical amounts in various foods are presented in Table 12.7.

The average daily exposure is estimated to be 0.9 µg/kg b.w in babies, 0.41 µg/kg b.w in children up to two years, and 0.26 µg/kg b.w in older children and adults.

Table 12.7 Typical levels of furan in foods in the European market.

Food	Furan (µg/kg)[a]		
	Mean value	Minimum	Maximum
Baby food			
Baby food with meat	4.0	3.0	6.0
Baby food with vegetables, including potatoes	40.3	4.0	153.0
Baby food with fruits	4.3	1.0	16.0
Vegetables			
Vegetables, canned	5.3	<2	12.0
Vegetables, fresh	<1	–	–
Fruits			
Fruits, canned	3.5	<1	6.0
Fruits, dry	2.6	<1	7.0
Meat and meat products			
Meat and meat products	3.0	<1	10.0
Meat, canned	9.0	4.0	14.0
Bread			
Whole (white)	25.7	<4	148.0
Crust	65.6	5.0	193.0
Snacks and nuts			
Potato chips, peanuts, almonds	51.1	<5	143.0
Sugar	–	<1	<2
Caramel	484.4	17.0	1956
Coffee and coffee surrogates			
Coffee, ground	2677	959.0	5938
Coffee, brewed from ground	78.0	13.0	199.0
Coffee, instant	929.0	44.0	2150
Coffee, brewed from instant	18.8	2.0	51.3
Milk and beverages			
Milk	nd	nd	nd
Chocolate	<2	<2	<2
Fruit juice	1.0	1.0	1.0
Coca-cola	<1	<1	<1
Beer (dark)	3.0	3.0	3.0

[a] nd, not determined (under limit of determination).

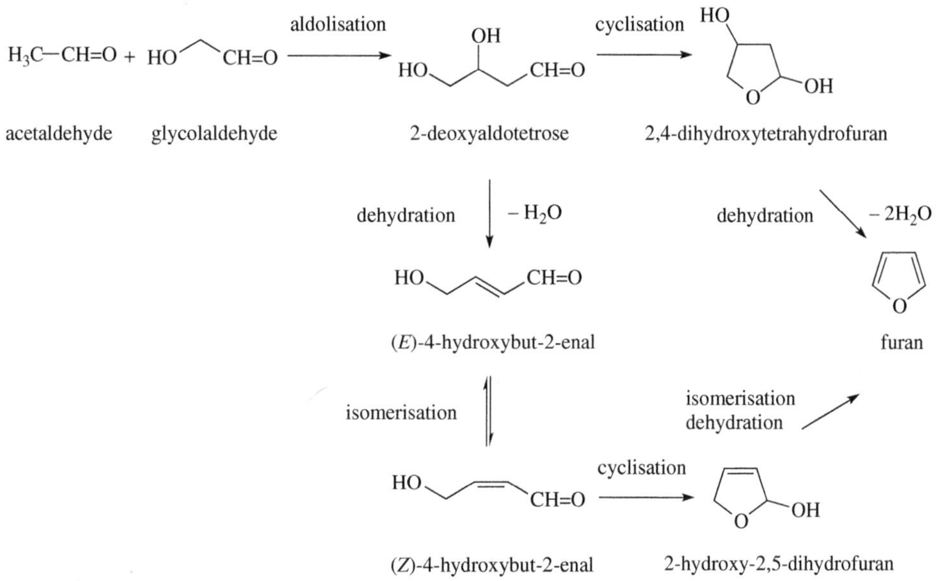

Figure 12.9 Formation of furan from amino acids.

12.3.3.3 Reactions

The presence of furan has been demonstrated in many foods of very different composition, which shows that there are multiple possibilities for its formation. It has been found that the precursors of furan in foods may be a number of food components, including amino acids, carbohydrates, and ascorbic acid, as well as unsaturated fatty acids and carotenoids.

The most important precursors of furan are amino acids that yield acetaldehyde and glycolaldehyde on degradation. Acetaldehyde is produced in the Strecker degradation of alanine (alanine is also the product of aspartic acid decarboxylation) or by decarboxylation and deamination of serine (see Section 2.5.1.3.2); glycolaldehyde results from the Strecker degradation of serine. Furan arises from these reactive intermediates by aldol condensation via 2-deoxyaldotetrose, which undergoes cyclisation and dehydration (Figure 12.9). Glycolaldehyde is additionally produced by fragmentation of reducing sugars.

Important precursors of furan are oxidation products of polyunsaturated fatty acids, in particular of α-linolenic acid and linoleic acid. In the case of linolenic acid, C—H bond cleavage at C-17 of (9Z,12Z,15Z)-17-hydroperoxyoctadeca-9,12,15-trienoic acid yields (9Z,12Z,15Z)-17-oxoheptadeca-9,12,15-trienoic acid. Subsequent autoxidation (primarily cleaved is the C—H bond at C-11, and the free radical isomerises) produces (9Z,11E,15Z)-13-hydroperoxy-17-oxoheptadeca-9,11,15-trienoic acid, which subsequently decays to form (E)-but-2-enal, which can isomerise to (Z)-but-2-enal (crotonaldehyde, Figure 12.10).

Autoxidation of linoleic acid gives rise to (10E,12Z)-9-hydroperoxyoctadeca-10,12-dienoic acid as one of the major products, which is degraded to give (2E,4Z)-deca-2,4-dienal. Its isomer (2E,4E)-deca-2,4-dienal, formed by isomerisation, decomposes to (E)-but-2-enal and other products (Figure 12.11). A likely intermediate in polyunsaturated fatty acids, as well as amino acids, is 4-hydroxybut-2-enal formed by oxidation of (E)-but-2-enal, which can provide furan after cyclisation, isomerisation, and subsequent dehydration.

Another possibility is the Maillard reaction. Furan was produced in the highest amounts in model mixtures of alanine with glycolaldehyde heated to 250 °C. Serine in mixtures with ribose and sucrose gave 30% furan, and in mixtures with ribose or glucose 10–25%. About eight times more was produced in the reaction mixture with erythrose than in mixtures with glucose or fructose, as tetroses (their degradation products) are immediate precursors of furan. In terms of their potential to form furan, the following order of sugars can be compiled: D-erythrose > D-ribose > sucrose > D-glucose = D-fructose. The expected reaction mechanisms are summarised in Figure 12.12.

Four-carbon furan precursors are formed from hexoses in essentially three ways: via 1-deoxyhexo-2,3-diulose and aldotetrose, via 2-deoxytetros-3-ulose, and via 3-deoxyhexos-2-ulose. The most common 1-deoxyhexo-2,3-diulose is 1-deoxy-D-*erythro*-hexo-2,3-diulose, which is produced by 2,3-enolisation of ketosamines derived from glucose as the major intermediate; furan is also produced in small amounts from fructose in the absence of amino compounds. Fragmentation of 1-deoxyhexo-2,3-diulose or hexose yields aldotetrose (carbons C-3, C-4, C-5, and C-6). Ribose gives rise to 1-deoxy-D-erythro-hexo-2,3-diulose, glucose, or fructose. This pathway seems to be the most important, since it provides approximately 50% furan. Another furan precursor, 2,3-deoxytetros-3-ulose, is produced as one of many products of the transformation (dehydration and retroaldolisation) of hexoses. The molecule of 2,3-deoxytetros-3-ulose contains the carbons C-1, C-2, C-3, and C-4 from the original skeleton of the hexoses. This pathway produces about 10% furan. From 2-deoxyaldotetrose, which is formed by oxidation, decarboxylation, and cleavage of hexoses, carbon atoms C-2, C-3, C-4, and C-5 are incorporated into the furan molecule. This pathway produces about 10% furan.

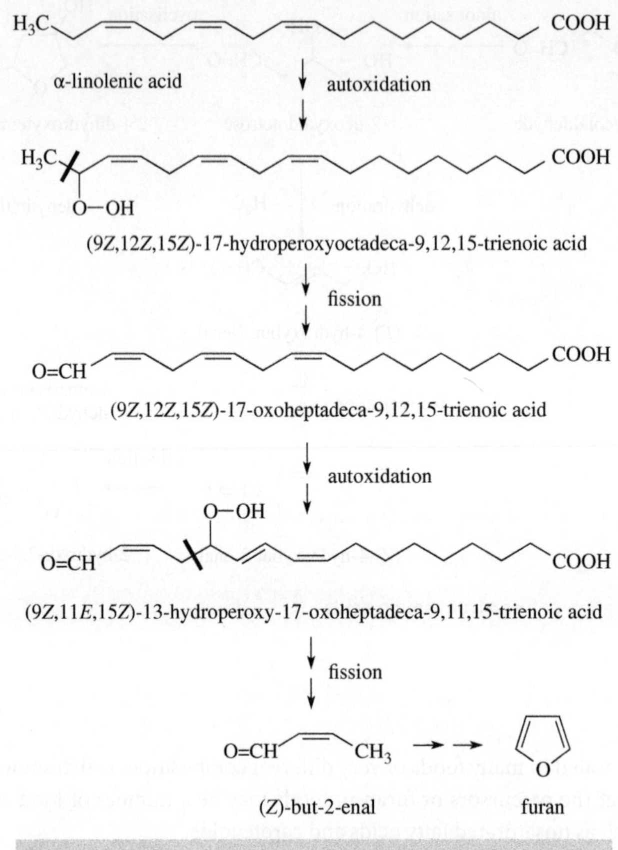

Figure 12.10 Formation of furan from linolenic acid.

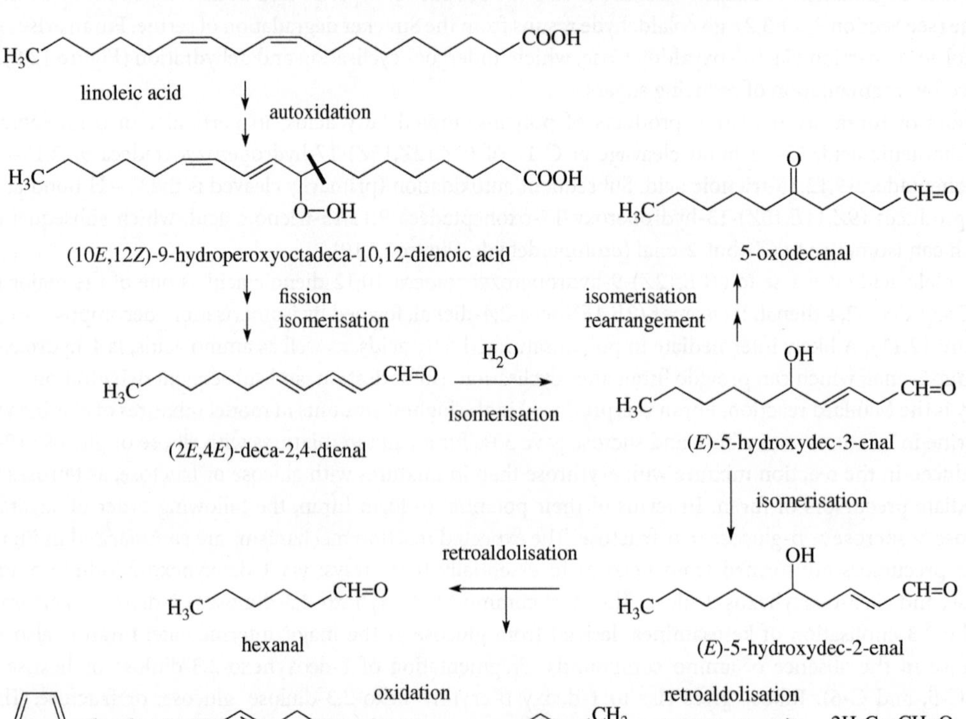

Figure 12.11 Formation of furan from linoleic acid.

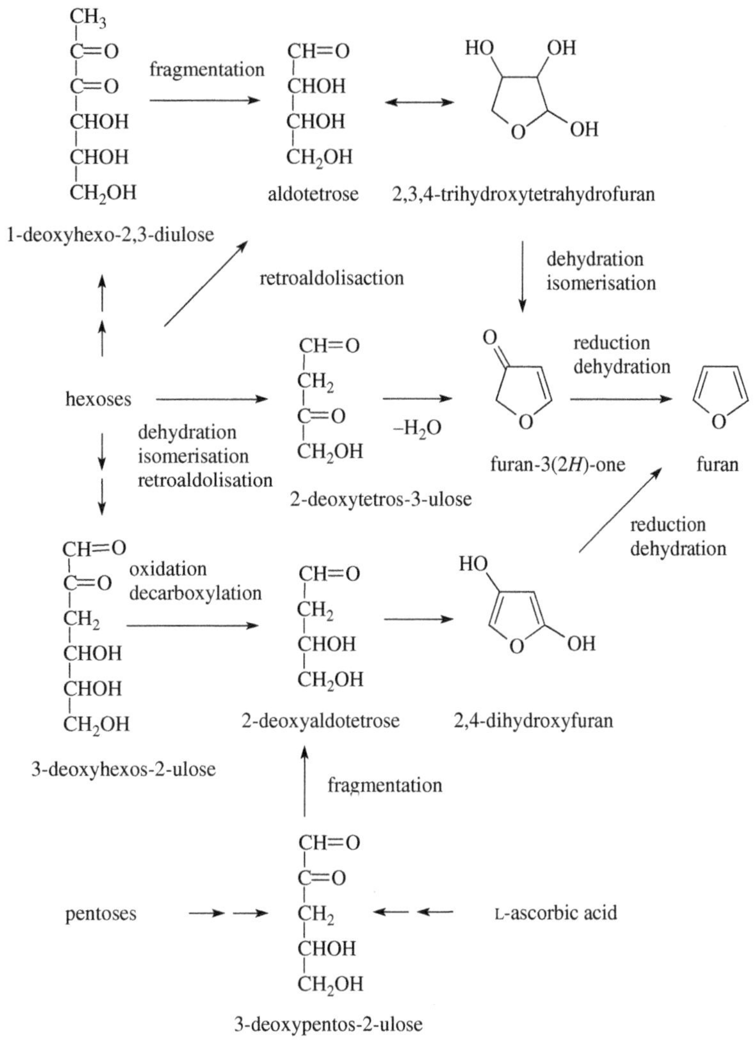

Figure 12.12 Formation of furan from saccharides and ascorbic acid.

Another natural furan precursor is L-ascorbic acid (the potential of dehydroascorbic and isoascorbic acids found in model experiments was still one order of magnitude higher). The reaction mechanism of furan formation from ascorbic acid is not yet fully understood. The expected precursor is 3-deoxy-L-*threo*-pentos-2-ulose (3-deoxy-L-glycero-pentos-2-ulose), also known as 3-deoxy-L-xylosone (see Section 5.14.6.1).

12.3.3.4 Health and toxicological evaluation

Furan is rapidly absorbed by the body due to its easy penetration across biological membranes. Its excretion is also relatively fast. It seems that it is first oxidised (epoxidation of one double bond), and the oxidation product is then rearranged, with opening of the cyclic structure. The reactive (Z)-but-2-ene-1,4-dialdehyde (maleic dialdehyde, **12-26**) formed is a strongly cytotoxic substance that can bind to proteins and nucleosides. An alternative metabolic pathway is its conjugation with glutathione. The resulting product is then expelled from the body through the urine.

12-26, maleic dialdehyde

Based on toxicological studies in rodents, the IARC has classified furan as a potential human carcinogen of Group 2B. The existing experimental data show the relationship between liver cancer and the exposure dose.

12.3.3.5 Mitigation

The dietary sources of furan are mainly two groups: cooked foods with relatively high water activity (products in cans or jars) and foods with low water activity processed by roasting or baking (such as coffee and bread).

The assumption that, on opening cans or glass containers, furan quickly escapes into the atmosphere due to its low boiling point has proved to be untrue. During normal culinary procedures, furan has a relatively high retention in food, and a decrease of its concentration occurs only during intense boiling, when it is removed with the large volume of water vapour. The amounts of furan in different types of heat treatment have also been compared. Furan concentrations in industrially manufactured products were considerably higher than in home-made products, probably due to a diminished possibility of vapour release from the equipment. The research also shows that the furan content in industrially produced vegetable meals is significantly higher than that in fruit mixtures.

12.3.4 Chloropropanols and their esters

12.3.4.1 Classification, structure, and properties

Chloropropanols (chlorinated propanols) are a group of three-carbon alcohols and diols with one or two chlorine atoms that are hypothetically derived from glycerol. Six compounds, one chloropropanol (3-chloropropan-1-ol), two dichloropropanols (1,3-dichloropropan-2-ol, 1,3-DCP for short, and 2,3-dichloropropan-1-ol, 2,3-DCP), two chloropropanediols (3-chloropropane-1,2-diol, glycerol-1-chlorohydrine, 3-CPD, also known as 3-MCPD (from 3-monochloropropane-1,2-diol), and 2-chloropropane-1,3-diol, glycerol-2-chlorohydrine, 2-CPD or 2-MCPD), and the structurally related 1,3-dichloropropane (trimethylenedichloride, **12-27**), arise as technological contaminants in acid protein hydrolysates (hydrolysed vegetable protein) and probably also in foods during culinary and technological processing. 3-CPD and 2,3-DCP are chiral compounds that occur in protein hydrolysates and foods as racemic mixtures of corresponding enantiomers (−)-(R)-3-chloropropane-1,2-diol, (+)-(S)-3-chloropropane-1,2-diol, (R)-2,3-dichloropropan-1-ol, and (S)-2,3-dichloropropan-1-ol (**12-27**), respectively.

1,3-DCP and 2,3-DCP were identified in hydrolysed vegetable protein in 1978, whilst the main chlorine-containing substance, 3-CPD, was found in 1982 and its 2-CPD isomer in 1987, in the Czech Republic. At the beginning of the 1990s, some CPD and DCP isomers were found in soy sauces, where hydrolysed vegetable proteins were used in their manufacture. A surprising finding was that these contaminants occur in a number of common foods where hydrolysed vegetable proteins are not used in their production.

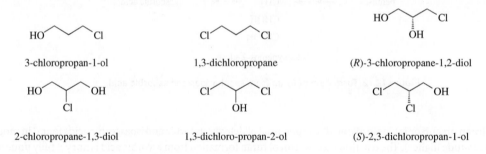

3-chloropropan-1-ol   1,3-dichloropropane   (R)-3-chloropropane-1,2-diol

2-chloropropane-1,3-diol   1,3-dichloro-propan-2-ol   (S)-2,3-dichloropropan-1-ol

12-27, structures of selected chloropropanols and chloropropanediols

In 1980, it was found that hydrolysed vegetable proteins contain esters of chloropropanols with higher fatty acids. As examples, structures of esters derived from 3-CPD are shown in formulae **12-28**. Monoesters (alkanoic acid 2-chloro-2-alkanoyloxymethyl ethyl esters) and diesters (alkanoic acid 2-chloro-1-alkanoyloxymethyl ethyl esters) of 3-CPD were found in one sample of goat's milk in 1984, but it was understood that they were exogenous contaminants. The current research, focusing on DCP and CPD esters in foods, began in 2004.

(R)-1-acyl-3-chloropropane-1,2-diol   (R)-2-acyl-3-chloropropane-1,2-diol   (S)-1,2-diacyl-3-chloropropane-1,2-diol

12-28, esters of chloropropanediols

12.3.4.2 Occurrence, main sources, and dietary intake

The major contaminant in traditional acidic protein hydrolysates is 3-CPD. Other contaminants are mainly 2-CPD, 1,3-DCP, and 2.3-DCP. Their ratio is roughly as follows: 3-CPD : 2-CPD : 1,3-DCP : 2,3-DCP = 1000 : 100 : 10 : 1. The hydrolysates contained 100–800 mg/kg of 3-CPD, 10–90 mg/kg of 2-CPD, 0.1–6 mg/kg of 1.3-DCP, and 0.01–0.5 mg/kg of 2,3-DCP. Their actual amounts depend primarily on the content of lipids in the raw proteinaceous material, the concentration and amount of hydrochloric acid, and the temperature during hydrolysis. Chloropropanols are also found in soy sauces made from mixtures of acid and enzymatic protein hydrolysates. Soy sauces produced exclusively by traditional fermentation do not contain chloropropanols.

Since the 1980s, new technological processes have gradually been introduced as substitutes for conventional ones, which have led to a decrease of chloropropanol content in acid protein hydrolysates and simultaneously in some soy sauces. This trend is documented in the results of controls carried out by the UK Ministry of Agriculture, Fisheries and Food (MAFF) in 2000: 3-CPD was found in amounts higher than 0.02 mg/kg in 25 cases and in amounts higher than 1 mg/kg in 16 cases. The highest amount found was 82.8 mg/kg. 2-CPD was found in amounts of 17.6 mg/kg in 26 samples. 1,3-DCP was detected in 17 cases (its concentration ranged from 0.006 to 0.345 mg/kg), 2,3-DCP in 11 (concentration 0.006–0.043 mg/kg). The same commodities analysed in 2002 contained 0.02 mg/kg or more of 3-CPD in only seven cases, and the highest concentration was 35.9 mg/kg. A concentration of 1,3-DCP of 0.017 mg/kg was found only in the sample with the highest amount of 3-CPD.

DCP and CPD esters of fatty acids (palmitic, stearic, oleic, linoleic, and linolenic acids) were detected only in the crude neutralised hydrolysates (Table 12.8). These lipophilic compounds (the main fatty acid was oleic acid) form a layer on the surface of neutralised hydrolysates that is easily removed by filtration and goes to the waste humins, so that in commercial products they may be present only in trace amounts.

Chlorinated propanols, especially 3-CPD and its positional isomer 2-CPD, have been identified in a number of common foods, generally in amounts lower than in acid protein hydrolysates. CPD isomers are present mainly in foods with lower water content that have undergone industrial and culinary processing at higher temperatures (such as bread crust, toasted bread, roasted malt, barley, and coffee). In other foods, such as processed and grilled cheese, meat products (e.g. fermented and smoked sausages), and fish, they are present in smaller amounts (Table 12.9).

CPD may not always be simply an endogenous contaminant: in smoked products, for example, it can be a component of liquid smoke preparations, in amounts of 200–760 µg/kg. It may also originate from acid protein hydrolysates used as condiments. High concentrations of 3-CPD (up to 1150 µg/kg) were found in chicken steaks. CPD can also migrate from some types of packaging that have been manufactured using chloromethyloxirane (also known as epichlorohydrin or 1-chloro-2,3-epoxypropane). In liquid smoke and garlic, CPD arises from unusual precursors. In the cellulose pyrolysates, acetol (3-hydroxyacetone) acts as a precursor, and in garlic, CPD arises from allyl alcohol, which is a degradation product of alliin.

The amounts of other chloropropanols (2-CPD, 1,3-DCP, and 2,3-DCP) are lower than those of 3-CPD. Higher levels are found sporadically; for example, the concentration of 1,3-DCP in a sample of ham was <3–21 mg/kg, in beef steak 70 µg/kg, and in various sausages <3–69 mg/kg.

Mono- and diesters of 3-CPD with fatty acids occurring in foods represent a new group of contaminants, because they may release 3-CPD by both chemical and enzymatic hydrolysis during food processing and possibly also by the action of lipases *in vivo*. In most cases, the content of 3-CPD esters is much higher than that of free 3-CPD (Table 12.10).

The highest amounts of 3-CPD esters (2500 mg/kg and more) undoubtedly occur in refined palm and olive oils, which are obtained from the pericarp of fruits (the outer and often edible layer in the fleshy fruits). The main proportion of 3-CPD esters arises in the deodourisation step, which takes place at temperatures exceeding 200 °C (see Section 3.4.3.7). Lower concentrations occur in refined vegetable oils obtained from oilseeds, and the lowest in virgin vegetable oils. Palm oil and mixtures of vegetable oils with palm oil are also used for frying foods, such as potato chips, crisps, cheeses, and meat. Products containing refined palm oil include substitutes for milk powder (creamers), whipped creams in sprays, various biscuits, sweet spreads, soup cubes (broth), and other products, including infant and follow-on formulae. For example, the maximum measured levels of 3-CPD esters in infant formulae were 4196 µg 3-CPD/kg fat content, which corresponds to a

Table 12.8 Contents of 3-CPD and 1,3-DCP esters in raw neutralised hydrolysate and humins.

Ester	Content (mg/kg) Raw neutralised hydrolysate	Humins
3-CPD diesters	4	35
3-CPD monoesters	35	205
1,3-DCP esters	8	65

Table 12.9 Contents of 3-CPD in selected foods and other products.

Food	3-CPD (µg/kg)	Food	3-CPD (p, g/kg)
Cereal products		*Meat products*	
Bread (white)	<10–55	Ground beef	<5
Bread crust	24–275	Regular patty (hamburger)	<10–71
Toasts	20–679	Ham (cured)	<5–22
Donats	11–24	Ham (smoked)	<10–47
Potato products		Salami	<10–69
Potato chips	<10–15	Sausages	<5–69
Potato crisps	15	Meat extracts	<10–14
Fish		*Various*	
Fish (baked)	<5–83	Coffee	<9–19
Fish (smoked)	<10–191	Malt	<10–850
Cheese	<10–95	Modified starches	<10–488
Fats and oils	<5–12	Garlic (thermally processed)	5–690

Table 12.10 Contents of 3-CPD esters in selected foods and other products.

Food	3-CPD esters (µg/kg)[a]	Food	3-CPD esters (µg/kg)[a]
Meat products		*Potato products*	
Salami	1399–1760	Potato chips	230–6100
Milk		Potato crisps	400[b]
UHT (3.5% of fat)	<3[b]	*Refined vegetable oils and other fats*	
Condensed (9% of fat)	67[b]	Palm oil	2821[b]
Milk, dry (26% of fat)	405	Palm kernel oil	1168[b]
Breast milk	<11–76	Coconut oil	1556[b]
Cereal products		Olive oil	<300–2462
Bread	6–85	Seed oils (mean value)	<300–1234
Bread crust	547[b]	Fats after frying	11206[b]
Toasts	86[b]	Margarines	500–1500
Cookies	200–1690	*Animal fats*	
Coffee, coffee surrogates, and malts		Butter, lard, tallow	<10–140
Coffee (roasted)	210–390	*Various*	
Rye (roasted)	145[b]	Sweet spreads	2300–10 300
Chicory root (roasted)	957[b]	Almonds, roasted nuts	433–1370
Barley (roasted)	1184[b]	Broths	380–670
Malt (Pilsener)	5–11	Cream in spray	50–730
Malt (roasted)	463–650	Baby and infant formulae	<72–8470

[a] Expressed in µg/kg of 3-CPD.
[b] Mean value.

concentration of 156 μg 3-CPD/l in ready-to-drink milk. Infants then could have a 3-CPD exposure of 25 μg/kg b.w per day, which is 12.5 times the TDI. During the examination of samples of follow-on formulae, a maximum level of 8467 μg 3-CPD/kg fat content was found, corresponding to 250 μg 3-CPD/kg in ready-to-drink milk. This could lead to intakes of 20 times the TDI. In adults, the daily intake could be up to five times the TDI in cases of high daily consumption of vegetable oils and margarines, which may contain up to 7356 μg 3-CPD/kg fat content.

Virgin oils from seeds (such as rapeseed, soybean, and sunflower oils) contain 3-CPD esters only at a level of <100–337 mg/kg, and virgin olive oils contain relatively very low amounts (<100 to <300 mg/kg). The content of 2-CPD esters in fats and oils represents about 20–60% of the 3-CPD amount (palm oils usually contain more than 1000 mg/kg and seed oils less than 150 mg/kg of 2-CPD esters). Oils with a high content of 3-CPD esters may rarely also contain 1,3-DCP esters (up to 11 mg/kg).

In addition to esters of chloropropanols, commercially refined palm oils with higher 3-CPD content contain esters of glycidol, also known as 2-(hydroxymethyl)oxirane or 2,3-epoxypropan-1-ol. Their amount increases with temperature during deodourisations. For instance, deodourisation of palm oil for 6 hours at 210 °C gives rise to 2800 μg/kg of 3-CPD esters and 300 μg/kg of glycidyl esters, whilst deodourisation at 250 °C over the same period yields 3300 μg/kg of 3-CPD esters of 2900 mg/kg of glycidyl esters. Glycidyl esters may become precursors of CPD esters and free CPD, respectively. Generally, the amount of glycidyl esters in refined oils and possibly also in other foods ranges from <100 to 4100 μg/kg, that in refined palm oils from 300 to 10 000 μg/kg, that in margarines from <150 to 5000 μg/kg, and that in the fat of infant formulae from <150 to 3000 μg/kg.

12.3.4.3 Reactions

The precursor of chloropropanols in acid protein hydrolysates is hydrochloric acid, which is used to hydrolyse the proteinaceous materials, as well as residual lipids of the raw material and glycerol arising by hydrolysis of lipids (acylglycerols) with hydrochloric acid. Wheat gluten, which is often used as a raw material, contains, for example, 0.5–3.0% residual lipids, of which 30–36% are neutral lipids, mainly triacylglycerols, and about 60% are phospholipids. Soybean meal contains 1.0–3.0% lipids, of which about 30% represent neutral lipids and 60% phospholipids. The main precursors of chloropropanols are triacylglycerols, followed by phospholipids and glycerol. The content of glycerol in acid protein hydrolysates ranges from 200 to 3000 mg/kg.

The precursors of chloropropanols in common foods are naturally present or intentionally added chlorides (sodium chloride) and lipids, and in some products also glycerol, which arises, for example, in dough as a result of the activity of yeast lipases. The primary reaction products are CPD diesters, which are hydrolysed to CPD monoesters. Esters of DCP are formed from CPD esters by reaction with chlorides. Free chloropropanols are then formed by hydrolysis of esters.

Exogenous sources of chloropropanols may be packaging materials containing epichlorohydrin copolymers with polyamines or polyamides that are used for the manufacture of tea bags, coffee filters, and absorbents of packaged meat juices. Other exogenous sources of chloropropanols may be starches modified with epichlorohydrin, copolymers of epichlorohydrin with dimethylamine (used during sugar refining as flocculants or for the immobilisation of glucose isomerase in manufacturing of fructose syrups); polyelectrolytes based on epichlorohydrin and polyamines (used as flocculating and coagulating agents in the production of drinking water) are only relevant in certain cases.

In acidic solutions, triacylglycerols are gradually hydrolysed to diacylglycerols, the diacylglycerols to monoacylglycerols, and finally the monoacylglycerols to glycerol. A simplified sequence of reactions that leads from triacylglycerols through partial glycerol esters and 3-CPD and DCP esters to free CPD and DCP is shown in Figure 12.13. The main reaction in acid protein hydrolysates and foods is the formation of reactive intermediates (cyclic acyloxonium cations), preferably from diacylglycerols (acyloxonium ions A and B in Figure 12.13) and monoacylglycerols. In particular, higher amounts of partial glycerol esters (diacylglycerols and monoacylglycerols) occur in oils obtained from seed pericarps (palm and olive oils), due to the higher activity of naturally occurring lipases. Ring opening of cyclic acyloxonium cations with chloride ions yields 3-CPD and 2-CPD diesters, which may split off fatty acid(s) and yield cyclic acyloxonium ions and the corresponding hydroxy derivatives, respectively. Acyloxonium ions can also be formed by the elimination of fatty acids from monoacylglycerols. Reaction with chloride ions then leads to 3-CPD (2-CPD) monoesters and 1,3-DCP and 2,3-DCP esters. In the absence or depletion of chloride ions, the pathway leading from diacylglycerols is believed to end at the stage of glycidol esters. The ratio of 3-CPD and 2-MCPD arising from acylglycerols in acid protein hydrolysates depends on steric and other effects (terminal ester group) and is approximately 10 : 1. During the refining of vegetable oils (and possibly in foods), 3-CPD diesters, 3-CPD monoesters, 2-CPD diesters, 2-CPD monoesters, and glycidol esters are produced at a ratio of about 49 : 14 : 28 : 7 : 7.

Glycidol esters are found in particularly high amounts in refined palm oil and palm oil-based fractions. Palm oil contains relatively high amounts of diacylglycerols (ranging from 2.3 to 4% in freshly extracted oils and from 4 to 7.8% in commercial oils) and monoacylglycerols (0.4–0.5%); therefore, these acylglycerols are supposed to be the main precursors of glycidol esters. The formation of glycidol esters proceeds at high temperatures by an intramolecular rearrangement, followed by elimination of a fatty acid molecule, which can be initiated by abstraction of the hydroxyl group's proton by the vicinal carboxyl group. The acyloxonium intermediate can then rearrange through charge migration, resulting in the release of a fatty acid and the oxirane ring formation by nucleophilic reaction of the alkoxide group.

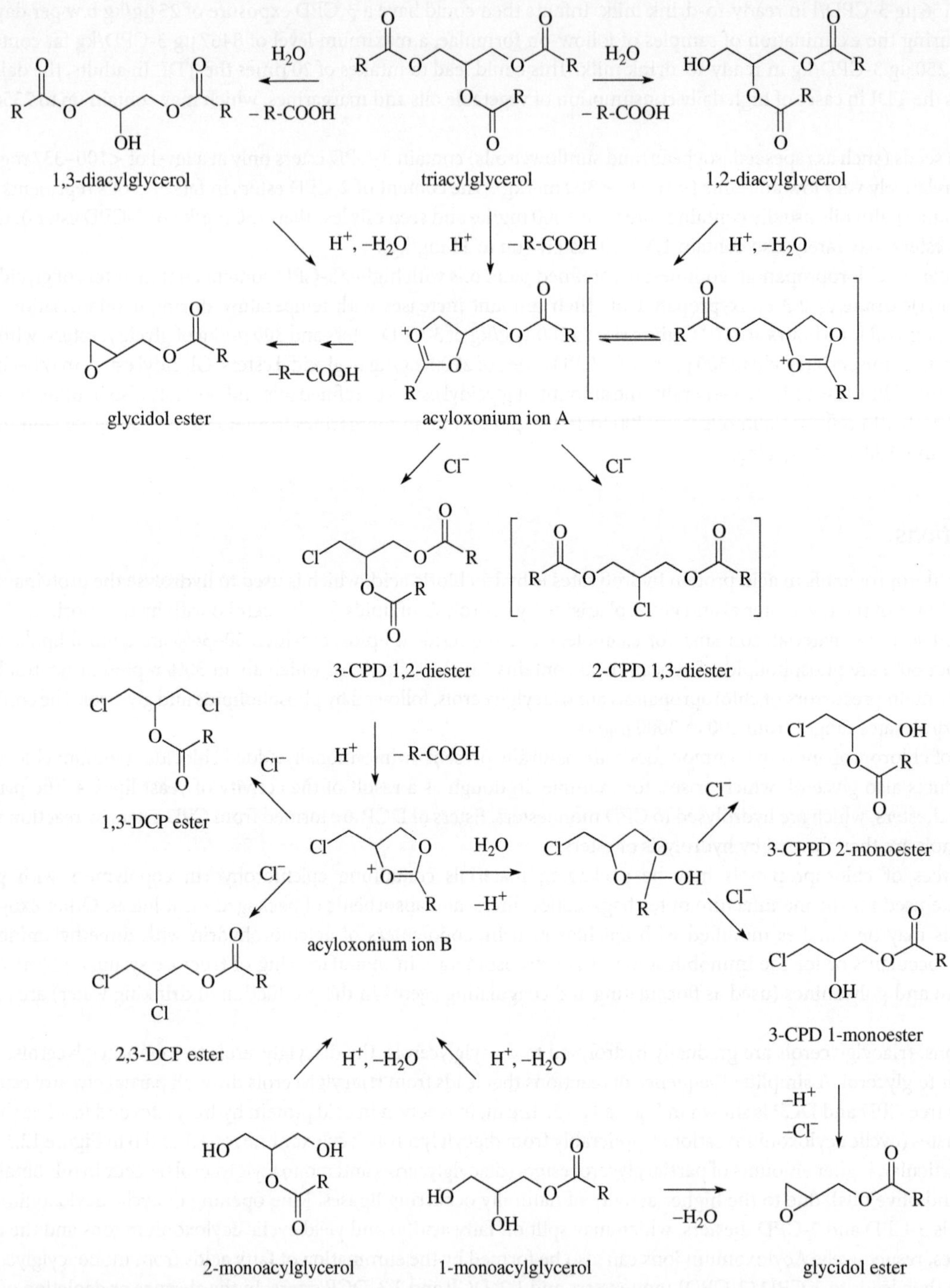

Figure 12.13 Formation of CPD and DCP esters from triacylglycerols and partial glycerol esters.

An analogous reaction (with elimination of water) may proceed with monoacylglycerols (Figure 12.13). In addition to glycidol esters, diacylglycerols heated at temperatures exceeding 140 °C in refined palm oil yield the corresponding oxopropyl esters at about 10% of glycidol esters levels (Figure 12.14).

Phospholipids are hydrolysed to deacylated products. For example, 1,2-diacyl-*sn*-glycero-3-phosphatidylcholine yields 1-acyl-*sn*-glycero-3-phosphatidylcholine, 2-acyl-*sn*-glycero-3-phosphatidylcholine, and *sn*-glycero-3-phosphocholine. By seemingly analogous mechanisms to CPD ester formation (Figure 12.13), partial esters yield chloroester derivatives (such as 1-chloroester), which appears to hydrolyse to CPD esters and free CPD. CPD also arises from *sn*-glycero-3-phosphocholine (Figure 12.15). The ratio of 3-CPD to 2-CPD formed from phospholipids is about 4 : 1.

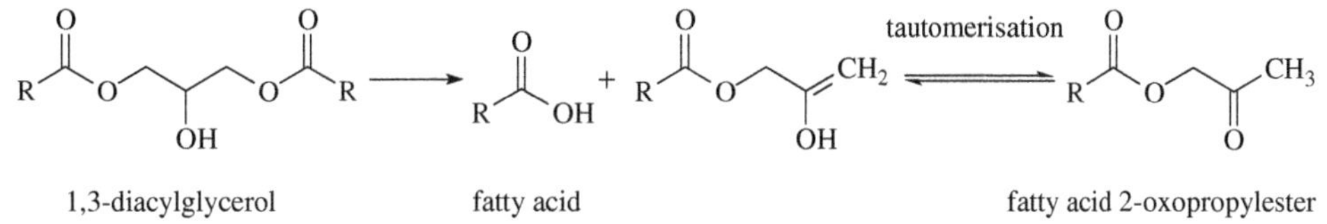

Figure 12.14 Formation of 2-oxopropyl fatty acid esters.

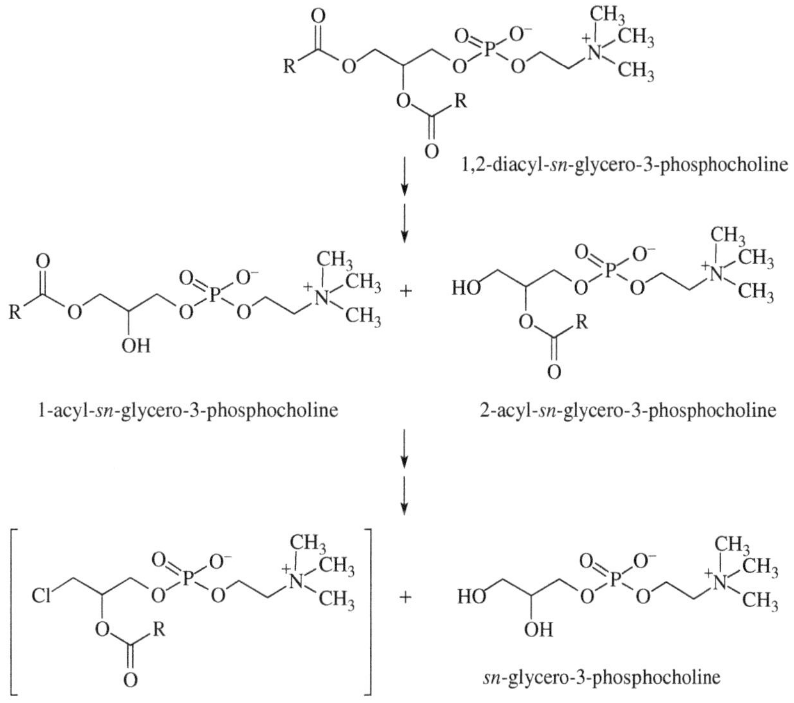

Figure 12.15 Formation of CPD from phospholipids.

Acid protein hydrolysates and many foods contain glycerol produced by hydrolysis of glycerolipids. Under acidic conditions, hydroxyl groups of glycerol can be protonated with the formation of alkyloxonium ions. In the case of the primary hydroxyl group, the next step is elimination of water, which is substituted by a chloride anion. This step is stereospecific and proceeds with inversion of the configuration at the carbon bearing the leaving group, which leads to a racemic mixture of 3-CPD. The alkyloxonium cation formed from the secondary hydroxyl group dissociates to a carbocation and water, and the reaction of the carbocation with chloride ion yields 2-CPD (Figure 12.16). The ratio of 3-CPD to 2-CPD formed from glycerol is about 2:1. Elimination of water from glycerol simultaneously gives rise to hydroxyacetone and isomeric 2-deoxyglyceraldehyde.

Another possibility, envisaged in foods with low water contents during heat treatment at higher temperatures, is dehydration of the carbocation to a cation of glycidol, from which arise 3-CPD and 2-CPD by the opening of the oxirane ring by chloride ions. In this case, the 3-CPD:2-CPD ratio is about 3:1 (Figure 12.17).

In garlic and garlic-containing foods, 3-CPD and 2-CPD can result from the addition of hypochlorous acid (occurring, for example, in chlorinated water) to allyl alcohol (Figure 12.18), which is one of the decomposition products of alliin.

12.3.4.4 Health and toxicological evaluation

Chloropropanols show various toxic effects. 3-CPD causes infertility in rats and suppression of the immune function, and has been shown to be genotoxic in several *in vitro* assays, but is not genotoxic and mutagenic *in vivo*. It is classified by the IARC as a chemical of Group 2B, probably carcinogenic to humans. In 2001, the Scientific Committee on Food (SCF) established a TDI for 3-CPD at 2 μg/kg b.w, and in 2002 the JECFA established a provisional maximal TDI of 2 μg/kg b.w. In the European Union, maximum levels of 0.02 mg/kg for free 3-CPD in hydrolysed vegetable proteins and soy sauces were established in 2001. The maximum levels have been applied since April 2002 and are integrated into the Commission Regulation (EC) No. 1881/2006. These limits were not designed to account for 3-CPD esters. Data for

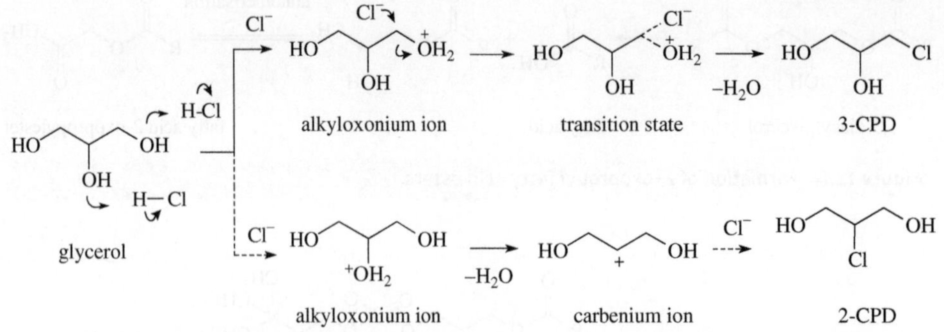

Figure 12.16 Formation of CPD from glycerol.

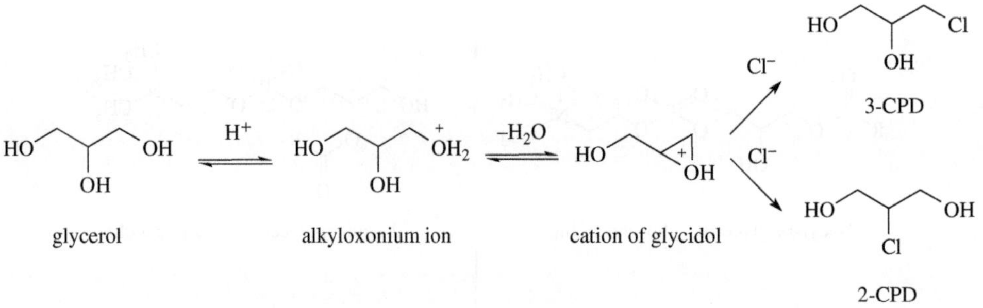

Figure 12.17 Formation of CPD from glycerol via glycidol.

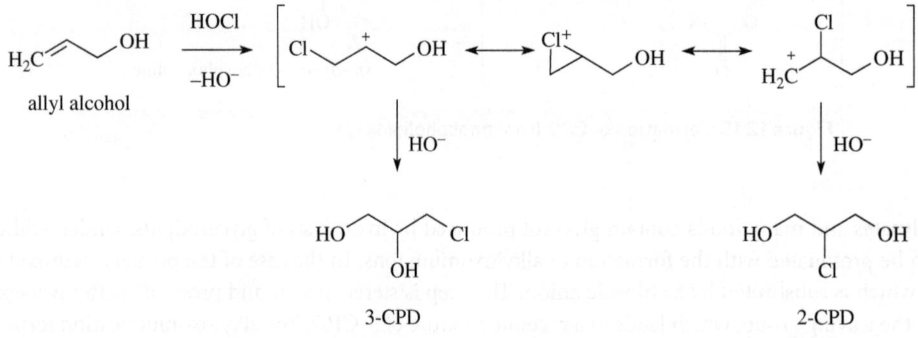

Figure 12.18 Formation of CPD from alliin.

2-CPD and its esters are lacking, and no toxicological data on esters of glycidol exist, but the degradation product glycidol and a metabolite 3-chlorolactic acid are genotoxic and carcinogenic compounds. In the human body, the glycidyl esters are metabolised into free glycidol, a compound classified by the IARC as Group 2A (possibly carcinogenic to humans), and to epichlorohydrin. 3-Chlorolactic acid, the main metabolite of 3-CPD in rats, was shown to be devoid of DNA-damaging effects *in vitro* in mammalian cells. 2-CPD is also a potential carcinogen.

The daily intake of 3-CPD in the UK population, which tends to consume foods with higher contents of 3-CPD, is estimated at 0.10 µg/kg b.w in adults, 0.18 µg/kg b.w in children aged 4–18 years, and 0.28 µg/kg b.w in children aged up to 4 years. In populations consuming higher amounts of hydrolysates, these estimates are higher (0.21, 0.38, and 0.58 µg.kg b.w, respectively). These values, of course, do not take into account the recent findings of high amounts of bound forms of 3-CPD (esters) in refined fats.

Genotoxic 1,3-DCP (a Group 2B carcinogen) also has other toxic effects (hepatotoxicity, nephrotoxicity, and thyreotoxicity), which appear to be linked to its toxic metabolites and degradation products (epichlorohydrin and 1,3-dichloroacetone). Its positional isomer 2,3-DCP is also potentially genotoxic and carcinogenic. The concentration of 0.005 mg/kg 1,3-DCP in hydrolysates and soy sauces was recommended based on the current limit of quantification. The daily intake of 1,3-DCP in the United Kingdom is estimated at 0.051 µg/kg b.w amongst adults and 0.136 µg/kg b.w in children.

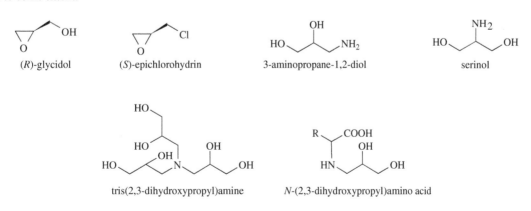

Figure 12.19 Dehydrohalogenation of vicinal chlorohydrins in alkaline media.

12.3.4.5 Mitigation

At the beginning of the 1980s, manufacturers of acidic protein hydrolysates started to implement practices that led to the reduction in the amount of chloropropanols. The stripping of crude hydrolysate with steam led to the removal of volatile chloropropanols, including 1,3-DCP and 2,3-DCP. Today, decontamination is performed on the crude hydrolysate in slightly alkaline medium, where dehydrohalogenation of chloropropanols proceeds. After decontamination, the hydrolysate is acidified with hydrochloric acid to have the same pH as the hydrolysate produced by traditional technology. Other decontamination methods were also considered, such as degradation of CPD by microorganisms and enzymes, CPD extraction with ethyl acetate, butan-1-ol, butan-2-ol, or isobutanol, and removal of residual solvent by steam stripping. Some manufacturers have stopped production of traditional acidic protein hydrolysates and replaced them with aromatised enzymatic hydrolysates.

The mechanism of dehydrohalogenation of vicinal chlorohydrins (α-chloroalcohols) is shown in Figure 12.19. Reaction in alkaline solutions is very rapid, and forms as its primary product the corresponding alkoxide (alcoholate). Oxiranes (1,2-epoxides) are important reaction intermediates. By opening of the oxirane ring, the corresponding vicinal diols appear as the final reaction products. In the case of 3-CPD and 2-CPD, the dehydrohalogenation intermediate is glycidol and the final product is glycerol. Enantiomers of 3-CPD are decomposed at the same rate, with (R)-CPD yielding (+)-(R)-glycidol and (S)-CPD yielding (−)-(S)-glycidol (**12-29**) by inversion of configuration, known as Walden inversion.

Glycidol, which is classified as a Group 2B carcinogen, accumulates under certain conditions in hydrolysates, because the rate of decomposition of CPD isomers is higher than that of glycidol. Glycidol is a very reactive compound, reacting not only with water, but also with many other nucleophilic reagents (alcohols, thiols, amines, and acids). Its reaction with glycerol yields polyglycols, whilst that with ammonia yields 3-aminopropane-1,2-diol (**12-29**). Both enantiomers occur in acidic protein hydrolysates at a level of about 30 mg/kg. Another product is the tertiary amine tris(2,3-dihydroxypropyl) amine (2–4 mg/kg, **12-29**). With ammonia, 2-CPD yields 2-aminopropane-1,3-diol (serinol, **12-29**) as the main product. The reaction with the amino group of amino acids produces N-(2,3-dihydroxypropyl)amino acids (**12-29**). These reactions paradoxically contribute to hydrolysate decontamination. Under alkaline conditions, L-amino acids isomerise to D-amino acids to some extent.

(*R*)-glycidol   (*S*)-epichlorohydrin   3-aminopropane-1,2-diol   serinol

tris(2,3-dihydroxypropyl)amine   *N*-(2,3-dihydroxypropyl)amino acid

12-29, glycidol, epichlorohydrin, and reaction products of glycidol

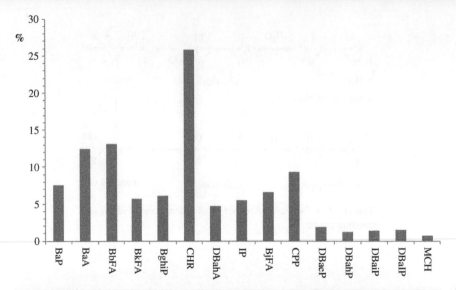

Figure 12.20 Relative average contributions of the EU priority PAHs to overall food contamination. BaP = benzo[*a*]pyrene, BaA = benzo[*a*]anthracene, BbFA = benzo[*b*]fluoranthene, BkFA = benzo[*k*]fluoranthene, BghiP = benzo[*ghi*]perylene, CHR = chrysene, DbahA = dibenzo[*ah*]anthracene, IP = indeno[1,2,3-*cd*]pyrene, BjFA = benzo[*j*]fluoranthene, CPP = cyclopenta[*cd*]pyrene, DbaeP = dibenzo[*ae*]pyrene, DbahP = dibenzo[*ah*]pyrene, DbaiP = dibenzo[*ai*]pyrene, DbalP = dibenzo[*al*]pyrene, MCH = 5-methylchrysene. Source: According to scientific opinion of the Panel on Contaminants in the Food Chain, on request from the European Commission on Polycyclic Aromatic Hydrocarbons in Food. *The EFSA Journal* **724**, 1–114 (2008).

12.3.5 Polycyclic aromatic hydrocarbons

12.3.5.1 Classification, structure, and properties

PAHs represent an important group of virtually ubiquitous environmental contaminants, some members of which have genotoxic and carcinogenic potential. They contain two to six fused aromatic rings. In total, more than 100 PAHs have been identified. Due to their carcinogenic activity, most attention is focused on the 16 that the US Environemental Protecting Agency (EPA) targeted in the 1970s. In 2005, the EU Commission recommended the monitoring of 15 EU priority PAHs highlighted by the JECFA. Compounds represented by formulae **12-30** are currently the most common, both in terms of their occurrence in the environment and in the context of health dietary risks for consumers. The relative average contributions of the EU 15 priority PAHs to the contaminantion of foods are shown in Figure 12.20. In addition, the JECFA recommended including benzo[*c*]fluorene as a further such compound. In formulae **12-30**, the symbol * indicates the 16 priority PAHs according to the EPA, whilst the symbol + indicates the 15 + 1 EU priority PAHs. Eight of these also appear in the EPA list.

Polycyclic systems of PAHs, in which the adjacent rings have common atoms, may be *ortho-*, *ortho–peri-*, or *peri-*condensed. If none of the common atoms are part of more than two rings, the system is *ortho-*condensed and can be either linear (e.g. anthracene) or angular (e.g. phenanthrene). The number of common atoms in this system is equal to twice the number of the common sides. A *peri-*condensed system exists if at least one of the common atoms belongs to three rings. The aggregate number of common atoms of several rings (at least three) is less than twice the number of common sides. An example of such a hydrocarbon is pyrene.

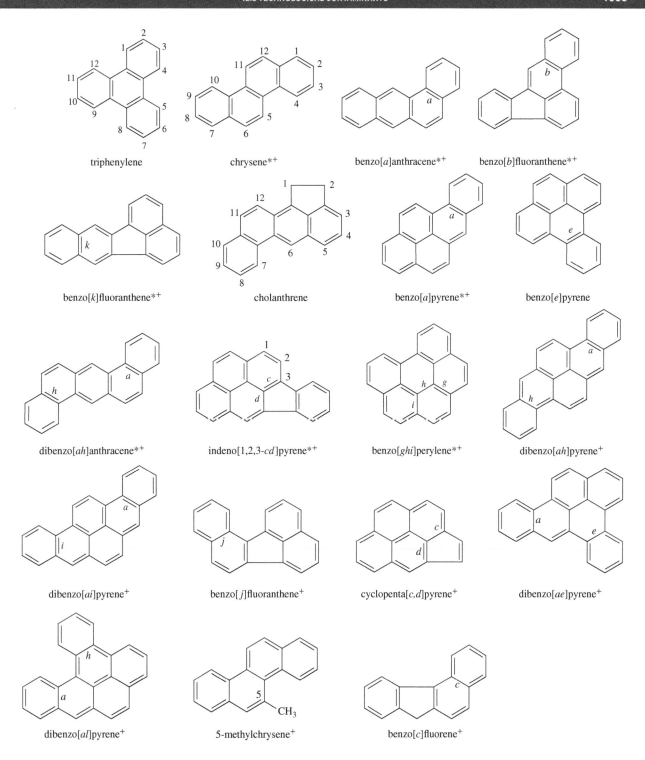

triphenylene

chrysene*+

benzo[*a*]anthracene*+

benzo[*b*]fluoranthene*+

benzo[*k*]fluoranthene*+

cholanthrene

benzo[*a*]pyrene*+

benzo[*e*]pyrene

dibenzo[*ah*]anthracene*+

indeno[1,2,3-*cd*]pyrene*+

benzo[*ghi*]perylene*+

dibenzo[*ah*]pyrene+

dibenzo[*ai*]pyrene+

benzo[*j*]fluoranthene+

cyclopenta[*c,d*]pyrene+

dibenzo[*ae*]pyrene+

dibenzo[*al*]pyrene+

5-methylchrysene+

benzo[*c*]fluorene+

12-30, structures of the most common polycyclic aromatic hydrocarbons

For selected hydrocarbons, trivial names are used. The names of *ortho-* and *peri-*condensed polycyclic hydrocarbons that have no trivial names are formed from the name of a hydrocarbon, for which there is an appropriate trivial name, and a prefix indicating the nature and location of other rings, for example benzo[*a*]pyrene. The individual isomers are distinguished by the letters *a*, *b*, *c*, and so on, by which all peripheral sides of the basic hydrocarbon are identified progressively from side *a* between C-1 and C-2. Examples of such isomers are benzo[*a*]pyrene and benzo[*e*]pyrene. The orientation of the formula should be such that the maximum number of benzene rings lie in a horizontal position, and as many as possible are lying at the top-right of this horizontal axis. The numbering of atoms is done in a clockwise direction, starting from the carbon atom of the ring of the far right and furthest from the horizontal axis (carbons that cannot carry a substituent are not numbered). Besides the already mentioned hydrocarbons, the names of other contaminants are derived from indene (**12-31**) and perylene (**12-32**). PAHs are generally substances of low reactivity, and the stability of linear condensed hydrocarbons generally decreases with the number of *ortho*-condensed rings. Angular and *peri*-condensed hydrocarbons are more stable. With increasing molecular weight, the volatility and water solubility of PAHs decrease, but the melting and boiling points and lipophilicity increase.

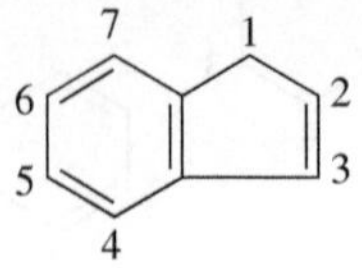

12-31, indene

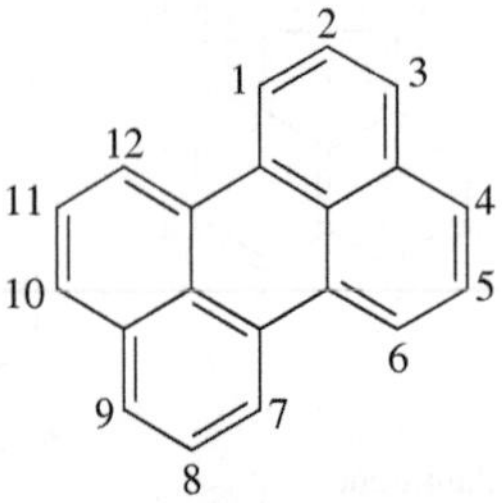

12-32, perylene

12.3.5.2 Occurrence, main sources, and dietary intake

PAHs have been part of the environment since time immemorial. They enter the ecosystem and the human food chain from anthropogenic sources (sources related directly or indirectly to human activities, such as pyrogenic sources) and non-anthropogenic sources (natural, especially geochemical sources). Some organisms are able to synthesise certain PAHs (Table 12.11).

Human exposure to PAHs occurs in three ways: inhalation, dermal contact, and consumption of contaminated foods.

Contamination of agricultural crops and food raw materials by exogenous PAHs is mainly due to polluted agricultural land (sorption from soil and water) and air (atmospheric) pollution (deposition from air). Plants receive PAHs by absorption from water, absorption

Table 12.11 Main sources of PAHs in the environment.

Anthropogenic sources	Non-anthropogenic sources
Industrial	*Geochemical*
Production of heat and electricity	Coal
Coke production	Natural oil seepage
Production and processing of coal tar	Sedimentary rocks
Production, processing and use of asphalt	Hydrocarbon minerals (such as curtisite and idrialite)
Catalytic cracking	Volcanic activity
Machines with internal combustion	
Production and use of carbon black	
Wastewater, sewage sludge	
Food technology	
Non-industrial	*Biological*
Fires of forests and prairies, tanker crash	Biochemical syntheses by macrophytes and microorganisms
Open burning of waste	
Incinerators	
Tobacco smoking	
Household heating	

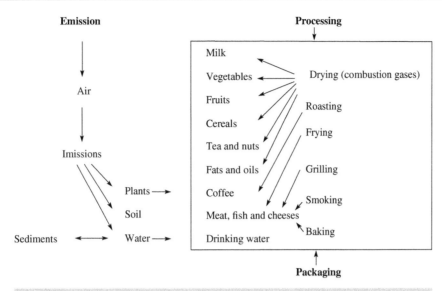

Figure 12.21 Possibilities of food contamination by PAHs (emissions are measured directly at source, e.g. chimney, whilst imissions are in the surrounding area).

from soil through the root surface, absorption through leaves from the vapour phase in air, and absorption through leaves from deposited particles. Farm animals mainly receive them from contaminated feed or water. A summary diagram of sources of PAHs in foods is given in Figure 12.21. It is obvious that many food commodities can be contaminated by several sources.

In the indoor environment, significant sources of PAHs are smoking, burning of candles or incense sticks, and heat treatment of food. Diet is the major source of human exposure to PAHs, accounting for 88–98% of contamination. For this reason, PAHs are classified as technological contaminants. Cooking and food processing at high temperature have been shown to generate PAHs that adhere to the food surface. The more intense the heat, the more PAHs are formed. Deep-fat frying, grilling, broiling, roasting, baking, stir-frying, braising, and smoking are common cooking methods that result in PAH formation. A particularly potent source is grilling of meat, fish, and other foods over a direct flame, as fat drips on the hot fire, yielding a flame containing high concentrations of a large number of PAHs.

Grilling may, under certain circumstances, lead to significant increases in both endogenous and exogenous PAHs in the product. The impact of the grilling process on the concentration of benzo[a]pyrene in barbequed frankfurters is shown in Table 12.12 as an example.

Benzo[a]pyrene, regarded for many years as a marker of food contamination by carcinogenic PAHs, was not found in all samples of a given commodity, although they contained high concentrations of other PAHs with similar toxic potential (Figure 12.20). This led scientists to consider the advisability of increasing the number of markers of contamination by carcinogenic PAHs. With regard to typical levels, three groups of PAHs were designed, representing 34, 60, and 80% of total PAH content:

Table 12.12 Contents of benzo[a]pyrene in grilled frankfurters under various conditions.[a]

Heat source	Mean content (µg/kg)	Range of values (µg/kg)
Frying pan	0.1	Not detected–0.2
Electric oven	0.2	0.1–0.3
Charcoal	0.3	Not detected–1.0
Pine cones	18	2–31
Logs	54	6–212
Embers	8	<1–25

[a]Original content of 0.2 µg/kg.

Table 12.13 Exposure of consumers to benzo[*a*]pyrene, PAH2, PAH4, and PAH8 through selected foods.

Food	Daily consumption (g)	Exposure (ng/day)			
		benzo[*a*] pyrene	PAH2	PAH4	PAH8
Cereals and cereal products	257	67	129	257	393
Sugar, sweets, and chocolate	43	5	13	25	39
Fats (vegetable and animal)	38	26	112	177	239
Vegetables, nuts, and legumes	194	50	124	221	378
Fruits	153	5	40	75	87
Coffee, tea, and cocoa (drinks)	601	21	55	106	156
Alcoholic beverages	413	4	12	25	74
Meat and sausages	132	42	107	195	279
Seafood	27	36	140	289	421
Fish and fish products	41	21	84	170	210
Cheeses	42	6	12	20	30

- PAH2: benzo[*a*]pyrene and chrysene;

- PAH4: PAH2, benzo[*a*]anthracene and benzo[*b*]fluoranthene;

- PAH8: PAH4, benzo[*k*]fluoranthene, benzo[*ghi*]perylene, dibenzo[*ah*]anthracene, and indeno[1,2,3-*cd*]pyrene.

Table 12.13 summarises some estimates of dietary exposure of EU consumers to benzo[*a*]pyrene and to the PAH2, PAH4, and PAH8 groups of PAHs. Generally, the most important dietary sources of PAHs to the European population are cereals, cereal products, and seafood.

In addition to PAHs, other polycyclic aromatic compounds (PACs), such as hydroxy- and oxy-PAHs, nitro-PAHs, amino-PAHs, cyano-PAHs, alkylated PAHs, and higher-molecular-weight PAHs, produced from various sources, are found in the environment. The term 'NSO-heterocycle' is used to denote compounds with nitrogen, sulfur, or oxygen atoms in an aromatic ring (e.g. indoles, acridines, dibenzothiophenes, and benzonaphthofuran). Information on the occurrence of derivatives of PAHs in food is very limited, but their concentrations are generally lower than those of the parent PAHs. As a biomarker to assess exposure to PAHs amongst workers in the petrochemical industry and in aluminum production plants, as well as ichthyofauna and ruminant exposure, the major pyrene metabolite 1-hydroxypyrene, which is excreted into the bile, urine, and milk, was employed. Petrogenic sources (mainly crude oil, coals, and similar materials) are rich in alkyl-substituted PAHs. High amounts of 1-nitronaphthalene, 2-nitrofluorene, and 1-nitropyrene were found in home-smoked meat, where their concentrations reached tens of µg/kg. Surprisingly high contamination by nitro derivatives of PAHs was demonstrated in some types of spices and tea. Typical concentrations did not exceed units of µg/kg.

12.3.5.3 Reactions

One of the most important routes of PAH formation is combustion (pyrolysis) of organic matter with limited availability of oxygen at temperatures of 500–900 °C, and especially above 700 °C. The mechanism of formation is **pyrosynthesis** from low-molecular-weight unsaturated aliphatic hydrocarbons. Figure 12.22 illustrates the formation of benzo[*a*]pyrene by recombination of free hydrocarbon radicals. This pathway, for example, produces PAHs during combustion of fossil fuels. Flame temperature and oxygen availability determine their composition. Many fuels already contain some PAHs, which can get into emissions directly.

Another important pathway leading to PAHs is the thermal elimination reaction of benzene derivatives (Figure 12.23). For example, the elimination of water from phenolic compounds (X = OH) yields, via 1,2-didehydrobenzene (often incorrectly called benzyn), another reactive intermediate benzobicyclo[2,2,2]triene and naphthalene as final products. A range of other natural compounds may be PAH precursors, such as plant steroids and other terpenoids. For example, benzo[*a*]pyrene and benzo[*a*]chrysene are the main decomposition products of stigmasterol. The contribution of endogenously generated compounds to the total dietary intake of PAHs is not too significant, as the contamination of food is largely exogenous.

Reactions of PAHs in the atmosphere are of particular interest, because the resulting products can contaminate plant or animal products cultivated in agriculture or aquaculture. The extent and nature of these reactions depend on the temperature, intensity of solar radiation,

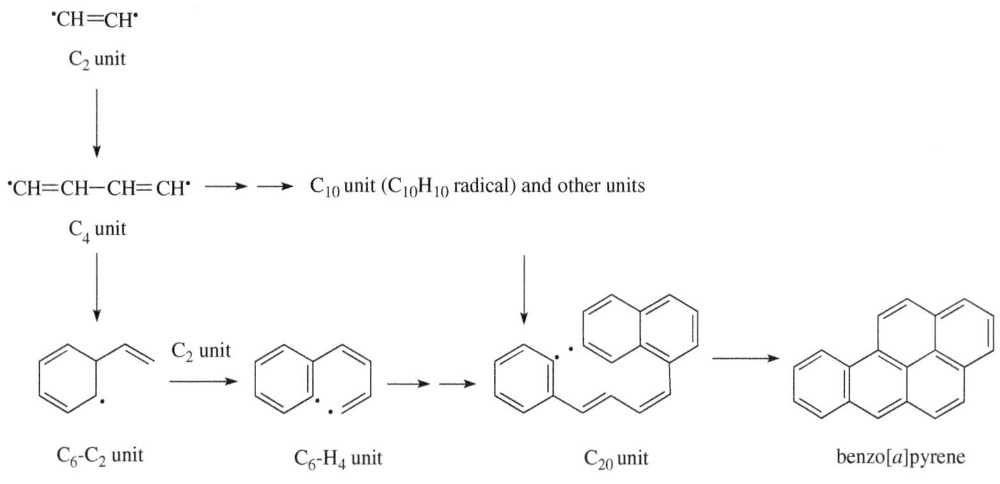

Figure 12.22 Formation of benzo[a]pyrene by pyrolysis of organic matter.

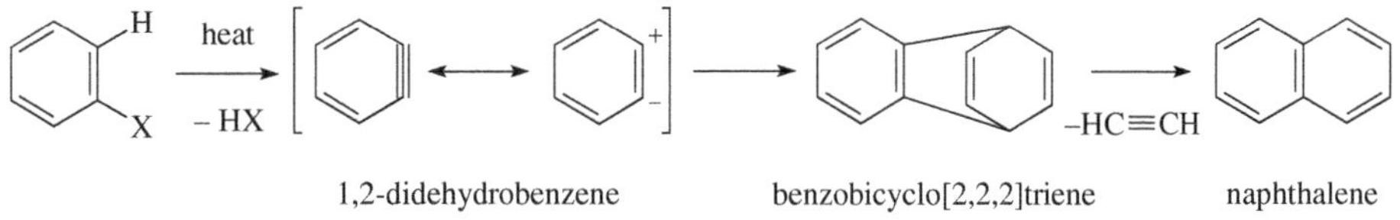

Figure 12.23 Formation of naphthalene from benzene derivatives.

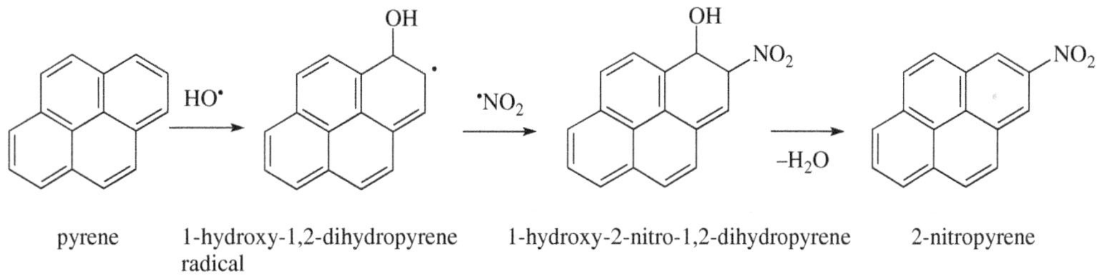

Figure 12.24 Formation of 2-nitropyrene in the atmosphere.

and concentration of substances, which are capable of electrophilic substitution on the aromatic nucleus. Particular attention is focused on reactions leading to nitro derivatives of PAHs, because of their high toxicity. Certain amounts of these contaminants are emitted directly into the atmosphere during organic matter combustion, but they may also arise by radical reactions in the gas phase. PAHs (or amino-PAHs) generally react with free hydroxyl radicals (HO•) to form reactive hydroxyl radicals of PAHs, which further react with •NO$_2$ radicals in the gas phase to form hydroxynitro derivatives of PAHs. Hydroxynitro derivatives eliminate water to form nitro derivatives. This reaction takes place in daylight. Another nitrating substance is nitric oxide (N$_2$O$_5$), which is cleaved to •NO$_3$ and •NO$_2$ radicals. The resulting radicals react with PAHs with elimination of nitric acid (HNO$_2$), yielding nitro-PAHs. Nitration of PAHs by nitric oxide radicals proceeds especially in the dark. Two different mechanisms of nitro-PAH formation can therefore be distinguished. Reactions of the first type, which produce 2-nitropyrene in daylight, are illustrated in Figure 12.24. Those of the second type, which produce 2-nitrofluoranthene in the absence of sunlight (in roughly the same amount as generated during daylight), are illustrated in Figure 12.25.

12.3.5.4 Health and toxicological evaluation

Commission Regulation (EU) No 835/2011 sets maximum levels for benzo(a)pyrene alone and the sum of benzo(a)pyrene, benz(a) anthracene, benzo(b)fluoranthene, and chrysene in a range of foodstuffs. For example, the maximum levels of benzo[a]pyrene and of the sum of the four PAHs in infant formulae and follow-on formulae (including infant milk and follow-on milk) and in processed cereal-based foods and baby foods for infants and young children are both 1 μg/kg. On the other hand, the maximum levels in smoked meat and smoked meat products are 2 μg/kg for benzo[a]pyrene and 12 μg/kg for the sum of the four PAHs.

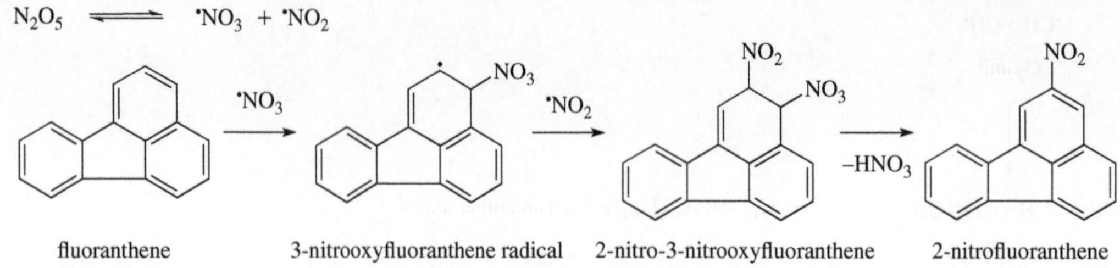

$$N_2O_5 \rightleftharpoons {}^{\bullet}NO_3 + {}^{\bullet}NO_2$$

fluoranthene   3-nitrooxyfluoranthene radical   2-nitro-3-nitrooxyfluoranthene   2-nitrofluoranthene

Figure 12.25 Formation of 2-nitrofluoranthene in the atmosphere.

PAHs generally have a low degree of acute toxicity to humans, but some have been linked with increased incidences of skin, lung, bladder, live, and stomach cancers. Certain PAHs can affect the haematopoietic and immune systems and produce reproductive, neurologic, and developmental effects. According to the IARC, benz(*a*)anthracene and benzo(*a*)pyrene are probably carcinogenic to humans. According to the EPA, probable human carcinogens include benz(*a*)anthracene, benzo(*a*)pyrene, benzo(*b*)fluoranthene, benzo(*k*)fluoranthene, chrysene, dibenz(*ah*)anthracene, and indeno(1,2,3-*cd*)pyrene. Their absorption in the gastrointestinal tract depends on food composition and varies over a range of about 12–99%. The bioavailability of PAHs from burnt food is surprisingly low, although their content is relatively high, due to the absorption of PAHs by charred particles. Absorption of PAHs is followed by a rapid distribution in the body, since they can easily penetrate the lipoprotein membranes. Thanks to extensive biotransformation, accumulation of PAHs in the human body does not occur. In the initial stage of biotransformation, PAHs are oxidised by oxidoreductases of the cytochrome P450 complex (located in endoplasmic reticulum or microsomes of cells in various tissues, especially the liver) to reactive metabolites. For example, benzo[*a*]pyrene metabolites, such as benzo[*a*]pyrene-7,8-diol and benzo[*a*]pyrene-7,8-dione, can yield covalent compounds with DNA (Figure 12.26). Pyrene is analogously transformed into 1-hydroxypyrene, which is the main biotransformation product. Oxidative DNA damage may similarly occur by reaction with superoxide radicals.

The metabolism of nitro derivatives of PAHs is different from that of the parent PAHs. Orally administered compounds are mainly reduced to amino derivatives by intestinal microflora. Their elimination from the body occurs after oxidation to hydroxy derivatives and acetylation of amino group in the liver. Hydroxynitro derivatives of PAHs are the major products that are resorbed unchanged from the gastrointestinal tract. The risks from dietary intake of nitro-PAH derivatives are not known in detail. Generally, these direct carcinogens require metabolic activation. Particularly intense mutagenic effects have been demonstrated in nitro derivatives of pyrene. The toxic potential of various PAH derivatives may often surpass that of the parent compounds.

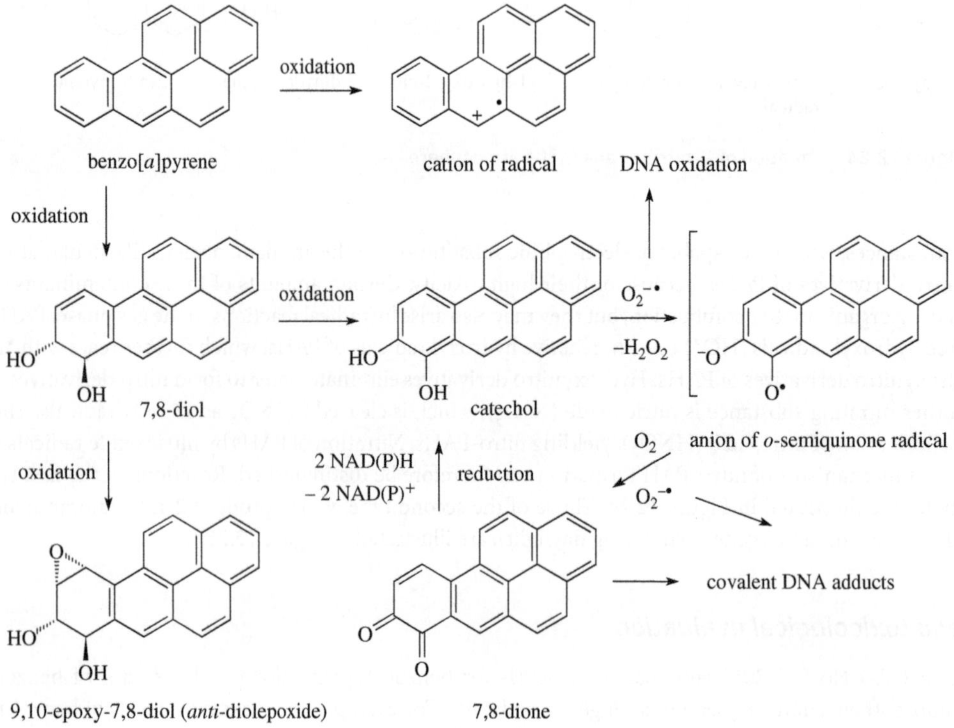

Figure 12.26 Metabolism of benzo[*a*]pyrene.

12.3.5.5 Mitigation

Possibilities for limiting the occurrence of exogenous PAHs are mainly connected with minimisation of anthropogenic and some non-anthropogenic sources and prevention of their dissemination into the food chain. Limitation of the formation of endogenous PAHs can often be achieved by modifying the process conditions or choosing less risky cooking processes.

12.3.6 Monocyclic aromatic hydrocarbons

Foods contain a variety of hydrocarbons, which arise in natural processes in the plant or animal organism. The main hazardous substances of particular importance are, however, PAHs. Some MAHs are also undesirable contaminants, particularly benzene, toluene, ethylbenzene, and xylenes. Compared with PAHs, MAHs are less hydrophobic and therefore somewhat more soluble in water.

The main sources of contamination of the food chain by MAHs are oil spills and a variety of coatings and plastics. MAHs may also arise as products of the combustion of various materials. Food contamination occurs primarily by absorption from the air and water, and in animals from the surrounding environment. Another type of food contamination is migration from packaging materials. In recent years, much attention has been paid to the contamination of yoghurt and biscuits by styrene (vinylbenzene) released from the polystyrene used to manufacture their packaging.

Findings of MAHs in certain foods in the UK market (see Table 12.14) illustrate the inputs of these substances from the external environment, because the food surface layers are generally more contaminated than the inner layers. In addition to foods, the MAH content is also monitored in drinking water. From a toxicological point of view, findings of benzene or styrene are particularly serious because of their carcinogenic potential.

Some MAHs may occur in foods as technological contaminants. A study carried out in the United States, Canada, and the United Kingdom in 2006–07 reported on the discovery of benzene in some soft drinks. In some samples, the benzene concentration exceeded $10\,\mu g/kg$ (the highest amount was $90\,mg/kg$), which caught the attention of food safety experts. In a follow-up study conducted in the Czech Republic, benzene levels higher than $10\,\mu g/l$ were found in cucumbers and caper berries containing benzoic and ascorbic acids, but only when packed in plastic pouches and after prolonged storage at room temperature, when ascorbic acid was partially or totally degraded during storage.

The formation of MAHs in soft drinks and foods can be explained by decarboxylation of aromatic acids, whether natural or used as additives. The second case corresponds to the formation of benzene in soft drinks and fermented vegetables preserved with benzoic acid, which in the presence of ascorbic acid can yield benzene by decarboxylation. The reaction is catalysed by traces of transition metals. Traces of benzene found in cranberry and mango products that were not preserved with benzoic acid resulted from a higher content of naturally present benzoic acid.

Table 12.14 Concentrations of monocyclic aromatic hydrocarbons in selected foods.

Food	Sampling site	Content (µg/kg)		
		Benzene	Toluene	Other hydrocarbons
Butter	Surface	<10–28	<10–275	<10–204
	Inner part	<10–15	<10–189	0–27
Cheeses	Surface	<10–18	<10–109	<10–158
	Inner part	<10–13	<10–55	<10–16
Pork lard	Surface	<10–12	10–216	<10–270
	Inner part	<10–14	<10–118	<10–18
Margarines	Surface	<10–21	12–977	<10–119
	Inner part	0–16	<10–274	0–13
Bacon	Surface	0–34	<10–180	0–44
	Inner part	<10–11	<10–10	0 <10
Sausages	Surface	0–34	<10–122	<10–19
	Inner part	<10	<10–56	<10–20

In comparison with other possible sources of exposure (emissions from motor vehicles, evaporation losses during handling, storage, and distribution of gasoline) the benzene intake via soft drinks is negligible. Nevertheless, in order to reduce total benzene intake, soft-drink recipes should be formulated to avoid its formation, which means avoiding or eliminating the simultaneous presence of benzoic acid (or its salts) and ascorbic acid.

Decarboxylation of cinnamic acid analogously yields styrene. Cinnamic acid is present as a minor constituent in cinnamon, where it arises by autoxidation of cinnamaldehyde. The concentration of styrene in cinnamon may even reach 40 mg/kg.

12.3.7 Nitroso compounds

12.3.7.1 Classification, structure, and properties

Since 1956, when the carcinogenicity of N-nitrosamines was proved, special attention has been paid to this group of contaminants. All nitroso compounds contain a nitroso group (N=O) in the molecule. The most common nitroso compounds are N-nitroso compounds (**12-33**), which include N-nitrosamines derived from secondary amines and N-nitrosamides derived from N-substituted amides of carboxylic acids. O-, S-, and C-nitroso compounds also exist.

12.3.7.2 Occurrence, main sources, and dietary intake

Nitroso compounds are formed from various organic compounds by the action of nitrosation reagents. These agents may enter into food during technological processing as food additives (e.g. nitrites or nitrates) or contaminants (e.g. nitrates from fertilisers) or during drying of food by direct heating using combustion gases containing nitrogen oxides. Nitroso compounds in foods may also occur as exogenous contaminants migrating from different sources, such as certain types of elastomers or latex toys, teats, and soothers. Another source of nitrosamines is tobacco smoke.

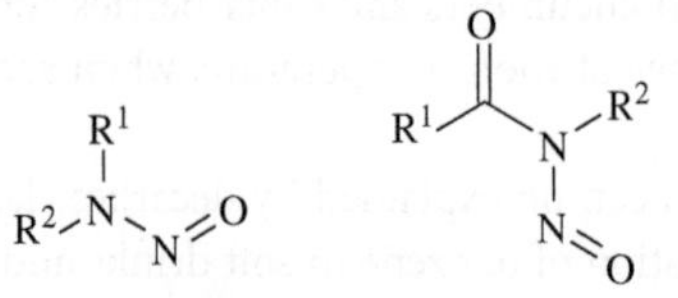

N-nitrosamines N-nitrosamides

12-33, N-nitroso compounds

N-Nitrosamines have been found in a number of foods. Greatest attention has been paid to their occurrence in smoked meats, cheeses (especially in smoked chesses), skimmed milk, fish, beer, and spirits, especially whisky. The contents of volatile and non-volatile N-nitrosamines in selected foods are given in Tables 12.15 and 12.16, respectively.

The formation of nitrosamines and their concentrations in foods depend on many factors, such as the presence and quantity of amines and their precursors, the type and quantity of nitrosation agents, pH, temperature, reaction time, the composition of the food (e.g. fat content), the method of heat treatment, the presence of substances that catalyse the nitrosation reaction (e.g. rhodanides and to a lesser extent chlorides), and the presence of inhibitors (e.g. ascorbic acid and its esters and salts, sulfur dioxide, and phenolic antioxidants such as gallates and melanoidins).

Table 12.15 Contents of volatile nitrosamines in selected foods.

Food	Nitrosamine	Content (µg/kg)	Food	Nitrosamine	Content (µg/kg)
Pickled meat	NDMA, NDEA, NPYR, NPIP	trace–55	Non-fat milk powder	NDMA	0.1–3.7
Fried bacon	NPYR, NDMA, NPIP	trace–200	Fermented vegetable	NDMA, NPYR	trace–5
Microwaved bacon	NDMA, NPYR	trace–1.2	Tea	NDMA	trace–1.2
Fish	NDMA	trace–10	Alcoholic beverages	NDMA	trace–4.9
Cheeses	NDMA	trace–15	Beer	NDMA	trace–68

NDMA = N-nitrosodimethylamine, NDEA = N-nitrosodiethylamine, NPIP = N-nitrosopiperidine, NPYR = N-nitrosopyrrolidine.

Table 12.16 Contents of non-volatile nitrosamines in selected foods.

Potravina	Content (μg/kg)				
	NPRO	NHPRO	NHMTCA	NTCA	NMTCA
Cured meat					
Cooked ham	0–40	0–100	–	–	–
Beef	70–100	240–250	130–255	328–570	0–28
Raw bacon	–	0–30	0–40	0–30	–
Fried bacon	0–20	0–80	0–100	0–50	–
Cured and smoked meats					
Mutton	230–360	350–560	160–320	960–1070	–
Sausages	0–70	–	110–410	180–210	
Raw bacon	0–20	0–60	0–1300	0–501	0–26
Fried bacon	0–40	0–90	0–2100	0–550	–
Ham	–	–	196–495	219–490	0–21
Cheeses	–	–	1062–1328	5–24	–

NPRO = *N*-nitrosoproline, NHPRO = *N*-nitroso-4-hydroxyproline, NHMTCA = *N*-nitroso-2-(hydroxymethyl)-4-thiazolidine carboxylic acid, NTCA = *N*-nitroso-4-thiazolidine carboxylic acid, NMTCA = *N*-nitroso-2-methyl-4-thiazolidine carboxylic acid.

12.3.7.3 Reactions

Volatile *N*-nitrosamines are formed by nitrosation of secondary amines and arise as intermediates through the nitrosation of primary and tertiary amines and quaternary ammonium salts in acidic media. Numerous precursors of nitrosamines, such as amino compounds containing a secondary amino group, are natural components of food (amines, amino acids, amino sugars, some vitamins, lipids, aromatic substances, and other compounds). For example, the amino acids proline, histidine, tryptophan, and sarcosine contain a secondary amino group in the molecule. Other secondary amines are formed from amino acids by decarboxylation. For example, histamine, tryptamine, putrescine, and agmatine arise by decarboxylation of the parent amino acids, spermidine and spermine are produced from putrescine, and tryptamine yields the hormone serotonine. Various products of the Maillard reaction include secondary amines, such as carbinolamines, glycosylamines, aminodeoxysugars (Amadori compounds), and heterocyclic compounds (pyrroles, imidazoles, oxazolines, and others). Choline bound in phospholipids yields dimethylamine by degradation. Secondary amines may also get into foods as contaminants.

The general mechanism of nitrosation of secondary amines is shown in Figure 12.27. Nitrosation agents are formed in acidic solutions from nitrous acid (nitrites) by a sequence of reactions indicated in Figure 12.28. Nitrites may be present in foods as additives or contaminants. In foods and beverages obtained by fermentation, nitrites may arise by reduction of nitrates by microbial reductases. In beer, for example, wild yeasts partly assimilate nitrates to give ammonia, and the activity of nitrate reductase yields nitrites which are further reduced to ammonia by the action of ammonia nitrite reductase. Nitrates are also reduced to nitrites by some contaminating denitrifying bacteria. Nitrites are reduced to nitrous oxide (dinitrogen monoxide, N_2O) and nitric oxide (nitrogen monoxide, NO) by cytochrome cd_1-nitrite reductase, whilst nitric oxide is reduced to nitrous oxide, from which elemental nitrogen arises by reduction, catalysed by reductases.

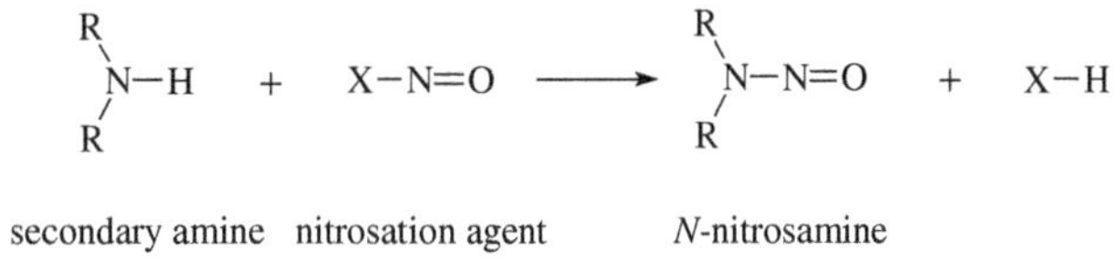

Figure 12.27 General mechanism of secondary amine nitrosation.

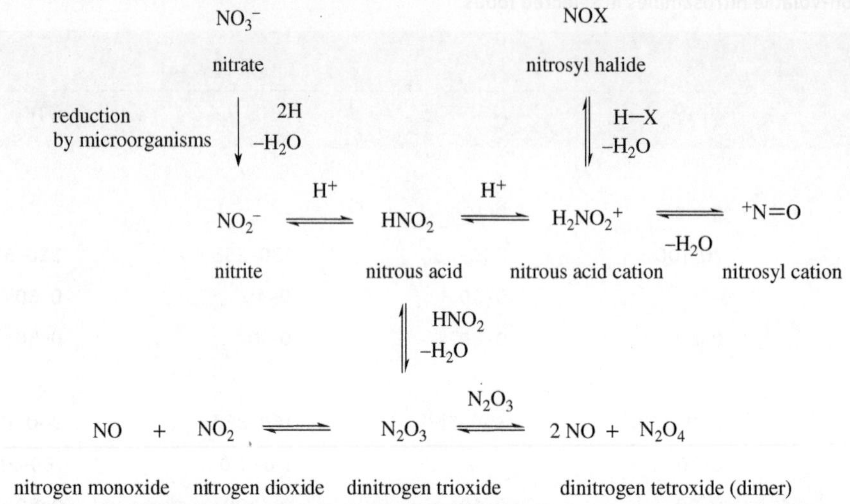

$$NO_3^-$$

nitrate

reduction by microorganisms $\Bigg\downarrow \begin{array}{c} 2H \\ -H_2O \end{array}$

$$NOX$$

nitrosyl halide

$\Bigg\updownarrow \begin{array}{c} H-X \\ -H_2O \end{array}$

$$NO_2^- \underset{}{\overset{H^+}{\rightleftharpoons}} HNO_2 \underset{}{\overset{H^+}{\rightleftharpoons}} H_2NO_2^+ \underset{-H_2O}{\rightleftharpoons} {}^+N{=}O$$

nitrite   nitrous acid   nitrous acid cation   nitrosyl cation

$\Bigg\updownarrow \begin{array}{c} HNO_2 \\ -H_2O \end{array}$

$$NO \;+\; NO_2 \rightleftharpoons \overset{N_2O_3}{N_2O_3} \rightleftharpoons 2\,NO \;+\; N_2O_4$$

nitrogen monoxide   nitrogen dioxide   dinitrogen trioxide   dinitrogen tetroxide (dimer)

Figure 12.28 Formation of nitrosation agents in acidic media.

Effective nitrosation agents are the nitrosyl cation (NO^+), nitrogen oxides (dinitrogen trioxide, N_2O_3, and nitrogen dioxide, NO_2), and nitrosyl halogenides formed in the presence of hydrohalogen acids. The relative ratio or nitrosation agent depends on the pH. In media with pH < 2, the predominant nitrosation agent is the nitrous acid cation ($H_2NO_2^+$) or the nitrosyl cation (NO^+). At pH > 3, the main nitrosation agent is dinitrogen trioxide (N_2O_3). Both reagents are present in media with pH 2–5. Dinitrogen trioxide (N_2O_3) readily decomposes to dinitrogen monoxide (N_2O) and nitrogen monoxide (NO) and its dimer (N_2O_2). For example, at 25 °C and under atmospheric pressure, solutions contain equal amounts of both oxides and about 10% undissociated dinitrogen trioxide (N_2O_3). Nitrogen monoxide (NO) is also produced, together with nitrates, by disproportionation of nitrous acid:

$$3\,HNO_2 \rightarrow 2\,NO + H^+ + {}^+NO_3^- + H_2O$$

Nitrogen monoxide (NO), formed by disproportionation of dinitrogen trioxide (N_2O_3), is easily oxidised by oxygen to nitrogen dioxide (NO_2):

$$2\,NO + O_2 \rightarrow 2\,NO_2$$

Nitrosation reactions may be catalysed by – and also inhibited in the presence of – other substances (Figure 12.29, Y—N=O is a stronger nitrosation agent than X—N=O). Catalytic effects are exhibited, for example, by thiocyanate (SCN^-) and halide (iodides, bromides and chlorides) anions. Inhibitory effects have been reported for some inorganic and organic compounds that preferentially react with and reduce nitrosation agents. In this way, ascorbic acid inhibits nitrosamine formation (at concentrations of 500–1000 mg/kg) and tocopherols (at concentrations of 100–500 mg/kg). These vitamins, if used in combination, show a higher inhibitory effect than when applied individually. Inhibitory effects are also exhibited by other substances, such as sulfur dioxide, cysteine, and glutathione.

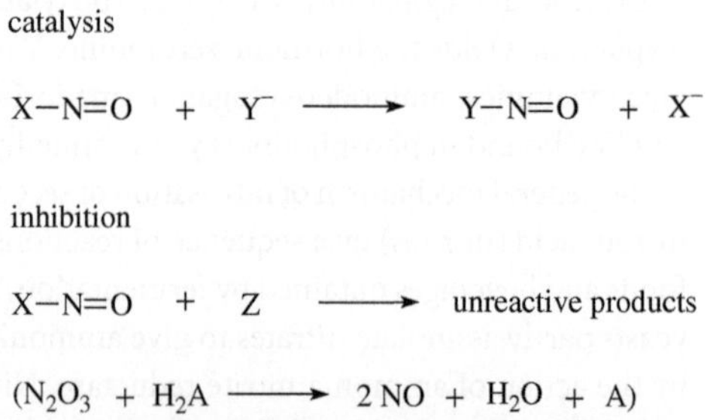

For practical reasons, *N*-nitrosamines are divided into:

- volatile nitrosamines;

- non-volatile nitrosamines.

catalysis

$$X{-}N{=}O \;+\; Y^- \longrightarrow Y{-}N{=}O \;+\; X^-$$

inhibition

$$X{-}N{=}O \;+\; Z \longrightarrow \text{unreactive products}$$

$$(N_2O_3 \;+\; H_2A \longrightarrow 2\,NO \;+\; H_2O \;+\; A)$$

Figure 12.29 Mechanism of catalysis and inhibition of nitrosation reaction. Y^- = catalyst (such as iodide), Z = reducing agent (H_2A = ascorbic acid, A = dehydroascorbic acid).

Volatile nitrosamines are a group of relatively non-polar low-molecular-weight compounds, whilst non-volatile nitrosamines are polar compounds. The most common and also the most toxic volatile nitrosamine is *N*-nitrosodimethylamine, which is produced from dimethylamine and, via dimethylamine, from other amino compounds, such as sarcosine, creatine, trimethylamine, and choline (Figure 12.30). In malt, the main precursors of dimethylamine are alkaloids hordenine and gramine, occurring in germinating barley. The concentration of dimethylnitrosamine in malt and beer depends on the conditions during barley germination and malt storage. Concentrations found in malt and beers in the past have been drastically reduced by the introduction of indirect heating during malt kilning.

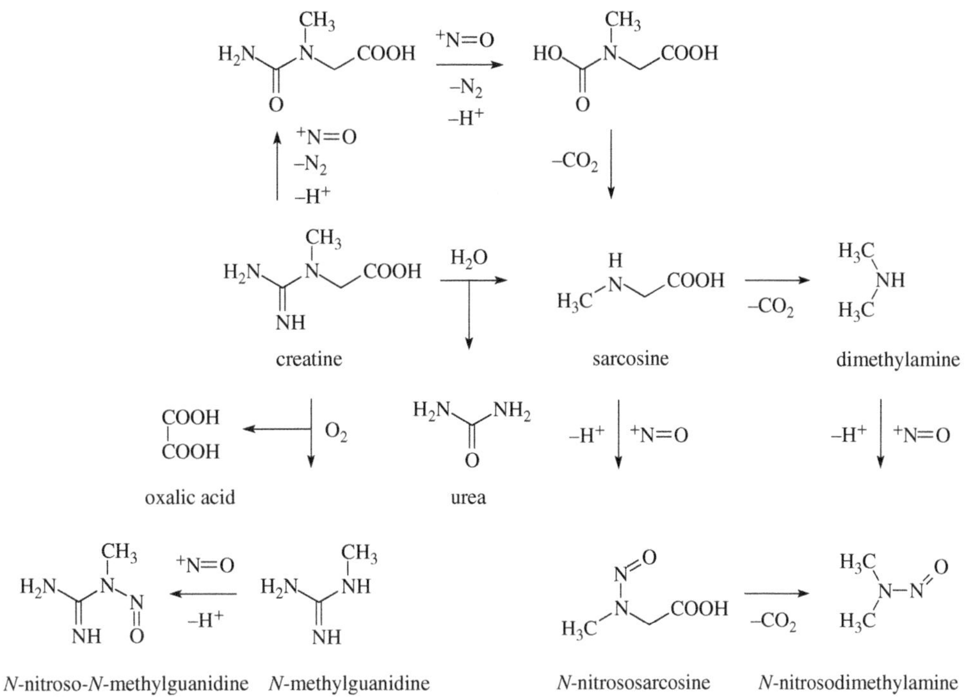

Figure 12.30 Formation of N-dimethylnitrosamine and other N-nitroso compounds from creatine and sarcosine.

Precursors of non-volatile – but also of certain volatile – nitrosamines are often amino acids and derived amines. It is estimated that N-nitrosoamino acids constitute about 1% of the total N-nitrosocompounds present in foodstuffs. For example, N-nitrososarcosine arises from sarcosine and from creatine (Figure 12.30), proline yields N-nitrosoproline, and the decarboxylation of N-nitrosoproline gives N-nitrosopyrrolidine. N-Nitrosoproline also arises from ornithine. Similarly, N-nitrosopyrrolidine is produced from putrescine and spermidine, and also from pyrrolidine, which is the product of proline decarboxylation (Figure 12.31). By the same sequence of reactions, N-nitroso-4-hydroxyproline arises from 4-hydroxyproline. Lysine and biogenic amine cadaverine give rise to N-nitrosopiperidine (**12-34**).

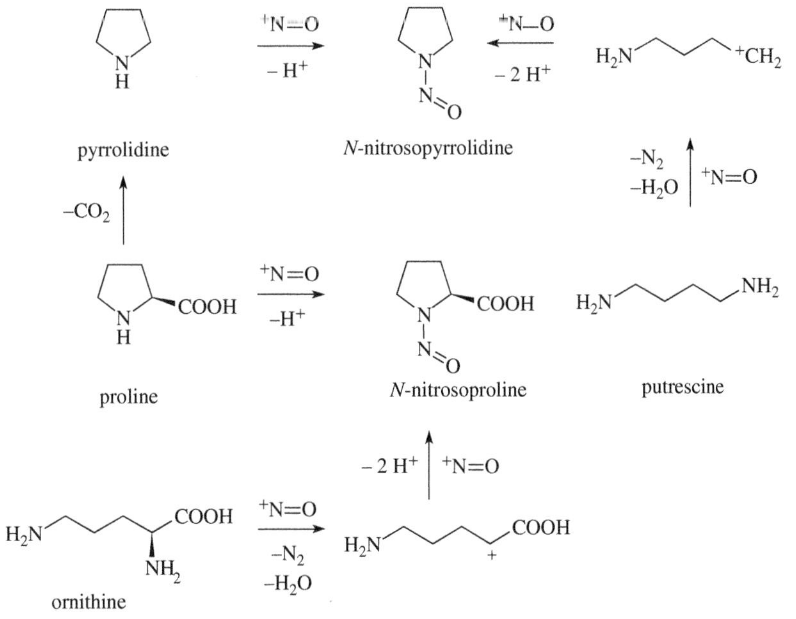

Figure 12.31 Formation of N-nitrosoproline and N-nitrosopyrrolidine.

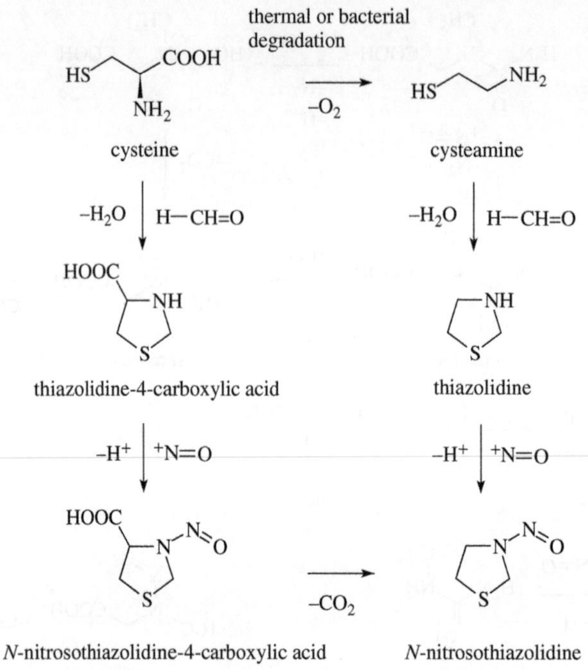

Figure 12.32 Formation of nitrosamines from cysteine.

N-Nitrosomorpholine (**12-34**) is formed from ethanolamine, which is a component of some phospholipids. Under certain conditions, nitrosation of diethanolamine, which is used in the manufacture of personal care products, yields *N*-nitrosodiethanolamine.

Precursors of some non-volatile nitrosamines are formed in the Maillard reaction. For example, 4-thiazolidine carboxylic acid arises from cysteine and formaldehyde (Figure 12.32), which is a component of combustion gases and liquid smoke preparations, and is also produced by the Strecker degradation of glycine and retroaldolisation of sugars. Nitrosation of 4-thiazolidine carboxylic acid then yields *N*-nitroso-4-thiazolidine carboxylic acid. 4-Thiazolidine carboxylic acids substituted at C-2 arise in reactions of cysteine, with higher aldehydes resulting from the Strecker degradation of certain amino acids or in degradation of sugars. Reaction with acetaldehyde, for example, yields 2-methyl-4-thiazolidine carboxylic acid, and 2-hydroxymethyl-4-thiazolidine carboxylic acid is produced in a reaction with glycolaldehyde.

Non-volatile nitrosamines are likewise produced by nitrosation of *N*-alkylsubstituted guanidines and *N*-alkylsubstituted ureas. *N*-Methylguanidine, for example, occurs in fresh meat at concentrations up to 10 mg/kg, and also arises together with *N*-nitrosoguanidine from the decomposition of creatine in the presence of nitrites (Figure 12.30). Another example of an *N*-substituted nitrosoguanidine is *N*-nitrosoagmatine (**12-34**), formed by nitrosation of agmatine, which is a product of arginine decarboxylation. In the meat of certain shellfish, agmatine occurs in concentrations up to 200 mg/kg. Methyl-, propyl-, and but-3-en-1-ylureas and the corresponding *N*-nitroso-*N*-alk(en)ylureas were found in fish, shellfish, and ham. The key *N*-methylurea precursor is creatinine, which is present in fairly high concentrations in meat, fish, and seafood and can be nitrosated to form traces of *N*-nitroso-*N*-methylurea (Figure 12.33) via 1-methyl-5-oxohydantoin-5-oxime. The optimum pH for the formation of *N*-nitroso-*N*-methylurea from creatinine ranges between pH 1 and 3. The step from 1-methyl-5-oxohydantoin-5-oxime to *N*-nitroso-*N*-methylurea is not clear, however; 1-methyl-5-oxohydantoin-5-oxime might be nitrosated directly to *N*-nitroso-*N*-methylurea or indirectly via the intermediate formation of methylurea. *N*-Nitroso-*N*-methylurea is a potent carcinogen that has been shown to induce cancer of various organs, mainly the forestomach, brain, and nervous system, in a wide variety of animal species. At gastric pH, small amounts of *N*-nitroso-*N*-methylurea (0–2.6 µg/kg) were formed from various cured meats, increasing with additional nitrite up to 17.6 µg/kg of meat. At gastric pH, *N*-nitroso-*N*-methylurea can also be produced in other foods, such as fish sauce, herring, shrimp, sardines, oysters, mussels, and pickled vegetables.

Some nitrosamines are formed only in certain commodities. For example, the nitrosation of goitrin present in cruciferous vegetables and rapeseed yields *N*-nitroso-5-vinyloxazolidin-2-one (**12-34**). During the burning of tobacco, in addition to common volatile nitrosamines (*N*-nitrosodimethylamine and its higher homologues), non-volatile nitrosamines are derived from nicotine and other tobacco alkaloids. Examples are *N*-nitrosonicotine, *N*-nitrosonornicotine, *N*-nitrosoanatabine, and *N*-nitrosoanabasine (**12-34**). Tobacco-specific nitrosamine

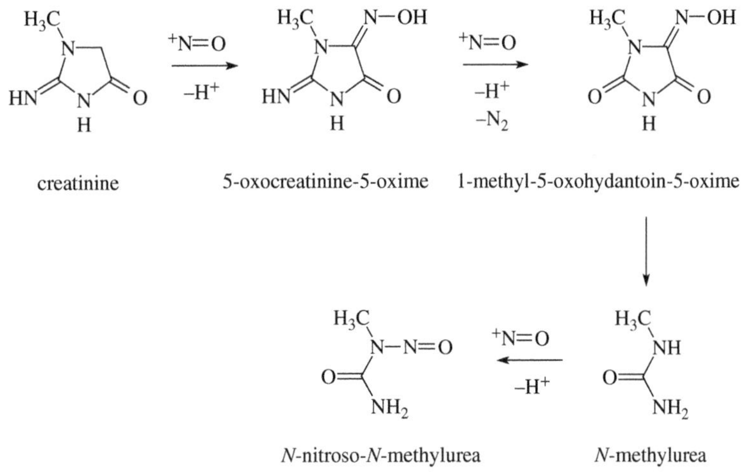

Figure 12.33 Nitrosation of creatinine.

formation in tobacco is influenced by alkaloid levels and the availability of nitrosating agents (NO$_x$).

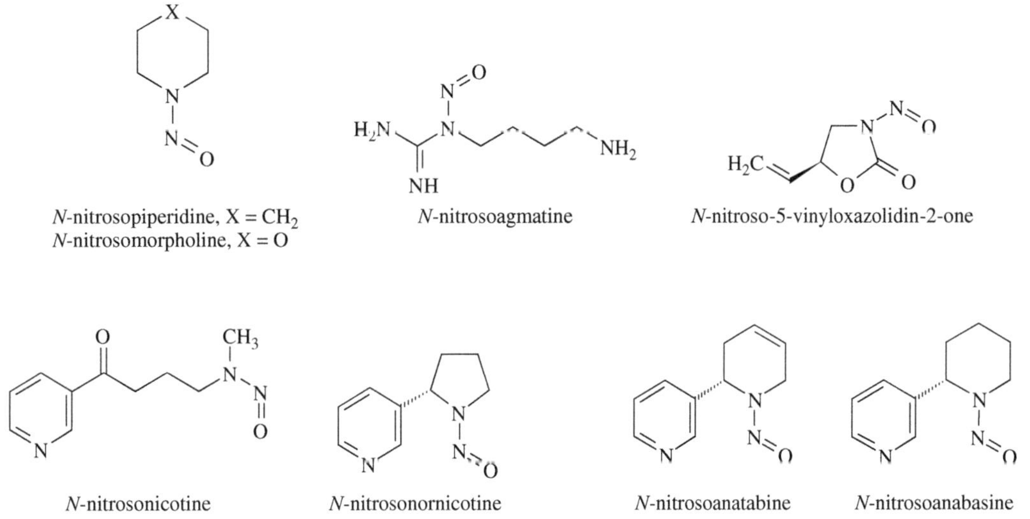

N-nitrosopiperidine, X = CH$_2$
N-nitrosomorpholine, X = O

N-nitrosoagmatine

N-nitroso-5-vinyloxazolidin-2-one

N-nitrosonicotine

N-nitrosonornicotine

N-nitrosoanatabine

N-nitrosoanabasine

12-34, less common non-volatile *N*-nitrosamines

Nitrous acid reacts with primary aliphatic amines in the presence of mineral acids to form volatile nitrosamines, which isomerise to stable diazohydroxides. Diazohydroxides are split into nitrogen and a primary alkyl cation, which may partly isomerise to a secondary cation, and both cations are eliminated, which yields the corresponding alkene. Substitution reactions then yield primary and secondary alcohols and symmetrical and unsymmetrical secondary amines. In the presence of hydrohalogen acids, both alkyl halides are produced (Figure 12.34). Aromatic primary amines, which should however not be natural food components, yield diazonium salts (Ar–N=N)$^+$ X$^-$ by nitrosation. In the absence of mineral acids, nitrous acid reacts with primary, secondary, and tertiary aliphatic amines with the formation of nitrites. Secondary amines, for example, yield salts, the structure of which is (R^{1}R^{2}NH$_2$)$^+$NO$_2^-$. Tertiary amines react with nitrous acid in the presence of mineral acids to form nitrosamine cations, which may be cleaved analogously to nitrosamines derived from primary amines to alkyl cations and the corresponding nitrosamines. Splitting of hyponitrous acid (H$_2$N$_2$O$_2$) yields the corresponding imine cation, which decomposes to an aldehyde and a secondary amine (Figure 12.35).

N-Substituted amides are produced in small amounts by heating of carboxylic acids with amines. They are formed relatively easily from fatty acids or their esters (e.g. triacylglycerols) and amines. The reactivity of esters of fatty acids is greater than that of free fatty acids. At

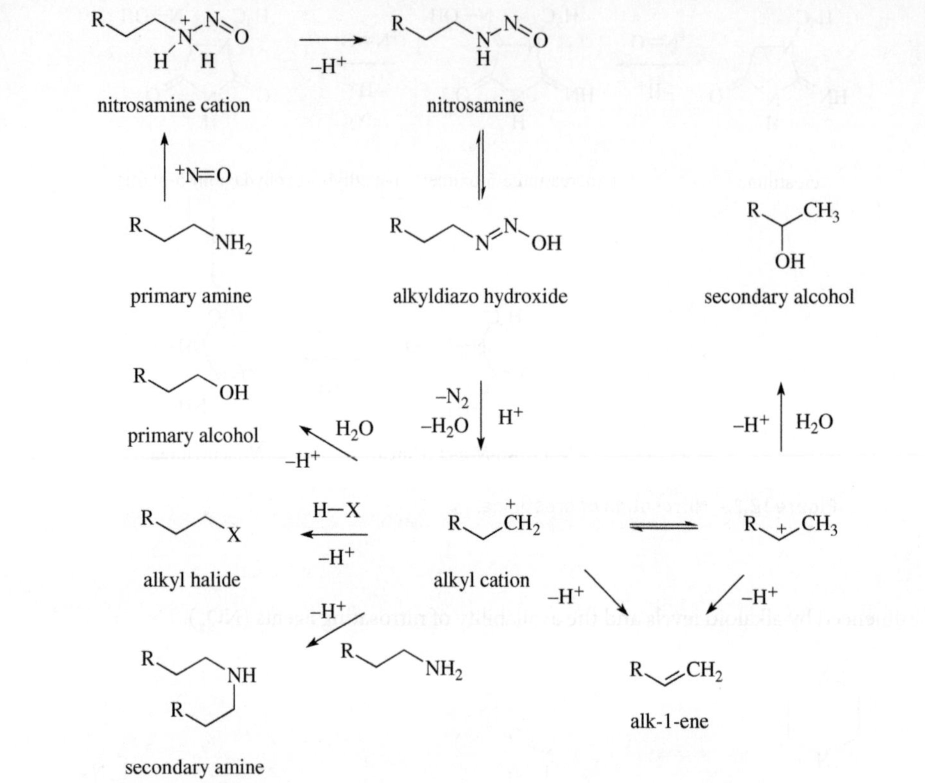

Figure 12.34 Reaction of primary amines with nitrous acid in acidic solutions.

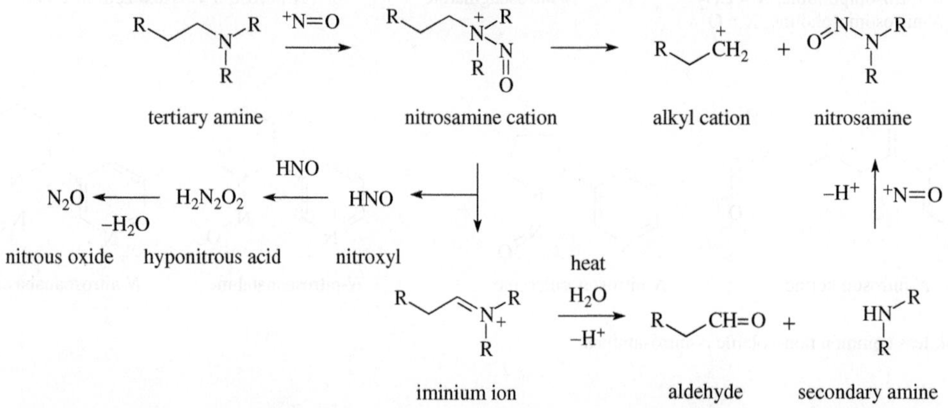

Figure 12.35 Reaction of tertiary amines with nitrous acid in acidic solutions.

temperatures above 150 °C, amino acids also react to form amides. Nitrosamides are formed by nitrosation of N-substituted amides through the same mechanisms as N-nitrosamines from secondary amines (Figure 12.36).

In addition to N-nitroso compounds, S-nitroso compounds are also formed in foods. The reaction of nitrite with the sulfhydryl group of free cysteine or protein-bound cysteine leads to the formation of mutagenic S-nitrosocysteine (**12-35**), which was found in meat and meat products containing nitrites. Nitrosothiols in meat, including S-nitrosocysteine, consume 3–12% of added nitrite. They show antimicrobial activity, participate in the development of the characteristic aroma of smoked meat, and take part in transnitrosation reactions. For example, they react with myoglobin to yield nitroxymyoglobin (see Section 9.2.1.5.3).

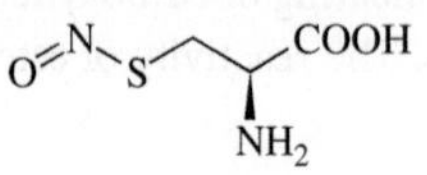

12-35, S-nitrosocysteine

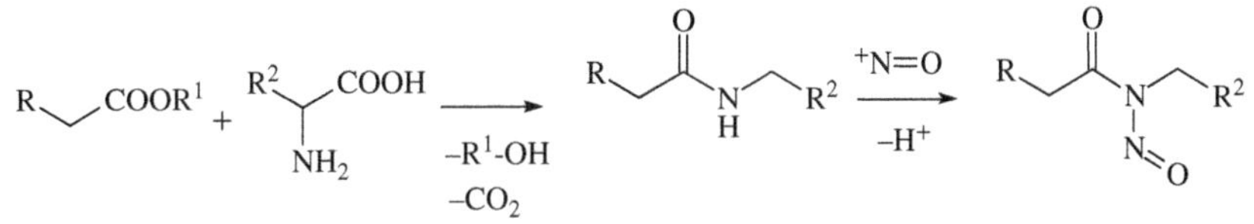

fatty acid ester amino acid fatty acid amide *N*-nitrosamide

Figure 12.36 Formation of nitrosamides from amino acids and lipids.

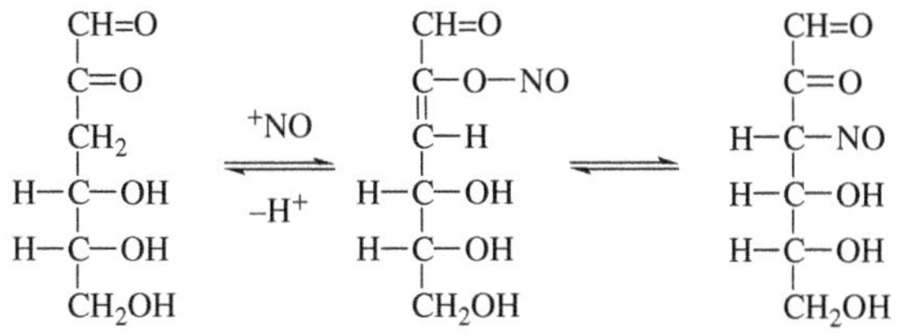

3-deoxy-D-*erythro*-hexosulose 3-deoxy-3-nitroso-D-*erythro*-hexosulose

Figure 12.37 Nitrosation of 3-deoxy-D-*erythro*-hexosulose.

Less information is available on the occurrence of *C*-nitroso-compounds in foods. Stable compounds are derivatives in which the nitroso groups are bound to the tertiary carbon. Such a compound is the nitrosation product 3-deoxy-D-*erythro*-hexosulose, which is one of the major degradation products of glucose and other sugars (Figure 12.37).

Compounds with a nitroso group on the primary or secondary carbon irreversibly isomerise to isonitroso compounds, which are oximes of aldehydes or ketones:

$$R{-}CH_2{-}N = O \longrightarrow R{-}CH = N{-}OH$$
$$R^1R^2CH{-}N = O \longrightarrow R^1R^2C = N{-}OH$$

Examples of such compounds are products of the nitrosation of ketones arising from autoxidation of lipids. Their oximes and oxime decomposition products are flavour-active components of smoked meat. The reaction begins at the methylene group adjacent to the carbonyl group. The elimination of a proton provides a ketone anion, which reacts with a nitrosyl cation to yield a *C*-nitroso compound, which isomerises to oxime (Figure 12.38).

Another example is the nitrosation of creatinine in meat, leading to the formation of 5-oxocreatinine-5-oxime and 1-methyl-5-oxohydantoin-5-oxime (Figure 12.33). Some phenols also react with nitrosation reagents, yielding nitroso compounds and oximes. In addition, in reactions with nitrosation agents, there also arise nitro, oxynitro, and nitronitroso compounds, as in the case of PAHs. The mechanisms of the formation of these compounds and other aspects of their presence in food are not yet sufficiently known. An example is the nitrosation of *p*-coumaric acid, which is dependent on pH. 4-Hydroxybenzaldehyde, 1′,4-dihydroxybenzeneacetaldehyde oxime, 4-hydroxy-1′-oxobenzeneacetaldehyde oxime, and 7-hydroxy-1,2(4*H*)-benzoxazin-4-one all form in acidic solutions (at 16, 59, 26, and 6%, respectively), whereas 4-(2-oxido-1,2,5-oxadiazol-3-yl)phenol forms in acidic (6%), neutral (1%), and alkaline (1%) ones (Figure 12.39).

12.3.7.4 Health and toxicological evaluation

Although food and tobacco products are important sources of external exposure to *N*-nitrosamines, exposure also occurs from nitrosamines produced internally in the digestive tract. About 5% of ingested nitrates (e.g. nitrates occurring in vegetables) are reduced to nitrites in saliva. These nitrites can subsequently react with secondary and tertiary amines, as well as *N*-substituted amides, carbamates, and other related

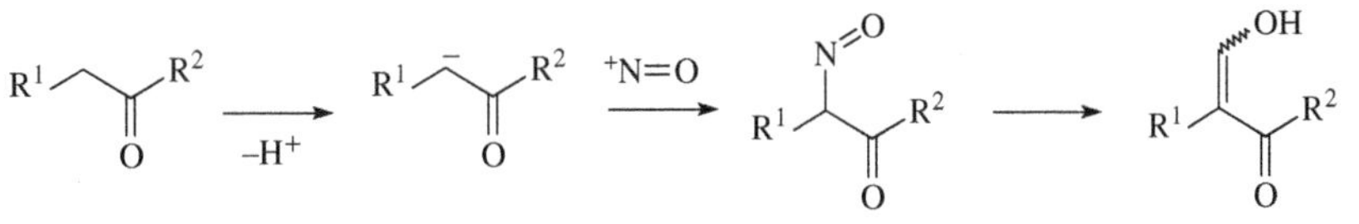

ketone ketone anion *α*-*C*-nitrosoketone *α*-ketone monoxime

Figure 12.38 Formation of oximes from ketones.

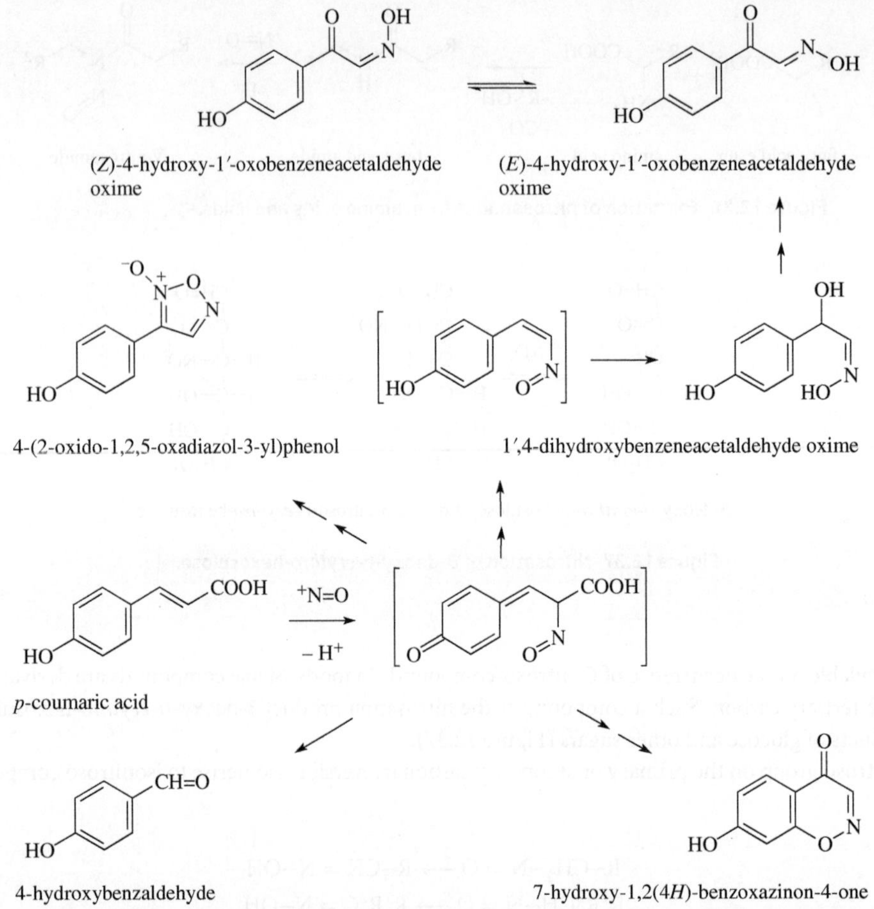

Figure 12.39 Nitrosation of *p*-coumaric acid.

compounds, to form *N*-nitroso compounds within the gastrointestinal tract. This internal formation is a major source of human exposure to *N*-nitrosamines.

N-Nitrosamines exhibit mutagenic, teratogenic, and carcinogenic effects. Carcinogenicity has been demonstrated in various organs of a number of animals. Many nitrosamines are classified by the IARC as Groups 2A or 2B carcinogens. The synergistic effect of *N*-nitrosamines taken in with other carcinogenic compounds has also been demonstrated. In general, reducing the levels of nitrites to the minimum necessary amount seems to be an appropriate strategy for limiting the exposure of consumers to nitrosamines. In terms of chemical safety, it is also necessary to monitor and possibly eliminate the migration of nitrosamines from food packaging and other materials.

At present, maximum levels of nitrosamines in foods are set only for beer in some EU member states, which stipulate that beer may contain up to 0.5 µg/l of *N*-nitrosodimethylamine and that the sum of nitrosamines including *N*-nitrosodimethylamine, *N*-nitrosodiethylamine, *N*-nitrosodibutylamine, *N*-nitrosopyrrolidine, *N*-nitrosopiperidine, and *N*-nitrosomorpholine must not exceed 0.0015 mg/l. Action levels for *N*-nitrosodimethylamine in barley malt and malt beverages range from 5 to 10 µg/l. In other foods, nitrosamines have not been limited since 2004. Within the European Union, legislation also exists for infants' teats and soothers (Commission Directive 93/11/EEC), which limits total *N*-nitrosamines to 10 µg/kg and nitrosatable materials to 200 µg/kg. European Cosmetic Directive 76/768/EEC specifies that nitrosamine r *N*-nitrosodiethanolamine contamination must not be present in health and beauty care products or cosmetics in amounts exceeding 10 µg/kg.

12.3.7.5 Mitigation

The formation of nitroso compounds in foods can be successfully controlled by reducing the concentration of added nitrite, drying foods via indirect heating, using substances that inhibit nitrosation (such as ascorbic acid and tocopherols), and making suitable changes to the technological processes (lowering the temperature during processing). For example, the technological modification of drying malt indirectly instead of directly by combustion gases leads to a significant reduction in the content of volatile nitrosamines in beer.

12.3.8 Ethyl carbamate

12.3.8.1 Classification, structure, and properties

The presence of toxic carbamic acid ethyl ester (ethyl carbamate, also called urethane, **12-36**) in fermented food products was first observed in 1976. Canadian authorities drew attention to this fact in 1985, when they carried out an extensive inspection of spirits imported from Europe. The ethyl carbamate level was restricted to 30 µg/kg in table wines, 100 µg/kg in fortified wines, 150 µg/kg in distilled spirits, and 400 µg/kg in fruit brandies and liqueurs. These guidelines have been used as a reference in other countries that do not have a specific legislation on this issue.

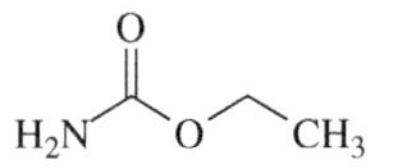

12-36, ethyl carbamate

12.3.8.2 Occurrence, main sources, and dietary intake

Concentrations of ethyl carbamate in bread, yoghurt, soy sauce, beer, and wine generally reach tens of µg/kg. Substantially higher amounts can be encountered in alcoholic beverages, especially spirits made from stone fruits, which in rare cases can contain thousands of µg/l of ethyl carbamate. The results of the latest EU study are presented in Table 12.17.

Table 12.17 Typical concentrations of ethyl carbamate in selected foods and alcoholic beverages.

Material	Total number of samples	Number of positive samples	Average amount (µg/kg)	Range (µg/kg)[a]
Foods and other products				
Bakery products	50	49	6	nd–20
Fermented dairy products	22	0	–	nd
Fermented olives	3	0	–	nd
Fermented sauces	44	28	3–4	nd–18
Sauerkraut	1	1	–	29
Vinegars	10	1	–	nd–33
Yeast extracts	1	1	–	41
Alcoholic beverages				
Beers	13	1		nd–1
Wines	17	11	10–11	nd–24
Rice wines (sake)	2	2	123	81–164
Liqueurs	4	2	45–47	nd–170
Gin	1	1	–	580
Whisky	210	196	41	nd–1000
Rum	11	10	325–328	nd–1020
Vodka	60	57	386–387	nd–2140
Brandy	42	19	123–129	nd–2100
Fruit brandy	328	281	663–667	nd–7920
Stone fruit brandy	3244	2912	848–851	nd–22 000

[a]nd, not determined (under the limit of detection).

12.3.8.3 Reactions

The high amounts of ethyl carbamate in some beverages were first explained by the use of diethyl dicarbonate, which was permitted as a preservative at that time (see Section 11.2.1.2.6). Findings of ethyl carbamate in products that were not stabilised with diethyl dicarbonate, however, highlighted a number of other mechanisms of its formation. Research in the follow-up period showed that the main precursors of ethyl carbamate in fermented foods were cyanides, urea, citrulline, and carbamoyl phosphate.

In distilled spirits, especially stone fruit spirits, ethyl carbamate is formed via reaction of ethanol with cyanides, and correspondingly with isocyanate ions that are byproducts of the degradation of cyanogenic glycosides. Examples of cyanogenic glycosides are amygdalin and prunasin, occurring in stone fruits (cherries, apricots, and plums), and epiheterodendrin, found in barley. Although it can be assumed that ethyl carbamate may be formed during the fermentation of stone fruits, its transition to the distillate is not very significant due to its low volatility (relatively high boiling point of 185 °C). A substantial proportion of ethyl carbamate in distillates arises by photochemical reactions. The most efficient light is that with wavelength 350–425 nm. In addition to the amount of precursors present, light exposure, temperature, pH, and the presence of certain other compounds are the main factors influencing the formation of ethyl carbamate in distilled spirits during storage and transport.

It is supposed that the main reaction is the oxidation of cyanides, released from cyanogenic glycosides, to cyanates, which may isomerise to isocyanates:

$$2\,N{\equiv}C^- + O_2 \rightarrow 2\,N{\equiv}C{-}O^- \leftrightarrow 2\,O{=}C{=}N^-$$

This reaction is effectively catalysed by divalent copper ions, which are often released from the distillation apparatus. By another reaction, copper cyanate is produced, and nucleophilic addition of water gives rise to copper carbamate, which provides ethyl carbamate by alcoholysis or decomposes to copper hydroxide, carbon dioxide, and ammonia:

$$2\,C{\equiv}N^- + 4\,HO^- + 4\,Cu^{2+} \rightarrow 2\,N{\equiv}C{-}O^- + 4\,Cu^+ + 2\,H_2O$$
$$Cu^{2+} + 2\,N{\equiv}C{-}O^- \rightarrow Cu(O{=}C{=}N)_2$$
$$Cu(O{=}C{=}N)_2 + 2\,H_2O \rightarrow Cu[O(O{=}C)NH_2]_2$$
$$Cu[O(O{=}C)NH_2]_2 + 2\,CH_3CH_2OH \rightarrow 2\,CH_3CH_2OC({=}O)NH_2 + Cu(OH)_2$$
$$Cu[O(O{=}C)NH_2]_2 + 2\,H_2O \rightarrow Cu(OH)_2 + 2\,CO_2 + NH_3$$

Ethyl carbamate can also be produced in small quantities by other reactions, for example from vicinal dicarbonyl compounds, such as methylglyoxal, biacetyl, and pentane-2,3-dione, which are fermentation byproducts.

Ethyl carbamate precursors in wine, beer, yoghurt, and some other fermentation products are mainly N-carbamoyl compounds, such as urea (carbamide) and citrulline, which result from catabolic processes in yeasts and bacteria from arginine.

Ethyl carbamate in wine is formed (mostly at the end of fermentation) from urea. The intermediates of its degradation are probably cyanates and cyanic acid (HO—C≡N), also known as hydrogen cyanate, which may isomerise to isocyanic acid (H—N=C=O). Isocyanic acid can also arise by protonation of the cyanate anion, and nucleophilic addition of ethanol to isocyanic acid yields ethyl carbamate. Isocyanic acid also reacts with other nucleophilic reagents, such as water (with the formation of ammonia and carbon dioxide), thiols, and amino groups of proteins. By catalysis with ornithinecarbamoyl transferase, citrulline is transformed into ornithine and carbamoyl phosphate, the ethanolysis of which yields ethyl carbamate (Figure 12.40).

Azodicarbonamide, used as a blowing agent in beer bottle cap liners and as a bread improver, has also been suggested as a possible ethyl carbamate precursor.

12.3.8.4 Health and toxicological evaluation

Ethyl carbamate is characterised by a broad spectrum of biological effects. Even in the 1940s, it was used as an anaesthetic substance; it was particularly suitable for children, as it acts on the central nervous system and has sedative effects similar to those of barbiturates. At the beginning of the 1940s, however, reports appeared of its carcinogenicity. In 2007, the IARC upgraded its classification from Group 2B to Group 2A. The current TDI is 20 ng/kg b.w, which corresponds (in the case of a person weighing 70 kg) to an annual intake of 520 μg. This value can easily be exceeded in some population groups (e.g. consumers of higher amounts of alcohol).

At present, some countries, such as Canada, Korea, and certain EU member states (e.g. France, Germany, and the Czech Republic), have established maximum levels for ethyl carbamate in alcoholic beverages. In the Czech Republic, for example, the limit in fruit spirits is 0.4 mg/l, whilst the limit for other spirits and wines is lower by one order of magnitude.

Toxicological studies have shown that nearly 90% of orally received ethyl carbamate is excreted as carbon dioxide by the action of cytochrome P450 oxidase complex. In the minor pathway, however, illustrated in Figure 12.41, ethyl N-hydroxycarbamate, vinyl carbamate, and oxiran-2-yl carbamate arise as minor products, which are responsible for ethyl carbamate's genotoxic and carcinogenic effects.

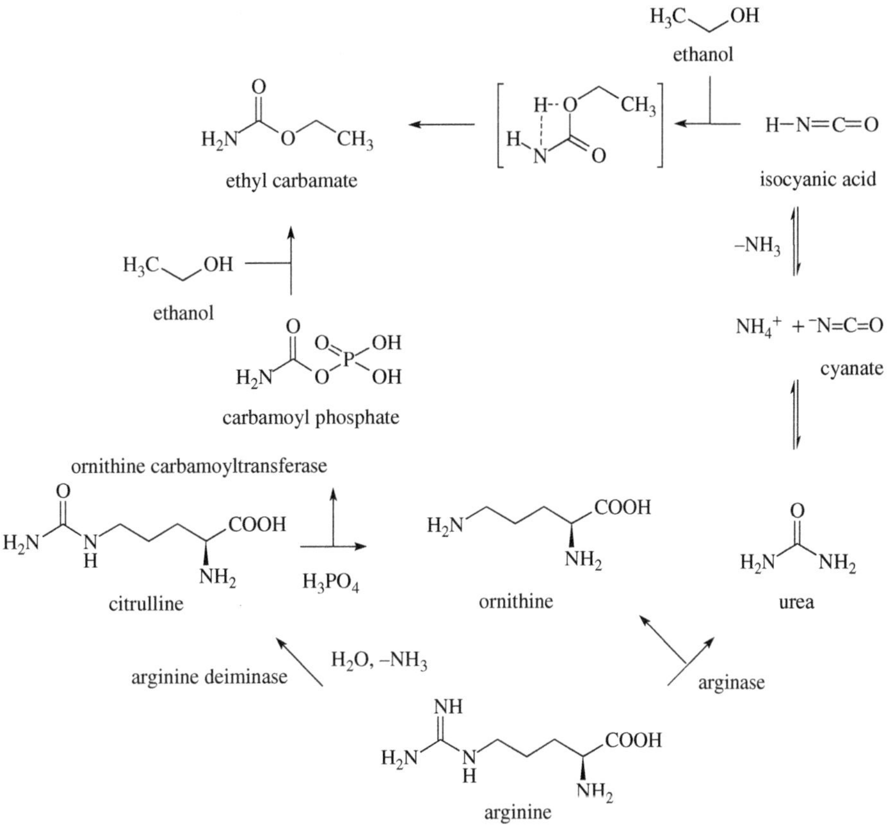

Figure 12.40 Formation of ethyl carbamate from *N*-carbamoyl compounds.

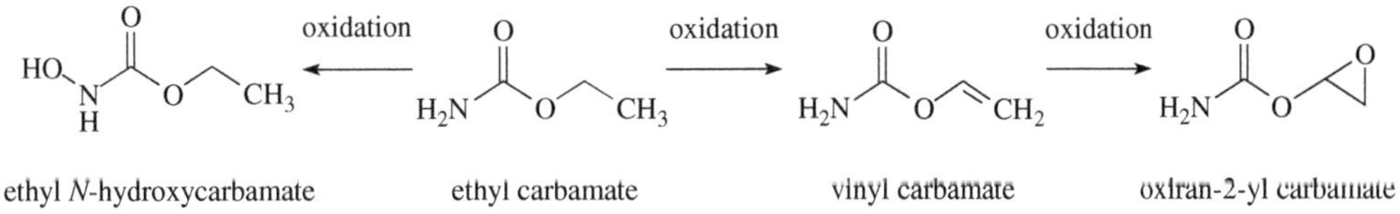

Figure 12.41 Metabolic activation of ethyl carbamate.

12.3.8.5 Mitigation

Preventive measures in the production of distilled spirits (such as whisky and stone fruit spirits) are based on reducing the cyanogenic glycoside content. It is therefore recommended to use low-cyanogen barley varieties or fruit that has had the stones removed and to minimise the mechanical damage to stones and seeds. Even for small distilleries, simple options like fruit destoning exist. It is also recommended that the first fraction of distillate, which contains higher amounts of isocyanic acid, not be used. After exposure of spirits in clear glass bottles to solar radiation, maximum concentrations of urethane are reached in about two days; therefore, measures should be taken to prevent light exposure, such as by using proper containers and covering boxes. Other possibilities for reducing ethyl carbamate levels include adding special copper catalysts to the mash or in front of the dephlegmators in conventional distillation equipment.

One possibility for the production of wines is the use of genetically modified *S. cerevisiae* yeast with low arginase activity, which can metabolise urea, or the enzyme urease, which catalyses the hydrolysis of urea into carbon dioxide and ammonia. In the case of malolactic fermentation, the use of pure *Oenococcus oeni* yeast cultures is recommended. Generally, it is recommended that temperatures be kept in the range of 18–25 °C during fermentation, or that a lower pH value be used, as this increases during fermentation due to the ammonia generated. Special care should be taken to minimise heat exposure by maintaining the temperature, preferably at or below 20 °C, and critically not above 38 °C, along the chain from production through shipment to storage and retail.

12.4 Persistent organohalogen contaminants

An undesirable consequence of some anthropogenic activities in the twentieth century was, and still is, a series of environmental contaminations caused by POPs. POPs are synthetic chemicals that are released either intentionally or non-intentionally. Examples of intentionally produced chemicals include PCBs, which have been useful in a variety of industrial applications, and DDT, which is still used in some parts of the world. An example of an unintentionally produced chemical is dioxins, which result from some industrial processes and via combustion (e.g. municipal and medical waste incineration and burning of trash). POPs are resistant to chemical changes; therefore, their **biotransformation** and **biodegradation** are very slow. They are persistent in the environment, and it may take them decennia or centuries to be degraded. The long-range transport of POPs through the air and ocean currents leads to global pollution. As a result of global transport and the low evaporation rates in cold climates, POPs tend to accumulate in Arctic regions. For example, in Canada's Arctic region, the level of POPs in the breast milk of Inuit women has been found to be up to nine times higher than in women living in southern Canada.

POPs are lipophilic substances that commonly enter organisms and deposit in their fat-rich tissues (an exception is perfluorinated contaminants, which bind to proteins). The terms describing the transfer of POPs and other contaminants from the external environment to organisms are **bioaccumulation** and **bioconcentration**. In terrestrial organisms, the contaminants enter the body through contaminated food and water (trophic transfer), by respiration, or dermally, whilst in aquatic organisms bioaccumulation mainly occurs via exposure to bottom sediment or through the food chain. Bioconcentration is the accumulation of contaminants by aquatic animals through non-dietary exposure routes. **Biomagnification** is an increase in contaminant concentration in an organism in excess of bioconcentration. It appears most significant for very hydrophobic contaminants resistant to biotransformation and biodegradation (such as PCBs).

To address the global concern over POPs, the UN treaty Exit, known as the Stockholm Convention, was signed in Sweden in May 2001. Under this treaty, the United States, the European Community, and 90 other countries agreed to reduce or eliminate the production, use, and release of 12 key POPs (called the 'dirty dozen').

The following sections present, from the perspective of human dietary exposure, the major groups of POPs, representing industrially manufactured products and intermediates and untargeted pollutants. Currently, 29 chemicals are designated as POPs (Table 12.18). Classical persistent organochlorine pesticides are discussed, together with modern pesticides, in Section 12.5.

12.4.1 Polychlorinated biphenyls

PCBs belong to a broad family of artificial organic chemicals known as chlorinated hydrocarbons. They are probably one of the world's most watched groups of POPs.

12.4.1.1 Classification, structure, and properties

PCBs are substances of molecular formula $C_{12}H_{10-(x+y)}Cl_{x+y}$ (**12-37**), where $x+y = 1$–10, x being the number of Cl atoms in ring 1 and y the number of Cl atoms in ring 2. Theoretically, 209 individual compounds, called congeners (members of the same group), each of which has its own systematic name, can be derived from the basic biphenyl skeleton (1 deca-CB, 3 mono- and nona-CBs, 12 di- and octa-CBs, 24 tri- and hepta-CBs, 42 tetra- and hexa-CBs, and 46 penta-CBs). Specification of the individual PCBs (number of chlorine atoms and their position in the aromatic rings of biphenyl) is presented in Figure 12.42. To simplify communication at the level of professional and legislative practice, a uniform nomenclature is used, proposed at the international level by the International Union of Pure and Applied Chemistry (IUPAC), where congeners have assigned numbers from 1 to 209, in order of increasing numerical index (2, 3, 4, 22′, 23, 23′, etc.). Three isomeric monochlorobiphenyls do not belong to the polychlorinated group of compounds but are commonly classified as PCBs for practical reasons. Indicated are **planar congeners** containing up to two substituents in the *ortho* position, which are further discussed in the context of their biological activity relative to dioxins. Also highlighted are **indicator congeners**, employed as contamination markers. Seven selected indicator congeners are PCBs 28 (2,4,4′-tetrachloro-), 52 (2,2′,5,5′-tetrachloro-), 101 (2,2′,4,5,5′-pentachloro-), 118 (2,3′,4,4′,5-pentachloro-), 138 (2,2′,3,4,4′,5′-hexachloro-), 153 (2,2′,4,4′,5,5′-hexachloro-), and 180 (2,2′,3,4,4′,5,5′-heptachlorobiphenyl). They are present in technical mixtures of PCBs in relatively high concentrations and are characterised by medium to high persistence. Sometimes, only six indicator congeners are followed, without congener 118. The impetus for the introduction of indicator congeners was the result of the so-called Belgian crisis in 1999, when a feed contaminated by Aroclors 1254 and 1260 was distributed to many poultry farms. Systematic monitoring often also includes PCBs 8, 18, 31, 44, 66, 70, 74, 99, 128, 149, 163, 170, 183, 187, and 194.

12-37, polychlorinated biphenyls

Table 12.18 Substances listed under the EU POPs regulation.

Substance	Main uses
Aldrin	Pesticide applied to soils
Alkanes C_{10}–C_{13}, chlorinated paraffins	Flame retardant
Chlordane	Insecticide on agricultural crops
Chlordecone	Historically used as an agricultural pesticide
DDT	Historically used as an insecticide
Dieldrin	Insecticide
Endosulfan	Insecticide
Endrin	Insecticide, also used as a rodenticide
Heptabromodiphenyl ether	Flame retardant
Heptachlor	Pesticide applied to soils
Hexabromobiphenyl	Flame retardant
Hexabromocyclododecane (HBCD)	In expanded and extruded polystyrene
Hexabromodiphenyl ether	Flame retardant
Hexachlorobenzene (HCB)	Fungicide
Hexachlorobutadiene	Solvent
Hexachlorocyclohexanes (HCH), incl. lindane	Broad-spectrum insecticide
Mirex	Insecticide
Pentachlorobenzene	Dyestuff carriers, fungicide, flame retardant
Pentachlorphenol	Pesticide and desinfectant
Perfluorooctane sulfonic acid and derivatives (PFOS)	Various industrial applications
Polybrominated diphenyl ethers (PBDEs)	Flame retardants
Polychlorinated biphenyls (PCBs)	Various industrial applications
Polychlorinated dibenzo-p-dioxins (PCDD)	Products of various industrial processes
Polychlorinated dibenzofurans (PCDF)	Products of various industrial processes
Polychlorinated naphthalenes	Various industrial applications
Toxaphene	Insecticide

In unsubstituted biphenyl, the phenyl residues can freely rotate around the single bond connecting the aromatic rings, and the planar conformation with the lowest energy content is preferred. The conformation of substituted biphenyls (PCBs) depends on the number and position of substituting chlorine atoms, which may prevent rotation. Substituents in the *ortho* position play a particularly important role. At a high degree of substitution, the relative position of aromatic rings is orthogonal and conformations of molecules determine the physico-chemical properties of individual congeners and their toxic effects.

The evaluation of health risks of PCBs is based on the WHO proposal from 1997, which divided PCBs into two groups based on their structure and related toxicity. The first group consists of PCBs able to bind to the aryl hydrocarbon receptor (AHR, a transcription protein that regulates gene expression). They include a total of 12 congeners, four non-*ortho* and eight mono-*ortho* PCBs, the so-called **dioxin-like PCBs** (DL-PCBs), which have toxic effects similar to the effects of to 2,3,7,8-tetrachlorodibenzo-p-dioxin. The most toxic is congener 126 (3,3′,4,4′,5-pentachlorobiphenyl). The second group consists of the **non-dioxin-like PCBs** (NDL-PCBs), which are characterised by different (not necessarily identical) toxicity profiles. It is assumed that the contribution of indicator congeners to the overall intake of NDL-PCBs is about 50%.

12.4.1.2 Occurrence, main sources, and dietary intake

PCBs were manufactured from 1929 until 1979, when their production was banned. Owing to their low acute toxicity, non-flammability, chemical stability, high boiling point, and electrical insulating properties, PCBs were used in hundreds of industrial and commercial

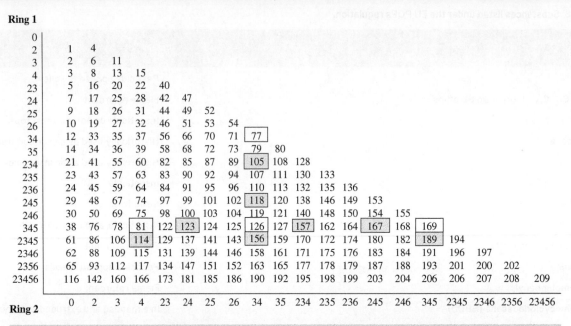

Figure 12.42 Structure of PCB congeners and their labelling according to the IUPAC (axes show the positions of chlorine atoms in the biphenyl aromatic rings; white rectangles = *non-ortho* PCB, grey rectangles = *mono-ortho* PCB).

applications, including electrical, heat-transfer, and hydraulic equipment, as lubricants, additives to pesticides, plasticisers in paints, plastics and rubber products, and in pigments, dyes, and carbonless copy paper. Table 12.19 summarises the history of their production and use.

The largest share of manufactured PCBs in the world falls to technical mixtures with the trade name Aroclor, manufactured by Monsanto (USA). A representation of different groups of congeners in these products is shown in Table 12.20 (where the chlorine content in % w/w specifies the last two digits in the Aroclor designation). The trademark Aroclor included not only technical PCB mixtures, but also a wide

Table 12.19 History of PCBs on a global scale.

Year	Events
1881	First synthesis of PCBs
1929	Start of industrial production of PCBs (Monsanto Chemical Company, USA)
1936	Negative health effects in persons who come into contact with the PCBs demonstarted
1966	Evidence of PCB occurrence in tissues of animals and humans (Sweden); PCBs classified as global environmental contaminants
1968	Mass poisoning of more than 1000 people by contaminated rice oil (Yushō disease, Japan)
1971	Voluntary restrictions on the distribution of products based on PCBs by Monsanto (only supplied for controlled applications)
1973	WHO working group charged with complex evaluation of PCBs; proposal of restrictions on use of PCBs; restrictions on the manufacture, sale, and use of PCBs in OECD countries
1976	Prohibition of the manufacture, processing, distribution, and use of PCBs by the US Congress (with the exception of closed systems)
1978	Restrictions on the use of PCBs (except for closed systems) in most advanced industrial countries
1979	Mass poisoning of more than 2000 people by contaminated rice oil (Yu-cheng disease, Taiwan); production of PCBs banned under US law
1983	Voluntary cessation of production PCBs by the German company Bayer
1987	Ban on all new applications of PCBs in 24 OECD countries; intensification of the solution of problems connected with the replacement of existing PCBs and provision of suitable means of PCB disposal

Table 12.20 Distribution of homologues in the technical PCB mixture Aroclor (% w/w) and their properties.

Number of chlorine atoms	Aroclor					
	1221	1232	1242	1248	1254	1260
0	10	–	–	–	–	–
1	50	26	1	–	–	–
2	35	29	13	1	–	–
3	4	24	45	22	1	–
4	1	15	31	49	15	–
5	–	–	10	27	53	12
6	–	–	–	2	26	42
7	–	–	–	–	4	38
8	–	–	–	–	–	7
9	–	–	–	–	–	1
Solubility in water[a]	15 000	1400	240	52	12	3
Lipophilicity[b] (log K_{ow})	0.65	1.7	3.5	6.4	12	22

[a] In µg/l at 25 °C.
[b] The octanol/water partition coefficient (K_{ow}, also known as P_{ow}) is the ratio of a chemical's concentration in the octanol phase to its concentration in the aqueous phase.

Table 12.21 Bioconcentration of PCBs in fish.

Congener groups	log K_{OW}	BCF	Congener groups	log K_{OW}	BCF
MonoCB	4.7	2500	HexaCB	6.7	250 000
DiCB	5.1	6300	HeptaCB	7.1	630 000
TriCB	5.5	16 000	OctaCB	7.5	1 600 000
TetraCB	5.9	40 000	NonaCB	7.9	4 000 000
PentaCB	6.3	100 000	DecaCB	8.3	1 000 000

range of products based on polychlorinated compounds, such as Aroclor 5460, which was a mixture of polychlorinated terpenes. Other products also existed on the market, with trade names including Clophen, Phenochlor, Kanechlor, Pyralene, and Fenclor.

The advantageous physico-chemical properties that led to the widespread use of PCBs were their thermostability, photostability (resistance to sunlight), non-combustibility, inertness (resistance to acids, bases, oxidation, and reduction), low volatility, high permittivity (dielectric material, electrical insulators), excellent heat-transfer properties, excellent miscibility with organic solvents, broad interval of melting point, and high boiling point. Due to these properties, PCBs are found globally in virtually all parts of the environment, and as lipophilic substances they have high affinity for the fat in organisms. Owing to the restrictive measures that have been adopted, their concentrations are declining gradually, but some particularly persistent congeners are doing so very slowly. As an example, Table 12.21 represents the average log K_{OW} values and bioconcentration factors (BCFs) for fish. K_{OW} is a very valuable parameter with numerous environmental applications. It can be related to BCF, the ratio of a chemical's concentration in an organism or biota to its concentration in water. Compounds with high potential for bioconcentration have log $K_{OW} > 5$ and BCF > 5000.

It is estimated that about 400 000 tons of technical mixtures were released into the external environment, which corresponds to about 31% of worldwide production. Approximately 65% of the total stock of PCBs is now rigorously recorded and used in so-called closed systems. The remaining 4% of PCB production has already been disposed of by safe incineration.

Transport between the non-living (abiotic) and living (biotic) parts of the environment is a dynamic process that is influenced by many factors. The most important transport medium is the atmosphere. The average atmospheric concentrations of PCBs in urban areas are usually

Table 12.22 Typical concentrations of PCBs in the abiotic environment at the beginning of the 1980s.

Part of environment	Location and environment part specification	Concentration	Units
Air	Urban agglomerations	0.5–30	ng/m^3
	Rural areas	0.1–5	
	Over the Atlantic Ocean	0.05	
	Around the factory transformers	20–6000	
Water	Rivers – background values	<0.5	ng/l
	– mean contamination	<50	
	– strong contamination	<500	
	Atlantic Ocean	0.3–8	
	Mediterranean Sea – coast	13	
	Rain – countryside	1–50	
	– urban agglomerations	10–250	
Soil	Background values	1–5	μg/kg
	Around factory transformers	15–18 000	
Sediments	Lakes – background values	0.1–5	μg/kg
	Rivers – normal load	8–20	
	– in the vicinity of PCB leaks	up to 60 000	

much higher than in rural areas or over oceans, and the highest concentrations can be found in the atmosphere of Arctic regions. Thanks to long-distance transmission in the atmosphere and dry and wet atmospheric deposition, PCBs are found in water, soil, on the surface of vegetation, and in animal tissues, even in outlying places far from the primary emission source. Oceans contain probably the most significant portion of the total quantity of PCBs emitted to the environment. PCBs can be found in virtually all abiotic and biotic components of the aquatic ecosystem, such as water, bottom sediments, plankton, arthropods, molluscs, fish, seabirds, and mammals. PCBs penetrate organisms mainly via food, but penetration from sediments also plays an important role. Other processes (such as diffusion into the surroundings, biodegradation, and excretion) may proceed simultaneously. Additionally, biomagnification takes place in predatory organisms. Table 12.22 lists the typical concentrations of PCBs in various abiotic ecosystems. Table 12.23 gives similar data for biotic ecosystems.

The main source of human exposure (except for professionally exposed persons) is food, which represents more than 90% of PCB intake. Exposure to PCBs in air, dust, and soil is relatively insignificant. Foods of plant origin contribute to the dietary intake of PCBs to a lesser extent; the degree of contamination varies at levels of ng/kg. Concentrations of PCBs in foods of animal origin (meat and milk, and in particular fish – the major dietary sources of PCBs) are generally considerably higher, typically μg/kg, and their content increases with lipid content. In foods of animal origin, there is lower amount of low-chlorinated PCBs (28, 52, but also 101), as living organisms gradually biotransform these substances and subsequently eliminate them. Table 12.24 summarises the levels of six indicator PCBs in foods in the European market in 2005.

Table 12.23 Typical concentrations of PCBs in the biotic environment at the beginning of the 1980s.

Organism	Concentration (mg/kg fat)	Organism	Concentration (mg/kg fat)
Vegetation	0.001–0.01	Marine mammals	0.1–1000
Aquatic zooplankton	0.005–2.0	Birds	0.1–1000
Aquatic invertebrates	0.005–10	Eggs of birds	0.05–500
Fish	0.01–25	People	0.1–50

Table 12.24 Average amounts of indicator PCB congeners in foods in the European market in 2005.

Food	PCB 28	PCB 52	PCB 101	PCB 138	PCB 153	PCB 180	Σ6(PCB)
Cereals and ceral products[a]	0.00835	0.00501	0.00189	0.00217	0.00246	0.00141	0.0213
Fruits and vegetables[a]	0.00632	0.00389	0.00262	0.00742	0.0105	0.0191	0.0495
Eggs[b]	0.59	0.41	0.70	1.81	2.00	1.05	6.60
Vegetable oils[b]	0.65	0.35	0.56	1.42	1.51	0.57	5.05
Animal fats[b]	0.13	0.11	0.13	0.63	1.16	0.46	2.61
Fish oils[b]	0.79	3.44	9.18	23.2	25.5	8.09	70.2
Fish and fish products[c]	0.29	0.63	1.64	3.88	4.41	1.63	12.5
Poultry[b]	0.77	0.96	0.89	5.34	2.31	2.03	12.7
Beef[b]	0.58	0.96	0.55	3.74	2.60	1.10	9.53
Pork[b]	0.45	0.66	0.58	1.54	2.34	1.23	6.80
Milk and diary products[b]	1.48	0.99	1.01	2.57	3.21	1.47	10.7

[a] In μg/kg.
[b] In μg/kg fat.
[c] In μg/kg edible portion.

12.4.1.3 Metabolism and degradation

The spectrum of PCB congeners observed in invertebrate organisms and fish is not too different from the composition of PCB congeners in the surrounding abiotic environment (due to the low metabolic activity). In warm-blooded animals, especially higher vertebrates, the spectrum of PCB congeners can be substantially different, as the biodegradable congeners are partially or completely eliminated, whilst in PCBs substituted in positions 4,4′ or 3,4′,5, preferential bioaccumulation and an increased tendency to bioaccumulate toxic DL-PCBs can be observed. Exposure of livestock to PCBs leads to their accumulation in meat, especially liver and adipose tissue. A significant proportion of accumulated PCBs are transferred to eggs and milk, and to offspring by the mother's milk. The largest transmission range, roughly 50–60%, can be observed in highly persistent PCB congeners 138 and 153. In humans, highly persistent PCB congeners 138, 153, and 180 of the group of NDL-PCB dominate.

Biodegradation is an important means of eliminating low-chlorinated PCBs from the environment. Biotransformation mechanisms and terminal metabolites are different in individual species. Certain strains of bacteria of the genera *Alcaligenes*, *Pseudomonas*, *Nocardia*, and others have the ability to metabolise PCBs, as do some eukaryotic microorganisms (such as fungi). The degradation rate decreases with the number of chlorine atoms and, starting from pentachlorobiphenyls, is very low. In the initial stage of aerobic degradation of PCBs by dioxygenases (and monooxygenases in some bacteria and fungi), dihydrodiols are produced, which are gradually transformed into chlorobenzoic acids, inorganic chlorides, and lower aliphatic acids. The use of bacteria, fungi, or green plants with biodegradation potential is the essence of bioremediation techniques (use of biological agents to remove or neutralise contaminants, as in polluted soil or water).

In vertebrates, the first reaction of PCB biotransformation by cytochrome P450 complex is the formation of epoxides and hydroxyl derivatives. Congeners with two unsubstituted carbons in the positions *meta* and *para* (such as PCB 52, PCB 95, and PCB 136) can be oxidised at multiple sites. The reactive electrophilic intermediates arene oxides give rise to hydroxylated products, such as dihydrodiol-PCBs and polychlorobiphenylols (hydroxy-PCBs), as well as conjugates with glutathione or adducts with macromolecules (DNA and proteins) and lipids. Dihydrodiols can become the precursors of metabolites with a catechol structure, and may be in equilibrium with hydroquinones and quinones. Hydroxy-PCBs may also arise by 1,2-rearrangement of chlorine atoms. Most of these compounds are excreted from the body in a free form or as conjugates (glucuronides or sulfates).

12.4.1.4 Health and toxicological evaluation

Toxicological evaluation of PCBs is an extremely difficult problem, as the considered organism is always exposed to mixtures of congeners with different properties. It is practically impossible to distinguish the effects of individual toxic congeners. Food of animal origin in particular may contain NDL-PCBs, DL-PCBs, and possibly also PCDD and PCDF simultaneously in different proportions. The acute toxicity of PCBs is low and depends on the degree of chlorination of the technical mixture. However, dioxin-like congeners (such as

Table 12.25 Comparison of toxic equivalency factors (TEFs) for PCB congeners (for humans and mammals in general).

Congener	Structural type	TEF–WHO	Congener	Structural type	TEF–WHO
77	Non-*ortho*	0.0001[a]	118	Mono-*ortho*	0.000 03[b]
81	Non-*ortho*	0.0003[c]	123	Mono-*ortho*	0.000 03[b]
126	Non-*ortho*	0.1[a]	156	Mono-*ortho*	0.000 03[d]
169	Non-*ortho*	0.03[e]	157	Mono-*ortho*	0.000 03[d]
105	Mono-*ortho*	0.000 03[b]	167	Mono-*ortho*	0.000 03[f]
114	Mono-*ortho*	0.000 03[d]	189	Mono-*ortho*	0.000 03[b]

[a]The same in 1998.
[b]0.0001 in 1998.
[c]0.0001 in 1998.
[d]0.0005 in 1998.
[e]0.01 in 1998.
[f]0.000 01 in 1998.

3,3′,4,4′-tetrachlorobiphenyl) elicit a much higher toxicity. Symptoms of PCB intoxication are manifested by wasting syndrome (progressive weight loss, not related to food consumption), skin disorders, lymphoid involution (immunosuppresion), and endocrine and reproductive effects (menstrual irregularities and reduced conception rate, early abortion, excessive menstrual haemorrhage, and, in males, testicular atrophy and decreased spermatogenesis); PCB mixtures did not cause mutation or chromosomal damage in a variety of test systems, although adducts with DNA, RNA, and proteins could be detected (prior metabolisation was necessary). According to the IARC's evaluation of PCB mixtures, there is sufficient evidence for carcinogenicity to animals; therefore, PCBs are classified as Group 2A agents, and are probably carcinogenic to humans.

The risk assessment resulting from exposure to PCDDs, PCDFs, and dioxin-like PCBs for humans, fish, and wildlife uses the system of toxic equivalents (TEQs). To obtain TEQ, the mass of each chemical in a mixture is multiplied by its toxic equivalency factor (TEF, relative toxicities determined using a database of relative effect potencies that meet WHO-established criteria) and then summed with all other chemicals to report the total toxicity-weighted mass (Table 12.25). The toxicity of a toxic substance is rated in relation to the compound with the greatest toxicity, 2,3,7,8-tetrachlorodibenzo-*p*-dioxin (2,3,7,8-TCDD, which has a TEF of 1.0; sometimes PCB 126 is also used as a reference chemical, with a TEF of 0.1). The TEF approach has also been used to assess the toxicity of other chemicals including PAHs and xenoestrogens.

Some hydroxy derivatives of PCBs can, through structural relationship with thyroxine, interfere with the metabolism of thyroid hormones by competitive inhibition of transthyretin (the serum and cerebrospinal fluid carrier of the thyroid hormone thyroxine and retinol-binding protein) and hormonal processes in the body. Toxicological studies have confirmed that sulfur metabolites of PCBs may bind to the secretoglobin called uteroglobin (progesterone-binding protein). Arylmethylsulfones accumulate mainly in the lungs and probably play an important role in the aetiology of toxic manifestations of the so-called Yushō disease, a mass poisoning by PCBs that occurred in northern Kyūshū, Japan, in 1968.

12.4.1.5 Mitigation

EC Recommendation 2006/88/EC was issued in order to facilitate the active reduction of dibenzodioxins, dibenzofurans, and PCBs in foods and feeds. With respect to risk-management evaluation, so-called action levels have been published for dioxins, dibenzofurans, and dioxin-like PCBs in foods and feeds, which may be used to take measures for their reduction or elimination.

Commision Recommendation No. 1881/2006/EU lists maximum levels for WHO-PCDD/PCDF-TEQ and WHO-PCDD/PCDF-PCB-TEQ. If dioxine-like PCBs are not included in the sum, the allowed values are approximately half the WHO-PCDD/PCDF-TEQ values. For example, for raw milk and dairy products (including butterfat, hen's eggs, and egg products), 3 pg/g fat of WHO-PCDD/PCDF-TEQ and 6 pg/g fat of WHO-PCDD/PCDF-PCB-TEQ are allowed. The highest values are set for meat and meat products (excluding edible offal) of bovine animals and sheep (3 and 4.5 pg/g fat), whilst lower values are set for poultry and pigs (2 and 4 pg/g fat and 1 and 1.5 pg/g fat, respectively). Higher values are set for muscle meat of fish and fishery products (4 and 8 pg/g f.w.). The highest allowable values are for liver of terrestrial animals (12 pg/g) and (by Commission Regulation No. 565/2008/EU) and for fish liver and its products (25 pg/g), which are close to the real concentrations in these products.

12.4.2 Polychlorinated dibenzo-*p*-dioxins and dibenzofurans

PCDD and PCDF are, from the toxicological point of view, amongst the most important groups of global environmental contaminants. Unlike all other halogen compounds, PCDD and PCDF do not have, and have never had, any practical use, and thus they have never specifically been synthesised in large quantities. They therefore represent the category of unintentionally produced POPs. The levels of these POPs that accumulate in food chains and are associated with health risks for living organisms (even in long-term exposure to very low concentrations) are the subject of special attention from experts, politicians, and the public.

12.4.2.1 Classification, structure, and properties

PCDD (**12-38**) and PCDF (**12-39**) are almost planar tricyclic compounds. Theoretically, the total number of PCDD derivatives is 75 (Table 12.26) and the total number of PCDF derivatives (containing one to eight chlorine atoms) is 135 (Table 12.27).

12-38, polychlorinated dibenzo-*p*-dioxins

12-39, polychlorinated dibenzofurans

The properties of PCDD and PCDF are similar to those of other POPs whose aromatic rings are substituted with halogens. PCDD and PCDF are characterised by relatively high lipophilicity, high values of log K_{OW} and very low solubility in water. High-chlorinated PCDD and PCDF generally have considerable bioaccumulation potential and high persistence. Their exclusion from the exposed organism is very slow. Typical ranges of half-life values for elimination of toxic congeners from human adipose tissue range from 3.2–6.6 years for PCDD 73 to 3.5–>70 years for PCDD 67. Elimination half-lives of PCDF are generally shorter.

12.4.2.2 Occurrence, main sources, and dietary intake

PCDD and PCDF arise, unlike most other organochlorine contaminants, almost exclusively as byproducts of anthropogenic activities. Recent research has also confirmed the formation of trace amounts of PCDD and PCDF in enzymatically catalysed reactions. For example, they may be biosynthesised by microbial peroxidases in sewage sludge contaminated with chlorophenols. The most important primary emission sources were – and perhaps still are – industrial syntheses, such as syntheses of certain pesticides (particularly hexachlorobenzene, pentachlorophenol, and phenoxyalkanoic acids), production of technical mixtures of PCBs, production of intermediates of some organic products (especially chlorophenols and chlorobenzenes), bleaching of wood pulp with chlorine dioxide (ClO$_2$), and production of chlorine using graphite electrodes. PCDD and PCDF also arise from the incineration of organic and inorganic compounds containing chlorine during burning of organic compounds containing chlorine (such as PVC). An important primary source is photochemical reactions in the atmosphere and reactions of components found in emissions.

PCDD and PCDF can be released into the environment, and subsequently to the food chain, not only from these primary sources, but also from secondary sources such as atmospheric deposition, point source emissions (various incinerators or industrial plants), and contaminated materials in the agroecosystem. For ruminants, the main route of dioxin intake is vegetation. For the aquatic ecosystem, the main source of contamination is atmospheric deposition. In the initial phase, PCDD and PCDF (as well as other POPs) accumulate in the benthos, which is fed to animals at higher trophic levels. As with PCBs, it is expected that dietary intake represents up to 95% of the total amount of PCDD and PCDF.

Under normal circumstances, the concentrations of PCDD and PCDF in food raw materials and products are very low, usually in the range of units to tenths of µg/kg lipids; only in exceptional cases, for example in fish from contaminated localities, do they reach values of an order of magnitude higher. Table 12.28 shows the levels of PCDD/PCDF and DL-PCB in different foods on the Dutch market in 2004. This table also illustrates the relatively high contribution of DL-PCBs to the total TEQ value, which even exceeded the contribution of dioxins in some cases. In countries with a high consumption of fish, like Japan, this commodity significantly contributes to the intake of PCDD/PCDF and related POPs. Increased levels of PCDD and PCDF in milk and dairy products are alarming, because a large part of the population may be exposed to PCDD and PCDF via these commodities, including children, who are particularly vulnerable to the toxic effects of xenobiotics.

Table 12.26 Nomenclature of PCDD according to International Union of Pure and Applied Chemistry (IUPAC).

Number	Structure	Number	Structure	Number	Structure	Number	Structure
MonoCDD		19	1,3,6	39	1,2,7,8		
1	1	20	1,3,7	40	1,2,7,9	59	1,2,4,6,9
2	2	21	1,3,8	41	1,2,8,9	60	1,2,4,7,8
DiCDD		22	1,4,6	42	1,3,6,8	61	1,2,4,7,9
3	1,2	23	1,4,7	43	1,3,6,9	62	1,2,4,8,9
4	1,3	24	1,4,8	44	1,3,7,8	*HexaCDD*	
5	1,4	25	1,7,8	45	1,3,7,9	63	1,2,3,4,6,7
6	1,6	26	2,3,7	46	1,4,6,9	64	1,2,3,4,6,8
7	1,7	*TetraCDD*		47	1,4,7,8	65	1,2,3,4,6,9
8	1,8	27	1,2,3,4	*48*	*2,3,7,8*	*66*	*1,2,3,4,7,8*
9	1,9	28	1,2,3,6	*PentaCDD*		*67*	*1,2,3,6,7,8*
10	2,3	29	1,2,3,7	49	1,2,3,4,6	68	1,2,3,6,7,9
11	2,7	30	1,2,3,8	50	1,2,3,4,7	69	1,2,3,6,8,9
12	2,8	31	1,2,3,9	51	1,2,3,6,7	*70*	*1,2,3,7,8,9*
TriCDD		32	1,2,4,6	52	1,2,3,6,8	71	1,2,4,3,7,9
13	1,2,3	33	1,2,4,7	53	1,2,3,6,9	72	1,2,4,6,8,9
14	1,2,4	34	1,2,4,8	*54*	*1,2,3,7,8*	*HeptaCDD*	
15	1,2,6	35	1,2,4,9	55	1,2,3,7,9	*73*	*1,2,3,4,6,7,8*
16	1,2,7	36	1,2,6,7	56	1,2,3,8,9	74	1,2,3,4,6,7,9
17	1,2,8	37	1,2,6,8	57	1,2,4,6,7	*OctaCDD*	
18	1,2,9	38	1,2,6,9	58	1,2,4,6,8	*75*	*1,2,3,4,6,7,8,9*

In virtually all cases, the total TEF value is contributed to mostly by 2,3,7,8-tetraCDD, 1,2,3,7,8-pentaCDD, 1,2,3,6,7,8-hexaCDD, 2,3,4,7,8-pentaCDF, and PCB 126. The contribution of DL-PCBs varies over a wide range, and in some cases exceeds 50%. There is a high risk of increased exposure during lactation, as the gastrointestinal tracts of nursed babies absorb virtually all lipids contained in breast milk, and these lipids represent a reservoir of PCDD, PCDF, and other POPs.

12.4.2.3 Metabolism and degradation

To date, incineration has been the only inexpensive method of destroying dioxins and furans, although many other physicochemical treatment procedures have been developed. From an economical point of view, bioremediation processes comprising fungi and bacteria can be used. PCDD and PCDF are subjected to reductive dehalogenations leading to less halogenated congeners, which can be attacked efficiently by oxidases and dioxygenases analogously to PCBs.

12.4.2.4 Health and toxicological evaluation

Many PCDD and PCDF congeners are characterised by considerable toxicity, and their relative effect is related (as in DL-PCB) to congener 2,3,7,8-tetrachlorodibenzo-*p*-dioxin (TCDD), which was selected on the basis of toxicological studies.

TCDD and related compounds have negative effects on several organs in different animal species. The most famous case where people were exposed to dioxins was the chemical plant accident in Seves, Italy that occurred in 1976. The most important harmful effects are immunotoxicity, behavioural disorders, and reproductive disorders. Subchronic effects in experimental animals include weight loss, pathological changes in the liver, thymus, and lymph glands, hair loss, and porphyria (decomposition of haemoglobin). Evidence of the carcinogenicity of 2,3,7,8-tetraCDD is also ambiguous; the IARC classifies this substance as a Group 1 carcinogen. TetraCDD and related

Table 12.27 Nomenclature of PCDF according to International Union of Pure and Applied Chemistry (IUPAC).

Number	Structure	Number	Structure	Number	Structure	Number	Structure
MonoCDF		34	1,4,8	69	1,3,6,8	104	1,2,6,7,9
1	1	35	1,4,9	70	1,3,6,9	105	1,3,4,6,7
2	2	36	1,6,7	71	1,3,7,8	106	1,3,4,6,8
3	3	37	1,6,8	72	1,3,7,9	107	1,3,4,6,9
4	4	38	1,7,8	73	1,4,6,7	108	1,3,4,7,8
DiCDF		39	2,3,4	74	1,4,6,8	109	1,3,4,7,9
5	1,2	40	2,3,6	75	1,4,6,9	110	1,3,6,7,8
6	1,3	41	2,3,7	76	1,4,7,8	111	1,4,6,7,8
7	1,4	42	2,3,8	77	1,6,7,8	112	2,3,4,6,7
8	1,6	43	2,4,6	78	2,3,4,6	113	2,3,4,6,8
9	1,7	44	2,4,7	79	2,3,4,7	114	2,3,4,7,8
10	1,8	45	2,4,8	80	2,3,4,8	HexaCDF	
11	1,9	46	2,6,7	81	2,3,6,7	115	1,2,3,4,6,7
12	2,3	47	3,4,6	82	2,3,6,8	116	1,2,3,4,6,8
13	2,4	48	3,4,7	83	2,3,7,8	117	1,2,3,4,6,9
14	2,6	TetraCDF		84	2,4,6,7	118	1,2,3,4,7,8
15	2,7	49	1,2,3,4	85	2,4,6,8	119	1,2,3,4,7,9
16	2,8	50	1,2,3,6	86	3,4,6,7	120	1,2,3,4,8,9
17	3,4	51	1,2,3,7	PentaCDF		121	1,2,3,67,8
18	3,6	52	1,2,3,8	87	1,2,3,4,6	122	1,2,3,6,7,9
19	3,7	53	1,2,3,9	88	1,2,3,4,7	123	1,2,3,6,8,9
20	4,6	54	1,2,4,6	89	1,2,3,4,8	124	1,2,3,7,8,9
TriCDF		55	1,2,4,7	90	1,2,3,4,9	125	1,2,4,6,7,8
21	1,2,3	56	1,2,4,8	91	1,2,3,6,7	126	1,2,4,6,7,9
22	1,2,4	57	1,2,4,9	92	1,2,3,6,8	127	1,2,4,6,8,9
23	1,2,6	58	1,2,6,7	93	1,2,3,6,9	128	1,3,4,6,7,8
24	1,2,7	59	1,2,6,8	94	1,2,3,7,8	129	1,3,4,6,7,9
25	1,2,8	60	1,2,6,9	95	1,2,3,7,9	130	2,3,4,6,7,8
26	1,2,9	61	1,2,7,8	96	1,2,3,8,9	HeptaCDF	
27	1,3,4	62	1,2,7,9	97	1,2,4,6,7	131	1,2,3,4,6,7,8
28	1,3,6	63	1,2,8,9	98	1,2,4,6,8	132	1,2,3,4,6,7,9
29	1,3,7	64	1,3,4,6	99	1,2,4,6,9	133	1,2,3,4,6,8,9
30	1,3,8	65	1,3,4,7	100	1,2,4,7,8	134	1,2,3,4,7,8,9
31	1,3,9	66	1,3,4,8	101	1,2,4,7,9	OctaCDF	
32	1,4,6	67	1,3,4,9	102	1,2,4,8,9	135	1,2,3,4,6,7,8,9
33	1,4,7	68	1,3,6,7	103	1,2,6,7,8		

Table 12.28 Mean PCDD/PCDF and DL-PCB concentrations (TEQ, ng/kg).

Foods	Average fat content (%)	Dioxins	Non-*ortho*-PCB	Mono-*ortho*-PCB	Total TEQ[a]
Butter	81	0.96	0.52	0.03	1.50
Cheese	31	0.14	0.14	0.01	0.29
Eggs	10	0.09	0.03	0.003	0.12
Milk	1	0.01	0.01	0.001	0.01
Beef	16	0.05	0.13	0.001	0.18
Pork	26	0.09	0.02	0.004	0.11
Poultry meat	9	0.04	0.01	< 0.001	0.05
Vegetable fats and oils	57	0.23	0.02	0.001	0.25
Cereals	1	0.09	< 0.01	0.001	0.14
Fruits and fruit juices	—	0.13	0.01	0.001	0.15

[a]For the calculations, updated values of TEF from 2006 were used.

compounds act at different levels, through initiation of hormonal mechanisms supporting carcinogenic processes, on the development of the intellect in the foetus just before or shortly after birth (doses received by breast milk may exceed the normal dietary intake of adults). As in the case of PCBs, PCDD and PCDF (depending on the structure of the individual congeners and their planarity, respectively) induce a hepatic microsomal system of oxidases (cytochrome P450) in exposed organisms. The most effective inducer is TCDD. Table 12.29 shows the relative toxicities of 2,3,7,8-tetraCDD and other monitored PCDDs and PCDFs for humans and other mammals. TEF values, which were estimated for other species and used in assessing ecotoxicological risks, may be more or less different. A particularly significant difference in the intensity of biological responses and the TEF values is for fish.

12.4.2.5 Mitigation

Measures aimed at reducing emissions of PCDD and PCDF into the environment and the contamination of food chains have undoubtedly influenced the decline in population exposure in the United States in recent years: whilst, in those born in the 1950s, levels of bioaccumulated pollutants were on average almost 15 ng/kg fat, those born in the 1980s were exposed significantly less.

Table 12.29 Comparison of TEF values for PCDD and PCDF congeners used to estimate exposure risk.

Chlorinated dibenzo-*p*-dioxins	TEF-WHO	Chlorinated dibenzofurans	TEF-WHO
2,3,7,8-TetraCDD[a]	1	2,3,7,8-tetraCD[a]	0.1
1,2,3,7,8-PentaCDD[a]	1	1,2,3,7,8-pentaCDF[b]	0.03
1,2,3,4,7,8-HexaCDD[a]	0.1	2,3,4,7,8-pentaCDF[c]	0.3
1,2,3,6,7,8-HexaCDD[a]	0.1	1,2,3,4,7,8-hexaCDF[a]	0.1
1,2,3,7,8,9-HexaCDD[a]	0.1	1,2,3,6,7,8-hexaCDF[a]	0.1
1,2,3,4,6,7,8-HeptaCDD[a]	0.01	1,2,3,7,8,9-hexaCDF[a]	0.1
OctaCDD[d]	0.0003	2,3,4,6,7,8-hexaCDF[a]	0.1
		1,2,3,4,6,7,8-heptaCDF[a]	0.01
		1,2,3,4,7,8,9-heptaCDF[a]	0.01
		octaCDF[d]	0.0003

[a]The same in 1998.
[b]0.05 in 1998.
[c]0.5 in 1998.
[d]0.0001 in 1998.

In 2000, the WHO calculated TDI values in the range from 1 to 4 pg WHO-PCDD/PCDF-PCB-TEQ per kg b.w. These exposure limits are based on non-carcinogenic effects, with the main ones to be evaluated being immunotoxicity, neurotoxicity, and interference with hormonal systems. In 2001, the EU SCF calculated a tolerable weekly intake (TWI) of 14 pg WHO-PCDD/PCDF-PCB-TEQ per kg b.w. In the same year, the JECFA calculated a provisional tolerable monthly intake (PTMI) of 70 pg WHO-PCDD/PCDF-PCB-TEQ per kg b.w per month. In 2006, the SCF confirmed the provisional TWI of 14 pg TEQ per kg b.w, which corresponds to the value of 2 pg TEQ per kg b.w per day proposed in 2001. Setting these limits causes serious consequences, with restrictive measures aimed at reducing PCDD and PCDF exposure. TDI must always protect all populations, and from this perspective the exposure of infants to breast milk is critical. Despite these facts, from the point of view of nutrition, breastfeeding is considered irreplaceable, and so the only way to reduce this risk is by a drastic reduction in population exposure. The target value is 1 pg TEQ per kg of b.w per day. Under normal circumstances, the intake of PCDD and PCDF is much lower. The risks that could arise from ordinary low concentrations of these contaminants are not yet sufficiently documented, so it is recommended to reduce the intake of PCDD and PCDF to a minimum.

12.4.3 Brominated flame retardants

The development of anthropogenic activities is accompanied by the growing use of combustible synthetic materials, which brings an increased risk of fire. One of the ways to prevent economic impacts and the possible loss of human lives is the use of flame retardants, chemicals that reduce the flammability of materials or delay their combustion. These synthetic chemicals are used in electronics, upholstery, carpets, textiles, insulation, vehicle and airplane parts, baby blankets, children's clothes, strollers, and many other products. According to the method of incorporation into polymers, flame retardants can be either reactive or additive compounds. Reactive flame retardants are added to the monomers; therefore, they are covalently linked in the polymer and do not significantly penetrate into the environment. Additive flame retardants are added into polymers, where they are not covalently bound.

Currently, nearly 180 different compounds that increase the fire safety of flammable materials are known. The most important are brominated flame retardants (BFRs). The first mention of the occurrence of BFRs in the environment was in Sweden in the late 1970s, when they were identified in fish. Greater attention began to be paid to BFRs during the 1990s when they were found in breast milk. Like PCBs and other POPs, residues of BFRs can be found in virtually all environmental components, biotic and abiotic: the air, water, soil, sediments, sewage sludges, fish, birds, and mammals, including humans. In addition to leaks during their production, their presence in the environment is related particularly to the use of materials containing them. An important source is the disposal of plastics, electronic equipment, and other wastes. These compounds can be leached from landfills; furthermore, they release toxic polybrominated dioxins and furans when burned.

The classic BFRs are polybrominated biphenyls (PBBs), whilst the most commonly used are mixtures of polybrominated diphenyl ethers (PBDEs), 2,2′,6,6′-tetrabromobisphenol A (TBBPA), and 1,2,5,6,9,10-hexabromocyclododecane (HBCD or HBCDD) (12-40). PBBs are primarily used as additive flame retardants, PBDEs and HBCD are used only as additive flame retardants, whilst TBBPA is primarily a reactive flame retardant; in only about 10% of cases is it used as an additive flame retardant. For example, the HBCD content in polystyrene foam ranges from 0.4 to 8%, and the TBBPA content in polyesters may reach 13–28%.

The mechanisms of action of flame retardants differ between groups. In the case of halogenated compounds, the burning process is slowed by removal of hydrogen and hydroxyl radicals from the flame. At higher temperatures, flame retardants release free bromine or chlorine radicals, which react with hydrocarbons and yield hydrogen bromide or hydrogen chloride, which in turn react with combustion products to form water and bromine or chlorine radicals.

polybrominated diphenyl ethers

polybrominated biphenyls

2,2′,6,6′-tetrabromobisphenol A

1,2,5,6,9,10-hexabromocyclododecane

12-40, common polybrominated flame retardants

12.4.3.1 Polybrominated diphenyl ethers

12.4.3.1.1 Classification, structure, and properties

PBDEs are structurally similar to PCBs. The numbering of 209 individual PBDE congeners is analogous to the IUPAC nomenclature used for numbering PCBs. The basic skeleton is diphenyl ether, with a different number and location of bromine atoms. The PBDE congeners include 3 mono-, 12 di-, 24 tri-, 42 tetra-, 46 penta-, 42 hexa-, 24 hepta-, 12 octa-, 3 nona-, and 1 decabromodiphenylether. In industrial production, four technical PBDE mixtures are applied: decaBDE (decabromodiphenyl ether), octaBDE, pentaBDE, and tetraBDE. Technical decaBDE contains about 97% of 209 BDE congeners and 3% of nonaBDE; octaBDE contains mainly heptaBDE and octaBDE, and to a lesser extent hexa-, nona-, and decaBDE; pentaBDE contains predominantly penta- and tetrabromodiphenyl ether congeners; and tetraBDE contains tetra-, penta-, hexa-, and partly other products.

PBDEs are very stable compounds with high boiling points, low volatility, high lipophilicity, and low solubility in water. They are very persistent, having a tendency to accumulate in different parts of the environment by bioaccumulation and biomagnification processes. The bioaccumulation potential of low-brominated congeners is very high (BCF value is higher than 5000).

12.4.3.1.2 Occurrence, main sources, and dietary intake

The most important congeners, which are monitored in the environment, include PBDE 28 (2,4,4'-tribromodiphenyl ether, 47 (2,2',4,4'-tetrabromodiphenyl ether), 99 (2,2',4,4',5-pentabromodiphenyl ether), 100 (2,2',4,4',6-pentabromo diphenylether), 153 (2,2',3,4,4',5'-hexabromodiphenyl ether), 154 (2,2',4,4',5,6'-hexabromodiphenyl ether), and 209 (2,2',3,3',4,4',5,5',6,6'-octabromodiphenyl ether). Concentrations of PBDE congeners and their amount in the various components of the environment depend essentially on their physico-chemical properties. The behaviour in the environment of low-brominated PBDE congeners (with 4–7 Br atoms) is similar to that of PCBs or PCDD and PCDF. They are relatively soluble in water and air and are transported over long distances, but their concentrations in sediments and soils are lower than those of high-brominated congeners. Elevated levels of soil contamination are often caused by sewage sludge from wastewater treatment plants, used in agriculture for soil enrichment by organic components.

PBDE penetrate into the tissues of animals, including humans, where they accumulate. As with other halogenated POPs, their concentrations increase with each trophic level in a food chain, ranging from µg/kg fat to mg/kg fat depending on the dietary habits of the species, age, and level of contamination. The highest concentrations are found in the tissues of predatory fish. The dominant PBDE is often tetraBDE 47, which may constitute more than 50% of the total amount. Other relatively frequently occurring PBDEs are congeners 99, 100, 153, and 154. Despite the extensive use and common occurrence in the environment, the amounts of BDE 209 are relatively low. However, recent research revealed the presence of BDE 209 in the eggs of raptors (20–430 µg/kg fat) and fish (about 50 µg/kg fat).

Human exposure to PBDEs occurs in various ways, and a human organism can be exposed to these contaminants and other halogenated compounds even before birth, through breast milk. An important source of human exposure to PBDE is dietary intake, particularly the consumption of contaminated fish, as well as other foods with a higher fat content, such as meat, eggs, dairy products, fats, and oils. Results obtained in Spain in 2000 are summarised in Table 12.30 as an example. The total daily intake of PBDEs was estimated to be 82–97 ng, including intake from fish and seafood at the level of 30 ng, from vegetable oils at 24 ng, and from meat and meat products at 20 ng, which corresponds to a daily intake of 1.2–1.4 ng/kg b.w. The dominant congeners were tetraBDE and pentaBDE.

Unlike for PCBs and other POPs, another important source of PBDEs is inhalation of contaminated dust and dermal intake. In recent years, much attention has focused on BFRs in dust, where high-brominated substances are mainly found, as they have a higher affinity for solid particles. By far the highest concentration of these substances is in indoor environments, via a number of sources (electronics, furniture, and various textiles).

12.4.3.1.3 Metabolism and degradation

The most abundant congener BDE 47 (2,2',4,4'-tetrabromodiphenylether) is absorbed at up to 95% in the gastrointestinal tract, distributed to the tissues (during lactation excreted to breast milk), and slowly metabolised by the action of cytochrome P450. Only a small amount is excreted via excrements. Similarly, as in the case of aromatic xenobiotics, complex epoxides are produced as the primary products, which, on hydrolysis, yield the corresponding hydroxy derivatives. The most persistent congener BDE 153 (2,2',3,3',4,4',5,5',6,6'-decabromodiphenyl ether) is absorbed only up to 10–25%. The half-life of its metabolic transformation is a few days or weeks.

Various hydroxylated and methoxylated PBDEs have been found in organisms of mammals, including humans; the source could be fish, marine mammals, or cyanobacteria, such as *Oscillatoria spongeliae*, living in symbiosis with marine sponges, algae, mussels, and other seafood organisms. Debrominated, hydroxylated, and methoxylated metabolites were also found in maize that had been exposed to low-brominated diphenyl ethers.

Table 12.30 Contents of PBDEs in selected foods.

Food	Content	
	ng/kg fat	ng/kg dry matter
Vegetables	–	5–8
Fruits	–	0–6
Cereals	–	0–36
Shellfish	2961–3140	83–88
White fish	2052–2359	37–88
Fish tins	1997–2117	246–260
Pork meat and products	565–597	166–172
Poultry meat	0–247	0–10
Beef and products	248–290	36–42
Eggs	482–530	58–64
Milk	525–630	20–24
Dairy products	557–677	34–48
Vegetable oils	795–805	794–804

12.4.3.1.4 Health and toxicological evaluation

The acute toxicity of commercial technical mixtures is relatively low ($LD_{50} < 1$ g/kg b.w in rats), but at high doses increased liver weights and changes in the structure of liver tissue have been found. A much greater risk is long-term exposure to low doses of PBDEs. Harmful effects include neurotoxicity and interference with the endocrine system (hormonal processes), as PBDEs act as endocrine disruptors. One of the most striking symptoms observed in animals was an increase of thyroxine concentration in the blood plasma, because some PBDEs (and especially their hydroxyl derivatives) have a structure similar to thyroid hormones. In addition to the adverse effects, there is also a risk of the formation of toxic polybrominated dibenzo-*p*-dioxins (PBDDs) and dibenzofurans (PBDF, **12-41**) during photolysis or pyrolysis.

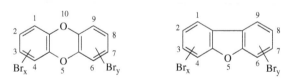

12-41, polybrominated dibenzo-*p*-dioxins and dibenzofurans

12.4.3.1.5 Mitigation

The European Union pushed for the elimination of BFRs in the 1990s. At present, there is a ban in the European Union and United States on the production of pentaBDE and octaBDE. Since 2009, the European Union has also prohibited the use of decaBDE for electronic equipment. Even so, a large number of products containing these BFRs are still in use, resulting in a continuous release of PBDEs into the environment.

12.4.3.2 Polybrominated biphenyls

PBBs (**12-40**) represent an older group of BFRs; 209 possible congeners can be derived, but only 101 individual ones have had registration numbers assigned. Commercial mixtures of PBBs, produced since the early 1970s, contained mainly hexa-, octa-, nona-, and decabromo-biphenyls.

Data on the occurrence of PBBs in foods are essentially unavailable. The most serious penetration of these POPs into food chains occurred in Michigan, United States in the 1970s. Contamination of feed by technical mixtures FireMaster BP-6 led to high concentrations in a variety of foods of animal origin.

The formation of degradation products of PBBs depends generally on the temperature, the amount of oxygen present, and certain other factors. Some PBBs are considered possible precursors of toxic PBDDs and polybrominated dibenzofurans (**12-41**), which can be formed during pyrolysis in the presence of oxygen.

No comprehensive toxicological evaluation of PBBs is available. It can be assumed, however, that due to the structural similarity of PBBs and PCBs, PBBs can act by the same mechanisms. The lethal dose (LD_{50}) in laboratory animals after oral administration of commercial mixtures indicates low acute toxicity of these substances (LD_{50} > 1 g/kg b.w), but prolonged use of PBB mixtures results in a range of adverse effects, including weight loss, bodily dehydration, reproductive disorders, morphological and histopathological changes in the liver and kidneys, thymus shrinkage, and swelling of the thyroid gland. PBBs also belong to endocrine disrupters, and some reports on their carcinogenicity suggest that whilst they are not mutagenic themselves, their presence supports the carcinogenicity of other substances, such as nitrosamines and PAHs.

The toxicities of individual PBB congeners may vary considerably; the most toxic substances are coplanar congeners, which lack a substituent in the *ortho*-position. Also, the degree of bromination can affect the toxicity. The most toxic substance is 3,3′,4,4′,5,5′-hexabromobiphenyl (BB 169), which is present in low concentrations in a commercial mixture under the trade name FireMaster. Conversely, the main component of this mixture, 2,2′,4,4′,5,5′-hexabromobiphenyl (BB 153), is relatively non-toxic.

Since the 1970s, there has been stagnation in the production of PBBs, and the last known commercial production of decaBB was completed in France in 2000. In 2006, the EFSA recommended monitoring of the PBB congener 153, a marker of contamination by this group of POPs.

12.4.3.3 Tetrabromobisphenol A

Tetrabromobisphenol A (TBBPA, **12-40**) is currently the world's most widely used BFR. It is produced industrially by bromination of bisphenol A. It has found use as a reactive flame retardant in the production of epoxy and polycarbonate resins added to circuit boards used in computers and many other devices. In about 10% of cases, it is used as an additive flame retardant, especially in the production of acrylonitrile-butadiene-styrene polymers, polystyrene, papers, textiles, televisions, and office equipment.

TBBPA is not subject to the process of biomagnification, and the main source of exposure is considered to be inhalation. It is assumed that it does not contribute to contamination of the environment as it is easily degraded by UV radiation and various bacteria. The main photodegradation product is 2,4,6-tribromophenol, which is accompanied by di- and tribromobisphenol A, dibromophenols, 2,6-dibromo-4-(bromoisopropylene)phenol, 2,6-dibromo-4-(dibromisopropylene)phenol, and 2,6-dibromo-1,4-hydroxybenzene.

Acute toxicity of orally administered TBBPA in laboratory animals is low (LD_{50} for rats is >5 g/kg b.w), with very little or no effects on b.w, organ abnormalities, or behaviour. By contrast, absorption from water by fish is very fast, and fish exposed to TBBPA (half-life of its degradation is up to several hours) often have jerking movements, seizures, darker colour, abnormal breathing, various inflammations, increased concentration of the thyroid hormone thyroxine in plasma, and, in some cases, reduced egg production and juvenile survival.

12.4.3.4 Hexabromocyclododecane

1,2,5,6,9,10-Hexabromocyclododecane (HBCD, **12-40**) has 16 different diastereomers with different physico-chemical properties and biological activities. The commercially manufactured product is a mixture of three of these: α (optical antipodes α_1 and α_2), β (β_1 and β_2), and γ (γ_1 and γ_2) (**12-42**), with traces of others. At temperatures higher than 160 °C, β- and γ-isomers rearrange to the most thermodynamically stable α-isomer. Commercially used HBCD is produced in two forms, as low-melting-point and high-melting-point products, which differ in their content of diastereomers. The low-melting-point mixture contains 70–80% γ-isomer and 20–30% α- and β-isomers, whilst the high-melting-point mixture may contain more than 90% γ-isomer.

12-42, hexabromocyclododecane diastereomers

HBCD has been used as an additive flame retardant for more than 20 years. Its main use is in the construction industry, where it is used in polystyrene foams. The second major application of HBCD is in the upholstery and textile industries. Products in which HBCD occurs include upholstered furniture, various textiles, car seats and upholstery, insulation in trucks and caravans, and many types of building materials. Unlike other flame retardants, HBCD is not used in electronic circuits.

Currently there is very little knowledge about the possible risks from the use of HBCD. Tests have found a very rapid absorption in the digestive tract of animals and subsequently elevated levels of HBCD in the blood and certain organs. The acute toxicity of HBCD is very low (LD_{50} for rats is $>10\,g/kg$ b.w by oral administration and $>20\,g/kg$ b.w for dermal application). At high doses, HBCD can damage the skin or eyes. Prolonged exposure in rats caused an increase in liver weight and reproductive disorders and affected the nervous and hormonal systems.

HBCD accumulates in soil, sediments, and airborne dust, is rapidly absorbed in the gastrointestinal tract, and accumulates in the fatty tissue of organisms, but has a very short half-life. For example, in fish, the half-life of HBCD is only one day. Representation of the individual diastereomers in the sediments reflects the compositions of commercial mixtures, whilst the tissues of aquatic animals, especially fish, have a quite different composition. The dominant substance is the α-isomer; the γ-isomer is present in much smaller amounts, and the content of the β-isomer remains virtually constant, which can be explained by the almost selective biotransformation of the γ-isomer to the α-isomer.

12.4.3.5 Other flame retardants

The alternative non-PBDE flame retardants (**12-43**) include bis(2,4,6-tribromophenoxy)ethane (BTBPE), octabromo-1-phenyl-1,3,3-trimethylindane (Br indane), decabromodiphenylethane (DBDPE), and certain other compounds. These substances have similar properties to PBDEs.

bis(2,4,6-tribromophenoxy)ethane   octabromo-1-phenyl-1,3,3-trimethylindane   decabromdiphenylethane

12-43, structures of alternative polybrominated flame retardants

BTBPE is used as an additive flame retardant, replacing octaPBDE, for polystyrene and acrylonitrile-butadiene-styrene products. It was found in the eggs of birds and especially in the dust and air of industrial plants for the liquidation of electronic waste. Animal studies indicate minimum BTBPE absorption in the digestive tract. The major part is probably excreted, partly as metabolites (e.g. hydroxylated BTBPE and 2,4,6-tribromophenol).

Br indane is an additive flame retardant that is manufactured under the name FR-1808. It is used in technical resins and polystyrene foams. Br indane is kinder to the environment than PBDEs.

DBDPE is an additive flame retardant with similar uses as decaBDE. It is employed in acrylonitrile-butadiene-styrene products, and is added to polystyrene and insulation materials. It has been detected in air, soil, and sludge at concentrations one order of magnitude higher than BTBPE. DBDPE enters the body in small quantities due to its high molecular weight.

12.4.4 Perfluorinated compounds

Perfluorinated or per(poly)fluoroalkyl compounds (PFC)s or per(poly)fluoroalkyl substances (PFASs) are a group of synthetic fluorinated compounds resulting from anthropogenic activities in the latter half of the twentieth century. Their presence has been confirmed in abiotic and biotic components of the environment around the world.

12.4.4.1 Classification, structure, and properties

PFCs comprise carbon chains of varied length, in which hydrogen atoms have been wholly or partly replaced by fluorine atoms. Attention was originally focused on volatile chlorofluorocarbons (CFCs), known by the DuPont brand name Freons (greenhouse gases), which have been widely used as refrigerants, propellants, and solvents and whose production has been phased out. PFCs are used in several industrial branches, but they also occur in a large range of consumer products. Because of their special physico-chemical properties, including chemical inertness, non-wetting, high fire resistance, very high temperature rating, and high weather resistance, they are applied in non-stick-coating cookware (polytetrafluoroethylene, PTFE, known as Teflon), surface treatment of carpets, fabrics, leather, paper, packaging materials, sports clothing, extreme weather-resistant military uniforms, medical equipment, motor oil additives, fire-fighting foams, paint, and ink, as well as water-repellent products.

Table 12.31 Overview of per(poly)fluoroalkyl substances.

Name	Abbreviation	Classification
Perfluoroctanoic acid	PFOA	1,3
Perfluorobutane sulfonate	PFBS	1
Perfluorohexane sulfonate	PFHxS	1
Perfluorooctane sulfonate	PFOS	1
Perfluorodecane sulfonate	PFDS	5
Perfluorooctane sulfonamide	PFOSA	5
Perfluorooctane sulfonyl fluoride	PFOSF	2
N-Methylperfluoroctane sulfonamide	N-MeFOSA	2
N-Ethylperfluoroctane sulfonamide	N-EtFOSA	2
N-Methylperfluoroctane sulfonamidoethanol	N-MeFOSE	2
N-Ethylperfluoroctane sulfonamidoethanol	N-EtFOSE	2
1H,1H,2H,2H-Perfluoroctanol	6:2 FTOH	2
1H,1H,2H,2H-Perfluorodecanol	8:2 FTOH	2
1H,1H,2H,2H-Perfluorododecanol	10:2 FTOH	2
1H,1H,2H,2H-Perfluoroctyl acrylate	6:2 FTA	4
1H,1H,2H,2H-Perfluordecyl acrylate	8:2 FTA	4
1H,1H,2H,2H-Perfluoroctyl methacrylate	6:2 FTMA	4
1H,1H,2H,2H-Perfluorodecyl methacrylate	8:2 FTMA	4

Criterion for classification: 1 = probable degradation product, 2 = important intermediate in the manufacture, 3 = important commercial product, 4 = important monomer for the production of polymers, 5 = data not available.

12.4.4.2 Occurrence, main sources, and dietary intake

An important category of fluorinated contaminants is **fluorosurfactants**, commercial products containing per(poly)fluorinated alkyl chains of varying lengths. Some PFCs are persistent compounds with bioaccumulation potential that have a negative impact on the environment. An overview of common PFCs is given in Table 12.31. Their structures are represented by formulae **12-44**. Some fluorosurfactants, such as perfluorooctane sulfonate (PFOS) and perfluorooctanoic acid (PFOA), are POPs.

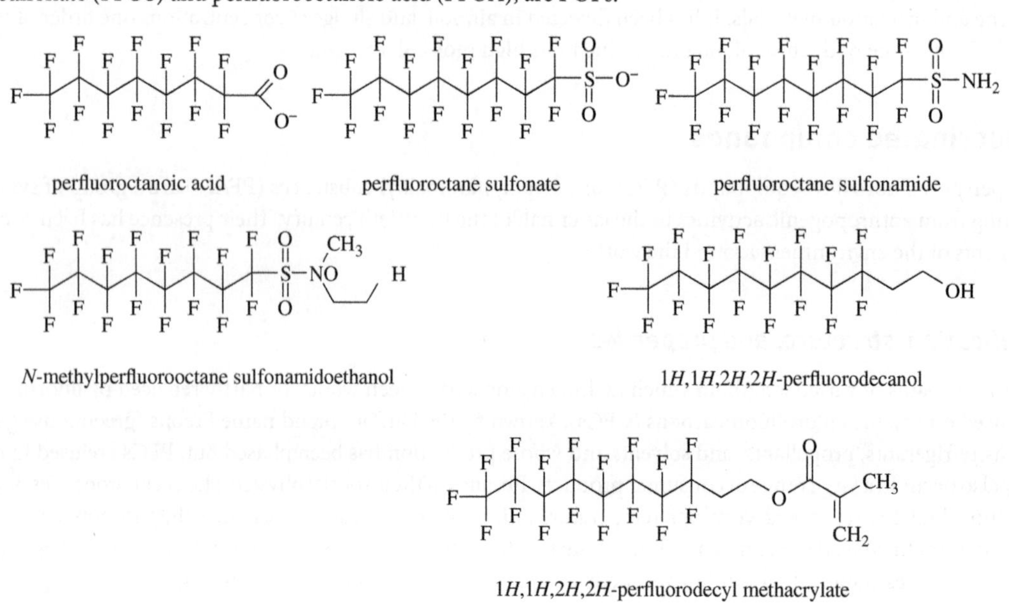

perfluoroctanoic acid

perfluorooctane sulfonate

perfluorooctane sulfonamide

N-methylperfluorooctane sulfonamidoethanol

1H,1H,2H,2H-perfluorodecanol

1H,1H,2H,2H-perfluorodecyl methacrylate

12-44, structures of selected per(poly)fluorinated compounds

Table 12.32 Amounts of perfluorooctane sulfonate (PFOS) and perfluorooctanoic acid (PFOA) in foods.

Foods	PFOS (µg/kg)	PFOA (µg/kg)	Foods	PFOS (µg/kg)	PFOA (µg/kg)
Fruits	<0.017	<0.036	Sea fish	0.407	<0.065
Vegetables	0.022	<0.027	Canned fish	0.271	<0.126
Legumes	<0.027	<0.045	Eggs	0.082	<0.055
Cereal products	<0.069	<0.080	Milk, full fat	<0.014	0.056
Pork meat	0.045	<0.053	Milk, reduced fat	<0.019	<0.028
Veal meat	0.028	<0.034	Dairy products	0.121	<0.040
Chicken	0.021	<0.067	Margarines	<0.034	<0.115
Seafood	0.148	<0.029	Vegetable oils	<0.099	<0.247

PFCs are emitted directly to the environment at the place of manufacture (mostly sorbed on dust particles) and during normal use of products containing them, both in industry and in the home. PFOS and related substances may also be emitted from wastewater treatment plants and landfills, where elevated concentrations have been observed. Increased concentrations of PFOS were found mainly in wastewater and water leaking from landfills. Fire foams are also a significant source of PFOS emissions. After penetration into the environment, PFCs are adsorbed to the organic fraction of the sediment or bioaccumulate in living organisms.

PFOS and related PFCs have been found in many organisms around the world. Generally, they are found in the highest amounts in the tissues of predators at the top of the food chain, which have fish as the major food component. In addition to PFOS, perfluorooctane sulfonamide (PFOSA) is usually present – one of its precursors. In fish, there are often higher PFOSA concentrations than PFOS concentrations, whereas in mammals, the opposite is true. For example, the concentration of PFOS in the liver of seals from the Baltic Sea was 240 µg/kg, whilst that in the liver of Canadian polar bears was 3100 µg/kg.

Human exposure to PFCs, including PFOA and PFOS, probably occurs orally, dermally, or by inhalation. Indirect factors such as place of residence, age, and type of PFC may also affect the extent of exposure. The most important sources of human exposure are fish, fish products, and drinking water. The PFCs occurring in the highest concentrations are PFOS and PFOA, which are the substances most commonly found in the environment. Table 12.32 summarises the amounts of PFOS and PFOA in selected foods in Spain, by way of example.

12.4.4.3 Metabolism and degradation

Owing to the high dissociation energy of the C—F bond, many fluorinated organic compounds, other than acrylates, are resistant to hydrolysis, photolysis, and biodegradation. The end products of microbial degradation of many PFCs in wastewater, such as N-ethylperfluorooctane sulfonamidoethanol (N-EtFOSE) and N-methylperfluorooctane sulfonamidoethanol (N-MeFOSE), are PFOS and PFOA, which are stable under aerobic and anaerobic conditions and are not further degraded.

12.4.4.4 Health and toxicological evaluation

After PFOS enters the human body orally, it is relatively easily absorbed and distributed mainly to the liver and blood serum (apparently through interactions with phospholipids), but is not metabolised. In general, the concentrations of PFOS in human plasma are higher than those of PFOA. PFOS is excreted in the urine and faeces, but its excretion is very slow. It is estimated that half of the received dose may be removed from the human body in approximately nine years. PFOA can enter the human body not only orally, but also (to a lesser extent) by inhalation and dermal absorption. It is not accumulated in the adipose tissue, but is covalently bound to macromolecules, especially proteins in the liver, plasma, and testes of males. It is not metabolised, and is excreted from the body in urine and faeces. It is estimated that half of the ingested amount can be removed from the human body in one to three years.

The mechanisms of the toxic effects of PFCs are not yet fully understood. Amongst the possible harmful effects are effects on the transport and metabolism of fatty acids in the biological membranes and on the bioenergetic processes in the mitochondria. PFOS is practically non-toxic (e.g. in evaluation of subchronic and chronic toxicities) for freshwater algae and higher plants, but shows a slight toxicity to invertebrates. Fish are more sensitive to the presence of this substance than algae, plants, or invertebrates (the lowest LC_{50} value was 7.8 mg/l for rainbow trout). PFOA is also virtually non-toxic for all tested species of freshwater organisms (bacteria, algae, and fish).

12.4.4.5 Mitigation

The fluorocompounds PFOA and PFOS have both been investigated by the EFSA and the EPA, which regard them as harmful to the environment. The EFSA has been involved with PFCs in food since 2004, when recommendations that they be monitored were first issued. In 2009, PFOS, its salts, and perfluorooctane sulfonyl fluoride (PFOSF) were subjected to restrictions on their production and use and became a part of the Stockholm Convention on POPs. For PFOS, the lowest No Observed Adverse Effect Level (NOAEL) was established at 0.03 mg/kg b.w per day, a TDI was derived at 150 ng/kg b.w, and an indicative figure of 60 ng/kg b.w per day for human exposure was selected. The estimated indicative exposure of high consumers of fish and fishery products is approximately three times as high (200 ng/g b.w per day). Non-food sources of PFOS were estimated to contribute in the order of 2% or less to average dietary exposure. Drinking water appears to contribute less than 0.5%. For PFOA, an indicative figure of 2 ng/kg b.w per day was established for average human exposure (6 ng/g b.w per day for the exposure of high consumers of fish and fishery products). At these estimated intakes, non-food sources could contribute up to 50% of the average dietary exposure, whereas drinking water would contribute less than 16%. The TDI value was estimated at 1.5 µg/kg b.w, and the average daily intake for the European population was estimated to be in the range of 2–6 ng/kg b.w.

12.5 Pesticides

Efforts to protect the food supply against attacks by various pests have been documented since ancient times. The ancient Chinese treated cereals and other plants seeds with soda ash and olive oil before storage and used preparations containing arsenic to protect crops during the growing season. The ancient Greeks and Romans used the antiseptic properties of combustion gases resulting from the burning of sulfur to ward off pests. Substances acting against insect pests are found in many plants that have been used in many countries. Traditional natural pesticides include the alkaloids nicotine and anabasine, so-called nicotinoids, contained in extracts from the leaves and roots of tobacco, mainly from the arborescent tobacco species *Nicotiana glauca* (Solanaceae), native to Argentina. Anabasine also occurs in the leaves of *Anabasis aphylla*, a plant belonging to the Amaranthaceae (formerly Chenopodiaceae) family, which grows in the steppes of Central Asia. Plants of the genus *Pyrethrum* (Asteraceae), in particular *P. cinerariaefolium* originating from the Balkans and the Adriatic Sea islands and *P. carneum* growing in the Caucasus and around the Black Sea, contain active components called pyrethroids (**12-45**), which have a strong insecticidal effect. Extracts (called pyrethrum) are a mixture of three structurally related esters of a monoterpenoid called chrysanthemic acid (pyrethrins I) and three esters of pyrethrinic acid (pyrethrins II). Pyrethrins later became prototypes of synthetic pyrethroids. One of the most significant natural insecticides and acaricides (pesticides that kill mites and ticks) is (5′*R*,6a*S*,12a*S*)-rotenone (**12-46**), from the group of isoflavones called rotenoids, which are highly toxic to all life forms. Rotenoids occur in some plants of the family Fabaceae, for example in the roots of *Derris elliptica* lianas, native to Indonesia, and liana species *Lonchocarpus utilis*, *L. urucu*, and *L. nicou*, native to South America. Powders of lianas are used as a fish poison, but their toxic effects on insects and mites have also been shown. The extract of quassia (also known as bitterwood, *Quassia amara*, Simaroubaceae) from Central and South America likewise has an insecticidal effect, the carriers of which are substances called quassinoids (see Section 8.3.5.1.3).

The beginning of a scientifically conceived, systematic study of the applications of chemical compounds for crop protection dates back to the mid-nineteenth century. In 1867, the so-called Paris Green, copper(II) acetoarsenite, $4\,Cu^{2+}(As_3O_6{}^{3-})_2(CH_3COO^-)_2$ (also known as Emerald Green, Schweinfurt Green, Imperial Green, Vienna Green, Mitis Green, and Veronese Green), began to be used as an insecticide and rodenticide. For the prevention of fungal infections in vineyards and potatoes, the Bordeaux mixture (copper sulfate, lime, and water) is used. This mixture was invented in the Bordeaux region of France in 1886 and is still in use today. A number of other inorganic compounds have since been tested, but a significant milestone was the introduction of organic mercury compounds for the protection of seeds in 1913. An expansion of the use of organic compounds with pesticidal effects began in the 1930s. Amongst the first herbicides designed to kill weeds in fields were dinitroderivatives of *o*-cresols, the fungicide thiram from the dithiocarbamates group, and pentachlorophenol, used to combat wood-destroying fungi. One of the first insecticides was the organophosphate tetraethyl pyrophosphate (TEPP).

In 1939, the efficient contact insecticide DDT (an abbreviation of the technically incorrect name DichloroDiphenylTrichloroethane) was introduced to the market. Its use, as well as the use of several other organochlorine compounds, spread worldwide in the following years. Important active substances that appeared on the market in the period after World War II include carbamate insecticides and herbicides. Also significant was the discovery of herbicidal phenoxyacetic acids, which represent the first group of the so-called hormonally active pesticides. Around the middle of the twentieth century, a number of other biologically active substances were discovered, many of which, such as herbicides based on substituted urea, *s*-triazines (1,3,5-triazines), quaternary ammonium salts, and insecticidal synthetic pyrethroids, along with many others, are still used in countries around the world.

Alarming signals regarding the occurrence of residues of DDT and other lipophilic organochlorine pesticides (such as aldrin and dieldrin) in virtually all parts of the global ecosystem and their adverse ecotoxicological effects, which began to emerge in the 1960s, substantially contributed to the documentation of the potential risks arising from the use of pesticides. DDT and related compounds were progressively banned in almost all countries around the world. This helped initiate research aimed at studying the possibility of environmental protection, including protection of human food chains, against continuing contamination by such persistent pollutants.

In general, the use of pesticides and related biologically active compounds has become an indispensable means of intensifying agricultural production. Besides agriculture, pesticides are also used in forestry and water management and in communal hygiene. Their use not only provides an increased harvest, but very often manifests in a higher nutritional, sensory, and technological quality of the treated crops. The FAO estimates that crop losses caused by various harmful agents worldwide, if protective agents were not used, could be as high as 30%, and in developing countries even higher. Of course, the extent and type of damage in a given region is subject to weather conditions and the character of the agroecosystem.

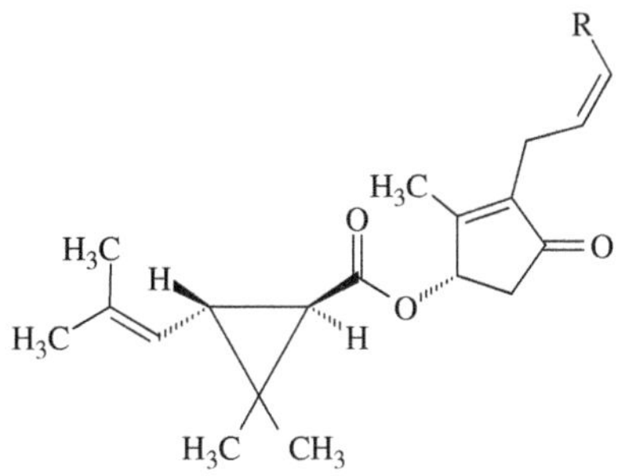

12-45, pyrethrins I

pyrethrin I, R = CH=CH$_2$
jasmolin I, R = CH$_2$CH$_3$
cinerin I, R = CH$_3$

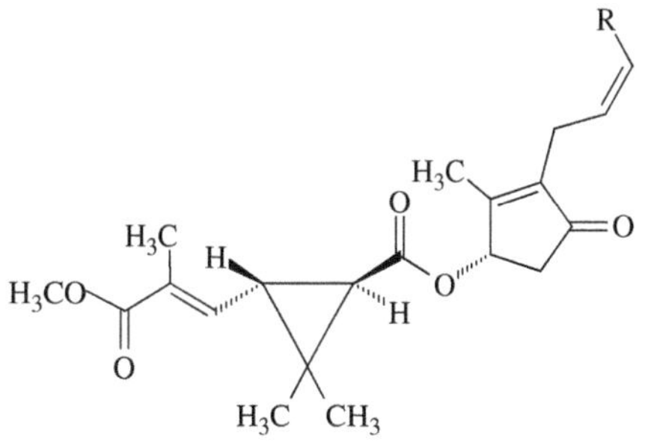

pyrethrins II

pyrethrin II, R = CH=CH$_2$
jasmolin II, R = CH$_2$CH$_3$
cinerin II, R = CH$_3$

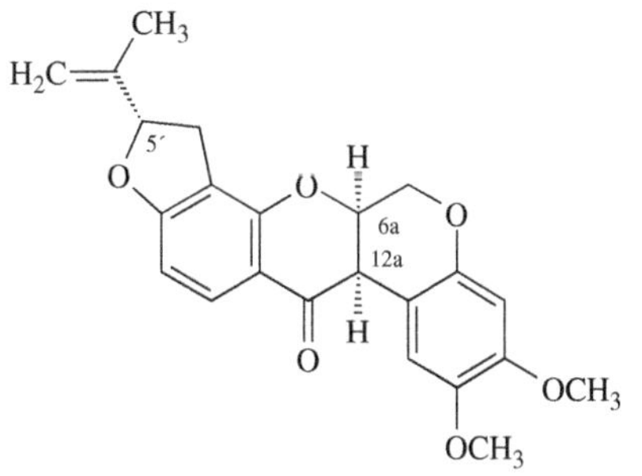

12-46, rotenone

In addition to the indisputable positive aspects arising from the use of pesticides, however, there are also negative impacts of the chemical processing of agriculture on the biotic components of the environment, including humans. According to the Stockholm Convention on POPs, 9 of the 12 most dangerous POPs are pesticides. Consumers are particularly sensitive to the potential presence of pesticide residues in food, which is one of the main incentives for the increasing interest in foodstuffs produced from so-called ecological and organic farming, which exclude the use of synthetic pesticides.

12.5.1 Classification, structure, and properties

A pesticide is any substance or mixture of substances specifically intended to prevent, repel, destroy, or lessen the effect of pests or undesirable microorganisms, plants, and animals during the production, storage, transport, distribution, and processing of agricultural commodities, foods, and feeds. The term also includes compounds administered to animals for the control of ectoparasites, growth regulators, desiccants, and germination inhibitors applied to crops either before or after harvest.

Pesticides are substances representing a wide range of chemical compounds, often of very complex structure. To facilitate communication, they are known by their common name, trade name, and chemical name. The common name is the name of the active ingredient: the biologically active substance that controls the pest. The trade name is the prominent brand name that the manufacturer gives to the product. Pesticides with different trade names can contain the same active ingredient or ingredients. In practice, the names predominantly used are the common and trade names. The chemical name is the name of the chemical structure of the active ingredient and is used by scientists. Common names of pesticides and other agrochemicals have been adopted by the Technical Committee of the International Organization for Standardization, created in 1953 (ISO/TC 81). Principles for the selection of common names of pesticides and other agrochemicals are explained in ISO 257:1988 and revised by ISO 257:2004. For example, fungicide containing captan as the active ingredient, the chemical name of which is *N*-trichloromethylthio-4-cyclohexene-1,2-dicarboximide, is called by the trade name Captan or Maestro. The insecticide carbaryl (1-naphthyl methylcarbamate) has the trade names Sevin and Vet-Tek, amongst others.

Table 12.33 Classification of pesticides by target harmful organisms.

Group of pesticides	Target harmful organism	Group of pesticides	Target harmful organism
Acaricides (miticides)	Mites	Ovicides	Insect eggs
Algicides	Algae	Larvicides	Insect larvae
Avicides	Birds	Adulticides	Adult insects (imago)
Bactericides	Bacteria	Molluscicides	Molluscs
Fungicides	Fungi, parasitic fungi	Nematicides	Worms
Herbicides	Weeds	Rodenticides	Rodents
Insecticides	Insect	Virucides	Viruses

At present, about 1000 compounds are registered globally that are used as pesticides or other agrochemicals (particularly as herbicides to protect cultivated plants). A common criterion used for their classification is the target (usually harmful) organisms. The most common types of pesticides in use are listed in Table 12.33. Table 12.34 summarises the most important groups of pesticides used over the past several decades, many of which are still in use today. They are classified according to their biological effects. Naturally, the list does not include all registered pesticides, but focuses mainly on the new generation, often referred to as modern pesticides. For completeness, substances that can generally be described as traditional pesticides or historical pesticides are listed. Their use globally is practically nil.

12.5.1.1 Persistent chlorinated hydrocarbons

Organochlorine pesticides are chlorinated hydrocarbons. They were used extensively from the 1940s in agriculture and pest control, but most were banned in 2004 when the Stockholm Convention went into effect. Measures to reduce or eliminate these POPs included elimination of their production and use (category A, which includes aldrin, chlordane, dieldrin, endrin, heptachlor, hexachlorobenzene, mirex, and toxaphene, and also PCBs and hexachlorobenzene, which also has use as an industrial chemical) and restriction and control of their production and use (category B, which includes DDT). Pentachlorophenol, formerly used as a herbicide, insecticide, and fungicide, is now used mainly as a fungicide to protect wood or as a biocide in masonry. Its use as a herbicide is banned. Lindane is currently being examined at an international level and its worldwide ban is being considered.

Organochlorine pesticides are mostly contact insecticides with neurotoxic effects that are characterised by their high lipophilicity and related high potential to accumulate in the adipose tissues of living organisms. These properties, along with their high chemical stability and limited biodegradability, are the reasons why many organochlorine pesticides were banned, but their residues are still commonly found in the environment, either as the parent insecticides or as some of their persistent metabolites. Some organochlorine pesticides are still registered for use in certain countries. The spectrum of the major contaminants in the food chain is a result of their persistence. They are mainly present in fish and other aquatic animals, but also in tissues of higher mammals, including humans.

12.5.1.1.1 DDT

The insecticide DDT, 1,1,1-trichloro-2,2-bis(4-chlorophenyl)ethane (technical-grade DDT), was composed of up to 14 chemical compounds, of which just 65–80% was the active ingredient, p,p'-DDT (**12-47**). The other components included 15–21% of the nearly inactive isomer o,p'-DDT (**12-47**), up to 4% of p,p'-DDD, dichlorodiphenyldichloroethane, up to 1.5% of 1-(4-chlorophenyl)-2,2,2-trichloroethanol, and lower amounts of o,p'-DDD and closely related compounds with low insecticidal properties, p,p'- and o,p'-isomers of DDE, 1,1-dichloro-2,2-bis(4-chlorophenyl)ethylene.

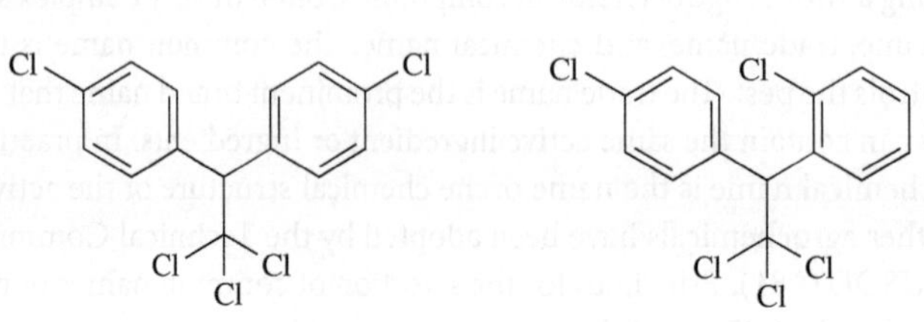

12-47, p,p'-DDT and o,p'-DDT

Table 12.34 Classification of pesticides into groups according to the target organism, application method, activity, and structure.

Pesticides	Names of pesticides representing a given group[a]
Non-systemic insecticides	
Organochlorine compounds	Aldrin, DDT, dieldrin, dicofol, endosulfan, endrin, heptachlor, chlordan, γ-HCH (lindane), methoxychlor, toxaphen
Organophosphorus compounds	Azinphos-methyl, diazinone, dichlorvos, ethione, etrimphos, fenitrothione, fomophos, phosalone, chlorfenvinphos, chlorpyriphos, chlorpyriphos-methyl, quinalphos, malathione, mecarbam, methidathione, mevinphos, parathione-ethyl, parathione-methyl, pirimiphos-methyl, sulfotep, terbuphos, tetrachlorvinphos, tolclophos-methyl, triazophos
Carbamates	Phenoxycarb, formethanate, carbaryl, methiocarb, methomyl, propoxur
Synthetic pyrethroids	Acrinathrin, allethrin, biphenthrin, bioresmethrin, cyfluthrin, cypermethrin, deltamethrin, fenvalerate, etofenprox, fenpropathrin, flucythrinate, fluvalinate, lambda-cyhalothrin, permethrin, piperonylbutoxide, tau-fluvalinate
Benzoylureas	Diflubenzurone, flucycloxurone, flufenoxurone, lufenurone, teflubenzurone, triflumurone
Others	Fipronyl, pyridabene, thiocyclam, thiodicarb, thiophanox
Systemic insecticides	
Organophosphorus compounds	Acephate, dimethoate, disulfotone, phorate, formothione, phosphamidone, heptenophos, methamidophos, mevinphos, thiometone, trichlorphon, vamidothion
Carbamates	Aldicarb, bendiocarb, benfuracarb, ethiophencarb, furathiocarb, carbofuran, carbosulfan, methomyl, oxamyl, pirimicarb
Others	Acetamiprid, bensultap, buprofezine, cyromazine, diaphenthiurone, hexaflumuron, imidacloprid, triazamate
Specific acaricides	
Without fungicide activities	
Organochlorin compounds	Chlorobenzilate, tetradifon
Organotin compounds	Cyhexatin
With fungicide activities	
Dinitro compounds	Binapicryl, dinocap
Molluscicides	
Aquatic	
Botanic	Endod
Others	Niclosamide, sodium pentachlorophenolate, tributyltin, triphenmorph, triphenyltin
Terrestrial	
Carbamates	Aminocarb, methiocarb, mexacarbate
Others	Metaldehyde
Non-systemic protective fungicides	
Dithiocarbamates	Maneb, mancozeb, metiram, propineb, thiram, zineb
Phthalimides	Carbendazim, dichlofluanide, folpet, captafol, captan
Dinitro compounds	Binapicryl
Organomercury compounds	Phenylmercury
Dicarboximides	Vinclozolin
Organotin compounds	Fentin
Chlorinated aromatic compounds	Dichlone, diclorane, chlorothalonil, quintozene (PCNB), tecnazene (TCNB)

(continued overleaf)

Table 12.34 (continued)

Pesticides	Names of pesticides representing a given group[a]
Cation-active tensides	Dodine, glyodine
Others	Iprodion, procymidone
Systemic curative fungicides	
Antibiotics	Blasticidine, cyclohexamide, kasugamycine, streptomycine
Benzimidazoles	Benomyl, thiabendazole, thiophanate-methyl
Morpholines	Dodemorph, tridemorph
Pyrimidines	Bupirimate, ethirimol
Piperazines	Triforine
Others	Metalaxyl, propiconazole, triadimefon
Herbicides applied to leaf	
Systemic or translocated	
Phosphonoamino acids	Gluphosinate, glyphosate
Benzoic acid derivatives	Dicamba, chlorophenprop-methyl, 2,3,6-TBA
Chlorinated aliphatic acids	Dalapon, TCA
Oxyphenoxyacid esters	Cycloxydim, diclofop-methyl, fenoxaprop-ethyl, fluazifop-butyl, haloxyfop-methyl, quizalofop-ethyl
Phenoxyalkanoic acids	2,4-D, 2,4-DB, dichlorprop, mecoprop, MCPA, MCPB, silvex, 2,4,5-T
Quaternary ammonium compounds	Diquat, paraquat
Applied to leaf, contact	
Benzonitriles	Bromoxynil, dichlobenil, ioxynil
Benzothiadiazoles	Bentazone
Carbanilates	Phenmedipham
Cyclohexenones	Cycloxydim, clethodim, sethoxydim
Dinitrophenols	Dinoseb
Diphenylethers	Acifluorfen, lactofen, nitrofen, oxyfluorfen
Applied to soil	
Acetanilides	Alachlor, butachlor, metolachlor, propachlor
Amides and anilides	Benzoylprop-ethyl, difenamide, naptalam, pronamide, propanil
Carbanilates and carbamates	Asulam, barban, bendiocarb, carbetamide, chlorpropham, propham, triallate
Dinitroanilines	Benephin, pendimethalin, trifluralin
Pyridazinones and pyridinones	Amitrol, dimethazone, fluridone, norflurazone, oxadiazone, pyrazone
Pyridineoxy- and picolinic acids	Fluroxypyr, clopyralid, picloram, triclopyr
Phenylureas or other substituted	Dimefurone, diurone, fenurone, fluometurone, chlorbromuron, chlorotoluron, isoproturon,
Ureas	linuron, metobromuron, metoxuron, monolinuron, siduron
Sulfonylureas	Amidosulfuron, flupyrsulfuron-methyl, chlorimuron-ethyl, chlorsulfuron, metsulfuron-methyl, nicosulfuron, primisulfuron-methyl, prosulfuron, rimsulfuron, sulfometuron-methyl, sulfosulfuron, thiameturon-methyl, triasulfuron, tribenuron, triflusulfuron-methyl
Thiocarbamates	Butylate, cycloate, EPTC, molinate, pebulate, thiophencarb, triallate
Triazins	Ametryne, atrazine, desmetryne, hexazinone, cyanazine, methoprotryne, metribuzine, prometone, prometryne, propazine, simazine, terbumetone, terbuthylazine, terbutrizine, terbutryne

(continued overleaf)

Table 12.34 (*continued*)

Pesticides	Names of pesticides representing a given group[a]
Uracils	Bromacil, lenacil, terbacil
Desiccants and defoliants	
Organophosphorus compounds	Merfos
Derivatives of phenols	Dinoseb
Quaternary ammonium compounds	Diquat, paraquat
Growth regulators and growth-promoting substances	
Auxins	2,4-D, MCPB, NAD
Cytokinins	Adenine, kinetin
Gibberellins	GA$_3$, giban
Substances producing ethylene	Ethephon
Growth inhibitors and retardants	
Quaternary ammonium compounds	Chlormequate, mepiquate
Hydrazides	Daminozide, maleinhydrazide
Triazoles	Paclobutrazola, uniconazole
Rodenticides	
Fumigants and anticoagulants	
Hydroxycoumarins	Brodifacoum, difemacoum, coumafuryl, coumatetralyl, warfarin
Indanediones	Diphacinone, chlorophacinone, pindone
Without coagulation activities	
Arsenic compounds	Sodium arsenite, arsenic trioxide
Benzeneamines (anilines)	Bromethaline
Thioureas	Antu, promurit
Natural compounds	Red squill (Drimia maritime, Asparagaceae), strychnine
Others	Fluoroacetamide, sodium fluoroacetate, zinc phosphide, sodium norbormide

[a]Abbreviated names are derived from chemical or common names: DDT = dichlorodiphenyltrichloroethane, HCH = 1.2.3.4.5.6-hexachlorocyclohexane, PCNB = pentachloronitrobenzene, TCNB = 1,2,4,5-tetrachloro-3-nitrobenzene, 2,3,6-TBA = 2,3,6-trichlorobenzoic acid, TCA = trichloroacetic acid, 2,4-D = (2,4-dichlorophenoxy)acetic acid, 2,4-DB = 4-(2,4-dichlorophenoxy)butyric acid, MCPA = (4-chloro-2-methylphenoxy)acetic acid), MCPB = (4-chloro-o-tolyloxy)butyric acid), 2,4,5-T = trolamine or 2,4,5-T-triethylammonium, EPTC = S-ethyldipropylthiocarbamate, GA$_3$ = gibberellic acid or gibberellin A$_3$.

DDT was first introduced in World War II, in 1940, as a very effective agent to wipe out malaria by killing the mosquitoes that carry the disease. Since then, DDT, DDE, and other components of technical-grade DDT have entered the environment. DDT stays in the environment for long periods and can travel long distances from where it was originally used. Both compounds, DDT and DDE, have even been found in Arctic and Antarctic animals, although DDT was never used in those regions. DDE also enters the environment as a DDT metabolite, produced by enzymatically catalysed biotransformation (dehydrochlorination) of DDT in living organisms. In adipose tissue of animals, the DDT dechlorination product DDD can also frequently be found. In air, DDT, DDE, and DDD are rapidly broken down by sunlight (the estimated half-life is about three days) and sorbed strongly to soil. Most DDT in soil is broken down slowly to DDE and DDD by microorganisms; half the DDT in soil will break down in 2–15 years, depending on the type of soil. Biodegradation of DDT, for example by white rot fungus *Phanerochaete chrysosporium*, yields polar, water-soluble metabolites including DDD, FW-152, and

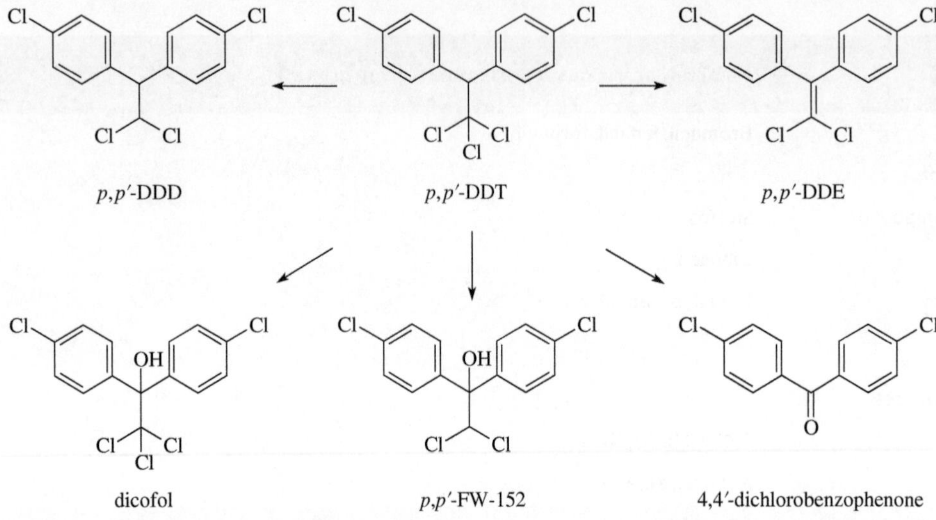

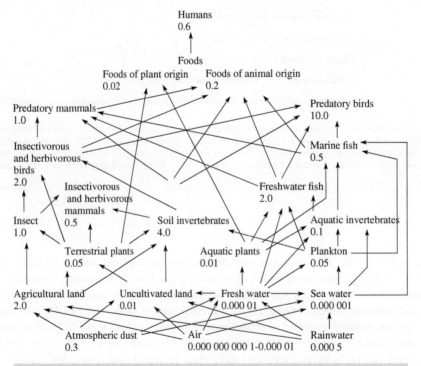

Figure 12.43 Transformation products of *p,p′*-DDT.

4,4′-dichlorobenzophenone (DBP), as well as 2,2,2-trichloro-1,1-bis(4-chlorophenyl)ethanol (dicofol), one of the intermediates of DDT synthesis, effective against red spider mite, which was formerly used as amiticide (Figure 12.43).

Typical levels of DDT and derived compounds in different parts of the ecosystem in the mid-1970s can be seen in Figure 12.44. As a result of biomagnification, in some clams and fish, concentrations of DDT are 10 times greater than they are in plankton, and the process of biomagnification goes up the food chain from one trophic level to the next. Gulls, which feed on clams, can accumulate 40 or more times the concentration of DDT that was in their prey. This represents a 400-fold increase in concentration along the length of this short food chain. The people applying the pesticides are exposed primarily to DDT, but nearly all of the general population is exposed to the metabolite DDE in the diet or drinking water. In 1991, the IARC reported that the mean concentration of DDT in the population had declined in much of the world, from 5–10 mg/kg to around 1.0 mg/kg of milk fat or even less over the last three decades. DDT residues (residues of *p,p′*-DDT,

Figure 12.44 Typical levels (in mg/kg) of DDT and derived compounds in different parts of the ecosystem.

o,p'-DDT, *p,p*'-DDE, and *o,p*'-DDE) in human fat in the United Kingdom in 1995–1997 were 1.0–9.3 mg/kg (47 cases), 0.1–0.9 mg/kg (135 cases), and 0.01–0.09 mg/kg (19 cases), which may reflect the measures taken since it was banned. DDE is mostly found in autopsy or biopsy samples of organisms, and also in the breast milk of mammals, often in much higher concentrations than the parent insecticide DDT. The ratio of these two substances (conversion rate) thus indicates the time from the primary organism load. The major metabolite detected in faeces and liver samples is *p,p*'-FW-152, 2,2-dichloro-1,1-bis(4-chlorophenyl)ethanol.

The amount of DDT in food has greatly decreased since it was banned, and should continue to decline. Between the years 1986 and 1991, the average adult in the United States consumed an average of 0.8 μg of DDT a day, differing slightly according to age and sex. Today, the largest fraction of DDT in a person's diet comes from meat, poultry, dairy products, and fish. Leafy vegetables generally contain more DDT than other vegetables, possibly because DDT in the air is deposited on the leaves. Infants may be exposed by drinking breast milk. The changes of concentrations of DDT and related compounds during industrial and culinary processing of foods are minimal due to their relatively low volatility and considerable resistance to degradation. Health standards have laid down very strict criteria for DDT content in food. For example, in the European Union, the maximum concentration of DDT in drinking water is 0.1 μg/l, whilst in the United States a more tolerant limit of 50 μg/l is set. The maximum residue level expressed as the sum of *p,p*'-DDT, *o,p*'-DDT, *p,p*'-DDE, and *o,p*'-DDE for fresh and frozen fruits and nuts is 0.05 mg/kg.

Potential mechanisms of DDT action on humans are genotoxicity and endocrine disruption. Acute toxicity is moderate, whilst chronic toxicity has been linked to diabetes. A number of studies have found a prevalence of diabetes in a population with increased serum levels of DDT or DDE. The IARC classifies DDT as a Group 2B agent.

12.5.1.1.2 Other chlorinated hydrocarbons

From a toxicological point of view, the most hazardous chlorinated insecticides are cyclodienes known as aldrin and aldrin epoxide dieldrin, named after the Diels–Alder reaction, which was used for the synthesis of aldrin. Aldrin, 1,2,3,4,10,10-hexachloro-1, 4,4a,5,8,8a-hexahydro-1,4 : 5,8-dimethanonaphthalene, was widely used until the 1970s, when it was banned in most countries. Dieldrin, 1a*R*,2*R*,2a*S*,3*S*,6*R*,6a*R*,7*S*,7a*S*)-3,4,5,6, 9,9-hexachloro-1a,2,2a,3,6,6a,7,7a-octahydro-2,7 : 3,6-dimethano-naphtho[2,3-*b*]oxirane, is an extremely persistent insecticide that tends to biomagnify as it is passed along the food chain. It was widely used from the 1950s until the early '70s. It is also produced by photooxidation of aldrin, as are another product, photodieldrin, a highly toxic stereoisomer of dieldrin known as endrin, and certain other compounds (Figure 12.45). Endrin was primarily used as an insecticide and also as a rodenticide.

Chlordane, 1,2,4,5,6,7,8,8-octachloro-3a,4,7,7a-tetrahydro-4,7-methanoindane (12-48), also known as chlorindane and chloro-tox, is another cyclodiene produced by chlorinating cyclopentadiene to form hexachlorocyclopentadiene and condensing the latter with cyclopentadiene. It is an insecticide used primarily in agriculture for the treatment of maize and citruses, and in households against termites. It was banned in the United States in 1988, but is still used in some countries today. The cyclodiene heptachlor (**12-48**), 1,4,5,6,7,8,8-heptachloro-3a,4,7,7a-tetrahydro-4,7-methano-1*H*-indene is about three to five times more active as an insecticide, and is also produced via the Diels–Alder reaction of hexachlorocyclopentadiene and cyclopentadiene. It was used similarly to chlordane in agriculture between the 1960s and '70s, and in households as an insecticide to kill ants, termites, and worms. It is still in use in some countries against ants and in the protection of underground electrical cables and around transformers. It is also a component of the

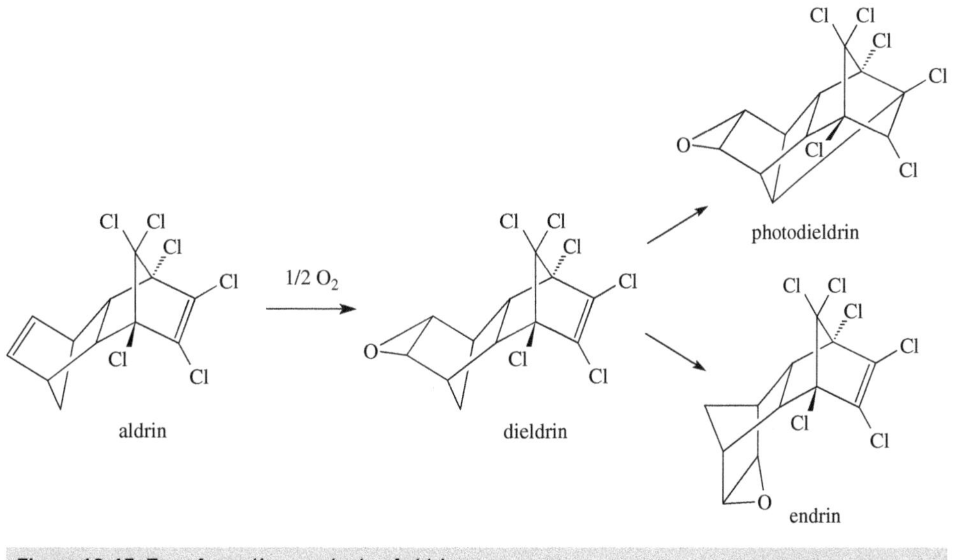

Figure 12.45 Transformation products of aldrin.

insecticide chlordane. Mirex, 1,1a,2,2,3,3a,4,5,5,5a,5b,6-dodecachlorooctahydro-1*H*-1,3,4-(methanetriyl)cyclobuta[*cd*]pentalene (**12-48**), produced by the dimerisation of hexachlorocyclopentadiene in the presence of aluminum chloride, was used in agriculture to control ants and termites. In the 1960s, it was also used as a flame retardant in plastics and building materials.

Another persistent chlorinated insecticide is toxaphene, a complex mixture of about 200 structurally related chlorinated terpenes (containing 6–11 chlorine atoms) derived from camphene by chlorination to an overall chlorine content of about 70%. Very little is known about the mode of action and metabolic fate of its constituent compounds. They differ not only in their toxicities, but also in their stabilities in the environment. In technical preparations, octachlorobornanes known as toxicants A-1, A-2, and B (**12-48**) are predominant. For example, toxicant A-1 has the chemical name 2,2,5,6-tetrachloro-1,7-bis-chloromethyl-7-dichloromethylbicyclo[2.2.1]heptane. It was used mainly in the 1970s; in 1982, it was banned for most situations, and in 1990 it was banned for all uses in the United States.

Important chlorinated insecticides of the group of alicyclic and aromatic chlorinated compounds are hexachlorobenzene (HCB), also known as perchlorobenzene or benzenehexachloride (BHC, **12-48**), pentachlorophenol (PCP, **12-48**), and hexachlorocyclohexane (HCH). Hexachlorobenzene is environmentally very dangerous, because as a result of its volatility, high stability, and bioaccumulation it can spread over large distances. For example, similarly to DDT, it has been demonstrated in the air, water, and tissues of animals and humans living in the Arctic region, although it was never used there. The technical product HCH is a mixture of eight stereoisomers, of which the most effective insecticide is γ-HCH, (1α,2α,3β,4α,5β,6β)-hexachlorocyclohexane, also called lindane, hexachloran, or gammexan (**12-48**). This isomer is accompanied by α-HCH, which is (1α,2α,3β,4α,5α,6β)-isomer, β-HCH, (1α,2β,3α,4β,5α,6β)-isomer, δ-HCH, (1α,2α,3α,4β,5α,6β)-isomer, ζ-HCH, (1α,2α,3α, 4β,5β,6β)-isomer, ε-HCH, (1α, 2α,3α,4α,5α,6α)-isomer, η-HCH, (1α,2α,3α,4α,5β,6β)-isomer, and ξ-HCH, (1α,2α,3α,4α, 5α,6β)-isomer.

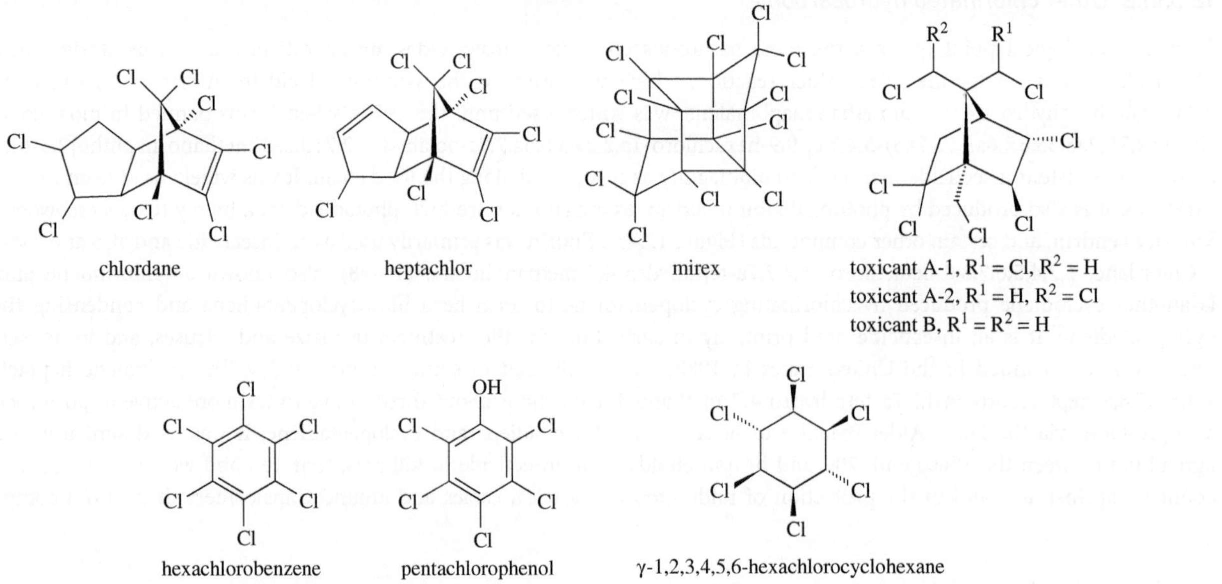

chlordane   heptachlor   mirex

toxicant A-1, R^1 = Cl, R^2 = H
toxicant A-2, R^1 = H, R^2 = Cl
toxicant B, R^1 = R^2 = H

hexachlorobenzene   pentachlorophenol   γ-1,2,3,4,5,6-hexachlorocyclohexane

12-48, structures of common chlorinated pesticides

The IARC classifies aldrin, dieldrin, and endrin as Group 3 chemicals not classifiable as to their carcinogenicity to humans, but all the other chlorinated hydrocarbons (chlordane, heptachlor, mirex, toxaphene, hexachlorobenzene, pentachlorophenol, and hexachlorocyclohexanes) are classified as Group 2B, probably carcinogenic to humans.

12.5.1.2 Modern pesticides

12.5.1.2.1 Classification, structure, and properties

This section includes commonly used compounds, residues of which have been and still are found relatively frequently in foods. In comparison with the previous group of persistent organochlorine compounds, modern pesticides are mostly substances that have higher solublity in water. Structures of important modern pesticides are given in formulae **12-49**.

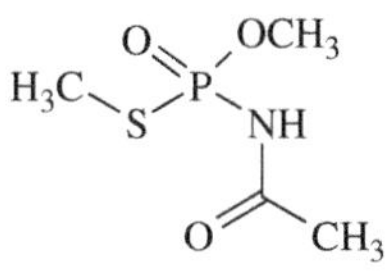

acephate

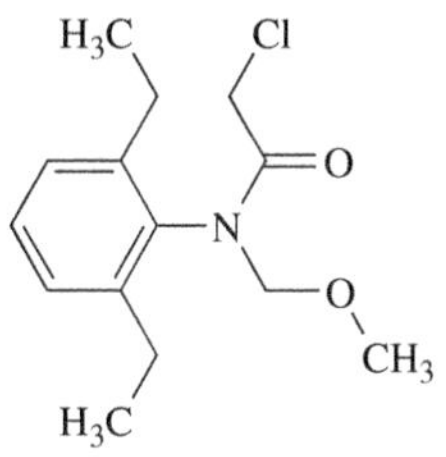

alachlor

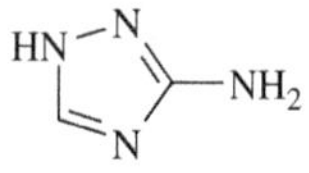

amitrol

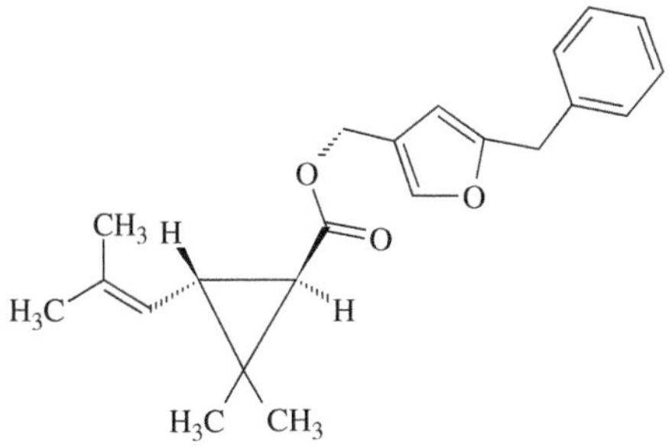

bioresmethrin

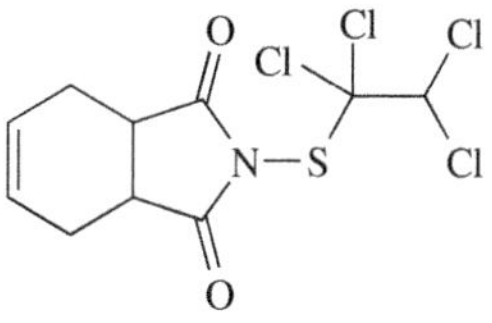

captafol

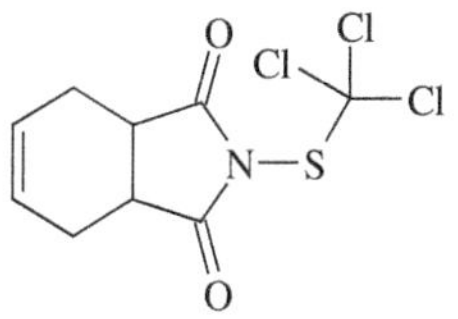

captan

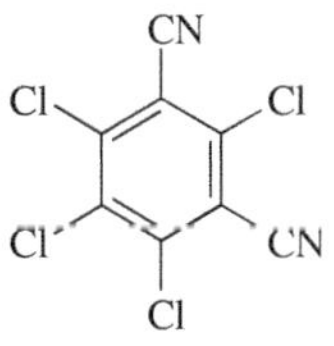

chlorothalonil

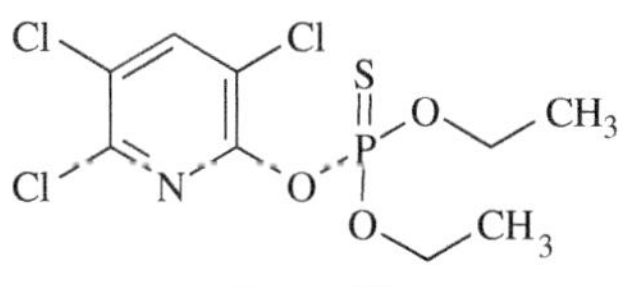

chlorpyrifos

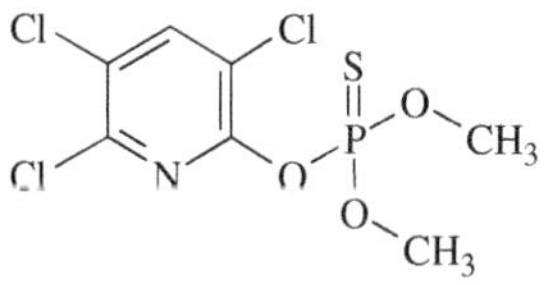

chlorpyrifos-methyl

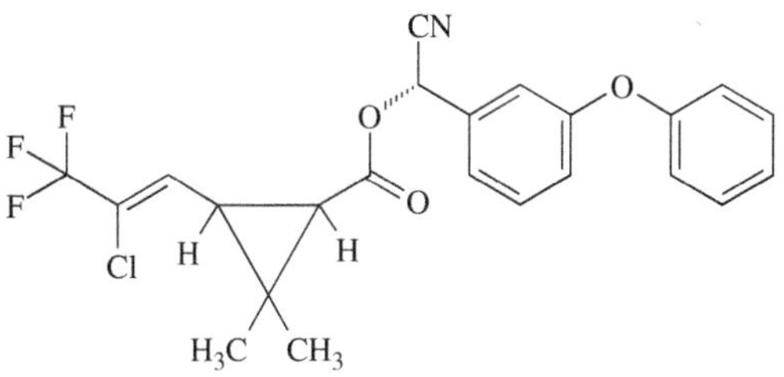

cyhalothrin, (S)-alcohol (Z)-(1R)-cis-acid

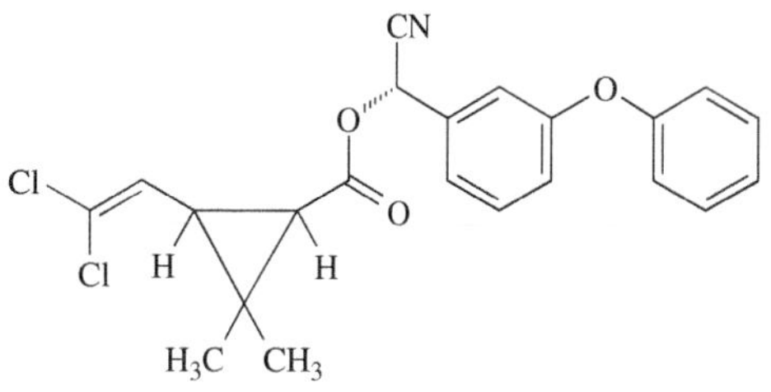

cypermethrin, (S)-alcohol (1R)-cis-acid racemate comprising (R)- and (S)-α-cyano-3-phenoxybenzyl (1S) and (1R)-cis-3-(2,2-dichlorovinyl)-2,2-dimethylcyclopropanecarboxylate

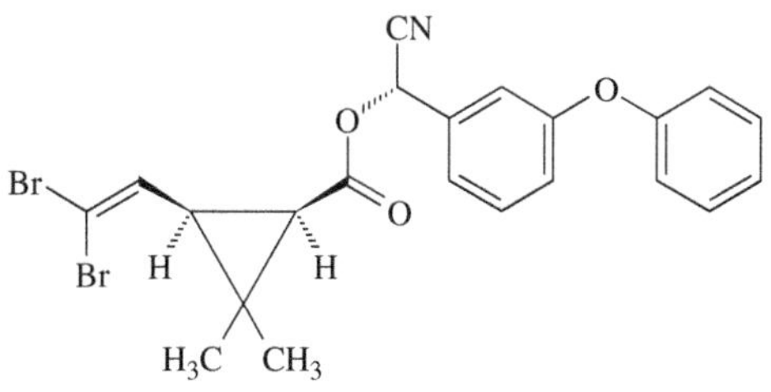

deltamethrin, (S)-alcohol (1R)-cis-acid

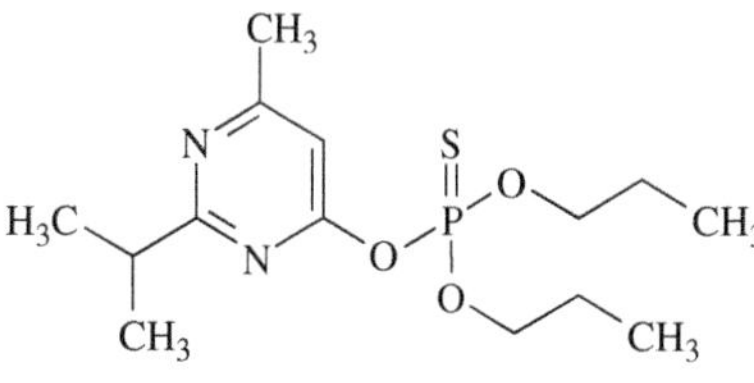

diazinone

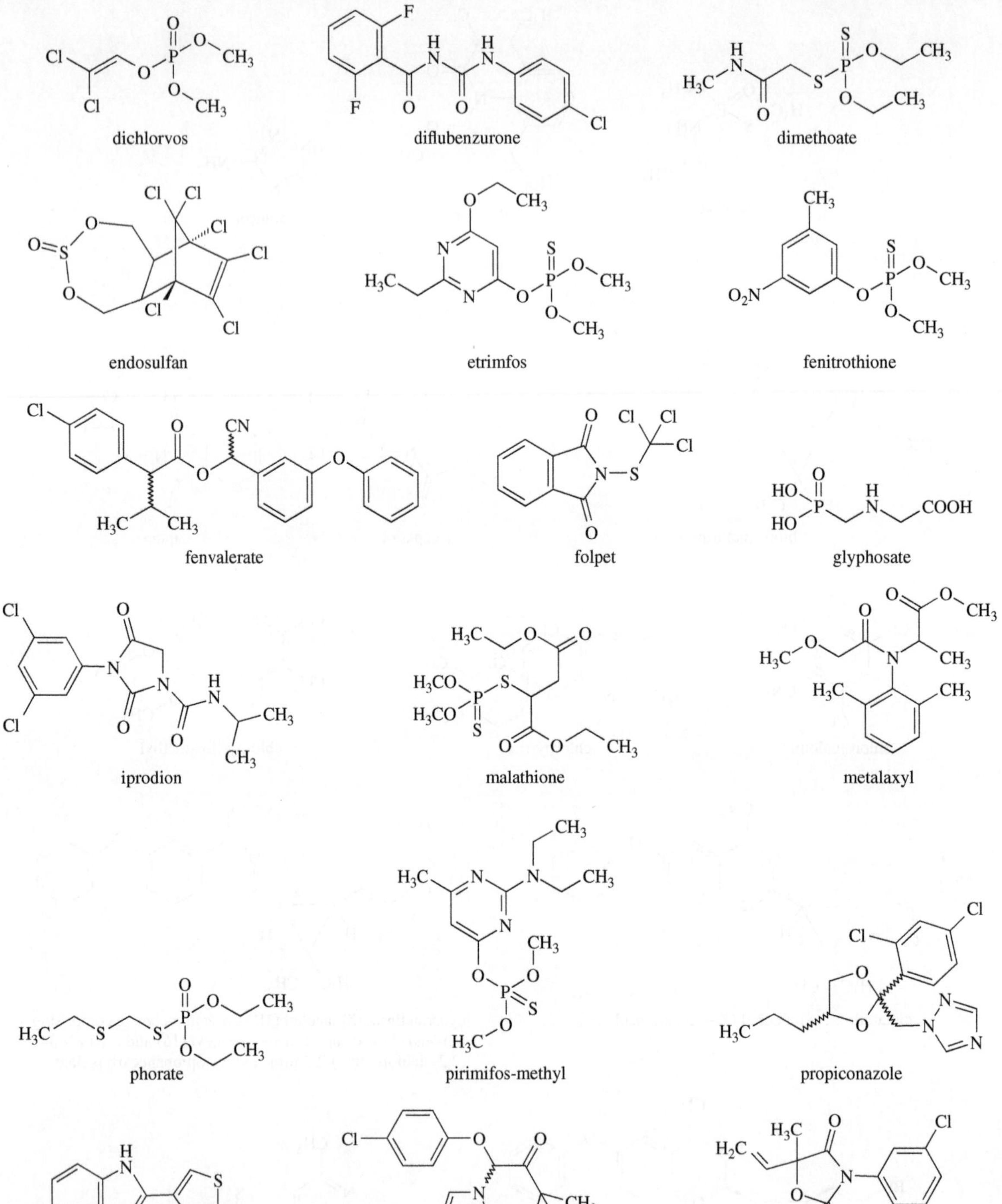

dichlorvos   diflubenzurone   dimethoate

endosulfan   etrimfos   fenitrothione

fenvalerate   folpet   glyphosate

iprodion   malathione   metalaxyl

phorate   pirimifos-methyl   propiconazole

thiabendazole   triadimefon   vinclozolin

12-49, structures of selected modern pesticides

The basis of insecticidal effects typical of classical organochlorine compounds, such as DDT, and pyrethrins, organophosphates, and carbamates is their neurotoxicity, but the mechanism of action is not identical for all pesticides. In insects, DDT and pyrethrins act on the voltage-gated sodium channel by prolonging the inactivation current. Other chlorinated hydrocarbons, including lindane and octachlorobornane toxicant A-2 (the principal active ingredients of hexachlorocyclohexane and toxaphene), as well as the cyclodienes, are neurotoxicants with a different target, which has finally been defined as the heteropentameric γ-aminobutyric acid (GABA)-gated chloride channel. Organophosphates and carbamates, such as carbaryl, ethephon, dichlorvos, dimethoate, and acephate, inhibit the enzyme acetylcholine esterase (act as serine hydrolase inhibitors) via phosphorylation or carbamoylation of the bound serine hydroxyl group in the active centre of the enzyme (acetylcholine is the only neurotransmitter employed in the motor division of the somatic nervous system and the principal neurotransmitter in all autonomic ganglia). Insecticidal benzoylureas, by contrast, are inhibitors of chitin synthesis, and thus inhibit the construction of insect cuticles.

The mechanism of action of fungicides is actually heterogeneous, as the attack targets may be the fungal hyphae and their spores. For example, the fungitoxic effect of widely used ethylenebisdithiocarbamates (EBDC) and phthalimides is based, amongst other things, on the inhibition of enzymes containing sulfhydryl groups. The toxicity of another important group of fungicides, the benzimidazoles, is based on their interference with the biosynthesis of DNA (its replication during cell division).

Herbicidal effects are based on an extraordinarily diverse range of mechanisms. For example, phenoxyalkanoic acids and derivatives of benzoic acids cause an atypical growth of sensitive plants (targeted weeds), which is based on disorders of nucleic acid metabolism caused by herbicide applied on the leaves, similar to the effects of plant hormones or plant growth regulators termed auxins. Total (non-selective) herbicides from the group of quaternary (bipyridylium) ammonium compounds interfere with the process of photosynthesis, mainly due to irreversible changes in cell membranes, in which they catalyse the oxidation of fatty acids. The effect of nitroaniline is based on retardation of weed seed germination, inhibition of root development, and cessation of mitosis. Herbicides, such as triazines, phenylureas, and uracils, inhibit electron transport during photosynthesis (the Hill reaction in chloroplasts). The Hill reaction is formally defined as the reduction of an electron acceptor by electrons and protons from water, with the evolution of oxygen, when chloroplasts are exposed to light.

12.5.1.2.2 Occurrence, main sources, and dietary intake

With the exception of certain fumigants, pesticides are rarely used as pure compounds. In addition to the active ingredients, commercial preparations (formulations) contain various other components (solvents, emulsifiers, and adhesives) that facilitate or improve their storage, handling, and biological effects in the period after application. The treatment of crops (or livestock) is achieved by pesticide formulations in the form of sprays, aerosols, powders, or granules. Unlike other groups of environmental contaminants, the entry of pesticides into the environment should be under controlled conditions and in accordance with the principles of good agricultural practice. The conditions under which pesticides are used must ensure effective and reliable pest control, but must also ensure that pesticide residues in treated products are kept to a minimum. Pesticides are often applied to the leaf surface, but the target organism is not necessarily the plant itself, but rather parasitic fungi or insects. Pesticides with **systemic effects** penetrate through the leaf cuticle and are translocated within the plant. The mobility of substances with **quasi-systemic effects** (some fungicides and insecticides) is lower, and only a limited amount of the active agent penetrates the cuticle. Finally, **contact pesticides** exhibit a local effect only in places where their surface deposits are found. Pesticides may also be applied directly to the soil or else get there during crop treatment; this accounts for about 35–50% of the pesticide used, depending on the type of plant and density of vegetation. Systemic pesticides in soil are taken up by the roots and transported to the aerial parts of the plant. The availability of the pesticide from the soil is mainly determined by its solubility and the organic carbon content of the soil.

Application of pesticide formulations, in particular in the form of sprays and powders, can lead to significant contamination of the atmosphere. On average, 10–20% of the pesticide used is thus in the form of vapour or droplets associated with the solids; this is transported by air flow to more or less remote locations, which may cause the pollution of agroecosystems. Pesticides with higher vapour pressure may be re-evaporated from the terrestrial or aquatic environment. Long-range transport of residues is particularly relevant in the case of persistent organochlorine compounds.

After application, pesticides are affected by several chemical, physical, and biological factors. Insecticides, herbicides, and fungicides to which harvested crops were exposed during the growing season have pre-harvest withdrawal times (the minimum interval between the last application and harvest), and following label directions will ensure that residues of these pesticides either do not remain on the plant material or remain there in amounts that do not exceed the MRLs (Maximum Residue Limits) when crops are harvested, eaten, or fed to livestock.

The operations used in the processing of farm crops or domestic culinary procedures usually have a significant impact on the residue levels in the final products. Generally, the following can occur:

- a decrease in the level of pesticide residue due to pesticide degradation, volatilisation, or selection of the edible portion with lower pesticide content;

- concentration of the pesticide residue in the edible portion due to the uneven distribution of residues in the commodity or higher affinity of the pesticide to the edible portion;

- formation of toxic degradation products from relatively non-toxic precursors.

Drying Under certain circumstances, drying may instigate an increase in pesticide concentration caused by an increase in the dry matter content. The level of volatile pesticides, on the other hand, decreases in the surface of the product. Drying in the sun can also cause photolysis of some pollutants. Freeze-drying does not usually result in lower levels of pesticide residues.

Removal of surface layers Removal of surface layers may, in crops such as bananas, pineapples, kiwi, citruses, and melons, result in a marked reduction in the levels of non-systemic pesticide residues applied directly to the crop before or after harvest. A similar effect can be achieved by removing the surface leaves of some vegetables (such as cabbage) and the surface layers of potatoes or cereals. For example, decontamination of potatoes grown in soil containing phorate was accomplished by peeling, which reduced the residue levels by 30%. Removing the outer husks of the rice grain treated with insecticide based on pirimiphos-methyl and, especially, removing the bran may lower the pesticide residue level to 70–90% of the original content. In the case of systemic pesticides, however, this process does not lead to product decontamination. Feeding livestock contaminated waste often leads to inadvertent infiltration of pesticide residues into the human food chain, especially in the case of lipophilic pesticides, which may bioaccumulate.

Washing and blanching Most of the technological processes used in fruit and vegetable processing include washing and blanching. This can decrease the levels of pesticide residues particularly in the case of pesticides that are soluble in water or of contact pesticides deposited on the plant surface, which can be mechanically removed. The overall effects resulting in a reduction of pesticide residues depend on a number of factors, the most important of which is localisation of residues.

Boiling Hydrothermal processes lead to a large reduction of residue levels of pesticides, which is associated not only with volatile substances that vaporise with water vapour but also with elevated temperatures that accelerate the hydrolysis of pesticides. For example, the loss of malathione can reach almost 100% in cooked spinach and 92% in cooked rice, whilst cooking tomatoes contaminated with carbaryl can decrease the pesticide concentration to 31%, and cooking meat contaminated with organophosphate insecticides similarly has a decontaminating effect.

The changes in concentration of some groups of pesticide residues during heat treatment are not significant. For example, the decrease in concentrations of synthetic pyrethroids during cooking, baking, or frying of foods of plant origin is negligible. Similarly, the decrease of thiabendazole concentration during the baking of potatoes is also negligible.

Milling The active ingredients of pesticide products used in the treatment of grains in the post-harvest period are found mainly in the surface layers of the grains. The levels of residues present in bran are therefore high. Concentration factors of frequently used organophosphorus insecticides and synthetic pyrethroids thus often reach values in the range of 2–4. For products made from wholemeal flour, however, the risk of residues is higher than for those made from white flour (Table 12.35). The husking process, involving removing the rice hull from the rough rice, removes up to 90% of pesticide residues.

Vegetable oils and animal fats The treatment of oilseed plants during the growing period leaves only very low or undetectable pesticide residues, but the risk of contamination is significantly higher in the case of post-harvest application of pesticides. In particular, relatively lipophilic pesticide residues pass to the crude oil during pressing or extraction. Concentration factors >1 were found primarily for the group of synthetic pyrethroids (such as cyhalothrin, cypermethrin, fenvalerate, and permethrin) and some organophosphates (diazinone, dichlorvos, etrimfos, malathione, parathione, pirimifos-methyl, and chlorpyrifos). Pesticide residues can be found especially in virgin oils, but only sporadically in refined oils (mostly traces of pyrethroids). A significant reduction of pesticide concentrations occurs in oil refining, especially during deodourising (Table 12.36).

Table 12.35 Influence of wheat processing on pesticide contents in cereal products.

Pesticide	Pesticide content (%)[a]				Concentration factor
	Wholemeal flour	White flour	Whole-grain bread	White bread	Bran/Wheat
Bioresmethrin[b]	5	64	43	80	4.0
Deltamethrin[c]	29	91	69	94	3.3
Diflubenzurone	31	83	22	66	2.2
Glyphosate	46	55	64	93	2.2
Permethrin	9	65	68	84	3.2

[a]Content in wheat = 100%.
[b]Bioresmethrin occurs in resmethrin and makes up 35–40% of it.
[c]Unlike other pyrethroids, deltamethrin consists of one pure compound.

Table 12.36 Distribution of and changes in selected pesticides in the production of vegetable oils.

Pesticide	Raw material	Concentration factor raw oil/seeds	Loss of pesticide during deodourisation (%)
Dichlorvos	Soybeans	5.3	100
Etrimfos	Rapeseeds	2.7	98
Chlorpyrifos	Soybeans	4.1	100
Malathione	Soybeans	3.9	100
Parathione	Olives	4.5	–
Pirimifos-methyl	Peanuts	1.7	–
Permethrin	Sunflower seeds	0.7	17

Table 12.37 Changes in levels of selected pesticides in wine production (%).

Pesticide[a]	Must	Clarified must	Wine	Pesticide[a]	Must	Clarified must	Pesticide
Benomyl	0	0	0.8	Iprodion	45–70	60–80	70–78
Captafol	50	95	100	Metalaxyl	0	30–50	66
Folpet	50	95	100	Vinclozolin	59–88	80	89–93

[a]Content in grapes = 100%.

Fruit juices The amount of pesticides passing from raw materials into fruit juices depends on the distribution of residues between the solid (skin or flesh) and liquid fruit portions. Moderately to highly lipophilic pesticides, such as parathion, captan, folpet, and synthetic pyrethroids, usually pass to the raw juices only to a very limited extent. A further decrease of their contents may occur during operations such as clarification and ultrafiltration. The risk of infiltration of residues into the human food chain, however, exists when the solid waste with higher concentrations of pesticides is used as an animal feed.

Alcoholic beverages The transition of residues of pesticides used to protect grapes depends on their solubility in water and the rate of their degradation, particularly hydrolysis. For example, a total transfer of polar benomyl residues to the final products can be expected. Conversely, a lower transition of frequently used fungicides such as folpet, captafol, propiconazole, triadimefon, and vinclozolin from grapes to wine can be expected, mainly due to sorption of these compounds to the solid waste (Table 12.37). In beer, pesticide residues are found only exceptionally, due not only to dilution, but also to degradation during technological operations and sorption to yeast cells.

Residues in foods The presence of pesticide residues in the human diet is mainly due to their targeted pre- and post-harvest applications. The source of contamination of agricultural crops by some more persistent compounds may be atmospheric transport from distant places and translocation of residues from soil contaminated in previous growing seasons; various agrochemicals may also contain water from rivers and reservoirs. Foods of plant origin contain mainly residues of modern pesticides or their metabolites. Foods of animal origin may become a source of persistent organochlorine pesticides that accumulate in fat tissues. Residues of modern pesticides occur rather rarely and their presence can be attributed, for example, to contamination of feeds or to pesticides used in protecting animals. By way of illustration, Table 12.38 brings together data obtained from reports on pesticide monitoring published in various countries. Only pesticides found in more than 10% of cases during the investigation of a given commodity are summarised. In particular, pesticide residues are frequently found in cereals, where all of them are used in post-harvest applications. Particularly frequent are findings of residues in fruits and vegetables with a waxy surface. The most commonly detected pesticides include acephate, benomyl/carbendazim, carbaryl, chlorothalonil, chlorpyrifos, cypermethrin, deltamethrin, dicofol, dimethoate, dithiocarbamates, endosulfan, fenitrothione, fenvalerate, malathione, parathione, permethrin, and vinclozolin. Only a small proportion of residues exceed the values set by the national health limits. The upper limit of incidence of residues above the limit stated is 5% in most sources, but mutual comparability of data is very difficult for many reasons.

12.5.1.2.3 Metabolism and degradation

Biological factors involved in the degradation of pesticides in soil and aqueous environments include various bacteria, fungi, and actinomycetes. In principle, there are two types of degradative processes:

Table 12.38 Combination pesticide/commodity with a frequent incidence of positive residue findings.

Pesticide	Commodity	Number of analysed samples	Samples with pesticide resudues (%)
Carbaryl	Barley	188	17.6
Carbaryl	Oat flakes	130	33.9
Carbaryl	Sorghum	156	96.8
Chlorpyrifos	Kiwi	127	20.5
Chlorpyrifos	Tomatoes	3613	27.8
Chlorpyrifos	Bell peppers	1428	27.7
Chlorthalonil	Celery leaves	375	65.6
Dicofol	Citrus fruits	1022	14.5
Dimethoate	Green peas	837	11.5
Dithiocarbamates	Apples	135	19.9
Dithiocarbamates	Grapes	285	17.9
Dithiocarbamates	Tomatoes	866	27.8
Dithiocarbamates	Cucumbers	613	11.9
Endosulfan	Spinach	183	21.9
Endosulfan	Melons	925	26.9
Endosulfan	Lettuce	2169	16.8
Endosulfan	Cucumbers	1659	19.8
Endosulfan	Green beans	574	11.7
Fenitrothione	Cereals	12759	74.5
Permethrin	Celery leaves	422	15.4
Permethrin	Lettuce	2451	18.3
Permethrin	Spinach	183	12.6
Vinclozolin	Strawberries	509	48.1
Vinclozolin	Kiwi	126	43.7

- co-metabolism, where the pesticide biotransformation proceeds by normal metabolic processes in a microorganism that is incapable of using the substrate as a source of energy or building material (the organism is actually growing on a second substrate and is transforming the pesticide without gaining the benefit);

- catabolism, where the pesticide becomes a substrate for a given microorganism (a source of carbon or nitrogen), adaptations can be found especially in bacteria during repeated exposure to xenobiotics.

Biodegradation of pesticides, however, also takes place in exposed plants and animals, whether they are target or non-target organisms, and the formed products can reside in the human food chain. In particular, vertebrates (especially birds and mammals) have active enzyme systems capable of effective degradation of xenobiotics. Activation of protective mechanisms occurs when the harmful substance penetrates into the body. The penetration rate of pollutants and their distribution in the body, however, as well as the range of detoxification or elimination mechanisms, is subject to anatomical, physiological, biochemical, and other factors. Phase I biotransformation typically involves changes catalysed by hydrolases and oxidases, in which polar functional groups are introduced to the molecule of the parent compounds, or are formed from the parent molecule by degradation. For the example of organophosphates, Figure 12.46 shows the places of enzymatic attack in the initial phase of the biotransformation. The resulting primary metabolites often additionally enter into secondary reactions of phase II, which leads to their conjugation with small polar endogenous molecules to yield products that can be easily excluded from the organism. The type of resulting secondary metabolites is characteristic of each species. For example, in plants and invertebrates, conjugation with glucose predominates. In plants, the waste products are apparently stored in lignin structures. In mammals, birds, and certain fish, conjugation with glucuronic acid and reduced glutathione (GSH) primarily takes place, whilst in other fish, conjugation with glycine occurs too. The metabolites in animals are transported by blood and excreted.

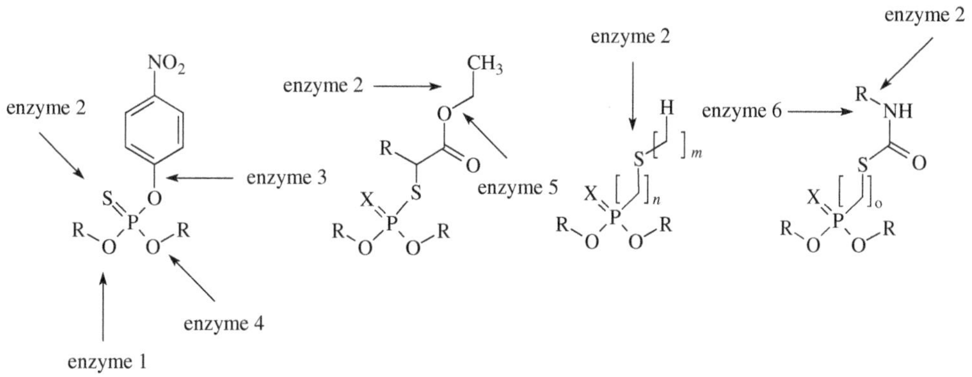

Figure 12.46 Enzymes involved in the biotransformation of organophosphorus compounds.

Physical factors applied during the degradation of pesticides are mostly the energy from light and heat. In particular, photolysis of residues occurring on the surface of leaves, in soil, in the atmosphere, or in the aquatic environment is one of the most important processes leading to elimination of a number of pesticides from the environment. Along with singlet oxygen, reactive hydroxyl and superoxide radicals and other free radicals generated by photochemical reactions play an important role. Direct photolysis by solar radiation is involved only in part of the photodegradation processes. An example of these reactions is photolysis of aldrin (Figure 12.45).

An important reaction to which pesticides are subjected in the environment is hydrolysis. An example is the hydrolysis of the synthetic pyrethroid insecticide permethrin to isomers of dichlorovinyl derivatives of chrysanthemic acid and 3-phenoxybenzyl alcohol (Figure 12.47). A relatively rapid hydrolytic cleavage of ester bonds also occurs in many organophosphates. Examples of pesticides that are easily degraded by heating are N-methyl carbamates. For example, carbaryl is degraded to naphthalen-1-ol (1-naphthol) and methylisocyanate (Figure 12.48). Decomposition of fungicide benomyl analogously yields butylisocyanate and carbendazim, which have similar effects to benomyl. Carbendazim is slowly hydrolysed and decarboxylated to 2-aminobenzimidazole. Decarboxylation of butylcarbamic acid produces butylamine as a minor product, which reacts with butylisocyanate to yield N,N'-dibutylurea (Figure 12.49).

The complexity of the degradation processes that take place in the biotic environment is demonstrated by a widely used herbicide, atrazine (Figure 12.50). Substances of this group (symmetrical 1,3,5-chlorotriazines known as s-chlorotriazines) may be hydrolysed

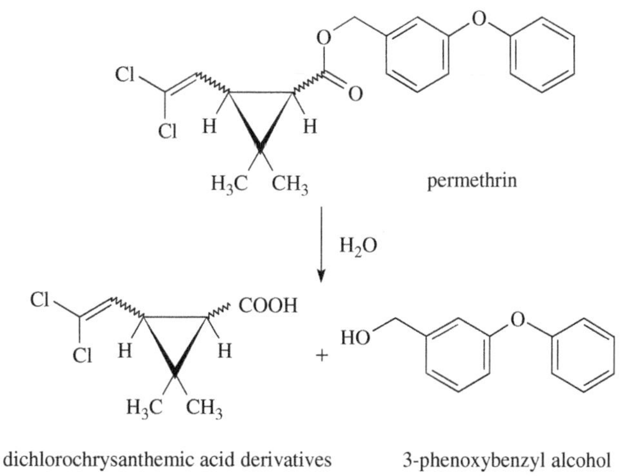

Figure 12.47 Hydrolysis of permethrin.

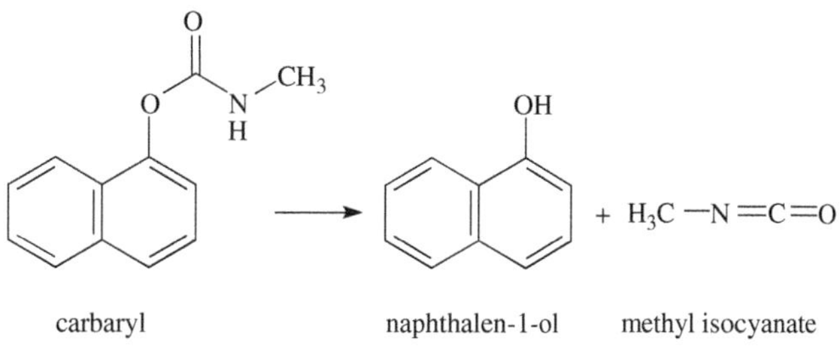

Figure 12.48 Thermal decomposition of carbaryl.

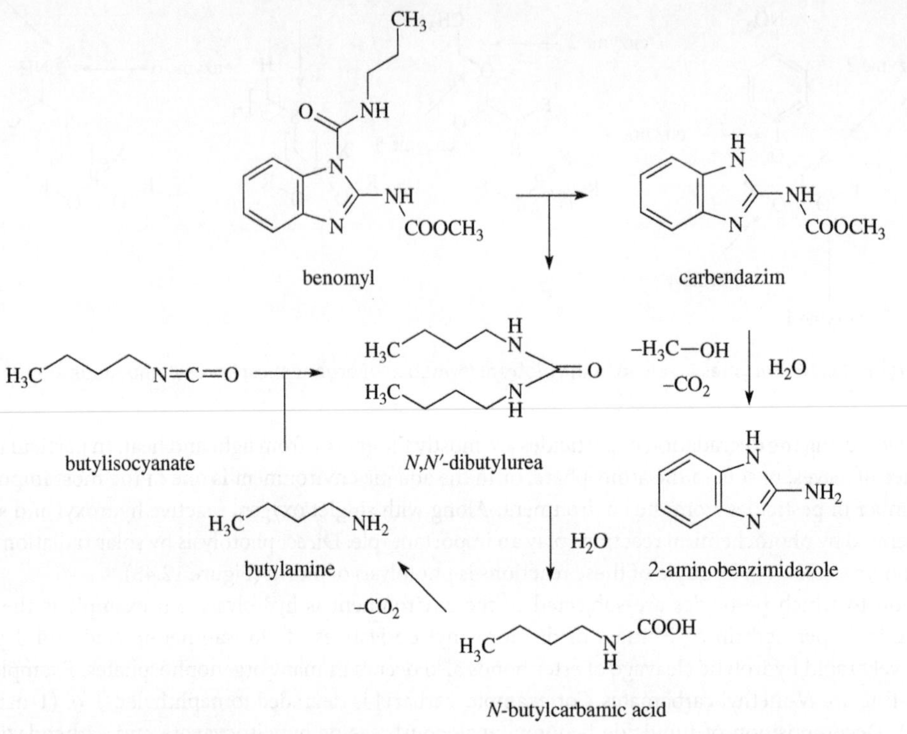

Figure 12.49 Decomposition of benomyl.

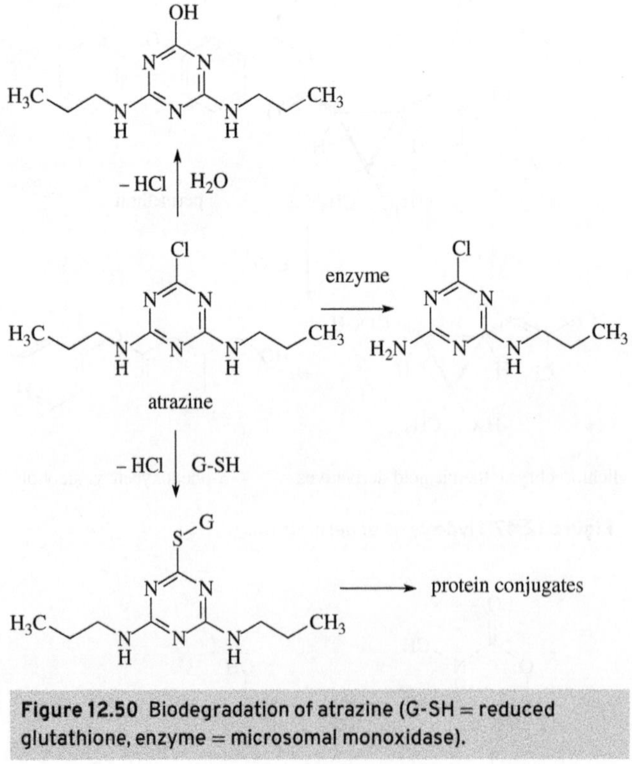

Figure 12.50 Biodegradation of atrazine (G-SH = reduced glutathione, enzyme = microsomal monoxidase).

non-enzymatically to hydroxytriazines, which no longer exhibit herbicidal activity. Cleavage of a chlorine atom from the carbon C-2 is catalysed by the naturally occurring benzoxazinoid hydroxamic acids, such as (R)-2,4-dihydroxy-7-methoxy-2H-1,4-benzoxazin-3(4H)-one, which is known under the acronym DIMBOA (see Section 10.3.12.8). Together with other derivatives, glucoside of this benzoxazinoid hydroxamic acid is present in plants tolerant to triazines, such as maize, wheat, rye, and other monocotyledonous plants. Alternative detoxification of atrazine is an enzymatically catalysed conjugation with glutathione. Resulting products are also inactive as herbicides. As is evident from Figure 12.50, another possibility may be atrazine dealkylation on the secondary amino group at carbons C-4 and C-6. Dealkylated metabolites partly retain the herbicidal activity of the parent compound.

Figure 12.51 provides another example of the diversity of metabolic processes to which pesticides are subjected after application. In the case of the carbamate insecticide carbaryl, the predominant processes are oxidation and hydrolysis, which may be followed by conjugation of primary metabolites with glutathione.

Transformation processes generally have a detoxifying nature, but can also lead to the formation of products that show more or less similar toxic effects on the target harmful organisms and on humans. Examples include the formation of dicofol from DDT (Figure 12.43), carbendazim from benomyl (Figure 12.49), and aldoxycarb from aldicarb (Figure 12.52). In some cases, the formed substances can even be significantly more toxic than the parent compound. For example, the *in vivo* activation (desulfuration) of the insecticide parathione leads to the formation of paraoxone (Figure 12.53), the active form of the insecticide and a very potent inhibitor of acetylcholinesterase. Atypical pro-insecticide is carbosulfan, which is hydrolysed in the insect body to biologically active carbofuran (Figure 12.54). In mammals,

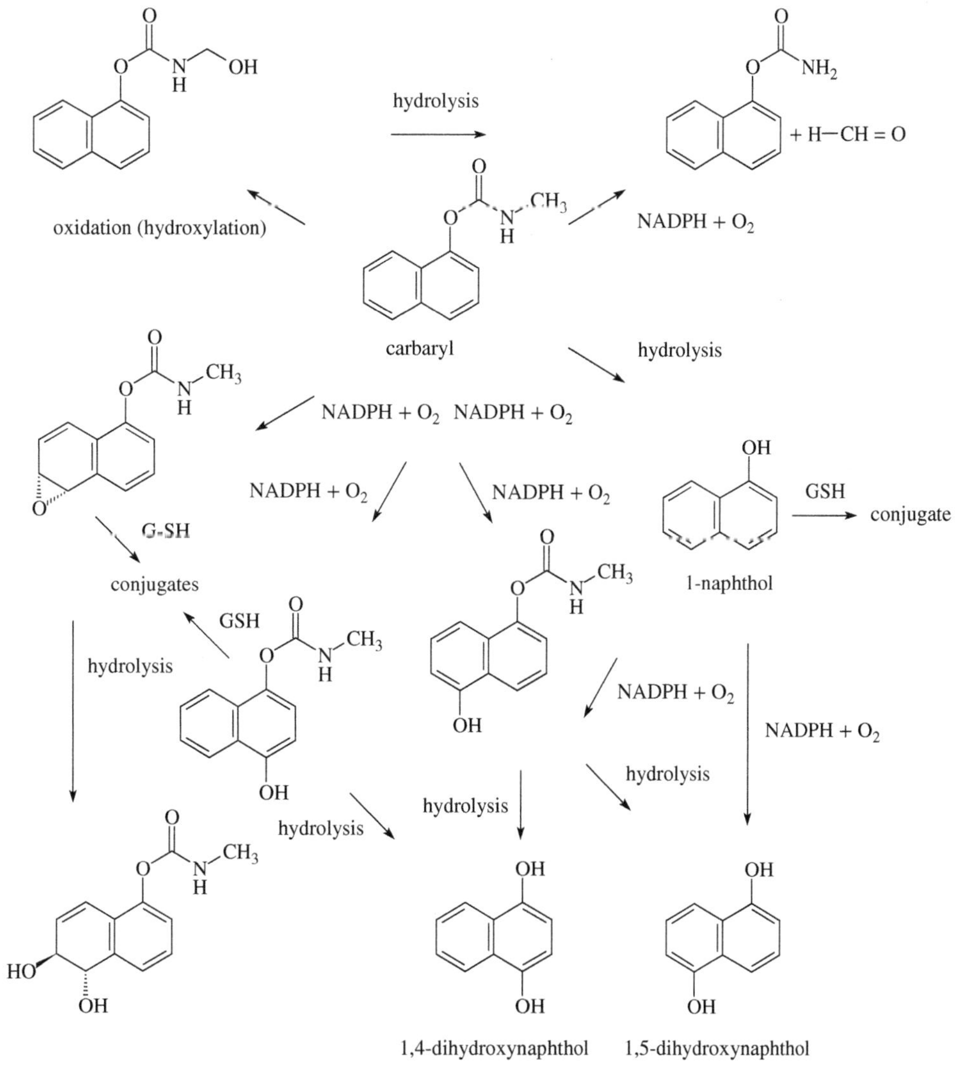

Figure 12.51 Biodegradation of carbaryl (G-SH = reduced glutathione).

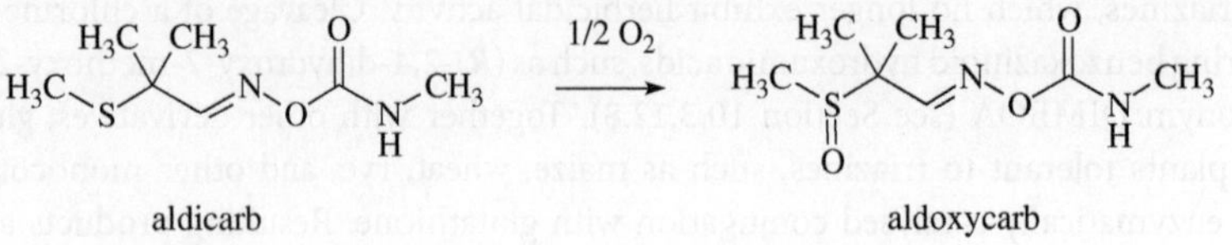

Figure 12.52 Transformation of aldicarb to aldoxycarb.

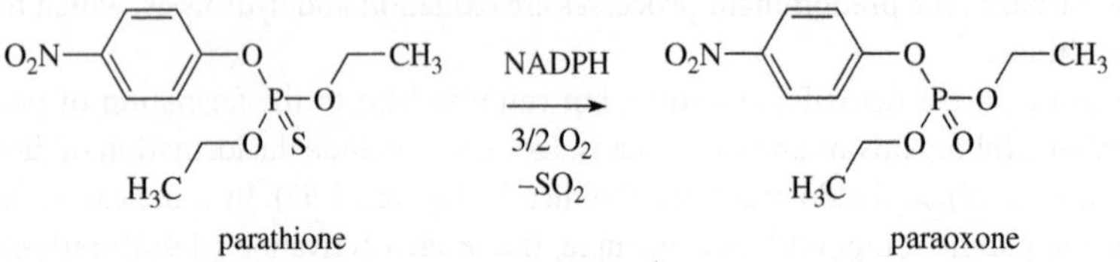

Figure 12.53 Transformation of parathione to paraoxone.

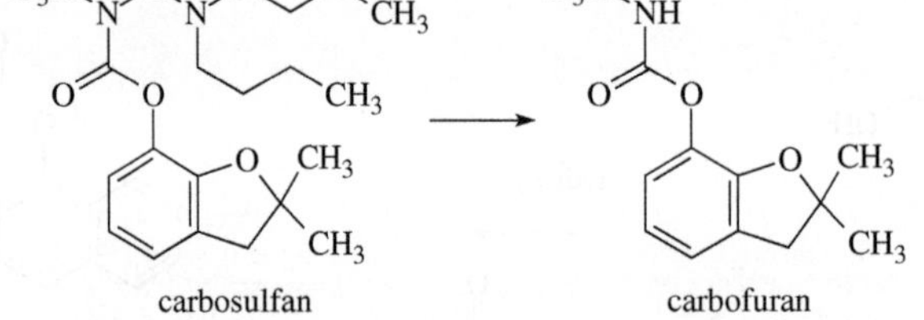

Figure 12.54 Transformation of carbosulfan to carbofuran.

this reaction does not take place (the case of a highly selective compound), which explains the very low acute toxicity of carbosulfan for mammalian organisms.

Toxic compounds may likewise be produced from relatively non-toxic precursors during storage or handling of contaminated materials. Probably the most significant example is the formation of ethylenethiourea (ETU), which is one of many degradation products of extensively used fungicides known as ethylenebisdithiocarbamates (EBDCs, Figure 12.55). According to the IARC, ETU is not classifiable as to its carcinogenicity to humans. In the late 1980s, special attention was devoted to the formation of the *N,N*-dimethylhydrazine from the popular growth regulator daminozide used for fruit ripening, which is a potential carcinogen of Group 2B (Figure 12.56). Phenoxyalkanoic acids can yield relatively persistent chlorinated phenols as decomposition products, which can accumulate in the human food chain. Chlorinated phenols also arise from other precursors, for example from the pesticides lindane and pentachlorophenol.

12.5.1.2.4 Health and toxicological evaluation

Pesticides can be naturally toxic to non-target organisms, including humans, in particular if the effect of a substance is based on interference with similar processes taking place in the human organism. In terms of the possibility of pesticide poisoning, the most discussed risks are associated with long-term exposure to residues obtained in the diet. Generally, the level of these risks, as well as other high-risk groups of food components, is derived not only from the dose received, but also from the mechanism of absorption, distribution, metabolism, and excretion of pesticides from the body. A general methodology of pesticide toxicity classification was issued by the WHO.

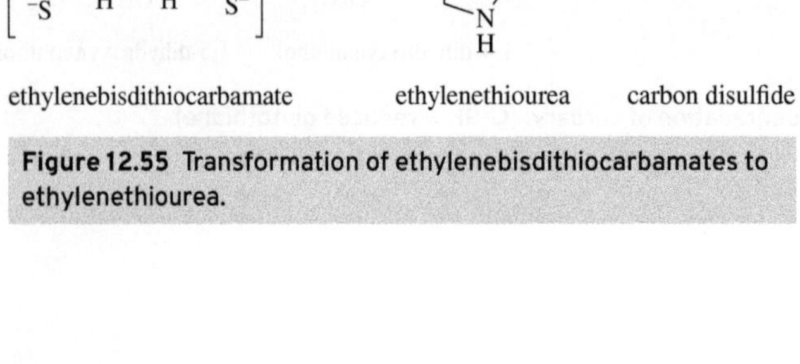

Figure 12.55 Transformation of ethylenebisdithiocarbamates to ethylenethiourea.

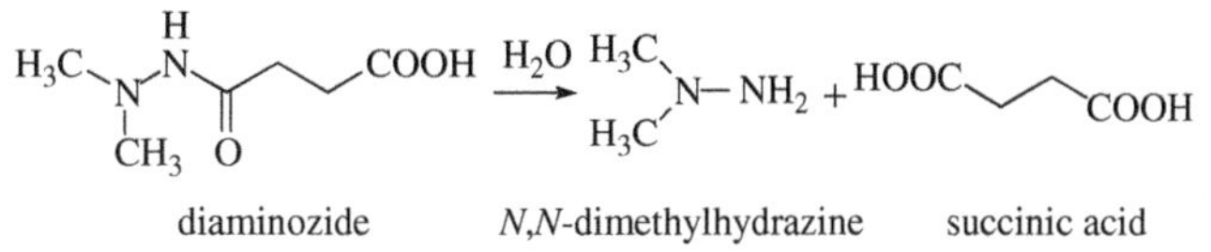

Figure 12.56 Transformation of daminozide into N,N-dimethylhydrazine.

The mechanism of effects toxic to mammals is accurately described for only a few groups of pesticides. As already mentioned, the toxicity of organophosphates and carbamates is due to inhibition of acetylcholinesterase; the toxic effects of dinitrophenols and polychlorinated phenols lie in the inhibition of oxidative phosphorylation. The increased toxicity of certain pesticides may proceed *in vivo* by mechanisms that have been previously listed. This mainly concerns the formation of oxoanalogues of organophosphates, which are more potent acetylcholinesterase inhibitors than the parent thionophosphates and thiophosphates. Pesticides that accumulate in human fat (DDT, lindane, and others), although not easily metabolised, may be mobilised under certain conditions (e.g. during starvation or breast feeding), which may lead to a significant increase in their concentration in blood serum with the manifestation of toxic effects.

None of the pesticides used today are classified by the IARC as a human carcinogen, yet some of them (particularly the classic pesticides DDT and toxaphene and modern pesticides amitrole and phenoxyalkanoic acids) are classified as potential human carcinogens. In recent years, in the context of assessing the exposure risks, the ability of certain pesticides to interfere with the hormonal processes of vertebrates, including humans, and to cause many side effects, is often discussed. Pesticides that show oestrogenic effects as xenooestrogens (anthropogenic oestrogens) include only some organochlorinated pesticides with high accumulation potential, such as toxaphene, dieldrin, heptachlor, DDT, and derived compounds, especially DDE, but also modern pesticides, represented by the herbicides 2,4-D, atrazine, and alachlor, the fungicides benomyl and ethylenebisdithiocarbamates, and the insecticides carbaryl, dicofol, permethrin, and parathione, amongst others.

12.5.1.2.5 Mitigation

Legislative measures concerning the presence of pesticide residues in food are generally set at the EU level. Limits introduced in the past in national regulations were harmonised with the relevant EU limits. The MRLs of pesticides in foods and crops and rules for their application are currently set at two different levels, global (Codex Alimentarius) and European, and in each case the limits for a large number of pesticide–commodity combinations are always given. Based on Regulation (EC) No. 396/2005 from September 2008, all responsibility for assessing the risk associated with the MRLs of pesticides lies with the EFSA, which is responsible both for assessing the risks of new MRLs and for periodically reviewing existing ones. Currently, the EFSA coordinates revision of MRLs for more than 300 substances. In the case of previously used persistent pesticides that are still found in foods due to contamination of the ecosystem, so-called extraneous residue limits (ERLs) are set. Unlike with other groups of contaminants, inputs of which into the human food chain are difficult to completely eliminate, in the case of pesticides alternative ways of protecting crops from harmful agents in order to minimise pesticide residues are increasingly being sought. In particular, emphasis is put on integrated protection incorporating complementary practices, including breeding of resistant plants and use of biological pesticides.

For foods listed in Annex I to Regulation (EC) No. 396/2005 (315 foods, including fruits, vegetables, cereals, spices, and foods of animal origin), if the relevant MRL (Maximum Residue Limit or Level) value for a particular pesticide is not declared in Annex II (list of MRL values) or III (temporary EU limits), then the default limit is 0.01 mg/kg. This of course does not apply to pesticides referred to in Annex IV (52 pesticides for which MRLs are not declared, referring to their low risk) or Annex V (pesticides with other limits than 0.01 mg/kg). Annexes to the Regulation (EC) No. 396/2005 are gradually amended by other extensive regulations. Owing to the vast number of MRLs given by combinations of several hundred pesticides with many types of food, and due also to the need for constant updating of the pronounced values, the legislation concerning pesticides is very confusing and difficult to use in practice.[2] Carcinogenic and mutagenic substances, or substances acting as endocrine disrupters, are banned. However, they may obtain exemption if there is no substitute and they are proved to be necessary for pest control. Immunotoxic and neurotoxic pesticides are included in the list of candidates that will be replaced with safer alternatives where such alternatives exist.

---

[2] Current MRLs can be found at https://ec.europa.eu/food/plant/pesticides/max_residue_levels_en.

12.6 Veterinary drugs

The production of high-quality and hygienically acceptable foods of animal origin in agriculture implies ensuring the overall health and welfare of livestock through good nutrition and reduction of stress. Livestock production is therefore not possible without the use of veterinary drugs. Veterinary drugs (pharmaceuticals) are pharmacologically and biologically active chemicals used for the treatment, prevention, and diagnosis of diseases in animals. With ever-increasing demands on productivity, growth in livestock production is manifested by the high concentration of animals in factory farms and their frequent geographic relocation, which promotes the spread of infectious diseases. As a result, the use of veterinary drugs in animal production has greatly increased over the last few decades. In pursuit of high productivity, inadequate sanitary conditions were in many cases replaced by increased use of pharmaceuticals, with the aim of preventing deterioration of the health of farm animals. An extreme way of increasing productivity can be the illegal use of pharmacologically active compounds as growth promoters, such as anabolic steroids, thyreostatics, and antibiotics in low doses.

12.6.1 Classification, structure, and properties

The veterinary products for a variety of animals, including pets, livestock, and numerous exotic species, comprise several hundreds of pharmaceutical, biological, diagnostic, feed medications, and parasiticide products. The most widely used groups of veterinary drugs include antimicrobial substances and antiparasitic agents. Also often used as veterinary drugs are anti-inflammatory agents called antiflogistics, drugs acting on the nervous system including anaesthetising agents or anaesthetics and sedative agents or tranquilisers, drugs acting on the kidney known as diuretics, preparations acting on the digestive tract (e.g. against diarrhoea) known as antidiarrhoics, hormones, vitamins, trace mineral elements, and others.

In the European Union, approximately 700 pharmacologically or biologically active chemical compounds are currently either lodged for registration or registered for use in the member states. Examples of the most important groups of pharmacologically active substances used in veterinary medicine are given in Tables 12.39 and 12.40.

Table 12.39 List of frequently used veterinary antimicrobial drugs.

Pharmaceutical	Name	Pharmaceutical	Name
Antibiotics		Macrolide	Tylosin
β-Lactam			Spiramycin (selectomycin)[a]
Penicillins	Benzylpenicillin (penicillin G)		Erythromycin[b]
	Ampicillin	Chloramphenicol group	Chloramphenicol
	Amoxicillin		Thiamphenicol
	Cloxacillin	Lincosamides	Lincomycin
Cephalosporins	Ceftiofur	*Sulfonamides*	Sulfadiazine
	Cephaloridine		
Tetracycline	Tetracyclin		Sulfadimidine
	Chlortetracycline		Sulfadoxine
	Oxytetracycline		Sulfadimethoxine (dinosol)
Aminoglycoside	Dihydrostreptomycin		Trimethoprim
	Gentamycin		
	Kanamycin[c]	*Quinolones*	Enrofloxacin

[a]Mixture of spiramycin I, II, and III.

[b]Erythromycin A is the main metabolite; other components are erythromycin B (without an OH group in the C-12 position of aglycone, called erythronolide) and erythromycin C (contains instead of cladinose (2,6-dideoxy-3-C-methyl-3-O-methyl-L-*ribo*-hexopyranose also known as 3-O-methylmycarose) saccharide, known as mycarose).

[c]Pharmaceuticals contain 95% of kanamycin A and 5% of kanamycin B; kanamycin C is produced from kanamycin B.

Table 12.40 Other important groups of veterinary drugs.

Pharmaceuticals	Name[a]
Antiparasitic agents[b]	
Coccidiostats[c]	Monensin, dicarbazim, narazin
Anthelmintics[d]	
Benzimidazoles[e]	Thiabendazole, albendazole
Probenzimidazoles[e]	Netobimin, febantel
Macrolide endectocides[f]	Ivermectin,[g] doramectin
Pyrethroids[h]	Permethrin
Organophosphates[h]	Foxim
Other agents	
Non-steroid anti-inflammatory agents (antiflogistics)	Carprofen, ketoprofen
Analgetics, antipyretics[i]	Salicylic acid
Substances effecting the autonomic nervous system[j]	Clenbuterol
Substances effecting the central nervous system	Pentobarbital Amphetamine[k]
Sedatives[l]	
Analeptics[m]	

[a]Structures of pharmaceuticals can be found, for example, in the National Center for Biotechnology Information. PubChem Compound Database: https://pubchem.ncbi.nlm.nih.gov

[b]Drugs against external (ectoparasitica) and internal (endoparasitica) parasites.

[c]Drugs used primarily to protect chickens against coccidiosis (enteritis or cachexia).

[d]Drugs used in the treatment of helmintoses, infections caused by lower parasitic worms.

[e]Endoparasitica.

[f]Endoparasitica and ectoparasitica.

[g]Contains about 80% of dihydroavermectin B_{1a} (22,23-dihydroavermectin) and 20% of dihydroavermectin B_{1b}.

[h]Ectoparasitica (insecticides).

[i]Analgesics are drugs that decrease or inhibit the perception of pain of various origins; antipyretics reduce abnormally elevated body temperature but do not affect the normal body temperature (they often also have an analgesic effect).

[j]Bronchodilator effect.

[k]Mixture of enantiomers.

[l]Drugs causing a decrease of increased excitability of the central nervous system.

[m]Drugs with a stimulating effect on the central nervous system, acting on circulatory and respiratory centra in the medulla.

12.6.2 Occurrence, main sources, and dietary intake

The supply of veterinary drugs into the body of livestock necessarily leads to the occurrence of residues in the muscle, various organs, milk, and eggs, so trace amounts of veterinary drugs may be found in meat, milk, meat, dairy, and other products.

In terms of food quality, veterinary drugs are considered extraneous, contaminating substances whose presence is undesirable but often inevitable. In this context, these substances are classified in the same group as environmental contaminants, such as mycotoxins, pesticides, PCBs, and other POPs. Like other types of contaminants, residues of veterinary drugs in foods present a potential health risk to the consumer. Therefore, the incidences and levels of veterinary drug residues in tissues have to be effectively and consistently controlled.

Levels of veterinary drug residues are subject to pharmacokinetic parameters: absorption, tissue distribution, metabolism, and excretion. The degree of absorption depends on the physico-chemical properties of the compounds and the method of administration, which may, for example, be **peroral** (through the gastrointestinal tract, such as in capsules, food, or drinking water), injected, **intramuscular** (within a muscle), **subcutaneous** (under the skin), or **dermal** (on the skin surface).

Generally speaking, the absorption from the gastrointestinal tract depends on the solubility of the drug in water (degree of ionisation at a given pH) and lipophilicity, which determines the transition across biological membranes. Lipophilic compounds, such as certain antiparasitic drugs (organophosphates and pyrethroids), exhibit a relatively high degree of absorption after application to the skin. On the other hand, injection preparations in the form of an oil suspension can lead to up to several months of persistence of residues at the injection site, with the risk of occurrence of extreme levels of residues in food (particularly meat) prepared from the given tissue.

Accumulation of residues in the liver and kidneys generally exceeds levels in muscle. Some compounds, such as benzimidazoles, show the highest concentration in the liver, whilst the majority of antibiotics are located in the highest amounts in the kidney. Lipophilic chemical compounds, such as some organophosphates, accumulate in adipose tissue. The most important factors affecting the transfer of veterinary drugs from the blood plasma into milk (mammary gland cells) are degree of ionisation, lipophilicity, binding to proteins, and relative molecular weight.

At the lower pH of milk, the ratio between milk and plasma for weak organic bases (such as antibiotics erythromycin and tylosin) is >1, whilst for organic acids (benzylpenicillin, sulfadimethoxin) it is ≤1. During mammary gland tissue inflammation (mastitis), the pH of milk increases, and this ratio of the organic bases decreases.

12.6.3 Metabolism and degradation

The elimination of most veterinary drugs from the body depends on the metabolic rate and mechanism of the excretion process. Examples of differences in the rate of elimination of veterinary drugs from the body of selected animal species are presented in Table 12.41. Metabolism proceeds in two phases. The initial one involves oxidation, reduction, and hydrolysis, whilst the second leads to conjugation with endogenous components, for example with D-glucuronic acid and glutathione, or acetylation or other transformation of the drug. In most cases, the metabolic process yields water-soluble, easily eliminated products that lack pharmacological activity. A summary of the most common biotransformation processes is given in Table 12.42.

The major excretory organs are the kidneys, but veterinary drug metabolites can also be excreted to the bile and milk. Renal excretion is the primary elimination for drugs that are ionised at physiological pH and compounds with low lipophilicity. Drugs excreted unchanged include many antibiotics, such as penicillin, cephalosporins, aminoglycosides, oxytetracycline, and most diuretics. The process of renal excretion of drugs is very complex, so in addition to the parent compound, pharmacologically active metabolites must always be analysed. Individual MRLs are set according to the occurrence of residues in the body for the meat and various edible organs, such as the liver and kidneys.

Until recently, there were surprisingly few studies dealing with the impact of processed raw materials of animal origin on residues in consumed foods. Knowledge of this area is still too limited to allow any generally valid conclusions. For example, cooking and baking meat containing residues of the antiparasitic drug ivermectin lead to reduced levels of this drug, on average down to 50%. Similarly, oxytetracycline residue levels decrease by 35–94% depending on temperature and type of heat treatment. On the other hand, under standard cooking or baking procedures, no changes were found in the concentration of residues of clenbuterol, which is sometimes used illegally as a growth

Table 12.41 Half-life of elimination (hours) of veterinary drugs from the body of livestock.

Veterinary drug	Cattle	Horse	Pig
Hepatic metabolism			
Chloramphenicol	4.2	0.9	1.3
Salicylic acid	0.8	1.0	5.9
Pentobarbital	0.8	1.5	–
Amfetamine	0.6	1.4	1.1
Hepatic metabolism and renal excretion			
Sulfadoxine	11.7	14.0	8.2
Sulfadimethoxine	12.5	11.3	15.5
Trimethoprim	1.5	3.2	2.3
Renal excretion			
Benzylpenicillin	0.7	0.9	–
Ampicillin	1.2	1.6	–
Oxytetracycline	9.1	10.5	–
Kanamycin	1.9	1.5	–

Table 12.42 Methods of biotransformation of veterinary drugs.

Functional group/skeleton	Biotransformation
Aromatic ring	Hydroxylation
Hydroxyl group Alcohols	Side-chain oxidation, conjugation with glucuronic acid and sulfate (to a lesser extent)
Phenols	Ring hydroxylation, conjugation with glucuronic acid and sulfate, methylation
Carboxylic group	Conjugation, side-chain hydroxylation
Alifatic and aromatic substances	Conjugation with glucuronic acid and glycine
Primary amines	Deamination
Alifatic	Side-chain hydroxylation, acetylation
Aromatic	Conjugation with glucuronic acid and sulfate, methylation
Sulfhydryl group	Conjugation with glucuronic acid, methylation, oxidation
Ester and amide bonds	Hydrolysis

stimulator. Frying of beef liver containing the antiparasitic drug oxfendazole (also known as fenbendazole sulfone, **12-50**) leads to changes in the relative ratio of oxfendazole residues and the main metabolites, such as oxfendazole sulfone, fenbendazole, and some others. Oxfendazole amine (**12-50**) is produced as a product of hydrolysis of oxfendazole.

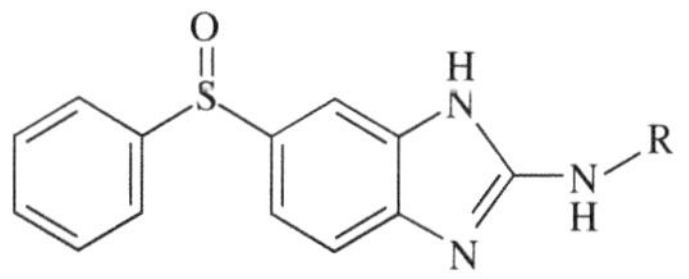

12-50, oxfendazole, R = COOCH$_3$
oxfendazole amine, R = H

12.6.4 Health and toxicological evaluation

In recent years, increasing attention has been devoted at both the national level and by international organisations to the risk assessment of veterinary drugs occurring in foods and the implementation of effective measures to reduce this risk.

The use of hormones as growth promoters is banned in the European Union. Other states, namely the United States, Australia, and Argentina, allow the controlled use of hormonal preparations. The drugs used include natural steroid hormones that have an anabolic effect, such as oestradiol, testosterone, and progesterone. Another group consists of synthetic hormones with corresponding effects, such as trenbolone (17β)-17-hydroxyoestra-4,9,11-triene-3-one (**12-51**), and non-steroidal substances having an oestrogenic effect, such as zearalenol and zearalanol (see Section 12.2.2.2.11).

The opinions of professionals as to the potential risks from hormone residues are not uniform. Their potential carcinogenic effects in humans (particularly in reproductive organs and mammary glands) is the area predominantly discussed. Altogether, it is considered that the residual amounts of used hormones are low and are within the normal, physiological values of endogenous hormones, so that there is no significant risk to consumer health. Their ban in European countries is mainly based on the negative attitudes of the general public to this type of increased production. In this context, other legitimate factors that are unacceptable from an ethical point of view are also discussed, such as the burden upon the animal of excessive growth stimulation or excessive production of milk or eggs.

Similar effects to those for hormones are seen in antithyroid substances (strumigens), which inhibit the synthesis of thyroxine in the thyroid gland. Heterocyclic compounds containing thiourea residues in their molecules, such as thiouracils and mercaptoimidazoles, are very potent strumigens. Examples of the main antithyroid drugs include propylthiouracil (6-propyl-2-sulfanylpyrimidin-4-one, **12-51**), which in the past was used in human medicine. In 2009, the US FDA published an alert notifying healthcare professionals of the risk of serious liver injury, including liver failure and death, related to the use of propylthiouracil. As a result, propylthiouracil is no longer recommended for non-pregnant adults or children. Examples of other antithyroid agents are methimazole (1-methyl-3H-imidazole-2-thione) carbimazole (ethyl 3-methyl-2-sulfanylideneimidazole-1-carboxylate, **12-51**), which acts as a pro-drug, because after absorption it is converted into the active form, methimazole.

trenbolone   propylthiouracil   methimazole, R = H
                                              carbimazole, R = COOCH$_2$CH$_3$

12-51, structures of synthetic hormones and antithyroid substances

There is also evidence that the excessive and indiscriminate use of drugs in veterinary medicine leads to some undesirable effects. Discussions occur at the international level, especially in the case of antibiotics, the excessive use of which may result in resistance of pathogenic microorganisms in the tissues of treated animals, reducing the effectiveness of the drugs. A similar phenomenon is known in human medicine, where effective antibiotics are very difficult to find for certain infections. A very important issue is the risk of transmission of resistance in livestock pathogens from one strain to another. Even more serious is the possibility of infection of food consumers with resistant strains of pathogenic microorganisms or induced cross-resistance to antibiotics used in human medicine only. A typical example is a finding from Denmark in 1995, when the preventive use of the antibiotic avoparcin (used as a feed additive to promote growth in chickens and pigs) led to the emergence of the resistant enterococci *Enterococcus faecium*. Based on the structural similarity, this strain, which is highly pathogenic to humans, also showed cross-resistance to vancomycin, which is only used against pathogenic enterococci in human medicine. *E. faecium* resistant to vancomycin was also isolated from raw chicken and pork from the retail network. Based on this information, the use of avoparcin was banned as an additive in animal feed. Under pressure (especially in Scandinavian countries), in 1999, the EU member states also prohibited the use of other antibiotics (bacitracin, tylosin, virginiamycin, and spiramycin) as growth stimulators in animal feed.

12.6.5 Mitigation

The actual production, distribution, and testing of veterinary drugs (including medicated feed) are summarised in the EU Rules Governing Medicinal Products (Eudralex). The approval process for a new veterinary drug in the European Union includes, according to Regulation (EC) No. 726/2004, the evaluation of its therapeutic efficacy, side effects in animals, environmental impact, and ensured good manufacturing practice.

MRLs (Maximum Residue Limits or Levels) of veterinary medicinal products in foods of animal origin established according to Council Regulation (EEC) No. 2377/90 represent internationally generally accepted thresholds, which indicate the amounts of drug residues that may occur in foods of animal origin. With regard to the character of veterinary drugs, MRLs are classified into four groups: Group I, where MRLs are set; Group II, where MRLs are not set (with respect to the identified harmlessness for human health); Group III, where MRLs are set only temporarily in the European Union; and Group IV, where drugs are dangerous for human health and cannot be used on animals intended for food production.

Group I includes numerous chemotherapeutics and antibiotics, non-steroidal anti-inflammatory drugs, corticosteroids, antiprotozoa agents, endo- and ectoparasitica, and drugs affecting the nervous and reproductive systems. Examples of MRL values (reflecting the diversity of species and distribution of residues between various tissues and organs of animals) are listed in Table 12.43. Group II contains mainly inorganic substances, such as chlorides, sulfates, phosphates, gluconates, carbonates, vitamins, other organic compounds (such as acetylsalicylate and alcohols), herbs, tinctures, homeopathics, and other substances generally recognised as safe for the intended use. The fact that a substance is included in Group II does not mean that its use is generally permitted. The legislation typically clearly defines the target organism and method of use. Great attention of control systems at the national and European levels is focused on agents assigned in Group IV, including, for example, chloroform, chloramphenicol, and chlorpromazine, which must not be present (detected) in animal products at all. In accordance with European legislation, prohibited substances are those with thyreostatic, oestrogenic, androgenic, or gestagenic activities, and some sympathomimetic agents (which stimulate the heart through activation of β-adrenoceptors; β-adrenoceptor agonists, also known as β-agonists, bind to β-receptors on cardiac and smooth-muscle tissues). Legislation also includes oestradiol, stilbenes, and all their derivatives. To allow the free trade of animal products within the European Union, each member state must carry out official control and monitoring of residues of pharmaceuticals and biologically active substances.

Table 12.43 Examples of maximum residue limits (MRL) for benzylpenicillin and ivermectin.

Animals	MRL (mg/kg)	Raw material
Benzylpenicillin		
All species	0.050	Meat, liver, kidneys, fat
	0.004	Milk
Ivermectin (dihydroavermectin B$_{1a}$)		
Cattle	0.100	Liver
	0.040	Fat
Sheep	0.020	Fat
Horse	0.015	Liver

12.7 Contaminants from packaging materials

Packaging has become an indispensible element in the food manufacturing process. It maintains the benefits of food processing after the process is complete, makes food more convenient, gives greater safety assurance against microorganisms and biological and chemical changes, and enables the food to travel safely for long distances from its point of origin and still be wholesome at the time of consumption.

Contamination of food by components of packaging materials is one of the most serious health problems relating to food packaging. In almost all cases, when the food is in direct contact with the packaging material, a reciprocal mass transfer proceeds between the food and the packaging components, even when relatively stable packaging materials are used. In order to maintain food quality, inert materials of high quality should be used for food packaging. Food packaging is therefore more expensive to produce than the packaging of the majority of other goods. The most important and commonly used packaging materials, which come in direct contact with the packaged food and can thus significantly affect its quality, are metals, glass and ceramics, paper, wood, and plastics.

12.7.1 Metals

Metals are the most versatile packaging materials as they offer a combination of excellent physical protection and barrier properties, decorative potential, and recyclability. Tin, steel, and aluminum have long histories of successful use in storing food. The basic problem of metals in food packaging materials is metal corrosion, which occurs due to the action of food (mainly acidic food) on the metal container, leading to partial dissolution of the package and resulting in a higher metal content in the food. The mechanisms of corrosion are basically twofold. Chemical and electrochemical corrosions can be distinguished. Methods of preventing corrosion are very diverse (e.g. tinning and passivation of the package, use of lacquered packages).

Just under one-third of the world's total tin production goes into the manufacture of tinplate, a sheet or strip of low-carbon steel (blackplate) coated with a thin layer of pure tin. Tinplate has been used to preserve food for well over a hundred years as it is an excellent barrier to gases, water vapour, light, and odours and can be heat-treated, hermetically sealed, and easily recycled. About 25 000 million food cans are produced and filled in Europe per annum, about 20% of which have plain internal (unlacquered) tin-coated steel bodies. Plain internal tinplate cans may be used, for example, for tomatoes and other tomato-based products, white fruits, and some vegetables (such as asparagus and mushrooms). Although tin provides steel with some corrosion resistance, tinplate containers are often lacquered to provide an inert barrier between the metal and the food product. Food-grade lacquers are based on natural resins and on a number of synthetic lacquers, such as phenol-formaldehyde resins, epoxy and vinyl resins, and polyesters. Sometimes, the lacquers used to protect the internal surfaces of metal packaging may be sources of food contaminants. Their effect on food contamination is analogous to the effect of polymeric packaging materials.

The tin-free steel used for food and beverage cans is chromium- or chrome oxide-coated steel, which was developed to meet economic requirements and surpass tinplate in paintability and paint adhesion. It requires a coating of organic material to provide complete corrosion resistance.

Aluminum's natural resistance to corrosion is advantageous for its role in packaging as, unlike iron, aluminum oxide forms a protective and not a destructive layer. Aluminum also provides an excellent barrier to moisture, air, odours, light, and microorganisms. One of the most common end uses for more than 20% of the aluminum manufactured worldwide is packaging, including drinks cans, foil wrappings, and bottle tops. Lamination of packaging involves the binding of aluminum foil to paper or plastic film to improve barrier properties. As well as recent health worries linking aluminum to Alzheimer's disease, the main disadvantages are its high cost compared with other metals (e.g. steel) and its inability to be welded.

12.7.2 Glass

Glass has an extremely long history in food packaging, dating back to approximately 3000 BCE. Glass is impermeable to gases and vapours, so it maintains product freshness for long periods without impairing its sensory properties and is reusable and recyclable. Glass is much more resistant to corrosion than most materials – so much so that it is easy to think of it as corrosion-proof. Nevertheless, glass may frequently be attacked by acidic solutions that remove the alkali from the surface and wash out ions of sodium, calcium, and other alkali metals and alkaline earth metals, leading to the formation of alkali silicates. These salts, if not removed by washing, can get into the food product. Water corrosion acts at a much slower rate than corrosion by acids, but may become important at high temperatures, especially during longer periods of rain or in case of condensation in glass containers that have been improperly stored before filling. In coloured glass, metal ions used for the colouring (such as iron, manganese, chromium, and some other metals) can be washed out, which may result in increased concentrations of these ions in packaged products.

12.7.3 Paper

The use of paper for food packaging dates back to the seventeenth century. Paper and paperboard are commonly used in corrugated boxes, milk cartons, bags and sacks, and wrapping paper. Plain paper has poor barrier properties and poor resistance to moisture, and therefore is not usually suitable for direct contact with food of higher water activity or in protecting food for long periods. When used in direct contact with food, paper is prepared from cellulose pulp by modified technological processes and is often coated, laminated, or impregnated with waxes, resins, or lacquers to improve its functional and protective properties. Such products include papers treated with sulfate (kraft paper, used for example to package flour and salt) and sulfite (sulfite paper, used for example to package confectionery), grease-proof paper (prepared by extended cellulose hydration), and parchment paper (prepared from sulfuric acid-treated pulp). Paper laminates (paper coated, for example, with polyethylene or aluminum foil) are based on kraft and sulfite papers.

When in contact with food raw materials and foods under low-moisture conditions, the migration of some paper components (such as fillers, binders, precipitation, fixation, and retention agents, drainage accelerators, dispersants, flotation and antifoaming agents, preservatives, lubricants, optical brighteners, and plasticisers) is possible. Contamination of food by paper laminates depends on the properties of these materials and is similar to the contamination caused by polymeric packaging materials. A special group is volatile substances, which may be present in the paper and can pass into the packaged food. Such substances can include various chlorinated phenols (2,4-dichlorophenol, 2,4,6-trichlorophenol, 2,3,4,6-tetrachlorophenol, and other substances), which arise, for example, in bleached papers and cork stoppers, and carbon disulfide, which may be found in kraft paper. Recycled paper, which is still only rarely used in contact with foods, may contain PCBs and PAHs. Benzophenone (diphenyl ketone) may be present in cartonboard food-packaging materials as a residue from UV-cured inks and lacquers used to print on the packaging. It may also be present if the cartonboard is made from recycled fibres recovered from printed materials. The main odour-active compounds of cardboard include vanillin, (E)-non-2-enal (R/S)-γ-nonalactone, (R/S)-δ-decalactone, 2-methoxyphenol, p-anisaldehyde, and 3-propylphenol. Paper exposed to higher temperatures may contain degradation products of cellulose, such as 5-hydroxymethyl-2-furancarbaldehyde, which may adversely affect the flavour of packaged foods.

12.7.4 Wood

Wood packaging is generally unsuitable for direct contact with food as the wood components, in particular resins and tannins, are easily released to food products by leaching or volatilisation. Before their first use, packaging materials and containers intended for direct contact with food have to be treated with water and solutions of sodium carbonate or sulfites, or else their inner surface must be covered with a protective layer of, for example, polymer-based materials. The possibility of food contamination is then dependent on the quality of the protective layer and the type of plastic. An exception is the storage of spirits and wines in oak barrels, where the extraction of wood components to the product is desirable.

Table 12.44 Basic types of polymeric packaging materials.

Group	Main types	Group	Main types
Polyolefins	Polyethylene (PE-HD, PE-LD, PE-LLD) Polypropylene (PP)	Vinyl polymers	Polyvinyl chloride (PVC) Polystyrene (PS)
Nitrogen polymers	Polyamide (PA) Polyurethane (PUR)		Polyvinyl acetate (PVAc) Polyvinyl alcohol (PVOH)
Thermosets	Phenol-formaldehyde resins Amino-formaldehyde resins Epoxy resins	Polyesters	Polyethylene terephthalate (PET) Polycarbonate (PC) Thermoset polyesters

12.7.5 Plastics

The production of plastic materials on an industrial scale began in the 1940s and '50s. In 2011, food packaging applications used 39.4% of the global plastics production. These materials included a range of organic polymers, **thermoplastics** (which soften upon exposure to heat and return to their original condition at room temperature), and **thermosets** (which solidify or set irreversibly when heated).

The advantage of thermoplastics is that they are low-cost materials, can easily be shaped into various products (e.g. bottles, jugs, and films), have some functional advantages (e.g. thermosealability and microwavability), and are mostly recyclable. Thermosets are mainly used for kitchen equipment and cookware, because of the demand for heat- and corrosion-resistant properties, which are needed to withstand exposure to acids in food and to cleaning materials. They are also used for their flame resistance around machined parts (in steamers, fryers, meat cutters, and other kitchen equipment). An overview of the main types of polymeric packaging materials is given in Table 12.44.

Mass transfer, which is commonly referred to as migration, is a typical interaction of food with packaging made from plastics. Low-molecular-weight substances are transferred to foods, especially residual monomers and several thousand different types of additives (antioxidants, plasticisers, thermal stabilisers, lubricants, condensation components, slip additives, and antistatic and antiblocking agents). Characteristically, the packaging material is not completely destroyed, but retains technologically important properties, whilst only certain components diffuse into the food. Migration is always a two-way process, as some food constituents migrate into food packaging materials. In terms of food quality, it is of course desirable to limit the transfer of substances in both directions. In principle, two basic types of migration are distinguished: **overall (global) migration** and **specific migration**.

Overall (global) migration refers to the transfer of all (in many cases, unknown) components from the packaging to the food. Table 12.45 shows the typical values for the overall migration of plastic materials used for food packaging. Specific migration is the transition of one or more substances that are particularly important from the hygienic and toxicological point of view. Because of the enormous complexity and number of potential contaminants (several thousand substances are used as additives) and the extent of the use of polymeric packaging materials, specific migration is one of the most significant problems of food contamination from packaging.

Table 12.45 Typical values of overall migration for plastic foils for food packaging.

| Polymer | Migration (mg/dm^2) | |
	Aqueous simulants	Olive oil
Low-density polyethylene (PE-LD)	0.1–1.5	4–20
Linear low-density polyethylene (PE-LLD)	0.1–1.0	1.5–5.5
Polyethylene terephthalate (PET)	<0.2	0.3–6.9
Polypropylene (PP)	0.1–1.5	0.5–5.0
Polystyrene (PS)	0.2–5	1.2–26
Polyamides (PA)	0.0–15	1.0–6,0
Polyvinyl chloride (PVC)	0.5–3	3.0–100

The most important potential food contaminants are listed in Table 12.46, but this list is far from exhaustive. It does, however, illustrate the types of contaminants and the complexity of the potential contamination of foods by plastic packaging materials. Structures of the most important compounds are given in formulae **12-52**.

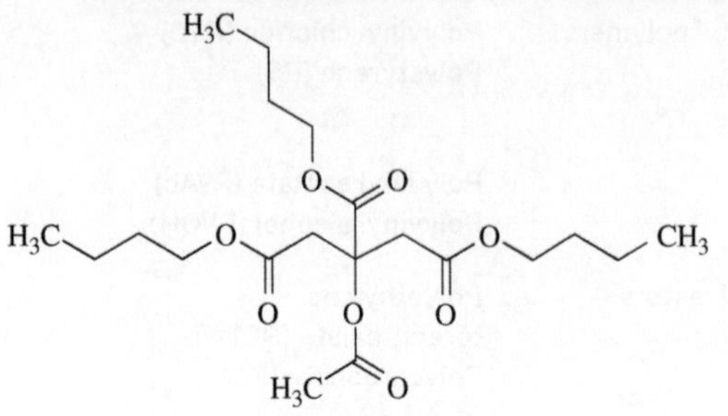

acetyl tributyl citrate

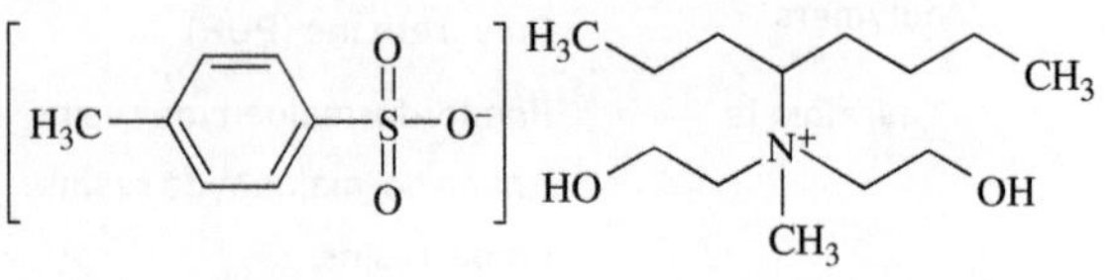

N,N-bis(2-hydroxyethyl)-*N*-(4-octyl)-*N*-methylammonium-
p-toluenesulfonate

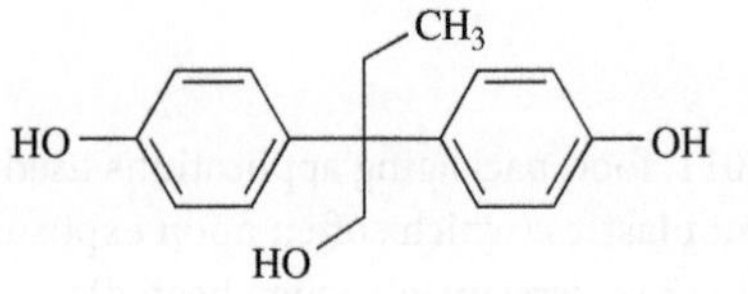

2,2-bis(4-hydroxyphenyl)butan-1-ol

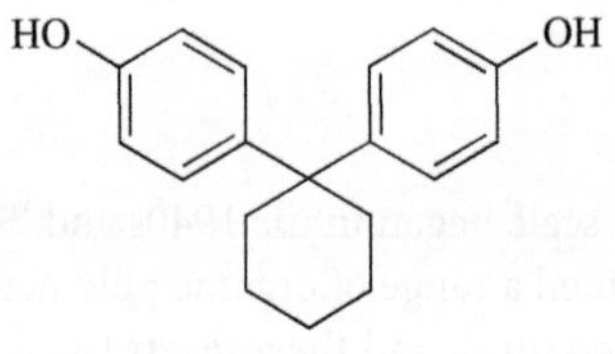

1,1-bis(4-hydroxyphenyl)cyclohexane

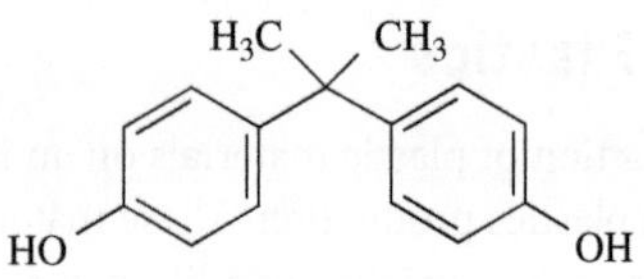

bisphenol A

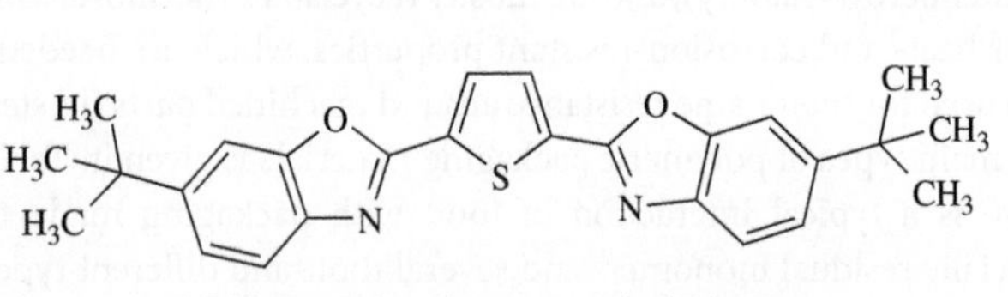

2,5-bis(5-*tert*-butylbenzoxazol-2-yl)thiophene

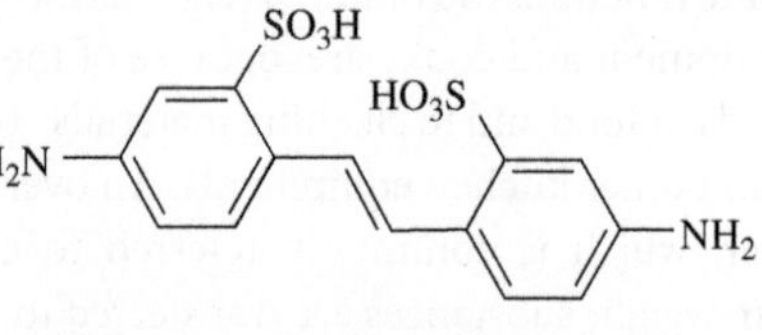

4,4′-diamino-2,2′-stilbenedisulfonic acid

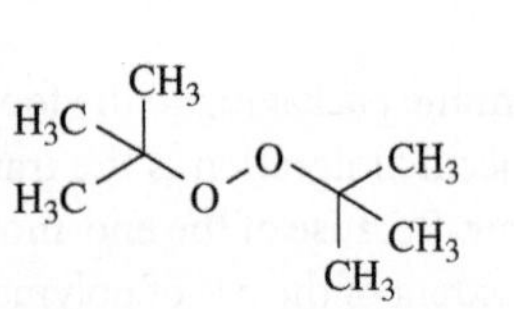

di-*tert*-butylperoxide

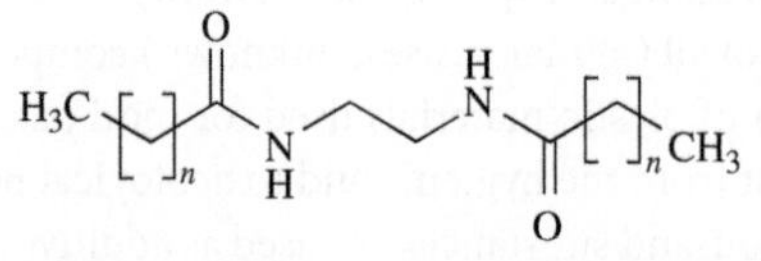

N,N′-dipalmitoylethane1,2-diamine, *n* = 1

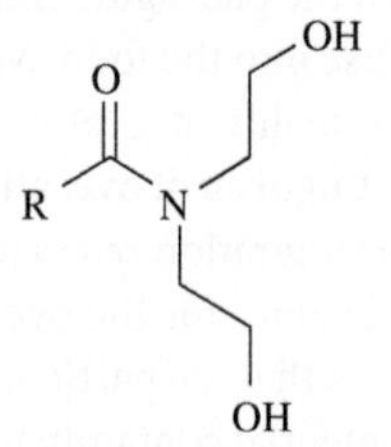

fatty acid diethanolamides

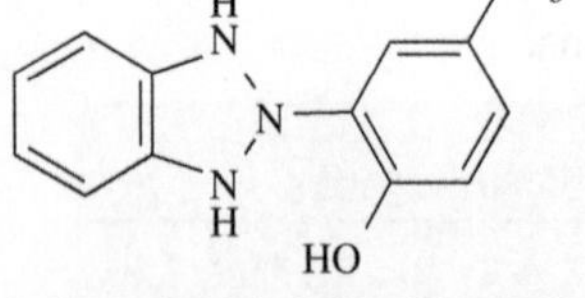

2-(2-hydroxy-5-methylphenyl)benzotriazole

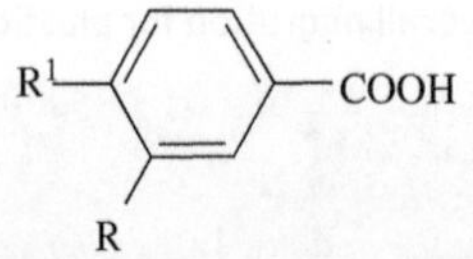

isophthalic acid, R = COOH, R^1 = H
terephthalic acid, R = H, R^1 = COOH

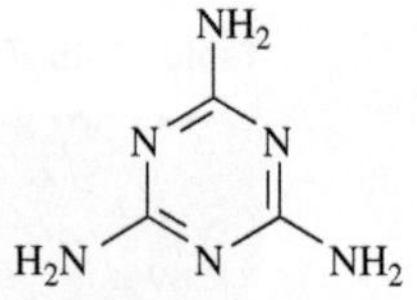

melamine

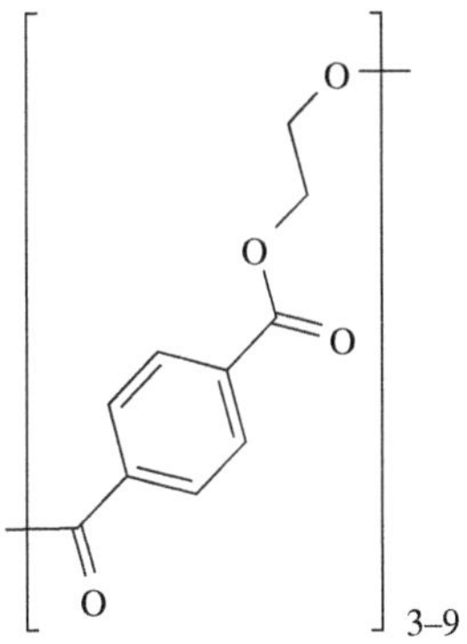

ethyleneterephthalate oligomers

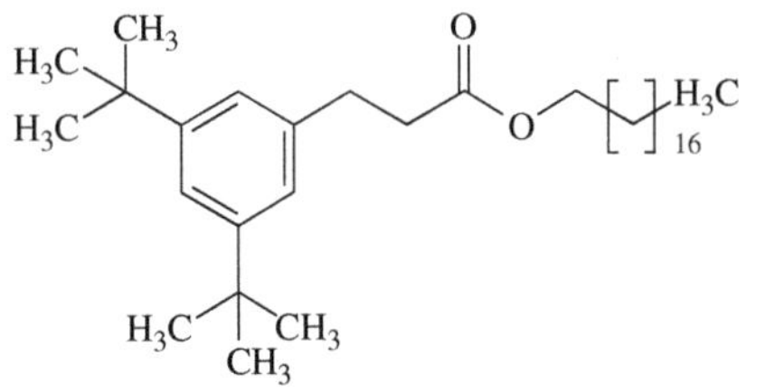

octadecyl-3-(3,5-di-*tert*-butyl-4-hydroxyphenyl) propionate

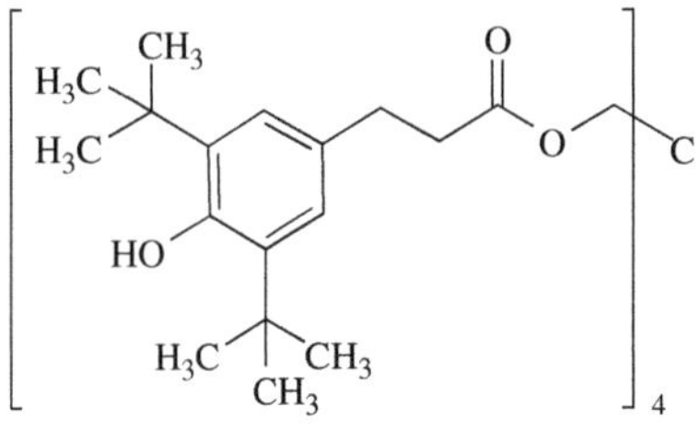

tetrakis[methylene-3-(3,5-di-*tert*-butyl-4-hydroxyphenyl) propionyl]methane

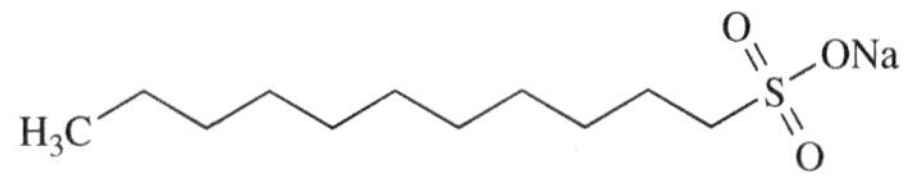

sodium 1-undecanesulfonate

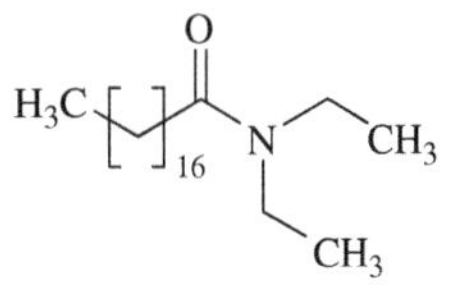

stearic acid *N,N*-diethylamide

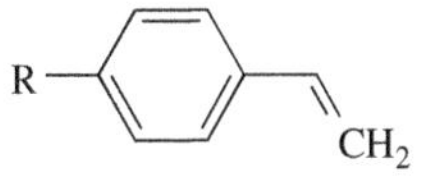

4-methylstyrene, R = CH₃

12-52, selected potential food contaminants from plastics

Table 12.46 Classification of contaminants from packaging materials.

Groups of contaminants	Compounds and examples of their occurrence[a]
Residues of monomers	Adipic acid dimethyl adipate (polyesters), acrylamide (acrylonitrile copolymers), acrylonitrile (acrylamide copolymers), acrylic acid (acrylic polymers), buta-1,3-diene (PS copolymers, elastomers), butan-1,4-diol (polyesters), 2,2-bis(4-hydroxyphenyl)butan-1-ol (polyesters), 1,1-bis(4-hydroxyphenyl)cyclohexane (polyesters), 4,4′-(propane-2,2-diyl)diphenol known as bisphenol A (polyesters), ethyleneglycol (polyester), formaldehyde (phenol-formaldehyde resins), isophthalic acid (PET), caprolactame and C_6-C_{12} aminocarboxylic acids and their lactames (polyamides), melamine (amino-formaldehyde resins), methacrylic acid and methyl methacrylate (acrylic polymers), methylstyrene (PS and copolymers), propyleneglycol (polyesters), sebacic acid and dimethyl sebacate (polyesters), styrene (PS and copolymers), terephthalic acid and dimethyl terephthalate (PET), vinyl acetate (vinyl acetate copolymers), vinyl chloride (PVC and copolymers), vinylidenechloride (PVdC and copolymers)
Residues of additives	
Initiators	Di-*tert*-butylperoxide, dibenzoylperoxide
Regulators	Dodecylmercaptan
Catalysts	Oxygen compounds of Al, Ti, Ca, Mg, Si, Cr, Sb, Mn, Li, Zn, Co, and others
Emulsifiers	Sodium 1-alkylsulfonates (C_{12}-C_{18})
Protective colloids	Polyvinyl alcohol, lecithin, gelatin, carboxymethylcellulose
Solvents	Acetone, butyl acetate, dichloromethane, ethanol, ethyl acetate, hexane, pentane, toluene
Hardeners	Formaldehyde, glutaraldehyde, glyoxal, and others
Stabilisers and antioxidants	2- and 3-*tert*-Butyl-4-hydroxyanisole (BHA), 3,5-di-*tert*-butyl-4-hydroxytoluene (BHT), 2-ethyl-di-2-ethylhexyl-monooctyltintrithioglycolate (Irgastab 17 MOK), 2-phenylindole, octadecyl 3-(3,5-di-*tert*-butyl-4-hydroxyphenyl)-propionate(trade name Irganox 1076), tetrakis[methylene-3-(3,5-di-*tert*-butyl-4-hydroxyphenyl)propionyl]-methane (trade name Irganox 1010)

(*continued overleaf*)

Table 12.46 (*continued*)

Groups of contaminants	Compounds and examples of their occurrence[a]
Plasticisers	Acetyl tributyl citrate, butylesters of soybean oil fatty acids, butyl stearate, dibutyl adipate, dibutyl sebacate, dioctyl adipate, dioctyl sebacate, epoxidised soybean oil, phthalates, glycerol, isobutyl stearate and others
Lubricants	Stearic acid diethylamide, *N,N*′-dipalmitoylethane-1,2-diamine, *N,N*′-distaroylethane-1,2-diamine, glycerol, stearic and palmitic acids, silicone oils, Ca, Mg, Al and Zn stearates
Antistatic agents	Fatty acid diethanolamides, *N,N*-bis(2-hydroxyethyl)-*N*-4-octyl-*N*-methyl-ammonium- *p*-toluenesulfonate
UV absorbers	2,4-Dihydroxybenzophenone, 2-(2-hydroxy-5-methylphenyl)benzotriazole
Optical lighteners	2,5-Bis(5-*tert*-butylbenzoxazol-2-yl)thiophene, 4,4′-diamino-2,2′-stilbenedisulfonic acid
Fillers	Silica, kaolin, graphite
Blowing agents	Isopentane or pentane, petroleum ether
Degradation products (of polymers and additives)	Acetaldehyde (PET), oligomers (PET), bisphenol A (polyesters), hydrochloric acid (PVC), dibutyl adipate (PVC), nitrosamines (elastomers, rubber)

[a]PS = polystyrene, PET = polyethylene terephthalate, PVC = polyvinyl chloride, PVdC = polyvinylidene chloride.

Packaging materials are not considered toxic, but they may contain toxic degradation products and many additives and plastics may occur as non-bound residual monomers, which have toxic properties. There is a risk with any packaging material that its components may be transferred in some way to the food. Fortunately, the level of contaminant transfer is extremely low in most cases, but there are instances where a real hazard exists and must be controlled. No official internationally agreed guidelines exist, but in the European Union there is a general requirement that food packaging components must not be transferred into food during its normal shelf-life, pose a health risk, or adversely affect the quality, flavour, texture, or appearance of the food. The major area of concern is the transfer of plastic material monomers and additives, especially plasticisers. The residual monomer content depends on the polymer type and polymerisation technique. The contents of monomers in various polymers vary from very low (mg/kg levels) up to 40 000 mg/kg. The amount of additives used is also highly variable. Polyvinylchloride (PVC), for example, requires the highest level of additives, and accounts for 73% of the world's production of additives by volume, whilst polypropylene and polyethylene account for 10% and styrene for 5%.

In 1974, the monomer of PVC, vinyl chloride (chloroethene), was first reported to cause angiosarcoma (a form of tissue cancer that arises in the lining of blood vessels) of the liver both in humans and in animals. Additional research has demonstrated its carcinogenicity to other organs. The target organs include the liver, brain, lung, and probably the lymphohaematic system. According to the IARC, vinyl chloride is a Group 1 agent carcinogenic to humans. Dietary exposure to vinyl chloride is controlled by limiting either its content in food-contact material (typically less than 1 mg/kg) or its migration to foods at the lowest technically achievable level (typically less than 0.01 mg/kg). Vinylidene chloride has some detrimental effects in mice and rats, but no carcinogenic effects have been seen. It is probably not carcinogenic to humans (a Group 3 agent).

Acrylonitrile, which is mutagenic after metabolic activation in the liver, is considered more toxic than chlorinated monomers and is possibly carcinogenic to animals and humans as a Group 2B agent. It is metabolised to cyanides, which are subsequently transformed into thiocyanates and excreted in the urine.

Reports of organ toxicity upon chronic exposure to styrene are rare; however, since the main intermediate in styrene metabolism is an epoxide (styrene-7,8-oxide), hepatotoxicity due to covalent binding at the site of formation appears to be a possibility. Both of these substances, styrene (a Group 3 agent) and its oxide (recently upgraded to a Group 2A as probably carcinogenic to humans), have been shown to produce chromosomal aberrations under certain conditions.

Amongst the additives used to modify the properties of polymeric packaging materials, plasticisers have raised much concern from the hygienic point of view. Butyl stearate, acetyltributyl citrate, alkyl sebacates, and adipates are important because they are types of plasticisers that typically have low toxicities. Materials such as epoxidised soybean oil are widely used in polyvinyl chloride, polyvinylidene chloride, and polystyrene as thermal stabilisers and lubricants at a level of 0.1–27%. Toxicity of epoxidised soybean oil is affected by the presence of oxirane, also known as ethylene oxide, which was upgraded to Group 1 as a carcinogenic agent to humans, based on mechanistic and other relevant data.

Bisphenol A occurs in polycarbonate plastics that have many applications, including use in some food and drink packagings, such as water and infant bottles, compact discs, impact-resistant safety equipment, and medical devices, as well as in epoxy resins used as lacquers (such as the resin with the trade name Araldite). It is an endocrine disruptor with a high oestrogenic potential and can mimic the natural hormone oestradiol, which may lead to negative health effects. Some studies have linked prenatal exposure to bisphenol A with later physical and neurological difficulties. Other studies have also presented a hypothesis that the current level of human exposure to growth hormones that fatten livestock, to pharmaceuticals that induce weight gain, and to bisphenol A can damage many of the body's natural weight-control mechanisms. Furthermore, it is posited that these effects, together with a wide range of additional, possibly synergistic factors, may play a significant role in the worldwide obesity epidemic. In 2010, Canada became the first country to declare bisphenol A a toxic substance. The European Union, Canada, and recently the United States have banned the use of bisphenol A in baby bottles.

The legislation is based on three principles: a positive list of substances, which may be used for the production of objects coming into contact with food; determination of specific migration limits (SMLs) for substances with proven adverse health effects; and compliance with the principles of good manufacturing practice for manufacturers of these items. For example, plastic materials and objects shall not transfer their constituents to foods in quantities exceeding 60 mg/kg of food, which is called the overall migration limit (OML). In some cases, this limit may be expressed as 10 mg/dm² of surface of the material or object. To determine whether a particular plastic formulation meets these criteria, legislative regulations describe in detail the terms and procedures of migration tests that are commonly performed not on individual foods, but on standardised food stimulants, which represent different conditions for leaching of undesirable substances. These are: distilled water, 3% aqueous solution of acetic acid, 10% aqueous solution of ethanol (or greater, if the alcoholic beverage in question has higher alcohol content), and olive oil. The regulations also specify which simulants should be used for each category of food. In general, there are no simulants listed for dried foods, which can be considered not to take up plastic constituents from contact materials. Very strict criteria in terms of the migration of harmful substances are also set for paper and cardboard, which are the traditional packaging materials coming into contact with food.

For substances with proven toxic effects, SMLs are declared in food simulants or the maximum amounts in a given material for unstable compounds are stated. An example might be the SML for melamine (2,4,6-triamino-1,3,5- triazine), which is set at 30 mg/kg. In sealing lids for jars intended for infant and baby food, the migration of epoxidised soybean oil is often controlled (SML = 30 mg/kg).

12.7.5.1 Phthalates

Phthalates (esters of benzene 1,2 dicarboxylic acid, known as phthalic acid) belong to the group of virtually ubiquitous organic environmental contaminants. The reason for their abundant spread to all components of the environment is their suitable physico-chemical properties, which is why they are used as plasticisers, substances improving the mechanical properties of plastics (mainly PVC); these include their flexibility, transparency, durability, and longevity.

The US EPA includes six phthalic acid esters (**12-53**) on its list of priority environmental contaminants. Of these, the most widespread are dibutyl phthalate and bis(2-ethylhexyl)phthalate.

12-53, phthalates

The release of phthalates into the environment occurs not only during their manufacture and the production of materials containing them, but also during the use and subsequent disposal of such materials. Phthalates occur in a wide variety of products, ranging from common household objects, such as floor coverings, rubber gloves, and children's toys, through to food packaging and specialised products used, for example, in health care such as bags for the storage of blood and blood plasma, syringes, and dialysis units. Plastic materials treated with phthalates may contain up to 40% plasticiser, which is not chemically bound to the polymer and may thus be released into any foods that come into contact with them. Contamination in industrial centres or enclosed areas can reach values of tens of mg/m³. Like other persistent contaminants, phthalates may occur in the form of vapour (esters with lower relative molecular weights), aerosols, or bound to dust particles and can be transported over long distances – especially bis(2-ethylhexyl) phthalate.

In the air over the oceans, concentrations of phthalates are at the level of ng/m³. Contamination of surface and ground water varies by location, differing within tenths to thousands of µg/l, but in sediments, concentrations of hydrophobic phthalates can reach hundreds of mg/kg. The main sources of soil contamination are industrial and municipal waste, agrochemicals, and emissions. In the soil, phthalates tend to adsorb to organic matter, where they accumulate.

The most important source of phthalates is food intake. Generally, food contamination by phthalates may be in the range of hundredths up to units of mg/kg, with occasional extreme values primarily in fatty foods, such as cheeses, cakes, and sandwiches, which may readily contain phthalates in amounts of tenths of mg/kg food. The occurrence of phthalates in food may be the result of contamination of raw materials, intermediates, or finished products. The migration of phthalates from packaging materials into food is generally influenced by many factors, such as the type of polymeric packaging material, type of food (the amount of fat is important), temperature, and length of contact with the food.

In Europe, the average maximum exposure to bis(2-ethylhexyl) phthalate from packaging is estimated at 0.02 mg per person per day, and exposure to all phthalates (expressed as dimethyl phthalate) is estimated at 4.37 mg per person per day.

Hydrolysis and photodegradation of phthalic acid esters in the abiotic components of the environment is extremely slow, and higher esters, such as bis(2-ethylhexyl) phthalate and dioctyl phthalate, are virtually stable. The major route of elimination of phthalates from the environment is their biodegradation. Almost all organisms have the biochemical means (non-specific esterases) to be able to catalyse the hydrolysis of phthalates. Microbial degradation of phthalates occurs mainly under aerobic conditions in soil and via aquatic bacteria and fungi. The degradation process begins through hydrolysis to phthalic acid monoesters and subsequently to free phthalic acid. Degradation of phthalic acid via pyruvate and succinate produces carbon dioxide and water.

Immediately after entering the human organism via the lungs (by respiration) and intestines (via food), phthalates get into the blood and advance to the liver, kidney, and testes. To a limited extent, they are stored in the fat tissue. Metabolism of bis(2-ethylhexyl) phthalate begins with hydrolysis in the gastrointestinal tract to 2-ethylhexyl phthalate and 2-ethylhexan-1-ol. In the next metabolic step, a small proportion of 2-ethylhexyl phthalate may be hydrolysed to phthalic acid and 2-ethylhexan-1-ol, but the major proportion of the monoester is oxidised in the aliphatic side chain to yield various metabolites (about 30 metabolites have been identified). In most mammals, 2-ethylhexyl phthalate and its oxidised derivatives react with D-glucuronic acid to form conjugates, which are excreted from the body. Metabolism of phthalates with shorter side chains is similar. Despite the relatively rapid metabolism and excretion of phthalates (60–90% are excreted in urine and faeces within 24 hours), a certain proportion of lipophilic phthalates are accumulated in the body, because the rate of their intake exceeds the rate of metabolic conversion.

The acute toxicity of phthalates is low and is manifested by gastrointestinal irritation, nausea, sleepiness, low blood pressure, dizziness, hallucinations, blurred vision, and tearing. Inhalation of vapours leads to coughing and irritation of the throat and oesophagus. In terms of subacute and chronic effects, phthalates with branched side chains, such as bis(2-ethylhexyl) phthalate, are dangerous. The primary target organ is the liver, which manifests by liver enlargement, increased pigmentation, fat sediments, increasing number of hepatic peroxisomes, and mitochondria. The kidneys and other organs may also be affected. Chronic intake of phthalates, particularly of bis(2-ethylhexyl) phthalate, may likewise have teratogenic and carcinogenic effects (liver cancer) and might affect the reproductive ability of the body (weight reduction of testes/ovaries, reduction in sperm count). With regard to possible carcinogenic effects, bis(2-ethylhexyl) phthalate is classified as a possibly carcinogenic agent to humans (Group 2B) and butybenzyl phthalate as an agent that is not classifiable as to its carcinogenicity to humans (Group 3). The SML (Specific Migration Limit) value for dibutyl phthalate is 0.3 mg/kg and that for bis(2-ethylhexyl) phthalate 1.5 mg/kg, but plasticised plastics should not be used for fatty foods.

Bibliography

Abdel-Aal, E.-S.M., Young, C.J., Rabalski, I. et al. (2007). Identification and quantification of seed carotenoids in selected wheat species. *J. Agric. Food Chem.* **55**: 787.

Abdel-Aal, E.-S.M., Young, J.C., and Rabalski, I. (2006). Anthocyanin composition in black, blue, pink, purple, and red cereal grains. *J. Agric. Food Chem.* **54**: 4696.

Adams, A. and De Kimpe, N. (2006). Chemistry of 2-acetyl-1-pyrroline, 6-acetyl-1,2,3,4-tetrahydropyridine, 2-acetyl-2-thiazoline, and 5-acetyl-2,3-dihydro-4*H*-thiazine: extraordinary Maillard flavor compounds. *Chem. Rev.* **106**: 2299.

Adams, J.B. (1997). Food additive-additive interactions involving sulphur dioxide and ascorbic and nitrous acids: a review. *Food Chem.* **59**: 401–409.

Adams, M.R. and Moss, M.O. (1997). *Food Microbiology*. Cambridge: The Royal Society of Chemistry.

Agerbirk, N., Olsen, C.E., Cipollini, D. et al. (2014). Specific glucosinolate analysis reveals variable levels of epimeric glucobarbarins, dietary precursors of 5-phenyloxazolidine-2-thiones, in watercress types with contrasting chromosome numbers. *J. Agric. Food Chem.* **62**: 9586–9596.

Agon, V.V., Bubb, W.A., Wright, A. et al. (2006). Sensitizer-mediated photooxidation of histidine residues: evidence for the formation of reactive side-chain peroxides. *Free Radic. Biol. Med.* **40**: 698–710.

Agyemang, D., Bardsley, K., Brown, S. et al. (2011). Identification of 2-ethyl-4-methyl-3-thiazoline and 2-isopropyl-4-methyl-3-thiazoline for the first time in nature by the comprehensive analysis of sesame seed oil. *J. Food Sci.* **76**: C385–C391.

Ahlborg, U.G., Hansberg, A., and Kenn, K. (1992). *Risk Assessment of PCBs*. Stockholm: Nord Report.

Ahmed, N., Smith, R.W., Aristizabal Henao, J.J. et al. (2018). Analytical method to detect and quantify avocatin B in hass avocado seed and pulp matter. *J. Nat. Prod.* **81**: 818–824.

Aizawa, H. (ed.) (1982). *Metabolic Maps of Pesticides*. New York: Academic Press.

Akoh, C.C. and Min, D.B. (1998). *Food Lipids: Chemistry, Nutrition, and Biotechnology*. New York: Marcel Dekker.

Aksnes, A. (1984). Methionine sulphoxide: formation, occurrence and biological availability. A review. *Fiskeridirektoratets Skrifter, Serie Ernæring* **2** (5): 125–153.

Allamy, L., Darriet, P., and Pons, A. (2017). Identification and organoleptic contribution of (*Z*)-1,5-octadien-3-one to the flavor of *Vitis vinifera* cv. Merlot and Cabernet Sauvignon musts. *J. Agric. Food Chem.* **65**: 1915–1923.

Allscher, T., Klüfers, P., and Mayer, P. (2008). Carbohydrate-metal complexes: structural chemistry of stable solution species. In: *Glycoscience* (eds. B.O. Fraser-Reid, K. Tatsuta and J. Thiem). Berlin, Heidelberg: Springer.

Al-Saleh, I.A. (1987). Pesticides: a review article. *J. Environ. Pathol. Toxicol. Oncol.* **13**: 151–161.

Alves, R.C., Casal, S., and Oliveira, B.P.P. (2007). Factors influencing the norharman and harman contents in espresso coffee. *J. Agric. Food Chem.* **55**: 1832–1838.

Amado, R. and Battaglia, R. (eds.) (1997). *Authenticity and Adulteration of Food – the Analytical Approach*. Winterthur, Switzerland: Sailer.

Amaglo, N.K., Bennett, R.N., Lo Curto, R.B. et al. (2010). Profiling selected phytochemicals and nutrients in different tissues of the multipurpose tree *Moringa oleifera* L., grown in Ghana. *Food Chem.* **122**: 1047–1054.

Amerine, M.A. and Ough, C.S. (1980). *Methods for Analysis of Musts and Wines*. New York: Wiley.

Amrani-Hemaimi, M., Cerny, C., and Fay, L.B. (1995). Mechanisms of formation of alkylpyrazines in the Maillard reaction. *J. Agric. Food Chem.* **43**: 2818–2822.

Andersen, R.M., Opheim, S., Aksnes, D.W., and Frystein, N.L. (1991). Structure of petanin, an acylated anthocyanin isolated from Solanum tuberosum, using homo- and hetero-nuclear two-dimensional nuclear magnetic resonance techniques. *Phytochem. Anal.* **2**: 230–236.

Anderson, J.A. (1996). Allergic reactions to foods. *Crit. Rev. Food Sci. Nutr.* **36** (Suppl. 001): 19–38.

Anderson, R.L. and Wolf, W.J. (1995). Compositional changes in trypsin inhibitors, phytic acid, saponins and isoflavones related to soybean processing. *J. Nutr.* **125** (Suppl): 581S–588S.

Andersson, J.T. and Achten, C. (2015). Time to say goodbye to the 16 EPA PAHs? Toward an up-to-date use of PACs for environmental purposes. *Polycycl. Aromat. Compd.* **35**: 330–354.

Andersson, S.C., Rumpunen, K., Johansson, E., and Olsson, M.E. (2011). Carotenoid content and composition in rose hips (Rosa spp.) during ripening, determination of suitable maturity marker and implications for health promoting food products. *Food Chem.* **128**: 689–696.

Andrawis, A. and Kahn, V. (1990). Ability of various chemicals to reduce copper and to inactivate mushroom tyrosinase. *J. Food Biochem.* **14**: 103–115.

Andreasen, M.F., Landbo, A.-K., Christensen, L.P. et al. (2001). Antioxidant effects of phenolic rye (Secale cereale L.) extracts, monomeric hydroxycinnamates, and ferulic acid dehydrodimers on human low-density lipoproteins. *J. Agric. Food Chem.* **49**: 4090–4096.

Andrea-Silva, J., Cosme, F., Ribeiro, L.F. et al. (2014). Origin of the pinking phenomenon of white wines. *J. Agric. Food Chem.* **62**: 5651–5659.

Angerosa, F., Mostallino, R., Basti, C., and Vito, R. (2000). Virgin olive oil odor notes: their relationships with volatile compounds from the lipoxygenase pathway and secoiridoid compounds. *Food Chem.* **68**: 283–287.

Antalick, G., Tempère, S., Šuklje, K. et al. (2015). Investigation and sensory characterization of 1,4-cineole: a potential aromatic marker of Australian Cabernet Sauvignon wine. *J. Agric. Food Chem.* **63**: 9103–9111.

Aparicio-Ruiz, R. and Gandul-Rojas, B. (2012). Thermal degradation kinetics of neoxanthin, violaxanthin, and antheraxanthin in virgin olive oils. *J. Agric. Food Chem.* **60**: 5180–5191.

Araújo, P.H.H., Sayer, C., Poco, J.G.R., and Giudici, R. (2002). Techniques for reducing residual monomer content in polymers: a review. *Polym. Eng. Sci.* **42**: 1442–1468.

Aresta, M., Boscolo, M., and Franco, D.W. (2001). Copper(II) catalysis in cyanide conversion into ethyl carbamate in spirits and relevant reactions. *J. Agric. Food Chem.* **49**: 2819–2824.

Arroyo-Abad, U., Pfeifer, M., Mothes, S. et al. (2016). Determination of moderately polar arsenolipids and mercury speciation in freshwater fish of the River Elbe (Saxony, Germany). *Environ. Pollut.* **208**: 458–466.

Arts, I.C.W., van de Putte, B., and Hollman, P.C.H. (2000). Catechin contents of foods commonly consumed in The Netherlands. 1. Fruits, vegetables, staple foods, and processed foods. *J. Agric. Food Chem.* **48**: 1746–1751.

Asano, N., Yamashita, T., Yasuda, K. et al. (2001). Polyhydroxylated alkaloids isolated from mulberry trees (Morus alba L.) and silkworms (Bombyx mori L.). *J. Agric. Food Chem.* **49**: 4208–4213.

Aucamp, J.P., Hara, Y., and Apostolides, Z. (2000). Simultaneous analysis of tea catechins, caffeine, gallic acid, theanine and ascorbic acid by micellar electrokinetic capillary chromatography. *J. Chromatogr. A* **876**: 235–242.

Aumann, D.C., Cloth, G., Steffan, B., and Steglich, W. (1989). Complexation of caesium 137 by the cap pigments of the bay boletus (Xevocomus badius). *Angew. Chem. Int. Ed.* **28**: 453–454.

Awika, J.M., Rooney, L.W., and Waniska, R.D. (2004). Properties of 3-deoxyanthocyanins from Sorghum. *J. Agric. Food Chem.* **52**: 4388–4394.

Baderschneider, B. and Winterhalter, P. (2000). Isolation and characterization of novel stilbene derivatives from Riesling wine. *J. Agric. Food Chem.* **48**: 2681–2686.

Baher, Z.F., Mirza1, M., Ghorbanli, M., and Rezaii, M.B. (2002). The influence of water stress on plant height, herbal and essential oil yield and composition in *Satureja hortensis* L. *Flavour Fragr. J.* **17**: 275–277.

Baillie-Hamilton, P.F. (2002). Chemical toxins: a hypothesis to explain the global obesity epidemic. *J. Altern. Complement. Med.* **8**: 185–192.

Bakker, J. and Timberlake, C.F. (1997). Isolation, and characterization of new color-stable anthocyanins in some red wines. *J. Agric. Food Chem.* **45**: 35–43.

Balasubramanian, K. (2006). Molecular orbital basis for yellow curry spice curcumin's prevention of Alzheimer's disease. *J. Agric. Food Chem.* **54**: 3512–3520.

Baldwin, E.A. and Nisperos-Carriedo, M.O. (1994). *Edible Coatings and Films to Improve Food Quality*. Lancaster, PA: Technomic Publishing Co.

Baldwin, J.E., Adlington, R.M., Bebbington, D., and Russell, A.T. (1994). Asymmetric total synthesis of the individual diastereoisomers of hypoglycin A. *Tetrahedron* **50**: 12015–12028.

Ballschmitter, K. (1992). Transport and fate of organic compounds in the global environment. *Angew. Chem. Int. Ed.* **31**: 487–515.

Baltes, W. (1993). *Lebensmittelchemie*, 3e. Berlin: Springer.

Balyaya, K.J. and Clifford, M.N. (1995). Individual chlorogenic acids and caffeine contents in commercial grades of wet and dry processed Indian green robusta coffee. *J. Food Sci. Technol.* **32**: 104–111.

Balz, M., Schulte, E., and Thier, H.-P. (1992). Trennung von Tocopherolen und Tocotrienolen durch HPLC. *Eur. J. Lipid Sci. Technol.* **94**: 209–213.

Barcenilla, J., Estrella, I., Gómez-Cordóves, C. et al. (1989). The influence of yeasts on certain non-volatile components of wine. *Food Chem.* **31**: 177.

Barlianto, H. and Maier, H.G. (1995). Acids in chicory roots and malt. II. Determination of acids derived from saccharides. *Z. Lebensm. Unters. Forsch.* **200**: 273–277.

Baroja-Mazo, A., Del Valle, P., Rúa, J. et al. (2005). Characterization and biosynthesis of D-erythroascorbic acid in Phycomyces blakesleeanus. *Fungal Genet. Biol.* **42**: 390–402.

Barrientos, L.G. and Murthy, P.P.N. (1996). Conformational studies of myo-inositol phosphates. *Carbohydr. Res.* **296**: 39–54.

Bartley, J.P. (1995). A new method for the determination of pungent compounds in ginger (Zingiber officinale). *J. Sci. Food Agric.* **68**: 215–222.

Bauer, K., Garbe, D., and Surburg, H. (1990). *Common Fragrance and Flavor Materials*, 2e. New York: VCH Publishers.

Bazemore, R., Harrison, C., and Greenberg, M. (2006). Identification of components responsible for the odor of cigar smoker's breath. *J. Agric. Food Chem.* **54**: 497–501.

Beecken, H., Gottschalk, E.M., von Gizycki, U. et al. (2003). Orcein and litmus. *Biotech. Histochem.* **78**: 289–302.

Beksan, E., Schieberle, P., Robert, F. et al. (2003). Synthesis and sensory characterization of novel umami-tasting glutamate glycoconjugates. *J. Agric. Food Chem.* **51**: 5428–5436.

Belitz, H.-D., Grosch, W., and Schieberle, P. (2009). *Food Chemistry*. 4th revised and extended edn. Berlin, Heidelberg: Springer-Verlag.

Belitz, H.-D. and Wieser, H. (1985). Bitter compounds: occurrence and structure-activity relationships. *Food Rev. Int.* **1**: 271–354.

Bell, E.A. (2003). Nonprotein amino acids of plants: significance in medicine, nutrition, and agriculture. *J. Agric. Food Chem.* **51**: 2854–2865.

Bell, E.A. and Charlwood, B.V. (eds.) (1980). *Secondary Plant Products*. Berlin: Springer.

Beltran, G., Ruano, M.T., Jimenez, A. et al. (2007). Evaluation of virgin olive oil bitterness by total phenol content analysis. *Eur. J. Lipid Sci. Technol.* **109**: 193–197.

Bendig, P., Maier, L., and Vetter, W. (2012). Brominated vegetable oil in soft drinks – an underrated source of human organobromine intake. *Food Chem.* **133**: 678–682.

Benkwitz, F., Nicolau, L., Lund, C. et al. (2012). Evaluation of key odorants in Sauvignon Blanc wines using three different methodologies. *J. Agric. Food Chem.* **60**: 6293–6302.

Bennett, J.W. and Klich, M. (2003). Mycotoxins. *Clin. Microbiol. Rev.* **16**: 497–516.

Benway, D.A. and Weaver, C.M. (1993). Assessing chemical form of calcium in wheat, spinach, and kale. *J. Food Sci.* **58**: 605–608.

Berg, H.E. and van Boekel, M.A.J.S. (1994). Degradation of lactose during heating of milk. 1. Reaction pathways. *Neth. Milk Dairy J.* **48**: 157–175.

Berhow, M.A., Vermillion, K., Jham, G.N. et al. (2010). Purification of a sinapine-glucoraphanin salt from broccoli seeds. *Am. J. Plant Sci.* **1**: 113–118.

Bernal, M.A. and Barceló, A.R. (1996). 5,5′-Dicapsaicin, 4′-O-5-dicapsaicin ether, and dehydrogenation polymers with high molecular weights are the main products of the oxidation of capsaicin by peroxidase from hot pepper. *J. Agric. Food Chem.* **44**: 3085–3089.

Berthiller, F., Schuhmacher, R., Adam, G., and Krska, R. (2009). Formation, determination and significance of masked and other conjugated mycotoxins. *J. Anal. Bioanal. Chem.* **395**: 1243–1252.

Bertini, I., Gray, H.B., Stiefel, E.I., and Valentine, J.S. (2007). *Biological Inorganic Chemistry – Structure and Reactivity*. Sausalito, California: University Science Books.

Bessiére, Y. and Thomas, A.F. (eds.) (1990). *Flavour Science and Technology*. Chichester: Wiley.

Bessonova, I.A., Yunusov, S.Y., Kondrat'ev, V.G., and Shreter, A.I. (1987). Alkaloids of *Aconitum coreanum*. I. Structure of acorine. *Chem. Nat. Compd.* **23**: 573–575.

Bezman, Y., Rouseff, R.L., and Naim, M. (2001). 2-Methyl-3-furanthiol and methional are possible off-flavors in stored orange juice: aroma-similarity, NIF/SNIF GC-O, and GC analyses. *J. Agric. Food Chem.* **49**: 5425–5432.

Bianchi, G. and Pozzi, N. (1994). 3,4-Dihydroxyphenylglycol, a major C_6-C_2 phenolic in *Olea europaea* fruits. *Phytochemistry* **35**: 1335–1337.

Bianco, A., Scalzo, R.L., and Scarpati, M.L. (1993). Isolation of cornoside from Olea europea and its transformation into halleridone. *Phytochemistry* **32**: 455–457.

Biemel, K.M., Bühler, H.P., Reihl, O., and Lederer, M.O. (2001). Identification and quantitative evaluation of the lysine-arginine crosslinks GODIC, MODIC, DODIC, and glucosepan in foods. *Nahrung/Food* **45**: 210–214.

Biemel, K.M., Friedl, D.A., and Lederer, M.O. (2002). Identification and quantification of major Maillard cross-links in human serum albumin and lens protein. *J. Biol. Chem.* **277**: 24907–24915.

Bilia, A.R., Bergonzi, M.C., Lazari, D., and Vincieri, F.F. (2002). Characterization of commercial kava-kava herbal drug and herbal drug preparations by means of nuclear magnetic resonance spectroscopy. *J. Agric. Food Chem.* **50**: 5016–5025.

Bin, Q. and Peterson, D.G. (2016). Identification of bitter compounds in whole wheat bread crumb. *Food Chem.* **203**: 8–15.

Birch, G.G. and Lindley, M.G. (eds.) (1985). *Interaction of Food Components*. London: Elsevier.

Blanc, P.J., Laussac, J.P., Le Bars, P. et al. (1995). Characterisation of monascidin A from Monascus as citrinin. *Int. J. Food Microbiol.* **27**: 201–213.

Blanco, C.A., Rojas, A., Caballera, P.A. et al. (2006). A better control of beer properties by predicting acidity of hop iso-α-acids. *Trends Food Sci. Technol.* **17**: 373–377.

Blank, I. and Grosch, W. (1991). Evaluation of potent odorants in dill seed and dill herb (*Anethum graveolens* L.) by aroma extract dilution analysis. *J. Food Sci.* **56**: 63–67.

Blank, I. and Grosch, W. (2002). On the role of (−)-2-methylisoborneol for the aroma of Robusta coffee. *J. Agric. Food Chem.* **50**: 4653–4656.

Blank, I., Lin, J., Fumeaux, R. et al. (1996). Formation of 3-hydroxy-4,5-dimethyl-2(5H)-furanone (sotolone) from 4-hydroxy-l-isoleucine and 3-amino-4,5-dimethyl-3,4-dihydro-2(5H)-furanone. *J. Agric. Food Chem.* **44**: 1851–1856.

Blank, I., Lin, J., Vera, F.A. et al. (2001). Identification of potent odorants formed by autoxidation of arachidonic acid: structure elucidation and synthesis of (E,Z,Z)-2,4,7-tridecatrienal. *J. Agric. Food Chem.* **49**: 2959–2965.

Blank, I. and Schieberle, P. (1993). Analysis of the seasoning-like flavour substances of a commercial lovage extract. *Flavour Fragr. J.* **8**: 191–195.

Blank, I., Sen, A., and Grosch, W. (1992). Potent odorants of the roasted powder and brew of Arabica coffee. *Z. Lebensm. Unters. Forsch.* **195**: 239–245.

Block, E., Birringer, M., Jiang, W. et al. (2001). Allium chemistry. Synthesis, natural occurrence, biological activity, and chemistry of Se-alk(en)ylcysteines and their γ -glutamyl derivatives and oxidation products. *J. Agric. Food Chem.* **49**: 458–470.

Block, E., Dane, A.J., Thomas, S., and Cody, R.B. (2010). Applications of direct analysis in real time mass spectrometry (DART-MS) in Allium chemistry. 2-Propenesulfenic and 2-propenesulfinic acids, diallyl trisulfane S-oxide, and other reactive sulfur compounds from crushed garlic and other Alliums. *J. Agric. Food Chem.* **58**: 4617–4625.

Block, E., Dethier, B., Bechand, B. et al. (2018). Ajothiolanes: 3,4-dimethylthiolane natural products from garlic (*Allium sativum*). *J. Agric. Food Chem.* **66**: 10193–10204.

Blunden, G. (2003). Betaines in the plant kingdom and their use in ameliorating stress conditions in plants. *Acta Hortic.* **597**: 23–29. http://www.actahort.org/books/597/597_2.htm (access date 4 August 2013).

Blunden, S. and Wallace, T. (2003). Tin in canned food: a review and understanding of occurrence and effect. *Food Chem. Toxicol.* **41**: 1651–1662.

Bockisch, M. (1998). *Fats and Oils Handbook*. Champaign: AOCS Press.

Bohn, J.A. and BeMiller, J.N. (1995). (1→3)-β-D-Glucans as biological response modifiers: a review of structure-functional activity relationships. *Carbohydr. Polym.* **28**: 3–14.

Bolarinwa, I.F., Orfila, C., and Morgan, M.R.A. (2015). Determination of amygdalin in apple seeds, fresh apples and processed apple juices. *Food Chem.* **170**: 437–442.

Bolarinwa, I.F., Orfila, C., Morgan, M.R.A. et al. (2014). *Food Chem.* **152**: 133–139.

Bordiga, M., Travaglia, F., Locatelli, M. et al. (2010). Histaminol: identification and HPLC analysis of a novel compound in wine. *J. Agric. Food Chem.* **58**: 10202–10208.

Borenstein, B. (1987). The role of ascorbic acid in foods. *Food Technol.* **41**: 98–105.

Borg-Karlson, A.-K., Englund, F.O., and Unelius, C.R. (1994). Dimethyl oligosul-phides, major volatiles released from Sauromatum guttatum and Phallus impudicus. *Phytochemistry* **35**: 321–323.

Bortolomeazzi, R., Berno, P., Pizzale, L., and Conte, L.S. (2001). Sesquiterpene, alkene, and alkane hydrocarbons in virgin olive oils of different varieties and geographical origins. *J. Agric. Food Chem.* **49**: 3278–3283.

Boselli, E., Grob, K., and Lercker, G. (2000). Determination of furan fatty acids in extra virgin olive oil. *J. Agric. Food Chem.* **48**: 2868–2873.

Boskou, D. (1996). *Olive Oil: Chemistry and Technology*. Champaign: AOCS Press.

Bost, M., Houdart, S., Oberli, M. et al. (2016). Dietary copper and human health: current evidence and unresolved issues. *J. Trace Elem. Med. Biol.* **35**: 107–115.

Bowen-Forbes, C.S. and Minott, D.A. (2011). Tracking hypoglycins A and B over different maturity stages: implications for detoxification of ackee (Blighia sapida K.D. Koenig) fruits. *J. Agric. Food Chem.* **59**: 3869–3875.

Bradley, D.G. and Min, D.B. (1992). Singlet oxygen oxidation of foods. *Crit. Rev. Food Sci. Nutr.* **31**: 211–236.

Braeuer, S., Goessler, W., Kameník, J. et al. (2018). Arsenic hyperaccumulation and speciation in the edible ink stain bolete (*Cyanoboletus pulverulentus*). *Food Chem.* **242**: 225–231.

Brambilla, G., Abete, M.C., Chiaravalle, E. et al. (2013). Mercury occurrence in Italian seafood from the Mediterranean Sea and possible intake scenarios of the Italian coastal population. *Regul. Toxicol. Pharmacol.* **65**: 269–277.

Brandl, W., Galensa, R., and Herrmann, K. (1983). HPLC-Bestimmungen von Sellerieeinhaltsstoffen (Hydroxyzimtsäureester, Zucker, Mannit, Phthalide). *Z. Lebensm. Unters. Forsch.* **177**: 325–327.

Branen, A.L., Davidson, P.M., and Salminen, S. (eds.) (1990). *Food Additives*. New York: Dekker.

Brash, A.R. (1999). Lipoxygenases: occurrence, functions, catalysis, and acquisition of substrate. *J. Biol. Chem.* **274**: 23679–23682.

Brauer, S., Borovička, J., Glasnov, T. et al. (2018). Homoarsenocholine – A novel arsenic compound detected for the first time in nature. *Talanta* **188**: 107–110.

Breksa, A.P. III, Dragull, K., and Wong, R.Y. (2008). Isolation and identification of the first C-17 limonin epimer, epilimonin. *J. Agric. Food Chem.* **56**: 5595–5598.

Brewer, S. (2004). Irradiation effects on meat color – a review. *Meat Sci.* **68**: 1–17.

Bridle, P. and Timberlake, C.F. (1997). Anthocyanins as natural food colours-selected aspects. *Food Chem.* **58**: 103–109.

Briggs, L.R., Miles, C.O., Fitzgerald, J.M. et al. (2004). Enzyme-linked immunosorbent assay for the detection of yessotoxin and its analogues. *J. Agric. Food Chem.* **52**: 5836–5842.

Brinkmann, M. (1987). Zur Produktion von Kochsalzreduzierten Fleisch- und Wursterzeugnissen: der Stellenwert des Kochsalzes. *Fleischerei* **38**: 187–192.

Britz, S.J., Prasad, P.V.V., Moreau, R.A. et al. (2007). Influence of growth temperature on the amounts of tocopherols, tocotrienols, and γ -oryzanol in brown rice. *J. Agric. Food Chem.* **55**: 7559–7565.

Brock, A., Herzfeld, T., Paschke, R. et al. (2006). Brassicaceae contain nortropane alkaloids. *Phytochemistry* **67**: 2050–2057.

Brown, A.J. and Jessup, W. (1999). Oxysterols and atherosclerosis. *Atherosclerosis* **142**: 1–28.

Brown, G.D. and Gordon, S. (2003). Fungal β-glucans and mammalian immunity. *Immunity* **19**: 311–315.

Bruhns, P., Kaufmann, M., Koch, T., and Kroh, L.W. (2018). 2-Deoxyglucosone: a new C6-α-dicarbonyl compound in the Maillard reaction of D-fructose with γ-aminobutyric acid. *J. Agric. Food Chem.* **66**: 11806–11811.

Brzeska, M., Szymczyk, K., and Szterk, A. (2016). Current knowledge about oxysterols: a review. *J. Food Sci.* **81**: R2299–R2308.

Bubník, Z., Kadlec, P., Urban, D., and Bruhns, M. (eds.) (1995). *Sugar Technologists Manual*, 8e. Berlin: Bartens.

Buettner, A. and Schieberle, P. (2001a). Evaluation of aroma differences between hand-squeezed juices from Valencia late and Navel oranges by quantitation of key odorants and flavor reconstitution experiments. *J. Agric. Food Chem.* **49**: 2387–2394.

Buettner, A. and Schieberle, P. (2001b). Evaluation of key aroma compounds in hand-squeezed grapefruit juice (*Citrus paradisi* Macfayden) by quantitation and flavor reconstitution experiments. *J. Agric. Food Chem.* **49**: 1358–1363.

Bumpus, J.A. and Aust, S.D. (1987). Biodegradation of DDT [1,1,1-trichloro-2,2-bis(4-chlorophenyl)ethane] by the white rot fungus Phanerochaete chrysosporium. *Appl. Environ. Microbiol.* **53**: 2001–2008.

Bunzel, M., Ralph, J., Kim, H. et al. (2003). Sinapate dehydrodimers and sinapate-ferulate heterodimers in cereal dietary fiber. *J. Agric. Food Chem.* **51**: 1427–1434.

Bunzel, M., Ralph, J., and Steinhart, H. (2004). Phenolic compounds as cross-links of plant derived polysaccharides. *Czech J. Food Sci.* **22**: 64–67.

Burapan, S., Kim, M., and Han, J. (2017). Curcuminoid demethylation as an alternative metabolism by human intestinal microbiota. *J. Agric. Food Chem.* **65**: 3305–3310.

Burdock, G.A. (1997). *Encyclopedia of Food and Color Additives*. Boca Raton, FL: CRC Press.

Buskov, S., Hansen, L.B., Olsen, C.E. et al. (2000). Determination of ascorbigens in autolysates of various Brassica species using supercritical fluid chromatography. *J. Agric. Food Chem.* **48**: 2693–2701.

Busquets, R., Puignou, L., Galceran, M.T. et al. (2007). Liquid chromatography-tandem mass spectrometry analysis of 2-amino-1-methyl-6-(4-hydroxyphenyl)-imidazo[4,5-b]pyridine in cooked meats. *J. Agric. Food Chem.* **55**: 9318–9324.

Butinar, B., Bučar-Miklavčič, M., Mariani, C., and Raspor, P. (2011). New vitamin E isomers (gamma-tocomonoenol and alpha-tocomonoenol) in seeds, roasted seeds and roasted seed oil from the Slovenian pumpkin variety 'Slovenska golica'. *Food Chem.* **128**: 505–512.

Buttery, R.G. and Takeoka, G.R. (2013). Cooked carrot volatiles. AEDA and odor activity comparisons. Identification of linden ether as an important aroma component. *J. Agric. Food Chem.* **61**: 9063–9066.

Buttery, R.G., Takeoka, G.R., and Ling, L.C. (1995). Furaneol: odor threshold and importance to tomato aroma. *J. Agric. Food Chem.* **43**: 1638–1640.

Cadenas, E. and Paker, L. (1996). *Handbook of Antioxidants*. New York: Dekker.

Calkins, C.R. and Hodgen, J.M. (2007). A fresh look at meat flavour. *Meat Sci.* **77**: 63–80.

Câmara, J.S., Marques, J.C., Alves, M.A., and Ferreira, A.C.S. (2004). 3-Hydroxy-4,5-dimethyl-2(5H)-furanone levels in fortified Madeira wines: relationship to sugar content. *J. Agric. Food Chem.* **52**: 6765–6769.

Cameleyre, M., Lytra, G., Tempere, S., and Barbe, J.-C. (2017). 2-Methylbutyl acetate in wines: enantiomeric distribution and sensory impact on red wine fruity aroma. *Food Chem.* **237**: 364–371.

Cannon, R.J., Curto, N.L., Esposito, C.M. et al. (2017). The discovery of citral-like thiophenes in fried chicken. *J. Agric. Food Chem.* **65**: 5690–5699.

Capone, D.L., Sefton, M.A., and Jeffery, D.W. (2011). Application of a modified method for 3-mercaptohexan-1-ol determination to investigate the relationship between free thiol and related conjugates in grape juice and wine. *J. Agric. Food Chem.* **59**: 4649–4658.

Caputi, L., Carlin, S., Ghiglieno, I. et al. (2011). Relationship of changes in rotundone content during grape ripening and winemaking to manipulation of the 'peppery' character of wine. *J. Agric. Food Chem.* **59**: 5565–5571.

Carlson, D.G., Daxenbichler, M.E., VanEtten, C.H. et al. (1985). Glucosinolates in radish cultivars. *J. Am. Soc. Hortic. Sci.* **110**: 634–638.

Carnegie, P.R., Ilic, M.Z., Etheridge, M.O., and Collins, M.G. (1983). Improved high-performance liquid chromatographic method for analysis of histidine dipeptides anserine, carnosine and balenine present in fresh meat. *J. Chromatogr.* **261**: 153–157.

Carpita, N.C. and Gibeaut, D.M. (1993). Structural models of primary cell walls in flowering plants: consistency of molecular structure with the physical properties of the walls during growth. *Plant J.* **3**: 1–30.

Casado, F.J., Sánchez, A.H., De Castro, A. et al. (2011). Fermented vegetables containing benzoic and ascorbic acids as additives: benzene formation during storage and impact of additives on quality parameters. *J. Agric. Food Chem.* **59**: 2403–2409.

Casado, F.J., Sánchez, A.H., Rejano, L., and Montaño, A. (2007). D-Amino acid formation in sterilized alkali-treated olives. *J. Agric. Food Chem.* **55**: 3503–3507.

Casida, J.E. (2011). Curious about pesticide action. *J. Agric. Food Chem.* **59**: 2762–2769.

Castillo, M.L.R., Flores, G., Herraiz, M., and Blanch, G.P. (2003). Solid-phase microextraction for studies on the enantiomeric composition of filbertone in hazelnut oils. *J. Agric. Food Chem.* **51**: 2496–2500.

Cataldi, T.R.I., Margiotta, G., and Zambonin, C.G. (1998). Determination of sugars and alditols in food samples by HPAEC with integrated pulsed amperometric detection using alkaline eluents containing barium or strontiumons. *Food Chem.* **62**: 109–115.

Cattoor, K., Dresel, M., De Bock, L. et al. (2013). *J. Agric. Food Chem.* **61**: 7916–7924.

Čechovská, L., Cejpek, K., Konečný, M., and Velíšek, J. (2011). On the role of 5,6-dihydro-3,5-dihydroxy-2-methyl-(4H)-pyran-4-one in antioxidant capacity of prunes. *Eur. Food Res. Technol.* **233**: 367–376.

Cejpek, K., Urban, J., Velíšek, J., and Hrabcová, H. (1998). Effect of sulfite treatment on allyl isothiocyanate in mustard paste. *Food Chem.* **62**: 53–57.

Cevc, G. and Paltauf, F. (1995). *Phospholipids: Characterization, Metabolism and Novel Biological Applications.* Champaign: AOCS Press.

Chadwick, M., Gawthrop, F., Michelmore, R.W. et al. (2016). Perception of bitterness, sweetness and liking of different genotypes of lettuce. *Food Chem.* **197**: 66–74.

Chahin, A., Guiavarc'h, Y.P., Dziurla, M.-A. et al. (2008). 1-Hydroxypyrene in milk and urine as a bioindicator of polycyclic aromatic hydrocarbon exposure of ruminants. *J. Agric. Food Chem.* **56**: 1780–1786.

Chai, W. and Liebman, M. (2005). Effect of different cooking methods on vegetable oxalate content. *J. Agric. Food Chem.* **53**: 3027–3030.

Chan, H.W.-S. (1987). *Autoxidation of Unsaturated Lipids.* London: Academic Press.

Chang, H.T., Cheng, H., Han, R.M. et al. (2017). Regeneration of β-carotene from radical cation by eugenol, isoeugenol, and clove oil in the Marcus theory inverted region for electron transfer. *J. Agric. Food Chem.* **65**: 908912.

Chang, S.-C., Lu, K.-L., and Yeh, S.-F. (1993). Secondary metabolites resulting from degradation of PR toxin by Penicillium roqueforti. *Appl. Environ. Microbiol.* **59**: 981–986.

Charlambous, G. (ed.) (1995). *Food Flavors: Generation, Analysis and Process Influence.* Amsterdam: Elsevier.

Charles, M., Martin, B., Ginies, C. et al. (2000). Potent aroma compounds of two red wine vinegars. *J. Agric. Food Chem.* **48**: 70–77.

Charve, J., Manganiello, S., and Glabasnia, A. (2018). Analysis of Umami taste compounds in a fermented corn sauce by means of sensory-guided fractionation. *J. Agric. Food Chem.* **66**: 1863–1871.

Cheeke, P.R. (ed.) (1989). *Toxicants of Plant Origin. Vol. III. Proteins and Amino Acids.* Boca Raton, FL: CRC Press.

Chen, B.H. (1997). Analysis, formation and inhibition of polycyclic aromatic hydrocarbons in foods: an overview. *J. Food Drug Anal.* **5**: 25–42.

Chen, C.-C. and Ho, C.-T. (1986). Identification of sulfurous compounds of shiitake mushroom (*Lentinus edodes* Sing.). *J. Agric. Food Chem.* **34**: 830–833.

Chen, C.-C., Kuo, M.-C., Wu, C.-M., and Ho, C.-T. (1986). Pungent compounds of ginger (*Zingiber officinale* Roscoe) extracted by liquid carbon dioxide. *J. Agric. Food Chem.* **34**: 477–480.

Chen, C.-N., Liang, C.-M., Lai, J.-R. et al. (2003). Capillary electrophoretic determination of theanine, caffeine, and catechins in fresh tea leaves and oolong tea and their effects on rat neurosphere adhesion and migration. *J. Agric. Food Chem.* **51**: 7495–7503.

Chen, H., Shurlknight, K., Leung, T.C., and Sang, S. (2012). Structural identification of theaflavin trigallate and tetragallate from black tea using liquid chromatography/electrospray ionization tandem mass spectrometry. *J. Agric. Food Chem.* **60**: 10850–10857.

Chen, J.S., Wei, C.-I., Rolle, R.S. et al. (1991). Inhibitory effect of kojic acid on some plant and crustacean polyphenol oxidases. *J. Agric. Food Chem.* **39**: 1396–1401.

Chen, L., Capone, D.L., and Jeffery, D.W. (2018). Identification and quantitative analysis of 2-methyl-4-propyl-1,3-oxathiane in wine. *J. Agric. Food Chem.* **66**: 10808–10815.

Chen, L., Yuan, P., Chen, K. et al. (2014). Oxidative conversion of B- to A-type procyanidin trimer: evidence for quinone methide mechanism. *Food Chem.* **154**: 315–322.

Cheng, W.-Y. and Yueh-Hsiung, K. (2007). Huang Ching-Jang: isolation and identification of novel estrogenic compounds in yam tuber (*Dioscorea alata* Cv. Tainung No. 2). *J. Agric. Food Chem.* **55**: 7350–7358.

Chetschik, I., Granvogl, M., and Schieberle, P. (2008). Comparison of the key aroma compounds in organically grown, raw West-African peanuts (*Arachis hypogaea*) and in ground, pan-roasted meal produced thereof. *J. Agric. Food Chem.* **56**: 10237–10243.

Cheyns, K., Waegeneers, N., Van de Wiele, T., and Ruttens, A. (2017). Arsenic release from foodstuffs upon food preparation. *J. Agric. Food Chem.* **65**: 2443–2453.

Chin, Y.-W., Jung, H.-A., Chai, H. et al. (2008). Xanthones with quinone reductase-inducing activity from the fruits of *Garcinia mangostana* (mangosteen). *Phytochemistry* **69**: 754–758.

Cho, J.-Y., Kim, C.M., Lee, H.J. et al. (2013). Caffeoyl triterpenes from pear (*Pyrus pyrifolia* Nakai) fruit peels and their antioxidative activities against oxidation of rat blood plasma. *J. Agric. Food Chem.* **61**: 4563–4569.

Choi, S.W., Lee, S.K., Kim, E.O. et al. (2007). Antioxidant and antimelanogenic activities of polyamine conjugates from corn bran and related hydroxycinnamic acids. *J. Agric. Food Chem.* **55**: 3920–3925.

Chowdhury, B., Rozan, P., Kuo, Y.-H. et al. (2001). Identification and quantification of natural isoxazolinone compounds by capillary zone electrophoresis. *J. Chromatogr. A* **933**: 129–136.

Christie, W.W. The lipid library. Lipid chemistry, biology, technology & analysis. https://lipidlibrary.aocs.org (accessed 28 October 2019).

Chrysanthou, A., Pouliou, E., and Kyriakoudi, A. (2016). Sensory threshold studies of picrocrocin, the major bitter compound of saffron. *J. Food Sci.* **81**: S189–S198.

Chu, F.L. and Yaylayan, V.A. (2008a). Model studies on the oxygen-induced formation of benzaldehyde from phenylacetaldehyde using pyrolysis GC-MS and FTIR. *J. Agric. Food Chem.* **56**: 10697–10704.

Chu, F.L. and Yaylayan, V.A. (2008b). Post-Schiff base chemistry of the Maillard reaction. *Ann. N. Y. Acad. Sci.* **1126**: 30–37.

Chung, S.K., Subedi, L., Kwon, O.W. et al. (2016). Wasabisides A–E, lignan glycosides from the roots of *Wasabia japonica*. *J. Nat. Prod.* **79**: 2652–2657.

Ciska, E., Drabińska, N., Narwojsz, A., and Honke, J. (2016). Stability of glucosinolates and glucosinolate degradation products during storage of boiled white cabbage. *Food Chem.* **203**: 340–347.

Clark, A.J. (1996). *Biodegradation of Cellulose*. Basel, Switzerland: Technomic.

Clifford, M.N. and Kazi, T. (1987). The influence of coffee bean maturity on the content of chlorogenic acid, caffeine and trigonelline. *Food Chem.* **26**: 59–69.

Clifford, M.N., Marks, S., Knight, S., and Kuhnert, N. (2006). Characterization by LC-MSn of four new classes of p-coumaric acid-containing diacyl chlorogenic acids in green coffee beans. *J. Agric. Food Chem.* **54**: 4095–4101.

Clifford, M.N., Stoupi, S., and Kuhnert, N. (2007). Profiling and characterization by LC-MSn of the galloylquinic acids of green tea, tara tannin, and tannic acid. *J. Agric. Food Chem.* **55**: 2797–2807.

Clydesdale, F.M. (1996). *Food Additives. Toxicology, Regulation, and Properties*. Boca Raton, FL: CRC Press.

Combs, F.C. and Combs, S.B. (1986). *The Role of Selenium in Nutrition*. New York: Academic Press.

Core Report Euro Diet. Nutrition & Diet for Healthy Lifestyles in Europe, Science & Policy, Implications. http://ec.europa.eu/health/archive/ph_determinants/life_style/nutrition/report01_en.pdf (accessed 28 October 2019).

Cornelis, R., Caruso, J.A., Crews, H., and Heumann, K.G. (eds.) (2005). *Handbook of Elemental Speciation, Handbook of Elemental Speciation II: Species in the Environment, Food, Medicine and Occupational Health*. Wiley.

Cornell, H.J. and Hoveling, A.W. (1997). *Wheat: Chemistry and Utilization*. Lancaster, PA: Technomic.

Cornell University Law School: Legal Information Institute, Pub. L. 93–523; 42 U.S.C. § 300f et seq. December 16, 1974. http://www.law.cornell.edu/uscode/text/42/300f (accessed 28 October 2019).

Coultate, T.P. (1989). *Food. The Chemistry of its Components*, 2e. London: The Royal Society of Chemistry.

Cramer, B., Königs, M., and Humpf, H.-U. (2008). Identification and in vitro cytotoxicity of ochratoxin A degradation products formed during coffee roasting. *J. Agric. Food Chem.* **56**: 5673–5681.

Cremer, H.-D., Hötzel, D., and Kühnau, J. (1980). *Biochemie und Physiologie der Ernährung*. Stuttgart: Georg Thieme.

Cremlyn, R. (1978). *Pesticides and Mode of Action.* Chichester: Wiley.

Crews, C., Hough, P., Brereton, P. et al. (2002). Survey of 3-monochloropropane-1,2-diol (3-MCPD) in selected food groups, 1999–2000. *Food Addit. Contam.* **19**: 22–27.

Crichton, R.R. (2008). *Biological Inorganic Chemistry – An Introduction.* Amsterdam: Elsevier.

Crittenden, R.G. and Playne, M.J. (1996). Production, properties and applications of food-grade oligosaccharides. *Trends Food Sci. Technol.* **7**: 353–361.

Crosby, D.G. and Aharonson, N. (1967). The structure of carotatoxin, a natural toxicant from carrot. *Tetrahedron* **23**: 465–472.

Cuadrado, C., Ayet, G., Robredo, L.M. et al. (1996). Effect of natural fermentation on the content of inositol phosphates in lentils. *Z. Lebensm. Unters. Forsch.* **203**: 268–271.

Cuevas, M.E., Rodriguez, A.M., Hillebrand, S., and Winterhalter, P. (2011). Anthocyanin composition of black carrot (Daucus carota ssp. sativus var. atrorubens Alef.) cultivars Antonina, Beta Sweet, Deep Purple, and Purple Haze. *J. Agric. Food Chem.* **59**: 3385–3390.

Cunha, S.C., Amaral, J.S., Fernandes, J.O., and Oliveira, M.B. (2006). Quantification of tocopherols and tocotrienols in Portuguese olive oils using HPLC with three different detection systems. *J. Agric. Food Chem.* **54**: 3351–3356.

Czepa, A. and Hofmann, T. (2004). Quantitative studies and sensory analyses on the influence of cultivar, spatial tissue distribution, and industrial processing on the bitter off-taste of carrots (Daucus carota L.) and carrot products. *J. Agric. Food Chem.* **52**: 4508–4514.

Czerny, M. and Buettner, A. (2009). Odor-active compounds in cardboard. *J. Agric. Food Chem.* **57**: 9979–9984.

Czerwińska, M.E. and Melzig, M.F. (2018). *Cornus mas* and *Cornus officinalis*-Analogies and differences of two medicinal plants traditionally used. *Front. Pharmacol.* **9** https://doi.org/10.3389/fphar.2018.00894.

D'Mello, J.P.F., Duffus, C.M., and Duffus, J.H. (eds.) (1991). *Toxic Substances in Crop Plants.* Cambridge: The Royal Society of Chemistry.

Da Silva Ferreira, A.C., Barbe, J.-C., and Bertrand, A. (2002). Heterocyclic acetals from glycerol and acetaldehyde in Port wines: evolution with aging. *J. Agric. Food Chem.* **50**: 2560–2564.

Da Silva, R., Rigaud, J., Cheynier, V. et al. (1991). Procyanidin dimers and trimers from grape seeds. *Phytochemistry* **30**: 1259–1264.

Dabrowski, W.M. and Sikorski, Z.E. (eds.) (2004). *Toxins in Food.* CRC Press – Taylor & Francis Group.

Damodaran, S., Parkin, K.L., and Fennema, O.R. (eds.) (2007). *Fennema's Food Chem*, 4e. Boca Raton: CRC Press.

Dao, L. and Friedman, M. (1992). Chlorogenic acid content of fresh and processed potatoes determined by ultraviolet spectrophotometry. *J. Agric. Food Chem.* **40**: 2152–2156.

Daramwar, P.P., Srivastava, P.L., Priyadarshinia, B., and Thulasiram, H.V. (2012). Preparative separation of α- and β-santalenes and (Z)-α- and (Z)-β-santalols using silver nitrate-impregnated silica gel medium pressure liquid chromatography and analysis of sandalwood oil. *Analyst* **137**: 4564–4570.

Das, M., Asthana, S., Singh, S.P. et al. (2015). Litchi fruit contains methylene cyclopropyl-glycine. *Curr. Sci.* **109**: 2195–2197.

Daskaya-Dikmen, C., Yucetepe, A., Karbancioglu-Guler, F. et al. (2017). Angiotensin-I-converting enzyme (ACE)-inhibitory peptides from plants. *Nutrients* **9** (316): 1–19.

Davídek, J. (ed.) (1995). *Natural Toxic Compounds of Foods Formation and Change During Food Processing and Storage.* Boca Raton, FL: CRC Press.

Davídek, J., Velíšek, J., and Pokorný, J. (1990). *Chemical Changes During Food Processing.* Amsterdam: Elsevier.

Davídek, T., Devaud, S., Robert, F., and Blank, I. (2006). Sugar fragmentation in the Maillard reaction cascade: isotope labeling studies on the formation of acetic acid by a hydrolytic α-dicarbonyl cleavage mechanism. *J. Agric. Food Chem.* **54**: 6667–6676.

Davidson, K.W., Booth, S.L., Dolnikowski, G.G., and Sadowski, J.A. (1996). Conversion of vitamin K_1 to 2′,3′-dihydrovitamin K_1 during the hydrogenation of vegetable oils. *J. Agric. Food Chem.* **44**: 980–983.

Davies, A.M.C., Wilkinson, C.C.L., and Jones, J.M. (1978). Carnosine and anserine content of turkey breast and leg muscles. *Br. Poult. Sci.* **19**: 101–103.

Davis, G.E., Garwood, V.W., Barfuss, D.L. et al. (1990). Chromatographic profile of carbohydrates in commercial coffees. 2. Identification of mannitol. *J. Agric. Food Chem.* **38**: 1347–1350.

Dawid, C., Dunemann, F., Schwab, W. et al. (2015). Bioactive C_{17}-polyacetylenes in carrots (*Daucus carota* L.): current knowledge and future perspectives. *J. Agric. Food Chem.* **63**: 9211–9222.

Dawid, C., Henze, A., Frank, O. et al. (2012). Structural and sensory characterization of key pungent and tingling compounds from black pepper (*Piper nigrum* L.). *J. Agric. Food Chem.* **60**: 2884–2895.

Dawid, C. and Hofmann, T. (2012). Structural and sensory characterization of bitter tasting steroidal saponins from asparagus spears (*Asparagus officinalis* L.). *J. Agric. Food Chem.* **60**: 11889–11900.

Day, L. (2013). Proteins from land plants – Potential resources for human nutrition and food security. *Trends Food Sci. Technol.* **32**: 25–42.

De Almeida, N.E.C., do Nascimento, E.S.P., and Cardoso, D.R. (2012). On the reaction of lupulones, hop β-acids, with 1-hydroxyethyl radical. *J. Agric. Food Chem.* **60**: 10649–10656.

De Freitas, O., Padovan, G.J., Vilela, L. et al. (1993). Characterization of protein hydrolysates prepared for enteral nutrition. *J. Agric. Food Chem.* **41**: 1432.

De Girolamo, A., Solfrizzo, M., Vitti, C., and Visconti, A. (2004). Occurrence of 6-methoxymellein in fresh and processed carrots and relevant effect of storage and processing. *J. Agric. Food Chem.* **52**: 6478–6484.

de Graaf, R.M., Krosse, S., Swolfs, A.E.M. et al. (2015). Isolation and identification of 4-α-rhamnosyloxy benzyl glucosinolate in *Noccaea caerulescens*. *Phytochemistry* **110**: 166–171.

De Keukeleire, D. (2000). Fundamentals of beer and hop chemistry. *Quim Nova* **23**: 108–112.

De Marino, S., Borbone, N., Gala, F. et al. (2006). New constituents of sweet Capsicum annuum L. fruits and evaluation of their biological activity. *J. Agric. Food Chem.* **54**: 7508–7516.

De Paula Lima, J., Farah, A., King, B. et al. (2016). Distribution of major chlorogenic acids and related compounds in Brazilian green and toasted *Ilex paraguariensis* (maté) leaves. *J. Agric. Food Chem.* **64**: 2361–2370.

De Rijke, E., Ruisch, B., Bakker, J. et al. (2007). LC-MS study to reduce ion suppression and to identify N-lactoylguanosine 5′ -monophosphate in bonito: a new umami molecule? *J. Agric. Food Chem.* **55**: 6417–6423.

De Rosso, V.V., Hillebrand, S., Cuevas, M.E. et al. (2008). Determination of anthocyanins from acerola (*Malpighia emarginata* DC.) and açai (*Euterpe oleracea* Mart.) by HPLC–PDA–MS/MS. *J. Food Compos. Anal.* **21**: 291–299.

De Taeye, C., Caullet, G., Eyamo Evina, V.J., and Collin, S. (2017). Procyanidin A2 and its degradation products in raw, fermented and rosted cocoa. *J. Agric. Food Chem.* **65**: 1715–1723.

De Taeye, L., De Keukeleire, D., and Verzele, M. (1979). Isolation and identification of the anti-isohumulones and the anti-acetylhumulinic acids. *Tetrahedron* **35**: 989–992.

De Vincenzi, M., Maialetti, F., and Silano, M. (2003). Constituents of aromatic plants: teucrin A. *Fitoterapia* **74**: 746–749.

De Voogt, P., Wells, D.E., Reutergardh, L., and Brinkman, U.A.T. (1990). Biological activity, determination and ocurrence of planar, mono- and di-ortho PCBs. *Int. J. Environ. Anal. Chem.* **40**: 1–46.

De Vos, R.H., van Dokkum, W., Schouten, A., and de Jong-Berkhout, P. (1990). Polycyclic aromatic compounds in Dutch total diet samples (1984–1986). *Food Chem. Toxicol.* **28**: 263–268.

Debolt, S., Melino, V., and Ford, C.M. (2007). Ascorbate as a biosynthetic precursor in plants. *Ann. Bot.* **99**: 3–8.

Degenhardt, A.G. and Hofmann, T. (2010). Bitter-tasting and kokumi-enhancing molecules in thermally processed avocado (*Persea americana* Mill.). *J. Agric. Food Chem.* **58**: 12906–12915.

Del Campo, G., Gallego, B., Berregi, I., and Casado, J.A. (1998). Creatinine, creatine and protein in cooked meat products. *Food Chem.* **63**: 187–190.

Del Prette, D., Millán, E., Pollastro, F. et al. (2016). Turmeric sesquiterpenoids: expeditious resolution, comparative bioactivity, and a new bicyclic turmeronoid. *J. Nat. Prod.* **79**: 267–273.

Delange, F., Bürgi, H., Chen, Z.P., and Dunn, J.T. (2002). World status of monitoring iodine deficiency disorders control programs. *Thyroid* **12**: 915–924.

Delaquis, P.J. and Mazza, G. (1995). Antimicrobial properties of isothiocyanates in food preservation. *Food Technol.* **49**: 73–84.

Deli, J., Matus, Z., and Tóth, G. (1996). Carotenoid composition in the fruits of Capsicum annuum cv. Szentesi Kosszarvú during ripening. *J. Agric. Food Chem.* **44**: 711–716.

Deli, J. and Ösz, E. (2004). Carotenoid 5,6-, 5,9- and 3,6-epoxides. *Arkivoc* **vii**: 150–168.

Dell'Agli, M., Giavarini, F., Ferraboschi, P. et al. (2007). Determination of aloesin and aloeresin a for the detection of Aloe in beverages. *J. Agric. Food Chem.* **55**: 3363–3367.

Delzenne, N.M. and Roberfroid, M.R. (1994). Physiological effects of non-digestible oligosaccharides. *LWT Food Sci. Technol.* **27**: 1–6.

Dembitsky, V.M., Abu-Lafi, S., and Hanuš, L.O. (2007). Separation of sulfur-containing fatty acids from garlic, Allium sativum, using serially coupled capillary columns with consecutive nonpolar, semipolar, and polar stationary phases. *Acta Chromatogr.* **18**: 206–216.

Department of Health and Human Services (1992). Toxicological profile for DEHP. Draft for public comment, U.S. Department of Health and Human Services (February 18).

Derksena, G.C.H., van Beeka, T.A., de Groota, Æ., and Capelleb, A. (1998). High-performance liquid chromatographic method for the analysis of anthraquinone glycosides and aglycones in madder root (*Rubia tinctorum* L.). *J. Chromatogr. A* **816**: 277–281.

Dermiki, M., Phanphensophon, N., Mottram, D.S., and Methven, L. (2013). Contributions of non-volatile and volatile compounds to the umami taste and overall flavour of shiitake mushroom extracts and their application as flavour enhancers in cooked minced meat. *Food Chem.* **141**: 77–83.

Desjardins, A.E. (ed.) (2006). *Fusarium Mycotoxins*. St Paul, MN: APS Press.

Destaillats, F., Craft, B.D., Dubois, M., and Nagy, K. (2012). Glycidyl esters in refined palm (*Elaeis guineensis*) oil and related fractions. Part I: Formation mechanism. *Food Chem.* **131**: 1391–1398.

Deutsch, J.C. (1998). Ascorbic acid oxidation by hydrogen peroxide. *Anal. Biochem.* **255**: 1–7.

Devlin, H.R. and Harris, I.J. (1984). Mechanism of the oxidation of aqueous phenol with dissolved oxygen. *Ind. Eng. Chem. Fundam.* **23**: 387–392.

Dewick, P.M. (2002). *Medicinal Natural Products: A Biosynthetic Approach*, 2e. Chichester: Wiley.

Dharmananda, S. (2004). Sweet fruit used as sugar substitute and medicinal herb. Institute for Traditional Medicine Online (2004). http://www.itmonline.org/arts/luohanguo.htm (accessed 28 October 2019).

Dharmaratne, H.R.W., Nanayakkara, N.P.D., and Ikhlas Akhan, I.A. (2002). Kavalactones from *Piper methysticum* and their ^{13}C NMR spectroscopic analyses. *Phytochemistry* **59**: 429–433.

Dias, R.C.E., Campanha, F.G., Vieira, L.G.E. et al. (2010). Evaluation of kahweol and cafestol in coffee tissues and roasted coffee by a new high-performance liquid chromatography methodology. *J. Agric. Food Chem.* **58**: 88–93.

Diawara, M.M., Trumble, J.T., Quiros, C.F., and Hansen, R. (1995). Implications of distribution of linear furanocoumarins within celery. *J. Agric. Food Chem.* **43**: 723–727.

Diaz-Maroto, M.C. and Pérez-Coello, M.S. (2002). Effect of different drying methods on the volatile components of parsley (*Petroselinum crispum* L.). *Eur. Food Res. Technol.* **215**: 227–230.

Dicenta, F., Martínez-Gómez, P., Grané, N. et al. (2002). Relationship between cyanogenic compounds in kernels, leaves, and roots of sweet and bitter kernelled almonds. *J. Agric. Food Chem.* **50**: 2149–2152.

Diem, S., Bergmann, J., and Herderich, M. (2000). Tryptophan-N-glucoside in fruits and fruit juices. *J. Agric. Food Chem.* **48**: 4913–4917.

Diem, S. and Herderich, M. (2001a). Reaction of tryptophan with carbohydrates: identification and quantitative determination of novel β-carboline alkaloids in food. *J. Agric. Food Chem.* **49**: 2486–2492.

Diem, S. and Herderich, M. (2001b). Reaction of tryptophan with carbohydrates: mechanistic studies on the formation of carbohydrate-derived β-carbolines. *J. Agric. Food Chem.* **49**: 5473–5478.

Dierkes, G., Krieger, S., Dück, R. et al. (2012). High-performance liquid chromatography–mass spectrometry profiling of phenolic compounds for evaluation of olive oil bitterness and pungency. *J. Agric. Food Chem.* **60**: 7597–7606.

Dihm, K., Lind, M.V., Sundén, H. et al. (2017). Quantification of benzoxazinoids and their metabolites in Nordic breads. *Food Chem.* **235**: 7–13.

Dimberg, L.H., Theander, O., and Lingnert, H. (1993). Avenanthramides – a group of phenolic antioxidants in oats. *Cereal Chem.* **70**: 637–641.

Din, Z., Xiong, H., and Fei, P. (2017). Physical and chemical modification of starches: a review. *Crit. Rev. Food Sci. Nutr.* **57**: 2691–2705.

Dini, I., Tenore, G.C., and Dini, A. (2006a). New polyphenol derivative in ipomoea batatas tubers and its antioxidant activity. *J. Agric. Food Chem.* **54**: 8733–8737.

Dini, I., Tenore, G.C., Trimarco, E., and Dini, A. (2006b). Seven new aminoacyl sugars in Ipomoea batatas. *J. Agric. Food Chem.* **54**: 6089–6093.

Diosady, L.L. (2005). Chlorophyll removal from edible oils. *Int. J. Appl. Sci. Eng. Technol.* **3**: 81–88.

Dobson, G., Christie, W.W., and Sebedio, J.L. (1996). The nature of cyclic acids formed in heated vegetable oils. *Grasas Aceites* **47**: 26–33.

Dreiucker, J. and Vetter, W. (2011). Fatty acids patterns in camel, moose, cow and human milk as determined with GC/MS after silver ion solid phase extraction. *Food Chem.* **126**: 762–771.

Drewes, S.E., Taylor, C.W., Cunningham, A.B. et al. (1992). Epiafzelechin-(4β→8, 2β→0→7)-ent-afzelechin from Cassipourea gerrardii. *Phytochemistry* **31**: 2491–2494.

Du, C.T. and Francis, F.J. (1977). Anthocyanins of mangosteen, *Garcinia mangostana*. *J. Food Sci.* **42**: 1667–1668.

Du, X., Song, M., and Rouseff, R. (2011). Identification of new strawberry sulfur volatiles and changes during maturation. *J. Agric. Food Chem.* **59**: 1293–1300.

Duetz, W.A., Fjällman, A.H.M., Ren, S. et al. (2001). Biotransformation of d-limonene to (+)-trans-carveol by toluene-grown Rhodococcus opacus PWD4 Cells. *Appl. Environ. Microbiol.* **67**: 2829–2832.

Dugasani, S., Pichika, M.R., Nadarajah, V.D. et al. (2010). Comparative antioxidant and anti-inflammatory effects of [6]-gingerol, [8]-gingerol, [10]-gingerol and [6]-shogaol. *J. Ethnopharmacol.* **127**: 515–520.

Dugrand, A., Olry, T., Duval, T. et al. (2013). Coumarin and furocoumarin quantitation in citrus peel via ultraperformance liquid chromatography coupled with mass spectrometry (UPLC-MS). *J. Agric. Food Chem.* **61**: 10677–10684.

Dunkel, A. and Hofmann, T. (2009). Sensory-directed identification of β-alanyl dipeptides as contributors to the thick-sour and white-meaty orosensation induced by chicken broth. *J. Agric. Food Chem.* **57**: 9867–9877.

Dutta, P.C. (1997). Studies on phytosterol oxides II. Content in some vegetable oils and in French fries prepared in these oils. *J. Am. Oil Chem. Soc.* **74**: 659–666.

Edge, R. and Truscott, T.G. (2018). Singlet oxygen and free radical reactions of retinoids and carotenoids – a review. *Antioxidants* **7**: 5.

Edwards, C.A. (1986). *Environmental Pollution by Pesticides*. New York: Plenum.

EFSA (2003). Opinion of the scientific panel on biological hazards on a request from the commission related to the effects of nitrites/nitrates on the microbiological safety of meat products. *EFSA J.* **14**: 1–34.

EFSA (2005). Opinion of the scientific panel of dietetic products, nutrition and allergies on a request from the commission related to the tolerable upper intake level of tin. *EFSA J.* **254**: 1–25.

EFSA (2008). Scientific opinion of the panel on contaminants in the food chain on a request from the European Commission on Polycyclic Aromatic Hydrocarbons in Food. *EFSA J.* **724**: 1–114.

EFSA (2009a). Panel on contaminants in the food chain: scientific opinion on arsenic in food. *EFSA J.* **7**: 1351.

EFSA (2009b). Scientific opinion of the panel on contaminants in the food chain on a request from the European Commission on cadmium in food. *EFSA J.* **980**: 1.

EFSA (2009c). Scientific opinion of the panel on food Additives and Nutrient Sources added to food (ANS) on orotic acid salts as sources of orotic acid and various minerals added for nutritional purposes to food supplements, following a request from the European Commission. *EFSA J.* **1187**: 1–25.

EFSA (2010). Panel on contaminants in the food chain: scientific opinion on lead in food. *EFSA J.* **8**: 1570.

EFSA (2011a). Panel on biological hazards. *EFSA J.* **9**: 2393–2486.

EFSA (2011b). Panel on contaminants in the food chain: scientific opinion on the risk for public health related to the presence of opium alkaloids in poppy seeds. *EFSA J.* **9**: 2405.

EFSA (2012a). Panel on contaminants in the food chain: scientific opinion on the risk for public health related to the presence of mercury and methylmercury in food. *EFSA J.* **10**: 2985.

EFSA (2012b). Cadmium dietary exposure in the European population. *EFSA J.* **10**: 2551.

EFSA (2008). Marine biotoxin shellfish-azaspiracid group. *EFSA J.* **723**: 1–52.

EFSA (2008). Safety of aluminium from dietary intake. Scientific opinion of the panel on food additives, flavourings, processing aids and food contact materials (AFC). *EFSA J.* **754**: 1–34, including Annex pp. 1–88.

Eggink, P.M., Maliepaard, C., Tikunov, Y. et al. (2012). A taste of sweet pepper: volatile and non-volatile chemical composition of fresh sweet pepper (*Capsicum annuum*) in relation to sensory evaluation of taste. *Food Chem.* **132**: 301–310.

El Ramya, R., Elhkimb, M.O., Lezmic, S., and Poul, J.M. (2007). Evaluation of the genotoxic potential of 3-monochloropropane-1,2-diol (3-MCPD) and its metabolites, glycidol and β-chlorolactic acid, using the single cell gel/comet assay. *Food Chem. Toxicol.* **45**: 41–48.

Eliasson, A.-C. (ed.) (1996). *Carbohydrates in Food*. New York: Dekker.

Elinder, C.G., Lind, B., Nilsson, B., and Oskarsson, A. (1988). Wine – an important source of lead exposure. *Food Addit. Contam.* **5**: 641–644.

Emura, M., Yaguchi, Y., Nakahashi, A. et al. (2009). Stereochemical studies of odorous 2-substituted-3(2H)-furanones by vibrational circular dichroism. *J. Agric. Food Chem.* **57**: 9909–9915.

Enamorado, S., Abril, J.M., Delgado, A. et al. (2014). Implications for food safety of the uptake by tomato of 25 trace-elements from a phosphogypsum amended soil from SW Spain. *J. Hazard. Mater.* **266**: 122–131.

Endlová, L., Laryšová, A., Vrbovský, V., and Navrátilová, Z. (2015). Analysis of alkaloids in poppy straw by high-performance liquid chromatography. *IOSR J. Eng.* **5**: 1–7.

Endo, H., Hosoya, H., Koyama, T., and Ichioka, M. (1982). Isolation of 10-hydroxypheophorbide as a photosensitizing pigment from alcohol-treated chlorella cells. *Agric. Biol. Chem.* **46**: 2183–2193.

Endo, Y., Thorsteinson, C.T., and Daun, J.K. (1992). Characterization of chlorophyll pigments present in canola seed, meal and oil. *J. Am. Oil Chem. Soc.* **69**: 564–568.

Engel, K.H. and Tressl, R. (1991). Identification of new sulfur-containing volatiles in yellow passion fruits (*Passiflora edulis* f. flavicarpa). *J. Agric. Food Chem.* **39**: 2249–2252.

Engel, W. and Schieberle, P. (2002). Structural determination and odor characterization of N-(2-mercaptoethyl)-1,3-thiazolidine, a new intense popcorn-like-smelling odorant. *J. Agric. Food Chem.* **50**: 5391–5393.

Engelhardt, U.H., Lakenbrink, C., and Pokorny, O. (2004). Proanthocyanidins, bisflavanols, and hydrolyzable tannins in green and black teas. Nutraceutical Beverages, Chapter 19, p. 254, ACS Symposium Series, Vol. 871, American Chemical Society.

Englyst, H.N. and Hudson, G.J. (1996). The classification and measurement of dietary carbohydrates. *Food Chem.* **57**: 15–21.

Englyst, H.N., Quigley, M.E., Englyst, K.N. et al. (1996). Dietary fibre. *J. Assoc. Public Anal.* **32**: 1–16.

Enman, J., Rova, U., and Berglund, K.A. (2007). Quantification of the bioactive compound eritadenine in selected strains of shiitake mushroom (*Lentinus edodes*). *J. Agric. Food Chem.* **55**: 1177–1180.

Eriksson, C. (ed.) (1981). *Progress in Food and Nutrition Science. Maillard Reactions in Food*. Oxford: Pergamon Press.

Esatbeyoglu, T., Wray, V., and Winterhalter, P. (2015). Isolation of dimeric, trimeric, tetrameric and pentameric procyanidins from unroasted cocoa bean (*Theobroma cacao* L.) using countercurrent chromatography. *Food Chem.* **179**: 278–289.

Escribano, J., Cabanes, J., Jiménez-Atiénzar, M. et al. (2017). Characterization of betalains, saponins and antioxidant power in differently colored quinoa (*Chenopodium quinoa*) varieties. *Food Chem.* **234**: 285–294.

Escribano-Bailón, M.T., Guerra, M.T., Rivas-Gonzalo, J.C., and Santos-Buelga, C. (1995). Proanthocyanidins in skins from different grape varieties. *Z. Lebensm. Unters. Forsch.* **200**: 221–224.

Escribano-Bailón, M.T., Gutiérrez-Fernández, Y., Rivas-Gonzalo, J.C., and Santos-Buelga, C. (1992). Characterization of procyanidins of Vitis vinifera variety Tinta del País grape seeds. *J. Agric. Food Chem.* **40**: 1794–1799.

Estrada, B., Bernal, M.A., Díaz, J. et al. (2002). Capsaicinoids in vegetative organs of *Capsicum annuum* L. in relation to fruiting. *J. Agric. Food Chem.* **50**: 1188–1191.

Etzbach, L., Pfeiffer, A., Weber, F., and Schieber, A. (2018). Characterization of carotenoid profiles in goldenberry (*Physalis peruviana* L.) fruits at various ripening stages and in different plant tissues by HPLC-DAD-APCI-MSn. *Food Chem.* **245**: 508–517.

Ewert, A., Granvogl, M., and Schieberle, P. (2014). Isotope-labeling studies on the formation pathway of acrolein during heat processing of oils. *J. Agric. Food Chem.* **62**: 8524–8529.

Ey, J., Schömig, E., and Taubert, D. (2007). Dietary sources and antioxidant effects of ergothioneine. *J. Agric. Food Chem.* **55**: 6466–6474.

Fahy, E., Subramaniam, S., Brown, A. et al. (2005). A comprehensive classification system for lipids. *J. Lipid Res.* **46**: 839–861.

Falcao, L.D., Lytra, G., Darriet, P., and Barbe, J.-C. (2012). Identification of ethyl 2-hydroxy-4-methylpentanoate in red wines, a compound involved in blackberry aroma. *Food Chem.* **132**: 230–236.

Fan, A.M. and Book, S.A. (1987). Sulfite hypersensitivity: a review of current issues. *J. Appl. Nutr.* **39**: 71–78.

Farre, R. and Lagarda, M.J. (1986). Chromium content in foods. *J. Micronutr. Anal* **2**: 201–209.

Fedrizzi, B., Pardon, K.H., Sefton, M.A. et al. (2009). First identification of 4-S-glutathionyl-4-methylpentan-2-one, a potential precursor of 4-mercapto-4-methylpentan-2-one, in Sauvignon Blanc juice. *J. Agric. Food Chem.* **57**: 991–995.

Feng, Z.-M., He, J., Jiang, J.-S. et al. (2013). NMR solution structure study of the representative component hydroxysafflor yellow A and other quinochalcone C-glycosides from *Carthamus tinctorius. J. Nat. Prod.* **76**: 270–274.

Fennema, O.R. (1985). *Food Chemistry*, 2e. New York: Marcel Dekker.

Fenwick, G.R., Heaney, R.K., and Mullin, W.J. (1983). Glucosinolates and their breakdown products in food and food plants. *Crit. Rev. Food Sci. Nutr.* **18**: 123–201.

Fergusson, J.E. (1990). *The Heavy Elements. Chemistry, Environmental Impact and Health Effects*. Oxford: Pergamon Press.

Ferland, G. and Sadowski, J.A. (1992a). Vitamin K_1 (phylloquinone) content of green vegetables: effect of plant maturation and geographical growth location. *J. Agric. Food Chem.* **40**: 1874–1877.

Ferland, G. and Sadowski, J.A. (1992b). Vitamin K_1 (phylloquinone) content of edible oils: effect of heating and light exposure. *J. Agric. Food Chem.* **40**: 1869–1873.

Fernández-Marín, M.I., Guerrero, R.F., García-Parrilla, M.C. et al. (2012). Isorhapontigenin: a novel bioactive stilbene from wine grapes. *Food Chem.* **135**: 1353–1359.

Ferrand, C., Marc, F., Fritsch, P. et al. (2000). Genotoxicity study of reaction products of sorbic acid. *J. Agric. Food Chem.* **48**: 3605–3610.

Ferreira-Lima, N., Vallverdú-Queralt, A., Meudec, E. et al. (2016). Synthesis, identification, and structure elucidation of adducts formed by reactions of hydroxycinnamic acids with glutathione or cysteinylglycine. *J. Nat. Prod.* **79**: 2211–2222.

Ferretti, A. (1973). Inhibition of cooked flavor in heated milk by use of additives. *J. Agric. Food Chem.* **21**: 939–942.

Festring, D. and Hofmann, T. (2010). Discovery of N2-(1-carboxyethyl)guanosine 5′-monophosphate as an umami-enhancing Maillard-modified nucleotide in yeast extracts. *J. Agric. Food Chem.* **58**: 10614–10622.

Festring, D. and Hofmann, T. (2011). Systematic studies on the chemical structure and umami enhancing activity of Maillard-modified guanosine 5′-monophosphates. *J. Agric. Food Chem.* **59**: 665–676.

Fibigr, J., Šatínský, D., and Solich, P. (2017). A UHPLC method for the rapid separation and quantification of anthocyanins in acai berry and dry blueberry extracts. *J. Pharm. Biomed. Anal.* **143**: 204–213.

Figge, K. (1996). *Plastic Packages for Foodstuffs*. Stuttgart: Wissenschaftliche Verlagsge-sellschaft.

Finot, P.A., Aeschbacher, H.U., Hurrell, R.F., and Liardon, R. (eds.) (1990). *The Maillard Reaction in Food Processing, Human Nutrition and Physiology*. Basel: Birkhäuser.

Fiorentino, A., Mastellone, C., D'Abrosca, B. et al. (2009). delta-Tocomonoenol: a new vitamin E from kiwi (*Actinidia chinensis*) früits. *Food Chem.* **115**: 187–192.

Firestone, D. (2005). Olive oil. In: *Bailey's Industrial Oil and Fat Products*, 6e (ed. F. Shahidi). Hoboken: Wiley.

Fischer, J. and Wüst, M. (2012). Quantitative determination of the boar taint compounds androstenone, skatole, indole, 3α-androstenol and 3β-androstenol in wild boars (*Sus scrofa*) reveals extremely low levels of the tryptophan-related degradation products. *Food Chem.* **135**: 2128–2132.

Fischer, U.A., Jaksch, A.V., Carle, R., and Kammerer, D.R. (2012). Determination of lignans in edible and nonedible parts of pomegranate (*Punica granatum* L.) and products derived therefrom, particularly focusing on the quantitation of isolariciresinol using HPLC-DAD-ESI/MSn. *J. Agric. Food Chem.* **60**: 283–292.

Floch, M., Shinkaruk, S., Darriet, P., and Pons, A. (2016). Identification and organoleptic contribution of vanillylthiol in wines. *J. Agric. Food Chem.* **64**: 1318–1325.

Flores, G., Castillo, M.L.R., Blanch, G.P., and Herraiz, M. (2006). Detection of the adulteration of olive oils by solid phase microextraction and multidimensional gas chromatography. *Food Chem.* **97**: 336–342.

Flores, P., Hellín, P., and Fenoll, J. (2012). Determination of organic acids in fruits and vegetables by liquid chromatography with tandem-mass spectrometry. *Food Chem.* **132**: 1049–1054.

Flynn, A. (1992). Minerals and trace elements in milk. *Adv. Food Nutr. Res.* **36**: 209–252.

Fondu, M. (1992). Food additives intake. *Food Addit. Contam.* **9**: 535–539.

Fox, D.L. (1976). *Animal Biochromes and Structural Colours*, 2e. Berekely and Los Angeles, CA: University of California Press.

Fraga, B.M. and Terrero, D. (1996). Alkene-γ -lactones and avocadofurans from *Persea indica*: a revision of the structure of majorenolide and related lactones. *Phytochemistry* **41**: 229–232.

Francis, G., Kerem, Z., Makkar, H.P.S., and Becker, K. (2002). The biological action of saponins in animal systems: a review. *Br. J. Nutr.* **88**: 587–605.

Francis, J.F. (1987). Lesser-known food colorants. *Food Technol.* **4**: 62–65.

Franitza, L., Granvogl, M., and Schieberle, P. (2016). Characterization of the key aroma compounds in two commercial rums by means of the sensomics approach. *J. Agric. Food Chem.* **64**: 637–645.

Franitza, L., Granvogl, M., and Schieberle, P. (2016). Influence of the production process on the key aroma compounds of rum: from molasses to the spirit. *J. Agric. Food Chem.* **64**: 9041–9053.

Frank, O., Blumberg, S., Krümpel, G., and Hofmann, T. (2008). Structure determination of 3-O-caffeoyl-epi-γ -quinide, an orphan bitter lactone in roasted coffee. *J. Agric. Food Chem.* **56**: 9581–9585.

Frank, O., Blumberg, S., Kunert, C. et al. (2007). Structure determination and sensory analysis of bitter-tasting 4-vinylcatechol oligomers and their identification in roasted coffee by means of LC-MS/MS. *J. Agric. Food Chem.* **55**: 1945–1954.

Frank, O., Heuberger, S., and Hofmann, T. (2001). Structure determination of a novel 3(6H)-pyranone chromophore and clarification of its formation from carbohydrates and primary amino acids. *J. Agric. Food Chem.* **49**: 1595–1600.

Frank, O., Ottinger, H., and Hofmann, T. (2001). Characterization of an intense bitter-tasting 1H,4H-quinolizinium-7-olate by application of the taste dilution analysis, a novel bioassay for the screening and identification of taste-active compounds in foods. *J. Agric. Food Chem.* **49**: 231–238.

Frankel, E.N. (1998). *Lipid Oxidation*. Dundee: Oily Press.

Franzke, C. (1990). *Lehrbuch der Lebensmittelchemie*. Berlin: Akademie-Verlag.

Frauendorfer, F. and Schieberle, P. (2006). Identification of the key aroma compounds in cocoa powder based on molecular sensory correlations. *J. Agric. Food Chem.* **54**: 5521–5529.

Fraústo da Silva, J.J.R. and Williams, R.J.P. (2001). *The Biological Chemistry of the Elements – The Inorganic Chemistry of Life*, 2e. Oxford: Oxford University Press.

Freccero, M. and Doria, F. (2009). Modelling properties and reactivity of quinine methides by DFT calculation. In: *Quinone Methides* (ed. S.E. Rokita). Hoboken: Wiley.

Freij, L. (ed.) (1994). *Phthalic Acid Esters Used as Plastic Additives*. Stockholm: PrintGraf.

Friberg, S.E. and Larsson, K. (1997). *Food Emulsions*. New York: Dekker.

Friedman, H.I. and Nylund, B. (1980). Intestinal fat digestion, absorption, and transport: a review. *Am. J. Clin. Nutr.* **33**: 1108–1139.

Friedman, M. (1996). Nutritional value of proteins from different food sources. A review. *J. Agric. Food Chem.* **44**: 6–29.

Friedman, M. (1997). Chemistry, biochemistry, and dietary role of potato polyphenols. A review. *J. Agric. Food Chem.* **45**: 1523–1540.

Friedman, M. (2006). Potato glycoalkaloids and metabolites: roles in the plant and in the diet. *J. Agric. Food Chem.* **54**: 8655–8681.

Friedman, M. and Cuq, J.-L. (1988). Chemistry, analysis, nutritional value and toxicology of tryptophan in food a review. *J. Agric. Food Chem.* **36**: 1079–1093.

Friedman, M., Levin, C.E., and McDonald, G.M. (1994). α-Tomatine determination in tomatoes by HPLC using pulsed amperometric detection. *J. Agric. Food Chem.* **42**: 1959–1964.

Friedman, M., McDonald, G.M., and Filadelfi-Keszi, M. (1997). Potato glycoalkaloids: chemistry, analysis, safety, and plant physiology. *Crit. Rev. Plant Sci.* **16**: 55–132.

Friedman, M. and Mottram, D. (eds.) (2004). *Chemistry and Safety of Acrylamide in Food*. Berlin: Springer.

Friedman, M., Roitman, J.N., and Kozukue, N. (2003). Glycoalkaloid and calystegine contents of eight potato cultivars. *J. Agric. Food Chem.* **51**: 2964–2973.

Friedrich, W. (1988). *Vitamins*. Berlin: Walter de Gruyter.

Friend, J., Threfad, D.R., and Overeem, J.C. (eds.) (1976). *Biochemical Aspects of Plant–Parasite Relationships*. London: Academic Press.

Fritsch, H.T. and Schieberle, P. (2005). Identification based on quantitative measurements and aroma recombination of the character impact odorants in a Bavarian Pilsner-type beer. *J. Agric. Food Chem.* **53**: 7544–7551.

Fujii, N. (2005). D-Amino acids in elderly tissues. *Biol. Pharm. Bull.* **28**: 1585–1589.

Fujii, T., Matsutomo, T., and Kodera, Y. (2018). Changes of S-allylmercaptocysteine and γ-glutamyl-S-allylmercaptocysteine contents and their putative production mechanisms in garlic extract during the aging process. *J. Agric. Food Chem.* **66**: 10506–10512.

Fujimaki, M., Namiki, M., and Kato, H. (eds.) (1986). *Amino-Carbonyl Reactions in Food and Biological Systems*. Amsterdam: Elsevier.

Fujita, T., Hada, T., and Higashino, K. (1999). Origin of D- and L-pipecolic acid in human physiological fluids: a study of the catabolic mechanism to pipecolic acid using the lysine loading test. *Clin. Chim. Acta* **287**: 145–156.

Fukushima, M. and Chatt, A. (2013). Rapid determination of silver in cultivated Japanese and South Korean oysters and Japanese rock oysters using the 24.6-s neutron activation product 110Ag and estimation of its average daily intake. *J. Radioanal. Nucl. Chem.* **296**: 563–571.

Gabelman, A. (ed.) (1994). *Bioprocess Production of Flavor, Fragrance, and Color Ingredients*. Hoboken: Wiley.

Gailer, J., George, G.N., Pickering, I.J. et al. (2000). A metabolic link between arsenite and selenite: the seleno-bis(S-glutathionyl)arsinium ion. *J. Am. Chem. Soc.* **122**: 4637–4639.

Galvan, T.L., Kells, S., and Hutchison, W.D. (2008). Determination of 3-alkyl-2-methoxypyrazines in lady beetle-infested wine by solid-phase microextraction headspace sampling. *J. Agric. Food Chem.* **56**: 1065–1071.

Gaman, P.M. and Sherington, K.B. (1986). *The Science of Food*, 2e. Oxford: Pergamon Press.

Gao, K., Ma, D., Cheng, Y. et al. (2015). Three new dimers and two monomers of phenolic amides from the fruits of *Lycium barbarum* and their antioxidant activities. *J. Agric. Food Chem.* **63**: 1067–1075.

Gaonkar, A.G. (ed.) (1995). *Ingredient Interactions*. Amsterdam: Elsevier.

García-Cruz, L., Dueñas, M., Santos-Buelgas, C. et al. (2017). Betalains and phenolic compounds profiling and antioxidant capacity of pitaya (*Stenocereus* spp.) fruit from two species (*S. pruinosus* and *S. stellatus*). *Food Chem.* **234**: 111–118.

Garrido, F.A., Adams, M.R., and Fernández-Díez, M.J. (1997). *Table Olives: Production and Processing*. London: Chapman and Hall.

Gasperotti, M., Masuero, D., Vrhovsek, U. et al. (2010). Profiling and accurate quantification of Rubus ellagotannins and ellagic acid conjugates using direct UPLC-Q-TOF HDMS and HPLC-DAD analysis. *J. Agric. Food Chem.* **58**: 4602–4616.

Gautschi, M., Schmid, J.P., Pappard, T.L. et al. (1997). Chemical characterization of diketopiperazines in beer. *J. Agric. Food Chem.* **45**: 3183–3189.

Gautschi, M., Yang, X., and Eilerman, R.G. (2012). Flavor chemicals with pungent properties. In: *Flavor Chemistry: Thirty Years of Progress* (eds. R. Teranishi, E.L. Wick and I. Hornstein), 199. Berlin: Springer.

Genovese, S., Fiorito, S., Locatelli, M. et al. (2014). Analysis of biologically active oxyprenylated ferulic acid derivatives in *Citrus* fruits. *Plant Foods Hum. Nutr.* **69**: 255–260.

Gerster, H. (1993). Anticarcinogenic effect of common carotenoids. *Int. J. Vitam. Nutr. Res.* **63**: 93–121.

Gilbert, J. and Senyuva, H.Z. (eds.) (2008). *Bioactive Compounds in Foods*. Oxford: Wiley Blackwell.

Gimeno, R.E. (2007). Fatty acid transport proteins. *Curr. Opin. Lipidol.* **18**: 271–276.

Ginz, M. and Engelhardt, U.H. (2000). Identification of proline-based diketopiperazines in roasted coffee. *J. Agric. Food Chem.* **48**: 3528–3532.

Glabasnia, A. and Hofmann, T. (2006). Sensory-directed identification of taste-active ellagitannins in American (*Quercus alba* L.) and European oak wood (*Quercus robur* L.) and quantitative analysis in bourbon whiskey and oak-matured red wines. *J. Agric. Food Chem.* **54**: 3380–3390.

Glässgen, W.E., Metzger, J.W., Heuer, S., and Strack, D. (1993). Betacyanins from fruits of *Basella rubra*. *Phytochemistry* **33**: 1525–1527.

Glicksman, M. (1982). *Food Hydrocolloids*. Boca Raton, FL: CRC Press.

Gody, H.T. and Rodriguez-Amaya, D.B. (1994). Occurrence of cis-isomers of provitamin A in Brazilian fruit. *J. Agric. Food Chem.* **42**: 1306–1313.

González, A.G., Irizar, A.C., Ravelo, A.G., and Fernández, M.F. (1992). Type A proanthocyanidins from *Prunus spinosa*. *Phytochemistry* **31**: 1432–1434.

González-Muñoz, M.J., Peña, A., and Meseguer, I. (2008). Role of beer as a possible protective factor in preventing Alzheimer's disease. *Food Chem. Toxicol.* **46**: 49–56.

Goosen, M.F.A. (ed.) (1996). *Applications of Chitin and Chitosan*. Basel: Technomic.

Gorman, J.E. and Clydesdale, F.M. (1983). The behavior and stability of iron-ascorbate complexes in solution. *J. Food Sci.* **48**: 1217–1220.

Gotoh, N., Matsumoto, Y., Nagai, T. et al. (2011). Actual ratios of triacylglycerol positional isomers consisting of saturated and highly unsaturated fatty acids in fishes and marine mammals. *Food Chem.* **127**: 467–472.

Goupy, P., Fleuriet, A., Amiot, M.-J., and Macheix, J.-J. (1991). Enzymatic browning, oleuropein content, and diphenol oxidase activity in olive cultivars (*Olea europea* L.). *J. Agric. Food Chem.* **39**: 92–95.

Govorushko, S.M. (2012). *Natural Processes and Human Impacts: Interactions Between Humanity and the Environment*. Berlin: Springer.

Gracanin, M., Hawkins, C.L., Pattison, D.J., and Davies, M.J. (2009). Singlet-oxygen-mediated amino acid and protein oxidation: formation of tryptophan peroxides and decomposition products. *Free Radic. Biol. Med.* **47**: 92–102.

Grajeda-Iglesias, C., Figueroa-Espinoza, M.C., Barouh, N. et al. (2016). Isolation and characterization of anthocyanins from *Hibiscus sabdariffa* flowers. *J. Nat. Prod.* **79**: 1709–1718.

Granvogl, M. (2014). Development of three stable isotope dilution assays for the quantitation of (*E*)-2-butenal (crotonaldehyde) in heat-processed edible fats and oils as well as in food. *J. Agric. Food Chem.* **62**: 1272–1282.

Granvogl, M., Beksan, E., and Schieberle, P. (2012). New insights into the formation of aroma-active Strecker aldehydes from 3-oxazolines as transient intermediates. *J. Agric. Food Chem.* **60**: 6312–6322.

Granvogl, M., Christlbauer, M., and Schieberle, P. (2004). Quantitation of the intense aroma compound 3-mercapto-2-methylpentan-1-ol in raw and processed onions (*Allium cepa*) of different origins and in other *Allium* varieties using a stable isotope dilution assay. *J. Agric. Food Chem.* **52**: 2797–2802.

Greger, V. and Schieberle, P. (2007). Characterization of the key aroma compounds in apricots (*Prunus armeniaca*) by application of the molecular sensory science concept. *J. Agric. Food Chem.* **55**: 5221–5228.

Griffith, D.W., Macfarlane-Smith, W.H., and Boag, B. (1994). The effect of cultivar, sample date and grazing on the concentration of S-methylcysteine sulfoxide in oilseed and forage rapes (*Brassica napus*). *J. Sci. Food Agric.* **64**: 283–288.

Griffiths, D.W. and Ramsy, G. (1992). The concentration of vicine and convicine in *Vicia faba* and some related species and their distribution within mature seeds. *J. Sci. Food Agric.* **59**: 463–468.

Griffiths, D.W., Shepherd, T., and Stewart, D. (2008). Comparison of the calystegine composition and content of potato sprouts and tubers from *Solanum tuberosum* group *Phureja* and *Solanum tuberosum* group *Tuberosum*. *J. Agric. Food Chem.* **56**: 5197–5204.

Gropper, S.S. and Smith, J.L. (2013). *Advanced Nutrition and Human Metabolism*, 6e. Belmont, CA: Cengage Learning.

Gros, J., Lavigne, V., Thibaud, F. et al. (2017). Toward a molecular understanding of the typicality of Chardonnay wines: identification of powerful aromatic compounds reminiscent of hazelnut. *J. Agric. Food Chem.* **65**: 1058–1069.

Gros, J., Peeters, F., and Collin, S. (2012). Occurrence of odorant polyfunctional thiols in beer hopped with different cultivars. First evidence of an *S*-cysteine conjugate in hop (*Humulus lupulus* L.). *J. Agric. Food Chem.* **60**: 7805–7816.

Grosch, W. (2001). Evaluation of the key odorants of foods by dilution experiments, aroma models and omission. *Chem. Senses* **26**: 533–545.

Grosch, W. and Schieberle, P. (1997). Flavor of cereal products-a review. *Cereal Chem.* **74**: 91–97.

Grosshauser, S. and Schieberle, P. (2013). Characterization of the key odorants in pan-fried white mushrooms (*Agaricus bisporus* L.) by means of molecular sensory science: comparison with the raw mushroom tissue. *J. Agric. Food Chem.* **61**: 3804–3813.

Grougnet, R., Magiatis, P., Laborie, H. et al. (2012). Sesamolinol glucoside, disaminyl ether, and other lignans from sesame seeds. *J. Agric. Food Chem.* **60**: 108–111.

Grúz, J., Pospíšil, J., Kozubíková, H. et al. (2015). Determination of free diferulic, disinapic and dicoumaric acids in plants and foods. *Food Chem.* **171**: 280–286.

Guebailia, H.A., Chira, K., Richard, T. et al. (2006). Hopeaphenol: the first resveratrol tetramer in wines from North Africa. *J. Agric. Food Chem.* **54**: 9559–9564.

Guerra, P.V. and Yaylayan, V.A. (2011). Thermal generation of 3-amino-4,5-dimethylfuran-2(5H)-one, the postulated precursor of sotolone, from amino acid model systems containing glyoxylic and pyruvic acids. *J. Agric. Food Chem.* **59**: 4699–4704.

Guerra, P.V. and Yaylayan, V.A. (2013). Cyclocondensation of 2,3-butanedione in the presence of amino acids and formation of 4,5-dimethyl-1, 2-phenylendiamine. *Food Chem.* **141**: 4391–4396.

Guerra-Hernández, E., Ramirez-Jiménez, A., and García-Villanova, B. (2002). Glu-cosylisomaltol, a new indicator of browning reaction in baby cereals and bread. *J. Agric. Food Chem.* **50**: 7282–7287.

Guillén, M.D. (1994). Polycyclic aromatic compounds: extraction and determination in food. *Food Addit. Contam.* **11**: 669–684.

Guillén, M.D., Sopelana, P., and Partearroyo, M.A. (1997). Food as a source of polycyclic aromatic carcinogens. *Rev. Environ. Health* **12**: 133–146.

Gunning, Y., Defernez, M., Watson, A.D. et al. (2018). 16-*O*-Methylcfestol is present in ground roast arabica coffees: implications for authenticity testing. *Food Chem.* **248**: 52–60.

Günther-Jordanland, K., Dawid, C., Dietz, M., and Hofmann, T. (2016). Key phytochemicals contributing to the bitter off-taste of oat (*Avena sativa* L.). *J. Agric. Food Chem.* **64**: 9639–9652.

Guo, D., Venkatramesh, M., and Nes, W.D. (1995). Development regulation of sterol biosynthesis in Zea mays. *Lipids* **30**: 203–219.

Gurr, M.I., Harwood, J.L., and Frayn, K. (2002). *Lipid Biochemistry*, 5e. Oxford: Blackwell Science.

Guth, H. (1996). Determination of the configuration of wine lactone. *Helv. Chim. Acta* **79**: 1559–1571.

Gutsche, B., Grun, C., Scheutzow, D., and Herderich, M. (1999). Tryptophan glycoconjugates in food and human urine. *Biochem. J.* **343**: 11–19.

Gutsche, B. and Herderich, M. (1997). High-performance liquid chromatography-electrospray ionisation-tandem mass spectrometry for the analysis of 1,2,3,4-tetrahydro-β-carboline derivatives. *J. Chromatogr. A* **767**: 101–106.

Guzmán-Maldonado, H. and Paredes-López, O. (1995). Amylolytic enzymes and products derived from starch: a review. *Crit. Rev. Food Sci. Nutr.* **35**: 373–403.

Hajfathalian, M., Ghelichi, S., García-Moreno, P.J. et al. (2018). Peptides: production, bioactivity, functionality, and applications. *Crit. Rev. Food Sci. Nutr.* **58**: 3097–3129.

Hakala, P., Lampi, A.-M., Ollilainen, V. et al. (2002). Steryl phenolic acid esters in cereals and their milling fractions. *J. Agric. Food Chem.* **50**: 5300–5307.

Hakk, H., Huwe, J.K., Murphy, K., and Rutherford, D. (2010). Metabolism of 2,2′,4,4′ -tetrabromodiphenyl ether (BDE-47) in chickens. *J. Agric. Food Chem.* **58**: 8757–8762.

Halstead, B.W. (1988). *Poisonous and Benomous Marine Animals of the World*, 2e. Penington, NJ: Darwin Press.

Hamilton, D.J., Holland, P.T., Ohlin, B. et al. (1997). Optimum use of available residue data in the estimation of dietary intake of pesticides. *Pure Appl. Chem.* **69**: 1373–1410.

Hamilton, R.J. (1995). *Waxes: Chemistry, Molecular Biology and Functions*. Dundee: Oily Press.

Hamlet, C.G., Asuncion, L., Velíšek, J. et al. (2011). Formation and occurrence of esters of 3-chloropropane-1,2-diol (3-CPD) in foods: what we know and what we assume. *Eur. J. Lipid Sci. Technol.* **113**: 279–303.

Hammer, M. and Schieberle, P. (2013). Model studies on the key aroma compounds formed by an oxidative degradation of ω-3 fatty acids initiated by either copper(II) ions or lipoxygenase. *J. Agric. Food Chem.* **61**: 10891–10900.

Hammerschmidt, C.R. and Fitzgerald, W.F. (2006). Bioaccumulation and trophic transfer of methylmercury in Long Island Sound. *Arch. Environ. Contam. Toxicol.* **51**: 416–424.

Hanbury, C.D., White, C.L., Mullan, B.P., and Siddique, K.H.M. (2000). A review of the potential of Lathyrus sativus L. and L. cicera L. grain for use as animal feed. *Anim. Feed Sci. Technol.* **87**: 1–27.

Hanft, F. and Koehler, P. (2005). Quantitation of dityrosine in wheat flour and dough by liquid chromatography-tandem mass spectrometry. *J. Agric. Food Chem.* **53**: 2418–2423.

Hanschen, F.S., Rohn, S., Mewis, I. et al. (2012). Influence of the chemical structure on the thermal degradation of the glucosinolates in broccoli sprouts. *Food Chem.* **130**: 1–8.

Harbowy, M.E. and Balentine, D.A. (1997). Tea chemistry. *Crit. Rev. Plant Sci.* **16**: 415–480.

Härmälä, P., Vuorela, H., Hiltunen, R. et al. (1992). Strategy for the isolation and identification of coumarins with calcium antagonistic properties from the roots of Angelica archangelica. *Phytochem. Anal.* **3**: 42–48.

Harsch, M.J., Benkwitz, F., Frost, A. et al. (2013). New precursor of 3-mercaptohexan-1-ol in grape juice: thiol-forming potential and kinetics during early stages of must fermentation. *J. Agric. Food Chem.* **61**: 3703–3713.

Hart, D.J., Fairweather-Tait, S.J., Broadley, M.R. et al. (2011). Selenium concentration and speciation in biofortified flour and bread: retention of selenium during grain biofortification, processing and production of Se-enriched food. *Food Chem.* **126**: 1771–1778.

Hartfoot, C.G. and Hazlewood, G.P. (1997). Lipid metabolism in the rumen. In: *The Rumen Microbial Ecosystem* (eds. P.N. Hobson and C.S. Steward), 382–426. London: Chapman & Hall.

Hasegawa, S., Bennett, R.D., and Verdon, C.P. (1980). Limonoids in citrus seeds: origin and relative concentration. *J. Agric. Food Chem.* **28**: 922–925.

Hasegawa, S. and Maier, V.P. (1983). Solutions to the limonin bitterness problem of citrus juices. *Food Technol.*: 73–75.

Hasegawa, S. and Miyake, M. (1996). Biochemistry and biological functions of citrus limonoids. *Food Rev. Int.* **12**: 413–435.

Haseleu, G., Intelman, D., and Hofmann, T. (2009). Identification and RP-HPLC-ESI-MS/MS quantitation of bitter-tasting β-acid transformation products in beer. *J. Agric. Food Chem.* **57**: 7480–7489.

Hassall, K.A. (1990). *The Biochemistry of Pesticides: Structure, Metabolism, Mode of Action and Uses in Crop Protection*, 2e. Verlag Chemie: Weinheim.

He, F., Liang, N.N., Mu, L. et al. (2012). Anthocyanins and their variation in red wines II. Anthiocyanin derived pigments and their color evaluation. *Molecules* **17**: 1483–1519.

He, J., Santos-Buelga, C., Silva, A.M.S. et al. (2010). Isolation and structural characterization of new anthocyanin-derived yellow pigments in aged red wines. *J. Agric. Food Chem.* **58**: 5664–5669.

He, J., Silva, A.M.S., Mateus, N., and de Freitas, V. (2011). Oxidative formation and structural characterisation of new a-pyranone (lactone) compounds of non-oxonium nature originated from fruit anthocyanins. *Food Chem.* **127**: 984–992.

Heartland Institute. Trends in dioxin levels in the environment and in humans. https://www.heartland.org/publications-resources/publications/trends-in-dioxin-levels-in-the-environment-and-in-humans (accessed 28 October 2019).

Heatherbell, D.A., Wrolstad, R.E., and Libbey, L.M. (1971). Carrot volatiles. 1. Characterization and effects of canning and freeze drying. *J. Food Sci.* **36**: 219–224.

Hefle, S.L. (1996). The chemistry and biology of food alergens. *Food Technol.*: 86–89.

Hefle, S.L., Nordlee, J.A., and Taylor, S.L. (1996). Allergenic foods. *Crit. Rev. Food Sci. Nutr.* **36**: S69–S89.

Heinrich, T., Willenberg, I., and Glomb, M.A. (2012). Chemistry and color formation during rooibos fermentation. *J. Agric. Food Chem.* **60**: 5221–5228.

Hellwig, M., Beer, F., Witte, S., and Henle, T. (2018). Yeast metabolites of glycated amino acids in beer. *J. Agric. Food Chem.* **66**: 7451–7460.

Hellwig, M., Degen, J., and Henle, T. (2010). 3-Deoxygalactosone, a 'new' 1,2-dicarbonyl compound in milk products. *J. Agric. Food Chem.* **58**: 10752–10760.

Hellwig, M., Kiessling, M., Rother, S., and Henle, T. (2016). Quantification of the glycation compound 6-(3-hydroxy-4-oxo-2-methyl-4(1*H*)-pyridin-1-yl)-l-norleucine (maltosine) in model systems and food samples. *Eur. Food Res. Technol.* **242**: 547–557.

Hempel, J. and Böhm, H. (1997). Betaxanthin pattern of hairy roots from *Beta vulgaris* var. *lutea* and its alteration by feeding of amino acids. *Phytochemistry* **44**: 847–852.

Henderson, D.E. and Henderson, S.K. (1992). Thermal decomposition of capsaicin I. Interactions with oleic acid at high temperatures. *J. Agric. Food Chem.* **40**: 2263–2268.

Hendriks, W.H., Moughan, P.J., Tarttelin, M.F., and Woolhouse, A.D. (1995). Felinine: a urinary amino acid of Felidae. *Comp. Biochem. Physiol. B: Biochem. Mol. Biol.* **112**: 581–588.

Hendry, G.A.F. and Houghton, J.D. (eds.) (1992). *Natural Food Colourants*. Glasgow: Blackie.

Hentschel, V., Kranl, K., Hollmann, J. et al. (2002). Spectrophotometric determination of yellow pigment content and evaluation of carotenoids by high-performance liquid chromatography in durum wheat grain. *J. Agric. Food Chem.* **50**: 6663–6668.

Herbach, K.M., Stintzing, F.C., and Carle, R. (2004). Impact of thermal treatment on color and pigment pattern of red beet (*Beta vulgaris* L.) preparations. *J. Food Sci.* **69**: 491–498.

Herbrand, K., Hammerschmidt, F.J., Brennecke, S. et al. (2007). Identification of Allyl esters in garlic cheese. *J. Agric. Food Chem.* **55**: 7874–7878.

Hernández-Hernández, O., Ruiz-Aceituno, L., Sanz, M.L., and Martínez-Castro, I. (2011). Determination of free inositols and other low molecular weight carbohydrates in vegetables. *J. Agric. Food Chem.* **59**: 2451–2455.

Herraiz, T. (1998). Occurrence of 1,2,3,4-tetrahydro-β-carboline-3-carboxylic acid and 1-methyl-1,2,3,4-tetrahydro-β-carboline-3-carboxylic acid in fruit juices, purees, and jams. *J. Agric. Food Chem.* **46**: 3484–3490.

Herraiz, T. (2007). Identification and occurrence of β-carboline alkaloids in raisins and inhibition of monoamine oxidase (MAO). *J. Agric. Food Chem.* **55**: 8534–8540.

Herraiz, T. and Galisteo, J. (2002). Identification and occurrence of the novel alkaloid pentahydroxypentyl-tetrahydro-β-carboline-3-carboxylic acid as a tryptophan glycoconjugate in fruit juices and jams. *J. Agric. Food Chem.* **50**: 4690–4695.

Herraiz, T., Galisteo, J., and Chamorro, C. (2003). L-Tryptophan reacts with naturally occurring and food-occurring phenolic aldehydes to give phenolic tetrahydro-β-carboline alkaloids: activity as antioxidants and free radical scavengers. *J. Agric. Food Chem.* **51**: 2168–2173.

Herzog, A., Herbert, E., Dehnhardt, M., and Claus, R. (1993). Vergleichende Messungen von Androstenon und Skatol in verschiedenen Geweben. *Mitteilungsblatt der Bundesanstalt für Fleischforschung, Kulmbach* **32**: 116–124.

Heydanek, M.G. (1981). Presence of 1,3-dioxolanes in commercial flavourings. *J. Agric. Food Chem.* **29**: 892–894.

Hidalgo, F.J., Delgado, R.M., and Zamora, R. (2013). Intermediate role of α-keto acids in the formation of Strecker aldehydes. *Food Chem.* **141**: 1140–1146.

Hidalgo, F.J. and Zamora, R. (2016). Amino acid degradations produced by lipid oxidation products. *Crit. Rev. Food Sci. Nutr.* **56**: 1242–1252.

Higashiguchi, F., Nakamura, H., Hayashi, H., and Kometani, T. (2006). Purification and structure determination of glucosides of capsaicin and dihydrocapsaicin from various Capsicum fruits. *J. Agric. Food Chem.* **54**: 5948–5953.

Hill, M.A. (1994). Vitamin retention in microwave cooking and cold-chill foods. *Food Chem.* **49**: 131–136.

Hillebrand, S., Schwarz, M., and Winterhalter, P. (2004). Characterization of anthocyanins and pyranoanthocyanins from blood orange [*Citrus sinensis* (L.) Osbeck] juice. *J. Agric. Food Chem.* **52**: 7331–7338.

Hillmann, H. and Hofmann, T. (2016). Quantitation of key tastants and re-engineering the taste of parmesan cheese. *J. Agric. Food Chem.* **64**: 1794–1805.

Hillmann, H., Mattes, J., Brockhoff, A. et al. (2012). Sensomics analysis of taste compounds in balsamic vinegar and discovery of 5-acetoxymethyl-2-furaldehyde as a novel sweet taste modulator. *J. Agric. Food Chem.* **60**: 9974–9990.

Ho, C.-T. and Manley, C.H. (eds.) (1993). *Flavor Measurement*. New York: Dekker.

Hofmann, T. (1998). Acetylformoin – a chemical switch in the formation of colored Maillard reaction products from hexoses and primary and secondary amino acids. *J. Agric. Food Chem.* **46**: 3918–3928.

Hofmann, T. (1999). Quantitative studies on the role of browning precursors in the Maillard reaction of pentoses and hexoses with L-alanine. *Eur. Food Res. Technol.* **209**: 113–121.

Hofmann, T. (2005). The 'tasty' world of nonvolatile Maillard reaction products. *Ann. N. Y. Acad. Sci.* **1043**: 20–29.

Hofmann, T., Bors, W., and Stettmaier, K. (1999). Studies on radical intermediates in the early stage of the nonenzymatic browning reaction of carbohydrates and amino acids. *J. Agric. Food Chem.* **47**: 379–390.

Hofmann, T., Münch, P., and Schieberle, P. (2000). Quantitative model studies on the formation of aroma-active aldehydes and acids by Strecker-type reactions. *J. Agric. Food Chem.* **48**: 434–440.

Hofmann, T. and Schieberle, P. (2000). Formation of aroma-active Strecker aldehydes by a direct oxidative degradation of Amadori compounds. *J. Agric. Food Chem.* **48**: 4301–4305.

Holland, P.T. (1996). Glossary of terms related to pesticides. *Pure Appl. Chem.* **68**: 1167–1193.

Holland, P.T., Hamilton, D., Ohlin, B., and Skidmore, M.W. (1994). Effects of storage and processing on pesticide residues in plant products. *Pure Appl. Chem.* **66**: 335–356.

Holm, R., Kennepohl, P., and Solomon, E.J. (1996). Structural and functional aspects of metal sites in biology. *Chem. Rev.* **96**: 2239–2314.

Holscher, W., Vitzthum, O.G., and Steinhart, H. (1992). Prenyl alcohol-source for odorants in roasted coffee. *J. Agric. Food Chem.* **40**: 655–658.

Holt, H.C., Demott, B.J., and Bacon, J.A. (1989). The iodine concentration of market milk in Tennessee, 1981–1986. *J. Food Prot.* **52**: 115–118.

Homer, J.A. and Sperry, J. (2017). Mushroom-derived indole alkaloids. *J. Nat. Prod.* **80**: 2178–2187.

Hong, C.M., Wendorff, W.L., and Bradley, R.L. Jr. (1995). Factors affecting light-induced pink discoloration of annatto-colored cheese. *J. Food Sci.* **60**: 94–97.

Hoppner, K. and Lampi, B. (1993). Pantothenic acid and biotin retention in cooked legumes. *J. Food Sci.* **58**: 1084–1085.

Horbowicz, M. and Obendorf, R.L. (2005). Fagopyritol accumulation and germination of buckwheat seeds matured at 15, 22, and 30°C. *Crop Sci.* **45**: 1264–1270.

Horie, H., Ito, H., Ippoushi, K. et al. (2007). Cucurbitacin C-bitter principle in cucumber plants. *Jpn. Agric. Res. Q.* **41**: 65–68.

Horníčková, J., Kubec, R., Cejpek, K. et al. (2010). Profiles of S-alk(en)ylcysteine sulfoxides in various garlic genotypes. *Czech J. Food Sci.* **28**: 298–308.

Horníčková, J., Kubec, R., Velíšek, J. et al. (2011). Changes of S-alk(en)ylcysteine sulfoxide levels during the growth of different garlic morphotypes. *Czech J. Food Sci.* **29**: 373–381.

Hornsby, A.G., Wauchope, R.D., and Herner, A.E. (1996). *Pesticide Properties in the Environment*. New York: Springer.

Horváth, G., Molnár, P., Radó-Turcsi, E. et al. (2012). Carotenoid composition and in vitro pharmacological activity of rose hips. *Acta Biochim. Pol.* **59**: 129–132.

Houessou, J.K., Maloug, S., Leveque, A.-S. et al. (2007). Effect of roasting conditions on the polycyclic aromatic hydrocarbon content in ground Arabica coffee and coffee brew. *J. Agric. Food Chem.* **55**: 9719–9726.

Howard, P.H. (1990). *Handbook of Environmental Fate and Exposure Data for Organic Chemicals*, vol. **1**. New York: Lewis Publishers.

Hrubec, J. (ed.) (1995). *Quality and Treatment of Drinking Water*. Berlin: Springer.

Hruska, A.J. (1988). Cyanogenic glucosinolates as defence compounds A review of the evidence. *J. Chem. Ecol.* **14**: 2213–1217.

Hu, Q., Yang, Q., Yamato, O. et al. (2002). Isolation and identification of organosulfur compounds oxidizing canine erythrocytes from garlic (*Allium sativum*). *J. Agric. Food Chem.* **50**: 1059–1062.

Hu, T.-Y., Liu, C.-L., Chyau, C.-C., and Hu, M.-L. (2012). Trapping of methylglyoxal by curcumin in cell-free systems and in human umbilical vein endothelial cells. *J. Agric. Food Chem.* **60**: 8190–8196.

Huang, A.-C., Burrett, S., Sefton, M.A., and Taylor, D.K. (2014). Production of the pepper aroma compound, (−)-rotundone, by aerial oxidation of α-guaiene. *J. Agric. Food Chem.* **62**: 10809–10815.

Huang, S., Wang, L.M., Sivendiran, T., and Bohrer, B.M. (2018). Review: amino acid concentration of high protein food products and an overview of the current methods used to determine protein quality. *Crit. Rev. Food Sci. Nutr.* **58**: 2673–2678.

Hubregtse, T., Kooijman, H., Spek, A.L. et al. (2007). Study on the isomerism in meso-amavadin and amavadin analogue. *J. Inorg. Biochem.* **101**: 900–909.

Hufnagl, J.C. and Hofmann, T. (2008). Quantitative reconstruction of the non-volatile sensometabolome of a red wine. *J. Agric. Food Chem.* **56**: 9190–9199.

Hull, G.L.J., Woodside, J.V., Ames, J., and Cuskelly, G.J. (2012). N^ε-(carboxy-methyl)lysine content of foods commonly consumed in a Western style diet. *Food Chem.* **131**: 170–174.

Hutzinger, O. (1984). *The Handbook of Environmental Chemistry*, vol. **3**, part C. Berlin: Springer.

Huwe, J.K. (2002). Dioxins in food: a modern agricultural perspective. *J. Agric. Food Chem.* **50**: 1739–1750.

Huyghues-Despointes, A. and Yaylayan, V.A. (1996). Retro-aldol and redox reactions of Amadori compounds: mechanistic studies with variously labeled D-[^{13}C]glucose. *J. Agric. Food Chem.* **44**: 672–681.

Hwang, H.-I., Hartman, T.G., and Ho, C.-T. (1995). Relative reactivities of amino acids in the formation of pyridines, pyrroles and oxazoles. *J. Agric. Food Chem.* **43**: 2917–2921.

Iacomino, M., Weber, F., Gleichenhagen, M. et al. (2017). Stable benzacridine pigments by oxidative coupling of chlorogenic acid with amino acids and proteins: toward natural product-based green food coloring. *J. Agric. Food Chem.* **65**: 6519–6528.

Ichikawa, M., Ide, N., Yoshida, J. et al. (2006). Determination of seven organosulfur compounds in garlic by high-performance liquid chromatography. *J. Agric. Food Chem.* **54**: 1535–1540.

Ichikawa, M., Ryu, K., Yoshida, J. et al. (2003). Identification of six phenylpropanoids from garlic skin as major antioxidants. *J. Agric. Food Chem.* **51**: 7313–7317.

Ichiyanagi, T., Kashiwada, Y., Shida, Y. et al. (2005). Nasunin from eggplant consists of cis-trans isomers of delphinidin 3-[4-(p-coumaroyl)-L-rhamnosyl(1→6)glucopyranoside]-5-glucopyranoside. *J. Agric. Food Chem.* **53**: 9472–9477.

Ieri, F., Innocenti, M., Andrenelli, L. et al. (2011). Rapid HPLC/DAD/MS method to determine phenolic acids, glycoalkaloids and anthocyanins in pigmented potatoes (*Solanum tuberosum* L.) and correlations with variety and geographical origin. *Food Chem.* **125**: 750–759.

Ikegami, F., Lambein, F., Kuo, Y.-H., and Murakoshi, I. (1984). Isoxazolin-5-one derivatives in *Lathyrus odoratus* during development and growth. *Phytochemistry* **23**: 1567–1569.

Ikegami, F. and Murakoshi, I. (1994). Enzymic synthesis of non-protein β-substituted alanines and some higher homologues in plants. *Phytochemistry* **35**: 1089–1104.

Imai, S., Akita, K., Tomotake, M., and Sawada, H. (2006). Identification of two novel pigment precursors and a reddish-purple pigment involved in the blue-green discoloration of onion and garlic. *J. Agric. Food Chem.* **54**: 843–847.

Impellizzeri, J. and Lin, J. (2006). A simple high-performance liquid chromatography method for the determination of throat-burning oleocanthal with probated antiinflammatory activity in extra virgin olive oils. *J. Agric. Food Chem.* **54**: 3204–3208.

Ina, K., Ina, H., Ueda, M. et al. (1989). Omega-methylthioalkyl isothiocyanates in wasabi. *Agric. Biol. Chem.* **63**: 537–537.

Institute of Medicine of the National Academies. Dietary reference intakes for energy, carbohydrate, fiber, fat, fatty acids, cholesterol, protein, and amino acids. https://www.nap.edu/catalog/10490/dietary-reference-intakes-for-energy-carbohydrate-fiber-fat-fatty-acids-cholesterol-protein-and-amino-acids (accessed 28 October 2019).

Intelman, D., Haseleu, G., Dunkel, A. et al. (2011). Comprehensive sensomics analysis of hop-derived bitter compounds during storage of beer. *J. Agric. Food Chem.* **59**: 1939–1953.

Intelmann, D. and Hofmann, T. (2010). On the autoxidation of bitter-tasting iso-α-acids in beer. *J. Agric. Food Chem.* **58**: 5059–5067.

Irmak, S. and Dunford, N.T. (2005). Policosanol contents and compositions of wheat varieties. *J. Agric. Food Chem.* **53**: 5583–5586.

Ishida, H., Wakimoto, T., Kitao, Y. et al. (2009). Quantitation of chafurosides a and b in tea leaves and isolation of prechafurosides a and b from oolong tea leaves. *J. Agric. Food Chem.* **57**: 6779–6786.

Ishii, T. and Ono, H. (1999). NMR spectroscopic analysis of the borate diol esters of methyl apiofuranosides. *Carbohydr. Res.* **321**: 257–260.

Isler, O. (ed.) (1971). *Carotenoids*. Basel: Birkhäuser.

Itoh, N., Kurokawa, J., Isogai, Y. et al. (2017). Functional characterization of epitheaflagallin 3-O-gallate generated in laccase-treated green tea extracts in the presence of gallic acid. *J. Agric. Food Chem.* **65**: 10473–10481.

Izawa, K., Amino, Y., Kohmura, M. et al. (2010). *Comprehensive Natural Products II, Chemistry and Biology. Volume 4: Chemical Ecology, 4.16 – Human–Environment Interactions – Taste* (eds. L. Mander and L. Hung-Wen), 631–671. Elsevier Ltd.

Izquierdo-Pulido, M., Hernández-Jover, T., Mariné-Font, A., and Vidal-Carou, M.C. (1996). Biogenic amines in European beers. *J. Agric. Food Chem.* **44**: 3159–3163.

Izydorczyk, M.S. and Biliaderis, C.G. (1995). Cereal arabinoxylans: advances in structure and physicochemical properties. *Carbohydr. Polym.* **28**: 33–48.

Jain, V., Garg, A., Parascandola, M. et al. (2017). Analysis of alkaloids in areca nut-containing products by liquid chromatography–tandem mass spectrometry. *J. Agric. Food Chem.* **65**: 1977–1983.

Jang, D.S., Cuendet, M., Fong, H.H.S. et al. (2004). Constituents of Asparagus officinalis evaluated for inhibitory activity against cyclooxygenase-2. *J. Agric. Food Chem.* **52**: 2218–2222.

Jang, D.S., Park, E.J., Hawthorne, M.E. et al. (2002). Constituents of Musa x paradisiaca cultivar with the potential to induce the phase II enzyme, quinone reductase. *J. Agric. Food Chem.* **50**: 6330–6334.

Jansman, A.J.M., Huisman, G.D.H., and van der Poel, A.F.B. (eds.) (1998). *Recent Advances of Research in Antinutritional Factors in Legume Seeds and Rapeseed.* Wageningen: Wageningen Pers.

Janssen, K. and Matissek, R. (2002). Fatty acid tryptamides as shell indicator for cocoa products and as quality parameters for cocoa butter. *Eur. Food Res. Technol.* **214**: 259–264.

Janusz, A., Capone, D.L., Puglisi, C.J. et al. (2003). (E)-1-(2,3,6-Trimethylphenyl)buta-1,3-diene: a potent grape-derived odorant in wine. *J. Agric. Food Chem.* **51**: 7759–7763.

Jautz, U., Gibis, M., and Morlock, G.E. (2008). Quantification of heterocyclic aromatic amines in fried meat by HPTLC/UV-FLD and HPLC/UV-FLD: a comparison of two methods. *J. Agric. Food Chem.* **56**: 4311–4319.

Jay, J.M. (1992). Microbiological food safety. *Crit. Rev. Food Sci. Nutr.* **31**: 177–190.

Jayaprakasha, G.K., Murthy, C.K.N., and Patil, B.S. (2011). Rapid HPLC-UV method for quantification of L-citrulline in watermelon and its potential role on smooth muscle relaxation markers. *Food Chem.* **127**: 240–248.

Jayaprakasha, G.K. and Sakariah, K.K. (1998). Determination of organic acids in Garcinia cambogia Desr. by high-performance liquid chromatography. *J. Chromatogr. A* **806**: 337–339.

Jedelská, J., Vogt, A., Reinscheid, U.M., and Keusgen, M. (2008). Isolation and identification of a red pigment from *Allium* subgenus *Melanocrommyum*. *J. Agric. Food Chem.* **56**: 1465–1470.

Jeffery, B., Barlow, T., Moizer, K. et al. (2004). Amnesic shellfish poisoning. *Food Chem. Toxicol.* **42**: 545–557.

Jensen, J. and Erickson, M.D. (1997). *Analytical Chemistry of PCBs*, 2e. Chelsea, Michigan: Lewis Publisher.

Jensen, S.R. and Nielsen, B.J. (1976). A new coumarin, fraxidin 8-O-β-D-glucoside and 10-hydroxyligstroside from bark of *Fraxinus excelsior*. *Phytochemistry* **15**: 221–223.

Jenske, R. and Vetter, W. (2008a). Enantioselective analysis of 2- and 3-hydroxy fatty acids in food samples. *J. Agric. Food Chem.* **56**: 11578–11583.

Jenske, R. and Vetter, W. (2008b). Gas chromatography/electron-capture negative ion mass spectrometry for the quantitative determination of 2- and 3-hydroxy fatty acids in bovine milk fat. *J. Agric. Food Chem.* **56**: 5500–5505.

Jezussek, M., Juliano, B.O., and Schieberle, P. (2002). Comparison of key aroma compounds in cooked brown rice varieties based on aroma extract dilution analyses. *J. Agric. Food Chem.* **50**: 1101–1105.

Jhoo, J.-W., Lin, M.-C., Sang, S. et al. (2002). Characterization of 2-methyl-4-amino-5-(2-methyl-3-furylthiomethyl)pyrimidine from thermal degradation of thiamin. *J. Agric. Food Chem.* **50**: 4055–4058.

Jiang, D. and Peterson, D.G. (2013). Identification of bitter compounds in whole wheat bread. *Food Chem.* **141**: 1345–1353.

Jin, J. and Wu, T. (2017). Anthocyanins from black wolfberry (*Lycium ruthenicum* Murr.) prevent inflammatiom and increase fecal fatty acid in diet-induced obese rat. *RSC Adv.* **7**: 47848–47853.

Jina, J.-S. and Hattori, M. (2011). A new mammalian lignan precursor, asarinin. *Food Chem.* **124**: 895–899.

Jordão, A.M., Ricardo da Silva, J.M., and Laureano, O. (2007). Elagitannins from Portugese oak wood (*Quercus pyrenaica* Willd.) used in cooperage: influence of geographic origin, coarseness of the grain and toasting level. *Holzforschung* **61**: 155–160.

Jordheim, M., Enerstvedt, K.H., and Andersen, Ø.M. (2011). Identification of cyanidin 3-O-β-(6″-(3-hydroxy-3-methylglutaroyl)glucoside) and other anthocyanins from wild and cultivated blackberries. *J. Agric. Food Chem.* **59**: 7436–7440.

Jorgensen, K. and Skibsted, L.H. (1993). Carotenoid scavening of radicals. *Z. Lebensm. Unters. Forsch.* **196**: 423–429.

Jorhem, L., Mattsson, P., and Slorach, S. (1986). Lead, cadmium, zinc and certain other metals in foods on the Swedish market. *J. Environ. Pathol. Toxicol. Oncol.* **6**: 195–256.

Jorhem, L. and Sundström, B. (1993). Levels of lead, cadmium, zinc, copper, nickel, chromium, manganese, and cobalt in foods on the Swedish market, 1983–1990. *J. Food Compos. Anal.* **6**: 223–241.

Joseph, A.I., Luis, P.B., and Schneider, C. (2018). A Curcumin degradation product, 7-norcyclopentadione, formed by aryl migration and loss of a carbon from the heptadienedione chain. *J. Nat. Prod.* **81**: 2756–2762.

Joubert, E., Viljoen, M., De Beer, D. et al. (2010). Use of green rooibos (*Aspalathus linearis*) extract and water-soluble nanomicelles of green rooibos extract encapsulated with ascorbic acid for enhanced aspalathin content in ready-to-drink iced teas. *J. Agric. Food Chem.* **58**: 10965–10971.

Jugdaohsingh, R., Anderson, S.H.C., Tucker, K.L. et al. (2002). Dietary silicon intake and absorption. *Am. J. Clin. Nutr.* **75**: 887–893.

Jung, C.H. and Wells, W.W. (1998). Spontaneous conversion of L-dehydroascorbic acid to L-ascorbic acid and L-erythroascorbic acid. *Arch. Biochem. Biophys.* **355**: 9–14.

Jung, H.-A., Su, B.-N., Keller, W.J. et al. (2006). Antioxidant xanthones from the pericarp of *Garcinia mangostana* (mangosteen). *J. Agric. Food Chem.* **54**: 2077–2082.

Jung, H.-P., Sen, A., and Grosch, W. (1992). Evaluation of potent odorants in parsley leaves [*Petroselinum crispum* (Mill.) *Nym.* ssp. *crispum*] by aroma extract dilution analysis. *Lebensm. Wiss. Technol.* **25**: 55–60.

Jung, K., Fastowski, O., Poplacean, I., and Engel, K.-H. (2017). Analysis and sensory evaluation of volatile constituents of fresh blackcurrant (*Ribes nigrum* L.) fruits. *J. Agric. Food Chem.* **65**: 9475–9487.

Jung, M.Y., Oh, Y.S., Kim, D.K. et al. (2007). Photoinduced generation of 2,3-butanedione from riboflavin. *J. Agric. Food Chem.* **55**: 170–174.

Juskelis, R., Li, W., Nelson, J., and Cappozzo, J.C. (2013). Arsenic speciation in rice cereals for infants. *J. Agric. Food Chem.* **61**: 10670–10676.

Kadrabova, J., Madaric, A., and Ginter, E. (1997). The selenium content of selected food from the Slovak Republic. *Food Chem.* **58**: 29–32.

Kaffarnik, S., Heid, C., Kayademir, Y. et al. (2015). High enantiomeric excess of the flavor relevant 4-alkyl-branched fatty acids in milk fat and subcutaneous adipose tissue of sheep and goat. *J. Agric. Food Chem.* **63**: 469–475.

Kahner, L., Dasenbrock, J., Spiteller, P. et al. (1998). Polyene pigments from fruit-bodies of *Boletis laetissimus* and *B. rufo-aureus* (Basidiomycetes). *Phytochemistry* **49**: 1693–1697.

Kalač, P. (2006). Biologically active polyamines in beef, pork and meet products: a review. *Meat Sci.* **73**: 1–11.

Kalač, P. and Abreu Glória, M.B. (2009). Biogenic amines in cheesess, wines, beers and sauerkraut. In: *Biological Aspects of Biogenic Amines, Polyamines and Conjugates* (ed. G. Dandrifosse), 267–309. Kerala: Transworld.

Kalač, P. and Krausová, P. (2005). A review of dietary polyamines: formation, implications for growth and health and occurrence in foods. *Food Chem.* **90**: 219–230.

Kalač, P. and Křížek, M. (2003). A review of biogenic amines and polyamines in beer. *J. Inst. Brew.* **109**: 123–128.

Kalaras, M.D., Richie, J.P., Calcagnotto, A., and Beelman, R.B. (2017). Mushrooms: a rich source of the antioxidants ergothioneine and glutathione. *Food Chem.* **233**: 429–433.

Kalio, H.P. (2018). Historical review on the indentification of mesifurane, 2,5-dimethyl-4-methoxy-3(2H)-furanone, and its occurrence in berries and fruits. *J. Agric. Food Chem.* **66**: 2553–2560.

Kallio, H., Yli-Jokipii, K., Kurvinen, J.a.-P. et al. (2001). Regioisomerism of triacylglycerols in lard, tallow, yolk, chicken skin, palm oil, palm olein, palm stearin, and a transesterified blend of palm stearin and coconut oil analyzed by tandem mass spectrometry. *J. Agric. Food Chem.* **49**: 3363–3369.

Kamal-Eldin, A. and Appelquist, L.-A. (1996). The chemistry and antioxidant properties of tocopherols and tocotrienols. *Lipids* **31**: 671–701.

Karametsi, K., Kokkinidou, S., Ronningen, I., and Peterson, D.G. (2014). Identification of bitter peptides in aged Cheddar cheese. *J. Agric. Food Chem.* **62**: 8034–8041.

Karleskind, A. and Wolff, J.-P. (1996). *Oils and Fats Manual*. Paris: Lavoisier.

Karlson, P., Gerok, W., and Gross, W. (1987). *Pathobiochemie*. Praha: Academia.

Katan, L.L. (ed.) (1996). *Migration from Food Contact Materials*. London: Blackie Academic & Professional.

Kataoka, H., Miyake, M., Saito, K., and Mitani, K. (2012). Formation of heterocyclic amine-amino acid adducts by heating in a model system. *Food Chem.* **130**: 725–729.

Kawakishi, S., Goto, T., and Namiki, M. (1983). Oxidative scission of the disulfide bond of cysteine and polypeptides by the action of allyl isothiocyanate. *Agric. Biol. Chem.* **47**: 2071–2076.

Kazuma, K., Takahashi, T., Sato, K. et al. (2000). Quinochalcones and flavonoids from fresh florets in different cultivars of *Carthamus tinctorius* L. *Biosci. Biotechnol. Biochem.* **64**: 1588–1599.

Kearley, M.L., Patel, A., Chien, J., and Tuma, D.J. (1999). Observation of a new nonfluorescent malondialdehyde-acetaldehyde-protein adduct by ^{13}C NMR spectroscopy. *Chem. Res. Toxicol.* **12**: 100–105.

Keller, R.F. and Tu, A.T. (eds.) (1983). *Handbook of Natural Toxins*, vol. **I**. New York: Dekker.

Kendrick, J.L. and Watts, B.M. (1969). Nicotinamide and nicotinic acid in color preservation of fresh meat. *J. Food Sci.* **34**: 292.

Kerney, P.C. and Kaufman, D.D. (1988). *Herbicides – Chemistry, Degradation and Mode of Action*, vol. **3**. New York: Dekker.

Khachik, F., Goli, M.B., Beecher, G.L. et al. (1992). Effect of food preparation and quantitative distribution of major carotenoid constituents of tomatoes and several green vegetables. *J. Agric. Food Chem.* **40**: 390–398.

Khandelwal, G.D. and Wedzicha, B.L. (1990). Nucleophilic reactions of sorbic acid. *Food Addit. Contam.* **7**: 685–694.

Kilcast, D. (1994). Effect of irradiation on vitamins. *Food Chem.* **49**: 157–164.

Killeit, U. (1994). Vitamin retention in extrusion cooking. *Food Chem.* **49**: 149–155.

Kim, C.S., Oh, J., Subedi, L. et al. (2018). Rare thioglycosides from the roots of *Wasabia japonica*. *J. Nat. Prod.* **81**: 2129–2133.

Kim, H.-O., Hartnett, C., and Scaman, C.H. (2007). Free galactose content in selected fresh fruits and vegetables and soy beverages. *J. Agric. Food Chem.* **55**: 8133–8137.

Kim, K., Kim, J.-J., Jung, Y. et al. (2017). Cyclocurcumin, an antivasoconstrictive constituent of *Curcuma longa* (turmeric). *J. Nat. Prod.* **80**: 196–200.

Kim, K.S., Park, S.H., and Choung, M.G. (2006). Nondestructive determination of lignans and lignan glycosides in sesame seeds by near infrared reflectance spectroscopy. *J. Agric. Food Chem.* **54**: 4544–4550.

Kim, O.K., Murakami, A., Nakamura, Y. et al. (2000). Novel nitric oxide and superoxide generation inhibitors, persenone a and b, from avocado fruit. *J. Agric. Food Chem.* **48**: 1557–1563.

Kimura, K., Nishimura, H., Iwata, I., and Mizutani, J. (1983). Deterioration mechanism of lemon flavor. 2. Formation mechanism of off-flavor substances arising from citral. *J. Agric. Food Chem.* **31**: 801–804.

Kiritsakis, A.K. (1998). *Olive Oil*, 2e. Trumbull: Food and Nutrition Press.

Kishimoto, T., Kobayashi, M., Yako, N. et al. (2008). Comparison of 4-mercapto-4-methylpentan-2-one contents in hop cultivars from different growing regions. *J. Agric. Food Chem.* **56**: 1051–1057.

Kizil, M., Fatih, O., and Besler, H.T. (2011). A review on the formation of carcinogenic/mutagenic heterocyclic aromatic amines. *J. Food Process. Technol.* **2**: 120.

Knödler, M., Reisenhauer, K., Schieber, A., and Carle, R. (2009). Quantitative determination of allergenic 5-alk(en)ylresorcinols in mango (*Mangifera indica* L.) peel, pulp, and fruit products by high-performance liquid chromatography. *J. Agric. Food Chem.* **57**: 3639–3644.

Knorr, D. (1984). Use of chitinous polymers in food. *Food Technol.* **38**: 85–89, 92–97.

Ko, J.A., Lee, B.H., Lee, J.S., and Park, H.J. (2008). Effect of UV-B exposure on the concentration of vitamin d$_2$ in sliced shiitake mushroom (*Lentinus edodes*) and white button mushroom (*Agaricus bisporus*). *J. Agric. Food Chem.* **56**: 3671–3674.

Kobata, K. (2006). Capsaicinol: synthesis by allylic oxidation and its effect on TRPV1-expressing cells and adrenaline secretion in rats. *Biosci. Biotechnol. Biochem.* **70**: 1904–1912.

Kobata, K., Mimura, M., Sugawara, M., and Watanabe, T. (2011). Synthesis of stable isotope-labeled precursors for the biosynthese of capsaicinoids, capsinoids, and capsiconinoids. *Biosci. Biotechnol. Biochem.* **75**: 1611–1614.

Kobata, K., Saito, K., Tate, H. et al. (2010). Long-chain N-Vanillyl-acylamides from *Capsicum oleoresin*. *J. Agric. Food Chem.* **58**: 3627–3631.

Kobata, K., Sugawara, M., Mimura, M. et al. (2013). Potent production of capsaicinoids and capsinoids by *Capsicum* peppers. *J. Agric. Food Chem.* **61**: 11127–11132.

Koehler, P. (2003). Effect of ascorbic acid in dough: reaction of oxidized glutathione with reactive thiol groups of wheat gluteline. *J. Agric. Food Chem.* **51**: 4954–4959.

Koge, T., Komatsu, W., and Sorimachi, K. (2011). Heat stability of agaritine in water extracts from *Agaricus blazei* and other edible fungi, and removal of agaritine by ethanol fractionation. *Food Chem.* **126**: 1172–1177.

Kohyama, N. and Ono, H. (2013). Hordatine a β-D-glucopyranoside from ungerminated barley grains. *J. Agric. Food Chem.* **61**: 1112–1116.

Koivu-Tikkanen, T.J., Ollilainen, V., and Piironen, V.I. (2000). Determination of phylloquinone and menaquinones in animal products with fluorescence detection after postcolumn reduction with metallic zinc. *J. Agric. Food Chem.* **48**: 6325–6331.

Kollmannsberger, H., Nitz, S., and Drawert, F. (1992). Über die Aromastoffzusam-mensetzung von Hochdruckextrakten. I. Pfeffer (*Piper nigrum*, var. *muntok*). *Z. Lebensm. Unters. Forsch.* **194**: 545–551.

Kolodziej, H., Sakar, M.K., Burger, J.F.W. et al. (1991). A-type proanthocyanidins from *Prunus spinosa*. *Phytochemistry* **30**: 2041–2047.

Kong, Q.J., Ren, X.Y., Hua, N. et al. (2011). Identification of isomers of resveratrol dimer and their analogues from wine grapes by HPLC/MSn and HPLC/DAD-UV. *Food Chem.* **127**: 727–734.

König, W.A. and Hochmuth, D.H. (2004). Enantioselective gas chromatography in flavor and fragrance analysis: strategies for the identification of known and unknown plant volatiles. *J. Chromatogr. Sci.* **42**: 423–439.

Kopsell, D.A. and Kopsell, D.E. (2010). Chapter 40 – Carotenoids in vegetables: biosynthesis, occurrence, impacts on human health, and potential for manipulation. In: *Bioactive Foods in Promoting Health. Fruits and Vegetables* (eds. R.R. Watson and V.R. Preedy), 645. Elsevier Inc.

Korky, J.K. and Kowalski, L. (1989). Radioactive cesium in edible mushrooms. *J. Agric. Food Chem.* **37**: 568–569.

Kotseridis, Y.S., Spink, M., Brindle, I.D. et al. (2008). Quantitative analysis of 3-alkyl-2-methoxypyrazines in juice and wine using stable isotope labelled internal standard assay. *J. Chromatogr. A* **1190**: 294–301.

Kovacs-Nolan, J., Phillips, M., and Mine, Y. (2005). Advances in the value of eggs and egg components for human health. *J. Agric. Food Chem.* **53**: 8421–8431.

Koyama, N., Aoyagi, Y., and Sugahara, T. (1984). Fatty acid composition and ergosterol contents of edible mushrooms. *Nihon Shokuhin Kōgyō Gakkaishi (J. Food Sci. Technol.)* **31**: 732–738.

Koyyalamudi, S.R., Jeong, S.-C., Cho, K.Y., and Pang, G. (2009a). Vitamin B_{12} is the active corrinoid produced in cultivated white button mushrooms (*Agaricus bisporus*). *J. Agric. Food Chem.* **57**: 6327–6333.

Koyyalamudi, S.R., Jeong, S.-C., Song, C.-H. et al. (2009b). Vitamin D_2 formation and bioavailability from *Agaricus bisporus* button mushrooms treated with ultraviolet irradiation. *J. Agric. Food Chem.* **57**: 3351–3355.

Kozukue, N., Han, J.-S., Lee, K.-R., and Friedman, M. (2004). Dehydrotomatine and α-tomatine content in tomato fruits and vegetative plant tissues. *J. Agric. Food Chem.* **52**: 2079–2083.

Kozukue, N., Yoon, K.-S., Byun, G.-I. et al. (2008). Distribution of glycoalkaloids in potato tubers of 59 accessions of two wild and five cultivated solanum species. *J. Agric. Food Chem.* **56**: 11920–11928.

Kramer, J.K.G., Parodi, P.W., Jensen, R.G. et al. (1998). Rumenic acid: a proposed common name for the major conjugated linoleic acid isomer found in natural products. *Lipids* **33**: 835–835.

Kramers, P.G.N. and van der Heijden, C.A. (1988). Polycyclic aromatic hydrocarbons (PAH): carcinogenicity data and risk extrapolations. *Toxicol. Environ. Chem.* **16**: 341–351.

Kreissl, J. and Schieberle, P. (2017). Characterization of aroma-active compounds in Italian tomatoes with emphasis on new odorants. *J. Agric. Food Chem.* **65**: 5198–5208.

Kreppenhofer, S., Frank, O., and Hofmann, T. (2011). Identification of (furan-2-yl)methylated benzene diols and triols as a novel class of bitter compounds in roasted coffee. *Food Chem.* **126**: 441–449.

Krogerus, K. and Gibson, B.R. (2013). Diacetyl and its control during brewery fermentation. *J. Inst. Brew.* **119**: 86–97.

Krzycki, J. (2005). The direct genetic encoding of pyrrolysine. *Curr. Opin. Microbiol.* **8**: 706–712.

Kuad, P., Schurhammer, R., Maechling, C. et al. (2009). Complexation of Cs+, K+ and Na+ by norbadione A triggered by the release of a strong hydrogen bond: nature and stability of the complexes. *Phys. Chem. Chem. Phys.* **11**: 10299–10310.

Kubec, R., Cody, R.B., Dane, A.J. et al. (2010). Applications of direct analysis in real time-mass spectrometry (DART-MS) in Allium chemistry. (Z)-Butanethial S-oxide and 1-butenyl thiosulfinates and their S-(E)-1-butenylcysteine S-oxide precursor from *Allium siculum*. *J. Agric. Food Chem.* **58**: 1121–1128.

Kubec, R., Curko, P., Urajová, P. et al. (2017). *Allium* discoloration: color compounds formed during greening of processed garlic. *J. Agric. Food Chem.* **65**: 10615–10620.

Kubec, R., Krejčová, P., Šimek, P. et al. (2011). Precursors and formation of pyrithione and other pyridyl-containing sulfur compounds in drumstick onion, Allium stipitatum. *J. Agric. Food Chem.* **59**: 5763–5770.

Kubec, R., Štefanová, I., Moos, M. et al. (2018). Allithiolanes: nine groups of a newly discovered family of sulfur compounds responsible for the bitter off-taste of processed onion. *J. Agric. Food Chem.* **66**: 8783–8794.

Kubec, R., Urajová, P., Lacina, O. et al. (2015). *Allium* discoloration: color compounds formed during pinking of onion and leek. *J. Agric. Food Chem.* **63**: 10192–10199.

Kubec, R., Velíšek, J., and Musah, R.A. (2002). The amino acid precursors and odor formation in society garlic (*Tulbaghia violacea* Harv.). *Phytochemistry* **60**: 21–25.

Kumar, P.P., Paramashivappa, R., Vithayathil, P.J. et al. (2002). Process for isolation of cardanol from technical cashew (*Anacardium occidentale* L.) nut shell liquid. *J. Agric. Food Chem.* **50**: 4705–4708.

Kumazawa, K. and Masuda, H. (1999). Identification of potent odorants in Japanese green tea (sen-cha). *J. Agric. Food Chem.* **47**: 5169–5172.

Kumorkiewicz, A. and Wybraniec, S. (2017). Thermal degradation of major gomphrenin pigments in the fruit juice of *Basella alba* L. (Malabar spinach). *J. Agric. Food Chem.* **65**: 7500–7508.

Kunert, C., Walker, A., and Hofmann, T. (2011). Taste modulating N-(1-methyl-4-oxoimidazolidin-2-ylidene) α-amino acids formed from creatinine and reducing carbohydrates. *J. Agric. Food Chem.* **59**: 8366–8374.

Kunz, C. and Rudloff, S. (1996). Strukturelle und funktionelle Aspekte von Oligosacchariden in Frauenmilch. *Z. Ernahrungswiss.* **35**: 22–31.

Kurihara, Y. and Nirasawa, S. (1994). Sweet, antisweet and sweetness-inducing substances. *Trends Food Sci. Technol.* **5**: 37–42.

Kurmann, A. and Indyk, H. (1994). The endogenous vitamin D content of bovine milk: influence of season. *Food Chem.* **50**: 75–81.

Kuroda, M., Kato, Y., Yamazaki, J. et al. (2013). Determination and quantification of the kokumi peptide, γ-glutamyl-valyl-glycine, in commercial soy sauces. *Food Chem.* **141**: 823–828.

Kuroda, M., Ohtake, R., Suzuki, E., and Harada, T. (2000). Investigation on the formation and the determination of γ-glutamyl-β-alanylhistidine and related isopeptide in the macromolecular fraction of beef soup stock. *J. Agric. Food Chem.* **48**: 6317–6324.

Kuru, R., Yilmaz, S., Tasli, P.N. et al. (2019). Boron content of some foods consumed in Istanbul, Turkey. *Biol. Trace Elem. Res.* **187**: 1–8.

Kusterer, J., Fritsch, R.M., and Keusgen, M. (2011). Allium species from Central and Southwest Asia are rich sources of marasmin. *J. Agric. Food Chem.* **59**: 8289–8297.

Kusterer, J. and Keusgen, M. (2010). Cysteine sulphoxides and volatile sulphur compounds from *Allium tripedale*. *J. Agric. Food Chem.* **58**: 1129–1137.

Kusterer, J., Vogt, A., and Keusgen, M. (2010). Isolation and identification of a new cysteine sulfoxide and volatile sulfur compounds from *Allium* subgenus *Melanocrommyum*. *J. Agric. Food Chem.* **58**: 520–526.

Kuwahara, H., Kanazawa, A., Wakamatu, D. et al. (2004). Antioxidative and antimutagenic activities of 4-vinyl-2,6-dimethoxyphenol (canolol) isolated from canola oil. *J. Agric. Food Chem.* **52**: 4380–4387.

Kuzuya, H., Tamai, I., Beppu, H. et al. (2001). Determination of aloenin, barbaloin and isobarbaloin in Aloe species by micellar electrokinetic chromatography. *J. Chromatogr. B* **752**: 91–97.

Kvasnička, F., Voldřich, M., Pyš, P., and Vinš, I. (2002). Determination of isocitric acid in citrus juice – comparison of HPLC, enzyme set and capillary isotachophoresis methods. *J. Food Compos. Anal.* **15**: 685–691.

Lachenmeier, D.W., Nathan-Maister, D., Breaux, T.A. et al. (2008). Chemical composition of vintage preban absinthe with special reference to thujone, fenchone, pinocamphone, methanol, copper, and antimony concentrations. *J. Agric. Food Chem.* **56**: 3073–3081.

Lachowicz, S., Oszmiański, J., and Pluta, S. (2017). The composition of bioactive compounds and antioxidant activity of Saskatoon berry (*Amelanchier alnifolia* Nutt.) genotypes grown in central Poland. *Food Chem.* **235**: 234–243.

Lagemann, A., Dunkel, A., and Hofmann, T. (2012). Activity-guided discovery of (*S*)-malic acid 1′-*O*-β-gentiobioside as an angiotensin I-converting enzyme inhibitor in lettuce (*Lactuca sativa*). *J. Agric. Food Chem.* **60**: 7211–7217.

Lahelma, O., Nuurtamo, M., Paaso, A. et al. (1980). Mineral element composition of Finnish foods: N, K, Ca, Mg, P, S, Fe, Cu, Mn, Zn, Mo, Co, Ni, Cr, F, Se, Si, Rb, Al, B, Br, Hg, As, Cd, Pb and ash. In: Koivistoinen P. (ed.): Miscellaneous Acta Agriculturae Scandinavica No. Suppl. 22 pp., 171 pp.

Lam, K.C. and Deinzer, M.L. (1987). Tentative identification of humulene diepoxides by capillary gas chromatography-chemical ionization mass spectrometry. *J. Agric. Food Chem.* **35**: 57–59.

Lambert, C., Richard, T., Renouf, E. et al. (2013). Comparative analyses of stilbenoids in canes of major *Vitis vinifera* L. cultivars. *J. Agric. Food Chem.* **61**: 11392–11399.

Lamberts, L. and Delcour, J.A. (2008). Carotenoids in raw and parboiled brown and milled rice. *J. Agric. Food Chem.* **56**: 11914–11919.

Landberg, R., Kamal-Eldin, A., Andersson, R., and Åman, P. (2006). Alkylresorcinol content and homologue composition in durum wheat (Triticum durum) kernels and pasta products. *J. Agric. Food Chem.* **54**: 3012–3014.

Lang, M.R. and Wells, J.W. (1987). A review of eggshell pigmentation. *World's Poult. Sci. J.* **43**: 238–246.

Lang, R., Bardelmeier, I., Weiss, C. et al. (2010). Quantitation of $^\beta$N-Alkanoyl-5-hydroxytryptamides in coffee by means of LC-MS/MS-SIDA and assessment of their gastric acid secretion potential using the HGT-1 cell assay. *J. Agric. Food Chem.* **58**: 1593–1602.

Lang, R., Klade, S., Beusch, A. et al. (2015). Mozambioside is an arabica-specific bitter-tasting furokaurane glucoside in coffee beans. *J. Agric. Food Chem.* **63**: 10492–10499.

Lang, V. (1992). Polychlorinated biphenyls in the environment. *J. Chromatogr. A* **595**: 1–43.

Langdon, T.T. (1986). Preventing of browning in fresh, prepared potatoes, without the use of sulfiting agents. *Food Technol.* **41**: 64–66.

Lapčík, O. (2007). Isoflavonoids in non-leguminous taxa: a rarity or a rule? *Phytochemistry* **68**: 2909–2916.

Lapierre, C. and Moutounet, M. (1993). Ellagitannins and lignins in aging of spirits in oak barrels. *J. Agric. Food Chem.* **41**: 1872–1879.

Largé, E., Zamora, S., and Gil, A. (2001). Dietary trans fatty acids in early life: a review. *Early Hum. Dev.* **65** (Suppl): S31.

Larisch, B., Pischetsrieder, M., and Severin, T. (1996). Reactions of dehydroascorbic acid with primary aliphatic amines including N^α-acetyllysine. *J. Agric. Food Chem.* **44**: 1630–1634.

Larsson, B.K. (1986). Formation of polycyclic aromatic hydrocarbons during the smoking and grilling of food, PAHs in Swedish food, aspects of analysis, occurrence and intake. Swedish Agricultural University.

Larsson, K. (1994). *Lipids: Molecular Organization, Physical Functions and Technical Applications*. Dundee: Oily Press.

Lätti, A.K., Riihinen, K.R., and Kainulainen, P.S. (2008). Analysis of anthocyanin variation in wild populations of bilberry (*Vaccinium myrtillus* L.) in Finland. *J. Agric. Food Chem.* **56**: 190–196.

Latza, S., Ganßer, D., and Berger, R.G. (1996). Carbohydrate esters of cinnamic acid from fruits of *Physalis peruviana*, *Psidium gujava* and *Vaccinium vitis-idaea*. *Phytochemistry* **43**: 481–485.

Lau, O.-W. and Wong, S.-K. (2000). Contamination in food from packaging material. *J. Chromatogr. A* **882**: 255–270.

Laub, E. and Woller, R. (1984). Über den Mannit- und Trehalosegehalt von Kulturchampignons. *Mitt. Geb. Lebensmittelunters. Hyg.* **75**: 110–116.

Lauber, R.P. and Sheard, N.F. (2001). The American Heart Association Dietary Guidelines for 2000: a summary report. *Nutr. Rev.* **59**: 298–306.

Lawrence, J.F., Michalik, P., Tam, G., and Conacher, H.B.S. (1986). Identification of arsenobetaine and arsenocholine in canadian fish and shellfish by high-performance liquid chromatography with atomic absorption detection and confirmation by fast atom bombardment mass spectrometry. *J. Agric. Food Chem.* **34**: 315–319.

Le Bourvellec, C., Gouble, B., Bureau, S. et al. (2013). Pink discoloration of canned pears: role of procyanidin chemical depolymerization and procyanidin/cell wall interactions. *J. Agric. Food Chem.* **61**: 6679–6692.

Lee, C.Y. and Whitaker, J.R. (eds.) (1995). *Enzymatic Browning and its Prevention*. Washington, D.C.: American Chemical Society.

Lee, E.J., Rezenom, Y.H., Russell, D.H. et al. (2012). Elucidation of chemical structures of pink-red pigments responsible for 'pinking' in macerated onion (*Allium cepa* L.) using HPLC-DAD and tandem mass spectrometry. *Food Chem.* **131**: 852–861.

Lee, H.-O. and Montag, A. (1992). Antioxidative Wirksamkeit von Tocochromanolen-Rangfolgebestimmung mit Response Surface Methodology (RSM). *Eur. J. Lipid Sci. Technol.* **94**: 213–217.

Lee, J., Zhang, G., Wood, E. et al. (2013). Quantification of amygdalin in nonbitter, semibitter, and bitter almonds (*Prunus dulcis*) by UHPLC-(ESI)QqQ MS/MS. *J. Agric. Food Chem.* **61**: 7754–7759.

Lee, J.Y., Lee, J.Y., Yun, B.-S., and Hwang, B.K. (2004). Antifungal activity of β-asarone from rhizomes of *Acorus gramineus*. *J. Agric. Food Chem.* **52**: 776–780.

Lee, K.D., Lo, C.G., and Warthesen, J.J. (1996). Removal of bitterness from the bitter peptides extracted from cheddar cheese with peptidases from *Lactococcus lactis* ssp. *cremoris* SKI. *J. Dairy Sci.* **79**: 1521–1528.

Lee, K.P.D. and Warthesen, J.J. (1996). Preparative methods of isolating bitter peptides from Cheddar cheese. *J. Agric. Food Chem.* **44**: 1058–1063.

Lee, S.H., Goto, T., and Oe, T. (2008). A novel 4-oxo-2(E)-nonenal-derived modification to angiotensin II: oxidative decarboxylation of N-terminal aspartic acid. *Chem. Res. Toxicol.* **21**: 2237–2244.

Leffingwell, J.C. Chirality & odour perception. http://www.leffingwell.com (accessed 28 October 2019).

Leffingwell, J.C. Cool without menthol & cooler than menthol. http://www.leffingwell.com/cooler_than_menthol.htm (accessed 28 October 2019).

Lehmann, D., Dietrich, A., Hener, U., and Mosandl, A. (2007). Stereoisomeric flavour compounds. LXX: 1-*p*-menthene-8-thiol: separation and sensory evaluation of the enantiomers by enantioselective gas chromatographyolfactometry. *Phytochem. Anal.* **6**: 255–257.

Lewis, R.S., Parker, R.G., Danehower, D.A. et al. (2012). Impact of alleles at the yellow burley (Yb) loci and nitrogen fertilization rate on nitrogen utilization efficiency and tobacco-specific nitrosamine (TSNA) formation in air-cured tobacco. *J. Agric. Food Chem.* **60**: 6454–6461.

Ley, J.P., Blings, M., Paetz, S. et al. (2006). New bitter-masking compounds: hydroxylated benzoic acid amides of aromatic amines as structural analogues of homoeriodictyol. *J. Agric. Food Chem.* **54**: 8574–8579.

Ley, J.P., Krammer, G., Reinders, G. et al. (2005). Evaluation of bitter masking flavanones from herba santa (*Eriodictyon californicum* (H. & A.) Torr., Hydrophyllaceae). *J. Agric. Food Chem.* **53**: 6061–6066.

Li, C. and Oberlies, N.H. (2005). The most widely recognized mushroom: chemistry of the genus Amanita. *Life Sci.* **78**: 532–538.

Li, F., Nitteranon, V., Tang, X. et al. (2012). *In vitro* antioxidant and anti-inflammatory activities of 1-dehydro-[6]-gingerdione, 6-shogaol, 6-dehydroshogaol and hexahydrocurcumin. *Food Chem.* **135**: 332–337.

Li, J., Li, N., Li, X. et al. (2017). Characteristic α-acid derivatives from *Humulus lupulus* with antineuroinflammatory activities. *J. Nat. Prod.* **80**: 3081–3092.

Li, J.-X., Schieberle, P., and Steinhaus, M. (2012). Characterization of the major odor-active compounds in Thai durian (*Durio zibethinus* L. 'Monthong') by aroma extract dilution analysis and headspace gas chromatography–olfactometry. *J. Agric. Food Chem.* **60**: 11253–11262.

Li, J.-X., Schieberle, P., and Steinhaus, M. (2017). Insights into the key compounds of durian (*Durio zibethinus* L. 'Monthong') pulp odor by odorant quantitation and aroma simulation experiments. *J. Agric. Food Chem.* **65**: 639–647.

Li, L., Shewry, P.R., and Ward, J.L. (2008). Phenolic acids in wheat varieties in the healthgrain diversity screen. *J. Agric. Food Chem.* **56**: 9732–9739.

Li, S., Lo, C.-Y., and Ho, C.-T. (2006). Hydroxylated polymethoxyflavones and methylated flavonoids in sweet orange (*Citrus sinensis*) peel. *J. Agric. Food Chem.* **54**: 4176–4185.

Li, Y., Zhao, M., and Parkin, K.L. (2011). β-Carboline derivatives and diphenols from soy sauce are *in vitro* quinone reductase (qr) inducers. *J. Agric. Food Chem.* **59**: 2332–2340.

Li, W.-X., Li, Y.-F., Zhai, Y.-J. et al. (2013). Theacrine, a purine alkaloid obtained from *Camellia assamica* var. *kucha*, attenuates restraint stress-provoked liver damage in mice. *J. Agric. Food Chem.* **61**: 6328–6335.

Liang, C.-H., Chan, L.-P., Ding, H.-Y. et al. (2012). Free radical scavenging activity of 4-(3,4-dihydroxybenzoyloxymethyl)phenyl-O-β-D-glucopyranoside from *Origanum vulgare* and its protection against oxidative damage. *J. Agric. Food Chem.* **60**: 7690–7696.

Liao, M.-L. and Seib, P.A. (1987). Selected reactions of L-ascorbic acid related to foods. *Food Technol.* **41**: 104–107, 111.

Li-Chan, E.C.Y. (2015). Bioactive peptides and protein hydrolysates: research trends and challenges for application as nutraceuticals and functional food ingredients. *Curr. Opin. Food Sci.* **1**: 28–37.

Lichtenthaler, F.W., Nakamura, K., and Klotz, J. (2003). (−)-Daucic acid: revision of configuration, synthesis, and biosynthetic implications. *Angew. Chem. Int. Ed.* **42**: 5838–5843.

Liégeois, C., Meurens, N., Badot, C., and Collin, S. (2002). Release of deuterated (E)-2-nonenal during beer aging from labeled precursors synthesized before boiling. *J. Agric. Food Chem.* **50**: 7634–7638.

Liener, I.E. (ed.) (1980). *Toxic constituents of food crops*, 2e. New York: Academic Press.

Liener, I.E. (1994). Implications of antinutritional components in soybean foods. *Crit. Rev. Food Sci. Nutr.* **34**: 31–67.

Liger-Belair, G. (2005). The physics and chemistry behind the bubbling properties of champagne and sparkling wines: a state-of-the-art review. *J. Agric. Food Chem.* **53**: 2788–2802.

Lin, Q., Gourdon, D., Sun, C. et al. (2007). Adhesion mechanisms of the mussel foot proteins mfp-1 and mfp-3. *Proc. Natl. Acad. Sci. U. S. A.* **104**: 3782–3786.

Lin, S.-P., Tsai, S.-Y., Lin, Y.-L. et al. (2008). Biotransformation and pharmacokinetics of 4-(3,4-Dihydroxybenzoyloxymethyl)phenyl-O-β-D-glucopyranoside, an antioxidant isolated from *Origanum vulgare*. *J. Agric. Food Chem.* **56**: 2852–2856.

Lin, Y.C., Chang, J.C., Cheng, S.Y. et al. (2015). New bioactive chromanes from *Litchi chinensis*. *J. Agric. Food Chem.* **63**: 2472–2478.

Lindenmeier, M., Faist, V., and Hofmann, T. (2002). Structural and functional characterization of pronyl-lysine, a novel protein modification in bread crust melanoidins showing in vitro antioxidative and phase I/II enzyme modulating activity. *J. Agric. Food Chem.* **50**: 6997–7006.

Linus, C.F. and Andersson, H.C. (2017). *Phytoestrogens in Foods on the Nordic Market: A Literature Review on Occurrence and Levels*. Nordic Council of Ministers.

Lipscomb, W.N. and Sträter, N. (1996). Recent advances in zinc enzymology. *Chem. Rev.* **96**: 2375–2434.

Little, M.C., Preston, J.F. 3rd, Jackson, C. et al. (1986). Alloviroidin, the naturally occurring toxic isomer of the cyclopeptide viroidin. *Biochemistry* **25**: 2867–2872.

Liu, H., Zheng, A., Liu, H. et al. (2012). Identification of three novel polyphenolic compounds, origanine a–c, with unique skeleton from *Origanum vulgare* L. using the hyphenated LC-DAD-SPE-NMR/MS methods. *J. Agric. Food Chem.* **60**: 129–135.

Liu, H.-L., Luo, R., Chen, X.-Q. et al. (2015). Identification and simultaneous quantification of five alkaloids in *Piper longum* L. by HPLC–ESI-MSn and UFLC–ESI-MS/MS and their application to *Piper nigrum* L. *Food Chem.* **177**: 191–196.

Liu, J., Zhu, X.-L., Ullah, N., and Tao, Y.-S. (2017). Aroma glycosides in grapes and wine. *J. Food Sci.* **82**: 248–259.

Liu, J.-K. (2006). Natural terphenyls: developments since 1877. *Chem. Rev.* **106**: 2209–2223.

Liu, K. (1989). *Soybean. Chemistry, Technology, and Utilization*. Gaithersberg, Maryland: Aspen Publishers, Inc.

Lo, C.-Y., Li, S., Tan, D. et al. (2006). Trapping reactions of reactive carbonyl species with tea polyphenols in simulated physiological conditions. *Mol. Nutr. Food Res.* **50**: 1118–1128.

Lo, Y.-H., Chen, Y.-J., Chang, C.-I. et al. (2014). Teaghrelins, unique acylated flavonoid tetraglycosides in chin-shin oolong tea, are putative oral agonists of the ghrelin receptor. *J. Agric. Food Chem.* **62**: 5085–5091.

Lodovici, M., Dolara, P., Casalini, C. et al. (1995). Polycyclic aromatic hydrocarbon contamination in the Italian diet. *Food Addit. Contam.* **12**: 703–713.

Loewus, F.A. (1999). Biosynthesis and metabolism of ascorbic acid in plants and of analogs of ascorbic acid in fungi. *Phytochemistry* **52**: 193–210.

Loo, J.V., Coussement, P., DeLeenheer, L. et al. (1995). On the presence of inulin and oligofructose as natural ingredients in the western diet. *Crit. Rev. Food Sci. Nutr.* **35**: 525–552.

Lopes, R.M., da Silveira Agostini-Costa, T., Gimenes, M.A., and Silveira, D. (2011). Chemical composition and biological activities of Arachis species. *J. Agric. Food Chem.* **59**: 4321–4330.

López, P., Pereboom-de Fauw, D.P.K.H., Mulder, P.P.J. et al. (2018). Straightforward analytical method to determine opium alkaloids in poppy seeds and bakery products. *Food Chem.* **242**: 443–450.

López-Olguín, J., De la Torre, M.C., Ortego, F. et al. (1999). Structure-activity relationship of natural and synthetic neo-clerodane diterpenes from Teucrium against Colorado potato beetle larvae. *Phytochemistry* **50**: 749–753.

Lorber, K. and Buettner, A. (2015). Structure–odor relationships of (E)-3-alkenoic acids, (E)-3-alken-1-ols, and (E)-3-alkenals. *J. Agric. Food Chem.* **63**: 6681–6688.

Lorber, K., Zeh, G., Regler, J., and Buettner, A. (2018). Structure–odor relationships of (Z)-3-alken-1-ols, (Z)-3-alkenals, and (Z)-3-alkenoic acids. *J. Agric. Food Chem.* **66**: 2334–2343.

Louche, L.M.-M., Luro, F., Gaydou, E.M., and Lesage, J.-C. (2000). Phlorin screening in various citrus species and varieties. *J. Agric. Food Chem.* **48**: 4728–4733.

Łuczyńska, J., Łuczyński, M.J., and Paszcyk, B. (2016). Assessment of mercury in muscles, liver and gills of marine and freshwater fish. *J. Elem.* **21**: 113–129.

Lund, E.D. and Bruemmer, J.H. (1992). Sesquiterpene hydrocarbons in processed stored carrot sticks. *Food Chem.* **43**: 331–335.

Lund, M.N. and Ray, C.A. (2017). Control of Maillard reactions in foods: strategies and chemical mechanisms. *J. Agric. Food Chem.* **65**: 4537–4552.

Lytra, G., Tempere, S., de Revel, G., and Barbe, J.-C. (2014). Distribution and organoleptic impact of ethyl 2-methylbutanoate enantiomers in wine. *J. Agric. Food Chem.* **62**: 5005–5010.

Ma, L., Watrelot, A.A., Addison, B., and Waterhouse, A.L. (2018). Condensed tannin reacts with SO_2 during wine aging, yielding flavan-3-ol sulfonates. *J. Agric. Food Chem.* **66**: 9259–9268.

Maarse, H. and van der Heij, D.G. (eds.) (1994). *Trends in Flavour Research*. Amsterdam: Elsevier.

MacLeod, A.J. and Cave, S.J. (1975). Volatile flavour components of eggs. *J. Sci. Food Agric.* **26**: 351–360.

Maga, J.A. (1981a). Pyridines in foods. *J. Agric. Food Chem.* **29**: 895–898.

Maga, J.A. (1981b). Pyrroles in foods. *J. Agric. Food Chem.* **29**: 691–694.

Maga, J.A. (1992). Pyrazine update. *Food Rev. Int.* **8**: 479–558.

Maga, J.A. (1996). Oak lactones in alcoholic beverages. *Food Rev. Int.* **12**: 105–130.

Magan, N. and Olsen, M. (eds.) (2004). *Mycotoxins in Food: Detection and Control*. Boca Raton, FL: CRC Press – Taylor & Francis Group.

Mai, F. and Glomb, M.A. (2014). Lettucenin sesquiterpenes contribute significantly to the browning of lettuce. *J. Agric. Food Chem.* **62**: 4747–4753.

Mai, F. and Glomb, M.A. (2016). Structural and sensory characterization of novel sesquiterpene lactones from iceberg lettuce. *J. Agric. Food Chem.* **64**: 295–301.

Majdalawieh, A.F., Fayyad, M.W., and Nasrallah, G.K. (2017). Anti-cancer properties and mechanisms of action of thymoquinone, the major active ingredient of *Nigella sativa*. *Crit. Rev. Food Sci. Nutr.* **57**: 3911–3928.

Maki, K.C., Shinnick, F., Seeley, M.A. et al. (2003). Food products containing free tall oil-based phytosterols and oat glucan lower serum total and LDL cholesterol in hypercholesterolemic adults. *J. Nutr.* **133**: 808–813.

Mall, V., Sellami, I., and Schieberle, P. (2018a). Identification and quantitation of four new 2-alkylthiazolidine-4-carboxylic acids formed in orange juice by a reaction of saturated aldehydes with cysteine. *J. Agric. Food Chem.* **66**: 11073–11082.

Mall, V., Sellami, I., and Schieberle, P. (2018b). New degradation pathways of the key aroma compound 1-penten-3-one during storage of not-from-concentrate orange juice. *J. Agric. Food Chem.* **66**: 11083–11091.

Manahan, S.E. (ed.) (1994). *Environmental Chemistry*, 6e. Boca Raton, FL: Lewis Publishers.

Manchand, P.S., Whalley, W.B., and Chen, F.-C. (1973). Isolation and structure of ankaflavin: a new pigment from *Monascus anka*. *Phytochemistry* **12**: 2531–2532.

Mann, D.A. and Beuchat, L.R. (2008). Combinations of antimycotics to inhibit the growth of molds capable of producing 1,3-pentadiene. *Food Microbiol.* **25**: 144–153.

Mann, J. and Truswell, A.S. (eds.) (2007). *Essentials of Human Nutrition*, 3e. Oxford: Oxford University Press.

Mariani, C. and Bellan, G. (1997). Presence of tocopherol derivatives in vegetable oils. *Riv. Ital. Sostanze GR.* **74**: 545–552.

Marks, H.S., Hilson, J.A., Leichtweis, H.C., and Stoewsand, G.S. (1992). S-Methylcysteine sulfoxide in *Brassica* vegetables and formation of methyl methanethiosulfinate from Brussels sprouts. *J. Agric. Food Chem.* **40**: 2098–2101.

Marnett, L.J. (1999). Lipid peroxidation-DNA damage by malondialdehyde. *Mutat. Res.* **424**: 83–95.

Marsh, K. and Bugusu, B. (2007). Food packaging-roles, materials, and environmental issues. *J. Food Sci.* **72**: R39–R55.

Marsili, R. (ed.) (1996). *Techniques for Analyzing Food Aroma*. New York: Dekker.

Marth, E.H. (1992). Mycotoxins: production and control. *Food Lab. News* **8**: 35–51.

Martin, R.E., Collette, R.L., and Slavin, J.W. (1997). *Fish Inspection, Quality Control, and HACCP*. Lancaster, PA: Technomic.

Marty, C. and Berset, C. (1986). Degradation of trans-β-carotene during heating in sealed tubes and extrusion cooking. *J. Food Sci.* **51**: 698–702.

Marty, C. and Berset, C. (1988). Degradation products of trans-β-carotene produced during extrusion cooking. *J. Food Sci.* **53**: 1880–1886.

Massey, L.K., Palmer, R.G., and Horner, H.T. (2001). Oxalate content of soybean seeds (Glycine max: Leguminosae), soyfoods, and other edible legumes. *J. Agric. Food Chem.* **49**: 4262–4266.

Massey, R. and Taylor, D. (1988). *Aluminium in Food and the Environment*. London: The Royal Society of Chemistry.

Masson, E., Baumes, R., Le Guernevé, C., and Puech, J.-L. (2000). Identification of a precursor of β-methyl-γ-octalactone in the wood of sessile oak (*Quercus petraea* (Matt.) Liebl.). *J. Agric. Food Chem.* **48**: 4306–4309.

Masuda, T., Inaba, Y., and Takeda, Y. (2001). Antioxidant mechanism of carnosic acid: structural identification of two oxidation products. *J. Agric. Food Chem.* **49**: 5560–5565.

Masuda, T. and Jitoe, A. (1994). Antioxidative and antiinflamatory compounds from tropical gingers: isolation, structure determination, and activities of cassumunins A, B, and C, new complex curcuminoids from *Zingiber cassumunar*. *J. Agric. Food Chem.* **42**: 1850–1856.

Masuda, T., Shingai, Y., Fujimoto, A. et al. (2010). Identification of cytotoxic dimers in oxidation product from sesamol, a potent antioxidant of sesame oil. *J. Agric. Food Chem.* **58**: 10880–10885.

Masuda, T., Toi, Y., Bando, H. et al. (2002). Structural identification of new curcumin dimers and their contribution to the antioxidant mechanism of curcumin. *J. Agric. Food Chem.* **50**: 2524–2530.

Matei, M.F., Jaiswal, R., and Kuhnert, N. (2012). Investigating the chemical changes of chlorogenic acids during coffee brewing: conjugate addition of water to the olefinic moiety of chlorogenic acids and their quinides. *J. Agric. Food Chem.* **60**: 12105–12115.

Matenga, V.R., Ngongoni, N.T., Titterton, M., and Maasdorp, B.V. (2003). Mucuna seed as a feed ingredient for small ruminants and effect of ensiling on its nutritive value. *Trop. Subtrop. Agroecosyst.* **1**: 97–105.

Mateo-Vivaracho, L., Zapata, J., Cacho, J., and Ferreira, V. (2010). Analysis, occurrence, and potential sensory significance of five polyfunctional mercaptans in white wines. *J. Agric. Food Chem.* **58**: 10184–10194.

Mateus, N. and de Freitas, V. (2001). Evolution and stability of anthocyanin-derived pigments during Port wine aging. *J. Agric. Food Chem.* **49**: 5217–5222.

Matsui, K., Takemoto, H., Koeduka, T., and Ohnishi, T. (2018). 1-Octen-3-ol is formed from its glycoside during processing of soybean [*Glycine max* (L.) Merr.] seeds. *J. Agric. Food Chem.* **66**: 7409–7416.

Matsui, T., Kudo, A., Tokuda, S. et al. (2010). Identification of a new natural vasorelaxatant, (+)-osbeckic acid from rutin-free tartary buckwheat extract. *J. Agric. Food Chem.* **58**: 10876–10879.

Matsuo, Y., Li, Y., Watarumi, S. et al. (2011). Production and degradation mechanism of theacitrin C, a black tea pigment derived from epigallocatechin-3-O-gallate via a bicyclo[3.2.1]octane-type intermediate. *Tetrahedron* **67**: 2051–2059.

Matsuo, Y., Okuda, K., Morikawa, H. et al. (2016). Stereochemistry of the black tea pigments theacitrins A and C. *J. Nat. Prod.* **79**: 189–195.

Matsutomo, T., Stark, T.D., and Hofmann, T. (2013). In vitro activity-guided identification of antioxidants in aged garlic extract. *J. Agric. Food Chem.* **61**: 3059–3067.

Mattila, P.H., Piironen, V.I., Uusi-Rauva, E.J., and Koivistoinen, P.E. (1994). Vitamin D contents in edible mushrooms. *J. Agric. Food Chem.* **42**: 2449–2453.

Mattison, C.P., Cavalcante, J.M., Gallão, M.I., and de Brito, E.S. (2018). Effects of industrial cashew nut processing on anacardic acid content and allergen recognition by IgE. *Food Chem.* **240**: 370–376.

Mattoli, L., Cangi, F., Maidecchi, A. et al. (2005). A rapid liquid chromatography electrospray ionization mass spectrometric method for evaluation of synephrine in Citrus aurantium L. samples. *J. Agric. Food Chem.* **53**: 9860–9866.

Matusheski, N.V. and Jeffery, E.H. (2001). Comparison of the bioactivity of two glucoraphanin hydrolysis products found in broccoli, sulforaphane and sulforaphane nitrile. *J. Agric. Food Chem.* **49**: 5743–5749.

Maugard, T., Enaud, E., Choisy, P., and Legoy, M.D. (2001). Identification of an indigo precursor from leaves of Isatis tinctoria (Woad). *Phytochemistry* **58**: 897–904.

Mavric, E. and Henle, T. (2006). Isolation and identification of 3,4-dideoxypentosulose as specific degradation product of oligosaccharides with 1,4-glycosidic linkages. *Eur. Food Res. Technol.* **223**: 803–810.

Mayer, A.M. (1987). Polyphenoloxidases in plants: recent progress. *Phytochemistry* **26**: 11–20.

McCance & Widowson (1998). *The Composition of Foods, Supplement: Fatty Acids*. London: Ministry of Agriculture, Food and Fisheries.

McCooey, A. Sugar and sweetener guide. http://www.sugar-and-sweetener-guide.com/index.html (accessed 28 October 2019).

McDougall, G.J., Gordon, S., Brennan, R., and Stewart, D. (2005). Anthocyanin-flavanol condensation products from black currant (Ribes nigrum L.). *J. Agric. Food Chem.* **53**: 7878–7885.

McDougall, G.J., Morrison, I.M., Stewart, D., and Hillman, J.R. (1996). Plant cell walls as dietary fibre: range, structure, processing and functions. *J. Sci. Food Agric.* **70**: 133–150.

McEvilly, A.J. and Iyengar, R. (1992). Inhibition of enzymatic browning in foods and beverages. *Crit. Rev. Food Sci. Nutr.* **32**: 253–273.

McGill, A.S., Moffat, C.F., Mackie, P.R., and Cruickshank, P. (1993). The composition and concentration of n-alkanes in retail samples of edible oils. *J. Sci. Food Agric.* **61**: 357–362.

McRae, R., Robinson, R.K., and Sadler, M.J. (1993). *Encyclopaedia of Food Science, Food Technology and Nutrition*, vol. **1**. London: Academic Press.

Medina-Meza, I.G., Aluwi, N.A., Saunders, S.R., and Ganjyal, G.M. (2016). GC–MS Profiling of triterpenoid saponins from 28 quinoa varieties (*Chenopodium quinoa* Willd.) grown in Washington state. *J. Agric. Food Chem.* **64**: 8583–8591.

Mehta, B.M. and Deeth, H.C. (2016). Blocked lysine in dairy products: formation, occurrence, analysis, and nutritional implications. *Compr. Rev. Food Sci. Food Saf.* **15**: 206–218.

Mekori, Y.A. (1996). Introduction to allergic diseases. *Crit. Rev. Food Sci. Nutr.* **36**: S1–S18.

Melough, M.M., Lee, S.G., Cho, E. et al. (2017). Identification and quantitation of furocoumarins in popularly consumed foods in the U.S. using QuEChERS extraction coupled with UPLC-Ms/MS analysis. *J. Agric. Food Chem.* **65**: 5049–5055.

Meng, X.-H., Zhu, H.-T., Yan, H. et al. (2018). C-8 N-Ethyl-2-pyrrolidinone-substituted flavan-3-ols from the leaves of *Camellia sinensis* var. *pubilimba. J. Agric. Food Chem.* **66**: 7150–7155.

Mennella, G., Rotino, G.L., Fibiani, M. et al. (2010). Characterization of health-related compounds in eggplant (Solanum melongena L.) lines derived from introgression of allied species. *J. Agric. Food Chem.* **58**: 7597–7603.

Merian, E. (1991). *Metals and Their Compounds in the Environment, Occurrence, Analysis, and Biological Relevance.* Wiley-VCH: Weinheim.

Mertz, W. (1987). *Trace Elements in Human and Animal Nutrition*, 5e. San Diego: Academic Press.

Mertz, W. (1993). Chromium in human nutrition: a review. *J. Nutr.* **123**: 626–633.

Metin, S. and Hartel, R.W. (2005). Crystallization of fats and oils. In: *Bailey's Industrial Oil and Fat Products*, 6e (ed. F. Shahidi). Chichester: Wiley.

Metzler, M., Kulling, S.E., and Pfeiffer, E. (1998). Genotoxicity of estrogens. *Z. Lebensm. Unters. Forsch.* **206**: 367–373.

Miínguez-Mosquera, M.I. and Hornero-Méndez, D. (1993). Separation and quantification of the carotenoid pigments in red peppers (Capsicum annuum L.), paprika, and oleoresin by reversed-phase HPLC. *J. Agric. Food Chem.* **41**: 1616–1620.

Milheiro, J., Filipe-Ribeiro, L., Vilela, A. et al. (2017). 4-Ethylphenol, 4-ethylguaiacol and 4-ethylcatechol in red wines: microbial formation, prevention, remediation and overview of analytical approaches. *Crit. Rev. Food Sci. Nutr.* https://doi.org/10.1080/10408398.2017.1408563.

Millane, R.P. and Hendrixson, T.L. (1994). Crystal structure of mannan and glucomannans. *Carbohydr. Polym.* **25**: 245–251.

Miller, A. and Engel, K.-H. (2006). Content of γ-oryzanol and composition of steryl ferulates in brown rice (Oryza sativa L.) of European origin. *J. Agric. Food Chem.* **54**: 8127–8133.

Miller, A.R., Kelley, T.J., and Mujer, C.V. (1990). Anodic peroxidase isoenzymes and polyphenol oxidase activity from cucumber fruit: tissue and substrate specifity. *Phytochemistry* **29**: 705–709.

Millward, D.J. (1999). The nutritional value of plant-based diets in relation to human amino acid and protein requirements. *Proc. Nutr. Soc.* **58**: 249–260.

Ming, D. and Hellekant, G. (1994). Brazzein, a new high-potency thermostable sweet protein from Pentadiplandra brazzeana B. *FEBS Lett.* **355**: 106–108.

Minguez-Mosquera, M.I., Garrido-Fernández, J., and Gandul-Rojas, B. (1989). Pigment changes in olives during fermentation and brine storage. *J. Agric. Food Chem.* **37**: 8–11.

Mittermeier, V.K., Dunkel, A., and Hofmann, T. (2018). Discovery of taste modulating octadecadien-12-ynoic acids in golden chanterelles (*Cantharellus cibarius*). *Food Chem.* **269**: 53–62.

Miyamoto, J. and Kerney, P.C. (eds.) (1985). *Chemistry: Human Welfare and Environment. Vol. 3, Mode of Action, Metabolism and Toxicology*. Oxford: Pergamon Press.

Miyazato, H., Nakamura, M., Hashimoto, S., and Hayashi, S. (2013). Identification of the odour-active cyclic diketone cis-2,6-dimethyl-1, 4-cyclohexanedione in roasted Arabica coffee brew. *Food Chem.* **138**: 2346–2355.

Moffat, C.F. and Whittle, K.J. (eds.) (1999). *Environmental Contaminants in Food*. Oxford: Wiley Blackwell.

Mogol, B.A. and Gökmen, V. (2013). Kinetics of furan formation from ascorbic acid during heating under reducing and oxidizing conditions. *J. Agric. Food Chem.* **61**: 10191–10196.

Molins, R.M. (1991). *Phosphates in Food*. Boca Raton, FL: CRC Press.

Möller, N.P., Scholz-Ahrens, K.E., Roos, N., and Schrezenmeir, J. (2008). Bioactive peptides and proteins from foods: indication for health effects. *Eur. J. Nutr.* **47**: 171–182.

Molnar-Perl, I. and Friedman, M. (1990). Inhibiton of browning by sulphur amino acids. III. Apples and potatoes. *J. Agric. Food Chem.* **38**: 1652–1656.

Momin, R.A. and Muraleeddharan, G.N. (2002). Pest-managing efficacy of trans-asarone isolated from *Daucus carota* L. seeds. *J. Agric. Food Chem.* **50**: 4475–4478.

Moon, J.-K. and Shibamoto, T. (2011). Formation of carcinogenic 4(5)-methylimidazole in Maillard reaction systems. *J. Agric. Food Chem.* **59**: 615–618.

Mora, L., Hernández-Cázares, A.S., Sentandreu, M.Á., and Toldrá, F. (2010). Creatine and creatinine evolution during the processing of dry-cured ham. *Meat Sci.* **84**: 384–389.

Mora, L., Sentandreu, M.Á., and Toldrá, F. (2008). Contents of creatine, creatinine and carnosine in porcine muscles of different metabolic types. *Meat Sci.* **79**: 709–715.

Morales, F.J., Somoza, V., and Fogliano, V. (2012). Physiological relevance of dietary melanoidins. *Amino Acids* **42**: 1097–1109.

Moreda–Piñeiro, J., Alonso-Rodríguez, E., Romarís-Hortas, V. et al. (2012). Assessment of the bioavailability of toxic and non-toxic arsenic species in seafood samples. *Food Chem.* **130**: 552–560.

Mosbach, K., Guilford, H., and Lindberg, M. (1974). The terphenyl quinone poly-poric acid: production, isolation and characterization. *Tetrahedron Lett.* **17**: 1645–1648.

Mossine, V.V. and Mawhinney, T.P. (2007). N^{α}-(1-deoxy-D-fructos-1-yl)-L-histidine: a potent copper chelator from tomato powder. *J. Agric. Food Chem.* **55**: 10373–10381.

Mottram, D.S. (1998). Flavour formation in meat and meat products: a review. *Food Chem.* **62**: 415–424.

Mukherjee, M. (2003). Human digestive and metabolic lipases – a brief review. *J. Mol. Catal. B Enzym.* **22**: 369–376.

Müller, C., Hemmersbach, S., Van't Slot, G., and Hofmann, T. (2006). Synthesis and structure determination of covalent conjugates formed from the sulfury-roasty-smelling 2-furfurylthiol and di- or trihydroxybenzenes and their identification in coffee brew. *J. Agric. Food Chem.* **54**: 10076–10085.

Muller, C.J., Joubert, E., Pheiffer, C. et al. (2013). Z-2-(β-D-glucopyranosyloxy)-3-phenylpropenoic acid, an α-hydroxy acid from rooibos (*Aspalathus linearis*) with hypoglycemic activity. *Mol. Nutr. Food Res.* **57**: 2216–2222.

Müller, D., Schantz, M., and Richking, E. (2012). High performance liquid chromatography analysis of anthocyanins in bilberries (*Vaccinium myrtillus* L.), blueberries (*Vaccinium corymbosum* L.) and corresponding juices. *J. Food Sci.* **77**: C340–C345.

Müller, H. (1993). Bestimmung der Folsäure-Gehalte von Getreide, Getreideprodukten, Backwaren und Hülsenfrüchten mit Hilfe der Hochleis-tungsflüssigchromatographie (HPLC). *Z. Lebensm. Unters. Forsch.* **197**: 573–577.

Mulsow, B.B., Jacob, M., and Henle, T. (2009). Studies on the impact of glycation on the denaturation of whey proteins. *Eur. Food Res. Technol.* **228**: 643–649.

Mumford, F.E., Stark, H.M., and Smith, D.H. (1963). 4-Hydroxybenzyl alcohol, a naturally occurring cofactor of indoleacetic acid oxidase. *Phytochemistry* **2**: 215–220.

Murkovic, M. (2004). Formation of heterocyclic aromatic amines in model systems. *J. Chromatogr. B* **802**: 3–10.

Murkovic, M., Mülleder, U., and Neunteufl, H. (2002). Carotenoid content in different varieties of pumpkins. *J. Food Compos. Anal.* **15**: 633–638.

Murthy Kishore Kumar, N.V. and Rao Narasinga, M.S. (1986). Interaction of allyl isothiocyanate with mustard 12S protein. *J. Agric. Food Chem.* **34**: 448–452.

Muso, H. (1979). The pigments of fly agaric, Amanita muscaria. *Tetrahedron* **35**: 2843–2853.

Mussinan, C.J. and Keelan, M.E. (eds.) (1994). *Sulfur Compounds in Foods*. Washington, D.C.: American Chemical Society.

Muto, N., Terasawa, K., and Yamamoto, I. (1992). Evaluation of ascorbic acid 2-O-α-glucoside as vitamin C source: mode of intestinal hydrolysis and absorption following oral administration. *Int. J. Vitam. Nutr. Res.* **62**: 318–323.

Myake, N., Kim, M., and Kurata, T. (1997). Formation mechanism of monodehydro-L-ascorbic acid and superoxide anion in the autoxidation of L-ascorbic acid. *Biosci. Biotechnol. Agrochem.* **61**: 1693–1695.

Myers, R.A., Fuller, E., and Yang, W. (2013). Identification of native catechin fatty acid esters in green tea (*Camellia sinensis*). *J. Agric. Food Chem.* **61**: 11484–11493.

Nagano, T., Oyama, Y., Kajita, N. et al. (1997). New curcuminoids isolated from *Zingiber cassumunar* protect cells suffering from oxidative stress: a flow-cytometric study using rat thymocytes and H_2O_2. *Jpn. J. Pharmacol.* **75**: 363–370.

Nagaoka, T., Nakatab, K., Kounoa, K., and Andoa, T. (2004). Antifungal activity of oosporein from an antagonistic fungus against Phytophthora infestans. *Z. Naturforsch.* **59c**: 302–304.

Nagata, T., Hayatsu, M., and Kosuge, N. (1992). Identification of aluminium forms in tea leaves by ^{27}Al NMR. *Phytochemistry* **31**: 1215–1218.

Nakai, S. and Modler, H.W. (1996). *Food Proteins. Properties and Characterization*. New York, Weinheim, Cambridge: VCH Publishers.

Nakai, T. and Ohta, T. (1976). β-3-Oxindolylalanine: the main intermediate in tryptophan degradation occurring in acid hydrolysis of protein. BiochImica et Biophysica Acta (BBA)-Protein. *Structure* **420**: 258–264.

Nakamura, T., Yoshida, A., Komatsuzaki, N. et al. (2007a). Isolation and characterization of a low molecular weight peptide contained in sourdough. *J. Agric. Food Chem.* **55**: 4871–4876.

Nakanishi, A., Fukushima, Y., Miyazawa, N. et al. (2017b). Identification of rotundone as a potent odor-active compound of several kinds of fruits. *J. Agric. Food Chem.* **65**: 4464–4471.

Nakanishi, A., Fukushima, Y., Miyazawa, N. et al. (2017c). Quantitation of rotundone in grapefruit (*Citrus paradisi*) peel and juice by stable isotope dilution assay. *J. Agric. Food Chem.* **65**: 5026–5033.

Narváez-Cuenca, C.-E., Kuijpers, T.F.M., Vincken, J.-P. et al. (2011). New insights into an ancient antibrowning agent: formation of sulfophenolics in sodium hydrogen sulfite-treated potato extracts. *J. Agric. Food Chem.* **59**: 10247–10255.

Nashalian, O. and Yaylayan, V.A. (2014). Thermally induced oxidative decarboxylation of copper complexes of amino acids and formation of Strecker aldehyde. *J. Agric. Food Chem.* **62**: 8518–8523.

Neilson, A.H. (ed.) (1998). *PAHs and Related Compounds. Vol. 3, Part I (Chemistry), Part J (Biology)*. Berlin: Springer.

Nelson, B.C., Putzbach, K., Sharpless, K.E., and Sander, L.C. (2007). Mass spectro-metric determination of the predominant adrenergic protoalkaloids in bitter orange (Citrus aurantium). *J. Agric. Food Chem.* **55**: 9769–9775.

Netz, N. and Opatz, T. (2015). Marine indole alkaloids. *Mar. Drugs* **13**: 4814–4914.

Neuser, F., Zorn, H., and Berger, R.G. (2000). Generation of odorous acyloins by yeast pyruvate decarboxylases and their occurrence in sherry and soy sauce. *J. Agric. Food Chem.* **48**: 6191–6195.

Ni, W., McNaughton, L., Lemaster, D.M. et al. (2008). Quantitation of 13 heterocyclic aromatic amines in cooked beef, pork, and chicken by liquid chromatography-electrospray ionization/tandem mass spectrometry. *J. Agric. Food Chem.* **56**: 68–78.

Nickavar, B., Mojab, F., Javidnia, K., and Amoli, M.A. (2003). Chemical composition of the fixed and volatile oils of *Nigella sativa* L. from Iran. *Z. Naturforsch. C* **58**: 629–631.

Nicolas, J., Richard-Forget, F.C., Goupy, P.M. et al. (1994). Enzymatic browning reactions in apple and apple products. *Crit. Rev. Food Sci. Nutr.* **34**: 109–157.

Nielsen, F.H. (1984). Ultratrace elements in nutrition. *Annu. Rev. Nutr.* **4**: 21–41.

Nielsen, K.F., Dalsgaard, P.W., Smedsgaard, J., and Larsen, T.O. (2005). Andrastins A-D, Penicillium roqueforti metabolites consistently produced in blue-mold-ripened cheese. *J. Agric. Food Chem.* **53**: 2908–2913.

Nielsen, K.F., Sumarah, M.W., Frisvad, J.C., and Miller, J.D. (2006). Production of metabolites from the Penicillium roqueforti complex. *J. Agric. Food Chem.* **54**: 3756–3763.

Niemeyer, H.M. (2009). Hydroxamic acids derived from 2-hydroxy-2H-1,4-benzoxazin-3(4H)-one: key defense chemicals of cereals. *J. Agric. Food Chem.* **57**: 1677–1696.

Nigg, H.N. and Seigler, D. (eds.) (1992). *Phytochemical Resources for Medicine and Agriculture.* New York: Plenum Press.

Nikolov, P.Y. and Yaylayan, V.A. (2012). Role of the ribose-specific marker furfuryl-amine in the formation of aroma active 1-(furan-2-ylmethyl)-1*H*-pyrrole (or furfuryl-pyrrole) derivatives. *J. Agric. Food Chem.* **60**: 10155–10161.

Nishino, A., Yasui, H., and Maoka, T. (2016). Reaction of paprika carotenoids, capsanthin and capsorubin, with reactive oxygen species. *J. Agric. Food Chem.* **64**: 4786–4792.

Nitz, S., Spraul, M.H., and Drawert, F. (1992). 3-Butyl-5,6-dihydro-4H-isobenzofuran-1-one, a sensorial active phthalide in parsley roots. *J. Agric. Food Chem.* **40**: 1038.

Noba, S., Yako, N., Sakai, H. et al. (2018). Identification of a precursor of 2-mercapto-3-methyl-1-butanol in beer. *Food Chem.* **255**: 282–289.

Noda, K., Masuzaki, R., Terauchi, Y. et al. (2018). Novel Maillard pigment, furpenthiazinate, having furan and cyclopentathiazine rings formed by acid hydrolysis of protein in the presence of xylose or by reaction between cysteine and furfural under strongly acidic conditions. *J. Agric. Food Chem.* **66**: 11414–11421.

Noonan, G.O., Begley, T.H., and Diachenko, G.W. (2008). Semicarbazide formation in flour and bread. *J. Agric. Food Chem.* **56**: 2064–2067.

Nörenberg, S., Kiske, C., Burmann, A. et al. (2017). Distributions of the stereoisomers of β-mercaptoheptanones and β-mercaptoheptanols in cooked bell pepper (*Capsicum annuum*). *J. Agric. Food Chem.* **65**: 10250–10257.

Nörenberg, S., Reichard, B., Andelfinger, V. et al. (2013). Influence of the stereochemistry on the sensory properties of 4-mercapto-2-heptanol and its acetyl-derivatives. *J. Agric. Food Chem.* **61**: 2062–2069.

Norisuye, T. (1996). Conformation and properties of amylose in dilute solution. *Food Hydrocoll.* **10**: 109–115.

Nriagu, J.O. and Simmons, M.S. (1990). *Food Contamination from Environmental Sources.* New York: Wiley.

Nykänen, L. and Suomalainen, H. (1983). *Aroma of Beer, Wine and Distilled Alcoholic Beverages.* Berlin: Akademie Verlag.

Nyman, P.J., Diachenko, G.W., Perfetti, G.A. et al. (2008). Survey results of benzene in soft drinks and other beverages by headspace gas chromatography/mass spectrometry. *J. Agric. Food Chem.* **56**: 571–576.

Nyström, L., Lampi, A.-M., Andersson, A.A.M. et al. (2008). Phytochemicals and dietary fiber components in rye varieties in the healthgrain diversity screen. *J. Agric. Food Chem.* **56**: 9758–9766.

O'Brien, J., Nursten, H.E., James, M. et al. (eds.) (1998). *The Maillard Reaction in Foods and Medicine.* Cambridge: The Royal Society of Chemistry.

O'Brien, R.D. (ed.) (1997). *Fats and Oils: Formulating and Processing Applications.* Lancaster, PA: Technomic.

Oakenfull, D.G. (1981). Saponins in food: a review. *Food Chem.* **6**: 19–40.

Obert, J.C., Hughes, D., Sorenson, W.R. et al. (2007). A quantitative method for the determination of cyclopropenoid fatty acids in cottonseed, cottonseed meal, and cottonseed oil (Gossypium hirsutum) by high-performance liquid chromatography. *J. Agric. Food Chem.* **55**: 2062–2067.

Oblath, S.B., Markowitz, S.S., Novakov, T., and Chang, S.O. (1981). Kinetics of the formation of hydroxylamine disulfonate by reaction of nitrite with sulphites. *J. Phys. Chem.* **85**: 1017–1021.

Oduho, G.W., Chung, T.K., and Baker, D.H. (1993). Menadione nicotinamide bisulfite is a bioactive source of vitamin K and niacin activity for chicks. *J. Nutr.* **123**: 737–743.

Ofuya, Z.M. and Akhidue, V. (2005). The role of pulses in human nutrition: a review. *J. Appl. Sci. Environ. Manag.* **9**: 99–104.

Ohloff, G. (1994). *Scent and Fragrances.* Berlin: Springer.

Ohnishi, M., Mori, K., Ito, S., and Fujino, Y. (1985). Chemical composition of steryl lipid in Maitake mushroom. *J. Agric. Chem. Soc. Jpn.* **59**: 1053–1059.

Ohta, T. and Nakai, T. (1978). The reaction of tryptophan with cysteine during acid hydrolysis of proteins: formation of tryptathionine as a transient intermediate in a model system. *Biochim. Biophys. Acta, Protein Struct.* **533**: 440–445.

Ohya, T. (2006). Identification of 4-methylspinaceamine, a Pictet-Spengler condensation reaction product of histamine with acetaldehyde, in fermented foods and its metabolite in human urine. *J. Agric. Food Chem.* **54**: 6909–6915.

Oka, Y., Omura, M., Kataoka, H., and Touhara, K. (2004). Olfactory receptor antagonism between odorants. *EMBO J.* **14**: 120–126.

Okamura, M. (1994). Distribution of ascorbic acid analogues and associated glycosides in mushrooms. *J. Nutr. Sci. Vitaminol.* **40**: 81–94.

Okuda, T., Yoshida, T., and Hatano, T. (1993). Classification of oligomeric hydrolysable tannins and specificity of their occurrence in plants. *Phytochemistry* **32**: 507–521.

Olds, S.J., Vanderslice, J.T., and Brochetti, D. (1993). Vitamin B$_6$ in raw and fried chicken by HPLC. *J. Food Sci.* **58**: 505–507.

Oliver, C.M., Melton, L.D., and Stanley, R.A. (2006). Creating proteins with novel functionality via the Maillard reaction. *Crit. Rev. Food Sci. Nutr.* **46**: 337–350.

Olson, A., Gray, G.M., and Ciu, M.-C. (1987). Chemistry and analysis of soluble dietary fiber. *Food Technol.* **41**: 71–80.

Osawa, T. and Namiki, M. (1982). Mutagen formation in the reaction of Nitrite with the food components analogous to Sorbic acid. *Agric. Biol. Chem.* **46**: 2299–2304.

Ostry, V. (2008). Alternaria mycotoxins: an overview of chemical characterization, producers, toxicity, analysis and occurrence in foodstuffs. *World Mycotoxin J.* **1**: 175–188.

Ottinger, H., Bareth, A., and Hofmann, T. (2001). Characterization of natural 'cooling' compounds formed from glucose and L-proline in dark malt by application of taste dilution analysis. *J. Agric. Food Chem.* **49**: 1336–1344.

Ottinger, H. and Hofmann, T. (2002). Quantitative model studies on the efficiency of precursors in the formation of cooling-active 1-pyrrolidinyl-2-cyclopenten-1-ones and bitter-tasting cyclopenta-[b]azepin-8(1H)-ones. *J. Agric. Food Chem.* **50**: 5156–5161.

Ottinger, H. and Hofmann, T. (2003). Identification of the taste enhancer alapyridaine in beef broth and evaluation of its sensory impact by taste reconstitution experiments. *J. Agric. Food Chem.* **51**: 6791–6796.

Ottinger, H., Soldo, T., and Hofmann, T. (2001). Systematic studies on structure and physiological activity of cyclic α-keto enamines, a novel class of 'cooling' compounds. *J. Agric. Food Chem.* **49**: 5383–5390.

Ottinger, H., Soldo, T., and Hofmann, T. (2003). Discovery and structure determination of a novel Maillard-derived sweetness enhancer by application of the comparative taste dilution analysis (cTDA). *J. Agric. Food Chem.* **51**: 1035–1041.

Ozga, J.A., Saeed, A., Wismer, W., and Reinecke, D.M. (2007). Characterization of cyanidin- and quercetin-derived flavonoids and other phenolics in mature Saskatoon fruits (Amelanchier alnifolia Nutt.). *J. Agric. Food Chem.* **55**: 10414–10424.

Paasikallio, A., Rantavaara, A., and Sippola, J. (1994). The transfer of ^{137}Cs and ^{90}Sr from soil to food crops after the Chernobyl accident. *Sci. Total Environ.* **155**: 109–124.

Paasivirta, J. (ed.) (2000). New types of persistent halogenated compounds. In: *The Handbook of Environmental Chemistry, Vol. 3. Part K*. Berlin, Heidelberg: Springer-Verlag.

Pais, P., Tanga, M.J., Salmon, C.P., and Knize, M.G. (2000). Formation of the mutagen IFP in model systems and detection in restaurant meats. *J. Agric. Food Chem.* **48**: 1721–1726.

Palani, K., Harbaum-Piayda, B., Meske, D. et al. (2016). Influence of fermentation on glucosinolates and glucobrassicin degradation products in sauerkraut. *Food Chem.* **190**: 755–762.

Panizzi, L. and Scarpati, M.L. (1954). Constitution of cynarine, the active principle of the artichoke. *Nature* **174**: 1062–1063.

Papageorgiou, V.P., Bakola-Christianopoulou, M.N., Apazidou, A.A., and Psarros, E.E. (1997). Gas chromatographic-mass spectroscopic analysis of the acidic triterpenic fraction of mastic gum. *J. Chromatogr. A* **769**: 263–273.

Papastamoulis, Y., Richard, T., Nassra, M. et al. (2014). Viniphenol A, a complex resveratrol hexamer from *Vitis vinifera* stalks: structural elucidation and protective effects against amyloid-β-induced toxicity in PC12 cells. *J. Nat. Prod.* **77**: 213–217.

Papiz, M.Z., Sawyer, L., Eliopoulos, E.E. et al. (1986). The structure of β-lactoglobulin and its similarity to plasma retinol-binding protein. *Nature* **324**: 383.

Parks, O.W., Schwartz, D.P., and Keeney, M. (1964). Identification of o-aminoaceto-phenone as a flavour compound in stale dry milk. *Nature* **202**: 185–187.

Parliment, T.H., Morello, M.J., and McCorrin, R.J. (eds.) (1994). *Thermally Generated Flavors*. Washington, D.C.: American Chemical Society.

Patai, S. (ed.) (1977). *Chemistry of Cyanates and their Thio Derivatives. Part 2*. Chichester: Wiley.

Patel, A., Mazzini, F., Netscher, T., and Rosenau, T. (2008). A novel dimer of α-tocopherol. *Res. Lett. Org. Chem.*: 1–4. https://doi.org/10.1155/2008/742590.

Patring, J.D.M., Jastrebova, J.A., Hjortmo, S.B. et al. (2005). Development of a simplified method for the determination of folates in baker's yeast by HPLC with ultraviolet and fluorescence detection. *J. Agric. Food Chem.* **53**: 2406–2411.

Patterson, H.B.N. (1994). *Hydrogenation of Fats and Oils: Theory and Practice*. Champaign: AOCS Press.

Pecháček, R., Velíšek, J., and Hrabcová, H. (1997). Decomposition products of allyl isothiocyanate in aqueous solutions. *J. Agric. Food Chem.* **45**: 4584–4588.

Pedersen, H.A., Laursen, B., Mortensen, A., and Fomsgaard, I.S. (2011). Bread from common cereal cultivars contains an important array of neglected bioactive benzoxazinoids. *Food Chem.* **127**: 1814–1820.

Pedreschi, R., Campos, D., Noratto, G. et al. (2003). Andean yacon root (Smallanthus sonchifolius Poepp. Endl) fructooligosaccharides as a potential novel source of prebiotics. *J. Agric. Food Chem.* **51**: 5278–5284.

Pennington, J.A.T. (1992). Total diet studies: the identification of core foods in the United States food supply. *Food Addit. Contam.* **9**: 253–264.

Pereira-Caro, G., Watanabe, S., Crozier, A. et al. (2013). Phytochemical profile of a Japanese black–purple rice. *Food Chem.* **141**: 2821–2827.

Perez, L.C. and Yaylayan, V.A. (2008). Isotope labeling studies on the formation of 5-(hydroxymethyl)-2-furaldehyde (hmf) from sucrose by pyrolysis-GC/MS. *J. Agric. Food Chem.* **56**: 6717–6723.

Pérez-Prior, M.T., Manso, J.A., Gómez-Bombarelli, R. et al. (2008). Reactivity of some products formed by the reaction of sorbic acid with sodium nitrite: decomposition of 1,4-dinitro-2-methylpyrrole and ethylnitrolic acid. *J. Agric. Food Chem.* **56**: 11824–11829.

Perkins, E.G. (1993). *Lipoproteins*. Champaign: AOCS Press.

Perkins, E.G. and Ericksson, M.D. (1996). *Deep Frying: Chemistry, Nutrition, and Practical Applications*. Champaign: AOCS Press.

Perkins, E.G. and Visek, W.J. (1983). *Dietary Fats and Health*. Champaign: AOCS Press.

Perpète, P. and Collin, S. (1999). Contribution of 3-methylthiopropionaldehyde to the worty flavor of alcohol-free beers. *J. Agric. Food Chem.* **47**: 2374–2378.

Persson, D.P., Hansen, T.H., Holm, P.E. et al. (2006). Multi-elemental speciation analysis of barley genotypes differing in tolerance to cadmium toxicity using SEC-ICP-MS and ESI-TOF-MS. *J. Anal. At. Spectrom.* **21**: 996–1005.

Peters, A.M. and van Amerongen, A. (1996). Sesquiterpene lactones in chicory (Cichorium intybus L.). Distribution in chicons and effect of storage. *Food Res. Int.* **29**: 439–444.

Peumans, W.J. and van Damme, E.J.M. (1996). Prevalence, biological activity and genetic manipulation of lectins in foods. *Trends Food Sci. Technol.* **7**: 132–138.

Pfeiffer, E., Hildebrand, A.A., Becker, C. et al. (2010). Identification of an aliphatic epoxide and the corresponding dihydrodiol as novel congeners of zearalenone in cultures of Fusarium graminearum. *J. Agric. Food Chem.* **58**: 12055–12062.

Pham, T.T., Guichard, E., Schlich, P., and Charpentier, C. (1995). Optimal conditions for the formation of sotolon from α-ketobutyric acid in the French 'Vin Jaune'. *J. Agric. Food Chem.* **43**: 2616–2619.

Phillips, K.M., Ruggio, D.M., Ashraf-Khorassani, M., and Haytowitz, D.B. (2006). Difference in folate content of green and red sweet peppers (Capsicum annuum) determined by liquid chromatography-mass spectrometry. *J. Agric. Food Chem.* **54**: 9998–10002.

Phillips, K.M., Ruggio, D.M., Horst, R.L. et al. (2011). Vitamin D and sterol composition of 10 types of mushrooms from retail suppliers in the United States. *J. Agric. Food Chem.* **59**: 7841–7853.

Picard, M., Lytra, G., Tempere, S. et al. (2016). Identification of piperitone as an aroma compound contributing to the positive mint nuances perceived in aged red Bordeaux wines. *J. Agric. Food Chem.* **64**: 451–460.

Picard, M., Revel, G., and Marchand, S. (2017). First identification of three *p*-menthane lactones and their potential precursor, menthofuran, in red wines. *Food Chem.* **217**: 294–302.

Piggott, J.R. (ed.) (1984). *Sensory Analysis of Foods*. London: Elsevier.

Pihlanto, A. and Korhonen, H. (2003). Bioactive peptides and proteins. In: *Advances in Food and Nutritional Research* (ed. L.T. Steve), 175–276. Academic Press.

Pinsky, P. and Lorber, M.N. (1998). A model to evaluate past exposure to TCDD. *J. Expo. Anal. Environ. Epidemiol.* **8**: 187–206.

Plastics Europe (2012). Plastics – the facts. An analysis of European plastics production, demand and waste data for 2011. https://www.plasticseurope.org/application/files/3715/1689/8308/2015plastics_the_facts_14122015.pdf (accessed 28 October 2019).

Pocock, K.F., Alexander, G.M., Hayasaka, Y. et al. (2007). Sulfate – a candidate for the missing essential factor that is required for the formation of protein haze in white wine. *J. Agric. Food Chem.* **55**: 1799–1807.

Podhradský, D., Drobnica, L., and Kristian, P. (1979). Reactions of cysteine, its derivatives, glutathione, coenzyme A, and dihydrolipoic acid with isothiocyanates. *Exp. Dermatol.* **35**: 154–155.

Pohland, A.E. (1993). Mycotoxins in review. *Food Addit. Contam.* **10**: 17–28.

Poisson, L. and Schieberle, P. (2008). Characterization of the key aroma compounds in an American bourbon whisky by quantitative measurements, aroma recombination, and omission studies. *J. Agric. Food Chem.* **56**: 5820–5826.

Pokorný, J. and Réblová, Z. (1995). Sinapines and other phenolics of Brassicaceae seeds. *Potravinářské Vědy* **13**: 155–168.

Pons, A., Allamy, L., Lavigne, V. et al. (2017). Study of the contribution of massoia lactone to the aroma of Merlot and Cabernet Sauvignon musts and wines. *Food Chem.* **232**: 229–236.

Pons, A., Lavigne, V., Darriet, P., and Dubourdieu, D. (2013). Role of 3-methyl-2,4-nonanedione in the flavor of aged red wines. *J. Agric. Food Chem.* **61**: 7373–7380.

Pons, A., Lavigne, V., Landais, Y. et al. (2008). Distribution and organoleptic impact of sotolon enantiomers in dry white wines. *J. Agric. Food Chem.* **56**: 1606–1610.

Pons, A., Lavigne, V., Landais, Y. et al. (2010). Identification of a sotolon pathway in dry white wines. *J. Agric. Food Chem.* **58**: 7273–7279.

Pons, M., Dauphin, B., La Guerche, S. et al. (2011). Identification of impact odorants contributing to fresh mushroom off-flavor in wines: incidence of their reactivity with nitrogen compounds on the decrease of the olfactory defect. *J. Agric. Food Chem.* **59**: 3264–3272.

Porta, H. and Rocha-Sosa, M. (2002). Plant lipoxygenases. Physiological and molecular features. *Plant Physiol.* **130**: 15–21.

Potter, N.N. and Hotchkiss, J.H. (1995). *Food Science*, 5e. New York: Chapman & Hall.

Poulton, J.E. (1990). Cyanogenesis in plants. *Plant Physiol.* **94**: 401–405.

Powell, J.J., McNaughton, S.A., Jugdaohsingh, R. et al. (2005). A provisional database for silicon content of foods in the United Kingdom. *Br. J. Nutr.* **94**: 804–812.

Preedy, V.R. (2013). *Isoflavones. Chemistry, Analysis, Function and Effects*. Cambridge: The Royal Society of Chemistry.

Preininger, M., Gimelfarb, L., Li, H.C., Dias, B.E., Fahmy, F., and White, J. (2009). Identification of dihydromaltol (2,3-dihydro-5-hydroxy-6-methyl-4H-pyran-4-one) in Ryazhenka kefir and comparative sensory impact assessment of related cycloenolones. *Journal of Agricultural and Food Chemistry*, **57**: 9902–9908.

Preston, C.M. and Sayer, B.G. (1992). What's in a nutshell: an investigation of structure by carbon-13 cross-polarization magic-angle spinning nuclear magnetic resonance spectroscopy. *J. Agric. Food Chem.* **40**: 206–210.

Pripis-Nicolau, L., de Revel, G., Bertrand, A., and Maujean, A. (2000). Formation of flavour components by the reaction of amino acid and carboxyl compounds in mild conditions. *J. Agric. Food Chem.* **48**: 3761–3766.

Przybylski, R. and McDonald, B.E. (1995). *Development and Processing of Vegetable Oils for Human Nutrition*. Champaign: AOCS Press.

Puah, C.W., Choo, Y.M., Ma, A.N., and Chuah, H.C. (2007). The effect of physical refining on palm vitamin E. (tocopherol, tocotrienol and tocomonoenol). *Am. J. Appl. Sci.* **4**: 374–377.

Puchľová, E. and Szolcsányi, P. (2018). Filbertone: a review. *J. Agric. Food Chem.* **66**: 11221–11226.

Qian, D., Zhao, Y., Yang, G., and Huang, L. (2017). Systematic review of chemical constituents in the genus *Lycium* (Solanaceae). *Molecules* **22**: 911.

Quemener, B. and Brillouet, J.-M. (1983). Ciceritol, a pinitol digalactoside from seeds of chickpea, lentil and white lupin. *Phytochemistry* **22**: 1745–1751.

Qureshi, A.A., Mo, H., Packer, L., and Peterson, D.M. (2000). Isolation and identification of novel tocotrienols from rice bran with hypocholesterolemic, antioxidant, and antitumor properties. *J. Agric. Food Chem.* **48**: 3130–3140.

Raab, T., Barron, D., Vera, F.A. et al. (2010). Catechin glucosides: occurrence, synthesis, and stability. *J. Agric. Food Chem.* **58**: 2138–2149.

Rang, H.P., Dale, M.M., and Ritter, J.M. (1996). *Pharmacology*, 3e. New York: Churchil-Livingstone.

Ranum, P. (1992). Potassium bromate in bread baking. *Cereal Food World* **37**: 253–258.

Rapior, S., Fons, F., and Bessièreb, J.-M. (2000). The fenugreek odor of *Lactarius helvus*. *Mycologia* **92**: 305–308.

Rassam, M. and Laing, W. (2005). Variation in ascorbic acid and oxalate levels in the fruit of *Actinidia chinensis* tissues and genotypes. *J. Agric. Food Chem.* **53**: 2322–2326.

Rauser, W.E. (1999). Structure and function of metal chelators produced by plants; the case of organic acids, phytin, and metallothioneins. *Cell Biochem. Biophys.* **31**: 19–48.

Ravber, M., Pečar, D., Andreja, G. et al. (2016). Hydrothermal degradation of rutin: identification of degradation products and kinetics study. *J. Agric. Food Chem.* **64**: 9196–9202.

Raymond, K.N., Cass, M.E., and Evans, S.L. (1987). Metal sequestering agents in bioinorganic chemistry: enterobactin mediated iron transport in *E. coli* and biomimetic applications. *Pure Appl. Chem.* **59**: 771–778.

Recheigl, M. Jr. (ed.) (1983). *Handbook of Naturally Occurring Food Toxicants*. Boca Raton, FL: CRC Press.

Reddy, N.R., Pierson, M.D., Sathe, S.K., and Salunkhe, D.K. (1989). *Phytates in Cereals and Legumes*. Boca Raton, FL: CRC Press.

Reglitz, K. and Steinhaus, M. (2017). Quantitation of 4-methyl-4-sulfanylpentan-2-one (4MSP) in hops by a stable isotope dilution assay in combination with GCxGC-TOFMS: method development and application to study the influence of variety, provenance, harvest year, and processing on 4MSP concentrations. *J. Agric. Food Chem.* **65**: 2364–2372.

Reilly, C. (1991). *Metal Contamination of Food*. Barking: Elsevier.

Reineccius, G. (ed.) (1994). *Source Book of Flavors*. New York: Chapman and Hall.

Reiners, J. and Grosch, W. (1999). Concentration of 4-methoxy-2-methylbutanethiol in Spanish virgin olive oil. *Food Chem.* **64**: 45–47.

Reischl, R.J., Bicker, W., Keller, T. et al. (2012). Occurrence of 2-methylthiazolidine-4-carboxylic acid, a condensation product of cysteine and acetaldehyde, in human blood as a consequence of ethanol consumption. *Anal. Bioanal. Chem.* **404**: 1779–1787.

Relkin, P. (1966). Thermal unfolding of β-lactoglobulin, α-lactalbumin, and bovine serum albumin. A thermodynamic approach. *Crit. Rev. Food Sci. Nutr.* **36**: 565–601.

Report of the DATEX Working Group on β-casomorphins. Review of the potential health impact of β-casomorphins and related peptides. EFSA Sci. Rep., 231: 1–107 (2009).

Requena, J.R., Fu, M.-X., Ahmed, M.U. et al. (1996). Lipoxidation products as biomarkers of oxidative damage to proteins during lipid peroxidation reactions. *Nephrol. Dial. Transplant.* **11** (Suppl. 5): 48–53.

Restania, P., Restellia, A.R., and Gallia, C.L. (1992). Formaldehyde and hexam-ethylenetetramine as food additives: chemical interactions and toxicology. *Food Addit. Contam.* **9**: 597–605.

Rice, S.L., Eitenmiller, R.R., and Koehler, P.E. (1976). Biologically active amines in food: a review. *J. Milk Food Technol.* **39**: 353–358.

Ricelli, A., Baruzzi, F., Solfrizzo, M. et al. (2007). Biotransformation of patulin by Gluconobacter oxydans. *Appl. Environ. Microbiol.* **73**: 785–792.

Rius Solé, M.A. and García Regueiro, J.A. (2001). Role of 4-phenyl-3-buten-2-one in boar taint: identification of new compounds related to sensorial descriptors in pig fat. *J. Agric. Food Chem.* **49**: 5303–5309.

Rizzi, G.P. (1997). Chemical structure of colored Maillard reaction products. *Food Rev. Int.* **13**: 1–28.

Robberecht, H., Van Dyck, K., Bosscher, D., and Van Cauwenbergh, R. (2008). Silicon in foods: content and bioavailability. *Int. J. Food Prop.* **11**: 638–645.

Roberfroid, M.R. (1993). Dietary fiber, inulin, and oligofructose: a review comparing their physiological effects. *Crit. Rev. Food Sci. Nutr.* **33**: 103–148.

Roberfroid, M.R., Gibson, G.R., and Delzenne, N.M. (1993). The biochemistry of oligofructose, a nondigestible fiber: an approach to calculate its caloric value. *Nutr. Rev.* **51**: 137–146.

Robin, J.P., Mercier, C., Charbonniere, R., and Guilbot, A. (1974). Lintnerized starches. Gel filtration and enzymatic studies of insoluble residues from prolonged acid treatment of potato starch. *Cereal Chem.* **51**: 389–406.

Robinson, D.S. and Eskin, N.A.M. (eds.) (1991). *Polyphenol Oxidase in Oxidative Enzymes in Foods*. London: Elsevier.

Roeber, M., Pydde, E., and Knorr, D. (1991). Storage time dependent accumulation of furanocoumarins in polysaccharide gel coated celery tubers. *Lebensm. Wiss. Technol.* **24**: 466–468.

Rojas-Garbanzo, C., Gleichenhagen, M., Heller, A. et al. (2017). Carotenoid profile, antioxidant capacity, and chromoplasts of pink guava (*Psidium guajava* L. Cv. 'Criolla') during fruit ripening. *J. Agric. Food Chem.* **65**: 3737–3747.

Rönner, B., Lerche, H., Bergmüller, W. et al. (2000). Formation of tetrahydro-β-carbolines and β-carbolines during the reaction of L-tryptophan with D-glucose. *J. Agric. Food Chem.* **48**: 2111–2116.

Roos, Y.H., Karel, M., and Kokini, J.L. (1996). Glass transitions in low moisture and frozen foods: effects on shelf life and quality. *Food Technol.* **50**: 95–108.

Roscher, R. and Winterhalter, P. (1993). Application of multilayer coil counter-current chromatography for the study of *Vitis vinifera* cv. Riesling leaf glycosides. *J. Agric. Food Chem.* **41**: 1452–1457.

Rose, M.D., Bygrave, J., Farrington, W.H.H., and Shearer, G. (1996). The effect of cooking on veterinary drug residues in food. 4. Oxytetracycline. *Food Addit. Contam.* **13**: 275–286.

Rose, M.D., Shearer, G., and Farrington, W.H.H. (1995). The effect of cooking on veterinary drug residues in food. 1. Clenbuterol. *Food Addit. Contam.* **12**: 67–76.

Rose, M.D., Shearer, G., and Farrington, W.H.H. (1997). The effect of cooking on veterinary drug residues in food. 5. Oxfendazole. *Food Addit. Contam.* **14**: 15–26.

Ross, A.B., Shepherd, M.J., Schüpphaus, M. et al. (2003). Alkylresorcinols in cereals and cereal products. *J. Agric. Food Chem.* **51**: 4111–4118.

Ross, A.B., Svelander, C., Karlsson, G., and Savolainen, O. (2017). Identification and quantification of even and odd chained 5-n alkylresorcinols, branched chain-alkylresorcinols and methylalkylresorcinols in quinoa (*Chenopodium quinoa*). *Food Chem.* **220**: 344–351.

Rossiter, K.J. (1996). Structure–odor relationships. *Chem. Rev.* **96**: 3201–3240.

Rössner, J., Velíšek, J., Pudil, F., and Davídek, J. (2001). Strecker degradation products of aspartic and glutamic acids and their amides. *Czech J. Food Sci.* **19**: 41–45.

Rotzoll, N., Dunkel, A., and Hofmann, T. (2005). Activity-guided identification of (S)-malic acid 1-O-D-glucopyranoside (morelid) and γ-aminobutyric acid as contributors to umami taste and mouth-drying oral sensation of morel mushrooms (*Morchella deliciosa* Fr.). *J. Agric. Food Chem.* **53**: 4149–4156.

Rotzoll, N., Dunkel, A., and Hofmann, T. (2006). Quantitative studies, taste reconstitution, and omission experiments on the key taste compounds in morel mushrooms (*Morchella deliciosa* Fr.). *J. Agric. Food Chem.* **54**: 2705–2711.

Roughead, Z.K. and McCormick, D.B. (1990). Qualitative and quantitative assessment of flavins in cow's milk. *J. Nutr.* **120**: 382–388.

Rouphael, Y., Bernardi, J., Cardarelli, M. et al. (2016). Phenolic compounds and sesquiterpene lactones profile in leaves of nineteen artichoke cultivars. *J. Agric. Food Chem.* **64**: 8540–8548.

Rowan, D.D., Allen, J.M., Fielder, S. et al. (1995). Identification of conjugated triene oxidation products of α-farnesene in apple skin. *J. Agric. Food Chem.* **43**: 2040–2045.

Roy, G. (1997). *Modifying Bitterness. Mechanism, Ingredients, and Applications*. Lancaster, PA: Technomic.

Rozan, P., Kuo, Y.-H., and Lambein, F. (2000). Free amino acids present in commercially available seedlings sold for human consumption. A potential hazard for consumers. *J. Agric. Food Chem.* **48**: 716–723.

Ruan, R.R. and Chen, P.L. (1997). *Water in Foods and Biological Materials. A Nuclear Magnetic Approach*. Lancaster, PA: Technomic.

Rudolph, S., Riedel, E., and Henle, T. (2018). Studies on the interaction of the aromatic amino acids tryptophan, tyrosine and phenylalanine as well as tryptophan-containing dipeptides with cyclodextrins. *Eur. Food Res. Technol.* **244**: 1511–1519.

Ruhl, I. and Herrmann, K. (1985). Organische Säuren der Gemüsearten. I. Kohlarten, Blatt- und Zwiebelgemüse sowie Möhren und Sellerie. *Z. Lebensm. Unters. Forsch.* **180**: 215–220.

Ruick, G. (1988). Untersuchungen zum Nickelgehalt von Lebensmitteln. *Food Nahrung* **32**: 807–814.

Ruisinger, B. and Schieberle, P. (2012). Characterization of the key aroma compounds in rape honey by means of the molecular sensory science concept. *J. Agric. Food Chem.* **60**: 4186–4194.

Ruiz Del Castillo, M.L., Flores, G., Herraiz, M., and Blanch, G.P. (2003). Solidphase microextraction for studies on the enantiomeric composition of filbertone in hazelnut oils. *J. Agric. Food Chem.* **51**: 2496–2500.

Rumpler, A., Edmonds, J.S., Katsu, M. et al. (2008). Arsenic-containing long chain fatty acids in cod-liver oil: a result of biosynthetic infidelity? *Angew. Chem. Int. Ed.* **47**: 2665–2667.

Ryan, L.A.M., Dal, B.F., Arendt, E.K., and Koehler, P. (2009). Detection and quantitation of 2,5-diketopiperazines in wheat sourdough and bread. *J. Agric. Food Chem.* **57**: 9563–9568.

Ryley, J. and Kajda, P. (1994). Vitamins in thermal processing. *Food Chem.* **49**: 119–129.

Ryu, K., Ide, N., Matsuura, H., and Itakura, Y. (2001). Nα-(1-deoxy-D-fructos-1-yl)-L-arginine, an antioxidant compound identified in aged garlic extract. *J. Nutr.* **131**: 972S–976S.

Saito, T. and Itoh, T. (1992). Variations and distributions of O-glycosidically linked sugar chains in bovine kappa-casein. *J. Dairy Sci.* **75**: 1768–1774.

Sakamaki, K., Ishizaki, S., Ohkubo, Y. et al. (2012a). Factors influencing the formation of medicinal off-flavor from ascorbic acid and α,β-unsaturated aldehydes. *J. Agric. Food Chem.* **60**: 12428–12434.

Sakamaki, K., Ishizaki, S., Ohkubo, Y. et al. (2012b). Elucidating the formation pathway of the off-flavor compound 6-propylbenzofuran-7-ol. *J. Agric. Food Chem.* **60**: 9967–9973.

Sampathu, S.R., Shirashankar, S., and Lewis, Y.S. (1984). Saffron (*Crocus sativus* L.)-cultivation, processing, chemistry and standardisation. *Crit. Rev. Food Sci. Nutr.* **20**: 123–157.

Sanders, R.A., Zyzak, D.V., Morsch, T.R. et al. (2005). Identification of 8-nonenal as an important contributor to 'plastic' off-odor in polyethylene packaging. *J. Agric. Food Chem.* **53**: 1713–1716.

Sandmann, G. (1994). Carotenoid biosynthesis in microorganisms and plants. *Eur. J. Biochem.* **223**: 7–24.

Sano, T., Okabe, R., Iwahashi, M. et al. (2017). Effect of furan fatty acids and 3-methyl-2,4-nonanedione on light-induced off-odor in soybean oil. *J. Agric. Food Chem.* **65**: 2136–2140.

Santos-Buelga, C., Escribano-Bailon, M.T., and Lattanzio, V. (eds.) (2011). *Recent Advances in Polyphenol Research*, vol. **2**. Oxford: Wiley Blackwell.

Sanz, M.L., Villamiel, M., and Martínez-Castro, I. (2004). Inositols and carbohydrates in different fresh fruit juices. *Food Chem.* **87**: 325–328.

Sapers, G.M., Hicks, K.B., Phillips, J.G. et al. (1995). Characterization of the coloured thermal degradation products of bixin from annatto and a revised mechanism for their formation. *Food Chem.* **53**: 177–185.

Sarwin, R., Laskawy, G., and Grosch, W. (1993). Changes in the levels of glutathione and cysteine during the mixing of doughs with L-threo- and D-erythroascorbic acid. *Cereal Chem.* **70**: 553–557.

Sasaki, K., Wright, J.L.C., and Yasumoto, T. (1998). Identification and characterization of pectenotoxin (ptx) 4 and PTX7 as spiroketal stereoisomers of two previously reported pectenotoxins. *J. Organomet. Chem.* **63**: 2475–2480.

Sato, K. and Ueno, S. (2005). Polymorphism in fats and oils. In: *Bailey's Industrial Oil and Fat Products*, 6e (ed. F. Shahidi), 77. Chichester: Wiley.

Sauvaire, Y., Girardon, P., Baccou, J.C., and Ristérucci, A.M. (1984). Changes in growth, proteins and free amino acids of developing seed and pod of fenugreek. *Phytochemistry* **23**: 479–486.

Sbihi, H.M. and Nehdi, I.A. (2014). Characterization of white mahleb (*Prunus mahaleb* L.) seed oil: a rich source of α-eleostearic acid. *J. Food Sci.* **79**: C795–C801.

Schaefer, O., Bohlmann, R., Schleuning, W.-D. et al. (2005). Development of a radioimmunoassay for the quantitative determination of 8-prenylnaringenin in biological matrices. *J. Agric. Food Chem.* **53**: 2881–2889.

Schaich, K.M. (1992). Metals and lipid oxidation. Contemporary issues. *Lipids* **27**: 209–218.

Scharbert, S. and Hofmann, T. (2005). Molecular definition of black tea taste by means of quantitative studies, taste reconstitution, and omission experiments. *J. Agric. Food Chem.* **53**: 5377–5384.

Scheidig, C., Czerny, M., and Schieberle, P. (2007). Changes in key odorants of raw coffee beans during storage under defined conditions. *J. Agric. Food Chem.* **55**: 5768–5775.

Scherb, J., Kreissl, J., Haupt, S., and Schieberle, P. (2009). Quantitation of S-methylmethionine in raw vegetables and green malt by a stable isotope dilution assay using LC-MS/MS: comparison with dimethyl sulfide formation after heat treatment. *J. Agric. Food Chem.* **57**: 9091–9096.

Schieber, A., Fügel, R., Henke, M., and Carle, R. (2005). Determination of the fruit content of strawberry fruit preparations by gravimetric quantification of hemicellulose. *Food Chem.* **91**: 365–371.

Schiff, P.L. Jr. (2002). Opium and its alkaloids. *Am. J. Pharm. Educ.* **66**: 186–194.

Schlutt, B., Moran, N., Schieberle, P., and Hofmann, T. (2007). Sensory-directed identification of creamness-enhancing volatiles and semivolatiles in full-fat cream. *J. Agric. Food Chem.* **55**: 9634–9645.

Schmarr, H.-G., Ganss, S., Sang, W., and Potouridis, T. (2007). Analysis of 2-aminoacetophenone in wine using a stable isotope dilution assay and multidimensional gas chromatography–mass spectrometry. *J. Chromatogr. A* **1150**: 78–84.

Schmid, C., Dawid, C., Peters, V., and Hofmann, T. (2018). Saponins from European licorice roots (*Glycyrrhiza glabra*). *J. Nat. Prod.* **81**: 1734–1744.

Schmidberger, P.C. and Schieberle, P. (2017). Characterization of the key aroma compounds in white alba truffle (Tuber magnatum pico) and burgundy truffle (*Tuber uncinatum*) by means of the sensomics approach. *J. Agric. Food Chem.* **65**: 9287–9296.

Schmidt, R.H. and Rodrick, G.E. (eds.) (2003). *Food Safety Handbook*. Chichester: Wiley.

Schmiech, L., Alayrac, C., Witulski, B., and Hofmann, T. (2009). Structure determination of bisacetylenic oxylipins in carrots (Daucus carota L.) and enantioselective synthesis of falcarindiol. *J. Agric. Food Chem.* **57**: 11030–11040.

Schneider, M., Klotzsche, M., Werzinger, C. et al. (2002). Reaction of folic acid with reducing sugars and sugar degradation products. *J. Agric. Food Chem.* **50**: 1647–1651.

Schoch, T.K., Manners, G.D., and Hasegawa, S. (2001). Analysis of limonoid glucosides from citrus by electrospray ionization liquid chromatography-mass spectrometry. *J. Agric. Food Chem.* **49**: 1102–1108.

Schoenauer, S. and Schieberle, P. (2018). Structure–odor correlations in homologous series of mercapto furans and mercapto thiophenes synthesized by changing the structural motifs of the key coffee odorant furan-2-ylmethanethiol. *J. Agric. Food Chem.* **66**: 4189–4199.

Schröder, H. and Netscher, T. (2001). Determination of absolute stereochemistry of vitamin E derived oxa-spiro compounds by NMR spectroscopy. *Magn. Reson. Chem.* **39**: 701–708.

Schröder, M., Lehnert, K., Hammann, S., and Vetter, W. (2014). Dihydrophytol and phytol isomers as marker substances for hydrogenated and rfined vegetable oils. *Eur. J. Lipid Sci. Technol.* **116**: 1372–1380.

Schuh, C. and Schieberle, P. (2005). Characterization of (E,E,Z)-2,4,6-nonatrienal as a character impact aroma compound of oak flakes. *J. Agric. Food Chem.* **53**: 8699–8705.

Schuh, C. and Schieberle, P. (2006). Characterization of the key aroma components in the beverage prepared from Darjeeling black tea: quantitative defferences between tea leaves and infusion. *J. Agric. Food Chem.* **54**: 916–924.

Schulz, G.E. and Schirmer, R.H. (1979). *Principles of Protein Structure*. Berlin: Springer.

Schwarz, B. and Hofmann, T. (2007). Isolation, structure determination, and sensory activity of mouth-drying and astringent nitrogen-containing phytochemicals isolated from red currants (Ribes rubrum). *J. Agric. Food Chem.* **55**: 1405–1410.

Schwarz, M., Wray, V., and Winterhalter, P. (2004). Isolation and identification of novel pyranoanthocyanins from black carrot (*Daucus carota* L.) juice. *J. Agric. Food Chem.* **52**: 5095–5101.

Schweizer, T.F. and Horman, I. (1981). Purification and structure determination of three α-D-galactocyclitols from soya bean. *Carbohydr. Res.* **95**: 61–71.

Schweizer, T.F., Horman, I., and Würsch, P. (1978). Low molecular weight carbohydrates from leguminous seeds; a new disaccharide: galactopinitol. *J. Sci. Food Agric.* **29**: 148–154.

Scott, P.M. (1995). Mycotoxins methodology. *Food Addit. Contam.* **12**: 395–403.

Seanes, K.T., Hohmann, S., and Prior, B.A. (1998). Glycerol production by the yeast Saccharomyces cerevisiae and its relevance to wine: a review. *S. Afr. J. Enol. Vitic.* **19**: 17–24.

Sébédio, J.L. and Christie, W.W. (1998). *Trans Fatty Acids in Human Nutrition*. Dundee: Oily Press.

Sedláčková, L., Kružíková, K., and Svobodová, Z. (2014). Mercury speciation from major Czech rivers and assessment of health risk. *Food Chem.* **150**: 360–365.

Seefelder, W., Knecht, A., and Humpf, H.-U. (2003). Bound fumonisin B_1:analysis of fumonisin-B_1 glyco and amino acid conjugates by liquid chromatography-electrospray ionization-tandem mass spectrometry. *J. Agric. Food Chem.* **51**: 5567–5573.

Seeram, N.P., Schutzki, R., Chandra, A., and Nair, M.G. (2002). Characterization, quantification, and bioactivities of anthocyanins in Cornus species. *J. Agric. Food Chem.* **50**: 2519–2523.

Sefidkon, F., Abbasi, K., and Khaniki, G.B. (2006). Influence of drying and extraction methods on yield and chemical composition of the essential oil of *Satureja hortensis. Food Chem.* **99**: 19–23.

Sekiwa, Y., Kubota, K., and Kobayashi, A. (2000). Isolation of novel glucosides related to gingerdiol from ginger and their antioxidative activities. *J. Agric. Food Chem.* **48**: 373–377.

Sell, D.R., Biemel, K.M., Reihl, O. et al. (2005). Glucosepane is a major protein cross-link of the senescent human extracellular matrix. *J. Biol. Chem.* **280**: 12310–12315.

Selman, J.D. (1994). Vitamin retention during blanching of vegetables. *Food Chem.* **49**: 137–147.

Sen Gupta, D., Thavarajah, D., Knutson, P. et al. (2013). Lentils (*Lens culinaris* L.), a rich source of folates. *J. Agric. Food Chem.* **61**: 7794–7799.

Sen, N.P., Seaman, S.W., Baddoo, P.A. et al. (2001). Pickled vegetables following incubation with nitrite under acidic conditions. *J. Agric. Food Chem.* **49**: 2096–2103.

Sen, N.P., Seaman, S.W., Burgess, C. et al. (2000). Investigation on the possible formation of n-nitroso-n-methylurea by nitrosation of creatinine in model systems and in cured meats at gastric pH. *J. Agric. Food Chem.* **48**: 5088–5096.

Sen, N.P., Seaman, S.W., Lau, B.P.-Y. et al. (1995). Determination and occurrence of various tetrahydro-β-carboline-3-carboxylic acids and the corresponding N-nitroso compounds in foods and alcoholic beverages. *Food Chem.* **54**: 327–337.

Senatore, F. (1992). Chemical constituents of some mushrooms. *J. Sci. Food Agric.* **58**: 499–503.

Seppanen, C.M. and Csallany, A.S. (2006). The effect of intermittent and continuous heating of soybean oil at frying temperature on the formation of 4-hydroxy-2-trans-nonenal and other α-, β-unsaturated hydroxyaldehydes. *J. Am. Oil Chem. Soc.* **83**: 121–127.

Servillo, L., Castaldo, D., Giovane, A. et al. (2017). Tyramine pathways in citrus plant defense: glycoconjugates of tyramine and its *N*-methylated derivatives. *J. Agric. Food Chem.* **65**: 892–899.

Servillo, L., Giovane, A., Balestrieri, M.L. et al. (2011a). Betaines in fruits of Citrus genus plants. *J. Agric. Food Chem.* **59**: 9410–9416.

Servillo, L., Giovane, A., Balestrieri, M.L. et al. (2011b). Proline derivatives in fruits of bergamot (Citrus bergamia Risso et Poit): presence of N-methyl-L-proline and 4-hydroxy-L-prolinebetaine. *J. Agric. Food Chem.* **59**: 274–281.

Servillo, L., Giovane, A., Balestrieri, M.L. et al. (2012). Occurrence of pipecolic acid and pipecolic acid betaine (homostachydrine) in citrus genus plants. *J. Agric. Food Chem.* **60**: 315–321.

Servillo, L., Giovane, A., Casale, R. et al. (2016). Homostachydrine (pipecolic acid betaine) as authentication marker of roasted blends of *Coffea arabica* and *Coffea canephora* (robusta) beans. *Food Chem.* **205**: 52–57.

Servillo, L., Giovane, A., Casale, R. et al. (2015). Serotonin 5-*O*-β-glucoside and its *N*-methylated forms in citrus genus plants. *J. Agric. Food Chem.* **63**: 4220–4227.

Servillo, L., Giovane, A., D'Onofrio, N. et al. (2014). *N*-methylated derivatives of tyramine in *Citrus* genus plants: identification of *N,N, N*-trimethyltyramine (candicine). *J. Agric. Food Chem.* **62**: 2679–2684.

Sessa, R.A., Bennett, M.H., Lewis, M.J. et al. (2000). Metabolite profiling of sesquiterpene lactones from *Lactuca* species. *J. Biol. Chem.* **275**: 26877–26884.

Seto, H., Okuda, T., Takesue, T., and Ikemura, T. (1983). Reaction of malonaldehyde with nucleic acid. I. Formation of fluorescent pyrimido[1,2-a] purin-10(3H)-one nucleosides. *Bull. Chem. Soc. Jpn.* **56**: 1799–1802.

Sevillano-Morales, J.S., Cejudo-Gómez, M., Ramírez-Ojeda, A.M. et al. (2015). Risk profile of methylmercury in seafood. *Curr. Opin. Food Sci.* **6**: 53–60.

Shamberger, R.J. (1983). *Biochemistry of Selenium*. New York: Plenum Press.

Sheard, N.F., Clark, N.G., Brand-Miller, J.C. et al. (2004). Dietary carbohydrate (amount and type) in the prevention and management of diabetes: a statement by the American Diabetes Association. *Diabetes Care* **27**: 2266–2271.

Shi, J. and Le Maguer, M. (2000). Lycopene in tomatoes: chemical and physical properties affected by food processing. *Crit. Rev. Food Sci. Nutr.* **40**: 1–42.

Shoulders, M.D. and Raines, R.T. (2009). Collagen structure and stability. *Annu. Rev. Biochem.* **78**: 929–958.

Shrestha, K., Stevens, C.V., and De Meulenaer, B. (2012). Isolation and identification of a potent radical scavenger (canolol) from roasted high erucic mustard seed oil from nepal and its formation during roasting. *J. Agric. Food Chem.* **60**: 7506–7512.

Siebert, T.E., Barker, A., Bartera, S.R. et al. (2018). Analysis, potency and occurrence of (*Z*)-6-dodeceno-γ-lactone in white wine. *Food Chem.* **256**: 85–90.

Siener, R., Hönow, R., Voss, S. et al. (2006). Oxalate content of cereals and cereal products. *J. Agric. Food Chem.* **54**: 3008–3011.

Sikorski, Z.E. (ed.) (1996). *Chemical and Functional Properties of Food Components*. Lancaster, PA: Technomic.

Silva Ferreira, A.C., Barbe, J.-C., and Bertrand, A. (2003). 3-Hydroxy-4,5-dimethyl-2(5H)-furanone: a key odorant of the typical aroma of oxidative aged Port wine. *J. Agric. Food Chem.* **51**: 4356–4363.

Simat, T.J. and Steinhart, H. (1998). Oxidation of free tryptophan and tryptophan residues in peptides and proteins. *J. Agric. Food Chem.* **46**: 490–498.

Simian, H., Robert, F., and Blank, I. (2004). Identification and synthesis of 2-heptanethiol, a new flavour compound found in bell peppers. *J. Agric. Food Chem.* **52**: 306–310.

Simonich, S.L. and Hites, R.A. (1995). Organic pollutant accumulation in vegetation. *Environ. Sci. Technol.* **29**: 2905–2914.

Simpson, R.F., Capone, D.L., and Sefton, M.A. (2004). Isolation and identification of 2-methoxy-3,5-dimethylpyrazine, a potent musty compound from wine corks. *J. Agric. Food Chem.* **52**: 5425–5430.

Sindt, L., Gammacurta, M., Waffo-Teguo, P. et al. (2016). Taste-guided isolation of bitter lignans from *Quercus petraea* and their identification in wine. *J. Nat. Prod.* **79**: 2432–2438.

Singldinger, B., Dunkel, A., Bahmann, D. et al. (2018). New taste-active 3-(*O*-β-D-glucosyl)-2-oxoindole-3-acetic acids and arylheptanoids in Cimiciato-infected hazelnuts. *J. Agric. Food Chem.* **66**: 4660–4673.

Singldinger, B., Dunkel, A., and Hofmann, T. (2017). The cyclic diarylheptanoid asadanin as the main contributor to the bitter off-taste in hazelnuts (*Corylus avellana* L.). *J. Agric. Food Chem.* **65**: 1677–1683.

Skibsted, L.H. (2012). Carotenoids in antioxidant networks. Colorants or radical scavengers. *J. Agric. Food Chem.* **60**: 2409–2417.

Skog, K. and Alexander, J. (eds.) (2006). *Acrylamide and Other Hazardous Compounds in Heat-Treated Food*. CRC Press – Taylor & Francis Group.

Skoglund, E., Carlsson, N.-G., and Sandberg, A.-S. (1997). Determination of isomers of inositol mono- to hexaphosphates in selected foods and intestinal contents using high-performance ion chromatography. *J. Agric. Food Chem.* **45**: 431–436.

Skouroumounis, G.K. and Winterhalter, P. (1994). Glycosidically bound norisoprenoids from Vitis vinifera cv Riesling leaves. *J. Agric. Food Chem.* **42**: 1068–1072.

Skrede, G., Wrolstad, R.E., and Durst, R.W. (2000). Changes in anthocyanins and polyphenolics during juice processing of highbush blueberries (*Vaccinium corymbosum* L.). *J. Food Sci.* **65**: 357–364.

Skurichin, I.M. and Volgarev, M.N. (1987). Chimiceskij sostav piscevych produktov, kniga 1. Spravocnyje tablicy sodìrzanija osnovnych piscevych vescestv i energeticeskoj cennosti piscevych produktov. Agropromizdat, Moskva.

Smarrito-Menozzi, C., Matthey-Doret, W., Devaud-Goumoens, S., and Viton, F. (2013). Glycerol, an underestimated flavor precursor in the Maillard reaction. *J. Agric. Food Chem.* **61**: 10225–10230.

Smart, G.A. and Sherlock, J.C. (1985). Chromium in foods and the diet. *Food Addit. Contam.* **2**: 139–147.

Šmidrkal, J., Karlová, T., Filip, V. et al. (2009). Antimicrobial properties of 11-cyclohexylundecanoic acid. *Czech J. Food Sci.* **27**: 463–469.

Smith, J.E., Lewis, C.W., Anderson, J.G., and Solomons, G.L. (1994). Mycotoxins in human nutrition and health. European Commission/Agro-Industrial Research Division.

Smith, K.T. (ed.) (1988). *Trace Minerals in Foods*. New York: Dekker.

Smith, L.L. (1996). Review of progress in sterol oxidations: 1987–1995. *Lipids* **31**: 453–487.

Smith, N.A. (1994). Nitrate reduction and N-nitrosation in brewing. *J. Inst. Brew.* **100**: 347–355.

Snowdon, E.M., Bowyer, M.C., Grbin, P.R., and Bowyer, P.K. (2006). Mousy off-flavor: a review. *J. Agric. Food Chem.* **54**: 6465–6474.

Sobolev, V.S., Neff, S.A., and Gloer, J.B. (2009). New stilbenoids from peanut (*Arachis hypogaea*) seeds challenged by an *Aspergillus caelatus* strain. *J. Agric. Food Chem.* **57**: 62–68.

Soldo, T., Blank, I., and Hofmann, T. (2003). (+)-(S)-Alapyridaine – a general taste enhancer? *Chem. Senses* **28**: 371–379.

Soleas, G.J., Yan, J., Seaver, T., and Goldberg, D.M. (2002). Method for the gas chromatographic assay with mass selective detection of trichloro compounds in corks and wines applied to elucidate the potential cause of cork taint. *J. Agric. Food Chem.* **50**: 1032–1039.

Söllner, K. and Schieberle, P. (2009). Decoding the key aroma compounds of a Hungarian-type salami by molecular sensory science approaches. *J. Agric. Food Chem.* **57**: 4319–4327.

Somoza, V., Lindenmeier, M., Wenzel, E. et al. (2003). Activity-guided identification of a chemopreventive compound in coffee beverage using *in vitro* and *in vivo* techniques. *J. Agric. Food Chem.* **51**: 6861–6869.

Sondey, S.M., Seib, P.A., and El-Atawy, Y.S. (1989). Control of enzymatic browning in apple with ascorbic acid derivatives, polyphenol oxidase inhibitors, and complexing agents. *J. Food Sci.* **53**: 997–1002.

Sones, K., Heaney, R.K., and Fenwick, G.R. (1984). An estimate of the mean daily intake of glucosinolates from cruciferous vegetables in the UK. *J. Sci. Food Agric.* **35**: 712–720.

Song, D., Liang, H., Kuang, P. et al. (2013). Instability and structural change of 4-methylsulfinyl-3-butenyl isothiocyanate in the hydrolytic process. *J. Agric. Food Chem.* **61**: 5097–5102.

Song, Y., Feng, Y., Zhao, S. et al. (2008). D-Amino acids in rat brain measured by liquid chromatography/tandem mass spectrometry. *Neurosci. Lett.* **445**: 53–57.

Sonntag, T., Kunert, C., Dunkel, A., and Hofmann, T. (2010). Sensory-guided identification of N-(1-methyl-4-oxoimidazolidin-2-ylidene)-α-amino acids as contributors to the thick-sour and mouth-drying orosensation of stewed beef juice. *J. Agric. Food Chem.* **58**: 6341–6350.

Sotheeswaran, S. and Pasupathy, V. (1993). Distribution of resveratrol oligomers in plants. *Phytochemistry* **32**: 1083–1092.

Soukupová, V., Čížková, H., and Voldřich, M. (2004). Evaluation of ketchup authenticity – chemical changes of markers during production and distribution. *Czech J. Food Sci.* **22**: 349–352.

Souquet, J.-M., Cheynier, V., Brossaud, F., and Moutounet, M. (1996). Polymeric proanthocyanidins from grape skins. *Phytochemistry* **43**: 509–512.

Sousa, A., Mateus, N., Silva, A.M.S. et al. (2010). Isolation and structural characterization of anthocyanin-furfuryl pigments. *J. Agric. Food Chem.* **58**: 5664–5669.

Spanier, A.M., Shahidi, F., Parliment, T.H. et al. (eds.) (2001). *Food Flavors and Chemistry: Advances of the New Millennium*, Special Publication No. 274. Cambridge: The Royal Society of Chemistry.

Spanos, G.A. and Wrolstad, R.E. (1992). Phenolics of apple, pear, and white grape juices and their changes with processing and storage-a review. *J. Agric. Food Chem.* **40**: 1478–1487.

Spiegel, J.E., Rose, R., Karabell, P. et al. (1994). Safety and benefits of fructooligosaccharides as food ingredients. *Food Technol.* **48**: 85–98.

Spiteller, P., Kern, W., Reiner, J., and Spiteller, G. (2001). Aldehydic lipid peroxidation products derived from linoleic acid. *Biochim. Biophys. Acta* **1531**: 188–208.

Sriwilaijaroen, N., Kadowaki, A., Onishi, Y. et al. (2011). Mumefural and related HMF derivatives from Japanese apricot fruit juice concentrate show multiple inhibitory effects on pandemic influenza A (H1N1) virus. *Food Chem.* **127**: 1–9.

St. Angelo, A.J. (1992). *Lipid Oxidation in Food*. Washington, D.C.: American Chemical Society.

Stadler, R.H. and Lineback, D.R. (eds.) (2009). *Process-Induced Food Toxicants*. Chichester: Wiley.

Stadler, R.H., Varga, N., Hau, J. et al. (2002a). Alkylpyridiniums. 1. Formation in model systems via thermal degradation of trigonelline. *J. Agric. Food Chem.* **50**: 1192–1199.

Stadler, R.H., Varga, N., Milo, C. et al. (2002b). Alkylpyridiniums. 2. Isolation and quantification in roasted and ground coffees. *J. Agric. Food Chem.* **50**: 1200–1206.

Stadler, R.H., Welti, D.H., Stämpfi, A.A., and Fay, L.B. (1996). Thermal decomposition of caffeic acid in model systems: identification of novel tetraoxygenated phenylindan isomers and their stability in aqueous solutions. *J. Agric. Food Chem.* **44**: 898.

Stadtman, E.R. and Levine, R.L. (2003). Free radical-mediated oxidation of free amino acids and amino acid residues in proteins. *Amino Acids* **25**: 207–218.

Stahl, W.H. (ed.) (1973). *Compilation of Odor and Taste Threshold Values Data*. Baltimore, MD: American Society for Testing and Materials.

Stamatopoulos, P., Brohan, E., Prevost, C. et al. (2016). Influence of chirality of lactones on the perception of some typical fruity notes through perceptual interaction phenomena in Bordeaux dessert wines. *J. Agric. Food Chem.* **64**: 8160–8167.

Stanley, W.L. and Vannier, S.H. (1967). Psoralens and substituted coumarins from expressed oil of lime. *Phytochemistry* **6**: 585–596.

Stark, T. and Hofmann, T. (2005). Isolation, structure determination, synthesis, and sensory activity of N-phenylpropenoyl-L-amino acids from cocoa (*Theobroma cacao*). *J. Agric. Food Chem.* **53**: 5419–5428.

Stark, T. and Hofmann, T. (2006). Application of a molecular sensory science approach to alkalized cocoa (*Theobroma cacao*): structure determination and sensory activity of nonenzymatically C-glycosylated flavan-3-ols. *J. Agric. Food Chem.* **54**: 9510–9521.

Stark, T., Keller, D., Wenker, K. et al. (2007). Nonenzymatic C-glycosylation of flavan-3-ols by oligo- and polysaccharides. *J. Agric. Food Chem.* **55**: 9685–9697.

Starkenmann, C., Cayeux, I., Decorzant, E. et al. (2009). Taste contribution of (*R*)-strombine to dried scallop. *J. Agric. Food Chem.* **57**: 7938–7943.

Starkenmann, C., Niclass, Y., and Troccaz, M. (2011). Nonvolatile S-alk(en)ylthio-L-cysteine derivatives in fresh onion (*Allium cepa* L. cultivar). *J. Agric. Food Chem.* **59**: 9457–9465.

Steele, D.H., Thorburg, M.J., Stanley, J.S. et al. (1994). Determination of styrene in selected foods. *J. Agric. Food Chem.* **42**: 1661–1665.

Steenackers, B., De Cooman, L., and De Vos, D. (2015). Chemical transformations of characteristic hop secondary metabolites in relation to beer proprties and the brewing process: a review. *Food Chem.* **172**: 742–756.

Steinbrecher, A. and Linseisen, J. (2009). Dietary intake of individual glucosinolates in participants of the EPIC-Heidelberg cohort study. *Ann. Nutr. Metab.* **54**: 87–96.

Steinhaus, M. (2015). Characterization of the major odor-active compounds in the leaves of the curry tree *Bergera koenigii* L. by aroma extract dilution analysis. *J. Agric. Food Chem.* **63**: 4060–4067.

Steinhaus, M. (2017). Confirmation of 1-phenylethane-1-thiol as the character impact aroma compound in curry leaves and its behavior during tissue disruption, drying, and frying. *J. Agric. Food Chem.* **65**: 2141–2146.

Steinhaus, M., Sinuco, D., Polster, J. et al. (2008). Characterization of the aroma-active compounds in pink guava (*Psidium guajava*, L.) by application of the aroma extract dilution analysis. *J. Agric. Food Chem.* **56**: 4120–4127.

Steinhaus, P. and Schieberle, P. (2000). Comparison of the most odor-active compounds in fresh and dried hop cones (*Humulus lupulus* L. variety Spalter Select) based on GC-olfactometry and odor dilution techniques. *J. Agric. Food Chem.* **48**: 1176–1183.

Steinhaus, P. and Schieberle, P. (2007). Characterization of the key aroma compounds in soy sauce using approaches of molecular sensory science. *J. Agric. Food Chem.* **55**: 6262–6269.

Stephen, A.M. (ed.) (1995). *Food Polysaccharides and their Applications*. New York: Dekker.

Stevens, J.F. and Maier, C.S. (2008). Acrolein: sources, metabolism, and biomolecular interactions relevant to human health and disease. *Mol. Nutr. Food Res.* **52**: 7–25.

Stevens, J.F., Miranda, C.L., Wolthers, K.R. et al. (2002). Identification and in vitro biological activities of hop proanthocyanidins: inhibition of nnos activity and scavenging of reactive nitrogen species. *J. Agric. Food Chem.* **50**: 3435–3443.

Stoddart, J.F. (1971). *Stereochemistry of Carbohydrates*. New York: Wiley.

Storkey, C., Pattison, D.I., Gaspard, D.S. et al. (2014). Mechanisms of degradation of the natural high-potency sweetener (2*R*,4*R*)-monatin in mock beverage solutions. *J. Agric. Food Chem.* **62**: 3476–3487.

Strålsjö, L., Alklint, C., Olsson, M.E., and Sjöholm, I.M. (2003). Total folate content and retention in rosehips (Rosassp.) after drying. *J. Agric. Food Chem.* **51**: 4291–4295.

Strålsjö, L., Witthöft, C.M., Sjöholm, I.M., and Jägerstad, M.I. (2003). Folate content in strawberries (*Fragaria* x *ananassa*): effects of cultivar, ripeness, year of harvest, storage, and commercial processing. *J. Agric. Food Chem.* **51**: 128–133.

Stratton, J.E., Hutkins, R.W., and Taylor, S.L. (1991). Biogenic amines in cheese and other fermented foods: a review. *J. Food Prot.* **54**: 460–470.

Strauss, C.R., Dimitriadis, E., Wilson, B., and Williams, P.J. (1986). Studies on the hydrolysis of two megastigma-3,6,9-triols rationalizing the origins of some volatile C_{13} norisoprenoids of Vitis vinifera grapes. *J. Agric. Food Chem.* **34**: 145–149.

Strecker, G., Fievre, S., Wieruczeski, J.-M. et al. (1992). Primary structure of four human octa-, nona-, and undeca-saccharides established by [1]H- and ^{13}C-nuclear magnetic resonance spectroscopy. *Carbohydr. Res.* **226**: 1–14.

Sugimura, T. (1986). Past, present, and future of mutagens in cooked foods. *Environ. Health Perspect.* **67**: 5–10.

Sugita, D., Inoue, R., and Kurihara, Y. (1998). Sweet and sweetness-inducing activities of new triterpene glycosides, strogins. *Chem. Senses* **23**: 93–97.

Sugiyama, H., Takahashi, M.N., Terada, H. et al. (2008). Accumulation and localization of cesium in edible mushroom (Pleurotus ostreatus) mycelia. *J. Agric. Food Chem.* **56**: 9641–9646.

Suksamrarn, S., Komutiban, O., Ratananukul, P. et al. (2006). Cytotoxic prenylated xanthones from the young fruit of *Garcinia mangostana*. *Chem. Pharm. Bull.* **54**: 301–305.

Sulser, H., Habegger, M., and Büchi, W. (1972). Synthese und Geschmacksprüfungen von 3,4-disubstituierten 2-Hydroxy-2-buten-1,4-oliden. *Z. Lebensm. Unters. Forsch.* **148**: 215–221.

Suto, D., Ikeda, Y., Fujii, J., and Ohba, Y. (2006). Structural analysis of amino acids, oxidized by reactive oxygen species and an antibody against N-formylkynurenine. *J. Clin. Biochem. Nutr.* **38**: 1–5.

Swallow, K.W. and Low, N.H. (1993). Isolation and identification of oligosaccharides in a commercial beet medium invert sugar. *J. Agric. Food Chem.* **41**: 1587–1592.

Sytar, O., Brestic, M., and Rai, M. (2013). Possible ways of fagopyrin biosynthesis and production in buckwheat plants. *Fitoterapia* **84**: 72–79.

Szefer, P. and Nriagu, J.O. (eds.) (2007). *Mineral Components in Foods*. Boca Raton, FL: CRC Press.

Szuhaj, B.F. (1988). *Lecithin: Sources, Manufacture Uses*. Champaign, IL: AOCS Press.

Szuhaj, B.F. and List, G.R. (1995). *Lecithins*. Champaign, IL: AOCS Press.

Szydłowska-Czerniak, A. and Rabiej, D. (2018). Octyl sinapate as a new antioxidant to improve oxidative stability and antioxidant activity of rapeseed oil during accelerated storage. *Eur. Food Res. Technol.* **244**: 1397–1406.

Tada, M., Shinoda, I., and Okai, H. (1984). L-Ornithine, a new salty peptide. *J. Agric. Food Chem.* **32**: 992–995.

Takahama, U., Yamauchi, R., and Hirota, S. (2013). Isolation and characterization of a cyanidin-catechin pigment from adzuki bean (*Vigna angularis*). *Food Chem.* **141**: 282–288.

Taleshi, M.S., Jensen, K.B., Raber, G. et al. (2008). Arsenic-containing hydrocarbons: natural compounds in oil from the fish capelin, Mallotus villosus. *Chem. Commun.* **39**: 4706–4707.

Tan, S.P., Kha, T.C., Parks, S.E., and Roach, P.D. (2016). Bitter melon (*Momordica charantia* L.) bioactive composition and health benefits: a review. *Food Rev. Int.* **32**: 181–202.

Tanabe, S. (1988). PCB problems in the future: foresight from current knowledge. *Environ. Pollut.* **50**: 5–28.

Tanaka, T., Nonaka, G., and Nishioka, I. (1986). Tannins and related compounds. Part 37. Isolation and structure elucidation of elaeocarpusin, a novel ellagitannin from Elaeocarpus silvestris var. elipticus. Journal of the Chemical Society. *Perkin Trans.* **1**: 369–376.

Tanaka, Y., Hosokawa, M., Otsu, K. et al. (2009). Assessment of capsiconinoid composition, nonpungent capsaicinoid analogues, in *Capsicum* cultivars. *J. Agric. Food Chem.* **57**: 5407–5412.

Tandon, K.S., Baldwin, E.A., and Shewfelt, R.L. (2000). Aroma perception of individual volatile compounds in fresh tomatoes (*Lycopersicon esculentum*, Mill.) as affected by the medium of evaluation. *Postharvest Biol. Technol.* **20**: 261–268.

Taniguchi, Y., Matsukura, Y., Ozaki, H. et al. (2013). Identification and quantification of the oxidation products derived from α-acids and β-acids during storage of hops (*Humulus lupulus* L.). *J. Agric. Food Chem.* **61**: 3121–3130.

Taniguchi, Y., Taniguchi, H., Matsukura, Y. et al. (2014). Structural elucidation of humulone autoxidation products and analysis of their occurrence in stored hops. *J. Nat. Prod.* **77**: 1252–1261.

Taniguchi, Y., Yamada, M., Taniguchi, H. et al. (2015). Chemical characterization of beer aging products derived from hard resin components in hops (*Humulus lupulus* L.). *J. Agric. Food Chem.* **63**: 10181–10191.

Tarantilis, P.A. and Polissiou, M.G. (1997). Isolation and identification of the aroma components from saffron (*Crocus sativus*). *J. Agric. Food Chem.* **45**: 459–462.

Tareke, E., Rydberg, P., Karlsson, P. et al. (2002). Analysis of acrylamide, a carcinogen formed in heated foodstuffs. *J. Agric. Food Chem.* **50**: 4998–5006.

Tatham, A.S. and Shewry, P.R. (1985). The conformation of wheat gluten proteins. The secondary structures and thermal stabilities of α-, β-, γ - and ω-gliadins. *J. Cereal Sci.* **3**: 103–113.

Tavčar, B.E. and Kreft, S. (2015). Fagopyrins and protofagopyrins: detection, analysis, and potential phototoxicity in buckwheat. *J. Agric. Food Chem.* **63**: 5715–5724.

Tavčar, B.E., Žigon, D., Friedrich, M. et al. (2014). Isolation, analysis and structures of phototoxic fagopyrins from buckwheat. *Food Chem.* **143**: 432–439.

Taylor, A.J. and Mottram, D.S. (eds.) (1996). *Flavour Science Recent Developments*. Cambridge: The Royal Society of Chemistry.

Taylor, S.L. (1992). Chemistry and detection of food allergens, in overview. *Food Technol.* **46**: 146–148.

Teixidó, E., Núñez, O., Santos, F.J., and Galceran, M.T. (2011). 5-Hydroxymethylfurfural-content in foodstuffs determined by micellar electrokinetic chromatography. *Food Chem.* **126**: 1902–1908.

Tenni, D., Martin, M., Barberis, E. et al. (2017). Total As and As speciation in Italian rice as related to producing areas and paddy soils properties. *J. Agric. Food Chem.* **65**: 3443–3452.

Teranishi, R., Buttery, R.G., and Sugusawa, H. (eds.) (1993). *Bioactive Volatile Compounds from Plants*. Washington, D.C.: American Chemical Society.

Teranishi, R., Takeoka, G.R., and Güntert, M. (eds.) (1992). *Flavor Precursors. Thermal and Enzymatic Conversions*. Washington, D.C.: American Chemical Society.

Ternes, W., Acker, L., and Scholtyssek, S. (1994). *Ei und Eiprodukte*, 90–135. Berlin, Hamburg: Verlag Paul Parey.

Terry, N., Zayed, A.M., de Souza, M.P., and Tarun, A.S. (2000). Selenium in higher plants. *Annu. Rev. Plant Physiol. Plant Mol. Biol.* **51**: 401–432.

Teshima, Y., Ikeda, T., Imada, K. et al. (2013). Identification and biological activity of antifungal saponins from shallot (*Allium cepa* L. Aggregatum group). *J. Agric. Food Chem.* **61**: 7440–7445.

Thakur, B.R., Singh, R.K., and Arya, S.S. (1994). Chemistry of sorbates-a basic perspective. *Food Rev. Int.* **10**: 71–91.

Thakur, B.R., Singh, R.K., and Handa, A.K. (1997). Chemistry and uses of pectin – a review. *Crit. Rev. Food Sci. Nutr.* **37**: 47–73.

Thavarajah, D., Vandenberg, A., George, G.N., and Pickering, I.J. (2007). Chemical form of selenium in naturally selenium-rich lentils (Lens culinaris L.) from Saskatchewan. *J. Agric. Food Chem.* **55**: 7337–7341.

The Chlorine Chemistry Council®: A Comparison of Dioxin Risk Characterizations. http://www.dioxinfacts.org/dioxin_health/public_policy/dr.pdf (accessed 28 October 2019).

The Medical Biochemistry Page. http://themedicalbiochemistrypage.org/index.php (accessed 28 October 2019).

Thiry, C., Ruttens, A., De Temmerman, L. et al. (2012). Current knowledge in species-related bioavailability of selenium in food. *Food Chem.* **130**: 767–784.

Thomas, B., Roughan, J.A., and Watters, E.D. (1974). Cobalt, chromium and nickel content of some vegetable foodstuffs. *J. Sci. Food Agric.* **25**: 771–776.

Thomas, M.J. (1995). The role of free radicals and antioxidants: how do we know that they are working? *Crit. Rev. Food Sci. Nutr.* **35**: 21–39.

Thomas, P.M., Flanagan, V.P., and Pawlosky, R.J. (2003). Determination of 5-methyltetrahydrofolic acid and folic acid in citrus juices using stable isotope dilution-mass spectrometry. *J. Agric. Food Chem.* **51**: 1293–1296.

Thorna, R.G., Tibell, L., Untereiner, W.A. et al. (2007). A higher level phylogenetic classification of the fungi. *Mycol. Res.* **111**: 509–547.

Thorpe, S.R. and Baynes, J.W. (2003). Maillard reaction products in tissue proteins: new products and new perspectives. *Amino Acids* **25**: 275–281.

Tian, Q., Konczak, I., and Schwartz, S.J. (2005). Probing anthocyanin profiles in purple sweet potato cell line (Ipomoea batatas L. cv. Ayamurasaki) by high-performance liquid chromatography and electrospray ionization tandem mass spectrometry. *J. Agric. Food Chem.* **53**: 6503–6509.

Tilley, K.A., Benjamin, R.E., Bagorogoza, K.E. et al. (2001). Tyrosine cross-links: molecular basis of gluten structure and function. *J. Agric. Food Chem.* **49**: 2627–2632.

Tils, M., Angenot, L., Poukens, P. et al. (1992). Prodelphinidins from Ribes nigrum. *Phytochemistry* **31**: 971–973.

Timbergen, B.J. and Slump, P. (1976). The detection of chicken meat in meat products by means of the anserine-carnosine ratio. *Z. Lebensm. Unters. Forsch.* **161**: 7–11.

Toelstede, S. and Hofmann, T. (2008). Quantitative studies and taste re-engineering experiments toward the decoding of the non-volatile sensometabolome of Gouda cheese. *J. Agric. Food Chem.* **56**: 5299–5307.

Togosaki, T. and Mineshita, T. (1988). Antioxidant effects of protein-bound riboflavin and free riboflavin. *J. Food Sci.* **53**: 1851–1853.

Tolba, M.F., Omar, H.A., Azab, S.S. et al. (2016). Caffeic acid phenethyl ester: a review of its antioxidant activity, protective effects against ischemia-reperfusion injury and drug adverse reactions. *Crit. Rev. Food Sci. Nutr.* **56**: 2183–2190.

Tominaga, T. and Dubourdieu, D. (2000). Identification of cysteinylated aroma precursors of certain volatile thiols in passion fruit juice. *J. Agric. Food Chem.* **48**: 2874–2876.

Tominaga, T. and Dubourdieu, D. (2006). A novel method for quantification of 2-methyl-3-furanthiol and 2-furanmethanethiol in wines made from Vitis vinifera grape varieties. *J. Agric. Food Chem.* **54**: 29–33.

Tominaga, T., Niclass, Y., Frerot, E., and Dubourdieu, D. (2006). Stereoisomeric distribution of 3-mercaptohexan-1-ol and 3-mercaptohexyl acetate in dry and sweet white wines made from Vitis vinifera (var. Sauvignon Blanc and Semillon). *J. Agric. Food Chem.* **54**: 7251–7255.

Tomlin, C.D.S. (ed.) (1997). *The Pesticide Manual.* Farnham, Surrey, UK: British Crop Protection Council.

Tomoda, M., Shimizu, N., Gonda, R., and Kanari, M. (1989). Anticomplementary and hypoglycemic activity of okra and Hibiscus mucilages. *Carbohydr. Res.* **190**: 323–328.

Tomomatsu, H. (1994). Health effects of oligosaccharides. *Food Technol.* **48**: 61–68.

Toyoda-Ono, Y., Maeda, M., Nakao, M. et al. (2004). 2-O-(β-D-Glucopyranosyl)ascorbic acid, a novel ascorbic acid analogue isolated from Lycium fruit. *J. Agric. Food Chem.* **52**: 2092–2096.

Tressl, R., Bahri, D., Holzer, M., and Kossa, T. (1977). Formation of flavor components in asparagus. 2. Formation of flavor components in cooked asparagus. *J. Agric. Food Chem.* **25**: 459–463.

Tressl, R., Helak, B., Kersten, E., and Nittka, C. (1993). Formation of flavor compounds by Maillard reaction. In: *Recent Developments in Flavor and Fragrance Chemistry* (eds. R. Hopp and K. Mori), 167–181. Weinheim: Wiley-VCH.

Tressl, R., Helak, B., Spengler, K. et al. (1985). Cyclo[b]azepine derivative-novel proline-specific Maillard reaction products. Liebigs Annalen der Chemie, 2017-2027.

Tressl, R., Holzer, M., and Apetz, M. (1977). Formation of flavor components in asparagus. 1. Biosynthesis of sulfur-containing acids in asparagus. *J. Agric. Food Chem.* **25**: 455–459.

Tressl, R., Kersten, E., Wondrak, G. et al. (1998). Fragmentation of sugar skeletons and formation of Maillard polymers. In: *The Maillard Reaction in Foods and Medicine*, Special Publication No. 223 (eds. J. O'Brien, H.E. Nursten, M.J.C. Crabbe and J.M. Ames), 9. Cambridge: The Royal Society of Chemistry.

Trevaskis, M. and Trenerry, V.G. (1996). An investigation into the determination of oxalic acid in vegetables by capillary electrophoresis. *Food Chem.* **57**: 323–330.

Tripathi, M.K. and Mishra, A.S. (2007). Glucosinolates in animal nutrition: a review. *Anim. Feed Sci. Technol.* **132**: 1–27.

Trucksess, M.W. (1997). Committee on natural toxins-mycotoxins General Referee Reports. *J. AOAC Int.* **80**: 119–138.

Truong, V.-D., Deighton, N., Thompson, R.T. et al. (2010). Characterization of anthocyanins and anthocyanidins in purple-fleshed sweetpotatoes by HPLC-DAD/ESI-MS/MS. *J. Agric. Food Chem.* **58**: 404–410.

Tsednee, M., Mak, Y.W., Chen, Y.R., and Yeh, K.C. (2012). A sensitive LC-ESI-Q-TOF-MS method reveals novel phytosiderophores and phytosiderophore–iron complexes in barley. *New Phytol.* **195**: 951–961.

Tsolakou, A., Diamantakos, P., Kalaboki, I. et al. (2018). Oleocanthalic acid, a chemical marker of olive oil aging and exposure to a high storage temperature with potential neuroprotective activity. *J. Agric. Food Chem.* **66**: 7337–7346.

Tsurumi, S., Tagaki, T., and Hashimoto, T. (1992). A γ-pyronyl-triterpenoid saponin from Pisum sativum. *Phytochemistry* **31**: 2435–2438.

Tweedy, B.G., Dishburger, H.J., Ballantine, L.G. et al. (eds.) (1991). *Pesticide Residues and Food Safety: a Harvest of Viewpoints.* Washington, D.C.: American Chemical Society.

US Environmental Protection Agency. Toxicological review of 2,2′,4,4′-tetrabromodiphenyl ether (BDE-47). http://www.epa.gov/iris/toxreviews/1010tr.pdf (accessed 28 October 2019).

Uchida, K., Kanematsu, M., Sakai, K. et al. (1998). Protein-bound acrolein: potential markers for oxidative stress. *Proc. Natl. Acad. Sci.* **95**: 4882–4887.

Ueda, Y., Tsubuku, T., and Miyajima, R. (1994). Composition of sulfur-containing components in onion and their flavor characters. *Biosci. Biotechnol. Biochem.* **58**: 108–110.

Ulrich, D., Kecke, S., and Olbricht, K. (2018). What do we know about the chemistry of strawberry aroma? *J. Agric. Food Chem.* **66**: 3291–3301.

Umano, K., Hagi, Y., Nakahara, K. et al. (1992). Volatile constituents of green and ripened pineapple (Ananas comosus [L.] Merr.). *J. Agric. Food Chem.* **40**: 599–603.

United States Department of Agriculture, USDA National Nutrient Database for Standard Reference. http://ndb.nal.usda.gov (accessed 28 October 2019).

Urban, J., Dahlberg, C.J., Carroll, B.J., and Kaminsky, W. (2013). Absolute configuration of beer's bitter compounds. *Angew. Chem. Int. Ed.* **52**: 1553–1555.

Ureno, D., Rombolà, A.D., Iwashita, T. et al. (2007). Identification of two novel phytosiderophores secreted by perennial grasses. *New Phytol.* **174**: 304–310.

Vámos-Vigyázó, L. (1981). Polyphenol oxidases and peroxidases in fruits and vegetables. *Crit. Rev. Food Sci. Nutr.* **15**: 49–127.

Van Der Merwe, J.D., Joubert, E., Manley, M. et al. (2010). In vitro hepatic biotransformation of aspalathin and nothofagin, dihydrochalcones of rooibos (Aspalathus linearis), and assessment of metabolite antioxidant activity. *J. Agric. Food Chem.* **58**: 2214–2220.

van der Spiegel, M., Noordam, M.Y., and van der Fels-Klerx, H.J. (2013). Safety of novel protein sources (insects, microalgae, seaweed, duckweed, and rapeseed) and legislative aspects for their application in food and feed production. *Compr. Rev. Food Sci. Food Saf.* **12**: 662–678.

Van Dokkum, W., de Vos, R.H., Muys, T., and Weestra, J.A. (1989). Minerals and trace elements in total diets in the Netherlands. *Br. J. Nutr.* **61**: 7–15.

Van Egmond, H.P. (1989). Current situation on regulation for mycotoxins. Overview of tolerances and status of standard methods of sampling and analysis. *Food Addit. Contam.* **6**: 139–188.

Van Egmond, H.P. (1995). Mycotoxins: regulations, quality assurance and reference materials. *Food Addit. Contam.* **12**: 321–330.

Vanderhaegen, B., Neven, H., Daenen, L. et al. (2004a). Furfuryl ethyl ether: important aging flavor and a new marker for the storage conditions of beer. *J. Agric. Food Chem.* **52**: 1661–1668.

Vanderhaegen, B., Neven, H., Verstrepen, K.J. et al. (2004b). Influence of the brewing process on furfuryl ethyl ether formation during beer aging. *J. Agric. Food Chem.* **52**: 6755–6764.

Vanderpump, M.P., Lazarus, J.H., Smyth, P.P. et al. (2011). Iodine status of UK schoolgirls: a cross-sectional survey. *Lancet* **377**: 2007–2012.

Varga, B. (2008). Regulations for radioisotope content in food- and feedstuffs. *Food Chem. Toxicol.* **46**: 3448–3457.

Varo, P., Alfthan, G., Ekholm, P. et al. (1988). Selenium intake and serum selenium in Finland: effects of soil fertilization with selenium. *Am. J. Clin. Nutr.* **48**: 324–329.

Vázquez, L., Fornari, T., Señoráns, F.J. et al. (2008). Supercritical carbon dioxide fractionation of nonesterified alkoxyglycerols obtained from shark liver oil. *J. Agric. Food Chem.* **56**: 1078–1083.

Vega-Mercado, H. and Barbosa-Cánovas, G.V. (1994). Prediction of water activity in food systems. A review on theoretical models. *Rev. Esp. Cienc. Tecnol. Aliment* **34**: 368–384.

Velíšek, J. and Cejpek, K. (2008). *Biosynthesis of Food Components*. Tábor: Ossis.

Velíšek, J. and Cejpek, K. (2011). Pigments of higher fungi – a review. *Czech J. Food Sci.* **29**: 87–102.

Velíšek, J., Kubec, R., and Davídek, J. (1997). Chemical composition and classification of culinary and pharmaceutical garlic-based products. *Z. Lebensm. Unters. Forsch.* **204**: 161–164.

Velíšek, J., Ledahudcová, K., Kassahun, B. et al. (1993a). Chlorine-containing compounds derived from saccharides in protein hydrolysates. II. Levulinic acid esters in soybean meal hydrolysates. *LWT Food Sci. Technol.* **26**: 430–433.

Velíšek, J., Ledahudcová, K., Pudil, F. et al. (1993b). Chlorine-containing compounds drived from saccharides in protein hydrolysates. I. 5-Chloromethyl-2-furancarbaldehyde. *LWT Food Sci. Technol.* **26**: 38–41.

Verano, A.L. and Tan, D.S. (2017). Family-level stereoselective synthesis and biological evaluation of pyrrolomorpholine spiroketal natural product antioxidants. *Chem. Sci.* **8**: 3687–3693.

Vetter, W., Schröder, M., and Lehnert, K. (2012). Differentiation of refined and virgin edible oils by means of the *trans*- and *cis*-phytol isomer distribution. *J. Agric. Food Chem.* **60**: 6103–6107.

Vichi, S., Guadayolb, J.M., Caixachc, J. et al. (2006). Monoterpene and sesquiterpene hydrocarbons of virgin olive oil by headspace solid-phase microextraction coupled to gas chromatography/mass spectrometry. *J. Chromatogr.* **1125**: 117–123.

Vieth, R. (2004). Why 'Vitamin D' is not a hormone, and not a synonym for 1,25-dihydroxy-vitamin D, its analogs or deltanoids. *J. Steroid Biochem. Mol. Biol.* **89–90**: 571–573.

Vincent, J.B. (2000). The biochemistry of chromium. *J. Nutr.* **130**: 4715–4718.

Vincent, J.B. (2003). Recent advances in the biochemistry of chromium (III). *J. Trace Elem. Exp. Med.* **16**: 227–236.

Vogt, R.N., Spies, H.S.C., and Steenkamp, D.J. (2001). The biosynthesis of ovothiol A(N^1-methyl-4-mercaptohistidine). *Eur. J. Biochem.* **268**: 5229–5241.

Voigt, M. and Glomb, M.A. (2009). Reactivity of 1-deoxy-D-erythro-hexo-2,3-diulose: a key intermediate in the Maillard chemistry of hexoses. *J. Agric. Food Chem.* **57**: 4765–4770.

Volker, K., Gutser, J., Dietrich, A. et al. (1994). Stereoisomeric flavour compounds LXVIII. 2-, 3-, and 4-alkyl-branched acids, Part 2: chirospecific analysis and sensory evaluation. *Chirality* **6**: 427–434.

Vollenweider, S., Grassi, G., König, I., and Puhan, Z. (2003). Purification and structural characterization of 3-hydroxypropionaldehyde and its derivatives. *J. Agric. Food Chem.* **51**: 3287–3293.

Wagner, R. and Grosch, W. (1997). Evaluation of potent odorants of french fries. *LWT Food Sci. Technol.* **30**: 164–169.

Wagstaff, D.J. (1991). Dietary exposure to furocoumarins. *Regul. Toxicol. Pharmacol.* **14**: 261–272.

Wakamatsu, J., Stark, T.D., and Hofmann, T. (2016). Taste-active Maillard reaction products in roasted garlic (*Allium sativum*). *J. Agric. Food Chem.* **64**: 5845–5854.

Walford, J. (ed.) (1980). *Development in Food Colours-1*. London: Applied Science Publishers.

Waller, G.R. and Feather, M.S. (eds.) (1983). *The Maillard Reaction in Foods and Nutrition*. Washington, D.C.: American Chemical Society.

Wallrauch, S. and Greiner, G. (2005). Differentiation of juices from wild and cultured blueberries. *Fluss. Obst.* **72**: 14–17.

Walther, B. and Sieber, R. (2011). Bioactive proteins and peptides in foods. *Int. J. Vitam. Nutr. Res.* **81**: 181–192.

Wan, X., Liu, R., Yang, Y., and Zhang, M. (2015). Isolation, purification and identification of antioxidants in an aqueous aged garlic extract. *Food Chem.* **187**: 37–43.

Wang, D., Nanding, H., Han, N. et al. (2008). 2-(1H-Pyrrolyl) carboxylic acids as pigment precursors in garlic greening. *J. Agric. Food Chem.* **56**: 1495–1500.

Wang, H.-J. and Murphy, P.A. (1994). Isoflavone content in commercial soybean foods. *J. Agric. Food Chem.* **42**: 1666–1673.

Wang, K., Maga, J.A., and Bechtel, P.J. (1996). Taste properties and synergisms of beefy meaty peptide. *J. Food Sci.* **61**: 837–839.

Wang, M., Zhao, J., Avula, B. et al. (2014). High-resolution gas chromatography/mass spectrometry method for characterization and quantitative analysis of ginkgolic acids in *Ginkgo biloba* plants, extracts, and dietary supplements. *J. Agric. Food Chem.* **62**: 12103–12111.

Wang, S., Zhang, S., Huang, H. et al. (2012). Debrominated, hydroxylated and methoxylated metabolism in maize (Zea mays L.) exposed to lesser polybrominated diphenyl ethers (PBDEs). *Chemosphere* **89**: 1295–1301.

Wang, T., Zhao, J., Li, X. et al. (2016). New alkaloids from green vegetable soybeans and their inhibitory activities on the proliferation of concanavalin A-activated lymphocytes. *J. Agric. Food Chem.* **64**: 1649–1656.

Wang, W., Zhang, L., Wang, S. et al. (2014). 8-C N-ethyl-2-pyrrolidinone substituted flavan-3-ols as the marker compounds of Chinese dark teas formed in the post-fermentation process provide significant antioxidative activity. *Food Chem.* **152**: 539–545.

Wang, X., Beckham, T.H., Morris, J.C. et al. (2008). Bioactivities of gossypol, 6-methoxygossypol, and 6,6′-dimethoxygossypol. *J. Agric. Food Chem.* **56**: 4393–4398.

Wang, Y. and Ho, C.-T. (2009). Polyphenolic chemistry of tea and coffee: a century of progress. *J. Agric. Food Chem.* **57**: 8109–8114.

Ware, G.W. (1989). *The Pesticide Book*, 3e. Fresno, CA: Thomson Publications.

Warren, M.W., Larick, D.K., and Ball, H.R. Jr. (1995). Volatiles and sensory characteristics of cooked egg yolk, white and their combinations. *J. Food Sci.* **60**: 79–84.

Watson, D.H. (ed.) (1987). *Natural Toxicants in Food, Progress and Prospects*. Chichester: Horwood.

Watson, D.H. (2001). *Food Chemical Safety, Vol. 1: Contaminants*. Boca Raton: Woodhead Publishers.

Watt, B.K. and Merrill, A.L. (eds.) (1975). *Handbook of the Nutritional Contents of Foods*. New York: Dover Publications.

Weber, G. (1985). The importance of tin in the environment and its determination at trace levels. *Fresenius' Z. Anal. Chem.* **321**: 217–224.

Wei, W., Lin, S., Chen, M. et al. (2017). Monascustin, an unusual γ-lactam from red yeast rice. *J. Nat. Prod.* **80**: 201–204.

Weiss, E.R., Pika, J., and Braddock, R.J. (2003). Isolation and identification of terpene chlorohydrins found in cold-pressed orange oil. *J. Agric. Food Chem.* **51**: 2277–2282.

Wellner, A., Huettl, C., and Henle, T. (2011). Formation of Maillard reaction products during heat treatment of carrots. *J. Agric. Food Chem.* **59**: 7992–7998.

Wen, J.-Q., Huang, F., Liang, W.-S., and Liang, H.-G. (1997). Increase of HCN and β-cyanoalanine synthase activity during ageing of potato tuber slices. *Plant Sci.* **125**: 147–149.

Wen, Y.-Q., He, F., Zhu, B.-Q. et al. (2014). Free and glycosidically bound aroma compounds in cherry (*Prunus avium* L.). *Food Chem.* **152**: 29–36.

Whitaker, B.D., Schmidt, W.F., Kirk, M.C., and Barnes, S. (2001). Novel fatty acid esters of p-coumaryl alcohol in epicuticular wax of apple fruit. *J. Agric. Food Chem.* **49**: 3787–3792.

Widder, S., Lüntzel, C.S., Dittner, T., and Pickenhagen, W. (2000). 3-Mercapto-2-methylpentan-1-ol, a new powerful aroma compound. *J. Agric. Food Chem.* **48**: 418–423.

Wiersma, D., van Goor, B.J., and van der Veen, N.G. (1986). Cadmium, lead, mercury, and arsenic concentrations in crops and soils in the Netherlands. *J. Agric. Food Chem.* **34**: 1067–1074.

Wijesundera, R.C. and Ackman, R.G. (1988). Evidence for the probable presence of sulfur-containing fatty acids as minor constituents in canola oil. *J. Am. Oil Chem. Soc.* **65**: 959–963.

Wijnena, B., Leertouwera, H.L., and Stavenga, D.G. (2007). Colors and pterin pigmentation of pierid butterfly wings. *J. Insect Physiol.* **53**: 1206–1217.

Wilkins, A.L., Rehmann, N., Torgersen, T. et al. (2006). Identification of fatty acid esters of pectenotoxin-2 seco acid in blue mussels (Mytilus edulis) from Ireland. *J. Agric. Food Chem.* **54**: 5672–5678.

Wilkinson, V.M. and Gould, G.W. (1996). *Food Irradiation: A Reference Guide*. Lancaster, PA: Technomic.

Williams, J.D., Yazarians, J.A., Almeyda, C.C. et al. (2016). Detection of the previously unobserved stereoisomers of thujone in the essential oil and consumable products of sage (*Salvia officinalis* L.) using headspace solid-phase microextraction–gas chromatography–mass spektrometry. *J. Agric. Food Chem.* **64**: 4319–4326.

Williams, P.J., Strauss, C.R., and Wilson, B. (1980). Hydroxylated linalool derivatives as precursors of volatile monoterpenes of muscat grapes. *J. Agric. Food Chem.* **28**: 766–771.

Winter, C.K., Segall, H.J., and Haddon, W.F. (1986). Formation of cyclic adducts of deoxyguanosine with the aldehydes trans-4-hydroxy-2-hexenal and 4-hydroxy-2-nonenal in vitro. *Cancer Res.* **46**: 5682–5686.

Wisutiamonkul, A., Promdang, S., Ketsa, S., and van Doorn, W.G. (2015). Carotenoids in durian fruit pulp during growth and postharvest ripening. *Food Chem.* **180**: 301–305.

Wittenauer, J., Falk, S., Schweiggert-Weisz, U., and Carle, R. (2012). Characterisation and quantification of xanthones from the aril and pericarp of mangosteens (*Garcinia mangostana* L.) and a mangosteen containing functional beverage by HPLC–DAD–MSn. *Food Chem.* **134**: 445–452.

Woldemichael, G.M. and Wink, M. (2001). Identification and biological activities of triterpenoid saponins from Chenopodium quinoa. *J. Agric. Food Chem.* **49**: 2327–2332.

Wollmann, N. and Hofmann, T. (2013). Compositional and sensory characterization of red wine polymers. *J. Agric. Food Chem.* **61**: 2045–2061.

Wolnik, K.A., Fricke, F.Z., Capar, S.G., and Brande, G.L. (1983a). Elements in major raw agricultural crops in the United States. 1. Cadmium and lead in lettuce, peanuts, potatoes, soybeans, sweat corn, and wheat. *J. Agric. Food Chem.* **31**: 1240–1244.

Wolnik, K.A., Fricke, F.Z., Capar, S.G. et al. (1983b). Elements in major raw agricultural crops in the United States. 2. Other elements in lettuce, peanuts, potatoes, soybeans, sweat corn, and wheat. *J. Agric. Food Chem.* **31**: 1244–1249.

Wolnik, K.A., Fricke, F.Z., Capar, S.G. et al. (1985). Elements in major raw agricultural crops in the United States. 3. Cadmium, lead and eleven other elements in carrots, field corn, onions, rice, spinach, and tomatoes. *J. Agric. Food Chem.* **33**: 807–811.

Wood, C., Siebert, T.E., Parker, M. et al. (2008). From wine to pepper: rotundone, an obscure sesquiterpene, is a potent spicy aroma compound. *J. Agric. Food Chem.* **56**: 3738–3744.

Wood, W.F., Sollers, B.G., Dragoo, G.A., and Dragoo, J.W. (2002). Volatile components in defensive spray of the hooded skunk, *Mephitis macroura*. *J. Chem. Ecol.* **28**: 1865–1870.

World Health Organization (1978). Principles and methods for evaulating the toxicity of chemicals, Environmental Health Criteria, No. 6. Geneva: World Health Organization.

World Health Organization (1990). *Principles for Toxicological Assessment of Pesticide Residues in Food*, Environmental Health Criteria, No. 104. Geneva: World Health Organization.

World Health Organization (1990). *Public Health Impact of Pesticides Used in Agriculture*. Geneva: World Health Organization.

World Health Organization (2006). *Evaluation of Certain Food Contaminants. Sixty-fourth report of the Joint FAO/WHO Expert Committee on Food Additives*, WHO Technical Report Series 930. Geneva: World Health Organization.

Wright, A., Bubb, W.A., Hawkins, C.L., and Davies, M.J. (2002). Singlet oxygen-mediated protein oxidation: evidence for the formation of reactive side chain peroxides on tyrosine residues. *Photochem. Photobiol.* **76**: 35–46.

Wright, D.A. and Welbourn, P. (2002). *Environmental Toxicology*. Cambridge: Cambridge University Press.

Wu, W., Tang, Y., Yang, J. et al. (2018). Avenanthramide aglycones and glucosides in oat bran: chemical profile, levels in commercial oat products, and cytotoxicity to human colon cancer cells. *J. Agric. Food Chem.* **66**: 8005–8014.

Wu, X., Gu, L., Prior, R.L., and McKay, S. (2004). Characterization of anthocyanins and proanthocyanidins in some cultivars of Ribes, Aronia, and Sambucus and their antioxidant capacity. *J. Agric. Food Chem.* **52**: 7846–7856.

Wu, X. and Priorm, R.L. (2005). Identification and characterization of anthocyanins by high-performance liquid chromatography-electrospray ionizationtandem mass spectrometry in common foods in the United States: vegetables, nuts, and grains. *J. Agric. Food Chem.* **53**: 3101–3113.

Wu, X. and Priorm, R.L. (2005). Systematic identification and characterization of anthocyanins by HPLC-ESI-MS/MS in common foods in the United States: fruits and berries. *J. Agric. Food Chem.* **53**: 2589–2599.

Wulfkuehler, S., Gras, C., and Carle, R. (2013). Sesquiterpene lactone content and overall quality of fresh-cut witloof chicory (*Cichorium intybus* L. var. foliosum Hegi) as affected by different washing procedures. *J. Agric. Food Chem.* **61**: 7705–7714.

Wybraniec, S. (2005). Formation of decarboxylated betacyanins in heated purified betacyanin fractions from red beet root (*Beta vulgaris* L.) monitored by LC-MS/MS. *J. Agric. Food Chem.* **53**: 3483–3487.

Wybraniec, S. and Mizrahi, Y. (2005). Generation of decarboxylated and dehydrogenated betacyanins in thermally treated purified fruit extract from purple pitaya (*Hylocereus polyrhizus*) monitored by LC-MS/MS. *J. Agric. Food Chem.* **53**: 6704–6712.

Xiang, C., Werner, B.L., Christensen, E.M., and Oliver, D.J. (2001). The biological functions of glutathione revisited in Arabidopsis transgenic plants with altered glutathione levels. *Plant Physiol.* **126**: 2564–2574.

Xiao, Z., Wu, Q., Niu, Y. et al. (2017). Characterization of the key aroma compounds in five varieties of mandarins by gas chromatography-olfactometry, odor activity values, aroma recombination, and omission analysis. *J. Agric. Food Chem.* **65**: 8392–8401.

Xiaolin, Z. and Simoneau, A.R. (2005). Flavokawain A, a novel chalcone from kava extract, induces apoptosis in bladder cancer cells by involvement of bax protein-dependent and mitochondria-dependent apoptotic pathway and suppresses tumor growth in mice. *Cancer Res.* **65**: 3479–3486.

Yabuta, G., Koizumi, Y., Namiki, K. et al. (2001). Structure of green pigment formed by the reaction of caffeic acid esters (or chlorogenic acid) with primary amino compounds. *Biosci. Biotechnol. Biochem.* **65**: 2121–2130.

Yamaguchi, F., Ariga, T., Yoshimura, Y., and Nakazawa, H. (2000). Antioxidative and anti-glycation activity of garcinol from Garcinia indica fruit rind. *J. Agric. Food Chem.* **48**: 180–185.

Yamashita, M., Yamashita, Y., Ando, T. et al. (2013). Identification and determination of selenoneine, 2-selenyl-N_α, N_α, N_α-trimethyl-L-histidine, as the major organic selenium in blood cells in a fish-eating population on remote Japanese islands. *Biol. Trace Elem. Res.* **156**: 36–44.

Yamashita, Y., Amlund, H., Suzuki, T. et al. (2011). Selenoneine, total selenium, and total mercury content in the muscle of fishes. *Fish. Sci.* **77**: 679–686.

Yamashita, Y. and Yamashita, M. (2010). Identification of a novel selenium-containing compound, selenoneine, as the predominant chemical form of organic selenium in the blood of bluefin tuna. *J. Biol. Chem.* **285**: 18134–18138.

Yamauchi, R., Miyake, N., Inoue, H., and Kato, K. (1993a). Products formed by peroxyl radical oxidation of β-carotene. *J. Agric. Food Chem.* **41**: 708–713.

Yamauchi, R., Miyake, N., Kato, K., and Ueno, Y. (1993b). Reaction of α-tocopherol with alkyl and alkylperoxyl radicals of methyl linoleate. *Lipids* **28**: 201–206.

Yamazaki, Y., Iwasaki, K., and Yagihashi, A. (2011). Distribution of eleven flavour precursors, S-alk(en)yl-L-cysteine derivatives, in seven Allium vegetables. *Food Sci. Technol. Res.* **17**: 55–62.

Yang, J., Wang, P., Wu, W. et al. (2016). Steroidal saponins in oat bran. *J. Agric. Food Chem.* **64**: 1549–1556.

Yang, X. and Eilerman, R.G. (1999). Pungent principal of *Alpinia galanga* (L.) Swartz and its application. *J. Agric. Food Chem.* **47**: 1657–1662.

Yassin, G.H., Koek, J.H., Jayaraman, S., and Kuhnert, N. (2014). Identification of novel homologous series of polyhydroxylated theasinensins and theanaphthoquinones in the sii fraction of black tea thearubigins using ESI/HPLC tandem mass spectrometry. *J. Agric. Food Chem.* **62**: 9848–9859.

Yasuda, K., Peterson, R.J., and Chang, S.S. (1975). Identification of volatile flavor compounds developed during storage of a deodorized hydrogenated soybean oil. *J. Am. Oil Chem. Soc.* **52**: 307–311.

Yasumoto, T. and Murata, M. (1993). Marine toxins. *Chem. Rev.* **93**: 1897–1909.

Yaylayan, V.A. and Kaminsky, E. (1998). Isolation and structural analysis of Maillard polymers: caramel and melanoidin formation in glycine/glucose model system. *Food Chem.* **63**: 25–31.

Yaylayan, V.A., Keyhani, A., and Wnorowski, A. (2000). Formation of sugar-specific reactive intermediates from ^{13}C-labeled L-serines. *J. Agric. Food Chem.* **48**: 636–641.

Ye, J., Wang, X.-H., Sang, Y.-X., and Liu, Q. (2011). Assessment of the determination of azodicarbonamide and its decomposition product semicarbazide: investigation of variation in flour and flour products. *J. Agric. Food Chem.* **59**: 9313–9318.

Ye, X.-S., He, J., Cheng, Y.-C. et al. (2017). Cornusides A–O, bioactive iridoid glucoside dimers from the fruit of *Cornus officinalis*. *J. Nat. Prod.* **80**: 3103–3111.

Yen, G.-C. (1992). Effect of heat treatment and storage temperature on the biogenic amine contents of straw mushroom (Vollvariella volvacea). *J. Sci. Food Agric.* **58**: 59–61.

Yılmaz, C., Kocadağlı, T.n., and Gökmen, V. (2014). Formation of melatonin and its isomer during bread dough fermentation and effect of baking. *J. Agric. Food Chem.* **62**: 2900–2905.

Yokokawa, H. and Mitsuhashi, T. (1981). The sterol composition of mushrooms. *Phytochemistry* **20**: 1349–1351.

Yong, Y.Y., Dykes, G., Lee, S.M., and Choo, W.S. (2017). Comparative study of betacyanin profile and antimicrobial activity of red pitahaya (*Hylocereus polyrhizus*) and red spinach (*Amaranthus dubius*). *Plant Foods Hum. Nutr.* **72**: 41–47.

Yoshikawa, K., Shimono, N., and Arihara, S. (1991). Antisweet substances, jujubasaponins I–III from *Zizyphus jujuba* revised structure of ziziphin. *Tetrahedron Lett.* **32**: 7059–7062.

Young, V.R. and Pellett, P.L. (1994). Plant proteins in relation to human protein and amino acid nutrition. *Am. J. Clin. Nutr.* **59** (Suppl): 1203S–1211S.

Yu, T.-H., Shu, C.-K., and Ho, C.-T. (1994). Thermal decomposition of alliin, the major flavor component of garlic, in an aqueous solution. In: *Food Phytochemicals for Cancer Prevention I. Fruits and Vegetables*, Vol. 546, Chapter 10 (eds. M.-T. Huang, T. Osawa, C.-T. Ho and R.T. Rosen), 144–152. American Chemical Society.

Yu, Z., Zhang, G., and Zhang, Y. (2005). Occurrence and analytical methods of acrylamide in heat-treated foods. *J. Chromatogr. A* **1075**: 1–21.

Yue, S., Tang, Y., Li, S., and Duan, J.-A. (2013). Chemical and biochemicl propertie of quinochalcone C-glycosides from the florets of *Carthamus tinctorius*. *Molecules* **18**: 15220–15254.

Yue, S.-H., Qu, C., Zhang, P.-X. et al. (2016). Carthorquinosides A and B, quinochalcone C-glycosides with diverse dimeric skeleton from *Carthamus tinctorius*. *J. Nat. Prod.* **79**: 2644–2651.

Yun, J.W. (1996). Fructooligosaccharides-occurrence, preparation, and application. *Enzym. Microb. Technol.* **19**: 107–117.

Zaki, M.M., El-Midany, S.A., Shaheen, H.M., and Rizzi, L. (2012). Mycotoxins in animals: occurrence, effects, prevention and management. *J. Toxicol. Environ. Health Sci.* **4**: 13–28.

Zamora, R., Delgado, R.M., and Hidalgo, F.J. (2012). Chemical conversion of phenylethylamine into phenylacetaldehyde by carbonyl–amine reactions in model systems. *J. Agric. Food Chem.* **60**: 5491–5496.

Zamora, R., Gallardo, E., Navarro, J.L., and Hidalgo, F.J. (2005). Strecker-type degradation of phenylalanine by methyl 9,10-epoxy-13-oxo-11-octadecenoate and methyl 12,13-epoxy-9-oxo-11-octadecenoate. *J. Agric. Food Chem.* **53**: 4583–4588.

Zamora, R. and Hidalgo, F.J. (2005). Coordinate contribution of lipid oxidation and Maillard reaction to the nonenzymatic food browning. *Crit. Rev. Food Sci. Nutr.* **45**: 49–59.

Zanardi, E., Battaglia, A., Ghidini, S. et al. (2007). Evaluation of 2-alkylcyclobutanones in irradiated cured pork products during vacuum-packed storage. *J. Agric. Food Chem.* **55**: 4264–4270.

Zanardi, E., Jagersma, C.G., Ghidini, S., and Chizzolini, R. (2002). Solid phase extraction and liquid chromatography-tandem mass spectrometry for the evaluation of 4-hydroxy-2-nonenal in pork products. *J. Agric. Food Chem.* **50**: 5268–5272.

Zanoli, P. and Zavatti, M. (2008). Pharmacognostic and pharmacological profile of Humulus lupulus L. *J. Ethnopharmacol.* **116**: 383–396.

Zapsalis, C. and Beck, R.A. (1985). *Food Chemistry and Nutritional Biochemistry*. New York: Wiley.

Zare, D., Muhammad, K., Bejo, M.H., and Ghazali, H.M. (2015). Determination of *trans*- and *cis*-urocanic acid in relation to histamine, putrescine, and cadaverine contents in tuna (*Auxis thazard*) at different storage temperatures. *Food Sci.* **80**: T479–T483.

Zarembski, P.M. and Hodgkinson, A. (1962). The oxalic acid content of English diets. *Br. J. Nutr.* **16**: 627–634.

Zelinková, Z., Doležal, M., and Velíšek, J. (2008a). Occurrence of 3-chloropropane-1,2-diol fatty acid esters in infant and baby foods. *Eur. Food Res. Technol.* **228**: 571–578.

Zelinková, Z., Novotný, O., Schůrek, J. et al. (2008b). Occurrence of 3-MCPD fatty acid esters in human breast milk. *Food Addit. Contam.* **25**: 669–676.

Zelinková, Z., Svejkovská, B., Velíšek, J., and Doležal, M. (2006). Fatty acid esters of 3-chloropropane-1,2-diol in edible oils. *Food Addit. Contam.* **23**: 1290–1298.

Zhang, L. and Peterson, D.G. (2018). Identification of a novel umami compound in potatoes and potato chips. *Food Chem.* **240**: 1219–1226.

Zhang, L.-Q., Yang, X.-W., Zhang, Y.-B. et al. (2012). Biotransformation of phlorizin by human intestinal flora and inhibition of biotransformation products on tyrosinase activity. *Food Chem.* **132**: 936–942.

Zhang, Q., Li, L., Lan, Q. et al. (2018). Protein glycosylation: apromising way to modify the functional properties and extend the application in food system. *Crit. Rev. Food Sci. Nutr.* https://doi.org/10.1080/10408398.2018.1507995.

Zhang, S., Balbo, S., Wang, M., and Hecht, S.S. (2011). Analysis of acrolein-derived 1,N^2-propanodeoxyguanosine adducts in human leukocyte DNA from smokers and nonsmokers. *Chem. Res. Toxicol.* **24**: 119–124.

Zhang, S., Yang, C., Idehen, E. et al. (2018). Novel theaflavin-type chlorogenic acid derivatives identified in black tea. *J. Agric. Food Chem.* **66**: 3402–3407.

Zhang, Y., Venkitasamy, C., Pan, Z. et al. (2017). Novel umami ingrediets: umami peptides and their taste. *J. Food Sci.* **82**: 16–23.

Zhao, S.-L. and Shen, J.-S. (2006). Enantiomeric separation and determination of *R,S*-salsolinol by capillary electrophoresis. *Chin. J. Chem.* **24**: 439–441.

Zhou, C., Zhao, D., Sheng, Y. et al. (2011). Carotenoids in fruits of different persimmon cultivars. *Molecules* **16**: 624–636.

Zhou, Y., Dong, F., Kunimasa, A. et al. (2014). Occurrence of glycosidically conjugated 1-phenylethanol and its hydrolase β-primeverosidase in tea (*Camellia sinensis*) flowers. *J. Agric. Food Chem.* **62**: 8042–8050.

Zhu, J., Jia, J., Li, X. et al. (2013). ESR studies on the thermal decomposition of trimethylamine oxide to formaldehyde and dimethylamine in jumbo squid (*Dosidicus gigas*) extract. *Food Chem.* **141**: 3881–3888.

Zhu, L., Lu, J.-G., Li, T. et al. (2012). Immunosuppresive decalin derivatives from red yeast rice. *J. Nat. Prod.* **75**: 567–571.

Zhuang, M., Lin, L., Zhao, M. et al. (2016). Sequence, taste and umami-enhancing effect of the peptides separated from soy sauce. *Food Chem.* **206**: 174–181.

Zidorn, C., Jöhrer, K., Ganzera, M. et al. (2005). Polyacetylenes from the Apiaceae vegetables carrot, celery, fennel, parsley, and parsnip and their cytotoxic activities. *J. Agric. Food Chem.* **53**: 2518–2523.

Index